ENCYCLOPEDIA
OF
FOOD MICROBIOLOGY

ENCYCLOPEDIA OF FOOD MICROBIOLOGY

Editor-in-Chief

RICHARD K. ROBINSON

Editors

CARL A. BATT

PRADIP D. PATEL

A Harcourt Science and Technology Company

San Diego San Francisco New York Boston
London Sydney Tokyo

This book is printed on acid-free paper

Academic Press
A Harcourt Science and Technology Company
24–28 Oval Road, London NW1 7DX, UK
http://www.hbuk.co.uk/ap/

Academic Press
525 B Street, Suite 1900, San Diego, California 92101-4495, USA
http://www.apnet.com

ISBN 0-12-227070-3

A catalogue for this encyclopedia is available from the British Library

Library of Congress Catalog Card Number: 98-87954

Access for a limited period to an on-line version of the Encyclopedia of Food Microbiology is included in the purchase price of the print edition.
This on-line version has been uniquely and persistently identified by the Digital Object Identifier (DOI)

10.1006/rwfm.2000

By following the link

http://dx.doi.org/10.1006/rwfm.2000

from any Web Browser, buyers of the Encyclopedia of Food Microbiology will find instructions on how to register for access.

Typeset by Selwood Systems, Midsomer Norton, Bath
Printed in Great Britain by The Bath Press, Bath

00 01 02 03 04 05 BP 9 8 7 6 5 4 3 2 1

EDITORIAL ADVISORY BOARD

FOREWORD

Public concern about food safety has never been greater. In part this is due to the ever increasing demand from consumers for higher and higher standards. But new food-borne pathogens like *E. coli* O157 have emerged in recent years to become important public health problems, and changes in production and manufacturing sometimes reopen doors of opportunity for old ones. A powerful reminder that food scientists have much unfinished business to attend to is provided by the succession of food scares that generate strong stories for the media.

Experience tells us that science must underpin all approaches to food safety, whether through the application and implementation of well-tried approaches or the development of new or improved methods. Microbiologists have had a central role in this since the high quality work of pioneers like van Ermengem on botulism and Gaffky on typhoid more than a century ago. The large amount of important data that has accumulated since then joins with the current rapid rates of technological and scientific advance to make the need for a structured and authoritative source of information a very pressing one. It is provided by this encyclopedia.

These are exciting times for food microbiologists. Expectations are high that as scientists we will soon provide answers to the many problems still posed by microbes – from spoilage to food poisoning. Approaches like HACCP are making everyone think hard about how best to apply the data we have to develop better ways for reducing and eliminating food-borne pathogens. The pace of scientific developments continues to accelerate and more and better methods are available for the detection and enumeration of microbes than ever before. The microbes themselves continue to evolve and so present moving targets. The solid foundation presented by the mass of information in this encyclopedia provides the launching pad and guide for meeting these challenges.

It could be said that a penalty of working in food microbiology is that because the subject is broad-ranging, mature and dynamic, its practitioners, teachers and students have to know about many things in breadth and depth. For most of us, of course, this is not a penalty but an attractive bonus because of its intellectual challenge. I am particularly pleased to be associated with the encyclopedia because it will help us all to meet this test with confidence. I wish it every success.

Professor H Pennington
Department of Medical Microbiology
University of Aberdeen

PREFACE

Although food microbiology and food safety have, in recent times, become major concerns for governments around the world, equally important is the fact that, without yeasts and bacteria, popular meals like bread and cheese would not exist. Consequently, a knowledge of the relationship between foodstuffs and the activities of bacteria, yeasts and mycelial fungi has become a top priority for everyone associated with food and its production. Farmers have concerns related to produce harvesting and storage, food processors have to generate wholesome retail products that are both free from pathogenic organisms and have a satisfactory shelf life and, last but not least, food handlers and consumers need to be aware of the procedures necessary to ensure that food is safely prepared and stored.

In order for these disparate groups to operate successfully, accurate and objective information about the microbiology of foods is essential, and this encyclopedia seeks to provide a source of such information. In some areas, introductory articles are provided to guide readers who may be less familiar with the subject but, in general, superficiality has been avoided. Thus, the coverage has been developed to include details of all the important groups of bacteria, fungi, viruses and parasites, the various methods that can be employed for their detection in foods, the factors that govern the behaviour of the same organisms, together with an analysis of likely outcomes of microbial growth/metabolism in terms of disease and/or spoilage. A further series of articles describes the contribution of microorganisms to industrial fermentations, to traditional food fermentations from the Middle or Far East, as well as during the production of the fermented foods like bread, cheese or yoghurt that are so familiar in industrialized societies. The division of these topics into 358 articles of approximately 4000 words, has meant that the contributing authors have been able to handle their specialist subject(s) in real depth.

Obviously, another group of editors might have approached the project in a different manner, but we feel confident that this encyclopedia will provide readers at all levels of expertise with the data being sought. A point enhanced, perhaps, by the inclusion at the end of each article of a list for further reading, comprising a selection of review articles and key research papers that should encourage further exploration of any selected topic. If this confidence is borne out in practice, then the efforts of the contributors, the members of the Editorial Board and the editorial team from Academic Press will be well rewarded, for raising the scientific profile of food microbiology is long overdue.

R.K. Robinson, C.A. Batt, P.D. Patel
Editors

INTRODUCTION

The advent of antibiotics gave the general public, and many professional microbiologists as well, the feeling that bacterial diseases were under control, and the elimination of smallpox and the control of polio suggested that even viruses posed few problems. However, this complacency has received a nasty jolt over the last decade, and the emergence of HIV and multiple-drug-resistant strains of bacteria has become a major concern for the medical profession. The food industry has been similarly shaken by the appearance of new, and potentially fatal, strains of *Escherichia coli*, a species that for over 100 years was regarded as little more than a nuisance. Equally unexpected was the devastating impact of BSE, and fresh reports of the activities of so-called 'emerging food-borne pathogens' are appearing with alarming regularity.

In some cases, it has been possible to understand, with the advantages of hindsight, why a particular species of bacterium, fungus or protozoan has become a major risk to human health while, on other occasions, the vagaries of nature have left the 'experts' totally bemused. However, even in these latter situations, control over the threat posed to food supplies has to be instituted, but the ability of the food industry, in conjunction with Public Health and other bodies, to develop effective responses can only be as good as the scientific knowledge available. In the case of food microbiology, this background has to be derived from a wide range of sources. Thus, agricultural practices may alter the biochemistry of a crop and, perhaps, its microflora as well; the microflora of any given foodstuff and/or processing facility will have specific characteristics that need to be understood before control is possible; techniques must be available to monitor a retail food for microorganisms that would pose a risk to the consumer. As the procedures necessary to monitor these various facets become ever more sophisticated, so fewer microbiologists can claim total competence, and the need for a specialist source of outside knowledge increases.

It is this latter need that the *Encyclopedia of Food Microbiology* seeks to satisfy for, within this work, a busy microbiologist can find details of all the important genera of food-borne bacteria and fungi, how the same genera may react in different foods and under different environmental conditions, and how to detect the growth and/or metabolism of the same organisms in foods using classical or modern techniques. In order to place this information into a broader context, the reader can explore the latest advice concerning food standards/specifications, or the role of monitoring systems like HACCP in achieving product targets for specific microorganisms; potential concerns over viruses and protozoa are also evaluated in the light of current knowledge. Readers interested in fermented foods will find the pertinent information in a similarly accessible form; indeed, purchasers of the print version of the encyclopedia will be entitled to register for access to the on-line version as well. This form allows the user the benefit of extensive hypertext linking and advanced search tools, adding value to the encyclopedia as a reference source, teaching aid and text for general interest.

It is inevitable, of course, that short articles written to a tight deadline may have omissions, but it is to be hoped that such faults are minimal and, in any event, more than compensated for through the careful selections of further reading. If this optimism is justified, then the major credit rests with the authors of each article. They are all recognized as experts in their fields, and their willing participation has been much appreciated by the editors. The role of the Editorial Advisory Board merits a special mention as well, for their constructive criticisms of the list of articles, their suggestions for authors and their expert refereeing of the manuscripts has provided a solid foundation for the entire enterprise.

However, the finest manuscripts are of little value to the scientific community until they have been published, and the editorial team at Academic Press – Carey Chapman (Editor-in-Chief), Tina Holland (Associate Editor), Nick Fallon (Commissioning Editor), Laura O'Neill (Editorial Assistant), Tamsin Cousins (Production Project Manager), Richard Willis (Freelance Project Manager), Emma Parkinson (Electronic Publishing Developer), Peter Lord (Publishing Services Manager), Emma Krikler (Picture Researcher) – have been outstanding in their support of the project. Obviously, each member of the team has made an important contribution, but it must be recorded that the role of Tina Holland has been absolutely invaluable. Thus, not only has Tina co-ordinated the numerous inputs from the editors, referees and authors, but even found time to help the editors with the location of authors; the editors acknowledge this unstinting assistance with much gratitude.

R.K. Robinson, C.A. Batt, P.D. Patel
Editors

GUIDE TO USE OF THE ENCYCLOPEDIA

Structure of the Encyclopedia

The material in the Encyclopedia is arranged as a series of entries in alphabetical order. Some entries comprise a single article, whilst entries on more diverse subjects consist of several articles that deal with various aspects of the topic. In the latter case the articles are arranged in a logical sequence within an entry.

To help you realize the full potential of the material in the Encyclopedia we have provided three features to help you find the topic of your choice.

1. Contents Lists

Your first point of reference will probably be the contents list. The complete contents list appearing in each volume will provide you with both the volume number and the page number of the entry. On the opening page of an entry a contents list is provided so that the full details of the articles within the entry are immediately available.

Alternatively you may choose to browse through a volume using the alphabetical order of the entries as your guide. To assist you in identifying your location within the Encyclopedia a running headline indicates the current entry and the current article within that entry.

You will find 'dummy entries' where obvious synonyms exist for entries or where we have grouped together related topics. Dummy entries appear in both the contents list and the body of the text. For example, a dummy entry appears for Butter which directs you to Milk and Milk Products: Microbiology of Cream and Butter, where the material is located.

Example

If you were attempting to locate material on Dairy Products via the contents list.

DAIRY PRODUCTS *see BRUCELLA*: Problems with Dairy Products; CHEESE: In the Market Place; Microbiology of Cheese-making and Maturation; Mould-ripened Varieties; Role of Specific Groups of Bacteria; Microflora of White-brined Cheeses; FERMENTED MILKS: Yoghurt; Products from Northern Europe; Products of Eastern Europe and Asia; PROBIOTIC BACTERIA: Detection and Estimation in Fermented and Non-fermented Dairy Products

At the appropriate location in the contents list, the page numbers for articles under *Brucella*, etc. are given.

If you were trying to locate the material by browsing through the text and you looked up Dairy Products then the following information would be provided.

DAIRY PRODUCTS *see* ***BRUCELLA***: Problems with Dairy Products; **CHEESE**: In the Market Place; Microbiology of Cheese-making and Maturation; Mould-ripened Varieties; Role of Specific Groups of Bacteria; Microflora of White-brined Cheeses; **FERMENTED MILKS**: Yoghurt; Products from Northern Europe; Products of Eastern Europe and Asia; **PROBIOTIC BACTERIA**: Detection and Estimation in Fermented and Non-fermented Dairy Products.

Alternatively, if you were looking up *Brucella* the following information would be provided.

BRUCELLA

Contents

Characteristics

Problems with Dairy Products

2. Cross References

All of the articles in the Encyclopedia have an extensive list of cross references which appear at the end of each article, e.g.:

ATP BIOLUMINESCENCE/Application in Dairy Industry.

See also: **Acetobacter**. **ATP Bioluminescence**: Application in Meat Industry; Application in Hygiene Monitoring; Application in Beverage Microbiology. **Bacteriophage-based Techniques for Detection of Food-borne Pathogens**. **Biophysical Techniques for Enhancing Microbiological Analysis**: Future Developments. **Electrical Techniques**: Food Spoilage Flora and Total Viable Count (TVC). **Immunomagnetic Particle-based Techniques**: Overview. **Rapid Methods for Food Hygiene Inspection**. **Total Viable Counts**: Pour Plate Technique; Spread Plate Technique; Specific Techniques; MPN; Metabolic Activity Tests; Microscopy. **Ultrasonic Imaging**: Non-destructive Methods to Detect Sterility of Aseptic Packages. **Ultrasonic Standing Waves**.

3. Index

The index will provide you with the volume number and page number of where the material is to be located, and the index entries differentiate between material that is a whole article, is part of an article or is data presented in a table. On the opening page of the index, detailed notes are provided.

4. Colour Plates

The colour figures for each volume have been grouped together in a plate section. The location of this section is cited both in the contents list and before the *See also* list of the pertinent articles.

5. Contributors

A full list of contributors appears at the beginning of each volume.

CONTRIBUTORS

Lahsen Ababouch
Department of Food Microbiology and Quality Control
Institut Agronomique et Vétérinaire Hassan II
Rabat
Morocco

D Abramson
Agriculture & Agri-Food Canada
Cereal Research Centre
195 Dafoe Road
Winnipeg
Manitoba
R3T 2M9
Canada

Ann M Adams
Seafood Products Research Center
US Food and Drug Administration
PO Box 3012
22201 23rd Drive SE
Bothell
WA 98041–3012
USA

Martin R Adams
School of Biological Sciences
University of Surrey
Guildford
GU2 5XH
UK

G E Age
PO Box 553
Wageningen
The Netherlands

M Ahmed
Food Control Laboratory
PO Box 7463
Dubai
United Arab Emirates

Imad Ali Ahmed
Central Food Control Laboratory, Ajman Municipality
PO Box 3717
Ajman
UAE

Peter Ahnert
Department of Biochemistry
Ohio State University
Columbus
OH 43210
USA

William R Aimutis
Land O' Lakes Inc.
PO Box 674101
St Paul
Minnesota
55164–0101
USA

J H Al-Jedah
Central Laboratories
Ministry of Public Health
Qatar

Cameron Alexander
Macromolecular Science Department
Institute of Food Research,
Reading Laboratory
Earley Gate
Whiteknights Road
Reading
RG6 6BZ
UK

Marcos Alguacil
Departmento de Genética
Facultad de Ciencias, Universidad de Málaga
Spain

M Z Ali
Central Laboratories
Ministry of Public Health
Qatar

M D Alur
Food Technology Division
Bhabha Atomic Research Centre
Mumbai 400085
India

R Miguel Amaguaña
US Food and Drug Administration
Washington, DC
USA

Vilma Moratade de Ambrosini
Centro de Referencia para Lactobacilos and Universidad Nacional de Tucumán
Casilla de Correo 211
(4000)-Tucuman
Argentina

Wallace H Andrews
US Food and Drug Administration
Washington, DC 20204
USA

Dilip K Arora
Department of Botany
Banaras Hindu University
Varanasi 221 005
India

B Austin
Department of Biological Sciences
Heriot-Watt University
Riccarton
Edinburgh EH14 4AS
Scotland, UK

Aslan Azizi
Iranian Agricultural Engineering Research Institute
Agricultural Research Organization
Evin Tehran
Iran

S De Baets
Laboratory of Industrial Microbiology and Biocatalysis
Department of Biochemical and Microbial Technology
Faculty of Agricultural and Applied Biological Sciences
University of Gent
Coupure links 653
B-9000
Gent
Belgium

Les Baillie
Biomedical Sciences
DERA
CBD Porton Down
Salisbury
Wiltshire
UK

Gustavo V Barbosa-Cánovas
Biological Systems Engineering
Washington State University
Pullman
Washington 99164–6120
USA

J Baranyi
Institute for Food Research
Reading
UK

Eduardo Bárzana
Departamento de Alimentos y Biotecnología
Facultad de Química
Universidad Nacional Autónoma de México
Mexico City 04510
Mexico

Carl A Batt
Department of Food Science
Cornell University
Ithaca
NY 14853
USA

Derrick A Bautista
Saskatchewan Food Product Innovation Program
Department of Applied Microbiology and Food Science
University of Saskatchewan
Canada

S H Beattie
Hannah Research Institute
Ayr KA6 5HL
UK

H Beck
Department for Health Service
South Bavaria
Veterinärstrasse 2
85764 Oberschleissheim
Germany

Reginald Bennett
FDA
Center for Food Safety and Applied Nutrition
Washington, DC
USA

Marjon H J Bennik
Agrotechnological Research Institution (ATO-DLO)
Bornsesteeg 59
6709 PD
Wageningen
The Netherlands

Merlin S Bergdoll (dec)
Food Research Institute
University of Wisconsin-Madison
Madison, WI
USA

R G Berger
Food Chemistry
University of Hannover
Germany

K Berghof
BioteCon Gesellschaft für Biotechnologische
Entwicklung und Consulting
Hermannswerder Haus 17
14473 Potsdam
Germany

P A Bertram-Drogatz
Mediport VC Management GmbH
Wiesenweg 10
12247 Berlin
Germany

Gail D Betts
Campden and Chorleywood Food Research Association
Chipping Campden
Gloucestershire
GL55 6LD
UK

R R Beumer
Wageningen Agricultural University
Laboratory of Food Microbiology
Bomenweg 2
NL 6703 HD Wageningen
The Netherlands

Rijkelt R Beumer
Wageningen University and Research Centre
Department of Food Technology and Nutritional Sciences
Bomenweg 2
NL 6703 HD Wageningen
The Netherlands

Saumya Bhaduri
Microbial Food Safety Research Unit
Eastern Regional Research Center
Agricultural Research Service
US Department of Agriculture
600 East Mermaid Lane
Wyndmoor
PA 19038
USA

Deepak Bhatnagar
Southern Regional Research Center
Agricultural Research Service
US Department of Agriculture
LA
USA

J R Bickert
Halosource Corporation
First Avenue South
Seattle
WA 98104
USA

Hanno Biebl
GBF – National Research Centre for Biotechnology
Braunschweig
Germany

Clive de W Blackburn
Microbiology Unit
Unilever Research Colworth
Colworth House
Sharnbrook
Bedford
UK

I S Blair
Food Studies Research Unit
University of Ulster at Jordanstown
Shore Road
Newtownabbey
Co. Antrim
Northern Ireland
BT37 9QB

G Blank
Department of Food Science
University of Manitoba
Winnipeg
MB
Canada

Hans P Blaschek
Department of Food Science and Human Nutrition
University of Illinois
488 Animal Science Lab
1207 West Gregory Drive
Urbana
IL 61801
USA

D Blivet
AFSSA
Ploufragan
France

R G Board
South Lodge
Northleigh
Bradford-on-Avon
Wiltshire
UK

Enne de Boer
Inspectorate for Health Protection
PO Box 9012
7200 GN Zutphen
The Netherlands

Christine Bonaparte
Department of Dairy Research and Bacteriology
Agricultural University
Gregor Mendel-Str. 33
A-1180 Vienna
Austria

Kathryn J Boor
Department of Food Science
Cornell University
Ithaca
NY 14853
USA

A Botha
Department of Microbiology
University of Stellenbosch
Stellenbosch 7600
South Africa

W Richard Bowen
Biochemical Engineering Group
Centre for Complex Fluids Processing
University of Wales Swansea
Singleton Park
Swansea
SA2 8PP
UK

Catherine Bowles
Leatherhead Food Research Association
Leatherhead
Surrey
UK

Patrick Boyaval
INRA
Laboratoire de Recherches de Technologie Laitière
65 rue de Saint-Brieuc
35042
Rennes Cedex
France

F Bozoğlu
Department of Food Engineering
Middle East Technical University
Ankara
Turkey

Astrid Brandis-Heep
Philipps Universität
Fachbereich Biologie
Laboratorium für Mikrobiologie
D-35032 Marburg
Germany

Susan Brewer
Department of Food Science and Human Nutrition
University of Illinois
Urbana
Illinois
USA

Aaron L Brody
Rubbright Brody Inc.
PO Box 956187
Duluth
Georgia
30095–9504
USA

Bruce E Brown
B. E. Brown Associates
328 Stone Quarry Priv.
Ottawa
Ontario
K1K 3Y2
Canada

G Bruggeman
Laboratory of Industrial Microbiology and Biocatalysis
Department of Biochemical and Microbial Technology
Faculty of Agricultural and Applied Biological Sciences
University of Gent
Coupure links 653
B-9000
Gent
Belgium

Andreas Bubert
Department for Microbiology
Theodor-Boveri Institute for Biosciences
University of Würzburg
Am Hubland
97074 Würzburg
Germany

Ken Buckle
Department of Food Science and Technology
The University of New South Wales
Sydney
Australia

Lloyd B Bullerman
Department of Food Science and Technology
University of Nebraska
PO Box 830919
Lincoln
NE 68583–0919
USA

Justino Burgos
Food Technology Section
Department of Animal Production and Food Science
University of Zaragoza
Spain

Frank F Busta
Department of Food Science and Nutrition
University of Minnesota
St Paul
Minnesota 55108
USA

Daniel Cabral
Departmento de Ciencias Biológicas
Facultad de Ciencias Exactas y Naturales
Pabellon II 4to piso – Ciudad Universitaria
1428 Buenos Aires
Argentina

María Luisa Calderón-Miranda
Biological Systems Engineering
Washington State University
Pullman
Washington 99164–6120
USA

Geoffrey Campbell-Platt
Gyosei Liaison Office
Gyosei College
London Road
Reading
Berks RG1 5AQ
UK

Iain Campbell
International Centre for Brewing and Distilling
Heriot-Watt University
Edinburgh
EH14 4AS
Scotland

Frédéric Carlin
Institut National de la Recherche Agronomique
Unité de Technologie des Produits Végétaux
Site Agroparc
84914
Avignon
Cedex 9
France

Brigitte Carpentier
National Veterinary and Food Research Centre
22 rue Pierre Curie
F-94709
Maisons-Alfort Cedex
France

Maria da Glória S Carvalho
Departamento de Microbiologia Médica
Instituto de Microbiologia
Universidade Federal do Rio de Janeiro
Rio de Janeiro 21941
Brazil

O Cerf
Alfort Veterinary School
7 Avenue du Général de Gaulle
F-94704
Maisons-Alfort Cedex
France

Lourdes Pérez Chabela
Universidad Autónoma Metropolitana-Iztapalapa
Mexico
Apartado Postal 55–535
CP 09340 Mexico DF
Mexico

Perng-Kuang Chang
Southern Regional Research Center
Agricultural Research Service
US Department of Agriculture
LA
USA

E A Charter
Canadian Inovatech Inc.
31212 Peardonville Road
Abbotsford
BC V2T 6K8
Canada

Parimal Chattopadhyay
Department of Food Technology and Biochemical Engineering
Jadavpur University
Calcutta-700 032
India

Yusuf Chisti
Department of Chemical Engineering
University of Almería
E-04071 Almería
Spain

Thomas E Cleveland
Southern Regional Research Center
Agricultural Research Service
US Department of Agriculture
LA
USA

Dean O Cliver
University of California, Davis, School of Veterinary Medicine
Department of Population Health and Reproduction
One Shields Avenue
Davis
California 95616–8743
USA

T E Cloete
Department of Microbiology and Plant Pathology
Faculty of Biological and Agricultural Sciences
University of Pretoria
Pretoria 0002
South Africa

Roland Cocker
Cocker Consulting
Bergeendlaan 16
1343 AR Almere
The Netherlands

Timothy M Cogan
Dairy Products Research Centre
Teagasc
Fermoy
Ireland

David Collins-Thompson
Nestlé Research and Development Center
210 Housatonic Avenue
New Milford
Connecticut
USA

Janet E L Corry
Division of Food Animal Science
Department of Clinical Veterinary Science
University of Bristol
Langford
Bristol
BS40 5DT
UK

Aldo Corsetti
Institute of Dairy Microbiology
Faculty of Agriculture of Perugia
06126 S. Costanzo
Perugia
Italy

Polly D Courtney
Department of Food Science and Technology
Ohio State University
2121 Fyffe Road
Columbus
OH 43210
USA

M A Cousin
Department of Food Science
Purdue University
West Lafayette
Indiana
47907–1160
USA

N D Cowell
Elstead
Godalming
Surrey
GU8 6HT
UK

Julian Cox
Department of Food Science and Technology
The University of New South Wales
Sydney
Australia

C Gerald Crawford
US Department of Agriculture
Agricultural Research Service
Eastern Regional Research Center
600 E. Mermaid Lane
Wyndmoor
PA 19038
USA

Theresa L Cromeans
Department of Environmental Sciences and Engineering
School of Public Health
University of North Carolina
North Carolina
USA

Kofitsyo S Cudjoe
Department of Pharmacology
Microbiology and Food Hygiene
Norwegian College of Veterinary Medicine
PO Box 8146 Dep
0033 Oslo
Norway

David Cunliffe
Macromolecular Science Department
Institute of Food Research
Reading Laboratory
Earley Gate, Whiteknights Road
Reading
RG6 6BZ
UK

Ladislav Čurda
Department of Dairy and Fat Technology
Prague Institute of Chemical Technology
Czech Republic

G J Curiel
Unilever Research Vlaardingen
PO Box 114
3130 AC Vlaardingen
The Netherlands

G D W Curtis
Bacteriology Department
John Radcliffe Hospital
Oxford
UK

Michael K Dahl
Department of Microbiology
University of Erlangen
Staudtstrasse 5
91058 Erlangen
Germany

Crispin R Dass
The Heart Research Institute Ltd
145 Missenden Road
Camperdown
Sydney
NSW 2050
Australia

E Alison Davies
Technical Services & Research Department
Aplin & Barrett Ltd (Cultor Food Science)
15 North Street
Beaminster
Dorset
DT8 3DZ
UK

Brian P F Day
Campden and Chorleywood Food Research Association
Chipping Campden
Gloucestershire
GL55 6LD
UK

J M Debevere
Laboratory of Food Microbiology and Food Preservation
Faculty of Agricultural and Applied Biological Sciences
University of Ghent
Coupure Links 654
9000 Ghent
Belgium

Joss Delves-Broughton
Technical Services and Research Department
Aplin & Barrett Ltd (Cultor Food Science)
15 North Street
Beaminster
Dorset
DT8 3DZ
UK

Stephen P Denyer
Department of Pharmacy
The University of Brighton
Cockcroft Building
Moulescoomb
Brighton
BN2 4GJ
UK

P M Desmarchelier
Food Safety and Quality
Food Science Australia
PO Box 3312
Tingalpo DC
Queensland 4173
Australia

Janice Dewar
CSIR Food Science and Technology
PO Box 395
Pretoria 001
South Africa

Vinod K Dhir
Biotec Laboratories Ltd
32 Anson Road
Martlesham Heath
Ipswich
Suffolk
IP5 3RD
UK

M W Dick
Department of Botany
University of Reading
Reading
RG6 6AU
UK

Vivian M Dillon
Department of Biology and Biochemistry
University of Bath
Bath
UK

Eleftherios H Drosinos
Department of Food Science and Technology
Laboratory of Microbiology and Biotechnology of Foods
Agricultural University of Athens
Iera Odos 75
Athens
Greece

F M Dugan
USDA–ARS Western Regional Plant Introduction Station
Washington State University
Washington
USA

B Egan
Marine Biological and Chemical Consultants Ltd
Bangor
UK

H M J van Elijk
Unilever Research Vlaardingen
PO Box 114
3130 AC Vlaardingen
The Netherlands

Hartmut Eisgruber
Institute for Hygiene and Technology of Foods of Animal Origin, Veterinary Faculty
Ludwig-Maximilians University
80539 Munich
Germany

Phyllis Entis
QA Life Sciences, Inc.
6645 Nancy Ridge Drive
San Diego
CA 92121
USA

John P Erickson
Microbiology – Research and Development
Bestfoods Technical Center
Somerset
New Jersey
USA

Douglas E Eveleigh
Department of Microbiology
Rutgers University
Cook College
76 Lipman Drive
New Brunswick
NJ 08901-8525
USA

Richard R Facklam
Streptococcus Laboratory
Respiratory Diseases Branch
Division of Bacterial and Mycotic Diseases
Centres for Disease Control and Prevention
Mail Stop CO-2
Atlanta
GA 30333
USA

M Fandke
BioteCon Gesellschaft für Biotechnologische Entwicklung und Consulting
Hermannswerder Haus 17
14473 Potsdam
Germany

Nana Y Farkye
Dairy Products Technology Center
Dairy Science Department
California Polytechnic State University
San Luis Obispo
CA 93407
USA

Manuel Fidalgo
Departmento de Genética
Facultad de Ciencias, Universidad de Málaga
Spain

Christopher W Fisher
Department of Food Science and Human Nutrition
University of Illinois
Urbana
IL 61801
USA

G H Fleet
CRC for Food Industry Innovation
Department of Food Science and Technology
The University of New South Wales
Sydney
New South Wales 2052
Australia

Harry J Flint
Rowett Research Institute
Greenburn Road
Bucksburn
Aberdeen
UK

Samuel Formal
Department of Microbiology and Immunology
Uniformed Services University of the Health Sciences
F Edward Hébert School of Medicine
4301 Jones Bridge Road
Bethesda
MD 20814
USA

Pina M Fratamico
US Department of Agriculture
Agricultural Research Service
Eastern Regional Research Center
600 E. Mermaid Lane
Wyndmoor
PA 19038
USA

Colin Fricker
Thames Water Utilities
Manor Farm Road
Reading
RG2 0JN
UK

Daniel Y C Fung
Department of Animal Sciences and Industry
Kansas State University
Manhattan
Kansas 66506
USA

H Ray Gamble
United States Department of Agriculture
Agricultural Research Service
Parasite Biology and Epidemiology Laboratory
Building 1040, Room 103, BARC-East
Beltsville
MD 20705
USA

Indrawati Gandjar
Department of Biology
Faculty of Science and Mathematics
University of Indonesia
Jakarta
Indonesia

Mariano García-Garibay
Departamento de Biotechnología
Universidad Autónoma Metropolitana
Iztapalapa, Apartado Postal 55–535
Mexico City 09340
Mexico

María-Luisa García-López
Department of Food Hygiene and Food Technology
University of León
24071-León
Spain

S K Garg
Department of Microbiology
Dr Ram Manohar Lohia Avadh University
Faizabad 224 001
India

A Gasch
BioteCon Gesellschaft für Biotechnologische
Entwicklung und Consulting
Hermannswerder Haus 17
14473 Potsdam
Germany

Michel Gautier
Ecole Nationale Supérieure d'Agronomie
Institut National de la Recherche Agronomique
65 rue de SrBrieuc
35042
Rennes cédex
France

Gerd Gellissen
Rhein Biotech GmbH
EichsFelder Str. 11
40595 Düsseldorf
Germany

N P Ghildyal
Fermentation Technology and Bioengineering
Department
Central Food Technological Research Institute
Mysore 570013
India

M Gibert
Institut Pasteur
Unité Interactions Bactéries Cellules
28 rue du Dr Roux
75724 Paris
Cedex 15
France

Glenn R Gibson
Microbiology Department
Institute of Food Research
Reading
UK

M C te Giffel
Wageningen Agricultural University
Laboratory of Food Microbiology
Bomenweg 2
NL 6703 HD Wageningen
The Netherlands

A Gilmour
Food Science Division (Food Microbiology)
Department of Agriculture for Northern Ireland
Agriculture and Food Science Centre
Newforge Lane
Belfast
BT9 5PX
Northern Ireland, UK

Giorgio Giraffa
Istituto Sperimentale Lattiero Caseario
Via A. Lombardo
11 – 26900 Lodi
Italy

R W A Girdwood
Scottish Parasite Diagnostic Laboratory
Stobhill Hospital
Glasgow
GL21 3UW
UK

Andrew D Goater
Institute of Molecular and Biomolecular Electronics
University of Wales
Dean St
Bangor
Gwynedd
LL57 1UT
UK

Marco Gobbetti
Instituto di Produzioni e Preparazioni Alimentari
Facoltà di Agraria di Foggia
Via Napoli 25
71100 Foggia
Italy

Millicent C Goldschmidt
Department of Basic Sciences
Dental Branch
The University of Texas Health Center at Houston
6516 John Freeman Avenue
Houston
Texas 77030
USA

Lorena Gómez-Ruiz
Departamento de Biotechnología
Universidad Autónoma Metropolitana
Iztapalapa, Apartado Postal 55–535
Mexico City 09340
Mexico

Katsuya Gomi
Division of Life Science
Graduate School of Agricultural Science
Tohoku University
Japan

M Marcela Góngora-Nieto
Biological Systems Engineering
Washington State University
Pullman
Washington 99164–6120
USA

S Gonzalez
Universidad Nacional de Tucumán, Argentina
Cerela–Conicet
San Miguel de Tucumán
Argentina

Silvia N Gonzalez
Centro de Referencia para Lactobacilos (Cerela) and
Universidad Nacional de Tucumán
Chacabuco 145 (4000)
Tucumán
Argentina

Leon G M Gorris
Unit Microbiology and Preservation
Unilever Research Vlaardingen
PO Box 114
3130 AC Vlaardingen
The Netherlands

Grahame W Gould
17 Dove Road
Bedford
MK41 7AA
UK

M K Gowthaman
Fermentation Technology and Bioengineering
Department
Mysore 570013
India

Lone Gram
Danish Institute for Fisheries Research
Department of Seafood Research
Technical University of Denmark Bldg 221
DK-2800 Lyngby
Denmark

AGE Griffioen
Stichting EFFI
PO Box 553 Wageningen
The Netherlands

Mansel W Griffiths
Department of Food Science
University of Guelph
Guelph
Ontario
N1G 2W1
Canada

C Grönewald
BioteCon Gesellschaft für Biotechnologische
Entwicklung und Consulting
Hermannswerder Haus 17
14473 Potsdam
Germany

Isabel Guerrero
Universidad Autónoma Metropolitana-Iztapalapa
Mexico
Apartado Postal 55–535
CP 09340 Mexico DF
Mexico

G C Gürakan
Middle East Technical University
Ankara
Turkey

Carlos Horacio Gusils
Centro de Referencia para Lactobacilos and Universidad
Nacional de Tucumán
Casilla de Correo 211
(4000)-Tucuman
Argentina

Thomas S Hammack
US Food and Drug Administration
Washington, DC 20204
USA

S A S Hanna,
48 Kensington Street
Newton
MA 02460
USA

Karen M J Hansen
Saskatchewan Food Product Innovation Program
University of Saskatchewan
Saskatoon
SK
S7N 5A8
Canada

J Harvey
Food Science Division (Food Microbiology)
Department of Agriculture for Northern Ireland
Agriculture and Food Science Centre
Newforge Lane
Belfast
BT9 5PX
Northern Ireland, UK

Wilma C Hazeleger
Wageningen University and Research Centre
Department of Food Technology and Nutritional Sciences
Bomenweg 2
NL 6703 HD Wageningen
The Netherlands

G M Heard
CRC for Food Industry Innovation
Department of Food Science and Technology
University of New South Wales
Sydney
New South Wales 2052
Australia

Nidal Hilal
Biochemical Engineering Group
Centre for Complex Fluids Processing
Department of Chemical and Biological Process Engineering
University of Wales Swansea
Singleton Park
Swansea SA2 8PP
UK

G Hildebrandt
Institute for Food Hygiene
Free University of Berlin
Germany

Colin Hill
Department of Microbiology and National Food Biotechnology Centre
University College
Cork
Ireland

A D Hitchins
Center for Food Safety and Applied Nutrition
Food and Drug Administration
Washington, DC
USA

Jill E Hobbs
George Morris Centre
345, 2116 27th Avenue NE
Calgary
Alberta
T2E 7A6
Canada

Ailsa D Hocking,
CSIRO Food Science Australia
Riverside Corporate Park
North Ryde
New South Wales 2113
Australia

Cornelis P Hollenberg
Institut für Microbiology
Heinrich-Heine-Universität Düsseldorf
40225 Düsseldorf
Germany

Richard A Holley
Department of Food Science
University of Manitoba
Winnipeg
Manitoba
R3T 2N2
Canada

Wilhelm H Holzapfel
Institute of Hygiene and Toxicology
Federal Research Centre for Nutrition
Bundesforschungsanstalt
Haid-und-Neu-Str. 9
D-7613 Karlsruhe
Germany

Rolf K Hommel
Cell Technologie Leipzig
Fontanestr. 21
Leipzig
D-04289
Germany

Dallas G Hoover
Department of Animal and Food Sciences
University of Delaware
Newark
DE 19717–1303
USA

Thomas W Huber
Medical Microbiology and Immunology
Texas A&M College of Medicine
Temple
Texas
USA

Robert Hutkins
Department of Food Science and Technology
University of Nebraska
338 FIC
Lincoln
NE 68583–0919
USA

Cheng-An Hwang
Nestlé Research and Development Center
210 Housatonic Avenue
New Milford
Connecticut
USA

John J Iandolo
Department of Microbiology and Immunology
University of Oklahoma Health Sciences Center
Oklahoma City
OK 73190
USA

Y Iimura
Department of Applied Chemistry and Biotechnology
Yamanashi University
Kofu
Japan

Charlotte Nexmann Jacobsen
Department of Dairy and Food Research
Royal Veterinary and Agricultural University
Rolighedsvej 3,0
1958 Frederiksberg C
Denmark

Mogens Jakobsen
Department of Dairy and Food Research
Royal Veterinary and Agricultural University
Rolighedsvej 3,0
1958 Frederiksberg C
Denmark

Dieter Jahn
Institute for Organic Chemistry and Biochemistry
Albert Ludwigs University Freiburg
Albertstr. 21
79104 Freiburg
Germany

B Jarvis
Ross Biosciences Ltd
Daubies Farm
Upton Bishop
Ross-on-Wye
Herefordshire
HR9 7UR
UK

Ian Jenson
Gist-brocades Australia Pty, Ltd
Moorebank
NSW
Australia

Juan Jimenez
Departmento de Genética
Facultad de Ciencias, Universidad de Málaga
Spain

Karen C Jinneman
Department of Veterinary Science and Microbiology
University of Arizona
Tucson
AZ 85721
USA

Juan Jofre
Department of Microbiology
University of Barcelona
Spain

Eric Johansen
Department of Genetics and Microbiology
Chr. Hansen A/S
10–12 Bøge Allé
DK-2970
Hørsholm
Denmark

Nick Johns
Independent Research Consultant
15 Collingwood Close
Steepletower
Hethersett
Norwich NR9 3QE
UK

Eric A Johnson
Department of Food Microbiology
Food Research Institute, University of Wisconsin
Madison
WI
USA

Clifford H Johnson
US Environmental Protection Agency
Cincinatti
Ohio
USA

Rafael Jordano
Department of Food Science and Technology
Campus Rabanales, University of Córdoba
E-14071 Córdoba
Spain

Richard Joseph
Department of Food Microbiology
Central Food Technological Research Institute
Mysore
570 013
India

Vinod K Joshi
Department of Post-harvest Technology
Dr YSP University of Horticulture and Foresty
Nauni
Solan-173 230
India

Vijay K Juneja
United States Department of Agriculture
Eastern Regional Research Center
600 East Mermaid Lane
Wyndmoor
Pennsylvania
USA

G Kalantzopoulos
Department of Food Science and Technology
Agricultural University of Athens
Greece

Chitkala Kalidas
Field of Microbiology
Department of Food Science
Cornell University
Ithaca NY 14853
USA

A Kambamanoli-Dimou
Department of Animal Production
Technological Education Institute
Larissa
Greece

Peter Kämpfer
Institut für Angewandte Mikrobiologie
Justus-Liebig-Universität Giessen
Senckenbergstr. 3
D-35390 Giessen
Germany

N G Karanth
Fermentation Technology and Bioengineering
Department
Mysore 570013
India

Embit Kartadarma
Department of Food Science and Technology
The University of New South Wales
Sydney
Australia

K L Kauppi
University of Minnesota
Department of Food Science and Nutrition
St Paul
USA

C A Kaysner
US Food and Drug Administration
22201 23rd Drive SE
Bothell
Washington 98021
USA

William A Kerr
Department of Economics
University of Calgary
2500 University Drive NW
Calgary
Alberta
T2N 1N4
Canada

Tajalli Keshavarz
Department of Biotechnology
University of Westminster
115 New Cavendish Street
London
W1M 8JS
UK

George G Khachatourians
Department of Applied Microbiology and Food Science
University of Saskatchewan
Saskatoon
Canada

W Kim
Department of Microbiology
University of Georgia
Athens
Georgia
USA

P M Kirk
CABI Bioscience UK Centre (Egham)
Bakeham Lane
Egham
Surrey
TW20 9TY

Todd R Klaenhammer
Departments of Food Science and Microbiology
Southeast Dairy Foods Research Center
Box 7624
North Carolina State University
Raleigh
NC 27695–7624
USA

Hans-Peter Kleber
Institut für Biochemie
Fakultöt für Biowissenschaften
Pharmazie und Psychologie
Universität Leipzig
Talstr. 33
Leipzig
D-04103
Germany

Thomas J Klem
Department of Food Science
Cornell University
USA

Wolfgang Kneifel
Department of Dairy Research and Bacteriology
Agricultural University
Gregor Mendel-Str. 33
A-1180 Vienna
Austria

Barb Kohn
VICAM LP
313 Pleasant Street
Watertown
MA 02172
USA

C Koob
BioteCon Gesellschaft für Biotechnologische
Entwicklung und Consulting
Hermannswerder Haus 17
14473 Potsdam
Germany

P Kotzekidou
Department of Food Science and Technology
Faculty of Agriculture
Aristotle University of Thessaloniki
PB 250
GR 540 06
Thessaloniki
Greece

K Krist
Meat and Livestock Australia
Sydney
Australia

Pushpa R Kulkarni
University Department of Chemical Technology
University of Mumbai
Matunga
Mumbai 400 019
India

Madhu Kulshreshtha
Division of Plant Pathology
Indian Agricultural Research Institute
New Delhi 11012
India

Susumu Kumagai
Department of Biomedical Food Research
National Institute of Infectious Diseases
Toyama 1–23–1
Shinjuku-ku
Tokyo 162–8640
Japan

G Lagarde
Inovatech Europe B.V.
Landbouwweg
The Netherlands

Keith A Lampel
US Food and Drug Administration
Center for Food Safety and Applied Nutrition HFS-327
200 C St SW
Washington
DC 20204
USA

S Leaper
Campden and Chorleywood Food Research Association
Chipping Campden
Gloucestershire
GL55 6LD
UK

J David Legan
Microbiology Department
Nabisco Research
PO Box 1944
DeForest Avenue
East Hanover
NJ 017871
USA

J J Leisner
Department of Veterinary Microbiology
Royal Veterinary and Agricultural University
Stigbøjlen 4
DK-1870 Frederiksberg C
Denmark

H L M Lelieveld
Unilever Research Vlaardingen
PO Box 114
3130 AC Vlaardingen
The Netherlands

D F Lewis
Food Systems Division
SAC
Auchincruive
Ayr KA6 5HW
Scotland
UK

M J Lewis
Department of Food Science and Technology
University of Reading
UK

E Litopoulou-Tzanetaki
Department of Food Science, Faculty of Agriculture
Aristotle University of Thessaloniki
54006
Thessaloniki
Greece

Aline Lonvaud-Funel
Faculty of Œnology
University Victor Segalen Bordeaux 2
351, Cours de la Libération
33405 Talence Cedex
France

S E Lopez
Departamento de Ciencias Biológicas
Facultad de Ciencias Exactas y Naturales
Pabellon II 4to piso – Ciudad Universitaria
1428 Buenos Aires
Argentina

G Love
Centre for Electron Optical Studies
University of Bath
Claverton Down
Bath
BA2 7AY
UK

Robert W Lovitt
Biochemical Engineering Group
Centre for Complex Fluids Processing
Department of Chemical and Biological Process Engineering
University of Wales Swansea
Singleton Park
Swansea
SA2 8PP
UK

Majella Maher
National Diagnostics Centre
National University of Ireland
Galway
Ireland

R H Madden
Food Microbiology
Food Science Department
Department of Agriculture for Northern Ireland and Queen's University of Belfast
Newforge Lane
Belfast
BT9 5PX
Northern Ireland

T Mahmutoğlu
TATKO TAS
Gayrettepe
Istanbul
Turkey

K A Malik
Chairman
Pakistan Agricultural Research Council
Islamabad
Pakistan

Miguel Prieto Maradona
Department of Food Hygiene and Food Technology
University of León
24071-León
Spain

Scott E Martin
Department of Food Science and Human Nutrition
University of Illinois
486 Animal Sciences Laboratory
1207 West Gregory Drive
Urbana
IL 61801
USA

L Martínková
Laboratory of Biotransformation
Institute of Microbiology
Academy of Sciences of the Czech Republic
Prague
Czech Republic

Tina Mattila-Sandholm
VTT Biotechnology and Food Research
Tietotie 2
Espoo
PO Box 1501
FIN-02044 VTT
Finland

D A McDowell
Food Studies Research Unit
University of Ulster at Jordanstown
Shore Road
Newtownabbey
Co. Antrim
Northern Ireland
BT37 9QB

Denise N McKenna
Microbiology – Research and Development
Bestfoods Technical Center
Somerset
New Jersey
USA

M A S McMahon
Food Studies Research Unit
University of Ulster at Jordanstown
Shore Road
Newtownabbey
Co. Antrim
Northern Ireland
BT37 9QB

T A McMeekin
School of Agricultural Science
University of Tasmania
Hobart
Australia

Luis M Medina
Department of Food Science and Technology
Campus Rabanales
University of Córdoba
E-14071 Córdoba
Spain

Aubrey F Mendonca
Iowa State University
Department of Food Science and Human Nutrition
Ames
Iowa
USA

James W Messer
US Environmental Protection Agency
Cincinnati
Ohio
USA

M C Misra
Fermentation Technology and Bioengineering Department
Central Food Technological Research Institute
Mysore 570013
India

Vikram V Mistry
Dairy Science Department
South Dakota State University
Brookings
South Dakota 57007
USA

D R Modi
Department of Microbiology
Dr Ram Manohar Lohia Avadh University
Faizabad 224 001
India

Richard J Mole
Biotec Laboratories Ltd.
32 Anson Road
Martlesham Heath
Ipswich
Suffolk
IP5 3RD
UK

M C Montel
Station de Recherches sur la Viande
INRA
63122 Saint Genès Champanelle
France

M Moresi
Istituto di Tecnologie Agroalimentari
Università della Tuscia
Via S C de Lellis
01100 Viterbo
Italy

André Morin
Imperial Tobacco Limited
3810 rue St-Antoine
Montreal
Quebec H4C 1B5
Canada

Maurice O Moss
School of Biological Sciences, University of Surrey
Guildford
GU2 5XH
UK

M A Mostert
Unilever Research Vlaardingen
PO Box 114
3130 AC Vlaardingen
The Netherlands

Donald Muir
Hannah Research Institute
Ayr
KA6 5HL
Scotland, UK

Maite Muniesa
Department of Microbiology
University of Barcelona
Spain

E A Murano
Center for Food Safety and Department of Animal Science
Texas A&M University
Texas
USA

M J Murphy
CBD Porton Down
Salisbury
SP4 0JQ
UK

K Darwin Murrell
Agricultural Research Service
US Department of Agriculture
Beltsville
Maryland 20705
USA

C K K Nair
Radiation Biology Division
Bhabha Atomic Research Centre
Mumbai 400 085
India

Motoi Nakao
Horiba Ltd
Miyanohigashimachi
Kisshoin
Minami-ku
Kyoto
Japan
601–8510

A W Nichol
Charles Sturt University
NSW
Australia

D S Nichols
School of Agricultural Science
University of Tasmania
Hobart
Australia

Poonam Nigam
Biotechnology Research Group
School of Applied Biological and Chemical Sciences
University of Ulster
Coleraine BT52 1SA
UK

M de Nijs
TNO Nutrition and Food Research Institute
Division of Microbiology and Quality Management
PO Box 360
3700 AJ Zeist
The Netherlands

S H W Notermans
TNO Nutrition and Food Research Institute
PO Box 360
3700 AJ Zeist
The Netherlands

Martha Nuñez
Centro de Referencia par Lactobacilos (Cerela)
Chacabuco 145 (4000)
Tucumán
Argentina

George-John E Nychas
Department of Food Science and Technology
Laboratory of Microbiology and Biotechnology of Foods
Agricultural University of Athens
Iera Odos 75
Athens
11855
Greece

R E O'Connor-Shaw
Food Microbiology Consultant
Birkdale
Queensland
Australia

Louise O'Connor
National Diagnostics Centre
National University of Ireland
Galway
Ireland

Triona O'Keeffe
Department of Microbiology and National Food Biotechnology Centre
University College
Cork
Ireland

Rachel M Oakley
United Biscuits (UK Ltd)
High Wycombe
Buckinghamshire
HP12 4JX
UK

Yuji Oda
Department of Applied Biological Science
Fukuyama University
Fukuyama
Hiroshima 729–0292
Japan

Lucy J Ogbadu
Department of Biological Sciences
Benue State University
Makurdi
Nigeria

Guillermo Oliver
Centro Referencia para Lactobacilos and Universidad Nacional de Tucumán
Casilla de Correo 211
(4000)-Tucuman
Argentina

Ynes R Ortega
Seafood Products Research Center
US Food and Drug Administration
PO Box 3012
22201 23rd Drive SE
Bothell
WA 98041–3012
USA

Andrés Otero
Department of Food Hygiene and Food Technology
University of León
24071-León
Spain

Kozo Ouchi
Kyowa Hakko Kogyo Co. Ltd
1–6–1 Ohtemachi
Chiyoda-ku
Tokyo 100–8185
Japan

Barbaros H Özer
Department of Food Science and Technology
Faculty of Agriculture
University of Harran
63040
Şanlıurfa
Turkey

Dilek Özer
GAP Regional Development Administration
Şanlıurfa
Turkey

J Palacios
Universidad Nacional de Tucumán, Argentina
Cerela-Conicet
San Miguel de Tucumán
Argentina

Ashok Pandey
Laboratorio de Processos Biotecnologicos
Universidade Federal do Parana
Departmento de Engenharia Quimica
CEP 81531-970 Curitiba-PR
Brazil

Photis Papademas
Department of Food Science and Technology
University of Reading
Whiteknights
Reading
Berkshire
RG6 6AP
UK

A Pardigol
BioteCon Gesellschaft für Biotechnologische Entwicklung und Consulting
Hermannswerder Haus 17
14473 Potsdam
Germany

E Parente
Dipartimento di Biologia, Difesa e Biotecnologie Agro-Forestali
Università della Basilicata
Via N Sauro 85
85100 Potenza
Italy

Zahida Parveen
University of Huddersfield
Department of Chemical and Biological Sciences
Queensgate
Huddersfield
HD1 3DH
UK

P Patáková
Faculty of Food and Biochemical Technology
Institute of Chemical Technology
Prague
Czech Republic

Pradip Patel
Science and Technology Group
Leatherhead Food Research Association
Randalls Road
Leatherhead
Surrey
KT22 7RY
UK

Margaret Patterson
Food Science Division
Department of Agriculture for Northern Ireland and The Queen's University of Belfast
Agriculture and Food Science Centre
Newforge Lane
Belfast
BT9 5PX
UK

P A Pawar
Fermentation Technology and Bioengineering Department
Central Food Technological Research Institute
Mysore 570013
India

Janet B Payeur
National Veterinary Services Laboratories
Veterinary Services
Animal and Plant Health Inspection Service
Department of Agriculture
1800 Dayton Road
Ames
IA 50010
USA

Gary A Payne
Department of Plant Pathology
North Carolina State University
Raleigh
North Carolina
USA

Ron Pethig
Institute of Molecular and Biomolecular Electronics
University of Wales
Dean St
Bangor
Gwynedd
LL57 1UT
UK

L Petit
Unité Interactions Bactéries Cellules
Institut Pasteur
28 rue du Dr Roux
75724 Paris
Cedex 15
France

William A Petri Jr
Department of Medicine, Division of Infectious Diseases
University of Virginia Health Sciences Center
MR4, Room 2115, 300 Park Place
Charlottesville
VA 22908
USA

M R A Pillai
Isotope Division
Bhabha Atomic Rsearch Centre
Mumbai 400 085
India

D W Pimbley
Leatherhead Food Research Association
Randalls Road
Leatherhead
Surrey
KT22 7RY
UK

J I Pitt
CSIRO Food Science Australia
Riverside Corporate Park
North Ryde
New South Wales 2113
Australia

M R Popoff
Institut Pasteur
Unité Interactions Bactéries Cellules
28 rue du Dr Roux
75724 Paris
Cedex 15
France

U J Potter
Centre for Electron Optical Studies
University of Bath
Claverton Down
Bath
BA2 7AY
UK

B Pourkomailian
Department of Food Safety and Preservation
Leatherhead Food RA
Randalls Road
Surrey
UK

K Prabhakar
Department of Meat Science and Technology
College of Veterinary Science
Tirupati 517 502
India

W Praphailong
National Center for Genetic Engineering and Biotechnology
Rajdhevee
Bangkok
Thailand

M S Prasad
Fermentation Technology and Bioengineering Department
Mysore 570013
India

J. C du Preez
Department of Microbiology and Biochemistry
University of the Orange Free State
PO Box 339
Bloemfontein 9300
South Africa

Barry H Pyle
Montana State University
Bozeman
Montana
USA

Laura Raaska
VTT Biotechnology and Food Research
PO Box 1501
FIN-02044 VTT
Finland

Moshe Raccach
Food Science Program
School of Agribusiness and Resource Management
Arizona State University East
Mesa
Arizona 85206–0180
USA

Fatemeh Rafii,
Division of Microbiology
National Center for Toxicological Research, US FDA
Jefferson
AR
USA

M I Rajoka,
National Institute for Biotechnology and Genetic Engineering (NIBGE)
PO Box 577
Faisalabad
Pakistan

Javier Raso
Biological Systems Engineering
Washington State University
Pullman
Washington 99164-6120
USA

K S Reddy
Department of Meat Science and Technology
College of Veterinary Science
Tirupati 517 502
India

S M Reddy
Department of Botany
Kakatiya University
Warangal
506 009
India

Wim Reybroeck
Department for Animal Product Quality and Transformation Technology
Agricultural Research Centre CLO-Ghent
Melle
Belgium

V G Reyes
Food Science Australia
Private Bag 16
Sneydes Road
Werribee
Victoria
VIC 3030
Australia

E W Rice
US Environmental Protection Agency
Cincinnati
Ohio 45268
USA

Jouko Ridell
Department of Food and Environmental Hygiene, Faculty of Veterinary Medicine
University of Helsinki
Finland

R K Robinson
Department of Food Science
University of Reading
Whiteknights
Reading
Berkshire RG6 6AP
UK

Hubert Roginski
Gilbert Chandler College
The University of Melbourne
Sneydes Road
Werribee
Victoria
3030
Australia

Alexandra Rompf
Institute for Organic Chemistry and Biochemistry
Albert Ludwigs University Freiburg
Albertstr. 21
79104 Freiburg
Germany

T Ross
School of Agricultural Science
University of Tasmania
Hobart
Australia

T Roukas
Department of Food Science and Technology
Aristotle University of Thessaloniki
Greece

M T Rowe
Food Microbiology
Food Science Department
Department of Agriculture for Northern Ireland and Queen's University of Belfast
Newforge Lane
Belfast
BT9 5PX
Northern Ireland

W Michael Russell
Departments of Food Science and Microbiology
Southeast Dairy Foods Research Center
Box 7624
North Carolina State University
Raleigh
NC 27695–7624
USA

G Salvat
AFSSA
Ploufragan
France

R Sandhir
Department of Biochemistry
Dr Ram Manohar Lohia Avadh University
Faizabad 224 001
India

Robi C Sandlin
Department of Microbiology and Immunology
Uniformed Services University of the Health Sciences
F Edward Hébert School of Medicine
4301 Jones Bridge Road
Bethesda
MD 20814
USA

Jesús-Angel Santos
Department of Food Hygiene and Food Technology
University of León
24071-León
Spain

A K Sarbhoy
Division of Plant Pathology
Indian Agricultural Research Institute
New Delhi 110012
India

David Sartory
Severn Trent Water
Shrewsbury
UK

Joanna M Schaenman
Department of Medicine
Division of Infectious Diseases
University of Virginia Health Sciences Center
MR4, Room 2115, 300 Park Place
Charlottesville
VA 22908
USA

Barbara Schalch
Institute of Hygiene and Technology of Food of Animal Origin
Ludwig-Maximilians-University Munich
Veterinary Faculty
Veterinärstr. 13
81369 Munich
Germany

P Scheu
BioteCon Gesellschaft für Biotechnologische Entwicklung und Consulting
Hermannswerder Haus 17
14473 Potsdam
Germany

Bernard W Senior
Department of Medical Microbiology
University of Dundee Medical School
Ninewells Hospital
Dundee
DD1 9SY
UK

Gilbert Shama
Department of Chemical Engineering
Loughborough University
UK

Arun Sharma
Food Technology Division
Bhabha Atomic Research Centre
Mumbai 400 085
India

M Shin
Faculty of Pharmaceutical Sciences
Kobe Gakuin University
Kobe
Japan

J Silva
Universidad Nacional de Tucumán, Argentina
Cerela–Conicet
San Miguel de Tucumán
Argentina

Dalel Singh
Microbiology Department
CCS Haryana Agricultural University
Hisar
125 004
India

Rekha S Singhal
University Department of Chemical Technology
University of Mumbai
Matunga
Mumbai 400 019
India

Emanuele Smacchi
Institute of Industrie Agranie (Microbiologia)
Faculty of Agriculture of Perugia 06126 S. Constanzo
Perguia
Italy

Christopher A Smart
Macromolecular Science Department
Institute of Food Research
Reading Laboratory
Earley Gate
Whiteknights Road
Reading R66 6BZ
UK

H V Smith
Scottish Parasite Diagnostic Laboratory
Stobhill Hospital
Glasgow
G21 3UW
Scotland, UK

O Peter Snyder
Hospitality Institute of Technology and Management
670 Transfer Road
Suite 21A
St Paul
MN 55114
USA

Mark D Sobsey
Department of Environmental Sciences and Engineering
School of Public Health
University of North Carolina
North Carolina
USA

Carlos R Soccol
Laboratorio de Processos Biotecnologicos
Departamento de Engenharia Quimica
Universidade Federal do Parana
CEP 81531–970
Curitiba-PR
Brazil

M El Soda
Department of Dairy Science and Technology
Faculty of Agriculture
Alexandria University
Alexandria
Egypt

R A Somerville
Neuropathogenesis Unit
Institute for Animal Health
West Mains Road
Edinburgh
EH9 3JF
UK

N H C Sparks
Department of Biochemistry and Nutrition
Scottish Agricultural College
Auchincruive
Ayr
Scotland

M Van Speybroeck
Laboratory of Industrial Microbiology and Biocatalysis
Department of Biochemical and Microbial Technology
Faculty of Agricultural and Applied Biological Sciences
University of Gent
Coupure links 653
B-9000
Gent
Belgium

D J Squirrell
CBD Porton Down
Salisbury
SP4 0JQ
UK

E Stackebrandt
DSMZ – German Collection of Microorganisms and Cell Cultures
Brunswick
Germany
Deutsche Sammlung von Mikroorganisem und Mascheroder
Weg 1 B
38124, Braunschweig
Germany

Jacques Stark
Gist-brocades Food Specialties
R&D
Delft
The Netherlands

Colin S Stewart
Rowett Research Institute
Greenburn Road
Bucksburn
Aberdeen
UK

G G Stewart
International Centre for Brewing and Distilling
Heriot-Watt University
Riccarton
Edinburgh
Scotland
EH14 4AS
UK

Gordon S A B Stewart (dec)
Department of Pharmaceutical Sciences
The University of Nottingham
University Park
Nottingham
NG7 2RD
UK

Duncan E S Stewart-Tull
University of Glasgow
Glasgow
G12 8QQ
UK

A Stolle
Institute of Hygiene and Technology of Food of Animal Origin
Ludwig-Maximilians-University Munich
Veterinary Faculty
Veterinärstr. 13
81369
Munich
Germany

Liz Straszynski
Alcontrol Laboratories
Bradford
UK

M Stratford
Microbiology Section
Unilever Research
Colworth House
Sharnbrook
Bedfordshire
MK44 1LQ
UK

M Surekha
Department of Botany
Kakatiya University
Warangal
506 009
India

B C Sutton
Apple Tree Cottage
Blackheath
Wenhaston
Suffolk
IP19 9HD
UK

Barry G Swanson
Food Science and Human Nutrition
Washington State University
Pullman
Washington 99164–6376
USA

Jyoti Prakash Tamang
Microbiology Research Laboratory
Department of Botany
Sikkim Government College
Gangtok
Sikkim 737 102
India

A Y Tamime
Scottish Agricultural College
Auchincruive
Ayr
UK

J S Tang
American Type Culture Collection
10801 University Blvd
Manassas
VA 20110-2209
USA

Chrysoula C Tassou
National Agricultural Research Foundation
Institute of Technology of Agricultural Products
S. Venizelou 1
Lycovrisi 14123
Athens
Greece

S R Tatini
University of Minnesota
Department of Food Science and Nutrition
1334 Eckles Ave
St Paul
MN 55108
USA

D M Taylor
Neuropathogenesis Unit
Institute for Animal Health
West Mains Road
Edinburgh
EH9 3JF
UK

John R N Taylor
Cereal Foods Research Unit
Department of Food Science
University of Pretoria
Pretoria 0002
South Africa

Lúcia Martins Teixeira
Departamento de Microbiologia Médica
Instituto de Microbiologia
Universidade Federal do Rio de Janeiro
Rio de Janeiro 21941
Brazil

Paula C M Teixeira
Escola Superior de Biotecnologia
Rua Dr António Benardino de Almeida
4200 Porto
Portugal

J Theron
Department of Microbiology and Plant Pathology
Faculty of Biological and Agricultural Sciences
University of Pretoria
Pretoria 0002
South Africa

Linda V Thomas
Aplin & Barrett Ltd
15 North Street
Beaminster
Dorset
DT8 3DZ
UK

Angus Thompson
Technical Centre
Scottish Courage Brewing Ltd
Sugarhouse Close
160 Canongate
Edinburgh
EH8 8DD
UK

Ulf Thrane
c/o Eastern Cereal and Oilseed Research Centre
K.W. Neatby Building, FM 1006,
Agriculture and Agri-Food Canada
Ottowa
Ontario K1A 0C6
Canada

Mary Lou Tortorello
National Center for Food Safety and Technology
US Food and Drug Administration
6502 South Archer Road
Summit-Argo
Illinois 60501
USA

Hau-Yang Tsen
Department of Food Science
National Chung Hsing University
Taichung
Taiwan
Republic of China

Nezihe Tunail
Department of Food Engineering
Faculty of Agriculture
University of Ankara
Dişkapì
Ankara
Turkey

D R Twiddy
Consultant Microbiologist
27 Guildford Road
Horsham
West Sussex
RH12 1LU
UK

N Tzanetakis
Department of Food Science
Faculty of Agriculture
Aristotle University of Thessaloniki
54006
Thessaloniki
Greece

C Umezawa
Faculty of Pharmaceutical Sciences
Kobe Gakuin University
Kobe
Japan

F Untermann
Institute for Food Safety and Hygiene
University of Zurich
Switzerland

Matthias Upmann,
Institute of Meat Hygiene
Meat Technology and Food Science
Veterinary University of Vienna
Veterinärplatz 1
A-1210 Vienna
Austria

Tümer Uraz
Ankara University
Faculty of Agriculture
Department of Dairy Technology
Ankara
Turkey

M R Uyttendaele
Laboratory of Food Microbiology and Food Preservation
Faculty of Agricultural and Applied Biological Sciences
University of Ghent
Coupure Links 654
9000 Ghent
Belgium

E J Vandamme
Laboratory of Industrial Microbiology and Biocatalysis
Department of Biochemical and Microbial Technology
Faculty of Agricultural and Applied Biological Sciences
University of Gent
Coupure links 653
B-9000
Gent
Belgium

P T Vanhooren
Laboratory of Industrial Microbiology and Biocatalysis
Department of Biochemical and Microbial Technology
Faculty of Agricultural and Applied Biological Sciences
University of Gent
Coupure links 653
B-9000
Gent
Belgium

L Le Vay
School of Ocean Sciences
University of Wales
Bangor
UK

P H In't Veld
National Institute of Public Health and the Environment
Microbiological Laboratory for Health Protection
PO Box 1
3720 BA Bilthoven
The Netherlands

Kasthuri Venkateswaran
Jet Propulsion Laboratory
National Aeronautics and Space Administration
Planetary Protection and Exobiology, M/S 89–2, 4800
Oakgrove Dr.
Pasadena
CA 91109
USA

V Venugopal
Food Technology Division
Bhabha Atomic Research Centre
Mumbai 400 085
India

Christine Vernozy-Rozand
Food Research Unit National Veterinary School
Lyon
France Ecole Nationale Véténaire de Lyon
France

B C Viljoen
Department of Microbiology and Biochemistry
University of the Orange Free State
Bloemfontein
South Africa

Birte Fonnesbech Vogel
Danish Institute for Fisheries Research
Department of Seafood Research
Technical University of Denmark Bldg 221
DK-2800 Lyngby
Denmark

Philip A Voysey
Microbiology Department
Campden and Chorleywood Food Research Association
Chipping Campden
Gloucestershire
GL55 6LD
UK

Martin Wagner
Institute for Milk Hygiene
Milk Technology and Food Science
University for Veterinary Medicine
Veterinärplatz 1
1210 Vienna
Austria

Graeme M Walker
Reader of Biotechnology
Division of Biological Sciences
School of Science and Engineering
University of Abertay Dundee
Dundee
DD1 1HG
Scotland

P Wareing
Natural Resources Institute
Chatham Maritime
Kent
ME4 4TB
UK

John Watkins
CREH Analytical
Leeds
UK

Ian A Watson
University of Glasgow
Glasgow
G12 8QQ
UK

Bart Weimer
Center for Microbe Detection and Physiology
Utah State University
Nutrition and Food Sciences
Logan
UT 84322–8700
USA

Irene V Wesley
Enteric Diseases and Food Safety Research
USDA, ARS, National Animal Disease Center
Ames IA 50010
USA

W B Whitman
Department of Microbiology
University of Georgia
Athens
Georgia
USA

Martin Wiedmann
Department of Food Science
Cornell University
Ithaca
NY 14853
USA

R C Wigley
Boghall House
Linlithgow
West Lothian
EH49 7LR
Scotland

R Andrew Wilbey
Department of Food Science
University of Reading
Whiteknights
Reading
UK

F Wilborn
BioteCon Gesellschaft für Biotechnologische
Entwicklung und Consulting
Hermannswerder Haus 17
14473 Potsdam
Germany

A G Williams
Hannah Research Institute
Ayr
KA6 5HL
UK

Alan Williams
Campden and Chorleywood Food Research Association
Chipping Campden
Gloucestershire GL55 6LD
UK

J F Williams
Department of Microbiology
Michigan State University
East Lansing
MI 48824
USA

Michael G Williams
3M Center
260–6B-01
St Paul
MN55144–1000
USA

Caroline L Willis
Public Health Laboratory Service
Southampton,
UK

F Y K Wong
Food Science Australia
Cannon Hill
Queensland
Australia

Brian J B Wood
Reader in Applied Microbiology
Dept. of Bioscience and Biotechnology
University of Strathclyde
Royal College Building
George Street
Glasgow
G1 1XW
Scotland

S D Worley
Department of Chemistry
Auburn University
Auburn
AL 36849
US

Atte von Wright
Department of Biochemistry and Biotechnology
University of Kuopio
PO Box 1627
FIN-70211 Kuopio
Finland

Chris J Wright
Biochemical Engineering Group
Centre for Complex Fluids Processing
Department of Chemical and Biological Process
Engineering
University of Wales Swansea
Singleton Park
Swansea
SA2 8PP
UK

Peter Wyn-Jones
Sunderland University
UK

Hideshi Yanase
Department of Biotechnology
Faculty of Engineering
Tottori University
4–101 Koyama-cho-minami
Tottori
Tottori 680–0945
Japan

Yeehn Yeeh
Institute of Basic Science
Inje University
Obang-dong
Kimhae 621–749
South Korea

Seyhun Yurdugül
Middle East Technical University
Department of Biochemistry
Ankara
Turkey

Klaus-Jürgen Zaadhof
Institute for Hygiene and Technology of Foods of Animal Origin
Veterinary Faculty
Ludwig-Maximilians University
80539 Munich
Germany

Gerald Zirnstein
Centers for Disease Control
GA
USA

Cynthia Zook
Department of Food Science and
University of Minnesota
St Paul
MN 55108
USA

CONTENTS

VOLUME 1

D

E

VOLUME 2

F

G

H

I

VOLUME 3

N

O

P

Q

R

S

T

NASBA *see* ***Listeria***: *Listeria monocytogenes* – Detection using NASBA (an Isothermal Nucleic Acid Amplification System).

Natamycin *see* **Preservatives**: Permitted Preservatives – Natamycin.

NATIONAL LEGISLATION, GUIDELINES & STANDARDS GOVERNING MICROBIOLOGY

Contents

Canada

Bruce E Brown, B. E. Brown Associates, Ottawa, Ontario, Canada

Introduction

The regulation of the microbiological safety and quality of foods in Canada operates in a complex jurisdictional context involving federal, provincial and municipal authorities and division of responsibilities between the federal departments of agriculture, fisheries and health. In addition, each of the 10 provinces have departments of agriculture and health that also regulate the microbiological safety and quality of food. In recent years under an initiative entitled 'The Canadian Food Inspection System' efforts are being made to harmonize legislation, regulations and guidelines and integrate inspectional services at the federal, provincial and municipal levels.

The Federal Acts and regulations involving the hygienic practices for production and manufacturing premises as well as the microbiological safety and quality of food products are administered by the departments of Health Canada, Agriculture and Agri-Food Canada, and Fisheries and Oceans Canada. The recent creation of the Canadian Food Inspection Agency which united the inspection resources of the three departments may well influence regulations, guidelines, etc. However, the mandate of each of the three departments to develop legislation and regulations and to designate enforcement policy remains unchanged.

Health Canada

Sections 4, 5, 6 and 7 of Part I of the Food and Drugs Act are the primary national legislation governing the overall safety and quality of food. The microbiological safety and quality regulations fall under Sections 4, 6 and 7.

Section 4 states that no person shall sell any article of food that: (a) has in or on it any poisonous or harmful substance; (b) is unfit for human consumption; (c) consists in whole or in part of any filthy, putrid, disgusting, rotten, decomposed or diseased animal or vegetable substance; (d) is adulterated; or (e) was manufactured, prepared, preserved, packaged or stored under unsanitary conditions. Foods containing pathogens in numbers that would constitute a direct hazard to health or their toxins would be considered not to be in compliance with Subsections 4(a) and 4(b) and possible 4(e). Spoilage (i.e. microbiological quality) can be considered to contravene Subsection 4(c). Subsection 4(e) and Section 7 deal with the hygienic conditions in which foods are processed, and opens the door for the sanitary inspection of premises where food is manufactured, prepared, preserved, packaged or stored. Subsection 6.1 permits the establishment of regulatory microbiological standards as being necessary to prevent injury to the health of the consumer or purchaser of the food. The term 'sell' is defined to include offer for sale, expose for sale, have in possession for sale and distribute, whether or not the distribution is made for consideration. 'Unsanitary conditions' means such conditions or circumstances as might contaminate with dirt or filth, or render injurious to health, a food, drug or cosmetic.

The Act also permits the establishment of regulations for carrying out the purposes and provisions of the Act, and examples of subject matter for regulations include:

- setting the sale or conditions of any food, drug, cosmetic or device
- prescribing standards of composition, strength, potency, purity, quality or other property of any article of food, drug, cosmetic or device
- respecting the importation of foods, drugs, cosmetics and devices in order to ensure compliance with the Act and regulations
- respecting the method of manufacture, preparation, preserving, packing, storing and testing of any food, drug, cosmetic or device in the interest of, or for the prevention of injury to, the health of the purchaser or consumer
- the keeping of books and records by persons who sell food, drugs, cosmetics or devices that are considered necessary for the proper enforcement and administration of the Act and regulations.

Microbiological Standards

The Food and Drug Regulations currently contain a number of regulatory microbiological standards. The standards have been developed on the basis of data gathered over the years as an aid to the administration of Sections 4 to 7 (inclusive) of the Act and relate to the microbiological safety and general cleanliness of food.

Most of the standards are specific to a microorganism or a group of microorganisms, while in others the organism is not specified but implied. There are two types of standards specific to microorganisms. One requires prohibition, that is zero tolerance, while the other permits some acceptable level. An example of a prohibition standard can be found in regulation B.08.014A which states that no person shall sell milk powder, whole milk powder, dry whole milk, powdered whole milk, skim milk powder or dry skim milk unless it is free from bacteria of the genus *Salmonella*, as determined by official method MFO-121, Microbiological Examination of Milk Powder, November 30, 1981. The official method is part of the regulation and specifies the method that must be used to establish compliance with the regulation, the sampling plan and compliance criteria (**Table 1**). Regulation B.08.016 which states that flavoured milks may contain not more than 50 000 total aerobic bacteria per cubic centimetre, is determined by official method MFO-7, Microbiological Examination of Milk, November 30, 1981 is an example of a standard in which an acceptable level is permitted.

The standards are classified with respect to three degrees of risk, referred to as Health 1, 2 and Sanitation. The degree of risk is reflected in the sampling plan and compliance criteria part of the official method. Two-class plans are used where there is a Health 1 risk (Table 1) and three-class plans for Health 2 and Sanitation risks (**Tables 2 and 3**). The sample size (n) designates the number of sample units to be taken and examined from a lot. The acceptance number (c) is the maximum allowable number of sample units that may exceed the level or concentration designated as acceptable, the m value. The lot is unacceptable and can be considered to be in violation of the respective regulatory standard when this number is exceeded. The acceptable concentration of microorganisms (m), expressed per gram or millilitre in a three-class plan separates sample units of acceptable microbiological quality from those classed as marginally acceptable, and in a two-class plan separates acceptable sample units from unacceptable. In a three-class plan unacceptable concentrations of microorganisms are represented by M which separates sample units of marginally acceptable quality from those of defective quality. The lot is unacceptable and in violation of the regulatory standard if one or more sample units exceed the M value.

Health 1 indicates that there is a direct risk to human health and appropriate action, for example a product recall, should be taken to limit exposure in the population. Follow-up action should ensure that the cause has been determined and appropriate corrective action has been taken. For Health 2 the hazard

Table 1 Microbiological standards for *Salmonella* considered as a Health 1 risk

Product	*Regulation*	*Method*	*Criteria*[a]		
			n	c	m
Chocolate	B.04.010	MFO-11	10	0	0
Cocoa	B.04.011	MFO-11	10	0	0
Milk powder	B.08.014	MFO-12	20	0	0
Egg products	B.22.033	MFO-6	6	0	0
Frog legs[b]	B.22.033	MFO-10	10	0	0

[a] *n*, sample size; *c*, acceptance number; *m*, acceptable concentration of microorganisms per gram.
[b] This regulation and hence standard is due to be repealed.

Table 2 Health 2 risk microbiological standards

Product	Microorganism	Regulation	Method	Criteria[a]			
				n	c	m	M
Cheese	*Escherichia coli*	B.08.048	MFO-14	5	2	10^2	2×10^3
made from pasteurized milk	*Staphylococcus aureus*			5	2	10^2	10^4
Cheese	*Escherichia coli*	B.08.048	MFO-14	5	2	5×10^2	2×10^3
made from unpasteurized milk	*Staphylococcus aureus*			5	2	10^3	10^4

[a] *n*, sample size; *c*, acceptance number; *m*, acceptable concentration of microorganisms per gram; *M*, unacceptable concentration of microorganisms per gram.

identified represents a risk to human health only if present in sufficient numbers and appropriate action, such as product recall, should be taken to limit exposure in the population to the product if the unacceptable level (*M* value) is exceeded. If the acceptance number (*c*) for levels between the acceptable level (*m*) and the unacceptable level (*M*) is exceeded, corrective action should be taken to bring about compliance. In Sanitation the hazard identified is an indication of a breakdown in hygienic practice. A review of the manufacturer's good hygienic practices (GHP) and/or the hazard analysis critical control point (HACCP) system is appropriate where *M* or *c*/*m* values are exceeded.

The sampling plans for *Salmonella* in Table 1, considered a Health 1 risk, are two-class plans. The lot from which the sample units were drawn to be classed is judged not to be in compliance with the specific regulation if *Salmonella* is found in any of the sample unit. Lots in violation of Health 1 risk standards are generally ordered to be destroyed and the legal owner could be prosecuted. Salvage operations may be permitted if it can be established that the treatment would decrease the hazard to acceptable levels. In such cases, the verification sampling and acceptance criteria may well exceed that of the particular regulation. In an investigation of a suspected outbreak of a food-borne illness, for example salmonellosis, sample numbers may well exceed the values designated in the regulation.

There are regulatory standards in which the microorganisms of concern are implied rather than stated. *Clostridium botulinum* is the microorganism of concern in regulation B.27.002 which requires a low-acid food packaged in a hermetically sealed container to be commercially sterile, unless it is kept refrigerated at a temperature not exceeding 4°C or frozen and is so labelled. Commercial sterility is defined in regulation B.27.001 as 'the condition obtained in a food that has been processed by the application of heat, alone or in combination with other treatments, to render the food free from viable forms of microorganisms, including spores, capable of growing in the food at temperatures at which the food is designed normally to be held during distribution or storage'. Under the regulation, a hermetically sealed container means a container designed and intended to be secure against the entry of microorganisms, including spores. A low-acid food is a food, other than an alcoholic beverage, where any component of the food has a pH greater than 4.6 and a water activity greater than 0.85.

Brucella and *Mycobacterium bovis*, and more recently *Salmonella* and *Listeria* are the organisms of concern in B.08.002.2. This regulation requires the normal lacteal secretion obtained from the mammary gland of the cow, genus *Bos*, or of any other animal, or a dairy product made with any such secretion, to be pasteurized by being held at a temperature and for a period that ensure the reduction of the alkaline phosphatase activity so as to meet the tolerances specified in official method MFO-3, Determination of Phosphatase Activity in Dairy Products, dated November 30, 1981.

Regulation B.21.025 deals with marine and freshwater animals, or marine and freshwater animal products, that are packed in a container that has been sealed to exclude air and that are smoked or to which liquid smoke flavour or liquid smoke flavour concentrate has been added. Under the regulation these products must be heat processed after sealing at a temperature and for a time sufficient to destroy all spores of the species *Clostridium botulinum* (i.e. commercially sterile). The only exception to the heat processing requirements are where the contents of the container comprise not less than 9% salt, the contents of the container are customarily cooked before eating, or the contents of the container are frozen and the product so labelled. The specific organism of concern is *C. botulinum* type E, which can grow – albeit slowly – at normal refrigeration temperatures.

Microbiological Guidelines

Guidelines take three forms – microbiological guidelines, codes of hygienic practice and field compliance guides. These have been developed after consultation with the Canadian food industry, are published and copies are available upon request. Microbiological guidelines were developed from surveys conducted

Table 3 Sanitary risk microbiological standards

Product	*Standard*	*Regulation*	*Method*	*Criteria*[a]			
				n	c	m[c]	M[c]
Flavoured milks	ACC[b]	B.08.016 B.08.018 B.08.026	MFO-7	5	2	5×10^4	10^6
Milk for manufacture	ACC	B.08.024	MFO-7	5	0	2×10^6	–
Cottage cheese	Coliforms	B.08.054	MFO-4	5	1	10	10^3
Ice cream	ACC	B.08.062	MFO-2	5	2	10^5	10^6
	Coliforms			5	2	10	10^3
Ice milk	ACC	B.08.072	MFO-2	5	2	10^5	10^6
	Coliforms			5	2	10	10^3
Mineral or spring water	Coliforms	B.12.001	MFO-9	10	1	0 per 1000 ml	10 per 100 ml
Prepackaged ice	Coliforms	B.12.005	MFO-15	10	1	< 1.8 per 100 ml	10 per 100 ml
Water in sealed containers	ACC	B.12.004	MFO-15	5	2	10^2	10^4
	Coliforms			10	1	< 1.8 per 100 ml	10 per 100 ml

[a] *n*, sample size; *c*, acceptance number; *m*, acceptable concentration of microorganisms; *M*, unacceptable concentration of microorganisms.
[b] ACC, Aerobic Colony Count.
[c] Per millimetre unless otherwise stated.

on specific products or groups of products across Canada. While guidelines are not regulatory standards, they are used in judging compliance with Sections 4 and 7 of the Act. Even though a given guideline may embody the same limiting criteria that would be employed in a standard, it is generally based on fewer data than are used in developing a standard. However, guidelines serve as useful indicators of levels that should be achievable using GHP. Guidelines can be readily modified, if necessary, as additional data become available.

The microbiological guidelines that are currently in force are given in **Table 4**. The same three levels of concern or risk (Health 1, 2 and Sanitation) are applied in the guidelines.

In addition to *Salmonella*, the microorganisms considered to be a Health 1 risk are *Campylobacter coli*, *C. jejuni*, *Yersinia enterocolitica*, *Pseudomonas aeruginosa*, and *Aeromonas hydrophila*. The microorganisms considered to be a Health 2 risk are *Escherichia coli*, *Staphylococcus aureus*, *Bacillus cereus*, and *Clostridium perfringens*. Aerobic colony count, coliforms and yeast and mould counts form the basis for a sanitation or hygiene hazard.

All the methods cited in the standards and guidelines are contained in the *Compendium of Analytical Methods*, published by Polyscience Publications for Health Canada. The compendium provides a ready reference of the food microbiological methods used by the Health Protection Branch (HPB) of Health Canada to determine compliance of the food industry with standards and guidelines, to assess the quality of foods with respect to their microbiological content, and to support investigations of food-borne diseases and consumer complaints.

Volume 1 of the compendium is devoted to the official microbiological methods (MFO). These are cited in the respective Food and Drug Regulations, are an integral part of the standard and must be used by the regulatory agencies to determine compliance. Health Protection Branch methods (MFHPB), used in the guidelines, are found in volume 2 of the compendium. Both the official and HPB methods have been fully validated by interlaboratory studies. Laboratory procedures (MFLP) are given in volume 3. These have been validated in at least one HPB laboratory, apart from the laboratory that originated the method. These methods include those undergoing development, newly developed rapid methods or methods for newly emerging pathogens. As for a regulatory standard, the sampling plans and compliance criteria are an integral part of each method and form part of the respective guideline.

The Code of Practice for the General Principles of Food Hygiene developed by the *Codex Alimentarius* Commission has been modified to reflect current Canadian good hygienic practices. The Canadian version of this code of practice is intended to provide guidance to the Canadian food industry on hygienic food handling practices in order to comply with Sections 4 and 7 of the Food and Drugs Act. The Code provides to both the regulatory inspector and the food industry a template for the sanitary or hygienic inspection of food processing and manufacturing premises.

The recommended Canadian Code of Hygienic Practice for Low-acid and Acidified Low-acid Foods in Hermetically Sealed Containers (Canned Foods) was adapted directly from the *Codex Alimentarius* Recommended International Code of Hygienic Practice for Low-acid and Acidified Low-acid Canned

Foods. The Canadian Code is a guideline for commercial processors who thermally process these products for compliance with the regulations B.27.001 to B.27.006, inclusively. While *Clostridium botulinum* is the primary microorganism of concern, the code also addresses microbiological spoilage.

A good example of a field compliance guide is that for ready-to-eat (RTE) foods contaminated with *Listeria monocytogenes*. An RTE food is defined as one requiring no further processing before consumption. Of primary concern are RTE foods that have been subjected to some form of processing not only to render them ready-to-eat but also to extend their shelf life. Such RTE foods can support the growth of *L. monocytogenes* even when maintained under conditions of commercial refrigeration. The field compliance guide combines inspection, environmental sampling and end product testing. The results of the inspection should show the inspector whether or not GHPs in place are adequate to control the potential for contamination of the product with *L. monocytogenes*. However, if the inspector does not consider them to be adequate, environmental sampling should be conducted. If the environmental sampling indicates that there is a probability of finished product becoming contaminated with pathogenic microorganisms, then the finished product should be sampled in accordance with **Table 5** and analysed. The results of that analysis will determine the choice of enforcement action as set out in **Table 6**. If environmental sampling indicates little or no probability for product contamination no further action is taken by the inspector other than to encourage strict implementation of GHPs, i.e. compliance with the Code of Practice for the General Principles of Food Hygiene.

For the purposes of this guide, an RTE food is considered capable of supporting growth of *L. monocytogenes* if, in a naturally contaminated lot of the RTE food under consideration, *L. monocytogenes* can be detected by direct plating after the food has been stored at 4°C or less until the end of its stated shelf life; *or* if, in an inoculated batch representative of the RTE food, *L. monocytogenes* increases in number by at least 1 log after it has been stored at 4°C or less until the end of its stated shelf life, as determined by the direct plating method. The guide encourages manufacturers to consider performing challenge tests not only under normal conditions of storage and distribution, but also under conditions of mild temperature abuse (e.g. 7–10°C). A challenge test involves the incubation of samples of the RTE food inoculated with a known concentration of a cocktail of at least five strains of *L. monocytogenes* for periods of time to reflect the desired product shelf life. A guide for the challenge testing of *L. monocytogenes* on refrigerated foods is included as an appendix to the Field Compliance Guide.

The definition for recall under the Food and Drugs Act with respect to a product, other than a medical device, means a firm's removal from further sale or use, or correction, of a marketed product that violates legislation administered by the Health Protection Branch. Three types or classes of recalls are designated. Class I is a situation in which there is a reasonable probability that the use of, or exposure to, a non-compliant product will cause serious adverse health consequences or death. Class II is a situation in which the use of, or exposure to, such a product may cause temporary adverse health consequences or where the probability of serious adverse health consequences is remote. Class III is a situation in which the use of, or exposure to, a product is not likely to cause any adverse health consequences.

Agriculture and Agri-Food Canada

This department administers a number of acts and associated regulations. Only the Canadian Agricultural Products Act, Health of Animals Act and the Meat Inspection Act and their associated regulations have microbiological standards or specifications directly applicable to foods. It should be noted that the administration of many of the Acts by this department is limited to foods that are imported, exported or traded interprovincially. The Food and Drugs Act and regulations have no such limitation.

Canadian Agricultural Products Act

The relevant regulations under the Canadian Agricultural Products Act are:

- Livestock and Poultry Carcass Grading Regulations
- Egg Regulations (upgraded 18/03/98)
- Processed Egg Regulations
- Dairy Regulations (upgraded 15/04/98)
- Fresh Fruit and Vegetable Regulations (updated 01/04/98)
- Honey Regulations (updated 01/04/98)
- Maple Products Regulations (updated 01/04/98)
- Processed Products Regulations (updated 15/04/98)
- Licensing and Arbitration Regulations (updated 04/03/98).

The Act stipulates that no person shall market a food product in import, export or interprovincial trade as food unless the food product, including every substance used as a component or ingredient thereof,

(a) is not adulterated
(b) is not contaminated
(c) is sound, wholesome and edible

Table 4 Microbiological guidelines

Food	*Method*	*Guideline*	*Risk*	*Criteria*[a]			
				n	c	m	M
Cocoa	MFHPB-18	ACC[b] includes aerobic spore-formers	Sanitation	5	2	10^5	10^6
	MFHPB-22	Yeasts and moulds[b,d]	Sanitation	5	2	2×10^3	10^4
	MFHPB-19	Coliforms[b]	Sanitation	5	2	< 1.8	10
Chocolate	MFHPB-18	ACC[b] includes aerobic spore-formers	Sanitation	5	2	3×10^4	10^6
	MFHPB-19	Coliforms[b]	Sanitation	5	2	< 1.8	10^2
Instant infant cereal and powdered infant formula	MFHPB-18	ACC[b]	Sanitation	5	2	10^3	10^4
	MFHPB-19	*Escherichia coli*[c]	Health 2	10	1	< 1.8	10
	MFHPB-20	*Salmonella*	Health 1	20	0	0	0
	MFHPB-21	*Staphylococcus aureus*[c]	Health 2	10	1	10	10^2
	MFHPB-42	*Bacillus cereus*[c]	Health 2	10	1	10^2	10^4
	MFHPB-23	*Clostridium perfringens*[c]	Health 2	10	1	10^2	10^3
Fresh and dry pasta	MFHPB-18	ACC	Sanitation	5	2	5×10^4	10^6
	MFHPB-22	Yeasts and moulds[e]	Sanitation	5	2	2×10^3	10^4
	MFHPB-19	*Escherichia coli*[e]	Health 2	5	2	< 1.8	10^3
	MFHPB-21	*Staphylococcus aureus*	Health 2	5	2	5×10	10^4
	MFHPB-20	*Salmonella*	Health 1	5	0	0	–
Bakery products[f]	MFHPB-18	ACC	Sanitation	5	2	5×10^4	10^6
	MFHPB-19	Coliforms	Sanitation	5	2	50	10^4
	MFHPB-19	*E. coli*	Health 2	5	1	< 1.8	10^3
	MFHPB-22	Yeasts and moulds	Sanitation	5	2	5×10^2	10^4
	MFHPB-21	*S. aureus*	Health 2	5	2	10^2	10^4
	MFHPB-20	*Salmonella*	Health 1	5	0	0	–
Heat-treated fermented sausage	MFHPB-19	*E. coli*[e]	Health 2	5	1	10	10^3
	MFHPB-21	*S. aureus*[e]	Health 2	5	1	50	10^4
Raw fermented sausage	MFHPB-19	*E. coli*[e]	Health 2	5	1	10^2	10^3
	MFHPB-21	*S. aureus*	Health 2	5	1	2.5×10^2	10^4
Heat-treated and raw fermented sausage	MFHPB-20	*Salmonella*	Health 1	5	0	0	–
	MFLP-46	*Campylobacter coli* or *C. jejuni*[g]	Health 1	5	0	0	–
	MFLP-48	*Yersinia enterocolitica*[g]	Health 1	5	0	0	–
Non-fermented meat products (ready-to-eat)[h]	MFHPB-19	*Escherichia coli*	Health 2	5	2	100	10^3
	MFHPB-21	*Staphylococcus aureus*	Health 2	5	2	100	10^4
	MFHPB-20	*Salmonella*	Health 1	5	0	0	
Deboned poultry products (precooked)	MFHPB-18	ACC	Sanitation	5	3	10^4	10^6
	MFHPB-19	*E. coli*[i]	Health 2	5	2	10	10^3
	MFHPB-21	*S. aureus*	Health 2	5	1	100	10^4
	MFHPB-20	*Salmonella*	Health 1	5	0	0	–
	MFLP-46	*Campylobacter jejuni* or *C. coli*[g]	Health 1	5	0	0	–
	MFLP-48	*Yersinia enterocolitica*[g]	Health 1	5	0	0	–
Dry mixes (gravy, sauce, soup) heat and serve	MFHPB-18	ACC	Sanitation	5	3	10^4	10^6
	MFHPB-19	Coliforms[j]	Sanitation	5	3	10	
	MFHPB-22	Yeasts and moulds	Sanitation	5	3	500	10^4
	MFHPB-19	*E. coli*[j]	Health 2	5	2	10	10^3
	MFHPB-21	*S. aureus*	Health 2	5	2	100	10^4
	MFHPB-23	*Clostridium perfringens*	Health 2	5	2	100	10^3
	MFHPB-20	*Salmonella*	Health 2	5	0	0	–
Soya bean products (ready-to-eat)	MFHPB-18	Psychrotrophic bacteria	Sanitation	5	2	10^5	10^7
	MFHPB-19	*Escherichia coli*[k]	Health 2	5	2	100	10^3
	MFHPB-21	*Staphylococcus aureus*	Health 2	5	2	100	10^4
	MFHPB-20	*Salmonella*	Health 1	5	0	0	–
	MFLP-48	*Yersinia enterocolitica*[l]	Health 1	5	0	0	–

Table 4 Microbiological guidelines *(Continued)*

Food	*Method*	*Guideline*	*Risk*	*Criteria*[a]			
				n	c	m	M
Spices (ready-to-eat only)	MFHPB-23	*Clostridium perfringens*	Health 2	5	2	10^4	10^6
	MFHPB-42	*Bacillus cereus*	Health 2	5	2	10^4	10^6
	MFHPB-19	*E. coli*	Health 2	5	2	100	10^3
	MFHPB-21	*S. aureus*[m]	Health 2	5	2	100	10^4
	MFHPB-20	*Salmonella*	Health 1	5	0	0	–
	MFHPB-22	Yeasts and moulds[m]	Sanitation	5	2	100	10^4
Bottled water and ice[n]	MFLP-61B	*Pseudomonas aeruginosa*	Health 1	5	0	0 per 100 ml	–
	MFLP-58B	*A. hydrophila*	Health 1	5	0	0 per 100 ml	–
Alfalfa and bean sprouts[o]	MFHPB-19	Coliforms[p]	Sanitation	5	2	10^3	10^5
	MFHPB-20	*Salmonella*	Health 1	5	0	0	–

[a] For definitions of *n*, *c*, *m* and *M* see previous tables.
[b] From HPB Data-Gathering Survey Results.
[c] From *Microorganisms in Foods 2. Sampling for Microbiological Analysis: Principles and Specific Applications*. International Commission on Microbiological Specifications for Foods (ICMSF).
[d] *M* adjusted for uniformity.
[e] *M* value adjusted for uniformity. For *E. coli* $M = 10^3$, for *S. aureus*, $M = 10^4$, and for yeasts and moulds $M = 10^4$.
[f] Products that are microbiologically sensitive, i.e. containing eggs or dairy products. This food category consisted previously of only cream pies but has been extended to other bakery products. Cream pies probably represent the worst case of this product type.
[g] Designates an optional analysis. It is not expected that these determinations will be done routinely.
[h] Guidelines for this food category are new. Only organisms indicating a health concern are provided. The limiting values are consistent with those for other products.
[i] *M* value adjusted for uniformity. For *E. coli* $M = 10^3$, for *S. aureus* $M = 10^4$.
[j] Values of *m* and *M* modified according to Health Canada monitoring results.
[k] *M* value adjusted for uniformity. For *E. coli* $M = 10^3$, for *S. aureus* $M = 10^4$.
[l] Designates an optional analysis. It is not expected that these determinations will be done routinely.
[m] Values are proposed by the International Commission on Microbiological Specifications for Foods (ICMSF).
[n] Includes mineral or spring water, or water in sealed containers, or pre-packaged ice. The microbiological standards for ACC and coliforms in these products (see Table 3) are under review.
[o] Based on data-gathering survey results.
[p] High coliform counts that do not confirm as faecal coliforms or *E. coli* should be investigated to determine if *Klebsiella pneumoniae* is present. Take action appropriate for a Health 2 risk if *K. pneumoniae* levels exceed those for coliforms.

(d) is prepared in a sanitary manner
(e) where irradiated, is irradiated in accordance with Division 26 of Part B of the Food and Drugs Act and the Food and Drug Regulations
(f) meets all other requirements of the Food and Drugs Act and the Food and Drug Regulations.

For the purposes of this Act, the term 'contaminated' means containing a chemical, drug, food additive, heavy metal, industrial pollutant, ingredient, medicament, microbe, pesticide, poison, toxin or any other substance not permitted by, or in an amount in excess of limits prescribed under, the Canadian Environmental Protection Act, the Food and Drugs Act or the Pest Control Products Act, or containing any substance that renders the food inedible. Paragraphs (b), (c), (d) and (f) address the microbiological safety and quality of foods. These general provisions are repeated in the regulations for each specific food group. The regulations for the various food groups may contain microbiological standards as well as directions with respect to sanitary preparation.

Processed Eggs Regulations The regulations contain a general stipulation that no processed egg shall be marked with a departmental inspection legend unless the processed egg tests negative for salmonellae and other pathogenic organisms of human health significance as determined by a method approved by the Minister. All establishments involved in the handling and processing of eggs and egg product for import, export or interprovincial trade are subject to inspection by Agriculture and Agri-Food Canada and the product packaging must bear the inspection legend. Unlike the microbiological standards under the Food and Drug Regulations the specifics of the method and sampling plan to be used are not given. In addition to this general stipulation, there are a number of microbiological standards for specific product types.

Frozen egg, frozen egg mix, liquid egg, liquid egg

Table 5 Sampling plans for analysing ready-to-eat (RTE) foods for *Listeria monocytogenes* (LM)

Food product category	*Sampling*	*Analysis*	*Type of analysis*
1. RTE foods causally linked to listeriosis (this list includes soft cheese, liver pâté, coleslaw mix with shelf life > 10 days, jellied pork tongue)	Five sample units (100 g or ml each) taken at random from each lot	5 × 10 g or 2 × 25 g analytical units[a] are either analysed separately or composited	ENRICHMENT *ONLY*
2. All other RTE foods supporting growth of LM with refrigerated shelf life > 10 days (e.g. vacuum-packaged meats, modified atmosphere-packaged sandwiches, cooked seafood, packaged salads, refrigerated sauces)	Five sample units (100 g or ml each) taken at random from each lot	5 × 5 analytical units[a] are either analysed separately or composited	ENRICHMENT *ONLY*
3. RTE foods supporting growth of LM with refrigerated shelf-life ⩽ 10 days and all foods not supporting growth[b] (e.g. cooked seafood, packaged salads, ice cream, hard cheese, dry salami, salted fish, breakfast and other cereal products)	Five sample units (100 g or ml each) taken at random from each lot	5 × 10 g analytical units[a] are analysed separately Where enrichment is necessary[c] 5 × 5 g analytical units[a] are analysed separately or composited	DIRECT PLATING ENRICHMENT

[a] The designated analytical unit is taken from each sample unit.
[b] Foods not supporting growth of LM include the following:
pH 5.0–5.5 and $a_w < 0.95$
pH < 5.0 regardless of a_w
$a_w \leqslant 0.92$ regardless of pH
frozen foods
The pH and a_w determinations should be done on three out of five analytical units. The food is presumed to support the growth of *L. monocytogenes* if any one of the analytical units falls into the range of pH and a_w values which are thought to support the growth of the organism.
[c] For the last category, if GMP is inadequate and *L. monocytogenes* has been found in the environment of the finished product area, or where examination of GMP status is not possible, the method to isolate *L. monocytogenes* from foods and environmental samples (MFHPB-30) and the method for enumeration of *L. monocytogenes* (MFLP-74) may be used as appropriate.

mix, frozen egg product or liquid egg product that is marked with an inspection legend shall, in addition to meeting the general requirements for salmonellae, have a coliform count of no more than 10 per gram, and a total viable bacteria count of no more than 50 000 per gram.

Dried egg, dried egg mix or dried egg product that is marked with an inspection legend shall, in addition to meeting the requirements for salmonellae, have a coliform count of no more than 10 per gram, and a total viable bacteria count of no more than 50 000 per gram in the case of whole egg, whole egg mix and yolk mix. In the case of albumen, the total viable bacteria count standard is reduced to 100 000 per gram.

The pasteurization of liquid egg products, while initially directed to reduce salmonellae to levels that do not represent a health hazard, will also have the same beneficial effect for other pathogens having the same or lower thermal resistance that may be present. The heating requirements are given in **Table 7**. Spray-dried albumen shall be pasteurized at 54°C (130°F) for 7 days, and pan-dried albumen at 52°C (125°F) for 5 days.

Dairy Products Regulations In addition to the general requirement as stipulated in the Act, the regulations specify the compositional standards (e.g. percentage of butterfat in various milk categories). The regulations reference the microbiological standards in the Food and Drug Regulations. As the production, processing, sale and distribution of milk and associated products for the most part are intraprovincial, they are also subject to regulation by each province. Each province has specific pasteurization requirements with respect to time and temperatures.

Processed Products Regulations The regulations require that a low-acid food product packed in a hermetically sealed container be thermally processed, until at least commercial sterility is achieved. A low-acid food product packed in a hermetically sealed container is exempt, if it is stored continuously under refrigeration and if the container in which it is packed, as well as the shipping container, is marked 'Keep Refrigerated', or kept continuously frozen and the container in which it is packed, as well as the shipping container, is marked 'Keep Frozen'. This duplicates the same requirement under B.27.002 of the Food and Drug Regulations.

Table 6 Compliance criteria for *Listeria monocytogenes* in ready-to-eat foods

Food product category	*Action level*	*GMP status*	*Immediate action*
1. RTE foods causally linked to listeriosis (see Table 5)	> 0 cfu per 50 g[a]	n/a	Class I recall to retain level, consideration of public alert
2. All other RTE foods supporting growth of LM with a refrigerated shelf life > 10 days	> 0 cfu per 25 g[a]	n/a	Class II recall to retain level, health alert consideration
3. RTE foods supporting growth of LM with a refrigerated shelf life ≤ 10 days and all RTE foods not supporting growth (see Table 5)	≤ 100 cfu g[b]	Adequate GMP	Allow sale
	≤ 100 cfu/g[b]	Inadequate or no GMP[c]	Consideration of class II recall or stop sale
	> 100 cfu/g[b]	n/a	Class II recall or stop sale

[a] Enumeration by enrichment only (MFHPB-30)
[b] Enumeration to be done by direct plating onto LPM and Oxford agar (MFLP-74)
[c] No information on GMP is considered as no GMP and the burden of proof remains with the legal agent for the product.
In all of the above cases where *L. monocytogenes* is detected, the processing establishment should be inspected to determine the source of the contamination and to ensure that corrective measures are taken.
cfu, colony forming unit; GMP, Good Manufacturing Practice; LM, *Listeria monocytogenes*; n/a, not applicable; RTE, ready-to-eat.

There is also a requirement that the water used to cool the containers after thermal processing shall be of an acceptable microbiological quality. The regulation does not specify what is an acceptable quality. Water used in a cooling canal system must contain a residual amount of a bactericide at the discharge end of the canal and records must be kept of all bactericidal treatments.

The specific hygiene requirements detailed in the regulations generally follow those in the Canadian Code of Practice for the General Principles of Food Hygiene for use by the food industry in Canada.

Meat Inspection Regulations All establishments involved in the slaughter, preparation, manufacture, storage, distribution and sale of meat and meat products in import, export and interprovincial trade must be registered by Agriculture and Agri-Food Canada. The regulations contain specific requirements:

- governing the design, construction and maintenance of registered establishments and of the equipment and facilities therein
- prescribing the equipment and facilities to be used, the procedures to be followed and the standards to be maintained in registered establishments to ensure humane treatment and slaughter of animals and hygienic processing and handling of meat products
- prescribing standards for meat products that are prepared or stored in registered establishments, for meat products that enter into interprovincial or international trade and for meat products in connection with which the meat inspection legend is applied or used.

Registered establishments have resident federal inspectors to ensure compliance with the regulations and to conduct product inspection sampling when required. All processes must meet departmental requirements. Premortem and postmortem inspections are carried out routinely in all registered slaughtering plants.

The requirements for low-acid meat products packaged in hermetically sealed containers duplicates the requirements found in Section B.27.002 of the Food and Drug Regulations The container cooling water must be of an acceptable microbiological quality and, in the case of water used in a cooling canal system, contains a residual amount of a bactericide at the discharge end of the canal, and the container be handled in a manner that ensures that the container remains hermetically sealed.

Fisheries and Oceans Canada

The microbial safety and quality of fish and fish products for export, import or interprovincial trade are regulated under the Fish Inspection Act and the Fish Inspection Regulations. The regulations:

- prescribe grades, quality and standards
- set the quality and specifications for containers and the marking and inspection of containers
- require the registration of establishments and the licensing of persons engaged as principals or agents in the export or import of fish or containers
- prescribe the requirements for the equipment and sanitary operation of establishments, of premises operated by an importer for the purpose of importing fish, and of any boats, vehicles or other equipment used in connection with an establishment or

Table 7 Pasteurization of liquid processed egg (see Processed Eggs Regulations – Schedule, Part I)

Liquid processed egg product	*Minimum temperature of the processed egg at the automatic diversion valve*		*Minimum heating time (mins)*
	°C	*°F*	
Whole egg with less than 24% milk solids	60	140	3.5
Whole egg with no less than 24% and no more than 38% egg solids	61	142	3.5
	60	140	6.2
Whole egg mix with less than 2% added salt or sweetening agent, or both	61	142	3.5
	60	140	6.2
Whole egg mix with no less than 2% and no more than 12% added sweetening agent	61	142	3.5
	60	140	6.2
Whole egg mix with no less than 2% and more than 12% added salt	63	146	3.5
	62	144	6.2
Yolk	61	142	3.5
	60	140	6.2
Yolk mix with less than 2% added salt or sweetening agent, or both	61	142	3.5
	60	140	6.2
Yolk mix with no less than 2% and no more than 12% added sweetening agent	63	146	3.5
	62	144	6.2
Yolk mix with no less than 2% and no more than 12% added salt	63	146	3.5
	62	144	6.2
Ova	63	146	3.5
	62	144	6.2
Egg product with less than 24% total solids[a]	61	142	3.5
	60	140	6.2
Egg product with no less than 24% and no more than 38% total solids[a]	62	144	3.5
	61	142	6.2
Egg products with more than 38% solids[a]	63	146	3.5
	62	144	6.2

[a] Regardless of the total solids content, egg product must be heated to 63°C (146°F) for 3.5 min or to 62°C (144°F) for 6.2 min if there is no less than 2% and no more than 12% added sweetening agent or salt, or both. The director, at the request of an operator, may designate a lower minimum temperature depending on the composition of the egg product.

in connection with fishing or the import or export of fish
- prescribe the manner in which samples of any fish may be taken.

The general requirement that no person shall import, export, sell for export or have in his or her possession for export any fish intended for human consumption that is tainted, decomposed or unwholesome has microbiological significance: 'decomposed' means fish that has an offensive or objectionable odour, flavour, colour, texture or contains a substance associated with spoilage; 'tainted' means fish that is rancid or has an abnormal odour or flavour; 'unwholesome' means fish that has in or upon it bacteria of public health significance or substances toxic or aesthetically offensive to humans. Any of these conditions can be the result of microbial growth, i.e. spoilage.

There are a number of regulations that are specific to the requirements for the equipment and sanitary operation of the fish processing establishments, for vessels used for fishing or transporting fish for processing, and for the storage of frozen fish.

Some microbiological guidelines have been designed to meet specific product risks (**Table 8**). The interpretation of the sampling plans and acceptance criteria is the same as that described for the microbiological standards and guidelines under the Food and Drug Regulations.

Table 8 Bacteriological guidelines for fish and fish products

Product	*Microorganism*	*Criteria*[a]			
		n	*c*	*m* (g^{-1})	*M* (g^{-1})
Cooked or RTE products	*Escherichia coli*	5	2	4	40
All other types	*E. coli*	5	2	4	4
All types	*Staphylococcus aureus*	5	1	10^3	10^4
All types	*Salmonella*	5[b]	0	0	
Cooked or RTE products	*Vibrio cholerae*	5[b]	0	0	

[a] *n*, sample size; *c*, acceptance number; *m*, acceptable concentration of microorganisms; *M*, unacceptable concentration of microorganisms.
[b] The analytical unit is 25 g. Each analytical unit must be negative for the microorganism. The analytical units may be pooled.
RTE, ready-to-eat.

The Canadian Shellfish Sanitation Program

The Canadian Shellfish Sanitation Program (CCSP) was developed in 1925 under the Fish Inspection Act as a result of a typhoid fever outbreak in the USA in 1924–25 involving 1500 cases and 150 deaths as a result of consuming contaminated oysters. The CSSP is jointly administered by the Canadian Food Inspection Agency (CFIA), Department of Fisheries and Oceans (DFO) and Environment Canada (EC) in cooperation with Health Canada. The parameters of interdepartmental cooperation between EC and DFO are established by a Memorandum of Understanding which is under revision to reflect the responsibilities of the Canadian Food Inspection Agency.

Environment Canada is responsible for carrying out shoreline sanitary and bacteriological water quality surveys of the shellfish growing areas according to the procedures, standards and protocols of the *Canadian Shellfish Sanitation Program Manual of Operations*. This includes the continuing evaluation of the level of faecal contamination in the water overlying shellfish growing areas, the identification of point and non-point pollution sources that have a negative impact on these areas, and classification of these areas based on sanitary quality and general sanitary conditions. In order for a shellfish area to be recommended for approval, the overlying waters must be free from hazardous concentrations of pathogenic microorganisms, poisonous or deleterious substances (or marine biotoxins and monitored by the CFIA) as outlined in the *Canadian Shellfish Sanitation Program Manual of Operations*.

The 1948 Canada–United States Bilateral Agreement on Shellfish governing trade in shellfish between the two countries required agreement on practices for sanitary control of the shellfish industry. The Canadian *Manual of Operations* is based on the protocols and procedures of the *American National Shellfish Sanitation Program Manual of Operations*, now called the *NSSP Guide*. The Agreement also required each country to facilitate inspections of each other's shellfish handling facilities and shellfish growing areas if requested. The United States Food and Drug Administration (USFDA) has routinely audited the CSSP (about every 2 years – most recently in 1996 on both Atlantic and Pacific coasts). Growing area classification based on water quality is the basis of the NSSP as it is for the CSSP. According to the NSSP 'the first critical control point in the sanitary control of shellfish is identifying harvesting areas of acceptable sanitary quality'. Water quality requirements in both programmes are the same for shellfish aquaculture as they are for wild harvest.

The sanitary surveys completed by Environment Canada are the basis for the classification of coastal harvesting areas for clams, oysters, mussels and whole scallops. Classifications are based on the sanitary conditions of the area as defined by the shoreline survey with supporting information from the microbiological evaluation of the area.

1. *Approved:* the area is not contaminated with faecal material, poisonous or deleterious substances or marine biotoxins to the extent that consumption of the shellfish might be hazardous. In these areas the median or geometric mean faecal coliform level must be less than 14 Most Probable Number (MPN) per 100 ml with no more than 10% of the samples in excess of 43 MPN per 100 ml.
2. *Conditionally approved:* this area must meet the same sanitary quality criteria as an approved area. However, under certain conditions which are predictable and verifiable, water quality can exceed approved area criteria. The quality can vary with: (a) the effectiveness of sewage treatment at a community, (b) rainfall or river flow, (c) seasonal changes in sanitary conditions (i.e. tourist or summer cottage activity, vessel traffic, seasonal industrial operation). Management plans which detail the criteria for opening and closing such areas, and the responsibilities of all parties are required for conditionally approved classifications.
3. *Closed:* direct harvesting from this area is prohibited owing to chemical or bacteriological contamination. Shellfish can be used only under specified permit conditions for depuration, relaying, experimental purposes or other approved processing. Depending on the level of contamination, harvesting may be prohibited for any purpose. The Closed classification includes the subclassifications of Restricted for Controlled Purification, Restricted for Relaying and Prohibited Area.

Regional Shellfish Classification Committees, composed of DFO, EC and provincial government representatives, are responsible for:

- the technical reviews of the sanitary and bacteriological surveys and evaluation of growing area classification recommendations
- reviewing the policies, procedures, criteria and regulations affecting the implementation and application of shellfish growing area classification
- recommendations to DFO (for closures under the Management of Contaminated Fisheries Regulations under the Fisheries Act) and CFIA (as required for approved areas under the Fish Inspection Regulations under the Fish Inspection Act
- implementation of the classification decisions, and recommending survey priorities.

It is the policy of Environment Canada either to use its own laboratories and personnel, or to audit laboratories and personnel under contract, so as to ensure impartiality of sample collection and integrity of the data. For example, Environment Canada currently contracts out sampling and analyses under QA/QC (Quality Assurance/Quality Control) controls to private laboratories in Quebec. In addition, EC has a cooperative agreement with the Department of the Environment of the Province of Prince Edward Island to carry out sampling and analysis for its growing areas. In all cases, the data are interpreted by Environment Canada which then makes classification recommendations to DFO at the regional classification committees. With reducing resources and increasing demand from the aquaculture industry for classified areas, EC distributed a discussion paper in May 1997 outlining cost-sharing approaches including stakeholder sampling and the use of third-party laboratories.

The programme is routinely evaluated by both internal and USFDA auditors to determine compliance with CSSP and NSSP protocols. The EC laboratories are evaluated by USFDA. In addition, EC participates in USFDA's Laboratory Evaluation Officer training programme. All laboratories participating in the CSSP must be evaluated by a Laboratory Evaluation Officer, have an internal QA/QC programme and participate in a split sample programme.

The Future

The preceding has summarized the present status of national legislation, standards and guidelines concerned with the microbiological safety and quality of food in Canada. There are five recent initiatives, the Canadian Food Inspection Agency Act and the Canadian Food Inspection System, that may well have a profound effect on the status quo.

Under the Canadian Food Inspection Agency Act, the Canadian Food Inspection Agency (CFIA) was created to consolidate all federally mandated food inspection and animal and plant health services. The CFIA legislation sets out its responsibilities, accountability, regimes, powers and reporting framework. The legislation also amended the enforcement provisions and penalty structures of the federal statutes relating to food and animal and plant health that are enforced and/or administered by the Agency. To meet its mandate, the Agency administers and/or enforces the following Acts:

- Canadian Agricultural Products Act
- Consumer Packaging and Labelling Act as it relates to food
- Feeds Act
- Fertilizers Act
- Fish Inspection Act
- Food and Drugs Act as it relates to food
- Health of Animals Act
- Meat Inspection Act
- Plant Breeders' Rights Act
- Plant Protection Act
- Seeds Act.

The Agency is a departmental corporation, reporting to Parliament through the Minister of Agriculture and Agri-Food. While the setting of standards and guidelines remains the prerogative of the mother departments that administer the underpinning legislation, these as well as the enforcement policy will certainly be affected by the feedback information from the inspectional services and associated laboratories.

A blueprint for the Canadian Food Inspection System (CFIS) was prepared in 1993 by the Joint Steering Committee of CFIS, the Federal/Provincial Agri-Food Inspection Committee and the Federal/Provincial/Territorial Food Safety Committee and revised in 1995. The goal of the system is to integrate the activities of all food inspection departments at all levels of government where appropriate. Priority will be placed on developing regulatory standards which will be outcome-based, and supported by nationally accepted guidelines. Effective, ongoing communication among stakeholders and government agencies is a requirement of integration.

Specifically, the process of integration will include one set of food safety standards which are nationally recognized and a common legislative base reflective of international developments. This will be extended to include the establishment of common standards for product identity, including grade, composition, net quantity and product description.

There will be a movement from prescriptive standards to standards which are outcome- or performance-based, where practical. Commonality will extend to the manner and environment under which food is produced, processed and distributed. Standard methods for laboratory testing and reporting that are reflective of the Canadian Food Inspection System and international developments will be established by the Joint Steering Committee as will an accreditation (certification) programme for government and private laboratories. Research into new methodologies is also critical to the success of this process.

See also: ***Brucella***: Characteristics. ***Clostridium***: *Clostridium botulinum*. **Eggs**: Microbiology of Egg Products. **Fish**: Spoilage of Fish. **Food Poisoning Outbreaks**. ***Listeria***: Introduction; *Listeria monocytogenes*. **Milk and Milk Products**: Microbiology of Liquid Milk. ***Myco-***

bacterium. **National Legislation, Guidelines & Standards Governing Microbiology**: European Union. **Rapid Methods for Food Hygiene Inspection**. ***Salmonella***: Introduction. **Sampling Regimes & Statistical Evaluation of Microbiological Results**. **Shellfish (Molluscs and Crustacea)**: Contamination and Spoilage.

Further Reading

Canadian Agricultural Products Act, RSC 1985 (4th Supp.), c. 20.

Compendium of Analytical Methods. Vols 1–3. Health Canada. Quebec: Polyscience Publications, PO Box 1606, Station St-Martin, Laval, Quebec, H7V 3P9, Canada.

Dairy Products Regulations, SOR/79–840.

Egg Regulations, CRC, vol. II, c. 284.

Fish Inspection Act (Canada), RSC 1985, c. F–12.

Fish Inspection Regulations, 1985, c. F–27.

Food and Drugs Act (Canada), RSC 1985, E. F–27.

Food and Drug Regulations (Canada).

Fresh Fruit Regulations, CRC, c. 870.

Meat Inspection Act, RSC 1985, (1st Supp.), c. 25.

Meat Inspection Regulations, 1990, SOR/90–288.

Processed Egg Regulations, CRC, vol. II, c. 290.

Processed Products Regulations, CRC, vol. III, c. 291.

European Union

B Schalch, Institute of Hygiene and Technology of Food of Animal Origin, Ludwig-Maximilians-University Munich, Germany

H Beck, Department for Health Service, South Bavaria, Oberschleissheim, Germany

Introduction

In the European Union many cultural, national and individual consumer groups exist with distinct consumption habits, preferences and specialities. Therefore, legislation is very complex and detailed for certain foods. Furthermore, microbial limit values vary according to the microbiological analysis technique used. Examination methods follow the International Standards Organization recommendations which imply appropriate techniques for different objectives. For example, if solid media are used for the microbiological analysis of minced meat, maximum limit values are lower than if liquid media are used, as most microorganisms show better growth in liquid media.

Drastic reduction and simplification of the subject is required to summarize the microbiological standards within the EU. For the full picture it is essential to read the original versions of the Council Directives and Decisions.

Most EU Directives containing microbiological standards imply the use of 'two-class' plans and 'three-class' plans. Two-class plans differentiate two categories of samples with the help of one limit value. Samples up to and including the limit value are judged as 'satisfactory', samples above the limit values are 'not satisfactory'. The number of obligatory samples is indicated by 'n'. With a three-class plan, samples can be divided into three categories:

- samples up to and including the criterion 'm'
- samples between the criterion 'm' up to and including the limit value 'M'
- samples exceeding 'M'.

Again, the number of obligatory samples is indicated by 'n'. The value 'c' indicates how many of the n samples may fall between m and M.

All the microbial standards in the following tables refer to colony forming units (cfu) per gram of the material examined, or cfu per millilitre for liquid foods, if no other unit (e.g. 25 g, 10 ml) is given.

Minced Meat and Meat Preparations

The Council Directive 94/65/EC of 14 December 1994 'laying down the requirements for production of, and trade in, minced meat and meat preparations' came into force in January 1994. Among other detailed regulations, it requires regular microbiological monitoring of such preparations from production plants. The Directive does not apply to minced meat and meat preparations that are produced in retail shops and sold directly to the final consumer. According to the Council Directive, the daily examination is compulsory for five samples of minced meat (100 g each) which must be representative of the day's production from one plant. All five samples have to be examined separately for the count of aerobic mesophilic bacteria, *Escherichia coli* and *Staphylococcus aureus* as well as the presence or absence of salmonellae (**Table 1**). Interpretation of the consignments is carried out as follows: the consignment is judged as 'satisfactory' when all five units are m or below and salmonella-negative. The consignment is

Table 1 Microbiological criteria of Directive 94/65/EC for minced meat ($n = 5$)

Microorganisms	m	M	c
Aerobic mesophile bacteria	5.0×10^5	5.0×10^6	2
Escherichia coli	5.0×10^1	5.0×10^2	2
Staphylococcus aureus	1.0×10^2	1.0×10^3	2
Salmonellae in 10 g	0	0	0

Standards are cfu g^{-1} unless otherwise specified.

Table 2 Microbiological criteria of Directive 94/65/EC for meat preparations ($n = 5$)

Microorganisms	m	M	c
Escherichia coli	5.0×10^2	5.0×10^3	2
Staphylococcus aureus	5.0×10^2	5.0×10^3	1
Salmonellae in 1 g	0	0	0

Standards are cfu g^{-1} or cfu ml^{-1} unless otherwise specified.

Table 3 Microbiological criteria of Directive 92/46/EEC for milk ($n = 5$)

Milk	*Microorganisms*	m	M	c
Packaged raw cow's milk for consumption	*Staphylococcus aureus*	10^2	5.0×10^2	2
	Salmonellae in 25 ml			0
Pasteurized milk on processing plant level	Pathogens in 25 ml	0	0	0
	Coliforms	0	5	0
	Total count after 5 days incubation at 6°C	5.0×10^4	5.0×10^5	1

Standards are cfu ml^{-1} unless otherwise specified.

judged as 'acceptable' when a maximum of two units (c) exceed m but not M and no salmonella is detected. If one unit of five exceeds M, or c exceeds 2 or is salmonella-positive, the whole consignment will be judged as 'unsatisfactory'.

For meat preparations (i.e. raw meat with spices, seasoning and/or additives) **Table 2** gives the microbiological criteria. Microbiological examination is compulsory once a week for meat preparations.

If the output from production plants does not meet these requirements, products can be excluded from trade within the EU.

Milk

Directive 92/46/EEC gives the detailed microbiological criteria for raw, pasteurized and sterilized cow, goat, sheep, and buffalo milk for human consumption at different stages of distribution. The maximum total count for raw cow's milk is 10^5 cfu ml^{-1}; milk of other ruminants has a maximum total count of 1.5×10^6 cfu ml^{-1}. Further limit values are laid down for milk intended for milk products. **Table 3** shows the limit values of milk for the consumer. Beyond the limit for packaged raw cow's milk given in Table 3, the Directive requires the absence of any pathogens and their toxins in counts that endanger human health.

Table 4 Microbiological criteria of Council Decision 93/51/EEC for cooked crustacea and molluscs without shells ($n = 5$)

Microorganisms	m	M	c
Staphylococcus aureus	10^2	10^3	2
Either: thermophil coliforms	10	10^2	2
or: *Escherichia coli*	10	10^2	1
Salmonellae in 25 g	0	0	0

Standards are cfu g^{-1} unless otherwise specified.

Table 5 Microbiological guidelines of Council Decision 93/51/EEC for on-line production control ($n = 5$)

Total count	m	M	c
Complete products	10^4	10^5	2
Products without carapace or shell (crab meat excluded)	5.0×10^4	5.0×10^5	2
Crab meat	10^5	10^6	2

Standards are cfu g^{-1} or cfu ml^{-1} unless otherwise specified.

Fish

Council Decision 93/51/EEC lists microbial limits for cooked crustaceans and molluscs. The Decision requires absence of any pathogens and their toxins in counts that endanger human health. **Table 4** gives further microbial limits for these cooked products without shells. Microbial guidelines intended as an on-line hygiene control for producers are listed in **Table 5**.

Drinking Water

Detailed standards for the organoleptic and physicochemical quality of drinking water for humans as well as undesirable components and toxic substances are laid down by Directive 80/778/EEC. Recommended total microbial counts for drinking water delivered directly to the consumer are:

- 10 cfu ml^{-1} (plate incubation at 37°C)
- 100 cfu ml^{-1} (plate incubation at 22°C).

Maximum colony counts for packed drinking water are:

- 20 cfu ml^{-1} (37°C)
- 100 cfu ml^{-1} (22°C).

Methods for further analysis of microbial parameters are either membrane filter technique or most probable number techniques. Whichever method is used, no coliforms, no *E. coli*, no faecal streptococci and no sulphite-reducing clostridia should be detectable in 100 ml of drinking water. Furthermore, the water should be free of all pathogens such as salmonellae, pathogenic staphylococci, faecal bacteriophages and enteroviruses. Neither should any parasites, algae or animalcules be detected.

Table 6 Microbiological criteria of Directive 89/437/EEC for egg products

Microorganisms	n	m	M	c
Salmonella	10	0	0	0
Total count	5	10^4	10^5	2
Enterobacteriaceae	5	10	10^2	2
Staphylococcus aureus	5	0	0	0

Standards are cfu g^{-1} or cfu ml^{-1}

Egg Products

Guidelines for hygiene requirements for egg products are listed in Directive 89/437/EEC. The microbiological standards are shown in **Table 6**.

See also: ***Aeromonas***: Detection by Cultural and Modern Techniques. **ATP Bioluminescence**: Application in Meat Industry; Application in Dairy Industry; Application in Hygiene Monitoring; Application in Beverage Microbiology. ***Bacillus***: Detection by Classical Cultural Techniques. **Biochemical and Modern Identification Techniques**: Food-poisoning Organisms. **Biosensors**: Scope in Microbiological Analysis. ***Campylobacter***: Detection by Cultural and Modern Techniques; Detection by Latex Agglutination Techniques. ***Clostridium***: Detection of Enterotoxins of *C. perfringens*. **Electrical Techniques**: Food Spoilage Flora and Total Viable Count (TVC). **Direct (and Indirect) Conductimetric/ Impedimetric Techniques**: Food-borne Pathogens. **Enrichment Serology**: An Enhanced Cultural Technique for Detection of Food-borne Pathogens. **Enterobacteriaceae, Coliforms and *E. coli***: Classical and Modern Methods for Detection/Enumeration. ***Escherichia coli***: Detection of Enterotoxins of *E. coli*. ***Escherichia coli* O157**: Detection by Latex Agglutination Techniques; Detection by Commercial Immunomagnetic Particle-based Assays. **Flow Cytometry**. **Fungi**: Food-borne Fungi – Estimation by Classical Culture Techniques. **Hydrophobic Grid Membrane Filter Techniques (HGMF)**. **Immunomagnetic Particle-based Techniques**: Overview. ***Listeria***: Detection by Classical Cultural Techniques; Detection by Commercial Enzyme Immunoassays; Detection by Colorimetric DNA Hybridization; *Listeria monocytogenes* – Detection by Chemiluminescent DNA Hybridization; *Listeria monocytogenes* – Detection using NASBA (an Isothermal Nucleic Acid Amplification System). **Molecular Biology – in Microbiological Analysis**. **National Legislation, Guidelines & Standards Governing Microbiology**: Japan. **Nucleic Acid-based Assays**: Overview. **PCR-based Commercial Tests for Pathogens**. **Petrifilm – An Enhanced Cultural Technique**. ***Salmonella***: Detection by Classical Cultural Techniques; Detection by Latex Agglutination Techniques; Detection by Enzyme Immunoassays; Detection by Colorimetric DNA Hybridization; Detection by Immunomagnetic Particle-based Assays. ***Staphylococcus***: Detection by Cultural and Modern Techniques; Detection of Staphylococcal Enterotoxins. **Total Viable Counts**: Pour Plate Technique; Spread Plate Technique; Specific Techniques; MPN; Metabolic Activity Tests. **Verotoxigenic *E. coli***: Detection by Commercial Enzyme Immunoassays. **Verotoxigenic *E. coli* and *Shigella* spp.**: Detection by Cultural Methods. ***Vibrio***: Detection by Cultural and Modern Techniques. **Water Quality Assessment**: Modern Microbiological Techniques.

Further Reading

Association Internationale de l'Industrie des Bouillons et Potages (1992) New microbiological specifications for dry soups and bouillons. *Alimenta* 31: 62–65.

Buchanan RL (1991) Microbiological criteria for cooked, ready-to-eat shrimp and crabmeat. *Food Technol.* 45: 157–160.

Baumgart J (ed.) (1993) Mikrobiologische Untersuchung von Lebensmitteln. Hamburg: Behr.

Bell RG (1990) Microbiological criteria in regulatory standards: reason or rhetoric. In: Nga BH and Lee YK (eds) *Microbiology Applications in Food Biotechnology.* P. 162. London: Elsevier Applied Science.

EC (1990) Proposal for a Council Regulation (EEC) laying down the health rules for the production and placing on the market of raw milk, of milk for the manufacture of milk based products and of milk based products. COM (89) 667. Brussels: Commission of the European Communities.

Garrett E (1988) Microbiological standards, guidelines and specifications and inspection of seafood products. *Food Technol.* 42: 90–93.

Gräf W, Hammes W, Hennlich G et al (1988) Mikrobiologische Richt- und Warnwerte zur Beurteilung von Lebensmitteln. *Bundesgesunhbl.* 31: 93–94.

Hygiene Alimentaire (1982) Journal Official De la Republique Francaise, No. 1488. Paris: 26 Rue Desaix, 75727 Paris Cedex 15.

Jay JM (1992) *Modern Food Microbiology.* New York: Van Nostrand Reinhold.

Jervis DI (1992) A manufacturer's view on how to achieve microbiological end product criteria. *Bull. IDF* 276: 36–43.

Mossel DAA, Morris GP, Struijk CB and Ehiri JE (1997) Shaping the new generation of microbiological food safety professionals: attitudes, education and training. *Int. J. Environ. Health Res.* 7 (3): 233–250.

Professional Food Microbiology Group of the Institute of Food Science and Technology (1995) Microbiological criteria for retail foods. *Lett. Appl. Microbiol.* 20 (6): 331–332.

Ramaswamy K, Reddy KS and Sivarami Reddy K (1990) Microbiological standards for processed cheese. *Indian Dairyman* 42: 205–208.

Robinson RK (ed.) (1990) *Dairy Microbiology.* Vols 1 and 2. London: Elsevier/Applied Science.

Snyder OP (1992) HACCP: an industry food safety self control program. Part 10. Derived overall microbiological standards for chilled food processes. *Dairy, Food and Environmental Sanitation* 12: 687–688.

Japan

Susumu Kumagai, Department of Biomedical Food Research, National Institute of Infectious Diseases, Tokyo, Japan

Legislative regulation for microbiological hazards in foods in Japan is based on the Food Sanitation Law, which is supplemented by an Enforcement Order (Cabinet Order) and Enforcement Regulations (Ministerial Ordinance). Prohibition of insanitary foods and food additives, restriction of sales of meat derived from animals with diseases, standards and specifications for foods and food additives, and standard procedures for manufacturing and processing are prescribed in the law, and by the enforcement order and regulation.

Food Sanitation Law

Articles 4, 5 and 7 of the Food Sanitation Law provide the basis of legislative regulation for microbiological hazards in foods. Article 4 prohibits the sale of insanitary foods and food additives, and restricts sales of newly developed foods. Article 5 restricts sales of meat derived from diseased animals. Article 7 establishes the standards and specifications for foods and food additives, and the approval system of the comprehensive sanitary manufacturing practice based on the hazard analysis critical control point (HACCP) system.

Article 4

Article 4 states that the food additives listed below shall not be sold (or given away), collected, manufactured, imported, processed, used, cooked, stored or displayed for the purpose of offering for sale:

1. Rotten, changed in quality or unripe food or food additives except for those approved by the Minister of Health and Welfare as safe and wholesome when used for food or drink.
2. Food or food additives that contain or are suspected to contain toxic or harmful substances. However, this clause shall not apply in cases where the Minister of Health and Welfare determines them to be safe and wholesome when used for food or drink.
3. Food or food additives that may be injurious to human health because they are contaminated or are suspected to be contaminated with pathogenic microorganisms.
4. Food or food additives that may be injurious to human health because they are insanitary or are mixed with extraneous substances.

Article 5

Article 5 states that meat, bone, milk, viscera and blood derived from animals (the term 'animals' includes cattle, horses, sheep, goats and other animals provided for in the Ministerial Ordinance of Health and Welfare) that have suffered or are suspected to have suffered from diseases as specified in the Ministerial Ordinance of Health and Welfare, or those derived from such animals that have died otherwise than by slaughter, or those derived from poultry ('poultry' means chickens, ducks, turkeys and other domestic fowls provided for in the Ministerial Ordinance of Health and Welfare) that have suffered from or are suspected to have caught diseases as specified in the Ministerial Ordinance of Health and Welfare, or those derived from such poultry that have died otherwise than by slaughter, shall neither be sold as food products nor be collected, processed, used, cooked, stored or displayed for the purpose of selling them as food products. However, this clause shall not apply to the meat, bone, and viscera derived from animals or poultry that have died otherwise than by slaughter, if they are determined to be harmless for human health and suitable for human consumption by the personnel of the competent authority.

The meat and viscera derived from animals or poultry, and their products described in the Ministerial Ordinance of Health and Welfare, shall not be imported for the purpose of selling them as food products, unless they are accompanied either by a certificate or by a copy thereof, which was issued by the competent authority of the exporting country and which specifies that they are neither derived from such animals or poultry as have suffered or are suspected to have suffered from such diseases as specified in the Ministerial Ordinance of Health and Welfare nor derived from such animals or poultry that have died otherwise than by slaughter. Article 5 also specifies other 'sanitation requirements'.

Provided, however, that this paragraph does not apply to the meat, etc. of animals imported from such countries as provided for in the Ministerial Ordinance of Health and Welfare, in cases where the sanitation requirements concerned were transmitted through electronic communication channels from the government organization of the country concerned to the computer provided for in paragraph 7 of Article 2 and recorded in the file of the computer concerned.

Article 7

The Minister of Health and Welfare may, from the viewpoint of public health, set the standards on the methods of manufacture, processing, use, cooking or storage or the specifications on the ingredients of food or food additives intended to be placed on the market.

In cases where standards or specifications have been established as provided by the preceding paragraph, any food or food additives shall not be manufactured, processed, used, cooked or stored otherwise than by the methods in compliance with such standards, any food or food additives manufactured by the methods other than by the methods in compliance with such standards shall not be sold or imported, or any food or food additives that do not meet such specifications shall not be manufactured, imported, processed, used, cooked, stored or sold.

The Minister of Health and Welfare may give approval to the person who manufactures or processes (including a person in a foreign country) food products, for which the standard procedures for manufacturing or processing are established in accordance with paragraph 1 of Article 7. These are also prescribed by the Cabinet Order, for each type of food product manufactured or processed whenever an application for such approval is filed by any person. The manufacture or process would take into account comprehensive sanitary manufacturing practice, hazard analysis critical control point system (HACCP) which means the HACCP system is implemented, to prevent (and control) the occurrence of food sanitation hazard during manufacturing or processing.

The Minister of Health and Welfare shall not give the approval under the preceding paragraph in cases where the procedures of manufacturing or processing and their sanitation control practice in the system for comprehensive sanitary manufacturing practice (HACCP) of the applicant field as specified in the same paragraph do not comply with the standards established in the Ministerial Ordinance of Health and Welfare.

Any person who files an application for approval under paragraph of Article 7 shall submit an application accompanied by the materials including the results of test performed on such food products manufactured or processed in conformity with their system for comprehensive sanitary manufacturing practice (HACCP).

The person who has obtained such approval (hereinafter referred to as the 'holder of approval') may request approval for the relevant change in cases where the holder wants to change any part of the system of comprehensive sanitary manufacturing practice (HACCP). In this case, the preceding two paragraphs may apply.

The Minister of Health and Welfare may withdraw in part or in whole the approval in any of the following cases:

- The method of manufacturing or processing and sanitation control procedures prescribed in HACCP of the relevant approval were modified without obtaining such approval as specified in the preceding paragraphs.
- The holder of approval in the foreign country (hereafter referred to as 'foreign holder of approval for manufacture') fails to report or falsely reports, in cases where the person is requested to report by the Minister of Health and Welfare.
- The inspection by the officers or employees of the competent authority of the food products, documents and other articles in the manufacturing or processing facility, office, warehouse or other places of the foreign holder of approval for manufacture is refused, hindered, or prevented.

The manufacture or processing of the food products through the system for comprehensive sanitary manufacturing practice (HACCP) approved under provision in paragraph 3 of Article 7 is regarded as that in compliance with the standards as provided in paragraph 1 of Article 7 and the provisions of this law or those of the order issued on the basis of this law shall apply.

Any person who files the application for approval or for the change in the approval shall pay the fee which is required by the Cabinet Order taking into consideration the cost required for the examination for such approval.

Enforcement Order and Enforcement Regulation of the Food Sanitation Law

Article 1 of the Enforcement Order (Cabinet Order) of Food Sanitation Law supplements Article 7 of the Law. It indicates that the following foods are those prescribed by Cabinet Order:

1. Cow's milk, goat's milk, skimmed milk and liquid milk containing reconstituted milk.
2. Cream, ice cream, condensed milk with no added sugar, condensed skim milk with no added sugar, fermented milk, lactic acid bacteria drinks and milk drinks.
3. Meat products.
4. Food products packaged and thermally processed under pressure.

The Articles of Ministerial Ordinance of Health and Welfare for Enforcement Regulation of the Food Sanitation Law that supplement Articles 4 and 5 are as follows:

Article 1 Such cases where food and food additive does not render it injurious to health as provided for in the proviso in subparagraph 2 of Article 4 of the Food Sanitation Law shall be as follows:

1. Any toxic or harmful substances are naturally contained in or borne on food or food additives, and the extent to which they are contained or the treatment thereof proves generally that such food or food additive is not injurious to the health of ordinary people.
2. Any toxic or harmful substances are inevitably mixed in or added to the food or food additive in the process of manufacturing such food or food additive and their mixing or addition proves generally that such food or food additive is not injurious to the health of normal people.

Article 2-1 Diseases of animals as provided for in paragraph 1 of Article 5 of the Law shall be those as listed in **Table 1** and **Table 2**. The whole carcasses and parts of the animal affected with the diseases in Tables 1 and 2 respectively, shall not be offered for human consumption.

2. Such an animal as prescribed by the Ministerial Ordinance of Health and Welfare as provided for in paragraph 1 of Article 5 of the Law shall be a buffalo.

Table 1 Animal diseases leading to prohibition of human consumption and prohibition of dressing in slaughterhouses

Rinderpest, contagious bovine pleuropneumonia, foot-and-mouth disease, epidemic encephalitis, rabies, influenza (exclusively for cattle), Q fever, anthrax, blackleg, haemorrhagic septicaemia, malignant oedema, leptospirosis, Johne's disease (exclusively for those with general symptoms), piroplasmosis, anaplasmosis, trypanosomiasis, leukaemia, glanders, pseudofarcy, equine infectious anaemia (exclusively for those with general symptoms), listeriosis, pox, hog cholera, swine erysipelas, African hog cholera, vesicular disease, toxoplasmosis, salmonellosis, tuberculosis (exclusively for those with systemic or serious disease, or accompanied by remarkable undernourishment, those with widespread lesions in not less than one organ or lymph node, or those with remarkably developed acute lesions), brucellosis (exclusively for those with general symptoms), tetanus, pyaemia, sepsis, uraemia, jaundice (exclusively for those affected seriously), oedema (exclusively for those affected seriously), tumour (exclusively for those with widespread lesions in meat, viscera, bones or lymph nodes), trichinosis, cysticercosis caused by *Cysticercus cellulosae*, cysticercosis caused by *Cysticercus bovis* (exclusively for those with widespread lesions in the whole body), various symptoms of intoxications (exclusively for those harmful to human health), various symptoms of fever (exclusively for those with remarkably high temperature), or reaction to injection (exclusively for those with remarkable symptoms in response to the injection of a biological preparation)

Table 2 Animal diseases leading to prohibition of human consumption of animal parts

Disease	*Part of the carcass*
Johne's disease (exclusively for those with lesions limited to within a part of intestine)	Intestine, mesentery and blood
Equine infectious anaemia (exclusively for those with the lesions, limited to within the viscera)	Viscera affected and blood
Tuberculosis (exclusively for those with the lesion localized in the udder or in other organs, or with the lesions localized in small part and do not manifest any acute symptom even in cases where the lesion is observed in not less than two organs or lymph nodes)	Udder or other organ affected, lymph node affected, lymph node covering udder or other organ affected, and blood
Brucellosis (exclusively for those with the lesions localized in the limited part of udder or genital organs)	Udder, genital organs and lymph node covering these organs, and blood
Jaundice (exclusively for those with the lesions localized in the limited part of meat or viscera)	Affected part involved and blood
Oedema (exclusively for those with the lesions localized in the limited part of meat or viscera)	Affected part involved and blood
Tumour (exclusively for those with the lesions localized in the limited part of meat, viscera, bone or lymph node)	Affected part involved and blood
Parasitosis (except for trichinosis, cysticercosis caused by *Cysticercus cellulosae* and *C. bovis* – exclusively for widespread lesions in the whole body	Part from which parasite cannot be separated, and blood in the case of *Sarcocystis* sp. infection
Actinomycosis	Affected part involved and blood
Botryomycosis	Affected part involved and blood
Trauma	Affected part involved
Inflammation	Affected part involved, part contaminated with inflammatory products, and blood for multiple purulent inflammation
Degeneration	Affected part involved
Atrophy	Affected part involved
Malformation	Noticeably affected part involved and blood

Table 3 Poultry diseases leading to prohibition of human consumption

Poultry pest, rabies, fowlpox (exclusively for those with general symptoms), infectious bronchitis (exclusively for those with general symptoms), infectious laryngotracheitis (exclusively for those with general symptoms), Newcastle disease, leukaemia of fowls, cell inclusion body hepatitis, Marek's disease, parrot disease, fowl cholera, tuberculosis, colibacillaemia (colibacillosis), infectious coryza (exclusively for those with general symptoms), swine erysipelas, pullorum disease and other salmonella disease, staphylococcia, listeriosis, toxaemia, pyaemia, sepsis, mycosis, protozoiasis (exclusively for those with general symptoms), except for toxoplasmosis, parasitosis (exclusively for those with a disease spread generally), degeneration (exclusively for those with generalized degeneration), uratosis (exclusively for those with general symptoms), oedema (exclusively for those affected seriously), hydroperitoneum, haemorrhage (exclusively for those with general haemorrhage), inflammation (exclusively for those with generalized inflammation), atrophy (exclusively for those with generalized atrophy), tumour except for Marek's disease or fowl leukaemia (except for those with such tumour as localized only in a limited part of meat, viscera, bone or skin), abnormal shape, size, hardness, colour or odour of viscera (except for those localized in the limited part of viscera), abnormal body temperature (except for those with remarkably high temperature (not lower than 43°C) or with remarkably low temperature (lower than 40°C), but including those caused by sunstroke or heatstroke), jaundice, trauma (exclusively for those with generalized trauma), various symptoms of intoxications (exclusively for those which may be harmful to human health), emaciation and arrested development (exclusively for those affected seriously), condition induced by the intense reaction against the administration of a biological preparation, contamination with inflammatory products, etc. (exclusively for those contaminated generally), or poor bleeding

Table 5 Microbial standards for milk and milk products

Product	*Bacterial count*	*Coliform*
Raw milk and raw goat's milk for manufacturing milk and milk products	$\leqslant 4 \times 10^6$ ml^{-1a}	
Normal liquid milk, partly skimmed milk, skimmed milk, liquid milk containing recombined milk	$\leqslant 50 \times 10^3$ ml^{-1a} 0^b	Negative
Pasteurized goat's milk	$\leqslant 50 \times 10^3$ ml^{-1a}	Negative
Certified milk	$\leqslant 30 \times 10^3$ ml^{-1a}	Negative
Milk products	$\leqslant 30 \times 10^3$ ml^{-1a} 0^b	Negative
Evaporated milk, evaporated skimmed milk	0	Negative
Sweetened condensed milk or skimmed milk, cream powder, whole or skimmed milk powder, whey powder, buttermilk powder, formulated milk powder, ice milk, lacto-ice	$\leqslant 50 \times 10^3$ ml^{-1a}	Negative
Ice cream, concentrated skimmed milk	$\leqslant 100 \times 10^3$ ml^{-1a}	Negative

[a] By standard plating method
[a] After 14 days at 30°C or 7 days at 55°C incubation for the product storable at ordinary temperature

3. The diseases of poultry as provided for in paragraph 1 of Article 5 of the Law shall be those as listed in **Table 3** and **Table 4**. The whole carcasses of the poultry affected with the diseases in Table 3 shall not be offered for human consumption. Parts of carcasses of the poultry affected with the diseases in Table 4 shall not be offered for human consumption.

Table 4 Poultry diseases leading to prohibition of human consumption of poultry parts

Disease	*Part of the carcass*
Fowlpox (except for those with systemic symptoms)	Meat, viscera, bone and skin of the affected part
Infectious bronchitis (except for those with systemic symptoms)	Meat, viscera, bone and skin of the affected part
Infectious laryngotracheitis (except for those with systemic symptoms)	Meat, viscera, bone and skin of the affected part
Infectious coryza (except for those with systemic symptoms)	Meat, viscera, bone and skin of the affected part
Protozoiasis except for toxoplasmosis (exclusively for those with a disease spread systemically)	Meat, viscera, bone and skin of the affected part
Parasitosis (except for those with a disease spread systemically)	Meat, viscera, bone and skin of the affected part
Degeneration (except for those with systemic degeneration)	Meat, viscera, bone and skin of the affected part
Uratosis (except for those with systemic symptoms)	Meat, viscera, bone and skin of the affected part
Oedema (except for those affected seriously)	Meat, viscera, bone and skin of the affected part
Haemorrhage (except for those with systemic haemorrhage)	Meat, viscera, bone and skin of the affected part
Inflammation (except for those with systemic inflammation)	Meat, viscera, bone and skin of the affected part
Atrophy (except for those with systemic atrophy)	Meat, viscera, bone and skin of the affected part
Tumour except for Marek's disease or fowl leukaemia (exclusively for tumour localized only in meat, viscera, bone or skin)	Meat, viscera, bone and skin of the affected part
Abnormal shape, size, hardness, colour or odour of viscera (exclusively for those localized in the limited part of viscera)	Viscera of the abnormal part
Trauma (except for those with systemic trauma)	Meat, viscera, bone and skin of the affected part
Contamination with inflammatory products, etc. (except for those contaminated systemically)	Meat, viscera, bone and skin of the affected part

4. Such cases where the relevant personnel find that the food derived from animals may not be injurious to human health and be suitable for eating and drinking pursuant to the proviso of paragraph 1 of Article 5 shall be the case where healthy animals died from and immediately after an unexpected accident.

Article 2-2 The products prescribed under the Ministerial Ordinance of Health and Welfare as provided for in paragraph 2 of Article 5 of the Law shall be meat products.

Article 2-3 The items to be described in accordance with the Ministerial Ordinance of Health and Welfare as provided for in paragraph 2 of Article 5 of the Law shall be as follows:

1. For the meat and viscera of animals or poultry, the species of animal or poultry and as to the products prescribed in the previous Article, the name and species of meat and viscera used as raw material.
2. Number and weight.
3. Name and address of the shipper (name and place of the corporation in cases where the shipper is a corporation).
4. Name and address of the consignee (name and place of the corporation in cases where the shipper is a corporation).
5. For the meat or viscera (excluding products derived from processing such as division or slicing), the name of the organization that carried out the examination and other related items including the following:

(a) For animals, name of the organization that carried out the inspection of slaughtered animals (including inspection of live animals before slaughter and the inspections carried out before and after the dissection of slaughtered animals) or the position and name of the employee of the competent authority who carried out the inspection of slaughtered animals.
(b) For poultry, the name of the organization which carried out the poultry inspection (including inspection of live poultry, and inspections carried out after the removal of feathers and after the removal of viscera of slaughtered poultry) or the

Table 6 Microbial standards of meat and meat products

Products	*Coliform*	E. coli	Staphylococcus aureus	Salmonella *spp.*	Clostridia[a]
Dry meat products		Negative			
Unheated meat products[b]		$\leqslant$100 per g	1000 per g	Negative	
Specified heated meat products[c]		$\leqslant$100 per g	1000 per g	Negative	$\leqslant$1000 per g
Meat products heated after packing[d]	Negative				$\leqslant$1000 per g
Meat products heated before packing[d]		Negative	$\leqslant$1000 per g	Negative	

[a]Gram-positive, spore-forming anaerobic bacilli reducing sulphurous acid.
[b]Meat products not heated at 63°C for 30 min nor heated under equivalent conditions.
[c]Meat products heated under conditions other than at 63°C for 30 min or equivalent conditions to 63°C for 30 min.
[d]Meat products other than dried, unheated and special meat products.

Table 7 Microbial standards of other foods

Products	*Bacterial counts (per g)*	*Coliform*	E. coli *(MPN per 100 g)*	*Enterococcus*	Pseudomonas aeruginosa
Oyster for raw eating	$\leqslant 50 \times 10^3$		$\leqslant$230		
Soft drink		Negative		Negative[a]	Negative[a]
Frozen food products					
Products not requiring heating	$< 100 \times 10^3$	Negative			
Products heated just before freezing	$\leqslant 100 \times 10^3$	Negative			
Raw, edible fresh-frozen fishery products	$\leqslant 100 \times 10^3$				
Products requiring heating before being served	$\leqslant 3 \times 10^6$		Negative		
Boiled octopus	$\leqslant 100 \times 10^3$	Negative			

[a]For mineral water (soft drinks made only from water) packaged in the container under the pressure of CO_2 at less than 1.0 kgf cm^{-2} at 20°C and that have not undergone pasteurization or removal of microorganisms (by filtration, etc.).
MPN, most probable number.

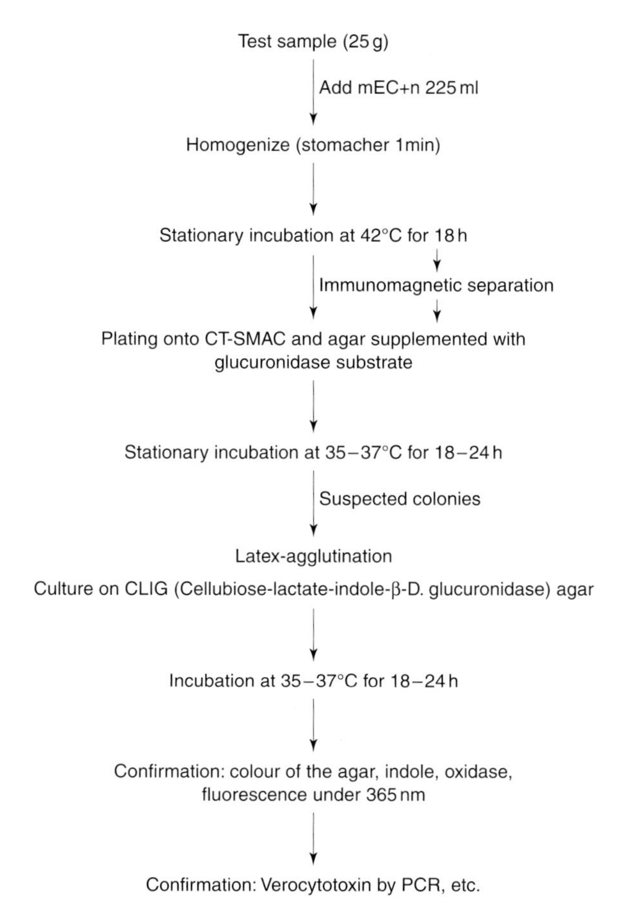

Figure 1 Protocol for detection of *Escherichia coli* O157:H7 in foods. Media: agar supplemented with glucuronidase substrate; e.g. CHROMagar O157, BCMO157; mEC+n, modified EC broth supplemented with 25 mg l^{-1} novobiocin; CT-SMAC; sorbitol–MacConkey agar supplemented with cefixime (0.05 mg l^{-1}) and potassium tellurite (2.5 mg l^{-1}). For investigation of causative food in *E. coli* O157:H7 infection, the following procedure should be taken in addition to the above procedure to promote sure detection of the organism: (1) Enrichment by incubation in trypticase-soy broth (TSB) and TSB supplemented with selective agents at 35–37°C and 42°C for 6 h and 18 h is recommended in addition to that in mEC+n at 42°C for 18 h. If 25 g sample is not available, use of an amount less than 25 g is recommended. (2) Enrichment culture and the fraction separated by immunomagnetic particles are recommended to be analysed by ELISA and other immunokits, and PCR. In any cases, isolation and confirmation of the organism may be needed. (3) Use of SMAC or other agars is recommended in addition to CT-SMAC and agar supplemented with glucuronidase substrate.

position and name of the employee of the competent authority who carried out the poultry inspection.

6. Name and place of the facility where procedures such as slaughter were carried out:

(a) For meat and viscera of animals (except for those undergoing processing as division, slicing, etc): the slaughterhouse where slaughter or dissection of animals was carried out.
(b) For meat and viscera of poultry (except for those which went through such processing as division, slicing, etc.): the poultry processing facility where the slaughter and removal of feathers and viscera were carried out.
(c) For meat and viscera of animals or poultry that underwent processing such as division, slicing, etc.: the facility where the processing was carried out.
(d) For the products as provided for in the preceding Article: the manufacturing facility where the products involved were manufactured.

7. The statement that the slaughter, dissection, removal of feathers and viscera, division, slicing, etc. specified in (a) through (d) of the preceding subparagraph were carried out in a sanitary manner in accordance with the national law of the exporting country.
8. Year and month when the slaughter, etc. listed in the following were carried out:

(a) For meat or viscera (except for those which went through such processing as division, slicing, etc.) of animals, slaughter and inspection of slaughtered animals.
(b) For meat or viscera (except for those undergoing processing such as division, slicing, etc.) of poultry, slaughter and inspection of poultry.
(c) For meat and viscera of animals or poultry which have undergone such processings as division, slicing, etc., processings concerned.
(d) For the products provided in the preceding Article, manufacture of the products concerned.

Article 2-4 For the certificate as provided for in paragraph 2 of Article 5, which certifies that meat or viscera of animals or meat or viscera of poultry underwent slaughter inspection in the country other than the exporting country, the copy of the certificate which was issued by the government agency of the country that had carried out the inspection of slaughtered animals or poultry and which specifies the items provided for in the preceding Article, shall be attached to the certificate provided for in the same paragraph.

Article 2-5 The countries which are provided by the Ministerial Ordinance of Health and Welfare under the proviso of paragraph 2 of Article 5 of the Law shall be the United States of America, Australia and New Zealand.

Abattoir Law, and Poultry Slaughtering Business Control and Poultry Processing Law

Slaughter and dressing of livestock is prohibited under the regulation of the Abattoir Law when the animals are recognized as suffering from the diseases listed in Table 1. When the diseases listed in Table 2 are recognized, the livestock slaughter must take disposal or other measures against the parts indicated in the right-hand column of the table.

The poultry slaughterer is prohibited from slaughtering animals suffering from the diseases listed in Table 3 under the regulation of the Poultry Inspection Law. When the diseases listed in Table 4 are recognized, the poultry slaughterer must take disposal or other measures against the parts indicated in the right-hand column of the table.

Microbial Standards

Microbial standards have been set up in Japan for 24 groups of foods including milk and milk products, meat and meat products. Examples of current microbial standards are summarized in **Tables 5–7**.

Development of Test Methods

A number of large outbreaks of *Escherichia coli* O157:H7 infection occurred in Japan in 1996. In order to facilitate surveillance of foods and investigation of food-borne *E. coli* O157:H7 infection, the Ministry of Health and Welfare recommended a test method for detection of the organism in foods (**Fig. 1**). This method was chosen on the basis of extensive studies including interlaboratory comparison of various methods.

See also: **Hazard Appraisal (HACCP)**: The Overall Concept; Involvement of Regulatory Bodies.

Further Reading

Japan Food Hygiene Association (1997) *Food Sanitation Law*. Tokyo: Japan Food Hygiene Association.

Japan Food Hygiene Association (1993) *Poultry Slaughtering Business Control and Poultry Inspection Law*. Tokyo: Japan Food Hygiene Association.

Food Sanitation Division, Ministry of Health and Welfare (1997) *Food Sanitation Administration*. Tokyo: Ministry of Health and Welfare.

Veterinary Sanitation Division, Ministry of Health and Welfare (1997) *Veterinary Sanitation Administration*. Tokyo: Ministry of Health and Welfare.

NATURAL ANTIMICROBIAL SYSTEMS

Contents

Preservative Effects During Storage

Vivian M Dillon, Department of Biology and Biochemistry, University of Bath, UK

Introduction

Interest in natural antimicrobial systems reflects objections to chemical preservatives by consumers, who nevertheless desire safe food products with convenient shelf lives. Only a few naturally occurring antimicrobial compounds have been tested in foods, and there are many others, of plant, animal or microbial origin, that could be exploited. In many cases, the mode of action of antimicrobial compounds, and their effectiveness in foods, is not known – to determine the preservative effect of a natural antimicrobial system during storage, the role of ecological factors needs to be understood.

Ecological Concept of Food Preservation

The microorganisms which contaminate food products are phylogenetically and phenotypically diverse, reflecting the origin of the raw materials and the conditions during transport, storage and processing. However, the intrinsic factors (food structure and composition) and the extrinsic factors (storage conditions) together select specific dominating microorganisms. Chemical, physical and natural preservation methods modify the food and so affect the microflora. Hence, ecological concepts must be considered when determining the suitability of a specific preservative, as well as the combined effects of pH; temperature; other microorganisms; treatments such as irradiation; vacuum packaging and modified-atmosphere packaging; interactions with food components; and the survival of the preservative in the food during storage. The significance of such factors is demonstrated by studies which show that some antibacterial compounds are more effective in broth cultures than in food products.

Traditional Methods

The traditional method of preserving food by fermentation depends on the natural microflora to produce the desired flavour and is associated with a lowered pH and other preservative characteristics. For example, the production of traditional oriental fermented foods depends on natural mixtures of yeasts, bacteria and moulds. Although the dominant microorganisms reflect the geographical location and seasonal variations, in many cases lactic acid bacteria play an important role. The processes of mincing, chopping and dense packing enhance the anaerobic conditions in the food, and ensure the even distribution of the nutrients and the fermentative microflora. Marination in salt and sugar, back slopping (the use of inoculum from previous fermentations) and the addition of substrates (usually cereals) encourage domination by the desired microorganisms.

When fermented foods were developed commercially, the fermentation was controlled by the inoculation of selected food-grade starter cultures. For example, the fermentation of milk gives products with either a short shelf life (e.g. yoghurt) or a long shelf life (e.g. cheese). Other fermented products which are produced commercially include buttermilk, fermented sausages, kefir, koumiss, sauerkraut, sourdough bread and tofu. A well-established history of the safe use of starter cultures made them readily acceptable to consumers, regulatory agencies and the food industry.

Growth of Desirable Microorganism

The traditional methods of fermentation of foods, to improve palatability and extend shelf life, were associated with a decline in pH: this encouraged the succession of microorganisms that tolerate acid conditions. The acid tolerance and production of antimicrobials by lactic acid bacteria enhanced their survival in these fermenting ecosystems. The identification of the important fermentative microorganisms led to the deliberate inoculation of starter cultures, to give foods desirable organoleptic properties. During the early use of starter cultures in the dairy industry, the antimicrobial bacteriocin, nisin, was discovered. This led to the addition of bacteriocins or bacteriocin-producing lactic acid bacteria to foods, to inhibit food spoilage bacteria and foodborne pathogens.

Inhibition of Undesirable Microorganisms

Microbial contaminants can be inhibited by the addition of preservatives, anaerobic atmospheres, low pH, low temperature or low water activity (a_w). The microorganisms that proliferate are those that can grow in the imposed conditions and can easily assimilate the simple nutrients available in the food. When physical and chemical factors are imposed, the remaining microflora may consist of only a few, or even one, species. In these conditions, a natural antimicrobial compound which specifically inhibits that species can be used. In less extreme conditions, several genera of microorganisms may persist, for example populations of lactic acid bacteria, yeasts and acetic acid bacteria which proliferate in acidic beverages.

In an environment which is mildly inhibitory to growth due to a combination of factors, only one genus or one species tends to reach numerical dominance and cause spoilage, although several might have been present originally. For example, in aerobically stored chilled meat, pseudomonads are the main spoilage microorganisms. They become predominant by converting glucose to gluconate, which may result in a lack of glucose limiting the growth of other microorganisms. This factor, combined with the ability of pseudomonads to grow rapidly at 4–6°C, decreases the shelf life of the meat. Modified-atmosphere packaging methods were developed in order to extend the shelf life of meat: CO_2 inhibits the growth of pseudomonads. Modified-atmosphere packaging favours slow-growing microorganisms such as species of *Carnobacterium*, *Weissella* and *Brochothrix thermosphacta*, but in many cases, lactic acid bacteria grow rapidly and dominate the microflora. Natural antimicrobial systems used in such cases would have to be effective against a cocktail of meat spoilage microorganisms.

The shelf life of meat can be extended by the use of sulphites. These impede the growth of Gram-negative bacteria, but favour that of yeasts and of Gram-positive bacteria such as *B. thermosphacta* and homofermentative lactobacilli. Pseudomonads are suppressed, resulting in growth of yeasts. These produce acetaldehyde, which binds to SO_2, neutralizing its preservative power. In such cases, a natural antimicrobial system that inhibits yeasts could be used in combination with a lower concentration of SO_2.

The inactivation of microorganisms by appertization is influenced by pH. Food with an acidic pH requires only a mild heat treatment to kill the same number of microorganisms as that achieved by higher temperatures for longer periods in foods with neutral pH. Natural antimicrobials that attack the cell envelope could be used in conjunction with appertization, to sensitize the microorganisms to heat treatment at high or low temperatures, or to high hydrostatic pressure.

Antimicrobial Systems Naturally Present in Foods

The preservation of traditional fermented foods by lactic acid bacteria is due mainly to: the acidic pH resulting from the metabolism of sugars to lactic and acetic acids; the increased concentration of undissociated acids, and the concomitant decrease in available carbohydrates; and the production of antimicrobial metabolites such as H_2O_2, diacetyl, reuterin and bacteriocins. Many of these compounds kill or suppress food-borne pathogens and spoilage microorganisms, by acting on specific targets (e.g. cell membranes or key enzymes). The activity of the antimicrobial compounds is affected by concentration, temperature, pH, contact time, food components and the type and growth stage of the bacterial contaminants. Factors such as nutrient limitation, although unlikely to occur in a food product, may result in the alteration of the structure of the outer and inner membranes of Gram-negative bacteria, permitting antimicrobial compounds to enter the cell. Other interacting factors may also come into play, for example, the lactic acid bacteria may multiply faster than the spoilage or pathogenic bacteria under the imposed conditions. In addition a lowered a_w or the exclusion of O_2 may have effects.

Crude or purified antimicrobial compounds, or antagonistic microorganisms themselves (probably in the form of safe 'food-grade' starter cultures), could be used to control pathogenic and spoilage bacteria. They would not, however, replace Good Manufacturing Practices: their role would be as an additional barrier in the 'hurdle' concept of food safety.

Organic Acids

The starter cultures in hard cheeses and fermented sausages produce acid, and the fall in pH ensures the stability of the food products and inhibits the growth of Enterobacteriaceae, coliforms, salmonellae, staphylococci, *Escherichia coli* and *Listeria monocytogenes*. The production of soy sauce involves a rapid fermentation with the concomitant production of acetic acid, and this inhibits yeasts and hence fermentation to alcohol. The effectiveness of this type of inhibition depends on the rate of fermentation, the duration of the processing time, the initial contamination level and the type of microorganisms present.

Acid production may be enhanced by the addition of sugar to a food product. *Lactobacillus plantarum* and *Pediococcus acidilactici*, inoculated into bacon supplemented with sucrose and a reduced amount of nitrite (the Wisconsin method), inhibit the growth of *Clostridium botulinum*, by means of acid production and hence a lowered pH. Similarly, staphylococci have been inhibited in country-style hams, and *C. botulinum* has been inhibited in chicken salads.

A fail-safe system, in which acidification occurs only at raised temperatures, can be used to extend the shelf life of non-fermented refrigerated foods such as red meat, poultry and seafood. Studies show that food-borne pathogens, in meat slurries supplemented with glucose, were inhibited by the acid produced by *Lactobacillus sake* at temperatures above 10°C. The efficacy of such a system would depend on the concentration of fermentable carbohydrate; the rates of growth and of acid production by the lactic acid bacteria at abuse temperatures; the initial pH and the buffering capacity of the food; the presence of other antimicrobial factors; and the type and concentration of the undesirable bacteria.

Organic acids disrupt the cell membrane and key enzymes. The lipophilic, undissociated acetate molecule penetrates the bacterial cell membrane and on entering the cell dissociates in the cytoplasm (where the pH is higher), thus releasing protons which must be exported from the cell to maintain a constant intracellular pH. This response disrupts the proton motive force and may uncouple oxidative phosphorylation and nutrient transport processes. The antimicrobial efficacy of organic acids increases as the pH decreases, and is a function of the undissociated molecules. Antimicrobial activity also increases in anaerobic conditions and with temperature. It is reduced by a high level of microbial contaminants and by the buffering capacity of the food product.

In theory, lactic acid bacteria could be genetically modified to produce more acetic acid – this would enhance their potential as food preservatives.

Hydrogen Peroxide

Hydrogen peroxide inhibits bacterial growth, respiration and viability. Its bactericidal effect is due to oxidation, caused directly or indirectly by a metabolite such as a hydroxyl radical (OH.), formed by the reaction between H_2O_2 and superoxides (compounds containing the O_2^- group). The hydroxyl radical is very reactive, and damages essential cell components such as membrane lipids and DNA.

As well as producing acid and H_2O_2, lactobacilli compete with other microorganisms for the limited amounts of vital nutrients available, (e.g. niacin, biotin). The combined effects of these three characteristics are important in inhibiting food-borne pathogens, such as *Salmonella typhimurium*, *Staphylococcus aureus*, enteropathogenic *Escherichia coli* and *Clostridium perfringens*.

Diacetyl

Diacetyl (2,3-butanedione) and its reduced forms (acetoin and 2,3-butanediol) are produced by the metabolism of sugars via pyruvate. However, diacetyl production is low unless there is an additional source of pyruvate, citrate or acetate. Several other factors affect the production of diacetyl by bacteria. The presence of lactate and an increase in temperature from 21°C to 30°C reduce diacetyl production. In contrast, production is increased with the presence of metal ions, particularly Cu^{++}, Mg^{++} or Mn^{++}, aeration or the addition of hydrogen and catalase to milk. The optimum pH for diacetyl production is pH 4.5–5.5, and its antimicrobial properties decrease with an increase in pH. Diacetyl is only antimicrobial at high concentrations,, and it has a greater effect on Gram-negative bacteria, yeasts and moulds than on Gram-positive bacteria.

Reuterin

Reuterin (3-hydroxypropionaldehyde) is produced and excreted by certain heterofermentative lactobacilli, e.g. *Lactobacillus reuterii*, during the anaerobic metabolism of glycerol or glyceraldehyde. It has a low molecular weight, is non-proteinaceous and is highly soluble in water, producing a solution with neutral pH. It also has broad-spectrum antimicrobial activity, being effective against yeasts, moulds, protozoa, and Gram-negative and Gram-positive bacteria.

Reuterin and/or reuterin-producing lactobacilli, with or without the addition of glycerol, could be used to control spoilage and pathogenic microorganisms in foods. Reuterin inhibits *Escherichia coli* and species of *Salmonella*, *Shigella*, *Clostridium*, *Staphylococcus*, *Listeria* and *Candida*.

Bacteriocins

Bacteriocins are protein-containing macromolecules of low molecular weight with either a narrow or a broad antimicrobial spectrum. *Lactococcus lactis* subsp. *lactis*, in particular, produces nisin, a widely used bacteriocin which is categorized as GRAS (generally recognized as safe).

Nisin is a cationic antimicrobial polycyclic peptide, which contains unusual residues – dehydroalanine, dehydrobutyrine, lanthionine and β-methyl-lanthionine. Nisin is nontoxic and stable to heat at a low pH, and its solubility in water decreases as the pH increases. Its activity is partially protected from heat damage by the large protein molecules of milk.

Nisin inhibits the vegetative cells of Gram-positive bacteria, and is bacteriostatic to the spores of *Bacillus* and *Clostridium* species. The effect of nisin on the outgrowth of *Bacillus* spores is affected by high pH, high spore loads and high incubation temperatures. Some species of *Bacillus* and lactic acid bacteria produce nisinase, a nisin-hydrolysing enzyme.

Nisin and nisin-producing starter cultures are added to processed Swiss-type cheeses, to inhibit the gas-forming *Clostridium butyricum* and *C. tyrobutyricum* and to suppress toxin production by *C. botulinum*. Heat-damaged bacterial spores are more sensitive to nisin than intact spores, so a gentler heat process can be used during canning to prevent the outgrowth of clostridial spores in canned vegetables. A combination of nisin and nitrite in fermented meat products such as frankfurters inhibits the outgrowth of *C. perfringens*.

Nisin is also used to prevent the spoilage of beer by diacetyl-producing lactobacilli, and to inhibit malolactic fermentation (the conversion of dicarboxylic malate to monocarboxylic lactate) in some white wines. Nisin and nisin-resistant *Leuconostoc oenos* are used to control malolactic fermentation in red wines. Nisin-producing *L. lactis* subsp. *lactis*, together with a nisin-resistant starter culture, can be used to control the fermentation of sauerkraut.

Nisin causes the formation of pores in the cytoplasmic membrane of susceptible cells, and the associated efflux of ATP, potassium ions and amino acids results in dissipation of the membrane potential. The sensitivity of *Salmonella* species and other Gram-negative bacteria to nisin can be enhanced by using disodium ethylenediaminetetraacetic acid (EDTA). This disrupts the lipopolysaccharide layer of the outer cell membrane by binding magnesium ions, enabling nisin to act on the cytoplasmic membrane.

Bacteriocin-producing lactic acid bacteria inhibit *Listeria monocytogenes*, *Staphylococcus aureus*, *Clostridium perfringens* and *Bacillus cereus*. An increase in antibacterial activity occurs when a combination of bacteriocins is used, e.g. pediocin AcH and nisin – this combination could be used as a natural preservative, to control *L. monocytogenes* in particular. The production and the effectiveness of bacteriocins are increased at low pH values. Either viable or non-viable producer bacteria, or the purified bacteriocins, can be added to foods. However, in complex foods, problems of solubility and distribution occur because bacteriocins migrate to the fat phase or bind to proteins.

Lactic acid bacteria could be genetically modified to generate strains that would produce great quantities of bacteriocins ('super-producers') or broader-spectrum bacteriocins. Bacteriocin production is plasmid-encoded, so can be transferred by genetic manipulation to a food-grade starter culture. Bacteriocin-producing starter cultures would ensure dominance of a fermentation by lactic acid bacteria. Ideally, bacteriocins would be effective against Gram-positive and Gram-negative spoilage and pathogenic bacteria, and also against vegetative cells and spores, but would not alter the organoleptic properties of foods. Bacteriocins should be effective at low doses, and stable during the storage of the product. Proteolytic enzymes in the human gastrointestinal tract should render them harmless.

Lactic Acid Bacteria

Food-grade lactic acid bacteria used as preservatives must have antagonistic mechanisms able to survive freezing, drying and storage methods. The physical and chemical characteristics, the antimicrobial mechanisms and the stability of the compounds when added to a food product must be determined. The bacteria must be fast-growing, phage-insensitive, salt-tolerant and genetically stable. Viable lactic acid bacteria, non-growing cells, spent medium containing antimicrobial compounds or the purified compound can be used to preserve a food product.

Yeasts

Saccharomyces cerevisiae, used to ferment bread, beer and wine, produces antimicrobial compounds including ethanol, sulphite and killer toxins. Sulphite is produced primarily by the reductive assimilation of sulphate. Its antimicrobial effect is pH-dependent, because only the undissociated sulphurous acid crosses the microbial membrane by passive diffusion. Sulphite is therefore more effective in acid foods, because at neutral pH, sulphite and bisulphite ions predominate. Sulphite is highly reactive – it reacts with aldehydes, ketones and thiamin and cleaves disulphide bonds. It decreases intracellular pH due to dissociation, and depletes ATP.

Killer Toxins Killer toxins are produced by yeasts, and are narrow-spectrum antifungal proteinaceous compounds. *Saccharomyces cerevisiae*, for example, produces four such killer toxins. The microorganisms sensitive to killer toxins belong to either the same genus or the same species as the producing organism. The toxin attaches to the cell wall or to receptors in the cell wall containing β-1,6-D-glucan, causing disruption of the plasma membrane, and consequently loss of ions and cell death. The toxins are effective at pH values within a narrow range (4.6–4.8), and are unstable at temperatures above 25°C. Killer toxins therefore have very limited potential for use in foods, particularly as they are not effective against bacteria.

Ethanol Ethanol, the end product of glycolysis, is inhibitory to all microorganisms. A concentration of 18–20% ethanol prevents the spoilage of beverages by lactobacilli. Ethanol migrates into the hydrophobic regions of cell membranes more effectively than water and may replace water bound to macromolecules, thus reducing membrane integrity. In acidic conditions, this increased membrane permeability permits the influx of protons, and the energy normally used for growth-related processes is used instead to maintain the intracellular pH. An associated leakage of ions, cofactors, magnesium and nucleotides occurs. The antimicrobial efficacy of ethanol is influenced by temperature and the associated reduction in a_w.

Antimicrobial Compounds Produced by Plants

Traditionally, herbs and spices have been used to enhance the flavour of foods and extend their shelf life. For example, the essential oils of oregano and thyme contain carvacrol and thymol, which inhibit the growth of *Aspergillus* species. Yeasts, Gram-negative and Gram-positive bacteria are also susceptible to the essential oils of plants. Phenolic compounds are the major antimicrobial components of the essential oils of spices, and affect the permeability of cell membranes. The antimicrobial activity of phenolic compounds is affected by binding to proteins, lipids or salts and by pH and temperature.

The uses of the essential oils of plants in foods are limited, because the oils would affect the aroma and flavour of the foods. Low concentrations could be used in a combined system, to enhance antimicrobial activity – for example, clove oil and sucrose have been shown to act synergistically. Alternatively, spices could be used with lactic acid bacteria – the manganese in spices (e.g. clove, cardamom, ginger, celery seed, cinnamon and turmeric) enhances the rate of acid production by lactic acid bacteria, which are used to ferment sausages. Sublethal heat treatment can be used to disrupt the cytoplasmic membrane of yeast cells, giving the anti-yeast components of essential oils access to the cytoplasm, where they inhibit the repair mechanisms.

Chelating Agents

Avidin Avidin is a basic tetrameric glycoprotein, representing 0.05% of the total protein in the albumen of hens' eggs. It combines with biotin, depriving microorganisms of this vitamin, an essential cofactor of several key enzymes. Consequently, it inhibits yeasts, Gram-negative and Gram-positive bacteria. Many microorganisms synthesize biotin, yet are inhibited by avidin – this may be due to avidin binding to the cell membrane and altering its permeability. The importance of avidin in the antimicrobial defences of hens' eggs is not proved.

Transferrins Transferrins, such as ovotransferrin (from hens' eggs) and lactoferrin (from milk), chelate iron, making it unavailable for microorganisms. However, this is probably not important in food products, where plenty of iron is usually available.

Lactoferrin Lactoferrin, an iron-chelating glycoprotein, competes well with bacterial siderophores. It also binds to the outer membrane of Gram-negative bacteria, and alters its permeability. This feature is modulated by Ca^{++} or Mg^{++}. Lactoferrin has been shown to inhibit *Listeria monocytogenes* in ultra-high temperature (UHT) milk. It has been used in infant formulae in developing countries, to combat enteritis.

Ovotransferrin Ovotransferrin has an enhanced antimicrobial effect on Gram-negative and Gram-positive bacteria when complexed with zinc. Synergism is also evident when ovotransferrin is combined with bicarbonate or citrate ions, or with EDTA. Salmonellae, *Clostridium botulinum* and *L. monocytogenes* are inhibited by ovotransferrin in combination with EDTA – EDTA may disrupt the cell wall, giving ovotransferrin access to the peptidoglycan.

Enzymes

The use of enzymes in foods is restricted owing to economic factors, their limited activity spectrum and their inactivation by endogenous food components.

Lysozyme Lysozymes are defined as 1,4-β-*N*-acetylmuramidases. They cleave the glycosidic bond between the carbon in position 1 of *N*-acetylmuramic acid and that in position 4 of *N*-acetylglucosamine in peptidoglycan. Most Gram-positive bacteria are extremely susceptible to lysozyme because their cell walls consist of 90% peptidoglycan. However in staphylococci, teichoic acids and other cell wall materials bind lysozyme and prevent its diffusion. *Bacillus cereus* is also resistant to lysozyme, because its glucosamine lacks *N*-acetyl groups. Gram-negative bacteria are usually resistant because the lipopolysaccharide of the outer membrane prevents the diffusion of lysozyme, but this protection can be disrupted by shifts in pH or temperature, or by the removal of Ca^{++} or Mg^{++} by EDTA.

Lysozyme (from hens' eggs) is used in semihard and hard brine-salted cheeses, to control the fermentation of lactate by *Clostridium tyrobutyricum*, which causes 'late blowing'. To improve the survival of lysozyme during cheese-making, the lysozyme gene from bacteriophage T4 has been modified, to make the enzyme more thermostable.

Lysozyme shows increased activity in combination with butyl-*p*-hydroxybenzoate, *p*-hydroxybenzoic esters, amino acids, organic acids, (e.g. ascorbate), salts or chelating agents (e.g. phytic acid or EDTA). Enhanced inhibition of *C. botulinum* and *L. monocytogenes* occurs when lysozyme is used in combination with EDTA. When bacterial cells (e.g. *Escherichia coli* are physically damaged (e.g. by freezing), they become more permeable to lysozyme.

Lactoperoxidase Lactoperoxidase is a glycoprotein (mol.wt 77 000) containing one haem group (protohaem IX) and is an important enzyme in bovine milk. It catalyses the oxidation of thiocyanate (SCN^-) by hydrogen peroxide (H_2O_2) to hypothiocyanite ($OSCN^-$), forming the lactoperoxidase system (LPS), which inactivates a broad range of microorganisms. The concentration of the thiocyanate anion, the principal electron donor in cow's milk, depends on the breed of cow and the type of feed consumed. Hydrogen peroxide is formed in milk by enzymic action, or is produced by lactic acid bacteria when dissolved oxygen is present. The LPS causes oxidation of the thiol groups of enzymes and affects the cytoplasmic membranes of sensitive microorganisms, causing the leakage of potassium ions, amino acids and polypeptides.

The LPS is harmless to mammalian cells, and also occurs in human saliva, milk and tears. It is, therefore, an ideal tool for extending the shelf life of milk products, particularly in the tropics where refrigeration is likely to be unavailable on farms. The LPS has antibacterial action on Gram-negative bacteria such as coliforms, pseudomonads, salmonellae, shigellae and *Campylobacter jejuni*. Its effectiveness depends on pH, the population size, species and the growth stage of the bacteria.

Future Developments

Natural antimicrobial compounds of plant, animal or microbial origin can be used to extend the shelf life of foods, but must meet the criteria of the regulatory agencies. The compound must remain active in the food product and must survive any heating, freezing or storage processes. It needs to be active in the appropriate type of food (e.g. liquid, semisolid, solid) at the correct pH range, and must be antagonistic to the specific microorganisms growing in the imposed storage conditions. For use as food preservatives, antimicrobial compounds must not cause organoleptic changes and must be economic to produce and toxicologically safe. They are likely to need to be used in additive or synergistic combinations, or alongside physical treatments (e.g. heating, freezing or irradiation), which can damage the microbial cells so that it becomes more sensitive to the antimicrobial compound. Such biopreservatives could also be used in combination with a lower concentration of an established preservative.

See also: **Bacillus**: *Bacillus cereus*. **Bacteriocins**: Potential in Food Preservation; Nisin. ***Clostridium***: *Clostridium perfringens*; *Clostridium botulinum*. **Fermentation (Industrial)**: Basic Considerations. **Fermented Foods**: Origins and Applications; Fermented Meat Products. **Fermented Milks**: Range of Products. **Good Manufacturing Practice**. ***Lactococcus***: *Lactococcus lactis* Sub-species *lactis* and *cremoris*. ***Leuconostoc***. ***Listeria***: *Listeria monocytogenes*. **Natural Antimicrobial Systems**: Antimicrobial Compounds in Plants; Lactoperoxidase and Lactoferrin; Lysozome and other Proteins in Eggs. ***Saccharomyces***: *Saccharomyces cerevisiae*. ***Staphylococcus***: *Staphylococcus aureus*. **Starter Cultures**: Moulds Employed in Food Processing. **Yeasts**: Production and Commercial Uses.

Further Reading

Banks JG, Board RG and Sparks NHC (1986) Natural antimicrobial systems and their potential in food preservation in the future. *Biotechnology and Applied Biochemistry* 8: 103–147.

Board RG, Jones D, Kroll RG and Pettipher GL (eds) (1992) *Ecosystems: Microbes: Food*. Society for Applied Bacteriological Symposium Series 21. *Journal of Bacteriology* 73: supplement.

Davidson PM and Branen AL (eds) (1993) *Antimicrobials in Foods*, 2nd edn. New York: Marcel Dekker.

Dillon VM and Board RG (eds) (1994) *Natural Antimicrobial Systems and Food Preservation*. Wallingford: CAB International.

Gould GW (ed.) (1989) *Mechanism of Action of Food Preservation Procedures*. London: Elsevier Applied Science.

Gould GW (ed.) (1995) *New Methods of Food Preservation*. London: Blackie Academic & Professional.

Gould GW (1996) Industry prospectives on the use of natural antimicrobials and inhibitors for food applications. *Journal of Food Protection* SS: 82–86.

Gould GW, Rhodes-Roberts ME, Charnley AK, Cooper RM and Board RG (eds) (1986) *Natural Antimicrobial Systems. Part 1: Antimicrobial Systems in Plants and Animals*. Bath: Bath University Press.

Lindgren SE and Dobrogosz WJ (1990) Antagonistic activities of lactic acid bacteria in food and feed fermentations. *FEMS Microbiology Reviews* 87: 149–164.

Ray B and Daeschel M (eds) (1992) *Food Biopreservatives of Microbial Origin*. Boca Raton: CRC Press.

Russell NJ and Gould GW (eds) (1991) *Food Preservatives*. Glasgow: Blackie.

Schillinger U, Geisen R and Holzapfel WH (1996) Potential of antagonistic microorganisms and bacteriocins for the biological preservation of foods. *Trends in Food Science and Technology* 7: 158–164.

Stiles ME and Hastings JW (1991) Bacteriocin production by lactic acid bacteria: potential for use in meat preservation. *Trends in Food Science and Technology* 2: 247–251.

Vandenbergh PA (1993) Lactic acid bacteria, their metabolic products and interference with microbial growth. *FEMS Microbiological Reviews* 12: 221–238.

Antimicrobial Compounds in Plants

C Umezawa and **M Shin**, Faculty of Pharmaceutical Sciences, Kobe Gakuin University, Japan

Introduction

Fungi and bacteria cause a wide range of plant diseases. Plants defend themselves from invading pathogens by a combination of 'physiological' and induced mechanisms. Physiological compounds are produced by way of a plant's normal metabolism. Induced defence mechanisms operate only in response to infection. In some cases, the mechanisms of resistance to pathogens are well understood: for example, phytoalexins are produced by plants in direct response to several pathogens. Phytoalexins constitute a chemically heterogeneous group of substances, such as isoflavonoids, sesquiterpenoids, polyacetylenes and stilbenoids. Most phytoalexins have been isolated from members of the Leguminosae. Virulent microorganisms usually tolerate higher concentrations of phytoalexins than avirulent strains; this difference is usually due to the ability of the virulent strain to degrade the phytoalexin.

Range of Antimicrobial Compounds in Plants

A variety of substances isolated from plants has been reported as having significant antimicrobial activity (**Table 1**).

Avocado

In unripe avocado, a long-chain diene, *cis,cis*-1-acetoxy-2-hydroxy-4-oxo-heneicosa-12,15-diene, has antifungal activity towards *Colletotrichum gloesosporioides*, which causes disease in a number of tropical crops. During ripening, the concentration of the diene falls rapidly, possibly due to the action of lipoxy-

Table 1 Antimicrobial compounds synthesized by plants

Antimicrobial compound	*Plant*
Cis,cis-1-acetoxy-2-hydroxy-4-oxo-heneicosa-12,15-diene	Avocado
Allicin	Garlic
Avenacin	Oat
Catechins:	Tea
Epigallocatechin gallate	
Epicatechin gallate	
Gallocatechin gallate	
Flavones:	Citrus
Nobiletin	
5,4'-dihydroxy-6,7,8,3'-tetramethoxy flavone	
Solanine	Potato
α-Tomatine	Tomato

genase, the activity of which increases. Lesions may then develop in the plant.

Garlic

The antibacterial and antifungal activity of garlic is attributed to allicin, a thio-2-propene-1-sulphinic acid *S*-allyl ester produced by the interaction of alliin, a non-protein amino acid, with alliinase. The major biological effect of allicin is attributed to its anti-oxidative activity and its rapid reaction with thiol-containing proteins. Allicin inhibits the SH-protease papain, $NADP^+$-dependent alcohol dehydrogenase from *Thermoanaerobium brockii* and the NAD^+-dependent alcohol dehydrogenase from horse liver. All three enzymes can be reactivated by thiol-containing compounds: papain by glutathione; alcohol dehydrogenase from *T. brockii* by dithiothreitol or 2-mercaptoethanol; and alcohol dehydrogenase from horse liver by 2-mercaptoethanol.

Oat

Avenacins, a group of triterpenoid saponins found in oat plants, are responsible for resistance to the fungus *Gaeumannomyces graminis* var. *tritici*, a major pathogen of wheat and barley.

Tea

The green tea plant, *Camellia sinensis*, has antibacterial and antiviral activity in vitro. This is attributed to catechins, polyphenol compounds. Tea polyphenols, especially epigallocatechin gallate, epicatechin gallate and gallocatechin gallate, inhibit the growth and adherence of *Porphyromonas gingivalis* onto buccal epithelial cells. Catechin oligomers are synthesized during the manufacture of oolong tea. These catechin oligomers markedly inhibit the glucosyltransferase of *Streptococcus mutans*. Catechins deactivate proteins and affect the membrane of the target cell. Catechins are much more active against Gram-positive than Gram-negative bacteria, probably owing to the protective outer membrane of Gram-negative bacteria.

Citrus

Several species of citrus trees, including mandarin and grapefruit, are resistant to *Deuterophoma tracheiphila*, a fungus that causes one of the most destructive diseases of citrus trees, mal-secco. Fungistatic flavones, such as nobiletin and 5,4′-dihydroxy-6,7,8,3′-tetramethoxy flavone, have been obtained from mandarins.

Potato and Tomato

Solanine from potato (*Solanum tuberosum*) and α-tomatine from tomato (*Solanum lycopersicum*) are both steroidal saponins. They contribute to the protection of the plants against attack by phytopathogenic fungi. In vitro, both solanine and α-tomatine caused the disruption of model membranes, possibly by the insertion of the aglycone section into the lipid bilayer.

Reduction of Spoilage by Antimicrobial Compounds

The antimicrobial activity of preformed defence compounds has been tested in vitro. It is unclear whether they are all important in conferring resistance to pathogens, but the roles of some physiological antimicrobial compounds are well understood.

α-Tomatine

α-Tomatine is a saponin found in tomato plants, in high concentrations. Successful pathogens of tomato are more resistant to α-tomatine in vitro, due to the ability to break down α-tomatine using the enzyme tomatinase. *Cladosporium fulvum* is sensitive to α-tomatine, and cannot break it down. When cDNA encoding tomatinase from the fungus *Septoria lycopersici* was expressed in *C. fulvum*, the tomatinase-producing transformants showed increased sporulation on the cotyledons of susceptible tomato plants and caused extensive infection of resistant tomato plants. Thus α-tomatine contributes to the ability of the tomato plant to restrict the growth of *C. fulvum*.

Avenacins

Avenacins are triterpenoid saponins found in oat plants. They confer resistance to the fungus *Gaeumannomyces graminis* var. *tritici*, but oat plants are susceptible to *G. graminis* var. *avenae*. The latter possesses the enzyme avenacinase, which detoxifies avenacin by the removal of the terminal D-glucose

Table 2 Phytoalexins and the plants that produce them

Phytoalexin	*Types of compound*	*Plant*
Betavulgarin	Isoflavonoid	Beet
Capsidiol	Sesquiterpenoid	Green pepper
Glyceollin	Isoflavonoid	Soybean
Lubimin	Sesquiterpenoid	Aubergine
6-Methoxymellein	Isocoumalin	Carrot
Momilactone	Diterpenoid	Rice
Oryzalexin A	Diterpenoid	Rice
Phaseolin	Isoflavonoid	French bean
Phytuberin	Sesquiterpenoid	Potato
Pisatin	Isoflavonoid	Pea
Rishitin	Sesquiterpenoid	Tomato
α-Viniferin	Stilbenedimer	Grape
Wyerone	Furanoacetylene	Broad bean

residues. Disruption of the gene encoding the enzyme, by homologous recombination, gives avenacinase-negative mutants, which are sensitive to avenacin-A1 and are not pathogenic to oat plants.

Response of Plants to Infection

Phytoalexins are antimicrobial metabolites of low molecular weight. Their structural diversity is shown by **Table 2**. They are not detectable in uninfected plant tissues, and are synthesized by plants in response to infection by a microbial pathogen. Phytoalexins accumulate at the sites of infection in concentrations which are inhibitory to the development of fungi and bacteria. There are similarities between the phytoalexins of plants within the same family. For example, over 80% of the phytoalexins reported in the Leguminosae family are isoflavonoid derivatives – plants in this family have not been reported to produce sesquiterpenoid phytoalexins, and those in the Solanaceae family have not been reported to produce isoflavonoid phytoalexins. Microbial infection can also induce other plant defence responses, for example the synthesis of proteinase inhibitors and the accumulation of hydroxyproline-rich glycoproteins. These defence responses can be induced by compounds known as 'elicitors'.

Role of Elicitors

Elicitors are fungal and bacterial extracellular metabolites, produced when plant cells are damaged by microbial infection. Their production is due to the action of plant hydrolases, released by necrotic plant cells, on the pathogen. The host plant then recognizes the elicitor molecule by means of a plasma membrane receptor, and this results ultimately in the synthesis of antimicrobial compounds by healthy tissue.

Infection of the soybean plant by the phytopathogenic fungus *Phytophthora megasperma* results in the attack of the fungal cell wall by β-1,3-glucanase, contained in the host tissue. As a result, an elicitor (β-glucan) is released and this initiates the accumulation of phytoalexin by the plant. Parsley leaves develop a reaction against *P. megasperma* which is typical of this plant species. The response comprises hypersensitive cell death, defence-related gene activation and the synthesis of the phytoalexin furanocoumarin. An oligopeptide of 13 amino acids, identified within a 42 kDa glycoprotein elicitor from *P. megasperma*, has been shown to stimulate phytoalexin formation in parsley.

Plant cells are capable of responding to elicitors in a species-specific manner. Elicitor-binding proteins function as receptors, transmitting extracellular signals which cause the activation of nuclear genes. **Figure 1** summarizes current knowledge about the perception and transduction of elicitor signals in parsley. When the fungal oligopeptide elicitor binds to its receptor in the plant plasma membrane, the ion channels for H^+, Ca^{++}, K^+ and Cl^- in the plasma membrane are affected. The absence of Ca^{++} from the culture medium, or the addition of calcium channel blockers, inhibits not only the flux of calcium ions but also phytoalexin production, indicating that calcium channels are involved in elicitor-mediated signal transduction. Elicitor-specific, calcium-dependent phosphorylation of several proteins has been demonstrated in vivo. The phosphorylated proteins may be involved in the expression of defence-related genes in parsley.

Phytoalexin Synthesis

The complete biosynthetic routes of several phytoalexins have been elucidated. Disease resistance is associated with increased levels of mRNA of the enzymes necessary for phytoalexin synthesis.

The major phytoalexin accumulating at sites of infection in soybean plants is glyceollin, as isoflavonoid. The biosynthetic pathway of glyceollin is shown in **Figure 2** . An important reaction in the biosynthesis of glyceollins is the formation of the essential intermediate 4,2′,4′-trihydroxychalcone from 4-coumaroyl-CoA and malonyl-CoA. The reductase, which requires NADPH as a cofactor, acts with chalcone synthase in the formation of 4,2′,4′-trihydroxychalcone. Chalcone synthase is involved in both the flavonoid and the 5-deoxyflavonoid pathways, and the reductase is a key enzyme in phytoalexin synthesis because it catalyses the essential first step which channels the metabolites into the biosynthetic pathway.

The induction of chalcone synthase and the reductase are stimulated and coordinated following the

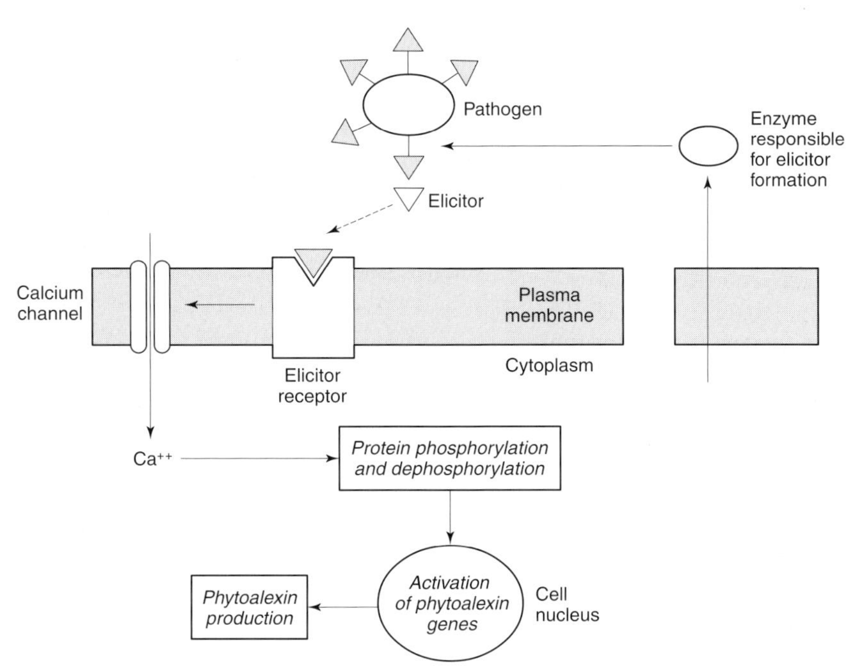

Figure 1 Hypothetical model of activation of phytoalexin synthesis by a pathogen.

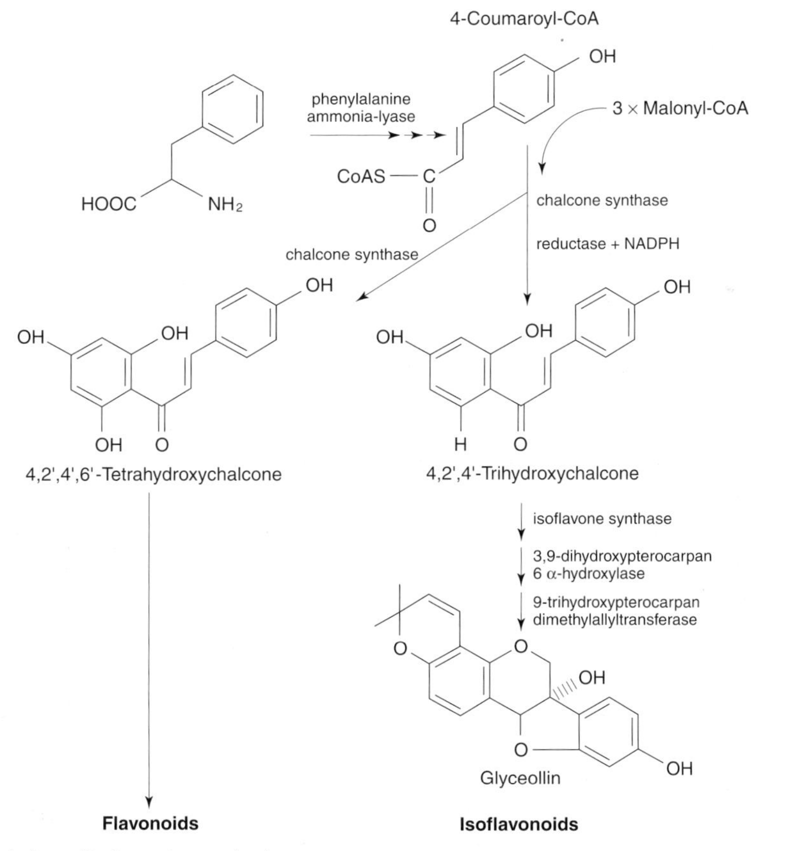

Figure 2 Biosynthesis of glyceollin in soybean plants.

treatment of soybean cell cultures with the elicitor of *Phytophthora megasperma* f. sp. *glycinea*, which induces glyceollin synthesis. Induction is preceded by increases in mRNA activity. The haploid genome of the French bean (*Phaseolus vulgaris*) contains six to eight genes of chalcone synthase. Marked differences were found in the pattern of accumulation of specific transcripts and encoded polypeptides in wounded

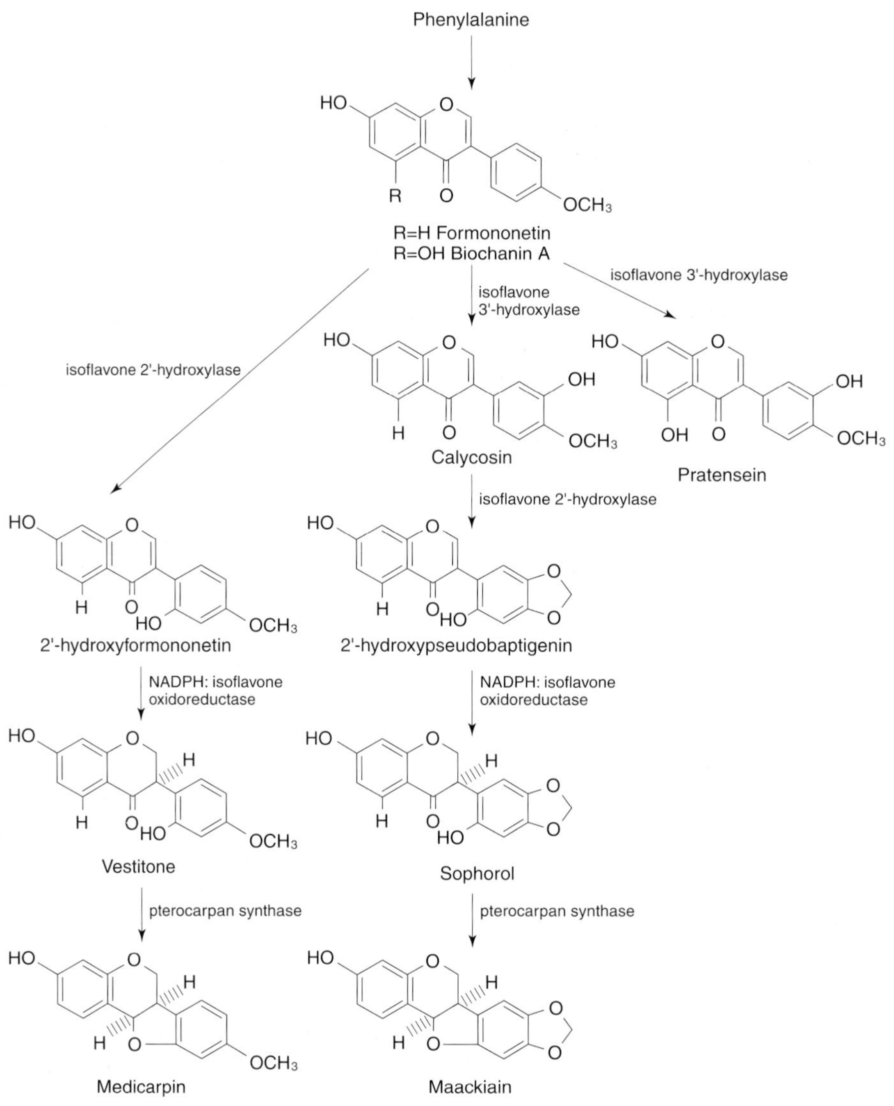

Figure 3 Biosynthesis of the pterocarpan phytoalexins, medicarpin and maackiain, in chickpea plants.

cells, compared to that in cells treated with fungal elicitor. This result suggests that the chalcone synthase genes are regulated differently in the bean in response to different environmental stresses. Compounds from the biosynthetic pathway of glyceollins also behave as signals in the interaction between the bean plant and the bacterium *Rhizobium*, and as regulators of the genes needed for the growth of nodules.

When the chickpea (*Cicer arietinum*) is infected with the fungus *Ascochyta rabiei*, the phytoalexins medicarpin and maackiain are produced (**Fig. 3**). Isoflavone 2′-hydroxylase and isoflavone 3′-hydroxylase both catalyse an early step in the pterocarpan-specific branches of biosynthesis and are induced by elicitors. However, isoflavone 3′-hydroxylase is also involved in the synthesis of the physiological isoflavone pratensein. NADPH:isoflavone oxidoreductase also catalyses pterocarpan-specific steps of the synthetic pathway, namely the reduction of 2′-hydroxyformononetin to vestitone and the reduction of 2′-hydroxypseudobaptigenin to sophorol. It too is induced by *A. rabiei* elicitor in chickpea cells in vitro.

Genetic engineering can be used to cause plants to produce 'foreign' phytoalexins, by the incorporation of genes that alter the phytoalexin biosynthetic pathway.

Stilbene-type phytoalexins are normally formed in a range of plants including grape, peanut and pine. Stilbenes are synthesized from malonyl-CoA and 4-coumaroyl-CoA, and the only enzyme specifically required for stilbene biosynthesis is stilbene synthase. The genes for stilbene synthase were isolated from grape and transferred into tobacco plants. In grape, stilbene synthesis can be induced by fungal infection or by treatment with elicitor. Similarly, when transgenic tobacco callus was treated with a preparation of fungal elicitor from *Phytophthora megasperma*, stilbene synthase mRNA accumulated and stilbene synthase activity followed. Thus it seems that defence-related genes can be transferred between grape and tobacco plants.

The level of phytoalexin-mediated resistance in plants is believed to depend on the ability to produce a high concentration of the compound within a short time after infection. In studies on transgenic tobacco plants containing stilbene synthase genes from grape, a correlation between resveratrol concentrations and disease incidence in the tobacco leaves could be demonstrated. Resveratrol is a phytoalexin formed in certain varieties of grape attacked by *Botrytis cincerea*. It was shown that for enhanced disease resistance, a high concentration of resveratrol 48 h after inoculation with *B. cinerea* was required.

The increased disease resistance of transgenic plants expressing the stilbene synthase gene raises the possibilities that the biosynthetic pathways of plants could be modified by recombinant DNA technology, and 'foreign' phytoalexin synthesis could be induced. The development of crop plants less prone to attack by pathogens would thereby be facilitated.

Mode of Action of Phytoalexins

Although the antimicrobial activity of phytoalexins is proven, the mechanism underlying their toxicity to pathogens have not been satisfactorily explained. One theory suggests that their toxicity is due to bacterial membrane disruption. This is supported by some studies.

Camalexin (3-thiazol-2′-yl-indole) is the phytoalexin formed by *Arabidopsis thaliana* infected with the virulent pathogen *Pseudomonas syringae* pv. *maculicola*. Studies have revealed that camalexin disrupts bacterial membranes.

Glyceollin, a soybean phytoalexin, was found to inhibit Ca^{++} transport in sealed plasma membrane vesicles isolated from the pathogenic fungus *Phytophthora megasperma* f. sp. *glycinea*. It also increased Ca^{++} leakage from *Phytophthora* membrane vesicles.

Detoxification of Phytoalexins

Phytoalexins accumulate in plants or cell cultures only transiently, because they are degraded or polymerized by extracellular peroxidases. It has been suggested that such detoxification of plant phytoalexins by fungi is a general mechanism for circumventing phytoalexin-mediated plant defences. Degradation may either be complete or be limited to one or a few reaction steps.

The degradation of phytoalexins may be important for the development of particular plant diseases. Examples include the degradation of the phytoalexin kievitone in beans, by *Fusarium solani* f. sp. *phaseoli*; the oxidation and reduction of the phytoalexins medicarpin and maackiain in chickpeas, by *Ascochyta rabiei*; and the degradation of the potato phytoalexins lubimin and rishitin by *Gibberella pulicaris*.

The phytoalexin pisatin helps to defend the pea (*Pisum sativum*) against *Nectria haematococca*, a fungal pathogen. Pisatin demethylase, a cytochrome P-450, detoxifies pisatin. In studies, the genes encoding pisatin demethylase have been transformed into and highly expressed in *Cochliobolus heterostrophus*, a fungal pathogen of maize but not of pea. The rates of pisatin demethylation by the transformants were equal to or greater than those of the highly virulent *N. haematococca* wild-type strain, and the recombinant *C. heterostrophus*, while remaining virulent on maize, also became virulent on pea. It appears that the production of high levels of pisatin demethylase, alone, enhances the ability of *C. heterostrophus* to attack pea. This suggests that pisatin demethylase is required by *N. haematococca* for pathogenicity on the pea, and hence demonstrates that pisatin is a plant defence factor. A knowledge of phytoalexin catabolism will underpin the development of compounds resistant to degradation by pathogens.

See also: **Spoilage of Plant Products**: Cereals and Cereal Flours. **Natural Antimicrobial Systems**: Lactoperoxidase and Lactoferrin; Preservative Effects During Storage; Lysozyme and other Proteins in Eggs.

Further Reading

Barz W, Bless W, Börger-Papendorf G et al (1990) Phytoalexins as part of induced defence reactions in plants; their elicitation, function and metabolism. *Ciba Foundation Symposium* 154: 140–156.

Denarie J, Debelle F and Rosenberg C (1992) Signaling and host range variation in nodulation. *Annual Review of Microbiology* 46: 497–531.

Ebel J and Grisebach H (1988) Defense strategies of soybean against the fungus *Phytophthora megasperma* f. sp. *glycinea*: a molecular analysis. *Trends in Biochemical Sciences* 13: 23–27.

Halverson LJ and Stacey G (1986) Signal exchange in plant-microbe interactions. *Microbiological Reviews* 50: 193–225.

Ingham JL (1982) In: Bailey JA and Mansfield JW (eds) *Phytoalexins*. Pp. 21–80. London: Blackie and Sons.

Kue R and Rush JS (1985) Phytoalexins. *Archives of Biochemistry and Biophysics* 236: 455–472.
Larson RA (1988) The antioxidants of higher plants. *Phytochemistry* 27: 969–978.
Phillips DA and Kapulnik Y (1995) Plant isoflavonoids, pathogens and symbionts. *Trends in Microbiology* 3: 58–64.
Rosen L, Davis EO and Johnston AWB (1987) Plant-induced expression of *Rhizobium* genes involved in host specificity and early stages of nodulation. *Trends in Biochemical Sciences* 12: 430–433.
Strange RN (1998) Plants under attack II. *Science Progress* 81: 35–68.

Lysozyme and Other Proteins in Eggs

E A Charter, Canadian Inovatech Inc., Abbotsford, BC, Canada

G Lagarde, Inovatech Europe BV, Zeewolde, The Netherlands

As the vessel that is designed for the embryogenesis of a bird or reptile, it is not surprising that an egg contains numerous antimicrobial proteins. By far the most widely studied of these is the small, but significant, lytic enzyme known as lysozyme. This enzyme has the ability to lyse specific bacteria. It is used in several food and pharmaceutical products. Three other egg proteins (avidin, ovotransferrin and yolk immunoglobulins) have antimicrobial properties of particular interest for application in foods and pharmaceuticals.

Occurrence

Lysozyme was first identified in 1921 in human nasal secretion by Alexander Fleming, who would later discover penicillin. Lysozyme has since been isolated from human tears, saliva and mother's milk, as well as viruses, bacteria, phage, plants, insects, birds, reptiles and other mammalian fluids. Commercially, the most readily available source of lysozyme is the egg white of the domestic chicken (*Gallus gallus*). Lysozyme is probably the most intensively studied of all proteins.

Lysozyme is not, however, the only antimicrobial protein in avian eggs. **Table 1** lists some of the proteins present in avian egg white. Most research to date has focused on the proteins of egg white, perhaps because they can readily be separated by ion-exchange chromatography.

Most yolk proteins, on the other hand, tend to be less water soluble, and because of their close association with lipids, they are somewhat less easily extracted and purified on a large scale. Nevertheless, there are antimicrobial proteins present in the yolk, including immunoglobulins (IgY – the chicken equivalent of IgG) and trace amounts of a biotin-binding protein. It has been proposed that antimicrobials in yolk work primarily to provide the developing chick with passive protection, as is the case with the immunoglobulins.

The four egg proteins that appear at present to have the greatest potential as natural antimicrobials in food and pharmaceutical applications are lysozyme, avidin, ovotransferrin and IgY. The remainder of this article focuses on the properties and applications of these particular egg proteins.

Table 1 Some of the major proteins in egg white and their antimicrobial functions

Protein	*Solids (%)*	*Antimicrobial function*
Ovalbumin	54	Unknown
Ovotransferrin	12	Binds multivalent cations, particularly iron
Ovomucoid	11	Inhibits trypsin and other proteases, antimicrobial properties
Ovoinhibitor	15	Inhibits trypsin, chymotrypsin and other proteases
Ovomucin	3.5	Increases viscosity of egg white preventing bacterial movement
Lysozyme	3.4	Lyses peptidoglycan layer of some Gram-positive organisms
Ovoflavoprotein	0.8	Binds riboflavin (vitamin B_2)
Ovomacroglobulin	0.5	Protease inhibitor
Ficin inhibitor (cystatin)	0.05	Inhibits cysteine proteases
Avidin	0.05	Binds avidin, making it unavailable to organisms

Data from: Ibrahim HR (1997) *Insights into the structure–function relationships of ovalbumin, ovotransferrin, and lysozyme*. pp. 37–56. In: *Hen Eggs Their Basic and Applied Science*. (1997) Yamamoto T, Juneja LR, Hatta H and Kim M (eds) Boca Raton, Florida: CRC Press. Li-Chan ECY, Powrie N and Nakai S (1995) The chemistry of eggs and egg products. In: Stadelman WJ and Cotterill OJ (eds) *Egg Science and Technology*. Pp. 105–175. New York: Food Products Press.

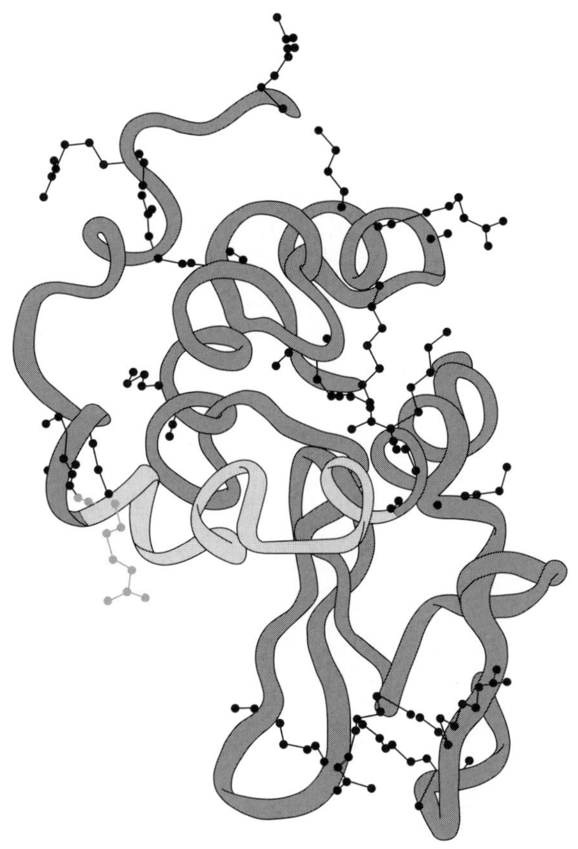

Figure 1 Three-dimensional structure of chicken lysozyme. The active site is highlighted by shading. (Courtesy of Canadian Inovatech Inc., Abbotsford, British Columbia, Canada.)

Structure

It is now recognized that lysozymes from diverse sources fall into several structural classes, with the three most common being type c (chicken), type g (goose) and type v (viral). Chicken lysozyme is composed of 129 amino acid residues with a molecular weight of approximately 14 000.

Figure 1 shows the three-dimensional structure of chicken lysozyme. Lysozyme is used extensively as a model enzyme, partly because it contains all of the twenty common amino acids. An α-helix links two domains of the molecule; one is mainly β-sheet in structure, and the other primarily α-helical. The hydrophobic groups are mainly oriented inward, with most of the hydrophilic residues on the exterior of the molecule.

It is proposed that the enzymatic action of the molecule is dependent on its ability to change the relative position of its two domains by hinge bending. Essentially, the small α-helical connection, or hinge, between the domains can bend sufficiently to cause large conformational changes in the molecule. This permits the enzyme to engage its substrate.

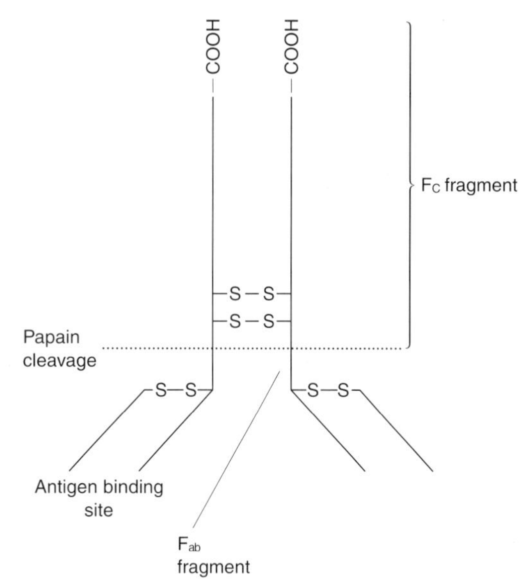

Figure 2 Schematic diagram of a typical immunoglobulin molecule. S-S represents disulphide bonds. F_{ab} is the antigen binding portion of the molecule. F_c is the heavy chain portion of the molecule.

Avidin is a glycoprotein composed of four identical subunits, each with 128 amino acid residues. The molecular weight of the entire molecule is around 67 000. A disulphide bridge links residues four and 83. There are four tryptophan residues per subunit.

Ovotransferrin, also known as conalbumin, is a glycoprotein with a molecular weight of 78 000. It contains two lobes connected by an α-helix. Each lobe is homologous, and can bind an Fe^{+++} ion. The iron-binding site in each lobe is situated between two subdomains. The presence of bicarbonate ion enhances the binding of iron to the molecule.

IgY is similar in structure to mammalian IgG, but with a higher molecular weight (170 000). **Figure 2** shows a schematic representation of the basic structure, with two heavy chains and two light chains, all connected by disulphide linkages, as indicated. The antigenic site is found in the F_{ab} fragment, which can be cleaved from the molecule by means of papain hydrolysis.

Properties

Lysozyme has an extremely high isoelectric point (> 10) and consequently is highly cationic at neutral or acid pH. In solution, lysozyme is relatively stable at pH 3–4 and can withstand near boiling temperatures for a few minutes. As the pH increases, however, its stability decreases. At pH 5.5–6.5 (the pH of many dairy products), lysozyme is stable up to about 65°C. Beyond this point, denaturation accelerates rapidly with temperature. Thus, many precooked food products, in which temperatures of 69–75°C are typically reached during processing, the enzyme may be partially or completely denatured.

Avidin is unique in that its complex with biotin results in an association coefficient (K_a) of about 10^{15}, the strongest such biological association known that does not involve covalent bonding. Avidin is inactivated at 85°C, but the avidin–biotin complex can withstand temperatures greater than 100°C for a brief period.

Ovotransferrin has an isoelectric point of 6.1. Each molecule is capable of binding two atoms of iron. Ovotransferrin will also bind aluminium and copper ions, with the order of strongest to weakest binding being $Fe^{+++} > Al^{+++} > Ca^{++}$.

IgY represents a heterogeneous population of molecules composed of several antibody subclasses, with an isoelectric point ranging from 5.0 to near neutrality. Yolk antibodies are more stable in the neutral pH range up to about 60°C, or above pH 4 up to 40°C. Below pH 4 or above 65°C, the antibody activity is greatly diminished.

Mode of Action

The primary mechanism by which lysozymes lyse microorganisms is by cleaving the bonds between C-4 of *N*-acetylglucosamine and the C-1 of *N*-acetylmuramic acid, the two repeating units of the peptidoglycan layer. It is fairly active against organisms with a relatively accessible peptidoglycan layer (some Gram-positive organisms), but against organisms where this layer is not as accessible (for example, Gram-negative organisms), the enzyme is not able to access its substrate and usually shows little or no antimicrobial effect. Compounds which help to destabilize the outer membrane of Gram-negative organisms (e.g. ethylenediaminetetra-acetate (EDTA)) appear to permit lysozyme to act on some of these otherwise unassailable targets. The lysing action of lysozyme can be quite dramatic when viewed by electron microscopy (**Fig. 3**).

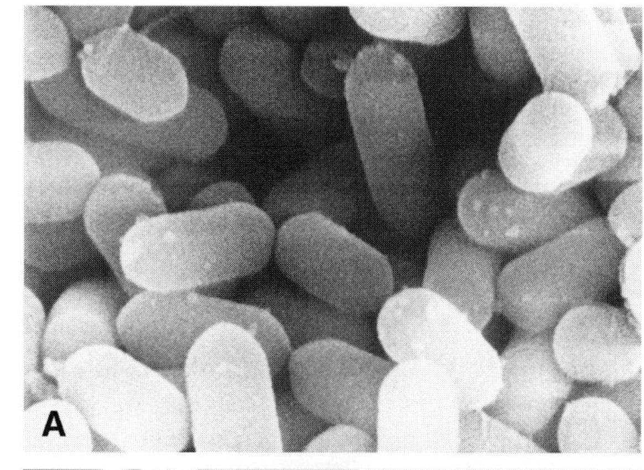

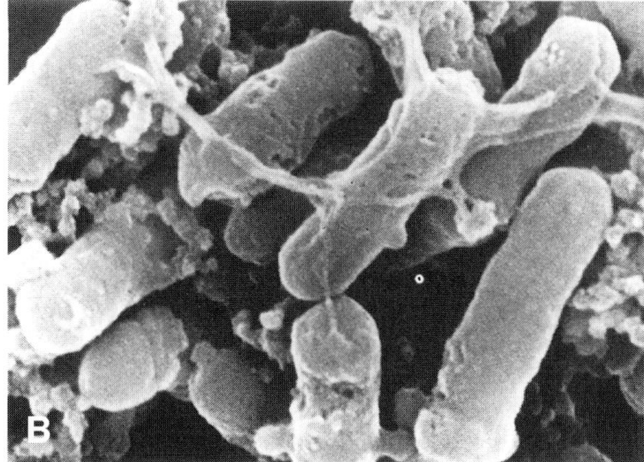

Figure 3 (**A**) Electron micrograph of *L. curvatus* grown in pork juice extract. (**B**) Electron micrograph of *L. curvatus* treated with 500 p.p.m. lysozyme. (Courtesy of Canadian Inovatech Inc., Abbotsford, British Columbia, Canada, and Dr. Frances Nattress, Agriculture & Agri-Food Canada, Lacombe Research Centre, Lacombe, Alberta, Canada.)

The antimicrobial properties of lysozyme are not only limited to its enzymatic action. It is also believed that by adhering to the exterior surface of some microorganisms (especially fungi), it can interfere with cell function and hinder growth and replication. Lysozyme also appears to cause agglutination of bacteria.

The antimicrobial effect of avidin is attributed to its ability to bind strongly with biotin. By depriving microorganisms of this essential nutrient, it has a bacteriostatic effect on biotin-reliant organisms.

For many years, the literature has described a bacteriostatic effect of ovotransferrin. This is attributed to the ability of ovotransferrin to bind iron, and thereby deprive iron-dependent organisms of an essential element. Recent studies have shown that there is also an antimicrobial effect that is relatively independent of the degree of iron saturation of the molecule. It is now believed that a peptide sequence embedded within the primary structure may, in fact, cause direct disruption of cell surfaces. This is based on knowledge of similarities in structure between ovotransferrin and lactoferrin, a milk transferrin containing a bactericidal peptide sequence known as lactoferricin. Recent research has identified a 92 residue sequence in the N-lobe portion of ovotransferrin that displays antimicrobial effect against *Staphylococcus aureus* and *Escherichia coli* K-12. Three disulphide bridges within this sequence contribute to a tertiary structure that is needed for the antimicrobial action.

IgY inhibits bacterial growth through a typical antibody precipitin reaction in which the antibodies coat the outer surface of target organisms and cross-link the cells into an insoluble mass. The optimal pH range for this effect is from pH 5 to 9.

Importance in Avian Eggs

Egg-white proteins protect the egg from invasion by foreign organisms by both physical and chemical defences. By forming a viscous layer between the shell and the yolk, egg white impedes the movement of organisms. Within the egg white, there is a network of fibres believed to be the result of interaction between lysozyme and ovomucin.

The chemistry of the defence in egg white is clearly linked to the antimicrobial proteins discussed earlier. The bactericidal effect of lysozyme is significant, but many organisms that gain access to the interior of avian eggs are Gram negative and therefore unlikely to be susceptible to the lytic action of lysozyme alone. At the basic pH of raw egg white ($pH > 9$), the iron sequestering ability of ovotransferrin is enhanced and it is also considered to be a major impediment to the growth of many organisms. The proteins presented in Table 1 present a relatively hostile environment to invading microorganisms by making biotin, riboflavin and iron relatively unavailable, inhibiting bacterial proteases and binding bacterial cells together through electrostatic interactions.

Importance in Food and Pharmaceutical Applications

Lysozyme has been used in pharmaceutical and food applications for many years, due to its lytic activity on the cell wall of Gram-positive microorganisms. These organisms are responsible for infections of the human body and for the spoilage of various foods.

Pharmaceutical Applications of Lysozyme

Hen egg white lysozyme is used in 'over-the-counter' drugs in order to increase the natural defences of the body against bacterial infections. Since lysozyme forms part of the human immune system, it has been proposed that supplementation with chicken lysozyme may have benefits. The pharmaceutical use encompasses applications such as otorhinolaryngology (lozenges for the treatment of sore throats and canker sores) and ophthalmology (eye drops and solutions for disinfecting contact lenses). Lysozyme is also added to infant formulae to make them more closely resemble human milk (cow's milk contains very low levels of lysozyme).

A recent publication has renewed interest in the antiviral properties of lysozyme by demonstrating anti-HIV activity of both human and chicken lysozyme in vitro. It is possible that these findings may lead to new means of treatment of this infection.

Food Applications of Lysozyme

Much research has been done on the use of lysozyme as a preservative in food products, particularly in the Far East and Japan. Several applications have been described and patented, including the treatment of fresh fruits, vegetables, seafood, meat, tofu, sake and wine.

The most important food application of lysozyme is the prevention of the problem known as 'butyric late blowing', which occurs during the ripening of certain European-type cheeses. This problem is due to contamination of milk by spores of *Clostridium tyrobutyricum*. The origin of this contamination lies in the widespread use of silage as a feed. The spores of *C. tyrobutyricum* are present in the soil and incorporated, together with soil particles, into the corn or hay used to make silage. The spores will proliferate in the silage if a rapid acidification does not take place. When the cows are fed the contaminated silage, the spores are excreted into the manure and, if the milking is not carried out under very strict hygienic conditions (thorough washing of the udder; elimination of the first drops of milk), the spores can subsequently contaminate the milk. It has been demonstrated that a very small amount of manure (less than one gram) is enough to contaminate a tank containing several thousand gallons of milk.

If cheese is made with milk contaminated with spores of *Clostridium tyrobutyricum*, the majority of the spores are retained in the curd. Here, the conditions (absence of oxygen and presence of large amounts of lactic acid) are favourable for the germination and for the development of vegetative forms during the ripening of the cheese. Lactic acid can be metabolized as the primary source of carbon by *C. tyrobutyricum* to produce butyric acid and a combination of two gases: hydrogen and carbon dioxide. The accumulation of butyric acid is responsible for organoleptic defects in the cheese due to the characteristic off-flavour caused by this short-chain fatty acid. The production of large volumes of hydrogen (totally insoluble in the water phase of the cheese curd), and of carbon dioxide (partly soluble), leads to an increase in the internal pressure of the cheese and, subsequently, to the formation of slits and cracks in the cheese during the ripening process. This has a dramatic and detrimental impact on the quality of the cheese and, consequently, on its commercial value. Cheese with a late blowing problem usually has to be downgraded or, in severe cases, cannot be sold at all.

Before the use of lysozyme, cheese makers developed a number of techniques to try to prevent butyric late blowing. There are two commonly implemented techniques: a physical process to eliminate the spores by centrifugation (known as 'bactofugation'); and the use of chemical inhibitors of *C. tyrobutyricum*, such as nitrates. Neither method can guarantee a complete solution to the problem.

Research begun in the late 1960s and 1970s, followed by cheese trials carried out in Europe in the early 1980s, has demonstrated the efficacy of lysozyme to prevent late blowing in different types of cheese. The principle of lysozyme action is based on its capacity to be retained in the cheese curd, through electrostatic attraction with the casein, and on the stability of its enzymatic activity throughout the ripening process. Lysozyme is active on the vegetative cells of *C. tyrobutyricum*, which appear during the ripening process. The usage level is usually 25 p.p.m. in the cheese milk. At this concentration, most of the lactic cultures used in the production of cheese, although Gram-positive bacteria, are not sensitive to the lytic action of lysozyme.

Lysozyme has been approved by the European Union, and has now been used with success for more than 15 years in several European countries (e.g. France, Italy, Spain, Portugal, Germany, Denmark, The Netherlands). Its use has also been successful in different types of cheeses, such as the hard cheeses (Parmesan, Swiss), the semihard cheeses (Gouda, Manchego), and the soft cheeses (Brie). Lysozyme recently received GRAS (Generally Recognized as Safe) status from the Food and Drug Administration of the United States, and is raising a lot of interest in North America for its application in speciality cheeses.

Because lysozyme is mainly effective against Gram-positive bacteria, recent research has focused on synergistic combinations of lysozyme and other antimicrobial compounds. These combinations often target the Gram-negative organisms against which lysozyme is relatively ineffective alone.

Applications of Avidin

Avidin, in conjunction with biotin, is used in a number of diagnostic and analytical applications, including biotinylated probes for a number of quantitative detection methods, affinity chromatography columns, immunoassays, immunohistochemistry and protein blotting.

Applications of Ovotransferrin

Although patents from the 1970s promoted ovotransferrin as a potential inhibitor of mildew in such Asian dishes as noodles, wonton and fried bean curd, it has not been used extensively for its antimicrobial properties in food applications. More recent patent applications propose the use of ovotransferrin to treat human immunodeficiency virus and to prevent periodontal disease. In Japan, immobilized ovotransferrin has been used to remove iron from drinking water, as well as water for brewing. Also, a company in the Netherlands has recently filed a patent for a nutraceutical drink containing ovotransferrin.

Applications of IgY

A number of recent patents have been issued for the use of IgY (often in the form of a crude extract or even administered directly with the egg yolk). Many include prophylactic or therapeutic use involving passive protection of fish, mammals or humans against pathogenic organisms or viruses. One interesting application involves the use of IgY to target food enzymes that cause deterioration of foods through discoloration, generating off-flavours or odours, or altering important physical properties.

The ability of domestic hens to generate antibodies to a large number of important antigens is likely to lead to the continued growth of applications for IgY.

See also: ***Bacillus***: *Bacillus cereus*. **Cheese**: Microbiology of Cheese-making and Maturation; Microflora of White-brined Cheeses. ***Clostridium***: *Clostridium tyrobutyricum*. **Eggs**: Microbiology of Egg Products. ***Listeria***: *Listeria monocytogenes*. ***Staphylococcus***: *Staphylococcus aureus*. **Starter Cultures**: Cultures Employed in Cheese-making.

Further Reading

Fennema OR (ed.) (1985) *Food Chemistry*, 2nd edn. New York: Marcel Dekker.

Ibrahim HR, Iwamori E, Sugimoto Y and Aoki T (1998) Identification of a distinct antibacterial domain within the N-lobe of ovotransferrin. *Biochimica et Biophysica Acta* 1401(3): 289–303.

International Dairy Federation (1987) *The Use of Lysozyme in the Prevention of Late Blowing in Cheese*. Bulletin 216. B-1040 Brussels, Belgium: International Dairy Federation.

Jolles P (ed.) (1996) *Lysozymes: Model Enzymes in Biochemistry and Biology*. Basle: Birkhauser Verlag.

Lee-Huang S, Huang PL, Sun Y, Huang PL, Kung H, Blithe DL, Chen H-C (1999) Lysozyme and Rnases as anti-HIV components in β-core preparations of human chorionic gonadotropin. Proceedings of the National Academy of Sciences 96: 2678–2681.

Mayes FJ and Takeballi MA (1983) Microbial contamination of the hen's egg: a review. *Journal of Food Protection* 46: 1092–1098.

Savage D, Mattson G, Desai S et al (1992) *Avidin–Biotin Chemistry: A Handbook*. Rockford, Illinois: Pierce Chemical Company.

Stadelman WJ and Cotterill OJ (eds) (1995) *Egg Science and Technology*, 4th edn. New York: Food Products Press.

Yamamoto T, Juneja LR, Hatta H and Kim M (eds) (1997) *Hen Eggs: Their Basic and Applied Science*. Boca Raton: CRC Press.

Lactoperoxidase and Lactoferrin

Barbaros H Özer, Department of Food Science and Technology, The University of Harran, Şanlıurfa, Turkey

Introduction

In general, milk is kept on farms for about 2–3 days before being transported to dairy plants, despite the fact that the storage of raw milk for several days at refrigeration temperatures diminishes the quality of dairy products. This deterioration is due to extracellular enzymes synthesized by psychrotrophic bacteria, and a reduction in the shelf life of the products is often observed. Some countries have insufficient cold storage systems for handling milk, leading to the excessive multiplication of bacteria and increases in the acidity of raw milk far beyond the level acceptable for processing. Chemical preservatives, such as H_2O_2 and sodium carbonate, are commonly used in many developing countries with inadequate refrigeration systems to maintain the acidity of milk at low levels, but their undesirable effects on human health are well known. The process of activating the natural antimicrobial systems in milk has been introduced into the dairy industry as an alternative way of preserving milk.

Natural Antimicrobial Systems

The specific immune system in mammals is suppressed during the period following birth, possibly for protection against hypersensitivity and allergic reactions. During this period, the animal's defence system is supported by nonspecific factors in milk.

A great deal of research has been carried out into the antimicrobial factors in milk, and three proteins have been found to be of particular importance: lactoperoxidase (LP), lactoferrin (LF) and lysozyme. There is a similarity between these proteins and leucocytes in terms of their antimicrobial effects. In most cases, the enzymes activate the natural antimicrobial system, resulting in a fatal effect on the target microorganisms. The most important characteristic of natural antimicrobial systems is the simultaneous attack on the oxidative and lytic mechanisms of the microorganism.

The Lactoperoxidase System

The lactoperoxidase system consists of lactoperoxidase, hydrogen peroxide (H_2O_2) and thiocyanate ions (SCN^-).

Components

Lactoperoxidase Lactoperoxidase (EC 1.11.1.7, donor H_2O_2 oxidoreductase) is one of the most abundant enzymes in bovine milk, constituting about 1% of whey proteins. Oxygen metabolism leads to toxic end products, such as oxygen free radicals and H_2O_2, in both prokaryotic and eukaryotic cells unless they are protected by enzymes such as superoxide dismutase, catalase and peroxidases. The primary function of superoxide dismutase is to convert superoxide radicals to H_2O_2. The H_2O_2 is then reduced to H_2O and O_2 by either catalase or peroxidase, the latter involving a variety of electron donors.

The oxidation of molecules by peroxidases leads to the formation of H_2O_2. As early as 1924, it was observed that freshly drawn milk was bactericidal to certain strains; this effect was ascribed to the presence of oxidizing enzymes. A few years later, the involvement of H_2O_2 and SCN^- in the antimicrobial properties of milk was demonstrated. Lactoperoxidase has been found in the milk of all species tested, and in many other types of secretion, e.g. saliva, tears, nasal fluid, uterine luminal fluid and vaginal secretions. Bovine milk contains high levels of LP, which can show activity even at very low concentrations.

Lactoperoxidase is a basic protein with one Fe^{3+}-containing haem group. The amino acid sequence of bovine LP was elucidated in the early 1990s. The molecule consists of 612 amino acid residues constituting a single peptide chain. The haem group is bound to the single peptide chain by a disulphide bond, but LP does not contain free thiol (–SH) groups. The complementary DNA (cDNA) sequences of bovine and human LP show that they are closely related. The predicted molecular weight of bovine LP is 77 500, and its isoelectric point (pI) is 9.8 (**Table 1**).

Some structural similarities between bovine LP, cytochrome *c* peroxidase and horseradish peroxidase have been demonstrated, using nuclear magnetic resonance (NMR). Similar arginine and histidine residues have also been found in these enzymes. Few metal ions are present in bovine LP at significant

Table 1 Specifications of lactoperoxidase and lactoferrin

Properties	*Lactoperoxidase*	*Lactoferrin*
Molecular weight	77 500	78 500
Isoelectric point (pI)	9.8	9.5
pH range	4–7	4–8

levels, except calcium: there is one calcium ion per iron atom, and the calcium has a high affinity for protein. The lactoperoxidase enzyme has an iron content of about 0.07% and a carbohydrate content of about 10%.

Using recently developed methods based on the oxidation of 2,2′-azino-di-(3-ethyl-benzthiazoline-6-sulphonic acid) (ABTS), average LP concentrations of about 1.4 units per millilitre in bovine milk and 0.9 units per millilitre in buffalo milk have been reported. Lactoperoxidase is resistant in vitro to levels of acidity as high as pH 3, and to human gastric juice. Bovine LP is relatively heat-stable, and high temperature short time (HTST) pasteurization at 72°C causes only partial inactivation: after pasteurization, the enzyme can still catalyse the reaction between SCN^- and H_2O_2.

Thiocyanate Animal tissues and secretions are the main sources of the thiocyanate (SCN^-) anion. Extracellular fluids contain large amounts of SCN^-, and it is concentrated by certain types of body cell. The concentration of SCN^- in blood is 0.1–0.3 mg kg^{-1} and in saliva it is 1–27 mg kg^{-1}. The anion is excreted in the urine and, given normal renal function, the half-life of elimination is 2–5 days. Bovine milk contains 1–10 mg kg^{-1} of SCN^-, although higher concentrations have been reported, particularly in milk with a high somatic cell count. However, the SCN^- concentration in milk does not normally exceed about 10 mg kg^{-1}, even if the animal feed is supplemented with SCN^-.

Glucosinolates and cyanogenic glucosides are the main dietary sources of SCN^-. Vegetables belonging to the genus *Brassica* (family Cruciferae), e.g. cabbage, kale, Brussels sprouts, cauliflowers, turnips and swedes, are particularly rich in glucosinates. On hydrolysis, these yield SCN^- and other reaction products. The hydrolysis of the glucosides releases cyanide, which is detoxified by conversion into SCN^- in a reaction catalysed by the enzyme rhodanase.

Hydrogen Peroxide It is widely believed that milk does not contain H_2O_2 at a level sufficient to activate the LP system, because the indigenous catalase and peroxidase enzymes reduce the H_2O_2 formed by mammary tissue throughout lactation. Therefore, the concentration of H_2O_2 is too low to exert an antimicrobial effect. However, H_2O_2 in milk, as well as being naturally formed, may be produced by lactic acid bacteria (mainly lactobacilli, streptococci and lactococci) that contaminate the milk or are added as starter cultures, and these can produce sufficient H_2O_2 to activate the LP system. H_2O_2 may also be added directly to milk, or an H_2O_2-generating system, e.g. sodium percarbonate, may be added. In general the latter is more efficient. Natural H_2O_2-generating systems include the oxidation of ascorbic acid; the oxidation of hypoxanthine by xanthine oxidase; and the manganese-dependent aerobic oxidation of reduced pyridine nucleotides by peroxidase.

Antimicrobial Effects of the LP System

Mechanism of Action The operation of the LP system leads to the oxidation of thiocyanate, to intermediates which may be further oxidized to sulphate, CO_2 and ammonia, or may be reduced back to SCN^-. The intermediate products of oxidation provide the antimicrobial effect of the LP system, by inhibiting growth, the uptake of O_2 and lactic acid production. The activities of some enzymes, such as hexokinase and glyceraldehyde-3-phosphate dehydrogenase, are also inhibited.

There is general agreement that the main intermediate product of oxidation is hypothiocyanate, $OSCN^-$. This inhibits the metabolic activity of streptococci as well as some enzymes, and the oxidation of SCN^-, catalysed by peroxidase, leads to the accumulation of $OSCN^-$. Two different pathways have been proposed for the production of $OSCN^-$ ions. In the first, the oxidation of SCN^- produces thiocyanogen $(SCN)_2$, which is rapidly hydrolysed to yield hypothiocyanous acid (HOSCN), and hence $OSCN^-$:

$$2SCN^- + H_2O_2 + 2H^+ \xrightarrow{peroxidase} (SCN)_2 + 2H_2O$$
$$(SCN)_2 + H_2O \longrightarrow HOSCN + SCN^- + H^+$$
$$HOSCN \longleftrightarrow H^+ + OSCN^-$$

However, the direct oxidation of SCN^- to $OSCN^-$ is also possible:

$$SCN^- + H_2O_2 \xrightarrow{peroxidase} OSCN^- + H_2O$$

The oxidation of the thiol (–SH) groups of enzymes and proteins is of crucial importance in the bacteriostatic and/or bactericidal effect of the LP system. The system may also inhibit SH-independent enzymes such as D-lactate dehydrogenase.

The LP system also causes structural damage to microorganisms, owing to a rapid escape of potassium ions, amino acids and polypeptides into the surrounding medium. Subsequently, the uptake of glucose, purines, pyrimidines and amino acids is suppressed, and the synthesis of proteins, DNA and RNA is inhibited. Thiocyanate ions may also be incorporated into protein substrates, under the catalysing effect of lactoperoxidase. The reaction of $(SCN)_2$ or $OSCN^-$ with proteins results in the oxidation of the protein –SH groups, to produce sulphenyl thiocyanate, usually followed by disulphide formation.

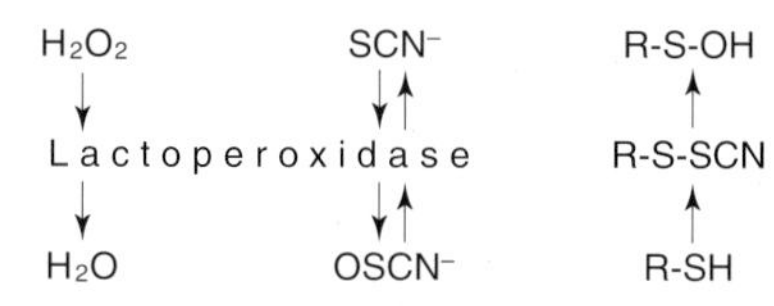

Figure 1 Oxidation of thiol (–SH) groups.

$$R\text{–}SH + (SCN)_2 \longrightarrow R\text{–}S\text{–}SCN + SCN^- + H^+$$
$$R\text{–}SH + OSCN^- \longrightarrow R\text{–}S\text{–}SCN + OH^-$$

Sulphenyl thiocyanate derivatives may be further modified, to yield mainly sulphonic acid:

$$R\text{–}S\text{–}SCN + H_2O \longrightarrow R\text{–}S\text{–}OH + SCN^- + H^+$$

The rate of release of SCN^- from sulphenyl derivatives is dependent on the concentration of SCN^-, and is higher at lower concentrations. The release of SCN^- from sulphenyls is important for the further oxidation of thiol (–SH) groups, in a form of chain reaction (**Fig. 1**).

Antibacterial Effects The antibacterial action of the LP system results from the oxidation of –SH groups by $OSCN^-/O_2SCN^-$ in vital enzymes such as hexokinase and glyceraldehyde-3-phosphate dehydrogenase, and/or depletion of reduced nicotinamide adenine nucleotides. The effect of the LP system on bacteria can be reversible or irreversible. Many Gram-positive bacteria, e.g. streptococci and lactobacilli, are inhibited by the LP system: there have been reports of the suppression of the growth of *Streptococcus pyrogenes*, *S. agalactiae*, *S. mutans*, *S. sanguis*, *S. mitis*, *S. salivarius* and *Lactococcus lactis* var. *cremoris*. The capacity of cells to recover from inhibition depends mainly on environmental conditions, e.g. temperature and pH, and on the particular strain.

In contrast, many Gram-negative bacteria are killed by an activated LP system, e.g. *Escherichia coli*, *Salmonella typhimurium*, *Staphylococcus aureus*, *Listeria monocytogenes*, *Campylobacter jejuni* and *C. coli*. Yeasts, including *Candida* species, are also strongly inhibited by the LP system. Its bactericidal effect is more pronounced on bacteria during the exponential growth phase than in the stationary phase.

Effects on Starter Bacteria

Mesophilic Starter Bacteria Many investigations have been carried out using *Lactococcus lactis* subsp. *cremoris* 972, which is very sensitive to the peroxidase system, and *L. lactis* subsp. *cremoris* 803, which is insensitive to it. Acid production by some of the mesophilic starter bacteria is enhanced by an activated LP system, but many starters used for making cheese are inhibited by the LP system to some extent. However, in practice LP activity depends on the severity of the heat treatment given to the milk. With skim milk pasteurized at a low temperature, which is LP-positive, *L. lactis* subsp. *cremoris* 972 is strongly inhibited, but with steamed milk, which is LP-negative, a high level of bacterial activity is found. Heat treatment denatures LP, and consequently the activity of the LP system decreases. The inhibitory effect of heat treatment is also related to an increased number of exposed sulphydryl groups, and to the redistribution of protein between the micellar and whey phases. Chromatographic analyses of heat-treated milk show that the inhibition is associated with the casein micelle fraction.

The inhibition of starter bacteria is stronger in winter milk than in summer milk. This is probably due to the increased concentration of compounds that contribute to the inhibitory effect.

Thermophilic Starter Bacteria Much less information is available about the inhibitory effect of the LP system on the activity of thermophilic starters in milk. Lactoperoxidase has no specific effect on *Streptococcus thermophilus*, which is used as a starter for yoghurt and some types of Swiss and Italian cheeses in combination with other thermophilic bacteria. However, when LP is used in association with H_2O_2 and SCN^-, clear inhibition of *S. thermophilus* is observed. The growth of yoghurt starter cultures containing *S. thermophilus* and *Lactobacillus delbrueckii* subsp. *bulgaricus* is clearly suppressed by an activated LP system – *L. delbrueckii* subsp. *bulgaricus* produces H_2O_2, as do *L. helveticus* and *L. delbrueckii* subsp. *lactis*. Thus the LP system in these cases is activated by H_2O_2 produced by the latter bacteria.

Lactoferrin

Lactoferrin (LF) is an iron-binding glycoprotein occurring in milk. It belongs to the transferrin family, and is a single-chain protein which is synthesized in the mammary gland and also in the lacrimal, bronchial and salivary glands. Lactoferrin has an antimicrobial function, as well as contributing to the nutritional content of milk.

Intensive studies on the purification and characterization of LF were carried out in the 1960s, and its structure and role are now well established. In human milk, LF is a major whey protein – constituting 10–30% of the total protein content. In general, LF concentrations in bovine milk and colostrum (about 0.2 mg ml^{-1} and 1 mg ml^{-1} respectively) are lower than those in human milk and colostrum (about 1.5 mg ml^{-1} and 5.0 mg ml^{-1} respectively). Diseases such as mastitis cause an increase in the LF level in milk. The level decreases sharply during the first few days of lactation. Lactoferrin requires bicarbonate ions for its activity.

The molecular weight of bovine LF is about 78 500 (see Table 1). Its structure is similar to that of transferrin, another iron-transporting protein, but LF shows a greater affinity for ferric ions. No differences between the biochemical properties of bovine, caprine and ovine LF have been reported. The amino acid sequence of bovine LF has been established, and DNA studies reveal similarities between bovine and human LF. Bovine LF contains α-1,3-linked galactose residues, and also glycans of the oligomannasidic type. Its molecular structure has been described in detail, following studies with X-ray crystallography. The molecule is composed of two domains, each binding 1 mol of iron. Each domain contains 125 residues at corresponding positions (with 37% homology), which suggests gene duplication.

The absorption of LF in the intestine does not lead to any structural deformation in the LF molecule. LF shows great stability towards proteases, especially chymosin, and hydrolysis with trypsin does not result in fragment separation. Although LF is considered to be a natural antimicrobial agent, in combination with α-lactalbumin it stimulates the growth of some strains of bifidobacteria – with a greater response from *Bifidobacterium infantis* and *B. breve* than from strains of *B. bifidum*.

Antimicrobial Effects of Lactoferrin

The antimicrobial effects of LF can be bacteriostatic or bactericidal.

Bacteriostatic Effect Bacteria require ferric ions for growth, and under certain conditions, LF can inhibit bacteria by competing effectively for available supplies. A number of factors affect the antibacterial activity of LF, the most important being the concentrations of calcium ions and citrate. An increased concentration of Ca^{++} results in a tetramization, which leads to a reduction in the iron-binding capacity of LF. Citrate competes with LF for Fe^{++}, and then makes it available to the bacteria. There is an inverse relationship between citrate and bicarbonate concentrations. Under normal conditions, citrate is absorbed rapidly from the intestine of the calf and as bicarbonate is the main intestinal buffer secreted, it appears that the conditions in the intestine should be favourable for inhibition by LF.

LF has been reported as having a bacteriostatic effect on cultures of *Escherichia coli*, *Salmonella typhi*, *Shigella flexneri*, *Shigella dysenteriae*, *Aeromonas hydrophila*, *Staphylococcus aureus* and *Listeria monocytogenes*, but its antibacterial effect in vivo is less clear. It is unlikely that LF plays a significant role in the defence of the bovine mammary gland during lactation, but in the non-lactating gland the conditions are more favourable for antibacterial activity – post-lactation, the LF concentration increases and the citrate concentration decreases. However, the results of in vivo studies are not always predictable on the basis of in vitro studies. For example, intramuscular challenge failed to inhibit the growth of *Escherichia coli*, *Salmonella typhimurium* and *Pseudomonas aeruginosa*, because siderophores produced by these bacteria take up LF-bound iron; this is in contradiction to the results obtained in vitro.

Lactoferrin alters the permeability of the outer membrane of Gram-negative bacteria by damaging it irreversibly – this action is regulated by Ca^{++} and Mg^{++} ions. The destruction of the membrane affects glucose uptake, and there is also a reduction in lactate production and a partial suppression of DNA and RNA synthesis.

Bactericidal Effect The partial hydrolysis of LF by heat, as well as by proteases, produces a hydrolysate with greater antibacterial activity than intact LF. The antibacterial peptide region, which is strongly basic, is 23 amino acids long and contains 18 amino acids in a loop formed by a disulphide bond, in part of the LF molecule which is distinct from the iron-binding region. The bactericidal activity of LF is not affected by the presence of iron.

Little is known about the inhibitory effect of LF on the activity of starter organisms, but it is unlikely to be of importance.

Effects of Antimicrobial Systems on Raw Milk and Dairy Products

Much more is known about the effects of the LP system on bacteria than about its effects on the quality of dairy products, in particular their firmness and consistency, organoleptic properties and yield – qualities of the utmost importance to cheese makers.

Yields of soft fresh cheeses (e.g. cottage cheese) or ripened hard cheeses (e.g. Cheddar) are generally increased by using milk preserved by an activated LP system instead of untreated milk. This is probably due to the lower proteolytic and lipolytic activity of treated milk – there is an inverse relationship between the concentration of SCN^- and H_2O_2 and lipoprotein lipase activity. Thus in treated milk, fewer proteins

Table 1 Differential sugar oxidation and polysaccharide formation for *Neisseria* spp.

Species	*Acid from*					*Polysaccharide from sucrose*
	Glu	*Mal*	*Suc*	*Fru*	*Lac*	
N. gonorrhoeae	+	–	–	–	–	None
N. kochii	+	–	–	–	–	None
N. meningitidis	+	+	–	–	–	None
N. lactamica	+	+	–	–	+	None
N. polysaccharea	+	+	–	–	–	Positive
N. cinerea	–	–	–	–	–	None
N. flavescens		–	–	–	–	None
N. mucosa	+	+	+	+	–	Positive
N. sicca	+	+	+	+	–	Positive
N. elongata	–	–	–	–	–	None

GLU = glucose; Mal = maltose; Suc = sucrose; Fru = fructose; Lac = lactose; + positive; – negative.

Table 2 Biochemical characteristics of *Neisseria* spp.

Species	*Acid from*					*Nitrate reduction*	*DNase*	*Superoxol*	*Pigment*	*Colistin resistance*
	G	*M*	*S*	*F*	*L*					
*N. gonorrhoeae**	+	–	–	–	–	–	–	Strong (4+) positive (explosive)	–	Resistant
N. meningitidis	+	+	–	–	–	–	–	Weak (1+) to strong (4+) positive	–	Resistant
N. lactamica	+	+	–	–	+	–	–	Weak (1+) to strong (3+) positive	–	Resistant
N. polysaccharea	+	+	–	–	–	–	–	Weak (1+) to strong (3+) positive	–	(Resistant)
N. cinerea	–	–	–	–	–	–	–	Weak (2+) positive	–	(Resistant)
N. flavescens	–	–	–	–	–	–	+	Weak (2+) positive	+	Sensitive
N. mucosa	+	+	+	+	–	+	–	Weak (2+) positive	–	Sensitive
N. sicca	+	+	+	+	–	–	–	Weak (2+) positive	–	Sensitive
N. elongata	–	–	–	–	–	–	+	–	–	Sensitive

G = glucose; M = maltose; S = sucrose; F = fructose; L = lactose; + positive; – negative.

mutation may occur, complicating this issue. There are a number of proteins divided into five classes. Class 2 and 3 function as porins resembling those found in *N. gonorrhoeae*, while class 4 and 5 proteins are analogous to Rmp and Opa proteins found in the gonococcal bacterium. Serotypes B and C have been further subdivided into other serotypes based on their class 2 and 3 proteins. Strain A group has the same protein serotype antigens in the outer membrane.

Biochemical Identification

Biochemical tests are used to identify different strains of *Neisseria* spp. based on their acid production activities and enzymatic production. **Table 2** shows some differential characteristics of biochemical properties of different *Neisseria* spp. Acid production from five sugars – glucose, maltose, sucrose, fructose and lactose – is detected as a result of the bacterial oxidative activity. This is an important contrast to most bacteria, which produce acid by fermentation; this distinction is important as more acid is produced by fermentation than by oxidation. Some *Neisseria* spp. such as *N. cinerea* produce acid from carbohydrates; however the acid is rapidly overoxidized to carbon dioxide, resulting in the non-accumulation of the acid in the accumulation tube even though carbohydrates have been used. This results in a misleading conclusion as *N. cinerea* is considered to be glucose-negative. *Neisseria* spp. produce ammonia from peptone, which neutralizes the already low acid production, complicating the test outcome.

The traditional test for determining acid production is cystine trypticase agar (CTA), but this is no longer recommended for detecting acid production by *Neisseria*, and should only be used for fermentative organisms. Alternative media are being used containing a lower ratio of peptone to carbohydrates. This is then inoculated with a heavy suspension of bacteria, as the action will be that of the enzymes present in the suspension.

Polysaccharide formation from sucrose is another

important feature used to identify *Neisseria* spp., as this property is found in a number of *Neisseria*, such as *N. perflavescens*, *N. mucosa*, *N. sicca* and *N. flavescens*. This test should not be used to distinguish between *N. meningitidis* and *N. polysaccharides*, as over 25% of organisms were found to be *N. polysaccharides* and were recorded as non-typable *N. meningitidis*. Changes to the test procedure were made, such as reducing the use of 5% w/v sucrose, which is inhibitory for some strains. *N. polysaccharea* can be detected on a medium containing 1% w/v sucrose and it will produce starch from sucrose as low as that. There is no need for carbon dioxide supplementation or incubation in a carbon dioxide environment if a heavy inoculum is used. Carbon dioxide can be used if needed to support growth; 5% atmosphere of this gas may be used in such a situation. *N. cinerea* does not have the ability to produce polysaccharides from sucrose; this is helpful to distinguish between *N. cinerea* and *N. polysaccharea*.

Neisseria Relationship to Humans

Neisseria spp. have a pathogenic and non-pathogenic relationship with humans, hence division can be made according to their effects on humans. The pathogenic species includes *N. meningitidis*, *N. gonorrhoeae* and *N. lactamica*. Between them they can cause a range of genital and meningococcal diseases. Meningococcal dissemination is an endemic disease caused by outbreaks of cases in geographically isolated areas. In developed countries true meningitis epidemics are rare; however, in developing countries this problem is somewhat more evident. Strains of *N. meningitidis* are carried as normal flora in the nasophrarynx in adults and children. The prevalence of this bacteria may occur more frequently in adults with gonorrhoea and in homosexual men.

N. lactimica was first described by Hollis in 1969. Its ability to produce acid from lactose was not used as a differential test for the identification of *Neisseria* spp. at that time. This bacteria had frequently been isolated from children but infrequently from adults. It often colonizes the throats of children. This has been attributed to the large volumes of milk children consume, due to the ability of this species to produce acid from lactose in milk. This is important concerning dairy products, as *Neisseria* spp. have been isolated from cheese which is not inoculated with lactic acid bacteria during processing.

Cheese made in Middle-eastern countries are more susceptible to contamination with *N. lactamica*, compared to fermented cheese made in other parts of the world. Processing what is known locally as white soft cheese does not involve the addition of lactic acid bacteria, thus its level of acidity and the presence of lactose in sufficient amounts enhance the growth of *N. lactamica*.

N. meningitidis and *N. gonorrhoeae* are two other important pathogenic types which infect humans, causing meningitis and gonorrhoea respectively (**Table 3**).

Pathogenicity of *Neisseria* Species in Humans

Meningococcal dissemination is an endemic disease in geographically isolated areas. True meningitidis epidemics are rare in developed countries; however cases are high in developing countries of high population density with bad hygiene or poor medical treatment. Endemic diseases appear to be caused by a single serotype. Serogroups B and C account for 46% in adults, while serotype C accounts for 69% of cases in children older than 2 years in the developed world. In China, Africa and other developing countries serotype A is the major causative agent.

N. meningitidis is only found in the human nasopharynx. This is the common reservoir. Transmission requires aspiration of infective particles attached to the non-ciliated columnar epithelial cells of the nasopharynx. This attachment is carried out with the aid of the pili and the outer membrane components. Meningococci disseminate from the upper respiratory tract into the blood stream; however it has been shown that serum antibody leads to complement-mediated bacterial lysis, thus preventing such dissemination.

N. gonorrhoeae is a sexually transmitted disease of worldwide importance (**Fig. 1**). The highest rate of incidence is in men and women between 15 and 29 years old. A number of factors lead to the onset of infection; these factors include number of sexual partners, use of contraceptives and sexual habits. This organism will cause the shedding of secretions and exudate of infected mucosal surfaces, which is transmitted through intimate sexual contact. Newborn babies may contract gonorrhoea during delivery if their mothers are infected with the bacterium. Gonorrhoea is not affected or destroyed by phagocytic vacuoles; however, it is not yet clear if it can replicate in the vacuole or not. Gonorrhoea is very effective and highly efficient in utilizing transferrin-bound iron for in vitro growth; many strains utilize lactoferrin-bound iron.

There are some host factors that play an important role in the susceptibility of a host to contract *Neisseria* infections. The age of the host female is an important factor in the case of gonococcal infection; oestrogen secretion will increase the host's resistance to the causative bacterium. The body's resistance will play

Table 3 Host range of *Neisseria* spp.

Species	*Host*	*Pathogen/commensal*
N. gonorrhoeae	Humans	Pathogen
N. meningitidis	Humans	Occurs in carrier state; some strains cause epidemics/pandemics
N. lactamica	Humans	Commensal; more frequent in children than adults
N. polysaccharea	Humans	Commensal
N. cinerea	Humans	Commensal
N. flavescens	Humans	Isolated from outbreak of meningitis. Apart from the original description, there are no reliable reports of this species; strains isolated as *N. flavescens* were *N. cinerea* or *N. polysaccharea*
N. subflava		
Biovar *sublava*	Humans	Commensal
Biovar *flava*	Humans	Commensal
Biovar *perflava*	Humans	Commensal
N. sicca	Humans	Commensal
N. mucosa	Humans	Commensal: similar strain isolated from dolphins
N. elongata	Humans	Commensal
N. canis	Cat	Commensal
N. ovis	Sheep	Causes keratoconjunctivitis in sheep

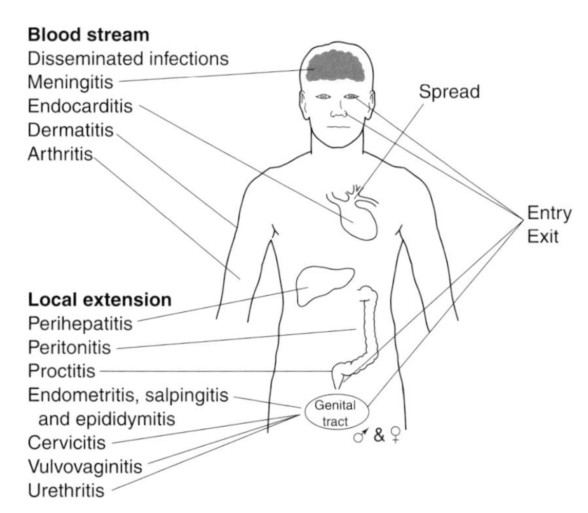

Figure 1 Clinical manifestations of *Neisseria gonorrhoeae* infection.

an important role in fighting off infections of *Neisseria* spp. (**Fig. 2**).

Host defence varies between individuals; likewise, there is variation in size or virulence of an infection due to the fact that the body's resistance will determine the size of the invading microbe showing clinical signs on the host. In males 1000 organisms will cause a gonococcal infection; in some cases 50% of this will be an infective dose (ID_{50}). Thus the ID_{50} for males is 50%. There is no ID_{50} for females concerning gonococcal infections.

There is not yet any vaccination against gonococcal infection. However there have been a number of attempts to produce such a vaccine, but with no effect. The evolution of antimicrobial-resistant bacteria, which resist antibiotics such as penicillin, tetracycline and erthromycin, has added to the problem of controlling this infection.

Meningitis vaccines have been produced based on the capsular polysaccharides for groups A, C, Y and W135. They are being used to control outbreaks with a high success rate. Chloramphenicol with sulphur drugs is effective in controlling and treating cases.

Neisseria spp. can gain entry to the human body by a number of routes, such as air, contact or food. A number of these species only cause infection if the conditions are suitable or if a state of mutation occurs. For example, *N. mucosa*, *N. sicca* and *N. flavescens* can cause meningitis or respiratory infections. The close genetic composition shared by members of this family makes mutations frequent, increasing the difficulties of producing vaccines, coupled with the detection of mutant strains resistant to the usual antibiotics. These factors enhance the need for accurate identification of reported cases of outbreaks or when isolated from food and dairy products (**Table 4**).

Conclusion

These recommendations update information regarding the polysaccharide vaccine licensed in the US for use against disease caused by *Neisseria meningitidis* serogroups A, C, Y and W135, as well as antimicrobial agents for chemoprophylaxis against meningococcal disease. This report provides additional information regarding meningococcal vaccines and the addition of ciprofloxacin and ceftriaxone as acceptable alternatives to rifampicin for chemoprophylaxis in selected populations. The incidence of meningococcal disease peaks in late winter to early spring. Attack rates are highest among children 3–12 months of age and then steadily decline among older age groups.

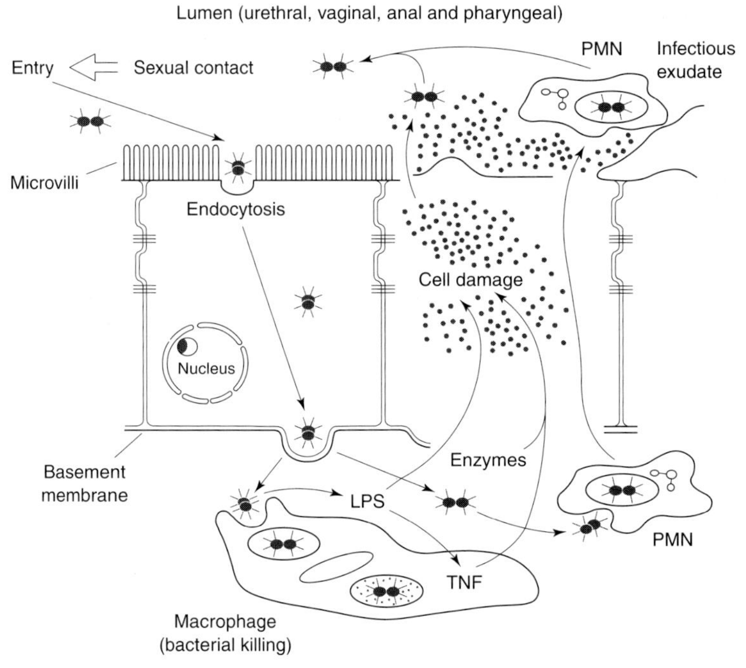

Figure 2 Pathogenesis of uncomplicated gonorrhoea. Gonococci can invade columnar epithelial cells, although they do not invade ciliated columnar epithelium of the genitourinary tract. PMN, polymorphonuclear cells; LPS, lipopolysaccharide; TNF, tumour necrosis factor.

Based on multistate surveillance conducted during 1989–1991, serogroup B organisms accounted for 46% of all cases and serogroup C for 45%; serogroups W135 and Y strains that could not be serotyped accounted for most remaining cases. Recent data indicate that the proportion of cases caused by serogroup Y strains is increasing. Serogroup A, which rarely causes disease in the US, is the most common cause of epidemics in Africa and Asia. In the US, localized community outbreaks of serogroup C disease and a statewide serogroup B epidemic have recently been reported.

Persons who have certain medical conditions are at increased risk for developing meningococcal infection. Meningococcal disease is particularly common among those who have component deficiencies in the terminal common complement pathway (C3, C5–C9); many of these people experience multiple episodes of infection. Asplenic persons may also be at increased risk for acquiring meningococcal disease with particularly severe infections. Those who have other diseases associated with immunosuppression (e.g. HIV and *Streptococcus pneumoniae*) may be at higher risk for acquiring meningococcal disease and for disease caused by some other encapsulated bacteria. Evidence suggests that HIV-infected persons are not at a substantially increased risk for epidemic serogroup A meningococcal disease; however, such patients may be at increased risk for sporadic meningococcal disease or disease caused by other meningococcal serogroups. Previously, military recruits had high rates of meningococcal disease, particularly serogroup C disease; however, since the initiation of routine vaccination of recruits with the bivalent A/C meningococcal vaccine in 1971, the high rates of meningococcal disease caused by those serogroups have substantially decreased and cases occur infrequently.

N. meningitidis is the leading cause of bacterial meningitis in older children and young adults in the US. The quadrivalent A, C, Y and W135 meningococcal vaccine available in the US is recommended for control of serogroup C meningococcal disease outbreaks and for use among certain high-risk groups, including those who have terminal complement deficiencies, those who have anatomical or functional asplenia and laboratory personnel who are routinely exposed to *N. meningitidis* in solutions that may be aerosolized. Vaccination may also benefit travellers to countries where the disease is hyperendemic or epidemic. Conjugate serogroup A and C meningococcal vaccines are being developed by methods similar to those used for *Haemophilus influenzae* type b conjugate vaccines, and the efficacies of several experimental serogroup B meningococcal vaccines have been documented in older children and young adults.

Antimicrobial chemoprophylaxis of close contacts of patients who have sporadic cases of meningococcal disease is the primary means of preventing meningococcal disease in the US. Rifampicin has been the

Table 4 Characteristics of pathogens of *Neisseria*

Pathogen	*Bacteriology*	*Pathogenesis*	*Disease*	*Treatment*
Neiserria spp.	High levels of cytochrome *c* oxidase; distinguishes *Neisseria* from other similar-appearing organisms, speciation done by fermentation	*N. sicca*, *N. catarrhalis*, *flavescens* = do not cause disease; *N. meningitidis* = do cause disease		
N. gonorrhoeae	Gram-negative diplococci, kidneys, pelvis-to-pelvis aerobic microaerophilic, CO_2 stimulates growth, won't grow at 22°C where commensal *Neisseria* will need selective medium to grow	Five different colony types (T1–T5), T1 & T2 are piliated and more virulent – seen on primary invasion	gonorrhea, salpingitis, PID & disseminated gonococcal infection, proctitis, pharyngitis	Bloodstream colour transparent
		No evidence of a capsule		epidemiologic treatment, potential carriers
		Pili in vivo determine disease site	Dermatitis-Arthritis-Tenosynovitis syndrome	many penicillinase producing *N. gonorrhoeae*
		Protein 1 (POMP) – jump to host: 16 types in GC	Ophthalmitis	
		Protein2: enhance attachment & entry into cell, opacity therefore in isolates	Perihepatitis Endocarditis	
		LPS = endotoxin, specific against human genital mucosa	Asymptomatic Women reservoirs	
		IgA1 protease		
		β-Lactamase		
		Peptidoglycan: induce TNF et al cytokines for septic shock		
	pilated or not have same shape	Capsule definitely present on disease-causing forms: 10 antigenic groups, antiphagocytic		same location appearance
N. meningitidis	Nasopharynx of humans is only reservoir, very common			
		antibodies are OPSONINS, vaccine against types A, C, Y & W135, not B		
		M-P1: 5 types, type 2 most serious (vac available)		
		MP2: apradoxically clear and in blood colonies		
		LPS: species profuse	rapidly fatal sepsis, meningitis; most likely for older children and young adults	antibodies are again and LPS
		IgA1 protease		Group B capsular polysaccharide as in *E. coli* K1.
		No β-lactamase plasmids	Especially groups B, C, Y	Multiple attacks should prompt search for complement deficiency
		capsule: groups A, B, C, Y, W135 antiphagocytic		state administration of penicillin species.
		Peptidoglycan: TNF		Vaccines for A, C, Y, W, not for B
				Rifampicin

drug of choice for chemoprophylaxis; however, data from recent studies document that single doses of ciprofloxacin or ceftriaxone are reasonable alternatives to the multidose rifampicin regimen for chemoprophylaxis.

See also: **Cheese**: In the Market Place; Microbiology of Cheese-making and Maturation. **Fermented Milks**: Products of Eastern Europe and Asia. **Milk and Milk Products**: Microbiology of Liquid Milk.

Further Reading

Bovre K (1984) Family VIII. *Neisseria* Prevot, In: Krieg DA (ed.) *Manual of Systemic Bacteriology.* Vol. 1, p. 288. Baltimore, MD: Williams & Wilkins.

Brooks GF and Donegan EA (1985) *Gonococcal Infection.* London: Edward Arnold.

Centers for Disease Control and Prevention (1995) *Neisseria. MMWR.* 42: 14–21.

Centers for Disease Control (1995) Serogroup B meningococcal disease – Oregon, 1994. *M.M.W.R.* 44: 121–124.

Centers for Disease Control (1996) Serogroup Y meningococcal disease – Illinois, Connecticut, and selected areas, United States, 1989–1996. *M.M.W.R.* 45: 1010–1013.

Centers for Disease Control (1997) Control and prevention of meningococcal disease and control and prevention of serogroup C meningococcal disease: evaluation and management of suspected outbreaks: recommendations of the Advisory Committee on Immunization Practices (ACIP) *M.M.W.R.* 46 (no. RR-5).

Conde-Glez CJ, Morse S and Rice P et al (eds) (1994) *Pathobiology and Immunology of Neisseriaceae.* Cuernavaca: Instituto Nacinonal de Salud Publica.

De Voe W (1982) The meningococcus and mechanism of pathogenicity. *Microbiol. Rev.* 46: 162.

Hanna SAS (1994) Neisseria *Species in Food Microbiology.* Baghdad, Iraq: Baghdad University Research Center Publications.

Holmes KK, Mardh PA and Sparling PF (eds) (1990) *Sexually Transmitted Diseases*, 2nd edn. New York: McGraw Hill.

Jackson LA, Tenover FC, Baker C et al (1994) Prevalence of *Neisseria meningitidis* relatively resistant to penicillin in the United States, 1991. *J. Infect. Dis.* 169: 438–441.

Jackson LA, Schuchat A, Reeves MW and Wenger JD (1995) Serogroup C meningococcal outbreaks in the United States: an emerging threat. *J.A.M.A.* 273: 383–389.

Knapp SA, Broome CV and Cannon AC (1989) Historical perspectives and identification of *Neisseria* and related species. *Clin. Microbiol. Rev.* 1: 415–431.

Leggiadro RJ (1987) Prevalence of complement deficiencies in children with systemic meningococcal infection. *Pediatr. Infect. Dis. J.* 6: 75.

Michele E (1995) *Meningococcal Infections, Infectious Diseases.* Pp. 768 and 771. New York: McGraw Hill.

Morse SA (1989) Perspectives on pathogenic *Neisseria. Clin. Microbiol. Rev.* 2: SI.

Riedo FX, Plikaytis BD and Broome CV (1995) Epidemiology and prevention of meningococcal disease. *Pediatr. Infect. Dis. J.* 14: 643–657.

Whintington WL (1988) Incorrect identification of *Neisseria gonorrhoeae* from infants and children. *Pediatr. Infect. Dis.* 7: 34.

Wong VK, Hitchcock W and Mason WH (1989) Meningococcal infection in children. *Rev. Pediatr. Infect. Dis.* 8: 224.

Nematodes *see* **Helminths and Nematodes**.

Nisin *see* **Bacteriocins**: Nisin.

Nitrate *see* **Preservatives**: Permitted Preservatives – Nitrate and Nitrite.

Nitrite *see* **Preservatives**: Permitted Preservatives – Nitrate and Nitrite.

Nitrogen Metabolism *see* **Metabolic Pathways**: Nitrogen Metabolism.

NUCLEIC ACID-BASED ASSAYS
Overview

Mansel W Griffiths, Department of Food Science, University of Guelph, Canada

Introduction

The elucidation of the genetic code by Watson and Crick in 1953 has had a profound influence on all the life sciences and spawned a scientific discipline, molecular biology, that is yet to realize its full potential. Food microbiology has been influenced by developments in molecular biology, and these techniques are particularly applicable to microbial detection. The inherent uniqueness of an organism's genetic make-up can be used to detect, identify and categorize the many microorganisms present in food. Of particular interest is the ability to identify the genetic determinants of pathogenicity and to devise tests that detect only the variants in foods with potential to cause human illness. Tests based on detection of unique sequences of nucleic acid have also made diagnosis of viral contamination of food a reality. Nucleic acid-based assays can be performed in one of two ways:

- direct detection of a unique nucleic acid sequence of the genome of the target organism
- detection of a unique sequence of nucleic acid following amplification of the sequence.

The principles of these technologies, and their advantages and limitations in detecting food-borne pathogens and spoilage organisms, are discussed below.

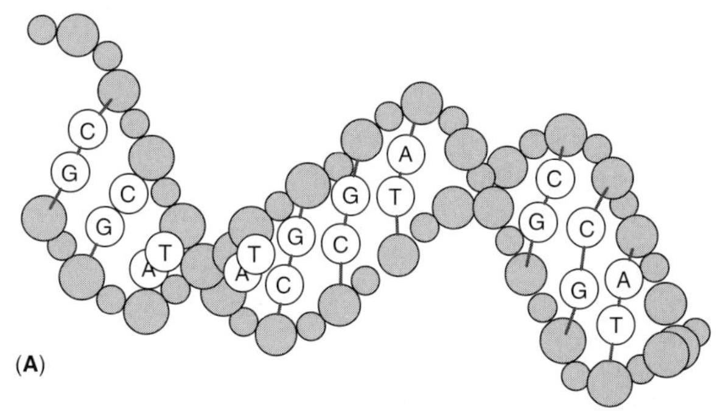

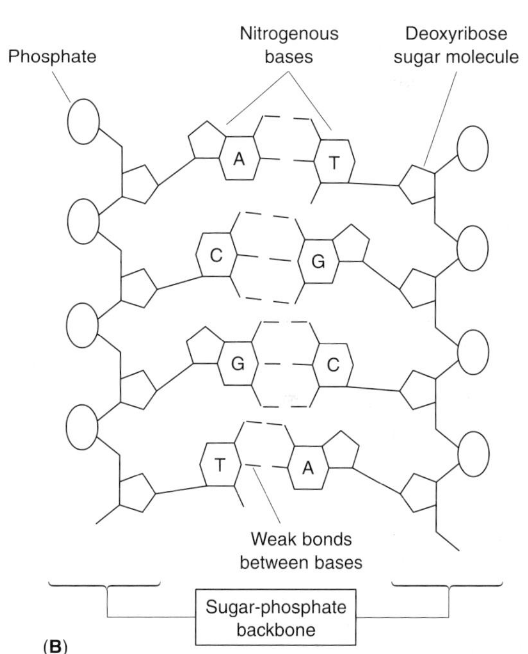

Figure 1 The structure of DNA.

Gene Probes

Principles

The DNA molecule had a double helical structure with the strands held together by hydrogen bonds between four repeating nucleotides – adenine, guanine, cytosine and thymine (**Fig. 1**). Bonds are only formed between guanine and cytosine, and between adenine and thymine. Thus, the complementarity of the two polynucleotide strands and the genetic information encoded by the sequence of the nucleotides forms the basis for detection by gene probes.

The strands of the DNA molecule can be separated by heat or alkaline pH and, on cooling or return to neutral pH, the strands will rejoin to form the double-stranded structure. A polynucleotide with a sequence complementary to single-stranded DNA or RNA will also be bound. This process is called hybridization. A gene probe is a short sequence of nucleotides that will hybridize with a target sequence unique to the organism to be detected. The sequence to be targeted can be:

- within the total DNA of the organism
- within the genomic DNA, such as a gene or part of a gene that encodes for a cell specific function (e.g. virulence factor, enterotoxin)
- part of a conserved region of genetic information, such as a portion of ribosomal RNA (rRNA).

The probes can be designed to detect genera, species or even strains. Many assays have been developed that target sequences within rRNA because it is present within the cell at high copy numbers. Compared with the detection of genomic DNA sequences, the use of an rRNA target increases the sensitivity of the assay several thousandfold. However, as rRNA tends to be highly conserved, identification of a sequence that enables detection at the species level is more difficult. Since viruses do not contain rRNA, unique sequences in their genomic nucleic acid (either DNA or RNA) must be used for detection. Once a suitable target sequence has been identified probes can be produced (1) by cloning the fragment into a host organism that will reproduce the probe sequence; (2) by chemical synthesis of the probe; or (3) by amplification of the probe sequence by the polymerase chain reaction (see below).

Properties of Probes

Probe Labelling In order to detect whether hybridization has taken place, the probes need to be labelled. Although radiolabelling with phosphorus 32 was the original preferred method, problems with waste disposal and health concerns have focused attention on non-isotopic labels. Several strategies have been investigated. Enzymes, such as horseradish peroxidase and alkaline phosphatase, can be chemically linked to nucleic acid probes. Following washing steps, the hybridized probe can be detected by using colorimetric or chemiluminescent substrates. An alternative approach is to incorporate detectable moieties directly into the probe. Biotin can be incorporated into nucleic acid through a biotinylated nucleotide analogue. The biotinylated probe can then be detected with enzyme-linked avidin which binds to biotin. As many avidin molecules are capable of binding to a single biotin molecule, this increases the sensitivity of the assay. Digoxigenin (DIG) can also be incorporated into probes and hybridization is detected using an anti-DIG antibody coupled to an enzyme. Hybridization to a digoxigenin-labelled probe can be detected very sensitively with a photon-counting charge-couple device (CCD) camera by measuring the chemiluminescence produced when an alkaline phosphatase-labelled anti-DIG antibody reacts with a chemiluminescent substrate. Using this technique, femtogram (10^{-15} g) amounts of DNA are detectable. More recently fluorescent probes have been developed.

Assay Format

Colony and Dot Blot Hybridization As with labelling techniques, there are several formats in which a nucleic acid probe can be used. Whichever format is used, the key step is to ensure that non-hybridized probes are effectively removed. The oldest technique is the colony hybridization method (**Fig. 2**). Colonies grown on plates are transferred to a solid support (usually a nylon or nitrocellulose membrane filter) by gently pressing on the agar surface. Alternatively, the cells can be grown directly on the surface of the membrane by laying it on the agar surface prior to incubation. The cells adhere to the surface of the membrane and can be lysed by alkaline treatment, heat or microwave irradiation. The DNA released is cross-linked to the membrane by exposure to ultraviolet light or heat. The probe is added and, after hybridization, excess probe is removed by washing. Hybridization is detected by a method appropriate to the probe labelling system used. This technique is extremely useful for screening presumptive colonies for the presence of virulence factors. An interesting variation to the colony hybridization assay has been described in which colonies are obtained after filtration through a hydrophobic grid membrane filter and incubation of the filter on the surface of an agar plate.

Dot blot hybridization is a variant of colony hybridization in which the target nucleic acid is extracted and denatured before being blotted onto a membrane through negative pressure. This allows a large number of samples to be processed and more dilute solutions of nucleic acid are concentrated during the filtration step.

A high level of background signal has been reported for colony and dot blot hybridization assays and so other formats have been investigated to improve specificity of the hybridization and reduce background signal noise.

Strand Displacement Hybridization A capture probe with a sequence that is complementary to the sequence to be detected is attached to a solid support. A short, labelled oligonucleotide is weakly hybridized to the capture probe. As the target sequence hybridizes with the capture probe, the labelled strand is displaced and the signal can be detected in the aqueous phase of the assay. Problems have been encountered because of the constant loss of the labelled probe into solution and the unavailability of the total capture probe sequence for hybridization with the target. The assay format is depicted in **Figure 3**.

Sandwich Hybridization The sandwich hybridization format also makes use of a capture probe that is linked to a solid support and is designed to hybridize with the target sequence. In addition, a signal probe with a sequence complementary to a sequence on the target nucleic acid adjacent to the hybridization site of the capture probe is used. The presence of a signal after washing indicates that the target sequence is

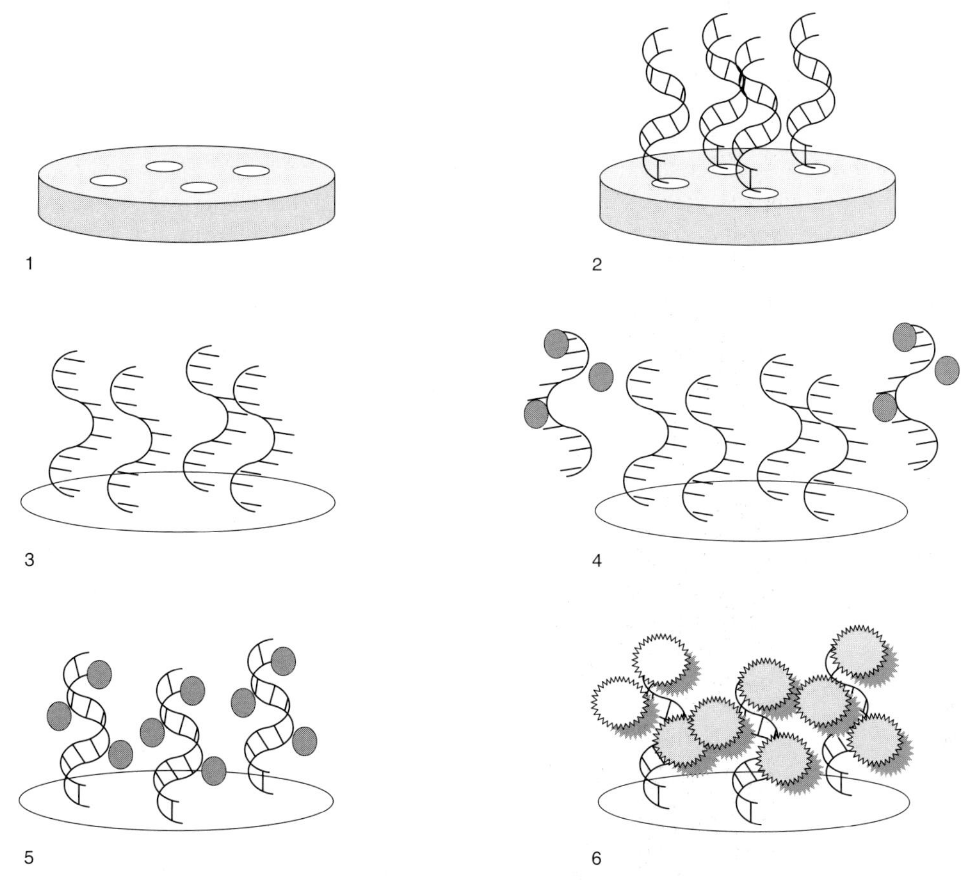

Figure 2 Colony hybridization assay. 1, isolation of colonies; 2, extraction of double-stranded DNA; 3, DNA is denatured and ssDNA is bound to membrane; 4, addition of labelled probe; 5, hybridization of probe and target; 6, washing followed by detection of signal.

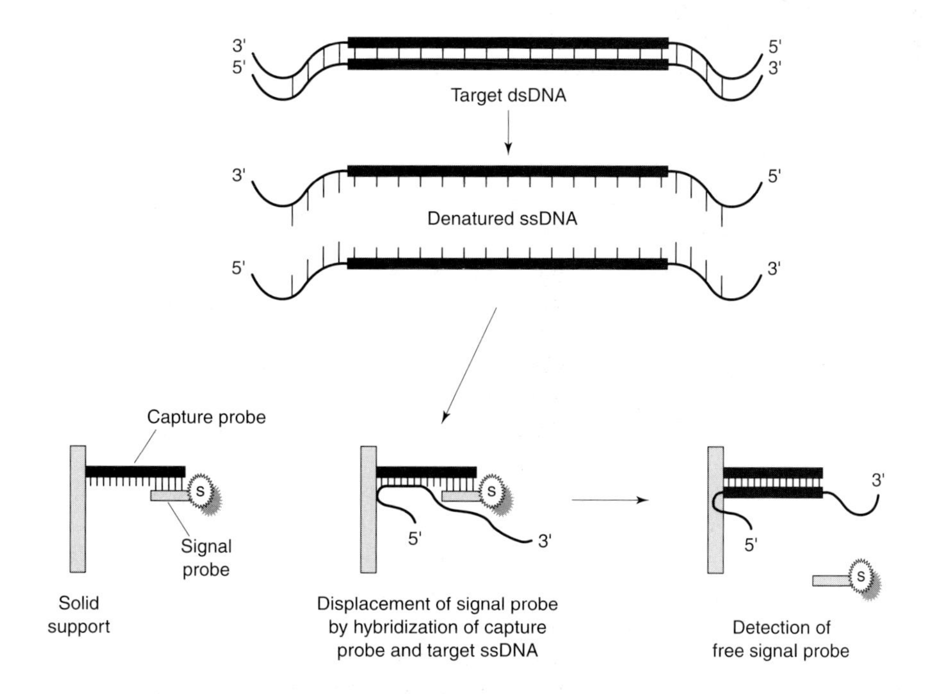

Figure 3 Strand displacement hybridization assay.

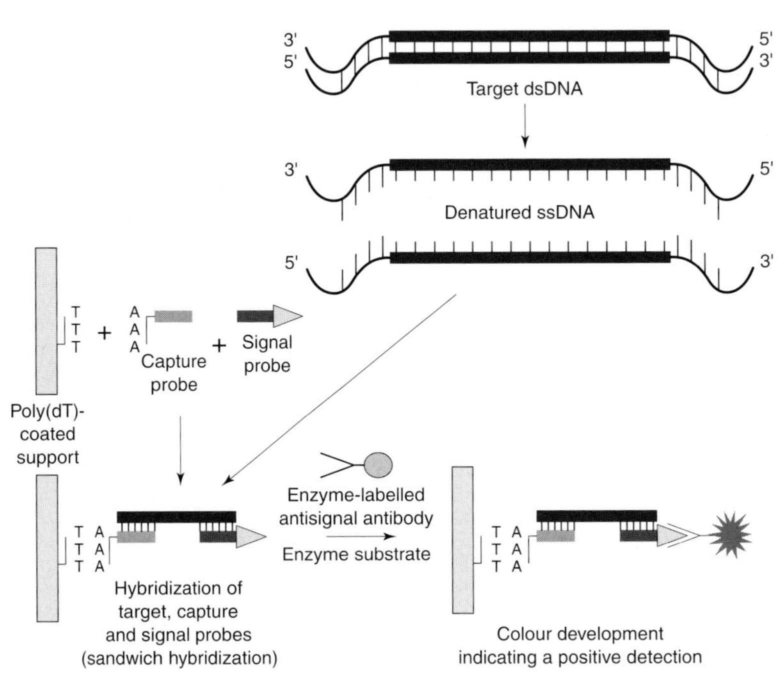

Figure 4 Sandwich hybridization assay.

present. A commercial system for the detection of *Salmonella* spp. and several other food-borne pathogens has been developed that uses a dipstick as the solid phase for the capture probe (**Fig. 4**). The commercial assay (GENE-TRAK®) is designed to detect unique sequences on either the 16S or 23S rRNA of the target organism. The capture probe is about 30 nucleotides long and has a polydeoxyadenylic acid (poly-dA) tail to link it with the solid support. The signal probe is about 35–40 nucleotides long and is labelled at both the 3 and 5 ends with fluorescein. The rRNA signal probe–capture probe complex is then detected with a polyclonal antifluorescein antibody conjugated to the enzyme horseradish peroxidase (HRP). The development of a blue colour (which subsequently changes to yellow following addition of sulphuric acid to stop the reaction) when hydrogen peroxide is added in the presence of the chromogen, tetramethylbenzidine, indicates the presence of the target sequence and, hence the organism. The GENE-TRAK® (Gene-Trak systems, Hopkinton, MA, USA) assay for *Salmonella* spp. has been tested on 1100 food samples representing 20 different food types and it attained 97.7% sensitivity and 100% specificity. The agreement between the commercial assay and the *Bacteriological Analytical Manual*/Association of Official Analytical Chemists culture method was 98.6%.

Other commercially available systems utilizing genetic probes are available from Gene-Probe, Gene Probe Technology, Gaithesburg, MD, USA, and Molecular Biosystems, Molecular Biosystems Inc., San Diego, CA, USA.

Riboprobes Short RNA probes directed against DNA target sequences can be produced from templates generated by polymerase chain reaction (PCR) incorporating the bacteriophage T7 promoter sequence. Hybridization of the RNA probe with the target can be detected immunoenzymatically using a monoclonal antibody raised against the RNA–DNA hybrid.

Specificity and Sensitivity The specificity of the assay is determined by the uniqueness of the target sequence, the probe sequence and the assay conditions. The probe should be designed so that self-hybridization does not occur. The size of the probe also has an effect on specificity. Short probes hybridize quickly but there is a greater risk of nonspecific hybridization and they are more difficult to label. Hybridization reactions with longer probes are much more stable but take longer to achieve. The hybridization reaction is also affected by ionic strength and temperature. At high ionic strengths and low temperatures (conditions of low stringency), more mismatches between the probe and the target sequence will be tolerated, resulting in a greater chance of nonspecific binding to other sequences.

Despite the improvement in labelling methods for the detection of hybridization, culture enrichment of the sample is required to obtain sufficient target nucleic acid to detect food-borne pathogens by hybridization assays, because about 10^5–10^6 cells are required to obtain a positive result. The enrichment can also serve to reduce problems associated with indigenous microflora. This increases the time required for the assay and, typically, a commercial test kit requires an overnight incubation of the sample followed by about 3 h to perform the hybridization and detection reactions.

Advantages and Drawbacks The greatest advantage of nucleic acid probe assays is their specificity. A properly constructed probe targeted against a well-defined sequence will be absolutely specific. The reaction is also less affected by the physiological status of the cell than other detection methods and the assay can be more robust than immunological techniques.

Difficulties can be encountered in extracting nucleic acid from the food matrix and that is another reason why enrichment cultures are performed. The assay will detect cells regardless of whether they are viable or not, but this is another argument in favour of prior enrichment as only viable cells will attain the levels required for detection.

DNA Array Like computer chips before them, biochips are expected to revolutionize the modern world. Also known as DNA arrays, biochips can hold thousands of gene probes that can hybridize with DNA, giving researchers the ability to analyze thousands of genes at a time. Hybridization can be detected by fluorescence and instruments are available that can 'read' the surface of the biochips. As more becomes known of the nucleic acid sequences of foodborne microorganisms, DNA arrayers will become an important detection tool for the food industry.

Amplification of Target Sequences

Much attention has been focused on nucleic acid amplification techniques to improve the sensitivity of gene probe assays. Several systems have been developed to amplify target nucleic acid sequences but only a few have found application for the detection of microorganisms in food.

Polymerase Chain Reaction

In 1983, Kary B. Mullis developed a method to amplify the number of specific DNA fragments in a sample, a technique for which he won the Nobel prize for chemistry 10 years later. The method, called the polymerase chain reaction, is a three-step process:

1. The target DNA is denatured at high temperature to yield two single strands.
2. Two synthetic oligonucleotides, termed primers, are annealed to complementary sequences on opposite strands of the DNA at a temperature that only allows hybridization to the correct target sequence.
3. The primers are extended by enzymatic polymerization with DNA polymerase using nucleotides present in solution to form a sequence complementary to the target DNA.

By repeating this cycle of events the amount of the target sequence is doubled; so if the cycle is repeated 20–40 times the number of targets increases exponentially, resulting in over a millionfold amplification of the original DNA sequence. The basic steps in PCR are shown in **Figure 5**.

The convenience of the technique was further increased with the isolation of a thermostable DNA polymerase (*Taq* DNA polymerase) from *Thermus aquaticus*. Thus, the DNA denaturation step can be performed without having an effect on the polymerase. This, in conjunction with the availability of instruments, has led to the development of simpler protocols. The routine application of PCR to the detection of food-borne pathogens has been brought even closer by the introduction of prepackaged reagents. Different primer sets directing the amplification of target sequences from more than one organism can be used in the same PCR test, resulting in a multiplex assay. Modifications of the PCR reaction have been made to broaden its application. Reverse transcriptase PCR (RT-PCR) can be used to amplify target sequences in RNA and is useful for the detection of certain viruses, such as hepatitis A virus.

Detection of Amplification Products The most common method of characterizing the PCR products is by agarose gel electrophoresis. The presence of a band of the expected size is indicative of the presence of the target nucleic acid sequence. However, apart from the amplicon having the expected size, there is no other basis for concluding that the product has the expected sequence. The amplified product can be sequenced to confirm its identity, but this is expensive and not suitable for routine analysis. Alternatively, because the sequence of the expected product should be known, internal restriction sites can be identified and endonuclease cleavage used to produce digestion products of known size. Electrophoresis can then be used to confirm that the cleavage products are of the expected size. A technique called 'nested PCR' can also provide confirmatory evidence that the amplified product has the expected sequence. An additional primer, homologous to a region located internally in the targeted sequence, can be used to amplify a several thousandfold dilution of the amplified product in conjunction with one of the original primers. The size of this second amplicon should correspond with the calculated size. Nested primers also form the basis of a test principle called 'detection of immobilized amplified nucleic acid' (DIANA). One of the inner primers is labelled with biotin and the other with a

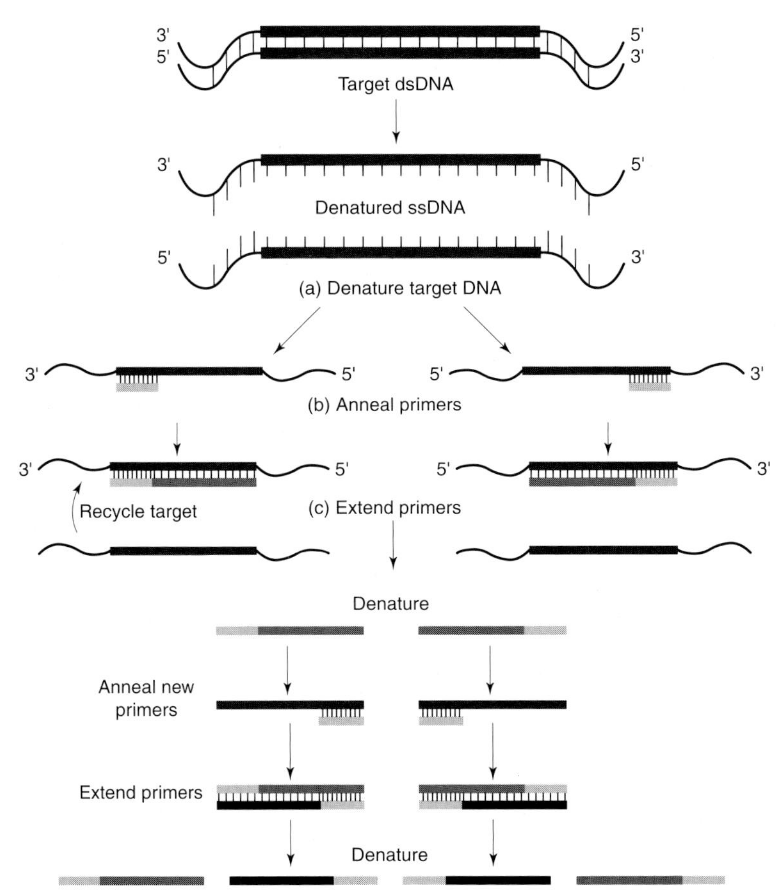

Figure 5 Polymerase chain reaction.

tail of a partial sequence of the *lac* operator gene (*lacO*) or DIG-dUTP. Steptavidin-coated magnetic beads can then be used to remove the labelled amplicons from solution and their presence detected by a suitable assay. The whole system can be automated by combining a PCR thermal cycler, a robotic work station and an automatic DNA sequencer.

A good method to confirm that the amplified product has the expected sequence is to use an internal hybridization probe on a Southern blot of a gel or on a dot blot of the final PCR products. Hybridization can be detected conveniently using the DIG-labelling system described previously. This method is about ten to a hundred times more sensitive than agarose gel electrophoresis.

Automated PCR Detection Systems Two automated PCR instruments have recently been described that may accelerate the routine application of the technology by the food industry.

The AG-9600 AmpliSensor Analyzer (Biotronics Corp, St. Lowell, MA, USA) is an automated system for the dispensing of PCR reagents and for the detection of PCR products using a microtitre plate format

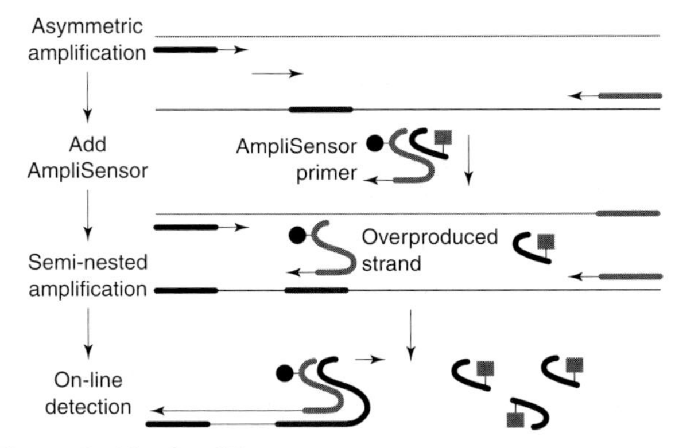

Figure 6 The AmpliSensor system.

and a single pipetting step. The AmpliSensor assay consists of two steps (**Fig. 6**).

1. An initial asymmetric amplification with normal primers to overproduce one strand of the target.
2. Subsequent semi-nested amplification and signal detection in which one of the outer primers and the AmpliSensor primer direct the amplification.

The AmpliSensor primer is a double-stranded single probe in which one strand is labelled with fluorescein isothiocyanate and the other with Texas red. During

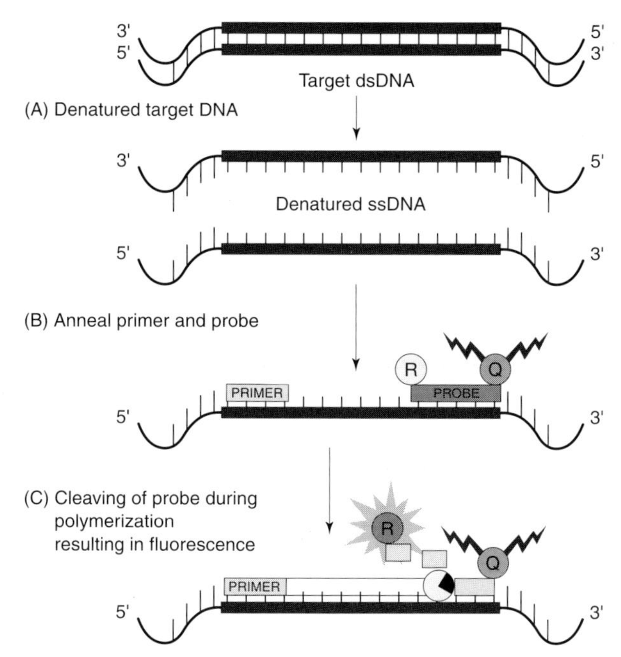

Figure 7 The TaqMan assay protocol. R, reporter; Q, quencher.

semi-nested amplification, one strand acts as a primer and the other as an 'energy sink'. Amplification results in strand dissociation of the AmpliSensor duplex and causes disruption of the fluorescence signal. The extent of this signal disruption is proportional to the amount of the AmpliSensor primer incorporated into the amplification product and can be used for quantification of the initial target. Using an AmpliSensor primer homologous to a target sequence within the *Salmonella invA* gene it has been possible to detect as few as 3 colony forming units (cfu) of *Salmonella typhimurium* per 25 g in chicken carcass rinses, ground beef, ground pork and milk following overnight enrichment in buffered peptone water. The detection limit was a hundred times more sensitive than ethidium bromide-stained agarose gel electrophoresis.

Another method that makes use of a fluorescence detection signal is the TaqMan LS-50B PCR Detection System (PE Applied Biosystems, Foster City, CA, USA). This method makes use of the 5′ nuclease activity of the AmpliTaq® DNA polymerase together with an internal probe that is labelled with a fluorescent reporter and a quencher dye (**Fig. 7**). If the target is present the probe anneals at a site between the forward and reverse primers during PCR amplification. As the primers are extended, the nucleolytic activity of the polymerase cleaves the probe hybridized to the target sequence, but the enzyme will not break down non-hybridized probe. Upon cleavage the reporter dye is separated from the quencher dye and the resulting increase in fluorescence of the reporter can be detected on a fluorescent plate reader. This process occurs during every PCR cycle and so the increase in fluorescence is a direct result of amplification of the target sequence. Three reporter dyes and two quencher dyes are available, which provide the opportunity for multiplex assays. A commercial test for *Salmonella* is available that can detect fewer than 3 cfu per 25 g of *Salmonella typhimurium* in a variety of foods following overnight pre-enrichment in buffered peptone water, and has over 98% correlation with cultural methods.

More recently, the LightCycler System (Roche Molecular Biochemicals, Indianapolis, IN, USA) has been developed for qualitative and quantitative PCR. Ultrarapid thermal cycling combined with a fluorescent detection system allows amplification and analysis to be performed in less than 20 minutes.

Some Drawbacks of the PCR Assay Problems may be encountered when applying PCR for the detection of organisms directly in foods owing to the presence of inhibitors of the PCR reaction. This necessitates extraction and purification of the target nucleic acid or pre-enrichment to allow growth of the organism of interest. The latter approach does enable the detection of viable bacteria and overcomes the criticism that PCR detects both live and dead cells. However, the use of a pre-enrichment step is not possible for the detection of microorganisms such as viruses and protozoa for which there is no convenient method of cultivation.

Several techniques have been applied for the extraction and purification of the target DNA including organic solvent extraction and the use of commercial resins. Many reports have also described the coupling of immunomagnetic separation with PCR. An interesting variation on the theme of immunocapture PCR was originally proposed to overcome the inhibitory effect of humic acid present in soil samples during PCR amplification, but has recently been applied to foods. Instead of coating magnetic beads with antibodies targeted to the organism of interest, nucleic acid probes specific for a sequence close to that to be amplified are coated onto magnetic beads. This can be carried out using biotin-labelled probes and streptavidin-coated beads. The target sequence is captured on the bead and can subsequently be amplified directly by conventional PCR. The technique, termed magnetic capture hybridization polymerase chain reaction (MCH-PCR), has been shown to be capable of detecting 1 cfu g^{-1} of verotoxigenic *E. coli* in ground beef within 15 h when biotin-labelled probes were used to capture specific regions of the genes for verotoxins (shiga-like toxins) 1 and 2.

Because of the sensitivity of the PCR technique it is important to combat contamination with extraneous

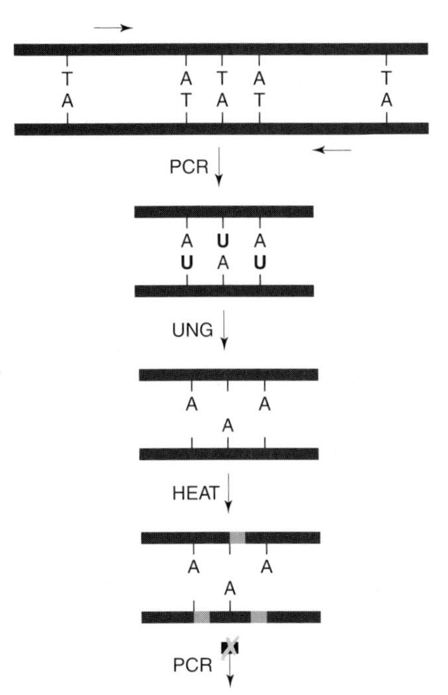

Figure 8 Polymerase chain reaction (PCR) carry-over prevention using uracil *N*-glycosylase (UNG) and dUTP to prevent false positives due to contamination.

nucleic acid fragments that may be amplified along with the target. There are methods to ensure that PCR products cannot be reamplified in subsequent PCR amplifications by using an enzymatic reaction to specifically degrade PCR products from previous PCR amplifications, in which dUTP has been incorporated, without degrading the target nucleic acid templates. The method used to make PCR products susceptible to degradation involves substituting dUTP for dTTP in the PCR mixture. Products from previous PCR amplifications are eliminated by excising uracil residues using the enzyme uracil *N*-glycosylase (UNG), and degrading the resulting polynucleotide with heat treatment prior to PCR amplification (**Fig. 8**).

Methods Utilizing PCR for Epidemiological Typing Random amplification of polymorphic DNA (RAPD) or arbitrarily primed PCR (AP-PCR) is a technique whereby primers having an arbitrary sequence are used to amplify sequences of genomic DNA generating amplicons that vary depending on the genus, species or even strain of organism under investigation. Thus, 'fingerprints' are generated that can provide valuable information for epidemiological studies. Further discrimination can be achieved by digesting amplicons with restriction endonucleases to obtain restriction fragment length polymorphism patterns (RFLP) on agarose gels.

In another typing strategy, amplification of enterobacterial repetitive intergenic consensus (ERIC)

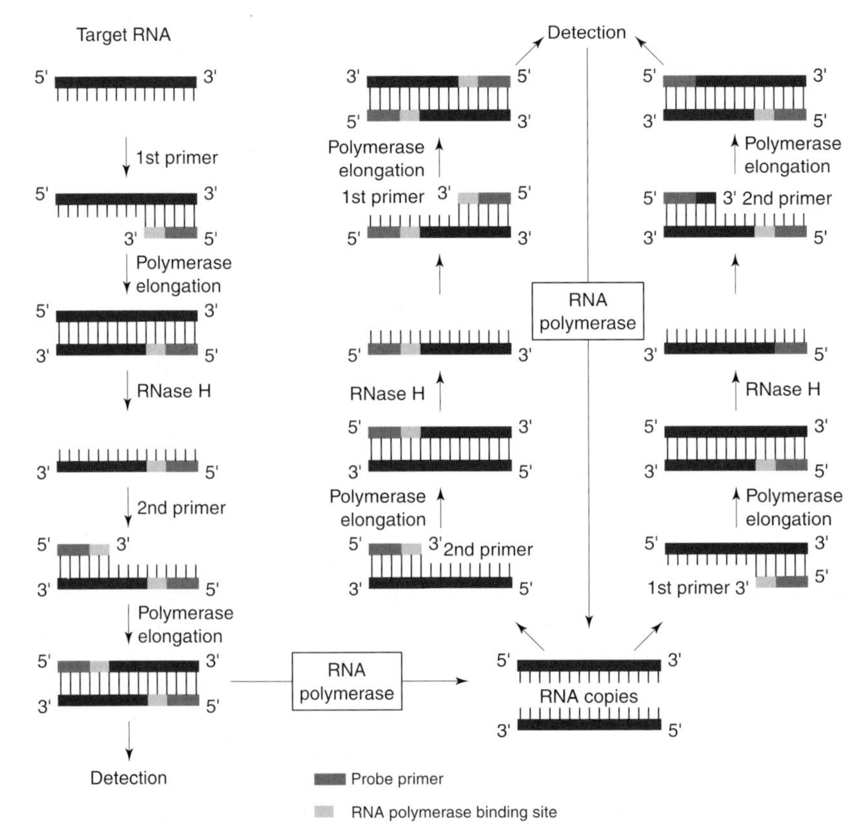

Figure 9 Nucleic acid sequence-based amplification (NASBA).

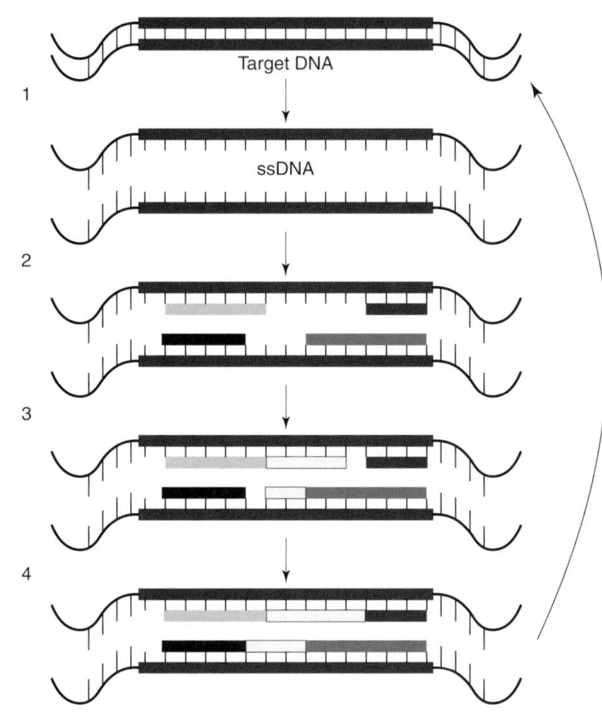

Figure 10 Ligase chain reaction. 1, heat denaturation; 2, annealing of four probes; 3, gap filling with thermostable polymerase; 4, ligation with thermostable ligase.

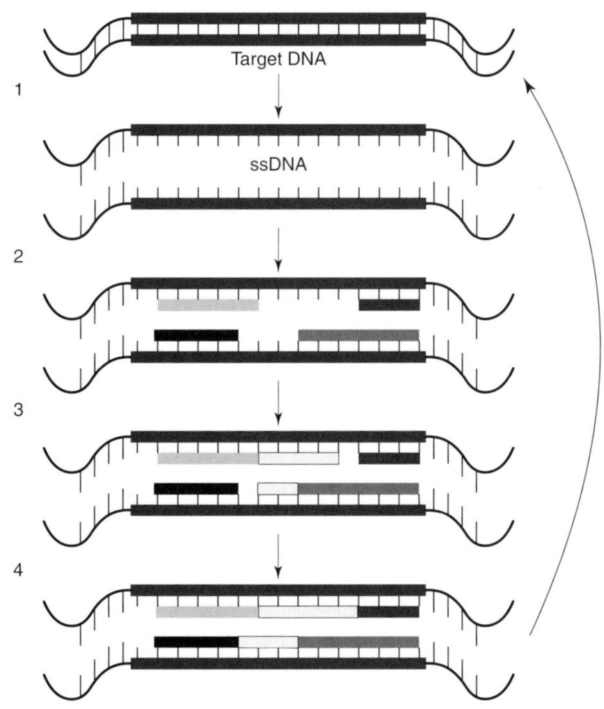

Figure 11 Q-β replicase assay.

motifs results in amplicons of different fragment lengths, depending on the number of repeat units which varies from strain to strain.

Other Amplification Reactions

Nucleic Acid Sequence-based Amplification The NASBA® process uses three enzymes (reverse transcriptase, RNase H and T7 RNA polymerase) and two target sequence-specific oligonucleotide primers (one carrying a bacteriophage T7 promoter sequence) to amplify an RNA target sequence (**Fig. 9**). The amplification is isothermal and therefore does not require a thermal cycler. The absence of a heat denaturation step also prevents the amplification of DNA sequences, although DNA targets can be amplified when a denaturation step is included. Because ribosomal RNA (rRNA) or the corresponding genes (rDNA) are highly conserved, it is very useful for detecting taxonomic groups. Ribosomal RNA is also present in cells at much higher copy numbers than DNA, and so the sensitivity of the assay is increased. There is also evidence that detection of RNA is a better indicator of viability.

Ligase Chain Reaction In the ligase chain reaction (LCR), two oligonucleotides with sequences complementary to adjacent regions on the target single-stranded DNA (ssDNA) are synthesized. When they hybridize with the target a thermostable *Taq* ligase joins them and these newly ligated oligonucleotides are used as a template in subsequent cycles (**Fig. 10**). Two commercial systems are available for detecting the amplified product. The AmpLiTek LCR kit (Bio-Rad Laboratories, Hercules, CA, USA) uses an oligonucleotide containing a biotin moiety, while the other oligonucleotide contains a sequence complementary to a detection oligonucleotide. After amplification, LCR products are placed in microtitre plates coated with streptavidin and the biotinylated products bind to the streptavidin. After washing the detection oligonucleotide is added and it hybridizes with the bound product. Hybridization is detected by a colorimetric change following addition of substrates for the enzyme conjugated to the detection probe.

In the second system, the Lcx Analyser marketed by Abbott, Abbot Laboratories, South Pasadena, CA, USA, the amplification products bind to microparticles to create an immune complex. This complex is then transferred to a reaction cell where it binds to an inert glassfibre matrix. An enzyme-labelled conjugate is added which in turn binds to the immune complex, and addition of a fluorogenic substrate results in the generation of a fluorescent product.

Q-β Replicase An RNA probe that contains a template region, MDV-1, for an RNA-directed RNA polymerase (Q-β replicase) is used to hybridize to the target sequence. The MDV-1 can then be amplified rapidly by the replicase (**Fig. 11**).

In Situ Hybridization and Amplification

In situ hybridization and *in situ* PCR are newly emerging techniques that allow the detection of minute quantities of DNA or RNA in intact cells or tissues. A number of applications of this technology has been reported recently, but despite its tremendous potential it has not been fully realized.

See also: **Hydrophobic Grid Membrane Filter Techniques (HGMF)**. **Immunomagnetic Particle-based Techniques**: Overview. ***Listeria***: *Listeria monocytogenes* – Detection using NASBA (an Isothermal Nucleic Acid Amplification System). **Molecular Biology – in Microbiological Analysis**. **PCR-based Commercial Tests for Pathogens**. ***Salmonella***: Detection by Colorimetric DNA Hybridization.

Further Reading

Blais BW, Turner G, Sookanan R and Malek LT (1997) A nucleic acid sequence-based amplification system for detection of *Listeria monocytogenes hlyA* sequences. *Applied and Environmental Microbiology* 63: 310–313.

Candrian U (1995) Polymerase chain reaction in food microbiology. *Journal of Microbiological Methods* 23: 89–103.

Chen S, Yee A, Griffiths MW et al (1997) The evaluation of a fluorogenic polymerase chain reaction assay for the detection of *Salmonella* species in food commodities. *International Journal of Food Microbiology* 35: 239–250.

Chen S, Yee A, Griffiths M et al (1997) A rapid, sensitive and automated method for the detection of *Salmonella* species in foods using AG-9600 Amplisensor Analyzer. *Journal of Applied Microbiology* 83: 314–321.

Fliss I, St-Laurent M, Emond E et al (1995) Anti-DNA:RNA antibodies: an efficient tool for non-isotopic detection of *Listeria* species through a liquid-phase hybridization assay. *Applied Microbiology and Biotechnology* 43: 717–724.

Griffin HG and Griffin AM (1994) *PCR Technology: Current Innovations*. Boca Raton: CRC Press.

Harris LJ and Griffiths MW (1992) The detection of foodborne pathogens by the polymerase chain reaction. *Food Research International* 25: 457–469.

Hill WE and Olsvik Ø (1994) Detection and identification of foodborne microbial pathogens by the polymerase chain reaction: food safety applications. In: Patel P (ed.) *Rapid Analysis Techniques in Food Microbiology*. P. 268. London: Blackie.

Marshall A and Hodgson J (1998) DNA chips: an array of possibilities. *Nature Biotechnology* 16: 27–31.

Olsen JE, Aabo S, Hill W et al (1995) Probes and polymerase chain reaction for detection of food-borne bacterial pathogens. *International Journal of Food Microbiology* 28: 1–78.

Swaminathan B and Feng P (1994) Rapid detection of food-borne pathogenic bacteria. *Annual Reviews in Microbiology* 48: 401–426.

Wiedmann M, Stolle A and Batt CA (1995) Detection of *Listeria monocytogenes* in surface swabs using a non-radioactive polymerase chain reaction-coupled ligase chain reaction assay. *Food Microbiology* 12: 151–157.

Wolcott MJ (1991) DNA-based rapid methods for the detection of foodborne pathogens. *Journal of Food Protection* 54: 387–401.

Zhai JH, Cui H and Yang RF (1997) DNA based biosensors. *Biotechnology Advances* 15: 43–58.

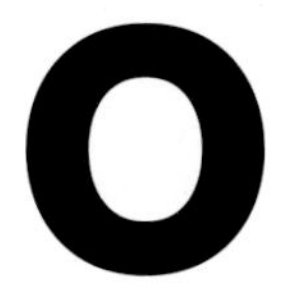

Oenology *see* **Wines**: Specific Aspects of Oenology.

Oils *see* **Fermentation (Industrial)**: Production of Oils and Fatty Acids; **Preservatives**: Traditional Preservatives – Oils and Spices.

Organic Acids *see* **Fermentation (Industrial)**: Production of Organic Acids, e.g. Citric, Propionic; **Preservatives**: Traditional Preservatives – Organic Acids

PACKAGING OF FOODS

Aaron L Brody Rubbright Brody Inc., Duluth, Georgia, USA

Packaging is intended to protect foods against environmental invasion. Among the many external variables that may adversely affect foods are an excess or deficiency of moisture, oxygen, dirt, humans (through tampering), dust, animals, insects and microorganisms. Packaging and processing are increasingly becoming integrated with each other; an example is canning, which is really a packaging and thermal preservation operation in which the can, its product contents, the filling temperature, air removal, closure, heating, cooling and distribution must be an uninterrupted continuum or else preservation is not effected.

More traditional preservation processes such as drying and freezing do not necessarily require close relationships between the product, process and packaging; the process and the packaging may be separate and the preservation effect will still be achieved. In contrast, in preservation processes such as thermal pasteurization, modified-atmosphere packaging, aseptic packaging, retort pouch and tray packaging, it is necessary to integrate all the elements to ensure the optimum preservation of the contained foods. For example, in aseptic packaging, preservation is achieved by sterilization of the product independently of the package, and the packaging equipment and assembly environment must therefore be sterile to exclude microorganisms from the ultimately hermetically sealed package. It is essential that the operations be connected by sterile linkages and that no microorganisms are permitted to contaminate any element.

For these reasons it has become increasingly important that the packaging be incorporated into the system if the objectives of delivering safe and high-quality food are to be achieved.

To understand fully the role of packaging in food preservation it is perhaps instructive to offer a few definitions. 'Packaging' is a term describing the totality of containment for the purpose of protecting the food contents and includes the package material, its structure and the equipment that marries the package structure to the food. Package materials are the components that constitute the structures usually known as packages or containers. Package materials are no longer single elements but rather composites of several different materials. In addition, new forms of packaging are increasingly replacing the traditional cans, bottles, jars, cartons and cases.

Preservation Requirements of Common Food Categories

Meats

Fresh Meat Most meat offered to consumers is freshly cut, with little further processing to suppress the normal microbiological flora present from the contamination received during the killing and breaking operations required to reduce carcass meat to edible cuts. Fresh meat is highly vulnerable to microbiological deterioration from indigenous microorganisms. These microorganisms can range from benign forms such as lactic acid bacteria or slime-formers, to proteolytic producers of undesirable odours and pathogens such as *Escherichia coli* O157:H7. The major mechanisms that retard fresh meat spoilage are temperature reduction to (or near) the freezing point and a reduced oxygen atmosphere during distribution to retard microbial growth. Reduced oxygen levels could provide conditions for the expression of pathogenic anaerobic microorganisms, a situation usually obviated by the presence of competitive spoilage organisms. Reduced oxygen levels also lead to the colour of fresh meat being the purple of myoglobin; exposure to air converts the natural meat pigment to the bright cherry-red oxymyoglobin characteristic of most fresh meat offered to and accepted by consumers in industrialized societies. Reduced oxygen packaging is achieved through the mechanical removal of air from the interiors of gas-impermeable multilayer flexible material pouches closed by heat-sealing the end after filling.

Ground Meat About 40% of fresh beef is offered in ground or minced form to enable the preparation of hamburger sandwiches and related foods. Ground beef was originally a by-product, that is, the trimmings from reducing muscle to edible portion size. The demand for ground beef is now so great that some muscle cuts are specifically ground to meet the demand. Grinding the beef further distributes the surface and below-surface microflora and thus provides a rich substrate for microbial growth even under refrigerated conditions. Relatively little pork is reduced to ground fresh form; however, increasing quantities of poultry meat are being comminuted and offered fresh to consumers, both on its own and as a cheaper substitute for ground beef. The major portion of ground beef is coarsely ground at abattoir level and packaged under reduced O_2 levels for distribution at refrigeration temperatures to help retard microbiological growth. The most common packaging technique is pressure-stuffing into chubs, which are tubes of flexible gas-impermeable materials closed at each end by tight-fitting metal clips. Pressure-stuffing the pliable contents forces most of the air out of the ground beef, and since there is no head-space within the package, little air is present to support the growth of aerobic spoilage microorganisms such as *Lactobacillus* and *Leuconostoc* spp. At retail level the coarsely ground beef is finely ground to restore the desirable oxymyoglobin red colour and to provide the consumer with the desired product.

In almost all instances, the retail cuts and portions are placed in expanded polystyrene (EPS) trays which are overwrapped with plasticized polyvinyl chloride (PVC) film. The tray materials are resistant to fat and moisture to the extent that many trays are internally lined with absorbent pads to absorb the purge from the meat as it ages and/or deteriorates in the retail packages. Because of the prognosis the PVC materials are not sealed but rather tacked so that the somewhat water vapour-impermeable structure does not permit loss of significant moisture during short refrigerated distribution. Being a poor gas barrier, PVC film permits the access of air and hence the oxymyoglobin red colour is retained for the short duration of retail distribution.

Case-ready Meat For many years, attempts have been made to shift the retail cutting of beef and pork away from the retailer's back room and into centralized factories. This movement has been stronger in Europe than in the USA, but some action has been detected in the latter country in the wake of the *E. coli* O157:H7 incidents. Case-ready retail packaging in the UK where the practice is relatively common, involves cutting and packaging meat under extremely hygienic conditions to reduce the probability of microbiological contamination beyond that of the indigenous microflora. Packaging is usually in a gas barrier structure, typically gas/moisture barrier foam polystyrene trays heat-sealed with polyester gas barrier film. The internal gas composition is altered to a high content of O_2 (up to 80%) and of CO_2 (up to 30%), with the remainder (if any) being nitrogen as a filler gas to ensure against package collapse arising from internal vacuum formation. The high O_2 concentration fosters the retention of the oxymyoglobin red colour preferred by consumers, while the elevated CO_2 level suppresses the growth of aerobic spoilage microorganisms. Using this or similar technologies, refrigerated microbiological shelf lives of retail cuts may be extended from a few days to as much as a few weeks, permitting long-distance distribution, e.g. from a central factory to a multiplicity of retail establishments. One thesis favouring the centralized packaging of ground beef is that the probability of the presence of *E. coli* O157:H7 is reduced. On the other hand, if the pathogen *is* present at the central location, the probability of its being spread among a number of retailers is greatly increased. Nevertheless, the use of central factories, which would probably be under federal government supervision in the USA, and certainly under technical supervision, would increases the probability of the emerging packaged meat being microbiologically safe.

Alternative packaging systems for case-ready beef and pork include the 'master bag' system used widely for freshly cut poultry (see below) in which retail cuts are placed in conventional PVC film overwrapped EPS trays and the trays are multipacked in gas barrier pouches whose internal atmospheres are enhanced with CO_2 to retard the growth of aerobic spoilage microorganisms. Another popular system involves the use of gas barrier trays with heat-seal closure using flexible gas and moisture barrier materials. Conventional non-gas-barrier trays such as EPS may be overwrapped with gas/moisture barrier flexible films subsequently shrunk tightly around the tray to impart an attractive appearance. Other systems, all of which involve removal of O_2, include vacuum skin packaging in which a film is heated and draped over the meat on a gas/moisture barrier tray. The film clings to the meat so that no head-space remains, with the result that the meat retains the purple colour of myoglobin. In one such system the drape film is a multilayer whose outer gas barrier layer may be removed by the retailer, exposing a gas-permeable film that permits the entry of air, which reblooms the pigment and restores the desired colour. Variations on this double film system include packaging systems in which the film is not multilayer but is composed of two independent flexible layers, the outer being impermeable to gas and moisture and the inner gas-permeable to permit air entry to restore the red colour. In all instances the microbiological shelf life is extended by reduced temperature plus reduced O_2 levels, which may incidentally or intentionally be enhanced by elevated CO_2 concentration.

Processed Meat Longer-term preservation of meats may be achieved by curing, using agents such as salt, sodium nitrite, sugar, seasonings, spices and smoke, and by processing methods such as cooking and drying; these treatments alter the water activity, add antimicrobial agents, provide a more stable red colour, and generally enhance the flavour and mouth feel of the cured meats. Cured meats are often offered in tubular or sausage form which means that the shape is dictated by the traditional process and consumer demand. Because of the added preservatives, the refrigerated shelf life of processed meat is generally several times longer than that of the fresh meat. Because cured meats are not nearly so sensitive to oxygen variations as fresh meat, the use of reduced O_2 atmospheres to enhance the refrigerated shelf life is quite common. The O_2 reduction may be achieved by mechanical vacuum, inert gas flushing or a combination of methods. Since the conditions have been changed to obviate the growth of anaerobic pathogenic microorganisms, reduced oxygen conditions are generally effective in retarding the growth of aerobic spoilage microorganisms.

The containers for reduced O_2 packaging of cured meats are selected from a multiplicity of materials and structures depending on the protection required and the marketing needs: frankfurters are generally sold in twin web vacuum packages in which the base tray is an in-line thermoformed nylon/polyvinylidene chloride (PVDC) web and the closure is a heat-sealed polyester (PET)/PVDC flexible material. Sliced luncheon meats and similar products are packed in thermoformed unplasticized PVC or polyacrylonitrile (PAN) trays, heat-seal closed with PET/PVDC. Sliced bacon packaging employs one of several variations of PVDC skin packaging (in contact with the surface of the product) to achieve the oxygen barrier. Ham may be fresh, cured or cooked, with the cooking often performed in the package. The oxygen barrier materials employed are usually a variation of nylon/PVDC in pouch form.

Poultry Poultry meat is most commonly chicken, but turkey is becoming an increasingly significant category of protein. Further, chicken is increasingly penetrating the cured meat market as a less expensive but nutritionally and functionally similar substitute for beef or pork. Since the 1970s poultry processing in westernized societies has shifted into large-scale, almost entirely automated killing and dressing operations. In such facilities the dressed birds are chilled in water to near the freezing point after which they are usually cut into retail parts and packaged in case-ready form: expanded polystyrene trays overwrapped with printed PVC or polyethylene film. The package is intended to appear as if it has been prepared in the retailer's back room, but in reality it is only a moisture and microorganism barrier. Individual retail packages, however, may be multipacked in gas-impermeable flexible materials to permit gas flush packaging, thus extending the refrigerated shelf life of the fresh poultry products.

Poultry is especially susceptible to infection with *Salmonella* spp., which are pathogenic in large quantities. Such organisms are not removed or destroyed by the extensive washing and chemical sanitation of current poultry processing plants, merely reduced in numbers. Modified-atmosphere packaging has relatively little effect on *Salmonella* and so refrigeration during distribution is critical in the drive to avoid increasing populations of this bacterium.

All meat products may be preserved by thermal sterilization in metal cans or, less frequently, glass jars. The product is filled into the container which is hermetically sealed, usually by double-seam metal end closure (see Fig. 2). After sealing, the cans are retorted to destroy all microorganisms present and cooled to arrest further cooking. The metal (or glass) serves as a barrier to gas, moisture, microbes etc. to ensure indefinite microbiological preservation. Cans or jars do not, however, ensure against further biochemical deterioration of the contents.

Fish

Fish is among the most difficult of all foods to preserve in its fresh state because of its inherent microbiological population, many organisms of which are psychrophilic, i.e. capable of growth at refrigerated temperatures. Further, seafoods may harbour a non-proteolytic, quasi-psychrophilic anaerobic pathogen, *Clostridium botulinum* type E. The need to prolong the refrigerated shelf life of fresh fish suggests the application of modified-atmosphere packaging in which reduced O_2 levels and elevated CO_2 levels are present (**Table 1**). However, a reduced O_2 atmosphere can permit the expression of type *E. botulinum*, and for this reason reduced O_2 packaging for seafood is discouraged in the USA. This is not the situation in Europe, where gas barrier flexible and semirigid plastic packaging similar to that described above for case-ready fresh beef is often applied.

Packaging for fresh seafood is generally moisture-resistant but not necessarily proof against microbial contamination. Simple polyethylene film is employed

Table 1 Pathogens of concern in modified atmosphere-packaged and vacuum-packaged foods

Psychrotrophs – growth at 3–4°C
Listeria monocytogenes
Yersinia enterocolitica
Bacillus cereus
Non-proteolytic *Clostridium botulinum*
Pseudopsychrotrophs – growth at 7–8°C
Escherichia coli O157:H7
Salmonella sp.
Mesophiles – growth at >10°C
Proteolytic *Clostridium botulinum*

Table 2 Ranges for bacterial growth

Organism	*pH range*
Gram-negative bacteria	
Escherichia coli	4.4–9.0
Pseudomonas fluorescens	6.0–8.5
Salmonella typhimurium	5.6–8.0
Gram-positive bacteria	
Bacillus subtilis	4.5–8.5
Clostridium botulinum	4.7–8.5
Lactobacillus sp.	3.8–7.2
Staphylococcus aureus	4.3–9.2

often as liners in corrugated fibreboard cases. The polyethylene serves not only to retain product moisture but also protects the structural case against internal moisture.

Seafood may be frozen, in which case the packaging is usually a form of moisture-resistant material plus structure such as polyethylene pouches or polyethylene-coated paperboard cartons.

Canning of seafood is much like that of meats since all seafoods have a pH above 4.6 and so require high-pressure cooking or retorting to effect sterility in metal cans (**Table 2**).

One variation unique to seafood is thermal pasteurization, in which the product is packed into plastic cans under reasonably clean conditions, achievable in contemporary commercial seafood factories. The filled and hermetically sealed cans are heated to temperatures of up to 80°C to effect pasteurization to permit several weeks of refrigerated shelf life. The system is usually effective because *Clostridium botulinum* type E spores are thermally sensitive and may be destroyed by temperatures of 80°C. To ensure against growth of other pathogens which may grow at ambient temperatures, however, distribution at refrigerated temperatures is dictated.

Dairy Products

Milk Milk and its derivatives are generally excellent microbiological growth substrates and therefore potential sources of pathogens. For these reasons, almost all milk is thermally pasteurized as an integral element of processing. Refrigerated distribution is generally dictated for all products that are pasteurized to minimize the probability of spoilage.

Milk is generally pasteurized and packaged in relatively simple polyethylene-coated paperboard gable-top cartons or extrusion blow-moulded polyethylene bottles for refrigerated short-term (several days to 2 weeks) distribution. Such packages offer little beyond containment and avoidance of contamination as protection benefits; they retard the loss of moisture and resist fat intrusion. Newer forms of milk packaging incorporate reclosure, a feature that was missing from the traditional gable-top cartons. Further, modern packaging environmental conditions have been upgraded microbiologically to enhance refrigerated shelf life by the use of pre-sterilization of the equipment, shrouding and use of clean air.

An alternative, popular in Canada, employs polyethylene pouches formed on vertical form/fill/seal machines and heat-sealed after filling. This variant has been enhanced by re-engineering into aseptic format, a system that has not become widely accepted. Pouch systems are generally less expensive than paperboard and semirigid bottles, but are less convenient for consumers. Little difference exists between the three packaging systems from a microbiological perspective.

In some countries, aseptic packaging is employed to deliver fluid dairy products that are shelf-stable at ambient temperatures. The most common processing technology is ultra-high temperature short time thermal treatment to sterilize the product followed by aseptic transfer into the packaging equipment. Three general types of aseptic packaging equipment are employed commercially: vertical form/fill/seal in which the paperboard composite material is sterilized by high temperature/high concentration hydrogen peroxide (removed by mechanics plus heat); erected preformed paperboard composite cartons which are sterilized by hydrogen peroxide spray (removed by heat); and bag-in-box, in which the plastic pouch is pre-sterilized by ionizing radiation. The former two are generally employed for consumer sizes while the last is applied to hotel, restaurant or institutional sizes, largely for ice cream mixes. Fluid milk is generally pasteurized, cooled and filled into bag-in-box pouches for refrigerated distribution.

Cheese Fresh cheeses such as cottage cheese fabricated from pasteurized milk are generally packaged in polystyrene tubs or polyethylene pouches for refrigerated distribution. Such packages afford little microbiological protection beyond acting as a barrier against recontamination, i.e. they are little more than rudimentary moisture loss and dust protectors, but are adequate because the distribution time is so short. Enhancement of refrigerated shelf life may be achieved by clean filling and/or the use of a low O_2, high CO_2 atmosphere, all of which retard the growth of lactic acid spoilage microorganisms.

Fermented Milks Fermented milks such as yoghurts fall into the category of fresh cheeses from a packaging perspective, i.e. they are packaged in polystyrene or polypropylene cups or tubs to contain and to protect minimally against moisture loss and microbial recontamination. Their closures are not hermetic and so gas passes through both the closures and the plastic walls, and microorganisms could enter after the package is opened. Because the refrigerated shelf life is short, however, few measures are taken from a packaging standpoint to lengthen the shelf life. Clean packaging is often used to achieve several weeks of refrigerated shelf life. Aseptic packaging is occasionally used to extend the ambient temperature shelf life of these products. Two basic systems are employed; one uses preformed cups, and the other is thermoform/fill/seal. In the former, the cups are sterilized by spraying with H_2O_2 and heating to remove the residue prior to filling and heat-sealing a flexible closure to the flanges of the cups, which are impermeable to gas and water vapour. In the thermoform/fill/seal method, a sheet of multilayer barrier plastic sheet (usually polystyrene plus PVDC) is immersed in H_2O_2 to sterilize it, air-knifed to remove the residual sterilant, heated to softening, and formed into cups by pressure. The web containing the connected cups is within a sterile environment under positive pressure of sterile air. The cavities are filled with sterile product and a flexible barrier material web, usually an aluminium foil lamination (also sterilized by H_2O_2 immersion), is heat-sealed to the cup flanges. Filled and sealed cups then pass through a sterile air lock. These aseptic dairy packaging systems may also be employed for juices and soft cheeses.

Recently, aseptic packaging of dairy products has been complemented by ultra-clean packaging on both preformed cup deposit/fill/seal and thermoform/fill/seal systems. In these systems, intended to offer extended refrigerated shelf life for low-acid dairy products, the microbicidal treatment is with hot water to achieve a 4D kill (i.e. four times the decimal reduction time) on the package material surfaces. The same systems may be employed to achieve ambient temperature shelf stability for high-acid products such as juices and related beverages.

Cured cheeses are subject to surface mould spoilage as well as to further fermentation by the natural microflora. These microbiological growths may be retarded by packaging under reduced O_2 atmospheres which may or may not be complemented by the addition of CO_2. To retain the internal environmental condition, the use of gas barrier package materials is commercial. Generally, flexible barrier materials such as nylon plus PVDC are employed on horizontal flow wrapping machines or on twin web thermoform/vacuum/seal machines. On twin web machines, the flat sealing web is usually a variant of polyester plus PVDC. One problem is that some cured cheeses continue to produce CO_2 as a result of fermentation, and so the excess gas must be able to escape from the package or else the package might bulge or even burst. Somewhat less gas-impermeable materials are suggested for such cheeses.

In recent years, shredded cheeses have been popularized. Shredded cheeses have increased surface areas which increase the probability of microbiological growth. Gas packaging under CO_2 in gas-impermeable pouches is mandatory. One feature of all shredded cheese packages today is the zipper reclosure which does not represent an outstanding microbiological barrier after the package has first been opened.

Ice Cream Ice cream and similar frozen desserts are distributed under frozen conditions and so are not subject to microbiological deterioration, but the product must be pasteurized prior to freezing and packaging. The packaging needs to be moisture resistant because of the presence of liquid water prior to freezing and sometimes during removal from refrigeration for consumption. Water-resistant paperboard, polyethylene-coated paperboard and polyethylene structures are usually sufficient for containment of other frozen desserts.

Fruit and Vegetables

In the commercial context, fruits are generally high-acid foods and vegetables are generally low-acid. Major exceptions are tomatoes, which commercially (not botanically) are regarded as vegetables, and melons and avocados, which are low-acid.

The most popular produce form is fresh, and increasingly fresh cut or minimally processed. Fresh produce is a living, 'breathing' entity with active enzyme systems fostering the physiological consumption of O_2 and production of CO_2 and water vapour. From a spoilage standpoint, fresh produce is more subject to physiological than to microbiological spoilage, and measures to extend the shelf life are designed to retard enzyme-driven reactions and water loss.

The simplest means of retarding fresh produce deterioration is temperature reduction, ideally to near freezing point but more commonly to about 4–5°C. Temperature reduction also reduces the rate of microbiological growth, which is usually secondary to physiological deterioration.

Since the 1960s, alteration of the atmospheric environment in the form of modified or controlled atmosphere preservation and packaging has been used commercially to extend the refrigerated shelf life of fresh produce items, such as apples, pears, strawberries, lettuce and now fresh cut vegetables. Controlled atmosphere preservation has been largely confined to warehouses and transportation vehicles such as trucks and seaboard containers. In this form of preservation, the O_2, CO_2, ethylene and water vapour levels are under constant control to optimize refrigerated shelf life. For each class of produce a separate set of environmental conditions is required for optimum preservation effect. In modified-atmosphere packaging, the produce is placed in a package structure and an initial atmosphere is introduced. The normal produce respiration plus the permeation of gas and water vapour through the package material and structure drive the interior environment towards an equilibrium gas environment that extends the produce quality retention under refrigeration. In some instances the initial gas may be air (passive atmosphere establishment). Produce respiration rapidly consumes most of the oxygen within the package and produces CO_2 and water vapour to replace it, generating the desired modified atmosphere.

The target internal atmosphere is to retard respiration rate and microbiological growth. Reduced O_2 and elevated CO_2 levels independently or in concert retard the usual microbiological growth on fruit and vegetable surfaces.

One major problem is that produce may enter into respiratory anaerobiosis if the O_2 concentration is reduced to near extinction. In respiratory anaerobiosis, the pathways produce undesirable compounds such as alcohols, aldehydes and ketones instead of the aerobic end products such as CO_2. To minimize the production of these undesirable end products, elaborate packaging systems are being developed. Most of these involve mechanisms to permit air into the package to compensate for the oxygen consumed by the respiring produce. High-gas-permeability plastic films, microperforated plastic films, plastic films disrupted with mineral fill, and films fabricated from polymers with temperature-sensitive side chains have all been proposed or used commercially.

The need for reduced temperature is emphasized in modified-atmosphere packaging because the dissolution rate of CO_2 in water is greater at lower temperatures than at higher temperatures. Carbon dioxide is one of the two major gases involved in reducing the rate of respiration and the growth of microorganisms.

Since the late 1980s, fresh cut vegetables, especially lettuce, cabbage, carrots, etc., have been a major product in both the retail trade and the hotel, restaurant and institutional markets. Cleaning, trimming and size reduction lead to a greater surface area to volume ratio and expression of fluids from the interior, increasing the respiration rate and offering a better substrate for microbiological growth than the whole fruit or vegetable. On the other hand, commercial fresh cutting operations generally are far superior to mainstream fresh produce handling in cleanliness, speed through the operations, temperature reduction and application of microbicides such as chlorine. Although some would argue, on the basis of microbial counts found in fresh cut produce in distribution channels, that uncut produce is safer, the paucity of its cleaning coupled with the rarity of adverse incidents related to fresh cut produce lead to the opposite conclusion – that fresh cut is significantly safer microbiologically. Another argument is that the low O_2 environment within most fresh cut produce packages plus the risk of soil contamination lead to ideal conditions for the proliferation of *Clostridium botulinum*. Further, distribution temperatures are often in excess of 10°C, well within the range of growth and production of spores. However, extensive testing has demonstrated that after responsible fresh cut processing, pathogenic spores are present in relatively small numbers, distribution temperatures prior to retail level are significantly lower than for uncut produce, and times are too short for pathogenic expression. These data indicate that while anaerobic pathogenic problems may occur, they are significantly less likely in fresh cut than in uncut fruit and vegetables.

Uncut produce packaging comprises a multitude of materials, structures and forms, ranging from traditional containers such as wooden crates, to inexpensive ones such as injection-moulded polypropylene baskets, to polyethylene liners within waxed, corrugated fibreboard cases. Much of the packaging is designed to help retard moisture loss from the fresh produce or to resist the moisture evaporating or dripping from the produce (or occasionally its associated ice), to ensure the maintenance of the structure throughout distribution. Some packaging designs recognize the issue of anaerobic respiration and incorporate openings to allow passage of air into the package, for example perforated polyethylene pouches for apples or potatoes. Almost none of the contemporary packaging for fresh uncut produce encompasses any specific microbiological barriers or countermeasures. That result is a direct extension of the observation that uncut produce 'processing' is virtually nonexistent. Packing house operations include collection and the removal of debris and gross dirt, and packaging is usually the least expensive structure that will contain the contents during distribution, often at sub-optimum temperatures.

For freezing, vegetables are cleaned, trimmed, cut and blanched, prior to freezing and then packaging (or packaging and then freezing). Blanching and the other processing operations reduce the numbers of microorganisms. Fruit may be treated with sugar to help retard enzymatic browning and other undesirable oxidations. Produce may be individually quick frozen (IQF) using cold air or cryogenic liquids prior to packaging, or frozen after packaging as in folding paperboard cartons. Frozen food packages are generally relatively simple monolayer polyethylene pouches or polyethylene-coated paperboard to retard moisture loss. No special effort is engineered to obviate further microbiological contamination after freezing, although the polyethylene pouches are generally heat-sealed.

Canning of low-acid vegetables to achieve long-term ambient temperature microbiological stability is the same as for other low-acid foods, with blanching prior to placement in steel cans (today all welded side seam tin-free steel, with some two-piece cans replacing the traditional three-piece type), hermetic sealing by double seaming, and retorting and cooling. Canned fruit is generally placed into lined three-piece steel cans using hot filling coupled with post-fill thermal treatment. Increasingly, one end is 'easy open' for consumer convenience. Newer techniques involve placing fruit hot into multilayer gas- and moisture-impermeable tubs and cups prior to heat-sealing with flexible barrier materials and subsequent thermal processing to achieve ambient temperature shelf-stability or extended refrigerated temperature shelf life. These plastic packages are intended to provide greater convenience for the consumer as well as to communicate that the contained product is not 'overprocessed' like canned food.

Tomato Products The highly popular tomato-based sauces and pizza toppings must be treated as low-acid foods if they contain meat, as so many do. For marketing purposes, tomato-based products for retail sale are commonly packed in glass jars with reclosable metal lids. The glass jars are often retorted after filling and hermetic sealing; major differences from the technique using metal cans include counterpressured retorting and longer times for heating and cooling, since the thick-walled glass is a thermal insulator.

Juices and Juice Drinks Juices and fruit beverages may be hot-filled or aseptically packaged. Traditional packaging has been hot-filling into steel cans and glass bottles and jars. Aseptic packaging, described above for paperboard composite cartons, is being applied for polyester bottles using various chemical sterilants to effect the sterility of the package and closure interiors. Much fruit beverage is currently hot-filled into heat-set polyester bottles capable of resisting temperatures of up to 80°C without distortion. Hermetic sealing of the bottles provides a microbiological barrier, but the polyester is a modest oxygen barrier and so the ambient temperature shelf life from a biochemical perspective is somewhat limited.

Since the 1970s high-acid fluid foods such as tomato pastes and non-meat-containing sauces have been hot-filled into flexible pouches, usually on vertical form/fill/seal machines. The hot filling generates an internal vacuum within the pouch after cooling so that the contents are generally shelf-stable at ambient temperature. Package materials are usually laminations of polyester and aluminium foil with linear low-density polyethylene (LLDPE) internal sealant; this resists the relatively lengthy exposure to the high heat of the contents during and immediately following filling. The heat seal is hermetic. Some efforts have been made to employ transparent gas/water vapour barrier films in the structures: polyester/ethylene vinyl alcohol laminations with the same LLDPE sealant. Transparent flexible pouches offer the opportunity for the consumer to see the contents, and for the hotel, restaurant or institutional worker to identify the contents without needing to read the label.

Other Products

A variety of food products that do not fall clearly into the meat, dairy, fruit or vegetable categories may be described as 'prepared foods', a rapidly increasing segment of the industrialized society food market during the 1990s. Prepared foods are those that combine several different ingredient components into dishes that are ready to eat, or simply require heating. If the food is canned, the thermal process must be suitable for the slowest heating component, meaning that much of the product is overcooked to ensure microbiological stability. If it is frozen, the components are separate but the freezing process reduces the eating quality. The preferred preservation technology from a quality retention or consumer preference perspective is refrigeration.

Incorporation of several ingredients from a variety of sources correctly implies many sources for microorganisms – aerobic, anaerobic, spoilage, benign and pathogenic. Where refrigeration is the sole barrier, microbial problems are minimized by reducing the time between preparation and consumption to less than 1 day (under refrigeration at temperatures above freezing) plus a nodding acknowledgment of cleanliness during preparation. As commercial operations attempt to prolong the quality retention periods beyond same-day or next-day consumption, enhanced preservation 'hurdles' have been introduced. These microbiological growth retardant factors include elevated salt or sugar concentrations, reduced water activity, reduced pH to minimize the probability of pathogenic microbiological growth, selection of ingredients from reduced microbial count sources, and modified-atmosphere packaging. The last is often suggested as a potential stimulus for the growth of pathogenic anaerobic microorganisms, since the multiple ingredient sources can almost assure the presence of *Clostridium* spores, and the reduced O_2 low-acid conditions are common to the types of products such as potato salad, pasta dishes, etc. Further, distribution temperatures may often be in the 5°C range or higher.

Packaging for air-packaged prepared dish products is generally oriented thermoformed polystyrene trays with oriented polystyrene dome closures snap-locked into position i.e. no gas, moisture or microbiological barriers of consequence. Refrigerated shelf life is measured in days. When the product is intended to be heated for consumption, the base tray packaging may be thermoformed polypropylene or crystallized polyester with no particular barrier closure. For modified-atmosphere packaging the tray material is a thermoformed, coextruded polypropylene/ethylene vinyl alcohol with a flexible gas/moisture barrier lamination closure heat-sealed to the tray flanges. Refrigerated shelf life for such products may be measured in weeks.

For several years, the concept of pasteurizing the contents, vacuum packaging and distribution under refrigeration has been debated and commercially developed in both the USA and Europe. The 'sous vide' technique is the most publicized process of this type. In sous vide processing the product is packaged under vacuum and heat-sealed in an appropriate gas/water vapour barrier flexible package structure such as aluminium foil lamination. The packaged product is thermally processed at less than 100°C to destroy spoilage microorganisms and then chilled for distribution under refrigerated or (in the USA) frozen conditions. The US option is to ensure against the growth of pathogenic anaerobic microorganisms. A similar technology is cook-chill in which pumpable products such as chili, chicken à la King and cheese sauce are hot-filled at 80°C or more into nylon pouches which are immediately chilled (in cold water) to 2°C and then distributed at temperatures of 1°C. The hot filling generates a partial vacuum within the package to virtually eliminate the growth of any spoilage microorganisms that might be present.

This listing is only a sampling of the many alternative packaging forms offered and employed commercially for foods subject to immediate microbiological deterioration. An entire encyclopedia would be required to enumerate all of the known options available to the food packaging technologist with the advantages and issues associated with each.

Package Materials and Structures

Package Materials

In describing package materials, different conventions are employed depending on the materials and their origins. The commercial conventions are used with some common indicator of quantitative meaning to establish relative values.

Paper The most widely used package material in the world is paper and paperboard derived from cellulose sources such as trees. Paper is used less in packaging because its protective properties are almost non-existent and its usefulness is almost solely as decoration and dust cover. Paper is cellulose fibre mat in gauges of less than 250 microns. When the gauge is 250 microns to perhaps as much as 1000 microns the material is known as paperboard, which in various forms can be an effective structural material to protect contents against impact, compression and vibration. Only when coated with plastic is paper or paperboard any sort of protection against other environmental variables such as moisture. For this reason, despite their long history as packaging materials, paper and paperboard are only infrequently used as protective packaging against moisture, gas, odours or microorganisms.

Paper and paperboard may be manufactured from trees or from recycled paper and paperboard. Virgin paper and paperboard, derived from trees, has greater strength than recycled materials whose fibres have been reduced in length by multiple processing. Therefore, increased gauges or calipers of recycled paper or paperboard are required to achieve the same structural properties. On the other hand, because of the short fibre lengths, the printing and coating surfaces are smoother. Paper and paperboard are moisture-sensitive, changing their properties significantly and thus often requiring internal and external treatments to ensure suitability.

Metals Two metals are commonly employed for package materials: steel and aluminium. The former is traditional for cans and glass bottle closures, but is

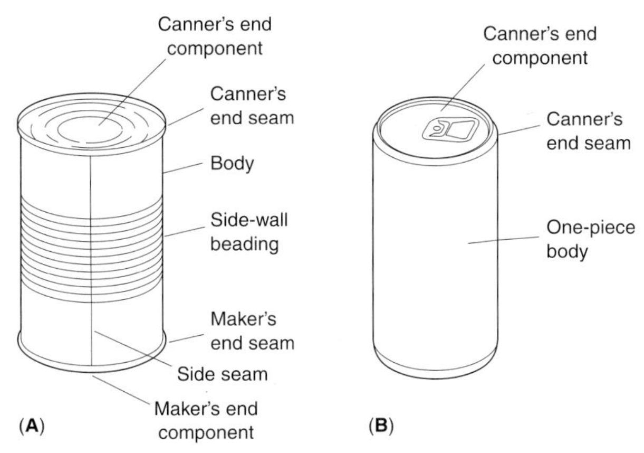

Figure 1 Metal can construction. (**A**) Three-piece steel can. (**B**) Two-piece steel or aluminium can. From Soroka (1995) with permission.

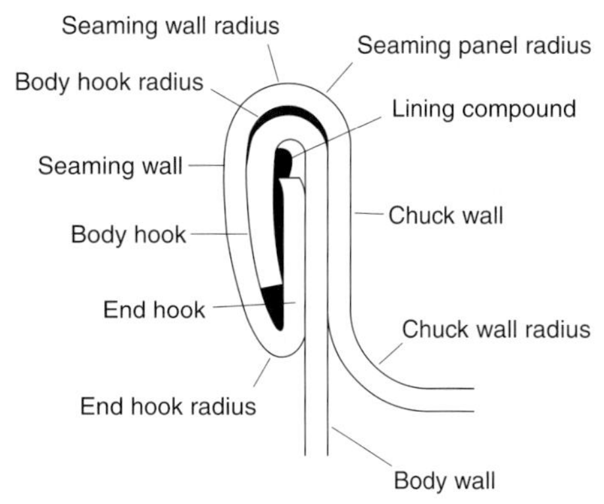

Figure 3 Double-seam closure on a metal can. From Soroka (1995) with permission.

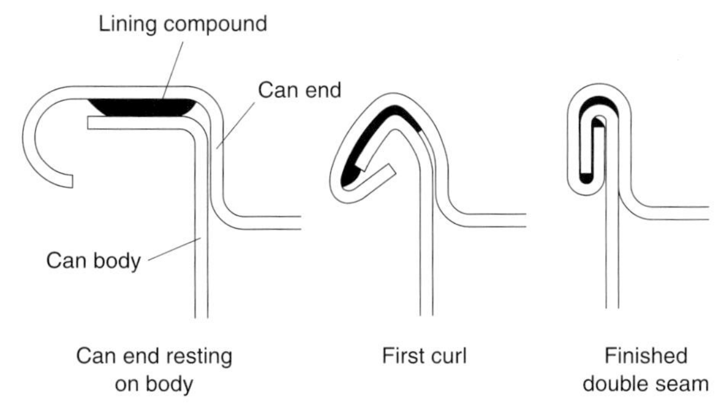

Figure 2 Operation of affixing or double-seaming a metal closure to a metal can body. From Soroka (1995) with permission.

subject to corrosion in the presence of air and moisture and so is almost always protected by other materials. Until the 1980s, the most widely used steel protection was tin, which also acted as a base for lead soldering of the side seams of 'tin' cans. When lead was declared toxic and removed from cans during the 1980s in the USA, tin was also found to be superfluous and its use as a steel can liner declined. The tin in 'tin-free' cans was chrome and chrome oxide. The construction and closure techniques of metal cans are shown in **Figures 1–3**.

In almost every instance the coated steel is further protected by organic coatings such as vinyls and epoxies which are really the principal protection.

Steel is rigid, a perfect microbial, gas and water vapour barrier, and resistant to every temperature to which a food may be subjected. Because steel–steel or steel–glass interfaces are not necessarily perfect, the metal is often complemented by resilient plastic to compensate for the minute irregularities.

Aluminium is lighter in weight than steel and easier to fabricate; it has therefore become the metal of choice for beverage containers in the USA and is favoured in other countries. As with steel, the aluminium must be coated with plastic to protect it from corrosion. It is the most commonly used material for can-making in the USA. However, aluminium cans must have internal pressure from CO_2 or N_2 to maintain their structure, and so aluminium is not widely used for food canning applications in which internal vacuums and pressures change as a result of retorting.

Aluminium may be rolled to very thin gauges (8–25 microns) to produce foil, a flexible material with excellent microbial, gas and water vapour barrier properties when it is protected by plastic film. Aluminium foil is generally regarded as the only 'perfect' barrier flexible package material. Its deficiencies include a tendency to pinholing, especially in thinner gauges, and to cracking when flexed.

In recent years, some applications of aluminium foil have been replaced by vacuum metallization of plastic films such as polyester or polypropylene.

Glass The oldest and least expensive package material is glass, derived from sand. Furthermore, glass is a perfect barrier material against gas, water vapour, microorganisms, odours, etc. The transparency of glass is often regarded by marketers and consumers as a desirable property. Technologists may view the transparency as less than desirable because visible and ultraviolet radiation accelerates biochemical (particularly oxidative) reactions.

Glass is energy-intensive to produce; it is heavy and vulnerable to impact and vibration even though it has excellent vertical compressive strength. For these reasons, glass is being displaced by plastic materials in industrialized societies.

Plastics The term 'plastics' describes a number of families of polymeric materials (**Table 3**), each with different properties. Most plastics are not suitable as package materials because they are too expensive or toxic in contact with food, or do not possess properties desired in packaging applications. The most commonly used plastic package materials are poly-

Table 3 Package plastic structures

Plastic	*Structure*	*Qualities*
Polyethylene (PE)	H H H H | | | | — C — C — C — C — | | | | H H H H	Three basic types: high-density linear low-density low-density Moisture barrier
Polypropylene (PP)	H CH_3 H CH_3 | | | | — C — C — C — C — | | | | H H H H	Higher temperature than polyethylene Low density high yield Very good moisture barrier
Ethylene vinyl alcohol (EVOH)	H H | | H O H O | | | | — C — C — C — C — | | | | H H H H	Excellent O_2 barrier resin Moisture sensitive, poor water barrier Used in coextrusion; expensive
Polyvinylidene chloride (PVDC)	H Cl H Cl | | | | — C — C — C — C — | | | | H Cl H Cl	Excellent O_2, moisture, flavour and fat barrier Dense
Polyvinyl chloride (PVC)	H Cl H Cl | | | | — C — C — C — C — | | | | H H H H	Stiff, clear – without plasticizer Soft with plasticizer No barrier
Polyamide (PA) (Nylon)	O O H H || || | | — C — $(CH_2)_4$ — C — N — $(CH_2)_6$ — N —	Temperature resistant Very good O_2 barrier Thermoformable
Polyethylene terephthalate (PET) (polyester)	O O H H || || | | — O — C — ⌬ — C — O — C — C — | | H H	High temperature after orientation
Polyacrylonitrile (PAN)	H H H H H | | | | | — C — C — C — C — C — | | | | | H C H C H ||| ||| N N	Very good O_2 barrier Not processable in extrusion unless copolymer
Polystyrene (PS)	H H H H H | | | | | — C — C — C — C — C — | | | | | H ⌬ H ⌬ H	Stiff, brittle, clear Very little barrier

Table 4 Properties of plastic package materials

Material	*Specific gravity*	*Clarity or colour*	*Water vapour transmission*[a]	*Gas transmission*[b]	*Resistance to grease*
Polyethylene					
high density	0.941–0.965	Semi-opaque	Low	High	Excellent
medium density	0.926–0.940	Hazy to clear	Medium	High	Good
low density	0.910–0.926	Hazy to clear	Good	High	Good
Polypropylene	0.900–0.915	Transparent	Good	High	Excellent
Polystyrene	1.04–1.08	Clear	High	High	Fair to good
Plasticized vinyl chloride	1.16–1.35	Clear to hazy	High to low	High	Good
Nylon	1.13–1.16	Clear to translucent	Varies	Low	Excellent

[a]Water vapour transmission rate is measured in gm^{-2} for 24h at 38°C and 90% relative humidity.
[b]Gas transmission is measured in $cm^3 ml^{-1} m^{-2}$ for 24h at 1atmosphere, 30°C and 0% relative humidity.

ethylene, polypropylene, polyester, polystyrene and nylon. Each has quite different properties (**Table 4**). Plastics may be combined with each other and with other materials to deliver the desired properties.

Polyethylene Polyethylene is the most used plastic in the world for both packaging and non-packaging applications. It is manufactured in a variety of densities ranging from 0.89 $g\,cm^{-3}$ (very low density) to 0.96 $g\,cm^{-3}$ (high density), and is lightweight, inexpensive, impact-resistant, relatively easily fabricated, and forgiving. Polyethylene is not a good gas barrier and is generally not transparent, but rather translucent. It may be extruded into film with excellent water vapour and liquid containment properties. Low-density polyethylene film is more commonly used as a flexible package material. Low-density polyethylene is also extrusion-coated onto other substrates such as paper, paperboard, plastic or even metal to impart water and water vapour resistance or heat sealability.

Although used for flexible packaging, high-density polyethylene is more often seen in the form of extrusion blow-moulded bottles with impact resistance, good water and water vapour barrier, but poor gas barrier properties. Any of the polyethylenes in proper structure functions as an effective microbial barrier.

Polypropylene Like polyethylene, polypropylene is a polyolefin, but it has better water vapour barrier properties and greater transparency and stiffness. Although more difficult to fabricate, polypropylene may be extruded into films that are widely used for making pouches particularly on vertical form/fill/seal machines. In cast film form, polypropylene is the heat sealant of choice on retort pouches because of its fusion sealing properties, and because in this form it is a good microbial barrier.

Polypropylene's heat resistance up to about 133°C permits it to be employed for microwave-only heating trays. Unfortunately microwave heating alone is insufficiently uniform to be a reliable mechanism for reducing microbiological counts or destroying heat-labile microbial toxins in foods.

Polyester A cyclical polymer that is relatively difficult to fabricate, polyethylene terephthalate polyester is increasingly the plastic of choice as a glass replacement in making food and beverage bottles. Polyester plastic is a fairly good gas and moisture barrier ; in bottle, tray or film form it is dimensionally quite stable and strong. Its heat resistance in amorphous form is sufficient to permit its use in hot-fillable bottles. When polyester is partially crystallized the heat resistance increases to the level of being able to resist conventional oven heating temperatures. For this reason crystallized polyester is employed to manufacture 'dual ovenable' trays for heat-and-eat foods ('dual ovenable' means that the plastic is capable of being heated in either conventional or microwave ovens).

The transparency of polyester makes it highly desirable from a marketing standpoint for foods that are not light sensitive.

Nylon Polyamide or nylon is a family of nitrogen-containing polymers noted for their excellent gas barrier properties. Moisture permeability tends to be less than in the polyolefin polymers and nylon is somewhat hygroscopic, meaning that the gas barrier may be reduced in the presence of moisture. Gas and water vapour barriers are enhanced by multilayering with polyolefins and high-gas-barrier polymers. Nylons are thermoformable and both soft and tough, and so are often used for thermoformed processed meat package structures in which the oxygen within the package is reduced to extend the refrigerated shelf life.

Polystyrene Polystyrene is a poor barrier to moisture or gas. It is, however, very machinable and usually highly transparent. Its structural strength is not good unless the plastic is oriented or admixed with a rubber modifier which reduces the transparency. Polystyrene is often used as an easy and inexpensive tray material for prepared refrigerated foods.

Polyvinyl Chloride Polyvinyl chloride is a polymer capable of being modified by chemical additives into plastics with a wide range of properties. The final materials may be soft films with high gas permeabilities, such as used for overwrapping fresh meat in retail stores; stiff films with only modest gas barrier properties; readily blow-mouldable semirigid bottles; or easily thermoformed sheet for trays. Gas and moisture impermeability is fairly good but must be enhanced to achieve 'barrier' status.

This material falls into a category of halogenated polymers which are regarded by some environmentalists as less than desirable. For this reason, in Europe and to a lesser extent in the USA, PVC has been resisted as a package material.

Polyvinylidene Chloride Polyvinylidene chloride (PVDC) is an excellent barrier to gas, moisture, fat and flavours, but is so difficult to fabricate on its own that it is almost always used as a coating on other substrates to gain the advantages of its properties.

Ethylene Vinyl Alcohol Ethylene vinyl alcohol (EVOH) is an outstanding gas and flavour barrier polymer which is highly moisture sensitive and so must be combined with polyolefin to render it an effective package material. Often EVOH is sandwiched between layers of polypropylene which act as water vapour barriers and thus protect the EVOH from moisture.

Package Structures

Currently, rigid and semirigid forms are the most common commercial structures used to contain foods. Paperboard is most common, in the form of corrugated fibreboard cases engineered for distribution packaging. In corrugated fibreboard three webs of paperboard are adhered to each other with the central or fluted section imparting the major impact and compression resistance to the structure. Folding cartons constitute the second most significant structure fabricated from paperboard. Folding cartons are generally rectangular in shape and often are lined with flexible films to impart the desired barrier.

Metal cans have traditionally been cylindrical (Figures 1, 2 and 3), probably because of the need to minimize problems with heat transfer into the contents during retorting. Recently, metal – and particularly aluminium – has been fabricated into tray, tub and cup shapes for greater consumer appeal, with consequential problems with measuring and computing the thermal inputs to achieve sterilization.

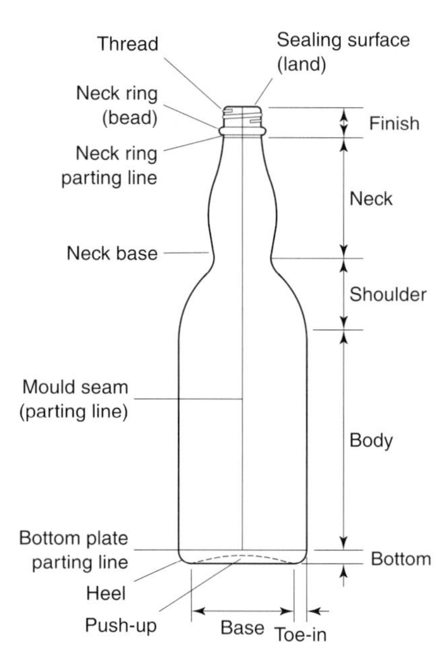

Figure 4 Glass bottle nomenclature. From Soroka (1995) with permission.

During the 1990s shaped cylinders entered the market again in efforts to increase consumer market share. Few have been applied for cans requiring thermal sterilization, but barrel and distorted body cans are not rare in France for retorted low-acid foods. Analogous regular-shaped cans are being used for hot filling of high-acid beverages.

Noted for its formability, glass has traditionally been offered in a very wide range of shapes and sizes including narrow-neck bottles (**Fig. 4**) and wide-mouth jars. Each represents its own singular problems in terms of fabrication, closure and – when applicable – thermal sterilization.

Plastics are noteworthy for their ability to be formed into the widest variety of shapes. Thin films can be extruded for fabrication into flexible package materials. These flexible materials may then be employed as pouch or bag stock or as overwraps on cartons or other structures, or as inner protective liners in cartons, drums, cases, etc. Thicker films (sheets) may be thermoformed into cups, tubs and trays for containment. Plastic resins may be injection- or extrusion-moulded into bottles or jars by melting the thermoplastic material and forcing it, under pressure, into moulds that constitute the shape of the hollow object, e.g. the bottle or jar.

See also: **Cheese**: In the Market Place. **Chilled Storage of Foods**: Use of Modified Atmosphere Packaging; Packaging with Antimicrobial Properties. **Fermented Milks**: Range of Products. **Fish**: Spoilage of Fish. **Heat Treatment of Foods**: Thermal Processing Required for Canning; Principles of Pasteurization. **Ice Cream**. **Meat and Poultry**: Spoilage of Meat; Curing of Meat; Spoilage of Cooked Meats and Meat Products. **Milk and Milk Products**: Microbiology of Liquid Milk; Microbiology of Dried Milk Products; Microbiology of Cream and Butter.

Further Reading

Brody AL (1989) *Controlled/Modified Atmosphere/Vacuum Packaging of Foods*. Trumbull, Connecticut: Food & Nutrition Press.

Brody AL (1994) *Modified Atmosphere Food Packaging*. Herndon: Virginia: Institute of Packaging Professionals.

Brody AL and Marsh KS (1997) *Wiley Encyclopedia of Packaging Technology*, 2nd edn. New York: John Wiley.

Jairus D, Graves R and Carlson VR (1985) *Aseptic Packaging of Food*. Boca Raton: Florida: CRC Press.

Paine FA and Paine HY (1983) *A Handbook of Food Packaging*. London: Blackie.

Robertson GL (1993) *Food Packaging*. New York: Marcel Dekker.

Soroka W (1995) *Fundamentals of Packaging Technology*. Herndon: Virginia: Institute of Packaging Professionals.

Wiley RC (1994) *Minimally Processed Refrigerated Fruits and Vegetables*. New York: Chapman & Hall.

Panary Fermentation *see* **Bread**: Bread from Wheat Flour.

PANTOEA

André Morin, Imperial Tobacco Limited, Montreal, Canada
Zahida Parveen, Department of Chemical and Biological Sciences, University of Huddersfield, UK

Introduction

The genus *Pantoea* is less than 10 years old and has been proposed to include some *Erwinia* and some *Enterobacter* species. The biochemical characteristics are still mainly being used to identify the various *Pantoea* spp. because the phylogenetic approach for this genus is still in its infancy. Cultural techniques and conditions used to grow *Pantoea* spp. are those used for typical members of the Enterobacteriaceae family, taking into account that some members of the *Pantoea* genus are facultative anaerobes. The genus *Pantoea* encompasses some species having beneficial effects on some edible plants, and others which are wound plant pathogens and that may be involved in plant decay. Some members of the genus *Pantoea* have been isolated from animals and human wounds and may be opportunistic pathogens.

Biochemical Characteristics of the Genus *Pantoea*

The genus *Pantoea* was proposed in 1989 to include the *Erwinia herbicola* group and one of the DNA hybridization groups of *Enterobacter agglomerans*. Analysis of DNA–DNA hybridization data and phenotypical characteristics revealed that these organisms were closely related to each other, and two species were originally established within this genus – *P. agglomerans* and *P. dispersa* (**Table 1**). Indeed, the genus *Pantoea* regroups species considered to be heterogeneous. For instance, *Enterobacter agglomerans* is synonymous with *Erwinia herbicola*, *E. uredovora* and *E. stewartii*. In 1992, three bacterial strains were identified as new species of the genus *Pantoea*. These were *P. punctata*, *P. citrea* and *P. terrea* (**Table 2**). Since then, three other bacterial strains belonging to the genus *Erwinia* have been placed in the genus *Pantoea* (**Table 3**): *P. ananas*, *P. stewartii* subsp. *stewartii* and *P. stewartii* subsp. *indologenes*.

The following describes the genus *Pantoea*. *Pantoea* (Pan.toe'a. Greek adjective *pantois*, of all sorts and sources): Gram-negative, non-capsulated, non-spore-forming, straight rods that are 0.5–1.3 × 1.0–3.0 μm. Cells are motile and peritrichous or non-motile. Cultures grow well on nutrient agar at 30°C. Colonies on nutrient agar are smooth, translucent and more or less convex with entire margins. Colonies are yellow, pale beige to pale reddish yellow or non-pigmented. *Pantoea* spp. are facultatively anaerobic and oxidase-negative. No gas is produced from D-glucose. No lysine decarboxylase, tryptophan deaminase or urease activity have been detected. Pectate is not degraded, and hydrogen sulphide is not produced from thiosulphate. Acid is produced from D-xylose, D-ribose, D-galactose, D-fructose, trehalose and *N*-acetyl-D-glucosamine, but not from D-arabinose, L-xylose, L-fucose or D-turanose. *Pantoea* spp. have been isolated from plants, seeds, fruits, soils and water, as well as from humans (wounds, blood, urine, internal organs) and animals in several parts of the world. The type species is *P. agglomerans*.

Identification Methods used to Characterize the Genus *Pantoea* and Related Microorganisms

Many commercial identification kits are widely available, such as API (BioMérieux), Vitek, Sensititre and Biolog. A number of these are automated and results obtained are compared to a database with profiles of

Table 1 Biochemical characteristics of *Pantoea agglomerans* and *P. dispersa*

	P. agglomerans *ATCC 27155*	P. dispersa *ATCC 14589*	*Acid production from:*	P. agglomerans *ATCC 27155*	P. dispersa *ATCC 14589*
Production of yellow pigment	+	+	L-Arabinose	+	+
Growth at 4°C	+(5)	−	D-Ribose	+	+
Growth at 41°C	−	+	D-Xylose	+	+
Growth at 44°C	−	−	D-Cellobiose	(+)(12)	(+)(8)
Motility	+	+	D-Galactose	+	+
KCN (growth)	+	−	Lactose	−	−
Gelatin liquefication	(+)(15)	(+)(16)	Maltose	+	+
Indole production	−	−	Melibiose	−	+
Voges–Proskauer reaction	+	+	Sucrose	+	+
Nitrate reduction	+	−	Trehalose	+	+
H_2S production	−	−	Melezitose	−	−
Hydrolysis of aesculin	+	−	Raffinose	+	−
Gas production from D-glucose	−	−	Glycerol	(+)(4)	+
Arginine dihydrolase	−	−	*meso*-Erythritol	−	−
Lysine decarboxylase	−	−	Adonitol	(+)(12)	−
Ornithine decarboxylase	+	+	Dulcitol	−	−
Phenylalanine deaminase	−	−	Inositol	(+)(15)	+
Tetrathionate reductase	−	−	D-Mannitol	+	+
Deoxyribonuclease	−	−	D-Sorbitol	−	−
β-Xylosidase	−	−	α-Methyl-D-glucoside	−	−
β-Galactosidase	+	+	Salicin	+	−
Urease	−	−	Glycogen	−	−
Utilization of:			Inulin	−	−
Citrate (Simmons)	+	+	Tartrate (Jordan)	−	−
Malonate	+	+	D-Fructose	+	+
D-Tartrate (Kauffman)		−	Mucate (Kauffman)	−	−
L-Tartrate (Kauffman)	−	−	*meso*-Tartrate (Kauffman)	−	+

+, Reaction present in at least 90% of strains within 24–48 h; −, reaction absent in at least 90% of strains after 2 days: (+), slow reaction (between 2 and 30 days: the number in parentheses indicates the number of days for reaction to occur); d, different reactions. Tests were done at 30°C unless otherwise indicated. (Gavini et al, 1989)

type species. The advantages of using kits such as these are the ease of use and production of rapid results. However, extreme caution should be exercised in the identification of *Enterobacter agglomerans*, *Erwinia herbicola* or *Pantoea* spp. when using commercial kits such as the API 20E. It is difficult to differentiate unequivocally between strains of *P. dispersa* and *P. agglomerans*, other named *Enterobacter agglomerans* and *Erwinia herbicola* strains and strains of other *Erwinia* species or of other genera. Other discriminating features are shown in Tables 1–3. Occasionally biochemical testing may not be sufficient for absolute identification, and DNA–DNA hybridization experiments and/or fatty acid analysis of bacterial cell walls may also be necessary. However, isolates identified as *P. agglomerans* by DNA–DNA hybridization experiments might not necessarily fit the biochemical profile of *P. agglomerans*. For instance, out of 65 identified isolates, 22 showed delayed acid production of acid from α-methyl-glycoside and none showed decarboxylated ornithine. Biochemical characteristics of 30 strains agreeing well with the *Enterobacter agglomerans–Erwinia herbicola* complex have been compared and divided into three phenotypic groups based on the difference of indole production and gas production from D-glucose on API 20E identification system (**Table 4**).

Thus, although the *Enterobacter agglomerans–Erwinia herbicola* complex has been proposed to regroup *Pantoea* species and related microorganisms, no consensus has been reached and these different taxonomic views have resulted in an ambiguous situation. For instance, phytopathologists often use the nomenclature of Dye, i.e. *Erwinia herbicola*, and clinical bacteriologists use the nomenclature of Ewing and Fife, i.e. *Enterobacter agglomerans*.

Cultural Techniques and Conditions of the Genus *Pantoea* and Related Species

Pantoea spp. have no special nutritional requirements and grow well on commercial media such as nutrient agar and trypticase soy agar. On media with a high carbon (usually sucrose) to nitrogen ratio, certain strains of *P. agglomerans* produce exocellular polymers. However, exopolysaccharide production by *P. agglomerans* can be repressed by initial carbon concentration higher than $60\,g\,l^{-1}$, and carbon limitation

Table 2 Differential characteristics of *Pantoea agglomerans*, *P. dispersa* and the species previously assigned to the *Erwinia herbicola–Enterobacter agglomerans* complex (Gavini et al, 1989)

	P. agglomerans	P. dispersa	Erwinia uredovora	Leclercia adecarboxylata	Escherichia hermanni	Escherichia vulneris	Enterobacter sakazakii	Rahnella aquatilis	Ewingella americana
Production of yellow pigment	+	d	+	+	+	d	+	–	–
Indole production	–	–	+	+	+	–	(–)	–	+
Voges–Proskauer reaction	+	+	+	–	–	–	+	+	–
Arginine decarboxylase	–	–	–	–	–	d	+	–	–
Lysine decarboxylase	–	–	–	–	–	(+)	–	–	–
Ornithine decarboxylase	d	–	–	–	+	–	+	–	+
Acid production from:									
D-Cellobiose	d	d	+	+	+	+	+	+	d
Lactose	d	–	+	+	d	(–)	+	+	+
Maltose	+	+	+	+	+	+	+	+	d
Melibiose	–	d	+	+	–	+	+	+	NT
Raffinose	d	–	+	+	d	+	+	+	–
Sucrose	–	+	+	+	d	–	+	+	–
Adonitol	–	–	+	+	–	–	–	–	–
Dulcitol	–	–	+	+	(–)	–	–	(+)	–
Sorbitol	–	–	–	–	–	–	–	+	–
α-Methyl-D-Glucoside	–	–	+	–	–	*d*	+	–	–
Salicin	+	–	–	+	*d*	*d*	+	+	+

+, Reaction present in at least 90% of strains; –, reaction absent in at least 90% of strains; (+), reaction present in 76–89% of strains; (–), reaction present in 11–25% of strains; d, reaction present in 26–75% of strains; NT, not tested. Data was obtained after 48 h of incubation at 30°C (*Pantoea* and *Erwinia* spp.) or 36 ± 1°C (other species).

might be the desired situation for improving exopolysaccharide production as observed for xanthan and alginate production. This is different from what has generally been reported in the literature, where exopolysaccharide production can be enhanced by an excess of carbon in the culture medium, as reported for *Enterobacter aerogenes* and by limited amount of nitrogen.

A medium using maple sap as the carbon source has been developed for the production of an exopolysaccharide by *P. agglomerans* (**Table 5**). The exopolysaccharide (EPS) production stage from *P. agglomerans* was observed mostly after 10 h of culture, when no further growth occurred. Sugar, yeast extract, urea (SYU) medium has been designed (Table 5) as an isolation medium for polymer-producing organisms. SYU was also used as the production medium for the EPS produced by *P. agglomerans*. Other media have been investigated as production media for EPS or lipopolysaccharide (Table 5). As a facultative anaerobe, *P. agglomerans* can grow under aerobic or anaerobic conditions but the resulting metabolites differ. In both SY and SYU, *P. agglomerans* was cultured under aerobic conditions at 25–30°C for up to 90 h. Anaerobic conditions led to reduced polymer production (**Fig. 1**) and an increase in the production of organic acids. The production of the EPS from *P. agglomerans* is also influenced by the concentration of yeast extract, the carbon/nitrogen ratio (C/N), the temperature, the pH of the medium and the agitation rate of the culture (**Table 6**).

When grown aerobically, *P. agglomerans* metabolizes sugars through the Embden–Meyerhof–Parnas pathway, where the end product, pyruvate, is completely oxidized into carbon dioxide and water through the tricarboxylic acid cycle. Intermediates of the Krebs cycle, such as succinate, α-ketoglutarate, fumarate and oxaloacetate, might be monitored in the broth resulting from the aerobic growth of *P. agglomerans*. α-Ketoglutarate was found to be invariably excreted along with the EPS during aerobic growth of *P. agglomerans*. Monitoring of metabolites has been used to detect oxygen deficiency when

Table 3 Differential characteristics of *Pantoea* species. Reproduced with permission from Mergaert et al (1993).

	P. ananas	P. stewartii *subsp.* stewartii	P. stewartii *susbsp.* indologenes	P. agglomerans	P. dispersa	P. citrea	P. punctata	P. terrea
API 20E tests								
Nitrate reduced to nitrate	d	–	–	+	–	+	+	+
Indole production	+	–	+	–	–	–	–	–
Citrate utilization	+	–	+	d	+	+	+	+
β-Galactosidase	+	+	+	+	+	+	–	–
API 50CHE tests: Acid produced from:								
Glycerol	+	–	(+)	(–)	d	+	+	+
D-*Arabitol*	+	–	+	d	+	(+)	–	–
Sorbitol	(+)	–	–	–	–	(–)	–	(–)
Cellobiose	+	–	+	d	+	–	–	–
Maltose	+	–	+	+	+	+	(–)	(–)
Lactose	+	–	+	(–)	(–)	+	–	–
*α-Methyl-*D*-mannosidase*	(+)	–	–	–	–	–	–	–
Arbutin	+	–	+	+	–	(–)	–	+
Salicin	+	–	+	+	(–)	(–)	–	+
Raffinose	(+)	+	+	–	–	–	d	d
D-*Turanose*	–		–	–	+	–	–	–
D-*Fucose*	–	–	–	–	–	+	+	+
Hydrolysis of aesculin	d	–	+	+	+	–	(–)	+
Other tests								
Motility	+	–	(+)	+	+	–	–	+
Malonate utilization	–	–	–	+	–	–	–	–
Phenylalanine deaminase	–	–	–	+	–	–	–	–
Growth on cis-aconitase	+	–	+	+	+	NT	NT	NT

+, At least 90% of strains are positive; –, at least 90% of strains are negative; (+), 76–89% of strains are positive; (–), 11–25% of strains are positive; d, 26–75% of strains are positive; NT, not tested. Data were obtained after 24 h (API 20E tests, motility), 48 h (API 50CHE tests) or after 3 days (other tests) of incubation at 30°C.

Table 4 Division of the *Enterobacter agglomerans–Erwinia herbicola* complex (based on Iimura and Hosono, 1996)

Strains (API 20E code)	*Indole production*	*Gas production from* D*-glucose*
Pantoea agglomerans	–	–
(1 205 133)		
(1 205 132)		
(1 005 333)		
(1 005 332)		
(1 005 133)		
(1 005 132)		
Erwinia ananas	+	–
(renamed *Pantoea ananas*)		
(1 245 773)		
(1 045 773)		
(1 045 573)		
Rahnella aquatilis	–	+
(1 205 573)		

growing continuous cultures of *P. agglomerans*. When grown in fermenter at dissolved oxygen concentration lower than 1% saturation, lactate, succinate and formate were formed by *P. agglomerans*, but neither EPS nor α-ketoglutarate was produced. Although EPS synthesis by facultative anaerobes was reported to occur only when the microorganism was grown aerobically, *P. agglomerans* could also produce EPS at dissolved oxygen concentration lower than 20% saturation. When *P. agglomerans* was cultivated at dissolved oxygen concentration greater than 14% saturation, EPS and a mixture of lactate, α-ketoglutarate and formate were produced. Thus, products from anaerobic metabolism might also be found during carbohydrate fermentation by *P. agglomerans* grown under partially aerobic conditions. This situation would occur towards the end of the fermentation due to the increase in viscosity as the EPS concentration increased. A reduction in oxygen tension occurs and anaerobic conditions would ensue.

Safety Guidelines for Working with the Genus *Pantoea*

Members of the genus *Pantoea* have been isolated from a number of sources, both animal and plant. *P. agglomerans* has been found in human wounds, blood and urine as an opportunistic pathogen, and other members of this genus are phytopathogenic, e.g. *P. ananas*.

P. agglomerans has been classified as a category II

Table 5 Suitable culturing media and conditions for *P. agglomerans*

SYU medium (Parveen, unpublished data)	*SY medium (Morin et al, 1993)*	*Lipopolysaccharide production medium (Karamanos et al, 1992)*	*SPY medium (Okutani and Kobayashi, 1991)*
10 g Sucrose[a]	80 g Sucrose[a] (as concentrated maple sap syrup)	Trypticase soya broth	30 g Sucrose[a]
0.5 g Yeast extract	1.5–3.4 g Yeast extract		1 g Yeast extract
0.5 g Urea[b]			5 g Peptone
5 g K_2HPO_4, 2 g KH_2PO_4	9 g K_2HPO_4, 3 g KH_2PO_4		
0.2 g $MgSO_4$, 0.1 g NaCl[a]	0.2 g $MgSO_4.7H_2O$[a]		1 l Sea water
20 g Agar	20 g Agar		15 g Agar
1 l Tap water	1 l Tap water		
Culture at 25°C	Culture at 30°C	Culture at 37°C	Culture at 25°C

[a]separate sterilization; [b]filter-sterilized.

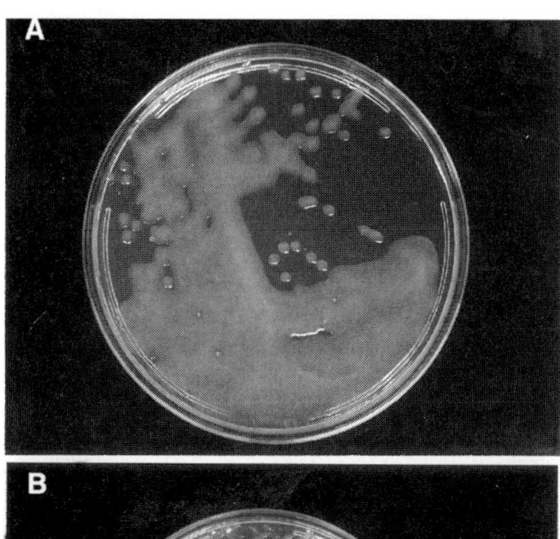

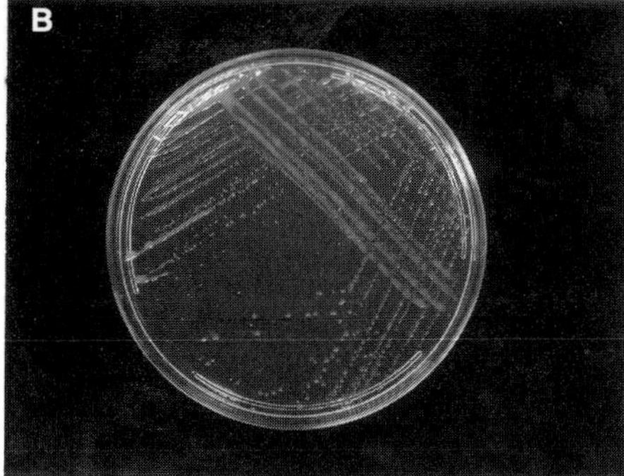

Figure 1 (**A**) Aerobic and (**B**) anaerobic growth of *Pantoea agglomerans* on sugar, yeast (SY) agar medium after 18 h of incubation at 30°C.

organism and as such certain precautions must be applied when working with it. Any work in open spaces is to be avoided and must be confined to a class II microbiological safety cabinet. The production of aerosols when grown in liquid culture is to be minimized. It is recommended that contact is avoided with those who are immunosuppressed. *Enterobacter sakazakii* and *P. agglomerans* have a tendency to cause life-threatening neonatal meningitis and sepsis, and as a consequence it is suggested that the measures mentioned above must be followed for *P. agglomerans*. General microbiological safety practice must also be followed and strictly adhered to.

Table 6 Range of values of components or parameters allowing the growth and exopolysaccharide production of *Pantoea agglomerans*

Components or parameters	*Values*
Sucrose	20–120 g l^{-1}
Yeast extract	1–10 g l^{-1}
Carbon/nitrogen ratio	131–218
Temperature	15–35°C
pH	5.0–8.0
Agitation rate	50–300 r.p.m.

Importance of the Genus *Pantoea* and Individual Species in the Food Industry

In the *Pantoea* genus, *P. agglomerans* is most often reported to have both beneficial and detrimental effects on food and feed, followed by *P. dispersa*, *P. ananas* and *P. citrea*.

Food- and Feed-related Beneficial Effects of *Pantoea* spp. (Table 7)

Pantoea agglomerans is a common inhabitant of soil and is often associated with plants, although its precise role has not always been elucidated. For instance *P. agglomerans* coexists with *Pseudomonas syringae* and various non-pathogenic epiphytic species in the phyllosphere of beans (*Phaseolus vulgaris*). *Pantoea agglomerans* was the most frequent endophytic species (59% of all isolates) recovered in foliage tissues of red clover plants (*Trifolium pratense*). *P. (Enterobacter) agglomerans* has been isolated from buckwheat (*Fagopyrum esculentum*) seeds and has been identified as one of the dominant contaminating species.

Some *Pantoea* spp. such as *P. agglomerans* can enter in a relationship that is beneficial to some host plants, such as wheat (*Triticum* spp.). *P. agglomerans* was shown to promote wheat growth by producing phytohormones, fixing atmospheric nitrogen and reducing nitrate.

Table 7 Food- and feed-related beneficial effects and/or uses of *Pantoea* spp.

Species	*Effect*	*Plant (Latin name); disease (pathogen)*	*Product*
P. agglomerans	Antifungal activity	Potato (*Solanum tuberosum*) Potato dry rot (*Fusarium sambucinum*)	NA
P. agglomerans	Production of phytohormones Nitrogen fixation Nitrate reduction	Wheat (*Triticum* spp.)	NA
P. dispersa	Albicidin detoxification enzyme	Sugar cane (*Saccharum officinale*) Lead-scald disease (*Xanthomonas albilineans*)	NA
P. agglomerans	Interestification of butter fats	NA	Lipase
P. agglomerans	Food stabilizer	NA	Exopolysaccharide
P. agglomerans	Vitamin supplement	NA	Nicotinamide
Pantoea spp.	Feed additive	NA	α-Hydroxy-4-methylthiobutyric acid
Pantoea spp.	Anticholesterolaemic, antiulcer, antiwithdrawing	NA	Lipopolysaccharide

NA = not applicable.

Some *Pantoea* species are associated with plant materials as saprophytes of pathogens. *P. agglomerans* has a beneficial effect because it can produce antimicrobial substances, antagonistic enzymic activity or detoxifying activity. For instance, the use of *P. dispersa* to detoxify albicidins, a family of phytotoxins and antibiotics, has been proposed to biocontrol sugarcane leaf-scald disease. *P. dispersa* produces an albicidin detoxification enzyme (ADE) capable of irreversibly inactivating albicidin and substantially reducing or inhibiting the development of leaf-scald disease caused by the xylem-invading bacterium *Xanthomonas albilineans* in sugar cane (*Saccharum officinale*). The enzyme can be introduced directly to the plant material by soaking or infiltrating the plant. A nucleotide sequence from *P. dispersa* encodes an ADE, and this could be used to transform sugar cane by microprojectile particle bombardment.

P. agglomerans has been isolated from soils in which potato (*Solanum tuberosum*) plants have been cropped and from which harvested potatoes exhibit a relatively low incidence of the potato disease to be controlled. *P. agglomerans* can be used as a bacterial antagonist of fungal species such as *Fusarum sambucinum* responsible for potato dry rot and other tuber diseases which occur in the field or in post-harvest storage.

P. agglomerans isolated from commercial starters suitable for the degradation of domestic wastes has been shown to be an efficient lipase-producing microorganism, and interesterification of butter fat triacylglycerols by enzymatic extracts of *P. agglomerans* has been demonstrated. Such a lipase could be used to perform a favourable interchange of hypercholesterolaemic fatty acids (12 : 0; 14 : 0 and 16 : 0) with those of hypocholesterolaemic fatty acids (18 : 0; 18 : 1) at the *sn*-2 position of triacylglycerol molecules. *Pantoea* sp. produces low-molecular-weight lipopolysaccharides that may be used as anticholesterolaemic, antidiabetic, antiulcer and antihaemorrhoidal, and also to prevent withdrawal symptoms in individuals addicted to alcohol. *P. agglomerans* also produced high-molecular-weight EPS when grown in maple sap. The estimated M_r of the EPS ranged from 5×10^5 to 2×10^6 Da and is close to that reported for xanthan (2.8×10^6). The polymer produced by *P. agglomerans* was investigated as an alternative to xanthan in the food industry and could find applicants for the stabilization of emulsions requiring low viscosity. The polysaccharide from *P. agglomerans* grown on SYU medium was investigated as a flocculant in conjunction with multivalent cations.

A process has been claimed for the production of industrially important amides by *P. agglomerans*. This process involves the conversion of a nitrile into a corresponding amide by the action of a *P. agglomerans* strain or an enzyme thereof, and the recovery of the resulting amide such as nicotinamide that is useful in the production of vitamins.

A species of *Pantoea* has also been claimed to produce α-hydroxy-4-methylthiobutyric acid. This is potentially useful as a feed additive for domestic animals, especially fowl, since it is a good source of sulphur-containing amino acids.

Food- and Feed-related Detrimental Effects of *Pantoea* spp. (Table 8)

Some species of *Pantoea* are associated with plant decay and plant diseases.

Table 8 Food- and feed-related detrimental effects of *Pantoea* spp.

Species	*Effect or disease*	*Causes*	*Plant (Latin name) or food*
P. agglomerans	Decay	NI	Celery (*Apium graveolens*)
P. agglomerans	Black spot necrosis	Production of cellulases and amylase	Beach pea (*Lathyrus maritimus*)
P. agglomerans	Stringiness of maple syrup	Production of exopolysaccharide	Maple syrup (*Acer saccharum*)
P. agglomerans	Necrosis at the leaf margins and tips	NI	Pearl millet (*Pennisetum glaucum*)
P. agglomerans *P. ananas*	Leaf blight, seed stable rot, bulb decay	NI	Onion (*Allium cepa*)
P. citrea	Pink disease	Production of glucose dehydrogenase	Pineapple (*Ananas comosus*)

NI, not identified.

Strains of *P. agglomerans* (*Erwinia herbicola*) were frequently isolated from the bacterial flora responsible for decay of pre-cut celery (*Apium graveolens*) and they appeared to be residents on the plants at harvest.

Some *Pantoea* spp. are wound pathogens indeed. *P. agglomerans* was identified as the aetiological agent for black spot necrosis on beach peas (*Lathyrus maritimus*). It is believed to penetrate the beach pea plant through mechanical injuries. The beach pea fixes nitrogen while growing in areas which are inhospitable to other plants. The seeds of beach peas can be eaten and may be a good source of protein, like the seeds of *L. sativus* for some populations in Bangladesh, India and Ethiopia. *P. agglomerans* has also been associated with a disease of pearl millet (*Pennisetum glaucum*) in Zimbabwe and India.

Pantoea agglomerans and *P. ananas* have caused severe leaf blighting, rapid collapse of tissue and rapid drying, seed stable rot and bulb decay of onion (*Allium cepa*) in South Africa and the US, respectively.

P. citrea causes pink post-harvest disease of pineapple, *Ananas comosus*. Symptoms are characterized by the formation of pink to brown discolorations of the infected portions of the pineapple fruit cylinder upon canning. The gene encoding for glucose dehydrogenase, one of the enzymes found to be essential for the disease, has been identified and characterized. Unfortunately, the pink disease is asymptomatic in the field, and its effects are only recognized when the ripe pineapple is cored and canned.

Maple sap is normally sterile in the maple tree. Spoilage of sap collected without sterile conditions may occur and microbial counts ranging from 0.3 to 1.4×10^7 cfu ml^{-1} have been reported. One of the most common defects reported by the maple sugar industry is the production of a ropy maple syrup resulting from contamination of maple sap. *P. agglomerans* has been isolated from soil able to support the growth of maple trees. When grown in maple sap, *P. agglomerans* renders it viscous. The stringiness was attributed to the formation of a microbial EPS. When each of the parameters described in Table 6 was studied separately, the EPS from *P. agglomerans* was best produced in medium containing 80 g l^{-1} sucrose, 1.5 g l^{-1} yeast extract, C/N ratio of 164, at 25 ± 5°C in a medium initially adjusted to pH 6.5 and agitated at 200 r.p.m. Under these conditions, EPS production reached a maximum value of about 20 g l^{-1} (2000 mPa s^{-1}) after 72 h of growth. These values are close to the ones reported for *P. agglomerans* (*Erwinia herbicola*), a microorganism producing levan. Using sucrose as the carbon source, maximal production of the EPS from *P. agglomerans* (*E. herbicola*) was observed at a C/N ratio of 140, an incubation temperature of 25°C, a pH value of 7.2, and a minimum agitation rate of 200 r.p.m. When the effect of agitation and yeast extract on EPS production was considered simultaneously, the maximum amount of EPS produced, corresponding to a viscosity of 2384 mPa s^{-1}, was predicted at a concentration of yeast extract of 3.4 g l^{-1}, and an agitation rate of 400 r.p.m.

The examples described above demonstrate the diversity of metabolites and enzymatic activity produced by members of the genus *Pantoea*. For the time being, because the era of nutraceuticals and probiotics is in its infancy, it is difficult to predict whether some of these examples will find applications in the food arena. Furthermore, because some findings have pointed out that some *Pantoea* species were opportunistic pathogens, the potential of using members of the *Pantoea* genus for health promotion and/or natural foods requires further investigations.

See also: **Enterobacter**.

Further Reading

Brenner DJ (1981) The genus *Enterobacter*. In: Starr MP, Stolp H, Trüper HG, Balows A, Schlegel HG (eds) *The Procaryotes*. P. 1173. New York: Springer-Verlag.

Dye DW (1969) A taxonomic study of the genus *Erwinia*. II. The 'carotorova' group. *N. Z. J. Sci.* 12: 81–97.

Ewing WH and Fife MF (1972) *Enterobacter agglomerans*

(Beijerinck) comb. nov. (the herbicolalathry bacteria). *Int. J. Syst. Bacteriol.* 22: 4–11.

Gavini F, Mergaert J, Beji A et al (1989) Transfer of *Enterobacter agglomerans* to *Pantoea* gen. nov. as *Pantoea agglomerans* comb. nov. and description of *Pantoea dispersa* sp. nov. *Int. J. Syst. Bacteriol.* 39: 337–345.

Holt JG, Krieg NR, Sneath PHA, Staley JT and Williams ST (1994) *Bergey's Manual of Determinative Bacteriology*, 9th edn. Pp. 184, 213, 237, 248. London: Williams & Wilkins.

Iimura K and Hosono A (1996) Biochemical characteristics of *Enterobacter agglomerans* and related strains found in buckwheat seeds. *Int. J. Food Microbiol.* 30: 243–253.

Karamanos Y, Kol O, Wieruszeski J, Strecker G, Fournet B and Zalisz R (1992) Structure of the O-specific polysaccharide chain of the lipopolysaccharide of *Enterobacter agglomerans*. *Carbohydrate Res.* 231: 197–204.

Mergeart J, Verdonck L and Kersters K (1993) Transfer of *Erwinia ananas* (synonym, *Erwinia uredovora*) and *Erwinia stewartii* to the genus *Pantoea* emend. as *Pantoea ananas* (Serrano 1928) comb. nov. and *Pantoea stewartii* (Smith 1898) comb. nov., respectively, and description of *Pantoea stewartii* subsp. *indologenes* subsp. nov. *Int. J. Syst. Bacteriol.* 43: 162–173.

Morin A, Moresoli C, Rodrigue N, Dumont J, Racine M and Poitras E (1993) Effect of carbon, nitrogen, and agitation on exopolysaccharide production by *Enterobacter agglomerans* grown on low grade maple sap. *Enzyme Microb. Technol.* 15: 500–507.

Okutani K and Kobayashi H (1991) The structure of an extracellular polysaccharide from a marine strain of *Enterobacter*. *Nippon Suisan Gakkaishi* 57: 1949–1956.

Starr MP (1981) The genus *Erwinia*. In: Starr MP, Stolp H, Trüper HG, Balows A, Schlegel HG (eds) *The Procaryotes*. P. 1173. New York: Springer-Verlag.

Parasites *see* ***Cryptosporidium***; ***Cyclospora***; ***Giardia***; **Helminthes and Nematodes**; ***Trichinella***; **Waterborne Parasites**: *Entamoeba;* Detection by Classic and Modern Techniques.

Pasteurization *see* **Heat Treatment of Foods**: Principles of Pasteurization.

Pastry *see* **Confectionery Products**: Cakes and Pastries.

PCR-BASED COMMERCIAL TESTS FOR PATHOGENS

P A Bertram-Drogatz, **F Wilborn**, **P Scheu**, **A Pardigol**, **C Koob**, **C Grönewald**, **M Fandke**, **A Gasch** and **K Berghof**, BioteCon Diagnostics GmbH, Potsdam, Germany

Comparison of PCR with Other Techniques

Since its invention in the early 1980s, the polymerase chain reaction (PCR) has become a routine method which is widely used, particularly in clinical diagnostic laboratories. It has a number of advantages over traditional techniques for the identification of microorganisms – enormous specificity and sensitivity as well as very high speed. In some applications PCR may have outpaced older ones, for example in virus diagnostics (e.g. human immunodeficiency virus) or diagnosis of some bacterial infections (e.g. *Chlamydia trachomatis*). Identifications by PCR are still being cross-checked with supplementary tests, e.g. immunological serum analyses (e.g. *Helicobacter pylori*) or culture (e.g. *Mycobacterium tuberculosis*) to reach a final result.

Today, in times of high-throughput screening, high cost of well-trained technical staff, and particularly the need to reduce processing and stock times, PCR and related techniques are revolutionizing other fields in addition to the clinical diagnostic laboratory. One such field is food diagnostics, which previously used test methods employing culture, impedance determination, bioluminescence determination, cytometry or immunoassay (e.g. enzyme-linked immunosorbent assay, ELISA) as the means to detect the presence of pathogenic or spoilage microorganisms in foodstuffs (**Table 1**). Thus far, in food diagnostics PCR is only available commercially for the identification of bacteria. However, further methods currently in development include systems for the identification of individual food sources in composite foods, e.g. meats, or for the identification of genetically modified organisms in all types of food.

Table 1 Comparison of test methods used in food analysis

Test method	*Advantages*	*Disadvantages*
Traditional culture	• Species- or group-specific identification • Quantitative tests available	• Moderately expensive • Time-consuming (3–6 days)
Rapid culture	• Useful for preliminary analysis, but biochemical testing is still required	• Few specific methods available • Moderately expensive • *Salmonella* spp.: no faster than traditional culture (3–4 days)
Impedance determination	• Direct determination of total viable cell number • Inexpensive	• Focus on sterility testing • Few specific methods available • Insufficient detection limit ($\geqslant 10^6$ cells per ml) • Moderate speed of evaluation (1–2 days)
Bioluminescence determination	• Direct determination of total viable cell number • Fast evaluation (<1 day)	• Focus on sterility testing • Few specific methods available • Insufficient detection limit ($\geqslant 10^3$–10^4 cells per ml)
Cytometry	• Direct determination of total viable cell number • Fast evaluation (<1 day) • Very low detection limit ($\leqslant 10$ cells per sample)	• Focus on sterility testing • Few specific methods available
Immunoassay	• Species- and group-specific identification • Automated systems available	• Insufficient detection limit ($\geqslant 10^5$ cells per ml) • Some systems may not be reliable[a]
Gene probe	• Species- and group-specific identification • Simple methods available (dipstick or luminometric assays)	• Limited number of specific tests available • Insufficient detection limit ($\geqslant 10^6$ cells per ml) • Time-consuming (2 days)
Polymerase chain reaction	• Species- and group-specific identification and identification of subspecies • Very low detection limit ($\leqslant 10^3$ cells per ml corresponding to $\leqslant 10$ cells per reaction) • Superior in sensitivity to culture methods • Fast evaluation (<1 day)	• Moderately expensive

[a]Refer to: Brett MM (1998) Kits for the detection of some bacterial food poisoning toxins: problems, pitfalls and benefits. *Journal of Applied Microbiology Symposium Supplement* 84: 110S–118S, and Mäntynen V, Lindström K (1998) A rapid PCR-based DNA test for enterotoxic *Bacillus cereus*. *Applied and Environmental Microbiology* 64: 1634–1639.

Comparison between microbiological food testing using traditional culture and new PCR technology shows that molecular biology carries a potential for high time savings (**Table 2**) and a potential for automated handling of large sample numbers above about 100 analyses per day. Thus, tomorrow's food production lines may no longer require initial long periods of stock-holding or risky delivery before final evaluation of microbiological safety, because PCR allows early clearance of products so that time-consuming analyses may no longer be required. To this end, analytical test methods employing molecular biology may directly lead to significant reduction in overall costs of food production.

With the exception of cytometry, in all methods enrichment by microbiological cultivation is recommended to achieve sufficiently high sensitivity when large quantities of food are being analysed, and in molecular testing to discriminate between viable and non-viable cells.

The following article is intended to introduce the reader to the new field of PCR analysis to aid with the first steps into molecular biology. Due to space limitations, it is not possible to cover every test method currently available, nor to explain the features of any test kit in great detail. Nevertheless, we hope to explain the various products in an unbiased and critical way. While describing more precisely those systems that target *Salmonella* spp., other systems are also mentioned. Thus, we present here an intro-

Table 2 Comparison of time-scales of detection methods targeting *Salmonella* employing traditional microbiology or PCR in food analysis

Step of analysis	*Microbiology*	*PCR*
Enrichment	Non-selective pre-enrichment in liquid media (16–20 h) followed by selective enrichment in liquid media (24–48 h) or rapid enrichment (16 h, some methods only)	Rapid enrichment (16 h)
Isolation/identification	Isolation of bacterial colonies on selective solid media (24 h) followed by identification by biochemical tests (24–48 h)	Sample purification, target amplification and detection (8 h)
Result	After 3–6 days	After 24 h

ductory review of commercial analytical PCR kit systems that can safely be applied in the rapid diagnosis of food contamination.

Principles and Types of Commercial PCR Tests

All PCR kit systems, as well as related molecular methods, follow a common principle of diagnosis. The underlying feature is the strain-specific uniqueness of genomic DNA sequences. Owing to differences between the genomes of various strains, one may distinguish bacteria at all taxonomic levels on the basis of appropriately variable segments in their DNA sequence. For instance, since the late 1970s the sequences of genes coding for ribosomal ribonucleic acid have been used to generate bacterial phylogenetic trees. This characteristic opens the possibility for a group-, species- and even subspecies-specific identification of bacteria by PCR.

PCR is an iterative amplification reaction that profits from the high stability of some DNA polymerase enzymes, e.g. from the bacterium *Thermus aquaticus*, during repeated cycles of heating and cooling between approximately 35°C and 95°C. Reaction mixtures used for PCR comprise a cocktail of reagents that includes, besides a putative target DNA provided with a sample, primer oligonucleotides, desoxyribonucleotide monomers, a DNA polymerase enzyme, buffer components and specifically adapted additives (to prevent inhibition by food matrix components carried over from the sample). In addition, an enzymatic system for the protection against false-positive contamination, e.g. through aerosols from the laboratory, may be included.

The PCR reaction proceeds by successive melting of the double-stranded DNA target (at 92–95°C); hybridization of each strand with the primer oligonucleotides (at 35–68°C); and polymerization by action of the supplied enzyme (at 72°C) before this cycle starts over again, usually through some 25–40 cycles in total (**Fig. 1**).

Theoretically, PCR over 25 cycles should produce a 3.4×10^7-fold increase in the number of target molecules. However, in practice one would expect yields in the range of approximately 65–70%, thus affording a 10^5-fold rise in target concentration throughout the process. Still, this would provide around 100 ng of product formed from 1 pg of target DNA. To afford precision in temperature and time that meets the exact requirements of any defined reaction, a thermal cycling device (thermocycler) is required.

In different commercial PCR systems, the detection step is accomplished by various techniques. The ELISA technique has the advantage of introducing one further level of discrimination through specific hybridization of a capture probe. Similarly, the automated TaqMan system employs the hybridization of amplified target DNA to a detection probe. By these mechanisms, the specificity of the overall reaction becomes greatly enhanced. In contrast, immediate evaluation of test results directly after PCR lacks this

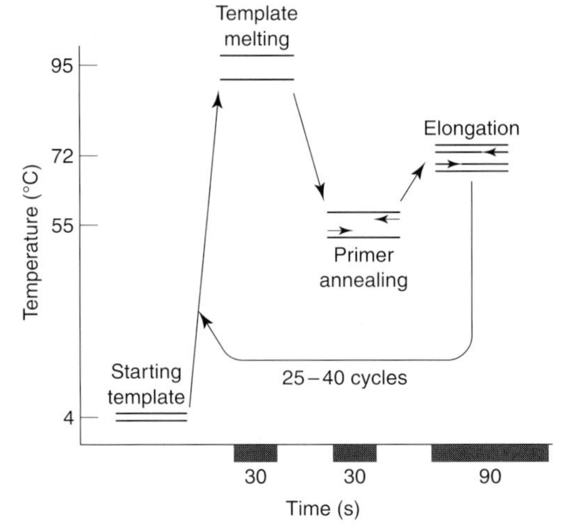

Figure 1 The polymerase chain reaction process.

property and, in the case of agarose gel electrophoresis, may be somewhat cumbersome because this latter technique is not amenable to automation. As an example, **Table 3** outlines a general protocol for a PCR-ELISA assay.

Due to the fact that the application of PCR is covered by proprietary rights, every professional laboratory is required to purchase licensed instruments and products. However, some companies have developed alternative methods to circumvent such restrictions. Although the resulting gene probe assays may be regarded as equally specific as PCR, they are carrying the trade-off that their sensitivity usually does not reach that of PCR. Thus, for reliable food testing such methods may not be sufficient because many contaminations which would be detectable by PCR or traditional methods may remain undetected by gene probe assays.

A number of commercial PCR systems are discussed here in order to shed light on their advantages and disadvantages (**Table 4**). To simplify, we focus on systems for the detection of *Salmonella* spp., since all manufacturers mentioned here produce kit systems that target this genus (**Table 5**).

BioteCon Diagnostics FoodProof Kit Systems

BioteCon Diagnostics (Berlin/Potsdam, Germany) is currently developing 13 kit systems for the detection of pathogenic and spoilage microorganisms in foods. By covering a broad spectrum of target bacterial species (or groups thereof) relevant to food contamination, industrial diagnostic laboratories of any size will be given a versatile tool for rapid analytical testing amenable to automation. The first two kit systems, aiming to detect *Salmonella* spp. and *L. monocytogenes*, are to be released in 1999. PCR-ELISA as the underlying principle enables a reliable detection of 1 cfu per 25 g food. The FoodProof *Salmonella* kit has been successfully tested following Food and Drug Administration (FDA)-II standards – analyses starting from 750 g food may be carried out on a routine basis.

Table 3 Analysis of foods using the PCR-ELISA technique

Step of analysis	*Laboratory action*
Sample preparation	• Determination of food sample weight • Homogenization
Enrichment	• Inoculation of rapid enrichment medium • Incubation of inoculated medium
Sample purification	• Sampling from enrichment broth • Extraction and purification of DNA
Amplification (PCR)	• Addition of sample DNA to reaction mixture • Thermal cycling
Detection (ELISA)	• Coupling of amplified DNA to sample and internal control wells in microtitre plate • Addition of ELISA reagents • Incubation at 37°C • Colorimetric detection in a microtitre plate reader
Result	• Evaluation of data printout

The BioteCon Diagnostics FoodProof kit systems follow a general protocol (**Table 6**) that, as an example, is explained in more detail below. This kit system is divided into three modules covering sample purification, amplification and detection procedures. Media and reagents for the enrichment of bacteria may also be obtained from the manufacturer. Additional services include extensive start-up consulting, training of laboratory personnel and a technical services hotline. Notably, BioteCon Diagnostics provides specifically adapted protocols for the detection of contaminants in foods from different product groups.

A validation of the FoodProof *Salmonella* kit investigating the test system's performance in the analysis of 75 different foods (including 51 inoculated food samples) showed no difference between in-house PCR-ELISA, in-house microbiology and additional independent ex-house microbiology testings, all of which were capable of detecting 46 positives. The remaining five samples were all spiked with *S. enterica* subsp. *enterica* serovar Agona. The authors supposed that the negative findings with this serovar originated from a low survival potential of these organisms in confectionery products in long-term storage. Interestingly, two cereal samples inoculated with serovar Agona gave a positive test result with all three detection modes, thus corroborating this interpretation. Tests with the FoodProof *Salmonella* kit involving 27 inoculated and 353 non-inoculated food samples gave 99.0% correct results. These validations have shown that this PCR system is capable of producing correct results even when inhibitory matrices (such as cocoa) were used.

In a collaborative study involving 12 independent laboratories, the sensitivity and specificity of this method were found to be 98.2% and 96.6%, respectively ($n = 240$) (sensitivity = no. of true positive test results/(no. of true positive test results + no. of false negative test results); specificity = no. of true negative test results/(no. of true negative test results + no. of false positive test results)). The assay's performance was thus found to equal that of traditional culture. To date, more than 470 strains of *S. enterica* (including all subspecies) and *S. bongori* have been successfully tested using the FoodProof *Salmonella* kit.

The BioteCon Diagnostics FoodProof *Salmonella*

Table 4 Properties of commercial PCR-based kit systems for food analysis

Target	*Manufacturer*	*Kit system*	*Method*	*Performance*	*Approval*[a]
Salmonella enterica					
	BioteCon Diagnostics	FoodProof *Salmonella*	PCR-ELISA	• 1 cfu per 25 g and 1 cfu per 750 g or 1E3 cfu ml^{-1} after enrichment • 100% correct results with 470 *S.*sp. strains and 51 non-*S.*sp. strains • 99% correct results with >1000 food samples	DIN, LMBG, (AOAC)
	PE Biosystems	TaqMan *Salmonella* PCR amplification/ detection kit[b]	Integrated PCR amplification/ detection system	• 1 cfu per 25 g • 99.7% specificity	(AOAC)
	Qualicon	BAX for screening/*Salmonella*	PCR	• 1E3 cfu ml^{-1} after enrichment • 99.7% specificity	AOAC
	Sanofi Diagnostics Pasteur	Probelia *Salmonella* sp.	PCR-ELISA	• 1 cfu per 25 g	AFNOR
	TaKaRa Shuzo	*Salmonella* (*invA* gene) one-shot PCR screening kit	PCR	• 100 cells per reaction	
Listeria monocytogenes					
	BioteCon Diagnostics	FoodProof *Listeria monocytogenes*	PCR-ELISA	• 1 cfu per 25 g • 100% correct results with 96 *L.m.* strains and 85 non-*L.m.* strains	(LMBG, AOAC)
	Qualicon	BAX for screening/*L. monocytogenes*	PCR	• 100% correct results with >90 *L.m.* strains and 100 non-*L.m.* strains	
	Sanofi Diagnostics Pasteur	Probelia *L. monocytogenes*	PCR-ELISA	• 10 cfu per 25 g • 98.4% correct results with 64 pâté samples spiked with *L.m.*	(AFNOR)
Escherichia coli					
	Qualicon	BAX for screening/*E. coli* O157:H7	PCR	• 1E5 cfu ml^{-1} after enrichment	
	TaKaRa Shuzo	O-157 (verocytotoxin genes) one-shot PCR screening kit	PCR	• 10 cells per reaction	

[a]Parentheses indicate that validation is in preparation or in progress.
[b]Follow-up product under development.
DIN = Deutsches Institut für Normung; LMBG = German Food and Consumable Goods Act; AOAC = Association of Official Analytical Chemists; AFNOR = Association Française de Normalisation.
S.sp. = *Salmonella* species; non-*S.sp.* = non-*Salmonella* species; *L.m.* = *Listeria monocytogenes*; non-*L.m.* = non-*Listeria monocytogenes*.

kit system currently serves as a model system for new German guidelines on molecular testing methods being prepared by the Deutsches Institut für Normung (DIN). It is expected that it will be shown to meet the requirements of the German Food and Consumable Goods Act (LMBG) as well.

Table 5 Comparison of features of various PCR-based kit systems for the detection of *Salmonella* spp. in foods[a]

Feature	*BioteCon Diagnostics FoodProof* Salmonella	*PE Biosystems TaqMan* Salmonella *PCR amplification/ detection kit*	*Qualicon BAX for screening/* Salmonella	*Sanofi Diagnostics Pasteur Probelia* Salmonella *sp.*	*TakaRa Shuzo* Salmonella (*invA gene) one-shot PCR screening kit*
Test principle	PCR-ELISA	Automated amplification/ detection system	PCR	PCR-ELISA	PCR
Detection mechanism	Colorimetric	Fluorimetric	Fluorimetric	Colorimetric	Fluorimetric
Dye employed for detection	Tetramethyl benzidine	6-Carboxy fluorescein	Ethidium bromide	Tetramethyl benzidine	SYBR green I
Additional level of probe-linked specific discrimination during detection	Yes	Yes	No	Yes	No
Exclusion of carry-over contamination	Yes[b]	No[c]	No	Yes[b]	No
Total number of strains tested positive by manufacturer	470	317	>1800	384	174
Total number of serovars tested positive by manufacturer[d]	192	121[e]	>1795	N.D.[f]	46
Maximum number of diagnostic tests per assay (excluding controls)[g]	94[h]	86	48[i]	93[h]	96[i]
Controls recommended by the manufacturer	Negative, positive (internal standard DNA), positive (wild-type DNA)	Autozero, negative, positive (internal control), positive (wild-type DNA)	Positive (control DNA)	Negative, positive (internal standard DNA), positive (wild-type DNA)	Positive (control DNA)
Total time required per run (h, post-enrichment)[j]	6	6	8	6	4
Additionally required laboratory equipment[k]	ELISA-reader, hybridization oven	ABI Prism sequence detector	None	ELISA-reader hybridization oven	None

[a]Numerical data may represent approximate values and may be subject to change.
[b]Enzymatic degradation of carry-over amplificates by uracil-*N*-glycosylase.
[c]Not required, since final fluorimetric evaluation occurs in the reaction vessel used for amplification (amplification is done in the presence of desoxyuracil triphosphate, however).
[d]*S. enterica* is currently subgrouped as follows (number of known serovars in brackets): *S. enterica* subsp. *enterica*/group I (1367), *S. enterica* subsp. *salamae*/group II (464), *S. enterica* subsp. *arizonae*/group IIIa (93), *S. enterica* subsp. *diarizonae*/group IIIb (309), *S. enterica* subsp. *houtenae*/group IV (64), *S. enterica* subsp. *indica*/group VI (10), *S. bongori*/group V (17).
[e]*S. enterica* subsp. *enterica* serovar Memphis is not detected.
[f]N.D. = No data available; *S. enterica* subsp. *enterica* serovars Wedding and Zürich are not detected. Some strains of *S. enterica* subsp. *houtenae* and *S. bongori* may not be detected.
[g]Minimum number of diagnostic tests is generally one single diagnostic determination.
[h]Microtitre plate format.
[i]Due to evaluation by agarose gel electrophoresis, only a limited number of diagnostic reactions may be analysed simultaneously.
[j]For the determination of *S. enterica* in foods, enrichment usually takes between 16 and 20 h. In all systems, hands-on times will amount to approximately 2 manhours per assay.
[k]Assuming presence of ordinary molecular biology laboratory equipment. Initial investments for setting up a molecular biology laboratory should amount to approximately £14 000–35 000; investment for a microtitre, plate reader and a hybridization oven should amount to approximately £10 000, and a sequence detector should cost approximately £30 000.

PE Biosystems TaqMan PCR Amplification/ Detection Kits

PE Biosystems (Foster City, CA, US) has developed an integrated amplification/detection system that allows for post-amplification data evaluation directly in the PCR reaction vessel. Hence, post-PCR procedures such as gel electrophoresis, hybridization or ELISA are not required. In addition to the reagents necessary for ordinary PCR, the reaction mixture of the TaqMan system includes an internal fluorigenic hybridization probe that is capable of binding to a target sequence present in specifically amplified products. This probe

Table 6 Outline of the BioteCon Diagnostics FoodProof *Salmonella* test protocol for the determination of *S. enterica* in foods[a]

Enrichment (16 h)

- 25 g food sample + 225 ml Salmosyst medium; incubation for 8 h at 37°C
- Sampling of 10 ml and addition of one tablet selective supplement; incubation for 8 h at 37°C
- Sampling of 1 ml for sample purification

or:

- 750 g food sample + 6750 ml UHT milk; incubation for 8 h at 37°C
- Sampling of 100 ml and addition of 10 tablets selective supplement; incubation for 8 h at 37°C
- Sampling of 1 ml for sample purification

Sample purification (0.5 h)

- Centrifugation at 10 000 ***g*** of 1 ml sample obtained from enrichment for 5 min at 4°C
- Resuspension of the pellet in 200 μl lysis buffer, incubation for 10 min at 98°C, mixing for 10 s at room temperature and centrifugation at 12 000 ***g*** for 2 min at room temperature
- Transfer of 100 supernatant into a new vessel, addition of 300 μl binding buffer
- Transfer of the sample into the upper reservoir of a pre-assembled filter/collection tube, centrifugation at 8000 ***g*** for 1 min at room temperature (discard flow-through)
- Addition of 500 μl washing buffer, centrifugation at 12 000 ***g*** for 1 min at room temperature (discard flow-through)
- Centrifugation at 12 000 ***g*** for 1 min at room temperature (discard flow-through)
- Elution of DNA by addition of 60 μl elution buffer (prewarmed to 70°C), incubation for 1 min at room temperature and centrifugation at 8000 ***g*** for 1 min at room temperature; if sample is positive the eluate contains DNA

Amplification by PCR (2.5 h)

- Addition of 2.5 μl sample eluate to pre-assembled reaction mix (plus control reactions)
- Transfer reaction vessel into thermocycler, run the PCR programme

Detection by ELISA (2 h)

- Denaturation of 13 μl sample by addition of 13 μl denaturation reagent
- Transfer of denatured sample from the PCR to microtitre plate: 10 μl each into a sample well and an internal control well
- Addition of 100 μl hybridization buffer to each well, incubation for 30 min at 50°C
- Four brief washes with 800 μl washing buffer I
- Addition of 100 μl antibody solution, incubation for 30 min at 37°C
- Four brief washes with 800 μl washing buffer II
- Addition of 100 μl staining solution, incubation for 15 min at room temperature
- Addition of 100 μl stop solution
- Determination of absorbance at 450 nm using a microtitre plate reader

[a]The protocol shown applies for foods such as cereals, cocoa, milk, milk powder or nuts. In the analysis of other food sources different protocols may apply that are provided by the manufacturer.

is labelled at the 5′-position with 6-carboxy fluorescein (reporter dye, FAM) and at the 3′-position with 6-carboxy-*N*,*NN*′,*N*′-tetramethyl rhodamine (quencher dye). During the polymerization process *Taq*-DNA polymerase may release the reporter dye due to the enzyme's 5′→3′-exonuclease activity, so that the reporter signal is no longer quenched. At the end of cycling the amount of free reporter product formed is used as a basis for qualitative data evaluation.

Future developments may include on-line detection of the amplification reaction, thereby allowing for quantification of the amount of target DNA present in a sample. The reaction mixtures also contain an internal standard control DNA, labelled with 6-carboxy-4,7,2′,7′-tetrachloro fluorescein that emits at a different wavelength from that of FAM, to monitor the amplification performance. The assay requires very elaborate high-end equipment produced by the manufacturer. Currently, a non-quantitative data output format is produced supplying a positive, retest or negative readout.

The *Salmonella* PCR amplification/detection kit has been on the market for several years. The development of new products targeting *S. enterica*, *Escherichia coli* O157:H7 (EHEC), *Listeria* spp. and *Campylobacter* spp. is currently underway. Initial investment costs for laboratory equipment required to run this new system are expected to be comparatively high.

In a validation of the TaqMan *Salmonella* PCR amplification/detection kit employing several meat sources and raw milk, a detection limit of 3–7 cfu per 25 g food was determined. Comparison between this kit system and a modified semisolid support Rappaport–Vassiliadis culture method using 110 naturally contaminated chicken and milk samples revealed that 32.7% of the samples were positive by PCR and 30.9% were positive by the culture method, thus exemplifying a good correlation between the two methods.

PE Biosystems' services include workshops and seminars on the TaqMan technology. Information on matrix-associated test inhibition and non-stereotype handling of specific foods are available on request.

Qualicon BAX Pathogen Detection Systems

Qualicon (Wilmington, Delaware, USA) was the first company to market a PCR system for the detection of pathogenic bacteria in foods. Its kits target *Salmonella* spp., *E. coli* O157:H7 and *L. monocytogenes*. Recently a genus-specific system for the detection of *Listeria* spp. has been released.

The BAX for screening kit systems are loaded with bacterial lysates obtained directly from enrichment cultures, i.e. without explicit DNA purification. The reaction vessels carry a tablet containing all PCR reagents required for the test. This means that the risk of cross-contamination is significantly reduced. Detection of amplified products with PCR is accomplished by unspecific intercalation by ethidium bromide. New developments aim to use the non-specific intercalating fluorescent dye SYBR green I in a system that permits detection inside the PCR reaction vessel, avoiding the time- and labour-consuming gel electrophoresis protocol.

The BAX for screening/*Salmonella* kit requires non-selective pre-enrichment (20 h) followed by short selective enrichment (3 h) in a brain–heart infusion broth-based medium. The protocol features a procedure that is easy to achieve, since the only steps required are a protease digest followed by brief high-temperature incubation and agarose gel electrophoresis before staining with ethidium bromide dye. It should be noted that the rather simple method of DNA extraction employed here may fail if applied to some food product groups rich in inhibitory components, e.g. cocoa and other confectionery products. Unfortunately, the manufacturer does not make suggestions concerning matrix-associated test inhibition or non-stereotype handling of specific foods, although some protocols are recommended, involving inclusion of polyvinylpyrrolidone in the analysis of certain foodstuffs.

More than 95% concordance of this method with traditional culturing methods has been demonstrated; for instance, in an evaluation employing 216 inoculated foods, the BAX for screening/*Salmonella* kit detected 59.7% positives as compared to 57.4% positives detected by culture. It has been approved by the Association of Official Analytical Chemists (AOAC). However, thus far in validations run on this system only, inoculant sample sizes of 25 g but not 750 g (FDA-II standard) were used, thus a verdict on its performance when using pooled large samples cannot be given here.

Sanofi Diagnostics Pasteur Probelia Kits

The Probelia kit systems produced by Sanofi Diagnostics Pasteur (Marnes-la-Coquette, France) rely on the PCR-ELISA method and employ a protocol similar to that described below. Starting with 25 g food of any product group, it is recommended to inoculate buffered peptone water for a nonselective enrichment of *Salmonella* spp. over 16 h. In addition to lysing bacterial cells by heating in the presence of an anionic Chelex®-resin, no further DNA purification is done, so the samples may proceed directly on to amplification and detection reactions. To overcome inhibitory effects resulting from certain food sources such as chocolate or raw eggs, high initial dilutions of the cell lysates are required. Further, an internal control template is included to verify correct function of the test.

Following information from the manufacturer, the Probelia *Salmonella* spp. kit has been tested on a large number of isolates from each group of *Salmonella*, and most serovars have been detected. However, some serovars or strains may not be amenable to detection (*S. enterica* subsp. *enterica* serovars Wedding and Zürich; some strains of *S. enterica* subsp. *houtenae* and *S. bongori*); it is possible that such isolates may not contain the target of this kit, the *Salmonella*-specific invasion gene *iagA*.

Sanofi Diagnostics Pasteur offers customers training courses and seminars on their technology.

TaKaRa Shuzo One-Shot PCR Screening Kits

TaKaRa Shuzo (Shiga, Japan) provides two types of PCR screening kits for food and environmental testing, one of which is directed against *E. coli* EHEC (O157) verocytotoxin genes and the other against the *Salmonella* spp. invasion gene *invA*. These kits contain a pre-mixed reagent solution aliquoted into reaction vessels, so that samples merely need to be added before the PCR is run. The kit targeting *Salmonella* spp. also includes a positive control as part of the supplied reagents. Detection is by molecular size discrimination, employing non-specific staining with the fluorescent dye SYBR green I of DNA fragments separated electrophoretically on an agarose gel, and thus bears a high risk of false-positive identification. Occasionally, ambiguous results may be obtained due to the fact that the positive control DNA is directly included in the diagnostic sample reaction; in the presence of large amounts of food sample-derived *invA*-DNA the positive control DNA reaction may be inhibited.

An outline for two alternative enrichment procedures is included in the kit manual: pre-enrichment in EE Broth Mossel for 18–24 h followed by enrichment in Rappaport–Vassiliadis broth for a further 18 h, or one-step enrichment for 18 h using Rappaport–Vassiliadis broth. However, the decision of which protocol to choose is left to the user. Extraction of bacterial DNA is done by heat shock in distilled water – a method that remains questionable as far as food samples are concerned. Unfortunately, further information is lacking regarding matrix-associated test inhibition and non-stereotype handling of specific foods. Data on collaborative testings, validations and the like have not been found available.

Outline of a PCR-ELISA test protocol

As an example of a PCR-based method, a detailed description of the BioteCon Diagnostics FoodProof *Salmonella* kit is given below (see Table 6). This method is characterized by a user-friendly design that leads the lab technician smoothly through successive steps of the procedure. Preceding installation of the method in the diagnostic laboratory, this company – like most other kit manufacturers – offers detailed consulting and laboratory training to the customer in preparation for correct laboratory set-up and a safe application of this technology.

To recognize and avoid major pitfalls common to routine food diagnostics, every reaction contains internal standard DNA to prove regular performance of the PCR protocol, and an enzymatic system for protection against carry-over contamination that may be picked up from aerosols in the user's laboratory.

Employing this method, enrichment of possible contaminants may start from food samples of 25 g or 750 g initial weight. After preliminary sample preparation, ordinary microbiological media recommended – and optionally supplied – by the manufacturer are inoculated to achieve a nonselective enrichment before selective growth in a round of enrichment. A sample from the resulting suspension is passed into a sample preparation which has been shown to remove inhibitory food matrix components effectively from DNA-containing samples recovered in the eluate. This brief DNA purification procedure is recommended for most types of foods and has proven to be superior to simple lysis-only-type protocols. Next, 2.5 μl of each purified sample is added each into one pre-assembled PCR reaction mix and the thermocycling protocol is initiated. Separate independent reactions comprise a negative control and a positive, wild-type DNA-containing control. Continuing with the ELISA, every sample from the PCR is split into two parts, one of which is added into a sample well and the other into an internal control well on a probe-coated microtitre plate. The protocol continues with hybridization of DNA from food-contaminating *Salmonella* that may be present in a sample (or hybridization of the corresponding control reagents) to a specific wild-type probe, and subsequent washing, staining and stopping. Then, absorbance at 450 nm is determined using a microtitre plate reader that may be interfaced to a personal computer system to handle the data output. A guide to interpretation of absorbance readings is presented in **Table 7**. Only in the rare event of an indeterminate result are additional follow-up testings recommended.

Table 7 Guide to interpretation of data obtained from PCR-ELISA using the BioteCon Diagnostics FoodProof *Salmonella* test protocol[a]

ΔA450 in sample well	*ΔA450 in internal standard well*	*Interpretation*
≥0.2	<0.2	Positive result
<0.2	≥0.2	Negative result
<0.2	<0.2	Indeterminate result, repeat determination

[a]Prerequisite readings: positive control well, ΔA450 ≥ 1.0; negative control well, ΔA450 < 0.2.

Validation of Commercial Tests

'Standardization is the unique, defined solution of a repetitively occurring task in the light of the scientific, technological and economical possibilities available at any given time'. This definition, made by Otto Kienzle, co-founder of the German Institute for Standardization, sets the stage for any standardized continuous manufacture of industrial products, to facilitate the trade and application of reliable products.

A number of national and international independent organizations are active in co-ordinating attempts to standardize rapid molecular test methods for application in microbiological food testing. Generally, such bodies supervise collaborative assessments of the technical performances of new procedures on the basis of comparison with existing official methods. The procedures usually involve evaluation of the specificity, sensitivity, precision and accuracy of the new method and include kit performance data, peer-verified laboratory study data and collaborative study data. Official validation of a new method reflects that it has been shown to comply with existing standards of the validating organization, and that it equals the performance of the existing standard method so that it may be safely used for commercial diagnostic testing. Some selected organizations responsible for official standardization and/or validation are described below.

The French Association for Standardization (Association Française de Normalisation, AFNOR, Paris, France) has five central departments linked to certifications: advice, implementation/management, networking of laboratories, representation of industrial interests and development/expansion of the range of available certifications. AFNOR as an independent, impartial non-government organization warrants that AFNOR-certified products conform to the AFNOR reference system. Thereby, the competitiveness and image of the manufacturers involved are strengthened and authorities and consumers are helped to distinguish quality products between seemingly identical offers on the basis of clear information. An AFNOR validation of an alternative new method follows a number of steps in a defined schedule over a period of at least 8 months and includes three types of studies. To achieve a first impression of the method's performance, in collaboration with AFNOR's expert representatives, the experimental design for the validation is worked out and a project study run to verify the validation protocol. After evaluation, tests continue with a preliminary study to compare the new rapid method and the reference method. Follow-up collaborative testings require the participation of at least eight independent certified laboratories and at least eight series of interpretable results from each participant.

AOAC International (Gaithersburg, MD, USA; formerly Association of Official Analytical Chemists) is an independent association of scientists from the public and private branches that promotes method validation and quality control. Among other tasks, this organization aims to co-ordinate the development and validation of microbiological analytical methods by experts from industrial, academic and governmental settings. In particular, the AOAC performance-tested methods and the AOAC official methods programmes are relevant to analytical food-testing products. Products subject to validation may run successively through both types of programmes. The performance-tested methods programme requires a period of at least 9 months and includes an on-site validation study at the manufacturer's premises and a collaborative study involving at least eight independent laboratories. After expert approval as a first-action AOAC official method, final-action status may be reached within 2 years (post-publication) after successful application. Accepted methods are published in the AOAC periodicals *Journal of AOAC International* and *Inside Laboratory Management* or *The Referee*.

Standardization organizations are active in the European states and in many other countries. As an example of important standardization bodies, two German organizations are described here. In Germany, production and trade with foodstuffs and consumables are regulated by the Food and Consumer Goods Act (Lebensmittel- und Bedarfsgegenständegesetz; LMBG). In Act 35 it is laid down that the German Federal Institute for the Protection of Consumer's Health and Veterinary Medicine (Bundesinstitut für gesundheitlichen Verbraucherschutz und Veterinärmedizin; BgVV, Berlin, Germany) publishes a collection of official methods for sampling and analysis of foodstuffs and co-ordinates the establishment of such nationally binding standard methods involving experts from the fields of monitoring, science and industry. As a non-governmental counterpart, the German Institute for Standardization (Deutsches Institut für Normung; DIN, Berlin, Germany) is active in serving citizens and public and private bodies by setting standards for technical and scientific applications. As a baseline, the DIN orients its work on the following 10 principles: voluntary participation, publicity, participation of every interested group, homogeneity/absence of contradictions, non-ideological action, consent, modern standards, economic settings, public usefulness and international relevance. The last principle reflects that much of DIN's work (currently about 80%) is devoted to preparing standard regulations for subsequent implementation into the European standardization network.

Similarly, the International Dairy Federation (IDF, Brussels, Belgium) is active in supporting the preparation of international standards by various bodies, in collaboration with several other non-governmental organizations. In particular, this organization has prepared a standard on the quantitative determination of bacteriological quality in milk (International IDF standard 16A) that gives guidance on the evaluation of routine methods. This text supplies useful information on the effective application of statistical methods in the interpretation of microbiological validation testings.

Internationally, organizations like AFNOR, AOAC and DIN are represented by working groups and committees of the European Committee for Standardization (Comité européen de normalisation, CEN; Brussels, Belgium) and the International Organization for Standardization (ISO, Geneva, Switzerland).

The CEN is an intergovernmental organization for voluntary standardization in the European Union (EU). EU member states are required to use standards set out by the CEN and are obliged to eliminate conflicting national standards to build a harmonized standards system. However, no other nation is obliged to follow CEN standards unless products manufactured by its internationally active companies are exported into the EU.

In contrast, the ISO forms a global board for the adoption of standards with currently 118 member nations. Following prior acceptance by the ISO's joint technical committee, a publication of draft standards as an international standard takes place after approval by at least 75% of the voting member's national bodies. As a general rule, the adoption of ISO standards is optional for its member nations.

CEN and ISO collaborate with a large variety of non-governmental organizations, industry, trade unions, consumers' representatives and with each other, to ensure general acceptability of new standard regulations.

New Technologies

Two major drawbacks continue to hinder the analysis of food samples with current commercial PCR kit systems: inability to quantitate in a sample the cell number of a bacterial species and inability to discriminate between viable and non-viable cells. In addition continuing pressure to reduce to a minimum the time lapse between sampling and data evaluation drives the search for even more sensitive technologies.

To date, quantification of microorganisms in a food sample may be done employing quantitative PCR as a non-automated stopped-time assay. While this technique may be a method that is very efficiently used in research laboratories, it is usually elaborate due to its time- and labour-intensive protocols. For instance, comparisons with properly standardized amounts of DNA must be included in each analysis to enable back-calculation to determine the starting amount of target nucleic acid from the amount of product obtained after a defined number of PCR cycles. The reliability of this method has been repeatedly questioned and currently no commercial system employing this technique is available. Hence, today's quantitative PCR may not be a method of choice for a routine food diagnostics laboratory. However, continuous monitoring as available in the PE Biosystems' TaqMan technology may be a future means of solving this problem.

Discrimination between viable and non-viable cells may be obtained by isothermal amplification systems (e.g. nucleic acid sequence-based amplification, NASBA) targeting ribonucleic acids (RNA), which often have a limited half-life and thus may be characteristic of viable cells. However, currently only clinical applications are available targeting human immunodeficiency virus (Organon Teknika, Turnhout, Belgium), *Mycobacterium tuberculosis* complex or *Chlamydia trachomatis* (bioMérieux, Marcy-sur-l'Etoile, France). Again, although this method may be useful to researchers, routine applications – particularly when starting from food samples that may be rich in hydrolase enzymes – should take into account that often the kinetics of decomposition of RNA in natural materials has been underestimated. Thus, the half-life of every single RNA species of interest may vary dramatically when isolated from various foods.

Similar to PE Biosystems' TaqMan, a new technology that may be both selective to viable cells (in the case of RNA detection) and quantitative is based on the LightCycler analytical system, recently released by Roche Diagnostics/Boehringer Mannheim (Mannheim, Germany). Relevant new features include a fast qualitative PCR within 30 min – accomplished by employing thin-walled glass capillaries as reaction vessels and a mass transfer thermal control unit – and a quantitative determination of PCR and reverse-transcribed PCR (of RNA targets) due to inclusion of an on-line real-time fluorimetric detector. Detection is accomplished either by non-specific staining with SYBR green I or by specific hybridization of an amplified product with two specific detection probes; one of the probes is labelled in the 3′ position with fluorescein as donor fluorophore and the other in the 5′ position with red 640 as acceptor fluorophore. Spatial proximity of the two probes allows for fluorescence resonance energy transfer, which can be detected by detector photohybrids. The LightCycler system currently carries 32 sample holding devices, at a capital cost of approximately £30 000.

It should be noted that, particularly for future applications in the clinical diagnostic laboratory, several companies are developing products for the nucleic acid analysis of multi-component systems (sequencing or identification of polyallelic genes) or for the detection of multiple targets in parallel (detection of DNA from a number of distinct species). For this purpose, various procedures employing silicon-based microchip technology are being developed. In these systems, numerous different DNA probes may be located on a gene chip microarray connected to a detection/data evaluation device. Amplification products may be passed across the gene chip area, and successful hybridization of an amplified DNA to one of the probes will trigger a positive signal. Today, this technology may be far from being applied in the food industry; however, it does have a potential for rapid and possibly quantitative analyses for identification of a variety of pathogenic or spoilage microorganisms in foods.

Conclusions

PCR has been used as a diagnostic tool for more than 10 years, but applications to the food industry are just beginning. It appears likely that the importance of this technology to the food industry may increase exponentially in the near future, particularly in view of the fact that national and international regulatory boards are currently preparing for implementation of PCR technology in official analytical standards. Not only in Germany such standards will be binding for the production industry (and thus to diagnostic laboratories as well), so that alternative new methods like PCR may drastically change the range of methods applied in such tasks. Traditional methods like culture and some more elaborate assays will remain important, especially when exact diagnosis of the origin of contamination is requested. However, new high-throughput production technologies and pressure to reduce overall production costs already require rapid, specific and sensitive diagnostic methods, the most powerful of which is PCR.

See also: **Biosensors**: Scope in Microbiological Analysis. ***Escherichia coli* O157:H7**: Detection by Latex Agglutination Techniques; Detection by Commercial Immunomagnetic Particle-based Assays. **Food Poisoning Outbreaks**. **Hydrophobic Grid Membrane Filter Techniques (HGMF)**. **Immunomagnetic Particle-based Techniques**: Overview. ***Listeria***: Detection by Classical Cultural Techniques; Detection by Commercial Enzyme Immunoassays; Detection by Colorimetric DNA Hybridization; *Listeria monocytogenes* – Detection by Chemiluminescent DNA Hybridization; *Listeria monocytogenes* – Detection using NASBA (an Isothermic Nucleic Acid Amplification System. **Molecular Biology – in Microbiological Analysis**. **Nucleic Acid-based Assays**: Overview. **Petrifilm – An Enhanced Cultural Technique**. **Polymer Technologies for Control of Bacterial Adhesion**. **National Legislation, Guidelines & Standards Governing Microbiology**: European Union; Japan. **Reference Materials**. ***Salmonella***: Detection by Classical Cultural Techniques; Detection by Latex Agglutination Techniques; Detection by Enzyme Immunoassays; Detection by Colorimetric DNA Hybridization; Detection by Immunomagnetic Particle-based Assays. **Sampling Regimes & Statistical Evaluation of Microbiological Results**. **Verotoxigenic *E. coli***: Detection by Commercial Enzyme Immunoassays. **Verotoxigenic *E. coli* and *Shigella* spp.**: Detection by Cultural Methods.

Further Reading

Bennett AR, Greenwood D, Tennant C, Banks JG and Betts RP (1998) Rapid and definitive detection of *Salmonella* in foods by PCR. *Letters in Applied Microbiology* 26: 437–441.

Biron MP, Moulay S, Thierry D and Le Guern M (1998) Rapid detection of *Listeria monocytogenes* in food samples using Probelia *Listeria monocytogenes*. 4th World Congress Foodborne Infections and Intoxications, Bundesinstitut für gesundheitlichen Verbraucherschutz, Berlin, Germany, 7–12 June, 1998.

Chen S, Yee A, Griffiths M et al (1997) The evaluation of a fluorogenic polymerase chain reaction assay for the detection of *Salmonella* species in food commodities. *International Journal of Food Microbiology* 35: 239–250.

Galan JE, Ginocchio C and Costeas P (1992) Molecular and functional characterization of the *Salmonella* invasion gene *invA*: homology of InvA to members of a new protein family. *Journal of Bacteriology* 174: 4338–4349.

Lücke FK and ten Bosch C (1998) Möglichkeiten und Grenzen für den Einsatz der Polymerase-Kettenreaktion (PCR) zum Nachweis und zur Typisierung lebensmittelhygienisch relevanter Bakterien. *Deutsche Lebensmittel-Rundschau* 94: 182–187.

Scheu P, Gasch A, Zschaler R, Berghof K and Wilborn F (1998) Evaluation of a PCR-ELISA for food testing: detection of selected *Salmonella* serovars in confectionery products. *Food Biotechnology* 12: 1–12.

PEDIOCOCCUS

Moshe Raccach, Food Science Program, School of Agribusiness and Resource Management, Arizona State University East, Arizona, USA

Characteristics of the Genus and its Species

The lactic acid bacterium *Pediococcus* has seven species. They have diverse and unique morphological, physiological, nutritional and genetic characteristics. A variety of methods were developed to detect the organism and its products. The genus is utilized in industrial fermentations of meat and vegetables and has potential as a biopreservative. Some genetic modifications have been successfully implemented to obtain tailor-made industrial strains. *Pediococcus* is also associated with spoilage of beer and wine.

The genus *Pediococcus*, a member of the Streptococcaceae, is a Gram-positive lactic acid bacterium, which uniquely divides in two planes to form tetrads. Cultures usually show cocci (0.4–1.4 μm in diameter) in pairs. The cross-linked peptide of the cell wall is L-Lys-L-Ala-D-Asp.

Based on comparisons of 16S ribosomal RNA (rRNA) catalogs and sequences of Gram-positive bacteria, the pediococcal evolutionary line of descent is within the clostridial lineage (which includes, among others, the spore-forming genera *Clostridium* and *Bacillus* and the non-spore-forming genera *Lactobacillus*, *Leuconostoc*, *Pediococcus* and *Strepto-*

coccus). The clostridial lineage is characterized by a low G+C ratio (< 50%). *Pediococcus* species have a G+C ratio in the range of 32–42% (T_m). The pediococci have a closer phylogenetic relationship to the lactobacilli than to the streptococci. Pediococci are found along with lactobacilli and *Leuconostoc* spp. in plant habitats. They have more in common physiologically with these organisms than with the streptococci that are more associated with animal habitats.

Eight species of the genus *Pediococcus* are listed in the 1984 edition of *Bergey's Manual of Systematic Bacteriology*: *P. damnosus*, *P. parvulus*, *P. inopinatus*, *P. dextrinicus*, *P. pentosaceus* (**Fig. 1**), *P. acidilactici*, *P. halophilus* and *P. urinaeequi*. In 1993 *P. halophilus* was reclassified in a new genus, *Tetragenococcus*, as the type species *T. halophilus*.

The DNA–DNA hybridization between *P. pentosaceus* and *P. acidilactici* showed a homology of 15–21%, justifying the classification of *P. acidilactici* as a distinct species. On the other hand, the study of aldolases suggests that the two species are closely related. *Pediococcus dextrinicus* did not show a close phylogenetic relationship (DNA homology of 4–8%) to other *Pediococcus* species.

The pediococcal colonies vary in size from 1.0 mm to 2.5 mm in diameter; they are smooth, round and greyish white. All species grow at 30°C but the optimum temperature range is 25–40°C. *Pediococcus pentosaceus* has a lower optimum temperature for growth (28–32°C) than *P. acidilactici* (40°C), but the latter grows at 50°C.

The optimum pH for growth is 6.0 to 6.5. Half the species grow at pH 4.2, and most of them (except *P. damnosus*) grow at pH 7.0.

Most *Pediococcus* species except *P. damnosus* can grow in the presence of 4.0% and 6.5% NaCl. None of the species can grow in the presence of 10% NaCl. Lactic acid production by *P. pentosaceus*, in a bacteriological medium at 27°C, was inhibited 36.0%, 42.0%, 44.0% and 51.0% by concentrations of NaCl of 3.0%, 3.3%, 3.6% and 3.9% (w/v), respectively.

Some strains of *P. acidilactici* and *P. pentosaceus* have proteolytic enzymes such as protease, dipeptidase, dipeptidyl aminopeptidase and aminopeptidase. *Pediococcus pentosaceus* showed strong leucine and valine arylamidase activities.

The pediococci are facultative anaerobic to microaerophilic, microorganisms. *Pediococcus damnosus* and *P. parvulus* require the most anaerobic conditions while *P. urinaeequi* is the more aerobic species. *Pediococcus acidilactici* and *P. pentosaceus* demonstrate good growth under both aerobic and microaerophilic conditions. Under aerobic conditions, pediococci produce acetic acid with less lactic acid in the culture medium.

The pediococci are usually catalase- and benzidine-negative. Catalase activity was detected in some pediococci when grown in low or high carbohydrate media. This pseudocatalase activity is insensitive to both cyanide and azide, suggesting the absence of a haem molecule. Some strains were able to incorporate exogenously provided haem into a catalase molecule. It is not clear whether pediococci contain cytochromes. The formation of hydrogen peroxide by some pediococci led to the consideration of a flavoprotein enzyme system as the electron transport chain. This system does not fully reduce oxygen to water but rather to the toxic hydrogen peroxide. Reduced nicotinamide-adenine dinucleotide (NADH) oxidase activity was detected in pediococci leading to the production of water. The pediococci may cause 'bleaching' to complete haemolysis of blood agar.

The pediococci are chemoorganotrophs requiring, among other things, a carbohydrate and an array of vitamins, amino acids and metals for growth. A monosaccharide such as glucose is probably transported into the pediococcal cell via the phosphoenolpyruvate : phosphotransferase system

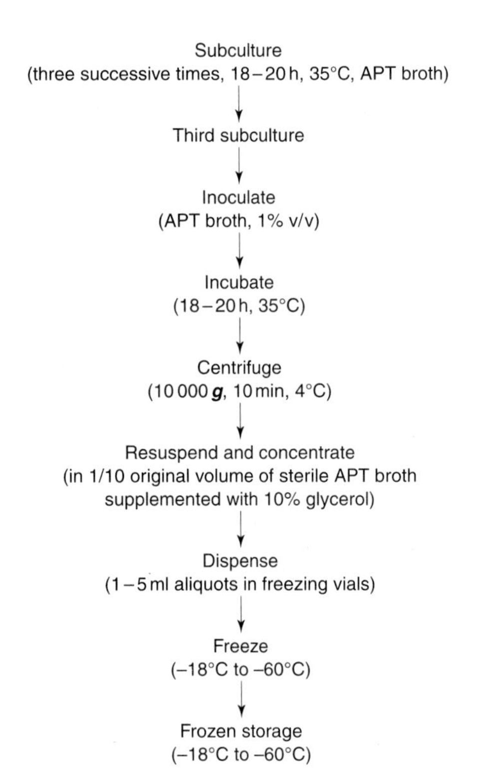

Figure 1 Inhibition of *S. aureus* ('lawn') by 'spots' of *P. acidilactici* (formerly *P. cerevisiae*).

(PEP : PTS) and undergoes glycolysis utilizing the Embden–Meyerhof–Parnas (EMP) pathway yielding pyruvate. The pyruvate is reduced to lactic acid with the coupled reoxidation of NADH to NAD^+. The overall pathway is homolactic fermentation because 90% or more of the end product is lactic acid (lactate). The process can be described as:

$$\underset{\text{(glucose)}}{C_6H_{12}O_6} + 2ADP + 2P_i \rightarrow \underset{\text{(lactic acid)}}{2CH_3CHOHCOOH} + 2ATP$$

The pediococci possess NAD-dependent D(−) and L(+) lactate dehydrogenase (LDH). They mainly form DL and L(+) lactic acid. The final pH in De Man Rogosa and Sharpe (MRS) broth is usually less than 4.0.

Lactate is also the major end product from the fermentation of fructose, ribose and arabinose, in addition to smaller amounts of ethanol and acetate. Equimolar amounts of lactate and ethanol were produced from the fermentation of xylose. Good growth of pediococci can be obtained on D-xylose under aerobic conditions and the presence of glucose. Acid was not produced by *P. pentosaceus* in the presence of D-arabinose and L- and D-xylose under anaerobic conditions. Glycerol utilization is favoured under aerobic conditions and by catalase-positive pediococci, which yield equimolar amounts of lactic acid, acetic acid and acetoin in addition to CO_2. Some strains of *Pediococcus pentosaceus* and of *P. acidilactici* had intracellular β-galactosidase activity when grown in the presence of lactose. Fewer strains of pediococci had such activity when grown in the presence of glucose. The synthesis of β-galactosidase by pediococci was inducible with lactose, galactose, maltose, melibiose, lactobionic acid and possibly cellobiose. Beta-galactosidase from a strain of *P. pentosaceus* had a molecular weight of 66 000 and its optimal activity is at pH 6.5 and at 45°C.

Pyridoxal is necessary for *P. acidilactici*. There are conflicting reports as to the requirement of biotin by pediococci. Some strains of *P. acidilactici* and of *P. pentosaceus* required biotin while other strains did not. Some strains of *P. pentosaceus*, *P. parvulus* and *P. dextrinicus* require folinic acid, which can be replaced by thymidine. Lactic acid bacteria including pediococci require vitamin B_{12}. Tween 80 stimulates the growth of pediococci and it is necessary for *P. parvulus* and *P. dextrinicus*.

Pediococci require 16–17 amino acids; the requirement is strain-specific. Methionine and lysine are stimulatory to the growth of pediococci. The requirement for amino acids by *P. acidilactici* stems from either a base substitution mutation or an extensive genetic lesion. They cannot grow on ammonium salts as a sole source of nitrogen and usually do not reduce nitrate. *Pediococcus pentosaceus* may have a haem-dependent nitrite reduction capability with ammonia as the sole product.

Pediococci require several inorganic ions, at trace level, for normal growth and metabolic activity. Some of these inorganic ions are potassium, phosphate, magnesium, calcium, zinc, iron and manganese.

Pediococci, as other bacteria, require potassium and phosphate. Rubidium may fully or partially substitute for potassium. Magnesium is used for cell division and is found in phosphorylation systems especially in the glycolytic pathway often acting as a link between substrate, enzyme and coenzyme. Magnesium stimulated the growth of pediococci, and is involved in the synthesis of intermediate metabolites, utilization of RNA, energy-yielding enzymes, assimilation of certain amino acids, decreasing the binding ability of heavy metals and protecting against metal toxicity. Calcium may be needed in the formation of some proteases. Some enzymes such as lactic dehydrogenases contain zinc. Iron in a preformed haem group is essential to catalase activity. Manganese is associated with ATP and can replace magnesium in many biological reactions. Manganese enhanced the activity of many enzymes. This metal is also necessary for the induction of enzymes participating in the fermentation of ribose. Manganese is also a component of the metalloflavin enzyme nitrite reductase. The catalytic active centre of non-haem catalase was found to contain a binuclear manganese centre, which undergoes oxidation (Mn^{+++}) and reduction (Mn^{++}) in the presence of H_2O_2. Manganese may substitute for superoxide dismutase in *P. pentosaceus* as a scavenger of the superoxide radical (O_2^-), and may be required for RNA polymerase. Ions such as Mg^{++}, Mn^{++}, Zn^{++} and Co^{++} stimulated the activity of β-galactosidase.

The minimal inhibitory concentrations (MIC) of neomycin, penicillin, chlortetracycline, streptomycin, erythromycin and chloramphenicol against *P. pentosaceus* and *P. acidilactici* are listed in **Table 1**. It is apparent that *P. pentosaceus* is more sensitive than *P. acidilactici* to all tested antimicrobics except for chlortetracycline and streptomycin. Pediococci are resistant to vancomycin.

Table 1 Minimal inhibitory concentrations of selected antimicrobics

Antimicrobic	P. pentosaceus (*μg ml⁻¹*)	P. acidilactici (*μg ml⁻¹*)
Neomycin	250.0	375.0
Penicillin	1.3	2.5
Chlortetracycline	10.0	10.0
Streptomycin	375.0	375.0
Erythromycin	0.24	0.12
Chloramphenicol	2.0	1.0

Pediococci harbour plasmids, some of which encode for bacteriocin formation and many metabolic traits including the fermentation of sucrose and the utilization of lactose.

All pediocin-producing strains of *Pediococcus acidilactici* contain a 9.4 kb (6.2 MDa) plasmid. Bacteriocins are proteinaceous antimicrobics; for example, class II bacteriocins, are small (30–100 amino acids), heat-stable and commonly not post-translationally modified. One of the most extensively studied bacteriocins is pediocin PA-1 produced by strains of *P. acidilactici* of meat origin. Pediocin PA-1 production, immunity and secretion are determined by an operon containing four genes. The production of a pediocin (PD-1) by *Pediococcus damnosus* starts during early growth and reaches a plateau at the end of the exponential growth. The size of this pediocin is approximately 3.5 kDa and its isoelectric point (pI) is approximately 3.5. Pediocin PD-1 is heat-resistant (10 min at 121°C) and remains active after 30 min of incubation at pH 2–10. Pediocin PD-1 is resistant to treatment with pepsin, papain, α-chemotrypsin and trypsin, but not proteinase K. Pediocin PD-1 was not active against other pediococci and differs in this respect from other pediocins produced by *P. acidilactici* and *P. pentosaceus*. Pediococci isolated from spoiled ciders produced ropiness and had resistance to oleandomycin, which was encoded by a plasmid DNA. These pediococci were tolerant to 10% ethanol and to 15 $\mu g\,ml^{-1}$, 25 $\mu g\,ml^{-1}$ and 50 $\mu g\,ml^{-1}$ total SO_2 (pH 3.8).

Genetic modification of meat starter cultures produced strains with desirable phenotypic characteristics. For example, curing of a strain of *P. pentosaceus* from a plasmid encoding for the fermentation of sucrose resulted in a sucrose-negative (Suc^-) industrial culture, which in a mix sugar fermentation, utilized glucose without affecting the sucrose. Some strains of pediococci were transformed by electroporation with the *Lactococcus lactis* lactose plasmids, pPN-1 or pSA3. The transformants rapidly produced acid and efficiently retained the plasmid in lactose broth, and were not attacked by bacteriophage in whey collected from commercial cheese facilities. Genetically modified strains of *P. pentosaceus* containing the MLS (macrolide lincosamide streptogramin B) R plasmid pIP 501 showed 16 000 and 32 times more resistance than the parent strain to erythromycin and chloramphenicol, respectively. The same was true for *P. acidilactici* except that the resistance to erythromycin increased about 32 000 times.

Cultures of pediococci can be stored for later use in refrigerated deeps of all-purpose Tween (APT) agar. They can also be lyophilized or frozen. Freezing and frozen storage are convenient, as no special equipment is required. A procedure for the preparation of a laboratory-frozen concentrate – 10^{10} colony forming units (cfu) per millilitre – is shown in **Figure 2**. Laboratory concentrates can be stored frozen for up to a year.

Testing the fermentative activity of pediococci can be done in quarter-strength MRS broth. *Pediococcus pentosaceus* produced about 0.4% lactic acid from glucose (0.5% w/v) of which 84% and 16% were L- and D-lactic acid, respectively. The same culture did not produce lactic acid from D-arabinose, D-lactose, or from L- and D-xylose.

Although pediococci are inherently resistant to vancomycin, these organisms were formerly considered harmless to humans and were often misidentified. Pediococci are being recognized with increased frequency as potential human pathogens: *P. acidilactici* caused septicaemia in a 53-year-old man and *P. pentosaceus* caused bacteraemia in a 64-day-old infant.

Methods of Detection

Pediococcal cultures inoculated in fermented meats, dough or vegetables can be detected and enumerated using lactobacilli MRS agar or APT agar supplemented with sodium azide. The incubation is up to 4 days at 30°C.

Pediococci can also be isolated from foods or drinks using selective media such as sucrose agar medium, Lee's multidifferential agar, lactobacillus selection agar (LBS), Rogosa SL agar (RNW agar), acetate agar (pH 5.6–5.8), universal beer agar (UBA), modified Wallerstein laboratory nutrient (MWLN) agar and homofermentative-heterofermentative differential

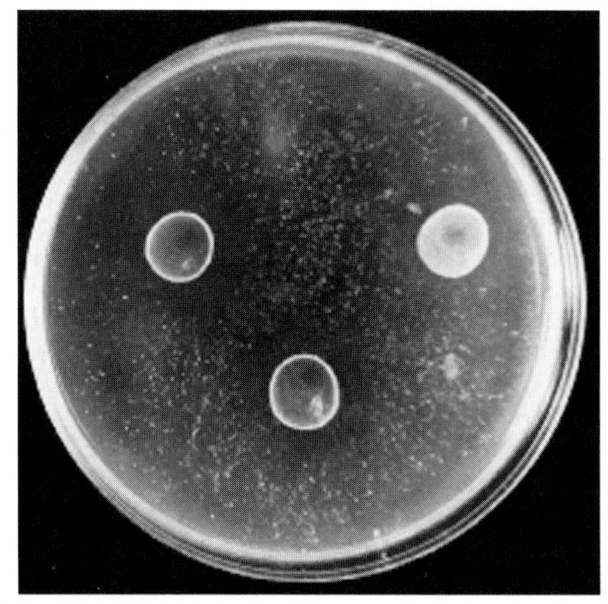

Figure 2 Preparation of a laboratory frozen concentrate of *Pediococcus pentosaceus*.

(HHD) medium. The pour plate method is recommended combined with aerobic or reduced oxygen incubation. Some surface-plated pediococcal cultures develop better if incubated under reduced oxygen conditions (a 'candle' jar, the microaerophilic Gas Pack system or a CO_2 incubator), while other strains do well aerobically.

The microbiology sub-committee of the analysis committee of the Institute of Brewing recommended sucrose agar (SA) for the routine isolation of pediococci and lactobacilli. This medium is normally used for surface plating. The incubation is under aerobic conditions for 3–6 days at 30°C. The medium yields larger colonies (2.5–3.6 mm in diameter) and more of them compared with other media.

In comparison with lactobacilli, pediococci usually develop smaller colonies on Lee's multidifferential agar. The pediococcal colonies (2.0–3.0 mm in diameter) are yellowish-green and have a halo limited to the edge of the colony. Prolonged incubation results in a larger halo with no change in the colour of the colony. The best differentiation among genera of bacteria is obtained with Petri dishes containing no more than 50 colonies per plate. It is recommended to incubate the plates under reduced oxygen conditions for up to 7 days at 30°C.

Either LBS agar or Rogosa agar (each at pH 5.4) can be used for the selective enumeration of pediococci. Either bromcresol green or brilliant green may be added as an enumeration aid. In the presence of bromcresol green the colonies are blue to green. Brilliant green may increase the selectivity of the medium. Cycloheximide may be added to suppress yeasts. Plates are incubated for 4 days (or until large enough colonies develop) at 32°C.

Differentiation of pediococci from lactobacilli can be done using MRS differential (MRSD) medium. For this purpose, MRS was modified to include 0.1 mol l^{-1} L-arginine–HCl, 0.0025% phenol red, 100 IU polymyxin B sulphate, by deletion of meat extract, use of only 1.2% (w/v) glucose and increase of Mn^{++} to 1000 µg g^{-1} and pH 5.5. In addition, the hydrophobic grid membrane filter system with 0.025% fast green FCF dye was used. After an anaerobic incubation (25°C) followed by staining (0.4% w/v bromocresol purple) the pediococcal colonies were blue, whereas colonies of homofermentative and heterofermentative lactobacilli were green.

As other lactic acid bacteria may develop colonies on the selective media for pediococci, it becomes necessary to further characterize the purified isolated colonies. **Table 2** shows a procedure for the presumptive identification of the genus *Pediococcus*. For further identification of the species of *Pediococcus*, it is recommended to consult *Bergey's Manual of Systematic Bacteriology*. Tests used for the identification of species include growth at different temperatures, pH values and concentrations of NaCl, as well as acid production from carbohydrates and sugar alcohols, hydrolysis of arginine and the type of lactic acid produced.

Table 2 A procedure for the presumptive identification of the genus *Pediococcus*

1. Purify the selected bacterial colony by isolation streak
2. Determine the Gram staining reaction. Proceed if the isolate is Gram-positive
3. Determine the cell morphology. Proceed if the cells are cocci
4. Determine the cell arrangement. Proceed if the cells are arranged in pairs or tetrads
5. Determine the catalase reaction. Proceed if the isolate is catalase-negative (most pediococci are catalase-negative)
6. Determine the oxygen need. Proceed if the isolate is either facultatively anaerobic or microaerophilic
7. Determine the type of lactic acid fermentation of glucose. Proceed if it is homofermentative
8. Determine whether lactose is fermented. Proceed if the isolate does not ferment lactose (most pediococci are lactose-negative)
9. Determine susceptibility to vancomycin. Proceed if the isolate is resistant
10. Determine the base ratio G+C%. For pediococci values are in the range 32–42%

Several media were developed for determining pediococcal proteolytic and nitratase activities, formation of acetylmethylcarbinol, gas, ammonia from arginine and the requirement of vitamins.

A simple, rapid test for the presumptive identification of catalase-negative non-haemolytic cocci such as pediococci has been developed, using disc tests for susceptibility to vancomycin (Van), production of leucine aminopeptidase (LAPase) and pyrrolidonylarylamidase (PYRase). The pediococci were unique in being vancomycin resistant (Vanr), PYRase negative and LAPase positive. The results indicate that together with Gram staining and the catalase test, the vancomycin, LPAase and PYRase disc tests can be used presumptively to identify Vanr strains of *Pediococcus* from human infections. Cell morphology, arginine dihydrolase and gas production in MRS broth allowed for the identification of vancomycin-resistant *P. pentosaceus* from humans.

Characterization of carbohydrate fermentation can be done using either the API *Lactobacillus* system or the Minitek system as used for lactobacilli. The API *Lactobacillus* system was tested for the reaction of pediococci towards 50 substrates. Increasing the inoculum size to 10^9 cfu ml^{-1} may help with slow-growing strains.

Viable cells of pediococci can be counted using

fluorescence microscopy and staining with two fluorochromes, erythrosine B (ERB) and 4′,6-diamidino-2-phenylindole (DAPI). Viable cells appear as bright blue or bright green fluorescence, whereas dead or heat-treated cells have only low-intensity fluorescence. Pediococci can be detected by electrical impedance and by a fluorescent antibody technique.

Different strains of the genus *Pediococcus* can be detected using nucleic acid probes. For example, probes exist for the detection of beer spoilers and for the identification of glucan-producing *P. damnosus* in wines. Strains of *P. acidilactici* and *P. pentosaceus* can be identified using a polymerase chain reaction assay; this is based on the amplification of a 16S rRNA gene fragment, specific for each species, and a D-lactate dehydrogenase gene fragment specific for strains of *P. acidilactici*.

An enzyme-linked immunosorbent assay was used to isolate *Pediococcus* species from fermented meat products by colony immunoblotting. The monoclonal antibody Ped-2B2 did not show any cross-reactions with other lactic acid bacteria or other Gram-positive or Gram-negative organisms. A membrane immunofluorescent antibody test was developed to detect diacetyl-producing *Pediococcus* contaminants of brewers' yeast. Specific precipitin was detected for pediococci, which can be used as an aid in the identification of these cultures. A monoclonal antibody-based enzyme immunoassay for pediocins of *P. acidilactici* has been developed.

Importance to the Food Industry

The pediococci are used as starter cultures in the commercial fermentation of meats (**Table 3**) and vegetables. They can also be used to sour wheat flour with no added sugar. The pediococci were tested as biopreservatives of refrigerated foods, but the refrigeration temperature did not favour their fermentative activity. *Pediococcus acidilactici* (NCIB 6990) is highly sensitive to pantothenic acid and can be used for the bioassay of this vitamin. The pediococci are also associated with the spoilage of beer, wine and juices.

Table 3 A generalized process for the preparation of fermented semidry sausage using *P. pentosaceus*

1. Grind or chop the meat
2. Add:
   a fermentable carbohydrate (0.7–1.0% w/w)
   salt (NaCl 2.8–3.0% w/w)
   nitrite (according to regulations)
   spice mix (according to manufacturer's recommendation)
   a *P. pentosaceus* starter culture (about 10^7 cfu g^{-1} sausage-mix)
3. Comingle the sausage mix
4. Stuff sausage mix into casings
5. Ferment to a pH of 5.3–5.0 (27–43°C, relative humidity 75–85%)
6. Smoke (optional, natural or liquid smoke)
7. Heat treat (to an internal temperature of at least 60°C)
8. Refrigerate (1–3 days)
9. Slice and vacuum package (optional)
10. Store under refrigeration (prolongs the shelf life)

Pediococcus is one of the main genera used in the fermentation of meats. The species commercially used are *P. acidilactici* and *P. pentosaceus*. The pH of salami products made with starter cultures containing no added manganese lagged behind that of products made with added manganese (5 $\mu g\, g^{-1}$) by 0.2 pH units. A level of 1.2 $\mu g\, g^{-1}$ of added manganese was sufficient to achieve optimal fermentation of the meat. A Mettwurst spice blend can provide the fermenting meat with 0.77 $\mu g\, g^{-1}$ manganese. Up to 50% of the NaCl in a fermented sausage formulation was substituted with KCl. Pediococci were inhibited by KCl and the inhibition was more pronounced when they were inoculated with a strain of *Lactobacillus plantarum*. The meat starter culture *Pediococcus pentosaceus* was immobilized in calcium alginate beads and then lyophilized. Upon inoculation into meat, the immobilized culture was found to ferment more rapidly than a comparable free cell culture. Chlortetracycline was most inhibitory and penicillin was least inhibitory to the fermentation of glucose by *P. pentosaceus* in meat; streptomycin and neomycin were in-between.

The use of mix starter cultures could be a problem as some strains of pediococci may inhibit the growth of other strains of pediococci, *L. plantarum* and *Leuconostoc mesenteroides*.

Pediococci were tested as biopreservatives to control the growth of *Salmonella typhimurium* (in pasteurized liquid whole eggs and cooked, mechanically deboned poultry meat), *Staphylococcus aureus* (cooked, mechanically deboned poultry meat), *Listeria* (milk), *Pseudomonas* sp. (pasteurized liquid whole eggs and cooked, mechanically deboned poultry meat). Pediococci also increased the shelf life of refrigerated, mechanically deboned poultry meat, ground beef and ground poultry breast.

The pediococci are found in spoiled beer (*P. inopinatus* and *P. dextrinicus*) and wine. *Pediococcus damnosus* is a major problem in the brewing of beers.

Some pediococci are able to form depressor amines such as tyramine (in beer) and histamine. The level of tyramine in beer was found to be a reliable indicator of the degree of contamination by *Pediococcus* sp. during beer fermentation.

Lactose-positive pediococci have potential as a replacement for *Streptococcus thermophilus* in Italian cheese starter blends and may facilitate development of new strain rotation schemes to combat *S. thermophilus* bacteriophage problems in mozzarella cheese plants.

There are conflicting reports as to the inhibition of *Clostridium botulinum* by pediococcal bacteriocin. One report claims that a pediococcal bacteriocin inhibited spores of *Clostridium botulinum*. Another report shows that when *Pediococcus* was co-inoculated with *C. botulinum*, in sous vide beef, it was capable of significantly delaying the appearance of toxin but not of preventing it. Pediocin was also effective in controlling *Listeria* in milk. Both *P. acidilactici* and *P. pentosaceus* controlled the growth of *Yersinia enterocolitica* serotype O:3 and O:8 in fermenting meat. Pediocins produced during the fermentation of turkey summer sausage provided an additional measure of safety against listerial proliferation.

See also: ***Clostridium***: Clostridium botulinum. **Fermented Foods**: Fermented Meat Products. **Starter Cultures**: Uses in the Food Industry.

Further Reading

Back W (1978) Zur taxonomie der gattung *Pediococcus*. Phanotypische und genotypische der bisher bekannten arten sowie beschereiburg einer neuen bierschadlichen art: *Pediococcus inopinatus*. *Brauwissenschaft* 31: 237–250, 312–320, 336–343.

Deguchi Y and Morishita T (1992) Nutritional requirements in multiple auxotrophic lactic acid bacteria: genetic lesions affecting amino acid biosynthetic pathways in *Lactococcus lactis*, *Enterococcus faecium* and *Pediococcus acidilactici*. *Biosci. Biotech. Biochem.* 56: 913–918.

Giraffa G, Gatti M and Beltrame A (1994) Antimicrobial activity of lactic acid bacteria isolated from fermented meat products. *Ann. Microbiol. Enzymol.* 44: 29–34.

Pederson CS (1979) *Microbiology of Food Fermentations*, 2nd edn. Westport: Avi Publishing.

Raccach M (1987) Pediococci and biotechnology. *CRC Crit. Rev. Microbiol.* 14: 291–309.

PENICILLIUM

Contents

Introduction

J I Pitt, Food Science Australia, North Ryde, NSW, Australia

It is difficult to overestimate the importance of *Penicillium* in nature and in the affairs of humans. *Penicillium* species are almost everywhere: ubiquitous, opportunistic saprophytes. Nutritionally, they are supremely undemanding, being able to grow in almost any environment with a sprinkling of mineral salts, any but the most complex forms of organic carbon, and within a wide range of physico-chemical parameters.

Many *Penicillium* species are soil fungi, and their occurrence in foods is more or less accidental and rarely of consequence. Others have their major habitat in decaying vegetation, or in seeds, or wood, for example, ecological niches which prepare them well for a role in food spoilage. Overall, *Penicillium* species are very important agents in the natural processes of recycling used biological matter. In consequence, they also play an important role in the spoilage of many kinds of foods.

Taxonomy

Penicillium is a large genus, with about 200 recognized species, of which 50 or more are of common occurrence. All common species grow and sporulate well on a wide range of laboratory media, producing small, circular colonies, low and usually profusely sporulating in grey–green or grey–blue colours. In consequence, most *Penicillium* species can be readily recognized at genus level.

Classification within *Penicillium* is based primarily on microscopic morphology of the fruiting structure, termed the penicillus (**Fig. 1**). The genus is divided into subgenera based on the number and arrangement of phialides (elements producing conidia) and metulae and rami (elements supporting phialides) which make up the penicillus, which is borne on the main stalk cells (stipes). The currently accepted classification includes four subgenera. In subgenus *Aspergilloides*, penicilli are monoverticillate, i.e. phialides are borne directly on the stipes without intervening supporting elements.

Figure 1 Fruiting structures (penicilli) characteristic of *Penicillium*, showing differences among subgenera. **A**, **B**, **D**, **E**, **G**, **H**, **J**, Nomarski interference contrast, ×750; **C**, **F**, **I**, **K**, scanning electron microscopy, **C**, **F**, ×3500, **I**, ×2000, **K**, ×2750. (**A–C**) Penicilli characteristic of subgenus *Aspergilloides*: (**A**, **C**) *P. glabrum*, (**B**) *P. restrictum*; (**D**, **F**) penicilli from subgenus *Furcatum*: (**D**) *P. corylophilum*, (**E**) *P. janczewskii*, (**F**), *P. citrinum*; (**G–I**) penicilli from subgenus *Biverticillium*: (**G**) *P. variabile*, (**H**) *P. purpurogenum*, (**I**) *P. verruculosum*; (**J**, **K**) penicilli from subgenus *Penicillium*, (**J**) *P. chrysogenum*, (**K**) *P. crustosum*.

In subgenera *Furcatum* and *Biverticillium*, penicilli are biverticillate, i.e. phialides are supported by metulae; and in subgenus *Penicillium*, both metulae and rami are usually present, producing terverticillate penicilli (Fig. 1). Separation of subgenus *Furcatum* from subgenus *Biverticillium* relies on small differences in phialide shape, metula length and some other features, not all obvious at first, but which reflect fundamental phylogenetic differences between subgenus *Biverticillium* and the other subgenera.

Identifying *Penicillium* isolates is not easy for the inexperienced. The species commonly occurring in foods are mostly similar in colour and general colony appearance. Reproductive structures are small and often ephemeral. However, identification of a high percentage of isolates to species level can be accomplished if isolates are grown under standardized conditions of medium and temperature, and examined after a standard time, so that important taxonomic attributes including colony diameters, colony colours and fruiting structures are reproducible. Identification is normally carried out from 7-day-old colonies on Czapek yeast extract agar (CYA) and malt extract agar (MEA) at 25°C, preferably supplemented with growth on 25% glycerol nitrate agar (G25N) at 25°C and on CYA at 5°C and 37°C (see Table 2).

A wide range of macroscopic morphological characters is used in taxonomy, including colony diameters, colours of conidia, mycelium, exudates and medium pigment, and to a lesser extent, colony texture. Microscopic observations are also essential, especially of penicillus type, conidial morphology and dimensions and the dimensions of the penicillus components. In recent times, secondary metabolite profiles have become a valuable aid to identification, though only rarely essential to differentiate closely related species.

Teleomorphs

Penicillium species are associated with two Ascomycete teleomorphs, *Eupenicillium* and *Talaromyces*. Only a few *Penicillium* species produce teleomorphs, and of these fewer still occur in foods, but those that do possess important properties.

Eupenicillium

The name *Eupenicillium* was applied to the very hard (sclerotioid) cleistothecial state produced by certain *Penicillium* species as early as 1892. However, for many years it remained common practice to name both teleomorph and anamorph by their *Penicillium* name. As well as conflicting with provisions of the International Code of Botanical Nomenclature, this practice ignored the influence of the teleomorph on cultural appearance, longevity, heat and chemical resistance, etc. These important characteristics are readily overlooked if the *Penicillium* name is used. It is current practice for use of the *Penicillium* names to be restricted to references to the anamorphs alone.

Eupenicillium is characterized by the production of macroscopic (100–500 μm diameter), smooth-walled, often brightly coloured cleistothecia, in association with a *Penicillium* anamorph. In many species cleistothecia become rock hard as they develop, and may remain so for many weeks or months, finally maturing from the centre to yield numerous eight-spored asci. Most of the 40 or so recognized *Eupenicillium* species are soil fungi, and of little interest to the food microbiologist. However, as the result of soil contamination of raw materials, *Eupenicillium* species have been isolated as heat-resistant contaminants of fruit juices on several occasions. No particular species appears to be responsible, and growth of the fungus in the product has been rare. As a cause of food spoilage, *Eupenicillium* species can be safely ignored unless an unusual set of circumstances leads to excessive contamination of some raw material or product with soil.

Talaromyces

The name *Talaromyces* is derived from the Greek word for a basket, which aptly describes the body in which asci are formed. Known as a gymnothecium, this ascocarp is composed of fine hyphae woven into a more or less closed structure of indeterminate size. *Talaromyces* is characterized by the production of yellow or white gymnothecia in association with an anamorph characteristic of *Penicillium* (or less commonly, *Paecilomyces* or *Geosmithia*).

As with *Eupenicillium*, until recently *Talaromyces* species with *Penicillium* anamorphs were commonly known by their *Penicillium* names. Again, it is current practice for use of the anamorph names to be restricted to references to the *Penicillium* state alone.

Talaromyces is a genus of about 25 species, mostly inhabiting the soil. However, heat-resistant ascospores are produced and consequently *Talaromyces* species are sometimes isolated from pasteurized fruit juices and fruit-based products. Of these, *T. macrosporus* is the most frequently isolated. *T. flavus*, the most common *Talaromyces* species in nature, is occasionally isolated from commodities such as cereals. *T. wortmannii*, similar in many respects to *T. flavus*, but readily distinguished by its slower growth, is the only other species likely to be isolated from foods.

Enumeration and Isolation

Enumeration procedures suitable for all common *Penicillium* species are similar. Any effective antibacterial enumeration medium can be expected to give useful results. However, some *Penicillium* species grow rather weakly or uncharacteristically on very dilute or carbohydrate-deficient media such as potato dextrose agar or plate count agar. Moreover, it is important that enumeration media for *Penicillium* species restrict growth of spreading fungi which would overgrow the slowly developing penicillia, and also inhibit bacteria. For these reasons, the media most often recommended for enumerating *Penicillium* species are dichloran rose bengal chloramphenicol agar (DRBC) for foods of high water activity (more than 0.95 a_w) and dichloran 18% glycerol agar (DG18) for foods of lower a_w.

Nearly all food-borne *Penicillium* species produce characteristic small, low and heavily sporulating, blue or green colonies on DRBC and DG18, and are readily recognizable to genus. Confirmation requires microscopic examination of a wet mount made from a sporing portion of the colony, where the fruiting structures (the penicilli) characteristic of the genus will be seen. A few species are more floccose, with fewer spores; again microscopic examination will provide confirmation to genus. Identification of *Eupenicillium* and *Talaromyces* colonies on primary isolation plates requires microscopic examination of colonies, with observation of developing cleistothecia or gymnothecia as well as penicilli.

Isolation

Isolation of *Penicillium* species is straightforward. Media such as Czapek yeast extract agar (CYA) or malt extract agar (MEA) are usually used for isolation and storage. Purity can usually be ensured by the use of a wet needle to select a discrete clump of spores from a colony on an antibacterial medium such as DRBC or DG18. Purity can be checked by inoculating a CYA plate at three points, incubating at 25°C for 7 days, and examining for sectoring or other indications of variation in growth rate such as might be caused by a mixed culture or bacterial contamination.

Preservation

Penicillium species survive well on slants at room temperature, but storage at refrigeration temperatures is preferable to eliminate the danger of mite infestations. Long-term storage at very low temperatures or by lyophilization is strongly recommended, and usually presents no problems.

Occurrence in Foods

As noted above, a wide range of *Penicillium* species can occur in foods. Many simply turn up as adventitious contaminants, and rarely cause serious losses. Some, however, have a clear ecological association with certain raw material types, or certain food processes, and play a major role in food spoilage. Only a few have serious implications for toxicity.

Several important food-borne species have natural habitats in cereal grains, appearing before or immediately after harvest. In particular, *P. aurantiogriseum*, *P. verrucosum*, *P. viridicatum* and other closely related species from subgenus *Penicillium* occur on wheat, barley and other small grains grown in temperate regions. *P. funiculosum* and *P. oxalicum* are closely associated with maize cultivation, and are universally found in maize grains. In contrast, *Penicillium* species are not associated with rice during cultivation. However, they may infect rice during storage, with *P. citrinum* being the most important species.

Other *Penicillium* species, notably *P. commune*, *P. roqueforti* and closely related species, are associated primarily with proteinaceous foods such as cheese and meats. A few, notably *P. brevicompactum*, *P. chrysogenum* and *P. glabrum*, are ubiquitous in foods, with no obvious preferred habitat.

Some *Penicillium* species are more specialized ecologically. Several are destructive pathogens on fruit. *P. expansum* is the major cause of post-harvest losses in pome fruits (apples and pears), whereas *P. digitatum* and *P. italicum* cause similar losses in *Citrus* fruits. *P. sclerotigenum* is known only from yams.

A few species have adapted to challenging physiological conditions. Some grow below 0.80 a_w, or under low oxygen tension or are resistant to preservatives. Many are psychrotrophic and capable of causing food spoilage at refrigeration temperatures.

Only a few *Penicillium* species are directly utilized in food production. *P. camemberti* (a domesticated *P. commune*) and *P. roqueforti* are used in the production of various mould-ripened cheeses, and *P. nalgiovense* is a preferred species for inoculation of meat surfaces during the manufacture of mould-ripened European sausages.

Physiology

Penicillium species have a highly evolved physiology, resulting in adaptation to a very wide range of habitats. All food-borne *Penicillium* species are capable of growth at low pH, certainly down to pH 3 and some to pH 2. All species studied have been capable of growth at pH 9, and some above pH 10.

Oxygen Tension

Some *Penicillium* species can grow in low oxygen tensions. *P. expansum* and *P. roqueforti* are able to grow normally in 2% O_2. *P. roqueforti* is capable of slow growth in 0.5% O_2, even in the presence of 20% CO_2. Growth and sporulation can occur in the gas combination 20% O_2 plus 80% CO_2. These species are exceptional, however; most *Penicillium* species require relatively high O_2 concentrations for normal growth.

Heat Resistance

Species of *Penicillium* with teleomorphs, i.e. *Eupenicillium* and *Talaromyces*, display notable heat resistance. Values around a D_{90} of 2–6 min with a z value of 5–10°C have been reported for ascospores of *Talaromyces macrosporus*. As these fungi do not produce ascospores under conditions prevailing in food factories, the presence of heat-resistant ascospores in foods is invariably the result of soil contamination of raw materials.

Water Activity

Many *Penicillium* species are marginally xerophilic. Nearly all studied species in *Penicillium* subgenus *Penicillium* are able to grow down to 0.82 a_w. A few species from subgenus *Penicillium* and subgenus *Furcatum* are capable of growth down to 0.78 a_w, including *P. brevicompactum*, *P. chrysogenum*, *P. implicatum*, *P. fellutanum* and *P. janczewskii*. In contrast, only one or two species from subgenus *Biverticillium* are capable of growth below 0.86 a_w.

Temperature

Most *Penicillium* species grow over lower temperature ranges, and none are thermophilic. Nearly all species in *Penicillium* subgenus *Penicillium* are capable of growth below 5°C, and some at 0°C, making these very important spoilage fungi in foods stored at refrigeration temperatures. A few common species, e.g. *P. citrinum*, *P. oxalicum* and *P. funiculosum* grow well at 37°C, but *Penicillium* species rarely compete with *Aspergillus* species at high temperatures.

Preservatives

A few *Penicillium* species are preservative resistant. Notable is *P. roqueforti*, which is a frequent source of spoilage of cereal products, especially rye breads, preserved with weak acids in Europe. *P. roqueforti* is also unusually tolerant of sorbic acid, which it degrades to produce a kerosene taint.

Penicillium Species Important in Foods

As many as 50 *Penicillium* species are of common occurrence in nature. Although many of these are soil fungi and irrelevant in food spoilage, a number are commonly isolated from foods. The major species are described briefly below, grouped by subgenus.

Subgenus *Aspergilloides*

Penicillium citreonigrum is the major source of citreoviridin, and is believed to have been very important as a contaminant of yellow rice in Japan 100 years ago. However, this species has occurred only rarely in foods in recent times, and it is mentioned here only because of its undoubted toxicity and historic importance. On CYA and MEA at 25°C, *P. citreonigrum* grows slowly, producing small, yellow-pigmented colonies 20–25 mm in diameter in 7 days, and diminutive monoverticillate penicilli.

The most important spoilage species in subgenus *Aspergilloides* is *P. glabrum*. On CYA and MEA, this species grows rapidly (40–55 mm diameter), with usually low and flat colonies, heavily sporing, grey–green, with little other pigmentation or sometimes a yellow or orange reverse on CYA. Penicilli are monoverticillate, swollen at the apices, and conidia are spherical and finely roughened. This species is of common occurrence in a wide range of foods and raw materials, and sometimes causes spoilage of cheese and margarine.

Subgenus *Furcatum*

The most important food-borne species in this subgenus is *P. citrinum*. This species is found in foods from all geographic areas, indeed it is among the most ubiquitous of fungi. It occurs universally in cereals and nuts. Spoilage due to *P. citrinum* appears to be rare, however. It forms relatively small colonies on CYA (25–30 mm diameter) and characteristically smaller ones on MEA (less than 20 mm diameter). Penicilli are distinctive, consisting of a cluster of divergent metulae and phialides, with conidia produced in columns. Sometimes the colony reverse and medium on CYA are coloured yellow from citrinin production.

Penicillium corylophilum produces rather similar penicilli to *P. citrinum*, but with less metulae, often of unequal length. Colonies on CYA and MEA are larger, 25–45 mm diameter after 7 days, flat, and with pale greenish colours often evident in the colony reverses. *P. corylophilum* causes spoilage of high fat foods, and sometimes jams. Occurrence in cereals and nuts is common.

Unlike the previous two species, *P. oxalicum* is widespread in tropical foods, and in maize. It grows rapidly on CYA at 25°C (colonies 35–60 mm diameter) and at 37°C (up to 40 mm diameter). Colonies are flat and profusely sporulating, so that after 7 days conidia will break off in crusts if the colony is jarred.

Subgenus *Penicillium*

A number of important food spoilage species are classified in subgenus *Penicillium*, and their taxonomy is especially difficult. One readily recognized species is *P. expansum*, the common apple rot fungus. In culture on CYA and MEA, it produces deep, dark green, rapidly growing colonies (30–35 mm diameter in 7 days) with brown exudate and reverse pigments. Microscopically, this species produces the closely appressed three-stage penicillus characteristic of species in subgenus *Penicillium*, and stipes are smooth walled. All apple and pear cultivars are more or less susceptible to growth of this fungus, which causes very large losses, especially in roughly handled or long-stored fruit. Indeed, *P. expansum* is a broad-spectrum fruit pathogen, capable of spoiling tomatoes, avocados, mangoes and grapes. It is the major source of the mycotoxin patulin in fruit juices.

The species that cause rots in *Citrus* fruits are also readily recognized. *P. italicum* produces colonies 30–40 mm diameter on CYA but often larger (up to 55 mm) on MEA. Colonies are dark-green, flat with brown pigmentation. Penicilli are terverticillate, with distinctive ellipsoidal to cylindroidal conidia. *P. italicum* causes destructive rots on all kinds of *Citrus* fruits, but is rarely found on other kinds of foods. *P. digitatum* produces flat and usually spreading colonies on both CYA and MEA. It is readily distinguished from other species by its olive colony colour, and by forming large penicilli, with two or three branching stages, and large (up to 8 μm or more long) ellipsoidal to cylindroidal conidia. Like *P. italicum*, *P. digitatum* causes destructive rots in *Citrus* fruits, and again is rarely isolated from other sources. Neither species produces mycotoxins.

P. chrysogenum, like *P. citrinum*, is a ubiquitous fungus, with no obvious preferred habitat. It is among the most common penicillia isolated from foods, but rarely causes spoilage. On CYA and MEA, it produces flat, large (usually 35–45 mm diameter after 7 days) yellow green colonies, often with yellow pigmentation in exudate or medium. Penicilli are terverticillate, but rather spindly by comparison with other species, and with smooth stipe walls. Conidia are small and ellipsoidal. Nearly all isolates produce the mycotoxin cyclopiazonic acid.

Although of rather less common occurrence than *P. chrysogenum*, *P. crustosum* is nevertheless a very important species, because it produces the potent neurotoxin penitrem A. Isolation of more than an odd colony of this species from spoiled foods is a warning signal. *P. crustosum* forms rapidly growing (30–40 mm diameter in 7 days) colonies on CYA and MEA, with heavy dull-green sporulation, and usually little other pigmentation. This species is most readily recognized on MEA, by the formation of crusts of conidia which break off when the plate is jarred. Penicilli are large, three- or four-stage branched, and stipe walls are rough. Conidia are smooth and spherical.

Used in cheese manufacture, *P. roqueforti* is also a common spoilage fungus in cheeses, breads and other cereal products, in preserved foods, or in foods stored under modified atmospheres where conditions have not been maintained stringently. Colonies on CYA and MEA are large (40–70 mm diameter in 7 days), flat and spore in dull green colours. Reverse shades may be green or brown. Penicilli are large and terverticillate, with very rough stipes. Conidia are large and spherical. This species produces a range of mycotoxins (**Table 1**).

A second important cheese spoilage species is *P. commune*, known to be the wild ancestor of the cheese mould *P. camemberti*. *P. commune* produces dull grey–green colonies on CYA and MEA (30–37 and 23–30 mm diameter, respectively), with terverticillate penicilli, rough-walled stipes and smooth spherical conidia. This species produces cyclopiazonic acid.

Penicillium verrucosum is distinguished by producing small (less than 25 mm diameter) yellow–green, usually deep colonies on CYA and MEA. Penicilli are usually three-stage branched, though sometimes with two or four stages evident, broad, with rough stipes and small smooth conidia. This species is found almost exclusively on cereals from cool temperate climates, and is unknown in warmer regions. It produces ochratoxin A in cereals whenever it grows. The selective medium dichloran rose bengal yeast extract sucrose agar (DRYS) was developed to assist in isolation of this species, as on that medium *P. verrucosum* (and *P. viridicatum*) produces a distinctive red–brown reverse colour.

Subgenus *Biverticillium*

Species classified in subgenus *Biverticillium* are relatively rare in foods. The most common species is *P. variabile*, which is most frequently isolated from cereals and flour. On CYA and MEA it produces small (15–22 mm diameter), grey–green colonies which are flat, and usually with some yellow pigment. Penicilli are two-stage branched, with a tight cluster of metulae supporting slender phialides and ellipsoidal conidia. This species makes the minor toxin rugulosin, but this is not of serious concern in food processing.

Endemic in maize, *Penicillium funiculosum* also

Table 1 Significant mycotoxins known to be produced by specific *Penicillium* species

Mycotoxin	*Toxicity, LD_{50}* [a]	*Species producing*
Citreoviridin	Mice, 7.5 mg kg^{-1} i.p.	*P. citreonigrum*
	Mice, 20 mg kg^{-1} oral	*Eupenicillium ochrosalmoneum*
Citrinin	Mice, 35 mg kg^{-1} i.p.	*P. citrinum*
	Mice, 110 mg kg^{-1} oral	*P. expansum*
		P. verrucosum
Cyclopiazonic acid	Rats, 2.3 mg kg^{-1} i.p.	*P. camemberti*
	Male rats, 36 mg kg^{-1} oral	*P. commune*
	Female rats, 63 mg kg^{-1} oral	*P. chrysogenum*
		P. crustosum
		P. griseofulvum
		P. hirsutum
		P. viridicatum
Ochratoxin A	Young rats, 22 mg kg^{-1} oral	*P. verrucosum*
Patulin	Mice, 5 mg kg^{-1} i.p.	*P. expansum*
	Mice, 35 mg kg^{-1} oral	*P. roqueforti*
Penitrem A	Mice, 1 mg kg^{-1} i.p.	*P. crustosum*
PR toxin	Mice, 6 mg kg^{-1} i.p.	*P. roqueforti*
	Rats, 115 mg kg^{-1} oral	
Secalonic acid D	Mice, 42 mg kg^{-1} i.p.	*P. oxalicum*

[a] i.p., intraperitoneal injection; s.c., subcutaneous injection; i.v., intravenous injection.

occurs in a wide range of other foods, and sometimes causes spoilage. It grows moderately rapidly on MEA and CYA at both 25 and 37°C, with colonies of 25–45 mm diameter. Colonies are pale grey, with a loose surface texture. Penicilli are biverticillate, with short stipes (less than 100 μm long) and ellipsoidal conidia.

Mycotoxins

Penicillium species possess exceptionally diverse metabolic capabilities, with reports of production of literally hundreds of compounds by one species or another. The profiles of such compounds have proven to be highly species specific; sometimes whole families of such metabolites are produced by a single species, but not at all by closely related taxa. Not surprisingly, then, a very wide range of potentially toxic compounds has been reported to be produced by *Penicillium* species. The situation is complicated by the fact that some important compounds are produced by more than one species and that the literature is cluttered with a great number of inaccurate reports of production of specific metabolites by particular species. For example, the well-known mycotoxin citrinin has been reported from no less than 20 species, but only three have been shown to be authentic producers. A further important fact is that some highly toxic compounds, known to be produced by particular *Penicillium* species, are not of practical importance because the species concerned very rarely enter the food chain. For example, verruculogen and rubratoxin A are both highly toxic compounds. However, the producers of verruculogen, *P. simplicissimum* and *P. paxilli*, are soil fungi, and very uncommon in foods, whereas rubratoxin A is known to be produced by only three isolates of an unnamed species.

The most important *Penicillium* mycotoxins are listed in Table 1 and are discussed below.

Citreoviridin

Acute cardiac beri beri, a disease often responsible for the deaths of healthy young people, was prevalent in Japan 100 years ago, as the result of consumption of mouldy 'yellow rice'. The role of citreoviridin in this disease has been well documented. It is principally produced by *P. citreonigrum* (synonyms *P. citreoviride*, *P. toxicarium*), a species usually associated with rice, less commonly with other cereals, and rarely with other kinds of foods or raw materials. Citreoviridin is also produced by *Eupenicillium ochrosalmoneum*, a relatively uncommon though widespread species associated with cereals, especially maize, and this is a cause for some concern.

Citrinin

Primarily recognized as a metabolite of *P. citrinum*, citrinin is also produced by *P. expansum* and some isolates of *P. verrucosum*. *P. citrinum* is among the more commonly occurring *Penicillium* species, and the toxin citrinin appears to be abundantly produced in nature. Citrinin is a significant renal toxin affecting monogastric domestic animals including pigs, dogs and poultry. It causes watery diarrhoea, increased water consumption, and reduced weight gain due to kidney degeneration. Its effects in humans are uncertain.

Cyclopiazonic Acid

Table 1 lists seven *Penicillium* species which produce cyclopiazonic acid. As this toxin is also produced by *Aspergillus flavus*, it must, therefore, be of common occurrence in the environment. It has been detected in naturally contaminated maize, peanuts and other foods. It is quite toxic to chickens, but appears to be of less concern in humans. Apart from *Aspergillus flavus*, *P. commune* appears to be the most common natural source of cyclopiazonic acid.

Ochratoxin A

Ochratoxin A (OA) is the most important toxin produced by a *Penicillium* species. Originally described as a metabolite of *Aspergillus ochraceus*, it was subsequently reported to be produced by *P. viridicatum* and this view prevailed for more than a decade. Eventually it became clear that isolates classified as *P. viridicatum* but producing ochratoxin (and citrinin) were more correctly classified in a separate species, *P. verrucosum*. OA has immunosuppressive and embryonic effects, and has recently been classified by the World Health Organization as a probable human carcinogen. Because OA is fat soluble and not readily excreted, it accumulates in the bodies of animals. Recent studies have shown OA to be present in the blood of most Europeans, but the consequences for human health remain uncertain. It has been suggested that OA is a causal agent of Balkan endemic nephropathy, a kidney disease with a high mortality rate in certain areas of Bulgaria, former Yugoslavia and Romania, but evidence for this connection remains elusive.

Patulin

The most important *Penicillium* species producing patulin is *P. expansum*, best known as a fruit pathogen, but also of widespread occurrence in other fresh and processed foods. The production of patulin in rotting apples and pears by *P. expansum* can be a problem. The use of such fruit in juice or cider manufacture can result in quite high concentrations of patulin (up to 350 $\mu g\,l^{-1}$) in the resultant juice. The acceptable level in foods is considered to be 50 $\mu g\,kg^{-1}$. Scrupulous attention to culling of diseased fruit is essential to maintain levels of patulin in commercial juices below this value; the use of high-pressure water jets for washing fruit used in juice manufacture is recommended.

Penitrem A

Chemicals capable of inducing a tremorgenic (trembling) response in vertebrate animals are regarded as rare, except for fungal metabolites, of which at least 20 such compounds have been reported. Tremorgens are neurotoxins; in low doses they appear to cause no adverse effects on animals, which are able to feed and function more or less normally while sustained trembling continues to take place. Several of these tremorgenic mycotoxins are produced by *Penicillium* species, the most important being penitrem A, a highly toxic compound (Table 1). The common source of this toxin is *P. crustosum*. Virtually all isolates of *P. crustosum* produce penitrem A at high levels, so the presence of this species in foods is a warning signal. Diagnosis of the mycotoxicosis caused by penitrems is difficult. However, reports of death or severe brain damage in sheep, horses, and dogs due to naturally occurring penitrems have been sufficiently frequent to indicate that these compounds are both potent neurotoxins and of widespread occurrence. The effect of penitrem A in humans is unknown, but circumstantial evidence suggests a powerful emetic effect, which may serve to limit toxicity.

PR Toxin

Cheese moulds, i.e. the moulds used to produce mould-ripened cheeses, which are staple human foods in many countries, have understandably come under intense scrutiny for potential mycotoxin production. The search for toxins has not gone unrewarded. As discussed above, *P. camemberti* produces cyclopiazonic acid, and *P. roqueforti*, the other major cheese mould, produces at least three toxins: PR toxin, roquefortine, and patulin. However, extensive studies indicate that neither *P. camemberti* nor *P. roqueforti* produces toxins at appreciable levels in cheese.

Secalonic Acid D

Secalonic acid D is produced as a major metabolite of *P. oxalicum* and has significant animal toxicity. It has been found in nature, in grain dusts, at levels of up to 4.5 $mg\,kg^{-1}$. The possibility that such levels can be toxic to grain handlers, especially in maize silos, cannot be ignored. However, the role of secalonic acid D in human disease remains a matter for speculation.

Media

The media recommended for enumeration and isolation of penicillia are listed in **Table 2**.

See color Plates 26 and 27.

See also: ***Aspergillus***: *Aspergillus flavus*. **Mycotoxins**: Occurrence. **Penicillium**: *Penicillium* in Food Production. **Spoilage of Plant Products**: Cereals and Cereal Flours. **Spoilage Problems**: Problems caused by Fungi.

Table 2 Media used for enumeration and isolation of *Penicillium* species

Medium	*Component*	*Amount*	*pH*	*Comments*
Dichloran rose bengal chloramphenicol agar (DRBC)	Glucose	10 g	5.5–5.8	After addition of all ingredients, autoclave at 121°C for 15 min. Store away from light (photoproducts of rose bengal are highly inhibitory to some fungi, especially yeasts). Medium is stable in dark for > 1 month at 1–4°C. Stock solutions of rose bengal and dichloran need no sterilization, stable for very long periods
	Peptone, bacteriological	5 g		
	KH_2PO_4	1 g		
	$MgSO_4.7H_2O$	0.5 g		
	Agar	15 g		
	Rose bengal (5% w/v in water, 0.5 ml)	25 mg		
	Dichloran (0.2% w/v in ethanol, 1 ml)	2 mg		
	Chloramphenicol	100 mg		
	Water, distilled	1 litre		
Dichloran 18% glycerol agar (DG18)	Glucose	10 g	5.5–5.8	Add minor ingredients and agar to ca. 800 ml distilled water. Steam to dissolve agar, then make to 1 litre with water. Add glycerol – final concn. is 18% w/w not w/v. Sterilize by autoclaving at 121°C for 15 min. Final a_w 0.955
	Peptone	5 g		
	KH_2PO_4	1 g		
	$MgSO_4.7H_2O$	0.5 g		
	Glycerol, AR	220 g		
	Agar	15 g		
	Dichloran (0.2% w/v in ethanol, 1 ml)	2 mg		
	Chloramphenicol	100 mg		
	Water, distilled	1 litre		
Dichloran rose bengal yeast extract sucrose agar (DRYS)	Yeast extract	20 g		Sterilize by autoclaving at 121°C for 15 min
	Sucrose	150 g		
	Dichloran (0.2% in ethanol, 1 ml)	2 mg		
	Rose bengal (5% in water, 0.5 ml)	25 mg		
	Chloramphenicol	100 mg		
	Agar	20 g		
	Water, distilled	to 1 litre		
Czapek yeast extract agar (CYA)	K_2HPO_4	1 g	6.7	Use refined table grade sucrose free from SO_2. Sterilize by autoclaving at 121°C for 15 min
	Czapek concentrate	10 ml		
	Trace metal soln	1 ml		
	Yeast extract, powdered	5 g		
	Sucrose	30 g		
	Agar	15 g		
	Water, distilled	1 litre		
Czapek concentrate	$NaNO_3$	30 g		Keeps indefinitely without sterilization. Shake before use to resuspend ppt of $Fe(OH)_3$
	KCl	5 g		
	$MgSO_4.7H_2O$	5 g		
	$FeSO_4.7H_2O$	0.1 g		
	Water, distilled	100 ml		
Trace metal solution	$CuSO_4.5H_2O$	0.5 g		Keeps indefinitely without sterilization
	$ZnSO_4.7H_2O$	1 g		
	Water, distilled	100 ml		
25% Glycerol nitrate agar (G25N)	K_2HPO_4	0.75 g	7.0	Glycerol should be of high quality with low (1%) water content (if lower grade is used, allowance should be made for the additional water). Sterilize by autoclaving at 121°C for 15 min
	Czapek concentrate	7.5 ml		
	Yeast extract	3.7 g		
	Glycerol, AR grade	250 g		
	Agar	12 g		
	Water, distilled	750 ml		
Malt extract agar (MEA)	Malt extract, powdered	20 g	5.6	Commercial malt extract used for home brewing is satisfactory as is bacteriological peptone. Sterilize by autoclaving at 121°C for 15 min. Do not sterilize for longer or the medium will become soft
	Peptone	1 g		
	Glucose	20 g		
	Agar	20 g		
	Water, distilled	1 litre		

Further Reading

Pitt JI (1979) *The Genus* Penicillium *and its Telemorphic States* Eupenicillium *and* Talaromyces. London: Academic Press.

Pitt JI (1988) *A Laboratory Guide to Common* Penicillium *Species*, 2nd edn. North Ryde, NSW: CSIRO Division of Food Research.

Pitt JI and Hocking AD (1997) *Fungi and Food Spoilage*, 2nd edn. London: Blackie Academic and Professional.

Pitt JI and Leistner L (1991) Toxigenic *Penicillium* species. In: Smith JE and Henderson RS (eds) *Mycotoxins and Animal Foods*. P. 91. Boca Raton, Florida: CRC Press.

Samson RA, Hoekstra ES, Frisvad JC and Filtenborg O (eds) (1995) *Introduction to Foodborne Fungi*, 4th edn. Baarn, Netherlands: Centraalbureau voor Schimmelcultures.

Penicillium in Food Production

G Blank, Department of Food Science, University of Manitoba, Winnipeg, MB, Canada

Various types of cheese, including Roquefort and Camembert, and meat sausages such as salami are manufactured using a dual fermentation procedure. Initially a primary fermentation is involved which converts available carbohydrate to acid using either natural microflora or selected strains of lactic acid bacteria. In the case of milk the principal carbohydrate is lactose while in raw meat (pork or beef) it is glycogen, glucose or added sugars such as glucose and sucrose. In a secondary fermentation, a succession of non-lactic microorganisms, including various adventitious yeasts and bacteria but primarily moulds, colonize the surface and/or matrix of the ripening product. Subsequent microbial growth yields a series of complex biochemical and enzymatic reactions which impact on the taste, aroma and texture of the food. *Penicillium*, which consists of a wide and diverse group of aggressive fungi, has been used to augment the ripening of such fermented foods for hundreds of years. In many cases its beneficial involvement occurs either by design or by contamination during manufacture and ripening.

Exploitation of *Penicillium* for the production of organic acids and enzymes used in food processing is also significant, especially in the area of biotechnology. However, interaction of *Penicillium* with food can result in spoilage, shortened shelf life and waste. The deleterious effects of *Penicillium* growth and metabolism are frequently observed on both fresh or perishable and stored or processed foods. In addition, the ability of many forms of *Penicillium* to synthesize potent mycotoxins, including patulin and citrinin, in staple foods such as cereals and cheese and in animal feeds has prompted serious health concerns. Although *Penicillium* performs various distinct roles as regards to food, only its direct involvement during manufacture will be discussed in this article.

Fermented Meat Sausages

Air-dried fermented meat sausages, which are surface-ripened by moulds, were initially produced in Italy ca. 250 years ago. Today they are consumed in various European regions. In Romania, Italy, Hungary, Switzerland, Spain and France, approximately 60–100% of all dry sausages produced, including salami and French *saucisson*, are mould-ripened. To a lesser extent, similar products are produced in Bulgaria, Austria, Belgium and Germany. Although the presence of mould on sausages is often regarded as a criterion of quality in these countries, in North and South America, the UK, Scandinavia, Greece and Russia such products are not embraced with the same degree of enthusiasm and therefore are not manufactured routinely. Indeed, in these countries observance of mould growth on the surface of meat products is usually interpreted as a sign of spoilage and therefore shunned.

Essentially two main categories of mould-ripened sausages are produced: those which are lightly cold-smoked during curing, which is mainly the case with Hungarian and German sausages, and those which are not smoked. The latter types primarily include Italian sausages, including Milano, Genoa and Varzi. Historically, the majority of moulds which participated in the ripening process (the chemical, physical and microbiological changes which occur between stuffing and the time the product is sold) originated as part of the natural mycoflora of the processing facility and belong to the genus *Penicillium*. In some cases they have been reported to constitute up to 95% of the surface mycoflora (**Table 1**).

The dominant species include *P. nalgiovense, P. aurantiogriseum, P. olsonii* and *P. chrysogenum* and these grow mainly on the surface of the casing. This growth is both desirable and necessary since it provides important sensory and preservative functions. For instance, during product ripening, *Penicillium* is known to secrete various proteolytic and lipolytic enzymes. Although diffusion of these enzymes into products may be slow and often incomplete, they are particularly important in small-diameter (< 50 mm) sausages and their actions result in the formation of important degradation compounds. These compounds not only contribute to uniform and pleasant aromas but also to enhanced flavour development. The moulds also perform other functions during growth, including the oxidation of lactic acid produced during carbohydrate (primary) fermentation.

De-acidification coupled with NH_3 production arising from proteolysis results in a less acid but more delicate-tasting product. In this regard moulds are extremely importance since they exert a profound

Table 1 *Penicillium* species and potential mycotoxin production associated with naturally fermented dry sausages

Species	*Mycotoxin capability*
P. nalgiovense[a]	No production reported
P. aurantio-griseum	Aurantiamine, auranthine, anacine, terrestric acid, penicillic acid, verrucosidin
P. olsonii	No production reported
P. spathulatum	No production reported
P. oxalicum	Oxaline, roquefortine C, secalonic acid D
P. capsulatum	No production reported
P. chrysogenum[a]	Chrysogine, emodic acid, roquefortine C
P. verrucosum	Citrinin, ochratoxin A
P. commune	Cyclopiazonic acid, isofumigaclavine, rugulovasine
P. viridicatum	Cyclopenin, penicillic acid, verrucofortine, viomellin
P. roqueforti	PR toxin, roquefortine C, isofumigaclavine A,B, mycophenolic acid, patulin, penicillic acid
P. brevi-compactum	Botryodiploidin, mycophenolic acid
P. polonicum	Penicillic acid, verrucosidin, verrucofortine
P. variable	Rugulosin
P. expansum	Roquefortine C, patulin, citrinin, chaetoglobosin C

[a]Also produces penicillin.

influence on the pH of the ripening sausage, especially in the small-diameter varieties. In these types the pH profile from the surface to the core may not differ appreciably, especially after 20–25 days of ripening. Additionally, the catalase produced by moulds may serve as an antioxidant by reacting with surface oxygen and surface peroxides, thereby decreasing the rate of oxidative rancidity (reaction of oxygen with fat) in the sausage, while synthesis of nitrate reductase may promote the development and stability of a red surface colour. Furthermore, the mould coat or mat that develops:

- can reduce moisture loss, facilitating uniform drying
- provides evidence that the ripening process is complete and that a quality product has been produced
- serves as positive recognition which is closely associated with certain sausage varieties.

In Italy, for example, some fermented salami are expected to have a dry, velvety ivory-white (not white) surface coating whilst in Hungary a light grey covering is preferred on the sausage. In other cases, however, the mould is brushed down and/or washed off at the end of the ripening period to reduce its visibility.

Increasingly there is a growing tendency in the sausage industry not to rely on natural or traditional fermentation processes but rather to use known, well-characterized starter cultures for both primary and secondary fermentation. The reasons for this divergence include the fact that during ripening of naturally fermented sausages the traditional mycoflora which becomes established is often unknown and tends to vary depending on the establishment and the season. This may create problems with respect to the proper processing conditions necessary to produce a quality product. In addition there is the possibility that house mycoflora which become established could produce mycotoxins and/or antibiotics during the ripening process, thereby creating potential health hazards (see Table 1). For example, fermented sausages experimentally inoculated with *Penicillium* have been confirmed to contain several mycotoxins, including ochratoxin A, cyclopianzoic acid, rugulosin and citrinin. Starter cultures such as *P. nalgiovense, P. expansum* and *P. chrysogenum* which are used for ripening raw fermented meat sausages on the other hand are expected to suppress the growth of such undesirable moulds which, depending on the processing conditions used, can develop spontaneously. In Italy, for example, the resultant pH (ca. 4.3–7.1) and available moisture (ca. 0.67–0.92) of salami vary significantly from region to region. In part the composition (spices, preservatives, meat type) and the traditional manufacturing regimes employed may account for these differences. In addition, a high relative humidity (> 70%) during curing and the lack of a smoking step in some cases facilitate mould colonization. In this regard the starter cultures are expected to compete aggressively and rapidly establish dominance. Therefore, through competitive exclusion unwanted mycoflora can be controlled. Also, the cultures should not produce mycotoxins or antibiotics at temperatures employed during ripening.

Commercially, products such as Italian salami or German sausage or rohwurst are manufactured using a mixture consisting of pork fat, lean pork and beef as well as selected spices and curing salts. Sugar may be included and serves as a substrate for the production of lactic acid by primary fermentation microorganisms (lactic starter cultures are frequently used). Following meat cutting, mixing and seasoning, the mixture is stuffed into either natural (intestine) or artificial casings (cellulose or collagen). The sausages can then be surface-inoculated with mould spores and hung in climate-controlled chambers or greening rooms adjusted to a minimum temperature of 20–22°C with a relative humidity over 90% (typically 95%) for 5–7 days. During this time a uniform dry fungal mat develops over the sausages. The temperature and relative humidity are then reduced to ca. 15°C and 75% respectively. During the curing or drying period, which may last between 1 and 2 months, ca. 40% by weight of product is lost as water. The final product appearance is dry and firm. When a natural fermentation process is utilized, it is especially important to regulate closely both the temperature and relative humidity in the curing rooms in an effort to control mycotoxin synthesis during fungal growth. For example, although many toxinogenic *Penicillium* spp. grow below 15°C, their ability to secret toxin is diminished, if not halted completely.

Ham

Penicillium is frequently (> 90% contaminated samples) encountered growing during the ripening of various speciality hams, including country-cured or country-style hams which are popular in the south-eastern US; also various European types such as Südtiroler Bauernspeck manufactured in the German-speaking area of northern Italy and Bindenfleisch or Bundnfleischer from Switzerland. Traditionally these types of raw ham are initially prepared by hand-rubbing their surfaces with a mixture of salt, sugar and sodium nitrate. Following a cool smoke (optional; 22–27°C) they are ripened or aged for a period of 6–10 months. It is during this time that the hams become covered with a thick layering of mould. *Penicillium* usually predominates during the early (3–6 months) ripening period, however, as the water activity (a_w) in the ham decreases (<0.70), and/or if the ambient temperature increases to greater than 30°C, which is often the case during the summer months, especially in the southern US, the more xerotolerant *Aspergillus* begins to predominate. *Penicillium* spp., including *P. expansum* (most frequently isolated), *P. commune* and *P. aurantiogriseum*, contribute to the typical appearance and aroma of the hams and are usually found as part of the house flora. Prior to sale, varying amounts may be removed by trimming, brushing and/or washing. Similar to salami, it has been concluded that some fungi (*P. aurantiogriseum* var. *viridicatum*) are capable of synthesizing mycotoxins including citrinin, especially when ripening temperatures greater than 15°C are used. It has also been suggested that the presence of mould on ham may be more potentially dangerous since these products are devoid of a casing. Ostensibly this could allow for greater mycelial penetration and diffusion of mycotoxins into the product if toxinogenic *Penicillium* was present and environmental conditions were favourable.

Soft White Cheese

Mould-ripened white-surfaced cheeses made from raw or pasteurized cow's, goat's or sheep's milk are particularly popular in France and primarily include Camembert, Coulommier and Brie (**Table 2**). Brie consists of a group of cheeses including Brie de Meaux, Brie de Montereau and Brie de Valois, which are similar to Camembert but are formed in larger-diameter wheels. Differences among these variants are subtle and are primarily based on size, weight and texture of the cheese wheel. In contrast, Camembert cheese is distinctive in having a nutty mushroom-like flavour and a soft yellow creamy texture. Following a primary lactic acid fermentation the surface of the cheese is colonized by a secondary flora of non-lactics which is primarily mould-dominated. As with salami this process occurs naturally with the participation of house fungi or by the use of starter cultures. The white mould *P. camemberti* Thom, which exhibits various growth forms (**Table 3**) and was formerly referred to as *P. caseicolum* and *P. album*, is primarily used for this purpose. After approximately 3 weeks (depending on strain or form) of ripening a confluent surface layer of *Penicillium* mycelium is observed with spores (also referred to as conidia or condidiospores) embedded in the cheese curd. Final products appear covered with a fabric-like white mat, the density and height of which are strain-dependent. It should be mentioned that strains will also differ in regard to their lipolytic and proteolytic activities as well as their tolerance to NaCl.

Table 2 Some cheese varieites ripened using *Penicillium*

Variety	*Country of manufacture*
Blue-veined	
Bleu d'Auvergne	France
Roquefort	France
Bleu de Bresse	France
Fourme d'Ambert	France
Gorgonzola	Italy
Stilton	England
Wensleydale	England
Lymeswold	England
Caledonian blue	Scotland
Danish blue (Danablu)	Denmark
Mycella	Denmark
Edelpilzkäse	Germany
Bayrisch blau	Germany
Gammelost	Norway
Blauschimmelkäse	Switzerland
Tulum	Turkey
Adelost	Sweden
Magura	Bulgaria
Grunschimmelkäse	Austria
Merinofort	Hungary
White-surfaced	
Carré de l'Est	France
Chaource	France
Brie	France
Camembert	France
Coulommier	France
Neufchâtel	France
Weissschimmeläse	Germany

Table 3 Forms of *Penicillium camemberti* used in surface-ripening cheese

- Neufchâtel form consisting of a thick mat of white-yellow mycelia; rapid growth. Produces the highest amounts of protease and lipases
- Short, dense white aerial mycelia; rapid growth
- Long, white loose aerial mycelia; slow growth
- Fluffy or floccose aerial mycelia; young cultures are white, turning greenish-grey with maturity. This form is considered to be the original and is sometimes referred to as *P. album* or *P. caseicolum*

In France, true Brie undergoes a lengthy ageing process which involves a complex succession of colonizing organisms. After approximately 12–14 weeks under 80–90% RH at 10–15°C the cheese surface is smeared or becomes overgrown with *Brevibacterium linens* and related coryneform bacteria. The latter microorganisms contribute to the formation of a red-brown surface and sulphur-like taste. In part the ability of the acid-sensitive *B. linens* to proliferate and further contribute to the ripening process is due to the *Penicillium* which assimilates some of the lactic acid (ca. 1%) produced during the primary fermentation. Raising the pH on the cheese surface, for example in freshly pressed blue cheese, from 4.5–4.7 to 7–7.5 increases the activity of enzymes involved in ripening. Although de-acidification takes place initially at the surface, a pH gradient is created extending to the interior of the cheese. Diffusion of lactate towards the surface also contributes to internal pH changes. Once the lactate has been totally metabolized, *Penicillium* begins to assimilate proteins. Resulting amino acids are deaminated with the formation of NH_3. This further contributes to an increase in both surface and internal pH. Additionally, the solubility of calcium phosphate at the surface begins to decrease with an increase in pH, resulting in the formation of a precipitate at the surface. This creates another gradient, resulting in additional calcium phosphate diffusing towards the surface. Diminishment of calcium phosphate from the interior helps to soften the cheese, giving it its creamy-like texture.

Blue-veined Cheese

These types of cheeses (see Table 2), including Roquefort and Gorgonzola, are ripened by *Penicillium* which grows throughout the cheese matrix. In the case of Roquefort the cheese curds are normally fashioned into 8–10 cm wheels and historically are ripened in a network of caves or grottos. The caves not only serve as a natural source for *Penicillium* but also provide for favourable ripening conditions which include uniform temperatures of 10°C with a relative humidity of 95%. Characteristically these cheeses develop a blue-veined appearance signifying mould growth. Traditionally during manufacture the curd is pierced or skewered and inoculated with *Penicillium* spores. The framework of holes created increases air contact or diffusion within the curd which allows for CO_2 escape. The hyphae tend to follow and grow in the cracks and crevices created. This also encourages spore development within the cheese matrix, which usually occurs 2–3 weeks post-inoculation.

Secondary Starter Cultures

Under suitable growth conditions, including temperature, pH, exposure to oxygen, available moisture and adequacy of growth medium (or composition of the food product), *Penicillium* spores will germinate to form hyphae. Some of the developing hyphae become specialized and produce fertile (spore-bearing) structures, referred to as conidiophores. Both hyphae and conidiophores contribute to the characteristic colour and/or veining of cheese and fermented meat sausages. In addition they provide an exogenous source of several enzymes, namely lipases and proteinases, which are important for the degradation of fat and protein in these cheeses. Pure spore cultures usually in the form of lyophilized (freeze-dried) powders are available from various culture collections and commercial laboratories.

Production of Fungal Spores

Spores are produced separate from the lactic acid bacteria since their cultural conditions are different. In many cases, however, commercial laboratories supply both types of culture. Individual processing companies may also propagate or carry their own stock spore cultures by maintaining them on specialized growth media such as Czapek agar slants. Basically, two methods are employed in the production of *Penicillium* spores for commercial use (bulk starters); these methods are primarily differentiated according to whether a liquid or solid substrate is used.

Liquid Substrate In this method, which is often used for large-scale production, spores are grown under strictly controlled aseptic conditions using a liquid medium from which the resultant spores are harvested and concentrated into a small volume. The choice of the growth medium may be controlled by factors such as cost, ability to produce high spore densities and the ease with which spores can be harvested. The culture vessels range from 1 l foam or cotton-stoppered sterile Roux and/or wide-mouth bottles to large-capacity (ca. 1000 l), pH and temperature-controlled stainless steel fermenters. After inoculation and incubation of the growth medium, typically at 20–25°C for 10–20 days, the resulting fungal biomass and associated conidia are concentrated, in most cases using centrifugation. The concentrate is usually suspended in sterile water with or without the presence of a surface active agent such as Tween 80 and agitated to dislodge spores which are separated from mycelia by filtration. The aqueous spore suspension can be concentrated by centrifugation and then pelletized and/or preserved by freeze-drying, deep-freezing or via the use of a fluid-bed drier.

Solid Substrate Typically, for *P. roqueforti* spores the following protocol is used: mycelium or spores are inoculated on to moistened 1 cm cubes of bread contained in wide-mouth containers or flasks and sterilized at 121°C for 1 h. During cooling the flasks are shaken to prevent the cubes from matting. Following inoculation the flasks are incubated at 20–25°C for 10–20 days or until confluent sporulation is observed. Shaking periodically during incubation ensures even mould growth. The mould-covered bread is harvested, dried and pulverized into a fine powder. Following screening or sifting the powder can be used as an inoculum. Alternatively, the mouldy bread can be gathered into cheesecloth which is then squeezed into the milk along with the addition of the lactic acid starter prior to rennetting. *P. camemberti* spores can be cultured in a similar fashion using sterile oyster or water crackers wetted with a growth broth such as Czapek. Resulting spores can be separated from the substrate and mycelium by filtration through cheesecloth or the entire contents can be dried and pulverized. The fungi can also be grown on Petri dishes or agar slants containing a suitable medium. Following incubation at 20–25°C for 7–10 days the surfaces of the growth media are washed with sterile water and the resulting spore suspensions are harvested, cleaned (separated from any accompanying mycelia by washing and centrifugation) and concentrated.

Addition of Spores

Cheese Depending on the preference of the cheese maker, the mould spores are either added to the bulk milk prior to rennetting or to the fresh curd before or after draining. Neither method has been reported to have any advantage as regards final product quality. When spores are added to the curds after draining, they are frequently sprayed on to the layers during hooping, using an atomizer. Most frequently, after hooping, the curds are dipped into sanitized tanks containing a spore suspension. Inoculum levels of at least 10^9 spores per 1000 l of cheese milk or per litre suspension for spraying or dipping are recommended. Spores also occur as natural residents in processing or ripening facilities located on shelves and in the air and are automatically inoculated on to cheese during handling and turning. In some instances the established house flora serves as the inoculum.

Sausages The mould culture, consisting of a liquid spore concentrate or lyophilized powder (the latter usually supplied in a foil pouch), is suspended in a clean container containing boiled and cooled non-chlorinated tap water. The suspension is subsequently stored for several hours or overnight, ideally at 5–8°C and on the day of production added to a tank containing tap water (20–22°C) and stirred. Resultant concentrations usually range from 10^6 to 10^8 spores per millilitre. Following stuffing, the sausages are immersed into the tank for 10–15 s, during which time a uniform inoculum is applied. Alternatively, the aqueous spore suspension may be sprayed on to the sausages during conveyor transport. Inoculation using these techniques ensures the development of a well-developed adherent fungal coat that covers the entire sausage surface. Hungarian, German and Romanian sausages which are often cold-smoked after stuffing should be washed down before mould inoculation in order to reduce the levels of surface mycostats deposited by the smoke.

Table 4 Some important flavour groups produced by *Penicillium* during cheese-ripening

Free fatty acids (FFA)
Released during the lipolysis of milk triglycerides. Typical concentrations: 20–50 and 30–60 g kg^{-1} in Camembert and Roquefort cheese, respectively. Higher levels of palmitic and oleic acid compared to most non-mould-ripened cheeses. High levels of caproic (C_6), caprylic (C_8), caproic and capric (C_{10}) acids contribute a sharp peppery taste
Methyl ketones
Dehydrogenases secreted by *Penicillium* (spores and mycelium) convert FFA to saturated fatty acids which are subsequently oxidized to β-ketoacids (R-CO-CH_2-COOH) including acetone, propylmethyl ketone, amylmethyl ketone, isobutylmethyl ketone, methylamyl ketone, 2-butanone, 2-pentanone, 2-hexanone, 2-octanone and 2-nonanone. Decarboxylation gives rise to CO_2 and methyl ketones (R-CO-CH_3) which impart a piquant flavour to the cheese. Saturated methyl ketones with 7, 9, 11 and 13 carbon atoms are especially aromatic. Levels range from 6 to 200 mg mg^{-1} and are much lower for Brie and Camembert compared to Roquefort. In Camembert and Brie, 2-nonanone followed by 2-heptanone are the most abundant. Also found are 2-undecanone, 3-octanone, 1,5-octadien-3-one and oct-1-en-3-ol. The latter two volatiles contribute to an earthy mushroom-like flavour. In Roquefort cheese, 2-pentanone, 2-nonanone (highest concentration) and 2-heptanone (usually second highest) are important flavour compounds
Secondary alcohols
Methyl ketones are reduced by dehydrogenases to their corresponding alcohols with similar but somewhat heavier flavour notes. 2-Pentanol, 2-heptanol (highest concentration) and 2-nonanol (second highest) are most common. In Camembert cheese, 1-octen-3-ol is responsible for the characteristic mushroom-like flavour. Levels typically found in blue cheese and Brie are 20 and 0.6 mg kg^{-1}, respectively
Small peptides and free amino acids (FAA)
This group represents compounds produced by proteolysis. They are non-volatile, water-soluble and primarily affect the taste of the cheese. Overall, FAAs contribute to background flavour and serve as substrates for biochemical-based, flavour-generating reactions, including decarboxylation and deamination. Representative (%) of the total major FAAs found in mould-ripened varieties are given as follows: glutamic acid (22), leucine (12) and lysine (9)

Fungal Enzymes

The primary contribution of *Penicillium* during cheese-ripening lies in its ability to produce a variety of hydrolytic enzymes. These enzymes result in the formation or liberation of a litany of volatile and non-volatile compounds, many of which serve as substrates for a succession of biochemical reactions (**Table 4**). In addition, through proteolysis milk proteins or caseins are hydrolysed, resulting in an increase in soluble nitrogenous products. The characteristic bouquet, taste, colour and indeed texture or body of these cheeses are intimately associated with the synthesis of enzymes which are capable of hydrolysing high-molecular-weight peptides and triglycerides. The two major types of enzymes produced by *Penicillium* during ripening are the lipases and proteases (proteinases).

Protease Proteolysis in mould-ripened cheese, especially in blue-veined varieties, is extremely intense and probably represents the most important biochemical event. Degradation products range from high-molecular-weight polypeptides to medium and small peptides and even individual amino acids. Although lactic acid bacteria and indigenous milk enzymes participate in proteolysis, it is generally considered that *Penicillium* is the main agent responsible for this activity. The nutty ammonical flavour of Brie, for example, is directly the result of proteolysis. Overall, the extent of proteolysis in mould-ripened cheeses, particularly in regards to β-casein, is much higher compared to other varieties such as Cheddar. In addition, the extent of proteolysis is greatest at the cheese surface where most growth takes place.

The extracellular proteolytic system of *P. camemberti* includes two extracellular endopeptidases and two exopeptidases. The primary endopeptidase is a metalloproteinase which is produced during growth at pH 6.5 but is stable at a pH range from 4.5 to 8.5; optimum activity on casein is at pH 5.5–6.0. Enzyme hydrolysis is primarily directed towards α_{s1}-casein rather than β or κ caseins. The other endopeptidase is an acid protease, also referred to as aspartate proteinase. It is produced during growth at pH 4.0 and is stable at a pH from 3.5 to 5.5, with optimum activity on casein at pH 5.0. The casein fractions: α_{s1} (highest), β and κ are degraded. These enzymes have wide specificity and noticeably contribute to an increase in both the level of pH 4.6-soluble and non-protein nitrogen; however, they hydrolyse short chain peptides poorly and do not liberate amino acids. The exopeptidases produced include an alkaline aminopeptidase and an acid carboxypeptidase, with optimum activities at pH 8–8.5 and 3.5–5.5 respectively. The latter enzyme has properties similar to serine carboxypeptidase and has been reported to reduce the bitterness in cheese via the hydrolysis of casein or peptide fractions. Overall the exopeptidases are responsible for the release of substantial levels of free amino acids in blue cheese varieties (26 000–57 000 mg kg^{-1} compared to 16 000 mg kg^{-1} in Cheddar).

The endopeptidases and exopeptidases produced by *P. roqueforti* are similar to those of *P. camemberti*. Some *P. roqueforti* strains, however, have also been reported to produce alkaline carboxypeptidases and proteinases.

Lipase The primary lipolytic agent in these types of cheeses is also *Penicillium*. Natural lipases, even in raw milk, are not very active. The degree of lipolysis in mould-ripened cheese is also much higher compared to other varieties such as Cheddar and, like proteolytic activity, is always higher near the surface. *P. camemberti* produces one lipase enzyme with an optimum activity from pH 8.5 to 9.5 at 35°C. Repeated studies have shown that the extent of lipolysis rises dramatically after approximately 6–10 days of ripening, which coincides with the initiation of mould growth. Activity increases concomitantly with growth and mycelial lysis. In contrast, *P. roqueforti* produces both an acid and an alkaline lipase with optimum activities at pH 7.5–8.0 and 6.5, respectively, at 37°C. Although the acid lipase is normally synthesized in higher amounts, its activity is lower on milk fat compared to its alkaline counterpart.

The degradation of milk fat via lipase enzymes results in the release of free fatty acids (FFA). These FFAs contribute directly to the flavour of the cheese since many are highly aromatic. Indirectly, many are oxidized to carbonyl compounds (methyl ketones) and subsequently reduced to secondary alcohols (alkanols). Esters, aldehydes and γ-lactones are also formed. The latter compounds are formed by intramolecular esterification of hydroxy acids through the loss of water to form a ring structure. Lactones possess a strong aroma which, although not cheese-like, may contribute to the overall cheese flavour. Mycelia and especially resting spores produce methyl ketones via oxidation of fatty acids at the β-carbon position, which takes place optimally at pH 5–7. Enzymatic activity varies among strains, therefore the nature and concentration of methyl ketones, which contribute the most recognizable flavour tones, are also expected to differ.

Characteristics of Some *Penicillium* Species Used as Starter Cultures

Various commercial laboratories supply strains of these starter cultures. They differ not only in their appearance but also in their enzymatic activity and growth characteristics, including tolerance to salt, CO_2 and growth at refrigeration temperature.

***P. roqueforti* (Thom)** This mould produces dark-green to deep blue-green, almost black, colonies. Conidia are spherical and usually dark green. Good growth occurs at 3°C but not at 37°C. Of all *Penicillium* species, *P. roqueforti* (also referred to as *P. glaucum* and *P. gorgonzola*) has the lowest oxygen requirement for growth (<4.2%). It can also grow slowly in atmospheres containing 80% CO_2. Tolerant of alkali environments, its pH growth range is from

3.0 to 10.5 with an optimum from 4.5 to 7.5. This mould is unusual in that it can also grow in the presence of weak acid preservatives such as 0.5% acetic acid. The minimum a_w for growth is 0.83 and it tolerates from 6 to 10% NaCl in cheese. It is a widely distributed spoilage microorganism primarily colonizing refrigerated foods, including meats and fresh vegetables, and cool-stored foods such as cereals. A white mutant of this mould has been developed for use in Nuworld cheese.

***P. camemberti* (Thom)** Four forms of this mould are recognized. It produces white, floccose colonies which become pale grey-green with age. Conidia are white or grey, large and spherical. Growth takes place at 5°C but not at 37°C. *P. camemberti* is sometimes referred to as *P. caseicolum*, *P. caseicola*, *P. candidum* and *P. album*. However, these species are now considered to be white mutants of this fungus. The pH range for growth is from 3.5 to 8.5. However, it grows slowly at the pH and salt concentrations (4.7–4.9 and 2–4% respectively) that are present on the surface of fresh cheese. *P. camemberti* is largely found within the confines of cheese-manufacturing plants and causes spoilage in cheeses not intended for ripening.

***P. chrysogenum* (Thom)** This species produces white to yellow to pale or grey-green colonies. Conida are ellipsoidal to subspheroidal and range from colourless to blue-green. Growth takes place at 5°C but is variable at 37°C. Being a xerophilic species it is able to germinate at 0.78–0.81 a_w. The pH range for growth is 3–8. It is a common contaminant in foods, including refrigerated meats and vegetables, cereals, grapes, spices and dried foods. This species grows relatively fast, is very aggressive and is therefore recommended as a starter for fermented sausages where high levels of house mycoflora are anticipated.

***P. nalgiovense* (Laxa)** This mould produces white or creamy-white colonies, which sometimes become pale green with age. Conidia are globose to subglobose, smooth and colourless. *P. nalgiovense* is recommended for fermented sausages where house mycoflora is low. Aroma qualities are superior to *P. chrysogenum*. Growth at 5°C is slow and with no growth at 37°C. *P. nalgiovense* tolerates 6–8% NaCl. Its pH range for growth is from 3.5 to 8.0. This species grows relatively fast and is also considered aggressive. It is a common contaminant on sausages, frequently isolated in processing facilities and is also used as a starter culture for some speciality cheeses.

See also: ***Brevibacterium***. **Cheese**: Mould-ripened Varieties. **Fermented Foods**: Fermented Meat Products. **Mycotoxins**: Classification. ***Penicillium***: Introduction. **Starter Cultures**: Uses in the Food Industry; Cultures Employed in Cheesemaking; Moulds Employed in Food Processing.

Further Reading

Arora DK, Mukerji KG and Marth EH (eds) (1991) *Handbook of Applied Mycology*. Vol. 3. New York: Marcel Dekker.

Early R (ed.) (1998) *The Technology of Dairy Products*. New York: Blackie Academic and Professional.

Pitt JI and Hocking AD (eds) (1985) *Fungi and Food Spoilage*. New York: Academic Press.

Samson R, Hoekstra E, Frisvad J and Filtenborg O (eds) (1995) *Introduction to Food-borne Fungi*. Baarn, the Netherlands: Centraalbureau voor Schimmelcultures.

Stiegbert P and Pederson PD (1998) Mold cultures for the food industry. *Danish Dairy and Food Industry* 6: 8–12.

Peronosporomycetes *see* **Fungi**: Classification of the Peronosporomycetes.

PETRIFILM – AN ENHANCED CULTURAL TECHNIQUE

Rafael Jordano and **Luis M Medina**, Department of Food Science and Technology, University of Córdoba, Spain

Application of the hazard analysis control critical point (HACCP) system and the use of statistic-based sampling plans require the analysis of a large number of samples and rapid availability of results. This system ensures the application of suitable corrective measures in the manufacturing process and prevents the food product from being marketed before the quality control results have been received. Advantages of conventional methods based on the development of colony-forming units (cfu) in solid media which have been used up to now include sensitivity and simplicity and the low cost of the equipment required for their application. However, the delay in obtaining data and the laboriousness of these methods have had a decisive influence on the investigation and development of new techniques for the daily task of controlling the microbiological quality of foods.

Petrifilm plate methods are an alternative to conventional agar plate methods for microbiological testing of food and beverages. As far as rapid methods are concerned, this system can be classified as improvement on traditional colony count methods. The automation of conventional methods permits a reduction in the use and preparation time of the material and increases the number of samples which can be analysed; however, in most cases, it does not significantly shorten the time necessary to complete each assay. At present, it is possible to have equipment that, amongst other functions, automates the procedures of culture medium sterilization, plate-filling, preparation of the food dilutions, plating of bacterial suspensions on agar and the cfu counts.

In this article the following sections are included:

- range of food product applications
- principle and standard protocol
- procedures specified in regulations, guidelines and directives
- advantages and limitations compared with conventional and other modern techniques
- interpretation and presentation of results
- list of available commercial products.

Range of Food Product Applications

Table 1 shows studies in which Petrifilm plate products have been evaluated or compared with conventional methods for the determination of aerobic bacteria, coliforms, *Escherichia coli* O157 : H7 and yeasts and moulds in a wide range of food products.

The main groups of foods sampled were:

- dairy products (raw milk, pasteurized milk, cheese and yoghurt)
- meat and meat products (meat, lamb carcasses, beef, pork, poultry, minced meat and sausages)
- eggs
- seafoods
- vegetables (fresh potatoes and beans and frozen green beans)
- fruit (apples and strawberries) and nuts (pecans)
- cereals and cereal products (corn meal, flour and pasta)
- spices (thyme and black pepper)
- selected dairy and high-acid foods (hard and soft cheese, cottage cheese, yoghurt, sour cream, fruit juice, salad dressings, relishes and sauces)
- frozen desert mixes and yoghurt ice cream (Table 1).

Principle

The dry film methods constitute a ready-to-use system consisting of two plastic films attached together on one side and coated with culture media ingredients (selective or nonselective depending on the microorganism researched) with an indicator dye, and a cold-water-soluble gelling agent. The calculation area has different squares depending on the product (see section on interpretation and presentation of results, below) to facilitate estimation. These methods are an alternative to standard plate count methods since it is possible to dispense with Petri dishes, with the making-up of culture media and with several of the steps of the traditional methodology.

Indicators provide a rapid interpretation of the colonies. Thus, there are plates for Enterobacteriaceae, coliforms and *E. coli* which use violet red bile nutrients as culture media and tetrazolium (2,3,5-triphenyl tetrazolium chloride) as an indicator dye. In the case of *E. coli*, there is a glucuronidase indicator (5-bromo-4-chloro-3-indolyl-β-D-glucuronide) and the *E. coli* colonies have a blue precipitate around them.

The methodology for plating on Petrifilm plates is simple and the steps to be followed in each product are identical, with the exception of some specific peculiarities concerning the type of spreader. The differences in incubation time and temperature of the plates are in line with what has been established for the standard plate methods. In general, a standard protocol could be drawn up as follows (**Figs 1–4**):

1. Place the Petrifilm plate on a flat surface.
2. Lift the top film and carefully dispense 1 ml of sample (or dilution as required) on to the centre of the bottom film. For a high-sensitivity coliform count (HSCC) plate, 5 ml of sample is required.
3. Carefully roll the top film down on to the sample to prevent air bubbles being trapped.
4. Distribute the sample evenly within the circular inoculating area, applying pressure on the centre of a plastic spreader (flat side down) provided with each pack. Do not slide the spreader across the film.
5. Remove the spreader and leave the film undisturbed until the gel solidifies (usually 1 min: 2–5 min for HSCC films).

The plates should be incubated horizontally with the clear side up, in stacks of less than 20. Current count standards should be followed for incubation temperature, depending on the microorganism being investigated and the method followed. Some examples are given in **Table 2**.

It is possible to isolate colonies for research or

Table 1 Selection of food applications of Petrifilm plate products

Food products	*Petrifilm plate products*	*References*
Dairy products:		
Raw milk	PAC, PEC, HEC	Chain and Fung 1991, Calicchia et al 1994, Ginn et al 1984
Pasteurized milk	PAC, PCC, PEC, PYM	Senyk et al 1987, Beuchat et al 1991, Jordano et al 1995
Cheese	PEC, HEC, PYM	Beuchat et al 1990, Matner et al 1994, Vlaemynck 1994, Calicchia et al 1994
Yoghurt	PYM	Vlaemynck 1994
Meat and meat products:		
Meat	PEC, HEC	Okrend et al 1990
Minced meat	PAC, PCC, PEC, PYM	Jordano et al 1995
Lamb carcasses	PCC	Guthrie et al 1994
Beef	PAC, PCC, PEC, HEC	Smith et al 1989, Restaino and Lyon 1987, Chain and Fung 1991, Calicchia et al 1994
Pork	PAC	Chain and Fung 1991
Poultry	PAC, PCC, PEC, HEC	McAllister et al 1988, Matner et al 1990, Chain and Fung 1991, Calicchia et al 1994
Sausages	PYM	Beuchat et al 1991
Seafoods	PEC, HEC	Calicchia et al 1994
Eggs	PAC, PCC, PEC, PYM	Jordano et al 1995
Vegetables	PAC, PCC, PEC, HEC, PYM	Matner et al 1990, Beuchat et al 1991, Calicchia et al 1994, Jordano et al 1995
Fruit and nuts	PAC, PCC, PEC, HEC, PYM	Beuchat et al 1991, Chain and Fung 1991, Calicchia et al 1994, Jordano et al 1995
Spices	PAC, PYM	Beuchat et al 1991, Chain and Fung 1991
Cereal products	PYM	Beuchat et al 1991
Selected dairy and high-acid foods	PYM	Beuchat et al 1990
Frozen dessert mixes and yoghurt ice cream	PAC, PCC, PEC, PYM	Smith et al 1989, Jordano et al 1995

PAC, Petrifilm aerobic count; PEC, Petrifilm *E. coli*; HEC, Petrifilm Kit HEC; PCC, Petrifilm coliforms count; PYM, Petrifilm yeast/mould.

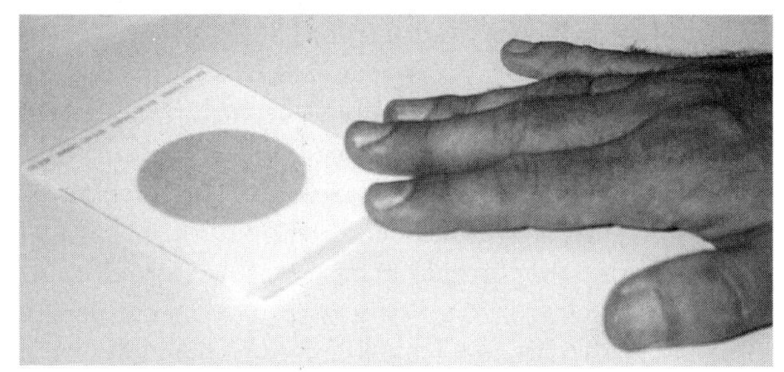

Figure 1 Protocol of inoculation.

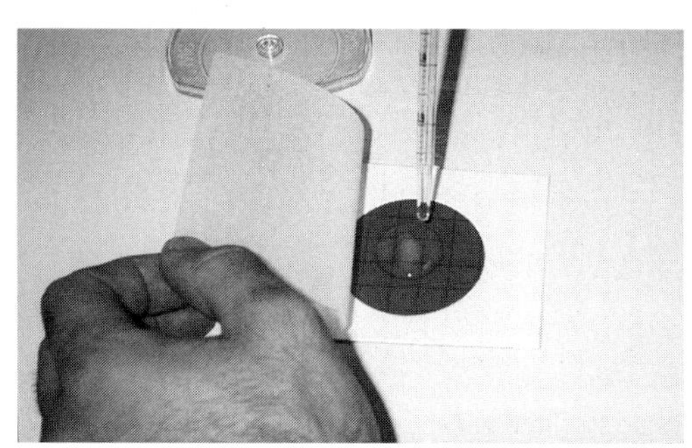

Figure 2 Protocol of inoculation.

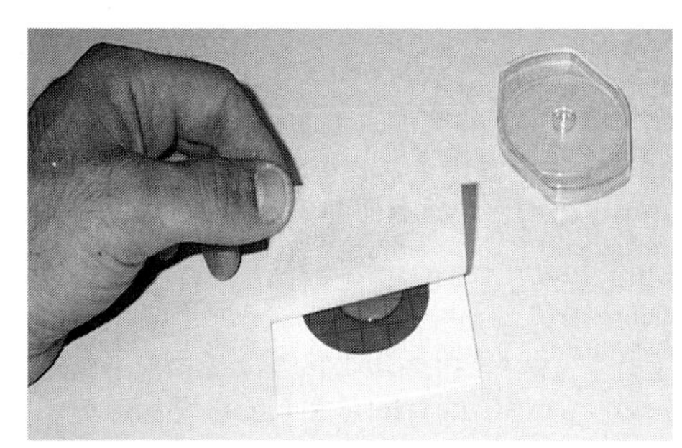

Figure 3 Protocol of inoculation.

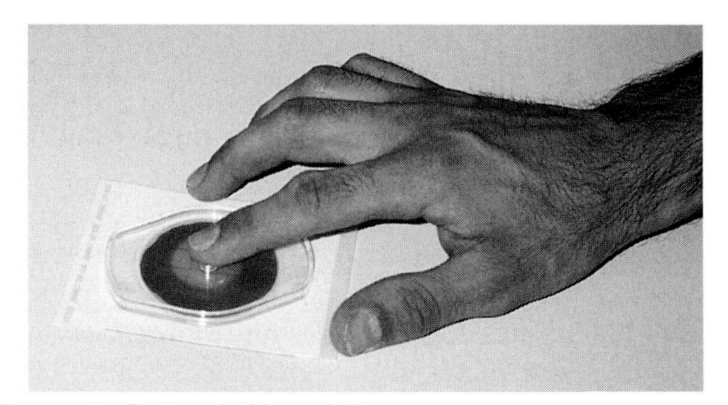

Figure 4 Protocol of inoculation.

Table 2 Time and temperature range for incubation of microorganisms

	Hours	*Temperature (°C)*
Petrifilm aerobic count plate	48 ± 2	30 ± 1
Petrifilm Enterobacteriaceae count plate	24 ± 2	35 ± 1 or 37 ± 1
Petrifilm coliform count plate	24 ±2	30 ± 1 or 37 ± 1
Petrifilm *Escherichia coli*/coliform count plate	According to each method followed	
Petrifilm yeast and mould count plate	72–120	25
Petrifilm high-sensitivity coliform count plate	24 ± 2	32–35
Petrifilm series 2000 rapid coliform count plate	24 ± 2	35 ± 1

identification by lifting up the top film and removing the colony from the gel.

Procedures Specified in Regulations, Guidelines and Directives

In order to be used in food analysis new microbiological methods must comply with the following criteria: rapid supply of data, suitable for routine analysis, precise and accurate, technically viable and internationally acceptable. This last is particularly interesting: various international organizations related to food control have responded by carrying out validation programmes in a collaborative study.

Petrifilm plate methods are subject to this acceptance process. They have already been validated by the French Standardization Association (AFNOR) and recognized as official methods by the Association of Official Analytical Chemists (AOAC) and the Nordic Committee on Food Analysis (NMKL), in specific cases (aerobic bacteria, coliforms and *E. coli*). Similarly, the Victorian Dairy Industry Authority (VDIA) in Australia and the Health Protection Branch in Canada have specified several Petrifilm methods (**Table 3**). Validation of Petrifilm plate methods by the International Dairy Federation (IDF) has not been completed up to July 1998, although the evaluation has been done (stage 2: proprietary methods).

These methods have been approved in Australia (New South Wales Dairy Test Manual), Belgium (Department of Agriculture), Czech Republic (Ministry of Health), Finland (Finnish Standard Union, Committee of Provisions), Japan (National Milk Quality Improvement Association), Romania (Ministry of Agriculture) and the US (Grade A Pasteurized Milk Ordinance and USDA Agricultural Marketing Service Dairy Grading Branch), among other countries.

The Standard Methods for the Examination of Dairy Products, an American Public Health Association compilation of microbiological methods for analysing dairy foods and dairy food substitutes, recognized in its 16th edition the Petrifilm aerobic count plate as an alternative method to standard plate count and the Petrifilm coliform count (PCC) plate for coliform organisms. In addition, the 8th edition of the *Bacteriological Analytical Manual* of the Food and Drug Administration has taken into consideration Petrifilm among other commercially available rapid methods, in relation to the determination of bacteria and total coliforms/*E. coli*.

Advantages and Limitations

Petrifilm plate methods have been compared with conventional methods (plate count agar (PCA), violet reed bile agar (VRBA), levine eosin methylene blue (EMB), and oxytetracycline glucose yeast extract (OGYE) agar) for the enumeration of mesophilic aerobic bacteria, coliforms, *E. coli* and yeasts and moulds in six different food groups (pasteurized milk, yoghurt ice cream, eggs, minced meat, fresh strawberries and frozen green beans). For all microbiological criteria, except for yeasts and moulds and mesophilic aerobic bacteria in frozen green beans, the mean values of counts made with Petrifilm plates were higher than those obtained with traditional methods. In the case of *E. coli*, this was only observed in frozen green beans with the Petrifilm plate methods and was not detected by the traditional method (**Table 4**). The correlation coefficient and slope of Petrifilm aerobic bacteria, coliforms and yeasts and moulds versus PCA, VRBA and OGYE agar for each microbiological criterion for six food products are shown in **Table 5**. The results obtained with the use of the Petrifilm plate method were satisfactory and it can therefore be considered an acceptable alternative to traditional methods for routine microbiological food control.

According to the bibliography consulted and in our own experience, the main advantages of Petrifilm plate methods compared with conventional techniques are:

- simple and rapid (ready-to-use system) application
- ease of transport
- storage and preparation of materials
- flexible films (for surface control)
- indicators and grids aid in interpretation
- accuracy
- excellent selectivity
- less space is required for storing and incubating
- less time is needed for plating samples.

The use of selective nutrients in the most available products means that other microorganisms, which

Table 3 Official validation method for Petrifilm plate products (up to November 1998)

Official organization	*Food products*	*Petrifilm plate products*	*Reference*
	Raw and pasteurized milk	Aerobic count, coliform count	986.33 method
	All dairy products	Aerobic count, coliform count	989.10 method
		High-sensitivity coliform count	996.02 method
AOAC	All foods	Aerobic count	990.12 method
		Coliform count, *E. coli*/coliform count	991.14 method
		Yeast and mould count	997.02 method
		Series 2000 rapid coliform count	Validation in progress (March 1999)
AFNOR	All foods[a]	Aerobic count	3M certification 01/1-09/89
		Coliform count	3M certification 01/2-09/89
		E. coli/coliform count	3M certification 01/4-09/92
		Series 2000 rapid coliform count	3M certification 01/5-03/97
		Enterobacteriaceae count	3M certification 01/6-09/97
NMKL	All foods	Aerobic count	146.1993
		Coliform count, *E. coli*/coliform count	147.1993
VDIA	All dairy	Aerobic count	Certificate number 9503
		Coliform count	Certificate number 9504
HPB	All dairy	Aerobic count, coliform count	MFLP-71 method
		High-sensitivity coliform count	MFLP-85 method
	All foods	Aerobic count	MFHPB-33 method
		Test kit-HEC	MFLP-83 method
		E. coli/coliform count	MFHPB-34
		Coliform count	MFHPB-35
		Yeast and mould count	MFHPB-32 method
	Environmental sampling		MFLP-41A

AOAC, Association of Official Analytical Chemists; AFNOR, French Standardization Association; NMKL, Nordic Committee on Food Analysis; VDIA, Victorian Dairy Industry Authority (Australia); HPB, Health Protection Branch, Compendium of Analytical Methods (Canada).

[a]No validation for raw shellfish in coliform count plates and thermotolerant coliforms for all foods.

could confuse the results, are unlikely to grow. In the case of the Petrifilm yeast/mould (PYM) plate, no tartaric acid or antibiotic addition is required but interpretation problems may arise in the identification and counts of yeast and mould colonies.

As far as the economic aspect is concerned, and compared to the traditional plate count techniques, Petrifilm plate methods can be considered to have an advantage as a greater number of samples can be processed, thus increasing the work capacity of the laboratory and reducing labour costs per sample or test.

In our opinion, the main limitation of this system in comparison with other rapid methods is that it does not significantly shorten the time required for each assay to be completed because of the incubation period necessary for each microbiological criterion. However, a recent product, Petrifilm series 2000 rapid coliform count (RCC) plate gives provisional results in 6–14 h and confirmation in 8–24 h.

Comparing Petrifilm plate methods with other modern techniques, it was ascertained the Petrifilm *E. coli*/coliform count plate method is as good as or better than the AOAC most probable number (MPN) method for the detection of *E. coli* in inoculated cheese, vegetables and naturally contaminated poultry in 24 h. In addition, the Petrifilm *E. coli*/coliform count plate method was equivalent to the confirmed MPN, VRBA and PCC methods when detecting coliforms in the same food groups. A comparative study has been made of Redigel tests, Petrifilm plate methods, Spiral Plate System, Isogrid and aerobic plate counts (APC) to determine the number of aerobic bacteria in selected foods. The results indicated that all five methods were highly comparable ($r = 0.97$ and higher, with the exception of Petrifilm compared to the Spiral Plate System, which was 0.88) and exhibited a high degree of accuracy and agreement. The four alternative methods were found to provide an accurate aerobic bacteria count of foods compared to the APC method.

Interpretation and Presentation of Results

The interpretation of the Petrifilm plates is based on the identification of the typical colonies considered in each case for each microorganism. The role of the indicators is an important one. For instance, in the Petrifilm plate method for *E. coli*/coliform, thanks to glucuronidase indicator, the colonies considered as *E. coli* can be distinguished from the rest of the coliforms which have also grown in the culture medium. In the

Table 4 Mean microbial counts detected in six food products using Petrifilm plates and conventional methods. From Jordano et al (1995)

Food products	*Microbiological criteria*	*Mean counts[a] recovered (log_{10} cfu g^{-1} or ml^{-1})*	
		Petrifilm plates	*Conventional methods*
Pasteurized milk	Aerobic bacteria	2.69	2.30
(*n* = 5)	Coliforms	1.30	1
	E. coli	ND	ND
	Yeasts and moulds	1.30	1.60
Yoghurt ice cream	Aerobic bacteria	5.17	5.08
(*n* = 5)	Coliforms	ND	ND
	E. coli	ND	ND
	Yeasts and moulds	ND	ND
Eggs	Aerobic bacteria	2.53	ND
(*n* = 5)	Coliforms	ND	ND
	E. coli	ND	ND
	Yeasts and moulds	1.77	2.30
Minced meat	Aerobic bacteria	5.81	5.69
(*n* = 5)	Coliforms	5.60	5.58
	E. coli	ND	ND
	Yeasts and moulds	5.47	5.48
Fresh strawberries	Aerobic bacteria	2.71	2.64
(*n* = 5)	Coliforms	2.35	2.02
	E. coli	ND	ND
	Yeasts and moulds	5.26	5.29
Frozen green beans	Aerobic bacteria	5.34	5.49
(*n* = 5)	Coliforms	2.31	1.55
	E. coli	1	ND
	Yeasts and moulds	2.82	3.86

[a] Comparisons of mean values of the enumeration methods for each microbiological criterion were made from six food products by Kruskal–Wallis one-way analysis by ranks test. In none of the cases was the difference significant ($P > 0.05$).
ND = not detected.
Source: Jordano et al (1995).

Table 5 Statistical comparison in six food products using Petrifilm plates and conventional methods. From Jordano et al (1995)

Measure of performance	*Microbiological criteria*	*Petrifilm plates methods vs conventional methods[a]*
Correlation coefficient	Aerobic bacteria	0.897
	Coliforms	0.861
	Yeasts and moulds	0.981
Slope	Aerobic bacteria	0.599
	Coliforms	1.051
	Yeasts and moulds	0.998

[a] Comparisons of mean values are made of six food products (pasteurized milk, yoghurt ice cream, eggs, minced meat, fresh strawberries and frozen green beans).
Source: adapted from Jordano et al (1995).

case of Petrifilm yeast and mould (PYM), the indicator can also help distinguish between both types of microorganisms, and between the microorganisms themselves, due to the phosphatase reaction. Similarly, the size of the colonies or the colony edges can help us interpret them and, in this case, it is important not to confuse the change of colour of the indicator with the reaction that can be caused by some live cells present in the foods. In each case, it is advisable to consult the guides provided which contain different interpretation examples and hypotheses.

For colony counts interpreted as being positive, a Quebec colony counter or any colony counter with a magnified light source can be used.

The recommended counting range on Petrifilm plates is 15–100 colonies for the Enterobacteriaceae plate, and 15–150 for PCC, *E. coli*/coliform, Petrifilm series 2000 rapid coliform count and yeast and mould plates. For an HSCC, which is especially indicated for a small number of this type of microorganism, a maximum of 150 is considered. For the Petrifilm Aerobic Count plate, the recommended counting range is 25–250.

Samples having higher counts than those cited above can be estimated by determining the average number of colonies in one square (1 cm^2). The researcher will take into account the inoculated area and multiply the average obtained from it. These inoculated areas are the following:

- Petrifilm Enterobacteriaceae count plate, Petrifilm *E. coli*/coliform plate, Petrifilm coliforms plate, Petrifilm series 2000 rapid coliform count plate and

Table 6 Guidelines for interpretation of Petrifilm results

Petrifilm aerobic count plate (**Fig. 5**) 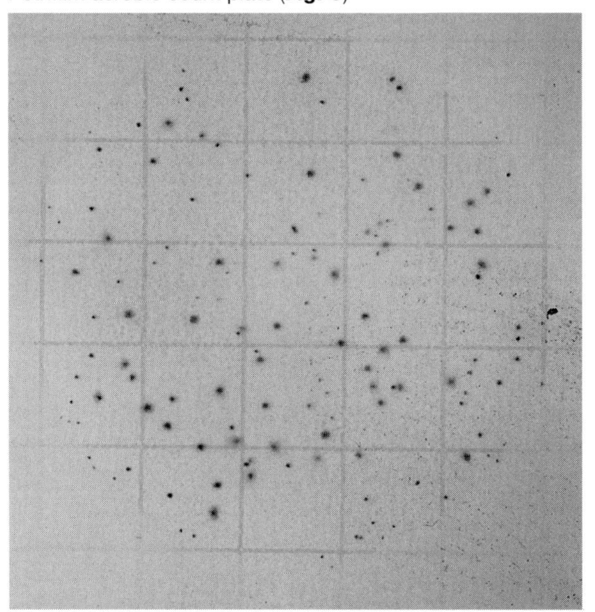 **Figure 5** Colonies on Petrifilm aerobic count plate.	Read all colonies coloured red by an indicator
Petrifilm aerobic count plate (for lactic acid bacteria: see list of commercial products available)	Colonies appear red to reddish-brown in colour and may be associated with a gas bubble
Petrifilm Enterobacteriaceae count plate (**Fig. 6**) 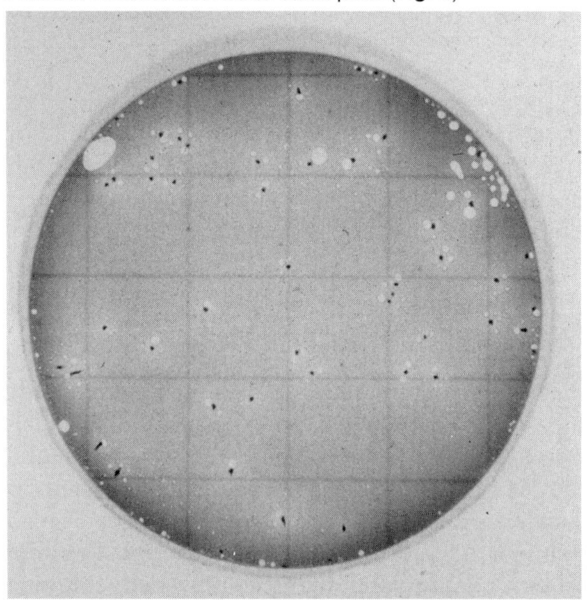 **Figure 6** Colonies on Petrifilm Enterobacteriaceae.	Production of acid and gas can be observed, enhanced by an indicator which turns yellow and the top film that traps the gas produced by substratum fermentation. Red colonies with a yellow halo, or with gas or red colonies are numerated as Enterobacteriaceae
Petrifilm coliform count plate	Colonies are red depending on their ability to reduce the tetrazolium indicator dye. The top film traps the gas produced by lactose fermentation

Petrifilm aerobic count plate: 20 cm^2

- Petrifilm yeasts and moulds count plate: 30 cm^2
- Petrifilm HSCC: 60 cm^2.

Petrifilm plates are also reliable for monitoring the environment. Obviously, if surface control is carried out (direct contact procedure), results will be expressed as counts per 20 cm^2 for aerobic count plates, *E. coli* plates and coliform plates and counts per 30 cm^2 for yeasts and moulds. If an air-sampling method is performed, results will be expressed as counts per

Petrifilm *E. coli*/coliform count plate (**Fig. 7**)	Colonies have a blue precipitate around them thanks to a glucuronidase indicator

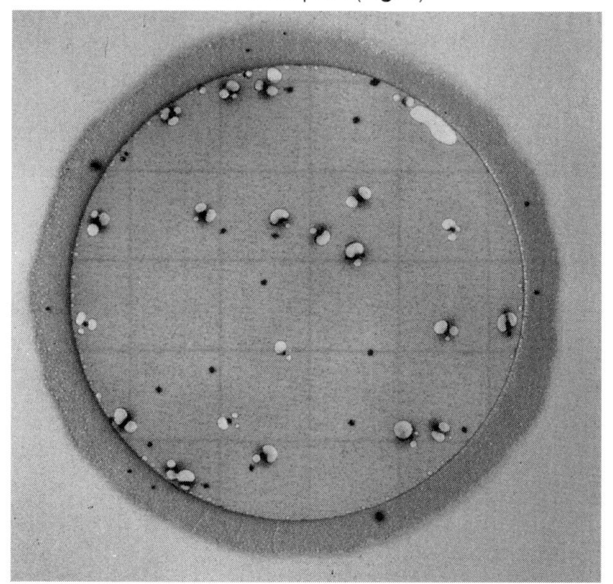

Figure 7 Colonies on Petrifilm *Escherichia coli*.

Petrifilm yeast and mould count plate (**Fig. 8**)	A very different type of colony. Yeast colonies are small, with defined edges and generally blue. Mould colonies are of a variable aspect, generally large, with diffuse edges and a focus in the centre of the colony

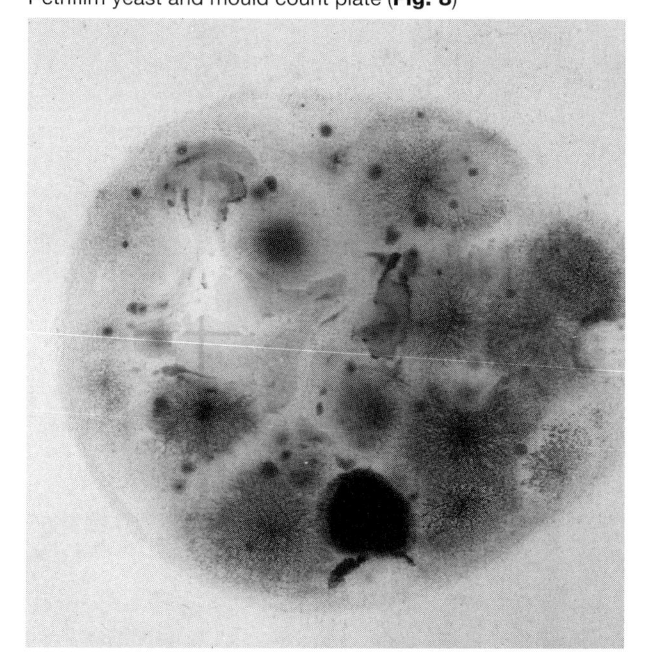

Figure 8 Colonies on Petrifilm yeast/mould.

Petrifilm high-sensitivity coliform count plate	Identical to Petrifilm coliform count plate, detecting small numbers of this type of microorganism

40 cm^2 for aerobic count plates, *E. coli* plates and coliform plates. Yeast, and mould, results will be expressed as counts per 60 cm^2 because a double surface is exposed to the air for 10–15 min.

Where may small colonies or gas bubbles develop, the count is expressed as TNTC (too numerous to count). This can also be indicated when the gel thins or sometimes changes colour. Many colonies close to the edges of the growth area should be considered TNTC.

Interpretation of the results in each case is carried out as shown in **Table 6**.

Available Commercial Products

Dry rehydratable films are marketed by 3M, and at April 1999 the following products are available:

- Petrifilm aerobic count plate: boxes of 100 and 1000 units

Petrifilm series 2000 rapid coliform count plate (**Figs 9 and 10**)

Colonies are red (with or without gas). Acid production is detected from the start of the formation of the colony with a yellow zone thanks to a very sensitive pH indicator. Provisional results are available at 6–14 h

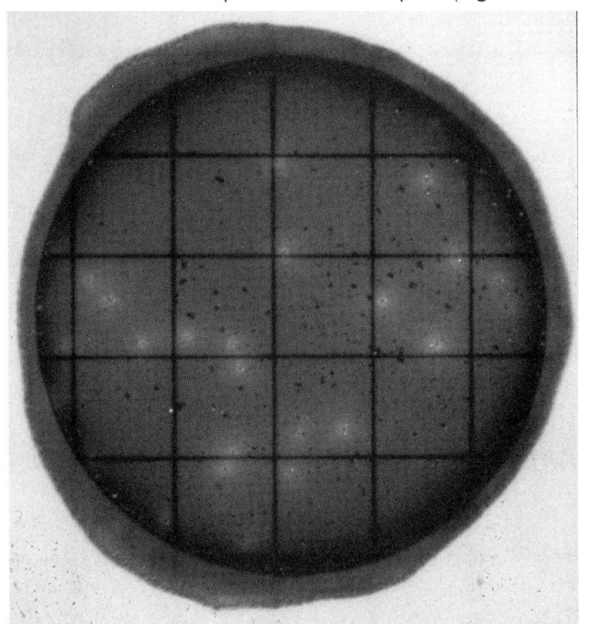

Figure 9 Colonies on Petrifilm series 2000 rapid coliform count plate. Results at 18 h.

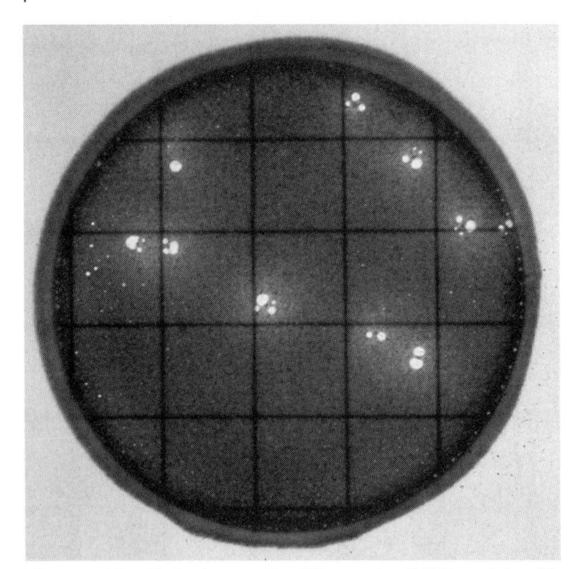

Figure 10 Colonies on Petrifilm series 2000 rapid coliform count plate. Results at 24 h.

- Petrifilm Enterobacteriaceae count plate: boxes of 50 and 1000 units
- Petrifilm coliform count plate: boxes of 50 and 1000 units
- Petrifilm *E. coli*/coliform count plate: boxes of 50 and 500 units
- Petrifilm yeast and mould count plate: boxes of 100 and 1000 units
- Petrifilm high-sensitivity coliform count plate: boxes of 50 and 500 units
- Petrifilm series 2000 rapid coliform count plate: boxes of 50 and 500 units.

The Petrifilm kit HEC (detection of enterohaemorrhagic *E. coli*) was previously available but has been withdrawn. The Petrifilm aerobic count plate with a De Man, Rogosa and Sharpe (MRS) broth diluent in combination with anaerobic incubation can be used to enhance the growth of homo- and heterofermentative lactic acid bacteria in processed meats and high-acid products.

The boxes are provided with a plastic spreader to assist in applying pressure in order to spread the inoculum over the culture media area.

Once a box has been opened, it should be sealed with tape and stored at room temperature and at less than 50% relative humidity. The remaining plates should be used within 1 month.

Acknowledgements

We wish to express our appreciation to Marie-Pierre Copin, Muriel Lienhardt, Emmanuelle Brunot and 3M Microbiology Products, Europe, for their willing collaboration, providing information and figures. We also thank Inmaculada Raventós and Carlos Sancho, Bioser, Spain, for their kind co-operation.

See also: **Enterobacteriaceae, Coliforms and *E. coli***: Classical and Modern Methods for Detection/Enumeration. ***Escherichia coli***: *Escherichia coli*; ***Escherichia coli* O157**. *Escherichia coli O157:H7*. **Fungi**: Food-borne Fungi – Estimation by Classical Cultural Techniques. **Process Hygiene**: Risk and Control of Aerial Contamination; Testing of Disinfectants. **Rapid Methods for Food Hygiene Inspection**. **Total Viable Counts**: Specific Techniques. **Water Quality Assessment**: Modern Microbiological Techniques.

Further Reading

Anonymous (1997) *Petrifilm Products Information*. Malakoff, France: Laboratories 3M Santé. Microbiology Products, Europe.

Beuchat LR, Nail BV, Brackett RE and Fox TL (1990) Evaluation of a culture film (Petrifilm YM) method for enumerating yeasts and molds in selected dairy and high-acid foods. *J. Food Prot.* 53: 869–874.

Beuchat LR, Nail BV, Brackett RE and Fox TL (1991) Comparison of the Petrifilm yeast and mold culture film method to conventional methods for enumerating yeasts and molds in foods. *J. Food Prot.* 54: 443–447.

Calicchia ML, Reger JD, Wang CIN and Osato DW (1994) Direct enumeration of *Escherichia coli* O 157: H7 from Petrifilm EC count plates using the Petrifilm test kit – HEC without sample pre-enrichment. *J. Food Prot.* 57: 859–864.

Chain VS and Fung DYC (1991) Comparison of Redigel, Petrifilm, spiral plate system, Isogrid, and aerobic plate count for determining the numbers of aerobic bacteria in selected foods. *J. Food Prot.* 54: 208–211.

Ginn RE, Packard WS and Fox TL (1984) Evaluation of the 3M dry medium culture plate (Petrifilm SM) method for determining numbers of bacteria in raw milk. *J. Food Prot.* 47: 753–755.

Guthrie JA, Dunlop KJ and Saunders GA (1994) Use of Petrifilm 3M to assess coliform numbers on lamb carcasses. *J. Food Prot.* 57: 924–927.

Jordano R, López C, Rodríguez V, Córdoba G, Medina L and Barrios MJ (1995) Comparison of Petrifilm method to conventional methods for enumerating aerobic bacteria, coliforms, *Escherichia coli* and yeasts and molds in foods. *Acta Microbiol. Immunol. Hung.* 42: 255–259.

McAllister JS, Stadtherr MP and Fox TL (1988) Evaluation of the 3M Petrifilm culture method for enumerating aerobic flora and coliforms in poultry processing facilities. *J. Food Prot.* 51: 658–659, 662.

Matner RR, Fox TL, Mciver DE and Curiale MS (1990) Efficacy of Petrifilm *E. coli* count plates for *E. coli* and coliform enumeration. *J. Food Prot.* 53: 145–150.

Okrend AJG, Rose BR and Matner R (1990) An improved screening method for the detection and isolation of *Escherichia coli* O157:H7 from meat, incorporating the 3M Petrifilm test kit – HEC – for hemorrhagic *Escherichia coli* O157:H7. *J. Food Prot.* 53: 936–940.

Restaino L and Lyon RH (1987) Efficacy of Petrifilm VRB for enumerating coliforms and *Escherichia coli* from frozen raw beef. *J. Food Prot.* 50: 1017–1022.

Senyk GF, Kozlowski SM, Noar PS, Shipe WF and Bandler DK (1987) Comparison of dry culture medium and conventional plating techniques for enumeration of bacteria in pasteurized fluid milk. *J. Dairy Sci.* 70: 1152–1158.

Smith LB, Zottola EA, Fox TL and Chausse K (1989) Use of Petrifilm to evaluate the microflora of frozen dessert mixes. *J. Food Prot.* 52: 549–551.

Vlaemynck GM (1994) Comparison of Petrifilm and plate count methods for enumerating molds and yeasts in cheese and yogurt. *J. Food Prot.* 57: 913–914.

Phages *see* **Bacteriophage-based Techniques for Detection of Food-borne Pathogens**; **Minimal Methods of Processing**: Potential Use of Phages and/or Lysins.

PHYCOTOXINS

Arun Sharma, Food Technology Division, Bhabha Atomic Research Centre, Mumbai, India

Phycotoxins are the toxins produced by algae or seaweeds (Greek phucos means seaweed and toxin means an organic poisonous substance). Algae are phototrophic eukaryotic microorganisms. A large part of the marine and aquatic flora comprises algae. The larger algae are also called macroalgae or seaweeds. However, the microscopic algae or microalgae dominate the marine environment. A group of marine algae, called dinoflagellates, are notorious for the production of certain potent toxic compounds. Whether the ability to produce toxins by algae and cyanobacteria is acquired during the course of evolution to avoid predation is not very clear. The toxins produced by algae are also known as phycotoxins.

Food poisoning caused by the consumption of certain fish species has been observed worldwide. Many of these fish are inherently non-poisonous but become poisonous after feeding on poisonous algae. Thus, humans may suffer poisoning either due to consumption of herbivorous fish and shellfish that feed on the toxigenic algae or through the consumption of those carnivorous fish that feed on these herbivorous fish. Human and animal algal poisoning can also occur by direct consumption of poisonous algae that contaminate drinking water and water used for the preparation of food and feed.

Toxins Produced by Dinoflagellates

Dinoflagellates are unicellular flagellated algae belonging to the Pyrrophyta. Their cells contain chlorophylls *a* and *c*. They occur in both freshwater and marine habitats. A typical representative is *Gonyaulax*. The carbon reserve material in Pyrrophyta is starch and the cell wall contains cellulose. Dinoflagellates produce two types of toxins. One causes gastrointestinal problems (diarrhetic poisoning) and the other causes respiratory paralysis (paralytic poisoning). Most cases of seafood poisoning include both gastrointestinal and neurological symptoms.

Paralytic Poisoning

The symptoms of paralytic poisoning involve tingling and prickly sensations around the face, fingers and toes; in extreme cases, there is muscular paralysis and respiratory difficulty that can lead to paralysis of skeletal muscles, respiratory paralysis and death. The onset of symptoms may start within a few minutes to a few hours of the consumption of contaminated seafood.

Saxitoxin Saxitoxin is one of the major causes of paralytic poisoning occurring after consumption of seafood such as mussels, clams or oysters. Saxitoxin is a neurotoxin named after the butter clam (*Saxidomus giganteus*). The toxin is produced by dinoflagellates such as *Protogonyaulax* sp., *Gymnodium catenatum*, *Alexandrium catenella* and *Alexandrium minutum*. Mussels, clams and oysters feed on these dinoflagellates or red algae with which they may be associated and thus become toxic. The carnivorous fish that feed on these organisms will also become toxic. Thus human consumption of seafoods harvested from areas where these dinoflagellates thrive in abundance, i.e. algal blooms, can lead to outbreak of paralytic poisoning. Saxitoxin is a sodium channel blocker. All paralytic shellfish toxins contain a guanidino group. The positively charged guanidino group interacts with a negatively charged carboxyl group at the mouth of the sodium channel on the extracellular side of the plasma membrane of nerve and muscle cells. Blocking of the sodium ion transport through nerve and muscle cell membranes results in paralysis. Saxitoxin is also produced by a genus of cyanobacteria *Aphanizomenon*.

Gonyautoxin Another toxin, related to saxitoxin, involved in paralytic shellfish poisoning is sulphocarbamoyl gonyautoxin. This toxin is also produced by *Gymnodium catenatum* and *Gonyaulax catenella*, now renamed *Alexandrium*. *Alexandrium* is one of the important species of toxic marine dinoflagellates responsible for reported shellfish poisoning from Australia and America. Shellfish that are generally safe to eat may acquire this poison as a result of the presence of this organism in the environment. After consuming the algal mass the shellfish may remain toxic for 1–3 weeks. Extracts of *Gonyaulax catenella* have been found to cause toxicity in mice. The shellfish escape poisoning as the algal toxin is bound by the hepatopancreas from where it is gradually excreted. The regions of the world where paralytic shellfish poisoning has occurred include areas around the North Sea, the North Atlantic coast of America, the North Pacific coast of America, the coastal area of Japan, South Africa and New Zealand. Gonyautoxin is a derivative of saxitoxin and, therefore, like saxitoxin it blocks sodium influx into the nerve and muscle cells. The symptoms of poisoning are similar to those of saxitoxin.

Diarrhetic Poisoning

Diarrhetic poisoning results in nausea, abdominal pain and discomfort, followed by diarrhoea after consumption of contaminated shellfish and other seafood. A number of toxins may cause diarrhetic poisoning.

Okadaic Acid Polyether compounds such as okadaic acid and dinophysistoxin-1 produced by the dinoflagellates *Dinophysis acuminata* and *Dinophysis fortii* may be responsible for diarrhetic shellfish poisoning. The structures of okadaic acid and dinophysistoxin are quite similar. Okadaic acid is named after the sponge *Halichondria okadai* from which it was first isolated, although the source of the toxin was probably the associated dinoflagellate *Prorocentrum lima*. These toxic compounds cause abdominal pain, nausea, vomiting and diarrhoea. Recent evidence suggests these compounds have a tumour promoting activity. These toxins are powerful inhibitors of certain protein phosphatases.

Neurotoxic Poisoning

Blooms of a halophilic dinoflagellate *Ptychodiscus brevis* are reported to be the cause of a neurotoxic factor in shellfish. Consumption of the toxic fish can cause tingling of facial muscles, cold and hot sensory reversal, bradycardia, dilation of pupils and a feeling of inebriation.

Ciguatera Poisoning

Ciguatera poisoning is caused by eating certain tropical fish. The ciguatera syndrome has been documented since the sixteenth century and is probably a leading cause of morbidity in tropical regions. About 300 species of fish and shellfish inhabiting shallow waters are known to cause ciguatera. These include fin fish such as reef and island fish, grouper and surgeon fish. The poisoning is a serious food-borne disease in some of the island nations in the Caribbean and the Pacific where fish form a major proportion of the human diet. It may also have economic and legal implications for the hotel and food industries dealing in tropical fish. The name ciguatera is derived from the Spanish name for a marine snail.

These fish feed on a marine macroalgae on which some dinoflagellates such as *Diplopsalis* sp. and *Gambierdiscus toxicus* are found as epiphytes. These dinoflagellates have been found in Hawaii, the Caribbean, areas of the Pacific and Australia. *Diplopsalis* sp. and *Gambierdiscus toxicus* have been identified as the source of ciguatoxin. The toxin is concentrated in the liver and viscera of the coral reef fish. Once contaminated, the toxic fish lose their toxicity rather slowly. The onset of poisoning takes 3–4 h after consumption of the toxic fish. Initially there are gastrointestinal symptoms, such as nausea, vomiting, diarrhoea and abdominal pain. These are followed by neurological symptoms including muscular paralysis starting with numbness, tingling of face, convulsions, respiratory arrest and may even lead to death. Skin rashes may develop after a few days. Cardiovascular symptoms including bradycardia, tachycardia, arrythmias and hypotension have also been noted in a number of cases.

Ciguatoxin is a polycyclic ether with extremely high toxicity (LD_{50} 0.045 mg kg^{-1} in mice). Its molecular weight is 1111. It is reported to act at the molecular level on voltage-dependent sodium channels and to increase the permeability of excitable membranes to sodium.

Gambierdiscus toxicus also produces another toxin called maitotoxin, which causes nausea and neurological deficits. Neurological dysfunction includes the reversal of the sensations of hot and cold, called dry ice sensation. Relapse may occur but death is rare.

Toxigenic Yellow-brown Algae

Brackish water ponds and estuarine water may also harbour toxic yellow-brown algae belonging to the Chrysophyta, also called phytoflagellates or diatoms. These are unicellular algae containing chlorophylls *a*, *c* and *e*. *Navicula* is the typical genus. The carbon reserve in the Chrysophyta is lipid and the cell walls contain components made of silica. Chrysophyta occur in soil, freshwater and marine environments. *Prymnesium parvum*, a common yellow-brown alga found in brackish water, is involved in toxicity of fish. It produces a toxin that inhibits transfer of oxygen across the gill membranes and has been a great problem in commercial farms in Israel. The toxin is a non-dialysable, thermolabile, saponin-like compound that is a potent haemolytic agent.

Domoic Acid Domoic acid is responsible for amnesic shellfish poisoning. The source of this toxin is a marine diatom *Nitzchia pungens*. The poisoning is caused by the consumption of contaminated shellfish such as mussels, scallops and clams. The toxin can cause gastrointestinal and neurological disorders including headache, dizziness, vomiting, diarrhoea, difficulty in breathing and coma. Short-term memory loss or amnesia is the characteristic symptom of domoic acid poisoning. Domoic acid is a potent agonist at receptors for excitatory amino acids, such as glutamic and kainic acids.

Toxigenic Blue-green Algae (Cyanobacteria)

The blue green algae or cyanobacteria are a very ancient and diverse group of microorganisms. Cyanobacteria are prokaryotic oxygenic phototrophs that contain chlorophyll *a* and phycobillins but not chlorophyll *b*. Although cyanobacteria do not belong to the algae, they have been commonly called blue-green algae. Therefore, traditionally the toxins produced by this group of microorganisms have been discussed under algal toxins. This may also be due to the fact that like dinoflagellates their toxins have been involved in food poisoning caused by the consumption of fish and other seafoods.

Like dinoflagellates many of the cyanobacteria produce compounds that are toxic to humans and animals. Toxicity of blue-green algae has been reported in livestock from a number of countries. The earliest reports involve *Nodularia spumigena*, which thrives in brackish water. The cause of death is reported to be hepatotoxicity and liver failure. A cyclic pentapeptide of molecular weight 824 is reported to be the cause of toxicity. The toxin has been named nodularin after the name of the producer organism. It is resistant to boiling at neutral pH, and has LD_{50} of the order of $70\,\mu g\,kg^{-1}$ in mice by intraperitoneal injection.

Another blue-green alga involved in poisoning of livestock from freshwater lakes is *Microcystis aeruginosa*. This produces a water-soluble cyclic polypeptide, microcystin. The compound is quite similar to nodularin. The compound is also hepatotoxic to many animals.

Some cyanobacteria also synthesize and accumulate certain alkaloidal toxins. The common species producing such toxins include *Anabaena circinalis* and *Anabaena flox-aquae*. These cyanobacteria are also found in shallow freshwater lakes and are toxic to many animals. *Anabaena circinalis* produces neurotoxins. The heat-stable alkaloid anatoxin *a* has been found to be a blocking agent for post-synaptic neuromuscular transmission. The clinical symptoms in test animals include leaping movements, abdominal breathing and convulsions occurring within a few minutes of an intraperitoneal injection. Another toxin from *Anabaena circinalis* is called anatoxin-*a*(s). This is a naturally occurring organophosphate neurotoxin related to guanidine. It is an irreversible cholinesterase inhibitor. The symptoms of toxicity include lacrimation, salivation, urination and diarrhoea. *Anabaena flox-aquae* produces a toxic substance called very fast death factor. It is toxic to many animals and is more potent than *Microcystis aeruginosa* toxin. Another cyanobacterium *Aphanizomenon flos-aquae* is known to produce saxitoxin, the cause of paralytic shellfish poisoning.

Several species of *Caulpera* found in the Western Pacific Ocean cause dizziness in humans. *Schizothrix calcicola*, found in the Pacific Ocean, causes poisoning similar to ciguatoxin.

Toxins of Bacterial Origin Associated with Algae

There are fishes that become poisonous due to the presence of particular bacteria in their body. These are tetrodons or puffer fishes. Though the toxin in these fishes is of bacterial origin, the bacteria responsible for production of these toxins may be associated with algae on which the fish feed.

Tetrodotoxin Tetrodotoxin is present in pufferfish also called blowfish, fugu or sea squab. Tetrodotoxin is actually produced by marine bacteria as well as by intestinal microflora of tetrodotoxin-producing fish. The bacteria include *Vibrio alginolyticus*, *V. damsela*, *Staphylococcus* sp., *Bacillus* sp., *Pseudomonas* sp., which lead a parasitic or symbiotic existence on puffer fish. These bacteria have been found to be epiphytic on the species of calcareous algae, *Jania* sp. and *Alteromonas* sp., on which these fish normally feed. Tetrodotoxin blocks the sodium pump. It binds to the sodium channel in nerve cells and blocks the propagation of nerve impulses. This causes elimination of electrical differential created by the influx of sodium and efflux of potassium ions. The onset of poisoning could be as soon as 10–45 min after consumption. It involves nausea, vomiting and diarrhoea followed by dizziness, tingling of lips and extremities, paralysis, respiratory arrest and death. The toxin can be avoided by thorough evisceration of the fish before consumption.

Controlling Phycotoxin Poisoning

Shellfish toxicity can be prevented only by effective management. The waters containing dinoflagellate blooms need to be identified and the propagation of algae in water bodies should be controlled. Release of effluents containing nutrients such as nitrogen and phosphates results in eutrification of water bodies, which also encourages the formation of toxic algal blooms. Soluble nitrogen and phosphates should therefore be removed from effluents before release into water bodies. The toxin content of the fish harvested from notified areas should be regularly monitored. While preparing fish for human consumption adequate care should be taken to eviscerate fish and particularly remove those organs that are known to concentrate phycotoxins.

See also: **Fish**: Spoilage of Fish. **Food Poisoning Outbreaks. Shellfish (Molluscs and Crustacea)**: Contamination and Spoilage.

Further Reading

Bidaud JN, Hank PM, Vijerberg C et al (1984) Ciguatoxin is a novel type of sodium channel toxin. *Journal of Biological Chemistry* 259: 8353.

Capra MF and Cameron J (1992) Ciguatera Poisoning. In: Watters D, Lavin M, Maguire D and Pearn J (eds) *Toxins and Targets*. Chur: Harwood Academic.

Falconer IR (1992) Poisoning by Blue Green Algae. In: Watters D, Lavin M, Maguire D and Pearn J (eds) *Toxins and Targets*. Chur: Harwood Academic.

Hashimoto Y (1987) *Marine Toxins and Other Bioactive Marine Metabolites*. Tokyo: Japan Scientific Societies Press.

Kelly GJ and Hallegraeff GM (1992) Dinoflagellate toxins in Australian shellfish. In: Watters D, Lavin M, Maguire D and Pearn J (eds) *Toxins and Targets*. Chur: Harwood Academic.

Kuenstner S (1991) *Seafood and Health Risks and Prevention of Seafood Borne Illnesses*. Boston: New England Fisheries Development Association.

Murata M, Legrand A-M, Ishibashi Y and Yasumoto T (1989) Structure of ciguatoxin and its congener. *Journal of the American Chemical Society* III: 8929.

Schantz EJ (1974) Shellfish, Fish, and Algae. In: Liener IE (ed.) *Toxic Constituents of Animal Foodstuffs*. London: Academic Press.

Yasumoto T and Yotsu M (1984) Biogenetic origin and natural analogs of tetrodotoxin. In: Keeler RF and Madeva NB (eds) *Natural Toxins: Toxicology, Chemistry and Safety*. Fort Collins: Alaken, Inc.

Phylogenetic Approach to Bacterial Classification *see* **Bacteria**: Classification of the Bacteria – Phylogenetic Approach.

PHYSICAL REMOVAL OF MICROFLORAS

Contents

Filtration

Patrick Boyaval, INRA, Laboratoire de Recherches de Technologie Laitière, Rennes, France

The main goal of a filtration operation is the separation of a continuous phase (liquid or gaseous) from a discontinuous phase (solid or liquid) initially mixed with it. The physical removal of microorganisms by filtration can be employed to recover the solid discontinuous phase (the cells) to produce, for example, concentrated starters, or to rid the continuous phase of microbial particles (air filtration, liquid product cold sterilization). In either case, separation of the two phases enables an upgrading of the product and facilitates downstream processing. Consequently, filtration is conducted early in the recovery or removal process. In general, media filtration is more expensive than thermal methods, but it can be used for heat-sensitive products (the cells or the continuous phase) or for fluids with low boiling points which cannot be thermally sterilized.

Principles of Filtration

Let us consider a particle-containing fluid passing through a filter. The pore dimensions of the filter are smaller than the average size of the particles, which then will form a cake at the filter's area, A. The pores are considered as small cylinders of length l. The velocity v of the fluid (laminar flow rate) in the pore is given by Poiseuille's law:

$$v = \frac{d^2}{32\eta} \cdot \frac{\Delta p}{\Delta l}$$

where d is the diameter of the pore, η is the fluid viscosity and $\Delta p/\Delta l$ is the pressure drop per length unit (with p_1 = upstream pressure and p_2 = downstream pressure of the filter, $\Delta p = p_1 - p_2$). Moreover, flow rate through a pore p (dV_p/dt) is given by:

$$\frac{dV_p}{dt} = \frac{\Pi d^2}{4} \cdot v$$

If the filter possesses n pores per surface area, then the instantaneous flow rate will be:

$$\frac{dV}{dt} = A \cdot n \frac{\Pi d^2}{4} \cdot \frac{d^2}{32\eta} \cdot \frac{p_1 - p_2}{l}$$

$$= k\frac{A \cdot n \cdot d^4 (p_1 - p_2)}{\eta \cdot l}$$

and if $R_s = \dfrac{1}{k \cdot n \cdot d^4}$ then

$$\frac{dV}{dt} = \frac{A(p_1 - p_2)}{\eta \cdot R_s \cdot l}$$

R_s is a characteristic of the filter. It increases when the resistance of the filter to the filtration flux increases. An important consideration in cell filtration is cake formation: because the cell deposit at the surface acts as a filter itself, R_s is the sum of the R_{s1} of the filter (considered to be constant during the operation) and the R_{s2} of the cell deposit, which increases during filtration as the thickness of the cake increases. Very often R_{s2} equals or exceeds R_{s1} during long periods of filtration. When the cake thickness increases or when the depth filter has retained a certain mass of solids, the energy required for filtration must be increased. A positive pressure must be applied between the upstream and the downstream sides of the filter. It increases the cost of pumping, and depends on the mechanical resistance of the filter.

This important problem of decreasing permeate flux rate may be overcome by:

- increasing the area of the filter
- discarding the cake.

Numerous techniques are used to remove the cells from the filter surface, e.g. gravity, a fixed knife on the rotating filters, back-flush of gas or permeate, and manual cleaning. An interesting approach to this problem is the use of tangential flow (cross-flow) filtration. The fluid containing the particles is pumped tangentially to the filter surface (**Fig. 1**). The erosion caused by the fluid velocity at the filter's surface counteracts the particle convection which tends to build a deposit. Among the different membrane filtration processes (**Fig. 2**) the most interesting for microflora removal is microfiltration (membrane pore size range 0.1–10 μm). Although microfiltration is not a new process, it attracted renewed industrial interest in the 1980s with the development of membranes with improved physicochemical properties. These inorganic membranes (aluminium oxide, titanium oxide or zirconium oxide over an agglomerated carbon or an alumina support) offered new opportunities for cell processing because of their stability to heat, acid and alkali, which rendered them steam sterilizable.

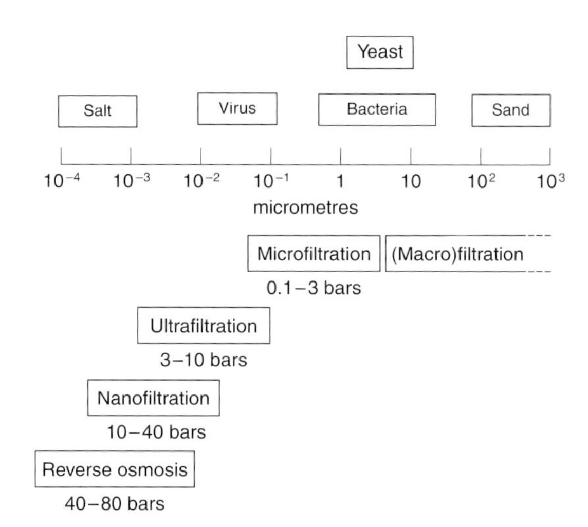

Figure 2 Pressure-driven membrane processes (1 bar = 10^5 Pa).

Factors Affecting Performance

For efficient removal of microorganisms by filtration, it is important to focus both on the physical characteristics of the cells and on the specificities and mode of operation of the filters.

Microorganism Characteristics The size of the cells is one of the main parameters to consider. Easiest to remove by filtration are mycelial organisms which are several millimetres in size. Yeast cells (typically 2–6 μm), bacteria (0.1–3 μm) and viruses (0.03–0.1 μm) are more difficult to remove. Many other parameters may be involved in the efficiency of the microflora removal: the cell concentration in the fluid; the morphology of the cells (rods, cocci, cells in 'Chinese characters'); plasticity and compressibility of the cell bodies, allowing leakage of cells through spaces that are smaller than the smallest dimension of the organism (especially under high pressure); global electric charge presented by the cells to the surrounding

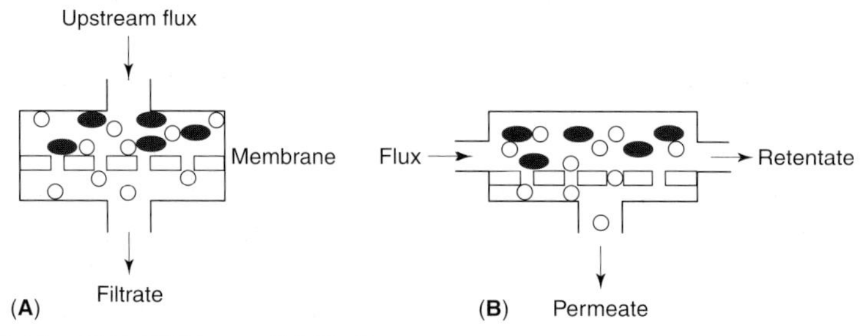

Figure 1 Principle of (**A**) dead-end and (**B**) tangential filtration.

medium; presence of extracellular polysaccharide (slime) which can stick the cells to the filter; the cell mode of association (length of the chains) which dramatically increases the apparent size of the particle; toxicity of the microflora, etc. The characteristics of the surrounding medium are also of prime importance: temperature; viscosity; Newtonian or non-Newtonian behaviour of the broth; pH; presence of particles other than the microbial cells of interest; toxicity; chemical aggressivity; volatility, etc.

Filter Characteristics The type of equipment selected for the filtration of media or gas depends on the desired flow rate, cost of the considered products and materials and on the 'expected characteristics' of the filtered product and/or retained material (complete cell removal is not always desired). The filters most commonly used are of two types: depth filters and absolute or screen filters.

Depth Filters The suspension of particles passes through a porous or fibrous mass made of packed materials such as fibreglass, asbestos fibre mats or cotton. These filters work on the high probability of collision between the microorganisms and the fibres; this probability increases with the length of the filter. This type of filter leads to pressure drops. They must be properly maintained (to avoid bacterial growth in the filter) and used. They fail to perform efficiently for gas filtration when the gas is wet (water or oil droplets) or when gas velocity is too high. This type of filter is not recommended if the cells must be efficiently recovered.

Absolute or Screen Filters The mechanism of filtration is absolute size exclusion: the maximum pore size is less than the minimum size of the particles to be removed. Membrane filters made with polymeric material (including cellulose nitrate, cellulose acetate, vinyl polymers, polyamides and fluorocarbons), having a fixed sub-micrometre pore structure of uniform size distribution, tend to be used in preference to depth filters. Most of them are steam sterilizable and are stable against aqueous solutions and many organic materials. A pleated structure is often used to increase the area of the filter in a small unit.

Mode of Operation of the Filter Pressure of filtration and velocity of the fluid are key factors affecting performance in microfiltration. The flow velocity at the membrane surface induces a high wall shear stress which is essential to avoid fouling phenomena. In order to master this essential operational control, several works focus on the study of transient and stationary operating conditions, of initial cell concentration (C_i), permeation flux (J), and ratio of permeation flux to wall shear stress (τ_w) on the performance of microfiltration of lactic acid bacteria and especially on *Lactobacillus helveticus*.

The effect of the pressure difference applied across the membrane depends on the filtration process considered (see Fig. 2).

Filter Aids In the classical filtration methods, the accumulation of cells on the filter surface decreases the filtration rate, with undesirable economic consequences. The addition of filter aids to the medium may resolve this problem, especially in fermentation broths. The major additives are diatomaceous earths and perlite, used in the fermentation industry. Their use can increase the relative flow rate by more than 20 times. The known binding properties of certain molecules (proteins, antibiotics) on these materials must be examined before scaling up the process, to avoid losses. Moreover, although these filter aids are cheap, they may necessitate expensive effluent treatments, and environmental laws relating to their use are becoming stricter.

Integrity Testing of Membrane Filters The integrity of the filter is of major importance to ensure the quality of filtration. Besides visual examination of the filter surface and the filtrate optical density in classical filtration, bubble point and forward-flow tests are used to ensure that the membrane filter is not defective and that it has been installed correctly. For particular uses, many other laboratory tests could be used to examine the selectivity of the separation – for example, the use of a filtrate sample for a bacterial growth test.

Materials and Systems for Food Uses

Classical Filtration

The type of filters, their geometry, size and chemical nature, their mode of operation, the cleaning procedures and the costs are almost as numerous as food products. Versatility of construction allows specific design of a filter based on the characteristics of the microbes and fluid to be separated, and the size and cost of the process considered.

Tangential Filtration

The choice for the size of membrane pores depends on the manufacturer. Different configurations of filtration membranes include plate and frame, spiral-wound, tubular and hollow-fibre membranes. They are assembled in modules. The term 'module' is used to define the smallest practical unit containing one or more membranes and supporting structures. The different geometries offer specific advantages and disadvantages with regard to filtration area, control of fouling, facility of cleaning, investment and operational costs. In most cases, the final choice of filtration unit will be made only after it has been tested on the real product in the plant, where a complete technical and economic evaluation can be done. Most manufacturers are conscious of the importance of such testing and offer the use of a pilot plant for assays.

Applications in the Food Industry

Cross-flow filtration, especially with ceramic membranes, can be used for most applications of traditional filtration: water potabilization for industrial or drinking uses, including removal of pyrogens, waste water treatment and oil–water separation. Most industries are now using this type of filtration process, e.g. the chemical and mechanical industries, paper and textile manufacturers, the petroleum industry and biotechnology and pharmaceutical industries, but the most heavily involved in the uses and development of filtration processes is the food industry, in which this technology is used chiefly to remove microfloras.

Milk

Raw milk is classically heat-treated to kill the natural microflora which can impair the taste and biochemical characteristics of the product. The time–temperature combination applied to the milk depends on the bactericidal effect expected (pasteurization, thermization, ultra-high temperature treatment, etc.). These treatments are efficient, but the shelf life of lightly treated milk is short, while more highly processed milk may have impaired organoleptic properties and be less desirable for cheese manufacturing. Clearly, there was a need for an efficient treatment that would not impair milk quality. Filtration technology has been in use in the dairy industry since the early 1970s, when the pressure-driven filtration processes, reverse osmosis and ultrafiltration, emerged as the perfect tools for the concentration of whey and standardization of milk proteins. The microbial purification of milk by microfiltration is a major recent advance.

The introduction of microfilters with pore sizes of 0.1–2 μm and high flux rates for the elimination of milk bacteria was proposed in 1986. Fouling problems were overcome by a new hydraulic concept: a recirculation loop of microfiltrate which permits a constant low transmembrane pressure all along a microfiltration ceramic membrane with a highly permeable support and a multichannel geometry in spite of a high retentate recirculation velocity (6–8 $m\,s^{-1}$). Filtration fluxes of 500–700 $l\,h^{-1}\,m^{-2}$ are obtained. The temperature of operation ranges between 35°C and 50°C. The retentate flow extracted (3–5% of the entering flow) contains the bacteria and may be heat-treated for animal feeding or mixed with cream for fat standardization (**Fig. 3**). The retentate-cream mix is blended with the permeate and finally homogenized, pasteurized, cooled to 6°C and packaged. This means that less than 15% of the milk is treated at high temperature, which is why this system has so little effect on the natural taste of the milk. The rate of casein (the main protein of milk) permeation is about 99%, while the rate of bacterial retention is above 99.5%. The average decimal reduction is 2.6 (or more if two microfiltration steps are coupled) and is independent of the initial level of the bacterial population. There is a non-negligible effect of soluble milk components which interact with the membrane support before favouring internal adsorption of microorganisms. Experimentation with various microbiological species or strains indicates that the morphology and cell volume affect the retention by the membrane. The first industrial use of this process was in Sweden to improve the shelf life of pasteurized milk, which was increased from 6–8 days to 16–21 days, with a notable improvement of the milk flavour. Several dozen industrial plants with capacities between 1000 $l\,h^{-1}$ and 20 000 $l\,h^{-1}$ are currently in use, mainly in Europe.

Whey, which is the main by-product of the cheese and casein industries, is clarified by cross-flow microfiltration (0.1 μm) in order to improve flux and hygienic conditions during subsequent ultrafiltration (for the recovery of whey protein concentrates).

Wine

The bottling of wines is one of the most important operations in wine-making. Filtration, together with adjustments of chemical and gas compositions, is a key step before bottling. The continuing trend towards the use of lower levels of chemical additives

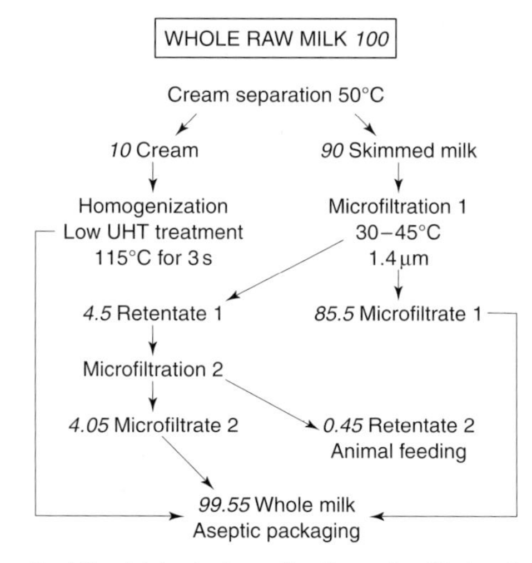

Figure 3 Microbiological purification of milk by tangential microfiltration. Italic figures indicate percentage by volume.

enhances the use of membrane filtration to prevent unwanted microbial action in bottles. Membrane filters are used to exclude not only large particles such as microorganisms but also crystals and partially soluble colloids such as polysaccharides, proteins or tannins. Generally wines are prefiltered with depth filters before membrane filtration. Most of these membranes are made from synthetic polymers (cellulose acetate, cellulose nitrate or polysulphone). Tangential microfiltration is now the most efficient method but it is expensive. Fouling of the filters is caused by colloids from the yeasts or grapes rather than by the microorganisms. The traditional use of cold temperature for that filtration decreases the permeation flux. All the bottling equipment (including the filters) is heat-sterilized after pre-treatment with detergents and sanitizing agents. The control quality tests for sterility in the bottling line include sampling of airborne microbes in the bottling area and of bottle and cork washings.

Beer

In beer manufacture, significant losses occur in fermentation and maturation tank bottoms after beer removal. The recovery of beer from tank bottoms after yeast fermentation with cross-flow microfiltration (ceramic membranes) appears technically feasible and is attractive in economic terms. The process is installed in order to save on polishing and clarifying agents and to reduce effluent problems. Nevertheless, membrane filtration is not effective for total beer filtering because low flux and the retention of protein and aroma compounds renders the process uneconomical.

Clarification of Beverages and Fluids

Tangential filtration is already used in vinegar, cider and fruit juice clarification and sterilization, brine regeneration, gelatin treatment and water purification. The animal blood valorization for foods must begin by complete flora removal which is currently done by tangential filtration with steam sterilizable membranes. The areas of interest in the food industry for complete microflora removal are growing as the industrial and consumer preference for food of high microbial quality increases (long preservation periods and very low risk of pathogenic or spoilage bacteria presence).

Fermentation Industry

The development of biotechnology has increased the use of membrane separation techniques. Major applications include media sterilization, the separation and harvesting of microorganisms and enzymes, the elaboration of continuous high-cell-density bioreactors, and tissue culture reactor systems.

Media Sterilization Media for animal cell multiplication or addition of heat-sensitive nutrients (e.g. vitamins and antibiotics) are mainly sterilized by filtration. However, most industrial fermentation media are heat-treated for technical and economical reasons. The ingredients could exhibit severe fouling problems, leading to losses of nutrients in the medium and decreasing the operating time of the filtration operation; numerous expensive cleaning procedures would then be required. The low cost of most of these industrial nutrients limits the use of filtration for such a purpose, with some exceptions. The production of clean water is also dependent on several filtration steps, some of which are designed for bacteria and virus removal.

Cell Processing Lactic acid bacteria are used extensively in many food fermentations to preserve, to retard spoilage and to improve flavour and texture. Cultures of lactic acid bacteria are being increasingly used in agriculture as inoculants in the preservation of fodder as silage and in probiotic feed supplements. They also produce bioantagonists such as antibiotics and bacteriocins. They find commercial applications in the dairy industry, in preservation of sausages and meats, in pickling vegetables and in preparing fermented beverages.

Physical concentration of starter cultures can be achieved using techniques such as centrifugation, spray-drying, freeze-drying and membrane processes. Greater recovery of biomass is one of the major advantages of filtration technology compared with centrifugation. Moreover, filtration (1) obviates the production of aerosols that may cause allergic reactions among employees; (2) can produce cell-free extracts not usually found in other techniques such as centrifugation; (3) provides a higher production rate unmatched by other techniques; and (4) is economically attractive. Besides, bioreactors coupled with cell recycling membrane process require high cell viability and high permeate flux to remove lactic acid from the fermentation broth in order to obtain increased lactic acid and cell mass production.

Nonetheless, few reports describe the optimization of cross-flow microfiltration (MF) and ultrafiltration (UF) of lactic acid bacteria. Higher MF flux was obtained with a 0.45 μm pore size membrane compared with 0.8 μm and 1.2 μm with a permeate enriched with a mixed culture of *Lactococcus lactis* subspp. *diacetylactis* and *cremoris* and *Leuconostoc mesenteroides* subsp. *citrovorum*. Recent work underlines the relevance of media composition, microbial strain and membrane roughness on cross-flow filtration of cell suspensions.

Available information is scarce about the optimization of operating conditions during UF and/or MF of lactic acid bacteria. A better management of the cell filtration operation is undoubtedly commercially important in improving membrane bioreactor performance.

Continuous High-cell-density Bioreactors Most biological conversion processes are conducted in batch mode. The advantages of such technologies are numerous, particularly simplicity and the low cost of the technology and materials. Nevertheless, there are disadvantages in comparison with continuous fermentation processes: low productivity (long periods for start-up and shut-down), batch-to-batch variations in quality, and high upstream and downstream processing costs. Researchers and manufacturers looking for an efficient continuous process with high volumetric productivity and simplified downstream processing technology therefore developed the concept illustrated in **Figure 4** for the continuous production of propionic acid. A continuous stirred tank reactor is coupled to a microfiltration unit in a closed loop configuration including a recirculation pump to achieve the flow velocity required by the filtration process. The total volume of the system is kept constant by adjusting the incoming new medium flow rate to the permeate flow rate. The membrane is chosen to allow a complete retention of the cells in the loop and to obtain a cell-free permeate which contains the desired biosynthesized molecules (which must be able to pass through the pores of the membrane). The downstream processing is simplified by the sterile nature of the effluent, allowing most separation methods to be used afterwards (electrophoresis, chromatography, etc.). One of the major advantages of this system is the increase of the cell biomass inside the loop. In such bioreactors cell densities as high as $100\,g\,l^{-1}$ (dry weight) can be obtained for bacteria and can reach more than $330\,g\,l^{-1}$ for yeasts. With such cell concentrations the volumetric productivities increase drastically: propionic acid productivity is more than 430 times higher than in the classical batch process.

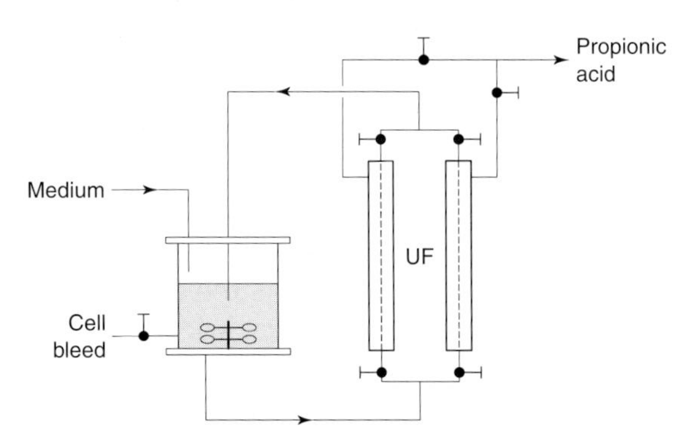

Figure 4 Membrane bioreactor for propionic acid production. UF, ultrafiltration.

Such membrane bioreactors could be used with one or several stages with the same or different microorganisms in each bioreactor. The concept of membrane bioreactors is also widely used with enzymes. The membrane is chosen to retain the enzyme (and very often, the macromolecular substrates such as proteins or starch) in the loop while the products pass through the membrane (in most of the cases, ultrafiltration membranes). Many such processes with microorganisms or enzymes have been developed on an industrial scale worldwide since the 1980s for the production of ethanol, sparkling wine, organic acids and peptides.

Filtration of Air

The filtration of air, employed in numerous industries such as fermentation and electronics, is being more and more used in food industry processes like cooked meals preparation. Filtration is the most practical and economical solution to the problem of removing dust and organisms from enclosed spaces. The quality of the unfiltered air could vary from 1.5×10^7 particles (size $\geqslant 0.1\,\mu m$) per cubic foot ($28.3\,dm^3$) for clean air, to 3×10^8 particles (size $\geqslant 0.1\,\mu m$) per cubic foot for dirty air. To obtain a reduction of 10^8 (to obtain 1–10 particles per cubic foot) filters with a minimum efficiency of 99.999% on particles (size $0.12\,\mu m$) must be used. The filters must have several properties: high retention efficiency, low cost, ease of use and simple cleaning procedure. Particle retention by air filters involves four mechanisms:

- direct interception
- inertial impaction
- diffusional interception
- electrostatic effects.

The quality of the air in 'white chambers' depends on the filtering system and also on the people working there, on the properties of the raw materials used in the room, the shape of the furniture, the quality of the manipulations and on the complete air distribution system. The most widely used classification comes from US Federal Standard 209 published in 1963 and subsequently modified (209A in 1973, etc.). The Class 100 corresponds to no more than 100 particles (size $\geqslant 0.5\,\mu m$) per cubic foot ($28.3\,dm^3$); class 10 000 corresponds to no more than 10 000 particles (size $\geqslant 0.5\,\mu m$) per cubic foot; class 100 000 corresponds to no more than 100 000 particles (size $\geqslant 0.5\,\mu m$) per cubic foot or 700 particles (size $\geqslant 5\,\mu m$) per cubic foot. The number of particles must be measured near to the working areas while the rooms are in use.

Air filters are often used in combination with other methods of eliminating microbial particles in the air: ultraviolet or thermal treatments before filtering could improve the efficiency of particle removal. Drying treatments must be applied to the air before depth filtration.

Filter system maintenance includes the performance of integrity tests at any time. These challenge tests, which have been developed by most of the filter manufacturers, are made with suspensions of uniformly viable microorganisms or phages.

The Future

Filtration technology is increasingly the focus of attention of food engineers. The availability of a wide range of pore sizes and hydraulic design for viscous liquids, together with high resistance and durability, make filter techniques appealing to the industry. Nevertheless, some current disadvantages are a barrier to widespread use: energy consumption and the price per unit area must decrease, and an increase of package density and filtration selectivity (sharp cutoffs) must be obtained.

See also: **Cheese**: Microbiology of Cheese-making and Maturation. **Fermentation (Industrial)**: Basic Considerations; Media for Industrial Fermentations. **Heat Treatment of Foods**: Ultra-high Temperature (UHT) Treatments; Principles of Pasteurization. **Hydrophobic Grid Membrane Filter Techniques (HGMF)**. **Starter Cultures**: Uses in the Food Industry. **Wines**: Microbiology of Wine-making; Specific Aspects of Oenology.

Further Reading

Boulton RB, Singleton VL, Bisson LF and Kunkee RE (1996) *Principles and Practices of Wine Making*. New York: Chapman & Hall.

Boyaval P and Corre C (1995) Production of propionic acid. *Lait* 75: 453–461.

Boyaval P, Lavenant C, Gesan G and Daufin G (1996) Transient and stationary operating conditions on performance of lactic acid bacteria crossflow microfiltration. *Biotechnology Bioengineering* 49: 78–86.

Cheryan M (1997) *Ultrafiltration and Microfiltration Handbook*. Basel: Technomic Publishing.

McGregor CW (ed.) (1968) *Membrane Separations in Biotechnology*. New York: Marcel Dekker.

Daufin G, Rene F and Aimar P (eds) (1998) *Procédés de Séparation sur Membranes dans les Industries Alimentaires*. Paris: Tec-Doc Lavoisier.

Maubois JL (1991) New applications of membrane technology in the dairy industry. *Australian Journal of Dairy Technology*, Nov: 91–95.

Tutunjian RS (1985) Cell separations with hollow fiber membranes. In: Moo Young M (ed.) *Comprehensive Biotechnology*. Vol. 2, p. 367. Oxford: Pergamon.

Centrifugation

Vikram V Mistry, Dairy Science Department, South Dakota State University, Brookings, USA

Spoilage of liquid food products is caused by many factors. Bacteria may play a significant role in such spoilage, and may also be detrimental to the quality of products (e.g. cheeses) that are manufactured from liquid foods such as milk containing these bacteria. It is a common practice to pasteurize fluids such as milk and juices to kill bacteria with the aim of improving shelf life and making the products safe for consumption without loss of flavour quality. Pasteurization kills over 99% of non-spore-forming bacteria, but most spore-formers are difficult to kill. Aerobic spore-formers such as *Bacillus cereus* and anaerobic spore-formers such as *Clostridium tyrobutyricum* are of particular concern in the dairy industry. Only very high temperatures (120°C for 15 min or equivalent) will damage or kill them. Such temperatures are not desirable for product processing because of their adverse effect on product quality. Milk treated with such high temperatures has a marked 'cooked' flavour. Temperature-dependent effects on proteins such as denaturation make highly heated milk undesirable for cheese-making. It is important, therefore, to be able to remove or kill bacteria and spores by employing processes that do not require large amounts of heat, such as centrifugation. These bacteria-removing processes are not only useful for improving the quality of milk, they are also employed in the starter culture industry for harvesting starter bacteria in large numbers for the production of cultures for fermented foods.

Background

Interest in the physical removal of bacteria from milk can probably be traced back to early Italian cheesemakers in the manufacturing of Asiago, Parmigiano-Reggiano and Grana Padano cheeses. As these cheeses are manufactured from raw milk and aged for extended periods (over 2 years in some cases), it is important that the milk be of excellent quality. The bacterial count of milk is lowered by allowing the milk to sit at room temperature in shallow trays for 8–10 h. During this period the fat rises to the surface, carrying with it bacterial cells. The low-bacteria milk is then decanted and used for cheese-making. This gravity separation of raw milk, a standard practice for these cheeses even today, lowers the total bacterial count by 90–95% of that in raw milk. Such a system would not be practical for large modern dairies, which commonly handle over a million kilograms of milk a day.

Breed reported in 1926 on the scientific comparison of gravity versus centrifugal separation of cream on bacterial cell numbers. It was observed that cream that was separated by gravity carried with it large numbers of bacteria. Subsequent work has demonstrated that the spores of some bacteria such as *Clostridium tyrobutyricum* attach to the fat globules and rise to the surface with cream.

The application of centrifugal force to remove bacteria from milk was first reported by Professor Simonart of Belgium in 1953. In his studies centrifugation of milk at 9000 *g* and at temperatures of 65–75°C removed more than 99% of the bacteria. This procedure, originally developed to improve the keeping quality of fluid milk, has since been perfected. The term 'bactofuge' is a trade name for the centrifuge manufactured by the Tetra Pak Company (Lund, Sweden) for removing bacteria. Other companies such as Westfalia (a division of GEA, Bochum, Germany) also manufacture such centrifuges and refer to them as bacteria-removing centrifuges. Commercial units with capacities of 15 000–25 000 l h^{-1} are now common in the dairy industry (**Fig. 1**). Applications include fluid milk treatment, processing of milk for cheese-making, dried milk and infant formula manufacture, and the production of highly concentrated direct vat set starter cultures. Other applications include bacterial clarification of fruit juices and production of bakers' and distillers' yeast.

Principle of Centrifugation

Centrifugation is used especially in dairy processing operations for continuous clarification, separation and standardization of milk, and removal of bacteria. These processes employ centrifuges fitted with a stack of conical discs assembled in a bowl which rotates at high speeds, and utilize differences in density of milk components, extraneous matter and bacteria to achieve the desired separation. The spaces between consecutive discs serve as channels for fluid flow and separation. The equipment design features that distinguish the centrifuges for these three processes include differences in the discs and the speed of centrifugation.

Figure 1 Bacteria-removing centrifuges at an Emmental cheese factory.

The purpose of clarification is to remove extraneous particles from milk without affecting milk composition, whereas that of separation is to remove milk fat. The removal of microflora by centrifugation is based on size and density differences between bacteria and fluid components. The specific gravity of milk (1.032 for whole milk, 1.036 for skim milk) and cream (0.993 for 40% fat cream) is less than that of bacteria which is in the range 1.07 to 1.13; that of heat-resistant spores is higher than that of other bacterial cells. Consequently, when milk is centrifuged under a high force (9000 to 10 000 *g*) the bacterial cells are removed from the milk. Under continuous application of such centrifugation, the bacteria form a concentrated suspension in a small amount of milk.

A bacteria-removing centrifuge is similar to a centrifugal milk clarifier but rotates at a much higher grativational force to enable removal of bacteria. Milk flows in an upward direction through a hollow spindle and into the bottom of the rotating disc stack via the centre. In some designs milk enters the centrifuge from the top but in either case milk moves in an upward direction through the disc stack via holes near the periphery of each disc, starting from the bottom disc. The heavy bacterial cells are thrown in an outward direction against the wall of the bowl, whereas the milk with a reduced bacterial load moves up to the top centre of the bowl and out. In a clarifier the gravitational force is lower ($< 4000\,g$) so bacteria are not removed. A cream separator operates at approximately 4000 *g*. Milk moves up the disc stack through holes in the discs located approximately halfway between the edge and the centre of the disc. As milk travels in the channels between the discs, milk fat in the form of cream (which has a lower density than skim milk) migrates towards the centre of the disc, and the skim milk (which is heavier) is thrown out towards the edge of the discs. Cream and skim milk emerge from the top of the centrifuge through separate outlets.

Modern bacteria-removing centrifuges may be of either one-phase or two-phase design (**Fig. 2**). In the one-phase design, the bacterial mass accumulates on the walls of the bowl with some milk solids as sludge, and is also known as the bactofugate. It is discharged intermittently (every 20–40 min) and forms approximately 0.15% of feed volume. The bowl of the cen-

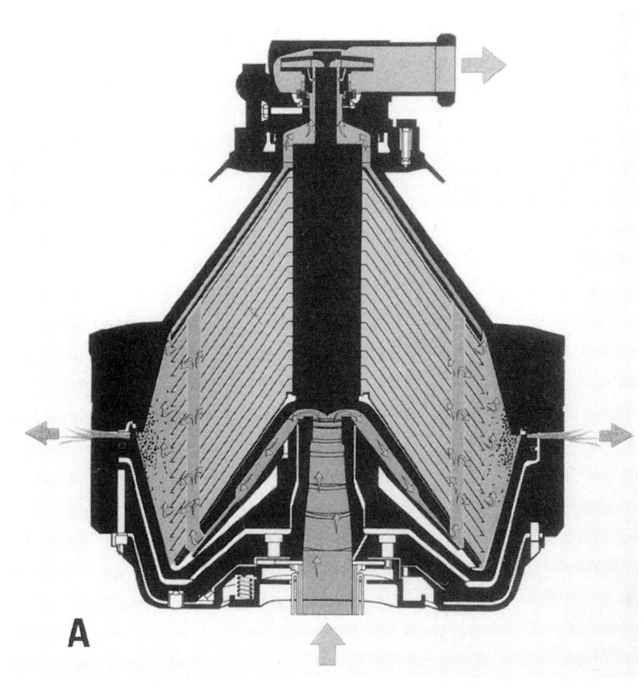

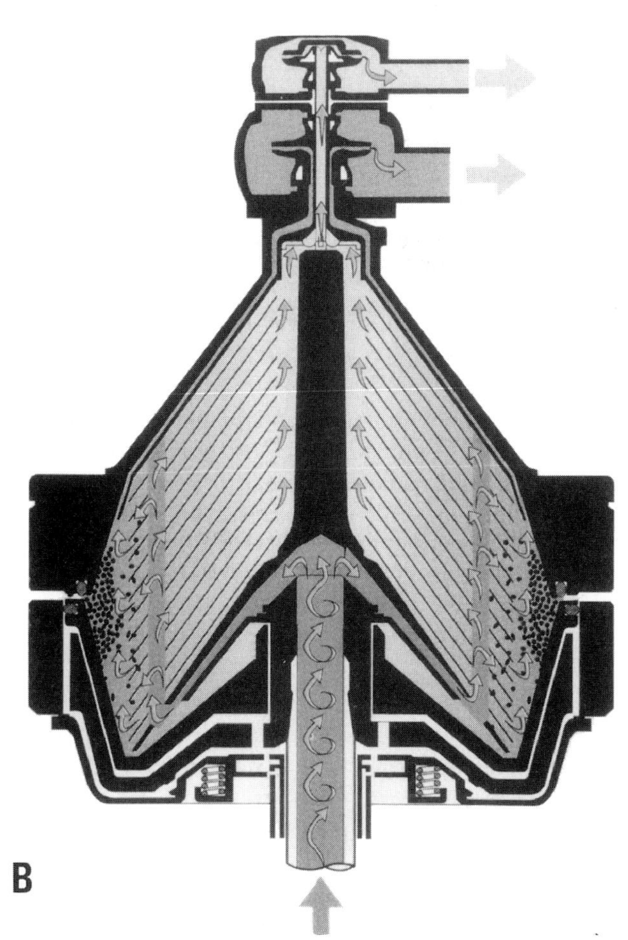

Figure 2 Bowl of (**A**) one-phase and (**B**) two-phase bacteria-removing centrifuge. Courtesy of Tetra Pak Processing Systems, Lund, Sweden.

trifuge, where the sludge accumulates, may be surrounded by cooling water to prevent it from baking on to the walls. In the two-phase design there are two outlets at the top of the centrifuge, one for the bacteria-reduced milk and the other for the continuous removal of the bactofugate. In both cases the aim is to remove the sludge so as to maintain the efficiency of bacterial removal during continuous operation for periods of more than 10 h. In early models the lack of timely removal of the sludge limited the efficiency.

In the two-phase system the bactofugate is approximately 3% by volume. This bactofugate, which is approximately 18% solids, may contain as much as 13% milk protein but practically no fat. This creates an imbalance in the composition of milk and causes a 5–10% loss in cheese yield, and also produces a softer cheese. Kosikowski in a 1970 patent proposed a new process in which the sludge is diverted to a sterilizer to kill the spores. Somatic cells are also concentrated in the sludge and destroyed by sterilization. The sterilized sludge is blended back with the bactofuged milk. The protein lost in the sludge is thus recovered and consequently there is no loss in cheese yield. Cheeses produced with such milk do not have the softness defect either. This process has the Tetra Pak trade name 'Bactotherm'. The process is also used to manufacture low-heat-treated milk powder.

The Bactofuge is connected in series with the pasteurizer and the cream separator (**Fig. 3**). Raw milk is first pre-heated in the regeneration section of the plate pasteurizer to the cream separation temperature. The milk is standardized to the required fat level with the help of the cream separator before flowing into the Bactofuge at the same temperature. If the plant is equipped with a Bactotherm unit, the bactofugate is sterilized and blended with the bacteria-reduced milk and routed back to the pasteurizer where it is heated to the pasteurization temperature (72°C) and held for 15 s prior to cooling.

Operating Parameters

At a constant *g* force the bacterial removal efficiency of a centrifuge depends on various factors. The efficiency differs between types of bacteria because of density differences. The higher the density the greater the efficiency. An important operating factor is the temperature at which milk is centrifuged. The optimum temperature is 55–60°C. At lower temperatures efficiency is low, whereas at higher temperatures it becomes difficult to ascertain whether the reduction of bacterial counts is because of temperature or centrifugation.

The impact of temperature on the removal of bac-

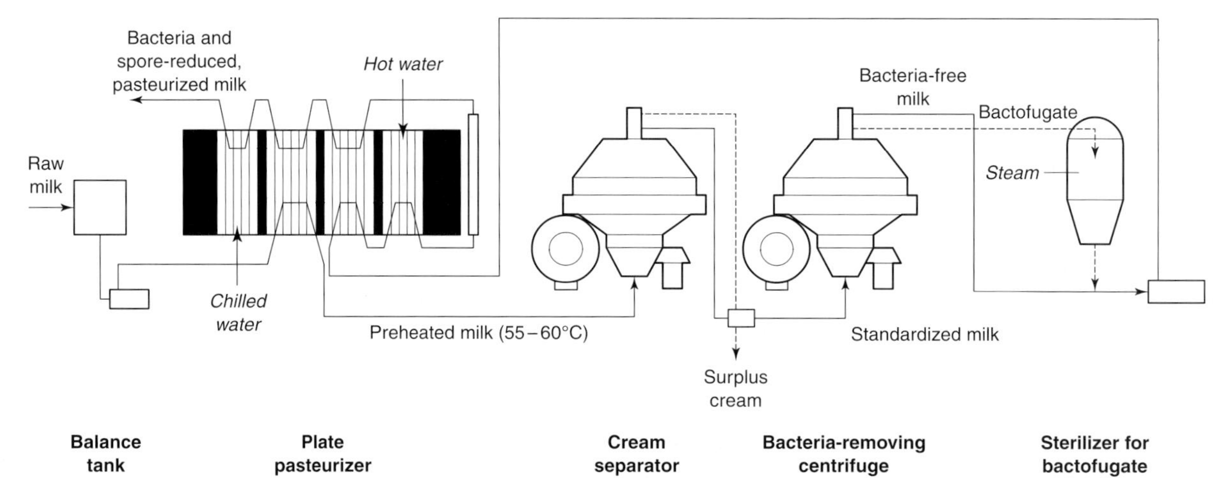

Figure 3 Process for removing bacteria from milk by centrifugation.

teria may be attributed to viscosity changes and agglutinins. As the temperature of milk is increased its viscosity decreases, making it easier for the bacterial cells to migrate through the fluid in the centrifuge. For example, removal is over 97% at 60°C but may be as low as 85% at 45°C. Another factor at high temperatures (>75°C) is the inactivation of agglutinins. When they are inactivated, the binding of spores of *Clostridium tyrobutyricum* to fat globules is limited, therefore removal efficiency during centrifugation increases. The effect of temperature is not as critical for all fluids. For example, in apple juice reductions of over 99% have been reported at 7°C. This is advantageous because the flavour of juice may deteriorate at high temperatures.

Microorganisms

The application of centrifugation for the removal of bacteria was originally developed to extend the shelf life of fluid milk. To this end the interest primarily was to significantly lower the total bacterial count. The major application today is the removal of the anaerobic spore-forming bacterium, *Clostridium tyrobutyricum* from milk intended for cheese-making. The spores of this microorganism are extremely heat-resistant, and normal pasteurization treatments are not sufficient. Fortunately the density of these organisms is high, therefore the efficiency of removal by centrifugation is also high.

Scientific studies have been conducted to determine the efficiency of removal by centrifugation of various microorganisms that include aerobic and anaerobic spore-formers, total bacteria counts, propionic acid bacteria, *Escherichia coli*, *Aerobacter aerogenes*, lactobacilli, yeasts and mould (in apple juice), and staphylococci and enteric bacteria in liquid egg whites. Furthermore, the removal of somatic cells from milk by centrifugation has also been studied.

Table 1 Removal of microflora by centrifugation

	Reduction (%)
Milk (54–65°C)	
Anaerobic spore-formers	96–99.6
Aerobic spore-formers	90–95
Total bacterial count	90–95
Escherichia coli *and* Aerobacter aerogenes	99.4
Staphylococci	98.5
Enteric bacteria	99.8
Apple juice (7°C)	
Total bacteria count	99.8
Yeast, mould	99.9

The efficiency of removal of these microorganisms varies because of varying physical properties. In raw milk it is possible to achieve at least a 98% reduction of *Clostridium tyrobutyricum* by centrifugation. Reductions of over 99.5% are possible by employing two centrifuges in series. Removal efficiencies are listed in **Table 1**.

Applications

Bacterial centrifugation is used for a wide range of commercial applications, especially in the dairy industry.

Fluid and Dried Milk Processing

With bacterial centrifugation of raw milk it is possible to extend the shelf life of pasteurized fluid milk by 3–5 days. This approach is used commercially by some dairies. In this application the bacteria-removing centrifuge usually supplements rather than replaces pasteurization or ultra-high temperature (UHT) treatment (135–150°C for 2–5 s). Enzymes that cause milk spoilage (e.g. lipases causing rancidity) are not removed by centrifugation and have to be inactivated by pasteurization. Pasteurization or UHT treatment also kills the small numbers of bacteria that are not removed by centrifugation. Minimal temperatures for UHT may be employed because of the low load of spores in milk after centrifugation. The ability to remove aerobic spore-formers (*Bacillus cereus*) is of particular value in the manufacture of UHT milk, dried milk and evaporated milk including infant formula. In these products spores of *Bacillus cereus* that survive heat treatment cause storage-related defects such as bitter flavours, and age gelation (in evaporated milk). Removal at low temperatures by centrifugation minimizes these defects.

Cheese and Whey Processing

The cheese industry has probably been the biggest beneficiary of bacterial centrifugation technology. After the introduction of corn silage in France in 1970 the large increase in *Clostridium tyrobutyricum* counts in raw milk caused problems, especially in Emmental cheeses. This has also been a problem in several other countries in Europe and affects other salt-brined cheese varieties such as Gouda, Edam and Gruyère. Spore counts in milk of as low as 300 spores per litre for Gouda cheese or more than 2000 spores per litre for Emmental cheese lead to vigorous butyric acid fermentation and consequently, late gas blowing in cheese.

The large amount of gas formed during this reaction within 2 months of cheese-making leads to cracking and in extreme cases explosion of the cheese (**Fig. 4**). This problem is of great economic significance because the cheese is unsaleable. Slow salt diffusion in cheeses such as Gouda and the low salt content of Maasdam cheese make them more vulnerable to this defect. Measures taken to prevent such defects include the addition of lysozyme, nitrates or formaldehyde to milk (not permitted in some countries) prior to cheese-making. In France and perhaps in other countries where high counts of *Clostridium tyrobutyricum* occur in milk from some farms, cheese-makers separate such milks by using compartmentalized milk tankers. Economic incentives are offered for milks with low anaerobic spore-former counts and penalties are assessed on a graded scale for high counts. Milks with high counts are bactofuged and blended with the rest of the milk supply prior to cheese-making.

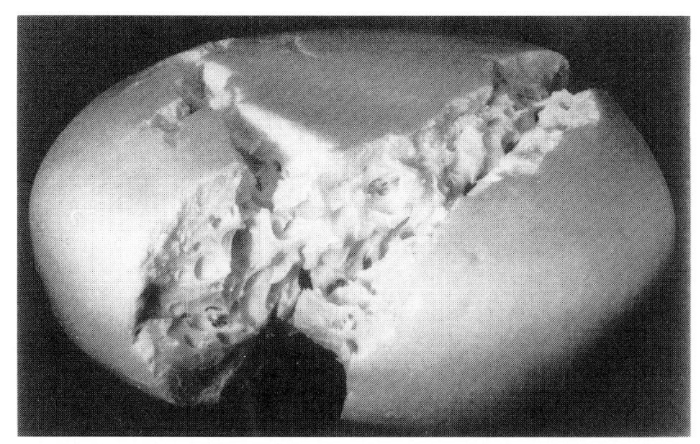

Figure 4 Late gas blowing in Gouda cheese. Courtesy of FV Kosikowski LLC, Great Falls, Virginia, USA.

When milk is not pasteurized, as in some cheese-making operations, centrifugation may be supplemented with treatments such as application of formaldehyde, hydrogen peroxide, nitrates or lysozyme where permitted. A two-stage bactofugation system is usually sufficient to lower the spore count to safe levels, but if the spore count is high, 2.5–5 g of sodium nitrate per 100 litres of milk in addition to bactofugation is effective. Without bactofugation, much larger amounts are needed: 15–20 g per 100 litres. Similarly, in the manufacture of Grana Padano cheese in Italy, formaldehyde at 25–40 mg per litre was traditionally added to milk. This practice in combination with natural creaming of milk lowered the anaerobic spore counts by 91%. Some manufacturers have now adopted bacteria-removing centrifuges and are able to lower spore counts by over 98% without using formaldehyde.

Other applications of bacteria-removing centrifuges in the cheese industry include whey processing. Whey is generally used for the manufacture of products such as whey protein concentrates, which in turn are used as ingredients in foods. It is important therefore to retain the functional properties of the whey proteins, which are adversely affected by heat. Whey is clarified to remove cheese curd fines and then bactofuged at low temperatures (35°C). It is possible to lower the bacterial population (which consists mainly of starter bacteria) by 95% or more. After minimal pasteurization the whey is concentrated by ultrafiltration and dried.

Harvesting Bacteria and Yeasts

Concentrated cultures of lactic acid bacteria are commonly used in the production of fermented milk products and cheeses. These concentrated cultures, often referred to as direct vat set cultures, contain 50–100×10^9 bacterial cells per gram. Selected bacteria are grown in specialized growth media containing the required nutrients to optimum numbers. The entire liquid mass is then centrifuged using a bacteria-separating centrifuge. Unlike the centrifugation of milk, in this case the bactofugate is the desired end product as it contains the concentrated mass of culture bacteria. Where automatic desludging is employed to recover the concentrated cell mass, the concentrate volume may be adjusted to a constant level, e.g. 5% of the original feed volume, to obtain consistency in the cell numbers. This is important to achieve reliable day-to-day starter performance among different batches of culture concentrates for cheese-making.

Figure 5 Pellets of frozen concentrated starter culture.

The concentrate thus prepared has a cell concentration approximately 20 times that of the feed. Cell recoveries are high during centrifugation; only approximately 1% of the total culture cells in the feed is lost during the centrifugation process. The cell concentrate is immediately frozen in liquid nitrogen typically at −196°C and packaged in aluminium cans or paperboard containers (**Fig. 5**). The culture can be added directly to milk for fermentation to proceed. Proper sanitary operation of the centrifuge is essential to ensure that other bacteria do not contaminate the culture concentrate.

Centrifugation is also applied to separate yeast cells from fermentation media during the manufacture of bakers' yeast and single-cell protein. In these processes, the fermentation medium is fed to the centrifuge when yeast cell numbers have reached the desired level. A force of 4000–5000 *g* is used, which is lower than that for bacteria because yeast cells are denser. The sludge consists of a concentrated mass of yeast cells, sometimes called 'yeast cream' because of its cream-like consistency. This mass is then processed further prior to packaging.

Since 1987 centrifugation technology for removing bacteria has faced stiff competition from a new technology for milk processing, namely microfiltration. This process enables the removal of bacteria, spores and somatic cells from milk at low temperature and with very high efficiency using specific membranes, and is already being used commercially for processing fluid milk and milk for cheese-making. The economics of operation will determine which technology is preferable for a given manufacturing plant.

See also: **Bacteriophage-based Techniques for Detection of Food-borne Pathogens**. **Cheese**: Microbiology of Cheese-making and Maturation. ***Clostridium***: *Clostridium tyrobutyricum*. **Heat Treatment of Foods**: Ultra-high Temperature (UHT) Treatments. **Milk and Milk Products**: Microbiology of Liquid Milk; Microbiology of Dried Milk Products. **Physical Removal of Microfloras**: Filtration.

Further Reading

Breed RS (1926) The number of cells in cream, skim milk, and separator and centrifuge slimes. *NY State Agricultural Experiment Station Circular* 88.

Burrows S (1970) Baker's yeast. In: Rose AH and Harrison JS (eds) *The Yeasts*. Vol. 3, p. 348. London: Academic Press.

Bylund G (1995) *Dairy Processing Handbook*. Pp. 110–111, 293–295. Tetra Pak Processing Systems AB, Lund Sweden.

Eck A (1986) *Cheesemaking Science and Technology.* P. 178. New York: Lavoisier.

Kosikowski FV (1970) Sterilization of milk. US Patent 3 525 629. 25 August.

Kosikowksi FV and Mistry VV (1997) *Cheese and Fermented Milk Foods*, Vol. 1. *Origins and Principles*, 3rd edn. P. 260. Great Falls: Kosikowski.

Porubcan RS and Sellars RL (1977) Preparation of culture concentrates for direct vat set cheese production. US Patent 4 115 199, 19 September.

Simonart P and Debeer G (1953) Researches concerning the improvement of microbiological quality of milk by ultracentrifuging. *Netherland Milk Dairy Journal* 7: 117–128.

Van den Berg G (1990) New technologies for hard and semi-hard cheeses. *Proceedings of the 23rd International Dairy Congress*, Montreal, Canada. Vol. 3, p. 1864.

PICHIA PASTORIS

Chitkala Kalidas, Department of Food Science, Cornell University, USA

Characteristics of the Species

Pichia pastoris is a methylotrophic yeast that belongs to the class Ascomycetes. It normally exists in the vegetative haploid state. Vegetative reproduction is by multilateral budding. Nitrogen limitation stimulates mating and leads to the formation of diploid cells. *P. pastoris* is considered to be homothallic since cells of the same strain can mate with each other. There may be more than one mating type in the population, that switches at a high frequency so that mating occurs between haploid cells of opposite mating type. Diploid cells maintained in standard vegetative growth medium remain diploid. If they are moved to nitrogen-limited medium, they undergo meiosis and produce haploid spores.

Physiological regulation of mating in *P. pastoris* facilitates its genetic manipulation. Since it is most stable in its vegetative haploid state, easy isolation and characterization of mutants are possible.

Currently, *P. pastoris* is one of the main expression systems for the production of heterologous proteins. A number of bacterial and mammalian proteins have been expressed in *P. pastoris* (**Table 1**).

History

Pichia pastoris was initially chosen for the production of single-cell proteins for feed stock. This was due to its ability to grow to very high cell densities in simple media containing methanol. The production of single-cell proteins from *P. pastoris* was considered a commercially viable option since the synthesis of methanol from natural gas (methane) was inexpensive in the late 1960s. Phillips Petroleum Company, Bartlesville, OK, developed protocols to grow *P. pastoris* on methanol in continuous cultures. However, with the increase in the price of methane due to the oil crisis in the 1970s, interest in *P. pastoris* for the production of single-cell proteins waned. In the 1980s, Salk Institute Biotechnology/Industrial Associates (SIBIA), located in La Jolla, CA, under contract with Phillips Petroleum Company developed the *Pichia* expression system for the production of foreign proteins. In 1993, Phillips Petroleum Company decided to release the *Pichia* expression system to research laboratories in academic institutions. Since then, *P. pastoris* has been widely used as an expression system even in laboratories not routinely working with yeasts. The *Pichia* expression system is now commercially available from Invitrogen (Carlsbad, CA, USA).

Advantages of Using *Pichia pastoris*

The use of *P. pastoris* as a protein expression system has gained rapid acceptance in the last decade. This can be attributed to high protein yields, very high levels of secretion with little or no secretion of native proteins, easy scale-up and ease of handling.

P. pastoris can be genetically manipulated using the same protocols as for *Saccharomyces cerevisiae*, one of the best-studied eukaryotes. It has a strong preference for respiratory growth and this trait enables it to be cultured to high cell densities compared to other fermentative yeasts. Apart from these benefits, *P. pastoris* is also capable of post-translational modifications of proteins such as proteolytic processing, glycosylation and disulphide bridge formation. Many proteins, which form inactive inclusion bodies in *Escherichia coli*, are expressed in their biologically active state in *P. pastoris*.

The yields from the *Pichia* expression system are generally better than those from higher eukaryotic systems such as insect and mammalian cell lines. It is also cost-effective and less time-consuming than the higher eukaryotic systems.

The *Pichia* Expression System

P. pastoris produces the enzyme alcohol oxidase that is required for it to metabolize methanol. The first step in the metabolism of methanol is the oxidation of methanol to formaldehyde, resulting in the formation of hydrogen peroxide. This reaction is catalysed by alcohol oxidase and takes place in a specialized membrane-bound organelle called the peroxisome. Strong proliferation of peroxisomes is seen

Table 1 Heterologous proteins expressed in *Pichia pastoris*[a]

Protein	*Expression levels* ($g\ l^{-1}$)	*Mode of expression and Mut phenotype*
Invertase	2.3	Secreted, Mut^+
α-Amylase	2.5	Secreted, Mut^S
Spinach phosphoribulokinase	0.1	Intracellular, Mut^S
Pectate lyase	0.004	Secreted, Mut^S
Bovine lysozyme C2	0.55	Secreted, Mut^+
Hepatitis B surface antigen	0.4	Intracellular, Mut^S
HIV-1 gp120	1.25	Intracellular, Mut^+
Carboxypeptidase B	0.8	Secreted, Mut^+/Mut^S
Bovine β-lactoglobulin	1.5	Secreted, Mut^+
Tumour necrosis factor	10.0	Intracellular, Mut^S
Human interferon (IFN)-α2b	0.4	Intracellular, Mut^S
Rabbit single chain antibody	>0.1	Secreted, Mut^S

[a]Invitrogen Corporation Pichia Expression Kit Instruction Manual. (Reproduced with permission from Invitrogen Corporation, 1996.)

during methanol utilization in *P. pastoris*. Peroxisomes sequester the toxic hydrogen peroxide from the rest of the cell.

Alcohol oxidase (AOX) has poor affinity for oxygen. *P. pastoris* compensates for this deficiency by expressing large amounts of this enzyme. Two genes, *AOX1* and *AOX2*, code for alcohol oxidase activity. The former accounts for more than 90% of the enzyme activity while the latter accounts for less than 5% of the activity. The *AOX1* promoter is tightly regulated and induced by methanol, but remains repressed under other conditions, including carbon starvation. This protein constitutes up to 5% of the total soluble protein during methanol induction in shake flask cultures. It can constitute more than 30% of the total soluble protein during growth on methanol in the fermenter. Thus, this promoter is specially suited for the controlled expression of foreign genes. By placing the foreign gene under the control of the *AOX1* promoter, it is possible to grow the culture on a non-inducing carbon source like glycerol until a suitable cell density is attained and then induce with methanol.

Other inducible and constitutive promoters have also been utilized for heterologous gene expression in *P. pastoris*, such as the promoter of the glutathione-dependent formaldehyde dehydrogenase gene (*pFLD1*) independently inducible by methylamine and methanol and the promoter of glyceraldehyde-3-phosphate dehydrogenase gene–constitutive expression.

Pichia Strains and Plasmids

All strains of *P. pastoris* are derivatives of the wild-type strain NRRL-11430 (Northern Regional Research Laboratories, Peoria, IL). Most strains are deficient in the enzyme histidinol dehydrogenase (**Table 2**). This aids in the selection of transformants which harbour the expression vector containing the *HIS4* gene. Auxotrophically marked strains are useful in the selection of diploid strains. Biosynthetic genes such as *arg4*-argininosuccinate lyase and *ura3*-orotidine 5′-phosphate decarboxylase are some of the other commonly used auxotrophic markers in the *Pichia* system.

Table 2 *Pichia pastoris* expression host strains

Strain name	*Genotype*	*Phenotype*
Y-11430	Wild-type	NRRL*
GS115	*his4*	Mut^+ His^-
KM71	*aox1Δ::SARG4 his4 arg4*	Mut^S His^-
MC 100–3	*aox1Δ::SARG4 his4 arg4 aox2Δ::Phis4 his4 arg4*	Mut^- His^-
SMD1168	*pep4Δhis4*	Mut^+His^-, protease-deficient
SMD1165	*prb1 his4*	Mut^+His^-, protease-deficient
SMD1163	*pep4 prb1 his4*	Mut^+His^-, protease-deficient

*Northern Regional Research Laboratories, Peoria, IL, US (reproduced with permission from Higgins and Cregg 1998).

There are three types of host strains according to the ability to utilize methanol, resulting from mutations in one or both *AOX* genes. The most commonly used strain is GS115. This strain has both the *AOX1* and *AOX2* genes and grows on methanol at wild-type rate. This phenotype is termed Mut^+ (*me*thanol *ut*ilization). KM71 is a strain in which the chromosomal *AOX1* gene is replaced with the *S. cerevisiae ARG4* gene. Therefore, this strain has lower alcohol oxidase activity from *AOX2*. It grows very slowly on methanol and this phenotype is termed Mut^S (methanol utilization slow). The strain MC 100–3 has a Mut^- phenotype as it has deletions at both the *AOX1* and *AOX2* loci. This strain is unable to grow on methanol but the *AOX1* promoter is inducible by methanol. Mut^- strains, therefore, require an alternative carbon source such as glycerol for growth. However, excess glycerol has a negative effect on expression and has to be fed at growth-limiting rates.

For the large-scale production of secreted proteins, Mut^+ strains are often used because they grow much faster on methanol compared to the *AOX*-defective strains. However, Mut^S and Mut^- strains are more tolerant to residual methanol in the fermenter than the Mut^+ strains. Due to this reason, the *AOX*-defective strains are sometimes preferred over Mut^+ strains for the production of secreted proteins. For intracellular expression, the *AOX*-defective strains are preferred since low levels of alcohol oxidase expression increase the specific yield of the heterologous protein.

Some secreted foreign proteins are unstable in the *P. pastoris* culture medium due to the action of endogenous proteases. The strain SMD1168 is similar to GS115 except that it lacks proteinase A activity. Other protease-deficient strains are SMD1163 and SMD1165. These strains are used in cases where proteolytic cleavage results in low yields of expressed proteins.

The schematic of a typical *Pichia* expression vector is shown in **Figure 1**. The vector, pPIC9, consists of the *AOX1* promoter fragment, the *AOX1* transcription terminator region and the 3′ *AOX1* region. The *AOX1* promoter is followed by the *S. cerevisiae* α-mating factor (α-MF) signal sequences and a multiple cloning site. This plasmid also carries the histidinol dehydrogenase gene (*HIS4*) which is used to select for recombinant *P. pastoris* clones. The ColE1 sequence and the gene for ampicillin resistance on the plasmid are useful for subcloning into *E. coli*

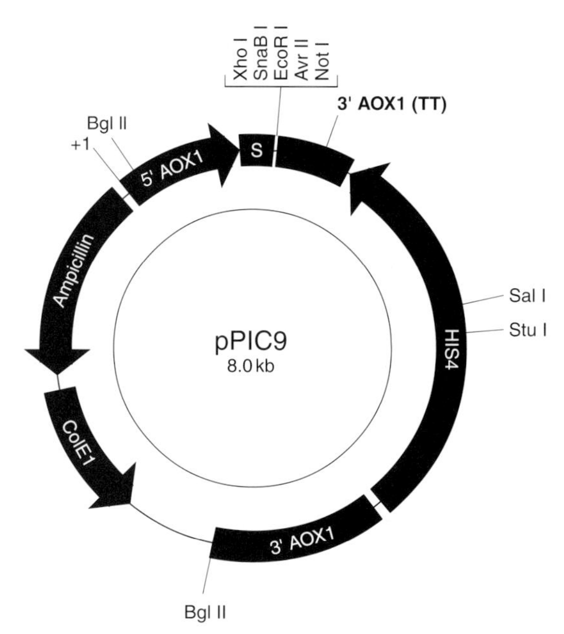

Figure 1 *Pichia* expression vector (reproduced with permission from Invitrogen Corporation, 1996).

Construction of Recombinant Strains

In order to obtain stable recombinant strains of *P. pastoris*, the expression vectors are integrated into the host genome. Linear DNA can generate stable transformants due to homologous recombination between the plasmid DNA and homologous regions within the genome. Selection of transformants is based on histidine prototrophy.

Integration of the expression vector can occur in three ways. This generates transformants with different methanol utilization phenotypes (Mut^+, Mut^S) depending on the host strain used (**Fig. 2**). It is possible to stimulate either single or double crossover events by linearizing the plasmid. If the plasmid is cut at one of the restriction sites within the *5′ AOX* sequences, it will stimulate single crossovers, leading to the insertion of the expression cassette at either the *AOX1* locus as in GS115 or *aox1::ARG4* locus as in KM71. The phenotype of such transformants would be His^+Mut^+ if the host strain is GS115 and His^+Mut^S if the host strain is KM71.

Insertion of the plasmid can also occur at the *his4* locus on the host genome. This results from a single crossover event between the *his4* on the chromosome and the *HIS4* on the plasmid. Since the genomic *AOX1* locus is not involved, the Mut phenotype of the His^+ transformant would be the same as the host strain used.

Double crossover events can be generated in the strain GS115 by cutting the plasmid at the *BglII* site. This leads to the formation of a fragment with the *AOX1* sequences at its termini and the gene of interest and *HIS4* in between. This would stimulate gene replacement events at *AOX1*. The resulting strains would lack *AOX1* and would have to depend on the weak *AOX2* gene for methanol utilization. The phenotype of such a transformant would be His^+Mut^S.

Multiple gene insertion events can occur at the *his4* or the *AOX1* loci, leading to the generation of multicopy recombinant strains. Such events have been found to occur spontaneously at a low frequency of 1–10% of all selected His^+ transformants. The Mut phenotype of such strains would be the same as the host strain used.

Even though high yields have been obtained from single-copy recombinants, yields from multicopy strains have often been found to be significantly higher. As a result, methods of generating multicopy strains have been developed. These are based on three different approaches. The first approach involves identification of multicopy strains that occur naturally within the population of transformants. A large number of transformants can be screened for protein expression using SDS-PAGE. Multicopy transformants can be identified based on their protein yields, which are typically higher than those from single-copy transformants. Alternatively, immunoblotting or colony hybridization can be carried out to detect multiple copies of the heterologous gene.

The second approach is to detect multicopy strains within a population based on their level of antibiotic resistance. For this, plasmids carrying the *Tn903kanR* gene are used. This gene confers resistance to G418. The level of antibiotic resistance depends on the number of copies of this gene. Multicopy strains will be resistant to higher concentrations of G418 and can thereby be identified.

The third approach involves transformation of the host cells with a vector carrying multiple copies of the expression cassette. By this method a single gene insertion event would be sufficient to generate a multicopy strain.

Protein Expression

Once stable recombinant strains are obtained, test tube cultures are used to screen for protein expression. In order to maximize protein yields, the cells are grown under non-inducing conditions until the culture reaches the log phase (OD_{600} 2–6). Then the culture is induced with methanol every 24 h until maximum production is obtained.

One of the advantages in using *P. pastoris* is the ease with which small cultures can be scaled up to larger volumes without any decrease in yield. Therefore, when optimum conditions for expression are

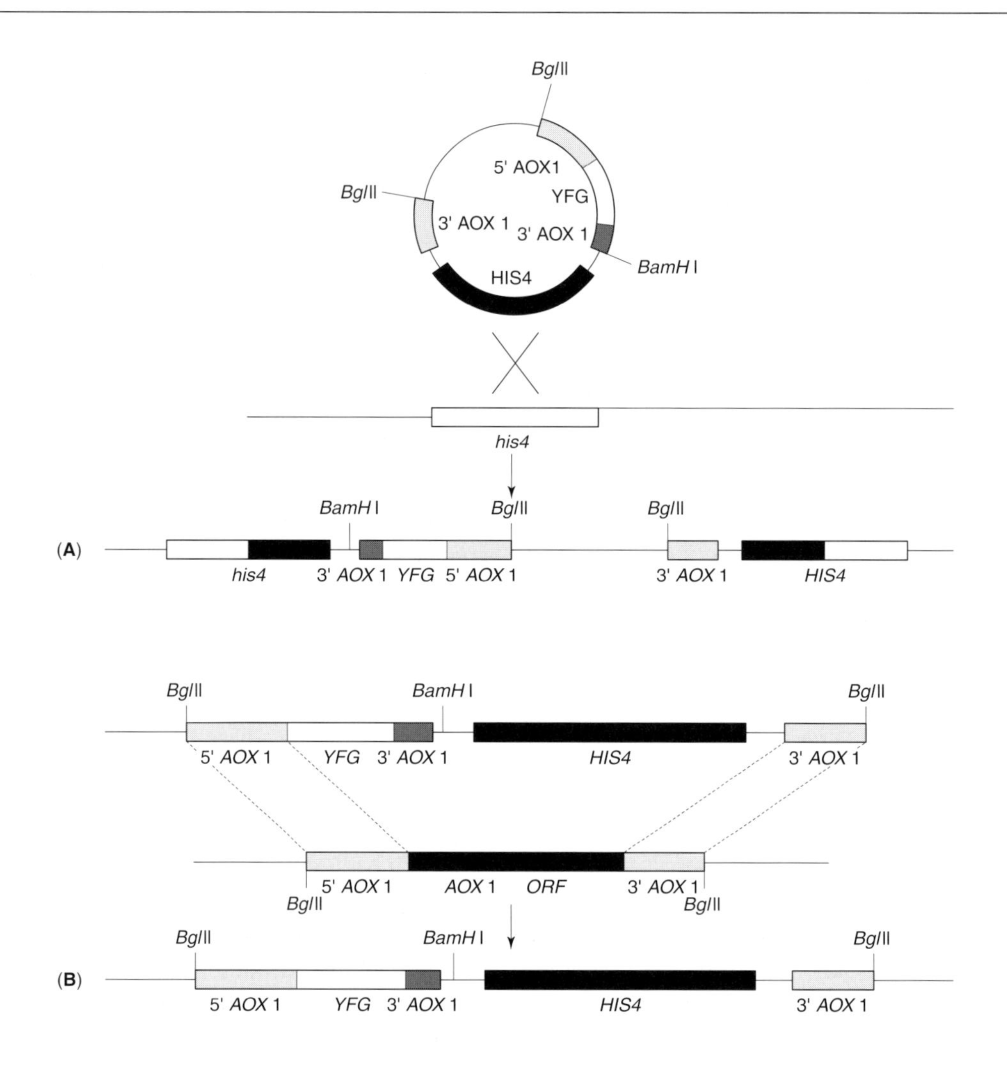

Figure 2 Integration of expression vectors into the *Pichia pastoris* genome. (**A**) Single crossover integration at the *his4* locus. (**B**) Integration of vector fragment by replacement of *AOX1* gene (reproduced with permission from Higgins and Gregg 1998).

determined, the culture volume can be scaled up using large shake flasks or by fermentation.

Protein yields are typically higher in fermenter cultures than in shake flasks. One of the reasons for this is that high cell densities (> 300 g l^{-1} dry cell wt) can be reached in fermenter cultures. The level of product yield is directly proportional to the cell density, especially in the case of secreted proteins. The other reason for higher yields in the fermenter cultures is that optimum levels of oxygen and methanol can be maintained. Both excess oxygen and methanol have a negative effect on protein expression.

A number of fedbatch and continuous culture schemes have been developed for the high-cell-density fermentation of expression strains. The process of fermentation using *P. pastoris* can be divided into three stages. The first is the continuous phase, using glycerol as the carbon source, which lasts for about 24 h. The second stage is the glycerol fedbatch phase. In this stage glycerol is fed at a growth-limiting rate. During growth on glycerol the cells multiply rapidly but remain strongly repressed. The third is the methanol fedbatch phase. The optimum period of induction and the time of harvest have to be determined empirically for each protein.

Post-translational Modification of Expressed Proteins

Signal Sequence Processing

One of the main attractions of the *Pichia* expression system is its ability to secrete heterologous proteins at high levels with little or no secretion of native proteins. As a result of this, downstream processing of secreted proteins becomes much easier.

A signal sequence is required for the proper secretion of the foreign protein. The native signal sequence of the foreign protein has been successfully used in some cases, e.g. bovine lysozyme. In some cases, however, native signal sequences have not worked well, e.g. invertase. Yeast signal sequences have been used as other alternatives. The most commonly used yeast signal sequence is the prepro α-MF sequence from *S. cerevisiae*. In *S. cerevisiae*, three enzymes are required for the proper processing of the α-MF signal sequence. They are a signal peptidase that cleaves the pre region, *KEX2* gene product that cleaves at the junction between α-MF prepro region and the foreign protein, and the product of *STE13* that removes Glu-Ala spacer residues from the amino terminal of the foreign protein. A number of foreign proteins have been expressed in *P. pastoris* at high levels and in the fully processed state using the α-MF. This indicates that the enzymes for processing the signal sequence are present in sufficient quantities in the *P. pastoris* secretory system as well. Incomplete processing of the signal sequence has also been observed in some cases. Conformational characteristics of the expressed protein could prevent access to the signal sequence processing enzymes.

Glycosylation

Glycosylation is another post-translational modification carried out by *P. pastoris*. Both O and *N*-linked glycosylation has been observed in proteins expressed in this yeast. *N*-linked glycosylation in *P. pastoris* is significantly different than in higher eukaryotes. Oligosaccharides from *P. pastoris* lack the *N*-acetylgalactosamine, galactose and sialic acid residues found in mammals. Carbohydrate side chains in mammals are composed of a mixture of different sugars (complex type) or of $Man_{5-6}GlcNAc_2$ (high mannose type) or both. In *Pichia*, the carbohydrate side chains are usually of the high mannose type. However, two distinct patterns of glycosylation are seen with regard to the number of mannose residues added. In some cases, 8–15 mannose residues are added, e.g. invertase from *S. cerevisiae*. In other cases hyperglycosylation has been observed.

Core structures of *P. pastoris* oligosaccharide molecules have been determined to be identical to those of *S. cerevisiae*. The structure of the oligosaccharide side chains from *P. pastoris* secreted invertase has been determined. The major species found were $Man_{8-11}GlcNAc_2$ and all except $Man_{11}GlcNAc_2$ were identical to *S. cerevisiae* core structures. The terminal mannose residues in *P. pastoris* were of the α-1,2 type, as opposed to the more common α-1,3 type in *S. cerevisiae*. Apart from the absence of α-1,3-linked mannose, little information is available about the structure of outer chain oligosaccharides. The mechanism underlying the addition of outer chains is also not well understood.

N-linked glycosylation in *P. pastoris* poses a problem with regard to the use of expressed proteins for therapeutic applications. The high mannose oligosaccharide could be highly antigenic and preclude therapeutic use. The other problem, due to differences in glycosylation pattern between *P. pastoris* and mammals, is that the long outer chains could interfere with the proper folding of proteins.

Importance to the Food Industry

Pichia pastoris was initially used to produce single-cell proteins, several unusual enzymes (e.g. alcohol oxidase, formate dehydrogenase) and metabolites such as ATP, aldehydes and amino acids. Since then, a number of food proteins and enzymes have been expressed in *P. pastoris* (Table 1). Recombinant amylases and sugar-converting enzymes from different sources have been expressed in *P. pastoris*. Examples of this class of proteins are α-amylases 1 and 2 from barley. These enzymes from *P. pastoris* were found to be similar in structure and function to those from malt extracts. β-Lactoglobulin, the major whey protein in bovine milk, has been expressed in *P. pastoris* at the level of $>1\,g\,l^{-1}$. The physical characteristics of this recombinant protein were found to be indistinguishable from the native bovine form.

P. pastoris has also been used as a biocatalyst in the conversion of glycolate and its derivatives to glyoxalate and other corresponding 2-oxo-acids. This reaction produces hydrogen peroxide that has to be metabolized for the efficient conversion of the substrate. Recombinant strains of *P. pastoris* carrying the glycolate oxidase gene from spinach and the endogenous catalase have been used as catalysts for this process.

Conclusion

Pichia pastoris has been successfully used for the production of a number of heterologous proteins. It can be used for both large-scale and small-scale production of proteins. Even though glycosylation in *P. pastoris* poses a big hurdle to the use of the expressed proteins for therapeutic purposes, glycoproteins such as vaccines that do not require authentic glycosylation to trigger a protective immune response can be produced in this system. It is also possible to use *P. pastoris* to produce non-glycosylated proteins at high levels. Secretion of these proteins into the culture supernatant would make protein purification much simpler. In vivo isotope labelling of proteins for nuclear magnetic resonance studies is possible in this system. Therefore *P. pastoris* serves as an excellent tool for the large-scale production of proteins for commercial purposes and also for the study of proteins. All these aspects make the *Pichia* expression system an attractive system for use by both industrial and academic research laboratories.

See also: ***Saccharomyces***: *Saccharomyces cerevisiae*. **Single Cell Protein**: Yeasts and Bacteria.

Further Reading

Cregg JM and Higgins DR (1995) Production of foreign proteins in the yeast *Pichia pastoris*. *Can. J. Bot.* 73 (suppl. 1): S981–S987.

Higgins DR and Cregg JM (eds) (1998) *Methods in Molecular Biology. Pichia* protocols. Vol. 103. Suite 808: Humana Press: Totowa, New Jersey.

Hollenberg CP and Gellissen G (1997) Production of recombinant proteins by methylotrophic yeasts. *Curr. Opin. Biotechnol.* 8: 554–560.

Invitrogen Corporation (1996) Pichia *Expression Kit. Instruction Manual*. Carlsbad, CA.

Romanos M (1995) Advances in the use of *Pichia pastoris* for high-level gene expression. *Curr. Opin. Biotechnol.* 6: 527–533.

Kim T, Gioto Y, Hirota N, Kuwata K, Denton H, Wu S, Sawyer L and Batt CA (1997) High level expression of bovine β-lactoglobulin in *Pichia pastoris* and characterization of its physical properties. *Protein Engineering* 10 (11): 1339–1345.

POLYMER TECHNOLOGIES FOR CONTROL OF BACTERIAL ADHESION

David Cunliffe, **Christopher A Smart** and **Cameron Alexander**, Macromolecular Science Department, Institute of Food Research, Reading Laboratory, Reading, UK

Introduction

Bacterial adhesion is important in many aspects of food microbiology. It profoundly influences the functioning of organisms in the human gut, the survival and growth of food-borne pathogens, and the uptake or transfer of microorganisms to different biological substrates. The attachment of cells to synthetic surfaces is also of considerable significance, with implications for the manufacture and processing of foods and, ultimately, the health and wellbeing of the consumer. Bacterial adhesion in this context is generally detrimental, for example where the attachment of pathogens and food spoilage organisms to foods and food-contacting surfaces leads to the contamination of the final products. However, cell adhesion can be used positively, as in the separation of microorganisms from foods by their adsorption to appropriate bacterial binding materials. The ability to control cell adhesion to artificial surfaces is therefore highly desirable, and one method by which this can be achieved is through the design and use of appropriate functional polymers. Although such materials are already used in a large variety of controlled separation and adsorption processes, it is only relatively recently that polymer technologies have been actively considered for similar applications in food microbiology.

Inhibition of Microorganism Attachment

Polymers for Retardation of Cell Adhesion

The development of large surface-bound communities of bacteria, known as biofilms, on materials used during food manufacture can lead to serious contamination of the final products. Biofilm removal often requires repetitive and harsh cleaning cycles and bactericides, which can themselves leave undesirable residues. The use of food contact materials capable of preventing the adsorption of bacteria is a potential method by which these problems can be avoided. In this context, a surface prepared from, or treated with, an appropriate ‘nonstick’ polymer is obviously of considerable interest to the food industry. While there has been some success recently in developing polymers that prevent cell attachment by surface display of bactericides (for example the introduction of Microban products by a UK Supermarket chain), there is nevertheless a need for a food contact material that intrinsically resists cell adhesion without the use of cytotoxic agents. However, the exact mechanisms of bacterial adhesion are not fully understood and accordingly the properties required of a surface to prevent attachment of cells are still the subject of intense study. Any investigations into nonstick surfaces must first consider the nature of the attaching material, in this case the bacterium. The specific roles of bacterial structures such as pili, cell wall components and extracellular lipopolysaccharides are of lesser significance in the initial stages of the attachment process than the thermodynamic factors involved and so are not discussed here; however, these specific cell structures are relevant to development of longer-term adhesion, particularly in the formation of biofilms, and indeed are the subject of intensive research. Nevertheless, to understand what makes a surface 'nonsticky' to bacteria in the vital first stages of adhesion, questions of thermodynamics must be posed.

Thermodynamic Treatments of Cell Adhesion

The attachment of a microorganism at a surface can be considered as a function of hydrophobicity or surface free energy, using the expression:

$$\Delta G_{ads} = \gamma_{SB} - \gamma_{SW} - \gamma_{BW} \qquad \text{(Equation 1)}$$

The free energy change of adhesion, ΔG_{ads}, is thus related to the free energies γ_{SB}, γ_{SW} and γ_{BW} of the solid–bacterium, solid–water and bacteria–water interfaces, respectively. These free energies cannot be measured directly, but estimates are readily made using wettability or contact angle measurements. Calculation of the interfacial energies of solids is possible using the equation:

$$\gamma_s - \gamma_{sl} = \gamma_l \cos\theta \qquad \text{(Equation 2)}$$

In this case γ_s is the surface free energy of the solid, γ_l is the surface free energy of the liquid and γ_{sl} is the solid–liquid interfacial energy. However, only two variables in Equation 2 can be ascertained directly: the surface tension of the liquid γ_l and the contact angle θ. To obtain the solid–surface free energy a further equation is required, linking the interfacial free energy γ_{sl} to the other variables. This can be derived empirically or calculated by a theoretical treatment, but in all cases a number of assumptions of how to describe interactions between different bodies are needed. However, if bacterial suspensions are regarded as living colloidal systems, the initial steps of adhesion can be approximated by colloid theories. These describe the free energies as a function of distance between two bodies. If steric factors are not important the total interaction energy can be obtained from a summation of the van der Waals and electrostatic interactions. Detailed considerations indicate that increasing the hydrophobicity of the bacterial or solid surfaces gives a larger van der Waals attraction, and correspondingly there is a considerable energy advantage for adsorption of bacteria to highly hydrophobic substrates. However, in vivo, polymeric organic materials such as proteins and polysaccharides are present; kinetic and thermodynamic arguments then suggest that (for hydrophobic substrates at least) a 'conditioning layer' of these components is adsorbed before bacterial attachment can take place. Therefore, in real situations the original substrate is likely to be wholly or partially masked by biological material which will interact rather differently with the cell surfaces.

Adsorbed biopolymers may significantly change the electrostatic and van der Waals interactions of the system, and since attached proteins or other polyelectrolytes can participate in charge–charge interactions, cell adhesion may also become strongly pH dependent. In some cases, both the microorganisms and the substrate may adsorb conditioning layers of macromolecules, which may hinder cell adsorption owing to steric chain–chain repulsion. Conversely, an attractive interaction may arise by charge or hydrogen-bonding between chains and this can represent a large binding energy for cell attachment. For chain–chain or chain–cell interactions to be significant the adsorbed layer must be partially flexible and extend from the surface, so that these forces can extend beyond the electric double layers associated with the bacteria and the substrate. Therefore, the thickness of the conditioning layer can also strongly affect the cell–surface interaction, and in practice the attachment of microorganisms does not always correlate simply with the physicochemical parameters of the adsorbed layer as obtained by contact angle measurements.

It is thus clear that bacterial adhesion in real systems is highly complex; although preparing a surface resistant to conditioning layer adsorption is an important step towards generating cell-repellent materials, the lack of correlation in some cases between protein and bacterial adhesion suggests that alternative strategies may also be needed. Accordingly, a large number of materials have been prepared and their properties investigated. A few examples are shown in **Table 1**, grouped in terms of the type of surface (hydrophilic, low-energy or mobile) they are designed to present to biomolecules and cells. However, these categories are not mutually exclusive, and even where design elements from more than one category have been employed, the 'perfect' cell-repellent surface remains an unattained goal.

Hydrophilic Surface Polymers

Hydrophilic materials are designed to resist the initial attachment of a protein to a surface, which is assumed to be the first step to bacterial adhesion in vivo. Thermodynamic considerations suggest that protein adsorption leads to a large increase in entropy as

Table 1 Polymer structures for prevention of cell adhesion

Hydrophilic surface	*Low surface energy*	*Mobile surface*
Poly(ethylene oxide), PEO	Poly(tetrafluoroethylene), PTFE	Poly(dimethylsiloxane), PDMS
Poly(vinyl alcohol), PVA	Poly[(3,3,3-trifluoropropyl)(methyl)siloxane]	Poly[(3,3,3-trifluoropropyl)(methyl)siloxane]
Poly(*N*-vinylpyrrolidone), PNVP	Poly(dimethylsiloxane), PDMS	Poly(ethylene oxide), PEO
Poly(hydroxyethylmethacrylate), PHEMA	Perfluoropolyethers (Fomblin, Krytox)	Poly(urethane-block-siloxane)

water molecules previously associated with the surface are displaced into the bulk aqueous solution. For hydrophobic substrates, there is also an enthalpic gain as hydrophobic segments of the protein adsorb, which is only partially offset by the loss in enthalpy as the poorly attached water desorbs. However, relative to water itself, all substrates can be considered as hydrophobic to some degree, which suggests that proteins will attach to almost any surface they can contact. Nevertheless, protein adsorption should be at least partially suppressed by reducing the energy gain on release of water from the substrate: this can be achieved by preparing a material with a very hydrophilic surface.

A number of synthetic polymers with such surfaces have been prepared by chemically cross-linking water-soluble precursor polymers: these are known as 'hydrogels'. The desorption of strongly bound water molecules at the surfaces of these polymers incurs an enthalpic penalty and does in practice reduce the adsorption of proteins. However, typical hydrogels are not mechanically tough enough to be used as food contact surfaces, and their porous nature means that with long-term exposure, microorganisms can eventually attach by physical entrapment at recesses in the substrate. As a result methods have been devised to attach materials with the functionality of hydrogels (i.e. those expressing a non-charged hydrophilic surface) to substrates of greater mechanical stability. These materials can be natural in origin, such as polysaccharides and their derivatives, or synthetic. Typical examples of the latter materials are poly(vinyl alcohol), poly(ethylene glycol), more accurately referred to as poly(ethylene oxide), polyacrylamides and poly(*N*-vinyl-2-pyrrolidone). Attachment to the appropriate robust substrate is carried out by adsorption, blending or surface grafting.

Adsorption of Hydrophilic Polymers Adsorption is the simplest method of attachment; it generally involves the preparation of an amphiphilic block or graft copolymer which physically bonds to the substrate by its hydrophobic segments, leaving the hydrophilic portions strongly solvated and projecting into the solution. Careful adjustment of the hydrophilic/hydrophobic balance in the copolymer is required for optimum performance, and these materials suffer from the disadvantage that repeated use or harsh cleaning cycles can remove the adsorbed layer.

Polymer Blends for Hydrophilic Surface Display The problem of durability can be alleviated by blending the amphiphilic block or graft material with a second substrate polymer which is miscible with the hydrophobic segments of the copolymer. During processing the hydrophilic segments migrate to the surface exposing the hydrogel functionality while leaving the hydrophobic portions bound into the bulk of the substrate. Materials of this type retain a hydrophilic surface longer than those prepared by adsorption, but the overall properties are crucially dependent on the processing conditions.

Surface Grafting of Hydrophilic Polymers The most versatile method of modifying a substrate is chemical grafting onto the surface, using polymers with functional endgroups which can react with chemical groups on the substrate. The resulting materials are considerably more resilient than those prepared by adsorption or blending, and any type of hydrophilic group can potentially be attached as long as the substrate and the graft polymer are appropriately functionalized. However, other factors can affect the performance of these materials, such as the difficulties of introducing the graft polymer to the substrate at concentrations sufficient to ensure complete surface coverage. Nevertheless, the technique has been widely used, and of the polymers mentioned above, surface-grafted poly(ethylene oxide) (PEO) has been found to be most effective in the prevention of protein and cell attachment. This is probably due to its unique solution properties, and considerable effort has been expended in determining the detailed structure of this polymer at solid–water interfaces.

Model studies of short-chain PEO self-assembled monolayers on crystalline substrates have shown the conformation of the polymer chain to be a critical factor in protein attachment. Tightly packed PEO chains are less resistant to protein adsorption owing to a reduction in the amount of water that can associate with the polymer. As a result, it has been suggested that it is the conformation-dependent degree of solvation – and consequently the stability of an interfacial water layer – that determines if a protein diffusing to a PEO surface can irreversibly adsorb.

Grafting PEO onto less-ordered substrates does not lead to the well-defined chain packing or surface density obtained with self-assembled monolayers, and in these cases PEOs with a greater molecular weight are required to obtain the same degree of protein rejection. For PEO grafted to substrates typically used as food contact surfaces such as glass or stainless steel, it is generally observed that protein adsorption decreases with increasing polymer chain length up to a molecular weight of about 2000; beyond that there is no further reduction in protein adsorption. This is often quoted as evidence of a steric mechanism for protein or cell repulsion, as it is assumed that above a certain chain length, the polymers must inevitably extend well into the solution, forming a solvated macromolecular barrier which covers the surface. However, for amphiphilic poly(ethylene oxide)–poly(propylene oxide) (PEO/PPO) copolymers adsorbed to glass substrates, it has been found that hydrophilic segments as short as six ethylene oxide units (molecular weight about 250) can be sufficient to suppress protein adsorption, and increasing the block length gives no further reduction. The same protein repelling ability by hexa(ethylene oxide) as pure self-assembled monolayers on gold has also been observed, whereas in mixed self-assembled monolayers longer ethylene oxide chains are required.

These somewhat confusing results have led to debate as to the mechanism of protein repulsion by adsorbed or grafted polymers. In addition, materials of differing chemistry to PEO, but designed to have similar interfacial properties, can resist protein adsorption, as shown by self-assembled monolayers of tri(propylene sulphoxide). Most recently a biomimetic approach to protein-resistant surfaces has been pursued, where a functional polymer is rendered amphiphilic by grafting hydrophilic dextran and hydrophobic fatty acid moieties to the chain. In this case, the copolymer presents a glycocalyx-like surface, with the oligosaccharide segments projecting into the solution and the hydrophobic parts anchoring the polymer to a hydrophobic substrate. As with the PEO/PPO block copolymers, the surface orders into a polymer 'brush' or 'comb' structure, and the resulting material is highly resistant to protein adsorption under conditions designed to mimic biological fluids.

In all the above cases, it is assumed that a surface that suppresses protein adsorption will provide the greatest resistance to bacterial adhesion at the initial stages owing to the absence of a conditioning layer to which the cells can attach. However, other methods that have been explored for controlling bacterial adhesion have been based on the premise that even if the cells do attach, if the bond between the microorganism and the substrate is weak the bacteria may be easily removed.

Low Surface Energy Polymers

A major strategy for designing 'nonstick' materials has involved the use of materials with low surface energies. The free energy at the surface is dependent on the additional energy associated with the groups at an interface compared with the same groups within the bulk of the material, and the magnitude of the surface free energy controls any interaction with an adherand. Materials possessing low surface energies have relatively low enthalpic gains on adsorption. A number of commercial polymers have suitably low surface energies and have been tested experimentally as nonstick substrates. Commodity polymers such as poly(ethylene) and poly(propylene) are potentially attractive for this reason but their optimum mechanical properties are reached at high molecular weights, at which the polymers are not soluble and are difficult to process. The corresponding unsaturated hydrocarbon polymers are more soluble but are much less chemically stable, being prone to oxidation which results in degradation of the surface and deterioration in their mechanical properties. However, hydrocarbon-based polymers with halogens substituted into the backbone or in the side chains are much more stable and possess even lower surface energies than the unsubstituted polymers. The properties of poly(tetrafluoroethylene) (PTFE) are well known, but this material does have disadvantages: it is difficult to process and is poorly resistant to bacterial colonization because it is highly porous in the pure state. Other commercial fluoropolymers have similarly low surface energies (15–31 $mJ\,m^{-2}$) and are as difficult to process as PTFE and the hydrocarbon polymers. To overcome the processing problems and to allow grafting to preformed substrates, a range of soluble and reactive fluorooligomers and polymers have been developed. Hydroxyl-functional fluorinated precursors can be reacted with isocyanates to form low surface energy polyurethanes, or fluoroacrylates can be copolymerized with conventional acrylic systems to produce the desired 'nonstick' material. This is currently an area of active research and new, ultra-low surface energy polymers are constantly being synthesized and tested for resistance to bacterial adhesion.

Mobile Surface Polymers

A third strategy for control of cell adhesion is the use of materials that present a constantly changing or highly mobile surface. It is believed that rapid molecular motion prevents the formation of strong adhesive

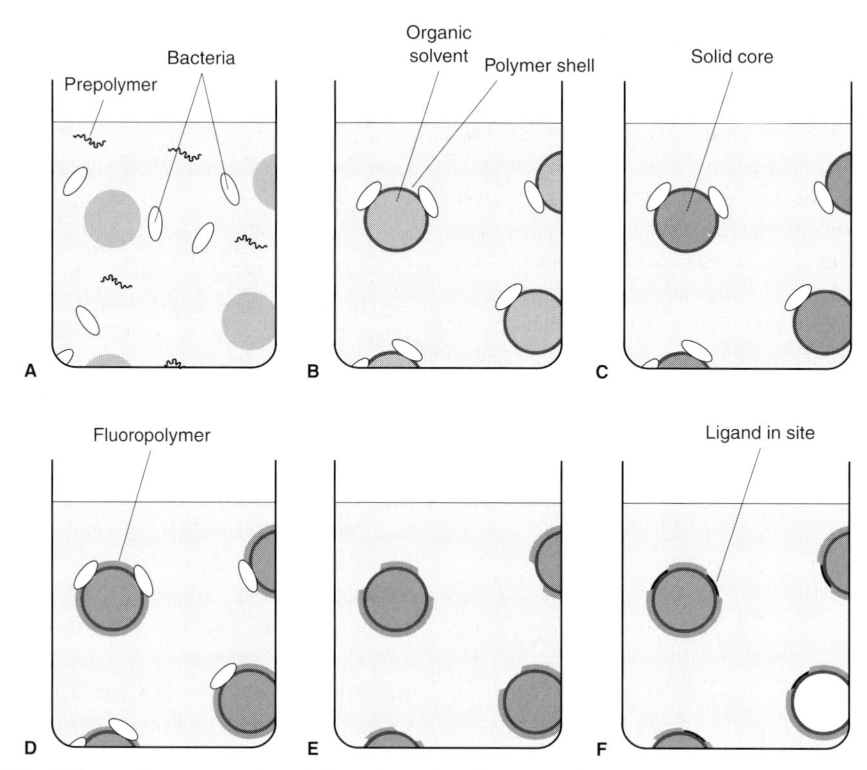

Figure 1 Cell-mediated lithography: preparation of the polymer beads (see text).

bonds between an adsorbing species and a substrate. Typical polymers with highly mobile backbones include silicones such as poly(dimethylsiloxane) (PDMS) which also has low surface energy. This polymer is not tough enough for use in food contact applications, but more robust silicone copolymers can be prepared which retain the highly mobile surface. Fluorinated poly(dimethylsiloxanes) are also available, combining surface mobility with the surface energy of materials such as PTFE. Although in aqueous environments the surface energies of these polymers increase owing to chain rearrangements exposing polar groups at the polymer–water interface, adhesion of biological moieties remains low, implying that the mobile surface is effective. In addition to the silicone-based polymers, poly(ethylene oxide) is also believed to present a highly mobile surface in water, owing to the flexibility of its hydrated chains. Adsorption of a bacterium to any mobile polymer surface results in the loss of mobility of the chains and an entropic penalty, which must be offset by an enthalpic contribution as surface–adherand bonds form. As stated earlier, the hydrophilic nature of PEO confers little enthalpic gain for adsorption, and this may be another reason for the ability of high-entropy, long-chain PEO to inhibit protein and bacterial adsorption.

There is still much debate as to the best method for preventing bacterial adhesion and much theoretical and experimental work is required. The selection of the best cell-resistant surface for a particular application is therefore still to some extent a matter of trial and error. The development of a truly bacteria-resistant food contact material would be of tremendous benefit to the food industry and to the consumer, and remains a challenging objective for researchers in polymer science and food microbiology alike.

Selective Adsorption of Bacteria

The control of bacterial adhesion is also of importance in the separation and concentration of bacteria from food prior to microbiological analysis. In particular, the ability to bind bacterial cells specifically to a surface as a means of removing microbial pathogens from foods remains a significant problem. The ideal material should adsorb bacterial contaminants rapidly, with high affinity, should be mechanically robust and capable of multiple assays, and should be adaptable to the particular end use required. Synthetic polymers offer a number of advantages in this context, as most are capable of withstanding higher temperatures and harsher conditions than natural materials, and are in general easy to manufacture in a variety of forms. A number of systems that use synthetic particles or polymer beads are already commercially available, and form the basis of the immunomagnetic separation technique. However, while this particular methodology has been highly successful, it nevertheless requires the use of cell-specific antibodies for adsorption of bacteria, which can increase the costs considerably if large amounts of antibody are required. As a result, new ways of preparing bacteria-specific adsorbents in which the need for antibodies is reduced through a technique termed 'cell-mediated lithography', are being explored. Although still in development, the methodology enables the preparation of highly durable polymer beads potentially capable of multiple use in microbiological assays.

Cell-mediated Lithography

The aim of this technique (which can be considered as a development of the imprinting method for preparing molecular-specific polymer adsorbents) is to allow a target bacterium such as a food pathogen to define its own binding site on the outer surface of a polymer bead as the polymer is formed. After removal of the microorganisms from the beads, the exposed binding sites on the polymer surface can act as an adsorbent for the type of cell used in its preparation. In this way, the methodology should be adaptable for the manufacture of a polymeric material as a specific adsorbent for any desired microorganism (**Fig. 1**).

A suspension of the bacteria of interest is stirred in a two-phase aqueous/organic medium containing prepolymers (Fig. 1A). The cells partition at the interface; at the same time, the prepolymer in the water phase reacts with a component in the organic layer to form a polymer microcapsule under and around the bacteria (Fig. 1B). The system is subjected to ultraviolet light to polymerize a third component in the organic phase, thus solidifying the core of the microcapsule (Fig. 1C). The resulting polymer beads with bacteria 'imprinted' in the surface are reacted with a perfluoropolymer of low surface energy, to block chemical groups in areas of the bead surface not covered by bacteria and to render the bulk of the polymer surface 'nonsticky' to microorganisms (Fig. 1D). The cells are then removed, exposing patches of the underlying reactive polymer, which are of the exact dimensions of the imprinted cells (Fig. 1E); these

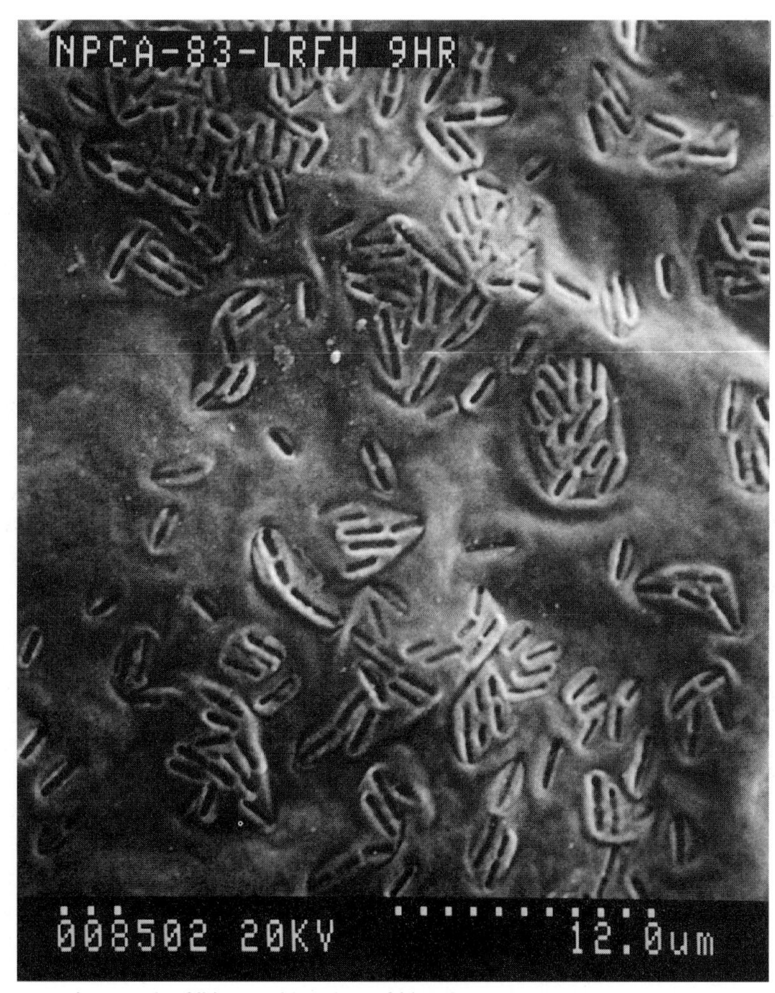

Figure 2 Scanning electron micrograph of lithographic prints of *Listeria monocytogenes* on a polymer bead.

are the 'lithographic prints' of the microorganisms, and their subsequent binding sites. The polymer beads can then be utilized as adsorbents for the bacteria used in their preparation, or undergo reactions with fluorescent labels (Fig. 1F) or specific biological ligands, to enhance further the binding affinities in the lithographed areas. Electron and confocal laser scanning microscopy enable the examination, respectively, of the physical structure and the chemistry of these binding sites.

As can be seen from the photomicrographs, the fluorescent areas (**Fig. 2**) are of dimensions exactly matching those of the lithographic prints of the bacteria (**Fig. 3**). The fluorescent label can react only in the printed sites as the remainder of the surface is rendered inert during the lithographic process. Similarly, specific cell-binding groups can be concentrated into these sites, enabling the precise positioning of functionality where it is needed, i.e. in the lithographic prints, rather than spread at random over the bead surface. In preliminary cell separation studies from pure cultures, beads lithographed using *Listeria monocytogenes* showed better adsorption of this microorganism than *Staphylococcus aureus* by a factor of 2, whereas the polymer beads 'printed' with *Staphylococcus* displayed preferential binding of *Staphylococcus* over *Listeria* by ratios of up to 8 : 1. The nonspecific lectin concanavalin A was incorporated into the lithographed prints for these experiments, indicating that expensive antibodies are not necessarily required for bacterial discrimination, although the selectivity would be greatly enhanced by filling the lithographed sites with target-specific affinity ligands.

Future Possibilities

Cell-mediated lithography is still in its early stages, and much greater selectivity for particular pathogens will be required before the technique can be adopted for routine use in food microbiology. However, it is clear that polymer technologies have much to offer in devising new ways of adsorbing bacteria to surfaces as well as preventing cells from attaching to substrates in the first place. The challenge for polymer scientists and food microbiologists is to develop methods for controlling or even 'tuning' cell adhesion to the exact

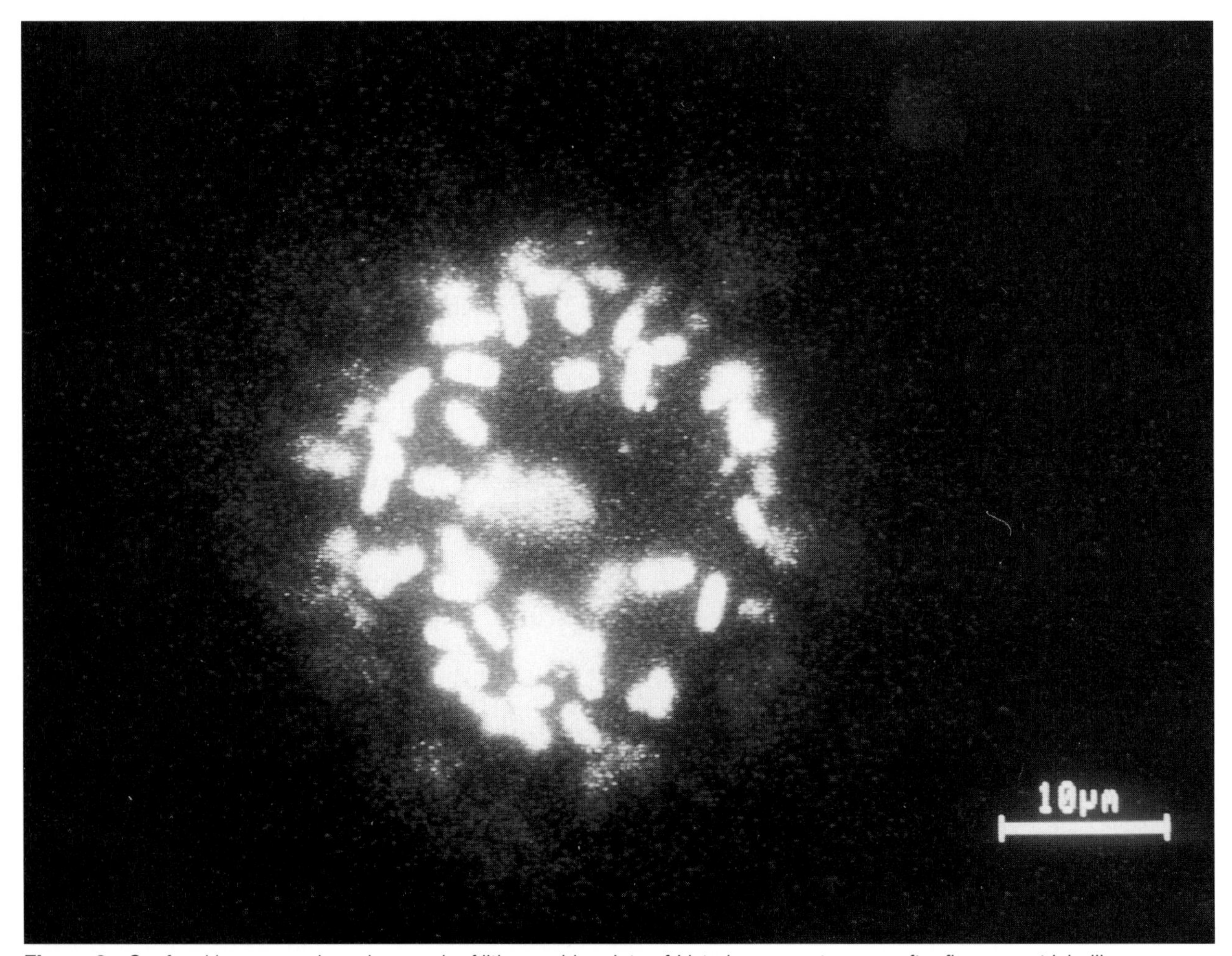

Figure 3 Confocal laser scanning micrograph of lithographic prints of *Listeria monocytogenes* after fluorescent labelling.

level required, and this continues to be a major research objective.

See also: **Biofilms**. **Biophysical Techniques for Enhancing Microbiological Analysis**: Future Developments. **Biosensors**: Scope in Microbiological Analysis. **Immunomagnetic Particle-based Techniques**: Overview. ***Listeria***: Detection by Classical Cultural Techniques; Detection by Commercial Enzyme Immunoassays; Detection by Colorimetric DNA Hybridization; *Listeria monocytogenes* – Detection by Chemiluminescent DNA Hybridization. **National Legislation, Guidelines & Standards Governing Microbiology**: European Union. **PCR-based Commercial Tests for Pathogens**. **Process Hygiene**: Overall Approach to Hygienic Processing. **Ultrasonic Standing Waves**.

Further Reading

Bitton G and Marshall KC (eds) (1980) *Adsorption of Microorganisms to Surfaces*. Chichester: John Wiley.

Bower CK, McGuire J and Daeschel MA (1996) The adhesion and detachment of bacteria and spores on food-contact surfaces. *Trends in Food Science and Technology* 7: 152–157.

Brady RF (1997) In search of non-stick coatings. *Chemistry and Industry* 6: 219–222.

Elbert DL and Hubbell JA (1996) Surface treatments of polymers for biocompatibility. *Annual Reviews in Materials Science* 26: 365–394.

Holland NB, Qiu YX, Ruegsegger M and Marchant RE (1998) Biomimetic engineering of non-adhesive glycocalyx-like surfaces using oligosaccharide surfactant polymers. *Nature* 392: 799–801.

Morra M and Cassinelli C (1997) Bacterial adhesion to polymer surfaces: a critical review of surface thermodynamic approaches. *Journal of Biomaterials Science: Polymer Edition* 9: 55–74.

Mrksich M and Whitesides GM (1996) Using self-assembled monolayers to understand the interactions of man-made surfaces with proteins and cells. *Annual Review of Biophysics and Biomolecular Structure* 25: 55–78.

Van Loosdrecht MCM, Lyklemam J, Norde W and Zehnder AJB (1990) Hydrophobic and electrostatic parameters in bacterial adhesion. *Aquatic Sciences* 52: 103–113.

Whitcombe MJ, Alexander C and Vulfson EN (1997) Smart polymers for the food industry. *Trends in Food Science and Technology* 8: 140–145.

Polysaccharides *see* **Fermentation (Industrial)**: Production of Polysaccharides, e.g. Xanthan Gum.

Poultry *see* **Meat and Poultry**: Spoilage of Meat; Curing of Meat; Spoilage of Cooked Meats and Meat Products.

Pour Plate Technique *see* **Total Viable Counts**: Pour Plate Technique.

PREDICTIVE MICROBIOLOGY AND FOOD SAFETY

T Ross and **T A McMeekin**, University of Tasmania, Hobart, Australia

J Baranyi, Institute for Food Research, Reading, UK

Introduction

Predictive microbiology may be considered as the application of research concerned with the quantitative microbial ecology of foods. In general, viruses and protozoa are inert in foods and have no 'ecology' as such, but may be inactivated. Thus, although predictive microbiology has been concerned almost exclusively with growth of bacteria and fungi in foods, the survival and inactivation of food-borne bacteria, fungi, viruses and protozoans have latterly begun to be modelled also.

Predictive microbiology is based on the premise that the responses of populations of microorganisms to environmental factors are reproducible and that, by characterizing environments in terms of those factors that most affect microbial growth and survival, it is possible from past observations to predict the responses of those microorganisms in other, similar, environments. This knowledge can be described and summarized in mathematical models which can be used to predict quantitatively the behaviour of microbial populations in foods, e.g. growth, death, toxin production, from a knowledge of the environmental properties of the food over time.

This article considers: the history, philosophy and impetus for development of the field; theory of mathematical modelling in general; types of models used in predictive microbiology; proposed uses and strategies for 'predictive microbiology' within the food industry; an assessment of the performance of 'predictive microbiology' models; and future research directions and anticipated outcomes.

Past and Present

Origins

In the 1980s, it was recognized that traditional microbiological end-product testing of foods was an expensive and largely negative science, and a more systematic and co-operative approach to the assurance of the safety of foods was advocated. The concept of 'predictive microbiology' was proposed within which the growth responses of the microbes of concern would be systematically studied, quantified and modelled mathematically with respect to the main factors controlling their growth in most foods, such as temperature, pH and water activity (a_w). It was suggested that models relevant to broad categories of foods would greatly reduce the need for *ad hoc* microbiological examination and enable predictions of quality and safety to be made quickly and inexpensively. The concept had been suggested as early as 1937, but was not seriously attempted until the early 1980s when the availability of funding in response to major food poisoning outbreaks, and the ready access to computing power, enabled its realization.

Although gaining acceptance, the concept continues to be viewed sceptically by some who consider that there are too many variables related to food structure and microbial physiology to enable reliable predictions to be made. These criticisms are discussed later. Nonetheless, mathematical models for the rate of death of spore-forming bacteria have formed the basis of canning for over fifty years, and modelling is well developed in the area of industrial fermentation. Modelling in fermentation is directed toward optimization, often in large scale, homogeneous, axenic, chemically defined, media. In predictive microbiology the interest is in growth minimization or arrest, in non-nutrient limited batch cultures, and at cell densities lower than is usual in fermentation or biotechnology.

Work in the 1970s had considered the limits to growth of microbial pathogens, but did not use mathematical models to summarize the data. The impetus in the 1980s led to research and data collection on growth rates, or in work dealing with *Clostridium botulinum*, data on the probability of toxin formation in foods within a certain time. Large programmes to develop models for the growth of food-borne microbial pathogens were initiated in the United Kingdom and United States, and were complemented by independent smaller programmes in other nations. A range of models for the growth rates of pathogens and spoilage organisms resulted, many of which are incorporated into software packages. Subsequently, the emergence of very low infectious dose pathogens (e.g. EHECs) in 'ready-to-eat' foods has refocused attention on describing limits to growth and rates of both thermal and non-thermal inactivation.

Impetus and Benefits

Consumers desire foods that are considered to be 'fresher', and more 'natural', i.e. less processed, while expecting food to be free from potentially harmful microbes, additives or contaminants. These competing demands require a better and more quantitative understanding of microbial physiology and ecology in foods and microbial responses to food preservation methods, so that food processing can be 'fine tuned' to minimize the processing while maintaining product safety and stability. Proponents claim that the development of a quantitative approach to microbial ecology and physiology, and made accessible through predictive microbiology, will enable:

1. prediction of the consequences (for product shelf life and safety) of changes to product formulation;
2. rational design of new processes and products to meet required levels of safety and shelf life;
3. objective evaluation of the microbiological consequences of processing operations and, from this, an empowering of the HACCP approach;
4. objective evaluation of the consequences of *lapses* in process and storage control and, from this, appropriate remedial action;
5. when combined with stochastic modelling techniques the ability to analyse systems to determine which steps in the handling of the product contribute most to the overall risk; and
6. development of teaching tools.

Thus, the quantitative approach to the microbial ecology of foods, and its application through predictive microbiology, are expected to become essential elements of modern approaches to microbial food safety and quality assurance.

Limitations

Objections to the utility of predictive microbiology have been raised and may be grouped into five interrelated areas:

1. assessment of initial conditions;
2. relevance of model systems to foods;
3. variability in responses;
4. provision of 'user-friendly' technology; and
5. empirical nature of the current generation of models.

Assessment of Initial Conditions Prediction of the absolute number of organisms present in a food at some time, requires knowledge of how many were present at some previous time, the physiological state of those organisms, and the environmental conditions the organisms had experienced subsequently.

If a food is produced to a consistent level of quality an initial level and initial physiological state (usually expressed as 'lag time') can be characterized, and used for subsequent calculations.

If the initial microbiological status of the product is unknown, useful information can still be derived by using the concept of relative rates. In this approach, the relative change in the microbial load can be calculated, i.e. whatever the initial population it will be n-fold greater or n-fold less, as a consequence of the environmental conditions to which the product has been exposed over time.

Theory and Philosophy of Mathematical Modelling

Considerations in Modelling

The essential purpose of mathematical models is to describe succinctly a set of acquired data. From a scientific perspective, however, it is more useful to consider a model as describing an underlying process, whether known or proposed, which generates data. Such models embody a hypothesis, i.e. the model is the mathematical expression of that hypothesis.

Predictive microbiology involves the systematic study and quantification of microbial responses to environments in foods. It first aimed, simply, at the collection and smooth representation of computerized microbial data, but mathematical modelling also provides a useful and rigorous framework for the scientific process. To develop a consistent scientific framework to interpret the microbial ecology of foods, it is desirable to integrate the patterns of microbial behaviour discerned with knowledge of the physiology of microbes.

Thus, two major types of model are recognized. Empirical models are derived from an essentially pragmatic perspective, and simply describe the data in a convenient mathematical relationship. When taken to its extreme, this approach has been described as 'curve fitting'. Mechanistic, or deterministic, models are built up from theoretical bases, and if they are correctly formulated, may allow the interpretation of the response in terms of known phenomena and processes. Mechanistic models are also easier to develop further as the quantity and quality of the information on the modelled system increases.

Although the fast development of the discipline has seen more and more mechanistic elements used in model construction, in practice most models are neither purely mechanistic nor purely empirical but lie between the two extremes. Nonetheless, even empirical models aid the food microbiologist in day-to-day decision making, and models have immediate use in improving food safety whether the underlying ecological, physiological and physicochemical processes are understood, or not.

Whatever its type, the mathematical expression of a model has several component parts, which are described and named in **Figure 1**.

Practical Model Building

A spectrum of needs and strategies exists for developing predictive models for food microbiology. These are summarized in **Table 1**.

Typically, a reductionist approach is adopted, and models are developed under well-defined and well-controlled conditions in laboratory broth media. The primary variable of interest may be growth rate, death rate, the time for some event to happen or some condition to be reached, or the probability that the event will happen within some predetermined time.

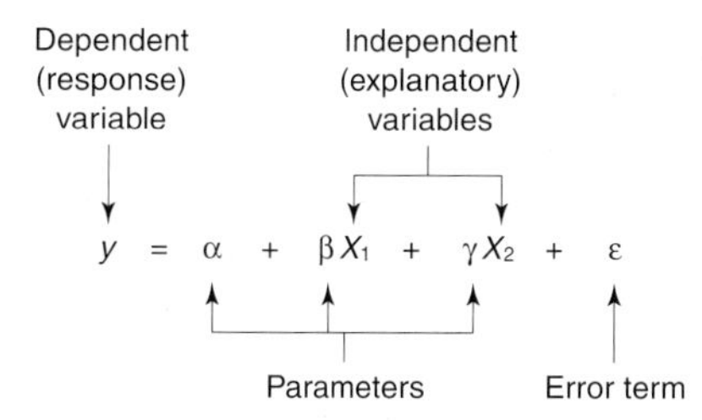

Figure 1 An example of a mathematical model showing the nomenclature of the component terms. The values of the independent variables (X_1, X_2) are known or set before the response (y) is observed. The values of the parameters (α, β, γ) are determined by the data, and are calculated, or 'fitted', to minimize the difference between the observed response and that predicted by the model. The stochastic term (ε) indicates the extent to which the predicted response differs, on average, from the observed response. (After McMeekin et al 1993).

Table 1 Diversity of problems and methods in predictive microbiology

Problem types	Toxin formation
	Shelf-life prediction (spoiler growth)
	Pathogen growth
	Pathogen survival
	Death or inactivation (pasteurization, canning, irradiation)
Model types	Death rate
	Probability of growth/toxin formation
	Growth rate
	Growth limits
Data collection methods	Turbidimetry
	Metabolite assays
	Viable counts
	Impedance/conductance
	Luminometry

Typically, this response will itself be determined by recourse to a model. For example, bacterial death rate cannot be measured directly; it must be derived from measurements of numbers of survivors over a period of time. The interpretation of this decrease in numbers is aided by a model, e.g. first order reaction rate model. Inherent in this approach would be the assumption that microbial death caused by high temperature is sufficiently well described by log-linear kinetics. Although this assumption is the subject of debate (other models for survivor curves include multi-hit and multi-target theory, energy-distribution, heterogeneous heat-resistance etc.), D-values could be used to summarize the reduction in numbers over time under different conditions of pH, temperature, water activity and so on.

Next, a model that relates the effect of those environmental conditions on D-values would be developed. In doing so, the log-linear death model has been 'embedded' into a more complex model for the effects of environmental conditions on the rate of thermal death. This process can continue, for example by including that model in a model for the fate of a pathogen during processing and subsequent distribution, storage and preparation for eating, such as would be done in an exposure assessment model in quantitative microbial risk assessment or the incorporation of that model into computer software applications.

Nomenclature was proposed for the different levels of models. Primary models include those that enable factors of primary interest to be determined, e.g. maximum specific growth rate (μ), death rate (D). The dependence of these factors on environmental conditions is then summarized in a secondary model. To enable practical application, i.e. to make predictions, that model could be incorporated into interactive computer software – a tertiary model. A more detailed description of the process of model building is given in the Appendix.

Variables Modelled

In the US and UK modelling programs, the key independent variables considered were temperature, pH, water activity (or concentration of a specific humectant), nitrate concentration, and gaseous atmosphere (see **Table 2**). In general, these same factors are modelled by most groups. Other factors modelled specifically include the concentration of specific organic acids or specific preservative compounds. In many practical situations, temperature has the most dominant effect on microbial growth rate, followed by water activity, then pH, with other factors playing lesser roles. Under certain circumstances, other factors will have a critical affect, but only when the dominant constraints have caused the organism to be near its limit for growth. Models for specific products, or packaging types, or microbial risks may also be developed and in some situations it is considered more practical to develop the model directly from observations made on the product of interest, or a model system closely based on the product of interest. The large-scale modelling projects represent a strategy that seeks to determine general patterns of response and from that base to work toward more and more specific cases. The former product-specific approach also aids the modelling initiative if all variables controlling growth in that situation are identified and quantified.

Modelling 'Rules'

Several factors dictate the choice of the model structure. Some of these are briefly described in **Table 3**, but a full explanation is beyond the scope of this article. Readers are referred to the suggested reading list for further details.

Pragmatically, two features of a model are critical to its utility. The first is the ability to predict accurately microbial responses under all conditions to which the model applies. Evaluation of this ability is loosely termed 'model validation', and is described later. Failure to address the issues listed in Table 3 may be revealed when the model is 'validated'. The second critical factor is the range of independent variables and variable combinations to which the model applies.

Interpolation

A fundamental principle is that unless a model is fully mechanistic, the model should *not* be used to make predictions of responses to conditions beyond the range of factors tested during the model's development, i.e. empirical models can be used for inter-

Table 2 Models available in software packages. From Ross T (1999) *Predictive Food Microbiology Models in the Meat Industry.* Meat and Livestock Australia.

Software product	*Type of model*	*Organism*	*Factors modelled*
Food MicroModel	Growth rate, lag time	*Aeromonas hydrophila*	temp, pH, NaCl
		Bacillus cereus	temp, pH, NaCl, CO_2
		Bacillus licheniformis	temp, pH, NaCl
		Bacillus subtilis	temp, pH, NaCl
		Bacillus thermosphacta	temp, pH, NaCl
		Clostridium botulinum	temp, pH, NaCl
		Clostridium perfringens	temp, pH, NaCl
		Escherichia coli	temp, pH, NaCl, CO_2
		Listeria monocytogenes	temp, pH, NaCl, CO_2, nitrite, lactate
		Staphylococcus aureus	temp, pH, NaCl
		Salmonella	temp, pH, NaCl, nitrite
		Yersinia enterocolitica	temp, pH, NaCl
Pathogen Modeling Program	Growth rate, lag time, non-thermal death rate	*Escherichia coli* O157 : H7	temp, pH, NaCl, nitrite, lactate, anaerobic
		Listeria monocytogenes	temp, pH, NaCl, nitrite, anaerobic, lactate
		Staphylococcus aureus	temp, pH, NaCl, nitrite, lactate
		Salmonella	temp, pH, NaCl, nitrite
	Growth rate, lag time	*Aeromonas hydrophila*	temp, pH, NaCl, nitrite, anaerobic
		Bacillus cereus	temp, pH, NaCl, nitrite, anaerobic
		Shigella flexneri	temp, pH, NaCl, nitrite, anaerobic
		Yersinia enterocolitica	temp, pH, NaCl, nitrite, anaerobic
	Time to toxigenesis	*Clostridium botulinum*	temperature, pH, NaCl
Food Spoilage Predictor	Growth under fluctuating conditions, remaining shelf life	Psychrotrophic pseudomonads	temperature, water activity
Delphi	Growth under fluctuating conditions	'Generic' *Escherichia coli*	temperature, anaerobic
Seafood Spoilage Predictor	Growth under fluctuating conditions, remaining shelf life	Various spoilage processes for temperate and tropical seafoods	temperature

Table 3 Some considerations in the selection of models

Subject	*Reasons*
Parameter estimation properties	Relates to the procedure of estimating the model parameters. In general, models should have parameters whose estimation properties are close to those of linear models, i.e. estimates should be 'iidn': (independent, identically distributed, normal)
Stochastic assumption	The form of the model, and choice of response variables, should be such that the difference between prediction and observations is normally distributed, and that the magnitude of the error is independent of the magnitude of the response, otherwise the fitting can be dominated by some data, at the expense of other data.
Parameter interpretability	It is useful if the parameters have biological interpretations that can be readily related to the independent and dependent variables. This can simplify the process of model creation and also aid in understanding of the model, though this may be less important initially than the behaviour and performance of the model.
Parsimony	Follows from the principle of 'Ockam's Razor': models should have no more parameters than are required to describe the underlying behaviour studied. Too many parameters can lead to a model that fits the error in the data, i.e. generates a model that is specific to a particular set of observations. Non-parsimonious models have better descriptive ability but poorer predictive ability.
Correct qualitative features	In mathematical terms, these are the analytical properties of the model function. They include convexity, monotony, locations of extreme and zero values. If biological considerations prescribe any of these, the model should satisfy that.
'Extendability'	When a model is developed further (such as to include more, or dynamically changing environmental factors) the new, more complex model should contain the old, simpler one as a special case.

polation but not extrapolation. Interpolation is currently the fundamental basis of predictive microbiology – predictions are made by interpolation between conditions at which the responses of microbes have been tested and recorded previously. So that model users do not attempt to use the model to make predictions for which it was never designed, users must be fully aware of the bounds of the model.

In cases where many variables are involved, the determination of the interpolation region is not self-evident – the region is sometimes unexpectedly small, but tools for its definition are available.

Model Types

Models fall into two main groups.

- *Kinetic models* are concerned with rates of response (e.g. growth, death).
- *Probability models* were originally concerned with predicting the likelihood that organisms would grow and produce toxins within a given period of time. More recently, probability models have been extended to define the absolute limits for growth of microorganisms in specified environments, e.g. in the presence of a number of stresses which individually would not be growth limiting but collectively prevent growth. This approach represents a quantification of the Hurdle Concept.

Kinetic Models

Modelling Death Rate

Thermal The killing of bacteria by lethal high temperature is the cornerstone of the canning industry. Models have been used in that industry since the 1920s. These models and their performance are discussed elsewhere in this volume.

Non-thermal The study of death kinetics of bacteria due to non-thermal factors, principally water activity, pH and organic acids is in its infancy and the patterns of behaviour are unclear. It appears, however, that complex multi-phase death curves result (**Fig. 2**).

Modelling Growth Rate The process of model development was discussed briefly earlier. A more detailed treatment is given in the Appendix.

Probability Models

'Probability' Models These models consider the probability of some event within a nominated period of time. The probability of detectable growth when plotted as a function of time, is a sigmoid curve with an upper asymptote representing the maximum probability of growth given infinite time. The probability is a function of the time required for germination or lag resolution, the rate of growth of the organism, and the number of cells initially present. As these models typically involve an element of time, the distinction that has traditionally been made between this kind of probability model and kinetic models is somewhat artificial. However, superimposed on the time dependency of probability, the probability of the response, even at infinite time, may not be '1' in all environments. Probability models incorporate stochastic elements, such as the variability in lag times and growth rates.

Growth/No Growth Interface Models The paradox of maintaining food safety while minimizing processing leads to a desire to define minimum combinations of preservative factors that prevent the growth of specific microorganisms. An example of this type of model is shown in **Figure 3**.

Model Performance

Each step in the model construction process introduces some error. **Table 4** indicates error sources relevant to predictive microbiology.

Model predictions can never perfectly match observations. To assess the reliability of models before they

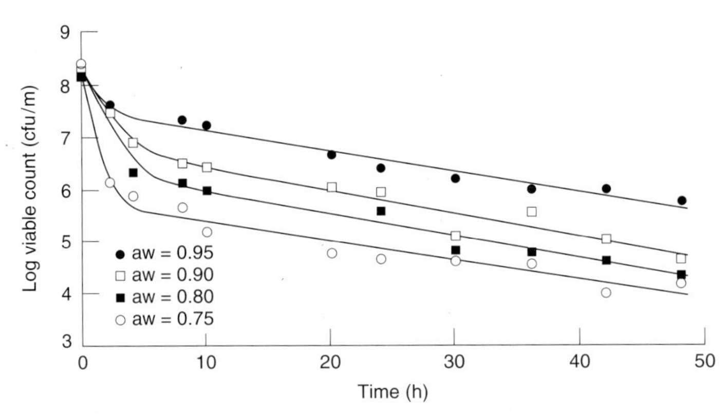

Figure 2 Example of non-thermal death curves. The curves represent the survival of *Escherichia coli* M23, grown to stationary phase in laboratory broth of $a_w = 0.996$, and then subcultured to fresh medium with lower water activity due to the addition of sodium chloride. In each case the water activity is below that which permits growth. The temperature of the incubations was 25°C which is, of itself, not lethal to *E. coli*. The solid lines are those from the same primary model fitted to each experimental dataset.

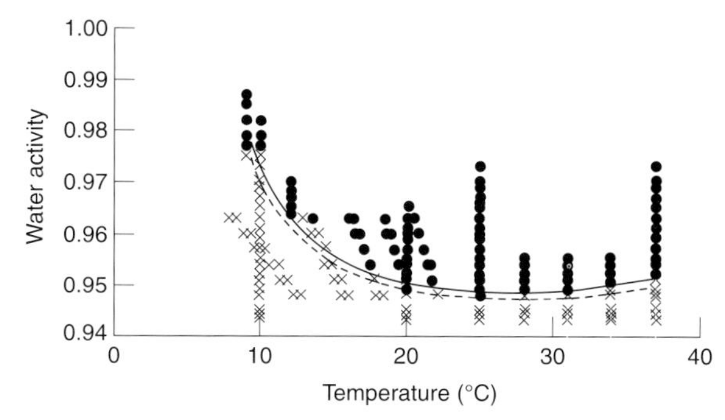

Figure 3 Growth limits of *E. coli* NT (R31) with respect to temperature and water activity and fitted by a 'generalized nonlinear regression' model. The solid line corresponds to those combinations of conditions at which growth is predicted in 50% of trials; the dashed line corresponds to combinations of conditions for which growth is predicted to occur only once in 10 trials. ×, no growth observed in 50 days; •, growth observed in ≤50 days.

Table 4 Sources of error in models in predictive microbiology

Error type	*Error source*
Homogeneity error	Arises because some foods are clearly not homogeneous and/or, at the scale of a microorganism, apparently consistent foods may comprise many different micro-environments. Current predictive models do not account for this inhomogeneity of foods.
Completeness error	Arises because the model is a simplification, i.e. only a limited number of environmental factors can be included in the model in practice.
Model function error	Arises mainly from the compromise made when using empirical models, i.e. that the model is only an approximation to reality.
Measurement error	Originates from inaccuracy in the raw data used to generate a certain model, i.e. due to limitations in our ability to measure accurately the environment and the microbial response.
Numerical procedure error	Includes all errors that are the consequences of the numerical procedures used for model fitting and evaluation, some of which are methods of approximation only. Generally, these are negligible in comparison with the other types of errors

are used to aid decisions a process termed 'validation' is undertaken. This typically involves the comparison of model predictions to analogous observations not used to develop the model. Two complementary, dimensionless, measures of kinetic model performance can be used to assess the 'validity' of models.

The bias factor is a multiplicative factor by which the model, on average, over- or under-predicts the response time. Thus, a bias factor of 1.1 indicates not only that a growth model is 'fail-dangerous' because it predicts longer generation times than are observed, but also that the predictions exceed the observations, on average, by 10%. Conversely, a bias factor less than one indicates that a model is, in general, 'fail-safe', but a bias factor of 0.5 indicates a poor model that is overly conservative because it predicts generation times, on average, half of that actually observed. Perfect agreement between predictions and observations would lead to a bias factor of 1.

The accuracy factor is also a simple multiplicative factor indicating the spread of observations about the model's predictions. An accuracy factor of two, for example, indicates that the prediction is, on average, a factor of two different from the observed value, i.e. either half as large or twice as large. The bias and accuracy factors can equally well be used for any time-based response, e.g. lag time, time to an *n*-fold increase, death rate, D-value, etc.

The indices may fail to reveal some forms of systematic deviation between observed and predicted behaviour in which case graphical methods can also be useful. The meaning of the bias and accuracy factors and examples of systematic deviations are illustrated in **Figure 4**.

The error in the estimate of maximum specific growth rate (or doubling time) of an organism determined from measurement of growth in laboratory media is ca. 10% per independent variable. As a 'rule of thumb', each additional environmental factor (pH, a_w, etc.) adds at least another 10% relative error to the model, assuming that the interpretation region of the model is comparable to the whole growth region. (Models with a small interpolation region have smaller error.) Thus, the best performance, that can be expected from a kinetic model encompassing the effect of three environmental factors on growth rate, is ca. 30%, or an accuracy factor of ca. 1.3.

The completeness error is still the greatest error source in predictive models due to other food and microbial ecology effects (structure, competition, etc.) which are difficult to quantify. The following scheme shows the relative contribution of error types to the overall error of predictive models when applied to microbial growth in foods:

homogeneity/completeness error
↓
model function error
↓
measurement error
↓
numerical procedure error

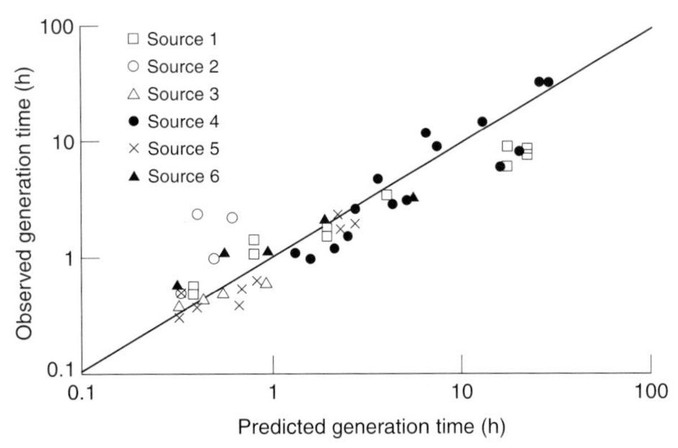

Figure 4 log ($GT_{observed}$) versus log ($GT_{predicted}$) for comparisons of the growth responses of *Staphylococcus aureus* in foods compared to that predicted by a kinetic model. Observed generation times and corresponding predictions for six data sets (sources 1–6) are shown. Each data set is derived from the published reports of independent researchers. The solid line is the 'line of equivalence', i.e. perfect agreement between prediction and observation. Points above the line of equivalence represent 'safe' predictions, i.e. when the predicted generation time is less than the corresponding observation. The scatter about the line of equivalence is reflected in the *accuracy factor*. The relative number of points above and below the line of equivalence is reflected in the *bias factor*. Thus, the pooled data are evenly distributed about the line of equivalence, and the model compared to the pooled data set has a bias factor of 1. The model, however, systematically over-predicts generation times in one region and systematically underpredicts in other regions for some data sets (e.g. data from Source 1). This highlights the need for supplementary tools for model validation. (After Ross 1996).

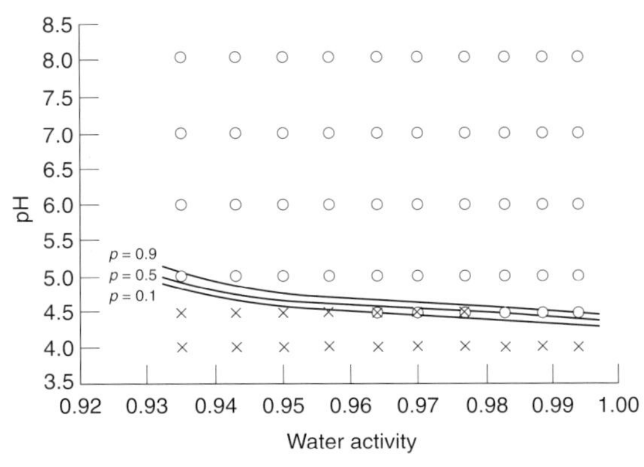

Figure 5 Published data for the effect of water activity and pH on the growth potential of *Listeria monocytogenes* NCTC 9863 in laboratory broth at 25°C (O, growth; ×, no growth) compared to the growth/no growth interfaces predicted by a probability model independently developed using other strains. The predictions of 90, 50 and 10% probability of growth are shown and reveal the abrupt transition from high ($p = 0.9$) to low ($p = 0.1$) probability of growth. There are three conditions (depicted ⊗) on the predicted boundary where growth was observed in some cases, but not in others.

Growth Limits Models

Quantitative indices of performance are not yet well developed for these models. A simple measure is the '% concordance' between the models predictions and observations. Alternatively, graphical comparisons are useful as shown in **Figure 5**.

Performance Evaluation: Applied A number of data sources for model validation exist. One is data from other modelling studies; another is 'inoculated pack studies' under well-controlled laboratory conditions. Another is to obtain analogous data from the literature. These sources represent different levels of data fidelity. The use of literature data is complicated because the data available were usually not designed for modelling studies, e.g. relevant data from which to make a matching prediction is often not supplied and has to be estimated and the growth rate data permit an approximation only of the growth rate. Conversely, in a practical situation in which a model was used, such data might not be available either, i.e. comparison to literature data, although underestimating the best possible performance of models, may be more indicative of the level of confidence that one can have in model predictions in the real world.

Models Compared to Other Models Models developed by different methods, by different workers, and using different strains, in different parts of the world are, nonetheless, usually consistent. Examples are given in **Table 5**. However, although models agree with one another quite well, in some cases they perform equally poorly when compared to the growth of pathogens in foods. Further examination of these 'failures' can reveal that there are deficiencies in the model, i.e. there is a 'completeness error'. The good performance of models in many situations however, provides confidence that the concept is sound but that models are far from being complete for all foods.

Fluctuating Conditions There are limited data available for the performance evaluation of models when applied to fluctuating storage conditions. In general, bacteria respond quickly to changed conditions and display growth rates characteristic of the new environment, i.e. there is little effect of prior history. However, when environmental changes are large and rapid, new lag phases may be induced. Existing models do not model such lag phases. The estimation of lag times remains a problem for the interpretation of fluctuating temperature histories.

Existing Technology and the Future

Technology

Numerous models have been published, or are available in software. One of the best-known software applications is the commercially available Food MicroModel. A similar suite of models, called the

Table 5 Evaluation of the performance of growth rate models

Model: organism and variables	*Data type*	*Number of data*	*Bias factor*	*Accuracy factor*
Staphylococcus aureus, temperature, water activity	Data used to develop model	212	1.00	1.20
	Inoculated foods, same strain	38	1.00	1.26
	Independent published data, various strains	49	1.01	1.53
Brochothrix thermosphacta, temperature, pH, water activity	Data used to develop model	44	1.00	1.26
	Independent data, various strains	102	0.73	1.83
Escherichia coli, temperature, pH, water activity, lactic acid	Data used to develop model (Model A)	240	1.00	1.30
	Independent data, various strains and foods[a]	178	0.84	1.43
	Independent studies in laboratory broths (Model A)[a]	75	0.78	1.61
	Independent studies in laboratory broths (Model B)[a]	75	0.73	1.56
	Model A cf. Model B	75	1.07	1.45
psychrotrophic pseudomonads, temperature, water activity	Data used to develop model	113	1.00	1.07
	Incubated foods, same strain	96	1.00	1.10
	Inoculated foods, same strain independent workers in industry	29	0.96	1.21
	Independent published data, various strains	266	0.87	1.30
	Independent model	integrated over the region 2–11°C, at a_w 0.995, pH 5.8.	⩾1.05	⩾1.10

[a]Model A and Model B were developed by independent research groups using different methods. They are compared to data published by yet other independent workers. The corresponding predictions of each model were then compared to each other. The models are more consistent with each other than either is with the available published data suggesting completeness errors.

Pathogen Modelling Program, was developed by the US Department of Agriculture and is distributed free of charge. Both of the suites of modelling software predict the increase of microbial populations under unchanging conditions. Other software integrates the effect of fluctuating environmental conditions over time to predict the change in microbial populations. An example of this is shown in **Figure 6**. Some of the available models and factors and organisms included were summarized in Table 2. Other computer-based applications including the development of expert systems have been described but are not generally available. The use of artificial neural networks to develop models from data has also been described.

New Zealand has used predictive models for regulatory purposes for several years, and other countries are beginning to explore their use. Large food processing and retailing organizations have also begun to adopt the philosophy and technology. This uptake was hastened, in part, by the endorsement of quantitative risk assessment by the World Trade Organization.

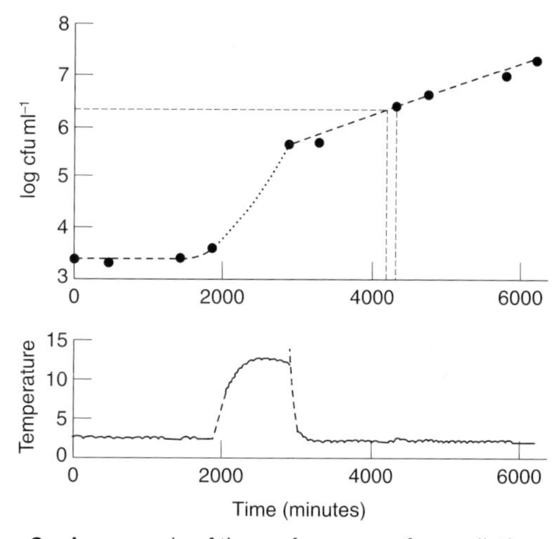

Figure 6 An example of the performance of a predictive model applied to microbial growth under fluctuating temperature conditions. The growth of pseudomonads in minced beef was monitored and compared to the predictions of an independent model derived previously. The solid line is the line of growth predicted by the model and ● is the observed growth. (N.B. the model did not include lag time predictions, thus the model predictions begin after the lag phase is resolved at ca 1500 min). The lower graph shows the temperature during the trial. Reproduced from Neumeyer K, Ross T, Thomson and McMeekin TA (1997) Validation of a model describing the effects of temperature and water activity on the growth of psychotrophic pseudomonads. *International Journal of Food Microbiology* 38(1): 55–64. With permission from Elsevier Science.

Future Developments

In the future, models will most likely become increasingly mechanistic. The ultimate aim is to develop a mechanistic and quantitative understanding of the factors that govern the microbial ecology of foods, i.e. to replace empiricism with quantitative data on microbial ecology, and physiological studies at the cellular and subcellular level, where gaps in knowledge remain.

Variability, in the initial status of a food, whether in terms of the initial load, the specific strains present, physiological condition (e.g. in lag, activated spores, stress responses invoked, etc.) or distribution within the food will always limit the ability of predictive models to predict the absolute safety and quality of a specific food item. When model inputs are variable, outcomes will also be variable. Thus, a probabilistic, or stochastic, approach is inevitable and the same software tools that facilitated the development of quantitative microbial risk assessment, namely Monte Carlo simulation software, make this possible.

This stochastic approach will complement further study and definition of bacterial responses close to the growth/no growth boundary and the further development of software-based tools that can answer questions such as 'What is the probability of growth and if growth occurs, at what rate?'

Predictive microbiology is a powerful tool to aid microbial food safety and shelf-life assurance, both in its own right and as a complementary tool for HACCP programmes, Hurdle Technology, and quantitative microbial risk assessment.

See also: ***Clostridium***: Detection of Neurotoxins of *C. botulinum*. **Enterobacteriaceae, Coliforms and *E. coli***: Classical and Modern Methods for Detection/Enumeration. ***Escherichia coli***: Detection of Enterotoxins of *E. coli*. ***Listeria***: Detection by Classical Cultural Techniques. **Quantitative Risk Analysis**. ***Staphylococcus***: Detection by Cultural and Modern Techniques; Detection of Staphylococcal Enterotoxins.

Further Reading

Baranyi J (1998) Comparison of stochastic and deterministic concepts of bacterial lag. *Journal of Theoretical Biology* 192: 403–408.

Baranyi J and Roberts TA (1995) Mathematics of predictive food microbiology. *International Journal of Food Microbiology* 26: 199–218.

Baranyi J, Robinson TP, Kaloti A and Mackey BM (1995) Predicting growth of *Brochothrix thermosphacta* at changing temperature. *International Journal of Food Microbiology* 27: 61–75.

Baranyi J, Ross T, Roberts TA and McMeekin T (1996) The effects of overparameterisation on the performance of empirical models used in Predictive Microbiology. *Food Microbiology* 13: 83–91.

Box GEP and Draper NR (1987) *Empirical Model-building and Response Surfaces*. New York: Wiley.

Buchanan RL (Guest Editor) (1993) *Journal of Industrial Microbiology*, vol 12 (3–5). (Entire Issue.)

Farkas J (guest editor) (1994). *International Journal of Applied Microbiology*, vol. 23 (3,4). (Entire Issue.)

McMeekin TA, Olley J, Ross T and Ratkowsky DA (1993) *Predictive Microbiology: Theory and Application*. Taunton, UK: Research Studies Press.

Roberts TA and Jarvis B (1983) Predictive modelling of food safety with particular reference to *Clostridium botulinum* in model cured meat systems. In: Roberts TA and Skinner FA (eds) *Food Microbiology: Advances and Prospects*, p. 85. New York: Academic Press.

Roels JA and Kossen NWF (1978) On the modelling of microbial metabolism. In: Bull MJ (ed.) *Progress in Industrial Microbiology*, vol. 14. Amsterdam: Elsevier.

Ross T (1996) Indices for evaluation of the performance of predictive models in food microbiology. *Journal of Applied Bacteriology* 81(5): 501–508.

Ross T and McMeekin TA (1994) Predictive microbiology – a review. *International Journal of Applied Microbiology* 23: 241–264.

Rubinow SI (1984) Cell kinetics. In: Segel LA (ed.) *Mathematical Models in Molecular and Cell Biology*, Chapter 6.6. Cambridge: Cambridge University Press.

Tsuchiya HM, Fredrickson AG and Aris R (1966) Dynamics of microbial cell populations. *Advances in Chemical Engineering* 6: 125–206.

Whiting RC and Buchanan RL (1996) Predictive Modeling. In: Doyle MP, Beuchat LR and Montville TJ (eds) *Food Microbiology: Fundamental and Frontiers*. Washington, DC: ASM Press.

Zwietering MH, Jongenburger I, Rombouts FM and van't Riet K (1990) Modelling of the bacterial growth curve. *Applied and Environmental Microbiology* 56: 1875–1881.

Appendix: Developing a Predictive Model for Bacterial Growth as a Function of Environmental Conditions

Let $x(t)$ be the concentration of a bacterial population at time t. If the concentration changes by a small amount, $\mathrm{d}x(t)$ in an infinitely small time, $\mathrm{d}t$, then $\mathrm{d}x(t)/\mathrm{d}t$, the 'derivative' of the $x(t)$ function at the time t, (i.e. the rate of change of cell numbers per unit time) is called the instantaneous, or absolute, growth rate. If growth is exponential, however, the instantaneous growth rate depends on the number of cells present. A more useful measure, independent of $x(t)$, is the relative growth rate, i.e. the proportional change in cell concentration per unit time. Proportional change is most easily understood by the familiar plot of logarithm of cell concentration against time.

Let $y(t)$ denote the natural logarithm of $x(t)$: i.e. $y(t) = \ln x(t)$. The derivative of $y(t)$, i.e. the slope of the $\ln x(t)$ vs. t plot, is the specific growth rate of the population, and can be considered as the number of divisions per cell per unit time (**Fig. 7**). It is denoted $\mu(t)$. Hence,

$$\mu(t) = \frac{1}{x(t)}\frac{\mathrm{d}x(t)}{\mathrm{d}t} = \frac{\mathrm{d}\,(\ln x(t))}{\mathrm{d}t} = \frac{\mathrm{d}y(t)}{\mathrm{d}t}$$

During exponential growth $\mu(t)$ is, in theory, constant.

Frequently the derivative (slope) of the $\log_{10}x(t)$ function is called the growth rate but, because the $\log_{10}$ scale is used it is 2.3 times smaller than the

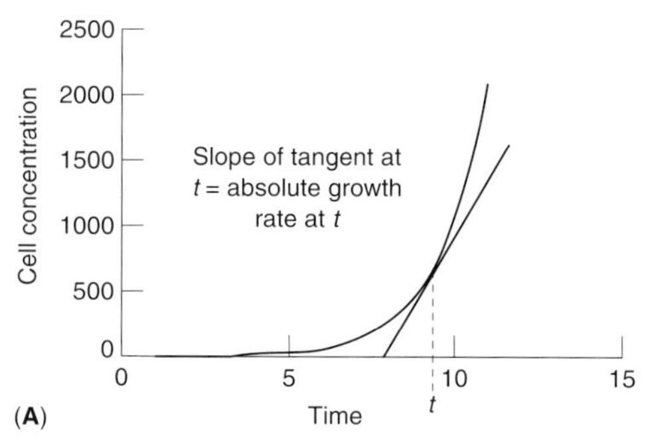

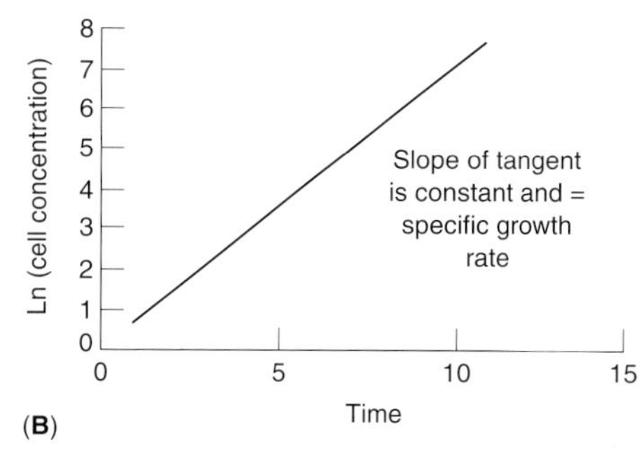

Figure 7 Comparison of (**A**) absolute and (**B**) specific growth rate. The absolute rate is measured on a linear scale, whereas the specific growth rate is measured on a logarithmic scale. Note the change in shape of the *Ln*-transformed growth curve.

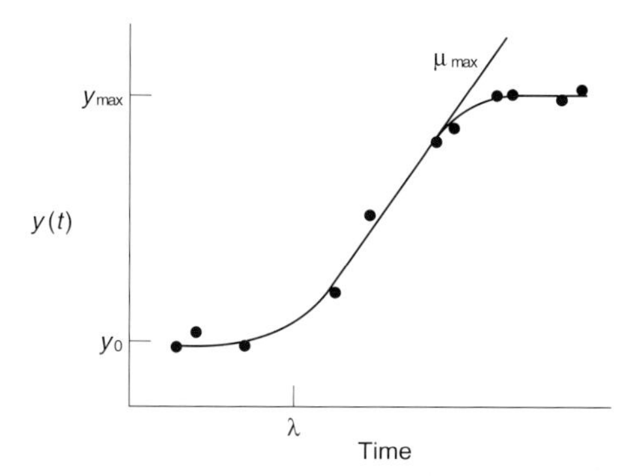

Figure 8 A classical bacterial growth curve showing three phases of growth (lag, exponential and stationary) and the parameters (λ, y_0, y_{max}, μ_{max}) needed to describe that curve mathematically.

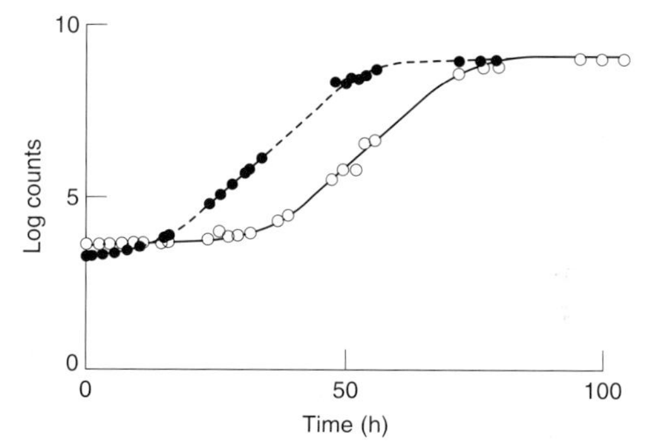

Figure 9 History-dependence of the maximum specific growth rate and the lag. These growth curves are from replicate experiments, except that the inoculation was prepared differently and led to different physiological states of the primary culture. The maximum specific growth rates are the same whereas the lag periods are different because the latter parameter depends on the history of the cells.

specific growth rate. This is not of itself a problem as long as it is used consistently and understood by all users.

In Figure 7 the graphs show exponential growth only. A typical bacterial growth curve has sigmoid shape, due to lag and stationary growth stages, and is commonly described by (at least) four parameters, as shown in **Figure 8**.

Following from the above, the steepest tangent to the curve is the maximum specific growth rate, μ_{max}, and occurs at the point of inflection. The end of the lag phase, denoted here by λ, is traditionally defined as the time when that steepest tangent crosses the level of inoculum.

Description of the dependence of these parameters on the actual physicochemical environment in foods (and analogous parameters describing survival and death) is the major thrust of predictive microbiology. Other parameters of the growth curve, like the maximum population density and the inoculum (y_{max} and y_0 in Fig. 8) are either less important, or not the subject of modelling. From the perspective of microbial safety and quality the stationary phase is of little interest and the inoculum level, y_0, does not depend on the environment but is specific to each situation.

A basic hypothesis is that, in the same environment, the maximum specific growth rate is a reproducible parameter, characteristic of the organism and does not depend on the history of the cells. It is considered to be an intrinsic characteristic, or intrinsic parameter, of the microbe in that specified environment. This does not hold for the lag parameter, as demonstrated in **Figure 9**.

Environment-dependence of Growth Parameters

From the above, μ_{max}, the maximum specific growth rate is the parameter that is primarily suitable for modelling. Frequently, for stochastic reasons, its square-root or logarithm is modelled as a function of temperature, pH, water activity, etc. either by a product or a sum of simple functions of the individual environmental variables, or by a second order multivariate response surface of the variables involved.

When lag-time models are considered, the raw data used for model creation should be collected in such a way that the history of the cells (expressed by the physiological state of the inoculum) is the same and also that the models are applicable only for cell populations with a similar history. The structure of most lag-models is similar to that of the growth rate models.

PRESERVATIVES

Contents

Classification and Properties

M Surekha and **S M Reddy**, Department of Botany, Kakatiya University, Warangal, India

Introduction

Fresh foods always contain microorganisms both on their surfaces and within. These microorganisms, if they are not destroyed, will spoil the food. The prevention of food spoilage by inhibiting or destroying the microorganisms is the basis of food preservation. This can be done by chemical treatment, freezing, curing, dehydration or thermal processing.

The chemicals used to prevent food spoilage have some antiseptic properties under the conditions of use, and are known as 'preservatives'. Broadly speaking, a preservative is a chemical substance capable of retarding or arresting the growth of microorganisms so as to prevent processes such as fermentation, acidification or decomposition, which cause deterioration of flavour, colour, texture, appearance and nutritive value. The main objectives of using preservatives are to extend the shelf life, retain nutritive value and ensure safety. Chemical preservatives are often used in combination with physical methods; such combinations may allow the preservatives to be used at lower concentrations, thus retaining the quality of the product.

The Need for Preservatives

The twentieth century witnessed radical technological advancement in the physical methods of food preservation. These developments include preservation of food by thermal processing, refrigeration, freezing, concentration, drying, and more recently the use of irradiation. In spite of this technological advancement, the worldwide population explosion has resulted in a crisis of food supply, which demands a reduction in losses to the minimum. The countries with the greatest nutritional need are the least developed, suffering from inadequate production, distribution, transportation, storage and preservation facilities. These countries are not in a position to afford the latest technologies for the preservation of food by physical methods, and thus depend on the use of chemical preservatives which are not only effective, but also safe and inexpensive. As physical methods are not suitable for all types of foods, these days even developed countries are making use of chemical preservatives.

Properties of Preservatives

The desirable properties of a chemical substance to serve as a preservative are as follows:

1. A preservative used for antimicrobial purposes should kill the microorganisms rather than inhibit their growth.
2. Any bacteriostatic preservative is most effective if it persists until the food is ready for consumption. If the food is undergoing processing, the bacteriostatic preservative should persist until the food is further processed.
3. A preservative should have an adequate degree of resistance to heat.
4. The specificity range of a preservative should correspond with the range of microorganisms that contaminate and develop on the food.
5. A preservative that is intended to supplant thermal processing should provide a degree of security against *Clostridium botulinum* similar to that given by the normal thermal processing.
6. The preservative should neither be destroyed by the miscellaneous reactions of the food nor be inactivated by the metabolic products produced by the microorganism.
7. Any antimicrobial preservative should not readily stimulate the appearance of resistant strains of microorganisms.
8. There should be a suitable procedure for determining the amount of the preservative in different foods.

Other desirable properties of a preservative are:

- it should have a practical value and be economical
- it should be non-irritant and have low (or no) toxicity
- it should not retard the activity of digestive enzymes or harm the consumer
- within the body, it should not decompose into substances more toxic than the preservative itself.

Classification

Preservatives include traditional (natural) preservatives, antibiotics and synthetic preservatives.

Traditional Preservatives

Compounds such as sugar, salt, vinegar, organic fruit acids, wood smoke, alcohol and various spices used in the preservation of food for centuries are regarded as traditional preservatives. Salts and sugars dissolve in the water of the food to form strong solutions in the process of curing and conserving. The difference between the concentration of the solution and that of the microbial cell cytoplasm causes dehydration of the cell, which leads to its inhibition or death. Salamis, hams, jams, and condensed and sweetened milk are examples of this principle. Smoking destroys bacteria on the surface of food.

Synthetic Preservatives

Apart from vinegar, some other acids and their salts are legally permitted preservatives (**Table 1**). The other synthetic preservatives used are nitrites and nitrates, sulphur dioxide and sulphites, carbon dioxide, phosphates and hydrogen peroxide.

Antibiotics

Antibiotics are chemicals, produced by microorganisms, that are able to destroy or prevent the growth of other microorganisms. Bacteriocin and nisin are the main antibiotics permitted as food preservatives. There is no legal restriction on the quantity of nisin in certain foods in some countries. It is used in creams, milk, cheese and some canned foods to prevent growth from spores. Pimaricin (natamycin) is used as an antimycotic.

Chemical preservatives are also classified based on their chemical nature and action. Based on their chemical nature, they are of two types, inorganic preservatives and organic preservatives. Nitrates, nitrites, sulphites, sulphurous acid, borates, hypochlorites and peroxide are inorganic preservatives. Benzoates, formic acid, sorbic acid, propionic acid and their sodium and calcium salts, and esters of *p*-hydroxybenzoic acid are classified as organic preservatives.

Antimicrobial Properties

Spectrum of Activity

None of the preservatives has a complete spectrum of action against all microorganisms that spoil foods. Most preservatives predominantly act against yeasts and moulds (**Table 2**). In general, most of the organic acids have the broadest spectrum of antimicrobial activity and are useful against many spoilage bacteria, fungi and yeasts. Benzoic acid is used primarily as an antimycotic agent and most yeasts and moulds are inhibited. The activity of benzoic acid against bacteria is very variable. Propionic acid and its salts are highly effective mould inhibitors, but yeasts and most bacteria are less affected. Inhibition of rope-forming bacteria in bread is a specific target for propionic acid. Acetic acid is more effective against yeasts and bacteria than moulds; *Acetobacter* spp., certain lactic acid bacteria and some yeasts are resistant to acetic acid. Lactic and citric acids have only moderate anti-

Table 1 Preservatives used in food

Traditional preservatives	*Synthetic preservatives*	*Antibiotics*
Sugar	**Organic:**	Nisin
Salt	acetic acid, acetates and diacetates	Pimaricin
Smoke	sorbic acid and its salts	Tylosin
Spices	benzoic acid and its salts	
Vinegar	*p*-hydroxybenzoic acid esters and their salts	
Alcohol	boric acid and borates	
	citric acid and its salts	
	formic acid and formates	
	lactic acid and its salts	
	propionic acid and its salts	
	Inorganic:	
	carbonic acid (CO_2)	
	sulphurous acid and sulphites (SO_2)	
	nitrites and nitrates	
	phosphates	
	hydrogen peroxide	

Table 2. Inhibitory action of sorbic acid, benzoic acid and sulphur dioxide on bacteria, yeasts and moulds

	Preservatives					
	Sorbic acid		*Benzoic acid*		SO_2	
Organism	*pH*	*MIC*[a]	*pH*	*MIC*[a]	*pH*	*MIC*[a]
Bacteria						
Escherichia coli	5.2–5.6	50–100	5.2–5.6	50–120		100–200
Serratia marcescens	6.4	50				50
Bacillus sp.	5.5–6.3	50–1000				
Clostridium sp.	6.7–6.8	100–1000				
Salmonella sp.	5.0–5.3	50–1000				
Lactobacillus sp.	4.3–6.0	200–700	4.3–6.0	300–1800		100
Pseudomonas sp.			6.0	200–400		
Streptococcus sp.			5.5–5.6	50–100		
Micrococcus sp.			5.2–5.6	200–400		
Yeasts						
Saccharomyces sp.	3.2–5.7	30–100			4.0	80–160
Hansenula anomala	5.0	500		200–300	5.0	240
Torulopsis sp.	4.6	400		200–500		
Candida krusei	3.4	100		300–700		
C. lipolytica	5.0	100				
Byssochlamys fulva	3.5	50–250				
Moulds						
Rhizopus	3.6	120	5.0	30–120		
Geotrichum candidum	4.8			1000		
Oospora lactis	3.5–4.5	25–200		300		
Penicillium sp.	3.5–5.7	20–100	2.6–5.0	30–280	5.0	160–400
Aspergillus sp.	3.3–5.7	20–100	3.0–5.0	20–300	4.5	220
Fusarium sp.	3.0	100				

[a] MIC, minimum inhibitory concentration, expressed in p.p.m.

microbial activity. These acids inhibit the formation of aflatoxin and sterigmatocystin. Sorbic acid and its salts have a wide spectrum of activity against catalase-positive bacteria, yeasts and moulds, and are highly active against osmophilic yeasts. Sulphur dioxide and sulphites also have broad spectrum of antimicrobial activity in acid foods. This preservative is more effective against bacteria than moulds and yeasts, Gram-positive bacteria being less susceptible than Gram-negative bacteria. Sulphites inhibit enterobacteria and *Salmonella*. Lactobacilli are highly sensitive to SO_2. Yeasts react differently to SO_2 depending on the strain. The practical importance of nitrite is in the inhibition of spore-forming bacteria; it also affects *Achromobacter*, *Aerobacter*, *Escherichia*, *Flavobacterium*, *Micrococcus* and *Pseudomonas*.

Mechanism of Antimicrobial Action

Food preservatives inhibit not only the general metabolism but also the growth of the microorganisms. Depending upon the type of preservative used, the final state at which the microorganisms are killed is reached within a few days or weeks, at the usual applied concentrations. The timescale for the killing of microorganisms under the influence of preservatives corresponds to the relationship

$$K = 1/t \cdot \ln Z_0/Z_t$$

or

$$Z_t = Z_0 \cdot e^{-Kt}$$

where K is the death rate constant, t_1 is the time period, Z_0 is the number of living cells at the time when the preservative begins to act, and Z_t is the number of living cells after time t.

The given formula is considered to be the basis for studying the action of preservatives in foods. However, this rule is valid only for relatively high dosages of preservatives and a genetically uniform cell material. A preservative added to a food when microbial counts are low inhibits microorganisms in the initial lag phase; the dosage of preservatives necessary in practice to inhibit microorganisms in the exponential log phase would be too high. Preservatives are not designed to kill microorganisms in substrates already supporting a massive germ population. In general, the action of preservatives includes physical as well as physicochemical mechanisms, especially the inhibitory action on enzymes.

The partial dissociation of weakly lipophilic acid food preservatives plays an important role in the inhibition of microbial growth. The undissociated lipophilic acid molecules are capable of moving freely through the membrane. They pass from an external environment of low pH (where the equilibrium favours the undissociated molecules) to the cytoplasm, which is of high pH (where the equilibrium favours the dissociated molecules). At the high pH level, the acid ionizes to produce protons which in turn acidify the cytoplasm and break down the pH component of the proton motive force. In order to maintain the internal pH, the cell then tries to expel the protons entering it. In doing so, it diverts the energy from growth-related functions and hence the growth rate and yield of the cell both fall. If the external pH is low and the extracellular concentration of the acid is high, then the cytoplasmic pH drops to a level where growth is no longer possible and the cell eventually dies. Some preservatives also exert specific efforts on metabolic enzymes. Sorbic acid is reported to react with the sulphydryl groups of enzymes such as fumarase, aspartase, succinic dehydrogenase, catalase and peroxidases in bacteria, moulds and yeasts. Antimicrobial activity of organic acids increases with chain length, but the limited water solubility of long-chain acids restricts their use.

Benzoic acid is effective only in acid foods. It inhibits enzymes of acetic acid metabolism, oxidative phosphorylation, amino acid uptake and various stages in the tricarboxylic acid cycle. It also alters membrane permeability of the microbial cell. Transport inhibition is the primary mode of action of parabens. Respiration of microbial cells is also inhibited.

Antimicrobial action of propionic acid is due to inhibition of nutrient transport and growth by competing with substances like alanine and other amino acids required by microorganisms. Antimicrobial action of formic acid is similar to any acidulant. Additionally, formic acid inhibits decarboxylase and haem enzymes, especially catalase. The antimicrobial effect of other acids, e.g. lactic, tartaric, phosphoric and succinic acids, is due to acidification of the microbial cell and inhibiting nutrient transport.

Sulphur dioxide is highly reactive, and therefore it interacts with many cell components. The sulphite ion acts as a powerful nucleophile, cleaving the disulphide bonds of proteins which changes the molecular configuration of enzymes, thus modifying active sites. It reacts with coenzymes (NAD^+), cofactors and prosthetic groups such as flavin, thiamin, haem, folic acid and pyridoxyl. In the case of yeast, the blocking of the oxidation of glyceraldehyde 3-phosphate to 1,3-bisphosphoglycerate is the salient feature. Sulphite treatment of yeast cells results in a rapid decrease in ATP content prior to cell death. This is attributed to inactivation of the enzyme glyceraldehyde-3-phosphate dehydrogenase. Sulphite also reacts with carbonyl constituents of the metabolic pool, to form hydroxysulphonates. Yeasts when treated with sublethal concentrations of sulphite tend to excrete increased amounts of acetaldehyde. This is due to the trapping of this metabolic intermediate as the stable hydroxysulphonate, thereby preventing its conversion to ethanol so that the reaction equilibrium shifts. Glycerol is formed instead of ethanol by reduction of glyceraldehyde 3-phosphate to glycerol 3-phosphate which is subsequently dephosphorylated. In *Escherichia coli* NAD-dependent formation of oxalacetate from malate is inhibited. Sulphite destroys the activity of thiamin by breaking the bond between the pyrimidine and thiazole portion of the molecule.

The antimicrobial action of nitrite is based mainly on the release of nitrous acid and oxides of nitrogen. Nitrite inhibits active transport of proline in *E. coli* and aldolase from *E. coli*, *Streptococcus faecalis* and *Pseudomonas aeruginosa*. Reaction between nitric oxide from the nitrite and iron of a cidophore compound involved in electron transport in clostridia accounts for the anticlostridial action. Nitrite reacts with haem proteins such as cytochromes and sulphydryl enzymes resulting in the formation of *S*-nitroso products.

Combination of Preservatives

No single preservative is active against all spoilage microorganisms. Attempts have been made to compensate for this by combining various preservatives with different spectra of action. In general, organic acids are compatible with other preservatives and many combinations are synergistic, for example:

- benzoate with SO_2, CO_2, NaCl, boric acid or sucrose
- propionate with CO_2 or sorbate
- sorbate with sucrose or NaCl
- lactic acid with acetic acid.

The combinations of sorbic acid, benzoic acid or esters of *p*-hydroxybenzoic acid with nisin and tylosin are useful because they extend the spectrum of action to cover *E. coli*, *Lactobacillus* and *Staphylococcus* strains. Cured meats are rarely involved in *Clostridium perfringens* food poisoning. This is a fine example of the 'hurdle' concept: individual preservatives such as salt content, nitrite and heat processing are insufficient to assure safety but effectively control growth of *C. perfringens* in combination. Contrary to general expectations, not all the combinations give better results than individual constituents. The presence of one preservative may sometimes weaken the effect of the other. For instance, boric acid has the tendency to weaken the effect of other preservatives in their action on *E. coli*. However, its action against fungi proved to be synergistic. On the other hand, presence of some chemical substances such as calcium chloride, which it is not a food preservative and has no antimicrobial effect individually, slightly weakens the efficacy of sorbic acid, benzoic acid and other preservatives. In general a beneficial effect will be obtained by using preservatives with substances that counter dissociation, such as acids, or those that reduce water activity, for example NaCl or sugar.

Degradation of Preservatives

In general, food preservatives are stable substances and are unlikely to decompose within the specified storage time. Occasionally, however, certain preservatives such as organic compounds are decomposed by microorganisms and are used as a source of carbon by them. Decomposition of this type of preservative is possible, if the preservative is ineffective against microbes and also if the food contains a large number of microbes. Therefore, it is impossible with such preservatives to arrest the spoilage of food and to maintain the food in an apparently fresh condition. The best example of this phenomenon is the conversion of sorbic acid to hexadienol by some strains of lactic acid bacteria. This product reacts with ethanol to form 1-ethoxy-2,4-hexadiene and 2-ethoxy-3,5-hexadiene, which give a geranium-type odour in wines.

Interaction of Preservative with Food Components

Chemical reaction between food preservatives, food components and microorganisms may lead to the formation of reaction products of toxicological importance and reduction in the concentration and the activity of the preservative. Some food preservatives such as sorbic acid, SO_2, sulphites and nitrites have an extensive reactivity with food components.

Sorbic acid reacts with low-molecular-weight thiols of food such as cysteine and glutathione to form the 5-substituted 3-hexenoic acid. Sorbic acid also undergoes autooxidation to malonaldehyde, acetaldehyde and β-carboxyacrolein.

Owing to its high chemical reactivity sulphur dioxide may be involved in a variety of interactions with food ingredients. The action of SO_2 in destroying thiamin in food is significant. An important nucleophilic reaction of the sulphite ion is its addition to the α-β-unsaturated carbonyl moiety of 3,4-dideoxyosulos-3-enes formed as reactive intermediates in Maillard and ascorbic acid browning, which causes a considerable depletion of the preservatives in foods susceptible to non-enzymatic browning.

The nitrite added to meat is converted to nitric oxide which combines with myoglobin to form nitric oxide myoglobin. The *N*-nitrosamines formed by the cooking of nitrite-cured meat are potent carcinogens. Nitrosophenols formed by C-nitrosation of phenolic components of food are readily oxidized to the corresponding nitro compounds; *S*-nitroso compounds are readily formed by the nitrosation of thiols and represent a reversibly bound form of the preservative.

Uses

Preservatives applicable to a particular need are determined by the composition of the food, the type of microbial spoilage and the desired shelf life. As well as the specific physical properties, cost is also an important consideration in the selection of preservative. Preservatives are incorporated directly into food products or developed during processing food. Traditional preservatives have been introduced through processes such as fermentation, salting, curing and smoking. Spices are commonly added in small amounts to the food as a preservative. Sugar is used in the preservation of jams, jellies, candied fruit and sweetened condensed milk. The main use of smoke is to preserve meat and fish products. Salt is used to preserve many foods, including butter, margarine, cheese, sausages, ham and fish.

Chemical preservatives may be applied directly, most often as an ingredient of manufactured foods, but also by dipping, spraying, gassing or dusting. Some preservatives are incorporated in the packing material rather than applied directly to the food itself. Vinegar or acetic acid is used in many foods, including mayonnaise, catsup, salad dressings, pickles and meat. Lactic acid is used as a flavouring agent in frozen desserts, and as an emulsifier in bakery products. It is also used for faster nitrite depletion and botulinal protection due to lowered pH in meat product processing.

Sodium benzoate is the most widely used preservative for acid foods, including carbonated and still beverages, salads, fruit desserts, fruit cocktails and margarine. Sodium benzoate is used at concentrations of 0.03% to 0.10%. Because of the astringent flavour of benzoates, they are often used in combination with sorbate or parabens. Parabens are used to preserve soft drinks, fruit products, jams, jellies, pickles, cream and pastes. The *N*-heptyl ester can be used in beer fermentation at a level of $12\,mg\,g^{-1}$. Parabens may be added as a dry or liquid ingredient to food.

Sorbic acid and its salts are frequently used because of their high solubility. Sorbates are used in cheese products, baked foods, fruits, fruit juices, vegetables, soft drinks, wines, jellies, jams, syrups, salads, margarine and fish products. Since sorbate inhibits yeasts, it is not used in yeast-raised bread. Sorbate may be added directly to the food or it may be applied by dipping, spraying, dusting or impregnating packing materials and wrappers. Recent studies showed the effectiveness of sorbate as an antibotulinal agent in meat products.

In food processing gases such as SO_2 and CO_2 may be used as antimicrobial agents or for other purposes. These gases may have direct or indirect antimicrobial effects. Sulphur dioxide is principally used in wine preservation. It is employed as a liquid under pressure or in aqueous solution. Additionally, various sulphite salts (sodium sulphite, sodium hydrogen sulphite, sodium metabisulphite, potassium metabisulphite and calcium sulphite) containing 52–68% active SO_2 are used to preserve a variety of foods such as fruit juices, soft drinks, dehydrated fruits and vegetables, pickles, syrups, meat and fish products. In wine, sulphite is also used as an equipment sanitizer, antioxidant and clarifier and to prevent bacterial spoilage during storage. A combination of 200 mg sorbic acid, 220 mg potassium sorbate and 20–40 mg SO_2 per litre provides a comprehensive protection to the wine. Proteolytic breakdown of meat may be prevented by sulphites. Sulphur dioxide is added to foods to prevent enzymatic reactions, notably browning.

Carbon dioxide is used to control psychrotrophic spoilage of meat and meat products, poultry, fish, eggs, fruits and vegetables. Carbon dioxide is generally used in the form of a liquefied gas, or as dry ice (solid CO_2) which sublimes to form CO_2 gas. Carbon dioxide applied under pressure with low temperature results in rapid biocidal action. Carbon dioxide is a major applicant in carbonized soft drinks, mineral water, wines, beers and ales. It functions as an antimicrobial and effervescing agent. It inhibits aerobic spoilage organisms when used in vacuum-packed meats at a concentration of 10–20%; higher concentrations may cause undesirable odours. The combination of O_2 and CO_2 in a controlled atmosphere delays the respiration, ripening and spoilage of stored fruits and vegetables.

Nitrites are added to cheese and meat products. Addition of nitrite not only prevents the growth of toxigenic microorganisms but also the production of toxins. Nitrite added to meat results in both chemical and antimicrobial effects. It reacts with haem proteins to form the characteristic cured meat colour and has a mild antioxidant effect that prevents rancidity and 'warmed-over' flavour. At low pH nitrite is depleted by increased formation of nitrous acid and nitric oxides, the reactive forms of nitrite. Because of this, addition of acids, acidulant or glucono-delta-lactone has a beneficial effect on the action of nitrite. For the positive chemical effect (colour and flavour) a nitrite concentration of $50\,\mu g\,g^{-1}$ is needed, while antibotulinal activity requires a concentration of $100\,\mu g\,g^{-1}$. Lower concentrations of nitrite (40–$80\,\mu g\,g^{-1}$) in combination with sorbate are more effective. In the USA the content of sodium nitrite in cured meat products is limited to $200\,\mu g\,g^{-1}$, with specific regulatory levels varying with the product. In the UK potassium and sodium nitrite are permitted in cured meats up to a maximum of $200\,\mu g\,g^{-1}$.

Antibiotics such as nisin, tylosin and pimaricin (natamycin) have been used for preservation of poultry, fish, meat, milk products, fresh fruits, vegetables and canned foods. Nisin is permitted in many countries chiefly for the preservation of processed cheeses or other dairy products, in which its action is directed against spore-forming bacteria, particularly butyric acid bacteria and clostridia. Nisin is used in canned products as a sterilizing auxiliary. Tylosin has a broader spectrum of action than nisin. Because of its activity against spore-forming bacteria, especially clostridia, tylosin is used as a thermal processing adjuvant. It is more heat-stable than nisin and has been tested in many canned foods including mushrooms. Tylosin 100 $\mu g\, g^{-1}$ in brine prevents botulinal toxin production in smoked fish. Natamycin (pimaricin) is permitted in some western European countries for surface preservation of cheese and as an additive to cheese coating. It has been used to retard yeast and mould spoilage of fruit, fruit juices, cottage cheese, poultry products and sausage.

Toxicology and Regulatory Status

Traditionally processed food in general finds ready acceptance by regulatory authorities. This is not the case for foods processed by the addition of chemical preservatives, where it is essential to ensure that the preservative used does not become a health hazard to human beings.

Benzoic acid and its salts have low toxicity in experimental animals and humans. Humans have a high tolerance to sodium benzoate because of a detoxifying mechanism, in which benzoate and glycine or glycuronic acid are conjugated and excreted as hippuric acid or benzoyl glucuronide. Benzoate is not mutagenic in *Drosophila* or *Salmonella* but interacts with nucleosides and DNA in vitro. Sodium benzoate and benzoic acid are generally recognized as safe (GRAS) at concentrations up to 0.1% in the USA. In the UK benzoic acid and its salts are permitted on a wide scale in accordance with the Preservatives in Food Regulations of 1979. Parabens toxicity is low, with an acute toxicity dose LD_{50} of 180–8000 $mg\, kg^{-1}$ of body weight in experimental animals, varying with the form of administration. The acceptable daily intake (ADI) is 10 $mg\, kg^{-1}$ of body weight of average man. The methyl and propyl parabens are GRAS in the USA, with a total addition limit of 0.1%.

Propionic acid and its salts are readily absorbed by the digestive tract owing to their high water solubility. This acid decomposes in mammals by linkage with coenzyme A via methylmalonyl-CoA, succinyl-CoA and succinate to yield CO_2 and H_2O. The ADI set by the Food and Agriculture Organization (FAO) and the World Health Organization (WHO) is not limited. These preservatives are permitted for use in many countries.

Sorbic acid is nontoxic and is metabolized by fatty acid oxidation, pathways common to both laboratory mammals and humans. The oral LD_{50} for rats is 7–10 $g\, kg^{-1}$ of body weight, and 6–7 $g\, kg^{-1}$ of body weight for the sodium salt. In highly sensitive individuals this preservative irritates the mucous membranes. Sorbates have no mutagenic, teratogenic or carcinogenic action. The FAO/WHO acceptable daily intake of sorbic acid and its salts is 25 $mg\, kg^{-1}$ of body weight, the highest ADI of the common preservatives. In the USA, sorbic acid and sorbates are GRAS. The maximum permissible level is 0.1–0.2%. These compounds are permitted in all countries for preservation of a wide variety of foods.

Sulphur dioxide and sulphites in the body are oxidized to sulphate and excreted in urine. Though vitamin B_1 deficiency, diarrhoea, organ damage and decreased usage of dietary protein and fat are some of the adverse effects of SO_2 in human beings, actual poisoning by SO_2 and sulphite is not possible because of vomiting. Sulphite is not a carcinogen, but SO_2 is mutagenic. Levels of application are restricted to 500 $\mu g\, g^{-1}$ owing to flavour problems. The FAO/WHO acceptable daily intake of SO_2 and sulphites is 0.7 $mg\, kg^{-1}$ of body weight per day for the average man. It is difficult to estimate an average human intake of SO_2 since consumption of treated foods is high. Sulphur dioxide intake may sometimes exceed the ADI value. For example, the consumption of about three glasses of wine per day alone leads to an SO_2 intake exceeding the ADI. Sulphur dioxide destroys the thiamin in foodstuffs and many of the reported toxicity problems are symptoms of thiamin deficiency. It also interacts with folic acid, vitamin K and certain flavins and flavoenzymes. This problem can be overcome by supplementing nutritionally adequate diet, which can withstand substantial intakes of SO_2 in terms of thiamin destruction. Humans ingesting up to 200 mg SO_2 per day showed no signs of thiamin deficiency or changes in urinary excretion.

Products of nitrite, which reduces haemoglobin and increases the methaemoglobin content of the blood, are highly toxic to humans. Methaemoglobinaemia may result in death due to oxygen shortage. Infants less than 6 months old are particularly susceptible. Neither nitrate nor nitrite have teratogenic action. The formation of potent carcinogenic compounds, nitrosamines, in cooked cured meat products can be reduced by the combination of nitrite with other preservatives such as sorbic acid or common salt. The LD_{50} of nitrite for human beings is 300 $mg\, kg^{-1}$ body weight. Sodium and potassium nitrite are permitted in many countries including the USA and the UK to preserve meat and fish products and cheese. Nitrate contributes little or no preservative action except as a source of nitrite, e.g. following reduction by *Micrococcus* species in curing or fermenting meats. The FAO and WHO acceptable daily intake of nitrate is 0–5 $mg\, kg^{-1}$ per day and for nitrite 0–0.2 $mg\, kg^{-1}$ per day.

See also: ***Clostridium***: *Clostridium botulinum*. **Preservatives**: Traditional Preservatives – Oils and Spices;

Traditional Preservatives – Sodium Chloride; Traditional Preservatives – Organic acids; Traditional Preservatives – Wood Smoke; Permitted Preservatives – Sulphur Dioxide; Permitted Preservatives – Benzoic Acid; Permitted Preservatives – Hydroxybenzoic Acid; Permitted Preservatives – Nitrate and Nitrite; Permitted Preservatives – Sorbic Acid; Permitted Preservatives – Natamycin; Permitted Preservatives – Propionic Acid; Traditional Preservatives – Vegetable Oils.

Further Reading

Adams MR and Moss MO (1996) *Food Microbiology*. New Delhi: New Age International.

Branen AL and Davidson PM (eds) (1983) *Antimicrobials in Food*. New York: Marcel Dekker.

Branen AL, Davidson PM and Salminen S (eds) (1990) *Food Additives*. New York: Marcel Dekker.

Busta FF and Foegeding PM (1983) Chemical Food Preservatives. In: Seymour RS and Block SS (eds) *Disinfection, Sterilisation and Preservation*. Philadelphia: Lea & Febiger.

Davidson PM and Branen AL (eds) (1993) *Antimicrobials in Foods*, 2nd edn. New York: Marcel Dekker.

Gould GW (ed.) (1989) *Mechanism of Action of Food Preservation Procedures*. London: Elsevier.

Hayes PR (1985) *Food Microbiology and Hygiene*. London: Elsevier.

ICMS (1980) *Microbial Ecology of Foods*. Vols 1 and 2. New York: Academic Press.

Lin JK (1990) Nitrosoamines as potential environmental carcinogens. *Manual of Clinical Biochemistry* 23: 67.

Lueck E (1980) *Antimicrobial Food Additives: Characteristics, Uses, Effects*. Berlin: Springer.

Norman N (1995) *Food Science*, 3rd edn. Westport: AVI Publishing.

Thorne S (1986) *The History of Food Preservation*. Carnforth: Parthenon.

Traditional Preservatives – Oils and Spices

George-John E Nychas, Department of Food Science and Technology, Laboratory of Microbiology and Biotechnology of Foods, Agricultural University of Athens, Greece

Chrysoula C Tassou, National Agricultural Research Foundation, Institute of Technology of Agricultural Products, Athens, Greece

Introduction

True spices are defined as the roots, bark, buds, seeds or fruits of aromatic plants, which usually grow in tropical and some temperate climates. Essential oils are defined as being a group of odorous principles, soluble in alcohol and to a limited extent in water, consisting of a mixture of esters, aldehydes, ketones and terpenes. These compounds are mainly responsible for the characteristic aroma and flavour of the spices. The use of volatile solvents (e.g. acetate, ethanol, ethylene chloride) could provide not only a more complete flavour profile than the essential oils (oleoresins) alone, but also potentially greater antimicrobial inhibition. The use of spices as antimicrobials in the food industry sector is popular for a number of reasons. New food introduced in the market requires a long shelf life and assurance of freedom from food-borne pathogens. The excessive use of chemical preservatives, many of which are suspect because of their potential carcinogenic and teratogenic attributes or residual toxicity, has resulted in increasing pressure on food manufacturers either to completely remove chemical preservatives from their products or to adopt more 'natural' alternatives for the maintenance or extension of a product's shelf life. Food microbiologists have investigated the antimicrobial properties of many herbs, spices and food plants.

Range of Extracted Oils and Spices Available

About 1400 herbs, spices and plants have been reported to be potential sources of antimicrobial agents. The antimicrobial spectra of some extracts, taken with steam, cold, dry or vacuum distillation or volatile organic solvents, have been studied in detail (**Table 1**).

Among the compounds having a wide spectrum of antimicrobial effectiveness are thymol from thyme and oregano, cinnamic aldehyde from cinnamon, and eugenol from cloves. Phenolic compounds are the

Table 1 Plant essential oils known to have antibacterial properties

Achiote, allspice, almond (bitter), angelica, anise, asafoetida (*Ferula* sp.)
Basil, bay, bergamot
Calmus, camomile, cananga, caraway, cardamom, celery, chilli, cinnamon, citronella, clove, coriander, cornmint, cortuk, cumin
Dill
Elecampane, tarragon, eucalyptus
Fennel
Gale (sweet), garlic, geranium, ginger, grapefruit
Laurel, lavender, lemon, lime, linden flower, liquorice, lovage
Mace, mandarin, marjoram, mastic gum (*Pistachia lentiscus* var. *chia*), melissa, mint (apple), musky bugle, mustard
Neroli, nutmeg
Onion, orange, oregano
Paprika, parsley, pennyroyal, pepper, peppermint, pettigrain, pimento
Rose, rosemary
Saffron, sage, sagebrush, sassafras, savory, spike-oil, spearmint, star anise
Tarragon, tea, thuja, thyme, turmeric
Valerian, verbena, vanilla
Wintergreen, wormwood

Table 2 Naturally occurring antimicrobial compounds from different plants

Apigenin-7-glucose
Benzoic acid, berbamine, berberine
Caffeine, caffeic acid, 3-o-caffeylquinic acid, carnosol, carnosic acid, carvacrol, caryophelene, catechin, cinnamic acid, citral, chlorogenic acid, chicorin, coumarin, *p*-coumaric acid, cynarine
Dihydrocaffeic acid, dimethyloleuropein
Esculin, eugenol
Ferulic acid
Gallic acid, geraniol, gingerols
Humulone, hydroxytyrosol, hydroxybenzoic acid, hydroxycinnamic acid
Isovanillic, isoborneol
Linalool, lupulon, luteoline-5-glucoside, ligustroside
Myricetin
Oleuropein
Paradols, protocatechuic acid, rutin
Quercetin
Resocrylic
Salicylaldehyde, sesamol, shogoals, syringic acid, sinapic
Tannins, thymol, tyrosol
Verbascoside, vanillin, vanillic acid

major antimicrobial components of the essential oils of spices (**Table 2**). This article reviews the antimicrobial action of essential oil extracts (produced by steam distillation or by volatile solvents) and of specific compounds, for example phenolics such as eugenol and thymol.

Foods to Which Oils and Spices May be Added

For millennia spices and herbs have been added to foods as seasoning because of their aromatic characteristics. The addition of these ingredients in fermented ethnic foods from the Middle East, the Balkans, the Indian subcontinent and the Far East gained popularity in other countries and spread to Europe, the USA and elsewhere. Nowadays the main consumer of spices is the meat industry, in particular for the manufacturing of sausages, while oils are used by the flavouring industry for flavour enhancement and antioxidant effect. The potential use of these ingredients as natural antimicrobial agents is less exploited than their use as flavouring compounds and antioxidants.

Antimicrobial Action

Although the mechanism of action of phenolics and essential oils on microorganisms has not been elucidated, it is generally accepted that these compounds not only attack the cytoplasmic membrane, destroying its permeability and thus releasing intracellular constituents, but also cause membrane dysfunction in respect of electron transport, nutrient uptake, nucleic acid synthesis and ATPase activity, etc. This may be due to impairment of a variety of enzyme systems including those involved in energy production and structural component synthesis. In other words, the bactericidal/bacteriostatic effect of phenolic compounds is shown by perturbations of the cytoplasmic membrane of sporulated microorganisms at two different levels: cell wall and membrane integrity, and the physiological status of the bacteria. In both cases the perturbation can be observed in several ways, such as by measuring the leakage of cellular materials, monitoring changes in the fluidity of the membrane and the variation of phospholipid content, changes in the membrane functions such as electron transport and nutrient uptake, and by monitoring the effect of these compounds on membrane-bound enzymes.

Cell Wall Integrity

The phenolic and essential oil compounds are membrane-active agents possibly because they affect membrane permeability. Indeed, these compounds attack the cytoplasmic membrane, releasing intracellular constituents. When *Escherichia coli*, *Staphylococcus aureus*, *Lactobacillus plantarum*, *Pseudomonas fragi* and *P. fluorescens* are exposed to phenolics or essential oils, there is a leakage of the sodium [3,4-^{14}C] glutamate, $NaH_2{}^{32}PO_4$, ultraviolet-absorbing material, ^{14}C-labelled compounds, nucleotides, glutamate, potassium and inorganic phosphate. The weakening or destruction of the permeability barrier of the cell membrane can account for this loss. Leakage of intracellular compounds is known to be a general phenomenon induced by many antibacterial substances. The increase in the permeability of the cell membrane could be attributed to the ability of the phenolics/essential oils to bind proteins, probably through hydrophobic interaction, or to react with the phospholipid component of the cell membrane of bacteria (e.g. *Pseudomonas aeruginosa*). These reactions may cause (a) significant changes in the fatty acid composition and phospholipid content of these organisms and (b) precipitation of proteins (e.g. cytoplasmic). In the former case, as most of the essential oils are highly hydrophobic, the cytoplasmic membrane is likely to be the main site of adsorption, since the cell wall of (for example) *Staphylococcus aureus* contains as little as 1–2% lipid material. The low water solubility of many phenolics and essential oils probably precludes any significant diffusion into the cytoplasm. Moreover, low temperature decreases the solubility and hence the concentration of the phenol in the cell membrane lipid. While this is one possibility, the effect of temperature on rates of reaction is probably more important. Lysis of bacterial protoplasts by phenolics and essential oils occurs at concentrations lower than those causing leakage of nucleotides from whole cells. This is an indication that the cell wall may play an important role in the relative resistance of whole cells to lysis by low concentrations of phenolic compounds. It has been suggested that such low concentrations affect the activity of enzymes associated with energy production, while higher amounts cause a precipitation of proteins. There is no definite answer to the question of whether the alleged damage caused to the cell membrane is quantitatively related to the amount of phenol derivative to which the cell is exposed, or whether the effect is such that, once a small amount of damage has been caused, the lesions enlarge and leakage proceeds continuously.

The binding with proteins is evident with the differences in high-performance liquid chromatography (HPLC) protein profile or sodium dodecyl sulphate polyacrylamide gel electrophoresis (SDS-PAGE) protein patterns in broths inoculated with *Staphylococcus aureus* or *Salmonella enteritidis* and supplemented with phenolics or essential oils. The extracellular material, as demonstrated with HPLC, is only a measure of a generalized loss of membrane function and it is more likely that these compounds interfere with energy metabolism, synthesis of macromolecules or the cell membranes of actively growing microorganisms. This accounts also for the differences found in the utilization of glucose and amino acids, as well as the production of metabolic products in broth inoculated with *S. enteritidis* when mint essential oil is added. Scanning electron microscopy shows that whole cells of untreated *Lactobacillus plantarum*, *Bacillus cereus* and *Staphylococcus aureus* are smooth compared with bacteria treated with phenolics for 24 h; in the latter the cell surfaces become irregular and rough.

Physiological Status

Phenolics and essential oils inhibit *Staphylococcus* enterotoxin B (SEB) and lactate production as well as the rate of glucose assimilation in well-buffered media, despite little or no effect on final cell mass. The decrease in the percentage of glucose and amino acid utilization, as well as the reduction in the formation of L-lactate, could be due to the inhibitory effect of phenolic/essential oil on substrate uptake, on specific enzyme(s) or on the electron transport chain. Similar results are obtained with *Salmonella enteritidis*. Inhibition of growth and enterotoxin A production by *Staphylococcus aureus* strain 100 by butylated hydroxyanisole (BHA) has also been noted. The effect of various phenolic compounds on the membrane-bound ATPase activity of *S. aureus* varies significantly. Some stimulate the activity while others either inhibit it to various degrees or are found to be neutral. These results suggest that there is no general overall effect of their activity on the membrane-bound ATPase of *S. aureus*.

As far as the effect of essential oils on *Lactobacillus plantarum* is concerned there is no influence on the rate of glycolysis, although a decrease in the ATP content of the cells has been observed.

Studies with spore-forming bacteria (e.g. *Bacillus*) shows that these may be more sensitive than non-spore-forming ones to the phenolic compounds. The activity of phenolics from olives could be decisive also for spores, as they may denature germination enzymes, inhibit the lytic enzyme subtilopeptidase A, or interfere with the use of L-alanine or other amino acids necessary for the initiation of the germination process.

The ability of phenolics and essential oils to affect many cell types could be explained if their mechanism of action was membrane perturbation, leading to cell dysfunction. These compounds probably do not share a common mechanism of action and there may not be a single target associated with the inhibition of microorganisms by these compounds. This mode of action could be beneficial in that it would be difficult for microbes to evolve resistance to phenolics and essential oils. This proposal would also explain the fact that phenolics, essential oils and phytoalexins generally cause static rather than outright toxic effects; cell membranes that leak or function poorly

Table 3. Antimicrobial spectrum of essential oils from herbs, spices and various plants

Gram-positive	*Gram-negative*	*Yeasts/fungi*
Arthrobacter sp.	*Acetobacter* spp.	*Aspergillus niger, A. parasiticus, A. flavus, A. ochraceus*
Bacillus sp., *B. subtilis, B. cereus*	*Acinetobacter* sp., *A. calcoacetica*	*Candida albicans*
Brevibacterium ammoniagenes, B. linens	*Aeromonas hydrophila*	*Fusarium oxysporum, F. culmorum*
Brochothrix thermosphacta	*Alcaligenes* sp., *A. faecalis*	*Mucor* sp.
Clostridium botulinum, C. perfringens, C. sporogenes	*Campylobacter jejuni*	*Penicillium* sp., *P. chrysogenum, P. patulum, P. roqueforti, P. citrinum*
Corynebacterium sp.	*Citrobacter* sp., *C. freundii*	*Malassezja ovale*
Lactobacillus sp., L. plantarum, L. minor	*Edwardsiella sp.*	*Rhizopus sp.*
Leuconostoc sp., L. cremoris	*Enterobacter sp., E. aerogenes*	*Saccharomyces cerevisiae*
Listeria monocytogenes, L. innocua	*Escherichia coli, E. coli O157:H7*	*Trichophyton mentagrophytes*
Micrococcus sp., M. luteus	*Erwinia carotovora*	
Pediococcus sp.	*Flavobacterium sp., F. suaveolens*	
Propionibacterium acnes	*Klebsiella sp., K. pneumoniae*	
Sarcina spp.	*Moraxella sp.*	
Staphylococcus spp., S. aureus	*Neisseria sp., N. sicca*	
Enterococcus faecalis	*Proteus spp., P. vulgaris*	
	Pseudomonas spp., P. aeruginosa, P. fluorescens, P. fragi, P. clavigerum	
	Salmonella spp., S. enteritidis, S. senftenberg, S. typhimurium, S. pullorum	
	Serratia sp., S. marcescens	
	Vibrio sp., V. parahaemolyticus	
	Yersinia enterocolitica	

would not necessarily be lethal, but would probably cause a slowing of metabolic process such as cell division.

Effect on Microbial Populations and Spoilage

The essential oils of spices or herbs show bactericidal activity against Gram-positive bacteria but only bacteriostatic activity against Gram-negative bacteria (**Table 3**). Their antimicrobial activity is influenced by the culture medium, the temperature and the inoculum size. A strong synergism with EDTA is noted with Gram-negative species.

Microbial Inhibition by Essential Oils

Gram-positive Species Active agents from linden flower, orange, lemon, grapefruit, mandarin, rosemary, oregano, thyme, cumin, caraway, clove, thyme, allspice, basil, sage, spearmint, mastic gum and onion retard the growth of *Staphylococcus aureus*, which is probably the most commonly used bacterium in studies of antimicrobial activity of essential oils.

The effects of different essential oils (rosemary, cloves and oregano) against *Listeria monocytogenes*, the psychrotrophic Gram-positive bacterium responsible for listeriosis, have been examined as well. The psychrotrophic and aciduric nature of this microorganism plays an important role in the final net effect. The essential oil of *Mentha piperita* var. *officinalis* against strains of *L. monocytogenes* – tested with the disc agar diffusion method – exhibits moderate inhibition, while the essential oils of cinnamon, cloves, oregano and thyme are the most inhibitory. The essential oil from clove at 0.5% and 1% concentrations is bactericidal to this organism when grown in laboratory media at 4°C and 24°C.

The inhibition of spore-forming bacteria by spices and their essential oils has also been studied. *Clostridium botulinum*, *C. sporogenes* and *C. perfringens* are inhibited by the essential oils of garlic, mace, achiote, onion, cinnamon, thyme, oregano, clove, pimento and black pepper. It was also observed that the effect of spice oils on spore germination is reversible. The oils of black pepper and clove have a greater inhibitory effect on vegetative cell growth than the other oils. None of the oils has a significant effect on outgrowth of spores, but there is a decrease in the rate and extent of germination of *Bacillus subtilis* spores in the presence of clove and eugenol. It is worth mentioning that the composition of the media used to test the oils affects the activity of essential oils.

Gram-negative Species The growth of *Salmonella* spp. is inhibited by essential oils of linden flower, basil, spearmint, thyme, oregano, orange, lemon, grapefruit, mandarin, almond, bay, clove, coriander, cinnamon, pepper and mastic gum, while growth of *Aeromonas hydrophila* is inhibited by clove, coriander and nutmeg.

Yeasts and Mycelial Fungi Clove is the strongest antifungal spice; cinnamon is also quite inhibitory, while mustard, garlic, allspice and oregano give smaller degrees of inhibition against *Penicillium* spp. and *Aspergillus* spp. Contradictory results have been obtained with oregano and thyme oil in tests with *Aspergillus niger*, *A. ochraceus*, *A. flavus*, *A. niger*, *Penicillium chrysogenum*, *Rhizopus* sp. and *Mucor* sp. Thyme and oregano oils can stimulate the growth of *Aspergillus flavus* and *A. parasiticus*, while at the same time acting as anti-aflatoxigens. *Aspergillus* and *Penicillium* isolates from black table olives are inhibited with methyl-eugenol and the essential oil of the spice *Echinophora sibthorpiana*.

Microbial Inhibition of Essential Oils in Food Model Systems

The effectiveness of essential oils decreases when experiments are conducted in vivo or in food model systems. The essential oils of basil and sage do not inhibit *Pseudomonas fragi* and *Salmonella typhimurium* in a meat gravy sauce, although their inhibition is observed in broth cultures when these compounds are present. There is a slight effect of these two essential oils on the flora of meat and on meat inoculated with *Salmonella enteritidis* for the first 3 days of storage. There is no additional inhibition of this bacterium in mayonnaise supplemented with essential oils.

The inhibition of growth of *Listeria monocytogenes* and *Salmonella enteritidis* in N-Z Amine A (NZA) broth supplemented with essential oil of mint confirms the consensus that the essential oils are more effective in laboratory media than in food systems. Thyme essential oil reduces the viable counts of *L. monocytogenes* in minced pork meat by about 2 $\log_{10}$ at 4°C and 1 $\log_{10}$ at 8°C, but its effectiveness is decreased compared with its performance in laboratory media. Moreover, cloves and oregano at 1% level fail to control the growth of *L. monocytogenes* in a meat slurry, while the same concentrations are bactericidal and bacteriostatic, respectively, in broth. There is evidence that the active substance is less stable in food systems than in a laboratory medium, a view that endorses the argument that essential oils are bound to food contents by addition and condensation reactions, leaving only a part of the total concentration of the essential oil free to exhibit the antibacterial effect.

The essential oil of mint loses its strong antibacterial activity against *Salmonella enteritidis* and *Listeria monocytogenes* in tzatziki salad (pH 4.2), in pâté (pH 6) and in fish roe salad (pH 5). Although in culture media supplemented with the essential oil, no growth is observed over 2 days at 30°C, *S. enteritidis* dies in tzatziki in all treatments and declines in the other foods, except pâté, as judged by viable counts. It seems that a low pH (e.g. tzatziki salad) favours inhibition, for *L. monocytogenes* populations show a declining trend towards the end of the storage period. The marked growth of *L. monocytogenes* in pâté confirms the fact that this food can harbour large populations of the microorganism even after the addition of antibacterial substances. The antibacterial action of mint essential oil depends mainly on its concentration, the food pH, composition, storage temperature and the nature of the microorganism.

A marked reduction of *Aeromonas hydrophila* occurs in inoculated samples of cooked, non-cured pork treated with clove or coriander and packaged either under vacuum or in air and stored at 2°C or 10°C. The lethal effect of these two oils is more pronounced in the vacuum-packaged samples.

Sublethal Effects, Strain Variability and Interaction with Other Factors

The ability of essential oils to act on different targets in bacterial cells, as well as the variation between organisms as to the relative sensitivity of these targets, needs to be noted. As far as the activity of naturally occurring compounds is concerned, Gram-positive bacteria are more sensitive, in general, to the abovementioned substances than are Gram-negative ones; variation in the rates of inhibition is evident also among the Gram-negative bacteria. A possible explanation for their generally reduced activity towards Gram-negative bacteria could be the lipophilic nature of the oils – they fail to diffuse across the outer membrane. However, the results with mint essential oil on *Salmonella enteritidis* and *Listeria monocytogenes* contradict the view that Gram-positive bacteria are more susceptible to essential oils than Gram-negative ones.

Factors Affecting Antimicrobial Action

Proteins, lipids, carbohydrates, water activity, pH, temperature and chelating compounds all affect the antimicrobial activity of phenolics. The antimicrobial effect of phenolics/essential oils against *Staphylococcus aureus*, *Pseudomonas fluorescens* and *Saccharomyces cerevisiae* is influenced by the presence of different amounts of casein and corn oil. The increase of proteins in the medium (broth culture, model food system or *in situ* food) affects the inhibitory action of these compounds on the survival and growth of the target microorganism. For example, the resistance to sage essential oil increases with decrease in water content and increase in the protein and fat content of the food. The loss of antimicrobial activity of phenolics has been attributed to binding of these compounds to food proteins (milk, meat, fish). Fats also influence the effectiveness of phenolics/essential oils and, in general, reduce their antimicrobial activity. Indeed, owing to their hydrophobic nature essential oils are much more soluble in the lipid ingredients than in the aqueous phase of the food. Because bacterial proliferation takes place in the latter, essential oils become ineffective towards the microbial flora. It has been suggested also that a fat coat could form on the surface of bacterial cells and possibly prevent the penetration of inhibitory substances. Addition of carbohydrates (e.g. glucose) has no effect on the inhibitory action of the essential oils in broth cultures inoculated with *Salmonella enteritidis* and *Staphylococcus aureus*. It can be concluded that the presence of a relatively high concentration of fats or protein in foods is more effective than the presence of carbohydrates in protecting microorganisms from the inhibitory action of essential oils.

The synergetic effect of salt in the inhibitory effectiveness of phenolics/essential oils is disputed. For example, the inhibitory effect of sage against *Bacillus cereus* in rice or in strained meat is more pronounced when salt is used in combination. Other research workers have not found this synergetic effect with *Salmonella enteritidis* or *Lactobacillus* spp., at least in broth cultures. In food model systems, such as fish roe salad (a very salty food), it seems that the essential oil of mint at concentrations of at least 1.0% (v/w) affects *S. enteritidis* as soon as it is inoculated into the food.

The effect of pH on the preservation capacity of phenolic is slightly more complex, as the effect depends on the microorganism tested and the form of carboxyl group of the compound. The synergism of pH and essential oil is evident especially in a food model system with low pH (tzatziki), where the greatest bactericidal effect is observed. On the other hand, in a food model system with neutral pH (pâté), addition of essential oil has little effect on the target organisms; pH modulates the rates of the inhibitory reactions as well as the chemical characteristics of the main substances that make up the essential oil. The increased antibacterial activity of the essential oil at low pH can be related to the fact that the essential oil constituents become more hydrophobic at low pH and dissolve better in the lipid phase of the bacterial membrane.

Temperature and the presence of EDTA can also play an important role in the inhibitory effect of essential oils on the death rate of microorganisms. For example, mayonnaise (an acidic food) stored at 18–22°C for 24 h exhibits a marked decrease in the populations of pathogens, but refrigeration temperatures protect *Salmonella* spp. Refrigeration temperatures enhance the inhibitory activity of sage, but not cloves and oregano, on *Listeria monocytogenes*; with cloves, the organism dies more rapidly in trypticase–soy broth (TSB) at 24°C than at 4°C. The essential oil in combination with EDTA is very inhibitory to *Pseudomonas fragi*, while a concentration of 0.01% (w/v) EDTA does not significantly increase the effect of the essential oil on *Salmonella enteritidis*. The chelating agent EDTA affects the surface of Gram-negative bacteria and in many cases causes the outer membrane to become more permeable to various molecules.

See also: **Bacteria**: The Bacterial Cell. **Fermented Foods**: Fermentations of the Far East. **Preservatives**: Traditional Preservatives – Sodium Chloride. ***Pseudomonas***: *Pseudomonas aeruginosa*. ***Staphylococcus***: *Staphylococcus aureus*; Detection of Staphylococcal Enterotoxins.

Further Reading

Billing J and Sherman PW (1998) Antimicrobial functions of spices: why some like it hot. *Quarterly Review of Biology* 73: 3–49.

Branen AL and Davidson PM (1983) *Antimicrobials in Foods*. New York: Marcel Dekker.

Deans SG and Ritchie G (1987) Antibacterial properties of plant essential oils. *International Journal of Food Microbiology* 5: 165–180.

Denyer SP And Hugo WB (1991) *Mechanisms of action of chemical biocides; their study and exploitation*. Society for Applied Bacteriology Technical Series 27. Oxford: Blackwell.

Dziezak JD (1989) Spices. *Food Technology* 102–116.

Farell KT (1990) *Spices, Condiments and Seasonings*, 2nd edn. New York: Van Nostrand Reinhold.

Fisher C (1992) Phenolic compounds in spices. In: Ho CT, Lee CY and Huang MT (eds) *Phenolic Compounds in Food and their Effect on Health: I, Analysis, Occurrence, and Chemistry*. Symposium Series 506. Washington: ACS.

Jay JJ and Rivers GM (1984) Antimicrobial activity of some flavouring compounds. *Journal of Food Safety* 6: 129–139.

Nychas GJE (1995) Natural antimicrobials from plants. In: Gound GW (ed.) *New Methods of Food Preservation*. P. 58. London: Blackie.

Shelef LA (1984) Antimicrobial effects of spices. *Journal of Food Safety* 6: 29–44.

Wilkins KM and Board RG (1989) Natural antimicrobial systems. In: Gould GW (ed.) *Mechanisms of Action of Food Preservation Procedures*. P. 285. London: Elsevier.

Traditional Preservatives - Sodium Chloride

M Susan Brewer, Department of Food Science and Human Nutrition, University of Illinois, Urbana, USA

Introduction

Sodium chloride occupies a unique place in human evolution. So abundant in nature and vital for life processes, it has been designated the fifth element, equated with earth, air, water and fire. Ancient records indicate that salt was used to cure meat in 3000 BC. By 850 BC, during the time of Homer, the use of salt and smoke were already old practices. In the Middle Ages, potassium nitrate (saltpeter) was added to the process to increase the preservative action of salt and to prevent botulism, particularly in sausage. Somewhat later, acid ingredients (fruit juice, wine) were added to salt-water brines used to preserve vegetables.

Preserving food means preventing spoilage and suppressing growth of pathogens, often by making the environment unfavourable for bacteria, yeasts and moulds. Microorganisms need relatively neutral pH and high water activity (a_w). The a_w of food can be lowered by removing water, by adding solutes (sugar, salt) or by freezing. Most fresh foods have a_w values of 0.95–0.99, allowing growth of most classes microorganisms (**Table 1**). The minimum a_w for most bacterial growth is 0.90–0.91, except for certain halotolerant and halophilic bacteria and osmophilic fungi. Often a higher a_w is required for toxin production. The growth rate of bacteria is greater than that of yeasts or moulds; therefore, in foods with high a_w, bacteria will generally outgrow the fungi to cause spoilage. Fruits and fermented foods are commonly spoiled by fungi due to the acidity of the product which restricts bacterial growth even at high a_w. Products with low a_w due to high salt concentrations (ham, salted fish) are often spoiled by halotolerant or halophilic bacteria.

NaCl Suppression of Growth

Salt has a variety of effects on both food tissues and microbial cells which are responsible for its preservative action. It can inactivate enzyme systems vital to the cell, slowing or stopping growth. It can draw water out of the cells due to osmotic pressure. It can have specific effects at the membrane level. In most cases, the effectiveness of NaCl as a preservative also depends on other environmental factors such as pH.

Prevention of pathogen growth is critical in preserved foods. Most spore-forming microorganisms can only grow at pH values of 4.6 or higher. To preserve them by tying up available water, low-acid foods (pH > 4.6) must have their a_w reduced to < 0.94 by adding solutes to prevent growth and toxin production. The endospore-forming rods (*Bacillus* and *Clostridium*) vary greatly in their salt tolerance, ranging from 2% to 25%. *C. botulinum*, like most pathogens, will not grow in an acid environment (pH < 4.6) (**Table 2**). The microorganisms found in the human gut, including enteric pathogens such as *Escherichia coli* and *Salmonella* spp., do not tolerate elevated salt levels growing at minimum a_w of 0.93–0.98. Some strains of *Vibrio* fail to grow without NaCl – the optimum concentration is about 3%. Skin flora, such as *Staphylococcus aureus*, because of their exposure to the salts found in sweat, are very salt-tolerant. Tolerance of an a_w of 0.86 induced by 10% NaCl classifies them as halophiles.

NaCl is more inhibitory than glycerol and sucrose to salt-sensitive bacteria (most spore-formers, Entero-

Table 1 Approximate water activity (a_w) values of selected foods and of sodium chloride and sucrose solutions. Adapted from Troller and Christian (1978) and Jay (1996).

a_w	*NaCl (%)*	*Sucrose (%)*	*Foods*
1.00–0.95	0–8	0–44	Fresh meat, fresh and canned fruit and vegetables, frankfurters, eggs, margarine, butter, low-salt bacon
0.95–0.90	8–14	44–59	Processed cheese, bakery goods, raw ham, dry sausage, high-salt bacon, orange juice concentrate
0.90–0.80	14–19	59–saturation (0.86 a_w)	Aged Cheddar cheese, sweetened condensed milk, jams, margarine, cured ham, white bread
0.80–0.70	19–saturation (0.75 a_w)		Molasses, maple syrup, heavily salted fish
0.70–0.60			Parmesan cheese, dried fruit, corn syrup, rolled oats, jam
0.60–0.50			Chocolate, confectionery, honey, dry noodles/pasta
0.40			Dried egg, cocoa
0.30			Dried potato flakes, potato chips, crackers, cake mix
0.20			Dried milk, dried vegetables, chopped walnuts
0.19			Sugar
0.10–0.20			Soda crackers

Table 2 Inhibitory water activity (a_w) values for growth of selected microorganisms. Source: Leistner and Rodel (1975) and Banwart (1981)

a_w	*Bacteria*	*Yeasts*	*Moulds*
0.98	*Clostridium* [a], *Pseudomonas*[b]		
0.97	*Clostridium*[c]		
0.96	*Flavobacterium*, *Klebsiella*, *Lactobacillus*[b], *Proteus*[b], *Pseudomonas*[b], *Shigella*		
0.95	*Alcaligenes*, *Bacillus*, *Citrobacter*, *Clostridium*[d], *Enterobacter*, *Escherichia*, *Proteus*, *Pseudomonas*, *Salmonella*, *Serratia*, *Vibrio*, *Clostridium* spores		
0.94	*Lactobacillus*, *Microbacterium*, *Pediococcus*, *Streptococcus*[b]		
0.93	*Lactobacillus*[b], *Streptococcus*, *Vibrio*[b], *B. sterothermophilus* spores		*Rhizopus*, *Mucor*
0.92		*Rhodotorula*, *Pichia*	
0.91	*Corynebacterium*, *Staphylococcus*[e], *Streptococcus*[b]		
0.90	*Micrococcus*, *Pediococcus*	*Saccharomyces*, *Hansenula*	
0.88		*Candida*, *Torulopsis*	*Cladosporium*
0.87		*Debaryomyces*	
0.86	*Staphylococcus*[f]		
0.85			*Penicillium*
0.81		*Saccharomyces*	
0.80	Most non-marine bacteria		*Aspergillus* mycotoxin produces aflatoxin
0.75	Halophilic bacteria		*A. flavus*, *A. ochraceus*
0.65			*Aspergillus*
0.61		*Xeromyces*, *Zygosaccharomyces*	

[a]*Clostridium botulinum* type C.
[b]Some strains.
[c]*C. botulinum* type E, some strains of *C. perfringens*.
[d]*C. botulinum* type A, B, *C. perfringens*.
[e]Anaerobic.
[f]Aerobic.

bacteriaceae, *Pseudomonas fluorescens*) and less inhibitory than glycerol to salt-tolerant bacteria (Micrococcaceae, *Vibrio*) at comparable a_w values. *Clostridium* growth is suppressed and spores are completely inhibited from germination when NaCl is used to lower a_w to 0.95; no inhibition occurs with glycerol, glucose or urea at this a_w (nor above 0.93). When a_w is maintained above the minimum for *P. fluorescens* growth, NaCl completely inhibits catabolism of glucose, and DL-arginine; glycerol is inhibitory at much lower a_w values. In general, NaCl is more inhibitory to respiring organisms than glycerol.

The interactive effects of pH and temperature with NaCl have been demonstrated for both lag phase and generation time of a mixture of six strains of (cold-tolerant) *Aeromonas hydrophila*. At 3°C, pH 7.0, increasing NaCl from 0.5% to 2.5% increased lag phase from 186 to 519 h and generation time from 9 to 28 h; when temperature was increased to 7°C, lag phase increased from 55 h (0.5% NaCl) to 128 h (2.5% NaCl) and generation time increased from 5 to 15 h. At 7°C, with pH reduced to 5.4, lag phase increased from 142 h (0.5% NaCl) to 449 h (2.5% NaCl) and generation time increased from 9 to 39 h. At pH 7, the effect of adding NaCl (0.5 or 2.5%) was proportionally approximately the same at 3°C and at 7°C. Decreasing pH to 5.4, under temperature abuse conditions (7°C), resulted in approximately the same lag and log phases (at 0.5 and 2.5% NaCl) as at 3°C indicating that pH is a factor under some conditions. However, it was reported that the minimum inhibitory concentrations of NaCl were independent of pH (5.7–7.0) for Gram-positive lactic acid bacteria and *S. aureus*. Toxin production by *S. aureus* is affected by both NaCl and pH. Even if growth is suppressed by addition of 10% salt, toxin production (per unit of growth) is unaffected between pH 5 and 7. Reduction of pH < 4.5 allows reduction of NaCl to about 4% to limit toxin production.

Mechanisms of NaCl Suppression of Growth

Several mechanisms of NaCl-induced suppression of microbial growth work in concert. The NaCl effect is partially the general effect of reduced a_w: cellular requirements that are mediated through an aqueous environment are progressively shut down, there is damage to the cell membrane (which must be maintained in a fluid state) and osmolysis, disruption of

N^+/K^+ balance, and in some cases, direct effects on specific enzymes and DNA. At low NaCl levels (0.5–2.0%), denaturation does not appear to be the mechanism by which NaCl affects microbial enzyme systems; however the contribution to the ionic strength and a_w of the system must be considered. Solutes such as NaCl that do not diffuse freely into the cell when concentration is high can affect processes that are occurring at the cell surface, such as transport. Most transport is active, exhibiting saturation kinetics at < 100 μmol concentrations of solute; the cell surface is exposed to much higher solute concentrations than what it has evolved to handle. However, microbial growth in the presence of NaCl concentrations that would inhibit enzymatic activity inside the cell requires that substrate (glucose, etc.) transfer should continue. NaCl can decrease enzyme activity by denaturing the enzyme, by reducing the catalytic activity, and by altering cofactors.

To damage microbial enzymes, NaCl must gain access to the intracellular pool. Enzymes can be damaged in a variety of ways by high ionic environments and perhaps by NaCl specifically. Many of those involved in the preservation effect of NaCl are involved with cellular recovery from stress. Enzymes important to microbial survival and recovery from injury which appear to be sensitive to NaCl concentration include the oxidoreductases – catalase, superoxide dismutase and peroxidase. The functions of these enzymes are to control the concentrations of hydrogen peroxide, lipid peroxides and superoxide anion, oxidation intermediates involved in formation of free radicals. *Bacillus subtilis* aldolase, which is involved in energy transfer, is inhibited by NaCl to the same degree as growth is suppressed; however, this relationship varies among genera and species. Protein decarboxylases appear to be inhibited in putrefactive microbes. Salt is a known pro-oxidant. Sufficiently high intracellular NaCl concentrations may disturb oxidative metabolism through its effects on non-haem iron. The iron cofactors required by enzymes in the tricarboxylic acid (TCA) cycle may ultimately back up the TCA cycle. Microbial iron sulphur enzymes may be inhibited. Restriction enzymes involved with DNA repair are also sensitive to NaCl. It has been suggested that the alteration in ionic strength of the environment results in a loss of ions from DNA molecules which destabilizes them sufficiently to allow conformational change, especially under heat stress.

NaCl can increase the activity of some enzymes. As NaCl increases (up to 8%), proteolytic and peroxidase activities of *Aspergillus parasiticus*, *A. flavus*, *A. ochraceus* and many *Penicillium* species increase markedly. Lipolytic activity in these fungi in the presence of NaCl increases to a lesser degree, but those without it are unable to develop it in the presence of NaCl. Above 8%, proteolytic, lipolytic and peroxidase activity decreases, and foods containing or coated with salt above this level are poor substrates for aflatoxin production by *A. flavus* and *A. ochraceus*. However, the increased activities up to 8% pose problems in some fermented products in which these fungi are selected for, alter the product, and potentially produce mycotoxins.

NaCl can promote the conversion of the microbial environment to a hostile one. The lactic acid-producing bacteria have inhibitory and lethal effects on a variety of food-borne pathogens and food spoilage microbes. The use of NaCl to select for these bacteria effectively concentrates their inhibitory effects whether or not their organic acid, bacteriocin, or hydrogen peroxide production is up-regulated. On the other hand, high salt brines select for acid-tolerant *Rhodotorula glutinis* varieties which produce both polygalacturonase and pectin methyl esterase, which softens texture by converting pectic substances into more soluble forms.

Salt Tolerance

The degree of tolerance of spoilage microbes for decreased a_w induced by NaCl depends on whether the required nutrients are present in optimal amounts. Salt tolerance, due to reduced a_w and possibly to NaCl-specific effects, also depends on whether factors such as pH, temperature and redox potential are optimal or suboptimal, and to some extent on their normal environment. Saline (0.85–0.90% NaCl) produces an isotonic condition for non-marine microorganisms. Higher salt concentrations increase osmotic pressure, resulting in net movement of water out of the cell; the result is plasmolysis which results in growth inhibition and possibly death. Most non-marine bacteria can be inhibited by a hypertonic solution of 20% NaCl. Some microbes, because their natural environment is high in NaCl or of high ionic strength, may be halophilic, requiring higher salt concentrations to live and grow. Others may be able to survive but not grow in high salt concentrations (halotolerant). The minimum a_w at which halophiles can grow is 0.75 (a saturated NaCl solution). Xerophilic fungi grow below a_w 0.85. Osmophilic yeasts do not have a general requirement for low a_w but tolerate it better than non-osmophiles for which the lower limit is about 0.87. Osmophilic and xerophylic yeasts and moulds such as *Zygosaccharomyces rouxii* and *Xeromyces bisporus* can grow at a_w values of 0.65–0.61. Osmotolerant yeasts include many members of the genera *Candida*, *Citeromyces*,

Debaryomyces, *Hansenula*, *Saccharomyces* and *Torulopsis*.

Mechanisms of Salt Tolerance

The primary mechanism by which (halotolerant and halophilic) microbes protect themselves against osmotic stress is through intracellular accumulation of compatible solutes such as K^+. This intracellular accumulation is the result of altered solute transfer and altered cellular synthesis of osmotically active species. Many halophilic bacteria require KCl from their environment; they accumulate and concentrate it to balance osmotic pressure. Many Gram-negative bacteria accumulate proline by enhanced transport. The imposition of osmotic stress that removes water from the cell, increasing the K^+ ion concentration, triggers accumulation of amino acids. Addition of L-proline (to growth medium) enhances growth of these bacteria in a high-osmotic-strength environment. In addition to transport of proline, synthesis of glutamine is used by *Staphylococcus aureus* under osmotic stress from 10% NaCl. *Listeria monocytogenes* accumulates glycine betaine and carnitine. Halotolerant and xerotolerant fungi tend to produce polyhydric alcohols such as glycerol and then alter their transport out of the cell. In non-tolerant fungi such as *Saccharomyces cerevisiae*, some tolerance can be produced by shifting energy into synthesis of protective polyhydric alcohols, resulting in an increased requirement for glucose. The transport of the polyhydric alcohols is then regulated to prevent their movement out of the cell. Other osmoprotectants synthesized or transported by microorganisms include glutamate, γ-aminobutyrate, alanine, sucrose and trehalose.

NaCl Effects on Heat Resistance

Heat resistance of some microbes increases when NaCl is used to decrease a_w. The enterococci, *Enterococcus faecium* and *E. faecalis*, often survive the pasteurization temperature (68°C) used for partially cooked, canned hams. These organisms are most heat-resistant at an a_w of 0.95 when salt is used. The heat resistance of *Lactobacillus* also increases when a_w is reduced with NaCl; it is maximal at 0.975–0.985, which also has implications for preserved meat products. NaCl (up to 10%) increases heat resistance of *Salmonella*.

Effect of NaCl on Spores

Lower than optimum a_w usually increases microbial resistance to heat – spores are most heat resistant at 0.20–0.40. *Bacillus stearothermophilus* spores, indicators of thermophilic spoilage, are quite heat-resistant at high a_w values; they pose a problem at pH values above 5.7. If foods containing these spores are stored at temperatures > 45°C, heat processing time at 121°C may be > 20 min to attain commercial sterility. When NaCl is the solute, these spores are strongly inhibited at a_w 0.93. However, not all solutes have this effect, and salt does not have this effectjyon all spores. *B. subtilis* spores are inhibited by 0.2 mol l^{-1} NaCl. *Clostridium botulinum* spores exhibit decreased ability to recover from heat stress and gamma radiation when grown out in the presence of 1–2% NaCl regardless of pH. NaCl has similar effects on the ability of *B. cereus* to form spores. Heat resistance of bacterial spores depends partly on maintenance of low water concentrations in the central protoplast, possibly via osmotic dehydration which is dependent on the presence of both anions and cations.

Salt-induced Microbial Selection

Often, preserving food is a matter of selection or suppression of specific microbes that alter pH. The mesophilic Gram-negative rods and psychrophiles are inhibited by 4–10% salt. The lactic acid-producing bacteria vary in salt tolerance from 4% to 15%. Spore-forming bacteria generally tolerate 5–6% salt.

Alteration of pH by production of organic acids from carbohydrate catabolism is widely used in the food fermentation industries. The fermentation process takes advantage of the fact that growth and activity of many undesirable microbes are inhibited by the presence of solutes. The groups which tolerate the higher salt concentrations found in brines and/or anaerobic environments (submerged in brine) using the sugar leached out of vegetables, such as cucumbers, as an energy source are primarily lactic acid-producing bacteria. These organisms predominate within 24–48 h in typical fermentation brines. The first 2 days of a fermentation process is a critical period during which effort must be directed towards encouraging growth of acid-producing bacteria, and inhibiting proteolytic bacteria such as pseudomonads that tend to raise pH by production of ammonia and other basic by-products.

A variety of vegetables are preserved by brining or fermenting. Fermented pickles are cured over a 3–6-week period to a pH of about 3.5. Most brined pickles are made in a low salt brine (3–5% salt) which contains some added acid in addition to the salt. After sufficient lactic acid is produced, acetic acid-producing bacteria take over to continue to lower the pH and alter the flavour of the product. Vegetables can be fermented in a high salt brine (10% salt) but must be de-salted by soaking in water before further processing. Pickled vegetables made in low salt brines need no de-salting. When cabbage is fermented to sauerkraut by lactic acid-producing bacteria, selection of the desired types of bacteria requires not less than 2% and not more than 3% salt. Sauerkraut is microbiologically stable without refrigeration at 3% salt and 1.50% titratable acidity.

Specific concentrations of salt favour the growth of *Leuconostoc mesenteroides*, *Lactobacillus brevis*, *Pediococcus cerevisiae* and *Lactobacillus plantarum* in the correct sequence during the fermentation process. The Gram-positive cocci include aerobic and facultative anaerobic bacteria in the family Micrococcaceae (genera *Micrococcus* and *Staphylococcus*) and the family Streptococcaceae (genera *Streptococcus*, *Leuconostoc*, *Pediococcus* and *Aerococcus*). *Micrococcus* are important spoilage bacteria which can grow in the presence of 5% NaCl. Most strains of *Staphylococcus* can grow in 7.5% salt; some tolerate 15% salt. Accurate salt concentration is critical: too little results in poor flavour and soft texture, too much selects for osmophilic yeasts.

Yeasts and moulds are able to grow at lower a_w than bacteria. If they dominate, they convert the lactic acid to non-acid products, raising the pH back up into the range where pathogenic organisms can grow. When the pH increases, the available proteins are used by proteolytic organisms such as *Bacillus*. Yeast species vary greatly in their tolerance to salt. They are more salt-tolerant between pH 3.0 and 5.0, but some yeasts will grow even in quite acid, salty brines. Several types of yeast can grow in pickle brines containing 19–20% NaCl. Sufficient salt prevents growth of *Saccharomyces rouxii* in foods with a_w below 0.81.

Cucumbers are usually submerged in a brine with a_w of about 0.87 using 18–20% NaCl but higher levels may be used. Cabbage is fermented in 2.5% brine. These salt concentrations may select for halotolerant yeasts such as members of the *Rhodotorula* genera which produce a softening and/or pink discoloration in sauerkraut. *R. glutinis* varieties produce polygalacturonase and pectin methyl esterase which softens texture. This yeast is selected for late in the fermentation period because of its acid tolerance.

In natural cheeses, salt retards the growth of undesirable bacteria, assures the predominance of the desired flora, controls the rate of lactic acid production, and aids in satisfactory development of flavour, body and texture during the ripening process as well as contributing salty flavour. Cheese may be rubbed with dry salt or brined in a 20–25% NaCl brine during ripening to limit the growth of proteolytic bacteria that require higher a_w. Salt is commonly used as a cheese component at 1.75–3.00%. In pasteurized process cheeses, and particularly in shelf-stable products that are not commercially sterile, salt plays a critical role in preventing growth of *C. botulinum*.

Direct Preservation by Salt

Salt in concentrated solutions exerts osmotic pressure sufficiently high to draw water from or prevent normal diffusion of water into microbial cells causing a preservative condition to exist. Between 18 and 25% salt in solution generally prevents all growth of microorganisms in foods. Products with high salt concentrations will keep indefinitely without refrigeration even if exposed to microbial contamination, provided they are not diluted above a critical (salt) concentration by moisture pickup. This has been the basis for the preservation of a wide variety of foods throughout history.

The art of curing meat with salt is very old. Salt crystals or corn (old Norse *korn* for grain) were applied dry or in brine to beef to produce corned beef. Finished products contain $\geqslant$6.25% salt. NaCl is used in combination with nitrite and other ingredients to delay growth of undesirable microorganisms and improve the product shelf life and safety of meat products. Cured meats generally contain sufficient NaCl to decrease a_w to between 0.88 and 0.95. Salt can prevent toxin production by *C. botulinum* type E in fish (under temperature abuse conditions), but the concentration needed is unacceptable to most consumers. Lipid oxidation in meat products is accelerated by added salt; the common usage level (1.5%) is particularly damaging. A 25% salt reduction (to 2.00–2.25%) does not adversely alter the shelf life or microbial characteristics of bologna, frankfurters, ham or bacon; however, further reduction significantly reduces both shelf life and predominant flora. Bacon with 0.7% NaCl spoils in <8 weeks, while that with 1.2–1.5% does not. Large reductions in NaCl levels in cured meat products are not recommended. Dry cured meat products depend on both salt and nitrite as well as smoking (dehydrating) for their shelf stability. Salt (3%) is applied directly to fresh pork belly together with sugar and a nitrite source. Dried beef is produced in a similar way, using about 7% salt.

Some meats are fermented or cured and fermented by submerging them in a brine (pickle) and allowing lactic acid-producing bacteria to lower the pH to <4.6. Gram-positive bacteria of the *Lactobacillus*, *Pediococcus* and *Micrococcus* genera, which are desirable for meat fermentation, generally tolerate a_w of 0.95 and sometimes less. On the other hand, yeasts and moulds of the genera *Debaryomyces* and *Pen-*

icillium are quite active at this a_w and below. Spore germination of members of the genera *Clostridium* and *Bacillus* is inhibited. Current brining techniques for hams allow reductions of salt; a 60–70% saturation level with a water activity of 0.87–0.82 is used for immersion. Additional solutes (sodium nitrite, various sugars and phosphates) add to the a_w-reducing effect of the brine.

Commercial salting using high levels of NaCl as a primary mechanism of preserving meat, fish and vegetables has become less important since the advent of refrigeration. Some types of fish (herring, anchovies) are still preserved by dry salting or in heavy brines. The fat in high-fat fish (cod, tuna) may oxidize as a result of the pro-oxidative effect of NaCl and growth of halophilic bacteria. Dipping fish in NaCl before storage in modified atmosphere packaging (MAP) significantly decreases bacterial counts during storage compared to holding in MAP alone. A 5% NaCl dip also decreases the extent of pH change and total volatile bases. The surface effect (osmotic) on bacteria subjected to dips is synergistic with the effect of CO_2, perhaps by increasing CO_2 (by fish tissue) absorption; CO_2 is converted to carbonic acid, lowering tissue pH.

Summary

NaCl is an effective antimicrobial operating in different ways. Halotolerant and osmotolerant bacteria, yeasts and moulds have developed compensatory metabolic processes that allow them to continue to live and in some cases grow and produce toxins and spores, even in relatively high salt concentrations. Use of salt alone is uncommonly practised today; however it is used in conjunction with a variety of other preservative conditions. Different preservation methods have developed to encourage the growth of acid-producing bacteria which, even in the presence of NaCl, are capable of suppressing the activity of many of these spoilage organisms. Fermentation in brines, addition of sugars, low-salt preservation in conjunction with refrigeration and modified atmosphere storage are but a few.

See also: ***Aeromonas***: Introduction. ***Aspergillus***: Introduction; *Aspergillus flavus*. ***Bacillus***: Introduction; *Bacillus stearothermophilus*; *Bacillus subtilis*. ***Clostridium***: Introduction; *Clostridium botulinum*. ***Escherichia coli***: *Escherichia coli*. **Fermented Foods**: Fermented Vegetable Products; Fermented Meat Products; Fermented Fish Products. **Heat Treatment of Foods**: Principles of Pasteurization. **Meat and Poultry**: Curing of Meat. ***Salmonella***: Introduction. ***Staphylococcus***: *Staphylococcus aureus*. ***Vibrio***: Introduction, including *Vibrio vulnificus*, and *Vibrio parahaemolyticus*. ***Xeromyces***. ***Zygosaccharomyces***.

Further Reading

Banwart GJ (1981) Control of Microorganisms by Retarding Growth. In: *Basic Food Microbiology*. P. 347. Westport, CT: AVI

El-Gazzar FE and Marth EH. Toxigenic and nontoxigentic strains of Aspergilli and Penicillia grown in the presence of sodium chloride cause enzyme-catalyzed hydrolysis of protein, fat and hydrogen peroxide. *J. Food Prot.* 49(1): 26–32.

Houtsma PC, DeWit JC and Rombouts FM (1996) Minimum inhibitory concentration (MIC) of sodium lactate and sodium chloride for spoilage organisms and pathogens at different pH values and temperatures. *J. Food Protect.* 59(12): 1300–1304.

Hutton MT, Koskinen MA and Hanlin JH (1991) Interacting effects of pH and NaCl on heat resistance of bacterial spores. *J. Food Sci.* 56(3): 821–822.

Jay MJ (1996) Intrinsic and Extrinsic Parameters of Foods that Affect Microbial Growth. In: *Modern Food Microbiology*, 5th edn. New York. Chapman & Hall.

Leistner L and Rodel W (1975) The Significance of Water Activity for Microorganisms in Meat. In: *Water Relations in Foods*. London: Academic Press.

McClure PJ, Cole MB and Davies KW (1994) An example of the stages of development of a predictive mathematical model for microbial growth: the effects of NaCl, pH and temperature on the growth of *Aeromonas hydrophila*. *Int. J. Food Microbiol.* 23: 359–375.

Pastoriza L, Sampedro G, Herrara JJ and Cabo ML (1998) Influence of sodium chloride and modified atmosphere packaging on microbiological, chemical and sensorial properties in ice storage of slices of hake (*Merluccius merluccius*). *Food Chem.* 61(12): 23–28.

Periago PM, Fernandez PS, Salmeron MC and Martinez A (1998) Predictive model to describe the combined effect of pH and NaCl on apparent heat resistance of *Bacillus stearothermophilus*. *Int. J. Food Microbiol.* 44: 21–30.

Troller JA and Christian JHB (1978) *Water Activity in Food*. New York: Academic Press.

Traditional Preservatives - Organic Acids

M Stratford, Microbiology Section, Unilever Research, Bedfordshire, UK

Introduction

Addition of acids to foods is, at first sight, an irresponsible action. Acids are often perceived as corrosive or toxic chemicals, capable of causing burns to skin or damaging clothing; clearly not the sort of materials to be added to food for human consumption. In reality, addition of acids to foods, beverages and water remains one of the most ancient, simple but effective methods of prevention of disease by food pathogens and preservation. Almost all bacteria, including the pathogenic species, cannot grow much below pH 4.0 (**Fig. 1**). Microbial growth at low pH causes only spoilage, notably by yeasts and moulds. Other benefits of acids in foods include improvements to flavour and antioxidant activity, maintaining the freshness of food taste and colour, over long periods.

Historically, organic acids have played a significant role in foods and beverages for thousands of years. Tartaric acid residues were found in Iranian wine jars, dated 5000–5400 BC. The natural acids from grapes, tartaric and citric acids, enhance the fruit and add freshness to wines, in addition to improving the antioxidant properties of the wine. In recent years, the addition of citric, malic or tartaric acids has become widespread, particularly to wines from riper, hot-climate grapes. It is no coincidence that the most long-lived wines are made from more acidic grape varieties.

An antimicrobial role for acids has been known for thousands of years. In areas where the water is suspect, it is common practice to add a little lemon juice or wine to the water before drinking. Pathogenic bacteria are rapidly inactivated by acidification to a low pH. It is recorded that Hannibal and the Carthaginian army, while crossing the Alps, were carrying substantial quantities of sour wine for this purpose.

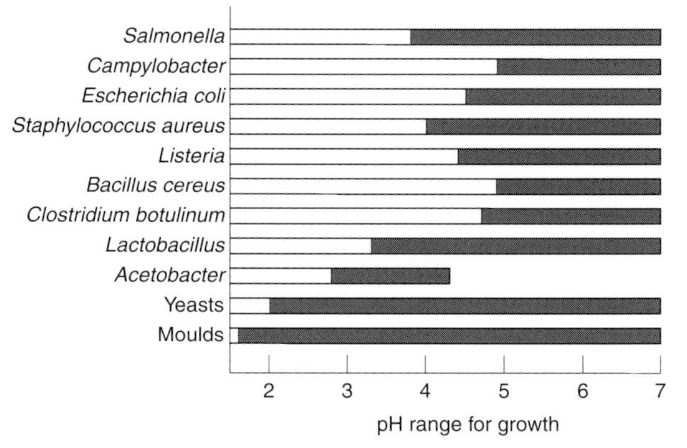

Figure 1 Minimum pH for growth of pathogenic bacteria commonly causing food poisoning, lactic and acetic acid bacteria, yeasts and moulds. Low pH restricts growth of pathogens but allows spoilage by yeasts and moulds.

Range of Acids Added to Foods

The range of foods containing acids is extensive, ranging from colas, fruit juices and wines, to spreads, mayonnaise and pickles. Some foods are preserved by artificial addition of acids, while acids form a natural, intrinsic part of other foods. This distinction between natural acids and artificial additions is somewhat arbitrary, since identical acids often are present in both, at similar concentrations. Traditional use does not involve chemically pure compounds, rather the addition of natural substances containing acids. Organic acids are entirely natural components of many foods, notably fruit juices. Grape juices and wine characteristically contain tartaric acid, malic acid is found in apples, and citric acid is found in lemons, grapefruit, oranges and other citrus fruit. Many fruit juices contain a mixture of acids, with citric and malic acids being most commonly found in substantial amounts (**Table 1**).

The concentrations of acids in fruit juices are not great, typically near 1%, although blackcurrant can contain up to 4% citric acid. Acid levels in fruits vary considerably with ripeness: unripe fruits contain acids but little sugar, and overripe fruits contain sugars but little acid. Unripe lemons may contain as much as 5–8% acid.

The legislation concerning food additives varies widely throughout the world. Acids may be found in several categories of additives to foods:

Table 1 Range of pH values and major acids present in various fruits, at a normal ripeness

Fruit	*pH range*	*Major acids*
Apple	2.9–4.5	Malic, citric
Cherry	3.7–4.4	Malic, citric
Grape	2.9–3.9	Tartaric, malic
Grapefruit	2.9–3.6	Citric
Guava	3.2–4.2	Citric, malic
Lemon	2.0–2.55	Citric
Lime	1.6–3.2	Citric
Mango	*c.* 4.35	Citric, tartaric
Orange	2.6–4.3	Citric, malic
Passionfruit	2.6–3.4	Citric, malic
Peach	3.6–4.0	Malic, citric
Pear	3.0–4.5	Malic, citric
Pineapple	3.1–4.0	Citric, malic
Plum	3.0–4.5	Malic, quinic
Raspberry	2.5–3.1	Citric
Strawberry	3.0–3.4	Citric

- acidulants – acids added to increase the acidity of a food, and/or impart a sour taste
- flavours – acids added as artificial flavours
- antioxidants – acids that preferentially combine with oxygen, thus preventing oxidation of the food
- preservatives – acids which protect foods against deterioration caused by microorganisms.

Since most acids have a variety of properties, it is possible for an acid to have an action in several categories. For example, acetic acid addition can increase the acidity of a food, as well as imparting a distinct flavour and having a definite action as a preservative. However, in the European Community, sorbic acid, benzoic acid and propionic acid are listed as preservatives (described fully elsewhere), ascorbic acid is listed as an antioxidant; and citric acid, malic acid, lactic acid, tartaric acid and acetic acid are recognized acidulants. In the USA, food additives are recorded on an approved list and have received pre-market clearance from the Food and Drug Administration (FDA).

The amount of acids added to foods depends on the acid, the food type, the taste and the purpose for which the acid is added. In general, acidulants are added in large quantity (several parts per hundred); preservatives are added in moderate amounts (100–500 parts per million), and flavours and antioxidants in small quantities.

Citric Acid

Citric acid is one of the most versatile and widely used acidulants, particularly in fruit-flavoured beverages, and is also used in jams, confectionery and candy, bakery products, cheese, wine, canned vegetables and sauces. Owing to its widespread use, citric acid has become the standard against which other acidulants are measured, in terms of taste, titratable acidity and acidification. Citric acid is much favoured because it has light fruity taste, dissolves readily, and is cheap and in plentiful supply.

Malic Acid

Malic acid is, like citric acid, a general-purpose acidulant. It is normally associated with apples, and while used in many products, is preferred in apple-containing foods such as ciders. Malic acid has a fuller, smoother taste than citric acid that is beneficial in low-energy drinks where malic acid masks the unpleasant flavours of some artificial sweeteners.

Tartaric Acid

Tartaric acid has a stronger, sharper taste than citric acid. It occurs naturally in grape juice and is used preferentially in foods containing cranberries or grapes, notably wines, jellies and confectionery. Tartaric acid is prepared from the waste products of the wine industry and is more expensive than most acidulants. When dissolved in hard water, insoluble precipitates of calcium tartrate can form.

Acetic Acid, Vinegar

Owing to its pungent odour and taste, acetic acid is used in substantial amounts but in very limited food applications: pickles, chutney, salad creams, mayonnaise, dressings and sauces. Acetic acid is almost always applied as vinegar. Vinegars contain some 4–8% acetic acid, formed by action of acetic acid bacteria on ciders, wines or yeast-fermented malt.

Lactic Acid

Lactic acid has a very smooth, mild taste, compared with other acidulants. It is used at substantial concentrations in fermented meat, dairy products, sauces brine-preserved pickled vegetables and salad dressings. It is also used in carbonated beverages, as a flavour modifier, and in some jams and jellies.

Phosphoric Acid

Phosphoric acid is an inorganic acid, used as an acidulant almost exclusively in cola-flavoured carbonated beverages. Phosphoric acid is cheap, with a characteristic flat sour taste. It is a relatively strong, dissociated acid, better able to acidify colas to the low desired pH (2.5) of these beverages.

Fumaric Acid

Fumaric acid has 'generally recognized as safe' (GRAS) status in the USA, but is not permitted in the EC. In the USA, fumaric acid is principally used in fruit juices, gelatin desserts and pie fillings. It is relatively cheap but has the disadvantages that it is stronger tasting than citric acid and difficult to dissolve.

Adipic Acid

Adipic acid may also be rarely encountered in the USA as an acidulant of fruit-flavoured beverages, jellies, jams and gelatin desserts. It is favoured in dry foods because it is not hydroscopic.

Behaviour of Different Acids in Foods

Chemical Properties of Acidulants

Acidification by acids requires release of protons. Acidulants vary in the number of acidic groups present on the molecule. Citric acid is a tricarboxylic acid; malic, tartaric, fumaric and adipic acids are dicarboxylic acids; lactic, ascorbic and acetic acids are monocarboxylic. Proton release depends on the

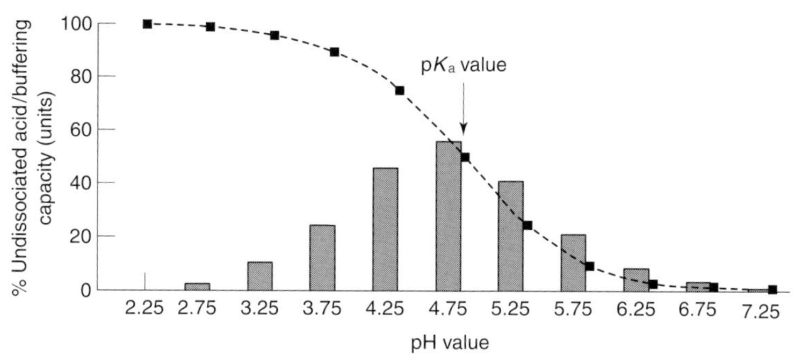

Figure 2 The effect of pH on the proportions of undissociated acetic acid and acetate anions. At the pK_a acid and anion are in equal proportion. Buffering capacity of acetic acid (histograms) was determined by titration. One unit of buffering capacity is the proton concentration (mmol l^{-1}) required to move the pH of 1 litre by one unit.

strength of the acid. Fully dissociated strong acids, such as hydrochloric acid, HCl, effectively release all protons at the pH range of foods. Acidulants of foods can be described as weak acids, being only partially dissociated. Weak acids in solution form equilibria between undissociated acid molecules and charged anions and protons.

$$\underset{\text{undissociated acid}}{HA} \rightleftharpoons \underset{\text{anion}}{A^-} + \underset{\text{proton}}{H^+}$$

Equilibria are pH dependent; at lower pH the proton concentration is higher, pushing the equilibrium towards more undissociated acid. **Figure 2** shows the dissociation curve for acetic acid. At a pH below 3.0, acetic acid exists almost entirely as undissociated molecular acid, whereas above pH 6.5, it is almost entirely dissociated into acetate anions. The pK_a value is the pH at which acid and anion coexist in equal proportion. Weak acids in solution form buffers, resisting changes in pH. Maximal buffering capacity occurs at the pK_a value (see Figure 2) with the effective buffering range extending 1 pH unit on either side of the pK_a.

Food acidulants have a variety of pK_a values (**Table 2**) and are thus dissociated to different extents at any given pH. Some acids contain several carboxylic acid groups, each with a different pK_a value (see Table 2). Citric acid, for example, contains three carboxylic acid groups and forms three anions, predominantly singly charged above pH 2.9, doubly charged above pH 4.3 and triply charged at above pH 5.7. This extends the buffering capacity of citric acid, to a buffering range from pH 1.9 to pH 6.7 (**Fig. 3**). Dicarboxylic acids also give a wide buffering range, effectively forming excellent buffers over the pH range of most acidic foods. Monocarboxylic acids have a limited buffering range. A food acidified with acetic acid is effectively unbuffered below pH 3.75, allowing easy movement of pH in this area.

Acidification and Taste of Acids in Foods

The primary effect of an acidulant in food is to lower the pH value. To what extent the pH falls depends on the buffering, fat content, acid type and concentration. The acidification power of different acids can be compared on a molar basis or by weight; food additions are determined as percentages by weight or parts per million (p.p.m.). On a molar basis, food acidulants are surprisingly similar in acidification power (**Fig. 4**). By weight, differences between acids become more marked, smaller acids with lower molecular weights being most effective.

Surprisingly, the taste of different acids does not always reflect acidification power. The human palate detects acidity more as concentration of acid at a constant pH value, rather than the pH itself. **Figure 5** shows the comparison between various acids in perceived acidity in relation to citric acid. More lactic acid has to be added to achieve an equivalent tart taste, and less malic, tartaric, acetic or fumaric acids.

Effects on Microbial Cells

A wide variety of acids occur naturally in, or added to, foods. These acids differ in structure and chemical properties (see Table 2) and different antimicrobial actions have been proposed for various acids. **Figure 6** shows the likely sites of action by traditional acid preservatives. These include action by low pH on the wall and plasma membrane; action lowering cytoplasmic pH; chelation of trace metal ions from the media and from the cell wall; and perturbation of membrane function by acid molecules. Some acids inhibit by a single mechanism, others may combine several actions.

Acidification of the Media

The primary function of acidulants is to lower the pH of foods; consequently the primary action of trad-

Table 2 Chemical properties and structures of commonly used food acidulants

Acid	*Structure*	*Mol. wt*	*pK$_a$*	log *P*$_{oct}$
Citric acid	COOH | $HOOCCH_2CCH_2COOH$ | OH	192.13	5.7; 4.3; 2.9	−1.222
Malic acid	$HOOCCHCH_2COOH$ | OH	134.09	4.7; 3.2	−1.984
Tartaric acid	OH | HOOCCHCHCOOH | OH	150.09	3.9; 2.8	−2.77
Fumaric acid	HOOCCH=CHCOOH	116.07	4.0; 2.8	−0.748
Lactic acid	$CH_3CHCOOH$ | OH	90.08	3.66	−0.186
Acetic acid	CH_3COOH	60.05	4.7	−0.168

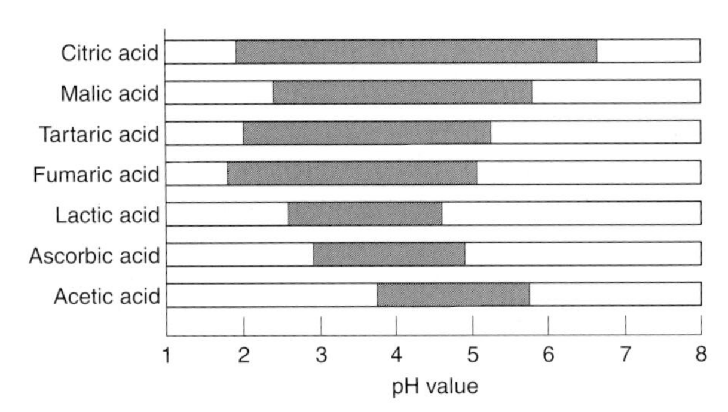

Figure 3 Effective buffering ranges of the acidulants commonly used in foods. Acids containing multiple carboxylic acids have broader buffering ranges than monocarboxylic acids.

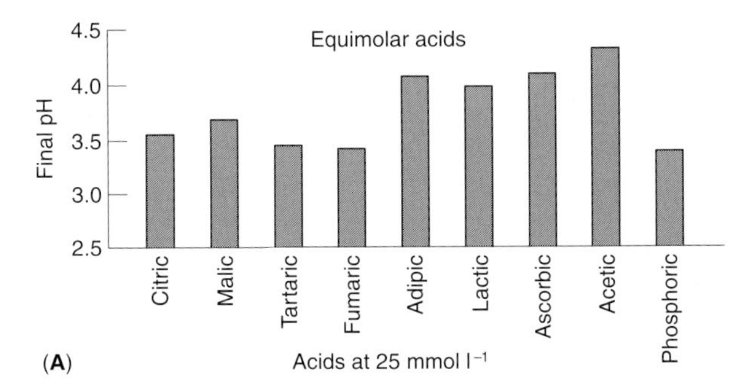

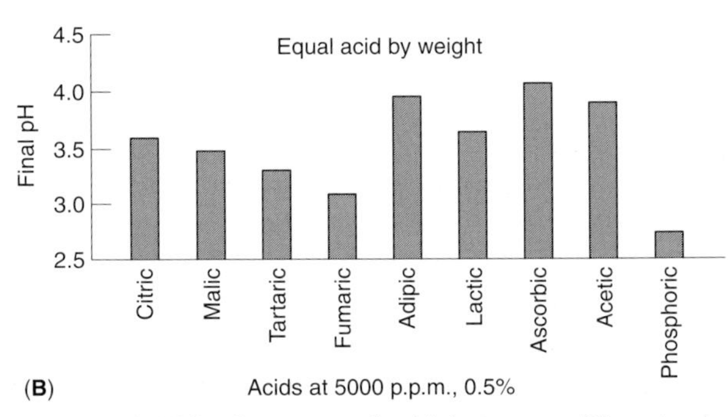

Figure 4 Acidification power of acidulants on an (**A**) equimolar or (**B**) weight basis. Acids were applied at 25 mmol l^{-1} or 5000 p.p.m. to a protein solution, 1% bacteriological peptone, initial pH 6.2.

itional acid preservatives on microorganisms involves direct action of protons on microbial cells. Protons are highly charged and pass slowly through lipid membranes. The action of pH on microbial cells is likely to involve cell walls, the outer faces of membranes and proteins protruding through the membrane. Of these, pH is most likely to affect proteins: enzymes, transport permeases and pumps. Charged proton association with proteins affects charge stability, altering conformation and folding. It has been shown that replacement of the H^+-ATPase proton pump in yeast membranes by an ATPase pump of plant origin prevented yeast growth in acidic conditions, demonstrating that the yeast pump could tolerate acidity but that the plant ATPase could not.

If acidulants inhibited microorganisms only via depression of media pH, at any given pH value all acids should be equally effective preservatives. **Figure 7** shows this not to be true. Acetic and citric acids inhibit microbes more effectively at substantially lower concentrations than other acids, showing that these acids possess additional mechanisms of action.

Cytoplasmic pH

Lipid membranes, by and large, are impermeable to charged ions, save by specific transport systems. Protons penetrate membranes poorly, as do charged

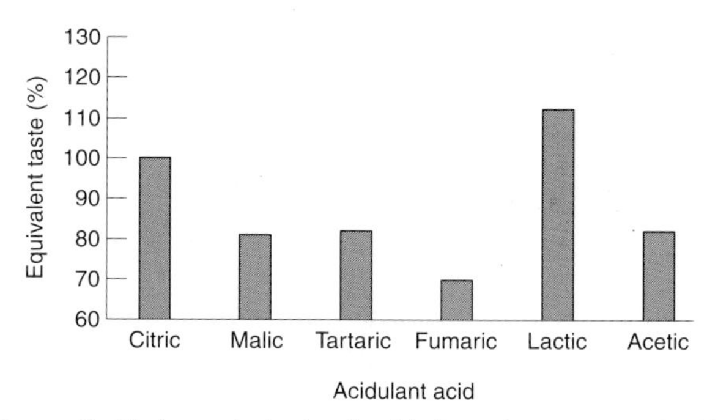

Figure 5 Taste equivalents of acids in water, compared with citric acid (100%). Fumaric acid, being stronger-tasting, required less addition for equivalent taste.

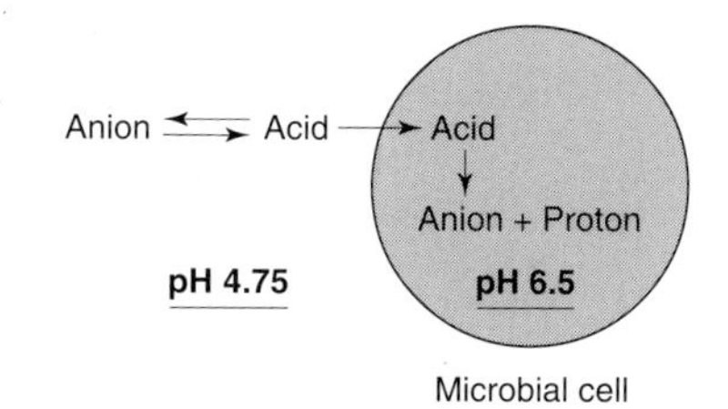

Figure 8 The 'weak acid' theory of microbial inhibition by acids. In media at pH 4.75, acid molecules and anions are in equimolar equilibrium. Lipophilic acid molecules diffuse rapidly through the membrane into microbial cells. The neutral cytoplasmic pH causes acids to dissociate into anions and protons, which being lipid-insoluble accumulate in the cytoplasm. Excess proton accumulation eventually lowers the cytoplasmic pH.

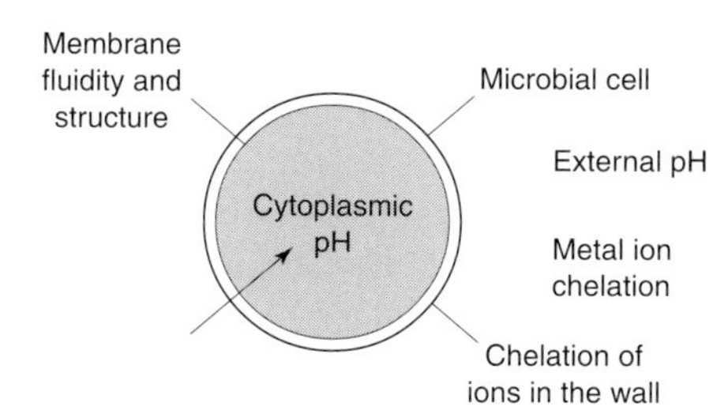

Figure 6 Potential sites of antimicrobial action by acids. Acids may act by lowering external pH, by affecting membrane structure and fluidity, depressing cytoplasmic pH, or chelating metal ions from the media or cell wall.

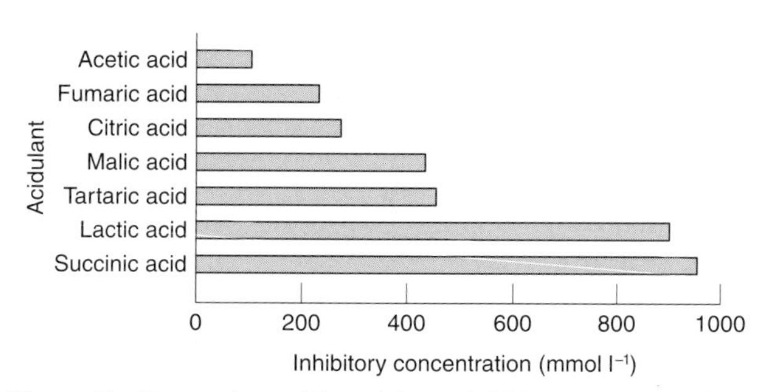

Figure 7 Comparison of the minimum inhibitory concentrations of acidulants, determined against *Saccharomyces cerevisiae* X180–1B in YEPD media, 30°C, at pH 4.0.

anions. However, uncharged acid molecules diffuse rapidly though the plasma membrane, if they are lipid-soluble. The 'weak acid' theory of microbial inhibition by lipophilic preservatives proposes that acid molecules in foods rapidly diffuse through the plasma membrane into the cytoplasm (**Fig. 8**). The neutral cytoplasmic pH causes acids to dissociate into charged anions and protons, both of which are unable to diffuse out of the cell. Diffusion continues until the acid concentration is equal on both sides of the membrane, by which time anions and protons have been concentrated in the cytoplasm. If preservatives are present at sufficient concentration, the accumulated protons overcome cytoplasmic buffering and lower internal pH. Low cytoplasmic pH (pH_i) causes denaturation of nucleic acids and enzymes, inhibiting metabolism and preventing active transport requiring a ΔH^+ gradient. Active pumping to remove protons from the cell by membrane-bound H^+-ATPases can raise internal pH but consumes excessive ATP and may cause inhibition by energy depletion.

The weak acid theory is often wrongly assumed to apply to all acids. For an acid to function as a weak acid preservative it must be:

- lipophilic
- able to diffuse rapidly through membranes
- concentrated within the cytoplasm, as a result of a low pK_a
- able to release sufficient protons in the cytoplasm at the minimum inhibitory concentration (MIC), to overcome cytoplasmic buffering and depress cytoplasmic pH.

Of the traditional acidic preservatives, it can be seen from the partition coefficient (see Table 2) that citric, malic, fumaric and tartaric acids are not lipophilic, and indeed are very lipophobic. It has been demonstrated these acids do not diffuse through lipid membranes. These impermeant acids cannot therefore act as weak acid preservatives, depressing cytoplasmic pH.

Acetic acid, in contrast, is lipid-soluble, diffuses rapidly through the plasma membrane, is accumulated in the cytoplasm and has been demonstrated to cause a rapid collapse in pH_i. A lowering of the pH of the medium greatly enhances the effectiveness of acetic acid (**Fig. 9**), not only increasing the undissociated acid concentration but also increasing the degree to which anions and protons are accumulated in the cytoplasm. Acetic acid appears therefore to inhibit as a classic weak acid preservative, in addition to action on external pH.

Lactic acid shows a degree of lipid solubility and has been shown to diffuse *slowly* through membranes. However, inhibition by lactic acid has not been correlated with decline in pH_i, and while weak acid action

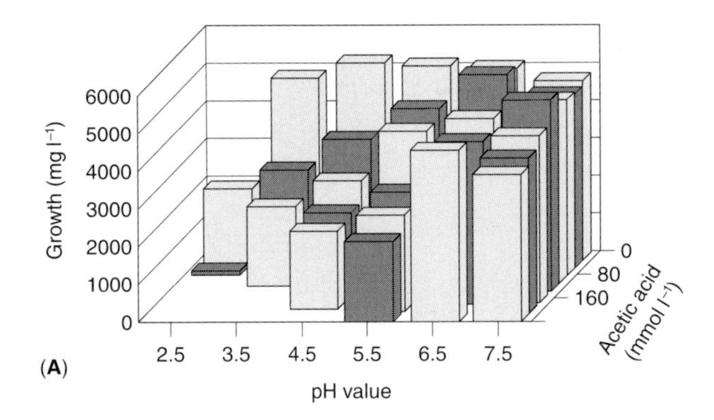

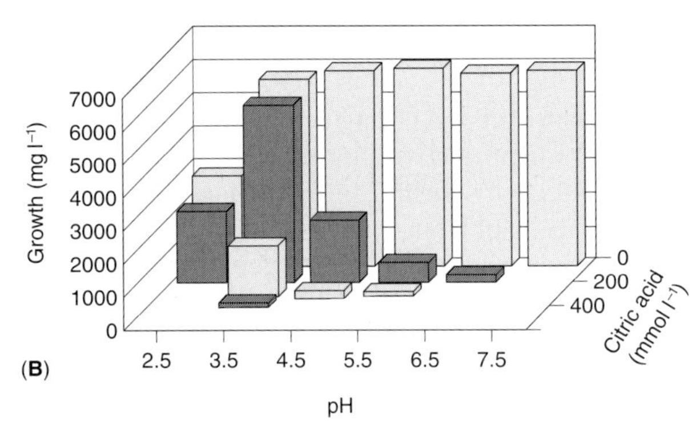

Figure 9 The effect of pH on inhibition of *Saccharomyces cerevisiae* X2180–1B by (**A**) acetic acid and (**B**) citric acid. Growth was examined in a matrix of flasks at pH 2.5–7.5 containing various acid concentrations, after 3 days shaken at 30°C. Acetic acid was more inhibitory at low pH, citric acid was more inhibitory at high pH.

may contribute to inhibition by lactic acid, it appears that other mechanisms of inhibition are also involved.

Chelation

Most acids form complexes with metallic ions, but for the majority, affinity of acids for metals is low and complexes are correspondingly unstable. However, certain acids, often with multiple carboxylic acid groups, form stable complexes and can chelate a substantial proportion of metallic ions, with greatest affinity for transition metal ions, e.g. Fe^{3+}. **Table 3** shows the affinity constants, K_1, for acid–metal complexes. Figures quoted are the log of the equilibrium constant. Stability of citric acid complexes are some 2–3 logs greater than those of malic, tartaric or lactic acid.

The antimicrobial action by citric acid is known to involve chelation, and is overcome by addition of metal ions, e.g. Mg^{++}, Ca^{++}. It appears probable that citric acid removes key nutrients from media, preventing microbial growth. Chelators have greatest affinity with Fe^{3+}, but the identity of growth-limiting nutrients depends on microbial ion requirements, concentrations of metal ions in media, and the affinity of acids for each ion. Action by citric acid via chelation is supported by the finding that inhibitory action is pH-dependent, greatest inhibition occurring at higher pH (see Fig. 9). Stability of complexes formed by multiply charged anions, predominating at high pH, are some 6 logs greater than those of undissociated acids (**Table 4**).

The inhibitory action by malic, tartaric and succinic acids may also involve chelation, given the affinity of these acids for metal ions, and the high concentrations of these acids used in food. Indeed the MIC of these acids against yeasts does appear to reflect the overall stability of acid/ion complexes.

In addition to chelation of nutrient ions, acids also act by damaging cell walls of bacteria by chelation of metal ions within the structure. It has been shown

Table 4 Stability constants, log equilibrium constant, of copper(II) complexes with malic acid, and malate ions. Undissociated malic acid predominates at below pH 3.2, malate⁻ predominates between pH 3.2 and 4.7, and malate²⁻ above pH 4.7

Acid		*Cation*	*Stability constant*
Malic acid	H_2L	Cu^{++}	2.00
$Malate^-$	HL^-	Cu^{++}	3.42
$Malate^{2-}$	L^{2-}	Cu^{++}	8.00

Table 3 Chelation properties of acidulants used in foods. EDTA, permitted in low concentration in the USA, is listed for comparison. Stability constant values, K_1 for acid–metal ion complexes quoted are the log of the equilibrium constant at 20–25°C

	Acid					
Cation	*EDTA*	*Citric*	*Malic*	*Tartaric*	*Succinic*	*Lactic*
K^+	0.96	0.59	0.18–0.23	–	–	–
Na^+	1.79–2.61	0.70	0.28–0.3	1.98	0.3	–
Mg^{++}	8.69	3.16–3.96	1.70	1.91	–	0.73
Ca^{++}	10.45–10.59	3.40–3.55	1.96	2.17	1.20	0.90
Mn^{++}	12.88–13.64	2.84–3.72	–	1.44–2.92	–	0.92
Zn^{++}	15.94–17.50	4.98	2.93	2.69–3.31	1.76–3.22	1.61
Cu^{++}	18.80–19.13	5.90	3.43–3.97	2.6–3.1	2.93	2.5
Fe^{3+}	23.75–25.15	11.40	7.1	6.49	6.88	6.4

that EDTA treatment makes Gram-negative bacteria sensitive to a variety of antibiotics. It is thought that EDTA removes metallic cations from the bacterial outer membrane, opening up the structure and allowing access to the cell by antimicrobial agents.

Effects on Microbial Populations and Spoilage

The effect of a preservative on a microbial population may be to cause cell death, stasis (viable but inhibited cells) or slow growth. Since traditional acid preservatives include many different acids, acting by different mechanisms against a variety of microorganisms, it is understandable that there is no single, unified effect on microbial populations.

Acidity – low pH – will cause loss of viability in most bacterial populations (**Fig. 10**). The kinetics of bacterial cell death depend to a great extent on the previous history of the population, notably phase of growth and prior exposure to acid. It is wrong, however, to assume that instant bacterial death occurs in foods at a pH lower than 4.0. As recent incidents of food poisoning caused by *Escherichia coli* O157:H7 and *Salmonella* have shown, bacteria can remain viable in apple juices at pH 4 for several weeks. Low pH values (2.5) per se have little or no effect on growth or viability of spoilage yeasts or moulds.

Weak acids able to diffuse through the plasma membrane, such as lactic, acetic or benzoic acids, greatly exacerbate the effect of pH. Bacterial populations are killed faster and at higher pH values when acidified with lipophilic acids. Growth of yeasts and moulds are inhibited by permanent weak acids, usually without loss of viability. Characteristically, weak acid preservatives cause prolonged lag phases, and at sub-inhibitory concentration, slow growth and reduced cell yield. Chelating agents also characteristically do not kill microorganisms but prevent growth by limiting metallic nutrient availability.

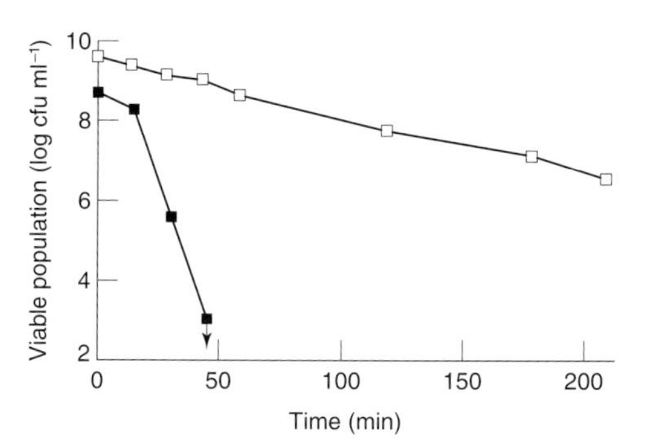

Figure 10 Growth-phase-dependent death of *Listeria monocytogenes* ScottA, in BHI media acidified to pH 3.0 with HCl. Open squares, stationary phase; solid squares, exponential phase. Courtesy of MJ Davis.

Sublethal Effects, Species/Strain Variability and Interaction with Other Factors

Acid Resistance and Sensitivity

There is considerable variation between microorganisms as to their sensitivity to traditional acid preservatives. This variation extends from the overall sensitivity to low pH of bacteria, as distinct from yeasts and moulds; to particular acid-resistant genera; to variation in sensitivity of strains within species; and even variation between individual cells in populations, caused by their phase of growth and previous history.

Acid-tolerant bacteria include acetic acid bacteria, *Acetobacter* and *Gluconobacter* spp., lactic acid bacteria, *Lactobacillus* and *Leuconostoc* spp., Lancefield group N streptococci, *Clostridium butyricum* and *C. pasteurianum*, *Alicyclobacillus* (ex. *Bacillus*) *acidoterrestris*, *Bacillus coagulans*, *B. macerans* and *B. polymyxa*.

Bacterial spoilage at low pH is most frequently associated with Gram-negative *Gluconobacter* (*Acetomonas*). These bacteria require oxygen for growth and are restricted by gas-impermeable packaging and minimal head space. Acetic acid bacteria are resistant to normal concentrations of preservatives. Lactobacilli and *Leuconostoc* spp., the lactic acid bacteria, are known to cause spoilage causing loss of astringency, slime, pressure, turbidity or buttermilk off flavours. Such bacteria can grow in products at pH 2.8 but are relatively heat-sensitive. Spore-forming *Alicyclobacillus* (formerly *Bacillus*) *acidoterrestris* can survive pasteurization, grows well at low pH and is capable of spoiling fruit juices, fruit juice blends and lemonade.

Preservative-resistant spoilage yeasts are selected using acetic acid agar. *Zygosaccharomyces bailii*, *Z. bisporus* and *Z. lentus* are highly preservative-resistant; *Z. rouxii*, *Saccharomyces cerevisiae*, *S. bayanus*, *S. exiguus*, *Schizosaccharomyces pombe* and *Torulaspora delbrueckii* are moderately to highly resistant. Individual strains show considerable variation in acid resistance (**Fig. 11**) despite genetic confirmation that strains are all of the same species. Moulds are generally more sensitive to weak acid preservatives, an exception being *Moniliella acetobutens*, an acetic acid-resistant mould causing spoilage in pickles and vinegar.

Habituation and Adaptation to Acids

Addition of acids to foods progressively lowers the pH value and increases the concentration of acid. Studies involving bacteria tend to focus on the effect of pH, this being the major bactericidal force; whereas

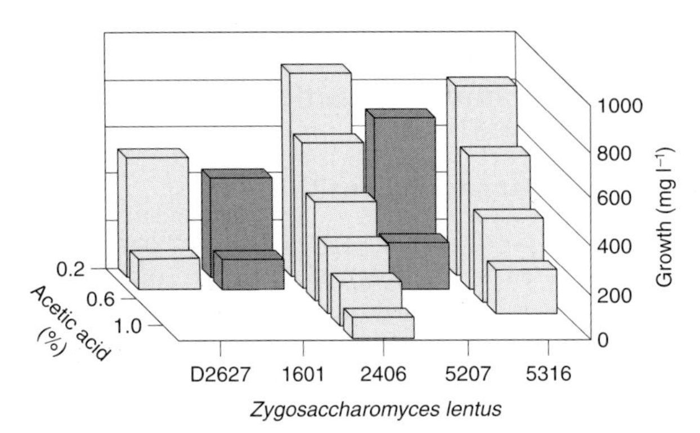

Figure 11 Variation in sensitivity to acetic acid by individual strains of *Zygosaccharomyces lentus*. Yeasts were grown in YEPD media corrected to pH 4.0 after acetic acid addition, at 25°C for 1 week. Courtesy of H Steels.

yeasts and moulds, which are substantially immune to low pH, have been more studied in relation to the acid concentration. Acid stress responses, sublethal adaptation and habituation to pH and acids, are not yet fully understood and this is an area of active research in bacteria and yeasts.

It is well established for many bacteria, including *Escherichia coli*, *Salmonella typhimurium*, *Listeria monocytogenes* and *Lactobacillus* spp., that survival at low pH is enhanced by prior exposure to mildly acidic pH. Bacteria transferred from neutrality to pH 3 die rapidly. An intermediate stage at pH 4.5 for some 20–60 min greatly increases the proportion of surviving cells. This acid tolerance response (ATR) is not yet fully characterized. There appear to be several components of the ATR, some requiring protein synthesis, parts of which are shared in stationary phase resistance. The ATR involves a set of some 50 gene products, acting to improve pH homeostasis, reducing energy dissipation, enhancing DNA repair, correction of protein mis-folding by chaperones and membrane biosynthesis. The *rpo*S gene encodes an alternative sigma factor s^s, a critical stationary phase regulator that is also involved in the ATR. The protein RpoSp is synthesized semi-constitutively and normally rapidly degraded. When growth rate is impaired, the regulatory s^s is stabilized and induces expression of a number of enzymes involved in protection and repair of DNA and proteins and in detoxification.

Adaptation by yeasts to acid preservatives has been known as an industrial problem for many years. Growth of yeasts on splashes of preserved products results in populations of adapted yeasts able to tolerate very high concentrations of preservatives. **Figure 12** shows that yeasts grown for a week in sub-inhibitory concentrations of acetic acid can subsequently grow in double this concentration of acetic acid. The explanation for adaptation by yeasts may involve mechanisms to conserve ATP, pdr12 drug-resistance pumps to remove acid, or simply that accumulation of acids within the cytoplasm creates buffering capacity (see Fig. 8), which resists further change in pH_i when cells are re-inoculated into higher levels of acetic acid.

Interaction of Acidulants with Other Factors

The most significant factor interacting with the preservative effect of acidulants is the pH and buffering capacity of the food. The primary antimicrobial action by acidulants is to lower pH value; foods with a higher pH or with substantial buffering will limit the fall in pH. Buffering may be caused by other acids and their salts in the food or by the presence of substantial quantities of proteins or amino acids. Lowering of pH substantially increases the effect of lipophilic acetic acid, acting as a weak acid preservative, but

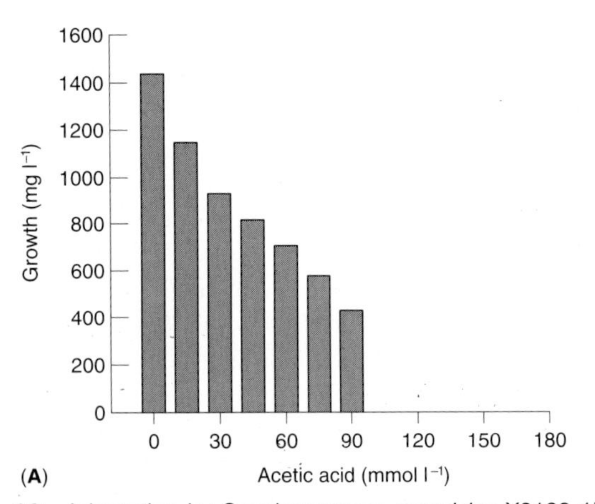

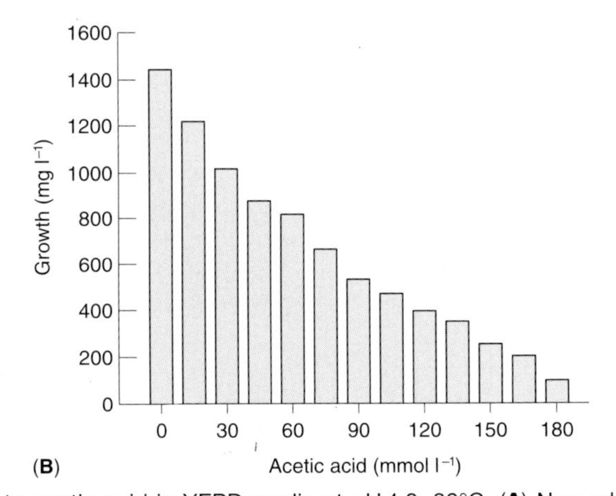

Figure 12 Adaptation by *Saccharomyces cerevisiae* X2180-1B to acetic acid in YEPD media at pH 4.0, 30°C. (**A**) Non-adapted yeasts were grown for 7 days in tubes of YEPD containing acetic acid. Adapted yeasts were taken from the highest concentration of acetic acid permitting growth (90 mmol⁻¹) and re-inoculated into a similar series of tubes. Growth of adapted yeast (**B**) was measured after a further 7 days.

may decrease the effect of chelating acids such as citric acid (see Fig. 9).

Many foods are preserved by high-temperature pasteurization. When acidulants are used to lower the pH, the pasteurization requirement is substantially reduced. This is partially due to the effect of acids on bacterial spores. Heat-resistant bacterial spores often require heating up to 121°C for many minutes to achieve sterilization; at pH values below 4.0, spore germination is inhibited and pasteurization at a much lower temperature is sufficient to kill vegetative bacterial and yeast cells. In addition, heat is a much more effective sterilant in acidic conditions, a fact used in the traditional preservation of fruit by bottling.

The behaviour of acids in foods will also depend on the nature and properties of the food itself. This may contain protein, composed of amino acids, in themselves buffers and resisting change in pH. Foods may also contain fats or lipids. Lipophilic acids may be removed from solution by partitioning into the lipid fractions. The partition coefficients for food acidulants are shown in Table 2, as log P_{oct} values, the log of the distribution between water and octanol. Negative values show that acids are preferentially soluble in water, rather than lipid. However, lactic acid and acetic acid are moderately lipophilic and a sizeable fraction of these acids may partition into the lipid phase in foods containing fats or oils, e.g. salad cream, mayonnaise and dressings, and be effectively reduced in concentration.

See also: **Ecology of Bacteria and Fungi in Foods**: Influence of Redox Potential and pH. **Spoilage Problems**: Problems Caused by Bacteria; Problems Caused by Fungi. ***Zygosaccharomyces***. **Preservatives**: Permitted Preservatives - Benzoic Acid; Permitted Preservatives – Sorbic Acid.

Further Reading

Baird-Parker AC and Kooiman WJ (1980) Soft drinks, fruit juices, concentrates, and fruit preserves. In: *Microbial Ecology of Foods*. Vol 2. *Food Commodities*. International Commission on Microbiological Specifications for Foods. P. 643. London: Academic Press.

Bearso S, Bearson B and Foster JW (1997) Acid stress responses in enterobacteria. *FEMS Microbiology Letters* 147: 173–180.

Beuchat LR and Golden DA (1989) Antimicrobials occurring naturally in foods. *Food Technology* 43: 134–142.

Booth IR and Stratford M (2000) Acidulants and low pH. In: Russell NJ and Gould GW (eds) *Food Preservatives*, 2nd edn. Glasgow: Blackie.

Chichester DF and Tanner FW (1972) Antimicrobial food additives. In: Furia TE (ed.) *Handbook of Food Additives*, 2nd edn. P. 155. Cleveland: CRC Press.

Corlett DA and Brown MH (1980) pH and acidity. In: *Microbial Ecology of Foods*. Vol. 1. *Factors Affecting Life and Death of Microorganisms*. International Commission on Microbiological Specifications for Foods. P. 92. London: Academic Press.

EEC (1989) *Council Directive on Food Additives other than Colours and Sweeteners*. (89/107/EEC). Brussels: EC.

Eklund T (1989) Organic acids and esters. In: Gould GW (ed.) *Mechanisms of Action of Food Preservation Procedures*. P. 161. London: Elsevier Applied Science.

Gardner WH (1972) Acidulants in food processing. In: Furia TE (ed.) *Handbook of Food Additives*, 2nd edn. P. 225. Cleveland: CRC Press.

Ingram M, Ottaway FJH and Coppock JBM (1956) The preservative action of acidic substances in food. *Chemistry and Industry* (London) 75: 1154–1163.

Kabara JJ and Eklund T (1991) Organic acids and esters. In: Russell NJ and Gould GW (eds) *Food Preservatives*. P. 44. Glasgow: Blackie.

Pitt JI and Hocking AD (1997) *Fungi and Food Spoilage*, 2nd edn. P. 439. London: Blackie Academic and Professional.

Somogyi LP (1996) Direct food additives in fruit processing. In: Somogyi LP, Ramaswamy HS and Hui YH (eds) *Processing Fruits: Science and Technology*. Vol 1. *Biology, Principles and Applications*. P. 293. Lancaster: Technomic.

Stratford M and Eklund T (2000) Organic acids and esters. In: Russell NJ and Gould GW (eds) *Food Preservatives*, 2nd edn. Glasgow: Blackie.

Taylor RB (1998) Ingredients. In: Ashurst PR (ed.) *The Chemistry and Technology of Soft Drinks and Fruit Juices*. P. 16. Sheffield: Sheffield Academic Press.

Traditional Preservatives – Wood Smoke

Lucy J Ogbadu, Department of Biological Sciences, Benue State University, Makurdi, Nigeria

Smoking as a method of food preservation is an age-old process which probably dates back to the start of human civilization. It is likely that the practice of smoking of meat and fish outdates the art of cooking in containers as open-fire contact (roasting) with food must have been the earliest form of cooking even before earthenware pots were made. The importance of wood as fuel, with wood smoke as an integral part of that, and perhaps later their utilization in food processing are all tied to the same aspect of civilization. Since wood smoke is generated by burning wood, both smoke and heat of wood combustion must have been discovered useful for food processing about the same time. The traditional practice of exposing food to wood smoke, which goes hand in hand with drying, is a food-preservative effort. The art must have assumed more importance and been adopted out of recognition of the desirable attributes of added flavour, odour and colour by the wood smoke.

The preservative role of food smoking is still of primary significance in developing countries because alternative or complementary methods for effective preservation require gadgets that depend on electricity or kerosene; all of which are costly and out of reach for the average citizen. It is no longer primarily a preservative effort in modern food industry but its importance lies more in the desirable flavours and odours that it imparts to food. This is such that efforts at simulating this form of food processing have been actualized in developing synthetic wood smoke in the form of liquid smoke. Food is simply dipped into the liquid smoke which contains essentially the chemical constituents of wood smoke. However, while food legislation in European Union countries defines food preservatives as excluding, among others, wood smoke or liquid solutions of smoke, the legislation of Canada, countries of Asia and Africa allow wood smoke to be used as a natural preservative together with other chemical preservatives. The legislative directive on food smoking generally only allows smoke or liquid solution of smoke obtained from wood or woody plants in their natural state and disallows those woods that have been impregnated, coloured, glued, painted or treated with any form of chemical.

Range of Foods which are Smoked

Quite a wide range of foods are subjected to smoking depending on the food culture of the people. Certain foods are commonly smoked across all food cultures where smoking is a culinary tradition. Animal flesh, which encompasses both meat and fish, forms the major category of foods that are processed by exposure to wood smoke. While meat covers mammalian flesh, other specific types are poultry and game. Fish includes both marine and non-marine water fish, all of which are commonly smoked. Other smaller marine animal life, like lobsters, shrimps and crustaceans, are equally processed by smoking. The list of specific types of each of these categories of animal flesh is as numerous as the number that exists in nature which are edible. And a variety of recipes exists for smoked products that are obtainable from each flesh type, again depending on the food traditions of the people. Some of the well-known animal flesh that are commonly smoke-processed both by traditional uncontrolled methods and by modern industrial techniques are listed in **Table 1**. Foods of animal flesh are highly perishable and require immediate preservative processing to halt microbial deterioration as soon as the animal is caught or slaughtered,

Foods other than those of animal flesh that are smoke-processed vary across the continents and are known by local names in different parts of the world. In many parts of Africa, as in Asia, different types of alcoholic and non-alcoholic beverages are smoke-produced by preparing the drink in smoked pots. A number of alcoholic beverages are smoke-processed in Ethiopia. The fermentation vessels for the beverages are densely smoked by inverting them over smouldering wood before being used for fermenting ground barley or wheat malt, as in the case for *talla* beer production or diluted honey fermentation, in the case of *tej* wine. These alcoholic beverages have a special smoky aroma and flavour. Non-alcoholic drinks are prepared in parts of Asia by collecting and preparing different plant saps in smoked pots. *Jaggeri* is a Sri Lankan drink prepared from coconut sap in this way. Copra meals are prepared by exposing coconut kernels to heat and smoke. In addition the oil obtained from smoked copra has a desirable smoky flavour. Smoked foods in other parts of the world include smoked plant products like nuts and seeds.

Active Antimicrobial Constituents of Wood Smoke

Thermal combustion of wood generates both smoke and heat simultaneously. Smoke generation results from the temperature gradient which exists between the outer combusting wood surface as it is being oxidized and the dehydrating inner core of the wood. The temperature rises from about 100°C to between 300 and 400°C as the internal moisture content of the wood approaches zero. It is this temperature gradient that favours wood degradation, giving rise to smoke generation. Wood smoke can be generated from any type of wood obtained from hewn dried green plants, but dicotyledonous plants give harder, more compact woody stems than monocots and are more suitable for steady smoke generation. Plants that have fibrous stems are not as good as hard woods which burn longer and with good-quality smoke. Different types of hard woody plants serve as sources for smoke generation and certain types are preferred over others.

Wood is used in the form of logs, chips, peat or sawdust for smoke generation. Hard woods commonly valued in the tropics for smoking are red mahogany, west African ebony, camwood and African white wood. Many other woody stems available in the vast tropical regions serve equally well and sometimes fibrous stems such as from the coconut plant and coconut shell are used where there is a scarcity of hardwood. The smoking industries make use of hard-

Table 1 Common types of smoked animal flesh and predominant forms of preservative treatment

Animal flesh/source		*Type of smoking/accessory treatments*			
		Controlled		*Traditional uncontrolled*	
Category	*Specific type*	*Method*	*Accessory*	*Method*	*Accessory*
Meat	Beef	H/s	Curing/refrign	H/s	Dry-salting/drying/refrign
	Mutton	H/s	Curing/refrign	H/s	Dry-salting/drying/refrign
	Pork			H/s	Curing/refrign
	Ham			–	–
	Bacon	H/s	Curing/refrign	–	–
	Frankfurter			–	–
	Bush (undomesticated)	–	–	H/s	Drying/refrign
Poultry	Chicken	–	–	H/s	Curing/refrign
	Duck	–	–	H/s	Curing/refrign
	Turkey	H/s	Curing/refrign	–	–
Game	Partridge	–	–		
	Pigeon	–	–	H/s	Drying
	Guinea fowl	–	–		
Fish: marine	Salmon				
	Herring				
	Mackerel	C/s	Brining/refrign	H/s	Drying
	Sardine				
	Cod				
	Haddock			–	–
	Trout	H/s	Brining/refrign	–	–
Fish: non-marine	Catfish	–	–		
	Mudfish	–	–		
	Cichlid	–	–	H/s	Drying
	Mullet	–	–		
Miscellaneous	Prawn	C/s	Brining/refrign	H/s	Brining/drying
	Shrimp			–	–
	Mussel	H/s	Brining/refrign	–	–
	Mollusc			H/s	Dry salting

H/s = Hot smoking; Refrign = chilling and freezing; C/s = cold smoking; – = uncommon method.

wood logs, chips or sawdust from hickory and oak. For a kiln of about 375 kg capacity, about 13 kg of wood is combusted per hour.

Smoke is made up of vaporized chemical compounds, most of which have been identified. They are mainly acidic compounds, terpenes, carbonyl compounds, phenolic compounds and the hydrocarbons, all of which vary in their roles (**Table 2**) and number about 200 different chemical compounds. Out of these, formaldehyde, a carbonyl compounds, is notable for its antimicrobial activity. It has a broad spectrum of activity and is the most active among the antimicrobial constituents of wood smoke. It is directly used as a bacteriostat in cheese milk to prevent *Clostridium* sp. from forming gas holes during the manufacture of Provolone cheese. It is equally used as a sterilizant in food-wrapping materials. Hexamethylene tetramine is a chemical compound commonly used for fruit-washing; this compound eventually breaks down to formaldehyde to give protection against microorganisms. Higher aldehydes are among the other carbonyl constituents and are produced in quantities five times greater than formaldehyde. The acidic constituents include acetic, formic, pyroligneous and higher acids, all of which are antimicrobial in activity. Organic acids generally have antimicrobial property: they lower the internal pH of microbial cells through easy passage across the membrane, as a result of their lipophilic property. They inhibit growth by interacting with the microbial cell membrane, thus neutralizing its electrochemical potential. Other antimicrobial constituents of note are the organic alcohols and the ketones.

The main function of phenolic compounds in wood smoke is their antioxidant activity. They prevent oxidative rancidity in smoked foods, thereby contributing to the preservation effect of smoking. Although phenol has long been known for its antimicrobial property (as used by Lister), it is not directly applied to foods as a preservative because of its toxicity even at low concentrations. About 60% of the phenolic fraction in wood smoke is made up of guaiacol (2-methoxyphenol), syringol (2,6-dimethoxyphenol) and their 4-substituted derivatives, namely eugenol (4-allylguaiacol), isoeugenol (4-propenylguaiacol), syringaldehyde (4-formylsyringol) and aceto-

Table 2 Major constituents of wood smoke of significance in food smoking

		Function/importance	
Group of compounds	*Chemical compounds*	*Specified(*) compound*	*The group*
Carbonyl compounds	Formaldehyde(*)	Antimicrobial	Antimicrobial
	Other aldehydes		Surface pellicle formation
	Alcohols		
Phenolic compounds	Phenol(*)	Antioxidant	
	Guaiacol	Antimicrobial	Antioxidant
	Syringol		Surface pellicle formation
	Eugenol		Aroma-enhancing
	Isoeugenol		Flavour-enhancing
	Acetosyringone		Colourant
	Syringaldehyde		
	Vanillin		
	Acetovanillone		
	Cathecol		
Acidic compounds	Formic acid(*)	Antibacterial	Antimicrobial
	Acetic acid(*)	Antibacterial	
	Higher acid	Antibacterial	
Hydrocarbons	Benzo(a)pyrene(*)	Carcinogenic	Colourant
	Tars		Surface pellicle formation
	Benzo(a)anthracene		
	Benzo(b)fluoranthene		
	Dibenz(ah)anthracene		
	Indeno(1,2,3-cd)pyrene		
Terpenes	Hemiterpenes		Aroma enhancing
	Sesquiterpenes		Surface pellicle formation
	Triterpenes		

(*), specific compound capable of exerting the specified function.

syringone (4-methylketone guaiacol). Phenols are responsible for the main desirable flavours and odours characteristic of smoked foods.

The rate and extent of smoke deposition on the food exposed during the process of smoking depend on a number of factors which are difficult to control with traditional kilns. These difficulties led to development of the modern techniques which place the important parameters under control within ranges that are best for safe and good-quality products. Such factors include smoke density and velocity, the concentration of certain smoke constituents in the air (particularly vapour), the location and distance of the food within the kiln/from the smoke source, the proportion of air and air velocity, the relative humidity and food surface moisture, among others. The rate of smoke absorption by smoking food under standard conditions of operation is proportional to the optical smoke density. Smoke density is measurable using a smoke density meter, which is incorporated in the kiln or smoke house. The cumulative smoke treatment applied to the food can also be measured. For normal commercial practice the smoke density value for fish is about $0.9\ m^{-1}$. The extent of smoke deposition on food can be directly estimated by measuring the phenol concentration as the phenol value.

Effect of Deposition on Microbial Cells/Microflora

Achievement of smoke deposition on products is visible: there is a characteristic dark-brown creosote coating appearance. Creosote is normally applied to wood for protection against deterioration; however, in spite of the known specific activity of some constituents of the smoke, the levels at which they are deposited seem to be more enhancing the aesthetic appeal rather than having a direct effect on the microorganisms. Barring the effects of all accessory preservative efforts such as brining and drying, the microflora of the food do multiply even in the presence of the smoke deposits. The concentration of the smoke constituents – particularly the antimicrobial constituents, formaldehyde and acidic compounds – seems not to be at levels which are inhibitory to the microflora. Smoked foods readily go mouldy when moisture levels are high enough to permit growth but aflatoxin production for instance is delayed. Smoke deposits exert an indirect effect; the glossy pellicle formed on the food surface by the phenol-coagulated surface protein presents the microbial cells with an unsuitable environment for optimal multiplication.

Likely Preservative Effect of Smoking

Given that the initial efforts at smoking foods was for preservative purposes, the achievement of that purpose can be attributed to a combination of accessory processes preparatory to smoking as well as smoke deposition on the food surface. These accessory processes include reduction of available moisture from the food tissues through brining or salting and evaporation due to heat generated by the thermal combustion of the wood. The effectiveness of any food preservation method has to do with the quality of the raw material or its sanitary state. Most foods that go for smoking have high levels of moisture, thus making them liable to fast development of microorganisms. The lower the initial microbial count, the better the raw material. Should there be any delay in processing, the food must be held at suitably low temperature until the process begins.

To start with, the raw materials for smoking are cleaned to rid the surface of dirt and slime (fish). This also helps reduce the surface microflora, particularly putrefying bacteria. Flesh foods like fish and meat are cured by direct application of salt to the surface or dipped in brine for about 5 min for small fillets or cuts and up to 15 min for large ones in 70–80% saturated salt solution. This leaves a desirable tissue salt concentration of about 2–4% or up to 8% in direct salt applications. The result is a preservative effect as it plasmolyses the microbial cells and is thus microbistatic or disruptive to the cells of contaminating microorganisms. The microflora associated with brined flesh foods are largely halophilic and osmotolerant microorganisms. These include both spoilage flora and pathogenic bacteria. The spoilage flora include *Lactobacillus*, *Micrococcus*, *Halobacterium*, *Halococcus*, *Pseudomonas* sp. and other Enterobacteriaceae while the pathogenic bacteria are *Staphylococcus aureus*, *Vibrio parahaemolyticus* and *Clostridium botulinum*. At the salt concentration attained during brining or salting, inhibition of growth of most of these microorganisms is achieved without making the product unpleasantly salty to taste. Brining reduces the microbial load by as much as 85–90% and it provides a surface gloss after smoking, in addition to serving as a condiment, thus improving the desirability of the product.

Brine strength is regularly checked using a brineometer and old brines are discarded at intervals; otherwise the microbiological quality of the product may be affected. As brines continue to be used, bacterial contamination builds up in them; this may increase the load on good-quality raw food entering the brine. In addition, reduction in brine strength results from either drip water from the subsequent flesh dips or absorption of salt by it. All these act on the effectiveness of the preservation process and must be prevented. After brining, the flesh food is drained and allowed to dry. Drying is done in chill rooms in industrial set-ups and can go on for several hours; with traditional processors in developing tropical countries, drying is achieved within an hour or two in a sunny, windy atmosphere. This step dries the surface water of the food, leaving a firm skin barrier against entry of microorganisms which is also suitable for smoke deposition.

The process of smoking requires arranging the food on racks or trolleys which separate the smoking compartments from the fire chamber. The wood is allowed to burn, producing smoke, but not allowed to burst into flame. This ensures a moderately low but steady heat which is allowed to rise within the first hour. The heat generated causes the food temperature to rise up to 55–80°C with the traditional uncontrolled system and sometimes up to 120°C for meat. For the more advanced techniques of smoking using the Torry kilns where smoke-influential parameters are controlled, cold smoking is done at about 30°C for about 2 h while hot smoking is done at temperatures of 70–80°C for 2–3 h after initial holding at 50°C for 30 min. At these temperatures most saprophytic, non-spore-forming microorganisms are killed. The food is thus pasteurized. Moreover, enzymatic activities are usually halted at about 60°C, preventing activity of the endogenous enzymes of the food. Furthermore the smoke generated in the process deposits its constituents on food surfaces as efforts are made to trap the smoke in the kiln to prevent its escape. The formaldehyde and phenols convert the brine-solubilized protein on the food surface into a coagulated, smooth, resinous pellicle on which other smoke constituents such as tars, aldehydes, alcohols, ketones, acidic compounds and phenols are deposited. These together serve as reinforcement of the food surface against development of microorganisms, thus helping to preserve the food. The phenols in their capacity as antioxidants also prevent oxidative rancidity which is a common spoilage feature of fatty foods.

In the uncontrolled system of smoking, even in tropical countries, the foods are smoke-dried until their water activity (a_w) falls within the range (a_w 0.6–8.5) for intermediate-moisture foods as they are shelf-stable without refrigeration. In such a form they are almost brittle and are retailed in the market for as long as 2 weeks. When moisture levels rise as a result of absorption from the atmosphere, they are returned for further smoking to check the development of microorganisms. All these factors together illustrate the efficacy of a combination of preservative treatments, termed the hurdle effect (**Fig. 1**) or synergism.

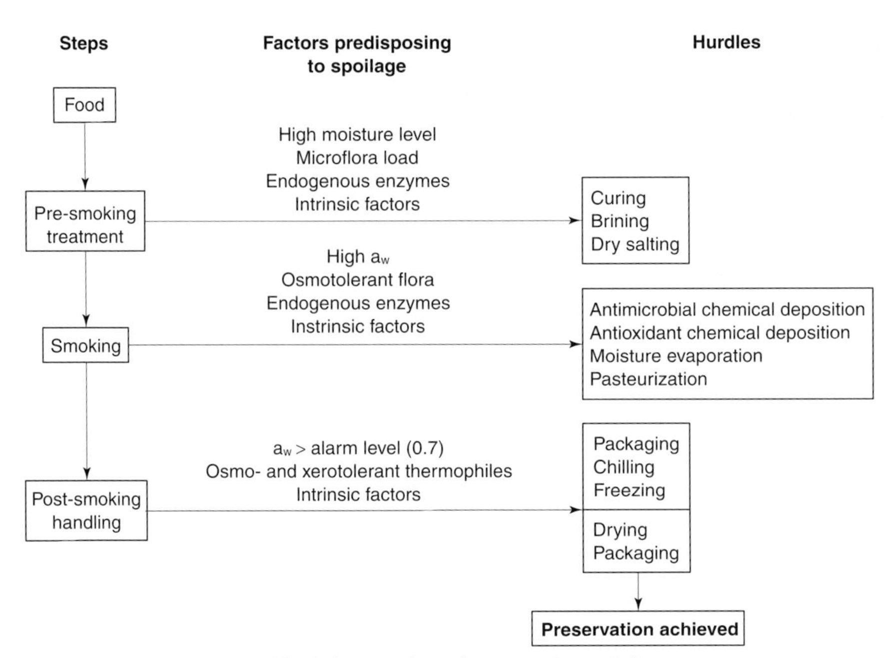

Figure 1 Hurdle effect in food preservation of flesh by wood smoke. a_w, water activity.

Smoked foods therefore can rightly be classified as hurdle-technology foods. By definition, these are products whose shelf life and microbiological safety are extended by several factors, none of which on its own would be sufficient to inactivate undesirable microorganisms.

Possible Risks to the Consumer

The possible hazards associated with smoking of foods began to attract attention in the 1950s when it was discovered that the polycyclic hydrocarbon (PAH) constituent of wood smoke could be carcinogenic. Equally hazardous are other chemicals generated as a result of reactions during exposure of food to smoke and heat. The following risks may be encountered by consumers of smoked products:

1. The hydrocarbon fraction of wood smoke consists of about 30 different PAHs as well as 40 other compounds. Among the PAHs, benzo(a)pyrene (b(a)P) is carcinogenic and has been detected in smoked foods in varying amounts. Levels ranging from 5.3 ng g^{-1} to 14.8 μg g^{-1} of foods have been detected in various species of fish as well as smoked ready-to-eat meats. Improved knowledge about food smoking allows control of smoking parameters so as to limit hydrocarbon generation in smoke. Important influential factors are temperature and air velocity. At temperatures above 400°C for wood decomposition and 200°C for oxidation, hydrocarbon generation from lignin and cellulose is highly favoured. Below these values, high concentrations of the desirable constituents (phenols, carbonyl compounds and acidic compounds) of wood smoke are produced with little or no carcinogenic aromatic hydrocarbons. Specific conditions at which formation of PAH is effectively suppressed while favouring production of desirable compounds are operated. The contents of B(a)P in fish smoked in traditional kilns are higher than in fish smoked in smoke houses under controlled parameters. Whether it is cold smoking (at 30°C) or hot smoking (at 80°C), temperatures are well below those established as favouring PAH generation.
2. Other hazardous chemicals closely associated with smoked animal proteins result from the effect of heat on the food components during the smoking process. A group of these are water-soluble, heat-stable polar substances with mutagenic activity. Mutagens such as amino-carbolines are formed when the surface temperature of meat rises to about 200°C or even 115°C for fish in the presence of high moisture content. The mutagens are formed as a result of pyrolysis of amino acids and proteins in heated meat and fish; heat is an integral part of smoking.
3. Maillard browning of food is another result of food heating. Reactions between sugars and amino compounds of foods subjected to heat and dehydration give rise to brown polymeric compounds – melanoidins – which are sugar–amino acid Amadori compounds. Maillard reactions occur readily in smoked products processed even under controlled conditions. Prolonged consumption of

Maillard browned diets causes physiological disorders of the liver and kidneys. Purified Maillard reaction products such as pyrazines, reductones, dicarbonyls and furan derivatives have shown mutagenic activities.

4. Heating of proteinaceous foods also leads to the formation of amino acid complexes that may be hazardous to health. Lysinoalanine is formed when stock fish (dried cod) is exposed to a temperature of 90°C at a pH above 13. This complex has been implicated in renal damage. Although there is a dearth of information regarding formation of lysinoalanine in foods, since conditions for smoked fish products are suitable for the formation of lysinoalanine, its importance cannot be overlooked.
5. Gizzerosine (2-amino-9-(4-imidazolyl)-7-azanonanoic) is another chemical substance associated with heat-exposed fish. This substance is a derivative of histamine or histidine. Gizzerosine causes stomach or gizzard erosion and black vomit in chicks.
6. While it is common for smoked foods to be viewed as ready-to-eat foods which do not need further cooking, certain risks are associated with this habit. Curing and smoking are not substitutes for cooking, particularly for flesh foods, given some of the temperatures attained during the process. Flesh foods that are known to harbour parasites require further cooking to remove any cysts. Pork (ham) retained in smoke houses until it reaches an internal temperature of 61°C will not be safe if it contains *Trichinella*. Ham labelled ready-to-eat is held at a temperature that raises the internal temperature to 68°C, at which point it is considered safe.

See also: **Dried Foods. Heat Treatment of Foods**: Synergy Between Treatments. **Hurdle Technology. Intermediate Moisture Foods**.

Further Reading

Burt JR (1998) *Fish Smoking and Drying: The Effect of Smoking and Drying on the Nutritional Properties of Fish*. Essex, UK: Elsevier Science Publishers Ltd.

Connell JJ (1979) *Advances in Fish Science and Technology*. Torry Research Station document. Aberdeen, Scotland: Fishing New Books.

Ikeme AI (1990) *Meat Science and Technology: A Comprehensive Approach*. Onitsha, Nigeria: Africana Fep.

Kordylas JM (1990) *Processing and Preservation of Tropical and Subtropical Foods*. London: Macmillan Education Ltd.

Russell NJ, Gould GW (1991) *Food Preservatives*. New York: Van Nostrand Reinhold.

Traditional Preservatives – Vegetable Oils

V Venugopal, Food Technology Division, Bhabha Atomic Research Centre, Mumbai, India

A number of vegetable oils are used as foods, food ingredients of frying media throughout the world. Fatty oils are employed traditionally to preserve a variety of pickled fish products. Several vegetable oils and their derivatives also have specific functions such as providing mouth feel, palatability and calorie content to foods. Fatty acid esters of sucrose and other polyhydric alcohols enhance the emulsifying properties of food formulations. In baked products, vegetable fats are added to improve plasticity, dough aeration, flavour, emulsifying properties and keeping qualities. Mono- and diglyceryl esters of fatty acids such as sodium steryl-2-lactate are important dough conditioners since they act as crumb softeners, emulsifiers and antistaling agents. Several vegetable oils are rich sources of unsaturated fatty acids, the nutritional significance of which is well recognized. The important vegetable oils used in foods are those from soybean, rape seed, coconut, corn, cottonseed, olive, palm, peanut, rice bran, safflower, sesame, sunflower and wheat germ. This article deals with the antimicrobial properties of vegetable oils.

Comparative Composition of Vegetable Oils

Vegetable oils and fats are triglycerides of fatty acids, the composition and nature of which vary depending on the source. The even-numbered, straight-chain fatty acids account for the largest proportion of fatty acids in plant and animal fats, although small amounts of odd-numbered carbon chains and branched and hydroxy acids are present. **Table 1** shows the nomenclature used for designating fatty acids.

Fatty acids with chain lengths of 10 carbon atoms or less are sometimes called short-chain fatty acids and they are all saturated. Fatty acids with 12 or 14 carbon atoms are medium-chain fatty acids and those with more than 14 carbon atoms are long-chain fatty acids. The fatty acids may be saturated or unsaturated. Seed fats have higher contents of unsaturated fatty acids compared with fats from red meats. Marine oils are rich in polyunsaturated fatty acids. Corn, soybean and sunflower seed oils contain less than 15% long-chain fatty acids and more than 55% are polyunsaturated fatty acids. Unprocessed vegetable oils contain only *cis* isomers, whereas processing may give rise to the formation of large quantities of *trans*

Table 1 Terms and symbols designating major fatty acids in food of animal and vegetable origin

Common name	*Chain length (C atoms)*[a]	*Double bonds*	*Symbol I*[b]	*Symbol II*[c]	*Symbol III*[d]	*Systematic name*
Butyric	4	0	$C_{4:0}$		4:0	n-Butanoic
Caproic	6	0	$C_{6:0}$		6:0	n-Hexanoic
Caprylic	8	0	$C_{8:0}$		8:0	n-Octanoic
Capric	10	0	$C_{10:0}$		10:0	N-Decanoic
Lauric	12	0	$C_{12:0}$		12:0	n-Dodecanoic
Myristic	14	0	$C_{14:0}$		14:0	n-Tetradecanoic
Palmitic	16	0	$C_{16:0}$		16:0	n-Hexadecanoic
Palmitoleic	16	1	$C_{16:1}$	$C_{16:1w7}$	16:1, *n*-7	*cis*-9-Hexadecanoic
Stearic	18	0	$C_{18:0}$		18:0	n-Octadecanoic
Oleic	18	1	$C_{18:1}$	$C_{18:1w9}$	18:1, *n*-9	*cis*-9-Octadecanoic
Linoleic	18	2	$C_{18:2}$	$C_{18:2w6}$	18:2, *n*-6	*cis, cis*-9,12-Octadecadienoic
γ-Linolenic	18	3	$C_{18:3}$	$C_{18:3w6}$	18:3, *n*-6	All *cis*-6,9,12-octadecatrienoic
α-Linolenic	18	3	$C_{18:3}$	$C_{18:3w6}$	18:3, *n*-9	All *cis*-9,12,15-Octadecatrienoic
Arachidic	20	0	$C_{20:0}$			n-Eicosanoic
Gadoleic	20	1	$C_{20:1}$	$C_{20:1w9}$	20:1, *n*-9	n-11-Eicosenoic
Arachidonic	20	4	$C_{20:4}$	$C_{20:4w6}$	20:4, *n*-6	All *cis*-5,8,11,14-Eicosatetraenoic
Behenic	22	0	$C_{22:0}$			n-Docosanoic
Erucic	22	1	$C_{22:1}$	$C_{22:1w9}$	22:1, *n*-9	*cis*-13-Docosenoic
Clupadonic	22	5	$C_{22:5}$	$C_{22:5w3}$	22:5, *n*-3	All *cis*-7,10,13,16,19-Docosapentaenoic

[a]Carbon atoms are numbered starting from the carboxyl group which is no. 1.
[b]In the case of unsaturated fatty acids, the symbol t is sometimes used to indicate points of unsaturation, e.g. $C_{18:2}$, $t^{9,12}$ for designating linoleic acid, the carboxyl atom being no. 1.
[c]Carbon atoms are numbered from the methyl group, w indicates the first carbon where a point of unsaturation is found.
[d]The letter 'n' gives the position of the first C atom where a point of unsaturation is found, starting from the methyl group.

isomers. **Table 2** shows the fatty acid contents and certain characteristics, such as iodine values, melting points and saponification values, of some important vegetable oils.

Antimicrobial Activities of Specific Components

Fatty Acids

Fatty acids are formed from triglycerides by the action of lipases which may be of varied origins such as muscle tissues and various microorganisms. Fatty acids with higher chain lengths, especially those that are unsaturated, may also be oxidized and hydrolysed to form short-chain acids. Studies reported so far, indicate that several fatty acids are inhibitory against bacteria and yeasts. Short-chain (C_2–C_6) fatty acids are the most biocidal, and have significant inhibitory activity against Gram-negative organisms. Acetic, lactic and citric acids can be acidulants as well as antimicrobial agents. Sodium salts of short-chain fatty acids (acetic, propionic, butyric and lactic) are inhibitory against *Streptococcus bovis* at concentrations of 0.05–0.3 M; inhibition of the bacterium being greater at lower pH. Medium- or long-chain fatty acids were less effective against Gram-negative organisms, but are more inhibitory against Gram-positive bacteria. For example, although fatty acids up to C_6 inhibited the growth of *Escherichia coli* and *Bacillus subtilis* to the same extent, about 50 times the concentration of C_{10} fatty acid was needed to inhibit *E. coli*. Some members of the Neisseriaceae and Enterobacteriaceae are susceptible to the antimicrobial action of some longer-chain fatty acids.

Chain length, degree of saturation and geometric configuration are all important determinants of antimicrobial activity of fatty acids. Saturated fatty acids reach their optimal inhibitory activity with a chain length of 12. Lauric, myristic and palmitic acids are effective inhibitors of bacteria; lauric acid depressed growth of *Lactobacillus casei* and tubercle bacilli. The inhibition was more effective by branched-chain α-methyl lauric acid. The inhibitory activity was limited to the *cis* form of the fatty acids. Myristic acid and its analogues have been shown to be inhibitory against the oral bacterium *Selenomonas artemidis* at a concentration of 100 μg ml^{-1} of growth medium. *cis*-Hexadecanoic and *cis*-octadecanoic acids were also inhibitory against another oral bacterium *Streptococcus sobrinus*. The weakness of long-chain fatty acids in their antimicrobial properties is due to their

Table 2 Fatty acid composition and analytical constants of major vegetable oils

Fatty acids	*Soy bean*	*Cotton seed*	*Corn*	*Palm*	*Sun-flower*	*Peanut*	*Safflower*	*Olive*	*Rape seed*	*Coco-nut*	*Palm kernel*
Caprylic	–	–	–	–	–	–	–	–	–	7.6	7.3
Capric	–	–	–	–	–	–	–	–	–	7.3	2.9
Myristic	0.1	1.0	0.8		1.2	0.8	0.1	–	–	16.6	18.4
Palmitic	10.5	25.0	11.5	46.8	7.0	11.0	6.7	16.9	4.3	8.0	8.7
Palmitoleic	–	0.7	–	–	–	–	–	1.8	1.3	1.0	–
Stearic	3.2	2.8	2.2	3.8	3.3	2.3	2.7	2.7	1.7	3.8	1.9
Oleic	22.3	17.1	26.6	37.6	14.3	51.0	12.9	61.9	59.1	5.0	14.6
Linoleic	54.5	52.7	58.7	10.0	75.4	30.9	77.5	14.8	22.8	2.5	1.2
Linolenic	8.3	–	0.8	–	–	–	–	0.6	8.2	–	–
Arachidic	0.2	–	0.2	0.2	–	0.7	0.5	0.4	0.5	–	–
Eicosenoic	0.9	–	–	0.3	–	0.5	0.1	–	–	–	–
Iodine value	120–141	97–115	124	48–56	124–136	84–100	143	80–88	114	81	14–23
Melting point (°C)	–23 to –20	10–16	–12 to –10	27–50	–18 to –16	–2	–18 to –16	80–88	114.3	23–26	24–26
Saponification value	189–195	189–198	187–193	196–202	188–194	188–195	190	–	–	250–264	–

ability to change to *trans* isomers during processing and their susceptibility to form complexes more easily with macromolecules such as starch and proteins.

In general, the unsaturated derivatives of long-chain fatty acids are more inhibitory than the corresponding saturated fatty acids. Thus, the inhibitory effect of linoleic acid is far greater than that of oleic acid. However, the minimum inhibitory concentration of arachidonic acid was about the same as that of linoleic acid. The antibacterial effect of unsaturated fatty acids on Gram-positive microorganisms increased as the number of double bonds in the molecule increased. The activity of fatty acids with chain lengths of C_{14} or longer are strongly influenced by the degree of unsaturation. The most effective monounsaturated fatty acid is $C_{16:1}$ (palmitoleic) and the most active polyunsaturated fatty acid is $C_{18:2}$ (linoleic). Germination of *Clostridium botulinum* spores has been reported to be inhibited by small amounts of unsaturated fatty acids.

A number of fatty acids present in vegetable oils may offer potential as microbial inhibitors for slightly acid foods. However, the effectiveness of fatty acids as food preservatives may be overcome by the presence of certain antagonists such as serum albumen, starch and cholesterol. Thus, foods containing these compounds would not be suitable for preservation by fatty acids. The antimicrobial activity of fatty acids as food additives is also related to their solubility. For example, capric acid was highly effective against film-forming yeasts in soy sauce, but this acid may not be a satisfactory additive because of its limited solubility. The microbial inhibitory effects of wheat bran and *Tetrahymena geleii* have been correlated to fatty acids. The antimicrobial properties of fatty acids may be enhanced when used in combination with multiple food additives.

The antifungal activity of fatty acids depends on chain length, concentration and pH of the medium. Unsaturated fatty acids are more fungicidal than saturated ones, acetylic derivatives being more active as compared with ethylenic fatty acids. Dicarboxylic acids are ineffective. Branched-chain fatty acids are less active than straight-chains fatty acids with an equal number of carbon atoms. Yeasts are affected by fatty acids with short chain lengths (C_{10}–C_{12}). Capric and lauric acids are most active against yeasts. A number of fatty acids up to a chain length of 20 carbon atoms were inhibitory to *Aspergillus niger*, *Trichoderma viride*, *Myrothecium verrucaria* and *Trichophyton mentagrophytes*, the maximum activity being exerted by those with chain lengths of 8–14 carbon atoms. For *A. niger*, the optimum chain length was 11 carbon atoms. The inhibitory activity was influenced by pH; the lower the pH the higher the extent of inhibition. Food ingredients such as starch, albumin, cholesterol etc. may, however, protect fungal organisms against growth inhibition by fatty acids. **Table 3** shows the antimicrobial activities of fatty acids.

Esters of Fatty Acids

Fatty acids esterified to alcohols have inhibitory activity against microorganisms, depending on their chemical nature. Esters of monohydric alcohol are inactive, whereas esterification to polyols increases activity. Esterification of lauric acid with methanol leads to loss of antimicrobial activity, although α-hydroxy lauric acid is active. The minimum inhibitory concentration of lauric acid required to inhibit *Staphylococcus aureus* is 2.49 mM, the corresponding value

Table 3 Inhibition of some microorganisms by fatty acids

Fatty acid	Staphylococcus aureus	Candida albicans	*Pneumococci*	*Streptococcus group A*
Caproic	NI	NI	NI	NI
Caprylic	NI	NI	NI	NI
Capric	2.9	2.9	1.5	1.5
Lauric	2.5	2.5	0.1	0.1
Myristic	4.4	4.4	0.2	0.5
Myristoleic	0.4	0.6	0.1	0.1
Palmitic	NI	NI	0.5	3.9
Palmitoleic	1.0	0.5	0.02	0.1
Stearic	NI	NI	NI	NI
Oleic	NI	NI	NI	1.8
Elaidic	NI	NI	NI	NI
Linoleic	NI	0.5	0.04	0.1
Linolenic	1.8	NI	0.18	0.4
Arachidonic	NI	NI	NI	NI

The values represent minimum inhibitory concentrations (mM. NI, not inhibitory at the concentrations tested. Adapted from Kabara (1993).

for methyl α-hydroxylaurate is 0.54 mM. Fatty acid esters of sucrose and other polyhydric alcohols have great potential for use as antimicrobial food additives in addition to their recognized role as emulsifiers. Sucrose dicaprylate and sucrose monolaurate have inhibitory activity against several Gram-positive bacteria and also *Escherichia coli*. Sucrose laurate can inhibit *E. coli* to a greater extent than lauric acid. Many of these compounds are industrially synthesized for use in the food industry, although a number of fatty acid esters of sucrose and other polyhydric alcohols occur naturally in plants.

Fatty acid mono-esters of glycerol have strong antimicrobial activities against a broad spectrum of Gram-positive bacteria. Di- and tri-esters are generally not active. The monoglycerides are more active than the fatty acids. C_{12} monoglyceride is more active than the lower (C_8, C_{10}) or higher (C_{14}) chain-length derivatives. The antimicrobial activity of mono-esters of glycerol and the diesters of sucrose compare favourably in activity with commonly used antiseptics, such as parabens, sorbic acid or dehydro-acetic acid.

The antimicrobial activity of glycerol monolaurate (monolaurin, lauricidin) has been well documented. It is more active than lauric acid, di- and triesters of glycerol. Monolaurin is more inhibitory to Gram-positive bacteria than Gram-negative microorganisms. It is one of the most effective bactericidal agents among different derivatives of lauric acid. Various species of *Streptococcus*, *Staphylococcus*, *Corynebacterium*, *Micrococcus*, *Sarcina*, *Saccharomyces* and also *Vibrio parahaemolyticus* are inhibited by monolaurin at concentrations of 5–100 μg ml^{-1}. It also enhances thermal inactivation of spores of *Bacillus* sp. Monolaurin is more effective against *V. parahaemolyticus* and *Salmonella typhimurium* than sodium benzoate or sorbic acid. Monolaurin has also been reported to be inhibitory to *Listeria monocytogenes* in tryptic soy broth in the presence of yeast extract at pH values ranging from 5.5–7.0. According to the US Federal Drugs Agency (FDA), lauricidin is a Generally Recognized as Safe (GRAS) chemical. For most applications a concentration of 500 p.p.m. or less is effective, which does not affect the taste of the food. Presence of ethylenediamine tetra-acetic acid (EDTA) can reduce the requirement of monolaurin to inhibit both Gram-positive and Gram-negative bacteria. The sensitivity of Gram-negative bacteria to monolaurin and other lipids can also be enhanced by other treatments such as cooling, mild heating, presence of citric acid or phosphate. Germinations of *B. cereus* and *C. botulinum* 62A spores in L-alanine have been reported to be inhibited by monolaurin, monomyristin and monolinolein at concentrations of 0.1–0.5 mM. The antifungal properties of monolaurin compared with some other known antifungal agents are given in **Table 4**. Like monolaurin, the glycerol ester of capric acid is inhibitory against Gram-negative bacteria. Like monolaurin, the antimicrobial property of glycerol monocaprate is enhanced by lactate, phosphate, citrate, sorbate, ascorbate and nisin, but may be reduced by starchy and proteinaceous compounds. A mixture of monolaurin and sorbic acid at 500–700 p.p.m. has been shown to be a potent inhibitor of several aerobic and anaerobic microorganisms.

Monolaurin and monocaprate (monocaprylin) also exhibit fungistatic action against *Aspergillus niger*, *Penicillium citrinum*, *Candida utilis*, and *Saccharomyces cerevisiae*. Monolaurin can also enhance the fungistatic activity of several antifungal compounds. Sucrose dicaprylate and sucrose monolaurate are inhibitory against fungi and yeast apart from Gram-positive and Gram-negative bacteria. Several

Table 4 Antifungal activity of lauricidin in comparison with some known antifungal compounds

Fungus	*Lauricidin*	*Sorbic acid*	*Propionic acid*	*Combination of lauricidin and sorbic acid*
Aspergillus flavus	28.6	83.5	84.1	100
A. niger	40.3	100	97.6	100
A. ochraceus	21.1	73.5	84.1	100
Candida albicans	17.5	100	11.7	100
Fusarium graminearum	20.1	100	100	100
Muco ciccinelloides	30.5	94.3	78.7	100
Pythium elongatum	10.0	100	100	100
P. patulum	18.9	75.2	65.7	100
Phytophthora citrophthora	41.2	100	100	100
Rhizoctonia solani	54.9	98.2	100	100
Rhizopus stolonifer	10.1	100	100	100
Saccharomyces cerevisiae	26.5	100	23.2	100
Sclerotium rolfsii	63.3	100	100	100
Sporobolomyces sp.	100	100	100	100

Results are expressed as percentage of mycelial growth inhibition at 1% concentration of the compounds. Adapted from Kabara (1993).

Table 5 Comparison of antimicrobial activities of fatty acid esters with some commonly used preservatives

Fatty acid ester	Bacillus subtilis	Bacillus cereus	Staphylococcus aureus	Aspergillus niger	Saccharomyces cerevisiae
Monolaurin	17	17	17	137	137
Monocaprin	123	123	123	123	123
Sucrose dicaprylin	74	74	148	–	–
Sodium lauryl sulphate	100	100	50	100	100
Butyl-*p*-hydroxybenzoate	400	200	200	200	200
Sorbic acid	4000	4000	4000	1000	1000

The values represent minimum inhibitory concentration (μ ml^{-1} medium). Adapted from Kato and Shibasaki (1975).

sucrose esters substituted to different degrees with a mixture of palmitic and stearic acids can inhibit the growth of *Aspergillus*, *Penicillium*, *Cladosporium* and *Alternaria* sp. in media at 1% concentrations each. The least substituted ester was the most active in reducing mould growth. Sucrose dicaprylate possessed the highest activity of the sucrose esters but was less active than monolaurin. The activity of monolaurin compares with that of sodium lauryl sulphate and is superior to butyl-*p*-hydroxybenzoate. Capric acid and monolaurin are the most active preservatives in soy sauce that may contain osmophilic yeast. **Table 5** shows comparative inhibitory properties of some esters of fatty acids and some other inhibitors.

Soaps

The earlier references on the germicidal effects of sodium or potassium salts of fatty acids date back to the beginning of the twentieth century. In 1908, potassium salts of myristate, palmitate, and stearate were found to inhibit growth of *Escherichia coli*. Although the sodium or potassium salts did not vary greatly in their germicidal action, the inhibitory effect depended on the nature of the fatty acids. Salts of lower members (C_4–C_{10}) had little or no germicidal effect, whereas soaps containing 14 carbon atoms had the maximum antimicrobial activity. Soaps of laurates, oleates, linoleates and linolenates can inhibit *Pneumococcus* spp., *Streptococcus* spp. and *Bacillus* spp. Of the saturated soaps, myristate and laurate were inhibitory against *Streptococcus faecalis*. Sodium oleate was more than 100 times as effective as sodium stearate or phenol in inhibiting certain microorganisms. The salient features of lipids as food preservatives are summarized in **Table 6**.

Mode of Action

The mechanism of antimicrobial action of lipids has not been well defined. However, it has been suggested that the antimicrobial action is related to the lipophilic properties of these compounds. As lipids have a balance of lipophilic and hydrophilic behaviour hydrophilic–lipophilic balance (HLB) helps them passively to transverse the relatively non-polar microbial membrane and yet to be sufficiently soluble in

Table 6 Some salient features of antimicrobial activities of lipids

Fatty acids	Generally more inhibitory against Gram-positive than Gram-negative bacteria Antimicrobial activity influenced by chain length, degree of saturation, branching and pH of the medium Antimicrobial activities of saturated fatty acids increase up to a chain length of 12. Inhibitory activity of unsaturated acids increases with degree of unsaturation Only *cis* isomers are active Antibacterial and antifungal activities enhance in combination with multiple food additives Compounds such as starch, cholesterol, albumin are antagonists
Fatty acid esters	Esters of monohydric alcohols inactive, those of polyhydric alcohols active Monoester of glycerol is active, di- and tri-esters are inactive Monoesters more active than the corresponding free acid. Glycerol monolaurate, a potent antimicrobial agent, occurs naturally in some foods Fatty acid esters of sucrose have good antimicrobial properties
Soaps	Sodium or potassium salts of fatty acids with 14 carbon atoms have good antimicrobial activities. Soaps of lower fatty acids are less active Microbial membrane is the major site of inhibitory activity.

the aqueous environment of the microorganism. The HLB scale is in the range of 1–20, a value of 10 indicating equal amounts of hydrophilic and lipophilic character. The HLB values for mono- and diglycerides are in the range of 2.8–3.5. The values for glycerol monostearate and monolaurin are 3.8 and 13, respectively. The ratio of solubility of compounds in the oil and aqueous phase is denoted by partition coefficient. Values of the coefficient are in the range of 0.17–2.5 for propionic acid and sorbic acid and 26–88 for ethyl and propyl esters of *p*-hydroxybenzoic acid, respectively. The values of several fatty acids are not, however, available.

The distribution of the preservatives in the cell causes a reduction in the proton motive force through the membrane and a reduction of pH affecting metabolic activity of the cell. The fatty acids may also induce lysis of protoplasts and stimulate uptake of oxygen. Growth, amino acid transport and oxygen consumption of Gram-negative organisms are mostly inhibited by short-chain (C_2–C_6) fatty acids. The stimulation of oxygen uptake at low bactericidal concentrations of fatty acids such as lauric and myristic acids and C_{18} unsaturated acids has been shown due to uncoupling of oxidative phosphorylation which produced disaggregation of the membrane. This also affects synthesis of ATP by the microbial cells. The interference on the membrane transport processes leads to cessation of cell growth or cell death. For example, high concentrations of oleic and linoleic acids inhibited uptake of amino acids by *Staphylococcus aureus*. Fatty acids at sublethal levels can also induce leakage of amino acids from *Bacillus megaterium*, suggesting interference of the compounds with amino acid uptake by bacteria. These observations suggest that the cellular membrane may be a major site of action of the preservatives.

Another theory is that the unsaturated fatty acids may undergo oxidation and the resulting free radicals could exert inhibitory action by attaching to critical sites on the cell membrane. The higher sensitivity to inhibition of Gram-positive bacteria as compared with Gram-negative organisms suggests that the lipopolysaccharide of the cell wall of the latter may be preventing the accumulation of fatty acids on the cytoplasmic membrane.

The mechanism of action of fatty acids and their esters depends on the concentration of both the ionized and unionized forms and the specific efficacy of each. In general, the undissociated acid, and not the anion, is the better antimicrobial agent. The minimum concentration of undissociated acid required to inhibit microorganism is generally one to two orders of magnitude smaller than that of the corresponding anion. The presence of the ionized form, however, cannot be neglected, particularly when at high pH, when its concentration may be much larger than that of undissociated acid. An increase from pH 6.5 to 7.5 increased the minimum inhibitory concentrations of the short-chain acids (caproic, caprylic and capric), but decreased the minimum inhibitory concentrations of the two medium-chain fatty acids (lauric and myristic). Fatty acids are also known to decrease glycolysis and stimulate gluconeogenesis.

The antimicrobial activity of monolaurin is produced through its ability to destabilize the function of the membrane. At lower concentrations it acts as a bacteriostatic by interfering with the uptake of nutrients. Monoglycerides have been shown to act on the oxygen side of the flavin of the NADH dehydrogenase. Enhancement of antimicrobial action of oils by cooling, mild heating, EDTA etc. has been shown to be due to increased permeability of the cells.

The germination of spores of *Clostridium pasteurianum* was inhibited by the oxidative rancidity in lard and corn oil. Since the inhibition was overcome by catalase, it is suggested that the activity was due to the peroxides formed due to oxidation. Peroxides have also been implicated in the antimicrobial and inhibitory activity of oleic acid.

Applications

Vegetable oils and their derivatives have several applications in a wide range of foods including their use as surfactants and stabilizers, as mentioned earlier. Recognition of their antimicrobial properties is resulting in increasing applications. The effectiveness of monolaurin as a conventional additive in deboned meat, chicken, sausages and minced fish and other foods has been studied. The natural occurrence of monolaurin in many food products, its low oral toxicity, its activity as anticaries and antifungal agent adds to its benefits as an antimicrobial food additive.

Staling of bread is characterized by loss of plastic properties, toughening and noticeable change of flavour. Monolaurin is known to prevent bread staling. The lower monocarboxylic acid (e.g. propionic acid) esters of a polyhydric alcohol with 2–6 hydroxyl groups are also effective mould inhibitors and softening agents. Typical polyhydric alcohols include glycol, glycerol and erythritol. Firming of bread can also be prevented by incorporating diglyceride (with one acyl radical derived from lactic acid and one acyl radical derived from a saturated fatty acid with 16–18 carbon atoms such as palmitic or stearic acid) into the dough before baking. Sodium, potassium or calcium salts of stearyl-2-lactylic acid are used for the purpose. Diethanolamides of lauric and myristic acids have found applications in controlling bacteria and yeast in beer. Monoglycerides derived from coconut oil have recently been shown to inhibit *Listeria monocytogenes* in beef frank slurries and seafood salads at a pH of 4.9 and temperature of 4°C.

Short-chain fatty acids have been used in coatings to several foods to prevent loss of quality such as dehydration and rancidity development. These coatings on animal carcasses can also reduce microbial proliferation and extend shelf life. Spraying to give a surface film of a mixture of mono-, di- and triglycerides of olefinic carboxylic acids containing 8–22 carbon atoms may be effective. The triglycerides may be obtained by using a mixture of olive, refined rape and peanut oils. Other compounds useful in this respect include uniform and/or mixed glycerol esters of caprylic, caproic, myristic, palmitic, stearic, arachidic, lauric, linoleic, linolenic or erucic acids.

Preservative systems consisting of multiple approved additives may have synergistic effects. Such systems hold promise for expanding the use of fatty acids as antimicrobial food additives. A preservative system consisting of lauricidin (500 p.p.m.), ethylenediamine tetra-acetic acid and phenol (100 p.p.m.) and butylhydroxyanisole (BHA) (100 p.p.m.) has been shown to preserve a wide range of foods. Another preservative system for meat, fish, vegetables and processed foods has been formulated using a combination of monolaurin, phosphates and sorbates. The formulation consists of one part of monolaurin, 20 parts of a mixture of pyro- and polyphosphate (1 : 1) and 40 parts of sorbic acid. The preservative mixture is added to the product to a final concentration of 0.3% sorbic acid. Lauric esters of propylene glycol, glycerol and polysorbate 80 (sorbitan) are able to preserve starch-containing foods. In summary, apart from the convention roles of lipids in food products, vegetable oils and their derivatives can also serve as antimicrobial agents in several food formulations.

See also: **Bread**: Bread from Wheat Flour. **Lager**. **Metabolic Pathways**: Lipid Metabolism.

Further Reading

Baldwin EA, Nisperos MO, Hagenmaier RD and Baker RA (1997) Use of lipids in coatings for food products. *Food Technology* 51(6): 56–62.

Beuchat LR and Golden DA (1989) Antimicrobials occurring naturally in foods. *Food Technology* 43(1): 134–142.

Branen AL, Davidson PM and Katz B (1980) Antimicrobial properties of lipids and phenolic antioxidants. *Food Technology* 34(5): 42–53.

Davidson PM and Branen AL (1993) *Antimicrobials in Foods*, 2nd edn. New York: Marcel Dekker.

Freeze E (1978) Mechanisms of growth inhibition by lipophilic acids. In: Kabara JJ (ed.) *The Pharmacological Effect of Lipids*, p. 123. Champaign, IL: American Oil Chemists Society.

Giese J (1994) Antimicrobials assuring food safety. *Food Technology* 48(6): 102–110.

Kabara JJ (1981) Food-grade chemicals for use in designing food preservative systems. *Journal of Food Protection* 44: 633–647.

Kabara JJ (1993) Medium-chain fatty acids and esters. In: Davidson P, Branen AL (eds) *Antimicrobials in Foods*, 2nd edn. P. 307. New York: Marcel Dekker.

Kato N and Shibasaki I (1975) Comparison of antimicrobial activities of fatty acids and their esters. *Journal of Fermentation Technology* 53: 793–797.

Kyzlink V (1990) *Principles of Food Preservation*, Development in Food Science Series no. 22. Amsterdam: Elsevier.

Marshall DL and Bullerman LB (1986) Antimicrobial activity of sucrose fatty acid ester emulsifiers. *Journal of Food Science* 51: 468–470.

Pintauro ND (1974) *Food Additives to Extend Shelf Life.* Park Ridge, Noyes: Data Corporation.

Shibasaki L and Kato N (1979) Combined effects on antibacterial activity of fatty acids and their esters against Gram-negative bacteria. In: Kabara JJ (ed.) *Pharmocological Effects of Lipids*, Monograph no. 5, p. 15. Champaign, IL: American Oil Chemists Society.

Permitted Preservatives – Sulphur Dioxide

K Prabhakar and **K S Reddy**, Department of Meat Science and Technology, College of Veterinary Science, Tirupati, India

Introduction

Sulphur dioxide is an important chemical extensively used in the processing and preservation of foods of both plant and animal origin. It has been known since ancient times as a sanitizing agent or antiseptic. It gained popularity as a preservative owing to its apparent lack of toxicity in mammals. Its use was widespread in the USA and other Western countries until the early part of the twentieth century, when incidents of abuses like masking the initial stages of spoilage in foods led to legislation to check indiscriminate and fraudulent commercial applications.

Sulphur dioxide is a colourless gas with a characteristic odour. It is highly soluble in water and liquefies at −10°C. It is used in gaseous or liquefied form, or as its neutral and acid salts.

Sulphur Compounds

The sulphur dioxide-generating compounds with application in the food industry are:

- sulphur dioxide as a gas
- sulphurous acid
- salts of sulphurous acid such as sodium sulphite, sodium bisulphite and potassium sulphite
- hydrosulphurous acid and its salt, sodium hydrosulphite
- pyrosulphurous acid and its salt, sodium pyrosulphite or metabisulphite.

The sulphur dioxide content of these compounds is listed in **Table 1**.

Table 1 Approximate theoretical available sulphur dioxide content of various sources. From Joslyn and Braverman (1954)

Compound	*Formula*	*Availability (%)*
Liquid sulphur dioxide	SO_2	100.00
Sulphurous acid (6%)	H_2SO_3	6.00
Potassium sulphite	K_2SO_3	33.00
Sodium sulphite	Na_2SO_3	50.8
Potassium bisulphite	$KHSO_3$	53.3
Sodium bisulphite	$NaHSO_3$	61.6
Potassium metabisulphite	$K_2S_2O_5$	67.4
Sodium metabisulphite	$Na_2S_2O_5$	57.7

Sulphur Dioxide Gas

The gas is obtained directly by burning sulphur from natural sources. It is the cheapest of all the sources of sulphur dioxide and very effective for purposes of disinfection. Grapes and cut fruits are exposed to fumes of burning sulphur before dehydration or transportation.

Salts of Sulphurous Acid

Sulphite, bisulphite and metabisulphite are extensively used in foods and beverages. They can be easily applied in dry form or as solutions. They are stable, economical and comparatively free from heavy metal impurities. Sulphite solutions are easily absorbed by fruits, which are dipped in the solution before freezing or dehydration.

Liquid Sulphur Dioxide

Liquid SO_2 is free from impurities, and is commonly used in wineries. Accurately measured quantities can be incorporated. Special steel containers are required for storage and transportation, making it a costly source of SO_2.

Range of Foods to Which Sulphites May Be Added

The range of foods into which sulphur dioxide is incorporated includes fruits, vegetables, fruit juices and concentrates, syrups, wines and jams, and to a lesser extent prawns, fish, minced meats and sausages.

The maximum permissible levels of SO_2 in some important foods as specified by the Preservatives in Food Regulation 1979 for the UK are listed in **Table 2**. There are only slight variations between the maximum levels permitted in various products in different countries, because of universal concern for the protection of consumers.

Antimicrobial Action of Sulphur Dioxide

Sulphur dioxide is highly soluble in water and forms sulphurous acid which dissociates into bisulphite or sulphite depending on the pH. Undissociated sulphurous acid is claimed to be the main antimicrobial agent inhibiting bacteria, yeasts and moulds.

The possible mechanisms of inhibition by sulphurous acid are attributed to:

- reaction of bisulphite with acetaldehyde in the cell
- reduction of essential disulphide linkages in enzymes
- formation of bisulphite addition compounds which interfere with respiratory reactions involving NAD.

Table 2 Maximum UK permitted levels of SO_2

Food, food product or beverage	*Maximum SO_2 level (mg kg^{-1})*
Fruits, fruit pulp, tomato pulp	350
Fruit spread	100
Grape juice products	70
Jams	100
Mushrooms, frozen	50
Pickles	100
Raw peeled potatoes	50
Salad dressing	100
Sauces	100
Soft drinks for consumption without dilution	70
Dehydrated potatoes	550
Dehydrated cabbage	2500
Yoghurt	60
Beer	70
Wine[a]	450
Flour for biscuits	200
Desserts, fruit-based milk and cream	100
Sausages or sausage meat	450
Hamburgers or similar products	450

[a]Levels vary for dry, sweet, sparkling, red or white wines.

Factors Influencing Antimicrobial Action

Initial Microbial Population and the Stage of Growth The initial level of bacterial contamination affects the preservative efficacy of SO_2. Minced meat samples containing 300 p.p.m. of sulphur dioxide during refrigerated storage revealed spoilage on the sixth day for samples with an initial contamination level of 7.6×10^7 colony forming units (cfu) per gram, compared with spoilage on the thirteenth day for samples with an initial microbial load of 6.9×10^5 cfu g^{-1}.

Type of Microorganisms Present Strains like acetic acid bacteria, yeasts and moulds are effectively eliminated through incorporation of SO_2. The inhibitory effect also depends on the levels of SO_2 incorporated and maintained. Coliaerogenous bacteria were not affected by 150 p.p.m. but at 450 p.p.m. their multiplication was totally inhibited.

Sulphur Dioxide-producing Sources Equilibrium between various forms of SO_2 – undissociated sulphurous acid, free sulphite or bisulphite ions and hydroxysulphonates – is determined by pH, temperature, composition and storage condition of foods.

Free and Bound Components of Added Sulphur Dioxide The free or unbound component of added SO_2 has the significant antimicrobial action. It is claimed that the inhibition power of the free component of added SO_2 is 30–60 times more effective than that of the bound component.

Composition of the Food and Food Products Foods containing higher levels of components that form inert complexes on reaction with SO_2 can not be effectively preserved with SO_2 alone, especially at room temperature.

Influence of pH The antimicrobial action of SO_2 is more effective in foods with acidic pH. Two to four times as much SO_2 is required to inhibit growth at pH 3.5 compared with pH 2.5. At higher pH values like 7.0, sulphites do not appear to have significant inhibitory action on yeasts and moulds and very high levels are required to control growth of bacteria. Acid is commonly added to lower the pH of foods, enabling preservation with lower levels of SO_2.

Effect of Heat Heating to high temperatures drives off SO_2 from foods and considerably reduces the antimicrobial effects. On heating, the sulphur compound decomposes and the free component escapes by volatilization. At pasteurization temperatures, it is reported to increase the thermal death rate of microorganisms present and enables more rapid destruction of microbes.

Temperature of Storage A synergistic action of lower temperatures and SO_2 addition is claimed by some investigators, as more pronounced bacteriostatic effects were observed in minced meat samples stored at lower temperatures than at higher temperatures (**Table 3**).

It is generally assumed that sulphite preservation of foods at room temperature competes with refrigerated storage of foods without any additives.

Behaviour of Sulphur Dioxide in Foods

Several reaction products are formed through reversible and irreversible reactions in SO_2-treated foods. The amounts of interaction products vary in different foods depending on the processing and storage conditions. Because of these reactions, SO_2 has multi-

Table 3 Approximate storable periods of minced meats at different storage temperatures, with or without SO_2 preservation

	Storage period	
Storage temperatures (°C)	*Without SO_2*	*With SO_2*
7	3–5 days	13 days
15	1–2 days	6–7 days
22	<20 h	1–2 days

farious functions in addition to its antimicrobial effects. It can act as an antioxidant, as a bleaching agent, as a colour fixative, and as an inhibitor of enzymic discolorations and non-enzymic browning. The interaction products of reversible reactions of sulphites do not pose serious problems as most of them are unstable. Addition of SO_2 to menadione, a water-soluble synthetic form of vitamin K, is reported to result in the formation of a reversible sulphonate adduct which readily dissociates in animals to become a source of vitamin K. However, irreversible reactions like cleavage of thiamin have toxicological significance.

Inhibition of Enzymic Discoloration

Enzymic browning is a result of processes involved in the production of pigments from enzymically oxidized phenolic compounds of natural origin. Sulphites form inactive complexes with enzymes or combine with breakdown products to form stable complexes, thus inhibiting enzyme-induced formation of abnormal colours in fruits and vegetables during processing and storage. Sulphite dips are used to control discoloration due to enzymic browning in frozen stored fruits and vegetables, as food enzymes are not destroyed by freezing.

Inhibition of Non-enzymic Browning

Non-enzymic browning involves reactions between amino groups and carbonyl groups leading to the formation of insoluble, dark-coloured compounds with a bitter taste. Sulphur dioxide is the most commonly used chemical for inhibition of non-enzymic browning in foods. Inhibition of browning reactions by SO_2 is attributed to the stabilization of the intermediate compounds formed. it combines reversibly with reducing sugars and aldehyde intermediates and irreversibly with certain unsaturated aldehyde intermediates.

The appearance of heat-processed and canned vegetables, fruits, fish and comminuted meat products like sausages can be improved through inhibition of non-enzymic browning. White wines are treated with SO_2 gas or metabisulphite to inhibit non-enzymic brown discoloration during storage. Sulphur dioxide also inhibits non-enzymic browning in dehydrated fruits and vegetables during storage at ambient temperatures.

Antioxidant Properties

Sulphur dioxide in the form of a gas or a sulphite dip during processing and storage of dehydrated vegetables, fruits and grape juice prevents loss of ascorbic acid. It is used in canned tomato sauce to prevent carotenoid oxidation and to preserve the bright colour. It is added to beer as a solution in water to inhibit adverse changes in flavour due to oxidation by dissolved oxygen. Lipids in sausages and comminuted meat products are protected from oxidation changes if sulphite or metabisulphite is included. It also prevents the oxidation of the essential oils and carotenoids and inhibits development of abnormal colour and flavour in citrus juices.

Reducing and Bleaching Actions

Sulphurous acid and the acid sulphites reduce many coloured compounds to colourless derivatives. Dried cut fruits with slight darkening can be almost completely restored to their original colour by treating with SO_2, probably owing to the formation of colourless compounds. In sugar processing, SO_2 bleaches the naturally occurring pigments such as anthocyanins and other coloured non-sugars and also reduces darkening during evaporation and crystallization owing to its combination with reducing sugars. As a reducing agent, it keeps reductones in the inactive reduced form rather than in the active dehydro form. The attractive bright pink colour of sulphited minced meat samples is maintained until spoilage during storage at 7–15°C. This colour fixation property of SO_2 is attributed to its ability to maintain haem iron in the reduced state. Studies revealed increased consumer preference for cooked sulphited minced meat samples. Sulphites also prevent grey discoloration in minced meats and raw sausages when they are exposed to air.

Losses Due to Binding to Food Constituents

Sulphur dioxide is highly reactive with other components in foods; hence it does not persist for long periods. A large part of the SO_2 added to foods remains fixed or bound. Glucose, aldehydes, ketone-like substances, pectin, etc. present in foods determine the extent of binding of added SO_2 in foods.

Glucose binds SO_2 in a reversible manner. The extent of binding is reported to be related to the total concentration of soluble solids in the food. Combination of bisulphites with sugars is much slower than with aldehydes and ketones, and the products formed are relatively more unstable. Sulphur dioxide after combination with sugars or aldehydes exercises very little antimicrobial action.

When increased levels of SO_2 are added to foods, the proportion of the free component increases. At low pH, combination of SO_2 with glucose is delayed, ensuring that more time is available for the SO_2 to act on the microorganisms present.

Levels of SO_2 decrease considerably during storage. Loss of SO_2 in sealed bottles of wine initially containing up to 400 p.p.m. ranges between 20% and 50%. In minced meat samples incorporating 450 p.p.m. of SO_2, levels started to decrease within a few hours. During storage at 7°C, levels of SO_2 decreased to around 295 p.p.m. after the first day, to 270 p.p.m. on the third day, to 240 p.p.m. on the fifth day, and stabilized at 200 p.p.m. on days 7–13, after which spoilage was observed. In samples stored at 15°C, residual SO_2 levels decreased to 350 p.p.m. on the first day, 280 p.p.m. on the third day and 220 p.p.m. on the fifth day. Spoilage was noticed on the sixth day of storage when the residual level was 120 p.p.m. Reduction in the concentration of SO_2 is faster at higher temperatures and it also coincides with increased microbial loads.

Importance of Species/Strain Tolerance

Sulphur dioxide is reported to have selective antiseptic action. Acetic acid bacteria, lactic acid bacteria and coliaerogenous bacteria are more sensitive than others. This compound is most effective against Gram-negative bacteria. Several studies indicate a general decline in the growth of spoilage organisms and also of added cultures of *Clostridium botulinum*, *Clostridium sporogenes*, *Clostridium perfringens* and *Salmonella typhimurium* in minced meats with SO_2 levels of 450 p.p.m. Bactericidal effect was found to be significant within 3 h of addition of *Salmonella enteritidis* and *Yersinia enterocolitica*. Germination of bacterial spores was also found to be affected. In minced meats without preservative, all groups of bacteria multiply throughout the storage period, whereas in sulphited samples only a portion of the microflora causes spoilage.

During storage of minced meat samples with 450 p.p.m. of SO_2 at 7°C, coliforms, salt-tolerant bacteria, and streptococci did not reveal significant changes in their numbers. However, lactobacilli were significantly inhibited by day 9 when spoilage was noticed. These organisms play a major role in the spoilage of vacuum-packaged meats during refrigerated storage. It is to be explored whether extension of refrigerated storage life of vacuum-packaged meats is possible with the addition of SO_2 or sulphites in a safe way. In minced meat samples with 450 p.p.m. of SO_2 stored at 15°C, lactobacilli, salt-tolerant bacteria and enterococci showed significant increases after a lag phase of 4–5 days.

Among yeasts, fermentative types are more resistant than true aerobic species. Certain desirable strains of yeasts required for fermentation are made sulphite resistant through gradual sensitization. Such resistant yeasts are utilized for fermentation in wine-making at levels of SO_2 at which other undesirable strains of yeasts and moulds do not develop.

Toxic Effects in Humans

The extensive use of SO_2 in the form of sulphites, bisulphites and metabisulphites in foods and beverages the world over indicates that allergic reactions and residual toxicity problems in consumers are almost nil in the normal pattern of human exposure. In spite of its high reactivity with biologically important molecules, SO_2 is oxidized to sulphate by sulphite oxidase enzyme and excreted in urine safely. The enzyme sulphite oxidase is reported to be present at higher than adequate levels in liver and other tissues of the human body. The capacity of the mammalian sulphite oxidase for sulphite oxidation is reported to be extremely high in relation to the normal sulphite load expected from both endogenous and exogenous sources.

Sulphites are known to destroy thiamin (vitamin B_1) in foods by cleavage of thiamin into 4-methyl-5-hydroxyethyl thiazole and the sulphonic acid of 2,5-dimethyl-4-aminopyrimidine. This cleavage is completed within 24–48 h at a pH of 5.0 and at room temperatures. Hence, sulphites are not used in foods that are major sources of thiamin. However, studies revealed that humans consuming up to 200 mg of SO_2 per day showed no signs of thiamin deficiency. This reaction need not be taken as a serious disadvantage since some nutrient losses are expected in almost all popular commercial methods of food preservation. Adverse effects were not observed even with chronic sulphite administration. Chronically ingested sulphite does not accumulate in the tissues or reach levels hazardous to human health because of its rapid metabolic removal. However, problems may occur in humans affected with sulphite oxidase deficiency disease.

The possibilities of undesirable interactions between SO_2 and other dietary components or cellular constituents leading to interference in metabolic processes or damage to the structural integrity of proteins have not been evidenced in human systems; hence SO_2 is considered as a safe preservative if used in permitted levels. A few cases of allergic reactions observed in asthma patients after consumption of sulphited foods such as pickled onions were found to be due to the presence of very high levels of sulphur dioxide. If foods are processed to permitted levels of SO_2, such problems may not arise.

Conclusion

The rapid strides made by the processed and convenience food industry would not have been possible without the use of traditional and chemical preservatives. In view of concerns about potential toxicity to the consumers in the long run, there is a worldwide trend to restrict the use of these preservatives to well below their legally permitted levels. No single permitted preservative fulfils the needed requirements of effectiveness and absolute safety. Sulphur dioxide is no exception to this in spite of its proven effectiveness and safety as indicated by its continued usage in a wide range of foods. Future development will lead to optimum utilization of combinations of permitted preservatives so that their individual levels of incorporation can be greatly reduced without compromising the safety and stability of food products. A combination of 50 p.p.m. of sorbate and 50 p.p.m. of SO_2 is reported to have inactivated yeasts such as *Saccharomyces cerevisiae* during heating, even in the presence of glucose. The food industry requires the continued use of preservatives like SO_2 in traditional ways until synergistic combinations have undergone detailed investigations on enhanced safety.

See also: ***Clostridium***: Introduction; *Clostridium perfringens*; *Clostridium botulinum*. **Heat Treatment of Foods**: Principles of Pasteurization. ***Salmonella***: Introduction. **Wines**: Microbiology of Wine-making.

Further Reading

Austin RK, Clay W, Phimphivong S, Smilanick JL and Henson DJ (1997) Patterns of sulfite residue persistence in seedless grapes during three months of repeated sulfur dioxide fumigations. *American Journal of Enology and Viticulture* 48 (1): 121–124.

Burke CS (1980) International legislation. In: Tilbury RH (ed.) *Developments in Food Preservatives*. Vol. 1, p. 25. London: Applied Science Publishers.

Cerrutti P, Alzamora SM and Chirife J (1988) Effect of potassium sorbate and sodium bisulfite on thermal inactivation of *Saccharomyces cerevisiae* in media of lowered water activity. *Journal of Food Science* 53(6): 1911–1912.

Duvenhage JA (1994) Control of post-harvest decay and browning of litchi fruit by sodium metabisulphite and low pH dips – an update. *Litchi Year Book. South African Litchi Growers Association* 6: 36–38.

Gray TJB (1980) Toxicology. In: Tilbury RH (ed.) *Developments in Food Preservatives*. Vol. 1, p. 53. London: Applied Science Publishers.

Gunnison AF (1981) Sulphite toxicity: a critical review of in vitro and in vivo data. *Food and Cosmetic Toxicology* 19: 667–682.

Joslyn MA and Braverman JBS (1954) The chemistry and technology of the pretreatment and preservation of fruit and vegetable products with sulphur dioxide and sulfites. *Advances in Food Research* 5: 97–154.

Roberts AC and McWeeny DJ (1972) The uses of sulphur dioxide in the food industry. A review. *Journal of Food Technology* 7: 221–238.

Sinskey AJ (1980) Mode of action and effective application. In: Tilbury RH (ed.) *Developments in Food Preservatives*. Vol. 1, p. 111. London: Applied Science Publishers.

Stammati A, Zanetti C, Pizzoferrato L, Quattrucci E and Tranquilli GB (1992) In vitro model for the evaluation of toxicity and antinutritional effects of sulphites. *Food Additives and Contaminants* 9(5): 551–560.

Studdert VP and Labuc RH (1991) Thiamin deficiency in cats and dogs associated with feeding meat preserved with sulphur dioxide. *Australian Veterinary Journal* 68(2): 54–57.

Taylor SL and Bush RK (1986) Sulfites as food ingredients. *Food Technology* 40 (6): 47.

Taylor SL, Higley NA and Bush RK (1986) Sulfite in foods, uses, analytical methods, residues, fate, exposure assessment, metabolism, toxicity and hypersensitivity. *Advances in Food Research* 30: 1.

Trenevey VC (1996) The determinations of the sulphite content of some foods and beverages by capillary electrophoresis. *Food Chemistry* 55 (3): 299–303.

Usseglio-Tomasset L (1992) Properties and use of sulphur dioxide. *Food Additives and Contaminants* 9(5): 399–404.

Permitted Preservatives – Benzoic Acid

Lucy J Ogbadu, Department of Biological Sciences, Benue State University, Makurdi, Nigeria

Introduction

Benzoic acid was among the first chemical preservatives allowed for use in foods by law. The major food legislation which dealt specifically with preservatives in foods was published in the UK in 1925 under the Public Health (Preservatives in Food) Regulations, with benzoic acid and sulphur dioxide as the permitted preservatives included therein. This early recognition of benzoic acid as an effective preservative may not be unconnected with the realization that it occurs naturally in certain foods such as cranberries. Its desirability as a preservative stems from its low toxicity and lack of colour, among other properties. One shortcoming, however, is its limited solubility in aqueous systems, which places preference for use on its sodium and potassium salts as these readily dissolve; the potassium salt is preferred in order to minimize sodium level in food.

Table 1 Important criteria of benzoic acid in food preservation

Criteria	*Description/values*
Chemical nature	Weak aryl carboxylic acid
Major purity criteria	White crystalline powder Not less than 99.5% content Melting point 121.5–123.5°C[a]
Forms in which used	Sodium benzoate Potassium benzoate Ammonium benzoate Calcium benzoate
Legislative status	Permitted artificial preservative GRAS
Maximum concentration allowed in food	0.1%
Acceptable daily intake (FAO/WHO)	0–5 mg kg^{-1} body weight[b]
pK_a	4.2
pH range of antimicrobial action	2.5–4.0
Combination of treatments with	Cold temperature Heat (pasteurization) Acidulants Other weak acid preservatives Anaerobiosis

[a] After vacuum drying in sulphuric acid desiccator.
[b] As the sum of benzoic acid and benzoates expressed as benzoic acid.

Benzoic acid has since been accepted by the European Community as well as by most countries of the world, and is listed wherever there are published food regulations. It is classified as a permitted preservative by the Food and Drugs Administration of the USA and is affirmed 'generally recognized as safe' (GRAS). Benzoic acid is a weak aryl carboxylic acid, recommended by the Joint Expert Committee of Food Additives (JECFA) of the Food and Agriculture Organization and World Health Organization (FAO/WHO) at an acceptable daily intake level of 5 mg kg^{-1} for humans. Other specifications regarding benzoic acid in food are listed in **Table 1**.

Foods to Which Benzoic Acid May Be Added

The range of foods to which benzoic acid can be added was realized almost as early as its preservative importance was appreciated. Benzoic acid is known to be effective at low pH, which serves as a pointer to the range of foods that can be well preserved by it. While most foods are close to neutral in pH and thus within a suitable pH range for microbial development and spoilage, some foods are acidic when harvested or before they are processed. Fruits, often with a pH between 2.5 and 4.0, fall into this category. Other foods develop acidity as a result of microbial growth (biological acidity), as in fermented foods like pickles.

Table 2 Major acid foods preserved by benzoic acid and permitted concentrations

Food product	*pH range*	*Concentrations used (mg kg^{-1})*
Fruit juices and concentrates	2.3–3.6	100–500
Canned fruits	< 4.6	250–500
Carbonated drinks	2.5–3.0	200–400
Non-carbonated drinks	2.8–3.3	500–1000
Pickles, relishes and sauerkrauts	2.7–3.0	250–1000
Vegetables and salads	3.5–4.0	1000–2000
Pie fillings	3.0–3.8	1000–2000
Mayonnaise and similar emulsified sauces	4.1–4.4	250–2000
Tomato purée and ketchup	4.0–4.3	250–1000
Sugar and flour-based confectionery	3.5–4.2	1000
Butter and margarines	6.2–6.5	100–1000
Fish semipreserves	6.0–6.5	1000–4000
Prawn and shrimp preserves	6.0–6.3	2000–4000
Tea and coffee liquid extract	4.5–5.4	250–1000

Although the hydrogen ion (H^+) concentration as released by the acid in itself is inhibiting to most microorganisms, some can tolerate the levels encountered in food. Such acid-tolerant microorganisms are the target of the added preservative.

High-acid foods well preserved by benzoic acid include fruit products, carbonated drinks, fermented vegetables, syrups and the like. Other similar products of low pH belong to the range of foods whose shelf life can be extended using the benzoates (**Table 2**). While the effectiveness of benzoic acid has been established for high-acid foods, other foods outside the H^+ concentration classification of high-acid foods now enjoy an extended shelf life through the use of benzoic acid in combination with other preservative methods (Table 2). Benzoic acid is used with ice and brine as an effective preservative for fish and other marine products.

Behaviour of Benzoic Acid in Food

In its pure form benzoic acid is a white crystalline powder. Salts of benzoic acid which are soluble in the aqueous phase of food are also white and are commonly referred to as benzoates. The effectiveness of the acid derives from its action within the aqueous phase of even compartmentalized foods consisting of both oily and aqueous phases, like butter, margarine and low-fat spreads. Even as benzoates, it is the benzoic acid molecule rather than the salt *per se* that is germicidal. The effectiveness of benzoic acid lies in the undissociated (non-ionized acid) molecules which increase in number as the pH of the food environment drops. The dissociation constant (pK_a) of benzoic acid is 4.2, at which pH the concentration of both the dissociated and undissociated fractions are equal:

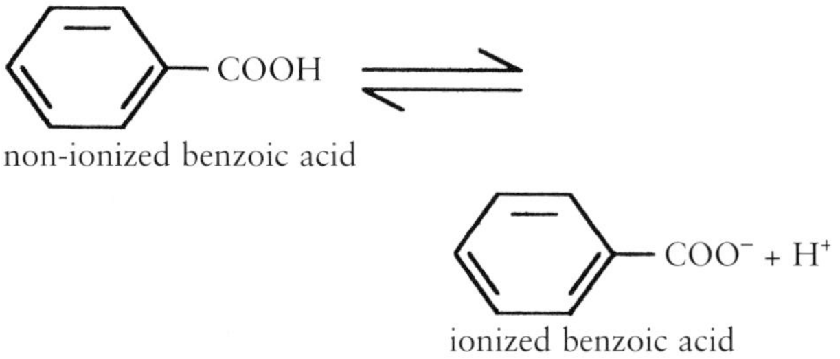

$$C_6H_5COOH \rightleftharpoons C_6H_5COO^- + H^+$$

non-ionized benzoic acid

ionized benzoic acid

A shift in the balance between the ionized and non-ionized acid is achieved exponentially in favour of the non-ionized molecule as the pH of the food drops below 4.2. It is the total effect of the concentrations of the benzoic acid, its anions and that of its H^+ at low pH that exert inhibitory action on the surviving microorganisms. The influence of food pH is such that the benzoates are relatively ineffective in foods like milk, meats, poultry, fish and butter with pH values of 6.0 or above. A rise in pH to 6.0 reduces the level of non-ionized molecules of benzoic acid to about 1.5%. A drop in pH from the pK_a value to 4.0 increases the level of the non-ionized molecules to 60%. The benzoates are most effective in foods of pH range 2.5–4.0. Even though benzoic acid has been used in combination with low temperature to preserve non-acid foods (e.g. fish), it is not able to control bacterial development effectively because the prevailing alkaline pH of such foods is outside its action range; in the case of fish, it is able to suppress the formation of trimethylamine, the compound responsible for the typical smell of spoilt fish. Unguarded addition of any chemical – even benzoic acid – to food would be objectionable and even poisonous. For this reason, the recommended maximum concentration of benzoic acid is 0.1% (w/v). No physiologic change has been observed in humans from this concentration of benzoic acid, although in fruit juices it imparts a disagreeable burning or peppery taste. Although benzoic acid is effective in acid foods, it does not permanently preserve apple juice in spite of its low pH (3.0) even at the maximum acceptable concentration, unless the initial contamination is low. In highly contaminated cider juice, 0.1% concentration of benzoic acid is unable to preserve the juice.

Antimicrobial Action of Benzoic Acid

For any chemical preservative to be effective at halting the development of microorganisms, it must interfere with at least one subcellular target such as the genetic material, the system for protein synthesis, metabolic enzymes, the cell membrane or the cell wall, since each of these components is essential for cell multiplication. The modes of action of most chemical preservatives including benzoic acid – even though it has a relatively simple structure – were not known until recently. This is because acceptance of chemical preservatives was simply based on observed effectiveness and presumed safety. The benzoates are active against a wide range of microorganisms, but at the low pH of 2.5–4.0 at which these compounds are most active, bacteria are generally unable to grow. Moulds and yeasts, however, can tolerate and grow within such a pH range. Benzoic acid is most effective against yeasts in foods, but it is also used to inhibit moulds and lactic and acetic acid bacteria.

As a weak acid, benzoic acid is lipophilic, and thus moves freely across the cell membrane. In a low pH food suitable for its antimicrobial action, the high concentration of undissociated molecules which predominates causes uncontrolled leakage of available H^+ across the membrane into the interior of the cell; because of its lipophilic nature, benzoic acid affects the cellular membrane processes, thus changing membrane fluidity. The resulting rise in the internal concentration of H^+ acidifies it causing protein denaturation and effectively inhibiting metabolic activities. Uptake of benzoic acid is closely linked with its interference with the membrane function and metabolic activities. In addition to acidification of the internal system of the cell, benzoic acid is inhibitory to a number of metabolic enzymes, namely those of the tricarboxylic acid cycle and the glycolytic pathway. Specifically, it inhibits the activities of the dehydrogenases of α-ketoglutarate and succinate and prevents their conversion to succinyl-CoA and fumarate respectively, as well as inhibiting formation of their respective reduced cofactors (NADH and $FADH_2$). It also inhibits the enzyme 6-phosphofructokinase and blocks oxidation of glucose and also pyruvate at the acetate level, thus further blocking formation of NADH. Inhibition of the reduced cofactors affects the efficiency of the oxidative phosphorylation reactions of the electron transport chain. By their modes of action the benzoates are generally antifungal as well as antibacterial when acid-tolerant fungi and bacteria are encountered.

Importance of Species/Strain Tolerance

Since high-acid foods naturally offer an unsuitable environment for growth of most bacteria, moulds and yeasts that grow in that range form their important spoilage flora. Bacterial inhibition in high-acid foods is even more significantly viewed in the light of the unfavourable pH for the growth of food poisoning species. A pH of 4.6 is the minimum for growth of most acid-tolerant food poisoning bacteria: *Clostridium botulinum*, *Staphylococcus aureus*, *Yersinia enterocolitica* and *Listeria monocytogenes*. It is in certain low-acid foods,some of which are preserved using benzoic acid (see Table 2), that these organisms are important, and care is needed in safeguarding against them. Reduction in the effectiveness of benzoic acid at higher pH levels is compensated for by its combination with other suitable preservative treatments. Important bacteria in high-acid foods are lactic acid bacteria and spore-formers, particularly *Bacillus coagulans*. Benzoic acid is effective at controlling these organisms.

Among the yeasts of importance in benzoic acid-preserved foods are the acid- and benzoate-tolerant species *Zygosaccharomyces bailii* and *Z. rouxii*, which can grow at food pH of 2–3 even in the presence of a high concentration of benzoic acid. These yeasts are a nuisance as contaminants of high-acid foods. Mycotoxigenic moulds, on the other hand, are inhibited by benzoic acid at low pH.

Interaction with Other Preservative Treatments

While the parahydroxybenzoates (closely related to the benzoates) have the advantage of being effective over a wide range of pH, the narrow pH range for the benzoates restricts their usefulness; for this reason they are used in combination with other treatments for effective stabilization of foods. Combinations of weak acid may be incorporated in foods where individual chemical concentrations in high proportions are known to affect palatability. Such combinations often act synergistically against microorganisms to enhance their effectiveness. Acidulants like acetic, citric and lactic acids are often used to lower the pH of foods, creating an environment conducive to the antimicrobial action of the benzoates.

Benzoic acid is also used in combination with temperature to effect food preservation. It is used with low temperatures to prepare germicidal ices in which fish, prawns and shrimps can be preserved. At a concentration of 0.1% solution of benzoic acid, the freezing point is only slightly more than 3°C. It is an eutectic mixture, and is sufficiently inhibitory to the growth of microorganisms. Benzoic acid alone in ice is more effective than in combination with other preservatives such as sorbate or EDTA with ice. Similarly, a combination of benzoates with higher temperatures (pasteurization) is required for effective preservation of most of the benzoate-preserved commercially manufactured packaged foods. Benzoic acid is not particularly good at stabilizing foods with high moisture content alone. It is effective in such foods in combination with pasteurization, anaerobiosis and low-temperature storage. While reduced and increased temperature treatments are microbistatic and microbicidal respectively, the benzoates act with either treatment synergistically to halt the growth of the microorganisms present or surviving in the food.

See also: ***Clostridium***: *Clostridium botulinum*. **Ecology of Bacteria and Fungi in Foods**: Influence of Redox Potential and pH. **Heat Treatment of Foods**: Synergy Between Treatments. **Hurdle Technology**. ***Listeria***: *Listeria monocytogenes*. **National Legislation, Guidelines & Standards Governing Microbiology**: European Union. ***Staphylococcus***: *Staphylococcus aureus*. ***Yersinia***: *Yersinia enterocolitica*. ***Zygosaccharomyces***.

Further Reading

Desrosier NW and Desrosier J (1987) *The Technology of Food Preservation*, 4th edn. New York: AVI Publishing.

Gould GW (1989) *Mechanism of Action of Food Preservation Procedures*. London: Elsevier.

Luck E (1980) *Antimicrobial Food Preservatives*. Berlin: Springer.

Tannenbaum SR (1979) *Nutritional and Safety Aspects of Food Processing*. New York: Marcel.

Russel NJ and Gould GW (1991) *Food Preservatives*. New York: Van Nostrand Reinhold.

Permitted Preservatives – Hydroxybenzoic Acid

Rekha S Singhal and **Pushpa R Kulkarni**, University of Mumbai, India

Esters of *p*-hydroxy benzoic acid or parabens differ from other commonly used preservatives by virtue of their activity in the neutral pH range, where most other preservatives fail. This article summarizes their physical properties, the range of foods to which they can be added, their permitted levels in various countries, antimicrobial profile, mechanism of action, metabolism and toxicity, and assay techniques used with parabens.

Esters of *p*-hydroxybenzoic acid, commonly termed parabens, are used as antimicrobial agents in pharmaceuticals, food products and also in cosmetics. The first report on the antimicrobial activity of parabens was published as early as in the 1920s. The relationship between the criteria prescribed by the International Association of Microbiological Societies

Table 1 Relationship between International Association of Microbiological Societies criteria for an ideal preservative and parabens

Criterion	*Compliance of parabens with the criterion*
Toxicological acceptability	+
Microbiological activity	–
Long persistence in foods	+
Chemical reactivity	Low
High thermal stability	Not relevant
Wide antimicrobial spectrum	BbYM
Active against food-borne pathogens	–
No development of microbial resistance	+
Not used therapeutically	+
Assay procedure available	+

+, Comply with criterion; –, do not comply.
B, Gram-positive bacteria; b, Gram-negative bacteria; Y, yeasts; M, moulds.

(IAMS) for ideal preservatives and parabens is given in **Table 1**.

Parabens were synthesized in order to replace benzoic and salicylic acids, which were limited to use within the acid pH range. The esters permitted for use in the US are methyl, ethyl, propyl and heptyl. They are available as ivory to white free-flowing powders, which are relatively non-hygroscopic and non-volatile compounds. The esters are stable in air and resistant to cold, heat and steam sterilization. Solutions of parabens at pH 3, 6 and 8 remain unchanged during storage at 25°C for 25 weeks, and at pH 3 and 6 when heated for 30 min at 120°C. With the exception of methyl ester, all parabens are odourless. Methyl ester has a faint characteristic odour.

When the number of carbon atoms in the ester chain is increased, their solubility in water decreases, while their solubility in oil, ethanol and propylene glycol increases. Some general properties of parabens are listed in **Table 2**.

Range of Foods to which Parabens may be Added

In order to take advantage of their solubility and antimicrobial profile, these parabens are generally used in combination at 0.05–0.10%. Common applications include the use of methyl and propyl parabens in the ratio of 2–3 : 1 in various food products. Applications have been used or tested in bakery products, cheeses, soft drinks, beer, wines, jams, jellies, preserves, pickles, olives, syrups and fish products.

A 3 : 1 combination of methyl and propyl paraben at 0.03–0.06% may be used to increase the shelf life of fruit cakes, non-yeast pastries, icings and toppings. A 2 : 1 combination of the same esters may be used in soft drinks and for marinated, smoked or jellied fish products (0.03–0.06%), flavour extracts (0.05–0.1%), for preservation of fruit salads, juice drinks, sauces and fillings (0.05%), in jams and jellies (0.07%), salad dressings (0.1–0.13%) and in wines (0.1%).

Parabens are effective at both acidic and alkaline pH. The pH range for antimicrobial activity of parabens is 3–8, compared to 2.5–4.0 for benzoate. Parabens are particularly useful in high-pH foods where other antimicrobials are rendered ineffective. This can be seen from **Table 3** which shows the minimum inhibitory concentration of parabens and other food additives against four types of moulds at pH 5 and 9. It is believed that parabens exert their antimicrobial action in the undissociated form; benzoic acid also operates in this way. Esterification of the carboxyl group retains the undissociated form of the parabens over a wide pH range. The weaker phenolic group provides the acidity rather than the carboxyl group, hence salt formations involve reactions with the phenolic hydroxyl group.

Parabens may be used in combination with benzoate, especially in foods which have slightly acidic pH values. Evidence of additive antimicrobial effects of parabens and benzoates is available. The compounds can be incorporated into foods by dissolving in water, ethanol, propylene glycol or the food product itself. Dry blending with other water-soluble food ingredients can also be done before addition to foods. Dissolution in water is generally carried out at room temperature, and if required, at 70–82°C. In ethanol or propylene glycol, a 20% stock solution of the parabens is prepared and then used in foods. The high cost of parabens limits their applications in food products. While some investigators believe parabens to have a definite taste at the concentrations used, other sources disagree.

The regulatory status in the US of different parabens is given in **Table 4**. Methyl and propyl parabens are permitted as antimycotics in food packaging materials. In the UK, methyl, ethyl and propyl parabens are permitted in food in accordance with the Preservatives in Food Regulations 1979. **Table 5** lists the maximum levels (mg kg^{-1}) in different foods according to these regulations, while **Table 6** gives the maximum permitted levels of parabens in foods in other selected countries. Many countries, including Japan, also permit the use of butyl ester in foods.

Table 2 Properties of parabens

Property	Prarben			
	Methyl	*Ethyl*	*Propyl*	*Butyl*
pK	8.47		8.47	
Molecular formula	$C_8H_8O_3$	$C_9H_{10}O_3$	$C_{10}H_{12}O_3$	$C_{11}H_{14}O_3$
Molecular weight	152.14	166.17	180.21	194.23
Melting point (°C)	131	116	96–97	68–69
Solubility (g per 100 g)				
in water at: 10°C	0.20	0.07	0.025	0.005
25°C	0.25	0.17	0.05	0.02
80°C	2.00	0.86	0.30	0.15
in ethanol at 25°C	52	70	95	210
in propylene glycol at 25°C	22	25	26	110
in olive oil at 25°C	2.9	3.0	5.2	9.9
in peanut oil at 25°C	0.5	1.0	1.4	5.0
Structure	$COOCH_3$ / OH	$COOC_2H_5$ / OH	$COOC_3H_7$ / OH	$COOC_4H_9$ / OH
LD_{50} (mg kg^{-1})[a]	2000	2500	3700	950

[a] The values are for the sodium salts of the corresponding ester.

Table 3 Minimum inhibitory concentration of parabens compared to other preservatives at pH 5 and 9

Minimum inhibitory concentration (%)	Chaelomonium globosum		Alternaria solani		Penicillium citrinum		Aspergillus niger	
	pH 5	*pH 9*	*pH 5*	*pH 9*	*pH 5*	*pH 9*	*pH 5*	*pH 9*
Benzoic acid	0.10	+	0.15	+	0.20	+	0.20	+
Methyl paraben	0.06	0.10	0.08	0.10	0.08	0.15	0.10	0.15
Propyl paraben	0.01	0.04	0.02	0.05	0.01	0.06	0.03	0.05
Propionic acid	0.04	+	0.06	+	0.08	+	0.08	+
Sorbic acid	0.06	+	0.02	+	0.08	+	0.08	+

Behaviour and Antimicrobial Action of Different Forms in Foods

Microbial inhibition due to parabens increases with an increase in the alkyl chain length. Branched chain esters have a low antimicrobial activity. **Table 7** shows the antimicrobial spectrum of esters of *p*-hydroxybenzoic necessary for total inhibition against selected organisms. Variations in the minimum inhibitory concentration spectrum for the same organism have been reported by different workers. This is due to the different strains of the organism, different incubation conditions with respect to time, temperature and pH, variations in media, assay techniques and in the analysis of results. Very little research is available on the activity of *n*-heptyl ester in foods, although it is known to be effective in inhibiting bacteria involved in malolactic fermentation of wines.

Table 4 Regulatory status of different parabens in the US

Compound	*Regulation*	*Limitation*
Methyl paraben (FEMA no. 2710)	FDA §182.1	GRAS, chemical preservative, up to 0.1%
	FDA §172.515	Synthetic flavour
Propyl paraben (FEMA no. 2951)	Same as above	Same as above
Butyl paraben (FEMA no. 2203)	FDA §172.515	Synthetic flavour
n-Heptyl paraben	FDA $121.1186	In fermented malt beverages to inhibit microbiological spoilage, 12 p.p.m. maximum

FEMA, Flavour and Extracts Manufacturers' Association; FDA, Food and Drug Administration; GRAS, generally regarded as safe.

Yeasts and moulds are more sensitive to parabens than bacteria. They are more effective against Gram-positive bacteria than Gram-negative species. Genera of microorganisms inhibited by parabens include *Alternaria*, *Aspergillus*, *Penicillium*, *Rhizopus*, *Saccharomyces*, *Bacillus*, *Staphylococcus*, *Streptococcus*, *Clostridium*, *Pseudomonas*, *Salmonella* and *Vibrio*. In combination with heating parabens are reportedly effective against *Salmonella* and yeast. Some organisms are not inhibited by parabens. For instance, *Alcaligenes viscolactis* in skimmed milk is not inhibited even by 600 p.p.m. of propyl paraben. Similarly, erratic behaviour is seen with some organisms. For example, 4000 p.p.m. of propyl paraben inhibits growth of *Pseudomonas fragi*, but 2000 p.p.m. actually stimulated the growth of the same organism.

Methyl and propyl parabens have also been shown to inhibit toxin formation by *Clostridium botulinum* at 100 μg ml^{-1}. The concentration for inhibiting toxin formation is much lower than the concentration of 1200 p.p.m. methyl and 200 p.p.m. propyl required for growth inhibition. Ethyl paraben is effective in inhibiting botulinal toxin in canned comminuted pork. The actual inhibition of *C. botulinum* is much lower in actual foods as compared to laboratory media. Propyl paraben is known to inhibit protease secretion by *Aeromonas hydrophila* at 200 μg ml^{-1}. Parabens are not very effective as replacements of nitrite in cured meats.

The antimicrobial action of parabens has been attributed to interference with nutrient transport functions, inhibition of germination of the bacterial spore, respiration, protease secretion, and DNA, RNA and protein synthesis. Parabens inhibit both membrane transport and the electron transport system.

Table 5 Foods permitted to contain parabens[a] according to the Preservatives in Food Regulations 1979

Food	*Maximum levels (mg kg^{-1})*
Beer	270
Beetroot, cooked and prepacked	250
Chicory and coffee essence	450
Colouring matter, except E150 caramel, if in the form of a solution of a permitted colouring matter	2000
Dessert sauces, fruit-based with a total soluble solids content less than 75%	250
The permitted miscellaneous additive dimethylpolysiloxane	2000
Aqueous solutions of enzyme preparations not otherwise specified, including immobilized enzyme preparations in aqueous media	3000
Flavourings or flavouring syrups	800
Freeze drinks	160
Fruit-based pie fillings	800
Fruit, crystallized, glacé or drained	1000
Fruit (other than fresh fruit) or fruit pulp, including tomato pulp, paste or purée	800
Glucose drinks containing not less than 23.5 lb of glucose syrup per 10 gallons of the drink	800
Grape juice products (unfermented, intended for sacramental use)	2000
Herring and mackerel, marinated, whose pH does not exceed 4.5	1000
Horseradish, fresh grated and horseradish sauce	250
Jam, diabetic	500
Olives, pickled	250
Pickles other than pickled olives	250
Preparations of permitted artificial sweeteners and water only	250
Salad cream, inlcuding mayonnaise and salad dressing	250
Sauces other than horseradish sauce	250
Soft drinks for consumption after dilution not otherwise specified in this schedule	800
Soft drinks for consumption without dilution not otherwise specified in this schedule	160
Tea extract, liquid	450
Yoghurt, fruit	120

[a] Parabens, methyl-, ethyl- and propyl-4-hydrozybenzoate and their sodium salts.

Inhibition of ATP production in the presence of parabens has been demonstrated with *Bacillus subtilis*.

Metabolism and Toxicology of Parabens

Parabens have an acute toxicity of low order. The oral LD_{50} value in mice for methyl and propyl esters in propylene glycol is greater than 8000 mg kg^{-1}.

The parabens are believed to be absorbed in the

Table 6 Maximum permitted levels (mg kg^{-1}) for parabens in foods in selected countries

Food	*Belgium*	*Denmark*	*Germany*	*Italy*	*Norway*	*Sweden*	*Canada*
Fish semipreserves	1000	300	1000	1000	500	500	1000
Fruit juice		200[a]				1000	1000
Fruit pulp		+				1000	+
Jam		300			900	1000	1000
Mayonnaise		300	1200	1000[b]			
Mustard, prepared		300	1500		900	1000	1000
Pickles		300			900	1000	1000
Sauces, spiced	1000[c]	300	1500		900	1000	1000
Soft drinks containing fruit juices		200			900	1000	1000
Soft drinks, flavoured, usually carbonated		200				1000	1000

[a] Not for direct consumption.
[b] Of the fat content.
[c] pH more than 5.
+ = No maximum permitted level.

Table 7 Antimicrobial spectrum of esters of *p*-hydroxybenzoic acid necessary for total inhibition against various organisms

Microorganisms	*Concentration (p.p.m.)*				
	Methyl	*Ethyl*	*Propyl*	*Butyl*	*Heptyl*
Bacteria					
Aeromonas hydrophila			>200		12
Bacillus cereus	2000	100–1000	10–400	63–400	
Bacillus megaterium	1000		320	100	
Bacillus subtilis	2000	1000	250	63–400	
Clostridium botulinum	1000–1200		200–400		
Enterobacter aerogenes	2000	1000	1000	4000	
Escherichia coli	2000	12–1000	30–1000	4000	
Klebsiella pneumoniae	1000	500	250	125	
Lactococcus lactis			400		12
Listeria monocytogenes	>512		512		
Proteus vulgaris	2000	1000	500	500	
Pseudomonas aeruginosa	4000	4000	8000	8000	
Pseudomonas fluorescens	1310		670		
Salmonella typhosa	2000	1000	1000	1000	
Salmonella typhimurium			>300		
Sarcina lutea	4000	1000	500	125	
Serratia marcescens	800	490	400	190	
Staphylococcus aureus	4000	1000	350–500	125–200	12
Vibrio parahaemolyticus			50–100		
Yeasts and moulds					
Alternaria spp.			100		50–100
Aspergillus niger	1000	50–500	10–250	125–200	
Aspergillus flavus			200		
Byssochlamys fulva			200		
Candida albicans	1000	500–1000	125–250	125	
Penicillium digitata	500	250	63	<32	
Penicillium chrysogenum	500	250	125–200	63	
Saccharomyces bayanas	930		220		
Saccharomyces cerevisiae	1000	80–500	40–200	200	100
Torula utilis			200		25

intestines and then travel to the liver and the kidney, where these are hydrolysed to *p*-hydroxybenzoic acid. This is excreted in urine unchanged or as *p*-hydroxyhippuric acid, glucuronic acid esters, or sulphates within 24 h. Parabens have been observed to have a local anaesthetic effect which increases with the increasing number of carbon atoms in the alkyl group. Ethyl and propyl parabens at 0.05% cause a local anaesthetic effect on buccal mucosa, while 0.1% methyl paraben has a similar effect as 0.05% procaine solution. Contradictory reports are available on the effect of parabens on the skin. While some reports suggest that parabens in foods cause dermatitis of unknown aetiology, other reports indicate no skin irritation, even at concentrations as high as 5%.

Assay Techniques for Parabens

Various techniques are available for quantitative and qualitative analyses of parabens. Thin-layer chromatography (TLC) on kieselguhr-silica plates using hexane–acetic acid solvent system is a simple qualitative technique. In an assay procedure accepted by Food and Agriculture Organization (FAO), the compounds are separated from an acidified food system using steam distillation. This is then followed by solvent extraction using diethyl ether. The extract is spotted on the TLC plates and separated with a 90 : 16 : 8 solvent system of toluene : methanol : acetic acid. After development, the plates are observed under ultraviolet light or Denige's reagent (mercuric oxide–H_2SO_4). Reversed-phase TLC on silanized silica gel using ether- or ethylacetate-saturated borate at pH 11 has been successfully used for the separation of methyl, ethyl, propyl and benzyl esters.

Millons reagent has also been used to detect the presence of parabens and free acids in foods. This reagent gives a rose red colour with neutral ammonium salt of *p*-hydroxybenzoic acid; hence, it is necessary to hydrolyse the esters with alcoholic potash before carrying out the tests. Another technique involves extraction of the compounds by the above method followed by saponification and spectrophotometric determination of *p*-hydroxybenzoic acid at 255 nm. Reaction of aminoantipyrine with *p*-hydroxybenzoic acid has also been claimed to be specific for qualitative detection.

Bromination of the samples enables separation of benzoic acid and parabens on the TLC plate. Benzoic acid does not brominate, while parabens give two spots after development.

Gas chromatography of the trimethylsilyl and silyl derivatives of parabens has shown good recovery rates of 92–100% from various foods such as ketchup, salad mixes, salad dressings, pickles and in fat-containing foods. High-pressure liquid chromatography (HPLC) techniques for determination of parabens are also available.

Conversion of the parabens into their 3-nitro derivatives and 2,4-dinitrophenyl esters followed by polarographic estimation has given less than 4% coefficient of variation in replicate determinations.

With the exception of the TLC method, no other collaborative studies have been recorded.

See also: ***Aeromonas***: Introduction. ***Alcaligenes***. ***Bacillus***: *Bacillus subtilis*. ***Clostridium***: *Clostridium botulinum*.

Further Reading

Branen AL and Davidson PM (ed.) (1983) *Antimicrobials in Foods*. New York: Marcel Dekker.
Branen AL, Davidson PM and Salminen S (eds) (1990) *Food Additives*. New York: Marcel Dekker.
Furia TE (ed.) (1972) *Handbook of Food Additives*, 2nd edn. Cleveland, OH: CRC Press.
Furia TE (ed.) (1980) *Regulatory Status of Direct Food Additives*. Boca Raton, FL: CRC Press.
Lewis RJ Sr (1989) *Food Additives Handbook*. New York: Van Nostrand Reinhold.
Maga JA and Tu AT (eds) (1995) *Food Additive Toxicology* New York: Marcel Dekker.
Tilbury RH (ed.) (1980) *Developments in Food Preservatives 1*. London: Applied Science Publishers.

Permitted Preservatives – Nitrate and Nitrite

Rekha S Singhal and **Pushpa R Kulkarni**, Department of Chemical Technology, University of Mumbai, India

Nitrites are used in a number of types of foods, principally in curing of meat, to deliver colour and flavour and to inhibit growth and toxin production by *Clostridium botulinum*. Nitrates, although ineffective as such, are reduced to nitrite by the microflora in some foods and in vivo by the gut microflora. Nitrite salts have been used in meat curing for many centuries as a part of the curing mixture. Their effectiveness was recognized by the Romans who used salt containing saltpetre or nitrite for curing and pickling meats including pork. Its antimicrobial role was, however, recognized only in the 1940s. Prior to this period, nitrate was thought to be the actual antimicrobial agent. Later, when the concept of nitrite being an antimicrobial agent was firmly established, nitrate was relegated to serving merely as a reservoir for generation of nitrite. The confusion that arose was due to ignorance about the other factors that affect the microbial profile of nitrite, especially the pH.

Some of the physical properties of nitrites and nitrates are given in **Table 1**. Both nitrites and nitrates present a fire hazard. When mixed with organic matter, these salts ignite with friction and at high temperatures may even explode. However, commercial curing mixes consist of mostly sodium chloride with only a small percentage of nitrites and nitrates, so this hazard is minimized.

Table 1 Properties of sodium and potassium nitrites and nitrates

Property	*Sodium nitrite*	*Sodium nitrate*	*Potassium nitrite*	*Potassium nitrate*
Molecular formula	$NaNO_2$	$NaNO_3$	KNO_2	KNO_3
Molecular weight	69.00	85.00	85.11	101.11
Colour	Slightly yellowish or white crystals, sticks or powder	Colourless transparent crystals	White or slightly yellowish deliquescent prisms or sticks	Transparent, colourless or white crystalline powder
Odour	–	Odourless	–	Odourless
Taste	Slightly salty	Saline, slightly bitter	–	Cooling, pungent, salty
Melting point (°C)	271	306.8	387	334
Boiling point (°C)	320 (decomposes)	380 (decomposes)	decomposes	400 (decomposes)
Density at 16°C	2.168	2.261	1.915	2.109
Solubility	Delinquescent in air, soluble in water; slightly soluble in alcohol	Deliquescent in moist air, soluble in water; slightly soluble in alcohol	Very soluble in water; slightly soluble in alcohol	Soluble in glycerol, water; moderately soluble in alcohol

Foods to Which Nitrate and Nitrite May Be Added

Nitrite in cured meats is known to be a multi-functional food additive. It produces a characteristic colour by reacting with the haem pigments in the muscle, contributes to flavour, inhibits *Clostridium botulinum* and the production of neurotoxin, and also retards rancidity by functioning as an antioxidant (this effect is brought about by reduction of the active ferric state to the ferrous state, which is inactive in lipid oxidation). Meat products, including bacon, bologna, frankfurters, corned beef, ham, various types of sausages and canned cured meats, benefit from the addition of nitrite. Nitrite may also be used in fish and poultry products, and in cheeses to prevent spoilage (very late fermentation and gas production) by *Clostridium tyrobutyricum* or *C. butyricum*.

The regulatory status of the sodium and potassium salts of nitrite and nitrate in the USA is given in **Table 2**. Maximum permitted levels of nitrites and nitrates in cured meat in various countries are given in **Table 3**. In bacon, rules for nitrite input were formed in 1986, and accordingly either of the following is allowed:

- 120 p.p.m. sodium nitrite (or 123 p.p.m. potassium nitrite) + 550 p.p.m. sodium erythorbate
- 40–80 p.p.m. sodium nitrite (49–99 p.p.m. potassium nitrite) + 550 p.p.m. sodium erythorbate + 0.7% sucrose + a lactic acid bacterial culture of *Pediococcus*.

Behaviour of Salts in Food

Ionic nitrite, although by itself unreactive in aqueous solutions, can be converted to a powerful nitrosating agent, N_2O_3, via nitrous acid under mildly acidic conditions (**Equations 1 and 2**). Additional nitrosating agents are produced in the presence of certain anions (Y) such as chloride and thiocyanate (**Equation 3**). These nitrosating agents are known to react with a wide range of nucleophilic substrates to form the corresponding nitroso derivatives (**Equation 4**).

$$NO_2^- + H_3O^+ \longrightarrow HNO_2 + H_2O \quad \text{(Equation 1)}$$

$$2HNO_2 \longrightarrow N_2O_3 + H_2O \quad \text{(Equation 2)}$$

$$HNO_2 + H^+ + Y^- \longrightarrow NOY + H_2O \quad \text{(Equation 3)}$$

$$\text{Substrate} + N_2O_3 \longrightarrow \text{Substrate-NO} \quad \text{(Equation 4)}$$

In cured meats, the levels of measurable nitrite decrease rapidly during curing and continue to decrease slowly during storage. Nitrite, being highly reactive, is reduced to nitric oxide which in turn binds to many components of the meat such as proteins, lipids and carbohydrates. It is possible that these chemical reactions which occur with meat may also occur with microbial cells.

Reaction with Proteins

The sites on which nitrosation reactions occur on proteins are as follows:

1. N-terminal amino or imino residues (**Equation 5**).

$$-CHRNH_2 + N_2O_3 \longrightarrow CHROH + HNO_2 + N_2 \quad \text{(Equation 5)}$$

2. Peptide linkages – *N*-nitrosated peptide bonds have been reported in cured meats, but are comparatively unstable and hence unlikely to persist. Hydrolysis of the *N*-nitrosated peptide bonds result in formation of N-terminal diazohydroxides which decompose rapidly with the elimination of nitrogen. This reaction is of significance in the formation of protein bound residues of *N*-nitrosoproline and *N*-nitrosohydroxyproline.

Table 2 Regulatory status of nitrite and nitrate in the USA

Compound	*Regulation*	*Limitation*
Potassium nitrite	MID[a]	To fix colour in cured products: 2 lb/100 gal pickle at 10% pump level; 1 oz/100 lb meat (dry cure); $\frac{1}{4}$ oz/100 lb chopped meat and/or meat by-product; 200 p.p.m. max. total nitrite in finished product; product specifications apply
Potassium nitrate	FDA §121.1132	Curing agent in processing of cod roe: 200 p.p.m. max.
	MID	A source of nitrite in cured products: 7 lb/100 gal pickle; $3\frac{1}{2}$ oz/100 lb meat (dry cure); $2\frac{3}{4}$ oz/100 lb chopped meat
Sodium nitrite	FDA §121.223	(1) 20 p.p.m. max. alone as preservative and colour fixative in canned pet food containing fish, meat and fish or meat by-products (2) In meat curing preparations for home curing of meat and meat products (including poultry and wild game); with sodium nitrate; directions must limit amount of sodium nitrite to 500 p.p.m. max. and sodium nitrite to 200 p.p.m. max.; product specifications apply
	MID	To fix colour in cured products: 2 lb/100 gal pickle at 10% pump level; 1 oz/100 lb chopped meat and/or meat by-product; 200 p.p.m. max. of nitrite in finished product, product specifications apply
Sodium nitrate	FDA §121.1063	(1) Preservative and colour fixative, with or without sodium nitrite; 500 p.p.m. max. in smoked and cured sablefish, salmon and chad; sodium nitrite in finished product 200 p.p.m. max. (2) In meat curing preparations for home curing of meat and meat products (including poultry and wild game): directions must limit amount of sodium nitrate to 500 p.p.m. max. and sodium nitrite to 200 p.p.m. max. in finished meat product; product specifications apply
	MID	Source of nitrite in cured products: 7 lb/100 gal pickle; 3.5 oz/100 lb meat (dry cure); 2.75 oz/10 lb chopped meat

[a]MID is the Meat Inspection Division of the US Department of Agriculture, which is responsible for clearing additives intended for use in all meats and meat products except poultry.

Table 3 Maximum permitted levels of nitrites and nitrates in cured meat in various countries

Country	*Nitrite (mg kg^{-1})*	*Nitrate (mg kg^{-1})*
Belgium	200[a]	
Denmark	75[b]	500[c]
France	150	None
Germany	None[a]	500
Italy	150	250
Netherlands	500	2000
UK	200	500
Norway	60[a,c]	
Sweden	200[a]	
Canada	200	200

[a]Must be mixed with NaCl before use.
[b]25 mg kg^{-1} in fully preserved products.
[c]Certain products only.

3. Amino acid residue side chains, which could be:

- *C*-nitrosation of tyrosine – presence confirmed by results obtained on enzyme hydrolysis
- *N*-nitrosation of the indole ring in tyrosine – unlikely to be present in cured meats owing to its unstable nature
- *S*-nitrosation of the thiol group of cysteine – this has been implicated as the major factor in the depletion of free nitrite in cured meats, accounting for 25% of the added nitrite in cured meats. The *S*-nitrocysteine formed is much less effective than nitrite in its anticlostridial action.

Reaction with Other Constituents

Examples of reactions of nitrite with other organic constituents are as follows:

1. Nitrite bound to lipids results from the formation of pseudonitrosiles (**Equation 6**).
2. With carbohydrates, reversible formation of nitrite esters results (**Equation 7**).
3. With other curing agents such as ascorbic acid, nitrite reacts rapidly and produces dehydroascorbic acid and nitric oxide; with sorbic acid it forms a number of products including ethyl nitrolic acid.
4. With the phenols generated during smoke curing, nitrite forms *C*-nitroso derivatives which being unstable are converted to *C*-nitro derivatives. A series of such *C*-nitroalkylphenols have been identified in bacon.
5. *C*-nitrosation of activated methylene groups such as that found in 3-deoxysuloses (formed as intermediates of ascorbic acid and Maillard browning reactions) are known to form oximes.
6. With alcohols, nitrite forms alkyl nitrites.
7. With thiols, nitrite forms thionitrites.
8. With primary amines, nitrite forms alcohols and unsaturated derivatives.

Table 4 Relationship between IAMS criteria for the ideal preservative and nitrite

Criterion	*Compliance of nitrite with the criterion*
Toxicological acceptability	?
Microbiological activity	–
Long persistence in foods	–
Chemical reactivity	High
High thermal stability	Not relevant
Wide antimicrobial spectrum	Gram-positive bacteria
Active against food-borne pathogens	+
No development of microbial resistance	+
Not used therapeutically	+
Assay procedure available	+

+, complies with criterion; –, does not comply; ?, debatable.

$$-CH{=}CH- + N_2O_3 \rightarrow -CHNO{-}CHNO_{2-} \quad \text{(Equation 6)}$$

$$ROH + N_2O_3 \longrightarrow RONO + HNO_2 \quad \text{(Equation 7)}$$

About 5–15% of the nitrite added to meat reacts with myoglobin; 1–10% is oxidized to nitrate; 5–10% remains as free nitrite; 1–5% becomes nitric oxide gas; 5–15% is bound to sulphydryl groups, 1–5% to lipids and 20–30% to proteins.

Analytical procedures for the analysis of such reaction products, and detailed knowledge of individual reactions and their toxicological concerns, have yet to be established.

Antimicrobial Action of Nitrite

Compliance of nitrite with the criteria laid down by the International Association of Microbiological Societies (IAMS) for an ideal preservative is summarized in **Table 4**.

It has long been recognized that nitrate at concentrations of up to 22 000 $\mu g\,g^{-1}$ cannot be relied upon to inhibit putrefactive anaerobes if other conditions are favourable for their growth. By the 1960s it was increasingly clear that nitrate, per se, had no antimicrobial properties. Sodium nitrite has been primarily used as an antimicrobial agent to inhibit the growth and toxin production by *Clostridium botulinum* in cured meats in association with other components of the curing mixture such as salt, ascorbate and erythorbate. Other factors that influence the antibotulinal efficacy of nitrite are the thermal process applied to the product, the growth of competitive flora, the level and type of phosphate that may be included in the formulation, and the temperature of abuse. This was made the basis of development of risk factors and is based on a notional value of 1 (unity) for pasteurized ham containing 3.5% salt + 200 p.p.m. nitrite at pH 6.0, with a process time of 6.7 min at 80°C, and stored for 6 months at 15°C. These risk factors can be limited or enhanced depending on the preservative interactions with other parameters. This is illustrated in **Table 5** which gives data for pasteurized ham.

Table 5 Risk factors for changes in the levels of nitrite and other components of the curing mixture, singly and in combination, in pasteurized ham

Components of the curing mixture	*Risk factors*
Components when used alone	
% Salt at (i) 2%, (ii) 3.5%, (iii) 4.5%, (iv) 5.5%	(i) 4, (ii) 1, (iii) 0.5, (iv) 0.1
Nitrite (mg kg^{-1}) at (i) 50 mg, (ii) 100 mg, (iii) 200 mg, (iv) 300 mg	(i) 10, (ii) 3, (iii) 1, (iv) 0.5
% Polyphosphate at (i) 0%, (ii) 0.3%	(i) 1, (ii) 0.2–1.0 depending on salt, pH and type of phosphate
pH value of (i) 5.7, (ii) 6.0, (iii) 6.4	(i) 0.3, (ii) 1, (iii) 4.0
Process time at 80°C for (i) 0.7 min, (ii) 6.7 min, (iii) 12.7 min	(i) 3, (ii) 1, (iii) 0.9
Storage temperature at (i) 15°C, (ii) 20°C, (iii) 25°C	(i) 1, (ii) 5, (iii) 7
Components in combination	
3.5% salt + 200 p.p.m. nitrite + processing time at 80°C for 6.7 min + storage at 15°C + pH 6.0	1
2% salt + 50 p.p.m. nitrite + storage at 20°C + pH 6.4	800
4.5% salt + 100 p.p.m. nitrite + storage at 20°C + processing time of 0.7 min at 80°C	23
4.5% salt + 100 p.p.m. nitrite + storage at 20°C + processing time of 0.7 min at 80°C + 0.3% polyphosphate	5

Botulinal toxin production can be inhibited by nitrite levels above 50 $\mu g\,g^{-1}$. Nitrite does not inhibit spore germination significantly. Nitrite is more inhibitory at an acidic pH where it exists as the undissociated nitrous acid, and under anaerobic rather than aerobic conditions. The disappearance of nitrite in meat or culture is related to pH and temperature and is given by an exponential equation as follows: $\log_{10}$ (half-life of nitrite) = 6.65 − 0.025 × temperature (in °C) + 35 × pH.

Bacterial growth and cell division of spore-formers are inhibited in the presence of nitrite. If the spore concentration is high, the inhibitory effect of nitrite and other curing adjuncts can be eventually overcome, leading to the formation of toxin within a short time.

Table 6 Conditions of inhibition of various organisms in the presence of nitrite

Organism	*Conditions of inhibition*
Clostridium perfringens	200 μg ml^{-1} nitrite + 3% salt (20°C, pH 6.2) 50 μg ml^{-1} nitrite + 4% salt (20°C, pH 6.2)
Salmonella	400 μg ml^{-1} nitrite + 4% salt (10–15°C, pH 5.6–6.2)
Escherichia coli	400 μg ml^{-1} nitrite + 6% salt (10°C, pH 5.6)
Achromobacter, Enterobacter, Escherichia, Flavobacterium, Micrococcus and *Pseudomonas*	200 μg g^{-1} and pH 6.0
Lactobacillus and *Bacillus*	Resistant to nitrite
Staphylococcus aureus	200 p.p.m. nitrite at pH 5.6 or less (inhibitory) 1.0% nitrite and pH 5.6 or less (bactericidal)

Temperature is another important parameter influencing production of clostridial toxin. A minimum of 10°C has been suggested in cured meats. Even in irradiated meat, limited amounts of nitrite are necessary to obtain an acceptable product (irradiation is not allowed in meat products in many countries). The antimicrobial effectiveness of nitrite alone and in combination with other ingredients is shown in **Table 6**.

Nitrite burn is a form of discoloration observed in cured meats and is believed to be due to a combination of excessive nitrite and reduced pH. Micrococci near the periphery of cured meat sausages or *Staphylococcus* spp. within the sausage convert nitrate to nitrite and may bring about this effect.

Heated nitrite-containing bacteriological media are more inhibitory to *Clostridium botulinum* than media heated prior to the addition of the preservative. This effect is due to substances called 'Perigo inhibitor', and has been modelled using mixtures of cysteine, nitrite and ferrous salts. The iron-sulphur bridge compounds which form on heating of such mixtures are believed to be effective inhibitors of clostridial spores. This is normally formed above 105°C which is higher than the temperatures used in the processing of cured meats. Perigo inhibitor is also effective against some *Bacillus* strains and *Enterococcus durans*. Certain organisms such as *Enterococcus faecalis* and *Salmonella* are known to be resistant to this factor. The antibacterial activity of Perigo inhibitor is neutralized by meat particles.

Mechanism of Action of Nitrite

The antimicrobial target of nitrite for the inhibition of spores has been recognized to be iron-containing cofactors or enzymes. Nitrite ion exerts its action on the phosphoroclastic system of enzymes that causes the conversion of pyruvate to acetate. Nitrite causes the ATP concentration in the microbial cell to decrease rapidly and leads to the excretion of pyruvate. The ion exists in equilibrium with non-ionic forms, including NO; the NO works as a ligand for non-haem iron, as well as haem iron of cytochrome oxidase in aerobic bacteria, and brings about inhibition. Inhibition of the iron-sulphur enzyme, ferredoxin and/or pyruvate ferrodoxin oxidoreductase, in *Clostridium botulinum* and *C. pasteurianum* is the ultimate mechanism of this antimicrobial action. Chelating agents such EDTA, erythorbates, sodium ascorbate and polyphosphates enhance the antibotulinal efficacy of nitrite by sequestering the iron. At high nitrite concentrations, S-nitrosation of sulphydryl enzymes are also a possible cause of inhibition.

Importance of Species/Strain Tolerance

Considerable variation is known to occur between proteolytic and non-proteolytic strains of *Clostridium botulinum* with respect to physical factors of inhibition such as temperature, water activity and acidity (pH). However, such variation in response to nitrite is not clearly documented. Since physical factors do influence the antibotulinal activity of nitrite, it can be said that strains differ indirectly in their tolerance to nitrite.

Toxicology

Toxicologically, nitrites and nitrates are considered together owing to the ease of conversion of nitrate to nitrite by organisms of the intestinal microflora. Estimates of acceptable daily intakes of sodium nitrate and sodium nitrite are, respectively, 0–5 mg and 0–0.2 mg per kilogram of body weight. The lethal dose of nitrite has been reported to be 300 mg kg^{-1} body weight. Nitrite shows acute toxic effects resulting from reduced oxygen transport by the bloodstream, which is mainly due to conversion of haemoglobin to methaemoglobin. Infants less than 6 months old are highly susceptible, with fatal cases of poisoning following consumption of vegetables or water containing 100–500 p.p.m. of nitrite. The possibility of nitrite itself being carcinogenic has been raised, but the lack of any dose–response relationship and the influence of nitrite toxicity at high dose levels complicated the safety evaluation of nitrite. Nitrite has also been reported to be a hypotensive agent and disruptor of thyroid function.

The toxicological safety of nitrite received considerable attention after the nitrosamine issue developed in the late 1960s.

Interaction of Nitrite with Other Preservatives

The joint effect of the cocktail of additives that is often consumed daily poses possible health risks. Nitrite is frequently used in combination with other additives, and an understanding of the chemistry of its interactions is therefore essential.

Nitrite is rarely used with sulphur dioxide, since the two react to form sulphonates of either hydroxylamine or ammonia, resulting in the loss of preservative action of both the additives. This has, however, been suggested as a method to remove excess nitrite in a patented process.

Reaction of ascorbic acid with nitrous acid has been considered as a means of preventing nitrosation in cured meats. This is due to the fact that the two compounds react to form dinitrosyl ascorbate, which in turn breaks down to dehydroascorbic acid and nitric oxide. The nitric acid so formed reacts with atmospheric oxygen and water to produce a mixture of nitric and nitrous acids. Since nitrate is relatively unreactive, this reaction works as a sink to remove excess nitrous acid from the system. Reaction between ascorbic acid and nitrous acid is also effective in eliminating the formation of *N*-nitrosopyrrolidine, a carcinogenic nitrosamine that can be generated in bacon during frying. Since the formation of nitrosamine takes place in adipose tissue, fat-soluble ascorbyl palmitate is more effective than ascorbic acid itself. Degradation products of ascorbic acid are also known to give as yet unidentified reaction products with nitrite.

Sodium chloride, a normal component of the curing mixture, facilitates nitrosamine formation through the formation of nitrosyl chloride, which reacts much more rapidly than nitrous acid.

Sorbic acid has been suggested as a partial replacement for nitrite in meat curing since it inhibits the growth of *Clostridium botulinum* and also reduces the formation of nitrosamines. However, this practice may lead to the formation of mutagenic reaction products: two such products, 1,4-dinitro-2-methylpyrrole (DNMP) and ethyl nitrolic acid, have already been identified, and can be formed in a few hours at 60°C. However, if ascorbic acid is used in conjunction, mutagenicity is completely abolished owing to the conversion of the C-4 nitro group in DNMP to a C-4 amino group.

Lecithin is used in meat processing as an emulsifying agent and as an anti-sticking agent. It is a source of choline which decomposes to trimethylamine and further to dimethylamine. The dimethylamine so formed can react with nitrite to form carcinogenic dimethylnitrosamine.

The toxicological threat posed by such interactions has not yet been completely established, and how far such interactions are relevant to real foods requires extensive investigation.

Transformation of Nitrite Nitrosamines

Nitrite interacts with various nitrogenous constituents of food, particularly amines and amides, to form highly potent carcinogenic nitrosamines and nitrosamides. The possibility of foods being contaminated with nitrosamines was realized in Norway in late 1950s when it was found that domestic animals fed nitrite-preserved fish died from severe liver disorders. The compound *N*-nitrosodimethylamine was later isolated and confirmed to be the causative factor. The *N*-nitrosamides decompose rapidly above pH 7.0.

Over 70% of the *N*-nitroso compounds tested so far have proved to be carcinogenic in animal feeding studies. The principal nitrosamines formed in foods include *N*-nitrosodimethylamine, *N*-nitrosodiethylamine, *N*-nitrosopiperidine and *N*-nitrosopyrrolidine. The *N*-nitrosation of a dialkylamine is as shown in **Equation 8**.

$$R_2NH + N_2O_3 \longrightarrow R_2N\text{-}NO + HNO_2 \quad \text{(Equation 8)}$$

The reaction rate is at a maximum in the pH range 2.25–3.4, and is catalysed by anions such as halides, acetates, phthalates, sulphur compounds, weak acids and thiocyanates. Thiocyanate is 15 000 times more catalytic than chloride. Nitrosation can also take place under basic conditions in the presence of carboxyl groups such as formaldehyde. Certain carbonyl-containing compounds can also work as catalysts for nitrosation, while simple and polyphenolic compounds may work as inhibitors or catalysts depending on their structures. Formation of *N*-nitrosamines during frying is oxygen-dependent; it is inhibited to the extent of 90% if oxygen is excluded from the environment. Processing conditions also affect the level of nitrosamine in foods. Higher temperatures for shorter times result in more *N*-nitrosamine formation than lower temperatures for longer times. Microwave processing gives undetected levels of nitrosamines.

Several amines present in animal tissues such as *N*-nitrosoproline, pyrrolidine, spermidine, proline and putrescine serve as precursors for the formation of *N*-nitrosopyrrolidine, of which proline is the major precursor. The majority of the *N*-nitrosopyrrolidine formation occurs in the fat portion of the edible meat, the fried-out fat and the vapour, but not in the lean edible meat. This finding led to the conclusion that the nitrosation reaction takes place mostly in adipose tissue; N_2O_3 partitions to the non-aqueous phase. This in turn is facilitated by surfactants and cell membranes. Since curing solutions are water-based, the use of effective lipid-soluble inhibitors of nitrosamine formation presents practical problems.

Nitrosamine formation is influenced by microorganisms that:

- reduce nitrate to nitrite or oxidize nitrite to nitrate
- oxidize ammonia to nitrite
- convert nitrate to amino acids or ammonia
- lower the pH
- produce substances that catalyse nitrosation.

Microorganisms can catalyse nitrosation enzymatically or non-enzymatically. While the enzyme nitrate reductase has been identified as the causative factor in *Escherichia coli* for promoting nitrosation, the same process is also brought about by *Neisseria* spp. which does not have nitrate reductase activity. Certain denitrifying bacteria such as *Pseudomonas aeruginosa*, *Neisseria* spp., *Alcaligenes faecalis* and *Bacillus licheniformis* bring about nitrosation much more rapidly than non-denitrifying bacteria.

Intestinal bacteria mediate the formation of nitrosamines in the body. Nitrate-reducing enterobacteria have been shown to produce nitrosamines from nitrate and secondary amines at neutral pH values. Several non-nitrate-reducing bacteria such as lactobacilli, group D streptococci, *Clostridium*, *Bacteroides* and *Bifidobacterium* can nitrosate secondary amines with nitrites at neutral pH. The mechanism is believed to be due to enzyme catalysis by the microorganisms.

Consumer Concerns about Amines

The *N*-nitrosamides produce tumours at the site of application, while *N*-nitrosamines produce them at a site different from the site of application. The median lethal dose (LD_{50}) of various nitrosamines is as low as 22 mg kg^{-1} for *N*-nitrosomethyl-2-chloroethylamine and as high as 7500 mg kg^{-1} for *N*-nitrosodiethanolamine.

It is believed that *N*-nitrosamines are metabolically activated to carcinogens by the hydroxylation of a carbon atom alpha to the nitrosamino nitrogen. The resulting α-hydroxyalkylnitrosamine is chemically unstable, hence it rearranges spontaneously to form an aldehyde and a primary alkyldiazohydroxide structure. The latter compound loses a hydroxide ion to form an alkyldiazonium ion. This ion further decomposes to molecular nitrogen and a carbonium ion which serves as an alkylating agent for reaction with proteins, nucleic acids, water or other nucleophiles. Symmetrical nitrosamines are primarily liver carcinogens, while unstable nitrosamines are oesophageal carcinogens. Such damage alone may not, however, be sufficient for tumour induction, since other factors such as site of damage at the molecular level and the extent of repair of DNA are also influential.

In order to overcome risks associated with the use of nitrite, investigators have attempted to reduce nitrite levels and simultaneously achieve inhibition of *Clostridium botulinum*. Use of nitrite at 40 μg g^{-1} along with sorbic acid or potassium sorbate has proved promising in bacon and canned comminuted pork as well as in chicken-meat frankfurters. However, it should be remembered that mixtures of sorbic acid and nitrite are mutagenic, and nitrous acid itself is a well-known mutagen. Besides, sorbates are reported to induce an allergic reaction, a chemical-like flavour, a 'prickly' mouth sensation and a sweet aroma in bacon, which are undesirable. Parabens have been investigated as substitutes for nitrite in meat products, but the results have not been encouraging. Use of a lactic acid starter culture along with dextrose decreases the pH to a point where even 50 μg g^{-1} nitrite is sufficient to inhibit growth and toxin production. Nisin at 75 p.p.m., hypophosphite at 3000 p.p.m., and 40 p.p.m. nitrite in combination with 1000 p.p.m. hypophosphite are other promising alternatives, but none has all the functional characteristics shown by nitrite in cured meats.

Conclusion

Exposure of humans to nitrite cannot be abolished, since nitrite can be formed by bacterial reduction of nitrate which is widely present in vegetables at levels of several thousand parts per million. Besides, nitrite is also present in saliva (1–10 p.p.m.) and drinking water (up to 45 p.p.m.). Dietary nitrate and nitrite act as precursors of nitrosamines in the mouth, stomach and urinary bladder. Attempts to minimize nitrite levels must not overlook the hazard posed by *Clostridium botulinum* toxin. This is especially true in the light of the fact that the hazard posed by the toxin is far greater than that posed by either nitrite or nitrate.

See also: **Clostridium**: *Clostridium tyrobutyricum*; *Clostridium botulinum*. **Meat and Poultry**: Curing of Meat. **Preservatives**: Traditional Preservatives – Sodium Chloride; Permitted Preservatives – Sorbic Acid.

Further Reading

Adams JB (1997) Food additive-additive interactions involving sulphur dioxide and ascorbic and nitrous acids: a review. *Food Chemistry* 59: 401–409.
Birch GG and Lindley MG (1986) *Interactions of Food Components* London: Elsevier Applied Science.
Branen AL and Davidson PM (eds) (1983) *Antimicrobials in Foods*. New York: Marcel Dekker.
Branen AL, Davidson PM and Salminen S (eds) (1990) *Food Additives*. New York: Marcel Dekker.
Furia TE (ed.) (1980) *Regulatory Status of Direct Food Additives*. Boca Raton: CRC Press.
Gomez RF and Herrero AA (1983) Chemical preservation of foods. In: Rose AH (ed.) *Food Microbiology*. London: Academic Press. Pp. 78–116.
Hotchkiss JH (1987) A review of current literature on N-nitrosocompounds in foods. *Advances in Food Research* 31: 53–115.
Lewis RJ (1989) *Food Additives Handbook*. New York: Van Nostrand Reinhold.
Maga JA and Tu AT (eds) (1995) *Food Additive Toxicology*. New York: Marcel Dekker.
Pierson MD and Smoot LA (1982) Nitrite, nitrite alternatives, and the control of *Clostridium botulinum* in cured meats. *CRC Critical Reviews in Food Science and Nutrition* 17(2): 141–187.
Tilbury RH (ed.) (1980) *Developments in Food Preservatives 1*. London: Applied Science Publishers.

Permitted Preservatives – Sorbic Acid

Linda V Thomas, Aplin & Barrett Ltd (Cultor Food Science), Beaminster, UK

Introduction

Sorbic acid derives its name from *Sorbus aucuparia*, because it was from berries of this tree that it was first isolated (**Table 1**). Seventy years later its potential as an antimicrobial agent was discovered, and sorbic acid and its salts (generally called sorbate) are now used as preservatives in a variety of foods in many countries.

Table 1 History of the use of sorbate as a food preservative

1859	Isolated from the oil of berries of the rowan (mountain ash) tree
1870–1890	Chemical structure formulated
1900	First synthesized by condensation of crotonaldehyde and malonic acid
1926	Synthesis of sorbic acid by oxidation of sorbaldehyde
1939–1940	Recognition of antimicrobial properties
1945	US patent for use as antifungal agent in foods
1940–1960	Industrial production. Use in dairy, fruit and vegetable products
1974	Potassium sorbate discovered to inhibit growth of bacteria

Table 2 Properties of sorbic acid and its potassium salt. Data from Sofos (1989)

	Sorbic acid	*Potassium sorbate*
EU number[a]	E200	E202
Molecular formula	CH_3-CH=CH-CH=CH-COOH	CH_3-CH=CH-CH=CH-COOK
Molecular weight	112.13	150.22
pK_a	4.76	
Melting range	132–137°C	Decomposes > 270°C
Solubility (%)		
in water at 20°C	0.15	58.20
in water at 100°C	4.00	64.00
in corn oil at 20°C	0.80	0.01
in 10% sucrose	0.15	58.00
in 10% NaCl	0.07	34.00

[a]Sodium sorbate E201, calcium sorbate E203.

Sorbic acid is an unsaturated aliphatic straight-chain monocarboxylic fatty acid, 2,4-hexadienoic acid. Salts and esters form by reaction with the carboxyl group; reactions also occur via its conjugated double bond. The acid and its sodium, calcium and potassium salts are used in food. The potassium salt is commonly used because it is more stable and easier to produce. Furthermore, its greater solubility extends the use of sorbate to solutions appropriate for dipping and spraying (**Table 2**). Other derivatives with antimicrobial capabilities (sorboyl palmitate, sorbamide, ethyl sorbate, sorbic anhydride) have limited use because they are more insoluble, toxic and unpalatable.

Sorbate has several advantages as a preservative in food. Initially thought to have only antimycotic activity, it is now known to also inhibit bacteria. Effective concentrations do not normally alter the taste or odour of products. In addition it has more activity at less acidic values (> pH 6.0) than propionate or benzoate. Sorbate is also considered harmless. Following thorough toxicological testing it was generally recommended as safe (GRAS). Metabolism of sorbate in the body is by β-oxidation (as for other fatty acids), forming CO_2 and water. It has a yield of 28 kJ g^{-1} (of which 50% is biologically usable) and a half-life in the body of 40–110 min. Its acceptable daily intake (ADI) of 25 mg kg^{-1} body weight is higher than that of other preservatives. It is considered less toxic than NaCl, with a median lethal dose (LD_{50}) of 10 g kg^{-1}, compared with 5 g kg^{-1} for NaCl.

Behaviour of Sorbate in Food

Sorbate levels may fall during storage because of microbial growth, oxidation or reactions with food constituents. Stability varies according to product type and depends largely on its composition (pH, organic acids, other additives, water activity, humectants, microbial numbers, etc.) as well as on storage temperature and packaging. For example, 10% loss was reported from cured meat after cold storage for 50 days, compared with nearly 50% loss in sliced cured meat stored at 22°C in a relative humidity of 70%.

Although the pure crystalline acid is stable, it undergoes autooxidation when dissolved in water, producing carbonyls such as malonaldehyde, acetaldehyde and β-carboxylacrolein. This process can occur in food, forming unsightly brown pigments by the reaction of β-carboxylacrolein with amino acids and proteins. This occurs more rapidly in light, with heat and acidity, and is influenced by irradiation, salts, trace metal ions, sugars, glycerol and amino acids. Autooxidation of sorbic acid in aqueous solutions decreases as pH rises. Degradation is generally more rapid at higher water activity levels, depending on the humectant. Glycine and EDTA, for instance, may enhance degradation. It has been suggested that an EDTA–Fe^{2+} complex can form by iron scavenging from packaging material, which catalyses autooxidation. Interactions with amino acids and proteins, particularly those containing free sulphydryl groups, may also affect stability.

Since oxygen causes sorbate decomposition, adding antioxidants or vacuum packaging in oxygen-impermeable material can reduce the problem. Tests on fruit and fish products found sorbate loss was proportional to the oxygen permeability of the packaging material (polypropylene > glass). Polyphosphate, propylgallate, dodecylgallate and nordihydroquaretic acid, and an ascorbic acid/nitrite combination have shown protection in vitro.

The fat to water partition coefficient of a preservative (its ratio of solubility in the fatty and aqueous phases) is an important consideration for products with a high lipid content. Sorbic acid has a fat to water partition coefficient of 3 and is consequently more efficient in these products than benzoate (partition coefficient 6.1) but less efficient than propionate (partition coefficient 0.17). The fat composition and concentration, soluble food components and pH also determine solubility. The solubility ratio, for example, is lowered by acetic acid, but rises with added NaCl, sucrose or glucose.

Foods to Which Sorbic Acid or Sorbate May Be Added

Sorbate is used as a preservative in a wide range of products (**Table 3**). It can be mixed with dry ingredients (e.g. flour, salt) or applied to surfaces by dipping, spraying or dusting. It can be incorporated within packaging material using organic carriers such as ethanol, vegetable oil or propylene glycol. Permitted levels depend on the product type and country of origin, but the maximum is generally 0.2%. Higher concentrations can be used in packaging or surface treatments. Sorbate use in the UK is covered by Schedule 2, Part A of the Miscellaneous Food Additives Regulations 1995 (Statutory Instrument 3187).

Sorbate is used in cheese products primarily as an antifungal agent and to prevent the formation of mycotoxin. It is more effective in this than propionate and benzoate. If sorbate is applied to the surface, the porosity and fat content of the cheese influence the rate and degree of absorption into the product. The complete transfer of sorbate from a wrapping into cheese can occur in 2 weeks. The calcium salt is least soluble in fat and water, and remains longer on the surface. This is the best sorbate to use on hard cheeses with long maturity periods. It is best not to use sorbate in cheeses whose flavour and appearance result from mould growth. It should not be added to wax coatings, because the temperatures used to melt the wax will cause volatization.

In the UK the use of sorbate in meat is restricted to the surface treatment of dried products. This is similar to the USA, where 10% sorbate solutions can be used to treat the surface or casings of dried sausages stored at room temperature. Sorbate is used more extensively in meat products in countries such as Japan and Korea.

Vacuum and modified-atmosphere packaging may extend the shelf life of fish products, but preservatives such as sorbate are needed to prevent the growth of anaerobic bacteria, which metabolize trimethylamine oxide, an osmoregulatory compound, to trimethylamine, causing off odours. Halotolerant organisms also cause fish spoilage. The mould *Sporendonema epizoum* causes 'dun', and sorbate is better than propionate in controlling this. Sorbate can be applied to the fish surface in various ways – by immersion, spraying, as a powder, in fat, packaging or ice. Sorbate is allowed only in packaging material in some countries, e.g. India. Fish can be preserved by treating with a solution of 0.5–2.0% sorbate and 15–20% NaCl, followed by refrigeration. A simple treatment of freshwater fish in Africa combined 1.5% NaCl with 1500 p.p.m. sorbate followed by 3 days drying in the sun. Similarly, sorbate used with lactic acid bacteria reportedly improved preservation. Shelf life was extended to 15 days by dipping in 5% sorbate, followed by packaging in 100% CO_2 and refrigeration.

Table 3 Food in which sorbate may be used as a preservative

Examples	*UK maximum in p.p.m. (typical levels in other countries)*
Dairy	
Prepacked slices	1000
Processed	2000
Desserts (not heat-treated)	300[a]
Mature cheese	(Surface treatment: 100–3000)
Cottage cheese	(500–700)
Meat	
Semimoist pet food	(1000–3000)
Dried meat	Surface treatment: no maximum
Pâté	1000[b]
Fish	
Semipreserved	2000[b] (500–2000)
Salted and dried	200[b] (5% immersion; 10% as spray)
Cooked shrimps	2000
Fresh	1–5% dip, or storage in ice with 1–5%
Vegetables	
In brine, vinegar or oil	2000[a] (500–2000)
Prepared salads	1500[a]
Fermented	(500–2000)
Olives	1000
Potato dough, prefried slices	2000
Fruit	
Sauces	1000 (500–1000)
Dried	(200–500 or 2–10% dip or spray)
Juice	(500–2500)
Candied or crystallized	1000[a]
Low-sugar jams and jellies	1000[a]
Bakery	
Cakes and mixes	(100–3000)
Prepacked sliced bread, partially baked goods and fine baked goods with $a_w > 0.65$	2000
Emusions	
Fat or sauces > 60% fat (excluding butter)	1000
Fat or sauces < 60% fat	2000
Margarine (unsalted)	(500–1000)
Mayonnaise	(1000)
Beverages	
Wine	200 (200–400)
Non-alcoholic drinks (not milk-based)	300 (100–1000)
Mead	200
Spirits < 15% alcohol by volume	200 or 400[a]
Liquid tea concentrates, fruit and herbal infusion concentrates	600[a]
Miscellaneous	
Chewing gum	1500[a]
Batters	2000
Confectionery (excluding chocolate)	1500[c] (500–2000)
Toppings and syrups	1000
Cereal or potato-based snacks, coated nuts	1000[b]
Mustards, seasonings, condiments	1000[a] (250–1000)
Liquid egg	5000[a]
Liquid soups and broths (not canned)	500[a]

[a] With benzoate.
[b] With *p*-hydroxybenzoate.
[c] With both benzoate and *p*-hydroxybenzoate.

Sorbate is often used to preserve fresh, fermented and pickled vegetables. Lower levels (0.025–0.05%) may be added during fermentation by lactic acid bacteria, with a subsequent increase in concentration to 0.1–1%, which prevents mould spoilage. Many fermented vegetable products are acidic. Low pH, NaCl, lactic acid and acetic acid increase sorbate effectiveness.

Spoilage of bakery goods can be better controlled using sorbate rather than propionate, and sorbate has less effect on flavour; it will, however, inhibit bakers' yeast. Products leavened with yeast can either be sprayed after baking or alternative products can be used: slow-release preparations such as encapsulated compounds or sorboyl palmitate. This anhydrous mixture of sorbic and palmitic acids will not inhibit yeasts, but during baking hydrolyses to form sorbic acid, which is active during storage. *Torulopsis holmii* (now reclassified as *Candida milleri* sp. nov), a sorbate- and propionate-resistant yeast, can be used to ferment sourdough bread. Sorbate can be used without problem in products leavened with baking powder, and added prior to baking as it normally withstands this process. It will control common spoilage organisms such as *Bacillus* (which forms 'rope') and osmotolerant yeasts (in sugary confectionery) as well as the pathogen *Staphylococcus aureus* (in cream-filled goods). Sorbate works well with citric acid, NaCl, propionate and sucrose. For instance, potassium sorbate in combination with calcium propionate extended the shelf life of tortilla dough at pH 5.8 to over 14 days.

Sorbates are used to protect fruit products (particularly intermediate-moisture fruits) from mould and yeast spoilage. Dipping (in 2–10% sorbate) is the recommended application method for fruits with irregular surfaces. Treatments such as heat processing, which affect product quality, may be reduced by use of sorbate. Additional pasteurization or sulphur dioxide treatment may be necessary to control enzymatic browning and oxidation. The storage life of cut apples and potatoes, for example, was extended using a cellulose-based edible coating containing antioxidants, acidulants and sorbate. Sorbate works well with sugar, and lower levels can be used in jams and jellies. It prevents spoilage fermentation by moulds and yeasts in fruit juices and wines, which are often acidic. Although benzoate is as effective as sorbate at low pH, the latter is a better preservative of beverages. It has less effect on flavour, but can cause turbidity. A sorbate/benzoate combination is often used in drinks, and sorbate is also effective with ascorbic acid. Higher levels of vitamin C remained in juice preserved using sorbate compared with pasteurization, but levels were lower compared with SO_2 treatment.

Not all countries permit sorbate in wine. In still wine it prevents unwanted yeast fermentation, particularly in young or sweet vintages. Potassium sorbate can precipitate as bitartrate. Sorbate is not recommended in sparkling wine since it may form ethyl sorbate, which has an unpleasant smell. The spectrophotometric analysis of bitter substances in beer may be affected by sorbate.

Sorbate is often used with benzoate in emulsions (e.g. butter, salad dressings). The shelf life of butter can be doubled with this combination. It prevents mould growth, reducing oxidation and the release of free fatty acids and thiobarbituric acid. In many countries, although not in the UK, sorbic acid is permitted in margarine.

More recently sorbate has been tested in African and Asian products. For example, it reportedly extended the shelf life of pinni, a traditional Indian sweet, from 10 days to 30 days in ambient temperature. Spoilage by *Lactobacillus*, *Bacillus* and *Saccharomyces* was similarly reduced in a Nigerian fermented rice product.

Antimicrobial Action of Sorbic Acid

Inhibition by sorbate may cause cell death, slowing of growth, attenuation of virulence and prevention of spore germination (**Table 4**). The extent of inhibition depends on the product composition as well as on environmental variables such as pH (**Table 5**).

Table 4 Examples of organisms reported to be inhibited by sorbate

Category	*Organisms*
Yeasts	*Brettanomyces, Byssochlamys, Candida, Cryptococcus, Debaryomyces, Hansenula, Kloeckera, Pichia, Rhodotorula, Saccharomyces, Torulaspora, Torulopsis, Zygosacchaomyces*
Moulds	*Alternaria, Aspergillus, Botrytis, Acremonium, Chaetomium, Cladosporium, Colleototrichum, Fusarium, Geotrichum, Helminthosporium, Heterosporium, Humicola, Candida, Mucor, Penicillium, Phoma, Pullularia, Rhizoctonia, Rhizopus, Sporotrichum, Trichoderma, Truncatella*
Gram-positive bacteria	*Arthrobacter, Bacillus, Clostridium, Lactobacillus, Listeria, Micrococcus, Mycobacterium, Pediococcus, Staphylococcus*
Gram-negative bacteria	*Acetobacter, Acinetobacter, Aeromonas, Alcaligenes, Alteromonas, Campylobacter, Enterobacter, Escherichia, Klebsiella, Moraxella, Proteus, Pseudomonas, Serratia, Vibrio, Yersinia*

Most yeasts are inhibited by 0.01–0.2% sorbate, and the induction of major heat-shock proteins has been reported in acidic conditions. Sorbate affects all stages of the growth cycle of moulds: spore germination, outgrowth and mycelial growth. It may interfere with transport mechanisms in conidia, causing depletion of ATP. Mycotoxin synthesis may be controlled by preventing nutrient uptake, although sub-inhibitory levels have, in certain circumstances, stimulated production. This is possibly due to accumulation of an intermediate in mycotoxin synthesis (acetyl-CoA) resulting from interference in the tricarboxylic acid cycle.

Bacterial inhibition has been shown in laboratory media, model systems and food products. Strains affected include catalase-positive and -negative species (*Lactobacillus*, *Clostridium botulinum*), mesophilic (*Salmonella*, *Staphylococcus*) and psychrotrophic strains (*Aeromonas*, *Listeria*, *Yersinia*). Conflicting reports on the efficacy of sorbate against *Listeria monocytogenes* may be due to variations in the media or food, pH or sorbate concentration. Suppression of listeriolysin production has been demonstrated. In laboratory media, high levels (500 p.p.m.) were required for complete inhibition of *E. coli* O157 at pH values greater than 4–5.

Several mechanisms for bacterial inhibition have been suggested. The cytoplasmic membrane is thought to be a major target. Like other organic acids, sorbic acid enters the cell in its undissociated form and dissociates in the more alkaline cytoplasm, making it more acidic. This reduces the electrochemical gradient across the membrane, dissipating the proton motive force causing, among other problems, depletion of ATP levels. Interference with electrochemical potentials across mitochondrial membranes has also been observed in *Penicillium crustosum*. Dissipation of the proton motive force may contribute to the inhibition of nutrient uptake that has been observed, but other mechanisms may be involved. It has been suggested that sorbate uncouples the nutrient transport system from the electron transport chain and, in addition, damages the structure and fluidity of the membrane.

Sorbate has been observed to disrupt cell walls. Cell division in *Clostridium*, for example, was completely inhibited, and at less inhibitory levels abnormal and elongated cells were formed. Divisional wall formation was also prevented in *Bacillus*, and *Listeria* was rendered more susceptible to NaCl. The cell wall of *Shewanellas putrefaciens* was observed to become more hydrophobic as well as experiencing damage to its outer membrane.

Endospore germination is prevented, as well as spore outgrowth after triggering of germination. This is possibly due to alteration of the spore membrane (increasing its fluidity) and/or inhibition of sporeolytic enzymes important for germination.

Sorbate is also thought to affect several enzymes, causing disruption to nutrient transport, metabolism, cell growth and division. Inhibition of fungi may partly result from interference with dehydrogenases, which are involved in fatty acid oxidation. Enzymes of the tricarboxylic acid cycle are inhibited, including malate dehydrogenase, isocitrate dehydrogenase, α-ketoglutarate dehydrogenase, succinate dehydrogenase, fumarase and aspartase. The active sites of certain enzymes (e.g. fumarase, aspartase, succinic dehydrogenase, ficin and yeast alcohol dehydrogenase) may be reduced by sorbate binding to their sulphydryl groups, possibly by addition to thiol in cysteine. Autooxidation of sorbic acid forms sorboyl peroxides, which affects catalase activity.

Table 5 Examples of minimum inhibitory concentrations (MICs) reported for sorbate. Data from Eklund (1989)

Yeasts	*Organism*	*pH*	*MIC (p.p.m.)*
Yeasts	*Byssochlamys fulva*	3.5	50–250
	Candida lipolytica	5.0	100
	Candida milleri	4.6	400
	Rhodotorula	4.0–5.0	100–200
	Saccharomyces	3.0	30–100
Moulds	*Aspergillus niger*	2.5–4.0	100–500
	Botrytis cinerea	3.6	120–250
	Cladosporium	5.0–7.0	100–300
	Fusarium	3.0	100
	Mucor	3.0	10–100
	Penicillium	3.5–5.7	20–1000
Gram-negative bacteria	*Escherichia coli*	5.2–5.6	50–100
	Pseudomonas	6.0	100
	Salmonella	5.0–5.3	50–1000
Gram-positive bacteria	*Bacillus*	5.5–6.3	50–1000
	Clostridium	6.7–6.8	100–10 000
	Lactobacillus	4.3–6.0	200–700

Sorbate also inhibits enolase and proteinase and may affect respiration by competing with acetate in acetyl-CoA formation.

Importance of Species/Strain Tolerance

It is generally observed that sorbate is more effective against catalase-positive and aerobic bacteria than against catalase-negative and anaerobic bacteria, particularly above pH 4.5. Reports vary as to the level of sorbate tolerated by lactic acid bacteria. For instance, the appearance and quality of a fermented cucumber product was impaired owing to inhibition of the starter culture by 0.1% sorbate. Lactic acid bacteria are capable of fermenting sorbate, producing ethyl sorbate, 4-hexenoic acid, 1-ethoxyhexa-2,4-diene and 2-ethoxyhexa-3,5-diene. Sorbate-containing wines with a high bacterial count can have a geranium-like smell.

Several strains of yeasts and moulds tolerate sorbate, including the yeasts *Brettanomyces*, *Candida*, *Saccharomyces* and *Zygosaccharomyces*, and the mould *Penicillium*. Preconditioning with sorbate induces and increases tolerance. Resistance may be due to an ability to expel the anions or to cell shrinkage creating smaller membrane pores. Sorbate effectiveness against *Z. rouxii*, for example, was influenced by changes in the lipid composition of the cytoplasmic membrane, altering its permeability. Enzymes affected by sorbate may be protected by the production of polyols and other such compatible solutes.

Sorbate metabolization by moulds in cheese and fruit products has been documented. Moulds capable of this include *Aspergillus*, *Fusarium*, *Geotrichum*, *Mucor* and *Penicillium*. Degradation may be by a decarboxylation reaction forming 1,3-pentadiene (which smells like kerosene), by esterification producing ethyl sorbate, or by reduction producing 4-hexenol and 4-hexenoic acid.

Interaction with Other Preservative Treatments

Food additives, storage conditions and processing treatments can all affect sorbate activity (**Table 6**). For example, more sorbate is usually required to preserve products of higher water activity and less acidity. This was demonstrated in grape juice, where 100 $\mu g\,ml^{-1}$ sorbate prevented *Talaromyces flavus* ascospore outgrowth at pH 3.5, but a higher concentration was needed at pH 5.4.

Sorbate can enhance heat inactivation of spores and diminish cell recovery from thermal injury. For example, treatment of kwoka (a Nigerian non-fermented maize meal) combined with 60 min steaming extended its shelf life by 2 days. The storage period of fruit products can be similarly extended by mild heat treatment with sorbate. Reduced water activity generally increases activity.

Sorbate is usually more effective at low pH (**Fig. 1**). Like other organic acid preservatives, its antimicrobial activity is greatest when it is undissociated. It has a pK_a of 4.75 at which it is 50% undissociated. This compares with 0.6% at pH 7, 6.0% at pH 6, 37% at pH 5 and 86% at pH 4. For example, repression of *Bacillus* spore germination was about five times greater at pH 6 than pH 7. The acidulant itself is influential; organic acids enhance activity more than inorganic acids. Mathematical modelling and growth investigations have indicated that the dissociated acid also shows antimicrobial activity. One study at pH levels over 6.0 reported that over 50% of observed inhibition was due to the dissociated form.

Nitrite is used not only to combat *Clostridium botulinum* growth in processed meat but also to enhance the product's colour and flavour. The possibility of replacing nitrite with sorbate has been investigated owing to concern about *N*-nitrosamine formation. In the USA the maximum nitrite level in bacon is 120 p.p.m. and 10 p.p.b. for nitrosamines. The use of 0.26% sorbate with reduced nitrite levels (40 p.p.m.) reportedly did not significantly impair the flavour or appearance of bacon and was effective against *C. botulinum*. Bacon inoculated with spores and stored for 60 days at ambient temperature developed toxin in 0.4% samples containing nitrite (120 p.p.m.), 58.8% with sorbate (0.26%), and none with 0.26% sorbate plus 80 p.p.m. nitrite. This sorbate-nitrite combination also significantly reduced nitrosamine levels and proved effective in other cured meats and against other bacteria. The treated bacon, however, had an unsatisfactory flavour, and some tasters experienced allergic reactions. This has not been repeated in other studies, and it was not proved that the reaction was due to sorbate. A further concern is that sorbate can react with nitrite to form potential mutagens including ethylnitrolic acid and 1,4-dinitro-2-methylpyrrole. These only form at very low pH and with high levels of nitrite, and are very unstable (particularly at pH above 5.0).

Incorporation of sorbate before irradiation can prevent off odours developing and generally improves product quality. Sorbate may also reduce vitamin C loss and browning, by reacting with hydrogen atoms and hydroxyl radicals.

Table 6 shows reported examples of synergy with sorbate. Synergy has also been observed with SO_2 fatty acids, sucrose fatty acid esters, betalains, propionate, ascorbate, amino acids, and polyphosphate as well as a range of antioxidants (butylated hydroxyanisole,

Table 6 Examples of sorbate interactions with other preservative treatments

Treatment	*Organism/product*	*Effect of combination*
NaCl	*Clostridium botulinum*	Synergistic reduction of toxin in meat slurries
Heat (49°C for 5 min)	Fresh fruit slices	Shelf life increased from 2 days to 90 days
Water activity of 0.87 and pH 3.7	Marmalade	Storage increased to 3 months
Irradiation (3 kGy)	*Apergillus flavus*	Synergistic inhibition of growth and aflatoxin production
	Fresh fish	Shelf life extended from 20 days to 35 days
High hydrostatic pressure	*Zygosaccharomyces bailii*	Reduced pressure required for inactivation in vitro
Isobutyric acid	*Aspergillus niger*	Synergy
Gluconic acid		
Cysteine-HCl		
$CaCl_2$	*Aspergillus niger*	Antagonism
Malonic acid		
Malic acid		
Formic acid	*Escherichia coli*	Addition
	Saccharomyces cerevisiae	Antagonism
	Aspergillus niger	Synergy
Acetic acid	Italian dry sausage	Enhanced inhibition of mould growth
Citric acid		
Nitrite	*Listeria monocytogenes*	Synergy
	Bacon and cured meat	Enhanced antibacterial activity
Nisin	*L. monocytogenes*	Synergy
	Vegetarian food	Shelf life extended

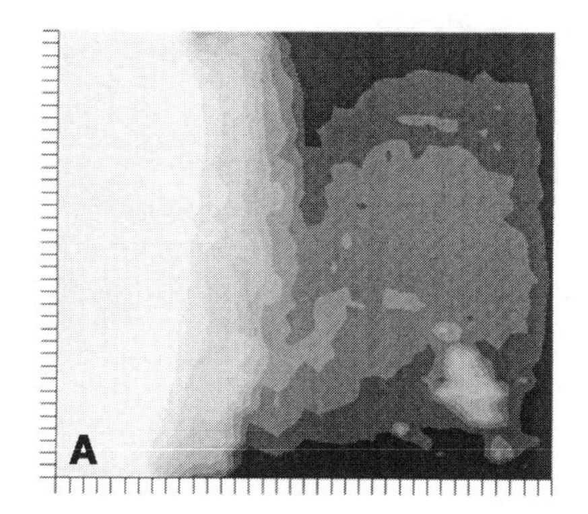

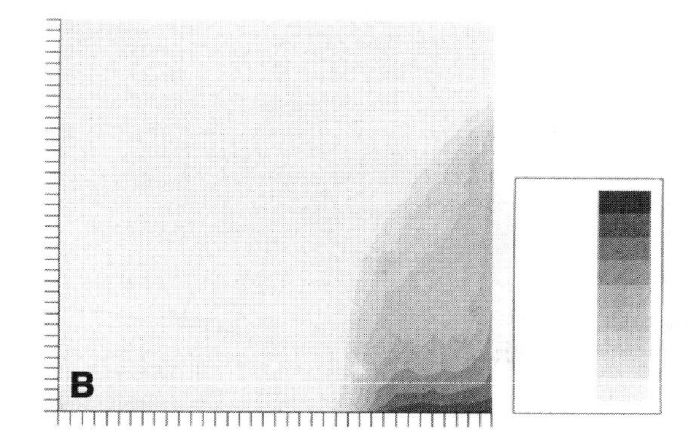

Figure 1 The effect of pH and NaCl concentration on sorbate activity. Contour maps showing growth (in optical density units) of *Salmonella typhimurium* after 48 h at 30 °C on brain–heart infusion agar plates with gradients of pH and NaCl concentration. (**A**) Control plate with no added preservative. (**B**) Plate containing 0.1% (w/v) sorbate.

butylated hydroxytoluene, tertiary butyl hydroxyquinone and propyl gallate). Antagonism has been observed with certain non-ionic surfactants.

See also: ***Aspergillus***: Introduction. ***Bacillus***: Introduction. **Bacteria**: Bacterial Endospores. **Bacteriocins**: Nisin. **Bread**: Bread from Wheat Flour. **Cheese**: Microbiology of Cheese-making and Maturation. **Chilled Storage of Foods**: Use of Modified Atmosphere Packaging. ***Clostridium***: Introduction. **Confectionary Products**: Cakes and Pastries. **Dried Foods**. ***Escherichia coli* O157**: *Escherichia coli O157:H7*. **Fermented Foods**: Fermented Vegetable Products; Fermented Meat Products. **Fish**: Spoilage of Fish. **Heat Treatment of Foods**: Synergy Between Treatments. **Intermediate Moisture Foods**. ***Lactobacillus***: Introduction. ***Listeria***: Introduction. **Meat and Poultry**: Spoilage of Meat. **Mycotoxins**: Classification. **Preservatives**: Traditional Preservatives – Sodium Chloride; Traditional Preservatives – Organic Acids; Permitted Preservatives – Sulphur Dioxide; Permitted Preservatives – Benzoic Acid; Permitted Preservatives – Hydroxybenzoic Acid; Permitted Preservatives – Nitrate and Nitrite; Permitted Preservatives – Propionic Acid. ***Rhodotorula***. ***Saccharomyces***: Introduction. ***Staphylococcus***: Introduction. **Starter Cultures**: Uses in the Food Industry. **Wines**: Microbiology of Wine-making. **Yeasts**: Production and Commercial Uses. ***Zygosaccharomyces***.

Further Reading

Eklund T (1989) Organic acids and esters. In: Gould GW (ed.) *Mechanisms of Action of Food Preservation Procedures*. Pp. 161–200. London: Elsevier.

Liewen MB and Marth EH (1985) Growth and inhibition of microorganisms in the presence of sorbic acid: a review. *Journal of Food Protection* 48: 364–375.

Robach MC and Sofos JN (1982) Use of sorbates in meat products, fresh poultry and poultry products: a review. *Journal of Food Protection* 45: 374–383.

Sofos JN, Busta FF and Allen CE (1979) Botulism control by nitrite and sorbate in cured meats: a review. *Journal of Food Protection* 42: 739–770.

Sofos JN and Busta FF (1993) Sorbic acid and sorbates. In: Davidson PM and Branen AL (eds) *Antimicrobials in Foods*, 2nd edn. Pp. 49–94. New York: Marcel Dekker.

Sofos JN (1989) *Sorbate Food Preservatives*. Boca Raton: CRC Press. [*Note*: this is the most extensive reference book on sorbate]

Thakur BR, Singh RK and Arya SS (1994) Chemistry of sorbates – a basic perspective. *Food Reviews International* 10: 71–91.

Thakur BR and Patel TR (1994) Sorbates in fish and fish products – a review. *Food Reviews International* 10: 93–107.

Permitted Preservatives – Natamycin

Jacques Stark, Gist-brocades Food Specialties R&D, Delft, The Netherlands

Introduction

For many years it has been shown that natamycin offers a good, safe solution for the prevention of fungal growth on food products.

Natamycin, also known as pimaricin, is a polyene macrolide antimycotic first isolated in 1955 in the Gist-brocades research laboratories from a culture of *Streptomyces natalensis*. The microorganism was isolated from a soil sample from the neighbourhood of Pietermaritzburg in the state of Natal in South Africa. Therefore the strain was called *S. natalensis* and the fungicide natamycin. Natamycin is produced on an industrial scale by fermentation using *S. natalensis*.

The potential applications of natamycin for food preservation were first described in the 1950s. In several early studies the effectiveness of natamycin against mould and yeast growth on cheese, sausages, fruit, berries and poultry was reported. Prevention of growth of yeasts and moulds in beverages was also described.

Today natamycin is widely used throughout the world as a preservative, mainly for the surface treatment of cheese and dry sausages. Natamycin is also used in human and veterinary medicine for the topical treatment of fungal infections such as fungal skin or eye infection and candidosis.

The major advantages of natamycin are its broad-spectrum activity and the fact that development of resistance does not seem to occur. Natamycin has no activity against bacteria, which makes it useful for application on products which require bacterial ripening processes, such as cheese and dry sausages.

Physical and Chemical Properties

Natamycin is a polyene macrolide antimycotic with a molecular weight of 665.75. Its empirical formula is $C_{33}H_{47}NO_{13}$. Its structural formula is shown in **Figure 1**.

In its common crystalline form, natamycin is present as a trihydrate. It is a creamy-white powder, which can be stored for several years with minimal loss of activity. Factors affecting the stability are extreme pH values, light, oxidants, chlorine and heavy metals.

Natamycin is poorly soluble in water and most organic solvents. In neutral aqueous systems, solubilities between 30 and 100 p.p.m. have been reported. Aqueous suspensions of natamycin will mostly have a pH value between 5.0 and 7.5. In these conditions most natamycin is present in crystalline form. Its solubility is increased in strong acid or alkaline solvents. Dissolved natamycin is less stable than the crystalline form. Where neutral aqueous suspensions are almost as stable as the dry powder, suspensions and solutions at pH values lower than 3 and higher than 9 are more susceptible to chemical degradation. At low pH values the mycosamine moiety is split off; at high pH values the lactone is saponified, with formation of the microbially inactive natamycoic acid.

Neutral aqueous suspensions of natamycin are quite stable to heat. Such suspensions can tolerate a temperature of 50°C for several days with only a slight

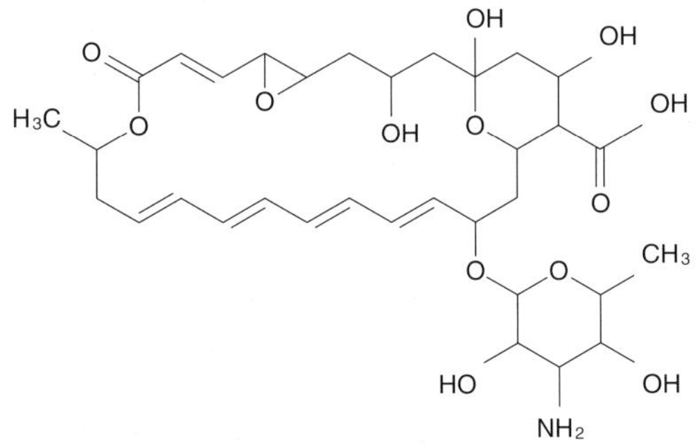

Figure 1 Structure of natamycin.

decrease in activity. It can even withstand sterilization conditions of 30 min at 116°C.

Natamycin is best stored in the dark since irradiation by ultraviolet light leads to decomposition of the molecule. Low concentrations of peroxides, oxidants or chlorine may quickly degrade natamycin. Contact of natamycin with such compounds must be avoided. This fact needs special attention, because some of these compounds are also used as cleaning agents in the food industry. It has occasionally been observed that hydrogen peroxide or chlorine inactivates natamycin, resulting in mould problems.

Natamycin solutions or suspensions are more sensitive to inactivation by ultraviolet light, peroxides, oxidants or chlorine than the solid form.

In addition to the stable trihydrate, other forms of natamycin are known. These forms are in general less stable. Barium and calcium salts are also known. These salts are stable in normal conditions.

In aqueous systems all forms tend to be converted to the more stable trihydrate, which leads to improved solution. This phenomenon leads to temporarily improved availability of active dissolved natamycin and consequently to enhanced antifungal activity. However, for practical use these preparations must be freshly prepared.

Production

Natamycin is produced on an industrial scale by fermentation processes using *S. natalensis*. Since natamycin has a low solubility in water, it is mainly present in the crystalline form in the fermentation broth. During fermentation insoluble particles are formed. These crystals may be in the form of needles or discs. Usually particles have a diameter of 0.5–20 μm. Needles longer than 40 μm have also been observed. For most applications a combination of larger and smaller particles is preferred. The smaller particles increase the density of the particle distribution on the product, which leads to an improved rate of solubilization. The larger particles ensure availability over a longer period of time.

The duration of fermentation depends on the quantity of cells in the inoculum, the composition of the medium and the desired yield – usually it varies between 48 and 120 h. Commercial fermentations may also be continued for longer if the increased yield thus obtained justifies the extra costs of a longer fermentation cycle. Fermentation is carried out at temperatures between 25 and 30°C and at pH 6–8.

After fermentation, the natamycin can be recovered using extraction, filtration and drying processes. First, the pH of the culture fluid is adjusted in order to dissolve the natamycin; then an organic solvent such as methanol or butanol is added. The biomass and other impurities are separated from the product by filtration. Precipitation of the crystals is achieved by readjusting the pH to about 7. Finally, after drying an almost pure white powder is obtained.

A recent patent application describes an alternative recovery method. It was found that when the particle size of the biomass is reduced, the natamycin can be separated from the fermentation broth without an organic solvent. The biomass of the *Streptomyces* strain used in the production generally consists of clusters of mycelia. After fermentation the biomass is disintegrated by homogenization or high shear mixing. The natamycin is then separated from the biomass using gravity gradient separation techniques. Particles of different densities or sizes are separated from each other subject to gravity or equivalent forces. After separation the natamycin suspension may be dried to obtain a powder.

Product Formulations

The pure natamycin powder thus obtained is the active compound of several commercially available products.

Delvocid® Instant, used in many food applications, is a powder containing 50% natamycin and 50% lactose. Delvocid® Instant can be added to plastic emulsions applied as a coating to cheese. Most cheese coatings, such as the Superdex® coating range, are polyvinyl acetate (PVA)-based.

Delvocid® Dip and Premi® Nat are special powder formulations developed as dipping or spraying solutions for surface treatment of cheese and dry sausages respectively. These formulations contain a food-grade thickening agent, which improves the adhesion of natamycin to the surface of the product.

Delvocid® Sol is a special formulation for large-scale use in PVA-coating manufacturing. The powder can be used to prepare suspensions of natamycin which have physical, chemical and microbiological stability over a longer period of time.

Delvocell® is a blend of food-grade powdered cellulose and natamycin. It combines the anticaking properties of powdered cellulose and the antifungal activity of natamycin. Delvocell® is mainly used in the production of shredded cheese.

Product Properties

Natamycin has proved to be an effective antifungal agent in food preservation. It has several advantages over other antifungal agents such as sorbic acid. Its

Table 1 Advantages of natamycin

Broad-spectrum activity against moulds and yeasts
No effect on bacterial starter or ripening cultures
No development of resistance
Effective at low concentrations
Prolonged working time
Remains on the surface
Easy to apply
Active at low, neutral and high pH
No negative effect on product quality
No colour, odour or taste
Safe for the consumer
Chemically stable

major advantages are summarized in **Table 1**. The combination of these properties makes it an extremely effective antifungal agent for food preservation.

Spectrum of Activity

The most important property of a preservative should be that it is an effective inhibitor of all target organisms without causing negative effects. In this respect, natamycin is unique in food preservation. It possesses a much broader spectrum of activity than any other fungicide allowed for food application. Natamycin does not inhibit bacteria, which makes it useful for application in products that require natural bacterial ripening processes.

Natamycin is active in small quantities against almost all moulds and yeasts occurring in food products. The sensitivity to natamycin of most moulds is lower than 10 p.p.m.; yeasts are even more sensitive (**Table 2**). Only dissolved natamycin has antifungal activity. Since the solubility of natamycin in aqueous food systems is around 40 p.p.m., usually sufficient dissolved natamycin will be available to prevent fungal growth.

Sometimes low solubility may be a limiting factor. The availability of active natamycin is determined by the solubility of the compound, diffusion of the dissolved fraction to the site of action and its elimination by light or binding to the target organism. Elimination of dissolved natamycin will usually be sufficiently compensated by dissolution from the crystals and diffusion to the site of action. Under less hygienic conditions, especially in combination with the presence of moulds which are not very susceptible to natamycin, solubility and diffusion may have a limiting effect on its inhibitory activity.

The availability of active natamycin can be enhanced by improving its solubility or by enhancing its solubility by using crystals with a smaller particle size. It is most likely that both effects can be achieved by using an aqueous solution prepared with either low or high pH. When such a solution is applied

Table 2 Sensitivity of fungi to natamycin

	Minimum inhibitory concentration ($\mu g\,ml^{-1}$)
Aspergillus chevalieri	≤ 5
Aspergillus clavatus	≤ 5
Aspergillus flavus	≤ 5
Aspergillus nidulans	≤ 5
Aspergillus niger	≤ 5
Aspergillus ochraceus	≤ 5
Aspergillus penicillioides	≤ 5
Cladosporium cladosporioides	≤ 5
Mucor racemosus	≤ 5
Penicillium camemberti	≤ 5
Penicillium chrysogenum	≤ 5
Penicillium digitatum	≤ 5
Penicillium expansum	≤ 5
Penicillium glabrum	≤ 5
Penicillium notatum	≤ 5
Wallemia sebii	≤ 5
Yeasts	
Aspergillus oryzae	≤ 10[a]
Aspergillus versicolor	≤ 10[a]
Fusarium spp.	≤ 10[a]
Penicillium roqueforti	≤ 10[a]
Rhizopus oryzae	≤ 10[a]
Scopulariopsis asperula	≤ 10[a]
Penicillium discolor	≤ 20[a]

[a]In general, the sensitivity of most mould strains is lower than 5 p.p.m.

on a cheese surface, most of the natamycin will be available directly; due to the pH of the cheese small crystals will probably form on the surface. A disadvantage is that such solutions must be used immediately after preparation. It has been demonstrated in practice that, even under unacceptable hygienic conditions, this method can prevent fungal growth. It must be borne in mind that a preservative should never become an alternative for proper hygiene and good manufacturing practice. From this point of view, mould growth where natamycin is in normal use is an indication of non-optimal hygienic or processing conditions.

Resistance

Most preservatives induce resistance. Resistance of moulds and yeasts to sorbate is a serious problem in the food industry. Some mould species can degrade sorbic acid to 1,3-pentadiene, a volatile compound which causes off flavours in food products such as cheese. The occurrence of resistant osmotolerant yeasts in beverages is a well-known problem.

Fungi cannot develop significant resistance to natamycin. Although natamycin has been used for decades, resistance has never yet been observed in the food industry.

Under laboratory conditions resistance to natamycin can only be induced with difficulty by mutagenesis in fungal cells containing either decreased ergosterol levels or no ergosterol at all. Ergosterol is a major compound in the fungal cell membrane. Mutant strains where ergosterol is replaced by another sterol cannot survive in nature.

Studies have indicated that the mode of action of natamycin is its irreversible binding to ergosterol, resulting in leakage of essential cellular material and, finally, lysis of the cell. Observations of resistance of *Aspergillus* and *Candida* species to natamycin have only been made by in vitro experiments. These mutants with no or reduced levels of ergosterol in their cell membrane have a slower growth rate and cannot survive in nature. Other studies, in which attempts were made to induce resistance under laboratory conditions, were not successful.

The most important reason why no resistance to natamycin has been observed is its mechanism of action – interference with an essential compound of the fungal cell membrane. An interesting additional explanation is the single-hit theory. It is suggested that, even in highly diluted aqueous solutions, polyene antimycotics always occur as micelles. In such a diluted solution fungal cells have a certain chance to come into contact with these micelles. If they don't, the cells survive. If they do, the local concentration of the polyene is always high and the cells do not survive. In both cases, selection of mutants is impossible.

Prolonged Working Time on the Surface

Natamycin protects food products against mould growth for a longer period of time. Mould development usually occurs on the surface of food products. In particular, foods ripened or stored in open air can be contaminated over the ripening or storage period. Such products may easily spoil and therefore need protection throughout this period.

If natamycin is applied for the surface treatment of cheese or sausages, only a small fraction (30–50 p.p.m.) is dissolved. Due to its low solubility in such food systems, natamycin is mainly present in the stable crystalline form. The natamycin slowly dissolves from the crystals. This slow release guarantees prolonged working time.

The natamycin crystals remain on the surface of food products such as cheese. Since the dissolved fraction hardly penetrates into the cheese, relatively low amounts per kilogram of product are sufficient to prevent fungal growth.

In several studies the penetration depth of natamycin in cheese has been examined. Natamycin could only be detected in the rind (about 1–4 mm thick). Also in the case of sausages the natamycin does not penetrate into the product. This means that the fungicide remains where it has to work – on the surface of the product. This is best illustrated by the application of natamycin on blue cheese. Growth of the blue mould and spoilage moulds on the surface of the cheese must be prevented without disturbing the development of the blue mould in the punch holes of the cheese. Where migration of sorbic acid cannot be controlled (it penetrates into the whole cheese and inhibits the interior growth of the blue mould), dipping such cheeses before punching in a natamycin solution results in a superior mould growth in the punch holes, while the surface of the cheese remains clean.

Use of Natamycin in Food Products

Fungal Spoilage of Foods

The prevention of fungal growth is an important issue in the food and agricultural industry. Worldwide economic losses due to spoilage of crops in the field are considerable. Particularly if humidity is too high, moulds easily grow on crops. After harvesting, most fruits, vegetables and cereals are even more susceptible to fungal growth.

Fresh berries, such as strawberries, must be consumed within a few days of harvesting or they will grow mouldy. Fruits such as apples and pears are often stored for longer. In general all products prepared from fruits can be spoiled by moulds or yeasts. Spoilage of fruit yoghurt, juices and marmalade is a well-known problem.

Cereals and animal feed are stored in silos or transported in ships where humidity locally can become high due to temperature fluctuations. Moulds can develop here.

Products stored or ripened in the open air are always susceptible to fungal contamination. Cheese and dry sausages are often ripened and stored in the open air at a high relative humidity of more than 80%. Under such conditions, moulds and yeasts can easily grow. Contamination with fungi and growth on the surface may occur throughout the ripening and storage period, which makes these products even more susceptible to fungal spoilage.

In spite of optimal hygiene and processing conditions, some products remain sensitive to fungal growth. Effective techniques such as sterilization or modified atmosphere packaging, cannot be used to prevent spoilage of all food products. Mostly an antifungal agent such as natamycin is the only way to prevent fungal spoilage.

Beverages

Natamycin is most effective in fluids. Due to its optimal availability, low concentrations are sufficient to eliminate all yeasts and moulds present. Since, after production, the packaging of products such as juices, lemonades, beer, wine and fruit yoghurt is well sealed, contamination before the packaging is opened by the consumer is unlikely. Usually just 1–5 p.p.m. of natamycin is sufficient to prevent spoilage of these products in the storage period.

Solid Foods

If natamycin is applied on solid food products, availability may be a limiting factor. The exterior structure of some fruits, vegetables and cereals forms a physical barrier. It is easy to understand that the active dissolved natamycin cannot reach all of the surface of some products, which may result in mould growth on unprotected places.

The best results are obtained on solid food products when the concentration of dissolved natamycin is not limited by external factors. In addition to the advantages described above, this is why natamycin is so suitable for the prevention of fungal growth on cheese and dry sausages.

Natamycin is usually applied via the plastic cheese coating. This coating covers the whole surface of the cheese by which the natamycin is homogeneously distributed over the surface. Most cheeses are treated three to five times with a coating containing 100–750 p.p.m. of natamycin. The concentration required depends on the type of cheese, time of storage and number of treatments.

Where dry sausages are concerned, the casings can be soaked for 2 h in a suspension of 1000 p.p.m. natamycin. This results in a homogeneous distribution of natamycin over the surface of the sausages.

Natamycin can also be applied by dipping or spraying. Both the distribution of natamycin on the surface, in particular with respect to different concentrations at the top and at the bottom of the product, as well as the final surface concentration can be optimized using the newly developed products Delvocid® Dip for cheese and Premi® Nat for dry sausages. Usually the dipping or spraying suspension contains 1–5 g natamycin per litre of water.

Natamycin can also be used to prevent fungal growth on fruit. Less spoilage is observed on strawberries, raspberries and cranberries which are sprayed in the field shortly before harvesting with a solution containing 50 p.p.m. natamycin. Dipping after harvesting is more effective than this. The shelf life of berries dipped in a solution of 10–100 p.p.m. natamycin can be prolonged by several days.

Apples and pears can be treated with an emulsion of lecithin. The lecithin coating affects the quality of the fruit during storage in a positive way. However, mould growth on the surface of the fruit may occur. Natamycin can be added to the lecithin emulsion applied to the fruit as a coating. The natamycin concentration in the coating varies from 20 to 100 p.p.m. When natamycin is added to the lecithin emulsion it will be distributed homogeneously over the surface of the fruit.

Regulatory Status

Although natamycin has proved to be an effective fungicide in many food applications, in most countries it is only permitted to be used in the surface treatment of cheese and dry sausages. In a number of countries natamycin is approved for other food applications. In some countries wider use is permitted, for example as a general food additive.

The acceptable daily intake (ADI) allowed for natamycin is $0.3\,\mathrm{mg\,kg^{-1}}$ body weight. As natamycin is mainly used for surface treatments, the ADI is not reached, even by consuming extreme quantities of treated products. In beverages low concentrations of natamycin are used.

The main application of natamycin is the surface treatment of cheese and dry sausages. In the European Union the maximum level of natamycin on the surface of hard, semihard and semisoft cheese and dry cured sausages is 1 mg per square decimetre surface, not present at a depth greater than 5 mm.

In Argentina and Venezuela natamycin is permitted for the surface treatment of certain cheese and sausages. In Australia natamycin is approved for the surface treatment of rinded cheeses and manufactured meat. In Brazil, Canada, Egypt, Israel, Oman, Qatar, Bahrein, Saudi Arabia, Turkey, Cyprus and the Philippines natamycin is permitted for the surface treatment of certain cheeses. Colombia allows its use on meat products only. Regulation concerning penetration depth and maximum levels varies in different countries.

In the US natamycin may be applied on cuts and slices of certain cheese varieties by dipping or spraying in aqueous suspensions containing 200–300 p.p.m. natamycin. In Estonia, Czech Republic, Slovakia, Hungary, Poland, Russia and former Yugoslavia its application is permitted in the dairy industry and sometimes also in the meat industry. South Africa approves the use of natamycin in a broad range of food products. In Mexico, Chile, Costa Rica, Dubai, Kuwait and the United Arab Emirates its use as a food additive is allowed. The details vary from country to country. Before use you should always check the situation in a particular region.

See also: **Cheese**: Microbiology of Cheese-making and Maturation. **Good Manufacturing Practice**. **Hurdle Technology**. **Meat and Poultry**: Spoilage of Cooked

Meats and Meat Products. **Preservatives**: Classification and Properties; Permitted Preservatives – Sorbic Acid. **Spoilage Problems**: Problems caused by Bacteria.

Further Reading

Brik H (1981) Natamycin. In: Florey K (ed.) *Analytical Profiles of Drug Substances*. Vol. 10, p. 513. New York: Academic Press.

Brik H (1994) Natamycin (supplement). In: Brittain HG (ed.) *Analytical Profiles of Drug Substances and Excipients*. Vol. 23, p. 399. San Diego: Academic Press.

Daamen CBG and Van den Berg G (1985) Prevention of mould growth on cheese by means of natamycin. *Voedingsmiddelentechnologie* 18: 26–29.

Davidson PM and Doan CH (1993) Natamycin. In: Davidson PM and Branen AL (eds) *Antimicrobials in Foods*. P. 395. New York: Marcel Dekker.

De Ruig WG and Van den Berg G (1985) Influence of the fungicides sorbate and natamycin in cheese coatings on the quality of cheese. *Netherlands Milk and Dairy Journal* 39: 165–172.

Gale EF (1984) Mode of action and resistance mechanisms of polyene macrolides. In: Omura S (ed.) *Macrolide Antibiotics. Chemistry, Biology and Practice*. P. 425. New York: Academic Press.

Hamilton-Miller JMT (1973) Chemistry and biology of the polyene macrolide antibiotics. *Bacterial Reviews* 37: 66–196.

Hammond SM (1977) Biological Activity of Polyene Antibiotics. In: Ellis GP and West GB (eds) *Progress in Medicinal Chemistry*. Vol. 14, p. 105. Amsterdam: Elsevier/North-Holland.

Morris HA and Castberg HB (1980) Control of surface growth on blue cheese using pimaricin. *Cultures Dairy Products Journal* 15: 21–23.

Raab WP (1974) *Natamycin (Pimaricin). Properties and Medical Applications*. Stuttgart: Georg Thieme Verlag.

Raghounath D and Webbers JJP (1997) Natamycin recovery. EP Patent Application no. 97/00588.

Struyk AP, Hoette I et al (1957–1958). Pimaricin, a new antifungal antibiotic. *Antibiotics Annual* 878–885.

Van Rijn FTJ, Stark J, Tan HS et al (1995) A novel antifungal composition. US Patent Application no. 08/446,782.

Permitted Preservatives – Propionic Acid

Rekha S Singhal and **Pushpa R Kulkarni**, Department of Chemical Technology, University of Mumbai, India

Propionic acid is a preservative developed especially to combat the rope-forming organism, *Bacillus mesentericus*, in bread. This compound is a member of the aliphatic monocarboxylic acid series containing from 1 to 14 carbon atoms, which all have antimicrobial activity. Propionates have been selected over other members of the series because the tastes and odours of the higher homologues become evident in baked goods. It is also a normal metabolite of the microflora in the gastrointestinal tract of ruminants. It is found to the extent of 1% in Swiss cheese due to the growth and metabolism of the genus *Propionibacterium*, where it also acts as a preservative.

Some general properties of propionic acid and its calcium and sodium salts are given in **Table 1**.

Foods Permitted to Contain Propionic Acid

Propionic acid and its salts are mainly used in baked products to suppress the bacteria causing 'rope' in the centre of bread, and the growth of moulds on bread and cakes. Most moulds are destroyed during baking, but surface contamination can occur under the wrapper during subsequent storage. This additive is also used as a mould inhibitor in cheese foods and spreads. Propionates do not inhibit yeasts, and hence do not interfere with the leavening in bread dough. Calcium propionate is the preferred salt, since it also contributes to the enrichment of the product. For chemically leavened products sodium propionate is preferred, since the calcium ions interfere with the leavening action. In processed cheese, propionates are added to the starting materials in the cooker. They may be added before or along with emulsifying salts in pasteurized processed cheese and cheese spread. Propionates may be mixed with other ingredients and used in a cold pack, or they be sprinkled onto the ground cheese base while is being worked and agitated.

Calcium and sodium propionates are also listed as antimycotics when migrating from food packaging materials. For butter, propionate-treated parchment wrappers give sufficient protection, although propionates are not used in butter itself. Maximum permitted levels of propionates in bread and cheese in selected countries are given in **Table 2**.

Concentrations of 0.2–0.4% sodium propionate are known to inhibit the growth of moulds on the surface of malt extract. Sodium propionate delays spoilage not only in fresh figs, syrup, apple sauce, berries and cherries, but also in neutral vegetables such as lima beans and peas.

The regulatory status of propionic acid and its calcium and sodium salts in the USA is summarized in **Table 3**.

Table 1 Properties of propionic acid and its calcium and sodium salts

Property	*Propionic acid*	*Sodium propionate*	*Calcium propionate*
Molecular formula	$C_3H_6O_2$	$C_3H_5O_2.Na$	$C_6H_{10}CaO_4$
Molecular weight	74.09	96.07	186.22
Appearance	Oily liquid	Transparent crystals or granules	White crystals
Odour	Slightly pungent, disagreeable	Cheese-like	Cheese-like
Melting point (°C)	−21.5		
Boiling point (°C)	141.1		
pK_a	4.87		
Solubility	Miscible in water, alcohol, ether and chloroform	150 g per 100 ml at 100°C in water 4 g per 100 ml in alcohol at 25°C	55.8 g per 100 ml at 100°C in water Insoluble in alcohol
LD_{50} (per kg body weight):			
oral ($mg\,kg^{-1}$)		5100 (rat)	3340 (rat)
intravenous ($mg\,kg^{-1}$)	625 (mouse)	1380–3200 (rat)	580–1020 (rat)

LD_{50}, median lethal dose.

Table 2 Maximum permitted levels of propionates in bread and cheese in selected countries

Country	*Baked goods ($mg\,kg^{-1}$)*	*Bread*[a] *($mg\,kg^{-1}$)*	*Cheese, including processed ($mg\,kg^{-1}$)*
Belgium		3000	
Denmark	3000	3000	
France		5000	
Germany		3000	
Italy	2000	2000	None[b]
Netherlands		3000	
UK	1000	3000	
Norway	1000	5000	
Sweden	3000	3000	
Canada	2000	2000	2000
USA	GMP[c]	3200	3000

[a]Certain types only, mostly sliced wrapped.
[b]Treatment of rind only.
[c]GMP is good manufacturing practice for which there is no specified level.

Behaviour of Propionic Acid in Food

The compliance of propionates with the criteria laid down by the International Association of Microbiological Societies (IAMS) for an ideal preservative is given in **Table 4**. Propionic acid has a low chemical reactivity and a long persistence in foods, two highly desirable qualities in a food preservative.

Antimicrobial Action of Propionic Acid

The use of propionic acid and propionates has been directed primarily against moulds, although some species of *Penicillium* can grow on media containing 5% propionic acid. Less calcium propionate by weight than sodium propionate is required to arrest the growth of moulds. It is especially active against the rope-forming bacterium *Bacillus mesentericus*; this spore-former being inhibited even at pH 6.0. Some Gram-negative bacteria are also inhibited. Sodium propionate at levels of 0.1–5.0% delays the growth of *Staphylococcus aureus*, *Sarcina lutea*, *Proteus vulgaris*, *Lactobacillus plantarum*, *Torula* and *Saccharomyces ellipsoideus* by 5 days. Organic acids including propionic acid are effective in combination with heating against *Salmonella* and yeast.

The propionate accumulates in the cell and competes with alanine and other amino acids necessary for the growth of microorganisms. It also inhibits the enzymes necessary for metabolism. The bacteriostatic action of sodium propionate can be overcome by addition of small amounts of β-alanine for *Escherichia coli*, but not for other organisms such as *Aspergillus clavatus*, *Bacillus subtilis*, *Pseudomonas* spp. and *Trichophyton mentagrophytes*. It is believed that the inhibitory action of sodium propionate against *E. coli* may be due to interference with β-alanine synthesis.

Metabolism and Toxicology of Propionates

Propionic acid is a normal human physiological metabolite, produced by the oxidation of certain fatty acids. Even after large doses, no significant amounts of propionic acid are excreted in the urine. In vitro, propionic acid is completely oxidized by liver preparations to carbon dioxide and water. Hence no acceptable daily intake has been prescribed by the Food and Agriculture Organization/World Health Organization, although technological limitations restrict the amount and scope of propionates as a food preservative. Sodium propionate is an allergen; when heated to decomposition it emits toxic fumes of Na_2O, and it is also reported to have some local antihistaminic activity.

Table 3 Regulatory status of propionates in the USA

Compound	*FEMA no.*	*Regulation*	*Limitation*
Propionic acid	2924	FDA §182.1	GRAS, chemical preservative
Sodium and calcium propionate		FDA §17	In bakery products, alone or with calcium propionate, up to 0.32% by weight of flour in bread, enriched bread, milk bread and raisin bread, and up to 0.38% of flour used in wholewheat bread
		FDA	In bread, up to 0.32% by weight of flour
		FDA §182.1	GRAS, chemical preservative
		MID	Alone or with calcium propionate, to retard mould growth in pizza crust, up to 0.32% by weight of flour; product specifications apply

FEMA, Flavour and Extract Manufacturers Association; GRAS, generally recognized as safe; MID, Meat Inspection Division of the US Department of Agriculture, which is responsible for clearing additives intended for use in all meats and meat products except poultry.

See also: ***Bacillus***: Introduction. **Bread**: Bread from Wheat Flour. **Penicillium**: Introduction. ***Propionibacterium***.

Further Reading

Branen AL, Davidson PM and Salminen S (eds) (1990) *Food Additives*. New York: Marcel Dekker.

Branen AL and Davidson PM (eds) (1983) *Antimicrobials in Foods*. New York: Marcel Dekker.

Furia TE (ed.) (1972) *Handbook of Food Additives*, 2nd edn. Cleveland: CRC Press.

Furia TE (ed.) (1980) *Regulatory Status of Direct Food Additives*. Boca Raton: CRC Press.

Lewis RJ (1989) *Food Additives Handbook*. New York: Van Nostrand Reinhold.

Maga JA and Tu AT (eds) (1995) *Food Additive Toxicology*. New York: Marcel Dekker.

Tilbury RH (ed.) (1980) *Developments in Food Preservatives 1*. London: Applied Science.

Table 4 Relationship between IAMS criteria for the ideal preservative and propionate

Criterion	*Compliance of propionate with the criterion*
Toxicological acceptability	+
Microbiological activity	–
Long persistence in foods	+
Chemical reactivity	Low
High thermal stability	Not relevant
Wide antimicrobial spectrum	Gram-positive bacteria, moulds
Active against food-borne pathogens	(+)
No development of microbial resistance	+
Not used therapeutically	+
Assay procedure available	+

Key: + complies with criterion; (+), complies to some degree; – does not comply.

PROBIOTIC BACTERIA

Detection and Estimation in Fermented and Non-Fermented Dairy Products

Wolfgang Kneifel, Department of Dairy Research and Bacteriology, Agricultural University, Vienna, Austria

Tiina Mattila-Sandholm, VTT Biotechnology and Food Research, Espoo, Finland

Atte von Wright, Department of Biochemistry and Biotechnology, University of Kuopio, Finland

Introduction

The role of probiotics in human nutrition has been increasingly recognized together with the growing public awareness of certain health benefits of fermented dairy products. Today, a broad variety of fermented and non-fermented dairy products containing microorganisms with probiotic function is available on the market. Historically, fermented milks have been regarded as valuable and sensorically attractive foods which positively influence the intestinal microbial balance of humans. Besides this long-term experience with this kind of traditional foods, numerous reports dealing with more or less proven evidence for certain prophylactic and therapeutic claims of probiotics have appeared in the last ten years and this has led to extensive discussions among experts from different disciplines. This development

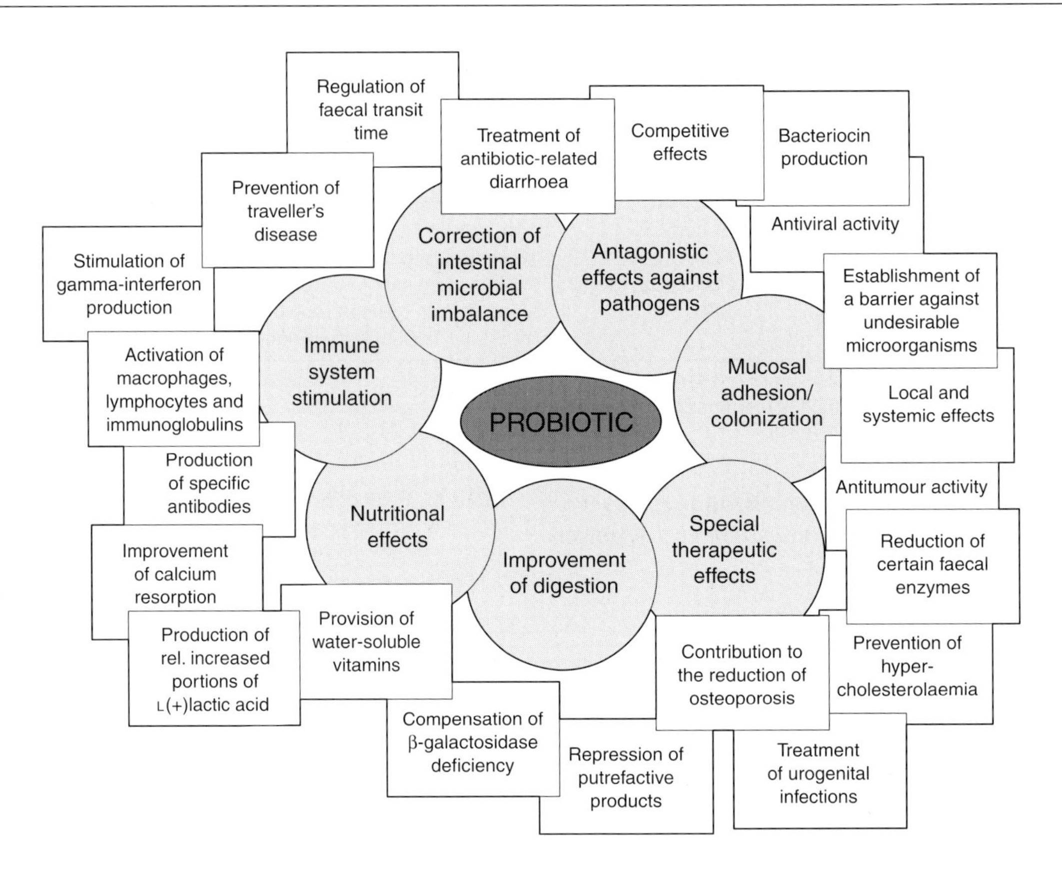

Figure 1 General survey of beneficial effects attributed to probiotics.

has also provoked research activities and scientific studies aiming at clarifying and a better understanding of the role and function of probiotics as an emerging branch of up-to-date nutrition.

Definitions and Nomenclature

The word probiotic has a Greek origin and can be translated as 'for life'. In 1965, Lilley and Stillwell used this word for the first time to describe the opposite effect of antibiotics. They defined probiotics as substances secreted by one microorganism that were able to stimulate the growth of another. Later on, in 1971, Sperti gave this name to tissue extracts promoting the growth of microbes. The definition by Parker in 1974 was more close to the most actual meaning: 'organisms and substances which contribute to the intestinal microbial balance'. Fuller in 1989 made a re-definition as 'a live microbial feed supplement which beneficially affects the host animal by improving its intestinal microbial balance'. This definition has also been extended to human nutrition, and probiotics have now been defined as live microbes, which when ingested, enhance the well-being of the host through their effect on the intestinal microflora. Probiotic strains must be able to survive in the intestinal tract, resist acid conditions during gastric passage and bile digestion. They should at least temporarily establish themselves among the natural microflora, and finally have various beneficial effects on the host (for general survey see **Figure 1**). Not in full agreement with the probiotics used in feeds, those for application in human nutrition usually belong to the group of lactic acid bacteria (LAB) only. The LAB 'family' can be generally described as a group of homo- and heterofermentative anaerobic, microaerophilic and facultatively anaerobic rods and cocci which produce lactic acid under meso- and thermophilic conditions. Originally, only four genera (*Lactobacillus*, *Leuconostoc*, *Pediococcus* and *Streptococcus*) were dealt with as LAB members, today others (*Carnobacterium*, *Aerococcus*, *Enterococcus*, *Lactococcus*, *Vagococcus*, *Tetragenococcus* and *Bifidobacterium*) are also assigned to LAB in a broad sense, although not all of them are linked to each other phylogenetically. At present, it has been shown that probiotic effects could be attributed primarily to certain species and strains of lactobacilli, bifidobacteria, enterococci and lactococci. It is important that this function must not be considered as a general property of a defined genus, but of defined strains.

Table 1 Positive and negative factors influencing the microbial balance of the intestine

Positive effects	*Negative effects*
Good state of health	Disease
Balanced nutrition	Stress
Probiotics/fermented milk products	Antibiotic therapy
Prebiotic foods	Malnutrition
Hygiene	Environmental pollution

Probiotic Function and Specific Properties

Role of Probiotics in the Human Intestine

The human gastrointestinal tract forms a very variable ecosystem with a wide range of different conditions. This naturally affects both the microflora and the physiological responses of the host in each particular intestinal location. Thus the natural microflora of the host forms the background for the action of each probiotic stain which may survive, colonize and act in the typical environment of the human host which may be exposed to various intrinsic as well as extrinsic factors (**Table 1**). Thus, both the qualitative and the quantitative composition of the intestinal microflora may vary to a considerable extent.

The microflora of the oral cavity is dominated by Gram-positive bacteria such as lactobacilli, actinomyces and streptococci. Members of genera such as *Bifidobacterium* and *Eubacterium* and Gram-negative species such as *Bacteroides* and *Fusobacterium*, all common in the lower parts of the intestinal tract, can also be found among the mouth microflora. The oral microflora present in the swallowed saliva might theoretically affect the rest of the intestinal ecosystem (**Fig. 2**).

The pH of an empty stomach (normally <3) very effectively eliminates most microbes. However, during meals foods have a buffering effect on the gastric lumen, and this allows for the survival of both salivary bacteria and of microbes present in the ingested foodstuffs. In the duodenum, the gastric content is neutralized by pancreatic secretions, thus further enhancing microbial survival. On the other hand, the presence of bile and pancreatic enzymes form an extra stress for bacteria. As a consequence, and because of the short luminal transit time in the small intestine, the total bacterial counts in the lower ileum are rather low ($< 10^6$ cells per gram of lumen).

By far the highest bacterial numbers are encountered in the colon. Bacterial densities up to 10^{11} cells per gram of lumen are frequently met. Anaerobes like *Eubacterium*, *Bifidobacterium*, *Bacteroides*, *Clostridium* and *Fusobacterium* form the bulk of this bacterial mass. The numbers of lactobacilli, streptococci and enterococci as well as those of coliforms and other enteric bacteria are generally several orders of magnitude lower, usually within the range of 10^5–10^9 cells per gram of faeces. The human mucosal microflora of the small and large intestine is not very well known. However, it appears that mucosa-associated bacteria differ qualitatively and quantitively from the luminar bacterial community.

Apparently, during the course of evolution, a certain balance has been established between the host and the intestinal microflora. Although the intestinal tract forms an ideal ecological habitat for numerous

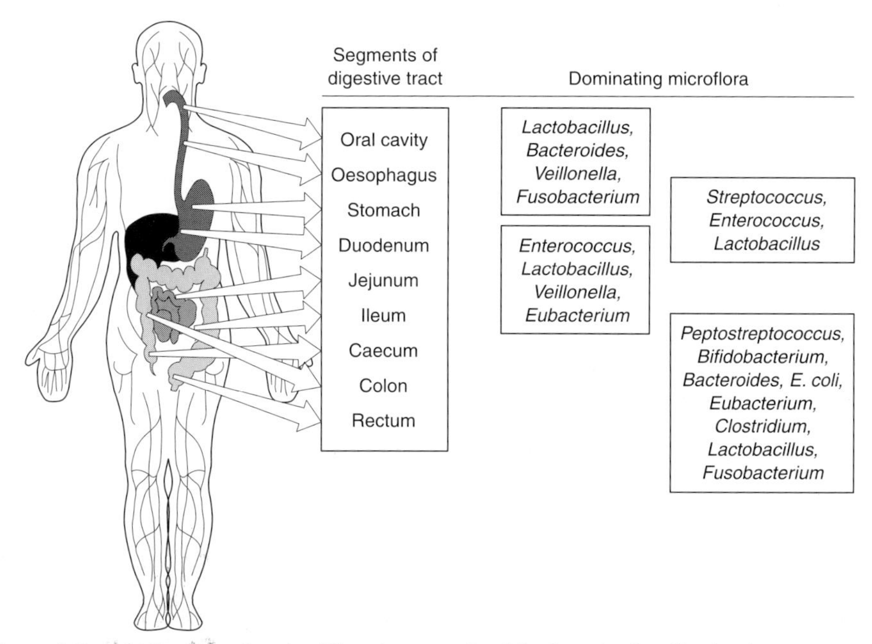

Figure 2 Qualitative variations in the microflora in different segments of the human digestive tract.

bacterial species and stains, their presence and metabolic activities have several important consequences – some beneficial, the others potentially harmful for the host. One of the most positive functions is the prevention of bacterial and viral infections. The presence of resident bacteria prevents pathogens establishing themselves or colonizing the intestinal tract (colonization resistance). It was demonstrated that germ-free guinea pigs could be killed by ten cells of *Salmonella enteritidis*, whereas 10^9 cells were required to kill guinea pigs with a complete intestinal microflora. The natural microflora also plays a role in the immunological response of the host as well as in the prevention of pathogens invading the enterocytes. Furthermore, the intestinal bacteria are involved in the metabolism of bile acids and xenobiotics. Here the balance between beneficial and potentially harmful effects can be rather delicate. Bacterial azo- and nitroreductases may lead to the formation of carcinogenic aromatic amines, or the bacteria may produce glucosidases and glucuronidases releasing toxic compounds in their active form from their glycoconjugates. Likewise the bacterial deconjugation of bile acids and the formation of 7-dehydroxylated secondary bile acids have been implicated as one of the factors in colorectal cancers.

An important phenomenon is the formation of butyrate and other short-chain fatty acids as products of carbohydrate fermentation by the strict anaerobes in the gut. Butyric acid appears to have a range of physiological functions in the gut epithelium both as a nutrient and as a controller of gene expression. Both epidemiological studies and laboratory tests on different cell lines indicate that butyrate prevents some of the cellular steps leading to cancerous growth.

Probiotic Effects

Given the many functions of the intestinal microflora the idea of keeping its microbial composition optimal by consuming bacteria with beneficial gastrointestinal effects is a natural one, and promoted by the notable immunologist Elie Metchnikoff more than 90 years ago. Most attention has been paid to lactobacilli and bifidobacteria. The reasons for this have been mainly intuitive, and the evidence of the beneficial effects largely anecdotal. However, rapidly accumulating reliable clinical data have confirmed at least some of the claims associated with these bacterial groups (see Fig. 1).

Most direct evidence for the beneficial probiotic action has been obtained from cases where there has been a severe disturbance in the normal intestinal microbial balance, i.e. bacterial and viral diarrhoeas, lactose malabsorption, antibiotic or surgical interventions or radiotherapy-induced intestinal disturbance. Anticarcinogenic properties associated with certain strains are based on animal experiments with model carcinogens. Clinical evidence of positive effects on certain food allergies have recently been obtained, whereas so far no consistent results are available on the possible lowering of the serum cholesterol as a result of probiotic consumption.

The mechanisms of probiotic action are still unclear. In case of alleviation of lactose malabsorption the lactase enzymes of the probiotic stains have an obvious function. Most of the successful probiotic strains have been shown to temporarily colonize the gastrointestinal tract, and thus they probably help to preserve and re-establish the normal composition of microflora. One of the results of colonization may be a general immunomodulation of the host, and this could be one of the factors behind the beneficial effects observed in some infections. The production of antimicrobial substances effective against pathogens or opportunistic pathogens may also play a role. The very low activities of glucosidases of glucuronidases typical for strains used in cancer prevention experiments may result in increased secretion of conjugated and detoxified carcinogens.

Probiotic Strains on the Market

A wide variety of human probiotic preparations are available commercially. Most of them are sold as pharmaceuticals (capsules, powders, tablets, suspensions), and very little information about their clinical efficiency is publicly available. Only a limited number of strains have been successful as components of functional foods (see below) with clearly shown beneficial effects. One of the oldest is *Lactobacillus casei* Shirota isolated already in the 1930s in Japan and forming the basis of the Yakult® products. *Lactobacillus rhamnosus* GG, isolated after a systematic search from human faeces in 1985, is currently probably the most thoroughly studied probiotic strain in the world, and food products based on it are licensed in at least 15 countries. Other clinically proven commercial strains include, among others, *Lactobacillus johnsonii* LA1, *Lactobacillus acidophilus* NCFB 1748, *Bifidobacterium longum* BB536 and *Bifidobacterium lactis* BB12. With the increased research effort, the number of successful strains is growing, and some of the marketed strains and potential candidates are listed in **Table 2**, together with their reported beneficial effects. Following the developments in probiotic research, the culture suppliers are now providing a constantly increasing collection of starter cultures with probiotic bacterial strains which can be applied to the manufacture of various dairy and non-dairy products.

Table 2 Reported effects and health benefits of selected probiotic bacterial strains (modified from Salminen et al 1996)

Strain	*Reported effects*
Lactobacillus acidophilus LA1	Immune enhancer, adjuvant, adherent to human intestinal cells, balances intestinal microflora
Lactobacillus acidophilus NCFB 1748	Lowering faecal enzyme activity, decreased faecal mutagenicity, prevention of radiotherapy related diarrhoea, treatment of constipation
Lactobacillus rhamnosus GG (ATCC 53013)	Prevention of antibiotic associated diarrhoea, treatment and prevention of rotavirus diarrhoea, treatment of relapsing *Clostridium difficile* diarrhoea, prevention of acute diarrhoea, Crohn's disease, antagonistic against cariogenic bacteria, vaccine adjuvant
Lactobacillus gasseri (ADH)	Faecal enzyme deduction, survival in the intestinal tract
Lactobacillus reuteri	Colonizing the intestinal tract, many animal studies, possibly an emerging human probiotic
Streptococcus thermophilus; Lactobacillus bulgaricus	No effect on rotavirus diarrhoea, no immune enhancing effect during rotavirus diarrhoea, no effect on faecal enzymes
Bifidobacterium bifidum	Treatment of rotavirus diarrhoea, balancing intestinal microflora, treatment of viral diarrhoea
Bifidobacterium lactis BB12	Prevention of traveller's disease, colonizing the intestinal tract
Bifidobacterium longum BB536	Decrease in levels of intraintestinal putrefactive products, prevention of diarrhoea, improvement of constipation

Characteristics of Probiotic Dairy Products

Fermented dairy products are the most commonly used vectors for probiotics. Conventional products (yoghurt, cultured milks) usually contain bacterial populations of around 10^8 living bacterial cells per gram. This is also in agreement with local product specifications and the legal situation. According to experience, probiotic bacteria are more sensitive against the influence of low pH and oxygen than the traditional, 'technologically proven' microflora as for instance used for yoghurt fermentation. Depending on their specific nature, the probiotic bacteria may either contribute to fermentation and acidification (e.g. *Lactobacillus acidophilus*) or they can be regarded as pure additives (e.g. some bifidobacterial strains or others incorporated in non-fermented products). Their viability and stability during fermentation and storage conditions as well as their availability in the fresh and the stored product can be regarded as cardinal points determining the specific quality of the food. From the sensory point of view, most of the probiotic products have a relatively mild taste and a reduced tendency for over-acidification (pH reduction caused by the metabolic activity of the living microorganisms despite cooling) during storage of the product in the retail shops. During manufacture in dairy technology, most probiotic bacterial cultures are applied to the products mainly as deep-frozen concentrates, pellets or freeze-dried powders which can be directly added to the process milk. In a few cases, conventional culture propagation via preparation of a mother culture is used. A novel and alternative way of incorporating certain probiotics in special foods is the application of encapsulated bacterial cultures. Because of its expensiveness this technology has so far only been used for the production of some infant formula and special pharmaceutical preparations.

Quality Parameters and Selection of Probiotic Bacteria

In addition to the above mentioned prerequisites, probiotic bacteria have to fulfil several requirements because of safety reasons and in order to ensure the desired effects. Most of the commercially used strains are the result of extensive systematic isolation and screening experiments. In general, probiotic bacteria have to meet six main criteria before they can be commonly applied.

1. They have to originate from the intestine of healthy humans. This specific origin implies a long-term familiarity and experience of the human organism with the bacterial strains.
2. They should have a clear identity based on up-to-date differentiation and on molecular-biological methodology. Furthermore, there has to be a guarantee that the strains do not undergo any biological change during conservation and industrial use.
3. They have to be 'generally regarded as safe' (*GRAS*) and should have proven apathogeneity, atoxicity and a defined antibiogramme pattern. Microorganisms attributed to security class 1 ('no health risk') according to the Official Grouping of Biological Agents (Sichere Biotechnologie: Merkblatt B006: Berufsgenossenschaft der chemischen Industrie) are preferred. Those attributed to class 2 ('single strains may possibly cause a disease') can be used only if there is proven evidence and long-term experience regarding their safety.

Table 3 Culture media for the selective detection and enumeration of the most commonly used probiotic bacteria in fermented dairy products

Bacteria	*Media*
Lactobacillus acidophilus	Aesculin cellobiose agar, TGV agar, X-Glu agar, MRS salicin agar
Lactobacillus rhamnosus	MRS agar containing vancomycin (50 μg ml^{-1})
Bifidobacterium spp.	MSB medium, MRS NNLP medium, Bif agar, RB agar
Streptococcus thermophilus, Lactobacillus delbrueckii subsp. *bulgaricus*	M17 agar and MRS agar, acidified

4. They have to constitute a high resistance against gastric juice, bile acids and digestion enzymes to facilitate their passage to the desired segments of the intestine.
5. They should exert defined beneficial effects to the humans which enable a clear distinction from 'conventional' starter cultures.
6. They have to meet several requirements which are necessary for a convenient, large-scale industrial production of foods (e.g. sensory properties, stability, viability). They need not necessarily contribute to the fermentation of the product.

Examination of Probiotic Dairy Products

Normally, probiotic bacteria exert their beneficial effects when ingested repeatedly at relatively high cell counts (> 10^6 cfu per gram of product of > 10^8 cfu per portion). Therefore, the manufacturers of probiotic products have to aim at making products which meet these requirements in order to guarantee the specific value. Because of these reasons there is a need for methods of quality control which enable reliable detection and enumeration of the various bacteria contained in the fermented products. Today, standard methods are being sought which are selective enough to distinguish qualitatively and quantitatively among the mixed microflora of fermented milk products. Based on the standard methodology of the International Dairy Federation (IDF) which has been established for routine examination of the yoghurt microflora, several culture methods and media have been proposed which can be applied for the assessment of yoghurt-related fermented milk products (**Table 3**). They are based either on inhibiting the growth of the accompanying bacteria using antibiotic additives to the medium or on specifically promoting growth of the special bacteria. Alternatively, the detection of defined bacteria by visualizing their specific enzymatic activities via chromogenic or fluorogenic dyes has been proposed. Moreover, the application of DNA- and RNA-based methods to detect the bacteria in various materials has become an area of growing importance, particularly for the examination of faecal material and in clinical studies.

Legal Aspects

Legally, microorganisms and cultures belong, in general, to the broad category of food additives which are part of an ingredients' list defined in national and multinational food law. Usually, they can be used in foods without any particular permission. It should be mentioned that this situation is different to the law valid for animal feeds which do require a special permission. According to EU law, it is also permitted to point out that the consumption of a probiotic food can be health-promoting if this statement is scientifically proven. On the other hand, it is not allowed to state that the particular probiotic food is able to prevent a defined illness. It is difficult for the marketing strategists to find the correct balance between the above-mentioned requirements. Primarily, in most European countries the statements for probiotic products have to meet two main criteria. (1) They must not be misleading to the consumers; and (2) they must not pretend to prevent or cure illness.

Whereas traditional yoghurt has to contain a typical microflora consisting of *Lactobacillus delbrueckii* subsp. *bulgaricus* and *Streptococcus thermophilus*, according to European quality standards, probiotic products are not so clearly defined regarding typical microorganisms and cell count levels. At present, several groups of experts are studying this topic, aiming at establishing a scientific basis for the evaluation of probiotics.

Future Trends in Probiotic Research

Probiotic foods can be ascribed to the novel branch of 'functional foods' which when ingested serve to regulate a particular body process. They can contribute to biological defence mechanisms, prevention of diseases, recovery from diseases, control of physical conditions and influence the ageing process. The incorporation of probiotic bacteria with scientifically supported health claims in foods may have great potential for improving the quality of life, and already constitute a rapidly growing market. There has been a great interest in the isolation and development of new probiotic strains during recent years. Results have shown that improved methodologies are urgently needed for screening new strains for functional food uses in order to facilitate rapid identification of specific properties suitable for this purpose. In current research projects targeted properties of successful probiotic strains are being validated, and important properties have been considered for future assessment and methodology development. Another research area related to that of probiotics is dealing with special carbohydrate sources (prebiotics), mainly oligosaccharides, which promote specifically the growth of beneficial and probiotic bacteria in certain regions of the intestine. Efforts have been made to combine pro- with prebiotics in a so-called synbiotic product which may exert a twofold beneficial effect to the human organism by providing not only the probiotic bacteria but also their food as a strong support for local activity and/or colonization in the colon.

Fermented dairy foods and cultures have a long history and large consumption in the diet of several countries. Industrial products, including cultured dairy products, and their probiotic properties have been studied for many years. Currently, a representative number of bacterial strains provided by European culture suppliers is being studied regarding in vitro cytokine release effects, adhesive properties, antimutagenicity, and behaviour in gastrointestinal tract models. Besides these specific effects, technological and production properties are also considered. New ways of predicting the probiotic properties and screening old and new strains will have to be developed. According to present knowledge, no single in vitro test is sufficient in selecting probiotics for targeted uses. There is some evidence that strains have to be tested on their inherent merit since even biochemically closely related strains may be very different. There is a clear need to validate the in vitro results with human clinical studies using the same probiotic strains to allow any firm conclusion on the predictive potential of the in vitro tests.

Europe has traditionally had a leading position on the probiotic market. However, there is considerable confusion and scepticism among consumers, consumer organizations, and the scientific community about the claims associated with probiotic products. To eliminate these hurdles, scientifically sound knowledge is needed. Topics of particular interest include the safety of probiotics, investigation of the role played by plasmids encoding for resistance factors, dose–response effects, health claims for probiotic products with regard to management of intestinal disorders and immune enhancement, pharmacokinetics of probiotics, tools for probiotic research and clinical testing of probiotics and, last but not least, technological suitability of probiotic strains. To speed up adaptation of the probiotic food technology, and to enhance the attractiveness of new probiotic foods, it is essential to demonstrate an up-to-date basis for marketable claims by presenting the health and nutritional benefits of probiotic bacteria and foods. This knowledge should be then made available to the extended audience consisting of industries, authorities and consumers.

See also: ***Staphylococcus***: *Staphylococcus aureus*.

Further Reading

Collins JK, Thornton G and Sullivan GO (1998) Selection of probiotic strains for human applications. *International Dairy Journal* 8: 487–490.

Conway P and Henriksson A (1994) Strategies for isolation and characterisation of functional probiotics. In: Stewart A and Gibson W (eds) *Human Health: The Contribution of Microorganisms*, p. 75. London: Springer-Verlag.

Fuller R (ed.) *Probiotics 2. Applications and Practical Aspects*, p. 1. London: Chapman & Hall.

Havenaar R and Huis in't Veld J (1992) Probiotics: a general view. In: Wood JB (ed.) *The Lactic Acid Bacteria in Health and Disease*, p. 151. London: Elsevier Applied Science.

Hill MJ (1995) The normal gut bacterial flora. In: Hill MJ (ed.) *Role of Gut Bacteria in Human Toxicology and Pharmacology*, p. 3. London: Taylor & Francis.

Kanbe M (1992) Uses of intestinal lactic acid bacteria and health. In: Nakazawa Y and Hosono A (eds) *Functions of Fermented Milk – Challenges for Health Sciences*, p. 289. London: Elsevier Applied Science.

Kneifel W and Bonaparte C (1998) Novel trends related to health-relevant foods: 1. Probiotics. *Ernaehrung/Nutrition* 22: 357–363.

Lee Y-K and Salminen S (1995) The coming of age of probiotics. *Trends in Food Science and Technology* 6: 241–245.

Salminen S and von Wright A (1998) Current probiotics – safety assured? *Microbial Ecology in Health and Disease* 10: 68–77.

Salminen S, Deighton M, Gorbach SL (1993) Lactic acid bacteria in health and disease. In: Salminen S and von Wright A (eds) *Lactic Acid Bacteria*, p. 199. New York: Marcel Dekker.

Salminen S, Isolauri E and Salminen E (1996) Clinical uses of probiotics for stabilizing the gut mucosal barrier: successful strains and future challenges. *Antonie van Leeuwenhoek* 70: 347–358.

Sanders ME (1994) Lactic acid bacteria as promoters for human health. In: Goldberg I (ed.) *Functional Foods, Designer Foods, Pharmafoods, Nutraceuticals*, p. 294. New York: Chapman & Hall.

Ziemer CJ and Gibson GR (1998) An overview of probiotics, prebiotics and synbiotics in the functional food concept: perspectives and future strategies. *International Dairy Journal* 8: 473–479.

Probiotics *see* ***Bifidobacterium***; **Microflora of the Intestine**: The Natural Microflora of Humans.

PROCESS HYGIENE

Contents

Designing for Hygienic Operation

G C Gürakan and **T F Bozoğlu**, Department of Food Engineering, Middle East Technical University, Ankara, Turkey

Design of hygienic operation is essential for food safety and successful implementation of Hazard Analysis and Critical Control Point (HACCP). In designing a hygienic operation, buildings, installations, equipment, nature of the product, scale of operation, reception, storage, transportation, air and water quality, cleaning and disinfection, personnel should also be on the checklist. Design and location of processing equipment within the plant, personnel hygiene, cleaning and disinfecting processes applied to the plant and equipment are all important factors for a safe product. The US Food Safety and Inspection Service (FSIS) proposed that sanitation Standard Operating Procedures (sanitation SOPs) are necessary because they clearly define each establishment's responsibility consistently to follow effective sanitation procedures and substantially minimize the risk of direct product contamination and adulteration. As FSIS proposed, both preoperational and operational sanitation procedures should be identified by each establishment. The preoperational sanitation programme includes cleaning of the general equipment and facility. All equipment is cleaned and disinfected before production begins. Each day the quality control (QC) manager should perform a sanitation inspection after preoperational equipment cleaning and disinfecting. The results of the inspection should be recorded. The QC manager should also perform daily microbial monitoring for total plate counts. If microbial counts are too high, the QC manager should notify the sanitation manager and try and determine the cause of this raised level. In this situation, all cleaning and disinfecting procedures and personal hygiene should be reviewed. Preoperational sanitation procedures should be distinguished from sanitation activities which are carried out during the operation. Processing operations should also be performed under sanitary conditions to prevent direct and cross contamination of food products. Employee hygiene practices, sanitary conditions and cleaning procedures should be maintained throughout the production shift (e.g. from slaughter to processing in a meat or poultry plant).

Hygienic practices for processing include the following:

1. Employees clean and disinfect all equipment, conveyer belts, tables and other product contact surfaces during processing to prevent contamination of food products
2. Employees take appropriate precautions when going from a raw product area to a cooked product area to prevent cross contamination of cooked products. Employees change their outer garments, wash and disinfect their hands with an approved hand-disinfectant, put on clean gloves for that room and step into a boot-disinfecting bath on leaving and entering the respective rooms
3. Raw and cooked processing areas are kept separate. There is no cross-utilization of equipment between raw and cooked products.

Nature of the Product and Scale of Operation

Buildings and operations are designed according to the nature of the product. The premises should be of sufficient size for the intended scale of operation and should be sited in areas which are free from problems such as a particular pest nuisance, objectionable odours, smoke or dust. The buildings should be large enough to maintain the necessary separation between processes to avoid the risk of cross contamination.

In food production, the raw materials are diverse. In some food-processing operations, cleaning and disinfection need to be done every few hours. The timing must be determined for the specific food and the nature of the operation. The soil type, which varies with the nature of the product, can be fat deposit, blood, milkstone and other organic or inorganic deposits such as metallic ones on processing equipment. It is important to identify the soil type and use the most effective detergents and sterilants.

The checks for raw material should include physical examination, microbiological tests, organoleptical assessment and inspection for evidence of pest infestation. These examinations are for high-risk category products. Depending on the nature of the product, the storage conditions for raw materials should be designed. The conditions should not be detrimental to the product; temperature control integrity is maintained and products should not be at risk from pests. Ingredients must be taken out of their boxes or bags or decanted from their outer packaging in a specific area. All unpacked materials should be transferred to the processing or production area in blue-coloured plastic bags or clean, inert containers, e.g. stainless steel tote bins or plastic trays. The type and level of microorganisms vary with the nature of the product. Besides natural microorganisms, a food can be contaminated with microorganisms from outside sources such as air, soil, water, humans, food ingredients, equipment, packages and insects.

Equipment, Vats, Pipework and other Plant Items

Equipment and its failings can also be the source of product contamination. Inadequate cleaning can be the source of cross contamination. In some equipment, small parts, inaccessible sections and certain materials may not be sufficiently cleaned and disinfected. Dead spots can serve as sources of pathogenic and spoilage microorganisms in food. When designing hygienic food-processing equipment, protection of food being processed from microbial contamination should be considered. Basic principles of hygienic design are given below:

1. All surfaces in contact with food should not react with the food and must not yield substances that might migrate to or be absorbed by food
2. All surfaces in contact with food should be microbiologically cleanable, smooth and nonporous to avoid particles being caught in microscopic surface crevices, becoming difficult to dislodge and a potential source of contamination
3. All surfaces in contact with food must be visible for inspection, or the equipment must be readily dismantled for inspection, or it must be demonstrated that routine cleaning procedures eliminate the possibility of contamination
4. All surfaces in contact with food must be readily accessible for manual cleaning, or if clean-in-place (CIP) techniques are used, it should be demonstrated that the results achieved without disassembly are equivalent to those obtained with disassembly and manual cleaning
5. All interior surfaces in contact with food should be arranged so that the equipment is self-emptying or self-draining. When designing equipment it is important to avoid dead space or other conditions which trap food and allow microbial growth to take place
6. Equipment must be designed to protect contents from external contamination. Product should not be contaminated by leaking glands, lubricant drips and the like, or through inappropriate modifications or adaptations
7. Exterior surfaces of equipment not in contact with food should be arranged to prevent the harbouring of soils, microorganisms or pests in equipment, floors, walls and supports. For example, equipment should fit either flush with the floor or sufficiently raised to allow the floor underneath to be readily cleansed
8. Where appropriate, equipment should be fitted with devices which monitor and record its performance by measuring factors such as temperature/time, flow, pH and weight.

Food-processing equipment becomes soiled with food residues in the course of use and can act as a source of microbiological contamination. Hygienic processing of food requires that both premises and equipment are cleaned frequently and thoroughly to restore them to the desired degree of cleanliness. The equipment should be arranged and located to permit easy access and cleaning, such as 90 cm off the floor, 45 cm from the ceiling and 90 cm from the wall and other equipment.

Microbial biofilm formation is one of the main

problems in food industry and it is very difficult to clean. Initial treatment with hydrolysing enzymes may be necessary. Following enzyme treatment, EDTA, together with quaternary ammonium compound treatment, can be used. In addition, design of processing equipment may be necessary to prevent or control biofilm formation.

After cleaning and disinfecting, tanks and containers should be inspected for cleanliness and for state of repair. Control is obtained through hygienic design of the equipment and properly evaluated cleaning and disinfecting regimes. Tanks, vessels, vats and pipes must be designed to prevent contamination, particularly by rodents and other pests, birds, dust and rain.

Tanks, plate heat exchangers, pipelines and homogenizers are examples of equipment which can be cleaned and disinfected by CIP systems. CIP systems are capable of cleaning storage tanks, vats and other storage containers by use of spray balls and provide the advantage of high hygienic standards with lower labour cost. Fixed or rotating spray balls produce a high-velocity jet of liquid to remove residual soil or other contamination. Pipelines and other plant items can be cleaned by high-velocity water ($>15\ m\,s^{-1}$) and appropriate detergents, which are recirculated.

Extreme temperature and abundant use of water and steam is one general problem of food processing. This creates a potential for water condensation on pipes and surfaces. This can also lead to microbiological contamination of exposed food. Proper vents, steam traps and guttering to void water are some of the control measures used. Dead ends or tees or low spots in pipework should be eliminated for internal surfaces to drain readily. Mixing vats should be covered, overhead pipes and exposed-beam ceilings should be eliminated. Pipework, light fittings and other services should be sited to avoid creating difficult-to-clean recesses or overhead condensation. Piping should not be exposed over the product stream. Horizontal surfaces (pipe hangers, beams, ductwork) over exposed product areas should be eliminated. Weld joints should be continuous-welded and ground smooth in food contact equipment, including pipelines.

Critical Points of High Contamination Risk

Foods can become contaminated during processing due to cross contamination via contaminated raw food or personnel, aerosols, malfunctioning or improperly disinfected equipment, misuse of cleaning materials, rodent and insect infestations and improper storage. Particularly, contamination from raw materials forms the major contamination source within the plant. For example, raw milk may be contaminated with pathogens such as *Salmonella*, *Brucella*, *Campylobacter*, *Listeria monocytogenes* and *Mycobacterium bovis* and psychrotrophic spoilage organisms such as pseudomonads. Hence control of raw milk delivery is considered to be a main point in reducing the risk of contamination. This control can be achieved by inspection after cleaning and disinfecting the tanks and containers, keeping the temperature constant during transportation and testing raw milk before use in the dairy factory. Similarly, the main source of potential hazard by pathogens in meat and poultry products is faecal contamination of carcasses during slaughtering. If insufficient care is taken when handling and dressing during slaughter and processing, the edible portions of the carcass can become contaminated with disease-causing bacteria. If these organisms are introduced into the environment, they may be transmitted from one carcass to another. Therefore, preventing moving faecal pathogens is vital and will also be a critical part of the HACCP plan in any slaughter establishment.

The overall layout of the plant should be designed to ensure a smooth flow from reception of raw materials and storage to product storage and dispatch. The production layout should include physical separation to prevent cross contamination via the operation or personnel. Areas may be designated as high-risk or low-risk depending on the sensitivity of the materials being handled and the process used. High- and low-risk areas should be physically separate, should use different sets of equipment and utensils, and workers should be prevented from passing from one area to the other without changing their protective clothing and washing their hands. The principal situation where such a separation would be required is between an area dealing with raw foods, particularly meat, and one handling the cooked or ready-to-eat product. It is essential that no cross contamination from raw/low-risk areas to high-risk areas occurs via drainage systems. The ideal system will provide two separate drainage patterns – one for raw/low-risk areas and another for high-risk areas, connecting via main below-ground drains via trapped inlets.

Chilling should also take place in high-risk areas. In chill rooms the major factors influencing the microbial growth are moisture and temperature. Psychrotrophic organisms need high humidity for growth. Psychrotrophic organisms growing in cold rooms are *Pseudomonas*, *Acinetobacter*, *Moraxella*, psychrotrophic Enterobacteriaceae, *L. monocytogenes* and psychrotrophic moulds. All such organisms prefer moisture to grow. Any excess moisture should be rapidly removed. By applying a bactericidal gel or detergent sanitizer, pipework can be cleaned.

The plant should be designed to restrict non-essential personnel from passing through high-risk processing and packaging areas. Food content packaging or primary packaging should be carried out in the high-risk area. The operation of outer packaging should be completed outside the high-risk area. Many types of packaging materials are used in the food industry. Since they are in direct contact with food and in some cases used in products which are ready-to-eat, the proper microbiological standards for packaging materials are essential.

Risks from the Environment

The environment in which food processing is conducted is an important factor in determining product quality. It is important that the buildings provide a comfortable and pleasant working environment conducive to good hygienic practices. The microflora of processing plants is composed of microorganisms that gain entry from the air and water and by animals, raw materials, dust, dirt and people. Improperly cleaned equipment or facilities may serve as vehicles of contamination. A sanitary food environment is free of insects, rodents, birds and other contamination sources.

Building

In processing areas, floors should be made of durable material which is impervious, non-slip, washable and free from cracks or crevices that may harbour contamination. Internal walls should be smooth, impervious, easily cleaned and disinfected and light-coloured. Ceiling areas should be designed to prevent the accumulation of dirt and debris, which could contaminate food products during processing. Floor–wall junctions should be coved to facilitate cleaning with a minimum 2.5 cm radius. Ceilings should be light-coloured, easy to clean, and constructed to minimize condensation, mould growth and flaking. Toilets should not open directly on to food-processing areas and must be provided with hand-washing facilities supplied with hot water, soap and hand-drying facilities.

Air

Collection of dust encourages infestation or bacterial growth. Microorganisms do not grow in dust, but are transient and variable depending upon the environment. Their level changes with the degree of humidity, size and level of dust particles, temperature and air velocity and resistance of microorganisms to drying. Generally dry air with low dust content and higher temperature has a low microbial level. The organisms which can be predominantly present in air are spores of *Bacillus* spp., *Clostridium* spp., moulds and some Gram-positive (e.g. *Micrococcus* spp. and *Sarcina* spp.) species as well as yeasts. If the surroundings contain a source of pathogens (e.g. animal farms or sewage treatment plant), different types of bacteria, including pathogens and viruses (including bacteriophages), can be transmitted via the air. Proper ventilation facilities to prevent dust entering the processing plant should be installed. Careful maintenance of ventilation and air filtration systems is needed to minimize airborne transfer of *Salmonella* in some plants such as a dried-milk-producing plant. Rooms designed for aseptic filling of foods, e.g. UHT milk, may require devices for sterile filtration of air. Moisture on the surface of unpackaged foods or ingredients enhances microbial growth. In warm areas effective ventilation can remove excessive heat, steam, aerosols and smoke. In addition to air filtration, using positive air pressure, reducing the humidity level and installing UV light can also be preventive measures for airborne contaminants including insects from the air.

Water

Water is used to wash and process foods, such as in canning and cooling of heated foods, and to wash and disinfect equipment. It is also used as an ingredient in many processed foods. Thus water quality can greatly influence microbial quality.

Water from unsafe sources has frequently caused enteric infections and intoxications, particularly *Salmonella enterica* serovar *typhi* infections and shigellosis. If there is no public supply or plant-owned well, adequate storage facilities made from non-corroding, nontoxic material must be provided. Chlorine-treated potable water (drinking water) should be used in processing, washing, disinfection and as an ingredient. Although potable water does not contain coliforms and enteric pathogens, it can contain spoilage bacteria. To overcome the problems of spoilage bacteria such as *Pseudomonas*, *Alcaligenes* and *Flavobacterium*, many food processors use water, especially as an ingredient, that has a higher microbial quality than potable water.

Soil

Bacillus aureus, *B. subtilis* and *B. licheniformis* are common in soil and can easily enter food premises. Many other bacterial genera as well as moulds and yeasts can get into foods from the soil. Soil contaminated with faecal materials may be the source of enteric pathogenic bacteria and viruses in food. Removing soil (and sediments) and avoiding soil contamination reduce microorganisms in foods from this source.

Personnel

All staff working in a food factory should be trained to be aware of the standards of personal hygiene. They are responsible for the quality and safety of the products they manufacture. Personnel handling food have been the source of pathogenic microorganisms in foods that later caused food-borne diseases, especially with ready-to-eat foods. In particular, high-risk staff should be aware of the medical screening procedures. All staff should be subject to health screening, including part-time and seasonal workers. *Staphylococcus aureus* is usually transferred to cooked food from personnel, and occasionally from pets and pests. Cross contamination within food-handling areas occurs by many vehicles, including personnel. Separation of clean from unclean sections and processes is crucial.

See also: **Hazard Appraisal (HACCP)**: The Overall Concept. **Process Hygiene**: Types of Biocides; Overall Approach to Hygienic Processing; Modern Systems of Plant Cleaning; Risk and Control of Airborne Contamination; Testing of Disinfectants; Hygiene in the Catering Industry.

Further Reading

Adams MR and Moss MO (eds) (1995) *Food Microbiology.* Science Park, Cambridge: Royal Society of Chemistry.

Food Safety and Inspection Service (1996) July 1996 Final rule. *Pathogen Reduction: Hazard Analysis and Critical Control Points (HACCP) Systems.* 9 CFR Parts 304, 308, 310, 320, 327, 381, 416 and 417; docket No. 93-016F, RIN 0583-AB69. Springfield, VA: NTIS, United States Department of Commerce.

Giese JH (1991) Sanitation: The key to food safety and public health. *Food Technology* 45: 74–80.

Gill CO, McGinnis JC and Badoni M (1996) Assessment of the hygienic characteristics of a beef carcass dressing process. *Journal of Food Protection* 59: 136–140.

Ray B (ed.) (1996) *Fundamental Food Microbiology.* New York: CRC Press.

Types of Biocides

J F Williams, HaloSource Corp., Seattle, WA, USA

S D Worley, Department of Chemistry, Auburn University, USA

Introduction

Industrial production and rapid distribution of foods have steadily increased the potential for amplifying hygienic inadequacies in processing, transport, storage and preparation of foods. Food safety measures introduced to counter these trends depend in part on the use of chemical antimicrobial agents, or biocides. However, their use in turn raises additional health and safety issues concerning food residues, and the environmental impact on waste streams at all levels of the production, transportation and consumption chain. Contemporary approaches to the incorporation of antimicrobial compounds into food products for purposes of preservation and spoilage avoidance are addressed elsewhere in this volume, as are physical methods such as temperature and irradiation. Our focus in this article is on the chemical agents used in food manufacturing and preparation, on food products themselves, in the immediate environment, and on the hands of food handlers. These measures are aimed at reducing opportunities for microbial contamination and spoilage, and enhancing prospects for delivery and consumption of safe food products, thereby minimizing risks to public health from food-borne diseases.

These use patterns obviously constrain the scope of antimicrobial compounds and formulations available and acceptable to the food industry. We review the nature of the major chemical biocidal classes below, and those characteristics and benefits particularly suited to food industry use are identified. The significance of these applications to waste streams, and the degradation pathways of the compounds, are also reviewed. Finally, we discuss the mechanisms of action and comparative features of the currently used biocides.

Chemical Classification of Biocides

The biocidal compounds used in so many applications in food production and processing and handling have to be able to bring about one paramount change: a reduction in total microbial numbers, ideally to a level that obviates risks of spoilage, extends shelf life and eliminates disease transmission. There are other benefits which accrue from their use, such as odour control, and a more attractive workplace (less attract-

Table 1 Ideal characteristics of food industry biocides

Rapid, broad-spectrum biocidal action
Odourless
Compatible with a wide variety of material surfaces (e.g. without bleaching, discolourant or corrosive effects)
Compatible with formulation components (e.g. cleaning agents, buffers, foaming gels)
Water-soluble and easily washed away
Stable on storage
Easy to use (e.g. to dilute, to apply)
Safe/non-irritant for the user
No safety hazards/residues that may harm consumers
No adverse effects on desirable features of food product (e.g. taste, appearance, texture)
Economically attractive

ive to vermin and flies, for example), but these are ancillary to the main goal of killing microorganisms. To meet this end, biocides, depending on their nature, may be applied to the food product surface itself, to the equipment (for food processing, transportation, preparation), the immediate environment (plant, kitchen, surfaces) and to the food handlers.

Food handlers are increasingly recognized as an important component of food hygiene maintenance, because their potential to disseminate pathogenic contaminants is enormous. Fewer than 5% of food-borne illness incidents are considered to come to light in the US, and estimates of annual cases run into tens of millions. However, when incidents are reported and investigated by public health authorities, one of the commonest findings is that there was a breakdown in food-handling hygiene. While the use of gloves as barriers is encouraged or required, the shortcomings of disposable gloves are well known; they frequently leak, and after a short time of wear the numbers of skin bacteria under the glove have increased rapidly. There is no substitute for personal hand care to improve safety levels, and the use of biocidal preparations is now commonly involved in these measures.

The most desirable features of biocidal agents for use in the context of the food production and preparation industry are shown in **Table 1**.

It is not possible for any one agent to display all these characteristics, and besides, the diversity of applications in the industry (e.g. for hands, for kitchen cutting boards and for carcasses) demands different combinations of these ideal attributes. Moreover, some of the most significant chemical agent groups in the broad categories of disinfectant/sanitizers commonly used today in medical, agricultural and consumer fields are unfortunately inappropriate even for consideration in this food production segment of industry. Compounds such as glutaraldehyde, formaldehyde, phenols, ozone and alcohol-based formulations, all widely used in other settings, often as sterilants as well as disinfectants, have little or no place in food production and processing. Strong oxidizers such as hydrogen peroxide and peroxide/peracetic acid mixtures can also be used for medical instrument sterilization and in industrial tubing disinfection, and skin and wound antisepsis, but have very limited utility in the food industry. These kinds of compounds and formulations are precluded by virtue of such traits as their powerful odours, corrosive natures, bleaching or discolouring actions, or because of their intrinsic toxicities to users, consumers, or the environment – or all three in the case of the aldehydes.

Table 2 Principal biocides currently used in the food industry

Environmental sanitizers:
Quaternary ammonium compounds
Iodophores
Chlorine-based agents
Hand sanitizers:
Quaternary ammonium compounds
Iodophores
Chlorohexidine
Polyhexamethylene biguanide
Parachlorometaxylenol
Triclosan
Carcass decontaminants:
Hypochlorites
Chlorate (source of ClO_2)
Organic acids

Table 2 lists the principal biocides currently accepted in the food industry: in the environment, on the handlers and on the food products. The molecular structures of these compounds are shown in **Figures 1–3**.

Physical and Chemical Properties of Biocides Acceptable to the Food Industries

Quaternary Ammonium Compounds (QACs or QUATS)

These cationic surfactant chemicals (Fig. 1) are stable to storage and heat, water-soluble, non-odorous and function as biocides over a wide range of pH (although they work optimally at slightly alkaline pH), and are little affected by water hardness. At use dilutions they are generally non-irritating to the skin, and are readily mixed and diluted from concentrates. Concentrates themselves are quite corrosive on skin and mucous membranes. QACs have evolved through several generations over the last few decades, but all tend to adhere in stable non-volatile forms to inanimate surfaces, providing for persistence of some degree of biocidal action and odour control for

Quaternary ammonium compound

$R^1 - N^+(CH_3)_2 - CH_2 - C_6H_5$, X^- $R^1 = C_{12}-C_{18}$, X = Cl, Br

Hypochlorites

$NaOCl$ $Ca(OCl)_2$

Sodium chlorate

$NaClO_3$

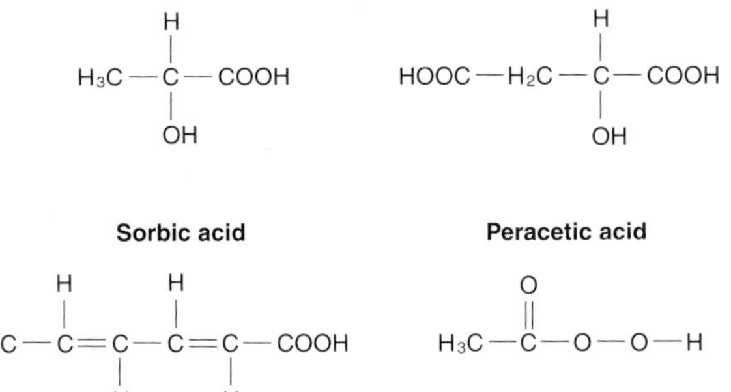

Figure 1 Molecular structures of biocides currently used in the food industry.

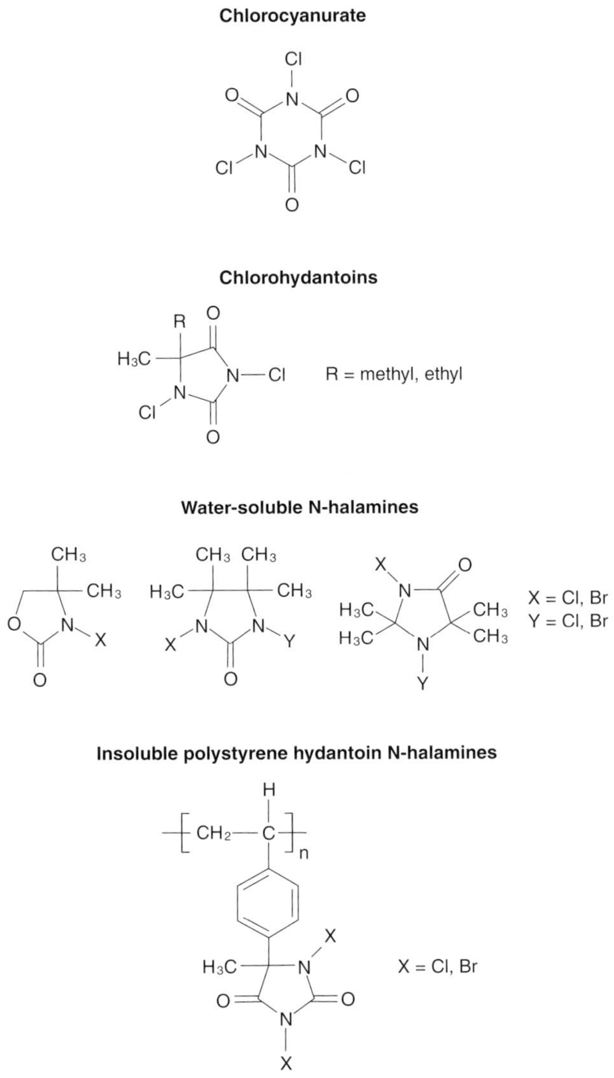

periods of hours to days, depending on temperature and moisture conditions.

There are limits to the extent to which they can be successfully formulated with detergents and soaps for cleaner/sanitizer preparations because of inherent incompatibilities with these compounds. Their spectrum of action, at least in the concentrations and formulations most suited to environmental sanitation in food manufacturing, is rather limited. This is especially the case for some of the Gram-negative pathogens and nuisance organisms that are so problematic in this context. They are effective against skin flora, particularly staphylococci, which can cause food poisoning. QACs are therefore used in some water-based hand rinse sanitizers for food industry personnel. Their ineffectiveness against *Pseudomonas* and other Gram-negatives renders these formulations subject to contamination, and these organisms may even grow luxuriantly in the nutrient-rich medium provided by the excipients in the formulations.

Iodophores

Iodine is a highly effective biocide with none of the spectrum of activity limitations of the QACs. While it enjoys wide use in alcoholic tinctures or aqueous solutions in medicine, the intense staining, irritant effects and strong metallic odour and taste of these

Chlorocyanurate

Chlorohydantoins

R = methyl, ethyl

Water-soluble N-halamines

X = Cl, Br
Y = Cl, Br

Insoluble polystyrene hydantoin N-halamines

X = Cl, Br

Figure 2 Molecular structures of biocides which stabilize chlorine for use in the food industry.

preparations preclude food industry applications in these forms. Instead, iodine complexed with organic carriers, as iodophores (Fig. 3), have become a mainstay of environmental sanitation in the food industry. Surfactant characteristics of some of these carriers offer advantageous features such as high solubility and detergent properties, along with the extremely rapid and potent biocidal effects of the iodine content. Water hardness and temperature have no serious deleterious effects on these attributes, and in the high-use dilutions permitted by this potency, there are no serious risks of iodine taint in odour or taste of the end product. Iodine does have corrosive oxidizing effects, however, even at these low concentrations, on many metallic surfaces, and it will discolour polymers used as food-processing and preparation surfaces too. Discoloration may be intensified by temperatures >42°C, when free iodine may be dissociated from the

Iodophor (providone-iodine)

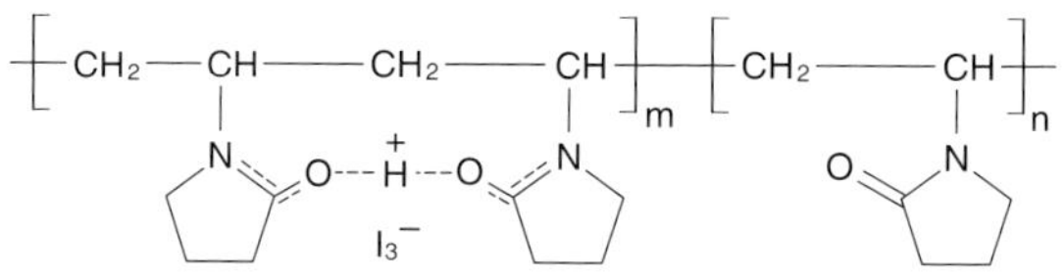

Chlorhexidine (a biguanide)

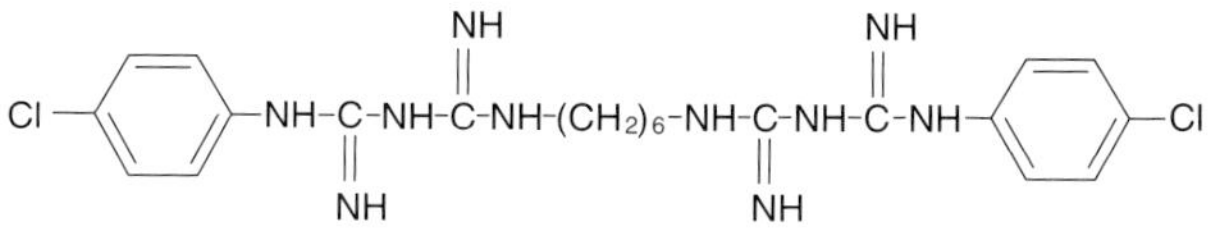

Polyhexamethylene biguanide (PHMB)

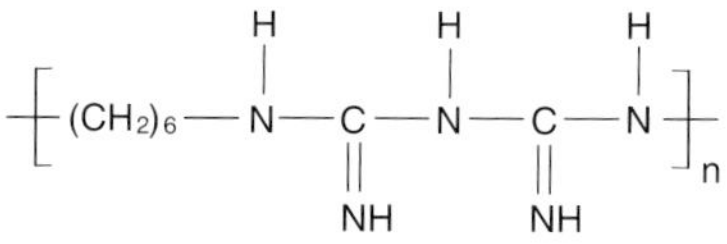

Parachlorometaxylenol (PCMX)

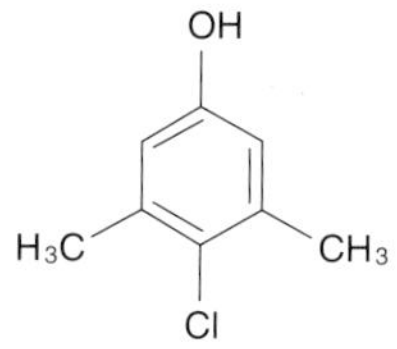

2,4,4'-Trichloro-2'-hydroxydiphenyl ether (triclosan)

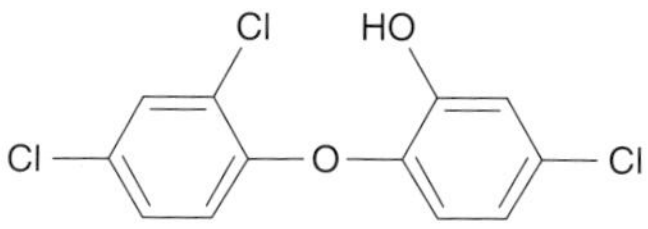

Figure 3 Molecular structures of other biocidal compounds used in the food industry.

iodophore complexes and deposited on the materials.

Despite these shortcomings, iodophors are amongst the most popular constituents of detergent sanitizers used in the industry for plants and for cleaning and decontaminating vehicles. They are also used in hand-washing solutions for food handlers. Iodine resistance is possible, and there are instances of *Pseudomonas* proliferation in concentrated iodophore formulations for hand use.

Chlorine Compounds

By far the most commonly used biocidal solutions in the food industry are those based on chlorine and chlorine-releasing/stabilizing compounds (Figs 1 and 2). Chlorine itself is used to generate highly biocidal hypochlorous acid, the most potent of the various constituents of chlorine compound formulations, such as sodium hypochlorite (liquid), calcium hypochlorite, chlorocyanurates, chloramine and chlorinated hydantoins (solids). Chlorine dioxide is a gas which can be produced on-site for certain equipment-sterilizing applications. It is short-lived and hazardous to users, so a variety of formulations have been developed to take advantage of the activity of oxides of chlorine of various species. Sodium chlorate especially, in mixtures with weak organic acids, can be used for on-site production of chlorine dioxide for environmental sanitation, and as a carcass rinse for meat and poultry to reduce the chances of contamination with *Salmonella* and *Escherichia coli* O157:H7 on the surface derived from faecal contamination during processing.

Chlorine as a gaseous element is very corrosive and dangerous on contact with mucous membranes of the eyes and nose. In solution, hypochlorous acid has remarkable penetrating ability resulting in rapid biocidal efficacy against virtually all forms of contaminant and pathogenic organisms, including bacterial spores. Hypochlorous acid rather than dissociated hypochlorite anions predominates in solutions that are on the acidic side of neutrality, but as the pH rises, chlorine solutions rapidly lose their effectiveness. Below pH 4 hypochlorous acid degrades, releasing free chlorine. Effective use therefore requires a balance between optimizing the biocidal efficacy and limiting the corrosiveness, which is maximal at pHs below neutral.

It is this powerful oxidizing effect of chlorine which leads to corrosion of many metallic surfaces and is so harmful to other components of equipment, such as rubber belts and O-rings. But the beneficial consequences of chlorine applications have sustained high rates of use in food-processing plants worldwide. Chlorine is the most commonly used means of disinfecting potable water, usually at about 1–2 p.p.m. However, in-plant chlorination in the food industry is often used to increase available levels up to 5 p.p.m. or more to ensure enhanced biocidal activity in washing and rinsing procedures. Even higher levels (approximately 20 p.p.m.) are used when corrosive surfaces are not exposed.

Liquid and solid forms of chlorine (as hypochlorites, chlorocyanurates, chlorohydantoins) and even *in situ* generation of chlorine by electrolytic means (so-called nascent chlorine) offer flexibility in use patterns in the food industry. Solid granules of calcium hypochlorite consist of almost two-thirds available chlorine, for example, and permit make-up of convenient formulations of sanitizers with detergent components for cleansing power. Loss of chlorine on liquid storage is an issue, so that when high dilutions are being made up routinely, a careful eye has to be kept on storage and preparation dates to ensure adequate levels. Another drawback is that the high reactivity of chlorine results in interactions with many organic substrates, leading to attenuation of the biocidal power. Chlorine-based food sanitizers are therefore most effective on clean surfaces. On the other hand, chlorine action is enhanced by higher temperatures, so that hot rinses generate extremely strong biocidal effects on environmental surfaces, tools and equipment.

Free chlorine in rinses for fish and poultry have become an important part of the processing, usually at about 10 p.p.m. Similar levels are used in canneries for food rinsing and canning equipment sanitation. However, vegetative stages of pathogenic and food-spoilage bacteria are most susceptible, and reliance on chlorine for decontamination where bacterial spores may be involved is unwise. Contact times for spores are too long to be practical.

It has also to be emphasized that the more recently recognized phenomenon of biofilm formation may undermine the effectiveness of sanitizers of all types, including chlorine. Biofilm and its significance are reviewed elsewhere in this volume. Suffice it to say here that the mechanisms whereby bacteria adhere to surfaces and rapidly establish complex multispecies consortia, embedded in a secreted organic matrix, can result in extraordinary declines in the biocidal efficacy of agents such as chlorine which interact so readily with organic molecules. Physical barrier effects also come into play, limiting the access of biocides to the target organisms, and physiological changes take place in the participants in the consortia that decrease their susceptibility to chemical agents of many kinds. Biofilms can form on porous and nonporous surfaces, on polymers, stainless steel, rubber and on many other surfaces in food-processing equipment. Their recent acceptance as commonplace, almost ubiquitous structural features of microbial populations in the environment provides an explanation for biocidal failures. It has also created an impetus for further research on methods on biofilm disruption and prevention, which will have a great impact on food sanitation practices in the future.

One of the most serious drawbacks to food industry dependence on chlorine has arisen in recent years with the demonstration that chlorine interaction with organic compounds leads to formation of so-called disinfection by-products (DBP). Among the DBP compounds are those commonly referred to as trihalomethanes (THM), including chloroform. They have acquired notoriety as evidence has accumulated of their carcinogenicity. This has led to increasing clamour for reductions in the use of chlorine despite its outstanding qualities and record in food and water safety and sanitation. It is for these reasons that new developments in chlorine-based biocide chemistry are so significant, since they address the public's concerns about DBPs to great advantage.

A series of heterocyclic ring compounds (N-halamines) in which N constituents are halogenated (variously by either chlorine or bromine, Fig. 2) display very attractive biocidal profiles. Depending on the configurations, compounds with extremely rapid actions and high potency can be prepared (e.g. a bromochloroimidazolidinone), or with a less potent, slower onset but extreme durability, and stability to heat and storage (e.g. monochloroimidazolidinone). Remarkably, these compounds exhibit spectra of efficacy which are comparable to chlorine, for example, but they do not liberate free chlorine and form hypochlorous acid to exert their effects. On the contrary, the biocidal effects demonstrably do not depend on this mechanism at all. The dissociations of chlorine atoms from the N atoms positioned between alpha carbons which are dimethylated are extremely low, yet the biocidal efficacy is high. This raises questions about the ways in which chlorine is working in the N-halamines, since these observations run counter to the generally accepted understanding of the role of free chlorine in biocidal events.

Most significantly, given the current concerns, the N-halamines do not generate THM and other DBPs at anything like the rate that other chlorine-based biocides do. They also exhibit much less corrosive capacity and less bleaching and odour than hypochlorites. They work at a much wider range of pHs than do other chlorine compounds such as chlorohydantoins (best at pH 7 or below).

In their most recently developed forms these N-halamine compounds can be synthesized as insoluble polymers, showing powerful biocidal effects on contact. These polymers release virtually no free chlorine, but in exercising their effects on bacteria and viruses, chlorine atoms are consumed in collisions with target-reducing groups on these organisms. The chlorine can then be restored and the polymers recharged by exposure to hypochlorite solutions. The possibilities are thereby raised for the creation of rechargeable polymeric biocidal surfaces for use in food processing, food preparation and storage, and as a complement to the likely emergence of water-soluble N-halamines as sanitizers for environmental applications and carcass-rinsing agents.

Chlorhexidine

Chlorhexidine (CHX; Fig. 3) belongs to the biguanide class of biocides. These compounds have a place in recreational water disinfection, mouthwashes, wound antiseptics and in antimicrobial rinses/dips for the hands of food-processing and preparation personnel. Their relatively benign toxicological profiles have expanded the use range in recent years, although they do not display a powerful or broad spectrum of activity across the entire range of organisms of importance to the food industry. Gram-negative species, such as *Pseudomonas*, are especially prone to resist CHX. Resistance as an acquired trait may appear in bacterial populations maintained under CHX pressure. Moreover, at use concentrations that are sufficiently effective, a proportion of users find CHX irritating. Incompatibilities between CHX or its salts and many detergents and soaps limit the extent to which formulation components can be added to counter this effect. CHX binds in non-volatile form to a wide range of substrates, including skin, so that its use in antimicrobial soaps results in a residue of bacteriostatic activity with beneficial consequences for food handlers. CHX has a bitter taste and odour so that to prevent contamination of sensitive food products, it is imperative to avoid transfer from user hands.

Polymerized biguanides (such as polyhexamethylenebiguanide, PHMB; Fig. 3) can also be formulated as components of antimicrobial soaps for food industry use. Their stability, low toxicity and non-sensitizing features make them attractive for this purpose, but the low efficacy traits of CHX against Gram-negative organisms are shared by the polymeric forms.

Chlorinated Phenolic Derivatives

Parachlorometaxylenol (PCMX, Fig. 3) has been used for many years in food service hand soaps and rinses. It is one of the phenolic derivatives most commonly used in household disinfectants and cosmetics, along with triclosan (2,4,4-trichlorohydroxyphenyl ether; Fig. 3), another representative of this group. Triclosan is a broad-spectrum, primarily bacteriostatic agent (at the concentrations commonly used), popular in consumer-level surface disinfectant and deodorizing formulations. Triclosan has a low level of toxicity to mammalian cells, and is a commodity chemical which is economically attractive for incorporation into many domestic and industrial use products.

Like all chlorinated phenolics, the efficacy spectrum of triclosan has an important gap in the Gram-negative bacteria area, and this factor, combined with its slow onset of action, limits its practical utility in food industry applications. Nevertheless, a large proportion of food industry soaps depend on the presence of triclosan for their claims of efficacy, though this reliance is perhaps imprudent. Even accidental contamination of antimicrobial soaps of this type can lead to luxurious growth of bacteria such as *Pseudomonas* (**Fig. 4**), the presence of which may not be apparent to users if the containers are opaque.

Figure 4 Popular antimicrobial soap heavily contaminated with growth of *Pseudomonas aeruginosa*, after accidental contamination during routine use. Gram-negative bacteria are often able to proliferate in antimicrobial formulations, resisting the weak biocidal effects and taking advantage of the rich supply of organic nutrients in these products.

Alcohol

An increasingly popular array of non-water-based formulations for hand sanitation has appeared, most of which depend on the presence of high concentrations of ethanol or isopropanol. These products incorporate emollients to counteract the inevitable drying effects of repeated exposure of hands to alcohols, and have a place in circumstances where hypersensitivities and idiosyncratic reactions to the common water-based formulations preclude their use by food handlers.

Organic Acids

Low molecular weight, weak organic acids such as lactic, citric, malic and sorbic find use as biocidal carcass rinses (Fig. 1). They have very low toxicity profiles because of their ubiquitous presence in virtually all eukaryotic cells; they can be cheaply and plentifully produced, and they can be conveniently applied due to their high solubility in water. Their utility in this role extends beyond efficacy against surface-contaminating Gram-negative bacteria on carcasses as a result of faecal contamination in the processing of meat and poultry. It includes beneficial inhibitory effects on spoilage organisms, spore-formers and moulds. Peracetic acid, although most useful in medical disinfection and sterilization as a concentrated and corrosive solution for instrument application, has been used in extremely diluted forms for carcass and seafood rinses as an effective decontaminant, as well as for container and surface sterilization in the beverage industry.

Effluent and Waste Stream Issues

Downstream consequences of biocidal use in food processing are an important consideration in their selection and use. Residues that may influence end product accessibility in terms of taste, odour or appearance have to be avoided. Contamination of food products by contact, aerosolization or transfer from biocidal preparations on hands cannot be permitted. Compounds currently used satisfy these kinds of needs and criteria provided they are applied correctly. More problematic can be issues raised by the large volumes of rinse and wash waters used in cleaning equipment in the food-processing industry, as well as for rinsing meat carcasses and seafood. The principal issues here relate to potential biocidal impacts on other non-target organisms in the waste streams, and chemical interactions that may lead to generation of undesirable compounds on route.

QACs and halogen-containing compounds are the most important in this regard because of their widespread use in food processing, meat packing plants, seafood sanitation, equipment and tool and pipe treatment. On the other hand QACs, iodine-containing compounds, chlorine and chlorine stabilizers have all been applied as microbiocides for commercial water treatment, either potable or waste water, so that the levels of concern about their environmental impact are limited, or were so until relatively recently.

QACs are ultimately degraded in the environment by biological mechanisms, and a number of bacterial species are known to degrade and use cationic detergent compounds as carbon sources. Halogenation of organic compounds, however, creates classes of disinfectant by products with known carcinogenic properties, albeit in very low qualities. It is this characteristic that has raised the spectre of limitations on the industrial use of chlorine, in particular, as a biocidal agent. Precisely because of these developments, the merits of N-halamine disinfectants, which are much less prone to generation of trihalomethanes, but show the same robust antimicrobial profile as the best chlorine-based disinfectants, loom large in the future of food service biocidal use. As water-soluble and environmental sanitizers, rinsers and as potential solid-phase biocidal surfaces, the N-halamines represent a new generation of biocidal compounds for applications in this industry, alleviating environmental concerns. N-halamines, chlorocyanurates and hydantoins are subject to environmental degradation by pathways which lead to innocuous amino acids, carbon dioxide, ammonia and water in the environment.

Effects of Biocides on Microorganisms

Chemical biocides are effective because they seriously disrupt central vitally important functions of all living cells. These characteristics may be expressed in modulated forms in commercial biocides depending on formulation constituents, temperature, water hardness, degree of contamination with other organic material and intrinsic susceptibility of the particular life forms targeted. Thus biocidal formulations may differ in their relative efficacies against bacteria and fungi for example, depending on the inherent biocidal efficacy of the active constituent, and the formulation and use pattern.

It is generally true that vegetative forms of bacteria, especially in their growth phase, are most susceptible to biocidal actions, while bacterial spores of certain species are the least susceptible, and are unlikely to be affected by most biocidal applications used in the food industry. Even more resistant infective agents have now been identified as a result of the emergence of the prion agents associated with bovine spongiform encephalopathy (BSE), but there are no practical chemical approaches to decontamination of these agents, nor are there likely to be in the near future. Mould-forming fungi also form spores, but these are generally less resistant than the food-spoiling or toxin-producing spores of *Bacillus* or *Clostridium*, for example, though commercial food preparation biocides vary in their fungicidal efficacy profiles.

QACs disrupt cell membranes, probably by intercalation into the lipid bi-layer structure, leading to breakdown of transport and control functions of membranes. These consequences are most likely to affect vegetative cells of bacteria and fungi, and membrane-bound enveloped viruses, but much less likely to kill non-enveloped viruses or spores. Unfortunately, Gram-negative pathogens, with thick outer membranes and glycolipid endotoxin components, are also less affected by QACs, though Gram-positive cocci (like food-poisoning staphylococci) are very susceptible.

Iodine is extremely effective against bacteria and fungi, including troublesome Gram-negative organisms that can grow at refrigeration temperatures. Iodine probably interacts with the same reducing group targets as other potent oxidizers like chlorine, and this accounts for the broad-efficacy spectrum. Many structural components of bacteria, viruses and fungi would be expected to display these susceptible groups, both on their surfaces and within. The rapid penetrating capacity of the halogen leads to speedy, potent biocidal activity, countered by the tendency for oxidative interaction with many organic agents in the environment.

Chlorohexidine and other biguanides bind membrane phospholipids disrupting membrane function and cytoplasmic membrane-associated enzymes of bacteria, yeasts and fungi. Surface charge changes of microorganisms are also brought about rapidly by biguanides, which are strongly cationic molecules.

PCMX and triclosan are chlorinated phenolics with distinct denaturing effects on proteins and nucleic acids precipitating functional molecules and causing cytoplasmic membrane leakage. They penetrate Gram-negative cell walls much less readily than Gram-positive, and this accounts for their widely recognized differential effects.

Ethanol and isopropanol function antimicrobially in part due to their lipid-solvent effects, disrupting lipid bi-layer membranes, and in part as a result of their denaturing influences on tertiary protein structure, thereby inactivating structural and functional protein constituents of all cells. These denaturing influences are enhanced in aqueous environments, so that pure ethanol, for example, is less bactericidal than lower concentrations in water.

Acidic conditions in general are inimical to survival and proliferation of bacteria and fungi, so that the organic acids used in the processing of meat or seafood are most likely affecting Gram-negative bacteria in this rather nonspecific way. Acidic conditions created by these acid washes prevent growth and delay sporulation of bacteria and fungi. At low pHs these weak acids remain unassociated, and this tends to enhance their inhibitory efficacy. Since many of the agents used are naturally occurring metabolic products in all cells, including mammalian cells, they have a high level of acceptance and are generally regarded as safe by food regulatory agencies worldwide. However, evidence is emerging that some of the notorious pathogens responsible for food-borne disease outbreaks such as *Salmonella* and *E. coli* O157:H7 are much more acid-resistant than was previously anticipated, possibly as a result of acquisition of enhanced proton pump mechanisms.

Clearly, the extraordinary capacity of microorganisms to adapt and resist adverse environmental conditions needs to be respected. Any relaxation of vigilance is likely to undermine the benefits reaped by the food production industry and the consuming public from the judicial selection and use of chemical biocides.

See also: **Biofilms**. ***Escherichia coli***: *Escherichia coli*. ***Escherichia coli* O157**: *Escherichia coli O157:H7*. **Food Poisoning Outbreaks**. **Heat Treatment of Foods**: Thermal Processing Required for Canning; Spoilage Problems Associated with Canning; Ultra-high Temperature (UHT) Treatments; Principles of Pasteurization; Action of Microwaves. ***Salmonella***: Introduction; *Salmonella enteritidis*; *Salmonella typhi*. **Preservatives**: Classification and Properties; Traditional Preservatives – Oils and Spices; Traditional Preservatives – Organic Acids; Traditional Preservatives – Wood Smoke; Traditional Preservatives – Vegetable Oils. Permitted Preservatives – Sulphur Dioxide; Permitted Preservatives – Benzoic Acid; Permitted Preservatives – Hydroxybenzoic Acid; Permitted Preservatives – Nitrate and Nitrite; Permitted Preservatives – Sorbic Acid; Permitted Preservatives – Natamycin; Permitted Preservatives – Propionic Acid. **Process Hygiene**: Hygiene in the Catering Industry.

Further Reading

Denyer SP and Hugo WB (1991) *Mechanisms of Action of Chemical Biocides: Their Study and Exploitation.* Society for Applied Bacteriology, technical series no. 27. Oxford, UK: Blackwell Scientific Publications.

Gravani RB (1986) Sanitation in the food industry. *Dairy Food Sanitation* 6: 250–251.

Kaysuyama AM and Shachan JP (1980) *Principles of Food Processing Sanitation.* Washington, DC: Food Processors Institute.

LaRocca MAK, LaRocca PT and LaRocca R (1985) Comparative study of three handwash preparations for efficacy against experimental bacterial contamination of human skin. *Advances in Therapeutics* 2: 169–274.

Lauten SD, Sarvis H, Wheatley WB, Williams DE, Mora EC and Worley SD (1992) Efficacies of novel N-halamine disinfectants against *Salmonella* and *Pseudomonas* species. *Applied and Environmental Microbiology* 58: 1240–1243.

Paulson DS (1996) A Broad-based Approach to Evaluating Topical Antimicrobial Products. In: Ascenzi JM (ed.) *Handbook of Disinfectants and Antiseptics*. New York: Marcel Dekker.

Paulson DS (1997) Developing effective topical antimicrobials. *Soaps, Cosmetics and Chemical Specialities* Dec. 50–58.

Truman JR (1977) The Halogens. In: Hugo WB (ed.) *Inhibition and Destruction of the Bacterial Cell.* New York: Academic Press.

Walker HW and LaGrange WS (1991) Sanitation in Food Manufacturing Operations. In: Block SS (ed.) *Disinfection, Sterilization and Preservation*, 4th edn. Philadelphia, PA: Lea & Febiger.

Williams DE, Elder ED and Worley SD (1988) Is free chlorine necessary for disinfection? *Applied and Environmental Microbiology* 54: 2583–2585.

Worley SD and Sun G (1996) Biocidal polymers. *Trends in Polymer Science* 4: 364–370.

Overall Approach to Hygienic Processing

M A Mostert and **H L M Lelieveld**, Unilever Research Vlaardingen, The Netherlands

Introduction

Most organisms in the environment, contaminating food products, do not present a health hazard. Regrettably there are exceptions, such as species of *Salmonella*, *Listeria* and *Staphylococcus*. From time to time new, sometimes potentially very dangerous, pathogenic microorganisms are identified (e.g. *Escherichia coli* O157). To ensure that food products are safe at the time of consumption, the manufacturing process must be hygienic. This requires the number of microorganisms in the raw materials to be reduced and reinfection prevented (**Fig. 1**).

Unfortunately, heat treatments which are intended to destroy unwanted microorganisms (pasteurization and sterilization) often adversely influence the characteristics of the product treated. Taste, flavour, colour and texture may be affected, as well as many nutrients.

Increasingly consumers demand fresh-like products which are less affected by processing. Therefore food processors look for milder processes. Such processes can become a realistic option only if hygienic processing requirements are correctly and consistently met.

Raw Materials

Most raw materials are contaminated on the outside only. The number of microorganisms can be minimized by proper treatment before processes such as cutting, milling and grinding. Such treatments include washing, blanching, flushing with water and a disinfectant (see below) and ultraviolet (UV) light treatment.

Storage and transport of raw materials should be free from residues of earlier batches. Storage temperature, time and humidity should be chosen such that growth of microorganisms and insects is halted or significantly slowed down. To this end, the atmosphere may be modified, for example by increasing its carbon dioxide content.

Insects, rodents and birds are significant sources of microbial contamination and therefore raw materials should be protected against such pests.

Decontamination Processes

Raw Materials and Products

The decontamination treatment depends on the product, its composition and parameters such as pH and water activity.

The design of the installation should ensure that all raw material or product receives the correct treatment. This means that the minimum residence time must be taken into account, and that sensors placed in line are in the correct position.

Equipment

To ensure that it can be freed from relevant microorganisms, equipment must be properly cleaned before the process begins. Equipment must be hygienically designed. Where crevices and/or dead areas cannot be avoided for functional reasons, equipment may have to be dismantled for cleaning. It is easy to decontaminate a clean surface. If this is done by heating, the coldest spot in the equipment must reach the required temperature for the treatment time

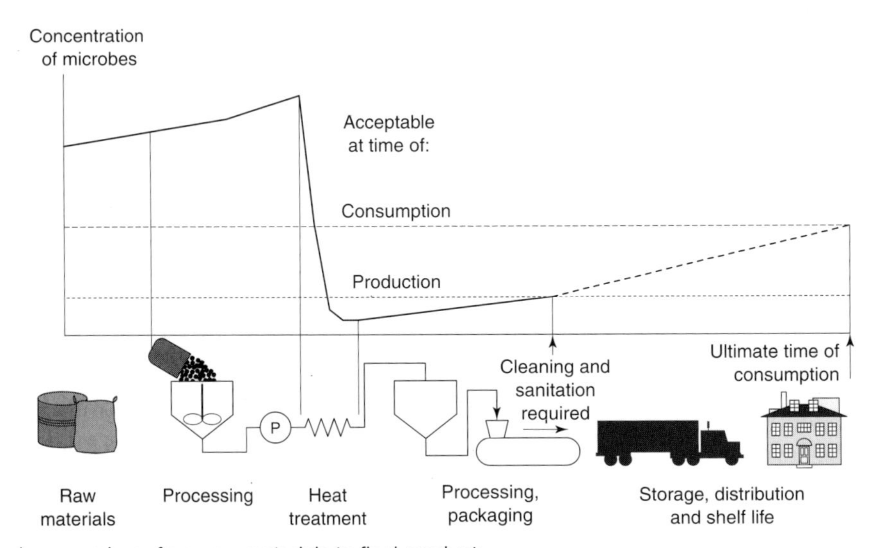

Figure 1 Fate of microorganisms from raw materials to final product.

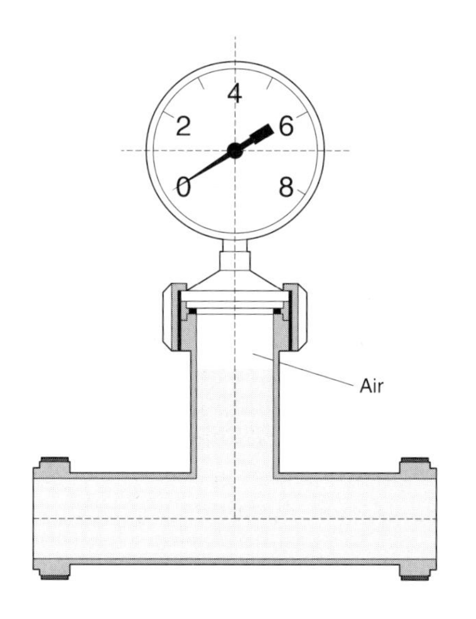

Figure 2 Upward-pointing dead leg, trapping air and therefore difficult to decontaminate.

needed. Usually heat treatment is based on a water activity (a_w) of 1.0. At lower a_w, more severe time/temperature combinations will be required. Therefore, the design of the process installation should either ensure that $a_w = 1.0$ or more severe heat treatment must be applied. Instead of heat, chemicals may be used, in which case it must be ascertained that the disinfecting liquid will be in contact with all surfaces (in particular, upward-pointing dead legs must be avoided, **Fig. 2**).

Packing Materials

Packing materials should not recontaminate the product. Most packing materials (e.g. glass and most polymers) are sterile at the point of manufacture. To preserve this condition production should take place in a clean environment. The secondary packing (in which the packing material is stored and transported) should protect and not contaminate the packing material. Special attention should be paid to cardboard because it may contain high microbial counts, in particular mould spores.

In some cases, packing materials may need to be decontaminated. Usually this is done before the product is filled on the packing machine. An example is packing of ambient stable UHT-sterilized milk. The most common method of decontaminating packing materials is treatment with H_2O_2 solutions at elevated temperatures ($>60°C$). All packing materials must receive the minimal treatment. Thus time, temperature and concentration should be adequately controlled.

Other methods are UV light, steam and gamma irradiation. Sometimes UV and H_2O_2 are combined to enhance effectiveness. Like H_2O_2, treatment with UV or steam can be applied on the packing machine.

The application of gamma irradiation is limited because it requires transport to and from a specialized facility, and this makes this method expensive. Moreover it requires the infection-free transfer of the sterilized material to the filling machine.

Buildings and Environment

Separation of Environment and Interior

The buildings where the raw materials and product are stored, processed and packed should not have an adverse influence on the microbiological quality of the product. Food factories should not be located in areas where the numbers of microorganisms, insects and/or rodents are high. They should not be close to waste disposal areas, animal farms or stagnant water. Insects, birds and rodents should be prevented from entering the building and care should be taken not to create suitable conditions for breeding – there should not be hidden places that are warm and damp and littered with edible materials. Trees, shrubs and flowers should be avoided as they attract animals.

The exterior of the building should not give the opportunity for birds to breed. Therefore the outer walls should not have ridges or other protrusions. If this is unavoidable, measures to discourage birds from settling should be taken, such as frequent cleaning.

The construction materials must be rodent-proof, especially the lower part of the wall. It should not be possible for rodents to gain access by digging. The contractor should guarantee that the building is rodent-proof.

To avoid the entrance of pests, doors and windows should be self-closing. They should fit tightly, limiting the space between the door or window and its frame so that insects cannot enter and breed. Staff should be required not to keep doors and windows open unnecessarily.

Design of Floor, Walls and Ceilings

To ensure cleanability, floors must be as smooth as possible and footwear should be adequate to compensate for any slipperiness. For different areas in the factory, different flooring materials can be used. For reception and many raw material storage areas, smooth concrete floors are normally acceptable. In areas where the product is exposed to the environment, ceramic tiles or resin floors are to be preferred. For high-care areas, seamless resin floors are best.

Where ceramic tiles are used, the foundation must be strong enough to avoid cracking of tiles and grouting. When cracking occurs, microorganisms will start to grow between and underneath the tiles, producing acid and dissolve the grouting.

If a seamless resin floor is chosen, the resin must be able to withstand the conditions to which the floor is exposed, such as cold and hot water, grease, cleaning and sanitizing agents, as well as the installation of heavy machinery.

Floors should have rounded corners where the floor meets the wall; this is also the case if tiles are used (for this purpose special tiles are usually available).

Walls should be cleanable. Many materials may be applied, including ceramic tiles and panels made from polymers, glass, aluminium or stainless steel. Rough base materials may be coated, provided the coating is firmly fixed and sealed to prevent ingress of moisture. When tiles are used, they must be properly grouted to ensure that the grouting does not come loose during normal use, including cleaning. Normal cement grouting will absorb moisture and spills and therefore support microbial growth. In wet areas cement mixed with synthetic resins should be used instead of normal grouting.

Ceilings should be hermetically sealed. Lighting should be built into or mounted against the ceiling, so that dust accumulation is avoided.

Ventilation and Air Conditioning

The risk of contamination of the product from the environment will always be proportional to the concentration of microorganisms in that environment. Therefore, multiplication of microorganisms in the production area must be prevented as far as possible. As microbial growth requires the presence of moisture, humidity should be kept low, preferably below 50%.

In processing areas where damp is generated (cooking, blanching) the damp air must be effectively extracted, e.g. by installing hoods. Similarly, where dust is generated (handling of dry powders, spices), care must be taken that the dust is extracted at that very spot. Dust-free emptying systems are available for both small and large bags. Usually such systems incorporate dust-free waste packing material compactors.

If located in an area with exceptionally high concentrations of airborne microorganisms, it may be necessary to install filters and to keep the interior of the plant slightly above atmospheric pressure.

Personal Hygiene

Proper hand-washing facilities should be available and used by whoever enters the food area. Such facilities must also be present in rest rooms (washrooms, toilets, flush lavatories). The facilities should be easy to clean, and be cleaned frequently.

Where contact with food is possible, access should not be allowed without a number of precautions to reduce the risk of product contamination. To prevent product contamination with foreign bodies, jewellery should not be worn in a food factory. Hair nets (and beard nets) or caps must be worn to prevent contamination with hair. Spectacles must be tied to a necklace-string. Eating, drinking and smoking must not be allowed in any food storage or processing area.

Other important measures include hand washing and sometimes disinfection, changing outdoor clothing for dedicated clean garments and footwear, and denying access to personnel with a health problem of microbial origin. If technical staff or visitors (including food inspectors) need to enter the food area, without exception they should comply with the rules in that area.

Prevention of Recontamination of Product

Logistics

The layout of the factory must be such that final products cannot be recontaminated by raw materials, intermediate product or packing materials. Thus, the flow of the product should be in one direction: from the raw materials entrance to the exit of the final product. Nowhere should intermixing be possible. As air and humans are major means of transport of microorganisms, they should move in the opposite direction only, i.e. from the end product (where the microbial load is lowest) to the raw materials side (where the load is highest). This must be taken into careful consideration when designing any ventilation or conditioning system. Staff who need to move with the stream of product must take careful preventive measures, such as washing hands and changing protective clothing, including shoes or boots and caps or hair nets.

Hygienic and Aseptic Equipment

To avoid recontamination from processing equipment, the equipment should be of hygienic design. This applies to both open and closed process lines and parts thereof. Open process lines are process lines where the product is exposed to the environment; examples are conveyor belts, tables and open vessels. Closed process lines consist of equipment which contains the product, such as pipelines, (most) pumps, valves and in-line mixers.

Hygienic equipment can be effectively cleaned and freed from microorganisms. To be and to remain cleanable, hygienic food-processing equipment must meet a number of essential requirements, such as smoothness, corrosion resistance (with respect to product and cleaning agents), absence of crevices, sharp corners and dead legs, smooth welds and correct use of sealants and lubricants (see Further Reading). If the cleaning frequency is correct, such equipment prevents the build-up of large numbers of microorganisms. Hygienic equipment, however, is not necessarily designed to prevent ingress of microorganisms and thus does not prevent recontamination of product completely.

For packing of products with a long shelf life at ambient temperature, such as UHT sterilized milk, after the UHT treatment both the process line and packing machine must be aseptic. Aseptic equipment is hygienic equipment which also prevents ingress of microorganisms and thus completely prevents recontamination of the product.

Clean Rooms

In areas where unprotected product is handled, which will not be subjected to a decontamination process, it may be necessary to keep the pressure above that of the adjacent areas and to do so by filtered air. In such cases stringent personal hygiene requirements apply. The area should be kept clean, and strict compliance is required with design criteria for building interiors. In some cases, it may be necessary to chill the area to reduce the growth rate of microorganisms. The measures required are product-dependent and should be discussed with specialists.

Who Can Help?

Several organizations are specialized in producing guidelines and standards for food-processing plants. Organizations in the US and Europe attempt to harmonize hygienic design standards and guidelines, facilitating international trade. Specifically, the US-based 3-A organization, NSF International and the European Hygienic Equipment Design Group (EHEDG) have established cooperation by exchanging staff and by participation in joint working groups. These organizations, where many of the publications listed under Further Reading are available, are listed below:

- 3-A Sanitary Standards Committees, St Paul, Minnesota, USA
- EHEDG Foundation, Brussels, Belgium
- NSF International, Ann Arbor, Michigan, USA.

See also: **Heat Treatment of Foods**: Ultra-high Temperature (UHT) Treatments. **Intermediate Moisture Foods**. **Packaging of Foods**. **Predictive Microbiology & Food Safety**. **Process Hygiene**: Designing for Hygienic Operation; Types of Biocides; Modern Systems of Plant Cleaning; Risk and Control of Airborne Contamination; Testing of Disinfectants; Involvement of Regulatory and Advisory Bodies; Hygiene in the Catering Industry.

Further Reading

International Association of Milk, Food and Environmental Sanitarians. *3-A Sanitary Standards*. US Public Health Service Dairy Industry Committee.

EHEDG Guidelines:

No. 1: Microbiologically safe continuous pasteurization of liquid foods (summarized in *Trends Food Sci. Technol.* 3, 1992, 303–307).

No. 3: Microbiologically safe aseptic packing of food products (summarized in *Trends Food Sci. Technol.* 4, 1993, 21–25).

No. 6: The microbiologically safe continuous-flow thermal sterilization of liquid foods (summarized in *Trends Food Sci. Technol.* 4, 1993, 115–121).

No. 8: Hygienic equipment design criteria (summarized in *Trends Food Sci. Technol.* 4, 1993, 225–229).

No. 9: Welding stainless steel to meet hygienic requirements (summarized in *Trends Food Sci. Technol.* 4, 1993, 306–310).

No. 10: Hygienic design of closed equipment for the processing of liquid food (summarized in *Trends Food Sci. Technol.* 4, 1993, 375–379).

No. 11: Hygienic packing of food products (summarized in *Trends Food Sci. Technol.* 4, 1993, 406–411).

No. 12: The continuous or semicontinuous flow thermal treatment of particulate foods (summarized in *Trends Food Sci. Technol.* 5, 1994, 88–95).

No. 13: Hygienic design of equipment for open processing (summarized in *Trends Food Sci. Technol.* 6, 1994, 305–310).

No. 14: Hygienic design of valves for food processing (summarized in *Trends Food Sci. Technol.* 5, 1994, 88–92).

No. 16: Hygienic pipe couplings (summarized in *Trends Food Sci. Technol.* 8, 1997, 88–92).

No. 17: Hygienic design of pumps, homogenisers and dampening devices.

No. 18: Passivation of stainless steel (summarized in *Trends Food Sci. Technol.* 9, 1998.

ISO/DIS 14159 (1998) *Safety of Machinery. Hygiene Requirements for the Design of Machinery.*

Lelieveld HLM (1999) Hygienic Design of Food Factories. In: *The Microbiology of Food*. London: Chapman & Hall.

NSF Standards. *Food Service Equipment and Related Products, Components, and Materials.*

Modern Systems of Plant Cleaning

Yusuf Chisti, Department of Chemical Engineering, University of Almería, Spain

This article details the principles of design, selection and operation of systems for cleaning of food and pharmaceutical processing plants. Automated clean-in-place (CIP) methods are an essential part of hygiene in large-scale processing where cleaning must be frequent, rapid and consistent. By reducing labour and improving the productive utilization of process plant, CIP makes economic production feasible.

Clean-in-place

A satisfactory standard of hygiene is a statutory requirement in food and pharmaceutical processing. Cleanliness is an essential component of process hygiene. Modern food and pharmaceutical processing plants are cleaned-in-place, i.e. the equipment is not dismantled during internal cleaning. CIP practice originated in the dairy industry in response to a need for frequent rapid and consistent cleaning. During CIP, various flushing, cleaning and sanitizing fluids are brought into contact with the wetted parts of the plant. Cleaning may be done on a once-through basis, or some of the cleaning agents may be recirculated to reduce consumption of water, chemicals and energy. Cleaning is achieved by the physical action of high-velocity flow jet sprays, agitation and the chemical action of cleaning agents enhanced by heat. While mechanical forces are necessary to remove gross soil and to ensure adequate penetration of cleaning solutions to all areas, most of the cleaning action is provided by chemicals – surfactants, acids, alkalis and sanitizers.

A CIP system consists of piping for distribution and return of cleaning agents, tanks and reservoirs for cleaning solutions, heat exchangers, spray heads, flow management devices (supply and return pumps, valves, sensors and gauges, recording devices) and a programmable control unit, as well as other items. Although a CIP system requires a significant initial capital investment, CIP is economical relative to manual cleaning in most large-scale processes. CIP greatly reduces labour demand. CIP methods often consume less water and chemicals relative to manual cleaning, especially if cleaning solutions are recycled. A properly designed, validated and operated CIP system ensures consistent and reproducible cleaning. Because there is little or no dismantling of equipment, the downtime is reduced and more time is available for productive use of machinery. Consistent cleaning eliminates product contamination and associated rejection of batches. As another important consideration, the CIP technology enhances safety by eliminating or minimizing operator contact with hazardous cleaning agents, bioactive products and potentially pathogenic biohazard agents. Other unsafe situations are avoided because worker entry into equipment is not required.

Although initially developed for cleaning liquid food-processing plant, CIP methods are now widely used in cleaning of all types of food and pharmaceutical process equipment and facilities. Machines for processing solids, e.g. those used in solid-state fermentation, can be effectively CIPd. Similarly, spray dryers, centrifuges, evaporators, chromatography columns and membrane filtration modules can be CIPd; however, CIP is only possible if the process equipment has been specifically designed to be CIP-capable. The specific design requirements are discussed later.

The Cleaning Problem

Various kinds of machinery are employed in food and pharmaceutical processing. A plant may have a variety of tanks, fermenters or bioreactors, pumps, valves, centrifuges, homogenizers, heat exchangers, evaporators, spray dryers, packaging machines, as well as other devices. A vessel may have internals such as agitators, ports for sensors, shafts and mechanical seals, mechanical foam breakers, baffles and gas spargers, all of which impact upon cleanability. Irrespective of the type of equipment, all plant components for food and pharmaceutical processing should be CIP-capable. Equipment design should ensure that all surfaces that in any way contact the product, including vapour, foam, and sprayed or splashed material, receive cleaning solutions during CIP. For example, a submerged culture fermenter may need to be supplied with CIP solutions at multiple points to ensure proper cleaning. In addition to being sprayed in the vessel, the CIP solutions may have to be sequenced through the submerged aeration pipe, the air exhaust lines that may be contaminated with fine culture droplets and foam, the mechanical foam breaker, and the various supply lines for the medium, inoculum, antifoam agents and pH control chemicals, as well as any harvest lines. Cleaning the sample valve and any retractable probes will require attention. Similar specifics need to be evaluated during design of other process items and in planning a CIP scheme.

Satisfactory cleaning standards are attained by closely matching the CIP devices and procedures to the specific configuration of the process equipment. In addition, the physicochemical nature of the cleaning problem needs to be evaluated in detail. The same type of equipment processing different foods or pharmaceutical broths will present cleaning problems of different difficulty. For example, a cream pasteurization unit processing a relatively viscous high-fat material may be more difficult to clean than a fruit juice pasteurizing plant. Similarly, fermenters used to culture yeasts and non-polymer-producing non-filamentous bacteria may be easier to clean than a bioreactor used to culture filamentous fungi, e.g. *Aspergillus niger* for making citric acid. Such considerations will affect the cleaning regimens as well as the choice of the specific cleaning agents. In one case a 1 min pre-rinse may be sufficient to remove gross soil; in another case, a 5 min pre-rinse may leave behind a lot of adhering debris, thus affecting the cleaning time, temperature and the strength of cleaning agents needed for subsequent cleaning steps.

CIP System Design Guidelines

General Aspects

Design and construction of a CIP system demand the same level of attention to hygienic practice as does engineering of the process equipment. Sanitary and sterile engineering standards vary somewhat among industries. Minimally, the process and CIP hardware should comply with the 3-A guidelines promulgated in the US and now widely followed. The specifications noted here conform to the best current practices, including the 3-A sanitation standards and the Good Manufacturing Practice (GMP) regulations established by the US Food and Drug Administration (FDA) and accepted in similar forms in Europe, Japan and Canada. Acceptable hygienic practices relevant to CIP capability require that:

1. Product contact surfaces should be durable, non-porous, non-absorbent and nontoxic. The surface should be non-shedding, and nothing should leach into the product. Suitable materials of construction are stainless steel types 304 and 316, certain types of glass and certain elastomers for gaskets
2. Because the surface finish affects cleanability, product contact surfaces should be smooth and free of crevices and pits. Generally, a 150 grit polished surface is the minimum acceptable for food contact surfaces, but significantly better finishes are commonly used in sterile bioprocess equipment
3. Product and solution piping and equipment should have CIP fittings with smooth surfaces and contours. Welded joints should be used as far as possible. Welds should be smooth and free from pits, cracks, inclusions and other defects. Welds should be ground flush with the internal surface
4. Removable fittings should be of hygienic type. The joint and gasket should be flush with the internal surface.
5. Lines should be relatively horizontal and sloped to drain points. Vessels and equipment should be similarly self-draining. Vessel drain nozzles should be flush with the internal surface and drains should be located at the lowest point. The base of a tank should be contoured to ensure complete emptying
6. The minimum acceptable CIP flow velocity through pipes and fittings is $\geqslant 1.5\ \mathrm{m\,s^{-1}}$ ($5\ \mathrm{ft\,s^{-1}}$)
7. The temperature of the CIP fluids should be controlled to within ±2.8°C (±5°F) and the return flow temperature should be recorded
8. Product and CIP lines joints require high-quality welding, as noted later in this article
9. The specific CIP methodology needs to be validated for satisfactory performance

The specific methods of complying with the various guidelines are discussed below.

Typically, stainless steel type 304 construction of CIP system components is satisfactory, although the equipment used directly in food and pharmaceutical processing is generally made of the higher-grade type 316L steel which resists corrosion better. Process vessels such as bioreactors intended for sterile operation require an easy-to-clean smooth surface finish: an electropolished surface with an arithmetic mean roughness (*Ra*) value of less than 0.3 μm is preferred. In contrast, the components of the CIP system, tanks, pipes, valves and so forth which do not come into contact with the product may have a lower level of finish at *Ra* of 0.4–0.5 μm, without electropolish. This level of finish allows a level of cleanability equivalent to that accepted in hygienically designed dairy product contact surfaces and is quite satisfactory for the CIP system. Further lowering of the surface finish is not recommended because the CIP system must be adequately self-cleaning; however, a 150 grit finish, equivalent to an *Ra* of about 0.8 μm, is the minimum accepted for food contact surfaces according to 3-A practices.

To ensure removal of gross soil and avoid its sedimentation, the minimum flow velocity through the CIP and transfer piping should be $1.5\ \mathrm{m\,s^{-1}}$, but a higher value of $2.0\ \mathrm{m\,s^{-1}}$ is recommended, especially in lines with obstructions. In addition, the Reynolds number of the flow must be well into turbulent regime to ensure good radial mixing, heat transfer (uniform heating), mass transfer (of cleaning chemicals and soil) and momentum (scouring action of eddies) transfer. A minimum Reynolds number of 10 000 has been

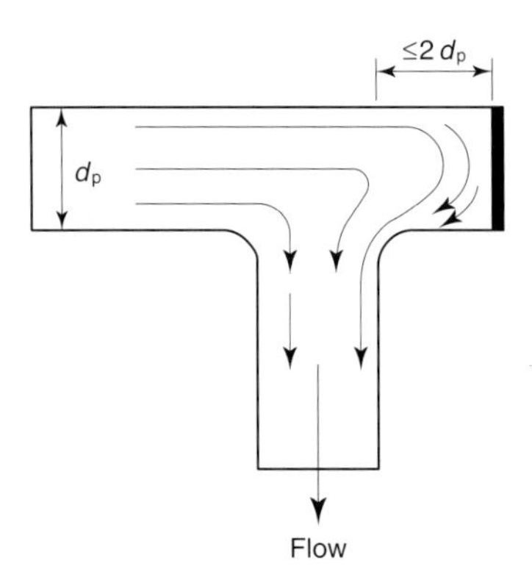

Figure 1 Preferred arrangement of any dead legs in pipework.

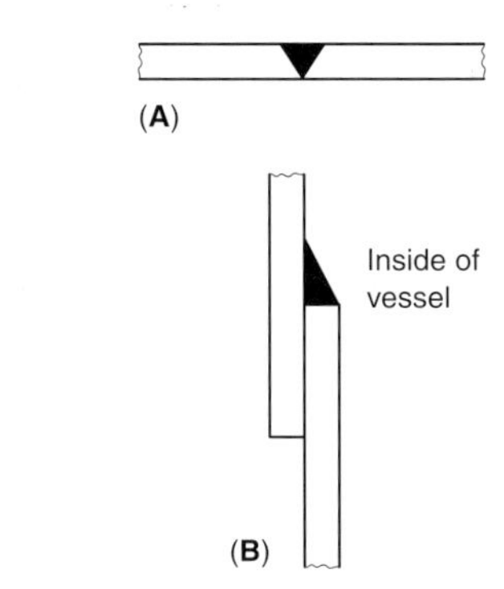

Figure 2 (**A**) Butt and (**B**) lap welds.

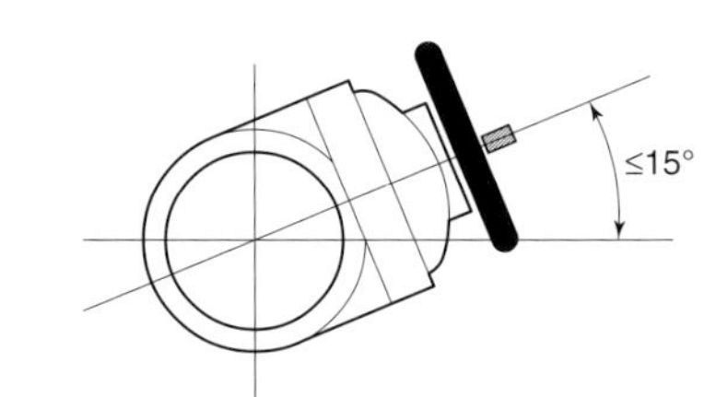

Figure 3 Self-drainage angle of diaphragm valves.

suggested, but a higher value of at least 30 000 is preferred. The Reynolds number for a volume flow rate Q ($m^3 s^{-1}$) through a circular pipe of diameter d_p (m) equals:

$$\frac{4Q\rho_L}{\pi\mu_L d_p}$$

where ρ_L ($kg\,m^{-3}$) and μ_L ($Pa\,s^{-1}$) are the density and the viscosity of the fluid. For cleanability, the CIP piping should be free of dead spaces and pockets as much as possible; if unavoidable, the depth of the dead zone must be less than two pipe diameters. When a dead leg occurs, it should not point vertically upward or downward where air may be entrapped or solids may settle. Any dead legs should face the oncoming flow (**Fig. 1**). The minimum radius of pipe bends should equal or exceed the outside pipe diameter. The pipework should be adequately supported to prevent sagging and consequent accumulation of stagnant pools. The lines and ducts should have a minimum slope of 1 cm per linear metre to drain points, but much higher inclinations are preferred.

Any corners and curves in tanks, vessels, bins and other items should be rounded with a minimum radius of 2.5 cm. Flat-bottomed tanks should have a minimum slope of 1-in-45 from the back to the outlet nozzle; the side-to-centre slope should be 1-in-24 minimum to ensure sufficiently rapid drainage that washes out any suspended solids. Vessels should be designed to withstand full vacuum, or suitably sized air vents should be provided to prevent tank collapse, e.g. while emptying rapidly during various CIP stages. Butt welded construction is preferred (**Fig. 2**). The welds should be ground flush with the internal surface and polished in the same way as the vessel. Welds are difficult to notice in high-quality construction. Lap welds contoured to ensure satisfactory drainage are acceptable in some cases (Fig. 2). Pipe welds should be of a similarly high quality. Product and CIP pipe welds require a shielded tungsten arc welding method or a technique capable of producing equally high-quality welds.

A CIP system and process machinery must use various kinds of valves for flow management. Only sanitary valves should be used in sterile process systems. Sanitary valve designs include valves with metal bellows-sealed stem, diaphragm valves and pinch valves. The operating mechanisms of these valves are fully isolated from the process stream. All other types of valves, even those commonly accepted in food processing plants, carry a significant risk of contaminating the process with accumulated debris or becoming a source of infection. Accumulation of debris at gaskets and valve spindles has been clearly documented for ball valves, butterfly valves and gate and globe valves, which are also difficult to clean using CIP methods. Except in vertical pipes, installation of a diaphragm valve requires attention to its self-drainage angle. The angle varies with valve size and manufacturer. In most cases, drainage through the pipe will not be impeded if the valve stem inclination from the horizontal (**Fig. 3**) is 15° or less. All valves and fittings should be of a self-cleaning design. Only sanitary types of flow measurement devices, e.g. vortex and magnetic flow meters, and mass flow meters relying on an oscillating flow-through tubular loop, are acceptable.

Design of the CIP system should consider cleanability of the system itself and attention must be given to drainage, elimination of crevices and stagnant areas, minimization of internals, arrangement of valves and pumps, piping welds, sanitary couplings, instrumentation and instrument ports. Spray devices are used in the CIP solution recirculation tanks to clean the tank at the same time as the process machinery. The CIP system must have a splash-resistant exterior of clean design which is easily washable by hosing or wiping. Smooth external contours and an absence of extensive ledges help ensure an easily cleanable exterior. Placement of the equipment should not interfere with thorough and easy cleaning of the area. In large facilities, the CIP equipment is installed in dedicated areas.

Depending on the process, the size of vessels and other process machines may range widely in a given plant. Thus, the volume requirements of cleaning solutions may vary tremendously, imposing difficult demands on the design of the CIP system. Preferably, the flow variation for cleaning different process circuits should be kept small: Sizing of CIP flow control valves, chemical dosage pumps, heat exchangers and other items becomes difficult if the flow variation exceeds 50%. A CIP system is designed to satisfy the flow requirements of the largest piece of equipment. The supply pump is sized for the pressure drop in the longest flow circuit. If a discharge flow control valve is used to control the supply rate of the CIP fluids, the supply pump needs to be sized for 1.2–1.3 times the head loss of the longest CIP circuit. The CIP solution tanks are sized to accommodate the volume needs of the largest flow circuit.

CIP Spray Devices

Tanks and other spaces are cleaned by spray of CIP solutions. Either static or dynamic spray devices (**Fig. 4**) are used to ensure that the CIP solutions reach all parts of the unit being cleaned. Static spray heads – usually spray balls, but also drilled tubes and 'bubbles' – have no moving parts and are considered self-cleaning sanitary devices. Dynamic self-cleaning and sanitary spray balls are also available. Dynamic spray nozzles, driven by liquid flow or mechanical means, are not self-cleaning or sanitary, hence, they are little used in hygienic applications.

Spray balls are designed to suit specific vessel volume and configuration. The balls are self-draining. Frequently, the balls are drilled to provide a directional spray pattern so that difficult-to-clean areas receive more spray (Fig. 4). Because of this directional flow, a removable spray ball needs to be correctly installed to ensure effective cleaning. Permanently installed spray devices are common in sanitary but non-sterile process equipment. Permanent installation is not recommended in fermenters and bioreactors because of potential difficulties with sterilization. Instead, a spray device should be mounted in a dedicated port on top of the fermenter just prior to cleaning. The CIP inlet pipe connecting to the spray ball and the region directly above the spray ball also needs to be cleaned. Typically, those areas are cleaned by an upward discharge of CIP fluids from engineered clearances between the spray ball sleeve and the inlet pipe (**Fig. 5**).

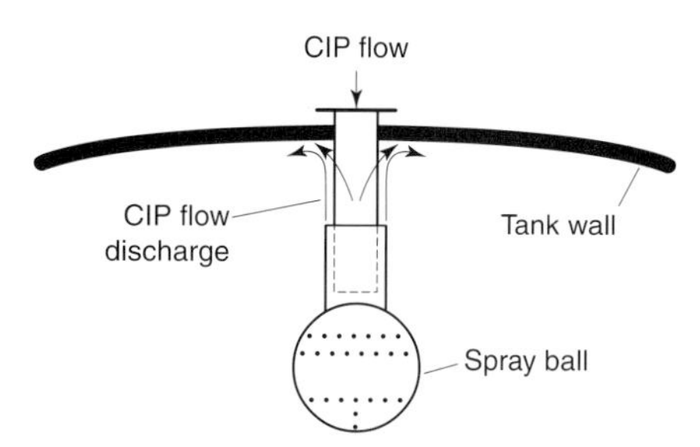

Figure 5 Upward discharge of CIP fluids from designed clearances for cleaning the spray ball supply pipe and the tank wall directly above the spray ball.

Spray balls typically require a flow of 4–12 $l\,min^{-1}\,m^{-2}$ of the internal surface in horizontal or rectangular tanks, and vessels with baffles, agitators and other projections. Similar CIP flow rates are used for spray dryers, cyclones, ductwork and other machinery. The balls are operated at 20–25 psig (1.4–1.7 bar gauge). Pressures greater than 30 psig (2.1 bar gauge) are not wanted because they lead to generation of a fine spray which is ineffective in cleaning. The CIP fluids should irrigate all parts of the vessel without leaving dead spots. Multiple spray balls are sometimes needed to ensure full coverage of the internal surface. Generally, the spray heads are positioned near the top of the tank, as close as possible to its vertical central axis. In vertical cylindrical tanks without internals, the recommended spray ball flow rate is 0.52–

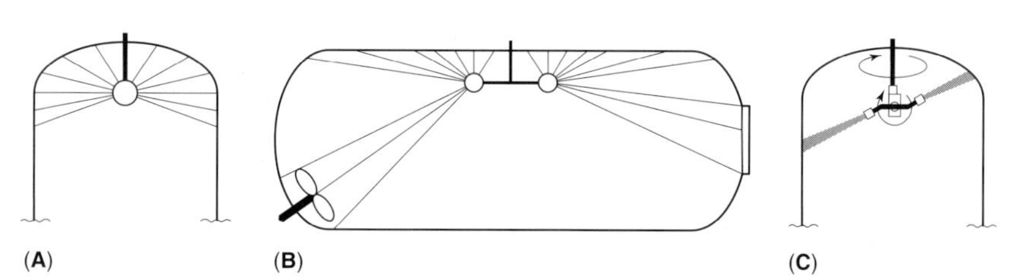

Figure 4 Spray devices: (**A**) spray ball CIP fluid distribution to upper parts of tank only; (**B**) directed spray pattern from a spray ball; (**C**) dynamic spray jets.

Table 1 Typical spray ball flow rates for horizontal cylindrical tanks

Tank volume (m^3)	*Flow rate ($l\,min^{-1}$)*
4.5	212
9.0	300
13.5	325

0.62 $l\,s^{-1}$ per m (2.5–3.0 gpm per linear ft) of the tank circumference. Often, the balls are designed to spray the upper third of the tank and the remaining surface is irrigated by the falling liquid film. The CIP solution flow should be sufficient to ensure a Reynolds number of at least 2000 in the irrigating fluid film. The Reynolds number (*Re*) is calculated as follows:

$$Re = \frac{\rho_L u d_f}{\mu_L}$$

which, for a tank with circular cross-section, approximates to:

$$\frac{\rho_L Q}{\pi \mu_L d_T}$$

Here d_f is the film thickness, d_T is the tank diameter, u is the liquid velocity in the falling film, and Q is the volume flow rate reaching the tank wall. Typical spray ball flow rates for horizontal cylindrical tanks are noted in **Table 1**.

High-pressure dynamic spray heads rely on the scouring action of the jet for cleaning. Because the jet continuously moves in three dimensions, the entire vessel wall is not irrigated at a given instance and, therefore, the flow rates can be lower, e.g. 19–38 $l\,min^{-1}$ (5–10 gpm). Movement of the jet is essential to effective functioning of the dynamic cleaning heads; hence, motion detectors may be needed to demonstrate satisfactory operation throughout the cleaning cycle. Also, a constant fluid supply pressure is essential to proper functioning. Jet spray cleaning requires higher pressures, typically 30–40 psig (2.1–2.8 bar guage). Extremely high-pressure jet blasting is employed in some industrial cleaning processes to remove hardened material, but this type of cleaning is uncommon in food and biopharmaceutical processing.

CIP Chemicals

CIP relies substantially on the action of chemicals on soil; hence, proper selection of cleaning agents is essential. Carbohydrate and protein-based soil is removed by alkalis. Fats and oils are insoluble in water and may protect other underlying soil. Fats are melted by heat and effectively solubilized by alkalis. Polyphosphates emulsify fats and oils, thus, increasing the rate of alkaline digestion. Mineral deposits are produced when hard water with or without alkali is heated. Similarly, calcium-containing deposits form when heating milk and spinach (calcium oxalate). Such deposits resist alkalis but are dissolved by acids.

Alkaline cleaners are especially useful as they digest most organics. Sodium hydroxide is a commonly used alkaline cleaner. Typically, a 0.15–0.5% (wt/wt) sodium hydroxide solution at 75–80°C (15–30 min) is used, but heat exchange surfaces with burnt-on protein deposits require treatment with 1–5% sodium hydroxide. Because sodium hydroxide is corrosive and difficult to rinse, silicates and wetting agents are added to inhibit corrosion and improve rinsing. Alkali may be supplemented with sodium hypochlorite (30–100 p.p.m.) to enhance protein and fat removal capability significantly. The damage to stainless steel normally associated with chlorine is insignificant in alkaline environment at the hypochlorite concentrations noted; however, at alkaline pH, chlorine does not act as a biocide. The amount of various additives, e.g. sodium metasilicate corrosion inhibitor, sodium tripolyphosphate sequestering or softening agent, wetting agents, and others, needed for a given volume of water is determined by the hardness of the water used in cleaning. High-quality soft water, e.g. reverse osmosis water, is used in many biopharmaceutical cleaning operations, but the potable-quality water used in cleaning of food-processing plant may be much harder. Generally, it is less expensive to use ion exchanger softened water than to add large amounts of softening chemicals.

An acid wash generally follows an alkaline wash in food-processing facilities. Acid neutralizes any alkaline residue and removes mineral deposits such as hardwater stone, milk stone, beer stone and calcium oxalate. Acid cleaners contain about 0.5% (wt/wt) acid. Formulations may have phosphoric acid; however, because mineral acids are extremely corrosive to steel, use of organic acid (e.g. lactic, gluconic, glyconic) cleaners is preferred. Routine acid washes of biopharmaceutical process equipment are not necessary if deionized water is used in production and cleaning, and the peculiarities of production (e.g. media high in Ca^{2+} and Mg^{2+}) do not contribute to build-up of acid-soluble deposits. An occasional acid wash – every 6 months – with 5 min recirculation of 0.5% (w/v) nitric acid at 60°C is sufficient. Nitric acid should not be used for routine cleaning.

Any wetting agents (surfactants) used in alkaline and acid CIP washes should be non-foaming or an antifoam agent may have to be included in the formulation. Typically, a cleaning formulation has

≈0.15% wetting agent. Depending on compatabilities, anionic, cationic or non-ionic wetting agents may be employed. Non-ionic ones (e.g. ethylene oxide–fatty acid condensates) are especially useful because they are poor foamers.

A sanitizing wash commonly follows acid treatment in food-processing facilities. A solution of quaternary ammonium salts (QATs) usually ≤200 p.p.m. is sometimes used. QATs are cationic wetting agents which have good bactericidal properties, especially against Gram-positive microorganisms. QATs are less effective against Gram-negative microbes such as *Escherichia coli* and *Salmonella* sp. QATs are incompatible with many minerals and soils; hence, they are used in the final treatment stages when all soil has been removed. QATs tend to foam. Other useful disinfecting agents are biguanides and peracetic acid. Peracetic acid should not be used in water containing excessive chloride, or corrosion could be promoted in stainless steel equipment. A sanitizing wash is essential in food processing, especially when equipment lines employ less sanitary devices such as butterfly and ball valves or other machines not intended for extended sterile processing.

The above-noted guidelines for selection of CIP chemicals are intended to provide only a general picture. Because the nature of soil varies tremendously, selection of suitable cleaning media requires consideration of relevant chemistry, microbiology, compatibility and safety issues, and cost.

Typical Cleaning Sequence

A typical cleaning sequence for food-processing plants consists of several steps: a water pre-rinse to remove gross soil; a hot alkaline detergent recirculation step to digest and dissolve away the remaining soil; a water wash to remove residual alkali; acid recirculation; a water rinse; a sanitizer recirculation; and a final water rinse. A 5 or 6 min pre-rinse or flush is usually sufficient for bacterial, yeast and animal cell culture reactors; often a 2 min pre-rinse is satisfactory. Usually, the pre-rinse is at ambient supply temperature, or at less than 45°C. Pre-rinsing should be on a once-through basis without recirculation. This ensures that the gross soil does not recirculate through the CIP system, thus reducing potential contamination. In biopharmaceutical plant, the pre-rinse liquid should be allowed to drain fully. In food-processing facilities, the time-saving practice of chasing the pre-rinse with subsequent wash solutions is common during CIP of certain equipment. For bioreactors for parenteral products and other biopharmaceuticals, potable-quality deionized water is recommended for all pre-rinsing and detergent formulations. Reuse of water from the intermediate rinses and the final rinse of the previous CIP cycle as the pre-rinse of the next CIP event should be avoided if cross contamination is an issue.

Following pre-rinse, an alkaline cleaner is circulated through the equipment so that all product contact surfaces are exposed to this solution, as specified earlier. Alkali is often reused for several cleaning cycles; however, reuse for next cleaning is not recommended when cleaning machinery for injectable products. Dilution, contamination with soil and microbial spores which can survive for long periods and loss of quality definition of the starting material for the next cleaning are some of the arguments against reuse of cleaning chemicals.

The alkali recirculation is followed by a short clean-water flush (e.g. 0.5 min) before the acid recirculation step. If acid recirculation is not used, a longer rinse with potable water or reverse osmosis water (25–35°C) is needed to remove all alkali. When acid washes are used routinely, a 5–10 min recirculation at ambient temperature is the norm. In biopharmaceutical processes for injectables, a hot water-for-injection (WFI) final wash ensures that all residual water complies with the requisite quality specifications. In food-processing facilities, a sanitization wash follows the acid treatment with or without an intervening clean-water flush. With some sanitizers, a final water flush may not be necessary, so long as all sanitizing solution is removed from the process circuit. A hot-air purge sometimes follows the final liquid rinse to aid emptying and drying. An air purge may also be used before the first water flush to remove all product liquid completely and reduce the soil load.

There may be additional specific considerations for CIP of various equipment. For example, for bioreactors, the sample valve may have to be manually or automatically repositioned during the various CIP steps to ensure cleaning. Similarly, sensors such as pH and dissolved oxygen probes may have to be manually or automatically retracted during processing. The cleaning programme should ensure – usually by a standard operating procedure – that such sensors are in the correct position in the fermenter during cleaning.

System Configuration and Layout

Two main types of CIP systems are available: single-use units, in which all cleaning solutions are used once and discarded; and solution recovery and reuse units which use cleaning media and rinse liquid more than once. The latter types of systems may recover the recirculated alkali which is stored for the next cleaning event. Similarly, the final rinse water is

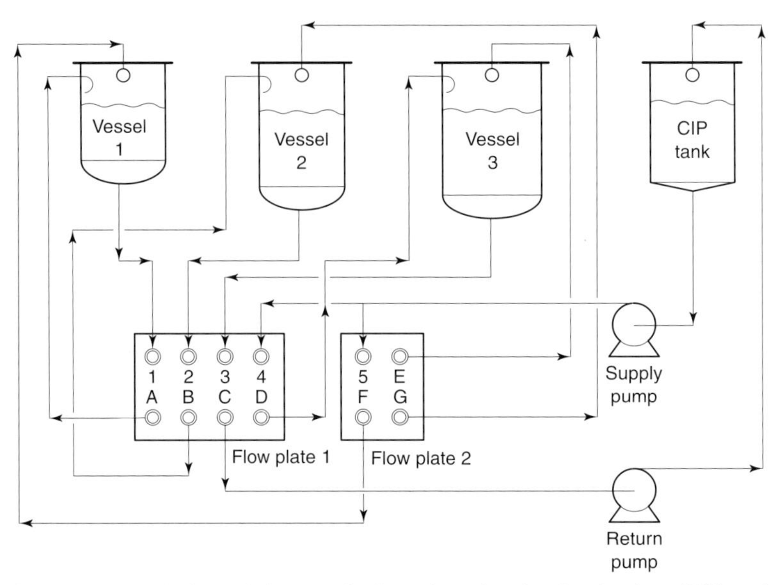

Figure 6 Arrangement of a train of vessels through the transfer flow plates for cleaning-in-place (CIP) and other transfer operations.

recovered and stored for use as pre-rinse or flush in the next cleaning event. If only the alkali is reused, the main supply tank may be sufficient for storage. Recovery and storage of rinse water may require increased storage capacity and a larger floor area for the CIP system. Consequently, the initial installation expense of reuse systems is greater than for single-use units, but the operating costs are lower. Recovery and reuse systems are the norm in large facilities, particularly in food-processing plants. Reuse of cleaning agents is feasible when soil loads are low and cleaning events are frequent, e.g. once a day. The quality of the cleaning agent is less well defined in multiple-use systems than in single-use ones. Single-use systems are preferred when cross contamination is a concern.

A single CIP system usually services all the process equipment in a facility. The CIP flows are directed to selected equipment or group of equipment by making appropriate pipe connections at one or more transfer flow plates (**Fig. 6**). The flow plates provide centralized locations for all the transfer piping inlets and outlets of the various process machinery in a given area of the plant. Usually one or two transfer plate locations are sufficient, but a complex facility may need several. A transfer system with two flow plates for a plant with three process vessels is shown in Figure 6.

During cleaning, or transfers between vessels, the inlets and outlets on a transfer plate must be connected by removable pipe sections which provide positive assurance against accidental mixing of the contents of various process vessels, or a vessel and the CIP fluids. In practice, flow plates are so configured that different, non-interchangeable, pipe sections are needed to connect specific inlets and outlets, thus eliminating the possibility of erroneous connection. Usually, either a standard operating procedure or a computer display instructs the operator to make the necessary connections on the flow plate. The 3-A sanitary practices require elimination of the possibility of accidental intermixing of the CIP flow with the in-process food. This is best achieved by physically disconnecting the CIP system from the process by removing the pipe sections at flow plates. Other than a few manual connections, the CIP system uses pneumatically operated automatic valves. The few manual connections, e.g. those at the flow plates, are made using easy-to-install sanitary couplings.

The simple flow plate scheme shown in Figure 6 allows for transfers between various process vessels (e.g. connection of 1 and B on the flow plate allows transfer of vessel 1 to vessel 2) and provides a means of circulating the CIP fluids through any of the tanks. Thus, during cleaning of vessel 2 (Fig. 6), the CIP supply point 4 is connected to the transfer inlet B and the outlet of the vessel (point 2) is connected to the CIP return line at point C. The spray ball CIP supply point 5 on the second flow plate is connected to the spray ball of the tank (nozzle G on flow plate 2).

A typical CIP system is shown in **Figure 7**. For acid or alkali recirculation, concentrated solutions are metered (pumps P1 and P2 in Fig. 7) into deionized water-filled alkali/acid tanks. The contents of the tank are mixed by recirculation to the tank through the CIP supply pump (P3) while the CIP supply valves (1 and 2) are closed. A heat exchange (E1) heats the solutions to desired temperature. The solutions now

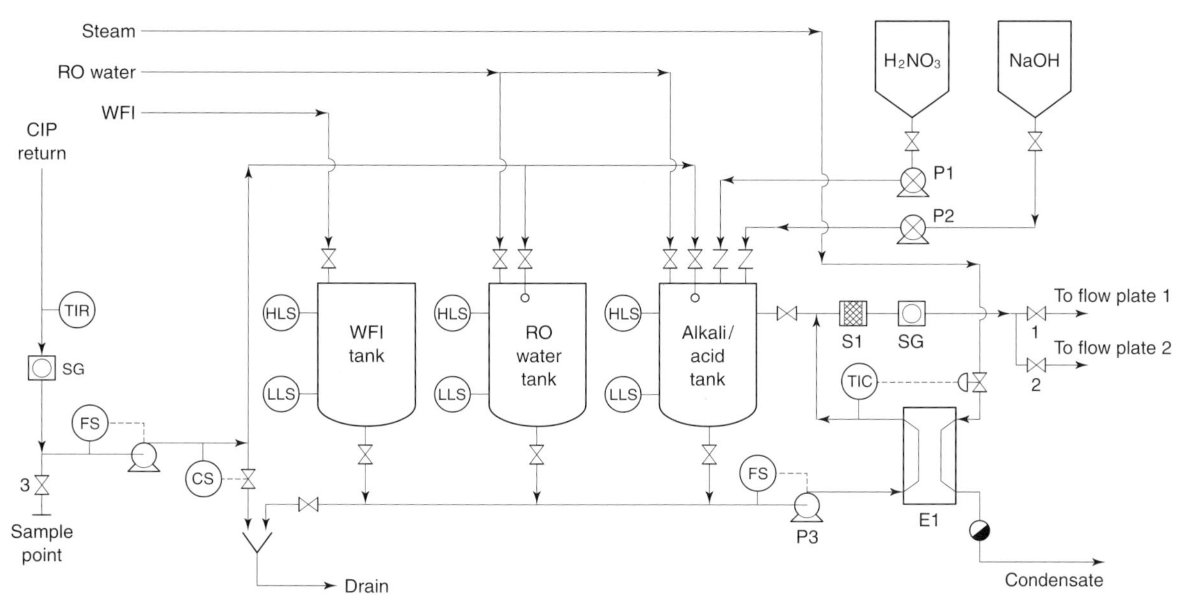

Figure 7 A typical multi-tank CIP system with pumped return. RO, reverse osmosis; HLS, high level sensor; LLS, low level sensor; see text for other symbols.

flow through the exchanger, strainer (S1) and sight glass (SG) to the flow plates for distribution to the spray devices and other areas of the process equipment and piping. The strainer may be a self-cleaning type that discharges accumulated debris to drain whenever the pressure drop across the device exceeds a preset value. Dry running of the supply pump is prevented by the no-flow sensor (FS). The CIP return line (and often the supply lines) has a sample point (valve 3) and the return pump, too, has no-flow protection (FS). A sight glass is also provided on the CIP return line. The return flow goes into one of the CIP tanks or to drain. The temperature of the return flow is monitored and recorded (TIR). The steam supply to the heat exchanger is controlled by the return flow temperature signal during recirculation of alkaline detergent. During final water wash, the conductivity sensor (CS) is used to monitor the return flow which is sent to drain until a preset low conductivity value has been reached indicating complete removal of acid or alkali from the system. Sensors can be provided to monitor the strength of cleaning solutions. Sometimes, the cleaning and sanitizing agents are dosed in-line.

The system shown in Figure 7 relies on a return pump to move the CIP fluids from the equipment back to the CIP tanks. Sometimes, the return pump is aided by an eductor. An eductor generates suction in the return line, thus ensuring that the return pump never air-locks. Whereas an eductor alone may be sufficient in CIP units with wide-bore and short-run return pipes, a return pump is usually necessary. To generate vacuum, an eductor requires a motive fluid. A small motive pump is used to circulate one of the CIP solutions – usually the same one as the returning solution – from the source tank, through the eductor, and back to the source tank. Thus, the motive fluid is different at different stages of CIP. Eductor-assisted return is shown schematically in **Figure 8**.

Automation

Manual operation of CIP systems is only feasible for relatively simple and smaller facilities. CIP of large or complex processing plant invariably requires automation and microprocessor-based control. Usually, programmable logic controllers (PLCs) are used to control the CIP operations. Different areas of a large facility may be in various stages of processing while other areas are being cleaned. Thus, selection of cleaning routes and operation of valves must eliminate any possibility of contaminating process streams with cleaning agents.

Typically, the PLC will be preprogrammed with

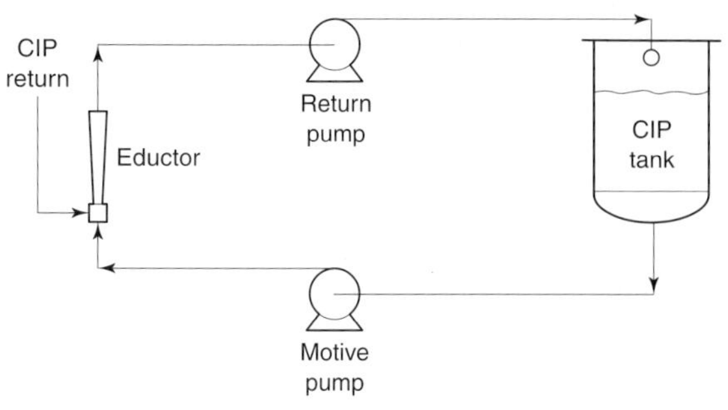

Figure 8 Schematic representation of eductor-assisted return of CIP solutions. The return pump may not be necessary.

several different cleaning programmes, one for each cleaning scheme required in a facility. A cleaning programme will specify the cleaning routes, the sequencing of cleaning chemicals, the flow rates and the temperature and strength of cleaning agents. The duration of each cleaning step will be programmed, and the reservoirs of the cleaning chemicals will be monitored to ensure that sufficient material is available. Upon completion of cleaning, complete removal of cleaning chemicals from process equipment will be monitored. The status of the many valves needed to direct the flow of chemicals will be continuously monitored using positive-feedback loops. In addition to using dedicated valves, pumps and other devices, a CIP system must make use of process valves, pumps and agitators during cleaning. Thus, control of a complex CIP operation must be closely integrated with control of the process machinery. In the event of a failure, a well-designed system shuts down in a safe state with the cleaning agents remaining confined to specified parts of the plant.

The automated cleaning sequence will stop for any steps that may require operator intervention; the sequence will recommence when the operator has acknowledged the execution of the manual step. For certain critical events, the controller will use sensor signals to verify the correct execution of the manual operation before proceeding with the CIP sequence. For example, installation of the correct pipe pieces at flow plates is detected using proximity switches.

While the internal CIP programme may be quite complex, the operator interface is kept simple. Simplified process flow diagrams or synoptic panels provide instant visual indication of the status of the CIP operations. Light signals indicate the status of each relevant valve, pump and agitator. The operator interface consists of a simple water-resistant key pad and display. Back-lit push buttons are commonly used to initiate the different cleaning programmes.

Validation of Cleaning Operations

Validation of a CIP system is essential to ensure consistent and acceptable cleanliness. Validation is a demonstration, to a reasonable degree of assurance, that cleaning according to a specified standard operating procedure will attain the required level of cleanliness, including removal of cleaning agents, in a reproducible manner. Cleanliness is regarded as having physical, chemical and microbiological characteristics. Physically clean equipment is visually free of soil and deposits that may be felt or smelt. Freedom from unwanted chemicals, including residues of cleaning and sanitizing agents, implies a chemically clean state. A microbiologically clean surface is free of unwanted and spoilage agents. In practice, an acceptable level of cleanliness may be taken as one which would eliminate such contamination as would alter the safety, identity, strength, quality or purity of the product.

Cleaning is influenced by time, temperature, degree of turbulence, type and strength of cleaning agents, and the type of soil and its load. Equipment configuration is another influencing factor and so is the quality of the water used in rising and formulation of the cleaning solutions. All these aspects should be considered in validating.

Validation can begin after the pre-validation steps – installation, qualification, performance qualification and operational qualification – have been successfully accomplished and suitable standard operating procedures have been developed. Validation should be carried out according to an internally reviewed and approved validation protocol specifying objectives, exact methods for achieving those objectives, and acceptance or rejection criteria. All validation and pre-validation must be clearly documented with piping and instrumentation diagrams, equipment checklists, calibration records and performance test results.

Suitably placed sampling points for the CIP flows, sensors (e.g. conductivity sensor) and sight glasses need to be designed into the CIP process for ease of validation later. In addition, manual overrides on supply and return pumps and positive-feedback switches for indication of valve positions are useful in validating.

Validation must document the correct sequencing of the various valves and pumps. Specified temperatures and flow rates should be attained for the specified times. The pre-rinse, alkali circulation and other steps must take place in the correct order. There should be checks on identity, strength and purity of the cleaning chemicals and on quality of water. Any CIP control hardware and software also needs to be validated using methods previously described for other computer-based control systems.

After CIP, the equipment should be visibly clean with no sign of adhering soil. Filling of a bioreactor vessel with clean water and intense agitation (or aeration in pneumatic reactors) for a few minutes should not release suspended solids into water. Residues of the CIP chemicals can be monitored by methods such as fluorometry, conductimetry and pH measurement. Measurements of total organic carbon, proteins, carbohydrates or some specific component such as an enzyme or antibody may be an indicator of residual soil. Validated analytical procedures must be employed in these measurements. When the residual concentrations in the final wash water are below the level of detection, concentrated samples may be used to prove reduction of soil to low levels. Wiping the cleaned surface with white tissue of appropriate material should not discolour the tissue. Extraction of the swab in solvent and analysis of the extract may indicate the level of residue on the equipment. The total residue may be calculated based on the surface area of the entire equipment and the area originally swabbed. Inspection of the cleaned surface under a 340–389 nm ultraviolet lamp should show no fluorescence.

Sometimes, simulated contaminants such as dyes are used to validate the CIP process. This approach may not give meaningful data because different substances have different rinsing kinetics. Under identical conditions, dyes such as sodium fluoresceinate may take significantly longer to rinse than a more realistic soil such as casein. Therefore, as far as possible, the CIP process should be validated with actual soil. Rinsing kinetics should be considered in designing and validating the CIP schemes. Surfactants are generally more difficult to rinse from stainless steel equipment than sodium hydroxide, nitric acid and phosphoric acid. Among surfactants, non-ionic ones are relatively easily rinsed, but many of these tend to foam a lot, which is undesirable in CIP systems. Satisfactory removal of pyrogens may be a consideration during cleaning of equipment producing injectables. Pyrogen removal issues are especially important for chromatography columns, membrane filtration modules, and reverse osmosis water systems that are operated in a non-sterile but bioburden-controlled manner.

Normally, the equipment is cleaned soon after use. If immediate cleaning is not possible, the validation exercise should demonstrate that satisfactory cleaning is achieved after the maximum allowable standing time of the soiled equipment. Also, if cleaned equipment is not immediately used for processing, validation should show that a satisfactorily clean state is maintained until next use. Alternatively, a second complete CIP sequence or at least a sanitization wash may be required just before use.

Some of the validation practices noted here are admittedly more rigorous than ones employed in the food industry, but scientifically based and thorough validation methods are well established in biopharmaceutical processing.

CIP in the Biopharmaceutical Industry

Cleaning demands and the acceptable practices in CIP of certain biopharmaceutical processing facilities can be very different from those encountered in hygienic processing of food and dairy products. Cross contamination between products needs to be rigorously prevented. Similarly, any level of contamination with cleaning agents is unacceptable. The quality of the cleaning agents may have to be controlled to significantly higher levels than in food processing. Sanitization washes, for example, with solutions of QATs are not needed for bioprocess equipment that is used sterile. The quality of the final rinse water needs to be especially high. The validation requirements can be more severe.

See also: **Process Hygiene**: Designing for Hygienic Operation; Types of Biocides; Overall Approach to Hygienic Processing; Risk and Control of Airborne Contamination; Testing of Disinfectants; Involvement of Regulatory and Advisory Bodies; Hygiene in the Catering Industry.

Further Reading

Adams D and Agarwal D (1988) Clean-in-place system design. *Biopharm* 2(6): 48–57.

Brennan JG, Butters JR, Cowell ND and Lilly AEV (1990) *Food Engineering Operations*, 3rd edn. p. 497. London: Elsevier.

Chisti Y (1992) Build better industrial bioreactors. *Chem. Eng. Prog.* 88(1): 55–58.

Chisti Y (1992) Assure bioreactor sterility. *Chem. Eng. Prog.* 88(9): 80–85.

Chisti Y and Moo-Young M (1994) Clean-in-place systems for industrial bioreactors: design, validation and operation. *J. Ind. Microbiol.* 13: 201–207.

Chisti Y (1998) Biosafety. In: Subramanian G (ed.) *Bioseparation and Bioprocessing: A Handbook*. Vol. 2, p. 379. New York: Wiley-VCH.

Harder SW (1984) The validation of cleaning procedures. *Pharmaceut. Technol.* 8(5): 29–34.

Hiddink J and Brinkman DW (1984) Cleaning in place in the dairy industry: some energy aspects. In: McKenna BM (ed.) *Engineering and Food*. vol. 2, p. 939. London: Elsevier.

Hyde JM (1985) New developments in CIP practices. *Chem. Eng. Prog.* 81(1): 39–41.

International Association of Milk, Food, and Environmental Sanitarians (1986) *3-A Accepted Practices for Permanently Installed Sanitary Product Pipelines and Cleaning Systems*. Ames IA: International Association of Milk, Food and Environmental Sanitarians.

Stewart JC and Seiberling DA (1996) Clean in place. *Chem. Eng.* 102(1): 72–79.

Risk and Control of Airborne Contamination

G J Curiel, H M J van Eijk and **H L M Lelieveld**, Unilever Research Vlaardingen, The Netherlands

Introduction

It is important to ensure that food products remain safe and wholesome throughout their shelf life. Thus, unacceptable contamination with harmful microorganisms, including airborne microorganisms during production and storage must be prevented. Where product is pasteurized or sterilized after packaging, this is relatively easy, as such treatment will take care of all relevant microorganisms. The success of in-line preservation treatments, such as high-temperature short-time pasteurization or ultra-high-temperature sterilization, depends on the successful prevention of recontamination of the treated product prior to packaging. This requires the use of aseptic process lines, sterile buffer tanks and internally sterile packing material downstream from the preservation treatment. In such cases it is essential that, where the product or the packing material is exposed to the air, the microbial quality of the air is well under control.

Where products are exposed to the environment during preparation and not subjected to a decontamination treatment, depending on the shelf-life conditions, control of the quality of the environmental air can be equally important.

If air (or other gases, such as nitrogen and carbon dioxide) is used as an ingredient for food products, such as whipped cream and ice cream, they must comply with microbiological product specifications.

This article will deal with the occurrence of microorganisms in the air, how they may contaminate food products, as well as methods of preventing such contamination by removal or inactivation. As it will enable one to calculate the risk of contamination by airborne microorganisms, methods of determining the concentration of microorganisms in the air are also discussed.

Presence, Transport and Sedimentation

Air carries many microorganisms. Concentrations of 100–10 000 microorganisms per cubic metre are quite normal. The concentration differs according to season and location. In agricultural areas during harvesting, the concentration of spores of bacteria and moulds in the outside air can be extremely high – close to a billion per cubic metre. In food factories, spaces between ceilings and so-called false ceilings as well as spaces between the ceiling and the roof must be carefully sealed off, as birds, rats and other pests may find this an attractive habitat. Unless access is denied, highly contaminated dust will accumulate here. Draughts carry such dust to the food-handling and processing areas. In contrast, in winter after snowfall, the concentration of microorganisms in the outside air may be very low. In enclosed spaces, the concentration usually varies less, between several hundred and a thousand per cubic metre. Most microorganisms are harmless; they do not cause illness but break down organic matter. If that organic matter is food, however, they cause spoilage. Some microorganisms can cause illness. They may be infectious, i.e. able to cause illness in relatively low numbers (between just a few and 100 000) or, when growing in the product, produce toxins which cause illness. Therefore, depending on the food produced or processed, control of airborne microorganisms may be necessary.

Ducts, in particular those for air conditioning, may, collect water and allow the growth of microorganisms, which are subsequently carried by the flow of air to the ventilated or conditioned areas. It is important therefore to ensure that air ducts are fully self-draining or that other measures are taken to prevent the accumulation of condensate.

Depending on type and origin, microorganisms may occur in a variety of states. As a result of their way of growth, conidia (mould spores of *Penicillium* and *Aspergillus* species) will often occur on their own, while bacterial spores are often clustered and surrounded by debris. Many microorganisms are associated with dust particles. Microorganisms may also be included in crystals of salt or sugar. In wet environments, microorganisms may be present in tiny droplets in aerosols. This will be the case, for instance, in the neighbourhood of sewage plants, where *Salmonella* species may be abundant. Rainfall will create aerosols as a result of relatively large drops impacting on contaminated solid surfaces. Similarly, contaminated aerosols will be created by the use of water in a factory. Water splashing in areas where product is exposed is unacceptable, even if its purpose is to flush away spilled food material. Contaminated aerosols also originate from refrigeration systems with automatic defrosters. Sometimes, the water collected supports selectively the growth of pathogenic microorganisms such as *Listeria monocytogenes*. Air contamination also happens as a result of sneezing, coughing and talking if people are suffering from an infectious disease. Fortunately, most vegetative bacteria die rapidly (in seconds) in dry air. Some, e.g. species of micrococci, may survive long enough to cause infection and disease. Spores survive in dry air and most airborne microorganisms consist of bacterial or fungal spores. How a microorganism exists has an important influence on its survival as well as on the effectiveness of inactivation treatments. Soil and crystals may retard inactivation and make some inactivation treatments useless. The appearance also influences the sedimentation rate of microorganisms, which is important as the sedimentation rate determines the rate of microbial recontamination of products during the time of exposure.

Risk of Contamination with Airborne Microorganisms

By exposing Petri dishes with a suitable agar medium to air for a known time, the sedimentation rate (*sr*) of airborne microorganisms can be determined. If at the same time the concentration (c) of microorganisms in the air is measured (as explained below), it is easy to calculate the settling velocity ($v = sr/c$). In the absence of vertical air currents, v is usually of the order of $3 \times 10^{-3}\,m\,s^{-1}$. Using these data, and assuming that the density of microorganisms and dust particles is approximately $1000\,kg\,m^{-3}$, it can be calculated using Stoke's law that the aerodynamic diameter of airborne microorganisms is effectively of the order of 9 μm.

As the diameter of spores of moulds and bacteria is in the range of 0.3–1.0 μm, it may be assumed that the larger apparent diameter is the result of clustering and association of the microorganisms with dust particles.

If v is measured in $m\,s^{-1}$ and c in m^{-3}, $sr = v \times c$ will be in $s^{-1}\,m^{-2}$. With a known exposed product surface area a measured in m^2 and a known exposure time t measured in s, the rate of infection of a product surface will be:

$$R = sr \times a \times s \text{ or } R = v \times c \times a \times s$$

(again in the absence of vertical air currents). Example: a sterilized product is poured into jars with an opening with a surface area of $3.5 \times 10^{-3}\,m^2$ ($35\,cm^2$). The time of exposure is 2 s. The concentration of microorganisms in the air (measured) is $800\,m^{-3}$. The contamination rate will then be $R = 0.0168$. In other words, on average, infection will take place approximately once per 60 containers.

If there are vertical air currents, as used in some laminar airflow systems, the sedimentation rate will be dramatically increased. The velocity of air in a laminar flow usually is $0.3\,m\,s^{-1}$, 100 times as high as the settling velocity. If laminar airflow is used to reduce the risk of contamination, the flow should be horizontal so that the sedimentation rate is not changed and hence the rate of infection during exposure remains low.

Methods to Determine the Concentration of Microorganisms in Air

Microorganisms can travel through the air in three ways; adhering to a dust particle, adhering to a droplet or as a single particle. There are in principle three methods of checking the number of microorganisms in the air.

The first and simplest method makes use of the sedimentation of particles, and for this purpose Petri dishes with a solid growth medium can be used: these are exposed to the air for a fixed time. After incubation the colonies can be counted and this number of microorganisms is expressed as counts per time and surface area. This method is not an accurate estimate of the amount of microorganisms in the air. Particle shape and size and air current patterns will influence the settling time of the microorganisms to be collected.

In the second method, air from the environment is directed to a growth medium at a high enough velocity to ensure that the microorganisms impact on the surface when the flow of air is diverted.

In the third method, air is passed through a filter which retains microorganisms and which can be examined by standard microbiological methods, such as by covering with a suitable agar medium. The advantage of the latter two methods is that the number of microorganisms can be expressed per volume of air sampled and is less dependent on the particle size.

There are also electrostatic and thermal precipitation samplers, but these are not widely used, as they are difficult to handle, and for this reason are not further discussed here.

Impact samplers are available in several models, which are described below.

Slit Sampler

The air to be examined is directed via a narrow slit to the surface of a solid growth medium (agar) in an open Petri dish (**Fig. 1**). To get an even distribution of microorganisms over the surface of the agar, the dish is rotated. Commercial slit samplers typically collect particles of 0.5 μm and above.

Andersen Perforated Disc Sampler

Impact is obtained by directing the air via small holes to the agar surface (**Fig. 2**). As this type of sampler can easily be extended by placing more units on top of each other, it is possible to discriminate in particle size as the diameter of the holes in the disc per stage can be varied. In this way it is possible to trap particles in a size range from 0.5 to 10 μm or larger.

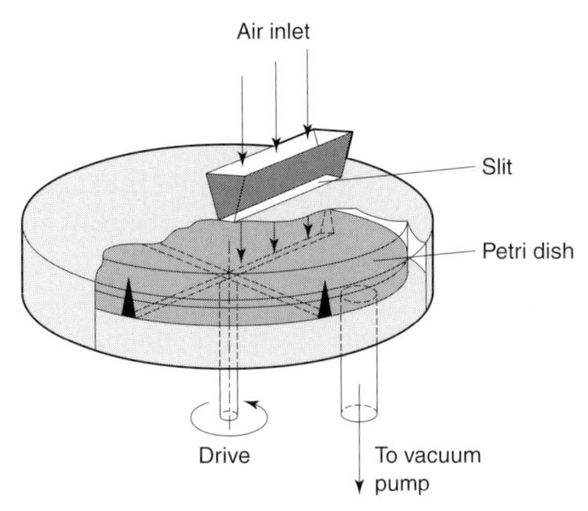

Figure 1 Principle of a slit sampler.

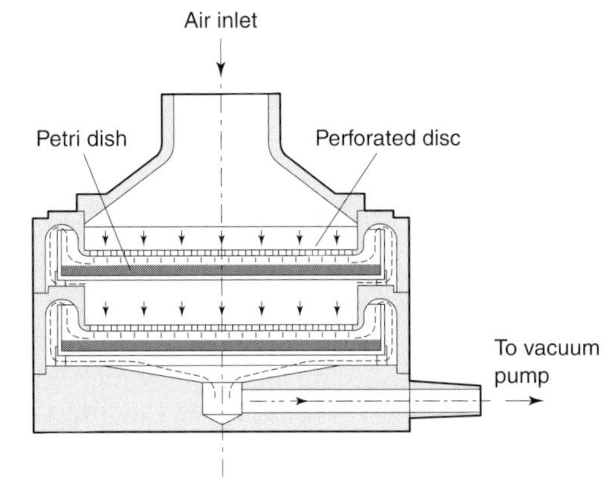

Figure 2 Andersen perforated disc sampler.

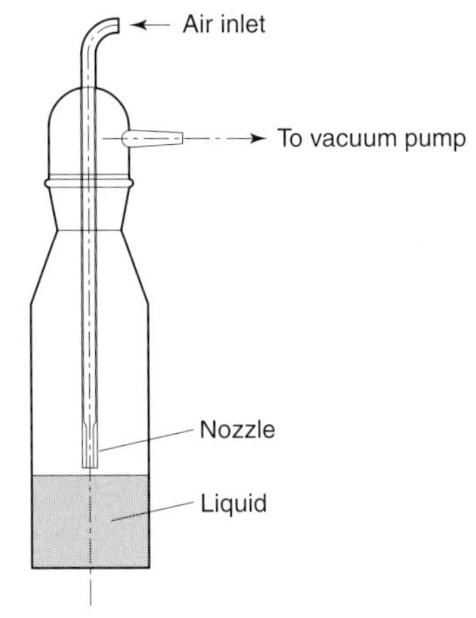

Figure 3 Impinger.

Impinger

The impinger makes use of a liquid medium in which the particles are to be trapped (**Fig. 3**). A tube with a narrow opening is placed just over the liquid surface to which the air stream is directed. Any particle from the air will be blown into this liquid which can afterwards be examined by standard microbiological methods. As particles may contain more than one microorganism, this method will give higher counts if those particles disintegrate in the process. In contrast, the high shear forces to which the microorganisms are subjected might inactivate them.

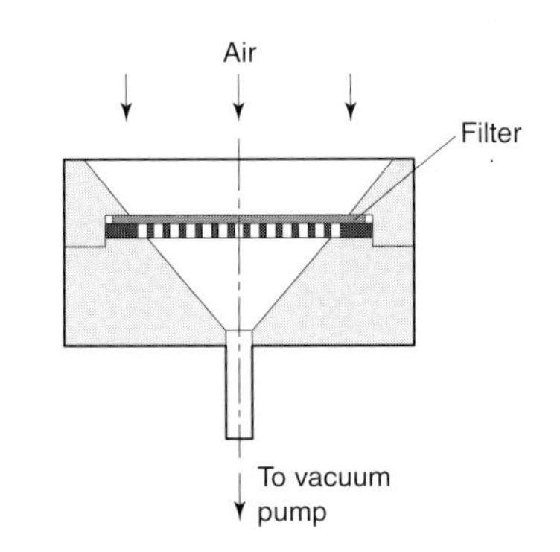

Figure 4 Bacteria air filter.

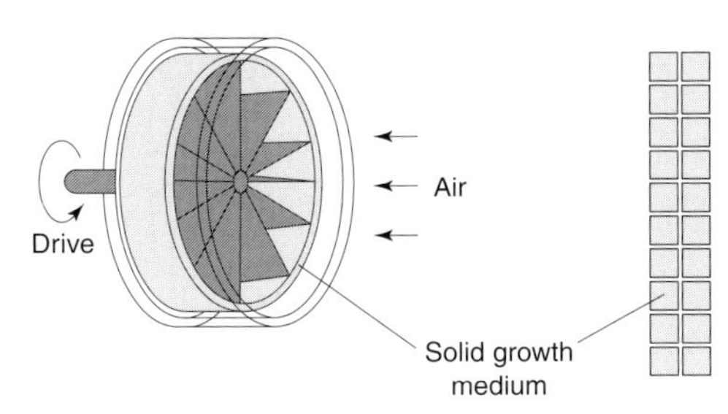

Figure 5 Centrifugal air sampler.

Filtration

Microorganisms can also be captured with a membrane filter through which the air is sucked, as shown in **Figure 4**. After sampling the filter can be put directly in a Petri dish with agar and incubated. The filter can also be soaked in saline solution and examined using standard microbiological methods. As not all particles will be released from the disc, liquid-soluble gelatine filters can be applied. Filtration of air is not always useful as vegetative cells may become dehydrated and die during sampling; the low flow is only suitable for very small volumes.

Centrifugation

A fan can be used to direct the air to an agar strip which collects the microorganisms (**Fig. 5**). After sampling the agar strip is removed and incubated, whereafter colonies can be counted. As no high velocities can be generated with this type of sampler, smaller particles are not trapped.

Reduction of Airborne Microbial Contamination

Air can be contaminated with microorganisms and may contain up to 10 000 cfu m^{-3}. Without proper treatment, the air may contaminate food products. Air that is intended to come in contact with sterile product or used to maintain a positive pressure in aseptic tanks and equipment must be sterile. The air to be treated must itself be of good quality; thus, removal of moisture, oil and particles is essential. Treatments such as inactivation or removal of microorganisms present in the air can then be applied.

Several methods may be used to reduce the number of microorganisms in the air. These include physical treatments and chemical agents or a combination of both. However, one of the most used methods of producing sterile air with higher assurance of sterility is filtration. Filtration is the removal of particles, including microorganisms, from the air. Other physical methods for removing air contamination are centrifugal (multi-cyclone), rotary flow collector, Venturi scrubber and electrostatic precipitator.

Once the number of microorganisms in the air is reduced or the air is sterile, recontamination should be prevented. The equipment should then be hygienically designed and easy to clean and to disinfect. Keeping the sterile area above atmospheric pressure will prevent recontamination.

Reducing the viable count in air can be achieved by several methods: physical, chemical or a combination of both. Physical means such as moist heat, dry heat (including incineration), ultraviolet (UV) radiation and ionizing radiation can inactivate microorganisms in air. Not all methods are equally reliable or effective, and differences in resistance between microorganisms must be taken into account. If spores must be inactivated, higher temperatures are required than for vegetative microorganisms. From the physical processes in practice only dry heat (including incineration) and UV are used for sterilization of air.

Inactivation by Heat

Dry heat (hot air or superheated steam) at a temperature of 160°C or higher, including incineration at higher temperatures, can be used to sterilize air. At lower temperatures, dry heat is much less effective than moist heat (saturated steam), which is applied at 121°C or higher. Incineration is an old method of destroying waste. Incineration has many desirable features; it destroys the structure and appearance of the waste; it reduces volume; it permits energy in the waste to be recovered as heat; and the heat can destroy all microorganisms in the air. Nowadays, incineration is used to treat a variety of air pollution control problems related to contamination in process exhaust gases. This includes the removal of volatile organic components, hydrocarbons, toxic chemicals and microbiological contaminants such as viruses and microorganisms. Air can be sterilized within less than a second by heating to a temperature of 350°C. Incinerators are probably more reliable than filters, provided that their design guarantees that all air is heated to the required temperature for the required time. Therefore an incinerator must be equipped with a reliable temperature and flow control system. Thermal incineration with regenerative heat recovery does not require much energy because of the low heat capacity of air. The energy required for incineration is derived from the preheat recovered in the regeneration section and from the heat from fuel or electricity in the combustion zone. Regenerative heat recovery efficiency can be up to 95%. For most applications the method is more expensive than filtration.

Inactivation by Ultraviolet Light

UV radiation can be used to kill microorganisms in air. UV radiation has a wavelength range between about 210 and 328 nm, with maximal bactericidal activity near the wavelength 260 nm of peak absorption of DNA. Bacterial spores are generally more resistant to UV light than vegetative cells. UV light systems are now available for treating airflow entering a sterile area. Air disinfection systems are fitted into duct work, and microorganisms present are deactivated as they are exposed to the UV source. UV can also be used to disinfect air for pressurizing tanks or pipelines. A disadvantage of UV is that it is not effective if microorganisms are protected by particles or embedded in dust, due to shade effects.

Inactivation Using Chemical Agents

Air in enclosed spaces can be decontaminated by fogging, a technique whereby a solution of antimicrobial chemicals is dispersed in the air. The aerosol consists of droplets typically between 5 and 50 μm diameter. These droplets may float around for some time before settling on horizontal or adhering to vertical surfaces. Often fans are used to improve the distribution of the aerosols. When the water evaporates dry residues of the chemicals may be left on the surface. Where chemically contaminated surfaces may – intentionally or unintentionally – come into contact with the food, the chemicals chosen should not be toxic. Corrosivity should also be considered in choosing a suitable chemical. The antimicrobial chemical agents may be several types, such as aldehydes, hypochlorite and hydrogen peroxide.

Toxicity of Chemicals The use of chemicals in concentrations toxic to microorganisms may also be toxic to humans. Obviously, this must be taken into consideration. Chemical vapours may be a hazard to the food product and may pose a danger to personal health, in particular with respect to the respiratory organs. The maximum allowed concentration in the

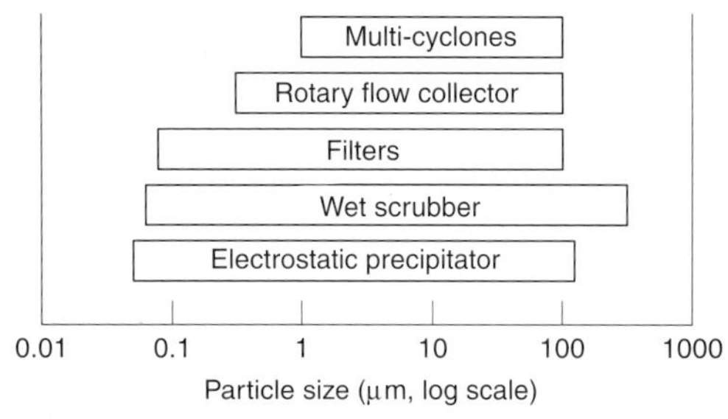

Figure 6 Particle size removal by various treatment methods.

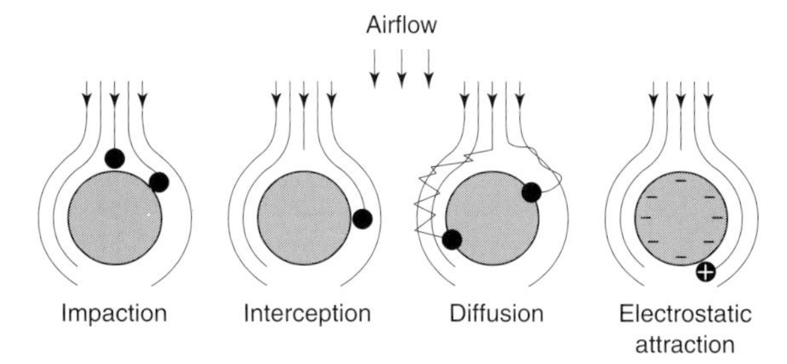

Figure 7 Mechanisms by which airborne microorganisms may be trapped in fibrous depth filters.

air and in the environment should be specified and monitored.

Removal of Particles from Air

Removal is a physical treatment method to reduce contaminants including microorganisms from air. The physical methods of removing air contamination are filtration, cyclone, rotary flow collector, wet-scrubber and electrostatic precipitator. Filtration is a reliable method of producing sterile air. **Figure 6** shows an outline of the particle size removal range for various physical treatment methods. Apart from filtration, the other techniques are mostly used as industrial dust collectors to clean exhaust gases.

Filtration Air filtration has the greatest practical potential of all the separation methods. Air sterilization is the physical removal of microorganisms from the air by filters of appropriate retention efficiency. Depth filters are made of cellulose, glass wool or glass fibre mixtures with resin or acrylic binders. The mechanism of filtration in depth filters can be interception, sedimentation, impaction, diffusion and electrostatic attraction (**Fig. 7**).

Depth filters are believed to achieve air sterilization because of the twisted passage through which the air passes, ensuring that any microorganisms present in the air are trapped not only on the filter surface, but also within the interior. Membrane filters consists of thin (10–100 μm) films of polymers such as polycarbonate polytetrafluoroethylene (PTFE). Membrane filters prevent microorganisms from passing

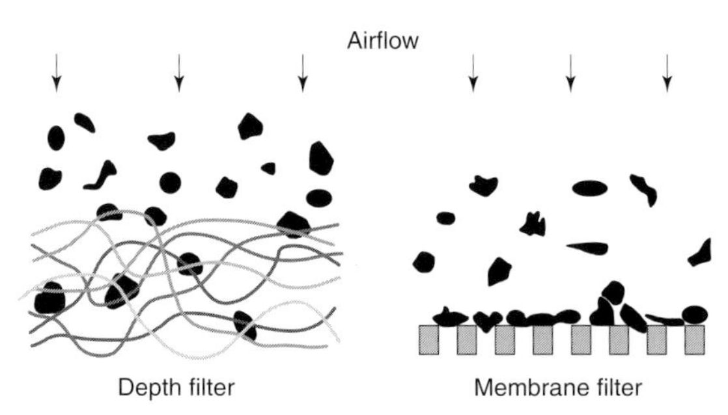

Figure 8 Mechanism of retention for depth and membrane filters.

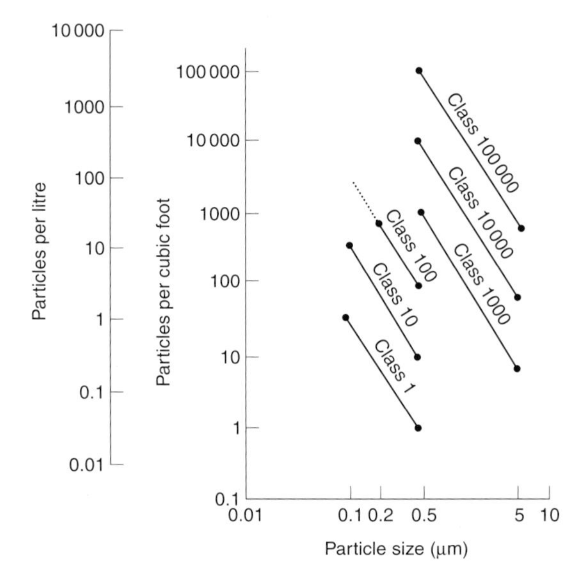

Note: The class limit particle concentrations shown in Figure 9 are based on ISO FS209f and defined for class purposes only. They do not necessarily represent the size distribution to be found in any particular situation.

Figure 9 Classes for the quality of air. Class limits in particles per cubic foot of size equal to or greater than particle sizes shown.

(straining effect) because the openings in the filter material are too small (average 0.2–0.5 μm). The mechanism for both materials is shown in Figures 7 and **8**.

The quality of air is laid down by the maximum level of contamination permitted. According to ISO FS209f, four classes are recognized: Class 100, Class 1000, Class 10 000 and Class 100 000. The maximum numbers of particles 0.5 μm or larger per cubic foot (litre) are, respectively, 100 (3.5), 1000 (35), 10 000 (350) and 100 000 (3500). **Figure 9** shows the class limits in particles per cubic foot of size equal to or greater than particle size shown. Class 10 and 1 are under discussion and not yet widely applied. Class 100 is most commonly used for aseptic applications, hence in sterile areas.

The fibrous sheet filters have a low resistance to airflow and a large surface area. Such a filter is required to provide air with an extremely low microbiological load to aseptic areas. High-efficiency particulate air (HEPA) filters meet those requirements and are able to remove particles of 0.3 μm or larger and may even remove particles much smaller than this. They have efficiencies of 99.97–99.997% retention for particles of 0.3 μm or larger. This type of filter is mostly used in laminar airflow (LAF) systems, such as LAF rooms and cabinets.

Multi-cyclones Cyclone collectors are one of the simplest dust collectors. They do not have moving parts and are easy to maintain. Particles are separated by centrifugal force. There are two types of cyclones, the axial inlet flow cyclone and the tangential inlet cyclone. In the axial flow cyclone, the air is rotated by guide vanes and in the tangential type of cyclone the air rotates by flowing from the tangentially connected inlet pipe. Both types can be applied in a multicyclone system to separate particles larger than 1 μm. This type can be used to treat air with a high content of pollen and spores and industrial exhaust gases containing powders, dust and fine particles.

Rotary Flow Collectors This type of collector can separate particles of 0.5 μm or below. The construction and mechanism are more complex than ordinary cyclones. Air enters a primary vortex chamber at the bottom where dust is separated by centrifugal force. Then the air flows upward through the exit nozzle as a swirling jet into the main cylindrical separation chamber. Here rotational descending pure airflow increases the magnitude of rotation of the primary vortex flow and the effect of the centrifugal separation of particles.

Wet Scrubber The wet scrubber washes particles off with water droplets. The collection mechanisms are approximately the same as infiltration – impaction, interception and diffusion. Four types – vortex, centrifugal, nozzle and Venturi scrubbers – can separate easily and cheaply down to a particle size of about 1 μm. The Venturi type is suitable for separating submicron particles down to about 0.1 μm, but is relatively expensive to operate.

Electrostatic Precipitator The electrostatic precipitator separates particles from air by the direct current (DC) high-voltage electric field. There is a collecting electrode shaped as a cylindrical pipe or a set of parallel plates connected to earth. A negative voltage is applied to a discharge electrode where a corona discharge from its surface is produced. Negatively charged particles are repelled by the electric charge to the surface of the collecting electrode. The particle separation is effective in the range of 0.1–10 μm. Electrostatic precipitators are mostly used as industrial dust separators.

Prevention of Recontamination

Once the air is clean or sterile, recontamination should be prevented. First the air distribution system (air lines and equipment) should be hygienically designed to prevent any recontamination of the air and growth of microorganisms in the system. Other areas, such as sterile tunnels, tanks and cabinets, should be kept at above atmospheric pressure to prevent recontamination. Sterile locks and air filtration units can be used for microbial free environments. Ultra clean rooms should also be kept at above atmospheric pressure of sterile LAF. Aseptic packing/processing requires the use of sterile air and LAF cabinets, rooms and tunnels. The building, rooms, cabinets, tanks and container (packaging) must then be kept at above atmospheric pressure.

Validation, Assurance and Maintenance of Air Filters

Filters are available in various grades. For sterile filtration membrane filters and HEPA depth filters are used. Where membrane filters are used the particles are retained by sieving; for depth filters they are retained by adsorption and interception. Sieve filters are so-called membrane or absolute filters; adsorption filters are depth filters and have no defined pore size. When sieving the pores are smaller than the particles to be trapped, whereas by adsorption the particles stick to the filter material and are captured within the fibre structure of the filter.

The efficiency of depth filters is commonly specified according to the dioctylphthalate (DOP) test (*DOP-smoke penetration and air resistance of filters*, MIL-STD 283). Using this test particles with an average diameter of 0.3 μm are sucked through the filter layer and the efficiency is expressed as a percentage of particles retained. HEPA filters have an efficiency of 99.97% and higher. Mechanical shocks or vibrations cause depth filters to release particles.

Sterilizing grades of membrane filters have pores ranging from 0.1 to 0.8 μm.

To protect HEPA filters against damage by particles of high density (with sufficient kinetic energy to perforate the membrane) and to increase the stand time, pre-filters are used. The stand time is influenced by:

- the concentration of dust in the air
- the humidity of the air
- air velocity
- chemicals

- temperature
- pressure differences across the element
- flow rate.

Too high a pressure drop across a filter element is normally a criterion for replacement. Pinholes and other small defects cannot be detected in this way. To check whether the filter is still within the manufacturer's specifications, special laser counters can be used. These are capable of detecting particles of 0.3 μm and above. Membrane filters, which are often used to supply sterile air to fermentation processes, can be checked by the water intrusion test. In this test the filter is completely wetted and a fixed pressure is applied which allows air to diffuse through the liquid, which can be quantified with a flow metre. This value is a measure of effective pore size.

To ensure filter quality, rely on the manufacturer's quality control system which guarantees that a product meets the specified requirements. Particles normally remain attached to the filter. This is not necessarily so with microorganisms. Moist air allows bacteria to grow on the filter surface and often through the filter. To check whether membrane filters are able to retain microorganisms, challenge tests have been developed during which the filter is exposed to large amounts of *Brevundimonas diminuta*, which is a very small and motile bacterium. It has been proven that this bacterium can grow through filters with pores of 0.3 μm within 24 h. This means that there should be no water in the pores and for this reason air filters should be hydrophobic.

Conclusion

Air carries many microorganisms which may make food unfit for consumption. There are ways of quantifying the risk of infection. There are also several effective ways of reducing the concentration of microorganisms in the air. With proper control, airborne contamination will not be a major food-poisoning concern.

See also: **Preservatives**: Classification and Properties. **Process Hygiene**: Types of Biocides; Involvement of Regulatory and Advisory Bodies.

Further Reading

Leahy TJ and Gabler R (1984) Sterile filtration of gases by membrane filters. *Biotechnology and Bioengineering* XXVI: 836–843.

Mostert MA (1993) Microbiologically safe aseptic packing of food products. *Trends in Food Science and Technology* 3: 21–25.

Ogawa A (1984) *Separation of Particles from Air and Gases*. Vols 1 and 2. Boca Raton, FL: CRC Press.

Stezenbach LD (1992) Airborne microorganisms. In: *Encyclopedia of Microbiology*. Vol. 1, p. 53.

US Atomic Energy Commission, US National Bureau of Standards (1956) *DOP-Smoke Penetration and Air Resistance of Filters*, MIL-STD 283.

Testing of Disinfectants

J F Williams, Department of Microbiology, Michigan State University, USA

J R Bickert, Halosource Corporation, Seattle, USA

Introduction

Public awareness in western countries of the hazards associated with food-borne pathogens is at an all-time high. National and international media coverage of incidents of contaminated meat and fruit causing outbreaks of *Escherichia coli* O157:H7, *Salmonella* and hepatitis A in recent years illustrates the potential for disastrous consequences of the breakdown of process hygiene. Hygienic measures in food processing involve many elements, including plant, equipment and process design, cleaning protocols and personnel training in the adoption and implementation of sanitary measures of behaviour and food handling. Intrinsic to many of these elements is the use of biocidal chemical agents to limit the scope and intensity of microbial contamination.

Characteristics of the chemical entities that find use in this context are reviewed elsewhere in this volume, as are the most important regulatory authorities and oversight mechanisms commonly put in place to control biocide use. Registration of chemical agents and formulations for applications in food process hygiene generally require the compilation of data on efficacy and safety testing, sufficient to justify confidence that the products match the demands of the proposed use pattern. Biocides used in food-processing applications fall into select categories based on features dependent on their chemistry. The specific uses of most concern are in environmental sanitation, especially for food-contact surfaces, topical application to the hands of food-processing and preparation personnel, and carcass or foodstuff decontamination by direct application. Types of protocols commonly applied are reviewed in this section, together with an account of their respective benefits and shortcomings. The suitability of these test systems in the fast-changing world of food preparation and distribution, emerging microbial food pathogens and conceptual shifts in overall understanding of the relevant microbial ecology is also considered.

Testing Protocols for Chemical Agents Used in Environmental Contamination Control

Principles of Efficacy Testing

The most important use patterns for biocides in process hygiene involve the application of chemical formulations to environmental surfaces. The principal microorganisms targeted are bacteria, and the test systems employed generally require controlled exposure of specified strains of disease-causing (pathogenic) bacteria to use dilution preparations of the agent or formulation under study. However, environmental sanitation in food processing is not entirely focused on the elimination or reduction in numbers of food-borne pathogens, but is aimed at the entire bacterial population, including microbes which may generate odours, spoil food products or decrease shelf life, but that have no disease-causing potential.

Bacteria are emphasized in the standards because mammalian viruses, although sometimes present in food-processing environments, cannot proliferate in the absence of animal host cells, and therefore have traditionally been less of a concern. This bias is beginning to be undermined by the realization that hardy enteroviruses, such as hepatitis A virus, can contaminate food products and processes via mechanisms that remain unclear, but which can lead to widespread dissemination of pathogens in packaged frozen fruit products. Likewise, efficacy-testing protocols for environmental biocides make no reference to protozoan pathogens, but the recognition that newly emerging disease agents such as *Cyclospora* can contaminate food surfaces and survive over extended storage and transportation periods to be distributed on fruits and vegetables has brought new attention to this group of microbes. *Cryptosporidium* outbreaks appear to have resulted from environmental contamination of fruits, and the high visibility of this pathogen in contemporary literature on gastroenteritis makes it likely that protozoan organisms will be incorporated into certain efficacy considerations in the near future. These are serious developments because such pathogens tend to show extraordinary durability in the environment and are not readily deactivated by any of the commonly used disinfectant agents used in the food industry at present.

Those disinfectant agents are generally formulations of quaternary ammonium compounds (QACs), iodophors and chlorine-based biocides, with the latter dominating in overall use frequency. Efficacy-testing protocols began to be defined for food sanitizers early in the 20th century as public health authorities devised preventive hygienic measures in the era of typhoid fever. Not surprisingly, demonstration of activity against *Salmonella typhi* became one of the hallmarks of these first testing protocols. *S. typhi* remained the gold standard organism in regulatory testing systems for many years, although it has now been supplanted by others. From the earliest days it was accepted that biological methods, rather than chemical assay methods, were necessary to assess the merits of food process disinfectant formulations, in recognition of the fact that quantitative determination of biocides in a product often provides less than optimal information about their efficacy in the practical formulations that are necessary for utility in practice.

Nowadays, an emphasis on speed of quantitation has produced a rash of chemical assays for monitoring microbial contamination levels, in preference to the more traditional approaches based on swabs, or the use of Rodac nutrient agar plates with an elevated agar surface, designed to be pressed on to the test surface area. Immunoassays for bacterial antigenic components and fluorometric assays that detect membrane-associated ATP and/or ADP in live bacterial contaminants are now widely used. Some can provide read-outs in seconds or minutes, depending on the sensitivity required. However, these systems have yet to find a place in the testing protocols used in determining the efficacy of sanitizing biocidal products for the food industry.

Disinfectant Test System Formats

Microbiological tests require exposure of the target organisms in one of two modes:

1. So-called suspension tests (**Fig. 1**), wherein suspensions of the organisms prepared from pure cultures are exposed to a specified volume of a suitable intended-use dilution of the disinfectant, for one or more contact periods, under carefully controlled conditions of temperature and pH. At the end of the contact period, a sample of the mixture is taken and subcultured into a liquid medium (to establish survival of any living target organisms) or plated on to appropriate media so as to be able to quantify the efficacy by measuring the numbers of survivors compared to the initial inoculum. End points for the liquid medium approach are usually the dilution yielding no growth, or the ratio of positive/negative tube cultures in a series of tubes at each dilution.
2. Carrier tests, traditionally favoured in the US, wherein target organisms exposed to the disinfectant are attached to the surface of inert carrier

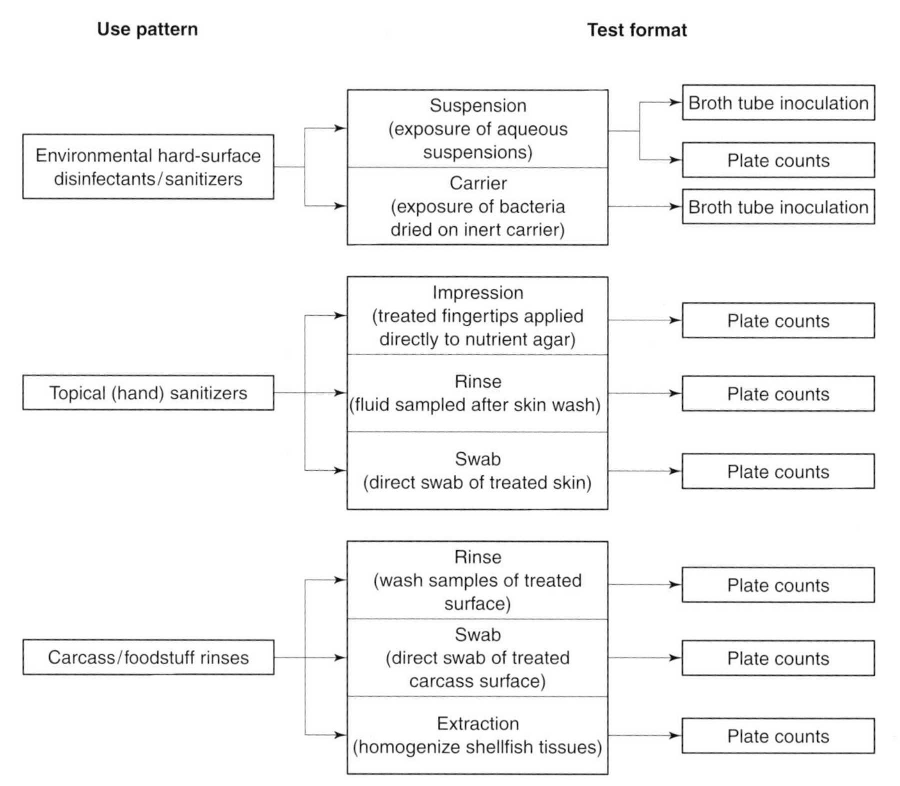

Figure 1 Bacteriological testing systems for biocides in food processing.

vehicles, usually made of glass, stainless steel or porcelain. Carriers are immersed in the biocide use dilution under test, and after the contact period is finished, the carriers are removed and transferred to liquid growth media and the numbers of carriers bearing live organisms are determined. Results are generally reported as the proportion of positive tubes out of a replicate set. Carrier tests are often considered to represent a more realistic measure of the efficacy, since in the real world of contamination, bacteria are usually stuck on to a surface rather than being exposed to disinfectants as suspensions of organisms in water. The tests are therefore regarded as tougher to pass than suspension tests.

Sources of Variation

On the face of it, both these approaches to evaluation have simple structures which would appear to permit ready standardization, and through appropriate selection of target organisms most relevant to food hygiene, reliable, reproducible test data for chemical disinfectants ought to be attainable. In reality, the number of variables that can be introduced into the matrices of these two protocols has turned out to be enormous. The reproducibility of test systems from one laboratory to another has been a constant problem that has plagued the industry for almost a century. Meticulous attention is required to the details of standardization of equipment, sources of consumable supplies, diluent water sources and quality, test organism history and strain maintenance protocols, inoculum preparation methods, temperature controls (both for the test conditions and the recovery media), neutralization of residual disinfectants, preparation and storage of test solutions, and benchmark chemical standards for positive control of target organism susceptibility. Benchmark standards were an essential component of the earliest test configurations. In fact, efficacy of new formulations was often referred to in terms of the antibacterial capacity of a certain number of units of the gold standard compound, such as phenol. That trend has declined in popularity in recent years, although there is still an acknowledged need to incorporate a reference biocide as an indicator of the expected behaviour of the test organism.

Food process hygiene test organism panels are now used instead of the original reliance on *S. typhi*. These generally include *E. coli*, *Staphylococcus aureus*, *Pseudomonas aeruginosa*, *Bacillus cereus*, and a species of *Salmonella*, usually *cholerae-suis*, plus a yeast organism, *Saccharomyces* or *Candida*, and a fungus, *Aspergillus*. The selection depends on the intended claims for the product. Strict culture handling and maintenance requirements are necessary, and the panel organisms are always tested separately, never as a mixture. Although the propagation procedures are well described in the literature and in the regulatory agency protocols, sources of variation still creep in to confound the comparability of test runs from site to site. Increasingly recognized as important in this regard are:

1. Factors related to organism injury and subsequent recovery under suitable conditions after exposure to chemical antimicrobial agents.
2. The influence of different amounts and types of bioburdens used to mimic the organic-rich environment in which food-processing disinfectants are expected to function. Commonly accepted are whole milk, meat extracts, dried faeces, yeast suspensions and serum.
3. Neutralization techniques, applied to the sample after disinfectant exposure, in an attempt to ensure that there is no continuation of the antimicrobial action during the recovery and enumeration of survivors.

Recovery from injury and the emerging concept of 'non-viable but culturable' organisms seem likely to generate additional controversy in the future. Disinfectant-injured organisms may have different optimal temperatures for growth and media requirements, for example, compared to unaffected survivors, and this may confound the reproducibility of assays. Chemical biocides of different classes cause different types of injury, requiring compound-specific recovery techniques. Disinfectant classes vary widely in the intensity of their interactions with bioburdens, so that the conditions of a given test protocol may furnish reproducible data only with specific classes of biocide, and perform unreliably with others.

Similar idiosyncracies arise in the success or otherwise of neutralization measures. This is a notoriously poorly studied aspect of test protocols, which has probably contributed to much of the confusion generated by laboratory-to-laboratory variations in results. The need for adequate neutralization is globally accepted by microbiologists to ensure that the efficacy evaluation is being done under conditions of contact time for the test only. The procedure avoids the possibility that efficacy is supplemented by residual persistent activity in the carry-over in samples either in the fluid phase of the sample, or as adherent residues on the bacteria themselves. However, there are still inadequate data on the quantitative influences of the different kinds of neutralizing agents on the biocidal chemicals, on the formulation components (i.e. the inactive constituents, which may influence neutralizer efficacy), and on the test organisms and their recovery success rates. Failure to appreciate the biocide-specific characteristics of neutralization and its importance has led in the past to serious errors in the overestimation of product potency of QACs used in the food-processing industry.

Process Hygiene Disinfectants

Regulatory agencies overseeing market entries in this product field generally act with specific statutory authority over the labels on disinfectants, and hence exercise full control over the claims made for every biocidal formulation currently sold in the regulated territory. Efficacy data requirements are detailed in published protocols, and microbiology laboratory data to meet these needs are increasingly produced by third-party contractors operating under prescribed conditions (such as Good Laboratory Practices (GLP) in the US), designed to enhance the prospects for reproducibility of the test data. Reference to these requirements, promulgated by national authorities in Canada and the US, and by the European Community, is recommended for an explicit account of the requisite experimental protocols, standard strains of target organisms, etc., rather than by exploration of the peer-reviewed literature on this subject. Regulatory authorities often arrive at designated protocols as a result of multi-laboratory comparative studies, and these data may ultimately be published in the mainstream food hygiene or microbiology literature, but the agencies' adherence to their own published stipulations takes precedence over reference to methods described in peer-reviewed journals. This fact must be borne in mind by those planning the assembly of data packs in support of the registration of products for commercial sale.

Product safety is also a responsibility of the regulatory authorities, and requires acute toxicological evaluation of the product as sold (i.e. the label specifies the toxicity of the concentrate in the bottle or container if the product is sold for dilution by the user). Acute toxicity testing for food hygiene disinfectants usually requires oral, dermal and aerosol exposure of rodents and rabbits, under stringently controlled conditions. For novel biocidal entities introduced into the market for the first time, subchronic exposures involving protracted periods of administration of the test article to animals are required.

The demands of the food industry ultimately determine the viability of disinfectant products in the marketplace. The requirements for simultaneous cleaning and disinfection have led to the introduction of a large number of product formulations containing surface-active non-biocidal constituents whose influence on the overall acceptability of the biocide may be profound and powerfully potentiating. Build-up of blood, serum and meat residues is a major impediment to the efficient exercise of biocidal properties, and detergent dispersal, often with high temperatures, is increasingly used to improve performance. Higher temperatures generally enhance antimicrobial efficacy, though instability of iodine-containing disinfectants can be a major deterrent to their use in this mode. Extremes of pH and water hardness can effectively counteract the effects of a biocide too, so that free chlorine-dependent biocides, for example, become non-functional outside a narrow pH range. Certain QACs are seriously adversely affected by both pH and the hardness of the water used to dilute them for rinses and sprays. For these reasons, novel halogen-based water-soluble disinfectants, the *N*-halamines, that function over wide temperature, hardness and pH ranges but do not liberate free chlorine, exemplify a new generation of disinfectants that should enjoy wide use in this industry in the future.

Label claims on environmental disinfectants encompass all the conditions of the recommended use patterns, and need to be adhered to diligently if the products are going to perform as expected, based on the laboratory testing. Misuse and abuse, inappropriate dilution, inadequate storage conditions and improper application techniques often turn out to be the sources of product failures when these occur in the real world of food-processing applications. In practice, cleaning agents in the formulations are especially important in compensating for these possible errors. However, cleaning capacities are not classified or tested by the regulatory agencies, and the characteristics of the so-called inert components are usually held as proprietary information by the manufacturers, and are not required to be revealed on the labels.

Sanitizing formulations properly used for food-contact surfaces are expected to exercise their biocidal effects extremely rapidly, against all bacterial types, so as to bring about a rapid knock-down of the contaminating microorganism numbers. They may not have the power to effect pathogen destruction over the longer term on a scale comparable to those products identified as 'environmental hard-surface disinfectants'. Adequate training of personnel is the key to compliance with the different types of designated use patterns for environmental biocides in food-processing facilities.

Testing Protocols for Hand-sanitizers

Effective hand-washing is a major factor in prevention of food-borne pathogen transmission. It has been increasingly recognized as a key element in the avoidance of hospital-acquired (referred to as 'nosocomial') infections in patients. Yet amongst highly educated, health-care professionals, compliance with the practice has been steadily declining to scandalously low levels. Pathogenic Gram-negative bacteria survive on human hands for up to several hours when deposited on skin, and there has been a growing appreciation of the desirability of incorporating effective biocides in hand-washing formulations in an effort to limit this problem. Testing protocols have been devised over the years for these 'topical sanitizers', as they are called in the food-handling industry, and there are convincing ways to demonstrate some degrees of efficacy (Fig. 1), but many problems remain and regulatory positions on this subject leave much to be desired.

The spectrum of biocides with properties that have utility in this use pattern has been described elsewhere in this volume. The requirements for biocidal expression and power but also for compatibility with frequent and prolonged contact with human skin impose constraints on the available choices. Biguanides, such as chlorhexidine (CHX), alcohols, chlorinated phenolic compounds (such as triclosan), and para-chlormetaxylenol (PCMX) and iodophors find use in this product category. Test systems for these products in the food industry are aimed at demonstrating significant reductions in populations of contaminating transient organisms, i.e. bacteria and viruses that do not take up residence on skin, but which are able to survive for long enough and in sufficient numbers to be transferable to food or food-contact surfaces, and serve as sources of infection. Resident microbial flora on skin are only removed by procedures and chemicals that effect surgical hand disinfection, such as prolonged scrubbing with iodine-containing formulations. Transients are typically Gram-negative faecal-derived bacteria and enteroviruses, but protozoan cysts such as those of *Cryptosporidium* and *Cyclospora* are now having to be considered in this context.

Standardization of protocols for registrable claims in this field has proven even more difficult to develop than in the case of environmental sanitizers. The variables of time and concentration are amplified in this instance by a wide individual variation in the extent to which inocula of microorganisms survive on human skin, and in the recoverability of challenge microbes in a reproducible fashion. The aim of the procedures is to determine the relationship between the numbers of organisms applied to the test skin and those recovered from hands exposed to the topical sanitizer, compared to control. Benchmark inocula include *E. coli* and *S. aureus*, and recovery techniques vary from direct application of finger tips to the surfaces of nutrient agar plates (impression tests), through swabbing of specified areas of skin post-treatment and challenge, to collection of rinses and quantitation of suspended survivors by plate counting techniques.

Biocide-containing and control non-biocide-containing formulations are compared to assess the merits of each proposed sanitizer product. Once again, however, non-biocidal formulations may have a profound potentiating effect on the efficacy of antimicrobial compounds in the mixture. Degreasing agents, for example, may be essential for reliable efficacy of formulations used for meat handlers.

The legitimacy of product claims is a responsibility of regulatory agencies, and permissible protocols for certain kinds of claims are published and revised periodically (e.g. in the Food and Drug Association Over The Counter (OTC) monograph). These need to be consulted in the process of product development for the evaluation component. On the whole, these protocols are simplistic and based on activity of the integral biocide in suspension or carrier tests against target bacteria, taking no account of the variables affecting efficacy *in situ*. Buyers and users should be aware of these limited claims of utility. Cleaning of food handlers' skin with soaps and detergents is a fundamental and highly desirable element of hygienic programmes. Removal of soil-containing contaminants is of paramount importance. However, soaps and detergents are not intrinsically biocidal and are readily contaminated by users, and as such they may serve as rich sources of nutrients for the growth of many bacteria unless biocides are present. Moreover, despite the disdain for antimicrobial sanitizer formulations amongst many in the health-care industry (who point to equivocal results on the proven benefits of the inclusion of biocides on skin bacterial counts), there are objective data supporting the usefulness of antimicrobial soaps in limiting contamination and transfer of pathogens.

Manufacturers trumpet such claims loudly. Especially strident are claims for 'instantaneous' efficacy, which almost assuredly requires the use of high concentrations of alcohols, themselves liable to produce desiccation and skin irritation if not properly formulated with appropriate emollients. Other popular claims are for the benefits of prolonged residual antimicrobial activity on the skin after use of the product. These effects are usually associated with CHX, which binds to skin cells for long periods, often many hours, although this compound has weak effects on the important Gram-negative bacteria. In the absence of a strong, science-based set of protocols and regulatory oversight for these sanitizers, encouragement of compliance with hand-washing requirements with a formulation that is easy and pleasant to use and well-tolerated by food handlers is probably the key measure in the food hygiene business.

Direct observational data on compliance suggest that those people required to wash their hands frequently on the job spend no more than 10 s on the procedure. The incorporation of a powerful biocidal agent therefore seems a sensible step, even if the proven advantages experimentally remain as yet unquantifiable, or not highly reproducible from site to site. Hand-sanitizers have to be considered a likely beneficial adjunct to the use of protective barrier gloves by food handlers, wherever possible in the process. Barriers leak with disturbing frequency and can lead to a false sense of security, especially since microbial proliferation on the skin under gloves can lead to enormous increases in numbers of contaminants.

Testing Protocols for Food Rinses

Antimicrobial food rinses represent a new use pattern for disinfectants/sanitizers in the food industry. Use of free chlorine at concentrations up to 10 p.p.m. has become the most widely accepted approach to concerns about carcass contamination in processing, but it is not without its shortcomings, and alternatives are sought. *E. coli* O157:H7 is the principal cause of the contamination problem on red meat, and *Salmonella* and *Campylobacter* on poultry carcasses. It has become clear in recent years that surface contamination of de-skinned beef carcasses and de-feathered broilers often arises through faecal material spread around during processing. Similar concerns are now arising about faecal contamination of vegetables and fruits (such as bean sprouts and apples) through the use of human sewage and animal manure for fertilizer, and from unhygienic practices by harvesters and food handlers.

Anomalously, the special cachet associated formerly with the designation of foods as 'fresh' has begun to be replaced with suspicions over their safety. Correspondingly, this is creating a rising demand for safe, effective biocidal rinses for a wider array of food products than ever. Shellfish, for example, collected from offshore sites often harbour sewage-derived bacteria and viruses, and in the absence of any suitable treatment methods they are sometimes marketed bearing warning labels identifying them as hazardous to eat. Clearly, chemical decontamination for this application must be water-based, and apart from the overriding needs for safety, the products must impart no taint or residues, must not discolour, or affect appearance adversely, and must exert their effects very rapidly.

Test protocols for this kind of performance are still in an exploratory stage (Fig. 1), but regulatory clearances are being issued. Poultry and red meat sanitizing washes are generally evaluated by performing post-rinse swab counts on treated surfaces, or by rinsing test surfaces with a wash solution which is then sampled for plate counting. Demonstration of significant declines after very short contact times is the goal of these products. Standardized test systems are beginning to be promulgated by regulatory agencies, such as the FDA and USDA (United States Department of Agriculture) in the US, and an array of new products is appearing, although they are not yet being taken up widely by the food-processing industry.

Dilute organic acids are popular because they fall into the category of being generally regarded as safe (GRAS) by the FDA, even though their low pH exerts a useful antimicrobial effect on food surfaces. However, acid-resistant strains of Gram-negative bacteria are rapidly appearing in response. Chlorine, most recently introduced in the form of chlorine dioxide (generated on-site from sodium chlorate by dilute organic acids), is still the most useful, but the formation of toxic disinfection by-products (DBP) is casting a shadow over this application. Trisodium phosphate (TSP) functions as a detergent perhaps more than as a biocide, but effectively displaces adherent bacteria from both beef and poultry carcasses. Ozone appears likely to find a place in this product field, especially because of its effects on parasite cysts, but it has a very short action and no residual effects, and may itself generate unwanted oxidative by-products. Water-soluble *N*-halamines, with their high safety profiles and broad spectrum of efficacy based on bound rather than free halogens, offer another alternative approach with some significant advantages.

Direct exposure of vegetables, fruits, fish and shellfish to sanitizing chemicals seems likely to increase as industry and regulators respond to the newfound visibility of microbial contaminants. Pressures to adopt safety measures will increase with the implementation of new bacterial contamination standards as part of the hazard analysis of critical control points (HACCP) system for improving food safety in the US, and elsewhere. But this field has not yet reached the state where experimental approaches has been widely agreed upon, let alone codified in regulatory agency protocols. Recent findings on the presence of *E. coli* O157:H7 within fruits and vegetables, rather than just adhering to their surfaces, suggest that bacteria may be able to penetrate and sequester in plant tissues and cells. Surface decontamination with biocidal rinses may prove to be ineffective if this phenomenon is more widespread amongst bacterial pathogens. Methods for reliably removing cysts of the protozoan pathogens from plant surfaces have yet to be devised, so this is likely to prove another problematic area for food hygiene for some time to come.

The hazards of shellfish contamination are also not readily dealt with through use of rinses. This is largely because oysters and clams need to be kept alive for distribution and sale, but they are too susceptible to biocides to be exposed to antimicrobially effective concentrations of most agents. Recent evidence that the *N*-halamines are very well tolerated by live shellfish is promising, especially if these exposures turn out to be effectively biocidal for the contaminating viruses and bacteria (such as *E. coli*, *Vibrio cholerae* and hepatitis A virus) that are filtered out of contaminated waters by these animals. Preparations of *N*-halamines in mineral oil have also been successfully employed to decontaminate shell eggs, commonly plagued by problems of *Salmonella* adherent to the shell surfaces. Aqueous biocides cannot confidently be used for this purpose because water passes readily through the shell and into the egg itself, whereas oil-based biocides do not pass through the shell pores. Protocols have not been developed for regulatory clearance of claims for this application.

Contemporary Issues in Disinfectant Testing

Testing protocols are under scrutiny as new research findings in epidemiology and microbial ecology continue to impact upon the field of food hygiene. Several developments are likely to bring about change:

1. Biofilms: No amount of rigorous test protocol standardization under defined laboratory conditions can mimic the full range of working circumstances in the real world of applications. While this has long been appreciated, the recently recognized significance of biofilm formation by mixed microbial populations on virtually all environmental surfaces is causing some re-evaluation of the *in situ* performance of disinfectant products. Testing protocols currently used for evaluation of process hygiene disinfectants take no account of the capabilities of approved formulations in dispersing or preventing biofilms – factors that can have great bearing on the ultimate effectiveness of the sanitizing/disinfecting process. Concerns over this shortcoming will probably result in new protocol developments aimed at assessing efficacy versus biofilms in the future.
2. Environmental resistance: Just as the formation of biofilms provides a new explanation of the observed resistance of bacterial organisms to disinfectant chemicals in real-world applications, so does emerging evidence that some food-borne pathogens have intrinsic durability profiles in the environment that are much more impressive than had been imagined previously. This is true whether they are dried on surfaces or suspended in organic

soil. Enteric pathogens are now known to survive for periods of up to many weeks in fully infective forms, rather than exhibit the rapid decays in viability expected from older, incomplete data sets. Environmental contamination has therefore assumed more importance in disease transmission, and this realization may also begin to influence protocol design for efficacy evaluation. Target strain selection for carrier tests may have to include organisms that exhibit these remarkable durability traits.

3. Resistance to disinfectants: Resistance of certain food-borne pathogens to therapeutic antibacterial agents has become a major concern, as evidence of multiple drug-resistant *E. coli*, *Salmonella* and *Staphylococcus aureus* has become commonplace. There is now evidence – albeit controversial – that some of these strains also show enhanced resistance to environmental disinfectants. Very specific, genetically based traits for resistance to the popular biocide triclosan have raised fears about comparable trends appearing with other biocides which formerly had been thought to be immune to this risk. These data will probably also be taken into account in the future selection of target strains in test protocols. It is already increasingly common for manufacturers of biocidal formulations to seek label claims of efficacy against multiple-resistant enteric pathogens, in response to marketplace concerns about this issue.
4. Surface residual disinfectant activity: Food-processing disinfectants are being developed with enhanced residual persistence on treated surfaces. These so-called self-sanitizing effects result from biocidal activity being retained in the dry, surface-bound state for periods of up to many weeks in some cases. New testing protocols will emerge that will address standardization of procedures for allowable claims for this new feature. It may prove to be a particularly attractive characteristic for food industry application. Certain QACs, the new silver-based polymeric biocides, and some of the water-soluble *N*-halamines all display this advantage, and are likely to enjoy wide use as a result.
5. New pathogen problems: Finally, pressures will undoubtedly rise on the food industry to reduce the risk associated with contamination with viral pathogens, and with the protozoan parasites *Cyclospora* and *Cryptosporidium*. These microbes will require special consideration in efficacy protocol improvements, because they are very tough adversaries for chemical biocide-based approaches.

It is always as well to remember, as this debate intensifies, that chemical disinfectants can never be more than a supplement to an entire array of hygienic measures based on well-established sound principles of cleanliness, and thoughtful facility and equipment design which can overcome both the old and the new microbial threats to food safety.

See also: **Biofilms**. ***Cryptosporidium***. ***Cyclospora***. **Hazard Appraisal (HACCP)**: The Overall Concept. **Process Hygiene**: Types of Biocides; Involvement of Regulatory and Advisory Bodies. ***Salmonella***: *Salmonella typhi*. **Shellfish (Molluscs and Crustacea)**: Contamination and Spoilage. **Viruses**: Introduction.

Further Reading

Barley PJ, Prince J and Finch JE (1981) The history and efficacy of skin disinfectants and skin bacteria assessment methods. In: Collins CH, Allwood MC, Bloomfield SF and Fox A (eds) *Disinfectants: Their Use and Evaluation of Effectiveness*. P. 17. New York: Academic Press.

Blood RM, Abbiss JS and Jarvis B (1981) Assessment of two methods for testing disinfectants and sanitizers for use in the meat processing industry. In: Collins CH, Allwood MC, Bloomfield SF and Fox A (eds) *Disinfectants: Their Use and Evaluation of Effectiveness*. P. 91. New York: Academic Press.

Haines KA, Klein DA, McDonnell G and Pretzer D (1997) Could antibiotic-resistant pathogens be cross-resistant to hard-surface disinfectants? *American Journal of Infection Control* 25: 439–441.

Lauten SD, Sarvis H, Wheatley WB, Williams DE, Mora EC and Worley SD (1992) Efficacies of novel N-halamine disinfectants against *Salmonella* and *Pseudomonas* species. *Applied and Environmental Microbiology* 58: 1240–1243

Nosken GA, Stosor V, Cooper I and Peterson LR (1995) Recovery of vancomycin-resistant enterococci on fingertips and environmental surfaces. *Infection Control and Hospital Epidemiology* 16: 577–581.

Paulson DS (1996) A broad-based approach to evaluating topical antimicrobial products. In: Ascenzi JM (ed.) *Handbook of Disinfectants and Antiseptics*. Pp. 17–42. New York: Marcel Dekker.

Russell AD (1981) Neutralization procedures in the evaluation of bactericidal activity. In: Collins CH, Allwood MC, Bloomfield SF and Fox A (eds) *Disinfectants: Their Use and Evaluation of Effectiveness*. P. 45. New York: Academic Press.

Springthorpe S and Sattar S (1996a) Handwashing product and technique comparisons. *Infection Control and Sterilization Technology* January: 19–22.

Springthorpe S and Sattar S (1996b) Improving the tests used to measure efficacy of chemical germicides. *Infection Control and Sterilization Technology* March: 26–31.

Walker HW and LaGrange WS (1991) Sanitation in food manufacturing operations. In: Block SS (ed.) *Dis-*

infection, Sterilization and Preservation, 4th edn. Pp. 791–801. Philadelphia, PA: Lea & Febiger.

Williams DE, Elder ED and Worley SD (1988) Is free chlorine necessary for disinfection? *Applied and Environmental Microbiology* 54: 2583–2585.

Involvement of Regulatory and Advisory Bodies

Roland Cocker, Cocker Consulting, Almere, The Netherlands

HLM Lelieveld, Unilever Research Laboratory, Vlaardingen, The Netherlands

Background

There is a steady increase in the involvement of regulatory and advisory bodies in the area of food process hygiene. Major programmes are underway to revise the nature of regulatory intervention, together with supporting educational, normative and accreditation programmes. There are a number of reasons for this, not least of which is the rate of illness and death stemming from poor process hygiene. For example, the US Food and Drug Administration talked in 1998 of: 'food-borne illnesses that annually claim 9000 lives and cause multimillion dollar economic losses'.

In addition, consumer concern about food safety is rising worldwide. In several countries, food scares have dented consumer confidence in legislation, scientific advisers and the food industry. Emotion, poor communication, misunderstanding of the principles of risk and commercial pressures add to the technical, governmental and scientific problems of assuring food safety. Controversy and dispute abound in these countries and regulatory matters can be highly charged, both politically and personally, leading to physical and legal skirmishes and threats.

We are starting to see 'farm to fork' approaches that consider the food process to include the whole chain from supply of animal feeds, to the farming of animals and crops, to industrial food processing to processing in the retail and restaurant outlets. This is supported by moves to unify and consolidate control strategies and agencies, in order to shift the balance towards prevention and to increase effectiveness.

Current Government Activities

An issue is the fragmented responsibility for regulation and control (see below). In response to these and other concerns, in 1998 the US government commenced its food safety initiative, with a budget of $43 million in 1998 and a further $101 million requested for 1999. In the EU, it led to the reorganization of the Commission services and calls by then Commissioner, Jacques Santer, to set up a centralized food safety authority. In the UK, the government announced proposals to set up a centralized Food Safety Authority.

Through their powers to control hazardous imports and also the fact that they have highly developed structures of legislation, the EU and the US may exert an influence which extends beyond their geographical boundaries. In the case of the EU, states aspiring to membership may adopt the EU directives as part of their commercial, legislative and political strategies. Other neighbours, such as Norway and East European Countries of the EU may do so for reasons of simple pragmatism and enlightened attitudes to harmonization, often as partners in the whole regulatory process.

Risk Management and HACCP

The most potent international trend has been towards methodologies based on risk management, such as hazard analysis and critical control points (HACCP). Legislative and regulatory implementation is at various stages around the world, with logistical problems of training for regulator and operator alike, in the switchover from prescriptive control to one based more on management of risk. There are signs of error in making conceptual change from one of fixed designs and threshold values to one of risk assessment and critical control point methodologies. For example, in the Netherlands, the application of HACCP in various food-processing sectors is supported by hygiene codes produced by industry associations under the control of the Ministry of Health, Welfare and Sport (VWS). In an investigation of recent hygiene codes, it was noted that key definitions such as *critical control point* did not agree between the various hygiene codes, leading to potential problems for operators who might be affected by a number of different codes. The EU has made the boldest move by making HACCP mandatory across the food industry, whilst Australia and New Zealand are moving in the same direction. In the US, the pattern has been one of introducing HACCP laws by industry sector, with considerable debate and discussion about how to ensure the best results.

Hierarchical Regulatory Structures

The structure of regulation generally is increasingly reflected in its phases of introduction. At the highest and earliest level, laws and regulations are introduced before being supported by subsequent standards.

Further support may be given in parallel by guidelines and standards that are produced in the first instance by voluntary bodies, but which may be promoted to the status of national or international standards. In the EU, the trend is towards guidelines for hygienic design linked to performance standards and tests. In the US, standard designs of equipment may be quoted in legislation, e.g. 3-A standards for equipment designs are quoted in the Pasteurized Milk Ordinance. The current moves in the EU towards harmonized standards for the interpretation and implementation of HACCP also reflect this pattern (see below).

Supporting Standards and Structure for HACCP

HACCP also requires to be supported by Good Manufacturing Practice and does not in itself facilitate food processors with regard to sourcing satisfactory equipment and process designs. To some extent, work to provide design standards is done by the European Hygienic Equipment Design Group (EHEDG), the 3-A Committee, International NSF (formerly National Sanitation Foundation) and the Comité Européen de Normalisation (CEN) *Safety in Biotechnology* standards and International Standards Organization (ISO).

An important principle of approaches based on risk management is that of verification and validation leading to equipment and process qualification – providing documented proof that they can achieve the required product safety. This requires much more development by legislators, inspectors, auditors and operators in the food industry, as it has consequences for the framing of supporting laws and standards. More recently, CEN standards and guidelines supporting directives based on risk management (90/219/EEC, 90/679/EEC) have provided optional methods in informative annexes, whilst providing for the use of *validated* alternatives. Fixed standards are reserved for reference activities such as measurement and testing. In contrast, laws are being passed elsewhere which prescribe fixed controls for food processing itself. For example, the recently enacted California law (California CURFFL section 113996(b)), intended to reflect most cooking requirements of the Food Code, specified cooking temperatures for foods of animal origin, microwave cooking of raw foods of animal origin, and re-heating of foods.

Some countries have seen the need for 'route maps' as exemplified by the UK *Industry Guide to Good Hygiene Practice*. This gives information about whether certain procedures are a legal requirement (in the UK) or just good practice.

Some EU member states such as the Netherlands have an accreditation scheme for HACCP. Independent auditors such as TNO, Bureau Veritas and SGS are themselves accredited as auditors and perform the accreditation services.

An overall international summary of the systems in the main trading blocs is given in **Table 1**.

As a necessary prerequisite to HACPP, food safety liability is already covered in EU countries and in the US by product liability laws and by general civil and criminal codes governing the behaviour of individual citizens. Worker safety (for example, exposure to bovine spongiform encephalopathy (BSE), *Escherichia coli* HO157, *Bacillus anthracis* or antibiotic-resistant strains of *Salmonella*) is usually covered by existing national industrial safety legislation such as the UK Control of Substances Hazardous to Health (COSHH) and Safety at Work laws. The EU directive 90/679/EEC controls work with pathogenic organisms, which affects especially (though not exclusively) food microbiology laboratories.

An important aspect of legislation is to define the scope of any new laws, especially to identify existing legislative instruments which may impact on hygiene aspects of food processing and to provide guidance or definition of what type of operation qualifies as a food-processing operation. In some jurisdictions such as the US, restaurants are currently treated differently from large food-processing operations, although moves are in place to require a form of HACCP.

A positive trend in the voluntary/industry sector has been the agreement between EHEDG, 3-A and International NSF jointly to develop standards for food-processing equipment. This is supplemented by efforts by EHEDG to involve Japanese bodies in this co-operation.

International Level

FAO/WHO Codex Alimentarius

The Food and Agriculture Organization/World Health Organization (FAO/WHO) Codex Alimentarius committee specifically concerned with food hygiene is the Codex Committee on Food Hygiene (CCFH), chaired by the US. It has produced the following standards:

- Draft Revised Recommended International Code of Practice – General Principles of Food Hygiene ALINORM 97/13, Appendix II; adopted with editorial changes, especially in the Spanish version
- Draft Revised Guidelines for the Application of the Hazard Analysis and Critical Control Point (HACCP) System ALINORM 97/13A, Appendix II: adopted with editorial changes, especially in the Spanish version.

The approved forward standards programme for

Table 1 The structure of regulatory systems in the main trading blocs

Jurisdiction	*Authority*	*Laws*	*Official standards*	*Voluntary standards*
International	World Trade Organization	SPS Code Agreement on Sanitary and Phytosanitary Measures		
	International Standards Organization		ISO/TC 199 Safety of Machinery (SC 2 Hygiene Requirements for the Design of Machinery)	ISO/DIS 15161 Guidance on the Application of ISO 9001/9002 to the Food and Drink Industry ISO/CD 14159 Hygienic Requirements for the Design of Machinery
	FAO/WHO Codex Alimentarius Commission			Codex Alimentarius (Alinorm 97/13 and Alinorm 97/13A)
	Codex Committee on Food Hygiene			
	Codex Committee on Meat Hygiene (CCMH)			
	Codex Committee on Milk and Milk products (CCMMP) International Dairy Federation			Code of Hygienic Practice for unripened cheese and ripened soft cheese (in preparation) Code of Hygienic Practice for Dried Milk (CAC/RCP 31:1983) Code of Hygienic Practice for Milk and Milk Products (in preparation)
Europe	European Council	93/43/EEC Food Hygiene 89/392/EEC Machinery Directive and its amendments 91/368/EEC, 93/44 and 93/68 EEC 92/59/EEC Council Directive Concerning General Product Safety EEC 93/465/EEC Conformity Assessment and Rules for Affixing the CE Mark EEC 93/68/EEC Amending Directives on CE Marking: 87/404/EEC, 88/378/EEC, 89/106/EEC, 89/336/EEC, 89/392/EEC, 89/686/EEC, 90/85/EEC, 90/384/EEC, 90/385/EEC, 90/396/EEC, 91/263/EEC, 92/42/EEC and 73/23/EEC EEC 94/62/EEC Packaging and Packaging Waste – amended by 97/129/EEC and 97/138/EEC 90/679/EEC Worker Safety Pathogenic Organisms 90/220/EEC Deliberate Release of Genetically Modified Organisms		
	Comité Européen de Normalisation CEN TC 153 Food Processing Machinery		EN 1672–1 and-2 and for specific machines: EN 453, EN 1673, EN 1974, EN 12505, EN 12505, EN 12331, EN 12853	
	European Hygienic Design Group (EHEDG)			Guidelines and standards (in association with 3-A, and International NSF: see below)

Table 1 (Continued)

Jurisdiction	*Authority*	*Laws*	*Official standards*	*Voluntary standards*
USA	Food and Drug Administration	Code of Federal Regulations Federal Food, Drug and Cosmetic Act		
	Federal Bureau of Investigation	Federal Anti-tampering Act		
	Department of Transportation	Sanitary Food Transportation Act		
	Department of Commerce National Oceanic and Atmospheric Administration			
	US Department of Agriculture (USDA)			
	Food Safety and Inspection Service (FSIS)	Clean Water Act (CWA)		
	Centers for Disease Control and Prevention (CDC)			
	Environmental Protection Agency (EPA)			
	3-A Organization			3-A standards
	International NSF			NSF and NSF/ANSI Standards
Australia and New Zealand	Australia New Zealand Food Authority	State and Territory legislation, i.e. Food Acts and associated food hygiene regulations Food Standards Code	Production Quality Arrangements (PQA) for meat processing Approved Quality Arrangement (AQA) for meat processing Meat Safety Quality Assurance (MSQA)	
Japan	Ministry of Health and Welfare			
	Environmental Health Bureau			
	Japan Food Hygiene Association 6-1 Chome, Jungumae Shibuya-ku Tokyo			
	Japan Food Machinery Manufacturers Association (Mr Sueichi Shimada, Managing Director) Fooma Building 3-19-20 Shibaura, Minato-ku Tokyo 108-0023			

the FAO/WHO Codex Alimentarius Committee on Food Hygiene (CCFH) committee includes:

- Code of Hygienic Practice for Milk and Milk Products
- Hygienic Recycling of Processing Water in Food-processing Plants
- Application of Microbiological Risk Evaluation to International Trade
- Revision of the Standard Wording for Food Hygiene Provisions (Procedural Manual)
- Risk-based Guidance for the Use of HACCP-like Systems in Small Businesses, with Special Reference to Developing Countries
- Management of Microbiological Hazards for Foods in International Trade.

Codex Committee on Milk and Milk Products

The Code of Principles concerning Milk and Milk Products was produced in 1958 at the initiative of the International Dairy Federation (IDF) by the Joint FAO/WHO Committee of Government Experts on the Code of Principles concerning Milk and Milk Products. At that time IDF was already active in drafting compositional standards for milk and milk products. The standards IDF elaborated as a non-governmental body missed official recognition by governments, as there was no structure to obtain government approval. To achieve regulatory status for compositional standards, IDF requested FAO and WHO to convene a meeting of government experts to initiate a Code of Principles and associated standards for milk and milk products. In 1993 the resulting Milk Committee was fully integrated into the Codex system as the Codex Committee on Milk and Milk Products (CCMMP).

IDF maintained its role as technical adviser to the new Codex Milk Committee and its formal status is specified in the revised *Procedural Manual of the Codex Alimentarius Commission* (9th edition, 1995): 'In the case of milk and milk products or individual standards for cheeses, the Secretariat distributes the recommendations of the International Dairy Federation (IDF)'.

Most of the standards concern composition of dairy products, but a few concerned hygienic practices:

- Code of Hygienic Practice for unripened cheese and ripened soft cheese (in preparation)
- Code of Hygienic Practice for Dried Milk (CAC/RCP 31:1983)
- Code of Hygienic Practice for Milk and Milk Products (in preparation).

The International Dairy Federation

The IDF is at 41 Square Vergote, 1030 Brussels, Belgium; Tel. +322 733 9888; Fax: +322 733 0413.

Europe

The laws applied by the national authorities have been harmonized at EU level by a framework directive. This lays down the law for general principles for the inspection, sampling and control of foodstuffs. It also provides for inspectors to be empowered to examine, record and seize or destroy foodstuffs which are unsafe or otherwise non-compliant. The framework directive requires the member states to inform the Commission of their control activities and provides for EU-wide co-ordination through annual control programmes. In addition, the Karolus programme provides for exchange of control officials.

Some controls are also undertaken at Union level. These are targeted at ensuring the adequacy and equivalence of the controls applied by the national authorities and involve teams of officials from the Commission in checking that the national systems are capable of meeting these goals. However, as in Australia, New Zealand and the US, much of the direct control is under the aegis of individual states.

The particular dangers arising from zoonotic diseases, like salmonellosis, tuberculosis and viral contaminants, have led the Commission's veterinary inspectorate to control and approve establishments in countries which produce food of animal origin for export to the EU. Such products are also controlled at the point of entry into the EU. However, in the main, food of non-animal origin has not been subject to this type of control, nor is the importation of these foodstuffs into the EU restrictive.

In recent years food policy at international level has been moving in a new direction, towards industry taking the responsibility for the control of the foodstuffs it produces, backed up by official control systems. The European foodstuffs industry has been at the forefront of the development of preventive food safety systems, in particular HACCP, which requires the industry itself to identify and control potential safety hazards. Control measures are decided and applied by industry, with a view to producing safe food. The national authorities check that the controls are adequate. Although initially introduced by industry and employed in a non-mandatory manner, the success of this approach has led to it being included in several directives.

Thirteen product-specific directives cover products of animal origin, from production to the point of distribution, and lay down detailed requirements. On the other hand, one horizontal hygiene directive covers all other products, with requirements based on goals, intended results, good hygiene practices and HACCP principles. This directive covers vegetal products throughout the chain and includes products of

Table 2 National standards supporting EU food safety directives

Member country	*Reference number*	*Title*
Ireland	IS 3219	Code of Practice for Hygiene in the Food and Drink Manufacturing Industry
	IS 340	Hygiene for the Catering Sector
	IS 341 (draft)	Hygiene for the Retail and Wholesale Sector
UK	ISO/DIS 15161	Guidance to the Application of ISO 9001 and ISO 9002 in the Food and Drink Industry
	Alinorm 97/13A	Draft Hazard Analysis and Critical Control Point (HACCP) system and Guidelines for its Application
Germany	DIN 10503	Food Hygiene – Terminology
	DIN 10514	Food Hygiene – Hygiene Training
	Draft	Food Hygiene HACCP System – Standardization of Flow Diagram Symbols
	DIN 10500, DIN 10500/A1, DIN 10501 supplement, DIN 10501-1, DIN 10501-2, DIN 10501-3, DIN 10501-3 supplement, DIN 10501-4, DIN 10501-5, DIN 10502-4, DIN 10504, DIN 10505, DIN 10507, DI 10510	Various standards for equipment, including testing
France	FD V 01-001	Hygiene and Safety of Foodstuffs – Methodology for drawing up of Guides to Good Hygiene Practice

animal origin even after the point of distribution. It imposes the responsibility for the safety of food and the prevention of unacceptable risks to the consumer on the industry. At the same time, it allows industry the flexibility to meet its obligations by the most appropriate means available, and to respond quickly to new pathogens or contaminants while providing a basis for innovation. This challenges industry, particularly smaller businesses, to maintain a good technical understanding of food safety. Voluntary business sector guidelines on hygiene practices and HACCP, produced by industry in conjunction with the competent authority, provides the basis for common understanding. Backed up by effective controls, this approach is intended to ensure a high level of health protection. However, some standardization of approach between sectors and states would be beneficial. A non-exhaustive list of national standards in support of EU directives is given in **Table 2**.

Countries known to be pressing for relevant unified CEN standards include Denmark, France, Ireland and the Netherlands.

EU directives which impact on food hygiene include:

- EEC 89/392/EEC: Council Directive on the Approximation of the Laws of the Member States Relating to Machinery – amended by 91/368/EEC
- EEC 91/368/EEC: Council Directive amending Directive 89/392/EEC on the Approximation of the Laws of the Member States Relating to Machinery – amended by 93/44 and 93/68
- EEC 92/59/EEC: Council Directive Concerning General Product Safety
- EEC 93/44/EEC: Amendment to 91/368 – Council Directive on the Approximation of the Laws of the Member States Relating to Machinery – Amended by 93/68
- EEC 93/465/EEC: Council Directive Concerning the Conformity Assessment and Rules for Affixing the CE Mark
- EEC 93/68/EEC: Amending Directives on CE Marking: 87/404/EEC, 88/378/EEC, 89/106/EEC, 89/336/EEC, 89/392/EEC, 89/686/EEC, 90/85/EEC, 90/384/EEC, 90/385/EEC, 90/396/EEC, 91/263/EEC, 92/42/EEC and 73/23/EEC
- EEC 94/62/EEC: Council Directive on Packaging and Packaging Waste – Amended by 97/129/EEC and 97/138/EEC.

The trend in the management of risk in the food-processing chain is increasingly towards 'farm to fork' initiatives. Amongst issues being addressed are:

- The exclusion of endemic animal disease which may affect humans, notably BSE, scrapie and *Salmonella*.

Sweden and Finland have laws and procedures which are aimed at eliminating *Salmonella* from the animal and human food chain. Sweden has been lobbying vigorously for adoption at EU level of their approach (see below).

- The control of antibiotic-resistant bacteria by banning the routine use of antibiotics in animal feedstuffs.

In short, the argument is that feeding antibiotics to animals will lead to an increased prevalence of bacteria-possessing resistance genes in the intestines of the animals. At slaughter, the carcass will inevitably be contaminated with bacteria containing these genes. The genes can be transmitted to human microbes when the food is prepared or consumed and in the end, humans can get infections with microbes harbouring these genes, causing treatment to fail. (It is ironic at a time when doctors are restricting the prescription of antibiotics to human patients in order to limit the development of resistant bacteria that some of the same or related antibiotics are being fed freely to farm animals.)

Several EU member states already ban routine feeding of certain antibiotics in addition to those not permitted at EU level. Some, such as Sweden, ban antibiotics entirely. The Swedish, Finnish and Danish governments have been taking a strong role in lobbying at EU level.

In late November 1998, the Commission proposed that four out of eight antibiotics should be removed from the list of authorized products. The four (spiramycin, tylosin, virginamycin and bacitracin) all belong to groups of antibacterials that are used in human medicine. For the remaining four, Sweden would have to apply Community legislation, i.e. authorize them in Sweden.

In Europe, three initiatives have been made which address deficiencies in hygienic food manufacture.

The EU Machinery Directive

The European Community Machinery Directive 89/392/EEC and its amendment 91/368/EEC made it a legal obligation for machinery sold in the EU after 1 January 1995 to be safe to use, provided manufacturer's instructions were followed. This requirement has vital implications for those supplying all types of machinery, including that described as suitable for food applications.

In cases of breaches of food safety legislation, inspectors in the EU can confiscate and destroy products and close down operations that threaten public health.

The European Hygienic Equipment Design Group

EHEDG develops design criteria and guidelines on factory design, including equipment, buildings and processing. They also develop equipment performance tests to validate compliance with the design criteria. This is in the spirit of avoiding prescriptive individual designs and specifications.

EHEDG is an independent group with currently 18 specialist subgroups dealing specifically with issues related to the design aspects of the hygienic manufacture of food products. Research institutes, equipment manufacturers, food manufacturers and government bodies are all represented.

The EHEDG has formed links with ISO, CEN, Japanese groups and, in the US, the 3-A and International NSF. The prime objective is to ensure that food products are processed hygienically and safely.

In the case of 3-A, the link is now a formal one. Standards are now being produced jointly and the Food and Drug Administration (FDA) and US Department of Agriculture (USDA) have an effective say via the 3-A input. The first result was a joint guideline on the passivation of stainless steel for hygienic use. The Executive Committee of EHEDG has a seat on the Steering Committee of 3-A and vice versa.

The work of developing guidelines is undertaken, via the subgroups, through the publication of clear recommendations for the hygienic and aseptic design and operation of equipment along with the principles and best methods to confirm that the equipment fulfils these requirements. These groups are drawn from equipment manufacturers, technical organizations and manufacturers, chiefly from the food and engineering industry. Whilst such a list will inevitably be incomplete because of the growth in membership, an impression of the composition of EHEDG is given in **Table 3**.

EHEDG was formed in response to a perceived need for higher standards in the design and testing of hygienic and aseptic equipment. In particular, participants have contributed considerable know-how in hygienic and aseptic design. The motivation has been to improve food safety and to reduce the complexity and cost of attaining satisfactory levels of safety in design. A series of guidelines have been or are being published in various languages. These are listed in **Table 4**.

Many items of equipment have by now been subject to the EHEDG tests, and this is always advertised by the suppliers.

An example of the contribution made by the participants in EHEDG has been the development of a new standard for hygienic/aseptic seals. Elastomeric seals are one of the more common sources of failure in aseptic processing. After a detailed study involving finite element analysis of the interaction of elastomeric components and different seal and housing geometries plus extensive cycles of testing for cleanability and sterilizability, two superior new designs have been produced and published via the German DIN standards organization as:

- DIN 11864–1, publication: 1998–07: Fittings for the food chemical and pharmaceutical industry – Aseptic connection – Part 1: Aseptic stainless steel screwed pipe connection for welding
- DIN 11864–2, publication: 1998–07: Fittings for the food chemical and pharmaceutical industry –

Table 3 Organizations represented in the European Hygienic Equipment Design Group

Research and government institutes	*Equipment manufacturers*	*Food manufacturers*
Biotechnological Institute, Denmark	Danfoss	BSN
Bundesanstalt fur Milchforschung, Germany	Sudmo	Cargill
Bundesgesundheitsamt, Germany	Tetra Laval	H.J. Heinz
Campden Food and Drink Research Association, UK	GEA Tuchenhagen	Italgel
College of Biotechnology, Portugal	APV (Seibe)	Kraft Jacobs Suchard
Institut National de la Recherche Agronomique, France	Clextral	General Foods
Ministry of Agriculture, Fisheries and Food, UK	Serac	Nestlé
Technical University of Munich, Germany	CMB	Rank Hovis MacDougall
TNO, Netherlands University of Lund, Sweden	Fristam	Unilever
	Gasti	
	Robert Bosch	
	Hamba	
	Huhnseal	
	KSB Amri	

Table 4 Current list of EHEDG guideline summaries

Title	*Reference*
European Hygienic Equipment Design Group (EHEDG)	3 (11) 1992, 277
The EC Machinery Directive and Food-processing Equipment	4 (5) 1993 153–154
Hygienic Equipment Design Criteria	4 (7) 1993 225–229
Welding Stainless Steel to meet Hygienic Requirements	4 (9) 1993 306–310
Hygienic Design of Closed Equipment for the Processing of Liquid Food	4 (11) 1993 375–379
Hygienic Pipe Couplings	8 (3) 1997 88–92
Hygenic Design of Valves for Food Processing	5 (5) 1994 169–171
Hygenic Design of Equipment for Open Processing	6 (9) 1995 305–310
A Method for Assessing the in-place Cleanability of Food-processing Equipment	3 (12) 1992 325–328
A Method for Assessing the in-place Cleanability of Moderately Sized Food-processing Equipment	8 (2) 1997 54–57
A Method for the Assessment of in-line Pasteurization of Food-processing Equipment	4 (2) 1993 52–55
A Method for the Assessment of in-line Steam Sterilizability Food-processing Equipment	4 (3) 1993 80–82
A Method for the Assessment of Bacteria-tightness of Food-processing Equipment	4 (6) 1993 190–192
Microbiology Safe Continuous Pasteurization of Liquid Foods	3 (11) 1992 303–307
Microbiologically Safe Continuous-flow Thermal Sterilization of Liquid Foods	4 (4) 1993 80–82
The Continuous or Semi-continuous Flow Thermal Sterilization of Particulate Food	5 (3) 1994 88–95
Hygiene Packing of Food Products	4 (12) 1993 406–411
Microbiologically Safe Aseptic Packing of Food Products	4 (1) 1993 21–25
Experimental Test Rigs are Available for the EHEDG Test Methods	6 (4) 1995 132–134
Passivation of Stainless Steel	9 (1) 1998 28–32
Hygienic Design of Pumps, Homogenizers and Dampening Devies	In press

Aseptic connection – Part 2: Aseptic stainless steel flanged pipe connection for welding.

See also the guidelines listed in **Table 5**.

CEN TC233 Safety in Biotechnology

The European Committee for Standardization (CEN) Technical Committee 233 on Safety in Biotechnology sets standards for equipment and procedures concerning the processing of recombinant and hazardous organisms. This is likely to benefit food process hygiene through the availability of type-approved components.

This committee has been funded by the European Community to produce new European standards relating to safety in biotechnology. The intention is to support and guide the (European) biotechnology industry in the implementation and regulation of activities governed by the European biotechnological safety directives 91/219/EEC, 90/679/EEC, 93/88/EEC and 90/220/EEC. (European directives are in effect laws applying to EU member states that have to be incorporated into their respective national legislatures.) Participants in the formulation of draft standards have included academics, equipment manufacturers, consultants and manufacturers from process industries, including pharmaceuticals, food and fine chemicals, research organizations and national standards bodies. Representatives have included European Free Trade Association (EFTA) countries, for example Switzerland. The emphasis has

Table 5 Supporting standards for food hygiene (US)

Food equipment	
ANSI/NSF 2-1996	Food equipment
NSF 2 Supplement	Descriptive Details for Food Service Equipment Standards
ANSI/NSF 3-1996	Commercial Spray-type Dishwashing and Glasswashing Machines
ANSI/NSF 4-1997	Commercial Cooking, Rethermalization, and Powered Hot Food Holding and Transport Equipment
NSF 5-1992	Water Heaters, Hot Water Supply Boilers and Heat Recovery Equipment
ANSI/NSF 6-1996	Dispensing Freezers (for Dairy Dessert-type Products)
ANSI/NSF 7-1997	Commercial Refrigerators and Storage Freezers
ANSI/NSF 8-1992	Commercial Powered Food Preparation Equipment
ANSI/NSF 12-1992	Automatic Ice Making Equipment
ANSI/NSF 13-1992	Refuse Compactors and Compactor Systems
ANSI/NSF 18-1996	Manual Food and Beverage Dispensing Equipment
ANSI/NSF 20-1998	Commercial Bulk Milk Dispensing Equipment
ANSI/NSF 21-1996	Thermoplastic Refuse Containers
ANSI/NSF 25-1997	Vending Machines for Food and Beverages
NSF 26-1980	Pot, Pan, and Utensil Commercial Spray-type Washing Machines
ANSI/NSF 29-1992	Detergent and Chemical Feeders for Commercial Spray-type Dishwashing Machines
ANSI/NSF 35-1991	Laminated Plastics for Surfacing Food Service Equipment
ANSI/NSF 37-1992	Air Curtains for Entranceways in Food and Food Service Establishments
ANSI/NSF 51-1997	Food Equipment Materials
ANSI/NSF 52-1992	Supplemental Flooring
ANSI/NSF 59-1997	Mobile Food Carts
NSF C2-1983	Special Equipment and/or Devices (Food Service Equipment)
3-A standards	
01-07	Storage Tanks for Milk and Milk Products
02-09	Centrifugal and Positive Rotary Pumps for Milk and Milk Products
04-04	Homogenizers and Reciprocating Pumps
05-14	Stainless Steel Automotive Milk and Milk Product Transportation Tanks for Bulk Delivery and/or Farm Pick-up Services
10-03	Milk and Milk Product Evaporators and Vacuum Pans
11-05	Place-type Heat-exchangers for Milk and Milk Products
12-05	Tubular Heat Exchangers for Milk and Milk Products
13/09	Farm Cooling and Holding Tanks
16-05	Milk and Milk Product Evaporators and Vacuum Pans
17-09	Formers, Fillers and Sealers of Single-service Containers for Fluid Milk and Fluid Milk Products
18-02	Multiple-use Rubber and Rubber-like Materials used as Product-contact Surfaces in Dairy Equipment
19-04	Batch and Continuous Freezers for Ice Cream, Ices and similarly frozen Dairy Foods
20-19	Multiple-use Plastic Materials used as Product-contact Surfaces in Dairy Equipment
22-07	Silo-type Storage Tanks for Milk and Milk Products
23-02	Equipment for Packaging Viscous Dairy Products
24-02	Non-coil Type Batch Pasteurizers for Milk and Milk Products
25-02	Non-coil Type Batch Processors for Milk and Milk Products
26-03	Sifters for Dry Milk and Dry Milk Products
27-04	Equipment for Packaging Dry Milk and Dry Milk Products
28-02	Flow Meters for Milk and Milk Products
29-01	Air Eliminators for Milk and Fluid Milk Products
30-01	Farm Milk Storage Tanks
31-02	Scraped Surface Heat Exchangers
32-02	Uninsulated Tanks for Milk and Milk Products
33-01	Polished Metal Tubing for Milk and Milk Products
34-02	Portable Bins for Dry Milk and Dry Milk Products
35-0	Continuous Blenders
36-0	Colloid Mills
38-0	Cottage Cheese Vats
39-0	Pneumatic Conveyers for Dry Milk and Dry Milk Products

been on performance rather than prescription and on an approach based on hazard assessment and risk management.

The agreement of standards between parties with such a wide group of perspectives and interests has taken considerable time and effort on the part of those involved. This in itself is of substantial potential value as a platform for advancement, for safety, and for greater freedom of trade and international activities in biotechnology and food processing. In many cases,

Table 5 Supporting standards for food hygiene (US)—*continued*

40-01	Bag Collectors for Dry Milk and Dry Milk Products
41-01	Mechanical Conveyors for Dry Milk and Dry Milk Products
42-01	In-line Strainers for Milk and Milk Products
43-0	Wet Collectors for Dry Milk and Dry Milk Products
44-02	Air, Hydraulically, or Mechanically Driven Diaphragm Pumps for Milk and Milk Products
45-0	Crossflow Membrane Modules
46-02	Refractometers and Energy-absorbing Optical Sensors for Milk and Milk Products
47-0	Centrifugal and Positive Rotary Pumps for Pumping, Cleaning and Sanitizing Solutions
49-0	Air Driven Sonic Horns for Dry Milk and Dry Milk Products
50-0	Level-sensing Devices for Dry Milk and Dry Milk Products
51-01	Plug-type Valves for Milk and Milk Products
52-02	Plastic Plug-type Valves for Milk and Milk Products
53-01	Compression-type Valves for Milk and Milk Products
54-02	Diaphragm-type Valves for Milk and Milk Products
55-01	Boot-seal Type Valves for Milk and Milk Products
56-0	Inlet and Outlet Leak-protector Valves for Milk and Milk Products
57-01	Tank Outlet Valves for Milk and Milk Products
58.0	Vacuum Breakers and Check Valves for Milk and Milk Products
59-0	Automatic Positive Displacement Samplers for Fluid Milk and Fluid Milk Products
60-0	Rupture Discs for Milk and Milk Products
61-0	Steam Injection Heaters for Milk and Milk Products
62-01	Hose Assemblies for Milk and Milk Products
63-02	Sanitary Fittings for Milk and Milk Products
64-0	Pressure-reducing and Back-pressure Regulating Devices Valves for Milk and Milk Products
65-0	Sight and/or Light Windows and Sight Indicators in Contact with Milk and Milk Products
66-0	Caged-ball Valves for Milk and Milk Products
68-0	Ball-type Valves for Milk and Milk Products
70-0	Italian-type Pasta Filata-style Cheese Cookers
71-0	Italian-type Pasta Filata-style Cheese Moulders
72-0	Italian-type Pasta Filata-style Cheese Moulded Cheese Chillers
73-0	Shear Mixers, Mixers and Agitators
74-0	Sensors and Sensor Fittings and Connections Used on Fluid Milk and Milk Products
75-0	Belt-type Feeders
78-0	Spray Devices to remain in Place
81-0	Auger-type Feeders
E-600	Egg-breaking and Separating Machines
E-1500	Shell Egg Washer
Drinking water treatment units	
ANSI/NSF 44-1996	Cation Exchange Water Softeners
ANSI/NSF 53-1997	Drinking Water Treatment Units – Health Effects
ANSI/NSF 55-1991	Ultraviolet Microbiological Water Treatment Systems
ANSI/NSF 58-1997	Reverse Osmosis Drinking Water Treatment Systems
ANSI/NSF 62-1997	Drinking Water Distillation Systems
Accepted practices (3–A)	
603-06	Sanitary Construction, Installation, Testing and Operation of High-temperature Short-time and Higher Heat Shorter Time Pasteurizer systems
604-04	Supplying Air Under Pressure in Contact with Milk, Milk Products and Product Contact Surfaces
605-04	Permanently-installed Product and Solution Pipelines and Cleaning Systems used in Milk and Milk Product Processing Plants
606-04	Design, Fabrication and Installation of Milking and Milk Handling Equipment
607-04	Milk and Milk Products Spray-drying Systems
608-01	Instantizing Systems for Dry Milk and Dry Milk Products
609-02	Method of Producing Steam of Culinary Quality
610-0	Sanitary Construction, Installation and Cleaning of Crossflow Membrane Processes
611-0	Farm Milk-cooling and Storage Systems
Food, Safety, and Quality Systems/HACCP-9000	
NSF HACCP-9000–1996	NSF Guidelines for the Application of ISO 9000 and HACCP Requirements to Global Food and Beverage Industries
ISO/DIS 15161	Guidance on the Application of ISO 9001 and ISO 9002 in the Food and Drink Industry

these standards have values beyond those connected solely with safety.

In the case of equipment, it will be possible for components such as valves, couplings, separators, pumps and sampling devices to be type-approved according to their cleanability, sterilizability and leak-tightness. These hygiene-related performance ratings will have to be obtained by recognized laboratories using documented test procedures and documented test conditions (e.g. for a mechanical seal: operating temperature, rotational speed, pressure, number of hours operation, sterilization conditions and frequency). Equipment that carries the CEN biosafety mark will have to be manufactured to a recognized quality management system. This has wider potential value than just for biosafety.

Again, there is an emphasis on type testing and certification of equipment, with similar control and documentation requirements to those of the EHEDG tests.

The idea of these tests is not to guarantee that a particular type of equipment will pass validation in every installed circumstance, but to give relative comparisons which can inform design choices.

The USA

General

The US maintains an interlocking monitoring system that watches over food production and distribution at every level: locally, state-wide and nationally.

Continual monitoring is provided by food inspectors, microbiologists, epidemiologists and other food scientists working for city and county health departments, state public health agencies and various federal departments and agencies. Local, state and national laws, guidelines and other directives dictate their precise duties. Some monitor only one kind of food, such as milk or seafood. Others work strictly within a specified geographic area. Others are responsible for only one type of food establishment, such as restaurants or meat-packing plants. Together they make up the US food safety organization.

The agencies listed below also work with other government agencies, such as the Federal Bureau of Investigation (FBI) to enforce the Federal Anti-tampering Act and the Department of Transportation to enforce the Sanitary Food Transportation Act.

US Department of Health and Human Services: Food and Drug Administration

The FDA enforces food safety laws governing domestic and imported food, except meat and poultry, by:

- Inspecting food production establishments and food warehouses and collecting and analysing samples for physical, chemical and microbial contamination.
- Monitoring safety of animal feeds used in food-producing animals.
- Developing model codes and ordinances, guidelines and interpretations and working with states to implement them in regulating milk and shellfish and retail food establishments, such as restaurants and grocery stores. An example published by the FDA is the Model Food Code, a reference for retail outlets and nursing homes and other institutions on how to prepare food to prevent food-borne illness.
- Establishing good food manufacturing practices and other production standards, such as plant sanitation, packaging requirements and HACCP programmes.
- Working with foreign governments to ensure safety of certain imported food products.
- Requesting manufacturers to recall unsafe food products and monitoring those recalls.
- Taking appropriate enforcement actions.
- Conducting research on food safety.
- Educating industry and consumers on safe food-handling practices.

US State and Local Governments

These work with the FDA and other federal agencies to implement food safety standards for fish, seafood, milk and other foods produced within state borders by:

- inspecting restaurants, grocery stores and other retail food establishments, as well as dairy farms and milk-processing plants, grain mills and food manufacturing plants within local jurisdictions.
- impounding (stopping the sale of) unsafe food products made or distributed within state borders.

US Department of Commerce: National Oceanic and Atmospheric Administration

Through its fee-for-service Seafood Inspection Program, this department inspects and certifies fishing vessels, seafood-processing plants and retail facilities for federal sanitation standards.

- Seafood Inspection Program, 1315 East–West Highway, Silver Spring, MD 20910, US. Tel: 800–422–2750

US Department of Agriculture: Food Safety and Inspection Service

This organization enforces food safety laws governing domestic and imported meat and poultry products by:

- inspecting food animals for diseases before and after slaughter
- inspecting meat and poultry slaughter and processing plants
- with USDA's Agricultural Marketing Service, monitoring and inspecting processed egg products
- collecting and analysing samples of food products for microbial and chemical contaminants and infectious and toxic agents
- establishing production standards for use of food additives and other ingredients in preparing and packaging meat and poultry products, plant sanitation, thermal processing and other processes
- making sure all foreign meat and poultry-processing plants exporting to the US meet US standards
- seeking voluntary recalls by meat and poultry processors of unsafe products
- sponsoring research on meat and poultry safety
- educating industry and consumers on safe food-handling practices.

- FSIS Food Safety Education and Communications Staff, Room 1175, South Building, 1400 Independence Ave SW, Washington, DC 20250, US. Media enquiries: Tel: 202–720–9113

Centers for Disease Control and Prevention

The Centers for Disease Control and Prevention (CDC) supports the work of the other US agencies involved in food hygiene by:

- investigating with local, state and other federal officials sources of food-borne disease outbreaks
- maintaining a nationwide system of food-borne disease surveillance
- designing and putting in place rapid electronic systems for reporting food-borne infections
- working with other federal and state agencies to monitor rates of, and trends in, food-borne disease outbreaks
- developing state-of-the-art techniques for rapid identification of food-borne pathogens at state and local levels
- developing and advocating public health policies to prevent food-borne diseases
- conducting research to help prevent food-borne illness
- training local and state food safety personnel.

- Centers for Disease Control and Prevention, 1600 Clifton Rd, NE, Atlanta, GA 30333, US. Media enquiries: Tel: 404–639–3286

The Food Safety Initiative

One of the initiative's major programmes got under way in May 1998 when the Department of Health and Human Services (which includes the FDA), the USDA and the Environmental Protection Agency signed a memorandum of understanding to create a Food Outbreak Response Co-ordinating Group, or FORC-G. The new group will:

- increase coordination and communication among federal, state and local food safety agencies
- guide efficient use of resources and expertise during an outbreak
- prepare for new and emerging threats to the US food supply.

Besides federal officials, members of FORC-G include the Association of Food and Drug Officials, National Association of City and County Health Officials, Association of State and Territorial Public Health Laboratory Directors, Council of State and Territorial Epidemiologists, and National Association of State Departments of Agriculture.

Whilst not strictly regulatory in nature, a powerful supporting capability is to be able to identify and pinpoint the source of outbreaks and incidents, especially those which cross state and other boundaries. A national computer network is being established which should help to identify outbreaks of food-borne diseases and help to issue alerts more quickly. The network, called PulseNet, will link public health laboratories and state health departments with investigators at the CDC, the FDA and the Agriculture Department. Using DNA fingerprinting to identify and match such food-borne pathogens as the *Escherichia coli* bacteria, it is intended to provide alerts via the Internet in as little as 48 h. At the outset, 16 states were connected to the system, and the CDC expected the rest to be added by 1999.

Supporting Standards

3-A, International NSF and the American National Standards Institute (ANSI) produce standards and guidelines relevant to food process hygiene (Table 5).

The Environmental Protection Agency and USDA are currently seeking comments on a Draft Unified National Strategy for Animal Feeding Operations.

Australia/New Zealand

Existing food hygiene regulations are contained within state and territory legislation, such as Food Acts and associated food hygiene regulations. The Australia New Zealand Food Authority (ANZFA) was formed in 1996 as a result of a treaty signed between the two countries to develop joint food standards. However, at this stage, food hygiene lies outside this treaty.

The Authority is currently developing national food safety standards for Australia. One of the proposed key requirements is for all food businesses (that can identify one or more hazards) to implement a food safety programme. However, these standards have yet to be approved in Australia by the Ministerial Council, which consists of state, territory and commonwealth Health Ministers. If they are approved (and the earliest possible time would be mid-1999), the Authority is recommending a 6-year implementation period for food businesses to implement food safety programmes, based on risk.

The national review of food regulation currently underway in Australia is seeking assistance through industry associations from a range of businesses interested in determining how much it costs them to comply. Hard data is being sought to assess how costly excessive and inefficient regulation is to the food industry, to consumers, and to government itself.

The aim of the review is to reduce the regulatory burden on the food sector and improve the clarity, certainty and efficiency of the food regulatory system, whilst protecting public health and safety.

Food safety programmes are currently voluntary in New Zealand, but if a food business chooses to develop a food safety programme it can be exempt from the current New Zealand food hygiene regulations.

The meat-processing sector has been in the vanguard of HACCP and as early as 1989 the federal inspection system introduced a voluntary system called Production Quality Arrangements (PQA). This covered sanitation, slaughter floor, boning room and offal room and small goods/canneries. Also in 1989 an Approved Quality Arrangement (AQA) was introduced for cold stores and transport of meat. Each system included HACCP. These systems allowed processors to take responsibility for many of the inspection duties traditionally undertaken by AQIS inspectors. At the end of 1994 the uptake was about 40%. In 1994 the Meat Safety Quality Assurance (MSQA) system was introduced gradually to replace the PQA system. It incorporated most of the ISO 9000 elements and used HACCP as the basis for process control. A second edition of MSQA, undated, has recently been published. It updates the previous MSQA and replaces the AQA system and may be used for all red and white meat operations, game meat and rabbits. MSQA is expected to be fully implemented in all export establishments by early 1999. About 50% of product from export plants enters the domestic market.

In early 1997, the various state, territory and commonwealth agencies adopted a set of common Australian standards for processing meat under ARMCANZ, whereas previously each had its own standard. Domestic processors use these standards whilst AQIS retains its equivalent Export Meat Orders. These standards are based on ISO 9000 and incorporate HACCP. Company staffs, with regulatory or external third-party auditing, now control most domestic production.

One of Australia's states, Victoria, has moved more quickly in implementing the proposed food safety reforms, and will begin requiring high-risk food businesses to have food safety programmes in place this year.

Canada

The Canadian Food Inspection Authority carries out enforcement of the Canadian Food and Drugs Act regarding food processing. Its inspectors have wide powers and may enter food preparation premises or conveyances and examine anything that the inspector believes on reasonable grounds is used or capable of being used for manufacture, preparation, preservation, packaging transport or storage of food products. They may also open and examine any receptacle or package that the inspector believes on reasonable grounds contains any article to which this Act or the regulations apply and also examine and make copies of, or extracts from, any relevant books, documents or other records.

They also have powers to impound materials and articles.

Scandinavia

Although Sweden and Finland are covered above as part of the EU, the Scandinavian group of Norway, Sweden and Finland are covered here specifically because of their distinctive and important approach to regulating the problem of *Salmonella* at source in the animal and human food chains. It is also vital for companies wishing to export animal or human feed to these countries to be aware of the compulsory controls that are involved, if they are not to incur a risk of substantial losses.

In many countries, the endemic presence of pathogens such as *Salmonella* and *Campylobacter* in domesticated animals and birds is accepted as inevitable. By adopting an approach combining unequivocal regulatory, educational, organizational and compensation measures, including compulsory intervention, it has been demonstrated that even in an area surrounded by countries where *Salmonella* is endemic, it has been possible to bring *Salmonella* to its knees.

This important approach may well spread to other jurisdictions, especially the EU, where the Scandinavian members have been fighting to be allowed to maintain their system and further to persuade the rest of Europe to do likewise.

In Sweden, *Salmonella* control was introduced for the first time in 1961, following a serious epidemic of *S. typhimurium* in humans in 1953, where some 90 people died and approximately 9000 were taken ill. The source was discovered to be contaminated meat and meat products from a slaughterhouse. This forced new legislation to be introduced.

Since 1961 notification of all kinds of *Salmonella* isolated in animals or animal feeding stuffs has been compulsory in Sweden. Continuous surveillance and control programmes were initiated and animals from infected herds were banned from sale. Food from which any *Salmonella* bacteria have been isolated is by law considered unfit for human consumption. Detection of *Salmonella* always triggers a number of compulsory measures regulated in the Swedish legislation, with the intent to trace and eliminate the infection and its sources. Norway and Finland have similar laws and systems.

Today much less than 1% of all animals and animal products for human consumption are contaminated with *Salmonella*. Detection in cutting plants and retail outlets is rare, in contrast to most other countries in Europe and in the USA, where it is not at all uncommon to find that, for instance, raw chicken, beef, pork and eggs host *Salmonella* bacteria.

Infection in Humans

In the case of Sweden, the *Salmonella* Control Programme in farm animals is the responsibility of the Swedish Board of Agriculture (SBA) and the National Food Administration (NFA), and if *Salmonella* is detected in animals or foodstuffs it must be notified. Specially appointed veterinarians are responsible for the official inspection and sampling.

In Sweden, Norway and Finland human infections account for only 0.04% of the population per annum, of which approximately 85% acquired the disease while travelling abroad. In other European countries the situation is reversed.

Salmonella Control in Sweden

The aim of the programme is to obtain animal products for human consumption free from *Salmonella*. The methods used to reach this aim are:

1. To monitor and control the feed and water used in all types of holdings where animals are kept, to prevent and exclude *Salmonella* contamination of all parts of the food-producing chain.
2. To monitor and control the animal breeding stock at all levels, to prevent *Salmonella* from being transmitted between generations in the food production chain.
3. To monitor and control all other parts of the food production chain from farm to retail outlets, at critical control points where *Salmonella* can be detected, and prevent *Salmonella* contamination in every part of the chain.
4. To undertake the necessary action in case of infection. This includes sanitation of infected flocks or herds.

Neither antibiotics nor hormones are permitted for use as prophylactic treatment or growth promotion in any farm animal, regardless of species. Such substances can only be used for treatment of specific diseases, after prescription by a certified veterinarian, and must be followed by a withdrawal period according to legislation, during which meat, milk and eggs are considered unfit for human consumption. In a survey carried out in 1997 no illegal substances were found, out of the 20 000 meat samples from cattle, swine, sheep and horses that were analysed from every slaughterhouse in Sweden.

Pigs and Cattle

The aim of the control is to monitor the animal population in order to identify *Salmonella*-infected herds, to minimize the spread of infection and to eliminate *Salmonella* from infected herds. The programme is officially supervised, and consists of two parts:

1. Monitoring the situation by official sampling in slaughterhouses and cutting plants: the number of samples is decided by the number of animals slaughtered.
2. Testing on the farms in health programmes monitored by the Swedish Animal Health Services, or when there is clinical suspicion of *Salmonella* in sick animals.

If *Salmonella* is detected on a farm the herd is put under official restrictions, which include specific hygienic measures in the herd, prohibition on moving animals to and from the farm, and prohibition on visiting the herd. Chronically infected animals are eliminated from the herd, and such slaughter may only take place after special permission and according to special rules. An official investigation to find the source of the infection is performed.

During 1997 close to 30 000 samples were collected and analysed in slaughterhouses and cutting plants. In slaughterhouses a total of only three *Salmonella*-positive lymph nodes from cattle and five from pigs were found, and none was found in cutting plants. That is a frequency of 0.08% for the country as a whole. In the cutting plants, surface swabs from the carcasses are analysed to detect if the plant has been contaminated by *Salmonella*. Only two positive, from pigs, were found in 1997.

Poultry

As practised in Scandinavia, the five basics of *Salmonella*-free production are:

1. The day-old chick has to be *Salmonella*-free.
2. Feed and water must be *Salmonella*-free.
3. The environment has to be, and must remain, *Salmonella*-free.
4. The entire production chain has to be checked regularly.
5. Immediate action has to be taken wherever *Salmonella* is detected, regardless of serotype.

There are two control programmes for birds while living on the farms, a voluntary and a mandatory one, with identical testing schemes. Both include production birds such as broilers, layer hens and turkeys, as well as breeder birds and egg production. The voluntary programme started in the 1970s, while the compulsory programme was started about 10 years later. Participation in the voluntary system is only possible if the higher levels of the production chain for that farm (parent and grandparent flocks) are also members. Farms not participating in the voluntary scheme are covered by the mandatory scheme. Participation is obligatory if producers are to deliver poultry to the slaughterhouse, or eggs to the packing centre.

The farms participating in the voluntary programme benefit from higher compensation in the case of an outbreak (up to 70% in the voluntary programme vs. up to 50% in the mandatory programme). In 1998 about 96% of the broiler farms (accounting for 98.5% of the produced poultry meat) and close to 25% of the layer farms were members. All breeder flocks are members today, except a few small ones. The high frequency of participation can be explained by the fact that the government no longer pays the costs associated with an outbreak of *Salmonella* in broiler flocks, and the insurance companies demand participation to compensate the farmers. The industry also makes demands on its members through the organization Svensk Fågel.

Sampling of the slaughter and cutting plants for poultry is a substantial element of the programme. The number and frequency of the sampling depend on the size of the plant. In broiler farms, sampling is organized in combination with an inspection on the farm, 2 weeks before slaughter. The birds are not admitted to normal slaughter procedures unless proven negative for *Salmonella*, to avoid contamination of the plant, but are destroyed if *Salmonella* is detected. From 1998 this was also compulsory for ostrich.

If *Salmonella* is found the infected flock, broilers and layer hens alike, as well as turkeys and ostriches, are immediately destroyed, strict hygienic measures are enforced on the farm and the source of infection is traced and eliminated. Eggs where an invasive (that is, transmitted within the eggs) serotype of *Salmonella* is detected are destroyed. On farms where a non-invasive *Salmonella* is present, the eggs can be heat-treated then sold. The layer hens where a non-invasive *Salmonella* is found can, after special permission from the NFA, be allowed to be slaughtered according to a special procedure, instead of being destroyed.

Out of nearly 4000 yearly samples of poultry taken from slaughterhouses and cutting plants during 1996 and 1997, only two were positive each year, indicating a detected frequency of *Salmonella* as low as 0.05%.

Feed Companies

Feed companies must apply strict testing for *Salmonella* both on raw materials and on finished feed stuffs, as well as a strict hygiene programme, the principles of which have existed for nearly 50 years. According to legislation it is compulsory to heat-treat all industrial poultry feed, including the concentrates. A strict separation between processed feed and unprocessed raw materials is compulsory in the plants. In 1996 *Salmonella* was found in only 0.5% and in 1997 in 0.6% of the 6000 analyses performed in the process control. This control system for animal feed is the strictest in Europe and probably in the world.

Does this Pay?

A cost–benefit analysis was made in 1994 by the National Veterinary Institute, the Swedish Board of Agriculture and the National Bacteriological Laboratory. It compared the annual costs arising from human salmonellosis and the annual cost of control measures in order to prevent and/or minimize the extent of *Salmonella* infection in domestic and imported animals (poultry, cattle and swine) and in animal products.

The analysis concluded that, should the control cease, the financial cost alone of treating human salmonellosis cases would exceed the cost of the prevention programme. Total annual costs, at 1992 prices, were estimated at between 112 and 118 million SEK with a control programme in effect, whereas the costs would be between 117 and 265 million SEK without one.

Costs for investigating outbreaks and control by local and regional authorities were not estimated. If these and other losses for pain and suffering, loss of leisure time and productivity losses in factories and establishments due to *Salmonella* outbreaks were included, the estimated benefits would increase considerably.

This information was provided by Dr Eva Örtenberg, DVM, Veterinary Inspector, National Food Administration, Uppsala, Sweden.

Japan

The Environmental Health Bureau of the Ministry of Health and Welfare is responsible for food hygiene. The Japan Food Hygiene Association (Shadan Hojin, Nihon Shokuhin Eisei Kyokai) can be found at 2–6–1 Gingumae, Shibuya-ku, Tokyo-150–0001. Tel: 03–3403–2111; Fax: 03–3478–0059.

See also: **Good Manufacturing Practice**. **Hazard Appraisal (HACCP)**: The Overall Concept; Critical Control Points; Involvement of Regulatory Bodies; Establishment of Performance Criteria. **National Legislation, Guidelines & Standards Governing Microbiology**: Canada; European Union; Japan. **Quantitative Risk Analysis**.

Hygiene in the Catering Industry

Nick Johns, Independent Research Consultant, Norwich, UK

The varied nature of most conventional catering output and the comparatively short time-scales of production place a different emphasis on the type of control that can practicably be used. **Figure 1** shows a simple systems model of generalized catering processes. The most effective approach to catering process hygiene is through the assessment of hazards and risks. In most cases cooking reduces pathogens in the food to safe levels, so the major risks are the hot and cold storage of prepared foods and the hygiene of premises, equipment and personnel. The relatively labour-intensive nature of catering operations also makes the training and motivation of personnel a critical factor.

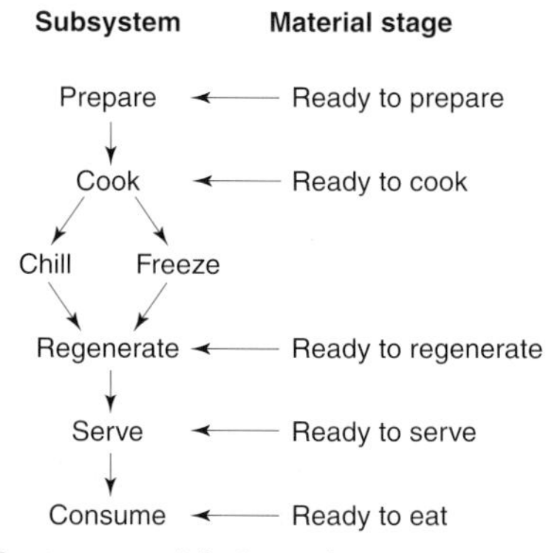

Figure 1 Systems model of catering processes.

Certain catering processes have much in common with food manufacturing, for example cook–chill and cook–freeze catering systems, which involve pre-cooking the food, rapidly chilling or freezing it and holding it refrigerated until use. It is then regenerated before serving. For example, many airline catering operations rely on chilled storage of food, portioned on to special trays. Manufacturing process conditions often apply, since large quantities of single lines may be produced during the course of a normal working day.

Holding Prepared Foods

Hot-holding relies on maintaining the temperature of food above 63°C, so that bacteria cannot multiply. Techniques available to the caterer include:

- bain-maries, which hold food and containers immersed in boiling water or steam
- hot tables which are heated from below, and used for displaying meats for carveries and hot buffets
- racks of infrared lamps, interspersed with ordinary lamps for holding thin food items such as bacon on breakfast buffets
- hot cabinets for holding pies, fish and chips and sometimes plated meats, prior to sale.

All of these systems have drawbacks as well as advantages. For instance, heat may be lost from the top or sides of bain-marie containers if they are not completely immersed in steam. Heat may also be inadequately distributed in solids and viscous liquids. On hot tables, food is not usually covered, so large joints can easily lose heat from the top and micro-organisms can multiply. Foods held beneath infrared lamps must be used up quickly as they tend to dry and shrivel, while items such as rare beef tend to go on cooking. Hot cabinets may contain cool spots, due to draughts. Temperatures should be checked regularly and stock rotated, to avoid both pathogen growth and the deterioration of the food. With hot-holding equipment it is generally more important to monitor the temperature of food items than that of the unit. This is best done using an electronic probe thermometer, which can be regularly sterilized and cannot contaminate food with glass or mercury.

The cold-holding of prepared foods is one of the most important risk factors in catering processes. Refrigerators, freezers, chilled display and vending units should have integral thermometers, with temperature displays on the outside of the cabinet, which should be checked regularly by a designated member of staff. Scheduled chilling and cold-storage operations should be prescribed in job cards, standard recipes or product specifications. Staff should be trained to keep strictly to time–temperature criteria. Standards and procedures for casual storage of over-produced items should also be included in standard recipes. For instance, recipes for sensitive left-over items such as large pieces of roast meat should contain chilling and cold-holding procedures in case of overproduction. Stock rotation and management are particularly important in the cold-holding of over-produced, left-over foods. These have not generally received any special hygiene care and are probably slightly contaminated so they cannot be held safely for long. All such foods should be used or discarded within 24 h. Items should be clearly labelled and dated, and more recent stock should be placed at the back of the refrigerator so that older stock is automatically taken first.

Capacity mismatches between different catering process stages may jeopardize the safety of chilled foods. For instance, a boiler may produce three of four times the amount of cooked potato that a mixer can mash at one time so that cooked potato must wait in bulk while smaller batches are mashed. Such foods should be adequately chilled and the process or menu design should anticipate potential process mismatch. Some items are too costly to throw away at the end of the working day. Often they can be used up satisfactorily the following day as long as the menu allows for this. Menu design requires clear decisions from the manager about what can or cannot be kept and these must be communicated to kitchen staff.

The inadequate thawing of frozen food items is a common problem in catering, because they may remain cool in the middle during cooking, so that pathogens can easily survive even after apparently adequate heat treatment. It is customary to thaw large food items in the refrigerator, to prevent micro-organism growth on the warming outer surface. This takes longer than at room temperature and thawing times should therefore be planned into the production process.

Hygiene of Premises and Plant

It is common to divide catering premises into 'clean' and 'dirty' areas. The former are defined as locations where cooked and ready-to-eat foods are handled and stored and any contamination will be harmful. Examples include hot-holding, plating-up, cooking and salad assembly areas. 'Dirty' areas of food operations are those where personnel, equipment or food might become contaminated. They include lavatories, washrooms, stores and preparation areas for raw meat and vegetables, areas for waste storage or collection, dish-washing and the restaurant (which often contains contaminants such as tobacco smoke, saliva and dirt from street clothes).

The flooring of food production areas must be safe, waterproof and easy to clean. It must be able to cope with the expected volume of traffic and to resist penetration by grease, oil and cleaning chemicals. It may also need to be resistant to heat or to steam cleaning. Suitable flooring materials include non-slip quarry tiles, thermoplastic, edge-welded vinyl (preferably with granules of garnet or corundum to reduce slip) and granolithic or terrazzo mineral matrix materials. Concrete can be coated with epoxy resin, rubber paint or light-grade asphalt to provide a serviceable floor. Unsuitable flooring materials are hard- and softwoods, cork, magnesite and granwood (a composite material of sawdust and cement). These floorings are porous, absorb bacteria and chemicals, and may also inactivate disinfectants. Floors should be coved to the wall without skirting to eliminate sharp edges and corners. Those which need frequent washing should be inclined slightly, towards a drain-hole located away from the walls.

Wall and ceiling surfaces must be smooth and non-porous for easy cleaning and to deter crawling insects. Finishes should be resistant to heat, moisture, grease and cleaning chemicals and should not crack, chip, peel or flake. The best general finishes are paints made of epoxy resin or rubber, in a light colour to show up dirt and to assist illumination. 'Dirty' areas such as those for raw food preparation, which must be regularly disinfected, may be surfaced with white, glazed tiles or with continuous thermoplastic vinyl.

Ceilings should be made from materials that are resistant to fire and to likely conditions of heat, steam and grease and should be coved to walls to assist cleaning. Suspended ceilings are often used to conceal pipes and other services, but the dead space above them may attract pests. Therefore there should be access to the area above a suspended ceiling to permit regular, thorough inspection.

Catering areas should always have artificial lighting and mechanical ventilation, so windows are not strictly necessary, and at best may only have psychological importance. If windows are present they should face north or east from food production and storage areas in order to avoid solar heat gain and glare. Windows should be screened with fine wire mesh to exclude flying insects, and all parts should be easily accessible for cleaning. Architrave should be round-section or absent altogether. Window sills should be absent or steeply canted so they do not collect dust and dirt. Reveals round windows should be tiled or rendered, not timbered. Polyvinyl chloride (PVC) or aluminium frames reduce maintenance costs and windows with condensation problems should be double-glazed.

Materials and finishes for doors should be smooth, non-porous, easy to clean and fire-resistant. Doors should allow easy transfer of personnel, trolleys and equipment between work areas and should not present an obstruction when they are open. Two-way swing doors should have transparent sight-panels and areas on either side should be well-lit. Doors should also have fittings to hold them open when required and to close them automatically in normal use. Doors in areas which need to be regularly hosed down should be of rubber or polypropylene, or have a skirt of these materials. Brushes or plastic strips fixed to the frames of outside doors prevent insects from entering.

Display equipment should be situated in a way that minimizes contamination. For example, low-fronted chilled cabinets in public areas should not be sited where children or adults can use the glass front as a temporary seat and eye-level displays should be protected with sneeze-guards. Chilled storage units should not be sited in hot or sunlit rooms and neither should hot-holding equipment be located in draughty areas. Operating or cleaning instructions should be fixed in a prominent place by food storage or processing machines. Inked, engraved signs on blue plastic plaques should be used, rather than signs printed on paper or cardboard, which may tear or disintegrate and contaminate food.

Equipment which comes in contact with raw foods should not be used with cooked items. The best way to prevent cross contamination is to colour-code equipment with a dot of paint. For example, green may be used to indicate raw vegetables and red, raw meat. Equipment for handling cooked food can be coded yellow or white. Paints used on food equipment should be non-toxic (e.g. guaranteed lead-free). Items to be marked should be properly prepared, i.e. by sanding and washing with sugar soap, and two or more coats should be applied. This makes cracking, flaking and wear less likely. Paint marks should be applied through a stencil. A circle cut from cardboard is adequate, but the outline should be smooth so that chipping, flaking and erosion will show up. Catering equipment should have integral hoppers and funnels, so that food materials can be fed in easily with the minimum of spillage. Their exteriors should be smooth, free of crevices and preferably glossy to aid the removal of split food. There should be no dips or dents in which moisture may collect and no unlagged cold water or coolant pipes to cause condensation. There should be the minimum of dents and recesses, and unavoidable recessed areas should be screened with fine wire mesh to exclude pests. Panels containing thermal insulation material should be sealed and water-tight to prevent the insulation from becoming water-logged and harbouring microorganisms.

Tools and utensils should also be easy to clean and in light colours that contrast with those of food. Items such as knife handles and chopping boards should not have decorative grooves or holes for hanging them up. Dishwashers corrode and darken aluminium items and may wear or crack on-glaze decorated crockery.

Surfaces with which food is in contact must not chip or crack, or absorb liquids. Otherwise pathogenic bacteria may grow within them and leak back into the food. Such surfaces should not contain constituents which may migrate into the food, such as the plasticizers in PVC, or metals other than stainless steel, which may react with food ingredients. The use of copper pans should be kept to a minimum, although they are sometimes necessary and appropriate in visual situations such as flambé work.

Chopping boards should be of plastic, rather than wood. The plastic must be food-grade, non-porous and must not yellow with use, as this indicates the absorption of constituents (and probably also of bacteria) from the food. Plastic chopping boards can be regularly planed smooth to avoid the deep scoring that is caused by prolonged use, which may harbour bacteria. Some makes of commercial board have layers of adhesive plastic which can be successively peeled off as they become scored. The handles of equipment such as knives and saucepans should also be of plastic, rather than wood.

Cleaning Systems

Process hygiene in catering establishments should adopt a *planned preventive* approach to maintenance, cleaning and pest control. In principle this involves estimating the time the next treatment will be needed, and scheduling the work, together with monitoring and documentation to ensure that it has been done to the correct standard. In practice, cleaning needs constant attention, so it is common to use two complementary approaches for cleaning catering premises and equipment: clean-as-you-go and deep cleaning. Clean-as-you-go is short-term, everyday cleaning and can be set up systematically by incorporating an element of cleaning into each culinary task. For example, the task of meat preparation can be designed so that it includes washing up the knife and chopping board and wiping over the work surface with sanitizer. Cleaning procedures can be written into job or recipe cards and should also be taught at training sessions. Clean-as-you-go systems make individual workers and groups responsible for the equipment they use and may promote team work and friendly rivalry through interdepartmental competition. However, clean-as-you-go systems cannot get equipment completely clean and should not be expected to do so.

Deep-cleaning operations are tasks in their own right. Separate job cards should be produced for them, rather than incorporating them into the paperwork for food productions tasks. Deep-cleaning should be carried out during slack periods, or better still, when production is closed down completely. Scrupulous monitoring and documentation of cleaning must be carried out in order to maintain standards, and any deterioration must be reported. Standards for cleaning can be defined in a number of ways. They can be established by microbiological analysis, e.g. by swabbing and plating or using commercial assay systems such as the Tillomed range, which give results quickly and do not require laboratory facilities or highly trained technicians. Day-to-day standards can be set on the basis of appearance and absence of dirt after cleaning, or sometimes the detail of the procedure itself may suffice as a standard. For example 'wipe down slicer blade with a damp cloth' is clearly an inferior standard (as long as it is followed faithfully) to 'dismantle slicer and soak components for 1 h in sterilant solution'. Routine and deep-cleaning operations should be scheduled so that residual accumulation of dirt or bacteria left after clean-as-you-go procedures is regularly removed, as shown in **Figure 2**.

Equipment may be cleaned manually or mechanically. Manual cleaning is generally more expensive and more susceptible to human error and hence quality fluctuations. However, it is more flexible and versatile and with well-trained staff can be more thorough. Manual cleaning is often the only way to achieve a satisfactory level of hygiene with premises or with large or awkward equipment items. Mechanical cleaning is available as clean-in-place systems with large, fixed food-processing or dispensing machines. In these systems cleaning is achieved by passing a solution of chemical cleaner through pipes and vessels; a commonly encountered example is provided by beer lines in bars. Clean-in-place systems are very effective, and give consistent results as long as staff are adequately trained and follow the correct procedures, and the equipment contains no design faults. Mechanical cleaning includes dish-washing machines, which usually clean to a more consistent standard than manual washing. However, food-handling equipment must be able to cope with it. For instance, components from food machines should fit into dishwashers and not be damaged in washing.

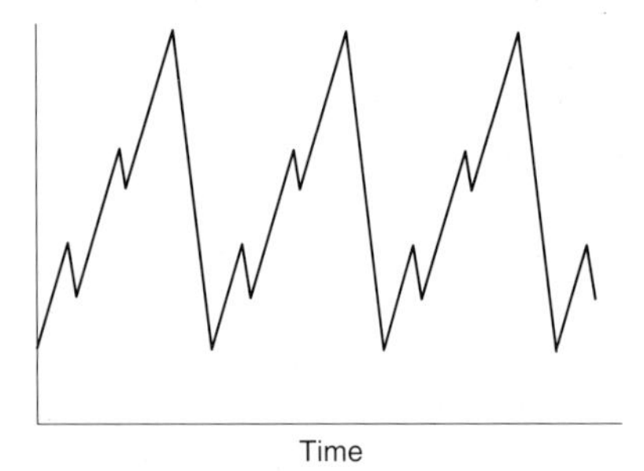

Figure 2 Deep cleaning and clean-as-you-go cycles.

It is important that the correct tools are available for manual cleaning and that they can be kept hygienic and stored conveniently for quick use between food-handling operations. The equipment required for manual kitchen cleaning is as follows:

- wiping cloths of heavy cotton for deep-cleaning operations, which must be boiled periodically to sterilize them
- semi-disposable cellulose wiping clothes, ready-impregnated with sanitizer, for cleaning food contact surfaces in clean-as-you-go operations
- abrasive nylon cloths for scouring dirty pots, which should be disinfected by boiling in water with a little detergent, preferably daily
- brushes with polypropylene or nylon heads and bristles for various cleaning jobs, which can be boiled, bleached and dried between use
- mops of the Kentucky or Foss type with string heads which use a specialized bucket and press. These are appropriate for general use and can be sterilized by boiling
- buckets of polypropylene, which are light, cheap and free from corrosion, mounted on a trolley, with a mop press for easy use.

Light mopping jobs and clean-as-you-go use single-solution mopping from a two-bucket assembly: one containing detergent solution and the other water. Deep cleaning requires double-solution mopping, with a three-bucket assembly: one with detergent, one with disinfectant and one with rinse water. Equipment used in kitchens should be colour-coded to distinguish it from that for lavatories or other use. Mop heads are sewn with coloured thread and handles and metal parts coded with coloured paint.

All cleaning processes are made up of five basic stages:

- preparation (or precleaning)
- cleaning
- rinsing
- disinfection
- drying.

The five stages vary in relative importance depending upon what is being cleaned and sometimes two stages may be accomplished at the same time. It is necessary to identify all five in order to understand any cleaning operation or to design a cleaning procedure.

Table 1 Accredited hygiene courses in the UK

	Minimum hours of study	*Course work (h)*	*Written exams*	*Oral exam*
IEHO Basic Course in Food Hygiene	6		1	
IEHO Intermediate Certificate Course in Food Hygiene	18		2	
IEHO Advanced Food Hygiene Certificate	36	3	3	3
RIPHH Primary Certificate in Hygiene for Food Handlers		7		1
RIPHH Certificate in Food Hygiene and the Handling of Food	16		2	3
RIPHH Diploma in Food Hygiene	42		4	3
RIPHH Diploma in Hygiene Management (supplement to Certificate in Food Hygiene and the Handling of Food; see above)			3	
RSH Certificate in the Hygiene of Food Retailing and Catering	22		1	3

IEHO, Institute of Environmental Health Officers; RIPHH, Royal Institute of Public Health and Hygiene; RSH, Royal Society of Health.

Personal Hygiene of Catering Operatives

Staff should be inspected before starting work to ensure the cleanliness of hands and forearms, absence of jewellery (apart from wedding rings) and cosmetics, correct protective clothing and absence of signs of illness.

Catering staff should wash their hands regularly with non-perfumed, antidermatitis, antibacterial liquid soap from a dispenser. Non-manual taps, operated by pedals or an infrared sensor, should be used and nail brushes should be available. Hot-air driers, paper towel dispensers and continuous roller towels should be used, rather than communal hand towels. Showers can also play a useful part in promoting hygiene.

Staff should handle food as little as possible and preferably with dry hands which are free of skin diseases, which can cause serious food contamination. Individuals showing symptoms of any illness or infection should be excluded from food-handling work and there should be a just and accepted system for compensating or redeploying them. Sick staff should see a doctor as soon as possible and where appropriate they should submit faecal samples for analysis.

Sufficient protective clothing should be supplied to catering operatives so that they can change regularly and always be clean and smart. Protective clothing must be laundered separately from table linen and sterilized by boiling at each wash. If clothes-washing facilities are not available on site, protective clothing must be sent to a commercial laundry, not taken home for washing. Work wear should include overalls, aprons, hair nets and hats to prevent hair, ornaments and grips from falling out, gloves (for work involving highly soiled conditions, chemicals or repeated immersion in water) and shoes or boots. Operatives who have to work with corrosive liquid chemicals (for example, chlorine bleach) should also be supplied with eye protection.

Training is sometimes a matter of sending scheduled staff on courses offered by outside bodies (**Table 1**). Basic courses in food hygiene include an outline of the bacteriology of food and food poisoning, together with practical hygiene, the prevention of contamination and pest control and cleaning. Alternatively, in-house courses make it possible to incorporate food hygiene into more general training programmes, or tailor the input to a company's specific needs. Training should be well organized, with classroom sessions which involve extensive use of visual aids, and handouts to support learning. Input by the trainer should be interspersed with activity by the trainees, e.g. in question-and-answer sessions. Learning is easier for trainees if the material is presented logically, with each topic leading naturally into the next. Training sessions should be introduced with a welcome and end with a summary of the session's learning. Food hygiene topics are taught (but not necessarily learned) on general catering courses and trainees may feel they have covered the material before. Training can therefore seem less repetitious and more relevant if it is given a company slant, i.e. 'this is the way we do it around here'.

Training should be reinforced, and this can be done by repeating some or all of the syllabus in greater detail, in a retraining session, or in a discussion group which translates the information imparted into more practical workplace terms. Food hygiene can be one of the areas included in company-wide total quality or customer care programmes. Training can also be reinforced by staff awareness schemes, with notices, posters or badges referring to hygiene.

The evaluation of training should focus on how well the original objectives have been achieved. Numbers of applicants, participants, drop-outs, examination successes and failure should all be recorded. It is also possible to assess less tangible aspects, such as improvements in staff morale (using questionnaires) or general hygiene standards (using independent inspections). Trainees and trainers should also be invited to comment on the programme content and delivery and on what they feel has been learned.

Training may improve staff attitudes in the short term but it cannot be relied upon to do so permanently. In order to achieve this, training needs to be supported by a conscious programme of motivation at the workplace. Motivation programmes should centre around fulfilling employees' workplace needs. Making sure that they have the correct equipment and that the jobs they must perform are clearly defined and communicated satisfies their *task needs*. In addition, managers and supervisors should aim to build work groups around themselves, so that employees feel that they are a part of a team which is taken seriously by management. Such groups can have a marked effect upon motivation if they are encouraged to discuss hygiene aspects of their work and to make recommendations and set codes of practice. This satisfies workers' *group needs*. Individuals can be motivated, and *individual needs* met, by recognizing and rewarding good practice.

In summary, catering processes are more complex and diverse than those of general food manufacturing, and this makes them more difficult to control through the rigorous control of hazards and risks. However, the most important risks can be identified as the hot and cold storage of foods, the cleanliness of surfaces and equipment and the performance of staff. All of these can be achieved through a systematic management approach that defines clearly what is to be done and controls, monitors and documents each activity.

See also: **Process Hygiene**: Designing for Hygienic Operation; Types of Biocides; Overall Approach to Hygienic Processing; Modern Systems of Plant Cleaning; Risk and Control of Airborne Contamination; Testing of Disinfectants; Involvement of Regulatory and Advisory Bodies.

Further Reading

Adair J (1968) *Training for Leadership*. Macdonald.

Allen DM (1983) *Accommodation and Cleaning Services*. Vols 1 and 2. Hutchinson.

Conning DM, Leigh L and Ricketts BD (1988) *Food Fit to Eat*. Sphere Books.

Cornwell PB (1973) *Pest Control in Buildings*. Hutchinson.

Eschbach CE (ed.) (1974) *Food Service Trends*. New York: Cahners Books.

Glew G (ed.) (1979) *Advances in Catering Technology*. London: Applied Science.

Knight JB and Kotschevar LH (1983) *Quantity Food Production Planning and Management*. CBI Books.

Longree K and Armbruster G (1987) *Quantity Food Sanitation*. Wiley.

Meehan AP (1984) *Rats and Mice: Their Biology and Control*. Rentokil Library.

Pannett A (1989) *Principles of Hotel and Catering Law*, 2nd edn. Cassell.

Pedderson RB et al (eds) (1973) *Increasing Productivity in Food Service*. Boston, US: Cahners Books.

Riemann H and Bryan FL (eds) (1979) *Food-borne Infections and Intoxications*. New York: Academic Press.

Skinner FA and Carr JG (eds) (1976) *Microbiology in Agriculture, Fisheries and Food*. London: Academic Press.

PROPIONIBACTERIUM

Michel Gautier, Ecole Nationale Supérieure d'Agronomie, Institut National de la Recherche Agronomique, Rennes, France

Taxonomic Approach

As early as 1906, Von Freudenreich and Orla-Jensen had isolated various bacteria from cheese, among them bacteria producing propionic acid which were named *Propionibacterium*. Orla-Jensen later specifically isolated propionibacteria and described them in more detail.

Genus Description and Phylogenetic Situation

The propionibacteria are classified within the Gram-positive bacteria group in the subdivision of Actinomycetes (the other subdivision being *Clostridium*) which groups together the numerous species with high G+C%. The G+C% content of the propionibacteria is in the range 65–67%, according to the species.

The classification of propionibacteria, as with numerous other bacterial genera, has advanced con-

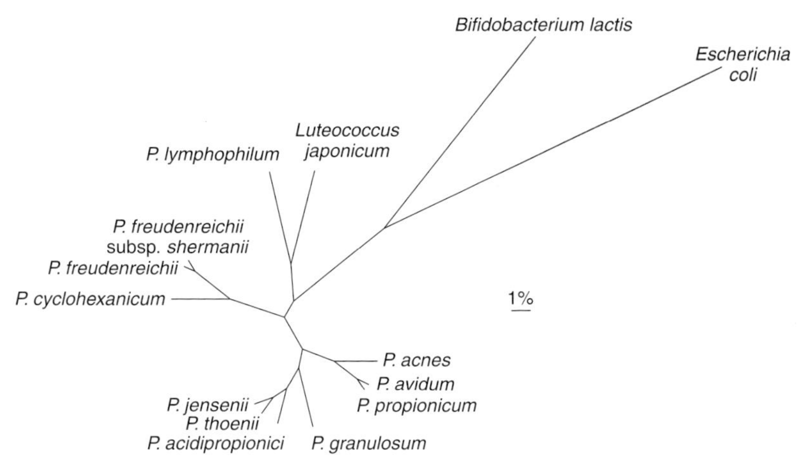

Figure 1 Phylogenetic tree of the *Propionibacterium* genus based on the analysis of the 16S rRNA. (Redrawn from Dasen et al. 1998.)

siderably during this century. In 1930 *Bergey's Manual* described eight species, in 1934 nine species and in 1939 11 species. The lastest modifications were not based on phenotypic characteristics, as previously, but rather on a genetic approach using DNA/DNA hybridization. Thus the species have been grouped together into eight species (*Bergey's Manual* of 1974 and 1986). The analysis of 16S rRNA allowed the construction of a phylogenetic tree (**Fig. 1**).

This genus is divided up into two groups of species:

1. The cutaneous species (*P. acnes*, *P. granulosum*, *P. lymphophilum*, *P. propionicum* and *P. avidum*) are commonly found on the skin and their role (*P. acnes*, *P. granulosum*) as possible causal agents of disease (acne vulgaris) is still unresolved.
2. The classical species, also called the dairy species (*P. freudenreichii*, *P. acidipropionici*, *P. thoenii* and *P. jenseinii*), are involved in the ripening of Swiss type cheeses. A new species, *P. cyclohexanicum*, belonging to the classical group but not found in the dairy products has been described recently. This species was isolated from spoiled orange juice and the 16S rRNA sequence shows that it is closely related to *P. freudenreichii*.

Only classical species involved in the manufacture of dairy products, are discussed here because of their implications in food microbiology.

P. freudenreichii groups together the classical species *P. globosum*, *P. orientatum*, *P. coloratum*, *P. freudenreichii* and *P. shermanii*.

P. freudenreichii includes two subspecies, *P. freudenreichii* subspp. *freudenreichii* and *shermanii*, but this division is currently under discussion because some authors have suggested the reintroduction of an additional *P. globosum* as a subspecies, whereas others think that DNA/DNA homology within the species is too high to justify a separation into two subspecies. This species is used as a starter for the manufacture of Swiss-type cheeses and is also used for the production of vitamin B_{12} and propionic acid.

P. jenseinii comprises the classical species *P. peterssonii*, *P. technicum* and *P. zeae*.

P. thoenii groups together the species *P. thoenii* and *P. rubrum*. However, recent studies based on numerical taxonomy and molecular biology (T_m comparison and sequence of 16S rRNA) showed that the strains of *P. rubrum* should be incorporated in with the *P. jenseinii* species. This error in classification has some repercussions on the value of the phenotypic keys of identification recommended in the last edition of *Bergey's Manual* (1986). In fact the keys used for the identification of *P. thoenii* strains were established, in part, from the phenotypic characteristics of *P. rubrum* strains.

P. acidipropionici includes the older species *P. pentosaceum* and *P. arabinosum*.

The taxonomy of propionibacteria is still evolving because, for some years, scientists have been looking for them ecological niches other than dairy products. It is probable that other new species will be isolated from the environment. A good illustration of this is the discovery of *P. cyclohexanicum*, a new species found in spoiled fruit juices.

Identification Methods

The propionibacteria are pleomorphic rods (**Fig. 2**) but the cells may be coccoid, bifid or branched. Cells may be single, in pairs forming a V or a Y shape, in short chains, and often grouped in a 'Chinese character' type pattern. Their morphology and arrangement varies according to the strain, the age of the culture and the culture medium. This pleomorphic characteristic means that contamination of all cultures of propionibacteria with other microorganisms is

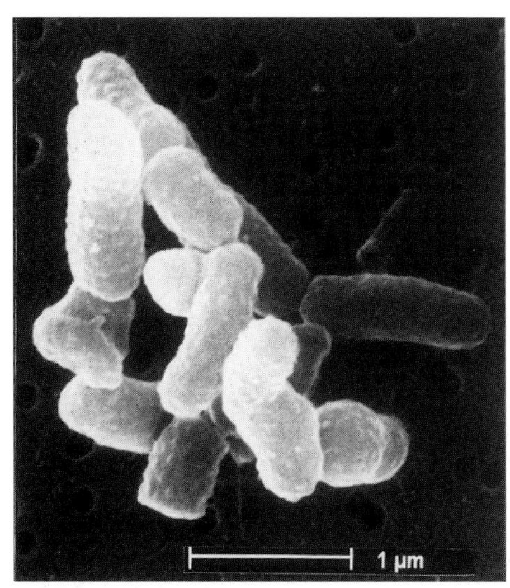

Figure 2 Scanning electron micrograph of a strain of *P. freudenreichii*. (Photograph J. Berrier.)

sometimes difficult to detect. These bacteria are also non-motile and non-spore-forming.

Propionibacteria are chemoorganotrophs and fermentation products include large amounts of propionic and acetic acids. They are anaerobic to aerotolerant and generally catalase positive. These characteristics, and especially the production of propionic acid, allow for relatively easy characterization of the genus.

Some phenotypic characteristics used in determining species have been established and are indicated in **Table 1**. However, the value of these identification keys has recently been called into question especially concerning the differentiation of *P. thoenii* and *P. jensenii*. Due to the great variety of strains, the variability of the phenotypic characteristics and the subjectivity relative to these identification methods, these phenotypic keys are not well adapted for species determination. Although the fermentation patterns allow for the differentiation of *P. freudenreichii* (which uses only a few sugars for fermentation) and *P. acidipropionici* (which uses a larger range of sugars), they cannot be used for the atypical strains or for differentiating between *P. thoenii* and *P. jensenii*.

Some classical methods, such as lysotyping, have been developed for the differentiation of *P. acnes* strains. However, concerning the dairy species, a lysotyping method could not be developed because, on the one hand, more suitable techniques of fingerprinting based on molecular biology were recently developed and, on the other hand, only bacteriophages infecting *P. freudenreichii* have been described and their variety is too limited to develop a lysotyping method. Other classical methods, such as serotyping, have not been well developed and, for the reasons previously outlined, are no longer available.

Methods Based on Molecular Biology The best identification methods are based on molecular biology. Some methods, based on the polymerase chain reaction (PCR) allowing efficient speciation, have been developed, the primers used having been obtained from the sequence of the 16S ribosomal genes. Two methods based on the polymorphism of the 16S rRNA genes have also been developed (namely the ribotyping and restriction analysis of the 16S rRNA genes).

Techniques leading to the differentiation of strains, particularly in order to follow the strains involved in fermentation processes and to evaluate biodiversity, have been developed using classical propionibacteria. These efficient methods, based on molecular biology, include the RAPD (randomly amplified polymorphism DNA) and the chromosome restriction pattern. In the latter technique the restriction endonuclease *Xba*I cuts the genome in rare sites and the large fragments obtained (about 20) are then separated with pulsed field gel electrophoresis (PFGE).

Ecological Niche

Milk and Dairy Products

Classical propionibacteria are found in raw milk. Some studies mention a concentration of 10–1.3×10^4 cfu ml^{-1} in French raw milk, and an average contamination of 7×10^2 cfu ml^{-1} and 2.5×10^2 cfu ml^{-1} in raw milk used for Italian Grana

Table 1 Main criteria for differentiation of lactic propionibacteria according to Cummins and Johnson (1986)

Organism	*Fermentation of sucrose and maltose*	*Reduction of nitrate*	*β-Haemolysis*	*Colour of pigment*	*Isomer of DAP in cell wall*
P. freudenreichii	–	d	–	*Cream*	*Meso*
P. jensenii	+	–	–	*Cream*	L-
P. thoenii	+	–	+	Red-brown	L-
P. acidipropionici	+	+	–	Cream to orange-yellow	L-

+, 90% or more strains are positive.
–, 90% or more strains are negative.
d, 11–89% of strains are positive.

cheese and in Swiss raw milk, respectively. This concentration is closely related to the hygienic quality of the milking line. Moreover, propionibacteria can grow in milk because, although they preferentially use lactate as a carbon substrate, they can also use lactose. However, their growth in milk is reduced due to their weak proteolytic activity.

The presence of propionibacteria in raw milk leads to their development in certain types of cheeses, where they can reach levels of 10^9 cfu g^{-1} of cheese. They are predominantly found in Swiss-type cheese because the temperature of the long ripening (several weeks at about 24°C), the low salt concentration and the relatively high pH (5.2) favour their growth.

Their development may also result in defects in other cheese varieties where propionic fermentation is not desirable (late blowing in Grana cheese-making, abnormal gas formation in mozzarella cheese).

Other Ecological Niches

Classical propionibacteria have been found in soil, silage and vegetable fermentations, but exhaustive studies have only been carried out on the habitat of propionibacteria. Propionibacteria have also been isolated from anaerobic environments. Within the rumen, *Propionibacterium* species are, among other species, responsible for urea breakdown and ammonia release and this bacterial genus has been found in anaerobic digesters.

Relations with Bacteriophages

Bacteriophages infecting a cutaneous species (*P. acnes*) have been described since 1970 and studied to develop a typing method for different strains. However, bacteriophages infecting the classical propionibacteria, and especially *P. freudenreichii* were only discovered in 1992. The presence of bacteriophages has been reported in a wide variety of French Swiss-type cheeses (16 cheeses of 32 analysed) at various levels ranging from 14 to 10^6 cfu g^{-1}. These bacteriophages present a classical morphology (they have an isometric head, a non-contractile tail and a tail plate) (**Fig. 3**) so they belong to the B1 group of Bradley's classification. Their genome consists of a linear double-stranded DNA molecule, 40 kb long, with cohesive ends.

Temperate bacteriophages have been described and it has been shown that some strains of *P. freudenreichii* are lysogenic since they harbour prophages inserted on their chromosome. The multiplication of bacteriophages was found to occur in cheese during the multiplication stage of propionibacteria in a warm curing room. Although propionibacteria bacteriophages are very common in cheeses, their impact on cheese technology and quality is probably limited. Bacteriophages coexist in cheese with an abundant population of phage-sensitive cells, indicating that destruction of propionibacteria is only partial. Because of the solid structure of Swiss-type cheese, bacteriophages cannot propagate throughout the cheese. Consequently, their multiplication occurs in separate sites and only partially hampers the propionibacteria development.

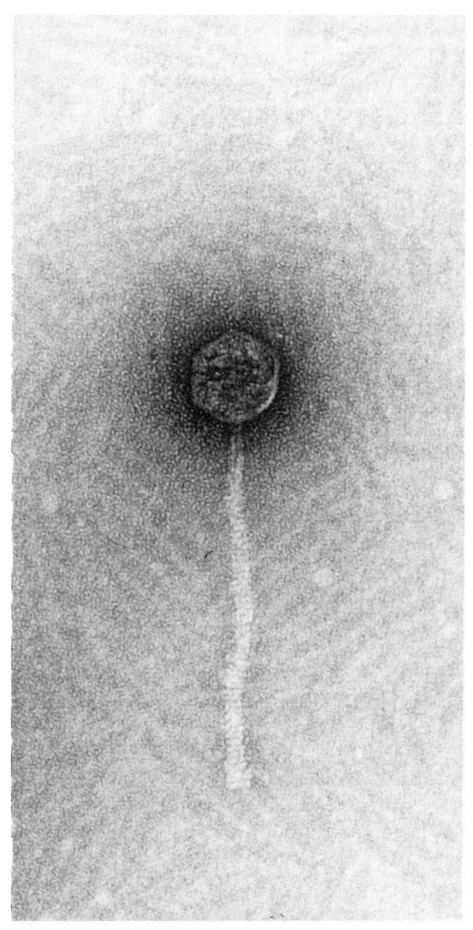

Figure 3 Electron micrograph of phage B22 infecting *Propionibacterium freudenreichii*. (Photograph F. Michel.)

No phage accident has so far been reported in other types of fermentation processes using propionibacteria.

Enumeration and Culture Procedures

Propionibacteria, being anaerobic to microaerotolerant, do not grow on solid media exposed to air and, consequently, obtaining colonies requires growth in anaerobic jars. However, probably because of their microaerotolerant character, growth in liquid culture media does not require anaerobic conditions.

They grow very well on complex media, such as brain heart infusion (BHI) broth, but because of their long generation time, contamination with other bacteria can occur, which is why more selective media are preferred. Yeast extract–sodium lactate (YEL), where the carbohydrate source is the lactate, is nor-

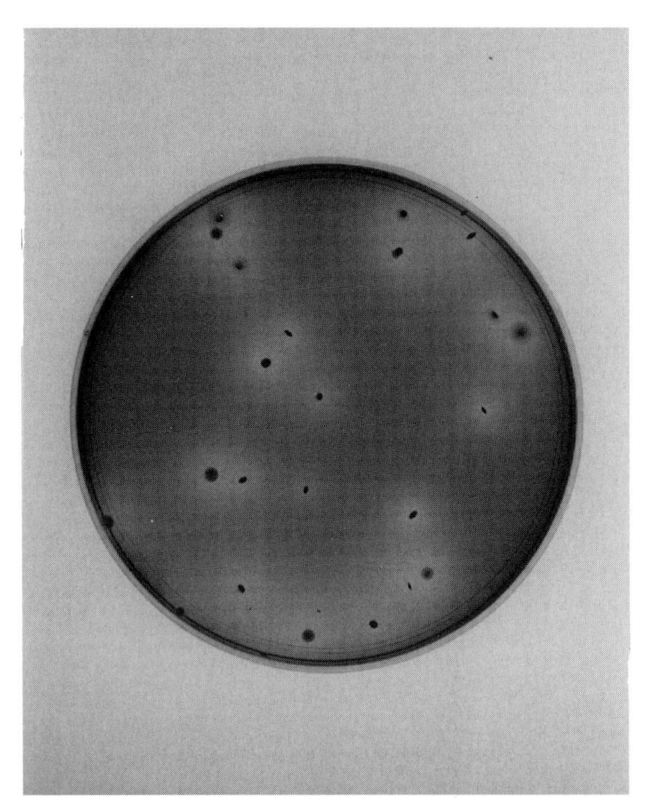

Figure 4 Colonies of *Propionibacterium freudenreichii* obtained on the selective medium Pal Propiobac. (Photograph A. Thierry.)

mally used. With this medium, to which agarose has been added (YELA), 5 to 6 days are required to obtain colonies of 2 mm in diameter which are cream coloured for *P. acidipropionici* and *P. freudenreichii*, orange or brick red for *P. jenseinii* and brick red for *P. thoenii*.

Selective Medium

Propionibacteria are usually a minority population in milk samples and, consequently, difficult to isolate with nonselective media. A recently developed medium now available commercially under the brand name Pal Propiobac® allows for improved isolation of propionibacteria from samples with a complex flora. In addition to classical nutritive elements, this medium contains glycerol as the fermentation substrate, lithium to inhibit various lactic acid bacteria, a cocktail of antibiotics active against Gram-negative bacteria, and bromocresol purple. The propionibacteria colonies appear on this medium as brown colonies larger than 0.5 mm in diameter, surrounded by a yellow area (**Fig. 4**) caused by the pH decrease resulting from glycerol fermentation.

Factors Interfering with the Growth of Propionibacteria

Temperature The optimal growth temperature is 25–32°C. The capacity to grow at low temperatures depends on the species and strain. In contrast to the three other species, the majority of strains of *P. freudenreichii* are able to grow at 7°C, probably due to modification of the fatty acids of the cytoplasmic membrane.

This ability to grow at low temperatures can present problems in cheese technology because when the ripened cheeses are stored in a cold room after ripening, the growth of propionibacteria can lead to excessive swelling of the cheese loaf.

pH It is difficult to define an optimum pH for growth of propionibacteria because this pH depends on the growth medium, the temperature and the water activity. In contrast to the lactic acid bacteria used in cheese manufacture, propionibacteria are rather sensitive to acidity.

In YEL medium, the optimal pH for growth of propionibacteria is between 6.5 and 7.0 but they may survive at much lower pH values. During the ripening of Swiss-type cheese, they grow at between pH 5.4 and 5.6 which is the pH of the cheese but they have much greater difficulty growing below pH 5. In these cheeses, at pH 5 to pH 6, a variation of 0.1 pH unit can have a significative effect on their growth.

Effect of Sodium Chloride As the manufacture of Swiss-type cheese requires brine salting of the cheeses just before ripening, the sensitivity of propionibacteria to sodium chloride has been studied. In cheese, salt is an important factor influencing the growth of propionibacteria and the inhibiting effect of sodium chloride on growth is all the more significant since the pH is low.

In YEL medium at pH 7, the great majority of propionibacteria strains are able to grow at a maximum concentration of 6–7% of salt. However, in cheese, their sensitivity is increased to pH 5.2–5.4, just prior to the beginning of ripening. Their growth decreases drastically at a concentration of 3% of salt in cheese, i.e. just beneath the cheese rind.

It has been shown that *P. freudenreichii* subsp. *shermanii* cells respond to changes in rich medium osmolarity by varying the concentrations of specific solutes and especially glycine-betaine, in order to maintain their constant turgor pressure. This property enables them to adapt to the rise in osmotic pressure due to the salting stage in the cheese process.

Use of Propionibacteria in Industry

Cheese Making

Propionibacteria are important in the development of flavour during the ripening process in the manufacture of Swiss-type cheese. It is difficult to determine the exact role of propionibacteria in the production of flavour compounds because the appearance of aromatic molecules results from the activity of various bacterial species developing in cheese. However, it can be stated that the characteristic flavour due to the propionibacteria results from the production of propionic acid, acetic acid and diacetyl. CO_2 is also produced which is responsible for the 'eyes' or gas vacuoles characteristic of such cheeses. Although their proteolytic activity is not significant, some studies emphasize their lipolytic capacities. It seems that this activity, which is responsible for the development of aromatic compounds, is most significant during ripening. However, it remains to be determined whether these bacteria are responsible for the production of aromatic amino acids or compounds resulting from the catabolism of amino acids.

These bacteria are naturally present in the raw milk used for cheese manufacture. However, this natural flora is becoming more and more depleted by the improvement of raw milk quality and the processes of microbiological purification of milk, such as microfiltration or bactofugation. For this reason, propionibacteria are added at a low concentration (10^5 cfu ml^{-1} of milk) by the cheese maker at the beginning of the process. In fact only the species *P. freudenreichii* is used as a starter, probably because it is the species most resistant to the heat treatment applied to the curd during cheese manufacture. However, indigenous propionibacteria are not completely eliminated by milk thermal treatment and, in spite of their low concentration, they are able to predominate during the ripening process because of the significant length of the ripening period.

Propionibacteria grow in the body of the cheeses after the development of thermophilic lactic acid bacteria, as they can obtain their energy for growth from the fermentation of lactate produced by lactic acid bacteria.

Production of Propionic Acid

Propionic acid, and its salt, are largely used as mould inhibitors in baking, as esterifying agents in the production of thermoplastics and in the manufacture of flavours and perfume bases. A large part of this production is via petrochemical pathways. Nevertheless, fermentation processes have been described since 1923. The increasing consumer demand for biological products and the more efficient performance of new fermentation processes have revived research and industrial interest for biological propionic acid production. Moreover, the association of propionic acid with lactic and acetic acids has been recommended for the preservation of foods. The US Food and Drug Administration (FDA) lists the acid, and the Na^{++}, Ca^{++} and K^{+} salts, as preservatives in their summary of generally recognised as safe (GRAS) additives and no upper limits are imposed except for bread, rolls and cheeses (0.30–0.38%).

The early work on propionic acid fermentation resulted in the formulation of the Fitz equation:

$$3 \text{ lactic acid} \rightarrow 2 \text{ propionic acid} + 1 \text{ acetic acid} + 1\ CO_2 + 1H_2O$$

or

$$1.5 \text{ glucose} \rightarrow 2 \text{ propionic acid} + 1 \text{ acetic acid} + 1\ CO_2 + 1H_2O$$

Theoretical maximum yields are 54.8% (w/w) as propionic acid and 77% as total acids. Formation of propionic acid is accompanied by the formation of acetate, for stoichiometric reasons and to maintain the hydrogen and redox balances.

Various processes of propionic acid production were described and are shown in **Table 2**.

Use of Propionibacteria as Probiotics

Although at present propionibacteria are not extensively commercialized as probiotics, it appears that they do have a probiotic effect. This probiotic action depends on the production of propionic acid, bacteriocins, nitric oxide, folacin, vitamin B_{12}, CO_2 and their stimulatory effect on the growth of other beneficial bacteria. Moreover, some strains resist the acid environment of the stomach and the bile salts of the intestine and reach a high population density within the digestive tract. Propionibacteria can be used as a probiotic for animals and for humans. They have been administered to piglets and the effect on the growth of the pigs was significant. In addition the fodder demand was clearly lower when compared with the control group. A mixture of propionibacteria, lactic acid bacteria and bifidobacteria have been used, with positive results, as probiotics for calves.

Propionibacteria have also been investigated with regards to their potential role as human probiotics, especially in curing certain intestinal disorders of children and elderly people. They are a source of beneficial enzymatic activities such as β-galatosidase and β-glucuronidase.

Table 2 Comparison of propionic acid fermentation processes (from Boyaval and Corre 1995)

Organism	*System*	*Carbon source*	*Cell concentration*	*Propionic acid concentration (g l^{-1})*	*Productivity (g l^{-1} h^{-1})*
P. acidipropionici	Batch	Lactose Glucose	–	2.37	0.033
P. acidipropionici (ATCC 25562)	CSTR	Glucose Xylose	1.5×10^9–9.5×10^{11} cells ml^{-1}	3.74	0.19 0.18
	Plug flow tubular reactor	Glycose Xylose	–	–	0.49 0.40
Propionibacterium sp.	Calcium alginate gel beads	Na-lactate 4	$10^{9\ cells}$ (free) ml^{-1}	5	2
P. acidipropionici (ATCC 25562)	CSTR + UF cell recycle	Xylose	95 (g $l^{-1)}$	18	2.2
P. acidipropionici	CSTR + UF cell recycle	Lactose	100 (g l^{-1})	25	14.3

Production of Vitamin B_{12}

Vitamin B_{12} is an important cofactor in the metabolism of carbohydrates, lipids, amino acids and nucleic acids. The vitamin is thus an important additive in animal feeds and is used in chemotherapy, in particular to prevent pernicious anaemia. Up to now vitamin B_{12} has been produced by fermentation on an industrial scale since chemical synthesis of the vitamin is very difficult. For a long time propionibacteria were used to produce vitamin B_{12} but *Pseudomonas denitrificans* strains have partly replaced propionibacteria in commercial vitamin B_{12} production because they grow faster and yields are higher. Actually two-thirds of the vitamin B_{12} produced is via *Pseudomonas denitrificans* and the remainder by *Propionibacterium freudenreichii*. As the vitamin produced by the two species is marketed at the same price, the advantages of using one process or the other seem negligible. The production rate of vitamin B_{12} is about 20 mg per litre of culture, the molecule produced by *Pseudomonas* being excreted into the culture medium whereas in the case of the propionibacteria it is intracytoplasmic.

Genetic Improvement of Propionibacteria

Very little work has been carried out on the genetics of propionibacteria. The development of a cloning system has been hampered by the lack of a DNA molecule able to replicate inside this genus. Such a molecule is necessary to develop a DNA transfer technique and an efficient DNA transfer technique is essential in order to select a DNA molecule able to replicate.

Propionibacteria have very few plasmids since about 30% of strains only harbour one to three plasmids and, moreover, the experiments concerning plasmid curing have not resulted in the connection of a phenotypic character to the presence of a plasmid. It was the discovery, in 1992, of a bacteriophage infecting *P. freudenreichii* which allowed the development of a DNA transfer technique. The phage chromosome has been used to optimize the conditions of electrotransformation and a transfer efficiency of 7×10^5 transfectants per microgram of DNA has been obtained. As the various vectors used in Gram-positive bacteria are inefficient in propionibacteria, a Japanese firm producing vitamin B_{12} has constructed a vector from a cryptic plasmid of propionibacteria. In addition to this plasmid, the patented vector consists of an *Escherichia coli* vector carrying a gene for chloramphenicol resistance which is under the control of a propionibacteria promotor. Consequently, due to this cloning system, it will now be possible to study the propionibacteria gene in *Propionibacterium*.

As regards the genetic studies of propionibacteria, no genes have as yet been cloned in propionibacteria and only a few genes have been cloned and studied in *E. coli*. These genes are involved in metabolic pathways, and particularly pathways involved in vitamin B_{12} production.

See also: **Bacteriophage-based Techniques for Detection of Foodborne Pathogens**. **Cheese**: Microbiology of Cheese-making and Maturation. **Genetic Engineering**: Modification of Bacteria. **Milk and Milk Products**: Microbiology of Liquid Milk; Microbiology of Dried Milk Products; Microbiology of Cream and Butter. **Preservatives**: Traditional Preservatives – Sodium Chloride; Permitted Preservatives – Propionic Acid. **Probiotic Bacteria**: Detection and Estimation in Fermented and Non-fermented Dairy Products. **Starter Cultures**: Cultures Employed in Cheesemaking.

Further Reading

Boyaval P and Corre C (1995) Production of propionic acid. *Lait*, 75: 453–462.

Charfreitag O and Stackebrandt E (1989) Inter- and intrageneric relationships of the genus *Propionibacterium* as determined by 16S rRNA sequences. *J. Gen. Microbiol.* 135: 2065–2070.

Cummins CS and Johnson JL (1986) Genus I, *Propionibacterium* Orla-Jensen 1909. In: Sneath PHA, Mair NS, Sharpe ME, Holt JG (eds) *Bergey's Manual of Systematic Bacteriology.* Vol. 2, p. 1346, Baltimore: Williams & Wilkins.

Dasen G, Smutny J, Teuber M and Meile L (1998) Classification and identification of propionibacteria based on ribosomal RNA genes and PCR. *System. Appl. Microbiol.* 21: 251–259.

De Carvalho A, Gautier M and Grimont F (1994) Identification of dairy *Propionibacterium* species by rRNA gene restriction patterns. *Res. Microbiol.* 145: 667–676.

Gautier M, Rouault A, Sommer P and Briandet R (1995) Occurrence of *Propionibacterium freudenreichii* bacteriophages in Swiss cheese. *Appl. Environ. Microbiol.* 61: 2572–2576.

Hettinga DH and Reinbold GW (1972) The propionic acid bacteria, a review. *J. Milk Food Technol.* 35: 295–301, 358–372, 436–447.

Langsrud T and Reinbold GW (1973) Flavor development and microbiology of Swiss cheese – a review. II. Starters, manufacturing process and procedures. *J. Milk Food Technol.* 36: 531–542.

Langsrud T and Reinbold GW (1973) Flavor development and microbiology of Swiss cheese – a review. III. Ripening and flavor production. *J. Milk Food Technol.* 36: 593–609.

Madec MN, Rouault A, Maubois JL and Thierry A (1994) Milieu sélectif pour le dénombrement des bactéries propioniques. French Patent no. 93 00823.

Mantere-Alhonen S (1995) Propionibacteria used as probiotics – a review. *Lait* 75: 447–452.

Riedel KHJ and Britz TJ (1993) *Propionibacterium* species diversity in anaerobic digesters. *Biodivers. Conserv.* 2: 400–411.

Riedel KHJ and Britz TJ (1996) Justification of the 'classical' *Propionibacterium* species concept by ribotyping. *System. Appl. Microbiol.* 19: 370–380.

Steffen C, Eberhard P, Bosset JO and Ruegg M (1993) Swiss-type Varieties. In: Fox PF (ed.) *Cheese: Chemistry, Physics and Microbiology.* Vol 2. *Major Cheese Groups.* 2nd edn. P. 83. London: Chapman & Hall.

Thierry A and Madec MN (1995) Enumeration of propionibacteria in raw milk using a new selective medium. *Lait* 75: 305–488.

Propionic Acid *see* **Fermentation (Industrial)**: Production of Organic Acids, e.g. Citric, Propionic; **Preservatives**: Permitted Preservatives – Propionic Acid.

PROTEUS

Bernard W Senior, Department of Medical Microbiology, University of Dundee Medical School, Ninewells Hospital, Dundee, UK

Bacteria of the genus *Proteus* are commonly encountered in nature and are frequently associated with humans where they act as opportunistic pathogens and cause a variety of infections. This article describes their characteristic features and virulence properties. The role of *Proteus* and other related organisms in food, food spoilage and as enteropathogens is considered and the methods to isolate, identify and, if necessary, type *Proteus* isolates are presented.

Characteristics of the Genus

Bacteria of the genus *Proteus* are named after the Greek deity Proteus who had the ability to assume different forms. The ability of isolates of *Proteus* spp. to do this and form a spreading growth ('swarming') over the surface of appropriate solid media is a characteristic of the genus. However, it should be remembered that this feature is not unique to *Proteus* and a similar type of growth may be given by isolates of other bacteria such as *Serratia marcescens* and *Vibrio parahaemolyticus*.

Bacteria of the genus *Proteus* belong to the tribe Proteeae (other genera include *Morganella* and *Providencia*) in the family Enterobacteriaceae. Therefore, they are Gram-negative, non-sporing, oxidase negative, facultatively anaerobic bacilli. Most isolates are actively motile, particularly when young, by peritrichate flagella. All *Proteus* strains form acid from glucose and most also form small amounts of gas. Unlike strains of other genera of the tribe, no strain of *Proteus* forms acid from mannose, mannitol, adonitol or inositol. Lactose fermentation is rare and when present is encoded by a plasmid acquired from outside the genus. All strains form catalase and an inducible urease. The G+C content of their DNA is 38–40 mol%, which is a much lower value than that of most other genera of the Enterobacteriaceae.

Strains of *Proteus* spp. (and also other members of the Proteeae) have one biochemical characteristic that distinguishes them from all other members of the Enterobacteriaceae (except the recently defined rare genera *Tatumella* and *Rhanella*) – the ability oxidatively to deaminate certain amino acids, such as phenylalanine and tryptophan to the corresponding keto acid and ammonia. The keto acid acts as a siderophore. Members of the tribe are also readily recognized by the red–brown, melanin-like pigment they form when cultured under aerobic conditions on media containing iron and an aromatic L-amino acid such as phenylalanine, tryptophan, tyrosine or histidine.

The genus *Proteus* has four species: *P. mirabilis* and *P. vulgaris* (formerly known together as *P. hauseri*), *P. penneri* and *P. myxofaciens*. The most frequently encountered species is *P. mirabilis* followed by *P. vulgaris*. *P. penneri* is rarely encountered and *P. myxofaciens* is not associated with humans.

Habitat

P. myxofaciens has been isolated only from the larvae of the gypsy moth (*Porthetria dispar*) and thus will not be considered further. The other *Proteus* spp., however, are widely distributed in nature and constitute an important part of the flora of decomposing matter of animal origin. They are constantly present in rotten meat and sewage and very frequently in the faeces of humans and animals and pests like cockroaches and flies. They are also commonly found in garden soil and on vegetables and fruit. In addition to their wide saprophytic existence, isolates of *Proteus* spp. are the cause of a number of septic infections in humans and animals.

Proteus and Food

All uncooked meats, fish, fruit, vegetables and that made with or from raw eggs or milk should be regarded as probably being infected with *Proteus* spp. Most isolates of *Proteus* spp. are proteolytic and lipolytic. In addition, *Proteus* spp. form a variety of inducible decarboxylases and aminotransferases depending on the proteinaceous nature of the food. Thus such foods that are rich in protein and fat may be spoiled if stored under inappropriate conditions.

Susceptibility to Physical and Chemical Agents

Cells of *Proteus* spp. are sensitive to heat and are readily killed by moist heat at 55°C for 1 h, by common disinfectants such as halogens, ozone and formaldehyde, and by UV and γ irradiation to which they are as sensitive as *Escherichia coli* and *Salmonella*. Exposure of *Proteus* to acid conditions (pH 3–4) for 24 h causes cell death. Some isolates of *Proteus* spp. can grow in 12–18% NaCl and survive in saturated NaCl for 5 days. *Proteus* strains have been shown to survive in frozen food at –20°C for 2–3 months although they do not grow below 4°C.

In view of these things, food which has been autoclaved correctly, or subject to high temperatures in cooking or effectively irradiated or disinfected is unlikely to bear viable *Proteus* cells.

Table 1 The major virulence factors of *Proteus* spp. and their functions

Virulence factor	*Role in pathogenicity*
Fimbriae	Enable bacteria to adhere to epithelial cells
MR/P	Adhesion to cells of upper urinary tract
MR/K	Adhesion to catheters
PMF	Adhesion to bladder cells
NAF (UCA)	Role unclear
Urease	Formation of alkaline pH, formation of bladder and kidney stones, anticomplementary activity, cytotoxic for kidney proximal tubular epithelial cells
Flagella	Movement of bacteria from bladder to kidneys
Haemolysins	Cytotoxicity and invasiveness
IgA protease	Limits effectiveness of immune response
Swarming	Coordinate induction of urease, haemolysin and IgA protease and cell invasion
Amino acid deaminases	Formation of α keto acids as siderophores
Capsular polysaccharide	Formation of biofilms and stones, translocation of swarmer cells
Endotoxin	Endotoxicity, serum resistance

Pathogenicity and Virulence

Nothing is known about the pathogenicity of *P. myxofaciens* but members of all the other *Proteus* spp. are pathogenic for humans. *P. mirabilis* is the most frequently encountered species and is responsible for *c.* 70–90% of human infections. *P. vulgaris* and *P. penneri* cause similar types of infection to *P. mirabilis* because their habitats and virulence factors are similar (see **Table 1**). However, they are isolated less frequently and they may be less virulent.

Urinary Tract Infections

The commonest site of *Proteus* infection is the urinary tract and *P. mirabilis*, the species most frequently implicated, is the organism after *E. coli* most frequently associated with urinary tract infections. *Proteus* urinary tract infections are common in young boys and the elderly. In the latter they are often associated in domiciliary patients with diabetes or structural abnormalities of the urinary tract, and in hospital patients, with various forms of urological instrumentation or manipulation. *Proteus* urinary tract infections tend to be more serious than those caused by *E. coli* and other coliforms because, although these are usually confined to the bladder, *Proteus* spp. have a predilection for the upper urinary tract where they may cause pyelonephritis.

The virulence of *Proteus* for the urinary tract arises through the interplay of several virulence factors of which the most important is urease. This nickel-containing cytoplasmic enzyme is induced solely by urea. The concentration of urea in urine is sufficiently high for the ureases of all *Proteus* spp. to be able to work at the maximum rate, that of *P. mirabilis* in particular giving the greatest rate of urea hydrolysis. Urease hydrolyses urea in urine to ammonia and carbon dioxide. This reaction may be important in supplying the bacteria with a source of usable nitrogen for growth in urine. However, the formation of ammonia also leads to the alkalinization of urine and at pH values above 8, calcium and magnesium ions are precipitated in the form of struvite and apatite crystals. These are bound by the polysaccharide slime formed by the cell to form bladder and kidney stones. The urease-induced ammonia also protects the bacterial cell from complement by inactivating it. Animal experiments have confirmed that urease is a critical virulence determinant for colonization of the urinary tract, stone formation and development of pyelonephritis. Urease together with haemolysin (see below) causes the death of human renal proximal tubular epithelial cells.

Proteus cells also form a number of different types of fimbriae (see Table 1) which play a significant but more subtle role in virulence for the urinary tract. Mannose resistant, *Proteus*-like (MRP) fimbriae are expressed in vivo, and though not essential for infection, appear to play a significant role in the colonization of both bladder and kidney and their presence correlates with the development of acute pyelonephritis. PMF (*P. mirabilis* fimbriae) are probably important for colonization of the bladder although mutants lacking them can still invade the kidney. Mannose-resistant *Klebsiella*-like (MR/K) fimbriae and uroepithelial cell adhesin (UCA) (alternatively known as NAF, non-agglutinating fimbriae) both bind to uroepithelial cells and the former also binds to Bowman's capsule of the glomeruli and tubular basement membranes of the kidney.

Most strains of *P. mirabilis* and some of the other *Proteus* spp. form a cell-associated, calcium independent haemolysin, HpmA. Some isolates of *P. vulgaris* and *P. penneri* form the calcium-dependent haemolysin, HlyA, that is very similar to the HlyA haemolysin of *E. coli*. Some *Proteus* isolates form both types of haemolysin. These haemolysins cause the lysis of a wide variety of cell types in addition to erythrocytes. Together with urease they play an important part in cell invasion and internalization and ultimately in cell death.

Most isolates of *Proteus* spp. produce a unique EDTA-sensitive metalloproteinase that cleaves at unique sites the heavy chain of immunoglobulin (Ig) A_1, IgA_2, IgG and both free and IgA-bound secretory component. The enzyme is formed in vivo and is active in patients with a *Proteus* urinary tract infection. The cleaved antibody fragments have defective immune effector functions and thereby the effectiveness of the immune response to the organism is limited. The proteinase may also play a role in generating products like glutamine which is important in inducing swarm cell formation. Experiments in vivo in mice have shown that proteinase-negative mutants can infect the bladder but have reduced ability to infect the kidney and form abscesses.

Motility and swarming are properties which are thought to be important, if not absolutely essential, in *Proteus* virulence. Antibodies to flagella have been shown to prevent infection of the kidney. The development of flagella is important in swarm cell formation. In this complex process, an environmental signal, such as that given by a viscous environment or a solid surface or glutamine, triggers the normally short (2–4 μm), sparsely flagellate, bacilli (referred to here as vegetative cells) to differentiate into swarmer cells which are multinucleate, densely flagellate, nonseptate, elongated (20–80 μm in length) cells whose enzymatic activity, antibiotic sensitivity, cell wall permeability and lipopolysaccharide composition are different from the vegetative cell. Moreover, in swarmer cell formation there is the coordinated expression of the virulence determinants, urease, haemolysin and protease. Swarmer cells have the ability, assisted by the secretion of a cell surface polysaccharide, to migrate over solid surfaces. Subsequently they cease migration and differentiate back into vegetative cells by division at several positions along the length of the cell. It has been shown that mutants lacking flagella are noninvasive to epithelial cells and motile, but non-swarming, cells are much less invasive than wild-type motile, swarming cells.

Infection of the urinary tract with *Proteus* can also give rise to hyperammonaemic encephalopathy and coma. In addition it frequently leads to bacteraemia.

Bacteraemia

Proteus bacteraemias are not uncommon and most are clinically significant. They are often hospital acquired and usually are associated with elderly people who have other underlying diseases. They can be difficult to treat and have a mortality rate of 15–88% according to the severity of the underlying disease.

Other Infections and Conditions

Proteus spp. have also been isolated in pure culture and with other organisms from various superficial lesions. Their presence in mixed culture favours the multiplication of pathogenic anaerobes. Occasionally they are isolated in pure culture from abscesses, from the meninges and from blood. Both *P. mirabilis* and *P. vulgaris* can cause osteomyelitis. In neonates, infection of the umbilical stump often leads to a highly fatal bacteraemia and meningitis. Patients with active rheumatoid arthritis often have raised antibody levels specific for *Proteus*. The reason for this may be because of shared epitopes between certain HLA types associated with the disease and particular *Proteus* antigens but the relationship between *Proteus* and rheumatoid arthritis remains unclear.

Gastroenteritis

Although there have been many reports in the past that strains of *Proteus* spp. cause diarrhoea, there is no strong evidence that isolates of *Proteus* are enteropathogens. Although some claim they are present more frequently in diarrhoeic stools than controls, the contribution of other bacterial and viral enteropathogens to the condition cannot be eliminated. Moreover, in some older reports associating *Proteus* with diarrhoea, the incriminating organism would now be classified as a species of *Morganella* or *Providencia*.

It is worth noting, however, that some species of other genera of the tribe Proteeae can be enteropathogenic. For example, *Providencia alcalifaciens* strains are found more frequently in the stools of British adults with diarrhoea who have travelled abroad than in those who have not. Recently, some *P. alcalifaciens* strains have been shown to have the ability to invade the intestinal mucosa, cause diarrhoea in rabbits, and bring about actin condensation in a manner similar to that caused by *Shigella flexneri*. The genetic determinants of invasiveness in *P. alcalifaciens* are different from those of invasive *Shigella* spp. and *E. coli* and are not plasmid-borne. In addition, *Morganella morganii* has been associated with scombroid food poisoning. Scombroid fish, such as mackerel, tuna, sardines, pilchards and anchovies may, through improper handling or storage or both, become contaminated with *M. morganii*. This organism is one of the best producers of histidine decarboxylase and it can be formed at temperatures as low as 7°C. As a result of the action of this enzyme on fish muscle, large amounts of histamine are formed and the critical amount (100 mg histamine per kilogram of food) can be formed in a short time. Ingestion of such food leads within 0.5–3 h to symptoms which may include nausea, vomiting, diarrhoea, headache, urticaria and a burning sensation in the mouth. The symptoms may persist for up to 8 h. The intoxication is not fatal and administration of antihistamines may be helpful. Cloves and cinnamon are spices that will reduce histidine decarboxylase production and decrease the likelihood of scombroid poisoning.

Detection and Isolation of *Proteus*

Proteus strains are frequently found in the intestinal tract of healthy people and animals. There is no evidence that ingestion of cells of *Proteus* spp. alone will give rise to gastroenteritis although ingestion of other organisms in the tribe such as *M. morganii* and *Providencia alcalifaciens* may. However, the presence of *Proteus* spp. in food suggests that it has been improperly prepared or stored or contaminated with faecal material after cooking. This may contain pathogens, such as *Salmonella*, *Shigella* or *Campylobacter* spp., which will give rise to gastroenteritis and dysentery. Such food should not be consumed.

Isolates of *Proteus* will grow readily on a wide variety of media in air over a wide temperature range below 42°C but optimally at 34–37°C. Culture at 22–30°C on a rich medium containing salt, such as blood agar, promotes swarming growth (**Fig. 1**). This feature is extremely strong evidence that *Proteus* is present.

To detect small numbers of *Proteus* in food, enrichment by overnight culture at 37°C in tetrathionate broth is recommended. In order to make colony counts of *Proteus* in food and subsequently to identify them, the food is emulsified and then diluted in a suitable sterile isotonic diluent, and then plated out on culture media with a dry surface that do not permit swarming growth. Swarming growth can be prevented by increasing the agar concentration of the medium to 3–4% (but this may alter the colonial morphology), by bile salts such as those in MacConkey agar or deoxycholate citrate agar (DCA) on which *Proteus* colonies appear pale as they do not ferment lactose, or

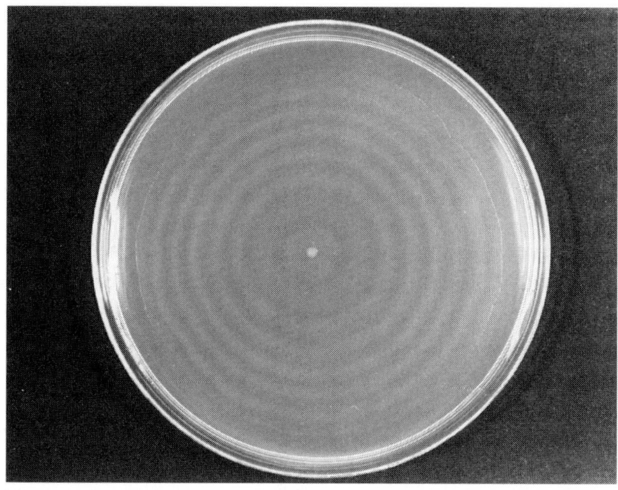

Figure 1 Typical swarming growth of *Proteus*. The culture medium was inoculated in the centre of the plate with *Proteus mirabilis* and incubated overnight at 30°C.

Urea–indole medium is prepared by supplementing aseptically, when cool, tryptone water (Oxoid CM 87) (1.5 g in 100 ml of distilled water) which has been sterilized at 121°C for 15 min with filtered sterile urea (40% w/v in water) to 2% w/v and with a 1 in 200 dilution of phenolphthalein 1% in isopropanol. The medium is dispensed aseptically in 1 ml volumes into sterile tubes.

Ornithine decarboxylase medium contains tryptone water (Oxoid CM 87) 0.5 g, L-ornithine.HCl 1 g and 2.5 ml of bromocresol purple dye 0.08% in 100 ml of distilled water. After sterilization at 121°C for 15 min and when cool, the medium is supplemented aseptically with sterile glucose (10% w/v in water) to 0.1% w/v and dispensed aseptically in 2.5 ml amounts into screw-capped bottles. The base control medium lacks ornithine but is prepared in an identical manner.

Maltose, mannose and xylose peptone water sugar media are made by supplementing aseptically, when cold, peptone water (Oxoid CM9) (1.5 g and 2.5 ml of bromocresol purple 0.08% in 100 ml of distilled water) which has been sterilized at 121°C for 15 min, with the sugar to a final concentration of 1% w/v, from a sterile (steamed for 1 h) stock solution of the sugar (10% w/v in water). These media are dispensed aseptically in 1 ml amounts into sterile tubes.

The media should be inoculated aseptically with a drop of the suspension. The inoculated ornithine decarboxylase medium and the base control medium should then be protected from the air by overlayering them with a small volume of sterile mineral oil. All the inoculated media should then be incubated at 37°C for 16–24 h in air and the reactions then read. A few drops of aqueous ferric chloride 10% are added to the phenylalanine deaminase medium and a few drops of Ehrlich's reagent to the urea–indole medium. The formation of a dark-green colour at the surface of the phenylalanine deaminase medium indicates deamination of phenylalanine and the formation of phenyl pyruvic acid. Formation of a pink colour in the urea medium indicates urease formation. Formation of a red colour in the Ehrlich's reagent above the medium indicates formation of indole. The devel-

Table 2 The important distinguishing biochemical reactions of *Proteus* spp.

Organism	*PAD*[a]	*Mannose*	*Urease*	*ODC*[b]	*Indole*	*Maltose*	*Xylose*
P. mirabilis	+	–	+	+	–	–	+
P. vulgaris	+	–	+	–	+	+	+
P. penneri	+	–	+	–	–	+	+
P. myxofaciens	+	–	+	–	–	+	–

[a]PAD, Phenylalanine deaminase test.
[b]ODC, Ornithine decarboxylase test.
+, formation of enzyme or product or acidification of sugar.
–, no formation of enzyme or product or acidification of sugar.

by the omission of salt as in cysteine lactose electrolyte deficient agar (CLED) on which *Proteus* colonies appear blue. The same method can be used to isolate other members of the Proteeae whose colonies will appear similar to those of *Proteus*.

Identification

After overnight incubation at 37°C, a single colony of pleomorphic Gram-negative bacilli that is oxidase negative and unable to ferment lactose should be picked and suspended in a small volume of sterile saline or nutrient broth. The key biochemical tests to identify the different species of *Proteus* are presented in **Table 2**. They can be made by preparation of the media as described below or use of commercially prepared media.

Media and Identifying Tests

Phenylalanine deaminase (PAD) medium contains tryptone water (Oxoid CM 87) 1.5 g, L-phenylalanine 1 g and agar 1.3 g in 100 ml distilled water. After sterilization at 121°C for 15 min, 2 ml amounts are dispensed aseptically into sterile tubes and left to solidify as a slope.

opment of a blue colour in the decarboxylase medium while the base control remains yellow indicates decarboxylation of ornithine. Fermentation of a sugar is denoted by a colour change of blue to yellow.

Results

The formation of phenylalanine deaminase is extremely strong evidence that the isolate belongs to the tribe Proteeae. The inability of *Proteus* spp. to ferment mannose distinguishes this genus from the other genera, *Morganella* and *Providencia*, of the tribe. All species of *Proteus* produce urease in large amounts. However, too much weight should not be put on this characteristic alone because some other members of the tribe and other organisms within the Enterobacteriaceae also produce urease. The ability to form ornithine decarboxylase is a property in the genus of only *Proteus mirabilis*. The ability to form indole from tryptophan is an important reaction because it is the definitive test that differentiates *Proteus vulgaris* (indole positive) from *Proteus penneri* (indole negative). *P. vulgaris* is the only *Proteus* spp. that forms indole. The term 'indole-positive *Proteus*' which is often seen in the literature, may be a misnomer for any lactose-negative indole- and urease-forming member of the Enterobacteriaceae and therefore could represent several different bacteria. The only true 'indole-positive *Proteus*' is *P. vulgaris*. Isolates of *P. vulgaris* can be divided into two biotypes. Biotype 2 strains acidify salicin and degrade aesculin whereas biotype 3 strains do neither of these reactions. Maltose fermentation is carried out by all species of *Proteus* other than *P. mirabilis* and all *Proteus* spp. except *P. myxofaciens* ferment xylose. Many isolates of *Proteus* spp. produce proteolytic and lipolytic enzymes.

Other important features characteristic of the different *Proteus* spp. and the features which distinguish them from other members of the Proteeae, some of which may act as enteropathogens, are presented in **Table 3**.

Most commercial identification systems give good (> 90% accurate) identification of *Proteus* spp. Most misidentifications label them as *Morganella morganii*, *Providencia rettgeri* or *Providencia stuartii*.

Proteus typing

If there is an outbreak of infection involving *Proteus* spp. it may be necessary to identify the strain(s) involved by typing. If only a few isolates are to be investigated, the simplest and most rapid method is Dienes typing. In this method, different strains of *Proteus* spp. are allowed to swarm towards each other. A line (called a Dienes line) of complete or partially inhibited growth forms where the spreading growths of incompatible strains meet. Such a line does not form between identical strains (**Fig. 2**). When larger numbers of isolates are involved, phage typing or serotyping methods can be used. Some 49 O antigens and 19 H antigens have been defined for *P. mirabilis* and *P. vulgaris*. Most O antigens are species specific but some are common to both species. However, the most discriminating typing method is that of bacteriocin typing (P/S typing) in which determinations are made of the type of bacteriocin (proticine) produced by a strain (the P type) and the sensitivity (the S type) of the strain to 13 different standard proticine preparations. Strains of the same P/S type, irrespective of their O and H serotypes, show compatibility in the Dienes test whereas strains of different P/S types are incompatible.

See also: **Bateriocins**: Potential in Food Preservation; Nisin. ***Campylobacter***: Introduction; Detection by Cultural and Modern Techniques; Detection by Latex Agglutination Techniques. ***Escherichia coli***: *Escherichia coli*. **Fermentation (Industrial)**: Basic Considerations. **Fish**: Spoilage of Fish. **Microflora of the Intestine**: The Natural Microflora of Humans. **Preservatives**: Traditional Preservatives – Oils and Spices; Traditional Preservatives – Sodium Chloride. ***Salmonella***: Introduction. ***Serratia***.

Further Reading

Albert MJ, Alam K and Ansaruzzaman M et al (1992) Pathogenesis of *Providencia alcalifaciens*-induced diarrhea. *Infection and Immunity* 60: 5017–5024.

Belas R (1996) *Proteus mirabilis* swarmer cell differentiation and urinary tract infection. In: Mobley HLT and Warren JW (eds) *Urinary Tract Infections. Molecular Pathogenesis and Clinical Management*. P. 271. Washington, DC: ASM Press.

Loomes LM, Senior BW and Kerr MA (1990) A proteolytic enzyme secreted by *Proteus mirabilis* degrades immunoglobulins of the immunoglobulin A_1 (IgA_1), IgA_2, and IgG isotypes. *Infection and Immunity* 58: 1979–1985.

Loomes LM, Kerr MA and Senior BW (1993) The cleavage of immunoglobulin G in vitro and in vivo by a proteinase secreted by the urinary tract pathogen *Proteus mirabilis*. *Journal of Medical Microbiology* 39: 225–232.

Mobley HLT (1996) Virulence of *Proteus mirabilis*. In: Mobley HLT and Warren JW (eds) *Urinary Tract Infections. Molecular Pathogenesis and Clinical Management*. P. 245. Washington, DC: ASM Press.

Rodriguezjerez JJ, Lopezsabater EI, Roigsagues AX and Moraventura MT (1994) Histamine, cadaverine and putrescine forming bacteria from ripened Spanish semi-preserved anchovies. *Journal of Food Science* 5: 998–1001.

Rozalski A, Sidorczyk Z and Kotelko K (1997) Potential

Table 3 The major distinguishing biochemical activities of members of the Proteeae

Organism	*PAD*[a]	*Mannose*	*ODC*[a]	*Urease*	*Indole*	*Trehalose*	*Maltose*	*Adonitol*	*Xylose*
Proteus mirabilis	+	−	+	+	−	+	−	−	+
Proteus vulgaris	+	−	−	+	+	V	+	−	+
Proteus penneri	+	−	−	+	−	V	+	−	+
Proteus myxofaciens	+	−	−	+	−	+	+	−	−
Morganella morganii	+	+	+	+	+	V	−	−	−
Providencia rettgeri	+	+	−	+	+	−	−	+	−
Providencia alcalifaciens	+	+	−	−	+	−	−	+	−
Providencia rustigianii	+	+	−	−	+	−	−	−	−
Providencia stuartii	+	+	−	(−)	+	+	−	(−)	−
Providencia heimbachae	+	+	−	−	−	−	V	+	−

[a]PAD, phenylalanine deaminase; ODC, ornithine decarboxylase.
+, formation of enzyme or product or acidification of sugar.
−, no formation of enzyme or product or acidification of sugar.
(), reaction of most isolates.
V, different isolates give different results.
In addition, the following properties are common to all members of the Proteeae: motility, inability to ferment dulcitol, lactose, sorbitol, raffinose and arabinose, lysine and arginine decarboxylase negative, malonate negative and mucate negative.

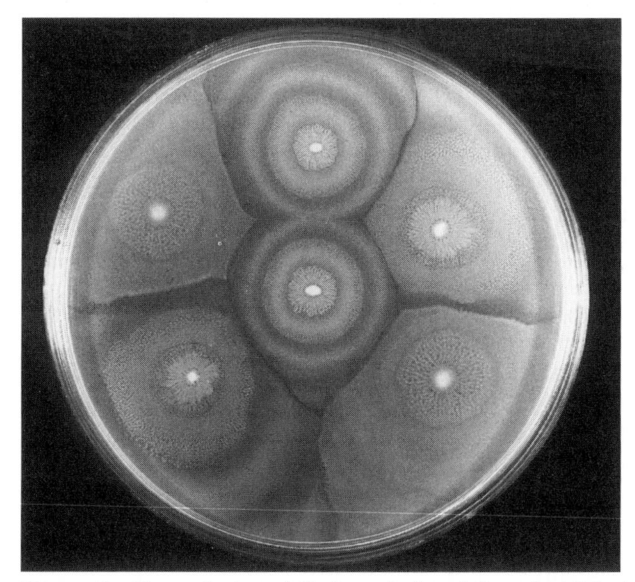

Figure 2 Dienes typing of *Proteus* strains. A line of inhibited growth forms only where the swarming growths of different strains meet. Therefore the strain at the top of the plate is identical to the one in the centre and each one of the remaining strains is different from its immediate neighbour.

virulence factors of *Proteus* bacilli. *Microbiology and Molecular Biology Reviews* 61: 65–89.

Senior BW (1977) The Dienes phenomenon: identification of the determinants of compatibility. *Journal of General Microbiology* 102: 235–244.

Senior BW (1997) Media and tests to simplify the recognition and identification of members of the Proteeae. *Journal of Medical Microbiology* 46: 39–44.

Senior BW (1998) *Proteus*, *Morganella* and *Providencia*. In: *Topley and Wilson's Microbiology and Microbial Infections*, 9th edn. Vol. 2, Balows A and Duerden BI (eds) *Systematic Bacteriology*. Ch. 43, p. 1035, London: Arnold.

Senior BW and Hughes C (1988) Production and properties of haemolysins from clinical isolates of the Proteeae. *Journal of Medical Microbiology* 25: 17–25.

Senior BW and Larsson P (1983) A highly discriminatory multi-typing scheme for *Proteus mirabilis* and *Proteus vulgaris*. *Journal of Medical Microbiology* 16: 193–202.

Senior BW, Loomes LM and Kerr MA (1991) The production and activity in vivo of *Proteus mirabilis* IgA protease in infections of the urinary tract. *Journal of Medical Microbiology* 35: 203–207.

Senior BW, McBride PDP, Morley KD and Kerr MA (1995) The detection of raised levels of IgM to *Proteus mirabilis* in sera from patients with rheumatoid arthritis. *Journal of Medical Microbiology* 43: 176–184.

Swihart KG and Welch RA (1990) The HpmA hemolysin is more common that HlyA among *Proteus* isolates. *Infection and Immunity* 58: 1853–1860.

PSEUDOMONAS

Contents

Introduction

M A Cousin, Department of Food Science, Purdue University, West Lafayette, USA

Characteristics of Genus

The Pseudomonadaceae is a family of Gram-negative bacteria that now includes four genera: *Frateuria*, *Pseudomonas*, *Xanthomonas* and *Zoogloea*. Species of these genera are common saprophytes in both fresh and marine waters and soil and are pathogenic to plants, animals and humans. The four genera are distinguished from one another by the characteristics listed in **Table 1**. *Pseudomonas* species are metabolically diverse aerobic bacteria that degrade a variety of low-molecular-weight organic compounds and hydrolytic products generated by actinomycete and fungal growth. They are aerobic with respiratory metabolism where oxygen is the terminal electron acceptor, motile by one or more polar flagella, and straight to curved rods that are 1.5–5.0 μm long with a 0.5–1.0 μm diameter. They are chemoorganotrophic with some species being chemolithotrophic using hydrogen or carbon dioxide for energy sources. They do not require organic growth factors and accumulate carbon reserves of poly-β-hydroxybutyrate. They are catalase-positive and usually oxidase-positive. They are not very acid tolerant and rarely grow below pH 5.0–6.0. The genus now has over 100 species; however, the use of genetic and molecular biological tools such as 16S and 23S rRNA gene sequences, rRNA–DNA and DNA–DNA hybridization, and cellular lipid and fatty acid profiles has resulted in some species being transferred into new genera, such as *Acidovorax*, *Burkholderia*, *Comamonas*, *Deleya*, *Flavimonas*, *Oceanospirillum* and *Ralstonia*. As more information about *Pseudomonas* species is learned from these molecular techniques, there will undoubtedly be new changes in both the genus and its species. Therefore, some names that appear in the literature as *Pseudomonas* species are no longer recognized in the genus.

Many papers on the taxonomy of *Pseudomonas* species have been published since the genus was first proposed over 100 years ago. In some cases, this genus became the home for any Gram-negative aerobic rod that had polar flagella. This has resulted in many misidentified species and difficulty in assigning new isolates to a given species. *Pseudomonas* species have plasmids that encode for enzymes that degrade many compounds that other bacteria cannot and for antibacterial resistance and related toxic factors. Some of these plasmids could have been transferred to other Gram-negative bacteria through conjugation; therefore, the gene pool probably has become less distinct over the years. The diversity of the species within the genus and the problems with creating a good classification scheme for *Pseudomonas* species meant that identifying isolates to the species level was difficult. Although an entire issue of volume 19 of *Systemic and Applied Microbiology* was devoted to the taxonomy of *Pseudomonas* species, the issue has not been completely resolved. Hence, the assignment of species names to the genus *Pseudomonas* by researchers in the food microbiology field may no longer be valid. The names as they appear in the literature with changes to new names that may be more appropriate are used in the following sections.

Species of *Pseudomonas* Identified from Foods

Pseudomonas species are very common in fresh foods because of their association with water, soil and vege-

Table 1 Characteristics that differentiate the genera of the Pseudomonadaceae

Genera	*Growth factors required*	*Growth at pH 3.6*	*Xanthomonadins*	*Dendritic flocs*	*Oxidase*
Frateuria	–	+	–	–	–
Pseudomonas	–	–	–	–	+
Xanthomonas	+	–	+	–	– (+)[a]
Zoogloea	+	–	–	+	+

[a] Weakly positive.

Table 2 Morphological and biochemical properties of some *Pseudomonas* species important in foods

Species	*Oxidase*	*Flagella*	*Grow at 4°C*	*Grow at 41°C*	*Denitrify*	*Arginine dihydrolase*	*Pigment production*	*Cell size (width × length) (μm)*
P. aeruginosa	+	1	−	+	+	+	Fluorescent, pyocyanin	0.5–0.7 × 1.0–3.0
B. cepacia[b]	+	>1	−	±	−	−	Diffusible colours	0.8–1.0 × 1.6–3.2
P. cichorii	(+)[a]	>1	−	−	−	−	Fluorescent	0.8 × 1.2–3.5
P. fluorescens	+	>1	+	−	±	+	Fluorescent	0.7–0.8 × 2.0–3.0
P. fragi	+	1	+	−				0.5–1.0 × 0.8–4.0
B. gladioli[b]	+	>1	−	+	−	−	Diffusible colours	0.8 × 2.0
P. maltophilia	−	>1	−	−	−	−	Yellow noncarotenoid	0.5 × 1.5
P. marginalis	+	1–3	+	−			Fluorescent green	
S. mephitica[c]	+	1	+	−				0.5–1.0 × 1.5–14.0
P. putida	+	>1	+	−	−	+	Fluorescent	0.7–1.1 × 2.0–4.0
P. syringae	−	>1	±	−	−	−		0.7–1.2 × 1.5
P. tolaasii	+	1–5	+	−			Fluorescent green or blue	0.4–0.5 × 0.9–1.7
P. viridiflava	−	1–2	−	−	−	−	Yellow to blue–green insoluble	0.8–1.5 × 1.5–3.2

[a] () weak reaction.
[b] Renamed *Burkholderia* to include plant or animal pathogens.
[c] Renamed *Stenotrophomonas*.

tation. They commonly contaminate eggs, meat, milk, poultry, seafoods and vegetables. Many species are psychrotrophic and are, therefore, important spoilage microorganisms in refrigerated foods. Since *Pseudomonas* species do not require organic growth factors, they will normally outcompete Gram-positive bacteria in these refrigerated foods. Also, they can use a variety of non-carbohydrate compounds for energy and degrade food components such as acids, amino acids, lipids, pectin, peptides, protein and triacylglycerols that are common in animal and plant foods.

Pseudomonas species have been isolated from many foods over the years; however, the species have not always been identified because of the difficulties in classification. There are many research papers that simply refer to unidentified *Pseudomonas* species rather than to specific species. The properties of some species that have been identified and that are important in foods are presented in **Table 2**. Most of these species names have been maintained in the latest reclassification of the pseudomonads; others have been renamed as designated in the table.

Methods of Detection

Generally, it is not important to detect *Pseudomonas* species in foods unless they are causing serious spoilage problems. The psychrotrophic count will detect many *Pseudomonas* species that cause spoilage of refrigerated foods. Agars that incorporate lipid, pectin or protein can be used to detect *Pseudomonas* or other psychrotrophs that are responsible for specific refrigerated food spoilage. Information on these methods can be found in the *Compendium for the Microbiological Examination of Foods*. If specific detection of *Pseudomonas* species is required, then an agar that has a broad-spectrum antimicrobial agent, such as Irgasan that does not affect *Pseudomonas* species, can be used.

Importance to the Food Industry

Pseudomonas species produce proteases, lipases and pectinases in the late logarithmic phase of growth and these enzymes are used to degrade food components (**Table 3**). *Pseudomonas* species produce pectinases that degrade pectin, the middle adhesive layer that holds the primary and secondary cellulosic walls of plants together. Because cellulose is not degraded by *Pseudomonas* species, there must be damage to the plant cell wall to allow bacterial entry to cause pectin degradation that results in soft rot (soft, mushy texture that looks water-soaked). Pectin is broken down into an intermediate that can enter the Entner–Doudoroff pathway to produce energy for the cell. By-products can result in off odours and flavours and breakdown of tissues in the vegetables.

Pseudomonas species need proteases, peptidases and related enzymes when growing in eggs, fish, meat, milk and poultry because there are very few residual or utilizable carbohydrates. Most *Pseudomonas* species cannot use lactose in milk; hence, they must be able to obtain carbon from protein. Many *Pseudomonas* species, such as *P. fluorescens*, *P. fragi* and *P. maltophilia*, can degrade the globular casein proteins in milk, especially the β-, α_{s1}- and κ-caseins, but not the whey proteins. Many of the proteases produced by *Pseudomonas* species are stable to heat and survive ultra-high temperature (UHT) processing; thereby, causing problems in long shelf-life dairy products (cheeses, 'sterilized' shelf-stable milks). Some *Pseudo-*

Table 3 Important properties of some *Pseudomonas* species involved in food spoilage

Species	*Levan*	*Lipase*	*Pectinase*	*Protease*	*Amylase*	*Psychrotroph*	*Presence in foods*
P. aeruginosa	–	+	–	+	–	–	Pathogen sometimes isolated
B. cepacia[b]	±	+	+	±	–	–	Plant associated
P. cichorii	–	–	+	–	–	–	Vegetable soft rots
P. fluorescens	±	±	+	+	–	±	Egg, meat, milk, vegetable
P. fragi		+		(+)	+	–	Meat, milk, seafood spoilage
B. gladiolli[b]	+	+	+	+	–	–	Onion soft rots
P. maltophilia		+		+	–	–	Proteases and lipases in milk
P. marginalis	–		+	+		+	Vegetable soft rot
S. mephitica[c]		(+)[a]	–	±	(+)	+	Skunk-like odour in butter
P. putida	–	±	–	–	–	±	Meat and milk spoilage
P. syringae	±	±	±	±	–	–	Vegetable soft rots
P. tolaasii				+	±	+	Mushroom spoilage
P. viridiflava	–	–	+	+	–	–	Vegetable soft rots

[a] () weak reaction.
[b] Renamed *Burkholderia* to include plant or animal pathogens.
[c] Renamed *Stenotrophomonas*.

monas species have generation times as low as 8–12 h at 3°C in milk. In meats, *Pseudomonas* species use the residual glucose, lactic acid, free amino acids and nucleotides for energy; therefore, they usually will not degrade the intact fibrous muscle proteins to any great extent. When populations reach 10^8 cfu cm^{-2} where spoilage is evident, growth can continue only to 10^{9-10} cfu cm^{-2}. Weakening of the protein bands in sarcoplasmic proteins, disruption of the actin/myosin myofibrillar proteins, and decrease in stromal proteins, especially elastin, have been observed after growth of *Pseudomonas* species; however, degradation of the fibrous structures is not evident. In finfish, *Pseudomonas* species use lactic acid, amino acids, nucleotides and trimethylamine oxide for carbon before they breakdown the protein at the amino- or carboxyl-terminal ends. This does not occur until the counts exceed 10^6 cfu cm^{-2}. Crustacean and molluscan shellfish contain 0.5% and 3.5–5.5% carbohydrate, respectively, and more free amino acids than finfish; therefore, fermentative bacteria usually cause spoilage before respiratory bacteria grow. Off odours and flavours (acids, amines, ammonia, hydrogen sulphide, mercaptans and other compounds) in meats and fish begin to appear at 10^6–10^8 cfu cm^{-2} and surface slime (coalescence of cells) begins at about 10^8 cfu cm^{-2}. Eggs have barriers to prevent microbial degradation, namely, the cuticle, shell and membranes plus chemical inhibitors (lysozyme, conalbumin, etc.). However, once they penetrate the pores of the eggshell, *Pseudomonas* species spoil eggs because they are motile and can swim through the egg white to the nutrient- and protein-rich yolk. The α-ketoacids (pyruvate, α-ketoglutarate, succinate, etc.) enter the tricarboxylic acid (TCA) cycle to yield energy for the continued growth of *Pseudomonas* species in these foods.

Pseudomonas species produce lipases that can be used to degrade fat in milk and in fish with high lipid contents (herring, mackerel, salmon); however, the adipose tissue in meat and poultry is insoluble fat that is not available for microbial growth. Lipases produced by *Pseudomonas* species can also be heat stable and cause problems similar to those discussed with proteases. The lipases selectively cleave the triglyceride at the 1 and 3 positions producing free fatty acids and 2-monoglycerides. Glycerol can enter the Entner–Doudoroff pathway of fatty acids and acetyl-CoA can go into the tricarboxylic acid (TCA) cycle to yield energy for continued growth. Rancidity from C_4 to C_6 fatty acids, soapy flavours from higher-molecular-weight fatty acids, fruity flavours from esterified free fatty acids, and cardboardy flavours from unsaturated fatty acids oxidized to ketones and aldehydes are some of the defects attributed to lipolysis.

Slime polymers can be produced by *Pseudomonas* species using a disaccharide, such as sucrose, because one monosaccharide is used for energy production and the other, for a polymer. Levans are polymers of fructose with β-(2→6) linkages that are formed as a defence or as food reserves. Levan formation results in food having a water-soaked appearance and feeling slimy to the touch.

Importance to the Consumer

The growth of *Pseudomonas* species in foods results in spoilage. Food that is spoiled is not harmful to eat; however, it will have a lower quality. Usually foods will have flavour, odour and textural changes that will cause consumers not to eat them. Although *Pseudomonas aeruginosa* is a medical pathogen, it is not generally associated with food-borne illness. Consumers can protect eggs, fish, meat, milk, poultry and vegetables from degradation by *Pseudomonas* species by properly refrigerating or freezing, preparing and cooking the food before consumption. Raw vegetables, once they are chopped, are subject to pectin degradation and should be properly packaged, refrigerated and used quickly to avoid spoilage. There will always be *Pseudomonas* species associated with fresh foods because they are found in many environments and degrade many organic compounds.

See also: **Heat Treatment of Foods**: Ultra-high Temperature (UHT) Treatments. **Meat and Poultry**: Spoilage of Meat. **Milk and Milk Products**: Microbiology of Liquid Milk. ***Pseudomonas***: *Pseudomonas aeruginosa*. ***Xanthomonas***.

Further Reading

Doyle MP, Beuchat LR and Montville TJ (eds) (1997) *Food Microbiology. Fundamentals and Frontiers*. Washington, DC: ASM Press.

Holloway BW (1996) *Pseudomonas* Genetics and Taxonomy. In: Nakazawa T, Furukawa K, Haas D and Silver S (eds) *Molecular Biology of Pseudomonads*. P. 22. Washington, DC: ASM Press.

Jay JM (1996) *Modern Food Microbiology*, 5th edition. New York: Chapman & Hall.

Kersters K, Ludwig W, Vancanneyt M et al (1996) Recent changes in the classification of pseudomonads: an overview. *System. Appl. Microbiol.* 19: 465–477.

Kraft AA (1992) *Psychrotrophic Bacteria in Foods. Disease and Spoilage*. Boca Raton: CRC Press.

Krieg NR and Holt JG (eds) (1984) *Bergey's Manual of Systemic Bacteriology*. Vol. 1. Baltimore: Williams & Wilkins.

McKellar RC (1989) *Enzymes of Psychrotrophs in Raw Foods*. Boca Raton: CRC Press.

Palleroni NJ (1992) Introduction to the Family Pseudomonadaceae. In: Balows A, Trüper HG, Dworkin M, Harder W and Schleifer KH (eds) *The Prokaryotes*, 2nd edn. P. 3071. New York: Springer-Verlag.

Palleroni NJ (1993) *Pseudomonas* classification. A new case history in the taxonomy of Gram-negative bacteria. *Antonie van Leeuwenhoek* 64: 231–251.

Vanderzant C and Splittstoesser DF (eds) (1992) *Compendium of Methods for the Microbiological Examination of Foods*, 3rd edn. Washington, DC: American Public Health Association.

Pseudomonas aeruginosa

Marjon H J Bennik, Agrotechnological Research Institute (ATO-DLO), Wageningen, The Netherlands

Characteristics of the Species

Pseudomonas aeruginosa is a Gram-negative rod (0.5–0.8 × 1.5–8 µm), which is mobile by polar flagella, and occurs singly, in pairs, or in short chains. This organism has a strictly respiratory type of metabolism with oxygen as the terminal electron acceptor, although nitrate can be used as an alternate electron acceptor. The optimum growth temperature of *P. aeruginosa* is 37°C. Growth occurs at temperatures as high as 42°C, but not at 4°C. It does not require organic growth factors, and is able to multiply on a wide range of substrates (⩾82 organic compounds). In addition to its nutritional versatility, this organism is able to produce a wide variety of extracellular enzymes and an extensive slime layer, which can confer resistance to various antimicrobial agents. *P. aeruginosa* has a relatively large chromosome (5.9 Mb; mol% G+C 67%), and it is known to harbour plasmids that can readily be exchanged.

P. aeruginosa is generally considered to be an ubiquitous bacterium that is common in moist environments with low nutrient availability and low ionic strength. It can be isolated from surface water, soil and vegetation, including vegetables and salads. Despite the presence of this bacterium in drinking water and on vegetables, it has rarely been associated with food-borne diseases. However, *P. aeruginosa* is a typical example of an opportunistic pathogen for humans; it does not attack normal healthy tissue, but can cause serious infection when tissue is damaged prior to exposure to virulent clones. Such clones, which constitute only a small percentage (1–2%) of environmental isolates, can synthesize extracellular products that are thought to play a role in its complex pathogenesis. These include a potent endotoxin or lipopolysaccharide, ADP-ribosyltransferase toxins (exotoxin A and exoenzyme S), haemolysins, phospholipase C (a haemolysin), cytotoxins, proteases, including two with elastase activity, and adherence factors (pili) (**Table 1**). In addition, the production of an alginate slime layer is a significant adaptation that protects *P. aeruginosa* against a wide range of challenges.

Methods of Detection

There is no standard detection method for *P. aeruginosa*. Selective isolation media which contain components to screen for this organism (such as cetrimide

Table 1 Extracellular products of *Pseudomonas aeruginosa*

Extracellular product		*Function*
Proteases	Metalloproteases E	Breakdown of proteins; release of substrates
	Metalloprotease AP	
	Elastase	Cleavage of transferrin: iron release
Phospholipase	Phospholipase C	Disruption of phospholipids in membranes and surfactants; haemolytic activity
Siderophores	Pyochelin	Growth in iron-limiting conditions
	Pyoverdin	
ADP-ribosyl transferase toxins	Exotoxin A	Inhibition of protein synthesis
	Exoenzyme S	
Mucoid exopolysaccharide	Alginate	Adhesion, protection of cells from antibiotics, disinfectants, biofilm formation
Endotoxin	Lipopolysaccharide (LPS)	Major component of bacterial cell wall
Phenazine pigment pyocyanin		Strong reducing potential
Rhamnolipid		Biosurfactant: non-enzymatic haemolysin and cytolysin; strong reducing potential

Table 2 Characteristics differentiating *Pseudomonas* species (percentage of strains positive in test)

	P. aeruginosa	P. fluorescens, P. chlororaphis, P. aureofaciens	P. putida	P. syringae	P. viridiflava	P. cichorii	P. stutzeri	P. mendocina	P. alcaligenes	P. pseudo-alcaligenes
Utilization of										
Glucose	+[a]	+	+	+	+	+	+	+	–	–
Trehalose	–[a]	+	–	–	–	–	–	–	–	–
2-Ketogluconate	+	+	+	–	–	–	–	–	–	–
meso-Inositol	–	+	–	v[a]	+	v	–	–	–	–
Geraniol	+	–	–	–	–	–	–	+	–	–
L-*Valine*	v	+	+	–	–	–	+	+	–	–
β-*Alanine*	+	+	+	–	–	–	–	+	v	v
DL-*Arginine*	+	+	+	v	+	+	–	+	+	+
Fluorescent pigments	v	+	+	+	+	+	–	–	--	–
Pyocyanin	v	–	–	–	–	–	–	–	–	–
Carotenoids	–	–	–	–	–	–	–	+	v	–
Growth at 41°C	+	–	–	–	–	–	v	+	+	+
Levan formation from sucrose	–	v	–	v		–	–	–	–	–
Arginine dihydrolase	+	+	+	–	–	–	–	+	+	v
Oxidase reaction	+	+	+	–	–	+	+	+	+	+
Denitrification	+	v	–	–	–	–	+	+	+	v
Gelatin hydrolysis	+	+	–	v	+	–	–	–	v	v
Starch hydrolysis	–	–	–	–	–	–	+	–	–	–
Number of flagella	1	>1	>1	>1	1–2	>1	1[b]	1[b]	1	1

[a] +, >90% of strains are positive in the test; –, <10% of the strains are negative in the test; v, variable results: 11–89% of the strains are positive in the test.
[b] Lateral flagella of short wavelength may also be produced.
Data from *Bergey's Manual of Systematic Bacteriology* (1984).

agar, King's agar, *Pseudomonas* CN agar, *Pseudomonas* CFC agar) each have their own sensitivity and selectivity, and typical colonies on these media need to be confirmed by additional testing. Characteristics differentiating *Pseudomonas* species are given in **Table 2**. The most distinguishing features of *P. aeruginosa* as compared with other *Pseudomonas* species are growth in the presence of nitrofurantoin (NF) and growth at 42°C. These characteristics form the basis of the detection method described by Mossel et al that is described in detail (see below). Another useful source describing the detection of *P. aeruginosa* is the APHA (American Public Health Association, 1981; Standard Methods for the Examination of Water and Wastewater).

Detection of *P. aeruginosa*

When high numbers of *P. aeruginosa* are present in foods, the organism can be detected by direct plating of food homogenates (tenfold serial dilutions in 0.1%

peptone water) onto selective agar media. However, when cells are injured, or when only low numbers are present, direct plating may not be suitable. Injured cells can be recovered in Tryptone Soy Yeast Broth (TSYB: 17 g tryptone, 3 g soy peptone, 3 g yeast extract, 2.5 g glucose, 5 g sodium chloride, 2.5 g dipotassium phosphate per litre; pH 7.3) by mixing the sample with TSYB (1 : 10 dilution) in an Erlemeyer that allows for sufficient aeration of the mixture, and incubation for 6 h at 30°C. Subsequent enrichment of *P. aeruginosa* can be achieved by adding equal amounts of twofold-concentrated nitrofurantoin (NF) broth (see below) and incubation at 42°C.

For detection in water, the sample can be concentrated by using hydrophobic grid membrane filters, and incubation of the membranes on selective agar medium (NF agar medium; see below). A representative number of colonies from each filter need subsequent confirmation by further tests.

NF broth (1×) is prepared by adding 5 ml of an NF stock solution (0.2 g NF in 100 ml polyethylene glycol 400) to 100 ml of sterile broth (15 g tryptone, 5 g NaCl per litre, pH 7.2). Turbid enrichments are streaked onto NF agar medium and again incubated at 42°C. NF agar medium is prepared by adding 5 ml of the NF stock solution to 100 ml of sterilized medium (20 g peptone, 10 g potassium sulphate, 1.4 g magnesium chloride, 15 g agar per litre, pH 7.0) after cooling to 48°C. Single colonies need confirmation by performing the tests described below.

Confirmation of Single Colonies as *P. aeruginosa*

1. Stab the colony in diagnostic tubes for gram-negative bacteria (GN tubes) to test for pigment formation, mobility, oxidative utilization of glucose and the absence of H_2S production. The medium in GN tubes consists of three different agar layers, each ca. 2 cm thick. The lower layer consists of violet red bile glucose (VRBG) agar, the middle layer of a 2% agar layer in water, and the upper layer of 20 g tryptone, 6.1 g soy peptone, 5 g sodium chloride, 0.2 g ferric ammonium sulphate, 0.2 g sodium thiosulphate, 3.5 g agar per litre, pH 7.3. After incubation for 1–2 days at 37°C, the upper layer should show green pigment formation, growth outside of the stabbing area (mobility positive) and absence of black discoloration (no H_2S production). Purple discoloration and precipitation, indicative of fermentative utilization of glucose, should be absent in the lower layer.
2. Stab the colony in VRBG agar to test for oxidative utilization of glucose. Purple discoloration in only the upper part of the medium after 1–2 days of incubation at 37°C indicates oxidative utilization of glucose. This test should be positive.
3. Inoculate the colony onto slants of cetrimide naladixic acid (CN) agar and incubate at 42°C for 1–2 days, after which growth should occur. CN agar is prepared by adding 1 ml of CN supplement (400 mg cetrimide, 30 mg naladixic acid per 8 ml of H_2O; filtersterilized) to 250 ml of *Pseudomonas* agar base (16 g gelysate peptone, 10 g caseinate hydrolysate, 10 g monopotassium phosphate, 1.4 g magnesium chloride, 10 ml glycerol, 15 g agar per litre, pH 7.1) after sterilization.
4. *P. aeruginosa* is oxidase positive; an oxidase reaction can be performed using commercially available oxidase strips.

P. aeruginosa can be further classified into several 'types'. This internal classification of the species is useful for epidemiological purposes to trace the origin of infections. Traditional methods are based on bacterial phenotypes and include lipopolysaccharide serotyping using polyclonal or monoclonal antibodies, flagellar antigen serotyping, antimicrobial susceptibility profiles, biotyping based on the utilization of various substrates and the production of certain enzymes, pyocin typing, and bacteriophage susceptibility. Molecular typing methods were developed more recently on the basis of more stable genetic markers. This typing results are not dependent on variable expression of a particular phenotype. These methods include: plasmid content and fingerprinting, DNA fingerprinting, Southern analysis using DNA probes hybridizing to the upstream region of endotoxin A or the pilin gene, and pulsed field electrophoresis.

Regulation

The normal risk to the general human population from *P. aeruginosa* is not significant. Food has no regulations limiting this species, and examination of all drinking-water types is not recommended as a routine procedure. Although *P. aeruginosa* is regulated in bottled water in the European Community (The Council of the European Community of 15 July 1980, Official Journal of the European Communities, No. 1, p. 229) this is not the case in tap water or bottled water in the US.

Because of the ubiquitous nature of the species, it is not practical to eliminate the species from food and drinking water. The presence of *P. aeruginosa* in foods and water does not justify specific attention by healthy people, but in hospital environments special precautions would be justified for this opportunistic pathogen, especially for immunocompromised individuals (see below).

Importance to the Consumer

P. aeruginosa rarely causes problems in healthy humans, but it has been implicated in a number of cases of epidemic diarrhoea of infants. However, the reported outbreaks were related to the consumption of infant food formulas that may have been prepared with contaminated water and it is not clear if this organism was the direct cause of the observed symptoms. Because of the ubiquitous nature of *P. aeruginosa*, humans are in constant contact with it. Furthermore this organism can be found and consumed in foods, particularly in vegetables. Many types of vegetables are contaminated with *P. aeruginosa*, including tomatoes, radish, celery, cucumber, onions and lettuce. *P. aeruginosa* has been isolated from vegetable salads with counts up to 10^3 cells per gram. In healthy volunteers, ingestion of more than 10^6 cells resulted in colonization of the gut and excretion of cells for up to 6 days, in the absence of clinical symptoms. Ingestion of up to 10^8 cells did not cause diarrhoea, indicating that an exceptionally large infectious dose is necessary to colonize the intestine.

P. aeruginosa can cause serious diseases, but only in well-defined groups of patients. In patients suffering from wounds caused by burning (third-degree burns), *P. aeruginosa* can rapidly colonize skin where blood vessels are exposed, possibly as a result of specific virulence factors, such as elastase production. A second group of patients that are at risk of *P. aeruginosa* infections are those with cystic fibrosis (CF). *P. aeruginosa* strains from CF patients are frequently mucoid, and capable of forming biofilms that coat the lung surfaces of such patients, leading to clogging of the airways by mucus and debris. This organism is the main cause of morbidity and mortality in CF patients as a result of progressive decline of pulmonary function. Other infections with *P. aeruginosa* are mainly associated with instilled medical devices that present a surface on which the organism can grow to high numbers (e.g. catheters, endotracheal tubes). Clearly, thorough decontamination and maintenance of medical devices is essential.

Importance to the Food Industry

P. aeruginosa is well known for its ability to form biofilms. A biofilm is a community of microbes embedded in an organic exopolymer matrix, adhering to a surface. The production of an alginate ('slime') layer is a key feature in biofilm formation by *P. aeruginosa*. Moreover, it was recently reported that cell-to-cell communication through a specific homoserine lactone plays a crucial role in differentiation of *P. aeruginosa* in biofilms. Cells embedded in a biofilm receive less oxygen and fewer nutrients than cells in suspension. This leads to reduced growth rates and an altered physiology, which in turn can result in a higher resistance to toxic agents, such as antibiotics and disinfectants.

In the clinical environment, *P. aeruginosa* biofilms can form on devices that are difficult to clean, for example on instruments, catheters, lung inserts, and several other surfaces (implants, contact lenses). In these situations, this organism is an important cause of hospital-acquired infection.

In the food industry, the attachment of bacteria and subsequent build up of biofilms can cause a variety of problems. Many bacteria, including *P. aeruginosa*, have been shown to be capable of attachment to materials commonly used in this industry such as stainless steel, rubber and Teflon®. Biofilm formation on food-processing equipment, such as stirrers in a mixing operation, can increase the frictional resistance and costs of the process, and may lead to corrosion of pipelines. Moreover, biofilms may pose a health hazard by harbouring pathogenic bacteria. Although *P. aeruginosa* is not considered a hazardous bacterium for healthy individuals, it can enhance the survival of pathogenic species in mixed culture biofilms. This can be a source of contamination by shedding these microorganisms into the surrounding foods.

Bacteria in biofilms will likely survive on surfaces of food-processing equipment if the cleaning, i.e. physical removal of biofouling deposits, and disinfection are inadequate. To avoid biofilm formation and build-up, materials should be used that are smooth, nonporous, and resistant to wear. Proper equipment design is essential to enable efficient mechanical cleaning, and areas should be avoided in which organic material can accumulate, such as inaccessible corners or cracks. The use of appropriate disinfectants, short intervals between cleaning routines, and thorough drying will further help to prevent biofilm formation.

See also: **Biofilms**. ***Cellulomonas***.

Further Reading

Bourion F and Cerf O (1996) Disinfection efficacy against pure-culture and mixed-population biofilms of *Listeria innocua* and *Pseudomonas aeroginosa* on stainless steel, Teflon® and rubber. *Sciences des Aliments* 16: 151–166.

Bower CK, McGuier J and Daeschel MA (1996) The adhesion and detachment of bacteria and spores on food-contact surfaces. *Trends in Food Science and Technology* 7: 152–156.

Boyd A and Chakrabarty AM (1995) *Pseudomonas aeruginosa* biofilm formation: role of the alginate exo-

polysaccharide. *Journal of Industrial Microbiology* 15: 162–168.
Carpentier B and Cerf O (1993) Biofilms and their consequences, with particular reference to hygiene in the food industry. *Journal of Applied Bacteriology* 75: 499–511.
Davies DG, Parsek MR, Pearson P et al (1998) The involvement of cell-to-cell signals in the development of a bacterial biofilm. *Science* 280: 295–298.
Fick RB (1993) Pseudomonas aeruginosa *the Opportunist: Pathogenesis and Disease*. Boca Raton: CRC Press.
Hardalo C and Edberg SC (1997) *Pseudomonas aeruginosa*: assessment of risk from drinking water. *Critical Reviews in Microbiology* 23: 47–75.
Mossel DAA, De Vor H and Eelerink I (1976) A further simplified procedure for the detection of *Pseudomonas aeruginosa* in contaminated aqueous substrata. *Journal of Applied Bacteriology* 41: 307–309.
Palleroni NJ (1984) Pseudomonadaceae. In: Kreig NR and Holt JG (eds) *Bergey's Manual of Systematic Bacteriology*. Baltimore: Williams and Wilkins.
Pier GB (1998) *Pseudomonas aeruginosa*: a key problem in cystic fibrosis. *ASM News* 64: 339–347.

Burkholderia cocovenenans

Julian Cox, **Embit Kartadarma** and **Ken Buckle**,
Department of Food Science and Technology, The University of New South Wales, Sydney, Australia

This article describes the essential features of the bacterium *Burkholderia cocovenenans* and the closely related *B. cocovenenans* biovar *farinofermentans*. Although these organisms are less well known than many food-borne pathogens, they are nevertheless significant in certain regions of the world, causing food-borne intoxication associated with high mortality.

Characteristics of the Organism

In 1932, a bacterium was isolated from the fermented food tempe bongkrek, implicated in an outbreak of food poisoning. The organism was named *Bacillus cocovenenans*, the specific epithet of the organism being derived from *cocos* (coconut) and *veneno* (to poison).

The organism is a pleomorphic Gram-negative bacterium, which is usually rod-shaped, but also appears as coccoid, vibrioid or filamentous, depending on cultural conditions. Rod-shaped cells are 0.4–0.5 × 0.8–1.5 μm in size. *B. cocovenenans* is motile by one to four polar flagella. It is catalase positive, and exhibits a very weak oxidase activity, originally described as negative. It grows at 30°C, but not at 4, 10 or 45°C. Further characteristics are given in **Table 1**. On ordinary nutrient culture media, colonies are round and slightly convex, smooth or rough in texture, and white to deep yellow in colour. Pigmentation reflects toxin production.

As the organism conforms to the general description of the family Pseudomonadaceae, it was relocated from the genus *Bacillus* to the genus *Pseudomonas*. Subdivision of *Pseudomonas* into groups and ultimately into several genera on the basis of rRNA analysis created the genus *Burkholderia*. Early biochemical test data suggested that species *cocovenenans* should remain in the genus *Pseudomonas*, but comprehensive biochemical and nucleic acid analysis of *P. cocovenenans* in 1995 led to reclassification of the species to the new genus.

An organism producing similar symptoms to *B. cocovenenans*, isolated in China from fermented cornflour, was described in 1980 as *Flavobacterium farinofermentans*. Further biochemical and physiological investigations and later nucleic acid and serological studies saw the organism first reclassified into the genus *Pseudomonas* and subsequently as a biovar of *B. cocovenenans*. Characteristics of the biovar are given in Table 1.

Toxins

Biochemistry

The major toxin produced by *B. cocovenenans*, bongkrek acid (**Fig. 1**), is a substituted glutaconic acid derivative of aconitic acid, with the formal chemical designation 3-carboxymethyl-17-methoxy-6,18,21-trimethyldocosa-2,4,8,12,14,18,20-heptenedioic acid, chemical formula $C_{28}H_{38}O_7$ and molecular weight of 486. A pure solution has an absorption maximum at 267 nm. The free acid is soluble in fat solvents but is soluble in water, although salts produced in alkaline solutions are soluble in the latter. In crude form, and in oil or solvent solutions, bongkrek acid is very heat stable, becoming less stable the more it is purified. Preparations of the toxin usually contain some fats or fatty acids, derived from the coconut-based media usually used for cultivation of *B. cocovenenans*.

Flavotoxin A is the major toxin produced by *B. cocovenenans* bv. *farinofermentans*. The compound has not been chemically defined, but mass spectral analysis has provided an empirical chemical formula of $C_9H_{13}O_3$ and a molecular weight of 169. Spectral analysis reveals absorption maxima at 232 and 267 nm. Although the molecular weight and chemical formula for flavotoxin A are somewhat different from those of bongkrek acid, similarities in mass spectra and absorption maxima suggest the two compounds

Table 1 Phenotypic and genotypic characteristics of *B. cocovenenans*, *B. cocovenenans* bv. *farinofermentans*, compared to *Pseudomonas aeruginosa* and *B. cepacia* (From Cox et al, 1997)

Test	Burkholderia cocovenenans	B. cocovenenans *bv.* farinofermentans	Pseudomonas aeruginosa	Burkholderia cepacia
Oxidase	(+)[a]	(+)	+	+
Production of poly-β-hydroxybutyrate	+	+	–	+
Anaerobic growth on nitrate	–	–	+	–
Hydrolysis of:				
gelatin	+	+	+	+
Tween 80	+	+	v[b]	+
Lecithin	+	+	–	v
Arginine (dihydrolase)	–	–	+	–
Utilization as sole carbon source of:				
Adonitol	–	+	–	+
α-Aminovalerate	–	+	–	–
m-Hydroxybenzoate	–	+	–	+
DL-*γ-Aminobutyrate*	–	v	+	+
DL-*α-Aminobutyrate*	–	v	na[b]	–
Mesaconate	+	–	+	–
l-phenylalanine	–	–	v	+
Hippurate	–	–	–	+
Benzoate	–	–	+	+
p-Phthalate	–	–	–	na
DNA relatedness[c]	100	97	23	55
DNA relatedness[d]	95	100	26	60
mol% G + C	69	69	67.2	67.4

[a] Weak reaction, originally considered negative.
[b] Variable; na, not available.
[c] To *B. cocovenenans*.
[d] To *B. cocovenenans* bv. *farinofermentans*.

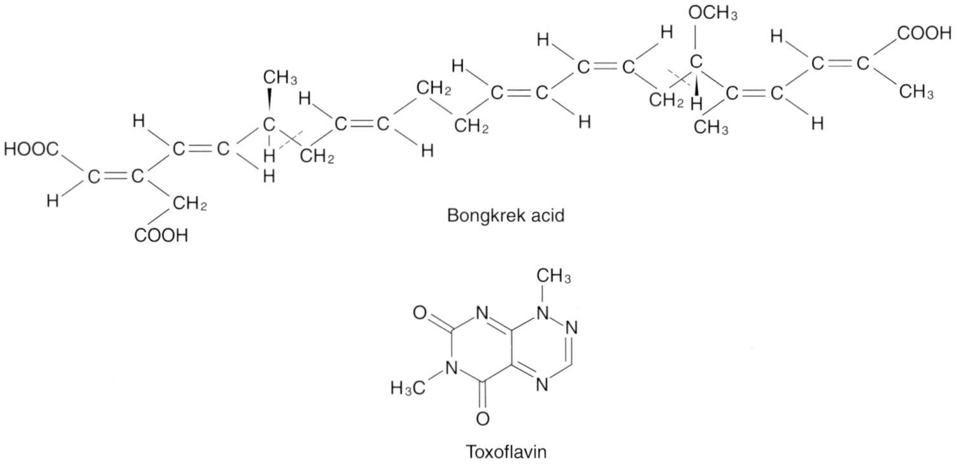

Figure 1 Structures of bongkrek acid and toxoflavin.

are related. If bongkrek acid was hydrolytically cleaved at the points indicated by dashed lines in Figure 1, compounds with a chemical composition similar to that reported for flavotoxin A would be generated.

B. cocovenenans and biovar *farinofermentans* produce a second toxin, known as toxoflavin because of its physico-chemical resemblance (yellow colour, green fluorescence, stability against oxidation, absorption spectrum) to riboflavin. The structure of the toxin is given in Figure 1. It can be extracted with chloroform and crystallizes into yellow flat needles with a melting point of 171°C. A pure solution has an absorption maximum at 258 nm. Toxoflavin is quite resistant to oxidizing agents, but discolours in the presence of sulphur dioxide.

Accurate determination of levels of bongkrek acid and toxoflavin requires solvent extraction, followed by purification and separation using paper, thin layer (TLC) or high pressure liquid chromatography

Table 2 Production (μg g^{-1}) of bongkrek acid and toxoflavin by different strains of *B. cocovenenans* grown in coconut culture medium incubated at 30°C (From Cox et al, 1997)

Strain	*Initial population of* B. cocovenenans *(cfu g^{-1} CCM)*	*Incubation time (h)*							
		24		*48*		*72*		*96*	
		BA	*TF*	*BA*	*TF*	*BA*	*TF*	*BA*	*TF*
ITB	1.2×10^6	80	nd[a]	1450	50	1550	50	1800	70
NCIB	2.7×10^6	20	nd	1700	nd	1150	nd	1400	nd
LMD	2.9×10^6	900	50	2600	20	2850	30	3150	50

CCM, coconut culture medium; BA, bongkrek acid; TF, toxoflavin.
[a] nd, not detected.

(HPLC). Once purified and separated, quantitation can be made using spectrophotometry. Routine analysis of simple extracts can be performed using TLC or HPLC, whereas spectrophotometry alone is unsuitable, due to the similar absorption maxima of the two toxins.

Production

Coconut culture medium (CCM, see below) has been used as a model system to study the production of toxins by *B. cocovenenans*. Growth of the bacterium and production of toxoflavin do not vary significantly between 30°C and 37°C, production of bongkrek acid is optimal at 30°C. Toxin production is low during the first 24 hours of growth, increasing substantially after 48 hours, suggesting that both toxins are secondary metabolites. Production of bongkrek acid and toxoflavin varies with strain (**Table 2**), and it appears that toxin production is attenuated during serial subculture.

It appears that the common link between *B. cocovenenans*, biovar *farinofermentans*, and their food associations is the presence of polyunsaturated fatty acids in vegetable matter, which putatively serve as the substrate for synthesis of bongkrek acid and flavotoxin A. Unlike bongkrek acid, production of toxoflavin by either organism occurs in simple bacteriological media, independent of a specific food matrix.

Action and Symptoms

Of bongkrek acid and toxoflavin, the former is the more severe. The main symptoms induced by bongkrek acid are an initial hyperglycaemia quickly followed by a marked hypoglycaemia, which exhausts the glycogen reserves in many tissues, particularly the liver and heart. These effects stem from inhibition of mitochondrial oxidative phosphorylation, in turn due to the inhibition of ADP/ATP translocation, as well as interference with the citric acid cycle in heart muscle.

Toxoflavin functions under aerobic conditions as an active electron carrier between NADH and oxygen, leading to production of hydrogen peroxide, and bypass of the cytochrome system. These characteristics, respectively, probably confer the strong antibiotic and poisoning properties associated with the toxin. Toxoflavin is inactive under anaerobic conditions. Although toxoflavin is lethal in small doses when administered intravenously to rats, little morbidity or mortality is observed in rats or monkeys when given orally, suggesting this toxin plays only a minor role, if any, in the symptoms observed during bongkrek food poisoning.

Approximately 4–6 h after ingestion of contaminated food, victims experience a range of symptoms, including malaise, abdominal pains, dizziness, extensive sweating and extreme tiredness, before lapsing into coma. Death usually occurs 1–20 h after the onset of the initial symptoms. After death, there is little evidence of cause, as no histological changes can be demonstrated on autopsy, no bacterial growth can be obtained from various organs, and laboratory animals do not succumb when fed such organ tissue. Although there are no precise figures, mortality is very high compared to many food-borne illnesses.

Although there is little precise information regarding the lethal dose of either toxin, it can be inferred that only milligram quantities of bongkrek acid, the more potent of the two toxins, are required to cause death. It is known that 1–3 mg of bongkrek acid can be produced per gram of food within 48 h when high numbers of the organism develop, and only a few grams of contaminated food, even after cooking, is sufficient to kill humans.

Significance in Foods

Unlike may other food-borne pathogens, *B. cocovenenans* and bv. *farinofermentans* are not associated with a wide range of foods, instead representing an almost unique ecology in food microbiology.

In many countries of the East and Far East, fermented vegetable products represent a major source of nutrients, particularly protein. These include many varieties of tempe, produced in Indonesia from a diversity of vegetable matter, as well as a range of

Table 3 Effect of onion extract on toxin production by *B. cocovenenans* in tempe bongkrek produced from coconut culture medium, after incubation for 48 h at 30°C (From Cox et al, 1997)

Onion extract (%)	*Treatment*		*Inoculum (cfu g^{-1})*		*Toxin production (μg g^{-1})*	
	pH	*NaCl (%)*	Rhizopus oligosporus	Burkholderia cocovenenans	*Bongkrek acid*	*Toxoflavin*
0	6.9	0	0	6.0×10^4	231	89
	6.9	0	0	2.1×10^7	2406	516
	6.9	0	3.5×10^5	2.1×10^7	1337	456
	5.5	0	0	7.5×10^6	753	491
0.6	6.9	0	0	1.9×10^5	nd[a]	nd
0.8	6.9	0	0	1.9×10^5	nd	nd
	6.9	0	0	2.1×10^7	788	330
	6.9	0	3.5×10^5	2.1×10^7	387	259
	6.9	1	7.0×10^4	7.5×10^6	nd	nd
	5.5	0	0	7.5×10^6	nd	nd

[a] nd, not detected.

cooked products made from fermented corn meal, used widely in poorer regions of China.

In parts of Java, tempe bongkrek is prepared from partially defatted coconut, either the presscake remaining after coconut oil extraction, or the material left after water extraction of coconut milk from shredded coconut meat. It is this form of tempe that has, to date, been exclusively associated with *B. cocovenenans* intoxication. The first deaths were reported in 1895, and since 1951, consumption of contaminated tempe bongkrek has resulted in approximately 10 000 cases of intoxication, including at least 1000 deaths.

In China, toxic products prepared from fermented cornflour derive from poor handling of the raw material. Corn is soaked in water for two to four weeks, rinsed, then ground into a wet flour and held at ambient temperature for an indefinite period before cooking and consumption. These conditions are conducive to growth of *B. cocovenenans* bv. *farinofermentans*. There were 327 cases in 23 outbreaks reported in China between 1961 and 1979, with 314 victims experiencing symptoms within 10 h and an overall mortality rate of 32.2%.

Although intoxication due to *B. cocovenenans* or biovar *farinofermentans* is at present geographically limited, migration to and adoption of foreign cuisine by other countries could lead to more widespread problems with these organisms.

Control

As intoxication due to either *B. cocovenenans* or bv. *farinofermentans* is associated with traditional fermented foods consumed by populations of low socioeconomic status, any approaches to control of these bacteria must be simple and inexpensive. The first measure, intrinsic to production of foods such as tempe, is inoculation of the substrate (e.g. coconut presscake) with a sufficient quantity of active fungal (e.g. *Rhizopus oligosporus*) culture. A rapid fermentation ensues, and growth of the pathogen is inhibited. Although this technique is suitable for fermented foods, it is obviously not applicable to non-fermented products. In addition, the inoculum at the village level is usually derived from a previous fermentation, which provides a culture of variable population and activity. If *B. cocovenenans* has already reached a high population in the substrate, the fungus, regardless of state, will be inhibited, and toxin production will follow.

Addition of up to 1.5% NaCl or acidification of the substrate to pH 4.5, serves to suppress synthesis of bongkrek acid, and eliminate toxoflavin production, whereas a combination of 2% salt and acidification to pH 5 prevents production of bongkrek acid. These amendments have no negative effect on the growth of *R. oligosporus*. Acidification is achieved through re-use of soak water or use of incompletely washed soaking vessels. Acidification has been attempted using acidic plant material, such as *Oxalis*, but this produces changes in the product that are unacceptable to the consumer.

Crude extracts of various spices, including garlic, onion, capsicum and turmeric, inhibit toxin production; their effects are enhanced by other amendments such as addition of salt or reduction of pH. However, efficacy is inversely proportional to the population of the pathogen.

The influence on toxin production of many of the abovementioned factors, including populations of fungal inoculum or pathogen, salt concentration, pH, or spice extract, is shown in **Table 3**.

Although the preceding techniques have proved effective in the laboratory or pilot studies, they have yet to be applied in the field. Other approaches have been taken. In order to curb the morbidity and mortality associated with consumption of contaminated product, the Indonesian government banned the production of tempe bongkrek in 1988. Such a legislative approach is unlikely to prove effective in the long term, as consumers see such products as a vital source of nutrition in the daily diet.

Isolation/Detection

At present there is no selective or differential medium specific for the isolation of *B. cocovenenans*. Coconut culture medium (CCM) is routinely used for culture of the bacterium, as it simulates the environment of tempe bongkrek, encouraging strong growth and stimulating toxin synthesis. CCM is usually prepared from fresh coconut meat by blending with water and pressing twice, although the pulp can also be produced from rehydrated desiccated coconut. The pulp is then shaped into a small round cake in a suitable vessel and sterilized by autoclaving. A more conventional medium can be prepared by mixing equal quantities of sterile (autoclaved) commercial coconut cream, and 3% agar. The latter medium permits simple cultivation of the bacterium and stimulates toxin production.

There are no routine non-cultural techniques for detection of the bacterium, although a monocolonal antibody specific for the lipopolysaccharide of both *B. cocovenenans* and biovar *farinofermentans* has been developed, and may prove useful in the rapid identification of food-borne disease outbreaks involving these organisms.

See also: **Fermented Foods**: Fermentations of the Far East.

Further Reading

Cox J, Kartadarma E and Buckle K (1997) *Burkholderia cocovenenans*. In: Hocking AD (ed.) *Foodborne Microorganisms of Public Health Significance*. P. 521. Sydney: Australian Institute of Food Science and Technology (NSW Branch) Food Microbiology Group.

PSYCHROBACTER

María-Luiga García-López and **Miguel Prieto Maradona**, Department of Food Hygiene and Food Technology, University of Leon, Spain

Introduction

The name *Psychrobacter* (*psychros*, cold; *bacter*, rod) refers to a group of bacteria which are psychrotrophic, i.e. they grow best at low temperatures. Currently the genus *Psychrobacter* is included in the family Moraxellaceae. Five species, *P. immobilis*, *P. urativorans*, *P. frigidicola*, *P. phenylpyruvicus* and *P. glacincola* are recognized although only *P. phenylpyruvicus* and *P. immobilis* appear to be of interest in food microbiology since both are among the spoilage flora of proteinaceous foods. *P. immobilis* has been implicated as an opportunistic pathogen. This article is a summary of the relatively scarce information on this genus covering mainly taxonomic aspects, its ecology, role in food spoilage and methods for isolation and identification.

History and Current Taxonomic Status

Relationship to Allied Genera

The genus *Psychrobacter* comprises Gram-negative, aerobic, non-motile, oxidase-positive cocci or coccobacilli (sometimes rods) which are psychrotolerant and halotolerant. Because of their typical microscopical appearance, psychrobacters resemble strains of *Moraxella* and *Acinetobacter*. The relationship has also been shown to be close from the compositional, biochemical and genetic viewpoints. **Table 1** summarizes the main characteristics of *Psychrobacter* and allied genera (*Moraxella*, *Acinetobacter* and non-motile strains of *Pseudomonas fragi*).

Psychrotrophic, saprophytic, non-pigmented aerobic Gram-negative bacteria isolated from food were originally classified as '*Achromobacter*'. This genus, which was not included on the Approved List of Bacterial Names, comprised both motile and non-motile strains. A first subdivision separated the non-motile strains into the genus *Acinetobacter*. Later, when the oxidase test became available, strains negative to this test were grouped together and labelled as *Acinetobacter*, whereas the oxidase-positive organisms were classified as *Moraxella* or *Moraxella*-like. By means of a genetic transformation assay, it was shown that part of this heterogeneous group was

Table 1 Characters that differentiate aerobic, catalase-positive, non-motile Gram-negative bacteria associated with foods

	Psychrobacter	Moraxella	Acinetobacter	*Non-motile* Pseudomonas fragi
Cell shape	CB/R[a]	CB/R	R	R
Oxidase	+	+	–	+
Tolerance to 6% NaCl	+	–	–	v[b]
Optimal temperature (°C)	20–25	33–35	33–35	28–30
Penicillium sensitivity (1 U ml^{-1})	(+)	(+)	–	–
Carbon sources for growth				
Aminovalerate	+		–	+
2-Keto-D-gluconate, glycerol, fructose and arabinose	–		–	+
Butyrate	+	+	+	–
Fatty acids				
10:0	+	+	–	
2-OH 12:0	–	–	+	
i17:0	+	–	–	
P. immobilis transformation	+	–	–	–
G+C (mol%)	41–46	40–47.5	38–47	58–60

[a] CB/R, coccobacillus; R, rods.
[b] v, variable percentages of positive strains; (+), most strains positive.

related to a single competent strain, making up a unique genospecies, which received the name *Psychrobacter immobilis*.

Until recently, moraxellae, acinetobacters and psychrobacters were accommodated in the family Neisseriaceae. The inclusion was based primarily on similarities in phenotypic properties and genetic interactions between *Neisseria catarrhalis* (*Moraxella* (*Branhamella*) *catarrhalis*) and some moraxellae, as well as the affinity between these and psychrobacters and acinetobacters. Later, it became clear that the above-mentioned members of the Neisseriaceae had to be grouped into a new family. Two proposals to group the species in a new family have been issued. Although it has been proposed to separate *Moraxella* and *Branhamella* from the Neisseriaceae and to accommodate them in a new family Branhamaceae, *Branhamella* is currently considered a subgenus of *Moraxella* comprising the coccoid forms. Other authors have proposed, on the basis of genetic studies, the exclusion of *Psychrobacter*, together with *Moraxella* and *Acinetobacter*, from the Neisseriaceae family, and to create a new family, Moraxellaceae, to accommodate these three genera.

The family Moraxellaceae would be placed in the class Proteobacteria together with other Gram-negative taxa, referred to as 'purple bacteria and their relatives'. Phylogenetically, Proteobacteria is divided, based on 16S rRNA sequence comparisons, into four RNA subclasses (α, β, γ and δ). The gamma (γ) subclass (which corresponds to the rRNA superfamily II) contains, among others, the highly interrelated genera *Moraxella*, *Acinetobacter* and *Psychrobacter*, and it is divided into two subdivisions, one for *Acinetobacter* and the other, which includes the moraxellae and psychrobacters, with four subdivisions:

- the authentic moraxellae;
- the generically misnamed taxon *Moraxella osloensis*;
- the generically misnamed taxon *Moraxella atlantae*; and
- the generically misnamed taxon *Moraxella phenylpyruvica/Psychrobacter*.

Psychrobacter species

The generically misnamed taxon *Moraxella phenylpyruvica/Psychrobacter* was formed by heterogeous strains. Also, *Moraxella phenylpyruvica* was considered taxonomically distant from other *Moraxella* species, not only because of 16S rDNA sequence analysis data, but also because of rRNA–hybridization data, phenotypic characters (psychrotrophy and halotolerance) and fatty acid composition. As a result, *M. phenylpyruvica* has been redefined as *Psychrobacter* (*Moraxella*) *phenylpyruvicus*. Furthermore, three new species have been described from the Antarctic habitat, *P. urativorans*, *P. frigidicola* and *P. glacincola*.

The relationships among the above species have been reported in a few papers. Phenotypic analyses carried out over selected groups of non-motile, oxidase-positive, Gram-negative strains showed a close similarity between *P. (Moraxella) phenyl-*

pyruvicus and *P. immobilis*. In the three clusters of strains from meat defined at the 80% (simple matching coefficient, Ssm) similarity level (unweighted pair group average linkage, UPGMA) two of them were assigned to *P. (Moraxella) phenylpyruvicus* and the third to *P. immobilis*. Overall similarity between the three clusters was 78% (Ssm). *P. (Moraxella) phenylpyruvicus* in particular reduced its participation as storage progressed.

When four species of *Psychrobacter* (all but *P. glacincola* isolated from meat and ornithogenic soils) were studied, analysis of phenotypical data also showed a high similarity (58% Jaccard coefficient, Sj) between *P. (Moraxella) phenylpyruvicus* and *P. immobilis*. *P. frigidicola* and *P. urativorans* exhibited a similarity level of 55% (Sj) with each other, but a similarity level of only about 37% with the two first species. Data from 16S rDNA sequence analysis and the DNA homology levels showed a different relationship, in which *P. immobilis* was slightly closer to *P. frigidicola* and *P. urativorans* than to *P. (Moraxella) phenylpyruvicus*. This last species showed overall levels of similarity to the other *Psychrobacter* species in the range 93–96%. *P. frigidicola* and *P. urativorans* were closer with overall similarity levels of 96% between them.

Overall, psychrobacters can be grouped into clearly defined species, and each one is isolated from delimited and specific environments, with the exception of *P. immobilis* (present in ornithogenic soils and foods) and the type strain of *P. urativorans*, isolated from pork sausage, whereas the rest of the *P. urativorans* strains described stem from ornithogenic soil. All members of a given species are very nearly genetically identical.

Habitats

Psychrobacter is considered to be a ubiquitous bacterium. Strains have been isolated from different marine and terrestrial environments, including foods, soil, sea water, sea ice and air. It is also considered an emerging and opportunistic pathogen because of its implication in clinical cases.

Food Isolates

Psychrobacter is commonly isolated from foods stored in air at low temperatures. Some of the isolates previously described in the literature and named as *Moraxella*-like or '*Achromobacter*' are in fact *Psychrobacter.* Two *Psychrobacter* species (*P. immobilis* and *P. (Moraxella) phenylpyruvicus*) have been isolated from fish (gills and skin), poultry, lamb, and several meat products (pork sausage, ground beef, Vienna sausage), some of them irradiated. Because of their presence in sea water, they contaminate fish. In the case of meat, the primary sources of contamination seem to be the soil or air, access to the meat being gained during slaughtering.

The presence and evolution of bacteria of the genus *Psychrobacter* in foods is greatly influenced by environmental conditions. Its presence is restricted to proteinaceous foods stored and kept in certain conditions. The ability to compete with other spoilage bacteria such as *Pseudomonas* is apparently low, and numbers of bacteria decrease during cold storage probably because of changes in the intrinsic conditions (a_w) of food.

Psychrobacter is highly radiation-resistant and constitutes the main residual flora in radurized proteinaceous foods (Vienna sausages, fish, poultry and ground beef). Certain strains show even greater resistance than some recognized spores, displaying some morphological and phenotypical (penicillin sensitivity) changes.

Incidence Problems in identifying members of this genus, misidentification of isolates before 1986 and the few studies from which their incidence in foods can be determined, make it difficult to establish whether these bacteria constitute a significant part of the aerobic spoilage population. In meat (beef and lamb) they often represent a low percentage of the total aerobic counts at spoilage (from <1 % to 5%), though higher figures (up to 50%) have been found on some fat surfaces. Their incidence among the spoilage flora of poultry, rabbit, milk and cheeses (soft and fresh) is often low. These organisms appear to be more prevalent in fish, particularly in sardine and cod, but they may also occur as a minor part of the microflora of other marine and freshwater fish species such as trout and pike.

Antarctic Isolates

Psychrobacter has been found in Antarctic ornithogenic soils, where it is the predominant culturable genus in this environment (66%). This soil stems from the deposition of faeces of various species of birds, and is characterized by a high content of uric acid. Besides *P. immobilis*, two new species (*P. urativorans* and *P. frigidicola*) have been described, which seem to be restricted to this ecosystem. Depletion of uric acid would reduce the population of psychrobacters to less than 1%. A new species, *P. glacincola*, has been recently isolated from the Antarctic sea ice.

Clinical Isolates

Little is known of the clinical significance of *Psychrobacter immobilis*. The occurrence of this species in clinical specimens such as blood, brain tissue, urethra, wounds, cerebrospinal fluid, vagina, eyes, etc. has been noted and several isolates from these sources formerly assigned to the CDC group EO-2 have been reclassified as *P. immobilis* on the basis of transformation studies. In recent years (since 1990), this bacterium has been involved in nosocomial infections (conjunctivitis, meningitis and bacteraemia) most often in infants and patients infected with human immunodeficiency virus. It appears that *P. immobilis* is an opportunistic pathogen and perhaps misidentification has occurred in the past with this organism being confused with related genera such as *Moraxella, Acinetobacter* or *Neisseria*.

Psychrobacter and Food Spoilage

It is not an easy task to determine whether a microorganism commonly isolated from a spoiled food plays a significant role in spoilage. Microbiological and chemical changes during storage must be studied, including the relative incidence and behaviour of the organism throughout the shelf life of that food, its spoilage potential and its spoilage activity. The latter is particularly important and qualitative and quantitative production of chemical compounds associated with spoilage should be investigated using the food and/or its components as substrates.

Spoilage Potential

In taxonomic studies, strains identified as *P. immobilis* are capable of forming acid from carbohydrates aerobically (**Table 2**) but incapable of using many compounds as carbon source. Proteins are not hydrolysed and trimethylamine (TMA), indole and H_2S are not produced. By contrast, most of them show lecithinase and lipolytic activity (Tween 20, 40, 60 and 80, triacyglycerols and cod liver oil).

Spoilage Activity

It is reported that *Psychrobacter* probably obtains energy from the oxidation of organic acids or amino acids; however information on the substrates used is not available.

On stored beef, one strain, identified as *Moraxella*-like, produced two nitriles and two oximes, but their relevance to spoilage odours was not clear. Some volatile compounds (methyl mercaptan, dimethyl disulphide, etc.) similar to those produced by *Shewanella putrefaciens* and *Pseudomonas fluorescens* were detected in sterile fish muscle inoculated with an '*Achromobacter*' strain showing characteristics of *P. immobilis*; however, the off-odours produced were not so intense and no marked or textural changes were observed. Ethyl acetate and other volatile compounds associated with spoiled chicken breast were also identified in sterilized poultry breasts inoculated with a strain of *Moraxella*.

The small relative proportion of psychrobacters within the total microflora of many proteinaceous foods, their tendency to decrease in most cases during chill storage and their low spoilage potential suggest that this bacterium is not an important contributor to the spoilage of proteinaceous foods although more detailed studies are required before definite conclusions can be reached. In fact, *Psychrobacter* constitutes the predominant flora of certain spoiled irradiated foods. Furthermore, it could be of significance in fresh products if present initially in high numbers because of its lipolytic activity and/or by restricting the availability of oxygen to other Gram-negative bacteria such as *Pseudomonas* or *Shewanella*. It appears that when oxygen limits growth, pseudomonads attack amino acids and *S. putrefaciens* generates H_2S.

Other Physiological and Structural Characteristics

Optimal growth temperature ranges are given in Table 2. *P. immobilis* strains of clinical origin are able to grow at 35°C. Isolation is possible from plates incubated at 7°C (10 days). Being oxidase-positive, they produce cytochrome *c*, although oxidase-negative *Psychrobacter* mutants have been reported. Enzymatic assays of cell-free extracts of strain ATCC 15174 (isolated from pork sausage and originally named *Micrococcus cryophilus*) have revealed the presence of all the enzymes of the tricarboxylic acid cycle. This strain is able to synthesize wax esters and phospholipids.

Covering and attached to the outer membrane of psychrobacters, there is an array made up of a single protein, which has phospholipase activity when released into the medium but not when arranged in the layered array. This additional surface layer, which is similar to those of other Gram-negative bacteria, appears to have a protective mission from disgenesic environments (Juni, 1992).

Recommended Methods of Detection and Enumeration in Foods

Isolation and Preservation

Psychrobacters usually grow well on standard media such as plate count agar (PCA), tryptone glucose extract agar supplemented with 0.5% NaCl (TGES), heart infusion agar (HIA), trypticase soy agar (TSA) or tryptone–yeast extract agar (LB agar). To improve the recovery and identification of strains of this genus as well as those of acinetobacters when they constitute a minor part of the flora, two media (medium M and medium B) have been proposed. The composition of

Table 2 Tests useful in differentiating *Psychrobacter* species

Characteristics	*P. immobilis*	*P. phenylpyruvicus*	*P. urativorans*	*P. frigidicola*	*P. glacincola*
Hydrolysis of Tween 80	+	+	−	−	+
Optimal growth temperature ranges[a]	25–31°C	32°C	17–19°C[a]	14–16°C	13–15°C
Growth stimulated by					
Na⁺	−	−	−	+	+
Bile salts	−	+	−	−	−
Simmons citrate test	−	+	−	−	−
Acid phosphatase activity	−	+	−	−	−
Acid produced from L-arabinose, D-xylose D-raffinose	+	−	−	−	−
Growth on					
DL-*lactate*	+	+	v[b]	−	v
3-*hydroxybutyrate*	+	+	+	−	v
Suberate and azelate	v	−	−	+	+
Sarcosine	−	−	−	+	
Caproate	+	−	+	+	+
Phenylalanine arylamidase activity	+		−	−	+
Deaminase activity					
Phenylalanine	+	+	−	+	−
Tryptophan	v	−	−	+	−
G+C (mol%)	44–46	42.5–43.5	44–46	41–42	43–44
Fatty acids: i17:0	+	T[c] or −	+		

[a] Data from Bowman *et al.* (1997).
[b] v, variable percentages of positive strains.
[c] T, trace (less than 0.1% of total fatty acids).

medium M is nutrient agar with 1 $\mu g\ ml^{-1}$ crystal violet and 0.1% bile salts no. 3 (Difco), whereas that of medium B is the Standard Mineral Base of Stanier et al (1966) plus 0.2% sodium butyrate, 0.05% bile salts and 1 $\mu g\ ml^{-1}$ crystal violet. These media show a good performance and surface colonies of both genera are recognized by their characteristics (convex, opaque and light blue). Prolonged incubation at 5°C improves the colony colouring. On nonselective media, colonies are mainly non-pigmented, small, circular, convex, entire, smooth, opaque and dull or glistening with a butyrous consistency.

Peptone water (0.1% w/v) and peptone or tryptone (0.1% w/v) plus NaCl (0.85% w/v) are used as diluents. Surface-inoculated plates should be incubated at 20–25°C (or lower) for 3–4 days though *P. immobilis* is often found on plates of psychrotrophic counts (7°C, 10 days). For short-term preservation (3 months), cultures may be kept on agar slopes (standard media) stored at 0–4°C. Lyophilization or storage at −20°C (mixed with sterile glycerol) are suitable for longer periods.

Characterization and Identification

Phenotypic Characters

Strains of *P. immobilis* are Gram-negative (often difficult to destain), aerobic, non-motile, non-pigmented, non-sporulating, oxidase-positive, catalase-positive rods or coccobacilli. Coccobacilli tend to be oval shaped and often occur in pairs, whereas rods, which are variable in length, appear swollen and chain formation can occur. Most strains are able to grow at low temperatures (4–10°C) with an optimal temperature near 20°C. They usually fail to grow at 35–37°C. Other characteristics are tolerance to salt (6% NaCl) and penicillin sensitivity (1 unit ml^{-1}). *P. immobilis* is able to produce acid aerobically from glucose; however, the incidence of glucose-positive strains among isolates obtained from foods may be as low as 27%. Acid is formed by most of the *P. immobilis* strains from mannose, melibiose, galactose, arabinose, xylose and rhamnose but is not formed from fructose, maltose, sucrose, trehalose, adonitol, xylitol, raffinose or sorbitol. Nitrate is usually reduced. Starch, gelatin, DNA and serum are not hydrolysed. Indole, H_2S and TMA are not produced. A high percentage show lecithinase activity (egg yolk reaction) and all show lipolytic activity on Tween 80. They are tolerant to 0.2% potassium tellurite and urease activity is present. Amino acids (lysine, ornithine and arginine) are not decarboxylated. Phenylalanine and tryptophan are deaminated. Phenethyl alcohol is formed from L-phenylalanine and ethanol. This property has been associated with an odour resembling that of a phenylethyl alcohol blood agar plate which was considered useful for differentiation of *P. immobilis* strains of human origin (formerly identified as CDC group EO-2). In our experience, the number of compounds used by *P. immobilis* as carbon sources is very limited. In other studies, DL-lactate, pyruvate, acetate, *n*-butyrate, *n*-valerate, δ-aminovalerate, DL-β-hydroxybutyrate, fumarate, glutarate, glutamate and histidine could be utilized by 90% or more of the strains isolated from foods. All the above characteristics can be tested for by using conventional procedures. Miniaturized systems such as API 20E, API 32ID and API ZYM (bioMérieux, Marcy l'Etoile, France) are also used to perform a number of tests. Isolates of clinical significance have been identified using an anaerobe identification (ANI) card (Vitek Systems, bioMérieux).

The three *Psychrobacter* species recently proposed (*P. urativorans*, *P. frigidicola* and *P. glacincola*) do not produce acid from carbohydrates. Table 2 lists the differential features of the five *Psychrobacter* species. It should be noted that most of the strains assigned to these novel species were isolated from ornithogenic soils and sea ice though one of them was obtained from irradiated foods (pork sausage).

Numerical Taxonomy

Numerical analysis of a large number of unit characters (morphological, cultural, physiological and biochemical) has proved to be useful for the differentiation of *P. immobilis* and other related Gram-negative bacteria from foods. The simple matching coefficient (Ssm) and/or the Jaccard coefficient (Sj) are usually employed and clustering is achieved by unweighted pair group average linkage (UPGMA). Since data obtained for each phenon allow the selection of the most discriminating characters, simple identification schemes have been derived from these studies. One of them based on five tests (oxidase and the ability to grow on four compounds – aminovalerate, 2-keto-D-gluconate, glycerol and fructose – as the sole source of carbon) has been proposed for the differentiation of *Acinetobacter*, *P. immobilis* and non-motile *Pseudomonas*.

In our experience, the best diagnostic characters to differentiate *Moraxella*, spp., *P. (Moraxella) phenylpyruvicus*, *P. immobilis* and non-motile strains of *Pseudomonas fragi* are acid production from melibiose, L-arabinose and cellobiose. Acid formation from other carbohydrates and the capacity to grow in simple mineral media at the expense of single carbon sources are also useful. Some authors say that psychrobacters are difficult to distinguish phenotypically from other moraxellae but this is not always the case. Variation is found in the patterns of carbon substrate utilization, but, in regard to the capacity to produce acid from carbohydrates, the strains show a large homogeneity. Phenotypically the major difference between *P. immobilis* on one side, and *P. phenylpyruvicus*, *P. frigidicola* and *P. urativorans* on the other, is the ability to produce acid from several carbohydrates (glucose, melibiose, L-arabinose, cellobiose, maltose, etc.), which is negative for the latter.

DNA Transformation Assay

Many psychrobacters are competent for genetic transformation. This fact made it possible to devise a transformation assay that permits definitive identification of isolates tentatively identified as *P. immobilis*. In this procedure, crude DNA of cultures to be assayed is obtained by suspending a loop of growth on HIA in 0.5 ml of lysing solution (0.05% sodium dodecyl sulphate in 0.15 M sodium chloride–0.015% trisodium citrate). The suspended cells are then heated in a water bath at 60°C for 1 h and refrigerated until tested. A hypoxanthine- and thiamine-requiring mutant *Psychrobacter* strain Hyx-7 (ATCC 43177) grown on HIA is the recipient strain used to assay the DNA samples. For the transformation assay, a cell paste of the recipient strain is placed on small areas of an HIA plate and a loopful of the crude DNA preparation from each test strain is added, mixed and spread over an area of about 5–8 mm in diameter. On the same plate, a loopful of each crude DNA is cultured for sterility and one section is reserved to culture the auxotrophic mutant strain without addition of DNA (non-DNA treated control). After overnight incubation at 20°C, no growth should be observed in the DNA control areas, but non-DNA treated samples and the DNA-Hyx-7 mixtures must grow. A loopful from each of the growth areas is then streaked onto a section of plates containing M9A medium or medium P96 (media composition given below). After incubation for 3 days at 20°C, all auxotrophic cells transformed to prototrophy appear as visible colonies whereas the mutant strain is unable to grow. The appearance of transformant colonies on one of the two latter media confirms that the tested strain belongs to the genus *Psychrobacter*.

Medium M9A is prepared by adding the following compounds to 150 ml of distilled water: casein hydrolysate (vitamin and salt free), 3.2 g; Na_2HPO_4, 1.12 g (or $Na_2HPO_4.7H_2O$, 2.12 g); KH_2PO_4, 0.4 g; NaCl, 2.0 g; and $MgSO_4$, 0.09 g (or $MgSO_4.7H_2O$, 0.18 g). The final volume is adjusted to 200 ml and sterilized by membrane filtration or by autoclaving for 20 min. The medium is mixed with 200 ml of hot sterile 3% agar in distilled water. Volumes of 3.2 ml of 50% glucose (filter sterilized) and 2.0 ml of 60% sodium lactate (sterilized by autoclaving) are finally added and after mixing thoroughly the M9A medium is distributed into Petri dishes. Plates can be stored at room temperature in plastic bags. Double-strength medium P96 contains per litre of distilled water: Na_2HPO_4, 11.2 g; KH_2PO_4, 4.0 g; NH_4Cl, 2.0 g;

$MgSO_4.7H_2O$, 0.4 g; $CaCl_2$ (1.0% solution), 1.0 ml; $FeSO_4.7H_2O$ (freshly prepared 0.1% solution), 0.5 ml; sodium DL-lactate (60% solution), 13.0 ml; monosodium L-glutamate, 10.0 g; Tween 80, 8.0 ml; L-methionine, 0.4 g; L-isoleucine, 0.4 g; L-leucine, 0.4 g; L-valine, 0.4 g. This medium (pH 7.2) is filter sterilized and may be stored at refrigeration temperatures for longer than one year.

Semisolid plates of medium P96 are prepared by mixing 200 ml of the above formulation with 200 ml of hot sterile water agar (3.0%), followed by addition of 2.0 ml of sterile 2.0% ferric ammonium citrate. The latter compound allows cells to produce colonies instead of confluent growth.

Other Genetic Methods

The DNA G+C base composition of psychrobacters is in the range 41–47 mol%. *P. immobilis* and *P. urativorans* have G+C values of 44–46 mol%, whereas *P. frigidicola* has G+C contents of 41–42 mol% and *P. phenylpyruvicus* 42–43 mol%. The DNA base composition is determined by the thermal denaturation method after extraction by the Marmur procedure.

Other DNA-based methods such as hybridization analysis and the 16S rDNA sequence analysis (Bowman et al, 1997) have proved to be useful for the identification of psychrobacter species as well as for the differentiation of *P. (Moraxella) phenylpyruvicus*.

Chemotaxonomy

The fatty acid profiles of *P. immobilis*, *P. urativorans* and *P. frigidicola* are very similar and characterized by large amounts (up to more than 50%) of oleic acid (18 : 1ω9c) and variable amounts of 16 : 1 ω7c (12–47%). These fatty acids, with the exception of i17 : 0, are also found in *P. phenylpyruvicus*, but this organism contains 18 : 2 as a major fatty acid and small amounts of two 11-carbon acids (i11 : 0 and 11 : 0). Moderate or small amounts of 16 : 0, i17 : 0, 17 : 1ω8c and 18 : 0 can also be detected. *P. immobilis* contains the relatively uncommon monounsaturated acid 12 : 1 ω9c. Clinical isolates of this species could be distinguished from closely related bacteria by the presence of 10 : 0 and i17 : 0.

P. urativorans and *P. frigidicola* are relatively rich in wax esters (11% of the total extractable lipids), which are probably constituent parts of their cell membranes. These compounds are also present in *P. immobilis*. The fatty alcohol patterns of the wax esters of these two species are quite similar with unsaturated fatty alcohols predominating. Saturated fatty alcohols are more significant in *P. immobilis*.

A common feature between *P. immobilis* and *P. phenylpyruvicus* is that both contain ubiquinone with eight isoprene units (Q-8). Lipid analysis has been performed by gas chromatography and gas chromatography–mass spectrometry. Quinones are examined by reverse-phase high performance liquid chromatography.

Applications

Cold-adapted enzymes produced by psychrophilic bacteria are of interest in biotechnology. Antarctic psychrophilic strains of *Psychrobacter* secrete a heat-labile class C β-lactamase able to perform catalysis at low temperatures though its level of specific activity is not higher than that of class C β-lactamases of mesophilic strains. It has been suggested that β-lactamase of *Psychrobacter* could be useful for analysis of structural factors influencing protein stability.

See also: ***Acinetobacter*****. Fish**: Spoilage of Fish. **Meat and Poultry**: Spoilage of Meat. **Milk and Milk Products**: Microbiology of Liquid Milk; Microbiology of Cream and Butter. ***Moraxella*****.** ***Neisseria*****.**

Further Reading

Bowman JP, Cavanagh J, Austin JJ and Sanderson K (1996) Novel *Psychrobacter* species from Antarctic ornithogenic soils. *Int. J. Syst. Bacteriol.* 46: 841–848.

Bowman JP, Nichols DS and McMeekin TA (1997) *Psychrobacter glacincola* sp. nov., a halotolerant, psychrophilic bacterium isolated from Antarctic sea ice. *System. Appl. Microbiol.* 20: 209–215.

Feller G, Sonnet P and Gerday C (1995) The beta-lactamase secreted by the Antarctic psychrophile *Psychrobacter immobilis* A8. *Appl. Environ. Microbiol.* 61: 4474–4476.

Gennari M, Parini M, Volpon D and Serio M (1992) Isolation and characterization by conventional methods and genetic transformation of *Psychrobacter* and *Acinetobacter* from fresh and spoiled meat, milk and cheese. *Int. J. Food Microbiol.* 15: 61–75.

Juni E (1992) The genus *Psychrobacter*. In: *The Prokaryotes*. Balows A, Trüper HG, Dworkin M, Harder W and Schleifer KH (eds). 2nd edn. Pp. 3241–3245. New York: Springer-Verlag.

Juni E and Heym GA (1980) Transformation assay for identification of psychrotrophic achromobacters. *Appl. Environ. Microbiol.* 40: 1106–1114.

Juni E and Heym GA (1986) *Psychrobacter immobilis* gen. nov., sp. nov.: Genospecies composed of Gram negative, aerobic, oxidase-positive coccobacilli. *Int. J. Syst. Bacteriol.* 36: 388–391.

Lozano F, Flórez C, Recio FJ, Gamboa F, Gómez-Mateos JM and Martín E (1994) Fatal *Psychrobacter immobilis* infection in a patient with AIDS. *AIDS* 8: 1189–1190.

Moss CW, Wallace PL, Hollis DG and Weaver RE (1988)

Cultural and chemical characterization of CDC groups EO-2, M-5, and M-6, *Moraxella (Moraxella)* species, *Oligella urethralis*, *Acinetobacter* species, and *Psychrobacter immobilis*. *J. Clin. Microbiol.* 26: 484–492.

Prieto M, García MR, García ML, Otero A and Moreno B (1992) Numerical taxonomy of Gram-negative, nonmotile, nonfermentative bacteria isolated during chilled storage of lamb carcasses. *Appl. Environ. Microbiol.* 58: 2255–2259.

Rossau R, Van Landschoot A, Gillis M and De Ley J (1991) Taxonomy of *Moraxellaceae* fam. nov., a new family to accommodate the genera *Moraxella*, *Acinetobacter* and *Psychrobacter* and related organisms. *Int. J. System. Bacteriol.* 41: 310–319.

Shaw BG and Latty JB (1988) A numberical study of non motile non fermentative Gram-negative bacteria from foods. *J. Appl. Bacteriol.* 65: 7–21.

Stanier RY, Palleroni NJ and Doudoroff M (1966) The aerobic pseudomonads: a taxonomic study. *J. Gen. Microbiol.* 43: 159–271.

Quality Assurance and Management *see* **Hazard Appraisal (HACCP):** The Overall Concept

QUANTITATIVE RISK ANALYSIS

S H W Notermans, TNO Nutrition and Food Research Institute, Zeist, The Netherlands

Introduction

Risk analysis is a structured, multidisciplinary approach to the identification and reduction of risk. Interest in risk analysis in the context of food-borne pathogens, contaminants and additives has increased due to the Sanitary and Phyto-Sanitary (SPS) Agreement of the World Trade Organization (WTO). The aim of the SPS Agreement is to endorse food safety objectives, such as microbial standards and guidelines, that are based on the application of risk analysis to sound scientific knowledge. **Figure 1** illustrates the use of risk analysis in the development of food safety objectives (e.g. end product specifications) from the food safety policy of the WHO/FAO Codex Alimentarius Commission. Risk analysis can also be used to determine criteria at the critical control points in HACCP (hazard analysis critical control point) processes.

Risk analysis involves the evaluation of risk in the context of science, an understanding of all the activities involved and a structured approach. The process of risk analysis consists of three essential components. These were recommended following FAO/WHO expert consultations, and later adopted by its Codex Alimentarius Commission. They are:

1. Risk assessment: the evaluation of known or potential adverse health effects resulting from human exposure to food-borne hazards. The outcome of the risk assessment is called the risk estimate.
2. Risk management: the control of risks associated with food-borne pathogens and contaminants, in order to protect consumers. Risks are controlled as effectively as possible through the selection and implementation of appropriate measures, as formulated by the World Health Organization and the Food and Agriculture Organization (FAO/WHO).
3. Risk communication: an interactive process of exchange of information and opinion between risk assessors, risk managers and other interested parties, such as consumers.

Figure 1 The use of risk analysis in the context of food safety.

The concept of risk analysis as adopted by FAO/WHO and several Codex Committees is primarily aimed at consumer protection, and involves the establishment of safety objectives for foods that are based on science. Risk analysis may also be used in selecting the most appropriate food processing and preservation methods for compliance with the food safety objectives set. In addition, it is possible to use risk analysis to set (sub)criteria at critical control points, as defined by the hazard analysis critical control point (HACCP) concept (see below). Thus risk analysis is used both in compliance with the food safety objectives set by the regulating bodies, and in meeting any additional objectives set by the producers themselves.

Components of Quantitative Risk Analysis

Risk Assessment

Risk assessment is the evaluation of known or potential adverse health effects resulting from human exposure to food-borne factors such as additives, contaminants and pathogenic microorganisms. Risk assessment involves the documentation and analysis of scientific evidence, the measurement of risk and the identification of factors that influence it. This information is used to produce the risk estimate. The process of risk assessment consists of four steps:

1. Hazard identification
2. Hazard characterization
3. Exposure assessment
4. Risk characterization.

There are several strategies for obtaining information about the factors which contribute to risk and their impact. One approach is a case–control study, in which unacceptable products are compared with acceptable ones.

Hazard Identification This is the identification of potential adverse health effects associated with exposure to, *inter alia*, additives, contaminants and pathogenic microorganisms. It is a qualitative approach. For example, the microbiological hazards present in food may be identified with reference to a list (based on published data) of pathogenic microorganisms able to cause food-borne disease. The likelihood is determined of each listed organism being present in the raw materials used and/or entering the food processing area. Organisms that have never been found in either location can be deleted from the list. Any organisms which are completely destroyed during processing can also be deleted from the list. The possibility of recontamination must then be considered. Any organisms which are not known to cause a food-borne disease involving either an identical or a related food product can be deleted from the list.

Hazard Characterization This is the qualitative and/or quantitative consideration of the nature of the adverse health effects associated with the biological, chemical and physical agents which may be present in food. If practicable, dose–response relationships should be assessed for all the adverse effects produced by the agents being considered, e.g. changes in organ function and clinical symptoms. In the case of additives and contaminants, epidemiological data are of value in verifying the dose–response relationships obtained in experimental animals. In the case of pathogenic microorganisms, such data may also be used to plot dose–response curves directly applicable to humans. Hazard characterization also involves consideration of the characteristics of a pathogenic microorganism in relation to factors such as the nature of the product, the processing conditions, and the storage conditions. This information is necessary, for example, to estimate the outgrowth of the organism in the food product of interest. There are a number of uncertainties in hazard characterization, and so the introduction of an uncertainty factor must be considered.

Exposure Assessment This is the qualitative and/or quantitative estimation of the likely intake of biological, chemical and physical agents via food. The ultimate goal of exposure assessment is the estimation of the hazardous agents in food at the time of consumption. This requires specific expertise and information, about food consumption (e.g. from intake surveys) and about the concentration and distribution of particular hazardous agents in foods. In the case of food-borne microbiological hazards, the estimated concentration of microorganisms may be based on product surveillance and testing, storage conditions and the use of mathematical models which predict the growth and death of microorganisms. There are many sources of uncertainty involved in exposure assessment, resulting in either underestimates or overestimates. These uncertainties should be reflected in the risk characterization. Although it is seldom possible to provide fully quantified assessments of uncertainties, the introduction of a negative or a positive bias should be made clear.

Risk Characterization This is the quantitative and/or qualitative estimation of the probabilities of occurrence and severity of known or potential adverse health effects in a given population, taking into account attendant uncertainties. It is the last step in risk assessment, and from it a risk management strategy can be formulated. Although the Codex Alimentarius document does not suggest that the identification and quantification of the factors contributing to a risk is part of the risk characterization, it is logical to include them.

Risk Management

Risk management is the process of evaluating alternative policies in the light of the risk estimate and, if required, selecting and implementing appropriate controls, including regulation. The purpose of risk management is the identification of acceptable risk levels and the development and implementation of control measures within the framework of public health policy. Risk management takes into account the factors contributing to a risk and their quantitative effect, and also a cost-benefit analysis of options.

The outcome of risk management is the derivation of food safety objectives, for example banning additives or reducing their usage; setting maximum levels for contaminants and pathogenic microorganisms; and the obligatory use of Good Manufacturing Practices (GMP) and controls at national level. In setting food safety objectives, risk managers should take into account the difficulties of control, the feasibility of monitoring, the availability of suitable methods of analysis and the economic importance of the food.

The FAO/WHO document on risk management formulates some general principles covering a structured approach embracing risk evaluation, the assessment of risk management options, decision implementation and monitoring and review. The document also emphasizes that the primary consideration in risk management should be the protection of human health, and that the decisions and practices associated with risk management should be clear.

Risk Communication

Risk communication is defined as an interactive process of exchange of information and opinion between risk assessors, risk managers, and other interested parties. Communication starts with the provision of information about food safety policy to all parties involved in the process of risk analysis, as the basis for the purpose and scope of risk assessment and risk management. Clear, interactive communication is necessary between all involved, including consumers, and at all stages of the processes, and is likely to assume increasing importance.

Framework for the Establishment of Food Safety Objectives

Food safety objectives reflect the food safety policy, which should present a general outline of what is acceptable or not acceptable, and quantitative risk analysis is used in the derivation process. There is increasing consensus that food safety policy is an international issue, and FAO and WHO are the international bodies responsible for setting this policy. They have delegated this task to the FAO/WHO Codex Alimentarius Commission (CAC). This was established in 1962 as an intergovernmental organization for developing food-related standards, guidelines and recommendations in order to protect the health of consumers and facilitate international trade. These standards are recognized by WTO, and provide a reference point for the safety of foodstuffs traded internationally.

Risk assessment in relation to additives and contaminants is carried out by the Joint FAO/WHO Expert Committee on Food Additives (JECFA). The outcome of its risk assessment, i.e. the risk characterization, is the starting point for the Codex Alimentarius Committee on Food Additives and Contaminants (CCFAC). This Committee, in which all member states are represented, is responsible for risk management and sets standards which are subsequently adopted by CAC.

Risk assessment in the context of additives and contaminants is a well-established activity. It involves, firstly, the determination of dose–response relationships for additives and contaminants which can cause an adverse health effect. This involves experiments on animals and the use of a data package obtained by testing several health parameters. From the dose–response relationship, the so-called 'no observed effect' level is estimated. This is used as the basis for determining acceptable daily intakes for additives, and provisional tolerable daily/weekly intakes for contaminants, through the application of uncertainty factors.

Risk analysis in relation to food-borne pathogens is a newly emerging activity. Criteria and guidelines are set by the Committee for Food Hygiene (CCFH), and are based on results obtained from the analysis of outbreaks of food-borne disease. At present, CCFH lacks the assistance of a JECFA-like body for risk assessment studies. WHO/FAO have recommended that such a body be established, because it is unacceptable that a single body should carry responsibility for both risk assessment and risk management. Both CCFH and the JECFA equivalent should then elaborate the criteria and guidelines regarding the risk assessment of food-borne pathogens. In addition, CCFH should clarify the criteria for the selection of pathogens for referral to the JECFA equivalent and should clearly identify the factors to be taken into account in its decision making, particularly in relation to the evaluation of risk management options.

The Use of Quantitative Risk Analysis in Food Production

The principles of quantitative risk analysis can also be applied to food production. Internationally established legal food safety objectives, together with safety objectives set by individual food production companies, focus on safe food production. In adhering to these objectives, food producers make use of general guidelines such as GMP. In addition, the use of hazard analysis critical control points (HACCP) is mandatory in most countries. The use of HACCP entails a systematic approach to the identification, assessment and control of hazards in a particular food operation. This approach aims to identify problems before they occur, and to establish measures for the control of stages in production that are critical in terms of food safety. The controls are thus preventive, remedial action being taken in advance of problems developing. The critical control points are defined as steps, points and procedures where control can be exercised. In relation to each point, criteria are specified such that if met, the food produced will be safe. The traditional, largely qualitative HACCP system can be converted into a quantitative system using elements of quantitative risk analysis, as indicated in **Figure 2**. In HACCP, a hazard is defined as it is in risk analysis: an agent with the potential to cause an adverse health effect. International standards have been established for most agents, which means that a risk assessment is not necessary. Critical control points may be defined as factors that contribute to the risk that a standard is not met. The effect of such a factor should preferably be quantified. In relation to each critical control point, risk managers set criteria, based on this quantification.

In most cases, the actual risk is the result of a combination of several factors. A knowledge of the effect of each individual factor enables optimization of the process, taking into account economic factors. This can be illustrated with a simple example described by Notermans, Zwietering and Mead in 1994. One of the legal standards for pasteurized milk is that *Bacillus cereus* must number < 10^4 organisms per millilitre at the time of consumption. The factors which determine the *B. cereus* count at the time of consumption are the spore load of *B. cereus* after pasteurization and the storage time and temperature. The effect of each factor can be calculated, as can the cost of control of each factor. Clearly, the wishes of the consumer, especially in relation to storage conditions, must be taken into consideration in reaching a final managerial decision.

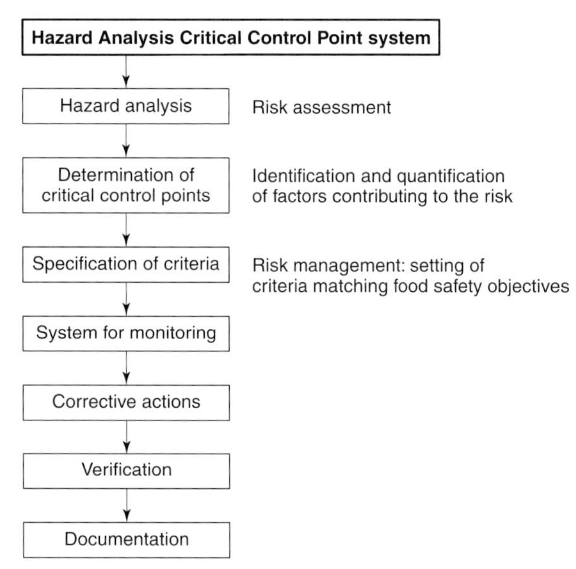

Figure 2 The hazard analysis critical control point (HACCP) concept and the possible use of elements of risk analysis.

See also: ***Bacillus***: *Bacillus cereus*. **Good Manufacturing Practice**. **Hazard Appraisal (HACCP)**: The Overall Concept; Critical Control Points; Involvement of Regulatory Bodies; Establishment of Performance Criteria. **International Control of Microbiology**. **National Legislation, Guidelines & Standards Governing Microbiology**: Canada; European Union; Japan. **Predictive Microbiology & Food Safety**. **Process Hygiene**: Involvement of Regulatory Bodies

Further Reading

Codex Alimentarius Commission (1996) *Terms and Definitions used in Risk Analysis*. Doc. CX/EXEC 96/43/6. Annex 1.

FAO/WHO (1995) *Applications of Risk Analysis to Food Standard Issues. Report of a Joint FAO/WHO Consultation.*

FAO/WHO (1997) *Risk Management and Food Safety. Report of a Joint FAO/WHO Consultation.*

Notermans S, Zwietering MH and Mead GC (1994) The HACCP concept: identification of potentially hazardous micro-organisms. *Food Microbiology* 11: 203–214.

World Trade Organization (1994) *The Results of the Uruguay Round of Multilateral Trade Negotiations: the Legal Texts. Agreement on the Application of Sanitary and Phytosanitary Measures*. MTN/FA II-A1A-4.

RAPID METHODS FOR FOOD HYGIENE INSPECTION

Matthias Upmann, Institute of Meat Hygiene, Veterinary University of Vienna, Austria
Christine Bonaparte, Department of Dairy Research and Bacteriology, Agricultural University, Vienna, Austria

Introduction

Supplying consumers with microbiologically safe products is a high priority with regulatory authorities worldwide. But, since recognizing that governmental supervision cannot assure absolute food safety, strong emphasis is placed on the manufacturer's responsibility for the hygienic and toxicological quality of foods, limiting the state's task to the 'control of the control'. To meet these product liability demands, the food industry increasingly relies on process control systems and longitudinally integrated quality and safety assurance programmes.

The underlying idea is that safety and quality of the products are controlled best through effective management of those processing areas where hazards may arise. After assessing the risks associated with the food, processing steps are selected where preventive measures will lead to the elimination of the hazard. Establishing critical limits within these processing steps and monitoring relevant parameters will result in its control. But, with respect to microbial hazards such a systematic approach known as hazard analysis critical control point (HACCP) system suffers from slow and cumbersome conventional methods in microbiology which neither allow rapid evaluation of raw materials on delivery nor 'on-line' control measures during processing. Even with end-product testing, they often permit only a retrospective assessment of the food's microbiological condition, since many foods are highly perishable. Therefore, much effort has been made to develop methods which enable a more rapid estimation of the microbiological quality of foods.

Microbiological Examination of Foods

General Considerations

To get reliable results from microbiological examination of foods many factors must be taken into consideration. Firstly, 'food' is an extremely varied matrix which contains infinite arrays of ingredients, shows a high variability in physical composition, is subjected to multifold processing technologies and is stored under many different conditions. Furthermore, its intrinsic flora may consist of high numbers of typical quality indicating microorganisms as in the case of fermented products. Also, they may contain varying amounts of shelf-life limiting or even hazardous microorganisms. On the other hand there are numerous sterilized products. In contrast to chemical and physical contaminants, microorganisms are mostly heterogeneously distributed in foods and their concentration seldom remains constant. Additionally, microbial cells may be injured sublethally due to food manufacturing processes or food ingredients, thus escaping detection if no preventive measures are taken. The same problem may occur when a high background flora prevents selective isolation of specific bacteria.

Methodological Requirements

Three main categories of analytical procedures can be distinguished. Firstly, analysis may be directed towards qualitative detection of specific microorganisms (presence/absence tests). Secondly, analysis may be performed in order to quantify the total microbial number, special indicative groups or specific microorganisms. Thirdly, characterization of isolated microorganisms may be desired.

Considering the broad range of analytical procedures available particular requirements were defined which an optimum method should meet. High sensitivity, which is defined as the lowest amount of microorganisms detectable, should be of primary importance. Likewise, high accuracy is essential. The analytical result should meet the true value and repetitions of the analytical procedure should ideally give the same results (i.e. high precision). As explained above, rapidity is another important factor. Under practical conditions economic considerations favour the use of simple, inexpensive, universally applicable and less laborious methods. Furthermore, the testing system must operate at a high level of hygienic safety, as for instance provided by self-contained units, especially if the user group consists of non-specialists.

Unfortunately, an optimum technique covering all requirements does not exist. In particular, the accuracy of different analytical techniques is quite different, hence validations by in-laboratory and/or inter-laboratory comparisons against commonly agreed standard methods are necessary. European standards for validation and official acknowledgement of alternative microbiological methods are now dealt with at the technical committee of the European Committee for Standardization (CEN/TC) in Brussels.

Improving Methodological Rapidity

Ideally, rapid methods should enable such a quick estimation of microbiological parameters that food manufacturers are able to take corrective actions immediately in the course of the manufacturing process. However, the majority of methods characterized as 'rapid' do not meet this demand. Nevertheless, they offer a more or less pronounced advantage in analytical time compared to their conventional equivalent by eliminating laborious and/or subjective elements through mechanization and automation.

Improved rapidity can be applied at each step of the analysis, i.e. the sampling process, sample treatment and detection/enumeration procedure. Although labour-saving and automated methods speed up the processes of sampling and sample treatment, thus improving the laboratory's output, the influence on the total analysis time is usually negligible due to the incubation time required for traditional culture-based methods. A real shortening of the analytical time can only be obtained if alternatives to the traditional incubation methods are developed.

Training of Inspection Staff

New inspection techniques make great demands on the qualifications of the inspection staff. The numerous analytical options can be confusing and overwhelming to the user. It is the user who decides whether a microbiological test is reasonable – the fact that it is applicable does not mean that it is necessary or useful – and which technique should be applied. Because new analytical procedures are based on various technologies and designs, their performances are highly variable. Moreover, many automated instruments exhibit a so-called 'black-box' phenomenon. The utmost caution is advised with such instruments; reliable results are only feasible when they are properly maintained and calibrated. Test results must never be accepted in an uncritical manner.

Therefore, incorporation of accelerated methods into the microbiological analytical repertoire must be accompanied by training of the inspection staff. By following the literature or attending occasional meetings, one is not likely to be able to keep abreast of the rapidly changing and developing field of inspection techniques.

Methods with Improved Rapidity

Sampling

The sampling method depends on the material (processing environment, solid, semisolid or liquid foods), the surface structure (smooth/rough, horizontal/vertical, flat/curved), and the expected microbial contamination level. Additionally, the practicability on the spot should be considered: the use of electrical sampling devices, for example in the processing areas, may be a problem due to lacking plug sockets.

With a few exceptions, such as ultrasonic sterility testing of heat-treated milk, food samples are taken destructively (excision, scraping) which destroys the integrity of the food. Sampling of foods is almost exclusively performed manually.

On the other hand, surfaces of the food processing environment are sampled by non-destructive methods. Mostly, contact slides or swabbing techniques are employed. The former is often regarded as 'rapid' as there is no necessity for further sample treatment and its simple application, although the incubation time remains unchanged. Other methods, such as manual or mechanical rinsing, do not have any practical importance.

Sample Treatment

During sample treatment, the sample is comminuted, liquefied and homogenized. Subsequently either dilution or enrichment steps may be necessary according to the expected level of microorganisms. Several semi- or fully automated dilution procedures which reduce the laboratory work have been developed (e.g. stomacher, pipetting instruments, gravimetric diluters).

Substituting for microbial enrichment procedures and enabling quantitative results at low microbial concentrations, several physical techniques for extraction and concentration of microorganisms are employed. Techniques such as filtration, centrifugation, ion exchange resins and the very promising area of magnetic separations are mentioned. Furthermore, the polymerase chain reaction (PCR), has become more applicable as a non-cultural means of target amplification for food analysis nowadays.

Microbial Detection, Enumeration and Characterization

New time-saving detection methods utilize principles originally belonging to disciplines such as chemistry, biochemistry, physics or immunology. This development was rendered possible because of major technological advances in data-processors, which allow rapid collection and interpretation of vast amounts of data. Since these methods often measure parameters which are different from the traditional ones, the correlation with traditional methods may be problematic.

The methods can be placed in two categories: (1) direct methods based on the detection of whole cells (single cells or colonies), and (2) indirect methods which measure cell components, metabolites, metabolic activities or changes caused by cell growth. **Table 1** gives a survey of rapid methods used for microbial detection, enumeration and characterization.

Direct Methods

Usually, colony-based techniques cannot be characterized as rapid due to the continuing necessity for incubation, although several devices (dehydrated nutrient pads, spiralplater, laser or image analyser etc.) can help to reduce the total laboratory work. Therefore, rapid direct methods are microcolony or single-cell based.

Microcolony and Single-cell Detection

Microscopical Techniques In order to visualize objects for microscopical examination, colouring agents are used to provide information on the total levels of microorganisms (e.g. methylene blue, acridine orange staining), special bacterial groups (e.g. Gram stains) or specific types of microorganisms (e.g. fluorescently labelled antibodies). In combination with different pre-treatments (e.g. membrane filtration, pre-incubation) and detecting principles (microscope, image analyser), microscopy has developed into a commonly used technique.

Epifluorescent techniques The direct epifluorescent filter technique (DEFT) was originally developed for rapid assessment of bacterial numbers in raw milk. However, the introduction of several pre-treatment techniques has considerably enlarged the range of successful applications.

Homogenized, prefiltered and subsequently enzyme-surfactant-treated food samples are passed through membrane filters and are stained, most commonly with acridine orange which binds to nucleic acids. On epifluorescence-microscopical examination, aggregates of orange fluorescing cells are counted. Another technique uses tetrazolium chloride, which is reduced to purple-coloured formazan by an active cellular respiration apparatus.

By using fluorescently labelled antibodies, specific types of microorganisms can be detected. This antibody-direct epifluorescent filter technique (Ab-DEFT) is especially useful in detecting pathogens. Since pathogens usually occur in foods at low numbers and microbial cell surface antigenicity has to be preserved,

Table 1 Rapid methods for microbial detection, enumeration and characterization in food microbiology: overview

Direct methods
Microcolony and single cell detection
Conventional microscopy
Epifluorescent techniques
Direct epifluorescent filter technique (DEFT)
Antibody direct epifluorescent filter technique (Ab-DEFT)
Membrane filter microcolony fluorescence technique (MMCF)
Flow cytometry
Indirect methods
Methods based on growth and metabolic activity
Optical methods
Colorimetry and fluorometry
Turbidimetry
Pyruvate determination
Thermal methods
Microcalorimetry
Electrical methods
Direct conductimetry/impedimetry
Indirect conductimetry/impedimetry
Radiometry
Immunological methods
Agglutination tests
Immunodiffusion tests
Immunoassays based on labelled antibodies
Immunofluorescent assays (IF)
Radioimmunoassays (RIA)
Enzyme immunoassays (EIA)
Immunomagnetic separation
Methods based on microbial cell components
Luminometry
ATP-bioluminescence
Bacterial bioluminescence ('in-vivo bioluminescence')
Limulus amoebocyte lysate test
Ergosterol determination
Nucleic acid-based methods
DNA probe hybridization
Polymerase chain reaction
Fingerprinting-like methods
Combined methods
Biosensors

special product preparation steps (enrichment, immunocapture, pre-incubation) are necessary.

Short-term incubation of the membrane filters before the staining process, results in the growth of microcolonies. Hence, only viable microorganisms are detected by this 'membrane filter microcolony fluorescence (MMCF)' technique.

DEFT and related techniques have been used for counting bacteria in milk, milk products, water, beverages, raw meat, fish, poultry and food contact surfaces. If large numbers of samples have to be analysed daily, an automated counting procedure linking the microscope to an image-analysing system is advisable. Due to its rapidity and broad applicability DEFT is recommended for quality control, shelf-life prediction, irradiation control, as well as hygiene monitoring. Further details are given in **Table 2**.

Flow Cytometry Flow cytometry enables both qualitative and quantitative analysis of microbial cells in liquids. The sample is injected in a thin, rapidly moving carrier fluid which passes through a light beam. The previously fluorescently labelled cells are detected one by one with a photoelectric unit. By using nonspecific and specific fluorochromes, different wavelengths and measuring at different angles, it is feasible to discriminate between bacteria in mixed populations.

The practical use of flow cytometry is still limited to few examples. However, since the possible applications are numerous, it should be considered as a promising technology in the future. Some methodological properties are given in Table 2.

Indirect Methods

Methods Based on Growth and Metabolic Activity Several promising analytical procedures are based on the detection of microbial growth during incubation. Detection times ranging from a few minutes to 30 h depend on many factors, including inoculum density, microbial growth rate and type of metabolic activity. Generally, detection time is related inversely to the bacterial number: the lower the initial bacterial content, the longer the detection time. According to the physico-chemical properties considered, optical, thermal, electrical, and radiometric methods are distinguished.

Optical Methods

Colorimetry and fluorometry: Specific physical or chemical changes (pH, oxidation/reduction potential, enzymatic transformations) associated with microbial metabolic activity can be indicated by changes in colour, fluorescence or colour intensity of an added reagent dye during sample incubation. Many chromogenic and fluorogenic dyes are used depending on the metabolic change to be shown. A multitude of miniaturized and computer-aided or even fully automated identification systems for pure cultures are based on this principle.

Colorimetry and fluorometry can also be used for quantitative purposes by measuring the required incubation time in order to produce a colour reaction or fluorochrome formation. Broadly known indicators are litmus and bromocresol purple for detecting pH shifts or resazurin, methylene blue, and triphenyltetrazolium chloride as oxidation/reduction indicators.

Fully automated procedures are now available which use reflectance colorimeters or fluorometers. These techniques are applicable for rapid estimation of total microbial numbers or specific microorganisms, for product shelf-life stability, starter culture activity, and antibiotic testing. Fur further details see Table 2.

Turbidimetry: Increasing cell numbers lead to an increase in optical density of liquid growth media. Therefore, a light beam will increasingly we weakened on transillumination when a liquid sample is incubated. By varying the sample dilution, the growth medium and the incubation temperature the result can be narrowed to specific bacterial species or numbers. Some methodological properties are given in Table 2.

Turbidimetry is widely applied in vitamin bioassays and disinfectant testing. It has been used for sterility testing in food quality control. Its application may be limited by background turbidity of foods (fat globules, blood cells, food particles).

Pyruvate determination: Pyruvate is a key compound in bacterial lactose metabolism and can serve as an indicator for milk quality monitoring. Pyruvate is measured indirectly by spectrophotometric detection of reduced nicotinamide–adenine dinucleotide (NADH) which is a cofactor in the enzymatic breakdown of pyruvate. Since somatic cells contribute to the pyruvate content of milk and not all bacteria produce pyruvate, the relation between this metabolite and total microbial count is limited (see Table 2).

Thermal Methods Bacterial growth is accompanied by heat production, which can be used for microcalorimetric estimation of the bacterial content. Highly sensitive calorimeters are necessary to detect the heat generated. Due to multiple interfering factors, microcalorimetry has thus far not assumed any practical importance.

Electrical Methods Measurement of electrical con-

Table 2 Methodological properties of selected rapid methods in food microbiology

Method	*Purpose*			*Detection limit (cells per ml or per g)*	*Rapidity*	*Selected instruments and suppliers*
	qual.	*quant.*	*char.*			
Direct methods						
Epifluorescence microscopy						
DEFT	–	+	–	ca. 10^4	< 1 h	Bio-Foss (Foss Electric, Denmark), COBRA (Biocom, France), Autotrak (A.M. Systems, UK)
Ab-DEFT	+	+	+	ca. 10^3	< 1 h	
MMCF	–	+	–	ca. 10^3	⩾7 h	
Flow cytometry	+	+	+	ca. 10^4	< 0.5 h	BactoScan (Foss Electric, Denmark), ChemFlow (Chemunex, France), Argus Flow Cytometer (Skatron, Norway)
Indirect methods						
Methods based on growth and metabolic activity						
Colorimetry, Fluorimetry	+	+	–	ca. 10^1	0.5–30 h	Omnispec (Wescor, USA), Fluoroskan (Labsystems Oy, Finland)
Turbidometry	+	+	–	ca. 10^2	0.5–30 h	Bioscreen analysing system (Labsystems Oy, Finland), AutoMicrobic System (Vitek Systems, USA), Cobas Bact Centrifugal Analyzer (Roche Diagnostica, Switzerland)
Pyruvate determination	(+)	(+)	–	ca. 10^2		
Conductimetry/impedimetry						
Direct	+[a]	+	–	ca. 10^1	0.5–30 h	Bactometer (Bio Merieux, Germany), BacTrac (Sy-lab, Austria), Malthus (Malthus Instruments, UK), RABIT (Don Whitley Scientific, UK)
Indirect	+	+	–	ca. 10^{-1}	0.5–30 h	Malthus (Malthus Instruments, UK), RABIT (Don Whitley Scientific, UK)
Radiometry	+	+	–	ca. 10^2	1–18 h	Bactec (Johnston Laboratories, USA)
Methods based on microbial cell components						
ATP bioluminescence	–	+	–	ca. 10^4	30 s–2 h	BactoFoss (Foss Electric, Denmark), Biocounter (Lumac/Perstorp Analytical, Netherlands), Luminometry System (Bio-Orbit Oy, Finland), AutoPICOLITE (Packard Instrument Company, USA), Biotrace Luminometer (Biotrace, UK)
ATP bioluminescence in hygiene monitoring	–	(+)	–		< 2 min	HY-LiTE (Merck, Germany), Uni-Lite (Biotrace, UK) Checkmate (Lumac/Perstorp Analytical, Netherlands), Lightning (Idexx, USA), Systemsure (Celsis, USA)
Bacterial bioluminescence	+	+	+	ca. 10^3	< 1 h	
Nucleic acid-based methods						
PCR	+	(+)	+	ca. 10^0 direct in food: ca. 10^4	< 3 days	
Dot Blot	+	(+)	+	ca. 10^4	< 3 days	
Colony hybridization	+	+	+	ca. 10^1	< 3 days	

qual., qualitative result (presence/absence); quant., quantitative result (enumeration); char., microbiological characterization.
+, applicable; (+), applicability restricted; –, not applicable.
[a]If special selective media are available.

ductivity changes have been applied quite successfully for microbial growth detection. Direct and indirect methods have been developed depending on the medium in which the changes are monitored. Using electrodes, impedance, conductance and/or capacitance on an alternating current, are measured continuously.

Conductimetric methods are well established in the dairy industry for monitoring total viable flora, indicator organisms and pathogens. Application to other foods is possible. Furthermore, conductimetry can be useful for producing data for predictive microbiology. Some further details are summarized in Table 2.

Direct conductimetry: Direct methods detect conductivity changes in specific growth media. Nutrient macromolecules are broken down to mostly charged particles, thus increasing the conductivity of the medium with incubation time.

Indirect conductimetry: Indirect conductimetry measures the conductivity reduction in a detection medium, which is caused by absorption of CO_2 produced during microbial metabolism. This method is suitable for food products with low contamination levels, since CO_2 formation can be detected much earlier. Specific growth media are not necessary.

Radiometry Radiometric measurement is also based on the detection of CO_2 release during microbial metabolism. However, in this case isotopically labelled carbon sources in the growth medium are converted into $^{14}CO_2$, the amount of which is measured in a ionization chamber. Table 2 contains more methodological information.

The major drawback of this technique is the use of radioactive material. Therefore, it has not been frequently used in Europe. Alternatively, CO_2 production may also be monitored by infrared spectrophotometry or volumetry. Both principles have been suggested for sterility testing of liquid samples.

Immunological Methods

Today, reactions between inducible animal-derived proteins (antibodies) and specific target molecules (antigens) are widely applied for rapid separation, identification, differentiation and quantification of microorganisms and their toxins.

Agglutination and Immunodiffusion Tests Bacterial agglutination ('clumping') due to the formation of antigen–antibody complexes are well introduced in microbiology. For example, the *Salmonella* differentiation scheme of Kauffmann and White is based on this technique. Similarly, it is used in several commercial latex agglutination kits; antibody coated latex particles show a macroscopically visible agglutination if the antigen is present.

Immunodiffusion has also been applied for a long time. A semisolid medium is inoculated at one spot with the sample and at another spot with a specific antibody. If the corresponding antigen is present, line-shaped immunoprecipitates will develop at the place where diffusing antibodies and antigens meet. Several test kits based on this technique are available.

Immunoassays Based on Labelled Antibodies In most immunoassays the primary antigen–antibody reaction is made detectable by means of labelling the antibodies with marker substances. Immunofluorescent (IF) assays, making use of antibodies labelled with fluorescent reporter molecules, do not enjoy widespread use. Equally, radioimmunoassays (RIA) are not used frequently, due to the inherent disadvantages of handling radioisotopes. Probably the fastest growing and most widely used formats are enzyme immunoassays (EIA) which employ enzyme markers in conjunction with a colorimetric or fluorometric substrate system. By immobilizing the capturing antibody on a solid matrix (microtitre trays, polystyrene or ferro-metal beads, dip-sticks) enzyme-linked immunosorbent assays (ELISA) were created. Other assay formats are conceivable depending on the number of antibodies involved.

Immunoassays are promising because of their sensitivity and rapidity. However, they normally require enrichment of the target bacterium to the level of the assay's detection limit. Disadvantageous false-positive or false-negative results have been significantly reduced by advances in antibody preparation.

Immunomagnetic Separation Immunomagnetic separation (IMS) has proved to be a very efficient method for separating target organisms from food materials and background flora. Antibody-coated paramagnetic particles are mixed with the sample. By exposure to a magnetic field, bound target cells are separated while the sample suspension is removed. A number of procedures may be used for subsequent final detection, such as conventional culturing, microscopy, impedance technology, ELISA, latex agglutination or DNA hybridization, partly involving amplification techniques. In addition to the short separation and concentration time, IMS technology also overcomes the problem associated with unwanted inhibition due to selective media components. Since IMS can be used in conjunction with many final detection technologies, it is expected that several automated analytical procedures will make use of this potent technique in the near future.

Methods Based on Microbial Cell Components

Luminometry

ATP bioluminescence: ATP is the universal intracellular energy carrier in living somatic and microbial cells. Although the level of ATP depends on the cellular type and is influenced by several external factors it seems to be a useful parameter in the estimation of the active microbial population.

The amount of ATP is detected by a bioluminescence reaction; in the presence of ATP and catalysed by the enzyme 'luciferase' the substrate 'luciferin' is oxidized to oxyluciferine. As a 'by-product' of this reaction photons are emitted – one photon of light for each molecule of ATP, provided all other reagents are present in surplus. Therefore, photometrical measurement of the light emitted gives rapid information on the amount of metabolically active cells. However, for microbiological food control additional steps prior to the assay are required in order to remove non-microbial (i.e. intrinsic and somatic) ATP which often greatly exceeds the microbial ATP content.

Nevertheless, ATP bioluminescence is a well established technique in the industry for quality control of raw milk, meat and fish. Further, it has been applied to end-product testing of beer, carbonated beverages, fruit juices, pasteurized or ultra-high temperature milk and dairy products, testing of starter cultures, as well as to shelf-life prediction. Manual and automated luminometers and standardized reagent kits are available from several suppliers.

In recent years, the application of ATP bioluminescence has become important for on-site hygiene control of surfaces in industrial plants. Various low-cost, portable luminometers are commercially available (see Table 2).

Bacterial bioluminescence ('in-vivo bioluminescence'): Bioluminescence also occurs naturally in several bacterial species using flavin mononucleotide as the energy source. Bacterial luciferase mediates the oxidation of aldehydes to fatty acids. Since the so-called *lux* genes encoding for bioluminescence are known, their bacteriophage- or plasmid vector-mediated transfer to other bacteria has been enabled. The resulting light emission is detected by standard luminometers, as with ATP-bioluminescence (see also Table 2).

The specificity of this sensitive and rapid technique depends on the host range of the bacteriophages used. Wide host ranges covering several bacterial species enable the detection of indicator organisms, whereas species- or subspecies-specific bacteriophages are used for the detection of pathogens. In addition, bacterial bioluminescence enables the detection of inhibitory substances (antibiotics, bacteriophages, antimicrobial compounds) by making use of starter cultures containing *lux* genes.

Limulus Amoebocyte Lysate Test The limulus amoebocyte lysate (LAL) test is a simple method for the detection of viable and non-viable Gram-negative bacteria. Certain cell-wall lipopolysaccharides (i.e. endotoxins) of this bacterial group lead to gelation of blood cell (amoebocytes) lysates of the *Limulus polyphemus* crab. Using a dilution row and determining the limit at which no more gel formation occurs, a semi-quantitative estimation of the Gram-negative content is possible.

Several test-kits are available. Mostly used for pyrogen control of pharmaceutical products the LAL-test is applicable for predominantly Gram-negative containing foods such as fresh meat, milk and eggs. Another field of application may be the retrospective assessment of the microbiological quality of heated products.

Ergosterol Determination Ergosterol is a steroidal component of fungal cell membranes. Therefore, the amount of ergosterol present can be related to fungal biomass. The chemical detection procedure includes high-pressure liquid chromatography and detection by ultraviolet absorption. Though yeasts also contain ergosterol in their cell wall, the rapid ergosterol assay seems to be a useful technique for food quality control purposes with respect to fungal invasion.

Nucleic Acid-based Methods The application of new methods based on nucleic acids has greatly stimulated the food microbiology field during recent years.

DNA/RNA Probe Hybridization With the classical DNA probe hybridization assay, microorganisms are collected on a filter, cells are lysed and the liberated DNA or RNA is immobilized in single-stranded form. The nucleic acid is identified by hybridization with radioactively or non-radioactively labelled DNA probes of defined origin.

Hybridization assays for several organisms are available commercially. However, their sensitivity depends on the initial amount of DNA/RNA; currently most of them do not allow detection of bacterial populations below 10^5 bacteria per gram. This disadvantage may be overcome by pre-enriched cultures and/or DNA amplification techniques such as PCR.

DNA hybridization with labelled DNA probes is mainly used for identification or confirmation of pure cultures thus providing qualitative results. It is sometimes considered as an indirect semi-quantitative method by comparison of the signal response intensity with that at the detection limit (for example Dot Blot, see Table 2). Quantitative results can be obtained by colony hybridization.

Polymerase Chain Reaction (PCR) In PCR technology specific DNA sequences are detected and multiplied in vitro. Small but specific DNA primers are added to the sample's liberated and denatured target DNA. If these primers meet a complementary nucleic acid base sequence, they will hybridize. Subsequently, a thermostable DNA polymerase will elongate the primers according to the complementary base sequence given by the target DNA. This cyclic process is usually repeated about 30 times, amplifying the target DNA by 10^5–10^7-fold.

The primers can be designed for different purposes. Specific primers are targeted to a known virulence factor of a single species and allow simultaneous detection and identification, whereas multiple primers with a broader spectrum are suitable for several species. Theoretically, several microorganisms in a sample can be detected at the same time with so-called multiplex PCR.

PCR is a sensitive, specific and rapid microbial detection (presence/absence testing) method, which provides qualitative results. Theoretically, a single molecule of target DNA can be detected within one working day. However, the PCR can be inhibited or its sensitivity severely reduced when applied to food samples (see Table 2). High amounts of fat and proteins in complex foods as well as certain components of selective culture media inhibit the enzymatic DNA amplification reaction. Additionally, the small PCR reaction volumes usually cannot be accommodated to large sample sizes dictated by low contamination levels (e.g. 25 g of product for *Salmonella* presence/absence testing).

Another problem in common with other rapid methods is the fact that PCR technology cannot distinguish between genetic material from dead and living cells. Therefore an mRNA-based modification of PCR (NASBA, nucleic acid sequence based amplification) has been developed which is indicative for living cells in the sample.

Routine application of PCR technology in food microbiology usually requires special sample pretreatment (e.g. purification, cell concentration, culturing or selective enrichment, immunomagnetic separation) adapted to each specific sample matrix.

Fingerprinting-like Methods Another group of DNA assays result in the determination of DNA patterns. These fingerprinting-like methods, for example restriction fragment length polymorphism (RFLP) analysis or random amplification of polymorphic DNA (RAPD), can be used for high-resolution characterization of microorganisms from pure cultures. They may gradually replace traditional serotyping systems. Such techniques may be useful for elucidating the contamination routes of pathogens in the food chain.

Combined Methods

Undoubtedly, the next major step will be the combination of advances from different technologies. One very promising example of combined methods is discussed here.

Biosensors Biosensors are a relatively new area in the automated food microbiology. They consist of immobilized, biologically sensitive material (e.g. enzymes, antibodies, antigens, nucleic acids) coupled with or in close proximity to a receiving transducer unit which gives an electrical, optical or thermal signal when the sensor reacts with its target. The intensity of the signal is proportional to the concentration of the target.

Due to problems with long-term stability, reusability and sterilizability, biosensors have so far been mostly used for detecting chemical substances. Nevertheless, their future potential is enormous, since they can offer a very sensitive and accurate 'on-line' control system for food manufacturing processes.

See also: **ATP Bioluminescence**: Application in Meat Industry; Application in Dairy Industry; Application in Hygiene Monitoring; Application in Beverage Microbiology. **Biochemical and Modern Identification Techniques**: Introduction; Food Spoilage Flora (i.e. Yeasts and Moulds); Food-poisoning Organisms; Enterobacteriaceae, Coliforms and *E. coli*; Microfloras of Fermented Foods. **Direct (and Indirect) Conductimetric/Impedimetric Techniques**: Food-borne Pathogens. **Direct Epifluorescent Filter Techniques (DEFT)**. **Electrical Techniques**: Introduction; Food Spoilage Flora and Total Viable Count (TVC); Lactics and other Bacteria. **Enzyme Immunoassays**: Overview. **Flow Cytometry**. **Hazard Appraisal (HACCP)**: The Overall Concept; Critical Control Points; Involvement of Regulatory Bodies; Establishment of Performance Criteria. **Immunomagnetic Particle-based Techniques**: Overview. **Molecular Biology – in Microbiological Analysis**. **Nucleic Acid-based Assays**: Overview. **PCR-based Commercial Tests for Pathogens**. **Predictive Microbiology & Food Safety**. **Total Counts**: Microscopy. **Total Viable Counts**: Metabolic Activity Tests; Microscopy.

Further Reading

Baumgart J (1996) Schnellmethoden und Automatisierung in der Lebensmittelmikrobiologie. *Fleischwirtschaft* 76: 124–130.

Blackburn C de W (1993) Rapid and alternative methods for the detection of salmonellas in foods. *J. Appl. Bacteriol.* 75: 199–214.

Hofstra H, van der Vossen JMBM and van der Plas J (1994) Microbes in food processing technology. *FEMS Microbiol. Rev.* 15: 175–183.

Karwoski M (1996) Automated direct and indirect methods in food microbiology: a literature review. *Food Rev. Int.* 12: 155–174.

Otero A, García-López M-L and Moreno B (1998) Rapid microbiological methods in meat and meat products. *Meat Sci.* 49 (Suppl.): S179–S189.

Patel P (ed.) (1994) *Rapid Analysis Techniques in Food Microbiology.* London: Blackie Academic & Professional.

Safarík I, Safaríková M and Forsythe SJ (1995) The application of magnetic separations in applied microbiology. *J. Appl. Bacteriol.* 78: 575–585.

Skovgaard N and Jakobsen M (eds) (1995) *Rapid Methods in Meat Microbiology; Advantages and Drawbacks.* Utrecht, NL: ECCEAMST Foundation.

Spencer RC, Wright EP and Newsom SWB (eds) (1994) *Rapid Methods and Automation in Microbiology and Immunology.* Andover, UK: Intercept.

Swaminathan B and Feng P (1994) Rapid detection of food-borne pathogenic bacteria. *Annu. Rev. Microbiol.* 48: 401–426.

Vanne L, Karwoski M, Karppinen S and Sjöberg A-M (1996) HACCP-based food quality control and rapid detection methods for microorganisms. *Food Control* 7: 263–276.

Wawerla M, Eisgruber H, Schalch B and Stolle A (1998) Zum Einsatz der Impedanzmessung in der Lebensmittelmikrobiologie. *Arch. Lebensmittelhyg.* 49: 76–89.

Wolcott MC (1991) DNA-based rapid methods for the detection of foodborne pathogens. *J. Food Protect.* 54: 387–401.

Redox Potential *see* **Ecology of Bacteria and Fungi in Foods**: Influence of Redox Potential and pH.

REFERENCE MATERIALS[1]

P H In't Veld, National Institute of Public Health and the Environment, Microbiological Laboratory for Health Protection, Bilthoven, The Netherlands

Introduction

Reference materials (RMs) are defined by the International Organization for Standardization (ISO) in Guide 30 as 'a material or substance, one or more properties of which are sufficiently well established to be used for the calibration of an apparatus, the assessment of a measurement method, or for assigning values to materials'. Besides RMs, there are also certified reference materials (CRMs). ISO Guide 30 defines a CRM as 'an RM one or more of whose property values are certified by a technically valid procedure, accompanied by or traceable to a certificate or other documentation which is issued by a certifying body'. An example of a certifying body is the Community Bureau of Reference (BCR) of the European Commission (EC).

RMs and CRMs have to fulfil a number of requirements, including:

- representative for its intended use
- homogeneity specified within defined limits
- stability specified within limits over a specified period of time.

'Representative' means that the RM has to resemble routine samples as much as possible. It is not possible to produce a stable and homogeneous RM for every type of food or environmental sample that is examined in practice, so compromises have to be made in order to fulfil the three requirements listed above.

Homogeneity is very important as heterogeneous RMs will lead to extra variation in results. Heterogeneity often exists in natural samples but has to be eliminated for RMs as far as possible if the performance of a method or laboratory is to be assessed. The distribution of microorganisms in a homogeneous sample (for example, a fluid) is described by a Poisson distribution. Due to the difference between this type of distribution and the normal distribution which can be applied to most other types of analysis (for example, chemical) a transformation of the counts is needed. For microbiological counts a $\log_{10}$ transformation was used as in most cases this yielded normally distributed data.

The inherent instability of microorganisms has long hampered the development of microbiological RMs. Depending on its use, an RM must remain stable for

[1] Whole article taken from *International Journal of Food Microbiology* 45 (1998): 35–41, Elsevier, by PH In't Veld, replacing Figs 2 and 3.

at least several months at a specified storage temperature.

The applications of (microbiological) (C)RMs include: testing accuracy in individual laboratories; comparing the performance of different laboratories; developing and validation of methods and media; testing the influence of matrix ingredients and competitive microorganisms; standard material for collaborative studies; use in first-line quality control.

Development of (C)RMs by the RIVM/EC

With financial support from the EC the RIVM (National Institute of Public Health and the Environment) had the opportunity to investigate further microbiological RMs. Contracts with the EC-BCR (presently called the Standard, Measurement and Testing Programme) started in 1986 and ended in 1995. The RMs investigated consist of gelatine capsules containing artificially contaminated spray-dried milk. The method of production of these materials is presented in **Figure 1**.

RMs have been developed for use in the field of water and food microbiology. In the framework of the BCR projects a total of nine different RMs have been evaluated in collaborative studies involving a minimum of 25 laboratories throughout the European Union. Based on the results of these studies, a number of RMs were subjected to the BCR certification procedure. In total, six RMs have been certified using this procedure. In **Table 1** the various RMs evaluated in collaborative studies in the field of water and food microbiology and those certified by the BCR are listed. The production process presented in Figure 1 was chosen, as in this way highly contaminated milk powders (HCMPs) are produced that can be used as stock for the production of batches of RMs. These HCMPs can be stored for many years. A decrease in the level of contamination of the HCMP can be compensated for by adjusting the mixing ratio.

The homogeneity will be affected by the ratio of HCMP to sterile milk powder. At first it appeared to be impossible to produce a homogeneous RM for *Listeria monocytogenes* (at a level of ca. 5 cfu per capsule) as the first HCMP produced needed to be diluted about 10^7 times to be used for presence/absence testing. A second HCMP was produced with a lower level of contamination and problems were no longer encountered in producing homogeneous batches of RM at a low level of contamination.

When the production of the RMs started, mixing was done with equipment containing 17 l drums. Later a new method was introduced. This procedure is used by pharmacists for the preparation of medicines, and uses a pestle and mortar to mix repeatedly small amounts (up to ca. 200 g of powder) with a ratio of 1 part contaminated powder to 1 part sterile powder. Using this procedure, homogeneity improved for all RMs. For most of the RMs produced at present no significant deviation from Poisson distribution can be detected.

For the production of stable RMs one needs to be patient. After the spray-drying process the HCMPs are stored, sometimes for several years, in order for them to stabilize. The time needed for stabilization

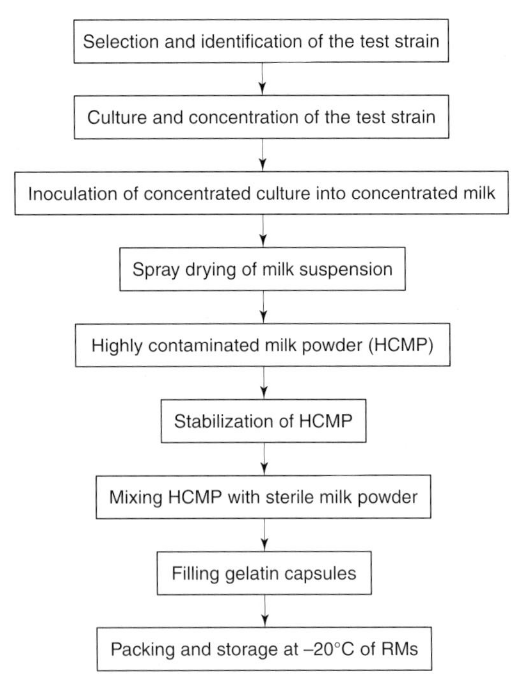

Figure 1 Production process for microbiological reference materials (RMs).

Table 1 Reference materials evaluated and certified by the BCR

Field of application	*Type of reference material*	*Certified by BCR*
Water microbiology	*Clostridium perfringens*	No
	Enterococcus faecium	Yes (CRM 506)
	Enterobacter cloacae	Yes (CRM 527)
	Escherichia coli	Yes (CRM 594)
	Staphylococcus warneri	No
	Salmonella typhimurium (p/a)[a]	No
Food microbiology	*Bacillus cereus*	Yes (CRM 528)
	Clostridium perfringens	No
	Escherichia coli	No
	Listeria monocytogenes (p/a)[a]	Yes (CRM 595)
	Salmonella typhimurium (p/a)[a]	Yes (CRM 507)
	Staphylococcus aureus	No

[a]Reference material for presence/absence testing.

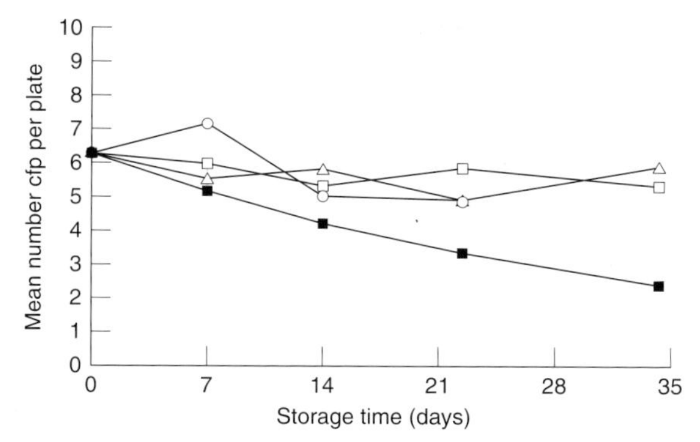

Figure 2 Influences on microbial growth in foods.

strongly depends on the type of organism dried. For example, *Bacillus cereus* spores needed no time for stabilization, while on the other hand it took ca. 2 years before a stable RM could be produced containing *Enterobacter cloacae*. The same was found for *Listeria monocytogenes*. The effect of the age of the HCMP for *E. cloacae* was tested at a storage temperature of 5°C. Lowering the storage temperature from 5 to − 20°C improves the stability of the RM. This was first shown for the *Salmonella typhimurium* RM. Stability at 5°C was about 6 months; lowering the temperature to − 20°C proved that the RM is stable for at least 4 years (the stability testing of CRM for *S. typhimurium* continues).

For shipment of RM and especially for CRM their stability at higher storage temperatures is important as shipment around the world at − 20°C is very expensive. Stability at higher temperatures is normally determined over a period of 4 weeks, examining at weekly intervals. Storage temperatures used are − 20°C (reference), 22°C, 30°C and 37°C. An example of the results of the stability test at higher temperatures is presented in **Figure 2** for the *S. panama* RM.

Routine use of reference materials

Routine use of RMs means that for each series of analysis RMs are examined. A series of analysis is defined as a number of measurements carried out by one technician in one part of a day, using one method, one batch of media etc. The counts obtained from examining the RM can be used for the preparation of a control chart. For this the following procedure is recommended. The first step in the preparation of a control chart is the determination of the (control) limits. This is done by analysing 20 capsules of one batch of RMs singly (i.e. no duplicate counts – although normal tests of real samples should, of course, always be done in duplicate), preferably on different days and by different technicians using the equipment normally used in the examination. It is

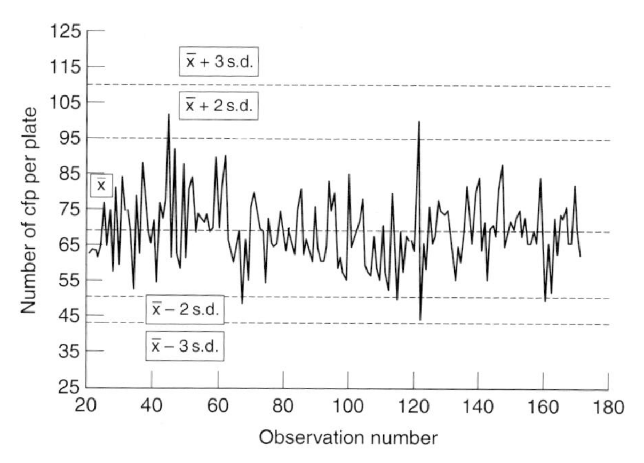

Figure 3 The HACCP concept.

also possible to construct a control chart from 10 measurements, in which case, a new control chart should be constructed as soon as 20 observations have been gathered. For calculation of the control limits the colony counts (c) are $\log_{10}$ transformed, such that $x = \log c$. The mean (x) and the standard deviation (s) are calculated as follows:

$$x = \sum x_i / n \text{ and } S = 0.8865R$$

where:

$$R = \sum (x_i - x_{i-1})/(n-1)$$

n = number of observations and x_i = the ith observation. The calculation of R (the so-called moving range) is based on the sum of the absolute differences between the first and the second count, the second and the third count and so on. This method of calculating the standard deviation is less influenced by the extreme counts (more robust) sometimes found in microbiology.

The control limits (based on the log-transformed data) are given by:

- warning limits: $\bar{x} \pm 2s$
- action limits: $\bar{x} \pm 3s$.

The control chart is constructed by back-transformation to the normal scale of the values for the mean (x), and the upper and the lower action and warning limits. Back-transformation will lead to asymmetrical distances for the upper and the lower warning and action limits around the mean. An example of a control chart is presented in **Figure 3**. The results for the 20 capsules analysed for the calculation of the limits should also be indicated in the control chart (this can be a separate chart). Whether or not these counts meet the criteria stated below (meaning that the analytical process is under control) should be checked. If results from one (or more) of these 20 capsules are out of control, the cause of the erroneous result should be identified and a decision about the validity of the result should be taken. If the result is not valid it should not be used to calculate the control limits.

Once the chart is constructed one capsule is analysed singly for each series of analyses and the result is indicated in the control chart. The result is out of control if:

- there is a single violation of the action limit ($\bar{x} \pm 3s$)
- two out of three observations in a row exceed the same warning limit ($\bar{x} \pm 2s$)
- there are nine observations in a row on the same side of the mean
- six observations in a row are steadily increasing or decreasing.

If the results are out of control, the cause should be identified and a decision should be taken about the validity of the results of the particular series of analysis.

A new control chart is constructed if necessary, for example when a shift in the mean level occurs. The decision to construct a new control chart is taken by the laboratory management.

Use of Certified Reference Materials

CRMs are used on an occasional basis as the availability of these materials is limited and the price high. CRMs are provided with a certificate and in most cases a certification report. The certificate states the certified values, including their 95% confidence limits. These certified values are obtained from a large number of examinations carried out by the participating laboratories during the certification study. They cannot be used directly by a laboratory examining only a few capsules. For use by a laboratory, user tables are prepared in which the 95% confidence limits are given for different combinations of capsules and replicates per capsule. The limits are calculated using the certified value and variance components calculated using $\log_{10}$-transformed counts. Back-transformation of the values obtained for the upper and lower limit will give the limits on the normal scale. An example of a user table is presented in **Table 2** for the *B. cereus* CRM enumerated on mannitol egg yolk polymyxin (MEYP) agar and incubated for 24 h at 30°C.

Before a recommendation can be made on the optimal number of capsules to be examined, a second aspect has to be taken into consideration. This relates to the difference between the certified value and the observed geometric mean value found by a laboratory (i.e. that can be detected). In statistical terms this is called the power of the analysis. This power analysis has been determined for the CRMs. The power of the test is defined as the chance that the experimental observations (based on the geometric mean count by a laboratory) do not conform to the expected (i.e. the certified) value when in fact no difference does exist; this power is denoted by $1 - \beta$. Similarly, the term $1 - \alpha$ refers to the chance that the observations conform to the expected value when this is true. The probability of $1 - \alpha$ is for most experiments set at 95%, meaning

Table 2 The 95% confidence limits of the geometric mean number of cfu of *Bacillus cereus* CRM per 0.1 ml on mannitol egg yolk polymyxin agar after 24 h incubation for different combinations of capsules and replicates

	24 h incubation			
	Lower limit		*Upper limit*	
Number of capsules analysed	*1*[a]	*2*[a]	*1*[a]	*2*[a]
1	38	40	79	74
2	42	43	72	69
3	43	45	70	67
4	44	46	68	66
5	45	46	67	66
6	45	46	66	65
7	46	47	66	65
8	46	47	66	65
9	46	47	65	64
10	46	47	65	64
11	47	47	65	64
12	47	47	65	64

[a]Replicate number.

Table 3 Lower and upper geometric mean values at a power of at least 0.8 in relation to the number of capsules and replicates examined using the *Bacillus cereus* certified reference materials on mannitol egg yolk polymyxin agar after 24 h incubation at 30°C.

Number of capsules examined	*Number of replicates examined*	*Lower geometric mean*	*Upper geometric mean*
1	1	32.5	87.7
	2	36.7	77.7
2	1	37.2	76.7
	2	40.4	70.6
3	1	39.4	72.4
	2	42.1	67.7
4	1	40.8	69.9
	2	43.1	66.2
5	1	41.7	68.4
	2	43.7	65.3
6	1	42.4	67.3
	2	44.2	64.5
7	1	42.9	66.5
	2	44.6	63.9
8	1	43.3	65.9
	2	44.9	63.5
9	1	43.7	65.3
	2	45.1	63.2
10	1	44.0	64.8
	2	45.3	62.9

that α is set at 5%. So, also the power of the test should be as high as possible, as one does not want to draw the wrong conclusion. For the power $(1 - \beta)$ a value between 0.8 and 0.95 is mostly used. For the power analysis of microbiological RMs a power of 0.80 is used.

The smaller the difference between the certified value and the observed geometric mean, the more capsules need to be examined in order to draw conclusions with the same power. The size of the difference has to be chosen in relation to what difference is acceptable in practice or what difference will have a consequence for the judgment of the results. In **Table 3** the lower and upper value for the observed geometric mean values for a power of ≥ 0.8 for various combinations of capsules and replicates using the *B. cereus* CRM is presented. Based on these observations it is recommended to examine five capsules in duplicate for this CRM.

See also: ***Bacillus***: Detection by Classical Cultural Techniques. **Biosensors**: Scope in Microbiological Analysis. **Direct (and Indirect) Conductimetric/Impedimetric Techniques**: Food-borne Pathogens. **Electrical Techniques**: Food Spoilage Flora and Total Viable Count (TVC). **Enrichment Serology**: An Enhanced Cultural Technique for Detection of Foodborne Pathogens. **Enterobacteriaceae, Coliforms and *E. coli***: Classical and Modern Methods for Detection/Enumeration. **Flow Cytometry**. **Hydrophobic Grid Membrane Filter Techniques (HGMF)**. **Immunomagnetic Particle-based Techniques**: Overview. ***Listeria***: Detection by Classical Cultural Techniques; Detection by Commercial Enzyme Immunoassays; Detection by Colorimetric DNA Hybridization; *Listeria monocytogenes* – Detection by Chemiluminescent DNA Hybridization; *Listeria monocytogenes* – Detection using NASBA (an Isothermal Nucleic Acid Amplification System. **Nucleic Acid-based Assays**: Overview. **PCR-based Commercial Tests for Pathogens**. **Petrifilm – An Enhanced Cultural Technique**. **Polymer Technologies for Control of Bacterial Adhesion**. ***Salmonella***: Detection by Classical Cultural Techniques; Detection by Latex Agglutination Techniques; Detection by Enzyme Immunoassays; Detection by Colorimetric DNA Hybridization; Detection by Immunomagnetic Particle-based Assays. ***Staphylococcus***: Detection by Cultural and Modern Techniques. **Water Quality Assessment**: Modern Microbiological Techniques.

Further Reading

Anonymous (1992) *ISO Guide 30*, 2nd edn. *Term and Definitions used in Connection with Reference Materials.* Geneva, Switzerland: International Organization for Standardization.

Beckers HJ, van Leusden FM, Meijssen MJM, Kampelmacher EH (1985). Reference material for the evaluation of the standard method for the detection of salmonellas in foods and feeding stuffs. *J. Appl. Bacteriol.* 59: 507–512.

Griepink B (1989) Aiming at good accuracy with the Community Bureau of Reference (BCR). *Quim. Anali.* 8: 1–21.

Heisterkamp SH, Hoekstra JA, van Strijp-Lockefeer NGWM et al (1993) *Statistical Analysis of Certification Trials for Microbiological Reference Materials.* Report EUR 15008 EN. Brussels: Commission of the European Communities, Community Bureau of Reference.

In't Veld PH, van Strijp-Lockefeer NGWM, van Dommelen JA, de Winter ME, Havelaar AH and Maier EA (1995) *The Certification of the Number of Colony Forming Particles of* Bacillus cereus *in a 0.1 ml Suspension of Reconstituted, Artificially Contaminated Milk Powder (CRM 528).* Report EUR 16272 EN. Brussels: Commission of the European Communities, Community Bureau of Reference.

In't Veld PH, van Strijp-Lockefeer NGWM, Havelaar AH and Maier EA (1996) The certification of a reference material for the evaluation of the ISO method for the detection of *Salmonella*. *J. Appl. Bacteriol.* 80: 496–504.

In't Veld PH, Havelaar AH and van Strijp-Lockefeer NGWM (1999) The certification of a reference material for the evaluation of methods for the enumeration of *Bacillus cereus*. *J. Appl. Bacteriol.* 26: 266–274.

Mooijman KA, in't Veld PH, Hoekstra JA et al (1992) *Development of Microbiological Reference Materials.* Report EUR 14375 EN. Brussels: Commission of the European Communities, Community Bureau of Reference.

Nelson LS (1985) Interpreting Shewhart $\bar{x}$ control charts. *J. Quality Technol.* 17: 114–116.

van Dommelen JA (1995) *Statistics for the Use of Microbiological (Certified) Reference Materials.* Report 281008008. Bilthoven, The Netherlands: National Institute of Public Health and the Environment.

Regulatory Bodies

see **Hazard Appraisal (HACCP)**: Involvement of Regulatory Bodies.

RHODOTORULA

Yeehn Yeeh, Institute of Basic Science, Inje University, Obang-dong, South Korea

Characteristics of the Genus

Rhodotorula belongs to a genus of imperfect yeasts within the family Cryptococcaceae, which are the anamorphic yeasts. Some species are of importance in food spoilage or in the fermentation industry. These yeasts are also considered to be medically important and have been reported to be allergenic. Positive skin tests in humans have been reported. The vegetative form of the organisms is usually a single spheroid, oval or elongated cell in malt extracts. The colonies are often reddish, orange or yellow as a result of synthesis of carotenoid pigment in malt agar cultures. Some of these organisms have a mucous appearance due to capsule formation, but others are pasty, or dry and wrinkled. Reproduction occurs by multilateral budding. *Rhodotorula* yeasts have an exclusively oxidative pathway of energy metabolism. They are all non-fermentative. They are considered to be unable to assimilate inositol as sole carbon source whereas most of organic acids and alcohols can be utilized. Strains of some species form pseudo- or true hyphae. Ascospores or ballistospores are not formed. Starch-like substances are not synthesized. Strains of some species such as *Rhodotorula glutinis* become sensitive to the presence of salt after mild heat treatment.

Rhodotorula can grow in a solid minimal synthetic medium without vitamins, grows well at 30°C and does not produce ethanol when grown on high-glucose media in aerobic conditions. Twenty-two species are generally known in the genus: *Rhodotorula acheniorum*, *R. acuta*, *R. araucariae*, *R. aurantiaca*, *R. aurantiaca B*, *R. bacarum*, *R. bogoriensis*, *R. flava* (*R. macerans*, *Cryptococcus flavus*), *R. glutinis* var. *glutinis*, *R. glutinis* var. *rubescens*, *R. gracilis*, *R. graminis*, *R. hylophila*, *R. lactosa*, *R. minuta* (*R. tokyoensis*), *R. minuta* var. *minuta*, *R. muscorum*, *R. pallida*, *R. philyla*, *R. pustula*, and *R. rubra* (*R. mucilaginosa*, *R. pilimanae*).

Occurrence

Rhodotorula species are isolated from bark-beetles, tree exudates, various types of plants and vegetables, soil, fresh water and clinical specimens including gastrointestinal tracts. They are also frequently encountered in coastal sediments and marine environments. *Rhodotorula* spp. are found in the omnivorous crabs *Aratus pisonii* and *Goniopsis cruentata*, and in the clam *Anomalocardia brasiliana*, but are rare in the clam *Tagelus plebeius* which thrives in mostly submerged, sandy sediments. *Rhodotorula* species are also associated with leaf-cutting ants, and the larvae and imago of mosquitoes; they are a component of the normal microflora of blood-sucking mosquitoes and have no pathogenic effect on the mosquito *Aedes aegypti*. In humans, *Rhodotorula* species occur rarely in the oral cavity in healthy individuals, but are found in the microflora of patients with oral cancer or psoriasis. They colonize terminally ill patients.

Identification of the Genus

Identification of species on the basis of morphological, nutritional and physiological tests often takes days to weeks to complete. Genetic methods are now preferred; these include the use of multiple segment-specific oligonucleotide priming of a region of the large subunit rDNA in a polymerase chain reaction. The 'hot start' reaction is used with one universal external delimiting primer and one internal species-specific primer. Two specific primers are tested (e.g. a primer for a biologically similar group of *Rhodotorula* species and a species-specific primer for *Rhodotorula mucilaginosa*). The universal rDNA segment is amplified in the absence of specific target DNA, while in the presence of target DNA the specific primer region is amplified. The technique is accurate within two base position differences when a 24 nucleotide-specific primer is used. The technique can be used to identify yeasts from single and mixed populations, including other marine eukaryotes.

Identification of *Rhodotorula* is also correct with Fongiscreen 4H and Rapidec albicans. Both tests can be rapid and easy-to-use tools in the clinical microbiology laboratory. Identification results of *Rhodotorula* species are available with MicroScan panels, Vitek cards and the API 20C strips, due to their high reliabilities.

For the differentiation and presumptive identification of clinically important yeasts such as *Rhodotorula*, CHROM-agar *Candida* medium is also recommended for recognition of mixed yeast cultures and differentiation of the common clinically yeast species (**Tables 1 and 2**).

Table 1 Characteristics of *Rhodotorula* species

	Assimilation									
	NO_3	*Me*	*Er*	*Ra*	*Mz*	*Ar*	*Tr*	*Xy*	*Ce*	*Su*
Rhodotorula acheniorum	+	+	+	+	+	+	+	+	+	+
Rhodotorula aurantiaca	+	–	–	–	+	v	v	+	v	+
Rhodotorula glutinis	+	–	–	+	+	v	+	+	+	+
Rhodotorula graminis	+	–	–	+	–	v	+	+w	v	+
Rhodotorula lactosa	+	+	–	+	+	+	+s	+	+	+
Rhodotorula minuta	–	–	v	–	v	+(–)	+	+(–)	v	+(–)
Rhodotorula rubra	–	–	–	+	v	v	+	+	v	+

Key: NO_3, nitrate; Me, melibiose; Er, erythritol; Ra, raffinose; Mz, melezitose; Ar, L-arabinose; Tr, trehalose; Xy, D-xylose; Ce, cellobiose; Su, sucrose; +, reaction positive; – reaction negative; v, some strains give a positive reaction, others a negative one; +w, reaction weakly positive; +s, reaction slow; +(–), reaction positive, seldom negative.

Table 2 Characteristics of *Rhodotorula* species

	Assimilation								
	Ga	*La*	*Ri*	*Rh*	*Mn*	*Ss*	*Rb*	*Ci*	*Sc*
Rhodotorula acheniorum	+	+w	+	–	+	–	+w	+	+
Rhodotorula araucariae	+	–	–	–	+	–	+	–	–
Rhodotorula aurantiaca	+	–	–	–	+	–	v	v	+
Rhodotorula glutinis	+	–	v	v	v	v	v	v	+
Rhodotorula graminis	+	–	v	v	+(–)	–	v	+w or–	+
Rhodotorula lactosa	–	+s or –	–	+	+	–	+s	+s	+
Rhodotorula minuta	v	v	v	–	+(–)	–	+(–)	–	+ or w
Rhodotorula rubra	v	–	v	v	v	–	+(–)	v	+w

Key: Ga, galactose; La, lactose; Ri, D-ribose; Rh, rhamnose; Mn, mannitol; Ss, soluble starch; Rb, ribitol; Ci, citric acid; Sc, succinic acid; +, reaction positive; –, reaction negative; v, some strains give a positive reaction, others a negative one; +w, reaction weakly positive; +s, reaction slow; +(–), reaction positive, seldom negative.

Biological Activity

Enzymes and Biochemicals

Phosphoric Monoester Hydrolases *Rhodotorula* species have potential application as a source of lipases. The purified enzymes from *R. glutinis* are all glycoproteins.

Lipolytic Enzyme Lipolytic yeasts occur in *Rhodotorula* and are involved in the breakdown of fats in dairy and meat products. *Rhodotorula rubra* shows phospholipase activity, which plays a significant role in damaging cell membranes and invading host cells. Since *R. rubra* has only rarely been demonstrated as a pathogen in humans, absence of dimorphism and low ability of adherence lessen the importance of high phospholipase activity in *R. rubra* as a pathogenicity determinant.

D-Amino Acid Oxidase D-Amino acid oxidase purified from the yeast *Rhodotorula gracilis* is a flavoenzyme which does not require exogenous FAD for maximum activity. The enzyme stability is influenced by 2-mercaptoethanol. The holoenzyme form of *R. gracilis* D-amino acid oxidase, an 80 kDa homodimer, reacts with general thiol reagents such as 2,2′-dithiodipyridine and 5,5′-dithiobis(2-nitrobenzoic acid), whereas the monomeric apoprotein is inactivated by these reagents. The enzyme is irreversibly inactivated by phenylglyoxal not influenced by the presence of FAD or benzoate, a pseudosubstrate. Incubation of the enzyme from *R. gracilis* with dansyl chloride may cause an irreversible inactivation of the enzyme. A DNA sequence to encode the enzyme from *R. gracilis* is known to have potential with regard to the treatment of uraemia.

ATP:Citrate Lyase ATP:citrate lyase purified from *Rhodotorula gracilis* is found to have a molecular weight of 520 000 and contains four identical subunits (M_r 20 000). It resembles the mammalian ATP: citrate lyase. The enzyme is stimulated by NH_4^+ ions and is inhibited by palmitoyl, lauroyl, oleoyl, myristoyl and stearoyl-CoA esters, glutamate and glucose 6-phosphate, but not by acetyl-CoA or shorter-chain fatty acyl-CoA esters.

Alpha-glucosidase *Rhodotorula javanica* ATCC 96991 (also known as *Candida javanica)* from alfalfa can produce α-galactosidase but lacks β-fructofuranosidase. The immobilization of yeast α-glucosidase purified from *Rhodotorula lactosa* causes a significant decrease in transglucosylation activity. The β-glucosidase is known to show a high transglucosylation activity.

Aspartate Aminotransferase Two enzymes are purified from *Rhodotorula minuta*. The yeast mitochondrial isoenzyme resembles animal mitochondrial isoenzymes in molecular weight, absorption spectrum and substrate specificity. The amino acid composition is closely similar to that of pig mitochondrial isoenzyme.

Cytochrome P-450 A cytochrome P-450 tentatively named P-450rm, catalysing the formation of isobutene from isovalerate, is purified from microsomes of *Rhodotorula minuta*. A P-450rm interacts with benzoate. The P-450rm-monooxygenase system of the yeast may function in the degradation of L-phenylalanine on the pathway to β-ketoadipate. The purified cytochrome is characteristic of a low-spin ferric haem protein.

Chorismate Mutase *Rhodotorula aurantiaca* contains only a single form of chorismate mutase, which is localized exclusively in the cytosol. The enzyme activity is activated by tryptophan.

Mitochondrial NADH Dehydrogenase *Rhodotorula minuta* and *R. mucilaginosa* possess the enzyme with significant features of the NADH: ubiquinone oxidoreductase.

Protease *Rhodotorula rubra* and *Rhodotorula glutinis* are able to bind collagen type I, fibronectin and laminin. They have extracellular protease activity as a capacity to bind lactoferrin as well as a capacity to excrete siderophores. These properties are correlated with virulence and with the capacity for colonization of other organisms.

Levanase Levanase, a slime-dissolving enzyme of *Rhodotorula* species, has been purified; its molecular mass of the enzyme is known to be 39 000 Da. The purified levanase shows substrate specificity towards inulin.

Other Enzymes All species within the genus *Rhodotorula* can synthesize urease, while α-ketoglutarate-dependent dioxygenase catalysing deoxyuridine hydroxylation occurs in the extract of *Rhodotorula glutinis*. An extracellular carboxylesterase from *Rhodotorula mucilaginosa* is also reported, while *Rhodotorula rubra* produces trehalase. Thymine hydroxylase from *Rhodotorula* species catalyses the oxidation of thymine to its alcohol, aldehyde and carboxylic acid, in three successive reactions. Each step involves stoichiometric consumption of O_2 and α-ketoglutarate and formation of CO_2 and succinate. Cells of *Rhodotorula glutinis* produce 6-phosphofructo-1-kinase, epoxide hydrolase and L-phenylalanine ammonia lyase, whereas mandelate dehydrogenase is purified from *Rhodotorula graminis*. A few species within *Rhodotorula* are able to produce pectinase associated with the decay and softening of fruits and vegetables, while potassium-dependent pyruvate kinase and malate dehydrogenase can be synthesized from *Rhodotorula rubra*.

Killer Activity

Rhodotorula glutinis and *Rhodotorula rubra* have 'killer' effects on other microorganisms. *Rhodotorula* species isolated from water or sediment samples are known to have killer activity against the majority of both ascomycetous and basidomycetous species. *Rhodotorula glutinis* isolated from phylloplane is found to produce antibacterial compounds inhibitory to both *Pseudomonas fluorescens* and *Staphylococcus aureus*.

Resistance and Sensitivity

Rhodotorula glutinis is fully resistant to an itraconazole, but sensitive to a pradimicin derivative, BMS-181184. Both 2-hydroxymethyl-1-naphthol diacetate and lovastatin also exhibit potent antimicrobial activity against *Rhodotorula* spp., including *Rhodotorula rubra*. A novel antibiotic, YM-47522, from *Bacillus* species presents potent in vitro antifungal activity, especially against *Rhodotorula acuta*. The C-terminal residue of syringomycin and syringotoxin from *Pseudomonas syringae* pv. *syringae* are active against *Rhodotorula pilimanae*, whereas a purified D-galactose-specific lectin Kb-CWL I extracted from *Kluyveromyces bulgaricus* is known to have marked antifungal effects on *Rhodotorula*. Moreover, both *Rhodotorula minuta* and *Rhodotorula rubra* infections are successfully treated with amphotericin, miconazole and 5-fluorocytosine.

Products

Fats and Lipids *Rhodotorula glutinis* is a strong synthesizer of lipids. *Rhodotorula* species can produce over 20% of their biomass as fat and the yields can approach 70% (dry weight) of cell mass under specialized culture conditions. The yeasts can produce a lipid yield of 40% from molasses and 67% from sugar cane syrup. *n*-Alkanes, starch, waste cellulose hydrolysates, molasses, peat moss hydrolysate, ethanol, glucose, lactose and xylose are all substrates for lipid synthesis. The major fatty acids synthesized are oleic, linoleic and palmitic acids. Arachidonic acid, a precursor of eicosanoid hormones, is also found in *Rhodotorula acheniorum*, *Rhodotorula aurantiaca* and *Rhodotorula bacarum*. Polyol fatty esters are produced by *Rhodotorula graminis*. The composition of extracellular, insoluble glycolipids can be influenced by the addition of precursors (long-chain lipids and hydrocarbons) to culture media. Fumonisin, a sphingolipid-like mycotoxin, inhibits the accumulation of free 8,9,13-trihydroxydocosanoic acid in *Rhodotorula* sp. YB-1502 cultures, leading to low trihydroxy acid production.

Carotene Carotene is potentially important as it is a precursor of vitamin A. Depending on the strain, various carotenoid pigments are detected, but the major carotenoid is always β-carotene. Carotenoproteins are formed by co-cultures of *Rhodotorula glutinis* 22P and *Lactobacillus helveticus* 12A in cheese whey containing lactose. Carotenoids from *Rhodotorula glutinis* 22P are identified as β-carotene, torulene and torulaoin. Carotenoid-containing yeasts of *Rhodotorula* may lose their ability to synthesize carotenoid pigments during growth in alkaline benzoate media. Fermentation by *Rhodotorula glutinis* is a method of producing torulaodin, a compound that enhances the carcinostatic activity of polysaccharides. The complex lipocarotenoid preparations from *Rhodotorula glutinis* have a normalizing effect on the lipid transport system and the peroxide oxidation of blood serum lipids of irradiated rats.

Exopolysaccharides Exopolysaccharides can be synthesized by *Rhodotorula glutinis*. The monosaccharide composition of the synthesized biopolymer is known to be predominantly D-mannose. A variety of sterols and steroid precursors are synthesized by *Rhodotorula* species.

Enzyme Inhibitor Lovastatin (monocolin K) is a competitive inhibitor of 3-hydroxy-3-methylglutaryl-CoA reductase. It influences the growth and ergosterol biosynthesis of a lovastatin-sensitive strain of *Rhodotorula rubra*.

Other Conversions Cephalosporane derivatives and chitin are synthesized by *Rhodotorula glutinis*. *Rhodotorula rubra* CBS 6469 can be used for the stereoselective reduction of substituted thiazolidines. The products may have application as antidiabetic, antihypertensive and eating disorder agents. The 2-halogeno-3-hydroxy-3-phenylpropionic acid ester compounds produced by *Rhodotorula* are useful as a coronary vasodilating agent. Transhydroxy sulphone intermediate is synthesized by both cells of *Rhodotorula rubra* and *Rhodotorula mucilaginosa* var. *mucilaginosa*, whereas *Rhodotorula minuta* is found to be an efficient producer of galactooligosaccharide from lactose. Species of *Rhodotorula* produce rhodotorulic acid. Ferrirhodotorulic acid in *Rhodotorula piliminae* exchanges the ferric ion with a membrane-bound chelating agent at the cell surface that in turn completes the active transport of iron into the cell. *Rhodotorula* species are potentially useful for the production of a wide range of other organic acids, particularly itaconic acid.

Pathogenicity

Human Disease

Rhodotorula is one of the most common yeasts associated with human infections. *Rhodotorula* is present in the oral cavity of patients with retrogenic defects. *Rhodotorula glutinis* infections are related to fungaemia in children and to granulomatous epididymitis, a disease characterized by hard connective tissue with several small abscesses and slight haemorrhage. Fungal infections can occur in neutropenic patients due to *Rhodotorula rubra* and cause respiratory allergy. *Rhodotorula* species are the fungal organisms associated with disruption of the natural barrier of the skin, including catheterization of urinary, venous and arterial systems.

Immunodeficiency increases the prevalence and severity of mycosis caused by fungi such as *Rhodotorula rubra*. Fungaemia in patients infected with human immunodeficiency virus (HIV) often presents as a community-acquired infection, which is frequently due to newly emerging opportunistic *Rhodotorula rubra*. *Rhodotorula minuta* central venous catheter infections with fungaemia are found in patients with advanced acquired immunodeficiency syndrome (AIDS), HIV nephropathy, end-stage renal disease requiring haemodialysis, and a permanent catheter. *Rhodotorula* is found in patients with acute myeloblastic leukaemia undergoing bone marrow transplant. *Rhodotorula rubra* is one of the major causative agents for fungal oesophagitis causing a haematemesis, while cells of *Rhodotorula glutinis* are responsible for infections during leukopenia. *Rhodotorula*-induced meningitis in HIV-infected patients can be characterized by severe headache and high body temperature. Extrinsic allergic alveolitis following exposure to *Rhodotorula rubra* is known, with an elevated titre of *Rhodotorula*-specific precipitating antibodies in the serums of patients.

Animal Disease

Rhodotorula spp. are often isolated from the gastrointestinal tract of reptiles and the oral cavity or cloaca of other living animals; some of these species are potential human pathogens. However, no sufficiently reliable criteria can be established to prove that positive culture results are associated with disease. In reptiles, dermatomycosis is detected. *Rhodotorula* is involved in bovine mycotic mastitis and can be isolated from milk samples from normal, clinical and subclinical mastitis quarters from dairy herds. It is known that serum IgG antibody concentrations against *Rhodotorula glutinis* in horses change seasonally.

Association with Foods

Mutagenic activity of the nutrient medium containing aflatoxin B_1 is reduced by a stain of *Rhodotorula mucilaginosa*. The technology is based on the use of particular strains for the treatment of food and feedstuffs. *Rhodotorula* species are preservative-sensitive yeasts. *Rhodotorula graminis* has emulsification ability: the bioemulsifier from this organism has potential for use in food. *Rhodotorula* species are sparse sources of yeast contamination of the curd during cheese-making. They are present in the factory environment, on working surfaces, the brine and on workers' hands and aprons.

Sugary Fruits

Colonization by *Rhodotorula* is usually associated with tropical fruits and ripe apples. The numbers of the yeasts increase if fruit is allowed to fall naturally, particularly if the skin is damaged. In apple processing, counts rise owing to the indigenous flora of the factory. The traditional rack and cloth press for juice production is also a major source of contamination. In pear juice, *Rhodotorula* is one of the main wild yeasts. Fermentation is not only affected by the species of wild yeasts but also by the tannin levels of the pear juice. *Rhodotorula aurantiaca* on apple suppresses grey mould caused by *Botrytis cinerea*; *Rhodotorula glutinis* controls blue mould, grey mould and side rot on pears. *Rhodotorula aurantiaca* Y1581 prevents or greatly reduces the frequency of decay and the decayed area in apples. The yeast cells are known to suppress the germination of *Botrytis cinerea* conidia. Treatment of wounds with yeasts reduces ethylene rates. Competition for carbon and suppression of ethylene production may be biocontrol mechanisms used by the yeasts.

Meat Products

Prevalence of *Rhodotorula* species is noted in eggs, fry and fingerlings of *Salmo gairdneri* Rich under artificial breeding; yeasts isolated from perished eggs and sick fry do not possess pathogenic properties. *Rhodotorula* plays an important role in the spoilage of frozen poultry carcasses. A large increase in the number of *Rhodotorula glutinis* on turkey carcasses is considered to be responsible for carcass off odour. *Rhodotorula* species predominate in nonpolluted waters. They are found on the skin, gills, mouth and faeces of fish. Red-pigmented species of *Rhodotorula* predominate among other isolates from seafoods such as oysters, quahogs, mussels and clams. It is known that at the end of storage of the shrimp on ice, only *Rhodotorula rubra* and *Rhodotorula minuta* are present. *Rhodotorula rubra* is a natural contaminant of oysters and a main species causing the pinking of both fresh and frozen oysters, but it appears to play little part in the spoilage of refrigerated seafoods. Yeasts associated with oysters are of special interest because of their potential to spoil these products, and because of their resultant public health significance, since oysters may be eaten uncooked. Unlike the human pathogenic yeasts such as *Candida* and *Torulopsis* in shellfish, these *Rhodotorula* species are considered not to pose a risk to public health.

Substrate Assimilation and Transformation

Rhodotorula species are capable of utilizing non-carbohydrate substrates as well as carbohydrates. Benzoic acid can serve as a sole source of carbon and energy for growth of *Rhodotorula*. Specific growth rate is strongly dependent both on the concentration of benzoate and the pH value of the cultivation media. The maximum specific growth rate on benzoate is obtained in alkaline cultivation media, whereas that on glucose occurs in mildly acidic media. *Rhodotorula glutinis* has the ability to grow on mandelate; utilization of this substrate is a widespread but not universal characteristic within the genus *Rhodotorula*. *Rhodotorula* isolates capable of biotransforming polycyclic aromatic hydrocarbons are known. Some esters of testosterone, acetate, propionate, enanthate, caprate, undecanoate, isobutyrate and isocaproate (some of them are used as drugs) are transformed by *Rhodotorula mucilaginosa*. Cysteate, taurine and isothionate can be utilized as sources of sulphur in *Rhodotorula glutinis*. Resting cells of *Rhodotorula rubra* convert transferulic acid to vanillic acid, then to guaiacol and protocatechuic acid under aerobic conditions. In an argon atmosphere, *Rhodotorula rubra* transforms ferulic acid to vanillic acid and 4-hydroxy-3-methoxystyrene. The biotransformation requires CoA, ATP and NAD^+.

Organic Acids and Alcohols

Tricarboxylic acid cycle intermediates such as citric acid, fumaric acid, pyruvic acid or succinic acid can act as carbon sources for *Rhodotorula* sp. Y-38. Substances such as lactic acid, acetic acid, malonic acid, propionic acid and tartaric acid can be also utilized. Other organic acids such as *n*-butyric acid, formic acid, glycolic acid and β-hydroxybutyric acid are not assimilated by the species. Among alcohols, ethyl alcohol and glycerol are good carbon sources for *Rhodotorula* species under aerobic conditions. *Rhodotorula* can also assimilate 2-amyl alcohol, butyl alcohol, propyl alcohol and polyvinyl alcohol. Other compounds such as ethylene glycol and mannitol can serve as a carbon source for *Rhodotorula* species, whereas amyl alcohol, 2-mercaptoethanol and benzyl alcohol are not metabolized.

Single-cell Protein Production

The utilization of C_2 compounds by *Rhodotorula* species is important in the production of single-cell protein. These yeasts are considered useful for single-cell protein (SCP) production owing to their high yields of biomass from both methyl alcohol and ethyl alcohol, and the favourable balances of amino acids in their SCPs. With a continuous supply of ethyl alcohol (1.0%), *Rhodotorula* sp. Y-38 is reported to have a cell yield of 64.4 g per 100 g ethyl alcohol, and the content of crude protein in the cells is found to be 50 g per 100 g biomass. The amino acid content is 42.6 g per 100 g biomass and that of nucleic acids is 9.4 g per 100 g biomass. *Rhodotorula glutinis* can produce SCP from methyl alcohol with the yield of 42 g per 100 methyl alcohol. Some *Rhodotorula* species are able to grow on acetic acid and on acetaldehyde. *Rhodotorula* Y-38 with a constant supply of acetic acid has a cell yield of 50.2 g per 100 g acetic acid; the crude protein value is 51.0 g per 100 g biomass, and the respective contents of amino acids and nucleic acids are found to be 42 g and 9.0 g per 100 g biomass. The isolate with acetaldehyde exhibits a cell yield of 33.8 g per 100 g acetaldehyde, giving a content of crude protein of 50.5 g per 100 biomass.

See also: **Fish**: Spoilage of Fish. **Meat and Poultry**: Spoilage of Meat. **Shellfish (Molluscs and Crustacea)**: Contamination and Spoilage. **Single Cell Protein**: Yeasts and Bacteria.

Further Reading

Barness EM, Impey CS, Geeson JD and Buhagiar RWM (1978) The effect of storage temperature on the shelf-life of eviscerated air-chilled turkeys. *Br. Poult. Sci.* 19: 77–84.

Berry DR, Russel I and Stewart GG (1987) *Yeast Biotechnology*. P. 42. London: Allen & Unwin.

Braun DK and Kauffman CA (1992) *Rhodotorula* fungaemia: a life-threatening complication of indwelling central venous catheters. *Mycoses* 35: 11–12.

Casorari C, Nanetti A, Cavallini GM, Rivasi F, Fabio U and Mazzoni A (1992) Keratomycosis with an unusual etiology (*Rhodotorula glutinis*). *Microbiologica* 15(1): 83–87.

Chand-Goyal T and Spotts RA (1996) Enumeration of bacterial and yeast colonists of apple fruits and identification of epiphytic yeasts on pear fruits in the Pacific Northwest United States. *Microbiol. Res.* 151(4): 427–432.

Carlile MJ and Watkinson SC (1996) *The Fungi*. P. 68. London: Academic Press.

Fattakhova AN, Ofitserov EN and Garusov AV (1991) Cytochrome P-450-dependent catabolism of triethanolamine in *Rhodotorula mucilaginosa*. *Biodegradation* 2(2): 107–113.

Frengova G, Simova E and Beshkova D (1997) Carotenoprotein and exopolysaccharide production by co-cultures of *Rhodotorula glutinis* and *Lactobacillus helveticus*. *J. Ind. Microbiol. Biotechnol.* 18(4): 272–277.

Fries N (1963) Induced thermosensitivity in Ophiostoma and *Rhodotorula*. *Physiol. Plant* 16: 415–422.

Gyaurgieva OH, Bogomolova TS and Gorshova GI (1996) Meningitis caused by *Rhodotorula rubra* in an HIV-infected patient. *J. Med. Vet. Mycol.* 34(5): 357–359.

Mansure JJ, Silva JT and Panek AD (1992) Characterization of trehalase in *Rhodotorula rubra*. *Biochem. Int.* 28(4): 693–700.

Pollegioni L, Ghisla S and Pilone MS (1992) Studies on the active centre of *Rhodotorula gracilis* D-amino acid oxidase and comparison with pig kidney enzyme. *Biochem. J.* 286(2): 389–394.

Spencer JFT and Spencer DM (1990) *Yeast Technology*. P. 109. Berlin: Springer.

Spencer JFT, Spencer DM and Tulloch AP (1979) Extracellular glycolipids of yeasts. In: Rose AH (ed.) *Economic Microbiology*. Vol. 3, *Secondary Products of Metabolism*. P. 523. New York: Academic Press.

Whitlock WL, Dietrich RA, Steimke EH and Tenholder MF (1992) *Rhodotorula rubra* contamination in fiberoptic bronchoscopy. *Chest* 102(5): 1516–1519.

Yeeh Y (1996) Single-cell protein of *Rhodotorula* sp. Y-38 from ethanol, acetic acid and acetaldehyde. *Biotechnol. Lett.* 18(4): 411–416.

Yeeh Y, Kim SH, Joo WH, Jun HK and Kwon OC (1996) Growth characteristics of *Rhodotorula* sp. Y-55 on ethanol, acetic acid and acetaldehyde. *Korean J. Biotechnol. Bioeng.* 11(3): 367–373.

Risk Analysis *see* **Quantitative Risk Analysis.**

SACCHAROMYCES

Contents

Introduction

Yuji Oda, Department of Applied Biological Science, Fukuyama University, Hiroshima, Japan

Kozo Ouchi, Kyowa Hakko Kogyo Co. Ltd, Tokyo, Japan

Yeasts are classified into three groups: ascosporogenous yeasts, basidiosporogenous yeasts and imperfect yeasts. *Saccharomyces* is the representative of ascosporogenous yeasts and historically the most familiar microorganism for human. This genus was first described by Meyen when he assigned beer yeast as *S. cerevisiae* in 1838, and redefined by Reess in 1870 from the observations of ascospores and their germination. The name is derived from the Greek words *sakcharon* (sugar) and *mykes* (fungus). The number of species has changed according to the criteria used to delimit species, and 16 species are now accepted in the genus *Saccharomyces* (**Table 1**).

Characteristics of the Genus

The vegetative cells of *Saccharomyces* species are round, oval or cylindrical and reproduce by multilateral budding. They may form pseudohyphae but not septate hyphae. These yeasts are predominantly diploid or occasionally of higher ploidy. Asci, which are persistent and usually transformed by direct change from the vegetative cells, may contain one to four ascospores. The ascospores are round or slightly oval, with smooth walls. Conjugation occurs on or soon after germination of the ascospores. Some strains of *S. cerevisiae* and its related species used in the brewing and baking industries hardly form ascospores at all; continuous selection with respect to practical properties seems to cause loss of sporulation ability in these strains.

The most famous physiological characteristic of *Saccharomyces* spp. is their capacity for vigorous anaerobic or semianaerobic fermentation of one or more sugars to produce ethanol and CO_2. These sugars include D-glucose, D-fructose and D-mannose, except in the case of certain mutants. Most strains of *Saccharomyces* can grow on D-galactose under aerobic or anaerobic conditions; however, none of them utilizes lactose, pentose, alditols and citrate as carbon sources, assimilates nitrate as a nitrogen source or hydrolyses exogenous urea. Among polysaccharides, starch and pectin are exceptionally utilized by certain strains of *S. cerevisiae*. They do not produce starch-like compounds. Their ubiquinone is exclusively Q-6, but this feature is common in the genera *Kluyveromyces Torulaspora* and *Zygosaccharomyces*.

Table 1 The species accepted in the genus *Saccharomyces*

Species	*Authority*
S. barnettii	Vaughan-Martini 1995
S. bayanus	Saccardo 1895
S. castellii	Capriotti 1966
S. cerevisiae	Meyen ex E. C. Hansen 1883
S. dairenensis	Naganishi 1917
S. exiguus	Reess ex E. C. Hansen 1888
S. kluyveri	Phaff, M. W. Miller and Shifrine 1956
S. kunashirensis	James, Cai, Roberts and Collins 1997
S. martiniae	James, Cai, Roberts and Collins 1997
S. paradoxus	Bachinskaya 1914
S. pastorianus	E. C. Hansen 1904
S. rosinii	Vaughan-Martini, Barcaccia and Pollacci 1996
S. servazzii	Capriotti 1967
S. spencerorum	Vaughan-Martini 1995
S. transvaalensis	van der Walt 1956
S. unisporus	Jörgensen 1909

Table 2 Discriminating characteristics of species of the genus *Saccharomyces*

Species	Fermentation					Assimilation: Carbon source									Assimilation: Nitrogen source			Resistance to cycloheximide		Growth temperature				Growth without vitamins	Active fructose transport system
	Malose	*Sucrose*	*Trehalose*	*Melibiose*	*Raffinose*	*Maltose*	*Sucrose*	*Melibiose*	*Raffinose*	*D-Ribose*	*Glycerol*	*D-Mannitol*	*α-Methyl-D-glucoside*	*Succinate*	*Ethylamine-HCl*	*Cadaverine*	*Lysine*	0.01%	0.1%	28ºC	30ºC	35ºC	37ºC		
S. cerevisiae	+	+	–	v	+	+	+	v	+	–	v	–	v	–	–	–	–	–	–	+	+	v	v	–	–
S. bayanus	+	+	–	v	v	+	+	v	+	–	v	v	v	–	–	–	–	–	–	+	+	v	–	+	+
S. paradoxus	v	+	–	–	v	+	+	–	+	–	–	+	+	–	–	–	–	–	–	+	+	+	+	–	–
S. pastorianus	+	v	–	v	v	+	v	v	v	–	v	–	v	–	–	–	–	–	–	+	+	v	–	–	+
S. barnettii	–	+	+	–	+	–	+	–	+	–	–	–	–	+	–	–	–	–	–	+	–	–	–	–	
S. castellii	–	–	–	–	–	–	–	–	–	v	–	–	–	+	–	–	–	–	–	+	+	+	v	–	
S. dairenensis	–	–	–	–	–	–	–	–	–	v	v	–	–	+	–	–	–	–	–	+	+	+	v	–	
S. exiguus	–	+	+	–	v	–	+	–	+	–	–	–	–	+	–	–	–	v	v	+	+	v	v	–	
S. kunashirensis	–	+	–	–	–	–	–	–	–	–	+	–	–	–	–	–	–	–	–						
S. martiniae	–	+	+	–	–	–	–	–	–	–	–	–	–	–	–	–	–	–	–						
S. rosinii	–	–	–	–	–	–	–	–	–	–	–	–	–	–	–	–	–	–	–	+	–	–	–	–	
S. servazzii	–	–	–	–	–	–	–	–	–	–	+	–	–	–	–	–	–	+	+	+	+	+	–	–	
S. spencerorum	–	+	+	–	–	–	+	–	–	–	+	–	–	+	+	+	+	–	–	+	+	+	+	–	
S. transvaalensis	–	–	–	–	–	–	–	–	–	–	–	–	–	–	v	–	v	–	–	+	+	+	+	–	
S. unisporus	–	–	–	–	–	–	–	–	–	–	–	–	–	+	+	+	+	+	+	+	+	+	v	–	
S. kluyveri	–	+	–	+	+	+	+	+	+	–	v	+	+	–	+	+	+	–	–	+	+	+	+	–	

Reactions: +, positive; –, negative; v, variable

Identification of *Saccharomyces* Species

Yeasts are usually classified by the characteristics of microscopic appearance, sexual reproduction and physiological features including (1) fermentation of certain sugars semianaerobically; (2) assimilation of various compounds each as sole carbon or nitrogen source; (3) growth without an exogenous supply of certain vitamins; (4) growth in the presence of 50% or 60% (w/w) glucose; (5) growth at 37°C; (6) growth in the presence of cycloheximide; (7) splitting of fat, production of starch-like polysaccharides, hydrolysis of urea; and (8) formation of acid.

The 16 species now accepted in the genus *Saccharomyces* are divided into three groups: *Saccharomyces sensu stricto*, *Saccharomyces sensu lato*, and *S. kluyveri*. **Table 2** compares the differences of physiological properties of the *Saccharomyces* species.

Saccharomyces sensu stricto

Saccharomyces sensu stricto species including *S. cerevisiae*, *S. bayanus*, *S. paradoxus* and *S. pastorianus* are phylogenetically closely related in the genus *Saccharomyces*. The species *S. cerevisiae*, *S. bayanus* and *S. pastorianus* are specifically found in the artificial environments of wineries and breweries. The relative genome sizes of these three species are estimated to be 1.00, 1.15 and 1.46, respectively. *Saccharomyces paradoxus* is exclusively isolated from natural sources such as tree exudates, soil and *Drosophila*. Cells of *S. paradoxus* are small in size and readily form asci compared with the other three species. Effective separation of the *Saccharomyces sensu stricto* species is complicated because these species often have apparently identical morphological, physiological and serological properties. The four species have been differentiated from each other by DNA reassociation study (**Fig. 1**). Strains with 80–100% overall homology of base sequences are considered as belonging to the same species, while the strains of distantly related taxa show homology of less than 30%. Among the four species, *S. pastorianus* reveals 53% homology to *S. cerevisiae* and 72% homology to *S. bayanus*,

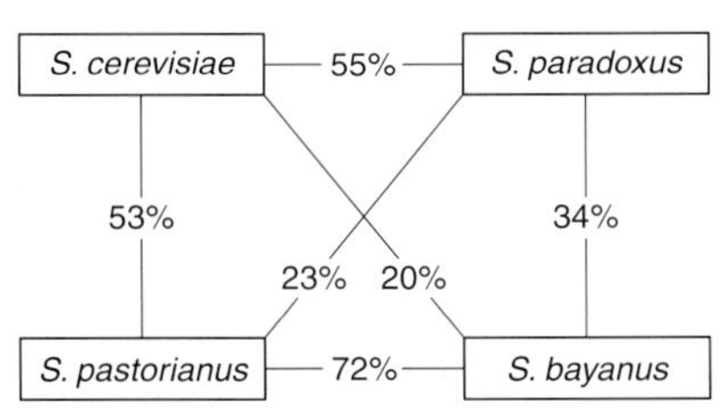

Figure 1 DNA relatedness among the type strains of *Saccharomyces sensu stricto* species.

suggesting an intermediate position between two unrelated species, *S. cerevisiae* and *S. bayanus*.

Since species division within *Saccharomyces sensu stricto* group were clarified on the molecular level, it became possible to determine those physiological responses necessary for the separation of the four taxa. *Saccharomyces bayanus* is the only species of the genus that can grow in the absence of vitamins. Maximum growth temperature immediately distinguishes *S. bayanus* and *S. pastorianus*, which never grow at above 35°C, from *S. cerevisiae* and *S. paradoxus*, which grow at 37°C and often at up to 40–42°C. An active fructose transport system is present in the group of *S. bayanus* and *S. pastorianus*, while absent in *S. cerevisiae* and *S. paradoxus*. *Saccharomyces cerevisiae* is distinguished from *S. paradoxus* with respect to assimilation of D-mannitol and fermentation of maltose.

Saccharomyces sensu lato

Saccharomyces servazzii and *S. unisporus* are unusual members of the genus as judged from narrow fermentative profiles and ability to grow in the presence of 0.1% cycloheximide. Assimilation of ethylamine, cadaverine and lysine can differentiate *S. unisporus* from *S. servazzii*.

Saccharomyces exiguus together with *S. servazzii* and *S. unisporus* is characterized by much lower G+C values (34.7–36.6%) than other *Saccharomyces* species (39.3–41.9%), but do not grow in the presence of 0.1% cycloheximide. *Saccharomyces barnettii* and *S. spencerorum* have been separated from *S. exiguus* and its anamorph, *Candida holmii*, on the basis of DNA relatedness. Separation of these three taxa is confirmed by conventional taxonomic tests. *Saccharomyces spencerorum* obviously differs from *S. exiguus* in the assimilation of glycerol as a sole carbon source and ethylamine, cadaverine and lysine as sole nitrogen sources.

Saccharomyces castellii, *S. dairenensis* and *S. rosinii* have not been distinguished by conventional physiological criteria although they represent separate taxa based on DNA relatedness. Physiological profiles of *S. dairenensis* and *S. transvaalensis* are similar, but *S. transvaalensis* has characteristics of larger vegetative cells and large refringent ascospores. The latter species shows intermediate levels of DNA homology to the type strains of *S. dairenensis* and *S. castellii*. Combination of physiological patterns and electrophoretic karyotypes permits with some difficulty the separation of most of these four species. *Saccharomyces dairenensis* is still a heterogenous taxon including at least two unrelated groups since three strains show less than 32% homology to the type strain of *S. dairenensis*.

Saccharomyces kunashirensis and *S. martiniae* are recently defined from the analysis of 18S ribosomal RNA (rRNA) gene sequences.

Saccharomyces kluyveri

Saccharomyces kluyveri is easily distinguishable from other species of the genus since it is characterized by a wide assimilative and fermentative profile including the ability to utilize ethylamine-HCl, cadaverine and lysine as sole nitrogen sources for growth as well as the ability to both assimilate and ferment melibiose. The distinct character was already anticipated by molecular taxonomic studies which showed no nucleotide homology between *S. kluyveri* with either *Saccharomyces sensu stricto* or *sensu lato* strains, where DNA homology values were never above 22%.

Molecular Methods to Differentiate Species

The conventional taxonomic tests to assess physiological features are fundamental for identification, but the results have shown to be insufficient for species delimitation and discrimination of interstrain variability. The genetic basis behind many of these characteristics is often either poorly understood or unknown. In the past DNA reassociation study has significantly contributed to molecular taxonomy of the genus *Saccharomyces*: this method was used to reestablish several species names, reduce other names to synonyms, describe new species, and raise the likelihood of the existence of additional species. However, the equipment used to measure DNA association is highly specialized and expensive and the amount of data obtainable in an average week's work is relatively small. Discrimination of the closely related species has been confirmed by other molecular techniques, such as whole-cell protein patterns, multilocus enzyme electrophoresis, fructose transport system, mitochondrial DNA restriction analysis, electrophoretic karyotypes, random amplified polymorphic DNA polymerase chain reaction (RAPD-PCR) and restriction fragment length polymorphism (RFLP) patterns, and rRNA gene sequencing. These methods are valid additions to the conventional taxonomic tests, and those applicable to the food industry are described below.

Electrophoretic Karyotypes

Chromosomal patterns resolved by pulse field gel electrophoresis are called electrophoretic karyotypes. By comparing the results from this method and DNA reassociation, it has been demonstrated that electrophoretic karyotypes of the two strains are identical when DNA sequence homology is over 85%, while

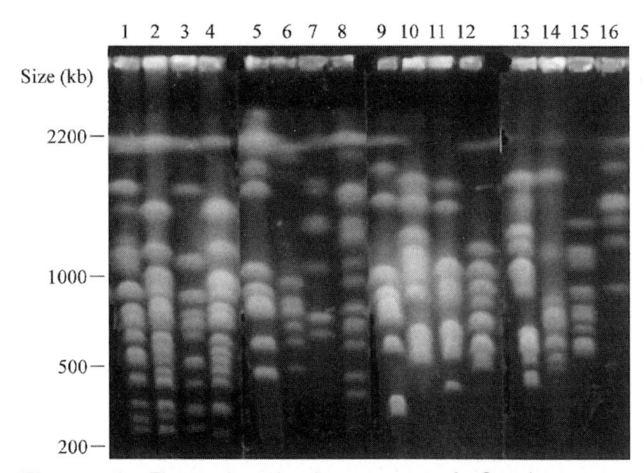

Figure 2 Electrophoretic karyotypes of *Saccharomyces* species. Chromosomal DNA of type strains was separated in 0.8% agarose gel in 0.5 × TBE (45 mmol l^{-1} Tris–borate, pH 7.3, 1 mol l^{-1} EDTA) at 125 V and 14°C. Pulse times were 3 min for 24 h followed by 5 min for 16 h. Lanes: 1, *S. cerevisiae*; 2, *S. bayanus*; 3, *S. paradoxus*; 4, *S. pastorianus*; 5, *S. barnettii*; 6, *S. castellii*; 7, *S. dairenensis*; 8, *S. exiguus*; 9, *S. kunashirensis*; 10, *S. martiniae*; 11, *S. rosinii*; 12, *S. servazzii*; 13, *S. spencerorum*; 14, *S. transvaalensis*; 15, *S. unisporus*; 16, *S. kluyveri*.

low DNA relatedness corresponds to completely different chromosomal patterns. Since similar but not identical karyotypes are not interpreted as either different species or polymorphisms in the same species, karyotyping is not so reliable as DNA base sequence comparisons, but is undoubtedly an important adjunct. Electrophoretic karyotypes can serve as a rapid, inexpensive and relatively easy first approach for the evaluation of a group of physiologically similar strains.

The general feature for ascosporogenous yeasts is the presence of one to five bands of chromosomal DNA larger than 1000 kb as in *S. kluyveri* (**Fig. 2**), whereas in most *Saccharomyces* species, chromosomes smaller than 1000 kb are observed. Chromosomes of *Saccharomyces sensu stricto* species were resolved into 12 to 16 bands in the range 200 kb to 2200 kb. None of the other species contains chromosomes smaller than 300 kb. The patterns of *Saccharomyces sensu stricto* species are similar, and are distinguishable from the other species at a glance. A multivariate analysis of the polymorphisms in the numbers and molecular weights of chromosomes has revealed that the *Saccharomyces sensu stricto* strains could be separated into four clusters that correspond to the four species.

RAPD-PCR

Analysis by RAPD-PCR involves the use of small random primers and low stringency primer annealing conditions to amplify arbitrary fragments of template DNA. The single primer will anneal at any point on the genome where a near-complementary sequence exists, and if two priming sites are sufficiently close, then PCR amplifies the fragment between them. A number of fragments of various sizes may be produced; formed patterns are specific for the particular DNA template used. This technique is suitable for typing and identification of microorganisms, but several problems are present. First, the whole patterns of electrophoresis are not always the same in independent experiments, and only the reproducible bands should be scored. Second, the results are affected by the nucleotide sequence of the primer used.

After the PCR products have been resolved, genetic distance is calculated manually as the number of different bands between two patterns divided by the sum of all bands in the same patterns. A value of 0 indicates that the two strains had identical patterns, and a value of 1 indicates that the two strains had completely different patterns. The dice matrix obtained from these data is used to construct an unrooted dendrogram.

Ribosomal RNA Gene Analysis

The analysis of rRNA genes, which can elicit exact data without pairing two samples, is one of the promising methods among these tools applied to the phylogenetic study of yeast. In *S. cerevisiae*, the co-transcribed genes for small (17S–18S), 5.8S, and large (25S–28S) rRNA and 5S rRNA genes occur as tandemly repeated units on chromosome XII. Sequence comparisons of the rRNA genes have shown a relatively high degree of evolutionary conservation and have been used as bases for inferring phylogenetic relationships. Almost the complete 18S rRNA gene sequence of *Saccharomyces* species has been sequenced and their relationships investigated in detail, but the entire region of 18S rRNA is not simply and rapidly determined by anyone.

The region spanning the internal transcribed spacers (ITSs) and the 5.8S rRNA gene entire is amplified by PCR using pITS1 (5′-TCCGTAGGTGAACCTGCGG-3′) and pITS4 (5′-TCCTCCGCTTATTGATATG-3′), which are derived from conserved regions of the 18S and 28S rRNA genes, respectively. The size is over 800 bp for *Saccharomyces sensu stricto* species and *S. transvaalensis*, and less than 800 bp for the other *Saccharomyces* species (**Fig. 3**). Furthermore, restriction analysis of ITS region allows the separation of *S. cerevisiae*, *S. bayanus* and *S. pastorianus*. These may be useful methods to identify *Saccharomyces* isolates.

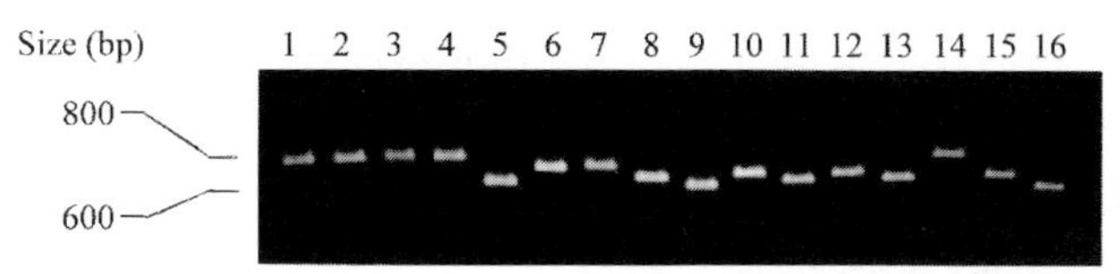

Figure 3 Size of ITS regions amplified from the type strains of *Saccharomyces* species. Each lane number corresponds to that of the strain in Figure 2.

Detection, Isolation and Cultivation

Composition of media for yeasts is shown in **Table 3**. The presence of yeasts in food is usually investigated using nutrient media such as YM agar which is principally composed of yeast extract and malt extract. Potato-dextrose agar is suitable for storage of cultures, but not satisfactory for detection because each species develops a less characteristic colony on this medium. Colonies of *Saccharomyces* and some species of *Hansenula* and *Pichia*, which ferment glucose vigorously, are simply discriminated on the agar plate: when medium containing 0.5% glucose, 0.05% 2,3,5-triphenyltetrazolium chloride and 1.5% agar is overlaid on the agar plate and incubated for 2–3 h at 30°C, the colour of the colony changes to pink or red.

Table 3 Composition of culture media

Medium	*Contents*	*Percentage w/v*
YM	Yeast extract	0.3
	Malt extract	0.3
	Peptone	0.5
	Glucose	1
	Agar (if required)	2
YPD	Yeast extract	1
	Peptone	2
	Glucose	2
	Agar (if required)	2
Potato-dextrose agar	Potato extract[a]	23 (volume)
	Glucose	2
	Agar	2
YNB glucose	Yeast nitrogen base without amino acids[b]	0.67
	Glucose[b]	2
	Agar (if required)	2
Fowell's acetate agar	Sodium acetate	0.5
	Agar	2
McClary's acetate agar	Glucose	0.1
	Potassium chloride	0.18
	Yeast extract	0.25
	Sodium acetate	0.82
	Agar	1.5
Gorodkowa agar (modified)	Glucose	0.1
	Peptone	1
	Sodium chloride	0.5
	Agar	2
Malt extract agar	Malt extract	5
	Agar	3

[a] The filtrate autoclaved for 1 h at 120°C after washed, peeled and finely grated potato (100 g) is soaked in 300 ml tap water for several hours in a refrigerator and filtered through cloth.
[b] Tenfold concentrated solution is filter-sterilized and added.

Selective isolation of yeasts and estimation of viable cell number require special techniques to repress the growth of bacteria and fungi. The use of acidified agar (< pH 5.0) is adequate for isolation of yeasts from samples containing bacteria. When acidified medium alone is inadequate to eliminate bacterial growth, chloramphenicol dissolved in ethanol (20 mg ml^{-1}) is added at a final concentration of 50 μg ml^{-1}. Unlike other antibiotics that depress bacterial growth, chloramphenicol can be added to the medium before autoclaving. Addition of either 0.2% sodium propionate or 2–3% ethanol to an acidic medium has only limited effect on prolonging the growth of fungi. However, it is possible to isolate yeasts selectively by the difference in growth rates in the presence of sodium propionate.

Saccharomyces sensu stricto species frequently appear after enrichment in YM broth in which glucose was added at 10–20%. The addition of a suitable indicator such as bromphenol blue permits the monitoring of changes in pH and periodic adjustments if desired. Glucose and other components should be autoclaved separately to avoid browning. Less than 0.1 g of sample is added to 10 ml of the modified YM broth containing chloramphenicol and sodium propionate in a test tube. After static incubation for the desired period, about 0.1 ml obtained from near the bottom is transferred to another medium. The fermentation rate can be followed by recording the weight loss of the medium. This procedure is repeated several times to enrich yeasts fermenting glucose vigorously. The culture broth is successively diluted and spread on a plate of acidified YM agar.

For cultivation, *Saccharomyces* yeasts are generally grown on YPD medium (see Table 3) rather than YM medium, at 25–30°C. The most common system for aerobic growth uses liquid medium in an Ehlenmeyer flask on an orbital shaker. Shaking at 150–200 r.p.m. provides efficient aeration for 100 ml volume of medium in a 300 ml flask. The growth rate varies depending on strain, medium and incubation temperature, typically within the range of 90 min to 3 h per doubling time. The cell concentration can be determined using an electronic particle counter or a haemocytometer and microscope. Alternatively, a spectrophotometer can be used to measure turbidity at 600 nm. Typically, a yeast culture in a logarithmic phase of growth of 0.1 will contain about 2×10^6 cells.

Ascospores are induced on the sporulation media, most of which have been developed for *Saccharomyces* species. Young cells grown on YM agar for 2–3 days are spread on Fowell's or McClary's agar based on sodium acetate, Gorodkowa agar or malt extract agar, and incubated at least 4–6 weeks. Freshly isolated cells sporulate on the isolation medium and ascospores can be observed after cultivation for about 1 month, while the cells cultured on the nutrient medium many times require certain sporulation media to convert asci.

Importance to the Food Industry

The genus *Saccharomyces* is the most extensively utilized group of yeasts for the benefit of humans. *Saccharomyces cerevisiae* and related species are employed in three main processes of the food industry (**Table 4**). The first is the production of industrial alcohol and alcoholic beverages, including wine, beer, saké and potable spirits. *Saccharomyces pastorianus* including *S. carlsbergensis* was initially recognized as a lager strain. *Saccharomyces bayanus* has been mostly associated with the wine industry. Second is the baking industry; originally, spent yeasts from the brewing and distilling industries were used for baking, but became insufficient as the baking industry expanded. Yeasts for dough leavening are now propagated to meet these growing needs. The third process includes the production of biomass, extracts, autolysates and flavouring compounds. The yeast used in such processes can be either purpose-grown or a by-product.

Saccharomyces exiguus and its anamorph, *Candida holmii*, are the predominant yeasts responsible for the leavening of sourdough which is usually prepared by adding a commercially produced culture containing lactic acid bacteria.

No other species of *Saccharomyces* is of commercial importance, although some strains of *Torulaspora* and *Zygosaccharomyces* spp., formerly accepted in the genus *Saccharomyces*, are used for baking and production of miso and shoyu, respectively.

Saccharomyces species are found in many foods and sometimes cause spoilage. Wild strains which contaminate the pure culture reduce the fermentation rate and diminish the quality of final products in the brewing process. Killer wild yeasts will dominate within a short period when inoculated strains are killer-sensitive. In saké brewing, killer saké strains were constructed by the methods of back-crossing and cytoduction to overcome these problems.

Table 4 Application of *Saccharomyces* species in the food industry

1.	Industrial alcohol and alcohol beverages
2.	Bakery products
3.	Biomass, extracts, autolysates and flavouring compounds

Most species of the genus *Saccharomyces* have 'generally recognized as safe' (GRAS) status from the fact that many strains have been applied to the food industry. There is no report of disease in healthy humans and other warm-blooded animals caused by *Saccharomyces sensu stricto* species.

See color Plates 28 and 30.

See also: ***Saccharomyces***: *Saccharomyces sake*; *Saccharomyces cerevisiae*; *Saccharomyces carlsbergensis (Brewer's Yeast)*.

Further Reading

Barnett JA (1992) The taxonomy of the genus *Saccharomyces* Meyen ex Reess: a short review for non-taxonomists. *Yeast* 8: 1–23.

Barnett JA, Payne RW and Yarrow D (1990) *Yeasts: Characterization and Identification*, 2nd edn. Cambridge: Cambridge University Press.

Benítez T, Gasent-Ramírez JM, Castrejón F and Codón AC (1996) Development of new strains for the food industry. *Biotechnol. Prog.* 12: 149–163.

Evans IH (1996) *Yeast Protocols*. Methods in Molecular Biology 52. Ottawa: Humana Press.

James SA, Cai J, Roberts IN and Collins MD (1997) A phylogenetic analysis of the genus *Saccharomyces* based on 18S rRNA gene sequences: description of *Saccharomyces kunashirensis* sp. nov. and *Saccharomyces martiniae* sp. nov. *International Journal of Systematic Bacteriology* 47: 453–460.

Kurtzman CP (1994) Molecular taxonomy of the yeasts. *Yeast* 10: 1727–1740.

Kurtzman CP and Fell JW (1997) *The Yeast – a Taxonomic Study*, 4th edn. Amsterdam: Elsevier.

Panchal CJ (1990) *Yeast Strain Selection*. New York: Marcel Dekker.

Reed G and Nagodawithana TW (1991) *Yeast Technology*, 2nd edn. New York: Van Nostrand Reinhold.

Rose AH and Harrison JS (1987) *The Yeast*, 2nd edn. Vol. 1, *Biology of Yeasts*. London: Academic Press.

Rose AH and Harrison JS (1993) *The Yeasts*, 2nd edn. Vol. 5, *Yeast Technology*. London: Academic Press.

Russell I, Jones R and Stewart GG (1987) Yeast – the primary industrial microorganism. In: Stewart GG, Russell I, Klein RD and Hiebsch RR (eds) *Biological Research on Industrial Yeasts*. Vol. 1, p. 1. Boca Raton: CRC Press.

Spencer JFT and Spencer DM (1990) *Yeast Technology*. Berlin: Springer.

Spencer JFT and Spencer DM (1997) *Yeast – in Natural and Artificial Habitats*. Berlin: Springer.

Vaughan-Martini A and Barcaccia S (1996) A reconsideration of species related to *Saccharomyces dairensis* (Naganishi). *International Journal of Systematic Bacteriology* 46: 313–317.

Vaughan-Martini A and Pollacci P (1996) Synonomy of the yeast genera *Saccharomyces* Meyen ex Hansen and *Pachytichospora* van der Walt. *International Journal of Systematic Bacteriology* 46: 318–320.

Vaughan-Martini A, Barcaccia S and Pollacci P (1996) *Saccharomyces rosinii* sp. nov., a new species of *Saccharomyces sensu lato* (van der Walt). *International Journal of Systematic Bacteriology* 46: 615–618.

Saccharomyces sake

Y Iimura, Department of Applied Chemistry and Biotechnology, Yamanashi University, Kofu, Japan

Saccharomyces sake is a species of yeast used for making saké, a traditional Japanese alcoholic beverage. Saké is produced from rice grains as the raw material by mean of both koji moulds (*Aspergillus oryzae*) and yeasts, called saké yeast. In saké-making, the hydrolysis of starch by koji, prepared by cultivating koji moulds on a rice substrate, is followed by alcoholic fermentation by saké yeasts. Such fermentation is called 'parallel combined fermentation', because the hydrolysis and the alcoholic fermentation proceed simultaneously in the same vessel. A number of useful saké yeast strains have been isolated from saké mash (*moromi*) in the saké-making factories of various regions of Japan. In addition, new types of saké yeast have been developed by methods such as mutagenesis.

Taxonomy and Characteristics

In 1897 the saké yeasts were first isolated from moto, the inoculum of saké mash, and named *Saccharomyces sake* Yabe. The morphological, physiological and biochemical characteristics of saké yeasts have mainly been established by Japanese researchers. The commercially used strains were isolated especially by the Brewing Society and the National Institute of Brewing in Japan. Successful strains, e.g. Kyokai numbers 7, 9, and 10, were supplied to saké makers not only for commercial production but also for research.

These strains belong to *Saccharomyces cerevisiae* Hansen according to Lodder's taxonomic study in 1970 after re-examination of the characteristics of the saké yeast strains. As a standard description, the vegetative cell of the saké yeasts is globose or oval, and transformed direct into asci containing one to three, occasionally more, globose or short oval ascospores. However, the ability to form ascospores is much lower compared with strains for laboratory use. The ability to assimilate and ferment various carbon compounds is similar to that of *S. cerevisiae*. Therefore, the name *Saccharomyces cerevisiae* var. *sake* was proposed for the saké yeasts. However, there are some differences in the taxonomic characteristics between the saké yeasts and *S. cerevisiae* strains such as brewers' and bakers' yeasts (**Table 1**).

Table 1 Differences in characteristics between saké yeasts and other industrial yeasts of *Saccharomyces cerevisiae*

Characteristic	*Saké yeasts*	*Other useful yeasts*
Fermentation ability for maltose	Weak	Moderate–strong
Growth in vitamin-free medium[a]	Growth	Variable
Growth in biotin-deficient medium[a]	Growth	Variable
Aggregation with *Lactobacillus casei*	Non-aggregation	Aggregation
Charge on cell surface at pH 3.0	Positive	Negative
Ethanol tolerance[b]	20–21%	16–19%

[a] Vitamin-free medium containing ammonium sulphate as nitrogen source.
[b] The highest ethanol concentration that permits cell survival.

In general, the saké yeasts show a weak fermentation ability for maltose, and have the ability to grow either in vitamin-free medium containing ammonium sulphate as the sole nitrogen source or in biotin-deficient medium, while brewers', wine and bakers' yeasts cannot grow in these media. The pantothenic acid auxotrophy of the saké yeasts, however, depends on the nitrogen source; most strains can grow in medium containing ammonium sulphate as the sole nitrogen source, but not in medium containing an organic nitrogen source such as casamino acid or asparagine, although the strain Kyokai 7 which has been supplied to saké makers as a seed yeast since 1962 requires pantothenic acid in both media.

The saké yeasts were reported to be resistant to yeastcidin, the product of koji mould; Kyokai strains 6, 7, 9 and 10 are resistant, although other useful strains such as wine and bakers' yeasts are inhibited in the presence of 50–200 $\mu g\,ml^{-1}$ yeastcidin. This resistance was suggested to be one of the key features for distinguishing saké yeasts from other useful strains.

The cells of wine, brewers' and bakers' yeasts – but not the saké yeasts – have the ability to aggregate with *Lactobacillus casei* in citric acid buffer (pH 3.0). These phenomena are due to the electric charge of the cell surfaces; because both the *L. casei* and the saké yeast cell surfaces have a positive charge in acidic solution, aggregation cannot occur between these cell surfaces. The other yeast cell surfaces have a negative charge, allowing them to aggregate with the *L. casei* cells. This physico-chemical property of the saké yeast cell surface is relevant to the formation of large amounts of froth in the early stage of saké moromi. The yeast cells showing the ability to form froth have the following characteristics:

1. Air bubbles are successfully adsorbed to the cell surface.
2. This adsorbing ability is lost following proteinase treatment.
3. The cells aggregate with the *Lactobacillus plantarum* cell, which has a negative charge in acidic solution (pH 3.0).
4. The cell is transferred to the benzene layer in a double-layer system of benzene and water.

From these facts, it was deduced that the froth-forming ability was due to the hydrophobic nature of the mannoprotein composing the yeast cell wall.

Another characteristic of the cell surface is its ability to adsorb to fibre such as hemicellulose from the rice in saké mash, moromi. To explain this phenomenon it was suggested that protein with the ability to adsorb to the fibre is exposed by the release of mannoprotein from the yeast cell wall. Generally saké yeasts with this characteristic show low fermenting ability, but not the strains suitable for saké fermentation.

Although saké-making is usually performed in open vessels, harmful contaminants from wild habitats are avoided by the addition of lactic acid to the yeast seed culture (*moto*), causing a reduction in the culture pH. Therefore, lactic acid tolerance seems to be essential for saké yeast growth.

The saké yeasts are tolerant of ethanol in concentrations as high as 19–20% in moromi. The following explanations for this characteristic have been suggested:

1. Both saccharification with koji enzymes and ethanol fermentation with yeast proceed simultaneously in the same vessel of saké moromi throughout parallel combined fermentation, which keeps the glucose concentration sufficiently high for yeast proliferation.
2. Saké moromi is kept at low temperature (e.g. 15°C) during the fermentation, decreasing the damage from ethanol.
3. Saké moromi contains a high proportion of solids, which are derived from the rice substrate and reduce the damage from ethanol.
4. Proteolipid derived from koji mould mycelia promotes yeast growth in the presence of ethanol at high concentration.
5. Oxidation–reduction potential is regulated in saké moromi, encouraging yeast proliferation.

On the other hand, it has been suggested that there is a genetic component in the alcohol tolerance of the saké yeasts, because the yeasts generally show high ethanol tolerance not only in saké moromi but also in conventional culture media.

Detection of Yeasts in Moromi

Since the alcoholic fermentation in saké moromi is generally performed in open vessels, wild yeasts from natural habitats may contaminate the moromi, damaging the product quality. Therefore, monitoring the population of saké yeast in the moromi and its starter culture is important for commercial saké-making as well as for ecological studies. The following methods are employed to distinguish seeded saké yeast strains from other wild contaminants.

TTC Agar Overlay Method

This method was first developed to detect respiratory deficient yeast mutant, which is unable to reduce 2,3,5-triphenyltetrazolium chloride (TTC); the mutants are distinguishable as white, small (petite) colonies, because colonies that reduce TTC are red in colour. This method can be applied to distinguish the yeasts in saké moromi as follows:

1. After dilution, the saké moromi sample is spread on TTC basal medium (**Table 2**) and incubated at 30°C for a few days.
2. After the colonies have grown on the basal medium, TTC agar is overlaid on the medium, and incubated at 30°C until the colour changes.
3. The colonies are counted according to the difference in the colour tones.

There are some differences in the colony colour tone among the yeasts; for instance Kyokai strains 6, 7, 9, 10 and 11 can reduce TTC, and appear red, whereas the petite yeasts and most wild yeasts including film-forming yeasts such as *Hansenula* spp. usually appear as white or pink colonies respectively. The colour tone changes depending on the kind of carbon source contained in the TTC basal medium. In practice, these

Table 2 The TTC basal medium and TTC agar

Basal medium		*TTC agar*	
Glucose	10 g	Glucose	0.5 g
Peptone	2 g	TTC[a]	0.05 g
Yeast extract	1.5 g	Agar	1.5 g
KH_2PO_4	1.0 g	Water	100 ml
$MgSO_4.7H_2O$	0.4 g		
Agar	30 g		
Water	1000 ml		
pH	5.5–5.7		

[a] 2,3,5-Triphenyltetrazolium chloride.

differences are very useful for distinguishing Kyokai yeasts from other yeasts, so this method should be widely used for monitoring the fermentation process. Although the colour changes seem to be due to differences in the ability to reduce TTC, the biochemical mechanism of the coloration is not yet clear.

β-alanine Method

The β-alanine method is used for distinguishing Kyokai strain 7, the most widely used strain for commercial saké production, from the other Kyokai strains. Kyokai 7 requires pantothenic acid in media containing an inorganic nitrogen source as well as in media containing an organic source, and also shows a temperature-sensitive requirement for β-alanine, a component of pantothenic acid; i.e. this strain can grow in a medium containing ammonium sulphate as the nitrogen source and β-alanine instead of pantothenic acid at lower temperature (20°C), but not at higher temperature (above 35°C). Using this physiological characteristic, the Kyokai strain 7 in saké moromi can be distinguished as follows:

1. After dilution, the saké moromi sample is spread on the β-alanine medium (**Table 3**) and incubated at 35°C for 2 days. The resulting colonies corresponding to yeasts other than Kyokai 7 are then counted.
2. The culture is then incubated at 20°C for 1–2 days. Newly appearing colonies will be only Kyokai 7.

Acid Phosphatase Stain Method

Kyokai yeasts 6, 7, 9 and 10 show no acid phosphatase activity on medium containing phosphate in high concentrations as the gene expression for this enzyme is repressed with excess of phosphate. The difference in the activity among yeasts can be determined by the diazo-coupling reaction. Using this method, Kyokai yeasts can be detected as follows:

1. The saké moromi sample is spread on TTC basal medium (see Table 2) containing enough phosphate to avoid the expression of phosphatase activity, and incubated at 30°C for a few days.
2. The upper layer agar, which is composed of 0.05% sodium α-naphthyl acid phosphate, 0.05% fast blue B, 0.01 mmol l^{-1} acetate buffer (pH 4.0) and 1.5% agar, is overlaid on the basal medium, and incubated at 30°C for 30–60 min.
3. The non-coloured colonies are Kyokai yeasts, and the coloured ones (red or dark red) are other yeasts.

Table 3 β-alanine medium

Glucose	20 g
$(NH_4)_2SO_4$	0.5 g
KH_2PO_4	1.5 g
$MgSO_4.7H_2O$	0.5 g
β-Alanine	40 μg
Thiamin	200 μg
Pyridoxine	200 μg
Nicotinic acid	200 μg
Inositol	1000 μg
Biotin	0.2 μg
p-Aminobenzoic acid	200 μg
Agar	30 g
Water	1000 ml
pH	5.0

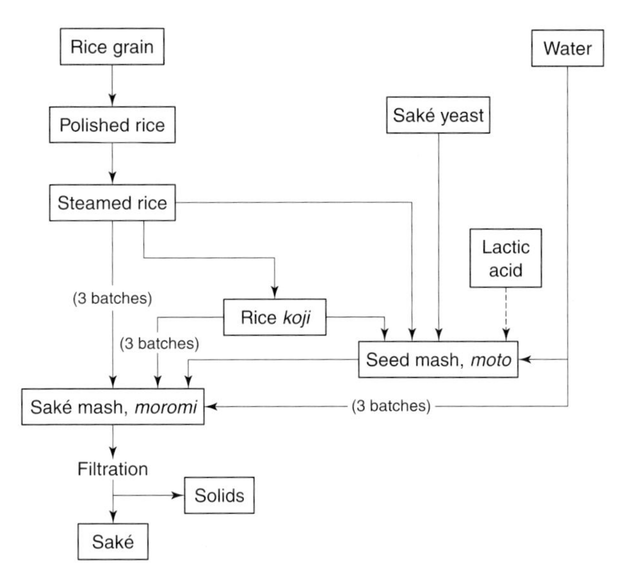

Figure 1 The saké-making process.

Importance in the Saké Industry

Outline of Saké-making

Saké is produced in most regions of Japan with the exception of the southern, warmer, parts, as saké fermentation requires low temperatures. The annual saké production in 1994 was about 1.2×10^9 litres. Saké has also been produced in countries other than Japan. However, there is little difference in the saké-making process (**Fig. 1**) between the regions.

The first stage in saké-making is the preparation of rice koji, a culture of *Aspergillus oryzae* on steamed rice. In the second stage, steamed rice, saké yeast and water are combined in a vessel to make the seed mash, moto, which is a yeast starter for the main saké mash, moromi. About 7–8% of the total amount of rice is used for moto preparation. Many microorganisms are involved in this preparation, and the changes in flora are well-defined. Since lactic acid is produced by lactic acid bacteria including *Leuconostoc mesenteroides* and *Lactobacillus sake*, acidifying the moto in the early stages of this preparation, harmful bacterial contaminants are successfully suppressed. In most cases lactic acid is initially added to the moto instead of depending on lactic acid bacteria.

During the moto preparation, the rice grains are gradually degraded and saccharified by the rice koji enzymes. By assimilating the produced components as well as glucose, the seeded saké yeasts are grown to a high cell density of about 3×10^8 cells per millilitre. In the final moto stage, the moto is usually cooled to prevent weakening or death of the grown yeasts, and rested for 3–7 days before its use for fermentation in saké mash, moromi.

In the next stage, saké moromi is subjected to parallel combined fermentation for about 25 days. In this stage, the rice koji, steamed rice and water are added successively in three batches to promote favourable yeast growth. When fermentation ceases, producing about 18% ethanol, the moromi is filtered to remove the solids, which are the undigested rice material, yielding fresh saké.

The Role of Yeast in Saké-making

In saké moromi the saké yeasts produce ethanol, organic acids, amino acids and other components that are essential for the saké flavour. Of the organic acids, lactate, succinate and malate are major components, the productivity of which is dependent both on the kind of saké yeast strains and on the conditions of the moromi fermentation. The productivity must be physiologically regulated; however, its regulation mechanism is not yet clear. Furthermore, the control of pyruvate production is important, because excess of this acid in the moromi results in the production of diacetyl, one of the components of off flavours in saké.

Generally, amino acids in the moromi are produced from the rice-grain protein by the proteolytic enzymes of koji. In the early moromi stage, amino acids are vigorously assimilated by the yeast, whereas in the late stage the amount of amino acids increases, because the yeast ceases growth owing to the increase of ethanol, and with cell death autolysis releases amino acids into the moromi. Excessive amounts of amino acids in the moromi cause discoloration, resulting in poor-quality saké. To prevent this spoilage saké moromi should be controlled to avoid yeast autolysis as much as possible; for instance keeping the fermentation temperature low is an effective measure.

Two kinds of ester, isoamylacetate and ethylcaproate, are major components giving fruity aromas to saké. Isoamylacetate is thought to be synthesized by alcohol acetyltransferase bonded to the yeast cell membrane. The enzyme is unstable at temperatures higher than 10°C. This enzyme activity is inhibited by incorporating unsaturated fatty acids such as linoleic acid to the yeast cell. This inhibition effect is greater as the number of fatty acid double bonds increases. As untreated rice grain contains a large amount of unsaturated fatty acid, especially in the outer layer, it is stringently polished to remove the fatty acids; 30–60% of the matter is usually removed. The most highly graded saké, *ginjyo-shu* is made from rice containing little fatty acid because of the high degree of polishing; above 40% of the matter is removed. In *ginjyo-shu* manufacture the fermentation is performed at lower temperature (usually below 12°C), which results in a long, slow fermentation. *Ginjyo-shu* contains more esters and less organic acid, amino acid and sugar than normally graded saké.

Table 4 Development of useful strains of saké yeasts

Useful character	*Isolation method*	*Practical effect*
Non-froth-forming	Non-adsorption to bubble	Effective utilization of fermentation vessel
Killer activity	Back cross or cytoduction	Prevention of wild yeast contamination
Ethanol tolerance	Enrichment culture	Prevention of yeast autolysis in saké mash
Adenine auxotrophy	Mutation and enrichment	Making pink saké
Fermenting ability for starch	Transfer of amylase gene	Increase of saccharifying ability in saké mash
Arginase deficiency	Mutation and canavanine resistance	Making saké free from urea
Ester production	Mutation and analogue resistance	Making saké containing aromatic compounds

Breeding of Saké Yeasts

Most of the saké yeasts in industrial use were isolated from moromi that produced high-quality saké. To develop new products and increase the efficiency of saké-making, these yeasts have been improved by mutagenesis and enrichment culture (**Table 4**).

Non-froth-forming Saké Yeasts

The usual saké yeasts such as the Kyokai strains form a large head of froth in the early stage of saké moromi, preventing vessels from being used to full capacity. Instances of absent froth formation were attributed to a characteristic of the yeast cells; froth-forming cells can be adsorbed to bubble, but not the non-froth-forming cells. This characteristic was applied to isolate non-froth-forming mutants from the saké yeasts as follows:

1. Sterilized air is blown vigorously into a saké yeast culture broth containing the non-froth-forming mutants which occur spontaneously at very low frequency, and the vessel overflows until most of the broth spills over (froth flotation method).

2. The mutants remain in the vessel, while most of the froth-forming yeast cells are removed with the air bubbles because of their adsorbing ability. By this procedure, the proportion of the mutants in the broth can be increased successfully.
3. Next, the residue is cultured in a fresh medium, followed by a repetition of the froth flotation method.
4. The proportion of mutants in the broth becomes higher as these procedures are repeated until the mutants can be finally isolated.

Because the froth-forming characteristic is a property of the yeast cell surface, the non-froth-forming mutant can be isolated not only by non-adsorption to bubble but also by non-adsorption to *Lactobacillus plantarum* cells, which have a negative charge in acidic solution (pH 3.0).

The gene involved in the froth-forming phenomenon has now been cloned and sequenced by a Japanese research group. A open reading frame of this gene seems to be for a protein composing the yeast cell wall. However, the relationship between the protein structure and the hydrophobicity of the cell surface is not yet clear.

Killer-resistant Saké Yeast

The 'killer' strain of *Saccharomyces* produces a protein that kills yeasts lacking immunity to this toxin. Because the killer strain has immunity to its own toxin, it is not self-destructive. However, since most of the saké yeasts have no such immunity, contamination by wild killer yeast in the fermentation process causes much damage to the seeded saké yeasts, resulting in slow fermentation and degradation of the product. Therefore, new killer yeasts that can produce high-quality saké are very effective against contamination, because these yeasts are not only resistant to wild killers, but also can destroy wild killer-sensitive yeast contaminants.

The killer activity and immunity depend on two kinds of killer plasmids, M-dsRNA and L-dsRNA, which are small and large double-strand, virus-like RNAs respectively. The gene M-dsRNA encodes the killer protein, whereas L-dsRNA encodes the envelope protein of these double-strand RNAs; i.e. both genes are essential for the killer expression. Therefore, a useful killer yeast can be bred by transferring these plasmids to the cytoplasm of the host yeast. The killer saké yeasts have been constructed by the transfer of killer plasmids to the cell of Kyokai strain 7 by cytoduction using a nuclear fusion-deficient killer mutant, which can form normal zygotes but is defective in nuclear fusion. By this means, it is possible to transfer the killer plasmids to useful yeast cells without crossing the nuclear genes.

Ethanol-tolerant Yeasts

In the late stage of saké moromi fermentation the ethanol concentration becomes higher (above 18%), leading to an increased content of dead yeast cells. The resulting increase in amino acids degrades saké quality. Yeast mutants that can tolerate ethanol at high concentrations have been isolated from Kyokai strain 7 by the enrichment culture using a medium containing 20% (v/v) ethanol. Saké produced by this mutant contains a smaller amount of amino acids than that produced by the usual saké yeasts, because the proportion of dead yeast is markedly less in the late moromi stage. The commercially used yeast with this characteristic has been developed as Kyokai strain 11. This mutation seems to be related to properties of the cell surface, because this mutant showed higher tolerance for digestion by the yeast-lytic enzyme, zymolyase, than the usual yeasts.

Adenine Auxotrophic Mutant

When oxidized, 5-aminoimidazole riboside (an intermediate of adenine biosynthesis) changes colour, giving a red pigmentation to the yeast cell. Of the adenine auxotrophic yeast mutants, *ade1* and *ade2* strains, which are deficient in phosphoribosylaminoimidazolesuccinylocarboxamide synthetase and phosphoribosylaminoimidazole carboxylase respectively, accumulate this pigment in the cells. To develop red saké as a new product by using this kind of yeast, the adenine auxotrophic mutant was isolated from saké yeast as follows. After the saké yeasts were treated by ultraviolet irradiation, the adenine auxotrophic mutants which occurred at very low frequency were enriched using medium containing nystatin; this antibiotic damages the growing cell membrane by bonding with its ergosterol, so that this antibiotic can depress prototrophic cells, but not non-growing auxotrophic cells in minimum medium. By this means, the mutant producing the red pigment has been successfully isolated from Kyokai yeasts, and commercially used. This mutant can also produce saké, in which the red yeast cells are suspended with white rice solids. Generally the biosynthesis of the pigment is repressed in medium containing excess adenine by the feedback inhibition effect. However, the successful coloration in saké shows that the moromi contains enough adenine for pigment production. The mutant has a slower fermentation ability than the usual yeasts because of the auxotrophy.

Other Useful Saké Yeasts

Strains of yeasts that can secrete glutathione or produce highly aromatic compounds, as well as non-urea-producing and starch-fermenting saké yeasts, have been developed and have been used commercially in saké manufacturer.

See color Plate 29.

See also: **Fermented Foods**: Fermentations of the Far East. ***Lactobacillus***: *Lactobacillus casei*. ***Saccharomyces***: Introduction; *Saccharomyces carlsbergensis* (Brewer's Yeast).

Further Reading

Fujii T, Nagasawa N, Iwamatu A, Bogaki T, Tamai Y and Hamachi M (1994) Molecular cloning sequence analysis, and expression of the yeast alcohol acetyltransferase gene. *Applied and Environmental Microbiology* 60: 2786–2792.

Fujii T, Yoshimoto H and Tamai Y (1996) Acetate ester production by *Saccharomyces cerevisiae* lacking the *ATF1* gene encoding the alcohol acetyltransferase. *Journal of Fermentation and Bioengineering* 81: 538–542.

Fukuda K, Kuwahata O, Kiyokawa Y et al (1996) Molecular cloning and nucleotide sequence of the isoamyl acetate-hydrolyzing esterase gene (*EST2*) from *Saccharomyces cerevisiae*. *Journal of Fermentation and Bioengineering* 82: 8–15.

Fukuda K, Yamamoto N, Kiyokawa Y et al (1998) Brewing properties of sake yeast whose *EST2* gene encoding isoamyl acetate-hydrolyzing esterase was disrupted. *Journal of Fermentation and Bioengineering* 85: 101–106.

Kitamoto K, Oda K, Gomi K and Takahashi K (1991) Genetic engineering of a sake yeast producing no urea by successive disruption of arginase gene. *Applied and Environmental Microbiology* 57: 301–306.

Kodama K (1993) *Sake*-brewing yeasts. In: *The Yeasts*. Vol. 5, 2nd edn, p. 129. London: Academic Press.

Magarifuchi T, Goto K, Iimura Y, Tadenuma M and Tamura G (1995) Effect of yeast Fumarase gene (*FUM1*) disruption on production of malic, fumaric and succinic acids in sake mash. *Journal of Fermentation and Bioengineering* 80: 355–361.

Mizoguchi H and Hara S (1997) Ethanol-induced alterations in lipid composition of *Saccharomyces cerevisiae* in the presence of exogenous fatty acid. *Journal of Fermentation and Bioengineering* 83: 12–16.

Mizoguchi H and Hara S (1998) Permeability barrier of the yeast plasma membrane induced by ethanol. *Journal of Fermentation and Bioengineering* 85: 25–29.

Ohbuchi K, Ishikawa Y, Kanda A, Hamachi M and Nunokawa Y (1996) Alcohol dehydrogenase I of sake yeast. *Saccharomyces cerevisiae* Kyokai No. 7. *Journal of Fermentation and Bioengineering* 81: 125–132.

Suizu T, Tsutsumi H, Kawado A, Murata K, Suginami K and Imayasu S (1996) Methods for sporulation of industrially used sake yeasts. *Journal of Fermentation and Bioengineering* 81: 93–97.

Watanabe M, Nagai H and Kondo K (1995) Properties of sake yeast mutants resistant to isoamyl monofluroacetate. *Journal of Fermentation and Bioengineering* 80: 291–293.

Saccharomyces cerevisiae

B C Viljoen, Department of Microbiology and Biochemistry, University of the Orange Free State, Bloemfontein, South Africa

G M Heard, Department of Food Science and Technology, University of New South Wales, Sydney, Australia

Characteristics of the Species

The yeast that has been most closely associated with humankind, *Saccharomyces cerevisiae*, has long been used for brewing, wine-making and baking bread. It is by far the most studied and best understood species of the yeast domain and an important model system for basic research into the biology of the eukaryotic cell. Indeed, the ability to rationally manipulate all aspects of its gene expression by in vitro genetic techniques offers *S. cerevisiae* a unique place among eukaryotes.

Saccharomyces cerevisiae is the type species of the genus *Saccharomyces*, introduced by Meyen in 1838 and defined by Rees in 1870, whereas Hansen described the beer yeast, *S. cerevisiae*, in 1888. The species is a member of the family Saccharomycetaceae and the subfamily Saccharomycetoideae, characterized as a unicellular fungus reproducing vegetatively by multilateral budding and sexually by means of ascospores. The asci are persistent and contain one to four globose ascospores. The vegetative cells are globose, ovoidal or cylindrical and appear butyrous and light cream-coloured, while the surface is smooth and flat.

The taxonomy of *S. cerevisiae* has undergone major changes, especially with the impact of molecular biology on classification. Over the years, the genus included a variable number of heterogeneous species, as many as 41 species in 1970. After extensive rearrangement of species and genera, seven species were recognized in 1984. *Saccharomyces cerevisiae* represented some 21 earlier taxa grouped in *Saccharomyces sensu stricto*, being taxonomically justified from physiological tests. Genetic analysis, however, contradicted the amalgamation and consequently the species were separated into four variably related species in 1985. The separation of the four species remains uncertain by physiological tests owing

Table 1 Key properties for the identification of *Saccharomyces cerevisiae*

Property	S. cerevisiae
Fermentation:	
Sucrose	+
Raffinose	+
Trehalose	–
Assimilation	
Sucrose	+
Maltose	+
Raffinose	+
D-*Ribose*	–
Ethanol	+
D-*Mannitol*	–
Nitrogen source:	
Cadaverine-2HCl	–
Ethylamine-HCl	–
L-*Lysine*	–
Growth:	
Cycloheximide (1000 p.p.m.)	–
30°C	+
37°C	v
Vitamin free	–
G+C	39–41%
Coenzyme	Q6

v, variable.

to exceptions that exist when larger groups of strains are involved. In 1990 seven species were proposed, listing 130 synonyms for *S. cerevisiae* with the inclusion of many subspecies and varieties encompassing bread-making, brewing, wine and cider yeasts, as well as naturally occurring species. At this time some taxonomists refrained from differentiating between closely related *Saccharomyces* species solely on the basis of DNA–DNA hybridization and consequently placed these species in *S. cerevisiae*. It was only in the 1990s that taxonomists agreed to accept DNA–DNA hybridization analyses as well as traditional genetic studies as a valid criterion, which led to the distinction between the four variably related species *S. bayanus*, *S. cerevisiae*, *S. paradoxus* and *S. pastorianus* previously grouped as *Saccharomyces sensu stricto*. As currently demarcated the genus *Saccharomyces* encompasses 16 species, while the species *S. cerevisiae* includes 95 synonyms. The accepted key characters for the identification of this species are listed in **Table 1**.

Physiological and Biochemical Properties

A substantial number of publications have accumulated over the years discussing the physiology and molecular biology of *S. cerevisiae* in detail. The significance of the yeast as a fermentative species, particularly for its role in alcoholic fermentations, has urged many scientists to study the factors governing the growth, survival and biological activities of this species in different food ecosystems.

The species is able to ferment the hexose sugars of D-glucose, D-fructose and D-mannose. The rate of D-glucose fermentation is normally the most aggressive. Other sugars that can be fermented by some strains of *S. cerevisiae* include sucrose, maltose and D-galactose, while dextrins and starch are rarely fermented, and lactose is not fermented by this species. The L-sugars and all pentoses are also considered non-fermentable, though xylulose can be fermented. Aside from the hexoses and their dimers and oligomers, the species readily metabolizes non-fermentable compounds such as lactic acid, other organic acids and polyhydroxy alcohols. Strains of *S. cerevisiae* differ in their ability to utilize nitrogen sources. Many inorganic ammonium salts have been found to promote growth, while strains exhibit different abilities to utilize free amino acids. Nitrates as well as L-amino acids such as L-lysine are not utilized. Conversely, some *S. cerevisiae* strains can utilize urea as a source of nitrogen.

Various growth factors are required by some *S. cerevisiae* strains, taken up during biosynthesis to relieve the cell of the need to synthesize the compound, thereby saving energy. Deficiency in inositol can lead to less effective cell division, and deficiency of thiamin in the absence of pyridoxine reduces growth. Some strains, however, require either thiamin or pyridoxine, the first stimulating growth. Biotin and pantothenate are also essential for all strains of *S. cerevisiae*.

With respect to their occurrence and survival in foods and beverages, the most distinctive characteristic of species of *Saccharomyces* appears to be their tolerance to high ethanol concentrations. The resistance of the different *S. cerevisiae* strains may vary, but in general most strains are able to grow in beverages containing 8–12% ethanol (v/v) and will survive concentrations of about 15%. Normal glucose fermentations may yield concentrations of about 12% ethanol, while saké fermentations may yield concentrations as high as 20% ethanol when the presence of unsaturated fatty acids promotes alcohol tolerance. The sensitivity of the species to ethanol, however, increases with temperatures above 30°C or lower than 10°C. During fermentations for ethanol production, temperatures of 5–10°C higher than the optimum result in a decrease of cell growth but increase ethanol productivity. Although most *S. cerevisiae* strains will grow at any temperature between 0°C and 40°C, optimum temperature for maximum growth rate is strain-dependent, generally in the region of 28–35°C. Optimum temperatures are lower, however, when cell yield is considered rather than growth rate. Alcoholic fermentation is further enhanced by the extraordinarily rapid growth capabilities of this species under oxygen limitation, resulting in high alcoholic fermentation rates. The species therefore has a selective advantage over strictly aerobic yeasts, thriving under low oxygen tension in beverages and in the inner layers of food. Oxygen is an essential nutrient for all yeasts, and although *Saccharomyces cerevisiae* can grow under microaerophilic conditions, oxygen is essential to maintain cell viability. Under anaerobic conditions the synthesis of certain cellular constituents such as fatty acids and sterols ceases and consequently the yeast cells stop growing.

The influence of water activity on yeast survival and growth is an important feature in respect to the growth of yeasts in foods. *Saccharomyces cerevisiae*, although not regarded as a xerotolerant species, responds to a decrease in the a_w value of a medium by synthesizing glycerol, thereby lowering the osmotic pressure difference across the yeast plasma membrane. However, the species leaks much of the polyol into the medium and therefore fails to adapt sufficiently to the stress. This xerotolerance is influenced by the nature of the solute, the temperature and other ecological factors, but minimum a_w values for growth of *S. cerevisiae* range between 0.89–0.91 with glucose, fructose and sucrose as the stressing solutes and values of 0.92 with salt (NaCl).

Saccharomyces cerevisiae, like all yeasts, prefers a slightly acidic medium with an optimum pH between 4.5 and 6.5. The species, however, shows a remarkable tolerance to pH, being capable of growth at pH values as low as 1.6 in HCl, 1.7 in H_3PO_4 and 1.8–2.0 in organic acids. It has a maximum tolerance to benzoic acid 100 mg kg^{-1} at pH 2.5–4.0, and to sorbic acid 200 mg kg^{-1} at pH 4.0. The inhibition of the growth of *S. cerevisiae* by these organic acids is due mainly to dysfunction of cell membrane permeability. Other acids reported to inhibit the growth include *p*-coumaric acid (100–250 p.p.m.) and ferulic acid (50–250 p.p.m.) as well as natural inhibiting compounds such as xylitol (0.5%), tuberine, the antioxidants butylated hydroxyanisole, tertiary butylhydroquinone and propylgallate (50–500 p.p.m.). Caffeine, conalbumin, lysozyme from eggs and high ethanol concentrations are all natural mycotic inhibitors of *S. cerevisiae* strains, although cinnamon and clove oils, which contain high levels of eugenol, stimulate pseudomycelium formation. With respect to resistance to inactivation by heat, vegetative cells have a decimal reduction time at 60°C (D_{60}) of 0.1–0.3 min, while ascospores are much more resistant, with a D_{60} of 5.1–17.5 min. Heat resistance can be enhanced when heating cells grown in media with a reduced water activity. In a medium based on fruit juice (pH 3.1, 0.99 a_w, 12° Brix) the D_{60} was 0.3–2 min, but at 0.93 a_w the D_{60} was 5 min or more.

The importance of *S. cerevisiae* in the food industry is strengthened by its ability to produce and secrete extracellular polygalacturonase enzymes which may have consequences for the fermentation of plant-derived substrates. The species also produces extracellular proteases for the breakdown of protein, while a biotype, *S. diastaticus*, produces glucoamylase initiating partial hydrolysis of starch.

Table 2 Significance of *Saccharomyces cerevisiae* in foods and beverages

Role of S. cerevisiae	*Examples*
Production of fermented beverages and breads	Wine, beer, cider, distilled beverages, bread, sweet breads, sourdough bread, cocoa, fermented juices and honey
Food spoilage	Processed fruit products – juices, purées, fruit pieces, bakery products containing fruit
	Fruit yoghurt, labeneh
	Minimally processed fruits and vegetables
	Cucumbers in brine
	Alcoholic beverages
Processing food wastes	Growth on vegetable by-products, citrus by-products, beet molasses and whey
Source of food ingredients	Flavour compounds, δ-decalatone, phenylethanol, yeast extract
	Fractionated yeast cell components – mannoproteins, glucomannans, yeast glycans, yeast protein concentrate, invertase, ergosterol, glucans
	Fructose syrup
	Probiotics (*Saccharomyces boulardii*)

Importance to the Food Industry

Saccharomyces cerevisiae is commercially significant in the food and beverage industries (**Table 2**) because of its role in the following:

- the production of fermented beverages and breads
- spoilage of foods and beverages
- processing food waste
- production of food ingredients
- as a probiotic.

Production of Fermented Foods and Beverages

Saccharomyces cerevisiae is best known for its domesticated role in the production of breads and fermented alcoholic beverages and for adding positively to the flavour and keeping quality of the final product. *Saccharomyces cerevisiae* converts hexose sugars, to ethanol, CO_2 and a variety of compounds, including alcohols, esters, aldehydes and acids that contribute to the sensory attributes of the food or beverage.

In the baking industry, *S. cerevisiae* is generally inoculated into bread dough as approximately 2% of the total ingredients. The primary role of *S. cerevisiae* is to convert the available carbohydrate into CO_2 gas. The yeast also assists in dough and flavour development, through the reduction of dough pH and production of reducing compounds such as glutathione, affecting dough rheology and strength, and volatile components. Growth and fermentation of the yeast is influenced by dough factors such as the inoculation rate, available carbohydrate (sucrose and maltose), the osmotic pressure and the water activity. Bakers' strains of *S. cerevisiae* are chosen for their ability to produce gas rapidly from the available carbohydrates in the dough, and because sucrose is preferentially fermented, strains with strong invertase activity are normally used to ensure rapid sugar utilization and rapid fermentation rate. *Saccharomyces cerevisiae* does not readily ferment maltose and it is necessary to select strains for baking that are adapted to maltose utilization, especially for use in doughs that contain little or no sucrose. Where high concentrations of sucrose are used (25% sucrose, a_w 0.92), strains are selected for osmotolerance. Osmotolerance of *S. cerevisiae* has been correlated with low invertase production and production and accumulation of low-molecular-weight metabolites include glycerol.

Saccharomyces cerevisiae also contributes to the acid fermentation of a wide range of bread and pancake doughs. The best known of these is the sourdough process, which uses rye and wheat flours. These fermentations are usually conducted by a mixed ecology of lactic acid bacteria and yeasts including *S. cerevisiae*, and may develop naturally or as a result of inoculation with a starter culture. *Saccharomyces cerevisiae* may be present as 60% of the yeast population together with other species of *Saccharomyces* and *Candida*. The activity of the yeast during acid fermentations is influenced by its ability to grow at low pH (< 4.5) and survive in the presence of organic acids. Further study is required to define the role of *S. cerevisiae* in these fermentations.

Saccharomyces cerevisiae is the principal yeast species involved in the production of many alcoholic beverages. Fermentation may be the result of spontaneous development of the microflora associated with the raw materials, or a pure culture of yeast may be used. Probably the best understood processes are the wine, beer and cider fermentations. Wine is fermented either by natural fermentation or by inoculation with a starter strain of *S. cerevisiae*. Starter cultures are added at approximately 10^6–10^7 colony forming units (cfu) per millilitre and achieve a maximum population of 10^8 cfu ml^{-1} at the end of fermentation. Fermentation is initially conducted by a complex flora. However, as the ethanol content increases *S. cerevisiae* becomes the dominant species. The final ethanol content of the wine varies within the range 8–14% and production of flavour compounds varies between strains. Of increasing interest is the ability of *S. cerevisiae* to produce glycosidases, which may enhance the flavour of wines by releasing flavour components from grapes, and pectin-degrading enzymes, which may be useful for clarification and filtration of grape juice. Cider fermentation has a similar ecology and biochemistry to wine fermentation, resulting in a beverage containing 1.5–8.5% ethanol. Although traditionally ciders have been fermented naturally, most commercial cider fermentations today are produced with an inoculated strain of *S. cerevisiae*. Other beverages are produced worldwide, from fruits, honey and tea, all involving yeast fermentation with contribution from *S. cerevisiae*. Less attention has been given to the ecology of these fermentations and more research is required to understand the role played by *S. cerevisiae*.

Beer is produced from alcoholic fermentation of extracts from cereal grains. Although there is a complex microflora associated with the process, in most commercial beer production the alcoholic fermentation is conducted by an inoculated strain of *S. cerevisiae*. The yeast is added to the wort (boiled, malted cereal extract) providing a starting population of $10–15 \times 10^6$ cfu ml^{-1}. Onset of fermentation is rapid, with a very short lag phase, and the fermentation is often complete within 2–4 days by which time the yeast population has increased to $6–8 \times 10^7$ cfu ml^{-1}, producing 2–6% ethanol. Brewing strains are selected for their influence on the flavour of beer, and for their ability to flocculate and sediment at the end of fermentation, assisting in clarification of the beer. As with wine and cider, *S. cerevisiae* produces a range of volatile flavour and aroma compounds, including higher alcohols and esters that influence the final quality of the beer. The yeast reduces compounds such as diacetyl to the more pleasant-tasting acetoin or 2,3-butanediol during beer maturation. Yeast autolysis products may also be beneficial to the flavour of the final product.

A variety of beer products are produced worldwide from cereals such as barley, maize and sorghum, most of them by natural fermentation. For example, *S. cerevisiae* is the primary yeast involved in the fermentation of Bantu beer, a sour, unhopped, unpasteurized African beer produced from sorghum by a natural fermentation. The process also involves other microorganisms such as lactic acid bacteria and results in a low pH product (pH 3.0–3.3) with 2–4% ethanol.

Saccharomyces cerevisiae is used in the production

of distilled alcoholic beverages such as rum and whisky and for saké fermentation. Distillery strains are inoculated into raw material extracts at approximately 10^6–10^7 cfu ml^{-1} to conduct the initial fermentation. For whisky production, *S. cerevisiae* ferments a slurry of cereals such as maize, wheat, rye or barley, to produce 6–9% ethanol. The resulting mash is distilled to produce whisky. Saké strains of *S. cerevisiae* are chosen for their ability to tolerate high ethanol concentrations. They ferment a steamed rice mash to produce 15–20% ethanol.

Saccharomyces cerevisiae is the most common species identified during the natural fermentation of cocoa. It contributes to the chocolate flavour by producing aroma compounds such as esters and higher alcohols including isoamyl alcohol and 2-phenylethanol. Strains of *S. cerevisiae* isolated from cocoa fermentation also exhibit pectinolytic activity which aids in the breakdown of the cocoa bean pulp.

Spoilage of Foods and Beverages

Saccharomyces cerevisiae occurs widely in foods but is infrequently designated as a causative agent for spoilage. It is mainly implicated in the fermentative spoilage of high-sugar foods and beverages. Because *S. cerevisiae* can tolerate ethanol concentrations of up to 15%, it may occasionally spoil alcoholic beverages, including wine and beer. *Saccharomyces cerevisiae* is also isolated from dairy products including milk, yoghurts and cheese, fermented vegetables and minimally processed vegetable products, although the significance of this species in the spoilage of these products is not clearly defined.

Saccharomyces cerevisiae is widespread in its occurrence in nature, on fruits, leaves and nectars, and although it is not commonly associated with spoilage of fresh fruits it is often implicated in the spoilage of processed fruit products. Contamination with yeasts may arise from the fruit, from insect vectors or from the processing environment. Final yeast populations of fruit juices may reach 10^7–10^8 cfu ml^{-1}. *Saccharomyces cerevisiae* has been reported to form approximately 25% of the yeast population of fruit juice concentrates. The species also causes spoilage of carbonated soft drinks and fruit drinks, sports drinks, purěed fruits and canned fruit products. Bakery products containing fruit are also susceptible to spoilage by *S. cerevisiae*. Yeast counts in products such as apple turnovers may reach up to 10^6 cfu ml^{-1} resulting in fermentative spoilage and blown packages. *Saccharomyces cerevisiae* has been isolated as part of the flora of dry or semidry dates and from figs and prunes. There is little quantitative data available on the occurrence of yeasts in minimally processed fruits; however, *S. cerevisiae* has been implicated as part of the spoilage flora of peeled oranges and commercially processed grapefruit sections. *Saccharomyces cerevisiae* is less commonly associated with vegetables, but it has been isolated from spoiled, softened cucumbers in brine. Softening was thought to occur as a result of yeast enzymatic activity. *Saccharomyces cerevisiae* has been isolated from minimally processed vegetable products such as processed lettuce and is implicated in the fermentative spoilage of low-pH, mayonnaise-based salads such as coleslaw. The extent of its contribution to salad spoilage requires further investigation.

Alcoholic beverages may be spoiled by undesirable strains of *S. cerevisiae*. For example, beer may be spoiled by strains that produce fruity flavours or sulphurous compounds. Haziness results from the presence of wild, non-flocculating strains in beer. During wine fermentation, indigenous strains of *S. cerevisiae* may produce undesirable characteristics. Killer strains of *S. cerevisiae* can prevent the growth of inoculated species, resulting in 'stuck' fermentation.

Saccharomyces cerevisiae may be isolated from a variety of dairy products, including milk, yoghurts and cheeses. Yeasts rarely grow in milk stored at refrigeration temperatures because they are outgrown by psychrotrophic bacteria. However, in sterilized milk in the absence of competition, *S. cerevisiae* exhibits weak lipolytic and proteolytic activity and is capable of growth to reach populations of 10^8–10^9 cfu ml^{-1}. In sweetened milks it can utilize the added sucrose, causing fermentative spoilage. *Saccharomyces cerevisiae* is also frequently isolated from fruit yoghurt, and can establish good growth (10^7 cfu g^{-1}) when inoculated into yoghurts stored at temperatures ranging from 5°C to 20°C. *Saccharomyces cerevisiae* is also present in labeneh, a strained yoghurt, as the predominant spoilage organism, reaching populations of 10^7 cfu g^{-1} during refrigerated storage. *Saccharomyces cerevisiae* is often present in soft and mould-ripened cheeses. It has been reported in Italian Stracchino cheese at 3.1% of the yeast population and in Camembert cheeses, when about a third of the samples tested contained 10^3–10^4 cfu g^{-1} *S. cerevisiae*. It is less frequently isolated from high-salt cheeses because of its inability to tolerate NaCl at concentrations greater than 5%. However, it may be present in semihard and hard cheese including Cheddar cheese. Growth of *S. cerevisiae* in cheeses is thought to be related to its ability to use lipid and protein products from other species and possibly its ability to utilize lactic acid present in the cheese. The significance of species such as *S. cerevisiae* as spoilage organisms in cheeses is not well understood and it has been suggested that rather than causing spoilage, it may play a role in flavour development during the maturation of cheeses.

Processing of Food Waste

Several reports describe the use of *S. cerevisiae* in the processing of food wastes to produce feedstocks and ethanol. *Saccharomyces cerevisiae* can be used to derive ethanol from substrates such as beet molasses, a by-product of the sugar industry. Depending on growth conditions, ethanol concentrations of 50–60 g l^{-1} have been achieved from an initial sugar concentration of 250 g l^{-1}. Other agroindustrial wastes may also be processed by *S. cerevisiae*. Whey and starch-fermenting strains of *S. cerevisiae* have been engineered for industrial production of ethanol. Ethanol yields as high as 50–70% have been reported. Plant leaf waste from vegetables such as cauliflower, mustard, turnip and radish plants have been fermented by a mixed culture of yeasts including *S. cerevisiae*, to produce animal feeds high in protein and vitamins. The biological oxygen demand was also reduced after fermentation. *Saccharomyces cerevisiae* has also been used to ferment citrus by-products in preparation of clouding agents for beverages. The soluble solids content of the peel was reduced by 50–60% after fermentation but the cloud stability was not sufficiently stable for addition to beverages.

Source of Food Ingredients and Processing Aids

Saccharomyces cerevisiae exhibits a wide range of biochemical properties of value as food ingredients. Approximately 90% of yeast and yeast-based products in the world originate from *S. cerevisiae*, and include yeast extracts, flavours and enzymes. The production of volatile flavour components such as 2-phenylethanol and δ-decalatone is of commercial importance. The rose flavour 2-phenylethanol is produced from phenylalanine during alcoholic fermentation and is produced during saké fermentation at levels of 20–70 p.p.m.. Overproducing mutants of *S. cerevisiae* are used to increase the yield by up to 40-fold. The compound δ-decalatone is produced by biocatalytic conversion by *S. cerevisiae* of 11-hydroxyhexadecanoic acid via the intermediate 5-hydroxydecanoic acid, which is lactonized at low pH. Its precursor, 11-hydroxyhexadecanoic acid, is present in large amounts in resin from the roots of the Mexican jalap plant. The process has been patented.

Saccharomyces cerevisiae is the main source of yeast extract, used for its flavour and nutrient-enhancing potential in the food industry. In foods, yeast extract is used to give a savoury, meaty flavour and as a source of vitamins. Strains of *S. cerevisiae* have been genetically engineered to overproduce glutathione (3–8%), which is used as a flavour enhancer and dough container. Other flavour enhancers produced by *S. cerevisiae* include disodium 5′-inositate and disodium 5′-guanylate. These compounds are obtained from the RNA of yeast cells via enzymic hydrolysis of inosine 5′-monophosphate (IMP) and guanosine 5′-monophosphate (GMP). Bakers' stains of *S. cerevisiae* contain approximately 8–11% RNA and concentrations of GMP and IMP of 1.5–6% may be achieved. The yeast extract market is growing annually at approximately 10–12% for food applications and also for feeds, aquaculture, fermentation substrates and cosmetics. Yeast extracts are usually produced from spent brewers' and bakers' strains of *S. cerevisiae*. The extract or autolysate is produced by heating the yeast cells to 40–60°C followed by addition of plasmolysing agents or hydrolytic enzymes. Cell debris, including cell walls, is removed by centrifugation after which the liquid extract is processed by filtration and spray-drying, to yeast extract powder or paste (> 70% dry matter).

Cells of *S. cerevisiae* may be completely fractionated. Cells are autolysed and the main components separated by a series of extraction and precipitation steps. Yeast protein concentrate (15% yield) and cell wall protein (15% yield) can be used in the food industry for their water-holding and oil-binding capacities. Primary yeast glycan is potentially useful as a food hydrocolloid, for thickening purposes, and glucomannans (up to 18% yield) show potential as emulsifying agents. Mannoproteins extracted from *S. cerevisiae* are similar in function to emulsifiers such as sodium caseinate. Other products include phospholipids, ergosterols, invertase and glucans. Glucans from cell wall extracts may be used as animal probiotics.

Several reports describe the use of *S. cerevisiae* for the production of fructose syrup from plant material such as deseeded carob pods and Jerusalem artichokes. Fermentation of carob with a glucophilic strain of *S. cerevisiae* produced a liquor containing up to 11.3% fructose which is useful as a sweetener in confectionery products. For the fermentation of Jerusalem artichokes, strains of *S. cerevisiae* are selected for their ability to ferment sucrose and small inulin polymers. Syrups containing up to 95% fructose are produced.

Saccharomyces cerevisiae is also potentially useful in the food industry as a probiotic organism. *Saccharomyces boulardii* (95% homologous with the genome of *S. cerevisiae*) has been shown to be of benefit for preventing antibiotic-associated diarrhoea in humans and rotavirus infections in children. Further evaluation of its benefits is required.

Methods of Detection and Enumeration in Foods

Saccharomyces cerevisiae grows on a wide range of media generally in use for isolation and enumeration of yeasts from foods. These media include malt extract agar, YM medium, glucose–yeast extract and tryptone glucose, yeast extract agar. In the brewing industry, brewing strains of *S. cerevisiae* are often enumerated on wort agar. Restriction of bacterial growth on these media may be achieved by adding antibiotics such as chloramphenicol or oxytetracycline ($100\,\mu g\,l^{-1}$), or by acidification to pH 3.5. However, acidification may prevent the growth of sublethally injured yeast cells. For the isolation of *S. cerevisiae* from food moulds (e.g. mould-ripened cheeses), biphenyl ($50\,mg\,l^{-1}$) can be incorporated into the medium as an antimicrobial agent.

Differential and selective media have been developed for enumeration of *S. cerevisiae*, for example, ethanol sulphite medium (ESY) was developed for the selective enumeration of *S. cerevisiae* in the presence of non-*Saccharomyces* species. The medium, containing 12% ethanol and 0.015% sodium metabisulphite, has been used to selectively enumerate *S. cerevisiae* during the early stages of wine fermentation. However, the performance of the medium in suppressing the non-*Saccharomyces* spp. is erratic because of the difficulty in maintaining consistent concentrations of ethanol in the agar. Biggy agar is used as a differential medium for enumeration of H_2S-producing strains of *S. cerevisiae*. Hydrogen sulphide-producing colonies react with bismuth sulphite present in the agar, causing the colonies to appear brown-black. Wallerstein Laboratories agar has been used to differentiate, by colony colour, between killer and sensitive strains of *S. cerevisiae*.

Identification of *S. cerevisiae* follows the morphological, physiological and biochemical tests described in *The Yeasts, a Taxonomic Study*, edited by C. P. Kurtzman and J. W. Fell (see Further Reading). The main characteristics of the species are listed in Table 1. Ellipsoidal or cylindrical cells, production of pseudohyphae and formation of smooth wall, globose to short ellipsoidal ascospores are factors assisting recognition of the species. Microscopic analysis and sporulation tests provide rapid presumptive diagnosis of the species. Vigorous fermentation of sugars is a key feature of the species. Several commercially available yeast identification kits give reproducible and reliable identification of this species. These include the API 20C and ID 32C systems (bioMérieux, Lyon, France) and the Biolog YT plate system (Biolog Inc., California, USA).

Pulsed-field gel electrophoresis is used to characterize strains of *S. cerevisiae*, based on the variability of their chromosomal constitution. The chromosomes are separated electrophoretically and visualized using staining techniques. A polymerase chain reaction–restriction fragment length polymorphism (PCR-RFLP) method for targeting single-stranded ribosomal DNA (ss rDNA) has been used to differentiate *S. cerevisiae* from species of *Zygosaccharomyces* and *Candida*. A 1200 base pair internal fragment of the ss rDNA is amplified and characteristic fragmentation patterns are generated by digestion with restriction endonucleases. The primers for DNA amplification were designed on the conserved region at the beginning and middle of the gene encoding ss rDNA for *Candida albicans* (primers: 5′-GTC-TCA-AAG-ATT-AAG-CCA-TG-3′ and 5′-TAA-GAA-CGG-CCA-TGC-ACC-AC-3′). The PCR product was digested using the restriction enzymes *Mse*I, *Ava*II, *Taq*I, *Scr*FI, *Hha*I, *Sau*3AI, *Msp*I and *Cfo*I. *Saccharomyces cerevisiae* was differentiated from the species of *Zygosaccharomyces*, *Candida valida* and *Candida lipolytica* by the *Mse*I digest.

As an alternative method, PCR fingerprinting primers for the microsatellite sequences may be used to identify *S. cerevisiae*. Microsatellite sequences are repetitious and occur randomly within the genome of yeasts, providing a 'fingerprint' of individual species. Primers consisting of the repeating oligonucleotides are synthesized and used in a PCR reaction to generate PCR fragments of differing lengths that are visualized using gel electrophoresis and staining techniques. The PCR patterns from a data base are used to identify and type yeast isolates. Random amplified polymorphic DNA (RAPD) assays are used to differentiate *S. cerevisiae* from other species. Randomly applied primers are used to generate a pattern of DNA fragments. Patterns are visualized and compared as described above.

See also: **Bread**: Bread from Wheat Flour. **Cocoa and Coffee Fermentations**. **Fermented Foods**: Beverages from Sorghum and Millet. **Lager**. **PCR-based Commercial Tests for Pathogens**. **Probiotic Bacteria**: Detection and Estimation in Fermented and Non-fermented Dairy Products. **Spoilage Problems**: Problems caused by Fungi. **Starter Cultures**: Uses in the Food Industry. **Wines**: Microbiology of Wine-making.

Further Reading

Benitez T, Gasent-Ramirez JM, Castrejon F and Codon AC (1996) Development of new strains for the food industry. *Biotechnological Progress* 12: 149–163.

Fleet GH (1992) Spoilage yeasts. *Critical Reviews in Biotechnology* 12: 1–44.

Fleet GH (1997) Wine. In: Doyle MP, Beuchat L and Montville T (eds) *Food Microbiology Fundamentals and Frontiers*. P. 671. Washington: American Society for Microbiology.

Kurtzman CP and Fell JW (eds) (1998) *The Yeasts, a Taxonomic Study*, 4th edn. Amsterdam: Elsevier.

Praphailong W, Van Gestel M, Fleet GH and Heard GM (1997) Evaluation of the Biolog system for the identification of food borne yeasts. *Letters in Applied Microbiology* 24: 455–459.

Roostita R and Fleet GH (1996) Growth of yeasts in milk and associated changes to milk composition. *International Journal of Food Microbiology* 31: 205–219.

Roostita R and Fleet GH (1996) The occurrence and growth of yeasts in Camembert and blue-veined cheeses. *International Journal of Food Microbiology* 28: 393–404.

Van der Vossen Jos MBM and Hofstra H (1996) DNA based typing, identification and detection systems for food spoilage microorganisms: development and implementation. *International Journal of Food Microbiology* 33: 35–49.

Wood BJB (1998) *Microbiology of Fermented Foods*, 2nd edn. Vols 1, 2. London: Blackie.

Saccharomyces: Brewers' Yeast

G G Stewart, International Centre for Brewing and Distilling, Heriot-Watt University, Edinburgh, Scotland, UK

The characteristic flavour and aroma of any beer is, in large part, determined by the yeast strain employed. In addition, properties such as flocculation, fermentation ability (including the uptake of wort sugars), ethanol tolerance and oxygen requirements have a critical impact on fermentation performance. Thus, proprietary strains belonging to individual breweries are usually (but not always) jealously guarded and conserved, however this is not always the case. In Germany, most of the beer is produced with only four lager strains and approximately 65% of the beer is produced with one strain.

Yeasts are of benefit to humanity because they are widely used for the production of beer, wine, spirits, foods and a variety of biochemicals. Yeasts also cause spoilage of foods and beverages, and some yeasts are of medical importance. At present, approximately 700 yeast species are recognized but only a few have been adequately characterized. No satisfactory definition of yeast exists, and commonly encountered properties such as alcoholic fermentation and growth by budding are not universal in yeast. However, all brewers' yeast strains multiply by budding. There are many definitions to describe the yeast domain, and the one that probably best describes the group is: 'Yeasts are unicellular fungi which reproduce by budding or fission'. Yeasts are both quantitatively and economically the most important group of microorganisms commercially exploited on this planet. The total amount of yeast produced annually, including that formed during brewing, wine-making and in distilling practices, is of the order of a million tonnes. Many microbiologists and fermentation technologists employ the term 'yeast' as synonymous with *Saccharomyces cerevisiae*. Although this yeast species is of critical economic and biochemical importance, and most (but certainly not all) of the research on yeast has been conducted on it, there are many exotic varieties of yeast species that offer advantages for experimental studies. Nevertheless, the genus *Saccharomyces* has often been referred to as 'the oldest plant cultivated by mankind'. Indeed, the history of beer, wine and bread-making with the fortuitous use of yeast is as old as the history of mankind itself. Many species of *Saccharomyces* are GRAS (generally regarded as safe) and produce two very important primary metabolites: ethanol and carbon dioxide.

As a group of microorganisms, yeasts are capable of utilizing a broad spectrum of carbohydrates and sugars. Nevertheless, none of the yeast species isolated to date from natural environments have been found capable of utilizing all of the readily available sugars. *Saccharomyces cerevisiae* has the ability to take up a wide variety of sugars, for example, glucose, fructose, mannose, galactose, sucrose, maltose, maltotriose and raffinose. In addition, the closely related species (or subspecies) *Saccharomyces diastaticus* and *Saccharomyces uvarum* (*carlsbergensis*) (lager yeasts) are able to utilize dextrins and melibiose, respectively. However, *Saccharomyces cerevisiae* and related species are not able to metabolize all sugars. Examples of carbohydrates and sugars in this category are pentose sugars (for example, ribose, xylose and arabinose), cellobiose (hydrolysis products of hemicellulose and cellulose), lactose (milk sugar), inulin and cellulose.

The taxonomic classification of yeasts is a complex and sometimes controversial matter. Early microbiologists used a system of classification employing morphological, physiological, reproductive and biochemical criteria by which many different types were distinguished as separate species. The systems more commonly used today tend to lump these types together as subspecies or variations of a few species. This is particularly true in the genus *Saccharomyces* where species names have come and gone with bewildering rapidity as successive taxonomic reviews appeared.

It is at the strain level that interest in brewing yeast

centres. There are at least 1000 separate strains of *Saccharomyces cerevisiae*. These strains include brewing, baking, wine-making, distilling and laboratory cultures. There is a problem classifying such strains in the brewing context; the 'minor' differences between strains that the taxonomist dismisses as inconsequential are of great technical importance to the brewer.

Also most brewing strains are polyphoid or aneuploid and as a consequence do not exhibit a haploid/diploid life cycle. These strains are usually genetically more stable and this has been attributed to the following characteristics: little or no mating ability, poor sporulation and low spore viability. As a consequence of this, the use of classical genetics employing the mating type and life cycle of the yeast are not possible to produce novel and improved strains of brewers' yeast. However, the new techniques of sphaeroplast fusion and recombinant DNA have been employed to produce novel yeast strains that possess industrial potential.

The following are some examples of successful fusions with commercial brewing yeast strains.

- The construction of a brewing yeast with amylolytic activity by the fusion of *Saccharomyces cerevisiae* and *Saccharomyces diastaticus*.
- A polyploid strain capable of high ethanol production by fusion of a flocculent strain with saké yeasts.
- Construction of strains with improved osmotolerance by fusion of *Saccharomyces diastaticus* and *Saccharomyces rouxii* (an osmotolerant yeast species).

Some successful examples of the use of recombinant DNA technology for improving brewers' yeast strains are:

- glucoamylase activity from the fungus *Aspergillus niger*
- β-glucanase activity from the bacterium *Bacillus subtilis*, the fungus *Trichoderma reesii* and barley
- α-acetolactate decarboxylase activity from the bacteria *Enterobacter aerogenes* and *Acetobacter* spp
- extracellular protease for chillproofing beer
- modification of the yeast's flocculation properties.

What are the future prospects for the use of recombinant DNA with brewers' yeast and their use in the brewing industry? At this time this is a difficult question to answer. It is quite surprising that there are not a number of recombinant brewers' yeasts commercially in use today. Permission has already been granted in the UK from the Ministry of Agriculture Foods and Fisheries Advisory Committee on Novel Foods and Processes for the use of a bakers' yeast strain that is genetically manipulated to enhance baking properties and for a brewing strain, cloned with DNA from *Saccharomyces diastaticus*, that secretes glucoamylase to produce low calorie beer.

Perhaps the availability of alternative inexpensive traditional solutions for many of the problems that it was hoped a cloned yeast could solve, such as inexpensive sources of glucanase and gluco- and α-amylase, has retarded implementation. Also in some cases recombinant DNA technology is ahead of the knowledge base in yeast biochemistry. There is also still concern over consumer acceptance. Although this is a difficult hurdle, it is thought that as people become accustomed to pharmaceuticals produced by recombinant DNA, and more plants with improved characteristics for farming/food gain regulatory approval and customer acceptance, the current reluctance to use the products of this technology in the brewing industry will slowly disappear.

The two main types of beer, lager and ale, are fermented with strains belonging to the species *Saccharomyces uvarum* (*carlsbergensis*) and *Saccharomyces cerevisiae*, respectively. Currently, yeast taxonomists have assigned to the species *Saccharomyces cerevisiae* all strains employed in brewing. Indeed, increasingly they are referred to in the scientific/technical literature as *Saccharomyces cerevisiae* (ale type) and *Saccharomyces cerevisiae* (lager type). However, there are several biochemical differences between these two types of yeast strains that warrant maintaining them as separate entities. For example, they have been distinguished on the basis of their ability to ferment the disaccharide melibiose (glucose–galactose). Strains of *Saccharomyces cerevisiae* (lager type) possess the *MEL* gene(s). They produce the extracellular enzyme α-galactosidase (melibiase) and are, therefore, able to utilize melibiose. However, strains of *Saccharomyces cerevisiae* (ale type) do not possess the *MEL* gene(s), do not produce α-galactosidase, and are, therefore, unable to utilize melibiose. Also, ale strains can grow at 37°C, whereas lager strains cannot grow above 34°C and this can be used as a distinguishing test.

There is genetic and molecular evidence indicating that modern lager yeast strains may have originated as a hybrid between an ale yeast (*Saccharomyces cerevisiae*) and an older type of bottom fermenting brewers' yeast formerly classified as *Saccharomyces monacensis* (which today is classified as part of the species *Saccharomyces pastorianus*). By using genetic analysis techniques such as restriction fragment length polymorphism (RFLP), pulsed-field gel electrophoresis and polymerase chain reaction (PCR), it has been shown that the genomes of these yeasts are highly dynamic. The genetic material is frequently added to, moved around or lost from the genome. It has been suggested that the ordinary species concept applied to the taxonomy of most other organisms may not be fully appropriate in the case of polyploid industrial yeasts and that lager yeast strains should be maintained as an independent taxonomic entity.

The objectives of wort fermentation are to consistently metabolize wort constituents into ethanol and other fermentation products in order to produce beer with satisfactory quality and stability. Another objective is to produce yeast crops that can be confidently re-pitched into subsequent brews. During the brewing process overall yeast performance is controlled by a plethora of factors. These factors include:

- the yeast strains employed and their condition at pitching and throughout fermentation
- the concentration and category of assimilable nitrogen
- the concentration of ions
- the fermentation temperature
- the pitching rate
- the tolerance of yeast cells to stress factors such as osmotic pressure and ethanol
- the wort gravity
- the oxygen level at pitching
- the wort sugar spectrum
- yeast flocculation characteristics.

These factors influence yeast performance either individually or in combination with others and also together permit the definition of the requirements of an acceptable brewers' yeast strain:

> In order to achieve a beer of high quality, it is axiomatic that not only must the yeast be effective in removing the required nutrients from the growth/fermentation medium (wort), able to tolerate the prevailing environmental conditions (for example, ethanol tolerance) and impart the desired flavour to the beer, but the microorganisms themselves must be effectively removed from the wort by flocculation, centrifugation and/or filtration after they have fulfilled their metabolic role' (Russell and Stewart, 1995).

It is worthy of note that brewing is the only major alcoholic beverage process that recycles its yeast. It is, therefore, important to jealously protect the quality of the cropped yeast because it will be used to pitch a later fermentation and will, therefore, have a profound effect on the quality of the beer resulting from it.

Over the years, considerable effort has been devoted in many research laboratories to the study of the biochemistry and genetics of brewers' yeast (and industrial yeast strains in general). The objectives of the studies have been twofold:

1. To learn more about the biochemical and genetic makeup of brewing yeast strains.
2. To improve the overall performance of such strains, with particular emphasis being placed on broader substrate utilization capabilities, increased ethanol production, and improved tolerance to environmental conditions such as temperature, high osmotic pressure and ethanol and finally to understand the mechanism(s) of flocculation.

When yeast is pitched into wort, it is introduced into an extremely complex environment due to the fact that wort is a medium consisting of simple sugars, dextrins, amino acids, peptides, proteins, vitamins, ions, nucleic acids and other constituents too numerous to mention. One of the major advances in brewing science during the past 25 years has been the elucidation of the mechanisms by which the yeast cell, under normal circumstances, utilizes, in a very orderly manner, the plethora of wort nutrients.

Although ethanol is the major excretion product produced by yeast during wort fermentation, this primary alcohol has little direct impact on the flavour of the final beer. It is the type and concentration of the many other yeast excretion products produced during wort fermentation that primarily determine the flavour of the beer. The formation of these excretion products depends on the overall metabolic balance of the yeast culture, and there are many factors that can alter this balance and consequently beer flavour. Yeast strain, fermentation temperature, adjunct type and level, fermenter design, wort pH, buffering capacity, wort gravity, etc., are all influencing factors.

Some volatiles are of great importance and contribute significantly to beer flavour, whereas others are important in building background flavour. The following groups of substances are found in beer: organic and fatty acids, alcohols, esters, carbonyls, sulphur compounds, amines, phenols and a number of miscellaneous compounds.

Traditionally, lager is produced by bottom-fermenting yeasts and fermentation temperatures of 7–15°C, and at the end of fermentation, these yeasts flocculate and collect at the bottom of the fermenter. This flocculated yeast is collected for re-use in a subsequent fermentation. Ale is produced by top-fermenting yeasts at fermentation temperatures of 18–22°C, and at the end of fermentation these form loose clumps of cells, which are adsorbed onto carbon dioxide bubbles, and are carried to the surface of the wort. Consequently, top yeasts are collected (a process called 'skimming') for reuse from the surface of the fermenting wort. With the advent of novel fermenting and yeast collection techniques, such as cylindroconical fermenters and centrifuges, the differentiation of lagers and ales on the basis of bottom and top cropping has become less distinct.

See also: **Lager**. **PCR-based Commercial Tests for Pathogens**. ***Saccharomyces***: *Saccharomyces cerevisiae*. **Yeasts**: Production and Commercial Uses.

Further Reading

Bamforth C (1998) *Beer – Tap into the Art and Science of Brewing*. New York: Plenum Press.

Hardwick WA (1995) *Handbook of Brewing*. New York: Marcel Dekker.

Kunze W (1996) *Brewing and Malting Technology*, International Edition. Berlin: VLB.

Lewis MJ and Young TW (1995) *Brewing*. London: Chapman & Hall.

Lyons TP, Kelsall DR and Murtagh JE (1995) *The Alcohol Textbook*. Nottingham, UK: Nottingham University Press.

Russell I and Stewart GG (1995) Brewing In: Rehm HJ and Reed G (eds) *Biotechnology*. Vol. 9, p. 419. Weinheim, Germany: VCH.

Spencer JFT and Spencer DM (1997) *Yeasts in Natural and Artificial Habitats*. Berlin: Springer-Verlag.

Walker GM (1998) *Yeast Physiology and Biotechnology*. Chichester, UK: Wiley.

Sake *see* ***Saccharomyces***: *Saccharomyces sake*.

SALMONELLA

Contents

Introduction

Julian Cox, Department of Food Science and Technology, University of New South Wales, Sydney, Australia

The role of *Salmonella* in food-borne disease was first documented in the late 1800s, although association with human clinical disease, in the form of typhoid, dates back to the beginning of that century. In 1885, an organism designated *Bacillus cholerae-suis* was isolated by a veterinary pathologist, D.E. Salmon, from pigs suffering hog cholera. Other similar organisms were isolated from outbreaks of food-borne disease and infected animals. To accommodate these organisms, the genus *Salmonella* was created by Lignières in 1900, in honour of Salmon.

Despite extensive efforts to understand and control members of the genus, *Salmonella* remains a major concern to food microbiologists throughout the world, due primarily to the ubiquitous association of the bacterium with food animals and their production environment.

Characteristics of *Salmonella*

The genus *Salmonella*, within the family Enterobacteriaceae, is comprised of facultatively anaerobic, oxidase-negative, catalase-positive Gram-negative rod-shaped bacteria; the rods are typically 0.7–1.5 × 2–5 μm in size, although long filaments may be formed. Most strains are motile and ferment glucose with production of both acid and gas. Further biochemical tests, commonly employed to distinguish *Salmonella* from other genera within the family Enterobacteriaceae, are given in **Table 1**.

Serology

Salmonella spp. are further characterized into serotypes or serovars within the Kauffmann–White antigenic scheme, based on differences in reaction with antibodies of two major and, in some cases, other minor types of cell-surface antigens.

Strains are divided into serogroups, based on differences in epitopes of lipopolysaccharide (LPS), a major component of the outer membrane of Gram-negative bacteria. LPS, designated the O or somatic antigen for serological purposes, is comprised of three components: lipid A, core polysaccharide and oligosaccharide side chain (which confers serogroup

Table 1 Biochemical tests used in the characterization and reactions of the genus *Salmonella*

Test	*Typical reaction*	*% Positive*
Decarboxylation of:		
Lysine	+	97.4
Ornithine	+	90
Production of:		
Hydrogen sulphide	+	95.3
Indole	–	1.1
Urease	–	0
Arginine dihydrolase	+	92.8
Phenylalanine deaminase	–	0
Metabolism of glucose:		
Fermentation	+	100
Methyl red	+	100
Voges–Proskauer	–	0
Gas production	d	89.4
Fermentation of:		
Arabinose	+	90.0
Xylose	+	94.6
Rhamnose	+	91.4
Maltose	+	97.3
Lactose	–	0.3
Sucrose	–	0.2
Raffinose	–	3.3
Mannitol	+	99.7
Sorbitol	+	94.5
Dulcitol	d	88.1
Inositol	d	38.5
Adonitol	–	0
Salicin	–	0.6
Liquefaction of gelatin	–	0.6
Utilization of citrate	d	86.9

+ = >90% positive; – = <10% positive; d = 10–90% positive. Some positive reactions may be delayed (3 or more days).

specificity). The LPS side chains are composed of repeating units of oligosaccharides, which in turn are composed of a range of sugars including rare heptoses, such as abequose and tyvelose. Strains lacking the side chain O antigen are known as rough strains, producing rough colonies on plate media, and failing to agglutinate with homologous antiserum.

Each serogroup is determined by a particular O antigen and each distinct antigen is denoted by a number; for groups A, B, C and D respectively, the O2, O4, O6 and O9 antigens are diagnostic. In addition, some serogroups contain subgroups, differentiated by other O antigens; O antigens 6,7 and 6,8 define subgroups C1 and C2 respectively. Many serovars possess other O antigens, such as O12, which are found in a number of serogroups and are thus of less diagnostic value (**Table 2**). The O antigens are numbered between 1 and 67, though non-contiguously as some antigens have been removed from the typing scheme since they were assigned to organisms which were originally but now are no longer within the genus *Salmonella*. These antigens may be encoded chromosomally, such as O12, or may be introduced into serovars by lysogenic bacteriophage. An example is O1 in many serovars of serogroups B and D. In the latter case, the antigen is underscored in the antigenic formula (Table 2) to denote its instability, as the antigen will be lost if a strain is cured either naturally or artificially of the encoding phage.

Occasionally, some chromosomally encoded antigens may not be produced, or only at levels not detected by agglutination. These antigens, such as O5 in serovar *typhimurium*, are denoted in the antigenic formula by square brackets (Table 2).

While the presence of lysogeny-derived O antigens is an accepted part of the antigenic formula for a number of serovars of long-standing, lysogenic variants of some serovars are designated as variants of the parent serovar, rather than affording them separate serovar status. This is particularly important in light of the identification of variants with more than one lysogeny-derived antigen, and the potential for discovery of many more such variants. This has led in some cases to the abolition of some serovar names (Table 2).

Within serogroups, strains are further differentiated into serovars, based on variation in flagellins or H antigens, the subunit proteins of flagella. Some serovars are easily defined with respect to H antigens as they produce only one form or phase; these serovars are termed monophasic. However, most serovars, termed diphasic, are capable of producing two distinct forms of flagella. Some strains are genotypically triphasic, capable of producing a third phase H antigen, which may be encoded chromosomally or more often extrachromosomally (plasmid-borne), although phenotypically these strains appear diphasic. An alternative H antigen produced in Phase 1 is referred to as a R-phase H antigen. Phase 1 H antigens are described by lowercase letters (a, b, c) or, beyond the 26th such antigen, by the letter z and a consecutive number (z1, z2 . . . z83 and so on). The first phase 2 H antigens identified are described, like O antigens, by numerals (1, 2, 3), although many serovars express H antigens typical of phase 1 as phase 2 antigens (e.g. e, n, x). In a culture of a multiphasic serovar, the population may be composed of a mixture of phase 1 and phase 2 cells, or a predominance of a single type. During serological characterization of a strain, alternative phase antigens are induced by culture of the strain in the presence of antisera to the first identified antigen.

As of 1996 there were 2435 serovars of *Salmonella enterica*, of which 58.9% belong to subspecies *enterica* (**Table 3**). On the basis of numerous characterized O and H antigens, the description of new antigens, and evidence for ongoing horizontal gene transfer,

Table 2 Antigenic formulae of selected *Salmonella* serovars, including those exhibiting normal, lysogeny-derived and weak O antigens, capsular (Vi) antigen, and R-, mono-, di- and tri-phase H antigens

Serovar (serogroup)	*Somatic antigens*	*Flagellar antigens*		
		Phase 1	*Phase 2*	*Phase 3*
paratyphi A	1,**2**,12	a	[1,5]	
typhimurium (B)	*1*,**4**,[5],12	i	1,2	
'sofia' (B)[a]	*1*,**4**,12,*27*	b	[e,n,x]	
virchow (C1)	**6**,7	r	1,2	
münchen (C2)	**6**,*8*	d	1,2	
typhi (D)	**9**,12[Vi]	d	–	
typhi (R-phase)	**9**,12[Vi]	j	z66	
dublin (D)	*1*,**9**,12[Vi]	g,p	–	
enteritidis (D)	*1*,**9**,12	g,m	–	
pullorum (D)	*1*,**9**,12	–	–	
orion (E)	**3**,10	y	1,5	
orion var. 15+ (*binza*)	**3**,15	y	1,5	
orion var. 15+,34+ (*thomasville*)	**3**,15,34	y	1,5	
rubislaw (F)	**11**	r	e,n,x	
rubislaw (triphasic)	**11**	r	e,n,x	d

[a] Subspecies II (*Salmonella enterica* subsp. *salamae*) serovars are no longer described by name. The serogroup-specific antigen is in bold and subgroup antigens are in italics for clarity, but usually appear in normal case.

Table 3 Number of serovars in each species and subspecies of *Salmonella*, 1992–1996

Salmonella enterica	*Year*				
subsp./spp.	*1992*	*1993*	*1994*	*1995*	*1996*
subsp. *enterica* (I)	1379	1405	1416	1427	1435
subsp. *salamae* (II)	466	471	477	482	485
subsp. *arizonae* (IIIa)	93	94	94	94	94
subsp. *diarizonae* (IIIb)	309	311	317	319	321
subsp. *houtenae* (IV)	64	65	66	69	69
subsp. *indica* (VI)	10	10	10	11	11
bongori (V)	18	19	19	20	20
Total number	2339	2375	2399	2422	2435

there is potential for the creation of many new serovars. Indeed, new serovars are regularly described (almost 100 during 1992–1996; Table 3), and it is almost certain more will continue to be discovered.

Strains of individual serovars can be subdivided using a range of phenotypic and genotypic methods, including biotyping, phage typing, antibiotic susceptibility testing or resistotyping, restriction endonuclease analysis, ribotyping, IS*200* typing, pulsed-field gel electrophoresis, plasmid profiling and fingerprinting. While it is beyond the scope of the present discussion to describe the technical details or advantages and disadvantages of many of these methods, or their applicability to specific serovars, it is worth noting that the efficacy of any technique varies with the serovar. The current method of choice for a growing number of serovars is phage typing; typing sets have been developed for serovars *enteritidis*, *heidelberg*, *paratyphi B*, *typhi*, *typhimurium* and *virchow*. Further strain analysis becomes necessary when a particular phage type (PT) of a serovar becomes dominant, such as *enteritidis* PT4 in the UK.

Taxonomic Considerations

Taxonomy

Based on biochemical and physiological tests, the genus *Salmonella* was originally differentiated into five and later seven subgenera, designated I, II, IIIa, IIIb, IV, V and VI (**Table 4**). Currently, after analysis by modern techniques, the genus is considered to consist of either one or two species. Evidence from DNA hybridization and numerical taxonomy supports the existence of only one species, *S. enterica*, while sequence analysis of the DNA encoding ribosomal RNA as well as multilocus enzyme electrophoresis suggests separation into two species, *S. enterica* and *S. bongori*, the former comprised of six subspecies, and the latter representing *Salmonella* subsp. V (Table 3). Most serovars of significance in humans, including those responsible for food-borne disease, belong to subspecies I (*S. enterica* subsp. *enterica*).

Table 4 Biochemical and physiological characteristics of the species and subspecies of *Salmonella* (note the old and new taxonomic designations)

Test	enterica *subspp.*						bongori
	enterica *(I)*	salamae *(II)*	arizonae *(IIIa)*	diarizonae *(IIIb)*	houtenae *(IV)*	indica *(VI)*	*(V)*
Fermentation of:							
Dulcitol	+	+	–	–	–	d	+
Sorbitol	+	+	+	+	+	+	–
Salicin	–	–	–	–	+	–	–
Lactose	–	–	d	d	–	d	–
ONPG	–	–	+	+	–	d	–
Utilization of:							
Malonate	–	+	+	+	–	–	–
Mucate	–	–	+	d	–	+	+

+ = >90% positive; – = <10% positive; d = 10–90% positive.
ONPG = *o*-nitrophenyl-β-D-galactopyranoside.

Nomenclature

Within the subgenus system, at the serotype level, *Salmonella* have historically been afforded species status. The name of the particular serotype appears in a typical italicized genus-species form; for example, serovar typhimurium is described as *S. typhimurium*. Currently, *Salmonella* are described in the scientific literature in two ways. While the historical form remains, in the second and more recent convention, subgenera I–IV and VI have been designated as subspecies of *S. enterica*, while subgenus V equates to *S. bongori* (Tables 3 and 4). Only serovars of subspecies I are given names and the serovar name is not afforded species status. Thus, serovar typhimurium is fully described as *S. enterica* subsp. *enterica* serovar *typhimurium*, which is conveniently abbreviated to *S. Typhimurium*. Since 1994, no new serovars of any subspecies have been assigned names. Retention of the serovar name, while distinguishing serovars from species, represents a satisfactory reconciliation of taxonomy and nomenclature on the one hand and practical microbiology on the other. Serovars of subspecies other than *enterica* are simply described by their antigenic formula, although some of the older and more common serovars of subspecies other than *S. enterica* are still often referred to by name; for example *S. enterica* subsp. *salamae* serovar 1,4,12,27:b:[e,n,x] is still commonly known as *S. sofia*.

Salmonella serovars were originally named for the disease syndrome in various hosts, examples being serovars *typhi*, *typhimurium*, *abortusovis* and *bovismorbificans* in humans, mice, sheep and cattle, respectively. Shortly thereafter, nomenclature based on species and syndrome became limiting, and names were assigned according to the first geographical site of isolation, examples including serovars *adelaide*, *dublin*, *london*, *miami* and *moscow*.

Physiology

A range of environmental conditions affects the growth, death or survival of *Salmonella*, forming the basis for control and preservation measures in the food-processing industry. These include temperature, pH and water activity (a_w), and combinations thereof.

Temperature

Salmonella can grow within the range 2–54°C, although growth below 7°C has largely been observed only in bacteriological media, not in foods, while growth above 48°C is confined to mutants or tempered strains. The optimum temperature for growth is 37°C, which is not surprising given that the natural ecology of most *Salmonella* strains of concern to public health is the gastrointestinal tract of warm-blooded animals.

Above the maximum growth temperature, *Salmonella* die quickly and are, in general, readily destroyed by mild heat processes, such as pasteurization. However, susceptibility varies with strain. Studies of many strains in model systems have demonstrated mean D values at 57°C and 60°C of 1.3 and 0.4–0.6 min respectively and z values of 4–5°C. *Salmonella senftenberg* 775W, the most heat-resistant strain of *Salmonella* thus far identified, had D values at the same temperatures of 31°C and 4–6 min. Exposure to adverse conditions, including sublethal temperatures and extremes of pH, increases resistance. Foods high in solid content, particularly protein or fat, and low in moisture (and a_w) are highly protective; survival in foods such as chocolate or peanut butter is measured in hours between 70 and 80°C. Increased heat resistance is less marked when solutes such as NaCl rather than sugars are used to reduce a_w.

Salmonella survive quite well at low temperatures. Although the time varies with substrate and the influence of factors such as pH and a_w, strains may survive for days to weeks at chill temperatures. During freezing, a population of *Salmonella* will be reduced in inverse proportion to the rate of freezing, further influenced by the degree of protection afforded by the matrix in which the organism is held and the physiological status of the cells. Log-phase cells are more susceptible to damage. After freezing, a population of *Salmonella* undergoes a slow decline which is inversely proportional to the storage temperature. In a protective matrix, and under commercial freezing conditions, *Salmonella* may survive for months or years.

pH

The optimum pH for growth of *Salmonella* is within the range 6.5–7.5. Strains grow at pH values up to 9.5 and down to 4.05, although the minimum varies considerably with the acidulant used to reduce pH. While growth occurs down to or close to the minimum pH with non-volatile organic acids such as citric acid or mineral acids such as hydrochloric acid, growth stops at higher pH values when volatile fatty acids (VFAs) are used (e.g. pH 5.4 in the presence of acetic acid). The inhibitory effect of VFAs is inversely proportional to chain length, and increases under anaerobic conditions, presumably due to a decrease in available energy (ATP) and a consequent decrease in ability to remove the acids from the intracellular environment.

Increasing temperature increases sensitivity to low pH, as does the presence of food additives such as salt or nitrite. Sensitivity to pH is a major preservative factor in foods to which acidulants are added, such as mayonnaise, or in which acids are produced by fermentation, such as salami or cheese. The effect of acidulant is exemplified by the increased survival of *Salmonella enteritidis* in mayonnaise made with lemon juice (citric acid) rather than vinegar (acetic acid).

Tolerance or adaptation to low pH is significant with respect to virulence, increasing the likelihood of surviving gastric acidity, or the acidic intracellular environment of phagocytic cells. *Salmonella* generally exhibit three distinct responses to acidity – a general pH-independent, RpoS-mediated stress response, and a pH-dependent response, both of which are activated in the stationary phase, as well as a pH-dependent log-phase response.

Water activity

Salmonella grow at a_w values between 0.999 and 0.945 in laboratory media, down to 0.93 in foods, with an optimum of 0.995. While there is no growth below 0.93, *Salmonella* survives; the time of survival increases as a_w decreases. In low-moisture foods, such as pasta, peanut butter and chocolate, survival is measured in months. Salt (NaCl), used as a solute to lower a_w and as a preservative in foods, is inhibitory towards *Salmonella* at concentrations of 3–4% tolerance increases with temperature between 10 and 30°C.

Virulence Factors

Lipopolysaccharide

Study of the LPS of strains of different serovars has shown that variation in the amount produced, the length of O side chains, and the degree of glycosylation all affect virulence, and virulence is enhanced when the former properties are increased. Long side chains sterically hinder the ability of components of the complement cascade system to bind to the surface of the *Salmonella* cell, preventing lysis.

In addition, the composition of the O side chains influences virulence, particularly the ability to cause invasive infection, as different serogroup determinants interact differently with components C5 to C9 of the complement cascade. Thus, the propensity to invade decreases respectively in serogroups B, D and C, regardless of other virulence characteristics. It is not surprising then that the majority of human *Salmonella* infections involve serovars of serogroups A–E.

Fimbriae

Many *Salmonella* produce one or more of the several distinct fimbrial structures described thus far. A major research model for the study of fimbriae is serovar *enteritidis*.

Type 1, mannose-sensitive fimbriae are expressed by strains of many *Salmonella* serovars including *enteritidis*, in which they are known as SEF21. Although they function similarly, the type 1 fimbriae of *Salmonella* are distinct from those of *Escherichia coli*.

Fimbriae composed of 14 kDa subunits have been described in *Salmonella enteritidis* and a small range of other serogroup D *Salmonella*, including serovars *blegdam*, *dublin* and *moscow*. The SEF14 fimbria is likely to be associated with virulence as it shares homology with an adhesin from enterotoxigenic *E. coli*.

SEF17 fimbriae, originally described in *S. enteritidis*, have been identified in a wider range of serovars as well as diarrhoeagenic strains of *E. coli*. These extremely hydrophobic and aggregative fimbriae bind fibronectin strongly, indicative of a role in adherence. SEF17 also activates tissue plasminogen, both directly and indirectly through induction of host tissue plasminogen activator, suggesting a role in dissemination.

SEF18, the latest of the fimbrial structures to be described, is manifested as a fibrillar structure in serovar *enteritidis*, but as a more amorphous matrix on the cell surface in the other *Salmonella* serovars in which it has been detected.

It is likely that other fimbrial structures will be identified among *Salmonella*, some of which may be widely distributed, while others, like SEF14, will be restricted to a narrow range of related serovars. At the same time, fimbrial structures currently associated with some serovars will be demonstrated in others.

Toxins

Enterotoxin is produced by many strains of *Salmonella*, representing the virulence factor responsible for the onset of diarrhoeal symptoms. While early studies suggested a serological relationship between the *Salmonella* enterotoxin, cholera toxin (CT) and the heat labile (HT) toxin of enterotoxigenic *E. coli*, more recent serological and nucleic acid studies indicate they are distinct entities. However, the *Salmonella* enterotoxin appears to be structurally similar to CT, consisting of A and B subunits that act respectively to stimulate host cell adenylate cyclase and produce a pore through which the former enters. Increased levels of cellular cyclic AMP (cAMP) lead to a net massive increase in concentration of sodium and chloride ions and a consequent accumulation of fluid in the intestinal lumen.

Salmonella also produce a membrane-bound proteinaceous cytotoxin, which is serologically and genetically distinct from Shiga toxins of *Shigella* and *E. coli*. The toxin, which may be released intracellularly as a consequence of limited bacterial lysis, inhibits protein synthesis, leading to host cell lysis and dissemination of the bacterium. Host cell lysis may also result from chelation of divalent cations by the toxin, causing disruption of host cell membranes.

Siderophores

Acquisition of iron is critical to survival and growth of microorganisms, and must be prised from the host during infection as iron is complexed in a range of proteins, or the little free iron available must be scavenged. Like many other members of Enterobacteriaceae, *Salmonella* produce two types of sequestering molecules, or siderophores, to acquire iron. The first, a high-affinity siderophore known as enterochelin or enterobactin, is a phenolate, composed of a cyclic trimer of dihydrobenzoic acid and L-serine, while the second, a hydroxamate called aerobactin, is synthesized from a citrate molecule, and two lysine derivatives. Enterochelin or aerobactin sequesters ferric ions from the environment (intestinal lumen, serum), and after binding to an outer membrane protein is translocated to the cytoplasm, where Fe^{+++} is reduced to Fe^{++}, which is released from the siderophore. Strains producing enterochelin are generally more virulent than those producing aerobactin.

Other Chromosomally Encoded Factors

A series of genes, the products or functions of which have not been fully characterized, occur within large gene loci, termed pathogenicity islands. A series of 15 genes, the *inv* region, occurs within one such island, and is necessary for epithelial cell invasion. Gene products are responsible for early stages of cell engulfment, including epithelial cell membrane ruffling and assembly and translocation to the bacterial cell surface of attachment appendages. The *inv* genes are expressed during late logarithmic or early stationary phase under conditions of high osmolarity and low oxygen tension – conditions encountered in many internal body sites. Other regions of the chromosome encode factors necessary for intracellular survival: the *oxyR* locus encodes proteins protective against the toxic oxygen products in macrophages, while the *phoP/phoQ* regulatory system is required for expression of factors permitting survival within phagocytic cells.

Regulatory gene products such as sigma factors also play a role in pathogenicity. Sigma factor RpoH, responsible for regulation of heat shock proteins (HSPs), is expressed during intracellular growth, while HSPs of 58 and 68 kDa are synthesized not only in response to heat shock, but constitutively at low levels at 37°C, and at increased levels during infection. The sigma factor RpoS is also strongly expressed intracellularly, and it is likely to regulate cytotoxic factors, as *rpoS*-negative mutants, while surviving intracellularly, causing less cell death than wild-type strains.

Plasmids

Strains of a number of *Salmonella* serovars, including *choleraesuis*, *dublin*, *enteritidis*, *pullorum* and *typhimurium*, harbour large serovar-specific plasmids (SSPs), between 30 and 60 MDa in size. Although these plasmids vary considerably in size, incompatibility group and overall homology, they contain an essentially identical piece of DNA, known as the *Salmonella* plasmid virulence or *spv* region. The region contains at least five genes, *spvRABCD*. Transcription is regulated by both the *spvR* gene product and the sigma factor RpoS, influenced in turn by factors such as the host intracellular environment, low pH, iron limitation and nutrient limitation concurrent with reaching the stationary growth phase. The gene products of *spvABCD* appear to enhance intracellular multiplication and systemic dissemination. Minor sequence differences in *spvR* markedly affect transcription of *spvABCD*, in turn influencing the invasiveness of different serovars.

For any serovar, the role in virulence played by the plasmid or, more specifically, the *spv* gene products varies with host. For example, the plasmid of serovar *enteritidis* affects invasion in cattle and mice, but not chickens.

Interestingly, some host-adapted serovars such as *typhi* and gastroenteric serovars with a predisposition to extraintestinal infection, such as *virchow*, do not harbour virulence plasmids; in some cases the homologous virulence genes have been located within the chromosome of these serovars.

Pathogenesis

Infectious Dose

Infection begins with ingestion of a dose of the bacterium sufficient to broach the first-line host defences and colonize the gastrointestinal tract. The dose required is influenced by the nature and physiological status of the strain itself, the matrix in which the strain is ingested and the status of the potential host. While the typical infectious dose is considered to be in the range of 10^6–10^8 cfu, epidemiological evidence from a number of outbreaks has demonstrated that the infectious dose may be substantially less – as little as a few cells (**Table 5**).

Different strains of *Salmonella* possess a diversity of virulence factors which can be brought to bear against the host defences; the number and type of these factors have a profound effect on pathogenesis. The physiological state of the organism, whether in active log phase or the stationary phase, may have an impact on survival, both in the food matrix and upon entry into the host.

The host also influences infectious dose. The very young have a poorly developed immune system and low gastric acidity, while the elderly and immunocompromised demonstrate only a weak immune response against infection.

Table 5 Infectious doses associated with outbreaks of salmonellosis

Food vehicle	Salmonella *serovar*	*Infectious dose (cfu ingested)*
Chocolate	*eastbourne*	100
Chocolate	*napoli*	10–100
Chocolate	*typhimurium*	<10
Cheddar cheese	*heidelberg*	100
Cheddar cheese	*typhimurium*	1–10
Peanut butter	*mbandaka*	10–100
Hamburger	*newport*	10–100
Ice cream	*enteritidis*	25–50
Paprika-flavoured potato chips	*saintpaul/javiana/rubislaw*	<45

A food matrix high in fat or protein offers significant protection to the organism within it, both in relation to the external environment, and that within the host. Such foods act as a barrier to gastric acidity, and those high in fat are also voided quickly from the stomach, both serving to transport *Salmonella* quickly and without injury to the lower gastrointestinal tract. Additionally, cells present in such a matrix will be in a dormant and thus more resistant physiological state.

The combination of these factors results in outbreaks involving low infectious doses, exemplified by an outbreak involving peanut butter in Australia. Small populations of serovar *mbandaka* were ingested in peanut butter by a largely young population, resulting in a widespread outbreak, with severe gastroenteritis experienced by many of those afflicted.

Disease

Salmonella are considered to cause two distinct disease syndromes, described simply as gastroenteritis and systemic disease, although it is now clear that some strains of typically gastroenteric serovars are capable of causing systemic disease.

Systemic disease is usually associated with strains of serovars that inhabit a narrow range of hosts, such as *Salmonella dublin* in cattle, *Salmonella pullorum* in poultry, and *Salmonella typhi*, *paratyphi*, and *sendai* in humans. The systemic syndrome is characterized by a long incubation period, a lower infectious dose than that generally associated with gastroenteric disease, a range of extraintestinal symptoms – particularly fever – and, commonly, establishment of an asymptomatic carrier state following resolution of acute symptoms. Further detail is provided in a subsequent discussion of *Salmonella typhi*.

The gastroenteric syndrome, most frequently associated with food-borne transmission in developed countries, begins after an incubation period of 8–72 h. After ingestion, organisms that survive the acidic environment of the stomach colonize and invade intestinal epithelial cells, and M cells at Peyer's patches. Both mannose-sensitive and mannose-resistant fimbriae and *inv* gene products facilitate attachment to these cells. Other products of *inv* genes stimulate intracellular accumulation of calcium ions, leading to membrane ruffling, actin filament formation and endocytosis of bacterial cells. *Salmonella* passively migrate within endocytotic vacuoles from the apical to the basal pole of the host cell, where they are released into the lamina propria. Production of enterotoxin stimulates cAMP production in host cells, leading to electrolyte imbalance and fluid accumulation. Inflammation at the lamina propria, resulting in release of prostaglandins, may exacerbate the influx of fluid to the intestinal lumen.

Disease, which usually only lasts several days, is characterized by mild fever, abdominal pain and diarrhoea, which may or may not be bloody, as well as, to a much lesser extent, nausea and vomiting. In most cases, the disease is self-limiting and symptoms resolve within 7 days. Mortality is low (0.1–0.2%), although this figure will vary dramatically with the nature of the afflicted population, rising among the very young, old or immunocompromised.

Strains of certain serotypes, such as *virchow*, are invasive in susceptible individuals or populations, causing an attenuated form of enteric fever, generalized septicaemia or localizing in any of a range of extraintestinal sites, including the appendix, gall bladder, spleen, liver, kidneys, urinary tract and brain, giving rise to a diversity of complications.

Following resolution of acute-phase symptoms and satisfactory management of any complications, as required, a complete recovery is made. Excretion of *Salmonella* ceases within several weeks, although in a small number of cases a carrier state may evolve. Additionally, chronic autoimmune conditions, such as reactive arthritis and Reiter's syndrome, may develop in some cases, which may be linked to incorporation of bacterial LPS molecules into cells at synovial joints, or HSP-induced proliferation of lymphocytes and a subsequent inflammatory response in those joints.

Significance to the Food Industry

Sources of *Salmonella*

Many biological entities, living and dead, act as reservoirs of *Salmonella*, and a diversity of foods have been implicated in outbreaks of food-borne disease (**Table 6**). As the natural habitat of *Salmonella* of significance to food-borne disease is the gastrointestinal tract of humans and other primarily warm-blooded animals, it is not surprising that the major food vehicles of transmission are animal-derived foods. However, plant foods may also act as vehicles, following environmental contamination.

Both animal and, to a lesser extent, plant-derived animal feed materials are important in the persistence of *Salmonella* in the food production environment. This is exemplified in the poultry industry, where stringent control of feed materials such as meat meal results in a significant decline in the carriage rate of *Salmonella* by poultry.

Incidence

Worldwide, food-borne disease associated with *Salmonella* is considered to be second only to that involving *Campylobacter*. The incidence of infection with *Salmonella* seems to be increasing. To some extent, this may be attributed to better reporting and surveillance rather than a real increase in disease. Nevertheless, a significant proportion of the reported cases represents an actual increase.

The case rate for human salmonellosis varies immensely, from < 1 to > 300 per 100 000 population, and is profoundly influenced by geographic, demographic, socioeconomic, meterological and environmental factors.

Concerning specific serovars, the dominant type associated with food-borne illness for many years in many parts of the world was the ubiquitous *typhimurium*. From the early 1980s, a major public health problem began to emerge, involving strains of serovar *enteritidis* capable of systemic colonization of poultry, leading to widespread food-borne disease associated with consumption of contaminated eggs and raw or lightly cooked foods containing them.

Since the early 1990s, a specific type of *Salmonella typhimurium* known as definitive type (DT) 104 has become a major problem in the UK and western Europe, and now in the US. Strains of *S. typhimurium* DT104 are extremely invasive, and many contain large plasmids, conferring resistance to a range of antibiotics, including ampicillin, chloramphenicol, streptomycin, sulphonamides and tetracycline. Newer strains have been found to be resistant to trimethoprim and ciprofloxacin.

Different regions of the world experience problems with specific serovars from time to time. Using Australia as an example, more than 80% of human infections with serovar *virchow* occur in the state of Queensland, while those involving serovar *mississippi* are largely confined to Tasmania.

Control of *Salmonella*

Control of *Salmonella* with particular regard to food-borne disease is problematic, given the close links between the environment, feeds, food animals and humans and requires vigilance at two levels – food production and food processing.

A range of management strategies have been developed or devised to control *Salmonella* in food production environments, particularly that for production of poultry, a major vehicle of transmission of *Salmonella*. These strategies include provision of *Salmonella*-free stock and feed, stringent biocontrol, particularly of rodents, vaccination with attenuated

Table 6 Major food-borne outbreaks of human salmonellosis

Year	*Country*	*Vehicle*	*Serovar*	*Cases*
1953	Sweden	Pork	*typhimurium* PT8	8845
1964	Scotland	Canned corned beef	*typhi* PT34	507
1967	US	Ice cream	*typhimurium* PT2a, *braenderup*	~1790
1968	Scotland	Raw pork	*typhimurium* PT32	472
1973	Canada/US	Chocolate	*eastbourne*	217
1973	Trinidad	Milk powder	*derby*	~3000
1974	US	Potato salad	*newport*	~3400
1976	Australia	Raw milk	*typhimurium* PT9	>500
1976	Spain	Egg salad	*typhimurium*	702
1976	US	Cheddar cheese	*heidelberg*	339
1977	Sweden	Mustard dressing	*enteritidis* PT4	2865
1981	Netherlands	Salad base	*indiana*	~600
1981	Scotland	Raw milk	*typhimurium* PT204	654
1982	England/Wales	Chocolate	*napoli*	245
1982	Norway	Black pepper	*oranienberg*	126
1984	Canada	Cheddar cheese	*typhimurium* PT10	2700
1984	England/Wales	Ham	*virchow*	274
1984	France/England	Pâté	*goldcoast*	756
1984	Various (airline)	Aspic glaze	*enteritidis* PT4	766
1985	US	Pasteurized milk	*typhimurium*	16 284
1987	China	Egg drink	*typhimurium*	1113
1987	Norway	Chocolate	*typhimurium*	361
1988	Japan	Cooked eggs	*Salmonella* spp.	10 476
1989	England	Cold meats	*typhimurium* PT12	538
1989	England	Roast pork	*typhimurium* PT193	206
1989	US	Cantaloupe melon	*chester*	295
1989	US	Mozzarella cheese	*javiana, oranienberg*	164
1990	US	Bread pudding	*enteritidis*	~1100
1990	US	Turkey meat	*agona*	851
1991	Germany	Fruit soup	*enteritidis*	600
1991	US/Canada	Cantaloupe melon	*poona*	>400
1991	US	Mexican fajitas	*heidelberg*	673
1993	France	Mayonnaise	*enteritidis*	751
1993	Germany	Potato chips	*saintpaul, javiana, rubislaw*	>1000
1994	Finland, Sweden	Alfalfa sprouts	*bovismorbificans*	492
1994	US	Ice cream	*enteritidis*	>200 000
1996	Australia	Peanut butter	*mbandaka*	>200
1997	Australia	Pork rolls	*typhimurium* PT1	>770
1998	US	Toasted oat cereal	*agona*	209

Salmonella strains and use of probiotic preparations (e.g. competitive exclusion).

Perhaps the most important control measure in food processing involves education, first of commercial food handlers in the areas of personal and food hygiene, particularly in the food service sector of the food industry, and second of consumers, who are the food handlers involved in food service at the domestic level.

While *Salmonella* may never be eliminated completely, significant reduction should be achieved through the application of appropriate control strategies within a well-developed and implemented hazard analysis and critical control point (HACCP)-based food safety plan from the commencement of production through to consumption.

Concepts in Detection

While it is not the purpose of this article to discuss at length the methods by which *Salmonella* are detected, it is worthwhile considering general issues that may affect the process.

The physiological state of *Salmonella* has a profound effect on culturability. For a clinical specimen, in which *Salmonella* is often present in high numbers and in a completely vegetative state, isolation by direct plating is feasible. The converse is generally true of a food sample, in that *Salmonella*, if present, will be in low numbers and often in a poor physiological state, suffering injury due to processes such as chilling, freezing, heating or extremes of pH. Nevertheless, such cells are still capable of recovery after ingestion, potentially causing disease, and thus must be detected. To aid recovery of *Salmonella* and facilitate the detection process, food samples are subjected to non-selective liquid pre-enrichment (resuscitation). This is followed by selective liquid enrichment, permitting further growth of the now vegetative *Salmonella*, while suppressing the background flora that develops during resuscitation. Finally, the selective enrichments are plated, and any isolates characterized.

Some foods may also influence the recovery of *Salmonella*. Many spices prove inhibitory toward *Salmonella* in culture, due in many cases to the antimicrobial activity of essential oils associated with odour and flavour, while the anthocyanins in chocolate and other cocoa-based products also inhibit growth. These must be neutralized to facilitate recovery, using strategies such as dilution, addition of neutralizing agents or use of an alternative enrichment medium.

Most *Salmonella* exhibit a common pattern of biochemical reactions and physiological traits, many of which are exploited in cultural methods for detection. However, some strains may display one or rarely more atypical reactions/traits, including fermentation of disaccharides, such as lactose and sucrose, failure to produce hydrogen sulphide, lack of lysine decarboxylation or lack of motility. In the case of such strains, cultural detection may fail. Atypical strains are rare in relation to the many thousands isolated annually, occurring at an incidence of less than 0.1% for any given trait. However, the incidence of atypical strains in relation to a specific food matrix may be much higher, due to selective pressure: an example is lactose-positive strains in dairy products.

Detection of *Salmonella*, including many atypical strains, can be performed using a range of non-cultural techniques, usually following some form of enrichment. These include serological techniques, such as latex agglutination and enzyme-linked immunosorbent assay, and nucleic acid techniques, such as polymerase chain reaction. These techniques are the subjects of subsequent chapters.

See also: **Food Poisoning Outbreaks**. **Hazard Appraisal (HACCP)**: The Overall Concept. **Microflora of the Intestine**: The Natural Microflora of Humans. ***Salmonella***: *Salmonella enteritidis*; *Salmonella typhi*; Detection by Classical Cultural Techniques; Detection by Latex Agglutination Techniques; Detection by Enzyme Immunoassays; Detection by Colorimetric DNA Hybridization; Detection by Immunomagnetic Particle-based Assays.

Further Reading

D'Aoust J-Y (1989) *Salmonella*. In: Doyle MP (ed.) *Foodborne Bacterial Pathogens*. P. 327. New York: Marcel Dekker.

D'Aoust J-Y (1997) *Salmonella* species. In: Doyle MP, Beuchat LR, Montville JJ (eds.) *Food Microbiology Fundamentals and Frontiers*. Ch. 8, p. 129. Washington: ASM Press.

Jay S, Grau FH, Smith K, Lightfoot D, Murray C, Davey GR (1997) *Salmonella*. In: Hocking AD et al. (eds.)*Foodborne Microorganisms of Public Health Significance*. Ch. 6, p. 169. Sydney: Australian Institute of Food Science and Technology (NSW Branch) Food Microbiology Group.

Salmonella enteritidis

Thomas S Hammack and **Wallace H Andrews**, US Food and Drug Administration, Washington, DC, USA

Characteristics of Species

Members of the International Subcommittee on Taxonomy of *Enterobacteriaceae* have designated *Salmonella choleraesuis* as the single species within the genus *Salmonella*. *S. choleraesuis* consists of five subspecies: *S. enterica* (subsp. I), *S. salamae* (subsp. II), *S. arizonae* (subsp. IIIa), *S. diarizonae* (subsp. IIIb), *S. houtenae* (subsp. IV) and *S. bongori* (subsp. V). *S. enteritidis* (SE) is a serovar in the subspecies *S. enterica*. The full name of SE is *S. choleraesuis* subsp. *enterica* serovar Enteritidis, although it is usually reported in the literature as *S. enteritidis*. It is a Gram-negative, non-spore-forming rod that is motile by peritrichous flagella; it is facultatively anaerobic, catalase-positive and oxidase-negative. As with most *Salmonella* serovars, it ferments carbohydrates with the production of acid and gas. It usually produces hydrogen sulphide, is urease-negative, indole-negative, methyl red-positive and Voges–Proskauer-negative. It is citrate-variable, KCN-negative, gelatinase-negative, lysine decarboxylase-positive and glucose-positive. It is usually lactose-negative, sucrose-negative, mannose-positive and malonate-negative. It is defined serologically by the Kauffmann–White scheme where numerical and alphabetical designations are given for its somatic (O), capsular (Vi), and flagellar (H) antigens. In the Kauffmann–White scheme, SE is a member of somatic group D_1 and its antigenic formula is 1,9,12:g,m:[1,7].

The optimum growth rate of SE occurs at 37°C when the water activity (a_w) of its growth medium is greater than 0.93. SE survives drying and has been isolated from many dried foods. Survival and growth increase with increasing a_w. The minimum pH necessary for growth for some strains of SE is 4. The optimal pH is 7.0. It can survive at a pH as low as 3, but will not survive prolonged exposure to this pH. It is heat-sensitive and is easily destroyed by pasteurization.

One means used to identify different strains of SE is to expose the organism to bacteriophages. Different phages will lyse different strains of *Salmonella* spp. A *Salmonella* sp. isolate is labelled as a particular phage type (PT) depending on the phage to which it is most sensitive. PT and invasiveness have been reported to be related, though different strains originating from different locations that are of the same PT can be variable for invasiveness. PT4 and 8 demonstrate the link between PT and invasiveness. SE PT4 strains generally demonstrate a greater degree of invasiveness in poultry than PT8 strains. PT4 strains are more virulent and produce a higher degree of morbidity and mortality in chickens than most other PTs. This PT is particularly well-adapted to the invasion of the intestinal tracts and reproductive organs of poultry. These characteristics contribute to the vertical transmission of SE from generation to generation and are the basis of many outbreaks: PT4-infected chickens produce contaminated eggs destined for human consumption. In the US, as of 1995, the main SE PTs involved in outbreaks were PT13A followed by PT8 and PT4. In 1991, PT8 was the dominant PT as PT4 had not yet been found in US layer flocks. In Europe the main PT involved in outbreaks is PT4.

SE is found mainly in avian species and is easily transmissible to mammals. Consequently, it is an agent of zoonotic infection between poultry and humans. Acute outbreaks among poultry occur most often in chickens under 1 month old and during periods of stress such as forced moults. The mortality rate in chicks under 2 weeks old, naturally infected with SE, has been reported to be as high as 20%. In chickens over 1 month of age, the mortality rates are low and many of the birds appear asymptomatic, although there may be hidden damage to the reproductive tract.

SE contains several virulence factors that enhance pathogenicity. SE produces a heat-labile enterotoxin that causes profuse intestinal fluid loss in infant mice. Moreover, it produces a heat-labile cytotoxin that is a protein component of its outer membrane, which has a molecular weight of 56 000–78 000. This cytotoxin affects human AV-3, CHO, Hela and Vero cells through the inhibition of protein synthesis. *Salmonella* cytotoxins also affect protein synthesis in intestinal epithelial cells and thereby account for damage to the intestinal mucosa, which is a hallmark of salmonellosis.

The lipopolysaccharide (LPS) of the bacterial cell wall is an additional virulence factor displayed by SE strains as well as by other *Salmonella* spp. The LPS consists of lipid A, an inner core, and O-side chain oligosaccharides. The O-side chain oligosaccharide of the LPS appears to be a key to the virulence of SE. For example, an SE PT4 strain was transformed to an SE PT7 strain after losing its ability to express the long chain LPS. This same SE strain became avirulent in mice. Changes in the structure of the O-side chain oligosaccharide affect the ability of SE to resist macrophage phagocytosis. *Salmonella* spp. organisms with defective O-side chains are less resistant to murine macrophage phagocytosis than *Salmonella* spp. organisms with intact O-side chains.

SE adheres to the intestinal epithelium with fimbriae and mannose-resistant haemagglutinin. SE utilizes two types of fimbriae: type 1 and/or type 3, to attach themselves to the intestinal cells. Type 1 fimbriae are found in most Gram-negative bacteria and allow the bacteria to agglutinate erythrocytes in the absence of D-mannose. Type 3 fimbriae agglutinate tannic acid-treated erythrocytes in the presence of D-mannose. Types 1 and 3 fimbriae are involved in the attachment of SE to human buccal and murine small intestinal epithelial cells.

Methods for Detection in Foods

There is a variety of conventional culture and rapid methods in use for the detection of SE in foods. Most of the culture methods consist of five steps:

- pre-enrichment in a nonselective medium to resuscitate injured organisms
- selective enrichment to allow *Salmonella* spp. organisms to proliferate to discernible levels while suppressing the growth of competitors
- selective/differential plating both to suppress the growth of competitors and to allow for the isolation of discrete suspect colonies
- biochemical testing
- serological testing.

These culture methods are definitive and result in a bacterial isolate. Methods such as the polymerase chain reaction (PCR) assay or the enzyme immunoassay (EIA) have been used for the rapid detection of this organism. Although rapid methods can be highly accurate (some are >98% in agreement with a reference culture method), normally they are not considered definitive because they usually do not produce an isolate. Rapid methods that exhibit both high specificity and high sensitivity can be used as a screening tool when they are performed in tandem with the culture method (especially when the rapid method and the culture method share the same pre-enrichment strategy). In cases where a validated rapid method does not detect *Salmonella* spp. in a pre-enrichment that it shares with the culture method, the culture method can be discontinued. In such a case, the rapid method has demonstrated, to a high degree of probability, that *Salmonella* spp. are not present in the test sample and no further work need be done. Thus, rapid methods can be used to save time and resources.

The methods discussed subsequently are recommended by standardization organizations for the detection of *Salmonella* spp. in foods for regulatory purposes by the US Food and Drug Administration (FDA), the US Department of Agriculture (USDA) and by the European Union (EU). In addition, some examples of rapid methods specifically designed to detect SE in foods are included. The FDA is mentioned most often in conjunction with the following culture methods because SE is isolated primarily from eggs or egg-containing foods. In the US, the FDA and USDA share regulatory responsibility for egg safety, but the primary regulatory responsibility for food safety in shell eggs and in egg-containing foods rests with the FDA.

Conventional Methods

Bacteriological Analytical Manual/Official Methods of Analysis

The FDA publishes the *Bacteriological Analytical Manual* (*BAM*) in conjunction with the Association of Official Analytical Chemists International (AOACI). The FDA relies on the *BAM* for the *Salmonella* spp. methods, including SE, used in its regulatory laboratories. In addition, it relies on the *Official Methods of Analysis* (*OMA*) published by the AOACI, which contains methods of analysis that have been verified through collaborative studies.

International Organization for Standardization

The International Organization for Standardization (ISO) is a worldwide federation of national standards bodies dedicated to standardizing microbiological and chemical methods worldwide. Its *Salmonella* methods are used in the EU for regulatory purposes. It recommends method ISO 6579:1993(E) for the detection of *Salmonella* spp. in foods other than milk and milk products, and method ISO 6785-1985 (E) for the detection of *Salmonella* spp. in dairy products. Its method for foods, other than dairy products, may not be appropriate under certain circumstances for all foods. The ISO recommends that, in such cases, deviations from the ISO method should only be allowed for established technical reasons. These methods, like the *OMA/BAM* method, can be used for the detection of SE in foods. The ISO *Salmonella* spp. culture methods are used for regulatory purposes in Europe, but not in the US.

Compendium of Methods for the Microbiological Examination of Foods

The *Compendium of Methods for the Microbiological Examination of Foods* is published by the American Public Health Association (APHA) and is used by the FDA when methods are not available in the *BAM* or *OMA*. The *Salmonella* spp. method referenced in the *Compendium* is the *BAM/OMA* method.

Standard Methods for the Examination of Dairy Products

The *Standard Methods for the Examination of Dairy Products* is published by APHA and is used by the FDA when methods are not available in the *BAM* or *OMA*. The *Salmonella* spp. method referenced in the *Standard Methods* is the *BAM/OMA* method.

Code of Federal Regulations

The *Code of Federal Regulations* (*CFR*) is a codification of the general and permanent rules published in the *Federal Register* (*FR*) by the executive departments and agencies of the US government. The CFR is kept up-to-date by the individual issues of the *FR*. The *CFR* and the *FR* must be used together to determine the latest version of any given rule. The SE problem has led to the publication of several methods for the detection of SE in foods in the *CFR* and *FR*. The *CFR* currently recommends methods to culture SE from the internal organs of poultry whose blood reacts to *pullorum*-typhoid antigens (9CFR147.11) and to culture SE from environmental, organ and intestinal test samples (9CFR147.12). These methods are used by both the FDA and the USDA for regulatory purposes.

Rapid Methods

Commercial

Currently, there is one commercially available rapid method for the detection of SE in foods. The *Salmonella* Screen/SE Verify manufactured by Vicam (Watertown, MA, USA) screens for all *Salmonella* spp. organisms and indicates the presence of SE. It has been approved for use by the AOAC Research Institute through its performance-tested program. In the procedure, a test sample is pre-enriched in a non-selective enrichment medium for 5, 6 or 16 h at 37°C. Foods with low levels of competitors are pre-enriched for 5–6 h. Foods with high levels of competitors or with expected low levels of injured *Salmonella* spp. organisms are incubated for 16 h. After incubation, 2 ml of the pre-enrichment is transferred to capture vials. Magnetic beads coated with monoclonal and polyclonal antibodies to SE somatic and flagellar antigens are added, and the pre-enrichment is rotated for 1 h at room temperature. The capture vial is placed into a magnetic rack and the supernatant is aspirated. The beads are washed twice in capture buffer before being resuspended in buffer. The beads are then spread on to XLD agar and incubated for 16–22 h at 37°C. Typical colonies are then confirmed as SE by latex agglutination. The drawback to this method is that the latex agglutination test used to confirm SE cross-reacts with *S. dublin*, *S. bledgam* and *S. moscow* so that additional serotyping, along with biochemical testing, must be performed to confirm SE. The manufacturer indicates that this test will detect *Salmonella* at 0.2 and 0.04 cfu g^{-1} after pre-enrichment periods of 5–6 and 16 h, respectively.

Generic

There are two basic techniques used by laboratories for the detection of SE in foods and other materials: PCR and EIA. PCR is used to amplify portions of the SE genome which are unique to SE. Samples are pre-enriched to increase the level of SE to what is necessary for PCR analysis. The SE cells are lysed in the pre-enrichment broth, cellular and food debris are removed by centrifugation, and the supernatant containing presumptive SE DNA is decanted. The DNA strands are then separated by heating and are annealed in the presence of a specific primer. The annealed strands are incubated with DNA polymerase and the 4-deoxyribonucleoside triphosphates. The strands are separated after incubation. Twenty to 30 cycles are necessary to amplify the DNA to an adequate level. After amplification, the amplified DNA is compared to a known standard using gel electrophoresis. Detection rates for SE and for *Salmonella* spp. have been reported to be as low as 1 cfu and 0.04 cfu per gram of chicken, respectively. A major disadvantage of this technique is that it is not easily automated and requires the use of gel electrophoresis which is a time-consuming process, especially when many samples must be analysed. Recently, there have been reports that automated PCR assay for *Salmonella* spp. can be performed in 96-well plates. A DNA probe tagged with a fluorescent marker is added after the DNA has been amplified. The amplified DNA is re-amplified with the tagged probe. Fluorescence readings are taken after the first cycle and every third cycle thereafter. If the test sample contains *Salmonella* spp., then the fluorescence will increase with time as the probe is incorporated into progressively larger numbers of DNA strands. With overnight pre-enrichment, the reported recovery rates of these methods were as low as 3 cfu *Salmonella* spp. per 25 g analyte.

There are several EIA-based methods for SE being used in regulatory and academic laboratories. These assays are mostly used to detect SE antibodies present in serum or eggs, although some have been used directly to detect SE in foods. Monoclonal antibodies to SE lipopolysaccharides, flagella, whole cell proteins, outer membrane proteins and fimbrial antigens have been used in SE EIA methods. The antibodies are bound to a solid surface (e.g. 96-well plate); a blocking agent such as bovine serum albumin is added to block any additional binding sites on the surface not occupied by the antibodies; the antigen is added and allowed to bind to the antibody; the plate is washed to remove any unbound antigens, and an antibody to the antibody–antigen complex that is linked to either horseradish peroxidase or alkaline phosphatase is added. A chromogenic (or fluorogenic) substrate specific to either horseradish peroxidase or alkaline phosphatase is added, and the absorbance (or fluorescence) is read. Increases in the absorbance indicate the presence of SE. A report by the World Health Organization in 1996 reported that the different generic EIA methods being used by 14 academic and regulatory laboratories were equivalent to one another for detecting antibodies to SE in commercial broiler breeder serum.

Importance in the Food Industry

In the food industry SE originates mainly within the poultry industry. In the US, between 1985 and 1991, eggs were implicated in 82% of SE outbreaks where the food vehicle was identified. Products as diverse as ice cream, custard, macaroni and cheese and hollandaise sauce have all been vehicles for SE outbreaks in which eggs have been implicated as the source of infection.

SE, along with other *Salmonella* serovars and other pathogens as well, is very important to the food industry. An outbreak of salmonellosis can cost millions of dollars in recall costs, charges and in product liability losses. There are also production losses for poultry and egg producers even when there are no outbreak-associated costs. SE increases losses to producers by increasing the morbidity and mortality of their flocks. Broiler and layer flocks infected with *Salmonella* suffer from low weight gain and increased mortality. Additional losses are sustained when contaminated eggs are traced back to a producer. In the US, a producer then becomes subject to the National Poultry Improvement Plan. The flock becomes a test flock which requires thorough testing of the birds and their environment. If SE is found, the flock's status becomes infected. Eggs from infected flocks in the US cannot be sold as table eggs and must be diverted to processors for pasteurization, boiling or export. An infected flock's status cannot be upgraded until further testing has demonstrated that the flock is no longer SE-positive. Furthermore, flocks cannot be replaced until the hen houses have been cleaned and disinfected. These measures provide a large incentive for egg producers to keep their flocks *Salmonella*-free.

Overall Economic Impact

The FDA estimates that the cost of salmonellosis from all causes in 1995 was $350 million to $1.5 billion in the US. These figures are based on 41 222 reported cases of salmonellosis and an estimated 20–100 unreported cases for each reported case. The Centers for Disease Control and Prevention (CDC) reports that from 1976–1994, the proportion of *Salmonella* spp. isolates that were SE rose from 5 to 26%. If these figures are extrapolated to include 1995, then it can be estimated that the total cost of SE in 1995 was $91–390 million. One can further extrapolate that the cost of SE contamination of eggs in 1995 was $75–320 million.

Contamination of Raw Foods

Eggs

As stated previously, eggs are a primary cause of SE-induced salmonellosis. In order to understand how eggs can be contaminated with SE, it is necessary to describe the egg itself and the process by which it comes to market. Eggs consist of a yolk, yolk membrane, albumen, an inner and outer shell membrane, the shell and the cuticle which covers the shell. The cuticle, shell and shell membranes present a physical barrier against contamination and the albumen contains bactericidal factors to defend against any organisms that penetrate the shell. The effectiveness of these protective features degrades with time. The cuticle and/or the shell may become cracked, thereby removing the barrier. The albumen thins because of moisture loss through the shell and, consequently, the yolk tends to move from the centre of the egg to the shell. The closer the yolk to the shell, the greater its chance of contamination. Eggs are stored point-down and upright to lessen this tendency.

Salmonella spp. can enter the egg through the shell or through transovarian contamination of the yolk membrane/albumen. The egg industry attempts to prevent external contamination by cleaning and candling. Conveyer belts transport the eggs from the pens to a washing machine that removes the manure and dust from the egg. Egg disinfection is a three-stage process in which the eggs are washed with hot soapy water and brushes or under high-power jets, rinsed with sanitizing agents and dried in a drying chamber. The wash cycle must be 20°F (11.1°C) above the egg temperature. If the wash water is cooler than the egg, the albumen will contract and pull the wash water into the egg along with any pathogens, spoilage organisms or other impurities present in the wash or on the surface of the egg. Disinfection, in addition to cleaning and disinfecting the egg, removes the cuticle. Many producers oil the egg as a way of replacing the barrier the cuticle provided. After eggs are cleaned and oiled, they are candled. Cracked eggs are sent to a breaker plant where they are pasteurized. Intact eggs are held in coolers at 45–55°F (7.2–12.8°C) and transported at 60°F (15.6°C) to retail outlets where the eggs are placed in cold boxes until sold. This process reduces the numbers of eggs that are contaminated with SE through mechanical means and reduces the multiplication of SE in contaminated intact eggs.

Transovarian egg contamination occurs in about 1 of 10 000 eggs produced in the US according to the USDA. This low level of contamination, however, results in approximately 4.5 million SE-contaminated eggs – hence, the exposure of large numbers of people to this pathogen.

There are two potential vectors of infection for poultry: contaminated feed and infected animals. Feed is made, in part, from rendered poultry and/or other livestock that may have been contaminated with SE. If the feed is not sufficiently heated, *Salmonella* spp. will survive and infect the flock. Sometimes, thorough heating may not be enough. *S. typhimurium* that has been habituated to dehydrated conditions has been reported to survive for 60 min at 100°C. Animals such as rodents, birds and cats can be carriers of *Salmonella* spp. and infect poultry with which they come into contact, either directly or indirectly. For example, a flock in California that was positive for SE PT4 was housed on a farm 1 km downstream from a sewage treatment plant. Effluent from the sewage plant, in a stream bordering the farm, contained the same SE PT4 strain. Moreover, 12.5% of the mice and 57% of the cats tested that were living at or near the farm and were believed to have drunk from the contaminated stream were also positive for the same strain. Rodent droppings in the feed bins contained as many as 2.3×10^5 SE PT4 organisms per faecal pellet. It is evident that the treatment plant, via feral animal life, was the source of the contamination.

The goal of the public health establishment is to reduce the risk of contracting salmonellosis from eggs to zero. The fact that 82% of the SE outbreaks, where the food vehicle has been identified, is the result of contaminated eggs, demonstrates that much work needs to be done. A two-pronged approach is necessary. First, every effort should be made to ensure that poultry are not infected initially. Second, every effort should be taken to ensure that any infected eggs that make their way into commerce are held under conditions that will not allow an increase in SE levels and that these eggs are subjected to a heat treatment sufficient to inactivate the pathogen, especially for those populations that are particularly at risk for infection.

Poultry

Chickens destined for consumption (broilers) become infected with SE in the same fashion that layer hens do: through feed, direct or indirect contact with feral animals and/or through transovarian infection. The broiler carcass can become contaminated during evisceration when it comes into contact with the contents of ruptured intestines or crop. Contaminated carcasses cross contaminate uncontaminated carcasses when they are chilled and rinsed in chiller water. The threat that a contaminated chicken presents can be mitigated through proper handling, as explained below.

Contamination of Processed Foods

Processed foods are contaminated with SE when the food production process fails or when the food preparer inadvertently cross contaminates a processed food with a contaminated foodstuff. Food production process failure occurs when, for example, the pasteurization process fails because of mechanical defect or when a finished ingredient comes into contact with a raw ingredient without further processing. An example of the latter case was an SE outbreak associated with ice cream in the midwestern US in 1994. In that outbreak, ice cream premix, which underwent no further processing, was transported in tanker trucks that had been used immediately before to transport non-pasteurized eggs. This outbreak resulted in 277 culture-confirmed cases of SE with an estimated 224 000 non-reported cases.

Importance to the Consumer

Salmonellosis often includes any or all of the following symptoms: nausea, vomiting, diarrhoea, fever, abdominal cramps and headache. It may also lead to arthritis 3–4 weeks after the acute disease in about 2% of infected persons. The time of onset is 6–72 h after inoculation and the illness may last up to 7 days. The illness may be more severe depending on the health of the person infected and the virulence of the strain. The infective dose may be as low as 1–10 cells depending, as before, on the health of the infected person and the virulence of the strain. The very young, the aged and the immunocompromised are particularly susceptible to the disease. Infections in these groups may lead to disorders such as septicaemia and meningitis, with sometimes fatal results. For example, in the US from 1985 to 1991, there were 59 SE outbreaks in hospitals or nursing homes, accounting for 12% of all outbreak-associated cases and 90% of all SE-related deaths. The case fatality rate in these settings was 70 times the rate in other settings. In 9 of 15 nursing home outbreaks, egg-containing dishes were the implicated vehicle. It is therefore advisable for these institutions to use only pasteurized egg products during meal preparation.

Epidemiology

SE has increased from 5% of *Salmonella* isolates in 1976 to 26% of isolates in 1994. In the years 1994 to 1996, SE was the most commonly isolated *Salmonella* serovar, followed by *S. typhimurium* and *S. heidelberg*. According to the CDC, the isolation rate of SE in the US increased from 0.5 to 3.9 per 100 000 population. There have been 660 outbreaks reported to the CDC during the years 1985 to 1996. This increase has been mainly associated with eggs, as noted above.

Prevention

Consumers can protect themselves from salmonellosis, as well as from several other food-borne enteric diseases when preparing food at home, both by cooking foods adequately to kill any possible pathogens and by handling food properly so as to reduce the chance of cross contamination of non-contaminated foods with contaminated foods. Proper storage of perishable foods is also important. SE grows exponentially in foods that are not properly refrigerated, which both extends the cooking time necessary for inactivation of the pathogen and increases the level of the pathogen in contaminated foods. Both circumstances increase the likelihood of contracting salmonellosis.

As a rule, raw animal products such as meat and eggs should be cooked until they reach an interior temperature of 60°C (140°F). This temperature is sufficient to kill most pathogens, including SE. There have been reported SE outbreaks where, after cooking, the centre of a dish was cool or actually cold while the periphery was hot. For example, in 1989 in New York state, 21 of 24 persons who attended a baby shower were stricken with SE-induced salmonellosis. The cause of the outbreak was a baked ziti pasta dish that consisted of one raw egg and ricotta cheese combined in a pan with cooked tomato sauce and refrigerated overnight. The dish was baked before serving at 350°F (176.7°C) for 30 min. Several of the participants reported that the centre of the dish was cold.

Because of the risk of salmonellosis, the FDA advises consumers to cook egg and egg-containing foods thoroughly. Eggs should be cooked until the egg has reached 60°C (140°F) throughout, i.e. until the yolk and white are firm. The following cooking times should be adhered to:

1. Scrambled eggs should be cooked at least 1 min at 121°C (250°F) in order to raise the temperature of the eggs to 73.9°C (165°F).
2. Poached eggs should be cooked for 5 min in boiling water.
3. Eggs fried sunny-side up should be cooked in a frying pan at 121°C (250°F) for a minimum of 7 min uncovered or a minimum of 4 min covered.
4. Eggs fried over-easy should be cooked with the frying pan at 121°C (250°F) for a minimum of 3 min on one side and 2 min on the other side.
5. Boiled eggs, in the shell, should be cooked while completely submerged in boiling water for 7 min.

The above guidelines should be sufficient to inactivate any *Salmonella* present in raw eggs. These guidelines should be considered as the minimum cooking times necessary for the inactivation of *Salmonella* spp. because the level of the pathogen is unknown. Higher levels of *Salmonella* spp. take longer to kill than lower levels. Also, in the case of boiled eggs, these guidelines assume near-sea-level conditions. High-altitude cooking requires longer boiling because water boils at a lower temperature than at sea level.

Once foods are cooked, there is a danger that these foods may still become contaminated with SE. This occurs through the re-use of unwashed cutting boards, dishes and other utensils after they have come in contact with raw foods of animal origin such as meat and eggs. A person who cuts up a chicken on a cutting board, cooks the chicken thoroughly and then cuts the cooked chicken on the same board without having first washed the board with hot soapy water endangers the intended diners. The consumer should always regard unpasteurized foods of animal origin as a source of infection and act accordingly. By doing so, the consumer will drastically reduce the chance of contracting salmonellosis.

See also: **Eggs**: Microbiology of Fresh Eggs; Microbiology of Egg Products. **Food Poisoning Outbreaks**. **Process Hygiene**: Involvement of Regulatory Bodies. ***Salmonella***: Introduction.

Further Reading

Cox JM (1995) *Salmonella enteritidis*: the egg and I. *Australian Veterinary Journal* 72: 108–115.

Ewing WH (1986) *Edwards and Ewing's Identification of Enterobacteriaceae*, 4th edn. P. 181. New York: Elsevier Science.

Holt JG, Krieg NR, Sneath PHA, Staley JT and Williams ST (1994) *Bergey's Manual of Determinative Bacteriology*, 9th edn. P. 186. New York: Williams & Wilkins.

Le Minor L (1984) *Bergey's Manual of Systematic Bacteriology*. Vol. 1, p. 427. Baltimore, MD: Williams & Wilkins.

Mishu B, Koehler J, Lee LA et al (1994) Outbreaks of *Salmonella enteritidis* in the United States, 1985–1991. *Journal of Infectious Diseases* 169: 547–552.

Susuki S (1994) Pathogenicity of *Salmonella enteritidis* in poultry. *International Journal of Food Microbiology* 21: 89–105.

Salmonella typhi

Julian Cox, Department of Food Science and Technology, University of New South Wales, Sydney, Australia

Unlike most *Salmonella*, which infect and often cause disease in a broad range of animal hosts, certain serovars are limited to a narrow range or perhaps a single animal host; these serovars are termed host-adapted. While some, such as *S. pullorum* and *S. gallinarum*, are of veterinary concern, others such as serovars *sendai*, *paratyphi* A, B and C and *typhi*, are significant human pathogens; *S. typhi* is the causative agent of enteric or typhoid fever. Although frequently disseminated through human-to-human contact or waterborne transmission, *S. typhi* does represent a food-borne pathogen of some significance.

Characteristics of the Organism

Salmonella typhi conforms generally, both genotypically and phenotypically, to the description of the genus *Salmonella*. However, it is biochemically distinct (**Table 1**), exhibiting weak or negative reactions in a number of tests normally employed in the characterization of *Salmonella*.

S. enterica subsp. *enterica* serovar *typhi* or *S. typhi* is a member of serogroup D within subspecies I of the genus *Salmonella*, and is represented by the antigenic formula 9,12 : d : -. Some strains of *S. typhi*, originally isolated in Java, Indonesia, and hence commonly known as var. Java, produce the R-phase 1 H antigens H-j and H-z66, rather than H-d. In addition to the O and H antigens, strains of serovar *typhi*, particularly

Table 1 Biochemical differentation of *Salmonella typhi* and other typical *Salmonella* strains

Test/substrate	S. typhi	*Typical* Salmonella
H_2S production	–/w	+
Utilization of:		
Citrate	–	+
Sodium acetate	–	d
Mucate	–	d
Ornithine decarboxylase	–	+
Gas from glucose	–	+
Fermentation of:		
Dulcitol	d	+
Inositol	–	d
Arabinose	–	+
Rhamnose	–	+

+ = >90% strains positive; – = >90% strains negative; w = weak reaction; d = variable reaction (+, – or w).

those recently isolated from clinical specimens, may also produce a capsular antigen, designated Vi.

Strains possess a genome of between 3.9 and 4.9 Mb and may also harbour large plasmids, many of which confer antibiotic resistance. Variation in either or both of these characteristics is associated with geographical distribution and differences in virulence.

Strains of serovar *typhi* may be further characterized by phenotypic or genotypic analysis; with respect to the former, the method of choice is phage typing. A set of bacteriophage specific for the Vi antigen can currently differentiate *S. typhi* into 108 phage types. Phage typing facilitates identification of outbreaks, and has demonstrated the geographical restriction of certain strains. For example, while phage types A and E_1 appear to be global types, others such as O, D_1, D_2 and the untypable H-j/H-z66 types are restricted to Papua New Guinea, Mediterranean countries, India and Indonesia respectively.

Strains of serovar *typhi* can also be differentiated using a range of molecular techniques. Restriction endonuclease analysis is effective in distinguishing strains of different H antigen type. Pulsed-field gel electrophoresis has been used successfully to subdivide strains of common phage types, and to examine differences among strains associated with sporadic cases versus outbreaks, and strains that appear to differ with respect to virulence. It has also highlighted genetic differences in isolates from blood and faeces during the course of a single infection. A range of restriction enzymes, including *Xba*I, *Avr*II and *Spe*I, produce useful and easily interpreted patterns. Ribotyping, using restriction enzymes including *Pst*I and *Cla*I to digest chromosomal DNA, has proven effective in tracing epidemic strains, and can subdivide strains of a range of phage types. Ribotyping, as well as typing with *fliC* probes, suggests that *S. typhi* evolved in South-East Asia, and the H-j genotype only occurs in Indonesia. The *Salmonella*-specific insertion sequence IS*200* has been used with some success in the typing of *S. typhi* strains, although typing may be confused by the occurrence of plasmid-borne sequences. *S. typhi* possesses an additional copy of IS*200* which, with respect to other *Salmonella* spp., occurs in a distinct region of the genome, facilitating both typing and detection.

Virulence and Pathogenesis

Virulence Factors

As with other *Salmonella*, virulence is complex and multifactorial. *S. typhi* produces an extensive and diverse array of factors that contribute to infection and disease, although the arrangement and location of some genes encoding such factors are distinct in the latter. In addition, *S. typhi* produces factors that facilitate systemic infection.

The chromosome of *S. typhi* contains three pathogenicity islands: the *inv* genes reside within the largest. A further locus has been described, composed of five genes designated *sipEBCDA*, and this is considered important for entry into epithelial cells, since it bears strong homology to the *ipa* genes present on the large virulence plasmid of *Shigella*, which confer the same function.

Like other *Salmonella*, *S. typhi* produces lipopolysaccharide (LPS) as an integral part of the outer membrane. The LPS of *S. typhi* is highly glycosylated, with side chains of considerable length. Both of these factors are strongly associated with virulence. In fact, variation in LPS exerts a greater effect on serum survival than variation in the amount of Vi antigen produced.

S. typhi is capable of producing both type I and type 3 fimbriae, and may produce others. Genes responsible for synthesis of fimbriae, including the *sef* operon, have been located on the chromosome of *S. typhi*, rather than on serotype-specific virulence plasmids in relevant serovars such as *typhimurium* and *enteritidis*.

Strains of *S. typhi* synthesize both types of siderophores common to enteric organisms, aerobactin and enterochelin. Production of the latter is much more common than the former in this serovar. The prevalence of enterochelin rather than aerobactin production is indicative of the invasive nature of this serovar.

Toxins are produced by *S. typhi*, including the enterotoxin produced by many *Salmonella*, which is structurally similar to cholera toxin, as well as a cytotoxin encoded by the *stpA* gene, similar to the enterotoxin produced by *Yersinia enterocolitica* and encoded by *yopE*. The cytotoxin was originally thought to be related to Shiga toxin, but this has been disproved by serological and nucleic acid studies.

The Vi antigen is a major and essentially distinct virulence factor of *S. typhi*. This polysaccharide capsular antigen, a homopolymer of α1,4-linked *N*-acetylgalactosaminuronic acid, variably acetylated at the C3 position, is expressed by most strains of *S. typhi*, some strains of *S. paratyphi* C and, rarely, by strains of *S. dublin*. The Vi antigen plays a role in pathogenesis, required for survival within macrophages, and in the blood stream, as it confers serum resistance, blocking C3b complement (opsonizing) activity against LPS. Interestingly, strains lacking Vi have been shown to be hyperinvasive, indicating that the antigen is not necessary during this phase of infection. Vi antigen synthesis is governed by two gene loci, designated ViaA and ViaB, the latter of which is only rarely found in other *Salmonella* spp. ViaB encodes at least eight polypeptides responsible for regulation, monomer synthesis, polymerization and translocation to the cell surface of the Vi antigen. While some *Salmonella* spp. as well as other organisms such as *Citrobacter freundii* possess the ViaB region, it is genetically distinct in the latter, and such organisms cannot synthesize Vi, due to a missing or dysfunctional ViaA region.

Pathogenesis and Disease

S. typhi causes typhoid fever, an enteric fever quite distinct from the typical gastrointestinal syndrome associated with most *Salmonella* spp. The infectious dose for typhoid is not well documented, and is influenced by the same factors that influence the dose required for infection by typical gastroenteric *Salmonella*. The required dose of *S. typhi* is considered to be 1–2 logs lower than that required for most *Salmonella*. Although infection begins in the gastrointestinal tract, the bacterium enters the lymphatic system, moving to the mesenteric lymph nodes, and begins to multiply within macrophages. After multiplication, *S. typhi* moves into the blood stream, circulating throughout the body, localizing in various organs, especially the spleen, liver and gallbladder. In the blood stream, the bacterial cells are ingested by macrophages, in which they continue to multiply, eventually killing the macrophages. At this stage, large numbers of *S. typhi* cells are released into the blood stream, leading to septicaemia, and onset of the symptoms typical of enteric typhoid fever. Progress to this first clinically overt phase of disease is slow. Onset takes anywhere between 3 and 56 days, although 10–20 days (on average 14 days) is more typical. The first symptoms include general malaise, headache, anorexia, abdominal tenderness and constipation, progressing to an undulant fever, and the appearance of red spots on the torso. During the early phase of disease, the bacterium may be isolated easily from blood and urine. The fever may persist for several weeks, during which time the organism reaches the gallbladder and multiplies in the bile. The organism-rich bile flows into the small intestine, causing enteritis and ulceration at Peyer's patches in the terminal ileum, caecum and ascending colon, leading to diarrhoea characterized by formation of loose to watery stools. The ulcers may haemorrhage, leading to blood in the stools and, potentially, perforation of the intestine and consequent peritonitis. During this second phase of disease, the organism is more readily isolated from faeces.

Other intra- and extraintestinal complications may also arise, including perforation of the terminal ileum or appendix, paralytic ileus, hepatitis, hepatic failure, bronchopneumonia, thyroid abscess, myocarditis, encephalopathy and meningitis, particularly in neonates. Even after resolution of acute-phase or chronic infection, other sequelae may develop, such as reactive arthritis in individuals of particular histocompatibility (human leukocyte antigen: HLA) types. There is evidence that such autoimmune conditions may involve cross-reactivity with either LPS or some stress (heat shock) proteins.

Mortality, compared to other *Salmonella*, is high – between 2 and 10% – whereas the rate for food-borne gastroenteric salmonellosis is 0.1–0.2%. Given the significance of the disease, intervention in the form of antibiotic therapy is usually warranted. Traditional drugs of choice are chloramphenicol, ampicillin, amoxycillin or sulfa compounds, such as trimethoprim or sulphamethoxazole. Multiple resistance is associated with the acquisition of large conjugative IncH1 R-plasmids of between 71 and 166 MDa in size, encoding between two and six different markers and conferring resistance to one or more of chloramphenicol, ampicillin, tetracycline, sulphamethoxazole and co-trimoxazole. In addition, resistance to gentamicin, kanamycin, streptomycin, piperacillin and ticarcillin has also been reported. The evolution of multiple drug-resistant (MDR) strains has caused concern over therapy, and newer drugs, including furazolidone, quinolones such as ofloxacin, norfloxacin, perfloxacin, ciprofloxacin and later-generation cephalosporins, such as cefixime, cefotaxime, ceftizoxime and ceftriaxone, are employed in such cases. Later-generation cephalosporins are favoured in paediatric therapy, given safety concerns over quinolones in this population. In addition, nalidixic acid-resistant strains have been identified which resist therapy with quinolones.

MDR strains appear to be more virulent than susceptible strains; the evidence ranges from higher blood stream populations to increased case fatality rates. Although there is strong evidence for enhanced virulence in MDR strains, investigations in underdeveloped countries suggest that the high fatality rate attributable to such organisms is overshadowed by fatality due to little or no intervention.

Usually, in milder cases and with appropriate therapy, the disease is completely resolved within several weeks, although in a significant proportion (10–15%) of cases, the organism localizes in the gallbladder or, more rarely, the kidneys, and a carrier state is established. While asymptomatic, an infected individual periodically sheds the bacterium from the sites of localization, especially the gallbladder, into the intestine and ultimately into faeces.

Significance to the Food Industry

Sources and Incidence of *Salmonella typhi*

A major vector in the dissemination of *S. typhi* is the infected food handler. Such individuals, in whom a chronic, subclinical carrier state is established, nevertheless continue to shed the bacterium over extended periods of time. A number of cases of carriers causing significant outbreaks have been recorded in the past, although perhaps the best documented is that involving Mary Mellon, better known as Typhoid Mary. She worked as a cook in several domestic and institutional settings over approximately 20 years, on occasion changing her name and moving to another location to continue her occupation once identified as a typhoid carrier at a particular place of employment. It is believed that Mary was responsible for more than 1000 cases of typhoid over her working life, and at least several deaths.

Beside human transmission, a range of foods have been implicated in the transmission of *S. typhi*, including desiccated coconut, unpasteurized liquid whole egg, raw milk, a range of soft cheeses made from raw milk, ice cream, cold cooked red meats and poultry, shellfish and salad vegetables. While the contamination of such foods may arise through human or waterborne transmission, such as irrigation or washing of salad vegetables with contaminated water, hazardous populations may develop in the food itself if internal and external environmental factors permit. Post-process intrusion of contaminated cooling water into canned foods, especially canned meats, has led to several outbreaks. Molluscan shellfish, including clams, oysters and mussels, have all been implicated in outbreaks and pose a significant risk, as filter feeding in contaminated water leads to concentration of the organism in the tissues of the animal, and often these products, especially oysters, are consumed raw. Depuration may not completely eliminate the organism from molluscan shellfish, and this is of particular concern.

Direct consumption of contaminated water has also led to outbreaks of disease. Water may become contaminated through seepage of sewage into natural sources, especially in areas where typhoid is endemic or, as in an outbreak in England, through contamination of well water by a carrier. Biofouling of reverse osmosis membranes has also led to a waterborne outbreak.

The rate of human infection with *S. typhi* in developed countries is very low. Most cases are associated with visitors, immigrants or residents returning from developing countries, where the serovar, as well as water and food contamination, is prevalent.

There are no regulations imposed on the food industry at any level (local, state, national or international) that relate specifically to *S. typhi*, although at the regulatory level the bacterium is considered like any other *Salmonella* serovar.

Control

As the host range of *S. typhi* is confined to humans and transmission is more commonly water- than foodborne, control measures differ somewhat from those advocated for broad-host-range *Salmonella*. While good personal hygiene and food-handling practices are still necessary, emphasis must be placed on the identification of carriers, and their exclusion from food handling and production. In addition, potable water should be used in food production and processing, and foods susceptible to and suspected of waterborne contamination, such as seafood and especially molluscan shellfish, must be purified before consumption.

In areas where typhoid is endemic, vaccination of susceptible populations, including the young, old and immunocompromised, represents an effective control strategy. First-generation vaccines comprised heat or solvent-killed whole cells and were not particularly effective. Newer vaccine strategies include the oral administration of live crippled strains such as single or multiple *aro* (deficient in aromatic amino acid biosynthesis) and temperature-sensitive mutants, as well as subunit vaccines, based on outer membrane proteins (porins).

Detection and Isolation

The long incubation period associated with the onset of typhoid fever generally precludes isolation of *S. typhi* from an implicated food. Detection or isolation is more likely from clinical specimens, given the common association of the bacterium with chronically infected food handlers (carriers). Isolation using cultural approaches commonly employed in routine food microbiology is problematic, while the techniques used in clinical microbiology prove effective. Generally, many of the liquid enrichment and agar media suitable for the isolation of most strains of *Salmonella* contain concentrations of selective agents inhibitory to *S. typhi*. Additionally, the elevated temperatures (⩾42°C) advocated for incubation of certain enrichment broth and plate media are unsuitable for this serovar.

Isolation from microbiologically clean acute-phase clinical specimens, such as urine and blood, is straightforward; the otherwise sterile state of the sample obviates the need for enrichment. In contrast, specimens containing low levels of *S. typhi* and/or a high background microflora, including faeces, food or soiled water samples, require enrichment prior to plating. Broth media containing selenite are not very selective, but support strong growth of *S. typhi*, and appear to enhance the production of H_2S on plate media, while tetrathionate broths vary in their ability to support growth as the Mueller–Kauffman formulation inhibits the organism. Rappaport-based broths, including Rappaport, Rappaport–Vassiliadis (RV), and RV with soy peptone (RVS) are inhibitory.

Despite the development of an extensive array of agars for the selective isolation of *Salmonella*, the medium of choice for *S. typhi* continues to be bismuth sulphite agar (BSA). Many media contain an inorganic system for the detection of H_2S production, eminently suitable for most *Salmonella*, but not *S. typhi*, which only produces H_2S from organic forms of sulphur. BSA responds better than any other medium to the low levels of H_2S production typical of *S. typhi*. It is also highly selective, especially when freshly prepared, at which time many *Salmonella* other than *typhi*, as well as other enteric organisms, are strongly inhibited. Detection is also satisfactory on media such as desoxycholate citrate agar (DCA) and lysine mannitol glycerol agar (LMG), but not on xylose lysine desoxycholate (XLD) medium, *Salmonella–Shigella*, most agars containing the dye brilliant green, such as mannitol crystal violet brilliant green agar (MLCB), or newer chromogenic media such as Rambach agar. Cultural detection of *S. typhi* is further complicated if atypical strains, including those that ferment lactose or fail to decarboxylate lysine, are encountered.

Detection of *S. typhi*, especially in subclinical carrier-state infections, may be possible using serology, detecting serum antibodies to a range of antigens, including O, H, Vi, as well as outer membrane proteins. Certain other *Salmonella* serovars, such as Dublin, are able to express the Vi antigen, although such strains are uncommon, thus false-positive reactions are unlikely. The simplest form of assay involves slide agglutination (the Widal test), although enzyme-linked immunosorbent assays for most of these antigens have been developed.

Polymerase chain reaction (PCR) assays for the detection of *S. typhi* have been developed. One assay, based on spacer region DNA between the 5S and 23S ribosomal RNA genes and yielding a 300 bp product, can detect small populations of the bacterium in foods. Further assays have been based on cell-surface antigens, including the *flicC* gene (which encodes the H-d flagellin subunit protein) and the *rfbS* gene (involved in LPS biosynthesis), while nested PCR protocols have been based on the H-d flagellin and the ViaB gene locus. In the case of the first two assays, specificity is questionable, given the distribution of these genes and their products among other *Salmonella* serovars, although there is variation in a specific region of the H-d*fliC* gene, which distinguishes it from the same gene in other serotypes expressing H-d. Additionally, the first assay may not detect the H variants H-j and H-z66.

See also: ***Salmonella***: Introduction; Detection by Classical Cultural Techniques; Detection by Latex Agglutination Techniques; Detection by Enzyme Immunoassays; Detection by Colorimetric DNA Hybridisation; Detection by Immunomagnetic Particle-based Assays. **Waterborne Parasites**: Detection by Classic and Modern Techniques. ***Yersinia***: *Yersinia enterocolitica*.

Further Reading

D'Aoust J-Y (1997) *Salmonella* species. In: Doyle MP, Beuchat LR, Montville TJ (eds) *Food Microbiology Fundamentals and Frontiers*. P. 129. Washington: ASM Press.

Franco A, Gonzalez C, Levine OS et al (1992) Further consideration of the clonal nature of *Salmonella typhi*: evaluation of molecular and clinical characteristics of strains from Indonesia and Peru. *Journal of Clinical Microbiology* 30: 2187–2190.

Liu SL and Sanderson KE (1995) Genomic cleavage map of *Salmonella typhi* Ty2. *Journal of Bacteriology* 177: 5099–5107.

Virlogeux I, Waxin H, Ecobichon C and Popoff MY (1995) Role of the *viaB* locus in synthesis, transport and expression of *Salmonella typhi* Vi antigen. *Microbiology* 141: 3039–3047.

Zhu Q, Lim CK and Chan YN (1996) Detection of *Salmonella typhi* by polymerase chain reaction. *Journal of Applied Bacteriology* 80: 244–251.

Detection by Classical Cultural Techniques

R Miguel Amaguaña and **Wallace H Andrews**, US Food and Drug Administration, Washington, DC, USA

Introduction

Salmonellosis is one of the leading causes of food-borne bacterial illness in the US and in other developed countries. Control measures for several of the food-borne pathogens have been implemented by public health agencies and seem to be successful in decreasing the incidence of food-borne illnesses caused by *Clostridium perfringens*, *Bacillus cereus*, *Vibrio parahaemolyticus* and *Staphylococcus aureus*. Control measures have also been implemented to reduce the incidence of *Salmonella* outbreaks, but have not met the same success as control measures for other pathogens. The incidences of food-borne salmonellosis, according to the Centers for Disease Control and Prevention, appear to be ascending.

The increase in salmonellosis in the US may be related to four factors:

1. Increase in the number of antimicrobial-resistant *Salmonella* spp. isolates.
2. Increase in individuals with immunodeficiency virus infection who are susceptible to *Salmonella* spp.
3. Increase in egg-associated *S. enteritidis* contamination due to increase in contaminated laying hens.
4. Increase in centralized food production facilities that may lead to large and widespread outbreaks.

The first factor merits additional comment. The use of antimicrobials in humans and animals is often followed by the appearance of resistant micro-organisms. This leads to treatment failure and the need for newer and costlier antimicrobials. Molecular biology techniques used for subtyping bacteria or identifying specific genes have provided scientists with a better understanding of the significance of anti-microbial-resistant *Salmonella*. The Centers for Disease Control and Prevention investigated 52 *Salmonella* outbreaks over a 12-year period. Seventeen of the 52 outbreaks involved antimicrobial-resistant organisms and affected 312 persons, 13 of whom (4.2%) died from salmonellosis. In contrast, the 19 outbreaks caused by antimicrobial-sensitive *Salmonella* resulted in only 4 (0.2%) fatalities in 1912 ill persons. In the 16 outbreaks caused by *Salmonella* of unspecified antimicrobial resistance, 4 (0.3%) of 1429 ill persons died.

All *Salmonella* spp. are Gram-negative, non-sporulating facultative anaerobic motile rods. The cells can survive under frozen and dried states for long periods of time. Moreover, they are capable of multiplying in many foods without affecting the general appearance of the foods.

Methods for recovering *Salmonella* spp. from foods involve the pre-enrichment of the food sample in a nutrient broth, followed by enrichment in selective enrichment media, streaking on selective agar media, and biochemical and serological confirmation. These five steps are used in the *Salmonella* method recommended by the US Food and Drug Administration's (FDA's) *Bacteriological Analytical Manual* (*BAM*), US Department of Agriculture (USDA), Association of Official Analytical Chemists International (AOACI), International Organization of Standardization, (ISO), International Dairy Federation and the American Public Health Association. Although the methods for detecting *Salmonella* spp. are basically the same for the aforementioned organizations, they differ in media used in the various steps. The media used also depend on the type of food being examined.

The *Salmonella* culture method is rather lengthy, requiring 4–5 days for a negative result. Rapid test kit methods have been developed to shorten the time needed to obtain a positive or negative result for *Salmonella* spp. The primary use of these rapid test kits is to screen out negative samples. Any negative results obtained with the test kit are considered definitive. Positive test kit results are always considered presumptive and must be confirmed by an approved culture method.

Pre-enrichment

Pre-enrichment, the first step in the culture method, is performed with a nonselective broth medium which provides for the non-inhibited growth of indigenous bacterial flora and the resuscitation and proliferation of stressed or injured *Salmonella* spp. cells to detectable levels. Cells are injured when exposed to adverse conditions during food processing, such as chilling, freezing or drying. This injury increases the sensitivity of the cell to selective agents in the various media. If a resuscitation step is not included in the method, then stressed cells that have not fully repaired cell injury may be missed.

Lactose broth and buffered peptone water are among the most widely used pre-enrichment media. Other pre-enrichment media for specific foods include distilled water with 1% brilliant green for instant non-fat dry milk, nutrient broth for frosting mixes and Trypticase soy broth for spices (**Table 1**). Studies have shown that the incubation time and temperature

Table 1 Analysis of foods by methods of the *Bacteriological Analytical Manual* and the Association of Official Analytical Chemists International

Food	*Pre-enrichment*[a]	*Preparation*[b]
Dried egg yolk, egg white and whole egg; liquid milk; powdered bread and pastry mixes; infant formulas; oral or tube feedings containing egg; coconut[c]; food dyes with pH > 6.0 (10% aqueous solution); gelatine[d]; guar gum[e]	Lactose broth	Mix
Pasta (noodles, macaroni, spaghetti); egg rolls; cheese; dough; prepared salads; fresh, frozen or dried fruits and vegetables; nut meats; crustaceans; fish; meats[c], meat substitutes[c], meat by-products[c], animal substances[c], glandular products[c], meals (fish, meat and bone)[c]; casein	Lactose broth	Blend
Shell eggs, liquid whole eggs (homogenized) and hard-boiled eggs	Trypticase (Tryptic) soy broth	Mix
Food dyes with pH < 6.0 (10% aqueous solution)	None[f]	Mix (25 g/225 ml TBG)
Non-fat dry milk (instant and non-instant)	Brilliant green water (add 2 ml 1% brilliant green dye solution per 1000 ml distilled water)	Soak
Dry whole milk	Distilled water; add 0.45 ml 1% brilliant green solution	Mix
Dried active[g] and inactive yeast; black pepper, white pepper, celery seed and flakes, chilli powder, cumin, paprika, parsley flakes, rosemary, sesame seed, thyme, and vegetable flakes	Trypticase (Tryptic) soy broth	Mix
Onion flakes and powder, garlic flakes and powder	Trypticase (Tryptic) soy broth containing 0.5% K_2SO_3 (final concentration)	Mix
Allspice, cinnamon, oregano	Trypticase (Tryptic) soy broth	Mix, using 1 : 100 sample/broth ratio
Clove	Trypticase (Tryptic) soy broth	Mix, using 1 : 1000 sample/broth ratio
Leafy condiments	Trypticase (tryptic) soy broth	Mix, using > 1 : 10 sample/broth ratio
Sweets (candy) and candy coating; chocolate	Reconstituted non-fat dry milk[h], add 0.45 ml 1% brilliant green solution	Blend
Frosting and topping mixes	Nutrient broth	Mix
Frog legs	Lactose broth	Immerse/rinse leg pairs (individual leg if ⩾25 g) and examine rinsings

[a] Unless specified otherwise, 25 g samples are pre-enriched in 225 ml of indicated medium.
[b] Unless specified otherwise, 1 ml volumes of incubated pre-enriched cultures of foods with a low microbial load are subcultured to 10 ml volumes of selenite cystine or tetrathionate brilliant green (TBG) broth prepared by adding 10 ml of 0.1% brilliant green dye solution and 20 ml of iodine potassium iodide solution to 1 l of tetrathionate broth base. For foods with a high microbial load, 0.1 ml and 1.0 ml volumes of pre-enriched cultures are subcultured to 10 ml volumes of Rappaport–Vassiliadis medium and TBG broth, respectively.
[c] Add up to 2.25 ml of Tergitol 7 or Trition X-100 to initiate foaming.
[d] Add 5 ml of 5% papain solution.
[e] Add 2.25 ml of 1% cellulase solution.
[f] Pre-enrichment in lactose broth may provide improved recovery.
[g] One millilitre volumes of pre-enrichment cultures of dried active yeast are transferred to 10 ml volumes of lauryl tryptose broth and TBG broth.
[h] Reconstituted non-fat dry milk is prepared by adding 100 g of non-fat dry milk to 1 l distilled water.

conditions applied at the pre-enrichment stage of the analysis are more critical than choice of nonselective pre-enrichment medium. Pre-enrichment such as lactose broth, for example, does not provide specific enrichment for *Salmonella* spp. This medium, provides a self-limiting environment for non-*Salmonella* spp. organisms caused by lowering the pH when microflora utilize the lactose.

Surfactants, such as Tergitol Anionic 7, can be added to a pre-enrichment broth medium to aid in recovering *Salmonella* spp. organisms from food with a high fat content. Surfactants disperse the lipid particles containing entrapped *Salmonella* spp. organisms. Other additives are sometimes incorporated into the pre-enrichment medium to neutralize toxic agents in foods, to reduce the viscosity of the food or to increase the selectivity of the media.

The recommended food/pre-enrichment broth ratio is most commonly 1 : 9, i.e. 25 g food to 225 ml pre-enrichment broth. The method of preparing the food pre-enrichment mixture varies, depending on food type (Table 1). Foods can be blended, soaked, swirled or stirred depending on the consistency of the food sample.

The incubation temperatures widely recommended for the pre-enrichment mixture are 35–37°C. Incubation time is normally 18–24 h. After incubation under these conditions, a portion of the pre-enrichment is subcultured to selective enrichment media.

Selective Enrichment

The second step in the culture method is selective enrichment. Selective enrichment broths are used to increase the population of *Salmonella* spp. organisms, while inhibiting the growth of other organisms in the food.

Three of the more commonly used selective enrichments are Rappaport–Vassiliadis (RV) medium, selenite cystine (SC) broth and tetrahionate (TT) broth. The amount of pre-enrichment subcultured into a selective enrichment varies depending on the medium used. Typically, a volume of 0.1 ml of the incubated pre-enrichment is subcultured to 10 ml of RV medium and 1.0 ml is subcultured to 10 ml of SC and TT broths. Studies have shown that the efficiency of subculturing smaller (0.1 ml for SC and TT broths) or larger (10 ml) volumes does not increase or decrease the method sensitivity when the pre-enrichment is incubated for 18–24 h.

Each of these selective enrichment media contains different selective agents. RV broth contains malachite green and magnesium chloride at a slightly reduced pH; SC broth contains sodium acid selenite; and TT broth contains brilliant green dye and bile. As no one selective enrichment is capable of recovering all of the 2400 recognized serovars, the various culture methods typically recommend the use of two selective enrichment broths.

The incubation temperature and period are critical to achieve optimal performance of the respective media. For foods with a low microbial load, SC broth is normally incubated at 35°C, TT broth at 35°C, and RV medium at 42°C. For foods with a high microbial load, the use of elevated temperatures enhances the recovery of *Salmonella* spp., increasing the inhibition of competitive microflora. Thus, TT broth at 43°C and RV medium at 42°C are recommended for foods such as shrimp, chicken and pork sausage. The authors are attempting to eliminate, by collaborative study, the use of SC broth for foods with a low microbial load, since SC broth has been found to be toxigenic. Moreover, SC broth, classified as a hazardous waste by the US Environmental Protection Agency, contains toxic levels of selenium, which increases the cost of disposal.

Pre-enrichments and selective enrichments can be refrigerated at 4°C for several days without affecting the recovery of *Salmonella* spp. in a variety of low-moisture foods. This gives the investigator flexibility as to when to begin the analysis for *Salmonella* spp. contamination.

The AOACI/*BAM* and USDA detection methods for *Salmonella* spp. in foods suggest that some raw or highly contaminated foods and poultry rinses can be directly enriched in selective enrichment media. The reason for not pre-enriching the food is that the *Salmonella* spp. organisms present in non-processed raw foods and rinses are in a relatively sound physiological condition and could withstand the rigor of a direct selective enrichment that would have adverse effects on competing microflora. Scientists have demonstrated that direct selective enrichment of highly contaminated foods is not significantly more productive for the recovery of *Salmonella* spp. than pre-enrichment of the food and may even lead to an increase in false-negative results.

Selective Plating

In the third step of the culture method, the selective enrichment broths are streaked on to selective solid media. The selective plating agars used to isolate and differentiate *Salmonella* spp. contain ingredients that restrict the growth of competing non-*Salmonella* spp. Among the more commonly used agars are brilliant green (BG) agar, xylose lysine Tergitol-4 (XLT-4), bismuth sulphite (BS) agar, Hektoen enteric (HE) agar and xylose lysine desoxycholate (XLD) agar. Since each agar uses one or more different selective agents and none is ideal for all types of *Salmonella* spp., it is recommended that two or more agars should be used in combination to ensure that atypical strains, such as lactose-utilizing strains, will not be missed.

These selective agars contain a source of carbohydrate, a protein and an indicator dye. They may also contain a hydrogen sulphide indicator and one or more inorganic salts. BG agar uses BG and phenol red dyes as carbohydrate utilization indicators. XLT-4 agar uses Tergitol-4 as a surfactant, and ferric ammonium citrate and sodium thiosulphate as H_2S indicators. BS agar uses bismuth sulphite and BG as inhibitory agents and ferrous sulphate as an indicator of H_2S production. HE agar uses bile salts for selectivity, bromothymol blue and acid fuchsin as carbohydrate utilization indicators, and ferric iron as an H_2S indicator. XLD agar uses sodium desoxycholate for selectivity, sodium thiosulphate and ferric ammonium citrate as H_2S indicators and phenol red as a carbohydrate utilization indicator.

Because the selectivity of the agars is critical, procedures for preparation and storage of the prepared plates must be adhered to closely. Freshly made BS agar plates inhibit growth of several *Salmonella* spp. serovars. These inhibitory properties decrease once the plates are stored for a few days, but selectivity also decreases if the plates are stored for an extended period of time. Freshly poured plates need a 48-h incubation period to yield *Salmonella* recoveries similar to that of the other selective plates incubated for 24 h. The other selective plates are stable once prepared, but some variability is possible in the media from lot to lot. Therefore, it is suggested that a single lot of the specific media be used for each examination. Also, the instructions for the preparation of the agar plates must be followed. If there are no specific instructions for media preparation in the method, then the media should be prepared as directed by the manufacturer.

Although other selective media, such as, EF-18 agar, Rambach agar and XLT-4 agar, have met moderate success in some methods, they have not been more effective than BG agar, BS agar, HE agar or the XLD agar. EF-18 agar was developed for use in the AOACI-approved hydrophobic grid membrane filter method for rapid identification of *Salmonella* spp. Rambach agar uses a unique phenotypic characteristic, the formation of acid from propylene glycol, and has mainly been used in clinical microbiology. The USDA has specifically designed XLT-4 agar to isolate *Salmonella* spp. from poultry in the presence of competitive organisms such as *Proteus*, *Pseudomonas* and *Providencia* spp.

The selective plating media are inoculated by streaking a loopful of the enrichment broth on to the surface of the prepared agar plates to obtain isolated colonies. The recommended incubation temperature is 35–37°C. The plates are usually incubated for 18–24 h. The appearance of typical *Salmonella* spp. colonies on the incubated plates varies with the medium. For example, on the BS agar plate, typical *Salmonella* spp. colonies may appear as black colonies, with or without a metallic sheen, and the surrounding medium changes from brown to black, producing a halo effect. On the HE agar plate, typical colonies appear as blue-green to blue colonies, with or without black centres, and may have large glossy black centres due to H_2S production. On the XLD agar plate, typical colonies appear as pink colonies, with or without black centres, and may have large glossy black centres. The BS agar plates should be recorded at 24 h but reincubated for an additional 24 h before making a final determination on the presence or absence of *Salmonella* spp.

Biochemical Testing

In the fourth step of the culture method, presumptive positive *Salmonella* spp. colonies from selective plating media are screened by biochemical testing. Suspect colonies from the selective agar plates are transferred to tubes of differential agar for the preliminary biochemical characterization of the isolates. A differential medium is one that provides certain biochemical information about the inoculated culture by one or more colour changes in that medium. Two or three suspicious presumptive positive colonies from each selective agar plate are selected for the primary differentiation procedure.

Two of the more commonly used differential agar tube media are triple sugar iron (TSI) agar and lysine iron (LI) agar. TSI agar measures the production of hydrogen sulphide and the utilization of glucose, lactose and sucrose. Sugar utilization changes the medium yellow and hydrogen sulphide production turns it black. The appearance of *Salmonella* spp. on TSI agar slant appears as red (alkaline) slant and yellow (acid) butt, with or without the blackening of the agar (H_2S production). LI agar demonstrates the production of hydrogen sulphide and the decarboxylation of lysine. The appearance of *Salmonella* spp. on LI agar slant appears as purple (alkaline) butt. Most *Salmonella* spp. cultures produce H_2S in LI agar. The TSI and LI agars are generally used in conjunction with each other. This combination is useful because they differentiate the cultures that require final confirmation from those that can be discarded. This test reduces the amount of cultures submitted to confirmatory serological testing. It should be stressed that a TSI culture should not be excluded if it appears to be non-*Salmonella* spp. when the accompanying LI agar is typical for *Salmonella* spp., because some lactose- and/or sucrose-positive *Salmonella* spp. produce atypical reactions (acid slant and acid butt with or without blackening). If the cultures appear to be contaminated on the agar tube media, they should be restreaked on to an appropriate selective plating agar and a small amount of a single colony exhibiting typical *Salmonella* spp. characteristics should then be transferred to the differential medium for retesting.

Many other nonselective differential media have been developed, including dulcitol lactose iron agar, malonate dulcitol lysine agar, DMS agar, and selective Padron–Dockstader agar. Although many biochemical tests can be performed to characterize *Salmonella* spp., urease, lysine decarboxylase, dulcitol broth, KCN (potassium cyanide) broth, malonate broth and indole reactions are usually sufficient to obtain presumptive identification of *Salmonella* spp. The results of these tests provide a relatively reliable indication of *Salmonella* spp. contamination in the food, but final confirmation is performed serologically because some strains of *Salmonella* spp. give atypical biochemical results.

Serological Confirmation

In the final step of the culture method, *Salmonella* spp. strains are characterized serologically by determining their antigenic composition. The antigens are classified as somatic (O) and flagellar (H). The somatic (O) antigens are determined by performing a slide agglutination test with somatic (O) antisera and growth from an agar culture of the food. The flagellar (H) antigens are determined by performing an agglutination test with the flagellar (H) antisera and a formalized saline infusion broth culture in a test tube. This test can also be performed using the Spicer–Edwards flagellar (H) antisera test. The test is positive if a specific somatic group (O) and Spicer–Edwards reactions are obtained in one or more TSI cultures.

Many culture methods are acceptable throughout the world for the detection of *Salmonella* spp. from foods; the most widely recognized and accepted of these methods are the ISO and AOACI/FDA methods. Although these culture methods are considered standard methods, advances in rapid test kit methods have proven to be a suitable alternative to the classical culture methods and have improved the microbiological testing of foods. Considering the recent advancement in rapid test kits, methods providing same-day test results seem to be within reach.

See also: **Bacteriophage-based Techniques for Detection of Food-borne Pathogens**. **Biochemical and Modern Identification Techniques**: Food-poisoning Organisms. **Biosensors**: Scope in Microbiological Analysis. **Direct (and Indirect) Conductimetric/Impedimetric Techniques**: Food-borne Pathogens. **Enrichment Serology**: An Enhanced Cultural Technique for Detection of Foodborne Pathogens. **Food Poisoning Outbreaks**. **Hydrophobic Grid Membrane Filter Techniques (HGMF)**. **Immunomagnetic Particle-based Techniques**: Overview. **Molecular Biology – in Microbiological Analysis**. **National Legislation, Guidelines & Standards Governing Microbiology**: European Union; Japan. **Nucleic Acid-based Assays**: Overview. **PCR-based Commercial Tests for Pathogens**. **Petrifilm – An Enhanced Cultural Technique**. **Polymer Technologies for Control of Bacterial Adhesion**. **Reference Materials**. ***Salmonella***: Detection by Latex Agglutination Techniques; Detection by Enzyme Immunoassays; Detection by Colorimetric DNA Hybridization; Detection by Immunomagnetic Particle-based Assays. **Sampling Regimes & Statistical Evaluation of Microbiological Results**.

Further Reading

Adams MR and Moss MO (1995) *Food Microbiology*. Cambridge, UK: Royal Society of Chemistry.

Association of Official Analytical Chemists International (1995) *Official Methods of Analysis*, 16th edn. Gaithersburg, MD: AOAC.

US Food and Drug Administration (1995) *Bacteriological Analytical Manual*, 8th edn. Gaithersburg MD: AOAC.

Detection by Latex Agglutination Techniques

Julian M Cox, Department of Food Science and Technology, University of New South Wales, Sydney, Australia

This article describes the application of latex agglutination (LA) reagents to the detection of *Salmonella*, both for the confirmation of isolates and as a screening technology for the likely presence or absence of *Salmonella* during the early stages of culture.

LA Reagents for *Salmonella*

A variety of reagents are available for the performance of LA of *Salmonella* (**Tables 1 and 2**). While some are targeted at the genus as a whole, others are designed to detect specific serogroups or serovars.

Genus-level reagents incorporate a mixture of antibodies raised against specific antigenic epitopes of various flagellar subunits (H antigens), the range considered sufficient to react with the most common serovars. For example, the Serobact *Salmonella* reagent incorporates antisera to the following flagellar determinants: a; b; c; d; eh; enx; enz$_{15}$; fg; gp; gpu; gms; gq; gst; i; k; lv; lw; mt; r; y; z; z$_4$; z$_6$; z$_{10}$; z$_{23}$; z$_{24}$; z$_{29}$; z$_{38}$; 1,2; 1,5; 1,6; and 1,7. Using flagella as a target, non-motile strains are not detected. However, many strains that have lost motility as a consequence of environmental stress usually regain motility during subculture. Even if a strain fails to regain motility, or is inherently non-motile, it may still agglutinate weakly as, according to one manufacturer, there is sufficient residual serological activity against somatic antigens.

In addition, the mixture of antibodies used in fla-

Table 1 Commercial latex agglutination reagents available for *Salmonella*

Reagent	*Manufacturer/supplier*	*Target antigens*	Salmonella *detected*
Microscreen super latex slide agglutination	Mercia Diagnostics	Flagella	Genus *Salmonella*
Serobact *Salmonella*	Medvet Science	Flagella (somatic)	Genus *Salmonella*
Salmonella Verify	Vicam	Flagella?	Genus *Salmonella*
SE Verify	Vicam	SEF14 fimbriae?	Serovar *Enteritidis*
Wellcolex Colour *Salmonella*	Murex Biotech	Somatic, capsular	Serogroups A, B, C, D, E/G, Vi

Table 2 Manufacturers of latex agglutination reagents for the detection/confirmation of *Salmonella*

Company	*Address*
Medvet Science	PO Box 79, Adelaide, South Australia 5000, Australia
Mercia Diagnostics	Surrey, UK
Murex Diagnostics	Central Road, Temple Hill, Dartford, Kent, DA1 5LR, UK
Vicam	Watertown MA 02472, USA

gellin-based reagents only detects the most common antigens; rare antigenic types will not be detected. Such rare antigens usually occur in only one of the H phases expressed by *Salmonella*; the likelihood of a rare antigen being present in both phases is very low. As a culture of a given *Salmonella* strain will contain a mixture of cells expressing either phase 1 or phase 2 flagella, the likelihood of detection is increased.

The Wellcolex reagents target the somatic antigens of the most common clinical *Salmonella* serogroups (A–E and G), as well as the capsular or Vi antigen of *Salmonella typhi* (and occasionally serovar *dublin*). Two reagents are supplied, each of which contains a mixture of three latex types, red, blue or green in colour, each latex-coated with a different antibody. The interpretation of reactions with these reagents is presented in **Table 3**.

The Vicam company produces two reagents, the first reacting with members of the genus *Salmonella*, the second reacting almost exclusively with *Salmonella enteritidis*. The second LA reagent relies on the use of a monoclonal antibody. As it cross-reacts with strains of serovars Blegdam, Dublin and Moscow, it is likely that the antibody recognizes the fimbrial structure SEF14.

LA reagents can be employed throughout the cultural procedure for the detection of *Salmonella*, as a confirmation test for isolates at the end of culture, or as a screening technology at earlier stages of culture (**Fig. 1**).

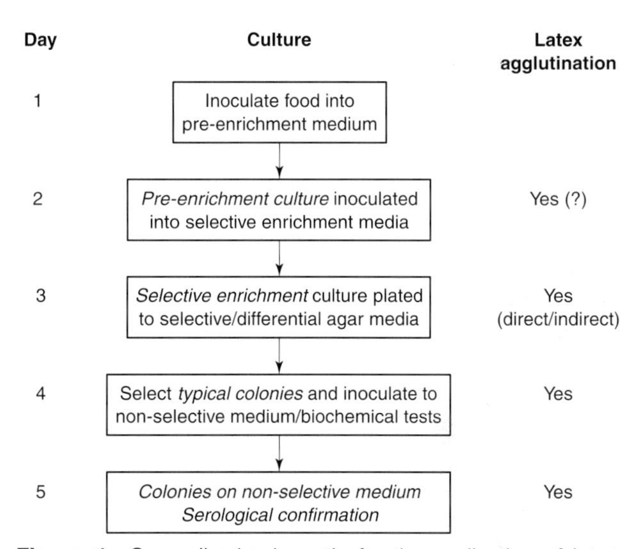

Figure 1 Generalized schematic for the application of latex agglutination during analysis of food samples for *Salmonella*. The sample used in the latex agglutination assay at each stage of culture appears in italic.

LA as a Confirmation Technology

Latex agglutination reagents, especially those designed to react at the genus level, have traditionally been employed as part of a protocol for the confirmation as *Salmonella* of typical colonies selected from selective-differential agar media. In practice, a part or all of a colony, depending on size, is removed from a plate and emulsified in a suspending medium, usually physiological (0.85%) saline. This is performed on a solid support such as coated paper, and the emulsion prepared with the aid of a toothpick or inoculating loop, both of which are supplied with the reagents. A drop of the LA reagent is added, and the support is rocked to mix the reagents together. If the isolate is a strain of *Salmonella*, agglutination will be observed as the development of discrete clumps or flocs, usually within one to several minutes. Controls must be employed to ensure the validity of the test. A reagent control is prepared using a mixture of one drop each of the LA reagent and the medium used to suspend the culture. If agglutination occurs, it may be due to contamination of either the suspending medium or the LA reagent, or deterioration of the latter due, for example, to freezing. A culture control is prepared by emulsifying growth in the suspending medium and observing for clumping of the culture alone; this phenomenon is referred to as autoagglutination. This is important if serologically rough strains are encountered, as they may yield false-

Table 3 Interpretation of reactions with the Wellcolex Colour *Salmonella* reagents

Colour of agglutinate	*Interpretation*	
	Reagent 1	*Reagent 2*
Green/olive	Serogroup D	Serogroup A
Blue	Serogroup C	Serogroup E or G
Red	Serogroup B	Vi antigen
Turquoise	Serogroups C and D	Serogroup A + E or G
Orange	Serogroups B and D	Serogroup A and Vi
Purple	Serogroups B and C	Serogroup E or G + Vi

positive results. A positive control is performed using a plate or broth culture of an authentic *Salmonella* strain. It is crucial that fresh cultures are used, as old cultures may yield false-negative results.

The Wellcolex reagents can be used to determine if an isolate belongs to any of serogroups A–E or G, and whether the isolate expresses the Vi capsular antigen. Rather than one, two suspensions of the isolate are prepared, and a drop of each of the latex reagents is added to one suspension. The combination of reactions (Table 3) determines the serogroup. Lack of reaction suggests that the isolate is either a less common type of *Salmonella*, or a non-*Salmonella*.

A suitable reaction may be observed using growth directly from some selective-differential media, such as xylose lysine desoxycholate (XLD), desoxycholate citrate, bismuth sulphite, *Salmonella–Shigella*, brilliant green or Hektoen agars. It is better practice, however, to subculture putative *Salmonella* isolates to a nonselective nutrient medium, and then perform LA, although this delays a definitive result by a day.

LA as a Screening Technology

While the primary use of LA reagents is in the confirmation of colonies at the end of a cultural procedure, it is possible to apply them at earlier stages to enrichment cultures as a tool to screen for the likely presence or absence of *Salmonella* (Fig. 1). In this way, the time, labour and materials associated with further processing of large numbers of *Salmonella*-negative cultures are reduced.

To minimize the time required to detect *Salmonella* in a food, LA can be applied after pre-enrichment, although use at this stage of culture is likely to prove problematic. Interference by sample matrix may be experienced, either masking a positive reaction or, for example with food powders, producing a false-positive reaction. In the latter case, performance of a sample control is critical. In addition, the dynamics of microbial growth may influence the efficacy of the LA test. *Salmonella* are often present in food samples in low numbers and in poor physiological condition. While such organisms recover during pre-enrichment, they may not reach a population sufficient for detection by LA. Nonselective pre-enrichment permits outgrowth of non-*Salmonella*, which may restrict the development of *Salmonella* to an undetectable level. Non-*Salmonella*, including other members of the Enterobacteriaceae or even *Pseudomonas* spp., may produce a false-positive reaction.

The stage of culture at which LA is best employed for screening is after selective enrichment, a cultural step that permits the outgrowth of *Salmonella* recovered in pre-enrichment, while suppressing growth of non-*Salmonella*. With respect to LA this step, in principle, will increase the detection of *Salmonella* and reduce the rate of false-positive reactions.

The balance between growth of *Salmonella* and suppression of other organisms will be influenced by the selective enrichment media used, of which there is an extensive range. It is recommended that two selective enrichment media are used, as in the conventional cultural procedure. Any single enrichment medium can prove inhibitory to certain strains of *Salmonella*, and vary in degree of selectivity.

In addition to their influence on microbial growth, enrichment media may interfere with the antigen–antibody reaction. For example, the high concentration of magnesium chloride and low pH of the various Rappaport–Vassiliadis (RV) broths appear to be factors responsible for poor reactivity with antibody-based assays in general. While selenite-based broths do not interfere with the antibody–antigen reaction, the lower selectivity of selenite compared to RV broths increases the likelihood of false-positive reactions. Selenite broths must also be used freshly prepared, as the presence of red precipitate may give rise to false-positive reactions. This is a particular problem with food samples high in solids, which may be carried over from pre-enrichment, as a red precipitate may form during incubation, even in freshly prepared medium.

The problem of inhibition or other interference by selective enrichment media can be overcome through the use of a short nonselective enrichment after selective enrichment. The nonselective medium generally recommended for this purpose is M-broth. This rich medium supports rapid growth and stimulates production of flagella, facilitating detection based on either O or H antigens. An example of a protocol accepted by regulatory authorities, which relies on post-enrichment of a selective enrichment medium, is presented in **Figure 2**.

While the time saved through use of LA at the selective enrichment stage is only 1 day, there are considerable cost savings in terms of labour and

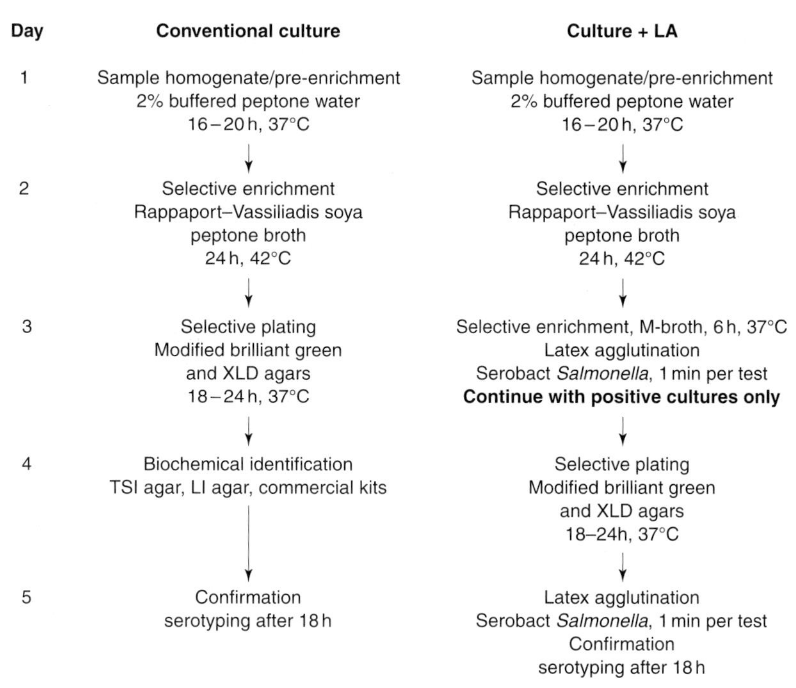

Figure 2 Comparison of conventional and latex agglutination (LA)-modified cultural methods approved by Ministry of Agriculture and Fisheries, New Zealand for detection of *Salmonella* in export meat products. XLD, xylose lysine desoxycholate; TSI, triple sugar iron; LI, lysine iron.

materials through screening and removal of negative enrichment cultures prior to plating.

O antigen-specific reagents, such as Wellcolex *Salmonella*, can potentially be used to screen enrichment cultures of foods for the presence of specific *Salmonella* serogroups. This approach has proven successful in the medical microbiology laboratory to screen enrichment cultures of samples such as stools. While optimal efficacy relies on the presence of only a single serogroup, it is possible to detect multiple serogroups in a mixed culture, as discrete clumps of each reacting serogroup-specific latex can be observed in suspension. Successful application of O-specific reagents to enrichment cultures of foods may be complicated in some matrices by the presence of multiple serotypes. Additionally, cross-reactivity with non-*Salmonella* may yield false-positive reactions, which have been observed during analysis of enrichment cultures of clinical specimens. This may be due to cross-reaction with somatic antigen epitopes shared with other Enteroacteriaceae, such as *Citrobacter freundii*, or the presence of Vi antigen on non-*Salmonella*. In the latter case, boiling can be used to destroy the Vi antigen, eliminating agglutination.

Advantages and Limitations of LA Technology

LA offers significant advantages over other technologies. As a confirmation technique, the use of latex particles greatly facilitates the visual observation of the agglutination reaction when compared to traditional slide agglutination using antisera on glass slides. As a screening technology, LA is simple and cheap. Other screening technologies such as enzyme-linked immunosorbent assay (ELISA), at least in manual format, require multiple manipulations. As the number of steps in the performance of a test increases, so too does the potential for error. Other technologies are still comparatively expensive, even in manual format. Technical errors can largely be overcome by automation, but cost increases markedly.

A potential and significant limitation on the use of LA for screening of enrichment broth cultures is lack of sensitivity. Using suspensions of known populations of *Salmonella*, a clear positive LA reaction requires 10^7–10^8 cfu ml^{-1}. This is 1–3 orders of magnitude lower in sensitivity than that of comparable technology, such as ELISA. This limitation may be overcome by using a centrifugation step to concentrate *Salmonella*.

Fewer positive samples have been detected in enrichment cultures of food samples when comparing LA with standard cultural procedures, as an insufficient population of *Salmonella* had developed in enrichment media; this problem can be considered a fault in the enrichment protocol as much as the LA test *per se*.

Use with other Technologies

LA reagents may also be used as a confirmation tool after pre-enrichment and enrichment in modified semi-solid Rappaport–Vassiliadis (MSRV) medium. After incubation, 1–3 drops of pre-enrichment is inoculated into the centre of a plate of MSRV. After incubation at 42°C for 24 h, a small trough is cut with a small spatula or scalpel in the MSRV medium within the outer zone of growth created by motile organisms. The liquid that accumulates in the trough is collected with a pipette and used directly to perform LA, rather than the recommended subculture to selective plate media or agglutination with polyvalent antisera.

Agglutination represents a rapid, specific and sensitive test for the confirmation of presumptive positive broth cultures following impedance analysis. Selective-differential broth media, many based on agar media for the detection of *Salmonella*, are inoculated with pre-enrichment culture. Analysis by impedance monitoring indicates which samples may contain *Salmonella*. Contents of presumptive positive impedance vessels are used directly in the LA assay. LA-positive cultures are then subcultured to plate media, and identified biochemically and serologically.

See also: **Bacteriophage-based Techniques for Detection of Food-borne Pathogens**. **Biochemical and Modern Identification Techniques**: Introduction; Food-poisoning Organisms. **Biosensors**: Scope in Microbiological Analysis. **Enrichment Serology**: An Enhanced Cultural Technique for Detection of Foodborne Pathogens. **Enzyme Immunoassays**: Overview. **Flow Cytometry**. **Food Poisoning Outbreaks**. **Hydrophobic Grid Membrane Filter Techniques (HGMF)**. **Immunomagnetic Particle-based Techniques**: Overview. **National Legislation, Guidelines & Standards Governing Microbiology**: European Union; Japan. **Nucleic Acid-based Assays**: Overview. **PCR-based Commercial Tests for Pathogens**. **Petrifilm – An Enhanced Cultural Technique**. **Rapid Methods for Food Hygiene Inspection**. **Reference Materials**. ***Salmonella***: Detection by Classical Cultural Techniques; Detection by Enzyme Immunoassays; Detection by Colorimetric DNA Hybridisation; Detection by Immunomagnetic Particle-based Assays. **Sampling Regimes & Statistical Evaluation of Microbiological Results**.

Further Reading

D'Aoust JY, Sewell AM and Greco P (1991) Commercial latex agglutination kits for the detection of foodborne *Salmonella*. *Journal of Food Protection* 54: 725–730.

Hansen W and Freney J (1993) Comparative evaluation of a latex agglutination test for the detection and presumptive serogroup identification of *Salmonella* spp. *Journal of Microbiological Methods* 17: 227–232.

Detection by Enzyme Immunoassays

Pradip D Patel, Science and Technology Group, Leatherhead Food Research Association, Leatherhead, Surrey, UK

Prelude

This article describes the application of a range of commercial enzyme immunoassays (EIAs), in particular enzyme linked immunosorbent assays (ELISAs), for the analysis of *Salmonella*, with a detailed example of a typical officially approved microtitre plate ELISA. The details concerning principles of EIAs, classical cultural methodology, non-enzyme based immunoassays and other techniques for *Salmonella* are covered in depth elsewhere.

The commercial markets for EIA techniques are changing continually to address the changing needs of the industry and it is expected that new technologies will undoubtedly be developed to replace or upgrade the current systems in the future. In this context, although a commercial EIA product may cease to exist, the basic principles and methodology can still be used to develop 'in-house' EIAs, if required. For most of the EIAs, the lower level of sensitivity for *Salmonella* is normally 10^4–10^6 cfu ml^{-1}, depending on the serotype. In order to detect low levels of salmonellae in foodstuffs (e.g. 1 cfu 25 g^{-1}) prior cultural enrichment of test samples is required to allow the *Salmonella* to resuscitate (if present in processed foods) and multiply to the levels that can be detected by an EIA. **Figure 1** gives an overview of a range of commercial EIAs reported for *Salmonella* and their requirements for prior cultural enrichment of a test food sample. At present, the earliest results (presumptive-positive) for a test sample can be obtained on day 2 of analysis (i.e. 24–30 h). No test is yet available that can give a result within a single working day (i.e. $\leq$8 h).

The ultimate analytical goal for low-level pathogen detection in foodstuffs is for analysts to be able to obtain a result within a single working day. This, in my view, is only likely using a combination of approaches involving enhanced cultural enrichment (e.g. Oxoid Selective Pre-enrichment and Rapid Isolation Novel Technology, SPRINT), physicochemical purification and concentration of the test pathogen (e.g. selective immunomagnetic and solid-phase

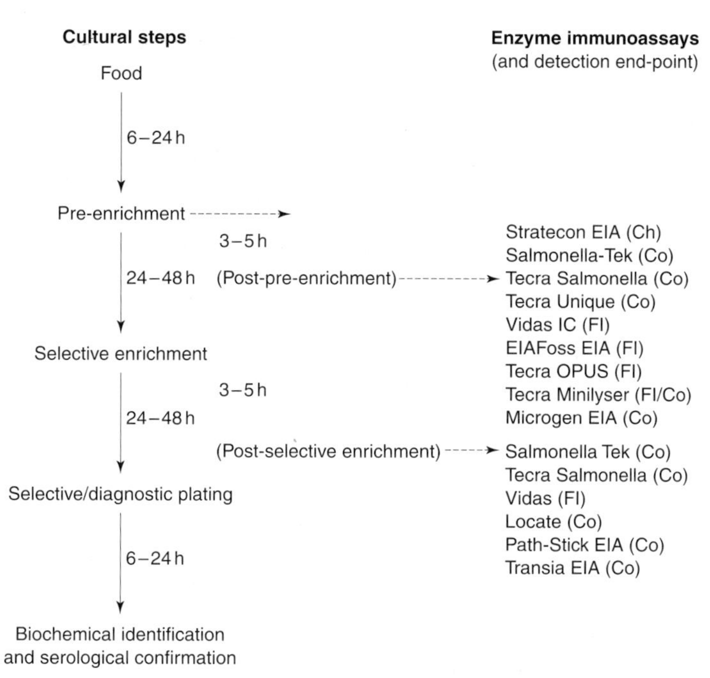

Figure 1 Overview of application of commercial EIAs for *Salmonella* showing the extent of prior cultural enrichment of a test sample. Co, colorimetric; Ch, chemiluminescent; Fl, fluorescent.

isolation) and a sensitive detection system (e.g. chemiluminescent end-point). There are several EIA products on the market currently available that have taken a step in this direction. In all cases, however, a 20–24 h cultural enrichment is still required. It is not within the scope of this article to describe all the techniques in depth. For further detailed information, the reader is directed to selected references in **Tables 1–3**.

Immunoconcentration Step

This solid-phase immunological step reported previously (Patel 1994) is introduced as a rapid alternative to the classical cultural selective enrichment of foodstuffs in order to specifically isolate and concentrate *Salmonella* by using immunomagnetic particles (e.g. Organon Teknika ELISA and EIAFoss) or plastic immunodipstick (e.g. Vidas IC, Tecra Unique). The concentrated *Salmonella* are allowed to multiply in a nonselective broth for a short period to increase their numbers for the subsequent detection by ELISA.

Membrane Filtration and Sensitive End-point ELISA

The Isoscreen *Salmonella* test (Stratecon Diagnostics Inc.) is an enzyme immunoassay that incorporates a disposable multiwell format, together with a portable instrument for the specific detection of a chemiluminescent signal produced in the presence of *Salmonella* concentrated on a filter. The combined filtration and the chemiluminescent signal allows as few as 10^4 *Salmonella* to be detected.

Similar levels of sensitivity have also been reported for some of the automated techniques (e.g. Vidas, EIAFoss and Tecra OPUS) in which the end points are based on fluorescent signal. Specifically, these tests incorporate methylumbelliferyl phosphate that is cleaved by the alkaline phosphatase (conjugated to a secondary antibody) to yield the fluorescent product, methyl umbelliferone, in the final reaction step of the assays.

Manual/Semiautomated ELISAs

Table 1 summarizes a range of current assay kits available for *Salmonella*, their official approval status, the marketing companies and selected references giving data on the performance characteristics based on collaborative trials according to approved procedures (where available). Many of these companies provide or recommend the use of automated microtitre plate washers, dispensers, incubators and readers for carrying out the various steps in the ELISA and, thus, these assays cannot be strictly classified as wholly manual. It should also be noted that most of the current generation ELISAs are based on a microtitre plate, simply because they are inexpensive, are capable of handling large numbers of samples and most of the developments over the past several decades have used this format for ELISA.

Table 1 Commercial manual or semi-automated microtitre based EIA techniques available for *Salmonella*

Test kit (Manufacturer[a])	*ELISA*		*Approval[b] status*	*Validation reference*
	Solid phase	*End point*		
Salmonella-Tek (Organon Teknika)	Microtitre well	Colorimetric	AOAC	Eckner et al (1994)
Assurance EIA (BioControl Systems)	Microtitre well	Colorimetric	AOAC	Feldsine et al (1992, 1993)
Path-Stik (Celsis-Lumac B.V.)	Dipstick	Colorimetric	AOAC[c]	Quintavalla et al (1996b), Brinkman et al (1995)
Bioline ELISA (Bioline)	Microtitre well	Colorimetric	AOAC[c]	Bolton (1996)
Tecra *Salmonella* VIA (Tecra Diagnostics)	Microtitre well	Colorimetric	AOAC, AFNOR DPIE/AQIS, VD	Hughes et al (1996)
Tecra Unique *Salmonella* (Tecra Diagnostics)	Dip-stick	Colorimetric	AOAC,[c] VDIA	Poppe and Duncan (1996)
Tecra *Salmonella* Immunocapture	Microtitre well	Colorimetric	VDIA	Flint and Hartley (1993)
Locate *Salmonella* (Rhône Poulenc)	Microtitre well	Colorimetric	AOAC, AFNOR	Moury et al (1994)
EQUATE EIA (Binax)	Microtitre well	Colorimetric	–	Holbrook et al (1989)
IsoScreen EIA (Stratecon)	Filter well	Chemiluminescent	–	–
Transia Plate *Salmonella* (Diffchamb)	Microtitre well	Colorimetric	AFNOR, AOAC	–
Mastzyme – *Salmonella* (Mast Diagnostics)	Microtitre well	Colorimetric	AOAC[c]	Bolton and Franklin (1995)

[a]Company addresses are shown in Appendix 1.
[b]AOAC, American Association of Official Analytical Chemists; AFNOR, Association Française de Normalisation, France; DPIE/AQIS, Dept. of Primary Industry & Energy Australia/Australian Quarantine & Inspection Service; VD, Veterinaerdirektoratet, Denmark; VDIA: Victorian Dairy Industry Authority, Australia.
[c]Performance tested certified test kits.

Typical Analytical Protocol

As an example, an AOAC official first action method No. 993.08, as detailed in the Organon-Teknika protocol 'Salmonella-Tek-ELISA test system' is described in detail. It should be noted that the example cited is for general reference only and does not indicate any preference over other commercial tests. The overall procedures of food sample preparation and subsequent analysis by ELISA on a microtitre plate are shown in **Figures 2 and 3**, respectively.

Food Sample Preparation Based on the food to be tested, prepare pre-enrichment media, selective enrichment media and other biochemical media according to instructions in the Food and Drug Administration's Bacteriological Analytical Manual (BAM). Further details of classical cultural enrichment procedures are reported elsewhere in this encyclopedia.

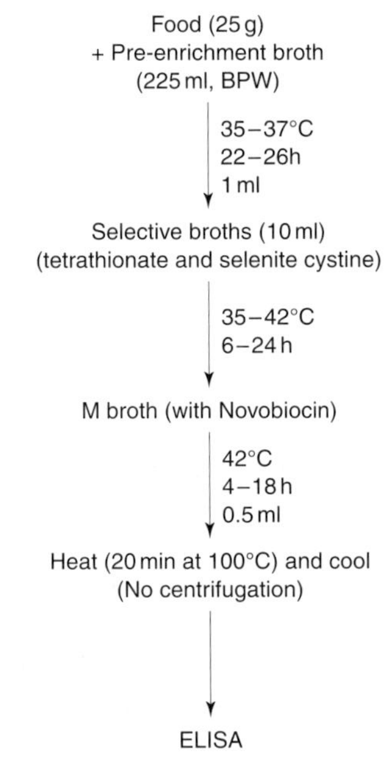

Figure 2 Food sample preparation protocol.

Raw Fleshy Foods (e.g. raw meats, poultry, fish, crustaceans, etc.)

Day 1: Pre-enrich samples for 24 ± 2 h at 35°C in the appropriate nonselective medium (lactose broth, nutrient broth, etc.) as indicated in BAM.

Days 2 and 3: Transfer aliquots of the pre-enrichment broth to selective enrichment broths (usually tetrathionate and selenite cystine) as described in BAM. Incubate the tetrathionate broth at 42 ± 0.5°C in a water bath for 18–24 h, and the selenite cystine broth at 35°C for 18–24 h. **Note**: It is critical that the 42 ± 0.5°C temperature range is maintained. After incubation, transfer 1.0 ml of tetrathionate mixture into a 10 ml tube of sterile M broth with 10 $\mu g\,ml^{-1}$ Novobiocin (see Appendix 2) which has been prewarmed to 42°C. Label as 'M broth Tet'. Carry out

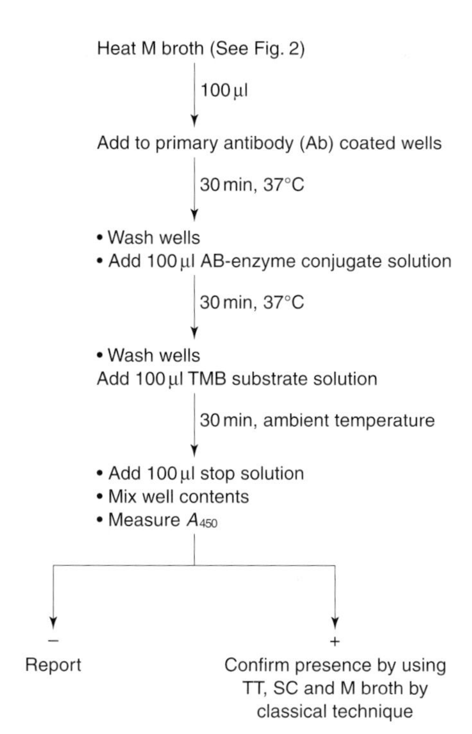

Figure 3 Detection of *Salmonella* by ELISA.

same procedure for the selenite cystine broth and label as 'M broth SC'. Incubate both broths in a 42 ± 0.5°C water bath for 4–6 h. Reincubate the tetrathionate and selenite cystine enrichments at 42°C and 35°C, respectively, and retain these broths for possible culture confirmation of enzyme immunoassay positive samples.

Following incubation, mix 'M broth Tet' and 'M broth SC' by hand or use a vortex mixer. Remove 0.5 ml from each M broth culture and place in the same screw-top test tube. Vortex to mix thoroughly and heat in boiling water or in an autoclave set at 100°C (flowing stream) for 20 min. Cool to 25–37°C prior to testing or store at 4°C for no more than 3 days.

Processed Foods (e.g. powdered egg, powdered cheese, no-fat dry milk, chocolate, spices, etc.)

Day 1: Pre-enrich samples for 18–24 h at 35°C in the appropriate nonselective medium as specified in BAM.

Day 2: Pre-warm tetrathionate broth to 42°C ± 0.5°C and selenite cystine to 35°C. Transfer 1.0 ml aliquots of the pre-enrichment broth to selective enrichment broths. Incubate the tetrathionate broth in a 42 ± 0.5°C water bath and the selenite cystine in a 35°C water bath for 6–8 h.

After incubation, transfer 1.0 ml of tetrathionate mixture to a 10 ml tube of sterile M broth with 10 μg ml^{-1} novobiocin; label as 'M broth Tet'. Also transfer 1.0 ml selenite cystine mixture to a second 10 ml tube of sterile M broth with 10 μg ml^{-1} novobiocin; label as 'M broth SC'. Incubate both M broths and remaining tetrathionate broth in a 42 ± 0.5°C water bath and selenite cystine broth at 35°C for 14–18 h.

Day 3: Following incubation, mix M broth Tet and M broth SC tubes by hand or use a vortex mixer. Remove 0.5 ml from each M broth culture and place in the same glass screw-top test tube. Retain remaining M broths, tetrathionate and selenite cystine broths for cultural confirmation of enzyme immunoassay positive samples. Heat the combined M broths in a boiling-water bath or in an autoclave set at 100°C (flowing stream) for 20 min. Cool to 25–37°C prior to testing or store at 4°C for no more than 3 days. **Note**: Alternatively, follow procedures outlined in the Official Methods of Analysis, 987.11 (*Salmonella* in Low-Moisture Foods).

ELISA Procedure For reagents, materials, equipment and storage instructions in relation to performing ELISA, see Appendix 3.

Reagent Preparation Carefully prepare the reagents before beginning the assay procedure. Reagents and samples should be at room temperature (20–25°C) before beginning the test and can remain at ambient temperature during testing. Return reagents to 2–8°C after use.

Control Antigens

1. Reconstitute the negative control by adding 2 ml of sterile distilled water to the vial.
2. Reconstitute the positive control by adding 1 ml of sterile distilled water to the vial.
3. Add 60 days to date of preparation and record expiration date on each vial label.

TMB Substrate

1. Calculate the total amount (ml) of TMB substrate needed by multiplying the number of wells used by 0.12.
2. Mix equal volumes (half of the total amount needed) of TMB solution A and TMB solution B in a clean glass test tube. **Caution**: Prepare TMB substrate just prior to use. Prepare only the amount needed for immediate use, any unused portion should be discarded.

Wash Solution Dilute wash concentration 1 : 50 with distilled water in a clean container (10 ml wash concentrate to 490 ml water). Wash solution is stable for 60 days.

Stop Solution Avoid contact with skin. If contact occurs, wash area with water.

Test Procedure

1. Fit stripholder with required number of Microelisa strips. Allow one well for each sample and incubate three wells for controls.
2. Pipette 100 µl of samples or controls into the assigned wells. Include two negative controls and one positive control per assay.
3. Cover the plate with a plate sealer and incubate at 37°C for 30 min.
4. Wash each well three times with wash solution (refer to Appendix 4, 'Microtitre plate wash procedure'). Go immediately to the next step. Do not allow the strips to dry.
5. Pipette 100 µl of conjugate into each of the wells containing samples and controls. **Note**: do not allow the pipette tip to touch the well.
6. Cover the plate with a plate sealer and incubate at 37°C for 30 min.
7. Wash each well six times with wash solution. Go immediately to the next step. Do not allow the strips to dry.
8. Pipette 100 µl of prepared TMB substrate (refer to 'Reagent preparation') into each well. Incubate at room temperature (20–25°C) for 30 min.
9. Pipette 100 µl Stop solution into each well. Plates should be read within 10–15 min.
10. Tightly close and return all unused reagents to 2–8°C storage.
11. Blank the Microelisa reader on air (without stripholder and strips) and then read the absorbance of the solution in each well at 450 nm.

Results

Calculations Calculations should be computed each time the assay is performed.

Qualification of Control Calculate the mean negative control (NC) value (NCX). This value should be less than 0.300. The positive control (PC) value must be greater than 0.700.

Note: A positive control with an absorbance that is greater than the upper limit of the reader is valid for use in the test.

All controls must be within the limits stated above for the test to be valid. If not, the run must be repeated.

Cutoff value The cutoff is calculated using the following equation:

$$NCX + 0.250 = 0.092 + 0.25 = 0.342$$

A test sample is presumed positive if sample absorbance is greater than or equal to the cutoff value. It is negative if the sample absorbance is less than the cutoff value.

Sample calculations absorbance: $NC = 0.090, 0.094$; $NCX = 0.092$; $PC = 1.35$.

Acceptance criteria These are: $NCX < 0.300$; $PC > 0.700$. All controls must be within these limits.

Interpretation of Results If the absorbance of a sample is greater than or equal to the cutoff value, it is presumed positive for the presence of *Salmonella* species. Positive results should be confirmed by culture. Samples of each broth (tetrathionate, selenite cystine, M broths) should be streaked onto selective agars, as indicated by BAM.

Negative ELISA readings indicate that the sample does not contain detectable levels of *Salmonella* antigens by this test method, and no further testing is required.

Automated ELISAs

It must be emphasized that automated ELISAs are exactly that, i.e. the various steps involved in an ELISA procedure are automated with results being analysed by a computer. The *Salmonella* analysis in foodstuffs still requires prior cultural enrichment procedures of 24–48 h. Table 2 summarizes the current commercial automated tests for *Salmonella* and, where available, their approval status and an example of a recent reference showing some validation data.

Advantages and Limitations of EIAs

The current EIAs have a sensitivity range of 10^4–10^6 cfu ml^{-1}, and generally show good specificity for *Salmonella* spp., although many of the antibodies show a degree of cross-reaction with *Citrobacter* spp. The results of EIA can be obtained rapidly, ranging from approximately 15 min (dipstick) up to 2.5 h (other ELISAs). Depending on the assay format, the equipment requirements range from virtually nothing (dipsticks) to highly automated ELISAs that can be quite expensive. Overall, the automated and 'manual' microtitre plate ELISAs allow high sample throughput with relatively little labour input compared with the disposable 'one-shot' dipsticks. Relatively little technical skill is required to perform an analysis depending on the assay format. Thus,

Table 2 Commercial automated ELISAs techniques for *Salmonella*

Test kit (Manufacturer[a])	*ELISA*		*Approval[b] status*	*Validation reference*
	Solid phase	*End point*		
VIDAS® *Salmonella* (SLM) (bioMerieux)	Solid phase receptacle	Fluorescence	AOAC, AFNOR	Curiale et al (1997), Keith (1997)
VIDAS® *Salmonella* ICS	Solid phase receptacle	Fluorescence	–	–
Minilyser® (Tecra Diagnostics)	Microtitre well	Fluorescence/ colorimetric	–	–
EIAFoss® (Foss Electric)	Magnetic particles	Fluorescence	–	Masso and Oliva (1997), Quintavalla et al (1996a)
PersonalLAB (Microgen)	Microtitre well	Fluorescence/ colorimetric	–	–
Transia ELISAMATIC (Microgen)	Microtitre well	Colorimetric	–	–

[a]Company addresses are shown in Appendix 1.

automated systems require lower skilled staff compared with the manual ELISAs. On the other hand, cultural step(s) are required prior to analysis by EIA and any positive results obtained must be confirmed by an approved cultural method. Additional drawbacks include the high cost of some kits and restricted commercial outlets for the kits.

Classical cultural techniques are highly labour-intensive and require skilled microbiologists to carry out the analysis. Negative results are obtained after approximately 4–5 days and confirmation of presumptive-positives require a further 2 days. In the case of ELISAs, negative results are obtained as early as 24–30 h and confirmation of presumptive-positive requires a further 2 days. Although the reagents and material costs in cultural techniques are generally lower than ELISAs, the labour costs are likely to be higher.

More recently, commercial tests for *Salmonella* based on polymerase chain reaction (PCR) (e.g. *Salmonella* Bax from Qualicon Inc., Wilmington, USA; Probelia PCR from Biocontrol Systems Inc. and Taqman system from PE Applied Biosystems, New Jersey, USA) have become available. These are generally an order of magnitude more sensitive and exhibit better specificity than the enzyme immunoassays, and the results obtained are usually definitive with no requirement for confirmation by classical cultural techniques. However, the techniques require high levels of technical skill, special laboratory facilities to avoid PCR contamination problems, generally high capital equipment costs and are prone to PCR inhibition depending on the matrix analysed. All the current generation of tests also require at least an overnight pre-enrichment step prior to PCR analysis.

Literature Update

It is not within the scope of this article to provide extensive literature data on EIAs for *Salmonella*. In particular, the aim is to list examples of some recent references on the applications of commercially available as well as research-based EIAs that are pertinent to *Salmonella* detection in food microbiology (**Table 3**).

See also: **Enzyme Immunoassays**: Overview. ***Salmonella***: Introduction.

Acknowledgement

The detailed methodology described in this article was reprinted with permission from Eckner et al (1994) with permission from the *Journal of AOAC International*. © 1994 AOAC International.

Further Reading

Anon (1984) *Food and Drug Administration: Bacteriological Analytical Manual for Foods*, 6th edn. Washington, DC: Association of Official Analytical Chemists.

Anon (1995) The IsoScreen *Salmonella* detection system. *Dairy, Food and Environmental Sanitation* 40.

Blais BW, Chan PL, Phillippe LM et al (1997) Assay of *Salmonella* in enrichment cultures of foods, feeds and environmental samples by the polymyxin-cloth enzyme immunoassay. *International Journal of Food Microbiology* 37: 183–188.

Bolton RE and Franklin J (1995) New immunoassay for the detection of *Salmonella* in foods. *International Food Hygiene* 6: 11–13.

Bolton RE (1996) AOAC research institute test performance testing program. Independent testing of the Bioline *Salmonella* ELISA test Cat. no. 96/1, by PHL, Royal Preston Hospital, Preston, UK.

Brinkman E, Van Buerden R, Mackintosh R and Beumer R (1995) Evaluation of a new dip-stick test for the rapid

Table 3 Examples of recent developments in EIAs for *Salmonella*

Solid phase	*End-point*	*Sensitivity (cfu ml^{-1})*	*Time to presumptive positive result*	*Reference*
Glass capillary tube	Fluorescent	$\approx 10^4$–10^5	30 h	Czajka and Batt (1996)
Immunomagnetic particles	Electrochemical	$\approx 10^4$–10^5	30 h	Gehring et al (1996)
Immunomagnetic particles	Electrochemical	10^3	ND[a]	Yu and Bruno (1996)
Sensor chip	Surface plasmon resonance	10^4–10^5	ND	Patel et al (1995)
Polymyxin-cloth	Colorimetric	–	50 h	Blais et al (1997)

[a]ND, Not evaluated in foods.

detection of *Salmonella* in food. *Journal of Food Protection* 58: 1023–1027.

Curiale MS, Gangar V and Gravens C (1997) VIDAS enzyme-linked fluorescent immunoassay for detection of *Salmonella* in foods: collaborative study. *Journal of AOAC International* 80: 491–504.

Czajka J and Batt CA (1996) Development of a solid phase immunoassay for the detection of *Salmonella* in raw ground turkey. *Journal of Food Protection* 59: 922–927.

Eckner KF, Dustman WA, Curiale MS, Flowers RS and Robison BJ (1994) Elevated-temperature, colorimetric, monoclonal enzyme-linked immunosorbent assay for rapid screening of *Salmonella* in foods: Collaborative study. *Journal of AOAC International* 77: 374–394.

Feldsine PT, Falbo-Nelson MT and Hustead DL (1992) Polyclonal enzyme immunoassay method for detection of motile and non-motile *Salmonella* in foods: Collaborative study. *Journal of AOAC International* 75: 1032–1044.

Feldsine PT, Falbo-Nelson MT and Hustead DL (1993) Polyclonal enzyme immunoassay method for detection of motile and non-motile *Salmonella* in foods: Comparative study. *Journal of AOAC International* 76: 694–697.

Flint SH and Hartley NJ (1993) Evaluation of the Tecra immunocapture ELISA for the detection of *Salmonella* in foods. *Letters in Applied Microbiology* 17: 4–6.

Gehring A, Crawford CG, Mazenko RS, Van Houten LJ and Brewster JD (1996) Enzyme-linked immunomagnetic electrochemical detection of *Salmonella typhimurium*. *Journal of Immunological Methods* 195: 15–25.

Holbrook R, Anderson JM, Baird-Parker AC and Stuchbury SH (1989) Comparative evaluation of the Oxoid *Salmonella* Rapid test with three other rapid *Salmonella* methods. *Letters in Applied Microbiology* 9: 161–164.

Hughes D, Dailianis A and Ash M (1996) Comparison of modified TECRA Visual Immunoassay with AOAC Method 989.14 and reference culture methods 967.25–967.28 for detection of *Salmonella* in foods and related samples. *Journal of AOAC International* 79: 1344–1359.

Keith M (1997) Evaluation of an automated enzyme-linked fluorescent immunoassay system for the detection of salmonellae in foods. *Journal of Food Protection* 60: 682–685.

Masso R and Oliva J (1997) Technical evaluation of an automated analyser for the detection of *Salmonella enterica* in fresh meat products. *Food Control* 8: 99–103.

Moury F, Fremy S and Lucon K (1994) *Detection of Salmonellae in Food by the Locate kit. Food Safety '94*, p. 319. Laval: ASEPT, Aseptic Processing Association.

Patel PD (ed.) (1994) *Rapid analysis techniques in food microbiology.* London: Blackie Academic and Professional.

Patel PD, Haines J and Walström L (1995) Analytical method (real-time detection of foodborne pathogens using the Biacore optical biosensor). Patent application filed.

Poppe C and Duncan CL (1996) Comparison of detection of *Salmonella* by the Tecra Unique *Salmonella* test and the modified Rappaport Vassiliadis medium. *Food Microbiology* 13: 75–81.

Quintavalla S, Bolmini L, Dellapina G, Pancini E and Barbuti S (1996a) Rapid methods for *Salmonella* detection in meat products. Part II – Evaluation of two immunochromatographic systems. *Industrial Conserve* 71: 306–314.

Quintavalla S, Bolmini L, Dellapina G, Pancini E and Barbuti S (1996b) Rapid methods for *Salmonella* detection in meat products. I. Evaluation of two automated immuno-enzymatic methods. *Industria Conserve* 71: 147–156.

Appendix I: Manufacturers of EIA kits and instruments for the detection of *Salmonella*

BioControl Systems, Inc., 12822 SE 32nd Street, Bellevue, WA 98005, USA.

Bioline, Hjulmaggervej 13c, DK-7100 Vejle, Denmark.

Binax, Portland, ME, USA.

bioMeriéux UK Ltd, Grafton House, Grafton Way, Basingstoke, Hants. RG22 6HY, UK.

Diffchamb, 8 rue St Jean de Dieu, 69007, Lyon, France.

Foss Electric, 69, Slangerupgade, DK 3400 Hillerød, Denmark.

Celsis-Lumac. Celsis Ltd., Cambridge Science Park, Milton Road, Cambridge CB4 4FX, UK.

Mast Diagnostics, Mast Group Ltd., Mast House, Derby Road, Bootle, Merseyside L20 1EA, UK.

Microgen Bioproducts Ltd., 1 Admiralty Way, Camberley, Surrey GU15 3DT, UK.

Organon Teknika N.V., Veedijk 58, 2300 Turnhout, Belgium.

Rhône Poulenc Diagnostics Ltd., Unit 3–03–3.06, West of Scotland Science Park, Maryhill Road, Glasgow G20 0SP, UK.

Stratecon Diagnostics Int'l, 12 Channel Street, Boston, MA 02210, USA.

Tecra Diagnostics, 28 Barcoo Street, Roseville, P.O. Box 20, Roseville, NSW 2069, Australia.

Appendix 2: Preparation of M-broth with 10 μg ml⁻¹ Novobiocin

1. Preparation of M broth: Medium formulation (per litre):
   Yeast extract 5 g
   Tryptone 12.5 g
   D-Mannose 2 g
   Sodium citrate 5 g
   Sodium chloride 5 g
   Dipotassium phosphate 5 g
   Mangenese chloride 0.14 g
   Magnesium sulphate 0.8 g
   Ferrous sulphate 0.04 g
   Tween® 80 0.75 g
2. Dispense 9.0 ml of M broth into test tubes (one for each sample to be tested) and autoclave at 121°C for 15 min. Prepare a 100 μg ml^{-1} novobiocin solution in distilled water and sterilize by filtration. Add 1.0 ml novobiocin solution to each sterile tube of M broth.

Appendix 3: Reagents, Materials, Equipment and Storage Instructions in Relation to Performing the ELISA

Components of the Test Kit

1 stripholder	Microelisa strips: eight per holder, each containing 12 wells coated with monoclonal antibodies to *Salmonella* species (murine); contained in a foil pack with silica gel desiccant
1 vial (2 ml)	Negative control: lyophilized antigen, non-reactive for *Salmonella* species; contains gentamicin sulphate as a preservative
1 vial (1 ml)	Positive control: lyophilized *Salmonella* antigen – reactive with antibodies to *Salmonella* species; contains gentamicin sulphate as a preservative
2 vials (6 ml)	Conjugate: peroxidase conjugated antibodies to *Salmonella* species. In a liquid diluent containing stabilizers and an antimicrobial agent in Tris buffer, pH 7.6
1 vial (7 ml)	TMB peroxidase substrate solution A: 3,3′,5,5′-tetramethylbenzidine
1 vial (10 ml)	Wash concentrate, 50×: contains 2.5% surfactant
1 vial (12 ml)	Stop solution: 2N sulphuric acid
10 sheets	Plate sealers: perforated, adhesive
1 each	Clamp and rod: Closure for foil pack

Additional Materials Required

Aspirator/wash system connected to a waste trap and vacuum source
Microelisa plate reader, capable of reading at 450 nm
Incubators, 35°C, 37°C
Water baths, 35°C, 42°C ± 0.5°C
Multichannel pipette, 50–300 μl
Single-channel pipette, 50–300 μl
Disposable micropipette tips
Reservoir troughs
Microbiological media and antibiotics for preparation of necessary enrichment broths
Hot plate, boiling-water bath, or autoclave (flowing stream)
Screw-top test tubes, glass
Appropriate containers for storage and disposable of materials potentially contaminated with infectious agents
Data record sheets

Instruments

Any ELISA reader (with appropriate filters), plate washers and plate incubators can be used.

Storage Instructions

1. Store all components at 2–8°C when not in use. Expiration date printed on the label of the kit indicates limit of stability.
2. The foil pack should be brought to room temperature (20–25°C) before opening to prevent condensation on the Microelisa strips. After the airtight foil pack has been opened, the strips are stable for 3 months at 2–8°C if the foil pack is resealed with the clamp and rod provided. The silica gel bag should not be removed.

Appendix 4: Wash Procedure

1. Incomplete washing will adversely affect the test outcome.
2. The wash procedure consists of an initial aspiration of well contents followed by filling of the wells (a minimum of 0.2 ml and maximum of 0.3 ml) with wash solution. Do not allow the wash solution to overflow into an adjacent well. This action should be performed several times as specified by the test procedure.

3. Routine maintenance of aspiration/wash system is strongly recommended to prevent carry over of antigen from samples containing a high concentration of antigen to non-reactive samples.
4. The aspiration/wash system should be flushed with copious amounts of distilled water upon completion of the final wash of the assay.

Detection by Colorimetric DNA Hybridization

Hau-Yang Tsen, Department of Food Science, National Chung Hsing University, Taichung, Taiwan, Republic of China

Conventional microbiological methods for the detection of *Salmonella* in foods may take 5–7 days since they include multiple subcultural, biotype and serotype identification steps. Rapid methods, such as immunoassay, DNA–DNA hybridization and polymerase chain reaction (PCR) methods have thus been developed in the past years. The DNA–DNA hybridization method requires a target gene (in target cells) specific probe. They are segments of DNA that have been labelled with radioisotopes, enzymes, antigenic substrates or chemiluminescent moieties and can bind specifically to the complementary sequences of nucleic acid. Probes can be directed to either DNA or RNA targets and may be oligonucleotide probes (often 20–50 bp) or DNA probes (50 to thousands of base pairs). Oligonucleotides can be synthesized chemically and easily purified while long DNA probes can be obtained by molecular cloning techniques.

In some cases, the sandwich hybridization system is used. This system requires the use of capture probes to capture the target cells on to the solid support. Then the captured target cells are detected by radioactively or non-radioactively labelled DNA probes, which are called the detector probes. Also, DNA hybridization can be performed on solid-phase nitrocellulose (NC) filters or in solutions. The applications of some of these probes for *Salmonella* detection in food samples are summarized in **Table 1**.

In the following section, a non-radioactive hybridization in solution system, the commercially available Gene-Trak system, is described as an example. This system is based on the detection of specific rRNA sequence unique to all the *Salmonella* serovars and thus allows the detection of all *Salmonella* cells. It has been reported that by comparing the sequences of variable regions in 16S rRNA genes for different bacterial species, specific oligonucleotide sequences can be derived and used to design the DNA probes or PCR primers to detect specific bacterial species.

Principles and Methods

In general, there are several steps in the DNA hybridization process. For the Gene-Trak system, these steps are shown in **Figure 1** and described here:

1. Culture of the sample with target cells in test tubes followed by cell lysis to release the target DNA or RNA (rRNA, for example).
2. Hybridization of the denatured target DNA or RNA with capture probe and detector probe. The detector probe has been labelled with fluorescein molecule. The hybridization reaction is carried out in solution.
3. Annealing of the polyadenine tail of the capture probe to polythymine which is linked to a dipstick. By this process the target DNAs with detector probes can be captured on the dipstick.
4. Detection of the captured target DNA which is annealed with the detector probe by the standard enzymatic/colorimetric approach commonly used in enzyme immunoassay. In the case of the Gene-Trak system, the target rRNA–probe complex captured on the dipstick is first reacted with a polyclonal antifluorescein antibody (anti-FL) conjugated to the enzyme horseradish peroxidase (HRP). This conjugate binds to the fluorescein molecules conjugated to the detector probe. The complex then reacts with a substrate of HRP, hydrogen peroxide, in the presence of a chromogen (tetramethylbenzidine; TMB) and a blue colour develops in proportion to the amount of enzyme conjugate bound to the complex and thus also in proportion to the amount of target DNA (or rRNA) captured. The reaction is stopped with sulphuric acid, and the developed colour changes from blue to yellow. The intensity of the colour is then measured by reading absorbance at 450 nm.

According to the instruction sheet for the Gene-Trak *Salmonella* assay system supplied by Gene-Trak, the assay protocol includes:

- Sample lysis: add lysis solution to sample tube and incubate to release rRNA target
- Hybridization: add probe solution and incubate. Both the capture probe and the detector probe hybridize to the rRNA target
- Hybrid capture: place dipsticks into tubes and incubate. Target : probe hybrid from the hybridization step will attach to the dipsticks
- Enzyme-label: wash dipsticks and place in tubes containing enzyme conjugate and incubate
- Colour development: wash dipsticks and place in

Table 1 DNA hybridization systems for the specific detection of *Salmonella* spp. with all serotypes

DNA probes and labelling	*Target gene*	*Food types*	*Sensitivity*	*Reference*
4.1 and 4.9 kb *Bam*HI fragment (P^{32})	Unknown[a]	Dairy products, soy flour, chocolate	10^6–10^7	Fitts et al (1983)
1.8 kb *Hind*III fragment (P^{32}, biotin)	Unknown[a]	Meat, fish, dairy	10^6–10^7	Tsen et al (1989, 1991a)
Oligonucleotide probes TS11, TS21, TS31 (P^{32})	1.8 kb *Hind*III fragment sequence	Meat, fish, feed, milk	10^6–10^7	Tsen et al (1991b)
Gene-Trak (first generation, P^{32})	RNA	Meat, fish, beef	10^6–10^7	Emswiler-Rose et al (1987)
Gene-Trak (second generation, fluorescein)	RNA	Dairy, soy flour, egg, pepper, turkey	10^6–10^7	Curiale et al (1990)
2.3 kb fragment (S^{35}, digoxigenin)	Unknown			Aabo et al (1992)
Oligonucleotide 16S1 and 16SIII	16S rRNA	Meat, fish, milk	10^6–10^7	Lin and Tsen (1995)

[a] Obtained by molecular cloning technique.

tubes containing substrate–chromogen. Incubate for colour development. Discard dipsticks and add stop solution to tubes

- Detection: read absorbance at 450 nm on the Gene-Trak photometer (Fig. 1).

In addition to *Salmonella* detection, similar principles and protocol can be used to detect *Listeria monocytogenes*, *Escherichia coli*, *Staphylococcus aureus*, *Campylobacter* and *Yersinia enterocolitica*. The DNA hybridization systems for these bacteria species are also available from Gene-Trak.

Protocol

(*Note*: for more details, see instructions from Gene-Trak.)

Sample Preparation and Enrichment

1. For raw meats (red meat, poultry, seafood) and raw milk, mix the homogenized food sample with 9 vol lactose broth. Incubate at 35°C for 22–24 h. For all other food products, pre-enrich the homogenized sample using methods described in Bacteriological Analytical Manual published by AOAC International, i.e., *BAM*/AOAC.
2. Mix 1 ml of the culture obtained to 10 ml tetrathionate broth (TT) and another 1–10 ml of selenite cysteine (SC) broth. Incubate each at 35°C for another 16–18 h.
3. Transfer 1 ml of TT culture and SC culture, respectively, to 10 ml Gram-negative (GN) broth. Incubate GN culture for 6 h at 35°C.
4. Take 0.25 ml of the GN culture for Gene-Trak assay. Save the rest for later confirmation, if required.

Gene-Trak Assay Procedure

1. Label the test tubes containing 0.25 ml GN culture of samples, and 0.5 ml each of the positive and negative controls.
2. Add 0.1 ml solution 1 (lysis solution) to each tube. Shake for 5 s and then incubate for 5 min.
3. Add 0.1 ml solution 2; shake for 5 s.
4. Cover the tubes with parafilm or aluminium foil and incubate the tube at 65°C for 15 min.
5. Add 0.1 ml *Salmonella* probe solution to each tube, shake for 5 s and incubate at 65°C for 15 min.
6. Prepare the 1× wash solution and rinse the dipstick for 2–3 min with 1× wash solution. Place the dipsticks into the sample tube in the 65°C bath. Incubate for 1 h.
7. Prepare 1× enzyme conjugate by mixing 100× enzyme conjugate concentrate (solution 5) and 1× wash solution. Note that enzyme conjugate is stable for 1 h after dilution.
8. Remove the dipsticks from the sample tube and wash them sequentially in 65°C 1× wash solution, then wash at room temperature.
9. Blot the dipstick on absorbent paper and place the dipsticks in tubes containing 1× enzyme conjugate solution. Incubate at room temperature for 20 min.
10. Prepare a substrate–chromogen solution by mixing 2 parts of substrate solution (solution 6) with 1 part of chromogen solution (solution 7).
11. When the 20 min incubation in step 9 is completed, remove the dipsticks from the tubes and wash them twice with 1× wash solution (1 min for each wash).
12. Blot the dipsticks on absorbent paper and place the dipsticks into tubes each containing 0.75 ml

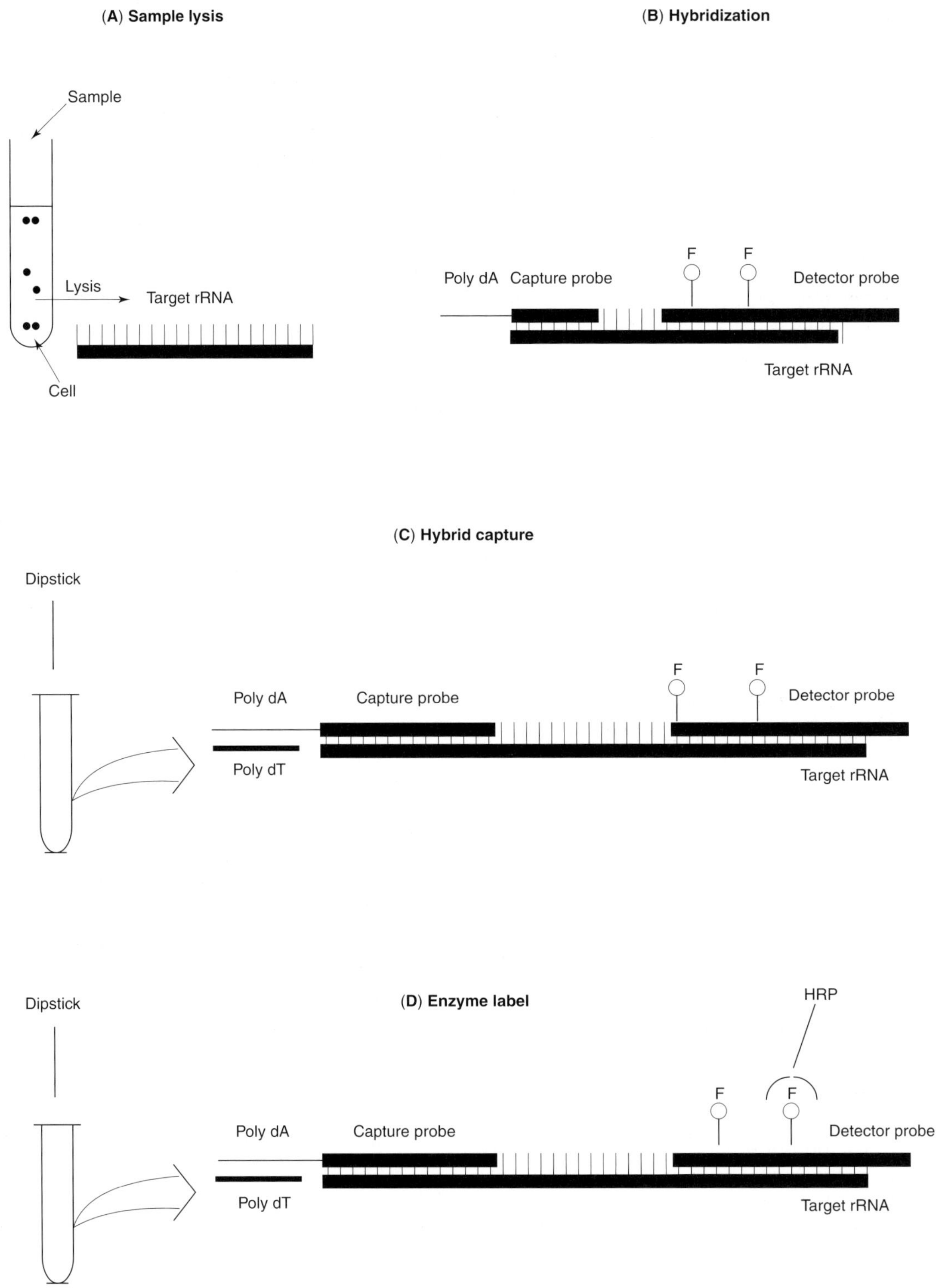

(**A**) **Sample lysis**

(**B**) **Hybridization**

(**C**) **Hybrid capture**

(**D**) **Enzyme label**

(**E**) **Colour development:** dipsticks are placed in substrate–chromogen solution

(**F**) **Detection:** Reading the OD at 450 nm

Figure 1 Protocol for Gene-Trak DNA hybridization assay of *Salmonella*. rRNA, ribosomal RNA; Poly dA, polyadenine sequence; Poly dT, polythymine sequence; F, fluorescein; HRP, horseradish peroxidase.

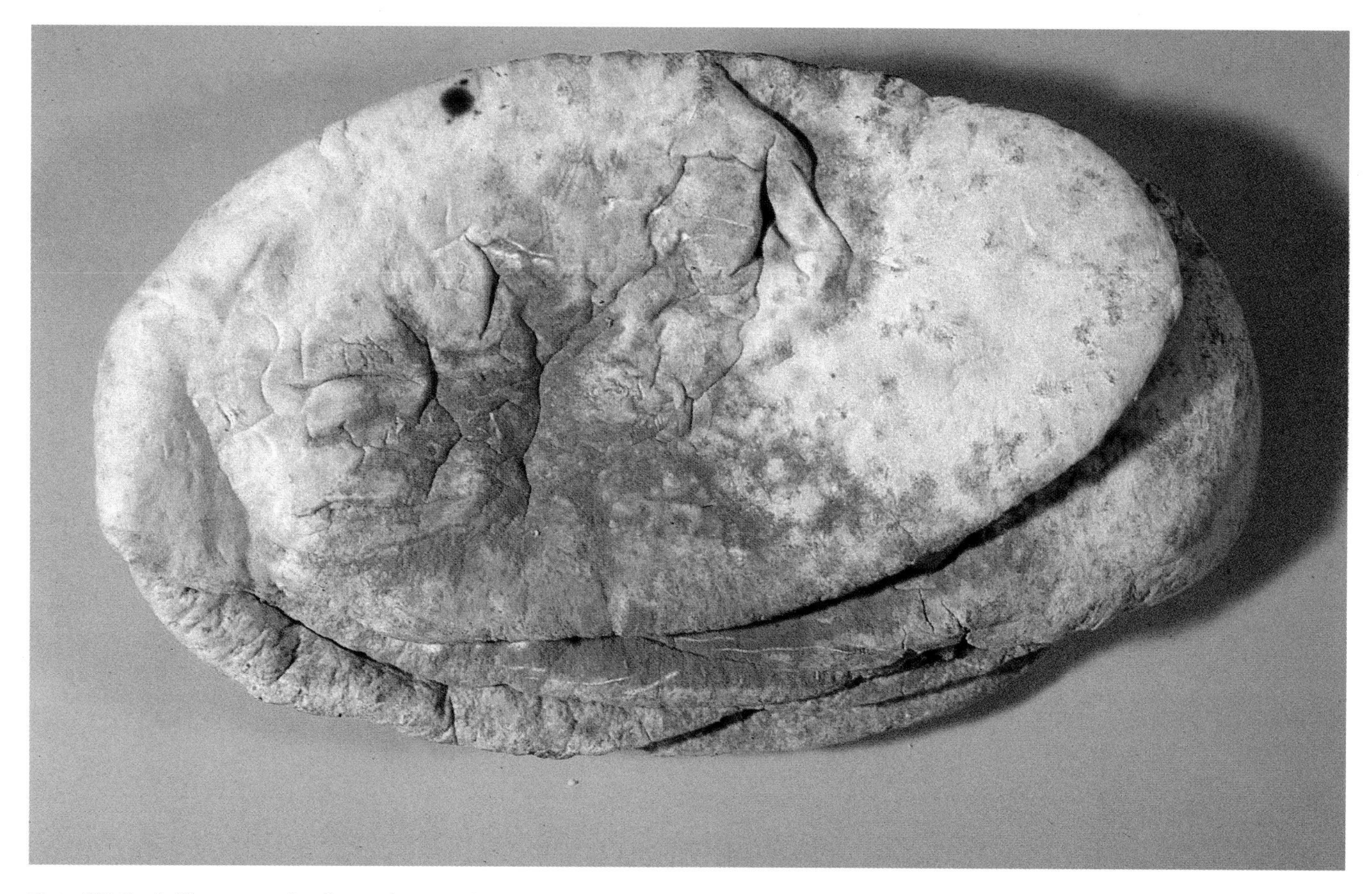

Plate 26 *Penicillium roquefortii* growing on pitta bread. (With permission from Dr Zofia Lawrence, CABI Bioscience, Surrey, UK.) See also entry **Penicillium: Introduction**.

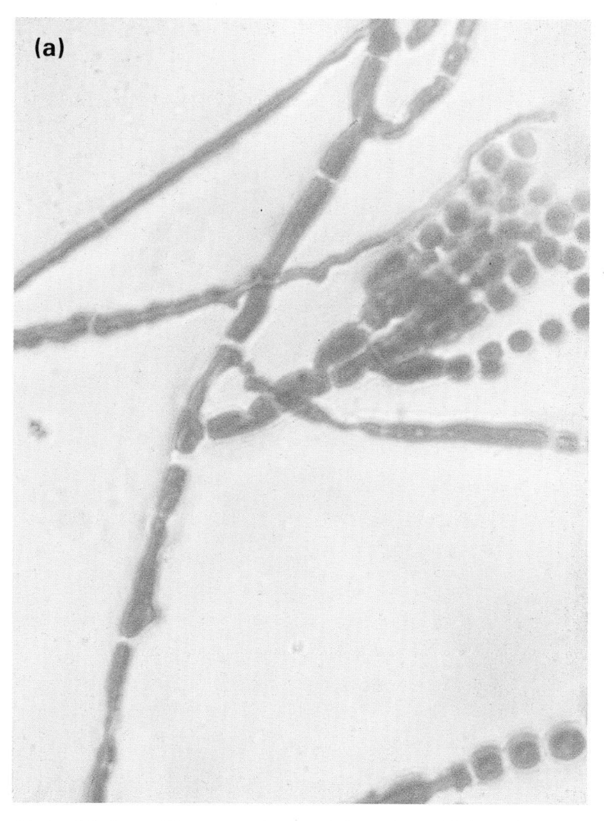

Plate 27 *Penicillium* sp. shown (a) under a light microscope with stain, and (b) using an electron micrograph. (With permission from Professor R. Davenport.) See also entry **Penicillium: Introduction**.

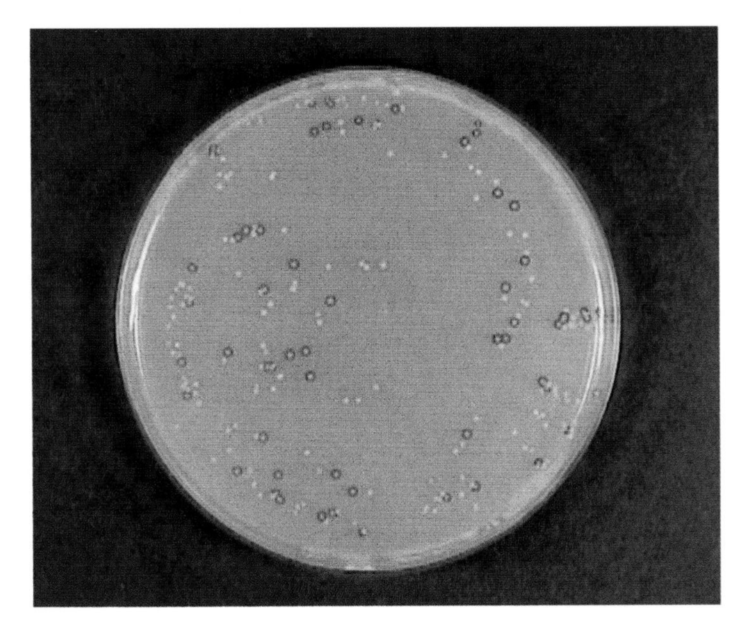

Plate 28 TTC (2,3,5-triphenyltetrazolium chloride)-agar overlay method to discriminate sake yeast strain. Sake yeast strain Kyokai 7 and a laboratory strain X2180-1A were mixed and spread on TTC-basal medium. Strain Kyokai 7 provided red colonies by the reduction of TTC. (With permission from Yuji Oda, Department of Applied Biological Chemistry, Fukuyama University, Japan.) See also entry **Saccharomyces:** *Saccharomyces sake.*

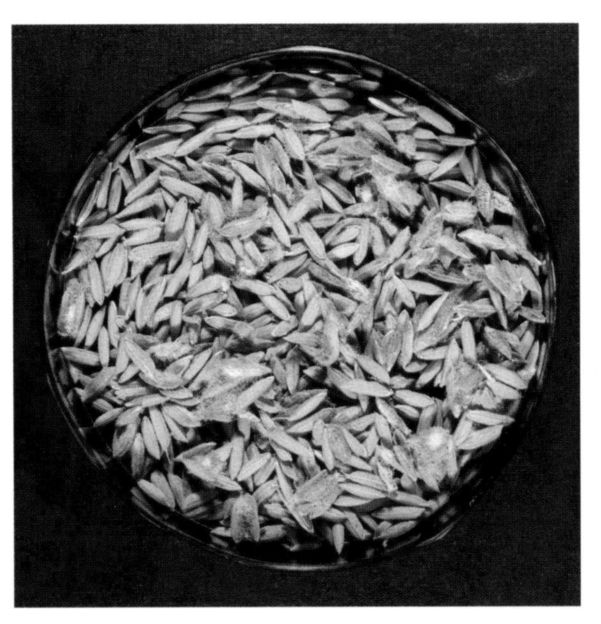

Plate 29 *Aspergillus flavus* growing on rice. (Courtesy of D. R. Twiddy and P. Wareing.) See also entry **Spolilage of Plant Products: Cereals and Cereal Flours.**

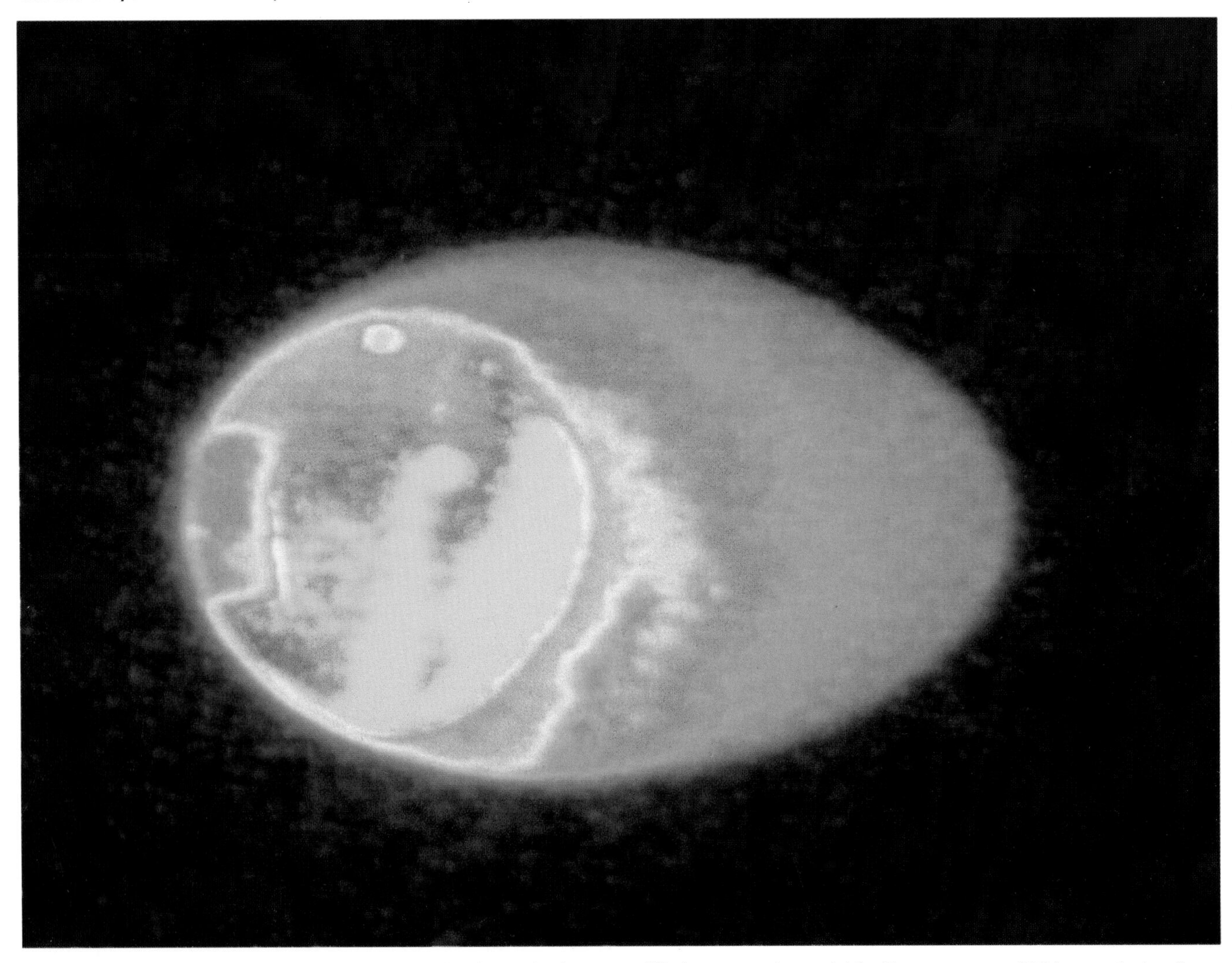

Plate 30 Salmonella detection in whole egg using bacteriophage modified to carry bacterial luciferase genes. (With permission from Dr Mansel Griffiths, Ontario Agricultural College, University of Guelph, Ontario, Canada.) See also entry **Salmonella: Introduction**.

Plate 31 (*right*) The patterns of colonies produced on Nutrient Agar plates using the Spiral Plater. By placing a special grid over a 'pattern' and counting the number of colonies in a prescribed segment, the number of colony-forming units ml^{-1} of the original sample can be calculated. (With permission from Dr Pradip Patel, Leatherhead Food RA, Surrey, UK.) See also entry **Total Viable Counts: Pour Plate Technique**.

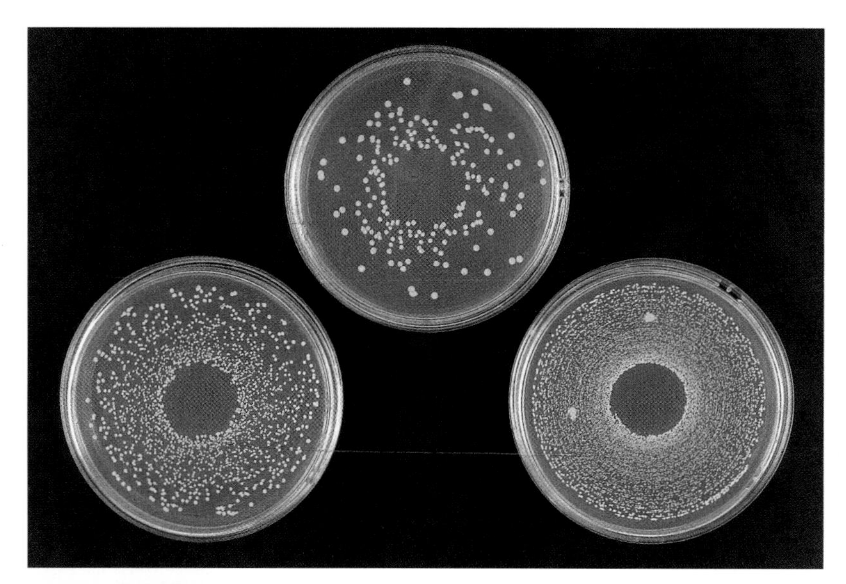

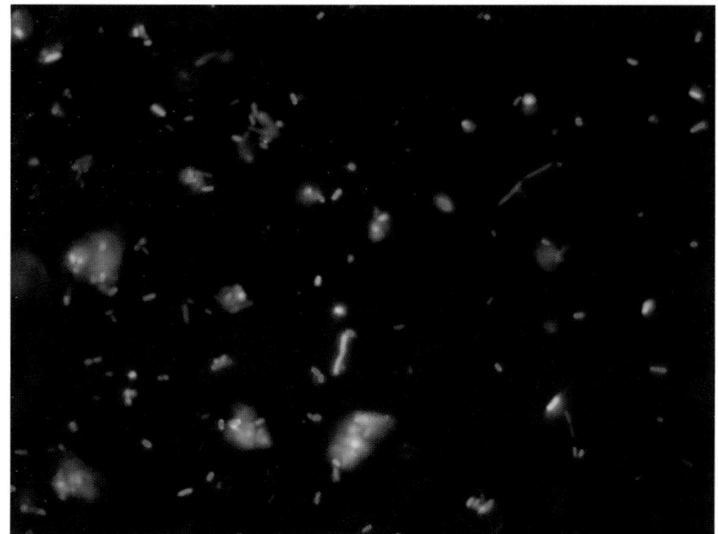

Plate 32 (*left*) Live/dead BacLight bacterial viability kit – the live cells fluoresce bright orange. (With permission from Haughland, 1996). See also entry **Total Viable Counts: Microscopy**.

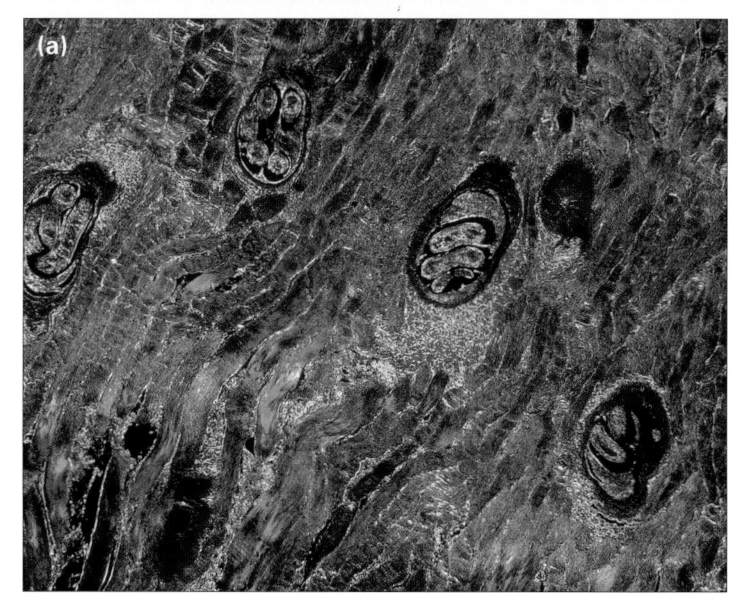

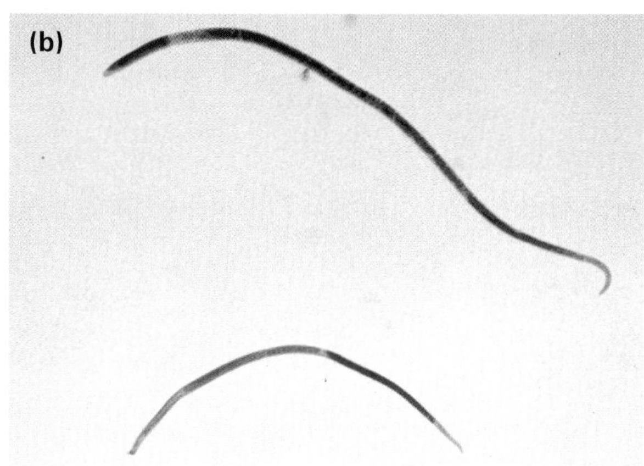

Plate 33 (a) Light micrograph of a section through cysts caused by *Trichinella spiralis* roundworms in pork muscle tissue. The pork muscle is stained blue. Several circular cysts are visible (dark blue, top right, lower right and lower left). Larvae are visible inside the cysts (purple). The cysts are being attacked by white blood cells (red). *T. spiralis* is a parasitic nematode worm. Human infection, usually caused by eating contaminated pork, causes trichinosis. The larvae mature in the intestines and reproduce, spreading around the body causing fevers and pain, and forming muscular cysts. Magnification: ×35 at 6×7cm size. (With permission from Science Photo Library.) **(b)** *Trichinella spiralis* causes trichinosis. Magnification of slide: ×16. (With permission from Custom Medical Stock Photo.) **See also entry Trichinella.**

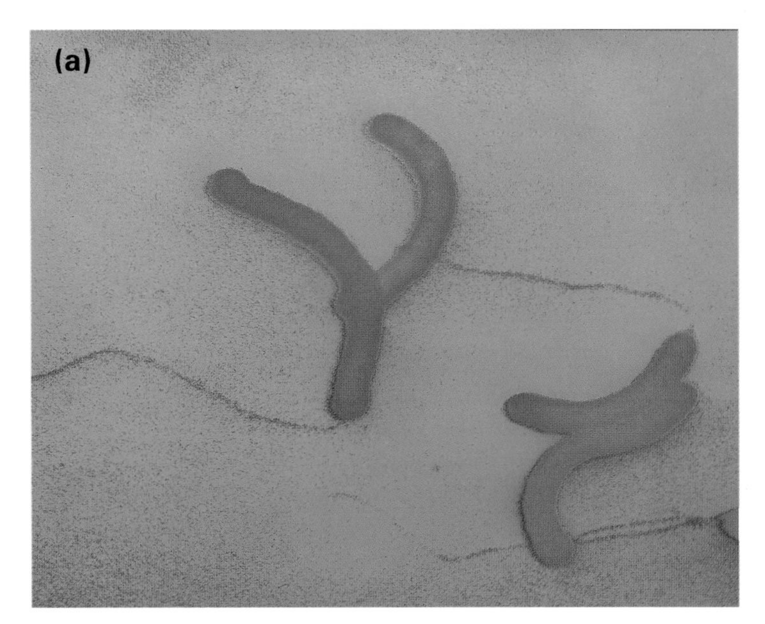

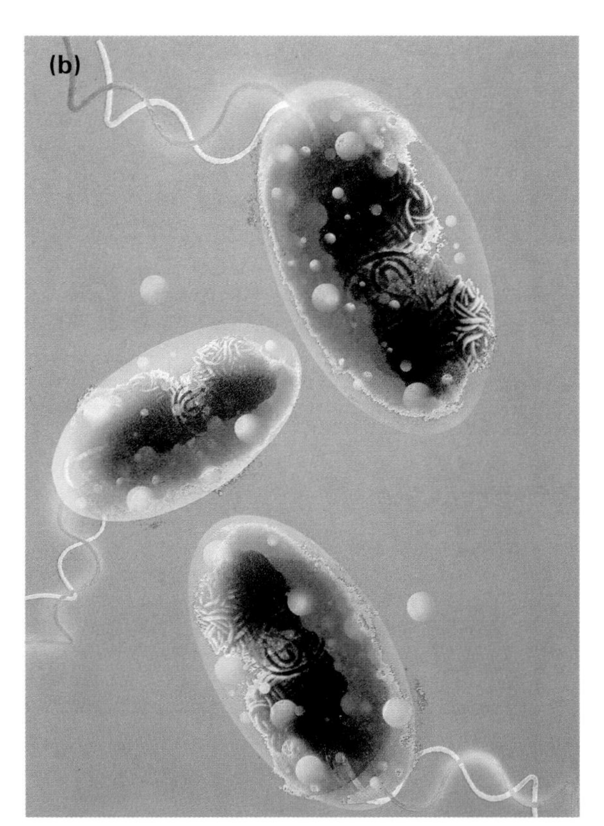

Plate 34 (a) (*above*) False-colour scanning electron micrograph of the bacterium, *Vibrio cholerae* – the cause of cholera. The bacterium moves by means of a single terminal flagellum (seen here). The spread of cholera is associated with poor sanitation and water supplies contaminated by faeces from a patient. The bacteria lodge primarily in the small intestines causing inflammation, vomiting, diarrhoea and, ultimately, dehydration. Magnification: ×6500 at 6×4.5cm size. (With permission from London School of Hygiene & Tropical Medicine/Science Photo Library.) **(b)** (*above right*) Illustration of three bacteria of the type *Vibrio cholerae*. These rod-shaped Gram-negative bacteria occur in colonies. Here, see within the bacteria, is genetic nucleic acid material (yellow) which lies freely in the cell. (With permission from Science Photo Library.) See also entry **Vibrio: *Vibrio cholerae***.

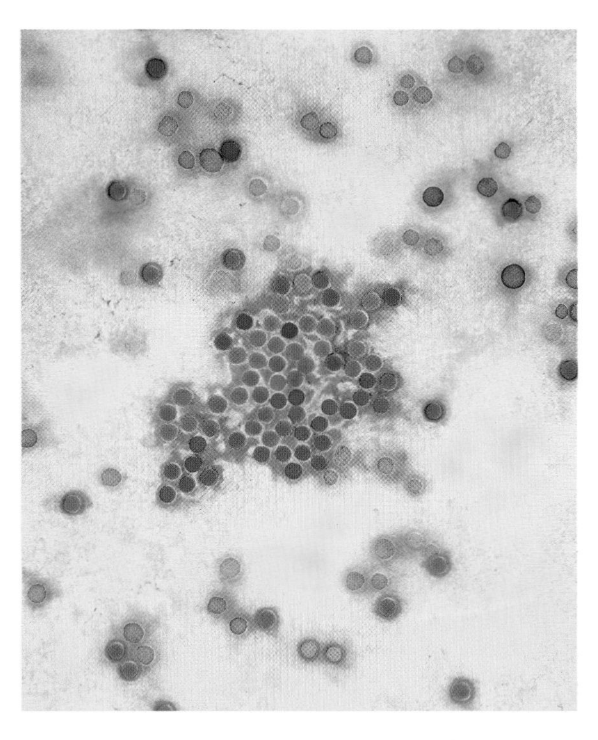

Plate 35 (*right*) Coloured transmission electron micrograph of a cluster of hepatitis A virus particles, the cause of infectious hepatitis. Virus particles are small red circles made of a protein coat encapsulating RNA-genetic material. These viruses are similar in structure to the Picornavirus group (which includes enterovirus and rhinovirus). During hepatitis A infection, the virus attacks the liver, producing liver inflammation, fever and jaundice. Hepatitis A is transmitted by water or food contaminated with infected faeces. The disease is more prevalent in warm countries with poor standards of hygiene. Magnification: ×56,000 at 6×7cm size. (With permission from Science Photo Library.) See also entry **Viruses: Hepatitis Viruses**.

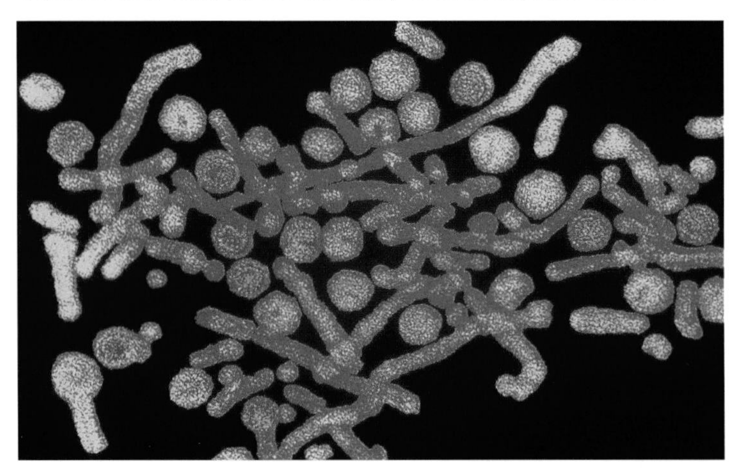

Plate 36 (***left***) Coloured transmission electron micrograph of the hepatitis B virus and its naturally occurring antigen. The pink spheres (called Dane particles) are the complete virus. Hepatitis B virus occurs typically in clusters of three types of particle (as seen here): the virus itself, with smaller spheres and rod-shaped forms made from the protein coat of the virus. The smaller non-infectious spheres are antigenically identical to the virus, and provide raw material for hepatitis B vaccines. Infection by this virus is more serious than hepatitis A, can be sexually transmitted, and causes chronic hepatitis. (With permission from Science Photo Library.) See also entry **Viruses: Hepatitis Viruses**.

tube is in the MPC. Remove the tube from the MPC and resuspend the beads in buffer D.

6. Wash the beads at 4°C for 30 min in buffer D four times using the MPC to collect the beads between each wash step.
7. After the last wash, resuspend the beads in original volume of storage buffer G. The coated beads are now ready for use. They can be stored for up to 6 months at 4°C.

Note that buffers A, B and C are used only for prewashing of beads and for solubilizing purified antibodies prior to coating. These buffers should not contain other proteins and sugars. Buffers D and E are used for washing all pre-coated beads and should never be used for washing uncoated beads. The TRIS component of buffer E will block free tosyl groups. The BSA in buffer D will prevent nonspecific binding of sugars and proteins onto the beads.

Commercial IMS Systems

Although anti-salmonella coated beads can be prepared in-house, two ready-prepared products are available commercially. Dynabeads anti-salmonella (Dynal, Oslo, Norway) uses magnetizable particles coated with specific antibodies to concentrate all *Salmonella* serovars selectively from foods, feeds and environmental samples. The technique takes 15–20 min when using traditionally pre-enriched samples, and may replace or enhance the performance of conventional selective enrichment, which takes 18–48 h. The other product, Salmonella Screen/Salmonella Verify (Vicam, Watertown, USA) detects only H_2S-producing salmonellae from a number of specific food and environmental samples. Both products have been validated independently, and are based on the same principles. Samples are pre-enriched before IMS. After two washes, the bead–bacteria complexes are resuspended in a small volume and plated onto selective *Salmonella* media. Isolated presumptive colonies are usually confirmed according to International Standards Organization protocols (ISO 6579) or methods specified in the *Bacteriological Analytical Manual*.

Dynal Protocols

The procedure for sample preparation is the same for all samples with the exception of shell eggs. Usually samples are blended or macerated in a 1:10 (w/v) dilution of buffered peptone water or a specific pre-enrichment broth. Certain foods, e.g. bony meat or pasta, should be blended to avoid the risk of perforation of the stomacher bag. After blending, the sample is transferred to a stomacher bag with a filter, which removes particulate matter as well as fatty components and allows easy pipetting of clear samples for examination.

The prepared samples are pre-enriched for 18–24 h at 37°C, and 1 ml of filtrate from the sample is used for IMS. Resuspended bead–bacteria complexes from all samples can be enriched selectively in Rappaport–Vassiliadis–soya (RVS) peptone broth for 18–24 h, before plating onto two media which are differential or selective for *Salmonella*. In the case of samples likely to have high levels of background flora, e.g. raw poultry or raw minced beef, the selective enrichment of the bead–bacteria complexes is essential. This approach is designated the 'core method', and is intended to replace or augment the direct plating of bead–bacteria complexes onto solid media, by virtue of its specificity and increased sensitivity. If direct plating of the complexes onto solid media (IMS–plating) is required, it is best achieved by using the swab-streak technique, rather than the standard technique of streaking with a loop.

The core method was developed because the IMS–plating of pre-enriched raw food samples occasionally resulted in the overgrowth of the target salmonellae by competitive enteric bacteria. These either bind non-specifically to the beads or combine with the antibodies coated onto the beads, and thus make the isolation of presumptive *Salmonella* colonies difficult. This problem arose most frequently in pre-enriched sample in which *Proteus* spp. and mucoid coliforms such as *Escherichia coli*, *Klebsiella aerogenes* and *Enterobacter* spp. were dominant. Without the meticulous and discriminate selection of suspect colonies to achieve purity, false negative results were sometimes obtained. The original approach of plating bead–bacteria complexes directly onto solid media using a swab-streak technique is now recommended only for processed foods or samples known to have a history of a very low level of resident flora. The most efficient approach, for all samples except shell eggs, is to swab-streak the bead–bacteria complexes and then selectively enrich the swabs. The following day, it is possible to plate out the RVS-enriched swabs corresponding to the plates on which no presumptive *Salmonella* can be detected. Either approach will also detect sublethally injured or stressed salmonellae. Both the core and the optional methods are designed to detect all current *Salmonella* serovars of importance in human and animal disease and occurring in food, feed and environmental samples, independent of their biochemical reaction pattern (**Table 3 and Table 4**).

The traditional selective enrichment method has now been combined with the use of an innovative timed-release capsule, containing the selective agents, and this approach is termed the SPRINT (Simple Pre-

Table 3 Performance testing[1] of Dynabeds anti-Salmonella for its ability to detect salmonellae from seeded foods compared to the conventional ISO 6579 method

Test sample	*Salmonella serovariant seeded*	*Level of Salmonella cells seeded per 25 g sample prior to pre-enrichment and detected (+) and not detected (–) by test method*			
		Low (1–5)		*Medium (10–50)*	
		ISO	*IMS*	*ISO*	*IMS*
Cake-mix	Hadar	+	+	+	+
	Braenderup	–	+	–	+
Casein	Java	–	–	–	–
	Kentucky	+	+	+	+
Cheese	Agona	+	+	+	+
	Bareilly	+	+	+	+
Chicken	Infantis	+	+	+	+
	Bredeney	–	–	+	+
Lasagne	Albany	–	+	+	+
	Blockley	–	+	+	+
Mayonnaise	Virchow	+	+	+	+
	Newport	+	–	+	+
Milk	Enteritidis	+	+	+	+
	Tymphimurium	+	+	+	+
Meat balls	Indiana	+	+	+	+
	Panama	+	+	+	+
Pepper	Heidelberg	+	+	+	+
	Mbandaka	+	+	+	+
Sausage	Montevideo	+	+	+	+
	Saintpaul	+	+	+	+
Total Positives		15	17	18	19
Actual Positives		20	20	20	20
False negatives		5	3	2	1
Concordance		90%		95%	

[1]a 100% concordance was achieved when the Dynal core and optional methods (IMS-Plating and IMS-RVS-plating) were tested on the same test samples.

enrichment and Rapid Isolation Novel Technology) salmonella system (Oxoid, Basingstoke, UK). This method ensures the recovery of stressed salmonellae while the timed-release selective agents are making the medium fully selective, and thus ensures the preferential growth of salmonellae over competing bacteria. The SPRINT system used with Dynabeads anti-Salmonella results in a significant reduction of the background flora compared with that achieved using the Dynal core method. This suggests that both the core and the optional methods could be unified, with significant savings in the time needed for analysis.

Shell eggs constitute the only food category with its own dedicated procedures for detecting the presence or absence of *Salmonella*. The level of *Salmonella* contamination in naturally infected eggs is generally low. Lysozyme, ovotransferrin and the alkaline pH of egg white kill or inhibit the growth of a wide variety of microorganisms, but salmonellae secrete chelating agents during extended storage, which compete with ovotransferrin, enabling these bacteria to survive and grow in eggs. If blended fresh eggs are supplemented with ferrous sulphate during an abbreviated enrichment for 6 h, any salmonellae present can multiply sufficiently to be concentrated by IMS and then be detected directly on plating media.

Sample Preparation

1. (a) Weigh 25 g of the sample test material, place into a stomacher bag (with filter) and add 225 ml of pre-enrichment broth *or* (b) for environmental samples using a swab, place the swab into 10–50 ml of pre-enrichment broth and incubate as described in step 2 below.
2. Incubate the prepared sample in the stomacher bag for 18–24 h at 35–37°C.

Table 4 Overview of the comparative evaluation of different IMS-related methods and standard methods for *Salmonella* detection. From Cudjoe and Krona (1997) with kind permission from Elsevier Science-NL, Sara Burgerhartstraat 25, 1055 KV Amsterdam, The Netherlands

Samples	*No. of samples*	*Number of salmonella-positive samples detected*[a]				
		IMS			*ISO*	*Total (all methods)*
		BPW (I)[b]	BPW (II)[b]	BPW-RVS[c]		
Chicken skin	50	6	10 (3)[d]	10 (3)	5	12
Chicken liver	15	4	6	6	7 (1)	8
Chicken meat	15	1	4 (2)	4 (2)	2	5
Faecal swabs	10	0	9 (1)	9 (1)	8	9
Poultry feed	10	9	10	10	9	10
Total	100	20	39 (6)	39 (6)	31 (1)	44

BPW, buffered peptone water; IMS, immunomagnetic separation; ISO, International Standards Organization; RVS, Rappaport–Vassiliadis–soya.

[a]All isolates from the respective methods have been confirmed.

[b]Bead–bacteria complexes from BPW-enriched samples were plated directly using the swab-streak technique (I), but the swabs were selectively enriched in RVS before plating (II).

[c]Bead–bacteria complexes from BPW-enriched samples were selectively enriched in RVS before plating.

[d]Numbers in parentheses indicate *Salmonella*-positive samples detected only by method.

3. Mix the pre-enriched sample thoroughly, by homogenizing once more.
4. Pipette a 1 ml sample of the filtered suspension for the IMS procedure below. For environmental samples, pipette a 1 ml sample for IMS. Use a new pipette for each new sample.

Immunomagnetic Separation using Dynabeads Anti-Salmonella

1. Remove the magnetic plate from the Dynal MPC-M and load the required number of 1.5 ml Eppendorf tubes.
2. Resuspend Dynabeads anti-Salmonella thoroughly using a vortex machine. Pipette 20 μl into each tube.
3. Add 1 ml of the pre-enriched filtered sample from above and close the tube. Use a new pipette for each new sample.
4. Invert the MPC-M rack five times. Incubate at room temperature for 10 min with gentle continuous agitation to prevent the beads from settling (e.g. in a Dynal MX3 sample mixer).
5. Insert the magnetic plate into the MPC-M. Allow 3 min for recovery of beads. During this period invert the rack several times in order to concentrate the beads into a pellet on the side of the tube.
6. Open the tube's cap using the opener provided and carefully aspirate and discard the supernatant as well as the remaining liquid in the cap. Use a new pipette for each new sample.
7. Remove the magnetic plate from the MPC-M.
8. Add 1 ml of wash buffer (PBS-Tween). Do not touch the tube with the pipette since this can cross-contaminate the samples as well as the wash buffer. Close the tube cap. Invert the rack several times to resuspend the beads.
9. Repeat steps 5 to 8.
10. Repeat steps 5 to 7.
11. Resuspend the particle–bacteria complex in 100 μl of wash buffer (PBS-Tween). Mix briefly using a vortex. Depending on the original sample type, the bead–bacteria complexes are now ready for further manipulation.

The Core Isolation Procedure The Dynal core method allows the detection of *Salmonella* from all food and environmental samples three days after receipt. The process of isolation after IMS is completed as follows:

1. Transfer the concentrated, resuspended bead–bacteria complexes into 10 ml of RVS–peptone broth and incubate at 42°C for 18–24 h.
2. Follow the standard procedure for isolation by spreading a loopful of RVS culture onto any *Salmonella*-selective or differential plating media.

The Optional Method The Dynal optional method is recommended for processed foods known to harbour little or no competing background flora. Presumptive *Salmonella*-positive results are available 2 days after receipt of samples. The process of isolation after IMS is completed as follows:

1. Transfer 50 μl of the resuspended bead–bacteria complex onto each of two *Salmonella*-selective agar plates (e.g. BGA, XLD agar, BSA, HE agar).
2. Spread the liquid over half of the plate with a

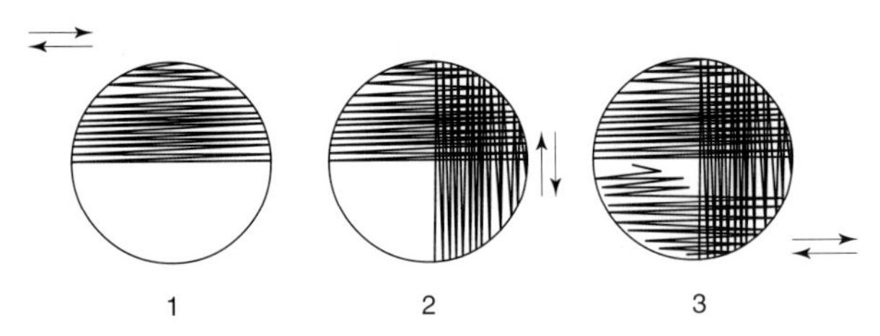

Figure 1 Dynal swab-streak plating technique. 1, Swab; 2, streak with a loop; 3, streak with a new loop.

sterile swab. Dilute further by streaking with a loop (see Fig. 1).

3. Incubate the agar plates upside down at 35–37°C for 18–24 h.

Method for Shell Eggs

1. Wash dirty eggs with a stiff brush under running water, and dry them with a paper towel. Dip the eggs into 70% ethanol for 5–10 s and allow to dry. Alternatively, follow any standard procedure for disinfecting shell eggs.
2. Aseptically crack open the eggs and blend the white and yolk thoroughly. Add ferrous sulphate ($FeSO_4$) solution to a final concentration of 35 mg l^{-1}.
3. Pre-incubate the egg mixture at 37°C for 6 h.
4. After pre-incubation, mix the egg mixture thoroughly. Dilute a sample fivefold with wash buffer or buffered peptone water. Use 1 ml of this dilution for IMS. Use a new pipette or a new pipette tip for each sample to avoid cross-contamination.
5. Incubate the remaining undiluted egg mixture overnight at 37°C.
6. Load the required number of Eppendorf tubes into the Dynal MPC-M with the magnetic strip removed.
7. Add 40 μl of Dynabeads anti-Salmonella into each of the Eppendorf tubes. Do not touch the tubes with the pipette since this can cross-contaminate the samples as well as the wash buffer.
8. Add 1 ml of diluted egg mixture. Use a new pipette for each new sample.
9. Incubate with constant rotation (e.g. Dynal MX3 sample mixer) for 30 min at room temperature (18–28°C).
10. Insert the magnetic plate into the MPC-M. Allow 3 min for recovery of the beads. During this period invert the rack several times in order to concentrate the beads into a pellet on the side of the tube.
11. Carefully aspirate and discard all of the supernatant. *Do not aspirate the beads*. If this occurs dispense the sample back into the tube and repeat step 10.
12. Remove the magnet and add 1 ml of wash buffer to each tube. Do not touch the tube with the pipette since this can cross-contaminate the samples as well as the wash buffer. Manually shake briefly to resuspend the beads.
13. Repeat step 11.
14. Add 100 μl of wash buffer and resuspend the bead–bacteria complexes using a vortex.
15. Add 50 μl of resuspended beads onto two different *Salmonella*-selective agar plates. Streak in a normal fashion.
16. Incubate the selective agar plates overnight at 37°C.
17. Steps 6 to 16 should be repeated on a fivefold dilution of the overnight incubated samples (step 5) that returned presumptive *Salmonella*-negative results after 6 h IMS examination.

Other End-point Detection Methods

Apart from giving cultural isolates for verification, Dynabeads anti-Salmonella allows the flexibility of applying different end point detection methods. An immunomagnetic particle-based enzyme-linked immunosorbent assay (IMP-ELISA) – for the detection of *Salmonella* in foods has been developed using Dynabeads anti-Salmonella. The use of appropriate procedures for sample preparation allows the rapid detection of *Salmonella* serovars in both processed foods (e.g. powdered egg products) and non-processed foods (e.g. raw chicken). Pre-enriched broths of heat-processed samples, which are likely to harbour only low levels of competitive enteric flora, can be boiled and used directly for IMP-ELISA. Non-heat-processed or raw samples are likely to contain higher levels of such competing organisms, and any live *Salmonella* cells must be isolated by IMS from the standard pre-enrichment broth, and then post-selectively enriched for 4 h in M-broth supplemented with novobiocin before boiling prior to IMP-ELISA. The total assay, including sample preparation, takes less than 26 h for these procedures, with a lower detection limit of 10^5 *Salmonella* cells per millilitre of sample. In an evaluation of naturally contaminated poultry samples, all 45 of 48 samples previously shown to contain salmonellae in a comparison of ISO, IMS-Plating, Salmonella-Tek ELISA (Organon Teknika, Durham, USA) and a modification of the latter based on IMS, were identified as positive. No other methods gave positive results for all 45 samples.

The demonstration that by replacing Organon Teknika's Salmonella-Tek ELISA conventional selective enrichment protocols with IMS significantly improved the *Salmonella* detection rate, led to the incorporation of IMS into that kit, now renamed Salmonella Capture-Tek. This 24 h screening procedure saves time, eliminates labour and costs of selective enrichment, and allows the early release of *Salmonella*-negative samples. Omission of selenite–cystine broth eliminates the biohazard disposal costs of selenite. Data demonstrate that this modified assay allows for recovery and detection of sublethally injured *Salmonella* and that IMS produces organisms unaffected by biochemical selection.

Salmonellae bound to Dynabeads have also been detected using conductimetric analysis, with improvement in the specificity of the conductance system and a reduced detection time. The improved ratio shift in favour of *Salmonella*, due to the reduction in the background flora, enables the target organisms to produce the desired conductivity rapidly in the growth media compared with samples not treated with IMS.

Inhibitory substances in food samples, which reduce the specificity and sensitivity of direct polymerase chain reaction (PCR) methods can be removed by the pre-treatment of samples using IMS. The resulting magnetic immuno-PCR assay (MIPA) combines the specific extraction of *Salmonella* by specific monoclonal antibodies with primer-specific PCR amplification to facilitate detection. A commercial sensor that combines IMS with electrochemiluminescence to detect *S. typhimurium* in foods and fomites, has demonstrated that the sensitivity is significantly improved, and the total processing time is reduced to less than 1 h. The detection of viable salmonellae in 20 min, using a combination of IMS and the measurement of intracellular microbial ATP by bioluminescence, demonstrates the effectiveness of this approach. At present it is applicable only to pure cultures, but the use of *Salmonella*-specific bacteriophages incorporating the *lux* gene should enable the approach to be used for the assay of any sample.

Vicam Protocols

Unlike the Dynal protocols, initial food sample preparation prior to using the Vicam Salmonella Screen/Salmonella Verify methods varies with the type of sample, but is normally based on the standard 1 : 10 dilution (w/v) in Vicam's own food enrichment medium or environmental enrichment medium before homogenization in a stomacher. Furthermore, depending on the sensitivity required, samples can be enriched for 5 h, 6 h or 16 h at 37°C. For shell eggs, raw steak and feeds (poultry meal), a 5 h pre-enrichment is required, whereas fish meal requires a 6 h pre-enrichment. For foods such as raw pork, raw ground beef, raw sausage, raw chicken, frozen egg yolk, frozen whole egg and frozen plain egg, a 16 h pre-enrichment is recommended. After pre-enrichment followed by filtration, 2 ml of the filtrate is used for IMS.

Immunomagnetic Separation using Vicam Salmonella Immunobeads

1. Add 100 μl of Salmonella immunobeads to each vial containing 2 ml of the sample to be analysed. Screw caps on firmly.
2. Place test vials on the capture vial rotator and rotate vials end-over-end at room temperature for 1 h on medium speed (15 rev min^{-1}).
3. Place the test vials in the magnetic rack with magnets in place and allow to stand for 10 min.
4. Gently aspirate the fluid from the test vials without disturbing the concentrated beads.
5. Immediately wash the beads by adding 2 ml of capture buffer to each test vial in the magnetic rack without the magnets, and resuspend the beads uniformly.
6. Return the magnet to the magnet rack and leave the test vials in place to concentrate immunobeads for 5 min.
7. Repeat steps 4 to 6 to achieve two washes.
8. Repeat step 4 to remove final wash liquid.
9. Immediately resuspend beads in 0.1 ml capture buffer and transfer all the immunobeads to the top half of the solid indicator medium plate (XLD) provided.
10. Spread the beads on the top half of the agar plates using a sterile glass spreader or loop.
11. Incubate plate at 37°C for 16–22 h.
12. Examine plates for colony growth. The absence of colony growth or presumptive positive colonies (black in colour) means that the result is negative and the test is complete.
13. If suspect positive colonies are observed an agglutination reaction is carried out with Salmonella Verify Kit followed by further biochemical reactions as detailed in the *Bacteriological Analytical Manual* by the AOAC if necessary.

Confirmation For both Dynal and Vicam protocols, representative numbers of isolated presumptive *Salmonella* are confirmed by standard biochemical and serological testing methods as defined by ISO 6579 and in the *Bacteriological Analytical Manual*.

Plating Techniques

Both Dynal and Vicam recommend special plating techniques for the isolation of target microorganisms bound to the paramagnetic beads. The Dynal swab-streak technique (**Fig. 1**), involves spreading the bead–bacteria complexes evenly onto half of the plating medium with a swab, and then streaking approximately 20 times from the second to the third quadrant and approximately 10 times from the third to the fourth quadrant with a sterile loop. Some of the *Salmonellae* bound to the beads are retained on the swab,

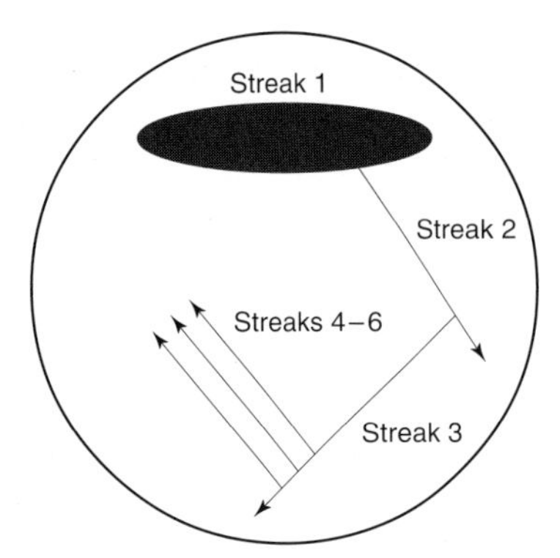

Figure 2 Vicam loop streaking technique.

but the swab-streak technique distributes the bead–bacteria complexes evenly over the plating medium, soaks up excess fluid and facilitates the examination and selection of clearly isolated colonies in the third and fourth quadrants of the plate.

The Vicam plating technique (**Fig. 2**) involves six steps. The first streak is broad and even, and is made with a spreader or loop. Fresh loops are used to make the second and third streaks, and then three more single streaks are made.

See also: **Bacteriophage-based Techniques for Detection of Food-borne Pathogens**. **Biochemical and Modern Identification Techniques**: Introduction. **Biosensors**: Scope in Microbiological Analysis. **Enrichment Serology**: An Enhanced Cultural Technique for Detection of Food-borne Pathogens. ***Escherichia coli***: Detection by Commercial Immunomagnetic Particle-based Assays. **Food Poisoning Outbreaks**. **Immunomagnetic Particle-based Techniques**: Overview. **National Legislation, Guidelines & Standards Governing Microbiology**: European Union. **Nucleic Acid-based Assays**: Overview. **PCR-based Commercial Tests for Pathogens**. **Reference Materials**. ***Salmonella***: Detection by Classical Cultural Techniques; Detection by Latex Agglutination Techniques; Detection by Enzyme Immunoassays; Detection by Colorimetric DNA Hybridization.

Further Reading

Andrews WA, June GE, Sherrod PS, Hammack TS and Amaguana RM (1995) Salmonella. In: *FDA Bacteriological Analytical Manual*, 8th edn. P. 5.01. Gaithersburg: AOAC International.

Cudjoe KS and Krona R (1997) Detection of *Salmonella* from raw food samples using Dynabeads® anti-Salmonella and a conventional reference method. *International Journal of Food Microbiology* 37: 55–62.

Food and Drug Administration (1995) *Bacteriological Analytical Manual*, 8th edn. Gaithersburg: AOAC International.

International Standards Organization (1990) *Microbiology: General Guidelines on Methods for the Detection of Salmonella*. Ref. 6579:1900 (E). International Standards Organization.

Kroll RG, Gilmour A and Sussman M (eds) (1993) *New Techniques in Food and Beverage Microbiology*. P. 1. Society for Applied Bacteriology Technical Series 31. Oxford: Blackwell Scientific Publications.

Salt *see* **Preservatives**: Traditional Preservatives – Sodium Chloride.

SAMPLING REGIMES & STATISTICAL EVALUATION OF MICROBIOLOGICAL RESULTS

G Hildebrandt, Institute for Food Hygiene, Free University of Berlin, Germany

In the past the acceptance or rejection of a food was ideally decided by inspection of 100% of the items of a lot. If the test was too laborious and slow or destroyed the unit, testing of representative samples had to be done instead. This concept is still used for lots of unknown history (port of entry situations). In contrast, quality and safety of a food cannot be retroactively tested in a product. Modern quality assurance systems, including hazard analysis and critical control point (HACCP), do not standardize the product but rather serve to standardize the production process. Sampling for surveillance of microbiological criteria is not superfluous but has shifted to the level of verification instead.

Sampling and Sample Preparation

Sampling

An unrestricted random sampling plan must be used to guarantee a valid estimate of the characteristic of interest. An organizational prerequisite is that all units of the population or lot are registered and available for sampling. In an ideal case, random sampling occurs by casting dice or pulling tickets. In view of the labour-intensive nature of these statistical procedures, random-number tables are often used instead. For process control, systematic sampling with a random starting point is recommended because of its great practicability. First, a starting point is randomly fixed and then additional samples are taken at prescribed intervals.

The random selection of a given number of random samples of a lot is referred to as a unitary (one-stage) procedure. This procedure is repeated from lot to lot without further division into subsamples, strata or phases. Occasionally, a two-stage or two-step strategy is recommended. This is where a second sampling must occur after an indifferent result in the first sampling to produce a clear acceptance or rejection result. However, the long time taken – often 1 day or longer – to obtain results in most microbiological procedures weighs against this strategy.

Sample Preparation, Bulk (Gross) and Pooled Samples

Sample collection, identification, shipment and preparation should follow the well-known rules. For microbiological analysis the test material must be carefully homogenized to minimize sampling error.

Instead of the usual procedure of analysing each sample separately, all samples can be combined into a composite sample and only an aliquot from the *bulk composite samples* is tested. This saves both time and labour. However, one consequence is that all information regarding the variability of the test characteristic is lost because only one result (realized arithmetic mean) is obtained and this has also been proven to be especially susceptible to outliers. For this reason, bulk (gross) sampling has only become accepted for determining the bacterial contamination on animal carcass surfaces with its unpredictable frequency distribution, or the evaluation of extremely skewed distributed characteristics such as the aflatoxin content of nuts or the histamine content of fish.

The term *pooling* describes when sample units are combined into one composite sample and the entire composite sample is then analysed. This procedure is useful with presence/absence tests where the random sample size is $n > 1$ and zero tolerance exists. This sample treatment is ideal for *Salmonella* testing where 25 g or 25 ml samples are combined and (pre-) enriched as a composite sample in a large container. A positive pool sample leads to rejection of the entire lot just as a single (or several) positive individual samples would be done. The theoretical threshold of detection can only be influenced by the total size of the pooled sample, whereas the size of the subsample is optional. Where there is heterogeneous distribution of microorganisms, it is recommended that the numbers of subsamples should be increased, with a corresponding reduction of the weight of the subsamples, while the total test sample quantity remains constant.

Sampling Plans

To simplify, each unitary sampling procedure is clearly defined once the number of analytical test units (= sample size) n and the acceptance number c have been established in a sampling plan for microbiological quality control. If the sample n has more than c defective units, the corresponding lot must be rejected. Otherwise (number of defective units $\leqslant c$), it must be accepted. The typical model of acceptance sampling runs as follows:

Lot → sample → sample plan → decision → accept/reject lot

'Intuitive' Sampling Plans

In view of the clarity and simple structure of the usual microbiological sampling design it is not surprising that plans of this kind are often drawn up intuitively. This expression characterizes a procedure in which the sample size is determined chiefly from the point of view of what is economically acceptable between the parties involved; these parties then agree on a plausible acceptance number. Although mathematical/statistical points of view do not enter into it, one can work with such intuitive plans of this kind and they may be found worthwhile in practice, even by national and international food legislation. The fact that pragmatic tests do not guarantee transparent quality assurance becomes evident when either the producer or the consumer has doubts about the reliability of a decision. In contrast to scientific sampling, neither the critical difference indicated by the acceptable negative deviations nor the probability of false decisions is concretely established or taken into account in the construction of a sampling plan.

Factors Influencing the Sample Size of a Sampling Plan

Those responsible for the introduction of sampling procedures are often only interested in finding out how many random samples must be drawn. To answer this question and to design a sampling plan for a specific requirement (What do we want the sampling plan to do for us?), the statistician needs to collate extensive data. A complex relationship exists between the sample size on the one hand and variance (homogenity of the lot), stringency and reliability on the other hand, this can be expressed by the equation:

Table 1 Minimum number of sample units required for evaluating infinitely large lots based on the number of tolerated percentage of bad units in the lot (reject quality level or RQL) as well as type 2 error[a]

RQL [%]	*1–β[%]*		
	95	*99*	*99.9*
25	11	17	25
10	29	44	66
5	59	90	135
1	299	459	688
0.5	598	919	1379
0.2	1497	2301	3451
0.1	2995	4603	6905

[a]Type 2 error [1 – β (in %)] is the reliability with which a bad lot should be rejected; rejection number $d \geqslant 1$.

$$\sqrt{\text{sample size}} = \text{reliability} \times \text{stringency} \times \text{variance}$$

Reliability (precision) denotes the probability of a correct decision at a given level of stringency, that means rejection of 'bad' lots and acceptance of 'good' lots. If a population with fewer defects than are normally tolerated is falsely rejected on the basis of a sampling result we call it producer's risk (type 1 error, α), whereas the consumer's risk (type 2 error, β) relates to a case where a 'bad' lot is wrongly accepted.

Stringency (discriminating power, critical difference, accuracy) in quality assurance is the degree of exceeding (microbiological) limits that can be detected with a given degree of probability. Smaller differences often go unnoticed. As with reliability, the extent of testing is positively correlated with stringency requirements.

The equation mentioned above shows that even plans where there is a high level of examination cannot be both reliable and stringent. For a given sample number, the two criteria are inversely proportional. The combined influence of reliability (1–β) and the stringency parameter called reject quality level (RQL) on the number of samples is demonstrated in **Table 1** for presence/absence tests.

The minimum number of sample units are shown where a bad lot can be detected under prevailing conditions such that at least one sample unit of the entire random sample yields a positive result (acceptance number $c = 0$, rejection number $d \geqslant 1$). Attention should be paid to the sample size $n = 60$ and $n = 300$. These ensure that, for RQLs of >5% and >1% respectively, the probability of (false) acceptance does not exceed 5%. Related to the detection of *Salmonella*, such high sample numbers should not be blamed on biometrics but are due to the statistical peculiarities of an alternative criterion.

In contrast to both the previously mentioned parameters, the *variance* cannot be stated. It is inherently related to the criterion itself.

Qualitative (alternative, Discontinuous or Discrete) Characteristics Generally manifest in the opposites good versus bad or present versus absent, these characteristics frequently appear in microbiology. The best-known example is probably the presence/absence test, where a bacterial cell is present or absent in a defined analytical volume or unit.

These qualitative characteristics usually follow the Poisson or binomial distribution. The variance in this case is determined by correlating good and bad units. This can be calculated as a direct mathematical relationship between variance and probability of occurrence. If one estimates the number of acceptable units in a random sample, the corresponding variance can be derived.

Occasionally there are also contagious alternative distributions with elevated variation (e.g. negative binomial distribution) for microbiological criteria. However, the sampling plans were not modified, because the clumping factor often only affects reliability.

Quantitative (Continuous) Characteristics These may be of any possible value in a defined distance. Almost all analytical data belong to this group, including bacterial counts in food samples. Such data points usually follow a normal distribution, at least after logarithmic (or square root) transformation. One of the characteristics of normal distributions is that no relationship exists at all between the mean and variance! Consequently, in order to construct an appropriate testing plan, the variation must be captured and determined independently of the average. It is either estimated simultaneous with the mean directly from random samples or is known from preliminary trials.

Unless the material being examined is very homogeneous, small sample sizes lead to imprecise and uncertain decisions. The single sample is the least informative test unit and should only be used for preliminary testing of freshly prepared fluids and powders. In contrast, the increase in information obtained from more than five samples ($n = 5$) is slight because of the square root function. Therefore, the addition of more random samples with complicated analysis is often not worth the effort, even for heterogeneous foods, especially when it involves a quantitative characteristic.

The importance of the *relationship between sample size n and population size N* is often overestimated. There are no interactions with reliability and accuracy in the extremely frequent cases where the relationship $n/N < 0.1$. Precision and stringency depend only on the sample numbers. A constant sampling fraction means that lots of smaller size are tested with a smaller margin of safety – a philosophy which is no longer followed within modern quality control.

A random sampling plan must not only fulfil theoretical statistical demands; it must also be practical and take *non-biometric factors* into account. For this reason the scope of random sampling should first be of a financially justifiable magnitude, that is, the increase in value of product quality through use of a statistical quality control programme should exceed the costs of this testing programme. Furthermore, the stringency and reliability of a random sampling plan are based on health risks, in addition to the testing costs within a microbiological quality control programme. In this category are included:

- type and extent of the hazard based on the target organism
- probability of contamination of the raw materials
- post-harvest process technology
- type of treatment which the product usually receives from the consumer
- consumption of the product by groups with lowered immune resistance (young, old, pregnant, immunosuppressive: YOPI).

The food products are placed into categories based on the number and importance of the risk factors and in turn these are placed into specific reliability and stringency requirements and finally into random sampling plans. The absolute number of random samples per lot is arbitrary and cannot be determined mathematically from non-biometric factors.

A compromise needs to be made between stringency, reliability, analytical effort and health risks to achieve a realistic level of testing. *There are no completely optimal plans, nor is there a single universal plan for all control situations.*

ABC-STD 105

A number of sampling plans and procedures have been developed for testing on the basis of qualitative characteristics. They are based on the generally recognized MIL-STD 105 D (1963), which is known internationally as ABC-STD 105. This was incorporated almost unaltered into ISO 2859–1974. The sampling plans were calculated on the model of binomial distribution (or hypergeometric distribution in the case of small lots) and also of Poisson distribution. However, the use of the plans in food microbiology has been impeded not only by the complicated variety of plans and inspection levels, which were not adapted to bacteriological needs, but also by the often unacceptable large sample sizes which are required.

Table 2 Product hazard characteristics and acceptance plans (Foster plan)

Category I	Non-sterile food for children, old people and the sick $n = 60$; $c = 0$ or $n = 95$; $c = 1$
Category II	Food with all three hazard characteristics – sensitive ingredient, no destructive step during manufacture, likelihood of growth if abused $n = 30$; $c = 0$ or $n = 48$; $c = 1$
Category III	Food with two hazard characteristics $n = 15$; $c = 0$ or $n = 24$; $c = 1$
Category IV	Food with one hazard characteristic; does not usually need to be controlled $n = 15$; $c = 0$ or $n = 24$; $c = 1$
Category V	Food with no hazard characteristic; does not usually need to be controlled $n = 15$; $c = 0$ or $n = 24$; $c = 1$

Unitary (one-step) Attributive Two-class Sampling Plans

As has already been explained, the simplest concept to control an alternative characteristic is to determine the number of random samples required, and then to fix how many sample units may exceed the limit m without rejecting the lot (acceptance number c). In the case of presence/absence tests, a result above m means a positive result. Regardless of whether the ABC-STD 105 also contains such sample designs, the first successful attempt at developing specific test plans appropriate for microbiological demands goes back to the *Salmonella* Committee of the National Research Council, US. The sampling procedures for monitoring *Salmonella* contamination have become known as the Foster plan. The problem-oriented approach to risk evaluation led to five categories with sample sizes between 15 and 60 25-g units for $c = 0$ or between 24 and 95 sample units for $c = 1$ (**Table 2**).

Later, the question arose as to how far tests of this kind can be applied, not merely to the control of *Salmonella* contamination, but also to other traits of microbiological quality assurance. In this situation the International Commission on Microbiological Specification for Foods (ICMSF) published sampling plans for microbiological analysis as an attempt at a universally applicable sampling strategy.

To improve the transparency of sampling plans and the decisions resulting from them, the operating characteristic curve (OC function, acceptance curve), a well-known function used in statistical decision theory, was introduced by ICMSF in describing a microbiological sampling plan. An example showing acceptable quality level (AQL) and lot tolerance per cent defectives (LTPD) is to be found in **Figure 1**.

The OC curve visualizes, for a particular sampling plan (n/c), the probability of acceptance as a function of the actual condition (e.g. number of *Salmonella*-

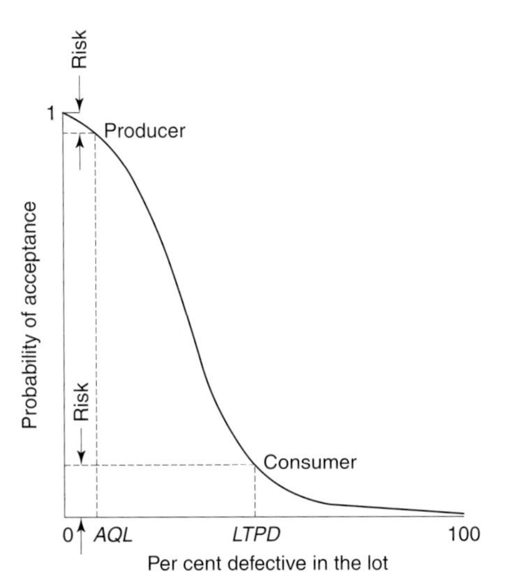

Figure 1 Acceptance curve (operating characteristic or OC function) showing producer's and consumer's risk correlated with acceptable quality level (AQL) and lot tolerance per cent defectives (LTPD). The AQL is the point on the horizontal axis measured from an OC curve such that a lot with that per cent of defectives has a high probability (e.g. 99%) of acceptance. This is also referred to as the producer's risk or low probability (e.g. 1%) that a good lot will be rejected. Similarly, the LTPD is referred to as the consumer's risk or the quality level at which a poor lot has a low probability (e.g. 5%) of being accepted.

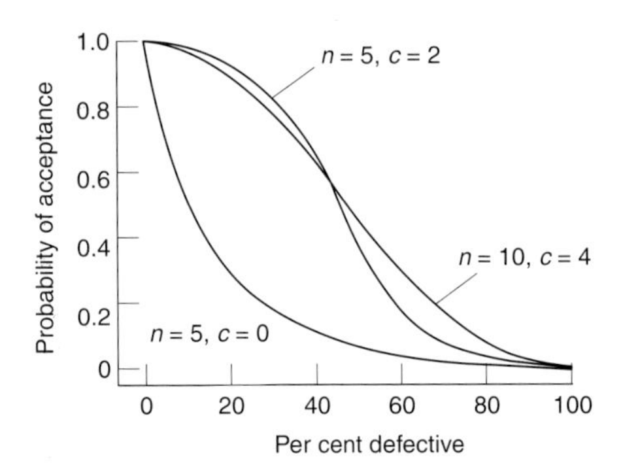

Figure 2 Operating characteristic (QC) function for three sampling plans with two different sample sizes and two different ratios of *c* to *n*.

positive 25 g samples) of the population. For each defective unit percentage the probability of achieving an acceptable sampling result is read from the OC function curve. The steeper the course followed by this function, the sharper the line drawn between bad and good populations. Only with the aid of the relevant OC function can one imagine how a sampling plan works. **Figure 2** shows the influence of two different sample sizes and acceptance numbers on the OC function curve.

The ICMSF sampling plans also work with risk categories, here called cases. The 15 possible cases derived from a combination of five health hazard classes with three forms of treatment are shown in **Table 3**.

The attributive two-class sampling plan can also be used for quantitative characters (e.g. colony counts) by transferring these quantitative data into attributive data. Such a transformation is achieved in two steps: first, a contamination limit *m* is fixed; then, as a second step, the transformation itself is carried out by assigning all test results to groups according to whether they exceed the previously determined limit or not. The result of this procedure is an attributive ± structure of the trait being analysed. Transforming a piece of quantitative information into an attributive criterion avoids the difficulties of having to determine the variability. In the case of quantitative criteria it is independent from the mean, whereas the mean and the variance of an alternative criterion are correlated by simple mathematical rules (**Fig. 3**).

Unitary Attributive Three-class Sampling Plans

The transformation of a quantitative into a qualitative criterion lowers the level of information considerably, because neither the actual amount of variation within the lot can be derived, nor how far away the single results are from the limit. In order not to lose the information contained in the cfu numbers or other quantitative data – as happens when applying the two-class plan – the three-class plan was developed and promoted. This sampling plan could be interpreted as a combination of two attributive two-class plans (**Table 4**).

One random sample of *n* units is tested for the two different microbiological limits *m* and *M* at the same time. The criterion *m* in common practice reflects the upper limit of a good manufacturing practice (GMP) and should rely on surveys on the microbiological condition of samples, drawn from consignments that were manufactured, stored and distributed under prescribed good conditions, which had previously been validated. In contrast, the criterion *M* marks the borderline beyond which the level of contamination is hazardous or unacceptable. Therefore, corresponding to its name, this plan discerns three classes of microbiological quality: the acceptable range from 0 to *m*, the marginally acceptable class between *m* and *M* and the so-called class of defective quality above *M*. An acceptance number *c* is assigned to each of the two limits *m* and *M*: the value assigned to *M* is always $C_M = 0$. A lot will be rejected if at least one unit in a sample exceeds the limit *M* and/or if more units range above the limit *m* than the acceptance number c_m would permit. In some guidelines c_m is defined as the number of samples that can fall between *m* and *M* without the lot being considered unacceptable. This modification leads to the same decisions, but appears somewhat confusing in consideration of the theoretical background.

Table 3 Suggested sampling plans for combination of degrees of health and conditions of use (International Commission on Microbiological Specification for Foods)

	Conditions in which food is expected to be handled and consumed after sampling, in the usual course of events[a]		
Degree of concern relative to utility and health hazard	*Conditions reduce degree of concern*	*Conditions cause no change in concern*	*Conditions may increase concern*
No direct health hazard Utility, e.g. shelf life and spoilage	Increase shelf life Case 1 3-class $n = 5$, $c = 3$	No change Case 2 3-class $n = 5$, $c = 2$	Reduce shelf life Case 3 3-class $n = 5$, $c = 1$
Health hazard	Reduce hazard	No change	Increase hazard
Low, indirect (indicator)	Case 4 3-class $n = 5$, $c = 3$	Case 5 3-class $n = 5$, $c = 2$	Case 6 3-class $n = 5$, $c = 1$
Moderate, direct, limited spread	Case 7 3-class $n = 5$, $c = 2$	Case 8 3-class $n = 5$, $c = 1$	Case 9 3-class $n = 10$, $c = 1$
Moderate, direct potentially extensive spread	Case 10 2-class $n = 5$, $c = 0$	Case 11 2-class $n = 10$, $c = 0$	Case 12 2-class $n = 20$, $c = 0$
Severe, direct	Case 13 2-class $n = 15$, $c = 0$	Case 14 2-class $n = 30$, $c = 0$	Case 15 2-class $n = 60$, $c = 0$

[a]More stringent sampling plans would generally be used for sensitive foods destined for susceptible populations.

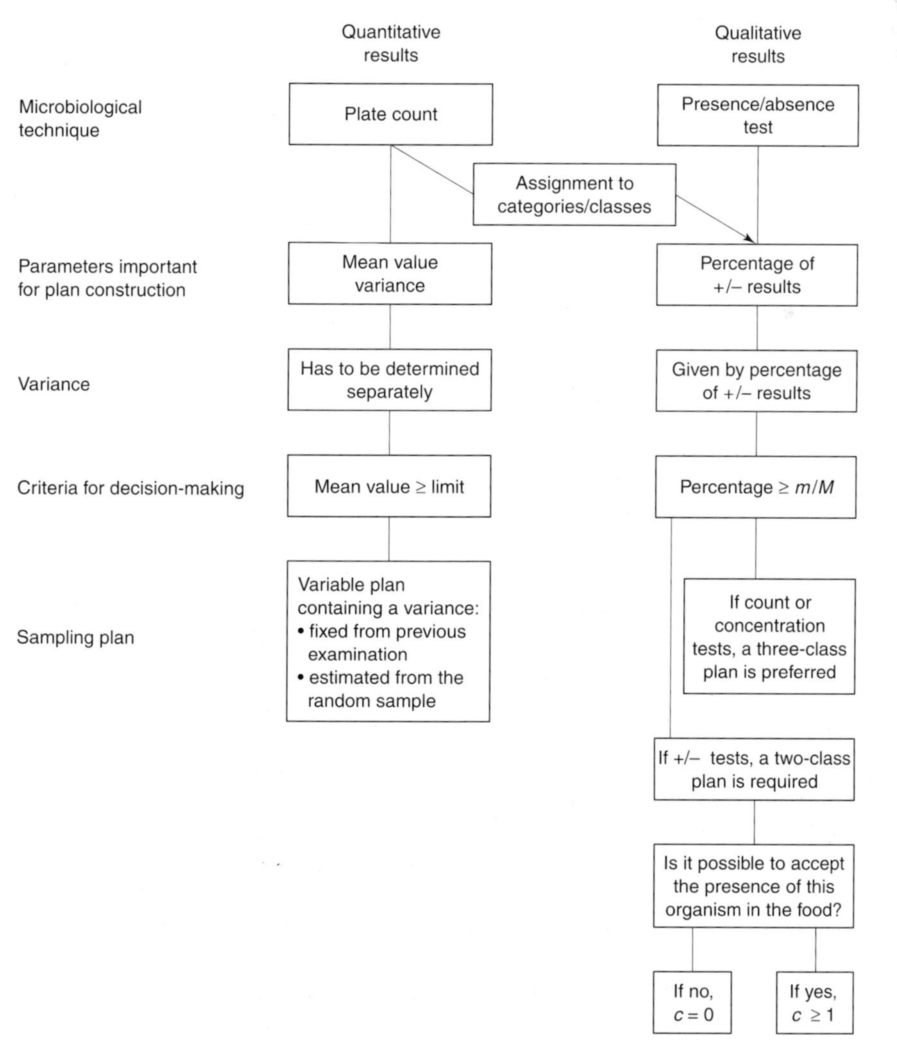

Figure 3 Design of variable and attributive sampling plans.

Table 4 Concept of the attributive two- and three-class plans

	Number of samples examined (n)	*Microbiological limit*	*Acceptance number (c)*
Two-class plan	$n = 5$[a]	m	$c_m \geq 1$
		⇩	
Additional limit		M	$C_M = 0$
		⇩	
Three-class plan	$n = 5$[a]	m	$C_M \geq 1$
		M	$C_M = 0$[b]

[a]$n = 5$ is common but not obligatory.
[b]Because C_M for M is generally 0, in most three-class plans only c_m is mentioned (given as c).

With the limit M and the corresponding acceptance number $C_M = 0$, the principle of zero tolerance is incorporated into the three-class plan. An acceptance number of $C_M = 0$ makes sense only if applied to pathogenic microorganisms or their toxins, whereas it seems inappropriate to reject a whole consignment when a single sample exceeds M for standard plate count or numbers of indicator organisms. A lot which still meets GMP conditions should not be condemned only because one outlier crosses the M border. In practice, the decision, do not accept the lot, seldom leads to real rejection. In most cases preventive measures are intensified.

In the three-class sampling plan where $n = 5$, $c_m = 2$ and $C_M = 0$, it was recently demonstrated that the additional risk of rejecting a lot with an acceptable, technological unavoidable standard deviation σ (in log units) solely due to a single sample lying above M is reasonable, if the difference between M and m does not fall within the distance $1.84 \times \sigma$. Results of surveys indicate that the usually chosen distances $(M - m)$ of 0.5 log units for recently homogenized foods and 1.0 log units for material with heterogeneous distributed microorganisms fulfil this condition. In addition to variation between samples these values include analytical errors of good laboratory practice (GLP) laboratories and sampling errors. Whether the limit M is the first to be fixed (for instance, for pathogenic bacteria) as the basis of calculating a GMP level with a suitable distance between m and M, or the lower GMP limit m is the one to start with (for instance, for total count or hygiene indicators) depends on the problem at hand. It should be mentioned that the original robust design of three-class sampling plans intentionally did not take into account the variance of the criterion.

After first publication the three-class attributes plans have undergone considerable changes in the hands of some users. The original design envisaged only two decisions for food – acceptance or rejection – in spite of being grouped into three classes. The design was amended to include three decisions for the three classes. Whether in the form of a traffic light system, red–yellow–green, or by tolerance intervals, the possibility of toleration with increased control (meaning retesting to verify the result) was included in the plan as a third decision in addition to immediate rejection or acceptance. Furthermore, some governmental regulations contain a fourth judgment – spoiled – this refers to a limit lying 1 or 2 log units above M. A statistical evaluation – not to mention a calculation of the OC function – of these modified three-class attribute plans has never been made. Thus the three or four-decision plans are pseudo-statistical procedures based on ICMSF instructions but which have lost their clearly defined working method and reverted to the intuition stage.

Variables Plans

If the result of a bacteriological examination is available in the form of a bacterial count (or as another quantitative characteristic content like lactate or histamine), the use of attribute plans always means a loss of information. The degree to which the results fall above or below the limit is not taken into account; also the true variability is not included in the construction of the plan. The use of the real standard deviation can thus lead to an improved decision-making process and even to economic advantages. But those plans require the data to follow a normal or log-normal frequency distribution. Apart from few investigations which expressed a preference for normal distribution, many surveys show that the log-normal distribution model is applicable to the behaviour of bacterial counts in parallel samples from a lot. The fact that microorganisms grow exponentially is itself already a plausible explanation for this assumption. There is no reason to forgo using variables plans and to prefer the robust working attributive three-class plan, which comprises no model for the real frequency distribution.

In the field of quality control, the so-called mean value plan is often chosen because it is the easiest to understand and apply. The general focus of this kind of test for significance is the decision as to whether the mean of the (log-) transformed colony counts exceeds the limit m with a given probability or not (Fig. 3). Incidentally, the mean value plan has a certain correspondence to an attributive two-class plan where $n = 5$ and $c = 2$, because in that plan acceptance/rejection depends only on the third highest value, which represents the median. In other words, a lot is to be condemned solely if the median in alternative plans or the arithmetic mean in variables plans lies above the limit m.

Unfortunately, the first recognized variables plans in food microbiology preferred an approach comparable to attributive three-class plans. As is the case with the attribute plans, a population gives cause for rejection if a certain percentage of samples P of the lot exceeds the critical bacterial count C. In statistics we call these limits *tolerance limits* (percentiles). The Kilsby plan works not only with two limits (rejection limit C and acceptance limit C_m) and two different corresponding percentiles (90% for C and 70% for C_m) but also estimates the variance from the sample. The strategy has rarely been used in microbiological food control, however, because of its complicated structure and it is difficult to understand.

Recent surveys realized a standard deviation of the total aerobic count of units of a lot often remained constant during the longer-term production sequence. This finding suggests that for variables plans a fixed variance which is established from experience (empirical standard deviation σ) can be used instead of estimating variation from the current sample from case to case. The advantage of working with a fixed variance is a definite reduction in the number of samples required. A variables plan of this kind where $n = 2$ can show an OC function comparable to a three-class attributes plan where $n = 5$ and $c = 2$, or to a variables plan using an estimated variance of $n = 5$.

Sometimes gauging offices use mean value plans with an additional 95% tolerance limit for single values. Working with a fixed empirical standard deviation σ and applying the design to microbiological food control, this modification of variables plans coincides with the philosophy and principles of attributive three-class plans ($n = 5$, $c_m = 2$, $C_M = 0$) where the difference between M and $m = 1.84 \times \sigma$. Altogether a mean value plan with an outlier trap has succeeded in the evaluation of quantitative microbiological data.

Control Charts

The random sampling plans discussed previously serve to evaluate uniform, discrete, defined lots. The information from a previous test does not influence the next decision. Such acceptance sampling plans for quality control of lots can also in principle be used for the continuous control of production stages. However, they represent a procedure which is more passive than active and with which the essential goals of quality control cannot be reconciled. The quality control strategy should:

- yield information about the characteristics of the process, especially average quality, as well as unavoidable variations
- maintain proper production as long as possible
- clearly show deviations from the quality standard so that production of faulty units is recognized early.

The task of lot testing, namely to accept or reject certain production units, becomes of less significance.

A control chart is employed for continuous evaluation of quality control to determine the agreement between the fixed standard and the reality of practice. The essential characteristic of a control chart is that it forms a continuous graphic representation of the quality status of production under consideration of prescribed tolerances.

Every control chart where random sampling results are recorded in chronological order requires the sample size to be established ($n \geq 2$), as well as sampling frequency, target organism, observed statistical parameter, critical limits and fixed corrective actions in case of rejection. Most quality control charts have 95% limits as attention limits and 99% limits as control limits, where a single event exceeding the attention limits – in practice the first warning step – provides a higher level of caution, while results outside the control limits demand immediate remedial action and countermeasures in the production process.

Although often proposed, a practical introduction of quality control monitoring charts within microbiological control has seldom succeeded to date, or at least there is scarcely any reference in the literature to its use. The wealth of data created by the internal quality control programme within the framework of quality control requirements presents the best situation of optimally exploiting the informational content of these results through quality control charts.

Future Scope of Sampling

In contrast to the well-developed and often-tested random sampling plans in the technical industry, many problems of quality control and risk control of food products still demand a satisfactory solution. Widely scattered characteristics due to raw materials and production processes which are hard to standardize, labour- and resource-intensive analytical methodology as well as controversial cut-off values make it difficult to fit the usual statistical concepts to the situation.

Therefore, there is a great demand for the simplest possible sampling plans whereby more or less intelligent solutions may be found for any problem. However, nobody can avoid observing objective mathematical rules. Only when it becomes possible to develop simple microbiological rapid methods suitable for on-line monitoring of the critical control points will one be able to forgo sampling plans. It is however a long (or endless) road to travel until this occurs.

See also: **Hazard Appraisal (HACCP)**: The Overall Concept.

Further Reading

Gould WA and Gould RW (1988) *Total Quality Assurance for the Food Industries*. Baltimore, MD: CTI Publications.

Harrigan WF and Park RWA (1991) *Making Safe Food: A Management Guide for Microbiology Quality*. London: Academic Press.

Hildebrandt G and Weiss H (1994) Sampling plans in microbiological quality control. *Fleischwirtschaft* 74: 56–59, 165–168.

IFST Professional Food Microbiology Group (1997) Development and use of microbiological criteria for foods. *Food Science and Technology Today* 11: 137–177.

International Commission on Microbiological Specifications for Foods (ICMSF) (1986) *Microorganisms in Foods 2. Sampling for Microbiological Analysis: Principles and Specific Applications*. 2nd edn. Toronto: University of Toronto Press.

International Commission on Microbiological Specifications for Foods (ICMSF) (1988) *Microorganisms in Foods 4. Application of the Hazard Analysis Critical Control Point (HACCP) System to Ensure Microbiological Safety and Quality*. Oxford: Blackwell Scientific Publications.

Jarvis B (1989) *Statistical Aspects of the Microbiological Analysis of Foods*. Amsterdam: Elsevier.

Kilsby DC (1982) Sampling schemes and limits. In: Brown HM (ed.) *Meat Microbiology*. P. 387–421. London: Applied Science Publishers.

Messer JW, Midura TF and Peeler JT (1992) Sampling plans, sample collection, shipment, and preparation for analysis. In: Vanderzant C and Splittstoesser DF (eds) *Compendium for the Microbiological Examination of Foods*, 3rd edn. P. 25–49. Washington, DC: American Public Health Association.

Niemelä S (1983) *Statistical Evaluation of Results from Quantitative Microbiological Examinations*. Uppsala: Nordic Committee on Food Analysis.

Tillett HE and Carpenter RG (1991) Statistical methods applied in microbiology and epidemiology. *Epidemiol. Infect.* 107: 467–478.

Scanning Electron Microscopy *see* **Microscopy**: Scanning Electron Microscopy.

SCHIZOSACCHAROMYCES

G H Fleet, Department of Food Science and Technology, The University of New South Wales, Sydney, Australia

Characteristics of the Genus

The genus *Schizosaccharomyces* represents a small group of yeast species that undergoes vegetative reproduction by fission. In this process, the cell elongates by polar growth and, after nuclear division is completed, a cross wall or septum is formed centripetally and the two cells separate. This method of division gives the cells a characteristic cylindroidal, sausage shape, depending on their age. The cytology and molecular dynamics of cell division in these yeasts have been extensively studied. The absence of budding, which is a typical feature of most other yeasts, is a distinctive characteristic of *Schizosaccharomyces*. Sexual reproduction within the genus occurs by the production of ascospores after conjugation of haploid vegetative cells. Growth of the yeasts on malt-, potato- or cornmeal agars at 25°C for 4–7 days usually induces ascospore formation which is readily seen by examination of the cells using light microscopy. The number of ascospores per ascus, spore shape and surface topography of the spores vary with the species. Sporulation is frequently observed during growth in liquid media and is often preceded by very obvious agglutination and flocculation of the cells.

Species of the genus are also distinguished from most other yeasts by the chemical composition of their cell walls. Whereas most ascomycetous yeasts possess cell walls composed of $\beta(1\rightarrow3)$ glucans, $\beta(1\rightarrow6)$ glucans, mannan and chitin, the walls of *Schizosaccharomyces* spp. are notable for the presence of $\alpha(1\rightarrow3)$ glucan in addition to the β-glucans, a galactomannan in place of the mannan, and the absence (or only trace amounts) of chitin. Other features of the genus include their strong fermentation of sugars, hydrolysis of urea, inability to utilize nitrate as a nitrogen source, and a negative reaction with diazonium blue to distinguish them from basidiomycetous yeasts. Nuclear staining and pulsed field electrophoresis show two or three large chromosomes

Table 1 Key properties of species within the genus *Schizosaccharomyces*

Property	S. pombe	S. japonicus	S. octosporus
Mol% G+C	42	32–36	40
Fermentation of			
Sucrose	+	+	–
Raffinose	+	+	–
Assimilation of			
Sucrose	+	+	–
Raffinose	+	+	–
D-Gluconate	+	–	–
Growth at			
32°C	v	+	–
37°C	–	+	–
Number of ascospores	4	6–8	6–8
True hyphae	–	+	–
Coenzyme	Q10	nd	Q9

v, variable; nd, not detected. From Kurtzman and Fell (1998)

in these yeasts, compared with the 16 chromosomes found in *Saccharomyces cerevisiae*.

Current taxonomy recognizes three species, namely, *Schizosaccharomyces pombe*, *Schizosaccharomyces japonicus* and *Schizosaccharomyces octosporus*, that are readily differentiated by the properties shown in **Table 1**. This classification has evolved through consideration of morphological, biochemical and physiological properties in conjunction with data on the composition of DNA, nuclear DNA reassociation studies and partial ribosomal RNA sequences. Earlier taxonomies accepted four species and two varieties in the genus, namely, *S. pombe*, *S. japonicus* var. *japonicus*, *S. japonicus* var. *versatilis*, *S. octosporus* and *S. malidevorans*. The two varieties of *S. japonicus* were originally differentiated on the basis of ascospore shape and ascospore reaction with iodine, but subsequent nuclear DNA reassociation (hybridization) data showed them to be 85–90% related. *S. malidevorans* was originally described as a separate species on the basis of a few differences in carbohydrate utilization tests but now it has been found to be 98% related to *S. pombe* from nuclear DNA reassociation studies.

Over the years, it has been proposed that *Schizosaccharomyces* should be separated into two or three genera, that have been named as *Hasegawaea*, *Octosporomyces* and *Schizosaccharomyces*. This classification is based on differences in properties such as composition of coenzyme Q, cytochrome spectra, fatty acid profiles, number and morphology of ascospores, and electrophoretic mobility of several metabolic enzymes. The genus *Hasegawaea* was proposed to represent the strains of *S. japonicus* and *Octosporomyces* was proposed to contain the strains of *S. octosporus*. A key criterion used in proposing the genus *Hasegawaea* was the distinct presence of higher amounts of linoleic acid in the fatty acid profiles of isolates of *S. japonicus* compared with the other species. Also, the sequence data on ribosomal RNA from these yeasts provided some justification for their inclusion in a new genus. Currently, yeast taxonomists are reluctant to establish new genera on the basis of a few phenotopic or genotypic data and, until further definitive evidence is provided, the accepted classification of *Schizosaccharomyces* is that given in Table 1.

Physiological and Biochemical Properties

The fission mode of cell division by *Schizosaccharomyces* spp. has attracted significant fundamental research aimed at understanding the physiology and molecular biology of this process. In contrast, very little information has been published on the factors which affect the survival, growth and biochemical activities of these yeasts in food ecosystems. The few data available are largely restricted to studies with *S. pombe*. A systematic and detailed study of these properties is needed.

These yeasts are facultative anaerobes and metabolize hexose sugars principally by fermentation through the glycolytic pathway to produce mainly ethanol and carbon dioxide, and a range of secondary metabolites (**Table 2**). Ethanol concentrations as high as 14% can be obtained. The concentrations of other volatile metabolites produced by *S. pombe* are not distinctive but compared with yeasts in other genera, high concentrations of hydrogen sulphide have been reported. Glycerol is catabolized, but the pathway is different from that in *Saccharomyces cerevisiae* and involves an NAD^+-linked glycerol dehydrogenase and a dihydroxyacetone kinase. Species of *Schizosaccharomyces* are notable among yeasts for their strong ability to metabolize L-malic acid to ethanol under anaerobic conditions. Malate is first decarboxylated to pyruvate by an NAD-dependent malic enzyme. The pyruvate is decarboxylated to acetaldehyde which is reduced to ethanol (**Fig. 1**). A proton–dicarboxylate symport has been demonstrated for the transport of malic acid into *S. pombe* and the presence of glucose is required for L-malic acid metabolism.

The absence of data suggests that extracellular amylases, pectolytic enzymes and proteases are not produced by *Schizosaccharomyces* spp. but more specific investigation is needed. However, *S. pombe* can be transformed to degrade starch with plasmids carrying the glucoamylase gene of *Saccharomyces diastaticus*. *Schizosaccharomyces octosporus* produces an extracellular lipase that is able to hydrolyse lard to

Table 2 Some relevant technological properties of *Schizosaccharomyces pombe*

End-product	*Concentration*[a]	*Preservative*	*Minimum concentration to inhibit growth*[b]
Ethanol	2–14%	Benzoic acid	600 mg l^{-1}
Propanol	1–15 mg l^{-1}	Sorbic acid	600 mg l^{-1}
Isobutanol	3–22 mg l^{-1}	Sulphur dioxide	125–200 mg l^{-1}
Isoamyl alcohol	5–40 mg l^{-1}	Methyparabenzoic acid	750 mg l^{-1}
Phenyl ethanol	13 mg l^{-1}	Acetic acid	16 g l^{-1}
Ethyl acetate	10–40 mg l^{-1}	Propionic acid	10 g l^{-1}
Isoamylacetate	< 0.1 mg l^{-1}	Carbon dioxide	6.6 g l^{-1}
Acetaldehyde	5–160 mg l^{-1}		
Acetic acid	0.1–0.3 g l^{-1}		

[a] As determined after fermentation of grape juice.
[b] Values measured at pH 3.5.

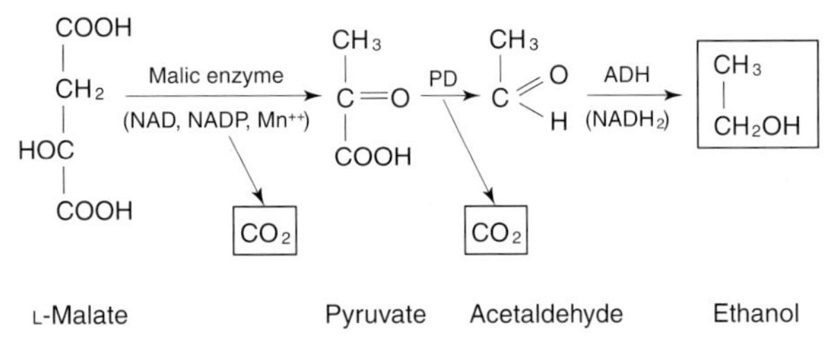

Figure 1 Biochemical mechanism for the fermentative decomposition of malic acid by *Shizosaccharomyces*. PD, pyruvate decarboxylase; ADH, alcohol dehydrogenase.

produce significant quantities of stearic acid, but the lipid degrading ability of other species of *Schizosaccharomyces* is not known.

The cardinal temperatures for the growth of *Schizosaccharomyces* species appear not to have been reported. The most recent taxonomy (Table 1) indicates *S. japonicus* as the only species able to grow at temperatures as high as 37°C, but earlier literature reports the growth of *S. pombe* and *S. octosporus* at this temperature. There is little doubt that these species grow well at temperatures of 15–30°C, but the minimum and maximum temperatures for growth need to be determined. It is also uncertain as to what pH values limit their growth. One study suggests that best growth occurs at pH values around 5.5, but it is significant that these yeasts are often isolated from the juices of acid fruits and readily grow in wines, at pH 3.0–3.5, although at slower growth rates than *Saccharomyces cerevisiae*.

With respect to occurrence and growth in foods, the most distinctive feature of these yeasts is their ability to tolerate low water activity environments as imposed by the presence of high sugar concentrations. In this context, they are widely recognized as xerotolerant or osmotolerant yeasts, capable of growth in the presence of 50% glucose (and possibly 60% glucose) at water activity (a_w) values as low as 0.78. Their xerotolerance depends on the solute and yeast species. For *S. pombe* and *S. octosporus*, minimum a_w values of 0.89–0.90 have been reported with glucose, fructose and glycerol as the stressing solutes, and for *S. japonicus* these values are 0.92–0.94. However, all these species are less tolerant of high salt (sodium chloride) concentrations and do not grow at water activity values less than 0.95 in the presence of this solute. There are isolates of *S. pombe* that are not able to grow in the presence of 3% NaCl, pH 5.5. Growth in low water activity environments is accompanied by the production of intracellular glycerol as a compatible solute.

Another notable property of *Schizosaccharomyces* spp. is their relatively high tolerance to preservatives, such as benzoate, sorbate, acetic acid and sulphur dioxide, that are commonly used in food processing (Table 2), although the data are mostly limited to observations with one species, *S. pombe*. *S. pombe* is generally 2–3 times more resistant to these preservatives than *Saccharomyces cerevisiae*. Resistance to inactivation by heat processing has been studied for *S. pombe*, where approximately 99% of the population suspended in phosphate buffer, pH 6.5, containing 48% sucrose (a_w 0.95), was killed at 65°C within 3 min (D_{65} 1.99 min). Faster death rates were obtained when sucrose was omitted from the buffer. However, greater thermotolerance is induced by pre-exposure of the yeast cells to mild heat (40°C), which induces the synthesis of intracellular trehalose as a thermoprotectant.

Significance in Foods and Beverages

There are negative and positive consequences of *Schizosaccharomyces* in foods and beverages. First, they have been implicated in the fermentative spoilage of several commodities and, second, they may contribute to fermentations in the production of alcoholic beverages and possibly other products.

Compared with other species, they are not commonly isolated as food spoilage yeasts. Nevertheless, the early and recent literature shows their repeated isolation from foods and beverages of high sugar content. All three species have, on different occasions, been isolated from honey, sugar syrups, sugar cane molasses and fruit juice concentrates where they have been responsible for fermentative spoilage, often with the production of unpleasant hydrogen sulphide odours. Growth of *S. octosporus* on the surfaces of dried prunes and figs produces a white sugar-like coating. In several cases, the addition to the product of sulphur dioxide, benzoate or sorbate at concentrations of 100–250 p.p.m. was insufficient to prevent these forms of spoilage. Clearly, their osmotolerance, resistance to preservatives and tolerance of acidic conditions are key properties which select for their occurrence and growth in these commodities. *Zygosaccharomyces rouxii* and *Zygosaccharomyces bailii* are also found in commodities with high sugar contents but, interestingly, at much higher frequencies than *Schizosaccharomyces* spp. Possibly, some form of antagonism exists between these species. Grapes, grape juice, apples, apple juice, cactus juice, wine and cider are other environments from which *S. pombe* has been frequently isolated.

Along with several other yeast species, *S. pombe* has been isolated from French natural sour doughs, but it was not clear whether they were involved in contributing to the dough fermentation. A similar situation exists with cocoa bean fermentations where, on some occasions, this species has been found but its role in the process is unknown. The Kombucha or tea fungus fermentation, which has attracted much popular press in recent years, appears to be a symbiotic involvement of *S. pombe*, *Saccharomyces* spp., *Pichia* spp. and acetic acid bacteria. However, the quantitative contribution of *S. pombe* to the ecology and biochemistry of this fermentation has not been defined.

Palm wines (toddy) are traditional alcoholic beverages produced in several African and south and southeast Asian countries. They are produced by natural fermentation of the sap collected from various species of palm trees. The fermentation develops from a mixed ecology of lactic acid bacteria, acetic acid bacteria, *Zymomonas* spp. and yeasts which comprise species of *Schizosaccharomyces*, *Saccharomyces*, *Candida* and *Pichia*. The alcohol content of the final product varies but is generally around 2%. Further study is required to define precisely the role of *Schizosaccharomyces* species in these fermentations. Duma is another traditional beverage in which *Schizosaccharomyces* spp. are reported to be involved in association with other yeasts and bacteria. It is produced in Sudan from the fermentation of honey. Rum is an alcoholic beverage produced by distilling fermented sugar cane molasses. As traditionally produced by countries in the West Indies, the fermentation is allowed to develop from the natural growth of *S. pombe* and *Saccharomyces cerevisiae* that originate from the molasses. The high sugar content of the molasses selects for growth of the osmotolerant *Schizosaccharomyces*. More modern processes aim to control the fermentation by inoculation of the molasses with cultures of *Saccharomyces cerevisiae* but it is reported that natural fermentations, where *S. pombe* makes a contribution, give a product with more complex and interesting flavour.

It is in the production of wine from grapes where the activity of *Schizosaccharomyces* spp. has received most attention. As mentioned above, *Schizosaccharomyces* spp. are sometimes isolated from grape and grape musts and, on rare occasions, they have been observed to successfully conduct a wine fermentation. However, their ability to rapidly degrade L-malic acid (Fig. 1) rather than conduct an alcoholic fermentation is their most desirable property. Many grape juices contain undesirable concentrations of L-malic acid which carries through into the final wine, giving it an unbalanced, acid flavour that detracts from its overall sensory quality. It is now common practice to put these wines through a secondary fermentation, called the malolactic fermentation, which is conducted by the bacterium, *Leuconostoc (Oenococcus) oenos*. In this process, L-malic acid is decarboxylated to L-lactic acid by the action of the malolactic enzyme, and the wine is sufficiently deacidified to give a marked improvement in its sensory profile. The technical hurdles and extra inconvenience of this operation have inspired the search for yeasts that might simultaneously degrade L-malic acid and conduct the alcoholic fermentation. The principal wine yeast, *Saccharomyces cerevisiae*, as well as many other species associated with wine fermentations, only partially degrade L-malic acid. However, there are strains of *S. pombe* that give complete degradation of this acid by action of the malic enzyme (Fig. 1) and various approaches have been investigated to utilize this yeast in wine fermentations. Unfortunately, there are two problems that have frustrated the success of these initiatives. First, strains of *S. pombe* can contribute undesirable concentrations of hydrogen sulphide, volatile acidity and other off flavours to the wine and, second, this species is readily overgrown and eliminated by *Saccharomyces cerevisiae*. In mixed culture fermentations with the two species, it is necessary to inoculate at the ratio of 75% *S. pombe* and 25% *Saccharomyces cerevisiae* to achieve effective L-malic acid degradation, but such combinations lead to off flavours. Increasing the temperature of fermentation to 30–35°C, and addition of higher concentrations of sulphur dioxide (> 150 p.p.m. total) to the juice will favour the growth of *S. pombe*, but these are not desirable wine-making practices. An alternative strategy includes pre-fermentation of the juice with *S. pombe* to achieve the desired deacidification before addition of *Saccharomyces cerevisiae*, but this runs the risk of leaving too little sugar available for fermentation by the latter. Another approach involves production of the wine with *Saccharomyces cerevisiae* followed by addition of *S. pombe* to conduct the deacidification. In this case, high populations (> 10^7 cells per ml) need to be inoculated since little nutrients are left for growth of the *Schizosaccharomyces*.

Non-proliferating cells of *S. pombe* at high densities can serve as a biocatalyst for the very rapid degradation of L-malic acid, but 10^9–10^{10} cells per ml are needed. Bioreactors with high densities of *S. pombe* cells immobilized in alginate, carrageenan, or in membrane reactors have been described for the continuous deacidification of wines or juice but their efficiencies are not commercially viable. In one system, a membrane reactor charged with 10^{10} cells per ml of *S. pombe* and operating at a flow rate of 2 ml min^{-1} and residence time of 15 min, gave only 5% degradation of malic acid in a wine containing 2.5 g l^{-1} of the acid. The reactor operating under the same conditions but charged with cells of *L. oenos* gave 95–100% degradation of the malic acid. Another system has exploited flocculation of *S. pombe* cells in an external loop reactor to produce the high density of biomass needed for biocatalytic deacidification but, again, practical efficiencies are questionable. In yet another initiative, the malolactic gene of *Lactobacillus lactis* has been expressed in *S. pombe* with the goal of improving its efficiency of acid degradation. To date, however, *S. pombe* has not been adopted into commercial practices for wine deacidification.

Enumeration and Identification

Schizosaccharomyces species grow on any of the media generally used for the isolation and enumeration of yeasts. These include malt extract agar, glucose yeast extract agar and tryptone glucose yeast extract agar. A recent international collaborative trial, suggests that *S. pombe* gives higher counts on tryptone glucose yeast extract agar than on malt extract agar. Bacterial antibiotics such as chloramphenicol, oxytetracycline, chlortetracycline, gentamicin and streptomycin can be added to these media at concentrations up to 100 μg ml^{-1} to suppress bacterial growth. Acidification of these media to pH 3.5 is another strategy often used to suppress bacterial growth but this approach is no longer recommended because the growth of sublethally injured yeast cells may be inhibited. Compared with other yeasts, *S. pombe* is particularly susceptible to sublethal injury after exposure to the stresses of mild heating (50°C for 5 min) or freezing and thawing (–196°C for 15 min). After such stresses, some 50% of the population is unable to form colonies on media such as acidified malt extract agar or yeast nitrogen base glucose agar, but produce colonies on malt extract agar. As noted above, *Schizosaccharomyces* species are osmotolerant and are most frequently associated with foods of high sugar content. Also, they are considered as somewhat tolerant to commonly used preservatives such as sorbic acid, benzoic acid, sulphur dioxide and acetic acid. Selective isolation of these yeasts can be achieved by plating on to malt extract, yeast extract, glucose agar where the glucose concentration has been increased to 50%, or where the particular preservative (e.g. acetic acid at 0.5%) has been incorporated into the medium. These media also allow the selective growth of other osmotolerant, preservative-resistant yeasts such as *Zygosaccharomyces bailii*, but the typical sausage–cylindroidal shape of the cells from colonies of *Schizosaccharomyces* species enables their rapid differentiation from the other yeast species. For the detection and enumeration of *S. pombe* in the presence of large populations of *Saccharomyces cerevisiae* as, for example, in cultures of brewing yeast, plating onto lysine agar or malt extract, yeast extract agar supplemented with 0.03% copper sulphate is recommended. Because *Saccharomyces cerevisiae* cannot utilize lysine as a nitrogen source, it will not form colonies on lysine agar whereas *S. pombe* is able to grow on this medium. Copper sulphate suppresses the growth of *Saccharomyces cerevisiae* but not that of *S. pombe*. As for all yeasts, the spread plate method rather than the pour plate method is recommended for isolation and enumeration, along with incubation at 25°C for 4–7 days.

The diluent in which food homogenates are prepared and diluted before plating can be significant in the enumeration of *S. pombe* and, presumably, other species of the genus. Diluents such as distilled water, saline or phosphate buffer can cause substantial death of the cells, leading to an underestimation of the original population. Incorporation of 0.1% peptone into the diluent minimizes this problem. When the organism is present in products with high sugar content, it is advisable to include about 20% of glucose or sucrose in the peptone diluent to minimize exposure of the cells to osmotic stress on dilution.

The identification of *Schizosaccharomyces* species follows standard morphological, biochemical and physiological tests and keys as outlined in Kurtzman and Fell, *The Yeasts, a Taxonomic Study* (1998) (Table 1). Observations of typical cell shape and the fission mode of reproduction are distinctive features of the genus. Analysis for ascospores can assist in speciation.

Consequently, simple microscopic examination of cultures may give a rapid diagnosis of these yeasts. This is probably the reason why molecular probe, polymerase chain reaction and immunodiagnostic tests have not yet been developed for them. Several commercial kits are available for conducting key fermentation and assimilation tests that facilitate yeast identification. These include the API20C, the ATB 32C, the MicroScan and the Biolog systems. However, the Biolog is the only system that incorporates *Schizosaccharomyces* spp. in its database and preliminary evaluation shows its ability to successfully identify isolates of *S. pombe* and *S. octosporus*.

See also: **Bread**: Sourdough bread. ***Saccharomyces***: *Saccharomyces cerevisiae*. **Spoilage Problems**: Problems caused by Fungi. **Wines**: Microbiology of Wine-making; The Malolactic Fermentation. ***Zygosaccharomyces***.

Further Reading

Ansanay V, Dequin S, Camarasa V, Schaeffer V, Grivet J, Blondin B, Salmon J and Barre P (1996) Malolactic fermentation by engineered *Saccharomyces cerevisiae* as compared with engineer *Schizosaccharomyces pombe*. *Yeast* 12: 215–225.

Ciani M (1995) Continuous deacidification of wine by immobilised *Schizosaccharomyces pombe* cells: evaluation of malic acid degradation rates and analytical profiles. *Journal of Applied Bacteriology* 79: 631–634.

Deák T and Beauchat LR (1996) *Handbook of Food Spoilage Yeasts*. Boca Raton, Florida: CRC Press.

de Jong-Gubbels P, van Dijken JP and Pronk JT (1996) Metabolic fluxes in chemostat cultures of *Schizosaccharomyces pombe* grown on mixtures of glucose and ethanol. *Microbiology* 142: 1399–1407.

Eriksson OE, Svedskog A and Landvik S (1993) Molecular evidence for the evolutionary hiatus between *Saccharomyces cerevisiae* and *Schizosaccharomyces pombe*. *Systema Ascomycetum* 11: 119–162.

Fleet GH (1992) Spoilage yeasts. *Critical Reviews in Biotechnology* 12: 1–44.

Fleet GH and Mian MA (1998) Induction and repair of sublethal injury in food spoilage yeasts. *Journal of Food Mycology* 1: 85–89.

Ganthala BP, Marshall JH and May JW (1994) Xerotolerance in fission yeasts and the role of glucose as a compatible solute. *Archives of Microbiology* 162: 108–113.

Gao C and Fleet GH (1995) Degradation of malic and tartaric acids by high density cell suspensions of wine yeasts. *Food Microbiology* 12: 65–71.

Hocking A (1996) Media for preservative resistant yeasts: a collaborative study. *International Journal of Food Microbiology* 29: 167–175.

Jeffery J, Kock JLF, Bota A, Coetzee DJ and Botes PJ (1997) The value of lipid composition in the taxonomy of the Schizosaccharomycetales. *Antonie van Leeuwenhoek* 72: 327–335.

Martini AV (1991) Evaluation of phylogenetic relationships among fission yeast by nDNA/nDNA reassociation and conventional taxonomic criteria. *Yeast* 7: 73–78.

Martini AV and Martini A (1998) *Schizosaccharomyces*. In: Kurtzman CP and Fell JW (eds) *The Yeasts, A Taxonomic Study*, 4th edn. Pp. 391–394. Amsterdam: Elsevier Science.

Praphailong W, van Gestel M, Fleet GH and Heard GM (1997) Evaluation of the Biolog system for the identification of food and beverage yeasts. *Letters in Applied Microbiology* 24: 455–459.

Tudor A and Board RG (1993) Food spoilage yeasts. In: Rose AH and Harrison JS (eds) *The Yeasts*, 2nd edn. Vol. 5, p. 435. London: Academic Press.

Secondary Metabolites *see* **Metabolic Pathways**: Production of Secondary Metabolites – Fungi; Production of Secondary Metabolites – Bacteria.

Sensing Microscopy *see* **Microscopy**: Sensing microscopy.

SERRATIA

Fatemeh Rafii, Division of Microbiology, National Center for Toxicological Research, Jefferson, USA

Characterization of the Genus

The genus *Serratia* is named after Serafino Serrati, an Italian physicist, and belongs to the family Enterobacteriaceae, tribe *Klebsiellea*. Nine species have been isolated from clinical samples: *S. marcescens*, *S. liquefaciens*, *S. rubidaea* (also called *S. marinorubra*), *S. ficaria*, *S. fonticola*, *S. odorifera*, *S. plymuthica*, *S. grimesii* and *S. proteamaculans* subsp. *quinovora*. *S. marcescens* is the best-characterized member of the genus. The classification of species is based on morphological, physiological, biochemical and carbon-source -utilization tests (**Table 1**). The differentiation within each species is based on biotyping, serotyping, phage typing, bacteriocin typing, whole-cell protein fingerprinting and DNA analysis.

All species are Gram-negative straight rods, with rounded ends, 0.5–0.8 μm diameter and 0.9–2 μm in length. They are facultative anaerobes, catalase-positive and motile with peritrichous flagella. In a minimal medium containing ammonium sulphate as the nitrogen source, they can use many different compounds as sole carbon sources, including D-glucose, D-fructose, D-ribose, L-malate, L-aspartate, citrate, *N*-acetylglucosaine, gluconate and mannitol (Table 1).

Colonies are most often opaque, somewhat iridescent and white, pink or red depending on the genotype and cultural conditions (e.g. amino acids, carbohydrates, pH, inorganic ions and temperature). The red colour is due to prodigiosin and pyrimine. Prodigiosin is a non-diffusible pigment produced by two biogroups (A1 and A2) of *S. marcescens*, and by most strains of *S. plymuthica* and *S. rubidaea*. Prodigiosin is best produced aerobically on peptone–glycerol agar at 12–36°C. Pyrimine is a water-soluble pigment that is produced by some strains of *S. marcescens* biogroup A4.

Some strains of *S. liquefaciens*, *S. odorifera*, *S. plymuthica* and *S. ficaria* can grow at 4–5°C; other strains of *S. marcescens*, *S. rubidaea* and *S. odorifera* can grow at 40°C. Almost all strains can grow from pH 5 to 9. The mol% G+C of the DNA is 52–60. *S. marcescens* responds to the environment with changes in shape and movement. In liquid media, the cells are short rods known as swimmers, with one or two flagella. On 0.70–0.85% agar, they transform to swarmers – aseptate, elongated cells with 10–100 lateral flagella. The flagellin proteins of swimmer and swarmer cells are identical.

Serratia species are important in food microbiology because not only are they involved in food spoilage but also they are opportunistic pathogens that can cause various diseases in humans and animals. The diseased food animals in turn may produce contaminated milk or meat, with further spread of the bacteria occurring through contaminated milking machines or carcasses.

Members of *Serratia* are distributed in soil, air and water. They are associated with large numbers of plants and animals (including insects, birds and their eggs). Some are insect pathogens and others produce antifungal and antimicrobial agents. They utilize a wide range of nutrients and have been shown to grow in disinfectant solutions and double-distilled water, and they resist some antiseptics. They colonize and survive on meat-packaging materials, hospital instruments and farm equipment, including milk pumps. *Serratia* spp. have been found in milk, ice cream from pasteurized milk, other dairy products, coffee from vending machines, frozen unpasteurized fruit juice, eggs and meats.

Many *Serratia* species produce haemolysins and have DNA that hybridizes to probes for the *S. marcescens* haemolysin DNA. *S. ficaria* has a smaller haemolysin that fails to hybridize to these probes. The restriction enzyme pattern of the haemolysin gene differs with the strains. The *Serratia* haemolysin causes pore formation in erythrocyte membranes. This results in the osmotic lysis of erythrocytes, leading to the release of haemoglobin.

S. marcescens produces extracellular enzymes that include a nuclease, a lipase, two chitinases and several proteases. The extracellular endonuclease of *S. marcescens* nonspecifically cleaves double-stranded or single-stranded DNA, as well as RNA.

Not all strains of *Serratia* are pathogenic. The virulence factors in *Serratia* are not well understood and may be a combination of several factors. Five strains of *S. marcescens* and two of *S. rubidaea*, out of 21 strains isolated from fruit juice and fish samples, killed mice upon parenteral inoculation but not through oral feeding. A heat-labile enterotoxin was detected in three strains of *S. marcescens* and one of *S. rubidaea* by the rabbit ligated ileal loop test, the mouse foot pad test and the vasopermeability factor test. Cytotoxic effects on a monolayer of Vero cells were found in the cell-free culture filtrates of two enterotoxigenic *S. marcescens* strains. Pathogenic strains had a different type of fimbriae from non-pathogenic strains. All pathogenic *Serratia* strains were agglutinated by comparatively lower salt concentrations than non-pathogenic strains and had multiple drug resistance. A strain of *S. marcescens* producing haemolysin colonized the urinary tract more and led to a stronger inflammatory response than an isogenic haemolysin-negative strain.

Methods of Detection

Serratia is one of the easiest genera to differentiate from others in the Enterobacteriaceae, even in the absence of pigments. A highly selective medium for the enrichment and isolation of all *Serratia* species is caprylate-thallous (CT) mineral salts agar. It contains 0.01% yeast extract, 0.1% caprylic (*n*-octanoic) acid as carbon source, and 0.025% thallous sulphate for inhibition of other organisms. CT agar supports the growth of all *Serratia* strains tested but few other bacteria. The efficiency of colony formation of known

Table 1 Biochemical reactions of different species of *Serratia*[a,b]

Biochemical reactions	*Species*							
	S. marcescens	S. marcescens *biogroup*[1]	S. liquefaciens *group*	S. rubidaea	S. odorifera *biogroup*[1]	S. odorifera *biogroup*[2]	S. plymuthica	S. ficaria
Indole production	1	0	1	0	60	50	0	0
Methyl red	20	100	93	20	100	60	94	75
Voges–Proskauer	98	60	93	100	50	100	80	75
Citrate (Simmons')	98	30	90	95	100	97	75	100
Hydrogen sulphide (TSI)	0	0	0	0	0	0	0	0
Urea hydrolysis	15	0	3	2	5	0	0	0
Phenylalanine deaminase	0	0	0	0	0	0	0	0
Lysine decarboxylase	99	55	95	55	100	94	0	0
Arginine dihydrolase	0	4	0	0	0	0	0	0
Ornithine decarboxylase	99	65	95	0	100	0	0	0
Motility (36°C)	97	17	95	85	100	100	50	100
Gelatin hydrolysis (22°C)	90	30	90	90	95	94	60	100
Growth in KCN	95	70	90	25	60	19	30	55
Malonate utilization	3	0	2	94	0	0	0	0
D-Glucose: acid	100	100	100	100	100	100	100	100
D-Glucose: gas	55	0	75	30	0	13	40	0
Lactose fermentation	2	4	10	100	70	97	80	15
Sucrose fermentation	99	100	98	99	100	0	100	100
D-Mannitol fermentation	99	96	100	100	100	97	100	100
Dulcitol fermentation	0	0	0	0	0	0	0	0
Salicin fermentation	95	92	97	99	98	45	94	100
Adonitol fermentation	40	30	5	99	50	55	0	0
myo-Inositol fermentation	75	30	60	20	100	100	50	55
D-Sorbitol fermentation	99	92	95	1	100	100	65	100
L-Arabinose fermentation	0	0	98	100	100	100	100	100
Raffinose fermentation	2	0	85	99	100	7	94	70
L-Rhamnose fermentation	0	0	15	1	95	94	0	35
Maltose fermentation	96	70	98	99	100	100	94	100
D-Xylose fermentation	7	0	100	99	100	100	94	100
Trehalose fermentation	99	100	100	100	100	100	100	100
Cellobiose fermentation	5	4	5	94	100	100	100	100
α-Methyl-D-glucoside fermentation	0	0	5	1	0	0	70	8
Erythritol fermentation	1	0	0	0	0	7	0	0
Aesculin hydrolysis	95	96	97	94	95	40	81	100
Melibiose fermentation	0	0	75	99	100	96	93	40
D-Arabitol fermentation	0	0	0	85	0	0	0	100
Glycerol fermentation	95	92	95	20	40	50	50	0
Mucate fermentation	0	0	0	0	5	0	0	0
Tartrate, Jordan's	75	50	75	70	100	100	100	17
Acetate utilization	50	4	40	80	60	65	55	40
Lipase (corn oil)	98	75	85	99	35	65	70	77
DNase at 25°C	98	82	85	99	100	100	100	100
Nitrate→nitrate	98	83	100	100	100	100	100	92
Oxidase, Kovac's	0	0	0	0	0	0	0	8
ONPG[c]	95	75	93	100	100	100	70	100
Yellow pigment	0	0	0	0	0	0	0	0
D-Mannose fermentation	99	100	100	100	100	100	100	100

[a] Each number gives the percentage of positive reactions after 2 days of incubation at 36°C.
[b] Data from Farmer JJ, Davis BR, Hickman-Brenner FW et al (1985).
[c] *o*-Nitrophenyl-β-galactopyranoside.

strains on CT agar is 80.7% as high as on a non-selective complex medium.

Another selective medium for differentiating non-pigmented *Serratia* strains from other Enterobacteriaceae is Tween 80 medium. The medium contains 3.3% tryptose blood agar base, 0.4% Tween 80 and 0.015% $CaCl_2$. *Serratia* hydrolyses Tween 80 via an esterase, resulting in the release of free fatty acids, which in the presence of calcium form an opaque zone around the colony. *Enterobacter sakazakii* also forms this zone around the colony but it is lecithinase-negative.

DNase agar, which contains DNA, and DTC agar, which contains cephalothin (to which most strains of *Serratia* are resistant), are also used for isolation.

On blood agar and some other media, *S. marcescens* produces red colonies. As the colonies grow, the colour intensifies. On CT agar, colonies of *Serratia* spp. are small and slightly bluish-white. On Tween 80 agar, colonies are large, pinkish and surrounded by a white zone of precipitate. The characteristics listed in Table 1 differentiate the members of this genus from other Enterobacteriaceae. The following biochemical reactions (Table 1) differentiate species of *Serratia* from each other: DNase (25°C), lipase (corn oil), gelatinase (22°C), lysine decarboxylase, ornithine decarboxylase, L-arabinose, D-arabitol, D-sorbitol, adonitol and dulcitol. Musty or vegetable-like odours in *S. odorifera* and *S. ficaria* and pigment formation in some strains of *S. rubidaea*, *S. marcescens* and *S. plymuthica* are also helpful in identifying these species.

Miniaturized biochemical detection systems may also be used to identify the species. Two of the systems are API 20E strips and Vitek GN1 Gram-negative identification cards (automated) from BioMérieux Vitek. The API 20E database contains information for all species of *Serratia* except *S. grimesii* and *S. proteamaculans*, which were previously classified under *S. liquefaciens*. Vitek GN1 does not include data for *S. grimesii*, *S. proteamaculans* or *S. ficaria*. The Biolog GN microplate (automated) system (Biolog) identifies all species of *Serratia* but considers *S. grimesii* as *S. liquefaciens*.

For isolation and identification of *Serratia*, food, milk or eluates from washed or suspended foods can be plated directly on blood agar and CT agar and incubated at 37°C for 24 h and 30°C for 48 h, respectively. To maximize the number of bacteria, the samples should be centrifuged. The pellet should be suspended in phosphate-buffered saline and either plated directly or enriched before plating by incubation with shaking at 37°C for 24 h. Suspected colonies of *Serratia* then are plated on Tween 80 agar and further incubated with shaking at 37°C for 24 h.

Table 2 Some reported associations of diseases with species of *Serratia* and types of food from which *Serratia* species have been isolated. The association of food with the aetiology of the disease is not implied

Species	*Reported cases of disease*	*Reported isolation from foods*
S. ficaria	Septicaemia	Figs
S. fonticola	Leg abscess	Fruit juice, coconut, coffee from vending machine
S. grimesii	Not reported, isolated from patient blood	Dairy products
S. liquefaciens	Nosocomial infections	Dairy and meat products
S. marcescens	Many nosocomial infections	Dairy and meat products, bread
S. odorifera	Infant fatal septicaemia	Unspecified food
S. plymuthica	Septic shock, bacteraemia, chronic osteomyelitis, cases of sepsis, septicaemia	Fish
S. proteamaculans subsp. *quinovora*	Pneumonia	Food source not reported
S. rubidaea	Infection of the bile and blood of a patient with a bile duct carcinoma	Frozen fruit juices, spoiled coconut, cheese

The bacteria that form precipitates around colonies on Tween 80 are identified to species by biochemical reactions.

The US Food and Drug Administration (FDA) has categorized *Serratia* under Miscellaneous Enterics and has no specific regulations. Both the US FDA and the US Centers for Disease Control and Prevention (CDC) consider infection with *Serratia* spp. among nosocomial infections.

Importance of Genus

Members of the genus *Serratia* do not cause infections in healthy individuals. Therapies, conditions and procedures that compromise patients immunologically or physiologically make them susceptible to colonization by opportunistic pathogens, including *Serratia*. Infants and very old patients are also susceptible. *S. marcescens* causes a variety of nosocomial infections; other species have been isolated from clinical specimens, but their involvement in pathogenicity was not clear until recently (**Table 2**). Almost all species of *Serratia* have been isolated from foods, including fruits and vegetables (Table 2), and infection by *S. ficaria* in a surgery patient appeared to be the result of eating contaminated figs.

The connection between the development of disease

and consumption of food contaminated with *Serratia* is not well established, except that contamination of a milk pump proved to be the source of an epidemic in a hospital nursery. *S. marcescens* isolated from the milk pump was identical to the isolates causing infections in the babies. It was resistant to disinfectants, but was eliminated by heat sterilization which ended the outbreak. Both *S. marcescens* and *S. liquefaciens* cause mastitis in dairy cattle, which may produce infected milk. Slaughter of animals with subclinical infections may result in meat contamination. *Serratia* spp. have been isolated from beef, milk, ham, chicken and fish.

Another important aspect of *Serratia* in food microbiology is involvement in spoilage of foods (eggs, butter, milk, coconut and bread and discoloration of cheeses). It also causes greening and malodour formation in meat. Dairy products could become contaminated by the use of milk from animals with mastitis or use of contaminated milking equipment. Contamination of ice cream and cheese can also occur during handling and processing at the retail market. *Serratia* spp. survive in foods unsuitable for the growth of other bacteria, such as smoked and dried fish. Strains of *Serratia* spp. were shown to have a mean minimum growth temperature of 1.7°C in beef or to resist pressure treatment during processing of ground chicken. *Serratia* spp. which cause red discoloration of cheese were relatively resistant to 9% salt and grew at pH 4–9.

See also: **Cheese**: In the Market Place. **Enterobacter**. **Milk and Milk Products**: Microbiology of Liquid Milk. **Spoilage Problems**: Problems caused by Bacteria.

Further Reading

Farmer JJ, Davis BR, Hickman-Brenner FW et al (1985) Biochemical identification of new species and biogroups of Enterobacteriaceae isolated from clinical specimens. *Journal of Clinical Microbiology* 21: 46–76.

Grimont PAD and Grimont F (1984) Family 1. Enterobacteriaceae, genus VIII. *Serratia* Bizio 1823, 288[A1] In: Krieg NR and Holt JG (eds) *Bergey's Manual of Systematic Bacteriology*. P. 477. Baltimore, MD: Williams & Wilkins.

SHELLFISH (MOLLUSCS AND CRUSTACEA)

Contents

Characteristics of the Groups

L Le Vay, School of Ocean Sciences, University of Wales, Bangor, UK

B Egan, Marine Biological and Chemical Consultants Ltd, Bangor, UK

Introduction

The term 'shellfish' covers the food species of molluscs and crustacea, which are mostly marine. Although shellfish are notorious sources of food poisoning outbreaks, and the consequences of consuming contaminated shellfish can be extremely severe, the overall risk of infection is relatively low.

Shellfish are harvested from the coastal marine environment, where they may be exposed to indigenous bacterial pathogens as well as to bacteria and viruses originating from sewage and other terrestrial drainage. Sewage-derived contamination can be avoided to a large extent, but not completely, by the monitoring of coastal waters for the presence of faecal coliform indicators and then depuration treatment in clean water if necessary. However, even without depuration the risk of infection is low if the shellfish are properly cooked immediately prior to eating – cross contamination and bacterial growth under inadequate refrigeration present the greatest risk in precooked products. The most likely route for the transmission of environmental contamination is via bivalve molluscan shellfish, due to their mode of feeding (by filtering large volumes of water) and to frequently being consumed raw or partly cooked. Filter feeding molluscs may also accumulate toxins produced by certain species of marine phytoplankton. The toxins are heat-stable and will survive normal cooking conditions, so control methods are focused on monitoring the presence of toxic species in shellfish waters and the occurrence of toxins in tissue samples.

Taxonomy of the Groups

Molluscan shellfish include lamellibranch (filter feeding) members of the Bivalvia, such as clams, mussels, oysters and cockles, and the prosobranch Gastropoda, such as abalone, whelks, winkles and conches (**Table 1**). The cephalopod molluscs, includ-

Table 1 Commercially exploited molluscan shellfish

Latin nomenclature	*Common name*
Bivalves	
Andara spp.	Blood cockles
Arca spp.	Arc clams
Arctica islandica	Quahog
Argopecten spp.	Scallops
Cardium edule	Cockle
Chlamys spp.	Scallops
Corbicula spp.	Clams
Crassostrea gigas	Pacific oyster
Crassostrea virginica	American oyster
Ensis spp.	Razor clams
Mactra spp.	Surf clams
Mercenaria mercenaria	Clams
Meretrix spp.	Clams
Modiolus spp.	Horse mussels
Mytilus spp.	Mussels
Ostrea spp.	Oysters
Panopea abrupta	Geoduck
Paphia spp.	Shortneck clams
Pecten spp.	Scallops
Perna spp.	Mussels
Protothaca spp.	Clams
Solen spp.	Razor clams
Spisula spp.	Surf clams
Tapes spp.	Clams
Tridacna spp.	Giant clams
Venus spp.	Venus clams
Gastropods	
Buccinum undatum	Whelk
Haliotis spp.	Abalone
Littorina littorea	Periwinkle
Strombus spp.	Conches
Turbo spp.	Turban shells

ing squid, cuttlefish and octopuses are not covered in this article.

Almost all the familiar crustacean shellfish species belong to the order Decapoda, within the class Malacostraca (**Table 2**). The few exceptions include minor species such as the stomatopod mantis shrimps and the goose barnacles. The decapods comprise one of the most diverse crustacean groups, represented by over 10 000 species worldwide and including the shrimps and prawns as well as the crabs (Brachyura), lobsters and freshwater crayfish (Astacidea) and spiny lobsters (Palinura).

The terms 'shrimp' and 'prawn' are confusingly applied in different countries, reflecting historical usage. The simplest guideline for international nomenclature is that 'shrimp' should refer to marine species and 'prawn' to freshwater species, replacing

Table 2 Commercially exploited crustacean shellfish

Latin nomenclature	*Common name*
Shrimps and prawns	
Crangon crangon	Brown shrimp
Macrobrachium spp.	Freshwater prawns
Metapenaeus spp.	Metapenaeid shrimps
Palaemon spp.	Common prawns
Pandalus boreali	Northern prawn
Pandalus spp.	Pink shrimps
Penaeus spp. (> 20 spp.)	Penaeid shrimps
Sergestidae spp.	Sergestid shrimps
Solenoceridae spp.	Solenocerid shrimps
Lobsters (clawed and spiny)	
Homarus homarus	European clawed lobster
Homarus americanus	American clawed lobster
Jasus spp.	Rock lobsters
Metanethrops challengeri	New Zealand lobster
Nethrops norvegicus	Norway lobsters
Palinurus spp.	Spiny lobsters
Panulirus spp.	Tropical spiny lobsters
Scyllarus spp.	Slipper lobsters
Crabs	
Callinectes sapidus	Blue crab
Callinectes danae	Swimming crab
Cancer pagurus	Edible crab
Cancer magister	Dungeness crab
Cancer borealis	Jonah crab
Carcinus maenas	Green shore crab
Chionectes spp.	Pacific snow crabs
Chionectes opilio	Queen crab
Geryon spp.	Deep sea geryons
Jacquinotia edwardsii	Southern spider crab
Lithodes spp.	King crabs
Maja squinado	Spider crab
Menippe mercenaria	Black stone crab
Paralomis spp.	Red stone crab
Paralomis spinosissima	Antarctic king crab
Portunus pelagicus	Blue swimming crab
Portunus spp.	Swimming crabs
Scylla spp.	Mud crabs
Mantis shrimps	
Squilla spp.	Mantis shrimps
Freshwater crayfish	
Astacus astacus	Noble crayfish
Astacus leptodactylus	Turkish crayfish
Cherax destructor	Yabby crayfish
Cherax quadricarinatus	Redclaw crayfish
Cherax tenuimanus	Marron crayfish
Pasifastacus leniusulcatus	Signal crayfish
Procambrus clarkii	Red swamp crayfish

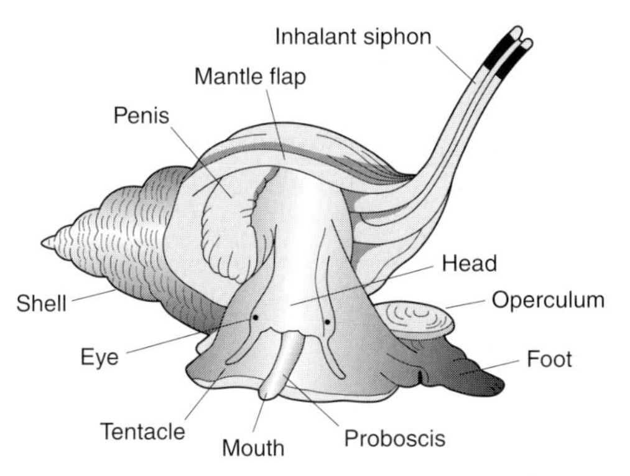

Figure 1 External anatomy of a marine prosobranch gastropod, the whelk *Buccinum undatum*. (Reproduced with permission from Barnes RD (1986))

previous attempts at standardization on the basis of taxonomy. Most large marine shrimps belong to the infraorder Penaeidea, which includes farmed species such as the tiger shrimp *Penaeus monodon* and the white-leg shrimp *Penaeus vannamei*. Apart from the related sergestid shrimp, most other marine shrimps and freshwater prawns fall within the Caridea, which can easily be differentiated because the side plates of the second abdominal segment overlap those of the first and third segments. In contrast, in the penaeid and sergestid shrimps the first abdominal plate overlaps that of the second segment.

Morphological and Anatomical Features

Molluscs

Molluscan shellfish typically have a hard shell, made up of calcium carbonate, as aragonite or calcite, deposited within an organic matrix. The shell is secreted by the mantle or epidermis, which underlies it and surrounds the body. The shell and mantle enclose a space, the mantle cavity, which houses the gills and through which water is circulated to promote efficient gaseous exchange. In the snail-like higher prosobranch gastropods, the visceral mass of the body is protected within a right- or left-handed spiral shell, with the head and muscular foot extending through the opening (**Fig. 1**). The body can be retracted into the shell, which in many species can be closed off by the operculum, a plate attached to the foot. The spiral shell structure is not apparent in some species, including the abalone and the limpet, which are characterized by low, flattened shells and a muscular foot adapted for clamping onto hard surfaces. These species feed by using a chitinous radula to scrape algal and other materials from hard substrata, as do some higher prosobranch gastropods such as the winkles. Other higher prosobranch gastropods may be carnivores, omnivorous scavengers or deposit feeders. Carnivores (e.g. the whelks) have a proboscis, allowing extension of the mouth, and a specialized radula.

The bivalve mollusc shell is formed from a pair of dorsally hinged valves, which usually enclose the body (**Fig. 2**). The valves are joined on the dorsal margin by a hinge ligament and, in most cases, interlocking teeth. The shell is closed by adductor muscles, attached internally to each valve anteriorly and posteriorly. In many species, a muscular foot can be extended beyond the open valves. Water is circulated through the mantle cavity, by cilia on the gills and pumping by the adductor muscles, entering and leaving via the inhalant and exhalant siphons respectively.

In the lamellibranch bivalves, including all food species, the gills are enlarged to form specialized lamellae made up of filaments which may be interconnected to varying degrees, in some cases forming continuous sheets. The lamellae filter food particles, such as phytoplankton, and fine sediments from circulating water. There is some evidence for the filtration of bacteria, which may be degraded by lysozyme and provide a source of nutrition for the animals. In large bivalves, more than 100 l of water may be filtered daily, indicating the potential for the concentration of waterborne or sediment-borne pathogens. Filtered particles are trapped in mucus and transported by ciliary action along the gill filaments to food grooves leading to the labial palps. The particles are sorted, both on the gills and at the palps – only small, light particles being taken into the mouth. Rejected particles are expelled from the mantle cavity through the exhalant siphon, as pseudofaeces. Digestive enzymes are released from the crystalline style, a proteinaceous rod retained in a sac which opens into the stomach. The style rotates as it winds in a mucus strand loaded with food material. Further, intracellular, digestion takes place within the digestive gland.

Crustaceans

The crustacean anatomy is extremely diverse but is based upon a generalized body plan of three sections – head, thorax and abdomen. The head carries two pairs of antennae, a feature that distinguishes crustaceans from other arthropods, including the insects. The thorax and abdomen carry biramous limbs, which are segmented and may be clawed or simple. The growth of crustacea is limited by the hard, calcified, chitinous exoskeleton. This is shed during moulting, after which the body expands rapidly before the underlying replacement skeleton hardens. Specialized shellfish markets exist for moulting crustaceans, particularly soft-shell crabs.

Figure 3 shows the external features of a generalized penaeid shrimp, with typical decapod

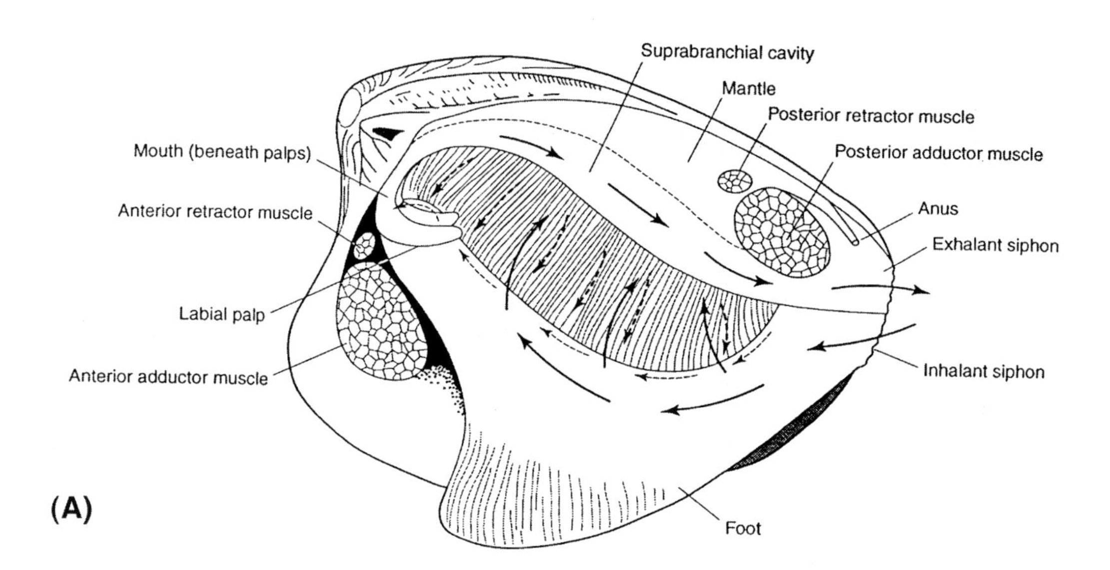

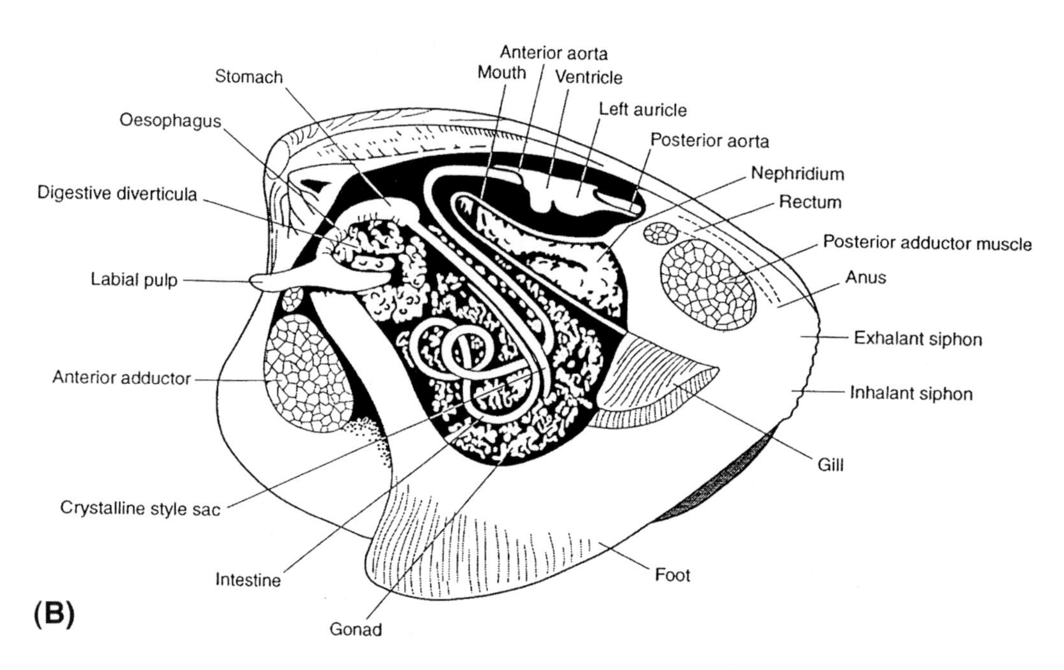

Figure 2 Internal anatomy of a bivalve mollusc, the clam *Mercenaria mercenaria*. Arrows show the direction of water flow. (**A**) Surface view showing gills, siphons and direction of water flow. (**B**) Dissected view showing the digestive tract. (Reproduced with permission from Barnes RD (1986))

anatomy. The head and thorax are fused dorsally and covered by a carapace, which encloses the gills. There are eight pairs of thoracic legs, of which three are adapted as mouth parts and five as walking legs (pereiopods). In crabs, lobsters and freshwater crayfish, the first pair of pereiopods is enlarged, to form the claws. In shrimps, prawns, lobsters and crayfish, the five pairs of abdominal legs (pleopods) are adapted for swimming, as well as more specialized functions such as copulation (in males) or carrying eggs (in females). In crabs (**Fig. 4**) the carapace is dorso-ventrally flattened, and the abdomen reduced and folded to fit closely underneath. The pereiopods are concealed by the abdomen, with only two present in males. In decapods, the digestive apparatus, a two-chambered stomach and secretory hepatopancreas, is located under the carapace. Food items are masticated by the mouthparts and passed into the stomach, where gastric teeth and setae mechanically break down the food into fine particles. The hepatopancreas secretes digestive enzymes into the stomach, and receives milled food particles from the posterior stomach chamber for final digestion and nutrient absorption. Waste material passes along the gut, which runs the length of the abdomen.

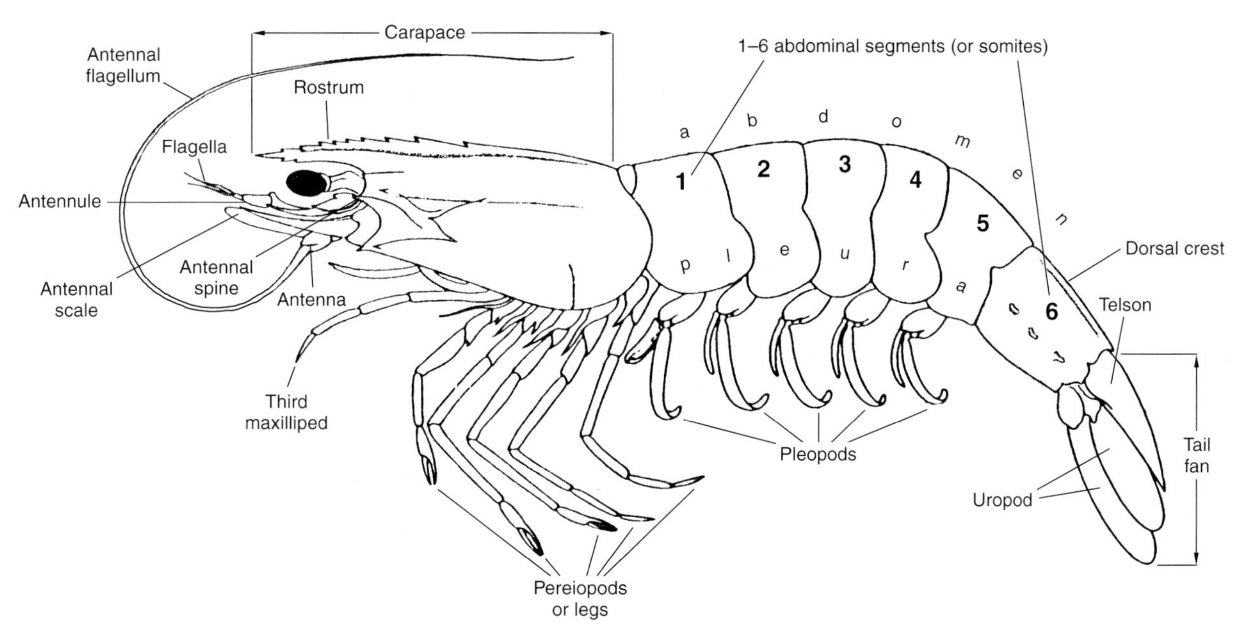

Figure 3 Schematic body plan of a generalized penaeid shrimp. (Reproduced with permission from FAO Species Identification Guide for Fishery Purposes)

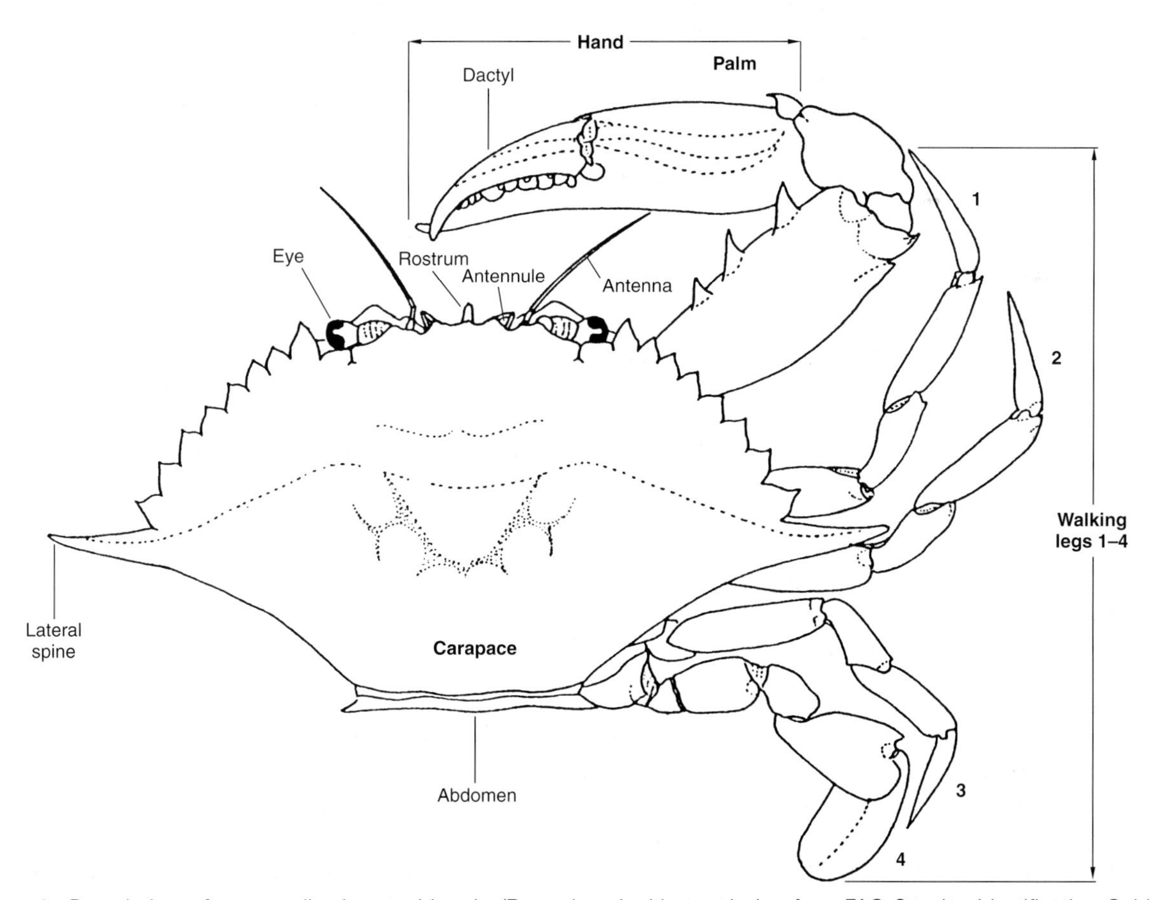

Figure 4 Dorsal view of a generalized portunid crab. (Reproduced with permission from FAO Species Identification Guide for Fishery Purposes)

Typical Habitats and Global Distribution

Molluscs

Most molluscan shellfish inhabit shallow coastal and estuarine waters, many species being cultured or harvested intertidally. Fisheries exist across a wide climatic range, but the high productivity of temperate coastal waters supports the highest yields of commercial bivalve species, which occupy a variety of habitats.

Clams and cockles generally live buried in soft sediments, filtering water through inhalant and exhalant siphons that reach to the sediment surface. In contrast, oysters and mussels live on the surface of the sea bed, oysters cementing themselves to the hard substrata, and mussels attaching themselves by secreting adhesive byssal threads. Scallops live unattached on the sea floor, and are able to swim short distances by using the adductor muscles to bring the shell valves together in a clapping action. Among the gastropod shellfish, some species (e.g. winkles and abalone) live in exposed habitats, typically in intertidal or shallow water, while other species (e.g. whelks and conches) can move over both hard and soft sediment sea beds.

Crustaceans

Most crustacean shellfish are captured from estuaries, shallow coastal waters or deeper offshore waters worldwide, from high latitudes to the tropics. A few minor crustacean shellfish species live in fresh water, including the *Macrobrachium* prawns, the astacid crayfish and some crabs. Most cold- and temperate-water species of shrimps, lobsters and crabs are the products of wild fisheries, but aquaculture in coastal ponds now accounts for close to 700 000 t of tropical penaeid shrimps annually.

Most crustacean shellfish species are either carnivores or omnivorous scavengers and detritovores. The crabs, lobsters and spiny lobsters are generally large, mobile, benthic species that walk or swim intermittently over the sea bed. They are usually caught in baited traps because they live at relatively low densities and often prefer rocky or heterogeneous substrates that provide shelter, although some species may bury themselves in soft sediments or excavate burrows. In contrast, shrimp species often occur in high densities, swimming over the sea floor, and are typically caught using trawl nets.

Habitats that Pose Special Risks

The primary environmental risk factor for humans in the context of shellfish is the presence of human pathogens in the waters in which the shellfish are living. Faecal coliform bacteria occur naturally in the intestines of mammals and birds, whose faecal material contains $1–10 \times 10^6$ coliforms per gram. In many coastal and estuarine locations around the world, the disposal of human sewage involves its release through offshore outfalls, with varying degrees of pre-treatment. The basic treatments do not, however, significantly reduce the bacterial or viral loading of the sewage, but merely address the aesthetic problems associated with the material.

In these circumstances, the filter-feeding and detritus-feeding habits of molluscan and crustacean populations are likely to lead to contamination of their external and internal tissues. The sessile nature of bivalve shellfish, and their method of feeding by filtering large volumes of sea water, increase their likely exposure in high-risk environments. In contrast, the foraging habits and selective feeding patterns of crustaceans may reduce the risk of exposure in this group. In many areas of the world it has become standard practice to monitor the water quality of shellfish beds. In sites free from contamination, the faecal coliform group of organisms is absent from the tissues of crustacean and molluscan shellfish, and its presence is a clear indication of the potential presence of other pathogens. The use of faecal coliform bacteria as indicators in drinking and recreational water has had almost a century of successful application in protecting public health, including about 50 years' use in shellfish regulation.

For water testing purposes, the detailed definitions of total and faecal coliforms vary from country to country, although there is general agreement that faecal coliform bacteria may be defined as coliforms which ferment lactose, with gas and acid formation, at 44–44.5°C. Total coliform bacteria are a relatively poorly defined group that includes genera not restricted to faecal origin, comprising aerobic and facultative anaerobic species that will grow in the presence of bile salts and produce acid and gas from lactose at 35–37°C, and as indicators they have been widely superseded by the more specific faecal coliforms. However, the bacterial tracers currently in use have been shown in a number of studies to be poor indicators of the distribution and survival of enteric viruses, and care is needed when interpreting data. There are also indigenous marine pathogens which are unrelated to human waste, and the standard indicators are inapplicable in assessing their distribution and the associated risk of contamination and disease. Despite these limitations, improvements in the control of water quality in shellfish harvesting areas in Europe and the US have significantly reduced the incidence of bacterial food poisoning outbreaks associated with shellfish in recent years. However, infections continue to result from the consumption of both raw and cooked shellfish products. In Japan and Southeast Asia the levels of infection remain relatively high, due to the relatively warm waters and the inadequate supervision of shellfish collection areas in some countries.

Microbial Contamination from the Environment

In general, the composition of the microflora associated with healthy invertebrates reflects the microbial populations of sea water and sediments, and may include Alcaligenes, Acinetobacter, *Bacillus*, Enterobacteriaceae, *Flavobacterium*, *Pseudomonas* and *Vibrio* species. The primary public health concern associated with both molluscan and crustacean shellfish is that they may be a point of transmission of pathogenic microorganisms. These may be indigenous to sea water (e.g. *Aeromonas hydrophila*, *Plesiomonas shigelloides*, *Vibrio cholerae* O1 and non-O1, *V. parahaemolyticus* and *V. vulnificus*), or may originate from sewage or other pollution (e.g. *Salmonella*, *Escherichia coli*, *Shigella* and enteric viruses including hepatitis A virus and small round structured viruses, including Norwalk virus).

Among the indigenous bacteria, the genus *Vibrio* contains 11 or more marine species that have been identified as pathogenic for humans and have the potential to cause extreme illness and sometimes death. Several *Vibrio* species pose a concern in coastal waters, the most common being *V. vulnificus*, which can cause rapid and devastating infection in humans. *V. vulnificus* is widespread, being found in waters with a temperature range of 10–30°C and a salinity range of 5–30 p.p.t. *V. parahaemolyticus* is also relatively common in temperate and warm-water coastal areas, and is a cause of gastroenteritis, particularly in India, Japan and Southeast Asia. This species has also been shown to be widely distributed in estuarine sediments in the US. *V. cholerae* occurs in tropical coastal waters and seasonally in warm and temperate climates, and the consumption of contaminated shellfish can be a significant factor in outbreaks of cholera (O1 and non-O1) infections. *Plesiomonas shigelloides* occurs naturally in sea water and may be implicated in outbreaks of gastrointestinal bacterial infection following the consumption of molluscan shellfish. *Aeromonas* species are low-risk indigenous marine species which may be transmitted to humans via shellfish, but documented evidence of food poisoning events is limited.

Although derived from human sewage and agricultural waste inputs, Enterobacteriaceae such as *Escherichia coli*, *Shigella* and *Salmonella* species are able to survive for extended periods in estuarine and coastal waters. *E. coli* contamination is a relatively low health risk, and is more important as an indicator of potential contamination by higher-risk sewage-borne pathogens. The environmental contamination of shellfish with *Salmonella* is relatively rare but outbreaks do occur, especially through the consumption of raw bivalves. Viral contamination of shellfish is derived from sewage inputs to the coastal environment. Enteric viruses are capable of surviving for extended periods in sea water, and infective doses are lower than those for bacterial pathogens. Consequently, infections with hepatitis A and small round structured viruses (e.g. Norwalk) represent major causes of seafood poisoning, usually being transmitted by the consumption of raw or improperly cooked bivalve molluscs.

Effects of Processing Shellfish

Molluscs

The standard method for reducing the pathogen content of bivalve molluscs from contaminated sites is depuration, either in recirculating clean water systems or by relaying in clean water areas prior to release for sale. Both of these procedures rely on the mollusc's ability to flush itself clear of pathogenic microorganisms during a prolonged period in clean water. However, although the overall bacterial count is very significantly reduced in the mantle fluids and body tissues, there is evidence that some pathogens, including viruses and algal toxins, are not effectively eliminated. The health risks are greatest when bivalve shellfish are consumed raw or partially cooked, and the potential for infection due to environmental bacterial and viral contamination is minimized in all shellfish products when thoroughly cooked immediately prior to consumption. The risk of eating contaminated tissue is much greater in the case of molluscs than crustaceans, because often the whole mollusc, including the gills and gut, is consumed and these can act as reservoirs for pathogens accumulated from the environment. Exceptions are gastropods such as the conch and the abalone, of which only the muscular foot may be consumed, and the scallops, in which only the adductor muscles and gonads are eaten.

In the consumption of crustacean shellfish, the muscle tissues of the abdomen and large pereiopods are usually eaten selectively, except in the case of small shrimps, which are often eaten whole. Consequently, the risk of consuming environmentally contaminated tissue is reduced because the viscera are largely avoided. The risk is further reduced compared to that associated with molluscs, because crustacean shellfish are usually prepared as cooked or pasteurized products. In cooking, adequate temperature and time are needed to reduce bacterial numbers. The bacterial quality of processed crustacean products, as measured

Table 3 Risk monitoring of algal biotoxins in molluscan shellfish

Syndrome	*Causative organism*	*Presence in shellfish waters*	*Tissue level of toxin*
Paralytic shellfish poisoning	*Alexandrium Gymnoclinium* and *Pyrodinium*	Presence	80 μg per 100 g
Diarrhoeal shellfish poisoning	*Dinophysis Aurocentrum* and *Prurocentrum*	100 cells per ml	Presence
Neurotoxic shellfish poisoning	*Ptychodiscus breve*	–	Presence
Amnesiac shellfish poisoning	*Pseudonitzschia* spp. *Nitzschia pungens*	150 000 cells per ml	20 $\mu g\, g^{-1}$ (domoic acid)

by aerobic plate counts, is also strongly dependent on plant sanitation and process management. Where machine picking is used, high standards of cleanliness of the equipment are required.

Spoilage

In products which are stored raw and cooked prior to eating, pathogenic bacteria may be less of a concern than in cooked products and products eaten raw. Indigenous sea water bacteria, such as *Vibrio* species, can grow rapidly at ambient temperatures and careful management of refrigeration conditions is required to control their development in uncooked products. Once mollusc meats have been shucked during raw processing, bacterial numbers tend to increase logarithmically, and may reach unacceptable levels after 2–3 days at abusive storage temperatures.

In cooked shellfish products, the reduction in bacterial contamination during cooking can be followed by increases to higher than initial levels during storage. In addition, cross contamination can occur during the handling and processing of cooked products, and pathogens may grow rapidly in the absence of competing bacteria. Pathogens that may be introduced include *Clostridium botulinum*, *Vibrio* species, *Listeria monocytogenes*, *Staphylococcus aureus*, *Salmonella*, *Shigella* and enteric viruses. In the case of products that are not further cooked before consumption, sanitation controls are required to minimize the risk of the introduction of pathogens, and temperature control is of great importance during post-processing storage and transport. Many shellfish products are pickled in brine, vinegar or spices. Although some pickling methods may reduce the overall bacterial loading and reduce certain specific pathogens, they cannot maintain the quality of the product if temperature control is inadequate. Recent evidence from the UK suggests that around 13% of ready-to-eat, pre-cooked shellfish products are of unsatisfactory quality, with *Vibrio* species (including *V. parahaemolyticus*), *Staphylococcus aureus*, *L. monocytogenes* and *Salmonella* species all being detected in the samples studied. Approximately 1% of products were deemed to have failed the published criteria sufficiently to constitute a threat to public health.

Toxin Production

The filter-feeding habit of bivalve shellfish can result in the accumulation of naturally occurring toxins that are produced by certain species of marine phytoplankton. The consumption of either cooked or raw shellfish contaminated in this way can lead to toxic syndromes in humans, because the toxins are generally heat-stable and will survive normal cooking conditions.

Toxic syndromes are often associated with so-called 'red tides', in which dinoflagellate populations grow rapidly to densities of $>1\times10^6$ cells per litre. The control of dinoflagellate populations is impossible, and the depuration of shellfish in clean water is not totally effective. Consequently, the monitoring of shellfish waters for the presence of causative species and for toxins in shellfish, by mouse bioassay and HPLC (high-pressure liquid chromatography), is the only means of avoiding poisoning incidents. **Table 3** shows the acceptable limits for risk monitoring.

The most common toxic syndrome, paralytic shellfish poisoning (PSP), occurs worldwide and is caused by neurotoxins (saxitoxins) produced by dinoflagellates of the genera *Alexandrium*, *Gymnodinium* and *Pyrodinium*. Symptoms include numbness, muscular weakness and incoherence, which may develop within 1–2 h of consumption. Fatalities can occur due to respiratory collapse, usually within 12 h of the onset of symptoms. PSP toxins have been reported at low levels in crabs as well as in bivalve shellfish. Diarrhoeal shellfish poisoning (DSP) can be caused by at least seven toxins, including okadaic acid, produced by dinoflagellates of the widespread genera *Dinophysis*, *Aurocentrum* and *Prorocentrum*. DSP is considered nonfatal, with symptoms that include diarrhoea and vomiting. Neurotoxic shellfish poisoning (NSP) is associated with bivalves exposed to the dinoflagellate *Ptychodiscus brevis*. NSP is similar to PSP but without paralytic symptoms, and is not usually fatal. Amnesiac shellfish poisoning (ASP) is caused by the toxin domoic acid, which is produced by the diatoms *Nitzschia pungens* and *Pseudo-nitzschia* spp. ASP, which can be fatal, has symptoms including nausea, vomiting and loss of balance as well as mental confusion and permanent short-term memory loss.

See also: ***Aeromonas***: Introduction. **Chilled Storage of Foods**: Principles. ***Clostridium***: *Clostridium perfringens*. **Enterobacteriaceae, Coliforms and *E. coli***: Introduction. **Fish**: Spoilage of Fish. ***Giardia***. **Phycotoxins**. ***Salmonella***: Introduction. **Shellfish (Molluscs and Crustacea)**: Contamination and Spoilage. ***Shigella***: Introduction and Detection by Classical Cultural Techniques. ***Staphylococcus***: Introduction. ***Vibrio***: Introduction, including *Vibrio vulnificus*, and *Vibrio parahaemolyticus*. **Viruses**: Introduction; Hepatitis Viruses. **Water Quality Assessment**: Routine Techniques for Monitoring Bacterial and Viral Contaminants.

Further Reading

Anderson DM, Kulis DM, Qi YZ et al (1996) Paralytic shellfish poisoning in southern China. *Toxicon* 34: 579–590.

Anonymous (1993) National shellfish sanitation program manual of operations. Part I: sanitation of shellfish growing areas. United States Food and Drug Administration, Washington, DC.

Austin B (1988) *Marine Microbiology*. Cambridge: CUP.

Barnes RD (1986) *Invertebrate Zoology*. San Diego: Harcourt Brace Jovanovich.

Cockey RR and Chai T (1991) Microbiology of crustacea processing: Crabs. In: Ward DR and Hackney C (eds) *Microbiology of Marine Food Products*. P. 41. New York: Van Nostrand Reinhold.

Colwell RR and Liston J (1960) Microbiology of shellfish. Bacteriological study of the natural flora of Pacific oysters (*Cassostrea gigas*). *Applied Microbiology* 8: 104–109.

Cook DW (1991) Microbiology of bivalve molluscan shellfish. In: Ward DR and Hackney C (eds) *Microbiology of Marine Food Products*. P. 19. New York: Van Nostrand Reinhold.

Council of the European Communities (1991) Directive No. 91/492 on shellfish hygiene: Classification and monitoring of shellfish harvesting water. *Official Journal of the European Communities* L268/1.

Gram L (1992) Evaluation of the microbiological quality of seafood. *International Journal of Food Microbiology* 16: 25–39.

Huss HH (1994) *Assurance of Seafood Quality. Fisheries technical paper 334*. Rome: Food and Agriculture Organization.

Little CL, Monsey HA, Nichols GL and de Louvois J (1997) The microbiological quality of cooked, ready-to-eat, out-of-shell molluscs. *PHLS Microbiology Digest* 14: 196–201.

Miget RJ (1991) Microbiology of crustacean processing: Shrimp, crab and prawns. In: Ward DR and Hackney C (eds) *Microbiology of Marine Food Products*. P. 65. New York: Van Nostrand Reinhold.

Richards GP (1987) Shellfish-associated enteric virus illness in the United States, 1934–1984. *Estuaries* 10 (1): 84–85.

Scoging AC (1991) Illnesses associated with seafood. *Communicable Disease Report* 1: R117. London: PHL.

Contamination and Spoilage

C A Kaysner, Seafood Products Research Center, Food and Drug Administration, Bothell, Washington, USA

Introduction

Numerous species of molluscs and crustaceans are harvested worldwide, for commercial supply and as private food. The bivalve molluscs (e.g. clams, oysters, mussels, scallops) and crustaceans, including crabs, shrimps, prawns, crayfish and lobsters are economically important in many countries. Species of gastropods, e.g. whelks and abalones, are also harvested, and are considered by some to be delicacies. Many of these products are exported worldwide.

Mechanisms of Spoilage and Contamination

Spoilage

Shellfish are highly perishable, and it is essential that they are handled appropriately after harvesting if their wholesomeness is to be maintained prior to consumption. Proper handling can prevent or delay spoilage, increase shelf life and prevent cross contamination by pathogenic organisms. Shellfish can be preserved at cold temperatures (freezing, chilling), by heat processing or by other preservation techniques, such as canning.

Contamination

Shellfish can become contaminated, toxigenic, or infectious to humans by one or more of several routes: by acquiring microorganisms and toxins that are naturally present in the aquatic environment; by exposure in their natural habitat to human sewage, animal faecal pollution and chemical pollution; by exposure to a contaminated environment, or contaminated equipment or plant workers, during processing; or during preparation in establishments serving food.

Species intended for human consumption are normally harvested from waters that are considered not to be polluted – areas that are affected by industrial pollution or are near sewage treatment outfalls are usually classified as prohibited for harvesting. Bacterial monitoring programmes have been established in many countries in response to public concern, to limit harvesting in areas which are suspected to be polluted. The principal causes of illnesses reported to be associated with seafood consumption are toxins, enteric viruses and human bacterial pathogens.

The bivalve molluscs are filter feeders, accumulating bacteria and other suspended particles that are present in the water in which they grow. They are sessile, and hence incapable of movement to cleaner waters. Several of these species are consumed raw, 'on the half-shell', so any contaminants are passed on to the consumer. Many documented outbreaks of human illnesses have resulted from the consumption of raw molluscan shellfish harvested from polluted waters. Crustaceans, in contrast, are generally cooked prior to consumption and so they tend to harbour a very different range of microorganisms.

Various procedures have been developed which attempt to 'purify' molluscs prior to marketing. One process involves 'relaying' molluscs, from one growing area to another that has better water quality. After a period of 'cleansing', the product is sent to market. A second depuration procedure involves placing the shellfish in man-made tanks with flowing sea water, usually sterilized by ultraviolet (UV) light. This procedure, over time, is effective at the purging of certain bacterial species accumulated by the molluscs. Both of these procedures, however, add handling steps which can affect the price of the product.

The display of live crustaceans and molluscs in tanks is popular at retail markets and restaurants, and live species can be exhibited in large tanks which use circulating artificial sea water. Often mixed species are displayed, which allows bacterial cross contamination between species. Concerns have been raised about the adequacy of the sanitation of these tanks, the maintenance of water quality during recirculation and the mixing of species. One study found that salmonella and *Vibrio cholerae* (non-O1) may be transmitted to consumers by oysters displayed in these tanks, although the study did not determine how the bacteria reached the display system.

Shellfish can have adverse effects on human health through a variety of mechanisms, described below. These may be categorized as:

- indigenous microorganisms
- non-indigenous bacteria
- non-indigenous viruses
- non-indigenous parasites
- marine toxins
- chemicals.

Indigenous Microorganisms

Numerous bacterial species which are indigenous to molluscs and crustaceans are subsequently involved in spoilage during storage, and several have been reported to be pathogenic to humans, including *Vibrio*, *Aeromonas* and *Plesiomonas*.

Vibrio

In the past few decades, illnesses caused by *Vibrio* species have increased in incidence: several mesophilic species are pathogenic to humans. Many *Vibrio* species are indigenous in coastal areas, particularly in brackish waters. They survive the winter in the marine sediment and 'bloom' during the warmer months of the year. As the water temperature rises above 15°C, vibrios become more abundant and are accumulated by molluscs during feeding. Crustaceans are also affected by the presence of vibrios, which are involved in chitin degradation – some species are able to attach to the shell of crustaceans.

The most notable *Vibrio* species in terms of public health are *V. parahaemolyticus*, *V. cholerae* and *V. vulnificus*. These account for most of those bacterial infections of humans which are associated with the consumption of molluscs and crustaceans. Most of these illnesses are reported during the summer months (June to September) in temperate regions, when estuarine water temperatures have risen. During the 14-year period up to 1994 in Florida, 95 infections caused by *V. vulnificus* and due to raw oyster consumption were reported to health officials; 82 of these cases involved primary septicaemia. *V. vulnificus* was responsible for 28% of the total vibrio infections reported in that time period, while together *V. parahaemolyticus* and *V. cholerae* (non-O1) accounted for 151 (45%) of the infections reported.

Vibrio parahaemolyticus This is a very common species in coastal and estuarine areas. The consumption of raw molluscs has caused a number of localized food poisoning outbreaks due to *V. parahaemolyticus*; and the post-processing contamination of crabs and shrimps has caused outbreaks involving large numbers of cases. However, not all strains cause human gastroenteritis. Approximately 5% of strains isolated from the environment produce a thermostable direct haemolysin (TDH) and are termed 'Kanagawa positive'. However, strains producing TDH are recovered from more than 95% of patients with gastroenteritis. A second haemolysin was produced by a strain isolated from patients during an outbreak in the late 1980s. This haemolysin is related to, but distinct from, TDH and is termed 'thermostable related haemolysin' (TRH). The genetic sequence of TRH was found to be only 60% homologous to that of TDH. Subsequently, a number of *V. parahaemolyticus* strains isolated from patients during outbreaks were found to contain both TDH and TRH sequences.

The exact mechanism of pathogenesis in humans is unknown, but the presence of TDH and/or TRH genes in *V. parahaemolyticus* is highly correlated with illness. In addition, recent studies in the Pacific Northwest, Asia and Japan have found that strains isolated from patients that contain the TDH and/or TRH genes also hydrolyse urea (urease positive). However, not all uh+ strains contain the haemolysin gene(s). Urease is thought to enhance the tolerance of *V. parahaemolyticus* to acidic conditions, and studies are in progress to determine whether acid-tolerance enhances virulence. A simple laboratory, production of urease, test can then be used to screen isolates from mollusc samples for further testing for potential pathogenicity, possession of TDH or TRH genes.

Vibrio cholerae *V. cholerae* has been known for centuries, causing epidemics of cholera involving large numbers of people and many deaths. Generally, this pathogen is endemic in areas where there is poor sanitation and sources of drinking water become contaminated. Cholera toxin was thought to be produced only by two biotypes of the O1 serogroup of *V. cholerae* until 1992, when another serogroup, O139, was identified as the cause of a large outbreak in Bangladesh.

The faecal–oral route is a common method of infection of humans, although various vectors have been involved, including seafood – usually molluscs. Although cholera can be a severe, life-threatening disease, it is easily treated and is preventable through good hygiene and the proper treatment of sewage and drinking water. In early 1991, epidemic cholera appeared in Peru and subsequently spread to countries in South America and other parts of the western hemisphere. This epidemic was of particular concern because estuaries became contaminated and in one instance in the US, crustaceans became vectors. Recently, outbreaks of cholera in the Americas have been linked with the extreme weather conditions associated with El Niño.

Other strains of *V. cholerae* are widely distributed in the estuarine environment, and include more than 150 somatic antigen designations. Some strains produce a cholera-like (CT-like) toxin and have been responsible for human illness, of a general type less severe than classical cholera. Molluscs and crustaceans were the vectors in many of the reported cases. Some non-O1 strains have been reported to cause a rare septicaemic condition in humans, from which several deaths have resulted. These strains did not produce CT or CT-like toxins: the mechanism of pathogenesis is still being investigated.

Vibrio vulnificus *V. vulnificus* was first identified in 1980 and is a particularly virulent human pathogen. This marine species can cause primary septicaemia, with a mortality rate exceeding 50%. Certain individuals are at particular risk, including patients with hepatic disorders (e.g. cirrhosis), diabetes, cancer or haemochromatosis; patients receiving long-term steroid treatment (e.g. for asthma or arthritis); patients with stomach problems (e.g. previous gastric surgery, low stomach acid due to the use of antacids) and the immunocompromised (e.g. patients with HIV infection). Such patients are cautioned to avoid eating raw or undercooked seafood. The majority of reported illnesses have been associated with the consumption of raw oysters – most often harvested in the US from estuaries communicating with the Gulf of Mexico. However *V. vulnificus* is also found in many other coastal areas of the world. Animal bioassay reveals both virulent and avirulent strains. The mechanism(s) of virulence have not been fully elucidated.

Other *Vibrio* species Several other species of *Vibrio* are found, in many coastal areas. *V. mimicus*, *V. hollisae*, *V. fluvialis*, *V. furnissii* and *V. metschnikovii* are all recently identified species. They have been isolated from seafood and from the stools of diarrhoeal patients, some of whose food histories point to molluscan shellfish as the probable vector. The illnesses caused by these species occur seasonally and are rare. These species may be opportunistic, attacking individuals with underlying medical conditions. The mechanisms of pathogenesis remain unclear, although some species produce protein toxins and proteases, which may have a role.

Aeromonas and *Plesiomonas*

Two other genera of the family Vibrionaceae – *Aeromonas* and *Plesiomonas* – contain species that are indigenous to molluscs and have been implicated in diarrhoeal illness associated with mollusc consumption. *A. hydrophila*, *A. caviae*, *A. sobria* and *P. shigelloides* have been isolated from human stools but have not yet gained acceptance as enteric pathogens. They appear to be opportunistic, attacking hosts with underlying medical problems, and the mechanisms of pathogenesis have not been elucidated. These species are widely distributed in the environment, and can be recovered from fresh and brackish water as well as shellfish. *Aeromonas* species have also been recovered from soil and from animal sources. Strains of *Aeromonas* can grow at refrigeration temperatures, and may be involved in the spoilage of seafood products during storage.

Prevention of Infection

There is much to learn regarding the control of indigenous bacteria in order to prevent human illness. Restricting the harvesting of molluscs to certain seasons, reflecting environmental temperatures, could minimize the levels of indigenous microorganisms to which consumers are exposed. All these bacteria are susceptible to mild heating, and are thus eliminated from seafood which is adequately cooked.

Non-indigenous Bacteria

A number of bacterial species considered non-indigenous to molluscs and crustaceans can cause human illness through the contamination of seafood products. Unhygienic practices during processing and direct faecal transmission from human or animal reservoirs are the main routes via which food becomes contaminated by these enteric pathogens.

Salmonella

Salmonellosis associated with the consumption of contaminated molluscs was reported as early as the 1900s in the US, when molluscs were harvested from waters polluted with human faeces. Controls on shellfish harvesting reflect these early outbreaks of illness, but *Salmonella* remains a major cause of food-borne illness. There are many reservoirs of *Salmonella* besides humans, including other mammals, birds and reptiles. Foods are often contaminated during handling, by the faecal–oral route.

Shigella

Shigella are enteric bacteria which contaminate foods via faeces. Although they are primarily thought of as waterborne pathogens, they have been associated with food-borne disease due to poor sanitation. Human waste in estuaries has resulted in the contamination of shellfish and subsequent outbreaks of shigellosis. Carriers of the organism who work in a food processing environment are a major concern, and good personal hygiene habits and sanitation practices are essential in preventing the spread of *Shigella*.

Yersinia and *Listeria*

Yersinia enterocolitica and *Listeria monocytogenes* are psychrotolerant bacteria which are widely distributed in the environment. They are associated with many animal species, and have been recovered from estuarine waters and shellfish. Not much is known about the association between seafood and the diseases yersiniosis and listeriosis, but the consumption of raw molluscs has been noted in the food history of some cases. Both species are capable of growth at refrigeration temperatures, and *L. monocytogenes* has been recovered from several seafood commodities, including frozen crustaceans, and from the processing environment.

Campylobacter

There are increasing reports of molluscan-borne human illness caused by *Campylobacter* species, enteric pathogens with many animal reservoirs. Recent evidence points to estuarine contamination by waterfowl and terrestrial run-off, and subsequent survival in the environment. Studies also indicate that this pathogen may survive in shellfish after harvesting, even when they are apparently handled adequately during storage. The infectious dose of this pathogen is suspected to be low, which elicits concern.

Staphylococcus

Enterotoxin-producing strains of *Staphylococcus aureus* have been responsible for outbreaks of intoxication with shellfish as the vehicle, and this pathogen is widely distributed, with human and animal reservoirs. Food handlers are the primary source of contamination, although equipment surfaces can be fomites. Heat-treated seafood which is subjected to further handling, such as hand-picking or sorting, can be a vehicle for *S. aureus*, which can predominate because competitive microorganisms have been destroyed. Temperature abuse (storage above refrigeration temperature) of the product can result in growth of the microorganism, and toxin production. The enterotoxins are resistant to heat, and may persist in the product even though the vegetative cells have been destroyed.

Clostridium and *Bacillus*

The spore-forming bacteria *Clostridium botulinum*, *C. perfringens* and *Bacillus cereus* have rarely been implicated in seafood-borne intoxications, although *C. botulinum* type E, one of seven types identified, is primarily of marine origin. These organisms are widespread in the environment, particularly in soil and not infrequently in the intestinal tract of animals. *C. botulinum* produces a potent neurotoxin, which causes the disease botulism. Safety concerns have been elicited by canning processes, vacuum-packaging, modified-atmosphere storage and the recently popular 'sous vide', involving the minimal processing of shellfish. Heat-processing must be adequate to destroy spores of the organisms prior to storage under anaerobic conditions, to prevent germination and subsequent toxin production. Several strains of C.

botulinum grow and produce toxins at temperatures as low as 3°C. The toxins are proteins and can be inactivated by heating at 60°C for 5 min. Therefore a product containing preformed toxin must be heated prior to consumption.

C. perfringens is commonly found in the intestinal tract of warm-blooded animals, and has been suggested for use as an indicator of faecal pollution in the estuarine environment. This organism, and also *B. cereus*, produce enterotoxins that can cause diarrhoeal illness in humans, but only if the organisms are present in high numbers and produce sufficient enterotoxin. The proper storage of shellfish at refrigeration temperatures will prevent the growth of these organisms, and they cause shellfish-borne illness only relatively rarely.

Non-indigenous Viruses

More than 100 enteric viruses can be found in human faeces, and so may reach the environment in sewage effluents. Seafood may then become contaminated, which is of concern because of the very low numbers of these viruses needed to cause human infection. However, only a few of the enteric viruses have been shown to be transmitted – these include hepatitis A, non-A and non-B hepatitis, Norwalk, Snow Mountain agent, astroviruses, caliciviruses and small round viruses. The shortness of this list probably reflects the lack of detection methods – some enteric viruses can survive in the marine environment longer than the bacteria used for monitoring water quality. However, recent advances in genetic techniques for the detection of viruses in samples will enable more rapid detection, either in shellfish associated with illnesses or for monitoring studies. Water temperature (< 10°C) appears to be most important for the survival of enteric viruses. The link between contaminated water, contaminated seafood and human infection has been demonstrated: in the US, all reported cases of seafood-associated viral infections have been linked with the consumption of raw or inadequately cooked molluscs, and the transmission of Norwalk and hepatitis A viruses to humans by contaminated molluscs has been documented. Enteric viruses were found in crabs sampled at a sewage sludge dump, and crustaceans can take up these viruses in experimental procedures (no field studies have taken place). No multiplication of these viruses has been demonstrated in molluscs, although within contaminated molluscs, some survived refrigerated storage for extended periods.

The transmission of enteric viruses may be prevented by heat-processing shellfish. A reduction in numbers of surviving viruses following commonly used cooking techniques (e.g. boiling, baking, frying, steaming) is expected, but in order to prevent low-level survival, it has been suggested that molluscs should be boiled for at least 20 min. The best way of preventing transmission to humans, however, is to ensure that shellfish are harvested from non-polluted waters, although in practice this is difficult.

Viruses may also be transmitted to humans via shellfish after contamination by infected food handlers, although hepatitis A is the only virus demonstrated to have been transmitted by this route. Nevertheless, the need for good hygiene practices is highlighted.

Non-indigenous Parasites

Recently the protozoan parasite *Cryptosporidium parvum* has become a concern in association with molluscan shellfish. The oocysts of *C. parvum* are quite resistant to chlorination, a common disinfection technique used for municipal drinking water systems. This parasite has been the cause of widespread infection carried by drinking water in the US, the most notable outbreak occurring in the State of Wisconsin in 1993: > 400 000 individuals were infected by water from a contaminated municipal supply, and during the following 4 years, 68 cryptosporidiosis-associated deaths were identified. In early 1998, four swimming pools in Melbourne, Australia were closed after the detection of *C. parvum* in the water systems.

This parasite is carried by many warm-blooded animals, including cattle, and could be washed via outfalls into estuarine areas where molluscs grow. Oocysts of *C. parvum*, which cause infection, have been recovered from the oyster *Crassostrea virginica* in Chesapeake Bay in the US and from the Asian freshwater clam *Corbicula fluminea*, introduced into the US in the early 1900s. Illnesses caused by *C. parvum* have not been traced to shellfish and as yet have not been determined to be parasitic to molluscs. However, the survival of the oocysts in an estuarine environment and their intake by filter-feeding molluscs indicate the possibility of bivalves being vectors.

Marine Toxins

Several natural or preformed toxins are found in, but not produced by, shellfish. The toxins are produced by other marine organisms, notably dinoflagellates, and then taken up by shellfish, either from the water or by ingestion of the whole toxin-forming organism. The bivalve molluscs are primarily involved, owing to their filter-feeding habit, which concentrates the toxins. The toxins are quite potent, may be present in minute quantities, and may undergo transformation within the shellfish, which may increase their toxicity to humans. The amount of accumulated toxin will not increase after harvesting, but the toxins cannot be removed from the shellfish, or destroyed. For example, they are far more stable to heat than are most bacterial toxins, and so the cooking of shellfish does not provide reliable protection. Most monitoring programmes involve animal bioassays, and have been effective in reducing human illness by prompting the closure of harvesting areas when toxins reach certain levels.

Paralytic Shellfish Poisoning

Paralytic shellfish poisoning (PSP) is a potentially fatal syndrome resulting from the ingestion of one or more of a family of potent neurotoxins called saxitoxins. PSP is generally associated with the consumption of bivalve molluscs, although crabs and whelks have also been implicated. At present 17 different saxitoxins are recognized, being derivatives of the parent saxitoxin molecule. These water-soluble toxins are produced by dinoflagellates, notably by species of *Alexandrium*. Low doses result in a tingling or burning sensation of the lips, but higher doses cause paralysis of the extremities, loss of motor coordination and even death by respiratory paralysis. No antidote is known.

Neurotoxic Shellfish Poisoning

The 'red tide' of the Gulf of Mexico is due to a bloom of *Ptychodiscus brevis*, a dinoflagellate which produces brevetoxins, which can cause neurotoxic shellfish poisoning (NSP) in humans. NSP is generally found along the west coast of Florida and other Gulf States, occasionally extending up the Atlantic coast of the US. Brevetoxins, of which nine are identified, consist of cyclic polyether backbones. They are lipophilic in nature and cause nausea and neurological symptoms. The syndrome usually lasts for a few days, and no deaths of humans have been reported.

Diarrhetic Shellfish Poisoning

Okadaic acid and its derivatives have caused diarrhetic shellfish poisoning (DSP), recognized in Japan and Europe. Bivalve molluscs are the known vectors. Acute diarrhoea is the primary symptom, with more rapid onset than that due to bacterial intoxications. DSP toxins have been isolated from molluscs implicated in illnesses, and from two genera of dinoflagellates.

Domoic Acid Poisoning

Domoic acid poisoning, also called amnesic shellfish poisoning (ASP), is a recently identified syndrome which seems to be restricted to Prince Edward Island, eastern Canada and the northern US west coast. The consumption of mussels and clams containing domoic acid has caused this rare form of poisoning, which can result in death. Symptoms include nausea, loss of equilibrium and neural involvement, including memory loss. It appears that brain damage caused by ASP may be irreversible, resulting in permanent loss of memory.

Tetramine Poisoning

The consumption of whelks and edible marine snails of the genus *Neptunea* has caused rare instances of tetramine poisoning. The symptoms caused by the tetramethylammonium ion appear approximately 30 min after ingestion, and recovery is usually complete within a few hours. The symptoms include headache, dizziness, and short periods of impaired vision. Tetramines are considerably less toxic than most other marine toxins, and there have been no documented cases of human mortality.

Chemicals

Industrial and agricultural chemicals can contaminate areas of shellfish, but risk assessment is difficult. Increasingly, waterways carry warnings against the consumption of fish and shellfish, because of chemical pollutants. Their effects on health are not readily obvious in the form of distinctive or acute illnesses. However, the effects of long-term exposure to low levels of these contaminants, on both shellfish and those who consume them, need to be studied further.

Chemical contaminants tend to be distributed unevenly in the environment and so are difficult to control. Examples include organic contaminants, such as polychlorinated biphenyls, dioxin, chlorinated hydrocarbon insecticides (e.g. DDT, Endrin, Chlordane) and petroleum hydrocarbons, and inorganic contaminants, including antimony, arsenic, cadmium, lead, mercury and selenium. Contaminants of aquatic environments may reach shellfish and ultimately humans, on whom they are known to cause serious health effects. The most effective way of protecting consumers is the prohibition of shellfish harvesting from areas subject to industrial pollution or excessive terrestrial run-off. The strengthening and enforcement of regulations aimed at minimizing chemical contamination of the environment is also necessary.

Spoilage

Microorganisms

The microflora of shellfish varies, reflecting that of their habitat. For example, warm-water crustaceans harbour mainly Gram-positive bacteria such as

Micrococcus or coryneforms, while cold-water species carry primarily Gram-negative organisms, including *Pseudomonas*, *Moraxella* and *Flavobacterium*. Warm-water shellfish generally have higher levels of (mesophilic) bacteria at harvest than those from cold water. However, it has been suggested that cold-water species spoil more quickly, since they contain bacteria more capable of growth at refrigeration temperatures. Studies have shown that certain organisms become predominant during chilled storage, as spoilage progresses. Species of *Pseudomonas*, *Flavobacterium*, *Acinetobacter* and *Moraxella* seem to be most commonly involved in spoilage. Yeasts can also be involved, for example *Rhodotorula*, *Candida* and *Torulopsis* species can cause discoloration of the product during storage.

The bacteria normally involved in spoilage are susceptible to heat, and normal cooking processes would eliminate them. Temperature abuse at some point between harvesting and consumption is the primary cause of rapid spoilage and the growth of pathogenic organisms.

Temperature

Molluscs and crustacea are highly perishable, and in order to maximize shelf life they must be subjected to reduced temperature immediately after harvesting and throughout processing and distribution. Temperature control is critical in delaying or retarding both decomposition by bacterial and autolytic enzymes, and the oxidation and hydrolysis of fats. A general rule is that for each Celsius degree increase in temperature from 0°C, shelf life is reduced by 2 days. Storage under or in ice affords maximum shelf life to products that are not stored frozen. Storage temperatures below 7°C inhibit the growth of most human enteric bacteria (except *Listeria*, *Yersinia* and *Aeromonas* spp.), and prevent toxin production.

Transport

Some species, particularly bivalve molluscs, are transported live. A low temperature and a damp environment must therefore be ensured throughout, i.e. for up to 2 weeks. Generally, decomposition does not occur as long as the animals remain alive. The shipment and storage of raw products must be kept separate from those of processed products, to prevent cross contamination. Some countries have strict regulations regarding the shipment and sale of certain live species.

Bioluminescence

The growth of bioluminescent marine bacteria in shellfish can result in products that 'glow in the dark', a phenomenon particularly noticeable in processed shrimps and crabs. These bacteria either survive processing or subsequently recontaminate the product, and grow during chilled storage to levels at which their luminescence is visible to the human eye. Bioluminescent bacteria include species of *Photobacterium*, *Xenorhabdus*, and *Vibrio*. These are generally thought to be non-pathogenic to humans, although a few species of *Vibrio* may prove otherwise.

See also: ***Aeromonas***: Introduction. ***Bacillus***: *Bacillus cereus*; Detection of Enterotoxins. **Biochemical and Modern Identification Techniques**: Food-poisoning Organisms. ***Campylobacter***: Introduction. ***Clostridium***: *Clostridium perfringens*; Detection of Enterotoxins of *C. perfringens*; *Clostridium botulinum*; Detection of neurotoxins of *C. botulinum*. ***Cryptosporidium***. **Food Poisoning Outbreaks**. ***Listeria***: Introduction; *Listeria monocytogenes*. ***Salmonella***: Introduction. **Shellfish (Molluscs and Crustacea)**: Characteristics of the Groups. ***Shigella***: Introduction and Detection by Classical Cultural Techniques. **Spoilage Problems**: Problems caused by Bacteria. ***Staphylococcus***: *Staphylococcus aureus*; Detection of Staphylococcal Enterotoxins. ***Vibrio***: Introduction, including *Vibrio vulnificus*, and *Vibrio parahaemolyticus*; *Vibrio cholerae*. **Viruses**: Introduction. **Water Quality Assessment**: Routine Techniques for Monitoring Bacterial and Viral Contaminants; Modern Microbiological Techniques. **Waterborne Parasites**: Detection by Classic and Modern Techniques.

Further Reading

Anderson DM, White AW and Baden EG (eds) (1985) *Toxic Dinoflagellates*. New York: Elsevier Science Publishing.

Colwell RR (ed.) (1984) *Vibrios in the Environment*. New York: Wiley.

Committee on Evaluation of the Safety of Fishery Products, Food and Nutrition Board, Institute of Medicine (1991) *Seafood Safety*. Washington: National Academy Press.

Fayer R, Graczyk TK, Lewis EJ, Trout JM and Farley CA (1998) Survival of infectious *Cryptosporidium parvum* oocysts in seawater and Eastern oysters (*Crassostrea virginica*) in Chesapeake Bay. *Applied and Environmental Microbiology* 64: 1070–1074.

Florida Dept. Agriculture and Consumer Services (1998) News about oysters. *website* http://www.fl-seafood.com/oystnews.htm.

Gerba CP (1988) Viral disease transmission by seafood. *Food Technology* 42: 99–103.

Green J, Henshilwood K, Gallimore CI, Browne DWG and Lees DN (1998) A nested reverse transcriptase PCR assay for detection of small round-structured viruses in environmentally contaminated molluscan shellfish. *Applied and Environmental Microbiology* 64: 858–863.

Hackney CR and Pierson MD (eds) (1994) *Environmental Indicators and Shellfish Safety*. P. 523. New York, London: Chapman & Hall.

Hlady WG (1997) *Vibrio* infections associated with raw oyster consumption in Florida, 1981–1994. *Journal of Food Protection* 60: 353–357.

International Committee on Microbiological Specifications for Food (1988) *Microorganisms in Foods 4, Application of the Hazard Analysis Critical Control Point (HACCP) System to Ensure Microbiological Safety and Quality.* Oxford: Blackwell Scientific.

Janda JM, Powers C, Bryant RG and Abbott SL (1988) Current perspectives on the epidemiology and pathogenesis of clinically significant *Vibrio* spp. *Clinical Microbiology Reviews* 1: 245–267.

Kaysner CA, Abeyta Jr C, Trost PA et al (1994) Urea hydrolysis can predict the potential pathogenicity of *Vibrio parahaemolyticus* strains isolated in the Pacific Northwest. *Applied and Environmental Microbiology* 60: 3020–3022.

Liston J (1990) Microbial hazards of seafood consumption. *Food Technology* 44 (12): 56–62.

Rippey SR (1994) Infectious diseases associated with molluscan shellfish consumption. *Clinical Microbiology Reviews* 7: 419–425.

Sado PN (1991) Isolation of luminescent bacteria from cooked seafood products. *Dairy, Food and Environmental Sanitation* 11: 361–363.

Update: Cholera – Western Hemisphere, and recommendations for treatment of cholera. *Morbidity and Mortality Weekly Report* 40: 562–565.

US Food and Drug Administration (1998) Foodborne Pathogenic Microorganisms and Natural Toxins Handbook. 'Bad Bug Book'. *website* http://vm.cfsan.fda.gov/~mow/chap8.html.

Ward DR and Hackney CR (eds) (1991) *Microbiology of Marine Food Products.* New York: Van Nostrand Reinhold.

World Health Organization (April 01, 1998) Cholera & El Nino, WHO Statement-Latin America. ProMED-Mail. <promed@usa.healthnet.org>.

SHEWANELLA

Lone Gram and **Birte Fonnesbech Vogel**, Danish Institute for Fisheries Research, Department of Seafood Research, Technical University of Denmark, Lyngby, Denmark

Introduction

Shewanella is a recently defined genus. The species of importance in the food industry, *Shewanella putrefaciens*, has been known since the early 1930s, although under changing names and taxonomic positions. Members of the genus are characteristic of the marine environment; however, *S. putrefaciens* has also been isolated from a variety of other environments, including freshwater lakes, sediments, oil fields and proteinaceous foods. The importance of *S. putrefaciens* in the food industry stems from its role in spoilage of low-temperature stored protein-rich foods of high pH, thus it is typical of spoiling marine fish, spoiling chicken and spoiling high-pH meat. The organism produces a variety of volatile sulphides, including H_2S, and in marine fish it reduces trimethylamine oxide (TMAO) to trimethylamine (TMA) which has a fishy smell. It has been identified as the specific spoilage bacteria of iced marine fish where levels of 10^8 *S. putrefaciens* per gram will cause unpleasant sensory changes. The organism has also been implicated in disease (bacteraemia and sepsis); however, it has recently been suggested that the clinical isolates probably belong to a different species, *S. alga*.

Characteristics of the Species

The taxonomic position of *S. putrefaciens* is not completely resolved. *S. putrefaciens* was formerly identified as *Achromobacter*, a group comprising various Gram-negative, non-fermentative, oxidase-positive, rod-shaped bacteria. The organism was transferred to *Pseudomonas* by Long and Hammer in 1941 and placed in the *Pseudomonas* group III/IV by the Shewan 1960 classification scheme. Due to the difference in guanine + cytosine (G+C) % of *Pseudomonas putrefaciens* (typically 43–53%) and other pseudomonads (typically 58–72%), the species was transferred to the genus *Alteromonas*. In 1985, MacDonnel and Colwell suggested that *Alteromonas putrefaciens* be transferred to a completely new species, *Shewanella*, so named in honour of Dr J. Shewan. This study also assigned two other marine species, *S. benthica* and *S. hanedai*, to the genus. The study was based on comparison of 5S rDNA gene sequences and suggested that the genus *Shewanella* be included in the Vibrionaceae family. More recently, 16S rDNA gene sequence comparisons have also suggested that *Shewanella* is more closely related to the Vibrionaceae than to the Pseudomonadaceae. However, this taxonomic position is controversial, as the Vibrionaceae includes only fermentative organisms and *Shewanella* is a strict respiratory genus. On the other hand, quite a number of phenotypic characteristics like association with the marine environment, the ability to use various electron acceptors, and the production of hydrolytic enzymes of the *Shewanella* spp. are quite similar to the genera of the Vibrionaceae. On a grander phylogenetic scale, the

Table 1 Major subgroups of *Shewanella putrefaciens* or different species of *Shewanella* as reported in the literature

Mol% G+C > 52; tolerance to NaCl > 6%; mesophilic	*Mol% G+C < 48%; tolerance to NaCl variable; mostly psychrotrophic*
Pseudomonas putrefaciens biovar 2	*Pseudomonas putrefaciens* biovars 1 and 3
P. putrefaciens species 2	*P. putrefaciens* species 1
P. putrefaciens group 2	*P. putrefaciens* group I
P. putrefaciens group VI	*P. putrefaciens* groups I, II, III
P. putrefaciens	*Alteromonas putrefaciens*
A. putrefaciens phenon A (A1, A2)	*A. putrefaciens* phenon B (B1, B2)
Shewanella putrefaciens group 4	*Shewanella putrefaciens* group 1, 2 and 3
S. putrefaciens	*S. putrefaciens* clusters 10, 11, 12 and 13
S. putrefaciens biovar 2	*S. putrefaciens* biovars 1 and 3
S. alga	*S. putrefaciens*

From Fonnesbech Vogel et al (1997).

species belongs to the β subclass of the Proteobacteria of the Bacteria kingdom.

It has been known for many years that *Shewanella* (formerly *Pseudomonas*, *Alteromonas*) *putrefaciens* is a very heterogeneous species, with several authors identifying a mesophilic, halotolerant group that has often been associated with warm-blooded animals and, occasionally, with disease in humans (**Table 1**). Recent research has shown that the vast majority of these mesophilic isolates are rightly members of a different species, *S. alga*. Classical phenotypic characterization will not distinguish the two species, but tolerance to 42°C and often to 10% NaCl as well as a high G+C% (52–56%) are all characteristics of *S. alga* and not of *S. putrefaciens*. The food spoilage strains are all psychrotrophic, with a lower G+C% (<48%) and identified as *S. putrefaciens*. Biochemically oriented analyses, e.g. protein profiling, ribotyping or 16S rDNA gene sequence confirms the existence of two separate species. However, *S. putrefaciens* with mesophilic traits can be isolated.

Phenotypic characterization, protein profiling and 16S rRNA gene sequencing have indicated the existence of subgroups of *S. putrefaciens*; however, this area needs further investigation. Along similar lines, little is known about clonal differences within the species, e.g. if particular clones are selected for during chill storage of foods. Recently randomly amplified polymorphic DNA (RAPD) analysis has been used to assess the genetic diversity of environmental isolates of *S. putrefaciens* and several genotypes were identified; however, the species appeared to be stable over time. Preliminary experiments with RAPD typing of isolates from fish show that, while the strains isolated from fresh fish are almost all genotypically different, some selection is seen during iced storage. However, a large variation is seen from fish to fish.

Also included in the genus *Shewanella* are the two species *S. benthica* and *S. hanedai*, which are both isolated from cold or actic marine waters. Very recently a number of new species (e.g. *S. colwelliana*) have also been assigned to the genus.

Biochemical and Physiological Attributes

Shewanella putrefaciens is a Gram-negative, motile rod with positive oxidase and catalase reactions. The organism will carry one polar flagella if cultured in liquid medium but on some solid media may contain several lateral flagella (**Fig. 1**).

The organism is strictly limited to respiratory

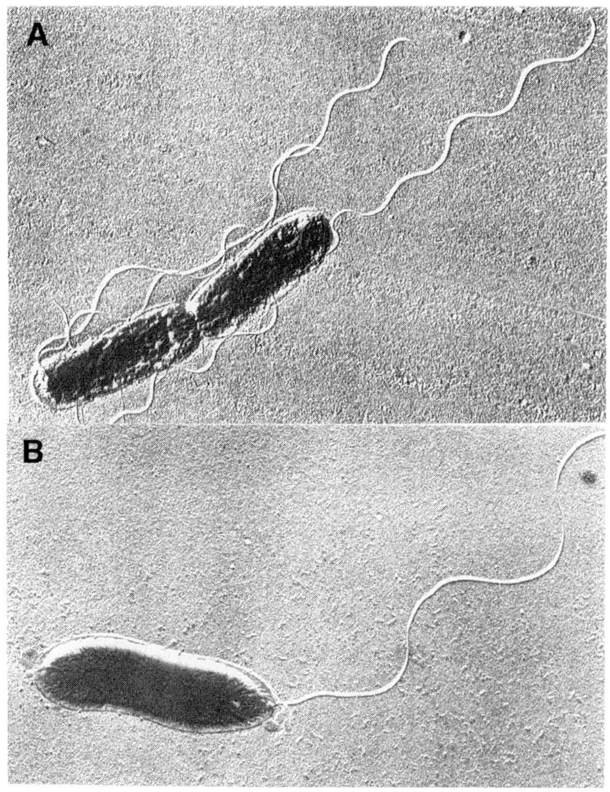

Figure 1 (**A**) Scanning electron microscopy of *Shewanella putrefaciens* cultured on nutrient agar (×21 000). From Gibson et al (1977). (**B**) Scanning electron microscopy of *Shewanella putrefaciens* cultured in nutrient broth (×28 000). From Gibson et al (1977).

metabolism; however, it is outstanding in its ability to use a variety of electron acceptors: oxygen, ferric iron (III), manganese (IV), TMAO, dimethyl sulphoxide, nitrate, nitrite, thiosulphate, fumarate, sulphite and elemental sulphur. Due to this versatility of electron sinks, *S. putrefaciens* may occur in many ecological niches and it is believed to be important in nature for the turnover of, for example, Fe and Mn in aquatic environments. Its ability to reduce Fe(III) and produce sulphides can be a cause of microbial corrosion of metals. During aerobic growth, the organism sequesters iron (for use in cytochromes) by use of low-molecular-weight iron chelators – so-called siderophores. Fish is an iron-limited substrate and facilitates siderophore production. Whether or not the organism chelates iron during anaerobic respiration is not known. During anaerobic respiration with Fe(III) as electron acceptor, some strains of *S. putrefaciens* are able to degrade aromatic hydrocarbons, e.g. benzene, and it may thus be a potential biodegrader in environmental clean-ups.

The ability of *S. putrefaciens* to use TMAO as electron acceptor is a major reason for its importance in fish spoilage, as the reduced compound, TMA, has a characteristic fishy smell. Respiration using TMAO or other compounds as electron acceptors is carried through a range of cytochromes of type *c* and to the final e^- carrier, the reductase. TMAO reductase has been localized to the periplasmic space, whereas some experiments have shown the Fe(III) reductase to be located in the outer membrane.

Experiments carried out in the 1980s indicated that *S. putrefaciens* uses the tricarboxylic acid (TCA) cycle during anaerobic respiration (**Fig. 2A**); however, recent experiments have suggested that it rather uses a fusion of the TCA cycle (a truncated TCA cycle) and the anabolic serine pathway (Fig. 2B). *S. putrefaciens* is a potent producer of volatile sulphides (see also Table 5, later) and produces H_2S from cysteine. Other sulphides probably originate from methionine metabolism. During storage of fish, *S. putrefaciens* also degrades ATP-related compounds and is capable of producing hypoxanthine (a bitter-tasting component) from inosine monophosphate.

S. putrefaciens produces a range of degradative enzymes. Apart from the ability to reduce TMAO and produce H_2S, the hydrolysis of DNA and the decarboxylation of ornithine are important tests in its classification. *S. putrefaciens* also hydrolyses proteins (casein and gelatin), RNA and some fatty acid esters of sorbitan (Tween compounds). By a number of simple biochemical tests *S. putrefaciens* may be distinguished from related Gram-negative organisms (**Table 2**).

Tolerance to Food-relevant Parameters

The food spoilage strains of *S. putrefaciens* are all psychrotrophic; all grow at 4°C and many at 0°C. Rarely does growth occur at 37°C. Although *S. putrefaciens* is associated with the marine environment, the bacterium does not tolerate high NaCl levels (e.g. 10%). The sensitivity of the psychrotrophic strains varies, with some being inhibited by 6% NaCl and others being tolerant. The organism, being non-fermentative, is relatively sensitive to low pH. Some strains are able to grow at pH 5.5 whereas others are not. In general, strains most tolerant to NaCl are also most tolerant to low pH. On a G+C% scale, the less tolerant strains have the lowest G+C% of the *S. putrefaciens*. This is consistent with the fact that *S. alga*, being much more tolerant to high temperature,

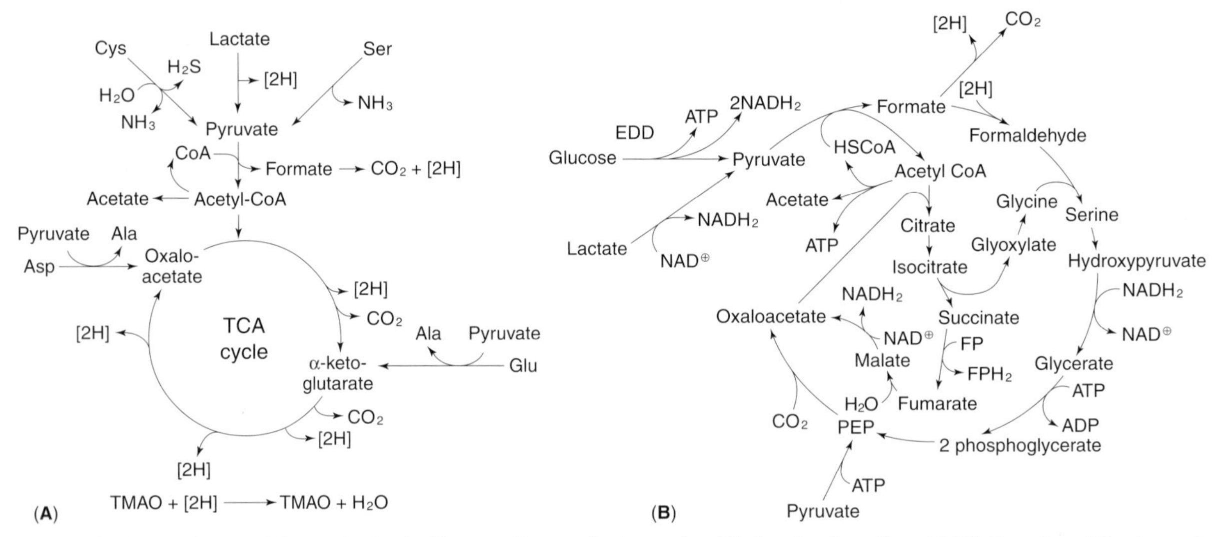

Figure 2 Pathway of anaerobic respiration in *Shewanella putrefaciens* using (**A**) the tricarboxylic acid (TCA) cycle or (**B**) a truncated TCA cycle merged with the serine pathway. (A) From Ringø et al (1984); (B) from Scott and Nealson (1994).

Table 2 Typical reactions of *Shewanella putrefaciens* and some closely related Gram-negative families and genera

Reaction	S. putrefaciens	S. alga	Pseudomonas *spp.*	*Vibrionaceae*	*Enterobacteriaceae*
Gram reaction	–	–	–	–	–
Shape	Rod	Rod	Rod	Rod	Rod
Motility	+	+	+	(+)	(+)
Cytochrome oxidase	+	+	+	+	–
Catalase	+	+	+	+	+
Glucose	–/O	–/O	O	F	F
TMAO reduction	+	+	–	+	+
H_2S production	+	+	–	(+)	(+)
Ornithine decarboxylase	+	+	–	(–)	+/–
DNAse	+	+	(–)	(+)	(–)
G+C%[a]	43–47	52–56	58–70	38–63	38–60

+ positive; (+) a few negative; +/– some positive, some negative; (–) a few positive; – negative.
TMAO = Trimethylamine oxide.
[a] May vary slightly depending on method of determination, e.g. high-performance liquid chromatography or T_m.

high NaCl and low pH, can be distinguished from *S. putrefaciens* by its high G+C%. Due to its ability to use various electron acceptors in anaerobic respiration, *S. putrefaciens* is not, in general, inhibited by vacuum-packing. The organism is sensitive to CO_2.

As *S. putrefaciens* is typical of chill-stored fresh foods, and is usually inhibited by low NaCl concentrations, little is known about its tolerance to chemical food preservation parameters. Unpublished results at our laboratory have shown that *S. putrefaciens* is sensitive to the food preservative sorbic acid. Even at relatively high pH (6.5), where most of the acid is not dissociated, some growth inhibition is evident and the minimal inhibitory concentration (MIC) is approximately 0.3%. Lowering pH to 6.0 reduces the MIC markedly to approximately 0.1%.

Methods of Detection and Enumeration in Foods

No selective method exists for the enumeration of *S. putrefaciens*; however, a number of indicative media and procedures have been described, e.g. relying on detection of H_2S production or reduction of TMAO. It should be borne in mind that all the methods described below are indicative of biochemical reactions characteristic of *S. putrefaciens*; however, other bacteria with the same abilities will obviously interfere.

Media

Several different media have been used for enumeration of *S. putrefaciens*, relying either on the ability of the organism to produce H_2S or its characteristic salmon-like pigmentation, or both (**Table 3**). Peptone iron agar is rich in peptones and contains ferrous sulphate. Bacteria producing H_2S from peptone degradation will appear with a black precipitate of FeS just below and around the colony. The iron agar, Lyngby, relies on the same basic principle; however, it has been modified in a number of ways: thiosulphate and L-cysteine are added to increase the pool of sulphur. *S. putrefaciens* degrades both the inorganic and organic sulphur sources. The pH of the medium is 7.4, which has a moderately stabilizing effect on the FeS. The medium is used as pour plate with a cover layer. This enhances FeS production. FeS is not stable and may be oxidized to $Fe(OH)_3$. Thus the black precipitate will be oxidized if the agar plates are left for too long and/or at too high a temperature. For routine purposes, incubation at 20–25°C for 3–5 days is suitable. *S. putrefaciens* produced pink/reddish/brownish colonies when screened on modified Long and Hammer medium, probably due to production of coloured cytochromes.

Table 3 Indicative agars for detection of *Shewanella putrefaciens* on fish

Principle	*Medium*	*Temp (°C)*	*Incubation (days)*
Pigment	Long & Hammer	21	2
Pigment +	Long & Hammer	15	7
H_2S production	Peptone iron agar (Difco)	20	3
H_2S production	H_2S medium	20	4
	Iron agar, Lyngby (Oxoid)	20–25	3–5

From Gram (1992).

Rapid Methods

Several attempts have been made to develop rapid methods for quantifying *S. putrefaciens* due to the importance of the organism in food spoilage.

Rapid Methods Based on Measurement of Reduction of TMAO While the agar methods described above rely on detection of H_2S, several rapid methods utilize the other main (fish) spoilage reaction, e.g. the reduction of TMAO to TMA. This causes three physical changes in the substrate: an increase in electrical con-

Table 4 Principles used for rapid incubation detection of reduction of trimethylamine oxide indicative of *Shewanella putrefaciens*

Physical parameter	*Physical change*	*Measured by*
Conductance	Increase	Automated conductance measurement (e.g. Malthus or Bactometer)
E_h	Decrease	E_h indicator read visually
pH	Increase	pH meter pH indicator read visually

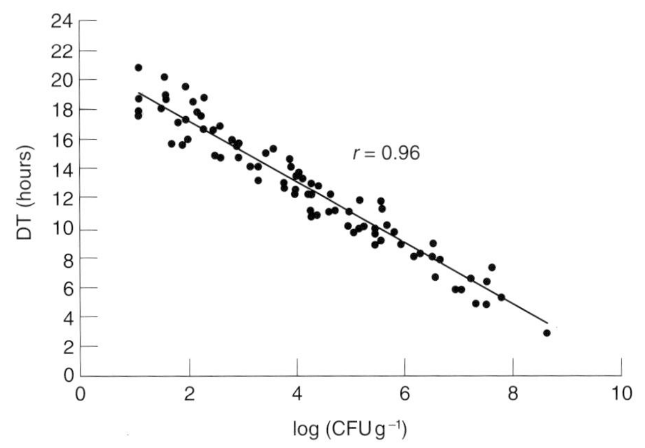

Figure 3 Comparison of H_2S counts (Iron Agar, Lyngby) in cod with Malthus detection times obtained in a trimethylamine oxide-containing broth. From Jørgensen et al (1988).

ductivity (at neutral pH), a reduction in the oxidation–reduction potential (E_h) and an increase in pH (**Table 4**). Typically, a fixed amount of sample with unknown amount of *S. putrefaciens* is incubated at 15–25°C with a fixed amount of TMAO-containing broth and the time taken to cause a significant change in any of the above mentioned parameters is measured. This so-called detection time is inversely proportional to the initial number of TMAO-reducing bacteria.

The reduction of TMAO, which is non-ionized, to TMA, which is positively charged at neutral pH, is an ideal reaction for conductometric measurements. The level of *S. putrefaciens* in iced cod, for example, can be estimated quite accurately by conductance measurement using a TMAO-containing broth in which 1 ml of a 10^{-1} suspension of the food sample is inoculated (**Fig. 3**). Changes in E_h are difficult to record automatically; however, the visual observation of the change of a redox indicator, e.g. resazurin, can be used. The changes in conductance are registered as soon as the ionized TMA is being produced, whereas changes in E_h are not observed until all TMAO has been reduced. Therefore, the E_h detection time will always be some hours longer than the conductance detection time. Increase in TMA can also be detected in a non-buffered environment by pH measurements. In these studies, it has been shown that addition of formate, which is a good electron donor for anaerobic respiration on TMAO, increases the TMA production and thus the sensitivity of the method. No attempts have been made to automate this method.

Immuno-based Methods Specific poly- and monoclonal antibodies have been produced against *S. putrefaciens* and employed in various enzyme-linked immunosorbent (ELISA) assays. However, a very high number (approximately 10^7 bacteria per gram) is required to detect the organism in these assays. Attempts to concentrate the bacteria, e.g. by immunomagnetic separation, have not been successful.

Molecular-based Methods With the elucidation of the 16S rRNA gene sequences of *S. putrefaciens*, the design and application of rRNA probes seems straightforward. A 16S rRNA probe has been designed and successfully applied in a study of Fe(III) reduction by *S. putrefaciens*, however, this – or any other – probe has not been tested in food-relevant situations.

As detection and quantification by gene probes typically rely on fluorochrome labelling and subsequent detection by fluorescence microscopy, 10^6–10^7 cfu g^{-1} is the minimum detectable level.

The problems of sensitivity may be overcome by amplifying the specific gene sequence by the polymerase chain reaction (PCR). Data have not been published using PCR to detect *S. putrefaciens* but work at our laboratory is underway to develop a procedure.

Cultural Techniques

No international standards or guidelines exist for the acceptable level of *S. putrefaciens* in foods. In Denmark the Veterinary and Food Directorate responsible for control of imported fish products uses a guideline of 10^7–10^8 H_2S-producing organisms per gram of fresh fish. However, as described below, the level of *S. putrefaciens* does not necessarily indicate the actual quality of the fish but can merely be used to predict the remaining shelf life of fresh iced fish (cod).

Importance of the Species in the Food Industry

Shewanella putrefaciens is an important spoilage organism, particularly of iced stored marine fish. Live fish accumulate virtually no glycogen and the decrease in pH postmortem (glycolysis) is limited. Rarely does pH fall below 6.0 and, as *S. putrefaciens* is slightly pH-sensitive, this is a suitable environment. The ability of

Table 5 Volatile compounds produced by *Shewanella putrefaciens* grown in foods and in model substrates

Substrate	*Compounds produced*			*Odour*
	Amines	*Sulphides*	*Aldehydes/ketones*	
High-pH beef	nd	Hydrogen sulphide Methane thiol Dimethyl disulphide Dimethyl trisulphide Methylthio acetate *bis*(methylthio) methan	Not different from sterile control	Eggs Putrid
Chicken	nd	nd	nd	Dishrag Wet dog Sulphur
Raw Jaira shrimp or banana prawn (*Penaeus merguiensis*)	Trimethylamine volatile bases	Methane thiol Dimethyl disulphide Methyl propyl disulphide Dimethyl trisulphide		Putrid Sulphidy
Marine fish (*Gadus morhua*)	Trimethylamine	Hydrogen sulphide Methane thiol Dimethyl sulphide		Putrid Cabbagey Fishy
Marine fish (*Sebastes melanops*)	Trimethylamine	Hydrogen sulphide Methane thiol Dimethyl disulphide Dimethyl trisulphide	Propionaldehyde 1-penten-3-ol 3-methyl-1-butanol	

nd = not done.

the organism to grow at low temperatures (generation time of approximately 24 h at 0°C), its reduction of TMAO to TMA and its production of volatile sulphides make it a significant fish-spoilage bacteria. Since the bacterium is relatively pH-sensitive it therefore does not normally grow on meat where post-mortem glycolysis causes a decrease in pH to approximately 5.5. However, on high-pH meat and poultry, *S. putrefaciens* is often isolated as part of the Gram-negative spoilage flora and when grown on these foods produces a range of volatile sulphides. Off odours associated with the growth of *S. putrefaciens* are described as putrid, sulphidy and rotten (**Table 5**) when numbers reach 10^8–10^9 cfu g^{-1}.

S. putrefaciens typically constitute only a minor fraction of the initial microflora on newly caught fish. During chilled (iced) storage, the microflora becomes dominated by a number of Gram-negative, psychrotolerant bacteria species, including *S. putrefaciens* which grows to levels of 10^7–10^9 cfu g^{-1} (**Fig. 4**).

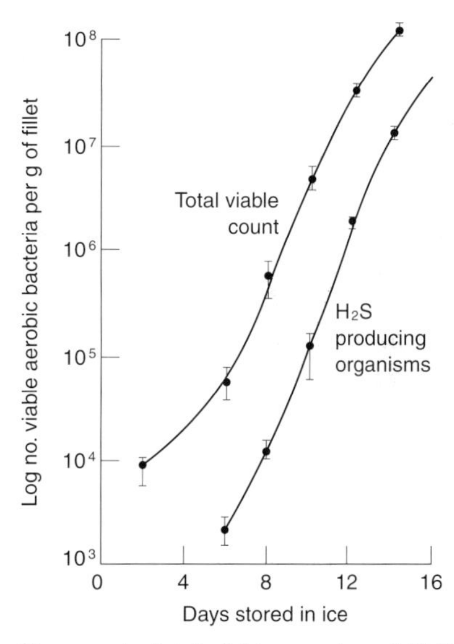

Figure 4 Changes in total viable count and H_2S-producing bacteria during iced storage of cod. From Herbert et al (1971).

The total count of bacteria carries no information about quality or spoilage of, for example, iced cod, as only few of these bacterial species produce the spoilage off odours and flavours. In iced cod and other marine fish, the biochemical changes resulting in spoilage off odours and off flavours can be explained by the growth of *S. putrefaciens*. However, the actual numbers of *S. putrefaciens* does *not* provide an indication of the eating quality of the fish, but the numbers of *S. putrefaciens* enables a prediction of the remaining shelf life (**Fig. 5**).

S. putrefaciens is, like many Gram-negative bacteria, capable of attaching to surfaces and when grown in relatively nutrient-rich environments forms a thick biofilm. Experiments with biofilms on steel surfaces have shown that the bacteria, like many others, produce a fibrous net of exopolysaccharides in which the bacteria proliferate (**Fig. 6**). This will enable the bacteria to adhere readily to surfaces in the food-processing industry and nothing is known about the effect of cleaning and disinfecting agents on the biofilm bacteria.

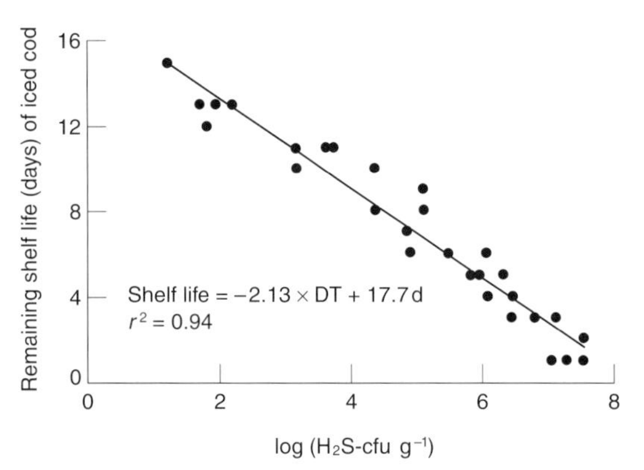

Figure 5 Comparison of remaining shelf life of iced cod and H_2S-producing bacteria counted on iron agar, Lyngby. From Gram and Huss (1999), modified from Jørgensen et al (1988).

Figure 6 Scanning electron micrograph of a mild steel coupon incubated for 9 weeks in B10 medium with *Shewanella putrefaciens* isolate 200. EP, Exopolysaccharide fibres; BC, bacterial cells. × 9200. From Obuekwe et al (1981).

Concluding Remarks

Shewanella putrefaciens is, due to its ability to grow at chill temperatures and produce amines and volatile sulphides, an important spoilage bacteria of certain proteinaceous foods, particularly marine iced fish. The organism is also important in the environment and is probably involved in the turnover of various metals in the geosphere. A better understanding of the role of the organism in food spoilage may be derived from studies of its importance in the environment. Further studies are needed to determine the genotypic variance within the species, e.g. if particular clones are involved in spoilage. Also the development of rapid sensitive methods, e.g. PCR-based, is an area that requires further studies.

See also: **Biochemical and Modern Identification Techniques**: Food Spoilage Flora (i.e. Yeasts and Moulds). **Biofilms**. **Chilled Storage of Foods**: Principles. **Fish**: Catching and Handling; Spoilage of Fish. **Meat and Poultry**: Spoilage of Meat. **Metabolic Pathways**: Release of Energy (Aerobic); Release of Energy (Anaerobic). ***Pseudomonas***: Introduction. **Rapid Methods for Food Hygiene Inspection**. **Spoilage Problems**: Problems caused by Bacteria.

Further Reading

Dalgaard P (1995) Qualitative and quantitative characterization of spoilage bacteria from packed fish. *Int. J. Food Microbiol.* 26: 319–333.

Fonnesbech Vogel B, Jørgensen K, Christensen H, Olsen JE and Gram L (1997) Differentation of *Shewanella putrefaciens* and *Shewanella alga* using ribotyping, whole-cell protein profiles, phenotypic characterization and 16S rRNA sequence analysis. *Appl. Environ. Microbiol.* 63: 2189–2199.

Gibson DM, Hendrie MS, Houston NC and Hobbs G (1977) The identification of some Gram negative heterotrophic aquatic bacteria. In: Skinner FA and Shewan JM (eds) *Aquatic Microbiology.* P. 135. London: Academic Press.

Gram L (1992) Evaluation of the bacteriological quality of seafood. *Int. J. Food Microbiol.* 16: 25–39.

Gram L and Huss HH (1996) Microbiological spoilage of fish and fish products. *Int. J. Food Microbiol.* 33: 121–138.

Gram L and Huss HH (1999) Microbiology of Fish and Fish Products. In: Lund BM, Baird-Parker AC and Gould GW (eds) *The Microbiology of Foods.* London: Chapman & Hall.

Herbert RA, Hendrie MS, Gibson DM and Shewan JM (1971) Bacteria active in the spoilage of certain sea food. *J. Appl. Bacteriol.* 34: 41–50.

Huss HH (ed.) (1995) *Quality and Quality Changes in Fresh Fish.* FAO Fisheries Technical Paper No. 348. P. 195. Rome: Food and Agricultural Organization of the United Nations.

Jørgensen BR, Gibson DM and Huss HH (1988) Microbial quality and shelf life prediction of chilled fish. *Int. J. Food Microbiol* 6: 295–307.

Lee JV (1977) *Alteromonas (Pseudomonas) putrefaciens.*

In: Russell AD and Fuller R (eds) *Cold Tolerant Microbes in Spoilage and the Environment*. P. 59. London: Academic Press.

Levin RE (1968) Detection and incidence of specific species of spoilage bacteria on fish. *Appl. Microbiol.* 14: 1734–1737.

MacDonnell MT and Colwell RR (1985) Phylogeny of the Vibrionaceae, and recommendation for two new genera, *Listonella* and *Shewanella*. *System. Appl. Microbiol.* 6: 171–182.

Obuekwe CO, Westlake DWS, Cook FD and Costerton JW (1981) Surface changes in mild steel coupons from the action of corrosion-causing bacteria. *Appl. Environ. Microbiol.* 41: 766–774.

Ringø E, Stenberg E and Strøm AR (1984) Amino acid and lactate catabolism in trimethyl-amine oxide respiration of *Alteromonas putrefaciens* NCMB 1735. *Appl. Environ. Microbiol.* 47: 1084–1089.

Scott JH and Nealson KH (1994) A biochemical study of the intermediary carbon metabolism of *Shewanella putrefaciens*. *J. Bacteriol.* 176: 3408–3411.

Stenström I-M and Molin G (1990) Classification of the spoilage flora of fish, with special reference to *Shewanella putrefaciens*. *J. Appl. Bacteriol.* 68: 601–618.

SHIGELLA

Introduction and Detection by Classical Cultural Techniques

Keith A Lampel, US Food and Drug Administration, Center for Food Safety and Applied Nutrition, Washington, DC, USA

Robin C Sandlin and **Samuel Formal**, Department of Microbiology and Immunology, Uniformed Services University of the Health Sciences, F Edward Hébert School of Medicine, Bethesda, USA

The Pathogen

Shigella are the causative agents of bacillary dysentery (shigellosis). These organisms have a worldwide distribution, and in Third World countries cause disease with high morbidity and mortality. *Shigella* are Gram-negative, non-motile, non-spore-forming, rod-shaped bacteria. The genus is a member of the family *Enterobacteriaceae* and is closely related to *Escherichia coli*. The genus *Shigella* consists of four species: *S. dysenteriae*, *S. flexneri*, *S. boydii* and *S. sonnei* (**Table 1**). The severity of disease varies depending on the particular species. *S. sonnei* causes the mildest form of disease, whereas *S. dysenteriae* produces the most severe disease; *S. flexneri* and *S. boydii* cause intermediate forms of the disease. Additionally, the geographic distribution and epidemiology of the disease caused by *Shigella* are different. *S. dysenteriae* is responsible for epidemics of the disease; shigellosis outbreaks in industrialized nations are usually caused by *S. sonnei*, whereas *S. flexneri* is found in developing countries.

Biochemical reactions and the O antigen type are the basis for classification of *Shigella* spp. into four major groups (Table 1). The groups are A–D, encompassing the different species *S. dysenteriae*, *S. flexneri*, *S. boydii* and *S. sonnei*. Each major group or species is further subdivided into serotypes (except *S. sonnei*) based on differences in the O antigen (**Table 2**). They are facultative anaerobes with relatively simple nutritional requirements. *Shigella* are typically lysine decarboxylase-negative and lactose non-fermenting (except for a slow reaction with *S. sonnei*) – characteristics which separate *Shigella* from closely related commensal *E. coli*. While typical *E. coli* strains are easily differentiated from *Shigella*, the enteroinvasive *E. coli* (EIEC) share many properties of *Shigella* and can cause a disease similar to bacillary dysentery caused by *Shigella* spp. This can complicate the accurate diagnosis of shigellosis.

Table 1 Characteristics of *Shigella* spp.

Species	*Serogroup*	*Serotypes*	*Geographic distribution*	*Distinguishing characteristics*
S. dysenteriae	A	15	Indian subcontinent, Africa, Asia	Produces Shiga toxin; Causes most severe dysentery High mortality rate if untreated
S. flexneri	B	6	Most common isolate in developing countries	Less severe dysentery
S. boydii	C	19	Rarely isolated in developed countries	Biochemically identical to *S. flexneri*; distinguished by serology
S. sonnei	D	1[a]	Most common isolate in developed countries	Mildest form of shigellosis

[a] Forms I and II are serotypically distinguishable.

Table 2 Tests to differentiate *Shigella* spp. from *Escherichia coli*

Test	*Reaction*	
	Shigella	E. coli
Motility	–	+[a]
Gas from glucose	–[b]	+[c]
Lysine decarboxylase	–	+[d]
Christensen's citrate	–	+
Acetate	–	+
Mucate	–	+

[a] Most positive, some negative.
[b] Some strains of *S. flexneri* 6 produce small amounts of gas from glucose.
[c] Some exceptions.
[d] Enteroinvasive *E. coli* are also negative.

An estimated 30 000 cases of shigellosis occur annually in the US. Although shigellosis accounts for less than 10% of the reported outbreaks of food-borne illness in the US, it can be a cause of significant morbidity in developing countries or in situations where sanitary waste treatment and water purification measures are disrupted, such as during war or following a natural disaster. *Shigella* is principally a disease of humans, although outbreaks of diarrhoeal disease in subhuman primates have been described. One outstanding characteristic of the disease is its high communicability, which is attributed to the low infectious dose required to cause disease. Ingestion of fewer than 100 organisms has been shown to result in clinical disease in volunteer studies. The low infectious dose is the primary reason for the ease that this organism can spread through a population living under crowded and unsanitary conditions. The incubation period is 1–7 days and onset of symptoms usually appears in 3 days. The clinical presentation of shigellosis can vary from mild diarrhoea to severe dysentery. The symptoms of bacillary dysentery include fever, cramps and frequent bowel movements containing blood, pus, and mucus. Infections are associated with mucosal ulceration, rectal bleeding and dehydration. Fatality may be as high as 10–15% with some strains, but the infection is usually self-limiting and resolves without treatment within 1–2 weeks. Reiter's disease, reactive arthritis and haemolytic–uraemic syndrome are possible sequelae that have been reported as complications of shigellosis.

These pathogens are not indigenous to any food. Most cases of illness are a result of close contact with infected individuals: transmission most frequently occurs via the faecal to oral route. The organism is frequently found in water polluted with human faeces and can be spread via flies, fingers, food and faeces. In highly endemic areas, shigellosis is associated with children who frequently lack stringent personal hygiene habits. The carrier state usually lasts for 1–4 weeks but long-term carriers have been described and can be a serious reservoir of infection. During the acute phase of the disease, 10^3–10^9 colony forming units (cfu) per gram of stool are shed, whereas in convalescing patients the number of recovered bacteria are 10^2–10^3 cfu. In order to control the spread of *Shigella*, infected patients, e.g. children in day-care centres, are monitored for the presence of the organism in stool until cleared.

Shigellosis is caused when virulent *Shigella* organisms attach to and penetrate M cells and epithelial cells of the large intestine. After invasion, the bacteria are released into the cytoplasm and multiply intracellularly. The infection is perpetuated by the spread of the organism to neighbouring epithelial cells through projections from the originally infected cell. This cycle results in damage and destruction of the intestinal epithelial cell which results in dysentery. Virulence factors which contribute to these processes are both chromosomally encoded and present on a large 220 kb virulence plasmid. The virulence plasmid is essential for many virulence phenotypes and in the absence of the plasmid, the organism is avirulent. A related virulence plasmid is also present in EIEC and is essential for this organism's ability to cause disease.

The genetic factors involved in the virulence phenotype can be loosely grouped into genes that encode proteins involved in uptake, genes that encode proteins involved in the intercellular spreading phenotype, and genes involved in the regulation of these processes. The virulence plasmid-encoded genes comprising the *ipaBCDA* (invasion plasmid antigens) operon encode proteins that are recognized by convalescent patient sera. IpaB, C and D are all essential for invasion and are thought to be involved in signalling the uptake of the bacteria by the host cell. Additionally, IpaB is hypothesized to be responsible for release from the phagocytic uptake vacuole and for induction of apoptosis in infected macrophages. The Ipas are secreted by a dedicated virulence protein secretion apparatus known as a type III secretory system. Related secretion systems have been described for a variety of plant and animal pathogens and share both functional and sequence homologues. The group of proteins that compose this secretion apparatus are encoded by the *mxi/spa* loci present on the virulence plasmid and are thought to comprise a pore that spans the inner and outer membrane to allow the release of the Ipa proteins from the host cell. Mutation in any of these structural components of the secretory apparatus results in a loss of virulence due to the inability to secrete the Ipa proteins. The *icsA* (*virG*) locus on the virulence plasmid is another essential virulence factor. This outer membrane protein is symmetrically localized at the old pole of the bacterial cell where it catalyses the polymerization of host cell actin. This results in the formation of an actin tail at one end of the bacterium that propels the organism through the host cell and allows the bacterium to enter the neighbouring cell without leaving the initially infected host cell and encountering the exterior environment.

The expression of these virulence proteins is regulated by a variety of environmental stimuli – temperature is the most notable. Temperature is an important factor in the regulation of virulence genes in a variety of pathogens, including *Salmonella typhimurium*, *Bordetella pertussis*, *Yersinia* spp. and *Listeria monocytogenes*. Virulence gene expression in *Shigella* is repressed at 30°C and induced at 37°C by the action of three gene products. VirR (H-NS) is a chromosomally encoded global negative regulator involved in a variety of regulatory pathways. The virulence plasmid-encoded VirF is a transcriptional activator that is a member of the AraC family of activators. H-NS acts to block VirF activity at the VirB promoter. VirB is a virulence-encoded gene that is thought to induce the expression of *ipa*, *spa* and *mxi* genes. The exact mechanism is unknown, but mutations in H-NS result in an absence of temperature regulation while mutations in VirF or VirB result in an inability to produce the Ipa, Spa and Mxi proteins.

The virulence plasmid-encoded genes are essential to the disease process. The ability of the pathogen to maintain the presence of the plasmid is critical for pathogenesis. In recent work it has been shown that the expression of the plasmid-encoded virulence genes is detrimental to the maintenance of the plasmid, presumably due to the high metabolic demand of synthesizing the virulence-encoded proteins. The virulence plasmid is lost from the bacterial cell at a higher frequency at 37°C when virulence protein expression is upregulated than when the bacteria are grown under repressing conditions at 30°C. This observation could affect the detection of the organisms from clinical, food and environmental samples using DNA-based or serological assays.

In addition to the mucosal damage which results from the process of epithelial cell invasion, *Shigella* also produce enterotoxins which are presumably responsible for the diarrhoeal component of the disease. One serotype, *S. dysenteriae*, elaborates a potent toxin (Shiga toxin) which inhibits protein synthesis and exhibits both cytotoxic and enterotoxic activity. Some *E. coli* strains also produce Shiga-like toxins and have been associated with outbreaks spread by contaminated meats and sprouts. The Shiga toxin is only produced in *S. dysenteriae*, which causes the most severe form of dysentery, and may significantly contribute to the severity of the disease.

Survival of *Shigella* in Foods

Since *Shigella* are not indigenous to any particular food, the contamination of foods usually occurs after processing by a food handler who did not take proper precautions such as hand-washing. Several types of foods have been associated with outbreaks of shigellosis; contaminated raw vegetables are a leading source. For example, in 1995, contaminated lettuce was responsible for outbreaks that occurred in a few European countries such as Norway and the UK. Other foods that are common vehicles for the spread of *Shigella* include tossed or potato salads, chicken and shellfish. But vegetables seem to be a popular food. In a recent study, the survivability of *Shigella* in packaged vegetables (sterile and unsterile) held at different temperatures (5°C, 10°C and room temperature) was reported. The greatest reduction in number of *Shigella* recovered was from over the first 24 h. However, the data indicate that *Shigella* do survive in the vegetables tested, albeit after a 3–7 log reduction, for at least 10 days (although the initial inoculum was high (10^{11}), the final numbers (cfu g^{-1}) were between 10^4 and 10^8).

In laboratory conditions, *Shigella* spp. can grow at temperatures as low as 6°C and up to 48°C and have a pH range of 4.8–9.3. *Shigella* can survive at room temperature up to 50 days in foods such as milk, flour, eggs, clams, shrimp and oysters; 5–10 days in acidic foods, e.g. orange and tomato juice and carbonated soft drinks; and 1–2 weeks in refrigerated fermented milk. *Shigella* do not survive well in acid foods, although these organisms are able to survive the acid environment of the stomach.

Detection of *Shigella* from Foods

Foods are not routinely examined for the presence of *Shigella*. Since humans and higher primates are the sole hosts, *Shigella* spp. are usually introduced into foods by infected food handlers with poor personal hygiene. An outbreak of shigellosis is usually initially identified from clinical findings and epidemiological investigations. Therefore 7–10 days may have passed after the first individual became ill. During this time, the fate and condition of the contaminated food are unknown and present a challenge to the laboratory responsible for isolating the aetiological agent. In a large outbreak at a mass meeting, over half the attendees (3000 people) developed shigellosis from consuming meals prepared at this gathering where one food handler was found to be ill. The need for rapid isolation and identification of *Shigella* in foods is necessary to reduce the spread of this pathogen.

Bacteriological

One protocol to isolate *Shigella* spp. from foods is detailed in the *Bacteriological Analytical Manual*. Initially, 25 g sample portions are added to 225 ml of *Shigella* broth supplemented with novobiocin (0.3 μg ml^{-1} for *S. sonnei*; 3 μg ml^{-1} for other *Shigella*; this antibiotic is added to suppress the natural bacterial flora of the food) and held for at least 10 min at room temperature. The broth is added to a 500 ml flask and incubated overnight at 37°C with shaking. In some cases, samples are added to other enrichment broths, such as Gram-negative broth, blended in a sterile stomacher bag and the contents are poured into a flask and incubated at 37°C with shaking for 16–20 h. At least two to three different ranges of selective agar plates should be used to increase the chance of recovery of *Shigella* from foods.

Since the process to recover *Shigella* from contaminated foods may not commence for a lengthy time period, the presence of injured cells is probable. Selective media without bile salts and desoxycholate are recommended since these compounds have an adverse effect on the growth of impaired cells. Food samples are added to 100 ml of tryptic soy broth, the pH is adjusted to 7.0, and then blended. After 8 h of incubation at 37°C, 125 ml of enrichment broth is added and the incubation extended for an additional 16–20 h.

A range of selective agar media can be used to plate overnight cultures. *Shigella* grown on MacConkey agar, a low-selectivity medium, produce colonies that are translucent and slightly pink (*Shigella* are lactose-negative) with or without rough edges. Eosin methylene blue (EMB) and Tergitol-7 agar are alternative low-selectivity agar containing lactose. Colourless colonies on EMB plates or bluish colonies on the yellowish-green Tergitol-7 agar are indicative of *Shigella*. Desoxycholate and xylose-lysine-desoxycholate (XLD) agars are intermediate selective media and are preferred media to isolate *Shigella* spp. *Shigella* colonies on XLD agar medium are translucent and red (alkaline). Although most *Shigella* do not ferment xylose, some species, e.g. *S. boydii* (variable) may be missed and therefore plating on XLD and MacConkey agar plates is recommended. *Shigella* spp. form reddish colonies on desoxycholate agar. Highly selective media include *Salmonella–Shigella* and Hektoen agars. Some *Shigella* spp., such as *S. dysenteriae* type I, are unable to grow on highly selective *Salmonella–Shigella* medium. On this agar medium, *Shigella* produce colourless, translucent colonies. Colonies on Hektoen agar appear to be green, as do colonies from *Salmonella* spp.; *E. coli* strains form yellow colonies. All presumptive colonies are inoculated into semisolid (motility) test agar. *Shigella* spp. are non-motile.

Biochemical tests are used to identify *Shigella* spp. Suspected colonies that are Gram-negative, non-motile rods are inoculated on to lysine iron or Kliger iron agar. *Shigella* produce alkaline slants, acid butt and no gas (Table 2) on these agars. Further biochemical characterizations show that *Shigella* spp. are negative for H_2S production, phenylalanine deaminase, sucrose and lactose fermentation, do not utilize citrate, acetate, KCN, malonate, inositol, adonitol, and salicin, and lack lysine decarboxylase. *Shigella* are negative for the Voges–Proskauer test (*S. sonnei* and *S. boydii* serotype 13 are positive); however, all *Shigella* are methyl red-positive and are unable to produce acid from glucose and other carbohydrates (acid and gas production occur with *S. flexneri* serotype 6, *S. boydii* serotypes 13 and 14, and *S. dysenteriae* 3). One *Shigella* sp., *S. dysenteriae*, is catalase-negative and has ornithine decarboxylase activity.

Table 2 and **Table 3** show key biochemical reactions to differentiate *Shigella* from *E. coli* and also to distinguish each *Shigella* sp. from each other. Growth on Christensen citrate, sodium mucate or acetate agar is one characteristic that discriminates between *E. coli* and *Shigella*; *Shigella* are unable to utilize citrate, acetate or mucate as sole carbon source.

Other biochemical tests are used to identify the serotypes of *Shigella*. The ability to utilize mannitol, dulcitol, xylose, rhamnose, raffinose, glycerol and indole and the presence of ornithine decarboxylase have been used to discriminate physiologically between *Shigella* spp.

Serological testing, using polyvalent antiserum, is used to identify the *Shigella* group (A–D). A note of caution should be addressed. EIEC causes the same disease – bacillary dysentery – as *Shigella*. In some cases, some EIEC strains share homology with antigenic structures of some *Shigella* serotypes. Several serotypes of *S. dysenteriae*, *S. flexneri* and *S. boydii* have reciprocal cross-reactivity with *E. coli* O antigens of the Alkalescens–Dispar bioserogroup or EIEC.

DNA and Antibody-based Assays

Several types of DNA-based assays have been used to detect *Shigella* spp. in foods. DNA probes, either DNA isolated containing segments of virulence genes or short oligonucleotides that are specific for a particular genetic marker, can be used in a colony hybridization format. The limitation of this method is similar to isolating *Shigella* from foods using conventional bacteriological techniques. Each food matrix and the physical state of the *Shigella* within the food can reduce the chance of isolating the pathogen.

Although DNA probes and the use of the poly-

Table 3 Biochemical differentiation of *Shigella* species

Test	S. dysenteriae	S. flexneri	S. boydii	S. sonnei
β-Galactosidase	–[a]	–	–	+
Ornithine decarboxylase	–	–	–	+
Indole production	+/–[b,c]	+/–[c,d]	+/–[c]	–
Acid from:				
Dulcitol	–[e]	–[f]	–	–
Lactose	–	–	–[g]	+[h]
Mannitol	–	+	+	+
Raffinose	–	+/–[c]	–	+[h]
Sucrose	–	–	–	+[h]
Xylose	–	–	+/–[c]	–

[a] *S. dysenteriae* type 1 are positive.
[b] *S. dysenteriae* type 1 is negative; *S. dysenteriae* type 2 is positive.
[c] Reaction is variable.
[d] *S. flexneri* 6 is negative.
[e] *S. dysenteriae* type 1 may be positive.
[f] *S. flexneri* 6 may be positive.
[g] *S. boydii* may be positive.
[h] Positive reactions may take 24 h or longer.
+/– = Variable reaction.

merase chain reaction (PCR) were initially developed to detect *Shigella* from clinical samples, these techniques have been applied to foods. PCR primers are selected that specifically target virulence genes and have been successfully used in PCR-based assays to detect the presence of *Shigella* in foods. Unlike other methods which may take several days to draw a conclusion about the presence of *Shigella* in foods, PCR assays can yield a result in less than 1 day.

A few commercially available latex agglutination tests are used to detect *Shigella* (and *Salmonella*) spp. but they were developed to be applicable to clinical specimens. Also an enzyme-linked immunosorbent assay (ELISA) uses antibodies raised to an EIEC virulence marker antigen. However, this ELISA has not been assessed for detecting *Shigella* spp. in foods.

Regulatory

Each year, there are a significant number of outbreaks of shigellosis from consumption of contaminated foods. Many of these outbreaks are confined to individual countries but international travel and global trade have been influential factors in the spread of the disease. For example, in 1994, contaminated lettuce from one country was the source of an outbreak that occurred in several European countries. Although the number of food-borne outbreaks has declined in the past few years in the US, *Shigella* are still the third or fourth leading cause of bacterial food-borne outbreaks.

In the US, as written in the Federal Food, Drug, and Cosmetic Act, Chapter IV–Food, if a food contains any poisonous or deleterious substances which may render it injurious to health, the food is considered adulterated. Pathogenic organisms are considered 'poisonous or deleterious substances' and therefore foods containing them are adulterated. The transmission of this pathogen in foods is very efficient and, because of the low infectious dose of *Shigella*, action to remove contaminated food products from commerce should be swift.

Impact on Food Industry

Unlike other bacterial food-borne pathogens, such as *Salmonella* spp., *Shigella* are not associated with any particular food. The contamination of foods with *Shigella* usually results from an infected food handler after the products are processed. Poor personal hygiene has caused shigellosis in institutions such as nursing homes and prisons. Instructions for proper handling of food in retail establishments in the US is described in the US Food and Drug Administration Food Code (FDA-1997).

Faecal contamination of food often results in the presence of unwanted microbes. Fresh and raw vegetables, whether grown for commercial distribution or available at local markets, have been implicated as the source of shigellosis outbreaks. Also, because of the trend of global markets, imported foods have become potential sources of contaminated foods. Another major factor that may lead to the spread of *Shigella* is the improper storage temperature of contaminated foods.

As a response to the increasing number of food-borne outbreaks due to contamination of fresh produce and fruits, the US Food and Drug Administration has made available a document entitled *Guidance for Industry – Guide to Minimize Microbial*

Food Safety Hazards for Fresh Fruits and Vegetables. Implementation of these voluntary guidelines addresses concerns with the following areas: agricultural and packing-house water use, manure management, worker hygiene, field and packing-house sanitation and transportation. These recommendations can be effective in reducing the transmission of *Shigella* and other enteric pathogens via fresh vegetables and fruits.

The food industry, in conjunction with regulatory agencies, has devised a programme (hazard analysis and critical control point: HACCP) to identify and control specific food-related practices. This analysis indicates where tests for the presence of food-borne pathogens at critical stages of food production should be implemented. HACCP is an alternative approach to the traditional method of analysing the final food product before marketing. Although HACCP may be a good monitoring system for other pathogens commonly known to be associated with foods, such as *Salmonella* spp. and *E. coli* O157 : H7, testing for the presence of *Shigella* in foods in this milieu may not be efficient. Contamination of foods with *Shigella* often occurs between the processing plant and the consumer through handling by infected employees. Proper adherence to good manufacturing process, such as hand-washing and temporary removal of employees with characteristic diarrhoeal illness, can not only reduce outbreaks of shigellosis, but also other forms of food-borne diseases.

Impact on Consumer

The consumer should be aware that, with respect to *Shigella*, many post-processed foods may have the potential of being contaminated. The source of many shigellosis outbreaks have been traced to the ingestion of raw or fresh vegetables, particularly in salads and seafood, bakery products, chicken and hamburgers. Outbreaks also occur in establishments that serve foods prepared on the premises, such as restaurants, where salads may be handled by infected food preparers. Additionally, there are examples of outbreaks caused by individuals who were either responsible for preparing meals for thousands of people at mass gatherings or from home-made products for local festivals. Consumers should obtain their fresh and prepared vegetables from reliable sources, wash produce, store them at suitable (refrigerated) temperatures, cook foods adequately, refrigerate leftovers promptly and cleanse cooking utensils and equipment. Many agencies, including the World Health Organization, are emphasizing adequate hand-washing by the consumer before and while preparing meals as an effective means of reducing food-borne outbreaks.

See also: ***Escherichia coli***: *Escherichia coli*. **Food Poisoning Outbreaks**. **Hazard Appraisal (HACCP)**: The Overall Concept; Involvement of Regulatory Bodies.

Further Reading

Bacteriological Analytical Manual (1998) *Shigella*. Ch. 6, p. 6.01. Gaithersburg: AOAC International.

Dupont HL (1995) *Shigella* species (bacillary dysentery). In: Mandell GL, Bennet JE and Dolin R (eds) *Principles and Practice of Infectious Diseases*. P. 2033. New York: Churchill Livingstone.

Ewing WH (1986) *Edwards and Ewing's Identification of Enterobacteriaceae*, 4th edn. P. 146. New York: Elsevier.

Ménard R, Dehio C and Sansonetti PJ (1996) Bacterial entry into epithelial cells: the paradigm of *Shigella*. *Trends in Microbiology* 6: 220–226.

Parson C and Sansonetti PJ (1996) Invasion and the pathogenesis of *Shigella* infections. *Current Topics in Microbiology and Immunology* 209: 25–42.

Smith JL (1987) *Shigella* as a foodborne pathogen. *Journal of Food Protection* 50: 788–801.

Smith JL and Palumbo SA (1982) Microbial injury reviewed for the sanitarian. *Dairy and Food Sanitation* 2: 57–63.

Parsot C and Sansonetti PJ (1996) Invasion and the pathogenesis of *Shigella* infections. *Current Topics in Microbiology and Immunology* 209: 25–42.

SINGLE-CELL PROTEIN

Contents

The Algae

Mariano García-Garibay and **Lorena Gómez-Ruiz**, Departamento de Biotecnologia, Universidad Autónoma Metropolitana, Mexico City, Mexico

Eduardo Bárzana, Departamento de Alimentos y Biotecnología, Universidad Nacional Autónoma de México, Mexico City, Mexico

Single-cell protein (SCP) refers to crude or refined protein of algal, bacterial or yeast origin which is used as either animal feed or human food. Microbial biomass is a convenient source of food proteins as it is rapidly propagated and it has been proposed as an accessible alternative source for proteins of agricultural origin. In addition to proteins, SCP contains other nutrients such as lipids and vitamins. Algae have been used for SCP production based on use of carbon dioxide and light energy (autotrophic growth). In contrast to other SCP-producing organisms, in many cases algae are grown by processes resembling traditional agriculture.

Suitable Organisms

The term algae is used generically for photosynthetic organisms, either microscopic or macroscopic, living largely in water habitats, growing as undifferentiated or little-differentiated tissues. The designation is applied to taxonomically unrelated species, and in some cases includes groups like cyanobacteria or *Euglena*. Cyanobacteria is included in the Prokaryotae kingdom, and therefore is more closely related to bacteria than to true algae. However, they are known as blue-green algae, and for SCP purposes are usually considered as such. *Euglena* is a genus of microorganisms belonging to the Protozoa kingdom; it is an unusual example of a unicellular animal with chlorophyll.

True algae belong to the Plantae kingdom, and are the simplest plants. There are unicellular and multicellular organisms, some reaching huge sizes.

Many algae have for a long time been used as food, including single-cell organisms as well as multicellular seaweeds, which are important in the diet of coastal communities. **Table 1** shows the main groups of algae used as food.

Macroscopic algae do not really fit the SCP definition, as they are multicellular and the final product contains little protein (from 6 to 30% on a dry-weight basis). They are mostly collected from the sea, and when cultivated, production is more like farming than a biotechnological process. The most widely consumed seaweed is *Porphyra*, particularly *P. tenera*, which has a widespread distribution. It is mainly eaten in Japan, and also in the Philippines, Wales and New Zealand. Among the most important species of seaweed used as food is *Ulva lactuca* (sea lettuce), which is used as a salad ingredient in western Europe. *Enteromorpha* is consumed in Hawaii and the Philippines in salads or as a flavour-enhancer for fish dishes. *Caulerpa* is also eaten in the Philippines.

Both true unicellular algae and cyanobacteria have been consumed for centuries. This practice dates back to the Aztecs in central Mexico, long before the discovery of the New World, when *Spirulina maxima* was harvested from natural habitats for human consumption. A similar species, *S. platensis*, is still used in Lake Chad in central Africa for the same purpose. Both species, *S. maxima* and *S. platensis*, have been classified interchangeably in either the genus *Arthrospira* or the genus *Spirulina*. **Figure 1** shows *S. maxima* harvested from Lake Texcoco in Mexico. Other ancient cultures have used microalgae as food but on a smaller scale. For instance, *Nostoc* is consumed in Mongolia, China, Thailand and Peru, *Oedogonium* and *Spirogyra* are eaten in Burma, Thailand and Vietnam, *Chlorella* is produced in Japan and Taiwan and *Scenedesmus* in China.

Production

In addition to the traditional harvesting of native algae in several regions, industrial and experimental efforts are being made in many countries in large, shallow ponds in areas of high isolation and with bioreactors for the production of unicellular organisms.

Macroalgae cultivation is started by providing anchorage to the initial culture in sheltered bays; once considerable growth of young plants is obtained, they

Table 1 Main groups of algae used as food

Genera	*Growth form*	*Cellular structure*	*Kingdom*	*Group*
Spirulina (Arthrospira)	Unicellular	Prokaryote	Prokariotae	Cyanobacteria
Nostoc	Unicellular	Prokaryote	Prokariotae	Cyanobacteria
Anabaena	Unicellular	Prokaryote	Prokariotae	Cyanobacteria
Tolypothrix	Unicellular	Prokaryote	Prokariotae	Cyanobacteria
Chlorella	Unicellular	Eukaryote	Plantae	Chlorophyceae
Scenedesmus	Unicellular	Eukaryote	Plantae	Chlorophyceae
Oedogonium	Unicellular	Eukaryote	Plantae	Chlorophyceae
Spirogyra	Unicellular	Eukaryote	Plantae	Chlorophyceae
Coelastrum	Multicellular	Eukaryote	Plantae	Chlorophyceae
Ulva	Multicellular	Eukaryote	Plantae	Chlorophyceae
Enteromorpha	Multicellular	Eukaryote	Plantae	Chlorophyceae
Caulerpa	Multicellular	Eukaryote	Plantae	Chlorophyceae
Porphyra	Multicellular	Eukaryote	Plantae	Rhodophyceae

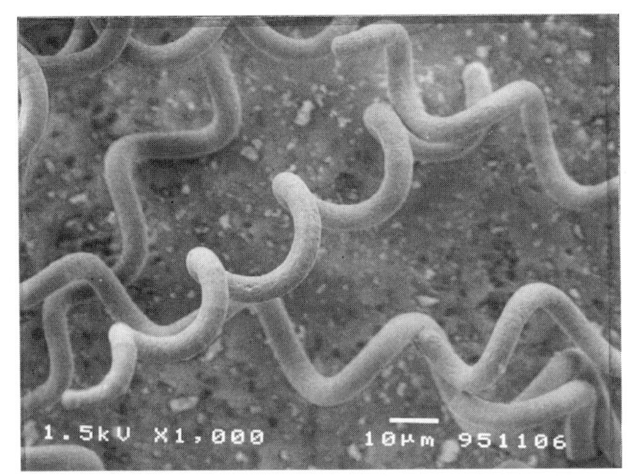

Figure 1 Cells of *Spirulina maxima* harvested from Lake Texcoco in Mexico, observed on scanning electron microscopy (courtesy of Dr Jorge Sepúlveda, Universidad Nacional Autónoma de México).

are transplanted to nitrogen-rich tidal estuaries with low salinity.

Single-cell algae are produced by a variety of methods, ranging from cultivation in lakes, earth ditches or ponds, to highly sophisticated fermenters and bioreactors. Intensive cultivation was initiated in the 1940s in Germany, followed by technological developments in Japan and Taiwan during the 1950s and 1960s. Later on, commercial production systems were developed in the US, Mexico, Thailand and Israel. Currently it is cultivated in many countries around the world, though on a modest scale.

Outdoor cultivation may be in open or closed systems. Growth occurs not more than 0.5 m from the water surface, as it is controlled by light penetration. In clean-culture systems a single species is inoculated and can be maintained over extended time periods; these are closed photobioreactors operating outdoors or indoors, ranging from large closed areas exposed to sunlight to smaller reactors illuminated with artificial light.

Substrate Requirements

Algae are autotrophic organisms which have the distinct advantage of using carbon dioxide as carbon source. Carbon dioxide represents the cheapest and probably most abundant carbon source for microbial growth. As their requirements are so inexpensive, these organisms are convenient for SCP production. Some species also grow heterotrophically using organic carbon sources.

The process is limited by the carbon source; the concentration of dissolved carbon dioxide is rather low owing to its low solubility in aqueous solution. Carbon dioxide can be injected from combustion gases. Some additional sources of carbon enhance cell growth; cheap organic materials such as manure, molasses and industrial wastes can be used. The added organic compounds are rapidly degraded by bacteria to make the carbon dioxide needed for algal reproduction. Lake Texcoco in Mexico has high concentrations of carbonate and bicarbonate which are efficiently consumed by *S. maxima*.

Nitrogen sources are generally nitrates, nitrites, ammonia and urea. Oxidized nitrogen compounds require energy to be reduced, therefore ammonia is a more convenient form. Many cyanobacteria are able to fix atmospheric nitrogen; species of the genus *Anabaena* are particularly active in fixing nitrogen.

Another important nutrient is phosphorus, which can be incorporated in its inorganic form. Micronutrients are needed in minimal amounts.

When water polluted with organic waste is directed into algal ponds, this has the advantages of using cheap raw material, promoting decontamination, while obtaining a good source of SCP. The macronutrients needed (ammonium and phosphate) are normally present in domestic sewage, animal waste and food-industry residual water.

Mass-culture Open Systems

For open systems, lakes, ponds and ditches are used. The floor can be of earth or be lined with concrete or plastic. Open systems show low cell densities with a large variation in productivity, depending on water properties and environmental conditions. In these systems weather conditions, especially concerning temperature and sunlight radiation, are critical. In contrast, the facilities needed are constructed at low cost and large areas are available.

It is a challenge to keep a monoalgal culture in an open-air system. The propagation of mixed populations and frequent problems of contamination by bacteria, fungi, protozoa and invertebrates usually disturbs productivity and lowers the quality of the product. Under such non-sterile, mixed-culture conditions some algal species tend to predominate. In ponds or ditches used for mass production the surface of the water is covered with polyethylene or other plastic material to reduce the risk of infection. Another way of avoiding contamination is the common practice of seeding a large inoculum to dominate the culture, at least during the first growth phase. Other approaches to promote single-species culture have been evaluated; for instance, the use of nitrogen-fixing species of the genus *Anabaena* in media with no other nitrogen source has proved useful. Similarly, the selection of thermophilic species, such as *Scenedesmus obliquus*, has shown some initial potential.

Gentle agitation is important to achieve high productivity. It prevents sedimentation, allows a more homogeneous exposure of algal cells to light and reduces nutrient and temperature gradients along the depth of the culture. To this end, several designs have been implemented in ponds and ditches, including paddlewheels, gravity flow and pump recycling, combined with special slop designs in oval ponds and horizontal raceways.

Productivities rarely exceed $30\,g\,m^{-2}$ per day and cell densities of $2\,g\,l^{-1}$; this is much lower than values for other industrial fermentation processes. Some experimental improvements have been achieved by optimizing the use of wastes through the addition of nitrogen sources, adding aeration ports and inoculating selected bacteria that efficiently degrade the diluted organic materials. By such means, the dry weight productivities reach about three and four times the average figures obtained for soya bean and maize respectively on an annual basis. With intensive modes of cultivation, algal cultures can produce up to 20–35 times more protein than soya bean for the same area of land.

An example of a typical mass-culture open system is that conducted by the Sosa Texcoco Company in Mexico, mainly during the 1970s and 1980s, where an alkaline (pH 9–11) lake of 900 hectares surface area produced up to 1000 metric tons per year of *Spirulina maxima* (equivalent to $0.111\,kg\,m^{-2}$). Weather conditions in central Mexico and the high alkalinity of the water favour effortless predomination of *S. maxima* on Lake Texcoco. Unfortunately, this factory stopped its production in the mid 1990s due to a long strike.

The most advanced system developed so far is the High-Rate Algal Ponds (HRAP), which combines the treatment of sewage with simultaneous massive production of algae. The project was developed by the Technion Research Centre at Haifa, Israel. Basic infrastructures are shallow canals that add up to a maximum of $1000\,m^2$, equipped with systems for gentle agitation and aeration. The process is operated continuously; retention times vary between 2 and 6 days depending on the season. A steady multiculture is established in the system within a few days of operation. This includes bacteria that degrade organic compounds, and well-defined algal species, with *Euglena*, *Chlorella* and *Scenedesmus* predominating. The maximum daily productivity reported at times of maximum solar radiation is $30\,g\,m^{-2}$, with an average annual production of 7 kg algae per square metre. To recover the cells, aluminium sulphate is added as a flocculant; the float is then dewatered by centrifugation and dried in a drum dryer to reach a final moisture content of 10%. The final product has been shown to have an excellent nutritional quality, containing 57.4 g crude protein per 100 g, and an amino acid profile superior to the average for soya bean protein. It has been used to complement at least 25% of fish diet and 10% of poultry diet with no toxic effects. The resulting effluent can be used directly for crop irrigation.

Chlorella ellipsoidea is produced in Taiwan for food use in open ponds with agitation or circulation. The product is recovered by filtration. In China pilot-scale open systems have been used for the production of *Spirulina*, mainly *S. platensis*. Ponds about $100\,m^2$, covered with polyvinyl chloride sheets, lead to a daily production of 10 g per square metre (equivalent to more than 3 kg per year per square metre if production could be maintained constant through the year). Other microalgae produced in China in open-air systems are *Scenedesmus*, *Chlorella* and *Anabaena*.

Photobioreactors

Photobioreactors are closed systems working either outdoors or indoors, in which a single species is inoculated to keep a clean culture operation. Closed cultivation systems offer better control in terms of contamination and cell physiology, leading to higher growth and better quality of the harvested product but increased manufacturing costs.

Large systems operating outdoors consist of tubes covering large areas exposed to sunlight and can be operated either in batches or continuously. Many designs have been constructed or proposed on a pilot scale. Tubes are made of either glass or plastic materials such as polyethylene. Since the tubes behave as solar collectors, overheating is a problem. Hence, tubular solar receptors must have a temperature control system, usually a water pool. Alternatively the use of thermotolerant strains has been proposed to avoid cooling facilities. Generally tubes are grouped in several modules to facilitate control and operation, and to offer flexibility to the system in terms of culture volume. Carbon dioxide supply systems such as carbonation towers, pumps for circulating the medium and tanks to mix nutrients are attached to the solar receptors. These photobioreactors have been used for the production of *Chlorella*, *Spirulina*, *Scenedesmus* and *Porphyridium*.

A design constructed in Chile up to 110 m^2 solar irradiation area consists of a cement pond lined with epoxic resin and covered with a polyethylene dome. The agitation system is a paddlewheel. It has been used to produce *Spirulina* biomass, reaching a growth density of 450–750 mg l^{-1}.

An innovative design, operated in the US, is based on oval plastic bags floating on thermal waters.

Photobioreactors operating indoors are smaller because artificial light is needed. Designs can be either plastic tubular systems or stainless steel fermenter-like reactors with internal illumination to allow maximum light incidence. Their use is rather limited for SCP production because of low throughputs; however, they are adequate for the production of algal metabolites with high added value, such as polysaccharides, carotenes and other pigments and polyunsaturated fatty acids.

Harvesting

Post-production recovery of microalgal biomass is difficult, particularly in large-area lakes, or when low concentrations occur. Some species, such as *S. platensis*, *S. maxima* and *Coelastrum probiscideum*, are easily skimmed off or harvested by filtration through cloths or screens. Filter presses can also be used. Owing to their small cell size (10 μm), other species are harvested by centrifugation or flocculation with flocculants such as lime, alum or polyelectrolytes.

After harvesting, the algal biomass must be dewatered by centrifugation and/or dried. Operations to dehydrate biomass are normally done by drum-drying, sun-drying or spray-drying; drum-drying is preferred most widely. In some cases, the product is not dried; for example, in China, *Scenedesmus* and *Chlorella* are fed to swine as fresh algae slurry.

Nutritional Value and Human Consumption

Algal SCP has a nutritional value similar to other SCP sources. Its crude protein content varies between 45 and 73% (N × 6.25), while its lipids and mineral contents are 2–20% and 5–10% respectively. The chemical composition of some algal species is shown in **Table 2**. The protein content of algae is higher than soya bean (40 g per 100 g).

The amino acid content of algal SCP is balanced except, as with any other microbial biomass, for the sulphur-containing amino acids methionine and cystine. It is rich in vitamins, especially water-soluble vitamins, and essential fatty acids. The content of some vitamins such as thiamin, riboflavin, folic acid and carotene is higher in algae than in many vegetable foodstuffs. The content of nutrients is highly dependent on cultivation and processing conditions. **Table 3** shows some parameters of protein quality of some algae. The protein efficiency ratio (PER), net protein utilization (NPU) and biological value (BV) of algal proteins are somewhat lower than casein. Nutritional tests have shown promise when algae supplemented with methionine and cystine are fed to broilers. However, monogastrics have problems digesting the whole cells and some processing is therefore needed.

The method of drying affects product bioavailability. In drum-dried algae compared to the air-dried product, NPU is increased to around 100%, while digestibility is increased to about 60%. This phenomenon may be due to the rupture of the algal cell walls when water is removed at controlled conditions. Algal cell wall is not readily digestible, therefore any treatment to disrupt the cell wall structure will increase digestibility and hence nutritional value. Many treatments have been suggested, such as mechanical disruptions, extractions with organic solvents and treatments with alkalis and/or acids.

Algal SCP have had many uses, including cultivation of daphnid and similar species that thrive on plankton as a food source in aquaculture. In addition to its use as feed for chicken and swine, algal biomass has been used as food worldwide. *Chlorella* is produced in Taiwan for food use; *Spirulina* has been

Table 2 Composition of algal species. Main components in g per 100 g dry wt. Amino acids and vitamins as specified

	Spirulina maxima	Chlorella	Scenedesmus obliquus	Scenedesmus acutus
Crude protein (N × 6.25)	55–71	40–58	50–56	46–64
True protein	48–61			44–48
Amino acids (g per 16 g N)				
Alanine	5.0–6.1	4.2–7.4		5.3–10.4
Arginine	4.5–9.3	5.8–10.2		4.6–7.1
Aspartic acid	6.0–15.2	6.9–8.8		6.5–11.1
Cystine	0.6–2.2	0.3–0.9		0.6–1.6
Glutamic acid	8.2–21.8	8.0		5.3–10.7
Glycine	3.2–4.0	4.9–5.5		3.4–7.0
Histidine	0.9–1.6	1.4–3.0		1.5–2.3
Isoleucine	3.7–4.5	3.1–6.4		2.2–4.9
Leucine	5.6–7.7	6.8–9.7		5.0–10.6
Lysine	3.0–4.5	4.9–9.4		5.0–6.4
Methionine	1.6–2.2	1.0–2.0		1.4–2.7
Phenylalanine	2.8–4.0	3.2–5.1		3.6–6.4
Proline	2.7–3.2	2.2–6.4		3.1–6.1
Serine	3.2–4.3	3.0–4.1		3.2–5.4
Threonine	3.2–4.5	3.6–4.7		3.0–5.8
Tryptophan	0.8–1.2	1.0–1.5		0.3–1.8
Tyrosine	3.9	2.6–4.1		2.0–4.6
Valine	4.2–6.0	4.8–6.0		4.7–7.4
Lipids	4–7	6–16	12–14	8–14
Carbohydrates	13–16		10–17	
Minerals	4–9	6–9	4–9	6–17
Vitamins (mg per 100 g)				
Thiamin	5.5	0.6–2.3		1.2–8.2
Riboflavin	4.0	2.0–6.0		3.4–36.6
Pyridoxine	0.3	0.1–3.2		1.1–2.5
Nicotinic acid	11.8	10–22		12–16.7
Pantothenic acid	1.1	1–10		1.5
Folic acid	0.05	0.1–4.0		0.7
Biotin	0.04	0.015–0.064		0.02–0.2
Cyanocobalamin	0.02	traces		0.04–0.44
Ascorbic acid		18–370		165–181
β-Carotene	0.17			
γ-Tocopherol	19.0	26–33		14–18.5

Table 3 Parameters of protein quality of some microalgae

Product	*NPU*	*PER*	*BV*
Spirulina platensis	52.7		68
S. platensis + methionine	62.4		82.4
Spirulina sp.	65	1.80	75
Spirulina sp. + methionine	73		82
Chlorella sp.	66		72
Chlorella sp. + methionine	78		91
Scenedesmus sp.	67	1.93	81
Casein	83	2.50	88

NPU = Net protein utilization; PER = protein efficiency ratio; BV = biological value.

produced commercially in the US, Mexico, Taiwan, Japan, Thailand and Israel with the same purpose.

Algal SCP has mainly been used for the preparation of tablets and other products to be sold as health foods, as protein and vitamin supplements, or to help people lose weight. However, few in-depth studies, if any, have been conducted to evaluate the nutritional or health-associated benefits of algal SCP. In countries such as the US, Mexico and Chile algal biomass is sold as tablets or powders in health food stores. In Japan and Taiwan it is sold either as dry powder or as pellets.

A wide range of food formulations has been prepared with algal biomass. It has been widely used as an additive or supplement to cereal foodstuffs, or as a garnish to salads. Experiences related to supplementation of cereal foods with algal SCP include mixtures with doughs for baked goods and pasta, such as bread, rolls, cookies and noodles. In Mexico, *S. maxima* has been used as a supplement for biscuits produced by a state company as part of a national breakfast programme for schoolchildren.

Concerning functional properties, it has been reported that *S. platensis* flour had similar emulsion and foam capacities, slightly lower water and fat absorption capacities, and a lower foam stability than

Table 4 Functional capacities of a protein concentrate obtained from *Spirulina platensis*

Water absorption	145 g per 100 g protein
Fat absorption	373 g per 100 g protein
Emulsifying capacity	113 ml oil per 100 g protein
Foam capacity	205%
Foam stability (1 h)	27%

soya bean meal. Using the protein concentrate obtained from the flour, improved functional characteristics are obtained, with the exception of water absorption. Some values for functional capacities of a protein concentrate obtained from *S. platensis* are given in **Table 4**.

The main problems of acceptability of algal biomass are its unfamiliar and sometimes bitter flavour, as well as the presence of dark green pigments which are difficult to mask. In addition to chlorophylls, other pigments such as carotenes, xanthines and phycocyanin are present in varying amounts. Flavour and colour may be improved if algal biomass is treated during downstream processing to remove undesirable components.

Toxicological Problems

Algae have been consumed as food for generations without ill effects. No toxic effects have been reported in animal evaluations. However, the following considerations must be taken into account.

A common problem for SCP from any microorganism is the high content of nucleic acids present in microbial cells. Consumption by humans of nucleic acids in amounts higher than 2 g per day can lead to accumulation of uric acid which develops into gout and kidney stones. The concentration of nucleic acids in algal biomass depends on several factors, such as species and growth conditions. Cyanobacteria have a nucleic acid content of 2.9–5 g per 100 g while microscopic plant algae contain 1–17 g per 100 g. These amounts are higher than in most other foodstuffs.

In order to reduce nucleic acid content, protein concentrates or isolate can be prepared by cell disruption and protein separation. **Figure 2** shows a process for the preparation of protein isolate from *Spirulina*. However, this increases the cost of the product. In general, algal biomass is consumed by humans in small amounts and so the intake of nucleic acids is below risk.

The cell wall of microalgal biomass represents about 10% of its weight. It is mainly composed of indigestible carbohydrates and some other compounds, e.g. murein in cyanobacteria. The bioavailability of protein from whole cells is therefore low.

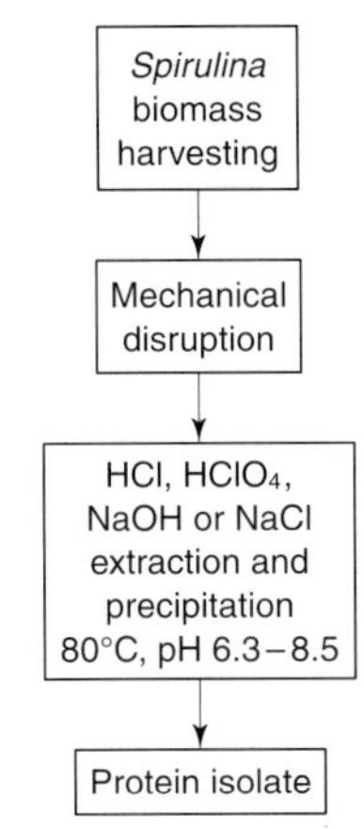

Figure 2 Process of *Spirulina* protein recovery (adapted from Litchfield 1991).

The preparation of protein concentrates or isolates produces products with a high nutritional value and free of undesirable pigments, although it represents a costly alternative.

Algae have the ability to remove heavy metals from polluted waters. Similar physiological phenomena account for the accumulation of pesticides and organochlorinated compounds. This is a problem when algae are intended for use as SCP. However, the causes are well identified and some steps can be implemented to maintain the final product composition within safety levels.

The origin of the water used in cultivation ponds determines the need for pre-treatment. In general, wastes generated in food industries carry low amounts of the contaminants mentioned. However, urban sewage and run-offs show high variability in the content of heavy metals and other toxicants. When these waste streams are subjected to standard secondary treatments, most organics are degraded, whereas metals remain associated with the activated sludge, rendering them safe as an input for algal growth. Furthermore, some studies have demonstrated that biological absorption of metals is a rather slow process, requiring more time than the usual retention time of the water within the bioreactor. It appears that fears concerning the presence of recalcitrant toxicants in algae might be excessive, although more research is needed to get a clear picture.

Another problem is the possibility of contamination by pathogenic microorganisms. Certainly, the culture practices in open systems increase the risk of pathogenic infections. The downstream processes of SCP products are designed to destroy most viable forms present, although some could survive in the product. Recommendations have been established for microbiological standards of SCP products for use in animal feeds.

See also: **Bacteria**: The Bacterial Cell; Classification of the Bacteria – Traditional. **Fermentation (Industrial)**: Basic Considerations. **Single Cell Protein**: Yeasts and Bacteria.

Further Reading

Goldberg I (1985) *Single-Cell Protein*. Berlin: Springer Verlag.

Litchfield JH (1991) Food Supplements from Microbial Protein. In: Goldberg I and Williams R (eds) *Biotechnology and Food Ingredients*. P. 65. New York: Van Nostrand Reinhold.

Moo-Young M and Gregory K (1986) *Microbial Biomass Proteins*. London: Elsevier.

Rolz C (1990) Características químicas y bioquímicas de la biomasa microbiana. *Archivos Latinoamericanos de Nutrición* 40: 147–193.

Rose AH (ed.) (1979) *Microbial Biomass. Economic Microbiology*. Vol. 4. London: Academic Press.

Stadler T, Mollion J, Verdus MC, Karamanos Y, Morvan H and Christiaen D (eds) (1988) *Algal Biotechnology*. London: Elsevier Applied Sciences.

Switzer L (1980) *Spirulina. The Whole Food Revolution*. Berkeley, CA: Proteus.

Wood A, Toerien DF and Robinson RK (1991) The Algae – Recent Developments in Cultivation and Utilization. In: Hudson BJF (ed.) *Developments in Food Proteins – 7*. P. 79. London: Elsevier Applied Science.

Yeasts and Bacteria

Mariano García-Garibay and **Lorena Gómez-Ruiz**, Departamento de Biotecnología, Universidad Autónoma Metropolitana, Mexico City, Mexico

Eduardo Bárzana, Departamento de Alimentos y Biotecnología, Universidad Nacional Autónoma de México, Mexico City, Mexico

The single-cell protein (SCP) concept is applied to the massive growth of microorganisms for human or animal consumption. SCP is a generic term for crude or refined protein whose origin is bacteria, yeasts, moulds or algae. The production and utilization of the bacteria and yeasts is discussed here. These two groups of microorganisms have been particularly important since yeasts and bacterial biomass have been consumed by humans since ancient times in fermented foods. The production of SCP has important advantages over other sources of proteins, such as its considerably shorter doubling time, the small land requirement and the fact that it is not affected by weather conditions. Much attention was focused on the use of petroleum derivatives as substrates for SCP production during the 1960s and 1970s when the price of this reserve was low; currently, the production of SCP is based on renewable resources, and its interest is also kept as a means of conferring value on waste materials. On the other hand, the organoleptic and functional properties of SCP are not always competitive, and its main drawback has been its high production costs.

Historical Developments and Implementation of SCP Production

SCP production was first developed during war time when conventional foods were in short supply. During World War I, *Saccharomyces cerevisiae* was massively produced in Germany from molasses to replace up to 60% of imported protein. A similar experience was repeated in World War II for the mass production of *Candida utilis* (previously known as Torula yeast or *Torulopsis utilis*) on sulphite liquor from paper manufacturing wastes. After the war, several plants were built in the US and Europe, mainly for *C. utilis* production.

Accelerated industrial development and general welfare expectancy led to renewed interest in SCP as an alternative to alleviate food shortages due to growing imbalance between food production and world population, mainly in developing countries. During the 1960s and 1970s several SCP production plants were built in the UK, France, Italy, USSR, Japan and Taiwan.

An important breakthrough took place when the production of SCP from hydrocarbons was demonstrated by several petroleum companies during the 1950s and 1960s. During the 1970s considerable research efforts resulted in the use of methanol and ethanol derived from petroleum as convenient substrates. In the early 1970s, the cost of *n*-paraffin was approximately US$80 per ton, while crude molasses cost approximately US$82 per ton. However, concern on substrate safety and increase in petroleum prices shifted interest back to the utilization of renewal sources, mainly food and agriculture by-products like molasses and whey, or industrial wastes rich in starch, cellulose and hemicellulose. The major SCP projects based on petroleum derivatives as substrates were abandoned in the 1980s.

In recent times, among the European communist countries the USSR had the largest capacity for SCP production with at least 86 plants in operation using different substrates. Other countries in eastern Europe like former Czechoslovakia were in a similar position. The Soviet delegation reported the production of 1 million tons of SCP per year in 1983 during an international symposium on SCP. They estimated that the

Table 1 Main industrial and pilot developments in yeast single-cell proteins

Substrate/yeast	*Process/plant*	*Country*
Sulphite liquor		
Candida utilis	St Regis Paper	US
Candida utilis	Boise Cascade	US
Candida utilis	State industry	Russia
Candida utilis	Attisholz	Switzerland
Hydrocarbons		
Yarrowia lipolytica	British Petroleum/Toprina	UK
Candida tropicalis	British Petroleum/Toprina	France
Candida tropicalis	Italprotein	Italy
Yarrowia lipolytica	Liquichimica/Liquipron	Italy
Yarrowia lipolytica	State industry	Russia
Yarrowia lipolytica	Swedt	Germany
Yarrowia lipolytica	Roniprot	Romania
Ethanol		
Candida utilis	Amoco/Torutein	US
Candida sp.	Mitsubishi	Japan
Methanol		
Pichia sp.	Mitsubishi	Japan
Pichia pastoris	Philipps Petroleum/Provesteen	US
Starch		
Saccharomycopsis fibuligera and *C. utilis*	Symba	Sweden
Saccharomyces cerevisiae	Brewers	Several
Molasses		
Candida utilis	Several industrial processes	Cuba
Candida utilis	Several industrial processes	Taiwan
Candida utilis	Cinvestav IPN	Mexico
Liquid sucrose		
Hansenula jadinii	Philipps Petroleum/Provesta	US
Whey		
Kluyveromyces marxianus	Wheast-Knudsen	US
Kluyveromyces marxianus	Amber Lab Universal Foods	US
K. marxianus, *K. lactis* and *C. pintolopesii*	Fromageries Bel	France
Kluyveromyces marxianus	SAV	France
Candida intermedia	Vienna	Austria
Candida utilis	Waldhof	US
Confectionary effluent		
Candida utilis	Bassett	UK

USSR would be producing 9 million tons at the end of the 1990s.

To date, a profusion of reports about SCP production has appeared in the scientific literature. Two main approaches have been followed: utilization of conventional substrates, and use of waste materials where SCP production brings about pollution control.

Many industrial processes have been developed worldwide; the most important are given in **Tables 1 and 2**. Countries with important industrial output are the US, UK, France, Russia and Cuba.

Suitable Organisms and Substrates

The most frequently used yeasts are the following: *Saccharomyces cerevisiae* (and related synonymous such as *S. carlsbergensis*, *S. uvarum* etc.), *Kluyveromyces marxianus* (including synonymous, subspecies and asexual forms such as *K. fragilis*, *K. lactis*, *K. bulgaricus*, *Candida kefyr* and *C. pseudotropicalis*), *C. utilis* (and its sexual form *Hansenula jadinii*) and *Yarrowia lipolytica* (formerly known as *Candida lipolytica* and *Saccharomycopsis lipolytica*).

All these yeasts have been widely used in the manufacture of human foods; *S. cerevisiae*, *K. marxianus* and *C. utilis* were granted GRAS status (generally recognized as safe for human consumption) by the US Food and Drug Administration. *S. cerevisiae* is also available as spent yeast from breweries and alcohol industries, but it is not commonly used for SCP production because it can easily deviate into alcoholic fermentation. *K. marxianus* and related species are widely used due to their capacity to assimilate lactose, the carbohydrate present in cheese whey, but can also grow on inulin, a fructose polymer found in some plants, and other simple sugars such as glucose, fructose and sucrose; therefore, it is also sometimes grown in molasses. Since it is able to grow at temperatures as high as 45°C, it has been used to produce biomass

Table 2 Main industrial and pilot developments in bacterial single-cell proteins

Substrate/bacteria	*Process/product*	*Country*
Methane		
Methylococcus capsulatus	Shell Oil	UK
Methanol		
Methylophilus methylotrophus	ICI/Pruteen	UK
Methylomonas clara	Hoechst-Uhde/Probion	Germany
Ethanol		
Acinetobacter calcoaceticus	Exxon-Nestlé	Switzerland
Cellulose		
Cellulomonas sp. and *Alcaligenes* sp.	Louisiana State University	US
Whey		
Lactobacillus bulgaricus and *Candida krusei*	Kiel	Germany

in tropical areas. *C. utilis* is used for a wide variety of substrates such as sucrose, ethanol and sulphite-spent liquor. It can also grow on wood hydrolysates because of its ability to assimilate pentoses. Starchy solids or water streams from potato and maize industries require previous hydrolysis for yeast growth, as in the case of *C. utilis*, or as in the Symba process, the utilization of an amylolytic yeast (*Saccharomycopsis fibuligera*). Yeasts able to assimilate hydrocarbons are *Yarrowia lipolytica*, *Candida tropicalis*, *C. rugosa* and *C. guilliermondii*, which can usually also be produced on lipids. Methanol is the preferred alcohol utilized as substrate by *Pichia* spp. (*P. pastoris*, *P. methanolica*, etc.), *Hansenula polymorpha*, *H. capsulata* and *C. boidinii*. The most important processes developed for yeast SCP production are shown in Table 1.

Bacteria have mostly been used for the production of animal feed. Commercially the most important species utilize methane and/or methanol as substrates, as shown in Table 2. Methanol is usually preferred over methane because it is water-soluble and less explosive. The Ministry of Agriculture, Fisheries and Food in the UK allows the use of Imperial Chemical Industries' (ICI's) product Pruteen in animal feed. For the production of bacterial SCP from ethanol, either *Acinetobacter calcoaceticus* or *Alcaligenes* sp. have been used in laboratory or pilot plant studies.

Other petroleum derivatives such as paraffins have generally not been considered for the production of SCP using bacteria; some studies have been done with *Acinetobacter calcoaceticus*.

To produce bacterial SCP from whey, several lactic acid and propionic bacteria have been investigated, frequently in mixed cultures with yeasts, as in the Kiel process. In this case *Lactobacillus delbrueckii* subsp. *bulgaricus* grows using the lactose, converting it to lactic acid; then *Candida krusei*, which is unable to ferment lactose, uses the lactic acid as carbon source. Both fermentations can be performed simultaneously, controlling the pH by adding ammonia, which is used as nitrogen source by the yeast.

Lactic acid bacteria proposed for the fermentation of whey include species such as *Lactobacillus delbrueckii* subsp. *delbrueckii*, *L. delbrueckii* subsp. *bulgaricus*, *Lactobacillus casei* subsp. *casei*, *Leuconostoc* sp., etc. It has also been proposed that fermentation of this substrate with rumen bacteria, producing biomass concentrated by ultrafiltration, should be used as feed.

Propionibacteria, such as *Propionibacterium freudenreichii* subsp. *shermanii* and *P. freudenreichii* subsp. *freudenreichii*, produce significant amounts of vitamin B_{12}, giving added value to their utilization in the production of SCP. A process using a mixture of *P. freudenreichii* subsp. *freudenreichii* and *K. marxianus* has been proposed for the production of SCP from whey rich in vitamin B_{12} and sulphur amino acids.

There has been little investigation into bacteria in the production of SCP from substrates such as starch and cellulose. A case in point is the project developed by Louisiana State University in the US to produce SCP from *Cellulomonas* sp. and *Alcaligenes* sp. from cane bagasse; this process was scaled up to pilot plant level.

Generally, yeasts have been preferred over bacteria for SCP production, especially for human consumption. It seems that yeasts are better accepted because they are more familiar to humans in foods like bread or beer. However, bacteria have some advantages over yeasts, such as higher protein content (**Table 3**), higher yields (carbon source to protein conversion) and faster growth rate, although higher nucleic acid content (**Table 4**) limits its uptake in the diet.

Production Processes

Since SCP must sustain a competitive price in the protein market, especially protein of vegetable origin, it is essential to guarantee efficiency in all stages of the process. In particular, the carbon source accounts for about 60% of operational costs. Therefore, high yields of substrate conversion are required, high productivity processes must be implemented, and utilization of an inexpensive but easy-to-assimilate carbon source must be guaranteed. This explains the generalized use of molasses, whey or industrial residues, depending on local availability, and attempts to implement fossil fuels as substrates. An important advantage in using hydrocarbons is the yield – 1 g of

Table 3 Nutritional parameters of single-cell protein products

	Kluyveromyces marxianus	Saccharomyces cerevisiae	Candida utilis	Methylophilus methylotrophus	Methylomonas clara	*Soya meal*
Protein (g per 100 g dry wt) crude (N × 6.25)	43–58	48	42–57	72–88	80–85	44–50
True protein	40–42	36	47	64	69–73	48
Essential amino acids (g per 16 g N)						
Isoleucine	4.0–5.1	4.6–5.5	4.3–5.3	5.2–5.4	3.6	5.4
Leucine	7.0–8.1	7.0–8.1	7.0	8.2–8.4	6.6	7.7
Phenylalanine	3.4–5.1	4.1–4.5	3.7–4.3	4.3–6.5	5.1	5.1
Tyrosine	2.5–4.6	4.9	3.3	3.5–3.8	5.1	2.7
Threonine	4.1–5.8	4.8–5.2	4.7–5.5	5.7–6.5	4.8	4.0
Tryptophan	0.9–1.7	1.0–1.2	1.2	1.1–1.6	1.5	1.5
Valine	5.4–5.9	5.3–6.7	5.3–6.3	6.3–6.5	4.8	5.0
Arginine	4.8–7.4	5.0–5.3	5.4–7.2	4.3–5.6	3.4	7.7
Histidine	1.9–4.0	3.1–4.0	1.9–2.1	2.2–2.3	2.8	2.4
Lysine	6.9–11.1	7.7–8.4	6.7–7.2	4.1–7.3	6.2	6.5
Cystine	1.7–1.9	1.6	0.6–0.7	0.8		1.4
Methionine	1.3–1.6	1.6–2.5	1.0–1.2	1.4–3.0	2.5	1.4
PER	1.8	2.0	1.7			1.4–2.2
NPU	67			84		64
Vitamins ($\mu g\ g^{-1}$)						
Thiamin	24–26	104–250	8–9.5	5		9.0
Riboflavin	36–51	25–80	44–45	40		3.6
Piridoxine	14	23–40	79–83	2		6.8
Nicotinic acid	136–280	300–627	450–550	57		24.0
Folic acid	6	19–30	4–21	15		4.1
Pantothenic acid	67	72–86	94–189	11		21.0
Biotin	2	1	0.4–0.8	3		
Vitamin B_{12}	0.015–0.05		0.0001			0

PER, Protein efficiency ratio; NPU, net protein utilization.

dry biomass is obtained per gram of hydrocarbon – compared to carbohydrates which typically yield around 0.5 g of dry biomass per gram of substrate.

The typical process stages for the production of SCP comprise raw material preparation, fermentation, biomass recovery, cell disruption and drying. **Figure 1** shows a successful process developed and established in Cuba for the production of *C. utilis* SCP from sugar-cane molasses, and is a good example of the SCP production processes currently in operation.

To maximize carbon assimilation the nutrients must be balanced. Sources of nitrogen, minor elements (P, K, S, Mg etc.) and trace elements (vitamins and minerals) are adjusted according to the general composition of the carbon source. This in turn is highly dependent on the strain used. In general, simple nitrogen sources such as urea, ammonia and nitrate are used to keep costs down. Phosphate is supplied as phosphoric acid or as soluble phosphate salts.

Fermentation variables like temperature, pH, ionic strength, level of oxygenation and, in continuous fermentations, dilution rate, have a strong influence on cellular yield. In particular, an abundant supply of oxygen promotes aerobic metabolism and higher growth rates. However, due to the low solubility of oxygen in aqueous media, the cost of aeration, through air sparging and agitation, increases rapidly with the scale of operation, resulting in an important technical challenge. Assimilation of *n*-paraffins requires high levels of oxygenation, and the growth of microorganisms on these water-insoluble substrates takes place on the hydrocarbon–water interface. The surface area becomes the limiting factor and heat production is about twice that using sugars as substrates.

In general, when yeast biomass is produced, alcohol accumulates due to oxygen limitation. Some alternatives proposed for SCP from whey are the production of alcohol as a by-product of fermentation or the use of *Kluyveromyces* in a mixed culture with *Candida pintolopesii*, where the latter consumes the alcohol produced by the former. The first approach has been followed by Amber Laboratories (Universal Foods, US) and the second by Fromageries Bel in France; *C. valida* has also been demonstrated to prevent ethanol accumulation when it is grown mixed with *C. kefyr* (the asexual form of *K. marxianus*), increasing the efficiency of whey conversion into biomass up to 20%. Limited oxygen supply when

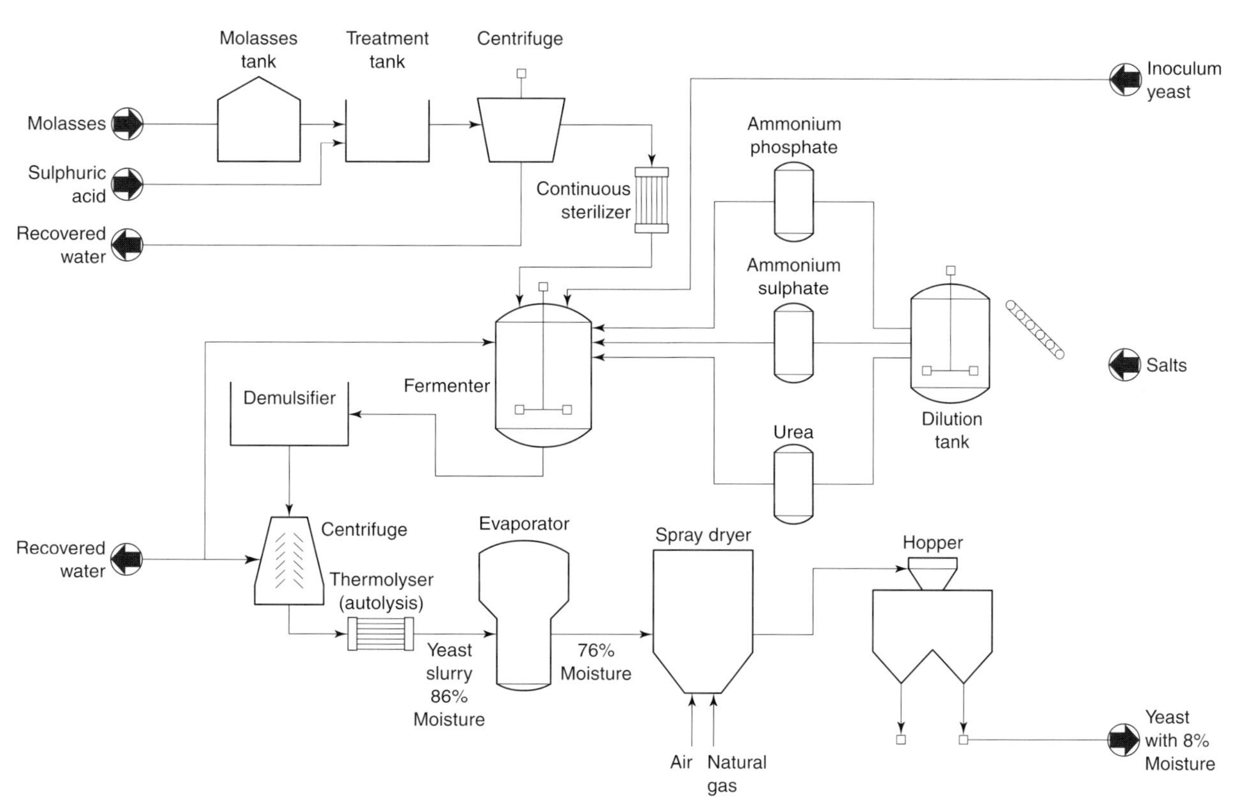

Figure 1 Industrial process for production of *Candida utilis* from sugar-cane molasses developed in Cuba. (Modified from de la Torre and Flores-Cotera 1993, with permission.)

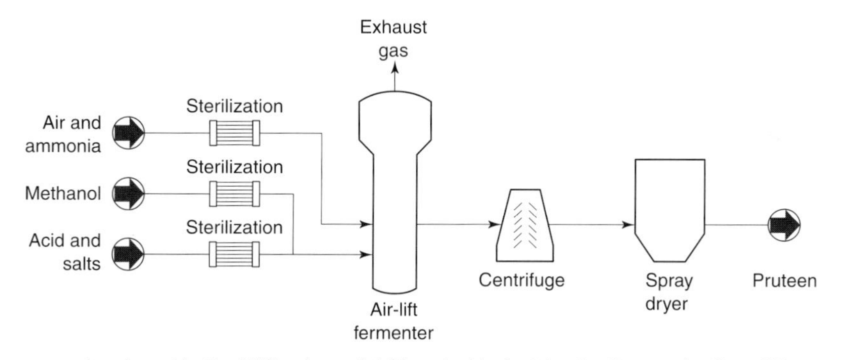

Figure 2 Industrial process developed in the UK by Imperial Chemical Industries for the production of Pruteen using *Methylophilus methylotrophus* grown in methanol.

hydrocarbons are used as substrates lead to the production of some metabolites such as mono- and dicarboxylic acids and ketones.

To sustain high oxygen transfer rates, large air volumes must be supplied with high agitation rates. Alternative fermenter designs include the air-lift type which achieves maximum oxygenation with minimum power, reducing aeration costs considerably. In fact, the largest fermenter ever operated is an air-lift (3000 m^3) for the aerobic production of *Methylophilus methylotrophus* by ICI. **Figure 2** shows a simplified diagram of the process developed by ICI; this is the most successful process developed for bacterial SCP production. Currently, the high-cell-density fermentation designs pioneered by Provesta (subsidiary of Philipps Petroleum) permits producing of up to 160 $g\,l^{-1}$ of yeast biomass while traditional fermentation techniques reach at most 30 $g\,l^{-1}$. These fermenters have efficient systems for heat removal and oxygen transfer.

Once obtained, the microbial biomass is recovered by filtration or centrifugation. The resulting cell suspension can be either spray-dried or cells broken down to obtain extracts, hydrolysates or autolysates. Finally, the protein can be concentrated, isolated or texturized. In the Philipps Petroleum process, owing to the high cell density, the spent medium is fed directly to the spray-drier without being concentrated.

Nutritional Value

The main nutritional contribution of SCP in either human food or animal feed is its high protein content. Bacterial have a protein concentration ranging from 50 to 85% and yeasts from 45 to 70%. Protein quality is also acceptable, compared to vegetable proteins. Generally, SCP from any source is limited in sulphur-containing amino acids; however, bacterial proteins tend to have better balances of essential amino acids than yeasts and other microorganisms, particularly with respect to sulphur amino acids. Some parameters of protein quality are given in Table 3. When methionine is added to SCP the protein quality increases considerably and reaches values similar to those of casein.

SCP is also an important source of vitamins; brewer's yeast has long been used as a vitamin supplement. Vitamin content is also shown in Table 3.

Toxicological Aspects

The safety of both microorganisms and substrates is an important consideration. Microorganisms used for SCP production must be subjected to extensive toxicological clearance. Those microorganisms, normally present in fermented foods, such as *Saccharomyces cerevisiae*, are free from suspicion. The main concern is in the use of petroleum derivatives in which residual alkanes must be removed with solvents. However, some residual hydrocarbons may still be present, and several reports have noted the presence of unusually high levels of odd-chain fatty acids and paraffin in tissues from animals fed with SCP from alkanes. These fatty acids, particularly unsaturated C17, have been suspected of having toxic effects, even though no evidence has been reported.

The high content of nucleic acids in microbial cells also has consequences. Human consumption of nucleic acids in amounts higher than 2 g day^{-1} could lead to the accumulation of uric acid, resulting in kidney stones and gout onset in susceptible people. The concentration of nucleic acids in the biomass is highly variable and depends on several factors. The first element is the nature, species and strain of the microorganisms; normally bacteria have higher concentrations than yeasts. Another factor is the growth conditions: the higher the growth rate, the higher the nucleic acid content in the cell. Nucleic acid content also changes with the growth phases. For instance, for *K. marxianus* grown on whey, the concentration of nucleic acids reached its peak at the middle of the exponential phase. Table 4 shows the nucleic acid content of several SCP products.

Table 4 Reported nucleic acid content of single-cell protein products (g per 100 g of biomass on a dry wt basis)

Kluyveromyces marxianus	2.7–10
Saccharomyces cerevisiae	2.5–15
Candida utilis	6.2–10
Methylophilus methylotrophus	16
Methylomonas clara	10–15
Isolated protein from *K. marxianus*	1.4–5.7

In order to reduce the nucleic acid content two approaches have been followed: growing the biomass at low rates, or isolating the protein by eliminating undesirable compounds. The latter is usually applied, to eliminate not only nucleic acids but also the cell wall. In recent years, research has been conducted on utilizing external (added) endonucleases to hydrolyse nucleic acids to obtain protein isolates free from these polymers. Immobilized endonucleases have been proposed for this purpose.

Yeasts and bacterial cell walls are difficult to digest, leading to poor protein bioavailability, flatulence, allergic responses and diarrhoea. Some cases of skin rash, nausea, vomiting and other allergic reactions have been reported, but they may be eliminated by reducing cell wall and nucleic acids.

Nucleic acid content in SCP for animal feed is not a problem, because animals have the enzyme uricase that prevents uric acid accumulation. However, cell wall digestibility in monogastric animals is poor.

Cell Disruption and Protein Isolation

Many processes have been developed to disrupt cells. A common one is autolysis, in which the biomass is exposed to heat shock or to chemical compounds, such as low-molecular-weight thiols. Yeast autolysis usually takes place when cells are heated to 45–55°C, and it is enhanced by the presence of NaCl. Further incubation for around 24 h induces cellular enzymes, leading to complete lysis of the cells. The process also activates endogenous ribonucleases that reduce nucleic acids. Lysis can also be facilitated by addition of exogenous enzymes such as proteases, β-glucanases or lysozyme. The disadvantages of these techniques are the high cost involved and extensive proteolysis, which reduces yield and functional properties of proteins. Chemical treatments with alkalis, organic solvents or salts which weaken cell walls are also used. Alkaline treatment may result in undesirable side reactions forming compounds such as lysinoalanine and off flavours.

Hydrolysis requires the use of hydrochloric acid at temperatures as high as 100°C, to treat a slurry with 65–85% of solids content, followed by neutralization with NaOH. This process has many drawbacks, such as the risk of accidents with concentrated acid, the use of anticorrosive equipment which is cost-intensive and the destruction of some amino acids and vitamins reducing the nutritive value of the product.

Physical methods to break cell walls are the most widely used. High shear rates are achieved by means of homogenizers or colloidal mills and have been used extensively for SCP processes.

Once the cells have been broken, the protein is extracted using water or alkaline conditions; cell wall debris is removed by centrifugation and the protein is further precipitated with acid, salt or heat treatment, while the nucleic acids remain in the supernatant. Usually microbial proteins have their maximum solubility at pH 12, and precipitation occurs at pH 4.5. The protein isolate is then obtained. Some chemical modification during protein extraction includes phosphorylation or succinylation which facilitates protein separation from nucleic acids and improves its functional properties.

Utilization for Human Food

For food applications, besides toxicity and nutritional quality, organoleptic acceptability and functional properties are important considerations.

SCP has been used as a protein supplement in baked goods, biscuits, snacks, soups, noodles and special foods like geriatric or baby meals. Its use as extender in sausages and other meat products has been important, mainly in eastern European countries. Despite the justified production of SCP in terms of world protein shortage and widespread malnutrition, real demand for protein is based on its chance of competing in terms of its functional properties such as solubility, water binding, emulsification capacity, gelation, whippability and foam stability. The successful supplementation of existing products and the replacement of traditional proteins with SCP depends on availability of proteins of equivalent functionality, price and organoleptic acceptability.

For food applications whole dried cells, disrupted cells and textured protein products are useful. From disrupted cells either protein concentrates or isolates can be obtained, which are better suited for the food industry. Moreover, SCP isolates can compete favourably with soya isolates from a functional point of view. However, isolation or concentration increases production costs dramatically.

Processes such as texturization by spinning and extrusion, and enzymatic or chemical modification can improve the functionality of SCP. For instance, protein fibres obtained by spinning can form textured protein products such as meat extenders.

Enzymatic modification includes partial proteolysis to improve solubility, emulsification capability and whippability, or the reverse reaction known as plastein (peptide bond formation) to improve nutritional value through addition of limiting amino acids. Promising chemical modifications include acetylation, which improves thermal stability, and succinylation or phosphorylation to increase solubility, emulsification and foaming capacities. However, such modifications tend to reduce the nutritive value of the proteins. Experiences with phosphorylated yeast proteins have demonstrated that protein recovery can be improved, reducing nucleic acids. Functional properties such as water solubility, water-holding capacity and thickening properties can be enhanced, while emulsifying activity is better than soya protein isolate and equivalent to sodium caseinate.

Although dried whole cells have limited functional properties, they are frequently used as flavour-carrying agents and food binders. Dried yeast cells can act as oil-in-water emulsion stabilizers.

The major market for microbial biomass is as a flavour enhancer for meat products, soups, gravies, barbecues, sauces, salad dressings, seasonings and any food with savoury, cheesy or meaty flavours (flavour notes associated with the fifth basic flavour, called umami), including pizzas, snacks and chips. In fact, yeast protein hydrolysates, autolysates and extracts have long been used as food flavourings.

Prospects

SCP has to compete with other protein sources such as soya bean, fish meal and milk proteins. It has been widely demonstrated that in proteins used as additives or incorporated in food, the most important factors are their functional properties and price; therefore, these are the main challenges that SCP has to face. Unfortunately, production and isolation of protein from microbial biomass is expensive because they are capital- and energy-intensive. Its broad utilization has been limited for economical reasons. However, autolysates or hydrolysates mainly prepared from yeasts have gained wide acceptability as functional food ingredients. In addition, recent biotechnological advances such as high-cell-density fermentations, more efficient downstream operations and the possibility of genetically improving microorganisms may lead to SCP being re-evaluated.

High-cell-density fermenters have made possible considerable reduction in equipment size, energy savings, high productivity and cheaper downstream processing. For instance, direct spray-drying from the fermenter is possible. These kinds of improvements may bring SCP into the competitive realm. Currently,

Philipps Petroleum is producing Provesta and Provesteen, SCP trademarks of different strains, using this process (Table 1).

The use of fed-batch fermentations has not been fully explored for SCP production. This technique is widely used to produce bakers' yeast, increasing yields and avoiding ethanol accumulation. Some reports applying this strategy to yeast SCP production have demonstrated increased yield in up to 70%. The possibility of implementing continuous fermentation technologies to improve productivity is currently also being considered for the production of SCP.

Genetic engineering has focused on the possibility of improving substrate utilization. The first modified microorganism utilized in an industrial process and the largest-scale application for genetic engineering is a strain of *Methylophilus methylotrophus*, developed by ICI in 1977. The improved strain growth on methanol is able to utilize ammonium ion as nitrogen source more efficiently than the wild strain, saving 1 mol of ATP per mole of ammonium assimilated, with an increase in efficiency of 4–7%.

Considerable research based on genetic engineering has been done to obtain yeasts able to utilize a broader range of carbon sources, such as lactose, starch, cellulose, xylose and chitin, in order to use cheaper and more available substrates.

Other interesting possibilities include selection of mutants with weaker cell walls, introduction of genes coding for lytic enzymes to facilitate cell disruption, and genes coding for nuclease activity to reduce nucleic acids content. Also, enhancement of nutritional value by modification of amino acid content, and proteins with better functional properties look promising. All these possibilities have been investigated but practical results will take longer to be implemented.

Another approach to improve the economics of SCP is to produce it as a low-cost by-product in multiproduct microbial processes such as during processing of food wastes to reduce biological oxygen demand, or as a by-product from high-added-value enzymes. Another possibility is the recovery of nucleic acids from bacteria and yeast biomass to produce 5′-nucleotides which can be used as flavour enhancers.

See also: **Bacteria**: The Bacterial Cell; Classification of the Bacteria – Traditional. ***Candida***: Introduction; *Candida lipolytica*. **Fermentation (Industrial)**: Basic Considerations; Media for Industrial Fermentations; Control of Fermentation Conditions. ***Kluyveromyces***. ***Saccharomyces***: Introduction; *Saccharomyces cerevisiae*. **Single Cell Protein**: The Algae; Mycelial Fungi. **Yeasts**: Production and Commercial Uses.

Further Reading

Batt CA and Sinskey AJ (1987) Single-cell Protein: Production, Modification and Utilization. In: Knorr D (ed.) *Food Biotechnology*. P. 347. New York: Marcel Dekker.

de la Torre M and Flores-Cotera LB (1993) Proteinas Unicelulares. In: García-Garibay M, Quintero-Ramirez R, López-Munguía A (eds) *Biotecnología Alimentaria*. P. 383. Mexico City: Editorial Limusa.

Goldberg I (1985) *Single-cell Protein*. Berlin: Springer Verlag.

Guzmán-Juarez M (1983) Yeast Protein. In: Hudson BJF (ed.) *Developments in Food Proteins 2*. P. 263. London: Elsevier Applied Science.

Harrison JS (1993) Food and Fodder Yeasts. In: Rose AH and Harrison JS (eds) *The Yeasts*, 2nd ed. Vol. 5, p. 399. London: Academic Press.

Litchfield JH (1983) Single-cell proteins. *Science* 219: 740–746.

Litchfield JH (1991) Food Supplements from Microbial Protein. In: Goldberg I and Williams R (eds) *Biotechnology and Food Ingredients*. P. 65–109. New York: Van Nostrand Reinhold.

Moo-Young M and Gregory K (1986) *Microbial Biomass Proteins*. London: Elsevier.

Reed G (1982) Microbial Biomass, Single Cell Protein, and Other Microbial Products. In: Reed G (ed.) *Prescott and Dunn's Industrial Microbiology*, 4th edn. P. 541. Westport, CT: AVI.

Reed G and Nagodawithana TW (1991) *Yeast Technology*, 2nd edn. New York: Van Nostrand Reinhold.

Rolz C (1990) Características químicas y bioquímicas de la biomasa microbiana. *Archivos Latinoamericanos de Nutrición* 40: 147–193.

Rose AH (ed.) (1979) *Microbial Biomass. Economic Microbiology*. Vol. 4. London: Academic Press.

Schlingmann M, Faust U and Scharf U (1984) Bacterial proteins. In: Hudson BJF (ed.) *Developments in Food Proteins 3*. P. 139. London: Elsevier Applied Science.

Mycelial Fungi

Poonam Nigam, Biotechnology Research Group, University of Ulster, Coleraine, Northern Ireland

The extent of shortfall in protein varies from country to country and must be considered within the framework of each national economy. The shift from grain to meat diets in developed and developing countries is of dramatic proportions and leads to a much higher per capita grain consumption, since it takes 3–10 kg of grain to produce 1 kg of meat by animal rearing and fattening programmes.

The experimental use of microbes as protein producers has been widely successful. This field of study has become known as single-cell protein or SCP production, referring to the fact that most micro-

Table 1 Time required to double the mass of various organisms

Organism	*Doubling time*
Bacteria and yeasts	20–120 min
Moulds and algae	2–6 h
Grass and some plants	1–2 weeks
Chickens	2–4 weeks
Pigs	4–6 weeks
Cattle (young)	1–2 months

Table 2 The advantages of using microbes for single-cell protein production

1. Microorganisms can grow at remarkably rapid rates under optimum conditions; some microbes can double their mass every 30–60 min
2. Microorganisms are more easily genetically modified than plants and animals; they are more amenable to large-scale screening programmes to select for higher growth rate and improved RNA content and can be more easily subjected to gene transfer technology
3. Microorganisms have a relatively high protein content and the nutritional value of the protein is good
4. Microorganisms can be grown in vast numbers in relatively small continuous fermentation processes, using a relatively small land area, and growth is independent of climate
5. Microorganisms can grow on a wide range of raw materials, including low-value agri-industrial residues and by-products

Table 3 Direct food uses of fungi

Fungal species	*Application*
	Edible macrofungi
Agaricus bisporus	Common edible mushroom
Lentinus edodes	Shiitake mushroom
Volvariella volvacea	Chinese or straw mushroom
Flammulina velutipes	Winter mushroom
Pleurotus spp.	Oyster mushroom
Tuber melanosporum	Truffle
	Cheeses
Penicillium roqueforti	Roquefort, Stilton, blue
Penicillium camemberti	Camembert, Brie, soft ripened cheeses
	Oriental food fermentations
Monascus purpurea	Ang-kak, anka koji or beni koki (red rice – culture grown on rice grains)
Aspergillus oryzae/*A. sojae*	Miso (fermented soybeans)
Aspergillus oryzae/*A. sojae*	Shoyu (soy) sauce
Rhizopus oligosporus	Tempeh or tempe kedele (fermented soybean cotyledons)

organisms used as producers grow as single or filamentous individuals rather than as complex multicellular organisms such as plants or animals.

Eating microbes may seem strange, but people have long recognized the nutritional value of the large fruiting bodies of some fungi, that is, mushrooms. Mushroom growing, because of its antiquity, can be

Table 4 Myco- and animal protein: conversion rates in protein formation

Producer	*Starting material*	*Product-protein*	*Total product*
Cow	1 kg feed	14 g	68 g beef
Pig	1 kg feed	41 g	200 g pork
Chicken	1 kg feed	49 g	240 g meat
Fusarium graminearum	1 kg carbohydrate + inorganic N	136 g	1080 g wet cell mass

considered a conventional type of food production. This article is concerned with novel processes for growing fungal mycelia which lend themselves to biotechnological processing.

Over the last two decades there has been a growing interest in using microbes for food production, in particular for feeding domesticated food-producing animals such as poultry. Use of SCP derived from low-value waste materials for animal feed may increase the food available to humans by reducing competition between humans and animals for protein-rich vegetable foods. Major companies throughout the world have long been involved in developing SCP processes, and many SCP products are now commercially available.

SCP may be used as a protein supplement, as a food additive to improve flavour and/or fat binding, or as a replacement for animal protein in the diet. Microorganisms have high DNA and RNA contents and human metabolism of nucleic acids yields excessive amounts of uric acid, which may cause kidney stones and gout. Because humans have a limited capacity to degrade nucleic acids, additional processing is required before SCP can be used in human foods. In animal feeding SCP may serve as a replacement for such traditional protein supplements as fishmeal and soya meal. The high protein levels and bland odour and taste of SCP, together with ease of storage, make SCP a potentially attractive component of manufactured foods. Also its high protein content makes it attractive for feeding farmed crustacea and fish.

Significance of Single-cell Protein

Microorganisms produce protein much more efficiently than any farm animal (**Table 1**). The yields of protein from a 250 kg cow and 250 g of microorganisms are comparable. Whereas the cow will produce 200 g of protein per day, microbes, in theory, can produce 25 tonnes in the same time under ideal growing conditions.

The advantages of using microbes for SCP production are outlined in **Table 2**.

Table 5 Maximum specific growth rates (μ_{max}) of filamentous fungi used for biomass production

Fermentation substrate	*Fungi*	*Temperature (°C)*	μ_{max} *(h^{-1})*
Cassava	*Aspergillus fumigatus* 121A	45	0.11
Carob extract	*Aspergillus niger* M1	36	ca. 0.16
Corn stover	*Chaetomium cellulolyticum*	37	>0.24
Glucose	*Fusarium graminearum*	30	0.28
Carob extract	*Fusarium moniliforme*	30	0.22
Not stated	*Fusarium* sp. M4	35	0.30
Whisky distillery spent wash	*Geotrichum candidum*	22	0.385
Sulphite liquor	*Paecilomyces variotii*	38	0.31
Milk whey	*Penicillium cyclopium*	28	0.20
Starch hydrolysate	*Penicullium notatum-chrysogenum*	30	0.20
Mung bean whey	*Rhizopus oligosporus*	32	0.16
Not stated	*Trichoderma album*	28	ca. 0.46
Coffee wastes	*Trichoderma harzianum, Rifai*	30	0.10

Table 6 Analysis of fungi for protein production (all values are in % dry weight)

Fermentation substrate	*Culture type (FP)*	*Fungi*	*Crude protein (total N × 6.25)*	*True protein nitrogen*	*Non-protein nitrogen*	*RNA*	*Lipid*	*Ash*
Cassava extract	b	*Aspergillus fumigatus* 121	40	31.5	21.3	ND	12.2	ND
Cassava	b	*Aspergillus fumigatus* 121A	37	27	27	ND	ND	ND
Ground barley	b	*Aspergillus oryzae* CMI 44242	39.4	30.2	23.4	2.5	ND	ND
Hydrolysed potato	b	*Aspergillus oryzae* NRRL 3483	40	22	45	ND	2	9
Hydrolysed potato	b	*Aspergillus oryzae* NRRL 3484	39	25	35.9	ND	2	9
Cassava starch	b	*Cephalosporium eichhorniae*	49.5	37.8	23.6	ND	ND	ND
Crop residues	c	*Chaetomium cellulolyticum*	45	ND		5	10	5
Glucose	c	*Fusarium graminearum* CMI 145425	60	42	30	10	13	6
Carob extract	c	*Fusarium moniliforme*	43	30	30	8	5	ND
Glucose	ND	*Fusarium semitectum* CMI 135410	48	34.5	28.1	3.2	ND	ND
Sulphite liquor	c	*Paecilomyces variotii*	55	ND		10	1.3	6
Hydrolysed potato	b	*Penicillium notatum-chrysogenum* CMI 138291	43	36	16.3	ND	1.6	5
Milk whey	c	*Penicillium cyclopium*	54	38	29.6	9	ND	ND
Cassava extract	b	*Rhizopus chinensis*	49	37	24.5	ND	ND	ND
Waste paper	b	*Scytalidium acidophilum*	45	36	20	6.2	2.6	3.5
Fibre board waste water	b	*Sporotrichum pulverulentum*	43	30	16.2	ND	10	6
Cassava extract	b	*Sporotrichum thermopile*	37	26	29.7	ND	6.6	ND
ND	b	*Trichoderma album*	64	54	16	4–6	6–12	6–9

FP, fermentation process; b, batch; c, continuous; ND, no data.

Choice of Mycelial Fungi in Biotechnology

Currently fungi are used for the production of secondary metabolites of medicinal and industrial importance (antibiotics, mycotoxins and fermented foods. Filamentous fungi also play a significant role in the food industry, for example, adding flavour to certain cheeses and in the production of oriental foods (**Table 3**); they are used as a major protein source in some food additives and extenders, and to improve the protein content of animal feeds.

In the examples given above the filamentous fungi, although playing an important role, are generally a minor component of the final product. However, it is possible to utilize the physical characteristics of these fungi to assemble structured food products whose sensory textures are similar to muscle tissue food products. An example of this approach is given below.

Texture and Flavour of Mycoprotein

In addition to the growth rates of organisms used for SCP, their conversion of substrate to protein is much more efficient than conversion of feed by farm animals. This is shown in **Table 4**.

The filamentous morphology of the fungi means

Table 7 Fermentative production of fungal protein

Carbon and energy source	*Fermenter*	*Microorganism*	*Temperature (°C)*	*pH*	*Specific growth rate (μ) or dilution rate (D : h⁻¹)*	*Culture density*[a] *(g l*	*Mycelial yield*[a] *g per g substrate*	
							Supplied	*Used*
Glucose, citrus press water, orange juice	1 l, 40 l bottle	*Agaricus blazei*	Ambient	3.5–7.5		15.2	30	42
Glucose	20 l fermenter	*Agaricus campestris*	25	4.5		20		
Malt syrup cane molasses			27	5.0–5.5		7.2	15.8	44.6
Carob bean extract	3000 l fermenter	*Aspergillus niger*	30–36	3.4	$\mu = 0.25$	31.5	45	
Coffee waste water	5000 gal. (18 927 l) tank	*A. oryzae*	28	4.0–4.5	$D = 0.037$			5
Glucose	300 ml flask	*Boletus edulis*	25	4.5–5.5	$\mu = 0.0017$	2.5		25
Brewery waste, grain press liquor	250 ml flask	*Calvatia gigantea*	25	6.0		6.25		24.8
Brewery waste, trub press liquor			25	6.0		27.72		74.9
Carob bean extract	14 l fermenter	*Fusarium, moniliforme*	30	5.5–6.5	$\mu = 0.18$	8.8		0.384
Corn and pea waste	37 854 l aeration pool	*Geotrichum* spp.	Ambient	3.7		0.75–1.0		
Corn and pea waste	189 270 l pool (continuous)	*Gliocladium deliquescens*	Ambient	4.6		1.2		
Soya whey	18 l fermenter		Ambient	4.6		3.2–3.5		
Glucose	30 l fermenter	*Lentinus elodes*	25	5.5	$\mu = 0.12$	7.5		26
Cheese whey, corn canning waste, sulphite liquor, pumpkin	10 l carboy, 7 l	*Morchella crassipes*	25	6.5		8.02	33.6	48.6
Canning waste	18.93 l carboy		Ambient	5.0–6.0		1.9		65
Sulphite waste liquor	18.93 l carboy	*M. deliciosa*	Ambient	6.0		10		32
Cheese whey, corn canning waste	37.85 l carboy	*M. esculenta*	25	6.5		7.85	32.8	48.1

[a] on dry weight basis; carboy-glass, plstic or metal round bottles

that the mycelial mass has a natural texture which can be used to impart a meat-like texture to the product, which may also be flavoured and coloured to resemble meat. The coarseness of the texture depends on the length of the hyphae, which can be controlled by adjusting the growth rate.

Commercial Exploitation of Mycelial Fungi

The following characteristics determine the choice of fungi as organisms to be used in a large-scale industrial fermentation process, producing a low-cost final product

- good at breaking down a wide range of complex substrates, e.g. cellulose, hemicellulose, pectin
- can tolerate low pH values, which helps in preventing contamination of the culture
- few nutritional requirements
- ease of recovery of biomass by filtration
- ease of handling and of drying the biomass
- structure conferred by hyphae allows the fabrication of textured foods.

The industrial production of SCP continues to excite attention, particularly in relation to the use of simple carbohydrates as feedstock for microbial growth and biomass production. Today, however, the economics of production have shifted the emphasis from the application of SCP to solve the problem of starvation to the production of novel foods for use in advanced economies.

Table 8 History of commercial mycoprotein

1965	The search is started for mycoprotein foods by Rank-Hovis-McDougall with ICI
1967	The microorganism used for production of mycoprotein is identified as *Fusarium graminarium*
1969	Initial work is begun into flavour and texture of mycoprotein
1975	Pilot development production facility is set up
1985	Ministry of Agriculture, Fisheries and Food acceptance in the UK
1986	Marlow Foods formed. The Quorn brand name launched. First ever mycoprotein retail product – a vegetable pie
1990	First home cooking product launched: Quorn pieces
1992	First European launch in Benelux countries
1996	Mycoprotein products launched in Switzerland
1997	Product range exceeds 50 items in UK supermarkets
1998	Expansion of product range in the UK and Europe. Development in other countries. Available in markets of UK, Belgium, Switzerland, Netherlands and Ireland

Table 9 Commercial food products made from mycoprotein which are available in UK supermarkets

Chilled products	Q pieces, Q mince, Q nuggets, Q chilled sausages, peppered Q steak, crunchy Q fillets garlic and herb, lemon and black pepper crisp crumb Q fillets, Q oriental fillets, Q steak Diane, Q lamb-flavour grills, Q en crûte, Q fillets, Q creatives Thai, Q creatives Italian, Q bolognese, Q fillets in a tomato, red wine and mushroom sauce, Q fillets in white wine and mushroom sauce, Q BBQ burger
Deli products	Roast beef-style, smoked chicken-style, honey roast ham-style, garlic sausages-style, turkey flavour with stuffing, new Q rashers
Ready meals	Q lasagne, Q tikka masala, Q oriental stir fry, Q cottage pie, Q mushroom pie, Q korma
Frozen products	Q burgers, Q quarter pounders, Q premium burgers, Q southern-style burgers, Q sausages, Q pieces, Q mince, new Q dippers, new Q lasagne, new Q chilli

Q, Quorn brand name.

Features of commercial exploitation of fungi

- rapid growth rate and high protein content compared to plants or animals
- can be produced in large amounts in a relatively small area, using biological by-products as sources of nutrient, such as the by-products from the confectionery and distillery, vegetable and wood-processing industries, although for human food application the use of food or reagent-grade nutrients is essential
- fungal cells contain carbohydrate, lipids and nucleic acids, and a favourable balance of lysine, methionine and tryptophan amino acids that plant proteins often lack.

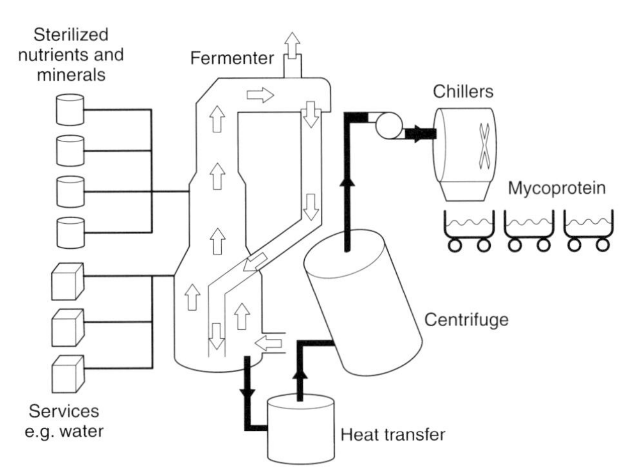

Figure 1 Schematic of the mycoprotein fermentation process.

For example, fungi can be used to improve the nutritional quality of food grains such as barley. Barley is deficient in lysine which is normally added to barley feed in the form of expensive proteins such as fish or soya meal. Fungal supplementation is achieved by adding a nitrogen source to a barley gruel and then inoculating it with an amylolytic (starch-decomposing) fungus such as *Aspergillus oryzae* or *Rhizopus arrhizus*. The barley starch is hydrolysed to glucose and the protein content increases as the fungus grows. Expensive sterilization steps are not required at any stage of the process and the product provides an ideal feed for use in pig production.

Growth Rates of Fungi

Although fungi usually grow more slowly than bacteria or yeasts, the data in **Table 5** show that, for the practical consideration of biomass production, the growth rates of fungi can be adequate.

Composition of Fungi and Nutritional Values

The nutritional value of fungal protein has been shown to be very satisfactory and compares well with protein from yeasts and bacteria. The compositions of some of the important fungi used for biomass production are shown in **Table 6**. The distinguishing feature of fungal composition lies in the distribution of the nitrogen content. Crude protein values based upon total nitrogen × 6.25 can be misleading for SCP, because of the RNA content of microbial cells and because fungi have substantial amounts of their nitrogen as *n*-acetylglucosamine in chitin of the cell wall. The protein content of the cells is approximately two-thirds of the total nitrogen, whilst RNA accounts for 15% and chitin for 10%. The chemical composition of single-cell biomass is not fixed; it varies with the limiting substrate, culture conditions, growth rate, temperature and pH. Biomass is grown for its protein content and is therefore never produced under nitrogen limitation; in consequence, the lipid content of the cells is almost invariably low, since fungal cells tend to synthesize maximum lipid content only under nitrogen limitation.

SCP Production Method

The central operation in SCP production is fermentation, for optimum conversion of substrate to microbial mass (**Table 7**). Any such operation requires the specification of the medium and the growth conditions, the design and operation of a suitable fermentation vessel and associated control systems, and the separation of the cell mass from the fermentation broth. On a commercial scale, SCP is invariably produced in submerged liquid culture. Batch or continuous culture techniques may be used. Continuous culture, which offers considerable advantages in terms of overall productivity of the fermentation processes, has been the chosen method of production in commercial SCP systems. On a commercial scale, this requires specialized plant, which is able to withstand initial sterilization procedures before each production run and which has sensors fitted into the vessel to monitor the parameters of the process.

RNA Reduction Processes

In order to meet the 1976 requirements of Food and Agricultural Organization/World Health Organization PAG (1976) to limit the ingestion of RNA from non-conventional food sources to 2 g per day, various methods of RNA reduction have been investigated:

1. Alkali extraction lowers the RNA levels of the mycelium. Treatment of proteins with alkali can lead to the formation of the dipeptide lysinoalanine which is undesirable in food materials, and care has to be exercised to prevent this occurring. It was claimed that the use of alkali extraction improved the consistency, colour and odour of Pekilo protein biomass if the alkali was neutralized with acid before washing. Care had to be taken to ensure that the pH did not fall below 6.0, as this caused RNA to be re-precipitated on to the biomass and hence increased the RNA level of recovered cells.
2. Endogenous enzymic hydrolysis reduces the RNA levels from 9% to less than 2%.
3. Heat shock at 64°C inactivates the fungal protease and allows the endogenous RNases to hydrolyse the disrupted ribosomal RNA.

Recovery of Biomass from Culture Broths

One of the major advantages possessed by the filamentous fungi over single-celled organisms is the ease with which the former can be separated from the culture medium. On a small scale, filtration using filter paper and Buchner funnels is usually adequate. For larger volumes, a low-speed, perforated-bowl centrifuge gives good results. On a large scale, rotary vacuum filters are the method of choice; nylon filter cloths of suitable retentivity can normally recover >99.9% of biomass mycelium and provision may be made for spray washing as part of the filtration operation; biomass removal is done by scraper blade. With a vacuum of 60–65 cmHg, filtration rates of around 70–80 $kg\,m^{-2}\,h^{-1}$ from a medium containing 20% total solids are achievable. In order to reduce subsequent costs of drying if required, various de-watering equipment can further reduce the 80% water content of filter cake. Continuous screw expellers of the type used in the brewing industry for de-watering spent grains can be used. High-volume throughput necessitates continuous equipment. In the Pekilo process mechanical de-watering produces a material of 35–45% total solids.

Drying

Fungal biomass is easy to dry since its structure does not tend to collapse and lead to case hardening, as does bacterial biomass. Using a continuous band drier with single-pass warm-air down-flow, an air temperature of 75°C is optimal for drying *Penicillium* mycelium ex-vacuum filter at 20% solids; a residence time of 20–30 min produces a product of 8–10% moisture. Heating at too high a temperature reduces the nutritional value of the product because of alteration in lysine availability. Other forms of simple driers such as rotary drum driers are also applicable.

General Product Specifications for Single-cell Protein as Human Food

Important aspects of the product quality of single-cell protein include:

- nutritional value
- safety
- production of functional protein concentrates.

Digestibility (*D*)

D is the percentage of total nitrogen consumed that is absorbed from the alimentary tract. The total quantity of microbial protein ingested by animals is measured and the nitrogen content (*I*) is analysed. Over the same period, faeces and urine are collected, and faecal nitrogen content (*F*) and urinary nitrogen content (*U*) are measured. Thus:

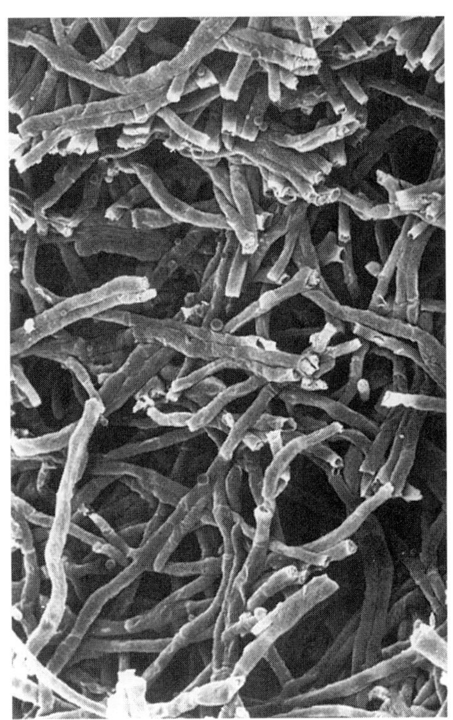

Figure 2 Micrograph of mycoprotein illustrating its filamentous nature.

$$D = \frac{I - F}{U} \times 100$$

Biological Value (*BV*)

BV is the percentage of total nitrogen assimilated that is retained by the body, taking into account the simultaneous loss of endogenous nitrogen through urinary excretion. Thus

$$BV = \frac{I - (F + U)}{I - F} \times 100$$

Protein Efficiency Ratio (PER)

PER is the proportion of nitrogen retained by animals fed the test protein compared with that retained when a reference protein, such as egg albumin, is fed.

Preservation of Mycoprotein

Preservation is by freezing or chill storage.

Testing mycoprotein for nutritive value and safety has been extensive and in 1985 resulted in permission being granted by the Ministry of Agriculture, Fisheries and Food for free sale of mycoprotein in the UK.

Commercial Production of Mycelial Protein

Pekilo Process

Pekilo is a fungal protein product produced by fermentation of carbohydrates derived from spent sulphite liquor, molasses, whey, waste fruits and wood or agricultural hydrolysates. It has a good amino acid composition and is rich in vitamins. Extensive animal feeding test programmes showed that Pekilo protein is a good protein source in the diet of pigs, calves, broilers, chickens and laying hens. Pekilo protein is produced by a continuous fermentation process. The organism, *Paecilomyces variotii*, a filamentous fungus, gives a good fibrous structure to the final product. The first plant was installed at the Jamsankoski pulp mill in central Finland in 1973. As an animal feed component, Pekilo protein is comparable to fodder yeast, which is also produced by fermenting spent sulphite liquor.

Mycoprotein Production

In the UK Rank-Hovis-McDougall, in conjunction with ICI (in 1993 ICI de-merged into Zeneca and became the new ICI) commercially marketed another fungal protein, mycoprotein (Quorn), derived from the growth of a *Fusarium* fungus on simple food-grade carbohydrates. Unlike almost all other forms of SCP, mycoprotein is produced for human consumption.

Technical Development of Mycoprotein and Quorn

Marlow Foods, based in the UK, is involved in the development, production and marketing of a range of Quorn consumer food products made with mycoprotein. Marlow Foods is a subsidiary of Zeneca Group PLC.

Mycoprotein is the generic name of the major raw material used in the manufacture of Quorn products. It comprises RNA reduced cells of the *Fusarium* species (Schwabe) ATCC 20334, grown under axenic conditions in a continuous fermentation process. **Table 8** shows the history of the development of commercial Quorn mycoprotein.

Quorn, the brand name of a range of meat-alternative products using mycoprotein as the principal component (**Table 9**), is registered to Marlow Foods. These products are sold throughout the UK and increasingly in western Europe.

Quorn products are a good source of protein, are lower in calorific value (89 kcal per 100 g) and have a higher dietary fibre content than their natural meat equivalents. They contain no animal fats or cholesterol and have a high level of dietary fibre. They can be eaten by anyone, although they are not rec-

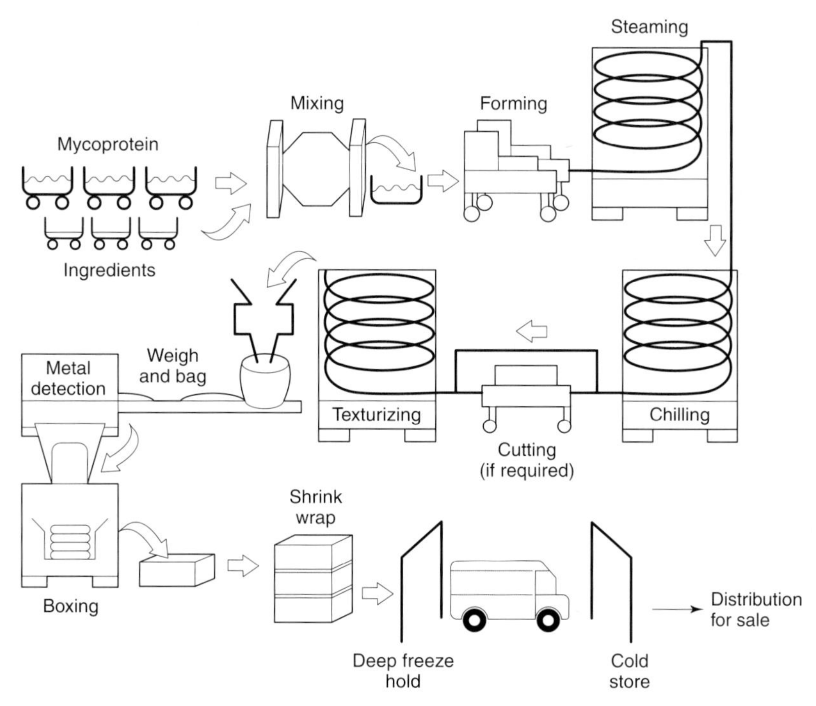

Figure 3 Schematic of the manufacturing process for Quorn products.

Table 10 Nutrition comparison: mycoprotein compared with other sources of dietary protein

Food type	*Measure* (g)	*Energy* (kcal)	*Protein* (g)	*Total carbohydrate* (g)	*Dietary fibre* (g)	*Total fat* (g)	*Fat breakdown (g)* Saturated	Mono	Poly	*Cholesterol* (mg)
Mycoprotein	85	72	9.4	7.7	5.1	2.6	0.6	0.4	1.6	0
Cheddar cheese	30	120	7.5	0.38	0	10	6.3	2.8	0.28	32
Whole eggs	50	75	6.3	0.6	0	5	1.6	1.9	0.68	212
Ground beef: regular, medium baked	85	245	19.6	0	0	17.8	7	7.8	0.66	74
Chicken: light meat, roasted	85	130	23.1	0	0	3.5	0.9	1.3	0.8	64
Fish: cod raw	85	89	19.4	0	0	0.7	0.1	0.1	0.25	47
Soy flour: raw, full fat	30	131	10.4	10.6	2.9	6.2	0.9	1.4	3.5	0

ommended for very young children because of their low energy density. Quorn products have a tender texture similar to that of lean meat. This makes them attractive to vegetarians who miss the taste of meat, as well as to consumers who are reducing their red meat intake.

The cells are grown by continuous aerobic fermentation (**Fig. 1**) for periods up to 1000 h of continuous operation. The plant is sterilized between operating runs by the use of steam under pressure.

The substrate, glucose syrup, and all the nutrients added to the fermenter are sterile and of food or reagent-grade quality. The water is purified before it is used in the fermenter. The pH in the fermenter is controlled by the injection of ammonia, which also provides part of the nitrogen source for the cells. A continuous spill from the fermenter carries away the biomass produced: the flow rate is such that the volume of the fermenter is displaced every 5–6 h. The cell suspension is then taken to a continuously stirred tank reactor held at approximately 65°C, to reduce the RNA content (dry weight) from 10% to less than

Table 11 Fatty acid profile of mycoprotein (fat content = 3 g per 100 g)

Fatty acid	*Grams per 100 g fat in mycoprotein*	*Grams per 100 g mycoprotein*
C16: Palmitic	9.3	0.3
C18:0 Stearic	2.0	0.1
C18:1 Oleic	9.6	0.3
C18:2 Linoleic	29.8	1.0
C18:3 α-Linolenic	13.5	0.4

Table 12 Protein, fat and calorie content of selected commercial mycoprotein food products compared to meat equivalents

Food (per 85 g cooked portion)	*Protein (g)*	*Total fat (g)*	*Saturated fat (g)*	*Calories (kcal)*	*Energy density (kcal)*
Q Pieces	10	2.5	0.5	70	0.9
Ground beef: regular (medium baked)	20	18	7	240	2.9
Ground beef: lean (medium baked)	20	16	6	230	2.7
Ground beef: extra lean (medium baked)	21	14	5	210	2.5
Q Burger	11	4	2	100	1.2
Ground beef patty (medium broiled)	21	17	7	240	2.8
Q Southern-style burger	9	7	1	150	1.8
Breaded chicken patties	14	20	4	280	3.3
Q Southern-style nuggets	10	6	1	150	1.8
Breaded chicken nuggets	14	20	4	280	3.3
Q Sausage	12	4	2.5	110	1.3
Pork sausage	17	27	9	310	3.7
Turkey sausage (smoked)	12	8	2.5	140	1.6

Q, Quorn.

Table 13 Essential amino acid content of mycoprotein compared to other foods that contain protein (g amino acids per 100 g edible portion)

Essential amino acids	*Mycoprotein*	*Cow's milk*[a]	*Egg*[b]	*Beef*[c]	*Soybeans*[d]	*Peanuts*[e]	*Wheat*[f]
Histidine	0.39	0.09	0.3	0.66	0.98	0.65	0.32
Isoleucine	0.57	0.20	0.68	0.87	1.77	0.91	0.53
Leucine	0.95	0.32	1.1	1.53	2.97	1.67	0.93
Lysine	0.91	0.26	0.90	1.6	2.4	0.92	0.30
Methionine	0.23	0.08	0.39	0.5	0.49	0.32	0.22
Phenylalanine	0.54	0.16	0.66	0.76	1.91	1.3	0.68
Tryptophan	0.18	0.05	0.16	0.22	0.53	0.25	0.18
Threonine	0.61	0.15	0.6	0.84	1.59	0.88	0.37
Valine	0.6	0.22	0.76	0.94	1.82	1.08	0.59

[a] Whole fluid milk (3.3% fat).
[b] Raw fresh egg.
[c] Ground beef (regular, medium baked).
[d] Mature raw soybeans.
[e] Raw peanuts (all types).
[f] Durum wheat.
Source: United States Department of Agriculture Nutrient Data Base for Standard Reference, March 12, 1998.

2%. The suspension is then heated to 90°C, and dewatered by centrifugation before being cooled. The harvested cells, collectively known as mycoprotein, are paste-like in consistency and contain around 75% moisture.

Commercial Production of Mycoprotein Food Products

The harvested cells have a similar morphology to animal muscle cells – they are filamentous with a high length/diameter ratio, length 400–700 μm, diameter 3–5 μm and branch frequency 1 per 250–300 μm (**Fig. 2**). The product assembly process seeks to reproduce the structural organization which exists in natural meats.

In meat, muscle cells are held together by connective tissue. To establish a similar product texture in Quorn products, the cells are mixed with a protein binder, together with flavouring and other ingredients, depending on the final product format, and then heated (**Fig. 3**). This causes the protein binder to gel and bind the cells together. The resultant structure is very meat-like in appearance, texture and formed products such as steaks or fillets.

Developing Texture

The ingredients added differ according to the product produced. The mixture is transported to a forming machine, set up for the product being produced. For Quorn pieces and mince the forming machine produces strips of Quorn which are then reduced in size. The next step is to steam the mixture. The high temperatures reached during steaming affect gelation of the albumen, which is added at the mixing stage; this in turn improves the texture of the product by increasing firmness. After steaming, the products are

Table 14 Protein digestibility corrected amino acid score (PDCAAS) of selected food proteins

Protein source	*PDCAAS*	*Data source*
Quorn pieces	1.0	d
Casein	1.0	a
Egg white	1.0	a
Chicken (light meat: roasted)	1.0	c
Turkey (minced: cooked)	0.97	c
Fish (cod: dry cooked)	0.96	c
Soybean protein	0.94	b
Beef	0.92	a
Mycoprotein	0.91	d
Pea flour	0.69	a
Kidney beans (canned)	0.68	a
Rolled oats	0.57	a
Lentils (canned)	0.52	a
Peanut meal	0.52	a
Whole wheat	0.40	a
Wheat gluten	0.25	a

a: Sources: [a] Food and Agricultural Organization/World Health Organization Joint Report (1989).
b: Sarwar and McDonough (1990).
c: Calculated from amino acid data in United States Department of Agriculture Data Base for Standard Reference, March 12, 1998 (assumes a digestibility equivalent to beef = 94%).
d: Calculated from Marlow Foods data.

cooled rapidly, before weighing, packing and storage.

Characteristics of Mycoprotein Products

Good Source of Dietary Fibre

Since Quorn food products contain 60–90% of mycoprotein there will be a corresponding significant dietary fibre content in the final product. Mycoprotein contains 5.1 g dietary fibre per 85 g (**Table 10**). Fibre comprises 65% β-glucans and 35% chitin and, of this, 88% is insoluble and 12% soluble.

No Interference with Mineral Absorption

Quorn products do not contain phytic acid or phytic salts that may interfere with mineral absorption. No significant effect on the absorption of calcium, magnesium, phosphorus, zinc or iron has been shown in comparison with a polysaccharide-free diet.

Lower in Fat, Saturated Fat and Cholesterol than Equivalent Meat Products

Mycoprotein contains only 2.6 g fat and 0.5 g saturated fat per 85 g (Table 10). It does not contain any *trans* fatty acids (**Table 11**). Quorn products, however, may contain slightly higher levels of fat and some *trans* fatty acids, as small amounts of fat may be added to enhance the taste and texture. These products contain between 2.4 and 8.4 g fat and 0.4–2.4 g saturated fat per 85 g cooked weight. In comparison, equivalent meat products contain 2.5–3.5 times more fat and saturated fat.

Rich in Protein Content

Quorn products typically contain between 10 and 13 g of protein per 85 g serving, most of which comes from mycoprotein. Small amounts come from egg albumen and milk proteins which are added in the manufacturing process. The nutritional advantages of Quorn products (**Table 12**) include the fact that they are excellent sources of high-quality protein but are significantly lower in fat, saturated fat and calories than many protein foods.

High-quality Protein

Dietary proteins contain a mixture of 20 amino acids, all of which are necessary to support growth. Although most amino acids can be made in the body, nine essential amino acids must be supplied by the diet: histidine, isoleucine, leucine, lysine, methionine, phenylalanine, threonine, tryptophan and valine. The quality of a dietary protein is based on its content of essential amino acids. **Table 13** compares the amino acid content of mycoprotein with other commonly consumed protein foods.

The PER for mycoprotein is 2.4, *BV* 84 and *D* 78. A recent development in the US required by the Food and Drug Administration is that the Protein Digestibility-corrected Amino Acid Scoring (PDCAAS) method must be used for most nutrition labelling purposes. This method takes into account the food protein's essential amino acid profile, its digestibility and its ability to supply essential amino acids in amounts required by humans. It compares the essential amino acid profile of a food, corrected for digestibility, to the Food and Agricultural Organization/World Health Organization 2–5-year-old essential amino acid requirement pattern. The 2–5-year-old pattern is used because it is the most demanding pattern of any age group other than infants.

The PDCAAS for mycoprotein is 0.91, based on a digestibility factor of 78% for mycoprotein. **Table 14** shows how mycoprotein compares to the PDCAAS of other food proteins.

Mineral and Vitamin Composition

Mycoprotein used in Quorn products compares well in mineral and vitamin composition with other sources of dietary protein (**Table 15**). However, the PDCAAS value for Quorn products (which all contain egg albumin) is 1.

See also: ***Aspergillus***: *Aspergillus oryzae*. **Fermentation (Industrial)**: Production of Oils and Fatty Acids. **Mycotoxins**: Classification. **Single Cell Protein**: Mycelial Fungi.

Table 15 Vitamin and mineral comparison: mycoprotein compared to other sources of dietary protein

Food type	*Measure (g)*	*Ca (mg)*	*Fe (mg)*	*Mg (mg)*	*P (mg)*	*K (mg)*	*Na (mg)*	*Zn (mg)*	*Vitamin A (μg)*	*Thiamin (mg)*	*Riboflavin (mg)*	*Niacin (mg)*	*Vitamin B_6 (mg)*	*Folic acid (mg)*	*Vitamin C (mg)*
Mycoprotein	85	38	0.5	38	215	85	4	8.5	0	0.01	0.2	0.3	0.1	0.1	0
Cheddar cheese	30	216	0.2	8.3	154	30	186	0.9	83RE	0.01	0.1	0.02	0.02	5.5	0
Whole eggs	50	25	0.7	5	89	61	63	0.6	95RE	0.03	0.25	0.04	0.07	0.24	0
Ground beef: regular medium baked	85	8.5	2.1	12.8	116.5	188	51	4.2	0	0.03	0.136	4	0.2	7.7	0
Chicken: light meat, roasted	85	11	0.92	19.6	185	201	43	0.66	7RE	0.05	0.08	8.9	0.46	2.6	0
Fish: cod raw	85	11.9	0.42	35.7	117	207	66	0.49	12RE	0.08	0.07	2.1	0.24	6.9	0.9
Soy flour: raw, full fat	30	62	1.9	129	148	755	4	1.0	4RE	17	0.35	1.3	0.14	104	0

RE, retinal equivalent

Further Reading

Anke T (ed.) (1997) *Fungal Biotechnology*. London: Chapman & Hall.

Atkinson B and Mavituna F (1991) *Biochemical Engineering and Biotechnology Handbook*, 2nd edn. New York: Macmillan.

Higgins IJ, Best DJ and Jones J (eds) (1988) *Biotechnology Principles and Applications*. Oxford: Blackwell Scientific Publications.

PAG ad hoc (1976) Working group meeting on clinical evaluation and acceptable nucleic acid levels of SCP for human consumption. PAG Bulletin. Vol. 5(3), pp. 17–26.

Sarwar G and McDonough FEC (1990) Journal of the Association of Official Analytical Chemists. Vol. 73, pp. 347–356.

Smith JE (1996) *Biotechnology*, 3rd edn. Cambridge: Cambridge University Press.

Solomon GL (1985) Production of filamentous fungi. In: Moo-Young M (ed.) *Comprehensive Biotechnology*. Vol. 3. Oxford: Pergamon Press.

Wainwright M (1992) Fungi in the food industry. In: *An Introduction to Fungal Biotechnology*. Chichester: John Wiley.

Sodium Chloride *see* **Preservatives**: Traditional Preservatives – Sodium Chloride.

Sorbic Acid *see* **Preservatives**: Permitted Preservatives – Sorbic Acid.

Sorghum *see* **Fermented Foods**: Beverages from Sorghum and Millet.

Sourdough Bread *see* **Bread**: Sourbough Bread.

Sous-Vide Products *see* **Microbiology of Sous-Vide Products**.

Spices *see* **Preservatives**: Traditional Preservatives – Oils and Spices.

Spiral Plater *see* **Total Viable Counts**: Specific Techniques.

SPOILAGE OF PLANT PRODUCTS

Cereals and Cereal Flours

D R Twiddy, Consultant Microbiologist, Horsham, West Sussex, UK, Formerly of the Natural Resources Institute.

P Wareing, Natural Resources Institute, Chatham Maritime, Kent, UK

Introduction

Cereals include wheat, oats, rye, barley, rice, maize, sorghum and millet. Worldwide, these are essential crops for human nutrition – rice is the staple food of more than 50% of the world's population. Cereals and cereal flours are used extensively in brewing, bread-making, animal feeds, pasta, biscuits, cakes and a wide variety of snack foods. Cereals can be stored very successfully if kept dry, but if they are inadequately dried before storage they are highly perishable.

Microflora

Cereals can be contaminated by a wide range of pathogenic and spoilage microorganisms during growth, harvest and storage. At the time of harvest, they typically harbour tens to hundreds of thousands of fungal propagules (e.g. spores and mycelia) and thousands to millions of bacteria per gram. Harvested grains retain part of their natural flora, and also become contaminated from the air, soil and insects. The microflora consists of a very wide variety of filamentous fungi, bacteria and yeasts, and some slime moulds and protozoa. However, many of these occur only as surface contaminants. Most microbial biodeterioration problems of cereals are caused by fungi. Although yeasts are numerous, they cause few spoilage problems.

The bacteria most commonly isolated from cereal grains are members of the Bacillaceae, Micrococcaceae, Lactobacillaceae and Pseudomonadaceae. Many of the bacterial human pathogens present are environmental contaminants, but the numbers of faecal organisms are low unless animal manure has been used during cultivation, as an organic fertilizer. Bacteria are not greatly involved in grain spoilage except in the case of very damp grain, in which they are particularly active during the final stages of moist grain heating. Moist grain heating occurs when grain is put into store with too high a moisture content for safe storage, or when it becomes wet during storage. A succession of fungi may then grow, producing metabolic heat and water, which eventually allow thermophilic bacteria to grow. Bacteria are generally unable to grow in dry grain due to its low water activity (a_w). However, they may persist in cereal flours even after milling. *Bacillus* species in flours can cause problems during bread-making, and *Salmonella* species in infant formulae containing cereal flours can be a risk to babies.

The fungi associated with cereal grains can be divided into two groups: the field fungi – plant pathogens which invade the grains before harvest; and the storage fungi, which invade the grains during drying and subsequent storage.

Spoilage Before Harvest

In the field, fungal invasion is most prevalent during humid and rainy weather. Field fungi invade the kernels of the grain while they are developing on the plant or after they have matured, but generally before harvesting. These fungi may cause blemishes, blights, discolorations and diseases of the kernels and, some also cause disease in plants grown from infected seed.

The principal invasive field fungi include members of the genera *Alternaria*, *Fusarium*, *Drechslera*, *Cladosporium*, *Botrytis* and *Phoma*, but many more genera are encountered. *Alternaria* species are particularly common: they are almost always present on wheat and barley kernels, where they cause blights and blemishes. Pre-harvest, rice carries a wide variety of mould species of the genera *Alternaria*, *Curvularia*, *Fusarium*, *Nigrospora*, *Chaetomium*, *Bipolaris*, *Acremonium*, *Aspergillus*, *Penicillium*, *Rhizopus* and *Trichoconiella*, and also bacteria of the genera *Pseudomonas*, *Enterobacter* and *Micrococcus*. Loose and covered smuts of wheat, barley and oats (caused by *Ustilago tritici*, *U. nuda*, *U. avenae*, *U. lordei*, *Tilletia caries* and *T. foetida*) may result in heavy crop losses pre-harvest.

The field fungi require equilibrium relative humidities (ERH) of 90–100% for growth (equivalent to >20% grain moisture content on a wet weight basis, or 28–33% on a dry weight basis). As the cereal grains mature, their moisture content decreases. Provided they are dried within a few days of harvest, significant mould growth will not normally occur. However, if the grain remains moist for a long period after harvest, mould growth will continue, discolouring the kernel, weakening or killing the embryo, and shrivelling the seed.

The main genera producing mycotoxins pre-harvest are *Alternaria* and *Fusarium*. However, *Aspergillus flavus*, which is normally considered as a storage fungus, can grow before harvest on maize cobs that have been damaged by insects, and it has the ability to produce aflatoxins under these conditions. Field fungi die slowly during storage as the relative humidity falls, so the damage they cause does not usually increase during storage. Grains which have been heavily invaded by field fungi have been shown to have increased resistance to the growth of other fungi during storage. Storage fungi, with a few exceptions (e.g. *Aspergillus flavus*, which can grow on corn before harvest), do not invade seeds before harvest to any significant extent.

Spoilage During Storage

During storage, the field fungi on grain are replaced by fungi capable of growing at a lower a_w. The two main genera causing spoilage during storage are *Aspergillus* and *Penicillium*. Many *Aspergillus* species are adapted to growth at low a_w, and some of the *Penicillium* species can grow at low temperatures, albeit at higher a_w. The types of fungi that can be isolated are indicative of the storage conditions of the grain. *A. penicillioides* predominates in grain at a_w of 0.68–0.75, causing blackening of the germs. The condition known as 'sick wheat' is due to *A. penicillioides* damaging the germs of grain which has a moisture content of 14–14.5% (grain ERH 68–73%) and which is being stored for several months. The germs are killed before fungal growth is apparent to the naked eye. At higher moisture contents, *A. penicillioides* can not compete with *Eurotium* species.

The fungi consistently associated with incipient deterioration are *A. penicillioides*, *A. restrictus* and *Eurotium* species. In grain with a_w less than 0.78–0.80, only *A. penicillioides* and *Eurotium* species can grow to any extent. However, with a_w above 0.80, a succession of fungi occurs. *A. penicillioides* and *Eurotium* species appear first, and these may be followed by *A. candidus*, *A. flavus*, *A. ochraceus*, *A. versicolor*, and *Penicillium* species. Each species has a lower limit of moisture level, below which it cannot grow (**Table 1**). *Rhizopus* and *Mucor* species require a higher a_w of 0.93–1.00.

Mites and insects, such as *Sitophilus granarius*, the granary weevil, can physically damage grain before and during storage and facilitate invasion by fungi. They also carry large numbers of fungal spores on their bodies, and help to spread infection.

Types of Spoilage

Spoilage will occur if the overall moisture content of the grain is sufficiently high to support fungal growth. However, moisture migration can occur within grain in bulk, due to temperature gradients or the activity of insects, and so apparently dry grain, with a moisture content as low as 10–12%, can be spoiled. This is particularly important in the tropics, where wide seasonal temperature changes occur. Temperature differences within the bulk grain of only 0.5–1.0°C can initiate storage problems, because moisture evaporates from the grain in a warmer section and condenses in a cooler section, increasing the moisture content to a level high enough to permit the growth of moulds.

The first fungus to grow is most commonly *A. penicillioides*, followed by *A. restrictus* and *Eurotium* species. These can raise the temperature of the grain to 35–40°C. Grain respiration and moisture generated locally by mould growth raises the a_w, permitting *A. candidus*, *A. ochraceus* and *A. flavus* to grow. When these moulds are growing on about 10% of the kernels in the moist mass, considerable heating occurs and the temperature may reach 50–55°C. *Penicillium* species, which require a relatively high a_w for growth, may also be present at this stage. The temperature may subside if the heat and moisture are allowed to dissipate, but thermophilic fungi (e.g. *Humicola lanuginosa*, *Thermoascus crustaceus*, *T. aurantiacus*) may take over and raise the temperature even more, to 60–65°C. Thermophilic bacteria then start to grow, and may raise the temperature to 75°C. At this point chemical processes take over, sometimes raising the temperature to a level which supports spontaneous combustion, but this phenomenon is common only in the case of commodities with a high oil content, such as soybeans, and rarely occurs in starchy cereals.

The invasion of cereal grains and their products by fungi can cause a number of undesirable outcomes, which are considered below. These are: loss of seed viability; loss of nutritional value; production of mycotoxins; deterioration of grain quality; and grain heating. The relative importance of the various kinds of damage is determined by the buyer or user of the grain.

Loss of Seed Viability and Germinability

The death of grain embryos frequently occurs before any visible discoloration, because the growth of fungi on cereals in their husks is often difficult to see until it is at an advanced stage. Fungal growth starts near the embryo of the grain and spreads from there to the endosperm. Many common storage fungi, such as *A. candidus* and *A. flavus*, rapidly decrease the germinability of grain because they preferentially invade the embryo of the seed.

Table 1 Approximate minimum, optimum and maximum temperatures and minimum water activity for the growth of common storage fungi on grain

Species of fungus	*Minimum temperature (°C)*	*Optimum temperature (°C)*	*Maximum temperature (°C)*	*Minimum water activity for growth*
Aspergillus restrictus	< 15	25–30	> 37	0.75
A. penicillioides	9	30	40	0.68–0.73[a]
Eurotium spp.	4–5[b]	25–27[b]	40–46[b]	0.69–0.71[b]
A. candidus	10–15	45–50	50–55	0.75
A. flavus	10–12	33	43–48	0.80
A. versicolor	9	25	35–40	0.76–0.80
Penicillium spp.	– 5–10[b]	20–31[b]	30–40[b]	0.79–0.99[b]

[a] Depending on the substrate, and species.
[b] Depending on the species.

Loss of Nutritional Value

Spoilage may render a commodity unfit for human or animal consumption. Losses of up to one third of the food grains stored every year have been reported in some tropical countries.

Production of Mycotoxins

Mould spoilage is undesirable from an organoleptic point of view, but more importantly, many fungi produce a variety of mycotoxins in cereals and cereal products which, if consumed, are potential hazards.

Deterioration in Grain Quality

Deterioration may take the form of visual discoloration, grain caking, loss of dry matter, off odours and other organoleptic changes. These represent advanced stages of grain spoilage. Direct loss results when, due to extensive visible mould growth, the product is recognized as being too mouldy to eat. Considerable fungal growth occurs before these stages are reached. Changes in colour, texture or flavour render the product unacceptable, and grain colour and odour are the principal factors used by traders for grading grain. For example, *Fusarium* species growing on maize cause a characteristic red streaking of the kernel, and 'blue eye' of maize is caused by the development of spores of *A. penicillioides* or *Eurotium* or *Penicillium* species. The caking of cereals (**Fig. 1**) can cause serious problems in silos and during the transport of grain, due to the blocking of transfer pipes and conveyor belts.

Grain Heating

This final stage of spoilage results in total or extreme loss. Heating of the grain by fungal activity can lead to considerable discoloration and blackening of the kernels, the loss of nutritional value and a reduction of germinability. 'Stack-burn' of maize has been reported in Africa, due to heating, sometimes caused by the growth of thermophilic fungi, to more than 43°C over a period of around 100 days. This results in brown discoloration of the pericarp and embryo of the stored maize, along with nutritional changes. The growth of mould on harvested rice can cause heating of the grain and the development of a yellow discoloration.

The invasion of germs by storage fungi often leads to brown or black discoloration, which can be mistaken for 'heat damage'. Although such damage is often associated with high levels of invasion by storage fungi, there is often no correlation between the extent of the damage and the temperature of the grain. For example, blackened 'heat-damaged' grain can be found even at storage temperatures as low as 5°C.

Control of Mould Growth

In the humid tropics, and in temperate regions when the harvest months are cool and damp, mould growth can be a serious problem. Major post-harvest losses are sustained worldwide due to the fungal spoilage of stored grains and seeds. Therefore, the control of fungal growth is important.

Moulds are capable of growing on a wide range of foodstuffs and other materials. Cereals typically contain in the region of 70–75% carbohydrate, 8–15% protein and smaller quantities of fat, fibre, vitamins and minerals, and are ideal substrates for fungal growth provided that sufficient water is also available. Temperature, O_2 and a_w are the most important factors influencing mould growth. Microbial activity on grain can be controlled by drying the crop to a 'safe' a_w, i.e. below 0.70 (**Table 2**). At this a_w, bacteria are rarely able to grow and moulds can grow only to a limited degree. This represents the maximum a_w to which the crop must be dried to enable satisfactory storage of up to 2 years, at a temperature of 21–27°C.

Cereals should be harvested at maturity, because grain harvested prematurely takes longer to dry. Cereals should be dried as quickly as possible, and protected at all stages from external moisture. Physical damage during harvesting and processing should

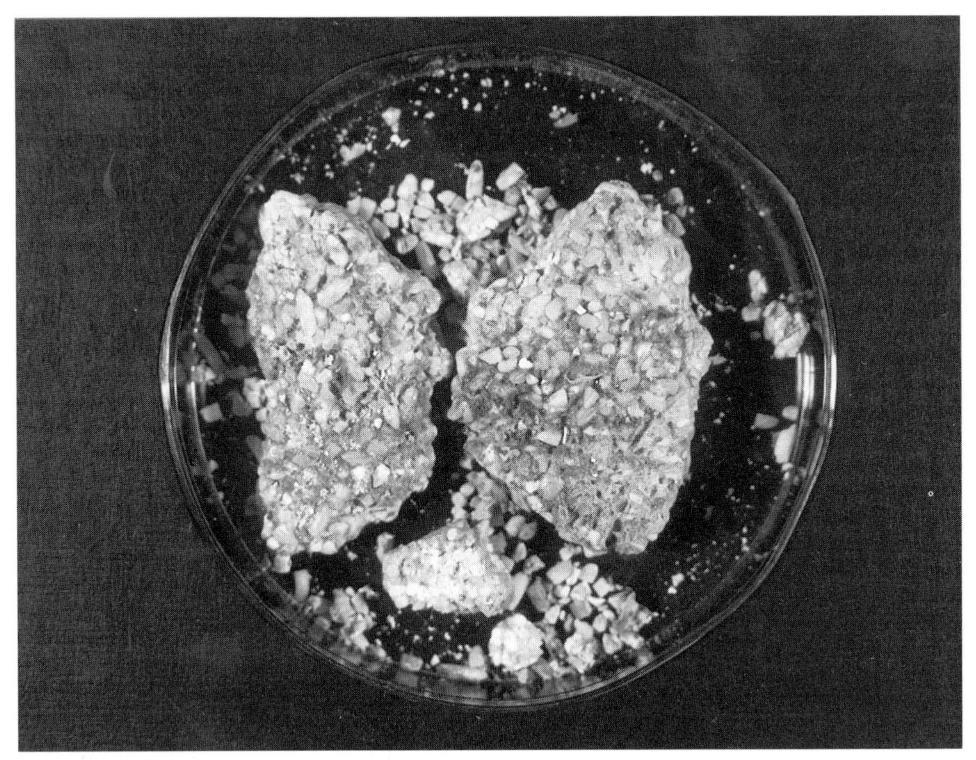

Figure 1 Severely caked milled rice.

Table 2 Equilibrium moisture content of barley, wheat, oats, rye, millet, sorghum and maize at different relative humidities at 25–30°C.

Moisture content (%)	*Relative humidity (%)*	*Corresponding water activity*
12.5–13.5	65	0.65
13.5–14.5	70	0.70
14.5–15.5	75	0.75
15.5–16.5	80	0.80
18.0–20.0	85–90	0.85–0.90

be avoided, because the natural physical barriers of the grain prevent rapid fungal invasion. Some grains, particularly maize, which is harvested with a high moisture content, must be artificially dried to prevent spoilage. High-temperature grain dryers can be used to rapidly reduce the moisture content by 5–10%.

The optimum temperature for the growth of many fungi is around 30°C, a common ambient temperature in tropical regions (see Table 1). The refrigeration of grains to about 5°C permits safe storage for prolonged periods, even when the water content of the grain is too high for safe storage at ambient temperatures. However, some species of *Penicillium*, e.g. *P. aurantiogriseum*, can grow slowly at temperatures as low as −2°C, and *P. expansum* can grow at −6°C – although they do require a relatively high moisture content.

Artificial aeration, involving the blowing or drawing of ambient air through the grain, reduces the temperature of the grain to 5–10°C – a temperature at which storage moulds grow slowly, and insects and mites are dormant. This results in a uniform temperature being maintained throughout the grain, which helps to prevent the transfer of moisture within the bulk grain and adds to its storage life. In cold climates, aeration is used to achieve a maximum moisture level of about 17% in the grain. In tropical climates, where the air is much warmer, the grain should not have a moisture content of >12.5% for long-term storage.

The storage of grain in controlled-atmosphere silos has been used to slow mould growth in the grain. Moulds are primarily aerobes, but some can grow in anaerobic conditions. An atmosphere of 20% CO_2 will inhibit the growth of some storage fungi, but higher percentages of CO_2 may be required for grain with a high moisture content, and some storage fungi will grow in atmospheres containing more than 80% CO_2 and less than 0.2% O_2. The depletion of O_2 in controlled atmosphere storage helps to reduce the activity of insects, before they become numerous enough to cause serious damage.

The fumigation of grain stores helps to reduce mould growth on cereal grains by killing the insects that damage the kernels. However, the effects of fumigants are not long-lasting. Methyl bromide destroys both moulds and insects, and phosphine at low levels ($0.1\ g\ m^{-3}$) can retard the development of storage fungi

Table 3 Mycotoxins produced in cereals

Mould species	*Mycotoxin produced*
Aspergillus parasiticus	Aflatoxins B_1, B_2, G_1, G_2
A. flavus	Aflatoxins B_1, B_2
Fusarium sporotrichioides	Trichothecenes: T-2 toxin, HT-2 toxin, neosolaniol, diacetoxyscirpenol
F. graminearum	Deoxynivalenol, nivalenol, zearalenone
F. moniliforme	Fumonisins, fusarin C
Penicillium islandicum	Islanditoxin, luteoskyrin
P. citreoviride	Citreoviridin
P. citrinum	Citrinin
Claviceps purpurea	Ergotamine
P. verrucosum, *A. ochraceus*	Ochratoxin A

and limit mycotoxin production. However, there are concerns about the toxicity of many of these fumigants, and methyl bromide is known to deplete ozone from the upper atmosphere.

The efficacy of fungicides depends on their dissociation in water, so under dry conditions the compounds may not be fungicidal. Dust fungicides, used for the protection of seeds against damping off, may not inhibit storage fungi – and concentrations which do so may also kill the seeds. Most fungicides are of little value, because of many factors including excessive cost, toxicity to humans and animals, effects on the suitability of the grain for processing, and their difficulty of application. However, systemic fungicidal seed treatments do control smuts of cereals effectively.

Propionic acid, and combinations of propionic and acetic acids, are effective preservatives of grain with a high moisture content destined for use as animal feedstuff, but the odour and flavour they impart make them unacceptable for use with foodstuffs for human consumption. Most of the chemicals used to preserve damp grain are fungistatic rather than fungicidal.

Mycotoxins

Mycotoxins are toxic secondary metabolites produced by some fungi. They can be produced at any stage during the processing or storage of cereal crops, provided that conditions are favourable for fungal growth. The principal mycotoxins in nature are found in cereal products and oilseeds, and they may be present in a food long after the moulds responsible for their production have died. The genera of fungi mainly associated with naturally occurring toxins are *Aspergillus*, *Penicillium* and *Fusarium*. Some examples of mycotoxins occurring in cereals are listed in **Table 3**.

Table 4 Visual characteristics of some storage and field fungi associated with cereal grains

Fungus	*Colour of sporing heads*	*Typical discoloration of grain*
Penicillium spp.	Dark blue-green	Blue coloration 'Blue eye' of maize
Aspergillus fumigatus	Dark blue-green	Discoloured germs
A. penicillioides	Dark blue-green	Discoloured germs 'Blue eye' of maize
A. candidus	Large, white	Powdery white patches
A. flavus	Large, yellow-green	Greenish discoloration
A. ochraceus	Large, ochre	Discoloured germs
Eurotium spp.	Small, grey-green	Discoloured germs 'Blue eye' of maize
Fusarium spp.	Rarely visible	Red streaking, particularly on maize
Alternaria spp.	Black	Darkening of grain
Cladosporium spp.	Dark brown to black	Darkening of grain
Curvularia spp.	Black	Darkening of grain

During the milling of grain, mycotoxins concentrated in the bran layers may be removed. The detoxification of aflatoxin-contaminated cereals, by ammoniation, has been used commercially for animal feeds. However, the control of mycotoxins is best addressed by controlling the growth of fungi.

Grain Sampling and Analysis

Once mycotoxins have formed, they are very difficult to remove effectively. Therefore it is vitally important to segregate mouldy grain, during visual inspection directly after harvest. Aflatoxin-contaminated maize grains will fluoresce under ultraviolet light, due to the presence of kojic acid along with the aflatoxins, and this enables rapid screening. Automatic sorting by colour can be used to segregate groundnut kernels of abnormal coloration or appearance. However, these two screening techniques are specific to these commodities.

Mould growth on cereal grains may be visible to the naked eye, as either mycelia, sporing structures or discoloration of the product. Grain should be inspected for any visible mould growth, and notes made on the presence and colour of any fungal sporing heads, the colour of any mycelia, the extent of growth, and any discoloration of the germ. Many of the more common storage and field fungi can be identified using a hand lens or the naked eye (**Table 4**). Detailed notes on the storage and transport conditions should also be made. The types of fungi identified on the grain may provide valuable information about its history and the likelihood of spoilage during storage.

Samples for mycological examination and analysis for mycotoxins should be taken in the field at the time of harvest, from different depths in bulk stores, and from individual grain sacks. The adequate sampling of large-volume bulk grain shipments is very difficult – mycotoxins may be concentrated in a small percentage of kernels, which may contain very high levels of toxin. Therefore the use of recognized sampling schemes is vital.

Cereal Flours

The microbial content of grain is reduced during processing into flour. The cleaning and washing of the grains and the bleaching and milling processes remove many of the extraneous microorganisms. A moisture content of less than 13% will prevent the further growth of almost all microorganisms. Heat generated by friction during the milling process raises the relative humidity inside the milling machinery, which results in mould growth in the flour adhering to surfaces. Fungal spores can then contaminate the flour being produced. Factory hygiene is important in preventing the build-up of contaminated flour. However, moderately high levels of storage fungi in flours are of no great importance.

The levels of moulds in correctly stored flours remain low but constant – in the region of thousands of spores per gram. Storage moulds such as *Penicillium*, *Aspergillus* and *Rhizopus* species predominate. If the a_w of the flour increases above normal levels, mould growth is likely to occur; if the a_w rises sufficiently, bacteria will multiply.

The endospores of *Bacillus* species, which may be present in large numbers, are able to survive baking and may outgrow in the finished product. *Bacillus subtilis* causes bread spoilage in the form of ropiness, and can lead to a mild form of food poisoning. Ropiness is caused by the hydrolysis of flour protein and starch, resulting in sticky, stringy bread with brownish spots and an off odour. It has largely been eliminated from commercially produced bread, by the addition of preservatives, such as propionates and sorbic acid, and the use of low-temperature storage. Red or 'bloody' bread is caused by the growth of the bacterium *Serratia marcescens* in the final product.

Moulds have been implicated in most cases of the spoilage of cakes incorporating cereal flours, although the role of other ingredients which may contain a variety of spoilage microorganisms, must not be overlooked. Preservatives such as sorbic acid have been effective in the control of fungal growth in many foods.

See color Plate 34.

See also: ***Alternaria***. ***Aspergillus***: Aspergillus flavus. ***Bacillus***: Bacillus subtilis. ***Fusarium***. **Mycotoxins**: Occurrence. ***Penicillium***: Introduction. ***Serratia***. **Spoilage Problems**: Problems caused by Fungi.

Further Reading

Beuchat LR (ed.) (1987) *Food and Beverage Mycology*, 2nd edn. New York: Van Nostrand Reinhold.

Coker RD (1997) *Mycotoxins and their Control: Constraints and Opportunities*. NRI Bulletin 73. Chatham: Natural Resources Institute.

Dendy DAV (ed.) (1995) *Sorghum and Millets: Chemistry and Technology*. St Paul: American Association of Cereal Chemists.

ICMSF (1996) *Microorganisms in Foods 5. Characteristics of Microbial Pathogens*. London: Blackie Academic & Professional.

Juliano BO (ed.) (1985) *Rice: Chemistry and Technology*, 2nd edn. St Paul: American Association of Cereal Chemists.

Lacey J (1986) Factors affecting fungal colonization of grain. In: *Spoilage and Mycotoxins of Cereals and other Stored Products*. P. 29. Biodeterioration Society Occasional Publication 2. Edited by B Flannigan, Commonwealth Agricultural Bureaux International, Farnham Royal, Slough.

MacGregor AW and Batty RS (eds) (1993) *Barley: Chemistry and Technology*. St Paul: American Association of Cereal Chemists.

Multon JL (ed.) (1988) *Preservation and Storage of Grains, Seeds and their By-Products*. New York: Lavoisier Publishing.

Pitt JI and Hocking AD (1997) *Fungi and Food Spoilage*, 2nd edn. London: Blackie Academic & Professional.

Pomeranz Y (ed.) (1988) *Wheat: Chemistry and Technology*, 3rd edn. St Paul: American Association of Cereal Chemists.

Sauer D (ed.) (1992) *Storage of Cereal Grains and their Products*, 4th edn. St Paul: American Association of Cereal Chemists.

Seiler DAL (1986) The microbial content of wheat and flour. In: *Spoilage and Mycotoxins of Cereals and Other Stored Products*. p. 35. Biodeterioration Society Occasional Publication 2. Edited by B Flannigan, Commonwealth Agricultural Bureaux International, Farnham Royal, Slough.

Watson SA and Ramstad PE (eds) (1987) *Corn: Chemistry and Technology*. St Paul: American Association of Cereal Chemists.

SPOILAGE PROBLEMS

Contents

Problems Caused by Bacteria

Karen M J Hansen and **Derrick A Bautista**,
Saskatchewan Food Product Innovation Program,
University of Saskatchewan, Saskatoon, Canada

Introduction

Spoilage of food can be described as a loss of qualitative properties in foods with regard to colour, flavour, texture, odour and/or shape. It is often the by-products of microbial metabolisms that make spoiled food offensive. However, spoiled food is a more subjective analysis rather than an objective one and is usually made by some organoleptic assessment. Most spoilage is caused by microorganisms such as bacteria, yeast and moulds. For the purposes of this article, spoilage of food products will be focused on those caused by bacteria only. For further information on characteristics of spoilage bacteria, the reader is encouraged to refer to other chapters within this book.

Meat Spoilage

In healthy animals, the combination of immune systems and the physical barrier of the skin adequately protect organs and muscle against microorganisms. Therefore, muscle tissues from freshly slaughtered carcasses should be relatively free of bacterial contamination. The surface of the skin and gastrointestinal tract are, however, heavily colonized with bacteria and provide a source of cross contamination during processing. For example, faeces and soil can harbour microorganisms such as *Micrococcus*, *Staphylococcus* and *Pseudomonas* spp. As faeces and soil can come into direct contact with animal surfaces, removal of hides during processing can contaminate tissues via the skinning knife or by handling. Fresh meat is an ideal source of nutrients (rich in nitrogenous compounds, minerals, water) and therefore bacterial spoilage of meat is greatly influenced by the sanitary conditions of the carcass and processing systems (**Table 1**).

Spoilage of Fresh Refrigerated Meat

Pseudomonas spp. are among the most common spoilage agents of refrigerated raw meats. This is

Table 1 Bacteria commonly associated with the spoilage of meat and meat products

Fresh beef (stored at approx. 4°C)	Surface slime and/or off odour	*Pseudomonas*, *Shewanella*, *Acinetobacter*, *Brochothrix*, *Leuconostoc*, *Moraxella* spp.
	Red spot	*Serratia marcescens*
	Blue discoloration	*Pseudomonas syncyanea*
	Yellow discoloration	*Micrococcus*, *Flavobacterium* spp.
Ham	Bone taint	*Clostridium* spp.
	Souring	*Lactobacillus*, *Leuconostoc*, *Alcaligenes*, *Pseudomonas* spp.
	Spongy consistency	*Bacillus* spp.
Vacuum-packed meats	Putrefaction and gas production	*Clostridium*, *Alcaligenes* spp.
	Souring	*Lactobacillus*, *Leuconostoc*, *Acinetobacter*, *Pediococcus* spp.
	Off odour and slime	*Pseudomonas*, *Brochothrix* spp.
Modified atmosphere packaging	Souring	*Lactobacillus*, *Leuconostoc*, *Acinetobacter* spp.
Cured meat (bacon)	Cheesy odour, sour taste, rancid	*Micrococcus*, *Lactobacillus*, *Alcaligenes*, *Bacillus*, *Clostridium* spp.
Sausage (pork)	Souring at 0–11°C	*Lactobacillus*, *Leuconostoc* spp.
	Souring at 22°C	*Microbacterium* spp.
	Surface slime	*Lactobacillus*, *Bacillus*, *Leuconostoc* spp.
Processed and cooked meats	Souring	*Lactobacillus*, *Brochothrix* spp.
Dried meats	Surface slime	*Micrococcus* spp.
	Discoloration	*Bacillus* spp.
Luncheon meats	Slime, greening	*Lactobacillus*, *Leuconostoc*, *Pseudomonas* spp.

especially true of raw meats stored over several days under aerobic conditions and at refrigeration temperatures (4°C). These psychrotrophic Gram-negative bacilli flourish at temperatures between 0°C and 20°C. Furthermore, the high humidity associated with domestic refrigeration systems can also increase the rate of spoilage. Excessive population of *P. fluorescens*, *P. fragi* and *Shewanella putrefaciens* (formerly *Pseudomonas putrefaciens*) will produce a green water-soluble slime and off odours.

Ground raw-meat products can spoil rather quickly due to the distribution of bacteria by the mechanical action of grinding and pooling of different meats to make one end product. The larger surface area of ground meat, in addition to cold temperatures (4°C) during storage, creates an environment favoured by *Pseudomonas*, *Acinetobacter*, *Moraxella* and *Aeromonas* spp. These organisms will produce discoloration and unpleasant odours in the product.

Growth of *Acinetobacter*, *Moraxella* and *Brochothrix* spp. can be favoured when *Pseudomonas* spp. are restricted by low oxygen conditions such as with modified atmosphere packaging (MAP). At high densities, *Acinetobacter* and *Moraxella* spp. can rapidly attack proteins to produce off flavours and odours whereas *Brochothrix thermosphacta* utilizes glucose and glutamate to produce off odours and slime. *Flavobacterium* spp. and *Serratia marcescens* are other common spoilage organisms of MAP and can produce greenish-yellow and red discolorations in meat, respectively.

Vacuum-packaged fresh meats tend to undergo considerably longer refrigeration than fresh meats without vacuum packaging. When these products spoil, the predominant spoilage agents are *Lactobacillus* spp. The prevalence of these bacteria is determined by a number of factors, including final pH and the level of available oxygen. In vacuum-packaged raw beef with a pH of about 5.6, *L. amylovorus*, *L. casei*, *Lactococcus lactis*, *Leuconostoc* spp., *Pediococcus* spp. and other lactic acid bacteria predominate. The sour taste and pungent odour associated with the growth of these organisms in meat are caused by their production of organic acid.

Homofermentative strains such as *Leuconostoc* spp. generate > 80% lactic acid from the fermentation of glucose. Heterofermenters, including *Pediococcus* spp., can produce at least 50% lactic acid as an end product of fermentation. In addition to souring, *Lactobacillus* spp. can create slime on the interior walls of the package and on meat surfaces. Often, a murky liquid is present in the packaging. Vacuum and MAP packaging can create an environment favourable for anaerobic and facultative anaerobes such as *B. thermosphacta*. In sufficient numbers, this organism produces diacetyl acetone and 3-methyl butane, resulting in a cheesy odour in meats.

Spoilage of Finished Meat Products

Due to differences in environmental conditions, moisture content and pH, the flora found in finished meat products is different from that of fresh raw meats.

Curing is a process where salt, nitrite and seasoning are added to meat to help develop unique characteristics and flavours. It was initially developed to extend the shelf life of meat products; however bacterial spoilage can still be a problem. Bacteria commonly isolated from spoiled cured products include lactic acid bacteria, *Bacillus*, *Micrococcus*, *Clostridium* and *Alcaligenes* spp. The high fat content and low water activity of products such as bacon can provide adequate conditions for the growth of *Lactobacillus*, *Lactococcus* and *Micrococcus* spp. Souring is often a result of these organisms utilizing sugar in the curing solution pumped into the product. It is of great concern when meat products have a sugar content ≥1% and are stored in MAP and vacuum packaging.

Processes employed during curing, such as smoking and brining, can reduce the susceptibility of finished meat products. Some exceptions are spore-formers such as *Bacillus* and *Clostridium* spp. Spores from these bacteria can survive the cooking process or can be introduced into the finished product during handling and packaging. *Bacillus cereus* and *Clostridium perfringens* can form gas in vacuum-packaged or canned meat products. They may also cause greening, odour, loss of texture and excessive liquid production.

Micrococcus and *Bacillus* spp. are common spoilage agents in sausages, especially when stored in MAP and a refrigeration temperatures (approx. 4°C). Excessive growth of these bacteria can result in slime formation and gas production.

Bacterial spoilage of luncheon meats such as frankfurters and bologna can result in sliminess, souring and greening. Slime spoilage usually occurs on the outer surfaces of sausages and frankfurters. It is normally caused by excessive growth of Gram-negative psychrotrophic bacteria (e.g. *Pseudomonas* spp.) Greening is also common on frankfurters and results from the action of peroxides produced by *Lactobacillus* and *Leuconostoc* spp. The lower brine levels and more neutral pH of these types of meat products can make them more susceptible to spoilage than traditional cured or cooked meats. In addition, luncheon meats are often sliced and kept in MAP or vacuum packaging. The increased surface area and humid storage conditions make surface contamination more noticeable.

A form of spoilage known as bone sour or bone

taint can be caused by *Clostridium* spp. As the name implies, a sour spoilage can occur between the flesh and bone of beef rounds and hams. Such internal spoilage of beef may be due to delayed chilling or prolonged storage at temperatures between 15 and 25°C. The low oxygen levels surrounding the bone allow *Clostridium* spp. to proliferate.

Due to a predominant lactic acid flora and the nature of fermented meat products, spoilage is minimal. Problems that occur include an over-production of organic acid by lactic acid bacteria. To rectify this problem, meat processors will cook the fermented meat product after the desired pH has been reached. This will stop any further production of lactic acid and souring.

Bacterial Spoilage of Milk and Milk Products

Spoilage in Raw Milk

Most sources of bacterial spoilage in milk can be traced to the hide and teats of the animal. Contamination may be due to infections of the udder, milk ducts or teats, but most often results from unclean or improper sterilization of equipment. Mastitis, an inflammatory disease of bovine mammary glands, is caused by a number of bacteria, including *Pseudomonas* and *Staphylococcus* spp. *Pseudomonas* spp. metabolize proteinaceous compounds to change the normal flavour of milk to a bitter or unclean taste. The production of ethyl butyrate by *Staphylococcus* spp. may give milk a fruity odour. Both *S. aureus* and *P. aeruginosa* can lipolyse milk lipids, resulting in rancidity of raw milk products.

Lactic acid bacteria can also contribute to spoilage of raw milk. Although end products of lactic acid bacteria can be desirable in many fermented milk products, they are considered to be a source of spoilage in raw milk. The lactic, formic, butyric acids and CO_2 produced by these bacteria result in souring, foaming and curdling of milk. *Alcaligenes* spp. can produce slime or ropiness in milk when left at ambient temperatures (22°C).

Spoilage in Pasteurized Milk

Most spoilage of pasteurized milk is the result of recontamination after thermal processing. Although pasteurization destroys many spoilage bacteria and attenuates growth of others, heat-resistant *Lactococcus* and *Lactobacillus* spp. can survive and grow to create spoilage problems. Their conversion of lactose to lactic acid lowers the pH of milk to about 4.5 and produces curdling. *Lactobacillus lactis* can metabolize leucine to produce 3-methylbutanol, which adds an undesirable malty taste. Normally, milk contaminated with *L. lactis* does not undergo a colour change. However, if this organism is grown in the presence of *P. syncyanea*, milk will turn bright blue.

Heat-stable proteinases and lipases of some psychrotrophic bacteria are not affected by pasteurization temperatures. These enzymes can cause proteolysis and lipolysis of casein and milk lipids, respectively, to produce flavour defects. Species of *Lactococcus*, *Lactobacillus* and *Clostridium* spp. may result in a sour taste, while *Proteus* spp. can give milk an undesirable sweet flavour. *Pseudomonas*, *Flavobacterium* and *Bacillus* spp. can give milk a bitter or off flavour when present in high numbers. Ropiness is a characteristic viscous and oily 'mouth feel' that can be produced by *Micrococcus* spp. and, especially, *Alcaligenes viscolactis*. This is particularly apparent when pasteurized milk is stored over long periods or kept at ambient temperatures (22°C).

Spores of *Bacillus* and *Clostridium* spp. can survive pasteurization temperatures. *Bacillus* spp. can cause bitty in cream, which is the result of lecithinase activity on phospholipids. It is a visual defect and appears as an aggregation of particles that adhere to the surface of milk cartons.

Ultra-high-temperature (UHT) products are commercially sterile milks which have been heated at or above 138°C for at least 2 s. Spores of thermophilic bacteria such as *Bacillus* and *Clostridium* spp. can survive these high temperatures and may result in rancidity or souring of pasteurized milk (**Table 2**).

Spoilage of Other Dairy Products

The microflora of whole milk tends to be present in the cream component. Since cream is the main ingredient of butter, microbial spoilage can be a problem. Microorganisms associated with the lipid hydrolysis of triglycerides to free fatty acids can produce increased acidity, rancidity and soapiness in butter. Causative agents include *Pseudomonas*, *Micrococcus* and *Serratia* spp. Surface taint or putridity results from the growth of *Shewanella putrefaciens*.

Bacterial contamination of cheese is usually the result of manufacturing with milk that has a high microbial content ($\geqslant$1000 cfu ml^{-1}). The undesirable fermentation of lactic acid bacteria such as *Leuconostoc* spp. and *Lactobacillus lactis* can cause an undesirable pink discoloration near the surface of cheese. The presence of *Bacillus*, *Leuconostoc* and *Pseudomonas* spp. can attack proteins and produce carbon dioxide. Production of large amounts of gas may result in the formation of undesirable holes in curd during cheese manufacturing. These bacteria are also responsible for bitter flavour and slime in soft and

Table 2 Bacteria associated with the spoilage of milk and milk products

Raw milk (at 10–37°C)	Souring	*Lactobacillus lactis*
(at 37–50°C)	Souring	*Lactobacillus*, *Staphylococcus*
(>50°C)	Souring	*Lactobacillus thermophilus*
	Unclean flavour	*Pseudomonas* spp.
	Fruity flavour	*Staphylococcus* spp.
Pasteurized milk	Bitter taste (proteolysis)	*Pseudomonas*, *Flavobacterium*, *Bacillus* spp.
	Souring (acid proteolysis)	*Micrococcus*, *Bacillus cereus*, *Lactobacillus*, *Clostridium* spp.
	Sweet proteolysis, curdling and slime	*Alcaligenes*, *Proteus* spp.
	Malty taste	*Lactobacillus lactis*
	Ropiness	*Micrococcus* spp., *Alcaligenes viscolactis*
	Blue colour	*Pseudomonas synceaes* with *Lactobacillus lactis*
Cream and butter	Surface taint	*Pseudomonas putrefaciens*
	Bitty	*Bacillus* spp.
Hard and soft cheese	Slime and off flavour	*Pseudomonas*, *Leuconostoc*, *Bacillus* spp.
	Pink discoloration	*Lactobacillus*, *Leuconostoc* spp.
	Holes in curd	*Bacillus*, *Pseudomonas*, *Leuconostoc* spp.
	Rancidity, soapiness	*Micrococcus*, *Serratia*, *Pseudomonas* spp.
Cottage cheese	Slimy curd, putrid odour	*Pseudomonas* spp.
	Unclean taste	*Escherichia coli*
	Discoloration	*Flavobacterium* spp.

Table 3 Bacteria commonly associated with the spoilage of vegetables and fruit

Various vegetables	Bacterial soft rot	*Erwinia* spp., *Erwinia carotovora*, *Pseudomonas marginalis*, *Bacillus* spp., *Clostridium* spp.
Potatoes	Black leg	*Erwinia* spp.
	Vascular ring and discoloration	*Corynebacterium sepedonicum*
	Scab of potatoes	*Streptomyces scabies*
Tomatoes	Bacterial canker	*Corynebacterium michiganese*
	Bacterial spot	*Xanthomonas vesicatoria*
	Bacterial speck	*Pseudomonas syringae*
Pears	*Erwinia* rot	*Erwinia* spp.

hard cheeses (e.g. Brie and Parmesan, respectively).

In cottage cheese, *Pseudomonas* spp., namely *P. fragi*, can alter the flavours leaving a putrid, rancid, bitter or fruity taste. Another problem is the growth of *Flavobacterium* spp., which can alter the colour of cottage cheese. *Escherichia coli* in high enough numbers (100 000 cfu ml^{-1}) can result in an unclean or barny taste, especially when cottage cheese is left at room temperature (22°C).

Spoilage of Fruits and Vegetables

Since most fruits and vegetables are usually harvested from or near the soil, these commodities can be subjected to a variety of flora that may cause spoilage. Losses in product and revenue associated with microbial spoilage have been estimated to be in excess of 20% of all fruit and vegetables. For the most part, biochemical composition of fruits and vegetables can be an excellent growth medium (**Table 3**). Carbohydrates which are present in high concentrations in these food products are easily utilized by a variety of bacteria, resulting in the production of various degradation by-products. The accumulation of these products can alter the appearance, texture and taste of fruits and vegetables. Unless certain precautionary measures are made, the shelf life of these products will be short.

Of the problems associated with bacterial spoilage, soft rot is of key importance in vegetables. Members of the *Erwinia* spp. (e.g. *Erwinia carotovora*), *Pseudomonas marginalis* and some *Bacillus* and *Clostridium* spp. are associated with this problem. These bacteria produce protopectinases that break down pectins found on the outer skin of vegetables. This results in softening and the production of off odours. Root crops, crucifers, cucurbits, solanaceous vegetables, onions and many other plants can be susceptible to these organisms.

In potatoes, *Erwinia* spp. are also responsible for black leg, which is a common rot of potatoes under poor storage conditions. Potatoes can be subjected to bacterial attack by *Corynebacterium sepedonicum* which creates a vascular ring and subsequently produces a creamy-yellow or light-brown discoloration and softening of the plant tissues. Scab of potatoes is caused by *Streptomyces scabies* and is seen as brownish spots that resemble enlarged corky areas.

C. michiganese causes bacterial canker in tomatoes. Random spotting occurs on the fruit, followed by decay. Bacterial spot is the production of small, black scabby fruit spots that is caused by *Xanthomonas*

Table 4 Chemical constituents of fruits and vegetables

Carbohydrates and related compounds	Polysaccharides, oligosaccharides, monosaccharides, sugar alcohols (e.g. sorbitol), sugar acids (e.g. ascorbic acid), organic acids (citric acid)
Proteins	Albumins, prolamines, peptides, amino acids
Lipids	Fatty acids, phospholipids, glycolipids
Nucleic acids and derivatives	Purines and pyrimidine bases, nucleotides
Vitamins	A, D, E (fat-soluble), thiamin, niacin, riboflavin (water-soluble)
Minerals	Sodium, potassium, calcium, magnesium, iron
Other components	Water, alkaloids, porphyrins, aromatics

vesicatoria. *Pseudomonas syringae* produces bacterial speck on tomatoes which appears as numerous small dark-brown spots.

Due to their low pH, bacterial spoilage is not a serious problem with fruits. If spoilage does occur, it is usually the acidophilic bacteria that causes the problem. Under normal storage conditions, *Acetobacter* and *Lactobacillus* spp. account for the reduced shelf life of fresh fruit products. One major exception is the *Erwinia* rot of pears. With a pH range of 3.8–4.6, it is believed that *Erwinia* spp. can initiate growth on the surface of the pear, where pH is suspected to be more neutral (**Table 4**).

Spoilage of Wines

A major problem of spoiled wine products is a significant increase in the acidity of wines due to the production of acetic acid and lactic acid by heterofermentative and homofermentative bacteria. In addition, wines with high amounts of these acids will often have high concentrations of histamine. Examples of the chemical breakdown of acidic substrates are depicted below.

Malolactic fermentation

$$\text{L }(-)\text{-malic acid} \xrightarrow{\text{Malolactic enzyme}} \text{L }(+)\text{-lactic acid} + CO_2$$

Tartaric acid decomposition

$$\text{Tartaric acid} \longrightarrow \text{acetic and lactic acids} + CO_2$$

Excessive malolactic fermentation will decarboxylate L-malic acid to produce pyruvic acid. The results are wines with reduced acid content and an unusual flavour. This spoilage problem is normally caused by *Leuconostoc*, *Pediococcus* and *Lactobacillus* spp. but the process may be instigated by *Oenococcus oenos*. The utilization of tartaric acid by *L. plantarum* will increase the acidity of wines. High levels of lactic acid bacteria can produce butandine 2,3,adikeytone that creates an undesirable buttery or whey-like aroma. Decarboxylation of amino acids by lactic acid bacteria is responsible for the production of off flavours and odours (e.g. phenethlyamine, tyramine, putrescine, cadaverine and spermidine).

Another common problem with wines is Tourne disease, caused by degradation of sugars by facultative and anaerobic bacteria under low alcohol content. It produces a silky cloudiness, mousy odour and unusual taste. Mousiness is described as an odour similar to mouse urine and this affliction is caused by *L. hilgardii*, *L. brevis* and/or *L. cellobiosos*.

Ropiness can be found in spoiled wines. There is a slimy, viscous, oily characteristic to the wine which is produced by *Streptococcus mucilaginosus*, *Pediococcus cerevisiae* and *Leuconostoc* spp. The problem occurs during wine manufacturing. It begins at the bottom of the fermentation vessel and slowly moves towards the top. *Clostridium butyricum* may give a rancid taint due to production of small amounts of butyric acid. *Bacillus circulans* and *B. subtilis* have been known to produce highly involatile acidity in wines.

Spoilage of Cereal Products and Bakery Goods

Due to the low water activity ($a_w \leqslant 0.60$) of grains and cereals, bacterial spoilage is not a serious problem. It is mostly of mould origin. However, a problem will arise if the a_w of the product increases by exposure to higher relative humidities.

With the production of baked goods, *Leuconostoc* and *Lactobacillus* spp. are common spoilage organisms for doughs, biscuits and rolls. After baking, *B. subtilis* may be a problem when baked goods are cooled slowly in a moist environment above pH 5. It can produce ropiness and a fruity aroma in the final product.

Concluding Remarks

Food spoilage may pose economic consequences if certain precautionary and preventive measures are not performed. The food industry has adopted methods to minimize spoilage with the use of natural preservatives, refrigeration, packaging material and, more recently, management systems. However, these techniques are for naught if incoming material are not of the highest quality and handled under good sanitary conditions. In all cases, shelf life of many foods can be extended if foods are prepared to minimize the level of bacterial contamination before final processing.

See also: **Confectionary Products**: Cakes and Pastries. **Hazard Appraisal (HACCP)**: The Overall Concept.

Heat Treatment of Foods: Principles of Pasteurization. **Meat and Poultry**: Spoilage of Meat; Curing of Meat. **Microflora of the Intestine**: The Natural Microflora of Humans. **Milk and Milk Products**: Microbiology of Liquid Milk; Microbiology of Cream and Butter. **Packaging of Foods**: Packaging of Solids and Liquids. **Spoilage of Plant Products**: Cereals and Cereal Flours. **Spoilage Problems**: Problems caused by Bacteria; Problems caused by Fungi. **Wines**: The Malolactic Fermentation.

Further Reading

Adams MR and Moss MO (1995) *Food Microbiology.* Cambridge: Thomas Graham House.

Brown MH (1982) *Meat Microbiology.* New York: Applied Science.

Davies A and Board R (1998) *The Microbiology of Meat and Poultry.* New York: Chapman & Hall.

Fleet GH (1994) *Wine Microbiology and Biotechnology.* Philadelphia: Hardwood Academic.

Frazier WC and Westhoff DC (1988) *Food Microbiology.* Toronto: McGraw Hill.

Jay JM (1992) *Modern Food Microbiology*, 4th edn. P. 185. New York: Van Nostrand Reinhold.

Kraft A (1992) *Psychrotrophic Bacteria in Foods – Spoilage and Disease.* Florida: CRC Press.

Lund BM (1971) Bacterial spoilage of vegetable and certain fruits. *Journal of Applied Bacteriology* 34: 9–20.

Radler F (1975) The metabolism of organic acids by lactic acid bacteria. In: Carr JG, Cutting CV and Whiting G (eds) *Lactic Acid Bacteria in Beverage and Food.* P. 127. New York: Academic Press.

Problems Caused by Fungi

Maurice O Moss, School of Biological Sciences, University of Surrey, UK

Introduction

By definition, foods are nutritious, which makes them suitable environments for the growth of microorganisms. The presence of nutrients and certain physico-chemical parameters influences the outcome of the competition between bacteria and fungi. The most important factors other than the presence of nutrients are water activity (a_w), pH, temperature and the presence of preservatives. The economic importance of fungal spoilage is illustrated by the fact that even in Australia, a country with a dry climate and advanced technology, losses of food due to fungal spoilage are estimated to cost in excess of Australian \$10 000 000 per annum: losses in the warm and humid tropical countries must be much greater. The two most important groups of fungi involved in food spoilage are those known colloquially as moulds and yeasts. (**Fig. 1**).

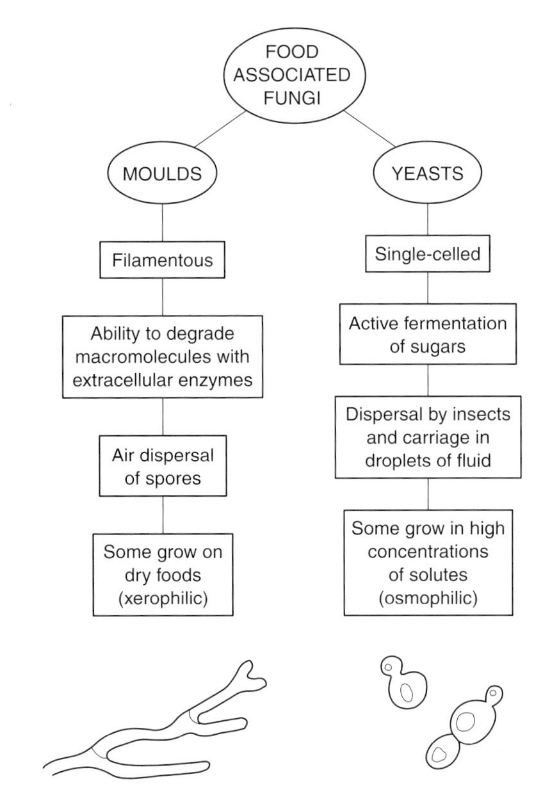

Figure 1 Contrasting characteristics of moulds and yeasts.

Moulds

Moulds are filamentous, with a growth habit that is well-adapted to colonizing the surfaces of, and penetrating into, solid substrates. They can, however, also grow in liquid environments. Many species can grow in environments with low a_w, or may even grow best in such environments, and are described as 'xerophilic'. Many species can also grow in acid conditions. Both of these types of environment offer conditions in which bacteria cannot normally compete. Associated with their growth on solid substrates and the absence of motile stages, the terrestrial moulds have evolved diverse mechanisms for dispersal by air. Some genera, including *Aspergillus* and *Penicillium*, produce vast numbers of asexual spores, which are small (2–3 μm), unicellular, dry, thick-walled and pigmented, and hence well-adapted for dispersal by air over long distances and for survival in the atmosphere for long periods. Such propagules are ubiquitous in the air, and pose a major threat to foods. Moulds have the ability to degrade macromolecules, by means of extracellular enzymes. The single-celled nature of the yeasts reflects their adaptation to growth in liquid environments rich in nutrients, e.g. the nectaries of plants. Many species are able to ferment carbohydrates, as a source of energy for growth under anaerobic conditions. In common with moulds, some species of yeasts can grow in environments with low a_w. However, in contrast to moulds, some yeasts grow optimally in the presence of high concentrations of solute, i.e. in conditions with high osmotic pressure, and are described as 'osmophilic'. Yeasts can also grow on solid surfaces, and there is overlap between the ranges of foods which yeasts and moulds can spoil and which would be considered safe from spoilage by bacteria. Yeasts are dispersed by insects and by droplets of fluid.

Fungi Associated with Food Spoilage

The fungi are eukaryotic, heterotrophic microorganisms from at least two biological kingdoms. The chytridiomycetes, zygomycetes, ascomycetes and basidiomycetes form the kingdom now considered to be the 'true fungi', and the oomycetes are more closely related to certain groups of algae.

The heterotrophic metabolism of fungi ranges from that of specialized plant pathogens, which require a living host to supply their nutrients, to an ability to utilize the simplest of organic compounds and inorganic salts. However, the most important aspect of the metabolism of moulds with respect to their growth on foods is the ability to break down complex macromolecules such as polysaccharides, proteins and lipids.

Plant pathogens have an impact on the earliest stages of food production, and in addition, some are pathogens in the field and develop as spoilage organisms after harvesting. For example, *Botrytis cinerea* is a pathogen affecting the blossom of a wide range of hosts in the field, including the strawberry plant, and is the serious grey mould associated with the spoilage in storage of many fruits and vegetables, especially strawberries.

Table 1 shows the diversity of fungi which may be associated with problems at all stages of food production, from the field to post-harvest storage and processing.

Until the discovery of mycotoxins, food spoilage by moulds was considered to be a relatively benign phenomenon, and foods were rejected for aesthetic reasons and because of the characteristic musty smell of mouldy foods. Many fungi have proteolytic activity, but they rarely produce the putrid odours frequently associated with bacterial spoilage. Rancidity, due to the hydrolysis of oils and fats and the subsequent degradation of the fatty acids so produced, is characteristic of some moulds, e.g. *Penicillium roqueforti*. This species is able to decarboxylate sorbic acid to *trans*-1,3-pentadiene, resulting in off flavours in foods in which sorbates have been used as preservatives (**Fig. 2**).

However, spoilage in one context may be the development of a desirable characteristic in a deliberately mould-ripened food: for example, *P. roqueforti* is the species used in the production of blue cheeses. Similarly, *Aspergillus niger* is used to manufacture enzymes and citric acid but is also involved in the spoilage of commodities, including oil seeds; *Fusarium graminearum* is used in the industrial production of mycoprotein but also causes mycotoxigenic spoilage of cereals; and the yeast *Saccharomyces cerevisiae* is widely used in the manufacture of fermented beverages, bread and food yeast but is also frequently implicated in the spoilage of fruit juices. In west Africa, a product of the fermentation of manioc, known as 'gari', involves bacteria and the yeast-like mould *Geotrichum candidum*. This species is implicated in the spoilage of dairy products and meat products (e.g. hams), and causes a sour rot of carrots, citrus fruits and tomatoes. *Geotrichum candidum* is also a very useful indicator organism for monitoring the overall hygiene of a factory or a food processing area: an increase in plate counts indicates a decline in the standards of hygiene and cleanliness.

Spoilage of Cereals and Cereal Products

Ten to twenty per cent of the world's estimated annual cereal crop of 2×10^9 t is lost through spoilage by moulds. The problem is particularly acute in the humid tropics, and is clearly important in view of overall worldwide food shortages.

Moulds can usefully be classified as either 'field fungi' or 'storage fungi'. As crops mature in the field, particularly as they progress into senescence, a mould flora develops which includes the genera *Cladosporium*, *Alternaria*, *Phoma*, *Drechslera*, *Epicoccum* and *Fusarium*. Once the crop is harvested, dried and put into storage, this flora becomes more difficult to isolate (with the possible exception of *Fusarium*), and is overgrown by other genera, including *Penicillium*, *Aspergillus* and *Eurotium* (which includes the teleomorph of *Aspergillus glaucus*).

The field fungi may be divided into two distinct groups: potentially pathogenic fungi (e.g. *Fusarium*), which can overcome the defences of living tissue and invade the growing plant; and fungi that do not normally invade living tissue but can grow on the senescent tissue which becomes straw, hay or chaff. The major genera in the latter category are *Alternaria* and *Cladosporium* and sometimes they can be isolated from 100% of the plants sampled from the field. The species which grow on senescent tissue have dark,

Table 1 Fungi associated with food production. (Classification based on *Ainsworth and Bisby's Dictionary of the Fungi* 8th edn.)

Species (Teleomorph)	*Phylum (Family)*	*Ecology*	*Nature of Spoilage*
Aspergillus niger	Imperfect fungus[a]	Saprophyte	Spoilage of oil seeds
Botrytis cinerea (*Botryotinia fuckeliana)*	Ascomycota (Sclerotiniaceae)	Saprophyte and weak pathogen	Grey mould rot of fruits and vegetables
Byssochlamys fulva	Ascomycota (Trichocomaceae)	Produces heat-resistant ascospores	Spoilage of heat-processed canned and bottled fruits
Cladosporium herbarum	Ascomycota	Minimum temperature for growth is −6°C	Black spot of chilled meat carcasses
(Mycosphaerella tassiana)	(Mycosphaerellaceae)		
Eurotium spp.	Ascomycota (Trichocomaceae)	Xerophilic ascospores	Spoilage of dried foods
Fusarium graminearum (*Gibberella zeae)*	Ascomycota (Hypocreaceae)	Facultative pathogen	Ear scab of cereals and spoilage mould in storage
Mucor racemosus	Zygomycotina (Mucoraceae)	Saprophyte	Spoilage of dairy products
Penicillium digitatum	Imperfect fungus	Saprophyte and weak pathogen	Green mould of citrus fruits
Penicillium expansum	Imperfect fungus	Saprophyte and weak pathogen	Soft rot of apples
Phytophthora infestans	Oomycota (Pythiaceae)	Obligate pathogen	Potato and tomato blight
Puccinia graminis	Basidiomycota (Pucciniaceae)	Obligate pathogen	Black stem rust of wheat
Rhizopus stolonifer	Zygomycotina (Mucoraceae)	Saprophyte	Post-harvest rot of fresh fruit and vegetables
Synchytrium endobioticum	Chytridiomycota (Synchytriaceae)	Obligate pathogen	Wart disease of potato
Zygosaccharomyces bailii	Ascomycota (Saccharomycetaceae)	Osmotolerant, preservative-resistant yeast	Spoilage of fruit juices, sauces and fermented beverages

[a]Widely referred to as 'Deuteromycotina'.

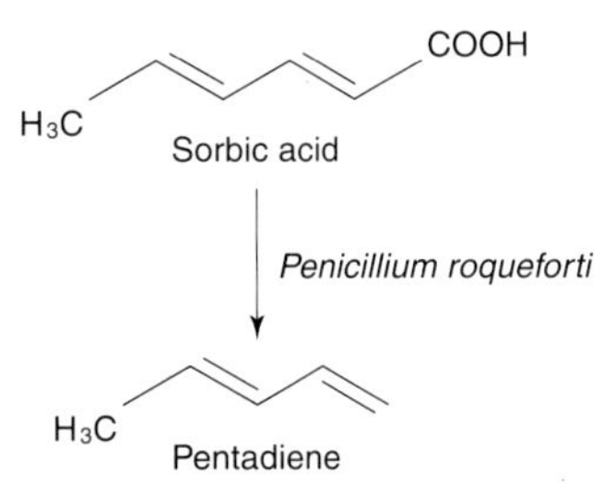

Figure 2 Decarboxylation of sorbic acid by *Penicillium roqueforti*.

pigmented, thick-walled hyphae and spores, and so are well-adapted to survive the rapidly changing environment of the surface of the aerial parts of plants, which involves rapid desiccation, rehydration and exposure to sunlight.

The most important factor influencing the overgrowth of field fungi by storage fungi post-harvest is the availability of water. Field fungi can survive desiccation on plant surfaces, but cannot grow in such conditions – in contrast, storage fungi can grow on desiccated surfaces (**Table 2**). The storage fungi are characterized by the production of small, resilient spores, which are widespread in the atmosphere and so have many opportunities to contaminate harvested crops.

Table 2 Minimum water activity (a_w) needed for the growth of some field and storage fungi in cereals

Species	*Minimum* a_w
Field fungi	
Cladosporium herbarum	0.85
Alternaria alternata	0.88
Fusarium graminearum	0.89
Fusarium moniliforme	0.87
Epicoccum purpurascens	0.99
Storage fungi	
Penicillium aurantiogriseum	0.83
Penicillium verrucosum	0.83
Aspergillus flavus	0.78
Aspergillus candidus	0.75
Eurotium amstelodami	0.73

Table 3 Fungal activity that may be associated with the spontaneous heating of cereals

Temperature	a_w	*Possible mould activity*	*Germination of grain (%)*	*Indications of spoilage*
Ambient	< 0.60	None	100	None
Ambient	0.75	Xerophiles, e.g. *Aspergillus restrictus*	90	None
Ambient, up to 25°C	0.87	Xerotolerant mesophiles, e.g. *Penicillium brevicompactum*, *Aspergillus versicolor*, *Penicillium aurantiogriseum*	45	Musty odours, discoloration
30–50°C	0.95	Thermotolerant mesophiles, e.g. *Aspergillus flavus*	15	Mycotoxin formation, caking
60°C	> 0.95	Thermophiles, e.g. *Aspergillus fumigatus*, *Talaromyces thermophilus*, *Malbranchea cinnamomea*	0	Caking, total spoilage

Table 4 pH values of a selection of fruits and vegetables

Fruits	*pH range*	*Vegetables*	*pH range*
Apple	3.1–3.9	Runner beans	5.0–6.0
Grape	3.0–4.0	Cabbage	5.2–5.4
Lemon	2.2–2.6	Carrot	5.2–5.4
Orange	3.3–4.0	Peas	5.8–6.5
Pear	3.7–4.6	Potato	5.4–6.0
Plum	3.2–4.0	Spinach	4.8–5.8
Raspberry	2.9–3.5	Tomato	4.0–4.4
Strawberry	3.0–3.9	Turnip	5.2–5.6

The terms 'fruits' and 'vegetables' are used here in the culinary rather than the botanical sense.

If grain is dried sufficiently and stored at a low, even temperature, it will be secure from post-harvest spoilage. The a_w which is safe (about 0.75) depends on the cereal – for example, for a typical cereal the safe water content is 14.0–14.5%; for a pulse (e.g. soybean) it is 12.5–13%; and for an oil seed (e.g. groundnut) it is 8.0–9.0%. However, even if the grain is dried sufficiently, problems can arise if it is subjected to fluctuating temperatures. Under these conditions, water can migrate through the gas phase of the bulk grain and water activity may temporarily increase locally, allowing the germination and growth of mould spores. Growth results in the production of water and heat, which can escalate into the phenomenon of spontaneous self-heating and spoilage. **Table 3** gives information about mould activity associated with post-harvest spoilage, one of the most sensitive symptoms being a decline in the ability of the grain to germinate.

Spoilage of Fruits and Vegetables

Fruits and vegetables have a high a_w and a plentiful supply of nutrients, but they often have a low pH (**Table 4**), which favours the growth of fungi over that of the majority of bacteria (with the exception of the lactic acid bacteria). In general, vegetables have pH values closer to neutral than do fruits, and this is reflected in the higher incidence of bacterial spoilage of vegetables.

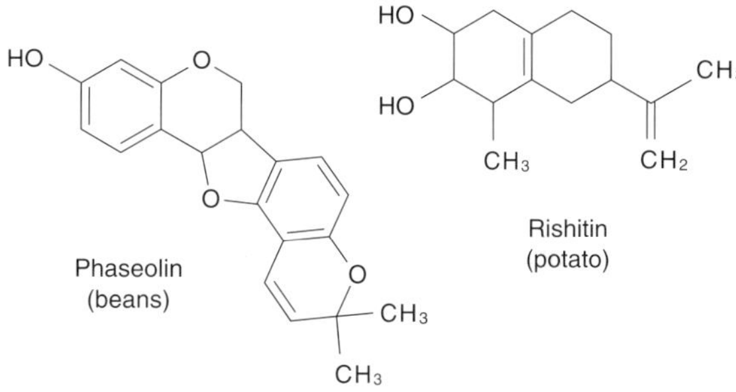

Figure 3 Examples of phytoalexins.

In order to attack the tissue of fruits and vegetables, fungi need the abilities to grow on the nutrients present and at the pH of the tissues; to produce enzymes capable of degrading the polymers responsible for the integrity of the tissues, particularly lignocellulose and pectin; to produce toxins lethal to plant cells; to overcome plant defence mechanisms, such as preformed antimicrobial metabolites and phytoalexins. Plants have various barriers against microbial invasion, including a tough, generally inpenetrable, cuticle. However, the cuticle has natural weak spots such as stomata, and its integrity can be damaged during harvesting. Phytoalexins (**Fig. 3**) are antimicrobial plant metabolites whose biosynthesis is normally repressed, but which are produced in response to infection, e.g. phaseolin, produced by the French bean, and rishitin, produced by the potato.

Plant tissues are generally held together by pectins, and their degradation is probably the most important activity in the fungal spoilage of fruits and vegetables. Pectins are α-1,4-linked polymers of galacturonic acid and its methyl ether. Their degradation involves a number of distinct enzyme activities, including demethylation, hydrolysis of the glycoside bond and *trans* elimination (**Fig. 4**).

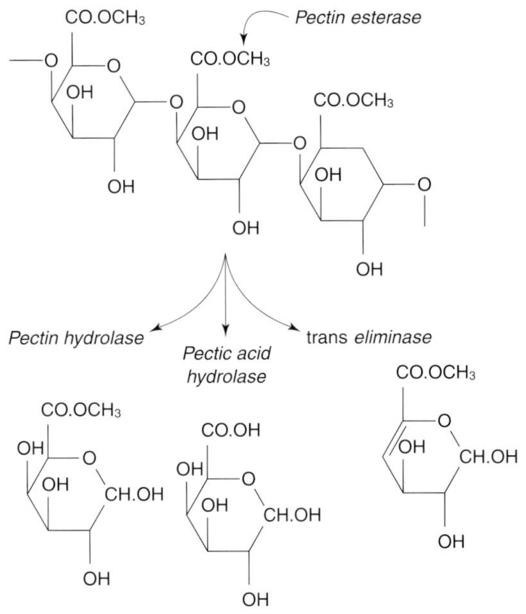

Figure 4 Enzymes involved in the degradation of pectin.

Table 5 Fungi causing post-harvest spoilage following harmless infection in the field

Fungus	*Fruit*
Gloeosporium perennans	Apple
Sclerotinia fructicola	Apricot
Colletotrichum gloeosporioides	Avocado pear
Gloeosporium musarum	Banana
Alternaria citri	Citrus fruits
Botrytis cinerea	Strawberry, raspberry, grape

Many of the fungi associated with the post-harvest spoilage of fruits and vegetables are very specific. For example, the ability of a species of *Penicillium* to cause rapid brown soft rot of apples is almost diagnostic for *P. expansum*. Once harvested, fruits and vegetables generally undergo changes which include ripening and some degree of senescence, damage to the skin and bruising of the tissues, all of which make them more susceptible to attack by moulds. Many plants carry fungi as endophytes during growth without adverse effects, but a progressive transition to pathogenicity may be triggered by the changes caused by harvesting. **Table 5** lists some fungi which may be present as latent infections of fruits in the field, and which can cause post-harvest spoilage in storage.

Fruits are particularly susceptible to infection at the site of the scar left when the fruit is separated from its stalk. For example, tomato rot caused by *Trichothecium roseum* is almost always initiated at the scar, but can be largely avoided by care during harvesting. All forms of damage can be a route of infection and subsequent spoilage. For example, cracks in the surface of carrots can lead to liquorice rot caused by *Mycocentrospora acerina*. Thus a major method of control of the post-harvest spoilage of fruits and vegetables is care during the harvest and subsequent storage.

Spoilage of Processed Foods

In general, moulds and yeasts are susceptible to heat treatment and are destroyed by the heat-processing of most foods. Fruits, however, because of their low pH and in order to retain an acceptable texture, are usually canned using a relatively mild heat-treatment, which the ascospores of some fungi, including *Byssochlamys fulva*, *Neosartoria fischeri* and *Talaromyces flavus*, can survive. Hence these fungi can grow in cans of fruit, causing off flavours and loss of structure.

Bakery products, such as cakes and biscuits, usually have a low a_w after cooking, but can be spoiled by xerophytic moulds such as *Wallemia sebi* following contamination by spores from the air. This species has also been implicated in the spoilage of other foods of low a_w, including jams and salted fish. Bread has a high enough a_w to allow the growth of moulds such as *Rhizopus stolonifer* (black bread mould) and *Chrysonilia sitophila* (the anamorph of *Neurospora sitophila*, red bread mould), as well as that of yeasts, e.g. *Endomyces fibuliger*, *Hyphopichia burtonii* (which causes chalky bread). The shelf life of bread and similar bakery products can be maximized by good hygiene and/or the addition of preservatives, e.g. sodium, potassium or calcium propionate.

Refrigeration usually offers good protection against the microbial spoilage of foods, but a number of fungi grow at refrigeration temperatures. For example, *Cladosporium herbarum* can cause black spot on chilled carcasses of meat, and can be seen on the walls of cold rooms which are not cleaned regularly. A number of species of *Penicillium* can grow at 4–5°C and *P. commune* is one of the most common spoilage organisms of cheeses kept in refrigerators.

Some confectionery, including chocolate and desiccated coconut, is protected from mould spoilage to some extent by its low a_w, but species of *Chrysosporium*, e.g. *C. farinicola* have been implicated in the spoilage of such products.

Pickles and sauces, fruit juice concentrates and fermented beverages are normally protected from spoilage by the antimicrobial activity of ethanoic acid, sorbic acid and ethanol respectively. However, one species of yeast, *Zygosaccharomyces bailii*, is remarkably resistant to all these compounds (see below). **Table 6** indicates the most common moulds and yeasts associated with the spoilage of processed foods.

Table 6 Moulds and yeasts implicated in the spoilage of processed foods and food products

Food product	*Moulds and yeasts responsible for spoilage*
Dairy products (generally)	*Geotrichum candidum*
Butter and margarine	*Cladosporium herbarum*
Yoghurts	*Candida parapsilopsis*
	Kluyveromyces marxianus
Cheeses	*Penicillium commune*
	Penicillium glabrum
Chilled meats	*Cladosporium herbarum*
	Thamnidium elegans
Cured meats	*Penicillium* spp.
	Debaryomyces hansenii
Pastas	*Eurotium* spp.
	Wallemia sebi
Bread	*Penicillium roqueforti*
	Chrysonilia sitophila
	Endomyces fibuliger
	Hyphopichia burtonii
Jams	*Eurotium* spp.
	Wallemia sebi
	Penicillium coryophilum
	Zygosaccharomyces rouxii
Dried fruit	*Eurotium herbarum*
	Xeromyces bisporus
	Chrysosporium spp.
Honey	*Zygosaccharomyces rouxii*
Salted fish	*Polypaecilum pisce*
Canned fruits	*Byssochlamys fulva*
	Byssochlamys nivea
	Neosartorya fischeri
	Tallaromyces spp.

Control of Spoilage of Foods

Preservatives such as SO_2, benzoic acid and sorbic acid in fruit juices and related products are usually effective except in the case of the remarkably preservative-resistant yeast *Zygosaccharomyces bailii.* This yeast is also resistant to the ethanoic acid present in sauces, mayonnaises and salad dressings, and is not inhibited by the levels of ethanol which occur in fermented beverages such as ciders and wines. A single viable propagule of this yeast can multiply to cause problems – in the presence of a fermentable substrate, *Z. bailii* produces CO_2 in amounts sufficient to cause the explosion of plastic containers and even glass bottles. However, it is not particularly heat-resistant, and can be eliminated by pasteurization, coupled with the avoidance of post-pasteurization contamination by means of good hygiene.

Fruits and vegetables are frequently transported over long distances and stored for extended times, to make them available throughout the year. This has necessitated the prevention of mould spoilage by the use of processes and preservatives. Temperature, relative humidity and the composition of the atmosphere can all be manipulated so as to increase the period of satisfactory storage, but there are limitations: some fruits and vegetables suffer physiological damage if the temperature is too low, and if the relative humidity is too low, many plant products lose water to the atmosphere and wilt. Modified atmospheres, for example those in which CO_2 levels are elevated and those of O_2 are reduced, inhibit many moulds but can also influence the metabolism of some fruits adversely, leading to the development of off flavours. The 'hurdle' approach to the use of antimicrobial tools works well in this instance: in combination, a reduced temperature, reduced relative humidity and modified atmosphere, all at levels which might not be effective on their own, can work synergistically to significantly increase the time for which undamaged fruits and vegetables can be stored.

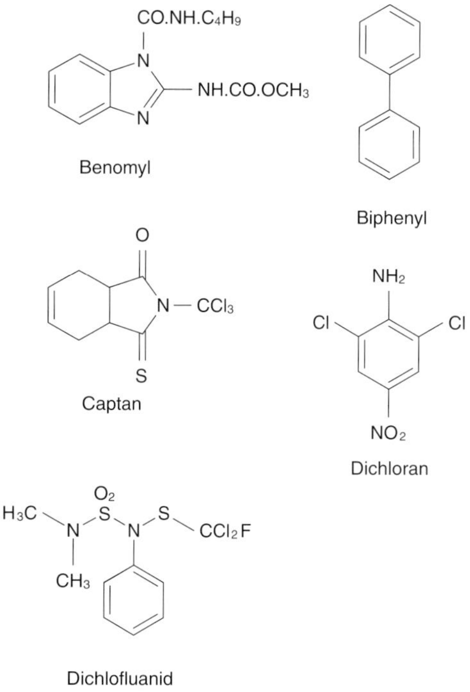

Figure 5 Some synthetic fungicides.

These methods of preservation, although effective, require sophisticated and expensive facilities, and so some chemical preservatives are permitted for use with certain products in different parts of the world (**Fig. 5**). Benomyl has been used on bananas and citrus fruits, on the assumption that their skins are not normally consumed, but there are now considerable problems of fungal resistance to benomyl, particularly in the case of *Penicillium digitatum*, which grows on citrus fruits. Biphenyl has been incorporated into the tissue paper used for wrapping high-value citrus fruits for transport. Captan has been used to control *Botrytis cinerea* in strawberries, but can only be applied until the fruit is developing. Dichloran has been used as a non-systemic dip for carrots in cold storage, and dichlofluanid has been used as a non-systemic, moderately persistent fungicide on soft fruits. Consumers are concerned, however, about the widespread use of such pesticides in foods, but their use reflects the considerable problems that fungi can cause in an increasingly complex international food industry.

See also: ***Alternaria***. ***Aspergillus***: Introduction. ***Botrytis***. ***Byssochlamys***. **Chilled Storage of Foods**: Principles. **Confectionery Products**: Cakes and Pastries. **Fungi**: The Fungal Hypha; Overview of Classification of the Fungi; Food-borne Fungi – Estimation by Classical Culture Techniques. ***Fusarium***. ***Geotrichum***. **Mycotoxins**: Classification. **Penicillium**: Introduction. **Preservatives**: Classification and Properties. ***Saccharomyces***: *Saccharomyces cerevisiae*. **Spoilage of Plant Products**: Cereals and Cereal Flours. ***Zygosaccharomyces***.

Further Reading

Beuchat LR (1987) *Food and Beverage Mycology*, 2nd edn. New York: Van Nostrand Reinhold.

Chelkowski J (1991) *Cereal Grain, Mycotoxins, Fungi and Quality in Drying and Storage*. Amsterdam: Elsevier.

Haard NF and Salunkhe DK (1975) *Post-Harvest Biology and Handling of Fruits and Vegetables*. Westport: AVI.

Hawksworth DL, Kirk PH, Sutton BC and Pegler DN (1995) *Ainsworth and Bisby's Dictionary of the Fungi*, 8th edn. Egham: Commonwealth Agricultural Bureau, International Mycological Institute.

Kozakiewicz Z (1989) *Aspergillus* Species of stored products. *Mycological Papers* 161. Egham: Commonwealth Agricultural Bureau, International Mycological Institute.

Pitt JI and Hocking AD (1997) *Fungi and Food Spoilage*, 2nd edn. London: Blackie Academic and Professional.

Samson RA, Hocking AD, Pitt JI and King AD (1992) *Modern Methods in Food Mycology*. Amsterdam: Elsevier.

Samson RA, Hoekstra ES, Frisvad JC and Filtenborg O (1997) *Introduction to Food-borne Fungi*, 5th edn. Baarn: Centraalbureau voor Schimmelcultures.

Samson RA and Pitt JI (1990) *Modern Concepts in* Penicillium *and* Aspergillus *Classification*. New York: Plenum Press.

Singh K, Frisvad JC, Thrane U and Mathus SB (1991) *An Illustrated Manual on Identification of some Seed-borne Aspergilli, Fusaria, Penicillia and their Mycotoxins*. Hellerup: Danish Government Institute of Seed Pathology for Developing Countries.

Tomlin C (1997) *The Pesticide Manual: A World Compendium*, 11th edn. Croydon: British Crop Protection Council.

STAPHYLOCOCCUS

Contents

Introduction

Scott E Martin, Department of Food Science and Human Nutrition, University of Illinois, Urbana, US

John J Iandolo, Department of Microbiology and Immunology, University of Oklahoma Health Sciences Center, Oklahoma City, US

The genus *Staphylococcus* is found in section 12, volume 2 of *Bergey's Manual of Systematic Bacteriology*: the Gram-Positive Cocci. The section includes aerobic genera (*Micrococcus*, *Planococcus* and *Deinococcus*), facultatively anaerobic genera (*Staphylococcus*, *Stomatococcus*, *Streptococcus*, *Leuconostoc*, *Pediococcus*, *Aerococcus* and *Gemella*) and anaerobic genera (*Peptococcus*, *Peptostreptococcus*, *Ruminococcus*, *Coprococcus* and *Sarcina*). The family Micrococcaceae consists of four genera: *Micrococcus*, *Stomatococcus*, *Planococcus* and *Staphylococcus*. Differential properties of the Gram-positive cocci include: arrangement of cells; strict aerobes; facultative anaerobes or microaerophiles; strict anaerobes; catalase reaction; cytochromes present; major fermentation products from carbohydrates anaerobically; peptidoglycan; teichoic acid in cell wall; and major menaquinones. Classification of *Staphylococcus* is as follows:

- Kingdom: Procaryotae
- Division: Firmicutes
- Class: Firmibacteria
- Family: Micrococcaceae
- Genus: *Staphylococcus*.

There are currently 32 species of *Staphylococcus* recognized. The G+C content of DNA from organisms

Table 1 Species of *Staphylococcus*

S. aureus	*S. epidermis*
S. capitis	*S. warneri*
S. haemolyticus	*S. hominis*
S. saccharolyticus	*S. auricularis*
S. saprophyticus	*S. caseolyticus*
S. cohnii	*S. kloosii*
S. xylosus	*S. simulans*
S. carnosus	*S. intermedius*
S. hyicus	*S. chromogens*
S. sciuri	*S. lentus*
S. gallinarum	*S. lugdenensis*
S. pasteuri	*S. caprae*
S. equorum	*S. arlettae*
S. felis	*S. piscifermentans*
S. schleiferi	*S. delphini*
S. muscae	*S. vitulus*

in the genus forms a narrow range of 30–39 mol%. Several other species are currently under investigation. **Table 1** lists the 32 genera found in the genus. **Table 2** lists those staphylococcal species of potential interest in foods. The name *Staphylococcus* is derived from the Greek *staphylo* (bunch of grapes) and *coccus* (a grain or berry), hence *Staphylococcus* – the grape-like coccus. The staphylococci are spherical Gram-positive cells, 0.5–1.5 μm in diameter, and can occur as single cells, in pairs, tetrads, short chains (three to four cells) or as irregular clusters. Cluster formation mainly occurs during growth on solid medium and results from cell division occurring in a multiplicity of planes, coupled with the tendency for daughter cells to remain in close proximity. The three-dimensional appearance is apparent with wet mounts; however, stained cells usually give the appearance of irregular sheets of cells. Staphylococci are non-motile and asporogenous; capsules may be present in young cultures, but are generally absent in stationary phase cells. Colony pigmentation on a nonselective medium such as tryptic soy agar can range from cream-white to bright orange. Most strains of staphylococci are considered by the American Type Culture Collection of Cultures on the Basis of Hazard as Class II (potential hazards). *Staphylococcus hyicus* is a Class IV (potential danger, extreme hazard) pathogen.

Staphylococcal species are aerobes or facultative anaerobes, and have both respiratory and fermentative metabolism. Staphylococci obtain energy via glycolysis, the hexose monophosphate shunt and the tricarboxylic acid cycle. They are catalase-positive and can utilize a wide variety of carbohydrates.

The cell wall contains peptidoglycan and teichoic acid. The peptidoglycan molecule is a glycan chain composed of regularly alternating *N*-acetylglucosamine and *N*-acetylmuramic acid residues linked by β-1,4 glycosidic linkages. The average chain length varies between 10 and 65 disaccharide units depending on the bacterial species. The carboxyl groups of all *N*-acetylmuramic acid residues are linked by amide bonds to the N-terminal L-alanine residues of an L-alanyl-γ-D-isoglutaminyl-L-lysyl-D-alanine peptide. Neighbouring peptides are interconnected by pentaglycine (occasionally hexaglycine) bridges extending from the carboxyl group of D-alanine of one peptide to the ε-amino group of the lysine residue of the next peptide. The peptidoglycan forms a rigid lattice surrounding the cell. The staphylococci are generally resistant to attack by the muramidase lysozyme but are sensitive to the lytic endopeptidase lysostaphin which attacks the glycine–glycine linkages present in the interpeptide bridges of the peptidoglycan. Teichoic acids are charged polymers in which repeating units of either ribitol or glycerol are linked through phosphodiester groups. Teichoic acids

Table 2 Staphylococcal species of interest in foods

Organism	*Coagulase*	*Nuclease*	*Enterotoxin*	*Haemolysis*	*Mannitol*	*Site*
S. aureus		+	–	+	+	Nasal membrane
S. intermedius	+	+	+	+	(+)	Nasal membrane, Carnivores
S. hyicus	(+)	+	+	–	–	Skin of pigs, milk
S. delphini	+	–		+	+	Dolphins
S. schleiferi	–	+		+	–	Human skin
S. caprae	–	+	+	(+)	–	Goat milk
S. chromogens	–	Weak	+	–	Variable	Cattle
S. cohnii	–	–	+	–	Variable	Human and primate skin
S. epidermidis	–	–	+	Variable	–	Human skin
S. haemolyticus	–	+	+	+	Variable	Human skin
S. lentus	–		+	–	+	Goat milk
S. saprophyticus	–	–	+	–	+	Human forehead
S. sciuri	–		+	–	+	Rodent skin
S. simulans	–	Variable		Variable	+	Human and primate skin
S. warneri	–	+	+	Weak	+	Human skin
S. xylosus	–	–	+	+	Variable	Human and primate skin

are found between the cell wall and the cytoplasmic membrane. They are proposed to function in maintaining the correct ionic environment for the cytoplasmic membrane. They also contribute to the surface charge of the staphylococcal cell.

In 1871, the German scientist von Recklinghausen first observed cocci in kidneys from a patient who died of pyaemia. Sir Alexander Ogston (a Scottish surgeon) and Louis Pasteur (both in 1880) demonstrated that certain pyogenic abscesses were caused by cocci that formed clumps. Ogston observed two types of cocci: chains, which he called *Streptococcus*, and groups or clusters, which he called *Staphylococcus*. Ogston is credited with the discovery of staphylococci and with giving the organisms their name. The staphylococci were first classified by Rosenbach in 1884, who adopted the Ogston name. Historically, efforts to classify the staphylococci have been both controversial and confusing. Because any classification scheme is arbitrary, continuous revisions will occur as new and better information and techniques are developed.

Most species of *Staphylococcus* are common inhabitants of the skin and mucous membranes. Some species have been found to have preferences for certain body sites. *S. capitis* is frequently found in large numbers on the scalp and forehead of humans. Species of *Staphylococcus* found on humans include: *S. aureus*, *S. epidermidis*, *S. hominis*, *S. haemolyticus*, *S. warneri*, *S. capitis*, *S. saccharolyticus*, *S. auricularis*, *S. simulans*, *S. saprophyticus*, *S. cohnii* and *S. xylosus*. *S. epidermidis*, coagulase-negative (see below) staphylococci, are natural inhabitants of the human skin, and comprise between 65 and 90% of all staphylococci isolated. Certain species of *Staphylococcus* are frequently found as aetiological agents of human and animal infections.

In the clinical laboratory, staphylococci are differentiated on the production of the enzyme coagulase. Two forms of coagulase are found to exist: one free and the other cell-bound. The two differ immunologically and have slightly different actions. Cell-free coagulase is protein in nature and different antigenic types have been identified. Coagulase-positive staphylococci have the ability to cause blood plasma to form a fibrin clot. This enzyme, produced by most virulent strains, has been suggested to facilitate staphylococcal pathogenicity by causing a fibrin barrier that appears to localize acute staphylococcal lesions. However, the importance of this barrier in infection is not entirely clear, as coagulase-negative staphylococci may also be pathogenic. Clotting of blood in vivo involves the interaction of a number of components. Very briefly, a clot is formed by the plasma protein fibrinogen which is present in a soluble form in the plasma, and which is transformed into an insoluble network of fibrous material called fibrin. Staphylococcal coagulase acts by converting (or by facilitating the conversion of) the fibrinogen to fibrin.

One of the reasons for the wide use of coagulase as a positive indication for the identification of staphylococci is that the test is easy to perform and very reliable. Cells of the suspect organism are mixed with commercially available human or rabbit plasma (with added citrate, oxalate or ethylenediaminetetraacetic acid, present to chelate calcium, required for in vivo clotting), using either the tube method, performed in a test tube, or the slide technique, performed on a microscope slide. The slide or test tube is incubated at 37°C, and read at 1 and 3 h. Previously, any degree of clotting, however slight, was considered positive. However, current procedures, as described in the *Bacteriological Analytical Manual*, require only a firm and complete clot that stays in place when the test tube is inverted to be considered positive. Results of the slide test, in which cell clumping is positive, correlate well with those of the test tube method.

Many foods (fresh and processed meats, poultry, some seafoods, dairy products) are contaminated with members of both genera *Staphylococcus* and *Micrococcus*. As a result the two genera may be confused. Differences between the two are found in **Table 3**.

S. aureus is the most common species in the *Staphylococcus* genus. It requires the presence of amino acids and vitamins for aerobic growth and, in addition, uracil and a fermentable carbon source for anaerobic growth. While *S. aureus* is capable of anaerobic growth, best growth occurs under aerobic conditions. The optimum temperature for growth is 35°C, although growth occurs over the range from 10 to 45°C. The pH range for growth is between 4.5 and 9.3, with the optimum between pH 7.0 and 7.5. As environmental conditions become more restrictive, so does the pH range for growth. *S. epidermidis* can cause human bacteraemia, endocarditis, infections of

Table 3 Differences between *Micrococcus* and *Staphylococcus*

Characteristics	Micrococcus	Staphylococcus
Irregular clusters	+	+
Tetrads	+	−
Capsule	−	−
Motility	−	−
Growth on furazolidone agar	+	−
Anaerobic fermentation of glucose	−	+
Oxidase and benzidine tests	+	−
Resistance to lysostaphin	+	−
Teichoic acid in cell wall	−	+
Mol% G+C of DNA	65–75	30–39

medical shunts, intravenous catheters, joint prostheses and genitourinary tract infections. *S. saprophyticus* is a common cause of urinary tract infections in young women. *S. intermedius* and *S. hyicus* are important pathogens in dogs and pigs, respectively. *S. chromogenes* is frequently isolated from the milk of cows suffering from mastitis.

Isolation and Detection

Direct microscopic examination of normally sterile fluids (blood, cerebrospinal fluid, etc.) may be useful in the clinical laboratory. Results should be reported as presumptive and the diagnosis of 'Gram-positive cocci resembling staphylococci' should be made. Most plating media for the detection of staphylococci have been developed specifically for *S. aureus*.

S. aureus-selective media utilize a number of different toxic chemicals to achieve selectivity. Some of the ingredients used include sodium chloride, tellurite, lithium chloride and various antibiotics. A number of media have been suggested for the isolation of *S. aureus* from food when >100 per gram may be present. Some of these include: staphylococcal medium 110, Vogel–Johnson agar, egg yolk–sodium azide agar, tellurite–polymyxin–egg yolk agar and Baird-Parker agar (**Table 4**).

Most selective media are suitable for the enumeration of normal or unstressed *S. aureus*. However, due to processing, preservation or other adverse conditions, sublethal stress may occur, resulting in the increased sensitivity of *S. aureus* to the selective agents. Because injured cells exhibit an increased sensitivity to selective agents, *S. aureus* may go undetected in conventional selective enumeration procedures. It has been demonstrated that the recovery of heated or dried cells of *S. aureus* may be lost or its activity reduced by heating or drying and that blood, which contains catalase, or the addition of pyruvate, helped in the enumeration by destroying H_2O_2 produced by recovering cells. Baird-Parker agar is most satisfactory in enumerating injured cells when compared with other staphylococcal selective media.

Table 4 Examples of selective media for *Staphylococcus aureus*

Agar medium	*Selective agents*	*Diagnostic agents*
Staphylococcus medium 110	Sodium chloride	Mannitol
		Gelatin
Vogel–Johnson	Lithium chloride	Mannitol
	Potassium tellurite	Tellurite
	Glycine	Phenol red
Egg yolk–sodium	Lithium chloride	Egg yolk
Azide	Potassium tellurite	Tellurite
	Polymyxin B sulfate	
Baird-Parker	Lithium chloride	Egg yolk
	Potassium tellurite	Tellurite

Some authors have suggested that most staphylococcal species of clinical significance can be identified on the basis of a few key characteristics. These include colony morphology, coagulase production, oxygen requirements, haemolysis, novobiocin resistance, acetylmethylcarbinol (acetoin) production, aerobic utilization of selected carbohydrates and certain enzyme activities. On a nonselective agar such as tryptic soy agar or nutrient agar, most staphylococcal species grow abundantly in 18–24 h when incubated at 35°C, with colony diameter generally 1–3 mm. Colony morphology may be an aid to species identification, and colony pigmentation is of importance. A number of commercial kit identification systems are available that permit identification of several staphylococcal species.

See also: ***Staphylococcus***: *Staphylococcus aureus*; Detection by Cultural and Modern Techniques; Detection of Staphylococcal Enterotoxins.

Further Reading

American Type Culture Collection (1982) Gherna RL and Pienta P (eds) *Catalogue of Strains I*, 15th edn. Rockville, MD: ATCC Publishers.

Bacterial Analytical Manual (1984) 6th edn, chs 13 and 14. Arlington, VA: Division of Microbiology, Center for Food Safety and Applied Nutrition, US Food and Drug Administration.

Baird-Parker AC and Davenport E (1965) The effect of recovery medium on the isolation of *Staphylococcus aureus* after heat treatment and after the storage of frozen or dried cells. *J. Appl. Bacteriol.* 28: 390.

Jay JM (1996) In: *Modern Food Microbiology*. P. 429. New York: Chapman & Hall.

Kloos WE and Jorgensen JH (1985) Staphylococci. In: Lennette EH (ed.) *Man. Clin. Microbiol.*, 4th edn. P. 143.

Mandell GL, Bennett JE and Dolin R (1995) In: *Principles and Practice of Infectious Diseases*, 4th edn. P. 1777. New York: Churchill Livingstone.

Minor TE and Marth EH (1976) *Staphylococci and their Significance in Foods*. New York: Elsevier.

Sneath PA, Mair NS, Sharpe ME and Holt JG (1986) *Section 12 Gram-Positive Cocci, Bergey's Manual of Systematic Bacteriology*. Vol. 2, p. 999. Baltimore, MD: Williams & Wilkins.

Staphylococcus aureus

J Harvey and **A Gilmour**, Department of Agriculture for Northern Ireland, Agriculture and Food Science Centre, Belfast, Northern Ireland, UK

Characteristics of the Species

The species *Staphylococcus aureus* is a member of the genus *Staphylococcus*, the natural reservoirs of which are the skin and mucous membranes of humans and animals. Many staphylococcal species have become adapted to life on particular animal species but in contrast *S. aureus* is present on most marine and terrestrial mammals, and may be present as a non-aggressive member of the normal skin microflora or may be associated with infectivity and disease. Up to 50% of humans may be healthy carriers of *S. aureus*; the body sites most often colonized are the nares, throat, hair and hands. The organism may also be isolated from healthy domestic and food animals as well as being associated with disease, particularly mastitis. As a result of adaptation to the extreme microenvironmental conditions which pertain on the surfaces of warm-blooded animals and humans, *S. aureus* has evolved a high degree of resistance to desiccation and osmotic stress. They are robust organisms which survive well outside their natural hosts in air, in dust and in water despite their lack of motility and their susceptibility to bacteriocins, bacteriophages and simple bacterial products of competitors – factors which would seem to mitigate against their survival.

S. aureus is one of more than 20 species which comprise the genus *Staphylococcus*, a member of the family Micrococcaceae. Staphylococci are Gram-positive, catalase-positive coccoid organisms, which grow within the temperature range 7–48°C with an optimum of 35–40°C and can metabolize glucose oxidatively or fermentatively. When staphylococcal cultures are examined microscopically, the cells are characteristically arranged in clusters as a result of their mode of division. Members of the *S. aureus* species produce a number of extracellular compounds including membrane-damaging toxins, epidermolytic toxin, toxic shock syndrome toxin, staphylococcal enterotoxins (SEs), pyrogenic exotoxin and exoenzymes such as coagulase and thermostable nuclease (TNase). Production of coagulase and TNase have been widely used as identification markers for *S. aureus*.

Extrachromosomal elements which may be carried by *S. aureus* include bacteriophages, plasmids and transposons. These elements increase the capacity of *S. aureus* cells to respond to changes in environmental conditions and facilitate the transfer of genetic information between cells. Studies have shown that *S. aureus* plasmids encode determinants for resistance to various antimicrobials, are involved in control of SE production and may also mediate in adherence processes. Some *S. aureus* plasmids are quite large (40–60 kb) and the genetic expression associated with these plasmids which has been identified to date probably constitutes a small proportion of the total genetic information encoded.

The SEs are a group of extracellular proteins, with molecular weights in the range 27 000–30 000. They are similar in composition and biological activity but are identified as separate proteins due to differences in antigenic structure. To date eight serologically distinct SEs (SEA, SEB, SEC_1, SEC_2, SEC_3, SED, SEE, SEH) have been identified, although analysis of food-borne staphylococcal outbreaks indicates that further unidentified SEs exist. A toxin identified as enterotoxin F (SEF) has been renamed toxic shock syndrome toxin-1 because it was found to lack emetic activity in monkeys and a gene for a new enterotoxin has been identified although the encoded protein (SEG) has not been purified. SEs are thought to act on both the intestinal epithelium and the vomiting centre of the central nervous system to cause typical food-poisoning symptoms. The possible role of SEs in other diseases is also being actively investigated.

Adherence of *S. aureus* to the skin surface is a major determinant for colonization and many studies have been carried out on different aspects of the adherence of the organism to various types of animal cells. From such studies it is known, for example, that *S. aureus* adheres preferentially to nasal cells from natural nasal carriers. More recently a few studies have been directed at investigation of the adherence of *S. aureus* to inert food contact surfaces. For example, studies have shown that colonization of poultry-defeathering machines by *S. aureus* appears to be partly due to mucoid growth that aids attachment of the cells to the equipment. It has also been reported that attachment of the microorganisms to surfaces results in increased resistance to cleaning and disinfecting agents.

Methods of Detection and Enumeration of *S. aureus* in Foods

Conventional Techniques

Enrichment When *S. aureus* are present in low numbers detection may require enrichment. Liquid media containing NaCl have been used but cannot be recommended since stressed cells are recovered poorly. The most suitable selective enrichment medium for *S. aureus* is Giolitti and Cantoni broth, as modified by Van Doone. Cultures with blackening after 24 h at 37°C are considered positive for *S. aureus*, although confirmatory testing is required. The medium is also suitable for enumerating low numbers of *S. aureus* using most probable number (MPN) determinations.

Selective Plating When numbers are sufficiently high, *S. aureus* may be isolated from foods by direct plating on selective media. Simple selective media containing NaCl or polymyxin have long been available but cannot be recommended in preference to Baird-Parker (BP) agar. BP agar is relatively complicated but this is outweighed by its greater selectivity and recovery of stressed cells. BP agar contains egg yolk plus tellurite for diagnositic purposes, pyruvate plus glycine as selective growth stimulators and tellurite plus lithium chloride as selective inhibitors. After inoculation and incubation of BP agar plates, presumptive *S. aureus* colonies are selected for confirmatory testing. It should be remembered that, although *S. aureus* colonies on BP agar plates are characteristically jet black surrounded by a white rim, an opaque zone and a zone of clearing, some strains of *S. aureus* are uncharacteristic in that they produce colonies which are lighter in colour than normal, whilst other strains lack a zone of opaqueness and/or clearing.

Confirmatory Testing Production of coagulase and/or TNase are the tests most commonly used to confirm the identity of presumptive *S. aureus* isolates, although it is now known that neither enzyme is unique to *S. aureus*. Slide agglutination test kits which detect clumping factor and protein A have recently become commercially available and are increasingly being used for identification of *S. aureus*. None of the aforementioned tests are 100% reliable and therefore none can be relied upon as the sole confirmatory test.

To eliminate the need for further testing, modifications of BP agar have been developed whereby coagulase or TNase activity of colonies are observed on selective agar media. However, difficulties in preparing the media and interpretation of results have resulted in many workers continuing to use unmodified BP agar, although this will probably change if the rabbit plasma agar now commercially available in complete dehydrated form is found to perform satisfactorily. Besides the selective media for *S. aureus*, there are also media available for selective isolation of all staphylococcal species. Although these media are suitable for isolation of *S. aureus* they may not be as successful as BP agar for the recovery of stress-injured cells.

Identification Since none of the commonly used confirmatory tests is unique to *S. aureus* species, it is prudent to carry out a fuller characterization of isolates obtained in this way. The tests shown in **Table 1** will confirm isolates as members of the genus *Staphylococcus* whilst those in **Table 2** will ensure they have been assigned to the correct species. The tests can be carried out using materials prepared in-house or using kits which are commercially available.

Biotyping *S. aureus* found on different hosts may vary in their physiological and biochemical properties and those differences form the basis of several systems which classify isolates into biotypes or ecovars. Thus one human ecovar (biotype A) is recognized, poultry and swine ecovars are classified in one biotype (biotype B) and cattle and sheep ecovars are classified in another biotype (biotype C). Biotyping has limitations and in some situations is inadequate to establish the origin of *S. aureus*.

Phage Typing Bacteriophages (phages) are highly specialized viruses that attack members of a particular bacterial species or strains of a species. Phage typing is based on the fact that surface receptors on phages bind to specific receptors on the bacterial cell surface. Only those phages that can bind to these cell-surface receptors can infect the bacteria and lyse the cells. Phage typing involves the examination of the sensitivity of microorganisms to the lytic action of selected phages. Since *S. aureus* may originate from different hosts, phages isolated from suitable host-adapted staphylococci will give the best results. This leads to the major disadvantage of the phage-typing method for this organism, namely, the difficulty of assembling a reasonably sized set of phages suitable for the purpose of typing a collection of isolates which originate from different hosts and ecosystems such as might be encountered in a collection of *S. aureus* food isolates.

Serotyping Methods for differentiating *S. aureus* on the basis of differences in their antigenic composition are long established and it has been suggested that antigens are more stable than phage markers. However, use of serological methods for typing *S. aureus* is restricted because of time and resources required for production of the antisera.

Plasmid Profiling Since many *S. aureus* isolates carry plasmids differing in size, plasmid profiling, whereby plasmid DNA is isolated from different strains and examined electrophoretically, has been used to type isolates from different sources. The use of this method is restricted to situations where there is a high per-

Table 1 Simple tests for differentiation of *Staphylococcus* from other genera of similar morphology

	Staphylococcus	Micrococcus	Streptococcus
Morphology	Coccoid	Coccoid	Coccoid
Gram reaction	+	+	+
Catalase reaction	+	+	−
Fermentation of glucose	+	−	*
Sensitivity to lysostaphin	S	R	*
Acid from glycerol in presence of erythromycin	+	−	*

* Test not applicable.

Table 2 Characteristics of selected *Staphylococcus* species

	S. aureus	S. hyicus	S. intermedius	S. epidermis
Coagulation of human plasma	+	−	+	−
Production of thermostable nuclease	+	+	+	−
Aetoin produced when grown on glucose medium	+	−	−	+
Pigment produced when grown on agar media	+	−	−	−

centage of plasmid carriage in the *S. aureus* population under investigation.

Enterotoxin Production Detection of SEs is routinely carried out using immunological methods, although biological assays using kittens, rhesus monkeys and chimpanzees have been described. Few laboratories have the facilities for handling these animals and such methods are only used for special purposes. Of the immunological methods available, gel diffusion, especially double gel diffusion with a reference toxin included to ensure the specificity of reactions, have been the methods of choice for years. However, these methods have been largely supplanted by reversed-phase latex agglutination (RPLA) and enzyme-linked immunosorbent assays (ELISAs) which are more sensitive and are commercially available as microtitre plate or polystyrene bead assay kits. Detection of SE production in pure cultures is straightforward and usually entails simultaneous growth/concentration using a sac-culture-broth or cellophane-over-agar technique followed by serological detection of toxin. The relative ease with which SE in pure culture can now be detected means that it is more useful to test *S. aureus* isolates directly for SE production with one of the new rapid methods (e.g. RPLA or ELISA) than to rely on any of the conventional identification tests as indicators of enterotoxigenicity. For detection of SE produced in foods using gel-diffusion methods it is necessary to carry out extraction, purification and concentration steps before assaying for toxin. Use of the much more sensitive RPLA and ELISA assays for SE detection in foods means that very simple extraction procedures are usually sufficient. The original ELISAs used polyclonal antibodies to detect SE but subsequently monoclonal antibodies have been used to increase the sensitivity of the assay.

Molecular Techniques

Isolation and Identification DNA-based techniques in combination with the conventional cultural techniques of enrichment have often been used to detect *S. aureus* in foods. Direct detection of *S. aureus* in foods using a DNA-based method is a more desirable approach but major problems remain to be solved before this can be routinely performed. For example, use of polymerase chain reaction (PCR)-based methods for the detection of *S. aureus* and other pathogens has been hampered by interference of the PCR reaction due to the presence of inhibitors in certain foods. Sample preparation methods continue to be improved and assay formats are becoming available which offer the possibility of more rapid and sensitive direct detection and identification of *S. aureus* in food compared to the use of cultural methods. Genes involved in the production of coagulase, protein A and TNase by *S. aureus* can be detected using PCR technology.

Typing The long-established conventional methods for typing *S. aureus* have been supplemented and even replaced in recent years by the introduction of methods such as multilocus enzyme electrophoresis (MEE), restriction fragment length polymorphism (RFLP), pulsed-field gel electrophoresis (PFGE) of macro-restricted DNA and PCR-based methods such as random amplification of polymorphic DNA (RAPD). Molecular typing methods have been shown to be valuable strain-specific discriminators for the epidemiological characterization of *S. aureus* and, because of their speed, ease of use and discriminatory power, allow for more detailed investigation of the epidemiology of this organism in the food chain.

Enterotoxins Much of the effort to develop rapid molecular methodology for *S. aureus* has been concentrated on the detection of SEs and methods using gene probes for the detection of enterotoxigenic strains have been developed. Since nucleotide sequences of genes coding for SEs are known, probes can be synthesized which hybridize with complementary sequences. Probe amplification methods can be used to produce highly sensitive detection methods to assay enterotoxigenic *S. aureus* in foods.

Procedures Specified in National/ International Regulations or Guidelines

The last 10 years have seen a continuation of co-operation between recognized scientific organizations both nationally and internationally through organizations such as the British Standards Institution and the US Food and Drug Administration, the Codex Alimentarius Commission, the International Dairy Federation, the International Organization for Standardization and the Association of Official Analytical Chemists. In line with the growing trend towards sharing of microbiological expertise on a global scale and as part of the continuing effort to validate and harmonize methods of microbiological analysis, the above organizations have issued recommendations on methodology for the detection and identification of *S. aureus* and its enterotoxins as well as recommending procedures more generally applicable to food safety assurance, such as the hazard analysis critical control point (HACCP) method.

Importance of *S. aureus* in the Food Industry

Although several staphylococcal species have been implicated in food-poisoning incidents, *S. aureus* is the predominant species. *S. aureus* is widespread in nature and many of the raw materials arriving at food establishments for processing and manufacture of foods will contain this organism. If materials containing *S. aureus* are not processed and handled properly during food manufacture, there is a risk of resulting staphylococcal food poisoning (**Table 3**). Although staphylococcal food poisoning is decreasing in many nations, the relative incidence in various countries varies substantially depending on geography and local eating habits. In the US, for example, it is one of the most economically important diseases, reported to cost $1.5 billion each year. Staphylococcal food poisoning is only one of a number of food-borne illnesses which have increasingly concerned the food industry, public health authorities and consumers and over several decades assurance of food safety has been a subject of growing interest. Individual procedures employed by those interested in assuring food safety have gradually been drawn together and a comprehensive procedure for food safety assurance known as the HACCP system has evolved which is now universally recognized and accepted for food safety assurance.

Essentially, HACCP is a three-component system consisting first of determining the hazard posed to the consumer; second identifying critical control points to ensure safe management of the hazard; and third, carrying out suitable monitoring to ensure critical control points are operating effectively. Many HACCP concepts are not new to the food industry and have been successfully employed by many sectors of the industry to ensure the safety of their products. However the HACCP system provides a systematic, uniform approach to food safety assurance. The hazard associated with the presence of *S. aureus* in food is staphylococcal food poisoning. To assure food safety with regard to this hazard, critical control points which should be considered are:

- use of raw materials containing the lowest practicable numbers of *S. aureus*
- use of treatments to reduce microbial load and eliminate *S. aureus*
- use of additives and/or low temperature to prevent multiplication of *S. aureus* during handling and storage
- use of hygienic handling to prevent re-introduction of *S. aureus*.

To ensure correct operation of critical control points, systematic monitoring should be carried out in accordance with a statistically based plan.

S. aureus is readily killed by the various processes (e.g. pasteurization) used to treat raw materials in the food industry. Therefore the challenge which faces the industry with regard to ensuring food safety with regard to staphylococcal food poisoning is the protection of products from contamination with *S. aureus* and the avoidance of conditions which allow multiplication of the organism. Contamination by humans is common in many food products, whilst other types of products have particular production problems which lead to contamination with *S. aureus*. For example, in poultry processing, defeathering equipment can become colonized with *S. aureus*, resulting in high numbers of the organism being deposited on carcasses being processed. Contamination problems can normally be effectively dealt with by paying careful attention to personal hygiene and to the cleaning and disinfection of equipment and food contact surfaces. In this regard it should be noted that *S. aureus* are not particularly heat-resistant and are

Table 3 Sources, risks and consequences of *Staphylococcus aureus* in the food chain

	Sources	*Risks*	*Consequences of failing to control risks*
Natural environment	Animals Humans Air, water, vegetation	Poor animal husbandry and poor human health care	Increase in animal and human *S. aureus* infections and occurrence in nature
Food-processing environment			
Raw materials	Animal carcasses Animal products Added ingredients	High numbers of *S. aureus*	*S. aureus* survive processing. Cross contamination between raw and processed materials
Processing	Food contact surfaces Air, water Human operatives	Processing failure, insufficient cleaning or disinfection, poor ventilation, insufficient water treatment, poor hygiene	*S. aureus* survive processing. Post-process contamination of product with *S. aureus*
Handling	Food product	Temperature abuse during storage, intrinsic growth control factors incorrectly adjusted	Multiplication of *S. aureus* and production of staphylococcal enterotoxin
Food preparation environment	Animals Humans Food contact surfaces Air, water	Contamination of processed food with enterotoxigenic *S. aureus*. Temperature abuse during holding of prepared food	Multiplication of *S. aureus* and production of staphylococcal enterotoxin

readily killed by disinfectants used in the food industry. Multiplication of *S. aureus* in food is prevented by storage at 7°C or less. If this temperature cannot be attained consistently it is necessary to limit the possibility of growth by the manipulation of intrinsic factors such as pH, water activity and NaCl concentration in the product.

When monitoring the effectiveness of critical control points, the results of analyses of samples for numbers of *S. aureus* and presence of enterotoxins should be interpreted with care. *S. aureus* cells compared to SE are less resistant to processes used in the food industry such as low pH, heat, irradiation and high-pressure treatments. Thus, a scenario is possible whereby prior growth of *S. aureus* occurs with SE production followed by reduction or elimination of the organism while biologically active SE remains. Although SEs are produced over a wide range of environmental and storage conditions, it is possible to have conditions of temperature, water activity and pH such that growth of the *S. aureus* organism occurs with little or no production of SE. If monitoring reveals the presence of *S. aureus* in the environment or product at unacceptable levels, then the use of a method with powerful discriminatory power to type isolates is invaluable for tracing the source of the organism.

Importance of *S. aureus* for the Consumer

Unlike the UK and US, many countries lack a comprehensive national surveillance system for food-borne disease and so the true extent of staphylococcal food poisoning in the world is unknown. A wide range of foods have been implicated, including meat, milk, fish, egg and vegetable products which have been processed by heating, fermenting or drying and concentrating. The most common factors in staphylococcal food poisoning outbreaks are:

1. Post-processing contamination of food with *S. aureus*, most often a human strain introduced directly by a person involved in food preparation or, less frequently, an animal strain of *S. aureus* introduced through cross contamination from raw foods.
2. Holding of contaminated food at a temperature and for a period sufficient to allow multiplication of *S. aureus* with concurrent production of SE. Small outbreaks may occur in the home, but those on a larger scale are usually associated with social occasions when catering practices such as preparation of food well in advance of eating and 'warm holding' of prepared food provide the opportunity for growth of *S. aureus*.

Ingestion of food containing preformed SE leads to the rapid development (1–6 h) of the symptoms of nausea, vomiting and diarrhoea that characterize staphylococcal food poisoning. All eight of the serologically identified SEs have been implicated in food-poisoning incidents, although SEA is the antigenic type most frequently found in cases of food poisoning. There is considerable variation in susceptibility to SE among normal adults and the precise dose required to cause illness in humans depends upon the susceptibility of the individual. The basis for this is not known, although prior exposure to SEs may confer a degree of immunity or tolerance. Susceptibility is greatest in young children; one study showed that as little as 0.5–0.75 ng ml^{-1} of enterotoxin A in chocolate milk was able to cause illness in schoolchildren. The elderly and generally unhealthy persons have a higher susceptibility than young or healthy persons, although there is no specific complaint which predisposes to staphylococcal intoxication. *S. aureus* food poisoning is associated with well-defined types of food, but there is no relationship with particular dietary habits and thus no cultural or other groups are at particular risk. In most cases of staphylococcal food poisoning, no treatment is required and complete recovery follows quickly after cessation of symptoms. In severe cases, rehydration and treatment for shock are necessary and hospitalization may be required. Death is rare and usually occurs only when the patient is elderly, very young or suffering from a debilitating disease.

See also: **Biochemical and Modern Identification Techniques**: Food-poisoning Organisms. **Biofilms**. **Chilled Storage of Foods**: Principles. **Direct (and Indirect) Conductimetric/Impedimetric Techniques**: Food-borne Pathogens. **Food Poisoning Outbreaks**. **Hazard Appraisal (HACCP)**: The Overall Concept. **International Control of Microbiology**. **Molecular Biology – in Microbiological Analysis**. **Predictive Microbiology & Food Safety**. **Process Hygiene**: Overall Approach to Hygienic Processing. ***Staphylococcus***: Introduction; Detection by Cultural and Modern Techniques; Detection of Staphylococcal Enterotoxins.

Further Reading

Anonymous (1989) *Microbiological Examination for Dairy Purposes*. Section 3.10.2 Detection of staphylococci. British Standard 4285. London: British Standards Institution.

Baird-Parker AC (1990) The staphylococci: an introduction. In: Jones D, Board RG and Sussman M (eds) *Staphylococci*. Society for Applied Bacteriology symposium series number 19. P. 1S. Oxford: Blackwell Scientific Publications.

Gilmour A and Harvey J (1990) Staphylococci in milk and milk products. In: Jones D, Board RG and Sussman M (eds) *Staphylococci*. Society for Applied Bacteriology symposium series number 19. P. 147S. Oxford: Blackwell Scientific Publications.

Grant KA, Dickinson JH and Kroll RG (1993) Specific and rapid detection of foodborne bacteria with rRNA sequences and the polymerase chain reaction. In: Kroll RG, Gilmour A and Sussman M (eds) *New Techniques in Food and Beverage Microbiology*. Society For Applied Bacteriology, technical series 31. P. 147. Oxford: Blackwell Scientific Publications.

ICMSF (1996) *Staphylococcus aureus*. In: Roberts TA, Baird-Parker AC and Tompkin RB (eds) *Microorganisms in Foods 5. Microbiological Specifications of Food Pathogens*. P. 299. London: Chapman & Hall.

Mossel DAA and Van Netten P (1990) *Staphylococcus aureus* and related staphylococci in foods: ecology, proliferation, toxinogenesis control and monitoring. In: Jones D, Board RG and Sussman M (eds) *Staphylococci*. Society for Applied Bacteriology symposium series number 19. P. 123S, Oxford: Blackwell Scientific Publications.

Tranter HS and Brehm RD (1990) Production, purification and identification of the staphylococcal enterotoxins. In: Jones D, Board RG and Sussman M (eds) *Staphylococci*. Society for Applied Bacteriology symposium series number 19. P. 109S, Oxford: Blackwell Scientific Publications.

Varnam AH and Evans MG (1991) *Staphylococcus aureus*. In: Varnam AH and Evans MG (eds) *Foodborne Pathogens: An Illustrated Text*. P. 235. London: Wolfe.

Wilson IG, Gilmour A and Cooper JE (1993) Detection of toxigenic microorganisms in foods by PCR. In: Kroll RG, Gilmour A and Sussman M (eds) *New Techniques in Food and Beverage Microbiology*. Society for Applied Bacteriology, technical series 31. P. 163. Oxford: Blackwell Scientific Publications.

Detection by Cultural and Modern Techniques

Sita R Tatini, Department of Food Science and Nutrition, University of Minnesota, St Paul, USA

Reginald Bennett, FDA, Center for Food Safety and Applied Nutrition, Washington, DC, USA

Introduction

Sir Alexander Ogston was the first to demonstrate the roles of cocci in purulent lesions in humans and he named this organism *Staphylococcus* (grape-like clusters) in 1882. Rosenbach studied this organism in pure culture and named the orange colony-forming coccus, *Staphylococcus pyogene aureus*, in 1884. The first association of staphylococci with food poisoning dates back to 1884. Cheese containing spherical organisms caused a large outbreak in Michigan. Cheese extracts consumed by humans caused illness. Partially cooked meat from a sick cow caused food poisoning in 1894 in France. Dried beef contaminated with staphylococci was associated with an outbreak in Kalamazoo, Michigan in 1907. In 1914, Barber, in the Philippines, clearly demonstrated that refrigerated stored milk from a mastitic cow caused staphylococcal food poisoning in humans.

In 1930, Dack rediscovered staphylococcal food poisoning by feeding sterile culture filtrates of yellow staphylococcal growth (*Staphylococcus aureus*) isolated from a Christmas cake involved in food poisoning to humans causing typical symptoms of staphylococcal intoxication. Isolation of *Staphylococcus aureus* from foods involved in food poisoning and later simultaneous isolation of *S. aureus* and direct detection of enterotoxins in foods became a reality. Development of sensitive (1 ng ml^{-1}) and practical in vitro methods in the mid 1980s established that coagulase-negative species also produce enterotoxins.

Taxonomy

Rosenbach described two species of *Staphylococcus* (*S. pyogenes aureus* and *S. pyogenes albus*) based on colony pigmentation. Subsequently, staphylococci and other Gram-positive, catalase positive cocci (micrococci) were placed in the family Micrococcaceae despite their distinct difference in their mol% G+C content of DNA (65–75% for micrococci and 30–39% for staphylococci (**Table 1**). Use of advanced molecular techniques and DNA–DNA hybridization techniques have now established that there are multiple species of staphylococci. Currently, there are 32 species (including six subspecies).

Sources of Staphylococci and Their Significance in Foods

Staphylococci are widely distributed in nature and are closely associated with warm-blooded animals (humans and food animals). *Staphylococcus aureus* is the type species of the genus *Staphylococcus* which is coagulase positive and, thus, potentially pathogenic to humans and animals. Even though *S. aureus* can cause a variety of infections, it is its capability to produce enterotoxins (see sections on enterotoxins) that is of concern in foods. When there was a lack of sensitive in vitro methods for the detection of the immunologically different enterotoxins, isolation and identification of large numbers of coagulase-positive staphylococci (*S. aureus*) have been used to signal the potential presence of enterotoxins in foods. Isolation of coagulase-positive *S. aureus* frequently from foods involved in food poisoning and the fact that 50% of *S. aureus* produce one or more of the known enterotoxins led to the use of coagulase positive *S. aureus* in specifications for foods. It is now clear that there are other species of coagulase-positive *Staphylococcus* and that there are several coagulase-negative species that also produce the known enterotoxins. The mere presence of enterotoxigenic species of *Staphylococcus* in foods is not of concern, it is their growth in foods to high levels (> 500 000 per millilitre or per gram) which results in preformed enterotoxins in foods. Further, enterotoxins are less likely produced in raw foods that are contaminated with enterotoxigenic staphylococci because of the competitive spoilage flora and the temperature of storage being below the minimum (10°C) required for production of enterotoxins. Raw milk from mastitic cows, sheep or goats is an exception to this in that they may contain enterotoxins produced in milk within the udder. Foods processed under conditions that allow growth of desirable organisms (fermented meats and cheeses) also allow limited growth of staphylococci. If the fermentation process fails then staphylococci could reach very high levels with the accumulation of enterotoxins. In these foods, processors monitor the ingredients and/or use pasteurized milk for cheese production to reduce and/or eliminate staphylococci and also monitor the number of coagulase-positive species (*S. aureus*) of staphylococci to screen out finished product likely to contain enterotoxins. Staphylococci (*S. carnosus* and *S. piscifermentans*) and some micrococci (*Micrococcus varians*) are used in fermented sausages, cheeses and fish to promote flavour and colour fixation of meat and also to reduce colour defects that may arise from hydrogen peroxide produced by lactic acid bacteria. Catalase production of staphylococci will produce this effect. In these products, speciation of staphylococci and their enumeration are important. Technological advances in the use of starter cultures in fermented foods and elimination of post-pasteurization contamination led to the current recommendation of < 10 000 coagulase-positive *S. aureus* per gram as a specification. In view of the fact that coagulase-negative staphylococci also produce enterotoxins, one must use < 10 000 of all

Table 1 Differentiating characteristics of *Micrococcus* and *Staphylococcus* genera

Characteristics	Micrococcus	Staphylococcus
Anaerobic glucose fermentation	–	+
Acid from glycerol in the presence of 0.4 μg erythromycin	–	+
Sensitive to:		
Lysozyme (25 μg ml^{-1})	S/R	R
Lysostaphin (200 μg ml^{-1})	R	S
Furazolidone (nitrofuran)	R	S
Oxidase	+	–[a]
Benzidine	+	–[a]
Mol% G+C content of DNA	65–75	30–39

S, sensitive; R, resistant.
[a]Three species of staphylococci are positive: *S. caseiolyticus*, *S. sciuri* and *S. lentus*.

Table 2 Differentiating characteristics of coagulase-positive *Staphylococcus*

	S. aureus	S. intermedius	S. hyicus	S. delphini	S. schleiferi *subsp*[a]	
					schleiferi	coagulans
Tube coagulase (Rabbit plasma EDTA)	+	+	V	+	–	+
Clumping factor	+	+	–	–	+	–
Protein A	+	–	+	–	–	+
TNase	+	+	+	–	+	+
Acid anaerobically from D-mannitol	+	–	–	–	–	–
Acetoin	+	–	–	–	+	–
Enterotoxin	+	+	+	?	?	?

V, variable.
[a]Pseudocoagulase reaction.

Table 3 Some characteristics of enterotoxigenic species of coagulase-negative staphylococci

Novobiocin resistant (≥1.6 μg ml^{-1}) strains:
- *S. cohnii*: subsp. 1 sucrose and ribose–, urease–, β-glucuroxidase$_+$ and β-galactosidase$_+$
- *S. cohnii*: subsp. 2 sucrose and ribose–, urease+, β-glucuronidase$_-$, and β-galactosidase$_-$
- *S. equarum*: optimum growth at 30°C, no growth at >37°C, urease+
- *S. lentus*: oxidase+, raffinose+, mannitol+
- *S. sciuri*: oxidase+, raffinose–, mannitol+
- *S. saprophyticus*: urease+, phosphatase–, trehalose–, mannitol–
- *S. xylosus*: xylose+, arabinose+, β-glucosidase+, β-galactosidase+

Novobiocin sensitive (≥1.6 μg ml^{-1}) strains:
- *S. caprae*: urease+, mannitol+, trehalose–, sucrose+, haemolysis+
- *S. capitis*: urease–, maltose and trehalose–
- *S. haemolyticus*: maltose and mannose–, sucrose and trehalose+, haemolysis+
- *S. epidermidis*: mannitol and trehalose–, maltose, mannose and sucrose+, urease+, some strains are β-glucosidase+
- *S. simulans*: urease and β-galactosidase+, sucrose and trehalose+, mannitol (aerobic and anaerobic)+
- *S. warneri*: urease+, phosphatase–, β-glucosidase+, sucrose+, trehalose+ and mannose–
- *S. chromogenes*: pigment, sucrose+, mannitol+, trehalose+, urease+ and hyaluronidase–

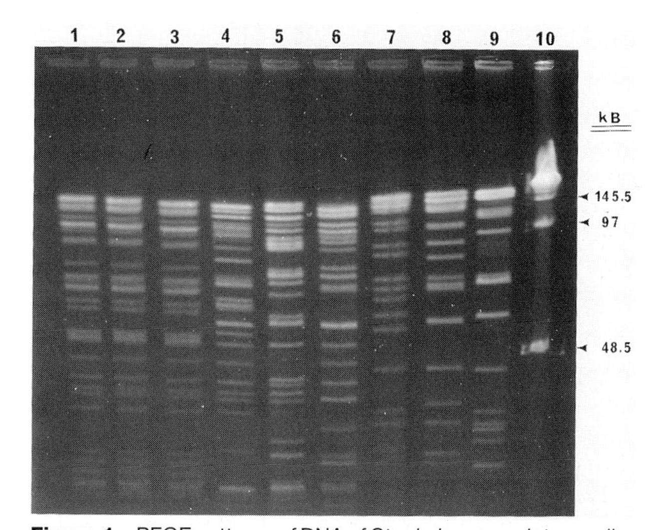

Figure 1 PFGE patterns of DNA of *Staphylococcus intermedius* and *S. aureus* (food and patient isolates). Lanes 1–3 *S. intermedius* isolates from a single outbreak (one from food, one from food-handling equipment and one from a patient).

staphylococci to assure safety from staphylococcal enterotoxins. Enumeration should be performed before numbers of staphylococci decrease during ripening of cheese or drying of sausage. Epidemiological investigations of staphylococcal food-borne outbreaks and establishment of sources of contamination of foods require subspecies or strain characterization. Use of bacteriophage sensitivity patterns was successfully used in the past for strains that were phage typable. Modern molecular typing techniques have more recently been used for subspecies or strain characterization and are currently used to establish sources of contamination in foods.

Enumeration of Staphylococci in Foods

Selective and differential enumeration media would permit direct enumeration of staphylococci. Baird-Parker Agar (BPA) medium contains lithium chloride, glycine and potassium tellurite to inhibit other organisms and allow staphylococci to form black colonies by reducing potassium tellurite. Egg yolk and sodium pyruvate would help resuscitate injured staphylococci, whilst in addition, egg yolk allows differentiation of lipolytic-positive *S. aureus* by forming a halo and/or a precipitate around the black colony. However, some strains of *Staphylococcus aureus*, especially those of bovine origin, do not show this reaction. Further, coagulase negative staphylococci grow poorly on this medium and only form pinpoint

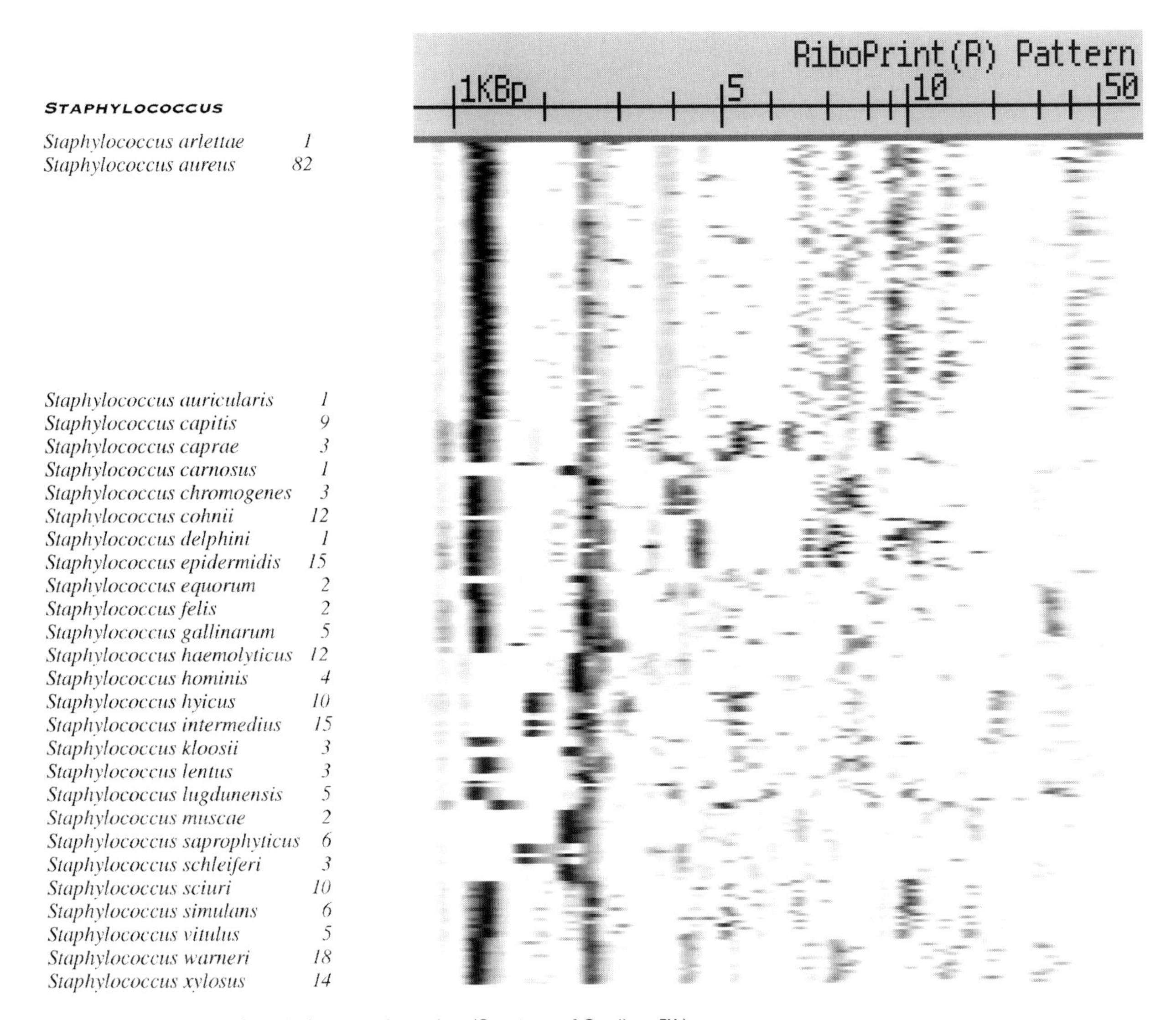

Figure 2 Ribotyping of staphylococccal species. (Courtesy of Qualicon™.)

or small colonies. Some strains of *S. intermedius* do not reduce tellurite and form white colonies. Some bacilli, micrococci, streptococci and *Proteus* may also grow on this medium. Modified Schleifer and Kramer agar is more selective and allows all staphylococci to grow on this medium. Micrococci, bacilli and streptococci are inhibited by sodium azide in this medium. Incorporation of rabbit plasma EDTA and fibrinogen into this medium allows direct enumeration of coagulase-positive (colonies with a precipitate) staphylococci. Modified Schleifer and Kramer medium is as follows:

1%	Tryptone or peptone
0.5%	Beef extract
0.3%	Yeast extract
1.0%	Glycerol
1%	Sodium pyruvate
0.05%	Glycine
2.25%	Potassium thiocyanate (KSCN)
0.06%	NaH_2PO_4
0.09%	NaH_2PO_4
0.2%	LiCl
1.3%	Agar

Add 10 ml of 0.15% filter-sterilized sodium azide per litre of sterile molten medium.

Details of the incorporation of bovine fibrinogen, rabbit plasma EDTA in this medium can be found in the literature.

Colonies on Baird-Parker agar are grouped on the basis of visual characteristics and one colony from each group is then tested by tube coagulase test. Rabbit plasma EDTA is used for this test because it is commercially available. Any degree of clotting rated as: 1+, small unorganized clots; 2+, organized but small clot; 3+, a clot as big as three-quarters of the volume; and 4+, a solid clot in 6–24 h is considered

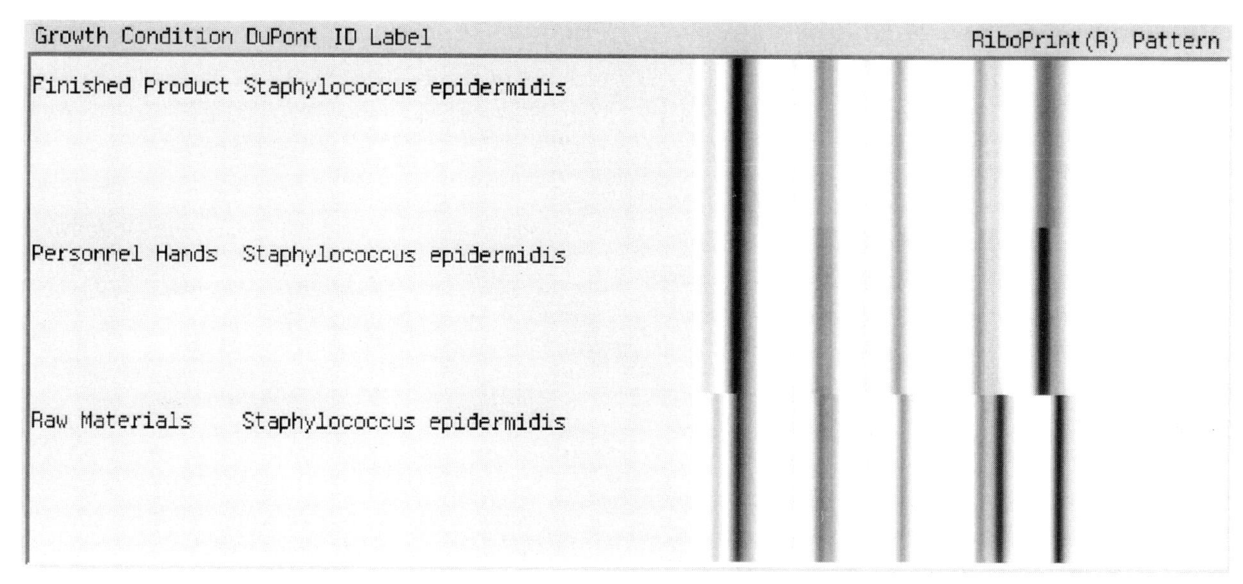

Figure 3 Staphylococcal strain discrimination by ribotyping. (Courtesy of Qualicon™.)

positive. Coagulase positivity is not an exclusive property of *S. aureus* and five other species show this trait. These are shown in **Table 2** (*S. intermedius*, *S. hyicus*, *S. delphini* and two subspecies of *S. schleiferi*). The former two species also produce enterotoxins. Cultures showing weak or doubtful coagulase reactions, especially those of 1+ or 2+ reaction, require additional testing and this is usually achieved by testing for heat-stable nuclease (TNase) which survives boiling. *S. schleiferi*, *S. hyicus* and *S. intermedius* are also TNase positive. Clumping factor (bound coagulase/fibrinogen affinity factor) and protein A in the cell wall of *S. aureus* are also found in other species (*S. hyicus* and *S. schleiferi*). However, acid is only produced anaerobically from D-mannitol by *S. aureus* from cultures shown in Table 2. Coagulase, clumping factor, protein A and TNase are not unique to *S. aureus* species. *S. aureus* can be definitively identified through use of gene specific rRNA probes (Accuprobe Systems, Genprobe, San Diego, CA or Gene Trak Systems, Hopkington, MA). These systems can identify *S. aureus* strains even if they are coagulase negative, clumping factor negative and/or protein A negative. Speciation of coagulase negative species requires testing for various enzymes, fermentation of various sugars, pigment, haemolysis and sensitivity to novobiocin. There are 13 coagulase-negative staphylococci that produce enterotoxins. Of these, seven are novobiocin sensitive (1.6 $\mu g\,ml^{-1}$) and six are novobiocin resistant ($\leqslant$1.6 $\mu g\,ml^{-1}$). Some of the diagnostic features are listed in **Table 3**. *S. equorum* can be differentiated from others by its optimum growth temperature (30°C) and it will not grow at > 37°C. This species may grow very poorly at 37°C. *S. lentus* and *S. sciuri* are oxidase positive. *S. caseiolyticus* which is also oxidase positive does not ferment D-mannitol whereas *S. lentus* and *S. sciuri* ferment mannitol. Two subspecies of *S. cohnii* do not ferment sucrose or ribose and one subspecies is negative and one is positive for urease, β-glucuronidase and β-galactosidase and both are negative for β-glucosidase.

S. saprophyticus is urease positive, phosphatase negative and does not ferment trehalose. *S. xylosus* ferments xylose and arabinose and is positive for β-glucosidase and β-galactosidase. Of the novobiocin-sensitive species, *S. caprae* is urease positive, ferments mannitol, does not ferment trehalose, ferments sucrose slowly and is weakly haemolytic. *S. capitis* does not ferment maltose or trehalose and is urease negative. *S. haemolyticus* does not ferment maltose or mannose and ferments arabinose, trehalose and sucrose. *S. epidermidis* does not ferment mannitol or trehalose but ferments maltose, mannose and sucrose. It is urease positive and some strains produce β-glucosidase. *S. simulans* is positive for urease and β-galactosidase and ferments sucrose and trehalose aerobically and ferments mannitol both aerobically and anaerobically. *S. warneri* is negative for phosphatase and positive for urease and β-glucosidase. It ferments sucrose and trehalose and does not ferment mannose. *S. chromogenes* is urease positive and negative for hyaluronidase, pigmented, ferments sucrose, mannitol and trehalose.

There are several commercially available systems to identify species of staphylococci which show a 90% correlation with conventional methods.

Establishing Sources of Staphylococcal Contamination

Both epidemiologists and food processors need to establish the source of contamination because species level identification is not adequate. Bacteriophage sensitivity has only limited application because there are strains that are not typable by this system. Genetic fingerprinting techniques are being applied to strain characterization of staphylococci. Two of these approaches are pulsed-field gel electrophoresis (PFGE) of DNA restriction fragment profiles and DNA restriction fragment polymorphism of rRNA genes (ribotyping). Pure cultures of staphylococci are isolated from the food, food processing environment, humans handling the food or food ingredients. These are then analysed by PFGE in which the culture is embedded in agarose and lysed with lytic agents. The released DNA is subjected to restriction endonuclease digestion. Fragments are resolved by gel electrophoresis (pulsed-field) stained with ethidium bromide and photographed. **Figure 1** shows PFGE patterns of DNA of *Staphylococcus intermedius* and *S. aureus*. Lanes 1–3 are *S. intermedius* isolates from a single outbreak (one from foods, one from food-handling equipment and one from a patient). Isolates of *S. intermedius* in lanes 1, 2 and 3 showed identical patterns and strains of the same species in lanes 4, 5 and 6 showed different patterns. Likewise *S. aureus* strains in lanes 7–9 showed different patterns within the species and differences from *S. intermedius*. This shows the discriminating power of this procedure.

Figures 2 and 3 show ribotyping of staphylococci using Riboprinter™ of Qualicon. In this system, pure cultures of staphylococci are introduced into a sample carrier and transferred to the machine. All the subsequent steps are performed by the machine: releasing DNA, digesting DNA to completion with restriction enzyme, transferring fragments to agarose gel cassettes, performing direct blot electrophoresis to a nylon membrane, hybridizing membrane with labelled rRNA operon from *Escherichia coli* and detecting hybridized fragments with chemiluminescence. Figure 2 shows ribotypes of staphylococcal species and Figure 3 shows strain discrimination.

Current advances in technology have brought to the investigator rapid, highly sensitive and molecular-based specific approaches in the identification of pathogens encountered in foods.

See also: **Nucleic Acid-based Assays**: Overview. ***Staphylococcus***: Introduction; *Staphylococcus aureus*; Detection of Staphylococcal Enterotoxins.

Further Reading

Beckers HJ, Van Ludsen FM, Bindschedler O and Guerraz D (1984) Evaluation of a power-plate system with a rabbit plasma-bovine fibrinogen agar for the enumeration of *Staphylococcus aureus* in food. *Canadian J. Microbiology* 30: 470–474.

Belkum et al (1998) Assessment of resolution and inter-center reproducibility of genotyping *Staphylococcus aureus* by pulsed-field gel electrophoresis of Sma I macrorestriction fragments: a multicenter study. *Journal of Clinical Microbiology* 36: 1653–1659.

Changing the Way Microbiology is Done. Principles and Application of the Riboprinter™ Microbial Characterization. Wilmington, DE: Qualicon Publication, Dupont Co.

Jones D, Board RG and Sussman M (1990) Society for Applied Bacteriology, Symposium Series number 19. *Journal of Applied Bacteriology* 69 (Suppl.): 1S–189S.

Schleiferi KH (1986) Gram-positive Cocci. In: Sneath PHA (ed.) *Bergey's Manual of Determinative Bacteriology.* Baltimore: Williams and Wilkins.

Detection of Staphylococcal Enterotoxins

Merlin S Bergdoll (dec), Food Research Institute, University of Wisconsin-Madison, USA

Types and General Properties of the Staphylococcal Enterotoxins

The staphylococcal enterotoxins are low-molecular-weight single chain proteins each having specific antigenic sites. Seven have been identified by their production of specific antibodies that identifies them as individual proteins. They are referred to as enterotoxins A (SEA), B (SEB), C (SEC), D (SED), E (SEE), G (SEG) and H (SEH). There is no enterotoxin F as this letter was given to a protein that proved not to be an enterotoxin. Several SECs have been identified with only three to five residue differences in their amino acid sequences, but all reacting with the same antibody.

The major structure that characterizes the enterotoxins is a cystine loop (**Table 1**) in the centre of the molecule. The significance of the loops is not known; however, it is assumed that they stabilize the molecular structure and may contribute to their resistance to proteolysis. An essentially common sequence of amino acids follows the cystine loop (**Table 2**). It was originally thought that this sequence was the active site, but amino acid substitution experiments were not confirmatory.

The enterotoxins are simple proteins that are hygroscopic and easily soluble in water and salt solutions.

Table 1 Cystine loops of the enterotoxins

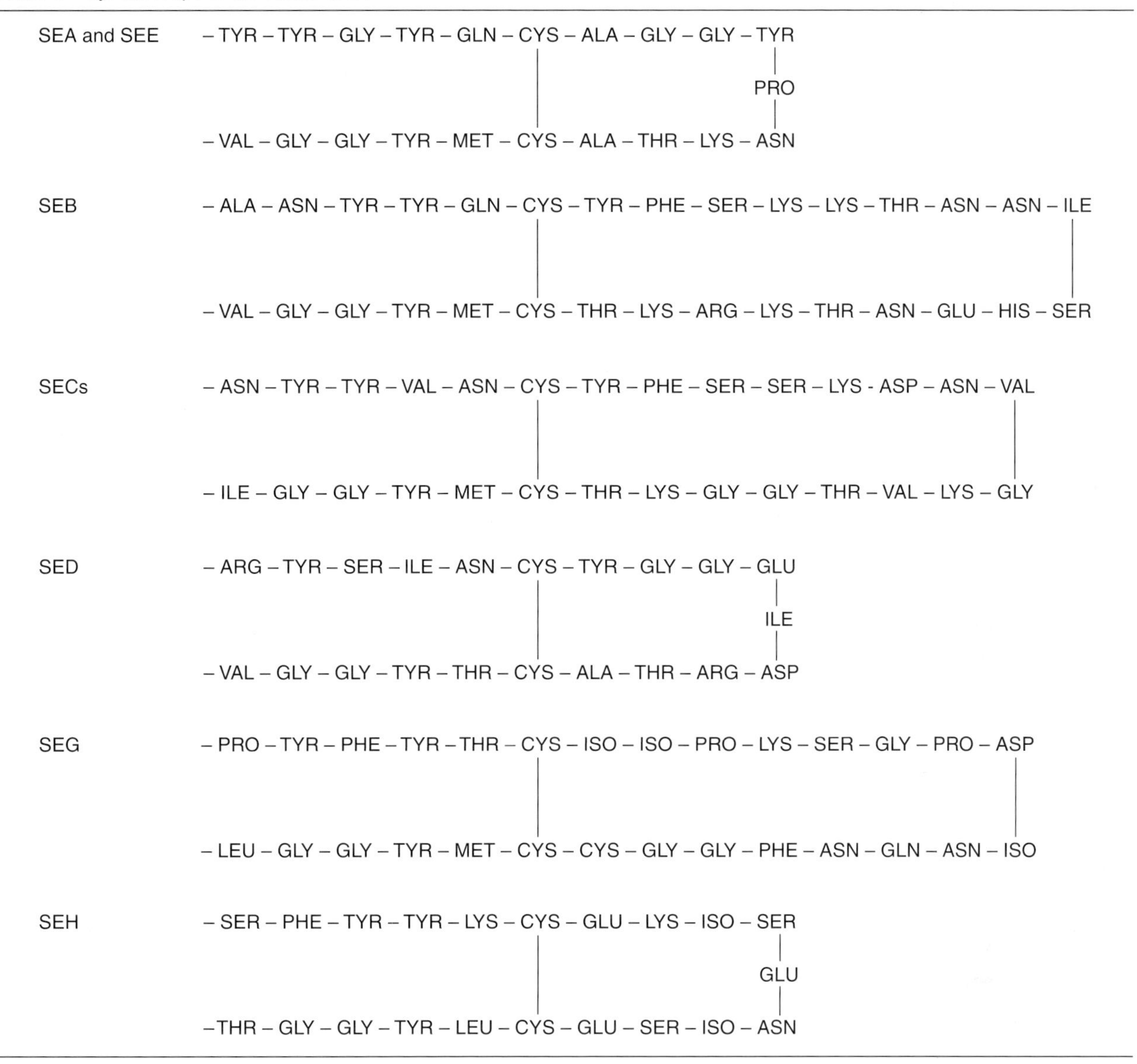

Enterotoxin	Cystine loop
SEA and SEE	– TYR – TYR – GLY – TYR – GLN – CYS – ALA – GLY – GLY – TYR
	PRO
	– VAL – GLY – GLY – TYR – MET – CYS – ALA – THR – LYS – ASN
SEB	– ALA – ASN – TYR – TYR – GLN – CYS – TYR – PHE – SER – LYS – LYS – THR – ASN – ASN – ILE
	– VAL – GLY – GLY – TYR – MET – CYS – THR – LYS – ARG – LYS – THR – ASN – GLU – HIS – SER
SECs	– ASN – TYR – TYR – VAL – ASN – CYS – TYR – PHE – SER – SER – LYS - ASP – ASN – VAL
	– ILE – GLY – GLY – TYR – MET – CYS – THR – LYS – GLY – GLY – THR – VAL – LYS – GLY
SED	– ARG – TYR – SER – ILE – ASN – CYS – TYR – GLY – GLY – GLU
	ILE
	– VAL – GLY – GLY – TYR – THR – CYS – ALA – THR – ARG – ASP
SEG	– PRO – TYR – PHE – TYR – THR – CYS – ISO – ISO – PRO – LYS – SER – GLY – PRO – ASP
	– LEU – GLY – GLY – TYR – MET – CYS – CYS – GLY – GLY – PHE – ASN – GLN – ASN – ISO
SEH	– SER – PHE – TYR – TYR – LYS – CYS – GLU – LYS – ISO – SER
	GLU
	–THR – GLY – GLY – TYR – LEU – CYS – GLU – SER – ISO – ASN

Table 2 Amino acid sequence homology in the enterotoxins[a]

SEA:	Asn-Lys-Thr-Ala-***Cys-Met-Tyr-Gly-Gly-Val-Thr***-Leu-***His***-Asp-Asn-***Asn***
SEB:	Lys-Arg-Lys-Thr-***Cys-Met-Tyr-Gly-Gly-Val-Thr***-Glu-***His***-Asn-Gly-***Asn***-
SEC:	Gly-Gly-Lys-Thr-***Cys-Met-Tyr-Gly-Gly***-Ile-***Thr***-Lys-***His***-Glu-Gly-***Asn***
SED:	Asp-Arg-Thr-Ala-***Cys***-Thr-***Tyr-Gly-Gly-Val-Thr***-Pro-***His***-Gly-Gly-***Asn***-
SEE:	Asn-Lys-Thr-Ala-***Cys-Met-Tyr-Gly-Gly-Val-Thr***-Leu-***His***-Asp-Asn-***Asn***-
SEG:	Phe-Gly-Gly-Cys-***Cys-Met-Tyr-Gly-Gly***-Leu-***Thr***-Phe-Asn-Ser-Ser-Glu-
SEH:	Asn-Ile-Ser-Glu-***Cys***-Leu-***Tyr-Gly-Gly***-Thr-***Thr***-Leu-Asn-Ser-Glu-Lys-

[a]Bold and italic common residues.
Residue numbers: SEA, 102–117; SEB, 109–124; SEC: 106–121; SED, 97–112; SEE, 99–114; SEG, 101–118; SEH, 88–103.

They have relatively low molecular weights of 25 000–29 000 daltons (**Table 3**). They are basic proteins, with pI values of 7.0–8.6, with the exception of SEG and SEH, which have pI values of 5.6 and 5.7, respectively. The maximum absorbance of the enterotoxins is 277 nm, which is higher than for normal proteins. The amino acid compositions of the enterotoxins are presented in **Table 4**. There is a degree of

Table 3 Properties of the enterotoxins

Molecular weight	25 000–29 000
Isoelectric points	5.7–8.6
Absorbance	Peak at 277 nm
Enzyme resistance	Trypsin and pepsin
Monkey emetic dose	
Intragastric	5–20 μg per animal
Intravenous	0.02–0.5 μg kg^{-1}

homology between the enterotoxins, but sufficient differences in the amino acid sequences result in each having different antigenic sites. Two of the enterotoxins, SEB and SEC_1 can be nicked in the cystine loop, without neutralizing the emetic reaction.

The enterotoxins are known primarily as food-poisoning agents. The most common symptoms of this disease are nausea, vomiting, retching, abdominal cramping and diarrhoea. When injected intravenously into animals, they are mitogenic, immunosuppressive, stimulate the production of interferon, interleukins-1 and -2, and tumour necrosis factor, and other activities. The enterotoxins have been labelled superantigens because they can activate as many as 10% of the mouse's T cell repertoire, whereas conventional antigens stimulate less than 1% of all T cells.

Foods Implicated and Major Outbreaks

Any food that provides a good medium for the growth of staphylococci may be involved in staphylococcal food poisoning. Large outbreaks have become relatively rare in recent years, particularly in the United States and other developed countries. Only two large outbreaks have been reported in the last decade in the United States, one from chocolate milk involving over 850 school children and the other in butter-blend spreads involving 265 cases. In the United States, pork, particularly baked ham, has been the food most frequently involved in outbreaks. Several large outbreaks have occurred in the United States that involved baked ham sandwiches served at picnics, with one of the largest outbreaks involving 1300 individuals. Cream-filled bakery goods, such as eclairs, are infrequently involved in the United States, but in 1990, 485 individuals became ill in Thailand with typical staphylococcal food poisoning symptoms from consuming eclairs. Cream-filled cakes have been implicated in several staphylococcal food poisoning outbreaks, one occurring in Spain resulted in 1800 illnesses with typical staphylococcal food poisoning symptoms. The cream filling contained more than 5×10^6 staphylococci per gram, which produced SEA. One of the largest outbreaks to occur in the US resulted in 1364 children becoming ill after consuming chicken salad. In Japan, rice balls is a common item taken on outings and has been a major cause of staphylococcal food poisoning, one outbreak involving 1500 individuals. In Brazil the two foods most frequently involved are cream-filled cake and a white cheese frequently produced on the farm or in small establishments.

Table 4 Amino acid residues of the staphylococcal enterotoxins

Amino acid	*SEA*	*SEB*	*SEC_1*	*SEC_2*	*SEC_{3a}*	*SEC_{3b}*	*SED*	*SEE*	*SEG*	*SEH*
Lysine	24	34	31	29	30	29	30	22	23	28
Histidine	6	5	5	5	5	5	6	6	3	2
Arginine	7	5	3	3	3	3	7	8	5	6
Aspartic acid	17	24	19	19	20	19	18	17	14	23
Asparagine	19	19	23	23	23	23	17	17	22	15
Threonine	17	13	15	16	15	16	12	17	16	10
Serine	13	14	15	16	16	16	16	13	14	14
Glutamic acid	15	12	14	14	13	13	16	14	17	18
Glutamine	11	8	5	5	5	5	9	9	6	6
Proline	4	6	6	6	6	6	5	5	11	2
Glycine	15	9	14	15	15	15	13	19	3	11
Alanine	7	5	7	6	6	6	6	7	3	9
1/2-cystine	2	2	2	2	2	2	2	2	3	2
Valine	13	16	18	17	18	17	11	11	15	8
Methionine	2	8	7	8	8	8	2	1	5	2
Isoleucine	10	9	10	10	9	10	15	10	14	22
Leucine	23	16	16	15	15	15	22	25	17	18
Tyrosine	18	20	16	17	17	17	12	17	13	14
Tryptophan	2	2	2	2	2	2	1	2	2	1
Phenylalanine	8	12	11	11	11	11	9	8	16	7
Total amino acids	233	239	239	239	239	239	228	230	232	218
Amides	30	27	28	28	28	28	26	26	28	21
pI	7.3	8.6	8.6	7.0	8.2	8.2	7.0	7.4	5.6	5.7

Overview of Classical Techniques for Enterotoxin Analysis

It was not possible to develop specific methods for the detection of the enterotoxins before the first enterotoxin was identified and purified in 1959. Until that time the only means of detecting the presence of the enterotoxins was with animals that gave emetic reactions to the toxin, either intragastrically or intravenously. The feeding of young rhesus monkeys provided the most reliable bioassay for enterotoxins because, of the biologically active substances produced by the staphylococci, only the enterotoxins caused emesis when administered by the oral route. Assays are made by administering solutions of the enterotoxins (up to 50 ml) to monkeys (2–3 kg) by catheter. The animals are observed for 5 h for emesis. A response in at least two animals is accepted as a positive reaction.

The intravenous injection of cats and kittens also proved useful for the detection of the enterotoxins. However, when culture supernatants are injected intravenously, it is necessary to inactivate other biologically active substances by treatment with trypsin or pancreatin. Cats are not as reliable as monkeys because they are subject to nonspecific reactions. Because antibodies are specific for each enterotoxin, it is necessary to continue the use of animal testing until each new enterotoxin is purified and antibodies produced against it.

All laboratory methods for detecting the enterotoxins are based on the use of specific antibodies to each of the enterotoxins. The first tests developed were based on the reaction of the enterotoxin with the specific antibodies in gels to give a precipitin reaction. Many types of gel reactions have been used in the detection of the enterotoxins, the most common being some form of either the Ouchterlony gel plate or the microslide. Modification of the Ouchterlony gel plate test used in the Food Research Institute for detection of enterotoxin-producing staphylococcal strains is the optimum sensitivity plate (OSP) method (**Fig. 1**). The antiserum is placed in the centre well and the control enterotoxin placed in outer wells 1 and 4 and the unknowns placed in wells 2, 3, 5 and 6. Incubation overnight at 37°C allows the enterotoxin and the antibody to diffuse toward each other and if enterotoxin is present in the unknown sample a precipitin line will join with the control line. The normal sensitivity is 0.5 μg ml^{-1} but can be increased to 0.1 μg ml^{-1} by a 5-fold concentration of the staphylococcal culture supernatant fluids.

The microslide method is the most sensitive of the gel-diffusion methods (0.05–0.1 μg ml^{-}), but care is needed in preparing the slides; however, the results are often difficult to interpret. Many things can go wrong with this method and experience is very important in using it successfully.

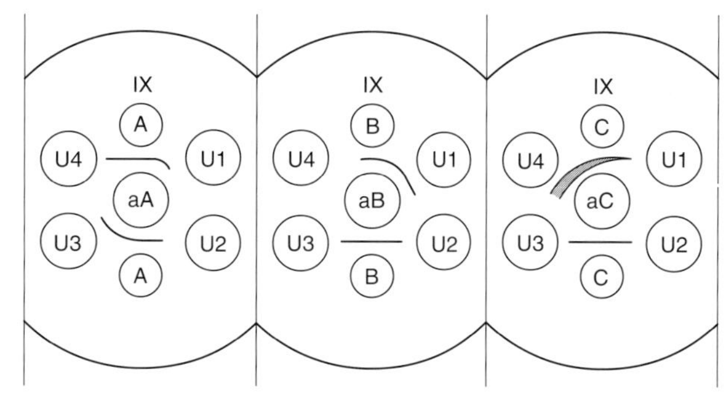

Figure 1 Enterotoxin analysis in optimum sensitivity plates. A, B, C, enterotoxins A, B, C, respectively (4 mg ml^{-1}). aA, bB, cC, antienterotoxins A, B, C, respectively. Unknown U1 contains approximately 0.5 μg of A and 6 μg of B per millilitre; unknown U2 contains no enterotoxins; unknown U3 contains approximately 2 μg A ml^{-1}; unknown U4 contains approximately 16 μg C ml^{-1}. IX indicates unconcentrated unknowns. (From Bergdoll.)

The single gel diffusion tube method developed by Oudin in 1952 was used for detecting enterotoxin, primarily for the detection of enterotoxin production by staphylococcal strains. Specific antibodies in gel were placed in the bottom half of a 4 mm diameter tube and the toxin solution was placed on top of the gel. Reaction of the enterotoxin with the antibody formed a band of precipitate that increased in length with time as the enterotoxin diffused into the agar. This method was used as an analytical method as there was a direct relationship between the amount of enterotoxin in the sample and the length of the precipitin band in the gel. The major problem with the gel-diffusion methods is the lack of sensitivity as the minimum amount that could be detected is about 50–100 ng ml^{-1} of enterotoxin with the microslide method. This was inadequate for the detection of enterotoxin in foods without time-consuming extraction and concentration of the extract of the food to be analysed. The gel diffusion methods are still used as screening methods for testing staphylococcal strains for enterotoxin production.

The first sensitive method developed for the detection of enterotoxin in foods was the radioimmunoassay (RIA) which involved the use of ^{125}I-labelled enterotoxin. The food extract was treated with specific antibody in a small plastic tube before the iodinated enterotoxin was added. The reaction product was precipitated with *S. aureus* cells containing protein A. The supernatant was removed and the radioactivity in the precipitate counted to determine the amount of enterotoxin in the food extract. This method is no longer used as newer methods do not require the use of radioactive materials.

Detailed Procedures for Modern Techniques

The detection of enterotoxin in foods requires methods that are sensitive to less than 1 ng g^{-1} of food. The quantity of enterotoxin present in foods involved in food-poisoning outbreaks may vary from < 1 to > 50 ng g^{-1}. Although little difficulty is usually encountered in detecting the enterotoxin in foods involved in food-poisoning outbreaks, outbreaks do occur in which the amount of enterotoxin is < ng g^{-1}, such as in the case of an outbreak among school children from chocolate milk (**Table 5**). Another situation in which it is essential to use a very sensitive method is determining the safety of a food for consumption. In such situations it is necessary to show that no enterotoxin is present by the most sensitive methods available.

The ELISA methods were applied to the detection of the enterotoxins in foods soon after they were originally developed for the detection of other proteins. The most common type of ELISA is the sandwich method in which the antibody is treated with the unknown sample before the antibody–enterotoxin complex is treated with the enzyme–antibody conjugate. This procedure is preferred because the amount of enzyme and, thus, the colour developed from the enzyme–substrate reaction, is directly proportional to the amount of enterotoxin present in the unknown sample. Both alkaline phosphatase and horseradish peroxidase enzymes have been used, with the former being easier to couple to the antibodies. However horseradish peroxidase gives a more sensitive reaction, primarily because the colour developed with the substrate is green, blue or orange; the colour developed with the substrate used with alkaline phosphatase, *p*-nitrophenyl phosphate (pNPP), gives a yellow colour.

The majority of users of ELISA methods use microtitre plates or strips to which the antibodies are attached. The samples are placed in the wells and allowed to react with the antibodies. After the reaction the samples are removed and the plate is washed.

Table 5 See analysis in 2% chocolate milk (Reproduced from Evenson et al (1988) *International Journal of Food Microbiology* 7: 311–116)

Sample	*ng ml^{-1}*	*ng per half pint*
1	0.40	94
2	0.75	177
3	0.45	106
4	0.63	149
5	0.78	184
6	0.65	153
Ave	0.61	144

The enzyme–antibody conjugate is added and allowed to react with the enterotoxin–antibody complex. The conjugate is removed and the plate washed before the substrate is added. The colour is allowed to develop for a specific time before it is read in a plate reader. Enterotoxin controls and blanks are included in the test. A positive reaction is recorded if the colour reading is higher than the blank reading.

An alternate procedure is the use of polystyrene balls to which the antibodies are attached. The ball method is more sensitive because 20 ml of food extract is used. The colour-coded balls are removed from the extract, washed, and each is added to the properly colour-coded tubes. The proper enzyme–antibody conjugate is added to each tube and allowed to react with the enterotoxin–antibody complex. The balls are washed in the tubes several times before 1 ml of the substrate is added. Colour development usually can be read visually because quantitation is not necessary for determination of a positive reaction. If the colour is to be read in a colourimeter development is halted after 30 min with sulphuric acid solution.

A dipstick method was developed by IGEN, Inc. (Rockville, MD, USA) with monoclonal antibodies developed in the Food Research Institute (Madison, WI, USA) and licensed to Transia (Lyon, France). The antibodies to each of the enterotoxins were adsorbed onto nitrocellulose paper attached to wells in plastic sticks, one well for each enterotoxin and a blank. The stick is placed in the food extract for adsorption of the enterotoxin. The stick is removed and washed before placing it in a solution of the conjugates for reaction with the antibody–enterotoxin complex. After removal and washing, the stick is placed in the substrate solution for colour development. Positive reactions are determined visually. This method is no longer available commercially.

Many investigators have developed ELISA methods for the detection of staphylococcal enterotoxins in their own laboratories, with slight differences in procedures. One difference involved the use of biotinylated antibodies and avidin–alkaline phosphatase instead of coupling the enzyme directly to the antibodies. The biotinylated antibodies apparently were more stable than the enzyme–antibody conjugates. A sensitivity of 0.1 ng ml^{-1} was possible. Most ELISA methods are sensitive to at least 0.5 ng of enterotoxin per millilitre.

A reversed passive latex agglutination (RPLA) method has been developed for the detection of enterotoxin in foods and also for the detection of enterotoxin-positive staphylococcal strains. In this method latex particles are coated with the enterotoxin antibodies before placing them in wells in microtitre plates. The sample is added and allowed to react. If

Table 6 Staphylococcal strain testing by RPLA method (From Bergdoll 1990)

Strain[a]	*Enterotoxins detected*	
	RPLA	*OSP*
188	SEA, SED	SEA, SED
311	SEA	SEA
228	SEA, SED	–
452	SEA	SEA
365	SEC	–
581	SEC	SEC
609	SEA, SED	–
754	SEA, SEB	SEA, SEB
802	SEB	SEB
887	SEA	–

[a]Strains were obtained from Dr James K. Todd, The Children's Hospital, Denver, CO, USA.

enterotoxin is present in the sample, reaction of the enterotoxin with the antibody results in agglutination, with the degree of agglutination determining the amount of enterotoxin present. If the latex particles settle to the bottom of the wells in a small dot, no enterotoxin is present. It is essential that the food extract is perfectly clear, but even so with an occasional food extract a nonspecific agglutination will occur. The method is adequately sensitive for the detection of enterotoxin in most foods that are implicated in food poisoning outbreaks. The method is useful for the detection of low enterotoxin-producing staphylococci that are not detectable by the OSP gel-diffusion method, approximately 10–20 ng ml^{-1} (**Table 6**).

The most recent technique developed for the detection of enterotoxigenic staphylococci is the polymerase chain reaction (PCR). This is dependent on synthesized DNA probes specific to the genes of the different enterotoxins. The technique is highly sensitive and specific, and allows the detection of enterotoxigenic staphylococci in a relatively short time with little sample preparation. One advantage of the method is that dead enterotoxigenic cells are also detectable, which is important in the analysis of heated foods implicated in food-poisoning outbreaks.

Commercial Products Available

An ELISA ball kit is available from Diagnostische Laboratorien, Bern, Switzerland and marketed in the United States by Toxin Technology Inc., Sarasota, FL. The enzyme used in this method is alkaline phosphatase, with pNPP as the substrate.

Another ELISA kit is the RIDASCREEN available from R-Biopharm GmbH, Darmstadt, Germany, and marketed in North America by Bioman Products Inc., Mississauga, Ontario, Canada. Horseradish peroxidase is the enzyme used, with urea peroxide and tetramethylbenzidine as the substrate. The substrate gives a blue colour which is changed to yellow when the stop reagent is added. Microtitre strips are used that are coated with the specific antibodies; there is one well for each enterotoxin and two negative controls.

An RPLA kit is produced by Denka Seiken Co. Ltd, Niigata, Japan, and marketed by Oxoid Limited, Hants, UK, and available in the United States from Oxoid, Inc., Ogdensburg, NY.

Another kit, VIDAS (Vitek ImmunoDiagnostic Assay System), is produced and distributed by bioMérieux Vitek, Inc., Hazelwood, MO. This method is based on a fluorometric reaction. However, an automated detection system is used which may not be readily available in many laboratories.

Two ELISA kits that include all the enterotoxin antibodies in one test are now available commercially. In some instances it may not be necessary to determine the type of enterotoxin, only if enterotoxin is present; for example, in examining the marketability of suspect foods. The food would not be marketable if any enterotoxin were present. For this purpose, including all of the enterotoxins in one test saves time as only one test is needed. However, this would not be useful in examining foods implicated in food poisoning outbreaks, because identification of the type of enterotoxin is valuable in tracing the source of the contamination. One kit utilizes small tubes coated with monoclonal antibodies to all of the enterotoxins (Transia-Diffchamb, Lyon, France). Horseradish peroxidase is used as the enzyme and the substrate is peroxidase substrate plus chromogen (Tetramethylbenzidine).

The second kit, TECRA, is produced by Bio-entraprises Pty Ltd, Roseville, New South Wales, Australia, and marketed in the United States by International BioProducts Inc., Redmond, WA. It utilizes microtitre plates with the wells coated with a mixture of the specific antibodies to SEA-SEE. The enzyme used is horseradish peroxidase, with 2,2-azino-di(3-ethylbenzthiazoline sulphonate), EDTA, and NaH_2PO_4 as the substrate.

Procedures Specified in Regulations, Guidelines, Directions or Collaboratively Validated Studies

It is difficult to specify a particular method as standard for the detection of the staphylococcal enterotoxins, primarily because all methods are based on the reaction of the enterotoxins with specific antibodies. The major problem with the screening kits is that equal quantities of each specific antibody may not be adsorbed to the active surface, such as the wells in the microtitre plates. It is difficult to specify any method because not all operators are equally efficient with any particular method, and each operator should be allowed to use the method with which they have the greatest success.

A collaborative study with the TECRA kit has been accepted by the Association of Official Analytical Chemists International as the official method for use in the United States and a collaborative study with the screening tube kit was validated by the French Normalization Agency for identification of the staphylococcal enterotoxins in foods and culture fluids in France.

Two other collaborative studies were done in Canada, one with the TECRA kit and one with the RIDASCREEN kit, but neither method has been accepted as official methods for use in Canada.

Table 7 Detection of enterotoxins in foods (From Wieneke (1991) *International Journal of Food Microbiology* 14: 305–312)

	ELISA			
Food	*RPLA*	*Ball*	*Dipstick*	*Tube*[a]
Ham	A	A	A, C[b]	+
Lasagne	A	A	–	+
Chicken	D	D	D	+
Ham	B	B	B	+
Rice	A	A	A	+
Chicken	–	A	A	+
Sandwich	–	A	–	–

[a]Small tube coated with all the antibodies.
[b]Non-specific reaction.

Advantages and Limitations of the Commercial Methods

All acceptable methods should have a sensitivity of at least 0.5 ng of enterotoxin per gram of food and all of the commercial methods claim to have this sensitivity. Although it may be possible to detect lower amounts with some of the methods, it is very unlikely that smaller amounts than was present in the chocolate milk would be encountered.

Those who have used the ELISA ball kit have found it to be a very good method for detecting enterotoxin in foods actually implicated in food poisoning outbreaks and was found to be superior when compared to the RPLA method, the dipstick method, and the screening tube method. Although the test can be completed in one day, recommendations are that the antibody-coated balls be shaken with 20 ml of food extract overnight to obtain the highest sensitivity. The method can be used quantitatively, although this is not necessary in checking foods for the presence of enterotoxins. If it is used quantitatively, the volume of the final solution is 1 ml which can be read in a colourimeter instead of an expensive microtitre plate reader as required for the ELISA methods. The main problem with the ball kit is that it is more cumbersome to use as each ball must be handled separately after the extraction process.

The TECRA, the RIDASCREEN, and the VIDAS methods have not been compared with other methods on foods actually implicated in food poisoning outbreaks. Tests of the TECRA, RIDASCREEN, and VIDAS methods experimentally on foods to which enterotoxin was added indicated that their sensitivity was adequate for the detection of the enterotoxin involved in essentially all outbreaks.

The advantage of the screening methods is that only one test is needed when it is not important to actually determine the type involved. Usually this is not of particular importance, hence, the TECRA kit has received wide usage. The RIDASCREEN method can be used in place of a screening method because the microtitre strips come already coated with antibodies to all of the enterotoxins, one in each well together with a positive and a control well. It proved to be equal to the TECRA method when tested collaboratively, with the advantage that the particular type of enterotoxin present could be determined in one test.

The RPLA method is relatively easy to use, particularly for the testing of culture supernatants, although it is recommended for examination of foods for enterotoxin. The major problem with the method for the examination of food extracts is that the extract must be perfectly clear or else a false coagulation reaction can occur. If the extract is not clear, it is necessary to filter the food extract through a nonprotein adsorbent filter. When it was compared with other methods for the detection of enterotoxin in implicated foods it was found to be less sensitive than the ball method and in at least one food a false coagulation reaction was encountered (**Table 7**). In the original method, the reaction was allowed to proceed overnight; however, with high-density latex particles the reaction can be completed in 3 h.

The cost of the reagents for research and the examination of foods involved in staphylococcal food poisoning is an important aspect. The reagents for use in gel diffusion methods are available from Toxin Technology, Inc, Sarasota, Florida, United States, but are expensive.

The cost of the different kits available commercially may limit their usefulness in research laboratories, however, most of the assays are being done in commercial laboratories. As the incidence of staphylococcal food poisoning has greatly decreased in the last few years in the developed countries, most of the research done today with the staphylococcal enterotoxins is as superantigens and not their involvement in food poisoning. The cost of the kits is influenced by the cost of preparing the specific antisera needed for each of the separate enterotoxins.

Presentation of Results of Collaborative Studies

Only the study with the tube kit used less than 1 ng of enterotoxin per gram of food, 0.5 ng g^{-1} for milk and cheese, 0.2–0.8 ng g^{-1} for meat, mushrooms, ravioli, and mayonnaise. The results of four collaborators were not considered because their duplicates were not in agreement. The nine remaining collaborators were able to detect 0.5 ng g^{-1} in mushrooms and ravioli, 0.8 ng g^{-1} in meat, 1 ng ml^{-1} in milk, and 1.5 ng g^{-1} in raw milk cheese.

The studies in Canada on the TECRA and RIDASCREEN kits used a minimum of 1 ng of enterotoxin per gram of food and all 13 collaborators were able to detect 1 ng of SEA per gram of ham and 1 ng of SEB per gram of salami. This would have been adequate to detect the enterotoxin in foods implicated in most staphylococcal food poisoning outbreaks. However, it might not be adequate to detect the minimum amount of enterotoxin (0.5 ng g^{-1}) present in foods suspected of being contaminated during manufacturing. The minimum amount required to cause food poisoning in the most sensitive individuals is not known, although in the chocolate milk outbreak, the amount of SEA detected was 0.61 ng ml^{-1} of milk and a total of 144 nanograms per half-pint, the amount that each student drank.

However, in the study with the TECRA kit conducted in the United States a combination of 4–10 ng of enterotoxin per gram of food was used in any of the five foods in the study. It is understandable that all 15 participants were able to detect the enterotoxin in all of the foods. It is doubtful that all of the participants would have been able to detect the enterotoxin in the chocolate milk involved in the outbreak mentioned above. This kit is widely used in the United States.

See also: **Food Poisoning Outbreaks**. **Immunomagnetic Particle-based Techniques**: Overview. ***Staphylococcus***: Detection by Cultural and Modern Techniques.

Further Reading

Bennett RW (1998) Current concepts in the rapid identification of staphylococcal enterotoxin in foods. *Food Testing and Analysis* 16–18: 310.

Bergdoll MS (1970) Enterotoxins. In: Montie TC, Kadis S and Ajl SJ (eds) *Microbial Toxins*, vol. III, *Bacterial Protein Toxins*, p. 266. New York: Academic Press.

Bergdoll MS (1972) The enterotoxins. In: Cohen JO (ed.) *The Staphylococci*, p. 301. New York: Wiley-Interscience.

Bergdoll MS (1977) Immunological aspects of staphylococcal enterotoxins. In: Catsimpoolas N (ed.) *Immunological Aspects of Foods*, p. 199. Westport CT: Avi Publishing.

Bergdoll MS (1979) Staphylococcal intoxications. In: Riemann H and Bryan FL (eds) *Food-Borne Infections and Intoxications*, 2nd edn, p. 443. New York: Academic Press.

Bergdoll MS (1983) Enterotoxins. In: Easmon CSF and Adlam C (eds) *Staphylococci and Staphylococcal Infections*, vol. 2, p. 559. London: Academic Press.

Bergdoll MS (1988) Monkey feeding test for staphylococcal enterotoxin. In: Harshman S (ed.) *Methods in Enzymology*, vol. 165, *Microbial Toxins: Tools in Enzymiology*, p. 324. New York: Academic Press.

Bergdoll MS (1989) *Staphylococcus aureus*. In: Doyle MP (ed.) *Foodborne Bacterial Pathogens*, p. 463. New York: Marcel Dekker.

Bergdoll MS (1990) Analytical methods for *Staphylococcus aureus* enterotoxins. *International Journal of Food Microbiology* 10: 91–100.

Bergdoll MS (1991) *Staphylococcus aureus*. *Journal of the Association of Official Chemists* 74: 706–710.

Bergdoll MS (1992) Staphylococcal intoxication in mass feeding. In: Tu AT (ed.) *Food Poisoning*, *Handbook of Natural Toxins*, vol 7, p. 25. New York: Marcel Dekker.

Bergdoll MS (1993) *Staphylococcus aureus*: *Food Poisoning*. In: Macraem R, Robinson RK and Sadler MJ (eds) *Encyclopaedia of Food Science, Food Technology, and Nutrition*, p. 4367. London: Academic Press.

Bergdoll MS and Bennett RW (1984) Staphylococcal enterotoxins. In: Speck ML (ed.) *Compendium of Methods for the Microbiological Examination of Foods*, p. 428. Washington, DC: American Public Health Association.

Bergdoll M and Wong AL (forthcoming) *Staphylococcus aureus*: Detection. In: Macraem R, Robinson RK and Sadler MJ (eds) *Encyclopaedia of Food Science*, *Food Technology and Nutrition*, p. 4362. London: Academic Press.

Bergdoll MS and Wong ACL (199?). Staphylococcal intoxications. In: Riemann H and Cliver DO (eds). *Food-Borne Infections and Intoxications*, 3rd edn, New York: Academic Press.

Su Yi Cheng and Wong ACL (1997) Current perspectives on detection of staphylococcal enterotoxins. *Journal of Food Protection* 60: 195–202.

Thompson NE, Bergdoll MS, Meyer RF et al (1985) Monoclonal antibodies to the enterotoxins and to the toxic shock syndrome toxin produced by *Staphylococcus aureus*. In: Macario AJL and de Macario EC (eds) *Monoclonal Antibodies Against Bacteria*, vol. II, p. 23. New York: Academic Press.

STARTER CULTURES

Contents

Uses in the Food Industry

R C Wigley, Boghall House, Linlithgow, Scotland

Introduction

Starter cultures can be defined as preparations of one or more strains of one or more microbiological species. They are subcultures, and depending on their concentration they are inoculated directly or require further subculture before use. Historically, a starter culture was simply a sample of fermented food, retained for inoculating the next batch of product. This process was known as 'back slopping' by sausage makers, and gave a very variable end result. Nowadays, a fermentation has to be predictable and consistent to ensure the maintenance of product quality, and this underpins the importance of commercial starter cultures in food production. Traditionally, starter cultures were associated with lactic acid fermentation, but commercial preparations of a wide range of species are now available.

Fermentation is one of the oldest methods of food preservation. It is also an attractive method, because it renders the food more digestible and more palatable, and is considered by consumers to be a harmless natural process. The bacteria most widely used in the fermentation of foods belong to the lactic acid group. These bacteria aid the preservation of foods by the utilization of carbohydrate and its conversion to lactic acid, the resulting drop in pH providing a measure of protection against some spoilage organisms. Some lactic acid bacteria produce compounds, including acetaldehyde, acetoin and diacetyl, which give rise to the characteristic flavour of dairy products. Other fermented products, e.g. kefir, koumiss, have a very complicated flavour profile, due to the by-products of multiple fermentations.

Starter cultures are most extensively used in the dairy industry, but are also used in the production of other fermented foods, including meat, wine, sourdough bread, vegetables and vegetable juices. Bacteria are selected for use in the food industry by virtue of their properties. For example, some bacteria produce organic acids from sugars by fermentation, and are used both to acidify foods and to produce aromatic compounds. The lactic, propionic and acetic acid bacteria fall into this category. Other bacteria produce primarily aromatic compounds, e.g. micrococci and surface bacteria. The lactic acid bacteria produce either solely lactic acid (homofermentative bacteria) or lactic acid and other organic acids (heterofermentative bacteria). The lactic acid bacteria include cocci and bacilli and belong to six genera: *Streptococcus*, *Lactococcus*, *Leuconostoc* and *Pediococcus* (cocci) and *Bifidobacterium* and *Lactobacillus* (bacilli).

Starter Cultures used in the Food Industry

Dairy Products

The range of cheeses and fermented milks is wide, and this is reflected in the variety of cultures used in the dairy industry (**Table 1**).

Cheeses In the past it would have been appropriate to specify the applications for each microorganism. However, there are now no conventions; for example, *Streptococcus thermophilus* is used in mixed direct-to-vat mesophilic cultures for Cheddar cheese to reduce the cost of inoculation through enhanced acidification. Improvements in farm hygiene have resulted in improvements in the quality of milk, and the use of defined strain cultures has led to the loss of character of some cheese varieties, e.g. Cheddar. Culture houses have devoted much effort to the selection of blends of cultures which recreate traditional flavours, for example producing mixed cultures of *Brevibacterium linens*, *Lactobacillus casei* and *Lactobacillus helveticus*, which would not normally have been associated with Cheddar manufacture.

Butter In mainland Europe, butter has traditionally been made from cream, ripened with a mixed culture containing *Lactococcus lactis* subspp. *lactis*, *cremoris* and *lactis* biovar *diacetylactis*, and *Leuconostoc mesenteroides* subsp. *cremoris*. A disadvantage was that the buttermilk was sour, and not suitable for further processing into other products. Technologists in Holland have developed an alternative method for the production of butter, which has been widely

Table 1 Organisms used in the production of fermented dairy products

Organism	*Application*
Mesophilic lactic acid bacteria	
Lactobacillus casei	Flavour production and texture improvement in cheeses and fermented milks
Lactococcus lactis subsp. *cremoris*	Acid production in cheeses and fermented milks
Lactococcus lactis subsp. *lactis*	Acid production in cheeses and fermented milks
Lactococcus lactis subsp. *lactis* biovar *diacetylactis*	Flavour production in cheeses and fermented milks
Leuconostoc lactis	Flavour production in cheeses and fermented milks
Leuconostoc mesenteroides subsp. *cremoris*	Production of eyes in Dutch cheeses; promotion of open texture, to aid growth of mould in Roquefort; flavour production in fermented milks
Other bacteria	
Micrococcus varians	Enhancement of activity of thermophiles
Propionibacterium freudenreichii subsp. *shermanii*	Production of eyes in Swiss cheeses
Thermotolerant lactic acid bacteria	
Lactobacillus delbrueckii subsp. *bulgaricus*	Acid and flavour production in yoghurts
Lactobacillus delbrueckii subsp. *delbrueckii*	
Lactobacillus delbrueckii subsp. *lactis*	Acid production in cheeses
Lactobacillus helveticus	Acid production in cheeses
Streptococcus thermophilus	Acid production in yoghurts and cheeses
Yeasts	
Candida utilis	Flavour production
Debaryomyces hansenii	De-acidification of curd to encourage growth of *Brevibacterium linens*
Kluyveromyces lactis	Flavour production and texture improvement in cheeses
Mycoderma spp.	Flavour production
Saccharomyces cerevisiae	Production of open texture in certain blue cheese varieties
Torulopsis spp.	Flavour production in soft cheeses, by proteases and lipases
Rhodosporidium infirmominatium	Pigmented surface coloration of cheeses
Surface-ripening bacteria	
Arthrobacter globiformis	Production of surface smear on cheeses; vigorous flavour production
Brevibacterium linens	Production of pigmented surface smear on cheeses; vigorous flavour production
Moulds	
Chrysosporium merdarium	Production of sulphur-yellow surface-ripened cheeses
Fusarium solani	Production of white surface-ripened cheeses
Geotrichum candidum	Production of white surface-ripened cheeses
Penicillium camemberti (or *album*)	Production of grey-blue surface-ripened cheeses
Penicillium caseicolum	Production of white surface-ripened cheeses
Penicillium cyclopium	Production of white surface-ripened cheeses
Penicillium nalgiovensis	Production of white surface-ripened cheeses
Penicillium roqueforti	Production of blue-veined cheeses
Scopulariopsis fusca	Production of beige surface-ripened cheeses
Sporendonema casei	Production of red surface-ripened cheeses

adopted and which produces not only butter indistinguishable from that made traditionally, but also valuable sweet buttermilk. This method involves the use of sweet cream instead of ripened cream, and a blend of two aromatic starter cultures for flavour production, together with NIZO-Lactic Starter Permeate to produce acidity at the final working stage.

Fermented Milk Products A trend in the production of fermented milk products is the use of 'probiotic' cultures, which have some connotation with the promotion of good health. The concept of a healthy life style, centred around improvements in digestion, originated in Japan in 1935. The concept was promoted by Dr Minoru Shirota, who developed Yakult, a milk fermented by *Lactobacillus casei*. Yakult is now manufactured in 15 countries worldwide, and an estimated 23 million bottles are drunk daily.

The benefits of the presence of the probiotic bacteria *Lactobacillus acidophilus* and *Bifidobacterium* in the stomach were studied in Germany, where the 'BIOgarde®' concept was developed. The most appropriate vehicle for these organisms was found to be milk. This led to the development of a fermented yoghurt-like product, which was sold as a health food. However, when the law changed in Germany in the 1970s, the claims associated with the health-promoting properties of BIOgarde® had to be reviewed and no epidemiological studies involving *L. acidophilus* or *Bifidobacterium* appeared to have been reported: hence no claims directly related to health have since been made for bio-yoghurt. Sales of BIOgarde® products, however, were not affected, because people preferred their smooth 'mouth feel' and mild taste to those of 'normal' yogurt. The legislation in Germany was changed in 1987, introducing a new class of 'mild yoghurt', defined as yogurt not containing *Lactobacillus delbrueckii* subsp. *bulgaricus*.

The first acidophilus milks were not popular, because they were unpalatable owing to their low pH. Sweet acidophilus milk, a low-fat pasteurized milk inoculated with a frozen concentrate of *Lactobacillus acidophilus*, was much more popular. Its production was made possible by the development in the early 1970s of techniques for the concentration of cultures, which led to the availability of deep-frozen cultures.

The most intensively studied and well-documented probiotic bacterium is *Lactobacillus rhamnosus* strain GG. Reports state that faecal recovery is possible following a daily dose of 1×10^9 cfu. This count can easily be reached in a fermented milk product, and it was concluded that the incidence of antibiotic-associated diarrhoea is significantly reduced when two pots of *Lactobacillus* GG yoghurt are eaten daily.

Some of the dairy products fermented with probiotic cultures are listed in **Table 2**.

Fermented Meat Products

Back slopping was used in the production of fermented meat products until the 1960s, when the first commercial starter cultures appeared. As is the case with all starter cultures, those first marketed contained strains of microorganisms that had been isolated from the best fermented sausages (**Table 3**). The advantages of standardized starter cultures soon became clear to manufacturers, and nowadays their use is universal in all but the smallest of plants. A summary of the advantages follows.

1. *Product safety.* The initial rapid population increase ensures dominance over spoilage or pathogenic bacteria early in the smoke cycle, when the pH has not dropped to a point where it inhibits the growth of contaminant organisms. The culture ensures that subsequently, the pH falls steeply, to restrict the further growth of contaminants.
2. *Product uniformity.* Consistent product quality can only be achieved when the quality of the raw materials and that of the manufacturing process are consistent. Fermentation parameters are crucial: hence reliable starter cultures ensure uniform flavour, colour and texture of the product.
3. *Drying time.* Meat protein loses its ability to retain moisture when it is acidified. Hence, acidification results in the quicker migration of water to the surface of the sausage and faster drying, with consequent energy cost savings.
4. *Increased plant efficiency.* Shorter fermentation and drying times allow a higher volume of product to be processed; fewer substandard batches result in lower costs; and reliable and predictable fermentations allow supervisory staff to be deployed elsewhere in the factory.
5. *Product body.* Sausages become firmer when the meat protein is acidified.
6. *Shelf life extension.* Ensuring the manufacture of a sausage with a low pH and low moisture content enables it to be sold unrefrigerated and with a long shelf life.

The addition of a can or sachet of commercially produced concentrated starter culture to the prepared meat is the preferred method of use. Lyophilized (i.e. freeze-dried) cultures are blended with the seasonings and additives, and added to the batch as early as possible in the process to ensure even dispersion throughout the mix. Such cultures can also be rehydrated and then poured over the mix.

The meat mix must contain added carbohydrate, in the form of a fermentable sugar such as dextrose, lactose or sucrose, in the proportion 0.38–1.00% w/w. The sugar is essential for bacterial growth, which results in the reduction of the pH to a point at which it has a preservative effect. The addition of 1.00% w/w sugar gives a final pH of 4.6, whereas addition of 0.38% sugar results in a pH of 5.2 at the end of fermentation.

The diversity of fermented meats is directly related to the wide differences in the process parameters employed, particularly the fermentation time and temperature combinations. The raw materials (pork or beef), the additives used (e.g. nitrate or nitrite) and the extent to which the meat is ground also affect the nature of the product. The fermentation conditions and the ingredients dictate the blends of culture strains best used in the starter. A typical production process is shown in **Figure 1**.

Starter cultures have two important roles in meat fermentation, in addition to acidification. These are performed by micrococci belonging to the genera *Staphylococcus* and *Micrococcus*. These bacteria reduce nitrates to nitrites, which combine with the myoglobin in meat to form the substrate for a series of reactions in which micrococci again participate, resulting in the formation of nitrosohemochrome, a stable compound which gives cured sausages their characteristic colour. In addition, micrococci generate aromatic compounds throughout the fermentation process, as a result of the actions of proteases, peptidases, esterases and lipases.

Table 2 Examples of probiotic organisms used in the production of fermented dairy products

Organism	*Product name*	*Producer*	*Country of manufacture*
Bifidobacterium bifidum *Lactobacillus acidophilus* *Lactobacillus casei*	Vita[b]	Tuffi	Germany
Bifidobacterium longum	Bl'AC[b]	Müller	Germany
Bifidobacterium longum	ProCult 3[a]	Müller	Germany
Bifidobacterium spp. *Lactobacillus acidophilus*	Daily Fit[b]	Ehrmann	Germany
Bifidobacterium spp. *Lactobacillus acidophilus*	PrioBiotic[b]	Bauer	Germany
Bifidobacterium spp. *Lactobacillus acidophilus* *Lactobacillus casei* *Lactobacillus reuteri*	SymBalance[a,b]	Toni AG	Switzerland
Bifidobacterium spp. *Lactobacillus acidophilus* *Lactobacillus casei* *Lactobacillus reuteri*	SymBalance[a,b] Reuteri	Chichiyasu	Japan
Bifidobacterium spp. *Lactobacillus acidophilus* *Lactobacillus reuteri*	RELA[a,b]	Ingman	Finland
Bifidobacterium spp. *Lactobacillus acidophilus* *Lactobacillus reuteri*	BRA[a,b]	Handlarnas	Spain
Bifidobacterium spp. *Lactobacillus casei*	Kindsköpfe[b]	Borgmann	Germany
Enterococcus faecium	Gaio[b]	MD Food	Denmark
Lactobacillus casei	Actimel[a]	Danone	Netherlands, Germany
Lactobacillus casei	BIO au casei actif[b]	Danone	France
Lactobacillus crispatus	Fysiq[b]	Campina AG	Netherlands
Lactobacillus johnsonii	LC1[b]	Nestlé	Switzerland, France, Germany
Lactobacillus paracasei subsp. *paracasei*	Yakult[a]	Yakult	Japan, Germany, Netherlands
Lactobacillus plantarum	ProViva[a]	Hansa Milch	Germany
Lactobacillus rhamnosus	ABC-Maelk[a]	Norske Meierier	Norway
Lactobacillus rhamnosus	Aktifit Plus[a]	Emmi	Switzerland
Lactobacillus rhamnosus	Gefilus[a,b]	Valio	Finland
Lactobacillus rhamnosus	La Serenisima[a,b]	Mastellone	Argentina
Lactobacillus rhamnosus	Next[a,b]	Soprole	Bolivia, Chile
Lactobacillus rhamnosus	Onaka He GG[a]	Takanashi Milk	Japan
Lactobacillus rhamnosus	Schärdinger[a,b]	Berglandmilch	Austria
Lactobacillus rhamnosus	Toni[a,b]	Toni	Ecuador

[a]Drinks.
[b]Spoonable products.

Cured Meat Products

'Curing', when applied to pork to produce ham, refers to the introduction into the meat of salt and also of nitrite, usually in the form of the sodium or potassium salt. The nitrite reacts with the meat pigment nitrosyl myoglobin to produce nitrosohemochrome, which gives the cured meat its characteristic purple-red colour. Prior to 1938, the addition of nitrites to curing systems was not permitted in Europe, and curing relied on the reduction of nitrate added to the brine and, in some cases, to the meat. Such conversion of nitrate to nitrite requires nitrate-reducing micro-organisms, and hence the need for a 'live' immersion brine. Freshly prepared immersion brines containing salt and nitrate would not cure pork because no conversion of nitrate to nitrite would occur. Such brines were therefore inoculated by using them to cure certain pork cuts, e.g. heads. Once a population of halophilic nitrate-reducing microorganisms was established, the brine was repeatedly reused, reaching an eventual microbial concentration of around 10×10^8 organisms per millilitre. Even after 1938, the practice of recycling the immersion brine was continued in the Wiltshire system, and is still in use. Commercial blends of starter culture offer fast and safe colour formation through the production of nitrate reductase; a reduced risk of the rancidity of fat through the production of catalase; and improved flavour in the cured product.

Table 3 Microorganisms used in the production of fermented meats

Microorganism	*Application*
Mesophilic lactic acid bacteria	
Carnobacterium piscicola, *Lactobacillus curvatus*, *L. sake*, *L. plantarum*, *L. pentosus*, *Pediococcus acidilactici*, *P. pentosaceus*	Acid production, to inhibit spoilage organisms, decrease water activity (a_w) and encourage drying of sausages
Micrococcaceae	
Micrococcus varians, *Staphylococcus carnosus*, *S. simulans*, *S. xylosus*	Flavour production; production of colour typical of the variety
Actinomycetes	
Streptomyces griseus	Flavour production
Moulds	
Penicillium chrysogenum	Production of characteristic appearance of certain sausage varieties; prevention of growth of surface contaminants
Penicillium nalgiovensis	Ripening, by production of lipases and proteases
Yeast	
Debaryomyces hansenii	Production of the characteristic appearance of certain varieties of sausage; prevention of growth of surface contaminants; ripening, by production of lipases and proteases

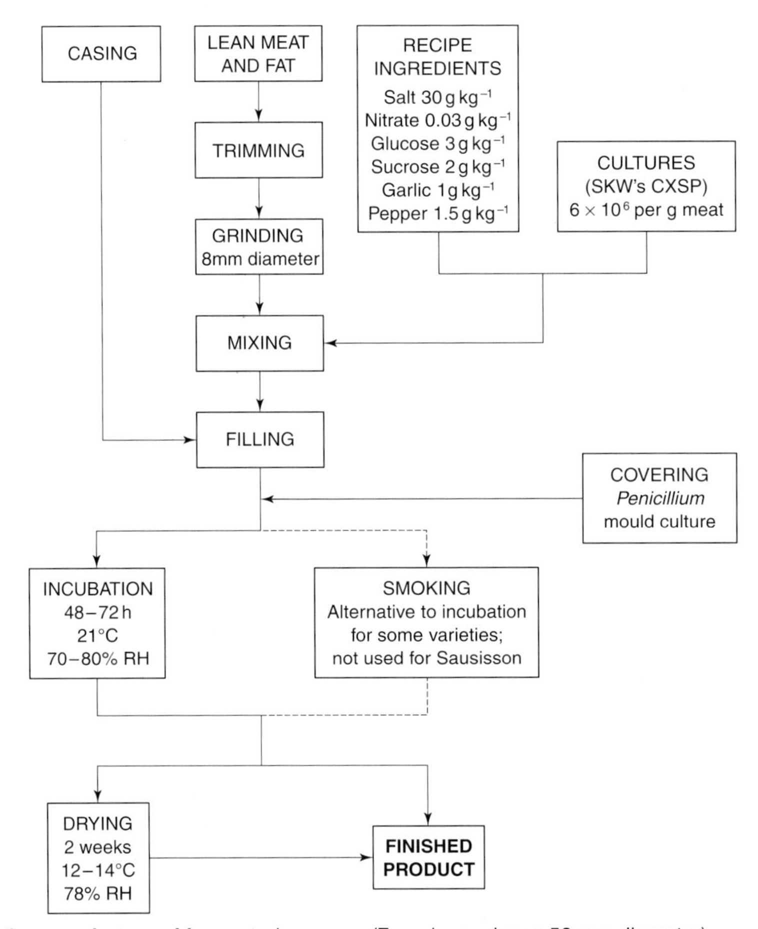

Figure 1 Flow diagram for manufacture of fermented sausage (French sausisson 50 mm diameter).

Sourdough Bread

Dough containing in excess of 20% rye flour is generally difficult to bake. Hence the properties of the flour must be altered by reducing the pH by fermentation: this is known as the 'sourdough' process. The starch of rye flour gelatinizes at a lower temperature than that of wheat flour, and as a result is susceptible to breakdown by α-amylases during baking. The reduction of the pH of the dough, to about 4.2, inhibits α-amylases, and this reduction in

Table 4 Microorganisms used in other food fermentation

Microorganism	*Application*
Mesophilic lactic acid bacteria	
Lactobacillus brevis	Sourdough bread; fermented vegetables (pickles) and vegetable juices
Lactobacillus casei	Fermented vegetable juices
Lactobacillus curvatus	Fermented vegetable juices
Lactobacillus delbrueckii	Sourdough bread
Lactobacillus hilgardii	Wine, to induce malolactic fermentation
Lactobacillus plantarum	Sourdough bread; olives; sauerkraut; fermented vegetable juices; wine, to induce malolactic fermentation
Lactobacillus rhamnosus	Fermented vegetable juices
Lactobacillus sake	Fermented vegetable juices
Lactobacillus sanfrancisco	Sourdough bread
Lactococcus lactis subsp. *cremoris*	Fermented vegetables (pickles) and vegetable juices
Lactococcus lactis subsp. *lactis*	Fermented vegetables (pickles) and vegetable juices
Leuconostoc oenos	Wine, to induce malolactic fermentation
Leuconostoc mesenteroides subsp. *cremoris*	Sauerkraut
Pediococcus spp.	Fermented vegetables (pickles)
Mould	
Geotrichum candidum	Beer (for malting barley)
Yeast	
Saccharomyces cerevisiae	Sourdough bread

pH can be achieved by starter cultures. The sourdough process also imparts a characteristic flavour to rye bread and retards staling. Concentrated cultures are used, and are generally lyophilized rather than deep-frozen. The strains used are listed in **Table 4**.

Fermented Vegetables and Vegetable Juices

Lactic acid bacteria are used in the production of pickled vegetables, the best-known perhaps being sauerkraut, the common name of pickled cabbage. Starter cultures are used extensively in the fermentation of the brine used for pickling cucumbers and olives, the salt limiting the growth of undesirable bacteria. Fermented vegetable juices are not widely available other than in Germany. The manufacturing process is not easily controlled, and the flavour of these products is variable. However, fermented vegetable juices could become popular as health foods in the future.

Wine

Starter cultures are used to effect the conversion of malic acid to lactic acid, which results in more mellow wine: red wines from northern wine-producing regions have been improved by malolactic fermentation since the early 1930s. Commercial starter cultures are reliable and produce wines of more predictable and uniform quality than those fermented naturally. However, wines from warmer regions contain less malic acid than those from colder regions, and there is a risk of excessive de-acidification if they are inoculated with starter culture.

Beer

The use of starters in the treatment of barley is relatively recent, and is a response to the growth of undesirable moulds and the production of mycotoxins during the malting of barley for beer production. A commercially produced, lyophilized culture of *Geotrichum candidum*, isolated by the French brewing institute, the Institut Français de la Brasserie et de la Malterie (IFBM), inhibits these moulds. It is claimed that the malting process is also more consistent, and that the physico-chemical properties of the malt are improved, as well as its safety.

Starter Culture Systems

Starter cultures are used either from a bulk starter vessel or inoculated directly into the mix (direct-to-vat cultures, DVC). The latter became available in the 1970s, but did not gain widespread acceptance until about 1980.

Daily Propagated Cultures

Daily propagated cultures are prepared from a mother culture, purchased at intervals of between two weeks and a year, depending on the success of the factory at culture maintenance. Mother cultures are either liquid or freeze-dried. Daily subculturing maintains the activity of the culture, which is also subcultured on through a feeder stage, for inoculation into the bulk starter vessel. The use of this system is now generally restricted to small creameries in the UK.

Deep-frozen Cultures

Frozen cultures eliminate the need for daily propagation by the food manufacturer. These cultures are supplied in bottles of reconstituted skimmed milk, which has been frozen after inoculation. The cultures are shipped in an insulated box and need to be transferred to a freezer immediately on arrival at the production site. Bulk starter is prepared by thawing the milk, clotting it overnight and inoculating the clotted culture into the bulk starter medium.

Deep-frozen Concentrated Cultures

Frozen concentrated cultures are packed in aluminium cans with ring-pulls, and generally a capacity of 70–

125 ml. The suppliers despatch the cans in insulated boxes containing solid CO_2 to maintain the temperature at about –70°C. On receipt the cans must be stored in a freezer operating at 50°C or lower. The shelf life of the cultures depends on the temperature of storage, but is typically 3 months at –50°C. Bulk starter is prepared by thawing the contents of a can and using them for inoculation.

Lyophilized Concentrated Cultures

Freeze-dried concentrated cultures have a significant advantage over deep-frozen concentrates: they can be stored in a conventional refrigerator and can be sent by post, reducing transport and storage costs. The cultures are packed in foil sachets, in quantities suitable for the direct inoculation of bulk starter medium. Bulk starter is prepared by sprinkling the contents over the surface of the medium.

Deep-frozen Direct Vat Cultures

These cultures must be stored at –50°C or below. They are supplied either in a 360 ml aluminium can with a ring-pull, or as granules in a cardboard box, polythene bottle or aluminium sachet, typically containing 500 g. Some suppliers offer cultures that are packed to reflect customer requirements, with the weight of culture tailored to the size of the batch to be fermented.

Lyophilized Direct Vat Cultures

These cultures are packed in aluminium foil sachets, and can be stored in the domestic refrigerator at 4°C. They are packed in quantities according to units of activity, with a specified number of units being required per 100 kg of food to be fermented.

Table 5 summarizes the merits of the different starter culture systems.

Handling and Use of Direct Vat Starter Cultures

Culture houses have refined the handling of starter cultures to the point at which it is simplicity itself. Frozen granules or freeze-dried powders are used directly from the freezer or refrigerator; the package is opened and the contents added to product. Cheese makers typically add the culture as soon as milk starts to flow into the vat. The time taken for the vat to fill is the ripening time, and as soon as the vat is full the coagulant is added. Wine-makers add the culture and agitate the wine to ensure it is dispersed homogeneously throughout the tank. The production of sourdough bread is essentially a bulk starter process where a fermented dough is made, but unlike in cheese-making the addition rate is high, up to 35% depending on the type of bread being produced, compared with about 1% in cheese-making. Culture houses provide detailed instructions regarding the storage, handling and use of cultures.

Types of Starter Culture

The three main types of starter culture used in cheese-making are listed in **Table 6**. Ideally, a cheese maker would use a single culture strain; this would give a product of consistent quality and eliminate any day-to-day variations in flavour. However, in practice pure single-strain starters are at risk from bacteriophage attack and for this reason single-strain cultures are grown separately but always used in pairs or triples. The suppliers of these starters have carried out extensive work on the phage sensitivity of the strains used, and usually the starters are sufficiently robust to enable the same blend to be used without rotation.

Culture systems must be monitored regularly, for bacteriophage infection, so that any strain being attacked by phages can be replaced. Although the stability of cultures can be maintained adequately in this way, some cheese makers prefer culture rotation, using different cultures for up to 3 days: in some factories, cultures are changed each time the vats are filled.

Multiple-strain cultures, sometimes known as 'defined multiple-strain (DMS) cultures', are blends of three or more compatible pure strains. They can be used in a non-rotational system but are more frequently used in rotation.

Mixed-strain cultures are 'naturally' occurring, and have been passed down from cheese maker to cheese maker, the less robust cultures falling into disuse. Those available today probably bear little resemblance to the culture originally put into a collection, owing to strain domination occurring progressively during subculture. This changing character of a culture leads to problems with phage typing, which do not arise with the other types of culture. However, traditional cultures are favoured by many manufacturers, because of the characteristic flavour and texture of the cheeses produced.

Culture suppliers are striving to develop DMS starters that perform more consistently than mixed-strain cultures, yet also produce cheese of the quality associated with such traditional cultures. The lack of definition of the strains in mixed cultures means that they are rarely used as DVC starters, whose predictability of performance is important, because there is no opportunity to obtain activity results prior to use.

Table 5 Summary of the characteristics of various starter culture systems

Characteristic	*Bulk starter cultures*				*DVC cultures*	
	Daily propagated cultures	*Deep-frozen cultures*	*Deep-frozen concentrated cultures*	*Lyophilized concentrated cultures*	*Deep-frozen direct vat cultures*	*Lyophilized direct vat cultures*
Cost of culture per vat	Low	Medium	Medium	Medium	High	Medium
Level of technical skill required	High	High	Medium	Medium	None	None
Cost of storage of cultures	Low	Low	High	Low	High	Low
Level of phage relationship data available	None	High	High	High	High	High
Period of planning required to manufacture starter	72 h	48 h	24 h	24 h	None	None
Data on culture performance available prior to use (fresh starter)	Fully tested	None	None	None	None	None
Data on culture performance available prior to use (stock starter)	Fully tested	Fully tested	Fully tested	Fully tested	None	None
Level of technical support provided for system	None	High	High	High	High	High
Range of cultures available	Good	Good	Good	Adequate	Good	Adequate

DVC, direct to vat culture.

Table 6 Mesophilic starter cultures used in the manufacture of cheese

Species of microorganism	*Composition of culture*	*Method of use*	*Applications*
Single-strain starters			
Lactococcus lactis subsp. *cremoris* *Lactococcus lactis* subsp. *lactis*	Pure culture	In pairs or triples, with or without rotation	Cheddar in New Zealand, Ireland, USA, Australia and UK
Multiple-strain starters			
Lactococcus lactis subsp. *cremoris* *Lactococcus lactis* subsp. *lactis* *Lactococcus lactis* subsp. *lactis* biovar *diacetylactis* *Leuconostoc mesenteroides* subsp. *cremoris*	Defined blends of three or more strains	Singly, with or without rotation	Hard cheeses in the USA and UK
Mixed-strain starters			
Lactococcus lactis subsp. *cremoris* *Lactococcus lactis* subsp. *lactis* *Lactococcus lactis* subsp. *lactis* biovar *diacetylactis* *Leuconostoc mesenteroides* subsp. *cremoris*	Undefined proportions of an unknown number of strains, which can vary on subculture	Singly or in pairs, always used in rotation	Most hard cheeses in the USA and Europe

Bacteriophage Infection

The most frequent cause of poor starter performance, apart from process failures such as low-temperature incubation, is infection by bacteriophages (phage). These are viruses that attack bacteria and have caused problems in the cheese industry for decades. Infection of the bulk starter tank may not be immediately apparent, the pH being unaffected. However, the culture will have no activity, i.e. there will be no growth. This situation arises when the bulk starter vessel is not grossly infected, and the inoculum passes through a number of growth cycles before it is killed completely. If the infection is in the cheese room, there is a slowing of acid development, culminating in a 'pack-up', a 'dead' vat, in which there is no acid development between starter inoculation and salting. At best, such cheese can be processed; at worst, it can only be used as pig feed. Thus uncontrolled phages have a catastrophic effect on the profitability of a plant.

Phages multiply by a process involving adsorption onto the surface of the bacterium, followed by the injection of genetic material into the cell. This material replicates, and up to 400 new virulent phages are released from each infected cell by lysis. This number is known as the 'burst size', and the cycle of replication is termed the 'lytic cycle'. The latent period, from initial contact between the phage and the bacterium to the liberation of progeny phages, varies between 10 min and 140 min. The burst size and the lytic cycle are highly dependent on a number of factors, including temperature, medium and pH.

One approach to the control of phages is a reduction in the time for which the cheese milk is ripened prior to the addition of coagulant. This period should ideally be 15 min and never longer than the typical lytic cycle of the phages, because following lysis, fewer phage particles can infect new cells in coagulated milk than in fluid milk.

Phages are ubiquitous, but infections can be limited by adopting appropriate working practices in the factory. It is important that the appropriate procedures are followed vigorously and consistently, because phages mutate quickly and easily. These procedures are summarized in **Table 7**. Technologists from the culture suppliers will help minimize the risk of phage attack by selecting a rotation of unrelated cultures, but the importance of maintaining the strictest hygiene procedures within the dairy cannot be overemphasized. A particularly important procedure is the disposal of whey, which is the primary source of phages.

Bacteriophage is not a culture problem as the majority (more than 80%) are virulent and not present in the chromosome of the strains. The DNA of temperate phages is present in the chromosome and the lytic cycle can be induced. Cultures supplied by culture houses are guaranteed phage-free when they are shipped, and it has been demonstrated that the activity even of single-strain cultures, used in pairs or triples on a non-rotational daily basis, can be maintained over many years. Problems become apparent only when there is a lapse in the strict hygiene regime necessary for single-strain cultures.

Table 7 Management of bacteriophage

Starter room
- Ensure area is physically separated from the cheese room and there is no direct access from cheese room
- Ensure staff are responsible for starter and no other part of the cheese process
- Ensure personal hygiene instructions are followed explicitly
- Provide chlorinated foot bath at door to room
- Use filtered air to pressurize tanks
- Use phage inhibitory medium
- Heat treat medium in the tank. If an external heat exchanger is used it must only be used to heat the medium partially, with the heat treatment completed in the tank
- Use aseptic inoculation techniques
- Sterilize contaminated starter before disposal
- Install a separate loss CIP system for the starter room
- Carry out full CIP prior to filling of starter vessels or vats if they have not been in use during the previous 24 h

Vat room
- Ensure air flow in room matches direction of product flow
- Provide a chlorinated foot bath at each door to the room
- Chlorinate the vats before the first fill of the day
- CIP and chlorinate the vats between the fills
- Monitor and control CIP procedures closely
- Ensure plant is cleaned in a logical sequence from milk intake to whey processing or install separate systems for each area
- Locate curd separator in the whey area physically separated from the vat room and pipe whey to separator in a closed system
- Locate raw milk handling and pasteurization plant in an area physically separated from the vat room
- Under no circumstances use whey to flush out curd from the vat at pitching
- Reduce ripening time of cheese milk prior to coagulant addition. Optimum is 15 min.

Whey processing area
- Process whey in a closed area separate from the vat and starter rooms
- Load tankers downwind of the starter and vat rooms
- Ensure whey processing and/or loading is done under the strictest hygienic conditions
- Eliminate spillages
- Never add whey cream back to the cheese vat or store in a silo used for cheese milk
- Ensure whey is not sold to farmers supplying milk to the dairy

Culture Rotation

It used to be thought that phages were very host-specific and could not survive without their host. Hence the longer the rotation of cultures, the greater the chance that any phages would die out. These beliefs led to the practice of using a different culture for each day of the week, or even changing cultures each time the vats were filled.

However, it is now known that phages can survive for several years without a host, and research has also shown that phages can attack more the one strain within a genus. Long rotations have therefore suffered a decline in popularity, additional reasons for this being that there is only a limited range of phage unrelated cultures, and most culture houses advise 'the shorter the rotation the better'. In practice, the length of the rotation depends on the confidence of cheese makers in their ability to control phage levels, and many use two different cultures, on alternate days.

Defined multiple-strain cultures reduce the risk of 'dead' vats, because it is unlikely that all the strains in the culture will be killed by phages. DMS cultures can be used in non-rotational systems, but are used more frequently in a short rotation. The use of DMS cultures is now common in large factories. It limits the population of potential phage hosts, making the problem of phages more manageable.

Most culture suppliers will analyse whey samples, to enable the identification of strains that will perform

Table 8 Major culture manufacturers

Company	Country	Starter cultures supplied[a]				
		Dairy products	*Meat products*	*Sourdough bread*	*Wine*	*Vegetables*
ABC Research Corp	USA		Yes			
ALCE Srl	I	Yes				
Australian Starter Culture Research Centre Ltd	AUS	Yes				
Biotec Fermenti Srl	I	Yes				
Centro Sperimentale del Latte	I	Yes				
Gewürzmüller	D		Yes		Yes	
Gist-brocades	NL	Yes	Yes	Yes	Yes	
Chr Hansen	DK	Yes	Yes	Yes	Yes	Yes
International Media & Cultures	USA	Yes				
NIZO	NL	Yes				
NZDRI	NZ	Yes				
Quest International	USA	Yes	Yes	Yes	Yes	Yes
Rhodia (R-P Texel)	F	Yes	Yes			
Rosell Institute Inc.	C	Yes				
SKW Biosystems	F	Yes	Yes			
Wiesby Biofermentation & Co. KG	D		Yes	Yes	Yes	Yes
Wiesby GmbH & Co. KG	D	Yes				

[a]Information believed to be correct at time of writing.
AUS, Australia; C, Canada; D, Germany; DK, Denmark; F, France; I, Italy; NL, Netherlands; NZ, New Zealand.

well in the factory, and also offer a service for monitoring the stability of the culture system. This is supplemented by a phage test kit, which enables the cheese maker to identify phage problems before cheese production is disrupted by a culture failure. The culture suppliers can then replace the infected strain or culture with an unrelated alternative. Phage test kits comprise a set of test tubes, each containing a genetically distinct strain or blend of starter culture in a sterile milk medium, together with an indicator which will change colour in the desired pH range. Each tube is inoculated with filtered whey from the same day's production, and incubated for 10 h. Cultures resistant to any prevailing phages exhibit the desired colour change, and develop a firm curd. This enables the selection of strains not susceptible to the prevailing phages.

Manufacture of Starter Cultures

The need for investment in research into isolating culture strains that perform better in a changing environment has resulted in specialization and rationalization within the industry. As a result a number of old manufacturing companies have been acquired by multinational corporations; these work closely with the food industries to whom they supply comprehensive technical and applications support (**Table 8**).

Bulk Starter Cultures

All cultures, other than direct vat cultures, must be subcultured in a bulk starter vessel to produce enough cells for addition to the food being fermented. Methods of bulk starter manufacture have evolved along with technology. For example, in the 1960s it was common for cheese makers to prepare bulk starter using whole milk in churns. Following the recognition that pretested reconstituted skimmed milk powder provided a more consistent substrate, its use was adopted as the industry standard in the 1970s, when there was also a move towards totally enclosed bulk starter vessels. Nowadays, the use of a phosphated medium is common. This type of medium chelates calcium, thereby making it unavailable to phages, which require it as a catalyst. Such media may therefore be described as 'phage-inhibitory' (**Table 9**).

Table 9 Bulk starter media

Substrate	*Comments*
Whole milk	Not widely used
Skimmed milk powder (SMP) used at 12% solids	Still used today
Phosphated media plus SMP at 12% solids total	First generation media
Phosphated media used at 12% solids	Isolated areas of use
Low solids phosphated media – 5–6% solids	Most popular media in use
Internal pH control media	Isolated areas of use
External pH control media	Popular in USA

Table 10 Comparison of activity of *Lactococcus lactis* subsp. *lactis* grown in Bactimedia® M55 and reconstituted skimmed milk powder (by kind permission of SKW Biosystems)

	pH after 6 h at 30°C	
Medium	*Control – no phage*	*4 × 10⁴ phage ml⁻¹*
RSM	5.60	6.54
Bactimedia® M55	5.16	5.22

Protection Against Bacteriophage Attack

Phage-inhibitory media, e.g. Bactimedia® (SKW Biosystems, France), are formulated so as to provide maximum protection against phages in the bulk starter vessel. The amounts of ammonium and sodium phosphates present ensure that all free calcium is chelated, thereby eliminating the risk of phage attack. **Table 10** shows that the activity of the culture grown in Bactimedia® M55 in the presence of 4.0×10^4 phage per ml at the end of the fifth culture cycle is at an equivalent level to the control without phage. The activity of the milk starter, on the other hand, is much worse than that of the control. In addition the bacterial strain inoculated at the time of the first cycle has completely disappeared, to be replaced by a *Lactococcus lactis* subsp. *lactis* mutant insensitive to the phage inoculum by the end of the fifth culture cycle. The total inhibition of phage on the Bactimedia® M55 medium is not synonymous with the total destruction of the phage. Its use, therefore, can only be considered as an additional precaution. The implementation of procedures to limit the risk of bacteriophage proliferation remains of critical importance.

Starter Room Precautions

Care in the preparation of bulk starter may be wasted if the starter room is poorly designed. There should be no direct access between the starter room and the cheese vat room, and the starter room should be situated so that it could not conceivably become a short cut to anywhere – especially the canteen. Ideally, starter room staff should not enter the cheese-making or whey processing area, and only authorized staff should be allowed into the starter production area. However, if this is impractical, then staff entering the starter area must change protective clothing (including footwear) immediately prior to entering the starter room.

Bulk Starter Quality Assurance

The critical characteristics of bulk starters are activity and phage sensitivity. Incubated bulk starter used to be chilled and stored until these were known, but nowadays the starter is often chilled and used immediately, because the performance of modern starter cultures is sufficiently reliable. However, activity and phage testing is still carried out in retrospect. Bulk starter is often used over long periods, and by chilling it to between 5°C and 7°C the activity can be maintained at a constant level.

Phage testing is part of the routine quality control of bulk starter. The traditional method involves comparing the growth of a sample of starter in reconstituted skimmed milk powder with its growth when

Table 11 Bulk starter quality control programme

Sampling regime
- Take samples of each of the starters used during the day
- Take whey samples at pitching from the last vat made with each culture in the rotation
- Take milk samples from each of the silos in use
- Chill all samples rapidly to 1–4°C and store until required
- Only retain samples at the end of each day if a culture does not perform satisfactorily
- Challenge the failed starter with whey samples from the failed and preceding vats and samples of milk used in the failed vat using the following procedure

Preparation of the whey sample
- Centrifuge 20 ml of whey at 3000 rev min⁻¹ for 10 min
- Filter whey through 0.45 μm filter
- Prepare 10^{-2} and 10^{-4} dilutions in maximum recovery diluent (MRD)
- If there is a serious phage problem include neat whey, 10^0, in the test
- Boil sufficient filtered undiluted whey for the control

Preparation of the culture
- Make 10^{-1} dilution in MRD
- Add 1 ml of the 10^{-1} dilution to 9 ml quantities of reconstituted skimmed milk autoclaved at 115°C for 10 min
- Five tubes are needed for each starter sample

The test
- *Activity control*: add 1 ml of MRD to the first tube
- *Phage test*: add 1 ml of the 10^{-2} and 10^{-4} whey dilutions to the second and third tubes
- *Phage control*: add 1 ml of the boiled whey to the fourth tube
- *Milk control*: add 1 ml of vat milk to the fifth tube
- Either incubate the tubes for 6 h at 30°C or use a stepped incubation – 4–5 h at 30°C followed by 1.5 h at 37°C to simulate the conditions in the vat

Interpretation
- *Activity control*: target for all tests is a developed acidity of not less than 0.50% LA
- *Phage test*: depression indicates the presence of phage. At 10^{-2} caution to be exercised. At 10^{-4} immediate removal of culture from rotation is advised
- *Phage control*: depression indicates methodology problem. Boiling will destroy phage. Repeat the test
- *Milk control*: depression indicates the presence of inhibitors in the milk which may or may not be confirmed with an antibiotic test

Notes
- Depression is defined as the negative difference between the test result and that of the activity control
- Depression of less than 10% – within limits of operator error
- Depression of 10% to 19% – cause for concern
- Depression of 20% or above – definite inhibition: take appropriate action

dilutions of a sample of whey are added. A smaller fall in pH in the tubes containing whey indicates the presence of phages (**Table 11**). Alternatively, the dilutions of whey can be used to inoculate a Petri dish containing agar that has been seeded with starter culture. A clear area or plaque indicates inhibition of the growth of the culture, and the possible presence of phages. However, a positive result in the plaque test is not always accompanied by problems in the factory. The reasons for this are not fully understood: phages are undoubtedly present, and have been termed 'non-disturbing' phages. Testing in tubes is therefore preferred.

Phage problems are mainly encountered in cheese factories, but phages of the thermophilic bacteria used in yoghurt production have been identified, and process failure due to infection is not unknown. It is important not to dismiss the points set out in Table 7 as being only relevant to cheese makers.

See also: ***Bifidobacterium***. **Bread**: Sourdough bread. **Cheese**: Microbiology of Cheese-making and Maturation. **Fermented Foods**: Fermented Vegetable Products; Fermented Meat Products. **Fermented Milks**: Range of Products; Yoghurt; Products from Northern Europe; Products of Eastern Europe and Asia. **Freezing of Foods**: Growth and Survival of Microorganisms. ***Lactobacillus***: *Lactobacillus acidophilus*. **Milk and Milk Products**: Microbiology of Cream and Butter. **Preservatives**: Traditional Preservatives – Sodium Chloride; Permitted Preservatives – Nitrate and Nitrite. **Wines**: Microbiology of Wine-making.

Further Reading

Bacus J (1984) *Utilization of Microorganisms in Meat Processing*. Letchworth: Research Studies Press.

Brown J (ed.) (1982) *The Master Bakers' Book of Breadmaking*. Ware: National Associations of Master Bakers, Confectioners and Caterers.

Campbell-Platt G (1987) *Fermented Foods of the World. A Dictionary and Guide*. London: Butterworth.

Campbell-Platt G and Cook PE (eds) (1995) *Fermented Meats*. Glasgow: Blackie Academic & Professional.

Cogan TM and Accolas JP (eds) (1996) *Dairy Starter Cultures*. New York: VCH Publishers.

Fugelsang KC (1997) *Wine Microbiology*. London: Chapman & Hall.

Lewis JE (1987) *Cheese Starters. Development And Application of the Lewis System*. London: Elsevier Applied Science.

Reed G (ed.) (1982) *Prescott & Dunn's Industrial Microbiology*, 4th edn. Westport: AVI Publishing.

Reed G (ed.) (1983) *Biotechnology*. Vol. 5. Weinheim: Verlag Chemie.

Rose AH (ed.) (1982) *Economic Microbiology*. Vol. 7. London: Academic Press.

Smith J (ed.) (1993) *Technology of Reduced-Additive Foods*. London: Blackie Academic & Professional.

Importance of Selected Genera

Crispin R Dass, The Heart Research Institute Ltd, Sydney, Australia

Introduction

Crude preparations of starter organisms have been used for thousands of years, to achieve variation in the aesthetic qualities and nutritive value of foods. Nowadays, starter cultures can be obtained in a relatively pure form, from the culture collections of research establishments. These commercial cultures are much superior to those utilized in the past in terms of consistency and reliability of performance. The safety of processes involving pure cultures can be ascertained and production losses from false fermentation are avoided. However, mixed cultures are usually more resistant than pure cultures to contamination by undesirable microbes.

Some contemporary societies store mixed cultures at home, for inoculating new batches of starting materials: for example, in certain regions in India cultures of microbes are maintained at home for the production of various types of yoghurt and salad dressings from cow, goat or buffalo milk.

Starter cultures perform reactions that chemically alter raw food material into a desirable edible product. By their growth and metabolism these microbes, either alone or in combination, also alter the physical and biological composition of food.

Important features of a microbe that ensures its use as a starter microbe include non-pathogenicity, lack of production of toxic chemicals, absence of production of chemicals reducing desirability of product, ease of handling, ease of storage and transport, period required for raw material to be converted to final edible product, genetic stability of the species, some degree of natural resistance to inhibitory factors (**Table 1**) present in the raw material and the accumulating product as fermentation proceeds, ability to grow symbiotically with other starter microbes if need be, production reproducibility between different batches of cultures grown and lack of spoilage of the product as it matures and during storage.

Starter Cultures in the Food Industry

Dairy Products

Lactic starter cultures impart particular features to various dairy products, as a result of the following actions:

Table 1 Inhibitory factors for starter cultures used in dairy product manufacture

Factors	*Comments*
Natural inhibitors present in milk	Some antibacterial systems, for example lactoperoxidase which converts thiocyanate and hydrogen peroxide to toxic products
Mastitis milk	High content of leukocytes lysing starter microbes, high level of lysozyme in milk, high antibody titre in milk
Antibiotics	Administered to lactating cows suffering from bacterial infections. Examples include streptomycin and erythromycin. Also produced by microbes present in milk, for example nisin and penicillin
Microbicides/sterilants	Usually from cleaning agents used for sanitation of fermentation equipment; also from detergents used on the farm, for example quaternary ammonium compounds
Bacteriophage	Presence of bacteria-destroying viruses in milk
Herbicides/insecticides	Used on the farm, for example benzene hexachloride

1. The conversion of lactose to lactic acid (glycolysis). Lactic acid coagulates milk, and so decreases the period required for the coagulation of milk proteins by rennet. The production of acid by the starter culture also inhibits the growth of contaminating microbes, and facilitates the separation of the liquid whey from the solidifying curd during cheese-making. If an inferior culture is used, the final product may have an undesirably high moisture content.
2. The secretion of lipolytic and proteolytic enzymes, which are important in degrading the lipids and proteins in milk. Lipids are degraded to ketoacids, ketones, esters and other substances, some of which give the final product its flavour and aroma. Proteins are degraded to peptones, peptides and amino acids.
3. The conversion of lactose to alcohol, which contributes to the unique flavour of fermented beverages such as koumiss and kefir.
4. The conversion of lactose to CO_2, which provides flavour and aerates dairy products.
5. The conversion of citric acid to aromatic compounds, which provide aroma and flavour.
6. The creation of the characteristic appearance of some cheeses, e.g. the blue veins of blue cheeses and the red hue of the surface of Limburger.

Table 2 lists some of the starter cultures used in the dairy industry.

Meat and Fish Products

The shelf life of meat may be substantially increased by the use of lactic starter cultures, which produce acid and/or antibiotics that inhibit the growth and metabolism of contaminating spoilage bacteria. For example: *Lactobacillus sake* inhibits the growth of *Listeria monocytogenes* in minced meat and sausages; *Leuconostoc gelidum* preserves fish meat by the production of antibiotics; and *Pediococcus acidilactici* is used against pathogenic species such as *Listeria monocytogenes* in sausages stored at 25°C. Another common pathogenic species present in meats is *Staphylococcus aureus*. Lactic starter cultures represent an alternative to salting or drying for the preservation of meat, the latter methods producing foods which are unsuitable for people on low-salt diets or concerned to avoid cancers of the digestive tract. The treatment of meat products with nitrates has now been largely replaced by other methods, because of the conversion of nitrogenous compounds to carcinogens when subjected to cooking temperatures.

The simultaneous injection of meats with lactic acid bacteria and sugar has been shown to lower the water content of the meat, and also to lower acidity, so inhibiting the growth of contaminating pathogenic microbes. These cultures are known as 'protective' rather than as starter cultures. In an ideal case, starter and protective functions are performed by the same microorganism.

The fermentation of meat is primarily a solid-phase process, with the inoculated bacteria growing in microcolonies. The important bacteria in the fermentation of sausages include the lactic cultures of *Lactobacillus plantarum*, *L. pentosus*, *L. curvatus*, *L. sake*, *Pediococcus pentosaceus* and *P. acidilactici*. These species form lactic acid and facilitate the drying of the sausages. Other bacteria are responsible for pigmentation, the synthesis of aromas and the degradation of excess nitrates, and include *Staphylococcus carnosus*, *S. xylosus* and *Micrococcus varians*. Fungal species used in sausage fermentation include *Debaryomyces hansenii* and *Penicillium nalgiovense*: both are responsible for aroma formation, and the latter is also involved in the production of pigmentation.

Lactococcus lactis subsp. *cremoris*, in combination with lactic acid, is used for adding flavour and aroma to catfish fillets and to control Gram-negative bacteria, which are sensitive to the acids and antibiotics produced by the lactic starter culture. Thus the shelf life of the fish is prolonged. Gram-negative bacteria

Table 2 Starter species used in the dairy industry

Starter	*Application*
Thermophilic cultures (optimum temperature range 30–42°C)	
Streptococcus thermophilus	Cooked cheeses and yoghurt
Lactococcus lactis	Production of slight aroma; substitute for *Lactobacillus bulgaricus*, for butter and cheese production
Lactobacillus helveticus	Strong acid producer and good proteolytic activity, used for production of butter, cheese, yoghurt
Lactobacillus lactis	Yoghurt
Lactobacillus acidophilus	Sweet acidophilus milk, yoghurt and other fermented products
Bifidobacterium bifidum	Fermented health products, yoghurt
Bifidobacterium longum	Fermented health products
Lactobacillus delbrueckii subsp. *bulgaricus*	Yoghurt
Lactobacillus delbrueckii subsp. *lactis*	Parmesan cheese
Propionibacterium shermanii	Emmental-type cheeses
Propionibacterium jensenii	Production of acetic acid from whey
Mesophilic cultures (optimum temperature range 22–27°C)	
Lactococcus lactis	Fermented products, including cheese
Lactococcus lactis biovar *diacetylactis*	Fermented products, including cheese
Leuconostoc cremoris	Cheeses with eyes, cottage cheese, sour cream, buttermilk
Enterococcus durans	Speciality cheeses
Enterococcus faecalis	Speciality cheeses
Mould and surface cultures	
Brevibacterium linens	Red hue for speciality cheeses
Geotrichum candidum	White mould-ripened cheeses
Penicillium candidum	Blue and white moulded cheeses
Penicillium roqueforti	Blue and white moulded cheeses
Penicillium camemberti	White moulded cheeses
Mucor spp.	Brown cheeses, vegetable-starch or cow-milk yoghurt products

are usually pathogens, and may also produce toxins harmful to humans. *Lactobacillus plantarum* is also used as a starter culture for fish fermentation, and species of the fungus *Monascus*, which produce pigments, have been used for the coloration of fish products.

Vegetable and Grain Products

Kimchi is a fermented vegetable product of South Korea, which for fermentation relies on the microbes naturally present in vegetables. The salt concentration, temperature, population of microbes, pH and air composition affect the final product. Production involves careful rinsing and brining, to reduce the number of microbes and also to create a more suitable environment for the growth of lactic acid bacteria. *Leuconostoc mesenteroides* is important during the early phases of fermentation, but *Streptococcus faecalis*, *Lactobacillus brevis*, *L. plantarum* and *Pediococcus cerevisiae* predominate during the later stages, *Leuconostoc mesenteroides* being inhibited by the increasing acidity as fermentation proceeds. These microorganisms, along with others, are being isolated and used in reproducible starter cultures, with the aim of improving the nutritional value of kimchi. For example, a starter culture of *Propionibacterium freudenreichii* subsp. *shermani* produces four times more vitamin B_{12} than normal fermentation. Kimchi is very similar to sauerkraut, the production of which requires starters for the fermentation of brined cabbage.

Vegetable fermentations in Nigeria have been attributed predominantly to *Leuconostoc mesenteroides. L. mesenteroides* is also used for the fermentation of soya milk, to produce a non-milk form of yoghurt. Alternatively, either *Lactobacillus casei* subsp. *rhamnosus* or *Leuconostoc mesenteroides* subsp. *cremoris* may be used: they both result in products with aromas that resemble those of true yoghurt, due to the conversion of citric acid to diacetyl.

Lactococcus lactis has been used for the fermentation-based preservation of gruel, made from cereal. The reduced growth of pathogenic microbes, including *Escherichia coli*, *Staphylococcus aureus*, *Clostridium perfringens*, *Shigella flexneri*, *Salmonella enteritidis* and *S. typhimurium*, is due to the increasing acidity of the gruel as fermentation proceeds. *Lactobacillus acidophilus* is also used for the preservation of gruel, and the amylase-producing yeast *Candida krusei* is included in the starter cultures used to degrade the starch in gruel which is to be used for weaning.

The rising of leavened bread is attributed to the

production of CO_2, by either *Saccharomyces cerevisiae* or *Candida krusei* in kneaded dough. *Lactobacillus fermentum* and *S. cerevisiae* are used in the fermentation of maize dough in the manufacture of 'koko' in Ghana; *Lactobacillus plantarum* is used for the fermentation of sorghum and maize into a porridge-type food called 'iogi' in Nigeria; and *Lactobacillus brevis*, with either *Propionibacterium acidipropionici* or *P. jensenii*, is used for the fermentation of rye dough for bread-making. Sections of cassava tuber are fermented with a mixed culture consisting of *Citrobacter freundii* and a species of each of *Saccharomyces*, *Candida* and *Geotrichum*.

Various oriental cooking sauces are produced by fermentation. For instance, *Aspergillus oryzae* and *A. sojae* are used in the production of soy sauce (liquid) and miso (paste). *Rhizopus microsporus* subsp. *oligosporus*, *R. chinensis*, *R. oryzae* and *Mucor indicus* are moulds used in the production of tempeh from soybeans. *Rhizopus microsporus* subsp. *oligosporus* is very effective against the growth of the pathogen *Clostridium perfringens*.

Sufu, a Chinese fermented product made from bean curd, is made with *Actinomucor taiwanensis* or *A. elegans*, by immersion in brine. In Indonesia, the solids remaining after the extraction of the oils from groundnuts are fermented using *Neurospora intermedia* or *Rhizopus* species, to produce a meat substitute. In Scandinavia, potato starch is fermented into a product used to feed animals. Starch is hydrolysed by *Saccharomycopsis fibuligera*, which is the starter used in the manufacture of sweet or sour alcoholic rice and tapioca products in the Orient. *Rhizopus* species and *Amylomyces rouxii* are used for the fermentation of tapioca starch and to produce semisolid snack foods.

Pigments produced by *Monascus* species, ranging from yellow to purple-red, have been used to provide colour for soybean curd products and wine. The fruits of the plant *Gardenia jasminoides* are fermented by *Bacillus subtilis* or *Aspergillus japonicus* to produce colours from the whole visual range. Fruit products are fermented using acetic acid bacteria, e.g. *Acetobacter mesoxydans* (also known as *A. pasteurianus*).

Numerous beverages, some alcoholic, are produced by mixed starter cultures. The list includes saké, wine produced from rice by *Aspergillus oryzae*, *A. japonicus*, *A. awamori*, *Pichia anomala*, *Zymomonas mobilis*, *Saccharomyces cerevisiae*, *Leuconostoc mesenteroides* subsp. *sake* and *Lactobacillus sake*. Wine is produced from grape juice by *Mucor indicus*, *Saccharomyces cerevisiae*, *Hanseniaspora guilliermondii*, *Kloeckera apiculata*, *Leuconostoc oenos*, *Pediococcus acidilactici* and *Lactobacillus casei*. Barley or wheat are fermented by *Saccharomyces cerevisiae* and species of *Dekkera*, *Brettanomyces*, *Lactobacillus* and *Pediococcus*, in the manufacture of beer. In wine-making, interactions between the yeasts are important for the development of the proper flavour and acidity. Initially in the fermentation, wild yeasts such as *Kloeckera apiculata* produce many of the compounds needed later in the production of flavour components. As the fermentation proceeds and the alcohol level rises, the wild yeasts are inhibited. *Saccharomyces cerevisiae* then predominates, and continues the production of ethanol.

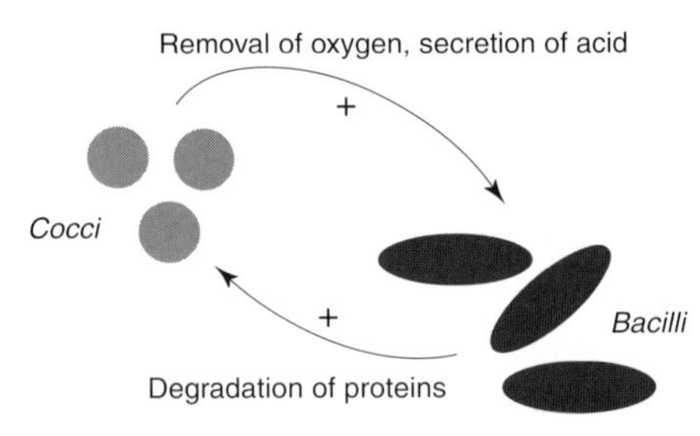

Figure 1 Production of yoghurt requires the symbiotic activity of *Streptococcus thermophilus* and *Lactobacillus delbrueckii* subsp. *bulgaricus*.

Composition of Starter Cultures

The first commercial starters usually comprised only one microorganism, but nowadays food manufacturers often use two or more microbes simultaneously. For example in the production of yoghurt, the symbiotic activity of *Streptococcus thermophilus* and *Lactobacillus delbrueckii* subsp. *bulgaricus* is involved (**Fig. 1**). However, a combination of starter culture microbes may also cause complications: an example is the combined use of *Lactobacillus delbrueckii* subspp. *lactis* and *bulgaricus* in Parmesan cheese manufacture, in which the latter subspecies is strongly inhibited.

Genetic Engineering of Species

Before the development of recombinant DNA technology, genetic variations in starter cultures were introduced by other methods, including exposure to mutagenic compounds or irradiation. However, these changes are by trial and error and entirely random, and mutated cultures could acquire undesirable attributes as well as showing enhancement. Recombinant DNA technology is now used to enhance the fermentative qualities of starter microbes. Conjugation involves the transfer of plasmids (bacterial DNA sequences) from one microbe to another. The donor and recipient microbes must be closely related. Chromosomal genes may also be transferred.

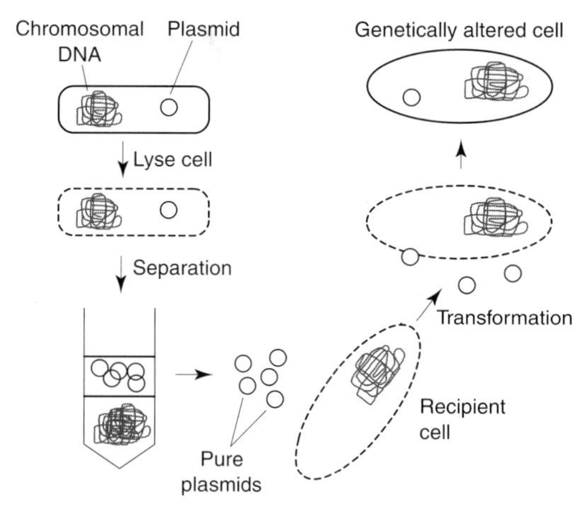

Figure 2 Transfer of plasmid-based gene for enhancement of starter microbe activity. The plasmid is isolated from a donor strain by cell lysis and centrifugational separation from chromosomal DNA. The pure plasmid preparation is then used to transform the recipient strain.

While genes are only transferable between closely related species by conjugation, transformation allows researchers to transgress the genetic barrier. In this method pure DNA, modified (**Fig. 2**) or unmodified, is introduced into recipient cells. The transferable elements may be plasmids that are not transferable through conjugation.

Transformed strains are subject to the laws governing genetic engineering, but conjugated strains are exempt from such regulation. On the basis of thousands of years of use in food manufacture, certain species of microbes are considered safe for consumption, and categorized as 'generally recognized as safe' (GRAS); genetic engineering should not affect the GRAS status of these microbes.

Microbes that have been altered genetically through protoplast fusion include *Aspergillus oryzae*, *Penicillium caseicolum*, *P. nalgiovense* and *Rhizopus niveus*. *Penicillium roqueforti*, *Mucor circinelloides* and *A. sojae* are starters that have been manipulated by transfection. Hybrids produced from the mesophilic *Saccharomyces cerevisiae* and the cryophilic *S. bayanus* have promising wine-making qualities at 10°C, releasing more flavour components than either of the parent species.

The limited ability of *S. cerevisiae* to degrade starch has been overcome by introducing the genetic elements for the synthesis of amylase and glycoamylase from the yeast *Schwanniomyces occidentalis*. Its inability to degrade maltodextrins has been overcome by introducing the genetic element for glucosidase from *Candida tropicalis*. *Aspergillus oryzae*, transformed with the glucoamylase genetic element from *A. shirousamii*, displays enhanced glucose synthesis and is used for saké production. *Saccharomyces cerevisiae* can be transformed with exogenous DNA genetic elements which express toxins against wild yeasts: this has allowed *S. cerevisiae* to dominate the initial stages of the wine-making process, leading to more reproducible results. Additionally, *S. cerevisiae* transformed with the *egl1* gene (encoding for an endoglucanase) can degrade all the glucans in wort, the final beer having similar organoleptic properties to that produced by the untransformed strain (**Table 3**). More properties endowed on *S. cerevisiae* via transformation are listed in Table 3.

Table 3 Transformation of *Saccharomyces cerevisiae*

DNA element	*Donor organism*	*Enhancement of function*
Inulinase	*Kluyveromyces marxianus*	Inulin converted to ethanol
Xylose reductase and xylitol dehydrogenase	*Pichia stipitis*	Xylose metabolism
Xylanase and glucosidase	*Bacillus pumilus*	Xylose metabolism
Galactosidase	*Kluyveromyces lactis*	Lactose metabolism
	Aspergillus niger	Lactose metabolism

Lactococcus lactis, transformed with a gene for lysozyme, causes lysis of *Listeria monocytogenes*. The gene for the bacteriocin helveticin J has been transferred from *Lactobacillus helveticus* to *L. acidophilus*, the recipient becoming resistant to pathogenic bacteria. Bacteriophage resistance can be conferred on starter microbes by plasmid-mediated transformation. This helps the fermentation industry and also facilitates the investigation of the relationship between the phages and their hosts. Genetically modified starter microbes used for fermented meat products include *Staphylococcus carnosus*, *Lactobacillus curvatus* and *L. sake*.

The activity of manipulated microbes must be carefully compared with that of parent strains, and also the food products must be analysed to ensure that there are no changes detrimental to consumer health or the aesthetic or organoleptic qualities of the product.

The undesirable properties of microbes may be removed using techniques such as gene disruption. In yeasts and moulds, homologous DNA fragments tend to integrate at the homologous site on the fungal genome, resulting in inactivation of the gene. This process, called homologous recombination, (**Fig. 3**), is being introduced for the development of starter strains without particular undesirable qualities.

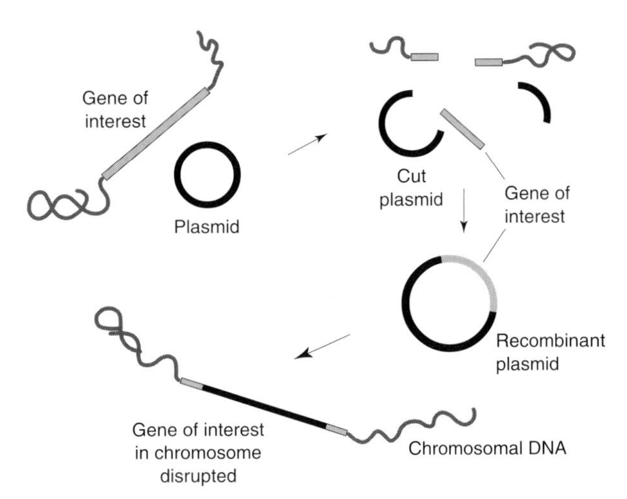

Figure 3 Gene disruption based on homologous recombination. A sequence of the gene (from the donor chromosome) to be disrupted is recombined into a plasmid. This plasmid is then used to homologously recombine with the undesirable gene in starter microbes, thereby 'silencing' expression of the gene.

Spoilage Problems caused by Starter Cultures

Spoilage can occur if the starter culture is impure or its activity has deteriorated over time: hence regular checks on starter activity are needed. Spoilage can be inhibited by a number of strategies. Lactic acid bacteria and the acid they produce inhibit the growth of numerous species of microorganisms. For example, *Lactobacillus acidophilus* inhibits the growth of pathogens such as *Staphylococcus aureus*, *Escherichia coli*, *Proteus vulgaris*, *Salmonella typhi* and *Yersinia enterocolitica*. Antibiotics such as nisin, produced by bacteria, also help to control the levels of spoilage microbes: for adjunct microbes able to produce antibiotics may be used together with the fermenting species. For example, *Enterococcus faecium* produces a bacteriocin against *Listeria monocytogenes*, and can be added to a mixed culture of *Streptococcus thermophilus* and *Lactobacillus delbrueckii* subsp. *bulgaricus*.

Some species of *Lactobacillus* inhibit the production of hazardous aflatoxins by fungi contaminating cheeses, but other microbes, e.g. *Lactococcus lactis*, stimulate aflatoxin production. Such characteristics must be taken into account in the selection of starter cultures for specific applications.

See also: **Cheese**: Microbiology of Cheese-making and Maturation. **Fermented Foods**: Origins and Applications; Fermented Vegetable Products; Fermented Meat Products; Fermentations of the Far East; Beverages from Sorghum and Millet. **Fermented Milks**: Range of Products. **Fish**: Spoilage of Fish. **Genetic Engineering**: Modification of Yeast and Moulds; Modification of Bacteria. ***Lactobacillus***: Introduction. ***Lactococcus***: *Lactococcus lactis* Sub-species *lactis* and *cremoris*. ***Leuconostoc***. ***Listeria***: *Listeria monocytogenes*. ***Pediococcus***. **Spoilage Problems**: Problems caused by Bacteria; Problems caused by Fungi. **Starter Cultures**: Uses in the Food Industry; Cultures Employed in Cheese-making. ***Streptococcus***: *Streptococcus thermophilus*.

Further Reading

Cheigh HS and Park KY (1994) Biochemical, microbiological, and nutritional aspects of kimchi (Korean fermented vegetable products). *Critical Reviews in Food Science and Nutrition* 34 (2): 175–203.

Geisen R and Holzapfel WH (1996) Genetically modified starter and protective cultures. *International Journal of Food Microbiology* 30: 315–324.

Lucke FK (1994) Fermented meat products. *Food Research International* 27: 299–307.

Cultures Employed in Cheese-making

Timothy M Cogan, Dairy Products Research Centre, Teagasc, Fermoy, Ireland

Introduction

Cheese-making is essentially a microbial fermentation of milk by specially selected microorganisms, which can be divided into two groups:

- those involved in both manufacture and ripening
- those involved in ripening only.

Only bacteria are involved in both manufacture and ripening. Their primary role is the production of lactic acid from lactose (milk sugar) which results in a reduction of the pH of the milk and curd. Invariably these are lactic acid bacteria (LAB), also they are called 'starter' or lactic cultures because they initiate (start) the production of lactic acid. The microorganisms involved only in ripening include other bacteria, yeasts and moulds (**Table 1**). All of the bacteria used in cheese-making are Gram-positive. Those in starter cultures are catalase-negative and hence essentially anaerobes, while those involved in ripening only are catalase-positive and, except for *Propionibacterium freudenreichii*, obligate aerobes.

Cultures Involved in Manufacture and Ripening

Starter Cultures

There are two types of starter cultures, mesophilic and thermophilic, with optimum growth temperatures of

Table 1 Lactic acid bacteria in commercial and natural (artisanal) cultures

Bacterium	*Type*	*Shape*	*Lactic acid produced in milk*[a]	*Metabolism of citrate*	*NH_3 from arginine*	*Growth at*				*Fermentation of sugar*	*Isomer of lactate produced*
						10°C	*15°C*	*40°C*	*45°C*		
Commercial starter cultures											
Streptococcus thermophilus	Thermophilic	Coccus	0.6	–	–	–		+	+	Homofermentative	L
Lactobacillus helveticus	Thermophilic	Rod	2.0	–	–	–	–	+	+	Homofermentative	D,L
Lactobacillus delbrueckii subsp. *bulgaricus*	Thermophilic	Rod	1.8	–	±	–	–	+	+	Homofermentative	D
Lactobacillus delbrueckii subsp. *lactis*	Thermophilic	Rod	1.8	–	±	–	–	+	+	Homofermentative	D
Lactococcus lactis subsp. *cremoris*	Mesophilic	Coccus	0.8	±	–	+	+	–	–	Homofermentative	L
Lactococcus lactis subsp. *lactis*	Mesophilic	Coccus	0.8	±	+	+	+	+	–	Homofermentative	L
Leuconostoc lactis	Mesophilic	Coccus	< 0.5	+	–	+	+	–	–	Heterofermentative	D
Leuconostoc mesenteroides subsp. *cremoris*	Mesophilic	Coccus	0.2	+	–	+	+	–	–	Heterofermentative	D
Natural (artisanal) starter cultures											
Bacteria involved in commercial starter cultures, plus:											
Lactobacillus casei		Rod		±	–		+		–	Homofermentative	L
Lactobacillus paracasei subsp. *paracasei*		Rod			–	+	+	+	±	Homofermentative	L
Lactobacillus paracasei subsp. *tolerans*[b]		Rod			–	+	+	+	–	Homofermentative	L
Lactobacillus rhamnosus		Rod			–	±	+	+	+	Homofermentative	L
Lactobacillus plantarum		Rod		±	–		+		–	Homofermentative	D,L
Lactobacillus curvatus		Rod			–		+		–	Homofermentative	D,L
Lactobacillus fermentum		Rod			+		+	+	+	Heterofermentative	D,L
Enterococcus faecalis		Coccus		+	+	+	+	+	+	Homofermentative	L
Enterococcus faecium		Coccus			+	+	+	+	+	Homofermentative	L

[a]Approximate values: individual strains vary.
[b]Survives heating at 72°C for 40 s.

about 28°C and about 42°C respectively. Each of these is further subdivided into defined and undefined cultures. Defined cultures are known strains, which have been carefully selected on the basis of:

- their ability to produce lactic acid rapidly in milk (e.g. lactococci should reduce the pH of milk below 5.3 in 6 h at 30°C from a 1% v/v inoculum)
- their salt tolerance
- their ability to withstand attack from bacteriophage
- their inability to inhibit other starter strains.

They are usually used in mixtures of two or three strains, known as multiple strain cultures. Bacteriophage-insensitive mutants are an inherent part of defined cultures (see below).

Undefined cultures consist of an unknown number of strains. They are subcultures of coagulated milks which produced good-quality cheese in the early part of the twentieth century when the scientific understanding of the microbiology of starter cultures was beginning. For this reason, undefined cultures are also called mixed-cultures. Many of them are also mixtures in the sense that they contain mixtures of different species of LAB. For example, many thermophilic cultures contain *Streptococcus thermophilus* and *Lactobacillus delbrueckii* subspp. *bulgaricus* and/or *lactis*, and many mesophilic cultures contain *Lactococcus lactis* subspp. *lactis* and *cremoris* and *Leuconostoc* spp. The exact species of *Leuconostoc* in these cultures is not known but is thought to be *L. mesenteroides* subsp. *cremoris* or *lactis*. The *Leuconostoc* species and some of the lactococci in these cultures also metabolize citrate. The citrate-positive (Cit^+) strains generally comprise only a small proportion of the bacteria present and their primary function is to produce flavour and aroma compounds, e.g. diacetyl and acetate, from citrate. Strains of lactococci unable to metabolize citrate (Cit^-) dominate these cultures and are responsible for acid production. Some distinguishing characteristics of the various organisms found in starter cultures are listed in Table 1.

Depending on the flavour producer, mixed-strain mesophilic cultures are classified into:

- L cultures containing only leuconostocs as flavour producers (L from ***L**euconostoc*)
- D cultures containing only Cit^+ *Lactococcus lactis* as flavour producers (D from *L. **d**iacetylactis*, an old name for this organism)
- DL cultures containing both *Leuconostoc* and Cit^+ *Lactococcus lactis* as flavour producers
- O cultures, no flavour producer.

The most common cultures are the DL type. Different cultures are used in different cheeses (**Table 2**).

Natural Cultures

In some countries, especially France and Italy, back slopping of starters is frequently practised. Whey or milk from one day's production is incubated under carefully controlled conditions for use in making cheese the next day. This system is generally used where there is a long tradition in cheese-making. These starters are called 'natural' or 'artisanal' cultures and usually contain both mesophilic and thermophilic organisms (Table 1), other lactobacilli and *Enterococcus* species. In some artisanal Spanish and Portuguese cheeses, e.g. La Serena and Serra da Estrela, no starter cultures are used; instead, the cheese maker depends on the adventitious LAB present in the milk for acid production.

Adjunct Cultures

Mixtures of mesophilic and thermophilic cultures are being used increasingly to make cheeses like Cheddar, which formerly were made only with mesophilic cultures. Facultatively heterofermentative (mesophilic) lactobacilli such as *Lactobacillus casei*, *L. paracasei* and *L. plantarum* are also used. These cultures are called adjunct cultures and are thought to produce flavours that are more rounded, less bitter, and sweeter. Thermophilic cultures also produce significant amounts of acid at the cooking temperature (~ 38°C) of Cheddar cheese.

Cultures Involved Only in Ripening

Cultures for ripening are generally added to the milk with the starter cultures, but in the case of smear-ripened cheeses, such as Tilsiter, Münster and Limburger, bacteria (*Brevibacterium linens*), moulds (*Geotrichum candidum*) and sometimes yeasts (*Candida utilis*) are inoculated onto the surface of the cheese after it has been removed from the brine and drained. *Brevibacterium linens* is unable to grow at low pH values. In smear-ripened cheeses, the yeast and moulds grow first, metabolizing the lactate to CO_2 and H_2O, which results in a distinct rise in the pH of the cheese e.g. from 4.8 to 5.8, at which point the acid-sensitive *B. linens* begins to grow and produce the typical red smear on the surface. This bacterium is unusual in that it undergoes a transformation from rods to coccal forms; rods dominate young, exponential cultures and cocci predominate in old, stationary-phase cultures. The surface organisms are very salt-tolerant and some strains can grow in the presence of NaCl concentrations as high as 15 g per 100 ml.

Growth of *Penicillium camemberti* is responsible for the white, fluffy surface growth characteristic of Brie and Camembert cheese, while *P. roqueforti* is

Table 2 Cultures used in dairy products

Product	*Starter cultures*	*Other cultures*	*Important compounds other than lactic acid*
Emmental	*Streptococcus thermophilus* and *Lactobacillus helveticus* Galactose-positive *Lactobacillus delbrueckii* subsp. *lactis* may also be used	*Propionibacterium freudenreichii*	CO_2, propionate and acetate
Mozzarella	*Streptococcus thermophilus* and *Lactobacillus helveticus* *or* a mixed-strain of thermophilic culture		
Cheddar	Defined strains of *Lactococcus lactis* subspp. *cremoris* and *lactis* *or* O, L or DL mesophilic mixed cultures Sometimes thermophilic cultures are included		
Edam and Gouda	Mainly DL mesophilic mixed cultures		CO_2 and acetate
Camembert and Brie	O, L or DL mesophilic mixed cultures	*Penicillium camemberti* *Geotrichum candidum* *Candida utilis*	
Tilsiter, Limburger and Münster	O, L or DL mesophilic mixed cultures	*Brevibacterium linens* *Geotrichum candidum* *Candida utilis*	Sulphur compounds, e.g. methional
Yoghurt	Mainly thermophilic mixed cultures Defined strains of *Streptococcus thermophilus* and *Lactobacillus delbrueckii* subspp. *bulgaricus* and *lactis* may also be used		Acetaldehyde
Fromage frais and quarg	O, L or DL mesophilic mixed cultures		Diacetyl and acetate
Lactic butter	DL mesophilic mixed cultures *or* L and D mesophilic mixed cultures		Diacetyl and acetate

responsible for the veins in blue cheeses. Moulds are aerobic organisms and blue cheeses are pierced to allow entry of sufficient oxygen into the cheese mass for the mould to grow. Growth of the *Penicillium* species also results in oxidation of the lactate to CO_2 and H_2O and a distinct rise in pH.

In surface-ripened cheeses, proteinases and lipases produced by the moulds and *B. linens* hydrolyse the casein and fat during ripening to the amino acids, peptides and fatty acids which are the precursors of many of the compounds responsible for cheese flavour. In addition, *P. roqueforti* produces ketones from the fatty acids which are mainly responsible for the strong flavour of blue cheese, and *B. linens* produces the sulphur-containing compounds (mainly methional from methionine) responsible for the 'smelly sock' odour of smear-ripened cheeses.

In the Emmental type of cheese *Propionibacterium freudenreichii* is used; although it is catalase-positive, it is essentially an anaerobe and sensitive to salt. Its main function is to ferment lactate:

$$3 \text{ lactate} \rightarrow 2 \text{ propionate} + 1 \text{ acetate} + 1\ CO_2$$

This fermentation occurs in the 2–3 weeks during which this cheese is held at 23°C. The CO_2 produced is responsible for the large holes (eyes) characteristic of Emmental cheese.

Bacteriocins

Almost all bacteria are capable of producing proteins, called bacteriocins, which inhibit the growth of other bacteria. Bacteriocins of LAB are being intensively studied because many of them inhibit pathogens such as *Listeria monocytogenes* and *Staphylococcus aureus*. The LAB are generally regarded as safe (GRAS), and so any bacteriocin they produce can be used in foods without the need for exhaustive testing to ensure its safety.

Nisin is the most intensively studied bacteriocin of LAB. It is a small (3.3 kDa), heat-stable peptide produced by some strains of *Lactococcus lactis* and was initially isolated in 1945. It contains several unusual amino acids like (β-methylanthionine, dehydroalanine and lanthionine, due to post-translational modification of the amino acids in the peptide. Nisin is soluble at pH 2 but its solubility decreases as the pH increases and it is virtually insoluble at pH 7.0. It inhibits numerous bacteria, including several Gram-positive food-poisoning organisms such as *Staphylococcus aureus*, *Listeria mono-*

cytogenes and *Bacillus cereus*, and also cheese spoilage bacteria such as *Clostridium tyrobutyricum* and *C. butyricum*. It has no effect on Gram-negative bacteria. Nisin acts by dissipating the proton motive force, which is an important component in transport of nutrients into cells. Other bacteriocins produced by LAB, e.g. pediocin produced by *Pediococcus pentosaceus*, are gaining in commercial importance.

Metabolism of LAB

Proteolysis

Proteolysis by LAB is important for their growth in milk and in the ripening of cheese. Starter LAB are auxotrophic and require several amino acids for growth. The level of free amino acids in milk is low and sufficient to sustain only 20% of the normal growth of LAB. Therefore, starter bacteria must have a proteinase system to hydrolyse the caseins to amino acids and peptides which are then transported into the cell. Proteolytic activity is low and not sufficient to cause visible hydrolysis or clearing of the milk.

The *Lactococcus* proteinase has been intensively studied. It is a large molecule of about 190 kDa and the amino acid sequence in all strains studied is similar. The slight differences that do occur are thought to be responsible for the different proteolytic specificities that have been observed with casein. The enzyme is a serine proteinase which hydrolyses the different caseins into numerous oligopeptides and amino acids which are then transported by various primary and secondary transport systems into the cell. Only peptides containing fewer than eight amino acid residues can be transported. Intracellular peptidases hydrolyse the peptides into the constituent amino acids for protein synthesis. Numerous peptidases, including aminopeptidases, endopeptidases and peptidases capable of hydrolysing proline-containing peptides, have been identified in LAB. The proteinase, peptide and amino acid transport systems and peptidases collectively comprise the proteolytic systems of LAB. Only limited studies have been carried out on the proteolytic systems of other starter LAB but the results suggest that they are similar.

Symbiotic relationships occur in both mesophilic and thermophilic cultures. In mesophilic cultures many isolates are proteinase-negative and rely on proteinase-positive strains to provide the amino acids necessary for growth. In thermophilic cultures, *Lactobacillus delbrueckii* subsp. *bulgaricus* is more proteolytic than *Streptococcus thermophilus* and hydrolyses casein to amino acids, particularly histidine, glycine, valine and isoleucine, which stimulate the growth of *S. thermophilus*. In turn, *S. thermophilus* produces formate (from lactose), which stimulates growth of *L. delbrueckii* subsp. *bulgaricus*.

Lactate Production

Lactose is the major sugar found in milk and is a disaccharide composed of glucose and galactose. *Lactococcus lactis* and *Lactobacillus helveticus* ferment lactose by the glycolytic (homofermentative) pathway almost completely to lactic acid, while *Streptococcus thermophilus* and *Lactobacillus delbrueckii* ferment only the glucose moiety and excrete galactose in proportion to the amount of lactose transported. This results in a build-up of galactose, which could act as an energy source for spoilage organisms. *Lactobacillus helveticus* ferments both sugars, and for this reason is used in Swiss cheese-making. *Leuconostoc* and the obligate heterofermentative lactobacilli (Table 2) ferment lactose by the phosphoketolase (heterofermentative) pathway to equimolar concentrations of lactate, ethanol and CO_2.

Two different systems are used to transport lactose. In most starter LAB, lactose is transported directly into the cell where it is hydrolysed by β-galactosidase. However, *Lactococcus* species use the phosphotransferase system, in which the energy in phosphoenolpyruvate is ultimately transferred to lactose to form lactose phosphate as lactose is transported across the cell wall in a complicated series of reactions involving several enzymes.

Diacetyl, Acetate and CO_2 Production

Diacetyl and acetate are important in determining the flavour of many fermented but unripened products such as cottage cheese, fromage frais, quarg and lactic butter, while CO_2 is responsible for the small number of holes or 'eyes' found in Edam and Gouda cheese. Citrate, which is present in low concentrations (about 10 mmol l^{-1}) in milk, is the precursor for each of these compounds. There is still considerable debate on what is the immediate precursor of diacetyl: some workers believe that diacetyl synthase condenses acetaldehyde-TPP, produced from pyruvate, with acetyl-CoA to form diacetyl directly; however, the majority believe that diacetyl is synthesized chemically from 2-acetolactate (AL) which is formed by condensation of acetaldehyde-TPP with a molecule of pyruvate. Acetolactate is unstable and is chemically decarboxylated, either oxidatively to diacetyl or non-oxidatively to acetoin; the latter reaction is also carried out by AL decarboxylase. Diacetyl is produced in much smaller amounts than acetoin. Diacetyl can be reduced to acetoin and acetoin to 2,3-butanediol by acetoin and 2,3-butanediol dehydrogenases respectively. These reactions are probably carried out by the same enzyme. An example of growth and product formation by a DL culture in milk is shown in **Figure 1**.

Acetaldehyde

Acetaldehyde is produced in small amounts by LAB and more is produced by thermophilic than by mesophilic starters. It is an important flavour component in yoghurt. It is generally believed to be formed from sugar (pyruvate), but in *Lactococcus lactis* it is formed from threonine in a reaction catalysed by threonine aldolase:

$$\text{Threonine} \rightarrow \text{Acetaldehyde} + \text{Glycine}$$

The physiological function of this reaction is to provide glycine for growth.

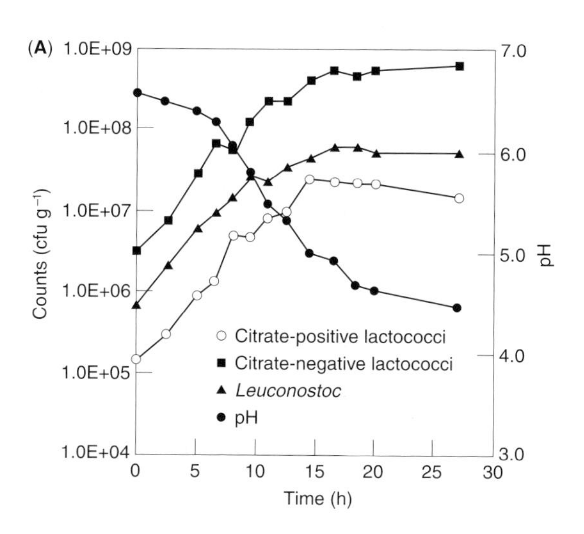

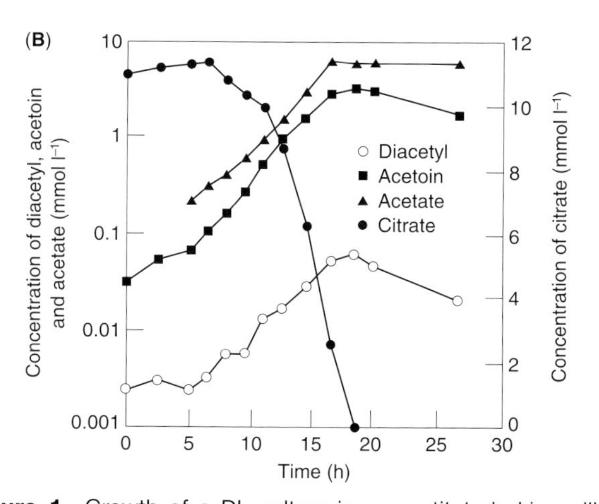

Figure 1 Growth of a DL culture in reconstituted skim milk ($100\,\text{g}\,\text{l}^{-1}$) at 21°C. (**A**) Growth of citrate-positive lactococci, citrate-negative lactococci and leuconostocs, and decrease in pH. (**B**) Production of diacetyl, acetate and acetoin, and utilization of citrate.

Plasmids

Many of the important properties of starter LAB – proteolytic activity, bacteriocin production and immunity, the first enzymes in lactose metabolism, the abilities to transport citrate and resist attack from bacteriophage – are encoded on plasmids. These are small, circular pieces of extrachromosomal DNA which replicate independently of the chromosome. They range in size from 2 kDa to 100 kDa. The number of plasmids found in starter LAB varies from none (most strains of *Streptococcus thermophilus* and *Lactobacillus helveticus*) to 11 (some strains of *Lactococcus*). The functions of most plasmids are unknown.

Plasmids have been the cornerstone of the major developments in starter genetics. Transformation, conjugation, transduction, transfection and protoplast fusion have all been identified in LAB, and numerous genes have been cloned. Several workers have shown that plasmid profiles are useful in distinguishing between strains in mixed cultures, but this approach does not give a permanent 'fingerprint' of a strain because plasmids are easily lost, for example on subculture.

Bacteriophage

Attack of starters by bacteriophage (phage) is the most serious problem in the commercial production of cheese. Once the cells have been attacked, their ability to produce lactic acid is impaired and, in severe cases, totally prevented. Phages are viruses that attack bacteria and can only multiply inside a bacterial cell. Generally, phages attack only closely related strains and phage-unrelated strains are an important component of defined cultures. Phages have a tail and a head, containing the DNA (**Fig. 2**).

There are two different mechanisms of phage multiplication: the lytic cycle and the lysogenic cycle. The lytic cycle involves four steps: phage adsorption, DNA injection, phage multiplication, and phage release. The phage tail attaches to special receptors on the bacterial cell and the phage DNA is injected into the host cell through the phage tail. The phage DNA then takes over the metabolic machinery of the host to synthesize more phage. Once the new phage matures, the phage lysin, present in the tail, hydrolyses the bacterial cell wall releasing new phage to attack new bacterial cells. The period from phage infection to phage release (the latent period) is usually quite short (about 30 min) and the number of phage produced per cell (the burst size) can be as high as 200. Assuming a latent period of 30 min and a burst size of 100, a single phage can multiply to 1 000 000 in about 1.5 h. In the same length of time one starter cell would

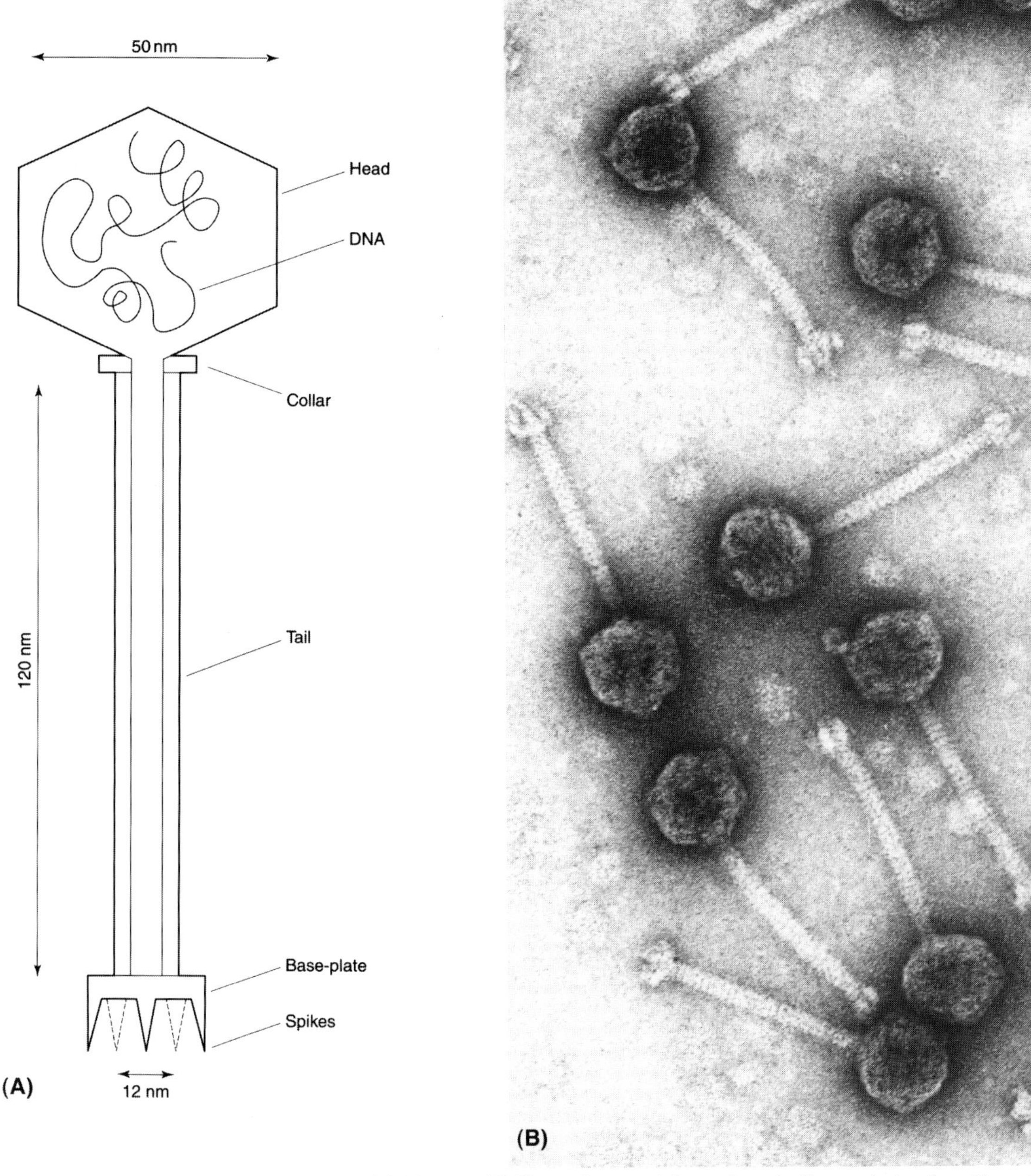

Figure 2 Bacteriophages of *Lactococcus lactis*. (**A**) Diagram. (**B**) Photomicrograph (reproduced with permission from Neve, 1996).

produce four to eight cells. These figures show the danger of contamination by just one phage. In the lysogenic cycle, the phage DNA is incorporated into the bacterial chromosome. Such phages are called 'temperate', and production of new phage particles does not occur. In commercial use of cultures lytic phages are obviously more important than temperate ones.

Lysogenic phage can be induced by ultraviolet light, H_2O_2 and mitomycin C, and will multiply if a suitable host, called an indicator strain, is present. The origin of phage has not been unequivocally demonstrated but lysogenic phage and raw milk have been implicated. However, lysogenic phage can only be a source if an indicator strain on which the induced phage can multiply is also present.

Many strains used in defined strain cultures are selected for their inherent resistance to phage. This is due to the presence of phage resistant plasmids in the cells. A large number of such plasmids have been identified in LAB, especially in *Lactococcus*. The mechanisms involved include the prevention of phage adsorption, probably through 'blocking' of the host's phage receptors, and restriction–modification systems involving endonucleases and methylases. Some forms of resistance involve neither of these mechanisms, and result in 'abortive infections'.

Virtually all starters will eventually succumb to attack by phage. When this happens, many cultures still contain a small number of phage-resistant cells which can be selected by growth in the presence of the phage. These bacteriophage-insensitive mutants, if they produce acid rapidly, can be used to replace the parent strain in multiple strain starters.

Control of Phage

The most important step in producing good-quality cheese is ensuring that the bulk starter is free of phage. This point cannot be overstressed, but it is often overlooked in cheese production. Determination of the phage levels in starters and cheese whey should be carried out daily in well-run cheese factories.

Many phages are heat-resistant and withstand heating to 85°C for several minutes. Therefore, the medium used for growing the starter must be heat-treated at high temperatures to inactivate any phage that may be present. The heat treatment also improves the nutritional value of the milk because it inactivates the natural inhibitors and causes partial hydrolysis of the proteins. Heat treatment should be carried out in the tank in which the culture is to be grown. Some cheese manufacturers heat-treat the milk in a pasteurizer and then fill the starter tank with the heated milk; this practice is not recommended because of the danger of phage being present in an otherwise clean starter tank. As already described, a single phage in a bulk tank is sufficient to cause problems.

Attachment of phage to the bacterial cell wall requires calcium ions. Phage-inhibitory media have therefore been formulated containing high concentrations of citrate and phosphate to chelate Ca^{++} and prevent phage multiplication.

During cooling, the air entering the starter tank should pass through a high-efficiency particulate air filter to prevent airborne phage entering the tank. A slight positive air pressure in the starter tank will also help to prevent entry of airborne phage.

Aseptic techniques should be used as far as possible to inoculate the medium. Most starter manufacturers advise that the package containing the inoculum be chlorinated before it is opened to prevent any phage on the outside from contaminating the culture. Chlorine is a very effective phagicide. Exposure to 100 $\mu g\,ml^{-1}$ of available or 'active' chlorine for 10 min is usually sufficient to inactivate all phage.

Rotation of phage-unrelated strains is also useful. In addition, many defined cultures are inherently phage-resistant and can be used daily without problems. This requires daily monitoring of the whey for phage, which is routine in any well-run quality control scheme in a modern cheese factory.

Addition of rennet to the milk as soon as possible after the addition of the starter causes the coagulum to physically separate phage-infected cells from noninfected cells. In addition, phages are unable to penetrate the curd particles to locate noninfected cells.

As little as 1 ml of residual whey in cheese vats can be a potent source of lytic phage. The final step in cleaning cheese-making equipment, such as vats and filling lines, should include chlorination. The residual chlorine should not be rinsed from the equipment, to prevent further contamination with phage from hoses or water. Any residual chlorine on the equipment is immediately inactivated once it comes in contact with the milk. Vats should be routinely chlorinated between fills, and the use of closed vats is recommended.

Genetic approaches have been used to improve the phage resistance of starter cultures. Many phage-resistant plasmids can be moved relatively easily into phage-sensitive strains. For example, plasmid pTR2030, which encodes a restriction–modification system and an abortive infection mechanism, has been introduced into several commercial strains.

Production of Bulk Cultures

The culture added to the milk in the cheese vat is called the bulk culture. The inocula used to produce the bulk cultures are purchased from specialized laboratories, as frozen or freeze-dried, concentrated suspensions, containing high numbers ($> 10^{10}$) of cells. These have been grown under controlled conditions (e.g. *Lactococcus* at pH 6.0 at 30°C) in proprietary media and are harvested, frozen in liquid N_2 and either stored at −80°C or freeze-dried in quantities sufficient to inoculate 300 l, 500 l or 1000 l.

Generally, the medium used for propagating starters in cheese factories is reconstituted skim milk powder (120 $g\,l^{-1}$), although proprietary phage-inhibitory media are also used. The medium is heat-treated to more than 85°C for at least 30 min, to inactivate phage and natural inhibitors in the milk. After cooling to the incubation temperature, the medium is inoculated and incubated for about 16 h at 21°C (mesophilic cultures) or about 10 h at 42°C (thermophilic cultures). Sometimes the pH of the cultures is controlled during growth, either externally, using a 'pH-stat' and a neutralizer, such as NH_4OH, or internally, using insoluble salts which solubilize as the pH of the culture decreases and help to maintain the pH above 5.5. This results in a more active culture because greater numbers of cells are produced per unit volume.

During incubation, the pH of milk-grown cultures decreases from its initial level of about 6.6 to 4.6 (mesophilic cultures and *Streptococcus thermophilus*) or to pH 3.5–4.0 (thermophilic lactobacilli). These pH values are equivalent to 0.7% and 1.2% lactic acid respectively.

After incubation, cultures are cooled, tested for their activity (i.e. their ability to produce acid) under defined conditions (e.g. a 1% inoculum followed by incubation at 30°C or 45°C for 6 h) and if satisfactory are used in cheese-making. Mesophilic bulk cultures can be stored at 40°C for 2–3 days and thermophilic ones at 4°C for 10–12 days without great loss of activity. Thermophilic cultures for cheese-making are generally grown separately, while those for yoghurt manufacture are grown together.

In small plants, bulk starters may not be produced. Instead, concentrated cell suspensions (produced as described above) are added directly to the milk in the vat. Generally a small but perceptible lag in acid production is observed. Such cultures are relatively expensive but they do away with the necessity for propagating and checking cultures for activity in plants.

Cheese Ripening

Each cheese is a unique ecosystem, and the reactions responsible for the production of the flavour compounds occur at different rates depending on the temperature, the pH, the presence or absence of a surface microflora, the method of salting (brine or dry), the rate of salt diffusion, and the shapes of brine-salted cheeses and the starters. Cheddar is an important cheese in world trade, and is used here as an example.

Cheddar cheese is made with mesophilic cultures and is dry-salted. The salt is added to the curd about 5.5 h after inoculation of the culture and the amount of salt controls the subsequent rate of lactose fermentation by the starter. The salt is dissolved in the moisture of the cheese and the percentage salt-in-moisture generally lies within the range 4–6%; the higher the level, the slower is the subsequent rate of fermentation. Generally all the lactose will be fermented within the first 2 weeks of ripening and the total amount of lactate at this stage will be about $13\,g\,kg^{-1}$.

The starter bacteria die and lyse during ripening at a rate that is an inherent property of the strain and also depends on the number of phage present that attack that particular strain. Lysis results in the release of intracellular enzymes. It is the activity of these enzymes, combined with the starter proteinase and the rennet, that results in the softening of the cheese texture and the increase in the levels of peptides and amino acids during ripening. The flavour compounds are produced from the peptides and amino acids by chemical rather than biochemical reactions. These transformations are thought to require the low oxidation–reduction potentials that are characteristic of all cheeses, and which are probably due to the metabolism of lactose by the starters.

In all cheeses – especially semihard and hard cheeses which are ripened for several months – growth of non-starter lactic acid bacteria (NSLAB) from low levels (about 10^2 per gram) to high levels (about 10^7 per gram) occurs. These strains include *Lactobacillus paracasei*, *L. plantarum*, *L. casei*, and *L. curvatus*, and are adventitious contaminants of the milk or the cheese-making equipment. The high salt levels, low pH, low redox potential and anaerobic conditions in the ripening cheese are ideal for their multiplication. Such high levels of NSLAB must have a role in formation of cheese flavour, although that role is not clear. However, these bacteria do transform L-lactate to D-lactate during ripening.

See also: **Bacteriocins**: Potential in Food Preservation; Nisin. ***Brevibacterium***. ***Candida***: Introduction. **Cheese**: Microbiology of Cheese-making and Maturation; Mould-ripened Varieties. ***Enterococcus***. **Fermented Milks**: Range of Products; Yoghurt; Products from Northern Europe; Products of Eastern Europe and Asia. ***Geotrichum***. ***Lactobacillus***: *Lactobacillus bulgaricus*. ***Lactococcus***: *Lactococcus lactis* Sub-species *lactis* and *cremoris*. ***Leuconostoc***. **Minimal Methods of Processing**: Potential use of Phages and/or Lysins. ***Penicillium***: *Penicillium in Food Production*. ***Streptococcus***: *Streptococcus thermophilus*.

Further Reading

Cogan TM and Accolas JP (1996) *Dairy Starter Cultures*. New York: VCH.

Cogan TM and Hill C (1993) Cheese Starter Cultures. In: Fox PF (ed.) *Cheese: Physics, Chemistry and Microbiology*, 2nd edn. Vol. 1, p. 193. London: Chapman & Hall.

De Vuyst L and Vandamme EJ (1994) *Bacteriocins of Lactic Acid Bacteria*. London: Blackie.

Gasson MJ and de Vos WM (1994) *Genetics and Biotechnology of Lactic Acid Bacteria*. London: Chapman & Hall.

Neve H (1996) *Bacteriophage*. In Cogan TM and Accolas JP (eds) *Dairy State Cultures*. New York: VCH. P. 157.

Moulds Employed in Food Processing

Tümer Uraz, Department of Dairy Technology, Ankara University, Turkey

Barbaros H Özer, Department of Food Science and Technology, Harran University, Şanhurfa, Turkey

Introduction

The multicellular filamentous fungi growing on foods that give them a cottony appearance are termed mould. The thallus, or vegetative body, is characteristic of thallophytes, which lack true roots, stems and leaves. The moulds are differentiated from yeasts due to the presence of hyphae, which are intertwined filaments, in the moulds. The branching and accumulation of these hyphae lead to the formation of mycelium. The hyphae may grow submerged within or on the aerial surface of foods.

Moulds are divided into two groups: septate, with cross walls dividing the hypha into cells (e.g. *Penicillium glaucoma*), and non-septate (coenocytic), with the hyphae apparently consisting of cylinders without cross walls (e.g. *Mucor muceda*).

Moulds do not contain chlorophyll and, therefore, in contrast to higher plants, they cannot synthesize the essential organic compounds for their growth. Instead, they use the organic compounds present in nature. Therefore, they show a saprophyte and parasite character.

The cell wall of moulds are thin, flexible and colourless. The cell cytoplasm includes vacuoles, cell core and some nutrients, including mainly glycogen and fat. In some species, the fat is the predominant component of total solids of cytoplasm.

Moulds can grow from a transplanted piece of mycelium. Reproduction of moulds is chiefly by means of asexual spores. Some moulds also form sexual spores. Such moulds are termed perfect and are classified as either Oomycetes or Zygomycetes if non-septate, or Ascomycetes or Basidiomycetes if septate. In contrast to imperfect moulds, no sexual spores have been found.

Physiological Characteristics of Moulds

Like other microorganisms, moulds require water, minerals and carbon and nitrogen sources for growth. Moulds are resistant to wide variations in environmental conditions, e.g. high salt and sugar content caused by chitin and cellulose in their cell walls. While, for example, moulds can grow in 50% sugar solution, the bacteria are lysed at the same sugar concentration due to high osmotic pressure. In general, moulds can utilize many kinds of foods, ranging from simple to complex. While some species use simple foods directly, others (e.g *Penicillium* spp.) reduce the complex molecules to simpler molecules enzymatically and then utilize them. The growth of moulds depends on the concentration and type of nitrogen salts, macro- and micronutrients such as P, K, Mg, Fe, Cu and Mn. In contrast to other microorganisms, the growth of moulds is stimulated by sodium chloride at moderate concentrations.

Table 1 Growth temperatures of moulds

Species	*Minimum*	*Optimum*	*Maximum*
Aspergillus candidus	10	28	55
Aspergillus clavatus	10	26	38
Aspergillus fumigatus	12	37	52
Aspergillus repens	5	30	38
Aspergillus flavus	10	30	45
Butrytis onerea	−4	22	37
Cladosporium herbarum	−6	25	40
Helicostylum pulchrum	0	5	28
Humicola lanuginosa	30	48	60
Mucor pusillus	20	40	55
Neurospora sitophila	5	36	45
Penicillium aurantiogriseum	−2	24	29
Penicillium brevicompactum	−2	22	28
Penicillium citrinum	12	27	34
Penicillium expansum	−2	24	29
Penicillium glabrum	3	24	29
Penicillium roqueforti	2	22	35
Penicillium viridicatum	−2	24	35
Rhizopus stolonifer	1	28	34

In general, most moulds are able to show activity at much lower water activity than most yeasts and bacteria. Most yeasts and bacteria cannot grow below 0.95 water activity (a_w), whereas Ascomycetes and Deuteromycetes can remain active when water activity is as low as 0.65. Salt is a determining factor for a_w and has a different effect on different species. For example, while *Mucor* spp. are sensitive to low salt concentrations, a slight stimulation of growth of *Penicillium* spp. is observed at the same NaCl concentration. Overall, below 14–15% total moisture in flour or some dried fruits, the growth of mould is prevented or considerably delayed.

The optimum growth temperature of moulds is around 25°C, ranging from 20 to 30°C. However, while some species require much lower temperature (5°C for *Helicostylum pulchrum*), some species show thermophilic character (optimum 48°C for *Humicola lanuginosa*; **Table 1**).

Moulds are aerobic – they need oxygen for growth. They also grow at low pH values. They can remain active over a wide range of hydrogen ion concentrations (pH 3–9), but the majority favour an acid pH (pH 5–6).

Some species, including *Aspergillus* spp. and *Penicillium* spp., are stimulated by CO_2 in air (up to 15% CO_2 concentration). *Penicillium roqueforti*, which is a starter used in the manufacture of Roquefort cheese, can remain active at very low oxygen concentrations (4.2% O_2 and 80% CO_2).

Compounds inhibitory to other organisms are produced by some moulds such as penicillin from *P. chrysogenum* and clavacin from *Aspergillus clavatus*. Certain chemical compounds are mycostatic, inhibiting the growth of moulds (sorbic acid, propionates and acetates are examples), or are specifically fungicidal, killing moulds.

Industrial Importance of Moulds

In general, foods on which moulds are grown are considered unfit to consume. In most cases, moulds are involved in spoilage of many kind of foods. On the other hand, some special moulds are useful in the manufacture of certain foods or food ingredients. Thus, some kind of cheeses are mould-ripened, e.g. blue-veined, Roquefort, Camembert, Brie, Gammelost etc. and moulds are used in making oriental fermented foods such as soy sauce, miso, sonti and tempeh. This article will focus on the moulds used as starters for the manufacture of foods rather than as spoilage agents.

Mould Cultures

Stock cultures of moulds are usually prepared on a suitable agar medium such as malt extract agar, and lyophilization (freeze drying) is the most common method of preserving the culture in a spore state for long periods. Mould cultures can be prepared in many different ways:

- surface growth on a liquid or agar medium in a flask
- surface growth on media in shallow layers in trays
- growth on wheat bran which is moistened and may be acidified or which may have liquid nutrient added, e.g. corn-steep liquor
- growth on previously sterilized and moistened bread or crackers
- growth by the submerged method in an aerated liquid medium, usually resulting in pellets composed of mycelium, with or without spores.

The efficiency of the recovery of mould spores varies depending on the method of production. Washing and drawing from dry surfaces, leaving in dry material that is ground up or powdered or incorporating in some dry powder, e.g. flour are the most common methods of recovering the spores. The pellets are used as such.

Growing spores of *P. roqueforti* on sterilized, moistened and acidified bread is a common practice, as well as growing on whole wheat or bread made of a special formula. Following the formation of mould spores, drying of bread and culture take place and the culture is used in powder form.

P. camemberti spores are prepared by growing the mould on moistened sterile crackers. A spore suspension is prepared for the surface inoculation of the Camembert, Brie or similar cheese.

When submerged fermentation is required, the pellets or masses of mycelium produced during the submerged growth of mould is used. In order to increase the yield of pellets, the culture should be shaken continuously during the submerged growth of culture. When surface growth is desired on liquid or agar medium or bran, mould spores produced by the methods listed previously ordinarily serve as the inoculum.

The koji, or starter, for soy sauce is usually a mixed culture carried over from a previous lot, although pure cultures of *Aspergillus oryzae*, together with a yeast and *Lactobacillus delbrueckii*, have been used. The mould culture is grown on cooked, sterilized rice.

Use of Substances and Enzymes Produced from Moulds in Food Industry

Moulds may be employed for the production of many kinds of substances such as lactic, citric, acetic, gluconic, malic, fumaric and tartaric acids. In similar manner, the moulds are successfully employed in the production of antibiotics, vitamins, ethyl alcohol, amino acids and hormones, and for the preparation of single-cell proteins and for the synthesis of fat.

Although it is not common, it has been reported that *Rhizopus oryzae* has been used to produce lactic acid from a glucose-salt medium in vitro. The production of citric acid by moulds is more common. In particular, *A. niger* is widely used in citric acid production. Additionally, other mould species, including *A. clavatus*, *A. wentii*, *P. luteum*, *P. citrinum* and *Mucor pyriformis* are also able to produce citric acid, mainly from molasses. There is a positive correlation between the concentration of metal ions and the yield of citric acid, and a negative correlation is reported between nitrogen concentrations and the amount of citric acid produced. Also, the production of citric acid is closely linked to the suppression of oxalic acid production. In modern citric acid production methods, the submerged growth of the mycelium is essential. A sugar concentration ranging between 14 and 20%, low pH, temperature around 25–30°C and

Table 2 Enzymes from moulds used in food processing

Enzyme	*Source*	*Industry*	*Application*
α-amylase	*Aspergillus niger*	Baking	Flour supplement
	Aspergillus oryzae	Brewing	Mashing
	Rhizopus spp.	Food	Syrup manufacture
	Mucor rouxii	Food	
Cellulase	*Aspergillus niger*	Food	Preparation of liquid coffee concentrates
	Trichoderma viride		
Glucose oxidase	*Aspergillus niger*	Food	Glucose removal from eggs
Lipase	*Aspergillus niger*	Dairy	Flavours in cheese
	Mucor spp.		
	Rhizopus spp.		
Pectinase	*Aspergillus niger*	Food	Clarification of wine and fruit juices
	Penicillium spp.		
	Rhizopus spp.		
Protease	*Aspergillus oryzae*	Brewing	Prevent chill haze in beer
	Mucor spp.	Food	Meat tenderizer
	Rhizopus spp.		
Rennin-like enzymes	*Mucor miehei*	Dairy	Curdling of milk for cheese-making
	Mucor pusilis		

a fermentation period of 7–10 days or lower are the optimum conditions of production of citric acid by moulds. In the food industry, citric acid is used to adjust the pH, to develop the flavour and to delay decolorization and browning in meat and fruits.

Some species, including *M. flavus*, *A. niger*, *Fusarium solani* and *P. notatum*, are able to bind some inorganic elements such as sulphur, calcium silicate and iron hydroxide and therefore contribute to the protection of environment.

The fungi *Aspergillus* spp., *Penicillium* spp. and *Fusarium* spp. may be used in the preparation of single-cell proteins.

Fats are synthesized in large quantities by yeasts, yeast-like organisms and moulds. *Geotrichum candidum* is primarily responsible for the synthesis of fat.

In addition to being used in the production of certain food additives, the moulds synthesize some enzymes which are widely used in the manufacture of foods. The enzymes produced by moulds are used in a wide area of the food industry, varying from baking to dairies (**Table 2**).

The use of moulds in the manufacture of some kinds of foods has been established for many years. For example, *Endothia parasitica*, *M. pusillus* and *M. miehei* are employed in the production of milk-clotting enzymes (rennin). In addition, lipases and proteases synthesized by *A. niger* and *A. oryzae* are often used to shorten the maturation period in cheese and to control the development of aroma/flavour. Certain proteases produced by *A. niger* have been used in cheese-making as an alternative to the traditional milk-clotting enzyme rennin.

Enzymes produced by *A. niger* and *A. flavus* such as α-amylase, β-galactosidase, β-gluconase, glucoamylase, glucosidase, catalase, pectinase and cellulase are employed in the manufacture of many kinds of foods. For instance, amylase is widely used in bread-making and in breweries. Pectinase and cellulase are used in the preparation of easily digestible foods.

α-Amylase is one of the commonest enzymes employed in the food industry. Moulds are considered as prime sources of amylases as well as other enzymes. Depending upon the method of production, different strains have been recommended in the production of amylase, e.g. *A. oryzae* for surface application and *A. niger* for submerged cultures. For the manufacture of amylases from *A. oryzae*, moistened steamed wheat or rice bran is widely used. The whole process takes about 40–48 h at 30°C in an atmosphere with high humidity. Amylases are employed in the clarification of beer, wines and fruit juices and removal of starch from fruit extracts. Also, in the development of consistency and gas retention of dough, amylases are used in combination with proteases.

Fungal types of pectinase are produced industrially by species of *Aspergillus* and *Penicillium*. Fungal pectin enzymes are used:

- to accelerate rates of clarification and filtration
- to remove pectin from fruit base prior to gel standardization in jam manufacture
- to prevent undesirable gel formation in fruit and vegetable extracts and purées
- to recover citrus oils
- to stabilize cloud in fruit juices.

Proteases are classified according to their optimal pH in acid, neutral or alkaline types. Acid proteases are mostly produced by fungi. *A. oryzae* is the main source of fungal proteases. Various media have been recommended for the production of proteinase, including wheat bran, soybean cake, alfalfa meal, middling, yeasts and other material. Acid fungal proteinases are used:

- to hydrolyse gluten to reduce mixing time, to make dough more pliable and to improve texture and loaf volume
- in meat tenderization
- to prepare liquid wheat products
- to reduce viscosity and to prevent gelation of concentrated soluble fish products
- to reduce the setting time for gelation without affecting the gel strength.

They are also actively used in the manufacture of oriental mould-fermented foods.

Glucose oxidase oxidizes glucose in the presence of oxygen to form gluconic acid. It is used to desugar – and hence to stabilize – egg products, and to increase the shelf life of bottled beer, soft drinks and other oxygen-sensitive foods. Glucose oxidase is primarily produced by the submerged growth of *A. niger*. Glucose oxidase and catalase are often employed together to remove residual oxygen from food packages.

Lipase from various species of moulds contributes to the development of flavour in mould-ripened cheeses and catalase from *A. niger*, *P. vitale* and *Mucor lysodelilaticus* is used in cake-baking, irradiated foods and hydrogen peroxide sterilization where the elimination of hydrogen peroxide is essential.

Use of Moulds in Cheese-making

Most cheese varieties are consumed as ripened. In the ripening process, the indigenous enzymes of milk, milk-clotting enzyme (rennin) and starter cultures play determining roles. Endo- and exoenzymes produced by starter bacteria (lipases and proteases) give a characteristic aroma/flavour to cheese. The type of microorganism used varies depending on the type of cheese. These microorganisms can be applied either alone or in combination with bacteria and yeasts.

In cheese-making, *Penicillium* spp. (e.g. *P. camemberti* and *P. roqueforti*) are widely used. *Geotrichum* spp. (e.g. *G. candidum*) are also applied, to a lesser extent. *P. camemberti* is employed as starter for French cheeses made from cow's milk, such as Camembert, Brie, Carré de l'Est, Neufchâtel and some other cheese produced from goat's milk. *P. roqueforti* gives a characteristic aroma/flavour to cheeses made from cow's milk, such as Bleu d'Auvergne, Bleu de Bresse, Gorgonzola, blue Danois, Stilton, Roquefort made from sheep's milk, and some cheeses manufactured from goat's milk. Although it is not common, *G. candidum* is used as a starter in some soft cheeses and, in some cases, Camembert, Pont l'Evêque and Saint-Nectaire cheeses.

Depending on the type of cheese, the moulds can be applied as follows:

- by adding to cheese milk with starter culture or with milk-clotting enzyme
- by adding to brine (*P. camemberti*)
- by spreading on to the surface of cheese in a cold room (*P. camemberti*, *G. candidum*)
- by inoculating into curd before pressing (*P. roqueforti*).

About 1000 species of *Penicillium* have been identified so far and only a few strains of the above-mentioned species are of industrial importance in terms of cheese production. *P. camemberti* has been known since 1906. The growth rate of *P. camemberti* is the same as other *Penicillium* species. On malt extract, it produces colonies of 25–35 mm diameter within 2 weeks at 25°C. The optimum growth temperature of *P. camemberti* is around 20–25°C. While it multiplies at 5°C, high temperatures (e.g. 37°C) have an inhibitory effect on *P. camemberti*. *P. camemberti* is moderately tolerant against salt and 20% salt concentration is accepted as the critical value at which growth ceases. Other *Penicillium* species are more acid-tolerant than *P. camemberti*. The growth of these species is favoured by low pH (pH 3.5–6.5). With the exception of selected strains, no strains are able to grow above pH 7.0.

The growth rate of *P. roqueforti* is higher than that of *P. camemberti*. On malt extract, it spreads within 7 days over a 40–70 mm diameter area at 25°C. The optimum growth temperature of *P. roqueforti* is around 35–40°C. On the other hand, due to its ability to grow at low temperatures (e.g. <5°C), it often spoils refrigerated foods. Although it is favoured by an acid environment (pH 4), it shows activity within a wider range of pH than *P. camemberti* (pH 3.0–10.5). Its growth is stimulated at low salt concentrations but, at 6–8% salt concentration, the growth rate decreases and, at 20% salt concentration, it stops completely.

Although it is not as common as *P. camemberti* and *P. roqueforti*, *G. candidum* is also used in the manufacture of some cheese varieties. It may grow at 5°C up to 38°C, but its optimum growth temperature is around 25°C. Although it remains active over a wide pH range, its growth is stimulated by an acidic environment (pH 5.0–5.5). Unlike *Penicillium* spp., *G. candidum* is sensitive to salt and even at 1% salt concentration its growth is slowed: at 5–6% salt concentrations an inhibitory effect is observed.

P. camemberti and *P. roqueforti*, which are both of

industrial importance, are aerobic. *P. roqueforti* can even remain active at oxygen concentration as low as 5%. Both *Penicillium* species (*P. camemberti* and *P. roqueforti*) are able to metabolize organic and inorganic compounds. Their lactose metabolism is of crucial importance in cheese-making.

Biochemical Activity of *P. camemberti* and *P. roqueforti* and their Role in Cheese-ripening

Both *Penicillium* species consume lactic acid and lactate present in cheese to cover their requirement for metabolites essential for growth. As a result of de-acidification occurring during the consumption of lactic acid and lactate, the pH of cheese increases, providing a suitable environment for the activity of enzymes. This leads to the development of aroma/flavour and cheese texture. In addition, the neutralization of lactic acid results in stimulation of bacteria and micrococci, which are acid-sensitive, at the end of the ripening period.

P. camemberti synthesizes two specific exocellular enzymes – acid protease, which is stable within the pH range 3.5–5.5 and has optimum activity at pH 5.0, and metalloprotease, which is stable at higher pH ranges (pH 8.5–9.5) and has optimum activity at pH 6.0. Also, carboxy-peptidase and amino-peptidase enzymes, whose optimum growth pH range is between 4.0–7.0 and 7.5–8.5, respectively, are produced by *P. camemberti*.

Proteolysis initiated by enzymes produced by *P. roqueforti* plays an important role in the development of aroma/flavour and the texture of blue-veined cheeses. Similar to *P. camemberti*, *P. roqueforti* also synthesizes acid protease (optimum pH 3.5–4.5), metalloprotease (optimum pH 5.5–6.0), carboxy-peptidase (optimum pH 3.5–4.0) and amino-peptidase (optimum pH 7.5–8.0) and some exoenzymes.

Proteolysis is an important feature of mould-ripened cheeses (both surface-smeared and blue-veined cheeses). At the end of the ripening period, the ratio of soluble nitrogen to total nitrogen is around 35% in Camembert cheese and 50% in Roquefort cheese.

These *Penicillium* species also synthesize lipases which are determining enzymes for the aroma/flavour of mould-ripened cheeses. Lipases produced by *P. camemberti* show activity within a pH range 1.0–10.0, with optimum pH at 8.5–9.5. The optimum temperature for lipase activity is 35°C; however, at 0°C they maintain 50% of maximum activity. *P. camemberti* causes oxidative degradation of lipids through its enzymes, especially degradation of caprilic acid and, secondarily, lauric acid. In cheeses where *P. camemberti* is used as a starter, a positive correlation between proteolytic activity and lipolytic and β-oxidative changes is established.

As with *P. camemberti*, the activity of lipases produced by *P. roqueforti* varies depending on the strains. However, an adverse relationship between proteolytic and lipolytic activity in cheeses made by *P. roqueforti* is present. *P. roqueforti* synthesizes two exocellular lipases, namely acid lipase (optimum pH 6.5) and alkali lipase (optimum pH 7.5–8.0). These lipases act on tricaproine and tributyrine, respectively. Further enzymatic conversions of free fatty acids lead to β-ketonic acids and methyl ketones, which give a characteristic aroma/flavour to mould-ripened cheeses. The level of free fatty acids in mould-ripened cheeses is an indicator for the level of lipolysis. For example, in Camembert cheese the level of free fatty acids ranges between 6 and 10%, in Roquefort cheese this falls to 7–12% and in blue cheese it is above 10%.

The Role of *P. camemberti* and *P. roqueforti* in the Aroma/Flavour of Cheese

The characteristic aroma/flavour of mould-ripened cheeses is a result of a series of enzymatic modifications of milk compounds. The enzymatic modifications of proteins and lipids, which depend on the technology applied, lead to the formation of many kinds of aroma/flavour compounds. The type and concentration of acids, primary and secondary alcohols, carbonyl compounds, esters and hydrocarbons determine the characteristic aroma/flavour of various cheeses. Such compounds, present in Camembert cheese, are shown in **Table 3**. The characteristic

Table 3 Volatile compounds isolated from Camembert cheese

Primary alcohols	C2, 3, 4, 6, 2-methylpropanol, 3-methylbutanol, oct-1-ene-3-ol, 2-phenylethanol
Secondary alcohols	C4, 5, 6, 7, 9, 11
Methyl ketones	C4, 5, 6, 7, 8, 9, 10, 11, 12, 13, 15
Aldehydes	C6, 7, 9, 2 and 3 methylbutanol
Esters	C2, 4, 6, 8, 10-ethyl, 2-phenylethylacetate
Phenols	Phenol, *p*-cresol
Lactones	C_9, C_{10}, C_{12}
Sulphur compounds	H_2S, methyl sulphide, dimethylsulphide, methanethiol, 2,4-dithiapentane, 3,4-dithiahexane, 2,4,5-trithiahexane, 3-methylthio, 2,4-dithiapentane, 3-methylthiopropanol
Anisoles	Anisole, 4-methylanisole, 2,4-dimethyl anisole
Amines	Phenylethylamine, $C_{2,3,4}$, diethylenamine, isobutylamine, 3-methylbutylamine
Miscellaneous	Dimethoxybenzene, isobutylacetamide

Table 4 Variation in the concentration of methyl ketones in blue and Roquefort cheeses during ripening (mg per 10 g of fat)

Cheese	*Age of cheese*	*($C_{15}+C_{13}$)*	*C_{11}*	*C_9*	*C_7*	*C_5*	*C_3*
Blue	2 months	1.1	1.7	7.6	30.2	20.0	5.1
Blue	3 months	0.9	0.7	2.5	3.4	0.9	Trace
Blue	4 months	1.3	1.7	8.5	12.4	7.2	Trace
Roquefort	2 months	0.5	0.7	7.4	15.6	11.0	2.4
Roquefort	3 months	0.4	1.6	12.4	9.2	1.2	0
Roquefort	4 months	0.3	0.3	2.6	4.2	0	0

aroma/flavour compound for this type of cheese is 1-octene-3-ol and the presence of this compound at high levels in Camembert cheese causes aroma defects.

In soft cheeses ripened by surface-smeared moulds, phenylethanol and its acetic and butyric acid esters can easily be recognized. Depending on the ripening coefficient of cheese, sulphuric compounds can also be detected organoleptically.

In Roquefort cheese, the level of fatty acids, methyl ketones and secondary alcohols determines the characteristic aroma/flavour. However, the level of such compounds in cheese varies greatly depending upon the technological variables (**Table 4**). In mature Roquefort cheese, for example, the methyl ketones having seven and nine carbon atoms are abundant. The methyl ketones are affected by technological applications. The stimulated lipolysis of milk fat also causes an increase in the level of methyl ketones. In contrast, the high level of fatty acids has an inhibitory effect on the growth of *Penicillium* spp. This eventually leads to the inhibition of the formation of methyl ketones. Similarly, salting influences the level of lipolysis and, as a result, retards the release of methyl ketones. It is known that at high salt concentrations, lipolysis is greatly delayed. Some methyl ketones are converted to secondary alcohols by *Penicillium* spp.

Compounds formed as a result of degradation of milk fat characterize the typical aroma/flavour of blue cheeses. Additionally, esters and phenylethanol present in cheese balance the typical aroma/flavour of mould-ripened cheeses.

Mycotoxins of *P. camemberti* and *P. roqueforti*

It has long been known that some moulds (e.g. *A. flavus*) produce mycotoxins which are harmful to humans and animals and which have carcinogenic effects on humans. These mycotoxins are secondary metabolites which are synthesized from amino acids and ketonic acids. Cyclopiazonic acid produced by *P. camemberti* was found to be fatal for mice. In mature Camembert cheese stored at 14–18°C, this compound was not found but the formation of cyclopiazonic acid in cheese stored at 25°C was reported. Additionally, some authors asserted that cyclopiazonic acid was present on the surface of Camembert cheese.

Table 5 Foods produced by fermentation of moulds

Food	*Mould*
Cheese	
Semihard blue-veined (Gorgonzola, Stilton, Roquefort, Danish blue etc.)	*Penicillium roqueforti*
Gammelost	*Penicillium roqueforti*, *Penicillium frequentas*, *Mucor racemosus*, *Rhizopus* spp.
Soft cheeses (Camembert, Brie)	*Penicillium camemberti*
Coffee	*Aspergillus* spp.
	Fusarium spp.
Soy sauce	*Aspergillus soyae* (*Aspergillus oryzae*)
First stage	*Mucor* spp.
	Rhizopus spp.
Tempeh	*Rhizopus oligosporus*
	Rhizopus stolonifer
	Rhizopus arrhizus
	Rhizopus oryzae
Tamari sauce	*Aspergillus tamarii*
Miso	*Aspergillus oryzae*
Ang-khak (Chinese red rice)	*Monascus purpureus*
Tou-fu-ju or tofu (soybean cheese)	*Mucor* spp.
Minchin	Various mould species
Fermented fish	*Aspergillus* spp.
Lao-chao	Various mould species
Poi	*Geotrichum candidum*

The effect of cyclopiazonic acid on humans has yet to be established.

In a study carried out in France, 30 commercial *P. camemberti* strains were tested both in vitro and in the cheese factories to determine their level of toxigenicity and it was reported that only three strains were found to be toxigenic (one very weak, one medium and one strong).

P. roqueforti produces PR toxin, PR imin and patulin, which are unstable in cheese.

Application of Moulds in the Manufacture of Oriental Foods

Most of the oriental foods mentioned below have moulds involved in their preparation (**Table 5**). In the starter named koji by the Japanese and chou by the

Chinese, moulds serve as sources of hydrolytic enzymes, such as amylases to hydrolyse the starch in grains, proteinases, lipases and many others. For the most part the starters are mixtures of moulds, yeasts and bacteria, but in a few products pure cultures have been employed.

Saké

Saké is a Japanese alcoholic beverage with 14–17% alcohol content manufactured by means of a mixed fermentation of moulds and yeasts. A starter named koji or chou is produced by growing *A. oryzae* on soaked and steamed rice mash until a maximal yield of enzymes is obtained. In the second stage of fermentation, *Saccharomyces* spp. carry out the alcoholic fermentation.

Sonti

Sonti is a rice beer or wine particular to India produced by mixed fermentation of *Rhizopus sonti* and yeasts.

Coffee

Pectinolytic enzymes from *Aspergillus* spp. and *Fusarium* spp. are used to remove the pulpy layers from the coffee beans and to prevent an off flavour.

Tempeh

The manufacture of tempeh, an Indonesian food, includes several steps. After the soybeans are soaked overnight at 25°C, the seed coats are removed; then the beans are boiled for about 20 min. Following drying, the beans are inoculated with a spore preparation including species of *Rhizopus* (*R. stolonifer*, *R. oryzae*, *R. oligosporus* or *R. arrhizus*). Incubation lasts at least 20 h at 32°C until mycelium is grown on the surface without any, or very little, sporulation. Finally, the sliced product is dipped in salt water and fried in vegetable oil until it turns light brown.

Soy Sauce

Soy sauce is manufactured by a two-stage fermentation process in which one or more fungal species are grown on an equal mixture of ground soybean. The starter used in the manufacture of soy sauce is grown on a mixture of soybeans, cracked wheat and wheat bran. This mixture is inoculated with spores of *A. oryzae* (*A. soyae*) and fermentation continues at 25–30°C until mould growth is observed on the surface of the mash. After preparation of koji, the autoclaved soybeans are inoculated with the starter and incubated for 3 days at about 30°C; then, the fermented soybeans are soaked in salt water containing 24% sodium chloride. The brined mash is then left from 2.5 months to a year or longer depending on temperature. The aroma/flavour of soy sauce is chiefly determined by proteinases, amylases and other enzymes synthesized by *A. soyae* (*oryzae*). On the other hand, the role of yeasts such as *Hansenula* spp. and *Saccharomyces rouxii* in the development of a tangy taste, and of bacteria such as *Tetragonococcus halophila* on the development of acidic aroma and clear appearance should also be noted.

Miso

The starter used in the manufacture of miso consists of *A. oryzae* and some yeasts, including *S. rouxii* and *Zygosaccharomyces* spp. In the preparation of starter (koji), *A. oryzae* is grown on a steamed polished-rice mash in shallow trays at 35°C, until the grains are completely covered without the formation of sporulation. Then, the starter is added into crushed, steamed soybeans, salt is added and fermentation continues for about 7 days at 28°C. Afterwards, the fermentation is extended for 2 more months at 35°C, and the mixture is matured for a few weeks at ambient temperature. Miso is usually consumed as a condiment with other foods.

Ang-khak

Ang-khak is a Chinese red rice which is produced by growth of *Monascus purpureus* on autoclaved rice. This oriental fermented product is usually used for colouring and flavouring fish and other food products.

Tamari Sauce

Tamari sauce is a special Japanese fermented product similar to soy sauce. In the manufacture of tamari sauce, the soybean mash is fermented, with or without rice, by *A. tamarii*.

Soybean Cheese

Soybean cheese or more familiarly, tofu or tou-fu-ju, is a well-known Chinese fermented product. The production of soybean cheeses involves several steps. First, soybeans are soaked and ground into a paste. Afterwards, ground soybeans are drained by means of a cheesecloth bag. The protein in the filtrate is coagulated with magnesium or calcium salts. After the coagulated product is pressed into blocks, fermentation takes place in fermentation rooms at 14°C for about a month. While the fermentation is continuing, the development of white moulds, probably *Mucor* spp., is observed. Finally, the fermented cheese blocks are kept in brine or in a special wine until ripening is complete.

Minchin

De-starched wheat gluten is the main ingredient of minchin. The raw gluten including some water is left in a closed jar and left to ferment for about 2–3 weeks, after which it is salted. Although the microbiology of minchin is not known with certainty, it has been reported that it contains seven or eight species of moulds as well as nine species of bacteria and three species of yeasts. Following fermentation, depending on personal choice, the product can be boiled, baked or fried.

Fermented Fish

Fermented fish is a common fermented product in China and Japan. Before fermentation, a cooking step takes place. Cooked fish is then fermented with species of *Aspergillus* and, after fermentation is complete, the product is dried. In China, fermented fish is preserved in a fermented rice product called lao-chao. Lao-chao is a result of mixed fermentation of moulds and yeasts. This product is slightly alcoholic.

See also: ***Aspergillus***: Introduction. **Cheese**: Mould-ripened Varieties. **Fermented Foods**: Fermentations of the Far East; Fermented Fish Products. **Fungi**: The Fungal Hypha. ***Geotrichum***. ***Hansenula***. ***Mucor***. **Mycotoxins**: Occurrence. ***Penicillium***: Introduction; *Penicillium* in Food Production. ***Saccharomyces***: Introduction.

Further Reading

Adda J (1984) Les propriétés organoleptiques du fromage: 2. Formation de la flaveur. In: Eck A (ed.) *Le Fromage*. P. 330. Paris: Lavoisier.

Akman AV (1956) *Microbiology*, No: 84. Ankara, Turkey: Ankara University Faculty of Agriculture Press.

Arora M, Mukerji P and Marthy M (1991) *Handbook of Applied Mycology*. Vol. 3, p. 621.

Choisy C, Gueguen M, Lenoir J, Schmidt JL and Tourneur C (1984) L'affinage du fromage: les phénomènes microbiens. In: Eck A (ed.) *Le Fromage*. P. 259. Paris: Lavoisier.

Fellows P (1988) *Food Processing Technology: Principles and Practice*. P. 505. Chichester: Ellis Harwood.

Fraizer W and Westhof D (1988) *Food Microbiology*. P. 539. Singapore: McGraw-Hill.

Gripon JC (1993) Mould-ripened cheeses. In: Fox PF (ed.) *Cheese: Chemistry, Physics and Microbiology*. Vol. 2, p. 121. Elsevier Applied Science, London.

Gueguen M (1992) Les moisissures. In: Hermier J, Lenoir J and Weber F (eds) *Les Groupes Microbiens d'Intérêt Laitier*. P. 325. Paris: CEPIL.

Lenoir J and Jaquet J (1969) Mécanismes intimes de l'affinage des fromages. *Economie et Médecine Animals* 10(1): 38–71.

Lenoir J, Gripon JC, Lambert G and Cerning J (1992) Les penicillium. In: Hermier J, Lenoir J and Weber F (eds) *Les Groupes Microbiens d'Intérêt Laitier*. P. 221. Paris: CEPIL.

Pamir H (1984) *Technical and Industrial Microbiology*, No. 681. Ankara, Turkey: Ankara University Faculty of Agriculture Press.

Statistical Evaluation of Microbiological Results *see* **Sampling Regimes & Statistical Evaluation of Microbiological Results.**

Sterilants *see* **Process Hygiene**: Types of Sterilant.

STREPTOCOCCUS

Contents

Introduction

Marco Gobbetti, Istituto di Produzioni e Preparazioni Alimentari, Facoltà di Agraria di Foggia, Foggia, Italy
Aldo Corsetti, Institute of Dairy Microbiology, Faculty of Agriculture of Perugia, Italy

After taxonomic revisions in recent years, the genus *Streptococcus sensu stricto* now includes species which may be grouped as 'oral', 'pyogenic' and 'other' streptococci. *Streptococcus* spp. make up a great part of the human and animal commensal flora. Some of the known species are important pathogens and some are food-related. *S. thermophilus* is used in food processing and other streptococci are indicators in the microbiological monitoring of foods.

Characteristics of the Genus

The genus *Streptococcus* consists of Gram-positive, spherical-ovoid or coccobacillary cells, that form chains or pairs. Cells in older cultures may appear Gram-variable, and some strains are pleomorphic. *Streptococcus* spp. are non-sporing and catalase-negative; they ferment carbohydrates to produce mainly lactic acid, but no gas, and have complex nutritional requirements. Under glucose-limiting conditions, formate, acetate and ethanol are also produced. Most are facultatively anaerobic or aerotolerant anaerobes; some are capnophilic (CO_2-requiring). Occasional strains synthesize peroxidases but none synthesize haeme groups. Some species produce capsules, either of hyaluronic acid (*S. pyogenes*) or a variety of type-specific polysaccharides (*S. pneumoniae*), but this is not a common feature of the genus. *Streptococcus* spp. generally grow within a temperature range of 20–42°C, ca. 37°C being the optimum in most cases. The mol% G + C of the DNA is 34–46. *Streptococcus pyogenes* is the type species of the genus.

Habitats and Taxonomy

The 39 currently classified species of the genus *Streptococcus sensu stricto* are grouped as 'oral', 'pyogenic' and 'other' streptococci.

Oral Streptococci

These streptococci are part of the normal body microflora found in the mouth and upper respiratory tract. 'Oral streptococci' comprise the viridans group of Sherman and show a variable type of haemolysis. Some are involved in human diseases as opportunistic pathogens. *S. salivarius* is mainly found on the dorsal surface of the tongue and in the saliva, whereas *S. sanguinis* colonizes tooth surfaces, and *S. vestibularis* favours the vestibular mucosa. The locations in the mouth mostly depend on the capacity to synthesize specific adhesins (lectin-like) which enable bacteria to bind to complementary host tissues.

Data from rRNA cataloguing and nucleic acid hybridization divide 'oral streptococci' into four main phylogenetic lineages: the *S. mutans* group, *S. mitis* group (formerly *S. oralis* group), *S. anginosus* group (formerly *S. milleri* group) and *S. thermophilus* group (formerly *S. salivarius* group).

***Streptococcus mutans* Group** Streptococci originally designated as *S. mutans* were isolated from dental caries and from bacterial endocarditis. In contrast to the initially perceived phenotypic homogeneity, subsequent serological, genetic and biochemical studies showed considerable heterogeneity (**Table 1**). The taxonomic position of *S. ferus* is less certain; it is included in this group on the basis of DNA homology but it appears to be related to the *S. mitis* group, based on multilocus enzyme electrophoresis.

***Streptococcus mitis* Group** *S. oralis* was the original name given to isolates from the human oral cavity. The initial phenotypic heterogeneity was later confirmed by cell-wall analysis, physiological data and nucleic acid hydridization; currently the name *S. oralis* is used to indicate a well-defined species belonging to the '*S. mitis* group' which includes several species (**Table 2**). *S. sanguinis* was the name given to mainly α-haemolytic, dextran- or non-dextran-forming streptococci isolated from patients with bacterial endocarditis. DNA-based studies have shown the existence of four DNA homology groups within strains designated as *S. sanguinis*. Two groups involve *S. sanguinis* and *S. gordonii* and the third was described as *S. parasanguinis*. The fourth group was initially described as the 'tufted fibril group' because

Table 1 Characteristics of species within the *Streptococcus mutans* group[a]

Characteristic	S. mutans	S. sobrinus	S. cricetus	S. ratti	S. macacae	S. downei	S. ferus
Acid from							
N-Acetylglucosamine	+	–	+	+	+	ND	+
Amygdalin	d	–	+	+	+	ND	+
Arbutin	+	–	+	+	ND	ND	+
Cellobiose	+	d	+	+	+	–	+
Dextrin	–	–	–	–	–	ND	+
Glycogen	–	–	–	–	ND	–	–
Inulin	+	d	+	+	–	+	– or +[b]
Lactose	+	d	d	d	ND	+	+
Maltose	+	+	+	+	+	+	+
Mannitol	+	d	+	+	+	+	+
Melibiose	d	d	+	+	–	– or +[b]	–
Raffinose	+	d	+	+	+	–	–
Ribose	–	–	–	–	–	ND	–
Sorbitol	+	d	+	+	+	–	+
Starch	–	–	–	–	–	–	+
Trehalose	+	+	d	+	+	+	+
Hydrolysis of							
Arginine	–	–	–	+	–	–	–
Aesculin	+	d	d	+	+	–	+
Glycogen	–	ND	ND	ND	–	–	+
Production of							
Acetoin	+	+	+	+	+	+	d
Alkaline phosphatase	–	–	ND	ND	ND	–	ND
Extracellular polysaccharide	+	+	+	+	+	–	+
α-L-Fucosidase	–	–	ND	ND	ND	ND	ND
β-D-Fucosidase	–	–	ND	ND	ND	ND	ND
α-Galactosidase	+	– or +[b]	ND	ND	ND	ND	ND
β-Glucosaminidase	–	ND	ND	ND	ND	ND	ND
H_2O_2	–	+	–	–	–	–	–
Hyaluronidase	–	–	ND	ND	ND	ND	ND
IgA protease	–	ND	ND	ND	ND	ND	ND
Neuraminidase	–	–	ND	ND	ND	ND	ND
Urease	–	–	–	–	ND	ND	–
Amylase binding	–	–	ND	ND	ND	ND	ND
Growth in/at							
4% NaCl	+	+	+	d	ND	ND	ND
6.5% NaCl	–	d	d	–	–	–	–
10% Bile	d	d	d	+	+	d	ND
40% Bile	d	d	d	d	+	d	ND
45°C	d	d	d	d	–	–	–
Cell wall components	Glucose, rhamnose	Glucose, galactose, rhamnose	Glucose, galactose, rhamnose	Galactose, glycerol, rhamnose	Glucose, rhamnose	Glucose, galactose, rhamnose	Glucose, rhamnose
Haemolysis	α, β, γ	α, γ	α, γ	γ	ND	ND	γ
Lancefield antigens	–, E	–	–	–	–	–	–
Mol% G+C	36–38	44–46	42–44	41–43	35–36	41–42	43–45
Murein type	Lys-Ala_{2-3}	Lys-Thr-Ala	Lys-Thr-Ala	Lys-Ala_{2-3}	ND	Lys-Thr-Ala	Lys-Ala_{2-3}
Serotype	c, e, f	d, g, h, –	a	b	c	h	c

[a] All species positive in glucose, mannose and sucrose. All species negative in adonitol, arabinose, arabitol, cyclodextrin, dulcitol, gluconate, rhamnose, sorbose or xylose. No species hydrolyses hippurate.
Symbols: +, 90% or more of strains are positive; –, 90% or more of strains are negative; d, 11–89% of strains are positive; ND, not determined.
[b] Proportion of strains reported as giving a positive result differs between studies.

the cells have fibrils arranged equatorially in lateral tufts. DNA homology studies indicated that they constitute a new species, called *S. cristatus*. 16S rRNA comparative sequencing showed that *S. pneumoniae* belongs to the *S. mitis* rRNA homology group. Due to its medical importance, it is considered within the *Streptococcus*-medical in *The Prokariots*.

***Streptococcus anginosus* Group** *S. constellatus*, *S. intermedius* and *S. anginosus* are included within the

Table 2 Characteristics of species within the *Streptococcus mitis* group[a]

Characteristic	S. mitis	S. oralis	S. gordonii	S. sanguinis	S. parasanguinis	S. pneumoniae	S. cristatus
Acid from							
N-Acetylglucosamine	+	+	+	+	+	d	+
Amygdalin	–	–	+	–	d	–	–
Arbutin	–	–	+	+	d	–	+
Cellobiose	d	–	ND	d	d	ND	ND
Glycogen	–	d	ND	ND	–	+	ND
Inulin	d	–	d	+	–	+	–
Lactose	+	+	+	+	+	+	d
Maltose	+	+	+	+	+	+	+
Mannitol	–	–	–	–	–	–	–
Melibiose	d	+	d	d	d	–	–
Raffinose	d	d	d	d	d	+	–
Ribose	d	d	–	–	ND	–	–
Sorbitol	d	–	–	d	–	–	–
Starch	–	d	–	d	–	d	ND
Trehalose	– or +[b]	d	+	+	d	+	+
Hydrolysis of							
Arginine	d	–	+	+	+	– or +[b]	d
Aesculin	– or +[b]	–	+	d	d	d	–
Production of							
Acetoin	–	–	–	–	–	–	–
Alkaline phosphatase	d	d	+	–	d	–	–
Extracellular polysaccharide	–	d	+	+	–	–	ND
α-L-Fucosidase	–	–	+	–	d	d	+
β-D-Fucosidase	d	+	–	d	d	d	–
α-Galactosidase	d	d	d	d	d	+	–
β-Glucosaminidase	d	+	+	–	ND	ND	ND
H_2O_2	+	+	+	+	+	+	+
Hyaluronidase	–	d	–	–	–	+	–
IgA protease	–	+	–	+	ND	+	ND
Neuraminidase	d	+	–	–	–	+	–
Urease	–	–	–	–	–	–	–
Amylase binding	+	–	+	–	d	ND	+
Growth in/at							
4% NaCl	–	–	ND	d	–	–	ND
6.5% NaCl	–	–	ND	–	–		ND
10% Bile	+	d	ND	d		ND	ND
40% Bile	d	–	ND	d	d	–	ND
45°C	d	d	ND	d	d	–	ND
Cell wall components	Rhamnose[c], ribitol	Glucose, galactose, N-acetyl galactosamine, rhamnose[c], ribitol	Glycerol, rhamnose	Glucose, rhamnose	ND	Glucose, galactose[c], N-acetyl galactosamine, rhamnose[c], ribitol	ND
Haemolysis	α	α	α	α	α	α	α
Lancefield antigens	–, K, O	–, K	–, H[d]	–, H[d]	–, F, G, C, B	–	–
Mol% G+C	40–41	38–42	38–43	46	41–43	36–37	42.6–43
Murein type	Lys-direct	Lys-direct	Lys-Ala_{1-3}	Lys-Ala_{1-3}	ND	Lys-Ala_2(Ser)	ND

[a] and [b] see Table 1.
[c] Trace amounts.
[d] Group H varies according to immunizing strain used.

S. anginosus group (formerly *S. milleri* group) (**Table 3**). They are found in the mouth, gastrointestinal and genitourinary tracts as part of the commensal flora, but are also recognized as pathogens associated with purulent abscesses. The taxonomy of the *S. anginosus* group has long been debated. The DNA–DNA hybridization levels among the three species are near the borderline of species delimitation, and the phenotypic differentiation between at least two species is not straightforward. Whole-cell protein electrophoretic

Table 3 Characteristics of species within the *Streptococcus anginosus*, *S. thermophilus* and nutritionally variant streptococci (NVS) groups[a]

Characteristic	S. anginosus	S. constellatus	S. intermedius	S. thermophilus	S. salivarius	S. vestibularis	S. adjacens	S. defectivus
Acid from								
N-Acetyl-glucosamine	d	d	+	–	+	d	ND	ND
Amygdalin	+	d	d	–	+	d	ND	ND
Arbutin	+	+	+	d	+	d	ND	ND
Cellobiose	d	d	d	–	+	d	ND	ND
Inulin	–	–	–	–	d	–	d	–
Lactose	+	d	+	+	d	d	–	d
Maltose	+	+	+	–	+	+	+	+
Mannitol	d	–	–	–	–	–	–	–
Melezitose				d	–	–		
Melibiose	d	–	–	d	–	–	–	–
Raffinose	d	d	d	d	d	–	–	d
Ribose	–	–	–	d	–	–	–	–
Sorbitol	–	–	–	–	–	–	–	–
Starch	ND	ND	ND	ND	–	–	–	+
Trehalose	+	d	+	–	d	d	–	+
Hydrolysis of								
Arginine	+	+	+	–	–	–	–	–
Aesculin	+	d	d	–	+	d	–	–
Production of								
Acetoin	+	+	+	d	d	d	d	d
Alkaline phosphatase	+	+	+	–	d	–	–	–
Extracellular polysaccharide	–	–	–	–	+	–	–	–
β-D-Fucosidase	–	–	+	ND	d	–	ND	ND
α-Galactosidase	d	–	–	–	d	–	–	+
β-Galactosidase	d	–	+	+	d	+	–	+
α-Glucosidase	d	+	+	–	+	d	ND	ND
β-Glucosidase	+	–	d	–	+	–	d	–
H_2O_2	d	–	d	–	–	+	ND	ND
Hyalurinidase	–	d	+	ND	–	–	ND	ND
N-Acetyl-β-glucosaminidase	–	–	+	–	–	–	d	–
N-Acetyl-β-galactosaminidase	–	–	+	ND	–	–	ND	ND
Pyrrolidonyl arylamidase	–	–	–	–	–	–	d	d
Urease	–	–	–	–	d	+	–	–
Sialidase	–	–	+					
Growth in/at								
4% NaCl	d	d	d		d	–		
6.5% NaCl					–			
10% Bile	d	d	d	–	+	d		
40% Bile	d	d	d	–	d	–		
45°C	d	d	d	+	–	–		
Cell wall components	Galactose, glucose, *N*-acetyl-galactosamine, rhamnose	Galactose, glucose, rhamnose	Glucose, rhamnose	ND	Galactose, glucose, *N*-acetyl-galactosamine, rhamnose		ND	ND
Haemolysis	α, β, γ	α, β, γ	α, β, γ	α, γ	α, β, γ	α	α	α
Lancefield antigens	–, F, A, C, G	–, F, A, C	–, F, G	–	–, K	–	–	–, H
Mol% G+C	38–40	37–38	37–38	37–40	39–42	38–40	36–37	46–47
Murein type	Lys-Ala$_{1-3}$	Lys-Ala$_{1-3}$	Lys-Ala$_{1-3}$	Lys-Ala$_{2-3}$	Lys-Thr-Gly	Lys-Ala$_{1-3}$	ND	ND

[a] See Table 1.

analysis supports the viewpoint that members of the *S. anginosus* group may be a single species.

***Streptococcus thermophilus* Group** The name *S. salivarius* was originally given to a streptococcus common in human saliva, present in the intestine, and occasionally isolated from patients with endocarditis, terminal septicaemia, and peritonitis. Extensive DNA–DNA hybridization studies, together with detailed phenotypic characterization, have shown that the level of DNA homology between *S. salivarius* and *S. thermophilus*, together with the newly described *S. vestibularis*, does warrant separate species status (Table 3).

Nutritionally Variant Streptococci Nutritionally variant streptococci (NVS) are referred to as satellite-forming, thiol-requiring, vitamin B_6-dependent, pyridoxal-dependent or nutritionally deficient streptococci. They are part of the normal flora of the human throat, as well as the genitourinary and intestinal tracts. By using DNA–DNA hybridization, NVS have been separated into *S. defectivus* and *S. adjacens* (Table 3). Based on recent 16S rRNA findings and phenotypic characterization, NSV should be placed in the new genus *Abiotrophia*, as *A. adjacens* comb. nov. and *A. defectiva* comb. nov.

Pyogenic Streptococci

Most pyogenic streptococci are grouped as the genus *Streptococcus*-medical in *The Prokariots*. They inhabit the skin and mucous membranes of the respiratory, alimentary and genitourinary tracts (**Table 4**).

S. pyogenes is the most common cause of streptococcal infections in humans. It produces an impressive array of erythrogenic and cytolytic toxins and non-suppurative sequelae infections include rheumatic fever and acute glomerulonephritis. It possesses the Lancefield group A carbohydrate antigen and the strains are divided according to the M, T and R surface protein antigens.

S. agalactiae, synonymous with Lancefield group B streptococci, was originally recognized as a bovine pathogen but has become increasingly important in human infections. Heavy colonization of the maternal genital tract is correlated with a high risk of infection in newborns; serious infections may also occur in adults. Due to colonization in udder tissues of cattle, *S. agalactiae* is also found in milk.

β-Haemolytic large-colony-forming streptococci with Lancefield group C or G antigen are isolated from human throat, skin, respiratory and gastrointestinal tracts, and are responsible for a variety of infections. They are also important animal pathogens. Group C strains were formerly divided into several species but chemotaxonomic and phenotypic examination suggests that *S. equisimilis* and *S. dysgalactiae*, together with other human isolates which produce group G or L Lancefield antigens, form the species *S. dysgalactiae*. Within *S. dysgalactiae* the name *S. dysgalactiae* subsp. *dysgalactiae* is used for strains of animal origin whereas *S. dysgalactiae* subsp. *equisimilis* is used for human isolates.

Animal isolates of *S. zooepidemicus* and *S. equi* are closely related and *S. zooepidemicus* has been proposed as a subspecies of *S. equi*. The name *S. canis* has been suggested for animal, but not human strains of group G streptococci.

The other pyogenic streptococci are of medical interest for animals. Some species such as *S. uberis, S. parauberis* and *S. porcinus* are frequently isolated from raw milk.

Other Streptococci

'Other streptococci' is a term used for a small group of mainly α-haemolytic streptococci (**Table 5**).

S. alactolyticus, S. bovis and *S. equinus* are currently the only Lancefield group D species included in *Streptococcus*. They are of medical importance at least for animals. Relatively close phylogenetic relationships have been shown by 16S rRNA comparative sequencing among *S. alactolyticus, S. bovis* and *S. equinus*. Although three DNA-homology groups were reported in *S. bovis*, further data from DNA–DNA hybridization showed that *S. bovis* and *S. equinus* belong to the same group and that the former should be reduced to synonymity. *S. bovis* has occasionally been isolated in large numbers from raw and pasteurized milk and cheese.

S. acidominimus was initially considered to be a variant of *S. uberis*, but 16S rRNA analysis has shown that it does not belong to the groups of other species with the possible exception of *S. suis*, an important pig pathogen. *S. caprinus* emerged from studies on the bacteria inhabiting the digestive tract of animals with tannin-rich diets. *S. intestinalis* is a relatively recent species present in the colon of pigs. Of particular interest is the ability of the strains to hydrolyse urea, an important aspect of nitrogen metabolism in animals.

Isolation and Cultivation

The nutritional needs of streptococci require the use of complex culture media that often contain meat extract. Rich agar-containing media (tryptic soy and heart infusion) supplemented with 5% animal blood (sheep or horse) are excellent for cultivating streptococci and determining haemolysis. An elevated CO_2

Table 4 Characteristics of species within the *Streptococcus pyogenes* group[a]

Characteristic	S. pyogenes	S. canis	S. agalactiae	S. dysgalactiae	S. parauberis	S. uberis	S. porcinus	S. iniae	S. equi	S. hyointestinalis
Acid from										
Amygdalin	–	ND	d	ND	+	+	d	ND	ND	d
Arbutin	–	ND	d	ND	+	+	d	ND	ND	+
Cellobiose	d	ND	d	ND	+	+	d	ND	ND	d
Cyclodextrin	d	–	–	ND	ND	ND	–	ND	+	ND
Glycerol	d	ND	d	d	–	–	d	–	–	–
Glycogen	d	–	–	d	–	d	–	ND	+	–
Inulin	–	ND	–	–	d	+	–	–	–	ND
Lactose	+	d	d	d	+	+	d	–	– or +[c]	+
Maltose	+	+	+	+	+	+	d	ND	+	+
Mannitol	–	–	–	–	+	+	d	+	–	–
Mannose	ND	ND	ND	ND	+	+	+	+	ND	+
Melezitose	–	–	–	ND	d	–	–	ND	–	ND
Methyl-D-glucoside	+	+	+	ND	–	d	d	ND	+	–
Methyl-D-xyloside	–	ND	d	ND	–	–	–	ND	ND	ND
Pullunan	d	+	+	ND	ND	ND	d	ND	+	–
Raffinose	–	–	–	–	d	–	–	–	–	d
Rhamnose	ND	ND	ND	ND	–	–	–	–	ND	–
Ribose	–	+	d	+	+	+	d	ND	– or +[c]	–
Salicin	+	ND	d	d	+	+	d	+	+	+
Sorbitol	–	–	–	d	+	+	+	–	– or +[c]	–
Starch	ND	ND	ND	+	ND	ND	V	ND	+	+
Sucrose	+	+	+	+	+	+	d	+	+	+
Tagatose	–	–	d	ND	d	d	–	ND	–	–
Trehalose	+	d	+	+	+	+	+	+	–	+
Hydrolysis of										
Arginine	+	+	+	+	+	+	+	ND	+	–
Aesculin	d	ND	–	d	+	+	d	+	d	+
Hyppurate	–	–	+	d	d	+	–	–	d	–
Starch	ND	ND	ND	ND	ND	ND	ND	+	ND	+
Production of										
Acetoin	–	–	+	–	+	+	d	ND	ND	+
Alkaline phosphatase	+	+	+	+	+	d	+	ND	+	+
α-Galactosidase	–	d	d	–	d	–	d	ND	–	d
β-Galactosidase	–	–	–	–	ND	–	–	ND	–	–
β-Glucosidase	–	–	–	ND	ND	ND	d	ND	–	ND
β-Glucuronidase	d	d	d	+	–	+	+	ND	+	–
Leucine arylamidase	ND	ND	ND	+	+	+	+	ND	+	+
Pyrolidonyl arylamidase	+	–	–	–	+	+	–	ND	–	–
Growth in/at										
4% NaCl	ND	ND	ND	ND	ND	+	ND	ND	ND	ND
6.5% NaCl	–	ND	d	ND	ND	–	ND	–	–	–
10% Bile	ND	ND	ND	ND	ND	ND	ND	ND	–	ND
40% Bile	–	ND	d	–	ND	d	ND	–	–	–
10°C	–	ND	d	ND	ND	+	–	+	–	ND
45°C	–	ND	–	ND	ND	–	–	–	–	ND
Cell wall components	Rhamnose	ND	Galactose, rhamnose, glucitol	*N*-acetylgalactosamine, rhamnose	ND	Glucose, rhamnose	ND	Galactose, glucose, rhamnose	*N*-Acetylgalactosamine, rhamnose or ND[c]	ND
Haemolysis	β	β	α, β, γ	β	α, β	α, γ	β	α, β	β	α
Lancefield antigens	A[b]	G	B	C, G, L	–, E, P	–, E, P, G	E, P, U, V	–	C	–
Mol% G+C	35–39	39–40	34	38.1–40.2	35–37	36–37.5	37–38	33	40–41 or 41.3–42.7[c]	
Murein type	Lys-Ala$_{2-3}$	Lys-Thr-Gly	Lys-Ala$_2$ (Ser)	Lys-Ala$_{1-3}$	ND	Lys-Ala$_{1-3}$	Lys-Ala$_{2-4}$	Lys-Ala$_{1-3}$	Lys-Ala$_{1-3\ (or\ 2-3)}$[c]	Lys-Ala (Ser)

[a] All species positive in *N*-acetylglucosamine, fructose, glucose and galactose. All species negative in adonitol, arabinose, arabitol, erythritol, gluconate, melibiose, methyl-D-mannoside, sorbose or xylose. Symbols: see Table 1. V, reported as 'variable'

[b] Further subdivision of Lancefield group A strains on the basis of M, T and R antigens.

[c] Depending on the subspecies.

Table 5 Characteristics of species within 'other streptococci'[a]

Characteristic	S. alactolyticus	S. bovis[b]	S. equinus	S. suis	S. acidominimus	S. intestinalis	S. caprinus
Acid from							
N-Acetylglucosamine	+	+; d	+	ND	ND	ND	ND
Amygdalin	d	d	+	ND	ND	ND	ND
Arbutin	d	+; d	+	ND	ND	ND	ND
Cellobiose	+	+; d	+	ND	ND	+	+
Gluconate	–	–	–	ND	ND	ND	ND
Glycerol	–	–	–	–	d	–	–
Glycogen	–	+; d	d	+	–	ND	Nd
Inulin	–	d	d	+	ND	ND	+
Lactose	–	+; d	d	+	d	–	+
Maltose	+	+	+	+	d	+	+
Mannitol	d	–; d	–	–	d	–	+
Melibiose	d	–; d	d	d	–	ND	ND
Melezitose	d	–; d	–	–	–	ND	ND
Methyl-D-glucoside	d	–; d	ND	ND	–	ND	–
Methyl-D-mannoside	–	–; d	–	ND	ND	ND	ND
Pullulan	–	ND	ND	+	–	ND	ND
Raffinose	+	+; d	d	d	–	–	+
Rhamnose	–	–; d	–	ND	ND	ND	–
Salicin	d	+; d	+	+	ND	+	ND
Sucrose	+	+	+	+	d	+	+
Tagatose	–	–; d	–	–	–	ND	ND
Trehalose	d	–; +; d	d	+	d	–	+
Hydrolysis of							
Arginine	–	–	–	+	–	–	ND
Aesculin	+	+; d	+	+	–	+	ND
Hippurate	–	–; d	–	–	ND	–	ND
Starch	ND	+; d	ND	+	ND	–	ND
Production of							
Acetoin	+	–; +	+	–	–	ND	ND
α-Galactosidase	+	–; d	d	+	–	ND	ND
β-Galactosidase	–	–; d	–	d	–	ND	+
β-Glucosidase	d	ND	ND	d	–	ND	ND
β-Glucuronidase	–	–; d	–	+	d	ND	ND
N-Acetyl-β-glucosaminidase	–	ND	ND	+	–	ND	+
Pyrrolidonyl arylamidase	–	–; d	–	d	–	ND	ND
Urease	d	ND	ND	–	–	+	ND
Growth in/at							
4% *NaCl*	ND	ND	–	ND	ND	ND	ND
6.5% *NaCl*	–	–	–	–	ND	–	ND
10% *Bile*	ND	ND	ND	ND	ND	ND	ND
40% *Bile*	ND	+	+	d	ND	–	ND
45°C	+	d	+	–	–	d	
Cell wall components	ND	Galactose, glucose, rhamnose	ND	Galactose[c], glucose, *N*-acetylgalactosamine, rhamnose	Galactose, rhamnose	ND	ND
Haemolysis	α, γ	α, γ	α	α, β	α	β	ND
Lancefield antigens	D	D	D	–, R, S, RS, T	–	–, G	–
Mol% *G+C*	40–41	ND	36.2–38.6	ND	40	39–40	ND
Murein type	ND	Lys-Thr-Ala	Lys-Thr-Ala	Lys-direct	Lys-Ser-Gly	ND	ND

[a] All species positive in fructose, galactose, glucose and mannose. All species negative in adonitol, arabinose, arabitol, cyclodextrin, dulcitol, erythritol, sorbitol, starch or xylose. No species produces alkaline phosphatase. Symbols: see Table 1.
[b] Variation of results obtained depending on DNA homology of groups.
[c] Trace amounts.

level (typically 5%) during incubation is essential for growing *S. mutans*, strains of the *S. anginosus* group and *S. pneumoniae*. Strains previously referred to as NVS also require the addition of ca. 0.001% pyridoxal HCl. Growth in liquid media must be buffered, otherwise the decrease in pH will soon become inhibitory. The Todd–Hewitt broth is the most widely used buffered medium.

If immediate laboratory processing is not possible, the reduced transport fluid (RTF) of Syed and Loesche is suitable for keeping streptococcal populations at room temperature. Commercially available silica gel or filter paper transport systems are also appropriate. Strains can usually be maintained for several days, or even weeks, on plates, at either room temperature or 4°C. Streptococci can also be kept in litmus milk containing glucose and yeast extract. Long-term preservation can be achieved by freezing at −70°C or in liquid nitrogen or by freeze-drying.

Oral Streptococci

Several species of oral streptococci give rise to characteristic colonial morphologies on trypticase–yeast extract–cystine (TYC) agar that are useful for very preliminary identification. Mitis salivarius (MS) agar is a widely used medium for the selection of oral streptococci (**Table 6**). Selective media for growing mutans streptococci are based on either MS agar or TYC agar with increasing amounts of sucrose and the addition of bacitracin (bacitracin susceptibility is a characteristic of group A streptococci). Selective compounds such as crystal violet, thallous acetate or sodium azide are also used.

Pyogenic and Other Streptococci

The medium colistin crystal violet sulphamethoxazole trimethoprim (CCSXT) agar (**Table 7**) is widely used for detecting group A streptococci together with some other confirmatory assays. Biochemical tests, easily performed in the laboratory, are an acceptable alternative to serological studies for identifying pyogenic streptococci (**Table 8**).

Lancefield group D streptococci may be determined by using the kanamycin aesculin agar medium (KAA) (**Table 9**).

Table 6 Composition of mitis salivarius (MS) agar

	$g\,l^{-1}$
Bacto tryptose (Difco)	10
Proteose peptone no. 3 (Difco)	5
Proteose peptone (Difco)	5
Bacto dextrose (Difco)	1
Bacto saccharose (Difco)	50
K_2HPO_4	4
Trypan blue	0.075
Bacto crystal violet (Difco)	0.0008
Bacto agar (Difco)	15

Dissolve by heating in deionized water to boiling. Sterilize at 121°C for 15 min. Cool to 50–55°C. Add 1 ml of 3.5% potassium tellurite per litre of medium.

Table 7 Composition of colistin crystal violet sulphamethoxazole trimethoprim (CCSXT) agar

Pancreatic digest of casein, powder	14.5 g
Papaic digest of soybean meal, powder	5 g
Sodium chloride	5 g
Crystal violet	0.2 mg
Colistin sulphate	10 mg
Sulphamethoxazole	24 mg
Trimethoprim	1.25 g
Agar	15 g
Distilled water	950 ml

Soak for 15 min, check and if necessary adjust pH to 7.3 ± 0.2, bring to the boil to dissolve the ingredients and sterilize for 20 min at 121°C. Cool rapidly to approximately 50°C, add 50 ml defibrinated sheep blood, mix with gentle rotation and pour into Petri dishes.

Streptococci in Foods

Except for *S. thermophilus*, streptococci *sensu stricto* are not used in food fermentation. However, some pathogens are introduced into humans and animals by foods, 'oral streptococci' are dependent on human diet for nutrition, 'pyogenic and other streptococci' are causes of mastitis and indicators of microbiological monitoring, thus all of them, at varying degree, may be considered food-related.

Streptococcus thermophilus

S. thermophilus is used as a starter to produce fermented milks, including yoghurt, and hard Italian and Swiss cheeses. It grows symbiotically with *Lactobacillus delbrueckii* subsp. *bulgaricus* during fermentation to produce lactic acid and acetaldehyde, responsible for the characteristic yoghurt flavour. In cheeses, *S. thermophilus* contributes to milk acidification and to flavour during ripening. *S. thermophilus* is discussed in detail in another article.

Foods as Delivery Agents of Pathogenic Streptococci

Raw milk may contain a wide variety of human pathogens including *S. pyogenes*. These usually reach the milk via the milker, from equipment either contaminated by humans or not adequately disinfected. In the past, infections by group A streptococci have

Table 8 Differentiation of streptococci with the use of biochemical tests

	Susceptibility to:							
	Bacitracin	*SXT*	*PYR*	*CAMP test*	*Hydrolysis of hippurate*	*Bile aesculin*	*Growth in 6.5% NaCl*	*Optochin and bile susceptibility*
S. pyogenes (group A)	+	−	+	−	−	−	−	−
S. agalactiae (group B)	−[a]	−	−	+	+	−	+[a]	−
Large colony (group C and G)	−[a]	+	−	−	−	−	−	−
S. pneumoniae			−	−	−	−	−	+
S. bovis	−	+[a]	−	−	−	+	−	−

+, positive; −, negative; SXT, sulphamethoxazole and trimethoprim; PYR, pyrrolidonyl arylamidase; CAMP test, test for enhancement of haemolysis by *Staphylococcus aureus* beta lysin.
[a] Exceptions occur occasionally.
Reproduced by the permission of Ruoff (1992).

Table 9 Composition of kanamycin aesculin (KAA) agar

Tryptone	20 g
Yeast extract powder	5 g
Kanamycin sulphate	0.02 g
Sodium chloride	5 g
Sodium citrate	1 g
Aesculin	1 g
Ferric ammonium citrate	0.5 g
Sodium azide	0.15 g
Agar	15 g
Distilled water	1 litre

Soak for 15 min, check and if necessary adjust pH to 7.0 ± 0.1 and bring to the boil to dissolve the ingredients completely. Sterilize for 15 min at 121 °C; cool to approximately 47 °C.

been spread by raw milk, but this risk has been eliminated by pasteurization. In recent years, outbreaks of streptococcal sore throats have occurred after the consumption of reconstituted milk and intensively manipulated foods such as salads, rice puddings and hams previously contaminated by handling by infected workers. If contaminated foods are left for several hours in a warm place, an explosive bacterial growth occurs causing outbreaks of pharyngitis.

It has long been disputed as to whether group B streptococci identical to *S. agalactiae*, a bovine pathogen, can cause serious diseases in humans. In one instance a correlation was established between the occurrence of group B streptococcal disease and consumption of raw milk. Recently, group B streptococci were identified as a possible cause of diseases transmitted by raw milk and dairy products made from unpasteurized milk. Consumption of raw milk or its incorporation into dairy products must be discouraged.

S. equi subsp. *zooepidemicus* causes outbreaks of severe infection when ingested with raw milk.

Oral Microflora

'Oral streptococci' constitute the dominant acidogenic population in supragingival dental plaque. Since the free sugar concentration in their natural habitat is often low, their main energy supply is from carbohydrates from dietary foods. Thus, 'oral streptococci' are transiently exposed to mixtures of various sugars and live under feast or famine conditions. The type of diet is one of the major factors in controlling the growth kinetics of 'oral streptococci'. Streptococci in dental plaque must survive cycles of acidification to pH ca. 4.0 and alkalinization to pH somewhat above 7.0. There is hierarchy to acid tolerance and mutans streptococci are inherently the most acid-tolerant bacteria. High levels of inherent tolerance is fundamental for cariogenicity and related to high levels of proton translocating F-ATPase activity, and to low pH optima for enzyme activity. Many of the less acid-tolerant bacteria in plaque, such as *S. ratti* and *S. sanguinis*, protect themselves by means of the arginine deaminase system. NH_3 produced within the cell combines with protons to yield NH_4^+, and this reaction raises the cytoplasmic pH value thereby protecting sensitive structures. NH_3 is also produced by urea hydrolysis.

Microbiological Monitoring of Foods

Faecal streptococci such as *S. bovis* and *S. equinus* are considered to be indicators of faecal pollution of food (i.e. pork, beef, poultry and sliced, vacuum-packed sausages and ham) because they have an advantage over coliforms in that they are more resistant to most environmental stress.

The most appropriate markers which indicate food contamination from the oral cavity and upper respiratory tract is the mitis-salivarius group. Their enumeration in food is also useful in assessing contamination in the food environment. This applies particularly in situation of mass feeding such as industrial catering and hospital food-preparation areas. The presence of the mitis-salivarius group on cutlery, glasses, etc, indicates inadequate elimination of contaminating organisms from the human oral cavity, and to a certain extent, from the respiratory tract.

Mastitis

S. agalactiae, *S. dysgalactiae*, *S. equinus*, *S. uberis* and *S. bovis* are some of the most important causes of mastitis. Most of them are spread between the udder quarters and cows primarily during milking and consequently, the milking clusters, the milker's hands, udder cloths, etc, become contaminated and may act as fomites transferring disease among the herd. The spread of *S. dysgalactiae* and *S. uberis* is less dependent on the milking process because they are more widely distributed in the environment. Mastitis is one of the most widespread and costly diseases affecting dairy herds. Nearly 50% of all cows suffer at least one outbreak of clinical mastitis per lactation. Disease control involves hygienic practices and the infusion of antibiotic drugs in the udder. When antibiotics are used, milk from treated cows cannot be sold for 3–5 days after treatment. This procedure is necessary to protect humans who show hypersensitivity to antibiotic drugs, as well as to protect starter cultures used in milk processing. Mastitis may be present in clinical form, in which macroscopic changes in the milk or udder are detectable, but subclinical conditions are more common. Subclinical mastitis can only be diagnosed by examining milk samples for the presence of pathogenic bacteria, an increased somatic cell count, or a variety of biochemical changes. Compositional changes are the result of an increased movement of blood components into the milk causing increased concentrations of bovine serum albumin, immunoglobulin and sodium and chloride ions, and decreased concentrations of caseins, lactose and potassium. Severe mastitis leads to the production of milk with a reduced casein/total N ratio, also due to the increased endogenous proteolytic activity of the milk. These compositional changes render the milk unsuitable for cheese-making due to the increased time needed for milk coagulation, unacceptable firmness of the coagulum and severe loss of whey.

Rumen Microflora

The population of microorganisms that inhabit the rumen of livestock is largely responsible for the digestion in these animals. *S. bovis* is a normal inhabitant of the rumen at moderate cell concentration (10^7 cfu ml^{-1}). Some strains are highly amylolytic and, in general, it is an essential proteolytic bacterium. If the animals' diet is radically altered by switching rapidly from a forage to a grain diet which is rich in readily fermentable carbohydrates, the rumen fermentation can become unbalanced, resulting in digestive disorders such as lactate acidosis. Although probably not the only cause, *S. bovis*, in this dietary condition, proliferates to ca. 10^{10} cfu ml^{-1} and becomes an important causative agent of this digestive disorder.

Since strains of *S. bovis* are amylolytic, they show a relatively short doubling time when grown in vegetables which lack readily fermentable carbohydrates. Since it is also a homolactic bacterium, it has been used as a substitute for *Enterococcus faecium* as a commercial inoculant in alfalfa silage.

Human Diseases

Some *Streptococcus* spp. are clinically relevant for humans, and a few of the most important diseases are briefly described here.

Dental caries in humans are related mainly to *S. mutans*. Members of the *S. anginosus* group have been mainly associated with purulent infections of the mouth and internal organs and possible virulence factors, such as hyaluronidase, gelatinase, collagenase, DNase, RNase and polysaccharide capsule, have been reported. *S. sanguinis*, *S. oralis* and *S. gordonii* cause endocarditis in patients with damaged heart valves and reach the bloodstream mainly from the mouth, as a result of dental procedures such as tooth extraction.

S. pyogenes has been reported as a rare cause of food-transmitted sore throats. Food-borne streptococcal pharyngitis is more severe and more confined to the pharynx than that caused by endemic air-borne streptococci. Other diseases related to Lancefield group A streptococcus are impetigo, scarlet fever, erysipelas, rheumatic fever, pneumonia, acute glomerulonephritis, toxic shock-like syndrome and septicaemia. Extracellular products, such as streptolysins O and S, hyaluronidase, DNase enzymes (A, B, C and D), streptokinase, NADase, three pyogenic toxins (A, B and C) and cell-associated proteins are the main virulence factors of *S. pyogenes*.

S. pneumoniae is involved in pneumonia, bacteraemia, otitis media, sinusitis and meningitis but it is also found as normal inhabitants in the upper respiratory tract from where it enters the host. In addition to the polysaccharide capsules, other virulence factors, such as the pneumococcal surface protein A, neuraminidase, pneumolysin and autolysin, have been detected. Isolates of pneumococci commonly resistant to penicillin and other β-lactam antibiotics, due to a reduced affinity of the high-molecular-mass penicillin binding proteins (PBPs), have arisen from horizontal gene transfer. Pneumococcal vaccines, made of capsular polysaccharides of the most prevalent serotypes isolated from infections, have been available since the 1970s. Unfortunately, the vaccines are often not effective in persons with poor immunological response.

S. agalactiae also causes serious diseases in humans. Both early and late onset neonatal sepsis, characterized by septicaemia and meningitis, and associations with invasive diseases in adults have been reported. Ribotyping and conventional epidemiological markers have been used to establish the epidemiological interrelationships between the bovine and human isolates of *S. agalactiae*. Several extracellular products, such as type-specific polysaccharide capsule, hyaluronidase and C5a peptidase, have been proposed as virulence factors. To prevent devastating perinatal group B streptococcal infections, intrapartum administration of antibiotics and, more recently, polysaccharide vaccines have been used.

S. bovis is the only Lancefield group D species of medical importance associated with human alimentary tract, faeces and tonsils. It causes septicaemia, endocarditis or meningitis in humans; the isolation of many of the polysaccharide-producing biotypes from the bloodstream may indicate the presence of colon cancer.

See also: **Milk and Milk Products**: Microbiology of Liquid Milk. **Starter Cultures**: Cultures Employed in Cheese-making. ***Streptococcus***: *Streptococcus thermophilus*.

Further Reading

Borne RA (1997). Oral streptococci . . . products of their environment. *Journal of Dental Research* 77: 445–452.

Bar-Dayan Y, Bar-Dayan Y and Shemer J (1997) Food-borne and air-borne streptococcal pharyngitis. A clinical comparison. *Infection* 25: 12–15.

Hardie JM (1986) Genus *Streptococcus*. In: Sneath PH, Mair NS, Sharpe ME and Holt JC (eds) *Bergey's Manual of Systematic Bacteriology*. Vol. 2, p. 1043. Baltimore: Williams and Wilkins.

Hardie JM and Whiley RA (1992) The Genus *Streptococcus*-Oral. In: Balows A, Trüper HG, Dworkin M, Harder W and Schleifer KH (eds) *The Prokariots*. P. 1421. New York: Springer-Verlag.

Hardie JM and Whiley RA (1995) The Genus *Streptococcus*. In: Wood BJB and Holzapfel WH (eds) *The Genera of Lactic Acid Bacteria*. P. 55. London: Blackie Academic and Professional.

Ruoff KL (1992) The Genus *Streptococcus*-Medical. In: Balows A, Trüper HG, Dworkin M, Harder W and Schleifer KH (eds) *The Prokariots*. P. 1450. New York: Springer-Verlag.

Vandamme P, Torch U, Falsen E, Pot B, Gossens H and Kersters K (1998) Whole-cell protein electrophoretic analysis of viridans streptococci: evidence for heterogeneity among *Streptococcus mitis* biovars. *International Journal of Systematic Bacteriology* 48: 117–125.

Streptococcus thermophilus

Gerald Zirnstein, Centers for Disease Control and Prevention, Foodborne and Diarrheal Disease Laboratory, Atlanta, USA

Robert Hutkins, Department of Food Science and Technology, University of Nebraska, Lincoln, USA

Introduction

The genus *Streptococcus* includes species of Gram-positive bacteria that have similar metabolic properties, but which live in diverse habitats and have many physiological differences. Some streptococci are pathogens to animals, others are part of the normal flora of humans, animals and foods. In the past two decades, advances in classification techniques have led to substantial reclassification efforts, and several important species have been reclassified as members of the recently named genera *Enterococcus* and *Lactococcus*. The latter genus now includes the well-studied dairy bacteria *Lactococcus lactis* subsp. *lactis* and *Lactococcus lactis* subsp. *cremoris*, organisms formerly classified as *Streptococcus lactis* and *Streptococcus cremoris*, respectively. The only 'dairy streptococci' remaining from those originally described by Sherman over 60 years ago is *Streptococcus thermophilus*. Even this organism has not been exempt from reclassification efforts. In 1984, researchers proposed that, based on DNA–DNA homology and membrane fatty acid profile studies, *S. thermophilus* should be reclassified as *Streptococcus salivarius* subsp. *thermophilus*. The adoption of this new nomenclature by the microbiology community, including industrial microbiologists in the business of marketing dairy starter cultures, was rather slow. There were some who actually resisted this name change, on the basis that *S. thermophilus* is not found in the same ecological niche as *S. salivarius* and that there are also significant physiological differences. In 1987, the results of more stringent DNA homology studies, the large number of phenotypic differences between *S. salivarius* and *S. thermophilus*, and their markedly different habitats, led investigators to propose that the name *Streptococcus thermophilus* be

Table 1 Properties and characteristics of *Streptococcus thermophilus*

Gram positive, non-motile coccus
Cell size about 0.7–0.9 μm, in pairs or long chains
Facultative anaerobe
Homofermentative (producing mainly L(+)-lactic acid)
Ferments lactose, sucrose, glucose and fructose, with preference for disaccharides; generally does not ferment galactose
Final pH in broth culture about 4.0–4.5
Catalase negative
Lacks cytochromes
Growth temperature maximum about 50–52°C, no growth at 10°C
Optimum growth at 40–45°C
Thermotolerant (survives 60°C for 30 min)
Weak or no growth with 2% NaCl
Weakly proteolytic
Ammonia not produced from arginine
Lacks group specific antigen
G + C mol ratio about 37–40%
Little DNA:DNA homology with *Lactococcus*

restored. Thus, it now appears that the more accepted nomenclature is indeed *Streptococcus thermophilus*.

General Properties

Like most lactic acid bacteria, *Streptococcus thermophilus* is non-spore-forming, catalase-negative, facultatively anaerobic and metabolically fermentative. Microscopically, *S. thermophilus* appears as spherical or ovoid cells (0.7–0.9 μm in diameter) in pairs or chains when grown in liquid media. Although its name would appear to indicate that *S. thermophilus* has a high optimum growth temperature, it actually grows best at the high end of the mesophilic range, about 42–45°C. Like other streptococci, *S. thermophilus* is heterotrophic and generally fastidious, requiring simple carbohydrates as an energy source, and preformed amino acids as a nitrogen source. Unlike other streptococci, however, *S. thermophilus* does not possess a group-specific antigen and is generally non-haemolytic. The structure of the peptidoglycan is identical to that of *Enterococcus faecalis*, an organism that shares other properties with *S. thermophilus*. However, it is distinguished from other enterococci and some lactococci by its sensitivity to salt; it does not grow in the presence of 4% salt, and some strains will not grow in as little as 2% salt. Its sensitivity to salt, and its inability to release NH_3 from arginine, or to grow at 10°C, at pH 9.6, or in the presence of 0.1% methylene blue, form the basis of various schemes used to distinguish between *Streptococcus* and related species. The main properties and characteristics of *S. thermophilus* are summarized in **Table 1**.

Ecology of *Streptococcus thermophilus*

The ecology and origins of *S. thermophilus* are enigmatic. Like other dairy lactic acid bacteria, *S. thermophilus* is well adapted to a milk environment. It can be isolated from milk, dairy utensils and especially from heated milk. Unlike *Lactococcus lactis* subsp. *lactis* and other lactic acid bacteria, *S. thermophilus* has never been isolated from green plant material or from any clinical sources. Other than milk, there is no known habitat for this microorganism. Because *S. thermophilus* can survive moderate thermal processes (e.g. 60°C for 30 min), it is often found in milk pasteurization equipment and in pasteurized dairy products.

Methods for Cultivation and Enumeration of *Streptococcus thermophilus*

Like other lactic acid bacteria, *S. thermophilus* has complex growth requirements. In addition to a source of fermentable carbohydrate, it also requires hydrolysed proteins as a source of amino acids. Most bacteriological media used to support growth of *S. thermophilus*, therefore, contain hydrolysed casein, tryptone or beef extract. Most other vitamin and nutrient requirements are satisfied by the addition of yeast extract. A number of common, commercially available, nonselective media have a long history of use, including Elliker medium, MRS medium, and M17 medium. The latter medium which contains β-glycerol phosphate as a buffer, has been advocated as providing optimum growth conditions for *S. thermophilus*. Other media, such as *S. thermophilus* agar (or ST agar) and reinforced clostridial agar, have also been used for recovery and enumeration of *S. thermophilus* in yoghurt.

A large variety of differential or selective solid media are also useful for enumerating and distinguishing between *S. thermophilus*, *Lactobacillus* spp. and other organisms ordinarily found in yoghurt and fermented dairy foods. Identification or differentiation of the lactic acid bacteria on these media is most often based on different colony morphologies that occur as a result of their fermentation properties. For example, Lee's agar includes lactose and sucrose, together with the acid–base indicator bromocresol purple (BCP). *Lactobacillus bulgaricus* ferments lactose but not sucrose and will produce only enough acid to form white or slightly yellow colonies. *S. thermophilus* forms colonies which are intense yellow because it ferments both lactose and sucrose. Yoghurt lactic agar is the recommended medium in the most recent edition of *Standard Methods for the Microbiological Examination of Dairy Products*. This medium is basically Elliker media containing Tween

80, skim milk, and triphenyltetrazolium chloride. *Lactobacillus bulgaricus* will form large white colonies, whereas *S. thermophilus* forms small, smooth colonies. L-S differential medium, which contains triphenyltetrazolium chloride, is also used to distinguish between *Lactobacillus* and *Streptococcus*. The former form irregular, red colonies surrounded by a white opaque zone, and the latter form round, red colonies surrounded by a clear zone. Hansen's yoghurt agar contains galactose, glucose and lactose (in addition to peptone and beef extract).

Carbohydrate Metabolism by *Streptococcus thermophilus*

Compared to other lactic acid bacteria, *S. thermophilus* is one of the more prolific producers of lactic acid. When grown in rich medium containing lactose, *S. thermophilus* has doubling times of less than 30 min and can produce more than 30 g of lactic acid per litre. According to most reports, *S. thermophilus* produces only L(+)-lactic acid. However, it ferments a rather narrow range of carbohydrates, leading one early investigator to conclude that '*S. thermophilus* is marked more by the things which it cannot do than by its positive actions'. It has a distinct preference for the disaccharides lactose and sucrose, and its growth on the constituent monosaccharides, glucose, fructose and galactose is usually slower than on the intact disaccharides. This observation led to the suggestion that the sugar transport systems used by *S. thermophilus* to accumulate monosaccharides were either absent or had low activity. In fact, it now appears that the ability of *S. thermophilus* to metabolize different sugars is, in part, dependent on the availability of the necessary transport systems.

Another early observation made by several investigators was that most strains of *S. thermophilus* are unable to ferment free galactose and are phenotypically galactose negative (Gal^-). Moreover, when cells were grown in medium containing excess lactose, galactose accumulation occurred in the medium. This phenomenon of galactose release during growth on lactose does not occur in *Lactococcus lactis* or *Lactobacillus helveticus*, dairy organisms that ferment the glucose and galactose moieties simultaneously. However, after extensive study, it now appears that most strains of *S. thermophilus*, as well as *Lactobacillus bulgaricus* and *Lactobacillus acidophilus*, share this unusual metabolic behaviour. Even those strains of *S. thermophilus* that ferment free galactose still do not do so if glucose is present, an observation that suggests that galactose metabolism is catabolite repressed by glucose.

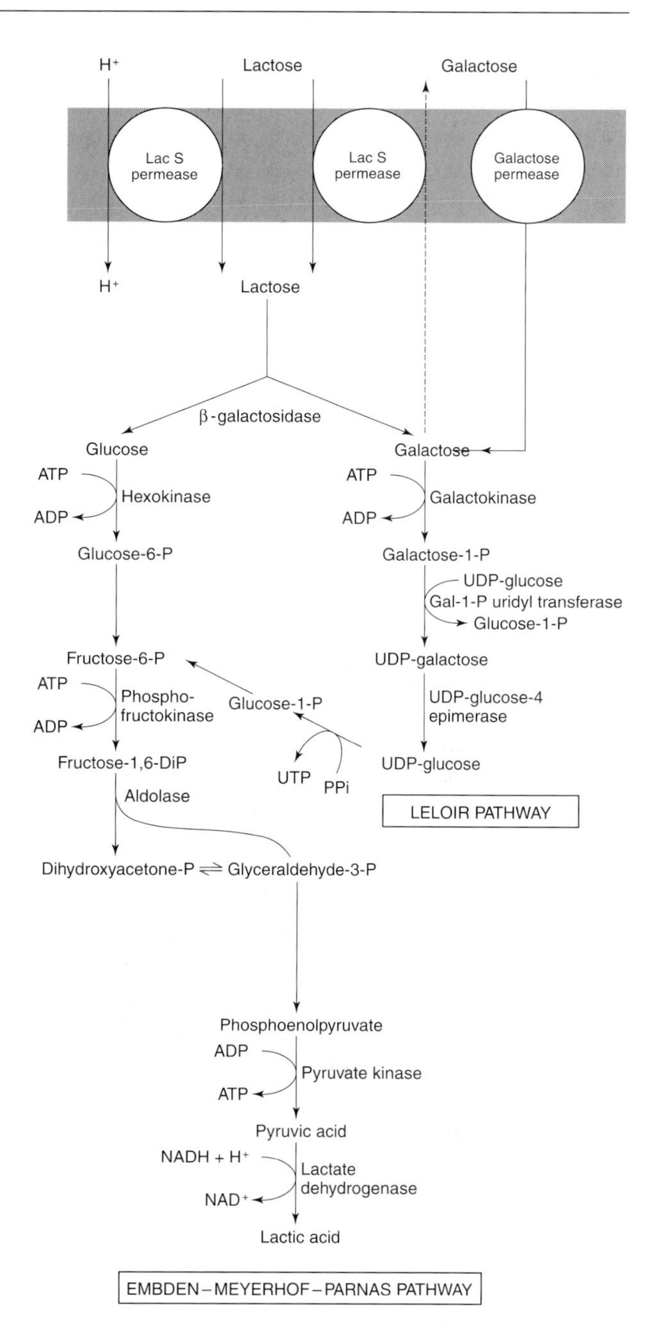

Figure 1 Pathways for lactose and galactose transport and metabolism by *Streptococcus thermophilus*.

Pathways for Sugar Metabolism

Pathways responsible for lactose and galactose metabolism in *S. thermophilus* are now well characterized (**Fig. 1**). Lactose is transported into the cell by a lactose transport protein, called LacS. The lactose reaches the cytoplasm as free lactose, and is then hydrolysed by a β-D-galactosidase to yield glucose and galactose. This system is in contrast to that in lactococci and many other lactic acid bacteria which phosphorylate lactose as it is transported into the cell (via a phosphotransferase system). The energy required to fuel the transport of lactose is provided by a proton motive force (pmf). Further evidence led to a broader model of the mechanism for lactose transport. Because galactose efflux occurred at rates nearly identical to that of lactose uptake, the notion that the two processes were linked was an attractive idea. Indeed, the experimental evidence has shown that lactose uptake and galactose efflux occurred via a common galactoside antiporter, the *lac*S transporter. Under laboratory situations, lactose uptake and galactose efflux may also occur independent of exchange, i.e. in symport with a proton (see Fig. 1). These alternate mechanisms of lactose uptake by the lactose transport protein could conceivably allow a degree of physiological flexibility for the organism. When only the glucose moiety of lactose is metabolized, then efflux of the accumulated intracellular galactose can serve as a driving force for lactose uptake by simultaneously exiting the cell via the *lac*S permease. However, when the galactose moiety of lactose is metabolized as rapidly as the glucose (most likely when the glucose supply is nearly exhausted), then the pmf-linked lactose-H^+ symport mechanism can be used to transport lactose.

Although the lactose transport protein can catalyse lactose-H^+ symport as well as lactose/galactose exchange, the latter reaction is much more rapid and appears to be responsible for most of the lactose uptake during ordinary growth on lactose. Researchers from The Netherlands have shown that maximal lactose-H^+ symport requires an alkaline internal pH, which implies that this uptake mechanism is slow when the medium pH is low (i.e. during fermentation). In contrast, the maximum rate of the lactose/galactose exchange reaction is at least 10 times greater than the maximal rate of the pmf-driven system. Thus, it would appear that the lactose/galactose exchange reaction is the predominant mechanism used by *S. thermophilus* under physiological conditions.

Despite the apparent utility of the lactose:galactose exchange system, the question remains: why do most strains of *S. thermophilus* not ferment galactose? One explanation is that the genes coding for the enzymes that comprise the galactose-catabolic pathway are absent from *S. thermophilus*. However, the enzymes of the Leloir pathway (Fig. 1) for galactose metabolism have been shown to be present in *S. thermophilus*, but their activities are exceedingly low. The activity of galactokinase, the first enzyme of the pathway, was essentially undetectable under ordinary growth conditions. More recently, it has been discovered that, although the genes coding for these enzymes are present, their transcription is strongly repressed. In fact, it appears that repression of these genes is the normal state. The few strains that ferment galactose do so only when the repressing sugars, glucose and lactose are absent, and the inducer, galactose, is present. Under these conditions, de-repression can occur and the operon is transcribed.

Recently, additional information was provided that further accounts for the Gal^- phenotype and explains the galactose-excretion system. According to this model, galactose mutarotase (the gene product of *gal*M, and part of the galactose operon in *S. thermophilus*) is responsible for converting β-D-galactose into α-D-galactose. Since the α-anomer serves as the substrate of galactokinase, phosphorylation may be dependent on the mutarotation rate. Although spontaneous mutarotation of galactose in water is very slow, for many years the intracellular mutarotation of β-galactose was believed to be spontaneous, despite biochemical evidence for the existence of mutarotase. It is now suggested that in *Escherichia coli*, the formation of α-D-galactose from β-D-galactose (generated by hydrolysis of lactose by β-galactosidase) is dependent on the presence of mutarotase activity. Thus, the mutarotase enzyme appears to link the metabolism of lactose and galactose in *E. coli*. Although *S. thermophilus* does have the *gal*M gene coding for mutarotase, it does not appear to transcribe the gene or express activity. Experiments have shown that the level of mutarotase activity was low whether cells were grown in glucose or lactose. Since *S. thermophilus* also uses β-galactosidase and the enzymes of the Leloir pathway to metabolize galactose, like *E. coli*, it is possible that a similar role of mutarotase in *S. thermophilus* explains, in part, the observed phenomenon of galactose metabolism in that organism.

Use of *Streptococcus thermophilus* in the Production of Dairy Foods

Due to its moderate thermophilic nature, *S. thermophilus* survives and produces acid at temperatures higher than can be tolerated by the mesophilic lactic acid bacteria. This characteristic makes *S. thermophilus* useful in the fermentation of dairy products, such as yoghurt and Swiss and Italian cheeses, that are ordinarily manufactured or incubated at elevated temperatures (**Table 2**). The benefits of this organism as a starter culture are due primarily to its ability to ferment lactose, in homolactic fashion, and to cause rapid reduction of the product pH. In the case of yoghurt, a decrease in pH to 4.7 or below causes casein to precipitate and the milk to coagulate. The acid formed during the mozzarella and Swiss cheese fermentation causes the pH to decrease to approxi-

Table 2 Applications of *Streptococcus thermophilus*

Swiss-type cheeses
Emmentaler
Gruyère
Pasta filata cheeses
Mozzarella
Provolone
Hard grating cheeses
Parmesan
Romano
Asiago
Other cultured milks
Yoghurt

mately 5.2, which is critical for successful manufacture of these products.

In nearly all fermented dairy products made using *S. thermophilus*, a compatible *Lactobacillus* spp., either *L. delbrueckii* subsp. *bulgaricus* or *L. helveticus*, is also added. The relationship between *S. thermophilus* and dairy lactobacilli has been the subject of much research interest, since it is well established that most strains grow better when grown together (as mixed genus cultures), compared to their growth as single species cultures. Several explanations have been proposed to account for this so-called symbiotic growth behaviour. The lactobacilli are generally more proteolytic than *S. thermophilus*; thus, in dairy fermentations, lactobacilli secrete extracellular proteinases that hydrolyse casein and other milk proteins and thereby provide *S. thermophilus* with necessary amino acids. In contrast, *S. thermophilus* is believed to produce small amounts of formic acid that may stimulate *L. delbrueckii* subsp. *bulgaricus*. The production of CO_2 from urea by *S. thermophilus* may also enhance growth of lactobacilli, due to the more reduced environment preferred by the latter. Finally, the excretion of galactose by most strains of *S. thermophilus* may provide galactose-fermenting lactobacilli with a source of fermentable carbohydrate when lactose is not available.

In addition to its role as an acid producer in various fermented dairy products, *S. thermophilus* can also synthesize other useful end products, most notably flavour compounds and polysaccharides. The production of acetaldehyde, which imparts 'green' or 'tart-apple'-like flavour notes in yoghurt, is due, in part, to metabolism by *S. thermophilus*. Acetaldehyde is synthesized by *S. thermophilus* via one of several possible pathways. Although glucose metabolism yields mostly pyruvate and some acetylphosphate, both of which can serve as substrates for acetaldehyde-yielding reactions, the relevant enzyme activities have not been reported. Rather, it appears more likely that acetaldehyde formation occurs via hydrolysis of the amino acid threonine by threonine aldolase. Whether *S. thermophilus* or *L. delbrueckii* subsp. *bulgaricus* is actually the primary producer of acetaldehyde, however, has not yet been established. Exopolysaccharides, produced by many lactic acid bacteria, including *S. thermophilus*, provide mouthfeel, body, viscosity, and other desirable rheological properties in yoghurt. The pathways and genes involved in these properties are discussed in more detail below.

Another important use of *S. thermophilus* in dairy foods has less to do with its technological applications and more to do with its putative nutritional role as a probiotic. Although there is evidence which indicates that *S. thermophilus* does not colonize the intestinal tract of humans, consumption of viable cells of *S. thermophilus* may enhance lactose digestion by otherwise lactose-intolerant individuals. Indeed, such individuals have been shown to tolerate yoghurt better than other dairy products containing equivalent amounts of lactose. Similarly, non-fermented fluid milk containing added *S. thermophilus* and *L. delbrueckii* subsp. *bulgaricus* is also better tolerated than the heat-treated fermented or non-fermented products. How does *S. thermophilus* alleviate symptoms of lactose intolerance? One possibility is that cells of *S. thermophilus* survive the stomach and are lysed in the gastrointestinal tract. Intracellular β-galactosidase is released by the lysed *S. thermophilus* and hydrolyses dietary lactose, ultimately preventing the lactose from reaching the large intestine and causing symptoms associated with lactose intolerance. Because sour cream, cultured buttermilk and other cultured dairy products are ordinarily made using lactococci, which do not produce β-galactosidase, these positive effects are not observed with those products. Although other health claims due to *S. thermophilus* consumption have been made, including cholesterol-lowering and antitumorigenic activities, such claims lack convincing evidence. Some strains of *S. thermophilus* do produce bacteriocins; however, it seems unlikely that these antimicrobial substances confer health benefits.

Production of Polysaccharides by *Streptococcus thermophilus*

The ability of some lactic acid bacteria to produce exopolysaccharides has important practical implications. In yoghurt and other cultured products, these polysaccharides impart desirable properties relating to viscosity. In general, the *S. thermophilus* polymers consist primarily of galactose and glucose monomers. Although many strains of *S. thermophilus* produce polysaccharide material, their composition and the amounts produced are often variable, even when growth conditions remain constant. Indeed, the 'ropy' trait is frequently diminished or lost altogether. It appears that growth temperature and culture conditions, especially the specific carbon source available, have profound effects on polymer composition and quantity. The unstable ropy character of *S. thermophilus* does not seem to be due to the loss of a plasmid, since many polysaccharide-producing strains do not contain plasmids. This is in contrast to the plasmid-carried ropy character of mesophilic lactic acid bacteria confirmed by several groups. Recently, researchers have reported the presence of a 15 kb gene cluster in *S. thermophilus* that consisted of 13 genes involved in exopolysaccharide synthesis. Most of these genes coded for proteins with homology to enzymes involved in synthesis of polysaccharides produced by other bacteria.

Bacteriophages and *S. thermophilus*

The presence of bacteriophages in cheese factories affects most lactic acid bacteria, and *S. thermophilus* is no exception. Until recently, phages that infect mesophilic lactic acid bacteria that are used in the manufacture of the most common cheese types attracted the most attention, and *S. thermophilus* phages were not thought to be much of a problem. However, due perhaps to the huge increase in the production of products made using *S. thermophilus* as a starter culture, mainly mozzarella and other Italian cheeses and yoghurt, problems with phage have become much more common.

Phages that infect *S. thermophilus* belong to the *Siphoviridae* group. In general, two (but perhaps as many as four) phage types are recognized, based on protein profiles, DNA homology and restriction patterns and other analyses. However, genetic analyses have shown little correlation between host range and gene-based groups. Morphologically, most *S. thermophilus* phages have isometric heads and long, noncontractile tails. All contain double-stranded DNA, with genomes in the range 30–45 kb. Most phages isolated from cheese plants and dairy environments are lytic; however, temperate *S. thermophilus* phages have also been characterized. Interestingly, the latter appear to be quite similar to lytic phages, thus leading investigators to suggest a common ancestral phage and that a single *S. thermophilus* phage species exists. Although little is known about natural phage defence systems used by *S. thermophilus*, the presence of restriction modification systems has been reported. Recently, the expression of a lactococcal restriction–modification system in *S. thermophilus* was shown to confer phage resistance.

Genetics of *Streptococcus thermophilus*

Although less well studied than the dairy lactococci, much information on the genetics of *S. thermophilus* has emerged in recent years. Physical maps of *S. thermophilus* have revealed a genome size of about 1800 kb. Most strains of *S. thermophilus* carry few, if any plasmids. Those plasmids that have been studied are small (< 10 kb) and cryptic. Despite the absence of plasmids by most strains of *S. thermophilus*, vectors have been constructed and used to express heterologous genes, such as the cholesterol oxidase gene from *Streptomyces*, in *S. thermophilus*. A number of gene transfer strategies have been used successfully in *S. thermophilus*, including transfection and transformation of sphaeroplasts, transduction, conjugative mobilization (with *L. lactis*) and electroporation.

Genes coding for peptidase utilization and DNA restriction and modification systems in *S. thermophilus* have been identified and sequenced. However, the most well-studied genes in *S. thermophilus* are those that code for enzymes involved in carbohydrate catabolism. **Figure 2** shows the order of the lactose genes in *S. thermophilus* A147. The *lac*S and *lac*Z genes in this strain form the lactose operon. The lactose transport protein (the *lac*S gene product) and β-galactosidase (the *lac*Z gene product) are expressed from a promoter located upstream of the *lac*S gene. The *lac*S gene of *S. thermophilus* A147 is 1902 bp long, and encodes a 69 454 Da protein with an NH_2-terminal hydrophobic region and a COOH-terminal hydrophilic region. The NH_2-terminal end is homologous with the melibiose carrier of *Escherichia coli*, whereas the COOH-terminal end is homologous to enzyme IIA proteins or IIA protein domains of phosphoenolpyruvate:sugar phosphotransferase systems. Since sugars are not phosphorylated during translocation by the *S. thermophilus* lactose transport system it was suggested that the enzyme IIA-like region serves a regulatory function in this protein.

Several genes coding for galactose metabolic enzymes have been identified in *S. thermophilus* (Fig.

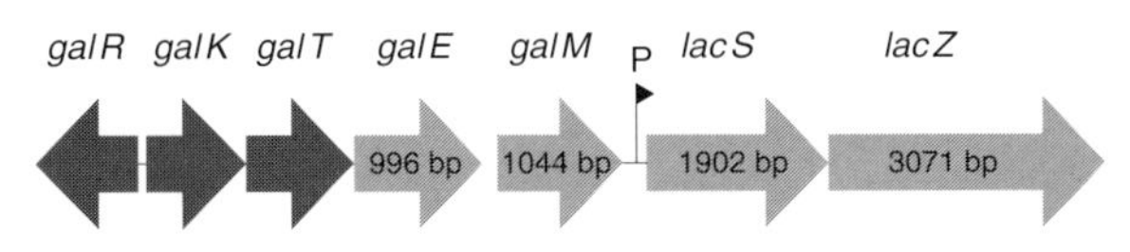

Figure 2 Arrangement of genes involved in lactose and galactose utilization by *Streptococcus thermophilus*. Genes encoding for lactose transport (*lacS*), lactose hydrolysis (*lacZ*) and a *lac* promoter region (P) and for galactose mutarotase (*galM*) and epimerase (*galE*) have been cloned and sequenced for strain A147. These genes are preceded by *galR* (repressor), *galK* (galactokinase) and *galT* (transferase). A gene coding for galactokinase (*galK*) has also been cloned in strains F410 and 19258, but appears to be located downstream of *gal*M.

2). In strain A147, the *gal*E and *gal*M genes are located 105 bp upstream from the lac operon. A putative promoter, P_{s1} is located 19 bp upstream of a possible start point for the *gal*M transcript. The *gal*M mRNA could not be reliably detected by Northern analysis after growth of the cells in glucose- or lactose-MRS, suggesting catabolite repression by glucose. Activities for β-galactosidase and lactose transport, as well as for enzymes involved in galactose metabolism (i.e. UDP glucose 4-epimerase, and UDP glucose-hexose-1-phosphate uridyltransferase) were 10- to 20-fold higher in lactose-grown cells than in glucose-grown cells. The gene coding for galactokinase (*gal*K) has also been cloned from *S. thermophilus*. This gene was also catabolite repressed by glucose. Collectively, these data indicate that expression of the *lac* and *gal* genes is regulated at the transcription level.

See also: **Cheese**: Microbiology of Cheese-making and Maturation. **Fermented Milks**: Range of Products; Yoghurt. ***Lactobacillus***: Introduction. **Probiotic Bacteria**: Detection and Estimation in Fermented and Non-fermented Dairy Products. **Starter Cultures**: Importance of Selected Genera; Cultures Employed in Cheese-making.

Further Reading

Auclair J and Accolas JP (1983) Use of thermophilic lactic starters in the dairy industry. *Antonie van Leeuwenhoek* 49: 313–326.

Cerning J (1990) Exocellular polysaccharides produced by lactic acid bacteria. *FEMS Microbiol. Rev.* 87: 113–130.

Farrow JAE and Collins MD (1984) DNA base composition. DNA–DNA homology and long-chain fatty acid studies on *Streptococcus thermophilus* and *Streptococcus salivarius*. *J. Gen. Microbiol.* 130: 357–362.

Grossiord B, Vaughan EE, Luesink E and de Vos WM (1998) Genetics of galactose utilization via the Leloir pathway in lactic acid bacteria. *Lait* 78: 77–84.

Hardie JM (1986) Genus *Streptococcus*. In: Sneath PHA, Mair NS, Sharpe ME and Holt JG (eds), *Bergey's Manual of Systematic Bacteriology*. Vol. 2, p. 1043. Baltimore: Williams and Wilkins.

Hutkins RW and Morris HA (1987) Carbohydrate metabolism by *Streptococcus thermophilus*: a review. *J. Food Prot.* 50: 876–884.

Hutkins RW and Ponne C (1991) Lactose uptake driven by galactose efflux in *Streptococcus thermophilus*: evidence for a galactose–lactose antiporter. *Appl. Environ. Microbiol.* 57: 941–944.

Mustapha A, Hutkins RW and Zirnstein GW (1995) Cloning and characterization of the galactokinase gene from *Streptococcus thermophilus*. *J. Dairy Sci.* 78: 989–997.

Ottogalli G, Galli A and Dellaglio F (1979) Taxonomic relationships between *Streptococcus thermophilus* and some other streptococci. *J. Dairy Res.* 46: 127–131.

Poolman B, Royer TJ, Mainzer SE and Schmidt BF (1989) Lactose transport system of *Streptococcus thermophilus*: a hybrid protein with homology to the melibiose carrier and enzyme III of phosphoenolpyruvate-dependent phosphotransferase systems. *J. Bacteriol.* 171: 244–253.

Poolman B, Royer TJ, Mainzer SE and Schmidt BF (1990) Carbohydrate utilization in *Streptococcus thermophilus*: characterization of the genes for aldose 1-epimerase (mutarotase) and UDP glucose 4-epimerase. *J. Bacteriol.* 172: 4037–4047.

Schliefer KH, Ehrmann M, Krusch U and Neve H (1991) Revival of the species *Streptococcus thermophilus* (ex Orla-Jensen 1919) nom. rev. *System. Appl. Microbiol.* 14: 386–388.

Schroeder CJ, Robert C, Lenzen G and McKay LL (1991) Analysis of the *lacZ* sequences from two *Streptococcus thermophilus* strains: comparison with the *Escherichia coli* and *Lactobacillus bulgaricus* β-galactosidase sequences. *J. Gen. Microbiol.* 137: 369–380.

Somkuti GA, Solaiman DKY, Johnson TL and Steinberg DH (1991) Transfer and expression of a *Streptomyces* cholesterol oxidase gene in *Streptococcus thermophilus*. *Biotechnol. Appl. Biochem.* 13: 238–245.

Stingele F, Neeser J and Mollet B (1996) Identification and characterization of the *eps* (exopolysaccharide) gene cluster from *Streptococcus thermophilus* Sfi6. *J. Bacteriol.* 178: 1680–1690.

Zourari A, Accolas JP and Desmazeaud MJ (1992) Metabolism and biochemical characteristics of yogurt bacteria. A review. *Lait* 72: 1–34.

STREPTOMYCES

Arun Sharma, Food Technology Division, Bhabha Atomic Research Centre, Mumbai, India

Introduction

Streptomyces is the type genus of the family Streptomycetaceae belonging to the order Actinomycetales of the class Schizomycetes. The bacteria belonging to this genus are mainly found in soil but are also occasionally isolated from manure and other sources. Though the *Streptomyces* are eubacteria, they grow in the form of filaments or as mycelium and do not show the usual bacterial bacillary or coccoid forms. They also form conidia, which are produced in chains from spore-bearing aerial hyphae. *Streptomyces* species show a Gram-positive reaction. They have DNA with a GC value of 69–78%. More than 500 species of *Streptomyces* are listed in *Bergey's Manual of Determinative Bacteriology.*

Streptomyces have a complex colony structure based on multinucleate, branching mycelia, with differentiation of the colony into vegetative and reproductive structures. This complex multicellular morphology led earlier microbiologists to believe that the actinomycetes were fungi or intermediate links between fungi and bacteria. However, it is now proved beyond doubt that Streptomycetes are prokaryotes. Their cell wall structure, genetic material and phages are similar to those of bacteria.

Characteristics of the Genus

The *Streptomyces* develop as fully mycelial organisms, and reproduce by the formation of immotile spores at the tips of the aerial hyphae. Nutritionally *Streptomyces* are not fastidious, and generally do not require special growth factors. Most isolates produce extracellular hydrolytic enzymes that permit utilization of polysaccharides, proteins and fats. Most species belonging to the genus are aerobic, psychrophilic or mesophilic saprophytes which are frequently found in soil. Some *Streptomyces* are thermophilic. A few of them are also parasitic on plants and animals. *Streptomyces* produce slender, branched hyphae, with or without cross walls, 0.5–2.0 μm in diameter. They grow with an extensively branched primary or substrate mycelium, and more or less abundant aerial or secondary mycelium. On nutrient media, colonies are small (1–10 mm diameter), at first with a rather smooth surface but later with a weft of aerial mycelium that may appear granular, powdery or velvety. The many species and strains of the organism produce a wide variety of pigments that colour the mycelium, spores and substrate. They also produce one or more antibiotics against bacteria, fungi, algae, viruses, protozoa or tumour tissues. All species are primarily soil inhabitants.

The type species of the genus is *Streptomyces albus*. *Streptomyces* species show differentiation which is uncommon in bacteria: two types of mycelia, spore-bearing structures, and spores are produced.

Pigment Production

Streptomyces species produce a range of pigments which are generally water-soluble. Pigment production has been used as a taxonomic marker for species identification. Pigment production is lacking in *S. albus*, whereas a faint brown pigment may be produced in protein media by species such as *S. longisporus* and *S. rochei*. Soluble blue pigment is produced by *S. coelicolor*, *S. pluricolor*, *S. cyaneus* and *S. violaceus*. The pigment may be initially red or yellowish red, and then change to blue. In other species such as *S. verne* and *S. viridans*, it may first appear green, turning to brown. Yellow or golden-yellow pigments are produced by *S. flaveolus*, *S. parvus*, *S. rimosus*, *S. aureofaciens* and *S. xanthophaeus*. The characteristic pigment is produced either in the substrate mycelium or aerial mycelium or the spores of the species.

Substrate Mycelium

The substrate or primary mycelium is a loose network of hyphae which grows by extensive branching on the surface of the substrate. Substrate mycelium may have a different colour in different species of *Streptomyces*. The characteristic colour of the substrate mycelium of some *Streptomyces* species is given below:

- orange turning to red – *S. aurantiacus*
- red turning to blue – *S. californicus*
- red turning to violet – *S. coelicolor*, *S. cyaneus*
- green turning to olive – *S. olivoviridans*
- grey turning to black – *S. hygroscopicus*.

Aerial Mycelium

The aerial mycelium bearing the characteristic pigment may be abundantly present in some species and scant in others. The aerial mycelium is more closely packed than the substrate mycelium. In those species that do not form conidia, the aerial mycelium may be altogether absent; *Streptomyces verne* usually does not possess aerial mycelium. Many species form spirals at the ends of aerial mycelium; these include *S. coelicolor*, *S. albus*, *S. longisporus* and *S. diastaticus*. Others, such as *S. globisporus*, *S. anulatus*, *S. vinaceus* and *S. cinnamonensis*, are not known to produce spirals. Aerial mycelium may be profuse in some species, for example *S. albus*. It may be produced in concentric zones as in *S. anulatus*, in tufts as in *S. viridoflavus*, or in whorls as in *S. reticuli*, *S. verticillatus* and *S. netropsis*. Morphogenetic mutants of *S. coelicolor* such as *bld* lack aerial mycelium. The colour of the aerial mycelium may vary from species to species:

- grey – *S. clavulogenes*, *S. violaceoruber*
- white – *S. somaliensis*
- red – *S. erythrogriseus*
- yellow – *S. fradiae*
- blue – *S. ipomoeae*
- violet – *S. mauvecolor*.

The genus can be divided into five main groups based on the structure of sporulating hyphae:

1. straight, sporulating hyphae without spirals
2. straight, spore-bearing hyphae arranged in clusters
3. spiral formation in aerial mycelium with long, open spirals
4. spiral formation in aerial mycelium with short, compact spirals
5. spore-bearing hyphae arranged on mycelium in whorls or tufts.

Spore Formation

Streptomyces forms chains of spores from the ends of the aerial mycelium, known as sporophores or conidiophores. At maturity the aerial mycelium shows characteristic modes of branching and transforms into sporophores which form chains of three or more spores. The spores are also called conidia, conidiospores or arthrospores. Spores contribute to the survival of species over long periods of drought and other unfavourable conditions.

Sporulation in *Streptomyces* is quite different from sporulation in spore-forming bacilli, and has been studied in detail. Sporulation in *S. coelicolor* is reported to have four stages. In stage 1 long cells of the aerial mycelium become coiled. In stage 2 sporulation septa are synchronously formed at regular intervals within these cells, each by the ingrowth of a double annulus continuous with the cell membrane and wall. This process is morphologically different from the formation of cross walls in substrate and aerial hyphae normally formed during cell division. After completion of the sporulation septum, the laying down of thick spore wall begins. In stage 4, which merges with stage 3, the cylindrical spore compartments become ellipsoidal. With the disintegration of the old cell wall external to the spore wall, the sporulation is complete. Ultimately the spore chains are joined at a very small interface and by the remnants of the fibrous sheath. The spores of *Streptomyces* are not very resistant to heat and do not contain dipicolinic acid. (Dipicolinic acid is found in association with the spores of bacilli and those of thermoactinomycetes, and imparts heat resistance.) The *whi* mutants lacking spore formation have been studied in detail.

In *S. coelicolor* after spore germination – which takes about 4 h at 30°C – a branched network of hyphae develops in the agar and on its surface, giving a colony of about 1 mm diameter after 48 h. At this stage the colony is somewhat bald and shiny. As the aerial mycelium develops it becomes hairy. The white colony gradually turns grey, the colour being the property of the mature spores. Regulation of morphogenesis in *S. coelicolor* has been studied in detail using molecular genetic methods. The genetics and life cycle of *Streptomyces* have been thoroughly studied by David Hopwood and his colleagues at the John Innes Institute, Norwich, UK. Spore colour may also vary from species to species:

- yellow or grey – *S. griseus*, *S. coelicolor*
- pink or light violet – *S. fradiae*, *S. toxitricini*
- brown – *S. fragilis*
- blue-green – *S. glaucescens*
- white – *S. albus*.

A few *Streptomyces* form sclerotia in which the cells are cemented together by a material that contains L-2,3-diaminopropionic acid. These organisms have been named *Chainia*. The ability to form sclerotia is reported to be lost during laboratory culture.

Thermophilic *Streptomyces*

Most species of *Streptomyces* are mesophilic. Thermophilic *Streptomyces* have an optimum growth temperature above 50°C; *Streptomyces thermophilus* has an optimum growth temperature of 50°C. Compost, manure heaps and fodders are common habitats of thermophilic *Streptomyces*. Some strains produce an antibiotic, thermomycin. *Streptomyces thermodiastiticus* and *S. thermofuscus* have an optimum temperature of 65°C. *Streptomyces casei* isolated from pasteurized cheese is reported to be highly resistant to high temperatures and to disinfectants; its thermal death point is 100°C.

Isolation of *Streptomyces*

Soil is the natural habitat of *Streptomyces*. These organisms are abundant in soil and are largely responsible for the odour of damp soil. This odour is due to the production of a number of volatile substances known as geosmins. The substances are sesquiterpenoid compounds, e.g. *trans*-1,10-dimethyl-*trans*-9-decalol. Geosmins are also produced by some cyanobacteria.

Isolation of *Streptomyces* from soil is relatively easy. A suspension of soil in sterile water is diluted and spread on a selective agar medium. The selective medium often contains the usual organic salts, a source of carbon such as starch, asparagine or calcium malate, and a source of nitrogen such as undigested casein or an organic nitrate. The medium can be enriched by the addition of calcium carbonate. Treatment of soil with phenol (1.5% for 10 min) can eliminate other bacteria and fungi. Alternatively antibiotics such as rifampicin, cycloheximide or nystatin can be added. Rose bengal can be used to curtail spreading growth of fungi. Composition of some suitable media is given in **Table 1**.

Trace elements may be added if required in the synthetic media. The plates are incubated at 25°C for 6–9 days. After incubation the plates are examined for characteristic colonies of *Streptomyces*. *Streptomyces* colonies on laboratory media often smell very strongly of earth. The colonies are typically firm and compact in early stages of development and may be difficult to subculture. Later, aerial hyphae showing loose cottony growth and bearing spores can be easily picked up. The spores of interesting colonies can be streaked and isolated. The dusty appearance, their compact nature and characteristic colour make detection of *Streptomyces* colonies on agar plates relatively easy. Identification of *Streptomyces* is also made easy by the zone of clearance generally found around the colonies.

Table 1 Selective agar media for the isolation of *Streptomyces*

Constituent	*Amount*
Glucose–yeast extract–malt extract agar medium	
Glucose	4.0 g
Yeast extract	4.0 g
Malt extract	10.0 g
Calcium carbonate	2.0 g
Agar	12.0 g
Distilled water	1.0 l
pH	7.2
Starch agar medium	
Starch	10.0 g
$(NH_4)_2SO_4$	2.0 g
K_2HPO_4	1.0 g
$MgSO_4 \cdot 7H_2O$	1.0 g
NaCl	1.0 g
Calcium carbonate	2.0 g
Distilled water	2.0 g
pH	7.0–7.4
Glycerol–asparagine agar medium	
Glycerol	10.0 g
L-Asparagine	1.0 g
K_2HPO_4	1.0 g
Agar	12.0 g
Distilled water	1.0 l
pH	7.0–7.4

Pathogenicity of *Streptomyces*

Plant Diseases

The group of diseases known as bacterial scab includes mainly diseases that affect the parts of plants below ground. The symptoms consist of more or less localized scabby lesions, affecting primarily the outer tissues of these parts. Scab bacterial diseases include common scab of potato and other below-ground crops caused by *S. scabies* and related species, and soil rot or pox of sweet potato by *S. ipomoeae* and related species. The scab bacteria survive in infected plant debris and in the soil, and penetrate tissues through natural openings or wounds. In the tissue these bacteria grow in the intracellular spaces of the parenchymal cells, but later these cells break down and are invaded. In a typical scab, the healthy cells below and around the lesion divide and form layers of corky cells. These cells push the infected tissues outward and give the scabby appearance. Scab lesions often serve as the point of entry for secondary and opportunistic pathogens and saprophytes, which may result in rotting of the crop.

Common Scab of Tuber and Root Crops Common scab of potato caused by *S. scabies* occurs throughout the world. It is most prevalent in neutral or slightly alkaline, light sandy and dry soils. The same pathogen can cause scab of garden beets, sugar beets, radish and other crops. The disease is generally superficial; reducing the value rather then the yield of the crop, unless the infection is severe. The symptoms of common scab of potato are observed mostly on tubers. At first they consist of small, brownish, slightly raised spots, but later they may enlarge, coalesce and become very corky.

The pathogen *S. scabies* is a hardy parasite that can survive indefinitely either in its vegetative mycelial form or in the form of spores in most soils except those that are extremely acidic. The vegetative form consist of slender (about 1 μm thick), branched mycelium without a few or no cross walls. The spores are cylindrical or ellipsoid, about 0.6 μm by 1.5 μm, and are produced on spiral hyphae that develop cross walls from the tip towards the base. As the cross walls constrict, spores are pinched off. The spores germinate by means of one or two germ tubes, which develop into mycelium. The pathogen is spread through soil and water and penetrates tissue through lenticels, wounds and stomata in young tubers.

The severity of scab infection worsens with increase in soil pH from 5.0 to 8.0. The optimum temperature for disease development is 20–22°C. Potato scab incidence is reported to be low in moist soil. Control of common scab of potato is through the use of certified scab-free seeds, or through seed treatment with pentachloronitrobenzene (PCNB) or maneb-zinc dust. Other *Streptomyces* reportedly associated with scab include *S. clavifer*, *S. fimbriatus*, *S. carnosus* and *S. craterifer*. The aetiological agent of the scab of sweet potato is *S. ipomoeae*; *S. poolensis* and *S. intermedius* are also reported to be associated with this disease. *Streptomyces tumuli* is found to be associated with the scab of sugar beets.

The phenomenon of antibiosis in soil has attracted the attention of plant pathologists for a long time. *Streptomyces* species have been used to control certain plant diseases owing to their antagonistic effects caused by the production of antibiotics against the soil pathogens.

Disease in Humans and Animals

Only a few *Streptomyces* species have been isolated from pathological material. However, their role as agents of infectious disease cannot be ignored. *Streptomyces somaliensis* is known to cause actinomycetomas in humans. The *Streptomyces* isolated from animal bodies and tissues include *S. listeri*, *S. galtieri* and *S. hortonensis*, which show limited proteolytic action in gelatin and milk. *Streptomyces somaliensis*, *S. kimberi* and *S. beddardii* are strongly proteolytic species of animal origin. *Streptomyces* species are also reported to be respiratory allergens in humans.

Spoilage of Food

Members of the genus are reported to cause undesirable odours and colour changes in foods. Typical musty or earthy odours and tastes in food could be attributed to the activity of *Streptomyces* in food or their growth in its vicinity. The presence of *Streptomyces* in freshwater environments and drinking water supplies is also common and is a source of contamination of foods.

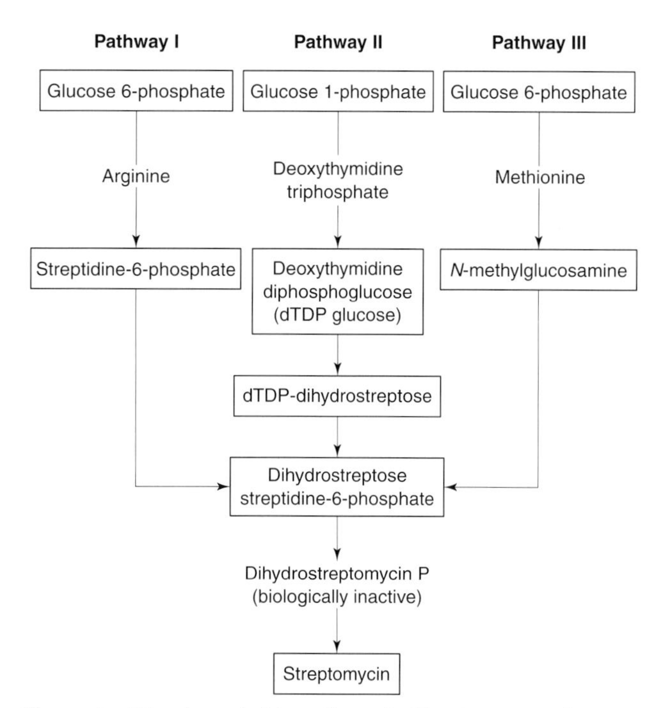

Figure 1 Streptomycin biosynthesis in *Streptomyces griseus*.

Antibiotics of *Streptomyces*

One of the most striking properties of *Streptomyces* is its capacity to produce antibiotics. This is often seen on agar plates from which the isolation is carried out. The surrounding clear zone around a colony points to its antibiotic-producing capacity. Owing to their great antibiotic-producing potential and the search for new antibiotics, much research effort has gone into the study of the genetics of antibiotic production. The production of antibiotics by *Streptomyces* is a complex process. For example, the three parts of the streptomycin molecule – streptidine, streptose and *N*-methylglucosamine – are formed by three different pathways and then joined together to form the antibiotic (**Fig. 1**). Production of this antibiotic in *S. griseus* is regulated by an inducer called factor A. The key enzymes in the pathway of streptomycin biosynthesis are not synthesized until the factor A concentration builds up. Thus factor A acts as a trigger. Similarly, the biosynthesis of the antibiotic tetracycline involves a large number of enzymatic steps. In the case of chlortetracycline as many as 72 steps and 300 genes are involved. Regulation of such an antibiotic would be very complex. More than 500 antibiotics are known to be produced by *Streptomyces* species and many more are likely to be discovered in the future. More than 50 antibiotics find practical application in human and veterinary medicine (**Table 2**). Besides antibacterial and antifungal products, *Streptomyces* species also form antiviral compounds such as ara A and tunicamycin; antitumour com-

Table 2 Common antibiotics produced by *Streptomyces*

Common name	*Producer organism*	*Antibiotic spectrum*
Streptomycin	*S. griseus*	Most Gram-negative bacteria
Spectinomycin	*Streptomyces* spp.	*Mycobacterium tuberculosis*, *Neisseria gonorrhoeae*
Neomycin	*S. fradiae*	Broad spectrum (topical application)
Tetracycline	*S. aureofaciens*, *S. rimosus*	Broad spectrum
Chlortetracycline	*S. aureofaciens*	Broad spectrum
Erythromycin	*S. erythreus*	Frequently used in place of penicillin
Lincomycin	*S. lincolnensis*	Obligate anaerobes
Nystatin	*S. noursei*	Antifungal
Chloramphenicol	*S. venezuelae*	Broad spectrum
Amphotericin	*S. nodosus*	Antifungal

pounds such as daunorubicin, mitomycin C, and actinomycin; antiparasitic compounds such as hygromycin, monensin and salinomycins; insecticidal compounds such as avermectins; and weed control compounds such as bialphos. They also produce several enzymes of industrial use, and enzyme inhibitors and immunomodifiers for therapeutic use.

The reasons for such profuse antibiotic production by *Streptomyces* are not yet clear. The antibiotics produced are typical secondary metabolites produced toward the end of the exponential phase of growth. Antibiotic production may be related to sporulation in *Streptomyces*.

Mode of Action of Some Antibiotics

The aminoglycoside antibiotics such as streptomycin and neomycin produced by *S. griseus* and *S. fradiae* inhibit 30S ribosome functions. Tetracyclines produced by *S. aureofaciens* inhibit binding of aminoacyl tRNAs to ribosomes. Polyenes such as amphotericin B and nystatin, produced by *S. nodosus* and *S. noursei* respectively, inactivate membranes containing sterols. Chloramphenicol produced by *S. venezuelae* inhibits the translation step of ribosome function. Macrolide antibiotics such as erythromycin and carbomycin, produced by *S. erythreus* and *S. halstidii* respectively, inhibit 50S ribosome function.

Phages and Plasmids of *Streptomyces*

All eubacteria are at risk of attack by bacteriophages: those that attack *Streptomyces* are called 'actinophages'. One actinophage may attack more than one species (polyvalent). Actinophages have been used to classify *Streptomyces* (phage typing). Industrial antibiotic-producing strains of *Streptomyces* should be resistant to actinophage attack. Several *Streptomyces* species have been found to contain large linear plasmids over 500 kb in length; some of the genes involved in antibiotic production are coded by these plasmids. Linear plasmids, although found in other bacteria, are widespread among *Streptomyces*.

See also: **Spoilage Problems**: Problems caused by Fungi.

Further Reading

Agrios GN (1988) *Plant Pathology*. London: Academic Press.

Breed RS, Murray EGD and Smith NR (eds) (1957) *Bergey's Manual of Determinative Bacteriology*, 7th edn. Baltimore: Williams & Williams.

Chater KF and Wibb MJ (1997) Regulation of bacterial antibiotic production. In: Kleinkoff H and von Dohren H (eds) *Products of Secondary Metabolism*. Biotechnology 6. Weinheim: VCH.

Felicitas K and Kutzner J (1991) The Family Streptomycotaceae. In: Balows A, Truper HG, Dworkin M, Hardr W and Schiefer KH (eds) *The Prokaryotes*. Berlin: Springer.

Goodfellow MM, Mordarski MM and Williams ST (eds) (1983) *The Biology of Actinomycetes*. London: Academic Press.

Holt JG (ed.) (1994) *Bergey's Manual of Determinative Bacteriology*. Williams and Wilkins, Baltimore.

Hopwood DA (1967) Genetic analysis and genome structure in *Streptomyces coelicolor*. *Bacteriological Review* 31: 373–403.

Kalaloutskii LV and Agre NS (1976) Comparative aspects of developments and differentiation in Actinomycetes. *Bacteriological Review* 40: 469–524.

Piepersberg W (1995) Streptomycin and related aminoglycoside antibiotics. In: Vining LC and Stuttard C (eds) *Biochemistry and Genetics of Antibiotic Biosynthesis*. Stoneham: Butterworth-Heinemann.

Shirling EB and Gottlieb D (1968) Cooperative description of type cultures of Streptomyces. II. Species description from first study. *Journal of Systematic Bacteriology* 18: 69–189.

Sulphur Dioxide *see* **Preservatives**: Permitted Preservatives – Sulphur Dioxide.

T

THERMUS AQUATICUS

C K K Nair, Radiation Biology Division, Bhabha Atomic Research Centre, Mumbai, India

The bacteria belonging to the genus *Thermus* are aerobic, thermophilic, Gram-negative and rod-shaped, and have attracted considerable attention because of the biotechnological potential of their thermostable enzymes, particularly their DNA polymerases which constitute the most important component of the polymerase chain reaction. These bacteria are isolated from neutral to alkaline hot springs of various geographical locations in the world.

Thermus aquaticus was first discovered by Thomas D. Brock in the 1960s during a long-term study of microbial life in hot springs of Yellowstone National Park in Wyoming, USA, as a pigmented rod-shaped Gram-negative bacteria capable of growing at temperatures as high as 79°C. After the discovery of this extremophile, search for microbes in hot springs and in environments around deep sea hydrothermal vents (called smokers, which are natural undersea chimneys through which super-heated mineral-rich fluid as hot as 350°C escapes) led to the isolation of a number of species belonging to the genus *Thermus*. Some of these hyperthermophiles are listed in **Table 1**. The cell walls of the genus *Thermus* has a characteristic peptide-glycan composition.

Some of the species have distinct features such as halotolerance (e.g. *T. thermophilus*), non-pigment formation (e.g. *T. scotoductus*), and filamentous nature (*T. filiformis*). However, the classification of the various species of the genus *Thermus* has been primarily based on whole genome DNA : DNA hybridization, as most biochemical and taxonomic parameters were found unsuitable. Recently, molecular biology techniques involving characterization of genomic macro-restriction fragment length polymorphism (RFLP) using pulsed-field gel electrophoresis (PFGE) and characterization of rDNA genes, ribotyping by restriction fragment length polymorphism, have been used in taxonomic studies and classification. Among the various species, *Thermus aquaticus* is the most widely studied and well characterized.

Table 1 Different species of hyperthermophiles belonging to the genus *Thermus* and their site of isolation

Isolate	*Site of isolation*
Thermus aquaticus	Yellowstone National Park, USA
T. brockianus	Yellowstone National Park, USA
T. filiformis	Rotorea, New Zealand; Takaanu, New Zealand
T. thermophilus	Mine, Japan; Izu-Atagawa, Japan; Hruni, Iceland; Savusavu beach, Fiji; Stream Is S. Miguel, Azores
T. scotoductus	Selfoss, Iceland; Bloomington, USA; London, UK; Calasde Vizele, Portugal
T. oshimai	S. Pedrodosal, Portugal; Hveresarli Hengill, Iceland

Characterization of *Thermus aquaticus*

A number of strains of *T. aquaticus* have been isolated from different thermal springs. This organism has also been isolated from artificial environments such as hot tap water in geographical locations quite distant from thermal springs.

T. aquaticus is a Gram-negative, non-sporulating, and non-motile rod of 0.5–0.8 μm diameter and 5.0–10.0 μm length. It forms long filaments of 20.0 μm or greater in stationary phase or at supra-optimal temperatures. The rods occur singly or in aggregates, by individual rods joining end to end or side to side. Large spheres of 10–20 μm in diameter are often formed in old cultures. Deoxyribonucleic acid base composition of *T. aquaticus* is 65.4–67.4 mol% G+C. The growth of this bacteria is inhibited by fairly low concentrations of cycloserine, streptomycin, penicillin, novobiocin, tetracycline and chloramphenicol. The growth of this organism is also inhibited by 100 $\mu g\ ml^{-1}$ sodium lauryl sulphate, 500 $\mu g\ ml^{-1}$ sodium azide, 200 $\mu g\ ml^{-1}$ oxgall or 2% NaCl. Nutritional studies show that *T. aquaticus* does not require vitamins or amino acids although the growth is considerably faster in enriched medium than in synthetic basal salts medium (**Table 2**). With 0.1–0.33% tryptone and yeast extract in basal medium, the organism

Table 2 Media used for the isolation of *Thermus aquaticus*

Basal salts medium	Nitrilotriacetic acid	0.1 g
	Micronutrient solution	0.5 ml
	$FeCl_3$ solution	1.0 ml
	$CaSO_4.2H_2O$	0.06 g
	$MgSO_4.7H_2O$	0.1 g
	NaCl	0.008 g
	KNO_3	0.103 g
	$NaNO_3$	0.689 g
	Na_2HPO_4	0.111 g
	Dissolve in 1 l distilled water, adjust to pH 8.2 with NaOH.	
Micronutrient solution	H_2SO_4, conc.	0.5 ml
	$MnSO_4.H_2O$	2.28 g
	$ZnSO_4.7H_2O$	0.5 g
	H_3BO_3	0.5 g
	$CuSO_4.5H_2O$	0.025 g
	$Na_2MoO_4.2H_2O$	0.025 g
	$CoCl_2.6H_2O$	0.045 g
	Dissolve in 1 l distilled water	
$FeCl_3$ solution	$FeCl_3$	0.2905 g
	Dissolve in 1 l distilled water	

grows well but at 1% tryptone and yeast extract there was inhibition of growth. The bacterial growth was good in 1% vitamin-free casein hydrolysate and 0.5% glutamic acid, as sole source of nitrogen and carbon. Ammonium as source of nitrogen and with acetate, succinate, sucrose, glucose or fructose as source of carbon supports the growth of this bacterium. *T. aquaticus* is an obligate aerobe; it has a pH optimum of 7.5–7.8 and does not grow below pH 6.0 and above pH 9.5. The optimum temperature for growth is 70°C, with a maximum at 79°C and minimum at 40°C. The generation time of this organism under optimum conditions is about 50 min. At temperatures 75°C and above the organism exhibits filamentous growth. Filamentous forms are also observed in stationary cultures after growth at 65–70°C. In older cultures large spherical bodies are often formed due to extrusion of the protoplasts or long rods or filaments from one of the poles, and also filaments with swollen ends can be seen frequently in these cultures. The rod-shaped forms have a tendency to aggregate either as linear arrays or as rosettes. The aggregation phenomenon is due to the presence of a slime on the surface of the organism.

Isolation and Enrichment

Thermus aquaticus can be isolated from soil or water from hyperthermal environments. Basal salts medium (Table 2) with 0.1% tryptone and 0.1% yeast extract in cap tubes is inoculated with the samples and incubated unshaken at 75°C for 1–2 days.

The growth of the bacterium is indicated by visible turbidity. When the turbidity is heavy the colour of the bacterial mass turns yellow or orange. Under the microscope the organism appears as rods, large filaments and/or large spheres. As this organism is a non-spore-former, cultures with spores are discarded. Pure cultures can be isolated by streaking on enrichment plates containing the same medium solidified with 3% agar. The plates are sealed with Saran Wraps and incubated for 1–2 days at 70°C to obtain yellow–orange colonies. These colonies are re-streaked and stock cultures are prepared as agar slants from the isolated colonies of the second plate. For longer storage the isolates are preserved by freeze-drying.

Biochemical Tests

Acid without gas is produced when *T. aquaticus* is grown on rich media with glucose, fructose, mannose, sucrose or maltose. Xylose is not utilized by the organism. *T. aquaticus* coagulates Brom Cresol Purple (BCP)-containing milk with slight acid formation and liquifies gelatin. Potassium nitrate is weakly reduced to nitrite and formation of ammonia is weakly positive.

Biotechnological Value of *Thermus aquaticus*

In the past 15–20 years the considerable value of enzymes produced by thermophilic organisms including *T. aquaticus* has been realized. Although their ability to survive and moreover flourish in environments that can exceed 95°C suggests a robust set of enzymes, not all of them are thermostable. Likewise not all enzymes recovered from mesophiles are thermolabile. The intrinsic properties that make a given enzyme thermostable are not fully understood. Efforts to classify enzymes based on structural motifs as well as other features including numbers of salt bridges, surface-to-volume ratio and disulphide linkages have had only limited success. Given the evolutionary diversity between enzymes isolated from thermophiles and mesophiles, it is difficult to correlate any particular amino acid difference to thermostability. Nevertheless, in general, thermophilic organisms tend to be a valuable source of thermostable enzymes. *T. aquaticus* is perhaps one of the best studied thermophiles or extremophiles and has been the source of a number of valuable enzymes. Thermostable enzymes are useful in a number of biotechnological processes since they typically have faster catalytic rates, they can be operated at high temperatures where contamination is suppressed and they have a longer half-life which makes them less expensive to use. In other applications, thermostable enzymes are a disadvantage as they cannot be inactivated and they can continue to affect the food product well after the optimal product composition is reached.

One of the most cited examples of a process whereby the inclusion of a thermostable enzyme dramatically improved the process is the use of the *T. aquaticus* DNA polymerase in the polymerase chain reaction (PCR). Although the original versions of PCR were accomplished using thermolabile *Escherichia coli* DNA polymerase, the process was compromised by the need to add enzyme after each step and the build up of inactive enzyme. The *T. aquaticus* DNA polymerase eliminated the need to add enzyme after each step. PCR is described in detail elsewhere and its application for the detection of food-borne pathogens holds the promise of revolutionizing food safety testing.

In addition to the *T. aquaticus* DNA polymerase, other enzymes from this organism have significantly improved a host of other biotechnological applications. For example, *T. aquaticus* DNA ligase is used in the ligase chain reaction (LCR). LCR is an amplification-based process in which a pair of adjacent oligonucleotides in a set of diametrically opposed oligonucleotide primer pairs are joined in a series of sequential ligation steps. The adjacent primers are only joined when the nucleotide at the junction matches the targeted 3′ end of the upstream primer. LCR is therefore diagnostic for that nucleotide and has been used to format assays to discriminate between two targets that differ only in a single nucleotide. Another *T. aquaticus* enzyme of value is *Taq*I, a restriction endonuclease that shares thermostable properties along with other DNA modifying enzymes from this organism. Aqualysin is a thermostable serine protease isolated from *T. aquaticus* and may have applications in food processing where its high heat stability is valuable. Other thermostable catalytic enzymes of potential value from *T. aquaticus* are superoxide dismutase and fumarase. Finally, a β-galactosidase has been isolated which may have value to hydrolyse lactose as well as being used for synthetic reactions to create oligosaccharides.

Regulation

In addition to hot springs, this bacterium has also been found in hot water taps and other thermal niches. Thus this organism may be considered as a predictor for thermal pollution. There have been no reports of illness due to *T. aquaticus* and no toxic material or poison has so far been isolated from this organism.

See also: **PCR-based Commercial Tests for Pathogens**.

Further Reading

Bej AK, Mahbubani MH and Atlas RM (1991) Amplification of nucleic acids by polymerase chain reaction (PCR) and other methods and their applications. *Critical Reviews in Biochemistry and Molecular Biology* 26: 301–334.

Brock TD (1967) Life at high temperature. *Science* 197: 1012–1019.

Brock TD and Freeze H (1969) *Thermus aquaticus* gen. n. and sp. n, a non-sporulating extreme thermophile. *Journal of Bacteriology* 98: 289–297.

da Costa MS, Duarte JC and Williams RAD (eds) (1989) *Microbiology of the Extreme Environments and Its Potential for Biotechnology*. London: Elsevier Applied Sciences.

Madigan MT and Marrs BL (1997) Extremophiles. *Scientific American* April 66–71.

Moreira LM, da Costa MS and Correia IS (1997) Comparative genomic analysis of isolates belonging to the six species of the genus *Thermus* using pulse field gel electrophoresis and ribotyping. *Archives of Microbiology* 168: 92–101.

Saiki T, Kimura R and Arima K (1972) Isolation and characterization of extremely thermophilic bacteria from hot springs. *Agricultural and Biological Chemistry* 36: 2357–2366.

Sato S, Hutchinson III, CA and Harris JI (1997) A thermostable sequence specific endonuclease from *Thermus aquaticus*. *Proceedings of the National Academy of Sciences USA* 74: 542–546.

Williams RAD and da Costa MS (1992) The Genus *Thermus* and Related Microorganisms. In: Balows A, Truper HG, Dworkin M, Harder W and Schleifer KH (eds) *The Prokaryotes*, 2nd edn. P. 3745. New York: Springer.

Williams RAD, Smith KE, Welch SG and Micallef J (1996) *Thermus oshimai* sp. nov., isolated from hot springs in

Portugal, Iceland, and the Azores, and comment on the concept of a limited geographical distribution of *Thermus* species. *International Journal of Systematic Bacteriology* 46: 403–408.

TORULOPSIS

Rolf K Hommel, CellTechnologie Leipzig, Germany

Hans-Peter Kleber, Institut für Biochemie, Fakultät für Biowissenschaften, Pharmazie und Psychologie, Universität Leipzig, Germany

Introduction

Torulopsis yeasts are a heterogenous collection of asporogenous species of the family Cryptococcaceae. They have diverse properties but have in common that they are imperfect (asporogenous) yeasts that do not form ascospores or basidiospores. Their reproduction is by multipolar budding and the haploid or diploid yeasts do not form ascospores, teliospores, ballistospores, endospores or arthrospores. The perfect state of most species is unknown. In general, the cells appear as globular or ovoid, more rarely elongated. Short fimbriae, about 0.5 μm long, may be present. There is no visible pigmentation. Some strains produce extracellular polysaccharides, which do not react positively with iodine. Inositol is not used as a carbon source. Weak alcoholic fermentation may occur.

Torulopsis spp. reveal many cross-connections to the genus *Candida*. The distinction is based on the pseudomycelium formation, which is well developed in *Candida* and absent or rudimentary in *Torulopsis*. Most strains of both *Candida* and *Torulopsis* cannot be resolved into natural taxa. The division into two genera is arbitrary and artificial. Both genera have been merged into *Candida*. *Candida* includes all yeast species that cannot be classified into other imperfect genera. **Table 1** surveys the taxonomic classification of selected former *Torulopsis* yeasts. In recent publications the term *Torulopsis* is still used; therefore, the old nomenclature is used here without reference to *Candida* or other genera.

Table 1 Taxonomic consideration of some examples of the former genus *Torulopsis*

Former taxonomic classification	*New taxonomic classification*
T. acris	*Cryptococcus albidus*
T. aeria	*Cryptococcus albidus*
T. apicola	*Candida apicola*
T. bovina	*Candia pintolopesii*
T. bombicola	*Candida bombicola*
T. candida	*Candida famata*
T. caroliniana	*Candida lactis-condensi*
T. cremoris	*Kluyveromyces marxianus*
T. colliculosa	*Torulospora delbrueckii*
T. dattila	*Kluyveromyces thermotolerans*
T. ernobii	*Candida ernobii*
T. glabrata	*Candida glabrata*
T. gropengiesseri	*Candida gropengiesseri*
T. halonitrophila	*Candida halonitrophila*
T. holmii	*Candida holmii*
T. lactis-condensi	*Candida lactis-condensi*
T. kefyr	*Candida kefyr*
T. mogil	*Zygosaccharomyces rouxi*
T. neoformans	*Cryptococcus neoformans*
T. shaerica	*Kluyveromyces marxianus*
T. stellata	*Candida stellata*
T. rasae	*Metschnikowia pulcherrima*
T. versatilis	*Candida versatilis*

Physiology

The temperature range of *Torulopsis* yeasts spans the whole scale from psychrophilic (upper limit for growth at or below 20°C), through mesophilic (temperature optimum of 20–50°C), to thermophilic (temperature optimum at or above 50°C). For example, *T. austromarina* and *T. psychrophila* are both obligate psychrophilics; examples of obligate thermophilics are *T. bovina* and *T. pintolopesii*. One mechanism of *T. bovina* to adapt to high temperature is a special membrane lipid composition: the lower the temperature, the higher the degree of lipid unsaturation. There is also a higher percentage (30–40%) of saturated fatty acids, as compared with mesophilic and psychrophilic yeasts. The latter contain approximately 90% unsaturated fatty acids, 55% of which is linolenic acid. There is differentiation with regard to b-type cytochromes. There are one, two and three b-type cytochromes in thermophilic, mesophilic and psychrophilic yeasts, respectively. *T. bovina* is characterized by a high cardiolipin (25% of the total phospholipid) and cytochrome oxidase content.

Most *Torulopsis* species are obligatory aerobic. Some can multiply under strict anaerobic conditions. The obligatory psychrophilic *T. psychrophila* and *T. austromarina* are obligate aerobes and are unable to grow anaerobically. In contrast, the obligatory thermophilic yeasts *T. bovina* and *T. pintolopesii* are facultative anaerobes. Both *T. bovina* and *T. pintolopesii* do not display conventional mitochondrial structures and may form respiration-deficient mutants spontaneously or drug induced. In *T. pintolopesii* and in its naturally occurring respiratory-deficient mutants, cytochrome oxidase, succinate oxidase and reduced nicotinamide adenine dinucleotide oxidase were absent. The spontaneous respiration-deficient mutants were similar to cytoplasmic petite mutants of *Saccharomyces cerevisiae*. Induction of respiration-deficient mutants, which lack cytochrome *c* oxidase, was also reported from *T. apicola* and *T. bombicola*. The mitochondria of drug or X-ray-induced respiratory-deficient mutants of *T. glabrata* did not contain mitochondrial DNA, indicating deficits in DNA repair.

Some *Torulopsis* yeasts can grow in environments that contain high concentrations of dissolved compounds and in low water activity media, and therefore are important in food spoilage. The water activity is defined as:

$$\ln a_w = -v\, m\varnothing / 55.5$$

where v equals the number of ions formed by each solute molecule, m is the molar concentration of the solute, and φ is the molar osmotic coefficient. The a_w value of pure water is 1.00 (a_w of foods is < 1.00) and decreases with increasing solute concentration. Below an a_w value of 0.95 xerotolerant (osmophilic/osmotolerant) yeasts, such as *T. candida*, *T. lactis-condensi* and *T. halonitratophila*, will grow. *T. halonitratophila* needs high salt concentrations to grow at elevated temperatures at which it normally cannot grow. One way of adaptation to low a_w values is by intracellular accumulation of polyols (glycerol, erythritol) on salt medium. Intracellular polyols reduce the difference of osmotic pressure across the plasma membrane keeping the intracellular enzymes functional. Extracellular accumulation of glycerol is known from *T. magnoliae* grown on high glucose medium.

The preferred pH-range for growth is at neutral pH and below. Specifically D-glucose, D-fructose and D-mannose as carbon and energy sources support growth of *T. pintolopesii* over the pH range of 2–8.

Biochemical Potency

In addition to the conventional sugar and polyol carbon sources, many *Torulopsis* yeasts accept unusual substrates like hydrocarbons as their sole source of carbon and energy, e.g. *T. bombicola*, *T. apicola*, *T. glabrata* and *T. gropengiesseri*. Subterminal oxidation of alkanes at position 2 by the latter species yields 2-hydroxy fatty acids. *T. candida* oxidizes the respective *n*-alkanes to dicarboxylic acids. The chain length of alkanes metabolized ranges from C_6 to C_{22} (except C_{11}). Other industrial paraffins, naphthenic acids and polycyclic aromatic hydrocarbons are also oxidized. Some of the enzymes involved in these metabolic pathways are inducible. Different types of cytochrome *P*-450 are described in *T. apicola*. Azole resistances of *T. glabrata* are closely related to the microsomal cytochrome *P*-450. Alcohol and aldehyde dehydrogenases have been characterized in *T. candida*, *T. bombicola* and *T. apicola*. Diterminal oxidation of alkanes is carried out by *T. gropengiesseri*; there is non-growth associated hydroxylation of oleic acid to 17-L-hydroxy fatty acid, e.g. fatty acids may be converted into triacylglycerol. *T. haemulonii* successfully grows on both fatty acids and animal lard. Co-cultured *T. holmii* and *Yarrowia lipolytica* gave a yield of 1 g biomass per gram of olein feedstock.

Some *Torulopsis* spp. excrete quantities of surface-active sophorose lipids into the medium if they are cultivated on hydrophobic substrates like alkanes, plant oils and fats. In these excreted lipids the sophorose is glycosidically linked to a long-chain ω- or (ω-1)-hydroxy fatty acid that mostly reflects the backbone of the hydrophobic carbon source. The free carboxylic residue may be lactonized with the 4″-hydroxyl group (**Fig. 1**). Similar compounds can be found in media with high concentrations of glucose. Sophorose lipids have antimicrobial and antiviral activities. In a human promyelocytic leukaemia cell line they induce differentiation into monocytes. Other products were obtained with some species, namely riboflavin from *T. candida*, D-glycero-D-manno-heptitol from *T. versatilis*, acetoin from *T. colliculosa* (up to 500 mg of acetoin per gram of substrate). Cane molasses, glucose and sucrose are suitable substrates for acetoin production.

Strains of *T. candida* and *T. glabrata* oxidize methanol. *T. molicshiana* assimilates methanol, ethanol and polyols.

Torulopsis yeasts have a weak fermentative capacity. In the stomach of neonates of the horse, dog, goat, sheep and newborn human babies ethanolic fermentation of glucose is carried out by *T. glabrata*. Extracellular enzymes, such as lipases, pectinases and cellobiose oxidase, have been reported from some species.

Habitats

Torulopsis yeasts have been isolated from a wide variety of environments. *T. austromarina* was isolated from the Antarctic ocean. *Torulopsis* yeasts have been detected in Guadeloupe coastal waters. Industrial

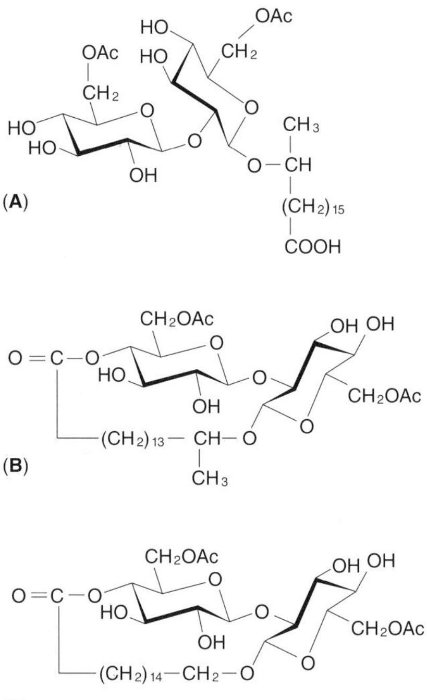

(**A**)

(**B**)

(**C**)

Figure 1 Sophorose lipids from *Torulopsis bombicola* (acidic form) (**A**) and lactonic sophorose lipids from *Torulopsis apicola* (**B**). Reprinted from Hommel and Ratledge (1993) with permission of Marcel Dekker Inc.

habitats are lubricants and fuels. *Torulopsis* spp. are present in a Russian water deposit with a low mineral content. *T. sorbophila* has been isolated from a guanosine monophosphate-manufacturing plant. Strains of *T. glabrata* with enhanced resistance to antibiotics have been found on bath surfaces, in sauna rooms, and in the water of swimming pools. This yeast was also present in the faeces and in the chlorinated water from bottle-nosed dolphins and in shellfish.

T. psychrophila has been found in the antarctic soil. Strains of *T. gaucheries* and *T. azyma* have been recovered from soil and rupicolous lichen in South Africa. *T. aeria* is present in the rhizosphere. Isolates from Amazonian soil samples contain 63% *Torulopsis* species, including *T. glabrata*.

Torulopsis strains have also been isolated from the air. *Torulopsis* yeasts have been found on plants, flowers and fruits; *T. pusula*, *T. bacarum* and *T. multis-gemmis* were isolated from wild berries, *T. candida* from apples and *T. molicshiana* from forest substrates.

Non-pathogenic microorganisms are found in specialized mycetocyte cells of many insects. *Torulopsis* yeasts are reported in Cerambycidae and Anobiidae (Coleoptera) and a few plant hoppers. Mycetocyte symbionts provide B vitamins and essential amino acids. In bees, fungi and yeasts are found in floral nectar, in honey stomachs, in honey and pollen, and in dead or moribund adults. *T. apicola*, first isolated from the intestinal tracts of bees in Yugoslavia, is common in bees in northern California. In other parts of California predominant species are *T. magnoliae* and *T. glabrata*, in Arizona *T. apicola*, *T. magnoliae* and *T. stella*. Adult bees acquire intestinal microflora by food exchange and the consumption of pollen. *Torulopsis* spp. are mainly involved in the conversion, enhancement and preservation of pollen stored as bee bread. *T. bombicola* is present in bumble bees or their honey in France and Canada.

T. insectorum, *T. dendrica*, *T. nemodendra*, *T. philyta* and *T. silvatica* are associated with South African ambrosia beetles and *T. dendrica*, *T. insectalens* and *T. torresii* with other beetles. *T. nitratophila* is present in lacewing insect adults. *T. buchnerii*, a symbiont from *Sitodrepa panicea*, obtains some nitrogenous compounds and carbohydrates, such as proline and trehalose from the host's haemolymph and delivers essential amino acids and vitamins, except biotin, to the host.

In reptiles, *Torulopsis* yeasts that were not causing organ mycosis, were most often isolated from the gastrointestinal tract, oral cavity and the cloaca. The number of yeast-carrying bodies was greatly increased in dead animals. *T. candida* and *T. glabrata* are present in the faeces of pigeons, considered to be an important vector of pathogens to humans and domestic animals. *Candida albicans* and *T. pintolopesii* have been isolated from the upper digestive tract of poultry from clinical thrush cases in Taiwan. *T. pintolopesii* is indigenous to the gastrointestinal tracts of mice and rats. It forms layers on the surface of the epithelium. This yeast is an opportunistic pathogen in guinea pigs: alterations of living circumstances may cause illness. In livers, spleens and lungs of bats from the Amazon basin identical fungal colonization or mixed colonizations in a single organ were partly constituent with *T. glabrata*. *Torulopsis* spp. were present in the organs of other free-living small mammals and in the cervix of mares. Different strains are known to cause bovine abortion. Cytopathic effects on renal tissue cultures of primates are exhibited by *T. apicola*.

Human Pathogens

Some *Torulopsis* yeasts are pathogenic to humans, causing opportunistic infections like candidiasis, thrush and moniliasis. Mild, severe, chronic and systemic diseases may be caused by *C. albicans* and *T. glabrata*, the most prevalent yeasts in humans. *T. glabrata*, present on the skin, for example, is usually non-pathogenic, living saprophytically with the normal flora and is of low virulence. This opportunistic pathogen has an effect only when the balance between the host and its flora is disturbed. It can become locally invasive and potentially disseminate. Risk factors are high age, pregnancy, a suppressed immune system, leukaemia, large interventions (operations, intravascular catheter), treatment with high doses of antibiotics and steroids, chemotherapy, etc.

T. glabrata may be involved in candidiases of the respiratory system, the urogenital system, the central nervous system, the eyes, bones and the mouth. Strains have been isolated from premature infants, suggesting transplacental passage. *T. glabrata* is the most common yeast species isolated from vaginal yeasts infections. It causes pancreatic infections, recurrent oral candidiasis and is associated with cancer. It is involved in haematological malignancies, fungal diarrhoea, fungaemia in children, etc. *T. candida* was characterized as a member of a fungal consortium causing mycosis of the heart. *T. neoformans*, living on soil contaminated with bird droppings, can cause a diffuse pulmonary infection. If inhaled, it may lead to meningitis and localized abscesses.

Strains of the species *T. glabrata* differ in many respects from the so-called albicans group (the filamentous *C. albicans*, *C. parapsilosis* and *C. tropicalis*). The absence of pseudomycelia formation might be an explanation for the mostly mild course of illness and for the weak response to cell wall-synthesis blocking antimycotics. Additional factors display the difference from the albicans group. Gene products of *T. glabrata* show preferred cross-reactivity with *Saccharomyces cerevisiae*. Sequence of analysis of both 5.8S rRNA and 18S rRNA showed a greater degree of evolutionary divergence of *T. glabrata* to the albicans group. It is specifically related to *S. cerevisiae* and more distantly related to *T. kefyr* and *T. cremoris*. Even the comparison of mitochondrial DNA revealed a closely related evolutionary origin of mtDNA with *S. cerevisiae*.

Regulation and Methods of Detection

General regulations and recommendations concerning *Torulopsis* yeasts have not been established. Methods for their detection and differentiation are based on standard assay procedures of carbohydrate assimilation and fermentation (biochemical and physiological characteristics), morphological considerations and estimates of reproduction characteristics. A general marker for detection and identification is not available except for the non-formation of pseudomycelium. The first steps of such a procedure include enumeration. Techniques used are the direct enumeration by cell counting on Petri plates, enumeration in liquid media and membrane filtering (for soluble sources).

The most common nonselective media for yeast separation containing glucose may be used to detect *Torulopsis* yeasts: Malt extract medium, tryptone glucose yeast extract medium, oxytetracycline glucose yeast extract medium, Sabouraud medium, and potato-dextrose medium (**Table 2**). Both acidification (pH 3.5) and additions of antibiotics increase the selectivity. Beer wort or grape must are used in brewing as liquid media. Media containing dyes, such as Schiff's reagent may be useful in detecting specific yeast spoilers through individual coloration.

Recommendations for microbial food analysis are focused on general procedures for typing bacteria, yeasts, and moulds by using agar media as above. Incubation conditions including those for xerotolerant and xerophilic strains are given in Table 2.

The medically important *T. glabrata* can be detected by semi-automated or automated yeast identification systems. Immunological methods, polymerase chain reaction–restriction enzyme analysis using a primer pair amplifying a cytochrome *P*-450 gene fragment, gas chromatography, restriction endonuclease digestion of chromosomal DNA, etc. are expensive modern methods to discriminate and type *Candida* species and *T. glabrata* in clinical samples. Different agar methods have been developed, like CHROMagar, comparison of blood agar and eosin methylene blue agar, pigmentation of colonies obtained by reduction of triphenyltetrazolium chloride added to Sabouraud medium, etc., all aiming to differentiate *T. glabrata* from other associated yeasts of the genus *Candida*.

Importance to the Food Industry

Due to their enzymatic capabilities and their ability to grow and survive in extreme environments, different species of *Torulopsis* are involved in food processing as acting agents and/or as spoiling microorganisms.

Food Processes

Kefir, a Caucasian fermented milk beverage, is produced from mare's milk. In addition to some bacteria and *Lactobacillus kefir* the yeasts involved in its production include *T. kefyr*, additional *Torulopsis* spsp. and *Saccharomyces fragilis*. The low fermentative capacity of these microorganisms results in low alcohol contents in the range 0.01–3%, usually 0.8–1.5%.

Natural fermentations result in enrichment of many carbohydrates with proteins and vitamins and in improvement of taste. Fermentation, initiated by spontaneous contamination or by starter cultures, results in soy being partly hydrolysed and therefore

Table 2 Examples of agar media for yeast enumeration

Media	*Carbon source*	*Incubation temperature (°C)*	*Incubation time (days)*	*pH*	*Modifications to enhance selectivity*
Potato-dextrose agar	Glucose 20 g l^{-1}	20–25	5	6–6.3	Acidified with tartaric acid, citric acid, hydrochloric acid, lactic acid or phosphoric acid to pH 3.5(–4)
Malt extract agar	Malt extract 20 g l^{-1}			5–6	As above or mixing with acetic to pH 3.2
Tryptone glucose yeast extract agar	Glucose 100 g l^{-1}			6	Addition of antibiotics (chloramphenicol, oxytetracycline 100 mg l^{-1}) is recommended acidification with acetic acid 3.5 – acidified TGY agar
Oxytetracycline glucose yeast extract agar	Gluose 20 g l^{-1}	20–25	2–5	7	Oxytetracycline (final conc. 100 mg l^{-1})
Sabouroud agar	Glucose 20 g l^{-1}	23–30	5	6–6.3	Addition of chloramphenicol (0.5 mg ml^{-1}) or gentamycin (0.04 mg ml^{-1}) or acidification
Dichloran rose bengal chloramphenicol agar	Glucose 10 g l^{-1}			5.6	Chloramphenicol (final conc. 100 mg l^{-1}); for enumeration in presence of moulds
Xerotolerant/xerophilic yeasts					
Glucose peptone yeast extract agar	Glucose 300–500 g l^{-1}			4.5	Incubation on rotary shaker to reduce growth of moulds
Dichloran 18% glycerol medium	Glucose 10 g l^{-1} glycerol 220 ml l^{-1}			5.6	Chloramphenicol 100 mg l^{-1}, final $a_w = 0.955$; intermediate a_w-foods
Malt extract yeast extract 10% salt 12% glucose agar	Glucose 120 g l^{-1}				$a_w = 0.88$; NaCl 100 g l^{-1}; growth at reduced water activity in presence of NaCl
Whalley's medium	Glucose 20 g, sucrose 400 g in 500 ml water		5–10/14–28		$a_w = 0.91$
Whalley's medium modified by Leveau and Bouix	Glucose 20 g, sucrose 20 g, fructose 400 g, glycerol 95 ml in 405 ml water		5–10/14–28		$a_w = 0.86$ (more selective vs. xerophilic yeasts)

more digestible. *Torulopsis* spp. are involved in the production of miso which is used as seasoning or as base for soups in Japan. The raw materials are soy and rice or barley. Starter cultures used today are mixtures of xerophilic yeasts (*Torulopsis* strains) and bacteria (lactic acid bacteria). The type of miso largely depends on the proportions of the ingredients (koji, salt, cooked soya beans) and also on the type of inoculum. *Torulopsis* spp. also contribute to the formation of shoyo, a soy sauce, and participate in the fermentation and ageing of other soy sauces.

Sake production is a multistage process in which various microorganisms and consortia of microorganisms are involved. One central step is the making of moto which corresponds to pitching yeast in beer brewing. One of three typical methods applied is Yamahai-moto. Steamed rice and rice koji are used. Starting at neutral pH values, nitrate-reducing bacteria and lactic acid bacteria are dominant. Due to the high concentration of sugars and acidification, the bacterial count decreases. In parallel, the proportions of wild yeasts change. Initially predominant film-forming yeasts, for example, *Hansenula anomala*, are replaced by non-film-forming yeasts, including *Torulopsis* spp., *Candida*, *Pichia* and *Debaryomyces*. These microorganisms are derived from rice-koji. Moto consists of yeast cells of high purity and their fermentative capacity may be retained over a long period of time. The subsequent fermentation process (moromi) proceeds slowly at constant activity.

Spontaneous aerobic fermentation of black olives is a typical yeast process. Appropriate sugar content, bitter factors, polyphenols, pH, salt concentration, and temperature favour a slow and mild fermentation. Under these conditions bacterial spoilage is largely prevented. The fermentation is carried out by a group of wild yeasts, including *Torulopsis* spp. The most representative yeast is *T. candida*, accounting in some cases for more than 20% of the yeast population. Salt concentrations below 7% also enable lactic acid-producing cocci and lactobacilli to settle. Aerobic fermentation has some advantages over anaerobic fermentation in which *T. candida* is also involved. Benefits are a lower incidence of gas-pocket spoilage, reduction of shrivelling in fruits, more homogenous brines and faster fermentation, and improvement of colour, flavour and texture.

Ripening of the coca fruit is accompanied by a succession of yeasts starting with the flower. Some of them are *T. candida*, *T. castelli*, *T. holmii* and *T. rasae*. *T. candida* belongs to those yeasts that are present from flower to ripe fruit. Large quantities of this yeast are found in naturally fermenting coca. *Torulopsis* yeasts are also involved in the fermentation of ginger.

The main processes during the fermentation of idli, an Indian food, are leavening and acidification. Besides bacteria, mostly lactic acid bacteria, *T. holmii* and *T. candida* are involved in this process. To shorten the fermentation time, soured buttermilk or yeast starters are added to the dough. That additionally alters the profile of the microbial consortium. By this process the nutritional value (increase of vitamin content) and sensory qualities are improved significantly.

Yeasts occur very regularly in sour dough starter cultures or in sour dough. *T. holmii* may sometimes be a minor component of sourdough starter cultures, dominated by lactobacilli, used in German bakeries. The yeast content in mixed fermentations is strongly reduced with the increase of acetic acid concentration. In San Francisco sourdough *T. holmii* is the dominant yeast. The ratio of *T. holmii* to lactobacilli is about 1 : 100. The two have established a unique symbiosis. The yeast cannot metabolize maltose released from starch by amylases, whereas lactobacilli have an absolute requirement for this sugar. All other sugars present in dough are metabolized by the yeast together with glucose released by lactobacilli. *T. holmii* is resistant to cycloheximide secreted by lactobacilli and moderately tolerant to acetic acid. Dead yeast cells serve as the source of several amino acids and fatty acids for the lactobacilli.

T. holmii is also present in rye flour and may contribute together with mainly lactobacilli and yeasts, for example *Candida* spp., *S. cerevisiae*, to establish a good sourdough culture from rye.

'Trosh', a starter for Sangak bread fermentation, is mainly composed of *T. colluculosa* and *T. candida*, which differ in their spectrum of carbohydrate assimilation. *T. cremoris* is a good utilizer of whey. *T. ethanolithrophicus* is used as industrial sulphite fodder yeast; it accepts ethanol as sole source of carbon.

Traditional preservation techniques, i.e. low water activity, low pH value, low temperature and presence of antibacterial agents, are conducive to yeast growth. Common preservatives, such as benzoate and sorbate, can be utilized by different *Torulopsis* spp., thus removing their protective effect. Otherwise, growth of *T. glabrata* is inhibited in the presence of fat emulsions.

Wild strains of *Torulopsis* act both as agents and spoilers in wine making. Grape skins are normally covered with bacteria, moulds and yeast. This population reaches its highest density at the time of maturity of the grapes. Wild yeasts, such as *Pichia*, *Kloeckera* and *Torulopsis*, are often more numerous than *Saccharomyces* and contribute to flavour, especially in the early stages of fermentation. For selected wines mixed inocula or yeasts other than *S. cerevisiae* are used. Bordeaux wine production is initiated by *Kloeckera* or *Torulopsis*.

Food Spoilage

A large number of spoilage yeasts is identified in the total microflora of wines and wineries. *T. stellata*, which tolerates 35°C and ferments fructose faster than glucose, is present particularly on overripe grapes invaded by *Botrytis cinerea*. The yeast appears during spontaneous fermentation. Its low alcohol tolerance (10%) accounts for its disappearance at the end of the fermentation. *T. stellata* is known from more than 70% of the best French wine-producing areas and occurs frequently in Italy. *T. dattila* and *T. kruisii* are found on wild grapes and *T. gropengiesseri* on grapes and in wines. *T. candida* is a known wine spoiler. *Torulopsis* yeasts do not belong to the virulent wine spoilers, like *Zygosaccharomyces bailii* and *S. cerevisiae* of which only one vital cell per bottle is sufficient for spoilage to occur.

Torulopsis yeasts as typical food spoilers have also occupied other niches. The xerophilic *T. lactis-condensii* and *T. versatilis* are found in concentrated juice. *T. candida* has been isolated from fruits and juices and even from concentrates of orange, apricot, peach and pear. The yeast flora of the fruit and of the product are always similar. Counts in juices document the hygiene maintained in the processing unit. Heat-tolerant *T. magnoliae* and *T. stellata* have been found in orange concentrates and drinks. *T. glabrata*, *T. versitalis* and *T. haemulonii*, all able to grow in 60% (w/w) glucose, have been reported from fruit juices and soft drinks. Many contaminations of soft drinks are due to *Torulopsis* spp. Even colas and lemonades are subject to growth of *T. glabrata* and *T. stellata*. The latter also grows in carbonated soft drinks.

T. lactis-condensi was isolated from sweetened condensed milk in China. Its low ability to ferment glucose and sucrose (and raffinose) resulted in gas formation. *T. apicola*, which spoiled and grew on refined crystalline sugar, was found in syrups and intermediary refinery samples together with *T. lactis-condensi*. *T. candida* was detected in raw cane sugar, and in syrups together with *T. globosa* and *T. kestoni*. Chocolate syrup may be a habitat for *T. etchellsii* and *T. versatilis*. Xerotolerant *Torulopsis* yeasts are present in floral nectars, which are infected by bees and other nectarophilus insects. This is an important source of spoilage for facilities processing sugar syrups, honeys etc. Beehives and mummified fruits are the most important niches for overwintering.

T. kefyr has been reported from kumyss leaven and *Candida* spp. and *Torulopsis* spp. have been described in brined fruits. The spoilage by cider yeasts reflects their population in the ground. Yeasts found in soil, the tree, the apple flowers and the fruits are *T. aeria*, *T. candida* and *T. nitratophila*. The handling of fruits is an additional source of spoilage.

Torulopsis spp., like *T. colliculosa*, are generally known spoilers of beer, causing unfinable hazes. *T. spandovensis* has been isolated from different samples of German Pilsner beer. One of the most predominant infectors in the UK is *T. inconspicua*. About 49% of the yeasts isolated from the air of a Belgium brewery were of *Torulopsis* origin.

The human-associated *T. glabrata* has been found in the bivalve shellfish. Plant oils or fat emulsions diminish its heat resistance and reduce its growth rate. Maintaining a high level of hygiene of any food is important in order to prevent oral applications of this yeast. *T. glabrata* causes ethanol production in infant food and cola drinks. This has been considered in connection with auto-brewery syndrome and sudden infant death syndrome. The ability to oxidize lactic acid and acetic acid increases the potential for putrefactive changes in food and feed. Under aerobic conditions *Torulopsis* yeasts may lower the acid concentration by assimilation and the rise in pH can allow other bacteria to spoil the fermented foods.

See also: **Biochemical and Modern Identification Techniques**: Introduction; Food Spoilage Flora (i.e. Yeasts and Moulds). **Bread**: Sourdough Bread. ***Candida***: Introduction; *Yarrowia (Candida) lipolytica*. **Fermented Foods**: Origins and Applications; Fermented Vegetable Products; Fermentations of the Far East. **Single Cell Protein**: Yeasts and Bacteria. **Spoilage Problems**: Problems caused by Fungi. **Wines**: Microbiology of Wine-making.

Further Reading

Cristiani-Urbina E, Ruiz-Ordaz N and Galindez-Mayer J (1997) Differences in the growth behavior of *Torulopsis cremoris* in batch and continuous cultures. *Biotechnology and Applied Biochemistry* 26: 189–194.

Deak T and Beuchat LR (1996) *Handbook of Food Spoilage Yeasts*. Boca Raton: CRC Press.

Douglas AE (1989) Mycetocyte symbiosis in insects. *Biological Reviews of the Cambridge Philosophical Society* 64: 409–434.

Fu XH, Li MX, Zhang TZ, Lu SG and Yang XM (1989) Studies on hyperglycophilic yeast causing gaseous fermentation of sweetened condensed milk. *Wei Sheng Wu Hsueh Pao – Acta Microbiologica Sinica* 29: 235–238.

Fujita S, Lasker BA, Lott TJ, Reiss E and Morrison CJ (1995) Microtitration plate enzyme immunoassay to detect PCR-amplified DNA from *Candida* species in blood. *Journal of Clinical Microbiology* 33: 962–967.

Gilliam M (1997) Identification and roles of non-pathogenic microflora associated with honey bees. *FEMS Microbiology Letters* 155: 1–10.

Hommel R and Ratledge C (1993) Biosynthetic mechanisms to low molecular weight surfactants and their precursor molecules. In: Kosaric N (ed.) *Biosurfactants: Production–Properties–Applications*. Pp. 3–63. New York: Marcel Dekker.

Kim SY, Oh DK, Lee KH and Kim JH (1997) Effect of soybean oil and glucose on sophorose lipid fermentation by *Torulopsis bombicola* in continuous culture. *Applied Microbiology and Biotechnology* 48: 23–26.

Kostka VM, Hoffmann L, Balks E, Eskens U and Wimmershof N (1997) Review of the literature and investigations on the prevalence and consequences of yeasts in reptiles. *Veterinary Record* 140: 282–287.

Lott TJ, Kuykendall RJ and Reiss E (1993) Nucleotide sequence analysis of the 5.8S rDNA and adjacent ITS2 region of *Candida albicans* and related species. *Yeast* 9: 1199–1206.

Morace G, Sanguinetti M, Posteraro B, Lo Cascio G and Fadda G (1997) Identification of various medically important *Candida* species in clinical specimens by PCR-restriction enzyme analysis. *Journal of Clinical Microbiology* 35: 667–672.

Morris JT and McAllister CK (1993) Fungemia due to *Torulopsis glabrata*. *Southern Medical Journal* 86: 356–357.

Pfaller MA, Houston A and Coffmann S (1996) Application of CHROMagar *Candida* for rapid screening of clinical specimens for *Candida albicans*, *Candida tropicalis*, *Candida krusei*, and *Candida (Torulopsis) glabrata*. *Journal of Clinical Microbiology* 34: 58–61.

Pitt JI and Hocking AD (1997) *Fungi and Food Spoilage*, 2nd edn. New York: Chapman & Hall (Gaithersburg, MD: Aspen Press).

Samson RA, Hoekstra ES, Frisvad JC and Filtenborg O (1995) *Introduction to Food-borne Fungi*, 4th edn. Baarn, The Netherlands: Centraalbureau voor Schimmelcultures.

TOTAL COUNTS

Microscopy

S R Tatini and **K L Kauppi**, University of Minnesota, St Paul, USDepartment of Food Science and Nutrition,

Microscopic examination of food permits rapid detection and enumeration of total numbers of microbial cells and in some instances somatic cells. Recognition of the morphology (rods and cocci) and arrangement of microbial cells allows for rapid investigation of problems and sources of contamination. Overall microbial count, regardless of viability, would be useful in quality control of foods. This chapter will provide information on the applications, procedures, advantages and disadvantages of the total count microscopy method.

Range of Applications

A variety of counting chambers are available, such as the Petroff–Hausser and haemacytometer for counting cells. These methods are applicable for transparent liquid systems with high microbial numbers ($>2\times10^7$ cells ml^{-1}). However, these methods should not be used for food samples as they do not distinguish between cells and food particles. In addition, the Howard Mold Count method uses direct microscopic examination for the detection and enumeration of mould fragments in canned tomato products such as ketchup and paste.

A more commonly used technique is the direct microscopic clump count (DMCC). In this method, a known volume (usually 0.01 ml) of liquid is placed on an etched or printed microscope slide. The liquid is dried to form a film, heat-fixed and defatted if necessary (in the case of whole milk). The film is then stained. A variety of stains can be used, but methylene blue, crystal violet and the Gram stain reaction are commonly used. Microorganisms (clumps and individual), both viable and dead, can be determined by microscopic examination. In using this method, the morphological characteristics such as rod or cocci can also be determined.

An application of the DMCC is in the examination of fluid raw milk and cream samples. The DMCC method can help determine microbial contamination by monitoring the quality of incoming raw milk in a dairy or cheese-processing plant. This method is not exclusive in determining microbial numbers as somatic cells can also be determined. If high numbers of somatic cells are found it could be an indication of udder problems or mastitis associated with the cows. When using the DMCC method for examining pasteurized milk or powdered milk products, care should be taken in interpreting the results as there should be low counts of bacterial cells in these products. However, this method can yield information on the past quality history of the product.

DMCC can also be used in the examination of liquid, frozen and rehydrated eggs.

Procedures for Slide Preparation

Preparation of Milk Samples

To disperse somatic cells in films efficiently, the samples to be tested for direct microscopic observation should be warmed to 40°C immediately before being transferred to slides. Samples to be used for direct microscopic observation of bacteria should not be warmed.

1. Shake the test sample vigorously 25 times.
2. Let the sample sit (but not longer than 3 min) to minimize the amount of foam.
3. Identify and label each sample.
4. Work carefully and rapidly to avoid bacterial growth while preparing the films.

Preparation of Film

1. Dip the tip of a Breed pipet 1 cm below the surface of the well-mixed sample. A calibrated metal syringe, microsyringe or platinum loop can be used to transfer the 0.01 ml sample. Before use, each instrument must be calibrated. Calibration is based on weight measurements. To calibrate the instrument, withdraw a test sample and wipe the outside of the tip. Weight the instrument with sample with an analytical balance. Record the weight. Discharge the sample and weigh the instrument without sample. A minimum of 10 trials should be averaged. The difference in weights between the instrument with sample and instrument without sample should average 0.0103 ± 0.0005 g.
2. Rinse the pipet by drawing in and expelling aliquots of the sample into a waste container.

3. Fill the pipet to above the calibration mark.
4. With clean tissue paper remove excess sample residue from the outside of the pipet. Touch the tip of the pipet to the tissue. This will draw the volume to the calibration line by capillary action.
5. Distribute the 0.01 ml mixed sample over the entire 1 cm^2 area of the slide using a flamed inoculating needle, not a loop. The slides should be cleaned with hot detergent solution or commercial alkaline or acid cleaners and rinsed with distilled water before use. Each slide should have one or more clear, etched or painted circles, 1 cm^2 in area (11.28 mm diameter).
6. Allow the film to dry thoroughly in a level position at 40–45°C for 5 min. The samples should be heated slowly to prevent cracking and peeling of the dried film. The drying surface should be clean and dust-free. The heat source can be a surface over an electric bulb, a substage microscopic lamp or a commercial drying box.
7. While your slide is drying, determine the microscopic factor of the microscope you are using (see section on microscopic factors, below).

Liquid and Frozen Eggs

1. For liquid and frozen eggs, place 0.01 ml undiluted egg material on a clean, dry microscopic slide.
2. Spread over a 2 cm^2 etched or painted area (a circular area with diameter of 1.6 cm is suggested). A drop of distilled water to the film aids in uniform spreading.
3. Let the film preparation dry on a level surface at 35–40°C for 5 min.
4. Immerse in xylene for at least 1 min.
5. Immerse in alcohol for at least 1 min.
6. Stain for at least 45 s in North's aniline oil methylene blue stain (10–20 min is preferred; exposure to up to 2 h does not overstain).
7. Wash slide by repeated immersions in distilled water and dry thoroughly before examination.
8. Express final result as number of bacteria per gram of egg material (see section on recording of results, below).
9. Note: The microscopic factor must be doubled since 2 cm^2 is used.

Dried Eggs

1. Place 0.01 ml of a 1 : 10 or 1 : 100 dilution of dried egg material on a clean dry microscopic slide. For samples that are relatively insoluble, 0.1 N LiOH may be used as a diluent.
2. The addition of a drop of water to the film aids in uniform spreading.
3. Spread over the 2 cm^2 etched or painted area.
4. Follow steps 3–9 of the method for liquid and frozen eggs (above). Because the dried egg sample was diluted, multiply each count by 10 or 100 depending on whether film was prepared from a 1 : 10 or 1 : 100 dilution.

Table 1 Examination of microscopic fields

Range of microscopic counts	*Number of fields to be examined if the field diameter measures:*	
	0.206 mm	*0.146 mm*
Under 30 000	60	120
30 000–300 000	30	60
300 000–3 000 000	20	30
Over 3 000 000	10	20

Stains and Staining Procedures

Staining Procedures

1. For products with fat, such as whole milk, cover the slide of the fixed, dried film with xylene for 1 min; otherwise, go to step 3, below.
2. Remove the solvent (xylene) and allow to drain dry.
3. Cover with 95% ethanol for 1 min. This will fix the sample to the slide.
4. Remove ethanol and allow to drain dry.
5. Cover with the appropriate stain for 1–2 min. Use the modified Levowitz–Weber for cow's milk and North's aniline oil methylene blue for eggs.
6. Pour off excess stain and rinse with distilled water.
7. Allow to air-dry thoroughly. The slide may be placed in the 45°C incubator or slide dryer to expedite drying.
8. Examine the stained film using the microscope. If the film is overstained, decolorize with a small amount of ethanol.
9. Count the number of organisms using the oil immersion objective. Select fields for counting in such a manner that they are representative and provide a general cross section of the entire film. Refer to **Table 1** to determine the number of fields to be counted.
10. Calculate the direct microscopic count per millilitre. This is accomplished by multiplying the average count per field by the microscopic factor by the reciprocal of the dilution factor. Counts should be rounded to two significant figures.

Stains

When preparing reagents, avoid using those that have suspended matter. After stain preparation, store the stain so as to minimize evaporation of solvents and/or precipitation of solvents.

Modified Levowitz–Weber stain (tetrachlorethane)

Methylene blue chloride (certified)	0.6 g
Ethyl alcohol (95%)	52 ml
Tetrachlorethane	44 ml
Glacial acetic acid	4 ml

Add 0.6 g certified methylene blue chloride to 52 ml of 95% ethyl alcohol and 44 ml tetrachlorethane in a 200 ml flask. Swirl to dissolve. Let the solution stand for 12–24 h at 4.4–7°C. Add 4 ml of glacial acetic acid and filter through fine filter paper (Whatman no. 42 or equivalent). Store prepared stains in a clean, tightly closed bottle with a cover that is not affected by stain reagents.

Modified Levowitz–Weber stain (xylene) Xylene is less toxic compared to tetrachlorethane.

Methylene blue chloride (certified)	0.5 g
Ethanol (95%)	56.0 ml
Xylene	40.0 ml
Glacial acetic acid	4.0 ml

Add 0.5 g methylene blue chloride to 56.0 ml of 95% ethanol and 40.0 ml xylene in a 200 ml flask. Swirl to dissolve, and then let stand for 12–24 h at 4.4–7.2°C. Add 4.0 ml glacial acetic acid, and filter through Whatman no. 42 paper or equivalent. Store the stain in a tightly closed bottle which has a cap that will not be effected by the reagents.

North's aniline oil–methylene blue stain

Aniline oil	3 ml
Ethanol (95%)	10 ml
Hydrochloric acid	1.5 ml
Methylene blue (saturated alcoholic) solution	30 ml

Mix 3.0 ml aniline oil with 10 ml of 95% ethanol, and then slowly add 1.5 ml hydrochloric acid with constant agitation. Add 30 ml saturated alcoholic methylene blue solution and then dilute to 100 ml with distilled water. Filter (Whatman no. 42 paper or equivalent) before use, and store the stain in a tightly closed bottle, using a cover that will not be affected by the reagents.

Counting Fields/Microscope Factors/ Recording of Results

Counting Fields

In using the direct microscopic method an appropriate number of fields (Table 1) must be examined in order to obtain a statistically accurate result. To obtain estimates of the bacteria or somatic cell count per millilitre, examine each film with an oil immersion objective. When counting bacteria, any two single cells or clumps of cells (of the same type) separated by a distance equal to or greater than twice the smallest diameter of the two cells nearest each other are considered separate clumps. Cells of different types are counted as a separate unit (**Fig. 1**). When determining somatic cells counts, count only those with an identifiable stained nucleus.

Microscopic Factor Determination

The microscope is an important piece of equipment when using the DMCC method. A basic compound light microscope can be used and should be equipped with a 1.8 mm oil immersion objective, substage, condenser, numerical aperture of 1.25 or higher, iris diaphragm, mechanical stage and oculars. For the DMCC method, it is important to determine the microscopic factor. The microscopic factor is a number by which the average number of bacterial clumps or somatic cells can be multiplied to determine the number of cells per millilitre.

1. Adjust the light source to provide maximum optical resolution.
2. Place the stage micrometer (ruled 0.1 and 0.01 mm divisions) on the stage of the microscope.
3. Use the oil immersion objective to focus on the lines on the stage micrometer. The oil used should have a refractive index of 1.51–1.52 at 20°C and be of a non-drying type. Use only one drop of oil.
4. Count the number of 0.01 mm intervals, which side by side reach across the field.
5. This number is the diameter (one-half of the diameter is the radius). Measure the diameter to three decimal places by counting the 0.01 mm intervals (e.g. 0.175 mm).
6. Calculate the area of the field. To do this, square the radius and multiply by π (3.1416). A microscopic field area determines the amount of milk that can be examined in any one field. Field diameters providing microscopic factors of 300 000–600 000 are recommended.
7. To convert the area of one field in mm^2 to cm^2, divide the field area in mm^2 by 100.
8. To determine the number of fields in 1 cm^2, divide 1 cm^2 (the usual area of the etched/painted portion of the slide) by the area (the number calculated in step 7) of one field in terms of cm^2.
9. Calculate the microscopic factor by multiplying the value obtained in step 8 by 100 (0.01 ml of sample was spread over the slide surface).

Alternatively, the microscopic factor can be calculated by:

Microscopic factor (MF) $= [(x)(y)]/\pi r^2$ or $10000/\pi r^2$

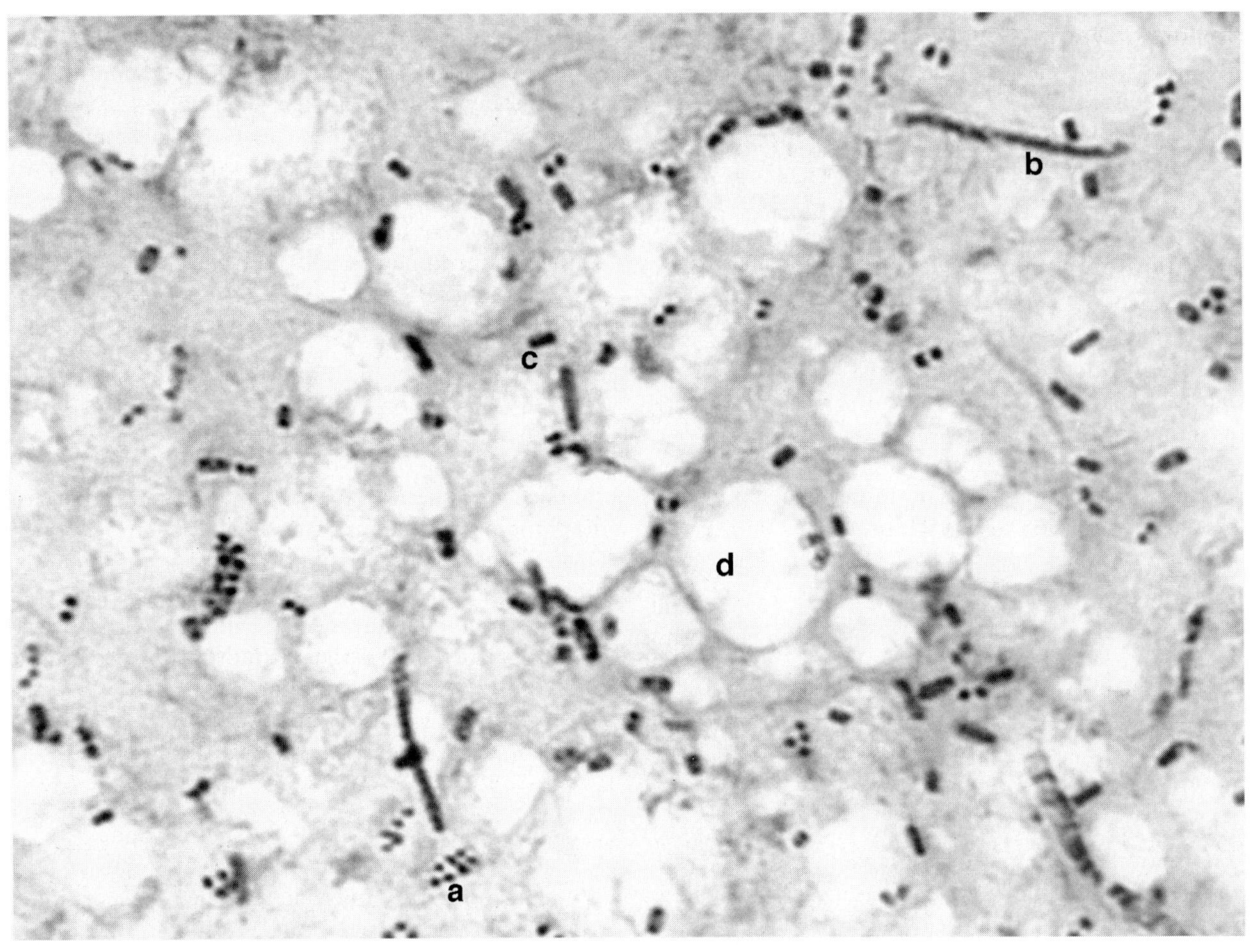

Figure 1 Levowitz Weber methylene blue stain of a whole, raw milk sample. (Light microscopy, 100× oil immersion.) (a) The cocci (round) cells could be representative of *Micrococcus*, *Staphylococcus* or *Streptococcus*. These organisms are commonly found in raw milk. High numbers are indicative of mastitis. (b) The rod (long) cells could be representative of the spore forming bacteria, *Bacillus* or *Clostridia*. These organisms can enter raw milk from unclean udders or teats or from the environment. (c) The rod (short) cells could be representative of coliforms. These organisms can enter the raw milk from the manure, soil, feed or water. (d) The large, clear areas represent fat globules found in the raw milk.

In this formula, x equals 100 and is the area in mm^2 covered by the 0.01 ml of food suspension on the slide. The value y also equals 100 and is the number of 0.01 ml portions of food suspension in 1 ml. The value r equals the radius in millimetres of the microscopic field, and r^2 equals the total area of one microscopic field. The radius, r, is the only unknown value in the above equation. This is obtained by means of the stage micrometer which consists of a glass slide with fine parallel lines placed at 0.01 mm intervals. The microscopic factor is dependent on the tube length and the objective and ocular lenses used. The reciprocal of the microscopic factor represents the amount of 1 ml of milk that is seen in one microscopic field.

Recording Results

The DMCC per millilitre or gram can be calculated by multiplying the average count per field by the microscopic factor by the reciprocal of the sample dilution. For example, if a 1 : 10 dilution was made, the reciprocal would be 10. Round off the counts to two significant figures. The DMCC value should be reported as DMCC per millilitre or gram of sample.

For samples requiring a $2\,cm^2$ area (egg products), the microscopic factor must be doubled.

Field-wide Single-strip Method

The field-wide single-strip method uses a single strip that covers the width of the microscopic field and across the diameter of the milk film. To make a single-strip count, focus on the edge of the film under oil immersion, cover the entire diameter of the milk film, and count cells within the strip and those cells touching one edge of the strip. Do not count bacteria or somatic cells that touch the other edge. During scanning of the strip, continually make fine focusing adjustments.

Single-strip Factor Calculation

1. Calculate the area of a single strip (mm^2) by multiplying the length of a strip (diameter of a 1 cm^2 circle) by the diameter of a microscopic field. The diameter can be determined by using a stage micrometer (0.01 mm divisions).
2. Determine the number of single strips in the area of the milk film (0.1 of milk) by dividing 100 mm^2 (the area of the 0.01 milk film) by the area of a single strip (calculated in step 1, above).
3. Convert the number of single strips on a 0.01 ml sample in terms of a 1 ml sample by multiplying the single number strips (0.01 ml) by 100. This value is the single-strip factor.
4. To calculate the number of cells per millilitre, multiply the single-strip factor by the number of cells (somatic or bacterial clump) in a single strip. This final value should be rounded to two significant figures.

Advantages and Disadvantages

Advantages

The DMCC method is rapid (individual samples can be analysed in 10–15 min), and a number of dried films may be made at one time, stained and read later. This method requires minimum equipment. Because a variety of stains can be used, different morphological and Gram types can be identified. The results obtained with the DMCC method are only estimates of the microbial levels of a food.

The count does not distinguish between viable and non-viable cells and usually represents both types. For this reason, direct microscopic counts will generally be higher than plate counts. However, organisms which do not normally grow at 32–35°C on agar plates, such as psychrophiles and thermophiles, will be detected by the direct microscopic count.

Disadvantages

The DMCC method requires high bacterial populations. Only a small volume of sample is examined, and this reduces the precision of the method, and there may be inaccuracy in measuring 0.01 ml aliquots. Debris and food particles may make it difficult to see the microorganisms, and a sufficient number of fields must be counted for the method to be accurate. Analyst fatigue is common if many samples are examined, and this could be a source of error or decreased precision. Another disadvantage is that viable and non-viable cells cannot be differentiated. In addition, sampling errors can come from an unrepresentative sample, failure to release microorganisms from solid foods during the blending process or irregular distribution of microorganisms in the film due to uneven spreading or not drying the film in a level position. Slide preparation, staining, counting (from improper illumination and focusing or counting too few fields) and calculations can also lead to errors.

See also: **Adenylate kinase**. **ATP Bioluminescence**: Application in Meat Industry; Application in Dairy Industry; Application in Hygiene Monitoring; Application in Beverage Microbiology. **Biophysical Techniques for Enhancing Microbiological Analysis**: Future Developments. **Electrical Techniques**: Introduction; Food Spoilage Flora and Total Viable Counts (TVC). **National Legislation, Guidelines & Standards Governing Microbiology**: European Union; Japan. **Sampling Regimes & Statistical Evaluation of Microbiological Results**. **Total Viable Counts**: Pour Plate Technique; Spread Plate Technique; Specific Techniques; MPN; Metabolic Activity Tests. **Ultrasonic Imaging**. **Ultrasonic Standing Waves**.

Further Reading

Packard VS Jr, Tatini S, Fugua R, Heady J, Gilman C (1992) Direct microscopic methods for bacteria or somatic cells. In: Marshall RT (ed.) *Standard Methods for the Examination of Dairy Products*, 16th edn. P. 309. New York: American Public Health Association.

TOTAL VIABLE COUNTS

Contents

Pour Plate Technique

James W Messer, Eugene W Rice and **Clifford H Johnson**, US Environmental Protection Agency, Cincinnati, Ohio, USA

The pour plate method (**Table 1**) is a technique designed to enumerate aerobic and facultative anaerobic bacteria in food, shellfish, water and dairy products that are capable of growth under the conditions employed (medium, time and temperature of incubation). There is no single colony formation method and set of conditions that will allow enumeration of all bacteria that may be found in or on a particular product.

The techniques differ from one another only in the manner in which they are carried out (**Table 2**). Once the optimum technique for a product is determined, repeated use on the product can provide significant public health information. Since the usefulness of the pour plate technique is highly dependent on its repetition, the competency and accuracy of the analyst performing the technique significantly affect the precision and accuracy of the bacterial count results.

Table 1 Bacterial pour plate techniques

Method	*Acronym*	*Source*
Aerobic plate count	APC	AOAC
Heterotrophic plate count	HPC	APHA
Standard plate count	SPC	APHA
Mesophilic aerobic bacterial count	MABC	APHA
Psychrotrophic bacterial count	PBC	APHA
Thermophilic bacterial count	TBC	APHA

AOAC = Association of Official Analytical Chemists;
APHA = American Public Health Association.

The pour plate method for estimating bacterial populations consists of:

1. mixing a measured volume of sample with a portion of sterile, melted and partially cooled agar medium in a Petri dish
2. allowing the mixture to solidify on a level surface
3. incubating the Petri dish containing the sample for the required time period
4. counting the bacterial colonies which develop in and on the agar medium
5. recording the colony count.

Pour Plate Method

Equipment and supplies required to perform the pour plate method must be carefully controlled to produce accurate bacterial counts.

Specifications for the most frequently used media and diluents are described in **Table 3**.

Table 2 Conditions of pour plate use

		Incubation		
Product	*Source*	*Temperature*	*Time*	*Medium*
Food	AOAC	35°C	48 ± 2 h	Plate count agar[a]
	APHA	7°C	10 days	Standard methods agar
Dairy	APHA	32°C	48 ± 3 h	Standard methods agar
	APHA	7°C	10 days	Standard methods agar
	APHA	55°C	48 h	Standard methods agar
Shellfish	APHA	35°C	48 ± 3 h	Standard methods agar
Water	APHA	35°C	48 ± 3 h	Plate count agar[a]
	APHA	28°C	5 days	R2A agar

[a] Or equivalent.
AOAC = Association of Official Analytical Chemists;
APHA = American Public Health Association.

Table 3 Specifications of the most frequently used media and diluents

Standard methods agar (plate count agar)[a]	
Pancreatic digest of casein (USP)	5.0 g
Yeast extract	2.5 g
Glucose (dextrose)	1.0 g
Agar, bacteriological grade	15.0 g
Reagent-grade water	1000.0 ml
Final reaction, after sterilization	pH 7.0 ± 0.1
R2A agar	
Yeast extract	0.5 g
Proteose peptone no. 3 or polypeptone	0.5 g
Casamino acids	0.5 g
Glucose	0.5 g
Soluble starch	0.5 g
Dipotassium hydrogen phosphate: K_2HPO_4	0.3 g
Magnesium sulphate, heptahydrate: $MgSO_4 7H_2O$	0.05 g
Sodium pyruvate	0.3 g
Agar, bacteriological grade	15.0 g
Reagent-grade water	1000.0 ml
Final reaction, before sterilization	pH 7.2

[a]Commercially prepared, dehydrated tryptone glucose yeast agar is equivalent.

Preparation of Culture Medium

Add correct weight of dehydrated culture medium to the specified volume of reagent-grade water and allow to soak 3–5 min with occasional agitation to aid wetting of the powder.

Heat the mixture in suitable containers (borosilicate glass or stainless steel) until ingredients are in solution and the agar is completely melted.

Adjust the pH of the medium if necessary using either a base (1 N sodium hydroxide) or an acid (1 N hydrochloric acid). Make pH determinations on undiluted agar at 45°C.

Restore water lost from evaporation, if necessary. If volume is checked, keep in mind that 1000 ml of water measured at 20°C will occupy 1027 ml at 80°C; 1034 ml at 90°C; and 1038 ml at 95°C.

Mix agar thoroughly and dispense in suitable containers. The volume of agar and the type of bottle used should be such that no part of the contents will be more than 2.5 cm from the glass or from the surface of the agar.

Sterilize bottled agar in the autoclave at 121°C for 15 min, allowing sufficient space between bottles to permit good circulation of steam.

The autoclave should reach 121°C slowly, but within 15 min. After sterilization, pressure in the autoclave should be gradually reduced to zero. No less than 15 min is recommended for this procedure. The maximum time from start-up to unloading the autoclave should not exceed 45 min.

Formulas of Buffers and Diluents

Formulas of buffers and diluents are given in **Table 4** and recommended diluents for use with various products are indicated in **Table 5**.

Table 4 Formulas of buffer/diluent

Stock phosphate buffer solution	
Monobasic potassium phosphate: KH_2PO_4	34.0 g
Reagent-grade water	500 ml
1N NaOH solution added to give pH 7.2 (about 175 ml is usually required)	
Add reagent-grade water to make 1000 ml	
Place in small screw-cap vials, sterilize at 121°C for 15 min. Store at 0–4.4°C after sterilization	
Stock magnesium chloride buffer solution	
Magnesium chloride: $MgCl_2 6H_2O$	81.1 g
Reagent-grade water to make 1000 ml	
Place in small screw-cap containers. Autoclave at 121°C for 15 min. Seal containers and store at 0–4.4°C	
Phosphate-buffered dilution water for food, dairy and shellfish product dilution	
Stock phosphate buffer solution	1.25 ml
Reagent-grade water to make 1000 ml	
Autoclave at 121°C for 15 min in volumes required for use	
After sterilization, tightly seal containers and store at room temperature	
Phosphate magnesium chloride-buffered dilution water for water dilution	
Stock phosphate buffer solution	1.25 ml
Stock magnesium chloride solution	5.0 ml
Reagent-grade water to make 1000 ml	
Dispense in quantities as required and autoclave at 121°C for 15 min. Tightly seal containers and store at room temperature. The addition of magnesium chloride improves recovery of metabolically injured organisms	
Peptone water (0.1%) for water dilution	
Peptone	1.0 g
Reagent-grade water to make 1000 ml	
Dispense in quantities as required and autoclave at 121°C for 15 min	
Tightly seal containers and store at room temperature	
Peptone water (0.5%) for shellfish dilution	
Peptone	5.0 g
Reagent-grade water to make 1000 ml	
Dispense in quantities as required and autoclave at 121°C for 15 min	
Tightly seal containers and store at room temperature	

Table 5 Recommended diluent

Product	*Phosphate buffer*	*Peptone water*	*Phosphate/magnesium chloride buffer*
Food	R		
Shellfish	R	R	
Dairy	R		
Water		R	R

R = Recommended.

Preparation of Sample for Dilution and Pour Plate

If additional tests are to be performed on the sample, first aseptically remove the portions to be used for microbiological analysis.

Food Transfer 50.0 g of food sample to a sterile tared laboratory blender jar. Add 450 ml of sterile phosphate-buffered dilution water and blend at 10 000–12 000 r.p.m. for 2 min. This is a 1 : 10 dilution. Stomaching has been suggested as an alternative to blending in the preparation of foods for microbial analysis. Stomaching consists of placing the food sample with the appropriate amount of diluent in a sterile plastic bag. The plastic bag with diluent is positioned within the stomacher, which is a metal box with metal paddles inside. The metal paddles, powered by a constant speed motor, move in a back-and-forth motion and pound the sample. The pounding removes the bacteria from the food particles, partly by the shearing forces of the liquid and partly by compression of the sample by the metal paddles. Samples with bones or other hard objects cannot be prepared by stomaching. A laboratory wishing to use stomaching in place of blending should compare the pour plate count of the food samples it analyses by stomaching and blending. If comparable plate count determinations are obtained, stomaching is a viable alternative.

Shellfish Transfer a suitable quantity of shelled shellfish and liquor from a sample jar to a sterile tared laboratory blender jar. Weigh the sample to the nearest gram. Add an equal amount by weight of sterile phosphate-buffered dilution water or 0.5% sterile peptone water. Blend at approximately 14 000 r.p.m. for 1–2 min. Excessive blending must be avoided to prevent heat build-up, which could result in injury or death of sensitive microorganisms. The blended sample should be cultured within 2 min of blending. This is a 1 : 2 sample dilution.

Dairy Before removal of the test portion, thoroughly and vigorously mix each sample until assured that a representative portion can be removed. Sample remains undiluted.

Water Thoroughly mix all samples before removing the test portion. Sample remains undiluted.

Sample Dilution

Select dilutions for the pour plate method so that the total number of colonies on the plate will be between 25 and 250.

Food Prepare all decimal dilutions with 90 ml of sterile phosphate-buffered dilution water plus 10 ml of previous dilution unless otherwise specified. Ordinarily dilutions 1 : 100–1 : 10 000 are sufficient.

Shellfish Use at least two dilutions per sample. For samples with unknown density, three or four dilutions might be needed. Prepare the 1 : 10 dilution by adding 20 ml of the blended sample (1 : 2) to 80 ml of sterile phosphate–buffered dilution water or 0.5% peptone water. Prepare all other decimal dilutions with 99 ml sterile phosphate-buffered dilution water or .5% peptone water. Ordinarily dilutions 1 : 10–1 : 1000 are satisfactory.

Dairy Dilute milk and dairy products having a viscosity similar to milk by transferring 1 or 11 ml of product to 99 ml of sterile phosphate-buffered dilution water; use to deliver (TD) pipettes and allow 2–4 s for the product to drain from the 1 or 11 ml graduation to the pipette tip. This produces a 1 : 10 or 1 : 100 dilution.

Dilute dairy products with a viscosity greater than milk by weighing 11 or 1 g of product into 99 ml of sterile phosphate-buffered dilution water which has been warmed to a temperature of 40–45°C. This produces a 1 : 10 or 1 : 100 dilution. If further decimal dilutions are required, transfer either 11 or 1 ml of product to 99 ml of sterile phosphate-buffered dilution water.

Water For most potable waters a 1 : 1000 dilution is suitable. Dilute sample types (sewage, turbid waters) requiring higher decimal dilutions by transferring 1 ml of sample or previous dilution into 99 ml of sterile phosphate magnesium chloride dilution water or 0.1% peptone water.

Pour Plate Procedure

Before starting the plating procedure, melt the required amount of agar plating medium in boiling water, flowing steam not under pressure or microwave oven. Do not melt agar more than once. Cool the medium to 45°C in a water bath that is operating in the range of 44–46°C. Do not depend on the sense of touch to indicate proper temperature of medium for use. Place a thermometer in a pilot bottle of agar (a 1.5% agar solution in a container identical to that used for medium) and expose it to the melting and cooling cycle with each batch of medium as an indicator of the temperature in the sterile bottles of medium in the tempering bath. Do not melt more medium than will be used within 3 h.

Sterility controls on agar, dilution blanks, Petri

dishes and pipettes used for each group of samples should be made.

The plating area should be free of dust and draughts. The microbial density of the air should be checked during plating by exposing for 15 min a freshly poured agar plate, cover removed, on the plating surface. After exposing the agar medium, replace the cover and incubate the plate with routine samples. Fifteen colonies or less is considered acceptable.

To ensure a dust-free laboratory bench top, wipe the plating area with clean paper towels moistened with any approved sanitizer.

Select the number of samples to be plated in any series so that all will be plated within 20 min after diluting the first sample. After depositing test portions in the plate, promptly pour the liquified cooled agar into each plate. Lift cover of Petri dish just high enough to pour medium. As each plate is poured thoroughly and evenly, mix the medium and test portion in the Petri dish. Allow the mixture to solidify on a level surface. Solidification should occur within 10 min. Invert plates and place in incubator within 10 min of solidification in stacks of not more than four high, with space at least 2.5 cm between walls of the incubator and culture dish stacks and between stacks to allow rapid equilibration of temperature. Arrange stacks over one another on successive shelves to permit circulation of air.

Check the incubator temperature in the areas where plates are incubated with not less than two thermometers. The entire thermometer bulb should be inserted through the stopper of a vial or bottle and completely immersed in water in order to obtain reliable readings of the average incubation temperature. Read and record the incubator temperature daily in the early morning and late afternoon when in use.

Avoid excessive humidity in the incubator to reduce spreader formation. Likewise, avoid excessive drying of plates.

Food

Seed duplicate Petri dishes in dilutions of 1 : 10, 1 : 100, 1 : 1000 etc. Place 1.0 ml of the appropriate dilution in each plate and add 10–12 ml of liquified cooled agar within 15 min from the time of original dilution. Incubate plates at 35 ± 1°C for 48 ± 2 h.

Shellfish

Plate 1.0 ml of appropriate dilutions (1 : 10, 1 : 100, 1 : 1000 and 1 : 10 000) in duplicate and add 10–12 ml of liquid cooled agar within 20 min from the time of original dilution. Incubate plates at 35 ± 0.5°C for 48 ± 3 h.

Dairy

Plate two decimal dilutions per sample. Plate 1.0 or 0.1 ml of undiluted or diluted sample and add 10–12 ml of liquid cooled agar within 20 min of original dilution. Incubate plates at 32 ± 1°C for 48 ± 3 h.

Water

Prepare replicate plates for each sample dilution used. Plate 1.0 or 0.1 ml of undiluted or diluted sample and add 10–12 ml of liquid cooled agar immediately. Incubate plate count agar plates at 35 ± 0.5°C for 48 ± 3 h and R2A plates at 28°C for 5 days.

Counting Pour Plates

Manual Counting

At the end of the incubation period, select spreader-free plates with 25–250 colonies. Using a dark-field Quebec colony counter or one with equivalent illumination and magnification and equipped with a guide plate ruled in square centimetres, count all colonies including those of pinpoint size. Avoid mistaking particles for pinpoint colonies. Use a hand or electronic tally in making counts. On laboratory data forms, for each dilution record the number of colonies counted on each plate.

Three types of spreading colonies are occasionally encountered on pour plates:

1. Chains of colonies appearing to be from a single source. Count each such chain as one.
2. A spreader that forms in the film of water between the agar and the bottom of the dish
3. A spreader that forms in the film of water at the edge of the surface of the agar.

The second and third forms usually result in distinct colonies and are counted as such.

Count and record spreader colonies and normal colonies under each dilution unless spreader growth plus area of repressed growth due to spreader growth exceeds 25% of the plate area. Record these counts as spreader.

For plates with no colonies, record the colony count on data forms as < 1. For plates with greater than 250 colonies, where the number of colonies per square centimetre is less than 10, estimate and record total colonies per plate by counting 12 representative square centimetre areas. Where the number of colonies per square centimetre is over 10, count four representative areas. Avoid recounting any square. In both instances, compute the average number of colonies per square centimetre and multiply the average number of colonies per square centimetre by the area of the plate. Record as estimated count. Each laboratory must determine the area of the plate used.

When a plate is known to be contaminated or unsatisfactory, record the count as 'laboratory accident' (LA).

When the colony count of a plate is significantly beyond the count range of 250 colonies, record as 'too numerous to count' (TNTC).

Automated Counting

Use the following American Public Health Association (APHA) guidelines. Automated colony counters, when determined in individual laboratories to yield counts that 90% of the time are within 10% of those obtained manually, may be used for counting plates. When using colony counting instruments, exercise the following precautions:

1. Align the Petri dish carefully on the colony counter stage.
2. Avoid counting the stacking ribs or legs of plastic Petri dishes.
3. Do not count plates having unsmooth (rippled) agar surfaces.
4. Avoid plates having food particles or air bubbles in the agar.
5. Do not count plates having spreaders or extremely large surface colonies.
6. Avoid scratched plates.
7. Wipe fingerprints and films off the Petri dish bottom before counting.

Computing and Reporting Pour Plate Counts

On the basis of the recorded sample data, compute the pour plate count by multiplying the total number of colonies (or average or estimated number) by the dilution used. Round off the count to two significant figures, by raising the second digit from the left to the next higher number when the third digit from the left is 5, 6, 7, 8 or 9 and dropping the third digit when it is 1, 2, 3 or 4. Examples:

1. 235 (number of colonies) × 100 (dilution) = 23 500. This is reported as 24 000.
2. 234 (number of colonies) × 100 (dilution) = 23 400. This is reported as 23 000.

Report counts or estimates as colony-forming units (cfu) per gram or millilitre, as applicable.

Counts from Duplicate Plates Compute arithmetic average of counts on duplicate plates of the same dilution. If only one plate of a certain dilution yields 25–250 colonies, average with the counts on other plates of the same dilution, unless excluded as a spreader or LA even though the count falls outside the 25–250 range. The arithmetic average is the cfu g^{-1} or cfu ml^{-1}.

Counts from Consecutive Dilutions If counts on plates from two consecutive decimal dilutions fall in the 25–250 colony range, unless excluded by spreader or LA, compute the counts per millilitre for each dilution by multiplying the number of colonies by the dilution used. The arithmetic average of the two dilution counts is the cfu g^{-1} or cfu ml^{-1}.

Counts from Plates With <25 Colonies Multiply actual number of colonies on lowest dilution by the lowest dilution. This is the cfu g^{-1} or cfu ml^{-1}.

Counts from Plates with >250 Colonies When plates from all dilutions yield greater than 250 colonies, estimate as directed and report as cfu g^{-1} or cfu ml^{-1}.

Counts from Plates Recorded as Spreader, LA or Unsatisfactory Report cfu g^{-1} or cfu ml^{-1} as demonstrated.

Counts from Plates with no Colonies If all plates from dilutions tested show no colonies, report the cfu g^{-1} or cfu ml^{-1} as < 1 times the lowest dilution.

See color Plate 35.

See also: **ATP Bioluminescence**: Application in Meat Industry; Application in Dairy Industry; Application in Hygiene Monitoring; Application in Beverage Microbiology. **Electrical Techniques**: Introduction. **National Legislation, Guidelines & Standards Governing Microbiology**: European Union; Japan. **Rapid Methods for Food Hygiene Inspection**. **Sampling Regimes & Statistical Evaluation of Microbiological Results**. **Total Viable Counts**: Spread Plate Technique; Specific Techniques; MPN; Microscopy. **Ultrasonic Imaging**. **Ultrasonic Standing Waves**.

Further Reading

American Public Health Association (1970) *Recommended Procedures for the Examination of Sea Water and Shellfish*. 4th edn. Washington, DC: APHA.

American Public Health Association (1992a) *Compendium of Methods for the Microbiological Examination of Foods*, 3rd edn. Washington, DC: APHA.

American Public Health Association (1992b) *Standard Methods for the Examination of Dairy Products*, 16th edn. Washington, DC: APHA.

American Public Health Association (1995) *Standard Methods for the Examination of Water and Wastewater*, 19th edn. Washington, DC: APHA.

Association of Official Analytical Chemists (1997) *Official Methods of Analysis*, 16th edn. Arlington, VA: AOAC International.

Niemela S (1983) *Statistical Evaluation of Results from Quantitative Microbiological Examinations*. Uppsala, Sweden: Nordic Committee on Food Analysis.

Spread Plate Technique

James W Messer, **Eugene W Rice** and **Clifford H Johnson**, US Environmental Protection Agency, Cincinnati, Ohio, USA

The spread plate method is a quantitative technique designed to enumerate mesophilic bacteria in food, water and dairy products that are capable of growth under the conditions employed (medium, incubation temperature, time and atmospheric conditions). The acceptable counting range for this method is 20–200 colonies, as surface colonies which are larger than subsurface colonies will result in crowding at lower levels than subsurface colonies of pour plates.

The spread plate method offers the following advantages over the pour plate method:

1. There is no danger of killing heat-sensitive organisms with hot media.
2. All colonies are on the surface where they can readily be distinguished from particles of food, dirt or residue.
3. Any non-selective or selective differential medium can be used, as translucence of the media is not essential for colony recognition because all colonies are on the surface.

The spread plate method for estimating bacterial population consists of:

1. Spreading a measured volume of sample on the surface of pre-dried agar medium in a Petri dish.
2. Letting the medium absorb the inoculum.
3. Incubating the Petri dish containing the sample for the required time.
4. Counting the bacterial colonies which develop on the agar surface.
5. Recording the colony count.

Because test volumes with the spread plate method are limited to 0.1–0.5 ml of sample, the spread plate method is not applicable to low count samples (**Table 1**).

Preparing Spread Plates

Equipment and supplies required to perform the spread plate method must be carefully controlled to produce accurate bacterial counts.

Specifications for the most frequently used media and diluents are described in the section on total viable counts – pour plate technique and in the specific sections of the three reference publications shown in Table 1.

Prepare the appropriate agar by adding the correct weight of dehydrated culture medium to the specified volume of reagent-grade water. Soak medium for 3–5 min with occasional agitation to aid wetting of the powder.

Heat the mixture in suitable containers (borosilicate glass or stainless steel) until ingredients are in solution and the agar is completely melted. The volume of agar and the type of container used should be such that no part of the contents will be more than 2.5 cm from the glass or from the surface of the agar.

Adjust the pH of the medium if necessary, using 1 N sodium hydroxide or 1 N hydrochloric acid. Make pH determinations on undiluted agar at 45°C. Replace water lost from evaporation, if necessary.

Sterilize in the autoclave at 121°C for 15 min, allowing sufficient space between containers to permit good circulation of steam.

The autoclave should reach 121°C slowly, but within 15 min. After sterilization, pressure in the autoclave should be gradually reduced to zero. No less than 15 min is recommended for this procedure. The maximum time from start-up to unloading the autoclave should not exceed 45 min.

Pour 15 ml of agar into each Petri dish. Lift the cover of the Petri dish high enough to pour medium. Use plastic dishes that are 12 mm deep and have bottoms of at least 80 mm inside diameter. Petri dishes may be made of glass or nontoxic sterilized plastic. Keep covers open slightly until agar has hardened. Close dishes, invert and store covered at room temperature for 3–5 days before use. Use of freshly prepared plates is not recommended.

Table 1 Validated spread plate methods

Organism	Section no.		
	APHA/ Food[a]	*AOAC*[b]	*APHA/ water*[c]
Aerobic bacteria			
Spread plate	4.52		9215C
Petrifilm plate count	4.53	17.2.07	
Petrifilm VRB count	24.53	17.3.02	
Petrifilm Escherichia coli count		17.3.04	
Staphylococcus aureus	33.53	17.5.02	
Aeromonas hydrophila	30.42		
Bacillus cereus	35.21		
Clostridium perfringens	37.72	17.7.02	
Listeria monocytogenes	38.516		

[a]American Public Health Association (1992).
[b]Association of Official Analytical Chemists (1997).
[c]American Public Health Association (1995).

Spread Plate Procedure

To use this procedure maintain a supply of suitable pre-dried, absorbent agar plates.

Sterility controls on agar plates, dilution blanks and

pipettes used for each group of samples should be made.

The plating area should be free of dust and draughts. The microbial density of the air should be checked during plating by exposing for 15 min a freshly poured agar plate, cover removed, on the plating surface. After exposing the agar medium, replace the cover and incubate the plate inverted with routine samples. Fifteen colonies or less is considered acceptable.

To ensure a dust-free laboratory bench-top, wipe the plating area with clean paper towels moistened with any approved sanitizer.

Prepare a series of decimal dilutions of the sample based on the estimated concentration of bacteria in the sample. Thoroughly mix the dilution and pipette 0.1 ml of each dilution onto the surface of the pre-dried agar plate. Sterilize a bent glass rod stick (shaped in the form of a hockey stick) by dipping in 95% ethanol and quickly flaming the rod to remove the alcohol. Cool the rod for several seconds. Test the glass rod on the edge of the agar in the plate to ensure a safe temperature for use. Spread the inoculum over the entire surface of the agar in the plate with the bent glass rod. Lift the glass rod from the agar surface and place it in 95% alcohol. Cover the plate and allow the inoculum to be absorbed before inverting the plate. Incubate as required for the specific test.

Check the incubator temperature in the areas where plates are incubated with not less than two thermometers. The thermometer bulb should be immersed in water in order to obtain reliable readings of the average incubation temperature. Read and record the incubator temperature daily in the early morning and late afternoon when in use.

Avoid excessive humidity in the incubator to reduce spreader formation. Likewise, avoid excessive drying of plates.

Counting, Computing and Reporting Spread Plate Counts

At the end of the incubation period select spreader-free plates with 20–200 colonies. Using a dark-field Quebec colony counter, or one with equivalent illumination and magnification, and equipped with a guide plate ruled in square centimetres, count all colonies, including those of pinpoint size. Use a hand or electronic talley in making counts. On laboratory data forms, for each dilution, record the number of colonies counted on each plate.

On the basis of the recorded sample data, compute the spread plate count by multiplying the total number of colonies (or average or estimated number) by the dilution used and the amount plated. Round-off spread plate counts using the guidelines described for the pour plate technique. Report counts or estimates as colony-forming units (cfu) per gram or millilitre, as applicable.

See also: **ATP Bioluminescence**: Application in Meat Industry; Application in Dairy Industry; Application in Hygiene Monitoring; Application in Beverage Microbiology. **Electrical Techniques**: Introduction. **Mycotoxins**: Toxicology. **National Legislation, Guidelines & Standards Governing Microbiology**: European Union. Japan. **Rapid Methods for Food Hygiene Inspection**. **Sampling Regimes & Statistical Evaluation of Microbiological Results**. **Total Viable Counts**: Pour Plate Technique; Specific Techniques; MPN; Microscopy. **Ultrasonic Imaging**. **Ultrasonic Standing Waves**.

Further Reading

American Public Health Association (1992) *Compendium of Methods for the Microbiological Examination of Foods*, 3rd edn. Washington, DC: APHA.

American Public Health Association (1995) *Standard Methods for the Examination of Water and Wastewater*, 19th edn. Washington, DC: APHA.

Association of Official Analytical Chemists (1997) *Official Methods of Analysis*, 16th edn. Arlington, VA: AOAC International.

Specific Techniques

Michael G Williams, 3M Center, St Paul, Minnesota, USA

Frank F Busta, Department of Food Science and Nutrition, University of Minnesota, St Paul, USA

Introduction

One of the most common tests in food microbiology is the enumeration of total viable microorganisms. Although there are a number of means to estimate the total viable count, each method is subject to limitations of accuracy and practicality. The aerobic plate count (APC) method is the most widely used method for enumerating total viable bacteria in food samples. Aerobic plate counts are also expensive and time-consuming. Therefore, a number of alternative methods have been developed to increase the speed or efficiency of the traditional agar plate count methods.

Most of the alternative methods described below have been approved for use in the dairy industry. In the case of commercial products, always refer to the package insert for detailed instructions for use. The advantages and disadvantages of each of these methods are discussed below. All the methods include an incubation period to allow for the growth and replication of the bacteria. The typical incubation temperatures for aerobic count plates are 32 ± 1°C for dairy products or 35 ± 1°C for non-dairy food and beverages.

Aerobic Plate Count (APC) Method

The aerobic plate count method is almost universally accepted for enumerating aerobic bacteria in food and dairy samples and is the standard to which other methods are compared. Diluted food samples are mixed with molten nutrient agar, poured into a Petri dish, and the mixture is allowed to solidify at room temperature before incubation. The agar does not contain dyes or indicators. Bacteria in the sample form opaque colonies that are visible in the essentially transparent agar matrix.

The APC method consists of the following steps.

1. Standard methods agar is prepared, autoclaved, and allowed to temper at 45°C.
2. The sample is prepared and diluted according to the standard methods used for each food type.
3. The Petri dishes are labelled and placed on a flat, level surface.
4. One millilitre of diluted food sample is pipetted into a sterile Petri dish.
5. Molten, tempered agar (12–15 ml) is dispensed into the Petri dish.
6. The dish is swirled gently to mix the sample into the molten agar. The agar is allowed to solidify at room temperature.
7. The dishes are incubated at the appropriate temperature for 2–3 days, depending upon the method used.
8. Colonies typically appear as white or cream-coloured ellipses on the surface or trapped inside the agar matrix. Plates with 25–250 colonies are used for enumerating aerobic bacteria in food or dairy samples.

3M™ Petrifilm™ Plate Method

The Petrifilm Aerobic Count plate method for the enumeration of aerobic bacteria in foods and dairy products has regulatory approvals, certification or official recognition in a number of countries (for example, AOAC International, AFNOR Certificate Number 3M 01/1 09/89, Belgium Department of Agriculture, Health Protection Branch – Canada – Method MFHPB-33, and Victorian Dairy Industry Authority, Australia). The Petrifilm plate consists of two plastic films coated with adhesive, powdered standard methods nutrients, and a dehydrated cold water-soluble gelling agent. An indicator dye, triphenyltetrazolium chloride (TTC), is included to help visualize the colonies for counting. This method is used globally in the food industry as a replacement for traditional agar pour plate methods.

The Petrifilm method consists of the following steps:

1. The sample is prepared according to the standard methods used for each food type.
2. The Petrifilm plate is labelled and placed on a flat, level surface.
3. The top film of the plate is lifted.
4. A 1 ml portion of the proper sample dilution is pipetted onto the centre of the bottom film (**Fig. 1A**).
5. The top film is released and allowed to drop onto the inoculum.
6. A plastic spreader with a concave surface is placed over the inoculum. Gentle pressure is applied over the centre of the spreader.
7. The spreader is removed and the plate is allowed to gel for 1 min before it is moved (**Fig. 1B**).

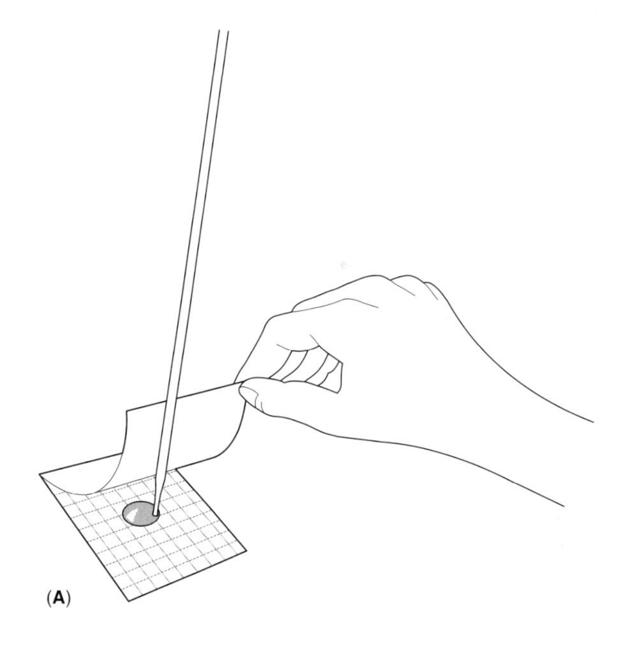

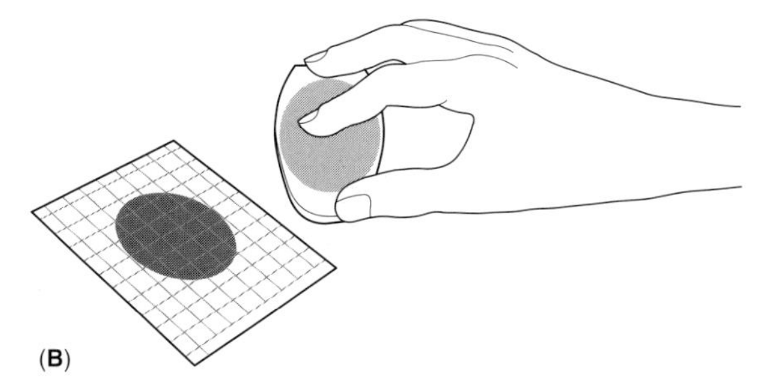

Figure 1 Petrifilm aerobic count plate method.

8. The plates are incubated for 2–3 days, depending on the method used. The plates are placed in stacks not exceeding 20 during incubation and are not inverted.
9. After incubation, all red spots, regardless of size, are counted as colonies. The counting range of the plate is 25–250 colony-forming units (cfu) per plate.

In addition to using the plate for enumerating bacteria in food, the Petrifilm Aerobic Count plate can also be used for environmental surface testing. In this application, the plate is hydrated with sterile diluent and allowed to gel for 1 h. The plate is then opened and the gel is pressed against the surface to be tested. The plate is closed and incubated under the appropriate conditions.

The Petrifilm Aerobic Count plate offers a number of advantages over traditional agar plate counts. The sample-ready Petrifilm plate significantly reduces the labour for total viable count tests. No media preparation is needed before the samples are plated. Another significant advantage is the size of the plates. A stack of 50 Petrifilm plates occupies approximately the same volume as two standard Petri dishes, which makes storage and disposal of the plates much more efficient than most methods.

Certain bacteria found in some food are able to hydrolyse the gelling agent in the Petrifilm plates. This hydrolysis may lead to large, spreading colonies. Frequently, these bacteria grow as 'spreaders' on agar, also making the enumeration of colonies on agar Petri plates difficult. In contrast to pour plates, where the sample is mixed into 12–15 ml of agar medium, the sample is not mixed into a larger volume in the Petrifilm plate method. Consequently, certain inhibitory components of the food (pH, salts, for example) may have a greater effect on the results in Petrifilm plates than they have in agar plates.

Spiral Plate Method

The spiral plate count method is accepted for total microbial enumeration by the US Food and Drug Administration and is an AOAC International Official Method for food testing. The spiral plate method has also been used to test milk samples. The spiral plate method is a variation of an agar spread plate method. In this method, a rotating agar plate is inoculated with a liquid sample that is dispensed at a constant rate. As the dispensing stylus moves away from the centre of the plate, the liquid sample is spread over a larger surface of agar. After incubation, the portion of the plate that contains a countable number of colonies is identified. The number of colonies is divided by the volume of liquid dispensed in this area to determine the cfu ml^{-1}.

The Spiral Plate method consists of the following steps.

1. Prepare and sterilize Plate count agar (standard methods agar).
2. Dispense the agar into sterile Petri dishes (on a level surface); allow it to solidify.
3. Prepare the food sample for plating, taking care to remove or minimize particulates that could plug the stylus. Label the Petri dishes.
4. Clean the stylus tip by rinsing with a sodium hypochlorite solution followed by a rinse with sterile distilled water.
5. Load the food sample into the Spiral Plater™.
6. Remove the cover and place an agar plate on the platform.
7. Place the stylus tip on the surface of the agar and start the motor.
8. After the plate is inoculated, replace the cover.
9. Incubate the plates for 48 ± 3 h.
10. Using the transparent plate overlay, choose any wedge and begin counting the colonies from the outer edge of the first segment toward the centre of the plate until 20 colonies have been counted. Complete the count of the colonies in the segment where the 20th colony is found.
11. Use the plate overlay to determine the sample volume in the region that was counted.
12. To estimate the count, divide the number of colonies by the sample volume that was counted.

The primary advantage of this method is that one inoculation can enumerate bacterial densities of 500–500 000 cfu ml^{-1}. Within that range of microbial densities, no additional materials (e.g. pipettes, dilution bottles) are needed.

The biggest disadvantage of the spiral plate method is the tendency of food particulates to plug the inoculating stylus. This limits the utility of the method. The small volume of sample plated also limits the sensitivity of this method. The spiral plate method can be affected by the consistency of the agar Petri dishes. Therefore, it is recommended that an automatic dispensing system be used to pour the plates that are used with the Spiral Plater machine.

3M™ Redigel™ Test Method

The Redigel test is a pectin gel method similar to agar pour plate techniques and is an AOAC International Official Method for aerobic count determinations in food. The method utilizes sample-ready reagent bottles and dishes to eliminate the media preparation necessary for normal pour plates. The reagent bottles contain sterile nutrient broth with pectin. The specially treated Petri dishes used in this method contain divalent cations to 'harden' the pectin. The gel hardens within about 40 min after the sample has been added to the plate and the colonies are enumerated using the same method as agar pour plates. The Redigel test has been compared to several other methods for the enumeration of aerobic bacteria in food.

The Redigel test consists of the following steps.

1. The food sample is prepared by blending or stomaching and diluted (if necessary).
2. The dishes are labelled with the sample identification.
3. The bottle of Redigel medium is opened and 0.1–1.0 ml of diluted food sample is added to the bottle.
4. The bottle is closed and swirled gently to mix.
5. The entire contents of the bottle are poured into a pretreated Redigel dish. The dish is swirled so that the medium completely covers the bottom.
6. The dish is placed on a level surface and allowed to solidify for approximately 40 min or the dishes with unsolidified Redigel medium may be placed (without inverting) into a level incubator.
7. The Redigel dishes are incubated at the appropriate temperature for 48 ± 3 h.
8. All colonies are counted. The counting range is 25–250 colonies. The total number of colonies is multiplied by the reciprocal of the dilution to obtain the cfu g^{-1} in the original food sample.

The primary advantage of the Redigel test is that the medium is premade and, thus, there is significantly less labour than with agar techniques. Since the procedures of this method are very similar to agar pour plate techniques, technicians can adopt the procedure with minimal training. Another advantage of the Redigel test is that the plates may be used as pour plates or spread plates. In the latter application, the sterile medium is added to the pretreated dishes and allowed to solidify before the sample is added to the dish.

The Redigel test is subject to the same interpretation issues as agar, such as occasional spreading surface colonies that may make it difficult to enumerate subsurface colonies. Even though the dishes used in the Redigel test appear similar to standard Petri dishes, the pectin gel will not harden when poured into a standard Petri dish. The specially treated dishes that are supplied with the Redigel test medium must be used.

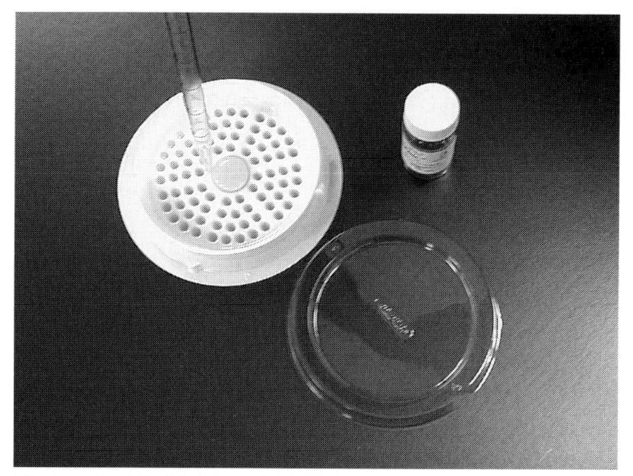

Figure 2 SimPlate total plate count method.

SimPlate Total Plate Count Method

The SimPlate total plate count method is a relatively recent development in aerobic count techniques. The method consists of mixing a food sample with nutrients and indicators and pouring the suspension into a plastic dish with microwells (**Fig. 2**). Fluorescent enzyme substrates in the medium are hydrolysed by the bacteria during the incubation period. After incubation, the number of fluorescent microwells is noted and a most probable number (MPN) table is consulted to obtain an estimate of the number of bacteria in the original suspension.

The SimPlate total count method consists of the following steps.

1. The food sample is prepared according to standard methods.
2. The plastic microwell dishes are labelled for sample identification.
3. A vial of dehydrated nutrients is hydrated with 9–10 ml sterile deionized water. The final volume of the water plus the food sample is 10 ml.
4. The food sample is added to the vial. The cap is replaced and the vial is shaken gently to mix the contents.
5. The contents of the vial are poured onto the centre portion of the plastic microwell dish.
6. The lid is placed on the microwell dish and the dish is swirled to distribute the liquid into the microwells.
7. Air bubbles are released from the microwells by tapping the dish gently on the counter top.
8. The notch in the lid is aligned with the spout on the microwell dish and the excess liquid is poured out of the plate.
9. The lid is rotated to close the spout and the plates are incubated at the appropriate temperature for 24 h.
10. The plates are examined under an ultraviolet light

source and the number of fluorescent microwells is recorded.

11. An MPN table is consulted and the result is multiplied by the reciprocal of the dilution to obtain the cfu g^{-1} in the original food sample.

The primary advantage of the SimPlate method is the shorter incubation time than most aerobic count techniques (24 vs. 48 h). Another advantage of the method is that you can obtain MPN estimates up to 738 cfu in the normal microwell plate and up to 1659 cfu in the high counting range plate.

The SimPlate method is relatively new and is not used broadly in the food industry. The technique is somewhat cumbersome. A number of foods produce endogenous enzyme activities that react with the indicators in the SimPlate medium and cause false positive reactions.

Hydrophobic Grid Membrane Filter Method

The hydrophobic grid membrane filter (HGMF) method has AOAC International Official Methods approval for aerobic plate counts in food. The method uses a membrane filter imprinted with a hydrophobic grid that divides the filter surface into 1600 equal compartments. When a sample is filtered through the HGMF, bacteria are randomly distributed into the filter compartments. The filter is placed onto a specially formulated agar plate and incubated. As colonies grow within the compartments of the HGMF, they absorb a green dye from the agar medium. The green dye facilitates the enumeration of bacteria in food samples. The number of compartments occupied by colonies is determined and a most probable number estimate is calculated to determine the cfu ml^{-1} in the original sample.

The hydrophobic grid membrane filter method consists of the following steps.

1. Tryptic soy–fast green agar is prepared, sterilized and dispensed into sterile Petri dishes.
2. The surfaces of the plates are dried before use. The plates are labelled with sample identifications.
3. The filtration apparatus is autoclaved and cooled. A sterile HGMF is aseptically placed into the filtration apparatus.
4. The food sample is prepared by blending.
5. If necessary, the blended sample is treated with various enzyme solutions to break down viscous or particulate matter in the food that may interfere with filtration of the sample.
6. The sample is filtered through the 5 μm mesh prefilter and the 0.45 μm HGMF.
7. The HGMF is aseptically removed from the filter apparatus and is laid onto the surface of a tryptic soy–fast green agar plate. Care is taken to avoid trapping air bubbles between the agar surface and the HGMF.
8. The plate is incubated at the appropriate temperature for 48 ± 3 h.
9. The number of (positive) compartments containing green colonies is enumerated.
10. The most probable number is calculated using the following formula:

$$\mathrm{MPN} = N \times \ln(N/(N - x))$$

where N = total number of squares and x = number of positive squares.

The HGMF method offers several advantages over the traditional agar methods. 1. The HGMF plates are easier to interpret because the colonies are stained a contrasting colour. 2. The sensitivity of the method can be greater than traditional methods because it is determined by the volume of sample filtered through the HGMF. 3. Samples containing up to approximately 10 000 bacteria can be enumerated on a single plate.

The primary disadvantage of this method is the labour involved in agar plate preparation, sample preparation and maintenance of the filtration apparatus. Each sample that is processed contaminates the filter device and, thus, it must be cleaned and sterilized before reuse. Another disadvantage of this method is the limitation imposed by food particulates that clog the prefilter or the HGMF. This limitation may be addressed by enzyme pre-treatments, which results in additional labour and material expenses for each test. Some food debris that is collected on the HGMF (e.g. corn, tuna) may be stained by the fast green dye. This may result in false-positives in some of the affected membrane compartments.

Plate Loop Count Method

The plate loop count method is used to reduce the time necessary to perform aerobic plate counts on milk samples. The method combines the use of a calibrated inoculating loop with a continuous pipetting syringe (**Fig. 3**) to facilitate the distribution of a small milk sample into a Petri dish. The sample is combined with molten agar and the colony counts are enumerated like other agar pour-plate techniques.

The plate loop count method consists of the following steps.

1. Sterile Petri dishes are labelled for sample identification.
2. The plate loop apparatus (Fig. 3) is sterilized by autoclaving prior to use.

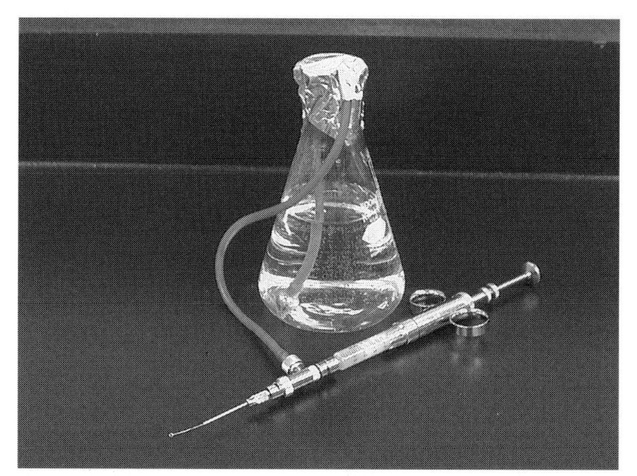

Figure 3 Plate loop count method.

3. After cooling, the filling tube of the continuous pipettor is placed aseptically into a bottle of sterile dilution media. The pipettor is adjusted to deliver 1 ml and the apparatus is rinsed several times.
4. The inoculating loop is passed briefly through a clean, high-temperature gas flame and allowed to cool for at least 15 s.
5. Several 1 ml portions of dilution media are used to rinse the inoculating loop.
6. The milk sample is shaken briefly to homogenize the sample.
7. The rinsed loop is dipped into the milk sample to the bend in the loop shank (ca. 3–4 mm above the loop). Care is taken to avoid passing the loop through foam on the surface of the sample.
8. The loop is moved through the sample several times before withdrawing the loop with the 0.001 ml sample (**Note**: The rate of withdrawal of the loop from the milk will affect the accuracy of the test.)
9. The loop is held over an open Petri dish and 1 ml of sterile dilution medium is discharged from the pipette to rinse the milk sample into the plate. The loop may be flamed between samples to prevent residue carryover from one sample to the next.
10. Molten, tempered plate count agar (12–15 ml) is poured into the plate and mixed with the sample.
11. The plates are incubated at 32 ± 1°C for 48 ± 3 h.
12. Colonies are counted, the number of cfu is multiplied by 1000 and the results are reported as 'plate loop count per mililitre'.

The primary advantage of this method is the speed with which large numbers of liquid samples can be processed. Another advantage of this method is that it minimizes the number of dilution bottles and pipettes that are used for plating large numbers of samples.

This method is not used routinely for food samples. The plate loop method relies on a very homogeneous, non-particulate sample. Particulate sample matter could have a large impact on the accuracy and reproducibility of the test. Another disadvantage of the plate loop method is the skill that is necessary to obtain an accurate volume with the calibrated loop. Lastly, the method includes the use of agar media, which is laborious to prepare.

Drop Plate Method

The drop plate method is similar to the spread plate method for aerobic plate counts. In this technique, the diluted sample is deposited over the surface of an agar plate using a calibrated pipette to deliver a predetermined number of drops. The drops are allowed to spread and dry over an area of the agar surface. The plates are incubated and the colonies are counted using the same procedures as spread plates. The colony count on each plate is divided by the total volume of sample deposited by the calibrated pipette to determine cfu ml^{-1}. This technique is not widely used for aerobic count determinations and it is not recommended for food samples containing less than 3000 cfu g^{-1}.

See also: **Adenylate Kinase**. **ATP Bioluminescence**: Application in Meat Industry; Application in Dairy Industry; Application in Beverage Microbiology. **Electrical Techniques**: Food Spoilage Flora and TVC. **National Legislation, Guidelines & Standards Governing Microbiology**: European Union; Japan. **Rapid Methods for Food Hygiene Inspection**. **Sampling Regimes & Statistical Evaluation of Microbiological Results**. **Total Viable Counts**: Pour Plate Technique; Spread Plate Technique. MPN; Metabolic Activity Tests; Microscopy. **Ultrasonic Imaging**.

Further Reading

Andrews WH (1998) AOAC Official Method 977.27. Bacteria in foods and cosmetics: spiral plate method. In: *AOAC Official Methods of Analysis* 16th edn, vol. 1, ch. 17, p. 5. Gaithersburg, MD: AOAC International.

Andrews WH (1998) AOAC Official Method 986.32. Aerobic plate counts in foods: Hydrophobic grid membrane filter method. In: *AOAC Official Methods of Analysis*, 16th edn, vol. 1, ch. 17, p.8. Gaithersburg, MD: AOAC International.

Andrews WH (1998) AOAC Official Method 988.18. Aerobic plate count: Pectin gel method. In: *AOAC Official Methods of Analysis*, 16th edn, vol. 1, Ch. 17, p. 10. Gaithersburg, MD: AOAC International.

Andrews WH (1998) AOAC Official Method 990.12. Aerobic plate counts in foods: dry rehydratable film method. In: *AOAC Official Methods of Analysis*, 16th edn, vol. 1, ch. 17, p. 10. Gaithersburg MD: AOAC International.

Andrews WH (1998) AOAC Official Method 989.10. Bacterial and coliform counts in dairy products: dry rehydratable film methods. In: *AOAC Official Methods of Analysis*, 16th edn, vol. 1, ch. 17, p. 12E. Gaithersburg MD: AOAC International.

Beuchat LR, Copeland F, Curiale MS et al (1998) Comparison of the SimPlate total plate count method with Petrifilm, Redigel, and conventional pour-plate methods for enumerating aerobic microorganisms in foods. *Journal of Food Protection* 61: 14–18.

Chain VS and Fung DYC (1991) Comparison of Redigel, Petrifilm, Spiral Plate System, Isogrid, and aerobic plate count for determining numbers of bacteria in selected foods. *Journal of Food Protection* 54: 208–211.

Curiale MS, Sons T, McAllister JS, Hallsey B and Fox TL (1990) Dry rehydratable film for enumeration of total aerobic bacteria in foods: Collaborative study. *Journal of the Association of Official Analytical Chemists* 73: 242–248.

Donnely CB, Gilchrist JE, Peeler JT and Campbell JE (1976) Spiral plate count method for the examination of raw and pasteurized milk. *Applied and Environmental Microbiology* 32: 21–27.

Entis P and Boleszuk P (1986) Use of fast green FCF with tryptic soy agar for aerobic plate count by the hydrophobic grid membrane filter. *Journal of Food Protection* 49: 278–279.

Houghtby GA, Maturin LJ and Koenig EK (1992) Alternative methods for standard plate counts. In: *Standard Methods for the Examination of Dairy Products*, 16th edn, ch. 6, p. 225. Washington, DC: American Public Health Association.

ICMSF (1978) Microorganisms in foods: their significance and methods of enumeration. In: *International Commission on Microbial Specifications for Foods*, 2nd edn. Toronto, Canada: University of Toronto Press.

Maturin LJ and Peeler JT (1995) Aerobic plate count. In: *FDA Bacteriological Analytical Manual*, 8th edn. Gaithersburg, MD: AOAC International.

Most Probable Number (MPN)

Michael G Williams, 3M Company, St Paul, USA

Frank F Busta, Department of Food Science and Nutrition, University of Minnesota, St Paul, USA

Introduction

The most probable number (MPN) technique is a method for estimating the number of bacteria in a food or water sample. In this technique, replicate portions of the original sample are cultured to determine the presence or absence of microorganisms in each portion. The replicate portions may be obtained from a serial dilution series. Alternatively, the replicate portions may be obtained by physically partitioning the sample, as in the hydrophobic grid membrane filtration system or in a microwell system. After subdividing the sample, each portion is incubated in a nutrient milieu that selects for the growth of certain organisms or groups of organisms. At the end of an appropriate incubation period, each portion is checked for the presence or absence of growth. A calculation is made or a table is consulted to determine the MPN of target organisms present in the original sample.

An MPN index number represents the MPN of bacteria in the original sample based upon the statistical probability of the coincidence of microorganisms in each sample replicate. A 95% confidence interval represents a range of actual counts in a sample, whereby there is a 95% probability that any sample containing a number of microorganisms within that range would yield the same result by MPN techniques. Although MPN techniques are commonly used to estimate the number of coliforms or *Escherichia coli* in food samples, they are not commonly used to estimate the number of total viable microorganisms in food samples.

The sensitivity of the MPN technique is dependent upon the total volume of sample tested in the smaller portions. For example, a typical nine-tube MPN consists of sets of three tubes, one set containing 1.0 ml per tube, another set containing 0.1 ml per tube, and another set containing 0.01 ml per tube, respectively. In this example, 3.33 ml of the original sample is tested and the sensitivity of the test is approximately 0.3 bacteria per ml.

The maximum counting range of the MPN technique is dependent upon two variables: the smallest volume of sample tested in the subdivisions and the number of portions of the smallest volume tested. In general, the smaller the volume of the replicates, and the more replicates of each volume you test, the greater the maximum counting range of the test.

The precision of the MPN procedure depends upon the number of replicates of each volume of sample tested. **Table 1** shows selected MPN index numbers and 95% confidence intervals for 3-tube, 5-tube, and 10-tube dilution series. It can be concluded from the data given here that the greater the number of replicates of each volume tested, the more precise the MPN index number.

Both the accuracy and precision of the MPN technique depend on the homogeneity of the sample. Sample non-homogeneity can lead to bias in the results. The technique relies on the assumption that the probability of any two organisms falling into the

Table 1 Examples of MPN estimates and 95% confidence intervals for MPN tube tests when three-tube, five-tube, and ten-tube series are used. The examples assume that a sample of food (11 g) has been homogenized in 99 mL of a suitable diluent

Example	*Series Type*	*Positive Tubes*	*MPN/g*	*95% Confidence Limits*	
				Lower	*Upper*
1	3-tube	1–0–0	4	< 1	18
	5-tube	1–0–0	2	< 1	10
	10-tube	1–0–0	1	< 1	5
2	3-tube	2–1–0	15	4	42
	5-tube	4–1–0	17	6	40
	10-tube	8–2–0	17	8	34
3	3-tube	3–2–1	149	37	425
	5-tube	5–3–2	141	52	402
	10-tube	10 6 4	141	70	278

same subdivision is purely due to random chance. The presence of particulates in a sample is an example of sample non-homogeneity, which may affect MPN results as well as other enumeration techniques.

The MPN technique was originally based on the Bayes theorem. A related technique, the maximum likelihood estimate, uses a different set of assumptions to derive an estimated bacterial count. The MPN and the maximum likelihood estimate yield the same result.

Range of Media for Aerobic Counts

The composition of growth medium used in an MPN procedure depends upon the method for detecting positive tubes. Formulations of common growth media for total viable counts are shown in **Table 2**. Typical methods for detecting bacterial growth in broth cultures include turbidity, pH indicators, oxidation–reduction indicators, chromogenic or fluorogenic enzyme substrates or stains. The most common detection method for total viable count MPN is based upon turbidity. Chemical indicators have been used to improve the speed and interpretation of MPN results. Chemical indicators that may be used for MPN total viable count procedures are listed in **Table 3**.

Table 2 General growth media for total viable counts

Medium	*Composition*
Standard methods broth	Pancreatic digest of casein, 10.0 g l^{-1}; yeast extract, 5.0 g l^{-1}; glucose 2.0 g l^{-1}; pH 7.0 ± 0.2 at 25°C
Nutrient broth	Bacto-beef extract, 3.0 g l^{-1}; bacto-peptone, 5.0 g l^{-1}; final pH 6.8 at 25°C
Trypticase soy broth	Pancreatic digest of casein, 17.0 g l^{-1}; sodium chloride, 5.0 g l^{-1}; papaic digest of soybean meal, 3.0 g l^{-1}; dipotassium phosphate, anhydrous, 2.5 g l^{-1}; glucose, 2.5 g l^{-1}; pH 7.3 ± 0.2 at 25°C

Table 3 Chemical indicators for detection of positive most probable number tubes

Detection method	*Indicator*
pH	Bromcresol purple
Oxidation–reduction	2,3,5-Triphenyltetrazolium chloride (TTC)
	Resazurin
	Methylene blue
Fluorogenic enzyme substrates	4-Methylumbelliferylphosphate, L-alanine 7-amido-4-methylcoumarin, 4-methylumbelliferyl-β-D-glucoside

The choice of a detection system used in an MPN procedure may be affected by the food samples. For example, foods that are relatively opaque or acidic may prevent the use of turbidity- or pH-based detection, respectively. Furthermore, some unprocessed foods contain enzymes that may interfere with detection systems based upon chromogenic or fluorogenic enzyme substrates.

Interpretation of Results

After a suitable incubation period, each portion of the original sample is examined for the presence of bacterial growth. The total sample volume demonstrating bacterial growth is proportional to the number of bacteria per millilitre in the original sample. The MPN in the sample is estimated using Thomas's approximation:

$$MPN/g = P / (N \times T)^{1/2},$$

where P is the number of positive tubes, N is the quantity of inocula (g) in the negative tubes and T is the total quantity of inocula (g) in all of the tubes. The results are reported as MPN/g.

See also: **Total Counts**: Microscopy. **Total Viable Counts**: Pour Plate Technique; Spread Plate Technique;

Specific Techniques; Metabolic Activity Tests; Microscopy.

Further Reading

Beuchat LR, Copeland F and Townsend DE (1998) Comparison of the SimPlate total plate count method with petrifilm, redigel, and conventional pour-plate methods for enumerating aerobic microorganisms in foods. *J. Food Protect.* 61: 14.

Eisenhart C and Wilson PW (1943) Statistical methods and control in bacteriology. *Bacteriol. Rev.* 7: 57.

Fisher RA (1922) On the mathematical foundations of theoretical statistics. *Philos. Trans. R. Soc. A.* 222: 309.

Garthright WE. Most probable number by serial dilutions. Appendix 2. In *FDA Bacteriological Analytical Manual*, 8th edn, Gaithersburg, MD: AOAC International.

Greenwood M and Yule GU (1917) On the statistical interpretation of some bacteriological methods employed in water analysis. *J. Hyg.* 16: 36.

McCarthy JA, Thomas Jr HA and Delaney JD (1958) Evaluation of the reliability of coliform density tests. *Am. J. Pub. Health* 48: 1628–1635.

Pearson ES (1978) *The History of Statistics in the 17th and 18th Centuries*. p. 361. New York: MacMillan.

Peeler JT, Houghby GA and Rainosek AP (1992) In: *Compendium of Methods for the Microbial Examination of Foods*. 3rd edn, Washington DC: American Public Health Association.

Sharpe AN and Michaud GL (1975) Enumeration of high numbers of bacteria using hydrophobic grid membrane filters. *Appl. Microbiol.* 30: 519–524.

von Mises R (1942) On the correct use of Bayes formula. *Ann. Math. Stat.* 13: 156–165.

Metabolic Activity Tests

Aubrey F Mendonca, Iowa State University, Department of Food Science and Human Nutrition, Ames, USA

Vijay K Juneja, USDA, Eastern Regional Research Center, Wyndmoor, Pennslyvania, USA

This article addresses a range of metabolic activity tests for indirectly assessing total viable counts of microorganisms in foods. Details of the technique, applications in food microbiology, correlation with plate counts and limitations of each test are discussed.

One of the most common of the microbiological diagnostic procedures used for assessing the overall quality of foods is the standard plate count. The major drawbacks of this method are that it is time-consuming, labour-intensive and becomes expensive when a large number of samples are to be analysed. This long testing procedure is not suitable for analysis of food with a limited shelf life. If food must be held until testing is complete, distribution is delayed and the storage cost can add considerably to the cost of processing. Moreover, it is of little value to food processors when microbial counts in food samples are quickly needed to facilitate early decision-making to accept or reject production lots. It is for this reason that interest remains in the application of metabolic activity tests for determining total viable microbial counts in foods. In order to be acceptable for routine analysis, metabolic activity tests must demonstrate direct correlation with the total viable microbial count. Therefore, standard curves that correlate measurements of microbial metabolic activity such as dye reductions, adenosine triphosphate (ATP) level, heat generation, detection time of electrical impedance and CO_2 production with viable counts in the test sample must be developed. The standard curves are used to estimate the total viable count in unknown samples. Generally, the greater the number of viable microorganisms in test samples, the shorter is the detection time of these tests.

Dye Reduction Tests

The dye methylene blue and indicators resazurin and tetrazolium salts are commonly used in dye reduction tests. Dye reduction tests give an estimation of viable microbial populations in a food sample based on the time taken by the dye or indicator to change colour. For example, bacterial growth utilizes dissolved oxygen in a liquid food sample such as milk, thereby reducing the oxidation–reduction (O/R) potential of the food. Under reduced conditions, certain commonly used dyes and O/R indicators can change colour.

Details of Technique

In the methylene blue reduction test, 1.0 ml of a freshly prepared solution of methylene blue thiocyanate is mixed with 10 ml of a liquid food sample, such as milk. Tubes of the blue-coloured mixture are usually held at 0–4.4°C if it is not immediately convenient to incubate them. The tubes of samples are placed in a thermostatically controlled water bath with sufficient water to heat the samples to 36°C within 10 min of incubation. The water level is maintained above the level of the tubes' contents and the samples are protected from light. During incubation the samples are observed for colour change. The time taken for colour change in the test sample from blue to colourless is inversely proportional to the number of metabolically active organisms in the sample. Basically, the same technique is used in reduction tests

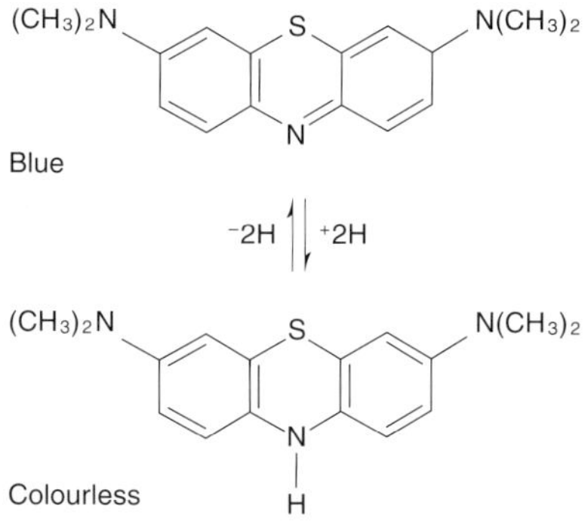

Figure 1 Reduction of methylene blue.

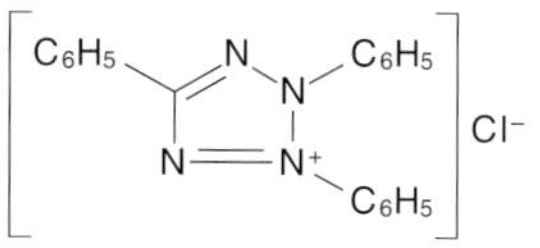

Figure 3 Triphenyl-tetrazolium chloride.

involving resazurin or 2,3,5 triphenyl tetrazolium chloride (TTC). Resazurin is used in two testing procedures to assess the microbial quality of milk: the 1 h test and the triple reading test. In the 1 h test the extent of colour change is observed after 1 h of incubation. The triple reading test involves measuring the time required (up to 3 h) for reduction of resazurin to a specified colour end point. In reduction tests involving tetrazolium salts, TTC is most often used because it is less toxic to bacteria.

Chemistry of Dye Reduction The reduction of methylene blue to its colourless form, leuko-methylene blue is shown in **Figure 1**. Resazurin can be used in place of methylene blue for assessing the number of viable microorganisms in raw milk. When added to milk this indicator undergoes two colour changes during reduction (**Fig. 2**). In the first change the indicator turns pink due to the formation of resorufin. This change results from the loss of an oxygen atom loosely bound to the nitrogen on the phenoxazine structure and is not reversible. In the second change, the pink resorufin is reduced to dihydroresorufin, which is colourless. This change is easily reversed in the presence of atmospheric oxygen. TTC (**Fig. 3**) is colourless when oxidized but gives a red colour when reduced.

Applications in Food Microbiology

Traditionally, methylene blue and resazurin reduction tests have been utilized to determine the microbial quality of milk and ice cream. These tests are used in the grading of raw milk and can be adapted for use at dairy processing plants, receiving stations, cheese factories and similar dairy operations. They are easy to perform and allow simultaneous testing of numerous food samples. Also, the methylene blue reduction test has been applied in assessing bacterial numbers in ground beef and predicting sterility of heat-processed foods. The resazurin reduction test is a rapid, inexpensive and objective test for determining excessive microbial contamination in foods. It has been applied to frozen meat, frozen poultry pies and vegetables, liquid and dried eggs and fresh scallops. In addition, it has been applied to heat-processed food products including frozen meals and precooked frozen shrimp, in which naturally occurring biological reducing agents have been inactivated.

The TTC reduction test has been applied in predicting shelf life of pasteurized milk and cream and for assessing contamination levels of food contact surfaces. The production of a red colour from reduction of TTC on areas of food contact surfaces represents sites of bacterial activity and/or soiled areas.

Correlation with Plate Counts

Dye reduction tests give a rough estimate of viable bacterial counts in food in a shorter time than that required by the standard plate count (SPC). The correlation between numbers of bacteria in milk and methylene blue and resazurin reduction time is tenuous.

Limitations

There are many limitations of the dye reduction tests that reduce their efficacy in accurately evaluating total viable microbial counts in foods. These limitations are:

- Naturally reducing substances in some foods can reduce the dyes
- Dissolved oxygen in food and oxygen absorbed from the atmosphere can prolong the dye reduction time
- Bacteria trapped in food particles, for example in

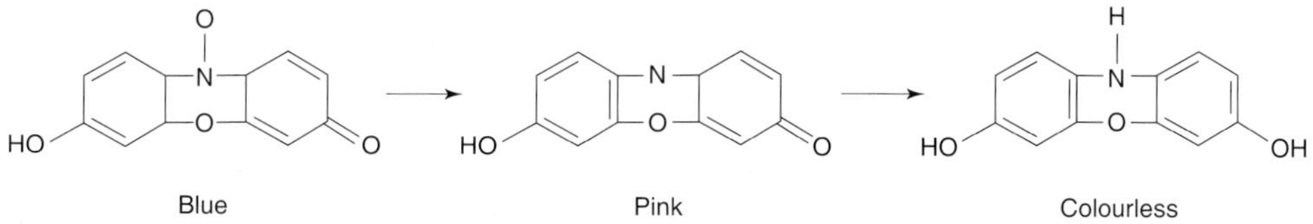

Figure 2 Reductin of resazurin.

the fat globules in the cream layer of raw milk, may not contribute to dye reduction

- Reduction rates differ among bacteria under the same test conditions
- Clumped microbial cells are not inhibited in their reducing activity but will cause a reduction in plate counts
- Presence of inhibitors such as antibiotic residues in milk can slow down the metabolic activity of organisms
- Some bacteria that reduce the dyes may be unable to grow on the agar medium and at the incubation temperature used

Adenosine Triphosphate Assay

The metabolic pool of ATP per bacterial cell is normally constant (about 0.5 fg per cell); therefore, the amount of viable microbial cells in a system can be determined by measuring cellular ATP. Bioluminescent measurement is a popular way of determining the amount of cellular ATP. This method is based on ATP measurement by use of the firefly luciferin-luciferase system. The firefly reaction is catalysed by the enzyme luciferase which uses energy contained in the ATP molecule to effect the oxidative decarboxylation of the substrate luciferin. This reaction results in the emission of light. Since the enzyme luciferase provides a means of testing for ATP, it permits the use of a rapid detection test for viable microbial populations.

Details of Technique

The example below for assessing the total viable bacterial count in raw milk by ATP bioluminescence briefly explains the technique. The raw milk sample is incubated with an extractant and apyrase to degrade somatic cells and the released ATP, respectively. The milk sample is then filtered through a positively charged nylon filter membrane (0.65 μm pore size) to concentrate bacterial cells. Residual ATP and apyrase are rinsed away from trapped bacterial cells by filtering sterile diluent. ATP is then released from the bacterial cells on the filter and the filtrate is tested for ATP. In the presence of luciferase, luciferin, oxygen and magnesium ions, the sample that yields ATP in the filtrate facilitates the light-producing reaction which is measured with a liquid scintillation spectrometer or a luminometer. The amount of light emitted from the sample is displayed as relative light units (RLU). A standard curve that relates the ATP assay (log RLU ml^{-1}) and plate count (log cfu ml^{-1}) is used to estimate the total viable bacterial count in the milk sample.

Applications in Food Microbiology

ATP bioluminescence has been applied for determining microbial quality of both raw and finished food products such as raw and pasteurized milk and cream, chicken, beef, pork, fish, beverages, fruit juices and ready-to-eat foods. In addition, this test has been applied in microbial biomass testing. As low as 10^2–10^3 viable bacteria and approximately 10 yeast cells per gram or millilitre of food can be detected by this method.

Correlation with Plate Counts

Since only living cells contain ATP, the quantity of this metabolite in a sample should be proportional to the numbers of viable cells present in that sample. The bioluminescent ATP assay can predict bacterial numbers within 0.5 $\log_{10}$ of the total viable counts in beef and chicken, thereby indicating a positive correlation.

Limitations

The limitations for reliably determining total viable microbial counts in foods by bioluminescent ATP assay are linked to the physiological state of viable cells, type of cells (e.g. yeasts, bacteria) the presence of non-microbial ATP and the type of food being tested. The amount of ATP in microbial cells may differ depending on physiological state and types of cells present. For example, injured or starved microbial cells may contain approximately 10–30% of ATP present in healthy cells, and yeast cells contain approximately 100 times more ATP than bacterial cells. Additionally, a major limitation is interference from non-microbial sources of ATP in food samples. Free ATP and ATP in cells of plant or animal origin must be removed from food samples before use of the ATP assay. Also, some intrinsic factors in foods such as pigments, extreme pH, inhibitors and certain enzymes can limit the reliability of the ATP assay. For example, red meat contains natural pigments that can quench the light produced; acidic pH of foods (fruit juices) and inhibitors can react with ATP or luciferase and interfere with the assay. Also, ATPases from somatic cells in milk can hydrolyse bacterial ATP and cause the ATP assay to underestimate the amount of viable bacteria present.

Catalase Test

The majority of microorganisms that negatively impact food quality and safety are catalase-positive. For example, the predominant spoilage bacteria in perishable foods aerobically stored at cold temperatures including *Pseudomonas* spp. and *Acinetobacter*/*Moraxella*/*Psychrobacter* spp. are catalase-positive. In addition, other aerobic spoilage organisms and the vast majority of bacterial genera in the Enterobacteriaceae family are catalase-positive. Since catalase is a constitutive enzyme in aerobic and many facultative anaerobic bacteria, its concentration increases as bacterial numbers increase. Therefore, catalase activity can be used to assess bacterial populations under certain conditions.

Details of Technique

The catalase test is based on measuring the amount of gas produced from a mixture of 3% hydrogen peroxide (H_2O_2) solution and a food sample in a closed system. Two methods have been developed to estimate microbial numbers via the detection of catalase: the catalase detection tube method and the catalasemeter.

The catalase tube method involves the use of a Pasteur pipet which is heat-sealed at the narrow end. A 0.05 ml aliquot of liquid sample is first dispensed into the wider open end of the pipet followed by 0.05 ml of 3% H_2O_2. The Pasteur pipet is swirled rapidly to mix its contents, then the mixture is forced into the narrow column of the pipet by a quick flip of the wrist. The pipet is inverted after 5 s and the mixture is held via surface tension in the narrow column of the pipet. Gas bubbles formed from the activity of catalase accumulate in the upper part of the narrow column. To lessen the impact of variation in diameters of Pasteur pipets, the amount of gas generated is expressed as a percentage of the length of the total column (gas plus liquid). For example:

$$\% \text{ Gas column} = (\text{gas column}/\text{total column}) \times 100$$

Generally, 10^5 cfu of catalase-positive bacteria per millilitre of test sample will form a column of gas which increases as the bacterial count increases.

The catalasemeter involves the use of the disc flotation method. A paper disc inoculated with an appropriate amount of liquid test sample is dropped into a test tube containing 3% H_2O_2 solution (5.0 ml) and 10^{-6} mol l^{-1} ethylenediaminetetraacetic acid (EDTA). EDTA prevents trace metal ions from decomposing the H_2O_2. The inoculated paper disc, held with forceps, is oriented perpendicularly to the surface of the H_2O_2, then released into the H_2O_2 solution. The time from the moment the disc contacts the solution to the time when it returns to the surface is recorded via interference of a light beam focused below the meniscus. This measured time is noted as the disc flotation time. The buoyancy of the paper disc is due to molecular oxygen generated from the reaction between catalase and H_2O_2 in the interstices of the disc. A short flotation time (in seconds) results from a high concentration of catalase which indicates the presence of a high population of catalase-positive microorganisms. Conversely, a long flotation time (100–1000 s) is due to low catalase concentration. When catalase is absent, the disc does not float. Data from plate counts and flotation time are used to prepare a standard curve for estimating the total microbial counts of unknown samples.

Applications in Food Microbiology

The catalase test gives a rough estimation of the level of microbial contamination of foods. This test has been applied in monitoring microbial contamination of raw materials, food samples from in-plant production lines and finished food products. It has been employed in assessing the bacterial quality of chicken and cod fillets and determining the sanitation level in meat-processing plants. In all instances, efforts are made to minimize the influence of non-microbial catalase on results of the tests.

Correlation with Plate Counts

The catalase test via use of the catalasemeter has shown good correlation with plate counts of psychotrophic bacteria in fish ($r = 0.95$) and chicken ($r = 0.93$).

Limitations

The major factors that limit the usefulness of the catalase test for estimating aerobic microbial populations are decomposition of substrate (H_2O_2), presence of naturally occurring non-bacterial catalase in test samples and poor sensitivity of the test for food samples with low microbial counts. Decomposition of H_2O_2 is caused by trace metal ions in test samples and can be minimized by addition of 10^{-6} mol l^{-1} EDTA to H_2O_2.

Also, naturally occurring non-microbial catalase in test samples interferes with the test. This interfering catalase can be inhibited by acidifying the test sample in pH 3.25 phosphate buffer. The catalase test is not sensitive enough to detect bacterial counts less than 10^4 cfu ml^{-1} or cfu g^{-1} food. Therefore, its use is limited to foods with relatively high bacterial populations. Another factor that can limit the effectiveness of the catalase test is the handling history of the food sample before testing. For example, freezing and thawing food samples can disrupt cellular membranes and reduce the total viable counts. A combination of lowered bacterial numbers and catalase activity in disrupted cytoplasmic membrane of non-viable cells can result in poor correlation between the catalase test and plate counts.

Electrical Impedance Test

Electrical impedance refers to the resistance to the flow of an alternating current through a medium. The use of impedance tests for estimating numbers of viable microorganisms is based on the association between microbial metabolic activity and electrochemical changes in a growth medium. During microbial growth, large, relatively uncharged molecules such as sugars, proteins and fats are metabolized to smaller, highly charged molecules such as lactic acid, amino acids and fatty acids. The production of these highly charged molecules results in a decrease in the electrical impedance of the growth medium which can be detected and measured before microbial colonies could become visible on agar media. Therefore, impedance testing is a relatively rapid way to estimate total viable microbial populations in food products. Several automated systems are commercially available and the manufacturers provide information of the technique and ways for interpreting results of impedance tests for estimating microbial populations in food.

Details of Technique

For impedance testing, 1 : 10 dilutions of solid food samples in 0.1% peptone water are prepared and 1.0 ml aliquots of the diluted samples are each added to 1.0 ml of an appropriate growth medium in sterile wells of the impedance detection instrument. Samples (1 and 0.5 ml) of liquid foods can be added directly to broth and agar media, respectively, in the wells of the instrument. The samples are incubated and impedance changes are measured over a 24 h period. The impedance detection time (IDT) is recorded and used to estimate viable microbial populations in the samples. The level of viable microbial populations is inversely related to the IDT.

The two methods used to determine if microbial numbers meet a set of specifications in a particular food sample are the calibration curve method and the sterility method. In the calibration curve method, a calibration curve is developed to relate IDT to a parameter of a comparison method such as the standard plate count. Data from analysis of approximately 80–100 food samples with microbial counts ranging over several $\log_{10}$ cycles are used to develop a meaningful calibration curve. Enough food samples with microbial numbers above and below a specified limit and representative of natural variation in microorganisms between different batches are used. Linear or quadratic regression analysis is used to analyse data from impedance and comparison methods. In developing a calibration curve, food samples with high (> 107 cfu ml^{-1}) and low (< 10 cfu ml^{-1}) contamination levels are carefully considered for the following reasons: first, at high contamination levels, it is impossible to distinguish microbial numbers because the detection threshold is rapidly achieved and second, at low levels, there is increased scatter in data points due to increased sampling errors. Food samples containing few microorganisms result in extended lag times and long, variable detection times. When a reliable calibration curve is developed, it can be used for estimating plate counts, determining generation times of contaminating microorganisms and classifying samples that are above or below an acceptable microbial level.

The sterility method was developed to detect impedance changes only if the test sample is contaminated above an acceptable level. In this method, the test sample is diluted to permit detection of higher numbers of microorganisms that might be present in the undiluted sample. Impedance detection in the diluted sample indicates a contaminated product. For example, if the food samples normally contain between 1 and 500 cfu ml^{-1} and the acceptable level is 1000 cfu ml^{-1}, a 10^{-2} dilution of the sample would be prepared for testing. This technique could be used for several levels of contamination by using appropriate dilution schemes. The usefulness of this technique depends on a large difference in counts between acceptable and unacceptable samples.

Applications in Food Microbiology

Impedance tests are applied in estimating total aerobic counts and selected groups of organisms, predicting shelf life of foods, determining sterility of UHT products, preservation challenge testing and hygiene monitoring. The most common application of impedance test is for estimating the aerobic plate count of food samples. The aim is to determine if the microbial population in a test sample falls above or below a set permissible level. This method has been employed for a variety of foods, including meats, fish, raw milk and frozen vegetables. A calibration curve must be developed before routinely applying this method.

A similar method involving the use of selective media is applied in detecting Gram-negative spoilage bacteria in pasteurized milk, coliforms in meat and dairy products, staphylococci in meat and yeasts in fruit juices and yoghurt. Impedance testing for shelf-life prediction involves a pre-incubation step whereby the food sample (with or without added growth media) is incubated at room temperature or slightly abusive temperature to permit multiplication of spoilage microorganisms. A specified volume of sample is then transferred to the measuring well for automatic impedance monitoring. Impedance tests are also used to monitor microbial growth rate in foods with added preservatives and to determine the inhibitory effects of pH, temperature and water activity on bacteria in foods. In addition, they are used to evaluate biofilms (from which it is difficult to release microbial cells in solution) and the efficacy of disinfectants against biofilms.

Correlations with Plate Counts

Impedance measurements for estimation of psychrotrophs on cod fillets, raw milk and pasteurized milk correlate well with plate counts. In testing of 200 samples of puréed vegetables for unacceptable levels of bacteria, 90–95% agreement is obtained between impedance measurements and plate counts. In comparing plate counts with rapid methods for estimating microbial shelf life of pasteurized milk and cottage cheese, impedance measurements are significantly correlated to shelf life.

Limitations

Natural inhibitors present in some foods, composition of microbial growth medium and differences in generation times of organisms limit the reliability of impedance tests. Natural inhibitors and/or added preservatives in foods extend the lag time and consequently delay production of an impedance signal. Inhibitors can also increase the generation time of organisms. The effects of inhibitors are minimized by diluting the food sample, neutralizing the inhibitor or using separation procedures. Variation in levels of inhibitors results in poor correlation between IDT and plate counts.

The nutrient composition of microbial growth media can cause microorganisms to utilize different metabolic pathways in different media. The addition of certain metal ions to liquid growth medium results in better impedance signals than others; therefore, it is assumed that certain metabolic pathways yield better impedance signals. Traditional growth media such as tryptic soy broth and brain–heart infusion might not always be useful for impedance testing. In many instances, specially formulated growth media produce better impedance signals than traditional media.

Microcalorimetry

Microcalorimetry involves the measurement of small amounts of heat that microorganisms produce during growth. The heat production is closely related to cellular catabolic activities and can be measured by very sensitive calorimeters such as the Calvet instrument. This instrument can detect 0.01 calories of heat per hour from a 10^{-1} sample. Microcalorimetry has good potential for use in the rapid determination of viable microorganisms in food.

Details of the Technique

Microcalorimetric tests conducted on food samples measure the exothermic heat production rate (HPR) of food-borne microorganisms. Results of these tests can be recorded as time to reach the peak HPR or as the minimum detection time. Samples of food homogenate enclosed in sealed disposable glass ampoules are temperature-equilibrated before inserting them in the ampoule holder in the thermophile area of the calorimeter. Control (sterile) samples are handled exactly the same way. The equilibration time is included in the times for achieving maximum HPRs obtained from the thermograms. The calorimeter's operation temperature is usually 30°C but other appropriate incubation temperatures can be used. Electric heaters (~ 50 ohm depending on type of instrument) are used for electrical calibration of the calorimeter. Proper calibration ensures that reactions common to both test samples and controls or minor temperature fluctuations do not contribute to background signals. The total exothermic HPR is represented by a microvolt output. The microvolt output is amplified and graphically recorded as a strip chart with microvolt values that correspond to appropriate HPRs in calories per hour.

Applications in Food Microbiology

Microcalorimetry has several applications in food microbiology. This technique has been applied in the study of microbial spoilage of ground beef and canned foods, and estimation of bacteria in milk and meat products. It has also been applied in differentiating genera in the Enterobacteriaceae family, detection of *Staphylococcus aureus* and characterization of commercially used yeast strains.

Correlations with Plate Counts

There is a linear relationship between initial bacterial levels (ranging from 10^1 to 10^9 cfu ml^{-1}) and peak HPRs. Accordingly, a significant correlation exists between initial viable counts of mesophiles or psychrotrophs and HPR peak times.

Limitations

Removal of aliquots of test samples from the calorimeter seriously interferes with the HPR signal. Therefore, replicate samples have to be prepared simultaneously and held under the same conditions as the calorimeter bath. It is assumed that numbers of viable microorganisms in replicate samples correspond to their counterparts in the calorimeter. Another limitation is that the operation temperature of the calorimeter can significantly influence the metabolic rate of microorganisms. At any particular temperature the generation times of microbial species vary considerably; for example, at 30°C mesophiles grow about five times faster than psychrotrophs. Consequently, the use of this temperature for determination of viable counts via calorimetry reduces the correlation between HPR peak times and viable counts. It is therefore necessary to explore the use of an operating temperature that could minimize the difference in generation times for these two groups of microorganisms. Correlation between HPR peak times and initial microbial counts is strong when monitoring of HPRs is carried out at 21°C rather than at 30°C. Also, the considerable variation that exists in heat production of bacteria influences the lowest number of bacterial cells necessary to produce detectable changes in HPRs.

Radiometry

Radiometry involves the measurement of radioactive CO_2 produced by bacterial metabolism of ^{14}C-labelled substrate incorporated in a growth medium. Labelled (^{14}C) glucose is used for microorganisms that utilize glucose. Other compounds such as ^{14}C glutamate and ^{14}C formate can be used for microorganisms that do not utilize glucose. The CO_2 liberated by the microorganisms is measured by a radioactivity instrument and the detection times for $^{14}CO_2$ are recorded. These results are used to estimate the viable microbial counts in test samples.

Details of Technique

Serum vials containing growth medium with radiolabelled metabolite are inoculated with test samples. Anaerobic conditions in the vials can be created by using reduced medium and sparging vial contents with appropriate gases such as N_2 or CO_2. Inoculated vials are incubated at an appropriate temperature and the head spaces of the vials are periodically tested for $^{14}CO_2$. The detection time for $^{14}CO_2$ is inversely related to the number of viable organisms in the test sample. The detection time can be shortened (within limits) by increasing the concentration of radioactive metabolite in the growth medium. Plate counts of samples are determined simultaneously and results are used to construct standard curves of $\log_{10}$ numbers of viable microorganisms versus $^{14}CO_2$ detection time. The standard curves are used to estimate total viable counts of unknown samples.

Correlation with Plate Counts

Results of radiometric tests for pure as well as mixed cultures show good correlation between viable microbial counts and $^{14}CO_2$ detection time. A high degree of correlation ($r = 0.97$) is observed between the concentration of microorganisms in cooked meat and the $^{14}CO_2$ detection time.

Limitations

First, the relatively high cost of radiolabelled substrate increases the cost of radiometric testing of food samples. Second, radiolabelled substrates are not accepted for use in the food industry.

Pyruvate Estimation

Pyruvate is a common intermediary metabolite in many microorganisms. It is formed during glycolysis via the Embden–Meyerhof–Parnas, the Entner–Doudoroff and Dickens–Horecker pathways. In addition, it is formed from deamination of amino acids and from free fatty acids. The vast majority of food-borne bacteria contain pyruvate in their metabolic pool and a portion of this metabolite is excreted into the surrounding medium. Determination of the amount of microbial pyruvate in a medium can be used to estimate the concentration of viable cells in that medium.

Details of Technique

The method for pyruvate estimation is based on an enzymatic reaction involving pyruvate, the enzyme lactate dehydrogenase (LDH) and reduced nicotinamide adenine dinucleotide ($NADH_2$). Pyruvate is enzymatically reduced to lactate by LDH and $NADH_2$.

$$\mathrm{CH3-CO-COO^- + LDH + H^+} \xrightarrow{\mathrm{NADH_2} \rightarrow \mathrm{NAD^+}} \mathrm{CH3-CHOH-COO^-}$$

The reduction in $NADH_2$ concentration is measured colorimetrically at 340 nm. The method is automated to facilitate instrumental analysis of large numbers of samples. Basically, the automated system consists of a sampler, pump, dialysers, a single-channel colorimeter, recorder and voltage stabilizer. Samples of liquid food are pumped in a split stream at a specified rate and pyruvate present in the samples is dialysed through special membranes. The dialysed pyruvate is passed into a stream of $NADH_2$ in Tris buffer or into a buffered mix of $NADH_2$ and LDH. The difference in absorbance (control versus pyruvate) is recorded. Standard curves that relate colorimetric (340 nm) values to the concentration of viable microorganisms are developed by using both colorimetric data and data from plate counts of test samples. The standard curves are used to determine the concentration of viable microbial counts of unknown samples.

Applications in Food Microbiology

Pyruvate estimation has been applied mainly to determine the adequacy of sanitation practices in milk production and the bacteriological quality of raw and pasteurized milk so as to obtain information on the keeping quality of milk.

Correlation with Plate Counts

Pyruvate concentration in skim milk correlates relatively well ($r = 0.64$) with direct microscopic counts of milk samples; however, pyruvate concentration in this milk type correlates poorly ($r = 0.31$) with standard plate counts and psychrotrophic plate counts of milk. Decreased viability of bacteria due to concentrating and drying of skim milk has been suggested as the probable cause for the poor correlations.

Limitations

The efficacy of pyruvate estimation for determining viable microbial counts in foods is limited by metabolic activities of microorganisms and the processing history of the product to be tested. Some bacteria, mainly psychrotrophs, can metabolize pyruvate excreted from cells. This microbial utilization of pyruvate in test samples can lead to gross underestimation of the microbial concentration of samples. Stresses imposed by food-processing operations such as heating, drying and freezing can severely injure or destroy microbial cells, resulting in decreased viability. Decreased viability of microbial cells can lead to poor correlations of pyruvate concentration with plate counts.

Nitrogen Reduction/Glucose Dissimilation Test

Among the many groups of food-borne microflora, those that reduce nitrate consist largely of psychrotrophic, non-fermentative, rod-shaped, Gram-negative bacteria. These organisms are the predominant spoilage microflora of raw protein-rich foods with high pH (>4.5) and high water activity. Therefore, nitrate reduction can be used to assess the bacteriological quality of certain foods by estimating the viable counts of nitrate-reducing bacteria. By combining the nitrate reduction test with a test for glucose dissimilation, food-borne microorganisms that do not attack nitrate are included. Since few food-borne bacteria lack the abilities to reduce nitrate and metabolize glucose as well, the nitrogen reduction glucose dissimilation (NRGD) test can be used for estimating total viable counts of bacteria.

Details of Technique

The technique of estimating total viable counts via the NGRD test is based on monitoring the rate of nitrate reduction and glucose dissimilation in test samples. Dilutions (1:10) of test samples in non-selective broth media are centrifuged to remove interfering levels of glucose and nitrite. The pellets are suspended in broth media containing sodium nitrate and glucose at concentrations of 5 and $0.5\,g\,l^{-1}$, respectively. The suspension is then incubated at an appropriate temperature and tested at set intervals for nitrite formation and glucose depletion via the use of urine analysis dipsticks. Plate counts of the suspension are conducted simultaneously. Nitrite formation results in a pink discoloration, whereas glucose depletion results in a colour change from green to yellow. The time taken for detection of nitrite or glucose depletion is inversely proportional to the numbers of viable microorganisms in the test samples. Standard curves relating nitrate and glucose depletion times to the concentration of viable microorganisms are developed by using depletion time data and data from plate counts of test samples. The standard curves are used to determine the concentration of viable microbial counts of unknown samples based on nitrate reduction and glucose depletion times. Separate standard curves for each different food product have to be developed because detection times can vary with the type of food being tested.

Applications in Food Microbiology

The NRGD test has several applications including selection of raw food materials with acceptable quality, microbiological monitoring of perishable foods, and detecting and correcting mistakes in good manufacturing practices in areas such as airline catering, meal preparation in restaurants, canteens, hospitals, nursing homes and production of precooked frozen foods. Results of NRGD testing can indicate sanitary failures or temperature abuse that lead to the loss of microbial quality of foods.

Correlation with Plate Counts

The NRGD test gives a rough estimate of viable bacterial counts in food in a shorter time than that taken for the SPC. Even though nitrate reduction used alone correlates poorly with total plate counts in food, nitrate reduction and glucose dissimilation used together correlate well with total plate counts.

Limitations

The effectiveness of the NRGD test for estimating total viable counts in food is mainly limited by variations in metabolic activities among groups of food-borne microorganisms, the presence of microbial inhibitors in food and the physiological state of the food-borne microorganisms. Certain food-borne bacteria such as lactobacilli can deplete nitrite via nitrite reductase activity to cause false-negative results of NRGD tests. Some bacteria that can metabolize nitrate and/or glucose may not form colonies on agar media at the incubation temperature used for test samples, thus underestimating the total viable count. Inhibitors in food such as antibiotics in milk can retard the metabolic activity of food-borne microflora and drastically reduce the rate of nitrate and glucose depletion. Various degrees of sublethal cellular injuries within the food-borne microbial population and variations in the make-up of microbial communities can result in pronounced differences in overall metabolic activities which are difficult to control.

See also: **Adenylate Kinase**. **ATP Bioluminescence**: Application in Meat Industry; Application in Dairy Industry; Application in Hygiene Monitoring; Application in Beverage Microbiology. **Electrical Techniques**: Introduction. **National Legislation, Guidelines & Standards Governing Microbiology**: European Union. Japan. **Rapid Methods for Food Hygiene Inspection**. **Sampling Regimes & Statistical Evaluation of Microbiological Results**. **Total Viable Counts**: Pour Plate Technique; Spread Plate Technique; Specific Techniques; MPN; Microscopy. **Ultrasonic Imaging**. **Ultrasonic Standing Waves**.

Further Reading

Bautista DA, Vaillancourt JP, Clarke RA, Renwick S and Griffiths MW (1995) Rapid assessment of the microbiological quality of poultry carcasses using ATP bioluminescence. *Journal of Food Protection* 58: 551–554.

Boismenu D, Lepine F, Thibault C, Gagnon M, Charbonneau R and Dugas H (1991) Estimation of bacterial quality of cod fillets with the disc flotation method. *Journal of Food Science* 56: 958–961.

Fung DYC (1994) Rapid methods and automation in food microbiology: a review. *Food Review International* 10: 357–361.

Gram L and Sogaard H (1985) Microcalorimetry as a rapid method for estimation of bacterial levels in ground meat. *Journal of Food Protection* 48: 341–345.

Jay JM (1996) *Modern Food Microbiology*, 5th edn. New York: Chapman & Hall.

Learoyd SA, Kroll RG and Thurston CF (1992) An investigation of dye reduction by food-borne bacteria. *Journal of Applied Bacteriology* 72: 479–485.

Marshall RT, Lee YH, O'Brien BL and Moats WA (1982) Pyruvate as an indicator of quality in grading nonfat dry milk. *Journal of Food Protection* 45: 561–565.

Mossel DAA, Corry JEL, Struijk CB and Baird RM (1995) *Essentials of the Microbiology of Foods: A Textbook for Advanced Studies*. Chichester: John Wiley.

Rule P (1997) Measurement of microbial activity by impedance. In: *Food Microbiological Analysis: new technologies, IFT Basic Symposium Series: 12*. New York: Marcel Dekker.

Russell SM (1998) Capacitance microbiology as a means of determining the quantity of spoilage bacteria on fish fillets. *Journal of Food Protection* 61: 844–848.

Microscopy

C D Zook and **F F Busta**, Department of Food Science and Nutrition, University of Minnesota, St Paul, USA

Introduction

The concentration of viable microorganisms is an important parameter of food quality and safety. The number of viable microbes could potentially relate to the concentration of toxins, enzymes and metabolites present, indicating the spoilage or pathogenicity potential of the food. Traditional enumeration techniques such as plating on solid media or multiple-tube most probable numbers (MPNs) are labour-intensive and require lengthy incubation before quantitative results are obtained. The food industry requires rapid, sensitive and accurate measures of viable microbial populations to assess food quality. Microscopic enumeration is a rapid quantification method that may also permit morphological and Gram reaction characterization. Techniques of direct total microbial counting will be addressed first since these methods are the foundation of direct viable counting methods. Total or viable counts can be obtained using light or fluorescence microscopy or flow cytometric cell sorting linked to microscopy. Traditional microscopic enumeration is limited by the inability to distinguish viable cells from inactivated cells. The emphasis of this article is to discuss the principles involved in microscopic discrimination and quantification of viable microbial cells.

Direct Total-cell Microscopic Counting

Direct total-cell microscopic counting techniques have traditionally been used as a quick estimate of microbial population. These techniques do not distinguish viable from non-viable cells; however, they are useful to evaluate microbial load, raw material quality and sanitation efficacy. Direct total-cell counting is the foundation upon which direct viable cell counting has been developed.

Counting Chambers

A counting chamber is a thick glass slide with an etched grid of defined area. A glass slip is placed over the grid, then a liquid sample is applied. The slide is allowed to settle for 1–5 min and then viewed with a 400×, high-dry or oil-immersion lens and a statistically relevant number of cells are counted (usually >500). The undiluted sample concentration is calculated using a grid volume factor. Procedures have been given for counting chamber enumeration and error minimization. This method is appropriate for foods that have high (10^8 cells ml^{-1}) populations and when food particles will not obscure or be mistaken for cells.

Howard Mould Count

The Howard mould count is used for quantification of mycelial fragments in tomato and fruit products. A positive field is identified as one containing mould filaments whose combined lengths are greater than one-sixth of the field diameter. For specific product applications, consult the Association of Official Analytical Chemists (AOAC) *Official Methods of Analysis* (1990).

Dried Films

The dried-film technique involves spreading a known quantity or volume of food over a known area on a microscopic slide. The smear is dried, defatted, stained and the number of microorganisms per field determined. The microscopic field area is used to convert raw numbers into concentration per gram or millilitre. This method is useful for food with high densities of organisms.

Flow Cytometry

In flow cytometry, aqueous cell suspensions are passed through a sensor where optical or electrical signals are produced and quantified. Flow cytometry can count or sort cells in suspension using a combination of impedence, scattering and absorption. Flow cytometry is not technically microscopy, but fluorescent-stained cells can be flow-sorted based on staining properties and collected and quantified using fluorescence microscopy described below. Flow cytometric enumeration is used in the dairy industry for somatic cell counts and pre-pasteurization bacterial counts (Bentley Instruments, Chaska, MN). Flow cytometry has also been used for bacterial enumeration of fermented milk products and frozen vegetables. Flow cytometry was coupled to immunofluorescent antibody labelling which allowed concurrent detection of two *Salmonella* serovars in milk.

Fluorescence Microscopy and Nonspecific Fluorochromes

Quantitative fluorescence microscopy measures fluorescence from stained cells in a defined area or number of specimens. The specimen is illuminated with short-wavelength light (UV or blue), absorbs part of the light and re-emits the rest as fluorescent light.

Fluorescent Dyes The fluorescent dye acridine orange (AO : 3,6-bis (dimethylamino) acridinium chloride) is used to count microorganisms directly with fluorescent illuminator-equipped microscopes. A spectrophotometric detector can also be used to quantify emitted light, which can be correlated to a microbial density standard curve. AO binds to phosphate groups in the backbone of DNA and RNA. High dye-to-nucleotide ratios produce orange-red fluorescence when bound to single-stranded nucleic acids. It is proposed that orange-red cells are viable due to a high concentration of ribosomes indicating active transcription. AO-stained double-stranded nucleic acids emit green fluorescence. Green fluorescing cells are considered inactive since DNA is double-stranded when transcription has ceased. Use of AO requires careful consideration of the organism's state, however. Degradation, starvation or heat inactivation can result in DNA denaturation causing a false-positive viable classification of non-replicating or inactive organisms. Although the AO direct count (AODC) is sometimes used as a viable count method, the validity of this has been questioned. Despite this, AO is used as a viable stain in combination with the direct epifluorescence filter technique (DEFT).

Direct Epifluorescence Filter Technique DEFT was developed for enumeration of bacteria in milk. Organisms are concentrated from the sample by vacuum filtration and stained with AO. Epi-illumination is used to examine opaque or thick specimens but can also be applied to quantification of transparent cells. Illumination light is reflected downwards through the objective by a beam splitter on to the sample. Microorganisms can be quantified in two ways: indirectly, by measuring the overall intensity of fluorescence by a fluorometric reader, or direct counting of individual fluorescent cells manually or by image analysis. This technique is useful for populations of 5×10^3 to 5×10^8 cells per millilitre. A DEFT/AODC procedure has been described for screening viable bacteria in raw or pasteurized milk. The antibody-DEFT (Ab-DEFT) is a modification that uses antibodies to detect specific bacterial types.

Direct Viable-cell Microscopic Counting

Viability may be traditionally defined as the ability to metabolize, grow and reproduce. Lack of reproduction does not necessarily indicate non-viability, however. Injured or fastidious cells may not reproduce on selective or nutrient-deficient media, respectively. The viable but non-culturable (VNC) state may also render cells temporarily unable to reproduce. Techniques to identify intact cells (those retaining membrane integrity and permeability) as well as biochemically active cells may result in a more useful measure of viability and impact on food safety and quality. The methods discussed consider membrane integrity and biochemical activity to be indicative of cell viability.

Fluorescence microscopy and flow cytometry, discussed above, can be used in conjunction with vital stains or indicators of membrane integrity to quantify viable cells. Stains for identification of viable cells should not decrease the viability of stained cells; however, many vital stains are cytotoxic. Specimens should be examined immediately after staining to prevent artifactual damage and to obtain accurate counts.

Membrane Integrity

Dye Uptake This method considers membrane integrity and enzymatic modification to be indicative of viability. Fluorescein diacetate (FDA) is an uncharged, non-fluorescent, lipid-soluble dye that is hydrolysed to fluorescein by nonspecific intracellular esterases after uptake. Free fluorescein is polar and is retained by intact cells; accumulation results in measurable fluorescence. The fluorescein ion diffuses out of damaged membranes; cells that do not fluoresce are considered non-viable. The limitation to this method is the assumption that membrane repair does not occur and no damaged cell can recover. This method is useful for evaluating both membrane integrity and intracellular enzyme activity. Carboxyfluorescein diacetate is retained better in Gram-negative cells where FDA may be cleaved by periplasmic enzymes.

Dye Exclusion This method considers the inability of certain dyes to cross the cell membrane as indicative of membrane integrity, thus viability. Some acid dyes such as trypan blue, eosin, erythrosin, nigrosin and primulin and the basic dye propidium do not cross intact cell membranes. As a general rule, cells that admit propidium or trypan blue are dead, but those that exclude these dyes are not necessarily viable. Dye exclusion is not to be confused with dye extrusion; this is a process by which intact membranes actively pump out dyes.

Two-fluorochrome Staining This method uses both SYTO®9 and propidium iodide stains. Alone, SYTO®9 stains both intact and damaged bacterial cell membranes. Propidium iodide enters only damaged cells and interacts with SYTO®9, modifying its fluorescence. When SYTO®9 and propidium iodide are combined, intact cells fluoresce green and damaged cells fluoresce red (**Fig. 1**). This method produces little background fluorescence and is useful for mixtures of different bacteria. As with other membrane integrity techniques, the disadvantages of this method include classification of all damaged bacteria as dead without regard to recovery, and classification of all intact bacteria as live even if they never reproduce.

Dielectrophoresis (DEP) This uses an external electrical field to impose a dipole moment in a cell. The field is non-uniform, therefore the cell is moved towards or away from the high field regions depending on several factors. Below 1 kHz, movement is associated with polarization due to surface charge. Membrane fluidity, dipolar relaxations at the cell wall or membranes and transmembrane ion transport are influential between 1 kHz and 1 MHz. Above 1 MHz, segregation is influenced by membrane capacitance and interfacial polarization. A frequency range can be used where viable yeast cells display positive DEP and non-viable yeast cells display negative DEP. In one study a 10 MHz signal was used to separate viable from heat-killed yeast cells. Suspensions were stained with methylene blue to observe distribution of cells in the dielectrophoretic chamber. Viable cells (unstained) collected at the electrode edges while dead (stained) cells aggregated on top of the electrodes in triangular formations. These investigators used direct light microscopy to determine a correlation ($r = 0.995$) between known viable concentration and dielectrophoresis. This method is limited by yeast cell concentration; above $10^7\,ml^{-1}$, separation was hindered by crowding at the electrode edges.

Synthetic Activity

Dye Uptake As discussed above, this technique considers enzymatic cleavage and retention as indicative of cell viability. Bacteria require an active enzyme system in order to convert the precursor molecule to a detectable fluorochrome.

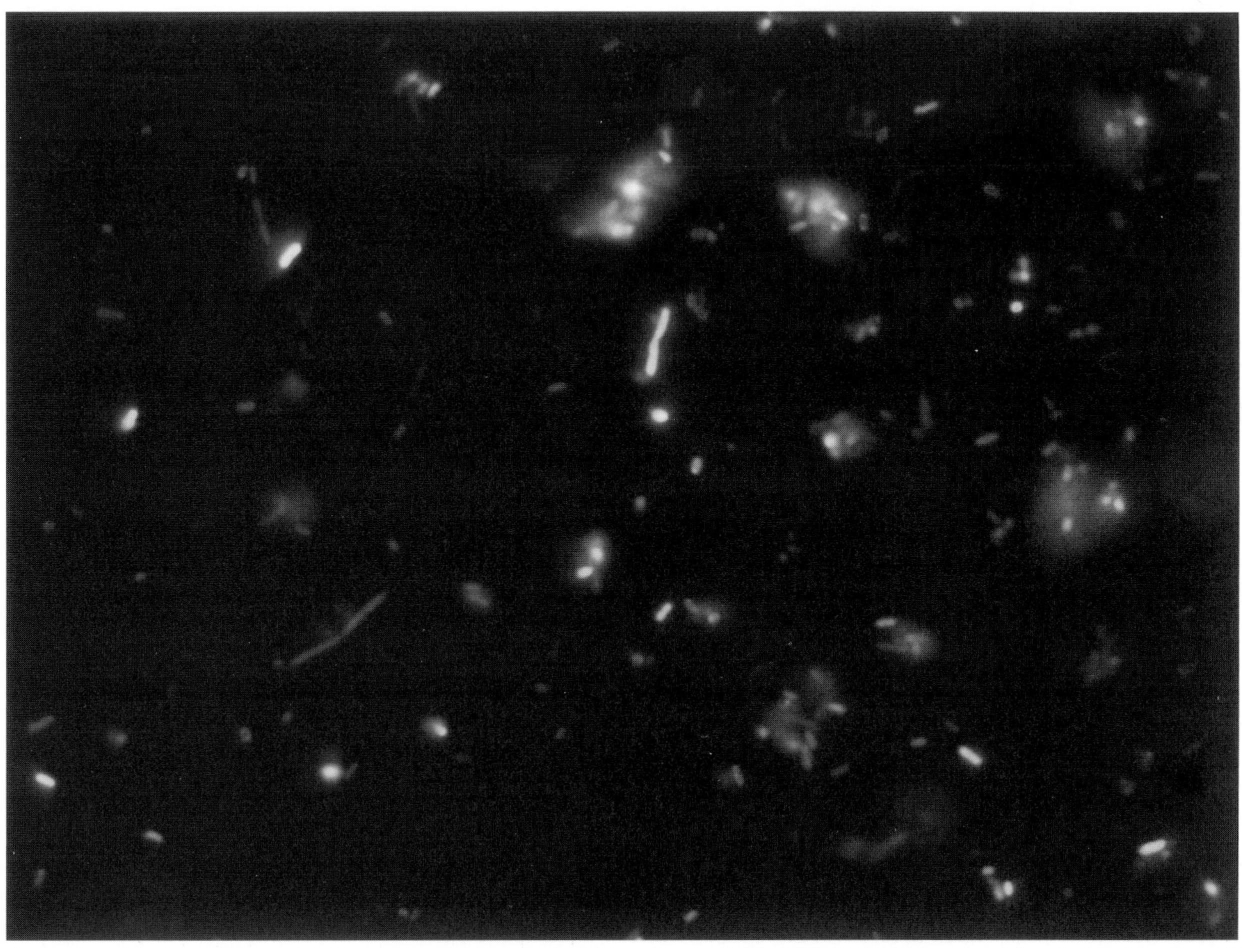

Figure 1 Mixed population of viable and inactivated *Escherichia coli* cells stained with SYTO™ and propidium iodide (Live/Dead™ Bac-Light, Molecular Probes, Inc.). Based on membrane permeability, green florescing cells are considered viable and red fluorescing cells are considered inactivated. See Haughland, 1996 reference for further information. Photomicrograph prepared by K. Kauppi, University of Minnesota. (See also color **Plate 36**.)

Microcolony Epifluorescence Microscopy (MEM) and Antibody-MEM The MEM technique combines the DEFT technique and hydrophobic filter method with a growth step, yielding an actual viable count. Cells are recovered from fluids and dilute homogenates by membrane filtration. The filter is transferred to agar media where microcolonies are allowed to develop. After incubating for 3–6 h, membranes are stained with AO then colonies are quantified by epifluorescence microscopy. This eliminates the concern over AO discrimination of cell viability, since microcolonies presumably only arise from viable cells. This technique was well correlated with total plate counting on nutrient agar when actual food samples were analysed. This technique has been used with selective media for microscopic enumeration of pseudomonads, coliforms, staphylococci and streptococci in food samples. Limitations include formation of visible microcolonies, recovery of injured bacteria on selective media and difficulty in filtering foods without using microbe-inhibitory detergents or enzymes. A further advance of MEM used *Salmonella*-specific fluorescent antibodies to label microcolonies: this could eliminate the need for stressful selective media.

Nalidixic Acid and Ciprofloxacin The nalidixic acid method compares formaldehyde-fixed cells with non-fixed cells to which yeast extract and nalidixic acid have been added. The non-fixed amended sample is incubated at a temperature specific for the sample and species present. Both samples are then filtered and stained with AO or fluorescein isothiocyanate and viewed by epifluorescence microscopy. Discrimination between viable and non-viable cells is based on growth without division; nalidixic acid inhibits DNA synthesis but not other metabolic activities. This results in elongation of viable cells. Elongated cells stained with AO that display orange-red fluorescence are counted as viable. Gram-positive bacteria are less sensitive to nalidixic acid and should be treated with ciprofloxacin, a cell division inhibitor in both Gram-negative and Gram-positive bacteria. Increased length and volume of rods and increased volume of cocci are used to quantify viable cells.

Microautoradiography Microautoradiography is used to detect uptake and incorporation of ^{3}H-labelled glucose, acetate, amino acids or thymidine. Samples are treated to yield 0.2–0.4 $\mu Ci\ ml^{-1}$ of specific activity, then fixed in 2% formaldehyde. Samples are filtered on 0.2 µm filters for dark-room application to an autoradiographic emulsion-coated slide. The slide is held in a light-tight cassette for a 3-day exposure and then removed. The bacterial emulsion is developed and fixed giving clusters of silver granules around labelled cells. The emulsion is stained with AO for viewing by epifluorescence microscopy.

Reporting

Regardless of direct viable count methodology, inclusion of conditions is critical to evaluating and comparing results. Reports should include preservative and stain types and concentrations, filter type and nominal pore size, counting strategy (number of cells or number of grids counted), total magnification and preparation and incubation protocols.

See also: **Direct Epifluorescent Filter Techniques (DEFT)**. **Flow Cytometry**. **Microscopy**: Light Microscopy; Confocal Laser Microscopy; Scanning Electron Microscopy; Transmission Electron Microscopy; Atomic Force Microscopy; Sensing Microscopy. **Total Counts**: Microscopy. **Total Viable Counts**: Pour Plate Technique; Spread Plate Technique; Specific Techniques; Metabolic Activity Tests.

Further Reading

Davies CM (1991) A comparison of fluorochromes for direct viable counts by image analysis. *Letters in Applied Microbiology* 13: 58–61.

Haughland R (1996) In: Spence M (ed.) *Handbook of Fluorescent Probes and Research Chemicals*, 6th edn. Eugene, OR: Molecular Probes 16: 365–377.

Kepner RL and Pratt JR (1994) Use of fluorochromes for direct enumeration of total bacteria in environmental samples: past and present. *Microbiological Reviews* 58: 603–615.

Kogure K, Simidu K and Taga N (1979) A tentative direct microscopic method for counting live bacteria. *Canadian Journal of Microbiology* 25: 415–420.

McClelland R and Pinder A (1994) Detection of *Salmonella typhimurium* in dairy products with flow cytometry and monoclonal antibodies. *Applied and Environmental Microbiology* 56: 4255–4262.

McFeeters G, Singh A, Byun S, Callis P and Williams S (1991) Acridine orange staining reaction as an index of physiological activity in *Escherichia coli*. *Journal of Microbiological Methods* 13: 87–97.

Murray RGE, Doetsch RN and Robinow CF (1994) Determinative and Cytological Light Microscopy. In: Gerhardt P, Murray RGE, Wood WA and Krieg NR (eds) *Methods for General and Molecular Bacteriology*. Washington, DC: American Society for Microbiology.

Nebe-von Caron G, Stephens P and Badley RA (1998) Assessment of bacterial viability status by flow cytometry and single cell sorting. *Journal of Applied Microbiology* 84: 988–998.

Pettipher GL, Kroll RG, Farr LJ and Betts RP (1989) DEFT: Recent developments for foods and beverages. In: Stannard CJ, Petitt SB and Skinner FA (eds) *Rapid Microbiological Methods for Foods, Beverages and Pharmaceuticals*. Society for Applied Bacteriology Technical Series Number 25. Boston: Blackwell Scientific Publications.

Pinder AC, Edwards C, Clarke RG, Diaper JP and Poulter SAG (1993) Detection and enumeration of viable bacteria by flow cytometry. In: Kroll RG, Gilmour A and Sussman M (eds) *New Techniques in Food and Beverage Microbiology*. Society for Applied Bacteriology Technical Series Number 31. Boston: Blackwell Scientific Publications.

Rodrigues U and Kroll RG (1989) Microcolony epifluorescence microscopy for selective enumeration of injured bacteria in frozen and heat-treated foods. *Applied and Environmental Microbiology* 55: 778–787.

Rodrigues U and Kroll RG (1990) Rapid detection of salmonellas in raw meats using a fluorescent antibody-microcolony technique. *Journal of Applied Bacteriology* 68: 213–223.

Toxicology *see* **Mycotoxins**: Toxicology.

Transmission Electron Microscopy *see* **Microscopy**: Transmission Electron Microscopy.

TRICHINELLA

H Ray Gamble, USDA, Agricultural Research Service, Parasite Biology and Epidemiology Laboratory, Beltsville, USA

Introduction

The nematode parasite *Trichinella*, which people most commonly associate with pork, has historically been a serious cause for concern as a food-borne pathogen. It is distributed worldwide and has been found in virtually all carnivorous and omnivorous animals. Under current methods of swine production, which emphasize good sanitary practices, *Trichinella* has virtually disappeared from the domestic pig in many countries. However, this parasite remains a problem where pigs are raised outdoors and are exposed to rodents and wildlife. *Trichinella* remains a common parasite in many game species and therefore poses a human health risk to hunters and others who do not prepare game meats properly to avoid infection.

Biology

Trichinella spiralis, and related species of *Trichinella*, have a direct life cycle, completing all stages of development in one host (**Fig. 1**). Transmission to another host can only occur by ingestion of muscle tissue infected with the encysted larval stage of this parasite. Once ingested, the muscle larvae are digested free from tissue in the stomach of the host and penetrate into epithelial cells of the small intestine, where they undergo four moults to the adult stage. Adult male and female worms mate and produce newborn larvae which leave the intestine and migrate, via the circulatory system, to striated muscle tissue. There, they penetrate a muscle cell, modify it to become a unique cyst called a nurse cell, and mature to become infective for another host. The time required for complete development takes 17–21 days. Adult worms in the intestine continue to produce larvae in most hosts for several weeks before they are expelled. Once adult worms are expelled and larvae reach and encyst in the musculature, no further contamination can occur. An animal that is infected with *Trichinella* is at least partially refractory to a subsequent infection due to a strong and persistent immunity. Contamination of pork, or other meat products, with *Trichinella* infective for humans, requires that an animal becomes infected a minimum of 17 days before slaughter; post-slaughter contamination with *Trichinella* is not a public health concern.

Classification and Distribution

Five species and two additional types of *Trichinella* are recognized. All species and types cause disease in humans. *Trichinella spiralis* (also called T-1) is distributed in temperate regions worldwide and is associated with a domestic pig cycle. It is highly infective for pigs, mice and rats. *Trichinella nativa* (T-2) is a cold climate-adapted species. It has limited infect-

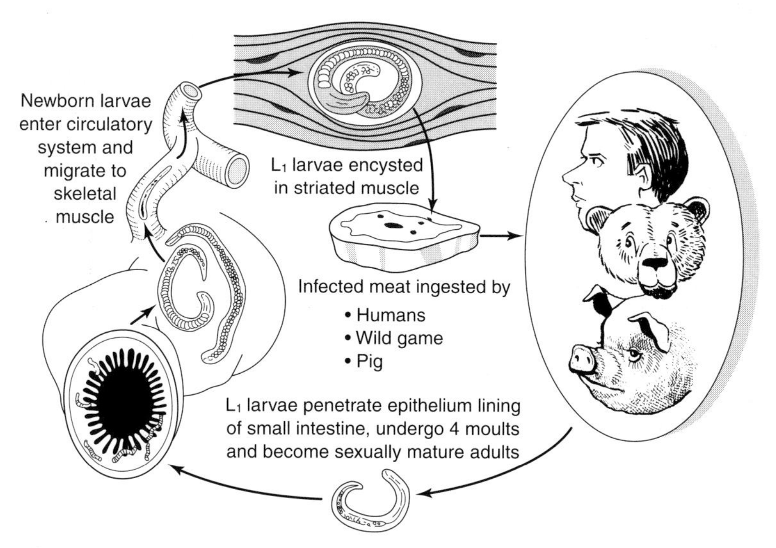

Figure 1 The life cycle and transmission patterns of *Trichinella spiralis*. Redrawn with permission from Gamble and Murrell (1988).

ivity for pigs, being found most commonly in wild canids, bear and walrus, and is further distinguished by its resistance to freezing. *T. nativa* causes human disease in arctic and subarctic regions. *Trichinella britovi* (T-3) is found predominantly in wild animals, although it may occasionally be found in pigs or horses. It occurs in temperate regions of Europe and Asia. *Trichinella nelsoni* (T-7) has been isolated sporadically from wildlife in Africa. It is characterized by greater resistance to elevated temperatures as compared to other species of *Trichinella.* Two types of *Trichinella*, designated T-5 and T-6 have been found exclusively in North America. *Trichinella* T-5 has been found in wildlife and pigs in temperate regions of the eastern United States. It appears to be closely related to *T. britovi* from Europe and Asia. *Trichinella* T-6 has been recovered from wild carnivores in mountainous regions of the continental US. It shares resistance to freezing with *T. nativa*, but does not extend as far north in its range. *Trichinella* T-5 and T-6 have lower infectivity for domestic pigs as compared to *T. spiralis.* The fifth species of *Trichinella* is *T. pseudospiralis.* This is the most divergent species, being the only one which does not form a cyst in muscle. It has been recovered from raptorial birds, wild carnivores, rats and marsupials in Asia, North America and the Australian subcontinent.

Human Trichinellosis

Humans acquire infection by ingesting raw or undercooked meat containing infective stages of the parasite. Trichinellosis (the disease caused by *Trichinella* in humans) is manifested by symptoms associated with worms developing in the intestine and in the musculature. Intestinal symptoms are only found in heavy infections and are characterized by abdominal pain and diarrhoea. Larvae invading the muscle cause fever, myalgia, malaise and periorbital oedema and an elevated eosinophil count is typical. Trichinellosis is generally not diagnosed until larvae reach the musculature. Diagnosis is based on a history of exposure to infected or suspect meat, symptoms consistent with trichinellosis, laboratory findings including positive serology, and in some cases, demonstration of parasites by biopsy. Treatment of trichinellosis generally consists of corticosteroids to reduce inflammation along with bedrest. Although some drugs can kill worms in muscle tissue, the resulting inflammation caused by dead worms often creates greater problems. Recovery is often complete, although muscle pain and weakness may persist. Occasionally, myocardial or neurologic complications may occur.

Prevalence in Animals and Humans

Swine The prevalence of *Trichinella* in swine varies from country to country, and regionally within countries. The lowest prevalence rates in domestic swine are found in countries where meat inspection programmes have been in place for many years (including countries of the European Union, notably Denmark and the Netherlands). In some instances, countries with long-standing inspection programmes consider themselves free from *Trichinella* in domestic swine. In countries of eastern Europe, higher prevalence rates of *Trichinella* in the swine population have been reported, which is supported by higher numbers of cases of human trichinellosis. Only sporadic information is available on the prevalence of trichinellosis in South America, Africa and Asia, but these limited reports suggest high infection rates in pigs. In the United States, the prevalence in pigs has changed dramatically. At the turn of the century, more than 2.5% of pigs tested had *Trichinella* infection. This number declined to 0.95% in the 1930s, 0.63% in 1948–1952, 0.16% in 1961–1965 and 0.12% in 1966–1970. Testing of sera drawn for a national swine survey in 1995 gave an infection rate of 0.013%. This dramatic decline in *Trichinella* infection in pigs is related to changes in the industry. Major inroads were made into *Trichinella* infection with the advent of garbage cooking laws passed for vesicular exanthema (1953–1954) and the hog cholera eradication programme (1962). Of equal importance has been the movement to high levels of biosecurity and hygiene under which most pigs are now raised. However, regional foci of *Trichinella* infection persist, as shown by studies conducted in northeastern US in 1985 and 1995, which revealed prevalence rates in pigs of 0.37–0.73%.

Horses Most evidence for infection in horses comes from the implication of horsemeat in human disease outbreaks. Despite widespread testing, detection of natural infections in horses has been rare. Recently, naturally infected horses have been identified from Mexico and Romania.

Wild Animals *Trichinella* infection in wildlife varies tremendously from region to region, but it is safe to say that no area is completely free from this parasite in nature. The highest rates of infection are found in foxes, wolves and bears, where infection rates can reach 85–90% of the population. It should be noted that infection rates in wildlife tend to increase in colder climates. In the domestic pig cycle, rats, skunks, raccoons and other small mammals play an important role and are often found to have high infection rates.

Humans Human infections resulting from pork vary from zero in some of the northern and western European countries to hundreds or thousands of cases annually in eastern European and Asian countries. The current rate of human infection in the US is about 25 cases per year (1991–1996) with only a portion of these infections attributable to pork. However, autopsy studies in the US and from other countries suggest that reported clinical cases of trichinellosis represent only a small fraction of actual infections in humans. A National Institute of Health report published in 1943 found 16.1% of the US population to be infected, although only about 400 clinical cases were reported each year between 1947 and 1950. This discrepancy, between reports of clinical disease and postmortem findings, probably results from a combination of subclinical infections and frequent misdiagnosis.

Epidemiology

Transmission of *Trichinella* in the sylvatic (wildlife) cycle relies on predation and carrion feeding. Generally, prevalence rates among carnivores increase up through the food chain. Sylvatic *Trichinella* infection affects public health in two ways. As a direct source, game meats pose a significant risk for human exposure to this parasite. Sources of documented human infection include wild boar, bear, walrus, fox and cougar. The reduction of exposure to *Trichinella* from these sources relies on education of hunters regarding the risks associated with eating raw or undercooked game meats. Sylvatic *Trichinella* infection also poses a risk as a source of infection to pigs. This is particularly true for *T. spiralis* and other species/types which can infect pigs. Limiting the contact of pigs with wildlife is part of an overall risk programme for control of *Trichinella* infection in the domestic pig cycle.

Exposure of domestic swine to *Trichinella* spp. is limited to a few possibilities including: (1) feeding of animal waste products or other feed contaminated with parasites; (2) exposure to rodents or other wildlife infected with *Trichinella*; or (3) cannibalism within an infected herd. The use of good production/management practices for swine husbandry will preclude most risks for exposure to *Trichinella* in the environment.

Legislation and Control

Legislation

Many countries require that pigs and horses sold as food animals be tested for *Trichinella* infection. These requirements are in the form of regulations governing slaughter inspection. The European Union (EU) outlines these requirements in Directives 77/96/EEC, 84/319/EEC and 91/447/EEC. Other countries have similar regulations and proof of freedom from *Trichinella* infection must accompany products sold for interstate commerce. The International Animal Health Code of the Office Internationale des Epizooties (OIE) specifies that importing countries should require that an international health certificate accompany imported pork products. The sanitary certificate should attest that the product has: (1) been tested for *Trichinella* infection at slaughter and was shown to be negative; (2) originated from a *Trichinella*-free country or territory; or (3) been processed to ensure destruction of *Trichinella* larvae. In the same document, it is specified that horsemeat sold for human consumption should be submitted to slaughter inspection or be processed by methods known to kill *Trichinella* larvae.

Some countries, including the United States, have no requirement for testing pigs at slaughter for *Trichinella* infection. Voluntary testing programmes, based on EU methods, allow the US to sell pork in international markets. Control of human exposure in the US relies on further processing of ready-to-eat products under methods described in the Code of Federal Regulations. Consumers of fresh product are advised to cook pork, and other meat products, to an internal temperature of 71°C (160°F).

Preharvest Control

Prevention of infection of pigs on the farm should be the first step in reducing the risk of human exposure. Prevention of infection requires implementation of sanitary management practices which: (1) prohibit feeding of animal products (without proper cooking); and (2) preventing exposure to rodents or other potentially infected mammals either directly, or through contamination of feed. Production practices which are free from risk or have minimal risk for exposure to *Trichinella* should be monitored periodically (by testing blood samples from live animals or by sampling slaughtered animals) to verify the absence of infection.

Testing pigs for *Trichinella* infection can be performed antemortem by detection of antibodies to the parasites in serum, blood or tissue fluid samples. An enzyme-linked immunosorbent assay (ELISA) has been used extensively for testing in both pre- and post-slaughter applications. Based on the use of an excretory–secretory antigen collected from short-term in vitro cultivation of *T. spiralis*, the ELISA has proven to be highly sensitive and specific; no known cross-reactions occur using this test. Since the ELISA is not in widespread use for the detection of trichinellosis in swine at slaughter, the reader is referred to the *OIE*

Manual of Standards for Diagnostic Tests and Vaccines for specific methodologies involving this test.

The International Animal Health Code of the OIE specifies that absence of *Trichinella* infection in pigs can be documented by declaring a country or territory as '*Trichinella*-free' based on certain criteria. These criteria include: (1) the disease is compulsorily notifiable; (2) waste food feeding is officially regulated; (3) a surveillance programme is in place to detect *Trichinella* infection at a very low prevalence in the disease-free area; (4) surveillance is intensified where infection was last reported; and (5) any outbreaks of infection in swine or humans are fully investigated to determine the source.

Postharvest Control

Prevention of human exposure to infected meat products, and in particular pork, is accomplished in a variety of ways. In many countries, inspection programmes are in place at slaughter for the detection of trichinellosis in pigs. Where fresh pork is not tested, alternative methods are used to prevent exposure of humans to potentially contaminated products. These include processing methods such as cooking, freezing and curing together with recommendations to the consumer concerning requirements for thorough cooking. Use of these processes renders pork free from infective *Trichinella spiralis* larvae.

Slaughter Inspection Many countries have approved methods for postmortem inspection of pork for *Trichinella.* As it is not possible to see *Trichinella* cysts within the tissue by macroscopic examination, it is necessary to perform one of several possible laboratory tests. These methods are outlined in detail in the *OIE Manual of Standards for Diagnostic Tests and Vaccines*.

The oldest method of direct detection of *Trichinella*, and one which is still frequently used, is the compression method. Small pieces of pork, or tissue from other animals, collected from the pillars (crus muscle) of the diaphragm, or alternative sites, are compressed between two thick glass slides (a compressorium) and examined microscopically. A minimum of 1 gram should be examined to allow a theoretical sensitivity of one larva per gram (a number frequently cited as the threshold of infection posing a public health risk). In practice, the compression method, using the trichinoscope, has an approximate sensitivity of more than three larvae per gram of tissue. The compression method is suitable for testing small numbers of samples and should be used to test carcasses of wild carnivores destined for human consumption.

An improvement in direct testing methods for *Trichinella* infection is the digestion method. Samples of tissue collected from sites of parasite predilection are subjected to digestion in acidified pepsin. Larvae, freed from their muscle cell capsules, are recovered by a series of sedimentation steps, then visualized and enumerated under a microscope.

Requirements for performing the digestion test are found in the Directives of the European Economic Community (77/96/EEC and 84/319/EEC), in the US Code of Federal Regulations, and in the *OIE Manual of Standards for Diagnostic Tests and Vaccines*. Because of the widespread use of these methods and their importance in the determination of *Trichinella* in pork, a brief outline of methodology is given below.

Pooled Sample Digestion Methods Post-slaughter samples are taken from the pillars (crus muscle) of the diaphragm or alternative sites including the tongue, neck, intercostals, or psoas. In pigs, the diaphragm and tongue accumulate considerably higher numbers of larvae compared to other tissues. In horses, the tongue, masseter and muscles of the neck contain the highest numbers of larvae. For samples from other mammals, tongue or diaphragm are good choices for testing. Samples size may be 1 g (as required by the European Union), 5 g (as specified by the US) or larger to increase sensitivity. Sensitivity of the method using a 1 g sample is approximately three larvae per gram (LPG) of tissue, whereas sensitivity using a 5 g sample is 1 LPG or less. Individual samples or pooled samples of up to 100 g of tissue should be ground or otherwise disrupted to allow for a thorough digestion. Ground, blended or minced samples are added to 1–2 l of digestion fluid containing pepsin and hydrochloric acid. Variations in these methods can be found in the literature. An effective digestion fluid uses 1 l per 100 g of tissue to be digested, containing 1.0% hydrochloric acid and 10 g of pepsin (1 : 10 000 National Standard Formulary). The digestion mixture should be agitated by mechanical stirring (magnetic stir bar or motorized stirring device) at a temperature of 42–46°C until digestion is visually observed to be complete (few pieces of musculature remaining). The time for digestion will be 30–60 minutes or longer depending on the method of sample preparation and the inherent digestibility of the tissue. When digestion is complete, stirring is stopped and the digestion mixture is allowed to settle for approximately 20 min. After settling, the top three-quarters of the fluid is decanted and the remaining fluid and sediment is poured through a sieve (180–350 μm mesh) into a settling vessel (conical bottom glass or sedimentation funnel). After settling for a period of 15–20 min, the top fluid is again decanted and the remaining sediment poured into a 50 ml conical bottom tube. After another 10 min of settling, the digest is siphoned to 10 ml and

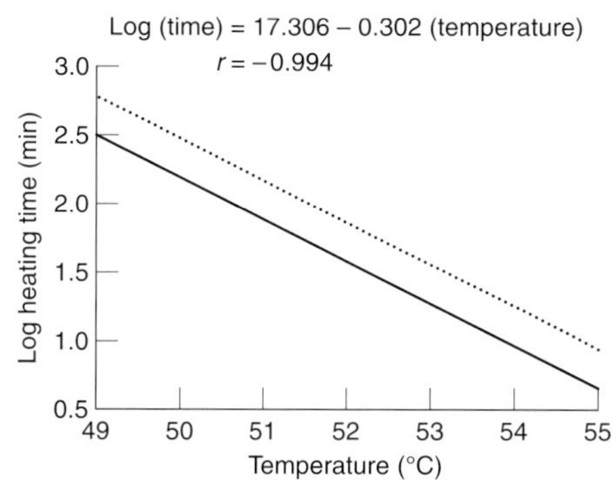

Figure 2 Linear regression (solid line) and the 99% upper confidence limits (dashed line) of the cooking time required at each temperature for the inactivation of *Trichinella spiralis* larvae. Reprinted with permission from Kotula et al (1983).

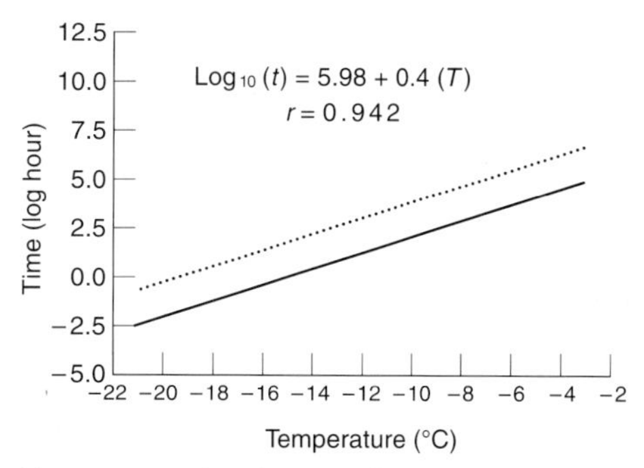

Figure 3 Linear regression (solid line), actual data points and the 99% upper confidence limits (dashed line) of the freezing time ($\log_{10}$) required at each temperature (–22 to –2°C) for the inactivation of *Trichinella spiralis* larvae. In the regression equation, t = required inactivation time and T = temperature in °C. Reprinted with permission from Kotula et al (1990).

this sample is poured into a gridded Petri dish for inspection. If the final settling step results in a cloudy sample, then additional settling steps may be performed using warm tap water. The final sediment is examined under 15–40× magnification for the presence of *Trichinella* larvae. These larvae measure 0.6–1.0 mm in length. Viable worms will be coiled or motile whereas dead worms will be in the characteristic C-shape.

In interpreting the results of direct methods for the detection of *Trichinella* in pork, the following should be considered. The sensitivity of testing is directly proportional to the amount of tissue examined. Using the most common methods of inspection testing, the sensitivity of the compression method and the digestion method is approximately 3 LPG. This level of detection is considered effective for identifying swine which pose a significant public health risk. Although there is insufficient information to determine the exact number of larvae which are necessary to cause clinical human disease (and these figures will be affected by the type of *Trichinella*, the amount of meat eaten, and the health of the individual), it is generally considered that infections > 1 LPG are a public health risk. Thus, most infections which could cause clinical human disease would be detected by currently employed direct testing methods.

Processing Specific parameters exist for the inactivation of *Trichinella* in pork products and these may be seen in depth in the US Code of Federal Regulations, and elsewhere. The following discussion is intended to provide a general overview of processing requirements.

Cooking Commercial preparation of pork products by cooking requires that meat be cooked to internal temperatures which have been shown to inactivate *Trichinella*. A thermal death curve for the interaction of temperature and time is shown in **Figure 2**. From these and other data, it was shown that *T. spiralis* is killed in 47 min at 52°C, in 6 min at 55°C and in < 1 min at 60°C. It should be noted that these times and temperatures apply only when the product reaches and maintains temperatures evenly distributed throughout the meat. Alternative methods of heating, particularly the use of microwaves, have been shown to give different results with parasites not completely inactivated when the product was heated to reach a prescribed end-point temperature. The US Code of Federal Regulations for processed pork products reflects these data, requiring pork to be cooked for 2 h at 52.2°C, for 15 min at 55.6°C, and for 1 min at 60°C.

The US Department of Agriculture recommends that consumers of fresh pork cook the product to an internal temperature of 71°C (160°F). Although this is considerably higher than temperatures at which *Trichinella* are killed (60°C), it allows for different methods of cooking which do not always result in even distribution of temperature throughout the meat. It should be noted that heating to 77°C or 82°C was not completely effective when cooking was performed using microwaves. The thermal death point for *Trichinella* is somewhat higher than temperatures recommended for inactivating bacteria in other types of red meat (i.e. heating to 54°C for ground beef products).

No information is available on the possible differential susceptibility of the different species and types of *Trichinella* to heating.

Freezing Thermal death curves have also been generated for the effect of cold temperatures on the viability of *T. spiralis* in pork. A thermal death curve for the interaction of temperature and time is shown in **Figure 3**. Based on these data, the predicted times required to kill *Trichinella* were 8 min at –20°C,

64 min at −15°C, and 4 days at −10°C. *Trichinella* were killed instantaneously at −23.3°C. The US Department of Agriculture's Code of Federal Regulations, requires that pork intended for use in processed products be frozen at −17.8°C for 106 h, at −20.6°C for 82 h, at −23.3°C for 63 h at −26.1°C for 48 h, at −28.9°C for 35 h, at −31.7°C for 22 h, at −34.5°C for 8 h and at −37.2°C for 0.5 h. These extended times take into account the amount of time required for the temperature to equalize within the meat along with a margin of safety.

It should be noted that other species and types of *Trichinella* are not susceptible to freezing in the same manner as *T. spiralis*. Both *T. nativa* (T-2) and T-6 can survive normal freezing temperatures and remain infective. Several outbreaks of human trichinellosis resulting from freeze-resistant species/types have been reported. Although *T. nativa* has low infectivity for pigs, other sources of infection, such as bears, are important in exposure of humans to this parasite.

Curing The wide variety of processes used to prepare cured pork products (sausages, hams, pork shoulder and other ready-to-eat products) makes it impossible to discuss standard requirements for inactivation of *Trichinella*. Most methods used have been tested on the particular process used to determine efficacy in killing parasites. In the curing process, the product is coated or injected with a salt mixture and allowed to equalize at refrigerated temperatures. After equalization, the product is dried, or smoked and dried, at various temperature/time combinations which have been shown to inactivate *Trichinella*. The curing process involves the interaction of salt, temperature and drying times to reach a desired water activity, percentage moisture or brine concentration. Unfortunately, no single or even combination of parameters achieved by curing has been shown to correlate definitely with *Trichinella* inactivation. All cured products should conform in process to one of many published regulations, such as the US Department of Agriculture's Code of Federal Regulations Title 9, Chapter III, 318.10. Products not produced in accordance with approved regulations should be subjected to testing by the manufacturer before sale to the consumer.

Irradiation Treatment of fresh pork with 30 krad of caesium-137 renders *Trichinella* completely non-infective.

See color Plate 37.

See also: **Meat and Poultry**: Curing of Meat. **National Legislation, Guidelines & Standards Governing Microbiology**: European Union. **Nucleic Acid-based Assays**: Overview.

Further Reading

Campbell WC (1983) *Trichinella and Trichinosis*. New York: Plenum.

Code of Federal Regulations (1990) *Animals and Animal Products*. Vol. 9, p. 212. Washington, DC: Office of the Federal Register, Government Printing Office.

European Economic Community (1977) Council Directive 77/96/EEC. *Official Journal of the European Communities* 26: 67–77.

European Economic Community (1984) Council Directive 84/19/EEC. *Journal of the European Communities* 167: 34–43.

European Economic Community (1991) Council Directive 91/433/EEC. *Journal of the European Communities* 268: 69–104.

Gamble HR and Murrell KD (1988) Trichinellosis. In: Balows W (ed.) *Laboratory Diagnosis of Infectious Disease: Principles and Practice*. P. 1018. New York: Springer-Verlag.

Gamble HR (1997) Parasites of pork and pork products. In: *OIE Scientific and Technical Review, Contamination of Animal Products: Risks and Prevention*. Vol. 16, no. 2.

Gamble HR (1998) Trichinellosis. In: *OIE Manual of Standards for Diagnostic Tests and Vaccines*. Ch. 3.5.3, p. 477.

Kotula AW, Murrell KD, Acosta-Stein L, Lamb L and Douglass L (1983) Destruction of *Trichinella spiralis* during cooking. *Journal of Food Science* 48: 765–768.

Kotula AW, Sharar A, Paroczay E, Gamble HR, Murrell KD and Douglass L (1990) Infectivity of *Trichinella* from frozen pork. *Journal of Food Protection* 53: 571–573.

Pozio E (1998) Trichinellosis in the European Union: epidemiology, ecology and economic impact. *Parasitology Today* 14: 35–38.

TRICHODERMA

Douglas E Eveleigh, Department of Microbiology, Rutgers University, New Brunswick, USA

Trichoderma spp. are cosmopolitan soil-dwelling moulds. They attack diverse organic materials and through their degradative activities produce a range of potentially useful enzymes and secondary metabolites. A few species are invasive and have recently become a scourge of the mushroom industry. On the more positive side, other species are being developed as biocontrol agents antagonistic to plant pathogens, while certain of them are being considered for use as stimulators of plant growth.

Enzymes

Trichoderma and the closely related *Gliocladium* species are metabolically versatile in attacking a diverse range of plant biomass – oligosaccharides melezitose, raffinose and sucrose and such polysaccharides as cellulose, chitin, inulin, laminaran, pectin, starch and xylan. *Trichoderma* spp. received particular attention when it was realized that degradative strains isolated during the Second World War, were particularly efficient in producing a complex of enzymes that attacked crystalline cellulose and in very high yields. Much of this realization came from the studies of E.T. Reese and M. Mandels at the US Army Laboratories, Natick, MA. One particular strain, *Trichoderma* sp. QM6a, later defined as *T. reesei* in honour of Elwyn Reese, produced up to 0.5% extracellular cellulase. In a bioenergy programme in which the production of ethanol was envisioned through the conversion of waste cellulosics, first to glucose, and followed by fermentation by yeast, hypercellulolytic mutants were created in order to facilitate the initial hydrolysis of cellulose. Such strains as *T. reesei* QM9414 and Rut C-30 routinely produced $20\,g\,l^{-1}$ extracellular protein, mainly cellulase. Higher productivities were gained through optimized fermentation protocols. As an outgrowth of the bioenergy programme, *Trichoderma* has since been considered a practical source of enzymes, including such food enzymes as cellulase, glucanase, xylanase, pectinase, chitinase, laminarinase and general hemicellulase enzymes – mannanase and arabinosidase. Commercial cell wall-degrading enzyme preparations were developed and initially used for the preparation of protoplasts from plants and also fungi; these wall-less cells were used for fundamental studies in cell fusion. These lytic enzyme preparations have been further developed and are now in use to enhance the efficiency of utilization of monogastric animal feeds. The Genencor International Multifect products are illustrative. Multifect xylanase is used principally in animal feeds. Mixes of xylanase and cellulase in Multifect XL and Multifect GC reduce the viscosity of plant materials, and are used to facilitate extraction of tea and coffee and production of fruit juice, besides being of general application in waste treatment and in agricultural silage. Though generally comprised of mixtures of lytic polysaccharases, they are available lacking protease, lipase and also amylase. Such special formulations (Multifect GGC) are used in the baking industry for modification of dough. A further *Trichoderma* cellulase-based preparation, Spezyme CP, is used in the corn milling, wheat starch and fuel alcohol industries, to enhance the separation of starch, gluten and fibre, thereby facilitating general operation by reducing viscosity, aiding filtration and reducing fouling of distillation equipment.

Trichoderma enzymes are also used in combination with enzymes from other microbial species. The classic mixed product is a cellulase–pectinase blend, the latter enzyme being from *Aspergillus niger*. Preparations include Superex Plus and also Vinemax C, respectively used for apple juice clarification and pigment extraction of red grapes. *Trichoderma* spp. produce a wide range of secondary metabolites, certain of which are potentially toxic. Thus it is emphasized that all these *Trichoderma* food-grade enzymes comply with the Food and Agriculture Organization/World Health Organization and Food Chemical Codex and fall into the US GRAS (Generally Recognized As Safe) category. Kosher grades of these enzymes are available.

Secondary Metabolites

Trichoderma produces a diverse range of secondary metabolites. Well over 100 have been characterized from *Trichoderma*, including polyketides, oxygen heterocyclic compounds, pyrones, terpenoids, polypeptides and derivatives of amino acids and fatty acids. The coconut odour associated with soils and due to the volatile 6-pentyl-α-pyrone is produced by *T. viride* and some *T. hamatum* strains. Pigments with unknown function include the anthroquinones *pachybasin* (1,8-dihydroxy-3-methyl-9,10-anthraquinone), *chrysophanol* (1,8-dihydroxy-3-methyl-9,10-anthroquinone) and *emodin* (1,6,8-trihydroxy-3-methyl-9,10-anthroquinone). Certain *Trichoderma* secondary metabolites are toxic to animals and plants, the more widely known ones falling into three mycotoxin groups: trichothecenes, cyclic peptides, and isocyanide-containing metabolites. Trichothecenes include trichodermin, which has been speculated to be produced in the soil and impairs plant growth. Cyclic peptides include the lipophilic alamethicin, suzukacilin, trichotoxins, trichopolyns and trichorianine, all of which attack membranes of bacterial and eukaryotic cells and promote lysis. Isocyanides such as trichoviridin are another class of toxic metabolites known to be produced widely by *T. hamatum* strains. This species occurs as a dominant soil microbe in certain sheep pastures, and has been implicated as a cause of ill thrift of sheep through their action in inhibiting their cellulolytic rumen microbes. As noted above, due to the potential toxicity of *Trichoderma* products, their application in the food industry is monitored extremely rigorously.

Green Mould of Mushrooms

In spite of their ability to produce large amounts of cellulases, most *Trichoderma* spp. attack wood quite poorly. They attack parenchymatous cells and bordered pits, but are not aggressive wood-destroyers, and appear as secondary colonizers sequentially following primary attack by true wood decay fungi. This has generated the theory that their ecological niche in wood decay is as necrotrophs: although they are wood saprophytes, they are also mycoparasites of the decay fungi. In this sense their role in attacking mushroom crops – the mushroom green mould – becomes apparent. *Trichoderma* spp. have been a periodic bane of mushroom producers worldwide, as their growth can cause major economic loss. The commercial production of shiitake (*Lentinus edodes*), oyster mushrooms (*Pleurotus ostreatus*) and button mushrooms (*Agaricus bisporus* and *A. bitorquis*) has sustained losses. Indeed, over the last decade, the occurrence of *Trichoderma* green mould has escalated, with button mushroom (*Agaricus* spp.) losses totalling millions of dollars. Though the disease was recognized as a disorder for mushroom production in 1953, over the last decade particularly aggressive strains of the mould have emerged in the UK, Ireland, Europe, Canada, the US and Australia. The outbreaks have been widespread, with crop yields being reduced up to 30%. The mould becomes predominant through infestation of the mushroom compost. A more minor disease, mushroom blotch (spot), in which dry, brown, sunken lesions appear on the stem and cap, has been proposed to be caused by *Trichoderma* toxins diffusing into the mushroom.

Speciation of *Trichoderma* by classical characterization requires sophistication and skill. The weed strains were initially all considered as taxons of *Trichoderma harzianum*. However, through molecular biological methodology new designations became apparent. The initial identifications have been qualified. There are four biotypes: *Th1* is the same as the *T. harzianum* neotype; *Th2* is a new species; *Th3* is in reality *T. atroviride*, while *Th4* is quite distinctive. The Irish *Th2* appears to be genetically related to the earlier Irish isolate *Th1* and may have evolved from the latter biotype at some earlier stage. The *Th2* and the North American *Th4* appear to have independent origins from other wild strains. Control of green *Trichoderma* mould is difficult but can be achieved through meticulous hygiene and sanitation, and ensuring that the composting phase attains full temperature. In addition to good microbiological aseptic technique, a further central control is to prevent the dissemination of green mould, whether by humans, insects or mites. Indeed, the use of alternative *Trichoderma* biocontrol species to control infestation by *Verticillium* mould has resulted in enhancement of red-pepper mites and spread of unwanted moulds. Modified composting strategies and the selective use of the rather costly fungicide benomyl are under evaluation as additional means of control.

Trichoderma in Biocontrol of Plant Fungal Pathogens

Trichoderma spp. also attack a range of pathogenic moulds and in such instances are being used to advantage to develop biocontrol agents active against soil-borne root diseases – cereal take-all (*Gaeumannomyces graminis*), damping off (*Pythium* spp.), root rot (*Rhizoctonia solani*) and wilts (*Sclerotinia sclerotiorum, Verticillium dahliae*) – besides leaf pathogens such as grey mould (*Botrytis* spp.). Superior biocontrol strains have been selected, larger-scale production of conidiospores and chlamydospores have been developed and specialized means for the delivery of these spores are under very active development. Strains of *T. harzianum* are particularly effective, together with strains of the closely related *Gliocladium*. Conidiospores and chlamydospores are considered more practical inocula as they survive in soil far longer than mycelia. The initial market niche for these biocontrol agents appears with greenhouse and ornamental plant production, where delivery methods and environmental conditions can be more effectively controlled. A corollary to the use of *Trichoderma* as biocontrol agents is that certain species can also promote plant growth. Effects include enhanced germination and growth of a range of agricultural plants including corn, tomato, radish and pepper. This intriguing concept is under development and again has to take full measure of the spread and persistence of *Trichoderma* growth-promoting agents in the soil.

See also: **Spoilage Problems**: Problems Caused by Fungi.

Further Reading

Claeyssens M, Nerinck W and Piens K (1998) *Carbohydrases from Trichoderma reesei and other Microorganisms*. London, UK: Royal Society of Chemistry.

Harman GE and Kubicek CP (eds) *Trichoderma and Gliocladium*. Vol. 2. *Enzymes, Biological Control and Commercial Applications*. London, UK: Taylor and Francis.

Kubicek CP and Harman GE (eds) *Trichoderma and Gliocladium*. Vol. 1. *Basic Biology, Taxonomy and Genetics*. London, UK: Taylor and Francis.

Sivasithamparam K and Ghisalberti EL (1998) Secondary metabolism in *Trichoderma* and *Gliocladium*. In: Kubicek CP and Harman GE (eds) *Trichoderma and Gliocladium* Vol. 1. *Basic Biology, Taxonomy and Genetics*. Ch. 7. London, UK: Taylor and Francis.

TRICHOTHECIUM

Arun Sharma, Food Technology Division, Bhabha Atomic Research Centre, Mumbai, India

Trichothecium Link ex Fr. is a small and still a heterogeneous genus of fungi. Some of the members of the genus are *T. polybrochum*, *T. cystosporium*, *T. pravicovi*, *T. luteum*, *T. parvum* and *T. roseum*. *T. roseum* (Pers.) Link is the type species of the genus. The conidial development in the type species has been extensively studied. *Trichothecium* Link was first reported in 1908. *Hyphelia*, *Cephalothecium* and *Trichodermia* are some of the synonyms.

Classification of *Trichothecium*

Many fungi that are known to have a septate mycelium reproduce by means of asexual reproduction by the formation of conidia. Since these fungi lack a sexual or perfect phase in the life cycle, they are grouped together as imperfect fungi or *Fungi Imperfecti* under a form class Deuteromycetes. Conidial stages of these fungi, however, are very similar to the perfect fungi that are placed in the class Ascomycetes, or Basidiomycetes. In the class Deuteromycetes and order Moniliales, there are four families namely, Moniliaceae, Dematiaceae, Tuberculariaceae and Stillbellaceae. The form family Moniliaceae is the largest of the Moniliales. All imperfect fungi that produce conidia on an unorganized, hyaline conidiophore or directly on hyaline hyphae are grouped in this family. Most species of the family are saprophytes, but many are plant parasites, animal predators or human pathogens. *Aspergillus*, *Penicillium*, *Oidium* (powdery mildews), *Monilinia* and *Botrytis* are some of the important genera of this family. *Trichothecium* is another small genus of this family.

Trichothecium has also been classified under the class Hyphomycetes which, under Deuteromycotina, is not a part of the main taxonomic classification of fungi. Like *Fungi Imperfecti*, it is also a part of the additional special purpose cross-classification in which different conidial forms of Ascomycotina, Basidiomycotina, and some Zygomycotina are grouped together. The conidial states of different conidiogenic fungi are grouped together into form genera for convenience in identification and nomenclature. The species included in the form genera are related to each other by the form of their conidia and conidiogenous apparatus. The various form genera comprising the different spore groups are Didymosporae, Basosporae, Hyalosporae and Ascosporae. Based on the conidiogenous cell the conidial group could be further classified as nonspecialized, ampulliform, raduliform, rachiform, annelliform, pluraliform and miscellaneous or nonspecific. *Trichothecium* belongs to miscellaneous group under the form genera Didymosporae. It produces arthrocatenate conidia from the apical meristem.

Characteristic Features

The morphological features of the *Trichothecium* conidia are shown in **Figure 1**.

Colony Morphology The fungus forms effused, velvety or powdery, whitish grey, yellowish or pink colonies on a solid medium.

Mycelium Upon germination *Trichothecium* conidia form creeping hyphae. The hyphae are septate, branched, smooth walled, hyaline or subhyaline.

Conidiophore The conidiophores arise singly or in loose groups, erect, straight or somewhat flexuous, mostly simple but occasionally branched, septate, scarcely swollen at the tip, with meristemic apices

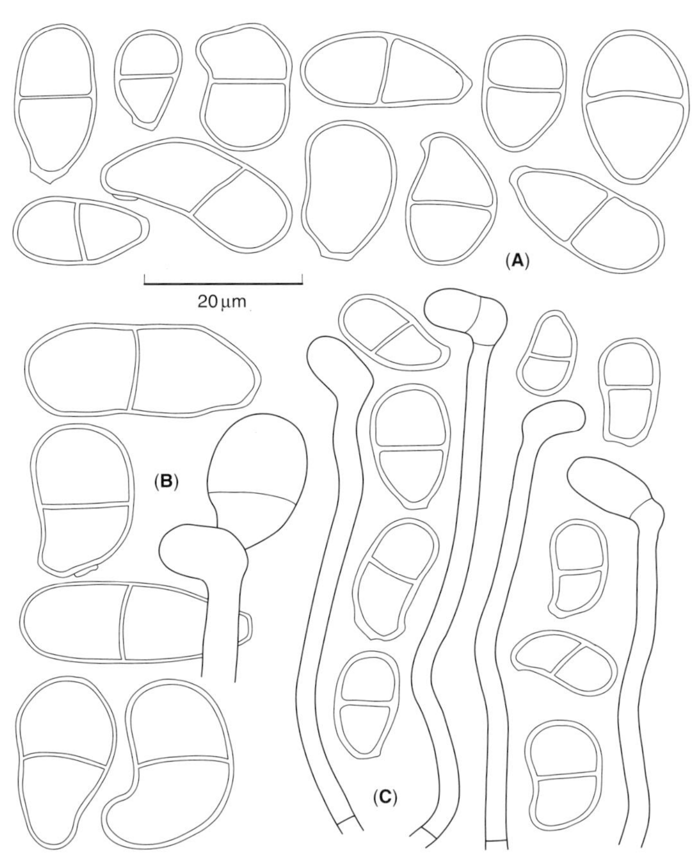

Figure 1 (**A**) *Trichothecium roseum* conidia; (**B**) *Trichothecium luteum*, conidia and conidiophores with young conidia; (**C**) *Trichothecium parvum,* conidia and conidiophores with young conidia. Reproduced from Rifai MA and Cooke RC (1966).

capable of producing conidia in basipetal succession to form characteristic chains. The meristematic apex of the conidiophore, which gives rise to meristem arthroconidia (arthrocatenate) apparently does not elongate significantly during spore formation. The characteristic basipetal catenulate conidial cluster and the asymmetric basal cell of the conidium are characters of great diagnostic value in assigning a species to the genus *Trichothecium.*

Conidia Conidia arise as blow outs from one side of the tip of the conidiophore, and after the first conidium has been put out before it fully matures, the next conidium is blown out from the opposite side. The conidia thus are pinched out one after the other and remain attached to each other on the shoulder to form zig-zag chains giving rise to a characteristic head. The conidia are ovoid or pear shaped, two celled, with the apical cell being larger and globose than the basal cell which is curved and conical. The conidia are hyaline or lightly coloured pink or pale, appear hyaline under a microscope, but pink in masses in culture or on the host. The conidia are attached to the conidiophore at the pointed end of their basal cell. The size of conidia is 12–18 μm long × 8–10 μm broad. Conidiophores and conidia of *T. polybrochum*, *T. cystosporium*, *T. pravicovi* are morphologically different from *T. roseum.*

Relatedness to Other Species

Trichothecium is closely related to a few other fungi, including *Cephalothecium roseum* Corda, *Hyphelia* and *Trichodermia* and *Spicellum roseum.* The first three are considered as synonyms. The genus *Cephalothecium* was proposed by Corda on the assumption that in *T. roseum* the conidia did not form chain-like clusters. It is probable that the two genera were based on the same species and the synonymy of the two was suggested. This suggestion was accepted without reservation by later authors. As far as *Spicellum* is concerned, a partial sequence analysis of the nuclear small 18S and nuclear large 28S ribosomal RNA subunits of the two species has been carried out. It has been suggested that the two species are from a monophyletic group. The colony characteristics and tricothecene production by the two species further strengthens this contention. However, there are dif-

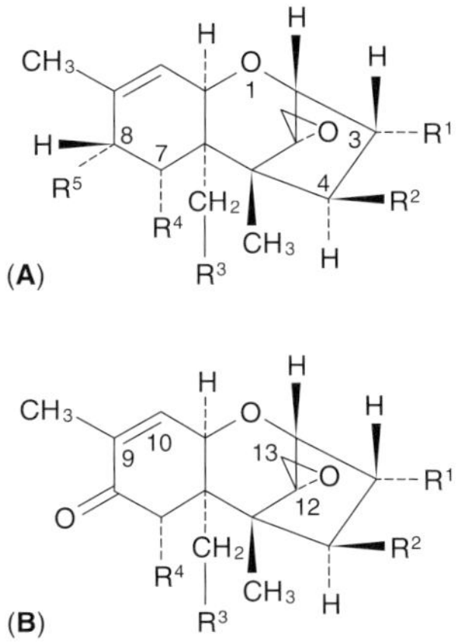

Figure 2 Naturally occurring trichothecenes. (**A**) Trichothecene: $R^1 = H$; $R^2 = H$; $R^3 = H$; $R^4 = H$; $R^5 = H$. T-2 toxin: $R^1 = OH$; $R^2 = OAc$; $R^3 = OAc$; $R^4 = H$; $R^5 = OCOCH_2CH(CH_3)_2$. (**B**) Trichothecin: $R^1 = H$; $R^2 = OCOCHCHCH_3H$; $R^3 = H$; $R^4 = H$. Deoxynivalenol: $R^1 = OH$; $R^2 = H$; $R^3 = OH$; $R^4 = OH$.

ferences in conidiophore branching and conidium ontogeny.

Phytopathogenic Potential

Trichothecium is basically a saprophyte, but it is being increasingly implicated as a secondary pathogen in fruits and vegetables. *Trichothecium* is known to cause pink rot of apples generally after apple scab infection. It can enter the fruit through the lesions caused by the primary pathogen of apple scab, *Venturia inaequalis*. Thus *Trichothecium* can be called an opportunistic pathogen in fruits and vegetables. Peach was earlier reported to be a host for the fungus. Recently, it has been reported to be associated with fruits, such as prunes (*Prunus persica* (L) Batsch), nectarines (*Prunus persica* (L) Batsch var. nactrine) and plums (*Prunus salicina*) where it has been shown to cause pink rot. It is also reported to grow on banana. *T. roseum* was found to sporulate better in the presence of *Monilinia* and *Cladosporium* infections. It is reported to cause pink rot of vegetables, such as tomato and cucurbits, and to be a pathogen on tea leaves and also to infect many forest trees. *T. roseum* is also reported to reduce the germination potential of seeds.

Mycotoxins Produced by *Trichothecium*

Trichothecium is a source of a number of secondary metabolites. These include toxins, antibiotics and other biologically active compounds. Trichothecenes comprise a group of closely related sesquiterpenoid compounds which contain a trichothecane nucleus (**Fig. 2**). The naturally occurring compounds possess an olefinic bond at C-9 and C-10, and an epoxy ring at C-12–13. Thus they are called 12,13-epoxytrichothecenes. Trichodermol is the simplest structure in the family. More than 30 derivatives of trichothecenes are known.

Acetyl-CoA
↓
Mevalonic acid
↓
Isopentyl pyrophosphate
↓
Dimethyl allyl pyrophosphate
↓
Geranyl pyrophosphate
↓
Farnesyl pyrophosphate
↓
Trichodiene
↓
Trichodiol
↓
Trichothecolone
↓
Trichothecin

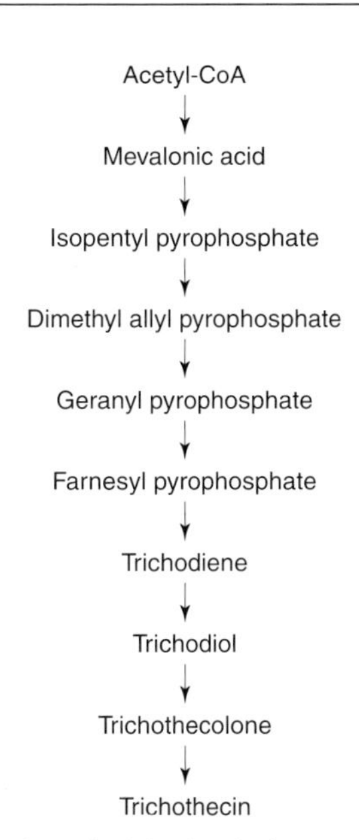

Figure 3 Formation of trichothecin from acetyl-CoA through the mevalonate pathway.

T. roseum produces a number of secondary metabolites including diterpenoids and sesquiterpenoids. The diterpenoids include roseolactone, rosololactone acetate, rosenonolactone, desoxyrosenonolactone, hydroxyrosenonolactone and acetoxyrosenonolactone. The sesquiterpenoids include 12,13-epoxytricothec-9-ene, crotocin, trichothecolone, trichothecin and trichodiols. Trichothecene mycotoxins are derived from the mevalonate pathway as shown in **Figure 3**.

The trichothecin skeleton is formed from farnesyl pyrophosphate. Trichodiene, a metabolite of *T. roseum*, was proposed as an intermediate in the biosynthetic pathway. Tritiated trichodiene was found to be incorporated into an epoxy alcohol, trichodiol which was also found to be an intermediary metabolite in the pathway for trichothecin formation. Incorporation of tritiated mevalonate in to both trichothecolone and trichodiol has been confirmed. Formation of trichodiene from farnesyl pyrophosphate in a cell-free extract has also been shown. *T. roseum* also produces roseotoxin, a cyclodepsipeptide. It consists of 1 mol each of L-isoleucine, N-methylvaline, N-methylalanine, *trans*-3-methyl-L-proline, β-alanine and 2-hydroxy-4-pentenoic acid joined together by amide and other bonds in a cyclic peptide lactone structure.

Apart from *T. roseum*, trichothecene compounds are produced by other fungi including *Fusarium*, *Myrothecium*, *Trichoderma*, *Cephalosporium* and *Stachybotrys*. In fact, the prevalence of trichothecenes in foods and feeds is considered next only to aflatoxin. The major source of trichothecenes in foods, however, is *Fusarium*. The first trichothecene compounds, verrucarins, were reported in 1946.

Detection of Trichothecenes

Trichothecenes are generally extracted from food and feed samples with aqueous methanol or acetonitrile, defatted with n-hexane, and partitioned with chloroform or dichloromethane. The chloroform or dichloromethane is evaporated from the extract and the residue is dissolved in methanol. A thin layer chromatographic (TLC) or gas chromatographic/mass spectrometric (GC/MS) analysis can be carried out on the extract. A TLC clean-up is usually recommended prior to GC/MS analysis. For TLC analysis a number of developing solvents are used. These include benzene : acetone (12 : 7 v/v), toluene : ethyl acetate :formic acid (6 : 3 : 1 v/v) and chloroform : methanol (9 : 1 v/v). Detection of trichothecenes can be carried out using various spray agents such as sulphuric acid 4-(*p*-nitrobenzyl)-pyridine, phluoroglucinol, aluminium chloride etc.

For GC/MS studies trifluoroacetate derivatives are obtained by reaction of the sample extract with trifluoroacetic anhydride at 60°C for 20 min. After evaporation of trifluoroacetate with nitrogen, the residue containing the derivatives is dissolved in chloroform. An aliqot is injected into a gas chromatographic/mass spectrometer. A fused silica capillary column coated with dimethyl silicone is used with a temperature programme of 80–300°C. Carrier gas (helium) is used with a flow rate of 90 ml min^{-1} at the inlet and 1.5 ml min^{-1} at the exit of the GC. Mass spectrometric measurements are performed in positive chemical ionization mode in methane plasma, with an electron energy of 200 eV, an ion source temperature of 150°C and a source pressure of 0.6 Torr. The quantitation is done based on area comparison of two selected ions for each trichothecene–triflouroacetate derivative with an appropriate external standard.

Toxicity

The fungus is a ubiquitous saprophyte. It is known to be associated with agricultural commodities, such as corn, sorghum, pearl millet, green gram and beans. It has also been found to be associated with poultry and cattle feed and oilseed cakes. Trichothecene mycotoxicosis can occur in cattle, swine, horse, sheep, fowl and humans. The problems associated with trichothecene exposure include poor feed conversion, feed refusal, diarrhoea, dermatitis, haemorrhage, immunosuppression, decreased egg and milk yield, haemorrhage and necrosis of rapidly proliferating tissues such as intestinal mucosa, bone marrow and spleen.

The International Agency for Research on Cancer (1993) has given a rating of 3 for trichothecenes, which means that the toxins are not classifiable as carcinogenic to humans since there is insufficient evidence of their carcinogenicity in humans and only a limited evidence in animals. However, general toxicity of trichothecenes may be very high. The first report of the toxicity of the extracts of *T. roseum* toward animals appeared in 1948 and was based on eye and skin tests in rabbits. Later a single dose of trichothecin, 250 mg/kg body weight given intravenously, was reported to cause death in rats. All trichothecenes are potent inhibitors of protein and DNA synthesis in eukaryotic cells. This property is, therefore, responsible for the potent cytotoxicity of trichothecenes. Inactivation of the initiation, elongation and termination steps in protein synthesis is reportedly caused by the conjugation of trichothecene molecules to protein, RNA and DNA. Trichothecenes have also been found to react to –SH groups of proteins. The toxicity of trichothecenes causes radiomimetic effects. Acute toxicity of trichothecenes to mice varies from 10 to 1000 mg kg^{-1} (p.o.) and in rats from 4 to 8 mg kg^{-1} (p.o.). The cyclodepsipeptide toxin from *T. roseum* (roseotoxin B) is toxic to ducklings and mice.

Other Metabolites of *Trichothecium roseum*

Reports concerning the metabolic products of the fungus *T. roseum* have appeared in the literature since 1946. Extracts of *T. roseum* have also been reported to reduce the infectivity of a number of plant viruses. Antagonism of *T. roseum* toward other fungi led to the identification of the antifungal activity of trichothecin. Since then several diterpenoid and sesquiterpenoid compounds have been isolated, purified and characterized from this fungus. Three new trichothecenes, trichothecinol A, B and C were isolated from *T. roseum* and unambiguously characterized on the basis of spectroscopic and chemical evidence. These exhibited potent inhibitory effects on Epstein–Barr virus early antigen (EBV-EA) activation induced by the tumor promoter 12-O-tetradecanoylphorbol-13-acetate (TPA). A new fungal antibiotic furanocandin has been reported from a *Trichothecium* species. *T. roseum* F1064 is also reported to produce an inhibitor of cholesteryl ester transfer protein (CETP).

See also: **Mycotoxins**: Classification.

Further Reading

Alexopoulos CJ and Mims CW (1979) *Introductory Mycology*. New York: John Wiley.

Engstrom GW, DeLance JV, Richard JL and Baetz AL (1975) Purification and characterization of roseotoxin B, a toxic cyclodepsipeptide from *Trichothecium roseum*. *Agriculture and Food Chemistry* 23: 244–253.

Gilman JC (1959) *A Manual of Soil Fungi*. Ames: Iowa State University Press.

Ingold CT (1956) The conidial apparatus of *T. roseum*, *Transactions of The British Mycological Society* 39: 460–464.

Kendrick WB and Carmichael JW (1973) Hyphomycetes. In: Ainsworth GC, Sparrow FK and Sussman AS (eds) *The Fungi*. New York: Academic Press.

Lida A, Konishi K, Kubo H et al (1996) Trichothecinol A, B and C potent antitumor promoting sesquiterpenes from the fungus *Trichothecium roseum*. *Tetrahedron Letters* 37: 9219–9220.

Moller TE and Gustavsson HF (1992) Determination of type A and Type B trichothecenes by gas chromatography using electron capture detection. *Journal of the Association of Official Analytical Chemists International* 75: 1049.

Rifai MA and Cooke RC (1966) Studies on some didymosporous genera of nematode-trapping hyphomycetes. *Transactions of the British Mycological Society* 49: 147–168.

Romer TR (1977) Analytical approaches to trichothecene mycotoxins. *Cereal Foods World* 22: 521–523.

Seifert KA, Lous SG and Savard ME (1997) The phylogenetic relationship of two trichothecene producing hyphomycetes, *Spicellum roseum* and *Trichothecium roseum*. *Mycologia* 89: 250–257.

Sinha KK and Bhatnagar D (eds) (1998) *Mycotoxins in Agriculture and Food Safety*. New York: Marcel Dekker.

Uraguchi K and Yamazaki M (1978) *Toxicology, Biochemistry and Pathology of Mycotoxins*. London: John Wiley.

Visconti A, Mirocha CJ, Bottalico A and Chelkowski J (1985) Trichothecene mycotoxins produced by *Fusarium sporotrichoides* strain P-11. *Mycotoxin Research* 1: 3–10.

UHT Treatments *see* Heat Treatment of Foods: Ultra-high Temperature (UHT) Treatments.

ULTRASONIC IMAGING
Non-destructive Methods to Detect Sterility of Aseptic Packages

Laura Raaska and **Tiina Mattila-Sandholm,** VTT Biotechnology and Food Research, Finland

Introduction

The safety criteria for aseptic foods are very important because or the long shelf life and unrestricted storage conditions of these foods. Microorganisms in aseptically processed food cause quality problems either by spoiling the product or by increasing the possible health risk. Aseptic products must be absolutely free of microorganisms, including their spores. When marketing ultra-high temperature (UHT) treated products, to ensure an acceptably low percentage of unsterile units it is necessary to check an appropriate number of packed samples for sterility from every lot. A sampling rate of about 1% is generally recommended when samples are evaluated for their microbiological safety and sensory quality. However, the system of destructive sterility testing by sampling currently in use by foodstuff producers does not guarantee consumer safety and causes large losses of ready-to-use food. At the commissioning stage of a new production line, tens of thousands of samples are tested by destructive microbiological methods, which are both uneconomical and a burden on the environment. Checking every single food container and its product content in a non-destructive way is expected to increase consumer safety and avoid losses of foodstuffs and packaging materials.

Traditionally, microbial quality control methods have focused on assessing specific food-borne pathogens. A wide range of kits and instruments are now commercially available for the detection of specific pathogens such as *Clostridium botulinum*, *Salmonella*, *Listeria* and *Escherichia coli*. However, in order to check the safety of commercial sterile products, methods that detect the growth of any microorganism are needed. To date, there are only two commercial non-invasive methods on the market. The Electester (TuomoHalonen Oy, Toijala, Finland) assesses viscosity changes in the product by first oscillating the product and then measuring the pattern of the subsequent damping of the induced motion in the fluid. Tap Tone (Benthos Inc., North Falmouth, MA, USA) uses an electric field to create a tone, by inducing vibrations in the aluminium foil of the package; the amplitude of the tone changes if, for example, gas is produced in a package without a head space. These indirect methods are often appropriate for detecting a number of different microorganisms, but as microorganisms produce different effects, the methods must be checked using a wide variety of microorganisms in order to establish the extent of their applicability.

Increasing consumer demand has accentuated the need for rapid, non-destructive on-line measurements of food quality. The ideal method, in addition to being non-destructive, should be nonspecific, to ensure that all types of microbial growth are detected. Preferably the method should measure factors that can be directly attributed to any organic growth in the product. A method with high sensitivity means that the necessary incubation time can be significantly less than that needed for standard microbial cultivation. Furthermore, in order to permit extensive testing the method should be rapid. A non-destructive sterility test method that shortens the incubation time, is more reliable than the standard methods used today and requires little labour input has great economic importance for companies producing aseptic food products.

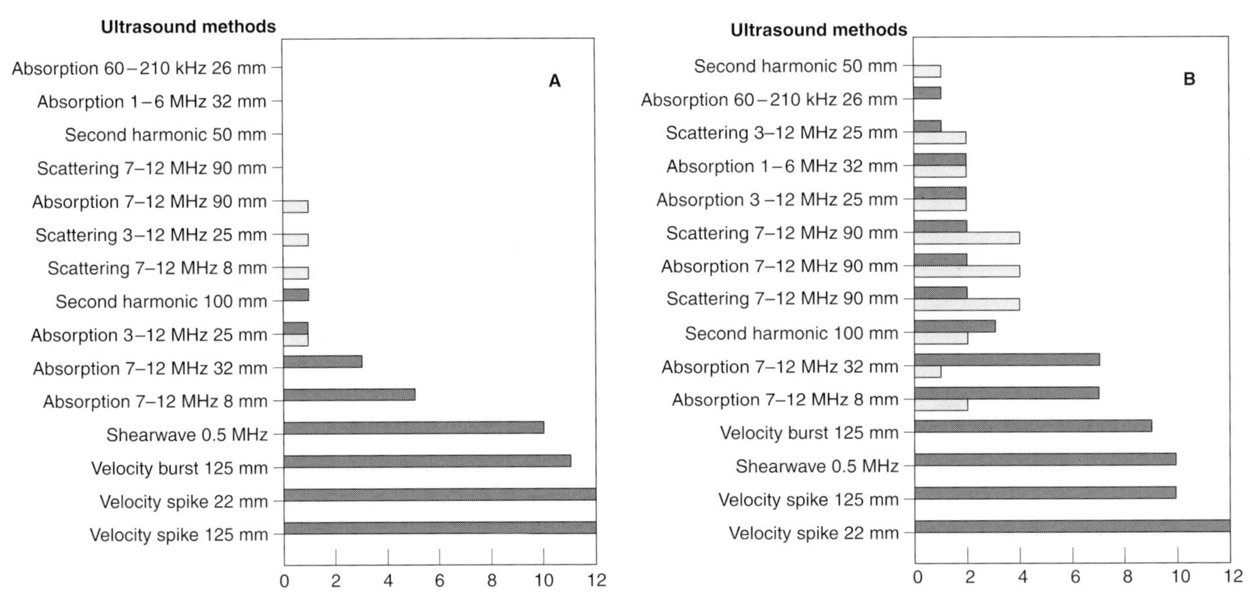

Figure 1 The sensitivity of different ultrasound methods to detect contamination. The shaded bars show the number undetected out of 12 samples; open bars are false positives. The ultrasound method, frequency and transducer distance shown are listed at the left of the graphs. (**A**) Inoculum level < 700 cfu ml^{-1} (**B**) Inoculum level < 10 cfu ml^{-1}.

Potential Non-destructive Sterility Test Methods

Contact Ultrasound Method

One interesting method for non-destructive testing is ultrasound technology. Ultrasonics can be used in food engineering for many purposes, for example to measure the different physical properties of foodstuffs. Its applicability to quality control of both fresh and processed foodstuffs has been studied. The stability of reconstituted orange juice, the skin texture of oranges, cracks in tomatoes and defects in husked sweetcorn have been investigated using ultrasonics. It is also possible to determine the presence of foreign materials such as metal and bone particles in food products with this method. Ultrasonic energy can be used for non-destructive measurement of the thickness of eggshells. It has also been shown that ultrasound imaging can be used for non-destructive testing of the microbiological quality of aseptic milk products.

The investigation was conducted with the primary idea that any change in properties compared with those of the controls is an indicator of a abnormality. The ability of some contact ultrasound techniques to detect contamination in UHT milk is shown in **Figure 1** . The sensitivity of the ultrasound methods was greater with a high inoculum level than with a low level. In addition, there were more false alarms when a low inoculum level was used. The most sensitive and accurate ultrasound techniques were second harmonic generation, absorption at 60–210 kHz, absorption at 1–6 MHz, and scattering at different frequencies and transducer distances. The measurements were clearly dependent on transducer distance and sound intensity. However, laser fluorescence and shearwave could not distinguish between contaminated products and controls, and in the case of velocity measurements, the number of undetected samples was high both with low and high inoculum levels. Repeatable and reliable ultrasound results were also shown to be dependent on the spreading and distribution of microbes in the package; shaking prior to measurement was significant, especially in the case of highly contaminated samples and more viscous products like Nantua fish sauce.

The potential of second harmonic generation and absorption was studied more thoroughly in an investigation using Tetrapak packages provided with 'windows', and water as a contact medium. The contact ultrasound measurement system was a semiautomatic system where the sample was manually changed (**Fig. 2**). The measurement vessel housed the transducers for the second harmonic measurement in addition to the ones used for the absorption measurement. An optical switch indicated when the package was positioned correctly. Small air bubbles which accumulated on the ultrasound window were cleaned off with brushes. Using this method, it was possible to record both the absorption and the second harmonic values at the same time. If only two 'windows' were used it was necessary to turn the package through 180°. One package was measured at a time; the change in absorption (1.1–5.6 MHz) and the generation of second harmonics (1 MHz → 2 MHz,

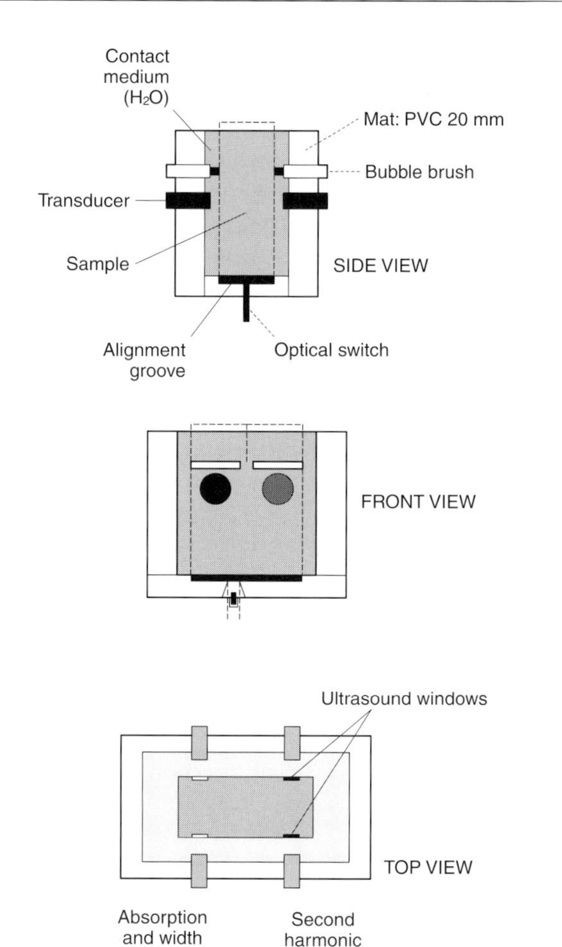

Figure 2 The measurement vessel used for assessing ultrasound techniques.

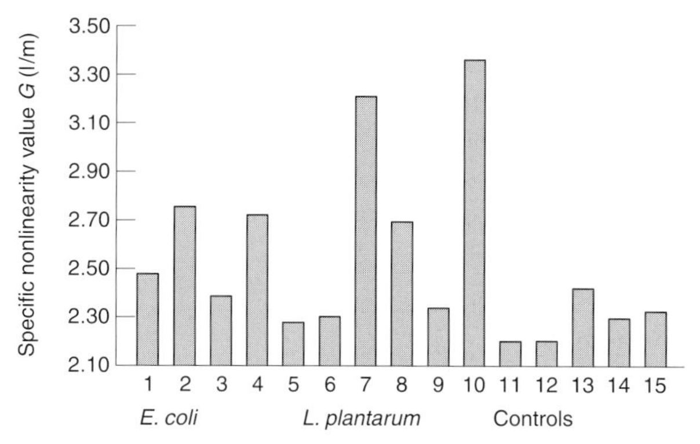

Figure 3 Detection ability of second harmonic (3 MHz) of *Escherichia coli* (samples 1–5) and *Lactobacillus plantarum* (samples 6–10) contaminated UHT milk packages from controls (samples 11–15) after 24 h. The measurement values have been corrected for the width of the package. The microbial levels were *E. coli* 6.9 log cfu ml^{-1}, *L. plantarum* 6.5 log cfu ml^{-1}.

3 MHz → 6 MHz) were measured; a powerful mathematical transform (partial least squares regression analysis) was used to extract the characteristic features from the data vector of the sample; and the characteristic vectors used for statistical inference and the samples differing from the controls were identified.

Absorption and second harmonic measurements were used to detect contaminated UHT milk packages. Measurements were performed through the 'windows' without breaking the package. All the contaminants could be detected after 24 h of incubation. During that time the microbial counts in UHT milk varied between 10^5 colony forming units (cfu) per millilitre and 10^6 cfu ml^{-1}. The detection threshold for second harmonic generation was 5% and for absorption 1.5%. The difference between ultrasound measurements of control and contaminated samples was greatest in the case of *E. coli* and smallest in the case of *Pseudomonas fluorescens*. Absorption appeared to be a more stable but less sensitive measurement technique than second harmonic generation. Although second harmonic (3 MHz → 6 MHz) and absorption could distinguish between control and contaminated packages, differences in their overall response could be detected. The ability of second harmonic (3 MHz) measurement to distinguish UHT milk packages contaminated with *E. coli* or *Lactobacillus plantarum* from control ones is shown in **Figure 3**.

The results of using second harmonic and absorption methods to detect contaminated UHT milk and Pursoup vegetable soup packages are presented in **Table 1** and **Table 2**. The packages were inoculated with several important contaminants and changes in ultrasound measurements compared to the controls were followed for 4 days. During the first day the measurements were performed 5–7 h after the inoculation, when the microbial counts were lower than 10 cfu ml^{-1}. From these results, absorption seems to be the most promising ultrasound measurement technique in detecting contamination in UHT milk. The best discrimination was obtained after 72 h of incubation when the microbial counts were 10^5–10^8 cfu ml^{-1}. However, *E. coli*, *L. plantarum* and *P. fluorescens* were detected 5–7 h after the inoculation. *Candida kefyr*, *Bacillus subtilis* and *Clostridium sporogenes* could be detected after 24 h of incubation owing to their slow growth rate. Second harmonic generation seems to be slightly better than absorption for detecting contamination in Pursoup vegetable soup. During the first incubation day 80% of the contaminated Pursoup vegetable soup packages would be detected by second harmonic generation. In the case of Nantua fish sauce, the second harmonic could detect only 20–40% and absorption 10% of the contaminated packages.

Variation between replicated measurements was slight but between samples was significant (see Fig. 3). The reliability of ultrasound measurements was

Table 1 The probability (%) of second harmonic generation and absorption detecting contaminated UHT milk samples. The measurement values have been corrected for the width of the package. The detection threshold level for second harmonic was 5% and for absorption 1.5%

	2nd Harmonic 3→6 MHz Incubation time			
Microorganism	*1 day*	*2 days*	*3 days*	*4 days*
Escherichia coli	20	20	40	nd
Lactobacillus plantarum	20	20	50	nd
Pseudomonas fluorescens	40	40	40	nd
Candida kefyr	nd	40	40	80
Bacillus subtilis	nd	40	40	40
Clostridium sporogenes	nd	100	100	100
Probability (day)	30	40	50	70

	Absorption Incubation time			
Microorganism	*1 day*	*2 days*	*3 days*	*4 days*
E. coli	40	80	90	nd
L. plantarum	60	40	100	nd
P. fluorescens	80	80	80	nd
C. kefyr	nd	100	60	100
B. subtilis	nd	60	80	100
C. sporogenes	nd	100	100	100
Probability (day)	60	80	90	100

nd, no data.

Table 2 The probability (%) of second harmonic generation and absorption detecting contaminated Pursoup vegetable soup samples. The measurement values have been corrected for the width of the package. The detection threshold level for second harmonic was 5% and for absorption 1.5%

	2nd Harmonic 3→6 MHz Incubation time			
Microorganism	*1 day*	*2 days*	*3 days*	*4 days*
Escherichia coli	nd	nd	nd	nd
Lactobacillus plantarum	80	80	60	80
Pseudomonas fluorescens	100	80	100	100
Candida kefyr	100	40	100	100
Bacillus subtilis	60	100	100	80
Clostridium sporogenes	80	100	100	80
Probability (day)	80	80	90	90

	Absorption Incubation time			
Microorganism	*1 day*	*2 days*	*3 days*	*4 days*
E. coli	60	60	60	60
L. plantarum	70	60	70	60
P. fluorescens	100	100	100	100
C. kefyr	80	60	80	100
B. subtilis	20	40	20	40
C. sporogenes	60	100	80	20
Probability (day)	70	60	70	60

nd, no data.

further improved by taking into consideration the effects of parameters such as temperature, air bubbles, amount of inoculum, changes in package width during incubation and differences in package weight. In particular, the effect of air bubbles in the package windows and changes in package width after inoculation and during incubation were shown to be important error factors. The use of brushes removed the air bubbles and also reduced the variation between replicate measurements and samples. The width was shown to vary between UHT milk packages and width also changed after the inoculation because of the needlestick. The possibility of measuring the width of the package using the same transducers used for interlocating the sample was investigated, and found to be feasible with the transducer pair used in absorption measurement for UHT milk and Pursoup vegetable soup. Nantua fish sauce, however, attenuates the signal too much for the pulse echo measurement to be successful.

Non-contact Ultrasound Method

The non-contact ultrasound method is based on the emission and reception of ultrasound by piezoelectric transducers. This includes three steps:

- generation of ultrasound with pulsed lasers
- detection of ultrasound with an interferometric probe
- data acquisition and signal processing.

Generation of ultrasound with pulsed lasers is achieved by electromagnetic radiation from the laser, which is absorbed in the surface region of the sample, causing heating; thermal energy then propagates into the specimen as thermal waves, the heated region undergoes thermal expansion, and thermoelastic stresses generate elastic waves (ultrasound) which propagate deep within the sample. The ultrasound waves are detected with an interferometric probe which is designed as a high-resolution optical spectrometer to detect changes in frequency of the scattered or reflected light. This method, which is being developed in France by SFIM-ODS and Technogram in cooperation with Danone, has shown some promising results in detecting contamination in aseptic food products.

Calorimetric and Volumetric Methods

The calorimetric and volumetric methods were developed in the Netherlands at the Delft University of Technology, in cooperation with the Unilever Research Laboratorium. Metabolically active and growing microorganisms consume energy and in turn generate small amounts of heat. This phenomenon is used in the calorimetric method which detects small

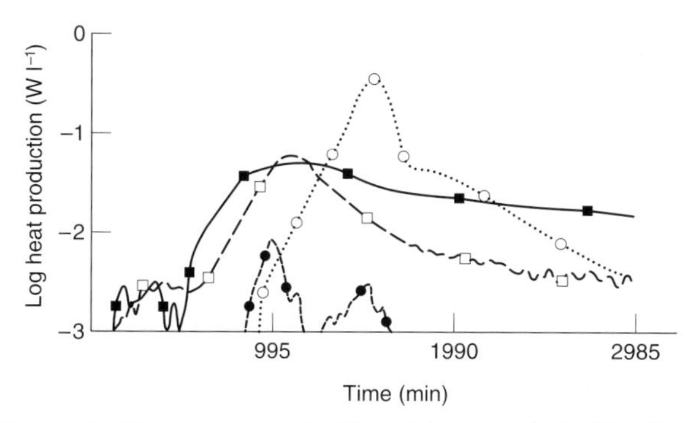

Figure 4 Thermograms of different bacteria in UHT milk at 30°C. Squares, *Escherichia coli*; triangles, *Salmonella arizonae*; circles, *Bacillus cereus*; crosses, *S. cremoris* H414.

temperature increases of a product caused by growing microorganisms. The method uses a specially designed calorimeter, smart temperature sensors and data processing equipment. The calorimeter contains 100 cavities for 1 litre packages. Each cavity is equipped with a smart sensor. The system is able to follow temperature changes of 100 packs simultaneously for an indefinite period. The system is microprocessor-controlled and is built in such a way that the influences of environmental temperature changes are minimized. The temperature changes in the products tested in this system are sensed using 'smart' sensor chips that are in direct contact with the package material of the product. With current technology, the heat production of certain organisms can be detected in the same time period as that needed for the destructive test method based on ATP bioluminescence. However, not all microorganisms produce enough heat during their growth cycle to be detected. In **Figure 4**, thermograms of some bacteria growing in UHT milk are shown.

In practice the implementation of the calorimetric and the volumetric methods is very similar, and by combining the two, the advantages of both non-specificity and sensitivity can be achieved. A prototype has been developed for simultaneous volume–temperature monitoring of two 1 litre Tetrapaks. Relative temperature changes of less than 10 mK and relative volume changes of less than 0.3 ml (0.03%) can reliably be distinguished. These results are achieved by ensuring a good thermal insulation and by applying smart sensor interfacing and smart data processing. This method has been shown to be very attractive for automated, non-destructive sterility testing of a number of food packs under laboratory conditions. It is not applicable for intensive, 100% sterility testing of the production lot, because the continuous testing procedure could last from a few hours up to a few days.

Impedimetric Method

For intensive, non-destructive sterility testing of 100% of the production lot a new impedance method has been studied in the Netherlands. The main problem to be solved in measuring the impedance of aseptically packaged food is how to pass an electric signal through the package. Most food containers are designed to prevent any contact between the food and the surrounding environment to ensure the highest food quality for the longest time possible. An intermediate aluminium foil layer acts as a Faraday's cage and does not allow electromagnetic fields to penetrate it. The small change in packaging technology required to apply the impedance method is already feasible and prototypes of new packages with one small electrode, fixed on the inside surface of the packaging material and reachable from outside, are being tested. The inside electrode can be in galvanic contact with the food or it can be isolated from the food by a thin thermoplastic layer. As a second electrode, the aluminium foil itself is used. In the impedance method, the changes in conductance and capacitance of the food are measured. The changes in impedance depend on the number of ions moving in the liquid – cations moving to the negatively charged electrode and anions to the positively charged electrode. The increase in conductance and capacitance caused by the metabolic activity of the microbes leads to a decrease in the impedance.

It appears that non-invasive sterility testing of the whole production lot guarantees high quality of the food immediately after it is produced, but not a few months or a year later, when it may actually be consumed. For such a guarantee, intensive quality checking of the package itself is also needed. For aseptically packed foods, this check principally focuses on the inner thermoplastic layer of the packaging laminate, since any possible leakage in this layer may result in the contents reaching the barrier layer (aluminium foil) or the fibre layer, at which time the other barrier properties of the laminate are lost, even if no actual liquid leakage occurs through the laminate. This procedure is destructive, time-consuming and makes significant demands on reliability and costs. At the same time, it does not ensure individual consumer safety as only a very small part of the production lot is tested.

The impedance method is easily applicable for simultaneous sterility testing and leakage detection. For this purpose, the small-surface electrode has to be in galvanic contact with the food. 'Internal' leakage has been simulated by making a hole in the thermoplastic layer, ensuring direct contact between the foil and the liquid. What is observed in case of leakage is an increase of the impedance components R_Σ and C_Σ that

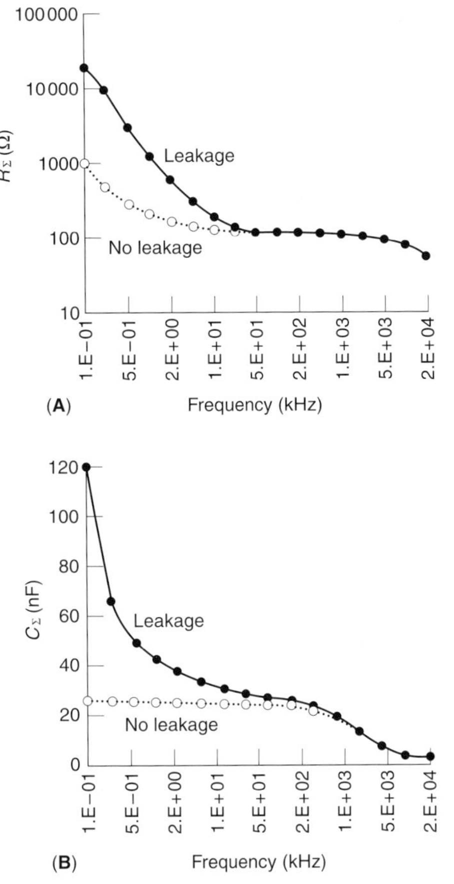

Figure 5 The increase of R_Σ (**A**) and C_Σ (**B**) results in decrease of the phase shift of the measured impedance $\omega = arctg\,(1/\omega CR_\Sigma C_\Sigma)$ at low frequency, which also can be used as an indication of leakage. R_Σ, resistance; C_Σ, capacitance. Delft University of Technology, Netherlands.

is easily detectable at frequencies below 10 kHz (**Fig. 5**).

Conclusions

The marketing study demonstrated the need to develop new non-invasive methods for detecting the growth of microorganisms and the spoilage of products; although new methods have been investigated and prototypes developed, none of these methods has yet succeeded in meeting the three 'ideal' criteria of nonspecificity (detects growth of any microorganism), high sensitivity (needs only a short incubation time) and rapidity (permits extensive on-line measurement) (**Table 3**). Changes in physical parameters are measured by ultrasound imaging, using Doppler techniques as well as contact and non-contact ultrasound methods. The impedance method detects changes in the conductance and capacitance of the food. The drawback of these non-destructive methods is that the presence of microorganisms is not detected directly but rather by a secondary parameter which changes with the presence and growth of the microorganism. These can be viscosity changes in the product or gas production by the microorganism. However, there are indications that these methods are also sensitive to physical parameters other than viscosity changes or gas production. Nevertheless, the effects of different microorganisms on the properties of liquid food products vary considerably and little is known about how different microbes change the physical properties of liquid food products; research in this area is certainly needed.

The smart temperature sensor method, which measures the minute temperature increases produced by growing microbes, is the only method that directly measures microbial growth. However, the measurements must be carried out before and during the exponential growth phase of the microbe, which makes the assessment time long and uncertain. In addition, not all microbes produce detectable amounts of heat during their exponential growth phase.

Several promising non-destructive sterility test methods have been developed and studied at the laboratory scale. Most of these methods presuppose some modifications in packaging technology. Research has shown, however, the potential for new industrial scale test methods. Optimization and on-line measurement tests at industrial scale are needed to verify the potential and applicability of these methods.

Acknowledgements

Non-destructive sterility testing methods have been investigated in an international European project called Endtest 'Development of non-destructive sterility testing equipment for aseptic products'. The project involved research institutes as well as com-

Table 3 Methods for non-invasive sterility contol in aseptically packaged foods

Method	*Type of changes registered*	*Nonspecificity*	*Sensitivity*	*Rapidity*
Ultrasonic imaging	Physical structures	++	++	++
Ultrasonic Doppler	Viscosity, physical structures	++	++	++
Contact ultrasound	Physical structures	++	++	+++
Smart temperature sensors	Temperature	+++	+++	+
Impedance	Electrical impedance	+++	++	+++

Key: +, low; ++, medium; +++, high.

panies from the Netherlands (Delft University of Technology, Unilever Research Laboratorium, Laboratory of Celsis-Lumac), Finland (Process Flow Ltd, VTT Biotechnology and Food Research, University of Helsinki), Sweden (Tetra Pak Research & Development AB) and France (SFIM-ODS, Technogram, Danone, INRA, CRSA).

See also: **ATP Bioluminescence**: Application in Dairy Industry. **Heat Treatment of Foods**: Ultra-high Temperature (UHT) Treatments. **Packaging of Foods**: Packaging of Solids and Liquids.

Further Reading

Ahvenainen R, Mattila T and Wirtanen G (1989) Ultrasound penetration through different packaging materials – a nondestructive method for quality control of packaged UHT milk. *Lebensm. Wissensch. Technol.* 22: 268–272.

Ahvenainen R, Wirtanen G and Manninen M (1989) Ultrasound imaging – a non-destructive method for monitoring the microbiological quality of aseptically-packed milk products. *Lebensm. Wissensch. Technol.* 22: 382–386.

Ahvenainen R, Wirtanen G and Mattila-Sandholm T (1991) Ultrasound imaging – a nondestructive method for monitoring changes caused by microbial enzymes in aseptically-packed milk and soft ice-cream base material. *Lebensm. Wissensch. Technol.* 24: 397–403.

Dubois M, Enguehard F and Bertrand L (1994) Analytical one dimensional model to study the ultrasonic precursor generated by a laser. *Physical Review E* 50: 1548–1551.

Gastagnede B, Deschamps M, Mottay E and Mourad A (1994) Laser impact generation of ultrasound in composite materials. *Acta Acoustica* 2: 83–93.

Gestrelius H (1994) Ultrasonic Doppler: a possible method for noninvasive sterility control. *Food Control* 5: 103–105.

Gestrelius H (1996) Aseptic packaging of food – non-destructive sterility testing. *Proceedings of the Fourth International Conference ASEPT – Food Safety '96*, 4–6 June 1996, Laval, France. P. 321.

Gestrelius H, Hertz TG, Nuamu M, Persson HW and Lindström K (1993) A nondestructive ultrasound method for microbial quality control of aseptically packaged milk. *Lebensm. Wissensch. Technol.* 26: 334–339.

Gestrelius H, Mattila-Sandholm T and Ahvenainen R (1994) Methods for noninvasive sterility control in aseptically packaged foods. *Trends in Food Science and Technology* 5: 379–383.

Haeggström E (1997) Ultrasound detection of microbe contamination in premade food. *Acta Polytechnica Scandinavica, Applied Physics Series* 214: 115.

Haus HA and Melcher JR (1989) *Electromagnetic Fields and Energy*. P. 260. Prentice Hall: New Jersey.

Javanaud C (1988) Applications of ultrasound to food systems. *Ultrasonics* 26: 117–123.

Margulies TS and Schwarz WH (1994) A multiphase continuum theory for sound wave propagation through dilute suspensions of particles. *Journal of the Acoustic Society of America* 96: 319–331.

Meijer GC, Kerkvliet HMM and Toth FN (1994) Non-invasive detection of micro-organisms using smart temperature sensors. *Sensors Actuators B Chemical* 18: 276–281.

Nihtianov SN (1996) Method for measuring the conductivity of fluids. Patent Application 96201096.3–2204, April 24 1996.

Nihtianov SN and Meijer GC (1995) Non-invasive impedimetric sterility testing of aseptically packed food products. *Proceedings of the Anniversary Scientific Conference: 'Fifty Years Technical University – Sofia', Fourth Edition of the National Scientific Conference 'Electronic Engineering'*, Sozopol 1995. Vol. 1, p. 52.

Nihtianov SN, Meijer GCM, Kerkvliet H and Demeijer E (1996) New methods for non-destructive sterility testing of aseptically packed food products. *Proceedings of the 1996 National Sensor Conference.* 20–21 March 1996, Delft, the Netherlands. P. 139.

Nihtianov SN, Kerkvliet HMM and Meijer GCM (1996) Non-invasive sterility-testing device of aseptically packed food products by simultaneous volume-temperature monitoring. *Fifth Edition of the National Scientific and Applied-Science Conference, Electronics ET '96*, Sozopol, Bulgaria, 27–29 September 1996.

Pless P, Futschik K and Schopf E (1994) Rapid detection of salmonellae by means of a new impedance-splitting method. *Journal of Food Protection* 57(5): 369–376.

Saito S (1993) Measurement of the acoustic nonlinearity parameter in liquid media using focused ultrasound. *Journal of the Acoustic Society of America* 93: 162–172.

Wirtanen G, Ahvenainen R and Mattila-Sandholm T (1992) Nondestructive detection of spoilage of aseptically-packed milk products: effect of frequency and imaging parameters on the sensitivity of ultrasound imaging. *Lebensm. Wissensch. Technol.* 25: 126–132.

ULTRASONIC STANDING WAVES
Inactivation of Food-borne Microorganisms using Power Ultrasound

Gail D Betts and **Alan Williams**, Campden and Chorleywood Food Research Association, Chipping Campden, Gloucestershire, UK
Rachel M Oakley, United Biscuits (UK Ltd), High Wycombe, Buckinghamshire, UK

Introduction

It has long been known that ultrasound is able to disrupt biological structures and much work has been done to investigate the mechanism by which it occurs. The killing potential of ultrasound was first demonstrated when it was discovered that sonar used in anti-submarine warfare was killing fish in the vicinity. Thereafter research into ultrasound as a method for inactivating cells has flourished. In the 1960s, research concentrated on understanding the mechanisms by which ultrasound interacted with microbial cells. This work investigated the effects of the cavitation phenomenon and associated shear disruption, localized heating and free radical formation. In the 1970s, it was found that brief exposure to ultrasound caused a thinning of bacterial cell walls, making them more susceptible to rupturing. In more recent times, application of ultrasound has been widely investigated for its potential to cause bacterial cell inactivation. In the food industry, ultrasound is being viewed as a potential food-processing tool, which can be used in combination with other treatments such as heat and chemicals to inactivate key target bacteria.

Many conventional methods of food processing involve the input of high levels of heat. Whilst this may be effective at inactivating food-borne pathogens and spoilage microorganisms, it can be detrimental to the overall quality of the food products in question. Techniques involving minimal processing are being investigated in an effort to ensure that the quality of a product is maintained and ultimately enhanced. Quality attributes that can be protected by the use of minimal processing techniques include appearance, flavour, nutritional value and absence of additives.

Power Ultrasound

Definition

Ultrasonic techniques are finding increasing use in the food industry for both the analysis and processing of foods. Normal human hearing will detect sound frequencies ranging from 16 Hz to 18 kHz and the intensity of normal quiet conversation is of the order of 10^{-11} W cm^{-2}. Low intensity ultrasound uses very high frequencies of the order of 2–20 MHz, with low power levels of 0.1–1 W cm^{-2}; this type of ultrasound is readily used for non-invasive imaging, sensing and analysis and is fairly well established in certain industrial and analytical sectors for measuring factors such as composition, ripeness, the efficiency of emulsification and the concentration or dispersion of particulate matter within a fluid.

Power ultrasound, on the other hand, uses lower frequencies, normally in the range of 20–100 kHz, and can produce much higher power levels of the order of 10–1000 W cm^{-2}. Low-frequency high-power ultrasound has sufficient energy to break intermolecular bonds. Energy intensities greater than 10 W cm^{-2} will generate cavitation effects, which are known to disrupt some physical systems as well as enhance or modify many chemical reactions.

Generation of Power Ultrasound

Whatever type of commercial system is used to apply power ultrasound to foods, it will consist of three basic parts: generator, transducer and coupler.

1. *Generator*: an electronic or mechanical oscillator that needs to be rugged, robust, reliable and able to operate with and without load.
2. *Transducer*: a device for converting mechanical or electrical energy into sound energy at ultrasonic frequencies. The three main types of transducer are:

- *Liquid driven transducers*: effectively a liquid whistle where a liquid is forced across a thin metal blade, causing it to vibrate at ultrasonic frequencies; rapidly alternating pressure and cavitation effects in the liquid generate a high degree of mixing. This is a simple and robust device, but because it involves pumping a liquid through an orifice and across a blade, processing applications are restricted to mixing and homogenization.
- *Magnetostrictive transducers*: electromechanical devices that use magnetostriction, an effect found in some ferromagnetic materials which change dimension in response to the application of a magnetic field. The dimensions of the transducer must

be accurately designed so that the whole unit resonates at the correct frequency. The frequency range is normally restricted to below 100 kHz and the system is not the most efficient (60% transfer from electrical to acoustic energy with losses mainly due to heat). The main advantages of these transducers are their ruggedness and ability to withstand long exposure to high temperatures.

- *Piezoelectric transducers*: electrostrictive devices that utilize ceramic materials such as lead zirconate titanate or barium titanate and lead metaniobate. This piezoceramic element is the most common of the transducers and is more efficient (80–95% transfer to acoustic energy) but less rugged than magnetostrictive devices; piezoelectric transducers are not able to withstand long exposure to high temperatures (normally not > 85°C).

3. *Coupler*: the working end of the system that helps transfer the ultrasonic vibrations to the substance being treated (usually liquid). The design, geometry and way in which the ultrasonic transducer is inserted or attached to the reaction vessel are crucial to its effectiveness and efficiency. For example, with ultrasonic baths, the transducer is bonded to the base or sides of the tank and the ultrasonic energy delivered directly to the liquid in the tank. However, with probes, the high-power acoustic vibration is amplified and conducted into the media by the use of a shaped metal horn; the shape of the horn will determine the amount of signal amplification.

There are several ultrasonic systems available; they differ mainly in the design of the power generator, the type of transducer used and the reactor to which it is coupled. Typical ultrasonic systems are:

1. *Ultrasonic baths*: transducers are normally fixed to the underside of the vessel, operate at around 20–40 kHz and produce high intensities at fixed levels due to the development of standing waves created by reflection of the sound waves at the liquid–air interface. The depth of the liquid is important for maintaining these high intensities and should not be less than half the wavelength of the ultrasound in the liquid. Frequency sweeping is often used to produce a more uniform cavitation field and reduce standing wave zones.
2. *Ultrasonic probes*: systems that use detachable horns or shapes to amplify the signal; the horns or probes are usually half a wavelength (or multiples) in length. The amount of gain in amplitude depends upon the shape and difference in diameter of the horn between one face (the driven face) and the other (the emitting face). If the probe is the same diameter along its length then no gain in amplitude will occur but the acoustic energy will simply be transferred to the media.
3. *Parallel vibrating plates*: opposing vibrating plates offer a better design for maximizing the mechanical effect of ultrasound than a single vibrating surface. Often plates vibrate at different frequencies (e.g. 20 and 16 kHz) to set up beat frequencies and create a larger number of different cavitation bubbles.
4. *Radial vibrating systems*: this is perhaps the ideal way of delivering ultrasound to fluids flowing in a pipe. The transducers are bonded to the outside surface of the pipe and use the pipe itself as a part of the delivery system. These are very good for handling high flow rates and high viscosity fluids. A cylindrical resonating pipe will help focus ultrasound at the central region of the tube, resulting in high energy in the centre for low-power emission at the surface; this can reduce erosion problems at the surface of the emitter.

Applications for Power Ultrasound in the Food Industry

Power ultrasound is already used to process food materials in a variety of ways, such as mixing, emulsification, tenderizing and ageing. Potential and interesting areas of applications for high power ultrasound as a processing tool for the food, pharmaceutical and chemical industries are: enzyme inhibition, hydrogenation of oils, crystallization control, extraction of proteins and enzymes and the inactivation of microorganisms. A more detailed list of potential applications for power ultrasound in the food industry is shown in **Table 1**.

Inactivation of Microorganisms

Mechanism of Action of Ultrasound

Microbial cell inactivation is thought to occur via three different mechanisms: cavitation, localized heating and free radical formation. Cavitation is produced when ultrasound waves pass through a liquid medium. The waves consist of alternate rarefactions and compressions and, if the waves are of sufficiently high amplitude, bubbles or cavities are produced. The bubbles collapse with differing intensities and this bubble collapse contributes to cell inactivation. There are two sorts of cavitation – transient and stable – which have been reported to have different effects.

Stable cavitation occurs due to oscillations of the ultrasound waves, which cause tiny bubbles to be produced in the liquid. It takes thousands of oscillatory cycles of the ultrasound waves to allow the

Table 1 List of current and potential applications for ultrasound in the food industry

Application	*Reported benefits*
Crystallization of fats and sugars	Enhances the rate and uniformity of seeding
Degassing	Carbon dioxide removal from fermentation liquors
Foam breaking	Foam control in pumped liquids and during container filling
Extraction of solutes	Acceleration of extraction rate and efficacy; research on coffee, tea, brewing; scale-up issues
Ultrasonically aided drying	Increased drying efficiency when applied in warm air, resulting in lower drying temperatures, lower air velocities or increased product throughput
Mixing and emulsification	On-line commercial use often using 'liquid whistle'. Can also be used to break emulsions
Spirit maturation and oxidation processes	Inducing rapid oxidation in alcoholic drinks; 1 MHz ultrasound has possible applications for accelerating whisky maturation through the barrel wall
Meat tenderization	Alternative to pounding or massaging; evidence for enhanced myofibrillar protein extraction and binding in reformed and cured meats
Humidifying and fogging	Ultrasonic nebulizers for humidifying air with precision and control; possible applications in disinfectant fogging
Cleaning and surface decontamination	On-line commercial use for cleaning poultry-processing equipment; possible pipe-fouling and fresh-produce-cleaning applications; can inactivate microorganisms in crevices not easily reached by conventional cleaning methods
Cutting	Commercial units available capable of cutting difficult products (very soft/hard/fragile) with less wastage, more hygienically and at high speeds
Effluent treatment	Potential to break down pesticide residues
Precipitation of airborne powders	Potential for wall transducers to help precipitate dust in the atmosphere; also removal of smoke from waste gases
Inhibiting enzyme activity	Can inhibit sucrose inversion and pepsin activity; generally oxidases are inactivated by sonication but catalases are only affected when at low concentrations; reductases and amylases appear to be highly resistant to sonication
Stimulating living cells	Lower-power sonication can be used to enhance the efficiency of whole cells without cell wall disruption, e.g. in yoghurt, action of *Lactobacillus* improved by nearly 50%; improved seed germination and hatching of fish eggs
Ultrasonically assisted freezing	Control of crystal size and reduced freezing time through zone of ice crystal formation
Ultrasonically aided filtration	Rate of flow through the filter medium can be substantially increased
Enhanced preservation (thermal and chemical)	Sonication in combination with heat and pressure has the potential to enhance microbial inactivation; this could result in reduced process times and/or temperatures to achieve the same lethality

bubbles to increase in size. As the ultrasonic wave passes through the medium it causes the bubbles to vibrate, creating strong currents in the surrounding liquid. Other small bubbles are attracted into the sonic field and this adds to the creation of microcurrents. This effect, which is known as microstreaming, provides a substantial force which rubs against the surface of cells, causing them to shear and break down without any collapse of the bubbles. This shear force is one of the modes of action which leads to disruption of the microbial cells. The effect of the pressures produced on the cell membrane disrupts its structure and causes the cell wall to break down.

During transient cavitation, the bubbles rapidly increase in size within a few oscillatory cycles. The larger bubbles eventually collapse, causing localized high pressures (up to 100 MPa) and temperatures (up to 5000° K) to be momentarily produced. The localized high temperatures can lead to thermal damage, e.g. denaturation of proteins and enzymes; however, as these temperature changes occur only momentarily and in the immediate vicinity of the cells it is likely that only a small number of cells are affected.

It is widely believed that cellular stress is caused by the cavitation effect, which occurs when bubbles collapse. The pressures produced during bubble collapse are sufficient to disrupt cell wall structures, eventually causing them to break, leading to cell leakage and cell disruption. The intensity of bubble collapse can also be sufficient to dislodge particles, e.g. bacteria from surfaces, and could displace weakly bound ATPase from the cell membrane – another possible mechanism for cell inactivation.

Free radical formation is the final proposed mode of action of ultrasound inactivation. Application of ultrasound to a liquid can lead to the formation of free radicals, which may or may not be beneficial. In the sonolysis of water, OH^- and H^+ ions and hydrogen peroxide can be produced and these have important bactericidal effects. The primary target site of these free radicals is the DNA in the bacterial cell. The action of the free radicals causes breakages along the length of the DNA, causing small fragments of DNA to be produced. These fragments are susceptible to attack by the free radicals produced during the ultrasound treatment and it is thought that the hydroxyl radicals attack the hydrogen bridges, leading to further fragmentation effects. The chemical environment plays an important part in determining the effectiveness of the ultrasound treatment and it may be possible to manipulate or exploit these conditions in order to achieve a greater level of inactivation.

Factors Affecting Cavitation

The frequency of ultrasound is an important parameter and influences the bubble size. At lower frequencies, such as 20 kHz, the bubbles produced are larger in size and when they collapse higher energies are produced. At higher frequencies, bubble formation becomes more difficult and at frequencies above 2.5 MHz cavitation does not occur. The amplitude of the ultrasound waves influences the intensity of cavitation; if a high intensity is required, a higher amplitude is used.

The intensity of bubble collapse also depends on factors such as temperature of the treatment medium, viscosity and frequency of ultrasound. As temperature increases, cavitation bubbles develop more rapidly, but the intensity of collapse is reduced. This is thought to be due to an increase in the vapour pressure, which is offset by a decrease in the tensile strength. This results in cavitation becoming less intense and therefore less effective as temperature increases. This effect can be overcome, if required, by the application of an overpressure to the treatment system. Combining pressure with ultrasound and heat increases the amplitude of the ultrasonic wave and it has been shown that this can increase the effectiveness of microbial inactivation. Pressures of up to 200 kPa (2 bar) combined with ultrasound of frequency 20 kHz and a temperature of 30°C have led to a decrease in decimal reduction time (D value; i.e. more effective microbial destruction) by up to 90% for a range of microorganisms.

Effect of Ultrasound on Microorganisms

Bacterial cells differ in their sensitivity to ultrasound treatment: some are more susceptible than others. It has been shown that, in general, larger cells are more sensitive to ultrasound. This may be due to the fact that larger cells have an increased surface area which is bombarded by the high pressure produced during cavitation, making them more vulnerable to sonication treatment. The effects of ultrasound have been studied using a range of organisms such as the Gram-positive *Staphylococcus aureus* and *Bacillus subtilis* and the Gram-negative *Pseudomonas aeruginosa* and *Escherichia coli*. Gram-positive cells have been found to be more resistant to ultrasound than Gram-negative cells and this may be due to the structure of the cell walls. Gram-positive cells have a thicker cell wall, which contains a tightly adherent layer of peptidoglycans, and it has been suggested that it is this that provides the cells with protection against sonication treatment. Meanwhile, other researchers have investigated this effect and found that there was no significant difference between the percentage of Gram-positive and Gram-negative cells killed by ultrasound. Cell shape has been investigated and it has been found that coccoid cells are more resistant to sonication than rods. Spore-forming bacteria such as *Bacillus* and *Clostridium* spp. have been found to be more resistant to sonication than vegetative bacteria and many of the bacteria known to be resistant to heat are similarly resistant to ultrasound.

Effect of Treatment Medium

The characteristics of the treatment substrate can influence the effectiveness of the ultrasound treatment applied. It has been found that the resistance of bacteria is different when treated in real food systems than when treated in microbiological broths. For example, foods that contain a high fat content reduce the killing effect of the treatment. Differences in effectiveness may be due to the intrinsic effect of the environment on the ultrasound action or a reflection of the changes in ultrasound penetration and energy distribution. In a liquid, the ultrasound waves will pass through relatively easily, causing cavitation to occur, but in a more viscous solution the ultrasound waves will have to be of a higher intensity to enable the same level of penetration to be achieved. Low-frequency, high-power ultrasound will be better at penetrating viscous products than higher-frequency ultrasound. This is because ultrasound waves with higher frequency will be more easily dispersed within the solution, causing a reduction in the overall intensity of the energy delivered.

Combination Treatments

Ultrasound applied on its own does not significantly reduce bacterial levels in systems that may be applied to foods. However, if it is combined with other preservation treatments such as heat or chemicals, the bacterial cells undergo a synergistic attack on their vital processes and structures.

Ultrasound and Heat The most commonly used combination treatments are the use of heat with ultrasound: this is known as thermosonication. It is thought that bacteria become more sensitive to heat treatment if they have undergone an ultrasound treatment. Increased cell death has been demonstrated in cells that have been subjected to a combined ultrasound and heat treatment compared with cells that were exposed to ultrasound treatment only or heat treatment only. Spore-forming bacteria have been

Table 2 Inactivation (D values) of a range of bacteria using heat and high-power ultrasound

Organism	*Heating temperature (°C)*	*Heat only (min)*	*Heat + ultrasound (min)*	*Ultrasound only (min)*
Bacillus subtilis	81.5	257	149	Not tested
B. subtilis	89	39.2	22.9	Not tested
B. licheniformis	99	5	2.2	No effect seen
B. cereus	110	12	1	No effect seen
Enterococcus faecium	62	11.2	1.8	30
Salmonella typhimurium	50	50	30	No effect seen
Staphylococcus aureus	50.5	19.7	7.3	Not tested

shown to have some degree of reduced resistance to heat if they are also treated with ultrasound. The increased heat sensitivity caused by sonication can be quantified in terms of a decimal reduction or D value (i.e. the time taken to achieve a 1 log reduction in cell levels) and **Table 2** shows the synergistic effect of heat and ultrasound for a range of bacterial species. Whilst these data show up to a 43% reduction in the heat resistance of the spore-formers tested, other studies have shown no effect or a limited effect for other spore-formers. This has been attributed to the fact that spores contain a highly protective outer coat which prevents the ultrasound waves passing through, thus limiting the amount of perturbation that occurs within the spore. During treatment with a combination of pressure and thermosonication, it has been shown that chemicals such as dipicolinic acid and low-molecular-weight peptides were released from spores of *Bacillus stearothermophilus*. In these combined treatments, spores are subjected to violent and intense vibrations due to increased cavitation effects. The loss of substances from spores during this combination of pressure, heat and ultrasound suggests that spore cortex damage and protoplast rehydration may account for the subsequent reduction in heat resistance.

There is evidence to suggest that the order in which heat and ultrasound are applied has a different effect on the inactivation observed. In work done at Campden and Chorleywood Food Research Association, samples of products such as orange juice and milk were inoculated with key pathogenic and spoilage organisms and subjected to ultrasound treatment. Some of the samples were presonicated at ambient temperature before application of a relatively mild heat treatment and sonication. Other samples were subjected to simultaneous heat and ultrasonic treatment only. These data are shown in **Table 3** for *Listeria monocytogenes* and in **Table 4** for *Zygosaccharomyces bailii*.

As previously discussed, the frequency of ultrasound used affects the type of cavitation response observed. **Fig. 1** shows the effect of treating *L. monocytogenes* at 20, 38 and 800 kHz in whole milk. It appears that 20 kHz was the most effective frequency whilst 800 kHz had little effect and resulted in a survivor tail.

Ultrasound in Combination with Chemicals Not only has ultrasound been used in combination with heat, but it has also been used in combination with chemical treatment. Chemicals such as chlorine are often used to decontaminate food products or processing surfaces and it has been demonstrated that chlorine combined with ultrasound enhances the effectiveness of the treatment. This theory was demonstrated using *Salmonella* attached to the surface of broiler carcasses. Bombardment with ultrasound caused the cells to become detached from the surfaces, making it easier for the chlorine to penetrate the cells and exert an antimicrobial effect (**Table 5**).

Ultrasound is able to disperse bacterial cells in suspensions, making them more susceptible to treatment with sanitizing agents. One of the possible advantages of combined treatments would be a reduction in the concentration of chemicals used in isolation for sanitation and disinfection or a reduction in the contact time required. This has additional advantages in that there is less likelihood of residual cleaning agents contaminating equipment after cleaning.

Ultrasound in Combination with pH A varying response of microorganisms to ultrasound treatment depending on the pH of the surrounding medium has been observed. In particular it has been found that if the microorganisms are placed in acidic conditions, this leads to a reduction in the resistance of the organisms to the ultrasound treatment. This may be due to the effects of the ultrasound on the bacterial membranes, which make them more susceptible to the antimicrobial effects of the acid or unable to maintain the essential internal pH conditions.

Conclusions

Ultrasound is currently used in the food industry for mixing, blending, speeding up the ageing processes in meats and wines and emulsifying fats and oils. It has the potential to be applied to the pasteurization of a range of low-viscosity liquid products. Ultrasound on

Table 3 D values (in minutes) for *Listeria monocytogenes*

Product	*Heat only (60°C)*	*Presonication (kHz)*			*Simultaneous (kHz)*		
		20	*38*	*800*	*20*	*38*	*800*
UHT milk	2.1	0.4	0.3	>10	0.3	1.3	1.4
Rice pudding	2.4	NT	NT	0.3–5.9	NT	NT	3.4–4.5

NT = Not tested.
Original data from Hurst et al (1995).

Table 4 D values (in minutes) for *Zygosaccharomyces bailii*

Product	*Heat only (55°C)*	*Presonication (kHz)*			*Simultaneous (kHz)*		
		20	*38*	*800*	*20*	*38*	*800*
Orange juice	10.5	2.4	0.9	1.4	3.9	1.8	>10
Rice pudding	11.0	2.3	0.5	NT	1.0	NT	>10
Orange juice + 5% starch	15.4	NT	NT	3.4	0.5–2.1	NT	NT

NT = Not tested.
Original data from Hurst et al (1995).

Table 5 Effect of high-power ultrasound and chlorine on *Salmonella* attached to chicken skin

Treatment	*Untreated*	*Ultrasound only for 30 min*	*Chlorine only (0.5 p.p.m. free residual) for 30 min*	*Ultrasound + chlorine (0.5 p.p.m. free residual) for 30 min*
Log_{10} reductions	1.19	2.59	2.08	4.07

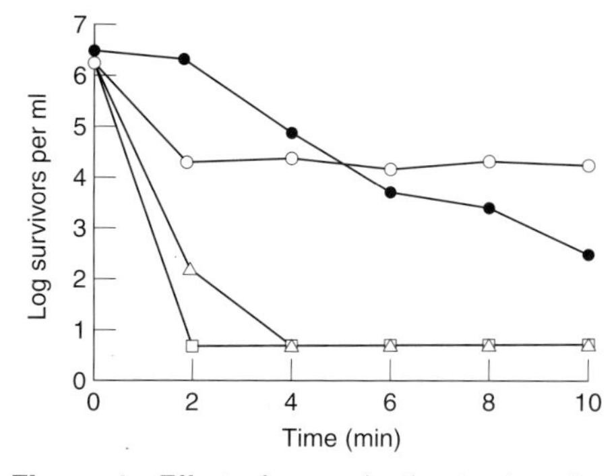

Figure 1 Effect of presonication treatment on the inactivation of *Listeria monocytogenes* in UHT whole milk at different power ultrasound frequencies. Filled circles, heat (60°C); open squares, heat + 20 kHz; open triangles, heat + 38 kHz; open circles, heat + 800 kHz.

its own needs high intensities and prolonged application to inactivate microorganisms and enzymes, and this may cause physical and sensory damage to foods. If, however, ultrasound is used as a combination treatment with mild pressure, heat or chemical preservatives, this technology has the potential to become a useful processing tool for achieving inactivation of food-borne pathogens or spoilage organisms. There are, however, a number of issues which need to be considered before it can be successfully employed. The reliability of the process needs to be more fully investigated in terms of assessing microbial inactivation and a wider range of bacteria needs to be investigated. The efficiency of the technology also needs to be assessed and food manufacturers must decide whether the ultimate benefits outweigh the costs of converting and maintaining the processing equipment.

See also: **Fermented Milks**: Yoghurt.

Further Reading

Ahed FIK and Russell C (1975) Synergism between ultrasonic waves and hydrogen peroxide in the killing of microorganisms. *Journal of Applied Bacteriology* 39: 31–40.

Alliger H (1975) Ultrasonic disruption. *American Laboratory* 10: 75–85.

Burgos J, Ardennes JA and Sala, FJ (1972) Effect of ultrasonic waves on the heat resistance of *Bacillus cereus* and *Bacillus licheniformis* spores. *Applied Microbiology* 24: 497–498.

Earnshaw RG (1998) Ultrasound: A New Opportunity for Food Preservation. In: Povey MJW and Mason TJ (eds) *Ultrasound in Food Processing*. P. 183. London: Blackie Academic and Professional.

Garcia ML, Burgos J, Sanz B and Ordonez JA (1989) Effect of heat and ultrasonic waves on the survival of two strains of *Bacillus subtilis*. *Journal of Applied Bacteriology* 67: 619–628.

Hughes DE and Nyborg WL (1962) Cell disruption by ultrasound. *Science* 138: 108–144.

Hurst RM, Betts GD and Earnshaw RG (1995) The antimicrobial effect of power ultrasound. CCFRA R&D

Report No. 4. Chipping Campden, Gloucestershire: CCFRA.

Kinsloe H, Ackerman E and Reid JJ (1954) Exposure of microorganisms to measured sound fields. *Journal of Bacteriology* 68: 373–380.

Lee BH, Kermala S and Baker BE (1989) Thermal ultrasonic and ultraviolet inactivation of *Salmonella* in films of aqueous media and chocolate. *Food Microbiology* 6: 143–152.

Mason TJ (1998) Power ultrasound in food processing – the way forward. In: Povey MJW and Mason TJ (eds) *Ultrasound in Food Processing*. P. 105. London: Blackie Academic and Professional.

Lillard HS (1993) Bactericidal effect of chlorine on attached salmonellae with and without sonification. *Journal of Food Protection* 56: 716–717.

Mason TJ, Paniwnyk L and Lorimer JP (1996) The uses of ultrasound in food technology. *Ultrasonics Sonochemistry* 3: S253–S260.

McClements DJ (1995) Advances in the application of ultrasound in food analysis and processing. *Trends in Food Science and Technology* 6: 293–299.

Ordonez JA, Sanz B, Hermandez PE and Lopez-Lorenzo P (1984) A note on the effect of combined ultrasonic and heat treatments on the survival of thermoduric streptococci. *Journal of Applied Bacteriology* 56: 175–177.

Palacios P, Borgos J, Hoz L, Sanz B and Ordonez JA (1991) Study of substances released by ultrasonic treatment from *Bacillus stearothermophilus* spores. *Journal of Applied Bacteriology* 71: 445–451.

Raso J, Codon S and Sala Trepat FJ (1994) Manothermosonication – a new method of food preservation? In: Leistner L and Gorris GM *Food Preservation by Combined Processes*. Final Report for FLAIR Concerted Action No. 7, Subgroup B. P. 37. Brussels: European Commission.

Roberts RT and Wiltshire MP (1990) High intensity ultrasound in food processing. In: Turner A (ed.) *Food Technology International Europe*. P. 83. London: Sterling Publications International.

Sala FJ, Burgos J, Condon S, Lopez P and Raso J (1995) Manothermosonication. In: Gould GW (ed.) *New Methods of Food Preservation by Combined Processes*. P. 176. London: Blackie.

Sanz P, Palacios P, Lopez P and Ordonez JA (1985). Effect of ultrasonic waves on the heat resistance of *Bacillus strearothermophilus* spores. In: Dring GJ, Elars DJ and Gould GW (eds) *Fundamental and Applied Aspects of Bacterial Spores*. P. 251. New York: Academic Press.

Scherba G, Weigel RM and O'Brien JR (1991) Quantitative assessment of the germicidal efficiency of ultrasonic energy. *Applied and Environmental Microbiology* 57: 2079–2084.

Scuhett-Abraham I, Trommer E and Levetzow R (1992) Ultrasonics in sterilisation sinks: applications of ultrasonics on equipment for cleaning and disinfection of knives at the workplace in slaughter and meat cutting plants. *Fleischwirtschaft* 72: 864–867.

Suslick KS (1988) Homogenous sonochemistry. In: Suslick KS (ed.) *Ultrasound. Its Chemical, Physical and Biological Effects*. New York: VCH Publishers.

Wrigley DM and Llorca NG (1992) Decrease of *Salmonella typhimurium* in skimmed milk and egg by heat and ultrasonic wave treatment. *Journal of Food Protection* 55: 678–680.

ULTRAVIOLET LIGHT

Gilbert Shama, Department of Chemical Engineering, Loughborough University, UK

Introduction

Ultraviolet light (UV) forms part of the electromagnetic spectrum. It can be harnessed to inactivate microorganisms associated with foods and food processing operations. The most effective wavelengths for direct inactivation of microorganisms lie in the so-called 'germicidal' region of the electromagnetic spectrum at approximately 260 nm. There are many different types of UV sources in current use, but low pressure mercury sources are efficient emitters of germicidal UV and these sources have been used to disinfect liquids, air and the surfaces of solids. UV has also been used in conjunction with other treatments, most notably peroxidation, to bring about synergistic disinfection. Intense irradiation of foods with short-wavelength UV may cause losses in food quality.

Nature of the Emission

'Ultraviolet light' (UV) is the term used for the portion of the electromagnetic spectrum which lies between visible light and X-rays. There is no universal agreement as to the precise boundaries of the UV spectrum. The upper limit is generally delineated with reference to the shortest wavelength detectable by the human eye (about 380 nm). The lower limit presents greater difficulties. UV is termed 'non-ionizing radiation' in contrast to 'ionizing radiation', e.g. X-rays. However, there is no clear cut-off and wavelengths of about 10 nm, classified by some as being in the UV range, do bring about some ionization. The exact location of the lower limit of the UV spectrum is in one sense irrelevant, because UV wavelengths < 100 nm are not routinely employed for inactivating microorganisms. The UV range below 100 nm is commonly referred to as the 'vacuum UV' region, because the radiation is strongly absorbed by passage through air and can only be transmitted through a vacuum. Biomedical interest in UV has given rise to subdivision of the upper end of the UV spectrum. The most commonly accepted boundaries are: UV-A for wavelengths of 315–380 nm, UV-B for wavelengths of 280–315 nm, and UV-C for wavelengths below 280 nm. This latter region has also been described as 'germicidal', because it includes the wavelengths most effective at inactivating microorganisms. Commonly used terms are: 'near UV', implying wavelengths of 300–380 nm; 'far UV' for wavelengths of 200–300 nm; and 'extreme UV' for wavelengths below 200 nm. The adjectives 'near', 'far' and 'extreme' refer to the proximity of the wavelengths to the visible region of the spectrum.

The amount of UV energy applied to a particular surface over a given interval of time is known as the UV 'dose' or 'fluence' (F) and is the product of the UV intensity (I) and the time of exposure (t)

$$F = I \times t$$

Doses are normally quoted in $mW\,s\,cm^{-2}$ or in $J\,m^{-2}$. The most convenient method of measuring dose is to use a UV radiometer to measure the UV intensity, from which the dose may be derived from the expression given above. Alternatively, dose may be determined directly by chemical actinometry. Actinometry relies on assaying the concentrations of the products of certain photochemical reactions which have well-characterized energetics, and relating these concentrations directly to the quantity of UV energy absorbed.

Sources of UV for Industrial Use

There are a number of commercially produced sources which emit energy in the UV range: these include mercury vapour, metal halide and xenon sources. As far as germicidal applications are concerned, the most widely used sources are of the mercury vapour discharge type. These may be broadly divided into two categories; low-pressure and medium/high-pressure sources or burners. Although use of the latter is by no means rare, low-pressure sources are generally favoured for most applications.

Low-pressure Sources

The low-pressure mercury discharge source is characterized by the conversion of a relatively high proportion of electrical energy to UV of wavelength 253.7 nm – typically, this efficiency is of the order of 50%. The source resembles a conventional fluorescent gas discharge tube, with electrodes at each end, but differs in that the glass envelope is made of UV-transmitting glass and does not contain a phosphor coating. Quartz gives high UV transmittance but is relatively expensive, and cheaper glasses with acceptable transmittances have been developed. The envelope is usually doped with titanium dioxide, to prevent the transmittance of wavelengths below 220 nm, which convert the O_2 of the surrounding air into O_3. This is a toxic and corrosive gas, but there are occasions when its production might be desirable (see below). The gas fill comprises a mixture of mercury vapour and argon. Maximum UV output is obtained by maintaining the temperature of the source at about 40°C – appreciable departures from this temperature result in a reduction of UV intensity.

The life of this type of source is measured in thousands of hours. Typically, after 10 000 h UV output will have fallen by approximately 50%. There has been a general reluctance in the food industry to use glass in areas where the contamination of food may occur, e.g. following breakages. This concern has been addressed by some manufacturers, who now use special grades of safety glass for the lamp envelopes or house the sources in sealed units. Alternatively, the source can be shrouded in tubes of clear grades of certain fluoropolymers (e.g. PTFE), which can be heat-shrunk over the lamp envelope and which provide high UV transmittance.

A fundamentally different design of low-pressure source was developed in the mid-1970s by the Brown Boveri company. This operates at very much higher current densities and at higher envelope temperatures (200°C) than conventional low-pressure burners. As a consequence, the UV intensities achieved are approximately three orders of magnitude greater. One of the principal applications of this type of source has been in surface disinfection.

Medium-pressure and High-pressure Sources

Medium- and high-pressure discharge sources are typified by a lower efficiency of conversion to germicidal wavelengths, typically 6–12%. They have a considerably broader emission range than that of low-pressure sources, making them suitable for a wider range of applications. They are more compact than low-pressure sources, and operate at temperatures up to 800°C.

Excimer Sources

Recently 'excimer' UV sources have been developed from excimer lasers. The UV output of these sources can be tailored, to produce essentially monochromatic wavelengths at high intensities. A particular advantage of these sources, it is claimed, is that they can be made in a wide range of configurations.

Biological Effects of UV

Molecular Level

The primary lethal effect of UV is due to its absorption by the DNA of the nucleus. UV also induces changes to many other functional and structural cell constituents, e.g. proteins. However all cell constituents other than DNA are present as multiple copies, so damage to even a large proportion of a single constituent has only a limited effect on cell function and integrity.

The high absorptance of DNA is entirely due to its pyrimidine and purine bases. Because the composition of DNA varies between species, the peak of UV absorptance occurs within a range, 260–265 nm. This corresponds very closely to the principal emission wavelength of the low-pressure mercury source (253.7 nm), and explains the efficiency of sources of this kind at inactivating cells (see **Fig. 1**). Germicidal effectiveness is measured relative to the lethality induced at 253.7 nm. The interaction of UV with DNA is complex, and the sequence of events following irradiation depending on a number of factors. The absorption of UV energy by the DNA bases promotes chemical reactions involving these bases, and the products if they persist, interfere with DNA replication and transcription. The products are commonly known as 'photoproducts'. A number of photoproducts have been identified and studied in detail, and most contain at least one pyrimidine base. They include pyrimidine dimers, other pyrimidine adducts and pyrimidine hydrates, and may involve cross-links with proteins. On rare occasions, breakages of DNA strands have been reported.

The formation of a photoproduct does not necessarily imply a lethal consequence, because of the existence of cellular DNA repair mechanisms. These play a significant role in cell survival, as demonstrated by work conducted with mutant organisms lacking repair capabilities. A number of quite different repair mechanisms exist, and organisms may possess more than one. Repair mechanisms detect and restore damaged sections of DNA, either by directly modifying photoproducts (e.g. the photoenzymatic monomerization of pyrimidine dimers) or by excising a damaged section of DNA and resynthesizing it with reference to the complementary strand. However repair mechanisms are prone to error and although DNA photoproducts may be removed successfully, the imperfect repair of a DNA strand may result in the formation of altered base sequences and the generation of mutants. These can vary in severity, from 'silent' to lethal mutations.

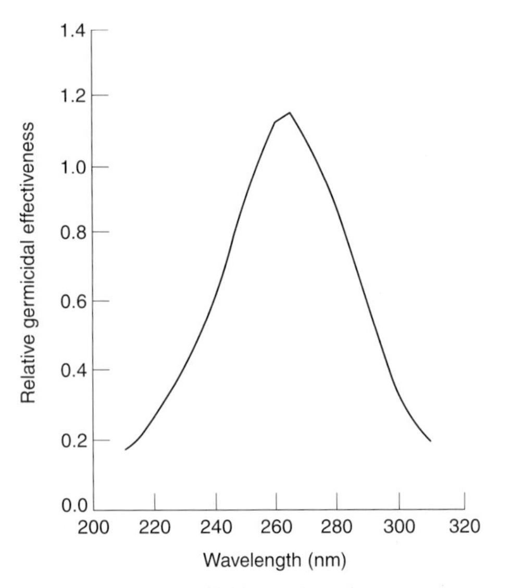

Figure 1 Direct lethality of UV wavelengths.

Cellular Level

The inactivation of a population of microorganisms or viruses is manifested by loss of the ability to form colonies or plaques. The inactivation of cells of a particular species is usually presented as a 'survival curve', also known as a 'dose–response curve'. This shows the fraction of the original population of cells surviving irradiation, as a function of dose. Data are often presented as a plot of the logarithm of the surviving fraction against UV dose, largely because this will produce a straight line if the inactivation follows simple first-order kinetics ((A) in **Figure 2**). This type of response is only strictly displayed by single-stranded DNA- or RNA-containing viruses and by certain repair-deficient mutants.

Most survival curves include a plateau region, corresponding to low UV doses. If the 'shoulder' is relatively small (curve B), first-order kinetics may be applied without serious error. However, the curves relating to some microorganisms display large plat-

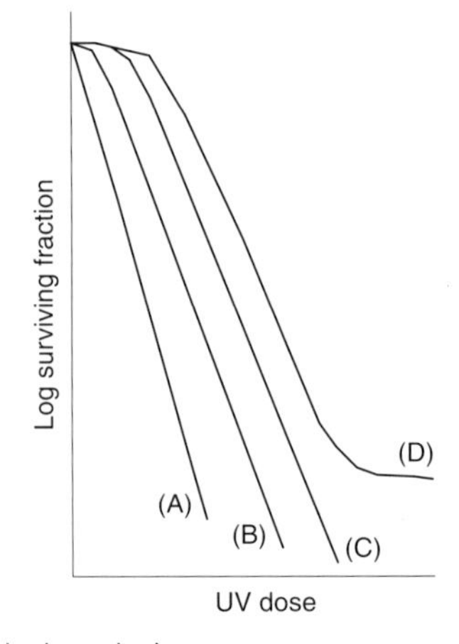

Figure 2 Typical survival curves.

eaus (curves (C) and (D)) and this requires more complex mathematical representation. This is available as a result of 'target theory', which was originally developed to account for inactivation by ionizing radiation. The theory is based on the fact that energy emitted from UV sources is quantized (in the form of UV photons) and that there is a discrete number of susceptible targets in a cell, which must be struck a finite number of times by UV photons for inactivation to occur. According to target theory, first-order kinetics are actually 'single hit–single target' functions. An alternative interpretation of the shouldered portion of these curves is that they represent a manifestation of cellular repair processes. At low UV doses a dynamic state can be visualized, in which the repair processes are able to repair damage to the DNA at a rate which exceeds that of infliction. As the UV dose is increased, the rate of repair lags behind that of photoproduct formation, resulting in progressively greater levels of inactivation.

Survival curves may display 'tailing' at high UV doses (curve (D) in Figure 2). This phenomenon has been attributed to the presence of UV-resistant cells in a heterogeneous population, or alternatively to the presence of aggregates of cells, in which the interior cells are protected by those cells at the exterior.

Compilations of the UV doses necessary to achieve a specified reduction in the viability of a population of organisms are often included in handbooks and in literature promoting the application of UV for disinfection. It is important that they be used only for general guidance, for a number of reasons:

1. Considerable variation in UV sensitivity has been reported within a single species.
2. The conditions of growth and the physiological state of an organism can have a bearing on its UV sensitivity, e.g. cells are more sensitive in the logarithmic phase of growth than in the stationary phase, and the UV resistance of some bacterial and fungal spores depends on whether they are wet or dry.
3. The conditions under which the cells are irradiated are important, because the penetration of UV into liquids other than clean water is usually low. Also, the presence of dense suspensions of microorganisms or high concentrations of solutes can significantly attenuate incident UV.
4. UV sources differ in spectral output, and individual wavelengths have different lethal effects (see Figure 1).

Table 1 shows UV doses typically required to inactivate different groups of microorganisms. The upper limit for bacteria is to some extent distorted by the inclusion of data for extremely UV-resistant micrococci (e.g. *Deinococcus radiodurans*). To place the data in context, a UV dose of $30\,\mathrm{mW\,s\,cm^{-2}}$ would theoretically result in 5 log reductions of the common enteric human pathogens *Campylobacter jejuni*, *Escherichia coli*, *Proteus vulgaris*, *Salmonella enteritidis*, *S. paratyphi*, *S. typhi*, *Shigella dysenteriae*, *Staphylococcus aureus*, *Vibrio parahaemolyticus* or *Yersinia enterocolitica*. However, a higher dose of $40\,\mathrm{mW\,s\,cm^{-2}}$ would be needed to bring about a similar theoretical reduction of *Listeria monocytogenes* or *Salmonella typhimurium*. The data for protozoa includes data for cysts of the common waterborne protozoan *Giardia lamblia*.

Table 1 D_{10} inactivation doses of UV (i.e. dose required to reduce population viability by one order of magnitude); measured at 253.7 nm

Microbial group	*UV dose (mW s cm⁻²)*
Bacteria (including spores)	0.4–30
Enteric viruses	5–30
Fungi	30–300
Protozoa	60–120
Algae	300–600

Practical Applications of UV

Liquids

The most successful application of UV disinfection processes has without doubt been in the field of water treatment. High-purity (e.g. drinking) water in particular represents an ideal medium for UV irradiation. However, UV can be used to disinfect other liquids, their treatability depending to some extent on their UV absorption characteristics.

As UV passes through an absorbing medium, its intensity is reduced, as a function of distance into the medium. Attenuation effects of this kind are described by the Lambert–Beer law. If monochromatic UV of intensity I_o is incident at one surface of a medium with absorption coefficient α, then the intensity (I) at depth d is given by:

$$I = I_o \times e^{\alpha d}$$

Table 2 shows values of α for various liquids.

Many compounds are strongly UV-absorbing in solution in water, and they adversely affect UV treatment, for example iron and all organic solutes to varying degrees. In addition, solid matter in suspension or emulsions can interfere with UV disinfection by scattering incident UV, resulting in attenuation. Larger particles may shield microorganisms from incident UV.

Another important factor influencing the efficacy

Table 2 Absorption coefficients of various liquids, for UV of wavelength 253.7 nm

Liquid	*Absorption coefficient* α *(cm*$^{-1}$*)*
Red wine	30
White wine	10
Beer	10–20
Clear syrup	2–5
Dark syrup	20–50
Milk	300
Distilled water	0.007–0.01
Drinking water	0.02–0.1

of UV treatment is the method used to bring the liquid into contact with UV. In most cases this will be a continuous flow-through device sometimes referred to as a contractor. The design of contractors should recognize that if a significant fraction of the liquid flow bypasses the UV sources, the effectiveness of disinfection will be reduced. The simplest enclosed configuration, used particularly in connection with low flow rates, is the annular contactor, in which water flows into the annular space between an axial tubular UV source and a cylindrical outer jacket. This configuration is quite common and is successful in the treatment of a wide variety of liquids. More sophisticated versions may feature additional UV sources at their circumference, and baffles to prevent the liquid from short-circuiting. It is possible to treat liquids which have high UV absorptance by presenting the liquid to the source of UV in the form of a thin film. The Lambert–Beer law given above indicates that attenuation is dependent on the product of the absorption coefficient (α) and the thickness of the liquid film (d), and so attenuation is reduced if d is decreased. A number of devices, of varying complexity, have been proposed for producing thin liquid films for irradiation, but they are mainly for low throughputs of liquid.

A feature of most contactors is that the liquid makes contact with the quartz sleeve which shrouds the UV source. The inevitable consequence is that over extended periods of operation, solid matter is deposited on the surface of the sleeve and attenuates the UV output from the source. Contactors used for liquids containing high concentrations of suspended matter are particularly prone to this type of fouling. Some manufacturers provide mechanical wipers, which maintain output intensity, or recommend the periodic interruption of treatment to enable cleaning of the sleeves to take place.

Air

UV has been successfully used to reduce the spread of microorganisms throughout buildings, including food processing facilities, by the irradiation of air. Most of the published evidence comes from studies aimed at reducing or controlling the spread of pathogenic organisms in hospitals and surgery waiting rooms. UV sources have been installed in heating and ventilation ducts, but they can also be placed directly in specified areas, usually being positioned so as to achieve 'upper air irradiation'. This is particularly important if personnel occupy the area, because it is essential to protect them from direct exposure to UV-C. Specially designed louvres are now available to adequately protect personnel at 'ground level' whilst achieving efficient upper air irradiation. Air irradiation is also routinely used in areas where aseptic conditions are necessary, e.g. for the transfer of starter cultures.

Surfaces

UV has very low penetrability into solid materials. Therefore UV irradiation is only effective for disinfecting the surfaces of solids. Furthermore, significant microbial inactivation is only achieved on surfaces which are relatively smooth and free of contamination. For example, crevices between ridges on the surface of a material may protect microorganisms from incident UV. Shadowing, by either surface irregularities or surface contamination such as dust, may also protect microbes.

Notwithstanding these limitations, UV is used to treat the surfaces of packaging materials, for drinks and foods. Materials such as packaging film may be irradiated using simple arrangements of sources, because the film only needs to be conveyed past a fixed, high-intensity, UV source. More complex geometries may require elaborate combinations of sources, and complex mechanical mechanisms for ensuring that all the surfaces of the object receive adequate irradiation. Surface disinfection of loaves of bread is being introduced in some large bakeries. Loaves emerging from the ovens are conveyed into 'UV tunnels' and irradiated. UV sources line the walls and ceiling of these tunnels. Treatment of bread in this way has been shown to significantly extend its shelf life.

Combined Treatments

In addition to its direct lethal effects on microorganisms, UV can exert a synergistic disinfectant effect in combination with other treatments. Synergism is the ability of combined treatments to produce a greater effect than the sum of the treatments applied separately. The most well-documented synergistic effect involving UV occurs with hydrogen peroxide (H_2O_2), which is thought to owe its disinfectant properties to the production of hydroxyl free radicals. Free radicals are very short-lived, but are highly reactive. They are capable of causing damage to microorganisms which is widespread, in contrast to that caused by UV, which is mainly restricted to nucleic acids. The rate of formation of lethal hydroxyl radicals can be greatly increased by irradiation with UV. H_2O_2 is strongly UV-absorbing, and studies have shown that maximum synergism is obtained using H_2O_2 solutions of concentration of about 1% over a wide intensity range. This synergistic disinfection process has been exploited commercially in the production of aseptic packaging for food. In some applications, an aqueous solution of H_2O_2 is sprayed onto the surface of the packaging prior to irradiation. However, spraying onto materials which are highly hydrophobic may result in incomplete surface coverage, and also in the presence of residual H_2O_2 on the surface. This H_2O_2 might come into contact with the food, and so some form of post-irradiation process, e.g. mild heat treatment, may be necessary to reduce the residual concentration to acceptable limits. A similar, although less widely investigated, synergistic effect is that between ozone and UV.

An interesting recent discovery is that of synergism between UV and treatment with a biological control agent, the yeast *Debaryomyces hansenii*, in preventing soft rot of fruits and vegetables.

A combined effect of a different kind is the irradiation of anatase, one of the crystalline forms of titanium dioxide, with UV-A. Irradiation with wavelengths in the vicinity of 400 nm in the presence of water leads to the formation of hydroxyl free radicals. Titanium dioxide has been incorporated into a variety of materials, e.g. ceramic tiles, to give what have become known as 'active surfaces'. These may have applications in food preparation areas helping to reduce the spread of pathogenic organisms.

Combined treatments thus effectively extend the range of UV wavelengths which is useful in inactivating microorganisms (see Table 1).

Hazards and Adverse Effects

Effects on Humans

Hazards to humans are associated with UV, particularly UV-C. The harmful effects generally depend on the amount of UV energy absorbed, and range from eye complaints (e.g. conjunctivitis) and erythema (reddening of the skin) to skin cancers. Adequate safety precautions (e.g. wearing safety goggles) must always be observed to prevent accidental exposure.

Mutants and Microbial Recovery

Fears that the irradiation of foods in commercial facilities may result in the production of highly UV-resistant mutants, possessing unquantifiable hazards to public health have proved groundless. Such mutants have only been generated under laboratory conditions.

The existence of repair mechanisms has been discussed above. The majority of studies investigating the impact of such mechanisms on industrial UV disinfection processes have focused on 'photoenzymatic' repair with reference to the irradiation of waste waters. Photoenzymatic repair involves the induction of enzymes capable of monomerizing pyrimidine dimers, by light in or near the UV-A region. The optimal wavelength varies, but it is in the range 310–480 nm. Although it has been shown that conditions conducive to photoenzymatic repair can arise, the extent of recovery has not generally been considered significant enough to warrant changes to existing practices. The extent to which recovery effects might be significant in other applications, e.g. on the surface of solids, remains largely unknown.

Effects on Foods

UV irradiation is known to destroy certain vitamins – vitamin C and the B vitamins are particularly susceptible. However, UV has long been known to promote the formation of vitamin D from its precursors. In the 1930s in the US milk was routinely irradiated to increase its vitamin D content as a measure to reduce the incidence of rickets in the population at large. The partial destruction of other nutrients was considered a price worth paying. The oxidative deterioration of oils and fats, leading to rancidity, has been reported following UV irradiation but there is some evidence to suggest that this can be reduced by conducting the irradiation under an inert gas blanket. This effectively limits the doses which can safely be applied to foodstuffs containing high levels of oils and fats. The intense irradiation of fish oils has been shown to be linked to the production of toxic by-products, e.g. aldehydes, but there have been no reports of the occurrence of these or similar compounds in foods which have undergone UV irradiation. A less well-understood phenomenon is the development of 'brown spot' on certain irradiated vegetables, lettuces apparently being particularly susceptible.

See also: **Minimal Methods of Processing**: Electroporation – Pulsed Electric Fields.

Further Reading

Bachman R (1975) Sterilization by Intense Ultraviolet Radiation. *Brown Boveri Review* 5: 206–209.

Gardner DWM and Shama G (1998) The Kinetics of *Bacil-*

lus subtilis Spore Inactivation on Filter Paper by UV Light and UV Light in Combination with Hydrogen Peroxide. *Journal of Applied Microbiology* 84: 633–641.

Harm W (1978) *Biological Effects of Ultraviolet Radiation.* Cambridge: Cambridge University Press.

Hatchard GC and Parker CA (1956) A New Sensitive Chemical Actinometer: II. Potassium Ferrioxalate as a Standard Chemical Actinometer. *Proceedings of the Royal Society* A235: 518–536.

Phillips R (1983) *Sources and Applications of Ultraviolet Radiation.* London: Academic Press.

Schenck GO (1987) Ultraviolet Sterilization. In: Lorch W (ed.) *Handbook of Water Purification*, 2nd edn. P. 530. Chichester: Ellis Horwood.

Shama G, Gardner DWM, Martin AP and Mason NL (1994) Disinfection of Particles using Ultraviolet Light. *Transactions of the Institution of Chemical Engineers Part C* 72: 197–200.

Shama G, Peppiatt C and Biguzzi M (1996) A Novel Thin Film Photoreactor. *Journal of Chemical Technology and Biotechnology* 65: 56–64.

Stevens C, Khan VA, Lu JY et al (1997) Integration of Ultraviolet (UV-C) Light with Yeast Treatment for Control of Postharvest Storage Rots of Fruits and Vegetables. *Biological Control* 10: 98–103.

Sunada K, Kikuchi Y, Hashimoto K and Fujishima A (1998) Bactericidal and Detoxification Effects of TiO_2 Thin Film Photocatalysts. *Environmental Science and Technology* 32: 726–728.

V

VAGOCOCCUS

Lúcia Martins Teixeira and Maria da Glória S Carvalho, Departamento de Microbiologia Médica, Instituto de Microbiologia, Universidade Federal do Rio de Janeiro, Brazil

Richard R Facklam, Streptococcus Laboratory, Respiratory Diseases Branch, Division of Bacterial and Mycotic Diseases, Centers for Disease Control and Prevention, Atlanta, USA

Introduction

The genus *Vagococcus* was established as a separate genus in 1989, to accommodate the motile cocci resembling lactococci, which were earlier referred to as motile lactic streptococci and were shown to be diverse from all known lactococci. Results of 16S rRNA sequencing studies demonstrated that such strains formed a separate line of descent within the lactic acid bacteria and represented a new species, which was named *V. fluvialis*. A subsequent molecular taxonomic investigation resulted in the description of a second species, in 1990, named *V. salmoninarum*. Although distinct, the genus *Vagococcus* has a close phylogenetic relationship with the genera *Enterococcus* and *Lactococcus*, and some species are difficult to differentiate solely on the basis of phenotypic characteristics. The significance of vagococci as agents of infections and their presence in food products of animal origin is still unclear. The scarcity of reports may be due, at least in part, to their recent recognition as a separate taxon.

Description of the Genus

The members of genus *Vagococcus* (wandering coccus) are facultatively anaerobic, catalase-negative Gram-positive cocci. Cells occur singly or arranged in pairs or as short chains. The colony morphology resembles that of enterococcal and streptococcal strains. Colonies are raised and grey-white and they are α- or non-haemolytic on agar media containing sheep blood. They have fermentative metabolism, with L-lactic acid being the predominant end product of glucose fermentation. They can react with Lancefield streptococcal groups D and N antisera. The cell wall peptidoglycan type is Lys-D-Asp. The DNA G+C content ranges from 33.6 to 36.5 mol%. Two species of *Vagococcus* have been described: *V. fluvialis* and *V. salmoninarum*. The type strain is *V. fluvialis* ATCC 49515 (NCFB 2497). The type strain for *V. salmoninarum* is ATCC 51200 (NCFB 2777).

The physiological tests for differentiating the vagococci and other catalase-negative Gram-positive cocci are listed in **Table 1**. The presumptive identification of a *Vagococcus* can be accomplished by demonstrating that the strain hydrolyses esculin in the presence of bile (bile-aesculin-test-positive), produces leucine aminopeptidase and pyrrolidonyl arylamidase (LAP and PYR test-positive), and is susceptible to vancomycin. Growth at 10°C, 45°C, and growth in broth containing 6.5% NaCl can be variable. Motility can also vary (only *V. fluvialis* strains are usually motile). The vagococci are non-pigmented, and are negative for production of gas, and arginine and hippurate hydrolysis. Delayed or weak reactions with anti-streptococcal group D serum as well as reaction with group N antiserum can be observed with some vagococcal strains when Lancefield hot-acid cell extracts are tested by the capillary precipitation method.

The difficulty in distinguishing these microorganisms, especially *V. fluvialis*, from other lactic acid bacteria by phenotypic criteria is widely recognized. On the basis of phenotypic characteristics, most isolates are initially classified as unidentified or atypical enterococci, because they resembled some of the less common atypical arginine-negative enterococcal species. Results of motility and arginine hydrolysis tests can be helpful in the differentiation from the lactococci, which usually give negative and positive reactions, respectively. All *Vagococcus* strains tested to date reacted with the AccuProbe *Enterococcus* culture confirmation test manufactured by Gen-Probe (San Diego, CA, US). This probe is based on a DNA oligomer having a structure complementary to a segment of enterococcal rRNA. Although testing with

Table 1 Phenotypic characteristics of facultatively anaerobic, catalase-negative, Gram-positive coccus genera

Genus	*Phenotypic characteristic*									
							Growth at:			
	VAN	*GAS*	*PYR*	*LAP*	*BE*	*NaCl*	*10°C*	*45°C*	*MOT*	*HEM*
Chains										
Vagococcus	S	–	+	+	+	+[a]	+	(–)[b]	V	α/n
Enterococcus	S[c]	–	+	+	+	+	+	+	V	α/β/n
Lactococcus	S	–	+	+	+	V	+	(–)[b]	–	α/n
Leuconostoc	R	+	–	–	V	V	+	V	–	α/n
Weissella	R	+	–	–	V	+	V	V	–	α/n
Abiotrophia	S	–	+	+	–	–	–	–	–	α
Streptococcus	S	–	–[d]	+	–[e]	–	V	V	–	α/β/n
Globicatella	S	–	+	–	–	+	–	–	–	α
Clusters and tetrads										
Pediococcus	R	–	–	+	+	V	–	+	–	α
Tetragenococcus	S	–	–	+	+	+	–	+	–	α
Aerococcus	S	–	+	–	V	+	–	+	–	α
Gemella	S	–	+	V	–	–	–	–	–	α/n
Helcococcus	S	–	+	–	+	+	–	–	–	n
Alloiococcus[f]	S	–	+	+	–	+	–	–	–	n
Dolosigranulum	S	–	+	+	–	+	–	–	–	n
Facklamia	S	–	+	+	–	+	–	–	–	n
Ignavigranum	S	–	+	+	–	+	–	–	–	α

VAN, susceptibility to vancomycin (30 μg disc); GAS, production of gas from glucose in *Lactobacillus* Mann, Rogosa, Sharpe (MRS) broth; PYR, production of pyrrolidonyl arylamidase; LAP, production of leucine aminopeptidase; BE, reaction on bile-aesculin medium; NaCl, growth in broth containing 6.5% NaCl; MOT, motility; HEM, haemolysis on blood agar containing 5% sheep blood; α, α-haemolysis; β, β-haemolysis; n, no haemolysis; S, susceptible; R, resistant; –, ⩾95% of strains with negative results; +, ⩾95% of strains with positive results; V, variable reactions (some strains positive, some negative).

[a]Strains are generally positive after long incubation (5 days or more).

[b]Some strains grow slowly at 45°C.

[c]Some enterococcal strains are vancomycin-resistant but still show a small inhibition around the disc; other strains grow right up to the disc and are vancomycin-resistant under the defined screening test criteria.

[d]Group A streptococci are PYR-positive and all other streptococci are PYR-negative.

[e]*Streptococcus bovis* and *S. equinus* are bile-esculin-positive as well as 5–10% of the viridans streptococci.

[f]*Alloiococcus* strains grow anaerobically only under defined conditions.

the enterococcal genetic probe does not allow differentiation between the enterococci and the vagococci, it can also be used as a tool to separate the vagococci from the lactococci, which are negative.

Characterization of the Species

Confirmation that a strain is a *Vagococcus* requires complete identification to the species level. It is generally accomplished by using a series of additional conventional physiological tests. In addition to the tests included in Table 1, the following tests are usually included: hydrolysis of arginine, hydrolysis of hippurate, pigment production, pyruvate utilization, tellurite tolerance, the Voges–Proskauer reaction, and acid production from L-arabinose, methyl-α-D-glucopyranoside, glycerol, inulin, lactose, maltose, D-mannitol, melibiose, raffinose, ribose, D-sorbitol, sorbose, sucrose and trehalose.

As already pointed out, even using extensive testing, the differentiation of vagococcal strains from enterococci is sometimes problematic. Tests for production of acid from L-arabinose and raffinose may be useful, since *V. fluvialis* strains are negative and the motile *Enterococcus* species, *E. gallinarum* and *E. casseliflavus*, are positive. The arginine test may also be a clue for such differentiation as most strains belonging to these enterococcal species are positive. However, uncommon arginine-negative variants of the physiological group II enterococcal species and *E. columbae* have biochemical characteristics which are very similar to those of the vagococci, especially *V. fluvialis*. **Table 2** lists some of the tests that can be used to differentiate between them.

Both vagococcal species *V. fluvialis* and *V. salmoninarum* share several phenotypic characteristics. They usually produce acids from maltose, ribose and trehalose, and are negative for the following tests: arginine, arabinose, inulin, lactose, melibiose and sorbose. Variable results are obtained for pyruvate utilization, tellurite tolerance and Voges–Proskauer tests. They can be distinguished on the basis of a few phenotypic characteristics, such as motility, and production of acids from mannitol, raffinose and sorbitol (Table 2). Tests for production of acids from

Table 2 Phenotypic characteristics used to differentiate between arginine-negative variants of physiological group II enterococcal species, *Enterococcus columbae* and species of *Vagococcus*

	Phenotypic characteristic												
Species	*MAN*	*SOR*	*ARG*	*ARA*	*SBL*	*RAF*	*TEL*	*MOT*	*PIG*	*SUC*	*PYU*	*MGP*	*GLY*
E. faecalis	+	–	–	–	+	–	+	–	–	+[a]	+	–	+
E. casseliflavus	+	–	–	+	V	+	V	+[a]	+[a]	+	V	+	V
E. gallinarum	+	–	–	+	–	+	–	+[a]	–	+	–	+	–[a]
E. columbae	+	–	–	+	+	+	–	–	–	+	+	–	–
V. fluvialis	+	–	–	–	+	–	–[a]	+	–	+[a]	V	+	+
V. salmoninarum	–	–	–	–	–	+	–	–	–	+	+	+	–

MAN, mannitol; SOR, sorbose; ARG, arginine; ARA, arabinose; SBL, sorbitol; RAF, raffinose; TEL, tellurite; MOT, motility; PIG, pigment; SUC, sucrose; PYU, pyruvate; MGP, methyl-α-D-glucopyranoside; GLY, glycerol; –, ⩾ 95% of strains with negative results; +, ⩾ 95% of strains with positive results; V, variable reactions (some strains positive, some negative).
[a]Occasional exceptions occur.

sucrose and glycerol, as well as pyruvate utilization and tellurite tolerance are also of some help. In the original description of *V. fluvialis* this microorganism was reported to be LAP-negative, PYR-variable and negative for growth in 6.5% NaCl broth, but all strains tested at the Centers for Disease Control have been both LAP- and PYR-positive and have demonstrated growth in 6.5% NaCl broth.

In addition to the determination of physiological characteristics, analysis of electrophoretic whole-cell protein profiles was shown to be a reliable method for the identification of vagococcal strains. *V. fluvialis* isolates correspond to a unique protein profile that is distinct from the protein profile characteristics of *V. salmoninarum* strains, enterococcal and lactococcal species. Genus- and species-specific oligonucleotide probes derived from 16S rRNA were also shown to facilitate the precise identification of vagococcal isolates.

Data on the antimicrobial susceptibility characteristics are only available for a few *V. fluvialis* isolates. Results of minimum inhibitory concentration determinations indicated that *V. fluvialis* strains are susceptible to ampicillin, cefotaxime, trimethoprim-sulphamethoxazole and vancomycin, and are resistant to clindamycin, lomefloxacin and ofloxacin. Strain-to-strain variation was observed in relation to 17 other antimicrobial agents tested, including cefaclor, cefazolin, cefixime, ceftriaxone, cefuroxime, chloramphenicol, ciprofloxacin, clarithromycin, erythromycin, gentamicin, meropenem, oxacillin, penicillin, piperacillin-tazobactam, rifampin, tetracycline and tobramycin.

Molecular characterization of *V. fluvialis* strains, using analysis of chromosomal DNA restriction patterns by pulsed-field gel electrophoresis (PFGE) after digestion with *Sma*I, resulted in distinctive PFGE patterns, suggesting the nonclonal nature of *V. fluvialis*, and indicating the potential ability of this typing technique to discriminate between *Vagococcus* isolates.

Importance of the Genus in Animal and Human Diseases: Importance for the Food Industry and Potential Hazard for the Consumer

As the members of the genus *Vagococcus* were only recently distinguished from other related bacteria, the involvement of the vagococci as infectious agents has been possibly hindered by difficulties in their precise identification, as they may have been misidentified or overlooked in diagnostic laboratories. Also, the role of *Vagococcus* in food spoilage, decreasing its shelf life, as well as the production of metabolic compounds important for the food industry, is still unclear.

The pathogenic role of these microorganisms was first recognized when they were identified among the agents associated with a complex of similar fish diseases known under the general denomination of streptococcosis, caused by different taxons of Gram-positive cocci. Such complex of diseases has long been considered a serious problem in cultured marine fish in the Far East. Nowadays, with the development of intensive aquaculture, fish infections caused by Gram-positive cocci, including vagococcosis, have become a major problem in various parts of the world, affecting different fish species destined for human consumption. Heavy economic losses caused by these infections are in some cases estimated to encompass 45% of total fish production. On the other hand, except for reports of the isolation of *V. salmoninarum* from diseased salmonid fish (brown trout, rainbow trout and Atlantic salmon) and of *V. fluvialis* from domestic animals (cats, horses and pigs), information on the significance of the vagococci as pathogens followed by proper identification of the microorganism is still very limited.

More recently, a report from our laboratories provided the first evidence of the possible connection of a vagococcal species as a cause of infections in human beings. Very little clinical information was submitted with the cultures identified. One strain was isolated from the peritoneal fluid of a nephrology patient, and another from an infected bite wound in a person who was bitten by a lamb. Two other human isolates were recovered from blood cultures, and no additional information on the clinical condition or associated illness was available. Two additional strains have been received for identification. One was isolated from a finger wound of a patient in Canada, and the other was recovered from the cerebrospinal fluid of a patient with meningitis in Argentina. All vagococcal strains isolated from humans until now are *V. fluvialis*. No human infection associated with *V. salmoninarum* has been documented.

Apart from the economic impact, concerns have been generated about the potential health hazard that handling or consumption of colonized or infected fish, as well as other animals or their products can represent for humans. The presence of lactic acid bacteria, including vagococci, in foodstuffs, is not usually considered to be a major concern, because they are widely distributed in nature. Therefore, they are not considered to be significant pathogens of humans, as they are probably only associated with opportunistic infections. However, recent evidence of acquisition of serious infections by people who had skin injuries during handling of *Streptococcus iniae* (another agent of fish streptococcosis that was not considered pathogenic for humans) -colonized or diseased fish grown by aquaculture, raises the question on the potential health hazard. During the investigation of *S. iniae* transmission from farm-cultured fish (*Tilapia*) to humans, many *V. fluvialis* strains were isolated from the surface of the fish. It is not known whether these isolates had any effect on the fish or whether they could be transmitted to humans. One additional example suggesting the possibility of animals or their food products as sources of transmission of infections, due to Gram-positive cocci, to humans, is given by *Lactococcus garvieae*, another agent of fish streptococcosis that has also been shown to cause disease in cattle and in humans.

In the light of the above-mentioned, the vagococci should be considered to be among the emerging zoonotic pathogens that have been isolated from various species of fish, as well as mammals, and finally humans. Diagnostic laboratories as well as those devoted to the analysis of food products of fish and other animal origin, especially fresh produces, must be aware of the methods for the precise detection of this microorganisms, as they may serve as vehicles for transmission of infections caused by this newly recognized pathogen. As more attention and accurate procedures are incorporated into the identification schemes to detect and characterize vagococcal strains in the diagnostic setting, more information will become available to help in answering the many questions raised about the significance of these microorganisms.

Suggested Laboratory Procedures for Isolation and Identification

Recommendations for Isolation and Preliminary Testing

Enriched infusion agar and broth, such as trypticase soy, heart infusion, Todd-Hewitt, *Lactobacillus* Mann, Rogosa, and Sharpe (MRS) or brain–heart infusion, support the growth of vagococci and 5% sheep blood agar plates are recommended to verify haemolytic activity. If the specimen to be processed for primary isolation is likely to contain other bacteria, such as food samples, bile-aesculin medium may be an option as a primary isolation-selective medium. Vagococci growth is better when incubated for 18–24 h in 3–10% CO_2 atmosphere. Special attention must be paid to the temperature requirements for optimal growth of each vagococcal species: *V. fluvialis* cultures should be incubated at 35–37°C and *V. salmoninarum* cultures at 25°C. The most consistent Gram stains can be prepared from growth in thioglycolate broth. The catalase test should be performed by flooding the growth of the bacteria on a blood-free medium with 3% hydrogen peroxide and observing for bubbling (positive reaction) or not (negative reaction).

Bile-aesculin Test The bile-aesculin medium can be used in agar slants or agar plates. Inoculate the bile-esculin medium with one to three colonies and incubate it at normal atmosphere for up to 7 days. A positive reaction is recorded when a black colour forms over one-half or more of the slant, or when any blackening occurs on the agar plate. No change in the colour of the agar indicates a negative reaction.

NaCl Tolerance Test Growth in broth containing 6.5% NaCl is determined in heart infusion broth base with an addition of 6% NaCl (heart infusion base contains 0.5% NaCl), 0.5% glucose and bromcresol purple indicator. When a frank growth occurs, the glucose is fermented and the broth colour changes from purple to yellow, but it is not necessary for a positive result: an obvious increase in turbidity without a change in colour is also considered to be a positive test. One or two colonies or a drop of an overnight broth culture is inoculated into the broth containing 6.5% NaCl. The inoculated broth is incubated up to 7 days.

LAP and PYR Tests The LAP and PYR disc tests are performed in the same manner, and the discs are available from several commercial suppliers. The strains to be tested are grown overnight on blood agar plates. The discs are placed on the blood agar plate in an area of little or no growth. The discs are inoculated heavily; two or more loopfuls of inoculum are necessary for satisfactory results. The plates with the discs are incubated at room temperature for 10 min, the detection reagent is added, and the reactions are read after 3 min. The development of a red colour is positive; no change in colour or a yellow colour is negative; and the development of a pink colour indicates a weak positive reaction. The test should be discarded after 10 min if still negative.

Tests for Growth at 10 and 45°C Growth at 10 and 45°C is determined in heart infusion broth base medium containing 0.1% dextrose and bromcresol purple indicator. The tests are performed by inoculating the broths with a single colony or a drop of an overnight broth culture and incubating at the respective temperatures for up to 7 days. The time between inoculation and placement at the proper temperature should not be longer than 10 min. When the test cultures are inspected for growth during the incubation period, the tubes should be returned to the proper temperature without being allowed to warm or cool. A positive result is indicated by frank growth, which may be accompanied by a colour change in the indicator. A colour change is not necessary to determine a positive reaction; an increase in turbidity indicates growth and a positive test. Be sure to rotate the tube vigorously after the incubation period. Some bacterial strains have a tendency to settle to the bottom of the tube, and turbidity will not be apparent until the contents of the tubes are mixed.

Vancomycin Susceptibility Identification Test To determine susceptibility to vancomycin, several colonies of the strain are transferred to one-half of a trypticase soy agar plate containing 5% sheep blood and spread with a loop or cotton swab to achieve confluent growth. The vancomycin susceptibility testing disc (30 μg) is placed in the heavy part of the streak. The inoculated plate is incubated at 5% CO_2 atmosphere for 18 h. Strains with any zone of growth inhibition are considered susceptible, and strains that exhibit growth up to the disc are considered resistant.

Carbohydrate Fermentation Test The carbohydrates that may be tested are L-arabinose, methyl-α-D-glucopyranoside, glycerol, inulin, lactose, maltose, D-mannitol, melibiose, raffinose, ribose, sorbitol, sorbose, sucrose and trehalose. Acid from carbohydrate is determined in heart infusion broth containing the specific carbohydrate (1%) and bromcresol purple indicator (0.0016%). The carbohydrate broth is inoculated with a drop or a loopful of an overnight broth culture or with several colonies taken from a blood agar plate. The inoculum should be from a fresh culture. The carbohydrate broth is incubated for up to 7 days. A positive reaction is recorded when the broth turns yellow.

Gas Production Production of gas from glucose is determined in *Lactobacillus* MRS broth. Two or more colonies from a blood agar plate or a drop of broth culture are used to inoculate the broth. The inoculated tube is overlaid with melted petroleum jelly and incubated for up to 7 days. Gas production is indicated when the wax plug is completely separated from the broth. Small bubbles that may accumulate over the incubation period are not read as positive.

Arginine Deamination The deamination of arginine is determined in Moeller decarboxylase medium containing 1% L-arginine. The medium is inoculated with a fresh culture and is then overlaid with sterile mineral oil (1–2 ml per 5 ml of arginine medium) and incubated for up to 7 days. A positive reaction is recorded when the broth turns deep purple, indicating an alkaline reaction. A yellow colour or no colour change of the broth indicates a negative reaction.

Motility Test Motility is determined in modified Difco motility medium. The medium is prepared by adding 16 g of motility test medium (Difco), 4 g of nutrient broth powder (Difco), and 1 g of NaCl to 1 l of distilled water. The medium is inoculated with a single stab (with an inoculating needle, not a loop) about 1 in (2.54 cm) into the centre of the medium in the tube. The inoculated tube is placed in a 25–30°C incubator. Motility is indicated by the spread of growth to the bottom and sides of the tube. Growth along and slightly away from the stab indicates negative motility.

Pigment Production Test Production of pigment is determined by examining a cotton swab that has been smeared across growth on a trypticase soy–5% sheep blood agar plate that has been incubated for 24 h in a 5% CO_2 atmosphere. Production of pigment is indicated by a yellow colour of the growth. A cream, white or grey colour does not indicate pigmentation.

Pyruvate Utilization Test A fresh culture is used to inoculate a tube of pyruvate broth. The broth is incubated for up to 7 days. A positive reaction is indicated by the development of a yellow colour. If the broth remains green or greenish yellow the test result is negative. A yellow colour with only a hint of green is interpreted as positive.

Tellurite Tolerance Test Tolerance to tellurite is determined on agar medium containing 0.04% potassium tellurite. The agar may be contained in a tube or plate. The medium is inoculated with a fresh culture of the strain to be tested and incubated up to 7 days. Tolerance (positive result) is indicated whenever black colonies form on the surface.

Voges–Proskauer Test The Colbentz modification of the Voges–Proskauer test is used to determine the production of acetylmethyl carbinol. The strains are grown on blood agar plates overnight. With a loop, all growth on the plate is transferred to 2 ml of Voges–Proskauer broth. The broth is incubated for 6 h or overnight and then tested. Ten drops of solution (A (α-naphthol) and 10 drops of solution B (sodium hydroxide and creatine) are added to 0.5 ml of the culture in the Voges–Proskauer broth. The tube is shaken vigorously, and a positive reaction is indicated when a red colour develops within 30 min. A pink or rust colour is interpreted as a weak positive reaction.

See also: ***Enterococcus***. ***Lactococcus***: Introduction. ***Streptococcus***: Introduction.

Further Reading

Carvalho MGS, Teixeira LM and Facklam RR (1998) Use of tests for acidification of methyl-α-D-glucopyranoside and susceptibility to efrotomycin for differentiation of strains of *Enterococcus* and some related genera. *Journal of Clinical Microbiology* 36: 1584–1587.

Collins MD, Ash C, Farrow JAE, Wallbanks S and Williams M (1989) 16S ribosomal ribonucleic acid sequence analyses of lactococci and related taxa. Description of *Vagococcus fluvialis* gen. nov., sp. nov. *Journal of Applied Bacteriology* 67: 453–460.

Facklam RR and Elliott JA (1995) Identification, classification, and clinical relevance of catalase-negative, Gram-positive cocci, excluding the streptococci and enterococci. *Clinical Microbiology Reviews* 8: 479–495.

Facklam RR and Teixeira LM (1997) *Enterococcus*. In: Collier L, Balows A and Sussman M (eds) *Topley & Wilson's Microbiology and Microbial Infections*, 9th edn. Vol 2, p. 669. London: Edward Arnold.

Pot B, Devriese LA, Hommez J et al (1994) Characterization and identification of *Vagococcus fluvialis* strains isolated from domestic animals. *Journal of Applied Bacteriology* 77: 362–369.

Schliefer KH and Kilpper-Balz R (1987) Molecular and chemotaxonomic approaches to the classification of streptococci, enterococci and lactococci: a review. *Systematic Applied Microbiology* 10: 1–19.

Schliefer KH, Kraus J, Dvorak C, Kilpper-Balz R, Collins MD and Fischer W (1985) Transfer of *Streptococcus lactis* and related streptococci to the genus *Lactococcus* gen. nov. *Systematic Applied Microbiology* 6: 183–195.

Schmidtke LM and Carson J (1994) Characteristics of *Vagococcus salmoninarum* isolated from diseased salmonid fish. *Journal of Applied Bacteriology* 77: 229–236.

Teixeira LM, Merquior VLC, Vianni MCE et al (1996) Phenotypic and genotypic characterization of atypical *Lactococcus garvieae* strains isolated from water buffalos with subclinical mastitis and confirmation of *L. garvieae* as a senior subjective synonym of *Enterococcus seriolicida*. *International Journal of Systematic Bacteriology* 46: 664–668.

Teixeira LM, Carvalho MGS, Merquior VLC, Steigerwalt AG, Brenner DJ and Facklam RR (1997) Phenotypic and genotypic characterization of *Vagococcus fluvialis*, including strains isolated from human sources. *Journal of Clinical Microbiology* 35: 2778–2781.

Wallbanks S, Martinez-Murcia AJ, Fryer JL, Phillips BA and Collins MD (1990) 16S rRNA sequence determination for members of the genus *Carnobacterium* and related lactic acid bacteria and description of *Vagococcus salmoninarum* sp. nov. *International Journal of Systematic Bacteriology* 40: 224–230.

Weinstein MR, Litt M, Kertesz DA et al (1997) Invasive infections due to a fish pathogen, *Streptococcus iniae*. *New England Journal of Medicine* 337: 589–594.

Williams AM and Collins MD (1992) Genus- and species-specific oligonucleotide probes derived from 16S rRNA for the identification of vagococci. *Letters in Applied Microbiology* 14: 17–21.

Vegetable Oils *see* **Preservatives:** Traditional Preservatives – Vegetable Oils.

VEROTOXIGENIC *E. COLI*

Detection by Commercial Enzyme Immunoassays

D W Pimbley, Leatherhead Food Research Association, Leatherhead, Surrey, UK

Introduction

The role of certain strains of *Escherichia coli* in human disease has been known for many years. More recently, strains of verotoxin-producing *E. coli* (VTEC) have been recognized that can produce severe illness in humans. Symptoms range from mild, non-bloody diarrhoea to haemorrhagic colitis and haemolytic–uraemic syndrome, resulting in death in severe cases. *Escherichia coli* O157 is the serogroup most commonly isolated in human infections, although other serogroups of VTEC have also been implicated. The verotoxigenic strains can possess the flagellar antigen (H7) or they can be non-motile (H–). Foods, particularly unpasteurized milk and raw meats, are thought to be an important source of infection, although person-to-person transmission may also occur.

Safety Considerations

At the time of writing, the UK Health and Safety Executive (HSE) and the European Union (EU) are reviewing the safety categorization of VTEC. In the UK it is likely that the Advisory Committee on Dangerous Pathogens (ACDP) will reclassify VTEC into ACDP hazard group 3, because of the low infective dose, the potential severity of the infection and the possibility of laboratory-acquired infections. In the meantime, it is recommended that a local risk assessment should be carried out by laboratories dealing with samples likely to contain *E. coli* O157. In this context, it is important to note that some immunoassays for *E. coli* O157 include a heat treatment stage that kills the organism, making subsequent stages in the assay less hazardous.

Commercial Immunoassays for *E. coli* O157

Commercial immunoassays for the detection of *E. coli* O157 and/or VTEC in foods are available in a variety of formats, including enzyme immunoassays (EIAs) and immunochromatographic and immunoblot-based assays. **Table 1** summarizes the main characteristics of the assays, including manufacturer, format, target antigen and approval status.

The majority of commercial immunoassays for *E. coli* O157 require a selective cultural enrichment of 18–28 h to increase the level of *E. coli* O157 to the minimum sensitivity of the test – typically 10^4–10^6 colony forming units (cfu) per millilitre. The total assay time for these tests (20–30 h) is still much shorter than the conventional cultural methods for *E. coli* O157, which take 3 days to give a presumptive positive result. Chemiluminescence-based EIAs (e.g. EC-LITE O157:H7) are claimed to be more sensitive (10^3–10^4 cfu ml^{-1}) and therefore require a shorter cultural stage (4 h), with an overall assay time of less than 7 h. However, to date no independent validation data have been published for this kit. **Figure 1** gives an overview of testing protocols and point of application for the commercial immunoassays for *E. coli* O157.

The examples of protocols given below are for reference only and their reproduction does not constitute an endorsement of these tests. For detailed information on specific assays for *E. coli* O157 refer to the publications listed under 'Further Reading', or contact the manufacturers listed in Appendix V.

Sample Preparation and Enrichment

Table 2 details variations in sample preparation and cultural enrichment between the different assays.

All the commercial immunoassays for *E. coli* O157 have similar sample preparation and cultural enrichment stages, which are based on those recommended by the US Department of Agriculture (USDA), and are also detailed in the Food and Drug Administration (FDA) *Bacteriological Analytical Manual* (BAM). All manufacturers of immunoassay kits recommend that presumptive positive results should be confirmed using conventional cultural, biochemical and serological techniques. Therefore, the selective enrichment culture must be retained and stored at 4°C until completion of the assay. The protocols for some immunoassays also require heat treatment of an aliquot of the

Table 1 Commercial immunoassays for the detection of *E. coli* O157 in foods

Manufacturer[a]	*Test*	*Format*	*End point*	*Target antigen*	*Approval status*
Organon Teknika	EHEK-Tek	Microwell	Colorimetric	O157:H7	Health Canada Method MFLP-91
Bioenterprises Pty Ltd	Tecra E.C. O157 VIA	Microwell	Colorimetric	All O157	
BioControl Systems Inc.	Assurance EHEC EIA	Microwell	Colorimetric	O157:H7	AOAC Official Method 996.10
Celsis-Lumac	PATH-STIK *E. coli* O157	Dip-stick	Colorimetric	All O157	
BioControl Systems Inc.	VIP EHEC	One-step	Colorimetric	O157:H7	AOAC Official Method 996:09
Diffchamb	Transia card *E. coli* O157	One-step	Colorimetric	All O157	
Neogen	REVEAL *E. coli* O157:H7	One-step	Colorimetric	O157:H7	
Meridian Diagnostics	ImmunoCard STAT! *E. coli* O157:H7	One-step	Colorimetric	O157:H7	AOAC Performance tested
3M	Petrifilm HEC	Immunoblot	Colorimetric	O157:H7	
GEM Biomedical	EC-LITE *E. coli* O157:H7	Tube	Chemiluminescent	O157:H7	
Difco Laboratories	EZ coli System	Detector tip	Colorimetric	All O157	
bioMérieux	VIDAS *E. coli* O157 (ECO)	Automated	Fluorescent	All O157	
Foss Electric A/S	EiaFoss *E. coli* O157	Automated	Fluorescent	All O157	

[a] See Appendix V for address details.
AOAC, Association of Official Analytical Chemists.

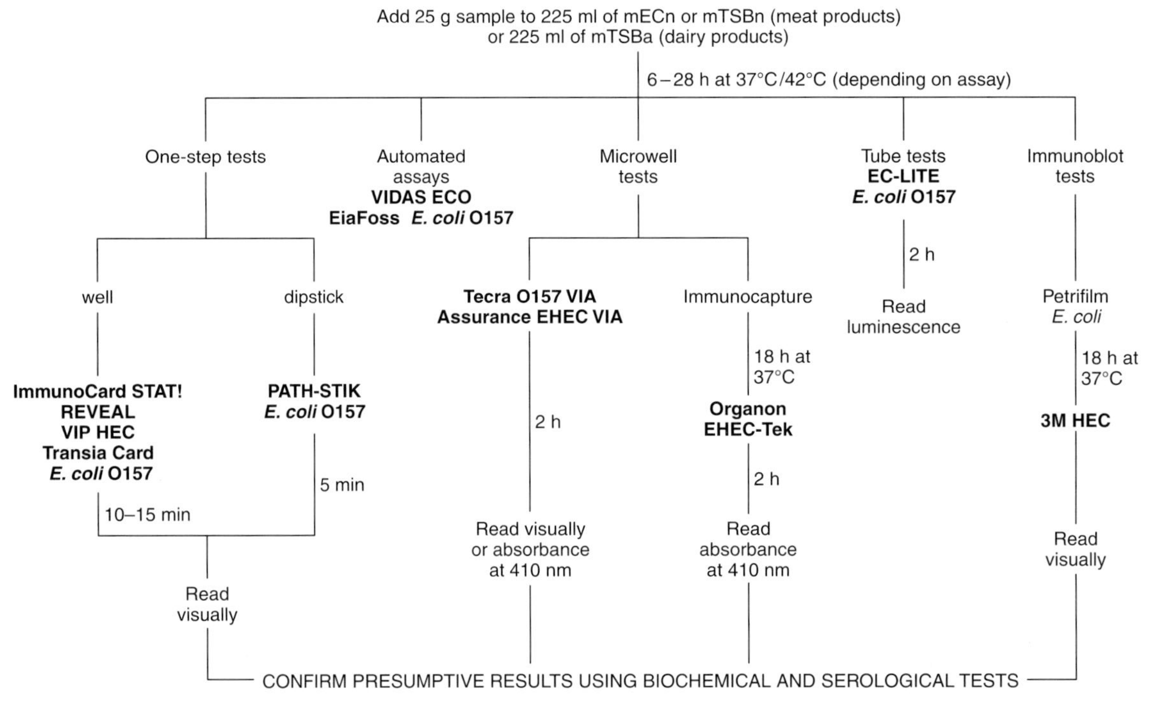

Figure 1 Protocols for commercial immunoassays for *Escherichia coli* O157 and point of application.

cultural enrichment to extract *E. coli* O157 antigens, prior to carrying out the assay (see Table 2).

Immunoseparation The Organon Teknika EHEK-Tek incorporates an immunomagnetic separation stage with the aim of improving the sensitivity and specificity of the assay. This reduces the incubation period for the selective enrichment stage from 18 h to 6 h, but a secondary 18 h selective enrichment stage is then required, so the overall assay length is similar to other EIAs. Immunoseparation can also be used with other EIAs for *E. coli* O157, since the immunomagnetic particles used in this process are available separately (Dynabead anti-*E. coli* O157, Dynal).

Table 2 Summary of sample preparation and enrichment methods for immunoassays for *E. coli* O157

Test	*Primary enrichment*[a]	*Blending instructions*	*Incubation time/temp.*	*Secondary enrichment*	*Heat treatment*[b]	*Minimum assay (h) time*
EHEK-Tek	mTSBa (dairy products) mECn (meat and meat products)	Stomach 1 min Stomach (filter bags)	6 h at 37°C (shaken)	18 h at 37°C in mTSBa (following IMS)	Yes	26
Tecra E.C. O157 VIA	mTSBa (dairy products) mECn (meat and meat products)	Stomach Stomach	18 h at 37°C (static)	None	Yes	20
Assurance EHEC EIA	mTSBn	Mix well	18–28 h at 37°C (static)	None	Yes	20
PATH-STIK *E. coli* O157	mECn	None	16–24 h at 37°C (static)	None	No	16.5
VIP EHEC	mTSBn	Mix well	18–28 h at 37°C (static)	None	No	18.5
Transia Card *E. coli* O157	BPW/mECn	Stomach (filter bag)	24 h at 36°C (static)	None	Yes	24
Reveal *E. coli* O157:H7	Reveal 8 h medium mECn/mTSBn	Stomach 2 min	8 h at 43°C (static)	None	Yes	8.5
			20 h at 37°C (static)			20.5
ImmunoCard STAT!	MacConkey broth	None	8–24 h at 43°C	None	No	8.5
Petrifilm HEC	mECn	Stomach 2 min	6–8 h at 37°C (shaken)	18 h at 42°C on Petrifilm EC	No	26
EC-LITE *E. coli* O157:H7	mECn	Mix thoroughly	4 h at 42°C (static)	None	No	6.5
EZ Coli	EZ coli broth	Blend or stomach	15–24 h at 42°C	None	No	15.5
VIDAS *E. coli* O157	mTSBa (dairy products) mECn (meat and meat products)	Stomach	6–7 h at 41°C	18 h at 35–37°C	Yes	25
EiaFoss *E. coli*	mECn	None	24 h at 41°C	None (IMS)	Yes	24

[a] See Appendix I.
[b] Heat treatment to extract antigen – boiling or steaming at 100°C for 15–20 min.
IMS, immunomagnetic separation.

Example of an Immunomagnetic Separation Protocol

1. Add 1 ml of enrichment broth to 20 μl of Dynabead anti-*E. coli* O157 in a 1.5 ml microcentrifuge tube and incubate for 10 min at room temperature.
2. Place tube in a magnetic particle collector (MPC, Dynal) and wait for 3 min.
3. With magnet still in place aspirate supernatant with a Pasteur pipette.
4. Add 1 ml of wash solution (provided), remove magnet from MPC and invert the tube several times to resuspend the beads.
5. Replace magnet and wait for 3 min.
6. With the magnet still in place, aspirate the supernatant.
7. Analyse the sample according to EIA kit instructions.

Manual Immunoassays for *E. coli* O157

Example of a Protocol for a Microwell-based EIA **Figure 2** shows a typical generic protocol for a microwell-based EIA for *E. coli* O157. The following detailed protocol is for the Assurance Polyclonal Enzyme Immunoassay for *E. coli* O157:H7, applicable to detection of *E. coli* O157:H7 in dairy foods, meats, poultry products, fruits, nutmeats, seafood, pasta and liquid egg.

Principle Assurance EHEC EIA is a microwell-based enzyme immunoassay that employs highly specific antibodies against *E. coli* O157:H7. Enriched test samples and positive controls are added to the microwells and any *E. coli* O157 present are captured. Nonreactive components of the sample are removed by washing. Alkaline phosphatase antibody conjugate is then added, which binds to the *E. coli* O157:H7 antigens and, after incubation, any unbound conjugate is washed away. The substrate (β-nitrophenylphosphate) is added and the absorbance of the resulting coloured product is read spectrophotometrically at 405–410 nm.

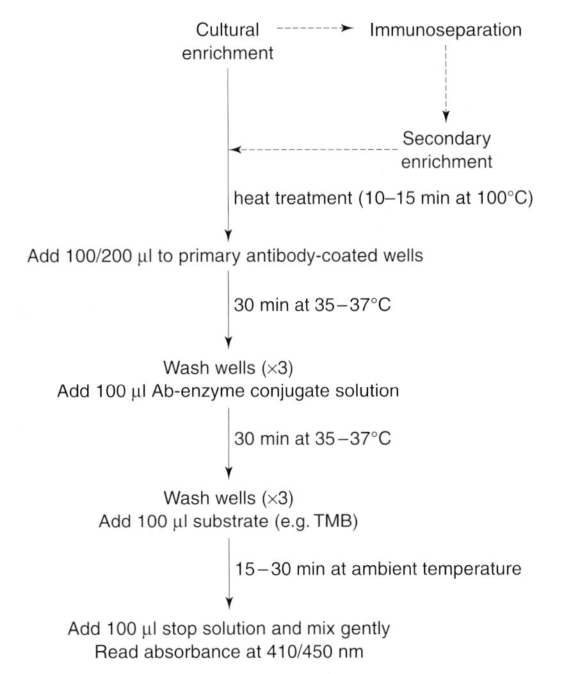

Figure 2 Typical protocol for microwell-based enzyme immunoassays for *E. coli* O157.

Reagents *see Appendix II.*

Apparatus *see Appendix II.*

Sample Preparation and Enrichment

1. Weigh 25 g sample into 225 ml of modified trypticase–soy broth (TSB) with novobiocin (mTSBn, see Appendix I). If larger sample sizes are analysed, proportionately increase the volume of mTSBn to maintain the 1 : 9 dilution ratio. Mix well. Incubate overnight (18–28 h) at 35–37°C.
2. After incubation, mix the sample and mTSBn well, allow particulate matter to settle for 1 min and then transfer 1 ml into a test tube. Retain the sample and mTSBn mixture for confirmation of presumptive positive samples.

Sample Inactivation Submerse the 1 ml aliquot in a boiling water bath or flowing steam for 20 min. Cool tubes to 25–37°C before testing. Tubes that have been heated can be refrigerated at 4–8°C up to 4 days prior to testing.

Test Procedure

1. Prepare wash and substrate solutions according to kit instructions.
2. Fit microwell reader with 405–410 nm filter.
3. Fit required number of microwell strips into the holder. Reseal unused microwells in foil pouch containing desiccant. In addition to samples, allow three extra wells for two positive controls and one blank. Carefully record positions of positive controls, blank and test samples in the holder.
4. Vortex mix samples and positive control before pipetting. A new pipette tip must be used for each sample. Pipette a 100 μl aliquot of sample into each well. Also pipette a 100 μl aliquot of reagent D (positive control) into each positive control well. Leave the blank well empty.
5. Cover and incubate for 30 min at 35–37°C. Do not stack anything on the microwell holder during incubation. Do not agitate the plate during any incubation step.
6. Following incubation, wash each well three times with wash solution, using a microwell washer or a wash bottle. Avoid underfilling wells to prevent ineffective washing.
7. Immediately following aspiration of the third wash invert the bottle of reagent E (conjugate) to gently mix, then add 100 μl to each well, including the control and blank wells. Cover and incubate for 30 min at 35–37°C.
8. Following incubation, aspirate and wash each well three times as in step (6).
9. Immediately following aspiration of the third wash, vortex mix diluted substrate thoroughly and add 100 μl to each well, including the control and blank wells. Cover and incubate for 30 min at 35–37°C. After incubation do not wash the wells, but proceed directly to the next step.

Reading To obtain valid results, the microwell plate reader must first be calibrated against the blank well. Standardize the reader by reading the blank well and adjusting the optical density (OD) to zero. Read the absorbance of each well, starting with the positive controls. *Note*: when the reader is standardized to the blank well, certain samples may read less than zero OD (a negative reading); this is not uncommon and indicates a negative result.

If reading of results is delayed, add 50 μl of stop solution to each well. Read within 1 h.

Interpretation of Results

1. Control value – the positive control absorbance values should be at least 0.8 OD units. Absorbance values that fall below this value may indicate problems with the washing procedure.
2. Cutoff value – calculate the mean of the two positive control absorbance readings and multiply by 0.25 to establish the cutoff value: $(PC1 + PC2)/2 \times 0.25$ where PC is the positive control absorbance value in OD units. Repeat positive controls for each run.
3. Confirmation of presumptive positive samples – samples with readings exceeding or equal to the cutoff value are presumptively positive, and must be confirmed culturally as described in current

BAM or USDA methods. Prepare appropriate dilutions from mTSBn. Spread plate to specified agars and confirm biochemically and serologically.
4. Negative results – samples with OD readings less than the cutoff value are negative.

Example of a Protocol for a One-step Immunoassay The protocol described below is for the Visual Immunoprecipitate assay (VIP for EHEC) from BioControl Inc. The protocols for other one-step immunoassays are similar, with the exception of the Lumac PATH-STIK *E. coli* O157, which is inoculated by dipping into the selective enrichment broth.

Principle The VIP for EHEC is a single-step device based on lateral flow technology. It employs highly specific *E. coli* O157:H7 antibodies which are bound to a chromogenic carrier and, separately, to a solid support matrix. During the initial rehydration of the test unit, *E. coli* O157:H7 antigens in the sample react with the antibody–chromagen complex to form an antigen–antibody–chromagen complex. This complex then flows laterally across the membrane and is bound by antibody immobilized on the membrane within the test sample window. A positive reaction is indicated by the presence of a detection line in the test sample window. Absence of a line in the test verification window invalidates the test.

Reagents *See Appendix III.*

Apparatus *See Appendix III.*

Sample Preparation and Enrichment Aseptically weigh 25 g sample into 225 ml mTSBn. If larger sample sizes are analysed, proportionately increase the volume of mTSBn to maintain a 1 : 9 dilution ratio. Mix well. Incubate overnight (18–28 h) at 35–37°C.

For viscous samples (e.g. powdered dairy products) add up to 2.25 ml steamed (15 min) Triton X-100 per 225 ml modified mTSBn at the time of the sample addition and prior to incubation.

Test Procedure
1. Open the sealed pouch and remove the required number of VIP units. Use one device for each sample. Reseal the unused VIP units in the pouch with a desiccant pack and store at ambient temperature in a cool, dark location.
2. Gently shake the incubated enrichment broth, then allow the food particles to settle.
3. Transfer 0.1 ml of incubated enrichment broth to the sample addition well.
4. Incubate 10 min at ambient temperature. Proceed immediately to the next step.

Reading and Interpretation of Results Examine the device immediately after 10 min incubation; otherwise faint lines may develop because of nonspecific colour development and should be disregarded.

1. Examine the VIP unit for the presence of distinct detection lines in the test verification windows. This line should be dark when contrasted with the white background and extend across the window. Absence of a control line indicates an invalid test result.
2. Observe the test sample window. The presence of a distinct line, as described above, indicates a presumptive positive sample. Absence of a line is a negative result. Intensities of test and control lines may differ as long as a procedural control line is present.
3. Positive and negative control cultures should be run to familiarize the analyst with results interpretation.

The VIP units may not be reused. Autoclave used units at 121°C for 15 min prior to discarding.

Confirmation of Positive VIP Samples Presumptive positive samples must be confirmed as described in current modified BAM or USDA methods. Prepare appropriate dilutions from mTSBn. Spread plate to specified selective agars and confirm biochemically and serologically.

Automated Immunoassays for *E. coli* O157

There are two types of automated system for carrying out immunoassays for *E. coli* O157. Fully automated systems (e.g. VIDAS and EIAFoss) carry out all stages in the assay (sample transfer, washing stages, reagent handling and interpretation) following cultural enrichment. The VIDAS system (bioMérieux) can perform a wide range of assays, including *E. coli* O157 (VIDAS ECO). Two instruments are available: the VIDAS is capable of running up to 30 assays and the mini-VIDAS up to 12 assays simultaneously. Results are obtained within 60 min, following a 24 h enrichment. An automated immunoconcentration kit (VIDAS ICE) is also available for use in conjunction with the VIDAS ECO test to confirm presumptive positive samples (see below).

Another fully automated system is the EiaFoss (Foss Electric), which includes an immunomagnetic separation stage and can carry out 27 determinations simultaneously. The EiaFoss *E. coli* O157 ELISA gives results in 2 h after a 22 h enrichment period.

There are several systems that automate the standard microwell-based EIAs and eliminate the manual pipetting of reagents, washing and reading stages (e.g. Minilyser, Tecra Diagnostics; Transia Auto EIA,

Diffchamb; PersonalLAB, Microgen Bioproducts Ltd). In general, these systems can be used with any standard format microwell-based EIA for *E. coli* O157.

Example of a Protocol for an Automated EIA The example given is for the VIDAS *E. coli* O157 automated EIA.

Principle The VIDAS ECO test is an enzyme-linked fluorescent immunoassay (ELFA) performed in the automated VIDAS instrument. A pipette tip-like disposable device coated with anti-*E. coli* O157 antibodies, the solid phase receptacle (SPR), serves as the solid phase as well as the pipette for the assay. All the reagents for the assay are contained in a sealed strip.

All the assay steps are performed automatically by the VIDAS instrument. An aliquot of the boiled enrichment broth is placed into the reagent strip and the sample is cycled in and out of the SPR for a specific length of time. The *E. coli* O157 antigens present in the sample bind to the anti-*E. coli* O157 antibodies coating the interior of the SPR. A final wash step removes the unbound conjugate. A fluorescent substrate (4-methyl-umbelliferyl phosphate) is introduced in the SPR. Enzyme remaining on the SPR wall will then catalyse the conversion of the substrate to the fluorescent product, 4-methyl-umbelliferone. The intensity of the fluorescence is measured by the scanner in the VIDAS. When the VIDAS *E. coli* O157 assay is completed, the results are analysed automatically by the computer, a test value is generated, and a report is printed for each sample.

Reagents *see Appendix III.*

Apparatus *see Appendix III.*

Sample Preparation and Enrichment

1. Aseptically add 25 g of sample to a stomacher bag. Add 225 ml of mTSBa (dairy products) or 225 ml of mTSBn (other foods) prewarmed to 41°C. Stomach the cultures for 2 min. Incubate the cultures for 6–7 h at 41°C.
2. Transfer 1 ml of the pre-enriched culture to 9 ml of MacConkey broth with cefixime and potassium tellurite (CT-MAC, see Appendix I). Incubate for 18 ± 1 h at 35–37°C.
3. After incubation, transfer 1–2 ml of the enrichment broth into a tube with a hermetic top. Homogenize and heat for 15 min at 100°C. Store the remaining enrichment broth at 2–8°C for confirmation of positive results.

Test Procedure

1. Remove the VIDAS ECO kit from the refrigerator. Remove the necessary components from the kit and allow them to reach room temperature (approximately 30 min). Return all unused components to storage at 2–8°C.
2. In the space provided label the ECO reagent strips with the sample identification numbers, without altering the bar code.
3. Enter the appropriate assay information into the instrument to create a work list.
4. Pipette 0.5 ml of standard, controls or boiled sample (after cooling) into the centre of the sample well of an ECO reagent strip. It is essential that the enrichment broth be heated for 15 min at 100°C prior to testing on the VIDAS.
5. Load the ECO reagent strips and SPRs into the positions that correspond to the VIDAS section indicated by the load list. Check to make sure the colour labels with the three-letter assay code on the SPRs and the reagent strips match.
6. Initiate the assay procedure as directed in the VIDAS Operator's Manual. The assay will be completed in approximately 45 min.
7. Dispose of all used SPRs and strips in appropriate biohazard containers.

Reading and Interpretation of Results Two instrument readings for fluorescence are taken for each sample tested. The first reading is a background reading of the cuvette and substrate before the SPR is introduced into the substrate. The second reading is taken after the substrate has been exposed to the enzyme conjugate remaining on the interior of the SPR. The background reading is subtracted from the final reading to give a relative fluorescence value (RFV) for the test result. A test value is generated for each sample by determining a ratio of the sample RFV to that of a standard. Test values from samples and controls are compared with a set of thresholds stored in the computer: values below the threshold are interpreted as a negative result, values equal to or above the threshold are positive.

A report is generated which records the type of test performed, the test kit lot number, date and time, and each sample identification, RFV, test value and interpreted result. Invalid results are reported when the background reading is above a predetermined cutoff. This indicates a problem with the substrate, in which case retest the sample with a new VIDAS strip.

Confirmation of Presumptive Positive Samples
Presumptive positive results should be confirmed using the VIDAS ICE kit followed by subculture onto an appropriate selective agar, e.g. sorbitol–

MacConkey agar (SMAC) or SMAC with cefixime and potassium tellurite (CT-SMAC). Presumptive *E. coli* O157 isolates from the selective agar should then be confirmed by standard biochemical and serological methods.

Automated Immunoconcentration of *E. coli* O157

The VIDAS ICE method is a fully automated procedure for the immunoconcentration of *E. coli* O157 from enrichment broths, following a positive VIDAS ECO test. It takes 40 min to perform and simplifies the confirmation of presumptive positive results by eliminating the multiple dilution and plating procedures used in the conventional confirmatory method for *E. coli* O157.

An aliquot of the enrichment broth is placed into the reagent strip and the sample is cycled in and out of the SPR for a specified time. The *E. coli* O157 antigens in the broth bind to anti-*E. coli* O157 capture antibodies coating the SPR. Unbound sample components are then washed away. A final enzymic step releases the captured *E. coli* O157 into a well in the reagent strip. The released organisms can then be plated onto selective agar media.

Advantages and Limitations of Commercial Immunoassays

In general, commercial immunoassays for *E. coli* O157 are simple, reliable and specific tests that give results within 15 min (one-step tests) to 2 h (microwell-based tests) from a sample enrichment culture. All commercial immunoassays for *E. coli* O157 require a cultural enrichment stage ranging in length from 4 h (EC-LITE *E. coli* O157) to 24 h (VIDAS ECO) to raise the level of *E. coli* O157 to the sensitivity of the test (typically 10^4–10^6 cfu ml^{-1}). The overall assay time including enrichment (6.5–26 h), is still much less than the conventional cultural method (CCM) for *E. coli* O157, which takes 3 days to complete, is labour intensive and lacks sensitivity and specificity. However, any presumptive positive results by immunoassays must be confirmed by conventional cultural, biochemical and serological methods. At present there are few commercially available alternatives to CCMs and immunoassays for the detection of *E. coli* O157. Nucleic acid-based techniques such as polymerase chain reaction (PCR) have been developed and are commercially available, but they require technical skills and are generally more expensive than immunoassays.

The choice of immunoassays will depend on a number of factors including sample throughput, staff skills and financial resources. Automated systems require higher capital investment, but a lower level of technical skill. For laboratories with large sample throughputs a manual or semiautomated microwell-based system may be the best choice, since the cost per test is relatively low. For lower volumes of samples the automated VIDAS or EIAFoss system might be more appropriate, although the cost per test is higher. With manual microwell-based EIAs, care must be taken to avoid cross contamination, so washing stages must be carried out thoroughly to avoid false positive results. Alternatively an automatic microwell washer can be used. Positive and negative controls are provided with the kits and these must give the appropriate readings, as recommended by the manufacturers. Some kits include optional colour comparator cards which require subjective interpretation. The use of a plate reader to measure the absorbance is therefore preferred. The Organon Teknika EHEC-Tek kit includes an immunomagnetic separation stage that is claimed to improve the sensitivity of the test, but also introduces a further series of manipulations. One-step assays are simple to perform, but they can be less sensitive than microwell-based assays, more labour intensive than semiautomated systems, and also require subjective interpretation.

It is important to note that some immunoassays for *E. coli* O157 only detect the O157:H7 serotype, which is the serotype most frequently isolated from cases of VTEC infection in Europe and North America. However, some O157:H– and non-O157 strains of *E. coli* also produce verotoxins (VT1 and VT2) and not all *E. coli* O157:H7 strains are verotoxin-positive. Thus, to detect VTEC other than O157 and phenotypic variants of *E. coli* O157 in foods the use of methods for the detection of verotoxin production and VT genes is recommended.

Validation Data

With a few exceptions detailed below, there is generally a lack of independent validation data on immunoassays for *E. coli* O157. Manufacturers usually provide some in-house performance data relating to sensitivity and specificity, but there is little information on the performance of the kits with food samples.

Two immunoassays for *E. coli* O157 (VIP *E. coli* O157 and Assurance EHEC EIA) have approved status from the Association of Official Analytical Chemists. The performance data from collaborative studies on these kits are summarized in **Table 3** and **Table 4**.

The TECRA Visual Immunoassay (VIA) was compared with the USDA cultural method for the detection of *E. coli* O157 in dairy foods; there was complete agreement between both methods when testing cheeses, milk powders, caseins and whey products

Table 3 Method performance for detection of *E. coli* O157:H7 in selected foods by Visual Immunoprecipitative assay (VIP). Adapted from Feldsine et al (1997a)

Sample	*MPN (cfu g⁻¹)*	*Sensitivity*[a]	*Specificity*[b]	*Agreement VIP/BAM*[c] *(%)*
Apple cider	0.043	100	99.4	100
	0.043	98.8	99.1	99.5
Liquid milk	0.043	100	100	100
	11.0	100	98.8	100
Ice cream	<0.003	100	95	100
	0.07	100	96.4	100
Raw ground beef	0.009	97.1	95.6	99.4
	0.44	100	92.4	100
Raw ground poultry (1)	0.03	100	95.7	100
	0.28	100	82	100
Raw ground poultry (2)	<0.003	100	96.1	100
	0.021	100	81.4	100

[a]Sensitivity is the total number of analysed positive test portions among 'known' positive test portions per laboratory divided by the total number of 'known' positive test portions per laboratory, where 'known' positive is defined as samples confirmed positive by reference method.
[b]Specificity is the total number of analysed negative test portions among 'known' negative test portions per laboratory divided by the total number of 'known' negative test portions per laboratory, where 'known' negative is defined as samples confirmed negative by reference method and negative controls.
[c]Rate reflects number of confirmed determinations that were equivalent between VIP and BAM culture methods.
BAM, *Bacteriological Analytical Manual*; MPN, most probable number.

Table 4 Method performance for detection of *E. coli* O157:H7 in selected foods by Assurance enzyme immunoassay. Adapted from Feldsine et al. (1997b)

Sample	*MPN (cfu g⁻¹)*	*Sensitivity*[a]	*Specificity*[b]	*Agreement EIA/BAM*[c] *(%)*
Apple cider	0.043	100	99.4	100
	0.043	97.6	99.1	98.9
Liquid milk	0.043	100	99.1	100
	11.0	100	98.8	100
Ice cream	<0.003	100	98.9	100
	0.07	100	98.7	100
Raw ground beef	0.009	96.7	97.7	99.4
	0.44	98.6	92	99.4
Raw ground poultry (1)	0.03	85.7	100	99.2
	0.28	88.6	95.3	96.1
Raw ground poultry (2)	<0.003	0	99.2	99.2
	0.021	94.9	85.2	98.3

[a] Sensitivity is the total number of analysed positive test portions among 'known' positive test portions per laboratory divided by the total number of 'known' positive test portions per laboratory, where 'known' positive is defined as samples confirmed positive by reference method.
[b] Specificity is the total number of analysed negative test portions among 'known' negative test portions per laboratory divided by the total number of 'known' negative test portions per laboratory, where 'known' negative is defined as a sample confirmed negative by reference method and negative controls.
[c] Rate reflects number of confirmed determinations that were equivalent between VIP and BAM culture methods.

inocuolated with *E. coli* O157. The Organon Teknika EHEC-Tek and Petrifilm HEC compared favourably with cultural methods when testing minced beef inoculated with low numbers of *E. coli* O157. The *E. coli* O157 Rapitest was found to be specific for all strains bearing the O157 and related antigens. With the exception of *Salmonella urbana* (group N), all the non-*E. coli* O157 bacteria tested gave negative results. The method detected levels as low as 0.4 cfu g^{-1} of *E. coli* O157:H7 inoculated into ground beef.

See also: **Biochemical and Modern Identification Techniques**: Introduction; Food-poisoning Organisms. **Enterobacteriaceae, Coliforms and *E. coli***: Classical and Modern Methods for Detection/Enumeration. **Enzyme Immunoassays**: Overview. ***Escherichia coli***: Detection of Enterotoxins of *E. coli*. ***Escherichia coli* O157**: Detection by Latex Agglutination Techniques; Detection by Commercial Immunomagnetic Particle-based Assays. **Food Poisoning Outbreaks**. **Hydrophobic Grid Membrane File Techniques (HGMF)**.

Immunomagnetic Particle-based Techniques: Overview. **National Legislation, Guidelines & Standards Governing Microbiology**: European Union; Japan. **PCR-based Commercial Tests for Pathogens**. **Petrifilm – An Enhanced Cultural Technique**. **Verotoxigenic *E. coli* and *Shigella spp.***: Detection by Cultural Methods.

Further Reading

US Food and Drugs Administration (1996) *Bacteriological Analytical Manual*, 8th edn. Arlington: AOAC International.

Bennett AR, MacPhee S and Betts RP (1995) Evaluation of methods for the isolation and detection of *Escherichia coli* O157 in minced beef. *Letters in Applied Microbiology* 20: 375–379.

Blais BW, Booth EA, Phillippe L, Pandian S and Yamazake H (1997) Polymacron enzyme immunoassay system for detection of *Escherichia coli* O157 inoculated into foods. *Journal of Food Protection* 60(2): 98–101.

Cohen AE and Kerdahi KF (1996) Evaluation of a rapid and automated enzyme linked fluorecent immunoassay for detecting *Escherichia coli* serogroup O157 in cheese. *Journal of the Association of Official Analytical Chemists International* 79(4): 858–860.

Feldsine PT, Falbo-Nelson MT, Brunelle SL and Forgey R (1997a) Visual immunoprecipitate assay (VIP) for detection of enterohaemorrhagic *Escherichia coli* (EHEC) O157:H7 in selected foods: collaborative study. *Journal of the Association of Official Analytical Chemists International* 80(3): 517–529.

Feldsine PT, Falbo-Nelson MT, Brunelle SL and Forgey R (1997b) Assurance enzyme immunoassay for detection of enterohaemorrhagic *Escherichia coli* O157:H7 in selected foods: collaborative study. *Journal of the Association of Official Analytical Chemists International* 80(3): 530–543.

Firstenberg-Eden R and Sullivan N (1997) EZ Coli Rapid Detection system: a rapid method for the detection of *Escherichia coli* O157 in meat and other foods. *Journal of Food Protection* 60(3): 219–225.

Flint SH and Hartley NJ (1995) Evaluation of the TECRA *Escherichia coli* O157 visual immunoassay for tests on dairy products. *Letters in Applied Microbiology* 21(2): 79–82.

Johnson JL, Rose BE, Shakar AK, Ransom GM, Lattuada CP and McNamara AM (1995) Methods used for detection and recovery of *Escherichia coli* O157:H7 associated with a food-borne disease outbreak. *Journal of Food Protection* 58(6): 597–603.

Johnson RP, Durham RJ, Johnson ST, MacDonald LA, Jeffrey SR and Butman BT (1995) Detection of *Escherichia coli* O157:H7 in meat by an enzyme-linked immunosorbent assay, EHEC-tek. *Applied and Environmental Microbiology* 61(1): 386–388.

Okrend AJ, Rose BE and Matner R (1990) An improved method for the detection and isolation of *Escherichia coli* O157:H7 from meat, incorporating the 3M Petrifilm test kit HEC – for haemorrhagic *Escherichia coli*. *Journal of Food Protection* 53(11): 936–940.

Padhye NW and Doyle MP (1992) *Escherichia coli* O157:H7, epidemiology, pathogenesis and methods for detection in food. *Journal of Food Protection* 7: 555–565.

Vernozy-Rozand C (1997) A review: detection of *Escherichia coli* O157:H7 and other verotoxin-producing *E. coli* (VTEC) in food. *Journal of Applied Microbiology* 82: 537–551.

Appendix I. Selective enrichment and agar media for *E. coli* O157

Modified EC broth (mECn)

Ingredient	*Quantity* ($g\ l^{-1}$)
Tryptone	20
Bile salts No. 3	1.12
Lactose	5
K_2HPO_4	4
KH_2PO_2	1.5
NaCl	5

Adjust pH to 6.9, autoclave at 121°C for 15 min. Add 20 mg l^{-1} novobiocin as filter-sterilized aqueous solution.

Modified trypticase–soy broth with novobiocin (mTSBn)

Ingredient	*Quantity* ($g\ l^{-1}$)
Trypticase–soy broth	30
Bile salts No. 3	1.5
K_2HPO_4	1.5

Dissolve ingredients in 1 l of distilled water. Mix thoroughly and dispense in 225 ml volumes. Sterilize by autoclaving at 121°C for 15 min. Cool to ambient temperature. Add 0.045 ml of filter-sterilized 10% novobiocin to each 225 ml of mTSB.

Modified trypticase–soy broth with acriflavine (mTSBa)

As above but substitute 10 mg l^{-1} acriflavine for novobiocin as filter-sterilized aqueous solution.

Modified buffered peptone water (BPW-VCC)

Buffered peptone water plus vancomycin 20 mg l^{-1}, cefixime 0.05 mg l^{-1} and cefsulodin 10 mg l^{-1}. Autoclave at 121°C for 15 min.

MacConkey broth with cefixime and potassium tellurite (CT-MAC)

Prepare MacConkey broth according to the manufacture's instructions. Sterilize by autoclaving at 121°C for 15 min. Cool and add 2.5 mg l^{-1} of filter-sterilized potassium tellurite and 50 µg l^{-1} of filter-sterilized cefixime.

Sorbitol–MacConkey with BCIG (SMAC-BCIG)

Prepare sorbitol–MacConkey agar (SMA) according to the manufacturer's instructions. Add 0.1 g l^{-1} of 5-bromo-4-chloro-3-indoxyl-D-glucuronide (BCIG). Autoclave at 121°C for 15 min.

Sorbitol–MacConkey agar with cefixime and potassium tellurite (CT-SMAC)

Prepare SMA according to the manufacturer's instructions. Autoclave at 121°C for 15 min. After autoclaving, add (per litre):

0.5 ml filter-sterilized 0.01% cefixime (0.05 mg l^{-1})
2.5 ml filter-sterilized 0.1% potassium tellurite (2.5 mg l^{-1}).

Appendix II. Materials and apparatus for Assurance EHEC microwell-based EIA

Media and Reagents

Items 1–7 are supplied with the test kit.

1. Wash solution concentrate – 2% polyoxyethylene 20 sorbitan monolaurate (Tween 20) in H_2O.
2. Substrate – 5 mg ρ-nitrophenylphosphate tablet.
3. Substrate diluent – 1 mol l^{-1} diethanolamine in H_2O.
4. Positive control – stabilized, inactivated *E. coli* O157:H7 antigen.
5. Conjugate solution – specific antibodies to *E. coli* O157:H7 conjugated to alkaline phosphatase.
6. Stop solution – 20% EDTA in H_2O.
7. Antibody-coated microwells – microwell strips, each well coated with *E. coli* O157:H7 antibody, 96-well holder, and cover.
8. Novobiocin solution (see Appendix I).
9. Modified trypticase soy broth with novobiocin (see Appendix I).
   Diagnostic reagents – necessary for confirmation of presumptive positive EIA tests (see BAM).

Apparatus

1. Incubators – capable of maintaining 35–37°C and 42–43°C (used in confirmation procedure only).
2. Water bath – capable of maintaining 100 ± 2°C. Alternatively, a flowing steam autoclave set at 100°C can be used.
3. Microplate washer – for washing microwell strips, or a plastic squeeze bottle can be used.
4. Microplate reader – photometer equipped with 405–410 nm filter, capable of reading microwell plates. May include optional printer.
5. Syringe – equipped with 0.45 µm or smaller porosity filter.
6. Micropipettes – capable of accurately dispensing 100 µl and 250 µl.
7. Vortex mixer.
8. Top-loading balance – for weighing test samples.

Appendix III. Materials and equipment for VIP EHEC one-step test

Media and Reagents

Item 1 is suplied with the kit.

1. One-step assay units – one per test.
2. Novobiocin solution.
3. Modified trypticase–soy broth with novobiocin (see Appendix I).
4. Diagnostic reagents – necessary for confirmation of presumptive positive EIA tests (see BAM).
5. Triton X-100 – used for preparing viscous foods.

Apparatus

1. Incubators – capable of maintaining 35–37°C and 42–43°C (used in confirmation procedure only).
2. Syringe – equipped with 0.45 µm or smaller porosity filter.
3. Micropipette – capable of accurately dispensing 0.1 ml.
4. Steam bath – used for steaming Triton X-100 when preparing viscous foods.
5. Top-loading balance – for weighing test samples.

Appendix IV. Materials and equipment for VIDAS ECO automated EIA

Media and Reagents

Items 1–5 are supplied with the kit.

1. ECO reagent strip – one per test.
2. ECO solid phase receptacle – one per test.
3. Standard – purified and inactivated *E. coli* O157 antigen.

4. Positive control – purified and inactivated *E. coli* O157 antigen.
5. Negative control – TRIS-buffered saline (TBS) – Tween.
6. Novobiocin solution (see Appendix I).
7. Modified EC broth with novobiocin (see Appendix I).
8. Sorbitol–MacConkey agar with BCIG – for confiration of presumptive positive results (see Appendix I).
9. Diagnostic reagents – necessary for confirmation of presumptive positive EIA tests (see BAM).

The *E. coli* O157 reagent strip

Wells	*Reagents*
1	Sample well: 0.5 ml of boiled enrichment broth is placed into the well
2	Prewash solution (0.4 ml): TBS–Tween with 0.1% sodium azide
3, 4, 5, 7, 8, 9	Wash solution (0.6 ml): TBS–Tween with 0.1 sodium azide
6	Conjugate (0.4 ml): alkaline phosphatase labelled polyclonal antibodies (gaot) with 0.1% sodium azide
10	Cuvette with substrate (0.3 ml): 4-methyl-umbelliferyl phosphate with 0.1% sodium azide

Apparatus

1. Vitek Immuno Diagnostic Assay System (VIDAS) instrument (bioMérieux).
2. Incubators – capable of maintaining 35–37°C and 42–43°C (used in confirmation procedure only).
3. Water bath – capable of maintaining 100° ± 2°C. Alternatively, a flowing steam autoclave set at 100°C can be used.
4. Syringe – equipped with 0.45 μm or smaller porosity filter.
5. Micropipette – capable of accurately dispensing 0.5 ml.
6. Stomacher or blender.
7. Top-loading balance – for weighing test samples.

Appendix V. Suppliers of immunoassay kits for the detection of *E. coli* O157

BioControl Systems Inc.
12822 SE 32nd Street
Bellevue, WA 98005
USA

Bioenterprises Pty Ltd
28 Barcoo Street
PO Box 20 Roseville
NSW 2069
Australia

bioMérieux
Chemin de l'Orme
69280 Marcy L'Étoile
France

Celsis-Lumac Ltd
Cambridge Science Park
Milton Road
Cambridge CB4 4FX
UK

Diffchamb
8 Rue St Jean de Dieu
69007 Lyon
France

Dynal
PO Box 158 Skoyen
0212 Oslo
Norway

Foss Electric A/S
69 Slangerupgade
PO Box 260
DK-3400 Hillerød
Denmark

GEM Biomedical Inc.
925 Sherman Avenue
Hamden, CT 06514
USA

3M Medical Products Group
3M Center
Building 275-5W-05
St Paul, MN 55144-1000
USA

Meridian Diagnostics Inc.
3471 River Hills Drive
Cincinnati, OH 45244
USA

Microgen Bioproducts Ltd
1 Admiralty Way
Camberley
Surrey GU15 3DT
UK

Neogen Corporation
620 Lesher Place
Lansing, Michigan 48912
USA

Organon Teknika nv
58 Veedijk
2300 Turnhout
Belgium

VEROTOXIGENIC *E. COLI* AND *SHIGELLA* SPP.

Detection by Cultural Methods

Christine Vernozy-Rozand, Food Research Unit National Veterinary School, France Ecole Nationale Vétérinaire de Lyon, France

Most *Escherichia coli* strains are harmless commensals in the human gut. However, some strains, such as *E. coli* O157:H7, can cause severe food-borne disease and are referred to as enterohaemorrhagic *E. coli* (EHEC). The EHEC strains produce toxins known as verotoxins or Shiga-like toxins because of their similarity to the toxins produced by *Shigella dysenteriae*. VT-producing *E. coli* (VTEC) have been shown to produce one or both of two immunologically distinct Vero cytotoxins (STX1 and STX2). Bloody diarrhoeal disease caused by serotypes of VT-producing *E. coli* (VTEC) other than O157:H7 is uncommon, and scarce outbreaks due to these other serotypes have been reported in the US, Canada, and the UK. Most outbreaks caused by *E. coli* O157:H7 have been food or water related. Likely vehicles of infection have been undercooked ground beef. Raw milk, cold sandwiches, vegetables and water have been implicated as sources of some outbreaks. These organisms may be present in low numbers in implicated foods containing high levels of competing microflora.

Many approaches have been taken in developing isolation and detection procedures for VTEC, and these can be generally divided into three categories: (1) the use of biochemical characteristics somewhat specific to strains of *E. coli* O157; (2) immunoblotting with antibodies to verotoxin (STX) or O157; and (3) the use of DNA probes for verotoxins or markers associated with serotype O157:H7 (**Fig. 1**).

The purpose of this chapter is to describe DNA methods for isolation and identification of *E. coli* O157:H7 and other verocytotoxin-producing strains of *E. coli* in food.

Throughout this report, the following convention has been adopted. When the term VTEC is used, it includes verocytotoxin-producing *E. coli* of all serogroups; the term *E. coli* O157 is used to refer to *E. coli* of that serogroup where the precise H antigen type is unknown or not specified; the term *E. coli* O157:H7 is used to refer to bacteria of that serotype only, which is frequently VT producing.

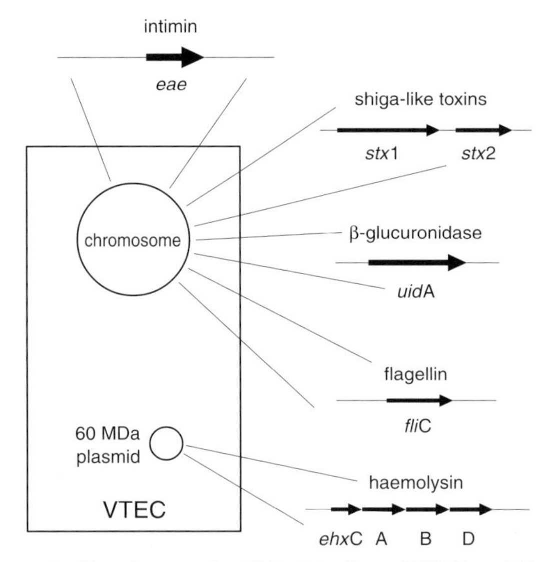

Figure 1 Target genes for DNA detection of VTEC and *E. coli* O157:H7.

Use of DNA Probes Specific for Verotoxin Genes (*stx* Genes)

Gene sequences for STX1 and STX2 demonstrate 58% gene sequence homology, and VTEC strains can therefore be differentiated by using DNA probes. In recent years, variants to STX2 have been described. STX2 is cytotoxic to both Vero and HeLa cells, whereas variants of STX2 usually show much-reduced or no cytotoxicity to HeLa cells.

Karch and Meyer examined four oligonucleotide probes of various lengths (20 and 40 bases) representing different regions of the *stx*1 structural genes and one oligonucleotide (41 bases) derived from the *stx*2 gene of *E. coli* O157:H7 strain for the identification of *E. coli* that produced cytotoxins for Vero or HeLa cells. The 20-bases probe appeared to be as valid as the 41-bases probe with regard to specificity and sensitivity of the hybridization reaction. Fifty isolates of five different serotypes of *E. coli* producing verotoxins were detected using a colony blot hybridization assay, whereas none of 416 non-verotoxinogenic *E. coli* strains was detected. *E. coli* strains that synthesized STX1 alone or *E. coli* O157:H7 isolates that coexpressed STX1 and STX2 were hybridized with all four probes that were complementary to the *stx* genes, suggesting that they had toxin genes with great homology in all the regions examined. The colony blot hybridization with the oligonucleotide probes described by Karch and Meyer could serve as a specific and sensitive test with potential diagnostic value.

Amplification of part of the *stx* gene, using the polymerase chain reaction (PCR), has also been used to test for the presence of VTEC. With this procedure, DNA is amplified to increase the level of target DNA when VTEC are present in very low numbers. The system first developed used 'degenerated primers' so that defined sequences of both *stx*1 and *stx*2 were amplified. PCR products were identified by hybridization using specific oligonucleotide probes complementary to part of the amplified sequence. It was possible to identify *stx*1 or *stx*2 sequences but variants of STX2 could not be distinguished from STX2.

In order to detect all types of VTEC isolated from animal and food sources, a PCR was developed using a pair of oligonucleotide primers, targeting conserved sequences found in *stx*1, *stx*2 and *stx*E genes. Supernatants of boiled broth cultures of VTEC (233 strains) isolated from ground beef, ground pork, raw milk, bovine faeces and porcine faeces, non-VTEC *E. coli* (72 strains), and other enteric and food bacteria (76 strains) were tested by PCR. The verocytotoxigenicity of these strains was verified by Vero cell assay. All 223 VTEC isolates, comprising over 50 different serotypes, were detected by the PCR procedure. *Shigella dysenteriae* type 1 was the only other bacteria that was positive in this assay. As little as 1 pg of VTEC DNA and as few as 17 cfu (colony forming units) of VTEC could be detected with this method. The results suggested that these primers detect VTEC over a wide range of serotypes. This method might be applicable as a screening procedure for the detection of VTEC in samples of foods and faeces.

More recently, a PCR technique has been adapted to make it suitable for the identification of VTEC directly from contaminated ground beef without isolation of the bacterium or purification of its DNA. Ground beef homogenates were diluted 1000-fold to reduce the concentration of components which inhibit the thermostable polymerase. As few as 30 VTEC per millilitre of a ground beef homogenate were detected using the PCR technique, although it was necessary to enrich six of the samples for positive detection. Assessment of four different ground beef samples using the PCR detection technique revealed that fat content was the major inhibitory component.

Masters *et al* examined the relationship between viability assessed by plate counts and detectability by PCR techniques with cells of *E. coli* previously exposed to a range of stress treatment. In all cases the organisms were detectable by PCR after plate counts had declined to zero. Treatment with acid or hydrogen peroxide caused loss of PCR soon after viability was lost, but strong PCR signals were obtained from starved or desiccated cells long after cells became nonviable. Exposure to temperatures up to 100°C had little effect on detection by PCR and even autoclaving cells at 121°C for 15 min failed to abolish PCR detection completely. There is thus no simple relationship between viability and detectability by PCR. Detection of pathogens by PCR in environmental monitoring requires additional evidence of viability before risk can be properly assessed.

Use of DNA Probe Specific for O157:H7

Most biochemical reactions of *Escherichia coli* O157:H7 are typical of *E. coli*, with the exception of sorbitol fermentation and β-glucuronidase activity. About 93% of *E. coli* isolates of human origin ferment sorbitol within 24 h, however, *E. coli* O157:H7 was reported as not fermenting sorbitol. However, in recent reports some O157 VTEC strains fermented sorbitol within 24 h. The prevalence of such strains is unknown but they would be discarded using the standard screening method described here. Additionally, 93% of *E. coli* possess the enzyme β-glucuronidase but the vast majority of O157 VTEC do not produce β-glucuronidase. Antibodies to the O157 antigen are also used in many assays to detect *E. coli* O157:H7 isolates in clinical and food samples. These tests, however, provide no information on the toxin types produced by the isolates and are not specific, since the O157 antigen is present on other *E. coli* species and anti-O157 sera often cross-react with *Citrobacter freundii*, *Escherichia hermanii*, and other bacteria. Analyses of food products with anti-O157 serum have recognized O157 isolates that neither produced verotoxin nor were of the H7 serotype. Similarly, antibody- or DNA-based assays for identifying verotoxin or bacteria carry *stx* genes will not discriminate *E. coli* O157:H7 isolates from the numerous other serotypes that also produce verotoxins.

Unlike biochemical differentiations such as sorbitol fermentation or β-glucuronidase activity, probe reactions do not rely on enzymatic activities and are therefore unaffected by media interference or the presence of bacteria, such as *E. hermanii*, which has similar phenotypes.

The *uid*A gene, the *eae*A gene and the *fli*C gene are specific for serotype O157:H7. They can be used solely or in combination with *stx* gene probes in multiplex PCR for example.

The *uid*A Gene

The *uid*A gene encodes for β-glucuronidase in *E. coli*. Although O157:H7 isolates do not exhibit β-glucuronidase activity, they carry the *uid*A gene. Earlier sequencing revealed that the *uid*A of O157:H7 had a G residue (rather than the T residue found in wild-type *E. coli*) at position 92 (**Fig. 2**). This highly conserved base change was used to identify O157:H7 isolates by allele-specific hybridization. Thus, although not directly involved in pathogenicity, the conserved base change in the *uid*A allele is a powerful, though coincidental, marker of O157:H7 strains.

An oligonucleotide probe, PF-27, directed to this region, containing a unique base substitution in the allele of the *uid*A gene has been used to identify isolates of *E. coli* O157:H7. Colony hybridization analysis of 239 bacteria, including *E. coli* and other enteric isolates showed that the probe reacted only with the 17 isolates of O157:H7 serotype. Interestingly, the probe did not hybridize with the 73 β-glucuronidase-positive *E. coli*, the 13-positive *Shigella*, or eight β-glucuronidase-positive *Salmonella* isolates analysed. Except for the single nucleotide base difference, the PF-27 sequence is identical to that *uid*A gene, which was present in almost all *E. coli* regardless of β-glucuronidase phenotype. Hence, the absence of probe hybridization with *E. coli* indicated that PF-27 could discriminate the single nucleotide difference between the 5′ region of the *uid*A gene of *E. coli* and its allele in the O157:H7 serotype.

The PF-27 probe appeared to be specific solely for serotype O157:H7 because it did not hybridize with isolates from the other VT-producing EHEC serotypes (O26:H11 and O111:NM) studied. The stringent specificity of the PF-27 probe may be valuable for clinical diagnosis and for the identification of O157:H7 isolates in foods. The stringency of PF-27 probe would

```
             +70             +80             +90
E.coli   ... ATCGCGAAAA  CTGTGGAATT  GATCAGCGTT ...
O157:H7  ... **********  **********  **G******* ...

             5'-GCGAAAA  CTGTGGAATT  GGG-3'
                       forward primer
```

Figure 2 Partial sequence of the *uid*A gene of *E. coli* and O157:H7, showing the position of the forward primer in mismatch amplification mutation assay. The consensus bases are shown with asterisks and mismatched bases in the primer are underlined.

also eliminate the need for serological confirmation, and thus, incidences of false-positive identification caused by antibody cross-reactivity with other organisms.

More recently, it has been shown that the probe also detected phenotypic variants of O157 serotype that were non-motile, β-glucuronidase negative and fermented sorbitol. These atypical pathogenic O157 strains were isolated from patients with haemolytic–uraemic syndrome in Germany and obtained from Gunzer *et al.* An isolate of *E. hermanii* examined by Feng did not hybridize with PF-27. By looking for several genes simultaneously in a multiplex PCR assay, several groups have reported assays that can specifically identify *E. coli* O157:H7. Cebula *et al.* described a mismatch amplification mutation assay (MAMA)-multiplex PCR assay in which primers for stx_1 and stx_2 are used in combination with primers specific for the *E. coli* O157:H7 allele of *uid*A, the gene encoding β-glucuronidase. In this assay, the authors exploited a minor sequence difference between the *uid*A genes of *E. coli* of other serotypes in designing the MAMA primers (Fig. 2). The MAMA primers amplified the *uid*A gene from only STEC of serotypes O157:H7 and O157:NM. An advantage of this method is that it allows the discrimination of *E. coli* O157:H7/NM strains from those of other STEC serotypes and at the same time provides information about the types of toxin encoded by the strain.

Multiplex PCR studies have largely concentrated on the identification of bacterial strains or toxins with DNA extracted from pure cultures. PCR amplification of enrichment broth was successful in foods only when an immunomagnetic separation step for *E. coli* O157:H7 was performed, but PCR methods for detection of bacteria directly from food samples need at least partially purified DNA.

Filtration greatly improved the sensitivity of the assay, such that it was possible to detect low numbers (10^2–10^3) of cells per gram after 6 h of enrichment and even 1 cfu g^{-1} after 18 h of enrichment. After 18 h of incubation and the two-step filtration, the PCR products became much more distinct and low-cell-number bands appeared. The absence of PCR product from 18 h, TSB-grown food homogenate clearly shows that metabolic by products remaining in the filtrate after 5 μm pore-size filtration posed a problem for PCR amplification. The sensitivity of PCR amplification was greatly enhanced once dissolved metabolites were eliminated from samples by 0.2 μm pore-size filtration.

The *eae*A Gene

Most EHEC isolates also possess a genetic locus associated with attachment to enterocytes and effacement of their microvilli. The latter property is shared with enteropathogenic *E. coli* (EPEC), and it has recently been shown that this activity is mediated by a 35 kb chromosomal region termed the 'locus of enterocyte effacement'. Although this locus contains several genes, one of these, *eae*A, encodes the protein intimin, which is thought to be responsible for the close association of EHEC and EPEC isolates to the cytoplasmic membrane of cell lines such as HEp-2 in vitro and enterocytes in vivo. The 5′ portion of the *eae*A gene appears to be relatively well conserved among EHEC and EPEC isolates; however, the 3′ one-third of the gene differs among EHEC and EPEC serotypes.

It has been shown that *stx*1 and *stx*2 genes and a 5′ conserved area of the *eae*A gene (*eae*A_{GEN}) can be used in PCR assays as targets for the identification of the majority of EHEC isolates and, in addition, that certain oligonucleotide primers with homology to the 3′ end of the *eae*A of *E. coli* O157:H7 (*eae*A_{O157}) are quite specific for this organism in PCR assays. In these studies, the *eae*A_{O157} oligonucleotide primers were shown to amplify DNAs from all *E. coli* O157:H7 (NM non-motile) strains, of serotype O55:H7 and an EHEC strain of serotype O145:NM but not DNAs from other *E. coli* strains.

A multiplex PCR method has been described for simultaneous amplification of three different DNA sequences of *E. coli* O157:H7: a specific fragment of the *eae* gene, conserved sequences of *stx*1 and *stx*2, and a fragment of the 60-MDa plasmid (all EHEC isolates contain a large (ca. 60-MDa) plasmid encoding factors which may contribute to their virulence, such as fimbria and entero-haemolysins). The detection limit for the plasmid, *eae*, and *stx* genes in the multiplex PCR was 1.2, 100 and 1000 cfu of *E. coli* O157:H7, respectively.

The assay gives a positive signal with all three primer pairs only for toxigenic O157 strains, whereas other *E. coli* strains are negative for at least one primer set. Like the MAMA multiplex PCR assay mentioned above, this multiplex assay allows the discrimination of *E. coli* O157:H7/NM strains from other VTEC strains and allows the differentiation of pathogenic groups of organisms possessing one or more target genes. In another study, a multiple digoxigenin (DIG)-labelled probe oligonucleotide hybridization (DLOPH) assay was developed for analyses of PCR products, the conditions for the multiplex PCR were optimized and the procedure was used for the identification of EHEC serogroup O157 isolated from foods. It was advantageous to include the multiple DLOPH assay in the procedure. First, the DLOPH assay was specific for confirming the identities of the amplification products. Second, the DLOPH assay was more sensitive than ethidium bromide staining of agarose gels. Furthermore, the DLOPH assay did not involve the use of radioactive materials.

The *fli*C Gene

Flagellar antigen group H7 is one of 53 flagellar antigen groups described for *E. coli*. The variability of the H antigen is found within the flagellar filament, which is a polymer of a single protein, flagellin, the product of the *fli*C gene. The N-terminal and C-terminal portions of flagellin are critical for the structure of the flagella and are highly conserved.

To further characterize non-motile isolates (designated NM), a PCR-restriction fragment length polymorphism (PCR-RFLP) test was developed to identify and characterize the gene encoding the H antigen (*fli*C) in *E. coli*. The entire coding sequence of *fli*C was amplified by PCR, the amplicon was restricted with *Rsa*I, and the restriction fragment pattern was examined after gel electrophoresis. A total of 280 *E. coli* isolates, representing serotypes O157:H7 and O157:NM, flagellar antigen H7 groups associated with other O serogroups, and all other flagellar antigen groups, were analysed. A single restriction pattern (pattern A) was identified for O157:H7 isolates, O157:NM isolates that produce verotoxins, and 16 of 18 O55:H7 isolates. Flagellar antigen group H7 isolates of non-O157 serotypes had one of three banding patterns distinct from pattern A. A wide variety of patterns was found among isolates of the other 52 flagellar antigen groups; however, none was identical to the O157:H7 pattern. Of 15 non-motile strains that did not produce the A pattern 13 had patterns that matched those of other known H groups. The PCR–RFLP (restriction fragment length polymorphisms) in conjunction with O serogroup determination will be useful in identifying *E. coli* O157:H7 and related strains that do not express immunoreactive H antigen and could be expanded to include other clinically important *E. coli* strains.

A multiplex PCR assay has been described with H7 specific primers (*fli*C gene) in combination with other primers which target the *stx*1 and *stx*2 genes and the *eae*A gene. In this study, *stx* and *eae*A PCR products were observed with DNAs from the majority of EHEC strains and *stx*, *eae*A, and *fli*C PCR products were observed with DNAs from *E. coli* O157:H7 or NM strains. Only *eae*A PCR products were present with DNA from enteropathogenic *E. coli*. Only *stx* PCR products occurred with STX-producing *E. coli* which are not EHEC. The multiplex PCR assays described allowed for the specific identification of *E. coli* O157:H7 or NM and other EHEC strains.

Acknowledgement: The author thanks Dr Jérôme Bouvet for his help in providing the two figures and the last review of this article.

See also: **Enterobacteriaceae, Coliforms and *E. coli***: Classical and Modern Methods for Detection/ Enumeration. ***Escherichia coli***: Detection of Enterotoxins of *E. coli*. ***Escherichia coli* O157**. Detection by Latex Agglutination Techniques; Detection by Commercial Immunomagnetic Particle-based Assays. **Food Poisoning Outbreaks**. **Hydrophobic Grid Membrane Filter Techniques (HGMF)**. **Immunomagnetic Particle-based Techniques**: Overview. **Molecular Biology – in Microbiological Analysis**. **National Legislation, Guidelines & Standards Governing Microbiology**: Japan. **Nucleic Acid-based Assays**: Overview. **PCR-based Commercial Tests for Pathogens**. **Petrifilm – An Enhanced Cultural Technique**. **Verotoxigenic *E. coli***: Detection by Commercial Enzyme Immunoassays.

Further Reading

Begum D and Jackson MP (1995) Direct detection of Shigalike toxin-producing *Escherichia coli* in ground beef using the polymerase chain reaction. *Molecular and Cellular Probes* 9: 259–264.

Cebula TA, Payne WL and Feng P (1995) Simultaneous identification of strains of *Escherichia coli* serotype O157:H7 and their Shiga-like toxin type by Mismatch Amplification Mutation Assay-Multiplex PCR. *Journal of Clinical Microbiology* 33: 248–250.

Deng MY and Fratamico PM (1996) A multiplex PCR for rapid identification of Shiga-like toxin-producing *Escherichia coli* O157:H7 isolated from foods. *Journal of Food Protection* 59: 570–576.

Feng P (1993) Identification of *Escherichia coli* serotype O157:H7 by DNA probe specific for an allele of *uid*A gene. *Molecular and Cellular Probes* 7: 151–154.

Feng P (1995) *Escherichia coli* serotype O157:H7: novel vehicles of infection and emergence of phenotypic variants. *Emerging Infections* 1: 16–21.

Fields PI, Blom K, Hugues J et al (1997) Molecular characterization of the gene encoding H antigen in *Escherichia coli* and development of a PCR–restriction fragment length polymorphism test for identification of *E. coli* O157:H7 and O157:NM. *Journal of Clinical Microbiology* 35: 1066–1070.

Gannon VP, Rashed M, King RK and Golsteyn-Thomas EJ (1993) Detection and characterization of the *eae* gene of Shiga-like toxin-producing *Escherichia coli* using polymerase chain reaction. *Journal of Clinical Microbiology* 31: 1268–1274.

Gannon VP, D'Souza S, Graham T et al (1997) Use of the flagellar H7 gene as a target in multiplex PCR assay and improved specificity in identification of enterohemorrhagic *Escherichia coli* strains. *Journal of Clinical Microbiology* 35: 656–662.

Gunzer F, Böhm H, Rüssmann H et al (1992) Molecular detection of sorbitol-fermenting *Escherichia coli* O157 in patients with haemolytic–uraemic syndrome. *Journal of Clinical Microbiology* 30: 1807–1810.

Karch H and Meyer T (1989) Evaluation of oligonucleotide probes for identification of Shiga-like-toxin producing *Escherichia coli*. *Journal of Clinical Microbiology* 27: 1180–1186.

Karch H and Meyer T (1989) Single primer pair for amplifying segments of distinct Shiga-like toxin genes by polymerase chain reaction. *Journal of Clinical Microbiology* 27: 2751–2757.

Masters CI, Shallcross JA and MacKey BM (1994) Effect of stress treatments on the detection of *Listeria monocytogenes* and enterotoxigenic *Escherichia coli* by the polymerase chain reaction. *Journal of Applied Bacteriology* 77: 73–79.

Read SC, Clarke RC, Martin A et al (1992) Polymerase chain reaction for detection of verocytotoxigenic *Escherichia coli* isolated from animal and food sources. *Molecular and Cellular Probes* 6: 153–161.

Thomas A, Smith HR and Rowe B (1993) Use of digoxigenin-labelled oligonucleotide DNA probes for VT2 and VT2 human variant genes to differentiate Vero cytotoxin-producing *Escherichia coli* strains of serogroup O157. *Journal of Clinical Microbiology* 31: 1700–1703.

Venkasteswaran K, Kamijoh Y, Ohashi E and Nakanishi H (1997) A simple filtration technique to detect enterohemorrhagic *Escherichia coli* O157:H7 and its toxins in beef by multiplex PCR. *Applied and Environmental Microbiology* 63: 4127–4131.

Vernozy-Rozand C (1997) Detection of *Escherichia coli* O157:H7 and other verocytotoxin-producing *E. coli* (VTEC) in food. *Journal of Applied Microbiology* 82: 537–551.

VIBRIO

Contents

Introduction, including *Vibrio vulnificus* and *Vibrio parahaemolyticus*

P M Desmarchelier, Food Science Australia, Cannon Hill, Queensland, Australia

Characteristics of the Genus *Vibrio*

The genus *Vibrio* includes Gram-negative, facultatively anaerobic non-spore-forming bacilli, which are often curved in shape and motile by means of a sheathed polar flagellum. The species, except *V. metchnikovii*, are oxidase-positive and ferment glucose, some with the production of gas. The growth of most vibrios is stimulated by the presence of sodium and some species have an obligate salt requirement for growth. *V. cholerae*, the type species of the genus, was first described almost 150 years ago and for many years this was the only species. *V. parahaemolyticus* was first isolated from food-borne infections in the 1950s and, since then, the number of species assigned to the genus has increased significantly to more than 20. This has resulted from extensive studies of microbial ecology in aquatic environments and the application of molecular techniques in determining the taxonomic position of bacterial isolates.

Identification of *Vibrio* spp.

Some of the *Vibrio* spp. are closely related and originally were considered to be biotypes of a single species until DNA homology studies were used to confirm their individual species status. The oxidase-positive *Vibrio* spp., particularly the human pathogens, can be divided in two groups based on their requirement for salt for growth (**Table 1**). *V. cholerae* and *V. mimicus* form one group and are non-halophiles, although their growth is stimulated by the presence of salt. These species are closely related and *V. cholerae* is primarily differentiated from *V. mimicus* by its fermentation of sucrose.

The other group includes the halophilic species. They vary in their salt tolerance and this can be used in their differentiation, together with other biochemical traits (Table 1). *V. vulnificus* and *V. parahaemolyticus* are related (40–50% DNA homology) and *V. vulnificus* was initially identified as a lactose-fermenting biotype of *V. parahaemolyticus*. This trait has since been shown to be inconsistent among *V. vulnificus* and additional phenotypic traits are required for their identification (Table 1). *V. parahaemolyticus* is closely related to *V. alginolyticus* (60–70% DNA homology) and is differentiated from *V. alginolyticus* by its inability to ferment sucrose and its lower salt tolerance. *V. furnissii*, *V. fluvialis* and *V. hollisae* produce arginine dihydrolase which differentiates them within the genus. *V. furnissii* and *V. fluvialis* share this trait with *Aeromonas* spp. from which they are distinguished by their salt requirement for growth. *V. furnissii* can be differentiated from *V. fluvialis* as it produces gas from glucose. Carbohydrate fermentation and carbon compound utilization further assist in species identification, although variations within species in these traits is often noted.

Importance of *Vibrio* spp. to the Food Industry

Most *Vibrio* spp. are autochthonous bacterial aquatic flora and about half have been associated with infections in human or aquatic animals. Human infection follows direct contact with aquatic environments or indirectly via contaminated food and water. Food-borne *Vibrio* infections do not occur frequently in most countries, although some specific countries have a higher incidence due to cultural food practices or because there are endemic disease foci associated with poor sanitation and hygiene. Some *Vibrio* infections are significant as they are listed as quarantinable diseases (e.g. *V. cholerae*) or are known to cause high mortality (e.g. *V. vulnificus*).

V. cholerae, the cause of cholera, is the most important human pathogen of the genus. The disease cholera continues as a major world health concern as it has emerged and re-emerges through a series of pandemics. Cholera is a notifiable disease in most countries and is also notifiable to the World Health Organization. The disease may be transmitted within a country and between countries via food, therefore food industries in cholera-endemic areas and their trading partners have to be aware of the risks and implement control measures.

Table 1 Traits for the differentiation of food-borne and closely related *Vibrio* spp

Trait	*Species*								
	1	*2*	*3*	*4*	*5*	*6*	*7*	*8*	*9*
Indole	+	+	+	–	–	±	–	–	+
ONPG	+	–	+	±	–	–	–	±	–
Voges–Proskauer	+/–[a]	–	–	–	–	–	+	+	±
Lysine decarboxylase	+	+	+	–	–	–	±	+	+
Ornithine decarboxylase	+	+	±	–	–	–	–	–	±
Arginine dihydrolase	–	–	–	+	+	–	+	–	–
Gas production from glucose	–	–	–	–	+	–	–	–	–
Fermentation of:									
Sucrose	+/–	–	–	+	+	–	–	+	+
Lactose	±	–	±	–	–	–	–	–	–
Salicin	–	–	±	–	–	–	–	–	–
Arabinose	–	+	–	+	+	+	–	+	–
Cellobiose	–	–	±	–	–	–	–	+	–
Utilization of:									
L-Leucine	–	+	–	–	–	–	–	–	+
L-putrescine	–	+	–	±	+	–	–	–	+
Ethanol	–	+	–	+	+	–	–	–	±
D-glucuronate	–/+	±	+	+	–	–	–	–	–
Growth in:									
0% NaCl	+	–	–	–	–	–	–	–	–
3% NaCl	+	+	+	+	+	+	+	+	+
8% NaCl	–	+	–	±	±	–	–	–	+
10% NaCl	–	–	–	–	–	–	–	–	+

1 = *V. cholerae* and *V. mimicus*; 2 = *V. parahaemolyticus*; 3 = *V. vulnificus*; 4 = *V. fluvialis*; 5 = *V. furnissii*; 6 = *V. hollisae*; 7 = *V. damsela*; 8 = *V. cincinnatiensis*; 9 = *V. alginolyticus*.
[a] Numerator and denominator are reactions for *V. cholerae* and *V. mimicus* respectively.

Two other important *Vibrio* spp. which are frequently food-borne are *V. parahaemolyticus* and *V. vulnificus*. *V. parahaemolyticus* usually causes uncomplicated gastroenteritis while *V. vulnificus* causes systemic infections associated with a high mortality. Other species which are infrequently food-borne or for which there is less conclusive evidence of food-borne transmission include *V. mimicus*, *V. fluvialis*, *V. furnissii* and *V. alginolyticus*.

Methods of Detection of *Vibrio* spp.

Specific methods of detection are only available for *Vibrio* spp. commonly associated with food-borne disease, e.g. *V. cholerae*, *V. parahaemolyticus* and *V. vulnificus*. Most of the other less commonly implicated species have also been isolated using the same media. The addition of 3% salt to the food diluent is necessary for the detection of the halophilic species. These species are often present in low concentrations, therefore selective enrichment of the food sample is usually necessary, followed by plating on selective agar. Colonies are primarily differentiated by fermentation of carbohydrates, e.g. sucrose. Further identification of isolates from raw seafoods can be difficult due to variability in their phenotypic traits and the close relationships between these species and other natural bacterial flora, which may be poorly classified. Commercial automated and rapid biochemical identification systems tend to perform better with the more common species, e.g. *V. cholerae*. Additional salt is required in biochemical test media used to identify halophilic species.

Characteristics of *V. parahaemolyticus* and *V. vulnificus*

V. parahaemolyticus and *V. vulnificus* have typical characteristics of the genus *Vibrio*, with the exception of *V. parahaemolyticus* that produces additional peritrichous flagella on solid media. The optimum conditions and limits for growth of *V. parahaemolyticus* and *V. vulnificus* are shown in **Table 2**. Both species

Table 2 Optimum growth conditions and growth limits for *Vibrio parahaemolyticus* and *V. vulnificus*

Growth parameter		V. parahaemolyticus	V. vulnificus
Temperature (°C)	Optimum	37	37
	Limits	5–43	8–43
pH	Optimum	7.8–8.6	7.8
	Limits	4.8–11	5–10
Water activity	Optimum	0.981	0.98
	Limits	0.940–0.996	0.96–0.997
NaCl (%)	Optimum	3	2.5
	Limits	0.5–10	0.5–5.0

are unable to grow in the absence of salt and require a minimum of 0.5% (w/v) NaCl for growth, while their optimum salt concentration for growth is approximately 3%. The salt tolerance range for *V. parahaemolyticus* is much wider than that for *V. vulnificus*. The optimum growth temperature for both species is 37°C and at this temperature the growth rate of *V. parahaemolyticus* is rapid, with doubling times of 9–10 min. The growth rate of *V. vulnificus* is slightly slower. These vibrios are mesophiles; however, the limiting temperatures for growth are affected by pH and salt concentrations. Both species have a high pH tolerance but are sensitive to acid conditions of pH < 5. Both are sensitive to reductions in water activity and ionizing radiation.

V. parahaemolyticus and *V. vulnificus* are autochthonous bacterial flora in coastal marine environments in tropical and temperate regions throughout the world. *V. parahaemolyticus* has occasionally been isolated from inland aquaculture or salt water environments. Their distribution and concentration at a particular location are influenced by interacting environmental factors, in particular water temperature and salinity, and their interaction with higher marine life. A distinct seasonal pattern occurs in temperate regions where the bacteria proliferate in the water during warm summer months and are not detectable during colder winters when water temperatures are lower than about 10°C. Contamination of fresh seafoods, e.g. crustaceans, shellfish and fish, with vibrios reflects the contamination of the environment from which the seafood was harvested. Their concentrations are higher in association with marine animals, e.g. on the exoskeleton of marine crustaceans or in filter-feeding shellfish, than in the water column. Vibrios are found in the gut and on the surface of marine animals and may be accumulated by filter-feeding molluscs. Concentrations of *V. parahaemolyticus* in seafood are generally lower than 10^3 per gram and may be higher when harvested from very warm waters. *V. vulnificus* is most commonly isolated from shellfish and is found in lower numbers than *V. parahaemolyticus* although, similarly, the numbers can be up to 10^5 per gram in very warm waters.

Food-borne disease outbreaks and cases caused by *V. parahaemolyticus* have been associated with the consumption of both raw and cooked seafood. In western societies, the foods most often implicated include cooked crustaceans (e.g. prawns, shrimps and crabs) and raw molluscs (e.g. oysters). Raw fish is an important food vehicle in societies where marine fish is traditionally eaten raw. *V. parahaemolyticus* is very sensitive to heat so outbreaks due to cooked products result from inadequate cooking or, alternatively, from poor handling and cross contamination with raw seafood or contaminated processing environments. The infectious dose of *V. parahaemolyticus* for healthy adults is generally much higher than that present on the freshly harvested and chilled product. To reach these levels, the contaminated raw or cooked product has probably been mishandled by holding at warm temperatures for significant periods.

In contrast, *V. vulnificus* infections are sporadic and associated with the consumption of raw shellfish, usually oysters. Although the species has been isolated from fish and crustaceans, these products have not been associated with infections. The infectious dose for this species is believed to be very low and most cases have underlying conditions predisposing them to infection. Such individuals may be susceptible to the numbers of *V. vulnificus* present in freshly harvested oysters; however, the bacterium is able to grow in oysters after harvest if not chilled and the concentrations consumed may be higher.

Food-borne infection with *V. parahaemolyticus* is typically an acute gastroenteritis that occurs after an incubation period of 4–30 h. Symptoms include diarrhoea, which may be bloody, abdominal pain, nausea and vomiting and are usually self-limiting. Fatalities are rare. Food-borne disease caused by *V. parahaemolyticus* is common in Japan where in the past it has been the cause of up to 70% of bacterial food-borne disease. In other countries the occurrence is much less frequent. Infections tend to follow a seasonal pattern in temperate regions in association with the proliferation of the bacterium in the marine environment. In contrast, the clinical symptoms of *V. vulnificus* food-borne infections are manifested extraintestinally, including fever, chills and nausea rather than gastroenteritis. The bacterium causes septicaemia and in many cases skin lesions (e.g. echthyma, vesicles and bullae, and necrotic ulcers) which often develop on the extremities and trunk. The incubation period is about 38 h, after which the patient's condition may deteriorate rapidly. Food-borne *V. vulnificus* infections are uncommon and patients frequently have chronic conditions such as liver dysfunction, alcohol abuse, diabetes and malignancies. The mortality rate for these individuals is high (40–60%).

V. parahaemolyticus strains are classified as Kanagawa-positive and -negative based on their ability to produce a thermostable direct haemolysin (TDH) that lyses fresh human red blood cells. The Kanagawa phenomenon is closely correlated with human pathogenic strains while in the environment most strains are Kanagawa-negative. Although the TDH has some enterotoxic activity, it is not proven to be the major virulence factor and TDH-negative strains have been isolated from cases of gastroenteritis. Human *V. para-*

haemolyticus isolates have been shown to produce several haemolytic substances as well as haemolysins related to TDH, cytotoxins related to Shiga toxins, and various adhesive factors facilitating cell attachment to the human gut. In some geographical areas, correlations have been observed between urease-producing strains and TDH production. The pathogenecity of *V. parahaemolyticus* and reliable virulence markers remain to be clearly determined.

V. vulnificus is an invasive bacterium with a variety of virulence properties which allow it to evade the host defence mechanisms. Virulent strains that have an opaque colonial morphology have a capsular-like surface layer which provides resistance to phagocytosis and complement activity. Other surface components also provide resistance to complement-mediated lysis of normal human serum. *V. vulnificus* is unable to grow in human serum with normal iron levels and increased serum iron levels are required. These factors help to explain the increased susceptibility to infection among the high-risk individuals described above whose medical conditions cause suppressed immunity or increased serum iron levels. All *V. vulnificus* produce a cytolysin that is believed to play a role in the tissue destruction manifested during these infections. As the cytolysin is common to the species it is a virulence trait and not a marker distinguishing virulent strains.

Methods of Detection of *V. parahaemolyticus* and *V. vulnificus*

Vibrios are often present in foods, especially in raw seafoods, in small numbers and in association with other marine bacteria that may outnumber them. Detection includes selective enrichment followed by plating on selective agar media and the bacteria can be enumerated by using these combinations and the most probable number (MPN) technique. When high numbers are anticipated, direct enumeration on selective agar is possible. The food sample is suspended and diluted in diluent containing 3% NaCl to maintain the integrity of the cells. A variety of media have been developed for the enrichment of these species but no single medium appears to perform optimally for all *Vibrio* spp. Alkaline peptone water (APW), consisting of 1% peptone and 1% NaCl, is a simple and inexpensive medium that has been widely used for the detection of *V. cholerae*. The salt concentration of the medium should be increased to 3% for the specific isolation of the halophilic *Vibrio* spp. The selectivity of APW is based on the high pH (8.5–9.0) at which vibrios are able to grow rapidly. As the pH can drop quickly after incubation and the selectivity is lost, a double enrichment is recommended with a short incubation period of 6–8 h and a secondary enrichment overnight.

Alternative enrichments tend to be specific for the individual species and include selective agents such as polymyxin, Teepol, ethyl violet and the addition of carbohydrates (arabinose) for *V. parahaemolyticus*, and starch and gelatin enhance isolation of *V. vulnificus*. Incubation at 42°C rather than 37°C increases selectivity for *V. vulnificus*, although some *V. parahaemolyticus* strains may not grow at this temperature. Stressed cells may be present in foods following processing or chilling and freezing of product during storage. These cells will be more susceptible to selective enrichment other than APW and can be resuscitated in non-selective low-salt media.

Thiosulphate citrate bile salts sucrose (TCBS) agar is commonly used as a selective agar medium for *V. cholerae* and other vibrios. TCBS is highly selective based on the addition of bile salts and a high pH. Sucrose is used for the differentiation of the yellow sucrose-fermenting *V. cholerae* and *V. alginolyticus* from the blue-green non-sucrose-fermenting *V. parahaemolyticus* and *V. vulnificus*. Additional media used for *V. parahaemolyticus* isolation have incorporated selective antibiotics and detergents and relied on the production of extracellular enzymes for the colony differentiation. None appears to be as widely accepted as TCBS. Alternative media recommended for the isolation of *V. vulnificus* include cellobiose polymyxin colistin agar (CPC agar; 4.0×10^5 U colistin methanosulphonate) and sodium dodecyl sulphate-polymyxin B sucrose agar. Incubation at 40°C also increases selectivity.

A variety of immunological assays and DNA-based tests have been developed for the detection of these species and can be used to screen enrichment broths before plating on agar media. Hydrophobic grid membrane filters can be used with colony hybridization as an alternative method for enumeration.

Identification of *V. parahaemolyticus* and *V. vulnificus*

Both *V. parahaemolyticus* and *V. vulnificus* are often primarily differentiated by their lack of sucrose fermentation on isolation media (e.g. blue-green colonies on TCBS) or by cellobiose fermentation for *V. vulnificus* (e.g. CPC agar). They are further tested for oxidase, motility and salt tolerance together with biochemical traits determined in media containing 3% NaCl. Key characteristics for both species are the production of lysine and ornithine decarboxylases (*V. vulnificus* may give slow reactions) and not arginine dihydrolase and no growth in 0% NaCl. Important individual characteristics include the maximum level of salt tolerated, the *o*-nitrophenyl-β-D-galactopyranoside (ONPG) reaction and the fermentation of cellobiose (Table 1).

V. parahaemolyticus isolates can be tested for the Kanagawa reaction on agar containing fresh human red blood cells, known as Wagatsuma's agar. Production of TDH causes haemolysis of the red blood cells within 24 h at 37°C. Alternatively, the TDH and related haemolysins can be detected using molecular probes. *V. parahaemolyticus* can be serotyped using O and K antisera.

Regulations

Food-borne infections due to *V. parahaemolyticus* and *V. vulnificus* occur throughout the world, although the occurrence varies greatly. Testing for these organisms in either food or clinical samples may be routine in countries with a high incidence, e.g. Japan; however, in other areas testing may be restricted to public health laboratories or laboratories specifically associated with seafood industries. The presence of these bacteria in food, processed in some way to provide a vibriocidal effect, is unacceptable. Such contamination indicates inadequate processing or post-process contamination and possible subsequent growth of the organisms. In contrast, both species may be present in raw seafood. Levels of *V. parahaemolyticus* up to 10^2–10^3 per gram appear usual for natural contamination of seafood and, from epidemiological evidence from outbreaks, are not associated with disease in the healthy adult population.

Acceptable levels of *V. vulnificus* in raw seafood are difficult to determine. Healthy consumers are able to tolerate the small numbers present in naturally contaminated seafood, while the same dose apparently causes infection in high-risk consumers. Due to the severity of the infection and the existing impaired health of the high-risk individuals, human volunteer studies cannot be performed. Small numbers are therefore considered hazardous for at-risk individuals. Oysters are the main product implicated and groups at risk should avoid eating this food. Some food authorities require restaurateurs and retailers of raw oysters from regions known to be naturally contaminated with *V. vulnificus* to have a health warning on the product notifying the risk to certain individuals.

Additional measures to health education are aimed at limiting the concentrations of *V. vulnificus* and controlling the growth of *V. vulnificus* in oysters after harvest. The US Interstate Shellfish and Sanitation Conference has recommended that any state whose oyster-growing waters have been confirmed as the original source of oysters associated with two or more *V. vulnificus* cases should be expected to require that the oysters be refrigerated within a specified time after harvest. The time is dependent on the average monthly temperature of the oyster-growing waters. For example, the time interval between harvest and refrigeration is 14 h when the water temperature is 18–23°C, 12 h at 23–28°C and 6 h at > 28°C. Another reduction strategy is to relay oysters from waters of low salinity to higher salinity. It should be noted that the depuration of oysters is effective in the reduction of introduced bacterial species but is ineffective in the removal of natural contamination with vibrios.

Importance to the Food Industry

Vibrios are natural contaminants of raw seafoods and have to be considered in any hazard analysis and critical control point plan for products containing seafood. Cooking is the most effective way of eliminating them from food, and products cooked to an internal temperature of > 65°C should be safe. Seafood, e.g. crustaceans, may present problems as the internal temperature may not reach these levels when the flesh appears cooked and of a desirable palatable texture. Prevention of growth of these bacteria is a critical control measure for all seafoods and is achieved by rapid chilling and refrigeration of stored product to temperatures < 5°C. Oysters should be rapidly chilled to < 15°C after harvest. Rapid chilling is essential as these species have a very short doubling time and most growth occurs in the first few hours of storage. Outbreaks of *V. parahaemolyticus* have frequently been attributed to inadequate processing or post-process contamination followed by poor temperature control during storage. Such outbreaks have resulted from food prepared at food service establishments and international airline caterers, and for picnics where the contaminated food has been stored for long periods at warm temperatures between preparation and consumption.

Oysters present a particular challenge as they are frequently eaten raw and thus for raw product there is no critical control point that will eliminate the hazard. Control includes health education and warnings for high-risk consumers, and pathogen reduction measures, neither of which offers absolute control. Mild heat treatments (e.g. 50°C for 10 min) and cold shock are proposed as alternative control measures. Given the potential high mortality and the impact of such an event on the industry, the risks are high, although the likelihood of occurrence low.

See also: **Food Poisoning Outbreaks**. **Hazard Appraisal (HACCP)**: The Overall Concept. ***Vibrio***: *Vibrio cholerae*.

Further Reading

Doyle M, Beuchat LR and Montville TJ (eds) (1997) *Food Microbiology Fundamentals and Frontiers*. Washington, DC: ASM Press.

Hocking AD, Arnold G, Jenson I, Newton K and Sutherland P (eds) (1997) *Foodborne Microorganisms of Public Health Significance*, 5th edn. Sydney, Australia: AIFST.

Roberts TA, Baird-Parker AC and Tompkin RB (eds) (1995) *ICMSF. Micro-Organisms in Foods 5*. London: Blackie Academic & Professional.

US Department of Health and Human Services (1995) *Revision. National Shellfish Sanitation Program, Manual of Operations*, Part II. Sanitation of the harvesting, processing and distribution of shellfish. Washington, DC: Shellfish Sanitation Branch, US Food and Drug Administration.

Vibrio cholerae

F Y K Wong and **P M Desmarchelier**, Food Science Australia, Queensland, Australia

Introduction

Vibrio cholerae is the aetiological agent of cholera, an epidemic disease of significant public health importance owing to its rapid spread in areas with poor sanitation and hygiene, and its severe consequences when access to health care is limited. A vibrio-like organism was first described as the cholera pathogen as early as 1854, although the *V. cholerae* bacillus was not successfully isolated until 30 years later. The modern history of cholera is characterized by seven recorded pandemics, during which *V. cholerae* spread globally in a series of disease upsurges. Cholera associated with the current seventh pandemic is endemic over much of Asia, Africa, and Latin America. In spite of improved public health controls in some of these regions, the disease remains a major public health menace. The emergence of novel epidemic strains has led to new disease upsurges, and greater challenges for cholera management and control.

Characteristics of the Species

Taxonomy

Vibrio cholerae is the type species of the genus *Vibrio*, which comprises a group of Gram-negative, mostly oxidase-positive, facultatively anaerobic bacteria. There are more than 30 described species within the genus, which include several known food-borne pathogens. Cells of *V. cholerae* are often curved, asporogenous rods that are motile by means of a single, sheathed polar flagellum. Strains are typically positive for nitrate reduction and sucrose utilization. *Vibrio* growth is stimulated by the presence of Na^+, although different species have varying requirements. A major feature that distinguishes *V. cholerae* from most of the other *Vibrio* species is the ability to grow in laboratory media without added NaCl. Constituents within the media are able to provide a sufficient source of NaCl for growth, although growth may be reduced under low Na^+ concentrations. *Vibrio cholerae* requires 5–15 mmol l^{-1} Na^+ for optimum growth. The bacterium is able to tolerate moderately alkaline conditions and growth occurs at up to pH 10; however, growth is inhibited at pH 6 or lower.

Vibrio cholerae is divided into two major serogroups based on the somatic O-antigen. The O-antigen is the outermost thermostable polysaccharide component of the lipopolysaccharide layer on the cell surface. The *V. cholerae* serogroup O1 includes the strains responsible for epidemic cholera and is further divided into two biotypes, classical and El Tor. The classical biotype is believed to be responsible for the first six cholera pandemics (1817 to 1923), which some researchers speculate to be a single pandemic highlighted by a series of major epidemic upsurges. The seventh cholera pandemic which began in 1961 resulted from the emergence of *V. cholerae* O1 biotype *eltor*. Biotype differentiation is based primarily on the haemolysis of sheep erythrocytes by El Tor strains, although this has become less reliable. El Tor strains were initially strongly haemolytic but as the seventh pandemic has progressed, non-haemolytic strains now predominate. The biotypes have no taxonomic significance but are epidemiologically useful and there appears to be an association with clinical severity. Cholera due to *V. cholerae* O1 *eltor* is often less severe and a larger number of asymptomatic cases occur compared with classical cholera.

Strains belonging to serovariants other than O1 are collectively known as *V. cholerae* non-O1. Non-O1 strains are associated with sporadic cases of gastroenteritis and milder forms of cholera-like illness, and were not thought to cause epidemic cholera. However, outbreaks of typical cholera have been attributed to strains of the novel non-O1 serogroup, *V. cholerae* O139 synonym Bengal. Since its emergence in India and Bangladesh in 1992, the rapid spread of O139-associated outbreaks to other regions has led some researchers to suggest that an eighth cholera pandemic may be in progress.

Numerical taxonomic studies using phenotypic criteria encompassing morphology, physiology, biochemical tests, antibiotic susceptibilities and substrate utilization show that no taxonomic distinction exists between the different biotypes and serogroups, or between clinical and nonclinical isolates of *V. chol-*

erae. Reported DNA–DNA homologies between O1 and non-O1 strains are greater than 80% at 60°C, and comparable to the homologies shown among O1 strains. *Vibrio mimicus* is phenotypically closely related to *V. cholerae*, and strains were previously referred to as atypical sucrose non-fermenting *V. cholerae* non-O1. These sucrose-negative strains were later designated as a separate species based primarily on their low DNA–DNA homologies (20–50% at 60°C) with the *V. cholerae* type strain. Analyses of 16S rRNA sequences have shown that *V. cholerae* and *V. mimicus* form a monophyletic grouping distinguishable from the core *Vibrio* group, which consists mainly of the halophilic vibrios.

Epidemiology

Cholera is a secretory diarrhoea typically caused by toxigenic strains of *V. cholerae* O1 and O139. Transmission is usually by ingestion of contaminated food or water, although in healthy persons the infectious dose is large (10^5–10^6 cells). *Vibrio cholerae* is sensitive to gastric acidity, which has a significant effect on the infectious dose. Patients with hypochlorhydria or predisposing gastric abnormalities are susceptible to a low infectious dose. The simultaneous ingestion of food may provide bacteria with a protective barrier against gastric acidity. Surviving cells attach to and colonize the walls of the small intestine. Following an incubation period of 12–72 h, clinical symptoms can progress to profuse, watery diarrhoea with the production of large volumes of distinctive, pale, opalescent stools likened to 'rice water'. Other symptoms may include nausea, abdominal cramps and fever. If not replaced by rehydration therapies, fluid loss can rapidly result in the loss of circulating blood volume, metabolic acidosis and vascular collapse, leading to death. Only a small percentage of the population infected with *V. cholerae* O1 *eltor* develop full-blown cholera, while a large proportion are asymptomatic or develop milder manifestations. There is no long-term carriage following infection.

In endemic areas where sanitation and hygiene are poor, the primary mode of transmission is the faecal-oral route via contaminated water. Acutely infected persons may individually excrete as many as 10^{13} *V. cholerae* cells per day, and epidemics can result in substantial pollution of the environment. Food contaminated with water, or directly with faeces, is an important secondary vehicle of transmission. Sporadic cases of cholera have been reported in some nonendemic countries, the USA and Australia in particular. *Vibrio cholerae* O1 exists autochthonously in environmental reservoirs (e.g. riverine and inshore coastal environments) which show no evidence of faecal contamination or disease. Ingestion of untreated river water was the main source of domestic cholera in Australia, while contaminated shellfish was the primary vehicle in the USA. *Vibrio cholerae* non-O1 is more prevalent in the environment than *V. cholerae* O1.

Ecology

The distribution of *V. cholerae* in the environment follows a distinct seasonal pattern. Temperature, salinity and nutrient concentration are considered to be the most important physical factors influencing the distribution and survival of *V. cholerae*. Cholera outbreaks and recovery of the bacterium from the environment often increase during warmer summer months. Optimum water temperatures for the isolation of *V. cholerae* are 20–35°C, and culturable cells may no longer be detectable when temperatures drop below 16°C. The bacterium is distributed in the environment over a range of salinities, from fresh water to marine areas with salinities up to 32 ppK. Higher organic nutrient concentrations may buffer the effects of sub-optimal salinity. Other factors such as pH, dissolved oxygen, and osmotic pressure also play a role in *V. cholerae* distribution.

The *V. cholerae* cells may enter a viable but non-culturable state when subjected to sub-optimal or stressful conditions, which may explain its apparent disappearance from the environment during colder months or nonseasonal periods. Viable but non-culturable cells retain virulence properties and remain metabolically active, but fail to grow on standard laboratory media. The inability to detect viable but non-culturable *V. cholerae* using standard culture techniques has implications for food and water testing. Some researchers believe that *V. cholerae* has an ecological role in the aquatic environment, through symbioses and associations with planktonic biota and higher organisms. There is evidence of an inter-relationship between global climate, phytoplankton and marine copepod cycles, and the distribution of *V. cholerae* in marine ecosystems.

Pathogenicity

Cholera pathogenesis is determined by a suite of virulence factors produced by toxigenic strains of *V. cholerae* O1 and O139. Key elements required for pathogenesis include the cholera enterotoxin (CT), and the toxin co-regulated pilus (TCP), which mediates colonization of the small intestine. The major subunit for the TCP is encoded in the *tcpA* gene within the *V. cholerae* chromosome, although a number of other accessory genes are needed for pilus assembly, function and regulation. The cholera enterotoxin is a multimeric toxin consisting of a single A subunit linked to five B subunits. The mature A subunit toxin

(27.2 kDa) is divided into two disulphide-linked fragments, A_1 and A_2, following proteolytic processing. The A_1 peptide is the enzymatic component involved with activating adenylate cyclase within the host cell. This causes an accumulation of intracellular cyclic AMP which disrupts the regulation of normal ion transport in the intestinal mucosal cells, leading to inhibition in sodium adsorption and increased chloride and water secretion into the lumen. The outcome is watery diarrhoea symptomatic of cholera. The B subunits (11.7 kDa) act to bind the CT to the host cell surface GM_1 ganglioside receptors, presenting the A_1 peptide for interaction with the host cell.

The cholera enterotoxin is encoded within a variable genetic element composed of a 4.5 kb core region that includes the *ctxAB* operon flanked by one or more copies of a 2.7 kb repetitive sequence called *RS1*. The core region also contains the zonula occludens toxin gene, *zot*, and the accessory cholera enterotoxin gene, *ace*. These accessory toxins increase permeability of intestinal mucosal cells, and may be responsible for diarrhoeal symptoms observed in human volunteers inoculated with *ctx*-deficient mutant strains. Other virulence factors produced by some *V. cholerae* O1 strains include the El Tor haemolysin and cytolysin, and a shiga-like toxin. Milder clinical symptoms are usually expressed when CT is absent. The enterotoxin is also expressed by some non-O1 non-O139 strains, although these strains occur less frequently than toxigenic *V. cholerae* O1. Non-O1 strains also produce haemolysins, and some produce a heat-stable enterotoxin (NAG-ST) similar to those produced by *Escherichia coli* and *Yersinia enterocolitica*; NAG-ST is rarely detected in *V. cholerae* O1.

Methods of Detection

The detection of *V. cholerae* in food and water may present difficulties, as the bacterium is often present in low numbers among significant concentrations of other contaminating flora. Bacteria in food may be stressed from the effects of heating, chilling, freezing and other sub-optimal conditions, and may fail to grow if inoculated directly into highly selective media. These cells have to be resuscitated using enrichment broths.

Enrichment and Isolation

Alkaline peptone water (APW) containing 1% peptone and 1% NaCl (pH 8.5–9.0) is the most effective and commonly used enrichment medium for *V. cholerae*; it relies on the ability of *V. cholerae* to grow rapidly in alkaline conditions. Since pH is the main selective factor, an incubation time of 6–8 h at 37°C is routinely recommended to minimize overgrowth by other bacteria. In samples containing stressed or low cell numbers, 6–8 h enrichments may give relatively poor recovery and a two-step enrichment can be used. This includes a secondary 18–21 h enrichment inoculated with culture from the primary 6–8 h enrichment. Recovery may also be improved by incubation at 42°C, which suppresses growth of competing organisms. The US Food and Drug Administration (FDA), and the *Official Methods of the AOAC International* (16th edn, 3rd revision, 1997) recommend preparation of a 1 : 10 dilution of oyster homogenate in APW and incubation at 42°C for 6–8 h. The use of a larger sample dilution (1 : 100) may give better recovery as this minimizes competing flora and the decrease in pH. Typically, a loopful of the incubated enrichment is taken from the top pellicle, and streaked onto solid selective medium for the isolation of *V. cholerae*.

Thiosulphate–citrate–bile salts–sucrose (TCBS) agar is a selective medium widely used for the isolation of *V. cholerae*. This medium is highly selective and colonies growing on it are easily differentiated. Overnight growth of *V. cholerae* on TCBS at 37°C will produce relatively large (2–4 mm diameter), circular, flattened, opaque yellow colonies. The sucrose non-fermenting species *V. mimicus* and *V. parahaemolyticus* are easily differentiated as they form blue-green colonies. The routine use of TCBS requires careful quality control as commercial preparations are subject to brand and batch variations. There is a misconception that only *Vibrio* species will grow on TCBS, as it can support the growth of other bacteria, particularly those from environmental samples. A disadvantage of TCBS is that *V. cholerae* forms sticky colonies that are difficult to emulsify and cannot be directly used for confirmatory tests, such as serotyping and the oxidase test.

An alternative medium is Monsur's tellurite–taurocholate gelatin agar (TTGA). This is easily prepared but is not available commercially, and unlike TCBS agar, its preparation requires autoclaving. It is commonly used for the isolation of *V. cholerae* from faecal samples where it is comparable to TCBS, but is less commonly used for food and environmental samples. *Vibrio cholerae* incubated on TTGA overnight forms relatively small (1–2 mm diameter), circular, opaque colonies with dark centres. Longer incubation results in gunmetal grey colonies owing to the reduction of tellurite, and the appearance of a halo around individual colonies owing to gelatinase production. Colonies can be directly used for serotyping and other confirmatory tests. Many *Vibrio* species have similar colonial morphology on TTGA, making differentiation difficult. A modified formulation which includes the substrate 4-methyl-umbelliferyl-β-D-galactoside improves differentiation. *Vibrio* species that exhibit β-D-galactosidase activity, such as *V. cholerae*, produce confluent colonies with bright blue fluorescence.

Cellobiose–polymyxin B–colistin (CPC) agar was developed for the isolation of *V. cholerae* and *V. vulnificus*. This agar relies on the resistance of the two species to the antibiotics colistin and polymyxin B. It also differentiates between the two species because *V. vulnificus* ferments cellobiose while *V. cholerae* does not, producing different coloured colonies. The medium is reported to be superior to TCBS and is widely used in combination with APW for the isolation of *V. cholerae* O1 *eltor*. However, CPC agar is not commercially available, has a complicated formulation, requires autoclaving, and may be too inhibitory to some *V. cholerae* strains. For example, strains of the classical biotype are sensitive to polymyxin B.

Identification

Presumptive isolates of *V. cholerae* have to be screened with a battery of morphological, biochemical, and serological tests for confirmation. When grown in liquid media such as nutrient broth, *V. cholerae* cells exhibit a characteristic rapid, darting motility. Triple sugar iron (TSI) agar, which contains glucose, lactose and sucrose, is a useful medium for the initial screening of isolates. *Vibrio cholerae* is able to ferment sucrose with no production of gas, resulting in an acid slant over an acid butt (A/A) reaction. Hydrogen sulphide is not produced by the organism. A number of other key phenotypic traits differentiate *V. cholerae* from related vibrios and other bacteria (**Table 1**). Most *Vibrio* species are oxidase-positive and this is the main test to distinguish them from the Enterobacteriaceae.

Vibrio cholerae is usually positive for lysine and ornithine decarboxylase but does not produce arginine dihydrolase. These test broths should be supplemented with 1% NaCl to stimulate growth. The ability to grow without added NaCl is a key trait distinguishing *V. cholerae* and *V. mimicus* from the other species. Salt tolerance tests can be performed using either 1% tryptone or nutrient broth, with the appropriate amount of NaCl added. Biochemically *V. cholerae* is similar to *V. mimicus* but can be differentiated on the basis of sucrose fermentation. In addition, many *V. cholerae* strains are positive for the Voges–Proskauer reaction and lipase, while most *V. mimicus* strains are negative for both traits. *Vibrio* species can be separated from the genus *Aeromonas* by their sensitivity to the vibriostatic compound 2,4-diamino-6,7-diisopropyl-pteridine phosphate (O/129). However, this reaction can be variable and should not be solely relied on. Many strains of *V. cholerae* O139 are reported to be resistant to 0/129. Confirmed *V. cholerae* isolates are usually typed to identify strains, and to establish epidemiological relationships.

Biotyping Biotyping is not a routine method, but may be useful for the epidemiological classification of strains newly introduced to a region. Biotyping includes testing for haemolysis on 5–10% sheep blood agar where positive activity is indicated by a zone of clearing around individual colonies after overnight incubation. Tube haemolysis using culture supernatants and 1% sheep erythrocytes have also been recommended. Because of the prevalence of non-haemolytic El Tor strains, other characteristics are also used for biotyping. The majority of El Tor strains differ from classical *V. cholerae* by a positive Voges–Proskauer reaction, resistance to polymyxin B (50 U), agglutination to chicken erythrocytes, and sensitivity to different classes of bacteriophage. Biotyping is only applicable to *V. cholerae* serotype O1.

Serotyping Currently, *V. cholerae* may be divided into 140 different serogroups based on seroagglutination to different O-antigens. Various research groups have developed serotyping schemes for non-O1 strains, although there is no standardized scheme and no commercial source of all antisera. The O1-antigen is further divided into three sub-serotypes designated Ogawa, Inaba and Hikojima. These are grouped according to the structure of the O-antigen, which consists of three antigenic fractions A, B and C. The A antigen is common to all three O1 serotypes, while expression of the B and C antigens allow differentiation into the Ogawa and Inaba serotypes, respectively. The Hikojima serotype expresses all three antigenic fractions but occurs rarely. *Vibrio cholerae* O1 strains have been observed to undergo seroconversion between the Ogawa, Hikojima and Inaba serotypes owing to spontaneous mutations of the *rfb*T gene involved with O-antigen biosynthesis.

Polyvalent O1 antiserum and respective monovalent antisera specific for the Ogawa and Inaba serotypes are available commercially. Specific antisera for the O139 antigen are also available. Serotyping is usually performed by slide agglutination. Non-emulsifiable or rough colonies cannot be used for the seroagglutination test. Isolates reacting with the polyvalent O1 antiserum should be checked against the Ogawa and Inaba monovalent antisera.

Molecular Typing Molecular typing provides highly discriminatory information on clonal and strain variations, origins, and ecology of *V. cholerae* isolates

Table 1 Key phenotypic traits for the identification of *Vibrio cholerae* and differentiation from related *Vibrio* species relevant to food and environmental samples

Test	*Reaction*[a]			
	V. cholerae	V. mimicus	V. parahaemolyticus	V. vulnificus
Oxidase	+	+	+	+
ONPG[b]	+	+	−	+
Arginine dihydrolase	−	−	−	−
Lysine decarboxylase	+	+	+	+
Ornithine decarboxylase	+	+	+	v
Fermentation:				
sucrose	+	−	−	−
lactose	−	−	−	+
arabinose	−	−	+	−
Growth in:				
0% NaCl	+	+	−	−
3% NaCl	+	+	+	+
8% NaCl	−	−	+	−
Voges–Proskauer	v	−	−	−
Lipase (corn oil)	+	−	+	+
O/129 reaction:				
10 μg	S	S	R	R
150 μg	S	S	S	S

[a] v, variable reaction; S, sensitive; R, resistant.
[b] ONPG, *O*-nitrophenyl-β-D-galactopyranoside.

when used in conjunction with epidemiological data during disease outbreaks. Methods such as multilocus enzyme electrophoresis, PCR-amplified DNA polymorphisms, and DNA hybridizations using specific gene probes have been effectively used with *V. cholerae* to expand on the limited discrimination provided by available serotyping and biotyping schemes. More recently, *V. cholerae* strains have been studied using whole-cell DNA restriction fragment length polymorphisms (RFLPs), analysis by pulsed-field gel electrophoresis (PFGE), and automated sequence analysis of specific genes (e.g. *ctx* genes). These techniques were used to confirm the existence of unique clones of toxigenic *V. cholerae* O1 from various global regions, including the El Tor strains indigenous to the USA and Australia.

Rapid Detection Methods

A variety of rapid detection techniques based on immunological and DNA methodology have been developed for *V. cholerae*, some of which have applications for food analysis. Isolates of *V. cholerae* from food can be tested for CT production to indicate potential pathogenicity. Commercial immunological test kits for CT are available and have incorporated reverse passive latex agglutination (RPLA) and enzyme-linked immunosorbent assay (ELISA) techniques.

Fluorescently labelled monoclonal antibodies (FA) targeting the O-antigen have been used in combination with epifluorescence microscopy for the detection and enumeration of *V. cholerae* cells in various samples. Direct detection also allows the identification of viable but non-culturable bacteria. Preincubation of samples with nalidixic acid and yeast extract causes metabolically active cells to become elongated and enlarged, making them easy to identify. Kits incorporating highly specific monoclonal antibodies to the O1-A antigen and the O139 antigen, and labelled with fluorescein isothiocyanate (FITC), have recently become commercially available for the detection of *V. cholerae* O1 and O139, respectively. If contaminating cells are present in low numbers, direct immunological labelling may lack sufficient sensitivity for detection. The detection limit in liquids can be improved by concentrating the sample using filtration, but this may not be possible with food samples. In most cases, it may be necessary to enrich samples before screening with these techniques.

Nucleic acid-based techniques applied to the detection of *V. cholerae* in foods have incorporated gene probes and PCR. The *ctx*AB genes have been the main targets. Whole gene probes for *ctx*A have been employed to screen isolates. However, other *Vibrio* species and bacteria from other genera (e.g. *Escherichia coli*) may possess genes with a high degree of homology to *ctx*, and cross-hybridizations with *V. cholerae* probes can occur. The use of shorter oligonucleotide probes derived from unique regions of *ctx*A has improved specificity. Direct hybridization in foods is limited by the low sensitivity of the technique, as approximately 10^5–10^6 copies of the target sequence are required to obtain a clear positive result.

This is normally achieved by cultural enrichment, which extends the time needed for analysis.

Polymerase chain reaction methods targeting both *ctx*A and *ctx*B provide higher sensitivity. This technique was successful in detecting *ctx*A from crabmeat implicated in a cholera outbreak, and demonstrated that the method could be used for direct testing of certain foods. In a PCR evaluation targeting *ctx*AB, positive detection was achieved in a variety of foods seeded with *V. cholerae* O1 cells. Enrichment of the samples in APW for 6–8 h improved the limit of detection, where 10 *V. cholerae* colony forming units (cfu) per gram of homogenized oysters and 1 cfu per 10 g of lettuce could be detected. The detection limit for inoculated crabmeat without enrichment was 4×10^4 *V. cholerae* cfu g^{-1}. Sensitivity of detection also depends on the food type as some food homogenates, particularly of oysters, inhibit the PCR. This can be improved by increasing the dilution of the sample during the pre-enrichment step.

Regulations and Importance to the Food Industry

Control of *V. cholerae* in foods differs between cholera endemic and non-endemic countries. There is a greater risk of contamination in endemic regions owing to higher levels of *V. cholerae* in the environment and cross contamination from the human reservoir. For endemic countries that export food, care must be taken to ensure proper food handling practices to minimize faecal contamination and human carrier transmission. Harvesting practices can also be monitored to minimize contamination.

Agricultural products irrigated or washed with sewage-contaminated waters should be avoided, or treated to destroy contaminating bacteria. Fish and shellfish products harvested from fresh water and inshore estuarine water which may be subjected to sewage pollution have an increased risk of contamination. Higher contamination levels may occur in seafood harvested during cholera seasonal periods or during months with elevated water temperatures, since *V. cholerae* is likely to be more prevalent in the environment. Normal depuration processes for commercially harvested oysters and other shellfish are ineffective in eliminating contaminating vibrios.

In non-endemic regions, contaminated foods may be imported. An incidence of a commercially imported product causing a small cholera outbreak involving four patients was reported in Maryland, USA in 1991. The implicated product was a single batch of frozen coconut milk manufactured in Thailand. However, commercially imported foods are not considered to be a major public health risk, and reports of product-related outbreaks are rare. The World Health Organization does not support the placing of embargoes on the importation of foods, including seafood and raw vegetables, from cholera-affected countries.

Vibrio cholerae survives for extended periods in certain foods, including vegetables, meat, fish and shellfish, dairy products and cooked cereals. Cooked food is particularly suited to survival, since growth inhibitors and competing flora may be eliminated during cooking. The vibrio can survive up to 14 days on raw vegetables and up to 21 days on cooked vegetables when stored under refrigeration. The organism is sensitive to heat (> 45°C) and irradiation, and shows better survival in foods kept under refrigeration (< 10°C) than at ambient temperatures (28–30°C). It is able to survive for up to 6 months in frozen seafood products. It is important to prevent contamination or to eliminate bacteria in foods intended for storage and export. Normal food manufacturing processes such as pasteurization, irradiation and adequate cooking should ensure that foods are safe. Additionally, *V. cholerae* is sensitive to low water activity and acidic conditions. Survival in dry cereals, grains and biscuits is poor compared with equivalent cooked products with a higher moisture content. Foods containing acidic sauces or marinades and acidic fruit juices may have a reduced level of contamination.

A number of disinfectants have been tested for effectiveness on *V. cholerae*, and may be applied for cleaning food processing surfaces. Peracetic acid was found to be most effective when compared against a quaternary ammonium compound and hypochlorite, a widely used disinfectant. Hypochlorite was effective provided that free proteinaceous material was minimal; 100 p.p.m. of free chlorine for 2 min was shown to reduce viable *V. cholerae* counts by more than $\log_6$. Studies have indicated that *V. cholerae* cells adsorbed to chitinous surfaces were relatively resistant to the action of disinfectants, and significantly higher concentrations for longer periods were required to achieve the same effect.

Importance to the Consumer

In non-endemic countries, cholera cases are often recently acquired infections imported from endemic countries among travellers. In cholera areas, contaminated domestic water sources provide the highest risk of transmission. Affected water sources can include wells, springs, household stored water, and even municipal systems that distribute improperly treated water. Contaminated municipal water was implicated in the spread of cholera in Peru. Beverages prepared and sold by street vendors containing ice have been incriminated as a source of infection. Ice is usually prepared under unhygienic conditions using water from high-risk sources. Fruit juices and other unheated beverages mixed with contaminated water have also been implicated, as have raw vegetables and fruits.

Raw and precooked seafood provide the most common source of food-borne infection. Almost all non-imported cholera cases due to the Gulf Coast O1 strain in the USA have implicated locally caught crabs as the primary vehicle. Adsorption into the chitinous shells of the hosts may protect some *V. cholerae* cells from brief cooking. Viable *V. cholerae* cells can be recovered from crabs boiled for less than 10 min or steamed for less than 30 min. Seafoods often have the appearance and texture of being cooked before internal temperatures are lethal for *V. cholerae*. Two case studies in the USA, in 1991, traced cholera outbreaks to hand-picked cooked crabmeat, and highlighted the risk of contamination from post-cooking processing. In both cases the crabmeat was bought in Ecuador, transported in personal baggage to the USA and eaten without reheating. Consumption of raw shellfish increases the risk of infection. Raw oysters are common vehicles of non-O1 gastroenteritis in the USA. Mussels and cockles have been reported as the sources of cholera in Europe. Consumption of raw or improperly cooked finfish has also been associated with cholera. Ceviche, a marinated raw fish salad, is a popular dish in Latin American countries and is a frequent source of infection.

In general, food-borne transmission occurs rarely in non-endemic areas where toxigenic *V. cholerae* is present in low numbers. Locally produced foods from these areas should be free of contamination, although there may be a higher risk in seafood due to indigenous strains. Adequate cooking and safe food handling practices provide sufficient protection, since a high infectious dose is normally required to cause illness in healthy individuals. Seafood should be consumed immediately after cooking or stored at refrigeration temperatures to prevent bacterial growth. Shellfish intended for raw consumption should be eaten immediately following purchase or harvest, and not left at ambient temperature. These shellfish should only be harvested from environments free of faecal contamination or other pollution. Individuals with predisposing illnesses and gastric abnormalities are more susceptible to *V. cholerae* infection and should avoid higher-risk foods, such as raw oysters.

See color Plates 38 and 39.

See also: ***Vibrio***: Introduction, including *Vibrio vulnificus*, and *Vibrio parahaemolyticus*; Standard Cultural Methods and Molecular Detection Techniques in Foods.

Further Reading

Colwell RR (1996) Global climate and infectious disease: the cholera paradigm. *Science* 274: 2025–2031.

DePaola A (1981) *Vibrio cholerae* in marine foods and environmental waters: a literature review. *Journal of Food Science* 46: 66–70.

Desmarchelier PM (1997) Pathogenic Vibrios. In: Hocking AD, Arnold G, Jenson I, Newton K and Sutherland (eds) *Foodborne Microorganisms of Public Health Significance*. P. 285. Sydney: Trenear.

Huq A and Colwell RR (1995) A microbiological paradox: viable but nonculturable bacteria with special reference to *Vibrio cholerae*. *Journal of Food Protection* 59: 96–101.

International Commission on Microbiological Specifications for Foods (ICMSF) (1996) *Vibrio cholerae*. In: Roberts TA, Baird-Parker AC and Tompkin RB (eds) *Microorganisms in Foods 5: Characteristics of Microbial Pathogens*. P. 414. London: Blackie.

Janda JM, Powers C, Bryant RG and Abbot SL (1988) Current perspectives on the epidemiology and pathogenesis of clinically significant *Vibrio* spp. *Clinical Microbiology Reviews* 1: 245–267.

Morris JG (1990) Non-O group 1 *Vibrio cholerae*: a look at the epidemiology of an occasional pathogen. *Epidemiologic Reviews* 12: 179–191.

Olsen JE, Aabo S, Hill W et al (1995) Probes and polymerase chain reaction for the detection of food-borne bacterial pathogens. *International Journal of Food Microbiology* 28: 1–78.

Pollitzer R (1959) *Cholera*. Monograph 43. Geneva: World Health Organization.

Wachsmuth IK, Blake PA and Olsvik Ø (eds) (1994) *Vibrio cholerae and Cholera: Molecular to Global Perspectives*. Washington: American Society for Microbiology Press.

West PA and Colwell RR (1984) Identification and classification of *Vibrionaceae*: an overview. In: Colwell RR (ed.) *Vibrios in the Environment*. P. 285. New York: John Wiley.

Standard Cultural Methods and Molecular Detection Techniques in Foods

Kasthuri Venkateswaran, Jet Propulsion Laboratory, National Aeronautics and Space Administration, Pasadena, USA

The Genus Vibrio

As presently defined in *Bergey's Manual of Systematic Bacteriology* the genus consists of 26 species, all of which have been isolated from or are associated with the aquatic environment. This genus received much attention in recent years; additional descriptions have been made, resulting in 41 currently recognized

Table 1 Differential characteristics of the genus *Vibrio* (Reproduced from Baumann and Schubert 1984)

Characteristics	*Reactions that are in the species of:*			
	Vibrio	Photobacterium	Aeromonas	Plesiomonas
Sheathed polar flagella	+[a]	–	–	–
Accumulation of poly-β-hydroxybutyrate couple with the inability to utilize β-hydroxybutyrate	–	+		
Sodium ion is required for growth or stimulates growth	+	+	–	–
Production of lipase	[+][b]	D[c]	+	–
Utilization of D-mannitol	[+][d]	–	[+]	–
Mol% G+C of DNA	38–51	40–44	57–63	51

[a]+, All strains positive; –, all strains negative.
[b] *V. nereis*, *V. anguillarum* and *V. costicola* are negative for this trait.
[c] Some species positive and some are negative.
[d] *V. nereis* and *V. anguillarum* are negative for this trait.

species. On the basis of rRNA phylogenetic data, the species that are pathogenic to fish or shellfish, such as *Vibrio anguillarum*, *V. aestuarianus*, *V. ordalii*, *V. pelagius* and *V. tubiashii* have been transferred into the genus *Listonella*. Likewise, luminous counterparts of vibrios, such as *V. damselae*, *V. fischeri*, *V. iliopiscarium*, and *V. logei*, are moved into the genus *Photobacterium*.

Definition

Vibrios are straight or curved rods, 0.5–0.8 µm in width and 2–3 µm in length. They do not form endospores or microcysts, are Gram-negative and motile by monotrichous or multitrichous polar flagella which are enclosed in a sheath continuous with the outer membrane of the cell wall. They are facultative anaerobes capable of both fermentative and respiratory metabolism. Sodium ions stimulate the growth of all species and are an absolute requirement for most. Most species are oxidase positive. They are very common in marine and estuarine environments and on the surfaces and in the intestinal contents of marine animals. The mol% G+C of the DNA is 38–51. Differential characteristics of the genus *Vibrio* from other related genera are shown in (**Table 1**).

Members of the genera *Vibrio* and *Listonella* can be categorized into six groups: (1) those that are capable of growth in 10% NaCl; (2) those that are pathogenic to humans; (3) those that are pathogenic to fish or shellfish; (4) those that are luminous; (5) those that are decarboxylase and dihydrolase negative; and (6) other species not listed above. *Vibrio* and *Listonella* species falling under these categories are listed in **Table 2**. Species that are originally described as members of the genus *Photobacterium* are not considered here.

Table 2 Categorization of known *vibrio* species

Category	Vibrio *species that are:*	*Taxon*
1	Capable of growing in 10% NaCl	*V. alginolyticus* *V. harveyi* *V. nereis*
2	Pathogenic to humans	*V. campbellii* *V. cholerae* *V. cincinnatiensis* *V. fluvialis* *V. hollisae* *V. metschinikovii* *V. mimicus* *V. parahaemolyticus* *V. vulnificus*
3	Pathogenic to fish or shellfish	*L. aestuarianus* *L. anguillarum* *L. ordalii* *L. tubiashii* *P. damselae* *V. carchariae* *V. pectenicida* *V. salmonicida* *V. scophthalmi*
4	Luminous	*P. fischeri* *P. logei* *V. orientalis*
	Decarboxylase and dihydrolase production negative	*L. pelagia* *V. gazogenes* *V. nigripulchritudo* *V. splendidus*
6	Not listed in the above group but of biotechnological importance	*P. iliopiscarium* *V. diabolicus* *V. diazotrophicus* *V. furnissi* *V. halioticoli* *V. mediterranei* *V. natriegens* *V. proteolyticus* *V. rumoiensis* *V. shiloi*

V, *Vibrio*; *L*, *Listonella*; *P.*, Photobacterium.

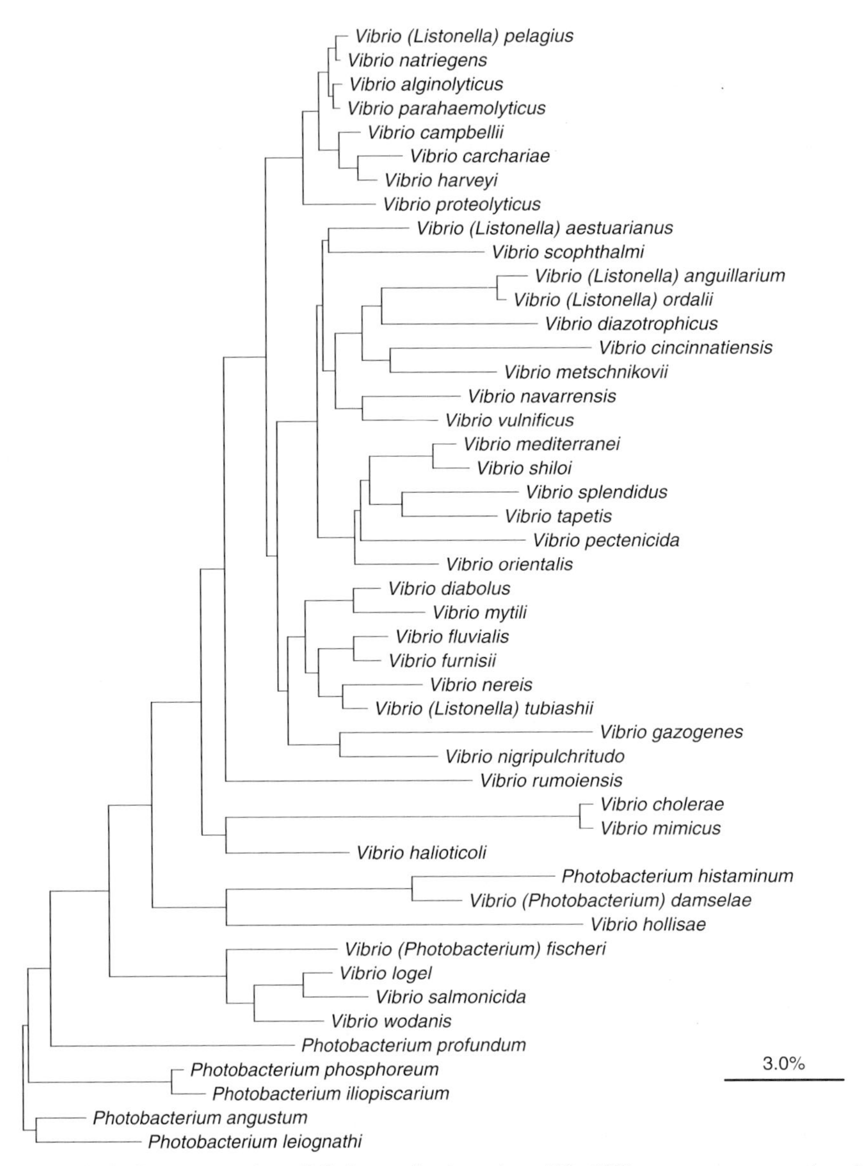

Figure 1 Phylogenetic tree including type strains of *Vibrio* species based on 16S rDNA sequence comparison by neighbour-joining method.

Phylogeny

A phylogenetic tree based on the 1.4 kb nucleotide sequences of the 16S rDNA of various *Vibrio* species covering base positions 11–1492 (*E. coli* numbering) is shown in **Figure 1**. The neighbour-joining and maximum-likelihood analysis indicates that the luminous counterpart *Photobacterium* (including luminous *Vibrio* species), form a group that is independent of the others. In this phylogeny, pathogenic strains did not group together; neither human pathogens nor fish pathogens formed separate identities. It is clear from these results that pathogenicity does not evolve with its species divergence. In addition, many of these phenotypically distinct species indeed showed close relationships with each other and they cannot be distinguished based on 16S rDNA sequences (see Molecular diagnosis of *V. parahaemolyticus* below).

Enrichment Media

Chemical composition and preparation of various enrichment media widely used in the isolation of vibrios are given (**Table 3**). High pH is used selectively to isolate vibrios and in some case, the resistance to polymyxin B is also exploited. High alkalinity might be harsh in the isolation of vibrios in some processed foods where the target organisms need to be resuscitated. In order to suppress the growth of competitive microflora, sodium cholate is used in the Monsur peptone water for the enrichment of *V. cholerae* from water and faecal samples. However, alkaline peptone

Table 3 Chemical constituents of various enrichment media used in the isolation of *Vibrio*

	Amounts (g l^{-1}) in the agar media of:			
Composition	*Alkaline peptone water*	*Polymyxin broth*[a]	*Monsur peptone water base*[b]	*AG medium*[a]
Yeast extract	5.0	3.0		1.0
Peptone	10.0	10.0	10.0	1.0
Soya peptone			2.0	
Sodium chloride	10.0	20.0		20.0
Sodium cholate			3.0	
Ammonium citrate				1.0
Arabinose				5.0
Sodium gluconate				3.0
Sodium carbonate			1.0	
Polymyxin B		250 000 U		20 p.p.m.
pH	8.6	7.4	9.2	8.5

[a]Heat to dissolve and sterilize at 121°C for 15 min. Avoid overheating.
[b]Heat to dissolve the medium and sterilize at 121°C for 15 min. Add 5 ml of 0.1% potassium tellurite solution to 1000 ml of the basal medium.

water (APW) is widely used in the isolation of vibrios in many different types of samples.

A simple, pad pre-enrichment technique to enumerate members of the family Vibrionaceae is also used. This procedure involves a short 6-h pre-enrichment incubation in APW to resuscitate starved or dormant cells followed by placement of the filtered membranes onto selective media. Briefly, after samples filtered in appropriate membrane filters (47 mm diameter; 0.2 μm pore size), the filter papers are placed onto sterile pads of the same size that are pre-soaked in APW and incubated for a 6-h period. Then the membrane filters are transferred onto a suitable agar medium for further incubation period. Using this resuscitation procedure, a 20% increase in the recovery of *Vibrio* sp. is reported when compared with the conventional membrane filtration procedure. Likewise, a 27% and 36% increase in *Vibrio* sp. recovery is noted for cold- (–20°C) and heat-shocked (48°C) cells, respectively.

Enumeration Media

Chemical composition and preparation of various agar media are tabulated (**Table 4**). In thiosulfate citrate bile salt (TCBS) agar, sucrose fermentative vibrios such as *V. cholerae* and *V. alginolyticus* form the turbid yellow colonies, whereas sucrose non-fermentative bacteria such as *V. parahaemolyticus* form green colonies with a blue centre. Faecal streptococcus and *Proteus* grew sometimes, but their colonies were very small and could be easily differentiated. Sodium cholate suppressed the growth of Gram-positive bacteria and bile salts inhibited most of the Gram-negative bacteria other than vibrios, thus allowing only vibrios to grow. In *Vibrio* agar, sucrose fermentative vibrios form blue colonies and non-fermentative vibrios form slightly reddish and translucent colonies. Sodium taurocholate and sodium lauryl sulphate suppressed growth of non-vibrios and the addition of the indicators such as cresol red and water blue indicated the colony colour change due to the fermentation of sucrose. In bromothymol blue (BTB) teepol agar, sucrose non-fermentative vibrios form blue–green colonies whereas fermentative vibrios form yellow colonies. The BTB teepol agar is the least sensitive among the three agar media described here, however non-vibrioid colonies form smaller size thus enumeration

Table 4 Chemical constituents of various agars[a] used in the isolation of *Vibrio*

	Amounts (g l^{-1}) in the agar media of:		
Composition	*TCBS agar*	*Vibrio agar*	*BTB Teepol agar*
Yeast extract	5.0	5.0	
Beef extract			5.0
Peptone	10.0	5.0	10.0
Sucrose	17.0	12.5	20.0
Sodium thiosulphate	10.0	8.5	
Sodium citrate	10.0	8.0	
Sodium taurocholate		5.0	
Sodium lauryl sulphate		0.2	
Disodium phosphate		7.5	
Sodium cholate	3.0		
Ferric citrate	1.0	3.0	
Sodium chloride	10.0	10.0	30.0
Oxgall (Bile salts)	5.0		
Teepol			2.0 ml
Bromo thymol blue	0.04		0.08
Thymol blue	0.04		
Cresol red		0.02	
Water blue		0.2	
Agar	15.0	15.0	15.0

[a]Do not autoclave. Heat to dissolve. Avoid overheating.
TCBS, thiosulfate citrate bile salts agar; BTB Teepol, bromothymol blue Teepol agar.

Table 5 Characteristics of pathogenic *Vibrio* species (Reproduced from Oliver and Kaper 1997)

	Reactions that are in the species of:									
Characteristics	V. alginolyticus	V. cholerae	V. fluvialis	V. furnissii	V. hollisae	V. metschnikovii	V. mimicus	V. parahaemolyticus	V. vulnificus	P. damselae
Oxidase	+[a]	+	+	+	+	–	+	+	+	+
Indole	±	+	–	±	+	±	+	+	+	–
Voges–Proskauer	+	±	–	–	–	+	–	–	–	+
Simmons citrate	±	±	+	+	–	+	±	+	+	–
Lysine	+	+	–	–	–	±	+	+	+	±
Ornithine decarboxylase	±	+	–	–	–	–	+	+	+	–
Arginine dihydrolase	–	–	+	+	–	±	–	–	–	+
Fermentation of:										
Sucrose	+	+	+	+	–	+	–	–	–	–
Lactose	–	(+)	–	–	–	±	(+)	–	+	–
L-Arabinose	–	–	+	+	+	–	–	+	–	–
D-Mannitol	+	+	+	+	–	+	+	+	±	–
Maltose	+	+	+	+	–	+	+	+	+	+
Cellobiose	–	–	±	±	–	–	–	–	+	–
Salicin	–	–	–	–	–	–	–	–	+	–
Gas from glucose	–	–	–	+	–	–	–	–	–	+
Nitrate to nitrite	+	+	+	+	+	–	+	+	+	+
Gelatinase	+	+	+	+	–	+	+	+	+	–
Growth in:										
0% NaCl	–	+	–	–	–	–	+	–	–	–
3% NaCl	+	+	+	+	+	+	+	+	+	+
6% NaCl	+	–	+	+	+	+	–	+	±	+
10% NaCl	+	–	–	–	–	–	–	–	–	–

+, Greater than 90% of strains positive; ±, variable reaction, predominant reaction is shown as the numerator; –, greater than 90% of strains negative; (+), positive reaction delayed ⩾3 days.

of vibrios is made easy. However, TCBS is widely used in the isolation of vibrios in many different types of samples.

Diagnosis of Pathogenic Vibrios

V. cholerae O1, *V. cholerae* non-O1, *V. mimicus*, *V. parahaemolyticus*, *V. fluvialis*, *V. furnisii*, *V. hollisae*, *V. vulnificus*, *V. alginolyticus*, *V. damselae* and *V. metschnikovii* are frequently implicated as etiological agents in human disease. Diagnosing such a wide variety of *Vibrio* species in a single given medium is impossible. But, all these pathogenic vibrios are able to grow in TCBS agar. Hence, stools and specimens collected from patients are directly streaked onto TCBS agar. If patients carried these pathogens, isolation of any of these vibrios is not difficult. Many commercially available kits would broadly classify these vibrios into groups such as *V. parahaemolyticus*-like or *V. cholerae*-like organisms. However, identification of the vibrios to the species level is time-consuming, labour-intensive and expensive. Phenotypic characteristics of pathogenic *Vibrio* species are given in **Table 5**.

Serology

The phenotypically similar species might be differentiated on the basis of their serological properties. Somatic 'O' antigens and flagellar 'H' antigens are widely used. The key confirmation for identification of *V. cholerae* O1 is agglutination in polyvalent antisera raised against the O1 antigen. Polyvalent antiserum for *V. cholerae* O1 is commercially available and can be used in slide agglutination or coagglutination tests. A monoclonal antibody-based coagglutination test suitable for testing isolated colonies or diarrhoeal stool samples has recently become available commercially. However antisera against the other 138 known serogroups of *V. cholerae* are not commercially available. Likewise, antisera against *V. parahaemolyticus*, *V. vulnificus*, *V. anguillarum* and *V. salmonicida* are available only in reference laboratories.

Commercial Kits and Collaborative Study

There is no commercial kit available for the detection of *Vibrio* species. Although some kits are available for the specific detection of *V. cholerae* O1 (Bead-ELISA, Nissui, Tokyo, Japan), its commercialization has stopped because of the recent advances in the polymerase chain reaction (PCR)-based methodologies. In addition, no collaborative study was carried out in the detection of *Vibrio* species except for the recovery of *V. cholerae* from oysters in 1988.

DNA Probes and PCR Techniques

The PCR assay is a useful detection method because of the demonstrated combination of speed and sensitivity, both of which are critical to any assay for the detection of bacteria. In addition to increased sensitivity, the use of unique oligonucleotide primers based on the sequence of the DNA probe also results in absolute specificity. A detailed description of the methodology can be found in the Further Reading list. Among human pathogenic vibrios, *V. cholerae*, *V. parahaemolyticus* and *V. vulnificus* are best described for their pathogenicity.

A number of DNA fragment probes and synthetic oligonucleotide probes have been developed to detect cholera toxin-encoding (*ctx*) gene. However, *ctx* gene is reported in *Vibrio* species other than *V. cholerae* such as *V. mimicus*, *V. cholerae* non-O1, *V. furnisii* and *V. hollisae*. *V. vulnificus* produces a large number of extracellular compounds, the most studied of which is a potent heat-stable haemolysin/cytotoxin. The gene-encoding the cytotoxin is used as a suitable molecular probe in the isolation and identification of *V. vulnificus*. However, cross-reaction of this gene in other *Vibrio* species is documented.

Pathogenic *V. parahaemolyticus*, which causes acute gasteroenteritis after the consumption of raw or partially cooked seafoods, has been known to produce thermostable direct haemolysin (TDH) or TDH-related haemolysin (TRH) or both TDH and TRH. The TDH produced by *V. parahaemolyticus* strains, has long been thought to be a virulence factor for humans because the preponderance of isolates from clinical samples do elaborate this molecule whereas the haemolysin is rarely produced by strains isolated from the environment. Of 214 clinical strains of *V. parahaemolyticus* tested by colony hybridization, 52% harboured only the *tdh* gene, 24% contained the *trh* gene, 11% encoded both genes, and about 17% had unknown virulence factors. This suggests that the role of *V. parahaemolyticus* haemolysins in human disease has not been firmly established. Hence, both the US Food and Drugs Administration (USFDA) and Japanese standards allows zero-tolerance of *V. parahaemolyticus* in foods. Thus, from the view of public health, it is necessary to identify *V. parahaemolyticus* irrespective of its virulence factors. A unique genetic marker described in detail later should recognize all vibrios identified as *V. parahaemolyticus* by conventional methods, but should not recognize the other bacteria.

The following section gives a specific example in detecting *Vibrio* species and compares the suitability of rapid methodologies in the detection of *V. parahaemolyticus* with conventional culture-based techniques.

Detection of *V. parahaemolyticus* in Foods

Conventional Assay

The procedure outlined here is detailed in *Bacteriological Analytical Manual* by USFDA (**Fig. 2**). A three-tube most-probable number (MPN) procedure using APW as an enrichment medium, incubated at 35°C for 18 h followed by streaking on TCBS agar is used for the enumeration of *V. parahaemolyticus*. All colonies appearing on TCBS can be considered as total vibrios. Green colonies should be transferred onto VP medium and strains that showed alkaline slant and acid butt in VP medium may be considered as *V. parahaemolyticus*-like organisms. Isolates that: showed positive reaction for oxidase, gelatinase, lysine and ornithine decarboxylases, indole production, motility, O/129 pteridine sensitivity, nitrate reduction; produced acid from D-mannose; grew in 3–8% NaCl and at 42°C; showed negative reaction for arginine dihydrolase, Voges–Proskauer reaction; did not grow in 0 and 10% NaCl; and did not produce acid from L-arabinose, myo-inositol, and sucrose are identified as *V. parahaemolyticus* (Table 5).

Fluorogenic Assay

In general, the detection of *V. parahaemolyticus* in foods involved a pre-enrichment step (APW or polymyxin broth) followed by streaking onto a selective agar (TCBS or BTB–teepol agars). Unfortunately, a number of other *Vibrio* species mimic *V. parahaemolyticus*, necessitating the use of additional biochemical tests for reliable identification. Hence, a more rapid and sensitive detection assay is warranted. A procedure that involves 6-h cultivation of cells in a specific medium (AG medium) followed by measuring the intracellular trypsin-like activity of *V. parahaemolyticus* is described (**Fig. 3**). Measurement of fluorescence intensity of the culture supernatant solution can be made with a fluorospectrometer. The excitation wavelength is 360 nm, and the fluorescence intensity is measured at 450 nm. Media, buffer and substrate controls should be included in the assay and fluorescence intensity is measured. One unit of trypsin-like activity is defined as the fluorescence intensity that is excited by 1 μg of trypsin (bovine pancreas origin). When the fluorescence intensity of the test sample is more than 1 trypsin unit, trypsin-like activity is recorded as positive for that sample. A typical negative control will range from 0 to 0.5; and

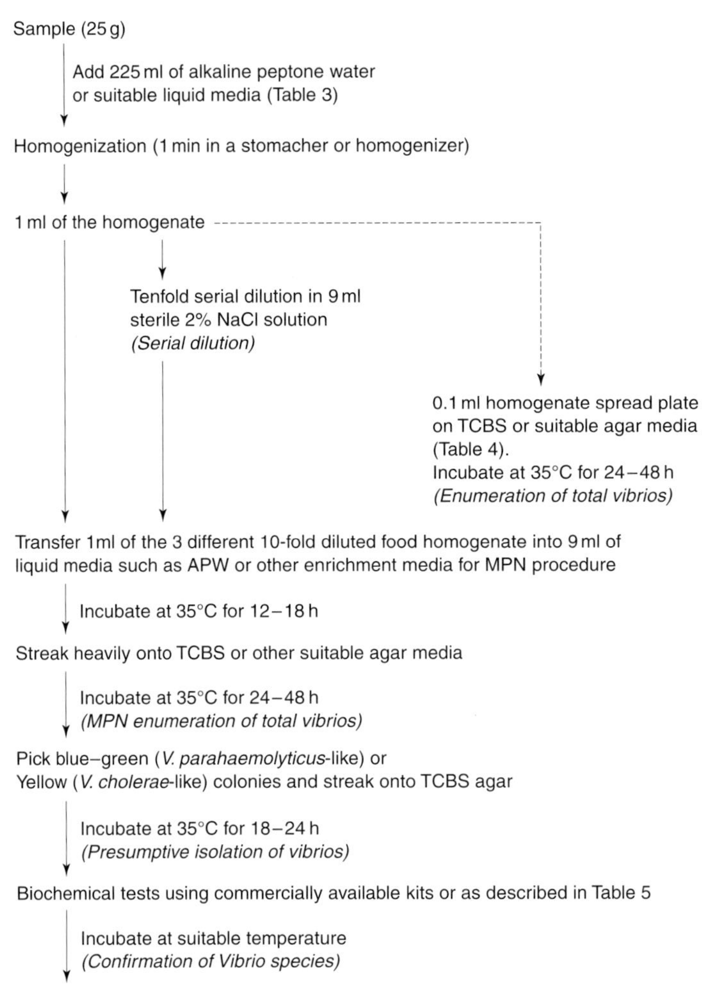

Figure 2 Conventional methodology in the isolation and identification of vibrios.

a positive control would exceed more than 4–6 trypsin units.

Molecular Detection of *V. parahaemolyticus*

Among *V. parahaemolyticus*, various phenotypes, serotypes and toxin-producing strains were reported. Specific haemolysin probes will not detect all types of *V. parahaemolyticus*, but to prevent *V. parahaemolyticus* food poisoning, it is necessary to detect all types of *V. parahaemolyticus* strains in foods and in their environment. As 16S rRNA sequences revealed 99.7% homology between *V. parahaemolyticus* and *V. alginolyticus* it is impossible to identify *V. parahaemolyticus* by any means except the time-consuming conventional methods. As no universal probe is available to differentiate *V. parahaemolyticus* from other related species, the use of *gyr*B gene that encode the subunit B protein of DNA gyrase (topoisomerase type II) as targets of highly specific probes was evaluated. A suitable PCR primer set has been designed that could amplify only the *gyr*B fragment of *V. parahaemolyticus* to specifically identify the pathogen irrespective of its phenotypes, serotypes and virulence factors. The procedure for application of *gyr*B primers in the detection of *V. parahaemolyticus* directly from food using a PCR protocol is shown in **Figure 4**. Briefly, the bacterial community from an APW-enriched sample is removed and washed in sterile 2% NaCl solution. Without extracting DNA from the samples, using a suitable primer set (**Table 6**), R is carried out to amplify the 285-bp *gyr*B fragment for the identification of *V. parahaemolyticus*.

Limitations of PCR

The sensitivity of detection in food samples has been low, because only small samples (10–100 μl) can be analysed, and many sorts of food contain substances that are inhibitory to PCR. Such PCR inhibitory substances have been reported in many clinical samples such as urine, blood, sputum, faecal specimens, food and environmental samples. However, with bacteria in food, the sensitivity of the PCR was far lower than that with bacteria in saline. With bacteria in food, the lower detection limit was higher than the number of

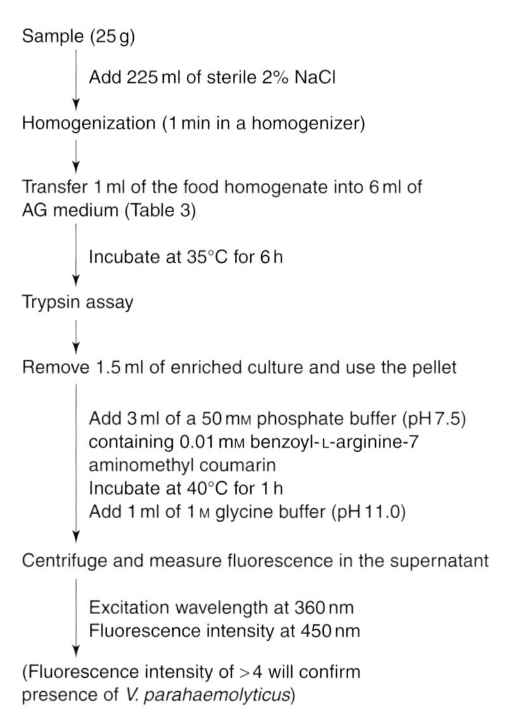

Figure 3 Fluorogenic assay in the identification of *V. parahaemolyticus.*

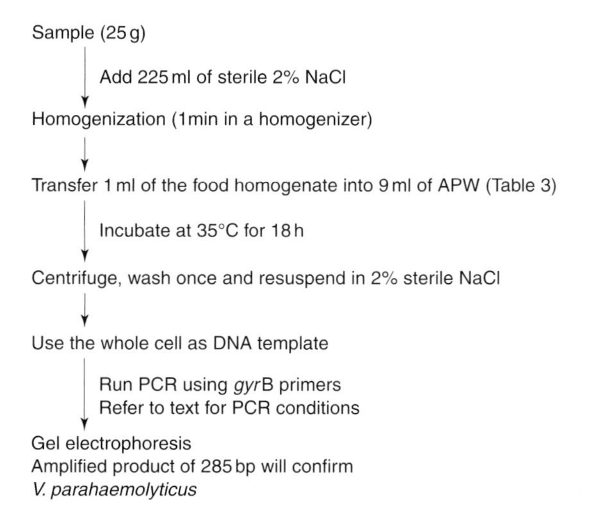

Figure 4 Molecular detection of *V. parahaemolyticus*.

Table 6 *gyr*B PCR primers in the identification of *V. parahaemolyticus*

Specificity	*Primer*	*Oligonucleotide sequence*	*No. of bp*
Universal primers	UP-1	GAA GTC ATC ATG ACC GTT CTG CAY GCN GGN GGN AAR TTY GA	41
	UP-2r	AGC AGG ATA CGG ATG TGC GAG CCR TCN ACR TCN GCR TCN GTC AT	44
V. parahaemolyticus	VP-1	CGG CGT GGG TGT TTC GGT AGT	21
	VP-2r	TCC GCT TCG GGC TCA TCA ATA	21

colony forming units per unit volume of food, which is usually found in processed food. Previous studies suggested extraction of DNA or application of chemicals and physical procedures to remove PCR inhibitory products. However, a simple pre-enrichment in a non-specific medium is applied to successfully remove any possible PCR inhibitory substances as well as proliferation of target bacteria.

Experimental Results

The bacterial strains (135 isolates) tested consist of the members of the genera *Vibrio* (36 species), *Alteromonas* (nine species), *Marinomonas* (two species), and others (nine species). Among 56 bacterial species tested for trypsin-like activity, *V. parahaemolyticus*, *V. alginolyticus*, *V. harveyi*, *V. mytili*, *V. carchariae* and *A. hydrophila* exhibited trypsin-like activity (4–5 trypsin units) when grown in trypticase–soy broth or marine broth. When bacteria were enriched in AG medium (Table 3), all *V. parahaemolyticus* (54 strains; 4–6 trypsin units), and 19 of 20 (95%) *V. alginolyticus* strains (4–5 trypsin units) showed both growth and trypsin-like activity whereas growth of *V. harveyi*, *V. mytili*, *V. carchariae* and *A. hydrophila* was suppressed. It could be inferred from this study that direct detection of *V. parahaemolyticus* from food was not possible because of the cross-reaction exhibited by *V. alginolyticus*. In TCBS, *V. parahaemolyticus* exhibit green coloration and *V. alginolyticus* show yellow coloration by producing acid from sucrose. Thus, after selective isolation of bacterial strains based on the colony coloration, *V. parahaemolyticus* could be identified by this simple and rapid method.

Log-phase *V. parahaemolyticus* ATCC 17802 cells grown in AG medium are resuspended at a concentration of $10^9 \times 1$ cfu ml^{-1} in 25 ml of tuna fish homogenate microcosms (1 g food ml^{-1}) and incubated at 35°C for 6 h by reciprocal shaking. The pre-enriched tuna fish homogenates are washed by centrifugation in sterile phosphate buffered saline and trypsin-like activity was measured (Fig. 3). The TCBS agar was used to count viable cells. The results showed that in AG medium a minimum of 2.7×10^4 cfu ml^{-1} *V. parahaemolyticus* cells was necessary as inoculum to attain a 10^6 cfu ml^{-1} density in 6 h to exhibit any detectable trypsin-like activity.

In summary, the fluorogenic assay is a simple and rapid methodology in the detection of *V. parahaemolyticus*. Trypsin-like activity was noticed in all *V. parahaemolyticus* but *V. alginolyticus* cross-reacted. Though there were some discrepancies, *V. parahaemolyticus* could be differentiated from other bacteria by fluorogenic assay after isolating the culture (sucrose-negative colonies) in a suitable selective agar medium. However, because of high false-positive results observed, direct detection of *V. parahaemolyticus* was not possible from seafoods using fluorogenic assay.

Designing *V. parahaemolyticus*-specific PCR primers

Whole sequences of about 1258 bp of *gyr*B fragment of *V. parahaemolyticus* ATCC 17802 and *V. alginolyticus* ATCC 17749 showed that the frequency of base substitutions in the published 16S rRNA was lower than that in *gyr*B. For example, between the sequences of *V. parahaemolyticus* and *V. alginolyticus*, 166 base substitutions were observed in *gyr*B whereas only five base substitutions were observed in 16S rRNA. The homology of the *gyr*B sequence between *V. parahaemolyticus* and *V. alginolyticus* is 86.8% as against 99.7% in the 16s rRNA sequence.

A species specific primer set to a length of 21 bp each was designed to specifically detect and differentiate *V. parahaemolyticus* from other bacteria. A forward primer with a nucleotide position of 75–95 (VP-1) and an antisense primer with a position of 321–341 were synthesized. When these primers were used to generate 267 bp PCR products, *V. parahaemolyticus* could be differentiated from *V. alginolyticus*. However, type strains of *V. harveyi*, *V. natriegens*, *V. mytili*, *Listonella pelagia* and *Pseudoalteromonas undina* also showed a positive amplification of 267 bp fragment. By comparing the *gyr*B nucleotide sequences of these five strains together with *V. parahaemolyticus* and *V. alginolyticus* sequences, a variable region between nucleotide position 339 and 359 of the *gyr*B fragment of *V. parahaemolyticus* comprising a 21 bp oligonucleotide was synthesized as an antisense primer (VP-2r). The nucleotide sequence of each primer is presented in Table 6. Amplification of *V. parahaemolyticus*-specific 285 bp was performed by using 30 cycles consisting of 1 min at 94°C, 1.5 min at 58°C, 2.5 min at 72°C with a final extention step at 72°C for 7 min.

Specificity of PCR Primers in the Detection of *V. parahaemolyticus*

Type strains of all 38 species of the genus *Vibrio* were examined for the *V. parahaemolyticus*-specific 285 bp PCR amplification. A specific band at 285 bp was noticed only for *V. parahaemolyticus* and no other bacterial species have showed any band. From this it could be said that the primers that have been developed and described are specific to *V. parahaemolyticus* and shall be applied to the molecular diagnosis of this pathogen. A total of 267 strains comprising 72 different species were checked for both 1.2 kb *gyr*B gene and *V. parahaemolyticus*-specific 285 bp fragments. Strains of the genera *Vibrio*, *Listonella*, *Photobacterium*, *Alteromonas*, *Pseudoalteromonas*, *Delaya*, *Marinomonas*, *Shewanella*, *Aeromonas*, *Escherichia*, *Pseudomonas*, *Salmonella*, *Shigella* and *Staphylococcus* showed the presence of 1.2 kb *gyr*B fragment but *V. parahaemolyticus*-specific 285 bp fragment was not observed. Twenty *V. alginolyticus* strains tested did not show any *V. parahaemolyticus*-specific fragment. *V. parahaemolyticus* strains isolated from various environments, food and clinical sources consisting of various phenotypes, serotypes and toxigenic properties were tested for PCR assay. All 117 *V. parahaemolyticus* strains exhibited 285 bp fragment, when tested under the PCR conditions described here.

Molecular Detection of *V. parahaemolyticus* in Artificially Contaminated Food

V. parahaemolyticus ATCC 17802 was grown in APW overnight at 35–37°C, serially diluted using food homogenate (1 g food per millilitre of APW) as diluent to get a final concentration in the range 0–10^9 cfu g^{-1}. These artificially contaminated food homogenate microcosms (10 ml each) were incubated at 35–37°C with shaking (140 r.p.m., rotary shaking). Subsampling (1 ml) was carried out after 0, 6 and 18 h (overnight) incubation. The subsamples were centrifuged (4°C; 10 000 ***g*** for 10 min) and resuspended in sterile 2% NaCl. A 10 μl sample suspension was used as template for PCR assay without extracting DNA. Suitable controls such as buffer, media and PCR reaction mixtures were used to check any false-positive reactions (Fig. 4).

The sensitivity of the PCR assay for detecting artificially contaminated *V. parahaemolyticus* in shrimp is presented in **Figure 5**. Absence of *V. parahaemolyticus*-like organism in the test sample was confirmed by both the conventional APW enrichment method (Fig. 2) and by PCR assay (Fig. 5, lane 1). When the food homogenate was incubated for 18 h in alkaline peptone water at 35–37°C, an initial inoculum of 1.5 cfu *V. parahaemolyticus* per gram of food homogenate amplified the desired PCR product (Fig. 5, lane 2). At time-zero, 1.5×10^5 cfu *V. parahaemolyticus* per gram of shrimp homogenate did not yield any PCR products (Fig. 5, lane 3). The detection of *V. parahaemolyticus* directly from food sample was possible by the combination of 18 h enrichment in APW and PCR assay even when *V. parahaemolyticus* and other competitive food microflora were present at a ratio of $1:10^2$.

A schematic diagram describing various steps involved, time taken in the conventional technique(s), fluorogenic assay and molecular methodologies for the detection of *V. parahaemolyticus* is given in

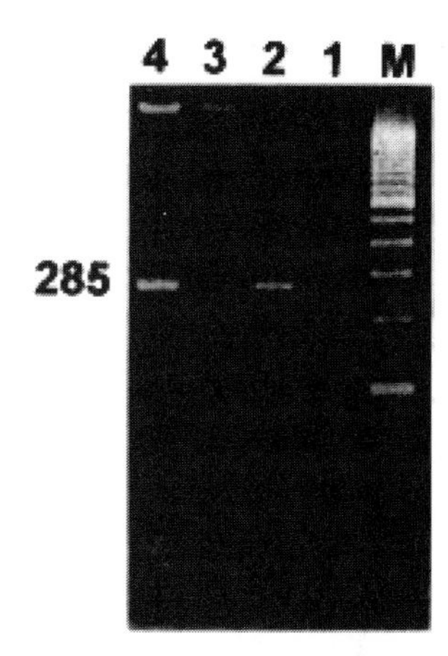

Figure 5 Detection of *V. parahaemolyticus* in artificially contaminated shrimp using VP-1 and VP-2r PCR primers (reproduced from Venkateswaran et al 1998). *V. parahaemolyticus* cells grown overnight in alkaline peptone water are serially diluted in shrimp homogenate (see details in text) to obtain appropriate dilutions. M, 100 bp DNA ladder; lane 1, shrimp homogenate having no *V. parahaemolyticus* cells; lane 2, an initial inoculum of 1.5 cfu g^{-1} *V. parahaemolyticus* is added in shrimp homogenate and incubated for 18 h at 35°C; lane 3, an initial inoculum of 1.5 × cfu g^{-1} *V. parahaemolyticus* is added in shrimp homogenate sampled at 0 h; and lane 4, 1.5 cfu *V. parahaemolyticus* cells prepared in PCR reaction mixture.

Figure 6. The *gyr*B-based PCR technique allows the isolation of the pathogen in 24 h compared to 3–7 days by the conventional technique. Although fluorogenic assay is rapid, cross-reaction shown by related species does not allow this procedure to be used for the direct detection of pathogens in foods.

The PCR-based procedure is a rapid, sensitive, specific, and reliable method for the detection of *V. parahaemolyticus* in foods. The facts that this technique allows detection of genetic potential and differentiation from related species may make it useful as both a screening test and a confirmatory test. The data provided could yield additional information useful to epidemiological studies.

See also: **Biochemical and Modern Identification Techniques**: Food-poisoning Organisms. **Enzyme Immunoassays**: Overview. **Hydrophobic Grid Membrane Filter Techniques (HGMF)**. **Immunomagnetic Particle-based Techniques**: Overview. **PCR-based Commercial Tests for Pathogens**. **Sampling Regimes & Statistical Evaluation of Microbiological Results**. **Water Quality Assessment**: Modern Microbiological Techniques.

Further Reading

Baumann P and Schubert RHW (1984) Family II. *Vibrionaceae* Veron 1965, 5245. In: Krieg NR and Holt JG (eds) *Bergey's Manual of Systematic Bacteriology*, vol. 1, p. 516. Baltimore: Williams & Wilkins.

Blake PA, Weaver RE and Hollis DG (1980) Diseases of humans (other than cholera) caused by vibrios. *Annual Review of Microbiology* 34: 341–367.

Honda T and Iida T (1993) The pathogenicity of *Vibrio parahaemolyticus* and the role of the thermostable direct hemolysin and related hemolysins. *Reviews in Medical Microbiology* 4: 106–113.

Joseph SW, Colwell RR and Kaper JB (1982) *Vibrio parahaemolyticus* and related halophilic vibrios. *Critical Reviews in Microbiology* 10: 77–124.

Kaper JB, Remmers EF and Colwell RR (1980) A medium for presumptive identification of *Vibrio parahaemolyticus Journal of Food Protection* 43: 936–938.

Oliver JD (1989) *Vibrio vulnificus*. In: Doyle MP (ed.) *Foodborne Bacterial Pathogens*, p. 569. New York: Marcel Decker.

Oliver JD and Kaper JB (1997) *Vibrio* species. In: Doyle MP, Beuchat LR and Montville TJ (eds) *Food Microbiology: Fundamentals and Frontiers*, p. 228. Washington, DC: ASM Press.

Popovic T, Fields PI and Olsvik O (1994) Detection of cholera toxin genes. In: Wachmuth IK, Blake PA and Olsvik O (eds) *Vibrio cholerae and Cholera: Molecular to Global Perspectives*, p. 41. Washington, DC: ASM Press.

Sambrook J, Fritsch EF and Maniatis T (1989) *Molecular Cloning: a Laboratory Manual*, 2nd edn. Cold Spring Harbor, NY: Cold Spring Harbor Laboratory.

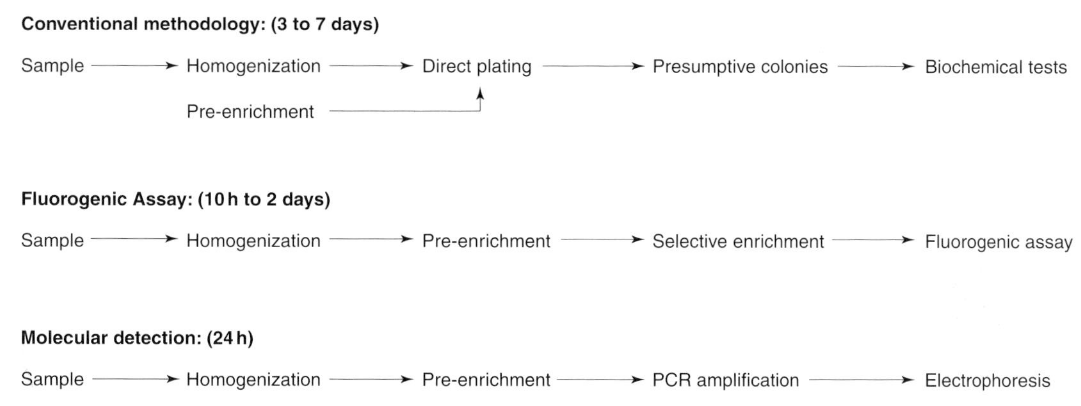

Figure 6 Schematic diagram in the detection of *V. parahaemolyticus* by various methodologies.

Twedt RM (1989) *Vibrio parahaemolyticus*. In: Doyle MP (ed.) *Foodborne Bacterial Pathogens*, p. 543. New York: Marcel Decker.

US Food and Drug Administration (1995) *Bacteriological Analytical Manual* 8th edn, Arlington, VA: AOAC International.

Venkateswaran K, Dohmoto N and Harayama S (1998) Cloning and nucleotide sequence of *gyrb* gene of *Vibrio parahaemolyticus* and its application in the detection of the pathogen in shrimp. *Applied and Environmental Microbiology* 64: 681–687.

VINEGAR

Martin R Adams, School of Biological Sciences, University of Surrey, Guildford, UK

Introduction

Vinegar is a dilute solution of acetic (ethanoic) acid in water. Acetic acid is produced by the oxidation of ethanol by acetic acid bacteria, and in most countries commercial production involves double fermentation, because the ethanol is produced from the fermentation of sugars by yeasts. The traditional vinegar and the traditional alcoholic beverage of a country are often made from the same raw material, for example wine vinegar in France, malt vinegar in the UK and rice vinegar in Japan, providing strong evidence that the first vinegar was the fortuitous result of a failure to produce an acceptable alcoholic beverage.

The long history of vinegar production testifies to the robustness of the fermentation steps involved. With only modest control measures and without the application of sophisticated microbiological expertise, the process will proceed reliably and reproducibly. The microflora necessary is generally part of the natural microflora of the raw materials used, and at each stage conditions restrict the growth of most other organisms.

Vinegar generally does not command the same high prices or esteem as do alcoholic beverages, but despite its somewhat down-market image it is an extremely important condiment and food ingredient. For centuries it has made an important contribution to the quality, safety and availability of foods – a role that shows no sign of diminishing, despite the advent of alternative methods of food preservation.

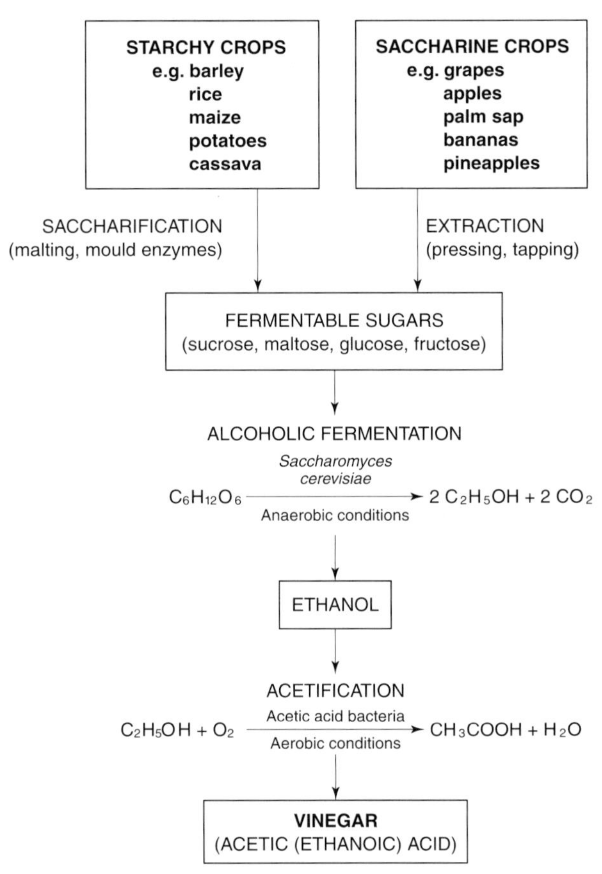

Figure 1 Vinegar production.

Industrial Output

In 1994, vinegar production in the countries comprising the EU was equivalent to 5.5 million hl (hectolitres) of vinegar of 10% acidity, and the annual turnover of the vinegar industry amounted to 280 million ECU. Similar volumes are produced in the US, and it is estimated that the average per capita consumption of vinegar in the industrialized world is equivalent to 1 l of vinegar of 10% acidity per year.

The most important single type of vinegar is that produced from purified ethanol, and is known as spirit vinegar in the UK and distilled vinegar in the US. It is widely used in food processing because it is water-white and can be produced at higher strengths than many other vinegars (up to around 18% acidity (w/v)). In developed countries, spirit vinegar accounts for 5–10% of ethanol usage (excluding potable spirits) and it accounts for 63% of all vinegar production in the EU.

The Production Process

The overall transformation of raw materials into vinegar is outlined in **Figure 1**. Both ethanol and acetic acid are primarily metabolites, the end products of energy-yielding pathways, and so yields are relatively high. Both the alcoholic fermentation and the acetification stages can proceed with efficiencies that are 90% of those predicted from a stoichiometric relationship (i.e. without substrate losses due to conversion into biomass and other materials). Consequently, a reasonable expectation would be the production of 1% (w/v) acidity in a vinegar from every 2% (w/v) fermentable sugar in the original substrate.

The minimum legal standard for the acetic acid content of table vinegar varies, but is generally between 4 and 6 % (w/v). Therefore any material that can supply sufficient carbohydrate to produce a solution containing 10–12% fermentable sugar has the potential to be used as a source of vinegar.

Alcoholic Fermentation

For many years, vinegar makers have exploited lapses in the brewer's or the wine maker's art as sources of relatively cheap raw materials. This is the exception nowadays, however, because for consistent and reliable supplies, raw materials must be produced specifically for vinegar production. Nevertheless, the alcoholic fermentation process used by vinegar brewers is broadly very similar to that used in the production of alcoholic beverages.

Starchy Crops

When starchy crops such as cereal grains are used, the starch must be converted into fermentable sugars. In the production of malt vinegar, the endogenous starch-degrading enzymes produced during the malting of barley achieve this, although in some cases commercially produced concentrates of microbial enzymes may be added. Other materials, such as unmalted cereals, can be used as a relatively cheap supplemental source of carbohydrate, up to levels of about 30%. This mix is converted to a fermentable solution known as 'wort' by the process of 'mashing'. The principal reaction within mashing is the hydrolysis of starch into fermentable sugars, principally maltose, by α- and β-amylases. The activities of other malt enzymes, such as proteases and glucanases are, however, important in achieving the fermentability and stability of wort. The α-amylase attacks the solubilized starch molecules randomly, cleaving α-1,4-glycosidic linkages and reducing the viscosity of the wort, and the β-amylase acts on the dextrins so produced by cleaving maltose from the non-reducing end of the molecule.

Two approaches to mashing are traditionally used: infusion and decoction. In the former, the malt and the starchy adjunct are milled together and heated with water at a constant temperature, usually around 65°C. In decoction mashing, the mix is taken through a range of temperatures by removing a proportion, heating it and returning it to the bulk. This procedure exploits the differing optimum temperatures of the different enzymes present, so maximizing their activity, and is particularly useful with less well-modified (degraded) malts. Nowadays, a hybrid of infusion and decoction mashing is frequently used.

Typically, mashing takes about 3 h: the sweet wort is then run off to the fermenters. It is at this stage that the production process of malt vinegar diverges from that of beer-brewing. In the latter, the sweet wort is boiled with hops before fermentation: this helps to stabilize the wort, extracts flavour components from the hops and inactivates the malt enzymes. None of these actions is necessary in the production of malt vinegar: hops are not necessary for flavour and it is advantageous that the activity of the starch-degrading enzymes continues as long as possible, to maximize the conversion of residual dextrins to fermentable carbohydrate, thus boosting the overall yield.

A different approach to mashing is applied in the production of rice vinegar in Asian countries. The rice starch is generally converted into fermentable sugars by the action of amylolytic enzymes produced by moulds growing on the substrate. In Japan, the mould *Aspergillus oryzae* is used, and *Rhizopus* species and *Mucor* species are used in China. The mould is grown on rice to produce 'koji', which is added to steamed rice in a ratio of about 1:3 and then mixed with 'moto' (a yeast inoculum: see below) and water. The amylolytic activity of the koji continues throughout the fermentation, although the further growth of the mould is prevented by the anaerobic conditions.

Saccharine Crops

In the case of saccharine materials such as fruits, the processes for preparing a vinegar stock are marked by their simplicity. Materials with a high sugar content, such as molasses or honey, have to be diluted to an appropriate strength, typically 10–15% sugar (w/v). Small amounts of nutritional supplements, e.g. ammonium salts, may be added and the pH may be reduced to 4.5–5.0 to favour yeast fermentation. Most fruits are simply crushed, to extract a juice which can be fermented directly, although pectinolytic enzymes may be used to facilitate the extraction of the juice.

Yeast Inoculum

Once a fermentable solution has been prepared, a suitable yeast inoculum, generally *Saccharomyces cerevisiae*, is normally added. The strains may be specially selected, or reconstituted active dried bakers' yeast may be used. In some regions a spontaneous yeast fermentation occurs, e.g. in the production of palm sap vinegars in countries including Sri Lanka and the Philippines. Here the palms are tapped to produce a saccharine juice, which is fermented by yeasts adhering to the inside of the collection vessels. The degradative fermentation that removes the sweet mucilage surrounding cocoa beans, to produce cocoa sweatings, is another spontaneous fermentation that has been exploited for the production of vinegar.

The prime objective at the alcoholic fermentation stage is to maximize the conversion of carbohydrate to ethanol: at this stage, organoleptic characteristics are far less important than they are in the production of alcoholic beverages. It is also, however, important to maintain adequate levels of hygiene, to ensure that premature acetification does not occur – this would inhibit yeast activity and lead to inefficiency in the overall conversion of sugar to acetic acid.

Acetic Acid Bacteria

The second main stage of vinegar production is acetification, which involves the oxidation of ethanol to acetic acid by acetic acid bacteria. Bacteria capable of oxidizing ethanol to acetic acid are described as acetic acid bacteria. They are Gram-negative, catalase-positive, oxidase-negative and ellipsoidal to rod-shaped. Depending on the genus, they can be non-motile, or motile with peritrichous or polar flagella. The cells can vary in length from $<1\,\mu m$ to $>4\,\mu m$ and can occur singly, in pairs and in chains.

Two genera, *Acetobacter* and *Gluconobacter*, are currently recognized. They are distinguished by the ability of *Acetobacter* species not only to oxidize ethanol to acetic acid, but to further oxidize the acetic acid (or alternatively lactic acid) to CO_2 and H_2O. This action is potentially detrimental to vinegar production, because it would convert product into CO_2, but it is inhibited at low pH and by ethanol. Therefore acetifications are run at low pH whenever possible, and are terminated when some ethanol is still present. Once a population of *Acetobacter* is over-oxidizing, reversion to a partially-oxidizing population cannot be achieved. Complete sterilization of the fermenter and restarting with a fresh culture are necessary.

Very early studies demonstrated that the oxidation of ethanol to acetic acid proceeds via acetaldehyde. In acetic acid bacteria, this appears to be largely a membrane-associated process. Membrane-bound alcohol dehydrogenase and aldehyde dehydrogenase, both using the prosthetic group pyrrolo-quinoline quinone (PQQ), have been described. Both enzymes donate electrons to a ubiquinone embedded in the membrane phospholipids, and the ubiquinol produced is then oxidized by a terminal oxidase, an activity associated with cytochromes a_1, b and d in different *Acetobacter* species. Thus the oxidation of ethanol results in the net translocation of protons across the cell's plasma membrane, generating a proton motive force that can be used to drive vital processes (**Fig. 2**).

Acetification Processes

Acetification is usually allowed to proceed directly in the clarified or partially clarified alcoholic stock, which retains sufficient nutrients to support the process. The production of spirit vinegar, however, generally uses the distillate from the fermented mash or, in the US, ethanol from petrochemical sources. The process of distillation removes nutrients, and these must be replaced for successful acetification to occur. Some companies have their own formulations for nutrient supplementation, but commercially produced preparations, which provide the necessary mixture of minerals, growth promoters and vitamins in a colourless form, are available.

From the stoichiometry of the acetification reaction, 1 l of ethanol should produce 1.036 kg of acetic acid and 0.313 kg of H_2O. Therefore, 1% (v/v) ethanol in the vinegar stock should yield approximately 1% (w/v) acetic acid on complete oxidation. This relationship allows the calculation of the expected strength of a vinegar and of the process efficiency, and enables monitoring of the extent to which losses may be occurring due to over-oxidation or evaporation. The composition of an acetifying liquid is expressed as its concentration sum or GK (from German, Gesammte Konzentration) (**Equation 1**):

$$\text{GK} = \%(\text{v/v})\ \text{ethanol} + \%\ (\text{w/v})\ \text{acetic acid} \qquad \text{(Equation 1)}$$

The GK indicates the ultimately attainable acetic acid content, although acetifications are usually terminated while some ethanol remains, in order to prevent over-oxidation. The GK should, in the absence of over-oxidation and evaporative losses, remain constant throughout the process. The extent to which this is achieved is expressed by the GK yield (**Equation 2**):

$$\text{GK Yield} = 100 \times \frac{\text{Final GK}}{\text{Initial GK}} \qquad \text{(Equation 2)}$$

The techniques used in commercial vinegar production differ in the way in which the three interacting components: ethanol, acetic acid bacteria and O_2 (air), are brought into contact.

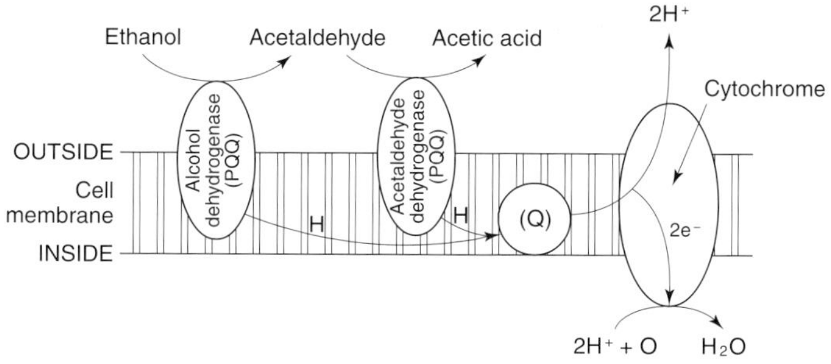

Figure 2 Oxidation of ethanol by *Acetobacter*. PQQ: pyrrolo-quinoline quinone; Q: ubiquinone.

Surface Culture

The simplest technique is surface culture, in which the acetic acid bacteria grow as a surface film on the acetifying stock. This is held in a partially full vessel, usually a barrel, in which holes have been drilled to improve air circulation. The process can be operated batchwise, the contents being allowed to acetify until the required level of acidity is reached and most of the ethanol is exhausted, at which stage the contents are removed and the vessel is refilled with fresh alcoholic stock. In a refinement of this technique known as the Orleans process, acetification is operated semi-continuously in order to reduce the delays and losses that result from the bacterial film having to re-establish itself after each batch. When the acetification is complete, a proportion of the stock is removed through a tap and replaced with fresh alcoholic stock without unduly disturbing the film. This approach has the additional advantages that the bacteria operate over a limited range of concentrations of acid and ethanol, and the acidity in the cask never drops to a level at which contamination may become a problem.

Quick Vinegar Process

Faster rates of acetification can be achieved by increasing the area of contact between the bacteria, the air and the acetifying stock. In the quick vinegar process, this increase is achieved by allowing the bacteria to grow as a surface film on an inert support material, such as wood shavings packed in a false-bottomed tower (**Fig. 3**). The acetifying wash is sprayed over the top of the wood shavings, and trickles down through the bed against a counter current of air passing up the tower. Theoretically, the oxidation of 1 l of ethanol at 30°C would require about 2150 l of air, but normally the passage of three to four times that volume is required. Air can be pumped in at the base of the tower, but in the most basic quick vinegar generators it is simply drawn up through the tower by the heat of reaction in the bed.

Acetification is a far more exothermic reaction than is alcoholic fermentation. Approximately 8.4 MJ are produced in the form of heat for every litre of ethanol oxidized, and the rates of acetification achieved by the quick vinegar process often necessitate cooling the generator to prevent the temperature from increasing to levels at which the bacteria would be inhibited. Cooling can be achieved by equipping generators with cooling coils or, more commonly, by passing the acetifying wash through a heat exchanger. In one type of automatic quick vinegar generator the wash temperature is adjusted to 28°C before passing through the packed bed and has increased to 35°C when it emerges at the bottom, but some generators have been run very efficiently at temperatures as high as 40–43°C.

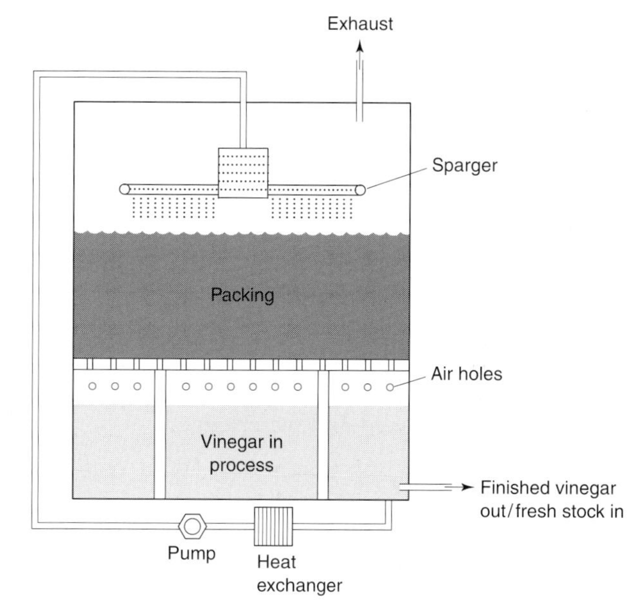

Figure 3 The quick vinegar process.

Following passage through the bed, the acetifying wash is collected in a sump at the base of the vinegar generator and is recirculated through the tower until acetification is complete. At this stage, a proportion of the wash is drawn off and replaced with fresh vinegar stock.

Problems may result either from a lack of homogeneity in the packed bed, which could lead to over-oxidation, or from the development of cellulose-pro-

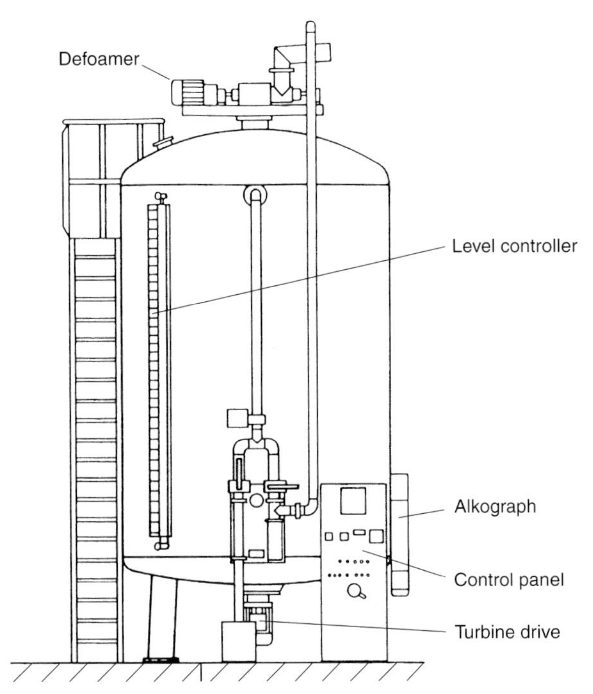

Figure 4 The Frings Acetator.

ducing strains of *Acetobacter*, which can cause the bed to become blocked and waterlogged.

Submerged Culture

Submerged acetification, the most technically advanced method of acetification, is essentially similar to other submerged fermentations, in which the organisms are suspended in the medium and an O_2 supply is maintained by bubbling air through the stirred suspension. The most successful submerged acetifier in commercial use is the Frings Acetator (**Fig. 4**). This has a number of special features, most notably the aerator, which consists of a hollow-bodied turbine surrounded by a stator, two static rings separated by a series of angled plates. The turbine is rotated at speeds of 1450 or 1750 rpm by a motor underneath the fermenter vessel. As the turbine rotates, it draws air down a pipe and creates an air–liquid emulsion (foam), which is ejected radially through the stator to ensure very efficient O_2 transfer to all parts of the vessel. Baffles in the vessel prevent its contents rotating as a whole. The Acetator is fitted with a mechanical defoamer and an Alkograph, which automatically monitors the concentration of ethanol in the vessel.

Like the Orleans and quick vinegar processes, submerged acetification is run semi-continuously, although continuous operation is possible with low vinegar strengths. Typically, the initial concentration of acetic acid is between 7 and 10% (w/v) and the ethanol concentration is about 5%. The cycle is terminated when the residual ethanol concentration falls to around 0.3%, after about 24–48 h. The Acetator will normally produce vinegar with a strength of up to 15% (w/v) acetic acid, which is comparable to that produced using other acetification techniques. However the Frings company has developed a technique using two fermenters in series, whereby vinegar strengths of up to 18.5% acidity can be attained.

Submerged acetification has the advantages of compactness, low requirement for labour and high efficiency. It does however have a high capital cost relative to other techniques, and it consumes cooling water and power at a higher rate. In addition, the process is particularly susceptible to power failures: if the O_2 supply is interrupted even briefly, the bacteria rapidly succumb to the high acidity of the medium and die. For example, in a wash with a GK of 11.35, an interruption in aeration of 1 min completely stopped acetification and there was no significant recovery when aeration was resumed. Also, the vinegar produced by submerged acetification is cloudy and requires fining and filtration.

Microorganisms

The organisms used in commercial vinegar production using the Orleans and quick vinegar techniques are generally not derived from pure cultures, but have been selected due to the efficiency of their performance in practice. Studies of isolates from commercial vinegar plants, using classical and modern molecular techniques, have shown that the microflora present on the wood shavings in quick vinegar generators is quite heterogeneous. This can be advantageous in terms of susceptibility to attack by bacteriophages.

Molecular studies have shown that the microflora in commercial submerged acetifiers in Germany is much simpler than that in quick vinegar generators, consisting of one or a few strains of the species *Acetobacter europaeus*. This may account for the greater susceptibility of submerged acetification to problems arising from phage infection. *Acetobacter*-specific phages have been demonstrated in both quick vinegar generators and submerged acetifiers, but in the former they may cause only a reduction in the acetification rate rather than complete cessation. Recommended precautions against phage infection include pasteurization of the vinegar stock; the use of sterile filtered air; and the physical separation of submerged acetifiers from quick vinegar generators, which may act as reservoirs of infection.

Post-fermentation Processing

After acetification, it may be necessary to store the vinegar while it stabilizes. During this period, previously soluble constituents such as tartrate (in wine vinegars) and proteins may precipitate out and chemical reactions may occur between constituents such as alcohols and acids, which contribute to the bouquet of the product. The period of storage necessary depends on the nature of the raw materials used and the acidity of the product – vinegars with high acidity are likely to stabilize more quickly.

Vinegars produced by the Orleans process are often quite clear when first removed from the cask, but most vinegars need to be filtered. Often this is preceded by fining with bentonite, particularly in the case of vinegars produced by submerged acetification.

Pasteurization or 'hot filling' will ensure microbial stability of the product and prevent the formation of turbidity or of a surface film due to the growth of *Acetobacter*. Sulphite is a permitted preservative in many countries, and is usually used in the concentration range 50–200 mg l^{-1}. In addition to its antimicrobial effect, its antioxidant properties help to prevent browning during storage. Colouring agents such as caramel may be added to ensure a consistent product colour, but in the case of spirit vinegar a water-white product is required and it may be necessary to decolorize it with activated charcoal.

Uses

The principal uses of vinegar are as a condiment and a food ingredient, to flavour and acidify foods. A large proportion of commercially-produced vinegar, up to 70% in the case of the US, is used in commercial food processing.

As a condiment, vinegar confers a sharpness to fatty foods or bland dishes. Its ability to extract the flavour components from herbs and spices has led to the production of a wide range of vinegars with flavourings such as garlic, chillies and tarragon and also of salad dressings incorporating flavoured vinegars.

The use of vinegar in pickling increases the shelf life and the safety of foods, due to the antimicrobial activity of acetic acid. The preservative action is not solely due to pH: at a given pH, the antimicrobial activity of acetic acid is far greater than that of mineral acids such as hydrochloric acid. This is because acetic acid is a weak organic acid (pK_a 4.75). As the pH decreases, so too does the proportion of the acid in the undissociated form. Table vinegars typically have a pH of 2.7–3.2 and contain 4–5% acetic acid, of which 98% or more is undissociated. (In pickling, vinegars containing more acid are often used, to allow for the dilution of the acid by moisture in the product.) In the undissociated state, acetic acid is moderately lipophilic and can pass freely through the microbial plasma membrane into the cytoplasm. Within the cell a higher pH is maintained, so some of the acetic acid dissociates, acidifying the cytoplasm. This imposes a metabolic burden on the cell, as it tries to maintain the transmembrane pH differential by diverting cellular energy away from growth-associated functions. The extent of this effect depends on the extracellular pH and the concentration of the acid, but a progressive decrease in growth rate results from an increased metabolic burden. In most pickles, this burden is great enough to prevent the growth of all but the most acid-tolerant organisms, and those which are unable to grow will die during storage.

See also: ***Acetobacter*****. Fermentation (Industrial)**: Production of Organic Acids. **Fermented Foods**: Fermented Vegetable Products; Fermentations of the Far East. ***Gluconobacter***.

Further Reading

Adams MR (1998) Vinegar. In: Wood BJB (ed.) *Microbiology of Fermented Foods*, 2nd edn. P. 1. London: Blackie Academic & Professional.

Ebner H and Follmann H (1983) Acetic acid. In: Rehm H-J and Reed G (eds) *Biotechnology 3b*. P. 387. Weinheim: Verlag Chemie.

Ebner H and Follmann H (1983) Vinegar. In: Rehm H-J and Reed G (eds) *Biotechnology 5*. P. 425. Weinheim: Verlag Chemie.

Kittelmann M, Stamm W, Follmann H and Trüper HG (1989) Isolation and classification of acetic acid bacteria from high percentage vinegar fermentations. *Applied Microbiology and Biotechnology* 30: 47–52.

Matsushita K, Toyama H and Adachi O (1994) Respiratory chains and bioenergetics of acetic acid bacteria. *Advances in Microbial Physiology* 36: 247–301.

Sievers M, Sellmer S and Teuber M (1992) *Acetobacter europaeus* sp. nov., a main component of industrial vinegar fermenters in Central Europe. *Systematic and Applied Microbiology* 15: 386–392.

VIRUSES

Contents

Introduction

Dean O Cliver, University of California, Davis, California 95616 8743, USA

In countries where causes of food-borne disease are regularly recorded and reported, viruses are consistently among the leading agents. This is true despite the special properties of viruses that complicate both diagnosis of viral illnesses in humans and detection of the agents as contaminants in foods. As an introduction to more specific sections on food-borne viruses, this section surveys the history and ecology of viruses, the common properties of food-borne viruses, the place of specific viruses among agents of food-borne disease, the role of foods as a means of virus transmission, inactivation of viruses in food and water, and approaches to the diagnosis of food-borne viral disease and detection of food-borne viruses.

History

The word '*virus*' meant a slimy liquid or poisonous substance in classical Latin; the word seems not to have been used in the plural, so no proper Latin plural is known. From the end of the nineteenth century the term was applied to infectious agents that did not grow on laboratory media and would pass through filters with pores small enough to retain bacteria.

Viruses were discovered that caused disease in plants, animals, and eventually humans. Bacteriophages – viruses that infect bacteria – were also identified; these served as powerful, but at times misleading, models in the study of other viruses. Eventually, a general definition emerged: viruses comprise genetic material (either DNA or RNA, but not both) inside a coat of protein. Some animal viruses also have a lipid-containing envelope outside the protein. In addition to the structural protein of the coat, some viruses contain enzymes needed early in their replicative cycles.

Properties and Replication of Viruses

All viruses selectively infect certain host cells: the viral nucleic acid, introduced into the host cell in various ways, directs the cell to produce the nucleic acid and protein components of progeny virus – often to the detriment of the host cell. Animal viruses are both species- and organ-specific: viruses transmitted to humans via foods, for example, infect only humans (and in some instances, a few other primates) and find entry by way of the digestive tract. A primary determinant of the specificity of infection appears to be the presence of suitable receptors on the surface (plasma membrane) of susceptible cells. If a viral particle contacts a complementary receptor on a cell's surface, the cell may be induced to engulf the viral particle and remove the viral coat protein, after which the viral nucleic acid begins the intracellular infection cycle.

The viruses most often transmitted via food are those that contain single-stranded RNA. The RNA codes for one or more structural (coat) proteins and for enzymes, required for viral replication, that are not inherent to the host cell. Initially, the entire viral genome is translated into one large protein; one or more enzymic portions of the protein catalyse its breakdown into smaller, functional units. An enzyme critical to viral replication that must be specified by the viral genome is RNA-dependent RNA polymerase. Complementary (minus-strand) RNA must be synthesized on the viral plus-strand template; and since cells do not synthesize RNA on an RNA template, the virus must specify this essential enzyme. There is no reverse transcription in the replicative cycles of these viruses, and DNA plays no role whatever. Essentially all of the viral replication occurs in the host cell's cytoplasm – the nucleus is not involved. One might suppose that the genomic segment that codes for RNA-dependent RNA polymerase would be highly conserved, but surprising diversity has been reported therein.

Viral components synthesized in the host cell assemble themselves spontaneously into progeny virus; hundreds or thousands of progeny may be produced per host cell. The progeny viruses exit the host cell in various ways: leaking through the plasma membrane, budding through the plasma membrane and taking a portion along, or bursting or lysing the cell. These progeny viral particles are quite inert: they cannot multiply, but only retain or lose their infectivity until another susceptible host cell is encountered.

If many cells or highly significant cells of a multicellular organism are subverted by virus infection, the host organism is likely to experience disease. Among animals, including humans, virus infection gradually induces an immune response by the body, which usually terminates the virus infection and prevents reinfection by the same virus. Typically, there are no drugs or antibiotics that specifically cure virus diseases. Some viral infections are now efficiently prevented by vaccines that use immunogenic products of viruses to evoke defences in the body before a real challenge occurs.

Vehicular Transmission

The hepatitis A virus and the Norwalk-like gastroenteritis (small, round) viruses have consistently ranked among the top ten causes of food-borne disease, as compiled by the US Centers for Disease Control and Prevention. A great deal of food-borne and waterborne gastroenteritis of unknown cause, some of which is probably viral, is also recorded. Hepatitis A is a reportable disease in the USA: although the statistics are highly incomplete, they suggest that vehicles play a limited role in the overall transmission of hepatitis A, even though the disease ranks high among food-borne illnesses. Rotaviruses and astroviruses are also transmitted by foods on occasion; and the hepatitis E virus causes waterborne, and possibly food-borne, disease outbreaks in Africa, Asia and Latin America.

Viruses transmitted to humans via food or water are typically transmitted by a faecal-oral cycle and are infectious only for humans. The viruses are shed in faeces after production in the intestinal lining or, in the case of the hepatitis viruses, production in the liver with subsequent drainage into the intestine via the common bile duct. Ingested with food or water, the virus particles enter the body by way of susceptible host cells in the lining of the small intestine. Some simply infect these lining (mucosal) cells; others infect the liver, central nervous system or other organs away from the digestive tract proper. Almost without exception, the progeny viruses are shed via the intestine in faeces, to infect other people by:

- person-to-person contact
- carriage by animal vectors such as flies or by inanimate fomites
- vehicular transmission by food and water.

Vehicular transmission, then, generally results from contamination of food or water by human faeces. One important exception is the Norwalk-like viruses, which are shed in vomitus that may contaminate food. A minor exception is the tick-borne encephalitis virus of Slovakia that infects goats, and perhaps other dairy animals, and sometimes infects humans via unpasteurized milk and milk products from these animals.

Because viral particles are totally inert outside the host cell, they cannot multiply in the environment in food or in water. Instead, the particles either remain capable of causing infection or are inactivated. Although viruses are most persistent at near-neutral pH, they are little affected by the acidity of foods such as strawberries or orange juice. Cooking of foods and disinfection of water with chemicals such as chlorine or ozone are common ways of inactivating viruses in vehicles. Chilling and freezing of food or water will preserve viruses. With the exception of the hepatitis A virus, the food-borne viruses are no more heat-stable than common food-borne vegetative bacteria, so thermal processes comparable to milk pasteurization will reliably inactivate these. Again, the hepatitis A virus may be an exception – it also withstands air-drying, which is strongly detrimental to most of the other food-borne viruses. Chemical disinfection is best accomplished by strong oxidizing agents, including halogens, ozone and hydrogen peroxide; these are useful only against viruses in water or exposed on food or food-contact surfaces. Water and surfaces are also the contexts in which viruses are accessible to inactivation by ultraviolet light. Ionizing radiation, such as gamma rays, would penetrate foods contaminated internally with viruses, but the small size of the virus target necessitates impractically large radiation doses to be effective. Viruses can be degraded slowly by bacterial action, but this effect is not very useful in foods.

Diagnosis and Detection

The submicroscopic size and inertness of viral particles make it extremely difficult to detect viruses in environmental samples, including food and water. For this reason, most of the record of virus transmission via food and water is based on diagnosis of viral illness in those affected. Unfortunately, even diagnosis is more difficult in viral than bacterial illnesses. Diagnostic methods have been developed based on the detection of virus or portions of virus in the faeces of infected persons – these tend to be difficult and are applicable only to stool specimens obtained during incubation, acute illness or early convalescence, depending on the virus that is sought. Diagnosis of hepatitis A is now based largely on the demonstration of IgM-class antibody against the virus in the blood of affected persons; the IgM response lasts a few weeks or months after the onset of illness and is demonstrable by means of commercial test kits. Test kits to detect antibody against other viruses, or viral nucleic acid, are evidently under development. A problem is that the development of such diagnostic aids is economically driven, and laboratory diagnosis of viral illness (other than hepatitis A) is seldom attempted because the illnesses are seen as self-limiting and not specifically treatable. As a result, a great deal of probably food-borne and waterborne disease goes undiagnosed, and thus unrecorded.

Especially when an outbreak of food-borne or waterborne disease is under investigation, it would be highly desirable to be able to detect virus in the suspect vehicle. This requires a suitable means of preparing the sample, followed by an effective method for detecting the virus. The infectivity of some types of viral particles may be detected in laboratory cultures of susceptible host cells, but no such cultures have been developed for detecting some of viruses most often transmitted in food or water. Alternative detection methods focus on the distinctive size and shape of the particle, as seen with an electron microscope, or on individual properties of the viral coat protein or nucleic acid. Highly sensitive and selective methods are now available for detecting specific portions of the genetic sequences of viral nucleic acids; adapting these methods to testing samples of food and water remains a challenge. This is because food and water samples often contain substances that interfere with the genetic detection procedures and because the volumes of processed sample that can be tested by these methods are extremely small; therefore, selective concentration of the virus from the sample, without enhancing the concentrations of interfering substances, is required. In sum, detection methods for viruses in food and water are exacting and expensive, when they work. Very few detections of viruses from vehicles implicated in outbreaks have yet been reported.

Meanwhile, the search continues for rapid, inexpensive monitoring methods, to detect indicator contaminants of food and water whose presence is correlated with the presence of human pathogenic viruses, especially for use in routine monitoring of water and shellfish. Because most of the viruses in question are shed in faeces, most indicators that have been studied are bacteria (especially *Escherichia coli*) associated with faecal contamination. However, these are not specifically of human origin and are affected differently from viruses by environmental influences outside the body, so the correlation of faecal bacteria with enteric viruses is weak or nonexistent. Human enteroviruses (e.g. the polioviruses) that are shed in faeces, but not transmitted in food and water in developed countries, have been studied for this purpose; they are slow and costly to detect and are thus not useful surrogates for the pathogens themselves. Coliphages – viruses that infect *E. coli* – may be good indicators of the possible presence of animal viruses of faecal origin, but are not necessarily related to *human* faecal contamination. Still, methods for detecting them are rapid and inexpensive, so their application continues to be studied. An indicator reliably related to the presence of human viruses in food or water has yet to be identified.

Conclusion

Human enteric viruses cause a good deal of the food-borne and waterborne illness recorded in developed countries. It is not possible to determine the role of food and water in transmitting these agents in parts of the world where food-borne and waterborne illnesses are not recorded. The agents involved are most often the hepatitis A and Norwalk-like viruses, less frequently the rotaviruses, astroviruses and the hepatitis E virus. All of these are transmitted via human-specific faecal-oral cycles in which food and water sometimes play a role. The virus particles in food and water are totally inert and cannot multiply, but may be inactivated before they are ingested by a susceptible human. Their extremely small size and inertness complicate the detection of viruses in food and water; the demonstration that an outbreak is caused by a virus often depends on laboratory diagnosis of human illnesses, which unfortunately is seldom attempted. Indicators of the presence of viruses, which could be detected quickly and easily, continue to be sought.

See also: **Viruses**: Environmentally Transmissible Enteric Hepatitis Viruses: A and E; Detection. **Water Quality Assessment**: Routine Techniques for Monitoring Bacterial and Viral Contaminants.

Further Reading

Cliver DO (1990) Viruses. In: Cliver DO (ed.) *Foodborne Diseases*. P. 275. San Diego: Academic Press.

Cliver DO (1997) Foodborne viruses. In: Doyle MP, Beauchat LR and Montville TJ (eds) *Food Microbiology: Fundamentals and Frontiers*. P. 437. Washington: American Society for Microbiology.

Cliver DO, Ellender RD, Fout GS, Shields PA and Sobsey MD (1992) Foodborne viruses. In: Vanderzant C and Splittstoesser DF (eds) *Compendium of Methods for the Microbiological Examination of Foods*. P. 763. Washington: American Public Health Association.

Hui YH, Gorham JR, Murrell KD and Cliver DO (eds)

(1994) *Foodborne Disease Handbook*. Vol. 2. New York: Marcel Dekker.

Environmentally Transmissible Enteric Hepatitis Viruses: A and E

Theresa L Cromeans and **Mark D Sobsey**, Department of Environmental Sciences and Engineering, University of North Carolina, USA

Introduction

The disease infection hepatitis can be caused by five different hepatitis viruses, but only two of these currently characterized viruses are potentially transmitted by contaminated food or water. Hepatitis A virus (HAV) and hepatitis E virus (HEV) are transmitted by ingestion of virus shed by infected individuals (faecal–oral route). In contrast, hepatitis B virus (HBV) and hepatitis C virus (HCV) are primarily transmitted by parenteral routes and intimate person-to-person contact. Hepatitis delta virus (HDV) is also parenterally transmitted, and is infectious only in the presence of acute or chronic HBV infection. Early studies in the 1960s established two different types of hepatitis based on the different modes of transmission, and two agents responsible for these two types were later identified as HAV and HBV. Initially unidentified hepatitis agents were referred to as non-A non-B (NANB) until subsequent studies established their identities and roles in hepatic disease. In developed countries most hepatitis infections of unknown aetiology were acquired after blood transfusions. When diagnostic assays were developed, HCV was identified as the major cause of NANB in developed countries. An antibody detection test for screening blood supplies was developed and applied in 1992 to eliminate donations contaminated with HCV. In the 1970s, the serologic test for HAV was applied to specimens collected in developing countries during previous large epidemics thought to be caused by HAV; these tests revealed that another agent was responsible for these epidemics, primarily in Asia. This agent was called enterically transmitted NANB; the causative agent was characterized in the late 1980s and designated HEV. Recently other viruses, named hepatitis F virus (HFV) and hepatitis G virus (HGV), have been isolated from patients with hepatitis, but these agents have not been established as causative agents of hepatitis disease. Hepatitis cases of apparent viral origin continue to occur that cannot be associated with any known hepatitis virus, or any of the other viral agents sometimes causing hepatitis, such as cytomegalovirus. At least two more unidentified viral agents exist, one transmitted parenterally and one transmitted by the faecal–oral route.

This article describes HAV and HEV, the only characterized hepatitis viruses transmitted through environmental sources, such as food and water. The basic characteristics of the viruses that are relevant to the modes of transmission and epidemiology of the diseases are described. Disease characteristics and unique patterns of transmission of both viruses worldwide are also described. Emphasis is placed on all characteristics that relate to the potential transmission via food, the protection of food samples and methods for detection in environmental specimens, including food and water.

Hepatitis A Virus

Structure, Genetics and Biology

Although HAV differs in important aspects from the other members of the Picornaviridae family, it is structurally similar. The biological differences have led to its classification in a separate genus, the *Hepatovirus*. HAV genomic organization is similar to other picornaviruses. A single-stranded RNA genome of positive polarity approximately 7500 nucleotides long contains a 5′ non-translated region (NTR) of 735 nucleotides, a 3′ NTR of 60 nucleotides and a single open reading frame (ORF) of 6500 nucleotides from which a single polyprotein is made. This polyprotein is cleaved by virus-specific proteases and can functionally be divided into three regions: P1 includes the viral capsid proteins, P2 and P3 both encode non-structural proteins, primarily enzymes involved in virus replication. Two small regions are exceptions: a region of P2 adjacent to P1, which may be cleaved with P1 to form structural proteins, and a region of P3, 3B which encodes the VPg, a genome-linked virus protein. Due to genome similarities, much of what is known about HAV is by analogy to other picornaviruses, such as poliovirus, that have been well studied. Since HAV replication in cell culture does not shut off host protein synthesis and the replication is slow, study of protein processing and replication has been difficult. Only very recently with the introduction of newer molecular tools have some of the slight differences of HAV from other picornaviruses been elucidated. Structural proteins are encoded by the P1 region and cleavages of the polyprotein yield three viral polypeptide (VP) precursors: VP0, VP1 and VP3. Final cleavages yield VP1, VP2, VP3 and VP4 in picornaviruses. In most picornaviruses, VP0 is cleaved to VP2 and VP4 in the assembled immature virion. During HAV morphogenesis, infectious HAV

Table 1 Persistence of HAV in Environmental Media and Foods

Environmental Media	*Temperature (°C)*	*Time*	*% Infectious*
Milk[1]	62.8	30 minutes	0.1
Cookies[2]	49	30 days	0.88
Shellfish[3]	85–95	>1 minute	None
Dried faeces[4]	25	1 month	Positive*
Septic tank effluent[5]	5	70 days	10
Human serum albumin, 5%[6]	60	3 hours	0.01
Underground water[7]	9	17 months	Positive*
Seawater[8]	5	2 months	0.02
Seawater + marine sediment[8]	5	2 months	100

[1] Parry et al., 1984. J. Med. Virol., 14:277.
[2] Sobsey et al., 1988. Viral Hepatitis and Liver Disease, p 121.
[3] Millard et al., 1987. Epidemiol. Infect., 98:397.
[4] McCaustland et al., 1982. J. Clin. Microbiol., 16:957.
[5] Deng and Cliver, 1995. Appl. Environ. Microbiol., 61:87.
[6] Barrett et al., 1996. J. Med. Virol., 49:1.
[7] Cromeans et al., 1998. Abstract. IAWQ Biennial Conference.
[8] Chung and Sobsey, 1993. Wat. Sci. Tech., 27(3–4):425.
* Indicates cell culture or animal infectivity without quantitation.

provirions are formed containing uncleaved VP0, and VP4 has not been isolated, although nucleotide sequence information suggests a 2.5 kDa protein. Five copies of the three structural proteins form a pentamer and 12 pentamers form the icosahedral, non-enveloped 27 nm HAV particle. VP1 is the major surface-exposed antigen. A single serotype exists worldwide and studies of neutralization-resistant mutants indicate a single conformational immunogenic epitope of sites on VP1 and VP3.

HAV is more resistant than other picornaviruses to environmental stresses, including temperature and pH, although the physiochemical basis is not known. Several of the known differences from enteroviruses such as poliovirus models may be factors that contribute to the virion stability of HAV. HAV is stable in water up to 60°C and to 80°C in aqueous solutions with high concentrations of divalent cations. A pH range of 1–10 is tolerated. Organic solvents, ether, alcohol, chloroform, fluorocarbons, most detergents, quaternary ammonium compounds and iodophores do not appreciably inactivate the virus. Selected experimental data on stability of infectious HAV in food items, environmental sources or sources relevant to transmission of virus are shown in **Table 1**. Disinfection of water with free chlorine, chlorine dioxide, ozone and UV radiation can achieve more than 99.99% inactivation under optimum conditions. If HAV is protected within organic matter or other particles, rates of inactivation can be dramatically reduced. For disinfection of contaminated surfaces, 5000 mg/l free chlorine with 1 min contact time has been shown to inactivate HAV.

Regions of the P1 segment of the genome have been targeted for detection of HAV by reverse-transcription polymerase chain reaction (RT-PCR) and subsequent nucleic acid sequence analysis. Nucleic acid sequence comparative analysis of 168 bases in the VP1 and adjacent P2A region (chosen for a high degree of sequence variability) of geographically diverse strains established seven genotypes of HAV with variation of 15–25% among genotypes. Four of the genotypes were isolated from humans and three from simian species. Subtypes within these genotypes differ by approximately 7.5% of the analysed base positions. However, there is considerably less genetic variability in comparison to a picornavirus such as poliovirus. Nucleic acid sequence analysis of the cloned and sequenced HAV genome demonstrated that the genome organization and translational strategy are similar to other picornaviruses. The 5′ NTR is the most highly conserved region of the genome, while the 3′ NTR has as much as 20% variability.

Geographical analysis of HAV sequence data suggests endemic circulation of HAV strains. Classical epidemiology combined with analysis of genomic sequence of HAV RNA isolated from patients makes determination of a common-source outbreak possible. If the virus can be isolated from the food or water source, epidemiologically implicated viral sequence can be compared with isolates from the patient specimens or any intermediate source of transmission to establish or verify the chain of transmission. As more nucleic acid sequence information is generated for HAV occurring in specific geographic locations or within specific patient groups, concurrent analysis of genomic sequences from environmental isolates for comparison with sequence from clinical samples will provide molecular epidemiological data for improved surveillance, prevention and control activities for hepatitis A.

HAV replication in cell culture was not achieved until the late 1970s, and cultivation from clinical specimens or environmental sources is difficult and not always possible. Repeated virus passages with incubation periods of weeks to months are required for growth of wild-type virus, and viral replication is detectable only by analysis for viral antigen or nucleic acid, since replication is usually not lytic. Persistent viral replication in the cytoplasm of infected cells occurs, and repeated passages of the infected cells can be made with continued production of HAV. After numerous passages in cell culture and/or primates, rapidly replicating lytic variants have been isolated and characterized and are useful in laboratory experiments. Cell culture-adapted HAV yields higher quantities of virus than initial isolates of wild-type virus, but the quantity of HAV produced is less than the quantity of poliovirus. Adaptation of HAV to human cells has given sufficient antigen production for inactivated vaccines.

The relatively slow replication in cell culture of HAV compared to many enteroviruses has not been fully explained. During adaptation in cell culture, certain mutations have been noted in the 2B and 2C and 5′ NTR regions of the genome. Mutations are different for adaptation to different cell lines or different isolates of HAV, indicating a role of host factors in virus replication and adaptation to cell culture. Adaptation in cell culture has been shown to coincide with loss of virulence for primates, the only non-human host. Factors such as protracted viral uncoating, rapid incorporation of new viral RNA into capsid, thus depleting the RNA pool for replication, and inefficient translation initiation due to the unusual internal ribosome entry site (IRES) in the 5′ NTR of HAV, have been suggested to contribute to the slow replication. Replication in vitro may reflect disease in vivo, since the liver damage appears to be immune-mediated and not a direct cytopathic effect of virus replication.

Epidemiology and Clinical Disease

Hepatitis A infection is common worldwide and in developing countries 90% of children are infected by age 6, usually asymptomatically. A significant decrease in childhood infections occurs with improved standard of living. These conditions create populations in which a large proportion of children and young adults are susceptible to infection. Since infection of older children and young adults is much more likely to result in clinical disease (50–90% above age 5 are jaundiced), there is an appreciable morbidity. The potential for large outbreaks of disease exists in areas where the standard of living has increased. Worldwide, the prevalence of HAV antibodies can be categorized into three general infection rates: high, intermediate and low. Western Europe and North America have low rates, Asia and Russia have intermediate rates, and South America, Africa and southern Asia have high rates. Rates within a region may fluctuate because periodic widespread epidemics can produce cyclical patterns of disease incidence. These cyclical patterns can result from several factors including asymptomatic infection in children, who serve as reservoirs; introduction into isolated populations; and introduction through contaminated food or water in areas where the standard of living has improved. In Europe and the US disease rates have declined since World War II and the epidemic cycles are of lower peak incidence with longer periods between epidemics. Community-wide outbreaks occur in which the highest attack rates are among children 5–14 years of age.

Epidemiologic data indicate that the faecal–oral route of transmission is the usual route of transmission, although blood-borne transmission has been documented. HAV is shed at up to 10^8 infectious units per gram of stool, while the level of viraemia in early clinical illness is 10^3–10^6 lower than in faeces. Infectious faeces is transferred from person to person or transferred by contaminated food or water. The most common risk factor for disease is personal contact with a person who has hepatitis; this accounts for approximately 25% of cases. Other known risk factors are drug use, day-care centre association, foreign travel or exposure to implicated food or water in an outbreak. Each of these contributes 15% or less to the total reported cases; therefore, about 45% of reported cases have no identified risk factor. Fifty per cent of those with no identified risk factor have a child under 5 years of age in the household. Transmission can occur readily in environments such as day-care centres, and in this age group infection is usually asymptomatic. Furthermore, children may shed virus longer than adults.

The incubation period of hepatitis A ranges from 2 to 6 weeks with an average of 28 days. Clinical symptoms of hepatitis A are the same as those of other hepatitis diseases. The degree of clinically apparent illness is related to the individual's age. Fewer than 10% of children under 6 develop jaundice, while in older children and adults, icterus occurs in 40–80% of infected individuals and most have some symptoms associated with viral hepatitis. Symptomatic infections are characterized by malaise, nausea, low-grade fever and headache, progressing to more severe symptoms such as vomiting, diarrhoea and right upper quadrant discomfort. The case fatality rate in the US is 0.3%, varying from extremely low in persons aged 5–14 years to 2.7% in persons over 49. Relapse of symptoms can occur in 3–20% of persons with associated reactivation of viral shedding.

Experimental infection of non-human primates has revealed patterns of virus replication and excretion. HAV can be detected in hepatocytes 1 week before liver enzyme elevations; several weeks after resolution of disease, HAV antigen is no longer detected. HAV is shed in faeces via bile entering the intestinal tract soon after it is detected in the liver. Once liver enzymes return to normal, excretion as measured by antigen detection methods cannot be detected. However, with RT-PCR, HAV RNA can be found up to several months after clinical infection is apparent. Molecular epidemiology in some outbreaks (see below) has been performed using acute sera from case-patients, rather than from faecal specimens, thus facilitating rapid analysis during an outbreak. Infants and children may shed virus for longer periods. The significance of this latter shedding in the spread of disease has not been determined, but it could contribute to the high rates among children and their contacts. Viral RNA can be detected in the serum of infected individuals by RT-PCR even after clinical infection is resolved, and studies are in progress to determine the duration of viraemia in model animal studied as well as infected humans.

Serological tests first detect immunoglobulin M (IgM) antibody 7–10 days after infection and usually do not detect IgM after 4–6 months. Immunoglobulin G (IgG) antibody appears soon after anti-HAV IgM and confers lifelong immunity. Competitive inhibition immunoassays to detect both immunoglobulins (total anti-HAV) are commercially available. Absence of anti-HAV IgM indicates a previous infection if the total is positive.

Although the liver is the primary target organ of the virus and the site of almost all viral replication, an initial low-level replication in the gastrointestinal tract is suggested by recent evidence in animal models and in vitro replication experiments. Hepatocellular damage is immune-mediated, and HAV may be noncytopathic in the liver cells, as in cell cultures.

Food-borne Disease, Detection and Prevention

Since hepatitis A is a reportable disease in many countries, data on the risk factors associated with reported disease can be compared. In this analysis 3–8% of reported cases in the US are due to contaminated food or water, and these are almost always associated with a common source outbreak. Sporadic cases associated with food would probably not be identified. Shellfish ingestion is the only food-related risk factor that is included in the questionnaire for disease surveillance of hepatitis by the Centers for Disease Control and Prevention (CDC), although a request for suspicion of a food- or waterborne outbreak is included in the questionnaire. Since data from other countries are compiled differently, no direct comparison of risk factors among western countries can be made. In countries where hepatitis A infection is low (such as Finland) and much of the population is non-immune, food-borne outbreaks have been suggested as potential significant sources of disease. A significant food-borne outbreak occurred in Shanghai in 1988 with 300 000 cases in 2 months due to contaminated clams. Shellfish were taken from a new harvest area that was later shown to have sewage pollution, and HAV was isolated from the clams. This outbreak demonstrates the potential for disease and outbreaks in a population where immunity is declining and sanitation is improving.

Without an outbreak to alert health officials, individual cases or cases seemingly unrelated to a common source, such as sporadic cases in large metropolitan areas, may not be identified as food- or waterborne cases. Molecular epidemiology has been used to identify multistate outbreaks of hepatitis A traced to frozen strawberries processed at a single plant. In 1997 a very large outbreak occurred among schoolchildren from several counties in Michigan, and the epidemiologically associated food was frozen strawberries served in school lunches. Analysis of temporally reported patient-cases from other states revealed strawberry-associated disease, and molecular epidemiology analysis was used to confirm the similarity of HAV from patient-cases in several other states. All strawberries associated with the cases were processed in the same plant in the US but imported from Mexico. All identified batches of contaminated strawberries were processed at a similar time; however, whether contamination occurred at the agricultural site or at the processing plant could not be determined.

HAV contamination of food may occur either at the production source, such as in cultivation fields and in large batch processing, or at the consumption site by infected food handlers who may directly contaminate food. Contamination may occur in the agricultural areas due to practices such as use of reclaimed waste water for irrigation or use of human waste as fertilizer. In addition to these direct methods, contamination of products may occur due to infected labourers. Lack of adequate sanitary facilities or lack of use will contribute to contamination. Fresh produce that is hand-picked and eaten unprocessed (or processed so that viral infectivity is retained) is particularly susceptible to contamination, and significant outbreaks have occurred due to this source. With increased consumption of fresh, uncooked foods imported to developed countries from areas with highly endemic disease, this source of disease could increase.

Contaminated shellfish eaten uncooked or inadequately cooked has been the source of outbreaks for decades. Sewage treatment does not completely remove HAV from waste water, and untreated sewage may also reach shellfish harvest areas. Filter-feeding shellfish concentrate any contamination entering the shellfish beds. Although controls are in place to regulate shellfish beds in most developed countries, these are not always adequate or effectively enforced. HAV may persist in the environment for months or years because the inactivation rates are slow, especially at lower temperatures such as encountered in shellfish beds or in waters in temperate climates. Depuration – holding of shellfish in clean flowing water to remove impurities during natural pumping – can be effective for HAV removal if time and conditions are adequate.

Since cultivation of wild-type HAV is extremely difficult, antigen and/or nucleic acid detection methods have been necessary for detection of virus particles in clinical and environmental samples. A radioimmunofocus assay (RIFA), utilizing radioactively labelled antibody to identify infectious foci in cell culture, analogous to the standard plaque assay, has been important in studies of virus replication, neutralization, virus stability, disinfection and clonal isolation of virus variants. Enzyme immunoassay (EIA) has been employed to detect antigen in stool samples and in analysis of replication in cell culture prior to development of the RIFA. The most sensitive detection method for environmental or clinical samples is RT-PCR, although only the HAV RNA is measured and infectivity is not evaluated. Food, water and other environmental samples usually require concentration of the virus for detection of the low levels typically present. HAV can be recovered and concentrated from water and other environmental samples by methods used for other enteric viruses. These virus concentrates may contain inhibitors of the enzymatic reactions for nucleic acid amplification, and numerous studies have sought to develop sensitive and specific methods for removal of inhibitors in the process of virus concentration. Polyclonal and monoclonal (immune or hyperimmune) antibody preparations may be used to select HAV from concentrates. Antibody may be bound to microfuge tubes or attached to magnetic beads for recovery of virus from samples for RT-PCR assays. This immunocapture RT-PCR method is selective for viral antigen and RNA and demonstrates the presence of immunoreactive viral antigen in addition to RNA. Alternatively, RNA may be extracted from the virus-contaminated concentrates, or the unconcentrated samples in some instances, for direct analysis in RT-PCR. If HAV RNA is not detected with RT-PCR alone, greater sensitivity and confirmation of specificity can be obtained with probe hybridization or a second round of PCR with internal primers. With laboratory modeling experiments, several methods for detection of HAV in shellfish with sensitivity of as low as 0.1 infectious units have been described, and HAV has been detected by RT-PCR in shellfish implicated in one US outbreak. Investigation of detection methodology for other foods has not been reported; however, several laboratories are developing recovery methods for types of fresh and frozen produce potentially contaminated. Although HAV detection by RT-PCR does not provide information on infectivity, it is an extremely useful method for indication of past or present contamination of environmental samples. In addition, methods have been developed and used successfully to recover and detect HAV in drinking water by RT-PCR or cell culture infectivity.

HAV is now a vaccine preventable disease and 2 vaccines are licensed for use in persons over 2 years of age. Modeling research suggests that routine childhood hepatitis A vaccination could greatly reduce HAV incidence levels within 3 to 10 years; however, routine childhood vaccination is not currently possible. Vaccination of persons at increased risk for disease, such as travellers to countries with high endemic rates of infection and persons with increased potential for fulminant disease (e.g., persons with chronic liver disease) is recommended. Another target group is children living in communities that have high rates of disease. Since children account for one-third or more of cases and frequently are a major source of infection, this strategy should result in reduction of hepatitis A disease in the United States. Persons who work in food service are at no greater risk than the general population, although they may have a significant role in common source outbreaks. Food-handlers have not been shown to contribute significantly to infection within communitywide outbreaks and routine vaccination of this group has not been shown cost-effective (CDC, unpublished data).

Immune globulin (IG) is recommended for persons recently exposed to HAV who have not been vaccinated at least 1 month before exposure to HAV. Confirmation of HAV infection in index patients by IgM anti-HAV testing is recommended before treatment of persons who may have been exposed. If a food handler is diagnosed with hepatitis A, other food handlers at the location should receive IG, although administration to patrons is usually not recommended. If the food handler handled uncooked foods or had poor hygienic practices, and patrons can be identified and treated within 2 weeks of exposure, consideration of IG use is justified. In the case of a common-source outbreak, IG should not be given to exposed persons after cases have occurred. The single most important factor in control of disease spread by infected food handlers is good personal hygiene.

Hepatitis E Virus

Structure, Genetics and Biology

HEV is currently unclassified, although it was previously considered a member of the Caliciviridae family based on nucleic acid type and length, general organization of the genome and lack of a lipid envelope. The genome is a positive-sense, single-stranded RNA molecule of 7.5 kb. Immunoelectron microscopy (IEM) studies of virus banded in sucrose gradients showed a 32–34 nm particle that is unstable upon exposure to high salt concentrations. The computed sedimentation coefficient is approximately 183 s, and the buoyant density 1.29 g ml^{-1}. These biophysical properties are consistent with the properties of calciviruses.

Although some recent cell culture systems have been developed in which viral RNA is detected by PCR in infected cells and some systems yield limited production of HEV, none of these are adequate for antigen production, detailed replication studies or infectivity studies. In the absence of any cell culture system to propagate virus, molecular cloning of material from bile of infected animals was used to characterize the virus. The genomic organization of HEV is substantially different from picornaviruses: the HEV genome codes for structural and non-structural proteins through discontinuous, partially overlapping open reading frames (ORF). The HEV genome is a positive-sense single-stranded 3′-polyadenylated (polyA) RNA molecule containing 3 ORF. Non-structural genes are located at the 5′ end and structural genes at the 3′ end. ORF 1 is approximately 5 kb and contains the sequence for several enzymes; ORF 2 contains the major structural proteins and is 2 kb, while ORF 3, overlapping ORF 1 by one nucleotide and ORF 2 by 328 nucleotides, encodes a phosphoprotein which is cytoskeleton-associated.

Although definitive comparisons cannot be made among all isolates due to variations in genome regions analysed, geographically distinct isolates of HEV have been identified. Mexican isolates vary from Asian isolates, of which three or four genotypes exist. Isolates from Nepal have greater identity with Indian and Burmese isolates than with African and Chinese isolates. Isolation of a novel HEV from swine in the US and comparison of that sequence with two US human isolates have identified a genotype unique among all previously described.

Antibody to HEV was first detected by IEM with virus isolates from different regions of the world. More recently, several immunoassays have been developed using recombinant expressed proteins from ORF 2 and ORF 3 for the detection of IgG and IgM activity. Synthetic peptides derived from proteins encoded by ORF 2 and ORF 3 and including at least four immunodominant epitopes have also led to the development of an EIA. Recombinant protein-based tests detect 90–95% of cases during outbreaks in HEV-endemic areas. The same tests however yield discordant results among blood donors from non-endemic countries, indicating that care must be taken in interpretation of results in developed countries. Further work with swine HEV and US-acquired human HEV may explain these discrepancies.

Clinical Disease and Epidemiology

The onset of symptoms of hepatitis E begins within 15 to 60 days of exposure, with a mode of 40 days. Symptoms include nausea, dark urine, abdominal pain, vomiting, pruritus, joint pains, rash and diarrhoea. About 20% of patients have fever and most have hepatomegaly. Information about the relationship of virological, immunological, clinical and histopathological events during infection has been obtained from study of non-human primate models, primarily the cynomologus macaque. In all animal models, liver enzyme elevation occurs 24–38 days after intravenous inoculation. HEV RNA can be detected by RT-PCR in the stool, beginning at 2–3 weeks and extending to 7 weeks. IgM and IgG antibody responses are detected about the same time as symptoms. IgM persists for 5–6 months while the persistence of IgG is not fully understood. Long-term persistence is suggested by the presence of IgG in adults 10 years after jaundice during an epidemic.

The demonstration of transmission of the swine HEV to non-human primates and the infection of swine with the US human isolates indicates the cross-species transmission potential of these similar viruses. In the US almost all cases of HEV have been related to travel to endemic areas; however, studies have demonstrated as many as 20% of persons (depending upon the test used) in non-endemic countries who are antibody-positive in one or more tests for HEV antigens.

Epidemics and acute cases of HEV have been reported throughout Asia and in several regions of Africa, while in the Americas, the only two reported outbreaks occurred in two rural villages in Mexico. Although the exact worldwide distribution of HEV infections has not been determined, the disease appears to be endemic in developing countries. In central Asia 7–10-year cycles have been documented. Where long-term surveillance has been conducted, a pronounced seasonal distribution associated with the local rainy season has been observed. A common source, such as contaminated water, is suggested in the epidemics, and HEV has been isolated from source water in some endemic areas. The highest rates of infection are in older children and young adults, and the mean age for symptomatic infection has been 29 years. Mortality rates of 17–33% among pregnant women has been an outstanding characteristic of all epidemics.

Although faecally contaminated water has been identified as the source of HEV in most outbreaks, secondary attack rates of 0.7–8% in case households have been observed. Several reports have suggested that HEV can be transmitted by food, however, no careful studies have been made. HEV has been detected in sewage, indicating that contaminated shellfish beds could be a source of infection; however, contamination of shellfish has not been demonstrated. HEV has been detected in raw and treated waste water in India by RT-PCR. The reservoir of HEV between epidemics is unknown. Disease could be maintained by serial transmission among susceptible individuals, occasional failures of water treatment plants or sporadic contamination of smaller untreated water supplies. Although epidemics are associated with contaminated water, the origin of the faecal contamination of the water is rarely determined. Another reservoir of HEV may be domestic or wild animals. In endemic regions, HEV has been detected by RT-PCR in faeces of caught wild pigs and anti-HEV antibody has been detected in serum of pigs, cattle and sheep. Contamination of natural water sources by local domestic and/or free-roaming animals is conceivable, especially during rainy seasons and floods. Cross-species transmission of HEV has not been demonstrated except in experimental infections. Further research is needed to clarify the potential relationships of animal HEV to human HEV in developing and developed countries.

Prevention and Control

HEV infection is likely to be controlled by adequate water and sewage treatment. Chlorination or boiling of water supplies have rapidly interrupted disease transmission. The removal and inactivation of HEV by water and sewage treatment processes and food sanitation processes has not been adequately studied due to the lack of a convenient infectivity assay system. The role that animals may have in transmission of this disease will have an impact on design of prevention and control measures. Although immune globulin has been used to prevent infection in outbreaks, no conclusions about efficacy can be made because these were not controlled trials. Immunization with recombinant ORF 2 proteins of HEV has demonstrated a potentially protective antibody response.

See color Plates 40 and 41.

See also: **Food Poisoning Outbreaks**. **National Legislation, Guidelines & Standards Governing Microbiology**: European Union. **Sampling Regimes & Statistical Evaluation of Microbiological Results**. **Viruses**: Introduction. **Water Quality Assessment**: Modern Microbiological Techniques.

Further Reading

Centers for Disease Control and Prevention (1996) Prevention of hepatitis A through active or passive immunization: recommendations of the Advisory Committee on Immunization Practices (ACIP). *Morbidity and Mortality Weekly Report* 45 (No. RR-15): 1–30. (Erratum appears *Morbidity and Mortality Weekly Report* 1997; 46: 588.)

Cliver DO (1997) Virus transmission via food. *World Health Statistics Quarterly* 50: 90–101.

Chang H and Sobsey MD (1993) Comparative Survival of indicator viruses and enteric viruses in seawater and sediment. *Water Science and Technology* 27: 425–428.

Cromeans T, Nainan OV, Fields HA, Favorov MO and Margolis HS (1994) Hepatitis A and E Viruses. In: Hue YH, Gorham JR, Murrell KD and Cliver DO (eds) *Foodborne Disease Handbook*. Vol. 2, p. 1. New York: Marcel Dekker.

DeSerres G, Cromeans TL, Levesque B et al (1999) Molecular confirmation of hepatitis A from well water: epidemiology and public health implications. *Journal of Infectious Disease* 179: 37–43.

Hollinger FB and Ticehurst JR (1996) Hepatitis A virus. In: Fields BN, Knipe DM, Howley PM et al (eds) *Fields Virology*, 3rd edn. P. 735. Philadelphia, PA: Lippincott-Raven.

Hutin YJF, Pool V, Cramer EH et al (1999) A multistate foodborne outbreak of hepatitis A. *New England Journal of Medicine* 340: 595–602.

Krawczynski K, McCaustland K, Mast E et al (1996) Elements of pathogenesis of HEV infection in man and experimentally infected primates. In: Buisson Y, Coursaget P and Kane M (eds) *Enterically-Transmitted Hepatitis Viruses*. P. 317. Tours, France: La Simarre.

Margolis HS, Alter MJ and Hadler SC (1997) Viral Hepatitis. In: Evans AS and Kaslow RA (eds) *Viral Infections of Humans Epidemiology and Control*. P. 363. New York, NY: Plenum Publishing.

Mast E and Krawczynski K (1996) Hepatitis E: an overview. *Annual Review of Medicine* 47: 257–266.

Meng XJ, Halbur PG, Shapiro MS et al (1998) Genetic and experimental evidence for cross-species infection by swine hepatitis E virus. *Journal of Virology* 72: 9714–9721.

Niu MT, Polish LB, Robertson BH et al (1992) Multistate outbreak of hepatitis A associated with frozen strawberries. *Journal of Infectious Diseases* 166: 518–524.

Robertson BH and Lemon SM (1998). Hepatitis A and E. In: Mahy BWJ and Collier L (eds) *Microbiology and Microbial Infections*. Pp. 693–716. London: Oxford University Press. Topley and Wilson's: *Microbiology and Microbial Infections*, Leslie Collier, Albert Balows, Max Sussman, 9th edition, Volume 1 *Virology*.

Robertson BH, Jansen RW, Khanna B et al (1992) Genetic relatedness of hepatitis A virus strains recovered from different geographical regions. *Journal of General Virology* 73: 1365–1377.

Sobsey MD (1994) Molecular methods to detect viruses in environmental samples. In: Spencer RC, Wright EP and Newsom SWB (eds) *Rapid Methods and Automation in Microbiology and Immunology*. P. 387. Andover: Intercept.

Sobsey MD, Shields FS, Hauchman AL, Davis VA, Rullman VA and Bosch A (1998) Survival and persistence of hepatitis A virus in environmental samples. In: Zuckerman AJ (ed.) *Viral Hepatitis and Liver Disease*. P. 121. New York: Alan R. Liss.

Detection

Dean O Cliver, University of California, School of Veterinary Medicine, Davis, USA

Every agent of food-borne disease should be detectable in food. Viruses are no exception. Still, given the challenges and difficulties of detecting food-borne viruses, it is well to reflect on why one is attempting to do so. None of the available procedures seems likely to become routine, so testing before release of a lot of food is not an option. Rather, testing is most likely to be undertaken as a basis for risk assessment or in the course of investigating an outbreak of apparently food-borne viral disease. In the former instance, one might select foods to be tested on the basis of the epidemiological record, and try to apply detection methods that would identify a wide variety of viruses. In the latter situation, one might be confronted with unexpected foods, but would at least have some idea of which virus (or viruses) was most likely to be present. Methods of processing food samples for virus detection may vary widely and are ultimately dependent on which viruses are sought, in that the sample extract must be quantitatively and qualitatively compatible with the test method that is to be applied. There is general consensus that an adequate method should enable detection of at least one peroral infectious dose for humans, in whatever size sample is to be tested; there is considerably less agreement as to what constitutes a peroral infectious dose of the various viruses in humans.

Viruses to be Detected

The viruses to be detected have generally been described in detail in preceding sections. Leading food-borne viruses are the Norwalk-like (small, round, structured) viruses and the virus of hepatitis A. Rotaviruses and astroviruses are sometimes transmitted via foods, and the hepatitis E virus (a calicivirus) is frequently transmitted via water in some parts of the world, so one suspects that it may be food-borne on occasion. For most purposes, it is well to consider drinking water a food in any case. The tick-borne encephalitis viruses that may infect dairy animals and be transmitted to humans drinking unpasteurized milk will not be considered here, in that they are seen as a highly local problem in Slovakia (and perhaps adjoining areas), and rarely there in recent years. Other than the tick-borne encephalitis viruses, none of the viruses transmitted to humans via foods is known to infect other animals (with the trivial exception that some primates, which are not a factor in food-borne disease, may also be susceptible to some of these agents). All of these human-specific food-borne viruses are shed in infected persons' faeces; the Norwalk-like viruses are also shed in vomitus.

Transmission of human enteroviruses (polioviruses, coxsackieviruses, echoviruses and numbered enteroviruses) via food is now rare in developed countries, but may still be occurring in some parts of the world. Because the world's poorest nations are unlikely to have the resources to test foods for viral contamination, detection of food-borne enteroviruses will be addressed here as historic, rather than current interest.

Available Detection Methods

A great deal of development work has been devoted to the detection of food-borne enteroviruses in cell cultures. Many viruses of this group are capable of replicating in cultured primate cells and producing cytopathic effects under fluid medium or plaques under agar overlay medium. Several established monkey kidney cell lines have been used for this purpose – the University of *B*uffalo [New York] Green *M*onkey (BGM) line (African green monkey: *Cercopithecus aethiops*) has an especially rich history in this application. Because the plaque technique yields quantitative results, the methods of extracting enteroviruses from foods could be evaluated for efficiency of recovery, using known quantities of experimentally inoculated virus. Unfortunately, none of the Norwalk-like viruses, nor the wild-type hepatitis A virus, replicates in cell cultures in a way that would be useful for detection from foods. The same is true of the rotaviruses and the hepatitis E virus, which leaves only the relatively rare astroviruses as potentially detectable on the basis of their infectivity in cell culture. In fact, detection in cell cultures of hepatitis A virus from field samples has occasionally been reported, but the tenacity required hardly recommends this as a practical approach.

The infectivity of viruses in food is the quintessential concern of detection; but if susceptible laboratory hosts are unavailable, other means of detection must be sought. Each virus particle has first, a specific, but not totally distinctive, morphology; second, a coat of protein that has specific, and sometimes group-specific, antigenic properties; and third, a genome encoded in a specific sequence of ribonucleotides (for whatever reason, all of the virus groups named above contain RNA rather than DNA).

Virus detection methods not based on infectivity must take advantage of one or more of these other characteristics.

Morphology

Groups of viruses are defined to some extent on the basis of their shared morphology. Norwalk-like (small, round, structured) viruses have a distinctive surface with characteristic depressions; these are now classified with the caliciviruses, which include some other gastroenteritis viruses and the hepatitis E virus that present a somewhat different appearance on the electron microscope. The hepatitis A virus is a member of the family Picornaviridae, as are the enteroviruses; all of these look identical on the electron microscope, with very little perceptible surface detail. The astroviruses, like the groups just mentioned, are at or near 30 nm in diameter, but present a different, star-like surface pattern. The rotaviruses are approximately 70 nm in diameter and have yet another surface pattern. Thus, electron microscopy can identify virus in a food extract to the group or family level, but this approach may be confounded by debris in the food extract and will not discriminate, say, between enteroviruses of human origin and those from cattle or swine. Immune electron microscopy, which obviates some of these problems, is discussed under combination methods below. It is noteworthy that electron microscopy, being nonspecific, is perhaps the only non-infectivity detection method that offers any chance of detecting a previously unknown virus in a food extract. On the other hand, the volume of sample extract that can be accommodated on an electron microscope grid is so small that the method is much less sensitive than most of the alternatives, and thus prone to yield false-negative results.

Serology

Particles (virions) of all of the food-borne viruses to be considered here have protein outermost.These coat proteins comprise repetitive subunits of one to a few polypeptides. Multiple antigenic sites (epitopes) occur on each viral particle surface: some of these are type-specific, and some may be group-specific. Some are related to the specificity with which viruses attach to susceptible host cells; all evoke antibody responses in infected persons and in inoculated laboratory animals. Methods based on serology have been developed for use in diagnostic virology. Some of these entail detection of the viral antigen in clinical specimens by means of laboratory antibody, but more often antibody produced by the patient is reacted with a reference viral antigen. This antibody response may not yet be detectable during the period of illness, but antibodies are highly reliable means of diagnosis, especially if the method specifically detects antibody of the IgM class, indicative of current or recent infection. However, methods for detecting viral particles serologically (e.g. enzyme immunoassay) are marginally sensitive enough for use with clinical (typically faecal) specimens, but unlikely to be sensitive enough to be applicable to detection of viruses in food extracts. Although all of the food-borne viruses are shed in faeces, the dilution factor involved in faecal contamination of food is too great to permit levels of virus detectable by serology to occur in foods. Nevertheless, antiviral antibody can be useful in combination detection methods, as discussed below.

Genome-based Methods

As was stated above, the viruses of interest contain RNA, rather than DNA. Detection methods based on oligonucleotide probes specific for a portion of the viral genome have been developed, but they have generally lacked sensitivity. Greater sensitivity is achievable by the polymerase chain reaction (PCR); this necessitates reverse-transcription (RT) of the viral genome, so that the PCR can be performed with complementary DNA (cDNA). RT-PCR offers an extremely sensitive and specific method for detecting viruses; however, components of food extracts often interfere with either RT or PCR, so a great deal of care is required in preparing food samples for testing by RT-PCR. Obviously, genomic test methods require knowledge of at least part of the nucleotide sequence of the viral genome, to permit synthesis of appropriate probes, PCR primers or both.

Combination Methods

The antigens of the viral coat can be used to collect viral particles specifically for subsequent detection on the basis of morphology or genomic specificity. Immune electron microscopy is based either on agglutination of the viral particles with homologous antiserum or on attachment of the particles to an electron microscope grid coated with homologous antibody. Type-specific antibody is typically used, but advantage may be taken of group cross-reactions.

Virus can also be collected specifically for RT-PCR. Antibody can be coated on to the interiors of microcentrifuge tubes in which the subsequent reactions are performed, or the antibody can be attached to paramagnetic beads that are mixed with the sample extract and collected by means of a magnet. In either case, the virus is specifically held by the antibody, while nonviral materials that may interfere with the test, as well as other viruses, are removed by washing. In most instances, these combination methods are of use only when the virus to be detected is already known and homologous antibody is available beforehand.

Foods to be Tested

By far the food most often implicated as a vehicle for viruses is bivalve molluscs (e.g. clams, oysters, mussels) which feed by filtration and are able to concentrate viruses from their growing waters, if the waters are subject to human faecal contamination. Although virus-contaminated shellfish are able to purify themselves during prolonged holding and siphoning of virus-free water, it is known that viruses are held more tenaciously than the bacterial indicators (e.g. thermotolerant coliforms) that are often used to determine the success of the purification process. This means that in situations that are perceived as likely to lead to virus contamination, testing is the only way to determine whether shellfish contain virus – short of feeding them to human volunteers, as has sometimes been done.

Shellfish are probably not the only foods subject to contamination via faecally polluted water; however, other foods contaminated via water are unlikely to concentrate the virus as shellfish do. Direct contamination of foods by infected persons is also an important possibility. During the incubation period of hepatitis A, the virus may be shed for 10–14 days before symptoms begin; shedding of the Norwalk-like viruses sometimes continues for some days after symptoms have ended. In both instances, the principal or exclusive route of shedding is via the intestines in faeces; contamination of food often takes place via unwashed hands that have faeces on them. Many outbreaks have been attributed to this mode of contamination, although the faecal deposits were not visible on the food, and presumably not on the workers' hands, either. Under such circumstances, there is no limit to the variety of foods at risk; but it is particularly important to consider foods that are handled just before serving or that will not be cooked after handling. Sandwiches and salads have most often been contaminated in handling and resulted in consumer infections, but many other foods have been vehicles in one-of-a-kind outbreaks. The recent US outbreak involving frozen strawberries, as unlikely as this may seem, was not unprecedented.

Sampling

There have been few studies of how viruses are likely to be distributed in contaminated food, and these have been done with model agents that, often, had not been produced in the human intestines and shed in faeces. Viruses may be shed at levels exceeding a million units (cell-culture infectious doses in most experimental situations, which are likely to comprise 100 times as many viral particles) per gram of faeces, so a milligram or less of faecal contaminant might cause infection. It is probably not safe to assume that virus is uniformly and randomly distributed in the suspect food. Therefore, a sample might best be a composite of several subsamples. The quantity of sample that is eventually tested after processing depends entirely on the test method applied and the resources of the laboratory. Most published methods have begun with 25–100 g of food. In that faecal contamination is unlikely to be visible, one can only sample on the basis of relative accessibility to external contamination. For solid foods that have specific form and perhaps an impervious skin or rind, there is little purpose in sampling the interior of an intact unit. Liquids and foods that have no specific form (e.g. comminuted meat products) can simply be stirred or mixed before sampling – there is no scientific basis for a sampling rationale beyond this, but see the section on interpretation of results later in this entry.

In addition to these theoretical considerations, it is important to note that representative samples are often unavailable in outbreak investigations. The long incubation period of hepatitis A – averaging 28 days – makes it unlikely that food eaten by the ill persons will remain. With other virus diseases, the desire of food laboratories to rule out all possible bacterial causes before considering a virus (and the reluctance to freeze samples that are to be tested for bacterial pathogens) may result in advanced decomposition of the samples before virus testing is proposed.

Sample Processing

A sample must be liquid in order to be processed. Foods that are solid or semisolid must be liquefied, by homogenization, washing, dilution or processing by other means. Almost inevitably, the fluid extract of the sample will need to be concentrated before testing; this requires that food solids be removed beforehand. Some testing methods will also require that the bacteria that were inevitably present in the sample (and any moulds or yeasts) be removed in some fashion or killed. Concentration of the sample consists either of selectively removing the virus from the aqueous food extract or of subtracting surplus water and leaving the virus in a smaller volume than formerly. Addition of substances to facilitate detection of the virus may also be required.

Liquefaction

At one time, the standard liquefaction apparatus for solid foods was a mortar and pestle. Now, the task is generally mechanized, using a high-speed rotary homogenizer or the milder Stomacher® or Pulsifier®. Soluble samples can be dissolved in water with stirring. Dissection of shellfish is sometimes done so as to include only the portions of the animal that are thought most likely to harbour the virus. What seems not yet to be generally available is an apparatus for washing the exteriors of foods that are likely to have only surface contamination.

Purification–extraction

Food solids have generally been removed from sample suspensions by centrifugation or by filtration. Inevitably, one runs a certain risk that a portion of the virus originally present in the sample is discarded with the solids. Although virus particles require extreme *g* forces to be sedimented efficiently, a portion will certainly by sedimented even by low-speed centrifugation. In one approach much of the food solids are collected in a bottom layer of trichlorotrifluoroethane in the centrifuge tube. In that this reagent is suspected of depleting the ozone layer, it is likely to become unavailable in the near future and require replacement with something else. Coagulation of food solids to enhance separation by centrifugation or filtration may be attempted by adding materials such as CatFloc®, a polycation sewage coagulant. A variety of prefilters and filters are available for clarification of food extracts in virus testing; care must be taken that they do not adsorb virus that is supposed to pass into the filtrate.

Decontamination

If bacteria, moulds and yeasts must be removed before the virus detection step, this may be possible with a chloroform treatment or by adding antibiotics. If clarification has been extremely effective, it may be possible to remove the microbial contaminants by passage through a filter of 0.45 or 0.20 μm porosity. Again, one must be careful to avoid virus loss by adsorption to the filter matrix. Antibiotics may serve the purpose, if only bacteria are expected to be present; some virus detection methods may not require prior microbial decontamination.

Concentration

Classic physical methods of virus concentration from dilute solution have included preparative ultracentrifugation and ultrafiltration (sometimes with dialysis membrane as the ultrafilter). Now, one has the option of precipitating virus with polyethylene glycol or proprietary products, or one may collect the virus by adsorption to special membrane filters (e.g. the Virosorb 1MDS®) and elute with beef extract or urea arginine phosphate buffer. These eluates may be treated to produce a second-stage concentration by co-precipitation of the virus. As was stated earlier, immunological capture of virus with antibody on the inner surface of a microcentrifuge tube or on paramagnetic beads offers a specific means of capturing virus and eliminating unwanted components of the sample extract by washing.

Additives

If molecular genomic testing is the intended method of virus detection, it may be found that the food extract contains substances that interfere with either RT or with PCR amplification. If rigorous purification cannot remedy this, it may be necessary to add substances such as cetyltrimethylammonium bromide to mitigate the interference. Neither the mode of action of the interfering substances nor those of the antidotes seems to be well understood.

Testing of Sample Extracts

The methods available for detection of viruses in food extracts have been reviewed above. No matter which is chosen, it is important that adequate controls be included. Positive controls should include the virus of interest, both in the food of interest and suspended in laboratory diluent. A negative control should be a sample of the food being tested, which is reasonably certain to be virus-free. As was noted earlier, foods to be tested for virus in outbreaks may be partially decomposed; foods that serve as controls in this instance should ideally be in a similar state of decomposition, although this will not always be possible. With genomic test methods, it is further necessary to add various positive and negative controls to demonstrate that the results of amplification, analysis or probe testing are specific for the agent sought.

Most test methods discussed are basically qualitative. One determines the presence or absence of the virus in whatever quantity of sample was actually tested. It is usually not possible to infer the level of contamination of the original food sample. A great deal of extra difficulty is involved in quantitative determinations, so these are ordinarily applied only to samples that are known or highly likely to be positive. Otherwise, the emphases in developing and evaluating test methods are on sensitivity (absence of false-positive results) and specificity (absence of false-negative results).

Interpretation of Results

One expects either a positive or a negative test result. Unfortunately, ambiguities are possible with some of the methods discussed. Even if the result appears to be clearly positive or negative, it is important to consider how the outcome should be interpreted.

Positive results may seem straightforward, especially if there have been ill persons and a sample of the food they ate appears to contain the virus that has been shown to have made them ill. One might well wish to be able to isolate some of the food-borne virus in cell culture, by way of verification, but this is seldom possible, for reasons stated previously. Verification by testing an additional portion of the same food sample, perhaps by another method or in a different laboratory, should be considered. Once the detection has been verified, the possibility of matching a portion of the viral genome, from the food sample and clinical isolates, should be considered.

A negative test result may have several explanations. The most likely is that virus was absent from the food or, more likely, from the sample tested. Given the probable heterogeneity of virus distribution in food, it is distinctly possible to draw a negative sample from a lot of food that contains virus somewhere. Additionally, it is possible that the food available for testing is not representative of the food eaten by the ill persons. Other possible explanations are loss of virus during extraction and concentration of the sample, presence of interfering food substances (especially when doing RT-PCR with a food that is unfamiliar) or simple failure of the detection system.

Alternatives

Given the difficulties described of detecting viruses in food and water, it is not surprising that indicators have been sought, the presence of which in a sample would suggest that a faecal, and possibly viral, contamination had occurred. Bacterial groups with various perceived relationships to faecal contamination (coliforms, faecal (thermotolerant) coliforms, enterococci, *Escherichia coli*) have been considered in this role, but in each case it has been shown that the presence of bacterial indicators is poorly correlated with the presence of human pathogenic viruses. The inference is that a proper indicator should be some kind of a virus that is certainly of human origin and is detectable on the basis of its infectivity. It should also be considered that an indicator of this sort is most likely to occur in conjunction with human pathogenic virus in faecally contaminated water or waste water and foods exposed to these, rather than in foods contaminated from the hands of a single infected individual. Human enteroviruses have been considered in this role; it is expected that vaccine polioviruses would predominate over coxsackieviruses and echoviruses in most situations, but doubt has been raised over the universality of poliovirus presence in community sewage. Alternatively, consideration has been given to viruses that infect enteric bacteria: these have included somatic coliphages that infect *E. coli* via its cell wall, male-specific coliphages that infect F^+ *E. coli* via the F-pilus and bacteriophages that infect *Bacteroides fragilis*. These share the advantage that infectivity tests in bacterial hosts are cheaper, and generally quicker, than in primate cell cultures. Each has some perceived disadvantages. Somatic coliphages can generally multiply in a range of ambient temperatures, and so might increase in number while pathogens were dying. Male-specific coliphages probably cannot multiply in the environment in most instances, but their numbers are lower than those of somatic coliphages in many instances where one would wish them to be plentiful. Further, coliphages strongly suggest faecal contamination (as does their host, *E. coli*), but the faeces may have come from any warm-blooded animal, rather than only from humans. Bacteriophages of *B. fragilis* have been considered because this host bacterial species is apparently more characteristic of human faeces than of faeces of other animals. However, *B. fragilis* is a strict anaerobe, which complicates its cultivation and use in seeking its phages in food extracts or in water samples. The incidence of *B. fragilis* phages in faecally contaminated water appears to be quite variable, which might mean that the phages would not be reliably present in faecally contaminated foods. Although it is reasonable to continue the quest for indicators of viral contamination, it should be recalled that these would be most applicable to monitoring water and shellfish, but probably of little use in monitoring food that is most likely to be contaminated (if at all) by infected individuals.

Summary

Because viruses are frequent causes of food- and waterborne disease, it is reasonable to seek methods of detecting them in food and water. Detection might be undertaken in routine monitoring of potential food and water vehicles, in risk assessment studies directed to specific vehicles or in investigation of suspected vehicles after an outbreak of human disease. The viruses reportedly most often transmitted via food and water are the hepatitis A virus and the Norwalk-like (small, round, structured) viruses; others include the astroviruses, rotaviruses, caliciviruses causing gastroenteritis or hepatitis E and enteroviruses. All of these agents are apparently human-specific and shed in faeces. The pathogenic viruses might be detected on the bases of their infectivity for cultured primate cells, their appearance as seen in the electron microscope, their specific coat protein antigens, the individual sequences of their ribonucleic acids or combinations of these. Each of these detection methods requires meticulous extraction and concentration of the sample in preparation for testing.

Because the most significant food- and waterborne viruses do not reliably infect cell cultures, the alternative approaches listed are most often applied. All are exacting and costly, and each has special problems that may lead to false-negative test results. Even so, these may serve specific purposes in risk assessment studies and in outbreak investigations. Indicators of probable viral contamination, such as coliphages, continue to be considered as means of monitoring water and shellfish; they are unlikely to be useful in monitoring foods (e.g. sandwiches) that would probably only be contaminated in handling by an infected person.

See also: **Food Poisoning Outbreaks**. **National Legislation, Guidelines & Standards Governing Microbiology**: European Union. **Sampling Regimes & Statistical Evaluation of Microbiological Results**. **Viruses**: Introduction; Environmentally Transmissible Enteric Hepatitis Viruses: A and E. **Water Quality Assessment**: Modern Microbiological Techniques.

Further Reading

Appleton H (1994) Norwalk virus and the small round viruses causing foodborne gastroenteritis. In: Hui YH, Gorham JR, Murrell KD and Cliver DO (eds) *Foodborne Disease Handbook*. Vol. 2, p. 57. New York: Marcel Dekker.

Atmar RL, Neill FH, Woodley CM et al (1996) Collaborative evaluation of a method for the detection of Norwalk virus in shellfish tissues by PCR. *Appl. Environ. Microbiol.* 62: 254–258.

Chung H, Jaykus LA and Sobsey MD (1996) Detection of human enteric viruses in oysters by in vivo and in vitro amplification of nucleic acids. *Appl. Environ. Microbiol.* 62: 3772–3778.

Cliver DO (1994) Other foodborne viral diseases. In: Hui YH, Gorham JR, Murrell KD and Cliver DO (eds) *Foodborne Disease Handbook*. Vol. 2, p. 137. New York: Marcel Dekker.

Cliver DO (1995) Detection and control of foodborne viruses. *Trends Food Sci. Technol.* 6: 353–358.

Cliver DO, Ellender RD, Fout GS, Shields PA and Sobsey MD (1992) Foodborne viruses. In: Vanderzant and Splittstoesser DF (eds) *Compendium of Methods for the Microbiological Examination of Foods*. P. 763. Washington, DC: American Public Health Association.

Cromeans T, Nainan OV, Fields HA, Favorov MO and Margolis HS (1994) Hepatitis A and E viruses. In: Hui YH, Gorham JR, Murrell KD and Cliver DO (eds) *Foodborne Disease Handbook*. Vol. 2, p. 1. New York: Marcel Dekker.

Cromeans TL, Nainan OV and Margolis HS (1997) Detection of hepatitis A virus RNA in oyster meat. *Appl. Environ. Microb.* 63: 2460–2463.

Grešíková M (1994) Tickborne encephalitis. In: Hui YH, Gorham JR, Murrrell KD and Cliver DO (eds) *Foodborne Disease Handbook*. Vol. 2, p. 113. New York: Marcel Dekker.

Herrmann JE (1994) Laboratory methodology. In: Hui YH, Gorham JR, Murrell KD and Cliver DO (eds) *Foodborne Disease Handbook*. Vol. 2, p. 177. New York: Marcel Dekker.

Le Guyader F, Neill FH, Estes MK, Monroe SS, Ando T and Atmar RL (1996) Detection and analysis of a small round-structured virus strain in oysters implicated in an outbreak of acute gastroenteritis. *Appl. Environ. Microbiol.* 62: 4268–4272.

Sattar SA, Springthorpe VS and Ansari SA (1994) Rotavirus. In: Hui YH, Gorham JR, Murrell KD and Cliver DO (eds) *Foodborne Disease Handbook*. Vol. 2, p. 81. New York: Marcel Dekker.

Vitamin Metabolism *see* **Metabolic Pathways**: Metabolism of Minerals and Vitamins.

WATER QUALITY ASSESSMENT

Contents

Routine Techniques for Monitoring Bacterial and Viral Contaminants

John Watkins, CREH Analytical, Leeds, UK
Liz Straszynski, Alcontrol Laboratories, Bradford, UK
David Sartory, Severn Trent Water, Shrewsbury, UK
Peter Wyn-Jones, Sunderland University, UK

Water is a basic requirement of living organisms. Throughout the world waterborne and water-associated diseases affect millions of people. In developed countries, clean water is readily available. It will have been subjected to treatment and disinfection, and will have been monitored for both chemical and microbiological qualities. In many areas of the world, although water is available and does not have to be paid for, the quality of the water is poor. Where there is no treatment or organized distribution of water, the incidence of gastroenteritis is high, particularly in young children. As well as being directly consumed, water is also used in the manufacture and preparation of food and drink, and is an important medium for recreation. Water quality standards must take account of all these uses, and the test procedures used must identify deterioration quickly and accurately.

Contamination of water with animal or human faecal material can introduce large numbers of pathogenic microorganisms, many of which have a low infective dose. Distribution of these organisms in drinking water can result in large numbers of people becoming infected in a short space of time, resulting in an outbreak of waterborne disease. Contamination may occur in the water before treatment, treatment may be ineffective or contamination may occur in the distribution system after effectively treated water has been distributed. Water and distribution systems may also contain relatively large numbers of saprophytic microorganisms. Increases in temperature during the summer months, or increases in temperature while water is stored and distributed in buildings, together with reduced amounts of disinfectants may result in growth of potentially hazardous species. These conditions may permit the growth of *Pseudomonas aeruginosa* or *Legionella* spp., and here testing for faecal contamination is of little value.

Rationale of Water Testing

Regular sampling and analysis provide data on the quality of raw water, the efficiency of water treatments and the integrity of distribution systems. The range of pathogenic microorganisms is extensive and therefore water is examined for microbiological indicators of contamination. The use of indicator organisms is based on the assumption that if they are present, the pathogens may also be present, but that if they are absent then the water is suitable for consumption or use in manufacturing. To be useful, indicators must always be present when contamination has occurred and must be present in higher numbers than pathogens.

The principal bacterial indicators are the coliforms, faecal coliforms, enterococci, *Clostridium perfringens*, the sulphite-reducing clostridia and total aerobic bacteria. The coliform group encompasses a wide range of genera of the family Enterobacteriaceae. Many of the group are common in soil and natural waters and can grow in water. Their presence in drinking water is undesirable, but does not necessarily indicate any health hazard. The faecal coliforms are a group of thermotolerant species which may indicate faecal contamination. The key species of the faecal coliform group is *Escherichia coli*, commonly found in the faeces of humans and animals and is thus a definitive indicator of faecal contamination. Other species of faecal coliforms (e.g. *Klebsiella pneumoniae* and *Enterobacter cloacae*), although associated with faeces, also occur naturally in the environment, on vegetation and in soils. The enterococci and the clostridia are able to survive in water for substantially longer periods and are therefore better indicators than the coliforms of remote pollution and serious contamination of raw water, when treatment and disinfection have removed coliforms and faecal coliforms. They are also used for determining the hygienic operation of new installations such as treatment works or boreholes and the proper laying, disinfection and repair of water mains. The total viable bacterial count (plate count) is a measure of the hygienic quality of the water and not necessarily a measure of its suitability for consumption. Counts in treated water are usually low, but they may increase dramatically if water becomes stagnant and the temperature increases. There have been many arguments about indicators reliably establishing the presence of pathogens, particularly those pathogens which are resistant to disinfection. Certainly, there have been a number of waterborne outbreaks of disease where indicator analysis has been satisfactory.

Current Legislation Governing Water Quality

The European Community Directive relating to the quality of potable water and water intended for food production (known as the Drinking Water Directive) was implemented in July 1980. It lists microbiological criteria, providing guide (G) values and maximum admissible concentrations (MACs), and details analytical methods and sampling frequencies. It does not require direct monitoring for specific pathogens but does state that water should be free from such organisms. The United Kingdom adopted these standards in the Water Supply (Water Quality) Regulations 1989, (amended 1990 and 1991). Private water supplies were similarly covered in the Private Water Supply (Water Quality) Regulations, 1991. In the USA, drinking water quality is covered by the National Primary Drinking Water Standards (1994) and in Canada by the Guidelines for Canadian Drinking Water Quality (1996).

Current Standards

The World Health Organization (WHO) Guidelines for Drinking Water Quality recommend that all 100 ml water samples intended for drinking must not contain *Escherichia coli* or thermotolerant coliforms. Where water is treated it should not contain coliforms on entry to the distribution system, and they should not be detected in 95% of samples from a distribution system. Under European legislation, treated waters should contain (per 100 ml) no coliforms, no faecal coliforms/*E. coli* and no enterococci. The current value for sulphite-reducing clostridia is not more than one in 20 ml. These are MAC values. For the aerobic plate counts there are guide values per millilitre of less than ten for colonies at 37°C and less than 100 for colonies at 22°C. There is an additional provision for the plate count that there should be no significant increase above those levels normally seen during routine analysis. If plate count levels normally exceed the guide values, this is acceptable providing that levels do not increase significantly. For distribution systems and at point of use, 95% of samples must be free from coliforms but faecal coliforms must be absent at all times. Standards for the United States are based on the monitoring of coliforms in distribution, where 95% of samples must be coliform free (or no more than one sample per month may be positive if less than 40 per month are taken). Isolation of coliforms requires that there be additional sampling and testing for *E. coli*. There are no standards for the other indicator organisms. In Canada the MAC for coliforms is stated to be zero, but there is compliance if no sample contains more than 10 coliforms and no faecal coliforms per 100 ml and no consecutive samples contain coliforms. In both the USA and Canada, aerobic plate counts are monitored, primarily as an adjunct to the coliform test as high numbers of heterotrophic bacteria (500–1000 ml^{-1}) can interfere with standard methods of coliform enumeration.

Frequencies of sampling and analysis for each indicator are set by each country, and are typically based on the sizes of the populations supplied.

Methods of Analysis

Early methods developed for water analysis were based on most probable number (MPN) or multiple tube techniques using liquid culture media. There was always the requirement for confirmation of any positive tests and therefore first results were always presumptive. Although relatively simple to perform, such test procedures were time consuming and of long duration, with up to four days required to obtain confirmed results. In the UK, a chemically defined medium, Minerals Modified Glutamate Medium, was introduced in 1969. This is still used by some laboratories. The introduction of enzyme-specific chromogenic and fluorogenic substrates has made the MPN test procedure popular again, and removed the need for confirmations. Results can now be available in 18 hours (**Table 1**).

Membrane filtration was introduced in the 1950s

Table 1 Isolation media for the detection of faecal indicators

Indicator organism	*Isolation medium*	*Incubation time and temperature*	*Confirmation*
Coliforms and faecal coliforms (UK)	Minerals modified glutamate medium	37°C 48 h	Oxidase, lactose peptone and tryptone water
UK	Membrane lauryl sulphate broth	30°C 4 h, 37°C 14 h or 44°C 14 h	Oxidase, lactose peptone and tryptone water
UK and USA	Colilert (Idexx)	37°C 18 h	None required
UK	Membrane lactose glucuronide agar	30°C 4 h, 37°C 14 h	None required
USA	m-ENDO Broth	35°C 20–22 h	Lauryl tryptose broth, brilliant green bile broth
USA	m-FC Medium	44.5°C 24 h	None required
Enterococci (UK)	Slanetz and Bartley agar	37°C 48 h	Bile aesculin azide or kanamycin aesculin azide agar
Clostridia	Tryptose sulphite cycloserine agar, OPSP agar	37°C 48 h	Litmus milk
Total viable bacterial count (UK, USA)	Yeast extract agar, plate count agar, R2A Agar	37°C 24 h, 22°C 72 h, 20°C 7 days	None required

and is currently the basis of the standard methods used in the UK for coliforms, faecal coliforms, enterococci and clostridia. The standard medium for coliforms and faecal coliforms is membrane lauryl sulphate broth. Coliforms are incubated at 37°C and faecal coliforms at 44°C, for 14 h, with both having a pre-incubation period of 4 h at 30°C. In the USA m-Endo medium with incubation at 35°C is employed for total coliforms and mFC agar with incubation at 44°C for faecal coliforms. More recently, methods based on defined substrate formulations have been widely used. Enterococci are enumerated on Slanetz and Bartley medium incubated at 37°C for 48 h for non-contaminated waters or at 44°C for 48 h for contaminated waters. Kanamycin aesculin azide medium may be used as an alternative. Clostridia present a more complex problem. Sample pasteurization may be required before incubation on tryptose sulphite cycloserine (TSC) agar or oleandomycin polymixin sulphadiazine (OPSP) agar. Chromogenic and/or fluorogenic substrates may be added to the above media to make confirmation unnecessary. Yeast extract agar is the medium of choice for plate counts in the UK, whereas in the USA plate count agar is used. Comparative trials suggest the two media give similar results. Many bacteria of water origin have simple nutritional requirements and the use of a nutritionally weak medium (for example R2A agar) together with a lower incubation temperature and longer incubation period can recover more bacteria from water samples than yeast extract or plate count agars.

Confirmation of Isolates

The media used for the isolation of faecal indicators may also isolate other bacteria which can give a typical positive reaction. Confirmation of positive MPN tubes or colonies on membranes is usually required. Coliform bacteria by definition ferment lactose at 37°C with the production of acid and do not possess the enzyme cytochrome oxidase. Faecal coliforms are coliforms that grow and can produce acid from lactose at 44°C. Some coliforms which are not faecal in origin may also grow at 44°C. For this reason, in the UK, *Escherichia coli* is considered to be the only true faecal coliform. It also produces indole from tryptophan at 44°C. Acid and indole production at 44°C can be found in coliforms other than *E. coli* and the test results should be viewed with caution. Experience has shown that coliforms other than *E. coli* can be isolated by some food microbiology methods. Biochemical test kits are commercially available for the identification of Enterobacteriaceae, but these can give widely differing results between kits and between single strains even when a single strain is tested in duplicate. The introduction of chromogenic and fluorogenic substrates has removed much of the ambiguity of confirmations although identification of coliforms can still be uncertain.

Coliforms are confirmed by subculture of typical colonies to nutrient agar and after growth, testing for oxidase and inoculation of lactose peptone water for the production of acid at 37°C within 48 h. Faecal coliforms are confirmed as oxidase negative and by the production of acid in lactose peptone water at 44°C and indole in tryptone water at 44°C. Enterococci are confirmed by their ability to hydrolyse aesculin when grown on bile aesculin azide agar or kanamycin aesculin azide agar. Sulphite-reducing clostridia can be confirmed as *Clostridium perfringens* by inoculation into litmus milk and incubation at 37°C for up to 5 days. Fermentation of the lactose causes acid production and clotting of the milk.

Quality Assurance

Legislation on water quality assessment in the water industry has included the requirement for laboratories to have internal quality control and participate in external quality assurance schemes. Because of the difficulties of preparing and keeping suspensions of bacteria for quality control, manufacturers have now developed dried suspensions which can be stored and rehydrated to give consistent suspensions for day to day quality assurance. These suspensions can also be used in some of the validation procedures for new methods and for the assessment of new analysts.

Interpretation of Results

Coliforms in water can derive from a number of sources of which faecal contamination is only one. In the absence of other indicators, they may be derived from vegetation, soil or surface water. In addition there is ample evidence that coliforms can grow in distribution systems in association with biofilms. Although their presence in water is undesirable, the water may be perfectly safe for drinking, but its use for other purposes may be unsatisfactory. The presence of faecal coliforms, especially *Escherichia coli*, and other indicators means that water is probably contaminated with faecal material and immediate remedial action must be undertaken. Aerobic plate counts above the guide values in the EC Directive do not mean that water is unfit for human consumption. High counts at 37°C are suspicious and may require further investigation. High counts at 22°C suggest stagnation and warming of water with the concomitant growth of normal water bacteria. Cleaning and disinfection are required in some circumstances, e.g. hospitals, where counts in excess of 1000 ml^{-1} at 22°C may require a water distribution system to be disinfected. Exceptional counts may be used as indications for the growth of other microorganisms in such as cooling towers and swimming pools. Pseudomonads, in particular *Pseudomonas aeruginosa*, aeromonads and *Legionella* spp. can colonize water systems where conditions are suitable, and high aerobic counts may signal the need to look for pathogens.

Other Pathogenic Microorganisms

There is a wide range of pathogens that can contaminate water, including bacteria, viruses and eukaryotic parasites. Analysis for bacteria such as *Campylobacter* spp., *Salmonella* spp., *Pseudomonas* spp. and *Aeromonas* spp. follow conventionally published methods. Immunological techniques such as enzyme-linked immunosorbent assays (ELISA) can also be adapted for water analysis. Methods for the isolation of *Legionella* spp. are currently detailed in a draft British Standard BSI 6068 (1994). The protozoan *Cryptosporidium parvum* is an important waterborne parasite that has been implicated in a number of waterborne disease outbreaks. Methods for detection have been published in the UK and USA and both methods are currently under revision. The significance and occurrence of this organism in water has led to the rapid development of faster and more efficient techniques for its detection, including the use of small volume processing, immunomagnetic separation and flow cytometry.

Escherichia coli O157:H7 has received much attention and can be transmitted by water. Evidence suggests that it is removed by conventional water treatment and is readily killed by conventional disinfectants. However, there have been several outbreaks where water was the vehicle of infection. In one instance, four people died. One of the main problems with such pathogens occurring at low levels in water is the opportunities for contamination of foods, where they may be able to multiply, or the contamination of food wash waters where they can be recycled.

Other Indicators

A wide range of other microorganisms have been described as useful indicators of faecal contamination. Coliphages are bacterial viruses which are specific to *Escherichia coli* and other coliforms. They can be divided into two groups. The somatic bacteriophages gain access to the cell through the cell wall and the F specific bacteriophages attach to the F or sex pilus. The latter have been described as useful indicators of the presence of enteric viruses in water.

Bacteriophages of *Bacteroides fragilis* have also been suggested as useful indicators, as have the bacteria themselves. Both are specific to the faeces of humans and farm animals. *Bifidobacterium* spp. can also be found in humans and some animals. Sorbitol-fermenting strains have been specifically linked to human faeces. Both groups of bacteria are strictly anaerobic. *Rhodococcus coprophilus* is a *Nocardia*-like actinomycete found in farm animals (cattle, sheep, pigs and horses) and may be a useful indicator of contamination with animal faeces.

Routine Techniques for Monitoring Viral Contamination

Viruses in Water

The aquatic environment (which includes fresh and marine water, raw and treated sewage, sludge and sediments) often provides conditions in which pathogenic viruses released from the body may retain infectivity, and so may cause disease on entry into a new host. Human enteric viruses enter the water cycle when faecal waste is discharged into the sewerage system; they may end up in water which, after treatment, is used for drinking. The purpose of sewage treatment is to remove microorganisms and other organic matter from sewage so that the resulting effluent can be discharged to receiving waters with minimal adverse effects. It also renders any receiving water fit for use as a raw water source for drinking water abstraction, as up to 90% of viruses are removed during sewage treatment. Viruses may rarely also gain access to raw water sources by percolation through the soil into groundwater used as a spring or a borehole supply. The treatment process to produce wholesome drinking water involves the removal of organic material by flocculation and sedimentation followed by disinfection with chlorine. Most viruses will be removed by flocculation and the remainder will be inactivated by disinfection. It is thus important to be able to monitor both raw and finished waters for viral contamination where there is likely to be a risk that these are present.

Human Enteric Viruses

Enteric viruses, i.e. those inhabiting the gastrointestinal tract, may be divided into two groups according to their effects on the gut lining. The enteroviruses cause little damage to the gut epithelium and thus do not usually cause gastroenteritis. Although they multiply in the gut and therefore may be pathogenic, in the majority of infections they do not migrate to other tissues. This group includes poliovirus, coxsackieviruses and echoviruses. They cause a variety of symptoms, mainly in children. They are shed into the sewage in large numbers and include poliovirus vaccine strains, which will be present all year round. The viral pathogens which do cause gastroenteritis include the rotaviruses, small round-structured viruses (SRSVs, Norwalk viruses) and astroviruses. It is important to distinguish between the enteroviruses, which is a taxonomic group, and enteric viruses, which is a term describing the habitat of all viruses in the gastrointestinal tract, and includes the enteroviruses.

Coincident with their pathogenesis is the ability of enteric viruses to grow in cell culture. Thus, most of the enteroviruses can be grown in BGM cells, a monkey kidney cell line used for many years in the detection of waterborne viruses. Enteroviruses are easily detected in this cell system after concentration. Because of this, and because enteroviruses are resistant to the pH changes carried out during the filtration process (see below), enteroviruses have been the group of waterborne viruses most investigated over the past 20 years. Conversely, the viruses of gastroenteritis cannot be grown in cell culture and are thus much more difficult to detect, as non-culture based systems are required.

The burden of disease imposed by enteric viruses varies, and in many cases is difficult to estimate, since many infections are subclinical or cause only minor symptoms. The enteroviruses cause relatively mild disease in the vast majority of infections, whereas rotaviruses and SRSVs cause moderate to severe diarrhoea with or without vomiting. The proportion of infections associated with waterborne spread is also difficult to gauge but will be very small compared with person to person spread. Most reports of waterborne disease are anecdotal. SRSVs have caused well-documented outbreaks of viral gastroenteritis when finished drinking water has been polluted by sewage, including contamination of a borehole in a tourist resort, water tanks at a mobile home park, well water used to make ice, and cross connections of drinking water supplies with water supplies used for irrigation.

Concentration and Detection of Enteric Viruses

Concentration All viruses are obligate intracellular parasites and so will not multiply outside the body. This, coupled with the diluting effects of water, means that direct detection of these agents in any kind of water (with the exception of raw sewage) requires the sample to be concentrated before virus can be detected. The degree of concentration will reflect the nature of the water and the potential quality of virus it contains. Thus raw waters will require less concentration than finished waters. Typically, volumes of about 10 l are taken from river waters, whereas it is necessary to take samples of 100–1000 l to detect any viruses in finished waters.

The most common concentration technique is adsorption/elution, where virus is adsorbed to an insoluble matrix then eluted in a smaller volume of fluid. Concentration from water samples usually involves passage of the sample through a filter which adsorbs the virus. Filters may be membranes or cartridges. In the UK, membranes are most popular, but in the US, where more potable water supplies are monitored, cartridge filters are used more frequently since they can handle greater volumes without clogging. Membrane materials include cellulose nitrate (commonest), cellulose acetate and glass fibre/epoxy resin, and are either 142 mm or 293 mm in diameter. Since retention of virus is by adsorption, not entrapment, the pore size used is determined by the turbidity of the sample. For membranes, the smallest pore size used is 0.45 μm and the largest, used for turbid waters, is 5 μm. A glass-fibre prefilter may be used for river waters but is not necessary for other types.

Viruses are negatively charged at neutral pH. Most filters are also negatively charged and it is necessary to precondition the water sample to increase the positive ion concentration on the virus particles. This is done by lowering the pH to about 3.5 and sometimes adding aluminium ions (as aluminium chloride) to the sample. Viruses will then adsorb efficiently to the filter.

Following filtration, adsorbed virus is eluted by passing through the filter 100–400 ml of a protein solution, either beef extract or skimmed milk at pH 9, which displaces the virus into the eluant. It is then further concentrated (secondary concentration) by lowering the pH to the isoelectric point of the protein, which results in the formation of a floc to which virus is adsorbed. This floc can be deposited by low-speed centrifugation, resuspended in a small volume (about 10 ml) of phosphate buffer and frozen until assayed. This is the method most popular in the UK and recommended by the Standing Committee of Analysts (SCA).

There are several variations on adsorption/elution. Some laboratories use glass fibre tube filters, which are more economic, but clog faster. Many EU and US laboratories use positively charged filters which avoid the need for preconditioning the water sample, but these are considerably more expensive. Novel techniques being evaluated include adsorption to positively charged nylon membranes, adsorption to glass powder and glass wool, and to immunomagnetic beads. The glass wool method has been used very successfully in France for analysis of drinking water samples for enteroviruses.

Entrapment techniques can be used to concentrate viruses from water; the most effective of these is ultrafiltration, which has been used in Italy and the US for analysis of drinking water supplies. Ultrafiltration can be effected either through hollow fibre systems or tangential flow membranes using a 100 000 M_r cut-off level. The advantages of this system are that a wide range of viruses may be concentrated, usually no secondary concentration is needed, and there is no need to condition the water before filtration. However, the capital costs are high and filtration of turbid waters can be a difficult and lengthy process.

Detection Detection of viruses in concentrates is best done by cell culture, which is the only way to detect infectious virus. As indicated above, of the viruses likely to be present in water, only the enteroviruses will grow well in culture. They are detected quantitatively by plaque assay under agar or by a liquid culture assay such as the MPN technique. Plaque assays are statistically more reliable in terms of the number of cultures usually used, but liquid culture assay will detect a wider range of viruses. Plaque assay may be done with BGM cells in a monolayer or with cells suspended in the agar. The latter approach is between three and 10 times more sensitive than the monolayer assay due to the larger number of cells available for virus infection, and is the SCA recommended method. The US Environmental Protection Agency (1984) recommends both methods. Plaques usually develop after 3–5 days.

To demonstrate the presence of infectious virus of the types which grow less well in culture, more complex procedures must be used. Rotaviruses and astroviruses will undergo limited replication in different cell lines and may be detected by immunofluorescence. Hepatitis A virus grows very slowly in a limited range of cells and may be detected by radioimmunoassay. At present there is no cell culture system for SRSVs.

Molecular biology methods (principally reverse-transcriptase–polymerase chain reaction, RT-PCR) offer an alternative approach to the detection of nonculturable viruses. Methods for the RT-PCR detection of several enteric virus types from water and related materials are available, and have proved as sensitive as culture where the techniques have been tested in parallel. The main drawback to RT-PCR is that it does not specifically detect infectious virus, but a positive result in such an analysis will nevertheless indicate viral pollution of the water. Routine analysis for these fastidious viruses is not practical at present.

Significance of Waterborne Viruses

The detection of viruses in water indicates faecal pollution. If animal viruses, which may gain access to water through soil run-off, are excluded, then it must be inferred that the water is polluted with sewage, since in the absence of virus multiplication in the environment, faeces from an infected individual are the only source of the virus. In this sense enteroviruses act as another indicator of pollution. Viruses are more resistant than many bacterial indicators and their infectious dose is lower than that for most bacterial infections. In the food context therefore, waterborne viruses are a significant, if rare occurrence and monitoring of water sources, and possibly supplies, for these agents as well as the normal bacterial indicators should be considered to maintain public confidence.

See also: **Enterobacteriaceae, Coliforms and *E. coli***: Classical and Modern Methods for Detection/Enumeration. **National Legislation, Guidelines & Standards Governing Microbiology**: European Union. **Viruses**: Introduction; Hepatitis Viruses; Detection. **Water Quality Assessment**: Modern Microbiological Techniques. **Waterborne Parasites**: *Entamoeba*.

Further Reading

Anon (1989) *Isolation and Identification of Giardia Cysts, Cyptosporidium Oocysts and Free-living Amoebae in Water etc.* London: HMSO.

Anon (1994) *The Microbiology of Water 1994 Part I – Drinking Water.* London: HMSO.

Anon (1995) *Methods for the Isolation and Identification of Human Enteric Viruses from Water and Associated Materials.* London: HMSO.

American Public Health Association (1995) *Standard Methods for the Examination of Water and Wastewater*, 19th edn. Washington DC, American Public Health Association.

Block JC and Schwartzbrod L (1989) *Viruses in Water Systems; Detection and Identification.* VCM Publishers.

Dutka B (ed.) (1981) *Membrane Filtration: Applications, Techniques and Problems.* New York: Marcel Dekker.

EC (1980) *Council directive relating to the quality of water intended for human consumption (80/778/EEC). Official Journal of the European Community* L229, (30.08.80) 11–29.

Gleeson C and Gray N (1997) *The Coliform Index and Waterborne Disease.* London: E & FN Spon.

Kay D and Fricker C (eds) (1997) *Coliforms and* E coli *Problems or Solution?* Cambridge: Royal Society of Chemistry.

Reasoner DJ and Geldreich EE (1985) A new medium for the enumeration and subculture of bacteria from potable water. *Applied and Environmental Microbiology* 49: 1–7.

WHO (1993) *Guidelines for Drinking-water Quality*, vol. 1 *Recommendations*, 2nd edn. Geneva: World Health Organization.

Modern Microbiological Techniques

Colin Fricker, Thames Water Utilities, Manor Farm Road, Reading, UK

Introduction

Many of the techniques currently employed for the microbiological examination of water have changed little in recent years. However, the recognition that organisms such as *Cryptosporidium parvum* can be transmitted by water, together with an increased public awareness of the potential for waterborne disease, has led to the development of new techniques for the examination of water. Some of these methods use molecular techniques, such as the polymerase chain reaction; some use instrumentation not normally associated with water microbiology; and others involve refinements in culture-based methods.

Escherichia coli and Coliforms

The tests performed most frequently are those for the presence of coliforms and *Escherichia coli* in drinking water. Most regulatory authorities expect both coliforms and *E. coli* to be absent from 100 ml samples of water intended for human consumption, thus most test formats are based around this volume. Traditionally, tests are carried out using membrane filtration or 'most probable number' procedures. However, these tests are time-consuming and can take several days to give a final result. There has been considerable effort in this area to develop tests that are simple to perform and which give more rapid results.

The majority of the newer tests for coliforms and *E. coli* are based upon the detection of the enzymes β-D-galactosidase and β-D-glucuronidase. Colorimetric or fluorimetric substrates for these enzymes have been incorporated into traditional membrane filtration media to enhance the specific detection of target organisms. However, such media have not become widely used, largely due to problems of interference by competing organisms growing on a membrane surface.

A number of other media have been produced commercially that do allow the specific and rapid detection of coliforms and *E. coli*. These media can be used either in a presence/absence or quantitative format. They are available from a number of commercial suppliers and are becoming widely used throughout the world. In particular, a procedure using 'defined substrate technology' has been widely adopted by many testing laboratories in countries such as the USA, UK, Japan, Canada and Brazil. The Colilert system incorporates a chromogenic substrate for the detection of β-D-galactosidase (orthonitrophenol galactoside) which turns yellow when the substrate is cleaved, and a fluorogenic substrate (methyl-umbelliferyl glucuronide) which fluoresces when cleaved by the enzymes β-D-glucuronidase. Thus, this medium detects coliforms and *E. coli* specifically. Defined substrate technology works on the simple principle of supplying these specific substrates as the sole carbon and energy source; thus if *E. coli* or other coliforms are to grow, they must use the specific substrates which serve as both the carbon and energy source and as an indicator of growth. Other essential growth factors and inhibitors to prevent the growth of competing organisms are also present. The system also uses a simple quantitation device, based on a most probable number system, to allow quantitation of organisms up to 2400 per 100 ml of water. The method has been approved for examination of drinking water by several regulatory authorities and is popular with water testing laboratories because of its simplicity and ease of use. A system for the detection of enterococci in water is also available.

Defined substrate technology and other similar media still require 18 h or more to obtain a result. Whilst this is an improvement over traditional techniques which may take several days, there is a need within the water industry to enable more rapid detection of indicator organisms. To this end several approaches have been investigated, but to date no system has been widely accepted. Many of these methods have been based on alternative, more sensitive methods for detecting the enzymes β-D-galactosidase and β-D-glucuronidase. Such methods have included the use of enhanced fluorescence, where the fluorescent signal is amplified by careful adjustment of pH and the use of an automated instrument for measuring the increase in fluorescence over time. This has allowed the time taken to detect coliforms and *E. coli* to be reduced to approximately 8 h.

Another procedure that has been suggested as being suitable for the rapid detection of indicators uses a laser scanning device to scan membrane surfaces and detect specific fluorescence events associated with bacteria. Organisms are concentrated by membrane filtration and the membranes are then incubated in the presence of fluorogenic substrates for β-D-galactosidase or glucuronidase for detection of coliforms or *E. coli* respectively. Compounds that stimulate the production of these two enzymes are also included in order to ensure that sufficient enzyme is produced to allow detection of the fluorescent product by the instrument. The fluorescent material remains associated with the cell, and analysis of the membrane surface by the laser scanning instrument, after an incubation period of 2–4 h, facilitates the detection of single bacterial cells which have cleaved the appropriate substrate. The instrument is also linked to a microscope with a motorized stage which allows particles to be visually validated, confirming that they are indeed bacteria (**Fig. 1**).

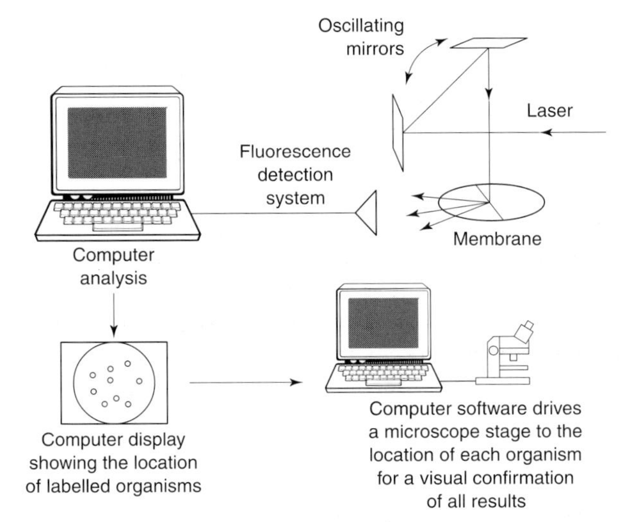

Figure 1 Detection of labelled organisms using the Chem*Scan* RDI instrument.

Molecular Methods

The development of the polymerase chain reaction (PCR) as a method of detecting microorganisms offered the potential to detect both indicator organisms and pathogens rapidly, specifically and with high sensitivity. This technique has been used by many workers in attempts to detect bacteria directly from water samples that have been concentrated by membrane filtration. Initial studies were promising, and suggested that single organisms could be detected in water samples rapidly using PCR. However, more recent reports have not confirmed these findings, and several issues have been identified that may prevent the use of PCR for the direct detection of bacteria in water. The first and most important of these is the issue of viability. Systems based on the detection of chromosomal DNA are unable to differentiate between viable and non-viable organisms. This has been studied most thoroughly using PCR systems targeting the *uidA* gene which codes for the production of β-D-glucuronidase in *E. coli*. Experiments using healthy, viable *E. coli* and cells that had been killed using chlorination showed that the PCR signals generated with viable and non-viable bacteria were similar and could not be distinguished. Further evidence of the ability of this PCR system to detect non-viable cells was presented when it was demonstrated that the nucleic acid from *E. coli* could be detected in human faeces that were 2000 years old. This inability to discriminate between viable and non-viable microorganisms means that PCR is not useful for the direct detection of organisms in water. Another obstacle is that even clean, wholesome drinking water may contain substances (notably iron and humic acids) inhibitory to PCR, significantly reducing its sensitivity, and the methods required to remove these substances would make the procedure too complicated for incorporation into a routine laboratory. In many countries there is a requirement to report the presence of coliforms and *E. coli* as the number of organisms present in 100 ml of water; while quantitative PCR procedures are being developed, they are not yet at a stage that allows accurate quantitation of bacterial cells in water.

An alternative to the use of PCR is the direct detection of bacterial cells using nucleic acid probes. While probing for single copy genes is unlikely to give the required sensitivity, targeting multiple copies of a gene could be successful. Most work in this area has concentrated on the use of 16S ribosomal RNA targets. The use of probes to the rRNA has distinct advantages since many copies of rRNA are present in living cells, and also because there is much sequence data available. Sequences of 16S rRNA are widely used in bacterial taxonomy and therefore should provide good targets for direct detection assays. A disadvantage, however, is that the coliform group of bacteria is not taxonomically defined and indeed is defined differently in different countries and even by separate industries within a country. Therefore, a single 16S rRNA probe cannot be defined for the coliform group, which consists of a wide number of genera within the Enterobacteriaceae. Probes specific to the 16S rRNA of *E. coli* have been developed and these have been shown to detect cells directly on a membrane surface. Direct labelling of the probe with a fluorescent label did not generate sufficient signal to be detectable in most cells, but the use of amplification systems such as biotin-avidin or enzymatic systems achieved the desired level of sensitivity, with single cells being detected microscopically. The use of a microscope to scan membrane surfaces is, however, extremely time-consuming and labour-intensive, and therefore laser scanning has been employed. The combination of 16S rRNA and laser scanning enabled quantitative detection of *E. coli* cells on membranes. However, the system has one major drawback – it is unable to discriminate between viable cells and those rendered non-viable by chlorine treatment. Experiments showed that detectable levels of rRNA remained within the cell in bacteria killed by chlorine treatment 20 days earlier. It is clear therefore that in order to use these sensitive and specific probes for the direct detection of bacteria, some other method of discriminating between viable and non-viable cells is necessary. This may be achieved by the detection of specific enzyme markers in the cell used prior to but in combination with a nucleic acid probe.

Advances in the methodologies for the detection of indicator organisms in water have been largely based on culture methods, although reports on methods of direct detection are becoming more frequent. However, systems that are robust, simple to perform and inexpensive are unlikely to be available for routine use for several years.

Cryptosporidium Oocysts

There is growing concern about the presence of pathogenic organisms in water, which may not be indicated by traditional indicator organisms. The organism that has raised most concern is the protozoan parasite *Cryptosporidium parvum*, which is present in many surface waters as an environmentally robust oocyst. This organism has presented many challenges to the water industry, in terms of both its removal and its detection in raw or potable water. The *Cryptosporidium* oocyst is unaffected by chlorine treatment and so conventional disinfection is not an adequate safeguard. The only reliable method of removing the oocysts from raw water is some form of filtration.

Detection of *Cryptosporidium* oocysts has also presented many problems to the water microbiologist. The low infectious dose of this parasite means that methods for detection need to be extremely sensitive if they are to be an effective measure of the likelihood of water causing disease. This means that it is necessary to concentrate large volumes of water for analysis. *Cryptosporidium parvum* is an intracellular parasite and to date it has not been possible to grow it in vitro, although initial growth of some cell culture lines has been demonstrated. The inability to grow *C. parvum* means that a direct detection system using microscopy is necessary.

Initial procedures for the detection of *C. parvum* used a yarn-wound filter of non-defined pore size for concentration, gradient centrifugation for separation of oocysts from extraneous particulates, and detection of the oocyst using epifluorescence microscopy following staining with a monoclonal antibody labelled with fluorescein isothiocyanate (FITC). These three stages of the procedure – concentration, separation and detection – continue to be used in current methods.

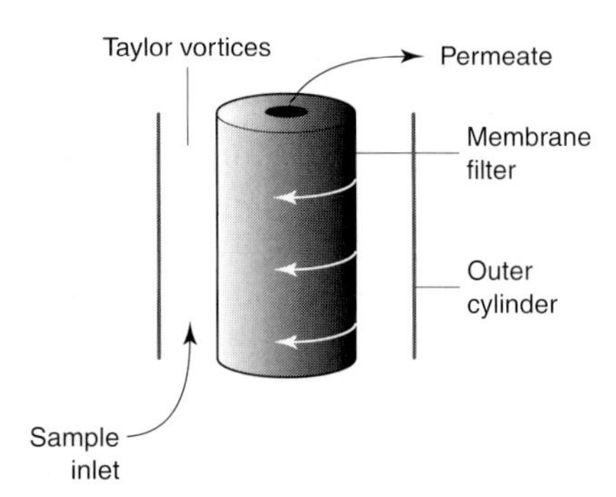

Figure 2 Principles of vortex flow filtration.

Concentration of Water Samples

Several concentration methods have been developed, mostly based on some form of membrane filtration. These procedures have generally better recovery efficiencies than the original yarn-wound filter method. Most of these procedure are only suitable for the concentration of relatively small (10–20 l) of water, although one procedure, vortex flow filtration, can be used with large volume samples and with highly turbid water. The procedure uses a cylindrical membrane cartridge with small pore size and water is pumped through the membrane which is continuously rotated at 4000 rev min^{-1}. This rotation causes a turbulence (Taylor vortices) within the filtration apparatus, which serves to 'scrub' the membrane clean and prevent blockage due to build-up of particulates (**Fig. 2**).

The various methods of concentration all concentrate particles non-specifically, and thus water concentrates contain many particles that are not *Cryptosporidium* oocysts. These extraneous particulates interfere with subsequent detection of the target organisms and therefore as many as possible are removed prior to microscopic examination. In the initial methods developed for *Cryptosporidium*, sucrose of Percol-sucrose density centrifugation was used to separate oocysts from more dense material. This procedure was not very efficient and many oocysts were lost. In the early 1990s, methods based on flow cytometry and fluorescence-activated cell sorting were developed, and these continue to be the most widely used methods in the UK. Essentially, water concentrates are stained with an FITC-labelled anti-*Cryptosporidium* monoclonal antibody and then analysed by the flow cytometer. The sample is broken into a series of very fine droplets and these are excited by a laser which causes the FITC to fluoresce. The droplets are then analysed and those containing particles of the correct size and fluorescence characteristics are sorted into one 'stream' while the remaining material runs to waste. The sorted particles can be collected on a glass slide or membrane filter for examination by epifluorescence microscopy. The procedure is effective in recovering oocysts but is labour-intensive, expensive and requires a considerable amount of operator training.

A more recent development is that of immunomagnetic separation (IMS), which has shown to be extremely useful in separating oocysts form background debris. The principle of IMS is that antibodies with high specificity to the oocyst wall are linked to paramagnetic beads using either a covalent link or an antibody–antibody linkage. The beads are then mixed with the sample and left to incubate for a period of time. After incubation, a magnet is used to separate the beads from background debris. The beads are then washed and treated (usually with acid) to release the beads from the oocysts. The beads are removed using a magnet and the oocysts are then placed on a slide or membrane filter before staining with an FITC-labelled antibody to allow visualization. The procedure is extremely simple to perform and requires little specialized equipment. Recovery efficiencies of oocysts using one of the modern sample concentration devices and IMS are often in excess of 50%, which is a considerable improvement over more traditional methods.

Whether oocysts are separated from background debris by density centrifugation, flow cytometry or IMS, they are confirmed by epifluorescence microscopy and where possible differential interference contrast microscopy to observe internal structures. This can be a time-consuming process and it has recently been simplified by the use of a laser scanning device. The device shown in Figure 1 reduces the time taken to examine a slide considerably, with examination usually taking no more than 5 min.

Molecular Methods

None of the antibody-based methods is able to discriminate between *C. parvum* and other species of *Cryptosporidium* which are not apparently infectious to humans. There is great interest therefore in developing tests to discriminate between species. Such discrimination requires molecular methods, and both fluorescent *in-situ* hybridization (FISH) and PCR-based approaches have been attempted. The FISH approach targets the 18S rRNA molecule, which like the 16S molecule in bacteria is present in multiple copies. Some workers have reported that a *C. parvum*-specific probe is available and that a positive signal means that the oocyst is alive and therefore potentially infectious. There is some disagreement as to whether there are sequences on the 18S molecule that are species-specific. Furthermore, the ability to detect the 18S rRNA may not correlate with viability. Nonetheless, the FISH approach is a useful one.

Direct PCR is not advocated for *Cryptosporidium* for the same reasons that it is not useful for the detection of bacteria. However, PCR has been used to detect the presence of oocysts after IMS, either with antibodies coated to the beads or nucleic acid probes complementary to the target sequence for PCR. Other methods have included the detection of messenger RNA (mRNA) coding for the heat shock protein in *Cryptosporidium* (Hsp70) after heat shocking. This procedure is claimed not only to be sensitive (able to detect one oocyst) but also to detect only live oocysts.

The Issue of Viability

Discrimination between viable and non-viable oocysts is considered by some to be essential, since only viable oocysts are potentially infectious to humans. Several methods have been examined for their ability to identify viable oocysts. Probably the most widely used procedure at present is the use of two dyes, 4,6-diamidino-2-phenylindole (DAPI; membrane-permeable) and propidium iodide (PI; membrane-impermeable). In this assay viable oocysts appear colourless with the nuclei of four sporozoites stained blue. Oocysts stained red are considered non-viable. The results of the assay correlate well with in vitro excystation but may not give a good indication of viability when oocysts have been treated with ozone or ultraviolet light. Other approaches include the use of a series of 'Syto' dyes which stain the nucleic acid of the oocysts and give results that correlate well with mouse infectivity tests. However, mouse infectivity assays are only useful for certain stains of *Cryptosporidium*, and thus further work is required to validate these dyes.

In Vitro Infectivity Tests

Many workers have attempted to determine the infectivity of oocysts by detecting infection of cell cultures. In these assays, samples are concentrated using normal procedures and separation techniques such as flow cytometry or IMS can be used. Contaminating bacteria can be removed by exposure of the concentrate to levels of chlorine which have no effect on the oocyst viability. Concentrates are then inoculated onto tissue culture monolayers, incubated for a period to allow attachment/infection to take place, and the extraneous debris is then removed by washing. Cell cultures are then incubated for 1–2 days and examined by one of several methods. Infectious foci can be detected using immunofluorescent techniques which may allow the number of infectious oocysts to be quantified. Other procedures that have been used include PCR or a reverse transcription PCR (RT-PCR) of the cell culture lysate. The benefit of the RT-PCR procedure is that DNA from non-infectious oocysts which may have adhered to the monolayer is not detected. A similar procedure can be used for the detection of infectious enteroviruses from water, which considerably reduces the time taken to obtain a result and may increase sensitivity.

The range of techniques available to the modern water microbiologist differs greatly from those used by traditional microbiologists. Molecular methods are becoming more frequently used, although there are still problems with these methods which need to be overcome before molecular methods become more widespread. However, as new challenges to the water industry become apparent, the demand will increase for more sensitive, rapid and specific tests to determine the presence of new organisms. Water microbiology continues to develop and the next decades will see a revolution in the type and range of techniques used. Routine detection of bacteria by culture is likely to become a thing of the past, and direct detection methods will become commonplace.

See also: **Biochemical and Modern Identification Techniques**: Enterobacteriaceae, Coliforms and *E. coli*. ***Cryptosporidium***. **PCR-based Commercial Tests for Pathogens**.

Further Reading

Fricker EJ and Fricker CR (1996) Alternative approaches to the detection of *Escherichia coli* and coliforms in water. *Microbiology Europe* 4(2): 16–20.

Grant MA (1997) A new membrane filtration medium for simultaneous detection and enumeration of *Escherichia coli* and total coliforms. *Applied and Environmental Microbiology* 63(9): 3526–3530.

Grant MA (1998) Analysis of bottled water for *Escherichia coli* and total coliforms. *Journal of Food Protection* 61(3): 334–339.

Iqbal S, Robinson J, Deere D, Saunders JT, Edwards C and Porter J (1997) Efficiency of the polymerase chain reaction amplification of the uid gene for detection of *Escherichia coli* in contaminated water. *Letters in Applied Microbiology* 24(6): 498–502.

Quinn CM, Archer GP, Betts WB and O'Neill JG (1996) Dose-dependent dialectrophorectic response of *Cryptosporidium* oocysts treated with ozone. *Letters in Applied Microbiology* 22(3): 224–228.

Wallis PM, Erlandsen SL, Isaac-Renton JL, Olson ME, Robertson WJ and van Keulen H (1996) Prevelance of *Giardia cysts* and *Cryptosporidium* oocysts and characterization of *Giardia* spp. isolated from drinking water in Canada. *Applied and Environmental Microbiology* 62(8): 2789–2797.

WATERBORNE PARASITES

Contents

Entamoeba

William A Petri Jr and **Joanna M Schaenman**,
Department of Medicine, University of Virginia Health Sciences Center, Charlottesville, USA

Entamoeba histolytica is the only *Entamoeba* species that causes amoebic colitis and liver abscesses. *E. histolytica* is endemic in the developing world and relatively uncommon in developed countries with temperate climates. Although diarrhoeal illnesses have been recorded in medical history since the time of Hippocrates, *E. histolytica* was not discovered until 1875 when F. Lösch described an abundance of actively motile 'Amöeben' isolated from the stool of a Russian peasant. The past few years have brought about an official reclassification of *E. histolytica* into two morphologically identical yet distinct species: the non-pathogenic *E. dispar* and the pathogenic *E. histolytica*. This separation is important because only *E. histolytica* requires antiprotozoal therapy, and it explains the long-standing mystery of why so few carriers of '*E. histolytica*' developed invasive disease. The trophozoite form of *E. histolytica* is the invasive form, whereas the cyst is the infectious form when ingested in faecally contaminated food or water.

Characteristics of the Species

Entamoeba are pseudopod-forming, protozoan parasites in the subphylum Sarcodina. The genus *Entamoeba* is composed of five species that infect humans: *E. histolytica*, *E. dispar*, *E. hartmanni*, *E. coli*, *E. polecki* and *E. gingivalis*. Only *E. histolytica* is known to cause disease in humans. *E. dispar*, *E. hartmanni*, *E. coli* and *E. polecki* are commensals found in the large intestine, and *E. gingivalis* is found in the oral cavity. Although *E. histolytica* and *E. dispar* are morphologically identical, the other species can be distinguished from one another by microscopy. The large trophozoite and cyst of *E. coli* can be identified by the splinter-like chromatid bodies in its cytoplasm, and by the fact that its cysts can possess up to eight nuclei. *E. hartmanni* is the smallest of the group, with cysts and trophozoites reaching no more than 10 µm. *E. polecki* is comparable in size to *E. histolytica* and *E. dispar*, but has a distinctive large karyosome and a mononucleate cyst. *E. histolytica* is endemic in the developing world, whereas in the industrialized world it is most commonly a disease of travellers and immigrants. Worldwide, it is believed to cause fifty million cases of invasive disease each year, resulting in approximately 100 000 deaths per annum.

Amoebiasis is the disease caused by *E. histolytica* and is composed of two main syndromes: dysentry and liver abscesses. The cycle of infection begins when the cyst form of the organism is ingested in faecally contaminated food or water. The acid-resistant cyst passes unharmed through the stomach until it reaches the small intestine where it excysts to form eight trophozoites, the motile and invasive form of the species (**Fig. 1A**). Trophozoites migrate to the large intestine, where they may either colonize the bowel lumen as commensal flora or invade into the colonic epithelium, causing inflammation and destruction of the bowel wall. What influences this decision to invade is as yet unknown, but potential factors include differences between strains of amoeba and host variations, such as genetic make-up, intestinal flora, nutritional state and immunocompetence. After invasion, amoebae can gain access to the portal circulation and be transported to various target organs including the liver, brain and lungs. The liver is the most common site of extraintestinal amoebiasis. The life cycle of the amoeba is completed when trophozoites in the large intestine encyst and are excreted in the faeces of the host (**Fig. 1B**).

The mononucleate trophozoite is 10–60 µm in diameter, but the quadrinucleate cyst is usually smaller, averaging 10 µm in diameter. Trophozoites passed in the stool are fragile and quickly die once outside the host. Cysts have a wall that makes them resistant to chlorine and desiccation and can live for weeks in a shady and moist environment at room temperature and up to months if refrigerated. Boiling or very high levels of chlorine, however, can destroy cysts. Although ingestion of contaminated food or water is the most common method of transmission, person to person transmission can also occur under settings of crowding and poor personal hygiene, for example in mental hospitals and day care facilities. Only 10–100 cysts are required to cause amoebic dysentery in animal models, an infectious dose comparable to the notoriously contagious *Shigella* sp., which can be transmitted by as few as 10–100 organisms.

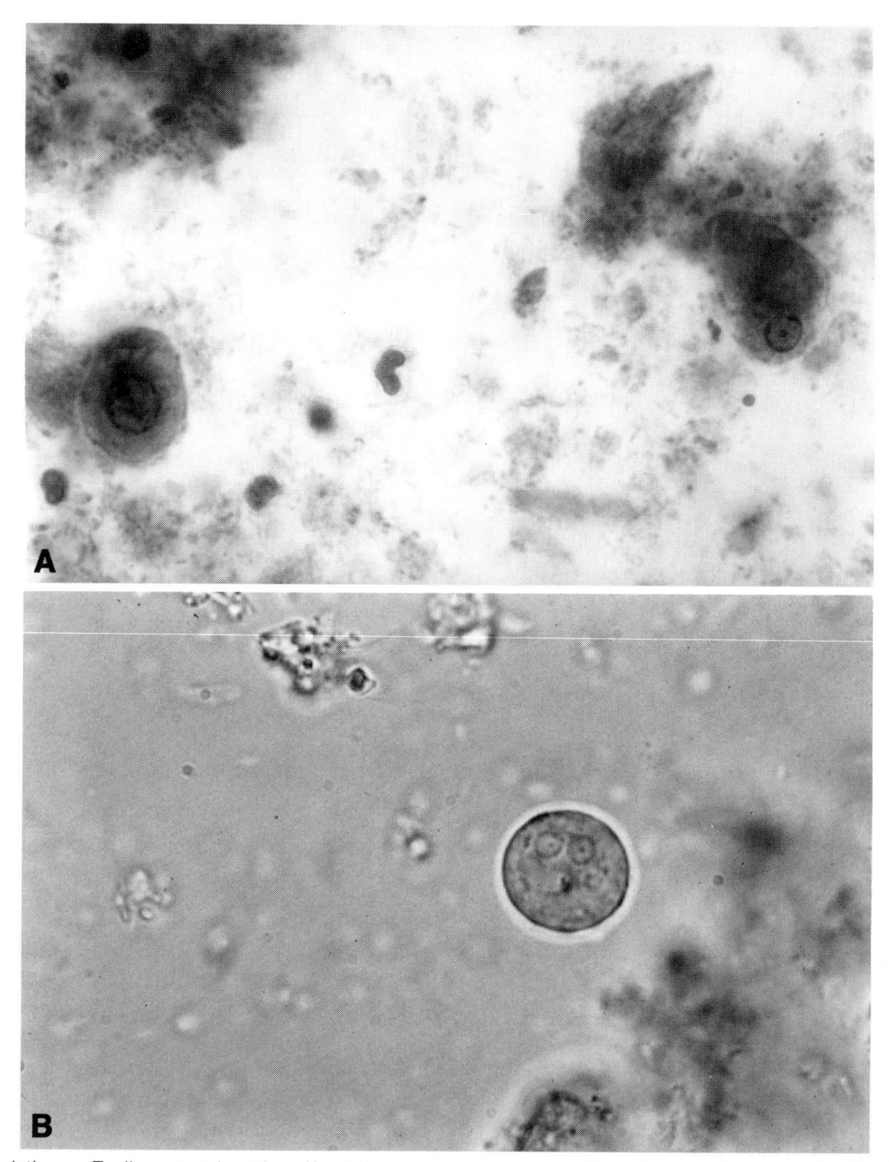

Figure 1 *E. histolytica* or *E. dispar* trophozoites (**A**, trichome stain) and cysts (**B**, iodine stain) in stool. Distinction between these two species is not possible by microscopy alone. (Photos courtesy of Centers for Disease Control.)

As its name implies, *E. histolytica* causes disease by lysing tissue. Cell killing is dependent on the amoeba's ability to adhere to target cells via a galactose-binding adhesin called Gal/GalNAc lectin. *E. histolytica* can kill a wide range of tissue culture cells as well as macrophages, lymphocytes and neutrophils. The Gal/GalNAc lectin is believed to play a central role in pathogenesis because in addition to its roles in adhesion and cell killing, it is also involved in complement resistance, an important survival skill when the trophozoite moves through the bloodstream. Other virulence factors include amoebopores, which create holes in target cell membranes, and cysteine proteases, which destroy extracellular matrix proteins.

When the amoeba invades into the intestinal wall it creates flask-shaped ulcers, and symptoms gradually develop from mild diarrhoea to frank dysentery. Amoebic dysentery is characterized by abdominal pain and tenderness and the production of bloody stools (up to 25 per day). The diagnosis can easily be missed in the countries of the developed world, where *Shigella*, *Campylobacter*, *Salmonella* rotaviruses and inflammatory bowel disease are common causes of bloody diarrhoea. More rarely intestinal amoebiasis can develop into acute necrotizing colitis. This complication is usually treated by partial or total colectomy and is often fatal.

Amoebic liver abscesses, which are more common in men than in women, manifest as abdominal pain and fever, sometimes accompanied by weight loss. Hepatic imaging studies such as ultrasound or computed tomography reveal an oval-shaped defect, most commonly in the right lobe of the liver. Ruptured liver abscesses can extend up through the diaphragm into

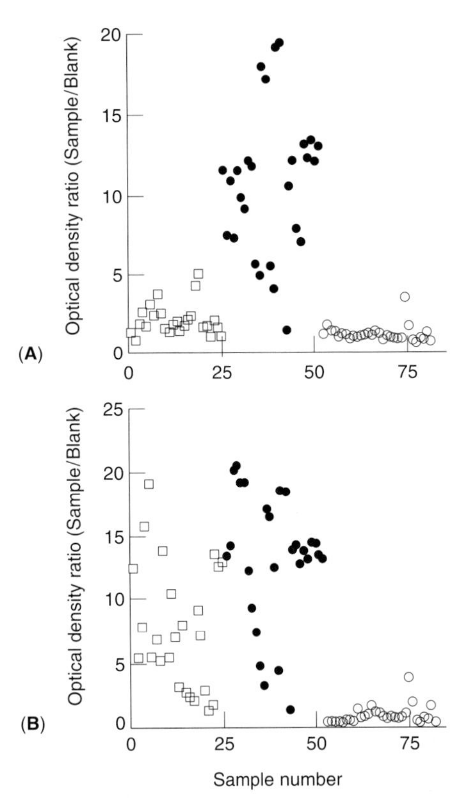

Figure 2 Antigen detection test for *E. histolytica* and *E. dispar* in stool. Stool specimens with culture-confirmed *E. dispar* (open squares) or *E. histolytica* (closed circles) infection, or stools with no detectable *Entamoeba* detected by microscopy (open circles) were assayed using an ELISA containing monoclonal antibody (mAb) specific for *E. histolytica* (**A**) or mAb cross-reactive with *E. histolytica* and *E. dispar* (**B**).

the lungs and pleural cavity. Less commonly, cerebral abscesses can occur when amoeba spread to the brain through the bloodstream. This uncommon complication of amoebiasis can lead to death in 12–72 h.

Asymptomatic colonization with *E. histolytica* can be treated with the luminal agents diloxanide furoate and paromomycin. Treatment for all forms of invasive disease is metronidazole or tinidazole, and treatment efficacy for typical drug regimens is greater than 90%.

Because humans are the only significant reservoir for *E. histolytica*, an effective vaccine could potentially eliminate the disease altogether. The ability of certain antigens, such as subunits of the Gal/GalNAc lectin, to provide protection against liver abscess formation in gerbils provides hope that a human vaccine will be developed in the future.

Methods of Detection

E. histolytica cysts can be detected in contaminated water by filtering a sample of 4 l or more through membrane filters with a pore size of 7–10 μm. After washing the filter with several millilitres of sterile distilled water into a small beaker, this concentrated sample can be examined directly by microscopy for the presence of cysts. Microscopy can also be used to identify cysts and trophozoites from stool samples, but this technique is not reliable for the diagnosis of intestinal amoebiasis for two reasons. First, the sensitivity of the test is very low, with detection rates estimated at only 33–50% for a single stool examination. Second, not only can *E. histolytica* be confused with neutrophils, macrophages and the commensal *E. coli*, which can resemble *E. histolytica* trophozoites to the untrained eye, the non-pathogenic species *E. dispar* is morphologically identical to *E. histolytica* and cannot be distinguished from its pathogenic relative by microscopy alone.

Additionally, serology is not a reliable technique for diagnosing amoebiasis because persons once infected can remain antibody positive for years.

A newly available method of detection is the stool antigen detection test, from TechLabs, Inc. (Blacksburg, USA), which uses monoclonal antibodies specific for the *E. histolytica* lectin in an enzyme-linked immunosorbent assay (ELISA). This test has been measured by the diagnostic gold standard, culture and isoenzyme analysis, to have a sensitivity of 93% and a specificity of 98%. Although this technique has been predominately used on stool samples, filtered material from suspect water supplies could probably be tested as well (**Fig. 2**).

The polymerase chain reaction (PCR) has shown promise as an alternative technique for the detection of *E. histolytica*-specific DNA from stool samples, but has not yet been standardized as a reliable and technically simple kit for clinical use. PCR could also be used to differentiate between different *E. histolytica* isolates, which could be helpful for epidemiological purposes.

Importance to the Food Industry

In the industrialized world, as agricultural production declines and population increases, countries are becoming more and more dependent on foods imported from the developing world. Thus crops grown in countries with contaminated water supplies can lead to faecal contamination of exported products, most importantly fruits and vegetables, such as salad greens which are not commonly boiled or cooked before eating. Non-potable water can be introduced to crops during irrigation and during spraying with pesticide or fertilizer dissolved in water, a practice which is sometimes performed only hours before harvesting occurs. More directly, crops can be fertilized with untreated human excrement, or 'night soil', again leading to faecal contamination. One study of strawberries imported from Mexico found that 20/21 samples were positive for parasites, 37%

of which were *E. histolytica*. Although a documented food-borne outbreak in imported foods has not yet occurred, outbreaks of other faecally transmitted protozoan pathogens such as *Cryptococcus* and *Cyclospora* suggest that such a route of transmission may be possible.

E. histolytica can also pose a threat to the food industry in the guise of infected food handlers, especially recent immigrants from areas where *E. histolytica* is endemic. An asymptomatic carrier of *E. histolytica* can excrete millions of cysts per day for years, and cysts have been shown to remain viable for almost an hour when present in faecal matter under fingernails. Therefore, persons known to be infected should be excluded from work in food processing or preparation.

Although cysts survive for long periods in moist locations, they are sensitive to desiccation, dying in 10 min when present on the surface of hands or when exposed to direct sunlight. Cysts can also be destroyed by heating to over 68°C and by iodine at 200 p.p.m. They are, however, resistant to acidic environments, such as that of the stomach.

Importance to the Consumer

As mentioned above, there is a possibility that *E. histolytica* and other enteric protozoa could be present on imported foods. Fruits and vegetables intended for raw consumption should be washed as well as possible, and care should be taken when travelling to avoid unboiled water and uncooked fruits and vegetables.

See also: **National Legislation, Guidelines & Standards Governing Microbiology**: European Union. **Nucleic Acid-based Assays**: Overview. **Water Quality Assessment**: Modern Microbiological Techniques.

Further Reading

Bray RS (1996) Amoebiasis. In: Cox F (ed.) *Illustrated History of Tropical Diseases*. P. 171. London: The Wellcome Trust.

Haque R, Ali IKM, Akther S and Petri WA Jr (1998) Comparison of PCR, isoenzyme analysis, and antigen detection for diagnosis of *Entamoeba histolytica* infection. *Journal of Clinical Microbiology* 36: 449–452.

Haque R, Neville LM, Hahn P and Petri WA Jr (1995) Rapid diagnosis of *Entamoeba* and *Entamoeba histolytica* stool antigen detection kits. *Journal of Clinical Microbiology* 33: 2558–2561.

Herwaldt BL, Ackers M-L and the Cyclospora Working Group (1997) An outbreak in 1996 of cyclosporiasis associated with imported raspberries. *New England Journal of Medicine* 336: 1548–1556.

Osterholm MT (1997) Cyclosporiasis and raspberries – lessons for the future. *New England Journal of Medicine* 336: 1597–1599.

Padilla y Padilla CA (ed.) (1974) *Amebiasis in Man*. Springfield, IL: Charles C. Thomas.

Petersen C (1995) *Cryptosporidium* and the food supply. *Lancet* 345: 1128–1129.

Petri WA Jr (1996) Recent advances in amoebiasis. *Critical Reviews in Clinical Laboratory Sciences* 33: 1–37.

Ravdin JI and Petri WA Jr (1995) *Entamoeba histolytica* (amoebiasis). In: Mandell G (ed.) *Principles and Practice of Infectious Diseases*. P. 2395. New York: Churchill Livingstone.

Ravdin JI (ed.) (1988) *Amoebiasis: Human Infection by* Entamoeba histolytica. New York: John Wiley.

Spindola FN, Rojas WG, de Haro AI, Cabrera BM and Salazar SPM (1996) Parasite search in strawberries. *Archives of Medical Research* 27: 229–231.

Detection by Conventional and Developing Techniques

H V Smith and **R W A Girdwood**, Scottish Parasite Diagnostic Laboratory, Stobhill Hospital, Glasgow, Scotland, UK

Introduction

Specific methods for the isolation and enumeration of the cysts and oocysts ((oo)cysts) of the intestinal protozoan parasites *Giardia duodenalis* and *Cryptosporidium parvum* from water are required for the following reasons: first, they can pass through physical and chemical barriers in water treatment, are chlorine-insensitive and have been detected in potable water supplies in the absence of indicator organisms; second, they can cause epidemic disease in consumers of contaminated potable water; third, being obligate parasites, their numbers cannot be augmented by conventional in vitro culture methods prior to identification; and fourth, the minimum infectious dose for human beings is low. Generally, with the exception of waste waters and filter backwash waters, (oo)cysts occur in low numbers in the aquatic environment and diverse techniques employing large volumes of sample followed by sample concentration have been employed in an attempt to concentrate them. A variety of methods from immunofluorescence to polymerase chain reaction (PCR) have been used to determine the presence of (oo)cysts. This article reviews current and developing methods for isolating and enumerating waterborne (oo)cysts.

Conventional Techniques

General Principles

A variety of methods have been developed which have specific advantages and disadvantages. Each has been used successfully to determine the occurrence of waterborne (oo)cysts. While there is no universally accepted procedure, all methods can be subdivided into the following elements:

1. sampling
2. elution, clarification and concentration
3. identification.

Standardized methods, which are continually evolving, are available in the UK and the US.

Methods for Sampling Water for the Presence of (Oo)cysts

As (oo)cysts tend to occur in low numbers in water, a system that enables their efficient recovery from large volumes of water is required. Two approaches to sampling have been adopted. In large-volume sampling, a sample is taken over a period of hours at a defined flow rate, whereas in small-volume sampling, a volume of 10–20 l, typically, is taken as a grab sample. In situations where little is known of the occurrence of (oo)cysts in the matrix tested, large-volume sampling can be useful as the sample is taken over a long time period. Grab samples generate occurrence data in smaller volumes, can provide higher recovery efficiencies than large-volume sampling and are easily collected. A compromise between both regimes is the collection of numerous grab samples over the large-volume sampling period in order to generate one composite sample.

Methods for Concentrating Cysts and Oocysts

Size-based concentration methods result in the accumulation of similar and larger-sized particles in the concentrate, and the use of filters, typically with a 1 μm (nominal) pore size, results in the accumulation of large volumes of extraneous particulate material. Such particles can interfere with organism detection and identification, and methods to reduce this interference have been adopted. Non-covalent interactions between the surfaces of (oo)cysts and other materials are reduced by the addition of detergents and surfactants, in an attempt to maintain them as individual organisms, and particles with a greater density than (oo)cysts can be clarified by centrifugation through a solution of a predetermined specific gravity on which (oo)cysts float. This clarification step is intended to separate (oo)cysts from the denser contaminants, which are pelleted in the flotation solution, and to concentrate the former. Many researchers believe this clarification step to be inefficient, leading to a loss of organisms, and prefer to omit it if the water concentrate is not too turbid. Non-viable (oo)cysts are more likely to penetrate the sucrose flotation interface than viable (oo)cysts, and viable organisms concentrate on the sucrose density interface. Thus the inefficiency of the sucrose flotation method may be a reflection of the numbers of non-viable (oo)cysts in a water concentrate. Clarification by sucrose flotation should be regarded as a method for the enrichment of viable organisms rather than as a method for concentrating and/or purifying all (oo)cysts present in a water concentrate.

The suspension of (oo)cysts, which now contains fewer contaminants, is washed free of flotation fluid and concentrated to a minimum volume by centrifugation which will vary according to the nature and turbidity of the sample. A proportion is examined microscopically for the presence of (oo)cysts. Because the infective dose to human beings is small (between 25 and 100 cysts for *G. duodenalis* and 30 and 132 oocysts for a bovine isolate of *C. parvum*), it is important to have a preparation as free as possible of inorganic and organic debris which might mask the presence of the organisms or interfere with their identification and to be able to identify small numbers of organisms accurately.

Methods for Identifying (Oo)cysts

(Oo)cysts are normally air-dried on to welled microscope slides (or deposited on to membranes) and genus-specific fluorescein isothiocyanate (FITC)-labelled monoclonal antibodies (FITC-mAbs), reactive with exposed epitopes on the (oo)cyst wall, are applied and used according to the manufacturer's instructions. A choice of commercially available FITC-mAbs is available for both *Giardia* and *Cryptosporidium*. Reagents are available as individual (e.g. *Giardia* only) or combination (e.g. *Giardia* and *Cryptosporidium*) kits, depending on requirement. Excess reagent is gently aspirated and the sample rinsed to remove unbound reagent. Application of the fluorogen 4′6-diamidino-2-phenyl indole (DAPI) enhances the visualization of nuclei in (oo)cysts. When possible, putative (oo)cysts should be examined under Nomarski differential interference contrast (DIC) microscopy in order to determine whether the putative organism contains identifiable organelles. Definitive criteria for epifluorescence and DIC microscopy (**Table 1**) are adopted to identify the presence of (oo)cysts.

Table 1 Characteristic features of *Giardia duodenalis* cysts and *Cryptosporidium parvum* oocysts by fluorescein isothiocyanate (FITC) epifluorescence microscopy and Nomarski differential interference contrast (DIC) microscopy

Appearance under the FTC filters of an epifluorescence microscope

The putative organism must conform to the following fluorescent criteria: uniform apple-green fluorescence, often with an increased intensity of fluorescence on the outer perimeter of an object of the appropriate size and shape (see below)

Appearance under Nomarski DIC microscopy

G. duodenalis *cysts*	C. parvum *oocysts*
Ellipsoid to oval, smooth-walled, colourless and refractile	Spherical or slightly ovoid, smooth, thick-walled, colourless and refractile
8–12 × 7–10 μm (length × width)	4.5–5.5 μm
Mature cysts contain four nuclei displaced to one pole of the organism	Sporulated oocysts contain four nuclei
Axostyle (flagellar axonemes) lying diagonally across the long axis of the cyst	Four elongated, naked (i.e. not within a sporocyst) sporozoites and a cytoplasmic residual body within the oocyst
Two claw hammer-shaped median bodies lying transversely in the mid-portion of the organism	

Large-volume Sampling

The procedure includes filtering a large volume (approximately 100–1000 l) of water through a depth filter (e.g. yarn wound (CUNO, flow rate 1.5 l min^{-1}), polypropylene (Filterite, flow rate 4 l min^{-1}), compressed foam (Crypto-Dtec, flow rate 1–2 l min^{-1})), which entraps cysts, oocysts and other particulates of similar or larger size. To release (oo)cysts entrapped in yarn wound and polypropylene cartridges, the filter matrix is cut, teased apart and eluted with large volumes (up to 4 l) of 0.01% Tween-80 in deionized water containing an antifoaming agent (antifoam A). For the compressed foam cartridge, the manufacturer's washing station with elution and concentrator tubes and associated filter membrane (3 μm, cellulose acetate) is a prerequisite for this method. Organisms concentrated on the membrane are eluted after placing the membrane together with a small volume of a dilute detergent solution (0.01% Tween-20 in phosphate-buffered saline) in a re-sealable bag and gently massaging the membrane, through the re-sealable bag, with finger and thumb.

Eluted organisms and particulates are concentrated, by large-volume centrifuge, to a small volume (about 20 ml) and for turbid samples a clarification procedure, utilizing a flotation medium such as sucrose (1.18 sp. gr.) or Percoll sucrose (1.1 sp. gr.), on which (oo)cysts float, is employed. Excess flotation medium is removed by further washing as it interferes with both the attachment of organisms on to slides and the binding of FITC-mAbs. A proportion of the final concentrate (at least 20%) is analysed by epifluorescence and DIC microscopy following the application of genus-specific FITC-mAbs and DAPI.

Small-volume Sampling

Between 10 and 20 l is sampled using one of two approaches: membrane filtration or flocculation.

Membrane Filtration The grab sample is filtered through either a flat-bed 142 mm, 1.2–2 μm cellulose acetate or polycarbonate membrane (flow rate ca. 150 ml min^{-1}) or a pleated membrane capsule (e.g. Gelman Environchek, flow rate 2 l min^{-1}) using a peristaltic pump and the (oo)cysts eluted by mechanical agitation. Entrapped organisms are eluted either by placing the flat-bed membrane together with a small volume of a dilute detergent solution (0.1% Tween-80 in deionized water) in a re-sealable bag and gently massaging the membrane with finger and thumb, through the re-sealable bag, or by placing the capsule containing the elution medium (Laureth 12-Tris-EDTA-antifoam A) in a wrist-action mechanical shaker. Eluted organisms are concentrated to a small volume (500 μl–5 ml) by centrifugation. A proportion of the final concentrate (at least 20%) is analysed by epifluorescence and DIC microscopy following the application of genus-specific FITC-mAbs and DAPI, as described above.

Flocculation Particulates in a water sample (normally 10 l) are flocculated following the production of a $CaCO_3$ floc generated by the addition of $CaCl_2$, $NaHCO_3$ and NaOH. As the floc settles by gravity, it causes cysts, oocysts and other particulates to settle with it, thus concentrating them. Once settled (≥4 h or overnight), the floc is dissolved in sulphamic acid and the particulates in the sample concentrated by centrifugation to a minimum volume of 0.5–1.0 ml. The concentrate, or an aliquot thereof, is analysed by epifluorescence and DIC microscopy as described above.

Identification

Currently, the only acceptable method for determining the presence of (oo)cysts in a sample is microscopy and this is dependent upon morphometry (the accurate measurement of size) and morphology. The morphological features of intact (oo)cysts are identified in Table 1.

Specificity and sensitivity are of paramount importance when attempting to detect small numbers of (oo)cysts in water concentrates. Organisms present in environmental samples can frequently be a mixture of both recently voided and aged organisms, often in low numbers. Many, if not all, may have been subjected to a variety of environmental pressures, including water treatment, which can alter their physical and chemical properties, making their appearance atypical. In contrast to the recently voided or intact organisms, where the typical morphological features in Table 1 can be seen readily, difficulties can arise in identifying (oo)cysts, in water concentrates, when internal contents have been lost. Physical changes associated with both survival and sample processing, such as distortion, contraction, collapse and rupture of the (oo)cysts, will inevitably reduce the number of organisms that conform to the criteria in Table 1, resulting in the under-reporting of positives. The antibody paratopes of commercially available FITC-mAbs, being of defined specificity and affinity, bind surface-exposed (oo)cyst epitopes and assist in consolidating and standardizing methodology. Thus, the fluorescence visualized defines the maximum dimensions of the organism, enabling morphometric analyses to be undertaken under near dark-field conditions.

The usefulness of mAbs for detecting (oo)cysts in water-related samples rests on their ability to react effectively with epitopes on (oo)cyst walls which are resistant to a variety of environmental conditions, including disinfection and ageing. Some anecdotal evidence is available which suggests that exposure of (oo)cysts to water treatment processes and/or the aquatic environment can affect epitope expression, and hence the intensity of FITC emission of the FITC-mAbs used for detection. It has been found that, whilst the levels of chlorine used for the disinfection of potable water do not appear to affect the integrity of the epitope on cysts, higher concentrations (5–500 mg l^{-1} free chlorine), whilst initially causing more FITC-labelled antibody to bind, enhancing fluorescence intensity (as determined by microfluorimetry), eventually eliminate antibody binding to *G. duodenalis* cysts, thus ablating cyst wall fluorescence. While exposure of *C. parvum* oocysts to similar concentrations of free chlorine does not affect FITC-mAb binding to chlorine-treated oocysts, exposure to more extreme concentrations of free chlorine (2% or 5% sodium hypochlorite on ice for 10–120 min) considerably reduces, but does not ablate, fluorescence intensity.

Recovery Efficiencies

Recovery efficiency will vary from matrix to matrix, with poor water quality, the presence of algae, suspended solids, clays and turbidity exerting a detrimental effect. In general, recoveries for *Giardia*, being larger organisms, are higher than those for *Crystosporidium*, therefore the recoveries for *Cryptosporidium* are identified below. A major disadvantage of large-volume (yarn wound and polypropylene) filtration is that the recovery efficiency can be low (< 1–40%). Recovery using the compressed foam cartridge, albeit on the small number of samples currently tested, is reported as 88–90%. Recovery using flat-bed membranes is approximately 5–60%, while recoveries for the pleated membrane capsule range from 58 to 81%, again on a small number of samples. Recovery using the $CaCO_3$ flocculation method can be variable (< 20 to > 70%) and, in addition, oocyst viability is reduced when using this method.

The large- and small-volume sampling methods described above are included in the current UK Department of the Environment Standing Committee of Analysts' (UKSCA) provisional recommended methods: *Methods for the Examination of Waters and Associated Materials: Isolation and Identification* of Giardia *cysts*, Cryptosporidium *oocysts and free living pathogenic amoebae in water.* Large-volume sampling is included in the US Environmental Protection Agency's (USEPA) *ICR Protozoan Method for Detecting* Giardia *Cysts and* Cryptosporidium *Oocysts in Water by a Fluorescent Antibody Procedure* and small-volume sampling is identified in the USEPA's draft *Method 1622*: Cryptosporidium *in Water by Filtration/IMS/FA*.

Alternative Concentration Methods

Whereas conventional nonspecific concentration methods rely primarily on biophysical parameters to affect the concentration of organisms, methods based on specific biological properties of (oo)cyst surfaces have also been developed recently. The availability of mAbs reactive with surface-exposed epitopes on (oo)cysts allows a variety of antibody-based technologies to be performed, which can either reduce processing time or increase the sensitivity of detection, or both. Two approaches, based on the immunoreactivity between (oo)cyst surface-exposed epitopes and mAbs to *Giardia* and *Cryptosporidium*, for separating (oo)cysts from contaminating debris in water concentrates, have been adopted.

Flow cytometers with cell-sorting (FCCS) capabilities which can concentrate particles of defined size and fluorescence intensity have been used to concentrate (oo)cysts. Fluorescent particles of the size of (oo)cysts can be separated readily from the myriad of contaminating debris on FCCS machines by analysis of immunofluorescence-stained samples and sorting of particles by both light scatter and pre-defined fluorescence characteristics. All fluorescent sorted objects are automatically deposited on a microscope slide or membrane, but require verification as being (oo)cysts by microscopy. FCCS makes subsequent microscopical analysis and identification easier because it produces a sample with maximum separation of fluorescent (oo)cysts from contaminating material. In a survey of 325 environmental samples, where equal volumes were analysed by both FCCS and microscopy, 92 samples contained oocysts (range 1–22 per volume analysed) by FCCS whereas 12 yielded oocysts (1 per volume analysed) by microscopy. With FCCS, larger volumes can be analysed more rapidly. A time of 5–15 min for the analysis of a sample equivalent to 1–10 l of raw or treated water has been reported by some workers. FCCS is included in the current UKSCA's *Isolation and Identification of Giardia cysts, Cryptosporidium Oocysts and Free Living Pathogenic Amoebae in Water.*

Immunomagnetizable separation (IMS) techniques, in which (oo)cysts are bound immunologically to beads coated with commercially available mAbs and which can be concentrated by magnet, can also concentrate (oo)cysts effectively from contaminating particulates. Both paramagnetic colloidal magnetite particles (40 nm) and iron-cored latex beads have been used to concentrate (oo)cysts selectively from water concentrates. Using antibody-coated paramagnetic colloidal magnetite particles, an average of 82% of mAb-coated *Giardia* cysts (seeded at 500 cysts per millilitre into water concentrates with turbidities from 6 to 6000 NTU) could be recovered. Turbidities of >600 NTU interfered with cyst recoveries, and significantly higher recoveries occurred with water samples of 600 NTU or less. Two benefits of using colloidal magnetite particles were identified. First, the small size of the colloidal paramagnetic particle is beyond the resolving power of a light microscope and does not interfere with the microscopic identification of an organism. Second, the surface area to volume ratio of these particles is larger than that for larger magnetizable particles and theoretically permits more antibody-binding sites per particle, producing a more reactive particle. Iron-cored latex beads coated with anti-*Giardia* mAb are recent commercial additions, but few comparisons of performance are available at present.

Recovery efficiencies, using IMS to concentrate *C. parvum* oocysts, have been assessed in the authors' laboratory, and range from 46 to 87% of oocysts seeded into raw, potable and mineral waters. Iron-cored latex beads coated with anti-*Cryptosporidium* mAb were used to separate oocysts from the above matrices which were seeded with 5, 50 or 100 oocysts per 1, 10 or 15 ml sample. Currently, two IMS kits are available commercially and kit performance is frequently better with less turbid samples. Both turbidity and the concentration of divalent cations affect the performance of IMS; turbidity is of greater significance in raw and potable water concentrates, with divalent cations being of greater significance in mineral water concentrates. In such instances, kits containing mAb paratopes with higher affinities for their epitopes can outperform lower-affinity mAbs. Such benefits of higher-affinity paratopes can be identified by comparison of commercial kits, whereby that using an IgG antibody isotype outperforms the IgM antibody isotype in very turbid (ca. 15 000 NTU) concentrates.

As for commercially available combination FITC-mAb kits for (oo)cyst detection, combination *Giardia* and *Cryptosporidium* IMS kits are recent additions to the current methods. IMS is included in both the current UKSCA's *Isolation and Identification of Giardia Cysts, Cryptosporidium Oocysts and Free Living Pathogenic Amoebae in Water*, and in the USEPA's draft *Method 1622: Cryptosporidium in Water by Filtration/IMS/FA*.

Developing Techniques

Antibody-based Detection Methods

Most effort has been directed towards the identification of organisms, although recent developments in sampling and concentration of organisms have been described earlier.

Electronic imaging using cooled charge-couple devices (CCDs) can provide a basis for the development of a sensitive, reliable and rapid technique for detecting fluorescent and/or luminescent emissions from mAb-labelled (oo)cysts. CCDs can detect fluorescent emissions from fluorogens which can enhance morphological detail, at low total magnification (×20–×30) and can reconstruct accurately the spatial organization of the intact fluorescent organism, thus providing the means for electronic measurement. A sensitive enhanced chemiluminescence (ECL) detection system, which can detect between two and six oocysts, using either X-ray film or a luminometer for imaging light emissions from mAb-coated *C. parvum* oocysts, has been developed, but has yet to be adopted. Laser scanning devices (e.g. Chem*Scan*, Chemunex) are designed to detect and enumerate fluorescently labelled microorganisms which have been captured by membrane filtration. FITC-labelled oocysts, captured on a 25 mm diameter membrane filter, are excited by a laser spot, directed across the membrane by an array of oscillating mirrors, and emissions are detected by a series of photo-multiplier tubes (detection channels), each recognizing fluorescence from different wavelength ranges. Signals generated undergo computer analyses to distinguish between labelled organisms and autofluorescent debris. A record of the location of each fluorescent event on the membrane is made and the membrane is then transferred from the instrument to an epi-fluorescence microscope fitted with a motorized stage. This stage can be driven to each documented event by the Chem*Scan* for visual verification of the findings. The inclusion of such automation should reduce the tedium associated with current methodology.

One definite advantage of emerging technologies is that they have the potential to be used together. The ability of CCDs to detect luminescent as well as fluorescent emissions makes them suitable candidates for consideration for a variety of the antibody and other probe-based novel technologies. Coupling of a CCD to an FCCS would provide a sensible solution by combining the advantages of both to produce a system which could not only sort fluorescent organisms, but which could also, with the inclusion of DNA-intercalating fluorogenic vital dyes and oligonucleotide probes, biotype, genotype and spatially reconstruct individual (oo)cysts.

The single-channel format of microscopy, FCCS and other developments identified dictates that samples have to be analysed consecutively, with sufficient quality assurance to prevent cross contamination of samples. Multichannel microplate systems, capable of analysing up to 96 samples simultaneously, using colorimetric or luminescence-based enzyme immunoassays, have been developed. They possess the advantage of being able to screen large numbers of samples but offer no significant increase in sensitivity and, as yet, have not been adopted.

Nucleic Acid-based Detection Methods

The potential for increased sensitivity and specificity using molecular techniques has led to the development of methods for detecting the presence of parasite nucleic acids liberated from or localized in (oo)cysts. These techniques also offer the potential for addressing some of the outstanding issues, such as the host specificity, infectivity and virulence and have generated research into the genomes of both *Giardia* and *Cryptosporidium*. Three approaches to detection have been investigated:

1. the detection of nucleic acids using a labelled oligonucleotide probe specific for the parasite in question
2. the use of PCR to amplify a specific region of parasite nucleic acid defined by oligonucleotide primers
3. the identification of individual organisms by fluorescence *in situ* hybridization (FISH) using species-specific oligonucleotide rDNA probes.

Although there is currently no consensus method, the increased sensitivity afforded by PCR is likely to reap benefits when attempting to detect small numbers of organisms. A disadvantage is that it is a destructive method, necessitating the rupture of the (oo)cysts in order to liberate nucleic acids for amplification. Of the primers published, some have been used to detect waterborne (oo)cysts. Far more advances have been made recently in the molecular detection of *Cryptosporidium* oocysts than in the detection of *Giardia* cysts.

A genus-specific (265 bp fragment, between base pairs 636 and 900; G+C content ca. 69%) probe was used to detect the presence of *Giardia* nucleic acid, following disruption of cysts with glass beads. While it was not capable of distinguishing between *G. duodenalis* and other *Giardia* species which might be present in the aquatic environment, it had a sensitivity between 1 and 5 cysts per millilitre of surface or waste water concentrates – similar to that obtained using a standard immunofluorescence method. PCR was used to discriminate *G. duodenalis* from *G. muris* and *G. ardeae*, under ideal laboratory conditions, by amplifying a 218 bp region of the *Giardia* giardin coding gene, and detecting the amplified product with a 28 mer oligonucleotide probe. As few as one cyst could be detected using this method; however, PCR analysis with this giardin probe, whilst being able to detect 250 cysts per litre of water, failed to detect species differences when 105 cysts were present in a concentrated 400 l sample.

In a comparative study, the sensitivity of optimized reactions using four pairs of previously published *Cryptosporidium* PCR primers ranged between 1 and 10 oocysts in purified preparations and 5–50 oocysts in seeded environmental water samples. The maximum sensitivity was achieved with two successive rounds of amplification followed by hybridization of an oligonucleotide probe to the PCR amplicon and chemiluminescent detection of the hybridized probe. Some primers amplified genus-specific amplicons whereas others amplified species-specific amplicons. Of the primers tested, the 'Laxer et al' primers provided the best combination of sensitivity and specificity. Primers which amplify a sequence of the 18S rDNA gene have also been used, with or without IMS, to amplify DNA from oocysts seeded into water, and possess a sensitivity similar to that described above. IMS helped reduce the inhibitors of PCR (humic and fulvic acids, organic compounds, salts and heavy metals) frequently found in water concentrates.

FISH has been used to identify both *Giardia* and *Cryptosporidium* spp. in water concentrates. In FISH, fluorescence-labelled oligonucleotide probes are hybridized to ribosomal RNA within organisms and detected using epifluorescence, confocal laser scanning microscopy or FCCS. Given sufficient discriminatory power of the target sequence, the technique can identify organisms to species level and the simultaneous detection of multiple organisms, such as different *Giardia* spp. or *G. duodenalis* and *C. parvum*, is possible by using probes labelled with different fluorescent markers. However, optimization of FISH can be time-consuming.

Determination of the species of individual *Giardia* cysts has been addressed using FISH. Species-specific oligonucleotide rDNA probes to the small subunit rDNA of *G. duodenalis*, *G. muris* and *G. ardeae* were developed and visualized with laser confocal scanning microscopy. In this approach, species-specific identification of cysts in environmental samples were performed utilizing the benefits of both FITC-mAb and FISH to determine both morphometry and species of the organism in question. Using 17–22-mer probes to the 16S-like rRNA of *G. duodenalis*, *G. muris* and *G. ardeae* linked to either FITC or high-quantum-yield carboxymethylindocyanine dyes (Cy3 or Cy5), together with an FITC-labelled mAb to the cyst wall of *Giardia*, individual *G. duodenalis* cysts present in a sewage lagoon concentrate were identified by laser confocal scanning microscopy.

In a laboratory-based approach, probes to unique regions of *C. parvum* rRNA were synthesized and used to detect oocysts on glass microscope slides. Two probes, with similar hybridization characteristics, were labelled with a novel fluorescent reporter, 6-carboxyfluorescein phosphoramadite (excitation 488 nm, emission 522 nm) and viewed by epifluorescence microscopy. The use of these probes to detect waterborne oocysts was advocated.

Biophysical-based Detection Methods

The composition of organisms is thought to be unique, based on the diversity, orientation, density, composition and fluidity of molecules and the interactions between them. When exposed to electrical fields, interactions between molecules produce identifiable electrical patterns which can be characterized. Dielectrophoresis and electrorotation (ROT) are both based on movement within AC electrical fields, and have been used to identify, concentrate and/or assess the viability of (oo)cysts. Both dielectrophoresis and ROT are dependent upon the relative conductive properties of the particle and the suspending medium and produce a fingerprint of the organism in question. The fingerprint can be used either to attract (collect) organisms to an electrode (dielectrophoresis) where their images can be captured on videotape, or to repel them from electrodes. Whereas dielectrophoresis can be described as the motion imparted on electrically neutral, but polarized particles, which are subjected to non-uniform electrical fields, ROT occurs as a result of rotational torque exerted on polarized particles subjected to rotating electrical fields.

The possibility of characterizing (oo)cysts by measuring relative rotational forces induced by AC electrical fields has been proposed. In this approach, *C. parvum* oocysts, when bound to antibody-coated beads and subjected to defined rotating AC electrical fields, can be differentiated from the background of unbound beads and other contaminants because they rotate at a speed different from the speed of the other contaminating particles. This is due to the interaction of the rotating field with the induced dipole of the bead-bound particle, which induces a physical torque on that particle, causing it to spin at a speed different from that of the background beads.

Viability

The conventional techniques of animal infectivity and excystation in vitro cannot be used for small numbers of organisms found in water concentrates; therefore surrogate techniques which can assess the viability of individual (oo)cysts are necessary.

Fluorogenic Vital Dyes

Insufficient information is available from phase contrast or DIC microscopy to determine the viability of the parasite within (oo)cysts consistently. Fluorogenic vital dyes have been used to develop rapid objective estimates of organism viability and are based upon the microscopic observation of fluorescence inclusion or exclusion of specific fluorogens as a measure of viability. The fluorogens fluorescein diacetate (FDA) and propidium iodide (PI) have been used to determine the viability of *G. muris* cysts. Cysts which included FDA and hydrolysed it to free fluorescein could cause infection in neonatal mice, whereas cysts which included PI were incapable of causing infection in neonatal mice. No correlation between the inclusion of fluorogens and the viability of *G. duodenalis* cysts as determined by in vitro excystation has been demonstrated, and a combination of PI inclusion/exclusion and assessment of morphology by DIC remains a currently accepted method. The role of PI in defining cyst death (loss of membrane integrity) has been questioned since cysts exposed to lethal levels of chlorine disinfectant failed to become PI-positive. Failure of such cysts to include PI is probably due to the gradual loss of membrane integrity following trophozoite death.

Two fluorogenic viability assays were developed for

G. muris (used as a surrogate for *G. duodenalis*, although less sensitive to disinfection) viability and compared to infectivity in neonatal CD-1 mice and in vitro excystation. For ozone and chlorine disinfection studies, SYTO®-9 (see below), which stained cysts bright yellow, was the best single stain for detecting dead cysts, while the Live/Dead BacLight™ kit (Molecular Probes, Eugene, OR, US), which stained viable cysts dark green and non-viable cysts light green-orange/yellow, also showed correlation with animal infectivity. The correlation between either SYTO®-9 or the Live/Dead BacLight™ kit was better with infectivity than with in vitro excystation.

A fluorogenic vital dye assay to determine viability of *C. parvum* oocysts, based upon the inclusion/exclusion of two fluorogenic dyes, DAPI and PI, has been developed. Discrimination between viable and non-viable oocysts is based upon the former including DAPI into the nuclei of the four sporozoites, but excluding PI, whilst the latter include PI either into the nuclei or cytoplasm as well as DAPI. A UV filter block is used to visualize DAPI (350 nm excitation, 450 nm emission) and a green filter block for PI (535 nm excitation, > 610 nm emission). Results correlated closely with optimized in vitro excystation. Its usefulness has been documented for assessing the survival of *C. parvum* oocysts under a variety of environmental pressures, although it overestimates the viability of oocysts exposed to chemical disinfectants used in water treatment.

Two further fluorogens (SYTO®-9, SYTO®-59; Molecular Probes) have also been developed as surrogates of oocyst viability. SYTO®-9 fluoresces light green-yellow under the blue filter (480 nm excitation, 520 nm emission) and SYTO®-59 fluoresces red under the green filter block (535 nm excitation, > 610 nm emission) of an epifluorescence microscope. Both are nucleic acid intercalators and have been shown to correlate closely with animal infectivity in neonatal CD-1 mice, but not with in vitro excystation. SYTO®-9 and -59 are permeable to oocysts with damaged membranes and, therefore, stain non-viable oocysts. SYTO®-9 stains dead oocysts green or bright yellow: viable oocysts display a green halo surrounding the oocyst wall but no internal staining. SYTO®-59 stains dead oocysts red: viable oocysts remain unstained. Both were used to determine the viability of *C. parvum* oocysts exposed to heat treatment (70°C, 30 min) and chemical disinfection (including chlorine, chlorine dioxide and ozone disinfection) of water and indicated $\log_{10}$ reductions similar to those produced when the same batches of oocysts were gavaged into CD-1 neonates.

In Vitro Infectivity

A combined viability and infectivity assay based on in vitro excystation and in vitro culture has been developed for *Cryptosporidium*. A variety of cell lines will support asexual development and are identified in Table 3 of the article on *Cryptosporidium*. Oocysts exposed to excystation stimuli are transferred to a cell culture which supports their invasion and asexual development in vitro. Development of the parasite is intracellular, but extracytoplasmic (see article on *Cryptosporidium*) and the developing asexual stages (primarily trophozoites and merozoites) are either highlighted by immunofluorescence using an anti-sporozoite/merozoite antibody or their DNA is amplified by PCR after extraction of *Cryptosporidium* nucleic acids. A period of up to 48 h is allowed to elapse before developmental stages are sought. Currently, the sensitivity and reproducibility of this method are unknown, although a single infectious oocyst was detected using a reverse-transcription PCR (RT-PCR) for extracted *C. parvum* mRNA, targeting the heat-shock protein 70 gene.

Nucleic Acid-based Methods

The detection of parasites which have undergone in vitro excystation, where parasites are released from their cyst or oocyst into a suspending matrix, has been used as the first stage in developing a PCR-based viability assay. Two approaches have been described for *C. parvum* oocysts. In one approach, sporozoites are excysted in vitro according to a standardized protocol, lysed and the DNA is amplified using previously published primers for a repetitive oocyst protein gene sequence. In the second approach, IMS is used to concentrate oocysts from the inhibitory matrix of faeces, the sporozoites are excysted and their DNA is released and amplified by PCR using the 'Laxer et al' primers in a nested PCR (IC-PCR). Whereas the sensitivity of the first approach was approximately 25 oocysts (100 sporozoites) in an experimental system, no information was available on its sensitivity in environmental samples. The sensitivity of the second (IC-PCR) approach was between 1 and 10 oocysts in purified samples and between 30 and 100 oocysts inoculated into stool samples.

The ability of an organism to respond to an external stimulus in its environment by producing increased amounts of messenger RNA (mRNA) has been used as a surrogate to determine viability, and mRNAs of heat-shock proteins have been especially targeted. Separate RT-PCR to amplify a sequence of the mRNA for *Giardia* heat-shock protein, and a sequence of the mRNA for *C. parvum* heat-shock protein 70 (*hsp 70*) have been developed. The *Giardia* heat-shock protein

RT-PCR using the Giardia heat-shock protein (GHSP) primers detects the human parasite *G. intestinalis* but has been reported as being inconsistent.

The *C. parvum* hsp 70 RT-PCR was tested in four different water types. Synthesis of hsp 70 mRNA was induced by 20 min incubation at 45°C followed by five freeze–thaw cycles (1 min liquid N_2, 1 min 65°C) to rupture the oocysts, and mRNA was hybridized on oligo $(dT)_{25}$-linked magnetizable beads. RT-PCR was undertaken according to the manufacturer's instructions using a primer specifically designed to prime *C. parvum* hsp 70. The RT reaction mixture was amplified by PCR and the amplicons detected visually on gels and by chemiluminescent detection of Southern blot hybridizations. An RNA internal positive control was also developed and included in each assay in order to safeguard against false-negative results caused by inhibitory substances. The sensitivity of *C. parvum* RT-PCR was good: it was able to detect one oocyst in each of the four water types.

The simultaneous detection of viable *Giardia* cysts and *C. parvum* oocysts using RT-PCR has also been described. For *Giardia*, primers, different to those identified above, were used to amplify heat-shock protein mRNA, whereas those described above were used for *C. parvum*. An internal positive control was developed to determine the efficiency of mRNA extraction and potential RT-PCR inhibition. Sensitivity for both organisms was reported in the range of a single viable organism and, in a comparison of the conventional immunofluorescence method for identifying (oo)cysts and RT-PCR in water concentrates ($n = 29$), the frequency of detection of viable *Giardia* cysts rose from 24% to 69% with RT-PCR. For *Cryptosporidium*, RT-PCR detected oocysts in one sample as compared with four samples by fluorescence microscopy. The authors suggested that the difference was due to the inability to distinguish oocysts of *C. parvum* from oocysts of other *Cryptosporidium* species with epifluorescence microscopy.

PCR has also been used to detect organisms in backwash waters and sewage influent/effluent. Fewer reports are available, but the protocols used are similar to those used in raw and potable water. In one study, both presence and reduction of (oo)cysts in sewage effluent were assessed by immunofluorescence and PCR. A nested PCR was used to detect *Cryptosporidium* oocysts whereas a double PCR was used to detect *Giardia* cysts. By immunofluorescence, there was a 3 $\log_{10}$ reduction of cysts and a 2 $\log_{10}$ reduction of oocysts through sewage treatment. The correlation between immunofluorescence and PCR detection of *Giardia* cysts was 100%, whereas for *Cryptosporidium* oocysts the correlation was slightly less. Because of the nature of these matrices, inhibitory substances are expected to be present in greater quantities, and this will decrease the sensitivity of detection. This demands more effective methods both to neutralize their effects and to extract nucleic acids more effectively. Methods for reducing inhibitory interactions in PCR have been reviewed and methods for preventing inhibitory substances from co-purifying with nucleic acids (such as concentrating (oo)cysts by IMS) have been identified elsewhere.

FISH has also been used to determine the viability of *C. parvum* oocysts. The fluorescently labelled oligonucleotide probe targets a specific sequence in the 18S ribosomal RNA of *C. parvum* and causes viable sporozoites (capable of in vitro excystation) to fluoresce. Neither dead oocysts nor organisms other than *C. parvum* organisms fluoresced following *in situ* hybridization. FISH-stained oocysts did not fluoresce sufficiently brightly to allow their detection in environmental water samples; however, simultaneous detection and viability could be undertaken when FISH was used in combination with a commercially available FITC-C-mAb.

Biophysical Methods

Dielectrophoresis can be used to discriminate between viable and non-viable oocysts given that their dielectrophoretic fingerprint spectra are sufficiently different. Ozone-treated, chlorine-treated and untreated oocysts produce different dielectrophoretic spectra; however, the viability of individual organisms cannot be determined readily by dielectrophoresis at present.

ROT has also been used to assess the viability of *Giardia* cysts and *Cryptosporidium* oocysts. An ROT chamber is manufactured on a microscope slide where particles are subjected to a uniform rotating electric field created by applying phase-shifted AC voltages (usually between 100 Hz and 10 MHz) to electrodes around the particle. The rotating field polarizes particles, which experience a torque, because of the finite time required to form the field-induced dipole, causing them to spin. The rates and directions of spin of the particle are dependent on the frequency of the applied field and on the passive electric properties of the particle and the surrounding medium. By observing the spin of a particle in an ROT chamber, the passive electrical properties (and therefore integrity, especially of the membranes of the organism can be determined.

ROT can demonstrate real-time assessment of viability of individual oocysts, which correlates closely with the fluorogenic vital dye technique and the maximized in vitro excystation assay. Viable *C. parvum* oocysts (i.e. include DAPI, exclude PI and are capable of excysting in vitro) can be differentiated from those non-viable oocysts (i.e. include DAPI and PI and which do not excyst in vitro) by ROT: viable oocysts

Table 2 Methodologies for detection of (oo)cysts in water, environmental and food samples

Technique	*Principle of technique*	*Advantages*	*Disadvantages*
Conventional	Identification by epifluorescence, Nomarski differential interference contrast (DIC) microscopy	Well tested, detects both empty and intact organisms, nuclei demonstrated with DAPI, reference standard for morphology and morphometry	Time-consuming, labour-intensive, expensive equipment, highly motivated and trained personnel necessary, DIC microscopy may be impossible in dirty samples, species-specific mAbs currently unavailable
Flow cytometry with cell sorting (FCCS)	Separation by size and intensity of fluorescence	Time- and labour-saving, detects both empty and intact organisms, dedicated software available, useful for concentrating organisms	Expensive equipment, sample often aerosolized, conventional microscopy necessary for identification, species-specific mAbs currently unavailable
Charge-couple devices (CCD)	Detection of fluorescent and luminescent low light emissions	Spatial reconstruction of individual organisms, detects both empty and intact organisms, time- and labour-saving, potential for automation and telemetry	Expensive equipment, microscope necessary, dedicated software in development phase, species-specific mAbs currently unavailable
Enhanced chemiluminescence (ECL)	Generation of light from ECL reaction	Time- and labour-saving, detects both empty and intact organisms, useful for primary screening	Low resolution, false positives may occur, confirmation by conventional microscopy necessary, species-specific mAbs currently unavailable
Polymerase chain reaction (PCR)	Detection of sporozoite nucleic acids	Sensitive, high-resolution technique with potential for species specificity and genotyping	Cannot detect empty organisms. Poorer resolution with organisms with damaged/fragmented nucleic acids, reduced sensitivity due to inhibitory substances in water concentrates
Laser scanning	Detects fluorescent emissions from organisms, memorizes organism location	Can be programmed for a variety of fluorogenic emissions and intensities, time- and labour-saving, readily automated	Expensive equipment, conventional microscopy necessary for identification, species-specific mAbs currently unavailable
Electrorotation	Assessment of passive electrical properties (integrity, especially of membranes) in an AC rotating field	Detects both intact and empty organisms, real-time assessment of viability	Requires added clean-up procedures and suspension in low-molarity solutions, further development and testing required

DAPI = 4′6-Diamidino-2-phenyl indole; mAbs = monoclonal antibodies.

will rotate clockwise and non-viable oocysts anticlockwise at a predetermined frequency. Automatic measurement of the rotation rate at one frequency or over a range of frequencies to create a full spectrum is also possible. As ROT is non-invasive, the organism can be further subjected to a variety of analytical methods.

Despite these developments, infectivity remains the reference standard as it is currently the only method that causes the parasites to multiply in a host. It demands both dedicated facilities and experienced personnel, which drastically limits its usefulness and applicability. Other surrogates, such as in vitro excystation, inclusion/exclusion of vital dyes, assessment of mRNA and in vitro infectivity act as separate standards in that they describe some of the series of events triggered in the parasite and the host which culminate in infection, but they have been used to good effect to determine survival in a variety of environments.

Regulations

The current European Union drinking water directive requires that 'water intended for human consumption should not contain pathogenic organisms' and 'nor should such water contain: parasites, algas, other organisms such as animalcules'. The proposed revision to this directive, in recognizing the impracticality of the current zero standard, will make it a general requirement 'that water intended for human consumption does not contain pathogenic micro-organisms and parasites in numbers which constitute a potential danger to health'. No numerical standard for *Cryptosporidium* (or *Giardia*) is proposed.

The UK government intends to propose a regulation for *Cryptosporidium* in drinking water. The regulation will set a treatment standard at water treatment sites determined to be of significant risk following risk assessment. Daily continuous sampling of at least 1000 l over at least a 22-h period from each point at which water leaves the water treatment works is required and the goal is to achieve an average density of less than 1 oocyst per 10 l of water.

The USEPA has issued several rules to address the control of *Cryptosporidium* and *Giardia*. One of the goals of the Surface Water Treatment Rule (SWTR), formulated to address the control of viruses, *Giardia* and *Legionella*, was to minimize waterborne disease transmission to levels below an annual risk of 10^{-4}. In order to accomplish this goal, treatment through filtration and disinfection requirements was set to reduce *Giardia* cysts and viruses by 99.9% and 99.99%, respectively. The SWTR also lowered the acceptable limit for turbidity in finished drinking water from a monthly average of 1.0 NTU to a level not exceeding 0.5 NTU in 95% of 4-h measurements. The Enhanced Surface Water Treatment Rule (ESWTR) includes regulation of *Cryptosporidium* but, in order to implement this rule, a national database on the occurrence of oocysts in surface and treated waters was required to be collected under the Information Collection Rule (ICR). Water authorities serving more than 10 000 individuals began an 18-month monitoring programme for *Cryptosporidium* under the ICR, which was issued in 1996 and is now completed.

Advantages and Limitations

The conventional and developing technologies are outlined in **Table 2**, together with their advantages and disadvantages.

Limitations of Isolation and Enumeration Methods and Conclusions

The occurrence of (oo)scysts in water is largely underestimated due to limitations in isolation and detection methods. Isolation and identification are influenced by the matrix, with water quality, the presence of biocontaminants, suspended solids, turbidity and clays reducing recovery efficiency. The standardization of filtration, concentration and identification procedures is imperative for the correct interpretation of results. Because of current limitations in our technologies, not only will an underestimation of (oo)cyst contamination continue to occur, but so will confusion arise from the detection of organisms which may have no significance to human health. Some current limitations have been overcome; however, others remain to be addressed. Specificity and sensitivity remain the main objectives. Novel technologies will address specific biological issues, but the ultimate goal remains on-line monitoring.

See also: **Biophysical Techniques for Enhancing Microbiological Analysis**: Future Developments. **Water Quality Assessment**: Modern Microbiological Techniques.

Further Reading

Anonymous (1990a) *Isolation and Identification of* Giardia *cysts*, Cryptosporidium *Oocysts and Free Living Pathogenic Amoebae in Water etc. Methods for the Examination of Waters and Associated Materials.* London: HMSO.

Anonymous (1990b) Cryptosporidium *in Water Supplies.* Report of the Group of Experts; chairman, Sir John Badenoch. Department of the Environment, Department of Health. London: HMSO.

Anonymous (1994) Proposed ICR protozoan method for detecting *Giardia* cysts and *Cryptosporidium* oocysts in water by a fluorescent antibody procedure. *Federal Register* 59: 6416–6429.

Anonymous (1995) Cryptosporidium *in Water Supplies.* Second Report of the Group of Experts; Chairman, Sir John Badenoch. Department of the Environment, Department of Health, London: HMSO.

Anonymous (1998a) Cryptosporidium *in Water Supplies.* Third Report of the Group of Experts; Chairman, Professor Ian Bouchier. Department of the Environment, Transport and the Regions, Department of Health. London: HMSO.

Anonymous (1998b) *Method 1622*: Cryptosporidium *in Water by Filtration/IMS/FA.* EPA 821-R-98-010. Washington: United States Environmental Protection Agency, Office of Water.

Craun GF (1990) Waterborne giardiasis. In: Meyer EA (ed.) *Giardiasis.* Vol. 3. p. 267. New York: Elsevier.

Rose JB, Lisle JT and LeChevallier M (1997) Waterborne cryptosporidiosis: incidence, outbreaks and treatment strategies. In: Fayer R (ed.) Cryptosporidium *and Cryptosporidiosis.* P. 51. Boca Raton, Florida: CRC Press.

Smith HV and Hayes CR (1996) The status of UK methods for the detection of *Cryptosporidium* sp. oocysts and *Giardia* sp. cysts in water concentrates and their relevance to water management. *Water Science and Technology* 35: 369–376.

Smith HV and Rose JB (1998) Waterborne cryptosporidiosis: current status. *Parasitology Today* 14: 14–22.

Smith HV, Robertson LJ and Ongerth JE (1995) Cryptosporidiosis and giardiasis: the impact of waterborne transmission. *J Water SRT – Aqua* 44: 258–274.

WINES

Contents

Microbiology of Wine-making

Graeme M Walker, School of Science and Engineering, University of Abertay Dundee, Scotland

Introduction

Wine-making is a biotechnology which is centuries old. It has become a global enterprise, which significantly affects the economic wellbeing of many countries. The basic activities of modern wineries are fundamentally the same as those undertaken traditionally; sugars from grapes or other fruits are physically extracted, and then fermented by yeasts to produce an alcoholic beverage. Yeasts and certain beneficial bacteria play additional roles, in the development of the flavour and aroma of the wine following the maturation of the liquid after primary fermentation.

Wines are generally classified according to colour (red, white or rosé) and alcohol content – table wines have ethanol concentrations of 7–14% v/v and in fortified wines, such as port and sherry, the ethanol concentration exceeds 14% v/v. Wines can also be categorized according to grape variety (there are hundreds of cultivars of *Vitis vinifera*), taste (dry, semidry, semisweet or sweet) and texture (still or sparkling). Important examples of wines based on *V. vinifera* cultivars are: Pinot Noir (red Burgundy), Chardonnay and Pinot Blanc (white Burgundy), Cabernet Sauvignon (red Bordeaux), Riesling, Müller-Thurgau and Sylvaner (German white), Zinfandel (Californian red), Palomino (Spanish sherry) and Sangiovese (Italian Chianti).

Wine Manufacture

The production process of red and white wines are outlined in **Figure 1**. Wine-making basically involves the extraction of grape juice ('must') by crushing the fruit, alcoholic fermentation by yeasts (endogenous or exogenous cultures), maturation, clarification and packaging. Crushing is followed by maceration, which facilitates the extraction of compounds from the seeds and skins. This extraction is initiated by the action of hydrolytic enzymes released from cells which have been ruptured during crushing.

Red wine is produced by the fermentation of the juice of black grapes, containing the skins. If the skins of black grapes are removed from the juice, or if white grapes are used, white wine is produced. Rosé wines are produced if the black grape skins are removed before all the pigment has been extracted. Malolactic fermentation may be carried out if desired, to decarboxylate L-malic acid to L-lactic acid, resulting in a decrease in the acidity of the wine. The organoleptic properties of wines depend primarily on the grape cultivar employed and on the metabolic activities of yeasts and bacteria. The actions of the enzymes of grapes and of microorganisms during maturation also contribute to the characteristics of the wine.

Microflora Involved in Traditional Wine-making

Numerous types of microorganism are involved in wine-making, and can be described as endogenous (from grapes or winery surfaces) or exogenous (from selected starter cultures). Yeasts and bacteria can make either beneficial or detrimental contributions to wine quality. Traditional, or natural, fermentations of wine exploit the wild microflora on the surface of grape skins, together with the yeasts indigenous to wineries (predominantly, strains of *Saccharomyces cerevisiae*). The principal microbial genera associated with grapes are:

- yeasts: mainly *Kloeckera* and *Hanseniaspora*, with lesser representations of *Candida*, *Metchnikowia*, *Cryptococcus*, *Pichia* and *Kluyveromyces* and very low populations of *Saccharomyces cerevisiae*
- lactic acid bacteria: *Lactobacillus*, *Leuconostoc*, *Pediococcus*
- acetic acid bacteria: *Gluconobacter*, *Acetobacter*
- fungi: *Botrytis*, *Penicillium*, *Aspergillus*, *Mucor*, *Rhizopus*, *Alternaria*, *Ucinula*, *Cladosporium*.

The skins of sound grapes typically harbour microbial populations of 10^3–10^5 cfu g^{-1}.

Table 1 shows the progression of the growth of yeasts in a typical natural wine fermentation. Although yeasts other than *Saccharomyces* grow well during the first few days, they are not very tolerant to the ethanol produced and generally start to die after about 4 days, when the ethanol concentration reaches

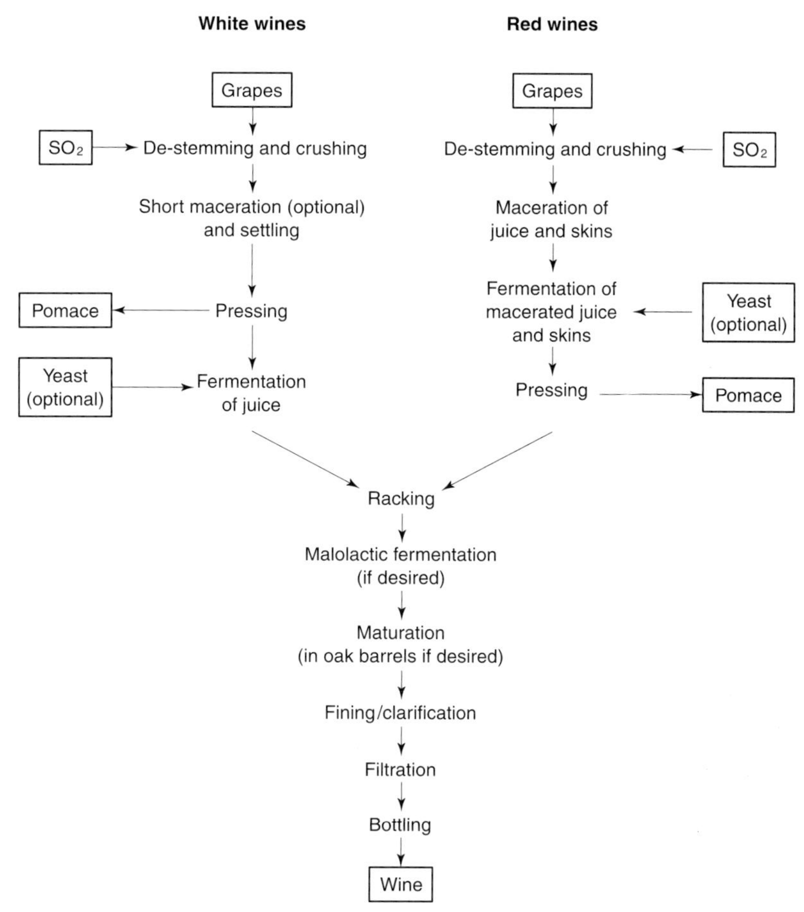

Figure 1 Wine production.

Table 1 Growth of yeasts in a typical natural wine fermentation

Stage of fermentation	*Ethanol content (% v/v)*	*Typical yeasts*
Early (days 1–3)	0–3	*Kloeckera apiculata* *Hanseniaspora valbyensis*
Middle (days 3–5)	3–5	*Candida stellata* *Torulaspora delbrueckii* *Kluyveromyces* spp. *Pichia* spp.
Latter (days 4–10)	5–12	*Saccharomyces cerevisiae*

about 5% v/v. Subsequently, fermentation is carried out predominantly by wine strains of *S. cerevisiae* which are able to tolerate high concentrations of ethanol (> 15% v/v).

Role of Yeasts

The composition of grape juice is summarized in **Table 2**. Yeasts utilize glucose and fructose, the principal sugars in grape juice, and metabolize them via the Embden–Meyerhof–Parnas glycolytic pathway, to pyruvate. This pathway furnishes the yeast cells with energy and with reducing power, for biosynthesis. Under anaerobic conditions, the yeasts decarboxylate pyruvate, in a reaction catalysed by pyruvate decarboxylase, to yield acetaldehyde and CO_2. The final step in alcoholic fermentation is catalysed by alcohol dehydrogenase and involves the reduced coenzyme NADH, and results in the reduction of acetaldehyde to ethanol. The conversion of glucose to ethanol by *S. cerevisiae* can be summarized as **Equation 1**:

$$\underset{\text{glucose}}{C_6H_{12}O_6} + 2Pi + 2ADP + 2H^+ \rightarrow$$
$$\underset{\text{ethanol}}{2C_2H_5OH} + 2CO_2 + 2ATP + 2H_2O \quad \text{(Equation 1)}$$

In addition to ethanol and CO_2, one of the quantitatively most important products of fermentation by wine yeasts is glycerol. Variable levels (generally in the range $2–10\,g\,l^{-1}$) of glycerol are found in wine, depending on the yeast strains and the fermentation

Table 2 Principal ingredients of grape juice

Component of grape juice	*Comments*
Carbon compounds	
Glucose	Typical concentration 75–150 g l^{-1}
Fructose	Typical concentration 75–150 g l^{-1}
Sucrose	Trace
Pentoses	Unfermentable by *Saccharomyces cerevisiae*
Pectins	Small amounts
Nitrogen compounds	
Free amino acids	0.2–2.5 g 1^{-1}
Ammonium ions	Small amounts, which may be limiting for yeasts
Proteins	Small amounts
Other organic acids	
Tartaric acid	2–10 g 1^{-1}: not metabolized by wine yeasts
Malic acid	1–8 g l^{-1}: partially metabolized by wine yeasts
Minerals	
Phosphorus, potassium, magnesium, sulphur, trace elements	Adequate supply of bulk minerals, and sufficient quantities of trace elements
Vitamins	Small amounts, but sufficient for growth of yeasts
Other compounds	Sulphite levels often high Decanoic and octanoic acids may inhibit yeast growth Sterols and unsaturated fatty acids may be limiting for yeasts

Table 3 Some biotechnological developments with wine yeasts

Development	*Comments*
Yeast strain identification	Genetic fingerprinting using RFLP and RAPD-PCR analyses, together with pulsed-field electrophoretic karyotyping (e.g. PFGE) have been used successfully to differentiate wine strains of *S. cerevisiae*
Genetic hybridization of yeast	Desirable flocculation (cellular aggregation and sedimentation) properties have been introduced into wine yeasts by this technique
Genetic engineering	Numerous genes from other organisms, including other yeasts, have been introduced into wine strains of *S. cerevisiae* using recombinant DNA technology, including genes encoding for: K1 (killer) toxin, to combat wild yeasts Pectinases, to increase the filterability of wine Glucanases, to release bound terpenes and increase fruity aromas Lactate dehydrogenase, to promote mixed fermentation and acidification Malolactic enzyme, to promote malolactic fermentation Malic enzyme, to promote maloethanolic fermentation Glycerol phosphatase, to increase glycerol levels
Fermentation technology	Sparkling wines can now be produced using yeasts immobilized in natural gels such as alginate and carrageenan

PFGE, pulse-field gel electrophoresis; PCR, polymerase chain reaction; RAPD, random amplification of polymorphic DNA; RFLP, restriction fragment length polymorphism.

conditions. It is produced by the following reaction (**Equation 2**):

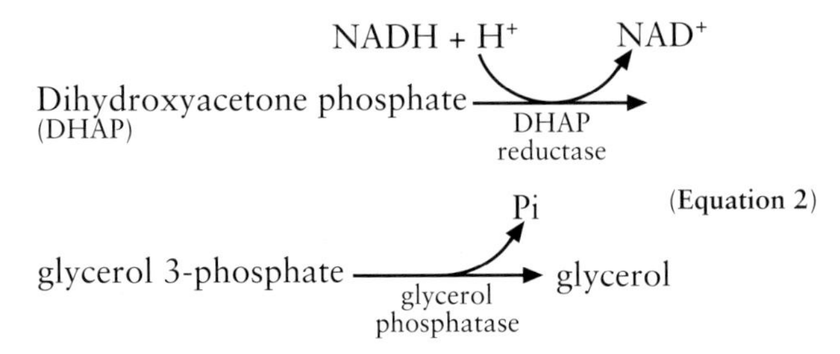

Glycolysis is the major mechanism for the catabolism of carbon compounds by yeasts, but other metabolic pathways also operate during grape juice fermentations, including the pentose phosphate pathway and the citric acid cycle. The limited operation of the latter pathway generates significant levels (typically around 0.5 g l^{-1}) of succinic acid in wine.

The metabolism of nitrogen and sulphur compounds yields, together with the main metabolites, hundreds of volatile and non-volatile minor metabolites, which collectively contribute to the flavour and aroma of wine. These include:

- higher alcohols: isoamyl alcohol, active amyl alcohol, isobutanol, propanol, 2-phenylethanol

Table 4 Spoilage of wine by microorganisms

Microorganism	*Spoilage effects*
Indigenous yeasts	Estery taints, due to high concentrations of ethyl acetate (> 200 mg l^{-1}) and methylbutyl acetate (caused especially by *Hanseniaspora* species Film formation on wine surfaces, caused by species of *Candida*, *Pichia* and *Metchnikowia*, which results in severe flavour taints
Exogenous yeasts	
Zygosaccharomyces bailii	Yeast growth and the re-fermentation of sugars in stored wine produce turbidity and off flavours
Brettanomyces spp.	Haze, turbidity, volatile acidity, mousy and phenolic taints in stored wine Occasionally, explosive CO_2 liberation in bottled sparkling wines
Schizosaccharomyces pombe	Osmotolerant yeast: may produce haze in stored wines
Saccharomyces cerevisiae	Re-fermentation of wines with residual sugars
Bacteria	
Acetic acid bacteria: *Gluconobacter oxydans*, *Acetobacter pasteurianus*, *Acetobacter aceti*	Vinegary taints, due to acetic acid in concentrations > 1.2 g l^{-1} Other acidic and estery taints, caused by aerobic metabolism of these bacteria Ropiness
Lactic acid bacteria	These bacteria have a low tolerance to alcohol, but may cause: increased acidity, due to acetic and lactic acids (produced especially by heterofermentative *Lactobacillus* and *Leuconostoc* spp.) mannitol taints (produced especially by *Lactobacillus brevis*) ropiness (caused by e.g. *Pediococcus cerevisiae*) diacetyl (produced e.g. by *Pediococcus* and *Lactobacillus* spp.) mousiness (e.g. *Lactobacillus* spp.) bitterness (e.g. the acrolein taint caused by *Leuconostoc* spp.) breakdown of tartaric acid (e.g. by *Lactobacillus* spp.) miscellaneous off flavours (e.g. esters, high alcohols) and odours
Clostridium spp.	Rarely cause spoilage, but occasionally produce *n*-butyric acid taints in low-acid wines
Bacillus spp.	Several species can grow in wine, to increase the volatile and total acidity
Actinomyces spp., *Streptomyces* spp.	Earthy, corky taints
Bacteriophages	Disruption of malolactic fermentations
Fungi	Earthy, corky taints due to fungal growth in wooden barrels and corks

- esters: ethyl acetate, ethyl lactate, phenylethyl acetate, isoamyl acetate, ethyl octanoate, ethyl hexanoate
- organic acids: succinic, tartaric, malic, lactic, acetic, citric
- aldehydes and ketones: acetaldehyde, diacetyl, acetoin
- sulphur compounds: H_2S, SO_2, dimethyl sulphide.

The relative concentrations of these compounds depend on the strains of yeast employed and the fermentation conditions, especially temperature. White wine fermentations are generally conducted at 10–18°C (for 7–14 days or longer), and red wine fermentations at 20–30°C (for around 7 days).

Starter Cultures

Modern, large-scale wineries generally use specially selected starter cultures of *Saccharomyces cerevisiae* in preference to relying on the fermentative activities of naturally occurring yeasts. Such cultures are available in dried form from several companies. The yeasts are normally inoculated (at 10^6–10^7 cells per ml) to which sulphite has been added to suppress natural yeasts and bacteria. Such starter yeasts may not completely prevent the growth and metabolism of indigenous yeasts, including winery-associated *S. cerevisiae* and grape-associated *Kloeckera apiculata*. However, because the starters experience shorter lag phases they are able to convert sugar to alcohol more rapidly in inoculated grape must than in must which has not been inoculated. Desirable characteristics of wine yeast starters are summarized below:

- genetics: homothallic diploid (usually polyploid)
- growth: minimal or no lag phase; 'killer' character; tolerance to SO_2
- metabolism: rapid and reproducible alcoholic fermentation; efficient conversion of grape sugars to

ethanol, CO_2 and desirable minor fermentation metabolites
- flavour: correct volatile acidity; appropriate character of aroma produced (e.g. esters, terpenes, succinic acid, glycerol); low acetaldehyde; correct balance of sulphur compound production
- others: low urea excretion, to minimize the production of potentially carcinogenic ethyl carbamate.

Biotechnological Developments

Advances in molecular biology and in fermentation technology are influencing the application of yeasts in wine-making. Some relevant developments are summarized in **Table 3**. Recombinant DNA technology offers the greatest potential for the improvement of wine yeasts, particularly in relation to performance in fermentation and final product quality.

Spoilage of Wine

The growth and metabolic activities of a variety of microorganisms, during wine-making and in finished wine, can spoil the organoleptic properties of the final product. Several stages of the production process can be affected by spoilage microorganisms, as can the grapes used. In particular, the fermentation step, stored or bottled wine and the corks used in bottling may be affected. **Table 4** lists the major spoilage microbes which affect wine, and summarizes their effects.

Hygiene is the key to preventing microbial contamination during wine-making, and hence is critical in spoilage control. The exclusion of O_2 and the appropriate use of SO_2 are additional measures, for the quality assurance of stored wines. Sulphur dioxide (added to wine as potassium or sodium metabisulphite or as gaseous SO_2) is an antimicrobial compound and antioxidant which has for centuries been used to preserve wine by preventing the growth of undesired microorganisms. It is added to grape must and wine at concentrations that inhibit the spoilage lactic acid bacteria and the undesired non-*Saccharomyces* yeasts, but do not prevent the growth of the malolactic bacteria and the desired fermentation yeasts (i.e. *S. cerevisiae*). Sulphur dioxide exerts its antimicrobial action by a combination of enzyme inhibition; coenzyme, protein and nucleic acid interactions; cleavage of vitamins (e.g. thiamin), and depletion of cellular ATP. It is used widely in all wine-producing countries – the maximum limits permitted vary, but generally the maximum total SO_2 is a few hundred $mg\,l^{-1}$. However in some asthmatic individuals it can cause allergic reactions which are mild to severe, and affect the skin and the respiratory and gastrointestinal tracts. Therefore some wines are now produced using low levels of exogenous SO_2, and alternative antioxidant flavour stabilizers and biocides, for use in wines and other alcoholic beverages, are being investigated.

See also: **Fermented Foods**: Fermented Vegetable Products. ***Leuconostoc***. **Preservatives**: Permitted Preservatives – Sulphur Dioxide. ***Saccharomyces***: *Saccharomyces cerevisiae*. **Starter Cultures**: Importance of Selected Genera. **Wines**: The Malolactic Fermentation; Specific Aspects of Oenology. **Yeasts**: Production and Commercial Uses.

Further Reading

Barre P, Vezinhet F, Dequin S and Blondin B (1993) Genetic improvement of wine yeast. In: Fleet GH (ed.) *Wine Microbiology and Biotechnology*. P. 265. Chur: Harwood Academic Publishers.

Boulton RB (1995) Red wines. In: Lea AGH and Piggott JR (eds) *Fermented Beverage Production*. P. 121. Glasgow: Blackie Academic & Professional.

Boulton RB, Singleton VI, Bisson LF and Kunkee RE (1994) *Principles and Practices of Winemaking*. New York: Chapman & Hall.

Dittrich HH (1995) Wine and Brandy. In: Reed G and Nagodawithana TW (eds) *Biotechnology*, 2nd edn. Vol 9, p. 404. Weinheim: VCH Publishers.

Ewart A (1995) White wines. In: Lea AGH and Piggott JR (eds) *Fermented Beverage Production*. P. 97. Glasgow: Blackie Academic & Professional.

Fleet GH (ed.) (1993) *Wine Microbiology and Biotechnology*. Chur: Harwood Academic Publishers.

Fleet GH (1997) Wine. In: Doyle MP, Beuchat L and Montville T (eds) *Food Microbiology Fundamentals and Frontiers*. P. 671. Washington: American Society of Microbiology.

Fleet GH (1998) The microbiology of alcoholic beverages. In: Wood BJB (ed.) *Microbiology of Fermented Foods*, 2nd edn. Vol. 1, p. 217. London: Blackie Academic & Professional.

Jackson RS (1994) *Wine Science. Principles and Applications*. San Diego: Academic Press.

Kunkee RE and Bisson L (1993) Winemaking yeasts. In: Rose AH and Harrison JS (eds) *The Yeasts*. Vol. 5, p. 69. London: Academic Press.

Steinkraus KH (1992) Wine. In: Lederberg J (ed.) *Encyclopedia of Microbiology*. Vol. 4, p. 399. San Diego: Academic Press.

Subden RE (1990) Wine Yeast: selection and modification. In: Panchal CJ (ed.) *Yeast Strain Selection*. P. 113. New York: Marcel Dekker.

Walker GM (1998) *Yeast Physiology and Biotechnology*. Chichester: J Wiley.

Malolactic Fermentation

T Faruk Bozoğlu, Middle East Technical University, Department of Food Engineering, Ankara, Turkey

Seyhun Yurdugül, Department of Biochemistry, Middle East Technical University, Ankara, Turkey

Introduction

Malolactic fermentation is a secondary wine fermentation, carried out by the malolactic bacteria. The word malolactic comes from the conversion of L-malic acid in to L-lactic acid, by the activity of these bacteria. During this conversion, CO_2 is also produced. This results in a natural decrease in total acidity and bacterial stability. Malolactic bacteria are capable of direct decarboxylation of malic acid to lactic acid by the enzyme malate carboxylase (EC 1.1.1.38), which is present in various lactic acid bacteria, but particularly in three genera: *Lactobacillus*, *Leuconostoc* and *Pediococcus*. This enzymatic activity serves only for a limited number of genera of lactic acid bacteria. The main interest of winemakers is in the few species that have high tolerance to acidity and ethanol. In some cases it is regarded as a spoilage activity, but under proper circumstances, malolactic fermentation, either naturally or artificially encouraged, can be a normal part of good winemaking practice, something to be appreciated and desired. For high-quality wines, the fermentation brings positive effects, such as deacidification, bacteriological stabilization, increased flavour and aroma complexity. A large percentage of premium red wines of the world have had a malolactic fermentation. Wines produced in cold regions, especially Germany, France and the eastern USA, have a high acid content and may benefit from deacidification by malolactic fermentation.

Malolactic bacteria are also involved in other kinds of fermentations, such as pickle, sauerkraut and soy sauce manufacturing and silage production. Most lactic acid bacterial isolates from wine would have malolactic conversion capabilities; many strains of *Lactococcus lactis*, the modern classification of several species of *Streptococcus* have this capability.

Intermediary Metabolism of the Malolactic Conversion

Malate carboxylase (the malolactic enzyme) is the enzyme which carries out the malolactic conversion: a direct decarboxylation of L(−)-malic acid into L(+)-lactic acid (**Fig. 1**). It is inducible and the activities found depend on the strain and the cultural conditions, e.g. glucose or a fermentable sugar in the medium is necessary for induction of the malolactic enzyme. This enzyme is Mn^{++} dependent. Manganese is replaceable by Co^{++} and Cd^{++}.

The decarboxylation of malic acid to lactic acid provides very little free energy ($\Delta G = -8.3\ kJ\ mol^{-1}$) if CO_2 is released in the hydrated form.

The presence of malic acid has a dramatic stimulatory effect on the initial growth rate of many strains of malolactic bacteria. A 0.3% concentration of L-malate is considered the normal range for the white wines of Albarinyo (northwest Spain) at the end of the alcoholic fermentation. Malolactic fermentation is progressively improved with *Lactobacillus curvatus* and *Lactobacillus plantarum* when the initial L-malate concentration is increasing whereas bacterial growth decreases if the initial L-malate concentration is decreased. On the other hand, the highest bacterial growth of *Lactobacillus plantarum* is observed when the initial L-malate concentration in the medium is 0.1% and when L-malate is added to the medium, even when no fermentable carbohydrate is present, it has a stimulatory effect on malolactic fermentation.

Due to a small amount of intermediate pyruvic acid, reduced NAD^+ occurs. Pyruvic acid helps to reoxidize the NADH and stimulate the growth of the bacteria out of the lag phase. Acetyl phosphate is formed which acts as a hydrogen acceptor. More NADH is then formed from acetaldehyde. Any acetyl phosphate formed could also result in the formation of ATP, which could have a positive effect either on bacterial growth rate or the formation of bacterial biomass.

Cellular efflux of lactic acid arising from the malolactic conversion could also provide extra energy, involving membrane ATPase and re-entry of protons through a contiguous symport with the resulting formation of one ATP for every three protons. Increased ATP formation can have a triggering effect on the formation of biomass toward the end of the growth cycle, precisely when the efflux effect would be minimal due to the increased extracellular concentration of lactic acid; 0.02% is the most frequent

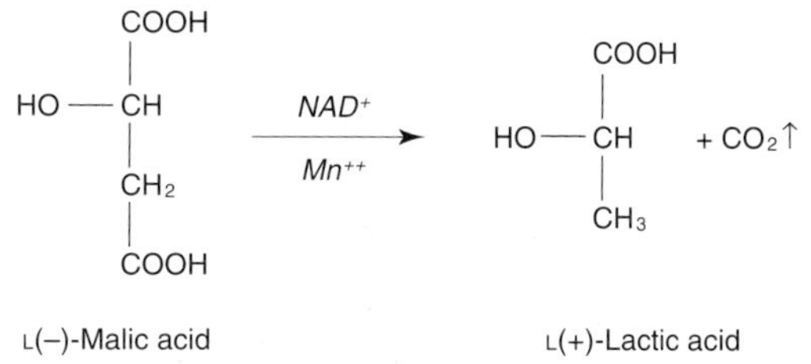

$$HOOC-CH(OH)-CH_2-COOH \xrightarrow[Mn^{++}]{NAD^+} HOOC-CH(OH)-CH_3 + CO_2\uparrow$$

L(−)-Malic acid → L(+)-Lactic acid

Figure 1 Malolactic conversion: a direct decarboxylation of L(−)-malic acid to L(+)-lactic acid. From Boulton et al (1996).

lactic acid concentration. The malolactic fermentation with *Lactobacillus plantarum* persists for two days in the absence of L-lactate. The addition of L-lactate increases the conversion of L-malate to L-lactate to 100%. It has been reported that in Spanish cider manufacturing, D-lactic acid increases markedly once the malolactic fermentation is completed. This D-lactate production is correlated with acetic acid production (volatile acidity). In France, this phenomenon is known as 'piqure lactique' when the concentration of D-lactic acid is greater than 2.2 mM. Traditional methods used for cider making in the Asturias region of Spain lead to an elevated production of acetic acid due to a high level of spoilage by lactic and acetic acid bacteria.

Lactobacillus curvatus displays a higher L-malate fermentative rate in the absence of additional D-glucose (0.1%) than in the presence of higher concentrations of this sugar, even though the growth of both *Lactobacillus plantarum* and *Lactobacillus curvatus* is improved in the presence of high D-glucose concentrations.

Volatile acidity and biogenic amines can be attributed to the metabolic activity of lactic acid bacteria in wines. Histamine, tyramine and putrescine are the major amines found in wines. The nitrogen content of wine is significantly altered during malolactic fermentation and the malolactic bacteria may also be involved in peptide and protein digestion.

Effect of Malolactic Fermentation on the Flavour and Aroma of Wines

The conversion of malic acid into lactic acid has been established as being a direct decarboxylation action by a single enzyme. In addition to the decarboxylation of malic acid, these kinds of bacteria have the capacity to metabolize citric acid. Citric acid is present in only small amounts in grapes and in wine, unless added. Although the metabolism of citric acid is not always seen in commercial winemaking, it can have an important influence on the formation of diacetyl, and thus on the flavour of the wine.

The two aspects of deacidification, the decrease in titratable acidity and the increase in pH, have important consequences for the winemaker. The perception of sourness comes essentially from the titratable acidity; thus wines too tart will benefit from having undergone malolactic conversion, but those already too flat will be further spoiled by this microbial activity. On the other hand, the microbial stability of the wine is primarily influenced by the pH. The winemaker needs to be alert to the increased susceptibility of wines to further microbial and chemical attack brought about by the elevation of pH resulting from the malolactic conversion.

The titratable acidity should be reduced to 50% of that represented by malic acid found before the malolactic fermentation. The titratable acidity may be altered by an additional formation of lactic acid from residual sugars, including pentoses, which the yeasts ignore during malolactic fermentation, by catabolism of malic acid by wine yeasts themselves, or by a loss of potassium bitartrate through precipitation if the wine is already saturated with this ion and if the pH, before the malolactic conversion, is below pH 3.56.

The increase in pH during malolactic fermentation depends on the buffering capacity of the medium (wine), and thus on the concentrations of the various weak acids before and after the decarboxylation, and also on the starting pH. Due to secondary effects of deacidification, a change in colour is found in red wine following the activity of these bacteria. A significant loss in red colour comes from equilibrium shifts in anthocyanin configuration, resulting from the increase in pH. If the conversion brings about an unusually high pH, the quality of the colour may also change from a full red to a bluish hue. Under situations of initially moderate or low acidities in grape musts or wines, the malolactic conversion may bring about a need to use acidulating agents to bring the titratable acidity and pH back to acceptable values. Unless the wine has been rendered completely free from malolactic bacteria, the addition of malic acid will bring about a continuation of bacterial activity, and considerable loss of the acid added.

Microflora Involved in Malolactic Fermentation

Malolactic fermentation is principally accomplished by three genera: *Lactobacillus*, *Leuconostoc* and *Pediococcus* (**Fig. 2**). They are generally grown in complex media, often containing peptone, tryptone or yeast extract at pH 5–6, and are usually stored in such media with agar, as stab cultures. The wine-related lactic acid bacteria will grow at much lower pH values, approaching 3.0 and low pH media are sometimes used by wine microbiologists. The cultivation of wine *Leuconostoc* spp. is difficult, as they need a special nutritional supplementation, including 4-O-(β-D-glucopyranosyl)-D-pantothenic acid, a tomato juice factor also found in apple and grape juices. A source of fatty acids is also helpful – the commercial mixture Tween 80 being commonly used. A yeast inhibitor, such as cycloheximide is also used for isolation. MRS broth has been used for the maintenance

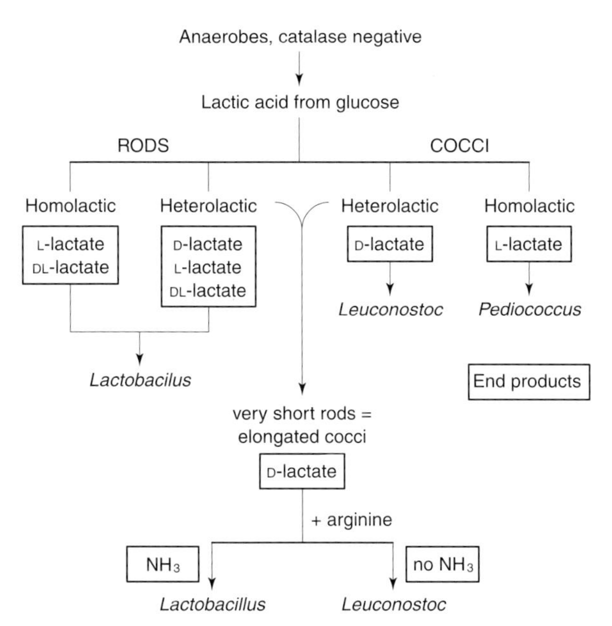

Figure 2 Genera of wine-related lactic acid bacteria. From Boulton et al (1996).

of lactic acid bacteria. The identification of these lactic acid bacteria is basically as follows.

Leuconostoc A simple identification procedure is applicable to *Leuconostoc* spp. If a *Leuconostoc* sp. is isolated from wine, it is automatically classified as *L. oenos*, as only strains of this species will grow in the presence of 10% ethanol at pH values less than 4.2. Cells are spherical, but lenticular when grown on agar, and usually occur in pairs or chains. Growth, compared to that of the non-acidophilic *Leuconostoc* spp., is slow and takes 5–7 days at 22°C.

Pediococcus Four species of *Pediococcus* that will grow at pH 4.2 or below are related to winemaking: *P. damnosus*, *P. parvulus*, *P. pentoseus* and *P. acidilactici*.

Lactobacillus Species classification of the wine-related lactic acid bacteria has not commanded great attention, since these organisms as a class show little resistance to the low pH values found in wine. Two species have been shown to be essentially important in the spoilage of high alcohol dessert wine: *L. hilgardii* and *L. fructivorans*.

The Influence of Malolactic Fermentation on Quality

Malolactic fermentation is said to improve the biological stability of wines as well as to increase the complexity of aroma and flavour. The effect of stabilization of the wine against further growth by other lactic acid bacteria is the most important consequence of the malolactic fermentation. The change in acidity is caused by infection of wines by spoilage strains of malolactic bacteria. The importance of strain selection for malolactic fermentation, not only for deacidification, but with respect of flavours, contrasts with the situation for the selection of wine yeast strains for the alcoholic fermentation. It is expected that desirable end metabolites, specifically associated with various non-spoilage strains of malolactic bacteria, are also made. In red wine the flavour effects of most of these malic fermenters is subtle and generally not perceived except under rigid taste panel conditions, in which case, one bacterial strain might be said to produce a wine with more complexity of flavour than another. An exception is diacetyl, an end product with a low sensory threshold.

Lactococcus lactis is able to ferment citrate that is converted into pyruvate by additional pathways, and this can lead to the formation of the flavour compound diacetyl (**Fig. 3**). Diacetyl is an important flavour compound in the dairy and winemaking industries. There is interest in the construction of *Lactococcus lactis* strains which can generate increased yields of diacetyl. Genetic manipulation of the biochemical pathway results in enzymes related to diacetyl production being induced. Several genes are involved: the *ldh* gene, encoding lactate dehydrogenase; the *ilvBN* gene encoding an α-acetoacetate synthase; the *aldB* gene encoding acetolactate decarboxylase; the *dar* gene encoding diacetyl–acetoin reductase; the *oad* gene encoding oxaloacetate decarboxylase; and the *pfl* gene encoding pyruvate formate lyase. By using genes *ldh*, *aldB* and *ilvBN* in metabolic engineering experiments, an increased yield of acetoin and diacetyl production has been obtained.

The benefits of malolactic fermentation include the lowering of acidity in high acid wines and enhancement of sensory characteristics through bacterial activity. Undesirable effects include the excessive reduction in acidity of high pH wines leading to the risk of spoilage, production of undesirable flavours, colour changes and formation of amines. It is necessary to seek better methods of controlling the occurrence and outcome of malolactic fermentation, including stimulation of indigenous flora, inoculation of bacteria, use of immobilized bacterial cells or enzymes and use of carbonic maceration for partial degradation of malic acid prior to natural malolactic fermentation.

The effect of malolactic fermentation also depends heavily on the grape variety. In a Chardonnay, for example, malolactic fermentation contributes strongly to aromas and compounds which have buttery, nutty, yeasty, oaky and sweet aromas. Malolactic fermentation slightly changes the aromatic profiles of wines. The Chardonnay wines are perceived

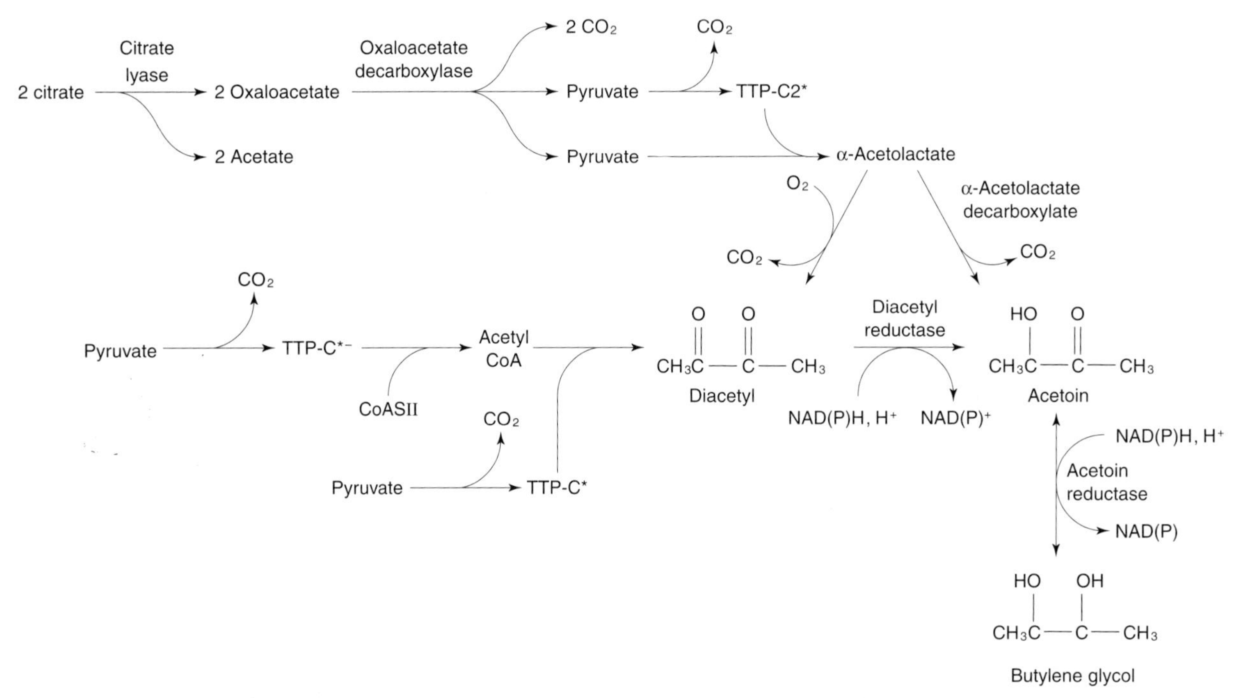

Figure 3 Pathways of diacetyl formation. *Active acetaldehyde. From Boulton et al (1996).

as higher in hazelnut, fresh bread and dried fruit notes, whereas Pinot Noir wines lose part of their berry notes in favour of animal and vegetable notes. On the contrary, the absence of malolactic fermentation retains specific aromas such as apple and grapefruit – orange in Chardonnay and strawberry–raspberry in Pinot Noir.

Factors Governing the Onset of Malolactic Fermentation

One of the tools for controlling malolactic fermentation could be the employment of the natural metabolic interactions between the yeast strain, used for the alcoholic fermentation, and the subsequent malolactic bacteria. The competition for nutrients, the formation and binding of sulphur dioxide by the yeast, the formation of inhibitory medium-chain fatty acids by the yeast, the strain variation in release of intracellular materials from the yeast during *sur lies* ageing are all important factors. The formation of a high concentration of ethanol by the yeast would be expected to have an important influence on secondary fermentation, depending on the strain of bacteria involved. Ethanol at low concentrations stimulates the growth of bacteria. On the other hand, the bacteria stimulate the death phase of the yeast.

The malolactic fermentation is stimulated either by making the conditions more favourable to the bacteria so as to speed up their growth rate or by supplying a high starting population of bacteria or both. Nominal additions of SO_2 inhibit bacterial growth and should be used as little as possible. During storage, the wine must be kept at a temperature above 18°C but not much higher because of potential damage to the wine. Before malolactic conversion, acid must not be added for stimulating the growth rate. Where the pH is low enough to prevent the onset of bacterial growth, winemakers can add calcium carbonate in order to raise the pH enough to allow bacterial growth. Some lactic acid bacteria can thrive at very high (>20%) concentrations of ethanol; the malolactic bacteria are generally sensitive to the ethanol concentrations in table wine, and the inhibition becomes more and more evident as the ethanol concentration reaches about 14% (vol./vol.). This inhibition is reported only for higher pH values. At low pH values it is a benefit for these bacteria.

In grape juice or laboratory media, the malolactic bacteria easily utilize glucose and fructose. In the absence of these sugars, as at the end of the alcoholic fermentation, it is not clear what the bacteria use as sources of energy. Malic acid has a dramatic stimulatory effect on the growth of these bacteria, but the bacteria can not use it as a sole energy source. All the malolactic bacteria can metabolize ribose, and some can metabolize other pentoses. These sugars are in low concentrations in wine, but sometimes there might be enough to provide sufficient energy for cell biomass to reach the limited amount typically found in wine. However, there may be a problem in induction of the permeases and enzymes for catabolism of these substrates, which are at such low concentrations. The micronutrient requirements for the lactic acid bacteria are extensive. They all require some amino acids and vitamins.

The lactic acid bacteria are microaerophilic, indicating a requirement for small amounts of oxygen. In heterolactic bacteria the presence of molecular O_2 should stimulate their growth by providing a direct means of reoxidizing NADPH. The latter is a required catalyst during the first steps in the heterolactic fermentation of hexoses, i.e. oxidative decarboxylation.

The use of bacterial starter cultures to stimulate the fermentation is becoming more and more popular and two routines have been devised to help overcome the difficulties of malolactic starter preparation. To allow better bacterial growth, the grape juice to be used as starter is initially soaked on the grape skins overnight and then the pressed juice is treated in the following commercially acceptable manner: dilution 1 : 1 with water; adjustment to pH 4.5 with calcium carbonate; and the addition of 0.1% yeast extract. To this is added a 1% starter of malolactic bacteria, and a 0.1 inoculum of wine yeast. The latter is added to allow an alcoholic fermentation to occur following the full growth of the bacteria. The resulting ethanol helps to protect the starter from unwanted invasive microorganisms. Another method of rapid induction of malolactic fermentation then becomes available; the wine which has just undergone malolactic conversion can now be employed as a starter culture. Here, there is essentially no loss of viability in transferring between such similar media, and also the consequent malolactic conversion rate is rapid and may occur in a few days and furnish even more starter material.

Several studies have demonstrated the possibility of controlling malolactic fermentation by using a reactor with immobilized bacteria. This type of reactor has numerous advantages: the possible reuse of the lactic acid bacteria; a lower formation of secondary products; and the possibility of carrying out fermentation at the right time by using selected bacteria. A new enzymatic reactor has been developed to improve the conversion of L-malic acid into L-lactic acid. Here the malolactic enzyme is produced by a strain of *Leuconostoc oenos*, a cell-free reactor for malic acid conversion. Good levels of enzymatic activity are feasible through the use of this reactor.

Inhibition of Malolactic Fermentation

The malolactic fermentation is inhibited by conditions that are less favourable to the bacteria. Sulphur dioxide inhibits bacterial growth, but the concentrations acceptable in good winemaking operations are not sufficient to assure complete absence of bacterial activity. To inhibit bacterial growth, the stored wine should be kept at normal cellar temperatures, at least around 18°C. The inhibitory effect of 12–13% concentrations of ethanol would able to prevent bacterial growth. Since the bacterial growth is inhibited by low pH, acid additions (preferably tartaric acid), can be used to control the fermentations. As with low temperature, low pH is not a cause of bacterial growth inhibition. If the pH is greater than 3.3, it will be difficult to prevent malolactic fermentation with the use of regular cellar operations.

Elimination of Viable Malolactic Bacteria

The need to block malolactic fermentation is usually most important in bottled white wines which are considered at risk, but which have not as yet undergone fermentation. This situation may arise from a blend of wines having and not having already undergone malolactic fermentation. With varietally distinct white wines, it is especially important to choose a method of eliminating bacteria which has minimal or no effect on sensory properties. With red wines, bottling is often delayed long enough for the wines to have already been bacteriologically stabilized by malolactic fermentation, or if not, it is reasonable to suppose that they do not support the growth of bacteria. Several methods are used for the absolute inhibition of malolactic fermentation: sterile filtration; treatment with chemical inhibitors; and heating the wine. When properly performed, sterile bottling assures complete removal of yeast to stabilize semi-dry wine. As with yeast removal, the essential removal of microbes may be done with depth filters, but the final filtration ought to be made with a membrane filter, which allows integrity testing (bubble point). Nisin at 100 U ml^{-1} can also be used.

Detection of Malolactic Conversion

Changes in titratable acidity and pH characterize the malolactic conversion, however these changes are variable in degree and may be masked by or arise from other reactions in wine. The measurement of disappearance of malic acid is the accepted means of determining whether malolactic conversion has occurred. On visual inspection of relatively clear and light coloured wines, an increase in turbidity can be seen coming from the increased concentration of bacteria. Increased effervescence may be evident from the formation of CO_2 and there may be a loss in colour due to the change in pH and available hydrogen ions, the latter, from NADH formed during the fermentation, can provide some indications of malolactic activity. The loss of colour in red wine can be measured spectrophotometrically.

In some cases, loss of colour is due to the increased pH but this is reversible; loss of colour may also come from reduction of anthocyanin pigments by net production of NADH by the bacteria. Paper chromatography, enzymatic analysis and liquid chromatography are used for determination of malic acid.

See also: **Bacteria**: The Bacterial Cell; Classification of the Bacteria – Phylogenetic Approach. **Fermentation (Industrial)**: Basic Considerations; Production of Organic Acids. ***Lactobacillus***: Introduction. ***Leuconostoc***. **Metabolic Pathways**: Production of Secondary Metabolites – Bacteria. ***Pediococcus***. **Nucleic Acid-based Assays**: Overview.

Further Reading

Battermann G and Radler FA (1991) A comparative study of malolactic enzyme and malic enzyme of different lactic acid bacteria. *Can. J. Microbiol.* 37: 211–217.

Britz TJ and Tracey RP (1990) The combination effect of pH, SO_2, ethanol and temperature on the growth of *Leuconostoc oenos*. *J. Appl. Bacteriol.* 68: 23–31.

Boulton RB, Singleton VL, Bisson LF and Kunkee RL (1996) Malolactic Fermentation. In: *Principles and Practices of Winemaking*. New York: Chapman & Hall.

Caridi A and Corte V (1997) Inhibition of malolactic fermentation by cryotolerant yeasts. *Biotechnol. Lett.* 19: 723–726.

Daeschel MA, Dong-Sun Jung and Watson BT (1991) Controlling wine malolactic fermentation with nisin and nisin-resistant strains of *Leuconostoc oenos*. *Appl. Environ. Microbiol.* 57: 601–603.

Formisyn P, Vaillant H, Lantreibecq F and Bourgois J (1997) Development of an enzymatic reactor for initiating malolactic fermentation in wine. *Am. J. Enol. Vitic.* 48: 345–351.

Gasson MJ, Benson K, Swindell S and Griffin H (1996) Metabolic engineering of the *Lactobacillus lactis* diacetyl pathway. *Lait* 76: 33–40.

Gerbaux V, Villa A, Monamy C and Bertrand A (1997) Use of lysozyme to inhibit malolactic fermentation and to stabilize wine after malolactic fermentation. *Am. J. Enol. Vitic.* 48: 49–54.

Kunkee ER (1991) Some roles of malic acid in the malolactic fermentation in wine making. *FEMS Microbiol. Rev.* 88: 55–72.

Salih AG, Le Quere J-M, Drilleau JF and Fernandez JM (1990) Lactic acid bacteria and malolactic fermentation in the manufacture of Spanish cider. *J. Inst. Brew.* 96: 369–372.

Sauvageot F and Vivier P (1997) Effects of malolactic fermentation of four Burgundy wines. *Am. J. Enol. Vitic.* 48: 187–192.

van Vuuren HJJ and Dicks LMT (1993) *Leuconostoc oenos*: a review. *Am. J. Enol. Vitic.* 44: 9–112.

Velazquez JB, Calo P, Longo E, Cansado J, Sieiro C and Villa TG (1991) Effect of L-malate, D-glucose and L-lactate on malolactic fermentation and growth of *Lactobacillus plantarum* and *Lactobacillus curvatus* wild strains isolated from wine. *J. Ferment. Bioeng.* 71: 363–366.

Volschenk H, Viljoen M, Grobler J et al (1997) Malolactic fermentation in grape musts by a genetically engineered strain of *S. cerevisiae*. *Am. J. Enol. Vitic.* 48: 193–197.

Specific Aspects of Oenology

Poonam Nigam, School of Applied Biological and Chemical Sciences, University of Ulster, Coleraine, UK

Introduction

This articles looks at various aspects of oenology, in particular the characteristics of sweet white table wines such as Sauternes; *Botrytis* infection and the metabolism of *Botrytis cinerea*; the role of 'noble rot' in botrytized wines; Champagne production and the characteristics of champagne yeasts; the role of yeasts in wine fermentation, including the potential of pure-culture wine fermentation as well as the merits of spontaneous versus induced fermentation.

Sweet White Table Wines

Sweet white table wines are characterized by a residual sugar concentration in excess of $30\,g\,l^{-1}$. There are two major categories: those produced by *Botrytis cinerea* infection and those produced by other techniques of sugar concentration. The former wines are the most complex, the classic examples being the Trockenbeerenauslese of Germany and the Sauternes of France. The German sweet white wines tend to have low alcohol levels (9–12% v/v) and high residual sugar concentrations ($120–150\,g\,l^{-1}$). Sauternes, in comparison, have a higher alcohol level (around 14% v/v) and a residual sugar concentration in the range $65–100\,g\,l^{-1}$, and have the distinctive aroma and flavour of new oak.

Production

The grape varieties used in the production of traditional white wines are Riesling, Semillon, Sauvignon blanc and Muscadelle. *Botrytis* infection results in the loss of the varietal characters of the grape, and any variety may be used with similar results. However, varieties differ in their susceptibility to infection and this influences the character of the wine. The sweet white table wines made without *Botrytis* are often made with Muscat-flavoured varieties, or contain some juice of Muscat gordo blanco (Muscat of Alexandria) grapes, to give the wine some distinctiveness. These wines are made by stopping the fermentation while some residual sugar remains, or by back-blending with conserved grape juice. They may also be made from grapes which have been partially dried, on the vine or on mats on the ground.

Sweet white tables wines are produced from high-sugar grapes. The high sugar concentration is achieved by either infection with *Botrytis cinerea* or by dehydration on the vine or on racks after harvest, followed by freezing of the bunches (to produce 'Eiswein'). Wines produced from *Botrytis*-infected grapes are the most complex and balanced. The epidermal cells of the grapes become more permeable to water, which is lost, resulting in the contents of the grape becoming more concentrated. However, because of the metabolism of some of the organic acids by *Botrytis*, their concentration is not significantly changed as a result of infection. Dehydration of the berries by simple drying concentrates both the sugar and the acids, resulting in a must with a high sugar concentration and high acidity.

Must produced from *Botrytis*-infected grapes contains laccase, a monophenol oxidase. In order to inhibit its activity, SO_2 is often added, in concentrations of about $150\,mg\,l^{-1}$. However, SO_2 becomes bound rapidly because of the generation of carbonyl compounds by the mould. It is therefore advisable to add $0.5\,mg\,l^{-1}$ of thiamin to the must, to reduce the level of pyruvic acid formed during fermentation and hence the overall sulphite-binding capacity.

The addition to the must of diammonium phosphate ($200\,mg\,l^{-1}$) and a selected yeast strain is desirable, using strains which can cope with high osmotic pressure and which produce low levels of volatile acidity. The yeasts are fermenting under high stress conditions, and so produce wines that are typically very high in volatile acidity.

Botrytis

Grapes that are heavily infected with moulds have an altered chemical composition, and the mould enzymes can adversely affect the flavour and colour of wine and the successful completion of the alcoholic and malolactic fermentations. Moulds found on grapes include species of *Botrytis*, *Penicillium*, *Aspergillus*, *Mucor*, *Rhizopus*, *Alternaria*, *Ucinula* and *Cladosporium*, of which *Botrytis cinerea* is particularly significant. Although *B. cinerea* can cause spoilage known as 'bunch rot', its controlled development on grapes is used to produce the distinctive sweet wines Sauternes, Trockenbeerenauslese and Tokay. Under certain climatic and viticultural conditions, this mould can parasitize healthy grape berries without disrupting the general integrity of their skin, causing 'noble rot'. This causes dehydration and the concentration of the chemical constituents of the berry. This concentration, along with the fungal metabolism of grape sugars and acids (especially tartaric acid), gives a juice with increased concentrations of sugar, glycerol, other polyols and gluconic acids, and less tartaric acid. The fermentation of such juices requires particular attention, because they are prone to become 'stuck', possibly due to nitrogen deficiency combined with the high sugar content. There is also evidence that the mould secretes anti-yeast substances. *Botrytis cinerea* also produces various phenolic oxidases and glycosidases, which affect the colour and flavour of the wine, and extracellular soluble glucans that block membranes during the processing of the wine by filtration.

In the unique climatic conditions of the Sauternes region grapes infected with *B. cinerea* produce one of the world's most exquisite white wines, the production of which appears to have been well-established in this region by 1830–1850. Nowadays similar wines are produced throughout much of Europe. The selective use of infected grapes for wine has been slow to become popular outside Europe, but these wines are now produced to a limited extent in Australia, Canada, New Zealand, South Africa and the USA. The reduction in berry volume due to dehydration, the risks of leaving the grapes on the vine to become overripe and the difficulties associated with fermentation and clarification result in the production of sweet white wines from *Botrytis*-infected grapes being expensive.

Metabolism of *Botrytis cinerea* Several hydrolytic enzymes are released by *B. cinerea*. Pectolytic enzymes degrade the pectins of the berry cell walls, causing the collapse and death of the affected tissues. The loss of physiological control causes the fruit to dehydrate in dry conditions, and because the vascular connections with the vine become disrupted as the fruit reaches maturity, the lost moisture is not replaced. Additional water is lost via evaporation from the conidiophores.

Drying grapes retards their invasion by *B. cinerea*, and appears to modify its metabolism. Drying also concentrates the juice, which is critical in the development of the organoleptic properties of the wine, and limits secondary invasion by bacteria and fungi. Saprophytic fungi are commonly present along with bunch rot caused by *Botrytis*, but the unpleasant phenolic flavour often associated with wines produced from grapes infected in this way is rarely observed in wines produced from grapes with noble rot. This may result from the low phenol content of their skins, and the separation of the juice from the skins before fermentation.

An indicator of *Botrytis* infection is the presence of

Table 1 Characteristics of juice obtained from *Botrytis*-infected grapes

	Sauvignon grapes		*Sémillon grapes*	
	Healthy	*Infected*	*Healthy*	*Infected*
Fresh weight of 100 grapes (g)	225	112	202	98
Component of juice (g l^{-1})				
Total sugar content	281	326	247	317
Acidity	5.4	5.5	6.0	5.5
Tartaric acid	5.2	1.9	5.3	2.5
Malic acid	4.9	7.4	5.4	7.8
Citric acid	0.3	0.5	0.26	0.34
Gluconic acid	0	1.2	0	2.1
Ammonia	0.049	0.007	0.165	0.025
pH of juice	3.4	3.5	3.3	3.6

gluconic acid. However, although *B. cinerea* produces gluconic acid, the acetic acid bacterium *Gluconobacter oxydans* is particularly active in its synthesis. Acetic acid bacteria frequently invade grapes infected by *Botrytis*, and are probably responsible for most of the gluconic acid found in diseased grapes. Acetic acid bacteria also produce acetic acid and ethyl acetate, and are the most likely source of the elevated content of these compounds in wines produced from *Botrytis*-infected grapes.

The by-products of *Botrytis* metabolism, and the concentration of the juice, are very significant in terms of the wine produced. The following factors are involved:

1. Some changes in the juice appear to be due to the concentrating effects of drying, e.g. the increased concentration of citric acid. Fungal metabolism results in a decrease in the concentrations of tartaric acid and ammonia. The selective metabolism by *Botrytis* of tartaric acid compared with malic acid is important in the avoidance of a marked decrease in pH or an excessive increase in perceived sweetness (**Table 1**).
2. An increase in the sugar concentration of the juice occurs in spite of a reduction in the total sugar content, from about 35% to about 45%. The decrease in osmotic potential induced by concentration of the sugar, enables *B. cinerea* to metabolize most of the sugars. Occasionally, glucose is selectively metabolized, resulting in a high fructose/glucose ratio.
3. The production and accumulation of glycerol during noble rotting may augment the smooth 'mouth feel' of these wines. This may be enhanced by the simultaneous synthesis of other polyols, including arabitol, erythritol, mannitol, *myo*-inositol, sorbitol and xylitol.
4. Noble rotting causes the loss of aroma. Several aromatic terpenes, including linalool, geraniol and nerol, are metabolized by *B. cinerea* to less volatile compounds such as β-pinene, α-terpineol, and various furan and pyran oxides. The latter may produce the phenolic and iodine-like odours reported in some wines produced using *Botrytis*.
5. *Botrytis* produces esterases, which degrade the esters that give many white wines their fruity character. Muscat varieties often lose more fragrance than they gain, but Riesling and Sémillon varieties generally gain more in aromatic complexity than they lose in varietal distinctiveness.
6. Fungal metabolism results in the synthesis of certain aromatic compounds, including sotolon, which contributes a distinctive honey-like fragrance to the wine. Infected grapes also contain the 'mushroom' alcohol 1-octen-3-ol. More than 20 terpene derivatives have been isolated from infected grapes.
7. *B. cinerea* affects the ease of grape-picking, the activity of yeasts and bacteria in the juice, and the filterability and ageing properties of the wine.
8. Laccases produced by *B. cinerea* inactivate antifungal phenols, pterostilbene and resveratrol in grapes. In wine, laccases can oxidize a wide range of important grape phenols, e.g. *p*-, *o*- and some *m*-diphenols, diquinones, anthocyanins and tannins, and a few other compounds, e.g ascorbic acid. The oxidation of 2(*S*)-glutathionylcaftaric acid may contribute to the golden colour of the wine.
9. *Botrytis* synthesizes a series of polysaccharides of high molecular mass, including polymers of mannose and galactose, and also small amounts of glucose, rhamnose and β-glucans. The rhamnose induces the increased production of acetic acid and glycerol during fermentation. The polysaccharides form strand-like lineocolloids in the presence of alcohol, and these can plug filters during clarification.
10. *Botrytis* produces an enzyme that oxidizes galacturonic acid (produced by the hydrolysis of pectin), to form mucic (galactaric) acid. Mucic acid slowly binds with calcium, forming a sediment in the bottled wine.

Botrytized Wines

Tokaji Aszú Tokaji was the first deliberately produced botrytized wine. Its most famous version, Aszú Eszencia, is derived from juice that spontaneously seeps out of the highly botrytized berries (*aszú*) placed in small tubs. About 1–1.5 l of *essencia* may be obtained from 30 l of *aszú*. Fermentation continues

Table 2 Chemical composition of botrytized Tokaji wines

Quality grade	*Total extract* ($g\,l^{-1}$)	*Extract residue* ($g\,l^{-1}$)	*Sugar content* ($g\,l^{-1}$)	*Ethanol content* (% v/v)
Two *puttonyos*	55	25	30	14
Three *puttonyos*	90	30	60	14
Four *puttonyos*	125	35	90	13
Five *puttonyos*	160	40	120	12
Six *puttonyos*	195	45	150	12
Eszencia	300	50	250	10

slowly, for weeks or months, and alcohol content reaches 14% (**Table 2**).

German Botrytized Wine *Auslese* wines are derived from specially selected clusters of late-harvested fruit. Beerenauslese (BA) and Trockenbeerenauslese (TBA) wines are similarly derived but come from individually selected berries or dried berries, respectively. Although the fruit is typically botrytized, this is not obligatory. The BA and TBA juices typically contain more sugar than is converted to alcohol during fermentation. The wines are correspondingly sweet and low in alcohol strength – commonly 6–8% (**Table 3**). Auslese wines may be fermented dry or may retain residual sweetness. The other main Prädikat wine categories, namely, Kabinett and Spätlese, may be derived from botrytized juice but seldom are.

French Botrytized Wine In France, the most well-known botrytized wines are produced in the Sauternes region of Bordeaux. Here, over a period of several weeks, noble-rotted grapes may be selectively removed from clusters on the vine. Because of the cost of multiple selective harvesting, most producers harvest but once and separate the botrytized berries from the bunches. Uninfected grapes are used in the production of a dry wine. Typically, only one sweet style is produced in Sauternes, in contrast to the many botrytized styles produced in Germany. French styles commonly exceed 11% alcohol. Sweet botrytized wines are produced particularly in the Alsace and Loire regions.

Induction of Noble Rot

If climatic conditions are unfavourable for the development of noble rot, the harvested fruit may be exposed to artificial conditions that favour its development. The fruit is sprayed with a solution of *Botrytis* spores, then placed in trays, at about 90–100% relative humidity for 24–36 h at 20–25°C. After the spores have germinated and the mould has penetrated the fruit, cool dry air is passed over the fruit to induce partial dehydration and restrict invasion. After 10–14 days, the infection has developed sufficiently.

Alternatively, the inoculation of juice with spores or mycelia of *B. cinerea*, followed by aeration, apparently produces many of the desirable organoleptic changes caused by vineyard infections.

Champagne

Champagne, the most northerly wine-producing region in France, is associated with the evolution of sparkling wine. Grapes from the region traditionally yielded poorly coloured, acidic juice which was low in sugar content and aroma – features now considered desirable in the production of most sparkling wines.

Primary Fermentation

The initial fermentation is carried out at 15°C – lower temperatures are considered to give a grassy odour, and higher temperatures yield wines lacking in finesse. Bentonite and/or casein may be added to aid fermentation. Inoculation with selected yeast strains avoids the production of perceptible amounts of SO_2, acetaldehyde, acetic acid and other undesired volatiles by indigenous yeasts.

If the pH of the juice is too low (3 or less), malolactic deacidification is commonly encouraged. This form of fermentation gives the wine a subtle bouquet. As the bacterial sediment produced is difficult to remove by riddling, it is important that the malolactic fermentation is complete before the second, in-bottle fermentation. Before preparing the blend (cuvée), the individual base wines are clarified and stabilized separately, according to cultivar, site and vintage.

Table 3 Composition of some Beerenauslese (BA) and Trockenbeerenauslese (TBA) wines

Wine type	*Total extract* ($g\,l^{-1}$)	*Sugar content* ($g\,l^{-1}$)	*Alcohol content* (% w/v)	*Total acidity* ($g\,l^{-1}$)	*Glycerol Content* ($g\,l^{-1}$)	*Acetaldehyde content* ($mg\,l^{-1}$)	*Tannin content* ($mg\,l^{-1}$)	*pH*
BA	163	74	7.9	8.7	12.0	73	250	3.2
BA	152	103	6.3	9.4	13.6	62	390	3.2
BA	119	78	7.9	7.9	10.9	139	390	3.0
TBA	299	224	5.3	11.4	13.0	56	291	3.6
TBA	303	194	6.4	10.5	40.0	163	446	3.5

Preparation of the Cuvée and Tirage

Blending improves the quality of sparkling wine and helps to minimize variations in supply and quality, and hence ensure consistency.

Tirage involves adding a concentrated sucrose solution (50–65%), also containing other nutrients, to the cuvée. The solution is added just before the inoculation of yeast, and results in a sucrose concentration of about $24\,g\,l^{-1}$. On fermentation, this produces a pressure considered appropriate for most sparkling wines, of about 600 kPa. If the cuvée contains residual fermentable sugars, less sucrose is added. About 4.2 g of sugar is required to generate 2 g of CO_2.

Thiamin and diammonium hydrogen phosphate, $(NH_4)_2HPO_4$, are often added to the cuvée, in concentrations of $0.5\,mg\,l^{-1}$ and $100\,mg\,l^{-1}$ respectively. Thiamin appears to counteract the alcohol-induced inhibition of the uptake of sugar by yeast cells.

Champagne Yeasts

The second fermentation requires the inoculation of the cuvée with a particular yeast strain – physiological variants of *Saccharomyces cerevisiae* are used. Desirable properties of such strains include the ability to effect complete fermentation under conditions of: low temperature (10–15°C); high ethanol concentration (8–12%); low pH (3.0–3.5); low nutrient availability; increasing pressure of CO_2; and in the presence of about $20\,mg\,l^{-1}$ of free SO_2, while producing a good flavour. Desirable strains should also undergo flocculation and sedimentation, to facilitate removal from the bottle, and should undergo autolysis during ageing. The yeast must also have a low tendency to produce H_2S, SO_2, acetaldehyde, acetic acid and ethyl acetate. The possession of an active proteolytic ability after fermentation aids the release of amino acids and oligopeptides during yeast autolysis.

Unfavourable conditions in the cuvée require the inoculum to be acclimated before addition, otherwise most of the inoculated yeast cells die, resulting in a prolonged latent period before fermentation commences. Acclimation usually involves the inoculation of a glucose solution with the yeast at about 20–25°C. The culture is aerated, to ensure the adequate production of the unsaturated fatty acids and sterols required for cell division and membrane function. Once growing actively, the culture is added to cuvée in the proportion 60 : 40. The bulk of cuvée is inoculated with the acclimated culture such that the concentration of yeast is about $3\text{–}4 \times 10^6$ cells per millilitre (about 2–5% of the cuvée volume). Higher inoculation levels are thought to increase the likelihood of H_2S production, and lower levels increase the risk of failed or incomplete fermentation. Over the next few days, more cuvée is added, achieving a mixture containing 80–90% cuvée. Simultaneously, the culture is cooled to the desired fermentation temperature.

Second Fermentation

Once the cuvée has been mixed with the tirage and yeast inoculum, the wine is bottled and sealed. The wine is kept at a stable temperature (10–15°C) for the second fermentation. At 11°C, a common fermentation temperature in the Champagne region, the second fermentation may last about 50 days.

During the early stages of fermentation, the yeast population goes through three or four cell divisions, reaching a final concentration of about $1\text{–}1.5 \times 10^7$ cells per millilitre. The rate of fermentation is largely dependent on the temperature, pH and SO_2 content of the cuvée.

An innovation in the production of sparkling wine involves secondary fermentation with the yeast cells immobilized in beads of alginate. Immobilization results in the cells being more readily removed from the bottle, giving significant cost savings. Immobilized cells produce satisfactory fermentation and ageing, and a product with organoleptic qualities comparable to those of sparkling wine made by the traditional process.

The amount of sugar added to the cuvée after the primary fermentation determines the amount of CO_2 produced. A sucrose concentration of about $24\,g\,l^{-1}$ produces a pressure of 600 kPa after fermentation, which is considered appropriate for most sparkling wines.

After fermentation, the bottles may be transferred to a different site, for maturation at about 10°C. Maturation lasts for 12 months; during this period, the number of viable yeast cells drops rapidly, falling below about 10^6 cells ml^{-1} after 80 days.

The production process is lengthy and labour-intensive, and this is reflected in the price of the finished product. Attempts have been made to shorten the maturation, for example by adding autolysed yeast cells or yeast extract at the end of the secondary fermentation. However, this is not allowed in France.

After the secondary fermentation, the yeast cells are sedimented into the neck of the bottle by a tedious and complicated process involving the manipulation of the bottle ('remuage'). The yeast is then removed by a process involving freezing, which results in the minimal loss of wine and CO_2 ('dégorgement'). The lost wine is replaced by base wine, which may contain a small amount of sugar to give the desired sweetness, the amount varying from about 0.5% for the production of very dry ('brut') champagne to about 10% for the production of sweet champagne.

Use of Yeasts in Fermentations

Fermentation can be conducted as either a 'natural' or a 'pure-culture' process. In the former, the yeasts present in the grape juice initiate and complete the fermentation. In 'pure-culture' fermentation, selected commercial strains of *Saccharomyces cerevisiae* are inoculated into the juice at population of 10^6–10^7 cfu ml^{-1}. *S. cerevisiae* is remarkably tolerant to high concentrations of sugar, ethanol and SO_2. Also, it grows and ferments rapidly at the low pH of grape must. Consequently, it generally metabolizes the fermentable sugars in must completely, primarily to ethanol and flavour products.

The pure-culture approach gives a more rapid and predictable fermentation. Natural fermentation has a more varied outcome, with the possibility of failures but also the prospect of wines with a more interesting character, due to contributions from a range of yeast species. As an alternative to spontaneous fermentations or those induced with a single yeast strain, the juice may be inoculated with a mixed culture of local and commercial yeast strains. The different species appear to influence the metabolism of one another, resulting in the production of wine with a uniform, although regionally distinctive, character.

Most wine fermentations are eventually dominated by *S. cerevisiae*. Consequently, this is widely recognized as the principal wine yeast, and has been commercialized in the form of starter cultures. However, wine fermentations induced by the inoculation of selected strains of *S. cerevisiae* are not devoid of contributions from the natural yeast flora, and hence are not strictly 'pure-culture' fermentations.

In inoculated fermentations, *S. cerevisiae* is usually added so as to achieve a population of about 10^5–10^6 cells per millilitre in the must. The indigenous *S. cerevisiae* of the grape flora often contributes an additional 10^4–10^5 cells per millilitre. The addition of SO_2, to a concentration of about 1.5 mg l^{-1} of free molecular SO_2, has generally been thought to inhibit, if not kill, most of the indigenous yeast population of the grapes. However, as the free SO_2 content falls rapidly during maceration and fermentation, the indigenous strains may grow and multiply.

Thus the addition of SO_2 to must reduces the total numbers of yeast cells present; it also results in qualitative changes in the yeast microflora, and after the initial stages of fermentation favours the resistant *S. cerevisiae* and *Saccharomycodes ludwigii* at the expense of more sensitive species such as *K. apiculata*. The patterns of microbial succession and dominance are modified by temperature. At 25°C, *Saccharomyces exiguus* and *Zygosaccharomyces bailii* may predominate. At 10°C, *S. cerevisiae* is dominant, but *K. apiculata*, *C. stellata* and *C. krusei* are also present in large numbers. Ideally, primary fermentation should take place at about 24°C for 3–5 days, for red wines and at 7–21°C, for 7 days to several weeks, for white wines.

The use of a commercial strain containing killer plasmids should prevent its elimination: the possession of killer plasmids enables the secretion of a glycoprotein toxin capable of killing sensitive strains, although the cell which secretes the toxin is protected from it.

Spontaneous Fermentations

Spontaneous fermentations may vary – from year to year, location to location, and according to the source and age of the cooperage. Occasionally, spontaneous fermentations generate higher concentrations of volatile acidity than those produced by induced fermentations. Nevertheless, those who favour spontaneous fermentation believe that indigenous yeasts donate desirable subtle characteristics to the wine, and possibly provide some of the distinctive characteristics of a regional wine.

In traditional European wine-making, fermentation depends on the naturally occurring yeasts in the must. Initially, *Saccharomyces* is not numerically significant. The dominant yeasts at this stage include *Aureobasidium pullulans*, *Candida stellata*, *Hanseniaspora uvarum*, *Issatchenkia orientalis*, *Kloeckera javanica*, *Metschnikowia pulcherrima* and *Pichia anomala*. *H. uvarum* is dominant in the early stages of fermentation in California and mid-Europe, while *H. osmophilia* dominates elsewhere. *Metschnikowia pulcherrima* is particularly prevalent in Europe. The climatic conditions during the growth of the vines affects the yeast microflora associated with a particular vineyard.

Induced Fermentations

Intentional inoculation is needed to initiate fermentation rapidly, for example after thermovinification; in the fermentation of pasteurized juice; to restart 'stuck' fermentations; and to promote the fermentation of juice made from significant numbers of mouldy grapes, which produce various inhibitors, such as acetic acid, that slow yeast growth and metabolism. Inoculation is also required to initiate the second fermentation in the production of sparkling wine. However, the predominant reason for using specific yeast strains is the avoidance of the undesirable flavours occasionally associated with spontaneous fermentation. In addition, the distinctive flavour characteristics generated by a particular yeast strain may influence its use.

Few wine makers in non-traditional areas rely on the adventitious inoculation of the must with *S. cer-*

evisiae: it is now common practice to add pure cultures at the start of fermentation. The use of pure cultures is also increasing in traditional regions. In some large wineries, the yeast is propagated from master cultures held in-house, but commercially produced liquid or dried cultures are more convenient in most cases. Dried cultures, which can be added directly to the vat without any propagation, are particularly useful in small wineries.

The use of pure cultures of *S. cerevisiae* is generally recognized as minimizing the problems of controlling the fermentation, and as producing wine of consistent quality. The addition of *S. cerevisiae* does not affect the presence of indigenous yeast, nor to a great extent the pattern of fermentation. However, it appears that during natural fermentation, each stage is characterized by the development of different strains of *S. cerevisiae*, and that the addition of pure cultures similarly influences the development of *S. cerevisiae*, rather than inhibiting other yeasts.

A molecular technique, mitochondrial DNA restriction analysis, has been used to study yeasts in inoculated and natural fermentations. Research shows that the inoculated strain is primarily responsible for fermentation, but that the indigenous strains are not suppressed during the first several days of fermentation, and hence may play a significant role. A great diversity of strains are present and, although only a few persist throughout the process, the same strains tend to be dominant in both natural and inoculated fermentations.

Although it is recognized that inoculated wines are generally of higher quality, there is some concern that the characteristics associated with particular natural microfloras may be lost. Studies have shown that some yeasts of low fermentation power, such as *K. apiculata*, produce significant quantities of volatile aroma compounds, particularly esters. Significant quantities of volatiles are also produced by some strains of *S. cerevisiae*, but other strains, including some used for inoculation of the must, produce less. Nevertheless, there is probably sufficient growth of natural aroma-producing strains in inoculated fermentations to cause the development of typical characteristics in wines. However, it has been suggested that an emphasis on the criterion of alcohol production in choosing strains of *S. cerevisiae* for inoculation may result in wine of low quality, and that mixed starter cultures, including yeasts which produce volatile aroma compounds, would be preferable.

See also: ***Botrytis***. ***Gluconobacter***. **Preservatives**: Permitted Preservatives – Sulphur Dioxide. ***Saccharomyces***: *Saccharomyces cerevisiae*. **Wines**: Microbiology of Wine-making; The Malolactic Fermentation.

Further Reading

Aylott RI (1995) Flavoured spirits. In: Lea AGH and Piggott JR (eds) *Fermented Beverage Production*. P. 275. Glasgow: Blackie.

Barre P, Vezinhet F, Dequin S and Blondin B (1993) Genetic improvement of wine yeasts. In: Fleet GH (ed.) *Wine Microbiology and Biotechnology*. P. 265. Chur: Harwood.

Berry DR (1995) Alcoholic beverage fermentations. In: Lea AGH and Piggott JR (eds) *Fermented Beverage Production*. P. 32. Glasgow: Blackie.

Degré R (1993) Selection and commercial cultivation of wine yeast and bacteria. In: Fleet GH (ed.) *Wine Microbiology and Biotechnology*. P. 421. Chur: Harwood.

Dittrich HH (1995) Wine and brandy. In: Reed G and Nagodawithana TW (eds) *Biotechnology*, 2nd edn. Vol. 9, p. 464. Weinheim: VCH Publishers.

Divies C (1993) Bioreactor technology and wine fermentation. In: Fleet GH (ed.) *Wine Microbiology and Biotechnology*. P. 449. Chur: Harwood.

Donèche RJ (1993) Botrytized wines. In: Fleet GH (ed.) *Wine Microbiology and Biotechnology*. P. 327. Chur: Harwood.

Ewart A (1995) White wines. In: Lee AGH and Piggott JR (eds) *Fermented Beverage Production* P. 97. Glasgow: Blackie.

Fleet GH (1998) The microbiology of alcoholic beverages. In: Wood BJB (ed.) *Microbiology of Fermented Foods*, 2nd edn. Vol. 1, p. 217. London: Blackie.

Fleet GH and Heard GM (1997) Wine. In: Doyle MP, Beuchat L and Montville T (eds) *Food Microbiology Fundamentals and Frontiers*. P. 671. Washington: American Society of Microbiologists.

Jackson RS (1994) *Wine Science Principles and Applications*. P. 467. London: Academic Press.

Ribèreau-Gayon J, Ribèreau-Gayon P and Seguin G (1980) *Botrytis cinerea* in enology. In: Coley-Smith JY, Vehoeff K and Jarvis WJ (eds) *The Biology of Botrytis*. P. 251. London: Academic Press.

Shimizu K (1993) Killer yeasts. In: Fleet GH (ed.) *Wine Microbiology and Biotechnology*. P. 243. Chur: Harwood.

Varnam AH and Sutherland JP (1994) Alcoholic beverages. II Wine. In: *Beverages Technology, Chemistry and Microbiology*. P. 362. London: Chapman & Hall.

Wood Smoke *see* **Preservatives**: Traditional Preservatives – Wood Smoke.

Xanthan Gum *see* **Fermentation (Industrial):** Production of Xanthan Gum.

XANTHOMONAS

Arun Sharma Food Technology Division, Bhabha Atomic Research Centre, Mumbai, India

The genus was created by Dowson in 1939 following a proposal by Burkholder in 1930, for a group of plant pathogens until then assigned to the genus *Phytomonas*. Bradbury in 1984 widened the definition of the genus to include characteristics such as the absence of denitrification, the nature of xanthomonadin pigment, and many more biochemical and physiological properties. The word *Xanthomonas* (Xan. tho'mo. nas or Xan. tho'. monas) is composed of the Greek adjective *Xanthus* meaning yellow and feminine noun *monas* meaning unit. In modern Latin it literally translates to 'yellow monad'. The bacteria belonging to the genus *Xanthomonas* are commonly associated with the diseases of plants. *Xanthomonas* species are seen as yellow-pigmented colonies on a nutrient agar plate and as Gram-negative, short and usually straight rods under a microscope. However, *Pseudomonas maltophila* has recently been included in the genus, which is neither a plant pathogen nor does it produce the characteristic yellow pigment.

Characteristics of the Genus

Cells of bacteria belonging to this genus are short, straight rods, but not vibrioid. The size is in the range 0.4–1.0 × 1.2–3.0 μm. The cells are monotrichous with a polar flagellum. The bacteria of the genus do not form spores, sheaths, appendages or buds. Xanthomonads are chemoorganotrophic, use low-molecular-weight compounds and some are able to depolymerize natural polysaccharides and proteins. The cells carry out aerobic respiratory metabolism and are non-fermentative. The members of the genus show oxidase negative (or weakly positive) and catalase positive reactions.

The major means of glucose catabolism is the Entner–Duodoroff pathway (**Fig. 1**). Acid is produced from mono- and disaccharides. In a weakly buffered medium acid is produced from many carbohydrates but not from rhamnose, inulin, adonitol, dulcitol, inositol or salicin, and rarely from sorbitol. Acetate, citrate, malate, propionate, and succinate are utilized

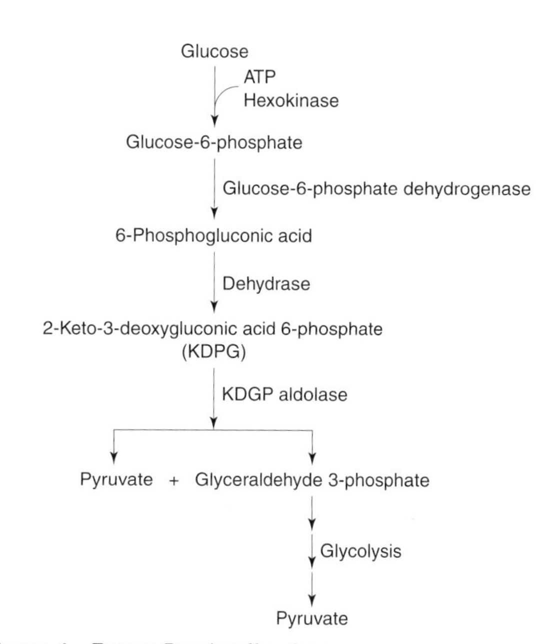

Figure 1 Entner–Doudoroff pathway.

but generally benzoate, oxalate and tartarate are not used. The tests for indole production from tryptophan, and acetoin production (Voges–Proskauer test) are negative. The growth on nutrient agar is inhibited by 0.1–0.02% triphenyltetrazolium chloride. Hydrogen sulphide is produced from cysteine and by most species from thiosulphate and peptone due to desulphurase activity. Proteins are usually readily digested and milk becomes alkaline with the growth of *Xanthomonas*. Asparagine is not sufficient as the only source of carbon and nitrogen. Most species hydrolyse starch and Tween 80 rapidly. Nitrates are not reduced. Some species hydrolyse cellulose and pectin. G+C content of the members of the genus is in the range 63–71 mol%. A majority of the species are plant pathogens.

The typical colonies on an LB agar plate (**Fig. 2**) are about 2–5 mm diam., mucoid or buttery, with a raised centre and entire margins. Minimum growth requirements are complex and include a need for

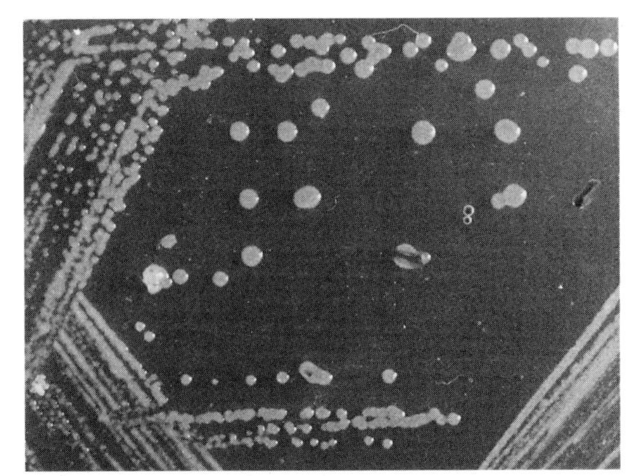

Figure 2 Typical colonies of *Xanthomonas* on LB agar plate.

methionine, glutamic acid and nicotinic acid in various combinations. Optimum temperature for growth is in the range 25–27°C. No growth is observed above 40°C and below 5°C.

Classification of *Xanthomonas*

Xanthomonas is one of the important genera of the family Pseudomonadaceae in the order Pseudomonadales. The genus was originally described as *Phytomonas* in the first edition of *Bergey's Manual*. The present description of *Xanthomonas* in relation to other genera of the family Pseudomonoadaceae as shown in *Bergey's Manual of Determinative Bacteriology* (7th edition) is:

Division: Protophyta
Class: Schizomycetes
Order: Pseudomonadales
Class: Pseudomonadaceae
Genus: *Xanthomonas*
Species: *campestris*

The general identification of *Xanthomonas* to species level is based on the ability of the organism to:

- hydrolyse gelatin and starch;
- produce nitrites and ammonia from nitrates;
- form yellow non-water-soluble pigment in nutrient agar;
- form brown pigment in beef extract agar.

Another key for the classification of *Xanthomonas* is based on the plant host that the bacterium attacks. The members of the genus are known to attack both monocotyledons and dicotyledons. A detailed classification of the genus is given in *Bergey's Manual of Determinative Bacteriology*. It is possible to classify a given species to subspecies level based on the pathogen race and host–cultivar relationship.

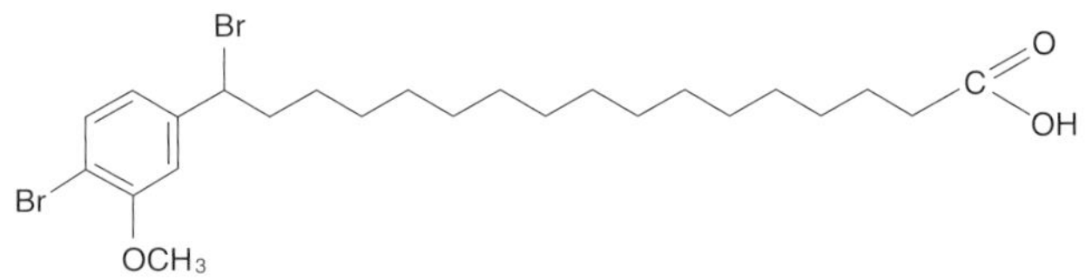

Br
Br
OCH_3
C
O
OH

Figure 3 Xanthamonadin: yellow pigment of *Xanthomonas*.

New Approaches to the Classification of *Xanthomonas*

Phenotypic properties are generally inadequate for differentiating xanthomonads because of overlapping characters among the strains. This has led to the development of the pathovar system of nomenclature for *Xanthomonas*. This system relies on the susceptibility of the host as the means of identifying the pathogen. However, many *Xanthomonas* pathovars have a wide host range. Therefore, new approaches to classification of xanthomonads have been developed.

Xanthomonadin Production The yellow pigment produced by most xanthomonads is a group of membrane bound, non-water-soluble, brominated arylpolyenes called xanthomonadins (**Fig. 3**), which are used as chemotaxonomic and diagnostic markers of the genus.

Nutritional Screening Carbon substrate utilization patterns have been found to be sufficiently uniform among the various genomic groups within xanthomonads to allow their differentiation.

Fatty Acid Composition The qualitative and quantitative differences in the whole-cell fatty acid methyl ester patterns have been used to determine the relationships between species and strains.

Phage typing Taxonomic classification based on bacteriophage sensitivity of the different strains of *Xanthomonas* has also been attempted.

Serological Typing A number of serological methods have been successfully used to identify and diagnose different species and strains of xanthomonads. Both polyclonal and monoclonal antibodies, produced to whole cells or flagellar extracts of the organism have been employed for the identification of xanthomonads and groupings of strains of different species. Techniques, such as immunofluorescence microscopy, enzyme immunoassays and dot-blot immunoassays, have been used for carrying out specificity tests. Based on studies with polyclonal antibodies three serovars of *X. albilineans* have been identified. Monoclonal antibodies have been used for the identification of xanthomonads and groupings of strains of *X. campestris* and *X. albilineans*. Serological studies using monoclonal antibodies have shown that *Pseudomonas gardneri* is also a xanthomonad.

Sodium Dodecyl Sulphate–Polyacrylamide Gel Electrophoresis (SDS–PAGE) Development of rapid high-resolution fingerprinting techniques such as protein electrophoresis in combination with the computer assisted processing of SDS–PAGE pattern of proteins make it possible to characterize and compare a large number of strains at infra-subspecies levels. The electrophoretic groupings corresponded in some, but not in all cases, to the existing pathovars. Numerical analysis on SDS–PAGE protein patterns of *Xanthomonas* strains has shown a good agreement between groupings obtained by gas chromatographic analysis of cellular fatty acids, and DNA–DNA hybridization.

DNA Based Techniques These techniques include DNA–DNA hybridization, DNA–RNA hybridization using 16S and 23S rRNA, restriction fragment length polymorphism (RFLP) analysis, plasmid profile analysis, and polymerase chain reaction (PCR) based detection using random or specific primers. Based on DNA–DNA hybridization experiments strain clusters have been identified that deserve independent status as species. As a result of such studies *Pseudomonas maltophila*, a member of the rRNA similarity group V of *Pseudomonas* together with *P. geniculata* and *P. gardeneri*, has been allocated to the genus *Xanthomonas* and is now known as *X. maltophila*. The organism is not a phytopathogen and it does not produce the xanthomonadin pigment. It is also reported to be an animal pathogen. On the other hand RFLP analysis has shown striking similarities among different *X. campestris* pathovars.

Conventional methods for detection and identification of *Xanthomonas* rely on pure culture isolation on selective media, followed by morphological and biochemical characterization of the isolates. Pathogenicity is ascertained by testing for Koch's postulates. A pathogenic microbe is further screened on a number of cultivars to identify the presence of race differences. Traditionally these methods have given reliable results. However, with the discovery of so many new strains it has been increasingly difficult to establish their relationship with each other and to classify them. Moreover, the traditional methods are cumbersome when quick results are required, for example in seed certification and quarantine procedures. Methods based on protein profiling and fatty acid analysis are also time consuming and need pure culture of each isolate for assessment. Polyclonal antisera often cross-react and fail to identify pathovars or races of the organism. Even monoclonal antisera may react with some or all the strains of a pathovar. Therefore, strain identification has become a difficult task. Development of probes based on more specific gene sequences such as those of *hrp* genes may provide a useful tool for the detection and identification of phytopathogenic xanthomonads.

Internal Subdivisions of *Xanthomonas*

In *Bergey's Manual of Determinative Bacteriology* five species of *Xanthomonas* were recognized. These species could be distinguished on the basis of biochemical reactions as shown in **Table 1**. *X. campestris* can grow at 35°C, hydrolyse aesculin (6,7-dihydroxycoumarin 6-glucoside), exhibits mucoid growth, and can liquefy gelatin. It causes proteolysis of milk and H_2S production from peptone. Salt tolerance of *X. campestris* is high (2–5%). It can utilize arabinose, glucose, mannose and cellobiose to produce acid. *X. fragariae* can not cause proteolysis of milk and does not produce H_2S. It can use glucose and mannose to produce acid. Its tolerance to salt is low (0.5–1.0%). *X. albilineans* does not show gelatin hydrolysis and mucoid growth, otherwise it is similar to *X. fragariae*. It has the least salt tolerance (< 0.5%). *X. axonopodis* differs slightly from *X. albilineans* in that it produces H_2S and can not use mannose to produce acid. *X. ampelina* differs from *X. albilineans* in that it has urease activity and use arabinose to produce acid. Two more species, *X. populi* and *X. maltophila*, have been recently added to the genus.

- *X. fragariae*, a pathogen of strawberries;
- *X. albilineans*, a pathogen of monocots;
- *X. axonopodis*, a pathogens of monocots;
- *X. ampelina*, a pathogen of grapes;
- *X. populi*, a pathogen of poplars;
- *X. maltophila*, recently transferred from the genus *Pseudomonas*;
- *X. campestris*, a pathogen of dicots.

Xanthomonas campestris

The determination of the genus *Xanthomonas* and its species is relatively easy, however, the characterization of *X. campestris* pathovars poses problems. An unambiguous identification of the pathovars of *X. campestris* can be of great use in plant pathology. The *X. campestris* pathovars that are defined by the host or disease symptoms are difficult to identify by other phenotypic characteristics. *X. campestris* group is the largest of all and causes diseases in many plant species. It is, therefore, classified into pathovars differentiated by the host reaction. However, application of the newer techniques of classification has been useful. A relationship of nutritional properties, host specificity

Table 1 Biochemical characters of different *Xanthomonas* species

Characteristic	X. campestris	X. fragariae	X. albilineans	X. axonopodis	X. ampelina
Growth at 35°C	+	+	+	+	−
Aesculin hydrolysis	+	+	+	+	−
Mucoid growth	+	+	−	−	−
Gelatin liquefaction	+	+	−	−	−
Milk proteolysis	+	−	−	−	−
H_2S from peptone	+	−	−	+	−
Urease	−	−	−	−	+
NaCl tolerance (%)	2–5	0.5–1.0	<0.5	1.0	1.0
Acid production from					
Arabinose	+	−	−	−	+
Glucose	+	+	+	+	−
Mannose	+	+	+	−	−
Cellobiose	+	−	−	−	−

and DNA homology groups has been observed. Genetic diversity among the strains of different pathovars of *X. campestris* has also been studied for a number of pathovars. It is believed that the variability could be more pronounced in the regions where the host plant originated. Ribosomal RNA and DNA probes could be useful tools for the epidemiological studies and in following the genetic evolution of the strains.

Methods of Detection and Enumeration in Foods

Starch hydrolysis is a major distinguishing feature of xanthomonads from other pseudomonads. Soluble starch has been used in several agar media for isolation for *X. campestris*.

Typical solid media for the general enumeration of xanthomonads may contain (per litre):

- Potato starch, 10.0 g; $K_2HPO_4.3H_2O$, 3.0 g; KH_2PO_4, 1.5 g; $(NH_4)_2SO_4$, 2.0 g; L-methionine, 0.5 g; nicotinic acid, 0.25 g; L-glutamic acid, 0.25 g; pH 6.8–7.0; with 15 g agar;
- Potato starch, 10 g; yeast extract, 5.0 g; $(NH_4)H_2PO_4$, 0.5 g; K_2HPO_4, 0.5 g; $MgSO_4.7H_2O$, 0.2 g; NaCl, 5.0 g, pH 7.4 with 15 g agar.

After autoclaving and cooling to 50°C the medium is fortified with one or more of the antibiotics such as cephalexin 20 μg ml^{-1}, kasugamycin, 20 μg ml^{-1}, chlorothalomine 15 μg ml^{-1}, gentamycin 2 μg ml^{-1}, and dyes such as brilliant cresyl blue 1 μg ml^{-1}, methyl green 1 μg ml^{-1} and methyl violet 1 μg ml^{-1}. Xanthomonads are generally resistant to these antibiotics.

After spread plating a given sample on the agar plates the typical yellow colonies of *Xanthomonas* are seen after incubating the plates at 26±2°C for 48 h. The starch hydrolysis is visualized as the zone of clearance around the colonies. In the absence of dyes, iodine solution (1%) is spread on the plate to visualize the zone of hydrolysis by starch.

Immunoassay for the Detection of *X. campestris*

Both polyclonal and monoclonal antibodies are employed for detection using radioimmunoassay or enzyme immunoassay. Production of antibodies, both monoclonal and polyclonal, has been described. These antibodies have been used for the detection of *Xanthomonas campestris* pv. *campestris* in crucifer seeds.

For assay the isolate is streaked on a solid medium. The plates are incubated at 26±2°C for 48 h. The putative *Xanthomonas* colonies could be directly tested by enzyme linked immunosorbent assay (ELISA).

A protocol for an indirect double antibody ELISA for the detection of *X. campestris* strains is as follows:

- Coat polystyrene micotitre plates with poly-L-lysine by adding 0.1 mg ml^{-1} solution in carbonate buffer (pH 9.6) and incubating for 30 min at 21°C;
- Wash microtitre plates three times with phosphate-buffered saline containing 0.01% Tween 20 (PBST);
- Add antigen (untreated or boiled bacterial cells 10^7 cells ml^{-1}) or bacterial protein extract (1–10 μg);
- Add PBST containing 0.5% BSA and incubate for 30 min at 37°C;
- Add antibody and incubate for 3 h at 4°C and wash the plates three times with PBST;
- Add goat anti-mouse alkaline phosphatase conjugate (1 : 1000 dilution in PBST containing 0.1% albumin) and incubate 2 h at 37°C;
- Wash three times with PBST and incubate with the enzyme substrate (0.75 mg ml^{-1} *p*-nitrophenyl phosphate in 10% diethanolamine buffer v/v, pH 9.8) for 0.5–1 h at 37°C;
- Read absorbance at 405 nm.

Phytopathogenic Potential of *Xanthomonas*

Among the bacterial diseases of plants, the most widespread and destructive losses are caused by the Gram-

Table 2 Some common plant diseases caused by *Xanthomonas campestris*

Plant	*Disease*	*Causative agent*
Cotton	Leaf spot	*X.c.* pv. *malvacearum*
Rice	Leaf blight	*X.c.* pv. *oryzae*
Cereals	Bacterial blight	*X.c.* pv. *translucens*
Walnut	Bacterial blight	*X.c.* pv. *juglandis*
Soybean	Bacterial pustule	*X.c.* pv. *glycines*
Sugarcane	Gumming disease	*X.c.* pv. *vascularum*

negative bacteria of the genus, *Erwinia*, *Pseudomonas* and *Xanthomonas*. The genus *Xanthomonas* is of great economic importance because of its broad host range. More than 145 species of plants are known to be infected, mainly by biotrophic pathogens, belonging to this genus. These pathogens cause many types of disease symptoms including spots, blights, cankers and vascular wilts. The type of physiological function that is affected first depends on the cells and tissues of the host plant that become infected. Thus the infection of xylem vessels interferes with the translocation of water leading to vascular wilts and cankers, whereas infection of foliage interferes with the photosynthetic process as in leaf spots, blights, and pustules. Some common plant diseases caused by *Xanthomonas* are listed in **Table 2**.

Angular leaf spot disease of cotton is caused by *X.c.* pv. *malvacearum*. The disease is present wherever cotton is grown. The bacterium attacks the leaves as well as young cotton bolls. In rice, *X.c.* pv. *oryzae* causes leaf blight disease. Bacterial blight or stripe of several cereals and streak of sorghum and maize is caused by *X.c.* pv. *translucens*. *X.c.* pv. *juglandis* causes blight of walnuts. In the case of soybean, bacterial pustule disease caused by *X.c.* pv. *glycines* is known to inflict considerable losses in yield. Gumming disease of sugarcane affecting yields of sugar is caused by *X.c.* pv. *vascularum*.

Bacterial pathogen–plant interactions involve an interplay of the various virulence factors, the hypersensitivity response and pathogenicity (*hrp*) and avirulence (*avr*) genes of the pathogen and the disease resistance genes in plants. The virulence factors comprise agents such as the hydrolytic enzymes, toxins, polysaccharides, and plant growth regulators secreted by the pathogen that damage or alter plant cells and provide optimal environment for the growth of the pathogen. On the other hand avirulence factors or the products of avirulence genes of the pathogen invoke hypersensitive response and death of the surrounding cells in the resistant host. This restricts the spread of the pathogen and in turn restricts its host range. *Hrp* genes in the pathogen regulate both the *avr*-induced hypersensitivity reaction as well as pathogenicity.

Table 3 Pre-harvest disease of fruits and vegetables caused by *Xanthomonas campestris*

Fruit/vegetable	*Name of disease*	*Causative agent*
Lima beans	Bacterial blight	*X.c.* pv. *phaseoli*
Tomato	Bacterial spot	*X.c.* pv. *tomato*
Pepper	Bacterial spot	*X.c.* pv. *vesicatoria*
Cabbage	Black rot	*X.c.* pv. *campestris*
Citrus	Canker	*X.c.* pv. *citri*

Food Spoilage Potential of *Xanthomonas*

Pre-harvest Spoilage

Xanthomonads are the cause of a number of pre-harvest diseases of fruits and vegetables. They cause blights, spots, cankers and rots of different fruits and vegetables (**Table 3**).

Lima bean pods can be spoiled by the common blight caused by *X.c.* pv. *phaseoli*. Bacterial spots of tomato and pepper caused by *X.c.* pv. *tomato* and *X.c.* pv. *vesicatoria* reduce quality and marketability of these important vegetables. One of the most feared diseases of citrus fruits, citrus canker, is caused by *X.c.* pv. *citri*. The commodity loses its aesthetic appeal, quality and marketability, thereby causing severe economic losses.

Post-harvest Spoilage

Being a part of the natural microflora xanthomonads are invariably associated with plants and plant products. The role of xanthomonads in the post-harvest spoilage of plant foods and food products has been poorly investigated. However, the association of various hydrolytic enzymes with the growth of xanthomonads suggests their high food spoilage potential. Many xanthomonads are known to possess proteolytic, amylolytic, cellulolytic, pectolytic and lipolytic activities. As fruits and vegetables provide ample substrate for these enzymes the food spoilage potential of xanthomonads is high. Xanthomonads may also bring about yellow discoloration of foods due to their pigment-producing potential. Xanthomonads also produce xanthan gum, which may cause undesirable gumminess in certain foods and fruit juices. Products formed from the infected plants and plant products are also likely to undergo spoilage by the action of hydrolytic enzymes.

Production of Xanthan Gum by *Xanthomonas*

Xanthan gum is a high-molecular-weight polysaccharide gum. It is produced extracellularly by strains of *Xanthomonas campestris* in a medium containing carbohydrate and protein. Industrially it is

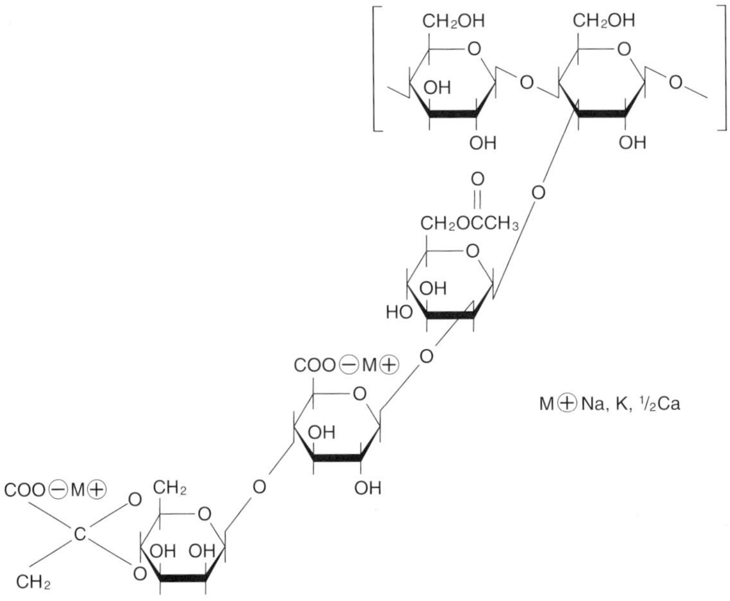

Figure 4 Structure of xanthan gum.

produced by a pure culture fermentation of a carbohydrate-containing medium with the strains of *X. campestris* developed for this purpose. Xanthan gum is a substituted cellulose. It is composed of β(1→4)-linked glucose backbone. Every alternate glucose residue in the backbone is connected to a mannose through an α(1→3) bond, which is connected to a glucoronic acid residue through an α(1→2) linkage. The glucoronic acid residue is in turn connected to another mannose unit through a β(1→4) linkage (**Fig. 4**). The repeating five-sugar unit thus comprises two glucose and two mannose molecules and one glucuronic acid molecule. The mannose units have acetyl and pyruvate groups. The contents of acetate and pyruvate are in the ranges 4–5% and 2–6%, respectively, in xanthan gum. The polymer has a molecular weight of $1–10 \times 10^6$ kDa.

Xanthan gum has the following useful properties:

- It gives high viscosities at low concentrations;
- It has remarkable emulsion stabilizing and suspending ability;
- Its viscosity is stable to changes in pH, salts and temperature;
- Xanthan gum solutions show pseudoplastic behaviour, i.e. with an increase in shear rate the viscosity of a xanthan solution decreases and vice versa.

Application of Xanthan Gum in Food Industry

Because of the above properties of xanthan gum it serves as an excellent stabilizing, thickening and emulsifying agent. Xanthan gum has a wide range of applications. It has several pharmaceutical, food and non-food uses. The thickening, stabilizing, jelling and emulsifying properties of this polysaccharide make it useful in food industry. It imparts good flavour release characteristics and sensory qualities to food. It can be pumped, sprayed and spread easily. Some of the applications of xanthan gum in the food industry are given in **Table 4**. Xanthan gum is used as a stabilizer in ice-creams and other milk-based products, in confectionery products and noodles, in salad dressings, and non-alcoholic beverages. It is used as thickener in soups, sauces, gravies, shakes, syrups, relishes and toppings. It is also used as a suspending agent in a number of foods including dressings, cake mixes and batter. In its non-food uses, xanthan gum finds application in the preparation of polishes, paints, adhesives, ceramics, and explosives. In the petroleum industry xanthan gum is used for enhanced oil recovery.

Table 4 Applications of xanthan gum in the food industry

Product	*Function*
Ice-cream	Stabilizer
Sauces and gravies	Thickener
Dressings	Stabilizing and suspending agent
Non-alcoholic beverages	Stabilizer
Cake mixes and batters	Suspending agent
Relishes	Thickener
Processed cheese	Stabilizer
Milk-based desserts	Thickener
Syrups and toppings	Thickener
Noodles	Stabilizer
Spring roll pastry	Stabilizer

See also: **Enzyme Immunoassays**: Overview. **Fermentation (Industrial)**: Production of Xanthan Gum.

Nucleic Acid-based Assays: Overview. ***Pseudomonas***: Introduction.

Further Reading

Alvarez AM and Lou K (1985) Rapid identification of *X.c.* pv. *campestris* by ELISA. *Plant Disease* 69: 1082–1086.

Alvarez AM, Schenk S and Benedict AA (1996) Differentiation of *Xanthomonas albilineans* strains with monoclonal antibody reaction patterns and DNA fingerprints. *Plant Pathology* 45: 338–366.

Bradbury JF (1984) Genus II, *Xanthomonas* Dowson 1939. In: *Bergey's Manual of Systematic Bacteriology*. Baltimore: Williams & Wilkins.

Breed RS, Murray EGD and Smith NR (1957) *Bergey's Manual of Determinative Bacteriology* (7th edn.). Baltimore: Williams & Wilkins.

Franken AAJM, Zilverentant JF, Boonekamp PM and Schots A (1992) Specificity of polyclonal and monoclonal antibodies for the identification of *Xanthomonas campestris* pv. *xampestris*. *Netherlands Journal of Plant Pathology* 98: 81–94.

Hildebrand D, Hendson M and Schroth MN (1993) Usefulness of nutritional screening for the identification of *Xanthomonas campestris* DNA homology groups and pathovars. *Journal of Applied Bacteriology* 75: 447–455.

Holt JG (ed.) (1979) *The Shorter Bergey's Manual of Determinative Bacteriology* (8th edn.) Baltimore: Williams & Williams.

Leite RP, Minsavage GV, Bonas U and Stall RE (1994) Detection and identification of *Xanthomonas* strains by amplification of DNA sequences related to *hrp* genes of *Xanthomonas campestris* pv. *vesicatoria*. *Applied and Environmental Microbiology* 60: 1068–1077.

Palleroni MJ (1991) Introduction to the Family *Pseudomonadaceae*. In: Balows A, Truper HG, Dworkin M, Harder W and Scheifer KH, vol 3, *The Prokaryotes*, New York: Springer-Verlag.

Sanderson GR (1981) Applications of xanthan gum. *The British Polymer Journal* 13: 71–75.

Vauterin L, Swings J and Kersters K (1991) Grouping of *Xanthomonas campestris* pathovars by SDS-PAGE of proteins. *Journal of General Microbiology* 137: 1677–1687.

XEROMYCES BISPORUS FRASER

Ailsa D Hocking and **J I Pitt** CSIRO Food Science Australia, North Ryde, New South Wales, Australia

The spoilage mould *Xeromyces bisporus* Fraser is able to grow at a lower water activity (a_w) than any other known organism. As a consequence it is capable of growth in dried and concentrated foods usually regarded as safe from microbial attack, and has been the cause of significant spoilage of such products. Cultural and microscopic features of this unique fungus are described below, together with its physiological properties and techniques for its isolation and maintenance.

Isolation Techniques

Because *X. bisporus* will not grow above 0.97 a_w, isolation on standard laboratory media is impossible. Moreover, unless significant numbers of mature ascospores are present, isolation by dilution plating will be unsuccessful because commonly used diluents such as 0.1% aqueous peptone will damage hyphae by osmotic shock. Dilution plating with low a_w diluents such as 50% glucose or 40% glycerol is possible but less practical because of their high viscosity. It is important therefore to attempt isolation of *X. bisporus* by direct plating.

Provided this species is present in effectively pure culture (as is often the case when it is growing on substrates of low a_w), the most suitable isolation medium is malt extract yeast extract 50% glucose agar (MY50G; 0.89 a_w; see formula below). This medium should be inoculated with small masses of mycelium or small pieces of substrate cut off with a sterile blade. Small amounts of powdered substrates may be sprinkled directly on the agar. Petri dishes should be enclosed in a polyethylene bag to stabilize the a_w of the medium and prevent drying out.

Isolation of *X. bisporus* is further complicated by the fact that it will not compete against the vigorous growth of *Eurotium* species (commonly known as the *Aspergillus glaucus* series) at a_w levels above about 0.75. In the presence of such fungi, identifiable by grey-green or yellow growth and the presence of *Aspergillus* heads, the most satisfactory isolation medium for *X. bisporus* is malt extract yeast extract 70% glucose fructose agar (MY70GF; see formula below), which is near 0.75 a_w. To prevent drying of this medium, Petri dishes should be incubated in the presence of a saturated solution of NaCl, in a sealed container such as a desiccator or plastic food box. Again, inoculation should be by direct plating.

Incubation should be at 25–30°C: on MY50G agar good growth should occur within 7–10 days; on

MY70GF, incubation times of 3 weeks or more are necessary. On both media, *Xeromyces* colonies are low and spreading: transfer peripheral hyphae to MY50G slants and again incubate at 25–30°C. Colonies of *Xeromyces* will remain low and white or translucent. Identity is confirmed by development of the characteristic cleistothecia (see description below).

Media

MY50G agar has the following composition: malt extract 10 g; yeast extract 2.5 g; glucose 500 g; agar 10 g; water 500 g; sterilize by steaming for 30 min. The final medium is about pH 5.5 and 0.89 a_w. If glucose monohydrate is used, add 550 g and reduce water to 450 g.

MY70G agar has the following composition: malt extract 6 g; yeast extract 1.5 g; glucose 350 g; fructose 350 g; agar 6 g; water 300 g; sterilize by steaming for 30 min. Final medium is about pH 5.5 and 0.75 a_w. If glucose monohydrate is used, add 385 g and reduce water to 265 g. Powdered light malt extract sold for home brewing is entirely satisfactory for the preparation of media such as these. Food-grade glucose monohydrate may be substituted for anhydrous glucose, provided that allowance is made for the water of hydration (see formulae). In the preparation of such concentrated media, it is desirable to melt the agar before adding the sugars, because in lowering a_w they also reduce agar solubility.

Maintenance

The most satisfactory medium for the maintenance of *X. bisporus* is MY50G agar. Allow 4–6 weeks' incubation at 25–30°C for ascospores to mature, and check for their presence by microscopy before storing cultures, which may be held at 5–25°C for several months.

Subculturing of *Xeromyces* isolates over long periods may cause delayed or diminished ascospore production. The most satisfactory method for long-term maintenance is lyophilization, provided adequate numbers of mature ascospores are present. An alternative method is to grow *Xeromyces* on commercially prepared fruit cake, which is commonly about 0.75 a_w. Place slices of cake in Petri dishes, enclose in foil or an oven-proof plastic bag, sterilize by steaming for 30 min, and then inoculate the upper surface of the cake with *Xeromyces*. Wrap the Petri dishes in polyethylene and store them at about 25°C. Cultures will be ready for use in about 2 months, and will provide inocula of viable ascospores for a long time.

Description

After incubation for 7 days on MY50G agar at 25°C, colonies are 3–6 mm in diameter, hyaline, low and sparse. After 14 days, colonies are 15–20 mm in diameter, with a low and dense texture, translucent with a glistening surface, and colourless or very pale red-brown; the reverse is uncoloured (**Fig. 1A**). After 4 weeks' incubation, colonies are 50–70+ mm in diameter, low, translucent and sometimes glistening, colourless or faintly red-brown, with contiguous layers of colourless cleistothecia visible under the low-power microscope; the reverse remains uncoloured.

Reproductive structures are cleistothecia and aleurioconidia, the latter solitary, developing only below 0.9 a_w, and usually measuring 15–20 × 12–15 μm. Cleistothecial initials are evident on MY50G agar at 2 weeks. They commence as three short cells, then develop finger-like processes from the bottom cell, which envelop the second and enlarge to form the cleistothecium (**Fig. 2**). Mature cleistothecia, often outnumbered by abortive ones, form within 4–6 weeks on MY50G agar. They measure 40–120 μm in diameter, and have thin structureless walls (Fig. 1B,C). Asci are inconspicuous and evanescent, and contain two ascospores only; ascospores are ellipsoidal, flattened on one side (D-shaped), measure 10–12 × 4–5 μm and have smooth walls (Fig. 1D).

Taxonomy

Xeromyces bisporus Fraser was originally described in 1953 by L. Fraser who isolated it from mouldy Australian liquorice. It is the only species in the genus. *X. bisporus* was transferred by von Arx to *Monascus*, but other taxonomists are agreed that *Xeromyces* is a distinct genus, probably related to *Monascus*.

Physiology

Water Relations

Xeromyces bisporus is the most xerophilic organism known. It can germinate at 0.61 a_w at 25°C after 120 days' incubation. It can form aleurioconidia in 80 days at 0.66 a_w and ascospores at 0.67 a_w within a similar time interval. The ability to complete a sexual life cycle below 0.7 a_w is a remarkable attribute, shared only with the xerophilic yeast *Zygosaccharomyces rouxii*.

The effect of a_w and solute on the germination of *X. bisporus* is shown in **Figure 3**. Ascospores of *X. bisporus* germinate remarkably slowly, taking at least 3 days even under optimal conditions, such as glucose/fructose medium at 0.90 a_w. More remarkable is the fact that germination on the same substrate at

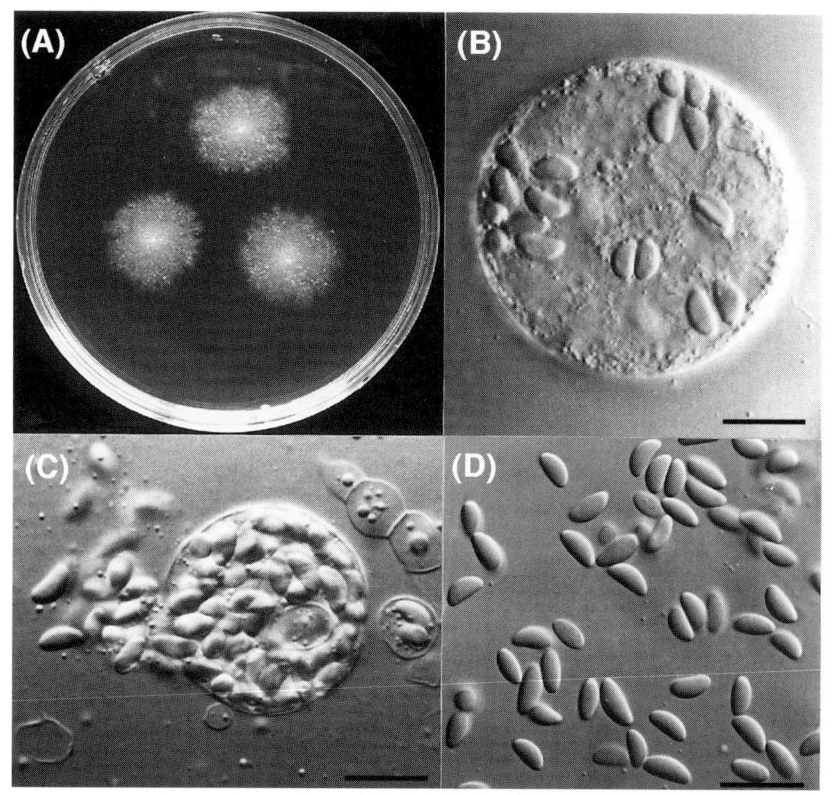

Figure 1 *Xeromyces bisporus*. (**A**) Colonies on malt extract yeast extract 50% glucose (MY50G) agar after 2 weeks at 25°C. (**B**) Developing cleistothecium showing ascospores forming in pairs. (**C**) Mature cleistothecium discharging ascospores. (**D**) Ascospores. (**B–D**) Bar =20 μg.

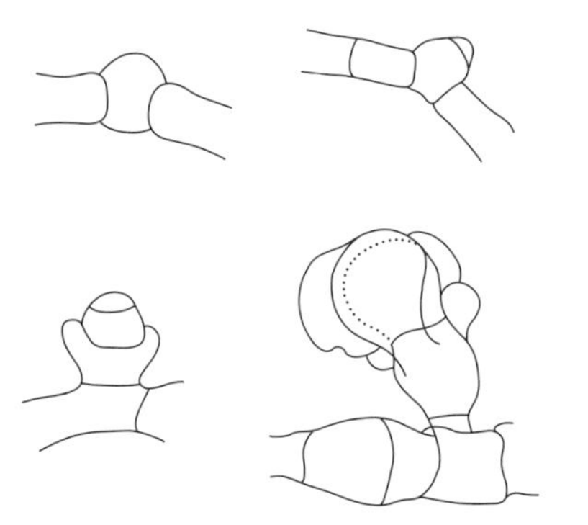

Figure 2 Stages in the development of cleistothecial initials of *Xeromyces bisporus*.

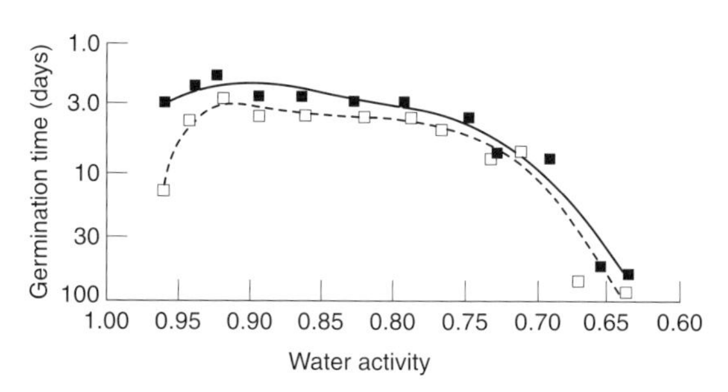

Figure 3 Effect of water activity on the time taken for germination of spores of *Xeromyces bisporus* on glucose/fructose-based media at pH 4.0 (open squares) and pH 6.5 (filled squares). Data from Pitt and Hocking (1977).

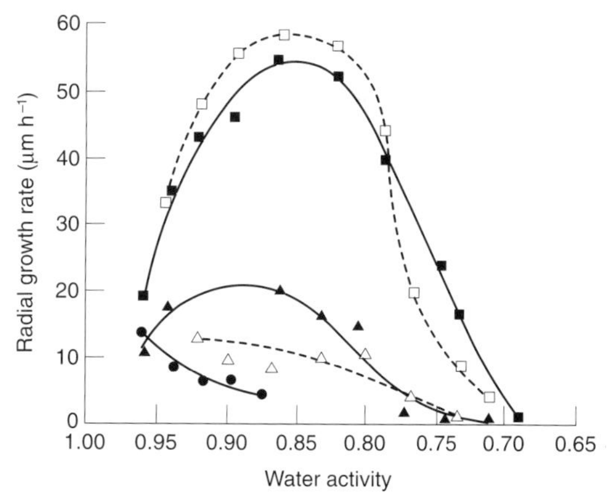

Figure 4 Effect of water activity and solute on the radial growth rate of *Xeromyces bisporus*. ■ = Glucose/fructose, pH 4.0; □ = glucose/fructose, pH 6.5; ▲ = glycerol, pH 4.0; △ = glycerol, pH 6.5; ● = NaCl, pH 4.0. Data from Pitt and Hocking (1977).

0.75 a_w requires less than twice this time. This very wide optimum range from 0.95 to 0.75 a_w in the germination curve for *X. bisporus* is unique among microorganisms. As illustrated in **Figure 4**, *X. bisporus* grows much more rapidly in media containing high concentrations of glucose or fructose than of glycerol or NaCl at the same a_w. It is thus to be expected that spoilage by *X. bisporus* will occur most

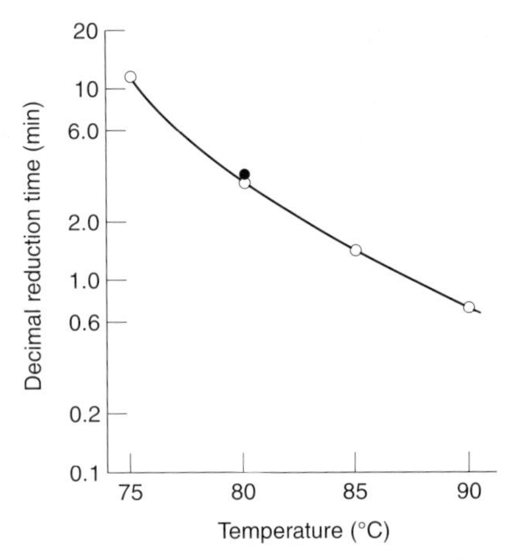

Figure 5 Relationship of decimal reduction time and temperature for *Xeromyces bisporus* ascospores, constructed using data of Dallyn and Everton (1969; open circles) and Pitt and Christian (1970; filled circle).

often in products containing sugars as the major solute.

Heat Resistance

The ascospores of *X. bisporus* are relatively heat-resistant. Two independent studies of the heat resistance of *X. bisporus* ascospores produced comparable results. The first study found 0.1% of 1000 ascospores survived heating at 80°C for 10 min; the second study found that when 2000 ascospores per millilitre were heated at 80°C, some survived after 9 min but none after 12 min.

By assuming survival of one spore per millilitre at the mean of the survival and death times given by these studies, decimal reduction times (DRT) can be estimated graphically, and these are plotted in **Figure 5**. Points at temperatures above 75°C can be connected by a straight line, which can be defined by a z value of 16.0°C (= 28.8°F) and a DRT at 82.2°C (F_{180}) of 2.3 min (Fig. 5). The test conditions for these heat runs (50% sucrose medium, 0.94 a_w) would be similar to a fruit cake mix, and the reduction in a_w as the cake dries during baking would confer increased heat resistance on the ascospores. From these data, the heat resistance of *X. bisporus* is sufficiently high to account for its presence in fruit cake as being due to survival of ascospores during baking.

Influence of Oxygen and Carbon Dioxide Tension

Early studies indicated that *Xeromyces* was capable of growth in an atmosphere containing only 1% O_2, even in the presence of 70–95% CO_2, so vacuum or gas packing of foods may not completely prevent growth of this mould. More recent studies indicate that *X. bisporus* is not quite so tolerant of modified atmospheres, and is unable to grow in an atmosphere containing $< 0.5\%$ O_2 and 20% CO_2.

Ecology

Originally isolated from spoiled liquorice, *Xeromyces bisporus* was later found to be a major cause of spoilage in Australian dried and high-moisture prunes and as an uncommon but widespread spoilage agent in the UK, where it was isolated from table jellies, dried prunes, tobacco, currants, chocolate and chocolate sauce. *X. bisporus* has been isolated in Australia from spice powders, Chinese dates (a_w 0.72), fruit cakes (a_w 0.75–0.76), sun-dried raisins (a_w 0.67), mixed dried fruit, diced dried apricots, cookies containing a high proportion of dried fruit pieces and its first known habitat, liquorice. It has been reported from honey in Japan, date honey in Israel, and tobacco and syrup in the Netherlands.

X. bisporus can become established in food-processing plants producing low a_w products providing a continuous low-level inoculum for foods such as packaged dried fruit, fruit cakes and cookies. Ascospores appear to be capable of surviving the baking process, leading to sporadic spoilage of baked goods which is often not manifested until many months after production. Tracing the source of the contamination can be difficult, if not impossible, particularly in view of the elapsed time between production and visual spoilage. Careful attention to cleaning of all product residues from production lines and equipment before sanitation is essential to eliminate this type of spoilage.

The natural habitat of *Xeromyces* is not known. It has never been isolated from soil or decaying vegetation, and its water relations would appear to preclude it from such habitats. Its physiological properties indicate that *Xeromyces* could grow in nature only on sugary materials such as drying, or dry, fruits and berries, honey or sugary exudates. All records to date have come from foods or dried tobacco, or food-processing equipment.

Prevention of Spoilage by *Xeromyces*

From ecological considerations and isolation data, the most likely sources of *Xeromyces* infections in foods or food plants appear likely to be dried fruits or spices. It is fortunate that such infections are rare: monitoring for *Xeromyces* is ineffective both because of its slow growth rate and because of its inability to compete with *Eurotium* species even on media specifically designed for isolation of xerophiles, such as dichloran 18% glycerol agar (DG18). If a processing or packaging plant does become infected with

Xeromyces, normal cleaning and sanitation procedures will usually be effective in removing it, as ascospores and aleurioconidia are easily wetted. For the same reason, the spores are apparently not aerially dispersed and recontamination is unlikely.

In summary, *X. bisporus* is a seldom recognized but significant cause of food spoilage, capable of destructive growth at exceptionally low levels of a_w and reduced O_2 tension. Its ascospores possess a high heat resistance, probably sufficient to withstand baking processes. Isolation and identification of *Xeromyces* are not difficult once it is realized that standard laboratory media and isolation techniques are completely ineffective.

See also: **Confectionary Products**: Cakes and Pastries. **Dried Foods**. **Spoilage Problems**: Problems Caused by Fungi.

Further Reading

Dallyn H and Everton JR (1969) The xerophilic mould, *Xeromyces bisporus*, as a spoilage agent. *Journal of Food Technology* 4: 399–403.

Fraser L (1953) A new genus of the Plectascales. *Proceedings of the Linnean Society of New South Wales* 78: 241–246.

Pitt JI and Christian JHB (1968) Water relations of xerophilic fungi isolated from prunes. *Applied Microbiology* 16: 1853–1858.

Pitt JI and Christian JHB (1970) Heat resistance of xerophilic fungi based on microscopical assessment of spore survival. *Applied Microbiology* 20: 682–696.

Pitt JI and Hocking AD (1977) Influence of solute and hydrogen ion concentration on the water relations of some xerophilic fungi. *Journal of General Microbiology* 101: 35–40.

Pitt JI and Hocking AD (1997) *Fungi and Food Spoilage*, 2nd edn. London: Blackie Academic and Professional.

YEASTS: PRODUCTION AND COMMERCIAL USES

Richard Joseph, Central Food Technological Research Institute, Mysore, India

Introduction

Commercial yeast production worldwide exceeds 1.8 million t per annum. The yeasts are used mostly by the baking industry, but also by the brewing and distilling industries. Yeast is also a commercial source of natural flavourings, flavour potentiators and the dietary supplements.

Yeasts are unicellular eukaryotes, and in several ways are akin biochemically to higher organisms. They have been shown to be suitable for the expression of valuable mammalian and plant proteins, and have therefore emerged as an important biotechnological asset in recent years.

The emphasis of this article is on the practical aspects of the production and preservation of bakers' yeast.

History

The practice of yeast husbandry can be dated to the Neolithic, i.e. long before scientific knowledge about microorganisms was available. More authentic evidence dates from 4–5 millennium BC when the arts of leavening, brewing and wine-making were well known. The excavation of Thebes in Egypt has revealed models of baking and brewing dating from the 11th dynasty (about 2000 BC).

The use of yeasts therapeutically is revealed in the Ebes Papyrus, one of the earliest known medical documents, dating from the 16th century BC. Hippocrates (4–5th century BC), the well-known Greek physician, also used yeasts therapeutically.

The first yeasts used for baking were obtained from the mashes produced in the manufacture of beer. The first compressed yeasts used for baking and brewing were made in England in about 1792, and by 1800 they were available throughout northern Europe. The large-scale commercial production of bread in the US was facilitated by the introduction of an improved strain of compressed yeast in 1868, by Charles Fleischmann. The vigorous research and development effort that ensued yielded yeast strains suited to each type of fermentation and leavening. For example, some of these strains were able to tolerate high sugar or salt concentrations, or the high temperatures used in fermentation and the proving of dough.

Anton van Leeuwenhoek (1632–1723) of Holland was possibly the first human to set eyes on a yeast, when he observed a droplet of fermenting beer with the aid of one of the first microscopes, capable of 250–270-fold magnification. Interest in yeasts, and in microorganisms in general, then lay almost dormant until Louis Pasteur (1822–1895) carried out extensive systematic studies which revealed the nature of yeasts and their extraordinary biochemical capabilities.

Classification

Yeasts are unicellular fungi reproducing asexually by budding or fission and sexually by spore formation. Emil Christian Hansen's studies, over a span of 30 years, provided insight into the biological features of yeasts and facilitated their differentiation and their characterization as species.

Currently more than 500 species of yeasts, belonging to around 50 genera, are known. Yeasts belong to the division Eumycota, within the subdivisions Ascomycotina, Basidiomycotina, and Deuteromycotina.

Bakers' yeast and the yeasts used in brewing, wine-making and distilling are strains of *Saccharomyces cerevisiae*, belonging to the family Saccharomycetaceae in Ascomycotina.

Saccharomyces cerevisiae

The genus *Saccharomyces* (translation 'sugar fungus') derives its name from its common occurrence in sugary substrates such as nectar and fruits. Strains of *S. cerevisiae* have been isolated from diverse sources, including breweries, wine, berries, cheese, pear juice and must, honey, eucalyptus leaves, kefir, *Drosophila*, soil and human skin, sputum and leg ulcers. *S. cerevisiae* has around 87 synonyms worldwide.

Morphology

After 3 days' growth in malt extract at 25°C, *S. cerevisiae* cells are either globose in shape (5.0–10.0) × (5–12.0) μm or ellipsoidal or cylindrical, measuring 3.0–9.5 μm × 4.5–21.0 μm. The cells may occur singly or in pairs, short chains or clusters.

Streak cultures on malt agar are butyrous and cream to brownish. They are either smooth and slightly raised with shallow striations, or raised, folded and (often) subdivided. They can be either glossy or dull.

Reproduction

Asexual reproduction usually occurs by budding. The buds arise on the 'shoulders' and at either pole of the cell. The vegetative cells are diploid or polyploid, and this phase predominates in the life cycle of the yeast.

Sexual reproduction involves the production of asci, within which ascospores develop directly following meiosis of the diploid nucleus. The sporulation of *S. cerevisiae* is encouraged by media containing acetate, such as acetate agar. Sporulation also occurs on potato–dextrose agar. True sexual reproduction is found in some strains, which exhibit heterothallism, of the bipolar physiological type. Compatibility is determined at one mating type locus, which may contain either of two alleles. Conjugation occurs either by the fusion of two ascospores or by the fusion of two haploid somatic cells of germinating ascospores. Haploid vegetative clones can be raised by the germination of isolated spores.

Hybrid Strains

Cells of opposite mating types can be fused to produce hybrid yeast strains. This technique can be used to combine industrially desirable traits such as high growth rates, high yields, resistance to drying and CO_2 production. A range of hybrids has been developed, suited to different needs. These include rapidly fermenting strains which produce high volumes of CO_2, for automated bakeries; strains with intermediate activity, for traditional bakeries; and strains which ferment more slowly, for in-store bakeries. Strains have also been developed which have improved resistance to drying, osmotolerance and tolerance to freezing.

If sexual mating is difficult to achieve owing to very low yields of viable spores, modern techniques for improving strains can be used. These include protoplast fusion and the construction of recombinant DNA, which can be achieved with relative ease using the tools and techniques of molecular genetics.

Commercial Production of Bakers' Yeast

Sources

Industries requiring yeast cultures can either obtain them from culture collection centres or isolate and develop their own cultures. In either case, the propagation and maintenance of cultures for long-term use ensures consistency of performance and quality. However, basic facilities and expertise within the industry are required.

Saccharomyces can be isolated from natural sources, and maintained in pure culture by conventional microbiological techniques. The source material (fermenting sugary materials, fruit juices or soil) is usually serially diluted and plated onto potato–dextrose agar (PDA) or yeast extract–peptone–dextrose agar (YEPDA). The growth of yeasts in preference to bacteria is achieved by the pH of the medium being below neutral (usually 4–6) and the incorporation of antibacterial antibiotics.

Enrichment culture is a technique by which strains with characteristics required by industry (e.g. tolerance to high temperatures) can be isolated from natural habitats. The required strains are selected either by gradually increasing exposure to the factors to which tolerance is required, or by cultivation with very high levels of the factors over a long period of time.

Maintenance of Cultures

If yeasts are used regularly, for example in a batch process with a constant periodicity, the simplest method of maintaining stock cultures is to use agar slopes or broth. For semi-continuous and continuous processes, fresh yeast cultures must be available because the overgrowth of less efficient and low-performing variants of the yeasts in continuous processes is a common problem.

Normally, slope and broth cultures are subcultured once every 2 months. After allowing for adequate growth at ambient temperature, they are kept at 4–8°C until use. The drawback of this simple technique is the risk of contamination and the development of genetic variants which are less efficient than the yeasts originally selected. These undesirable effects can be prevented by the incorporation of a 'selection pressure' to ensure the retention of strains which perform well in preference to any variants which perform less well. Examples include the incorporation of a high sugar or salt concentration in the maintenance medium, to retain yeasts which will be tolerant to high concentrations in the fermentation.

Preservation of Cultures

Yeasts that are sensitive to dehydration in slope or stab cultures may be maintained by overlaying with mineral oil. It is generally observed that microbes in soil culture retain their original characteristics over a relatively long period of time. This is a simple and inexpensive method, in which sterilized garden soil containing 60% moisture is inoculated with the culture and, after allowing for growth at ambient temperature for a week, is kept at 4–8°C. All micro-organisms except non-sporulating bacteria sporulate in soil, and in this form remain viable and functional for up to 2 years.

Cultures can also be frozen, keeping them functionally intact for long periods. The inclusion of glycerol (5–20%) in the suspension medium and storage at −20°C are recommended. Well-equipped culture collection centres have facilities for the extended storage of cultures in liquid N_2 at −196°C or by lyophilization. It is recommended that lyophilized cultures be stored at 4–8°C, but they can withstand the ambient temperatures imposed during transit by post. Frozen cultures stored in liquid N_2 and lyophilized cultures maintain their genetic constitution and functional characteristics for long periods of time.

Growth Requirements

S. cerevisiae is a heterotroph, i.e. it requires preformed organic compounds for growth. It is also a mesophile, growing best in the temperature range 25–40°C. In common with other living organisms, bakers' yeast has basic nutritional requirements for carbon and nitrogen sources, minerals and vitamins.

Carbon

A limited range of sugars is utilized as a carbon source by *S. cerevisiae*. Glucose and fructose are readily utilized, and of the disaccharides, sucrose and maltose are preferred. Other malto oligosaccharides can also be utilized, but less readily. Notably, *S. cerevisiae* cannot utilize pentoses, other hexoses, the disaccharides lactose or cellobiose or the polysaccharides.

In industry, the preferred carbon sources are cane or sugar-beet molasses, which have a fermentable sugar concentration of 50–55% and around 80% total soluble solids. The composition of typical molasses is shown in **Table 1**. However, it should be noted that being a by-product of the sugar industry, molasses is considerably variable in composition.

However, improvements in the processes for the recovery of sugar have resulted in the production of molasses containing a lower concentration of fermentable sugars. In addition, molasses is also the substrate for the production of ethanol, and so in many countries is subject to excise. Alternative substrates for the production of bakers' yeast have therefore had to be considered. Starchy substrates, derived from low-grade grains and tubers, are the logical alternative to molasses, but suitable manufacturing processes need to be developed. Sugar-cane juice and sugar-beet extract can also be used for the production of bakers' yeast, if the process is economically viable.

Table 1 Composition of molasses

	Percentage of total weight of molasses	
Component	*Range*	*Average*
Water	17–25	20
Sucrose	30–40	35
Glucose	4–9	7
Fructose	5–12	9
Other reducing substances	1–5	3
Other carbohydrates	2–5	4
Ash	7–15	12
Nitrogenous compounds	2–6	4.5
Non-nitrogenous acids	2–8	5
Waxes, sterols and phospholipids	0.1–1	0.4
Vitamins	Trace	Trace

Nitrogen

S. cerevisiae can utilize inorganic nitrogenous compounds such as ammonium sulphate, ammonium chloride or even ammonia. Urea can be used to provide N_2 in the commercial production of yeasts.

Minerals

The major minerals required for growth by *S. cerevisiae* are phosphorus (an important component of nucleic acids), potassium, calcium, sodium, magnesium and sulphur. Iron, zinc, copper, manganese and cobalt are required as trace elements. These requirements are largely met by the molasses, with the exceptions of phosphorus and magnesium.

Phosphorus is supplied in the commercial production of yeasts as phosphoric acid or as phosphate of sodium, potassium or ammonium. Magnesium is supplied as magnesium sulphate in the growth medium.

Vitamins

S. cerevisiae requires biotin, pantothenic acid, inositol and thiamin for growth. These, except for thiamin, are usually available from molasses in adequate quantities. Sugar-beet molasses, however, is deficient in biotin and hence requires supplementation with synthetic biotin. The biotin requirement of yeast is reported to be increased when urea is used as the N_2 source in the growth medium. A mixture of L-aspartate and oleic acid was found to completely eliminate the requirement for biotin.

Molasses must be supplemented with thiamin to enable maximum growth of the yeast.

Oxygen

S. cerevisiae possesses a remarkable ability to adapt to thrive in varying levels of available O_2. In very low levels of O_2, its metabolism responds by shutting off the respiratory enzymes. The yeast then leads a fermentative life, in which sugar is partially and non-oxidatively utilized for energy and the 'waste' product is ethanol. In contrast, when adequate O_2 is available, sugar is converted by the respiratory enzymes to CO_2 and H_2O, as well as to intermediates needed for the cell biomass.

Therefore in microaerophilic conditions (often erroneously termed 'anaerobic' conditions), the yeast grows significantly more slowly than in aerobic conditions. The maximum theoretical yields of yeast solids under microaerophilic (anaerobic) and aerobic conditions have been calculated as 7.5 kg and 54.0 kg respectively per 100 kg of sugar utilized. In the case of *S. cerevisiae*, the concentration of the substrate (sugar) influences the availability of O_2, with consequent effects on the growing culture. If the medium contains >5% glucose, glucose utilization via the TCA (tricarboxylic acid) cycle is almost completely blocked. Even if the culture is aerated, glucose can only be fermented, this phenomenon being known as the 'glucose effect' or the 'Crabtree effect'. If the glucose level is then lowered to 0.1% and aeration continued, the metabolism of the yeast shifts from fermentation to respiration.

For aerobic growth, therefore, the sugar has to be supplied incrementally so that the rate of growth of the yeast (μ) does not exceed 0.2 and the respiratory quotient (RQ) is maintained at the value of 1. The production of bakers' yeast should therefore take place in aerobic growth conditions, facilitating the efficient oxidation of glucose to CO_2 and H_2O and the concomitant formation of ATP, required for cellular metabolism and the build-up of biomass. In aerobic conditions, as much as a third of the available sugar is metabolized via the hexose monophosphate pathway, generating NADPH, which is mainly utilized in synthetic reactions. Thus, under aerobic growth conditions the yeast's metabolism is elegantly balanced, generating chemical energy for cellular metabolism and precursor molecules for cell growth and proliferation. The TCA cycle is particularly important, being involved in the production of both the biomass precursor molecules and the chemical energy. *S. cerevisiae* also possesses the enzymes involved in the 'glyoxylate shunt'. This replenishes the TCA intermediates taken up for biomass formation.

Value pH

S. cerevisiae grows optimally at pH 4.5–5.0, although it can tolerate a pH range of 3.6–6.0. At higher pH values, the yeast's metabolism shifts, producing glycerol instead of ethanol in microaerophilic conditions.

In the production of bakers' yeast an initial pH at the lower end of the range inhibits the growth of bacterial contaminants. As the process progresses towards harvesting, the pH is raised slightly so that any colouring matter taken up by the yeast from the molasses is desorbed.

Temperature

S. cerevisiae has one of the shortest generation times amongst yeasts, 2.0–2.2 h at 30°C. Bakers' yeast production is optimal when the temperature of the cultivation medium is maintained at 28–30°C. However, productivity, in terms of grams of yeast solids produced per litre of cultivation medium per hour, depends on the feed rate in fed-batch fermentations. Despite growth at diminished rates being non-exponential, high productivity is still achieved.

Manufacturing Processes used for Bakers' Yeast

The overall process is summarized in **Figure 1**.

Preparation of Medium

The molasses (containing around 50% fermentable sugars and 80% soluble solids) is usually diluted with an equal weight of water, and the pH is adjusted to 4.5–5.0 with sulphuric acid. The diluted molasses is clarified using a desludger centrifuge. Clarification by filtration is recommended for beet molasses, but is not necessary for cane molasses. The clarified molasses is then sterilized by the high temperature short time (HTST) process. Other sterilization methods, which involve prolonged heating at low temperatures, cause caramelization and hence a decrease in the fermentable sugar content.

Medium supplements are added, these typically being: a nitrogen source (e.g. urea, $2.5\,g\,l^{-1}$); potassium orthophosphate (KH_2PO_4), $0.96\,g\,l^{-1}$; hydrated magnesium sulphate (epsomite, $MgSO_4.7H_2O$), $0.1\,g\,l^{-1}$; and a defoamer (silicone, fatty acid derivatives or edible oil), $0.01\,g\,l^{-1}$.

The molasses is then transferred to the fermenter, filling it to about two-thirds of its total volume, and pitching yeast is added. More clarified molasses is added incrementally in the fed-batch cultivation.

A substrate of starch (from corn, sorghum or tubers), hydrolysed by acids or enzymes, or of sugar cane juice, does not require elaborate clarification. Supplementation with a nitrogen source, minerals and

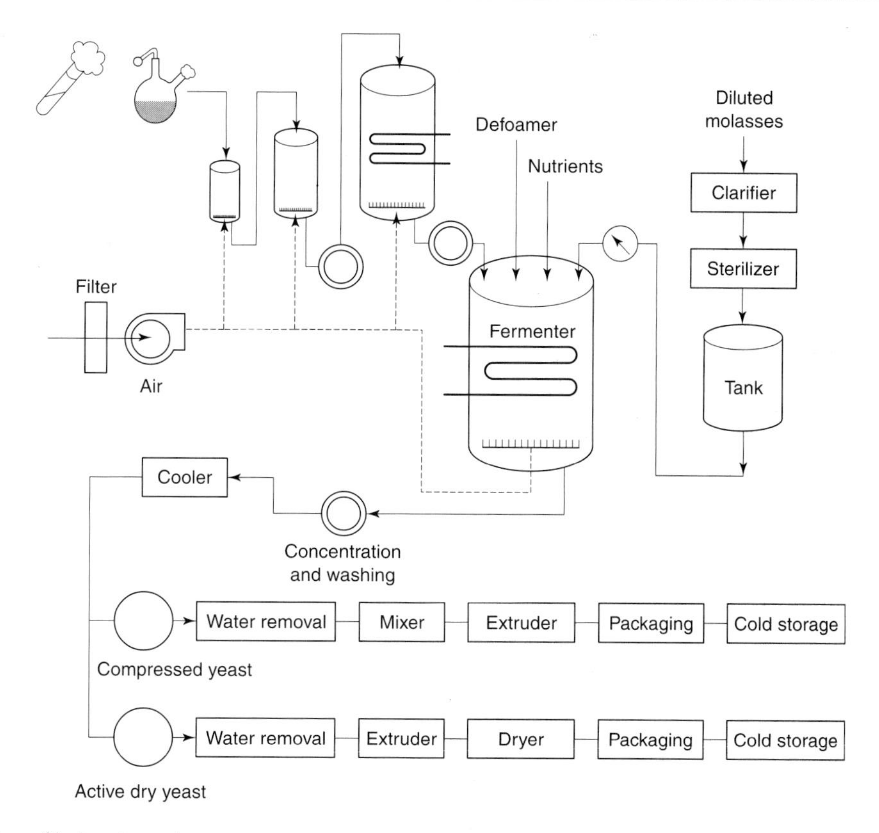

Figure 1 Production of bakers' yeast.

vitamins is, however, necessary. With suitable supplements, these raw materials are well-suited for the production of bakers' yeast, although economic considerations may dictate otherwise.

Cultivation

In industry, bakers' yeast is produced in fermentation tanks with a capacity of 200 m^3 or more. Tanks, and all connecting tubes, should preferably be made of stainless steel.

The design and operation of fermenters for the production of bakers' yeast must be carefully considered given the interrelationship between aeration, the specific growth rate of yeast and the substrate concentration. The facility for bubbling compressed air (as a source of O_2) into the medium, to ensure effective aeration, is therefore particularly important. In practice, O_2 transfer in a fermentation system can be manipulated by adjusting the bubble size and by the dispersion of air in the cultivation medium, by using mechanical agitation close to the point of entry of air into the medium. The level of dissolved O_2 during fermentation can be determined by using O_2 electrodes.

The features of a typical cultivation tank used for the production of bakers' yeast, an airlift fermenter, are shown in **Figure 2**. It consists of a cylindrical vessel, provided with a sparger (aerater) at the bottom for producing air bubbles in the medium. Aerobic growth results in the generation of almost 14 650 J of heat per gram of yeast solids, and because cultivation has to be carried out at 28–30°C, cooling is necessary. This is achieved by cooling coils, either within or

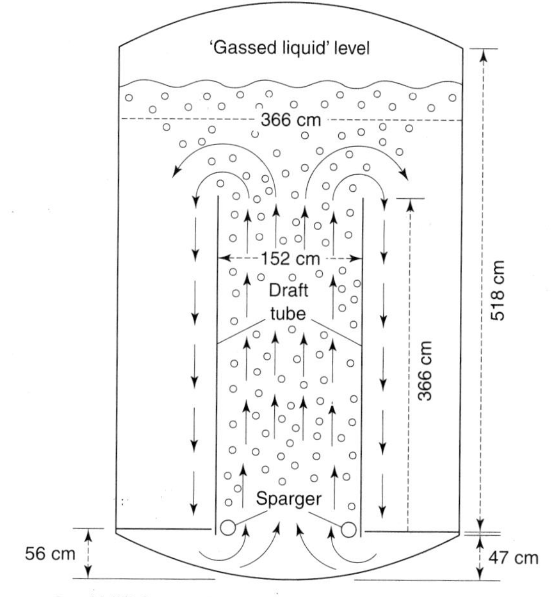

Figure 2 Airlift fermenter.

outside the vessel. Directional flow of the cultivation medium is achieved by pumping air in at the bottom of a 'draft tube'. The ratio between the depth of the cultivation medium and the diameter of the vessel has been shown to influence O_2 transfer. If the depth of the broth exceeds 3 m, the use of compressed air, to overcome the hydrostatic pressure of the liquid, is recommended. The design of the air sparger is also important for effective O_2 transfer. The inclusion of a motor-driven impeller further increases the effectiveness of O_2 transfer, but the additional investment and energy consumption involved may undermine their cost-effectiveness.

The manufacture of bakers' yeast begins with a number of stages which build up production, and involve the inoculation of the medium with pitching yeast. This development process is usually divided into eight stages, in which the yeast solids are gradually built up from the slope or flask culture of yeast. Over the eight stages, 0.2 kg of yeast solids give a final yield of about 100 000 kg of yeast. In the first two stages, sterilized medium and pure yeast cultures are employed in pressurized tanks, but the subsequent stages are operated in open tanks. The entire process is known to involve 24 generations of the yeast. Cultivation may be terminated before the normal final stage, in which case a yeast cream is obtained by centrifugation and used for pitching as required.

Maturation

At the end of the final stage of yeast cultivation, the feed rate is greatly reduced. This allows the yeast cells to mature and results in a low proportion of budding cells, which confers higher stability on compressed yeast in storage.

Finishing Stages

Yeast Cream The culture broth can be centrifuged in a continuous centrifuge (with a vertical nozzle) at 4000–5000 *g*, leading to almost complete recovery of the yeast cells. In the first run, around two-thirds of the fluid can be removed and in subsequent runs further concentration of the cells is achieved, producing a slurry called 'yeast cream', which contains about 20% yeast solids. Yeast cream can be stored at 4°C for a number of days, with good retention of viability.

Compressed Yeast This is prepared from yeast cream by filtration or by pressing in a filter press. Rotary continuous vacuum filters can also be used. The pressed cake thus obtained is mixed with 0.1–0.2% of emulsifiers such as monoglycerides, diglycerides, sorbitan esters and lecithin, and then extruded through nozzles. The extruded material, in the form of thick strands, is cut into suitable lengths and packaged (usually in packs of about 500 g) in wax paper or polythene sheet. The compressed yeast must be rapidly cooled, and stored at 5–8°C.

Active Dry Yeast Active dry yeast is useful in situations (e.g. homes) where storage at low temperatures is not possible. It is prepared by spreading out the pressed yeast cake to produce thin strands or small particles, which are then dried. Generally, a tunnel drier is used, taking 2–4 h with the air inlet temperature maintained at 28–42°C. However, more modern equipment, achieving either continuous drying or fluidized-bed drying (airlift drying), is also available. Emulsifiers such as sucrose esters or sorbitan esters (0.5–2.0%) are mixed with the dried yeast to facilitate rehydration. Antioxidants, such as butyl hydroxyanisole at 0.1%, are also added, to prevent undesirable oxidative changes. Active dry yeast has a moisture content of 4.0–8.5%.

Yeast Products and Uses

Nutritional Yeast

Yeast which has been heat-killed and dried is a source of protein and the B vitamins, and useful for supplementing foods and animal feeds. Currently the yeast used for these purposes is derived from brewing, wine-making or distilling, but the cultivation of yeast exclusively for food/feed supplementation is a possible future development. The official definitions and standards for such products are laid down by the International Union of Pure and Applied Chemistry (IUPAC), the National Formulary (NF XII) of the American Pharmaceutical Association and the Food and Drug Administration (FDA) of the US.

Lysine-enriched Yeast

Strains of bakers' yeast which can convert precursor molecules such as 5-formyl-2-oxovalerate or 2-oxoadipate to lysine, with high efficiency, have been reported. The use of such strains for the fortification of lysine-deficient cereals has potential.

Vitamin-enriched Yeast

The addition of thiazole and pyrimidine to the cultivation medium has been shown to cause bakers' yeast to synthesize high levels of thiamin (around 600 $\mu g\ g^{-1}$).

The irradiation of bakers' yeast with ultraviolet light has been shown to convert ergosterol to calciferol (vitamin D_2), with a vitamin potency reaching as much as 180 000 USP units per gram of yeast. Such strains will be useful for the fortification of food, feed and pharmaceuticals.

Yeast Lysates and Yeast Extract

Yeast autolysates are prepared by imposing conditions which trigger the yeast's native hydrolytic enzymes. These digest the yeast proteins and nucleic acids, converting them into soluble substances with an acceptable flavour and taste. The process involves the addition of ethyl acetate and NaCl (1–3%) to a yeast slurry containing 14–16% yeast solids. The mixture is maintained at 45°C for about 20 h, and then the whole autolysate is concentrated to a paste or dried to a powder. The soluble fraction can be separated out by centrifugation, and then concentrated.

Yeast hydrolysates can also be obtained, by adding hydrochloric acid to a yeast slurry containing 65–80% yeast solids and subjecting the mixture to reflux for about 12 h. It is then neutralized with sodium hydroxide solution, filtered, decolorized, concentrated and dried. Concentrates of hydrolysates, containing 42% solids, 18% NaCl and 3% N_2 can be obtained.

Biochemicals

S. cerevisiae is a good and economical source of several biochemicals which are in great demand by biochemists and molecular biologists. *S. cerevisiae* is a source of the pyridine nucleotides NAD and NADP and their reduced forms and of several ribonucleotides and deoxyribonucleotides, including AMP, ADP and ATP.

Heterologous Gene Expression in *S. cerevisiae*

With the advent of the recombinant DNA technology a major application has been to extract a desired plant or mammalian gene and introduce it into a suitable microorganism with the aid of a vector for expression. As microorganisms can be grown to huge numbers in a short period of time in fermenters, copious amounts of the 'foreign' gene product can be produced. The Human Genome Project makes great use of *S. cerevisiae* and its vectors.

Future Developments

Yeast has come to occupy a unique place in science and technology: being a unicellular microorganism with a short life span, it is readily amenable to cultivation and to manipulation to reflect process needs. It is also amenable to traditional and modern methods of genetic engineering, using its natural recombination processes as well as in vitro techniques. Yeasts are eukaryotes and their biochemistry has much in common with that of higher organisms, including glycosylation and cell sorting. Yeasts are therefore poised to be major players in biotechnology in the future.

See also: **Fermentation (Industrial)**: Basic Considerations; Media for Industrial Fermentations. **Genetic Engineering**: Modification of Yeast and Moulds. ***Saccharomyces***: *Saccharomyces cerevisiae*; *Saccharomyces carlsbergensis* (Brewer's Yeast). **Single Cell Protein**: Yeasts and Bacteria. **Wines**: Microbiology of Wine-making.

Further Reading

Chapman JW (1991) *Trends in Food Science and Technology*. P. 176. Elsevier Sci. Pub. Ltd (UK).

Collar C (1996) *Food Sci. Technol. Int.* 2(6): 349–367.

Doran PM (1995) *Bioprocess Engineering Principles*. London: Academic Press.

Edelmann K, Stelwagen P and Oura E (1980) In: Stewart G and Russell I (eds) *Current Develop. Yeast Research*. P. 51. Toronto: Pergamen Press.

Oura E, Soumalainen H and Viskari R (1982) In: Rose AH (ed) *Economic Microbiology*. P. 87. New York: Academic Press.

Peppler HJ (1979) In: Peppler HJ and Perlman D (eds) *Microbial Technology*, 2nd edn., vol 1. P. 157. New York: Academic Press Inc.

Reed G (1982) In: Reed G (ed) *Prescott & Dunn's Industrial Microbiology*, 4th edn. P. 593. Westport: AVI Publishing.

Reed G and Peppler HJ (1973) *Yeast Technology*. Westpor: AVI Publishing.

Rose AH and Harrison JS (eds) (1993) *The Yeast*, 2nd edn., vol 5. *Yeast Technology*. London: Academic Press.

Sato T (1966) *Bakers' Yeast*. Tokyo: Korin-Shoin Pub.

Trivedi NB and Jacobson G (1986) In: Adams MR (ed) *Progress in Industrial Microbiology*. P. 45. Amsterdam: Elsevier.

White J (1954) *Yeast Technology*. London: Chapman & Hall.

YERSINIA

Contents

Introduction

Peter Kämpfer Institut für Angewandte Mikrobiologie, Justus-Liebig-Universität Giessen, Germany

Introduction

Bacteria of the genus *Yersinia* belong to the family Enterobacteriaceae and can be isolated from various habitats (live and inanimate). Some species are adapted to specific hosts. Based on 16S rRNA sequence analysis, the genus *Yersinia* forms a distinct group within the family Enterobacteriaceae. Genus-specific signature nucleotides within the 16S rDNA allow differentiation from all other genera of this family. Eleven species and five biogroups are recognized, among them the phenotypically homogeneous species *Y. pestis*, the causal agent of plague, and *Y. pseudotuberculosis*, which is an important cause of sporadic zoonotic disease, including pseudotuberculosis. *Y. enterocolitica* and *Y. enterocolitica*-like strains represent a phenotypically heterogeneous assembly of strains, isolated from human clinical specimens, animals, environmental samples or various foods. They have been shown to represent several distinct species on the basis of DNA–DNA hybridization studies (*Y. enterocolitica sensu stricto*, *Y. intermedia*, *Y. mollaretii*, *Y. bercovieri* and *Y. frederiksenii*). At present five biogroups are still recognized within *Y. enterocolitica*. In further taxonomic studies, *Y. ruckeri* was described for organisms causing an inflammation around the head and mouth of rainbow trouts (redmouth bacterium), *Y. aldovae*, a species mainly from aquatic environments, and *Y. rohdei* isolated from water as well as from human and dog faeces. Beside *Y. pestis*, the species *Y. pseudotuberculosis* and *Y. enterocolitica* (*sensu stricto*) cause the overwhelming majority of human infections. *Y. enterocolitica* is primarily a gastrointestinal tract pathogen acquired through the oral route and epidemiologically linked to porcine sources. It is the species of major importance in food microbiology.

Characteristics of the Genus

The genus *Yersinia* contains Gram-negative, facultatively anaerobic, asporogenous oxidase-negative, catalase-positive straight rods (sometimes coccobacilli), 0.5 μm in diameter and 1–3 μm in length, having a DNA G+C content of 46–50 mol%. Pleomorphism is reported and depends on the type of medium used and the temperature of incubation. The enterobacterial common antigen is present in all species. With the exception of *Y. pestis*, from which the occurrence of a cell envelope is reported, *Yersinia* produces no capsules. *Yersinia* spp. are non-motile at 37°C, but motile with peritrichous flagella when grown at 30°C. (*Y. pestis* is always non-motile.) They have an optimum temperature of 28–29°C; most strains have a minimum at 4–5°C and a maximum of 45°C; however, growth at 0–1°C has been observed for *Y. enterocolitica* strains isolated from pork and chicken, milk and raw beef. *Y. pestis* and *Y. enterocolitica* grow over a pH range of 5.0–9.4, and the other *Yersinia* spp. tolerate a pH range of 4.0–10.0. *Y. pestis* and *Y. pseudotuberculosis* grow at salt concentrations up to 3.5%, while the other species can tolerate up to 5% NaCl. In composition and their fine structure of cell walls they do not significantly differ from other Enterobacteriaceae. Lipopolysaccharides (O antigens) have been isolated and characterized. *Yersinia* species grow well on nutrient agar, without any enrichment procedure. In contrast to most other members of the family, often a small colony diameter (1.0–1.5 mm after 24–30 h, and 2.0–3.0 mm after 48 h) differentiates members of the genus from other Enterobacteriaceae. The colonies are translucent, smooth and round with irregular edges. For *Y. pestis* and *Y. pseudotuberculosis* variable (often no) growth responses on MacConkey agar are reported. For all the other species this medium supports growth and the colonies reach a size similar to that observed on nutrient agar. *Y. pestis* hardly grows at all on *Salmonella–Shigella* agar incubated at 25°C, whereas all the other species grow and produce small colonies in 24–30 h.

In their general metabolism, *Yersinia* spp. do not differ significantly from other members of the family Enterobacteriaceae. Most of the species ferment carbohydrates (including glucose) without or with only trace gas production. Some biochemical activities

Table 1 Differentiation of *Yersinia* species (Wauters et al, 1988)

Test			Y. enterocolitica									
	Y. aldovae	Y. bercorvieri	*Biogroups 1–4*	*Biogroup 5*	Y. frederiksenii	Y. intermedia	Y. kristensenii	Y. mollarettii	Y. pestis	Y. pseudo-tuberculosis	Y. rohdei	Y. ruckeri
Indole production	−	−	+/−	−	+	+	+/−	−	−	−	−	−
Voges–Proskauer test	+	−	+/−	+	+/−	+	−	−	−	−	−	−
Citrate (Simmons)	+/−	−	−	−	+/−	+	−	−	−	−	+	−
L-Ornithine decarboxylase	+	+	+	+/−	+	+	+	+	−	−	+	+
Mucate (acid)	+/−	+	−	−	+/−	+/−	−	+	+/−	−	−	−
Pyrazinamidase	+	+	+/−	−	+	+	+	+	−	−	+	+/−
Acid from cellobiose	−	+	+	+	+	+	+	+	−	−	+	−
L-Fucose	+/−	+	+/−	−	+	+/−	+/−	−	+/−	−	+/−	+/−
Melibiose	−	−	−	−	−	+	−	−	+/−	+	+/−	−
L-Rhamnose	+	−	−	−	+	+	−	−	−	+	−	−
L-Sorbose	−	−	+/−	+/−	+	+	+	+	−	−	+/−	+/−
Sucrose	−	+	+	+/−	+	+	−	+	−	−	+	−

are temperature-dependent (e.g. cellobiose and raffinose fermentation, ornithine decarboxylase reaction, *o*-nitrophenyl-β-D-galactopyranoside (ONPG) hydrolysis, indole production, and the Voges–Proskauer reaction). These reactions are more constantly expressed at the optimum temperature (28–29°C) rather than at 37°C. *Yersinia* spp. are neither haemolytic nor proteolytic, with the exception of *Y. ruckeri*, for which gelatin liquefaction is reported and some strains of *Y. pestis*, which have fibrinolytic and coagulase activity.

Y. pestis, the aetiological agent of plague, is considered to be an infectious parasite of rodents. A definite proof of a saprophytic existence of the species in soil has not been obtained, although some reports support this. *Y. pseudotuberculosis* is widespread in mammalian and avian hosts and is a common causal agent of zootic disease in avian and mammalian species. Members of the remaining species, listed in **Table 1**, among them *Y. enterocolitica*, are ubiquitous, free-living saprophytes which can be isolated from soil, water and various foods, and which may be sources for warm-blooded animals.

Beside *Y. pestis*, pathogenesis is best studied for *Y. pseudotuberculosis* and *Y. enterocolitica*. Experimental intraduodenal inoculation of rabbits with *Y. enterocolitica* has shown that pathogenic strains (serogroups O:3, O:9) penetrated the epithelial linings of the intestinal mucosa, and entered the reticuloendothelial tissue. Subsequently intracellular multiplication within mononuclear cells could be observed. The resultant enterocolitis could be characterized by small focal ulcerations in the mucous membrane and granulomatous lesions (pseudotubercles) in the mesenteric lymph nodes, liver and spleen. In severe infections, the granulomas in lymph follicles underwent necrosis and ulceration. Based on these observations, i.e. the penetration through epithelial linings and subsequent multiplication in reticuloendothelial tissues, *Y. enterocolitica* can be considered as an invasive enteropathogenic species like *Salmonella* and *Shigella* and some *Escherichia coli*. For *Y. pseudotuberculosis* it was also shown that, as with *Y. enterocolitica*, pathogenesis is intimately associated with penetration of epithelial linings, survival and multiplication within host cells. Although experimental enteral infections are nearly coincident for both species, for *Y. enterocolitica* this capability is apparently restricted to certain serogroups. *Y. pseudotuberculosis* serotypes III and V are rarely encountered in human infections, but are pathogenic for rabbits and invaded, survived and multiplied within HeLa cells and macrophages in vitro.

For *Y. pseudotuberculosis* and *Y. enterocolitica* the initial event subsequent to ingestion of contaminated foods is invasion of intestinal epithelial cells with tracking to the lamina propria. Having entered the lymphatic system, macrophages are invaded but the microorganisms are still able to survive. A systemic spread is then possible to the liver, spleen and mesenteric lymph nodes.

The promotion of the epithelial cell invasion is connected with the presence of chromosomal genes termed the *inv* and *ail* loci in *Y. pseudotuberculosis* and *Y. enterocolitica*. *Inv* encodes for a 103 kDa protein, called invasin, located on the outer surface of the bacterium. The *ail* (attachment invasion locus) accounts for host specificity with regard to cultured cell lines, and seems to be uniquely associated with strains of serotypes causing human disease. In addition, the presence of plasmid-encoding determinants that play a major role in the overall virulence of pathogenic *Yersinia* is of major importance. For *Y. enterocolitica* it could be shown that both plasmid-bearing and plasmid-free strains of serogroup O:3 were able to penetrate the intestinal mucosa. But only the virulence plasmid-containing strains were subsequently capable of proliferating and surviving in the host tissue. These data advance the concept that plasmid-encoded surface constituents synthesized intracellularly in an environment of low calcium content, subsequent to epithelial cell penetration, act as either antiphagocytic factors, e.g. V (protein) antigen, W (lipoprotein) antigen, outer membrane proteins, or prevent intracellular killing within phagolysosomes.

All pathogenic strains of *Y. enterocolitica*, as well as the pathogenic species *Y. pestis* and *Y. pseudotuberculosis*, possess a 70 kbp plasmid carrying essential virulence genes, e.g. genes for the temperature-inducible (37°C) Yersinia outer membrane proteins (YOP), and the gene for the adhesin YadA. When present, YOP resists opsonization by inhibiting the deposition of complement C3b on the surface of the bacteria in the absence of specific YOP antibodies. The outer membrane proteins are then synthesized and these prevent complement-mediated opsonization and phagocytosis. The surface protein YadA mediates binding to diverse extracellular matrix proteins, adherence to epithelial cell lines, resistance to complement lysis and agglutination, and has been reported to be necessary for mouse virulence.

The virulence plasmid pYV is well conserved among the pathogenic species, although the routes of infection and the observed diseases may be different. However, loss of the pYV plasmid results in the loss of pathogenicity.

Despite the numerous virulence-associated plasmid-encoded attributes (**Table 2**), virulence in *Y.*

Table 2 Determination of virulence in *Yersinia enterocolitica* and *Y. pseudotuberculosis* (after Bottone 1992)

Temperature-dependent antiphagocytic surface factors (expressed at 37°C)
Outer membrane proteins, V antigen, W antigen
Intracellular survival and multiplication
Animal virulence
Mouse lethality, Sereny test
Cytotoxicity for cells of tissue cultures
Other characteristics
Calcium-dependence and autoagglutination (37°C), Congo red binding

enterocolitica and *Y. pseudotuberculosis* seems to be multifactorial.

By defining the two chromosomal loci (*inv* and *ail*), encoding for a protein invasin may in fact account for the true delineation of virulent from avirulent strains, with the plasmid playing an additional role. In *Y. pseudotuberculosis* the presence of invasin seems to be a constant property independent from serotype. Although often postulated, the acquisition of the virulence plasmid by a non-virulent strain of *Y. enterocolitica* need not result in a virulent phenotype if the recipient strain lacks a functional chromosomal locus for the synthesis of invasin. Unless both chromosomal and plasmid-associated virulence genes are co-transferred or coexist, virulence as defined by initial epithelial cell invasion does not exist.

With respect to pathogenicity, a further determinant was detected in *Y. enterocolitica*. A heat-stable (ST) enterotoxin which was more prevalent (90%) among human isolates of *Y. enterocolitica* as compared to *Y. intermedia* isolates (10%) with biological and physical characteristics similar to those produced by enterotoxigenic strains of *E. coli* can be produced during growth below 30°C (but is absent in culture filtrates of cells grown at 37°C). The enterotoxin (9000–9700 Da) survives at 100°C for 20 min and is not affected by proteases and lipases. Similar to the *E. coli* ST, it gives positive responses in suckling mice and rabbit ileal loop assays and negative responses in the CHO (chinese hamster ovary) and Y-1 mouse adrenal cell assays. ST (heat-stable enterotoxin) is methanol-soluble and stimulates guanylate cyclase and the cAMP response in intestines but not adenylate cyclase. Production is favoured in the pH range 7–8. It could be shown that, of 46 milk isolates, only three produced ST in milk at 25°C and none at 4°C. In a synthetic medium, enterotoxin production was favoured by aeration but inhibited by high iron content. Regarding the production of ST by species other than *Y. enterocolitica*, none of 21, 8 and 1 investigated *Y. intermedia*, *Y. frederiksenii* and *Y. aldovae* strains, respectively, was positive in one study of species from raw milk, whereas 62.5% of *Y. enterocolitica* were ST-positive. In two further studies some *Y. intermedia* and *Y. kristensenii* strains were positive for ST production. The pathogenic role of this compound in vivo, where body temperature approaches 37°C, remains questionable. In contrast to *E. coli*, where enterotoxin production is plasmid-encoded, enterotoxin production in *Y. enterocolitica* is chromosomally encoded. Perhaps it can be argued that *Y. enterocolitica* enterotoxin production is not significant in inducing diarrhoeal disease but toxin produced in food products, often incubated at lower temperatures, is active. It has been shown, however, that enterotoxin production may be repressed at refrigeration temperature (4°C), and that enterotoxin-negative strains of *Y. enterocolitica* O:3 can still induce diarrhoeal illness. Enterotoxin production has not been detected in *Y. pseudotuberculosis*.

Table 3 Selection of common and rarely observed clinical manifestations of *Yersinia enterocolitica* and *Y. pseudotuberculosis* infections (after Bottone 1992)

Often found
Enterocolitis (young children and old people)
Pseudoappendicitis (acute mesenteric lymphadenitis, terminal ileitis)
Acute typhoid-like septicaemia
Non-suppurative sequelae (erythema nodosum, arthritis, myocarditis, glomerulonephritis)
Rarely encountered
Cutaneous infections (cellulitis, wound infections, pustules)
Focal abscesses (liver, spleen, kidney)
Septicaemia (especially in immunosuppressed patients)
Pneumonia
Meningitis

The main clinical manifestations of human infection with *Y. pseudotuberculosis* and *Y. enterocolitica* are shown in **Table 3**. Although very similar, there are clear distinctions in frequency, age- and sex-related attack rate, pathology, diagnosis and epidemiology.

Within the genus *Yersinia*, siderophore production has only been described for *Y. pestis*, not for *Y. enterocolitica* and *Y. pseudotuberculosis*. Despite this, however, these species may use iron through low-affinity iron assimilation systems, or by transport across their membrane iron complexed in siderophores produced by other microbial species and both species are able to use desferroxamine B, a siderophore used therapeutically to control iron overload, for in vitro and in vivo growth. It has been reported that systemic infection with *Y. enterocolitica* and *Y. pseudotuberculosis* often occurs in patients with iron overload.

The host adaptation of *Y. enterocolitica* seems to be still underway, in contrast to its more staid counterpart *Y. pseudotuberculosis*. There seems to be a

Table 4 Differentiation of *Yersinia enterocolitica* biogroups including biogroups 3A and 3B, now described as *Y. mollaretii* and *Y. bercovieri*, respectively (after Bottone 1992)

Test	*Biogroup*							
	1A	*1B*	*2*	*3*	*3A*	*3B*	*4*	*5*
Indole production	+	+	–	–	–	–	–	–
Esculin hydrolysis	+/–	–	–	–	–	+	–	–
Voges–Proskauer test	+	+	+	+	–	–	+	+
L-Ornithine decarboxylase	+	+	+	+	–	–	+	+
Pyrazinamidase	+	–	–	–	+	+	–	–
Lipase	+	+	–	–	–	–	–	–
Nitrate reduction	+	+	+	+	+	+	+	–
Acid from:								
Inositol	+	+	+	+	+/–	–	+	+
Salicin	+	–	–	+	+/–	+/–	–	–
L-Sorbose	+	+	+	+	+	–	+	–
Trehalose	+	+	+	+	+	+	+	–
D-Xylose	+	+	+	+	+	+	–	+/–

close correlation between the serogroups of *Y. enterocolitica*, its biogroup designation and its ecological and pathogenic behaviour. The physiological heterogeneity of typical *Y. enterocolitica* isolates was the reason for devising several biotyping schemes. The importance of a stable biochemical typing scheme in this context is that it allows for the recognition of potentially human pathogenic strains which possess the numerous plasmid and chromosomally encoded virulence traits. In addition, one can couple this to serogrouping of strains comprising the various biogroups (**Table 4**).

In general, the antigenic structure of *Yersinia* spp. is complex. Some antigens are shared by *Y. pestis*, *Y. pseudotuberculosis* and *Y. enterocolitica*. *Y. pestis* produces the fraction 1 envelope antigen (F1) when cultures are incubated at 37°C on protein-rich media. This antigen is characterized as heat-labile (10 min at 100°C) and water-soluble. It contains a carbohydrate protein (F1A) and a carbohydrate-free protein (F1B). Passive haemagglutination with F1 antigen is used for serological surveys in plague foci. In *Y. pseudotuberculosis*, this antigen has also been demonstrated. Production of plasmid-encoded V and W antigens have been described in *Y. pestis*, *Y. pseudotuberculosis* and in *Y. enterocolitica*. The somatic antigen of *Y. pestis* is rough (R antigen) and therefore no serogroups have been described in this species. This R antigen is also detected in *Y. pseudotuberculosis*. Both species share at least 11 out of 18 antigens. *Y. pestis* and *Y. enterocolitica* also express some common protein antigens. The antigenic scheme for *Y. pseudotuberculosis* is based on six main thermostable serogroups (I–VI) with subgroups (A, B, 2–15) and five thermolabile flagellar H antigens (a–e). Antigenic relationships are recognized between *Y. pseudotuberculosis* (serogroups II, IV, IVA and VI) and *Salmonella* serogroups B and D, *E. coli* serogroups 017, 055 and 077, and *Enterobacter cloacae*. For *Y. enterocolitica* 34 different O antigen and 20 H antigen serogroups were described (including some serogroups defined by strains belonging to *Y. intermedia* (017) and *Y. kristensenii* (011, 012, 028)).

The main serogroups and their distribution among the *Y. enterocolitica* biogroups are summarized in **Table 5**.

The important human pathogenic strains are mainly included among biogroup 2 (serovar O:9; O:5, 27), biogroups 4 (serovar O:3) and 5 (serovar O:2,3); biogroup 3 contains two human pathogenic serovars (O:1,2,3; and O:5,27). Biogroups 3A and 3B, now described as *Y. mollaretii* and *Y. bercovieri*, show numerous serogroups; however, they have not been found as human pathogens.

The most problematic isolates are those comprising biogroup 1, which contains several human pathogens (e.g. serovar O:8) and numerous non-pathogenic isolates. Although esculin hydrolysis and salicin fermentation can be useful to differentiate between non-pathogenic environmental isolates (positive) from potential pathogens (negative; Table 4), the introduction of tests for pyrazinamidase activity has been used to delineate biogroup 1 isolates into avirulent 1A strains (positive) and virulent 1B strains (negative for all three characteristics). In this schemata, biogroup 1 strains of pathogenic serovar O:4; O:8; O:13a, 13b; O:18, O:20 and O:21 are clearly identified (Table 5). Strains comprising the latter serovars have been incriminated in human infections, especially in the US, and are known as the American strains. Given the considerable number of reports about the isolation of biogroup 1A strains from clinical specimens, one can suppose that these strains possess a certain pathogenic potential. They may

Table 5 Compilation of pathogenic potential *Yersinia enterocolitica* biogroups, serogroups and distribution, including biogroups 3A and 3B, now described as *Y. mollaretii* and *Y. bercovieri*, respectively (after Bottone 1992)

	Biogroup							
	1A	*1B*	*2*	*3*	*3A*	*3B*	*4*	*5*
Associated with human infections	(see text)	+	+	+	–	–	+	+
Serogroups	O:5; O:6,30; O:7,8 and further ones	O:8; O:4; O:13a,13b; O:18; O:20; O:21	O:9; O:5,27	O:1,2,3; O:5,27	Numerous	Numerous	O:3	O:2,3
Distribution	Environment Pigs	Environment, mainly in the US	Pigs	Chinchilla	Animal and human faeces	Animal and human faeces	Pigs	Hare

induce clinical symptoms similar to those produced by enteropathogenic *E. coli*. One potential virulence factor of a biogroup 1A strain was an enterotoxin in pathogenic O:3 strains. An investigation of 111 strains of *Y. enterocolitica* biogroup 1A for virulence-associated characters by DNA hybridization studies resulted in the detection of the chromosomal virulence-associated genes *ail* (7.2%), in addition to some others. The *Yersinia* virulence plasmid pYV could not be detected in any strain but about 30% of the strains produced an enterotoxin which was reactive in infant mice.

Throughout the world, the most commonly occurring *Y. enterocolitica* serovars (serotypes) in human infections are O:3; O:5, 27; O:8 and O:9. The majority of pathogenic strains in the US are O:8 (biogroup 2 and 3) and, except for occasional isolations in Canada, it is rarely reported from other continents. In Canada, Africa, Europe and Japan, serovar O:3 (biogroup 4) is reported to be the most common. The second most common in Europe and Africa is O:9, which has also been reported from Japan. Serovar O:3 (biogroup 4, phage type 9b) was practically the only type found in the province of Quebec, Canada. The next most common were O:5,27 and O:6,30. From all *Y. enterocolitica* human infections reported in Canada, O:3 represented 85% of 256 isolates, whereas for non-human sources, O:5,27 represented 27% of 22 isolates.

Susceptibility to the following antimicrobial agents has been reported for *Yersinia* species: tetracycline, chloramphenicol, aminoglycosides (streptomycin, gentamicin, kanamycin and neomycin), sulphonamides (alone or in combination with trimethoprim) and nalidixic acid. To a certain degree, *Yersinia* is susceptible to colistin and resistant to erythromycin and novobiocin. *Y. pestis* and *Y. pseudotuberculosis* are usually susceptible to β-lactam antibiotics but their susceptibility to penicillin is in the range of sensitive to intermediate. Resistance to ampicillin and to streptomycin has been described for *Y. enterocolitica*, *Y. intermedia* and *Y. pseudotuberculosis*. *Y. frederiksenii* and *Y. kristensenii* are resistant to penicillin and slightly susceptible or resistant to other β-lactam antibiotics (ampicillin, carbenicillin, cephalothin). *Y. enterocolitica* strains that are resistant to tetracycline, chloramphenicol, streptomycin and kanamycin have been reported.

Characteristics of Species

Yersinia pestis is the cause of plague. Although plague is primarily described as a disease of wild rodents, *Y. pestis* can be transmitted by fleas in which the bacteria multiply and block the oesophagus and the pharynx. The fleas can further transmit the organisms to humans when they take their next blood meal. After a bite by an infective flea, the typical bubonic form of plague is produced in humans. The multiplication of *Y. pestis* proceeds intracellularly in the host and by the lymphatic system. The lymph nodes then become rapidly enlarged when they are inflamed (bubos) and the infection process is usually so rapid that no characteristic lesions are found in the spleen or liver at autopsy. The disease evolves in 5–10 days to septicaemia and sometimes to a secondary pneumonia. From pneumonia, droplets can be transmitted from human to human (pneumonic plague).

Six factors are described in association with the virulence of *Y. pestis*:

- F1 antigen production
- V,W antigen productions
- a pigment (incorporation of Congo red dye or hemin)
- Pesticin I production (a bacteriocin active on *Y. pseudotuberculosis*)
- a toxin (murine toxin, whose activity is not clearly established)
- the ability to synthesize purines.

For mice, the LD_{50} dose inoculated with virulent strains is 1–10 cells. In avirulent strains of *Y. pestis* the production of V,W antigens is not observed (with the exception of the vaccine strain EV7). The lack of the other virulence factors does not completely diminish the virulence of *Y. pestis* strains.

Y. pseudotuberculosis is also a human and animal pathogen and described as the causal agent for epizootics in nearly all animal species, especially in rodents. The contamination occurs in most cases by the oral route. After 1–2 weeks of incubation, the bacteria are detectable in the mesenteric lymph node. Chronic diarrhoea, mesenteric lymphadenitis and septicaemia are observed as the main symptoms. Humans can be orally contaminated by animals. The incidence of this infectious disease varies with the season and is highest during the cold seasons. This is due to the fact that all *Yersinia* spp. multiply even at low temperatures (4°C) and therefore have a selective advantage over other bacteria. Like *Y. pestis*, *Y. pseudotuberculosis* is an intracellular parasite.

Y. enterocolitica is distributed worldwide in the terrestrial and aquatic environment, which can be sources of the organisms to warm-blooded animals. The species is also found among human isolates, more often than the other species shown in Table 1 (except *Y. pestis* and *Y. pseudotuberculosis*). Animals from which *Y. enterocolitica* has been isolated include beavers, birds, camels, cats, cattle, chickens, chinchillas, deer, fish, dogs, guinea pigs, horses, lambs, oysters, rats, raccoons and swine. Swine are an important reservoir for food-borne infections of humans. Of 43 samples of pork obtained from a slaughterhouse and examined for *Y. enterocolitica*, *Y. intermedia*, *Y. kristensenii* and *Y. frederiksenii*, 8 were positive and all four species were found.

The different biogroups are also ubiquitous, having been found in a wide range of animal and environmental sources. Some biogroups and serovars, however, are frequently associated with a specific host. Biogroup 5 strains have mainly been found from hares in Europe; biogroup 4, serovar O:3 strains and biogroup 3, serovar O:5,27 strains are often found to be responsible for human gastrointestinal infections in Europe, Canada and the republic of South Africa. *Y. enterocolitica* serogroup O:8 strains are often associated with various syndromes in the US. *Y. enterocolitica* is the species of major importance with respect to incidence of the genus *Yersinia* in foods. It has been found in cakes, milk, vacuum-packaged meats, seafood, vegetables and other food products. *Y. enterocolitica* has also been isolated from beef, lamb and pork.

Y. intermedia and *Y. frederiksenii* have been identified in Europe, the US, Australia and New Zealand, Israel and Japan. These two species have mainly been isolated from fresh water and foods and only rarely from non-irrigated soil, humans or animals other than fish. For *Y. frederiksenii* two genomic species have been described which are phenotypically indistinguishable. *Y. intermedia* has been detected in raw milk. *Y. kristensenii* has been found in Europe, the US, Japan and Australia. Strains of this species have mainly been isolated from soil, foods and asymptomatic animals; isolates from other environmental sources and from human infections are rare. Like *Y. enterocolitica*, *Y. kristensenii* produces a heat-stable enterotoxin.

Y. mollaretii and *Y. bercovieri* were first isolated from a terrestrial ecosystem and were initially described as *Y. enterocolitica* biogroups 3A and 3B, respectively. Most strains were isolated from environmental sources and foods (raw vegetables), but some were of human origin, mainly from stools of both healthy individuals and patients with diarrhoea. There is no evidence that these organisms are pathogenic for humans. Their frequent occurrence in soil, water, food and environmental samples suggests a saprophytic character. This is supported by a lack of virulence factors. Phenotypically, these two species most closely resemble each other and are next closest to *Y. enterocolitica* biogroups 3 and 4 (Tables 1 and 4). As in other *Yersinia* spp., higher metabolic activity was evident at 25°C than at 36°C, especially in tests for ornithine decarboxylase, and acid production from mucate, when reactions were sometimes delayed or negative at 36°C. In some strains of *Y. mollaretii* and *Y. bercovieri*, O antigens identical to those of *Y. enterocolitica* have been detected. Similar or identical antigens have previously been described for *Y. enterocolitica*, *Y. intermedia*, *Y. frederiksenii*, *Y. kristensenii*, *Y. aldovae* and *Y. rohdei*.

Y. rohdei was isolated from the faeces of dogs and humans, but also from surface water. At present it is not known whether *Y. rohdei* was the cause of human infections, but it was suggested that the natural habitat is water and that isolation from dogs and humans is only occasional. No reports on isolation from foods are available at present.

Like *Y. rohdei*, *Y. aldovae* is a species found in aquatic habitats. No human pathogenicity has been documented for this species either.

Y. ruckeri has been detected in freshwater ecosystems only in the US and Canada. The red mouth disease caused by this organism only appears when fish are exposed to large numbers of bacteria. The disease is usually enzootic and is occasionally epizootic in fish hatcheries.

General Aspects of Detection and Isolation Procedures

From non-contaminated samples (blood, lymph nodes) *Yersinia* strains can be isolated by using blood agar or nutrient agar incubated for 48 h at 28°C, or 24 h at 37°C followed by 24 h at room temperature. The isolation of *Y. pestis* from contaminated samples requires inoculation (subcutaneously or per-cutaneously) of animals (guinea pigs, mice or rats). The organism can be cultured from the spleen, liver or lymphatic nodes of the inoculated animals. All other *Yersinia* spp. will usually be isolated from stools or food samples by inoculating standard or special selective bile-salt media such as MacConkey agar, DCL (deoxycholate) agar, *Salmonella–Shigella* agar, cellobiose, arginine and lysine (CAL) medium, CIN (cefsulodin-irgasan novobiocin) medium, YVE (virulent *Y. enterocolitica*) agar, modified Rimler–Shotts agar and oxalate medium. The ability of *Y. enterocolitica*, *Y. pseudotuberculosis* and *Y. pestis* to digest polypectate (polygalacturonic acid) incorporated into an agar medium produced depressed colonies. Several formulations of the pectin agars have been devised. The addition of cellobiose, arginine and lysine (CAL), along with other ingredients to an agar base, is often used to enhance recovery. CIN agar was developed for the direct discrimination of *Y. enterocolitica* from most other Gram-negative species capable of growth on this medium. This agar contained cefsulodin (4 or 15 mg l^{-1}), irgasan (2,4,4′-trichloro-2′-hydroxy diphenyl ether; 4 mg l^{-1}) and novobiocin 2.5 mg l^{-1}), and the differential qualities derive from the ability of *Y. enterocolitica* to produce acid from mannitol, resulting in a deep red centre to the flat yersinial colonies. CIN agar is also effective in the isolation of *Y. pseudotuberculosis*. The incorporation of sodium oxalate into a medium consisting of casein hydrolysate, peptone, bile salts and lactose can be selective and differential for the recovery of *Y. enterocolitica*.

Preferably all these media should be incubated for 48 h at 28–29°C or for 24 h at 37°C, followed by 48–72 h at room temperature.

For maintenance, stab inoculations of *Yersinia* strains can be stored in the dark at 4°C for a long time. Lyophilization and deep-freeze storage in 10% glycerol are also recommended. If the virulence of a strain should be retained, it should never be subcultured at 37°C, but always at 25–28°C.

Yersinia spp. are distinct from other members of the family Enterobacteriaceae because of the poor growth response on enteric agar after 24 h incubation at both 22 and 37°C. Under both incubation conditions, colony sizes range from barely perceptible to pinpoint.

The recovery of *Yersinia* spp. from foods usually requires enrichment of the sample before cultivation on agar media. Several enrichment or selective techniques have been applied for the recovery, especially of *Y. enterocolitica* from meats and other foods. Enrichment broths can use three selective agents – irgasan, ticarcillin and potassium chlorate (ITC). In combination with coupling to direct plating on to SS-deoxycholate calcium agar, MacConkey agar, or to CIN agar, this may be superior for the recovery of *Yersinia* strains. Various cold enrichment techniques improve the recovery of *Yersinia* strains from contaminated samples. Furthermore, subsequent post-alkali treatment was reported to enhance isolation efficacy.

The development of DNA hybridization techniques combined with the application of polymerase chain reaction (PCR) are common techniques to detect pathogenic bacteria. With respect to food control, DNA techniques have most often been used to confirm culture-based techniques, i.e. bacteria are enriched and sometimes even purified by traditional culture procedures and thereafter identified by the use of DNA-based methods. With respect to the genus *Yersinia*, a genus-specific signature sequence within the 16S rRNA has been detected, which is promising for application in food microbiology. To detect non-pathogenic or pathogenic *Y. enterocolitica* strains, several approaches have been made. A PCR-amplified, digoxigenin-labelled yst probe (specific for pathogenic *Y. enterocolitica* heat-stable enterotoxin yst gene) was used for the detection and differentiation of *Y. enterocolitica* in naturally and artificially contaminated pork and milk samples following enrichment in ITC enrichment broth. Overall, the hybridization results were in good agreement when compared with those from standard biochemical and serological tests. The lowest concentration detectable by a colony hybridization technique in the milk samples inoculated with *Y. enterocolitica* O:8 PA-A corresponded to 10 cfu ml^{-1} in the original sample.

In a further study PCR was used for the detection of pathogenic *Y. enterocolitica*. Primers were selected for nested PCR directed at the attachment invasion locus, *ail*, on the bacterial chromosome, as well as at a sequence on the pathogenic marker plasmid, termed virulence factor, virF. Some advantages, i.e. higher sensitivity and specificity of the PCR method, compared with the conventional method were demonstrated. Another approach was the design of a 24-base oligonucleotide probe specific for a region of the *Y. enterocolitica* virulence plasmid. This probe, which is highly specific and sensitive for virulent *Yersinia*, detected pathogenic *Y. enterocolitica* isolates in foods which were artificially inoculated. The most desirable approach is, however, to detect organisms directly in the food, but at present several problems remain to be solved before this can be routinely performed.

See also: **Yersinia**: *Yersinia enterocolitica*.

Further Reading

Bottone EJ (1992) The genus *Yersinia* (excluding *Y. pestis*). In: Balows A, Trüper HG, Dworkin M, Harder W and Schleifer K-H (eds) *The Prokaryotes*. P. 2863. New York: Springer-Verlag.

Bottone EJ (1997) *Yersinia enterocolitica*: the charisma continues. *Clin. Microbiol. Rev.* 10: 257–276.

Jay MT (1996) *Modern Food Microbiology*. New York: Chapman & Hall.

Olsen JE, Aaabo S, Hill W et al (1995) Probes and polymerase chain reaction for detection of food-borne bacterial pathogens. *Int. J. Food Microbiol.* 28: 1–78.

Wauters G, Janssens M, Steigerwalt AG and Brenner J (1988) *Yersinia mollarettii* sp. nov. and *Yersinia bercovieri* sp. nov., formerly called *Yersinia enterocolitica* Biogroups 3A and 3B. *International Journal of Systematic Bacteriology*. 38: 424–429.

Yersinia enterocolitica

Saumya Bhaduri, Microbial Food Safety Research Unit, Agricultural Research Service, Wyndmoor, US

Yersinia enterocolitica is a relatively large, Gram-negative, cold-tolerant, facultatively anaerobic coccobacillus belonging to the genus *Yersinia* in the Enterobacteriaceae family. The organism is recognized as a food-borne pathogen and a large number of food-associated outbreaks of yersiniosis have been reported. In developed countries, *Y. enterocolitica* can be isolated from 1–2% of all human cases of acute enteritis. Since *Yersinia* can grow at low temperatures, refrigerated foods are potential vehicles for the growth of these organisms.

There is considerable confusion in the literature because not all *Y. enterocolitica* strains can cause intestinal infections. Unlike intrinsic pathogens, such as *Shigella* and *Salmonella*, strain-to-strain variation has been observed in the pathogenicity of *Y. enterocolitica*. The pathogenicity of *Y. enterocolitica* is correlated with the presence of a 70–75 kbp plasmid which is directly involved with the virulence of the organism. A number of temperature-dependent phenotypic characteristics, including mouse virulence, have been associated with the virulence plasmid and have been used to differentiate between virulent and avirulent strains of *Y. enterocolitica*. However, the physiological traits associated with the virulence plasmid are only expressed at 37°C, which also fosters the loss of the virulence plasmid and the concomitant disappearance of the associated phenotypic virulence characteristics. The plasmid is stable in cells maintained at 25–28°C. Because of the instability of the virulence plasmid at 37°C, it is difficult to isolate plasmid-bearing virulent strains after initial detection. As a consequence, detection has been hampered in clinical, regulatory and quality control laboratories that employ an incubation temperature of 37°C for isolation/detection of the organism. For example, such difficulties were reported in 1992 by the California Health Department and US Food and Drug Administration (FDA) in cases of yersiniosis in Los Angeles County, CA. The instability of the plasmid can lead to confusion concerning whether one is dealing with virulent or non-virulent strains. This chapter describes the application of plasmid-associated virulence determinants of pathogenic *Y. enterocolitica* as a tool for isolation of plasmid-bearing virulent (YEP^+) serotypes from foods.

Detection of YEP^+ Strains

A number of virulence plasmid-mediated phenotypic characteristics, including colony morphology, autoagglutination (AA), serum resistance, tissue culture detachment, hydrophobicity (HP) and low-calcium response (Lcr), have been applied to the determination of virulence in strains of *Y. enterocolitica*. These methods require specific reagents and conditions and do not give clear-cut results. In addition, most of these procedures are costly, time-consuming, complex and impractical for routine diagnostic use, particularly in field laboratories. Although the virulence can be effectively demonstrated using laboratory animals, this test is not suitable for routine diagnostic use. Molecular techniques such as DNA colony hybridization, DNA restriction fragment length polymorphisms and polymerase chain reaction (PCR) have also been successfully applied to the detection of virulent strains. However, these techniques are complex and time-consuming. These methods detect only the presence of a specific gene and not the presence of the organism. Although virulence is plasmid-mediated in all strains examined, the plasmids involved differ in molecular weight. Thus, in epidemiological studies it is not sufficient to search for plasmids of a particular molecular weight as an indicator of *Y. enterocolitica* virulence.

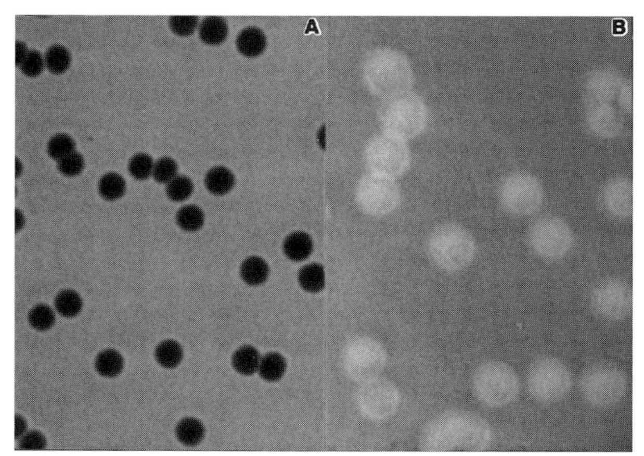

Figure 1 Crystal violet binding of colonies of *Yersinia enterocolitica* grown on brain heart infusion agar for 24 h at 37°C. After incubation, plates were flooded with 100 μg ml^{-1} crystal violet solution. (**A**) YEP$^+$ cells showing small dark-violet colonies. (**B**) YEP$^-$ cells showing large white colonies. (Bhaduri et al. (1987) *J. Clin. Microbiol.* 25: 1039–1042.)

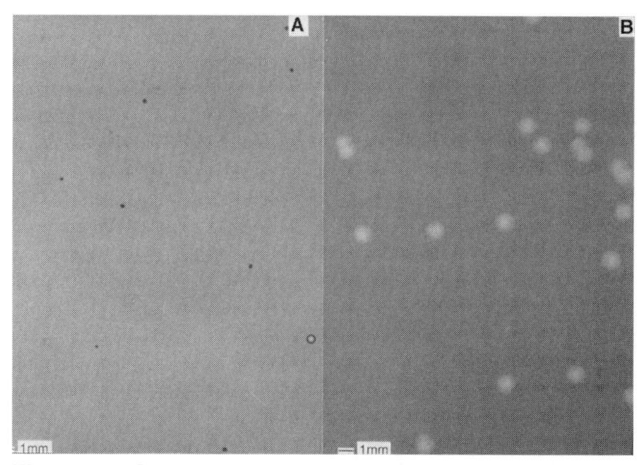

Figure 2 Congo red binding of colonies of *Yersinia enterocolitica* cells grown on Congo red brain heart infusion for 24 h at 37°C. (**A**) YEP$^+$ cells showing pinpoint red colonies. (**B**) YEP$^-$ cells showing large white or light-orange colonies. The concentration of congo red used in the binding assay was 75 μg ml^{-1}. (Bhaduri et al. (1991) *J. Clin. Microbiol.* 29: 2341–2344.)

Dye-binding Techniques for Detection

Virulence determinants provide a rapid, reliable and simple method for the detection and isolation of YEP$^+$ clones of *Y. enterocolitica*. The main disadvantage of virulence determinants for the isolation of YEP$^+$ strains is maintenance of the virulence plasmid since these plasmid-associated phenotypes are only expressed at 37°C, which fosters the loss of the virulence plasmid, resulting in plasmidless avirulent (YEP$^-$) strains. The ability of *Yersinia* to absorb hemin from agar media is correlated with the virulence plasmid. This fact led to the postulate that dye binding may be an indication of the presence of the virulence plasmid in YEP$^+$ strains. Several dyes, including crystal violet (CV), were included in brain–heart infusion agar (BHA) for the detection of YEP$^+$ cells but all of the dyes bound to both virulent and avirulent strains after incubation at 37°C for 24 h. An alternative procedure of flooding pregrown colonies of *Y. enterocolitica* at 37°C with CV solution at a concentration of 100 μg ml^{-1} indicated that YEP$^+$ cells bound CV and produced dark-violet colonies (**Figure 1A**). YEP$^-$ colonies did not bind CV and remained white (**Figure 1B**). The CV flooding and binding assay takes about 3–5 min. This technique quantitatively differentiates YEP$^+$ cells from YEP$^-$ cells from mixed culture with 95% efficiency (**Table 1**). The CV binding was also effective with other serotypes and correlated with mouse virulence and virulence-associated characteristics (**Table 2**).

Even though the CV binding provided a simple and efficient means of screening *Y. enterocolitica* for virulence and identification of individual plasmid-bearing clones, there were disadvantages to this technique: first, it required the extra step of flooding, and second, the CV solution kills the cells in the colonies. Thus, it was not suitable for isolation of viable YEP$^+$ strains. So an alternative approach was taken. Congo red (CR) had been used unsuccessfully to screen *Y. enterocolitica* for virulence, so more specific conditions for the binding of CR to YEP$^+$ cells were evaluated. Because agarose is a purer form of agar, it has been found that its calcium level is lower than in agar. Both agar and agarose were used as gelling agents in brain heart infusion (BHI) to attain low and high levels of calcium in the media. Taking advantage of the non-inhibitory nature of CR on bacterial growth, the dye was added at a concentration of 75 μg ml^{-1} before autoclaving BHI containing either

***Table 1** Efficiency of crystal violet (CV) binding in mixed cultures of virulent and avirulent strains[a]

Sample	*Estimated no. of colonies in the mixture*		*No. (%) of colonies bound to CV*[b]
	Avirulent	*Virulent*	
A	172	0	0
B	141	16	16 (100)
C	131	31	29 (93)
D	85	56	56 (100)
E	72	98	92 (93)
F	53	124	103 (83)
G	22	130	124 (94)
H	0	175	173 (98)

[a]Virulent cells of *Yersinia enterocolitica* were mixed in various ratios with cells from the plasmidless strain and surface-plated on brain heart infusion agar. The mixed colonies were incubated at 37°C for 24 h. The number of virulent colonies was determined by the CV-binding technique at a concentration of 100 μg CV per ml.
[b] Average binding was 94%.
*Bhaduri et al. (1987) *J. Clin. Microbiol.* 25: 1039–1042.

Table 2 Correlation between dye-binding techniques, virulence and virulence-associated properties of original and recovered[a] plasmid-bearing strains of *Yersinia enterocolitica*

Strains	*Serotype*	*CM*[b]	*CV binding*[c]	*Lcr*[d]	*CR binding*[e]	*AA*[f]	*HP*[g]	*Plasmid (70–75 kbp)*	*Diarrhoea in mice*[h]
GER	O:3	+	+	+	+	+	+	+	+
GER-RE	O:3	+	+	+	+	+	+	+	+
GER-C	O:3	–	–	–	–	–	–	–	–
EWMS	O:13	+	+	+	+	+	+	+	+
EWMS-RE	O:13	+	+	+	+	+	+	+	+
EWMS-C	O:13	–	–	–	–	–	–	–	–
PT18-1	0.5, O:27	+	+	+	+	+	+	+	+
PT18-1-RE	0.5, O:27	+	+	+	+	+	+	+	+
PT18-1-C	0.5, O:27	–	–	–	–	–	–	–	–
O:TAC	O:TACOMA	+	+	+	+	+	+	+	+
O:TAC-RE	O:TACOMA	+	+	+	+	+	+	+	+
O:TAC-C	O:TACOMA	–	–	–	–	–	–	–	–
WA	O:8	+	+	+	+	+	+	+	+
WA-RE	O:8	+	+	+	+	+	+	+	+
WA-C	O:8	–	–	–	–	–	–	–	–

[a]Recovered strains are designated as RE.
[b]CM = colony morphology. In a calcium-adequate agar medium (brain–heart infusion agar: BHA) YEP$^+$ cells appeared as small colonies (diameter 1.13 mm) as compared to larger YEP$^-$ colonies (diameter 2.4 mm).
[c]CV binding = crystal violet binding. YEP$^+$ cells appeared as small dark-violet colonies on BHA.
[d]Lcr = low calcium response. Calcium-dependent growth at 37°C: YEP$^+$ cells appeared as pinpoint colonies of 0.36 mm diameter compared to the larger YEP$^-$ colonies of diameter 1.37 mm on low-calcium Congo red brain heart infusion agarose medium (CR-BHO).
[e]CR binding = Congo red binding. YEP$^+$ cells appeared as red pinpoint colonies on CR-BHO.
[f]AA = autoagglutination.
[g]HP = hydrophobicity.
[h]Faecal material consistency was liquid; diarrhoea was observed, starting on days 3 and 4 post-infection.

Table 3 Estimation of calcium in the media to define low-calcium and high-calcium media

Medium	*Calcium concentration in µmol l^{-1a}*	*Medium*
BHO	238 (low)	CR-BHO
BHA	1500 (high)	CR-BHA
BHI broth	245 (low)	BHI broth

[a]The concentration of calcium in each medium was determined by atomic absorption analysis.
BHO = brain heart agarose; BHA = brain heart infusion agar; BHI = brain heart infusion; CR = Congo red.

agarose or agar. The addition of CR in the media did not change the concentration of calcium in the respective medium (**Table 3**). CR media containing low and high calcium levels were used for the determination of CR uptake by YEP$^+$ strains. When YEP$^+$ and YEP$^-$ strains were cultivated at 37°C for 12–24 h on these two media, only CR containing low-calcium BHI agarose medium (CR-BHO) demonstrated two types of readily discernible colonies. The YEP$^+$ cells absorbed CR and formed red pinpoint colonies (CR$^+$; **Fig. 2A**). The YEP$^-$ cells failed to bind the dye and formed much larger white or light-orange colonies (CR$^-$; **Fig. 2B**). The size and colony morphologies of the YEP$^+$ strain in CR-BHO also showed Lcr.

Another characteristic feature of the CR binding technique for YEP$^+$ strains is the appearance of a white opaque circumference around the red centre after 24 h of incubation (**Fig. 3A and 3B**). It is important to note that cells in the red centre contain the virulence plasmid whereas cells in the white surrounding border do not contain the plasmid since

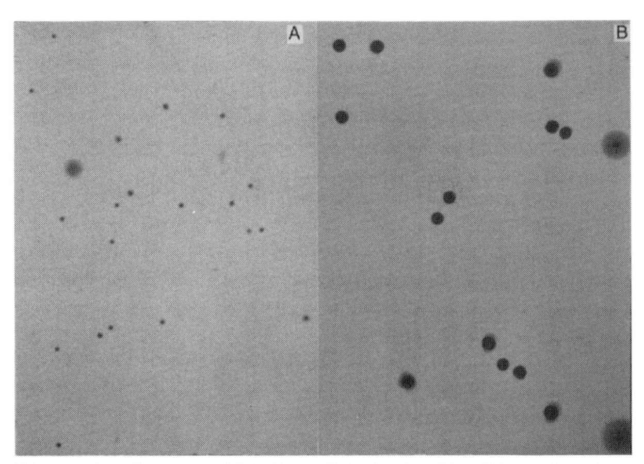

Figure 3 Congo red binding of virulent cells of *Yersinia enterocolitica* grown on Congo red brain heart infusion for 24 and 48 h at 37°C. (**A**) White border around the red centre of the colony after 24 h incubation. (**B**) Wide white border around the red centre of the colony after 48 h incubation. The white border around the red centre of the colony appears darker in the figure because of the red background of the agar plate, red-pigmented centre and reflected light source during photography of the colonies. (Bhaduri et al. (1991) *J. Clin. Microbiol.* 29: 2341–2344.)

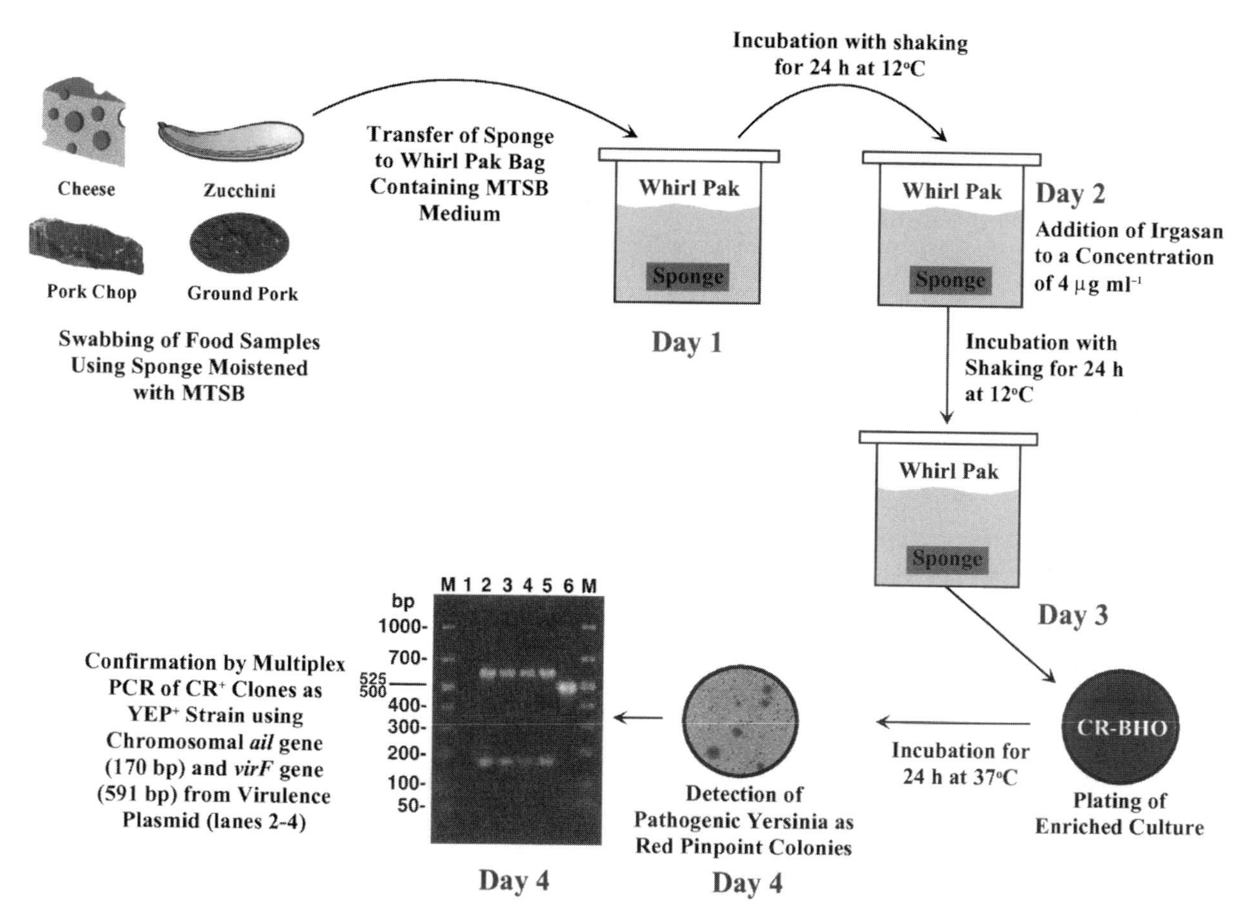

Figure 4 Diagrammatic scheme for direct detection and isolation of pathogenic *Yersinia enterocolitica*. MTSB, modified trypticase soy broth; CR-BHO, low-calcium (238 μm) Congo red brain heart infusion agarose; PCR, polymerase chain reaction.

they have lost the CR binding property. This observation was also confirmed by hybridization with a specific DNA probe. Initially cells do not lose the virulence plasmid during growth at 37°C but do so on prolonged incubation. This colonial characteristic is another parameter that can be used for the identification of YEP+ strains.

Identical results were obtained with other *Y. enterocolitica* serotypes (Table 2). A positive response in the CR binding test was correlated with the presence of the virulence plasmid as well as with a number of virulence-associated properties including mouse virulence (Table 2). The binding of CR to virulent strains consistently allowed the ready differentiation of virulent and avirulent strains of *Y. enterocolitica*. These techniques do not require special equipment and can be used to screen a large number of cultures. These tests allow small and large laboratories to detect a cluster of yersiniosis cases in a short period of time. This assay can effectively detect the presence of virulent YEP+ cells in cultures containing predominantly YEP− cells with 100% efficiency (**Table 4**). Such cultures are not uncommon in clinical laboratories where incubation at 37°C is the standard procedure. This has been demonstrated by an FDA investigation to assess a *Yersinia* outbreak in Los Angeles County, CA, in 1992, where the CR binding technique was able to detect only 0.3% plasmid-bearing virulent cells in clinical samples. Thus, it confirms that this technique is highly sensitive and can be used to detect a low level of YEP+ cells in a mixture of YEP− and other types of bacteria. The tests have also made it possible to study the effects of food-processing conditions on the stability of the virulence plasmid, including temperature, salt, pH level and atmosphere.

Recovery of YEP+ Clones by CR Binding

An additional value of the CR binding technique is that it can be used to isolate viable YEP+ cells since, unlike CV, the use of CR does not lead to death of the cells. This permits the recovery of YEP+ cells even in cultures which have grown at 37°C. This recovery technique has been successfully applied to five serotypes of *Y. enterocolitica* and the success rate varied from 5 to 95% (**Table 5**), indicating strain variation

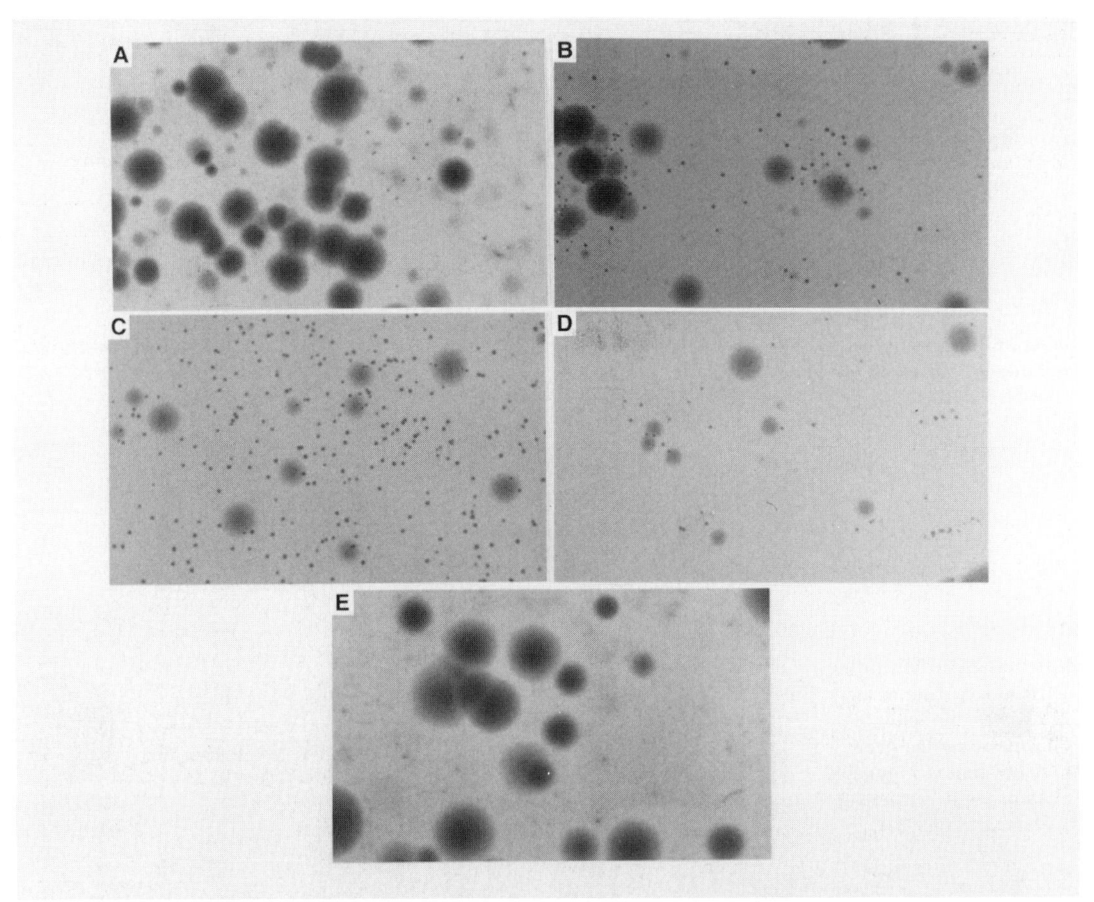

Figure 5 Recovery of YEP⁺ strains as red pinpoint colonies on CR-BHO from (**A**) artificially contaminated pork chops, (**B**) ground pork, (**C**) cheese, (**D**) courgettes and (**E**) naturally contaminated porcine tongue. (Bhaduri and Cottrell (1997) *Appl. Environ*. 63: 4592–4599.)

Table 4 Efficiency of Congo red (CR) binding in mixed cultures of virulent and avirulent strains[a]

Sample	*Estimated no. of colonies in the mixture*		
	Avirulent	*Virulent*	*No. (%) of colonies bound to CR*[b]
CR-BHO : A	67	4	4 (100)
B	63	10	10 (100)
C	49	30	30 (100)
D	35	50	50 (100)
E	21	70	70 (100)
F	7	90	90 (100)

[a]Virulent cells of *Yersinia enterocolitica* were mixed in various ratios with cells from the plasmidless strain and surface-plated on Congo red brain–heart infusion (CR-BHO). The mixed colonies were incubated at 37°C for 24 h. The number of virulent colonies was determined by the appearance of red pinpoint colonies.

[b]Average efficiency was 100%.

Bhaduri et al. (1991) *J. Clin. Microbiol*. 29: 2341–2344.

Table 5 Recovery of plasmid-bearing virulent (YEP⁺) strains after detection by Congo red (CR) binding test

Strain	*Serotype*	*Percentage recovery*
GER	O : 3	90–95
EWMS	O : 13	3–5
PT18-1	O : 5, O : 27	90–95
O : TAC	O : TACOMA	3–5
WA	O : 8	50–60

YEP⁺strains were recovered on Congo red brain heart infusion. The initial detection and percentage recovery of plasmid-bearing cells was determined by crystal violet and CR binding techniques.

in the stability of the plasmid. The recovered YEP⁺ strains show all the plasmid-associated properties, including virulence in mice (Table 2). By using the same recovery technique, FDA investigators were able to recover and enhance the level of plasmid carriage from 0.3% to over 92% from clinical samples obtained during the 1992 outbreak of *Y. enterocolitica* in Los Angles County, CA. Thus, the recovery technique is useful for isolating and enriching viable YEP⁺ cells even if they are present at low levels in mixtures of cells. This further confirms the presence of the virulence plasmid in these pathogenic strains for subsequent investigation. Thus, dye-binding assays offer distinct advantages over currently available commercial tests **Table 6**).

Table 6 Comparison of tests for detection of plasmid-bearing virulent strains of *Yersinia enterocolitica*

	Comparison criteria				
Detection method	*Detection principle*	*Specificity*	*Recovery*	*Convenience*	*Cost*
CV binding technique	Uses virulence determinants based on dye binding	Virulent strains	No	Simple and rapid	Inexpensive
CR binding technique	Uses virulence determinants based on dye binding and low calcium response	Virulent strains	Yes	Simple and rapid	Inexpensive
API	Biochemical tests	*Yersinia* spp.	No	No	Expensive
CIN agar	Medium-based	*Yersinia* spp.	No	No	Expensive
SYS	Enzyme-based	No	No	No	Expensive
Vitek	Medium-based	No	No	No	Expensive
Progen	Uses antibody	*Yersinia* spp.	No	No	Expensive

CV = crystal violet; CR = Congo red; API = Analytical Products, Plainview, NY; CIN = cefsulodin–irgasan–novobiocin; SYS = Analytical Products, Plainview, NY; Vitek = Hazelwood, MO; Progen = Heidelberg, Germany.

In conclusion, the CV and CR binding techniques permit rapid and accurate identification of *Y. enterocolitica* bacterial colonies harbouring the virulence plasmid. The detection methods are based on four virulence determinants:

1. CV binding and small colony size.
2. Appearance of pinpoint colonies (Lcr).
3. CR binding.
4. The appearance of a white border around the red centre of the colony on continued incubation at 37°C.

The combined use of CV binding and CR-BHO techniques provides a method for accurately differentiating between pathogenic and non-pathogenic *Y. enterocolitica*. In addition, the CR binding technique allows the isolation of viable YEP$^+$ strains from foods based on the expression of this virulence determinant.

Isolation of Pathogenic YEP$^+$ Strains from Foods

Common food vehicles in outbreaks of yersiniosis are meat (particularly pork), milk, dairy products, powdered milk, cheese, tofu and raw vegetables. Most strains isolated from these foods differ in biochemical and serological characteristics from typical clinical strains and are usually called non-pathogenic or environmental *Yersinia* strains. The increasing incidence of *Y. enterocolitica* infections and the role of foods in some outbreaks of yersiniosis have led to the development of a wide variety of methods for the isolation of *Y. enterocolitica* from foods. Since the population of *Y. enterocolitica* in food samples is usually low and the natural microflora tend to suppress the growth of this organism, isolation methods usually involve enrichment followed by plating on to selective media. Food matrices can also inhibit the enrichment of YEP$^+$ strains.

The efficiency of *Y. enterocolitica* enrichment techniques varies with serotype and depends on the type of food being tested. Different enrichment procedures have been described to recover the full range of YEP$^+$ serotypes from a variety of foods. The unstable nature of the virulence plasmid complicates the isolation of plasmid-bearing virulent *Y. enterocolitica* by causing the overgrowth of virulent cells by plasmidless revertants, eventually leading to a completely avirulent culture. Traditional methods employ prolonged enrichment at refrigeration temperatures to take advantage of the psychrotrophic nature of *Y. enterocolitica* and to suppress the growth of background flora. Due to the extended time period needed for this method, efforts have been made to devise selective enrichment techniques employing shorter incubation times and higher temperature, making them more practical for routine use. However, high levels of indigenous microorganisms can overgrow and mask the presence of YEP$^+$ strains, including non-pathogenic *Y. enterocolitica* strains. Enrichment media containing selective agents such as Irgasan, ticarcillin and potassium chlorate are effective in recovering a wide spectrum of *Y. enterocolitica* strains from meat samples. However, no single enrichment procedure has been shown to recover a broad spectrum of pathogenic *Y. enterocolitica*.

Since there is no specific plating medium for the isolation of YEP$^+$ strains, cefsulodin–irgasan–novobiocin (CIN) and MacConkey agars are commonly used to isolate presumptive *Y. enterocolitica* from foods. The initial isolation of presumptive *Y. enterocolitica* from enriched samples on CIN and MacConkey agars adds an extra plating step and picking presumptive *Y. enterocolitica* requires skilled recognition and handling of the colonies. The unstable nature of the virulence plasmid complicates the detection of YEP$^+$ strains, since isolation steps may lead to plasmid loss and the loss of associated phenotypic characteristics for colony differentiation. Since colonies of *Y. enterocolitica* are presumptive on the plating media, these isolates should be verified as YEP^+ strains. Biochemical reactions, serotyping, biotyping and virulence testing are essential for differentiation between YEP^+, plasmidless avirulent *Y. enterocolitica* (YEP^-), environmental *Yersinia* strains and other *Yersinia*-like presumptive organisms. Biochemical tests using systems such as API 20E give similar reactions among these organisms and are not conclusive. Serotyping involving major O and H factors differentiate between pathogenic and environmental *Y. enterocolitica* but fail to discriminate between YEP^+ and YEP^- strains. Biotyping involves biochemical tests which do not detect the presence of the virulence plasmid. Thus it does not identify YEP^+ strains. Several plasmid-associated phenotypic virulence characteristics, including colony morphology, AA, serum resistance, tissue culture detachment, HP, Lcr, CV binding, CR binding, isolation of the virulence plasmid and colony hybridization techniques have been described to determine the potential virulence of *Yersinia* isolates. Unfortunately, methods described in the literature do not treat confirmation of virulence in presumptive or known *Y. enterocolitica* isolates recovered from selective agars as an integral part of the detection method. The fastest enrichment procedure available for the isolation of a wide spectrum of *Y. enterocolitica* strains does not include the verification of isolates as YEP^+ strains.

Direct Detection, Isolation and Maintenance of Pathogenic YEP^+ Serotypes

No single procedure has been described in the literature for simultaneous detection and isolation of various YEP^+ serotypes from a variety of foods. Therefore, this section describes a technique to detect and isolate YEP^+ serotypes directly by enriching swabs of artificially contaminated foods such as pork chops, ground pork, cheese and courgettes in a single enrichment medium and applying CR binding, Lcr and PCR for confirmation. Since food surfaces are often the primary site of contamination, artificially contaminated foods are sampled by swabbing the surface with a sterile $5 \times 5 \times 1.25$ cm cellulose sponge moistened with modified trypticase soy broth (MTSB) containing 0.2% bile salts (**Fig. 4**). Swabs of each food type were enriched in MTSB containing 0.2% bile salts at 12°C for direct detection and isolation.

The addition of Irgasan plays a critical role in the enrichment of YEP^+ strains. Since its presence suppresses the growth of pure YEP^+ cultures grown in MTSB when added at the onset of growth but not when added after the lag phase, it should be eliminated from the initial enrichment medium. The addition of Irgasan after 24 h (day 2) and incubation for an additional 24 h (day 3) at 12°C allows the growth of YEP^+ strains while effectively inhibiting growth of competing microflora. Thus, YEP^+ strains were able to grow to a detectable level even in the presence of competing microflora. It was also determined that sampling should be done at 48 h total incubation to avoid sampling after competing microflora begin to predominate. This enhances the isolation of YEP^+ strains in the presence of competing microflora through the selection of incubation temperature, sampling schedule and timing of antibiotic addition. Since the actual food sample was not used and there was a low level of competing microflora, there was optimal growth of YEP^+ strains during enrichment. Thus, this technique allowed enhanced recovery of YEP^+ strains. The YEP^+ strains could be detected and isolated when directly plated on CR-BHO after 48 h total incubation at 12°C (day 3).

The YEP^+ colonies from artificially contaminated pork chops (**Fig. 5A**), ground pork (**Fig. 5B**), cheese (**Fig. 5C**), and courgettes (**Fig. 5D**) all appeared as CR^+ (red pinpoint) colonies (day 4). Thus, YEP^+ strains from each food sample were identified as harbouring the virulence plasmid. The CR^+ clones were further confirmed as YEP^+ strains by multiplex PCR using primers directed to the *ail* and *virF* genes amplifying a 170 bp product from the chromosome and 591 bp product from the virulence plasmid respectively (**Fig. 6:** pork chops, lanes 2–4; ground pork, lanes 8–10; cheese, lanes 14–16; courgettes, lanes 20–22). Thus, the YEP^+ virulent strains were identified by both virulence plasmid-associated phenotypic expression and the presence of specific virulence genes.

This method can be completed in 4 days from sample enrichment to isolation, including confirmation by multiplex PCR, and can recover YEP^+ strains from pork chops, and ground pork spiked with 10, 1 and 0.5 cfu cm^{-2} (**Table 7**). Cheese and courgettes required an additional day of enrichment for detection of samples within an initial inoculum of 0.5 cfu cm^{-2} (Table 7). YEP^+ strains could not be recovered from any of the samples at an initial contamination level of 0.1 cfu cm^{-2} regardless of the length of enrichment. This technique has been successfully applied in the recovery of different YEP^+ strains of five serotypes from artificially contaminated pork chops, ground pork, cheese and courgettes (**Table 8**). The successful isolation of YEP^+ strains from naturally contaminated porcine tongue verified the effectiveness of this method (Fig. 5E). PCR analysis confirmed the presence of a 170 bp product from

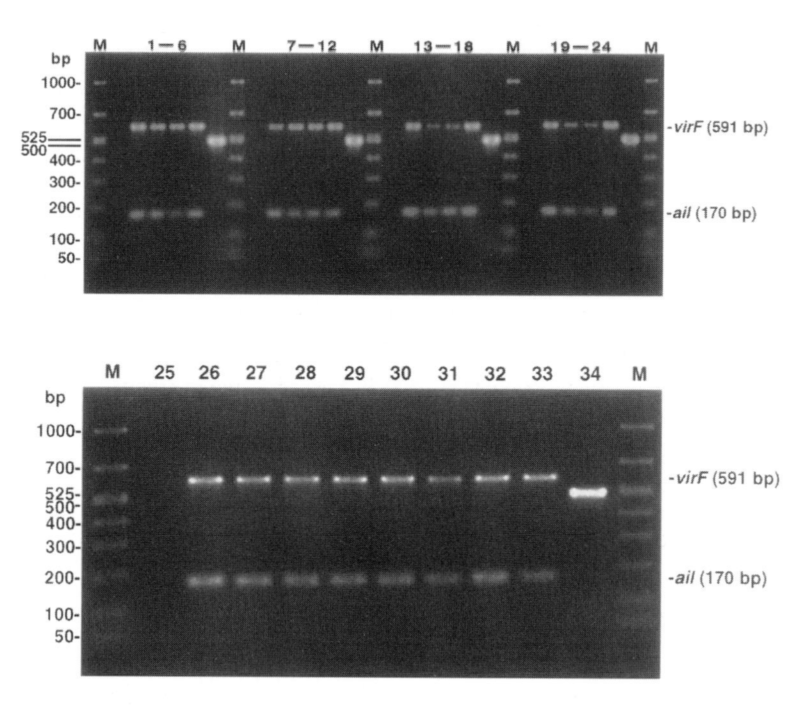

Figure 6 Confirmation of CP$^+$ clones isolated from various artificially contaminated foods and from naturally contaminated porcine tongue as YEP$^+$ strains by multiplex polymerase chain reaction (PCR) using chromosomal *ail* gene and *virF* gene from virulence plasmid. Lane M = 50–1000 bp ladder marker. Negative control with no template (lanes 1, 7, 13, 19 and 25). CR$^+$ colony showing the presence of 170 and 591 bp products with mixture of both *ail* and *virF* primers from chromosome and virulence plasmid respectively: pork chops (lanes 2–4), ground pork (lanes 8–10), cheese (lanes 14–16), courgettes (lanes 20–22) and porcine tongues (lanes 26–32). Positive control with purified DNA from YEP$^+$ strain showing the presence of 170 and 591 bp products with mixture of both *ail* and *virF* primers from chromosome and virulence plasmid respectively (lanes 5, 11, 17, 23 and 33). Positive control for PCR assay with λ as DNA template (lanes 6, 12, 18, 24 and 34). (Bhaduri and Cottrell (1997) *Appl. Environ.* 63: 4952–4955.)

Table 7 Sensitivity of recovery for YEP$^+$ strain from artificially contaminated foods

	YEP$^+$ strain confirmation[a]			
Concentration of YEP$^+$ strain (cfu cm^{-2})	*Pork chop*	*Ground pork*	*Cheese*	*Courgettes*
10	+	+	+	+
1	+	+	+	+
0.5	+	+	+	+
0.1	–	–	–	–

YEP$^+$ = plasmid-bearing virulent strain.
[a]Confirmed by Congo red binding (YEP$^+$ cells appeared as red pinpoint colonies on CR-BHO) and polymerase chain reaction on day 4 (pork chops and ground pork) and day 5 (cheese and courgettes) for 0.5 cfu cm^{-2} concentration, respectively.
+ Detected; – not detected.

the chromosome and a 591 bp product from the virulence plasmid (Fig. 6, lanes 26–32).

The virulence of YEP$^+$ strains recovered from both artificially contaminated food samples and naturally contaminated tongues was confirmed by plasmid-associated virulence characteristics and mouse virulence testing (**Table 9**). These results demonstrate that YEP$^+$ strains recovered using this method retain virulence plasmid, phenotypic characteristics and pathogenicity after isolation from pork chops, ground pork, cheese, courgettes and porcine tongue.

Table 8 Isolation and confirmation of YEP$^+$ strains from artificially contaminated foods

Serotypes	*Number of strains tested*	*YEP$^+$ strains confirmed by CR binding and PCR*[a]
O:3	5	+
O:8	5	+
O:TACOMA	4	+
O:5, O:27	4	+
O:13	3	+

YEP$^+$ = plasmid-bearing virulent strain.
[a]Confirmed by Congo red (CR) binding (YEP$^+$ cells appeared as red pinpoint colonies on Congo red brain heart infusion and polymerase chain reaction (PCR).

In conclusion, the method described here has the following advantages:

1. It requires only a single enrichment medium for a wide range of serotypes, including a large number of different strains from a variety of foods.
2. It eliminates a presumptive isolation step.
3. It uses a single medium (CR-BHO) for direct detection and isolation.

Table 9 Evaluation of plasmid-associated characteristics and mouse virulence of YEP⁺ strains recovered from artificially contaminated foods and naturally contaminated porcine tongue

Strain	*CM*[a]	*CV binding*[b]	*Lcr*[c]	*CR binding*[d]	*AA*[e]	*HP*[f]	*Plasmid (70–75 kbp)*	*Diarrhoea in mice*[g]
GER (O : 3) YEP⁺ strain as + ve control	Small	+	+	+	+	+	+	+
GER (O : 3) YEP⁺ strains from spiked foods	Small	+	+	+	+	+	+	+
SB (O : 3) from tongue	Small	+	+	+	+	+	+	+
GER (O : 3) YEP⁺ strain	Large	–	–	–	–	–	–	–

[a]CM = colony morphology. In a calcium-adequate agar medium (brain–heart infusion agar) YEP⁺ cells appeared as small colonies (diameter 1.13 mm) compared to larger YEP⁻ colonies (diameter 2.4 mm).
[b]CV binding = crystal violet binding. YEP⁺ cells appeared as small dark-violet colonies on BHA.
[c]Lcr = low calcium response. Calcium-dependent growth at 37°C: YEP⁺ cells appeared as pinpoint colonies of diameter 0.36 mm compared to the larger YEP⁻ colonies of diameter 1.37 mm on low calcium Congo red brain heart infusion agarose medium (CR-BHO).
[d]CR binding = Congo red binding. YEP⁺ cells appeared as red pinpoint colonies on CR-BHO.
[e]AA = autoagglutination.
[f]HP = hydrophobicity.
[g]Faecal material consistency was liquid; diarrhoea was observed starting on days 3 and 4 post-infection.

4. It preserves the virulence plasmid and the pathogenicity.

This procedure is a practical alternative to many other recovery methods which require significantly more time for completion.

Acknowledgement: I thank Journal Department of American Society Microbiology for permission to use the figures and tables from my papers published in Journal of Clinical Microbiology and Applied Environmental Microbiology.

See also: **PCR-based Commercial Tests for Pathogens**.

Further Reading

Bhaduri S and Cottrell B (1997) Direct detection and isolation of plasmid-bearing virulent serotypes of *Yersinia enterocolitica* from various foods. *Appl. Environ. Microbiol.* 63: 4952–4955.

Bhaduri S and Cottrell B (1998) A simplified sample preparation from various foods for PCR detection of pathogenic *Yersinia enterocolitica*: a possible model for other food pathogens. *Mol. Cell. Probes* 12: 79–83.

Bhaduri S, Conway LK and Lachica RV (1987) Assay of crystal violet binding for rapid identification of virulent plasmid-bearing clones of *Yersinia enterocolitica J. Clin. Microbiol.* 25: 1039–1042.

Bhaduri S, Turner-Jones C, Taylor MM and Lachica RV (1990) Simple assay of calcium dependency for virulent plasmid-bearing clones of *Yersinia enterocolitica. J. Clin Microbiol.* 28: 798–800.

Bhaduri S, Turner-Jones C and Lachica RV (1991) Convenient agarose medium for the simultaneous determination of low calcium response and Congo red binding by virulent strains of *Yersinia enterocolitica. J. Clin. Microbiol.* 29: 2341–2344.

de Boer E (1992) Isolation of *Yersinia enterocolitica* from foods. *Int. J. Food Microbiol.* 17: 75–84.

Doyle MP and Cliver DO (1990) *Yersinia enterocolitica*. In: Cliver DO (ed.) *Foodborne Diseases*. P. 223. San Diego: Academic Press.

Kapperud G (1991) *Yersinia enterocolitica* in food hygiene. *Int. J. Food Microbiol.* 12: 53–66.

Kwaga JKP and Iversen JO (1991) Laboratory investigation of virulence among strains of *Yersinia enterocolitica* and related species from pigs and pork products. *Can. J. Microbiol.* 38: 92–97.

Ravangnan G and Chiesa C (1995) Yersiniosis: present and future. 6th International Symposium Volume on *Yersinia*, Rome, September 26–28, 1994. Contribution to Microbiology and Immunology, Vol. 13, J.M. Cruse and R.E. Lewis, (Series Eds.), Karger, New York.

Toora S, Budu-Amoako E, Ablett RF and Smith J (1994) Isolation of *Yersinia enterocolitica* from ready-to-eat foods and pork by a simple two step procedure. *Food Microbiol.* 11: 369–374.

Weagent SD, Feng P and Stanfield JT (1992) *Yersinia enterocolitica* and *Yersinia pseudotuberculosis*. In: Tomlinson L (ed.) *Food and Drug Administration Bacteriological Analytical Manual*, 7th edn. P. 95. Arlington, VA: AOAC International.

Yoghurt *see* **Fermented Milks**: Yoghurt.

Z

Zygomycetes *see* **Fungi: Classification of the Zygomycetes.**

ZYGOSACCHAROMYCES

John P Erickson and **Denise N McKenna** Bestfoods Technical Center, Somerset, New Jersey, USA

The yeast genus *Zygosaccharomyces* contains eight recognized species, of which three pose serious economic spoilage risks to processed food manufacturers. They are *Z. bailii*, *Z. rouxii* and *Z. bisporus*. *Z. bailii* is particularly troublesome in mayonnaise, salad dressings, pickled vegetables, teas and various fruit drinks. In contrast, *Z. rouxii* spoilage is more closely associated with high sugar or salt-based ingredients and finished products like fruit concentrates, syrups, candied fruit pieces and confectioneries such as marzipan, chocolate candy fillings, etc. The *Z. bisporus* spoilage profile is similar to *Z. rouxii*, but occurs less frequently. The types of food ingredients and processed foods spoiled by *Zygosaccharomyces* are generally considered shelf-stable in that they readily inactivate a broad spectrum of food-associated microorganisms ranging from bacteria to moulds. Hence, *Z. bailii*, *Z. rouxii* and, to a lesser extent, *Z. bisporus* possess several intrinsic physiological resistance factors which allow them to survive and thrive in normally hostile environments. The most prominent resistance factors are acetic acid (vinegar) tolerance, chemical antimycotic preservative resistance and growth under high osmotic pressure. Effective contamination control is dependent upon ingredient ecology knowledge, stringent/consistent sanitation standards, tight formulation specifications and focused quality assurance strategies. Equally important, finished product microbiological stability strengths and weaknesses must be well documented and fully understood.

Taxonomic and Ecological Characteristics

The general taxonomic properties of *Zygosaccharomyces* are identical to ubiquitous foods yeast genera such as *Saccharomyces*, *Candida* and *Pichia*. Macroscopic and microscopic morphology observations cannot differentiate *Zygosaccharomyces* from other yeast or individual species within the genus. On various mycological agars, colonies are smooth, round, convex and cream-coloured. Microscopic observation shows large ovoid elongated cells and multilateral budding. All eight species vigorously ferment glucose and produce asci with one to four globose or ellipsoidal ascospores inside. The three food and beverage spoilage species can be distinguished by sucrose fermentation properties, presence/absence of growth at 37°C and acetic acid resistance (**Table 1**). However, the incubation time and laborious media preparation required to conduct these tests make them unsuitable for routine quality control applications. From a practical perspective, the physiochemical attributes of the ingredient or finished product can furnish useful information regarding species identification. For example, if yeast was recovered from a highly acidified food with a ≥0.90 water activity (a_w), there is a strong likelihood *Z. bailii* was

Table 1 Key taxonomic, biochemical and physiological tests required to differentiate three *Zygosaccharomyces* – food spoilage species

	Results		
Tests	Z. bailii	Z. rouxii	Z. bisporus
Glucose fermentation	Positive	Positive (slow)	Positive
Sucrose fermentation	Variable (slow)	Variable	Negative
Growth at 37°C	Variable	Variable	Negative
Growth in presence of 1% acetic acid	Positive	Negative	Positive
Water activity (a_w) tolerance	0.80–0.85	<0.80	<0.80

Adapted from Kreger-Van Rij NJW (1984) *The Yeasts, A Taxonomic Study*, 3rd edition. Amsterdam: Elsevier Science.

detected. Conversely, a low acid food with high sugar or salt content is more likely to be contaminated and spoiled by *Z. rouxii*.

Zygosaccharomyces isolates from alcoholic beverages are more difficult to discern since *Z. bailii* and *Z. rouxii* exhibit comparable ethanol and low pH tolerance, and the a_w of most wines is $\geqslant 0.95$. Also, with the exception of wines that contain sulphur dioxide, very few wines and fruit cordials are preserved with antimycotic acids such as acetic, benzoic and sorbic.

The ecological distribution of *Z. bailii*, *Z. rouxii* and *Z. bisporus* is not well defined. The most frequently cited natural habitats are mummified fruits, tree exudate, and at various stages of raw sugar refining and commercial syrup production. It appears that *Zygosaccharomyces* prefers specialized, selective ecological niches. But technical limitations have strongly influenced ecological investigations in two ways: first, the lack of reliable and simple-to-use selective recovery media prevented detection in highly competitive and dynamic environments, and second, slow, complex identification methods deterred screening large numbers of yeast isolates. Recent advances in selective plating media combined with powerful new genetic assays such as polymerase chain reaction (PCR) have greatly enhanced the ability to conduct in-depth and long-term ecological studies in both natural and food-processing plant environments.

Physiological and Preservative Resistance Characteristics

Z. bailii

Among the three *Zygosaccharomyces* food/beverage spoilage species, *Z. bailii* possess the most pronounced and diversified antimicrobial resistance attributes. Extensive research conducted over four decades has repeatedly documented resistance to high concentrations of monocarboxylic acids, chemical antimycotic preservatives and ethanol. *Z. bailii* also grows over wide pH and a_w ranges – 2.0–7.0 and 0.85–0.99, respectively. Even more significant for the food industry, the yeast can survive and defeat potent broad-spectrum synergistic preservative combinations (hurdles) that impart microbiological shelf-stability protection to acidified processed foods. Specifically, *Z. bailii* has been shown to be innately resistant to commonly used food and beverage preservatives, including acetic acid, lactic acid, propionic acid, benzoic acid, sorbic acid, sulphur dioxide and ethanol. Optimum preservative resistance is mediated by glucose levels, with 10–20% sugar concentrations producing maximum resistance responses. Depending upon accompanying formulation factors such as pH, it is not atypical to encounter *Z. bailii* strains spoiling fruit drinks and pourable salad dressings preserved with >1000 p.p.m. of potassium sorbate. Intrinsic preservative resistance mechanisms are extremely adaptable and robust. Their functionality and effectiveness are unaffected or marginally suppressed by physiochemical environmental conditions such as low pH, low a_w, high osmotic pressure and sparse nutrients. Interestingly, there is strong evidence that *Z. bailii* preservative resistance is stimulated by the presence of multiple antimicrobial constituents. Cellular acetic acid uptake was inhibited when sorbic acid, benzoic acid or ethanol was incorporated into yeast culture medium. Similarly, ethanol levels up to 10% did not adversely alter *Z. bailii* intrinsic sorbic acid and benzoic acid resistance at pH 4.0–5.0. Sugar substrate investigations also revealed negligible effects on preservative resistance. Comparable sorbic and benzoic acid resistance was observed regardless of whether *Z. bailii* cells were grown in culture medium containing glucose, fructose or sucrose as the fermentable substrate. Conversely, synergistic interaction between salt and acetic acid generated antagonistic effects against *Z. bailii*. As salt levels increased, the yeast was inactivated by lower amounts of acetic acid.

Sugar fermentation within the *Zygosaccharomyces* genus is unique. Unlike the vast majority of yeast genera, fructose is metabolized more rapidly than glucose. For species like *Z. bailii* and *Z. rouxii* this produces a phenomenon known as fructophily, in which yeast growth rates are greatly accelerated when a food's fructose level approaches and exceeds 1% of product composition. The slow and delayed fermentation of sucrose is directly linked to fructose metabolism: sucrose, a disaccharide composed of glucose and fructose, is hydrolysed by food acids, i.e. low pH conditions. Hence, in acidic processed foods and beverages there is a steady accumulation of glucose and fructose during storage. If sucrose is the primary carbohydrate ingredient and *Z. bailii* contamination is present, cell growth can be impeded for several weeks until sufficient quantities of fructose and glucose are available to support reproduction and proliferation. This is usually preceded by a 2–4-week lag before visible spoilage defects are noticeable. In toto, overt product quality deterioration does not surface until 2–3 months after manufacturing. *Z. bailii*'s key preservative, physiological and metabolic resistance properties are summarized in **Table 2**.

Z. rouxii and *Z. bisporus*

These two species differ from *Z. bailii* in their inferior resistance to acetic acid and chemical antimycotic

Table 2 Key preservative and physiological resistance factors of *Zygosaccharomyces bailii*

Factor	Resistance properties
Acetic acid (pH 4.0)	≥3%
Benzoic acid (pH 4.0)	≥1000 p.p.m.
Sorbic acid (pH 4.0)	≥1000 p.p.m.
Sulphur dioxide	≥500 p.p.m.
Ethanol	≥10–15%
Sodium chloride	≤10%
Sugar (glucose)	≥60%
pH	≤2.0
Water activity (a_w)	≥0.80–0.85
Temperature range	≥8–37°C
Atmosphere	≥Facultative (aerobic-stimulatory)

preservatives. In contrast, both yeasts grow at lower a_w than *Z. bailii*. *Z. rouxii* is capable of spoiling honey and related high-sugar foods at a_w as low as 0.62, whereas *Z. bailii* growth potential ceases in the 0.80–0.85 range. Comparative salt tolerance further illustrates exceptional capacity of *Z. rouxii* to survive high osmotic stresses. It can grow in liquid sugars at > 75° Brix and salt concentrations close to 20%, corresponding to 0.75 and 0.85 a_w respectively. With the exception of a few rare strains, the growth of *Z. rouxii* and *Z. bisporus* is completely inhibited by ≥500 p.p.m. of sorbic and benzoic acids within a 4.0–4.5 pH range. They are especially sensitive to acetic acid, rapidly dying off at ≥1% levels.

Physiochemical environmental effects on sugar and salt-tolerant yeast have not been studied as comprehensively as *Z. bailii*. This is most likely due to the latter's broader economic impact regarding food and beverage categories vulnerable to spoilage. Comparative studies of sugar- and salt-tolerant *Z. rouxii* strains indicated that distinct physiological differences existed. The sugar-tolerant strains grew over a 1.8–8.0 pH range in high sugar media, whereas salt-tolerant strains were more sensitive to pH conditions. At 1 mol sodium chloride concentration, growth was detected over a 3.0–6.6 pH range. When sodium chloride molarity was doubled, growth was restricted to a narrow 4.0–5.0 range. *Z. rouxii* optimum growth temperature increased as a_w decreased. Surprisingly, the optimum temperature reached 35°C at ≤0.96 a_w levels, which is more typical of mesophilic bacteria incubation requirements.

Zygosaccharomyces Heat Resistance

The heat resistance profiles of *Z. bailii*, *Z. rouxii* and *Z. bisporus* are comparable to other ascospore-forming yeasts. *Z. bailii* asci were significantly more heat-resistant than *Z. rouxii* at 0.963 and 0.858 a_w in liquid medium adjusted to pH 4.5. *Z. bailii* vegetative cells were also more heat-resistant than *Z. rouxii*. As expected, asci and vegetative cell heat resistance increased as a_w decreased. Six log reductions ($D_{64°C}$) were calculated as 1.2 min/0.963 a_w and 5.4 min/0.858 a_w. Commercially processed fruit drinks and sugar syrups are normally pasteurized at 75–85°C. This provides a large quality assurance margin with respect to *Zygosaccharomyces* thermal destruction efficacy.

Zygosaccharomyces Preservative and Environmental Resistance Mechanisms

Due to the pioneering and elegant research efforts of Dr A.D. Warth the *Z. bailii* preservative resistance mechanism has been elucidated and thoroughly understood. The organism utilizes an inducible, active transport pump to counteract the toxic effects produced by undissociated preservative molecule build-up inside individual cells. The pump provides two levels of protection. First, it physically expels preservative molecules from the cell, which assists in maintaining low, non-injurious preservative levels within the cell. Second, the rapid and efficient purging of undissociated molecules prevents deleterious cytoplasm pH changes that could disrupt or shut down critical metabolic pathways. Because the pump requires energy to function optimally, high sugar levels enhance *Z. bailii* preservative resistance. It is equally effective in excreting monocarboxylic organic acids and lipophilic straight chain fatty acid preservatives such as sorbic acid and benzoic acid. Additionally, the active transport successfully operates across a wide pH and a_w range, and broad nutritional conditions.

The resistance to osmotic stress of *Z. rouxii* is primarily associated with internal synthesis of polyols, mainly glycerol and arabitol, which raises intracellular pressure, bringing it in balance with the external osmotic gradient. Cell membrane and wall composition as well as ATPase enzyme activity may be important in augmenting polyol formation, thus regulating osmotolerance.

Zygosaccharomyces Spoilage in Food and Beverages

Z. bailii is the most troublesome and persistent spoilage yeast confronting acidified food and beverage manufacturers. Early reports of inexplicable fermentation spoilage in mayonnaise and salad dressing date back to the 1920s. Several incidents described violent fermentation coupled with the recovery of a few yeasts. More detailed investigations in the 1940s and 1950s confirmed *Z. bailii* spoilage in cucumber pickles, sundry pickled vegetable mixes, acidified

Table 3 Food and Drug Administration yeast fermentation – spoilage recalls in dressings and related acidified processed foods (1978–1996)

Year	Product
1978	Imitation mayonnaise
1981	Mayonnaise (single serving pouch)
1984	Low-calorie Roquefort dressing
1985	Homestyle salad dressing, real mayonnaise
1986	Real mayonnaise
1987	Carbonated beverages (soda and fruit concentrate)
1988	Ketchup
1990	Reduced-calorie mayonnaise, fat-free French dressing
1991	Light mayonnaise
1992	Lite Caesar dressing, lite creamy Parmesan dressing, olive oil vinaigrette dressing
1993	Salad dressing, tartar sauce, coleslaw dressing, burger tartar sauce, thousand island dressing, Parmesan pepper dressing, lite Parmesan pepper dressing
1995	White salad dressing
1996	Salad dressing

sauces, mayonnaise and salad dressings. Spoilage invariably occurred in acidic shelf-stable foods which relied upon acetic acid (vinegar) to negate microbiological growth risks. Around the same time, sporadic gaseous fermentation spoilage incidents suddenly appeared in high-acid/sugar fruit syrups and beverages preserved with moderate benzoic acid levels of 400–500 p.p.m. Again, *Z. bailii* was indisputably identified as the spoilage culprit. The near simultaneous emergence of *Z. bailii* spoilage in two divergent processed food categories was probably due to improved laboratory and field evaluation techniques, better communication channels, movement towards consolidated, mass-production manufacturing facilities and large-scale complex distribution networks. Remarkably, 50 years later, *Z. bailii* spoilage risks and root causes closely parallel the processed food industry's situation in the late 1940s to early 1950s. Despite quantum leaps in formulation control, food process equipment design and construction, and sanitation technologies (automated clean-in-place, sanitary transfer valves, etc.), *Z. bailii* remains problematic in mayonnaise, salad dressings, tomato ketchup, pickled/brined vegetables, low to moderate Brix fruit concentrates, and various non-carbonated fruit drinks. This is mute testimony to the organism's adaptability, resilience and overall hardiness. Furthermore, *Z. bailii* spoilage is expanding into new food categories. Two recent examples are spoilage incidents in prepared mustards and fruit-flavoured carbonated soft drinks containing citrus, apple and grape juice concentrates. The specialized but persistent nature of *Z. bailii* spoilage problems is exemplified in **Table 3**, which summarizes the US Food and Drug Administration yeast fermentation – spoilage recalls in high acetic acid and/or chemically preserved foods and beverages covering an 18-year period (1978–1996). Most likely, Table 3 represents a small fraction of economic losses produced by *Z. bailii* contamination. Obviously, company rejections, prolonged holding times and spoilage risks caught before finished production lots reached retail distribution channels are not accounted for.

In most cases it is difficult to pinpoint the specific cause of spoilage. This is because the problem does not show up until 2–4 months after production and the finished product conformed to applicable formulation, processing and microbiological specifications at the time of manufacture. *Z. bailii* spoilage manifestations are readily recognized by both customers and consumers. Overt physical and organoleptic decomposition signs include product oozing from jars or bottles, emission of pungent yeast and alcoholic odours, occasional emulsion breakage (dressings), sediment formation (beverages) and brown surface film development on product surfaces.

Z. bailii and *Z. rouxii* spoilage must never be taken lightly. Under extreme circumstances, internal CO_2 pressure increases inside glass jars or bottles to the level where on-the-shelf explosions may take place. Although the possibility is remote, personal injury could result from flying debris. Plastic containers and polyfilm pouches burst open rather than explode. If slip and fall injuries were caused by spilled product residue, the company is exposed to contentious liability issues. A plausible scenario is that individuals become angry after being soiled by high-velocity product expelled from jars or bottles immediately upon opening. More often than not, these types of incidents result in unwanted government involvement, or costly consumer complaint investigations.

Z. rouxii and *Z. bisporus* exclusively spoil high-sugar/low a_w foods and food ingredients. The most prevalent types are sugar syrups, fruit syrups, molasses, honey, fruit concentrates, sweetened wines and cordials and confectionaries such as marzipan and candy fillings. These yeasts ferment the food product, leading to effervescence, alcohol odour or taste, and turbidity which is more easily noted in clear sugar syrups. However, *Z. rouxii* and *Z. bisporus* ferment slower and less aggressively than *Z. bailii*. Thus, the potentially serious ramifications generated by vigorous *Z. bailii* fermentative metabolism are not a concern with osmotolerant yeast.

As mentioned previously, it is generally impossible to ascertain the exact reason why *Z. bailii* or *Z. rouxii* contamination occurred in processed food lots made 2–3 months earlier. Circumstantial and inferential evidence is readily attainable from additional finished product and environmental sample testing, but accurate and concrete information gathering is highly suspect when the investigation focuses on reconstructing and interpreting events that happened weeks ago. However, trend analysis of past spoilage incidents suggests that certain commonalties exist among diverse *Zygosaccharomyces* contamination failures. They include undetected introduction of the offending *Z. bailii* or *Z. rouxii* strain into the plant environment from a low-level heterogeneous contaminated ingredient lot. This is followed by yeast build-ups inside key processing equipment because of inadequate or poorly executed sanitation procedures. In tandem, routine quality assurance/quality control (QA/QC) monitoring protocols missed or overlooked contamination risks. Eventually, finished product cross contamination occurred during production runs, and QA/QC standard operating procedures (SOPs) lacked appropriate detection sensitivity, discrimination abilities and sampling discipline/focus to discover the problem. These shortcomings and deficiencies are compounded in product formulations which are excessively sensitive to yeast growth. Certain fruit beverages can be spoiled by *Z. bailii* contamination as low as one viable cell in ≥10 l of finished product. No sanitation or microbiological QA/QC programme can cope with this degree of risk. The only viable alternatives would be reformulation to increase stability and/or application of high-lethality thermal-processing parameters.

Zygosaccharomyces – Practical Food/Beverage Industry Aspects

Recovery and Enumeration – QA/QC Microbiology

Several media were developed for the selective detection and quantification of preservative-resistant and sugar/salt-tolerant yeast in susceptible foods and beverages. As regards the former yeast group, standard non-differential yeast and mould plating media (malt extract or potato dextrose agars) were supplemented with ≥0.5% glacial acetic acid and/or 0.05% sodium benzoate. Recommended incubation times ranged from 5 to 14 days. Common incubation conditions were 20–25°C, aerobic atmosphere and acidifying media to 4.0–4.5 pH range to ensure proper selectivity. The addition of 0.5% glacial acetic acid was usually sufficient to produce a final pH of 4.0–4.5. Specialized yeast enrichment broths were co-developed to increase recovery sensitivity below the typical ≥10 per gram or millilitre detection threshold of direct microbiological plating methods. Up to 1000-fold sensitivity increases were reported. The enrichment broths operate on three principles:

- Rich nutrient content and absence of acidulants concomitantly resuscitate debilitated cells and maximize growth rates.
- Use of antibiotics to eliminate bacterial growth.
- Plating or streaking on selective media to separate spoilage from innocuous yeast strains.

The detection and enumeration of sugar- and salt-tolerant yeast (*Z. rouxii*) rely upon high solute concentrations in plating media and enrichment broths. Typically, 40–60% glucose is added to the basal medium to lower a_w and establish a high osmotic pressure environment. The most popular basal plating medium is unacidified total plate count agar, which is often supplemented with antibiotics that obviate bacterial growth. Incubation conditions and length are identical to preservative-resistant yeast. One controversial aspect is whether plating diluent must contain 40–60% glucose in order to prevent osmotic shock (transfer from low to high a_w environment) that dramatically decreases yeast recovery rates. Recent observations suggest that 30°C is the best incubation temperature for yeast detection. Faster growth rates and larger surface colonies occurred at 30°C vs 25°C incubation for 3–5 days.

It is also widely accepted that all yeast and mould plating assays should employ surface plating techniques which simplify counting procedures, improve discrimination between food particles and yeast colonies, and enhance recovery rates due to high oxygen tension. Surface plating effectiveness is further optimized when the sample aliquot is analysed by membrane filtration methods (for example, hydrophobic grid membrane filtration). Membrane filtration physically separates viable yeast cells from food and beverage ingredients that may inhibit or slow yeast growth. The filtration step is superior to manual spreading and streaking regarding depositing and distributing individual yeast cells over the entire agar surface, which increases enumeration accuracy and improves analytical reliability and simplicity. Recommended preservative-resistant and sugar-tolerant *Zygosaccharomyces* microbiological QC plating and enrichment media are summarized in **Table 4**.

Zygosaccharomyces Identification

As discussed earlier, *Zygosaccharomyces* colonies on non-selective and semi-selective media are morphologically similar to many common yeasts, and traditional biochemical and physiological tests used to identify *Z. bailii* and *Z. rouxii* do not lend themselves to routine microbiological QC applications. Yeast identification kits have been available for over 10 years. However, they have limited QC benefits

Table 4 Recommended enumeration and recovery media for preservative-resistant and sugar/salt-tolerant *Zygosaccharomyces*

Medium	*Composition (per litre)*	*Preparation*	*Incubation time/temperature*	*Special comments*
Acidified tryptone glucose yeast extract agar (ATGYE)	100 g Glucose monohydrate 5 g Tryptone 5 g Yeast extract 15 g Agar 5 ml Glacial acetic acid	1. Mix and boil dry ingredients 2. Autoclave at 121°C for 15 min at 15 lb steam 3. Cool to 45–52°C 4. Add 5 ml glacial acetic acid and mix thoroughly 5. Pour plates, let solidify. Store at 4–6°C for up to 2 weeks	5 days 25–30°C	1. 0.5% glacial acetic acid in final medium: pH ca. 3.9 2. Best total recovery of preservative-resistant yeast 3. Not selective for *Z. bailii*
Z. bailii selective agar medium (ZBM)[a]	30 g Sabouraud's dextrose broth 30 g Fructose 25 g Sodium chloride 15 g Agar 5 g Tryptone 2.5 g yeast extract 0.1 g Trypan blue dye (optional) 5 ml Glacial acetic acid 1 ml Potassium sorbate 10% solution	1. Same as ATGYE except autoclave 121°C for 5 min at 15 lb steam 2. Add potassium sorbate and glacial acetic acid after cooling to 45–52°C	3–5 days 30°C	1. 0.5% glacial acetic acid in final medium: pH 3.7–4.0 2. Filter sterilize 10% potassium sorbate solution before use: 100 p.p.m. in final medium 3. Highly specific for *Z. bailii* 4. Targeted to pickled vegetables, fruit concentrates and salad dressings 5. Requires validation in alcoholic and acidified beverages
Dichloran-18% glycerol agar (DG18)	Commercially available	As directed by manufacturer	5 days 25°C	1. Recovered *Z. bailii* and *Z. rouxii* 2. Best yeast recovery when mixed with xerophilic mould
Malt–yeast extract–50% glucose agar (MY50G)	500 g Glucose 10 g Malt extract 10 g Agar 2.5 g Yeast extract	1. Mix and boil dry ingredients 2. Autoclave at 121°C for 15 min at 15 lb steam 3. Cool to 45–52°C 4. Pour plates, let solidify. Store at 4–6°C for up to 2 weeks	5 days 25°C	1. Does not recover *Z. bailii* 2. Colonies small at 5 days
Plate count 30% sugar agar (PCAS)	200 g Glucose 100 g Fructose 23 g Plate count agar 0.1 g Trypan blue dye (optional) 20 ml Sterile antibiotic solution	1. Mix and boil dry ingredients 2. Cool to 45–50°C 3. Add 20 ml antibiotic solution and mix thoroughly 4. Pour plates, let solidify. Store at 4–6°C for up to 2 weeks	3 days 30°C	1. Used with hydrophobic membrane filtration method 2. Leverages fructophily response of *Z. rouxii* to speed up growth rate

[a]Modified from Erickson JP (1993) Hydrophobic membrane filtration method for the selective recovery and differentiation of *Zygosaccharomyces bailii* in acidified ingredients. *J. Food Prot.* 56: 234–238.

because of labour requirements, complexity and length of testing, which can take ≥7 days for a final result. A trained laboratory staff is needed to reduce human error. These kits are useful plant investigation, trouble-shooting and problem-solving tools.

Various genetic DNA probe techniques have been successfully applied to identify spoilage yeast, including *Z. bailii* and *Z. rouxii*. Three technologies mentioned were random amplified polymorphic DNA (RAPD), RAPD-polymerase chain reaction, and microsatellite polymerase chain reaction assays. All have been described as extremely precise and faster than traditional cultural identification methods. Whether there is enough industry demand and volume to support commercialization and future technology development remains to be seen.

Genetic assays, direct epifluorescence, flow cytometry and impedance/conductance automated microbiological detection instrumentations are microbiological QC technologies with the potential to rapidly (24–48 h) detect and enumerate *Z. bailii* or *Z. rouxii* under high-volume sampling situations. However, some major technical barriers exist, such as compatibility with selective media, exclusion of an enrichment step and consistent performance in highly acidified and/or salted processed foods.

See also: **Nucleic Acid-based Assays**: Overview. **PCR-based Commercial Tests for Pathogens**. **Spoilage Problems**: Problems caused by Fungi.

Further Reading

Beuchat LR (1987) *Food and Beverage Mycology*, 2nd edn. New York: AVI–Van Nostrand Reinhold.

Deak T and Beuchat LR (1996) *Handbook of Food Spoilage Yeasts*. Boca Raton, Florida: CRC Press.

Erickson JP (1993) Hydrophobic membrane filtration method for the selective recovery and differentiation of *Zygosaccharomyces bailii* in acidified ingredients. *J. Food Prot.* 56: 234–238.

Hocking AD (1996) Media for preservative resistant yeast: a collaborative study. *Int. J. Food Microbiol.* 29: 167–175.

Kreger-Van Rij NJW (1984) *The Yeasts, A Taxonomic Study*, 3rd edn. Amsterdam: Elsevier Science.

Makdesei AK and Beuchat LR (1996) Performance of selective media for enumerating *Zygosaccharomyces bailii* in acidic foods and beverages. *J. Food Prot.* 59: 652–656.

Pitt JJ and Hocking AD (1985) *Fungi and Food Spoilage*. London: Academic Press.

Sand FEMJ (1980) *Zygosaccharomyces bailii* – an increased danger for soft drinks. *Brauwelt* 20: 418–425.

Smittle RB (1977) Microbiology of mayonnaise and salad dressings: a review. *J. Food Prot.* 40: 415–422.

Thomas DS and Davenport RP (1985) *Zygosaccharomyces bailii* – a profile of characteristics and spoilage activities. *Food Microbiol.* 2: 157–169.

Tokuoka K (1993) A review: sugar- and salt-tolerant yeasts. *J. Appl. Microbiol.* 74: 101–110.

Warth AD (1977) Mechanism of resistance of *Sacharomyces bailii* to benzoic, sorbic, and other weak acids used as food preservatives. *J. Appl. Bacteriol.* 43: 215–230.

ZYMOMONAS

Hideshi Yanase, Department of Biotechnology, Tottori University, Japan

Characteristics

The name of this bacterium, *Zymomonas*, originated from its unique metabolism: yeast-like, ethanogenic fermentation (= *zymo*) utilizing the Entner–Doudoroff glycolytic pathway found in Pseudomonads (= *monas*). On the basis of the characteristics listed in **Table 1**, all strains of *Zymomonas* classified to date belong to a single species, *Z. mobilis*, with two subspecies: *Z. mobilis* subsp. *mobilis*, the organism isolated from pulque or palm wine, and *Z. mobilis* subsp. *pomaceae*, the organism discovered in beer and cider spoilage.

Taxonomy

The following features are characteristic of *Zymomonas*:

- they are rods, 2–6 μm long and 1–1.4 μm wide
- they are Gram-negative
- they produce no spores
- they are usually non-motile, but if motile, movement is driven by one to four lophotrichous flagella, and motility may be spontaneously lost
- they do not grow on nutrient agar or in nutrient broth
- they are anaerobic but tolerate some oxygen (facultative anaerobic)
- they ferment glucose
- they ferment fructose
- they yield almost equimolar amounts of ethanol and CO_2
- they are oxidase-negative
- their genome contains 47.5–49.5% guanine plus cytosine (G+C).

These 11 features should be considered as the minimal description of the genus *Zymomonas*.

Culture Characteristics

On standard medium (SM), colonies are glistening, regularly edged, white to cream coloured, and 1–2 mm in diameter after 2 days at 30°C. Deep colonies in solid SM are lenticular, regular, entire-edged, butyrous, white or cream coloured, and 1–2 mm in diameter after 2–4 days at 30°C. Anaerobic surface colonies are spreading, entire-edged, convex or

Table 1 Characteristics of *Zymomonas mobilis*

Characteristics	*Reaction or result*
Morphological characteristics	
Shape	Rods, 2–6 μm long and 1.0–1.4 μm wide Rounded ends
Motility	Usually non-motile; if motile, movement driven by one to four lophotrichous flagella
Spore	Not formed
Gram stain	Negative
Cultural characteristics	
Colonies on standard medium	Glistening, regularly edged, white to cream coloured 1–2 mm in diameter after 2 days at 30°C
Deep colonies on standard medium	Lenticular, regular, entire-edged, butyrous, white or cream coloured 1–2 mm in diameter after 2–4 days at 30°C
Nutrient agar	No growth
Physiological characteristics	
Parasitic on warm blood	Negative; animals and/or humans
Pathogenic for humans	Negative
Causes donovanosis	Negative
Causes one form of rat-bite fever in humans	Negative
Causes vaginitis	Negative
Temperature for growth	Optimum between 25 to 30°C Growth at up to 38°C
Survival at 60°C for 5 min	Negative
Initial pH for growth	pH 3.85–7.55
Salt tolerance	1.0% NaCl
Ethanol tolerance	13%
Glucose tolerance	40%
Catalase	Positive
Oxidase	Negative
Voges–Proskauer test	Weak
Nitrate reduction	Negative
Indole production	Negative
H_2S production	Positive
Methyl red reduction	Positive
Thionin reduction	Positive
Triphenyltetrazolium reduction	Positive
Characteristics	*Reaction or result*
Physiological characteristics	
Gelatin hydrolysis	Negative
Tween 60 hydrolysis	Negative
Tween 80 hydrolysis	Negative
Urease	Positive
L-Ornithine decarboxylase	Positive
L-Arginine decarboxylase	Positive
L-Lysine decarboxylase	Positive
Vitamin requirements	Pantothenate, biotin
Mol%G+C of DNA	47.5–49.5
Carbohydrate metabolism	Fermentative and respiratory
Gas from glucose	Positive

Table 1 (cont) Characteristics of *Zymomonas mobilis*

Characteristics	*Reaction or result*
Major product from glucose	Ethanol; 1 mol of glucose fermented to 2 mol of ethanol and 2 mol of CO_2
Physiological characteristics	
Carbon sources for growth	
Sugars utilized	D-Glucose, D-fructose, sucrose
Sugars not utilized	D-Mannose, L-sorbose, D- and L-arabinose, L-rhamnose, D-xylose, D-ribose, D-sorbitol, salicin, dulcitol, D-mannitol, adonitol, erythritol, glycerol, ethanol, D-galacturonate, D,L-malate, succinate, pyruvate, D-lactate (D,L-lactate), tartarate, citrate, starch, dextrin, raffinose, D-trehalose, maltose, lactose, D-cellobiose
Antimicrobial agents (amount per disc)	
Resistant to	Ampicillin (10 μg), bacitracin (5 μg), cephaloridine (10 μg), erythromycin (10 μg), gentamicin (10 μg), kanamycin (10 μg), lincomycin (10 μg), methicillin (10 μg), nalidixic acid (30 μg), neomycin (10 μg), penicillin (5 U), polymyxin (300 U), streptomycin (10 μg), vancomycin (10 μg)
Sensitive to	Chloramphenicol (30 μg), fusidic acid (10 μg), novobiocin (30 μg), sulfafurazole (500 μg), tetracycline (10 μg)

umbonate and 1–4 mm in diameter after 2–7 days at 30°C. When incubated aerobically, colonies reach a maximum diameter of 1.5 mm or appear as microcolonies.

Physiological Characteristics

All *Zymomonas* strains are harmless to humans and are known to be used as natural inocula for making traditional alcoholic beverages. Most (90%) *Zymomonas* strains are able to grow at pHs between 3.85 and 7.55; 43% of the strains can grow at pH 3.5, but the organism does grow at pH 3.05 or below. Because *Zymomonas* is somewhat thermolabile, the organism grows best at temperatures between 25 and 30°C; 74% of strains grow at 38°C, but growth is rare at 40°C. Indeed, growth at 36°C is the best phenotypic index to classify the *mobilis* (positive) from the *pom-*

aceae (negative) subspecies. *Zymomonas* is killed by exposure to 60°C for 5 min.

Nitrate reduction and indole production are negative in *Zymomonas*. Catalase is positive, but oxidase is negative. Reduction of methylene blue, thionin and triphenyltetrazorium are positive. Hydrolysis of gelatin, Tween 60 and Tween 80 are negative. The *Zymomonas* genome is approximately half as large as that of *Escherichia coli* and contains about 1500 cistrons. Electropherograms of the soluble protein from 43 *Zymomonas* strains exhibit similar profiles, and their G+C contents range from 47.5 to 49.5%. Comparisons of rRNA cistrons, phenylalanine biosynthetic enzymes and the organization of the *trp* genes all indicate that *Z. mobilis* is most closely related to species of *Acetobacter*, *Rhizobium* and *Agrobacter*.

Zymomonas utilizes and easily ferments D-glucose and D-fructose, forming large quantities of CO_2 after 1–2 days at 30°C. Half of the strains will grow in glucose concentrations up to 40%, and many *Zymomonas* strains are able to ferment and grow on sucrose. The ability to ferment sucrose appears to be inducible, since this property is sometimes lost when *Zymomonas* is subcultured on D-glucose. In some instances, sucrose is converted to levan (β-(2,6)-fructose polymer), fructo-oligosaccharide or sorbitol. None the less, the range of carbohydrates utilized by *Zymomonas* is restricted to glucose, fructose and sucrose.

The nitrogen required for growth can be supplied as peptone, yeast extract, nutrient broth, beer, palm juice, apple juice or a mixture of 20 amino acids. *Zymomonas* strains require biotin and pantothenate as enzyme co-factors. They are somewhat tolerant to a number of antibiotics including bacitracin, gentamicin, kanamycin, lincomycin, methicillin, nalidixic acid, neomycin, novobiocin, penicillin, polymyxin, streptomycin and vancomycin.

Metabolism

Saccharomyces metabolize glucose to pyruvate via the Enbden–Meyerhof–Parnas pathway (EMP); ethanol is then formed from the pyruvate. In contrast, *Zymomonas* anaerobically ferments sugars via the Entner–Doudoroff pathway, forming pyruvate from gluconate (**Fig. 1**). As in *Saccharomyces*, the liberated pyruvate is decarboxylated yielding acetaldehyde and CO_2, and the acetaldehyde is in turn reduced to produce ethanol. Up to now, the Entner–Doudoroff pathway has only been observed in aerobic bacteria such as *Pseudomonas* and *Gluconobacter*; its identification in *Zymomonas* represents the first time the Entner–Doudoroff pathway has been seen in an anaerobe. In *Z. mobilis*, glycolytic enzymes account for

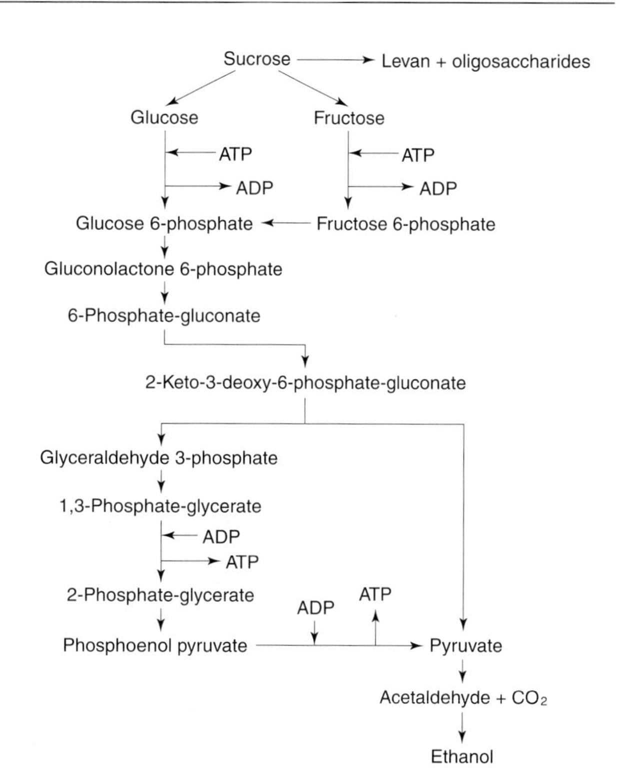

Figure 1 The Entner–Doudoroff pathway for carbohydrate metabolism.

30–50% of the soluble protein. Of these enzymes, glyceraldehyde 3-phosphate dehydrogenase and phosphoglycerate kinase are key regulators in the Entner–Doudoroff pathway. Furthermore, the presence of pyruvate decarboxylase and alcohol dehydrogenase isozymes that are tolerant to high ethanol concentrations enable *Zymomonas* to perform a pure ethanol fermentation.

When ATP formation via the EMP and Entner–Doudoroff pathways were compared, it was found that the EMP pathway yielded 2 mol of ATP per mol of glucose, whereas the Entner–Doudoroff pathway yielded only 1 mol of ATP per mol glucose. Thus, the cell yield of ATP from glucose is less in *Zymomonas* than in yeast. The equation describing the molar fermentation is as follows:

$$1\ C_6H_{12}O_6 = 1.93\ C_2H_5OH + 1.8\ CO_2 + 0.053\ CH_3CHOHCOOH$$

The kinetic parameters of ethanol fermentation by *Zymomonas* and *Saccharomyces carlsbergensis* have been compared in 25% glucose under anaerobic conditions. Ethanol productivity and glucose uptake by *Zymomonas* were three to four times faster than by *S. carlsbergensis*. The higher production resulted in an ethanol yield that was larger than was traditionally seen with yeast, despite the fact that the cell yield was smaller in *Zymomonas* than in *S. carlsbergensis*.

As mentioned previously, *Z. mobilis* utilizes only

glucose, fructose and sucrose for growth. Although, only trace amounts of by-product are produced during the fermentation of glucose, significant quantities of levan may be formed during the fermentation of sucrose. Three sucrose-hydrolysing enzymes are involved in the initial step in sucrose fermentation: intracellular invertase, extracellular invertase and extracellular levansucrase. Of them, extracellular levansucrase participates in the production of levan and some fructo-oligosaccharides. When both glucose and fructose are present in the culture medium, the sugar alcohol by-product, sorbitol, is accumulated. The formation of this product is due to the action of glucose–fructose oxidoreductase which catalyses the intermolecular oxidation–reduction of glucose and fructose to form gluconolactone and sorbitol. Gluconolactone is hydrolysed by a gluconolactonase and is then metabolized to form ethanol via the Entner–Doudoroff pathway. Although the physiological role of the oxidoreductase is still unclear, the accumulation of sorbitol in cells may be important for the acquisition of osmotic tolerance. Genes encoding glycolytic and related enzymes that have been cloned from *Zymomonas* are listed in **Table 2**.

Another characteristic peculiar to *Zymomonas* is its tolerance of high concentrations of both substrate and product. As shown in Table 1, some strains of *Zymomonas* will tolerate up to 30–40% glucose and 13% (wt/vol) ethanol. Such a high tolerance for ethanol is exceptional among bacteria; the growth of most bacteria is inhibited by ethanol concentrations of only 1–2% (wt/vol). By way of explanation, a major protective function has been ascribed to the hopanoids, pentacyclic tri-terpenoids that are present in large quantities in the cell membrane of *Z. mobilis*. Most likely, these amphiphilic, sterol-like substances stabilize the cell membrane of *Z. mobilis* against solubilization by ethanol.

Table 2 Genes cloned from *Zymomonas mobilis*

Gene[a]	*Enzyme*	*Reference*
gap	Glyceraldehyde 3-phosphate dehydrogenase	1
pgk	Phosphoglycerate kinase	2
pdc	Pyruvate decarboxylase	3–5
zwf	Glucose 6-phosphate dehydrogenase	6
edd	6-Phosphogluconate dehydratase	6
glk	Glucokinase	6
frk	Fructokinase	7
pgi	Phosphoglucose isomerase	8
gfo	Glucose–fructose oxidoreductase	9
adhA	Alcohol dehydrogenase I	10
adhB	Alcohol dehydrogenase II	11, 12
invA	Intracellular invertase	13
sacA	Intracellular invertase	14
invB	Extracellular invertase	15
levA	Extracellular levansucrase	15
zliE	Stimulator for production of invertase and levansucrase	16
zliS	Stimulator for secretion of invertase and levansucrase	16
phoC	Acid phosphatase	17
phoA	Alkaline phosphatase	18
phoD	Alkaline phosphatase	19
dnak	DnaK protein	20
groESL	GroES and GroEL proteins	21

[a]The gene having been sequenced.
1 = *J. Bacteriol.* 169: 5653 (1987); 2 = *J. Bacteriol.* 170: 1926 (1988); 3 = *J. Bacteriol.* 170: 3310 (1988); 4 = *Nucleic Acids Res.* 15: 1753 (1987); 5 = *J. Bacteriol.* 169: 949 (1987); 6 = *J. Bacteriol.* 172: 7227 (1990); 7 = *J. Bacteriol.* 174: 3455 (1992); 8 = *J. Bacteriol.* 173: 3215 (1991); 9 = *J. Bacteriol.* 174: 1439 (1992); 10 = *J. Bacteriol.* 172: 2491 (1990); 11 = *J. Bacteriol.* 169: 2591 (1987); 12 = *Nucleic Acids Res.* 18: 187 (1990); 13 = *Agric. Biol. Chem.* 55: 1383 (1991); 14 = *J. Bacteriol.* 172: 6727 (1990); 15 = *Biosci. Biotech. Biochem.* 59: 289 (1995); 16 = *Biosci. Biotech. Biochem.* 58: 526 (1994); 17 = *J. Bacteriol.* 171: 767 (1989); 18 = *Fems Microbiol. Lett.* 77: 3455 (1992); 19 = *Fems Microbiol. Lett.* 125: 2 (1995); 20 = *J. Bacteriol.* 175: 3328 (1993); 21 = *Gene* 148: 51 (1994).

Methods of Detection and Isolation

Detection

Detection of *Zymomonas* is based on its characteristic ability to produce CO_2 from glucose and its ethanol tolerance. The detection medium, which was originally designed in breweries, has the following composition ($g\,l^{-1}$): 3 g malt extract, 3 g yeast extract, 20 g D-glucose, 5 g peptone and 0.02 g actidione; pH is adjusted to 4.0. The medium is dispensed into 25 ml screw-capped bottles with a Durham tube (20 ml per bottle) and sterilized. Ethanol is then added to 3% (v/v). The presence of *Zymomonas* is indicated by abundant gas production after 2–6 days at 30°C. Because false-positive results may occur if lactobacilli or wild yeast are present, a Gram stain is recommended.

Isolation

It has been reported that in 1923 Lindner isolated *Zymomonas* from the sap of the maguey plant and Shimwell isolated the same bacterium from spoiled beer. Actually, *Zymomonas* can easily be isolated from palm sap and palm wines. None the less, it is our experience that many types of wine do not allow the growth of *Zymomonas*, presumably because of the presence of preservatives (e.g. potassium metabisulphite, $K_2S_2O_5$). Several protocols for the isolation of *Zymomonas* can be found in the literature. The media used for isolation of the organism are classified

Table 3 Details of media

Standard medium for identification (SM)	
Yeast extract (Difco)	0.5%
D-glucose	2.0%
For solid medium, agar is added to the concentration of 2.0%	
Apple juice–yeast extract medium	
Apple juice from a culinary or dessert apple variety	
Yeast extract (Difco)	1.0%
pH is adjusted to 4.8 with NaOH. For solid medium, agar is added to a concentration of 3.0%	
Apple juice–gelatin medium	
Four-times-diluted apple juice	
Yeast extract (Difco)	1%
Gelatin	10%
pH is adjusted to 5.5 with NaOH	
Beer–glucose medium	
Sterile beer	
Glucose	2%
MYPG medium	
Ethanol	3.0%
Glucose	2.0%
Malt extract	0.3%
Yeast extract	0.3%
Peptone	0.5%
Cycloheximide	0.002%
pH is adjusted to 4.0 After autoclaving, ethanol and cycloheximide are added	
Synthetic medium of Goodman *et al*	
Basal medium	
Glucose	2.0%
KH_2PO_4	0.1%
K_2HPO_4	0.1%
NaCl	0.05%
$(NH_4)_2SO_4$	0.1%
pH	6.0
Metal solution	
$MgSO_4{\cdot}7H_2O$	200 mg l^{-1}
$CaCl_2{\cdot}2H_2O$	200 mg l^{-1}
$Na_2MoO_4{\cdot}2H_2O$	25 mg l^{-1}
$FeSO_4{\cdot}7H_2O$	25 mg l^{-1}
Vitamin solution	
Calcium pantothenate	5 mg ml^{-1}
Thiamin hydrochloride	1 mg ml^{-1}
Pyridoxine hydrochloride	1 mg ml^{-1}
Biotin	1 mg ml^{-1}
Nicotinic acid	1 mg ml^{-1}
After autoclaving, the filter-sterilized metal solutions are added to the indicated concentrations, and 1 ml of the filter-sterilized vitamin solution is added to 1 l of basal medium. The medium is solidified with 1.5% of agar (Difco)	

roughly into two groups (**Table 3**): media based on fruit juice, saps and alcoholic beverages, which are supplemented with glucose and/or nutrients, and synthetic media used to isolate strains with higher ethanol productivity, which contain yeast extract as the essential additive.

Isolation of *Zymomonas* is carried out in two steps. First the organism is enriched in a liquid medium at 30°C under static conditions using screw-capped or cotton-plugged bottles. Then it is isolated by forming colonies on solid medium in Petri dishes at 30°C under either anaerobic (e.g. in a BBL® Gas Pak anaerobic system) or aerobic conditions. By transferring the cultures to complex media, the bacteria can be kept alive for 2–3 months at room temperature. Moreover, when lyophilized or frozen at −80°C in the presence of 12.5–25% glycerol, *Zymomonas* may be kept for several years.

Importance to the Food Industry

Although in the tropics various strains of *Zymomonas* are used to make alcoholic beverages, in Europe they are recognized as a causative agent in the spoilage of beer and apple cider.

Fermentative Agents

Palm Wine These alcoholic beverages, which are popular in Africa, South America and Southeast Asia (e.g. Indonesian Tuwak), are obtained from the spontaneous fermentation by *Zymomonas* of the sugary sap of various species of palm tree. The specific characteristics of palm wines depend on many factors including, among others, the length of time the palm trees are tapped, the palm species, the duration of storage and the season. Palm wine may contain 0.1–7.1% ethanol, depending upon the stage at which it was collected. A typical palm wine, however, contains approximately 4–5% ethanol, and it has a pH of 3–4 due to the presence of tartaric, malic, pyruvic, succinic, lactic, *cis*-aconitic, citric and acetic acids.

The main constituent of palm tree sap is sucrose with small amounts of glucose, fructose, maltose, raffinose and malto-oligosaccharides. Palm wines have rather complex microflora, among which *Saccharomyces*, *Zymomonas*, lactobacilli and *Acetobacter* are present. There is a hypothesis that palm wine may be the result of mixed alcoholic, acetic and lactic fermentation. None the less, it is certain that *Zymomonas* is a very important bacterial constituent. *Zymomonas* strains are largely responsible for the alcohol content and for the frothing that results from CO_2 formation. CO_2 and small amounts of lactic and acetic acid also contribute to the acidity. The production of some acetaldehyde and the characteristic fruity odour of *Zymomonas* probably positively influence the odour and the taste of palm wine.

Pulque This is a popular drink in Mexico. It is a milky and viscous beverage with about the same alcohol content as beer (approximately 4–6%). It is produced from the sap of the maguey plant; the most common species used is *Agave atrovirens*. It was the fermentation of Agave sap that Lindner was studying in 1923 when he discovered that the organism mediating fermentation was a bacterium that he called *Termobacter mobile* (now *Zymomonas mobilis* subsp. *mobilis*). Over the years this organism has been referred to by many names, including *Pseudomonas lindoneri* (1931), *Z. mobile* (1936), *Achromobacter anaerobium* (1937), *Saccharobacter* sp. (1937), *Saccharomonas anaerobia* (1950), *S. lindneri* (1950), *Z. anaerobia* (1963), *Z. mobilis* var. *recifensis* (1970) and *Z. congolensis* (1972). Like palm wine, pulque has a complex microflora constituted of *Z. mobilis*, *Saccharomyces*, lactobacilli and *Acetobacter* spp. Pulque is distinct from tequila which, although also derived from maguey, is produced by yeast fermentation and subsequently distilled.

Caldo-de-cana-picado This alcoholic beverage, popular in Northeast Brazil, is obtained from the spontaneous fermentation of sugar-cane juice. *Zymomonas* strains have been isolated from caldo-de-cana-picado that were very motile with mono- or lophotrichous flagella and could ferment at 42°C with a flocculent deposit. Based on these properties, a new taxon *Z. mobilis* var. *recifensis* was created for these organisms.

Spoilage Agents

Beer *Zymomonas* has been isolated from beer, from the surface of brewery equipment and from the brushes of cask-washing machines. When cask and keg beers are infected with *Zymomonas*, the bacteria cause heavy turbidity and an unpleasant odour of rotten apples due to traces of acetaldehyde and H_2S. In warm weather, beer can spoil within 2–3 days due to the growth of *Zymomonas* and the resultant accumulation of acetaldehyde. On the other hand, *Zymomonas* has not been reported in lager beers; the low temperature at which lagers are processed (8–12°C) is unfavourable for the growth of *Zymomonas*.

Cider and Perry In recent years, consumers have tended to prefer sweet ciders over drier farm-made ciders. A potential problem for sweet cider, however, is a second fermentation called cider sickness which produces heavy turbidity, high gas pressure and a reduction of the sweetness, aroma and flavour of the cider. To prevent the development of cider sickness, it is recommended that the acidity of the cider should be kept high and the storage temperatures low. It has been suggested that the *framboisement* of cider is caused by two different agents: the anaerobic cider sickness bacillus *Zymomonas* and aerobic acetic acid bacteria. *Zymomonas* strains have also been isolated from perry, a cider-like beverage made from pears.

Other Applications

As mentioned in the section on metabolism, *Zymomonas* accumulates levan as a by-product during the fermentation of sucrose. This characteristic has enabled the development of an effective one-step enzymatic process for synthesizing levan. By mixing a solution of sucrose with levansucrase derived from *Z. mobilis*, excellent quality, uniquely colloidal (viscous-like paste) levan is produced. The levans have a wide range of potential applications, including as food thickeners and as glazing agents in cosmetics. Interestingly, it has recently been reported that *Z. mobilis* levans exhibit anti-tumour activity against sarcoma 180 and Ehrlich carcinoma in Swiss albino mice.

Levansucrase also catalyses the fructosyl transfer from sucrose to various mono- and disaccharides to form hetero-fructo-oligosaccharides. These oligosaccharides have attracted special attention for their physiological and physical usefulness as intestinal flora conditioners and as non-digestible sugar.

Improvement by Genetic Manipulation

The specific rates of sugar uptake and ethanol productivity are three to four times faster in *Zymomonas* than in yeast. Consequently, many research groups worldwide have been attempting to develop industrial processes for producing ethanol using *Z. mobilis*. Unfortunately, the organism has some disadvantages as compared to yeast. In particular, its narrow spectrum of utilizable carbohydrates severely limits its commercial use. Genetic manipulation broadening this spectrum would substantially increase the industrial utility of *Zymomonas*.

Host Vector System

A variety of cloning vectors have been used to engineer *Z. mobilis* genetically; they can be classified into two groups: first, hybrid plasmids constructed from native *Z. mobilis* plasmid and an *E. coli* vector plasmid, because most *Z. mobilis* strains contain cryptic plasmids, and second non-conjugative plasmids with a broad host range. With respect to the former, an expression vector has been constructed that will insert *Z. mobilis* promoter and terminator genes into a shuttle vector. An example for such a shuttle vector between *Z. mobilis* and *E. coli* is shown in **Figure 2.** As for the latter, the only genetic markers that can be

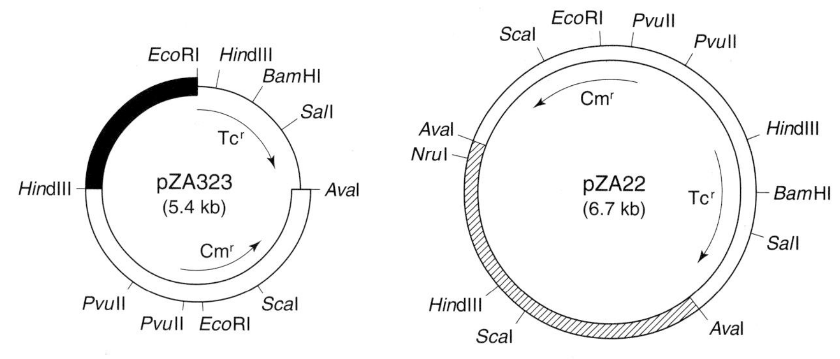

Figure 2 Shuttle vectors pZA323 and pZA22. Solid box, endogenous cryptic plasmid pZM3 from *Zymomonas mobilis* ATCC29192; hatched box, endogenous cryptic plasmid pZM2 from *Z. mobilis* ATCC29192; open box, pACYC184 DNA; line, pBR322 DNA.

used are resistance to tetracycline and chloramphenicol.

Gene Transfer

The conjugal method of transferring genes has proved suitable for the introduction of large, broad host-range plasmid vectors into *Zymomonas*. For example, it has been possible to use the kanamycin-resistant helper plasmid, pRK2013, to transfer a plasmid vector from *E. coli* to *Z. mobilis*. A conjugation has been reported in which a broad host-range plasmid vector containing the *tra* gene from RP4 was transferred from *E. coli* S17-1 to *Z. mobilis*; the *tra* gene was then integrated into the chromosomal DNA using the biparental mating method. Using a $CaCl_2$ heat-shock method originally devised for *E. coli*, a group of authors were able to transform *Z. mobilis* at a rate of 100–1000 cells per microgram of DNA. Our group have used partially spheroplasted cells induced with D-cycloserine and penicillin G to obtain *Z. mobilis* transformants at the rate of 10^4–10^5 cells per microgram of DNA. An electroporation method for reproducible transformation was also developed based on another modification. At a field strength of $10\,kV\,cm^{-1}$, transformants were obtained at the rate of 10^6 cells per microgram of DNA.

Because cell fusion may also be a useful method to improve *Zymomonas*, conditions for the formation, regeneration and fusion of protoplasts have been examined and established. In our experience fusants (prototrophic recombinants) could be obtained at a frequency of 10^{-5} per spheroplast through intraspecific fusion of mutants lacking the ability to ferment sucrose or fructose. This method may be useful for breeding salt-tolerant or thermotolerant *Zymomonas*.

Strain Improvement

The basic recombinant DNA techniques that would be used to improve *Zymomonas* strains are well established and include the host-vector system, the methods for gene introduction, and the methods for using *Zymomonas* promoters to drive gene expression. These techniques could potentially be used to confer on *Zymomonas* the ability to convert all of the major sugar components of plant biomass into ethanol fuel. For instance, lactose is the major organic constituent of whey, a waste material of the dairy industry. Production of ethanol from lactose or whey by *Z. mobilis* has been improved by introducing into the organism the *E. coli* lactose operon or the *lac* transposon. Almost all *Z. mobilis* bred this way can ferment lactose to form ethanol, although they still will not grow on lactose. *Z. mobilis* carry the *gal* operon derived from *E. coli* produced a small amount of ethanol from galactose. The recombinant strain did not grow on galactose, however, probably because of the inhibitory effects of galactose metabolites.

Beet molasses, one of the potent starting materials for ethanol fermentation, contains 3–4% raffinose. The *Z. mobilis* raf^+ recombinant, which was bred by introducing the α-galactosidase and lactose permease genes from *E. coli*, produced ethanol from raffinose. This suggests that raffinose is hydrolysed to fructose and melibiose by the native extracellular invertase and levansucrase, and then liberated melibiose is taken up into the recombinant cells by lactose permease and hydrolysed by α-galactosidase. The amount of ethanol produced this way was nearly equivalent to the amount of glucose and fructose liberated from raffinose.

Cellulosic materials are very important as potential fuel sources for production of ethanol, because they are the most abundant organic substances on earth. To breed *Zymomonas* that are able to ferment cellulose and produce ethanol directly, it was necessary to introduce and express both the endoglucanase and β-glucosidase genes. Bacterial cellulase genes, including endoglucanases from *Cellulomonas uda*, *Acetobacter xylinum*, *Erwinia chrysanthemi*, *Pseudomonas fluorescens* var. *cellulosa* or *Bacillus subtilis*, have been introduced into and expressed in *Z.*

mobilis. However, production of ethanol from cellulose by these recombinants has not as yet been reported. Recombinants carrying the β-glucosidase gene from *Xanthomonas albilineans* or *Ruminococcus albus* can ferment cellobiose to produce small amounts of ethanol.

Xylose is the principal constituent of hemicellulose, large quantities of which are found in agricultural waste such as rice straw, corncobs and parts of hardwood. Two operons encoding xylose assimilation and the enzymes of the pentose phosphate pathway from *E. coli* were constructed and transfected into *Z. mobilis*. Anaerobic fermentation of a pentose to ethanol was thereby achieved through combination of the pentose phosphate and Entner–Doudoroff pathways. This strain efficiently fermented both glucose and xylose, which is essential for economical conversion of lignocellulose to ethanol.

Z. mobilis has been improved by transforming it with a thermostable amylase gene from *Bacillus licheniformis*. The recombinant strain produced amylase that was immunologically identical to that of *B. licheniformis*, and virtually all of the enzyme activity could be recovered from the culture medium. The level of expression of α-amylase in this *Zymomonas* strain was still low, but this was a first step towards the engineering of a *Z. mobilis* strain able to convert starch into ethanol in a single step.

Many approaches aimed at broadening the spectrum of utilizable substrates through transfer of appropriate hydrolase genes have been tried. However, the recombinant strains were unable to produce ethanol directly from starch or cellulose. To breed the desired strains, it will be necessary for the cells efficiently to express and secrete cellulolytic enzymes or amylases. It has been proposed that the signal peptide-dependent general export pathway made up of the *sec* gene products and signal peptidases would be a good secretion mechanism in Gram-negative bacteria. In addition, we have found that extracellular levansucrase and invertase are secreted via a signal peptide-independent pathway in *Z. mobilis*; we have cloned the gene (*zliS*) that stimulates secretion of these enzymes.

Although there are obstacles that remain to be overcome, the unique and valuable characteristics of *Z. mobilis* mean that this organism will almost certainly be used as a biocatalyst for the industrial production of ethanol from plant biomass in the near future.

See also: **Cider (Hard Cider)**. **Genetic Engineering**: Modification of Bacteria. ***Saccharomyces***: *Saccharomyces carlsbergensis* (Brewer's Yeast). **Spoilage Problems**: Problems caused by Bacteria. **Wines**: Microbiology of Wine-making.

Further Reading

Carey VC and Ingram LO (1983) Lipid composition of *Zymomonas mobilis*: effects of ethanol and glucose. *Journal of Bacteriology* 154: 1291–1300.

Ingram LO, Eddy CK, Mackenzie KF, Conway T and Alterthum F (1989) Genetics of *Zymomonas mobilis* and ethanol production. *Journal of Industrial Microbiology* 30: 53–69.

Ingram LO, Gomez PF, Lai X et al (1997) Metabolic engineering of bacteria for ethanol production. *Biotechnology and Bioengineering* 58: 204–214.

Rogers PL, Lee KJ, Skotnicki ML and Tribe DE (1982) Ethanol production by *Zymomonas mobilis*. *Advance of Biochemical Engineering* 23: 37–84.

Saham H, Bringer-Myer S and Sprenger G (1992) The genus *Zymomonas*. In: Balows A, Truper HG, Dworkin M, Harder W and Schleifer K-H (eds) *The Prokaryotes*. P. 2287. New York: Springer-Verlag.

Swings J and De Ley J (1977) The biology of *Zymomonas*. *Bacteriological Reviews* 41: 1–46.

Swings J and De Ley J (1984) Genus *Zymomonas*, Kluyver and van Niel, 1936. In: Krieg NR and Holt JG (eds) *Bergey's Manual of Systematic Bacteriology*. P. 576. Baltimore: Williams & Wilkins.

Swings J, Kersters K and De Ley J (1976) Numerical analysis of electrophoretic protein patterns of *Zymomonas* strains. *Journal of General Microbiology* 93: 266–271.

Yanase H, Kato N and Tonomura K (1994) Strain improvement of *Zymomonas mobilis* for ethanol production. In: Murooka Y and Imanaka T (eds) *Recombinant Microbes for Industrial and Agricultural Applications*. P. 723. New York: Marcel Dekker.

Zhang M, Eddy C, Deanda K, Finkelstein M and Picataggio S (1995) Metabolic engineering of a pentose metabolism pathway in ethanologenic *Zymomonas mobilis*. *Science* 267: 240–243.

APPENDIX I: BACTERIA AND FUNGI

The genera listed here are those associated with food, agricultural products and environments in which food is prepared or handled.

Abiotrophia
Acinetobacter
Actinobacillus
Actinomyces
Aerococcus
Aeromonas
Agrobacterium
Alcaligenes
Alloiococcus
Anaerobiospirillum
Arcanobacterium
Arcobacter
Arthrobacter
Aureobacterium
Bacillus
Bacteroides
Bergeyella
Bifidobacterium
Blastoschizomyces
Bordetella
Branhamella
Brevibacillus
Brevibacterium
Brevundimonas
Brochothrix
Brucella
Budvicia
Burkholderia
Buttiauxella
Campylobacter
Candida
Capnocytophaga
Cardiobacterium
Carnobacterium
CDC
Cedecea
Cellulomonas
Chromobacterium
Chryseobacterium
Chryseomonas
Citrobacter
Clostridium
Comamonas
Corynebacterium
Cryptococcus
Debaryomyces
Dermabacter
Dermacoccus
Dietzia
Edwardsiella
Eikenella
Empedobacter
Enterobacter
Enterococcus
Erwinia
Erysipelothrix
Escherichia
Eubacterium
Ewingella
Flavimonas
Flavobacterium
Fusobacterium
Gardnerella
Gemella
Geotrichum
Gordona
Haemophilus
Hafnia
Hansenula
Helicobacter
Kingella
Klebsiella
Kloeckera
Kluyvera
Kocuria
Kytococcus
Lactobacillus
Lactococcus
Leclercia
Leptotrichia
Leuconostoc
Listeria
Malassezia
Methylobacterium
Microbacterium
Micrococcus
Mobiluncus
Moellerella
Moraxella
Morganella
Myroides
Neisseria
Nocardia
Ochrobactrum
Oerskovia
Oligella
Paenibacillus
Pantoea
Pasteurella
Pediococcus
Peptococcus
Peptostreptococcus
Photobacterium
Pichia
Plesiomonas
Porphyromonas
Prevotella
Propionibacterium
Proteus
Prototheca
Providencia

Pseudomonas
Psychrobacter
Rahnella
Ralstonia
Rhodococcus
Rhodotorula
Rothia
Saccharomyces
Salmonella
Serratia
Shewanella
Shigella
Sphingobacterium
Sphingomonas
Sporobolomyces
Staphylococcus
Stenotrophomonas
Stomatococcus
Streptococcus
Suttonella
Tatumella
Tetragenococcus
Trichosporon
Turicella
Veillonella
Vibrio
Weeksella
Weissella
Xanthomonas
Yarrowia
Yersinia
Yokenella
Zygosaccharomyces

APPENDIX II: LIST OF SUPPLIERS

The suppliers below are mentioned in the text as main sources of specialist equipment, culture media or diagnostic materials. This list is not intended to be comprehensive.

3M Microbiology Products
3M Center
Building 260–6B-01
St Paul
MN 55144–1000
USA

ABC Research Corporation
3437 SW 24th Avenue
Gainesville
FL 32607
USA

Adgen Ltd
Nellies Gate
Auchincruive
Ayr KA6 5HW
UK

Agi-Diagnostics Associates
Cinnaminson
New Jersey
USA

ANI Biotech OY
Temppelikatu 3–5, 00100
Helsinki
Finland

Applied Biosystems
The Perkin-Elmer Corporation
12855 Flushing Meadow Drive
St Louis
MO 63131 1824
USA

Becton Dickinson Microbiology Systems
7 Loveton Circle
Sparks
MD 21152–0999
USA

bio resources
9304 Canterbury
Leawood
KS 66206
USA

BioControl Systems
19805 North Creek Parkway
Bothwell
WA 98011
USA

BioControl Systems, Inc
12822 SE
32nd Street
Bellevue
WA 98005
USA

Bioenterprises Pty Ltd
28 Barcoo Street
PO Box 20 Roseville
NSW 2069
Australia

Biolog, Inc
3938 Trust way
Hayward
CA 94545
USA

Bioman Products, Inc
400 Matheson Blvd
Unit 4
Mississauga
Ontario
LAZ 1N8
Canada

bioMérieux
Chemin de l'Orme
69280 Marcy L'Étoile
France

bioMérieux (UK)
Grafton House
Grafton Way
Basingstoke
Hants RG22 6HY
UK

bioMérieux Vitek, Inc
595 Anglum Drive
Hazelwood
MO 63042 2320
USA

Bioscience International
11607 Mcgruder Lane
Rockville
MD 20852 4365
USA

Biosynth AG
PO Box 125
9422 Staad
Switzerland

Biotecon
Hermannswerder haus 17
14473 Potsdam
Germany

Biotrace
666 Plainsboro Road
Suite 1116
Plainsboro
NJ 08536
USA

Celsis
2948 Old Britain Circle
Chattanooga
TN 37421
USA

Celsis International plc
Cambridge Science Park
Milton Road
Cambridge
CB4 4FX
UK

Celsis-Lumac Ltd
Cambridge Science Park
Milton Road
Cambridge
CB4 4FX
UK

Charm Sciences Inc
36 Franklin Street
Malden
MA 02148 3141
USA

Chemunex Corporation
St John's Innovation Centre
Cowley Road
Cambridge
CB4 4WS
UK

Crescent Chemical Co, Inc
1324 Motor Parkway
Hauppauge
NY 11788
USA

diAgnostix, Inc
1238 Anthony Road
Burlington
NC 27215
USA

DIFCO
PO Box 331058
Detroit
MI 48232
USA

Diffchamb (UK)
1 Unit 12 Block 2/3
Old Mill Trading Estate
Mansfield Woodhouse
Nottingham NG19 9BG
UK

Diffchamb SA
8 Rue St Jean de Dieu
69007 Lyons
France

Digen Ltd
65 High Street
Wheatley
Oxford OX33 1UL
UK

DiverseyLever
Weston Favell Centre
Northampton
NN3 8PD
UK

Diversy Ltd
Technical Lane
Greenhill Lane
Riddings
DE55 4BA
UK

Don Whitley Scientific Ltd
14 Otley Road
Shipley
West Yorkshire
BD17 7SE
UK

DuPont/Qualicon
E357/1001A
Rouote 141 & Henry Clay Road
PO Box 80357
Wilmington
DE 19880 0357
USA

Dynal
PO Box 158 Skoyen
0212 Oslo
Norway

Dynal (UK) Ltd
Station House
26 Grove Street
New Ferry
Wirral
Merseyside L62 5AZ
UK

Dynal (USA)
5 Delaware Drive
Lake Success
NY 11042
USA

Dynatech Laboratories Inc
14340 Sulleyfield Circle
Chantilly
VA 22021
USA

Ecolab Ltd
David Murray John Building
Swindon
Wiltshire
SN1 1NH
UK

Envirotrace (BioProbe)
675 Potomac River Road
McLean
VA 22100
USA

Foss Electric (UK)
Parkway House
Station Road
Didcot
Oxon OX11 7NN
UK

Fluorochem Ltd
Wesley Street
Old Glossop
Derbyshire
SK13 9RY
UK

Foss Electric A/S
69 Slangerupgade
PO Box 260
DK-3400 Hillerod
Denmark

GENE-TRAK Systems
94 South Street
Hopkinton
MA 01748
USA

Gist-Brocades Australia
PO Box 83
Moorebank
NSW 2170
Australia

Gist-Brocades BV
PO Box 1345
2600 M A Delft
The Netherlands

I.U.L.
1670 Dolwick Drive
Suite 8
Erlanger
KY 41018

IDEXX Laboratories, Inc
One IDEXX Drive
Westbrook
ME 04092
USA

Industrial Municipal Equipment Inc (ime, Inc)
1430 Progress Way
Suite 105
Ridersburg
MD 21784
USA

Innovative Diagnostic Systems
2797 Peterson Place
Norcross
GA 30071
USA

International BioProducts Tecra Diagnostics
14780 NE 95th Street
Redmond
WA 98052
USA

Lab M Ltd
Topley House
52 Wash Lane
Bury
Lancashire
BL9 6AU
UK

Launch Diagnostics Ltd
Ash House
Ash Road
New Ash Green
Longfield
Kent DA3 8JD
UK

Lionheart Diagnostics Bio-Tek Instruments, Inc
Highland Park
Box 998
Winooski
VT 05404 0998
USA

M. I. Biol
BioPharma Technology Ltd
BioPharma House
Winnall Valley Road
Winchester SO23 0LD
UK

Malthus Instruments Ltd
Topley House
52 Wash Lane
Bury
Lancashire
BL9 6AU
UK

Merck (UK) Ltd
Merck House
Poole
Dorset BH15 1TD
UK

Meridian Diagnostics Inc
3741 River Hills Drive
Cincinnati
OH 45244
USA

MicroBioLogics
217 Osseo Ave N
St Cloud
MN 56303 4455
USA

Microbiology International
10242 Little Rock Lane
Fredrick
MD 21702
USA

Microgen Bioproducts
1 Admiralty Way
Camberley
Surrey GU15 3DT
UK

MicroSys, Inc
2210 Brockman
Ann Arbor
MI 48104
USA

Minitek-BBL
BD Microbiology Systems
7 Loveton Circle
Sparks
MD 21152
USA

Mitsubishi Gas Chemical America, Inc
520 Madison Avenue
25th Floor
New York
NY 10022
USA

M-Tech Diagnostics
49 Barley Road
Thelwall
Warrington
Cheshire WA4 2EZ
UK

National Food Processors Assoc
1401 New York
NW
Washington
DC 20005
USA

Neogen Corporation
620 Lesher Place
Lansing
MI 48912
USA

New Horizons Diagnostic Corp
9110 Red Branch Road
Suite B
Columbis
MD 21045 2014
USA

Olympus Precision Instruments Division
10551 Barkley
Suite 140
Overland Park
KS 66212
USA

Organon Teknika AKZO NOBEL
100 Akzo Avenue
Durham
NC 27712
USA

Oxoid, Inc
217 Colonnade Road
Nepean
Ontario
K2E 7K3
Canada

Oxyrase Inc
PO Box 1345
Mansfield
OH 44901
USA

Perkin Elmer Corporation
50 Tanbury Road
Mail Station 251
Wilton
CT 06897 0251
USA

Pharmacia Biotech
800 Centennial Avenue
PO Box 1327
Piscataway
NJ 08855 1327
USA

Prolab Diagnostics
Unit 7 Westwood Court
Clayhill Industrial Estate
Neston
Cheshire L64 3UJ
UK

QA Life Sciences Inc
6645 Nancy Ridge Drive
San Diego
CA 92121
USA

Radiometer Ltd
Manor Court
Manor Royal
Crawley
West Sussex
RH10 2PY
UK

R-Biopharm GmbH
Dolivostr 10
D-64293
Darmstadt
Germany

RCR Scientific Inc
206 West Lincoln
PO Box 340
Goshen
IN 46526 0340
USA

Remel
12076 Santa Fe Drive
Lenexa
KS 66215
USA

Rhone-Poulenc Diagnostics Ltd
3.06 Kelvin Campus
West of Scotland Science Park
Maryhill Road
Glasgow G20 0SP
UK

SciLog, Inc
14 Ellis Potter Ct
Madison
WI 53711–2478
USA

Silliker Laboratory Inc
1304 Halstead Street
Chicago Heights
IL 60411
USA

Spiral Biotech
7830 Old Georgetown Road
Bethesda
MD 20814
USA

Tecra Diagnostics
28 Barcoo Street
PO Box 20
Roseville
NSW
Australia

Tecra Diagnostics (UK)
Batley Business Centre
Technology Drive
Batley
W Yorkshire WF17 6ER
UK

Unipath, Oxoid Division
Wade Road
Basingstoke
Hampshire
RG24 8PW
UK

Unipath, Oxoid Division (USA)
800 Proctor Avenue
Ogdensburg
NY 13669
USA

Vicam
29 Mystic Avenue
Somerville
MA 02145
USA

Wescor, Inc
1220 E
1220 N
Logan
UT 84321
USA

INDEX

NOTE

Page numbers in **bold** refer to major discussions. Page numbers suffixed by T refer to Tables; page numbers suffixed by F refer to Figures. *vs* denotes comparisons.

This index is in letter-by-letter order, whereby hyphens and spaces within index headings are ignored in the alphabetization. Terms in parentheses are excluded from the initial alphabetization.

Cross-reference terms in *italics* are general cross-references, or refer to subentry terms within the same main entry (the main entry is not repeated to save space).

Readers are also advised to refer to the end of each article for additional cross-references – not all of these cross-references have been included in the index cross-references.

A

B

C

D

F

G

H

M

N

O

P

Q

R

S

T

U

W

X

Y

Z

Related Journal

Food Microbiology

Food Microbiology publishes primary research papers, short communications, reviews, reports of meetings, book reviews, and news items dealing with all aspects of the microbiology of foods. The editors aim to publish manuscripts of the highest quality which are both relevant and applicable to the broad field covered by the journal. Although all manuscripts will be considered, authors are encouraged to submit manuscripts dealing with the novel methods of detecting microorganisms in foods, especially pathogens of emerging importance, for example, *Listeria* and *Aeromonas*. Papers relating to the genetics and biochemistry of microorganisms that are either used to make foods or that represent safety problems are also welcomed, as are studies on preservatives, packaging systems, and evaluations of potential hazards of new food formulations. The editors make every effort to ensure rapid and fair reviews, resulting in timely publication of accepted manuscripts.

Research Areas Include

Food spoilage and safety

Predictive microbiology

Rapid methodology

Application of chemical and physical approaches to food microbiology

Biotechnological aspects of established processes

Production of fermented foods

Use of novel microbial processes to produce flavors and food-related enzymes

Database coverage includes Biological Abstracts (BIOSIS); Chemical Abstracts; Current Contents/Agriculture, Biology, and Environmental Science; Dairy Science Abstracts; Food Science and Technology Abstracts; Maize Abstracts; and Research Alerts

For further information: www.academicpress.com/foodmicro

V5ØØ M 8Ø B 25 DP 8 71521:5 P 5

EFM

ISBN 0-12-227070-3